W9-AVT-478

# TYPOGRAPHY

## AN ENCYCLOPEDIC SURVEY OF TYPE DESIGN AND TECHNIQUES THROUGHOUT HISTORY

Published by
Black Dog & Leventhal Publishers, Inc.
151 West 19th Street
New York, NY 10011

Distributed by
Workman Publishing Company
708 Broadway
New York, NY 10003

| | |
|---|---|
| Concept | Peter Feierabend, Friedrich Friedl, Nicolaus Ott, Bernard Stein |
| Project manager and editor | Sally Bald |
| Contributing editor | Sabine Gerber |
| Translation into English | Ruth Chitty |
| Translation into French | Cathérine Métais-Bührendt |
| Rights | Barbara Köthe-Löhausen |
| Research | Ute Brüning |
| Production manager | Detlev Schaper |
| Design | Ott + Stein |
| Jacket design | Jonette Jakobson |
| Text layout and typesetting | Wilhelm Schäfer |
| Picture layout | Claudia Grotefendt, Jennifer Neidhardt |
| Image processing | Christine Berkenhoff |
| Reproductions | Omniascanners, Milan |
| Printing and binding | Amilcare Pizzi, Milan |
| | Printed in Italy |

ISBN: 1-57912-023-7

hgfedcba

# TYPOGRAPHY

## AN ENCYCLOPEDIC SURVEY OF TYPE DESIGN AND TECHNIQUES THROUGHOUT HISTORY

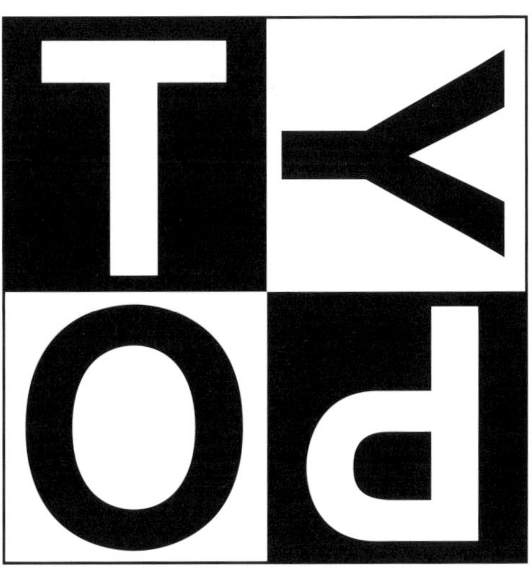

By Friedrich Friedl,
Nicolaus Ott and Bernard Stein

Black Dog
& Leventhal
Publishers

Information and assistance during the production of this publication was generously provided by many people.
In particular we would like to mention the following:

Carol Twombly, Adobe Systems; Stephanie Allen & Rick Grafe, American Institute of Graphic Arts, AIGI, SF; Michael Arlt; Olga Grisaitis, Art Directors Club ADC, NY; Jacqueline van Kimmenaede, Association of Dutch Design; Associazione italiana progettazione per la comunicazione; Phil Baines, Central Lettering Record; Ellen Lupton, Cooper-Hewett, National Design Museum; Antje Malec, documenta Archiv; Wolfgang Hartmann, Fundición Tipográfica Neufville; H. Gülgün, Eidgenössische Technische Hochschule, Zürich; Franco Grignani; Heinke Jenssen, Graphis Editorial Office, Zürich; Dietrich Helms; Sue Hofton; Cynthia Hollandsworth; Michael Hardt, Icograda; Icograda London; Joyce R. Kaye, International Type Corporation, ITC; Uwe Pflüger, Istituto Italiano di cultura Colonia; Toru Hoshuyama, Japan Graphic Designers Association Inc., JAGDA; H. Ernstberger, Kleukens-Archiv; Olaf Leu; Robin Nicholas, Monotype Typography Ltd; Dr. Thalmann, Museum für Gestaltung, Basel; Grazia Schenoni, Nebiolo, Paz Informazione; Frau Neuroth; Jennifer Loncar & Justus Oehler, Pentagram; David Quay; David Pankow, Rochester Institute of Technology, Cary Collection, Wallace Library; Helmut Schmid; Elizabeth Cromer, Society of Newspaper Design; James Mosley, St Bride Printing Library; Tonny Daalder, Stedelijk Museum; Sumner Stone, Stone Type Foundry; Carol Whaler, Type Directors Club, TDC; Visiva aiap; Prof. Hermann Zapf.



Photographic material
was kindly loaned to us by:

action press (H. Frankenfeld)
Anthony-Verlag
Astrofoto
Bavaria Bildagentur GmbH
Günter Beer
Bildarchiv Preußischer Kulturbesitz, Berlin
Bilderberg Archiv der Fotografen
(Rainer Drexel, Klaus D. Francke, Eberhard Grames)
Christoph Büschel
Central Order
Deutsches Historisches Museum, Berlin
Deutsches Plakatmuseum, Essen
Saša Fuis
Getty Images (Frank Tewkesbury)
Peter Greenaway Office
Haags Gemeentemuseum, Den Haag
Hochschule für Gestaltung, Offenbach am Main
Internationales Bildarchiv
Jan Kaplan
Kestner Museum, Hannover
Lorcán Köthe
Kunsthistorisches Museum Wien
Alexander Lavrentiev
Michael Löhausen
Lotos-Film
Middelhauve Verlag GmbH
Museum für Gestaltung, Zürich
Pelizaeus Museum (Sh. Shalchi)
Rheinisches Bildarchiv, Köln
Rijksmuseum van Oudheden, Leiden
Sammlung des Instituts für Papyrologie,
Universität Heidelberg
Schapowalow/Bahner
Staatliche Museen zu Berlin, Preußischer Kulturbesitz:
Antikensammlung
Kunstbibliothek
Museum für Indische Kunst
Museum für Islamische Kunst
Museum für Ostasiatische Kunst
Vorderasiatisches Museum
Stadtbibliothek Trier
Stiftung Buchkunst, Frankfurt am Main

A large number of the works illustrated come from the collection of Professor Friedrich Friedl. However, only the active support of most of the persons included in this book, or of their heirs, has made it possible. We would like to express out sincere thanks to all of them.

## Acknowledgements

We thank Peter Feierabend for giving us the idea for this book.
In collating the list of names of people and institutions contained in this book Ken Cato, Wim Crouwel, Olle Eksell, Adrian Frutiger, Takenobu Igarashi, Oswaldo Miranda, Morteza Momayez, Dan Reisinger, Paula Scher, Todor Vardjiev, Zdenêk Ziegler and Günter Gerhard Lange have advised us from the start. We owe them a debt of gratitude.
We would like to offer particular thanks to Jürgen Siebert of FontShop and Otmar Hoefer of Linotype Library. They provided us with the type samples which CitySatz & Nagel, Berlin generously prepared for the "who" section.
Our special thanks go to Sally Bald, Philipp Luidl and Wilhelm Schäfer.
In particular we would also like to thank Ute Brüning for her research and work on the index as well as Christine Berkenhoff, Claudia Grotefendt, Jennifer Neidhardt and Maximilian Meisse.
Those who have supported us in word and deed are Annette Beil, Ben Buschfeld, Jan Carell, Stephanie Ehret, Jeannine Fiedler, Matthias Gubig, Petra Jentschke, Gisela Limburg, Regina Lindenlauf, Alexander Nagel, Ingrid Ott, Friederike Ottnad, Wolfgang Rasch, Hubert Riedel, Elke Rauer, Christian Scheffler, Christa Scheld, Erik Spiekermann, Eva Stein, Heike Straßburger, Jürgen Thiede and especially Christa Schwarzwälder.

The Authors

## Danksagung

Für die Anregung zu diesem Buch danken wir Peter Feierabend.
Beim Erstellen der Namensliste der Personen und Institutionen, die in diesem Buch enthalten sind, haben uns von Anfang an Ken Cato, Wim Crouwel, Olle Eksell, Adrian Frutiger, Takenobu Igarashi, Oswaldo Miranda, Morteza Momayez, Dan Reisinger, Paula Scher, Todor Vardjiev, Zdenêk Ziegler und Günter Gerhard Lange beraten. Ihnen sind wir zu großem Dank verpflichtet.
Einen besonderen Dank möchten wir Jürgen Siebert von FontShop und Otmar Hoefer von Linotype Library aussprechen. Sie stellten uns die Schriften für die Figurenverzeichnisse zur Verfügung, die uns CitySatz & Nagel, Berlin dankenswerterweise für den „Wer-Teil" aufbereitet haben.
Unser außerordentlicher Dank gebührt Sally Bald, Philipp Luidl und Wilhelm Schäfer.
Als Mitarbeiter möchten wir vor allem Ute Brüning für Recherche- und Registerarbeiten sowie Christine Berkenhoff, Claudia Grotefendt, Jennifer Neidhardt und Maximilian Meisse danken.
Mit Rat und Tat geholfen haben Annette Beil, Ben Buschfeld, Jan Carell, Stephanie Ehret, Jeannine Fiedler, Matthias Gubig, Petra Jentschke, Gisela Limburg, Regina Lindenlauf, Alexander Nagel, Ingrid Ott, Friederike Ottnad, Wolfgang Rasch, Hubert Riedel, Elke Rauer, Christian Scheffler, Christa Scheld, Erik Spiekermann, Eva Stein, Heike Straßburger, Jürgen Thiede und in besonderem Maße Christa Schwarzwälder.

Die Herausgeber

## Remerciements

Nous remercions Peter Feierabend qui a eu l'idée de réaliser ce livre.
Pour établir la liste des personnes et institutions citées dans cet ouvrage, nous nous sommes adressés dès le début à Ken Cato, Wim Crouwel, Olle Eksell, Adrian Frutiger, Takenobu Igarashi, Oswaldo Miranda, Morteza Momayez, Dan Reisinger, Paula Scher, Todor Vardjiev, Zdenêk Ziegler et Günter Gerhard Lange qui nous ont dispensé leurs conseils. Nous les en remercions vivement.
Nous remercions aussi Jürgen Siebert de FontShop et Otmar Hoefer de Linotype Library qui ont mis à notre disposition des polices et hypothèques que CitySatz & Nagel Berlin ont eu la gentillesse de traiter pour le chapitre «Qui».
Nous exprimons aussi notre gratitude à Sally Bald, Philipp Luidl et Wilhelm Schäfer.
Nous exprimons aussi notre reconnaissance à nos collaborateurs, surtout Ute Brüning pour ses travaux de recherche de classement, ainsi qu'à Christine Berkenhoff, Claudia Grotefendt, Jennifer Neidhardt et Maximilian Meisse.
Nous remercions aussi Annette Beil, Ben Buschfeld, Jan Carell, Stephanie Ehret, Jeannine Fiedler, Matthias Gubig, Petra Jentschke, Gisela Limburg, Regina Lindenlauf, Alexander Nagel, Ingrid Ott, Friederike Ottnad, Wolfgang Rasch, Hubert Riedel, Elke Rauer, Christian Scheffler, Christa Scheld, Erik Spiekermann, Eva Stein, Heike Straßburger et Jürgen Thiede pour leurs conseils et leur aide, et très particulièrement à Christa Schwarzwälder.

Les auteurs

## Typography – when who how

## Why?
## A Preface

In all stages of the historical development of the world, the written word and typography have been a fundamental ingredient in human culture. Only a few specialists are, however, familiar with their background and effect. Whereas the names of the innovators in the "classical disciplines", such as painting, music and literature, are common knowledge, those of type designers and typographers of all historical periods are largely unknown; only a few appreciate the significance of their work. Yet the effects of text and typography on all spheres of human activity are constantly present. They influence the fields of aesthetics and technology, of the arts and economics. Without them the rapid exchange of information we take for granted in our contemporary world would be inconceivable.

The written word is universally present and accepted. Its origins lie in symbols and pictures containing certain elements which recur time and again. These repeated elements have at least one meaning or message attached to them. We are no longer able to prove for certain what they originally stood for. Yet we know that the connection between a symbol and the information it imparted is based on cultural consensus. Just as in this sense there are no pure pictures there are also no pure messages in the symbols; there is always something left over. The tension which arises between the information and the image is the essence of typography. Typography is to the written word what articulation is to the spoken word. Neither is conceivable without the other; together they make up what we perceive.

The world's systems of writing developed over lengthy, but necessary periods of time. In the vari-

ous corners of the globe notation evolved as a way of preserving and explaining a cultural inheritance. The needs which called for and further developed such writing systems were always existential. That we have any knowledge at all about past epochs is due to the fact that the content and form of written records give us information as to the striving of past civilizations for knowledge. All aesthetic and social factors apart, being able to convert thought into language and language into writing was always a symbol of power. It also meant being able to influence the typography of a piece of information. The essence of this process remains largely unaltered although democratic developments and technological innovation today permit us to approach our typographic inheritance in a manner which is almost playful.

Over the course of time writing, type and typography have undergone many additions, reforms and refinements. Subtle differences in alphabetic symbols and typographic forms provide information as to the technical and aesthetic innovations of a particular era. They also show how little and how slowly this cultural asset has actually changed in the space of a thousand years. And how little it had to change. The strength of typography lies in the fact that it fires the imagination and creates identities through manifest application; texts appear modern or historical, serious or amusing, weighty or light. Usually, the typography matches the text so perfectly that on reading the words the typography goes unnoticed. Long periods of learning and practice were necessary to master this use of text and typography. Respected professions evolved, practiced by proud, clever writing masters, craftsmen and designers. They were constantly afraid that the medium should suffer a drop in quality. This fear was, however, (and still is) often the fear of any kind of modification being made to familiar, much-loved forms.

The 20th century brought change to all areas of art and culture. The legacy of past centuries was consciously forgotten to make way for the new. Art saw the transformation from representational to abstract painting. Unfamiliar images were greeted with vehement enthusiasm and rejection alike before they were accepted as an expression of a society undergoing radical change. This in turn changed ideas of harmony, form and proportion. Typography, which had changed little since Gutenberg and then only in conformity with a rigid pattern of rules, was also embraced by these new concepts. In the past it had been a steady medium which served reading and writing; now it suddenly began to move. Type designers increasingly departed from the purely functional capacities of their work and created typefaces which were no longer merely for reading but which existed for their own sake.

There is plenty of information on typography. Yet this is often widely dispersed, difficult to access or dogmatically biased. In order to illustrate as many past and present developments in typography as possible this book is divided into three sections: *When*, *Who* and *How*. The *When* section starts in the present and works back to the origins of writing via the stylistic periods we have experienced in their initial stages. The *Who* section names the people and institutions in alphabetical order who have contributed to typography as designers and inven-

tors, teachers and calligraphers, entrepreneurs and theorists, publishers and artists, and schools and archives. The third section explores *How* text and typography have been produced, with which tools and aids. The relationships which make up the field of typography and which are necessary for the understanding of its past and present are made clear.

This volume is not a typography textbook. It is the beginning of a reference work which demonstrates the interaction between craft and technology, theory and practice, between functional and experimental ideas. By viewing the subject in its entirety it perceives typography as an important integral part of our culture. This book can be consulted during future typographic work and help to perceive type and typography more actively and understand them more consciously. Yet as the scope of this subject is so enormous, here we can only speak of the beginning of a beginning.

Friedrich Friedl
Nicolaus Ott
Bernard Stein

## Epilogue

Over the past decades typography has crept more and more into the consciousness of a public which asks how the things we take for granted and surround ourselves with come about. For many, typography was and is synonymous with writing and is at best the set of conditions which govern (or should govern) the use of writing and type.

The aim of the authors is to create a distinction here. To this end we began sorting existing material on writing and typography in accordance with the title of the book. In doing so we came to the painful realization that any form of order would mean omitting certain things and that after years of work something must be brought to a close which in reality can never be finished. Our beginning still leaves much to be done. This affects the When section with regard to further examples of writing and type from non-European cultures; the How section with regard to the rules of technique, classification and design, for which Philipp Luidl has already written the basic texts; and the Who section with regard to all those our limitations did not allow us to include. Together with the publishers, for whose constant support we are extremely grateful, we apologize for any shortcomings and exclusions in this first edition of *Typography – When Who How.*

## Typographie – wann wer wie

### Warum?<br>Ein Vorwort

In allen Epochen der geschichtlichen Weltentwicklung waren und sind Schrift und Typographie ein wesentlicher Teil der Kultur. Ihre Hintergründe und Auswirkungen sind jedoch nur wenigen Spezialisten bekannt, Zusammenhänge bleiben unklar. Während die Namen der Neuerer aus den ‚klassischen Disziplinen' wie Malerei, Musik und Literatur zum Allgemeinwissen gehören, sind die Namen der Schriftentwerfer und Typographen aller Epochen weitgehend unbekannt, und nur wenige wissen von der Wichtigkeit ihrer Werke. Dabei sind die Auswirkungen von Schrift und Typographie auf allen Gebieten menschlicher Betätigung stets präsent. Sie liegen im Ästhetischen und Technischen genauso wie im Kulturellen und Wirtschaftlichen. Ohne Schrift und Typographie wäre der rasante Austausch von Informationen, der uns heute selbstverständlich erscheint, nicht denkbar.

Die Verbreitung und Akzeptanz von Schrift ist universell. Ihr Ursprung liegt in Zeichen und Bildern, die immer wiederkehrende Elemente haben. Ihnen ist jeweils mindestens eine Bedeutung bzw. Mitteilung zugeordnet. Diese ursprüngliche Zuordnung ist für uns nicht mehr eindeutig beweisbar. Aber wir wissen, daß die Zusammengehörigkeit eines Zeichens und seiner Mitteilung auf einer kulturellen Vereinbarung beruht. Ebenso wie es in diesem Sinn keine reinen Bilder gibt, gibt es für die Schriftzeichen auch keine reinen Mitteilungen: Immer bleibt etwas übrig. Die Spannung, die zwischen Mitteilung und Bild entsteht, umschreibt das Wesen der Typographie. Die Typographie verhält sich zur Schrift wie die Artikulation zur Sprache. Beide sind nicht ohne einander vorstellbar und ergeben das, was wir wahrnehmen.

Die Entwicklung der Zeichensysteme vollzog sich in langwierigen, aber notwendigen Zeiträumen. In verschiedenen Regionen der Erde entstanden Zeichensysteme zur Bewahrung und Erklärung eines kulturellen Erbes. Immer waren es existenzielle Bedürfnisse, die solche Zeichensysteme erforderten und weiterentwickelten. Daß wir von vergangenen Epochen heute noch Kenntnis haben, liegt daran, daß Inhalt und Form schriftlicher Überlieferungen uns über das damalige Streben nach Erkenntnis Aufschluß geben. Neben allen ästhetischen und gesellschaftlichen Faktoren war es immer ein Zeichen der Macht, Gedanken in Sprache und Sprache in Schrift umsetzen zu können. Es bedeutete auch, eine Mitteilung typographisch mitbestimmen zu können. Am Wesen dieses Vorgangs hat sich kaum etwas geändert, auch wenn demokratische Entwicklungen und technische Neuerungen uns heute erlauben, beinahe spielerisch an unser typographisches Erbe heranzugehen.

Schrift und Typographie haben im Laufe der Zeit zahlreiche Ergänzungen, Neuerungen und Verfeinerungen erfahren. Subtile Differenzierungen an den alphabetischen Zeichen und typographischen Formen geben über die technischen und ästhetischen Neuerungen einer Epoche Auskunft. Sie zeigen auch, wie langsam und gering sich dieses Kulturgut in tausend Jahren eigentlich verändert hat und wie wenig es sich verändern müßte. Die Kraft der Typographie liegt darin, daß durch ihre sinnfällige Anwendung Imaginationen und Identitäten entstehen: Texte erscheinen historisch oder gegenwärtig, ernst oder lustig, schwer oder leicht. Meist ist die typographische Entsprechung eines Textes so stimmig, daß man sie beim Lesen gar nicht bemerkt.

Um den Umgang mit Schrift und Typographie beherrschen zu können, waren lange Lehr- und Übungszeiten notwendig. Sie führten zu geachteten Berufen, den stolze und kluge Schreibmeister, Handwerker und Gestalter ausgeübt haben. Ihnen war stets die Angst vor dem Qualitätsverfall des Mediums gegenwärtig. Diese Angst war und ist jedoch oftmals die Angst vor jedweder Veränderung liebgewonnener Formen.

Das 20. Jahrhundert brachte Veränderungen – auf allen Gebieten der Kunst und Kultur gleichermaßen – mit sich. Das Erbe vergangener Jahrhunderte wurde bewußt verlernt, um Neuem Platz zu machen. In der freien Kunst fand der Wechsel von gegenständlicher zur abstrakten Malerei statt. Bisher unbekannte Bildinhalte wurden gleichzeitig vehement begrüßt und abgelehnt, um schließlich als Ausdruck eines gesellschaftlichen Umbruchs akzeptiert zu werden. Mit ihm veränderten sich ebenfalls die Vorstellung von Harmonie, Form und Proportion. Auch die Typographie, die sich seit Gutenberg nur wenig und innerhalb eines festen Regelwerks verändert hatte, wurde von den Neuerungen erfaßt. War sie bis dahin ein ruhiges Medium, das dem Lesen und Schreiben diente, kam sie nun in Bewegung. Schriftentwerfer verabschiedeten sich mehr und mehr von ihrer rein zweckgebundenen Arbeit und kreierten Schriften, die nicht mehr nur gelesen werden sollten, sondern nun um ihrer selbst willen existierten.

Über Typographie gibt es reichhaltige Informationen. Diese sind jedoch häufig weit verstreut, schwer zugänglich oder dogmatisch gefärbt. Um eine möglichst große Zahl der vergangenen und gegenwärtigen Entwicklungen in der Typographie zu zeigen,

ist dieses Buch in die drei Teile *Wann, Wer* und *Wie* gegliedert. Der „Wann"-Teil beginnt in der Gegenwart und führt über die Stile, die wir noch in ihren Anfängen miterlebt haben, zurück zu den Ursprüngen der Schrift. Der „Wer"-Teil nennt in alphabetischer Reihenfolge die Personen und Institutionen, die als Entwerfer und Erfinder, als Lehrer und Kalligraphen, als Unternehmer und Theoretiker, als Verleger und Künstler, als Schulen und Archive ihren Beitrag zur Typographie geleistet haben. Der dritte Teil befaßt sich damit, „Wie" Schrift und Typographie erzeugt wurden und werden, mit welchen Hilfsmitteln und Werkzeugen. Die Zusammenhänge werden deutlich, die den Bereich der Typographie ausmachen, und die zum Verstehen ihrer Geschichte und Gegenwart gehören.

Das vorliegende Buch ist kein Lehrbuch der Typographie. Es ist der Beginn eines Nachschlagewerks, das die Wechselwirkungen zwischen Handwerk und Technik, Theorie und Praxis, zwischen dienender und experimentierender Auffassung zeigt. In seiner ganzheitlichen Betrachtungsweise nimmt es Typographie als wichtigen Bestandteil unserer Kultur wahr. Es kann bei künftigen typographischen Arbeiten zu Rate gezogen werden und dazu verhelfen, Schrift und Typographie offensiver wahrzunehmen und bewußter zu verstehen. Wegen des großen Umfangs des Themas sollten wir jedoch vom Beginn eines Beginns sprechen.

Friedrich Friedl
Nicolaus Ott
Bernard Stein

Ein Nachwort

Typographie ist in den vergangenen Jahrzehnten immer stärker in das Bewußtsein einer Öffentlichkeit gerückt, die danach fragt, wie die Dinge entstanden sind, mit denen wir uns wie selbstverständlich umgeben. Für viele war und ist Typographie gleichbedeutend mit Schrift und handelt im besten Fall von den Bedingungen, unter denen Schrift Verwendung findet oder finden sollte.

Die Absicht der Herausgeber ist es, hier eine Differenzierung vorzunehmen. Aus diesem Grund haben wir begonnen, vorhandenes Schrift- und Typographiematerial – im Sinne des Buchtitels – zu ordnen. Dabei ist uns schmerzlich bewußt geworden, daß jedes Ordnen ein Weglassen beinhaltet, und daß nach jahrelanger Arbeit etwas beendet werden mußte, was im eigentlichen Sinn nie fertig sein kann. Vieles bleibt nach diesem Anfang noch zu tun. Das gilt im „Wann-Teil" für die Schriftbeispiele von weiteren außereuropäischen Kulturen, im „Wie-Teil" für das Regelwerk von Technik, Klassifikation und Gestaltung, für das Philipp Luidl uns bereits die grundlegenden Texte erarbeitet hat, und im „Wer-Teil" für all die Personen, die aus unseren Begrenztheiten heraus nicht berücksichtigt wurden. Zusammen mit dem Verlag, für dessen andauernde Unterstützung wir sehr dankbar sind, bitten wir um Ihr Verständnis für Mängel in dieser ersten Auflage von *Typographie – Wann Wer Wie.*

## Typographie – quand qui comment

## Pourquoi ?
## Un avant propos

A toutes les périodes de l'Histoire, l'écriture et la typographie ont été et sont un aspect essentiel de la culture. Toutefois, seuls quelques rares spécialistes en connaissent les origines et les effets; et le contexte de leur apparition est souvent peu clair. Alors que les noms des novateurs des disciplines artistiques classiques comme la peinture, la musique et la littérature sont entrés dans le patrimoine des connaissances générales, à toutes les époques, les noms des concepteurs de polices et typographes sont souvent tombés dans l'oubli; et seules quelques rares personnes connaissent l'importance de leur oeuvre. Pourtant l'influence de l'écriture et de la typographie est constamment présente dans tous les domaines de l'activité humaine. Elle intervient au niveau esthétique et technique tout comme sans les secteurs culturels et économiques. Sans l'écriture et la typographie les échanges rapides d'informations qui nous semblent naturels aujourd'hui, ne seraient pas pensables.

L'écriture est universellement répandue et adoptée. Elle puise son origine dans des signes et images qui présentent des traits récurrents. A chacun d'eux est attribué au moins un sens, ou un message. Il n'est plus guère possible pour nous de fournir une démonstration sans équivoque de leur signification d'origine. Mais nous savons que l'association entre le signe et le contenu qu'il véhicule repose sur des conventions culturelles. Tout comme il n'existe en ce sens, aucune image à l'état pur, il n'existe aucun message univoque transmis par un signe écrit; il reste toujours une inconnue. La tension qui naît du rapport entre le message et l'image définit la typographie. Le lien qu'elle entretient avec l'écriture est le même que celui de l'articulation avec le langage. L'une est inimaginable sans l'autre, et toutes deux ensemble produisent ce que nous percevons.

Les systèmes de signes ont évolué sur des périodes d'une longueur nécessaire. Plusieurs systèmes sont apparus dans diverses régions du globe dans le but de conserver et de transmettre un héritage culturel. Ce furent toujours des besoins existentiels qui amenèrent à la formation et au développement de ces systèmes de signes. Si de nos jours nous avons encore connaissance des époques révolues, c'est parce que le contenu et la forme des textes écrits qui nous sont parvenus, nous éclairent sur les aspirations au savoir qui caractérisaient ces périodes. Parallèlement à tous les facteurs esthétiques et sociaux, pouvoir traduire sa pensée en langage et le langage en texte a toujours été un facteur de pouvoir. Ceci signifie aussi que la typographie codétermine le message. Or rien n'a changé dans la nature du processus, même si le développement de la démocratie et les innovations techniques nous permettent aujourd'hui d'aborder notre héritage typographique d'une manière presque ludique.

Avec le temps, l'écriture et la typographie ont subi de nombreux ajouts et des innovations, elles se sont affinées. De subtiles différences dans les signes de l'alphabet et dans les formes typographiques nous renseignent sur les progrès techniques et esthétiques de chaque époque. Ils montrent aussi que ce patrimoine culturel n'a subi que de lents et moindres changements en mille ans, et qu'au fond il ne lui suffisait que de quelques transformations. L'utilisation intelligente de la typographie est source d'imagination et d'identité, et fait sa force. Elle signale les caractéristiques du texte : contemporain

ou historique, sérieux ou drôle, difficile ou facile. Le plus souvent la concordance entre la typographie et le texte est telle qu'on ne la remarque même pas à la lecture. Pour maîtriser le maniement de l'écriture et de la typographie, il fallait un long apprentissage et beaucoup de pratique. Elle donna naissance à des métiers nobles exercés par des scribes, artisans et designers intelligents et fiers de leur rôle. La crainte de voir ce médium décliner en qualité était toujours présente à leurs yeux. Mais cette crainte correspondait souvent à la peur de voir toutes sortes de modifications apportées à des formes que l'on avait appris à aimer.

Le vingtième siècle fut celui des changements – dans tous les domaines, autant artistiques que culturels. L'héritage des siècles passés fut délibérément laissé de côté pour faire place neuve. Dans les arts libéraux, ce changement se manifesta par le passage de la peinture figurative à la peinture abstraite. Des thèmes picturaux jamais vus jusqu'alors rencontrèrent une adhésion aussi passionnée que la réprobation, avant qu'on les accepte comme étant l'expression d'une rupture. Parallèlement à cela, l'idée qu'on se faisait de l'harmonie, de la forme et des proportions se modifiait. La typographie qui, depuis Gutenberg, n'avait que peu évolué, et était restée dans le cadre de règles fixes, connut aussi un renouveau. Si jusqu'alors, elle était un médium paisible qui servait à lire et à écrire, elle se mit elle aussi en mouvement. Les concepteurs de polices se démarquaient d'un labeur purement pragmatique pour créer des alphabets dont le but n'était plus seulement d'être lus, mais qui désormais, voulaient exister par eux mêmes.

Nous disposons d'une foule d'informations sur la typographie. Pourtant celles-ci sont souvent dispersées, difficilement accessibles ou teintées de dogmatisme. Afin de présenter le plus grand nombre d'aspects des changements survenus en typographie dans le passé comme de nos jours, nous avons divisé cet ouvrage en trois parties: *Quand, Qui* et *Comment.* Le chapitre consacré au «Quand» commence par la période contemporaine et les styles que nous avons connus à leurs débuts, pour remonter le temps jusqu'aux origines de l'écriture. Le chapitre «Qui» présente en ordre alphabétique les hommes, femmes et institutions qui ont contribué au développement de la typographie en qualité de concepteurs et d'inventeurs, d'enseignants et calligraphes, de chef d'entreprises et théoriciens, d'éditeurs et artistes, d'écoles et archives. Le troisième chapitre traite du «Comment» l'homme produisait et produit écriture et typographie, avec quels moyens et quels outils. Ceci nous permettra peut-être de mieux aborder les contextes qui définissent le domaine de la typographie et qui contribuent à nous faire comprendre son histoire et sa présence.

Ce livre n'est pas un manuel de typographie. C'est la première pierre d'un ouvrage de référence qui ex-

pose les interactions entre artisanat et technologies, théories et pratique, entre les conceptions utilitaires et expérimentales. Vu globalement, il traite la typographie comme une des composantes majeures de notre culture. Il pourra être consulté dans la perspective de nouveaux travaux typographiques et contribuer à appréhender et à comprendre écriture et typographie de manière plus offensive et plus consciente. Compte tenu de l'étendue du sujet, ceci ne saurait être qu'un premier pas.

Friedrich Friedl
Nicolaus Ott
Bernard Stein

Postface

Au cours des dernières décennies, le public a pris de plus en plus fortement conscience de la typographie; il se demande comment sont nées ces choses que nous manions très naturellement. Pour bon nombre, la typographie était et est assimilée à l'écriture, et ne concerne, dans le meilleur des cas, que les conditions dans lesquelles on l'utilise ou doit l'utiliser.

L'intention des auteurs de cet ouvrage est ici de différencier. C'est pourquoi nous avons commencé par classer les documents écrits et typographiques existants en fonction du titre de cet ouvrage. Ce faisant, nous avons constaté à notre grand regret que toute classification entraînait des choix, donc des absences, et qu'après plusieurs années de recherche, il fallait mettre un terme à un travail qui, en vérité, ne prendra jamais fin. Après ce début, beaucoup reste à faire. Ceci vaut pour le chapitre «Quand» et les écritures de certaines cultures extérieures à l'Europe, pour le chapitre «Comment» et le répertoire des techniques, classifications et mises en forme, pour lequel Philipp Luidl nous a fourni les textes fondamentaux, ainsi que pour le chapitre «Qui», où, compte tenu du caractère limitatif de cet ouvrage, bon nombre de noms ont été laissés de côté. De concert avec la maison d'édition, que nous remercions pour le soutien constant qu'elle nous a apporté, nous vous prions de bien vouloir nous pardonner les lacunes que comporte ce premier volume de *Typographie – Quand Qui Comment.*

wann

quand

when

| The Chronology | Die Chronologie | La Chronologie |
|---|---|---|
| Multistylistic typography | Multistilistische Typographie | Typographie multistyles |
| Encyclopedic typography | Enzyklopädische Typographie | Typographie encyclopédique |
| New Wave | New Wave | New Wave |
| New Functionalism | Neuer Funktionalismus | Nouveau fonctionnalisme |
| Punk Typography | Punk Typographie | Typographie Punk |
| Pop and Psychedelia | Pop und Psychedelic | Pop et psychédélique |
| Experimental typography | Experimentelle Typographie | Typographie expérimentale |
| Fluxus | Fluxus | Fluxus |
| Functional typography | Sachlich-funktionale Typographie | Typographie objective fonctionnelle |
| Pictorial typography | Bildhafte Typographie | Typographie imagée |
| International typographic style | Internationaler typographischer Stil | Style typographique international |
| Traditional typography | Traditionsverbundene Typographie | Typographie traditionaliste |
| "elementare typographie" | Elementare Typographie | Typographie élémentaire |
| Individual written form | Individuelle skripturale Form | Formes scripturales individuelles |
| Art Deco | Art Deco | Art Deco |
| Expressionism | Expressionismus | Expressionnisme |
| Constructivism | Konstruktivismus | Constructivisme |
| Dada | Dadaismus | Dadaïsme |
| Futurism | Futurismus | Futurisme |
| Informative Functionalism | Informative Sachlichkeit | Objectivité informative |
| Art Nouveau | Jugendstil | Art nouveau |
| The Arts and Crafts Movement | Arts and Crafts | Arts & Crafts |
| Historicism | Historismus | Eclectisme |
| Classicism | Klassizismus | Néo-classicisme |
| Baroque | Barock | Baroque |
| Renaissance | Renaissance | Renaissance |
| The Gothic period | Gotik | Gothique |
| The Middle Ages and Romanesque | Mittelalter und Romanik | Le moyen-âge et la période romane |
| Rome and the Early Middle Ages | Rom und frühes Mittelalter | Rome et le haut moyen-âge |
| Phoenicia and Greece | Phönizien und Griechenland | Phéniciens et Grèce |
| Islam | Islam | Islam |
| China and the Indus Valley | China und Indus-Tal | Chine et Vallée de l'Indus |
| Egypt | Ägypten | Egypte |
| Mesopotamia/Sumer | Mesopotamien/Sumer | Mésopotamie/Sumer |

**Multistilistische Typographie**
**Typographie multistyles**
**Multistylistic typography**

Immer wieder erscheint der jeweils neueste gestalterische Ausdruck einer Zeit als ultimativ und nicht überbietbar; um bald wieder von neuen Überraschungen verdrängt zu werden. Aber kein gestalterischer Stil ist jemals zu Ende, immer wieder können richtungsweisende Impulse von vergangenen Formen ausgehen und zu anderen Qualitäten führen. Dehalb kann die beispielhafte Darstellung der Reichhaltigkeit in der Entwicklung der Typographie Voraussetzungen zum Verständnis des Vergangenen und Zukünftigen schaffen. Ihr Repertoire ist die Basis für eine neue praktische Typographie.
Um 1990 traten Schrift und Typographie mit der Einführung der Computer in die Welt der Gestaltung in eine neue Phase. Die neuen Apparate veränderten den gesamten Gestaltungsprozeß. Wurde früher nach einer Skizze der Satz bestellt und das Layout geklebt, war auf dem Bildschirm jetzt jeder Entwurfseinfall in einer Qualität zu sehen, die dem gedruckten Ergebnis sehr ähnlich war. So konnten Gestaltungsschritte schnell verworfen oder akzeptiert, konnten Ausdrucke im Originalformat gefertigt und Druckvorlagen ohne Zeitverlust hergestellt werden. Was bisher Spezialisten machten, erledigte jetzt der Gestalter mit Computer und Programmen selbst.
Zunächst war DTP für typographische Laien interessant, später auch für Profi-Gestalter. Denn schnell wurden die Programme verbessert, um Detailanforderungen erfüllen zu können. Die fast spielerische Herangehensweise an das Entwerfen war in den Händen von Fachleu-

A chaque époque, les moyens d'expression et de mise en forme les plus récents semblent avec un caractère ultime et insurpassable; et chaque fois, de nouvelles surprises les relèguent à l'arrière plan. Pourtant aucun style formel ne disparaît jamais complètement; sans cesse de nouvelles impulsions s'inspirant de formes passées ouvrent de nouvelles perspectives et conduisent à de nouvelles qualités. C'est pourquoi représenter de façon exemplaire la diversité de l'évolution de la typographie est la condition préalable pour mieux comprendre le passé et le futur. Son répertoire constitue une base pour la nouvelle pratique de la typographie.
Vers 1990, avec l'avènement de l'ordinateur, l'écrit et la typographie entrent dans une ère nouvelle. Les nouveaux outils transforment l'ensemble du processus de mise en forme. Si autrefois, le texte était composé d'après un schéma et la mise en page montée à la colle, désormais toutes les idées de maquette peuvent être visualisées sur l'écran avec une qualité presque égale à celle du résultat imprimé. Ainsi peut-on adopter ou rejeter rapidement une mise en page. Ceci permet de réaliser des impressions en format original et de fabriquer des épreuves sans perte de temps. Le maquettiste équipé d'un ordinateur et de logiciels réalise désormais les tâches des spécialistes d'autrefois, les typographes, les réprophotographes et les lithographes.
La DTP a d'abord attiré les néophytes; puis les maquettistes professionnels s'y sont intéressés à cause de ses nouvelles possibilités. Les logiciels se sont alors rapidement perfectionnés pour répondre aux exigences de détails. Entre les mains des spécialistes, cette manière presque ludique d'aborder la maquette évolue dans un sens positif. Mais le fait de pouvoir ac-

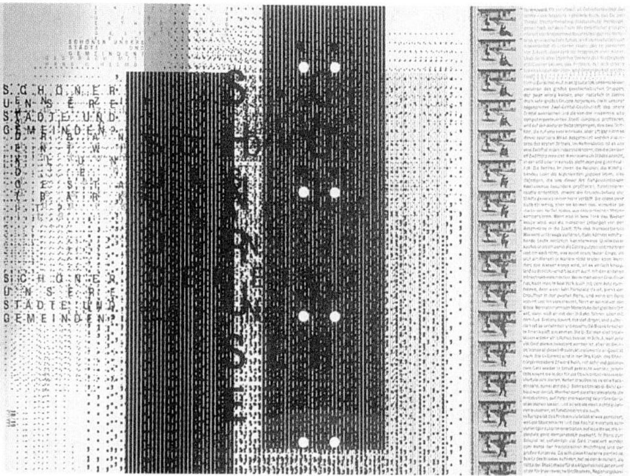

Cyan

Venezky *Speak*

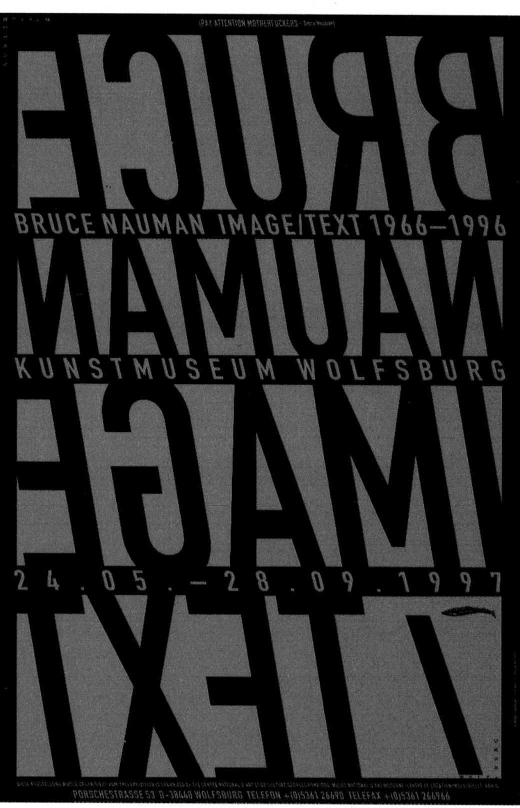
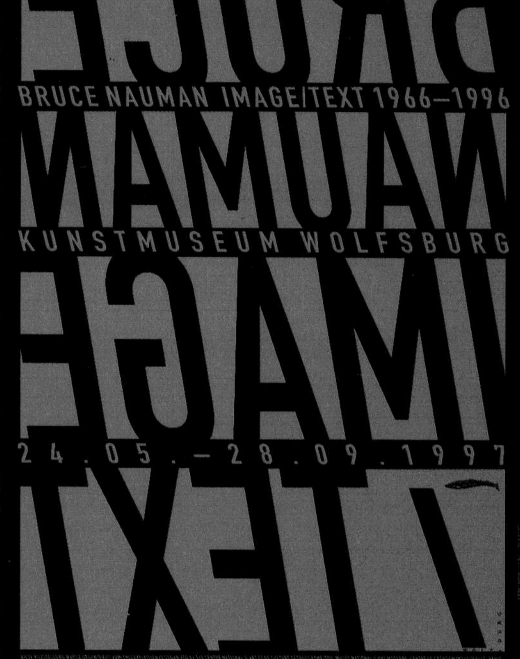


Hadders

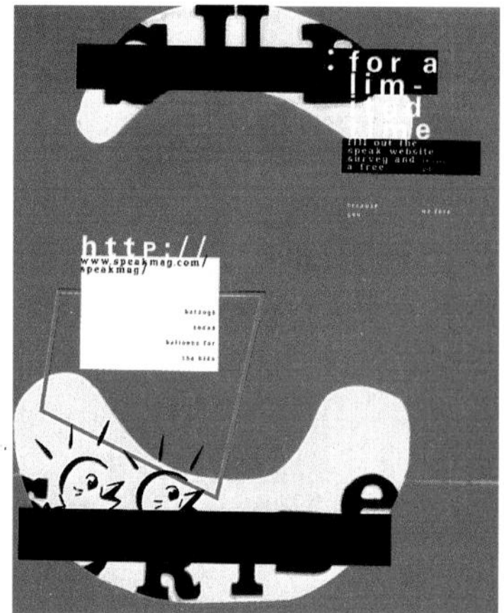


Venezky *Speak*

Time and again the most recent artistic expression of an age appears ultimate and unsurpassable – only to soon be superseded by new surprises. Yet no artistic style is ever at an end; new impetus can always be provided by past forms, pointing the way and leading to other qualities. Thus the exemplary presentation of variety in the development of typography creates conditions under which the past and the future are understood. Its repertoire is the basis for a new, practical typography.
Around 1990 type and typography entered a new phase with the introduction of the computer. This new machine altered the entire process of design. Whereas type matter used to be ordered and layouts pasted according to sketches, it was now possible to view each idea in the design process on the monitor in a quality representative of the final printed result. Stages in design could now be speedily accepted or discarded, print-outs could be made in the original format and printer's copies produced without loss of time. The work of specialists such as setters, repro photographers or lithographers could now be done by the designer himself with computers and appropriate software.
DTP was initially of greater interest to amateur typographers, yet its wealth of new possibilities soon began to appeal to the professionals. The programs were rapidly improved to meet more detailed requirements. The almost playful methods of design underwent a positive development at the hands of the experts. Yet the fact that any user was able to access

Multistilistische Typographie
Typographie multistyles
Multistylistic typography

ten eine positive Entwicklung. Aber der Zugriff auf das gesamte typographische Repertoire, auf die weitschweifigsten Modifikationsmöglichkeiten der Schrift durch jedweden Anwender ließ ein typographisches Horrorszenarium entstehen, das ähnliche Probleme aus der Zeit der Einführung des Fotosatzes in den Schatten stellte. Schriften, die in sorgfältigsten Prozessen mühsam entworfen und hergestellt wurden, konnten durch einfachste Handgriffe in die unwahrscheinlichsten Formen und Schnitte verwandelt werden. Ein enormer wirtschaftlicher Faktor war, daß Originalschriften einfach kopiert werden konnten, oder mit kleinsten Veränderungen und neuen Namen selbst produziert wurden. Firmen mit langer Tradition und großen Verdiensten um die Qualität der typographischen Kultur kamen in Schwierigkeiten oder verschwanden ganz, während Billiganbieter boomten.

Die DTP-Typographie brachte frische formale und inhaltliche Ansätze. Texte wurden experimentierfreudiger strukturiert und in ungewöhnlichen Schriftmischungen gesetzt. Programme wie Freehand, PageMaker, QuarkXPress führten zu formaler Offenheit. Die zunächst holprig und ungekonnt aussehenden Zacken, die an Buchstaben durch die Digitalisierung entstanden, wurden durch die Verfeinerung des Rasterstandards behoben, aber als gestalterisches Element sogar oft dominant eingesetzt. DTP förderte, außer bei technischer Dokumentation, nicht die funktionale, sondern die emotionale, bildhafte und ornamentale Typographie.

Designers Republic

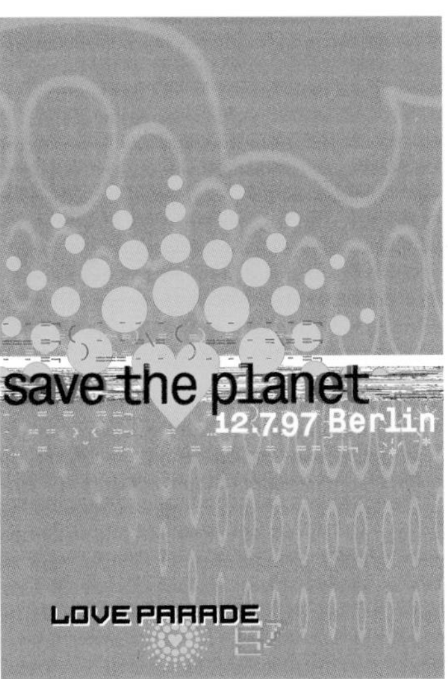

xplicit

Oliver

Glauber

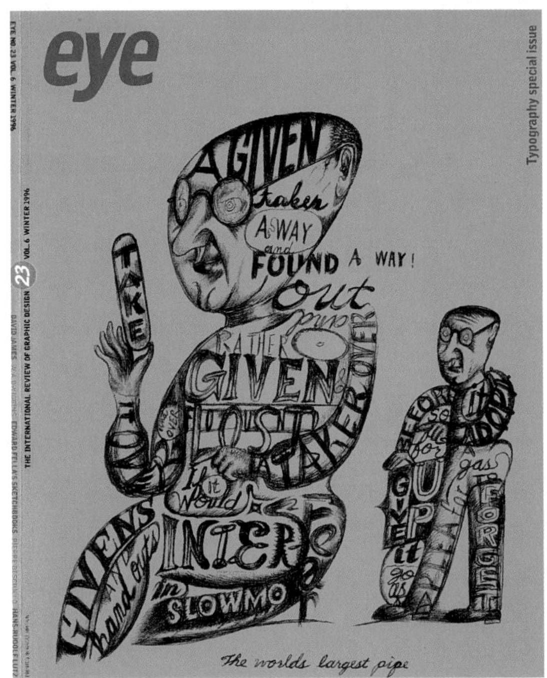


Fella *Eye*

céder à l'ensemble du répertoire typographique, et que n'importe quel utilisateur soit en mesure de modifier une police de la manière la plus extravagante, suscite des visions d'horreur, qui relèguent à l'arrière-plan les problèmes rencontrés au moment de l'instauration de la photocomposition. Les caractères qui, jadis, étaient soigneusement élaborés et fabriqués au cours de processus laborieux prennent les formes et les tailles les plus invraisemblables en un simple tour de main. Un facteur économique majeur intervient alors : dorénavant, les fontes originales sont faciles à copier ou à produire soi-même avec quelques modifications infimes et un nouveau nom. Les sociétés possédant une longue tradition et un grand mérite culturel et typographique connaissent des difficultés ou disparaissent tandis que la concurrence bon marché prospère.

La typographie DTP apporte de la fraîcheur dans les conceptions formelles. Les textes se structurent en expérimentant et sont composés au moyen de mélanges inhabituels de polices. Des logiciels comme le Freehand, PageMaker, QuarkXPress permettent une ouverture formelle. Les dentelures maladroites et irrégulières qui apparaissent sur les lettres en raison du traitement numérique, sont éliminées en affinant les trames, ou parfois même utilisées comme élément formel dominant. Hormis pour les documentations techniques, la DTP ne favorise pas le fonctionnalisme, mais une typographie émotionnelle, imagée et ornementale.

the entire typographical repertoire and undertake the most elaborate modifications to typefaces, resulted in a typographic horror scenario, overshadowing the similar problems experienced with the introduction of filmsetting. Fonts which had been laboriously designed and manufactured could be transformed into the most incredible forms and designs with speed and ease. An enormous economic factor was that original typefaces could be copied effortlessly, or "new" fonts manufactured at home with minimal alteration and a new name. Companies with a long tradition, whose quality typography had guaranteed their financial security, were plunged into crisis and in some cases vanished completely, while cheap suppliers boomed.

DTP typography offered fresh departures in form and content. Texts were experimented with and set with unusual mixtures of type. Formal openness was achieved with software such as Freehand, PageMaker and Quark XPress. Clumsy, inept-looking excrescences which appeared on letters through digitalization soon disappeared as grid systems were refined, although they were often used intentionally as design features. Except in technical documentation, DTP supported not functional but emotional, pictorial and ornamental typography.

## Enzyklopädische Typographie
## Typographie encyclopédique
## Encyclopedic typography

**Um 1985** Die zahlreichen stilistischen Nuancen unseres Jahrhunderts wurden alle mit großem Ernst vertreten und ausgearbeitet, aber immer wieder waren sie nach relativ kurzer Zeit überholt. Diese rasche Stilfolge führte zu immer kühneren Formvorstellungen.
Ausgelöst durch eine große publizistische Tätigkeit weltweit entstand bei interessierten Entwerfern eine umfangreiche Kenntnis der gestalterischen Entwicklung. Selbst die entlegensten Experimente und die außergewöhnlichsten Mittel wurden für die eigene Vorstellung und die praktische Arbeit verfügbar.
Es entstand eine nie geahnte Reichhaltigkeit des Ausdrucks und höchste Qualität der artifiziellen Formen. Die Gestaltung unterlag keinen festen Regeln, aber es gab auch keine einfachen Lösungen: Gebrauchsgrafik und Typographie waren endlich ein nicht endendes visuelles Abenteuer geworden, wie es undogmatische Freigeister schon immer wünschten. Obwohl der funktionale Ablauf der zu kommunizierenden Inhalte nicht immer klar nachzuvollziehen war, löste die Akzeptanz der neuen Formen, vor allem bei der jungen Generation, auch dieses Problem.
Der enzyklopädische Stil verband und zitierte Gestaltungselemente aus den unversöhnlichsten Epochen: wurde früher immer nach „reinen" Lösungen gesucht, wurde jetzt selbst das Unmögliche und der „schlechte" Geschmack herangezogen. Konstruktive Schriften wurden mit Fraktur, expressiven oder handschriftlichen Zeichen kombiniert. Ein ernstzu-

**Vers 1985** Bien que les innombrables nuances stylistiques de notre siècle aient été défendues et élaborées avec un grand sérieux, toutes sont devenues obsolètes en relativement peu de temps. La succession rapide des tendances conduit à des idées de plus en plus audacieuses.
L'énorme activité de publication amène les créateurs intéressés à mieux connaître l'évolution des formes. Les expériences marginales et les moyens les plus extraordinaires peuvent désormais être mis au service des idées de chacun et du travail pratique.
On voit apparaître une richesse insoupçonnée de modes d'expression et de formes artificielles d'une excellente qualité. Aucune recette ne semble trop difficile pour la création, mais par ailleurs, les explications simples n'existent plus. Le graphisme utilitaire et la typographie sont enfin devenus une aventure visuelle infinie, telle que l'avaient toujours souhaitée les esprits libres et non dogmatiques. Bien que les contenus devant être véhiculés ne soient pas toujours reconstituables de manière transparente, l'acceptation des nouvelles formes résout le problème, surtout dans la jeune génération.
Le style encyclopédique allie des éléments de mise en forme datant d'époques révolues tout en les citant. Si autrefois, on cherchait toujours des solutions «pures», dorénavant, on tire parti de l'impossible et même du «mauvais goût». Les caractères construits se combinent avec les gothiques et même avec des signes manuscrits. Les polices hybrides qui contiennent des alphabets entiers de sans sérifs et d'égyptiennes sont un produit intellectuel de cette époque qui doi être pris au sérieux. Dans ce contexte, la Stone de S. Stone, la Rotis de O. Aicher et la Thesis de L. de Groot sont essentielles.

abcdefghijklmnopqrstuvwxyz[äöüßåøæœç]
ABCDEFGHIJKLMNOPQRSTUVWXYZ
1234567890(.,;:?!$&-*){ÄÖÜÅØÆŒÇ}

abcdefghijklmnopqrstuvwxyz[äöüßåøæœç]
ABCDEFGHIJKLMNOPQRSTUVWXYZ
1234567890(.,;:?!$&-*){ÄÖÜÅØÆŒÇ}

abcdefghijklmnopqrstuvwxyz[äöüßåøæœç]
ABCDEFGHIJKLMNOPQRSTUVWXYZ
1234567890(.,;:?!$&-*){ÄÖÜÅØÆŒÇ}

Aicher

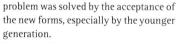

Wild Plakken

Barnbrook

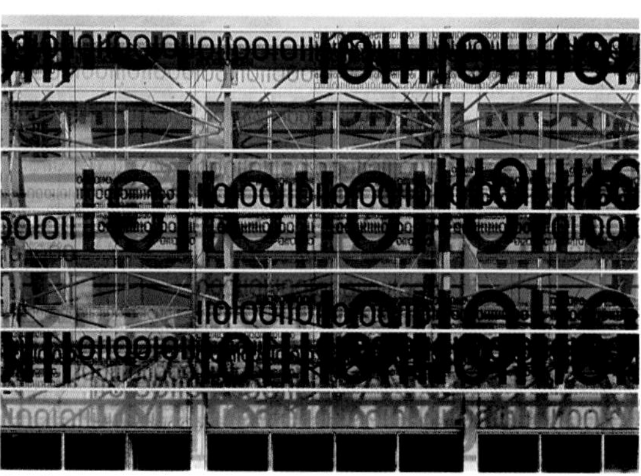

Baur, Kubiny

Tomato

**c. 1985** The many stylistic nuances of our century were all represented and developed with the utmost seriousness, yet were always overtaken within a relatively short time. This rapid succession of styles resulted in the creation of ever-bolder formal concepts.
Inspired by fervent journalistic activity throughout the world, interested designers gradually built up a comprehensive knowledge of developments in the field of design. They made use of even the oddest experiments and the most unusual materials for their own ideas and practical work.
The result of this was an unforeseen wealth of expression and inventiveness of the highest quality. Designers were unencumbered by any fixed formulas, but simple solutions were also few and far between: applied graphics and typography had finally become a never-ending visual adventure, just what undogmatic free spirits had always longed for. Although the function of the contents to be conveyed was not always easy to follow, this problem was solved by the acceptance of the new forms, especially by the younger generation.
Encyclopedic typography united and quoted elements of design from the most irreconcilable of epochs: whereas in the past, solutions had been chosen for their "purity", now even the impossible and "bad" taste were acceptable. Constructivist typefaces were combined with black letters, expressive or script characters. A major product of this period was and is a series of hybrid fonts, which offered

nehmendes Gedankenprodukt dieser Zeit waren und sind die Hybridschriften, die komplette Alphabete von Serifenloser bis Egyptienne anboten. Wichtig waren hier die Stone von S. Stone, die Rotis von O. Aicher und die Thesis von L. de Groot.
Die verwendeten Farbphantasien gingen weit über die Anwendung der reinen Grundfarben hinaus. Durch Farbdissonanzen und Kitschsymbolik entstanden bizarre Formen und ungezügelte Prächtigkeit. In dieser Zeit entstand auch die Vorstellung von der Autonomie des angewandten Gestalters. Man stellte sich vor, die dienende Funktion im Kommunikationsprozeß zwischen Sender und Empfänger könnte zu Ende sein, der Gestalter selbst sei der Sender. Diese Auffassungen führten zu der einen oder anderen interessanten Publikation, durch die Gestalter auch Verleger wurden. Aber eine tragfähige Autonomie wurde wegen ausbleibender Aufträge nie erreicht.
Mit all diesen Modellen der siebziger und achziger Jahre wurde die abstrahierende Moderne, wurden Bauhaus, De Stijl und Funktionalismus in allen kulturellen Bereichen theoretisch und praktisch in Frage gestellt. Das geschah nicht wegen der Qualität der Originale, sondern wegen der mangelnden Qualität bei den epigonalen Ergebnissen. Hochhaussiedlungen, Gleichförmigkeit der Produkte, langweilige typographische Textformen brachten nicht die Befreiung des Individuums, wie es die Avantgardisten erhofften.

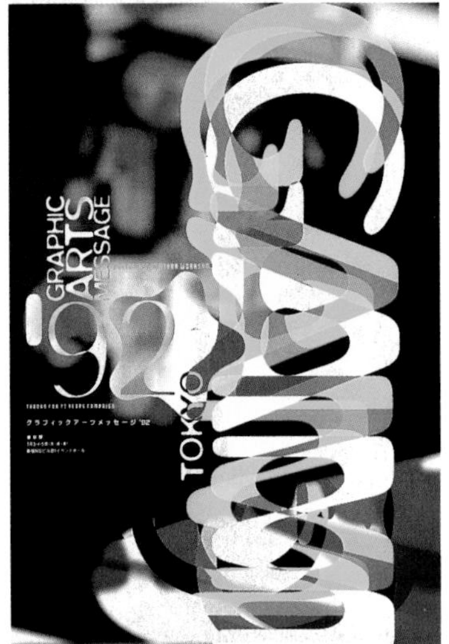

Brody

abcdefghijklmnopqrstuvwxyz[äöüßøæç]
ABCDEFGHIJKLMNOPQRSTUVWXYZ
1234567890(.,;:?!$&-*){ÄÖÜÅØÆŒÇ}

abcdefghijklmnopqrstuvwxyz[äöüßøæç]
ABCDEFGHIJKLMNOPQRSTUVWXYZ
1234567890(.,;:?!$&-*){ÄÖÜÅØÆŒÇ}

abcdefghijklmnopqrstuvwxyz[äöüßøæç]
ABCDEFGHIJKLMNOPQRSTUVWXYZ
1234567890(.,;:?!$&-*){ÄÖÜÅØÆŒÇ}

Sumner Stone

Les jeux de couleurs vont bien au-delà de l'emploi des couleurs primaires. Les dissonances chromatiques et le symbolisme kitsch génèrent des formes bizarres et une opulence débridée. C'est à cette époque que naît l'image du créateur autonome. On imagine la fin possible d'une fonction qui, dans le processus de communication, se situe entre l'émetteur et le récepteur, le créateur est alors lui-même émetteur. Cette conception conduit à quelques publications intéressantes pour lesquelles le maquettiste est lui-même éditeur. Mais l'absence de commandes ne permet pas d'arriver à une autonomie rentable.
Les modèles des années 70 et 80, l'abstraction moderne, le Bauhaus, De Stijl et le fonctionnalisme sont remis en question d'un point de vue théorique et pratique dans tous les domaines culturels. Ceci n'est pas dû à la qualité des originaux mais au manque de qualité des résultats obtenus par les épigones. Les tours des cités dortoirs, l'uniformité des produits, la typographie ennuyeuse des textes n'ont pas apporté à l'individu la libération que les avant-gardistes espéraient.

Carson *Ray Gun*

Uilen

complete alphabets from sans-serif to slab-serif. Some of the important ones are Stone by S. Stone, Rotis by O. Aicher and Thesis by L. de Groot.
Pure primary colors were no longer sufficient for the new fantasies of color. Dissonances of color and kitschy symbolism produced bizarre forms and unbridled splendor. It was also during this period that the idea of the autonomous applied designer arose. There arose the idea that the service function in the process of communication between transmitter and receiver was at an end; the designer himself was now the transmitter. This philosophy produced one or two interesting publications, where designers were also the publishers. Yet the lack of commissions meant that a workable autonomy was never achieved.
With the models of the seventies and eighties, Abstract Modernism, the Bauhaus, De Stijl and Functionalism were questioned in theory and in practice in all cultural fields. It was not the quality of the originals which was queried, but the lack of quality of later imitations. High-rise estates, uniformity of products, and boring typographic text structures did not bring freedom to the individual, as the avant-garde had hoped.

## New Wave

**Um 1980** Fernab von europäischer Enge, aber in voller Kenntnis der europäischen Entwicklung, entstand um 1980 an der Westküste der USA ein Stilgemenge ohne regionale oder dogmatische Starrheit.

Ausgelöst wurde diese Stilvariante durch anhaltende Kritik am erstarrten funktionalen Gestalten, aber vor allem durch die Anregung und Ausbildung, die viele Studierende aus den USA in der Hochburg interessanten Gestaltens, in Basel/Schweiz, erhalten hatten. Elemente aus der Lehre von Armin Hofmann und Wolfgang Weingart, die eher experimentell und modellhaft konzipiert waren, verbanden sich mit strebsamer wirtschaftsfördernder Ausrichtung der Studierenden.

Bei den Arbeiten von New Wave ging es mehr um unkonventionelles Äußeres als um substantielle visuelle Forschung. Elemente der Typographie der zwanziger Jahre wurden ausgeschöpft und aktualisiert. Aus Kreis, Dreieck und Quadrat entstanden mit skripturalen Anreicherungen buchstabenähnliche Formen und Signets. In vielen Entwürfen kamen zu horizontaler Textanordnung Zeilen in unterschiedlichen Winkelstellungen. Lesbarkeit war nicht mehr die wichtigste Aufgabe der Typographie. Pastellfarbige Phantasien, verschiedene Schriftschnitte und Größen wurden in einem Wort verwendet, ornamentierte Oberflächen ließen ungewöhnliche dreidimensionale Objekte eine wichtige Rolle im öffentlichen Kommunikationsprozeß spielen. Es war die Zeit, in der die postmoderne

**Vers 1980** Sur la côte ouest des Etats-Unis, loin de l'exiguïté du vieux continent, mais là où l'on a pleine connaissance de l'évolution de la pensée européenne, on assiste à la naissance d'un style disparate, dépourvu de tout dogmatisme ou régionalisme.

La critique permanente des formes fonctionnelles figées, mais surtout les idées de nombreux étudiants américains venant des bastions américains de la création, ou ayant été formés à Bâle en Suisse, sont à l'origine de cette variante stylistique. Certains aspects de la doctrine d'Armin Hofmann et de Wolfgang Weingart, conçus plutôt à titre expérimental ou comme modèles, concordent alors avec les ambitions des étudiants et leurs orientations économiques.

Dans ces travaux New Wave, il s'agit davantage d'obtenir un rendu non-conventionnel que des recherches visuelles substantielles. On puise au sein des éléments typographiques des années 20 en les actualisant. On enrichit les formes scripturales, on crée des structures évoquant les lettres et les signets à partir du cercle, du triangle et du carré. Sur beaucoup de maquettes, des lignes à inclinaison variable viennent s'ajouter à l'agencement horizontal du texte. La lisibilité cesse d'être la mission essentielle de la typographie. Des teintes pastel fantaisie, des lettres de fonte et de corps divers se retrouvent dans un même mot, des surfaces ornées font jouer à d'étranges objets tridimensionnels un rôle important dans le processus de communication. C'est l'époque où l'architecture postmoderne apparaît et où les produits du Memphis-Design italien attirent l'attention. En fait, le but de ces extensions dans les domaines bi- et tridimensionnels est de mêler un langage formel fonctionnel à un mode d'expression non-conventionnel.

1982 Skolos, Wedell

1982 Vanderbyl

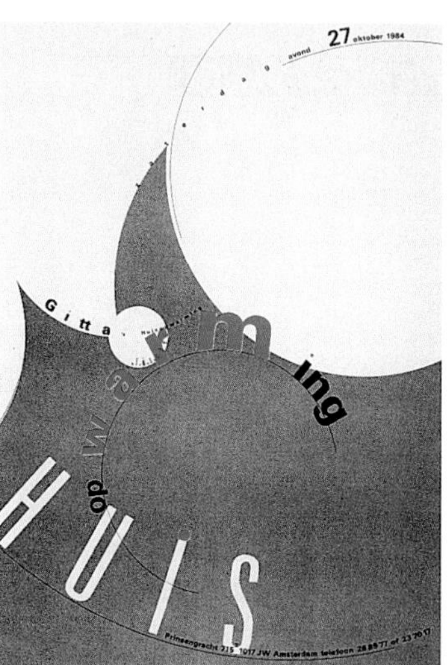

1984 Lieshout

1983 Greiman

1983 Brody *The Face*

**c. 1980** Far away from European restrictions, yet fully aware of European developments, a jumble of styles developed around 1980 on America's West Coast which were devoid of regional or dogmatic inflexibility.

These stylistic variations were the result of persistent criticism of ossified functional design, but also grew out of the stimulus and education students from America had received in that stronghold of interesting design, the Swiss city of Basel. Ideas from the teachings of Armin Hofmann and Wolfgang Weingart, conceived with a view to experiment, mingled with the industrious business orientation of the students.

New Wave was more concerned with unconventional appearances than with a substantial research of the visual. Elements of 1920s' typography were exploited and updated. With scripted embellishments, circles, triangles and squares became letter-like shapes and signets. Many designs included not only horizontal text arrangements, but also

Architektur entstand und die Memphis-Designprodukte aus Italien Aufsehen erregten. Bei all diesen Entwicklungen im zwei- und dreidimensionalen Bereich ging es um eine Vermischung von funktionaler und unkonventioneller gestalterischer Ausdrucksweise.
New Wave-Typographie war eine vieldiskutierte Stilrichtung, die von Gestaltern ausging und schnell viele junge Sympathisanten gewann. Diese machten Typographie in der aufkommenden Konkurrenz zu neuen Medien wieder zu einem attraktiven gestalterischen Bereich.
Die Arbeiten führten zu einer Öffnung verhärteter Stilformen und zu vielen hübschen verkaufsorientierten Anregungen. Durch radikalere nachfolgende Gestaltungskonzepte, denen sich die Hauptvertreter des New Wave anschlossen, verlor die Bewegung rasch an ernstzunehmendem Einfluß.

Vivement discutée, la typographie New Wave est une tendance stylistique qui émane de créateurs, et gagne rapidement la sympathie des jeunes adeptes. Face à la concurrence croissante des nouveaux médias, ils rendent la typographie de nouveau attrayante et créative.
Ce mouvement se libère des formes stylistiques crispées et produit de nombreuses idées charmantes destinées à favoriser la vente. Pourtant, il perd rapidement de son influence et de son sérieux avec l'apparition de concepts formels plus radicaux auxquels adhèrent les principaux adeptes de la New Wave.

1987 Scher

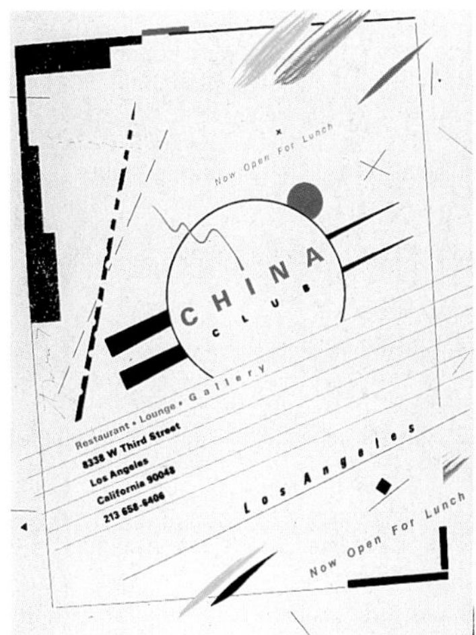


1981 Greiman

1985 Jones, McCabe, Thompson *i-D*

1986 Huff

1987 Hersey

lines positioned at various angles. Legibility was no longer the main task of typography. Pastel-colored fantasies and different font varieties and sizes were used in the space of one word. Ornamental surfaces allowed unusual three-dimensional objects to play a major role in the public communication process. It was the era of Post-Modernist architecture, when Memphis Design products from Italy caused a sensation. With all of these two- and three-dimensional developments, the main emphasis was on mixing functional and unconventional expressions of design.
New Wave typography was a much-discussed style which originated at the hands of designers and soon won many young adherents. In the face of rising competition from new media, they turned typography into an attractive branch of design once more.
New Wave led to an opening up of hardened stylistic forms and produced many appealing sales-oriented ideas. However, the numerous more radical subsequent design concepts, which the main representatives of New Wave endorsed, quickly deprived the movement of any serious influence.

**Neuer Funktionalismus**
**Nouveau fonctionnalisme**
**New Functionalism**

**Um 1980** Die Gestaltung im 20. Jahrhundert war bestimmt von einem dialektischen Auf und Ab verschiedenster Stile: rational-irreal, abstrakt-realistisch, surreal-konkret und so weiter. Anlaß für den Wechsel war, neben der reinen Spekulation, meist ein Verbrauch der ästhetischen Information in den praktischen Ergebnissen. Dabei bleibt die Innovation der wichtigste Antrieb und die größte Notwendigkeit kreativen Schaffens.
So war es nicht überraschend, daß nach Jahren des eruptiven und esoterischen Gestaltens um 1980 ein Neuer Funktionalismus sichtbar wurde, der die Elementare Typographie (1925) und die sachlich-funktionale Typographie (1960) aufnahm und weiterentwickelte. Denn allzuoft war hinter der Fassade interessanter expressiver Form der Inhalt nicht entschlüsselbar, fand Kommunikation gar nicht statt. Dazu kam, daß ganze Aufgabenbereiche durch eine subjektive grafische Form überhaupt nicht gelöst werden konnten: Informationssysteme, Gebrauchsanweisungen, Lehrprogramme brauchen einfach zu entschlüsselnde Formen.
War die sachlich-funktionale Gestaltung im Laufe der siebziger Jahre einfach langweilig geworden, glückte in vielen Zentren der Welt ein Neuansatz. Rastersysteme, die oft stur angewendet wurden, erlebten eine größere Variationsbreite, eine fast spielerisch-entspannte Anwendung bei der Text-Bild-Zusammenführung. Die Spalteneinteilung wurde schmaler, die Textspalten wurden nach oben und unten variabel plaziert und

**Vers 1980** Au 20e siècle, la dialectique de l'essor et du déclin des styles les plus divers, rationnel ou irrationnel, abstrait et réaliste, surréaliste et concret, etc. a souvent déterminé les formes. Outre la spéculation, la raison de ces changements est le plus souvent l'érosion de l'information esthétique. Pourtant l'innovation reste l'un des mobiles majeurs et l'une des nécessités absolues de l'activité créatrice.
Aussi, nous ne sommes pas étonnés d'assister au cours des années 80, après des années de création effervescente et ésotérique, à l'émergence d'un nouveau fonctionnalisme qui s'inspire de la typographie élémentaire (1925) et de la typographie objective et fonctionnelle (1960) et les développe. En fait ici, comme trop souvent, derrière une façade de formes intéressantes et expressives, le contenu est indéchiffrable et la communication ne se fait pas. A cela s'ajoute qu'on se trouve dans l'incapacité de résoudre toute une série de tâches au moyen de graphismes subjectifs. A titre d'exemple nous citerons les systèmes d'information, les modes d'emplois et les programmes d'enseignement qui requièrent des formes faciles à décoder.
Le style objectif et fonctionnel des années 70 est certes devenu ennuyeux, mais son renouvellement réussit en plusieurs endroits du monde. Les systèmes de grille, souvent utilisés par routine, s'enrichissent de variantes et leur utilisation devient plus ludique et plus souple dès qu'il s'agit d'associer le texte et l'image. Les colonnes plus étroites sont placées vers le haut ou le bas et cessent de s'aligner le long d'une bordure. Sur la page, les espaces vierges interviennent avec plus de raffinement qu'autrefois.
De nouvelles polices confèrent un caractère particulier à ce style. On trouve

1982 Odermatt + Tissi

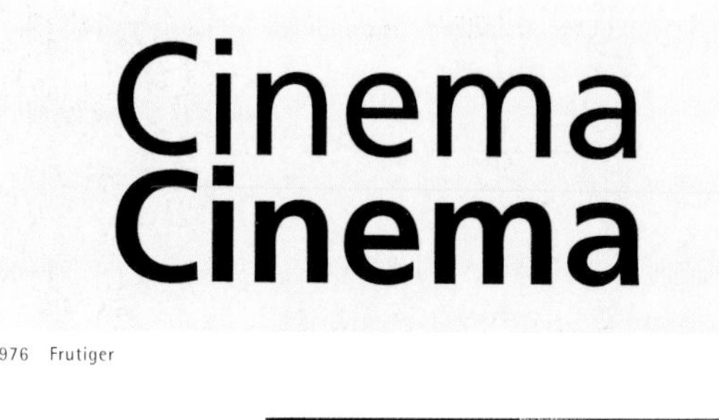


1976 Frutiger

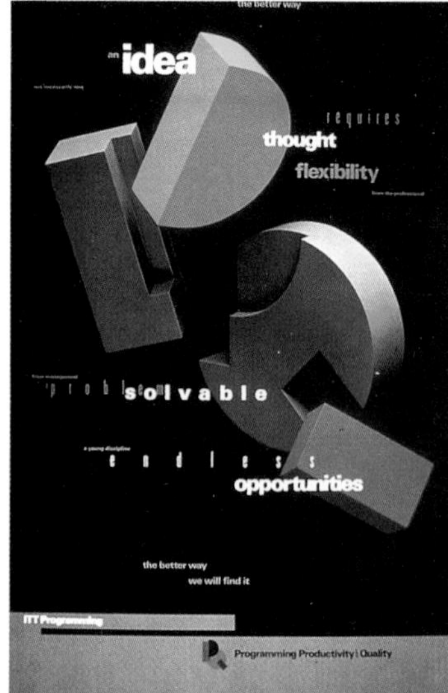


1983 Armstrong

1986 Ott + Stein

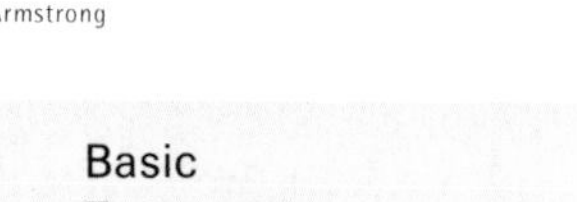

Ruedi Rüegg
ABC-Verlag
Zürich
Basic
Typography:
Design
with
Letters.
Typografische
Grundlagen:
Gestaltung
mit
Schrift.


1989 Rüegg

**c. 1980** Twentieth-century design was molded by a dialectical rise and fall of different styles, of the rational and unreal, the abstract and realistic, the surreal and concrete, etc.
The reason for these changes, apart from pure speculation, was usually that the conveying of aesthetic information was exhausted in practical results. Innovation remains the most important impetus of these changes and the greatest requirement of creativity.
It was thus hardly surprising that in 1980, after years of eruptive, esoteric design, a New Functionalism became apparent which absorbed and further developed "elementare typographie" (1925) and the functional typography of 1960. For often, hidden behind the façade of an interesting expressive form, the content of the text was not actually decipherable and no communication took place. Added to this was the fact that there were whole areas of printed matter which simply could not be represented by subjective, graphic structures: information systems, manuals and teaching programs needed easily-decodable forms.
Functional design had simply become boring during the course of the seventies; a new start brought success to many of the world's creative centers. Grid systems, which had often been used unthinkingly, experienced a new breadth of variation and an almost playful, relaxed employment in the union of words and pictures. Columns became narrower; they were placed sometimes higher, sometimes lower, and not all along one edge;

nicht mehr nur an einer Kante aufgereiht, die Leerräume auf den Seiten wurden raffinierter eingesetzt als früher.
Neue Schriftformen gaben dem Stil ein eigenes Gepräge. Zunächst war Adrian Frutigers Frutiger als fein und sensibel entwickeltes Alphabet charakteristisch für die achtziger Jahre.
Diese Schriften wurden in wenigen Graden und Schnitten verwendet, aber nicht nur in mager und halbfett in zwei Größen wie früher. Die vielfältigen Nuancen der Schriften wurden eingesetzt, um der Monotonie und Gleichförmigkeit der Textseiten zu entkommen. Im Gegensatz zu den lauten Stilformen wurden Linien, Flächen, Farben oder Diagonalen eher sparsam eingesetzt. Farbe in der Schrift selbst wurde sensibel und pointiert verwendet. Die ausgewählten Gestaltungselemente wurden nicht schmückend verwendet, sondern dazu, Inhalte besser zu strukturieren oder sichtbar zu machen.
Begleitet wurde der Neue Funktionalismus von wichtigen theoretischen Publikationen, die die rationalen Strukturen der Typographie im 20. Jahrhundert aufzeigten. Wichtigste Zeitschrift in diesem Zusammenhang war *octavo*, ein Projekt der Designgruppe 8vo, London. Dieses Projekt, das auf acht Ausgaben terminiert war, zeigte die Vielfalt des Funktionalismus der vergangenen Jahrzehnte und gipfelte in einem Ausblick auf die kommende mediale Entwicklung: die letzte Ausgabe der Zeitschrift erschien als CD-ROM.

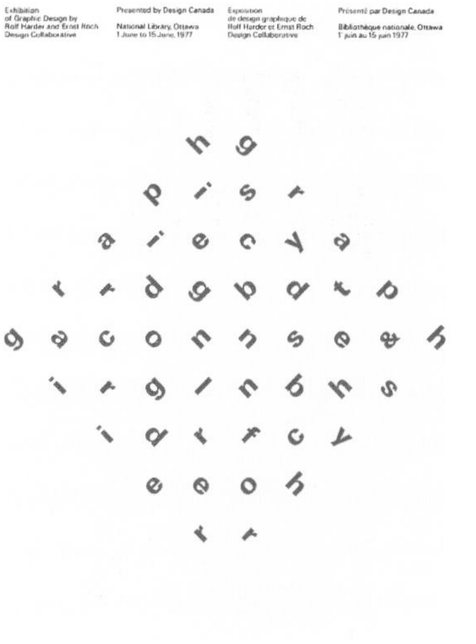
1977 Roch

1984 Mendell + Oberer

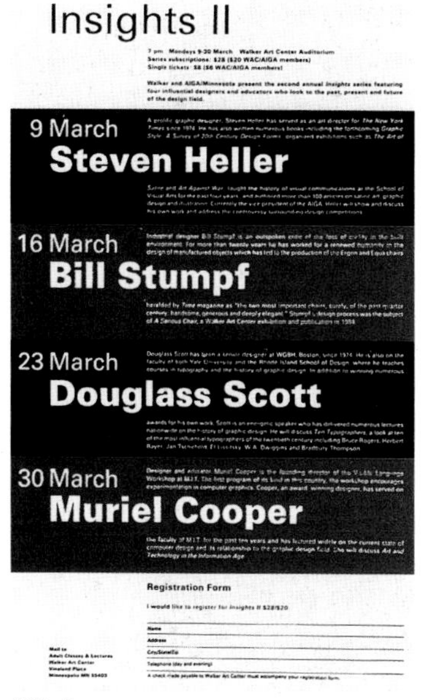

1989 Ferguson

d'abord la Frutiger d'Adrian Frutiger, un alphabet sensible et caractéristique des années 80.
L'écriture cesse d'être systématiquement maigre ou en deux corps comme autrefois, en revanche on utilise quelques fontes et diverses intensités. On tire parti de la multitude de nuances des caractères pour rompre la monotonie et l'uniformité des pages de texte. Contrairement aux formes stylistiques criardes, les lignes, surfaces, couleurs ou diagonales interviennent avec parcimonie. La couleur des lettres imprimées est employée avec sensibilité et pertinence. Les éléments formels choisis n'ont pas fonction de décor, mais servent à mieux structurer le contenu et à le rendre plus visible.
Ce nouveau fonctionnalisme s'accompagne d'importantes publications théoriques qui traitent des structures rationnelles de la typographie du 20e siècle. Dans ce contexte, la revue la plus importante *octavo* est l'émanation du groupe de design londonien 8vo. Ce groupe qui s'est fixé huit tâches, montre la diversité du fonctionnalisme des décennies précédentes et conclut en ouvrant une perspective sur l'évolution des médias du futur: la dernière édition de la revue paraîtra sous forme de CD-ROM.

1989 Baumann + Baumann

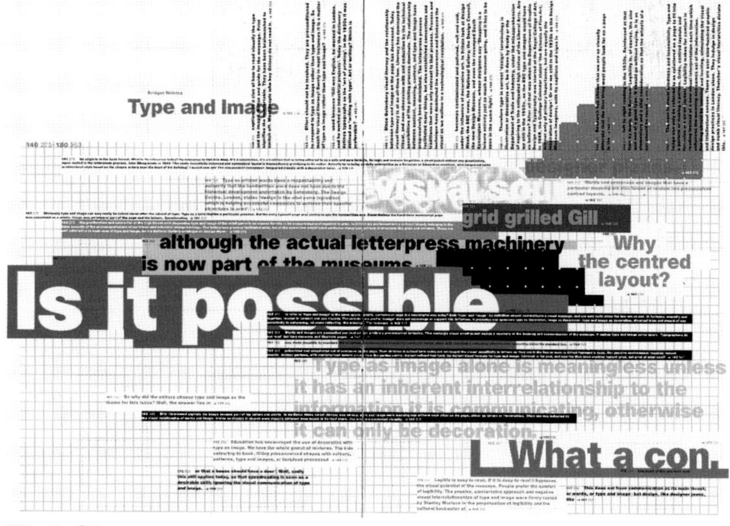

1990 8vo *Octavo*

empty spaces on the pages were used with a greater finesse than before.
New fonts left their mark on this New Functionalist style. Adrian Frutiger's Frutiger was a finely and sensitively developed alphabet characteristic of the eighties.
Type was employed in varying sizes and varieties, and not just in light or bold and in two sizes as in the past. The diverse nuances of the fonts were used to escape from the monotony and uniformity of the pages of text. As opposed to louder stylistic forms, lines, surfaces, colors and diagonals were used more sparingly; color in the type itself was used delicately and trenchantly. The selected design elements were employed with a view to better structuring or defining the content, and not as decoration.
New Functionalism was accompanied by significant theoretical publications demonstrating the rational structures of typography in the twentieth century. The most important journal in this respect was "octavo", a project run by the design group 8vo in London. The eight scheduled issues of this project illustrated the variety of Functionalism in the preceding decades and culminated in a preview of future developments in the media: the last issue of the magazine appeared as a CD-ROM.

**Punk Typographie**
**Typographie Punk**
**Punk typography**

**Um 1975** In der Mitte der siebziger Jahre entstand in London, der Stadt, die schon oft neue kulturelle Strömungen anregte, aus der Jugendszene die Punkbewegung. Mit agressiver Häßlichkeit wollten die Punks ihre Inhalte und Botschaften in die Welt schreien: Auflehnung gegen versöhnliche Lebensabläufe, Kampf gegen jedes Establishment, Ablehnung des guten Geschmacks. Es war eine anarchistische Richtung, die die gesamte Lebensform umfassen wollte. Sie äußerte sich in ungekonnt gespielter Rockmusik, in zerrissener Kleidung, in Schmuck aus Utensilien wie Ketten und Sicherheitsnadeln, bunten Irokesenfrisuren und gezähmten Ratten. Mit diesen Attributen inszenierte man die neue Variante des Bürgerschrecks.

Nur wenige namhafte Gestalter gingen aus der Punkbewegung hervor, denn der nihilistische Part der Bewegung gewann meist Oberhand. Die Plakate, Schallplattenhüllen, Pamphlete oder Handzettel wollten alle der guten Form widersprechen. Alles sah mehrmals recycled und verwittert aus, verziert mit chaotischer Typographie und collagierten Bildzitaten. Diese fotokopierten Collageninhalte waren aber oft schockierende Visualisierungen der real existierenden Welt. Es ist bemerkenswert, daß man trotz täglicher Sensationspresse diesen Schock erzielen konnte.

Neben verschiedenen Schallplattenfirmen entstanden 1980 als weitere kommerzielle Projekte Mode-Szenezeitschriften, die aus der Punkbewegung kamen. Auch hier waren zerzaustes Layout und

**Vers 1975** Au milieu des années 70, la mouvance punk voit le jour dans la jeunesse londonienne, la ville qui avait donné naissance à tant de courants culturels. Exhibant une laideur agressive, les punks hurlent leurs messages à la face du monde. Ils se révoltent contre un mode de vie conciliateur, luttent contre l'establishment, refusent le bon goût. Ils s'inscrivent dans un mouvement anarchiste qui tend à englober toutes les formes de la vie. Le punk se manifeste dans une musique rock endiablée et maladroite, des vêtements déchirés, des bijoux faits de divers ustensiles comme les chaînes et les épingles à nourrices, des crêtes de cheveux à l'iroquoise et des rats apprivoisés. Se servant de ces attributs, on se met en scène en affichant une nouvelle variante de l'épouvantail à bourgeois.

Très peu de créateurs de renom sont issus du mouvement punk puisque la partie nihiliste l'emportera le plus souvent. Les affiches, pochettes de disques, pamphlets et tracts sont délibérément contraires aux formes correctes. Tout semble avoir subi plusieurs recyclages et dégradations, être orné d'une typographie chaotique et de collages de citations picturales. Pourtant, ces collages photocopiés sont souvent la représentation choquante du monde réel. Il est toutefois remarquable qu'on parvienne à choquer de la sorte malgré la presse quotidienne à sensations.

Outre diverses sociétés de production de disques, d'autres projets commerciaux touchant la mode et les revues naissent du mouvement punk en 1980. Ici aussi, on trouve des maquettes ébouriffées comportant des éléments formels caractéristiques de la technique de la photocopie, mais qui n'égalent pas pour autant le radicalisme de la période des pionniers. Ces travaux n'ont qu'une valeur de modèle pour des dizaines de revues américaines

1977 Reid

1977 Anonymous

1980 Anonymous

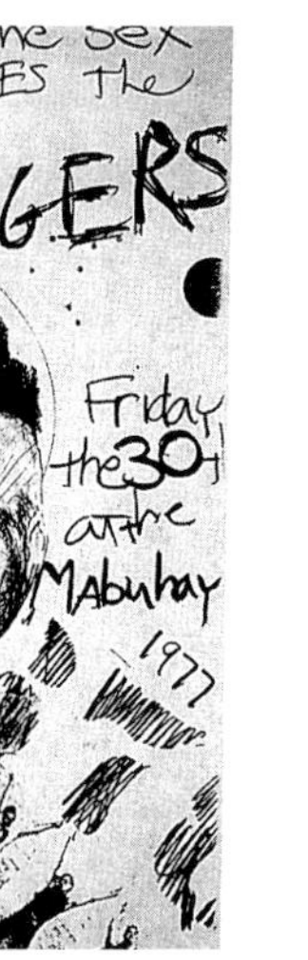


1977 Houston

1976 Anonymous

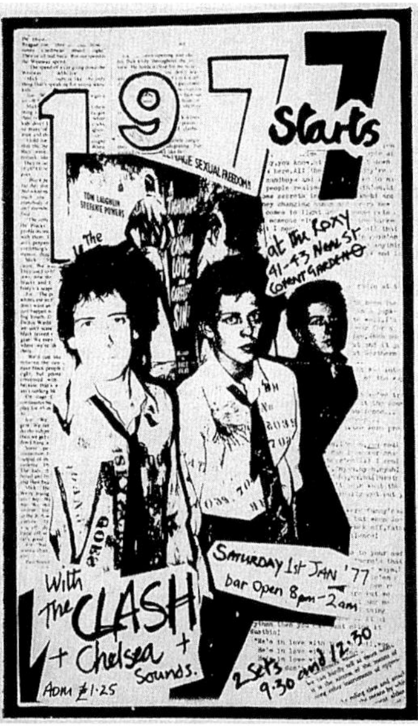


1977 Anonymous

1981 Suttle

**c. 1975** In the mid-seventies in London, the city which had often acted as a stimulus for new cultural trends, the youth movement produced Punk. With aggressive ugliness the Punks yelled out their messages to the world: reject good taste, revolt against conciliatory lifestyles, fight the establishment. It was an anarchy that wanted to include all of life. It was expressed in badly-played rock music, in ripped clothes, in bodily decoration made from chains and safety pins, in brightly-colored, spiked Mohican hairstyles and tame rats. These attributes were a new variant on the theme of *épater les bourgeois.*

There were few well-known Punk designers, as the nihilistic part of the movement usually gained the upper hand. Posters, record covers, pamphlets and handbills aimed to fly in the face of "good design". Everything looked flyblown and as if it had been recycled several times over, splattered with a chaotic typography and collages of words and pictures. These photocopied collages, however, were often shocking visualizations of the real world. It is remarkable that it was still possible to shock despite the daily sensationalism in the press.

In 1980, commercial projects born of the Punk scene included not only various record companies, but also fashion magazines. These also sported disheveled layouts and photocopying techniques as characteristic design elements, but no longer had the explosive force of the pioneering days of Punk. They merely became models for dozens of similar

die Kopiertechnik charakteristische Gestaltungselemente, aber die Brisanz der Pionierzeit des Punk erzielten sie nicht mehr: sie waren nur noch Vorbilder für dutzende ähnlicher Lifestyle-Zeitschriften in den USA und Europa. Es dauerte nur wenige Jahre, bis Punk zur Touristenattraktion mutierte: die Kreativen zogen sich zurück und entgingen so der Korrumpierung. Die einst beißenden Bildanklagen fanden sich auf unzähligen T-Shirts und Postern als Mitbringsel. Schnell wurden die Stilinnovationen von einst in das herrschende Grafik-Design der achziger Jahre integriert.

Durch Punk ist der Begriff Freiheit in der Gestaltung neu entdeckt worden. Die Radikalität dieses Stils entstand außerhalb bewährter und akzeptierter Ausbildungs- und Arbeitsszenen. Die Anregungen zu dieser Revolte wurden durch die Theorie und Praxis der philosophisch-künstlerischen Bewegung „Situationistische Internationale“ bestimmt. Aber auch die typographischen Vorstellungen aus dem vorrevolutionären Rußland trugen zu der Formzertrümmerung, aus der sich die neue Bewegung speiste, bei.

et européennes. Il suffit de quelques années pour que les punks deviennent une attraction touristique. Les plus créatifs d'entre eux se replient pour échapper à toute forme de récupération. Les dénonciations picturales provocantes de jadis commencent à figurer sur d'innombrables T-shirts et affiches que l'on rapporte en guise de souvenirs. Les innovations stylistiques d'autrefois s'intègrent rapidement dans le design graphique des années 80.

La mouvance punk a redécouvert le concept de liberté créative. Le radicalisme de ce style s'est développé en marge des normes établies et des milieux éducatifs et professionnels. La théorie et la pratique issues d'un mouvement philosophico-artistique tel que «L'Internationale situationniste» ont exercé une influence déterminante sur cette révolte. De même, les conceptions typographiques de la Russie pré-révolutionnaire ont contribué à l'explosion des formes dont se nourrissait cette nouvelle tendance.

1979 Suttle

1979 Suttle

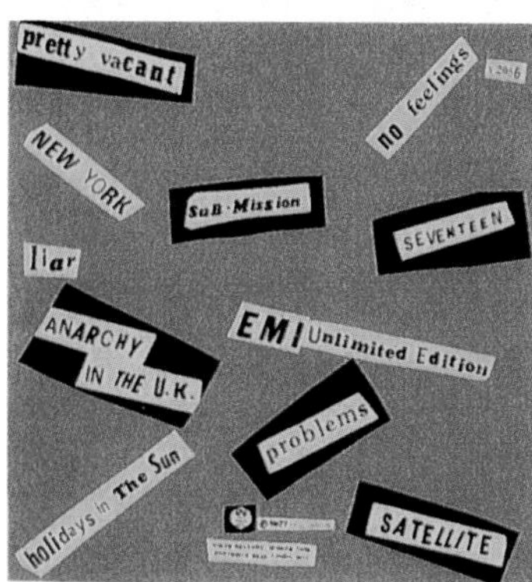

1977 Reid

1987 Garrett + Reid

lifestyle magazines in the USA and Europe. It was only a few years before Punk was demoted to a mere tourist attraction: creative minds retreated and so spared themselves the corruption. The formerly caustic pictorial accusations found themselves printed on numerous T-shirts and posters as souvenirs. The stylistic innovations of Punk were rapidly swallowed up into the overall graphic design of the eighties.

Through punk the concept of freedom in design was rediscovered. The radicalness of this style evolved apart from the well-established, accepted centers of work and education. The ideas behind this revolt were determined by the theory and practice of the philosophy and art movement "Situationist International". Typographic ideas from pre-Revolutionary Russia also contributed to the destruction of forms which fed the new movement.

**Pop und Psychedelic**
**Pop et psychédélique**
**Pop and Psychedelia**

**Um 1965** Jede Generation versucht ihre eigene visuelle Sprache zu entwickeln, wenn auch oft zaghaft und an bekannte Formen angelehnt. In den 60er Jahren waren die psychedelische Gestaltung und die Pop Art über einige Jahre eine sehr heftige Jugend- und Kunstbewegung mit starker Wirkung. Durch clevere Indienstnahme degenerierten sie jedoch schnell zu modischem Beiwerk und Verkaufsförderung. Die Pop Art bezog ihre Formen nicht aus einem fernen oder nahen Kunststil, sondern aus den Werbeformen des Konsumismus und den Comic-Strips. Auch dies war als eine Gegenbewegung zu der glatten, wohlgeordneten funktionalen Gestaltung gedacht. Es sollte eine Typographie des Profanen und des Massengeschmacks der Konsumwelt sein. Dies klappte so lange, bis die Pop-Formen nach Jahren selbst als akzeptabel integriert waren und die Innovation verbraucht war.
Die psychedelische Richtung entstand (wie der Jugendstil um die Jahrhundertwende) als eine ganzheitliche, ornamentale und anti-technische ästhetische Auffassung. Alle Aspekte des Lebens sollten mit diesen Formen romantisiert und verändert werden. Die gestalterischen Elemente waren der Welt der Comics, asiatischen Symbolen und Jugendstilformen entlehnt. Die Schriftformen und die Illustrationen waren halluzinativ verzerrt, was aus dem Drogenhintergrund und den dadurch veränderten Wahrnehmungsmöglichkeiten entsprang. Inhalt und Erscheinung waren jedoch untrennbar mit der Rock- und Popmusik verbunden. Die

**Vers 1965** Chaque génération tente de développer son propre langage visuel, parfois avec des hésitations et en s'inspirant de formes connues. Dans les années 60, le style psychédélique et le Pop art se font l'écho d'un mouvement qui a un impact considérable sur la jeunesse et les arts. Toutefois une récupération habile le fera rapidement dégénérer en un corollaire à la mode et en incitation à la vente. Les formes du Pop art ne se réfèrent pas à un style artistique lointain ou proche, mais au langage publicitaire et de consommation ainsi qu'à la bande dessinée. Il s'affirme comme contre-courant aux formes lisses et bien ordonnées de la création fonctionnelle. La typographie doit donc être profane et traduire les goûts des masses dans une société de consommation. Ceci fonctionnera plusieurs années jusqu'à ce que les formes pop soient admises et intégrées et que les innovations s'érodent.
Quand le mouvement psychédélique se constitue (comme le style Art nouveau à la fin du siècle dernier), ses conceptions globales, ornementales et esthétiques s'opposent aux technologies. Les formes visent à transformer tous les aspects de la vie et à lui donner une tournure romantique. Les éléments formels sont empruntés à l'univers de la bande dessinée, aux symboles asiatiques et au style Art nouveau. Les lettres et les illustrations présentent des déformations hallucinatoires suggérées sous l'effet de drogues et de la modification perceptive qu'elles provoquent. Toutefois, le contenu et l'apparence restent indissociables de la musique rock et pop. Cette musique est le prétexte et le moteur d'un renouveau social et typographique. Ce mouvement est le fruit du mode de vie libertaire dans les grandes métropoles occidentales et les campagnes environnantes. Los Angeles,

1970 Warhol

1968 Ruscha

1968 Max

1968 Indiana

ca. 1967 English

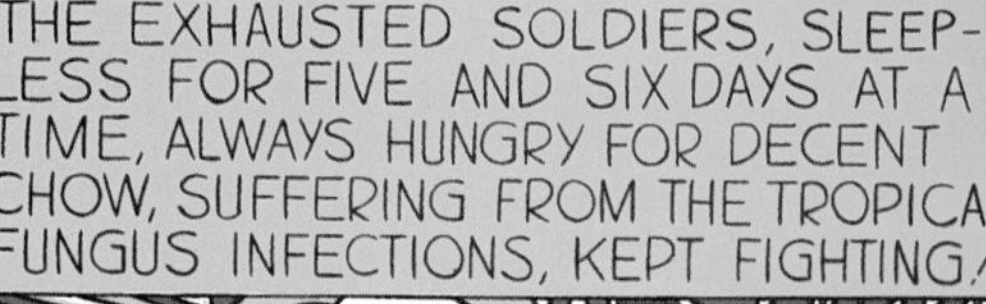

ca. 1961 Lichtenstein

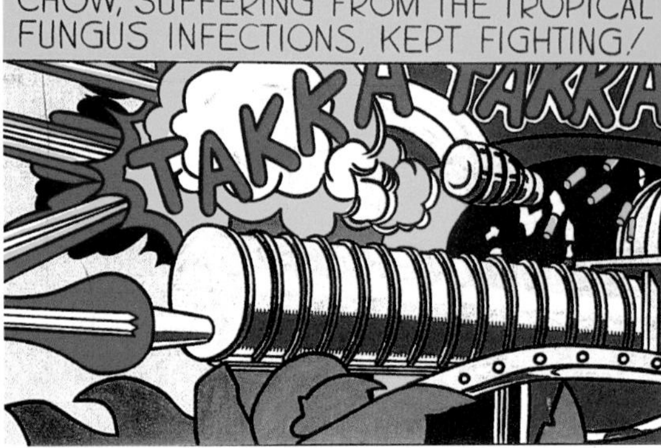

1962 Lichtenstein

**c. 1965** Every generation tries to develop its own visual language, even if the attempt is often timid and falls back on familiar forms. For several years during the sixties, Psychedelic design and Pop Art were intense youth and art movements, with powerful effects. Yet as a result of smart deployment, these forms soon degenerated to nothing more than fashionable accessories and aids for sales promotion. Pop Art took its form not from the artistic styles of the past or present, but from commercial advertising ploys and from comic strips. It, too, was born as a counter-movement to slick, well-ordered, functional design. It was to be a typography of the mundane and of the taste of the masses caught up in the net of consumerism. Yet the notion was successful until Pop Art itself became accepted and its inventive powers were spent.
Psychedelia, like Art Nouveau at the turn of the century, grew out of a holistic, ornamental and anti-technical aesthetic conception. The idea was to romanticize and transmute all aspects of life. Design elements were borrowed from the world of cartoons, Asiatic symbols and Art Nouveau. Typefaces and illustrations were distorted in hallucinatory fashion, a style which had its origins in the world of drugs and the altered perceptions they induced. Content and appearance were inextricably linked to rock and pop music. This music was the reason and drive behind social and typographical innovation. The whole movement was the result of the permissive lifestyle possible

Musik war Anlaß und Motor gesellschaftlicher wie typographischer Erneuerung. Die gesamte Richtung entsprang dem möglichen libertären Lebensstil in den großen Metropolen der westlichen Welt und derem freizeitlichen Hinterland. Zentren dieser romantischen Bewegung waren Los Angeles, New York, Chicago und London, aber auch in Japan gab es durch die schnelle internationale Kommunikation Übernahmen dieser Formen.
Das wichtige an beiden Bewegungen war das Experiment, eingefahrene praktische und theoretische Positionen in Frage zu stellen. Beide Richtungen wurden nicht von grafischen oder typographischen Fachleuten entwickelt, sondern von Gestaltern und Künstlern, die seismografisch den Zeitgeist erahnten und in Neues umsetzten. Die herrschende funktionale Gestaltung wurde international immer weniger kreativ eingesetzt, sondern immer mehr als bürokratisches Rezept in den immer gleichen Ritualen der Werbung verwendet.
Beide Richtungen entstanden am Beginn der Fotosatzzeit, in der Headline-Schriften einfacher, schneller und billiger produziert werden konnten als in der Bleisatzzeit. So lösten die meist handgezeichneten, skizzierten Alphabete des kreativen Anfangs eine Flut von ähnlichen Alphabeten aus, die aber sehr selten die Kraft der Originale hatten.

New York, Chicago, Londres font figure de centres de cette mouvance néoromantique. Puis, avec la rapidité des communications internationales, le Japon adopte ces formes à son tour.
L'expérience de la remise en question de positions théoriques et de pratiques routinières constitue l'aspect essentiel de ces deux mouvements. Ce ne sont donc pas les spécialistes en typographie ou en arts graphiques qui œuvrent, mais des créateurs et des artistes qui appréhendent l'esprit du temps tels des séismographes, et l'interprètent en innovant. En fait, le style fonctionnel alors dominant perd de sa créativité pour devenir une recette bureaucratique utilisant constamment les mêmes rituels publicitaires.
Ces deux mouvements apparaissent en même temps que la photocomposition qui permet de produire des gros titres plus facilement, plus rapidement et à meilleur prix qu'à l'époque de la composition au plomb. Ainsi les alphabets de la phase créative, souvent dessinés à la main, cèdent bientôt la place à une foule de caractères semblables possédant rarement la force des originaux.

1967 Weller

1968 McConnell

1966 Cleveland

1967 Griffin

1966 Wilson

1965 Yokoo

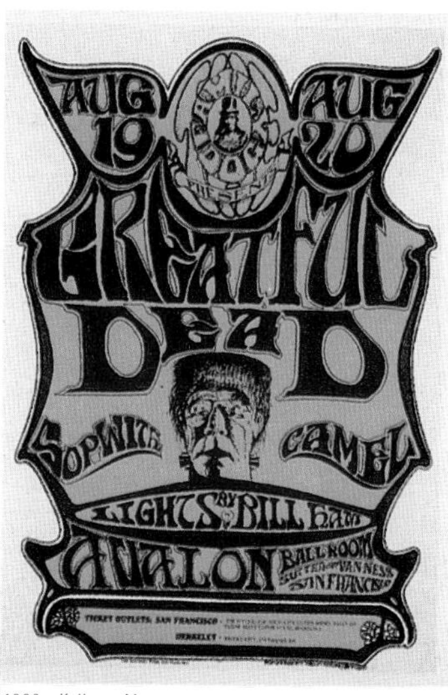

1966 Kelley + Mouse

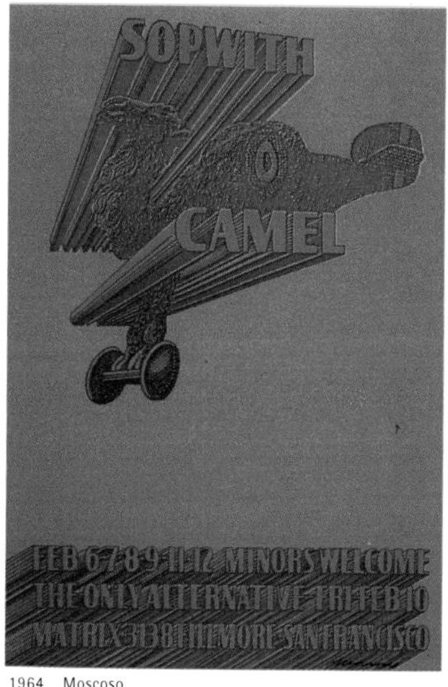

1964 Moscoso

in the great urban centers and amusement hinterlands of the Western world. The focuses of this romantic movement were Los Angeles, New York, Chicago and London, and forms even spread to Japan through the fast modes of international communication.
The essence of Pop Art and Psychedelia was the experiment of questioning tired practical and theoretical approaches. Both styles were developed not by graphics or typography experts, but by designers and artists who seismographically sensed the new spirit of the age and transformed it into something new. Worldwide, the prevalent functionalist style was employed with ever-decreasing creativity, becoming a bureaucratic recipe in the same old advertising rituals. The development of both styles coincided with the onset of filmsetting, where headline fonts could be produced more simply, more quickly and more cheaply than with metal composition. Thus the mostly hand-drawn, sketched alphabets of the creative initial stages produced a flood of similar alphabets, which, however, rarely captured the energy of the originals.

**Experimentelle Typographie**
**Typographie expérimentale**
**Experimental typography**

**Um 1965** Zur gleichen Zeit wie Fluxus entstanden in den klassischen typographischen Bereichen zunehmend Arbeiten mit experimentellem Charakter. Dazu trug der belanglos gewordene sachliche Stil bei, der die Notwendigkeit zur Innovation durch Form immer mehr außer acht ließ. Die ersten theoretischen Arbeiten zur Gestaltung mit Computern erfuhren Aufmerksamkeit und Diskussion.
Schrift und Typographie wurden durch die mathematisch-exakten Werke der Informationstheorie (z. B. von Max Bense, Abraham Moles und Kurd Alsleben) nicht mehr als traditions- und emotionsbelastete Medien beschrieben, sondern als „Zeichenvorrat". Diese aufgeklärte Sicht ließ die junge Generation immer stärker Fragen stellen nach dem Warum von Dogmen.
In der Literatur entstanden nun experimentelle Texte durch Permutation, Kombination oder Variation und diese wiederum führten zu einer angemessenen experimentellen Typographie.
Bei den Arbeiten zur Visuellen Poesie entstanden Formen zwischen Text und Bild. Diese Arbeiten wurden schnell als Anregungen für die angewandte Typographie erkannt.
Charakteristisch für alle Versuche war das Aufgeben der gewohnten waagerechten Zeilenanordnung, die jahrhundertelang die Form der Wörter und Bücher bestimmte. Es waren interessante Fortsetzungen experimenteller Alphabete der 20er und 30er Jahre.

**Vers 1965** A la même époque que Fluxus apparaissent des travaux de caractère expérimental dans les domaines de la typographie classique. La banalisation d'un style objectif devenu anodin et qui ignore de plus en plus la nécessité d'innover par la forme a contribué à l'émergence de ce phénomène. Les premiers travaux théoriques de mise en forme assistée par ordinateur attirent l'attention et font l'objet de vastes débats.
Dans les œuvres mathématiques et exactes de la Théorie de l'information (par ex. de Max Bense, Abraham Moles et Kurd Alsleben), l'écrit et la typographie cessent d'être considérés comme des médias traditionnels et chargés d'émotions, mais deviennent des «réserves de signes». Cette vision rationnelle permet à la jeune génération de poser des questions de plus en plus insistantes sur le pourquoi des dogmes.
En littérature, on voit apparaître des textes expérimentaux issus de permutations, combinaisons ou variations qui conduisent vers une typographie expérimentale adaptée.
Avec la poésie visuelle, on assiste à l'avènement de formes situées entre le texte et l'image. Ces travaux deviennent rapidement une source d'inspiration pour la typographie appliquée.
L'abandon de la disposition habituelle et horizontale des lignes qui, des siècles durant, avait déterminé la forme des mots, constitue l'une des caractéristiques de ces recherches. Ceci permet des prolongements intéressants des alphabets expérimentaux créés pendant les années 20 et 30.

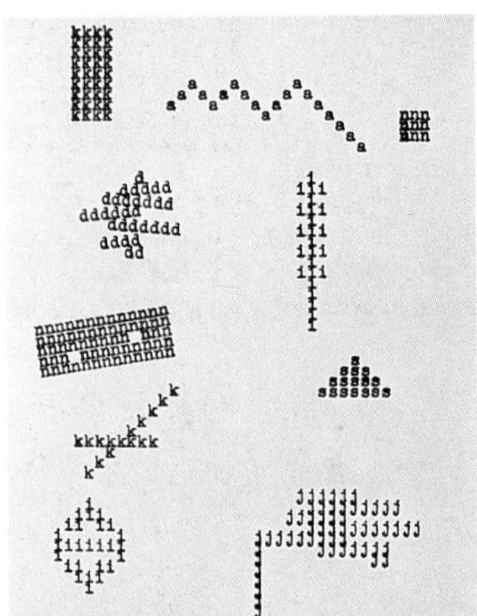
1966 Kolar

1969 Phillips

1973 Schmidt

1964 Massin

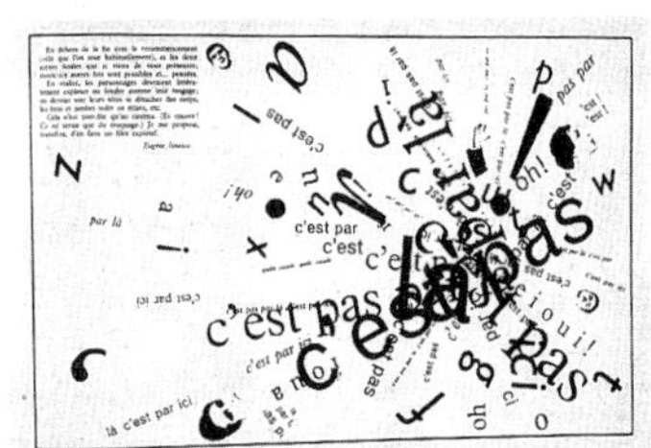
1964 Massin

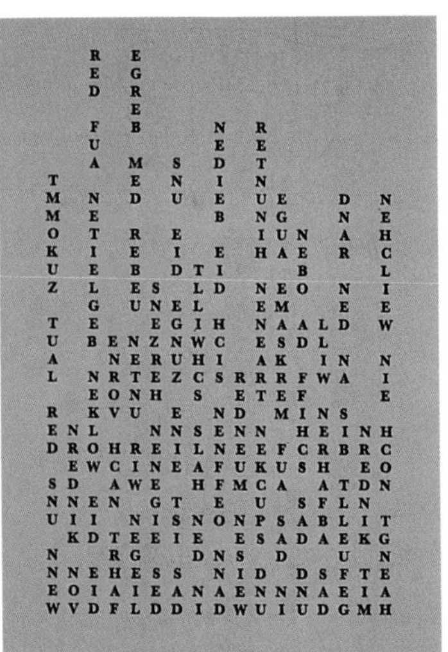
1967 Roth

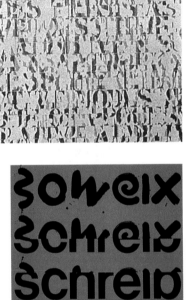

la poétique sémantique
bilder objekte graphik
10. juli bis 15. august 1965
zur eröffnung am 10. juli 20 uhr spricht william e. simmat
galerie gunar düsseldorf schützenstraße 63 an der kölner str.
1965 Schmidt

**c. 1965** At the same time as Fluxus, more and more work of experimental character was being produced in the classic fields of typography. This was due to the trivialization of the functional style, which increasingly ignored the necessity for innovation through form. The first theoretical pieces of work produced by computers in the field of design were the topic of discussion and attention.
The mathematically precise concepts of information theory (by Max Bense, Abraham Moles and Kurd Alsleben, for example) described typefaces and typography as a "set of signs", and no longer as media burdened by tradition and emotion. This enlightened view caused the younger generation to ask more and more questions as to the "why" of dogmas.
In literature experimental texts were put together by permutation, combination or variation, which in turn required suitable experimental typography.
Visual Poetry came up with forms between words and pictures. These literary works were soon recognized as providing a stimulus for applied typography. A major characteristic of all of these experiments was the abandonment of the familiar horizontal arrangement of lines, which for centuries had determined the shape of words and books. These alphabets were interesting continuations of the experimental fonts of the twenties and thirties.

**Um 1965** Durch kreatives Überschreiten und Zerstören formaler und inhaltlicher Normen entstanden im 20. Jahrhundert neue Anregungen für die Gestaltung. Diese Neuerungen wurden von der Fachwelt entweder nicht erkannt oder strikt abgelehnt. Nach dem Dadaismus (1915) war die radikalste ästhetische Opposition eine Bewegung, die sich Fluxus nannte. Dada war durch die kulturelle und politische Realität in der Welt lange vergessen, als in den 50er Jahren in den USA, in Japan und in der Bundesrepublik die Wiederentdeckung dieser Bewegung begann: Ausstellungen, Aktionen und Buchpublikationen zeigten, wie ungebrochen aktuell diese ästhetische Einstellung war. Was in Bewunderung der alten Vorbilder als Neodadaismus begann, entwickelte sich zu einer eigenständigen und weitergehenden Bewegung. Fluxus war gegen die herrschende Ästhetik von abstraktem Expressionismus und Neokonstruktivismus (Op Art) in der Malerei gerichtet. Zur kulturkritischen Einstellung kamen gesellschaftskritische Überzeugungen. Die Fluxusanhänger mokierten sich über die funktionierenden, aber leeren Staatsrituale ebenso wie über den exzessiven kopflosen Konsumismus. Wie in allen Medien suchte Fluxus auch in der typographischen Form neue Wege, und wieder half dabei handwerkliche Unbeschwertheit. Bei Broschüren, Plakaten oder Handzetteln wurden Schriften und Anordnungen erprobt, die dem herrschenden Geschmack dramatisch widersprachen. Keine dieser Arbeiten entstand für kommerzielle Zwecke.

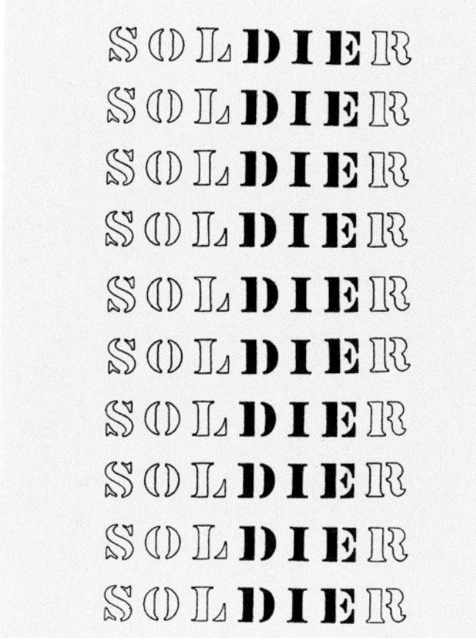


Williams

1963 Vautier

1964 Higgins

1964 Køpcke

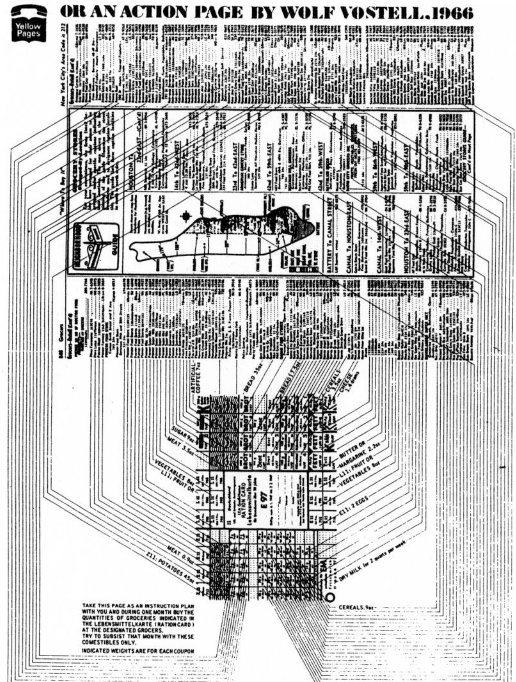


1966 Vostell

Maciunas

Maciunas

Brecht

1964 Maciunas

**c. 1965** During the twentieth century, the various artistic disciplines were able to contrive new ideas by creatively going beyond and even destroying structural and contextual norms. These innovations were either not recognized by the experts or immediately disregarded. The most radical, aesthetic oppostional movement after Dada (1915) was Fluxus. Cultural and political world reality had long forgotten Dada, when in the fifties it experienced something of a revival in the USA, in Japan and in Germany. Exhibitions, promotions and book publications illustrated the topicality of this aesthetic attitude. What began as Neo-Dada out of admiration for the old model soon grew into an independent and progressive movement.

Fluxus was against the ruling aesthetic of abstract Expressionism and Neo-Constructivism (Op Art) in painting. This culturally critical outlook was accompanied by socio-critical convictions. The followers of Fluxus mocked the functioning, empty rituals of state, and excessive, mindless consumerism. Fluxus looked for new roads in all forms of the media, and also in typography, employing a technical, carefree independence. In brochures, posters and handbills, fonts and arrangements of type were experimented with which dramatically flew in the face of prevailing tastes. None of this work was commissioned for commercial advertising.

**Vers 1965** Dans tous les domaines, le 20e siècle a été marqué par l'émergence de sources d'inspiration nouvelles générées par des transgressions créatives et la destruction des teneurs et des normes formelles. Or les milieux spécialisés ont toujours refusé de reconnaître ces innovations, ou même, les ont rejetées. Après le dadaïsme (1915), l'opposition esthétique la plus radicale se cristallisera autour d'un mouvement appelé Fluxus. Les réalités culturelles et politiques ont fait oublier Dada depuis longtemps quand, dans les années 50 on commence à le redécouvrir aux Etats-Unis, au Japon et en Allemagne fédérale. Des expositions, des manifestations et des publications montrent alors la continuité et l'actualité de cette attitude esthétique. Ce qui, au début, traduit une certaine admiration pour les modèles d'antan sous forme d'un néodadaïsme, évolue rapidement pour devenir un mouvement autonome.

Fluxus s'élève contre l'esthétique dominante de l'expressionnisme abstrait et du néoconstructivisme (Op Art). Une attitude sociale critique se greffe sur la critique culturelle. Les partisans de Fluxus se moquent autant du fonctionnement transparent mais vide de sens des rituels étatiques en vigueur que des excès d'une consommation effrénée.

Comme il le fait dans les nouveaux médias, Fluxus recherche aussi de nouvelles voies en typographie, et cette fois-ci encore, l'insouciance technique lui vient en aide. On se sert de brochures, affiches et tracts pour essayer des caractères et des mises en page radicalement contraires au goût dominant. Aucun de ces travaux n'est réalisé à des fins commerciales.

**Sachlich-funktionale Typographie**
**Typographie objective fonctionnelle**
**Functional typography**

**Um 1960** Nach all den längeren und kürzeren stilistischen Ausformungen der Vergangenheit war die Typographie um 1960 von einer Phase des Suchens in eine Phase der Stabilität gelangt. Die uneingeschränkte Mode der Zeit war ein sachlich-funktionaler Stil, der für die internationale Kommunikation bestens geeignet war. Die klaren Formen der serifenlosen Schriften von Akzidenz-Grotesk und Franklin Gothic und die Formen der neuen Generation Folio, Helvetica und Univers standen in asymmetrischen Kompositionen auf den Papierformaten. Als wichtiges Gestaltungsmittel wurde viel unbedruckter Freiraum in die Typographie integriert. Die Textspalten waren streng in Rastersysteme gegliedert, große halbfette Headlines standen über kleinen, mageren Body-Texten. Kein Ornament, keine Emotion, nur klare Information durch Text und Bild. In den besten Beispielen zeigte sich eine neue visuelle und typographische Intelligenz.
Diese sachlich-funktionale Typographie entstand durch die Weiterentwicklung der elementaren Typographie (1925) und des internationalen typographischen Stils (1945). Die Probleme der Anfangszeit (Blocksatz, Versalsatz, ungenügende Schriftauswahl und Schriftmischung), waren gemeistert; das Bild der Texte war ebenbürtiger Ausdruck der größten Präzision in der Technik, und oft visualisierte man sie mit den kongenialen Formen.
Schnell breitete sich dieser formal reduzierte Stil in ganz Europa, den USA, aber auch in Japan aus. Sowohl in der Ausbildung, als auch in der Industrie wurde

**Vers 1960** Dans les années 60, après les phases plus ou moins longues de tâtonnements stylistiques des périodes précédentes, la typographie entre dans une période de stabilité. La mode de l'époque va résolument dans le sens d'un style objectif et fonctionnel qui convient parfaitement à la communication internationale. Les formes clairement définies des caractères sans sérif de la Grotesk de ville et du Franklin Gothic ainsi que de la nouvelle génération, la Folio, l'Helvética et l'Univers se retrouvent sur les divers formats de papier dans des compositions asymétriques. Les espaces blancs, non imprimés, s'y intègrent et deviennent un outil important de mise en page. Les colonnes de textes s'organisent avec rigueur selon un système de grille, et les titres en semi-gras chapeautent de petits corps de texte imprimés en maigre. Aucun ornement, aucune émotion, uniquement l'information claire par le texte et l'image. Les meilleurs exemples témoignent d'une nouvelle forme d'intelligence visuelle et typographique.
La typographie objective fonctionnelle s'est constituée à partir de la typographie élémentaire (1925) et du style typographique international (1945). Les problèmes des débuts (composition par blocs, alinéas avec lettrines, choix insuffisant de caractères et mélanges de polices) avaient été résolus; l'image des textes est l'expression même d'une extrême précision technique que l'on visualise souvent en ayant recours à des formes apparentées.
Ce style caractérisé par une réduction formelle se répand dans toute l'Europe, aux Etats-Unis et au Japon. L'objectivité formelle est très bien acceptée autant par le secteur de l'éducation que dans l'industrie. De nombreux manuels définissent la typographie comme devant obéir à des

1962 Müller-Brockmann

1962 Kapitzki

1962 Wyss

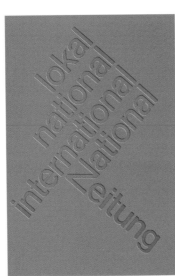

1960 Gerstner + Kutter

1962 Lohse

**c. 1960** By 1960, after all the periods of stylistic molding – of shorter or longer duration – typography had ceased searching and had entered a phase of stability. The universal fashion of the age was functional, well suited for international communication. The clear design of sans-serif fonts, such as Akzidenz-Grotesk, Franklin Gothic, and the newer generation of Folio, Helvetica and Univers, formed asymmetrical compositions on the page. An important typographical design feature was a large amount of unprinted space. Columns of text were arranged in rigid grid systems; large, bold-faced headlines towered over small, light-faced bodies of text. There was no ornamentation, no emotion, merely lucid information, conveyed in words and pictures. The best examples demonstrate a new, visual, typographical intelligence. This functional typography grew out of the further development of "elementare typographie" (1925) and the international typographic style of 1945. Teething problems (justification left and right, block capitals, insufficient choice and mixture of type) had been overcome; this new textual image was an expression of great technical precision, and was often visualized with congenial forms.
This formally reductionist style quickly spread through all of Europe, but also to the USA and Japan. This formal matter-of-factness was readily accepted by industry and education alike. Numerous new textbooks presented typography as a clear, easily comprehensible set of rules. These examples also prompted less ex-

**Sachlich-funktionale Typographie**
**Typographie objective fonctionnelle**
**Functional typography**

diese formale Sachlichkeit bereitwillig akzeptiert. In zahlreichen neuen Lehrbüchern wurde Typographie als klares, nachvollziehbares Regelwerk dargestellt. Diese Beispiele konnten auch wenig geschulte Kreise zu brauchbaren gestalterischen Ergebnissen anregen.
Im Laufe der Jahre führte die einfache Wiederholung und Wiederholbarkeit des klaren Groteskstils zu einer Abnutzung der Ausdruckskraft und zu einer Bürokratisierung der Gestaltung. Es kam zu einer Krise des Funktionalismus, die ihre Ursachen in der Typographie, vor allem auch in den immer wiederkehrenden Schriftformen und immer gleichen Konzepten hatte. Viele einflußreiche Unternehmen und öffentliche Einrichtungen verwendeten für ihr Erscheinungsbild als Hausschrift die Helvetica. Dies kam einer Uniformiertheit gleich. Lesbarkeitsanalysen und Gestaltungsmanuals waren an die Stelle einer kreativen Auseinandersetzung getreten. Dieser Verlust an Spontanität überschattete vorläufig die bedeutenden Fortschritte, die gerade in dieser Periode in bezug auf eine allgemein visuelle Kommunikation gemacht wurden. Die „Helvetica-Krise" spornte die Gestalter zur Suche nach Neuem an und führte bald zu völlig undogmatischen Erscheinungsformen in der Typographie.

ABCDEFGHIJK
LMNOPQRSTU
VWXYZ
abcdefghijklmn
opqrstuvwxyz
1234567890

1958 Miedinger

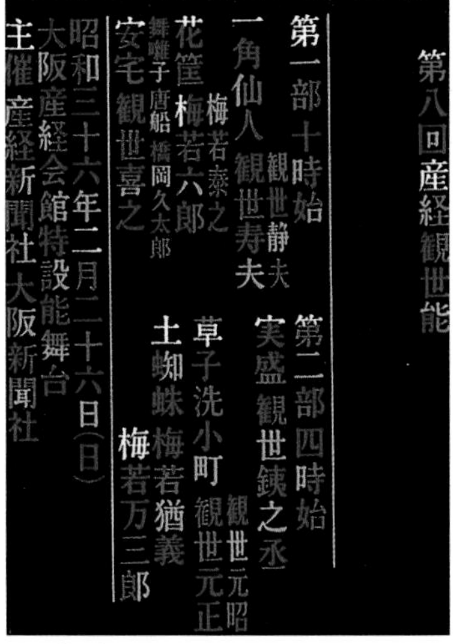

1961 Tanaka

1958 Odermatt

1962 Gerstner, Gredinger+Kutter

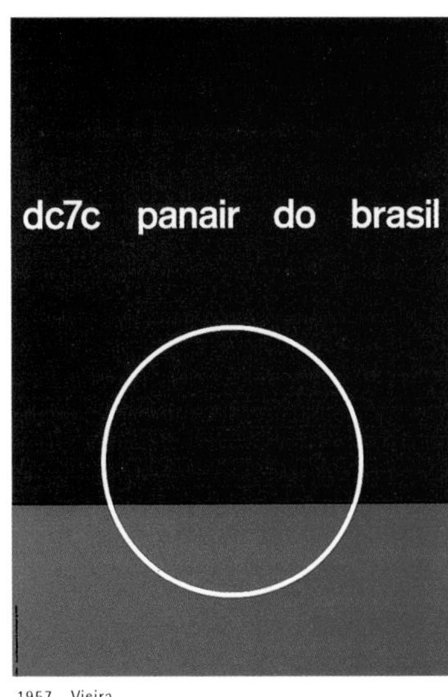


1957 Vieira

1963 Piatti

règles claires et logiques. Les exemples qu'ils donnent peuvent susciter des créations formelles utiles, même chez les non-initiés.
Au cours des années, la répétition et la monotonie d'un style basé sur la Grotesk avaient provoqué l'érosion de son expressivité et une bureaucratisation de la mise en page. Le fonctionnalisme typographique connaît alors une crise dont les causes principales sont la réutilisation permanente de certaines polices et concepts. De nombreuses entreprises et institutions publiques de renom utilisent l'Helvética pour leurs logotypes et comme police maison, tant et si bien qu'on arrive à une sorte d'uniformité. Les analyses de lisibilité et les manuels de mise en page prennent la place de la réflexion créative. Cette perte de spontanéité jette provisoirement une ombre sur les progrès importants qui, simultanément, touchent la création visuelle en général. La «crise de l'Helvética» incitera les créateurs à innover, et les amènera à inventer des formes de typographie contraires aux dogmes.

perienced circles to produce useful, structural results.
Over the years, the simple repetition and reproducibility of this clear sans-serif style led to a reduction in expressiveness and a bureaucratizing of design. Functionalism was hit by crisis, which had its roots in a typography of ever-repetitive alphabets and ever-identical concepts. Many influential companies and public institutions used Helvetica as a standard in their corporate literature, until it became akin to a uniform. Legibility analyses and design manuals had replaced creative discussion. This loss of spontaneity temporarily overshadowed the major progress in general visual forms of communication in this period. The "Helvetica crisis" spurred designers on to undertake a further quest for new ideas and soon led to completely undogmatic manifestations in typography.

**Bildhafte Typographie**
**Typographie imagée**
**Pictorial typography**

**Um 1955** Die ständige Zunahme der werblichen Kommunikation in den westlichen Industrieländern nach dem Zweiten Weltkrieg führte zu der Notwendigkeit, sich immer auffälliger durch visuelle und verbale Zeichen von den Mitbewerbern abzusetzen. Dies war für kreative Gestalter eine großartige Möglichkeit, außergewöhnliche Einfälle nicht nur zu entwerfen, sondern sie auch gedruckt und bald prämiert zu sehen. Für die Entwicklung neuer typographischer Ausdrucksformen waren veränderte technische Voraussetzungen von großer Bedeutung. Ausgehend von den USA bot die Verbreitung des Fernsehens eine unbelastete Herangehensweise an visuelle Kommunikation. Dazu kamen der Beginn des Fotosatzes, der Transfer-Technik und die Kopiertechnik. Publikationen und Ausstellungen verbreiteten Kenntnisse über die gestalterische Avantgarde, die mit allen Ausdrucksmöglichkeiten experimentierte, die seit den zwanziger Jahren das visuelle Spektrum bereichert hatten.
Durch die Praxis der Werbeagenturen und Grafik-Design-Studios wurde die gesamte Herangehensweise an das Entwerfen verändert. Immer seltener variierte man bestehende Gestaltungselemente. Immer häufiger gab es eine ganzheitliche, oft in kleinen Gruppen erarbeitete Text-Bild-Konzeption, bei der alle Gestaltungsteile als gleich wichtig erkannt wurden und untereinander verbunden waren. Ein Gestaltungsmittel gab nur mit dem anderen zusammen einen überraschenden Sinn. Bilder entstanden aus

**Vers 1955** Après la Seconde Guerre mondiale, l'explosion de la publicité dans les pays occidentaux industrialisés oblige à se démarquer des concurrents en utilisant des signes visuels et verbaux de plus en plus accrocheurs. Pour les créateurs, ceci représente une formidable opportunité non seulement de jeter sur le papier les idées les plus extraordinaires, mais aussi de les voir imprimées et même bientôt honorées de prix. Les changements survenus dans les technologies ont eu une grande influence sur le développement de nouveaux modes d'expression typographiques. Partant des Etats-Unis, la télévision se répand et favorise une approche de la communication visuelle presque vierge de tous préjugés. A cela vient s'ajouter l'avènement de la photocomposition, des techniques de transfert et de copie. Les publications et expositions propagent des informations sur cette avant-garde, laquelle expérimente en tirant parti de tous les moyens d'expression qui ont enrichi le spectre de la création visuelle depuis les années 20.
Grâce à l'expérience pratique des agences de publicité et des ateliers de design et de graphisme, on aborde la maquette d'une tout autre manière. Les graphistes déclinent de plus en plus rarement les éléments formels existants. Et l'on trouve de plus en plus souvent des concepts texte-image homogènes, fréquemment élaborés par de petits groupes, et où toutes les composantes formelles sont traitées à égalité, et cohérentes. La forme ne produit un sens surprenant que combinée à d'autres. Les images naissent du texte ou des mots. Pour cela, on modifie souvent la forme des lettres en ayant recours à diverses techniques ou au dessin, on intervient sur les surfaces, les chasses, au moyen du découpage et du collage. Or,

1957 Lubalin

1958 Silverstein

1955 Yamashiro

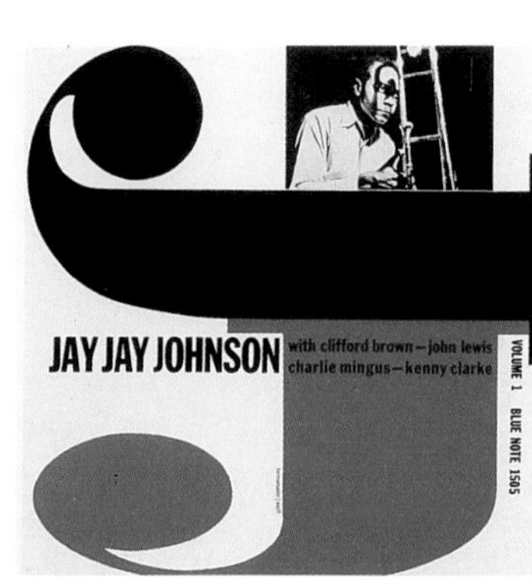

Hemansader

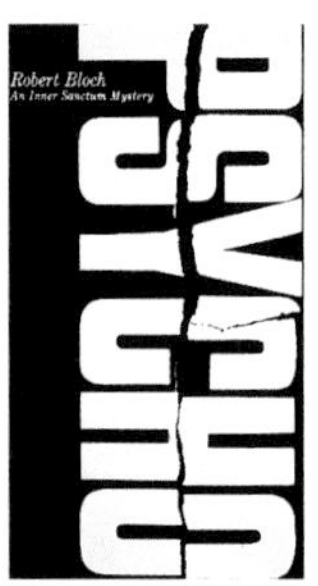

Palladino

Miles

1961 Beltrán

Grignani

**c. 1955** The steady increase of promotional communication in Western industrialized countries after the Second World War meant that companies had to distinguish themselves, visually and verbally, more and more from their competitors. This gave creative designers a splendid chance not only to think up unusual solutions but see them in print and quickly rewarded. Technological change was of great importance for the development of new forms of typographic expression. Starting in the USA, the growth of television offered a fresh medium of visual communication. The fifties also saw the birth of filmsetting, transfer technology and copying techniques. Publications and exhibitions disseminated knowledge about an avant-garde who were experimenting with all the forms of expression which had enriched the visual spectrum since the twenties.
The practices of advertising agencies and graphic-design studios changed the whole approach to design. Existing design elements were varied less and less. Integrated text-picture concepts, often worked out in small groups, became more and more common, where all the design elements were recognized as equally important and integrated. An element of surprise was created only by the link between one element and another. Pictures emerged from texts and words. Letters were often technically or graphically altered, extended two- or three-dimensionally, cut out or made into a collage. The age-old rules of the trade were ignored if it improved visualization of the

Texten oder Wörtern. Dabei wurden Buchstabenformen häufig technisch oder zeichnerisch verändert, flächig oder räumlich ergänzt, ausgeschnitten oder collagiert. Wenn es der Visualisierung der Inhalte diente, wurden tradierte Handwerksregeln negiert. Nie zuvor gab es so viele Erweiterungen und Erfindungen in den Bereichen des Buchstabenbildes, des Wortbildes und des Satzbildes. Durch fehlende Gestaltungsdogmen wurde Schrift nicht reduktiv ausgewählt (zum Beispiel nur serifenlose Schriften), sondern in einer reichhaltigen Formenpalette aussagekräftig eingesetzt. Manche Argumente gegen diesen gestalterischen Stil entstanden durch die Ähnlichkeit zu dem verpönten typographischen Formsatz, der im 19. Jahrhundert mit ornamentaler und abstruser Textanordnung Bilder nachzuahmen versuchte. Waren diese Arbeiten aber formalistischer Kitsch wurden jetzt inhaltlich-konzeptionelle Vorgänge sichtbar gemacht.
Diese expressiv-bildhaften typographischen Arbeiten entstanden zunächst hauptsächlich in den USA. Sie wurden durch die Wettbewerbe des Type Directors Club of New York unterstützt, dadurch international bekannt und bald in anderen Teilen der Welt mit Erfolg übernommen.

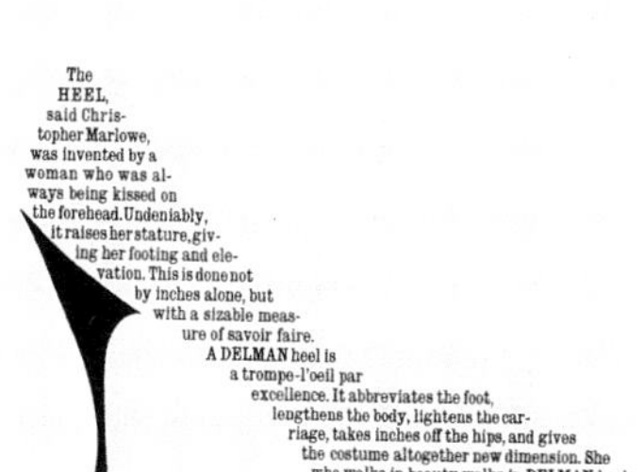


Sochis

NOW ARROW ANNOUNCES DECTON UNDERSHORTS AT 2 FOR 2.95 TO GIVE YOU A COMPLETE 2-FOR-2.95 DECTON UNDERWEAR PROGRAM. YES, T-SHIRTS, ATHLETIC SHIRTS, KNITTED BRIEFS AND NOW UNDERSHORTS. EVERYTHING IS DECTON. EVERYTHING PRICED COMPETITIVELY AT 2 FOR 2.95. EVERYTHING PUT UP IN BRIGHT NEW PACKAGING, TOO. THINK OF WHAT YOU'VE GOT GOING FOR YOU: THE QUALITY OF ARROW. THE DEPENDABILITY OF DECTON. ATTRACTIVE NEW PACKAGING AND A NEW ACROSS-THE-LINE PRICE. MORE REASONS THAN EVER TO STOCK THE NAME THEY KNOW: ARROW.

1964 Graham

1961 Engelmann

Oh
I know
such
a
lot
of
things,
but
as
I
grow
I know
I'll
know
much
more.

1956 Rand

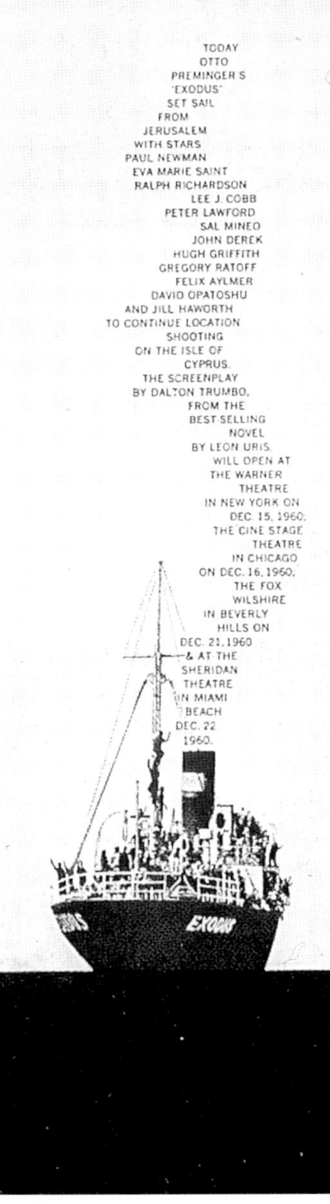


1960 Bass

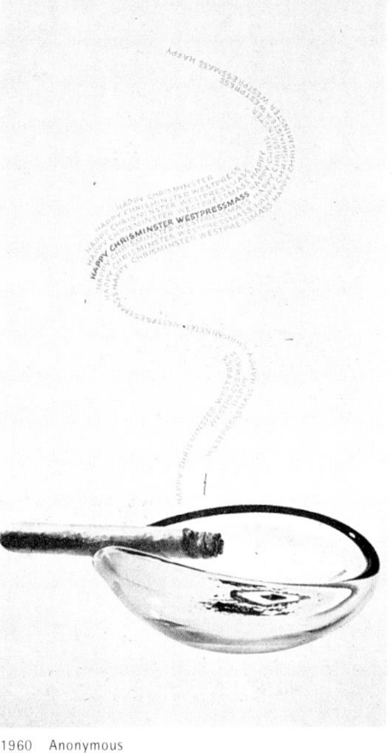

1960 Anonymous

1964 Engelmann

même si cela contribue à visualiser les contenus, on fait fi des règles traditionnelles de l'artisanat. Jamais auparavant, des aspects tels que l'image de la lettre, du mot et de la phrase n'ont connu autant de variations, d'inventions. En l'absence de dogmes formels, les polices ne sont pas sélectionnées de manière restrictive (par ex. ne choisir que des polices sans sérif), mais utilisées pour leur capacité expressive en puisant dans une palette très riche. Certains arguments opposés à ce style se fondent sur les ressemblances avec cette forme de composition typographique alors honnie qui, au 19è siècle, tentait d'imiter l'image par le biais d'ornements et de mises en pages obscures. Mais, même si ces travaux relèvent du kirsch formel, la démarche conceptuelle est désormais visible.
Les premières réalisations de cette typographie expressive et imagée ont surtout vu le jour aux Etats-Unis. Les concours du Type Directors Club of New York ont soutenu ce mouvement et l'ont fait connaître au niveau international et bientôt dans le monde entier.

content. Never before had there been such inventiveness regarding the appearance of letters, words and layout. The lack of design dogmas meant that typefaces were not selected on reductionist criteria (for example, only sans-serif fonts) but used expressively in an abundant palette of forms. Challengers of this type of typographic design argued that the new style shared too many similarities with the formal composition of the 19th century, frowned upon by critics because it attempted to imitate pictures with ornamental and abstruse arrangements of text. Yet whereas the older works were formalist kitsch, the newer style made context and conception visible.
This expressive, pictorial typography was initially employed chiefly in the USA. It was supported by competitions staged by the Type Directors Club of New York, through which it became internationally known and rapidly and successfully adopted by other countries.

**Internationaler typographischer Stil**
**Style typographique international**
**International typographic style**

**Um 1945** In zwei weit voneinander entfernten und sehr verschiedenen Gebieten, in der Schweiz und in den USA, wurde in den vierzigern Jahren ein typographischer Stil entwickelt, der für die neuen Bedürfnisse der Kommunikation und Werbung ein weiterführender Höhepunkt war. Die neuen Formen beruhten auf den vorhergegangenen Neuerungen des Konstruktivismus und vor allem der Elementaren Typographie, sie bezogen aber die zunehmende Notwendigkeit der internationalen Kommunikation und der Entwicklung des Mediums Fotografie mit ein. Während in vielen Ländern Europas Veränderungen und Erneuerungen stattfanden, war die Schweiz in den zwanziger und dreißiger Jahren im kulturellen und sozialen Bereich ein auf der Grundlage politischer Neutralität prosperierendes traditionelles Gebiet. Gestaltung im allgemeinen und Typographie im besonderen wurden nach lang erprobten Handwerksregeln gelehrt und entworfen. Grafische Arbeiten wie Anzeige, Prospekt und Plakat wurden in künstlerischem Malstil mit heimatbezogenen Motiven gestaltet.
Nach 1929 entstanden hier typographische Arbeiten, die nicht eine einfache Übernahme der Gedanken von Bauhaus und De Stijl waren, sondern eine Weiterentwicklung des sachlichen funktionalen Gedankens darstellten.
Diese neue Gestaltung, oft „Schweizer Stil" genannt, verzichtete auf Elemente, die formal durch den Konstruktivismus erklärbar waren, für einen universellen Einsatz bei der visuellen Kommunikation

**Vers 1945** Dans les années 40, deux pays pourtant très éloignés l'un de l'autre, la Suisse et les Etats-Unis, élaborent un style typographique qui répond parfaitement aux besoins nouveaux de la communication et de la publicité. Les nouvelles formes s'inspirent des innovations constructivistes et surtout de la typographie élémentaire, mais elles tiennent aussi compte des nécessités croissantes de la communication internationale ainsi que du développement d'un nouveau média, la photographie. A l'époque où des transformations et des innovations ont lieu dans de nombreux pays d'Europe, la Suisse des années 20 et 30 était restée un Etat traditionnel autant dans le domaine culturel que social, et prospérait sur la base d'une neutralité politique. La création en général, et la typographie en particulier, y étaient enseignées et pratiquées selon des règles artisanales ayant fait leurs preuves. Les annonces publicitaires, prospectus et affiches comportaient des motifs peints dans un style plutôt inspiré des arts plastiques et émaillés de références nationales.
Après 1929, on y réalise des travaux typographiques qui cessent d'être de simples emprunts aux idées du Bauhaus et de De Stijl, mais qui correspondent à des conceptions sobres et fonctionnelles. La nouvelle mise en page, souvent appelée «style suisse», renonce aux éléments formels qui rappellent le constructivisme et entravent toute utilisation universelle en matière de communication visuelle. Les titres en capitales, la composition en pavés, des caractères expressifs de divers types et fontes cèdent la place à un agencement clair de l'image et du texte, à des caractères et polices dont le corps correspondent au sujet, à une conception exempte d'arbitraire et à un rendu sobre

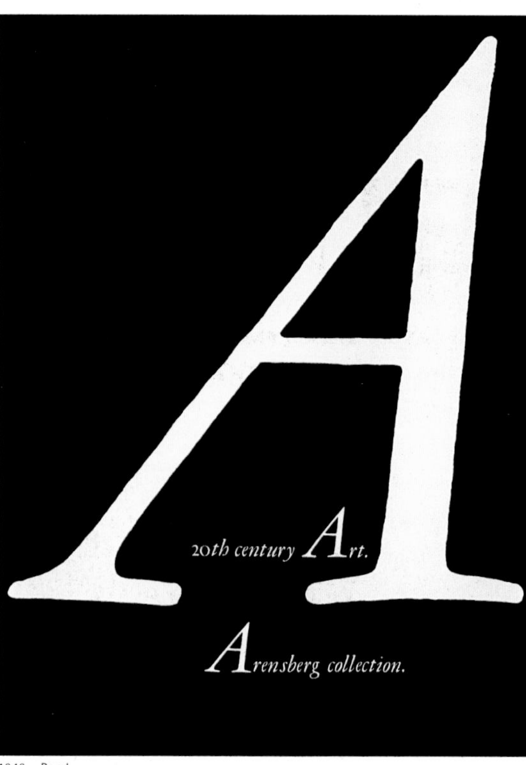

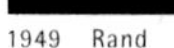
1949 Rand

1947 Federico

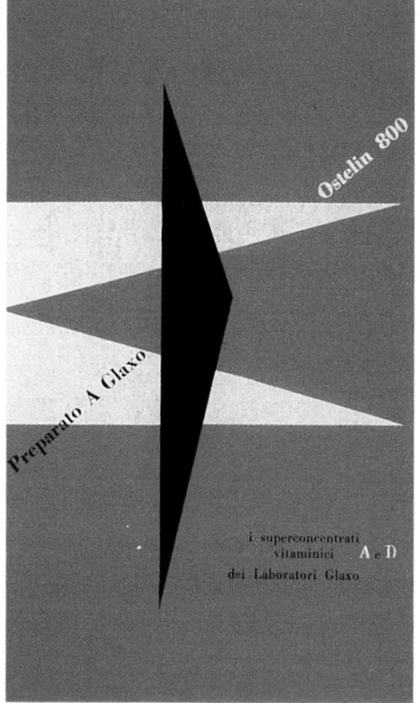

1946 Vivarelli

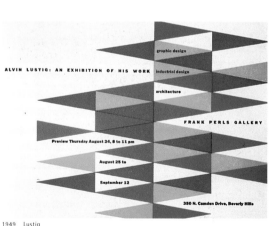

1949 Lustig

1948 Sandberg

1945 Bühler

1950 Matter

**c. 1945** In the forties, two very different and very distant parts of the world, namely the USA and Switzerland, saw the development of a typographical style which represented a forward-looking high point for the new requirements of communication and advertising. The new styles were founded on the previous innovations of Constructivism and more especially of "elementare typographie," while taking account of the increasing need for international communication and the rise of photography. Whereas in the twenties and thirties the accent in many European countries was on change and innovation, politically neutral Switzerland was, socially and culturally, a prospering area of tradition. Design in general, and typography in particular, were taught and practised according to tried and tested craft rules. Graphics, such as advertisements, brochures and posters, were in the style of artistic paintings and depicted motifs from the Swiss homeland.
After 1929, typographic work emerged here which did not just simply adopt the ideals of Bauhaus and De Stijl, but which represented a further development of functionalist ideas.
This new style, often called "Swiss Style," rejected elements which while formally explicable in Constructivist terms, were nevertheless unsuitable for universal use in visual communication. Lines of capitals, justification right and left, and expressive fonts in many varieties, gave way to a clear arrangement of illustration and text; font size and design were cho-

**Internationaler typographischer Stil**
**Style typographique international**
**International typographic style**

jedoch hinderlich waren. Versalzeilen, Blocksatz, expressive Schriften in vielen Arten und Schnitten wichen einer klaren Zuordnung von Bild und Text, einer inhaltlichen Schriftgrößen- und Schriftschnitteauswahl, einem willkürfreien Aufbau und einer sachlichen, informierenden Wirkung.

Die Zentren, in denen Auftraggeber und Gestalter den Mut zu Neuem aufbrachten und das Publikum dazu fanden, waren zunächst Zürich und Basel.

In den USA gab es schon lange eine starke Druckindustrie und Werbeszene, die mit patriotisch-traditionellen Motiven im Stil der Plakatmalerei und lautmalerischer Typographie eher zufällig arbeitete. Ähnlich wie in der Schweiz erfuhr die Szene in den USA durch Emigranten aus vielen Ländern Europas, die durch Weltkrieg und Ideologien vertrieben wurden, Erneuerung.

Geschaffen wurde ein Stil ohne regionale Zeichen, aber mit grenzüberschreitender Qualität. Es waren elementare Typographie, abstrakte Gestaltung, Fotografie und konzeptionelles Denken, die diesen „Internationalen typographischen Stil" zu der großen Bewegung nach 1945 werden ließen.

Während in der Schweiz zunächst nur serifenlose Schriften (Groteskschriften) für den neuen Stil eingesetzt wurden, kam später zögernd die klassische Renaissance-Antiqua, besonders die Garamond, zur Anwendung. In den USA war man weitaus weniger dogmatisch und verwendete beide Schriftgruppen gleichberechtigt.

et informatif. Zurich et Bâle sont les premières villes où commanditaires et créateurs osent innover et trouvent un public.

Aux Etats-Unis, il existe depuis longtemps déjà des imprimeries prospères, et un secteur publicitaire qui a recours à des motifs patriotiques et traditionnels s'inspirant de l'affiche peinte et présentant des typographies criantes d'un aspect assez hasardeux. Tout comme en Suisse, les milieux américains commencent à innover au contact des émigrants que la guerre et les idéologies ont chassés d'Europe.

On crée un style dépourvu de particularisme régional et capable de passer les frontières. La typographie élémentaire, la mise en page abstraite, la photographie, l'importance de la maquette font de ce «style international» l'un des mouvements décisifs de l'après-guerre.

Si, en Suisse, on avait tendance à n'utiliser d'abord que des caractères sans sérif (Grotesks), l'Antique Renaissance, une classique, et surtout la Garamond connaissent peu à peu un regain d'intérêt. Aux Etats-Unis où l'on est moins dogmatique, ces deux familles de polices se côtoient sans discrimination.

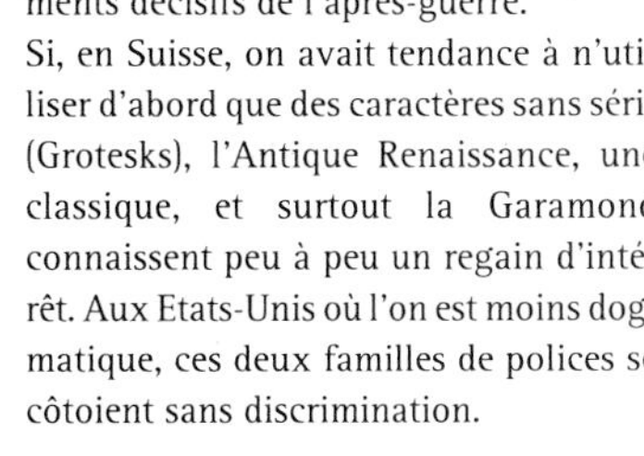

1948 Pintori

1949 Wittkugel

7
congrès international
d'architecture
moderne 22-31 juillet 1949
ciam
bergamo palazzo della ragione

1949 Huber

1950 Lohse

1951 Stankowski

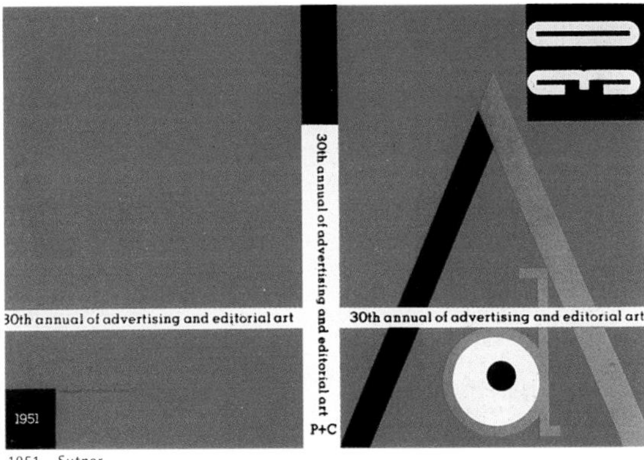

30th annual of advertising and editorial art
1951
30th annual of advertising and editorial art
P+C

1951 Sutnar

1951 Huber

sen according to textual content; the page was free of caprice in its layout; and the whole impression was one of straightforward informativeness. Zurich and Basle were the first centers where designers and their patrons encouraged innovation and found an audience for these new ideas.

The USA had long had a flourishing printing and advertising industry, whose products with patriotic, traditional motifs in the style of painted posters and onomatopoeic typography seemed, if anything, fortuitous. As in Switzerland, new life was pumped into the American scene by emigrants displaced by war and political ideology from numerous European countries.

A style was created with no regional characteristics but international quality. "elementare typographie," abstract design, photography and conceptional thought made this international typographic style a major force in the postwar years.

In Switzerland, only sans-serif fonts were initially employed in the new style, but later classic Renaissance roman faces, in particular Garamond, were hesitantly introduced. The USA was far less dogmatic, using both groups in equal measure.

**Traditionsverbundene Typographie**
**Typographie traditionaliste**
**Traditional typography**

**Um 1935** Während die eruptive Phase der politischen und ästhetischen Neuerungen zu Ende ging, kamen in den dreißiger Jahren die konservativen und reaktionären Tendenzen in allen Bereichen verstärkt zur Wirkung. Von Beginn an gab es starke Angriffe und Auseinandersetzungen um die „neue Kunst", die „neue Architektur" und die „neue Typographie". Es waren nur zum Teil ästhetische Gründe, die zu ihrer Ablehnung führten. Wesentlich einflußreicher war die Ablehnung der Avantgarde durch die politischen Mächte. Dies führte zum Beispiel in der Sowjetunion zu einem offiziellen Verdikt gegen die neue Kunst, in Deutschland zu dem Begriff „Entartete Kunst" und zu ihrem Verbot.
Neben der radikalen kulturellen Avantgarde gab es auch immer eine parallele moderate Entwicklung. So entstanden zum Beispiel in England, wo keine sozialen Revolutionen in Sicht waren, in einer Weiterentwicklung klassischer Werte Schriftformen, die für die Alltagstypographie von größtem Wert waren. 1932 erschien als neue Schrift für die Tageszeitung *Times* die Antiqua Times New Roman von Stanley Morison, die zu einer der meistgebrauchten Schriften weltweit werden sollte. Schon die Schrift für das Orientierungssystem der Londoner U-Bahn aus dem Jahr 1916, eine Groteskschrift von Edward Johnston, zeigte die Besinnung auf alte Formen für ein Alphabet des zwanzigsten Jahrhunderts: die Proportionen gehen auf die römische Capitalis zurück. Bei den i- und j-Punkten ist die kalligrafische Herkunft zu sehen.

**Vers 1935** Alors que la phase de renouvellement et d'explosions politiques et esthétiques tire à sa fin, les années 30 voient la résurgence de tendances conservatrices et réactionnaires qui influencent tous les domaines. Dès le début, on assiste à de violentes attaques, à des conflits autour de «l'art moderne», de la «nouvelle architecture» et de la «nouvelle typographie». Les raisons de ce refus ne sont que partiellement d'ordre esthétique, car en fait, les pouvoirs politiques ont un impact bien plus important en matière de rejet de l'avant-garde. En Union soviétique, par ex., ceci conduit à certaines condamnations de l'art contemporain, tandis qu'en Allemagne, le concept «d'art dégénéré» entraîne son interdiction.
Des tendances plus modérées coexistent toujours parallèlement aux avant-gardes extrémistes. En Angleterre par ex., là où la perspective d'une révolution est absente, on voit apparaître de nouvelles fontes qui ont une grande valeur pour la typographie quotidienne, mais qui se situent dans la lignée des classiques. En 1932, paraît une nouvelle police destinée au *Times*, une Antique de Stanley Morison appelée «Times New Roman» qui sera l'une des plus employées au monde. Déjà, les caractères utilisés pour le système d'orientation du métro de Londres en 1916, une Grotesk d'Edward Johnston, signalent un retour à des formes anciennes pour l'alphabet du 20e siècle, un alphabet dont les proportions renvoient aux capitales romaines. On remarque les origines calligraphiques des points sur les «i» et sur les «j».
L'œuvre de Johnston, mais aussi celle de Stanley Morison, d'Eric Gill et d'Oliver Simon indiquent une autre voie, plus traditionnelle de la modernité. En Italie, aux Pays-Bas, aux Etats-Unis en Alle-

ABCDEFGH
IJKLMNOP
QRSTUVWX
YZ&£
abcdefghijkl
mnopqrstuv
wxyzæœ?
1234567890

1931 Morison

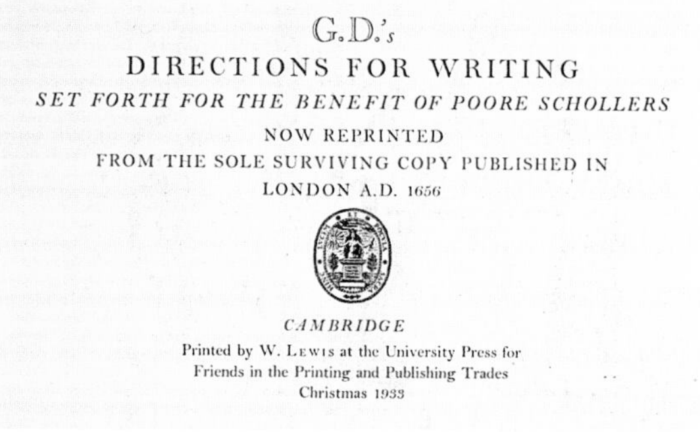


1933 Morison

1932 Ehmcke

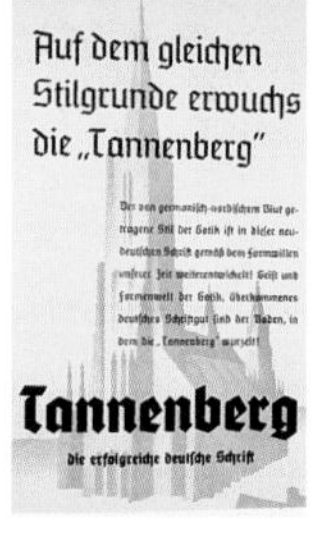


1934 Meyer

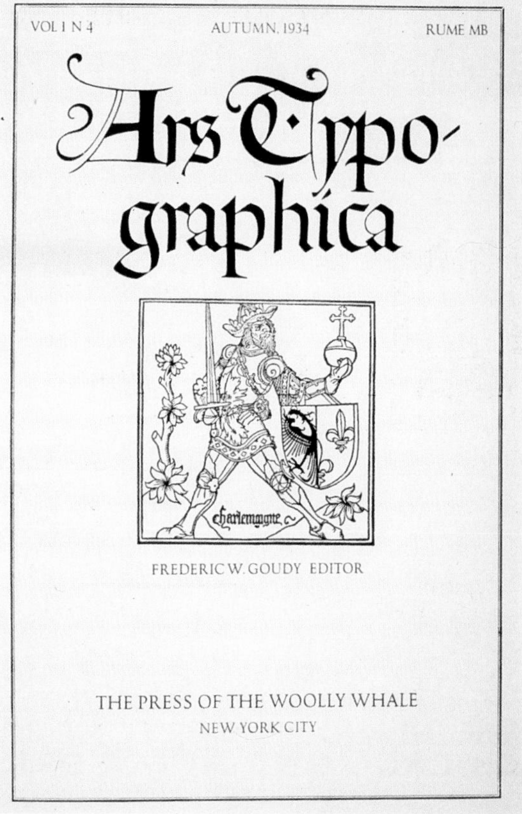


1934 Goudy

**c. 1935** As the turbulent phase of political and aesthetic innovation drew to a close, conservative and reactionary tendencies became stronger in all areas of creativity in the thirties. Right from the start, "new art", "new architecture" and "new typography" were the subject of heated arguments and critical attacks. They were rejected, but only partly on aesthetic grounds; contemporary political powers were extremely influential in suppressing the avant-garde. In the Soviet Union this led to official condemnation of the new art; in Germany, it was labeled "entartete Kunst" (degenerate art) and banned.
Yet there was always a more moderate form of development running parallel to the radical, cultural avant-garde. In England, for example, with not a single social revolution in sight, typefaces evolved as an expansion of classical values and became significant in everyday typography. In 1932, Stanley Morison produced a new roman font for the *Times* newspaper, known as Times New Roman, which was to become one of the most widely-used typefaces in the world. As far back as 1916, the lettering used on the London Underground system, a sans-serif by Edward Johnston, showed how a return to old forms could form the basis for a twentieth-century alphabet: the proportions go back to roman Capitals. Its calligraphic origins can be seen in the dots on the i and j. Johnston's work, and also that of Stanley Morison, Eric Gill and Oliver Simon, demonstrated a different, traditional, component of modernism.

**Traditionsverbundene Typographie**
**Typographie traditionaliste**
**Traditional typography**

Das Werk Johnstons, aber auch das von Stanley Morison, Eric Gill und Oliver Simon zeigen einen anderen, traditionelleren Weg der Moderne. Ähnliche Anschauungen führten in Italien, Holland, den USA oder Deutschland zu neuen Alphabeten und zu einer zeitgemäßeren Typographie des feinen Maßes: Mittelachse, Versalzeilen in Antiqua-Charakter, moderater Schmuck in neuer Bibliophilie ließen selbst die Vertreter der radikalen Typographie an ihren Entwürfen zweifeln. So wandelte sich zum Beispiel unter dem Eindruck dieser Entwicklung Jan Tschichold zu einem ebenso vehementen Vertreter des Traditionellen, wie er es vorher als Propagandist des Elementaren war.
Diese traditionsverbundene Entwicklung wurde von den Typographen mit ihren Entwürfen eingeleitet, und von den internationalen Schriftgießereien ermöglicht. Es erschienen interessante neue Alphabete, es wurde eine große Anzahl neuer Bücher zur Typographie herausgegeben, und in vielen neuen Fachzeitschriften wurden die handwerklichen Entwicklungen und die ästhetischen Probleme erörtert.

1936 Gill

ABCDEFGH
IJKLMNOP
QRSTUVW
XYZ&
abcdefghijk
lmnopqrstu
1234567890

1936 Gill

magne, des conceptions semblables conduisent à de nouveaux alphabets et à une typographie mieux adaptée à l'époque. L'axe central, les retraits avec lettrines en Antique, la sobriété décorative de la nouvelle bibliophilie, amènent les adeptes des typographies radicales à douter de leurs maquettes. Ainsi par ex. influencé par cette nouvelle tendance, Jan Tschichold, change d'optique et se met à défendre la tradition avec la même véhémence qu'il prônait la typographie élémentaire.
Ce retour à la tradition est donc provoqué par des typographes. En revanche, au niveau international, les fondeurs de caractères en permettent le développement. Des alphabets intéressants voient le jour, on publie un grand nombre d'ouvrages traitant de la typographie, et de nouvelles revues spécialisées débattent de l'évolution des techniques artisanales et des problèmes esthétiques.

1935 Wolpe

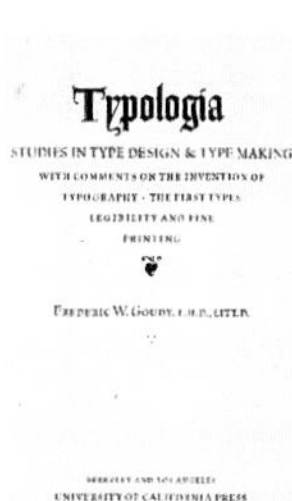


1940 Goudy

ZUR 650-JAHR-FEIER
DER SCHWEIZERISCHEN EIDGENOSSENSCHAFT

Goethes
Briefe aus der
Schweiz
1779

NACH DER AUSGABE LETZTER HAND 1832
HOLBEIN-VERLAG · BASEL

1941 Tschichold

1941 Mardersteig

1935 Freedman

Similar ideas in Italy, the Netherlands, the USA and Germany produced new alphabets and a more up-to-date typography with fine dimensions; center axes, lines of capitals with a roman character, and moderate decoration in a new bibliophily, caused even the representatives of radical typography to question their designs. The influence of these features was such that Jan Tschichold, for example, became as vehement an advocate of the traditional as he had previously been a propagandist for "elementare typographie".

This re-discovery of the traditional was sparked off by the designs of the typographers and made possible through the production facilities of international type foundries. Interesting new alphabets were produced; many new books were published on the subject of typography, while technical developments and aesthetic problems were debated in the many new specialist journals.

**Elementare Typographie**
**Typographie élémentaire**
**"elementare typographie"**

**Um 1925** Das Ende des Ersten Weltkriegs im Jahr 1918 ermutigte progressive Kräfte, neue Wege zu gehen. Dies zeigte sich in revolutionären sozialen Bewegungen ebenso wie in radikalen kulturellen Erneuerungsbestrebungen. Solche Versuche traten in vielen europäischen Ländern gleichzeitig auf und brachten variantenreiche theoretische und praktische Ergebnisse hervor.
Neben dem Expressionismus, der individuelle Emotion und Gestik in den Mittelpunkt seines Schaffens stellte, entwikkelte sich aus der klaren, reduzierten Formensprache des Konstruktivismus im Laufe der zwanziger Jahre die elementare Typographie, die für eine zweckgebundene, informierende Form der visuellen Kommunikation des ganzen Jahrhunderts die Grundlage bilden sollte.
Im Gegensatz zu einer rückwärtsgerichteten, denkmalartigen Typographie mit mittelachsialer Anordnung und mittelalterlicher Schrift besannen sich die Pioniere der Moderne auf die Substanz der neuesten technischen Entwicklungen und auf die Notwendigkeit, für diese neue Wirklichkeit eine kongeniale gestalterische Sprache zu entwickeln.
Für dieses Ziel war die serifenlose Linear-Antiqua (Groteskschrift) die konzeptionell und formal geeignetste Schrift. Die Qualitäten dieser schon um 1810 in England erstmals entworfenen und produzierten völlig neuen Schriftgruppe war fast hundert Jahre unterschätzt und wirtschaftlich wie ästhetisch ein großer Mißerfolg. Deshalb gab es am Beginn des Groteskzeitalters nur wenige Schriften,

**Vers 1925** En 1918, la fin de la Première Guerre mondiale encourage les forces progressistes à s'engager dans de nouvelles voies. Ces tendances émergent autant dans les mouvements sociaux-révolutionnaires qu'à travers des aspirations à un renouveau radical de la culture. Elles se manifestent dans plusieurs pays européens en même temps et donnent naissance à diverses variantes théoriques et pratiques.
Outre l'expressionnisme, qui place l'émotion et le geste individuels au cœur de la création, on assiste, dans les années 20, à l'avènement de la typographie élémentaire fondée sur un langage formel, clair et réducteur lié au constructivisme. Il constituera la base d'un modèle informatif et pragmatique qui marquera la communication visuelle de ce siècle.
Contrairement à la typographie rétrograde et édifiante, où des caractères médiévaux sont agencés en fonction d'un axe central, les pionniers du modernisme réfléchissent sur la nature des nouvelles technologies et sur la nécessité de développer un langage formel qui traduirait le génie de la nouvelle réalité.
Pour satisfaire à ces exigences l'Antique linéaire et sans sérif (Grotesk) semble être le caractère dont la forme et le concept sont le mieux adaptés. Les qualités de ce nouveau type de caractère, créé et produit en Angleterre dès 1810, avaient été sous-estimées pendant presque un siècle, et ce type de fonte avait connu un échec autant économique qu'esthétique. Ainsi ne possède-t-on au début de l'ère du Grotesk que quelques caractères capables d'être directement utilisés et répondant aux idées des maquettistes. Les ateliers de composition disposent surtout de Grotesk de ville (conçue dès 1898 en plusieurs fontes) et de Venus (à partir de 1907). Quand les hauts de casse

ca. 1925 Bayer

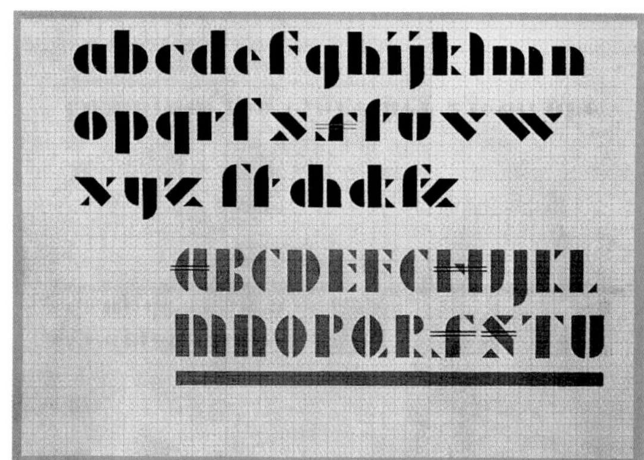
1923–1926 Albers

1929 Dexel

1927 Renner

1932 Schuitema

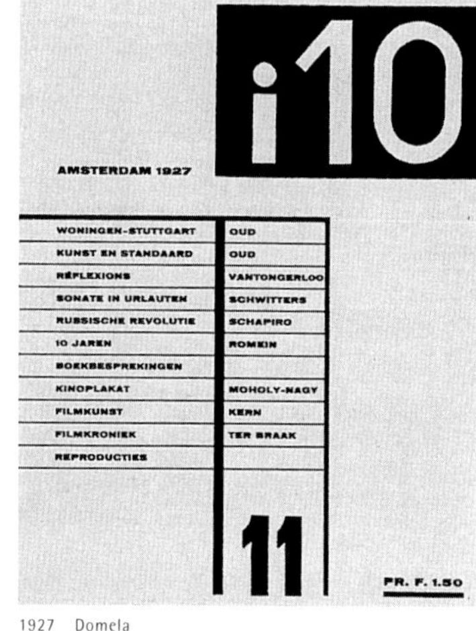

1927 Domela

Burchartz

**c. 1925** The end of the First World War in 1918 encouraged progressive minds to tread new paths. This was manifested by the revolutionary social movements of the time and by radical, cultural endeavors towards change. Such attempts were made simultaneously in many European countries, with a broad spectrum of theoretical and practical results.
Alongside Expressionism, whose creative focus was on the emotion and gestures of the individual, "elementare typographie" developed from the clear, reductionist forms of Constructivism during the twenties. This new style was intended to form the basis of a practical, informative form of visual communication for the rest of the century.
Rejecting backward-looking, monumental typography with medieval lettering and arrangements around a central axis, the pioneers of the modern age pondered the assets of recent technological advance and the need to develop a congenial structural language for this new reality. Conceptually and formally, sans-serif types seemed most appropriate. The qualities of this new group of fonts, first designed and produced in England in 1810, had been underestimated for almost a hundred years; the fonts had been an economic and aesthetic failure. At the beginning of the sans-serif age there were, thus, only a few available typefaces which were suitable for immediate use and which matched designers' ideas, in particular Akzidenz-Grotesk (developed in many styles from 1898 onwards) and

die für die unmittelbare Verwendung geeignet waren, und die den Vorstellungen der Gestalter entsprachen. Es war vor allem die Akzidenz-Grotesk (ab 1898 in vielen Schriftschnitten entworfen) und die Venus (ab 1907), die in den Setzereien vorrätig waren. Wenn von einzelnen Buchstaben wie E, F, H, I, L, T, nicht die benötigte Menge im Setzkasten vorrätig waren, wurden sie kurzerhand mit Linienelementen zusammengesetzt, was für unser geübtes Auge heute oft merkwürdig aussieht.

Die Elementare Typographie wollte dem Stilwirrwarr der vorhergehenden Zeit ein Ende machen. Es ging um formale Klarheit und um das Ausschalten ornamentaler Zutaten. Die Form der Typographie sollte sich aus dem jeweiligen Text ergeben, und nicht in eine vorgegebene Anordnung gepreßt werden. Man benutzte die Asymmetrie und nur wenige verschiedene Schriftgrößen oder Schriftschnitte. Da die Vorzüge der elementaren Typographie schnell bekannt wurden, und der Mangel an guten Schriften erkannt war, ging schon 1927 die Futura von Paul Renner in Produktion. Beruhend auf Vorarbeiten aus dem Atelier Ferdinand Kramer wurde die Futura schnell die Schrift, in deren einzelnen Zeichen sich die neue Typographie am besten verstanden sah. Danach enstanden in kurzer Zeit die Gill von Eric Gill (1928) und die Erbar von Jakob Erbar (1929). Mit diesen bis heute weitverbreiteten Alphabeten war die Pionierphase beendet und die Basis für nachfolgende Entwicklungen gelegt.

1923 Zwart

1928 Gill

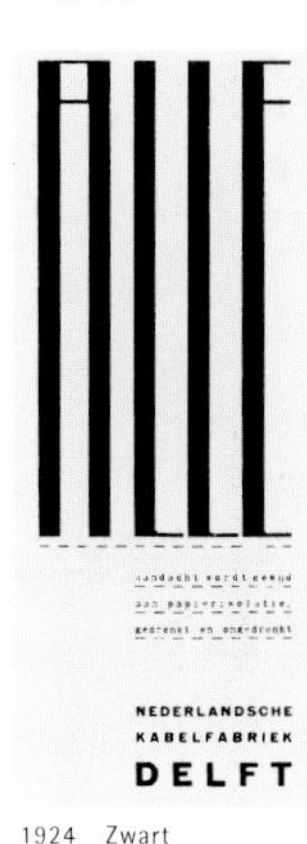


1924 Zwart

1932 Tschichold

1931 Trump

1931 Stankowski

1929 Anonymous

contiennent des E, F, H, I, L et T en quantité insuffisante, on les reconstitue en un tour de main à l'aide de fragments de filets, ce qui aujourd'hui étonne souvent notre œil exercé.
La typographie élémentaire veut mettre un terme à la confusion stylistique des périodes précédentes. L'enjeu est la précision formelle et l'élimination des ingrédients ornementaux. La forme typographique doit résulter du texte et ne pas être pressée dans un moule préalablement donné. On utilise l'asymétrie en variant peu les corps et les graisses.
Comme on prend rapidement conscience des avantages de la typographie élémentaire, mais aussi du manque de caractères adéquats, on commence à produire la Futura de Paul Renner dès 1927. Reposant sur des travaux préalables de l'atelier de Ferdinand Kramer, la Futura ne tarde pas à devenir le caractère qui convient le mieux à la nouvelle typographie. Puis, apparaissent en peu de temps la Gill d'Eric Gill (1928) et l'Erbar de Jakob Erbar (1929). Ces alphabets, encore très répandus de nos jours, marquent la fin de l'ère des pionniers et constituent la base des recherches ultérieures.

Venus (1907 onwards). If there were not enough of the letters E, F, H, I, L and T in the case, they were compiled from linear elements, which today often appears strange to our trained eyes.
"elementare typographie" aimed to put an end to the stylistic confusion of the previous years. It concentrated on formal clarity and the elimination of ornamental ingredients. Typographical form was to evolve from the text and not be squeezed into a given arrangement. Asymmetry was used along with a minimum of different font sizes and font varieties.
The merits of "elementare typographie" and the lack of good typefaces were quickly recognized, and in 1927 production of Paul Renner's Futura started. Based on preliminary work from Ferdinand Kramer's studio, Futura soon became the type whose individual characters the new typography felt best at home with. In a relatively short space of time, Futura was followed by Gill by Eric Gill (1928), and Erbar by Jakob Erbar (1929). These alphabets, still widely used today, ended the pioneering phase and formed the basis for future typographical developments.

**Individuelle skripturale Form**
**Formes scripturales individuelles**
**Individual written form**

**Um 1925** Viele Anregungen zur Typographie gingen immer wieder von Gestaltern aus, die nicht in den traditionellen Handwerksauffassungen oder gebrauchsgrafischen Berufen ausgebildet waren. Dabei entstanden Arbeiten, die bewußt die erwartete Form von Text und Schrift negierten, was natürlich eine weitgehende Kenntnis des Traditionellen vorraussetzte.
Es waren Maler, Zeichner, Holzschneider, aber auch Pressendrucker, die in für sie ungesicherte und unbekannte Bereiche vordrangen. Nach der „Befreiung" der traditionellen typographischen Form durch Futurismus, Dadaismus und Konstruktivismus entstanden in den zwanziger Jahren einflußreiche Arbeiten, die sowohl Auswirkungen auf neue einzelne Buchstabenformen hatten, aber auch auf die Auffassung von der Gestaltung ganzer Textseiten und der Textanordnung. Es waren Ergebnisse, die keinen funktionalen Interpretationsansatz darstellten, sondern einen sinnlich-individuellen Assoziationsspielraum in die Typographie brachten. Meist wurden keine vorgefertigten Buchstaben in diesen Entwürfen verwendet, sondern direkte skripturale Phantasien, und meist wurden die waagerechten Textabläufe vernachlässigt oder aufgehoben. Der Anteil von Schrift und zeichenhaften Elementen wird in ein ausgewogenes Gestaltungsverhältnis gebracht. Texte erscheinen in einem Zwischenreich von lesbar und sichtbar, was viele Jahre später z. B. in der Visuellen Poesie zu kultureller Blüte entwickelt wurde.

**Vers 1925** De nombreuses idées viennent sans cesse de graphistes qui n'ont pas été formés dans le respect des conceptions artisanales traditionnelles ou du graphisme utilitaire. On voit alors apparaître des travaux qui remettent consciemment en question la forme courante du texte et du signe, ce qui présuppose évidemment une ample connaissance des traditions.
Ce sont les peintres, les dessinateurs, les graveurs sur bois mais aussi les imprimeurs à la presse qui s'engouffrent dans ces domaines incertains et inexplorés. Après la «libération» des formes typographiques traditionnelles par le futurisme, le dadaïsme et les constructivistes, on assiste au cours des années 20 à la naissance de réalisations décisives qui auront des conséquences autant sur le graphisme des diverses lettres que sur la conception, la maquette, et l'agencement du texte. Ces travaux ne correspondent pas à une approche fonctionnelle et interprétative, mais introduisent dans la typographie une liberté permettant des associations personnelles et sensibles. Pour réaliser ces maquettes, les graphistes n'emploient pratiquement aucune lettre fabriquée d'avance, mais laissent libre cours à leur imagination scripturale. Le plus souvent, ils négligent la présentation horizontale du texte, ou même la suppriment. La proportion de caractères et d'éléments graphiques s'équilibre au sein de la mise en page. Les textes se situent à la limite entre lisibilité et visibilité, une technique qui, des années plus tard, connaîtra son aboutissement culturel dans la poésie visuelle.
Les expérimentations stylistiques se développent dans cette atmosphère de renouveau, propice aux grands changements, qui succède à la Première Guerre mondiale. Elles émergent du domaine de

1921 Itten

1920 Dicker

1922 Itten

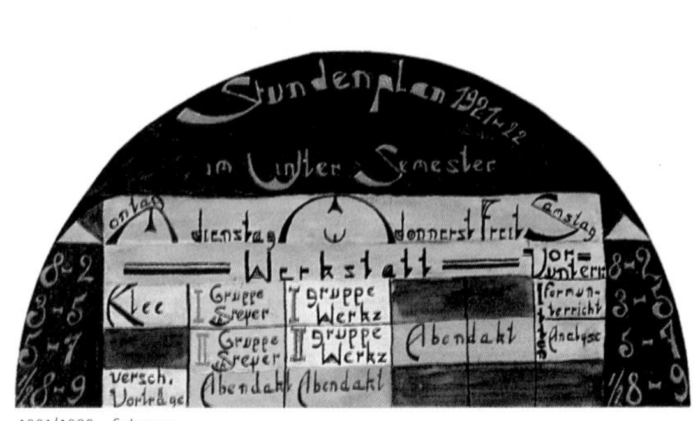

1921/1922 Schreyer

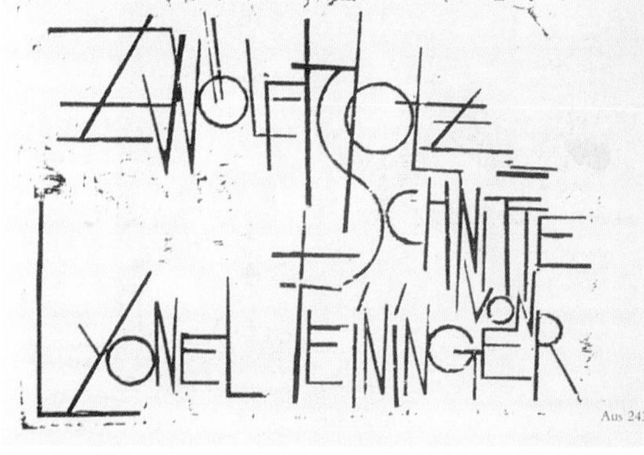

1918–1920 Feininger

**c. 1925** Many typographical ideas came from designers who had neither been trained in the traditional methods of the trade nor in commercial art. Work was thus produced which consciously negated expected forms of text and type, which of course presupposed broad familiarity with the traditional.
It was painters, draftsmen, wood cut artists and also press printers who explored fields new and unfamiliar to them. After the "liberation" from traditional typographical structures through Futurism, Dada and Constructivism, the twenties saw the appearance of influential works which had their effect not only on new, individual letter-forms, but also on attitudes concerning page layout and the arrangement of text. The outcome was not a functional approach but rather the introduction to typography of a sensuous, individual, associative latitude. Prefabricated letters were seldom used in these designs, the accent being on scripted fantasies, and horizontal lines of text were neglected or abandoned com-

Diese stilistischen Experimente entsprangen der allgemeinen veränderungsfreudigen Aufbruchsstimmung nach dem Ersten Weltkrieg. Sie entstanden im freien gestalterischen Bereich, wurden aber auch zur Identitätsfindung in Grundsemestern von Schulen wie dem Bauhaus erarbeitet. Hier entstanden gestische, zeichnerische und collagierte Arbeiten, die das individuelle Suchen nach Neuem und die Entwicklung der eigenen gestalterischen Identität der Studierenden einleiten sollte. Wichtigster Anlaß für diese Richtung blieb es jedoch, methodisch eine Alternative und Gegenbewegung zu den strengeren, parallel entstandenen Stilauffassungen dieser Zeit zu erarbeiten. Es ging bei diesem individuell-skripturalen Vorgehen nicht um Schönschrift und nicht um Kalligraphie. Beide Ausdrucksweisen wurden im akademischen Bereich und in privaten Zirkeln praktiziert. Durch eine stark traditionelle Ausrichtung ging von ihnen jedoch kein Impuls auf eine neue Typographie aus.

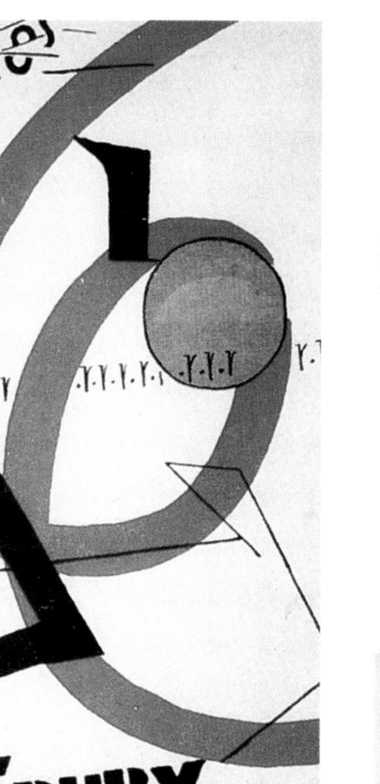

1925 Strzeminski

1920 Molzahn

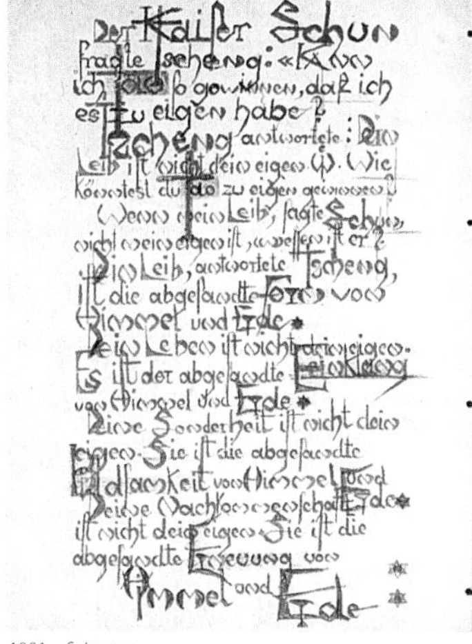
1921 Schreyer

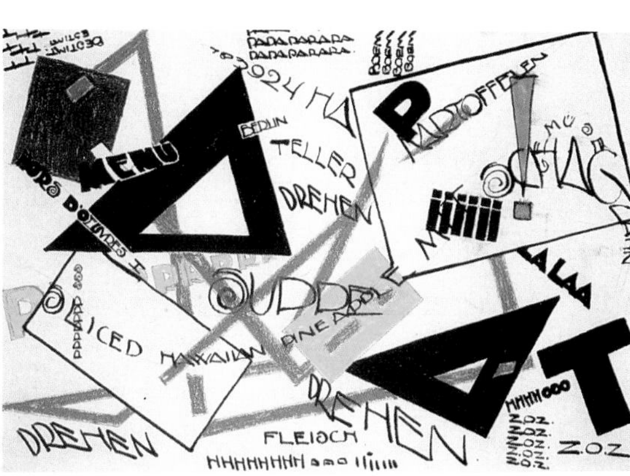

1928 Mendelsohn

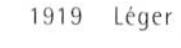
1919 Léger

la création indépendante, et se retrouvent dans des écoles telles que le Bauhaus où les étudiants des premiers semestres les élaborent dans le cadre d'une recherche de leur identité. Ils réalisent des travaux où le geste, le dessin et le collage jouent un rôle et doivent induire une recherche individuelle de la nouveauté et développer l'identité plastique des étudiants. Pourtant l'un des mobiles les plus importants qui pousse dans cette direction, est de développer une méthode, une alternative et un contre-courant aux conceptions stylistiques rigoureuses qui existent simultanément et parallèlement. L'enjeu de cette démarche scripturale individuelle n'est ni la belle écriture ni la calligraphie. Ces deux modes d'expression sont réservés aux pratiques des académies et des cercles privés. Mais en raison d'une orientation fortement traditionnelle, ils auront peu d'impact sur la nouvelle typographie.

pletely. The proportion of writing to symbolic elements was brought into balance. Texts hovered in a limbo between legibility and visibility, a technique which developed many years later to reach its cultural peak with the genre of visual poetry, for example.

These experiments with style arose from the general desire for change following the First World War. They originated among unaffiliated designers, but were also among the components for foundation courses in schools such as the Bauhaus, searching for an identity. Here gestural, graphic and collage works were produced which were supposed to promote the individual's search for new ideas and encourage students to develop their own stylistic identity. The most important reason behind this artistic departure was, however, the need to work out the methods of an alternative counter-movement to the more rigid contemporaneous stylistic conception of the age. The aim of this individual scriptual form was not to reinvent beautiful handwriting or calligraphy. Both forms of expression were practiced in the academic world and in private circles. Yet the strong adherence to tradition of both styles failed to provide stimulus for a new typography.

## Art Deco

**Um 1925** Der als Art Deco bezeichnete typographische Stil ging nicht von einem reduzierten Formenrepertoire aus, sondern, trotz des konstruktivistischen Ansatzes, von einer Anreicherung der elementaren Formen mit Ornamenten und mit vielfältigem Dekor. Neben dem geometrischen Ansatz gehörten florale und figurative Motive, die in kubischen Abstraktionen verwandt wurden, zu den Hauptmotiven des Art Deco. Es entstanden modisch-zeitgeistige Chiffren, die in den Niederlanden, Frankreich, England, in der Schweiz, in den osteuropäischen Ländern und den USA zu großer Blüte und Verbreitung kamen.
Die Gestalter des Art Deco-Stils wollten keineswegs (wie die Anhänger der elementaren Typographie) die Gesellschaft, in der sie lebten, verändern, sondern sie wollten die Konsumwelt verschönern. Diese Bestrebungen zeigten sich im grafischen Bereich bei Zeitschriften, Plakaten, Anzeigen oder Büchern ebenso wie im dreidimensionalen Bereich bei Architektur, Möbeln, Verpackungen, Mode oder Bühnenbild. Das Schriftschaffen der Art-Deco-Zeit brachte neben bizarr dekorierten Schriftzügen und schwer verwendbaren Alphabeten einige ausgearbeitete Schriftfamilien, die größere Verbreitung erlebten, und zum Teil bis heute in Anwendungen zu sehen sind.
Es waren Alphabete, bei denen die Geometrie, die Abstraktion und die Elementarformen nicht pur eingesetzt wurden (wie zum Beispiel bei der elementaren Futura), sondern die eine unausgeglichene Wechselwirkung von breit zu schmal,

**Vers 1925** Le style typographique que l'on appelle Art déco ne découle pas d'une réduction du répertoire formel, mais prône, en dépit de la démarche constructiviste, l'enrichissement des formes élémentaires au moyen d'ornements et de décors divers. Outre les éléments géométriques, certains motifs floraux et figuratifs issus des abstractions cubistes comptent parmi les thèmes majeurs de l'Art déco. Ceci donne naissance à des figures à la mode, reflétant l'esprit du temps et qui connaissent leur apogée aux Pays-Bas, en France, en Angleterre, en Suisse, dans les pays d'Europe centrale et aux Etats-Unis, où ce style a un grand retentissement.
Les créateurs Art déco ne cherchent en aucun cas (comme les adeptes de la typographie élémentaire) à transformer la société dans laquelle ils vivent, mais plutôt à embellir l'univers de la consommation. Leurs aspirations se reflètent dans les arts graphiques, les revues, affiches, publicités et livres, tout comme dans le domaine des volumes, dans l'architecture, le design de mobilier, d'emballages, ainsi que dans la mode et la scénographie. Outre des paraphes bizarrement ornés et des alphabets d'emploi difficile, il reste de la période Art déco quelques familles de caractères aboutis, dont certains sont encore utilisés de nos jours. Ces alphabets ne font pas intervenir la géométrie, l'abstraction et les formes élémentaires à l'état pur (comme la Futura), mais présentent une alternance asymétrique et délibérée de pleins et de déliés, de structures construites et d'éléments gestuels, de clair et de sombre. Ce sont des formes qui se veulent toujours plus riches en associations et en sensualité que les autres alphabets conçus dans un autre esprit stylistique. De nombreux caractères sont construits par assemblage de plans géo-

1927 Hlavaček

1929 Cassandre

1926 Renner

1925 Magritte

1927 McKnight Kauffer

1927 Cassandre

**c. 1925** The typographical style of Art Deco was not based on a reduced repertoire of forms, but, despite its Constructivist beginnings, on an enrichment of elementary forms through ornamentation and a variety of decorative elements. Alongside the basic geometric approach, the main motifs of Art Deco included floral and figurative patterns which were given a cubist abstract interpretation. Fashionable contemporary designs were created which met with great popularity in the Netherlands, France, England, Switzerland, eastern Europe and the USA.
Unlike the followers of "elementare typographie", the designers of Art Deco on no account wanted to change the society in which they lived; instead, they wished to beautify the world of consumerism. This is evident in the magazines, posters, adverts and books of the age, and also in the three-dimensional world of architecture, furniture, fashion, packaging and stage design.
Art Deco's typographical creations mainly produced crazily-decorated designs and barely-usable alphabets, yet there were also several families of type which were more widely employed at the time and can in some cases still be found today. These were alphabets where geometry, abstraction and elementary forms were not employed in the raw (as with Futura, for example), but which concentrated on a disharmonious interplay of broad and narrow, of the constructed and the gestural, of light and dark. These were forms which aimed at being more

von konstruiert und gestisch, von hell und dunkel bewußt einsetzten. Es waren immer Formen, die assoziationsreicher und sinnlicher sein wollten als die unter anderen stilistischen Gedanken entwickelten Alphabete. Viele Schriften wurden aus geometrischen Flächen zusammengesetzt, und nicht aus der Tradition der geschriebenen Schriften. Häufig wurden die Buchstaben räumlich-zeichnerisch ergänzt. Stilistische Anleihen wurden aus früheren Epochen genommen, zum Beispiel aus der aztekischen Kunst und aus der Formenwelt der nordamerikanischen Indianer. Bei den Alphabeten stand nicht die Lesbarkeit im Vordergrund, sondern die unmittelbar dekorative Wirkung.

Die Art Deco-Zeit und ihr gestalterischer Stil sind untrennbar verbunden mit der gleichzeitig eingeführten Gesamtelektrifizierung, der Stromlinienform und dem Beginn des Jazz.

Nach ihrer Blütezeit um 1925 erlebt die reichhaltige formale Pracht des Art Deco immer wieder Aufmerksamkeit und Aktualisierung, die jedoch keine neue Richtung hervorbrachte, sondern immer auf die ursprünglichen Stilelemente zurückgriff.

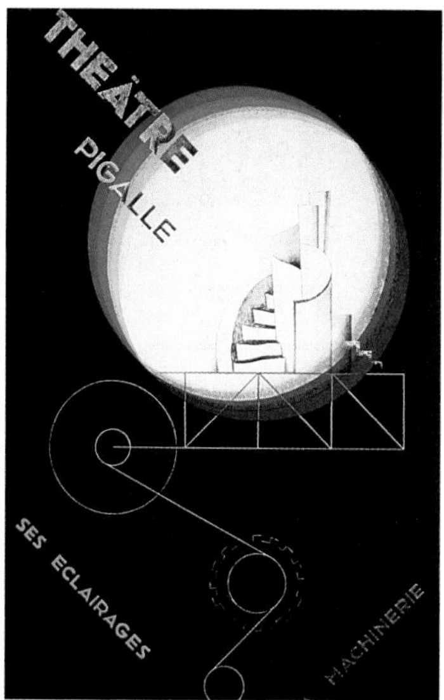

1929 Carlu

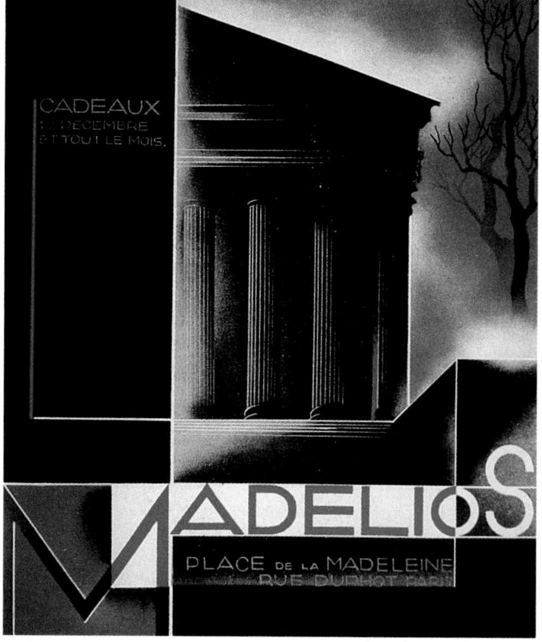

1928 Brodovitch

métriques et sans référence à la tradition de l'écriture manuscrite. Des extensions tridimensionnelles sont souvent ajoutées à la main aux lettres. On puise dans les styles antérieurs, avec des emprunts stylistiques à l'art aztèque et à l'univers formel des Indiens d'Amérique du Nord. Concernant ces alphabets, l'aspect primordial n'est pas la lisibilité mais l'effet décoratif.

La période Art déco et son style sont indissociables de la généralisation simultanée de l'électrification, des lignes aérodynamiques et du jazz. Après avoir vécu une période d'apogée vers 1925, la splendeur et la richesse formelle du style Art déco connaîtra à maintes reprises un regain d'intérêt et d'actualité sans pour autant que ceci donne naissance à une nouvelle tendance artistique, mais plutôt à des reprises des éléments stylistiques initiaux.

1922 Bernath

1930 McKnight Kauffer

1929 Csillac

associative and sensuous than alphabets developed under other stylistic ideals. Many of the typefaces were compiled from geometric surfaces and not based on traditional calligraphy. Three-dimensional extensions were often added to the letters by hand. Stylistic traits were borrowed from earlier epochs, such as Aztec art and the artistic forms of North American Indians. Alphabets were decorative in the first instance; legibility was a secondary factor.

The creativity of Art Deco and its designs is inextricably linked to other manifestations of the age: universal electrification, streamlining and the beginnings of jazz. Following its peak in around 1925, the rich, formal splendor of Art Deco enjoyed recurring phases of attention and revision yet no new direction was forthcoming; time and again designers returned to the original stylistic elements.

## Expressionismus
## Expressionnisme
## Expressionism

**Ca. 1910–1925** Als eine der ausgeprägten Ausdrucksformen am Beginn des 20. Jahrhunderts war der Expressionismus in den verschiedenen Gestaltungsbereichen ein besonders aufrüttelnder Zeitabschnitt.
Eruptive Schriftkaskaden, meist dem Medium Holz abgerungen, sind ohne den Zwang der waagerechten Schriftanordnung laut und bizarr über die Seitenformate gelegt. Dieser Stil war, wie so oft im 20. Jahrhundert, von freien Künstlern in die angewandten Gestaltungsbereiche eingebracht worden. Es war ein weiterer Versuch, eine bildhafte Opposition gegen die immer stärkere Anpassung und Gleichförmigkeit in der Gesellschaft zu schaffen.
Für die angewandte Typographie brachte der Expressionismus wenige Aufträge aus der Wirtschaftswerbung. In Buch, Zeitschrift und Plakat wurden allerdings durch die kleinen Auflagen, durch die original grafische Wirkung bei bibliophilen Publikationen und bei radikalen sozialkritischen Pamphleten überzeugende und innovative Ergebnisse erarbeitet. Der gesamte formale Anspruch der Schriftformen ging nicht in die Richtung einer allgemeinen, informierenden Sachlichkeit. Vielmehr entsprach er der durch die Industrialisierung und Politik verletzten individuellen Seele, die durch den expressionistischen Stil einen bildgewordenen Protest einlegte.
Obwohl der Expressionismus nur kurze Zeit überzeugende Ergebnisse hervorbrachte, sind bis heute immer wieder Anregungen von ihm auf die nachfolgende

**Vers 1910–1925** L'expressionnisme est l'un des styles les plus marquants de ce début de siècle, il reflète une période particulièrement agitée qui touchera divers secteurs de la création.
Des éruptions et des cascades de caractères, le plus souvent extraites d'un support bois, sont disposées de manière provocante et étrange sur les pages, sans tenir compte des contraintes de l'agencement horizontal des lettres. Comme souvent au 20e siècle, ce sont des artistes indépendants qui ont introduit ce style dans le domaine des arts appliqués. Il est l'expression d'une tentative visant à générer une sorte d'opposition plastique à l'uniformisation et l'assimilation croissante de la société.
Sur le plan des applications typographiques, l'expressionnisme ne recueillera que peu de commandes de publicités commerciales. Toutefois, les secteurs du livre, des revues et affiches produiront des petits tirages innovateurs et convaincants par leur originalité graphique, surtout en ce qui concerne les publications destinées aux bibliophiles et les pamphlets politiques extrémistes contenant des critiques sociales.
D'un point de vue formel, l'impression d'ensemble des caractères ne va pas dans le sens d'une objectivité informative et universelle. Au contraire, elle reflète cette âme individuelle, blessée par l'industrialisation et la politique, et qui, par le biais de l'expressionnisme, donne libre cours à la contestation picturale.
Bien que l'expressionnisme n'ait donné que brièvement des résultats convaincants, ses impulsions se retrouvent même de nos jours, et il exercera une influence sur la création artistique des périodes qui lui succéderont. On reconnaît et apprécie la tentative réussie de s'exprimer individuellement sans tenir compte des règles

1918 Schiele

1926 Schulz-Neudamm

1919 Fuchs

1919 Pechstein

**c. 1910–1925** Expressionism, one of the most distinctive early twentieth-century forms of artistic enunciation, was a particularly exciting stylistic period in the various fields of design.
Eruptive cascades of letters, mostly taken from woodcuts, spill across the pages, loud and bizarre, unconstrained by any horizontal arrangement of type. As with so many other twentieth-century movements, it was unaffiliated, individual artists who introduced this style to the applied creative disciplines. Expressionism was a further attempt at building up an artistic resistance to the ever-increasing conformity and uniformity of society.
The field of commercial advertising offered Expressionism few commissions for applied typography. In books, magazines and on posters, however, the small editions, and the original, graphic effect of bibliophile publications and of radical, socio-critical pamphlets produced convincing, innovative results.
The alphabets did not make any overall formal claim to general informative func-

künstlerische Gestaltung ausgegangen. Man erkannte und schätzte den gelungenen Versuch, einen persönlichen Ausdruck ohne Rücksichtnahme auf Regeln des Handwerks oder des herrschenden Stils zu schaffen. Durch individuelle Bearbeitung des Lithographie-Steins oder des Holzes konnten die Gestalter ihren Bekenntnissen Form geben.
Die Ergebnisse gliederten sich ohne Brüche ein in das gesamte Werk der expressionistischen Künstler. Sie mußten keine Zugeständnisse an die Welt der Wirtschaftswerbung machen, die in dieser Zeit extrem zunahm.
Das Individuelle war Stärke einerseits, aber Schwäche andererseits. Denn Wiederholungen und die Nicht-Übertragbarkeit der Formen auf alle Gestaltungsaufgaben begrenzten die Möglichkeiten und verhinderten, daß aus dieser Richtung eine umfassende Erneuerungsbewegung werden konnte.

que dicte le métier. En traitant la pierre à lithographie ou le bois à leur manière, les artistes peuvent formuler leurs désirs.
Ces travaux s'inscrivent sans rupture dans toute l'oeuvre des artistes expressionnistes. Ils ne sont plus obligés de se compromettre avec l'univers de la publicité commerciale qui gagne en importance à l'époque. Cet individualisme représente à la fois une force et une faiblesse car les reprises et l'impossibilité de transposer les formes dans les autres domaines de la création limitent et empêchent l'avènement d'un vaste mouvement novateur parti de cette tendance.

Kirchner

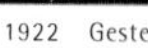

1922 Gestel

1928

1924 Schwab

1924 Kollwitz

1913 Kirchner

tionalism. They appeared instead to mirror the individual soul, damaged by industrialization and politics, which used the Expressionist style to make its protest in pictorial form.
Although Expressionism enjoyed only a brief boom, it has exerted a strong influence on artistic forms to this day.
This successful attempt at a personal expression which disregarded contemporary forms and the rules of the trade was recognized and appreciated. Free to work the litho stone or wood as they wished, designers were able to give form to their ideas.
Expressionist artists immediately integrated the products of these typographic experiments into their general repertoire. They were not obliged to make concessions to the world of commercial advertising which greatly expanded in this period.
This individualism was both the strength and weakness of this style. Repetition and the fact that forms could not be transferred to all areas of design was extremely limiting and prevented Expressionist typography becoming a new, all-embracing movement.

**Konstruktivismus**
**Constructivisme**
**Constructivism**

**Um 1920** Um 1920 entstand in Europa der Konstruktivismus. Der Begriff bezeichnete zunächst die Gestaltungsprinzipien, die am Beginn des Aufbaus der Sowjetunion entstanden. Hier sollte nach der Revolution von 1918 eine umfassende Erneuerung der Gesellschaft eingeleitet werden, und die Gestalter sollten als „Künstler-Ingenieure" an diesem Prozeß beteiligt sein. Dabei ging es nicht mehr um traditionelle und subjektive künstlerische Äußerungen, sondern um die Entwicklung einer Darstellungsform, die kollektiven und objektiven gesellschaftlichen Prozessen angemessen war. Dieses Ziel sollte mit einer allgemeinen, neuen Formensprache erreicht werden: geometrische Elemente, technische Klarheit und architektonischer Aufbau ergaben Bilder von kühler Harmonie, aber auch von dynamischer Vitalität.
In dieser Aufbruchszeit wurde in allen gestalterischen Bereichen mit konstruktivistischen Formen experimentiert.
Es entstanden Kinderbücher mit konstruktivistischen Formen und Inhalten, Plakate mit montierten Fotokonstruktionen, Fotoillustrationen für Bücher und konstruktivistische Filmsequenzen, Gedichtbände mit konstruktivistischer Gesamtgestaltung oder einzelnen Typokonstruktionen.
Die Blüte des Konstruktivismus in der Sowjetunion war von kurzer Dauer. Da die politische Führung die formalisierte und oft abstrakte Bildsprache für Agitation und Propaganda nicht geeignet fand, wurde ein realistischer Stil mit kraftvollen und fröhlichen Menschen als Motiv

**Vers 1920** Le constructivisme voit le jour en Europe vers les années 20. Ce concept définit d'abord les principes formels qui apparaissent au début de l'édification de l'Union soviétique. En 1918, juste après la révolution, le but est d'amorcer une profonde mutation de la société, et les créateurs doivent participer à ce mouvement en qualité d' «artistes ingénieurs». Il ne s'agit donc plus d'une expression artistique traditionnelle et subjective, mais d'élaborer des formes de représentation adaptées au processus de développement collectif et objectif de la société. Un nouveau langage formel doit permettre d'atteindre ce but; des éléments géométriques, la clarté technique et des compositions architecturées servent à produire des images possédant une harmonie froide, mais aussi une vitalité dynamique.
En cette période de bouleversement, on expérimente le constructivisme dans tous les domaines de la création.
On voit apparaître des livres pour enfants avec des formes et des sujets constructivistes, des affiches comportant des photomontages, des livres illustrés de photos, des films avec des séquences constructivistes, ainsi que des recueils de poèmes présentant une maquette et divers éléments typographiques de la même veine.
En Union soviétique, l'apogée du constructivisme fut de courte durée. Les dirigeants politiques considérant ce langage pictural formaliste et souvent abstrait comme inapte à l'agitation et à la propagande, ils lui préférèrent un style plus réaliste, montrant des hommes radieux et vigoureux; le constructivisme fut alors diffamé et interdit en tant que déviation formaliste de la vraie ligne prolétarienne.
Aux Pays-Bas, le mouvement De Stijl donna des impulsions fondamentales au constructivisme. Ce courant était égale-

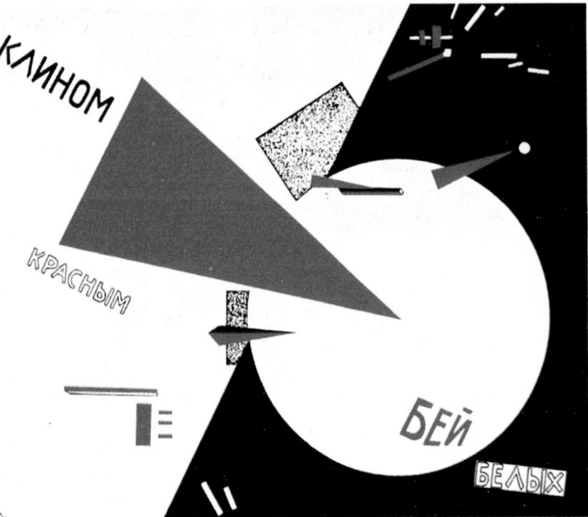

1920 Lissitzky

1924 Zwart

1926 Schuitema

1924 Kassak

1923 Schmidt

1922 Kassak

**c. 1920** Constructivism started in Europe in about 1920. The term originally described the style which originated during the formative years of the new Soviet Union. After the Revolution in 1918, a thorough restructuring of society was planned, where designers would participate as "artist engineers". Traditional and subjective artistic statements were no longer important; a form of presentation was to be developed which would be appropriate to the current collective and objective social processes. A new, general formal language was to be developed to achieve this: geometric elements, technical clarity and architectural constructions produced images in which cool harmony and dynamic vitality were both present.
In this time of social upheaval, Constructivist forms were experimented with in all areas of art, architecture and design. Children's books were produced with Constructivist form and content, posters with a montage of photographic constructions, photographic illustrations for books and Constructivist film sequences, volumes of poetry with an overall Constructivist design or individual Constructivist elements.
The heyday of Constructivism in the Soviet Union was short-lived. The political leaders found the formalized and often abstract pictorial language an unsuitable medium for agitation and propaganda purposes, preferring a realistic depiction of strong, happy people. Constructivism was attacked and banished, branded a formalist divergence from the correct proletarian path.

bevorzugt, der Konstruktivismus als formalistische Abweichung von der richtigen proletarischen Linie diffamiert und verboten.
In den Niederlanden stand die De Stijl-Bewegung für konstruktivistische Gestaltung. Auch in Ungarn, Tschechoslowakei, Polen und Jugoslawien gab es bedeutende konstruktivistische Strömungen. In Deutschland zeigte sich am Bauhaus nach einem eher mystischen Beginn eine starke konstruktivistische Ausrichtung. Aber auch der „ring neuer werbegestalter", die „Novembergruppe" und einzelne unabhängige Gestalter arbeiteten in diesem Sinne. Durch den konstruierten Ansatz wirken viele der angewandten typographischen Arbeiten heute steif und schwerfällig. Dies liegt daran, daß Texte im Blocksatz und Versalien gesetzt waren, daß positiv-negativ-Effekte, Flächen, Balken und Linien verwendet wurden, aber keine elementaren Alphabete. Die Fraktur- und Antiqua-Alphabete schienen den Konstruktivisten zu altertümlich, und geeignete Alphabete wie die Futura wurden erst ab 1927/28 entworfen und produziert.
Im Konstruktivismus wurden dennoch zahlreiche experimentelle Alphabete entworfen, die aber trotz klarer Konzepte nie für Mengensatz geeignet waren. Auch hierbei waren der radikale Neuerungswille und die Untersuchungen der Möglichkeiten des Entwerfens mit geometrischen Grundformen das eigentliche Ziel.

1928 Kluçis

1923 Rodtschenko

1926 Majakowski

1925 Schuitema

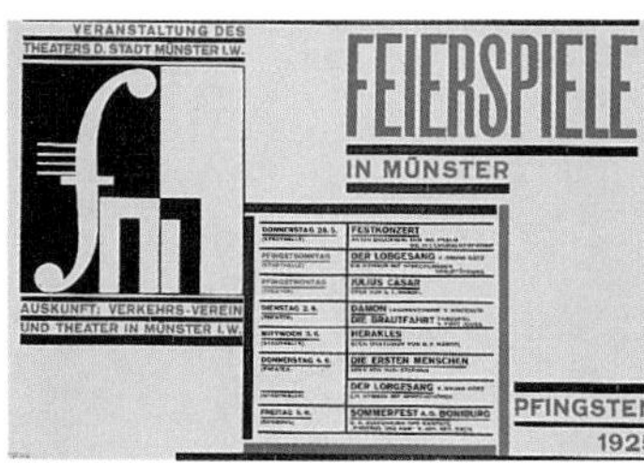

1925 Burchartz

1922 Moholy-Nagy

1922 Kluçis

ment présent et important en Hongrie, en Tchécoslovaquie, en Pologne et en Yougoslavie. En Allemagne, après des débuts plutôt mystiques, le Bauhaus opta pour une direction fortement empreinte de constructivisme. Par ailleurs, le «ring neuer werbegestalter» (cercle des nouveaux graphistes publicitaires), le «Novembergruppe» et plusieurs autres créateurs indépendants œuvraient également dans cette optique. Guidés par le principe de construction, de nombreux travaux de typographie donnaient une impression de raideur et de lourdeur. Ceci est dû au fait que les textes étaient composés par pavés et en capitales, et que l'on utilisait des effets positif-négatif, des barres et des filets, mais aucun alphabet élémentaire. Même si les constructivistes considéraient les caractères de type gothique allemand et les Antiques comme obsolètes, il fallut pourtant attendre les années 1927/28 pour que l'on crée et produise des alphabets mieux adaptés comme la Futura.
Ce mouvement a néanmoins permis l'émergence de nombreux alphabets expérimentaux qui, en dépit de concepts clairs, n'ont jamais été adaptés à la composition massive. Là aussi, le véritable objectif était l'innovation radicale et l'exploration des possibilités d'une création à base de formes géométriques fondamentales.

Constructivism also spread to western Europe. Right from its beginnings, the Dutch De Stijl movement had a fundamental, Constructivist impetus. There were major Constructivist trends in Hungary, Czechoslovakia, Poland and Yugoslavia. After a rather mystical start, the Bauhaus in Germany developed a strong Constructivist orientation. The "ring neuer werbegestalter", the "Novembergruppe" and independent designers also worked in this vein. The rather artificial approach, however, makes much of the applied typographical work today seem rigid and awkward. This is due to texts being set in justification left and right and in capitals, to the use of positive-negative effects, bars and stripes, but all in the absence of basic alphabets. The Gothic and roman alphabets were too old-fashioned for the Constructivists, while appropriate alphabets, such as Futura, were not designed and produced until 1927/1928.
The Constructivists drew up many experimental alphabets, which despite their conceptual clarity were never suitable for mass production. Here too, the primary aim was to satisfy the desire for radical innovation and explore the possibilities of designing with basic geometric forms.

## Dadaismus
## Dadaïsme
## Dada

**Um 1915** Der Dadaismus war eine literarische und gestalterische Bewegung, die sich um 1915 in Zürich sammelte und von hier aus in kurzer Zeit über ganz Europa wirkte. Er war zu Beginn eine stark anarchistisch und gesellschaftskritisch motivierte Bewegung, die die absurden Vorgänge des Ersten Weltkriegs und die fragwürdigen herrschenden gesellschaftlichen Werte aufnahm, und in der friedlichen Zone der Schweiz zu einem lauten, ästhetischen Protest umwandelte. Als Alternative zu den sinnlosen Ritualen wollte man Dinge ohne Sinn schaffen. Dabei ließen die Dadaisten soviele Regeln und Dogmen außer acht wie nur möglich. Sie akzeptierten zufällig entstandene Ergebnisse, die sie als Ausdruck des Unbewußten begrüßten.
Die Dada-Typographie war für die Werbegestaltung oder die Informationstypographie völlig ungeeignet. Auf individuelle gestalterische Versuche wirkte diese befreite und befreiende Gestaltung seit damals immer wieder sehr anregend und erneuernd.
Handwerkliche Setzerregeln wurden völlig mißachtet. Es wurden unfachliche Schriftmischungen in Schräg- und Rundsatz collagiert, die Buchstaben und Wörter tanzten wie in freiheitlicher Entrückung über die Formate. So wurden Einzelbuchstaben zu Bildelementen in einer Gesamtkomposition. Gewohnte Textabläufe wurden zugunsten optischer Bildüberlegungen zerteilt. Zu den Buchstaben kamen Linien in verschiedenen Stärken und Mustern dazu, Bildfetzen und Fotocollagen mit und ohne inhaltli-

**Vers 1915** Dada est un mouvement littéraire et plastique qui naît à Zurich vers 1915, et se répand à travers toute l'Europe en peu de temps. L'avènement de Dada dans un pays aussi paisible que la Suisse, marque le début d'une mouvance critique à l'égard de la société et fortement imprégnée d'anarchisme, qui se saisit de l'absurdité du drame de la Première Guerre mondiale ainsi que des valeurs sociales dominantes, jugées douteuses, pour les tourner en dérision par le biais d'une contestation esthétique et bruyante. Comme alternative aux rituels absurdes, on cherche à créer des choses dépourvues de sens. Et les créateurs de négliger délibérément les règles et les dogmes. Ils approuvent les réalisations nées d'un hasard qu'ils saluent comme étant l'expression de l'inconscient.
La typographie Dada est totalement inadaptée au graphisme publicitaire et à l'information. Pourtant, depuis cette époque, ces conceptions libérées et libératoires n'ont cessé d'influencer l'expérimentation individuelle qui y a puisé inspiration et renouveau.
Dada dédaigne toutes les règles de la composition artisanale. Au mépris de tout professionnalisme, on mélange les caractères. Montés en composition diagonale ou en arc de cercle, les lettres et les mots dansent sur la page comme s'ils cherchaient à s'échapper. Des lettres isolées deviennent des éléments picturaux au sein de la composition d'ensemble. Les textes courants sont fractionnés au profit de la recherche visuelle. Aux lettres on ajoute des filets et lignes présentant des épaisseurs et des motifs variés, on intègre à la maquette des fragments d'images et des collages photographiques avec ou sans lien avec le contenu de la page.
Le dadaïsme et sa typographie expriment un refus catégorique du compromis. Un

1917 Janco

1923 Schwitters

1920 Hausmann

1922 Zdanewitsch

**c. 1915** Dada was a literary and design movement, founded in Zurich in 1915 and quickly spreading to the rest of Europe. It started as a strongly anarchist and socially critical movement which took in the absurd processes of the First World War and the questionable social values of the time, transforming the peaceful haven of Switzerland into a raucous zone of aesthetic protest. As an alternative to pointless ritual, Dadaists aimed to create things without a purpose. In doing so, as many rules and dogmas as possible were flouted. Chance events were accepted, welcomed as an expression of the subconscious.
Dada typography was totally unsuitable for advertising or information purposes. But this liberated and liberating form had, and still has, a very stimulating and refreshing effect on individual design experiments.
Technical typesetting rules were completely disregarded. Collages were made of unprofessionally-jumbled alphabets, set obliquely or in rounded lines; letters and words danced in rapturous freedom across the margins. Single letters became illustrative elements in an artistic composition. Accustomed textual layouts were split up in favor of pictorial interpretations. Lines of various thickness and pattern were added to the letters; bits of picture and photo collages, whether related to the content of the text or not, were integrated into the designs.
Dada and Dada typography were bursting with a radical refusal to compromise. This had many consequences, particu-

che Bezüge wurden in die Entwürfe integriert.
Der Dadaismus und die dadaistische Typographie waren voll radikaler Kompromißlosigkeit. Besonders folgenreich war dies bei der Visualisierung literarischer Neuerungen. Selten in der Geschichte waren der inhaltliche Anspruch und die umgesetzte Form so eng zusammen und bedingten sich geradezu wie bei den Lautgedichten und akustischen Collagen, die eine neue Poesie hervorbrachten.
Wie keine andere gestalterische Neuerung im 20. Jahrhundert zog der Dadaismus (wie gewollt) die Kritik der gegensätzlichsten politischen Kreise und gesellschaftlichen Gruppen auf sich. Er wurde als die Inkarnation des Aufruhrs und der Verhöhnung verständnislos und bedrohend rezipiert.
Ausgehend von Zürich gab es Dada-Zentren in Berlin, Paris und New York. Überall entstanden in kurzer Zeit zahlreiche Buchpublikationen, Zeitschriften, Plakate und andere Drucksachen. Aber nach wenigen Jahren erlahmte der subversive Spott der Dada-Anti-Form. Protest wurde auf andere Weise artikuliert.

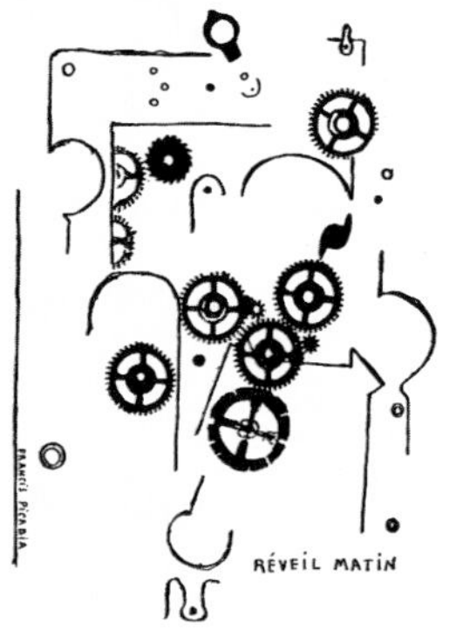


1919 Picabia

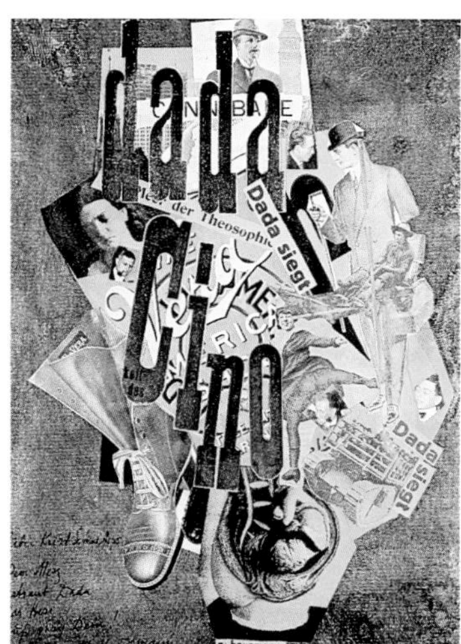


1920 Hausmann

1920 Schwitters

aspect qui aura un impact particulier sur la visualisation des innovations littéraires. Rarement dans l'histoire, les exigences de contenu et leurs applications formelles ont été aussi étroitement liées, se conditionnant même mutuellement comme dans ces poésies sonores et collages acoustiques qui donneront naissance à une nouvelle forme poétique.
Comme aucune autre innovation du 20e siècle, le dadaïsme attira à lui (comme voulu) la critique des milieux politiques et des groupes sociaux de l'autre bord. On l'interpréta comme étant l'incarnation de la révolte et de la dérision, mais aussi comme une menace.
A partir de Zurich, des groupes Dada se formèrent à Berlin, Paris et New York. De nombreux livres furent publiés un peu partout en un très bref laps de temps. Mais au bout de quelques années, l'ironie subversive de l'antiformalisme dada s'assoupit. La contestation s'exprime d'une autre manière.
A partir de Zurich, des groupes Dada se formèrent à Berlin, Paris et New York. De nombreux livres furent publiés un peu partout en un très bref laps de temps. Mais au bout de quelques années, l'ironie subversive de l'antiformalisme dada s'assoupit. La contestation s'exprime d'une autre manière.

1918 Janco

1922 Doesburg

larly in the visualization of literary innovations. Seldom had history experienced such closeness of contextual demand and actual form, mutually dependent as in sound poems and acoustic collages, producing a new poetic form.
No other formal artistic innovation of the 20th century attracted as much (probably intended) criticism from opposing political and social groups as Dada. It was misinterpreted, threateningly denounced as the incarnation of turmoil and ridicule.
With Zurich as its initial base, Dada centers sprang up in Berlin, Paris and New York. Within a short space of time numerous publications appeared: books, magazines, posters and other printed matter. Yet after only a few years the subversive mockery of Dada's anti-form waned. Protest was made in other ways.

**Ca. 1910–1920** Das neue Jahrhundert sollte durch die Literaten, Maler, Formgeber und Typografen schnell eine neue Ausdrucksweise erhalten. Der Futurismus war eine der ersten radikalen ästhetischen Bewegungen, die einen totalen Bruch mit traditionellen Formen der Gestaltung vollzog. Die allgemein gegenwärtige und überraschende Entwicklung der Technik brachte die Futuristen zu einer Mythologisierung der Maschinen und zu einer Religion der Geschwindigkeit. Bücher aus Metall zeigten den totalen Bruch mit Traditionellem und die Bewunderung für Auto oder Flugzeug. Die Gestalter setzten die Energie der Geschwindigkeit in dynamischen Bögen und Aufbauten auf der Fläche um. In gezeichneten, collagierten und konstruierten Formen wurden Bildgedichte und Texte über die Fläche gestreut. Erstmals wurde der lineare Lesevorgang, der seit Jahrhunderten das übliche war, in vielen Beispielen radikal in Frage gestellt. Die Ergebnisse zwischen Sehen und Lesen, zwischen Bild und Text, wurden von den Autoren selbst projektiert. Auffällig für diese Zeit ist auch, daß Schrift bzw. Buchstaben ein häufig verwendetes autonomes Gestalt-Element in der freien Kunst wurde.
Neben den gebrauchsgrafischen Medien wurden bald andere Bereiche erobert: Mode, Formgebung, Architektur und Film. Trotz der krassen Anti-Haltung gab es überraschenderweise in Italien und später in den USA viele Anwendungen des futuristischen Stils in der Werbung für Produkte. Auch hier wurden, wie bei

**Vers 1910–1920** Dès le début du siècle, écrivains, peintres, designers et typographes vont très vite chercher de nouveaux modes d'expression. Le futurisme est l'un des premiers mouvements esthétiques de caractère radical à accomplir un réel divorce par rapport à la création formelle traditionnelle. La généralisation et le développement étonnant des technologies portent les futuristes à mythifier les machines et à faire de la vitesse une religion. Des livres en métal témoignent de cette rupture brutale avec les traditions, et de l'admiration pour l'automobile et l'avion. Les maquettistes traduisent sur la page les notions d'énergie et de vitesse par des courbes dynamiques et des constructions. Textes et poésie visuelle se répartissent sur la surface au moyen de graphismes, de collages et de formes construites. Pour la première fois, la lecture linéaire que l'on pratiquait depuis des siècles est remise en question dans bon nombre de travaux. Le rendu visuel à mi-chemin entre le voir et le lire, entre l'image et le texte, est le produit de projections des auteurs. Ce qui saute également aux yeux à cette époque, est que l'écriture, ou les caractères, deviennent un élément formel souvent utilisé dans les arts plastiques.
Outre le domaine du graphisme utilitaire, d'autres secteurs ne tardent pas à être conquis : la mode, le design, l'architecture et le cinéma. En dépit de son attitude ostensiblement contestataire, nous remarquons avec étonnement que le style futuriste trouve aussi des applications dans la publicité de produits, d'abord en Italie, puis aux Etats-Unis. Ici aussi, on choisit, comme pour d'autres imprimés, des caractères d'une grande spatialité, en variant les corps et en créant des contrastes formels. On a recours à des dispositions en diagonale, à des doublons de

1914 Carrà

1915 Marinetti

1915 Soffici

1920 Depero

1919 Depero

**c. 1910–1920** Through contemporary literati, artists, designers and typographers, the new century soon assumed new forms of expression. Futurism was one of the first radical aesthetic movements to achieve a complete break with traditional form. The incredible and ubiquitous technological advance of the age caused the Futurists to adopt machines as their mythology and speed as their religion. Books made of metal demonstrated their total discarding of tradition and their admiration for the car and the airplane. Designers tried to capture the energy of speed with dynamic arches and superstructures on the page. Visual poetry and texts were strewn across the page in collage, in hand-drawn and in contrived forms. For the first time the linear method of reading text, the standard for centuries, was radically questioned in many Futurist works. The resulting mixture of seeing and reading, of pictures and text, was often planned by the authors themselves. Another feature of this period was that type and letters became autonomous elements of design often employed in free art.
Futurism soon spread to other areas beyond the functional graphic media, namely to fashion, design, architecture and film. Despite its blatant "anti" attitude, in Italy and later in the USA, Futurist styles surprisingly began to appear in advertisements for various products. Here too, as with other printed matter, three-dimensional fonts and contrasting forms in many sizes were employed. Diagonals were used and texts were dou-

den anderen Drucksachen, räumlich ergänzte Schriften in vielen verschiedenen Größen und Formkontrasten eingesetzt. Es wurden Diagonalstellungen und Textverdoppelungen verwendet, so daß ein abwechslungsreiches Verbalzeichen-Gemenge als Typographie entstand.
Niemand hatte in dieser schnellen Zeit Interesse an der Entwicklung gebrauchsfähiger neuer Alphabete in futuristischem Stil. Außer einigen signetartigen Schriftzügen für Zeitschriften und gezeichneten räumlichen Wörtern entstanden nur Arbeiten mit bekannten Buchstabenformen, die jedoch grafisch stark verändert oder skizzenhaft eingesetzt wurden. Bei der Schriftauswahl gab es keine formalen Einschränkungen; es trafen sich serifenlose, serifenbetonte und andere Antiquaschriften in einem Ensemble.
Viele Gestalter des Futurismus stellten die neue ästhetische Qualität in den Dienst des aufkommenden politischen Faschismus, der diese Akzeptanz durch eine ästhetische Bewegung begrüßte. Für nachfolgende Betrachter wurde eine unbefangene Bewertung der Qualitäten des Futurismus erschwert, vor allem weil auch eine offene Bewunderung für den Krieg ausgesprochen wurde.

1915 Balla

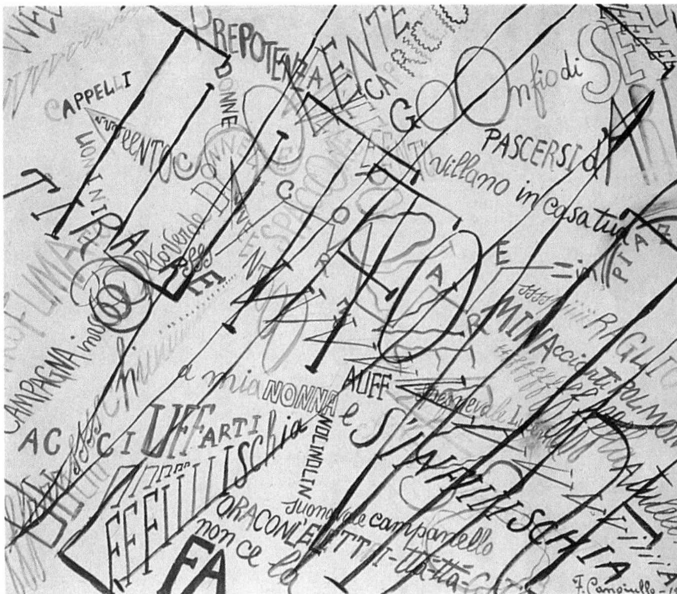

1914 Cangiullo

séquences de textes, de façon à obtenir un mélange très diversifié de signes verbaux d'images.
Personne, en cette période d'accélération, ne songe à dessiner des alphabets futuristes destinés à l'usage courant. Hormis quelques paraphes évoquant des signets utilisés dans les revues, et des mots dessinés pour leur conférer de la spatialité, les travaux de l'époque ne comportent que des caractères connus employés après de fortes déformations de leur graphisme ou une schématisation. Il n'existe plus aucune restriction formelle dans le choix des caractères, les lettres sans sérif côtoient les sérifs des Antiques sur une même page.
De nombreux maquettistes futuristes mettent ces nouvelles qualités esthétiques au service du fascisme montant qui se félicite de voir un courant esthétique adhérer à ses idées. Par la suite, le public aura peine à apprécier le futurisme sans préjugés, surtout parce qu'il a exprimé ouvertement son admiration pour la guerre.

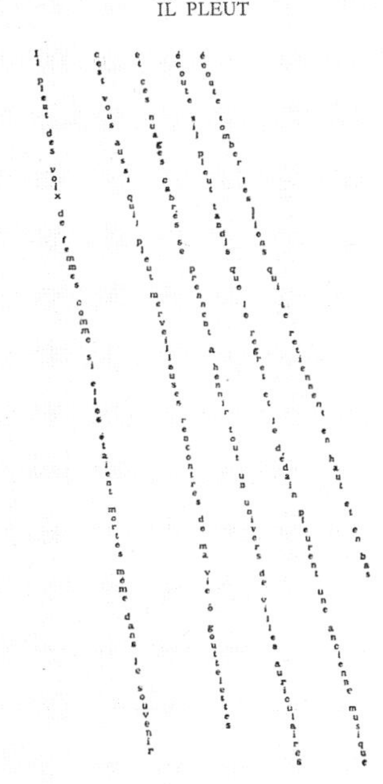


1916 Apollinaire

1919 Marinetti

1923 Balla

bled, creating an assorted, jumbled typography of verbal symbols.
In this era of speed, no-one was particularly interested in developing new usable alphabets in the Futurist vein. Apart from a few signet-type flourishes for magazines and the occasional hand-drawn, three-dimensional words, most printed material favored familiar letter forms which had either undergone radical graphic alteration or which had been reduced to sketched outlines. There were no formal restrictions dictating the choice of print; sans-serif, slab-serifs and other roman typefaces were used together in muddled ensemble.
Many Futurist designers placed this new aesthetic quality and, thus, themselves at the service of the rising political Fascist movement, which for its part welcomed this acceptance by an aesthetic faction. Later observers therefore found it hard to remain impartial when judging the qualities of Futurism, especially as the movement openly exalted war.

**Informative Sachlichkeit**
**Objectivité informative**
**Informative Functionalism**

**1910** Die individuelle künstlerische Gestaltung hat dort ihre Grenzen, wo neue Notwendigkeiten eine allgemeine, sachliche Form erfordern. Durch die Entwicklung industrieller Produktionsweisen, durch neue internationale Märkte, aber auch einfach durch eine Veränderung des Zeitgeistes nach der Jahrhundertwende, reichten die Formen des Jugendstils oder gar des Historismus nicht mehr aus, um eine zeitgemäße visuelle Kommunikation zu entwickeln.
Um 1910 gab es eine Hinwendung zur informativen Sachlichkeit in der Gestaltung. Dies war sowohl bei den ersten umfangreichen Erscheinungsbildern für Unternehmen zu sehen, als auch bei einzelnen Plakaten oder Plakatserien.
Vorbei war die Zeit, in der mit Abbildungen von Engelchen und Heiligen als Blickfang für die obskursten Inhalte geworben wurde. Erstmals entstanden Arbeiten, die konzeptionelle Form- bzw. Bildfindungen erzielten. Knappe Texte, aus denen klare Visualisierungen abgeleitet wurden, ergaben komprimierte Ergebnisse, Inhalte wurden auf den Punkt gebracht. Anders als im Jugendstil war die Typographie leicht lesbar. Es gab weniger Kalligraphie und künstlerische Geste, dafür großzügig in die Gesamtkonzepte integrierte, gemalte oder gesetzte Zeilen in klaren Schriften. Die serifenlose Schrift setzte sich neben den klassischen Antiquaschriften als gleichberechtigt durch.
Vor allem auch bei Zweckdrucksachen wurde dieser sachliche Stil stärker als bisher eingesetzt. Der Sinn verschiedener

**1910** Les limites de la création individuelle et artistique commencent là où de nouvelles nécessités requièrent des formes objectives et universelles. Or, du fait de l'évolution des modes de production industrielle, de l'internationalisation des marchés, mais tout simplement aussi à cause des changements survenus dans les esprits après le tournant du siècle, les formes de l'Art nouveau tout comme celles de l'éclectisme s'avèrent insuffisantes au développement d'une communication visuelle adaptée à l'époque.
Vers 1910, on assiste à une tendance à l'objectivité informative dans la mise en forme. Ceci apparaît autant sur les premiers logotypes présentant l'image d'entreprises que sur certaines affiches isolées ou en série.
Le temps où l'on tentait de capter le regard avec des publicités d'une teneur obscure, illustrées d'angelots et de saints, est révolu. Pour la première fois, apparaissent des réalisations témoignant d'une réelle conception formelle et picturale. Des textes concis, dont découle un agencement visuel clair, permettent de condenser le rendu et de synthétiser les contenus. A l'inverse de la période de l'Art nouveau, la typographie est aisément lisible. Le geste calligraphique et artistique y est moins présent, mais en revanche, les pages comportent des lignes dessinées ou composées en caractères nets qui s'intègrent généreusement dans le concept d'ensemble. Les caractères sans sérif s'imposent aux côtés des Antiques classiques comme ayant une valeur égale.
Mais c'est surtout sur les imprimés utilitaires que ce style sobre intervient davantage que par le passé. On porte plus d'attention qu'autrefois à la signification des divers éléments de la mise en page tels que le corps et la forme des caractères, l'aspect des lignes. On les choisit même

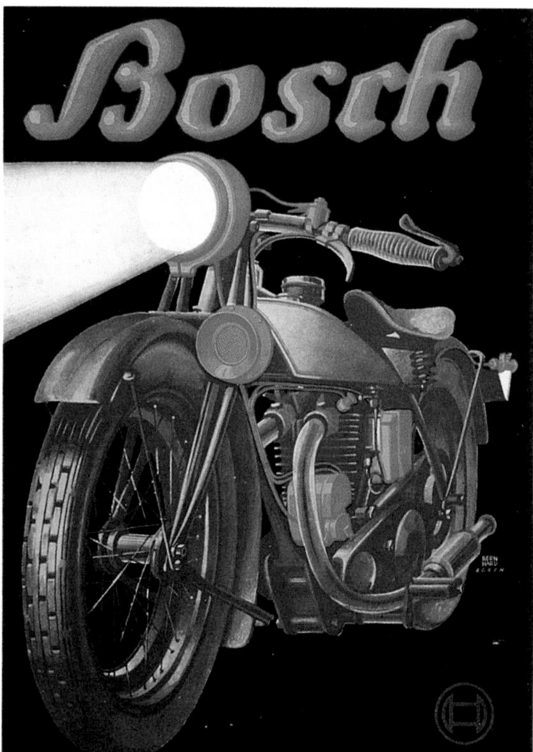

ca. 1920 Bernhard, Rosen

ca. 1912 Behrens

1912 Bernhard

1912 Behrens

1908 Hohlwein

**1910** Individual, artistic design reaches its limitations when new necessities require a general, functional form. The development of industrial manufacturing techniques, new international markets, and also simply the change in the Zeitgeist following the turn of the century meant that the forms of Art Nouveau, let alone Historicism, were no longer suitable for contemporary visual forms of communication.
By about 1910, there was a rise in informative functionalism in design. This

Gestaltungselemente wie Größe und Form der Schrift, das Aussehen von Linien, Flächen und Ornamenten wurde stärker beachtet als vorher: sie wurden nach ihren Funktionen ausgewählt. Insgesamt ging es nicht mehr um die Gestaltung einzelner genialischer Drucksachen, sondern um die formale Schlüssigkeit im Gesamterscheinungsbild.
Diese gestalterische Richtung stellt einen lange unterschätzten Ansatz einer neuen Herangehensweise an Gestaltung für wirtschaftliche Kommunikation dar.

Obwohl die Motive meist im Stil der klassischen Plakatmalerei ausgeführt waren, sind sie von inhaltlicher Reduktion und Klarheit geleitet, die Maßstäbe für die visuelle Kommunikation des 20. Jahrhunderts setzte.

1912 Oppenheim

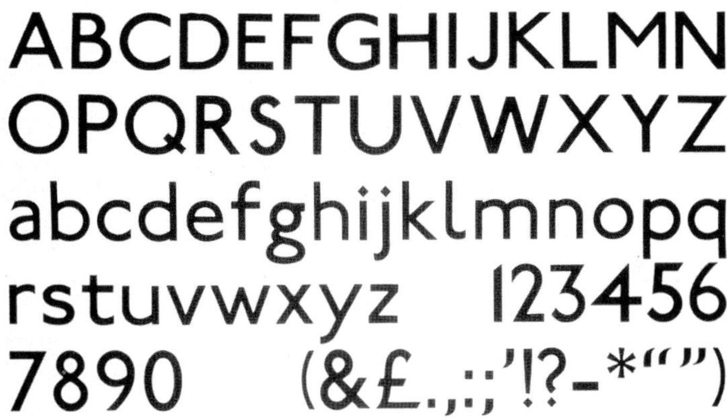

1916 Johnston

en tenant compte de leurs fonctions. Globalement, il ne s'agit plus de la mise en page de quelque imprimé génial, mais de l'homogénéité formelle du rendu visuel.
Cette tendance formelle correspond à une démarche longtemps sous-estimée en vue d'une nouvelle approche de la mise en forme au service de la communication des entreprises.
Bien que les graphismes rappellent le style classique de l'affiche peinte, ils sont commandés par un désir de clarté et de réduction des contenus qui fixe la norme de la communication visuelle du 20e siècle.

1910 Klinger

1909 Ehmcke

1914 Graz

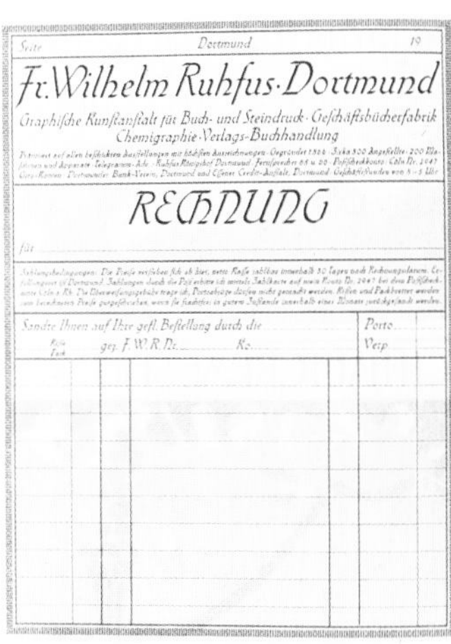
Fr. Wilhelm Ruhfus·Dortmund
Graphische Kunstanstalt für Buch- und Steindruck · Geschäftsbücherfabrik
Chemigraphie · Verlags-Buchhandlung

RECHNUNG

ca. 1910 Ehmcke

could be observed in the first manifestations of corporate identity and on individual posters or series of posters.
Gone was the age where cherubs and saints could be used to draw attention to obscure content. For the first time, the aim was to produce conceptional designs. Short texts with a clear visual message condensed the material while contents were brief and to the point. As opposed to Art Nouveau, the typography was easy to read. There was less calligraphy and artistic gesture; instead, drawn or printed lines of text in clear fonts were generously integrated into the overall concept. Sans-serif alphabets attained equal status alongside classic roman fonts.
Increasingly, this functional style was employed primarily for utilitarian printed matter. Greater attention was paid to the purpose of the various elements of design, such as the size and shape of the letters, the appearance of the lines, of the surfaces and ornaments. These elements were selected according to their function.

Typographers were no longer concerned with the creation of printed materials that displayed individuality and brilliance, but with a formal consistency in the overall appearance of the page. This avenue in design represents a long-underestimated attempt at a new approach to design for economic communication.
Although the motifs were often executed in the style of classic poster art they were marked by reduction in content and clarity, setting standards for the visual communication of the 20th century.

## Jugendstil
## Art nouveau
## Art Nouveau

**1890–1914** Am Beginn des 20. Jahrhunderts entstand der Jugendstil als eine bemerkenswerte Bewegung, die die ganze Sehnsucht dieser Zeit nach Schönheit und Individualität im Industriezeitalter zu einem Stil werden ließ.
Diese Bewegung fand in vielen Ländern Europas eigenständige Ausprägungen. Im Zuge des umfassenden Neuerungswillens gab es von der Architektur bis zum Gebrauchsgegenstand, von der Schrift über Plakat, Zeitschrift und Buch in allen Gestaltungsbereichen Ergebnisse, in denen sich der Jugendstil manifestierte. Seine Stilelemente wurden aus abstrahierten Formen und Vorgängen der Natur abgeleitet. Im Bereich der visuellen Kommunikation wurden diese symbolischen Bildformen ornamental mit den Schriftteilen zu einer kompositorischen Einheit zusammengefügt. Um dies in bester stilistischer Symbiose zu bewältigen wurden einzelne Wörter, Zeilen und ganze Alphabete im Sinne des Jugendstils neu entworfen. Für den Bleisatz gab es neben neuen Schriften vielfältigste florale Ornamentik. So entstanden Bücher in durchgängig formaler und künstlerischer Einheit.
Überall in Europa erschienen neugegründete prunkvolle Zeitschriften. Die wichtigsten waren „The Studio", „Revue blanche", „Pan", „Die Woche", „Insel" und „Jugend". Der Formenreichtum ist bis heute überraschend. Durch zahlreiche junge Illustratoren, Poeten, Schriftsteller und typographische Gestalter wurden diese periodisch erscheinenden Hefte schnell zu weitverbreiteten Publikatio-

**1890–1914** Le tournant du siècle voit l'avènement de l'Art nouveau, un mouvement remarquable incarnant, au cœur de l'ère industrielle, la nostalgie de la beauté et de l'individualité pour en faire un véritable style.
Ce mouvement trouve un langage singulier dans de nombreux pays européens. Dans le sillage des aspirations au renouveau total, l'Art nouveau se manifeste partout, dans l'architecture jusqu'aux objets usuels, des lettres d'imprimerie jusqu'aux affiches, revues et livres, et dans tous les domaines de la création artistique. Ses éléments distinctifs dérivent de formes abstraites ou empruntées à la nature. Concernant la communication visuelle, les formes symboliques et ornementales se fondent avec des corps de textes composés de manière homogène. Pour mettre en œuvre cette symbiose stylistique entre les mots et les lignes, des alphabets entiers sont conçus dans l'esprit de l'Art nouveau. On dessine même de nouveaux caractères et des fontes florales reproductibles destinées à la composition au plomb. C'est ainsi que paraissent des livres présentant une cohérence formelle et artistique constante.
Partout en Europe naissent des revues somptueuses dont «The Studio», la «Revue blanche», «Pan», «Die Woche», «Insel», et «Jugend», pour ne citer que les plus célèbres. La richesse des formes nous étonne encore aujourd'hui. La participation de jeunes illustrateurs, de poètes, d'écrivains et de typographes fait bientôt de ces périodiques des publications très répandues qui contribuent à la propagation d'idées autant culturelles qu'esthétiques.
Durant cette période, l'affiche compte parmi les moyens d'expression les plus importants et les plus répandus. L'affiche utilitaire à fort tirage et placardée partout

ABCDEFGHIKLM
NOPQRSTUVWXYZ
abcdefghijklmnopqr
sſtuvwxyz123456789

1899 Eckmann

1902 Moser

1900 Guimard

1902 Roller

1903 Klimta

**1890–1914** The turn of the century produced Art Nouveau, a remarkable movement which rapidly became a major artistic style, spurred on by a yearning for beauty and individuality in the industrial age.
It was a movement that enjoyed various independent forms of expression in many European countries. In the wake of an allround desire for innovation, Art Nouveau manifested itself in all forms of design, from architecture to household commodities, from writing to posters, from magazines to books. Its stylistic components were derived from abstract forms and the processes of nature. In the sphere of visual communication, these symbolic pictorial elements were combined with the written text to form a compositional unity. For optimal stylistic symbiosis, single words, lines and whole alphabets were redesigned. There were not only new fonts, but also diverse floral ornaments for lead composition. Thus books were created with a formal and artistic consistency throughout.
All over Europe, splendid new magazines appeared, the most important being "The Studio", "Revue blanche", "Pan", "Die Woche", "Insel" and "Jugend". The wealth of forms still amazes the beholder today. The many young illustrators, poets, authors and typographers working on these magazines helped make them readily available to a wide readership, elegantly promoting the aesthetic and cultural ideas of Art Nouveau.
The age of Art Nouveau saw the rise of the poster as one of the most widespread

nen, die zur Propagierung der ästhetisch-kulturellen Ideen vornehmlich beitrugen. In der Zeit des Jugendstils wurde das Medium Plakat zu einem wichtigen und verbreiteten gestalterischen Ausdrucksmittel. Sowohl das Gebrauchsplakat mit hohen Auflagen, das überall an den Anschlagtafeln, Litfaßsäulen oder Schaufenstern hing, als auch das Plakat als limitiertes druckgrafisches Sammelobjekt erfuhr eine große Verbreitung und Wertschätzung. Es wurden Geschäfte eröffnet, die ausschließlich Plakate verkauften. Gleichzeitig erschienen Bücher und Broschüren, in denen das Plakat dargestellt und analysiert wurde. In Sammelalben konnten faksimilierte Plakate aufbewahrt werden.

Die hohe Zeit des Jugendstils dauerte nur ca. zehn Jahre. Sie war dadurch beendet, daß die einst so visionären und euphorischen Zeichen nur noch oberflächlich wiederholt und industriell vervielfältigt auf geschmacklosen Produkten oder belanglosen grafischen Entwürfen reproduziert wurden. Die Rezeption des Jugendstils war nicht einheitlich, von Kitsch bis Offenbarung lauteten die Einschätzungen von damals bis heute. Aber es wurde im Laufe der Zeit auch klar, daß es ein Aufbruch war, der als Suche und Infragestellung immer wieder auftauchen und zu einer Erneuerung festgefahrener Stilformen führen kann.

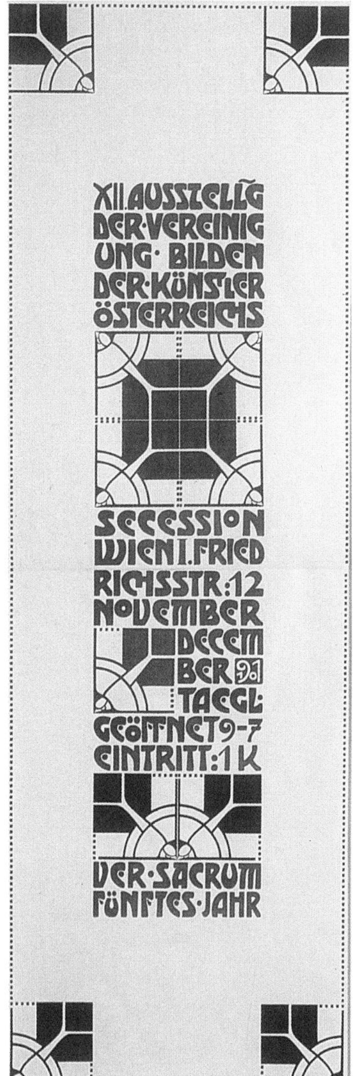


1901 Roller

1900 Guimard

sur les panneaux d'affichage, les colonnes Morris et les vitrines, mais aussi l'affiche d'art, objet de collection à tirage limité, connaissent une grande expansion et gagnent en estime. Des boutiques spécialisées dans la vente d'affiches ouvrent leurs portes en même temps que paraissent des livres et des brochures qui traitent de ce média et l'analysent. On publie même des albums contenant des facsimilés.

L'apogée de l'Art nouveau ne durera qu'une dizaine d'années. La raison de son déclin est que son esprit jadis visionnaire et euphorique s'épuise dans une répétition superficielle, une production industrielle de mauvais goût ou à cause de graphismes dénués d'intérêt. L'Art nouveau n'a pas connu de réception uniforme; aujourd'hui comme à l'époque les appréciations divergent, on parle tantôt de kitsch tantôt de révélation. Pourtant avec le temps, il apparaît que cette période a marqué une brèche qui devait ouvrir sur une quête, une remise en question incessante, et le renouvellement de formes stylistiques figées.

1898 Van de Velde

ca. 1905 Hoffmann

and important means of expression and design. Posters became esteemed and popular objects, whether mass produced as commercial posters, clinging to every billboard, advertising column or shop window, or as limited designer editions for collectors. Special poster stores opened. Books and brochures were published in which the poster was presented and analyzed. Facsimile posters could be collected and kept in special albums.

Art Nouveau's heyday lasted for only about ten years. Its end was brought about by the superficial, industrial mass production of tasteless products and by trivial graphical designs, devaluing what were once visionary and euphoric ideas. The movement has not always met with enthusiasm; now and in the past, it has been classed as anything from kitsch to revelation. With time it became clear, however, that it was a new departure, ever questioning and searching, refreshing stagnant stylistic forms.

## Arts and Crafts
## Arts & Crafts
## The Arts and Crafts Movement

**Um 1880** Die meisten Typographen und Kritiker waren sich einig, daß der Stil des Historismus nicht die formalen und inhaltlichen Möglichkeiten seiner Zeit wiederspiegelte, sondern einem Verfall an Qualität gleichkam. Dies zeigt unter anderem ein Blick auf das Gebiet der Architektur, wo 1851 als bahnbrechende Neuerung der Kristallpalast zur Weltausstellung in London eröffnet wurde: ein Haus aus Glas und Stahl, das die technischen Möglichkeiten und die ästhetische Utopie in eine revolutionäre, gültige Bauform umsetzte. Die Notwendigkeit zur Neubesinnung und der Drang zu einem angemessenen zeitgenössischen Stil in der angewandten Gestaltung wurden ebenfalls in England vorangetrieben. In diesem stark industrialisierten Land wurde der Widerspruch zwischen individuellem künstlerischen Anspruch und billiger Massenproduktion früh erkannt und in die Forderung nach Wahrheit und Aufrichtigkeit in der Gestaltung umgesetzt.
Die Versuche einer Neubesinnung orientierten sich an den Formen des Mittelalters, an den Drucken der Frühzeit, den Handschriften und den Kathedralbauten. Der Begriff des Handwerks erfuhr in diesem Zusammenhang eine erneute Wertschätzung. Die Trennung zwischen Kunst und Handwerk sollte aufgehoben werden. Sozialrevolutionäre Thesen, z. B. von John Ruskin, forderten, daß nicht die Maschinen, sondern der Einzelne die Produkte des Alltags erschaffen sollte.
In der Typographie war durch die mittelalterlichen Vorbilder immer noch eine rei-

**Vers 1880** La plupart des typographes et critiques s'accordaient à penser que le style éclectique ne reflétait ni les possibilités formelles ni les idées de l'époque, mais équivalait à une baisse de qualité. Ceci semble entre autres évident, quand on songe à l'architecture, et en particulier au Palais de cristal, une innovation radicale inaugurée en 1851 à Londres à l'occasion de l'Exposition universelle. Ce bâtiment tout de verre et d'acier était l'application même des possibilités technologiques; il représentait les utopies esthétiques traduites en formes révolutionnaires et valables. La nécessité d'une remise en question et l'aspiration à un style contemporain et approprié dans les arts appliqués faisaient aussi l'objet des préoccupations en Angleterre. Dans ce pays fortement industrialisé, on constata de bonne heure une contradiction entre les prétentions artistiques individuelles et la production de masse et à bon marché; celle-ci se traduisait par des revendications de vérisme et de sincérité de la mise en forme.
Pour se redéfinir, on s'orientait pourtant aux formes médiévales, aux imprimés anciens, aux manuscrits et à la construction des cathédrales. Dans ce contexte, l'artisanat connut une revalorisation. Le fossé entre art et artisanat devait être comblé. Les thèses socio-révolutionnaires d'un John Ruskin, par exemple, préconisaient que l'individu, et non les machines fabriquât les produits quotidiens.
En typographie, on trouvait encore des caractères riches en ornements, hérités du modèle médiéval. Mais des concepts tels que la «lisibilité de l'écriture» ou «l'homogénéité de la page» inspiraient des maquettes harmonieuses et claires.
William Morris, le fondateur du mouvement «Arts & Crafts» comptait parmi les théoriciens et créateurs les plus éminents

1891 Morris

1896 Bradley

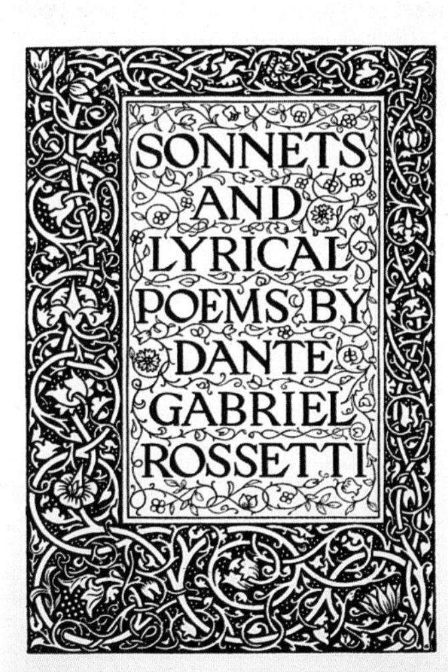

1894 Morris

ca. 1900 Pissarro

1909 Pissarro

**c. 1880** Most typographers and critics were united in their view that Historicism did not reflect contemporary possibilities of form and content but, instead, represented a drop in quality. This is demonstrated by the world of architecture, for example, where in 1851 the Crystal Palace was opened for the Great Exhibition in London. Heralded as a pioneering innovation, it was built from glass and steel, translating technical potential and aesthetic utopia into a revolutionary, valid form of construction. It was also heavily industrialized England which saw the need for a re-think and the pressure for a suitable contemporary style in applied design. Early on, it was recognized here that there was a contradiction between the artistic requirements of the individual and cheap mass production; the result was a growing demand for truth and sincerity of design.
Attempts in this direction were modeled on the forms of the Middle Ages, on early prints, on manuscripts and cathedrals. The concept of "craft" came to enjoy a new esteem. The division between art and craft was to be removed. Socio-revolutionary works, such as those of John Ruskin, demanded that everyday commodities be created by individuals, not machines.
In typography, the nature of the medieval models meant that book pages still boasted rich ornamentation. Yet concepts such as "legibility of the typeface" or "unity of the page" prompted the creation of designs full of harmony and clarity.

che Ornamentierung der Buchseiten zu finden. Aber Begriffe wie „Lesbarkeit der Schrift" oder „Geschlossenheit der Buchseite" ließen Entwürfe voller Harmonie und Klarheit entstehen.
William Morris, der Gründer der „Arts and Crafts"-Bewegung, gehörte zu den wichtigen Theoretikern und Praktikern der Zeit. Er entwarf für die Bücher seiner 1888 projektierten, 1891 erstmals aktiven „Kelmscott Press" Schriften, die aus den neuen Idealen entsprangen: die Golden Type geht auf Vorbilder von Nicolaus Jenson aus dem 15. Jahrhundert zurück, die Troy Type war eine Mischung aus gotischen Formen und Antiqua-Elementen. In diesem Sinne entwarf er auch Linien, Schmuckornamente und Initialen, die seine Alphabete unterstützten und ergänzten. „Arts and Crafts" wollte das schöne Buch in der industriellen Zeit wiederbeleben. Der Ansatz war ganzheitlich und ließ kein Detail unbedacht: neben den Schriften waren Papier, Illustration, Druck und Bindung der Bücher von Bedeutung. Auch in den USA und Deutschland wurden die Ziele von „Arts and Crafts" und die Forderungen nach Qualität im Industriezeitalter kreativ umgesetzt. Diese Bewegung war für die Moderne des 20. Jahrhunderts ausschlaggebend.

1894 Beardsley

1895 Beardsley

1903 Goudy

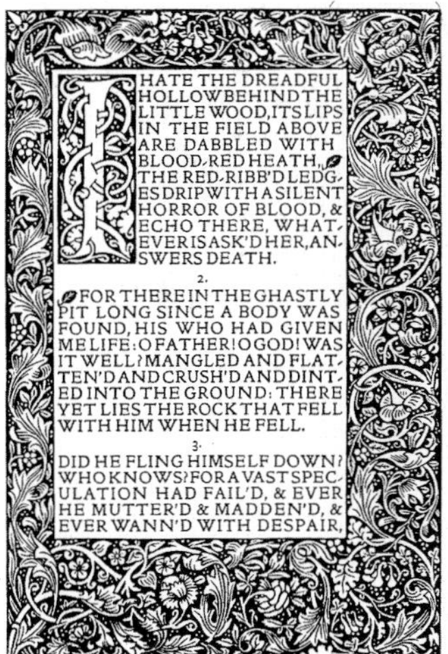
I HATE THE DREADFUL HOLLOW BEHIND THE LITTLE WOOD, ITS LIPS IN THE FIELD ABOVE ARE DABBLED WITH BLOOD-RED HEATH, THE RED-RIBB'D LEDGES DRIP WITH A SILENT HORROR OF BLOOD, & ECHO THERE, WHATEVER IS ASK'D HER, ANSWERS DEATH.

2.

FOR THERE IN THE GHASTLY PIT LONG SINCE A BODY WAS FOUND, HIS WHO HAD GIVEN ME LIFE: O FATHER! O GOD! WAS IT WELL? MANGLED AND FLATTEN'D AND CRUSH'D AND DINTED INTO THE GROUND: THERE YET LIES THE ROCK THAT FELL WITH HIM WHEN HE FELL.

3.

DID HE FLING HIMSELF DOWN? WHO KNOWS? FOR A VAST SPECULATION HAD FAIL'D, & EVER HE MUTTER'D & MADDEN'D, & EVER WANN'D WITH DESPAIR,

1894 Morris

de son temps. Pour les ouvrages de sa «Kelmscott Press» prévue pour 1888, mais qui n'était entrée en activité qu'en 1891, il créa des caractères inspirés des nouveaux idéaux. Le «Golden Type» renvoie aux modèles dessinés par Nicolaus Jenson dès le 15e siècle; le «Troy Type» est un amalgame entre les formes gothiques et des éléments empruntés à l'Antique. Dans le même esprit, il élabora des filets, des ornements et des lettrines pour soutenir et compléter ses alphabets. «Arts & Crafts» voulait redonner vie au beau livre en pleine ère industrielle. A la base, le mouvement obéissait à un principe de globalité en n'omettant aucun détail : outre les caractères, le papier, les illustrations, l'impression et la reliure des volumes étaient importants. Aux Etats-Unis comme en Allemagne, les objectifs d' «Arts & Crafts» et les exigences qualitatives de l'ère industrielle trouvaient des applications dans la création. Ce mouvement donna des impulsions qui furent décisives au 20e siècle.

1895 Beggarstaffs

ca. 1885 Anonymous

1895 Beggarstaffs

PARADISE LOST
THE AUTHOR
JOHN MILTON

OF MANS FIRST DISOBEDIENCE, AND THE FRUIT
OF THAT FORBIDDEN TREE, WHOSE MORTAL TAST
BROUGHT DEATH INTO THE WORLD, AND ALL OUR WOE,
With loss of Eden, till one greater Man
Restore us, and regain the blissful Seat,
Sing Heav'nly Muse, that on the secret top
Of Oreb, or of Sinai, didst inspire
That Shepherd, who first taught the chosen Seed,
In the Beginning how the Heav'ns and Earth
Rose out of Chaos: Or if Sion Hill
Delight thee more, and Siloa's Brook that flow'd
Fast by the Oracle of God; I thence
Invoke thy aid to my adventrous Song,
That with no middle flight intends to soar
Above th' Aonian Mount, while it pursues
Things unattempted yet in Prose or Rhime.
And chiefly Thou O Spirit, that dost prefer
Before all Temples th' upright heart and pure,
16

1902 Milton

William Morris, the founder of the Arts and Crafts Movement, was one of the most important theoreticians and practitioners of his time. For the books of his Kelmscott Press, conceived in 1888 and first active in 1891, he invented fonts which arose out of the new ideals of the era. His Golden Type goes back to Nicolaus Jenson's models from the 15th century; Troy Type was a mixture of Gothic forms and roman elements. He created lines, decorative ornaments and initials to support and complement his alphabets. The Arts and Crafts Movement aimed at reviving the beautiful book in the industrial age. It was a fully holistic attempt, leaving no detail unattended; not only the type but also the paper, illustrations, print and binding of their books were of great importance. In the USA and Germany, the aspirations of the Arts and Crafts Movement and the demand for quality in the industrial age were also translated into creative activity. The movement was to be the major stimulus for the modernism of the 20th century.

**Historismus**
**Eclectisme**
**Historicism**

**2. Hälfte 19. Jahrhundert** Um 1860 war die Gestaltung durch die Übernahme von Formen aus der Vergangenheit bestimmt. Was in der freien Kunst zu akademischer Strenge und zum Stillstand führte, bedeutete in den Bereichen der angewandten Kunst, daß stilistische Zitate und Verzierungen von der Antike über die Gotik bis zum Rokoko in der täglichen Arbeit verwendet wurden, und eine ganze Epoche charakterisierten. Formal trat der Historismus in malerischen und konstruktiv-architektonischen Elementen auf. Im Bleisatz wurden Bogen, Denkmale und Tempel aus dem Linienmaterial gesetzt, Einrahmungen, die an Gemäldegalerien mit klassischen Bildern erinnerten, schmückten alles Gedruckte.
Dies alles geschah rein formal, weil die vorfabrizierten Verzierungen ohne jeglichen inhaltlichen Bezug verwendet wurden: in der Architektur sahen Fabriken aus wie griechische Tempel, im gedruckten Bereich sahen Briefbogen und Rechnungen aus wie Ankündigungen zum Ritterschlag, Geschäftsanzeigen wie Denkmale. Zeilen und Texte waren auf Mittelachse ausgerichtet in weitschweifigsten Schriftphantasien gesetzt. Die Alphabete waren reichhaltig verziert: es gab industriell hergestellte Initialen wie aus der Inkunabelzeit, illustrative Ergänzungen einzelner Buchstaben, die als Wort wie ein Signet wirkten. Es wurden schattierte, räumliche, handgezeichnete und modellierte Alphabete eingesetzt, ornamentierte Flächen und Linien schlossen das ganze typographische Aufgebot ab. Die exzessiven Dekorationen aller

**Seconde moitié du 19e siècle** Vers 1860, les emprunts à des formes issues du passé déterminent la création. Cette attitude caractéristique de l'époque qui, dans les arts traditionnels, débouche sur un strict académisme puis à l'immobilisme, signifie pour les arts appliqués qu'on a constamment recours aux citations stylistiques et à l'utilisation de principes ornementaux antiques, gothiques mais aussi rococo. D'un point de vue formel, l'éclectisme se manifeste surtout dans le dessin et dans sa construction architecturée. Lors de la composition au plomb, on introduit des cintres, des accroches, des frontispices faits de filets et d'entrefilets; des encadrés, qui rappellent les galeries de peintures et les tableaux classiques, ornent tous les imprimés.
Ce procédé est purement formel puisque les éléments décoratifs sont fabriqués d'avance et interviennent sans aucune référence au contenu. En architecture, les usines ressemblent à des temples grecs; dans l'imprimerie, les papiers à lettres et les facturiers imitent des faire-part d'adoubement; les publicités commerciales rappellent les annonces de commémorations. Les lignes et le corps du texte sont disposés en fonction d'un axe central et composés en caractères fantaisistes et exubérants. Les alphabets se chargent de fioritures. On trouve des initiales fabriquées industriellement qui s'inspirent des lettrines des premiers temps de l'imprimerie, et des compléments illustratifs pour diverses lettres qui évoquent à la fois le mot et le signet. On utilise des alphabets présentant des pleins et des déliés, comme tracés à la main et modelés; les surfaces décorées et les lignes circonscrivent entièrement la partie typographiée. Le décor excessif de tous les imprimés, les estampages pompeux en or et en argent, des lignes composées en

1888

Der Waffenschmied von Dortmund.

Erzählung aus der Zeit des Faustrechts

von Rudolf Scipio.

(Fortsetzung.)

In der Thür desselben trat ihm ein alter Diener des Burgherrn entgegen. Dieser erschrak nicht wenig, als er eine so große Anzahl Bewaffneter inmitten der Burg erblickte. Schnell trat er zurück, offenbar in der Absicht zu fliehen.

„Du bleibst hier und begleitest mich zu dem Schloßherrn," gebot der Graf, der es verhindern wollte, daß dieser vor der Zeit etwas von seinem gewaltsamen Eindringen in die Burg erfahre.

„Darf ich nicht, wie es Brauch ist, den Herrn Grafen zuvor bei dem gnädigen Herrn anmelden?" fragte der Diener, der wohl noch immer die Hoffnung nicht aufgegeben hatte, den Feind seines Herrn, als welchen er den fremden Grafen erkannt hatte, zu täuschen und so den Burgherrn rechtzeitig warnen und ihm wenigstens Zeit zur Flucht geben zu können. Doch der Graf war ein zu erfahrener Kriegsmann, als daß er sich durch eine so plumpe List hätte sollen täuschen lassen.

„Es bedarf keiner vorherigen Meldung," versetzte er, „ich werde mich schon selber deinem Herrn anmelden."

Der erfahrene Diener mochte aus dem sicheren, hoheitsvollen Wesen des Fremden wohl erkennen, daß derselbe zu befehlen gewohnt und berechtigt sei, und schritt, ohne einen weiteren Einwand zu wagen, gesenkten Hauptes voran, dem Gemache des Burgherrn zu. Es war dieses dasselbe, in welchem auch Berthold am Abend zuvor ihn gesehen hatte.

Vor der Thüre angelangt, gebot Graf Engelbert seinen Leuten, hier stehen zu bleiben und trat dann, nur von Berthold gefolgt, in das Gemach ein, welches heute nur von einem einzigen Wandleuchter erhellt ward, unter welchem drei Männer an einem Tische beim Becher und Würfelspiel saßen.

Graf Engelbert bot ihnen, die seinen Eintritt gar nicht beachtet hatten, mit lauter Stimme einen guten Abend.

Während zwei von den Spielenden nur flüchtig auffahen, warf sich der dritte, eine mächtige Gestalt mit weingerötetem Gesicht und groben, plumpen Zügen, bei der Anrede des Grafen im Sessel herum und maß den Eintretenden mit erstauntem finsteren Blick vom Kopf bis zu den Füßen.

27

ca. 1885

ca. 1880 ca. 1880 ca. 1885

1820

**Second half of the 19th century** In c. 1860, design was marked by the adoption of past forms. Although in fine art this led to academic rigidity and stagnation, for applied art forms it meant that stylistic quotations and ornamentation from Antiquity, via Gothic to the Rococo, became a feature of everyday work, characterizing an entire epoch. Formally, Historicism was manifested in pictorial and architectural allusions. Using the lines of lead composition, arches, monuments and temples were set; frames and borders reminiscent of art galleries full of classical pictures decorated all printed matter.
This was all purely formal, as this prefabricated ornamentation was employed without any regard for the actual content. Just as in architecture factories looked like Greek temples, printed notepaper and invoices resembled royal warrants, and business advertisements small monuments. Lines and texts were centered, and set in fantasy scripts. Alphabets were richly decorated; there were industrially manufactured initials, like something from an incunabulum, and illuminated embellishments of individual letters, which as a whole word had the effect of a signet. Alphabets were shaded, three-dimensional, modeled or hand-drawn; while typographical offerings were rounded off with ornamental surfaces and lines. The excessive decoration of printed matter with magnificent gold and silver embossing, lines set obliquely or in arches, linear fantasies to complete the picture: it all had a narrative richness.

Druckwerke mit prunkvollen Gold- und Silberprägungen, Zeilen in Schrägsatz, Rund- oder Bogensatz und füllenden Linienphantasien hatten eine erzählende Reichhaltigkeit. Es waren Bilderbogen in einer Zeit, in der die Fotografie und das bewegte Bild noch nicht verbreitet waren. Für die Schriftgießereien war die Zeit des Historismus wirtschaftlich sehr lukrativ, denn alle Setzereien und Druckereien wollten die Gestaltungsmittel kaufen, um modisch auf der Höhe der Zeit zu sein. Das galt für alle Länder, in denen die Industrialisierung weit fortgeschritten war, und die Wirtschaftskommunikation stark zunahm: die USA, England, Frankreich und Deutschland. Von da aus wurden die Mittel in andere Länder exportiert.
Es ist überraschend, daß die gestalterische Kultur des Historismus, die hundert Jahre lang bekämpft, bespöttelt und überwunden wurde, in modernisierter Form durch die neuen Techniken in der DTP-Typographie (1990) und in der multistilistischen Typographie (1995) erneut aktuell ist. Der Stil war damals und ist heute ein ängstliches Aufbegehren des bewegten künstlerischen Individuums vor befürchteter normierter Enge.

1884

1889 Brünner

ca. 1885

ca. 1880

ca. 1890

diagonale, en accolades ou en forme de cintres, le remplissage au moyen de filets fantaisie présentent une grande richesse narrative. Ce sont des planches illustrées à une époque où la photographie n'est pas encore répandue.
Pour les fonderies de caractères, la période éclectique est très lucrative puisque tous les ateliers de typographie et les imprimeries désirent se procurer de tels moyens de mise en page pour répondre aux exigences de la mode. Ceci vaut pour tous les pays déjà industrialisés où la communication moderne est en plein essor, les Etats-Unis, l'Angleterre, la France et l'Allemagne, qui bien sûr exportent ces procédés à l'étranger.
Il est surprenant de voir que cette culture combattue, tournée en dérision pendant un siècle, et finalement considérée comme obsolète, soit actuellement réhabilitée sous une forme modernisée avec l'avènement de nouvelles technologies telles que la DTP (1990) ou les caractères multistyles (1995). Ce style était jadis, comme aujourd'hui, l'expression de l'indignation angoissée de l'individu ému par l'art, face à une normalisation réductrice et redoutée.

These were the picture books of an age where photographs and moving images were yet unborn.
The era of Historicism was a very lucrative period for type foundries, as compositors and printers all wanted to buy their own design materials to keep up with the latest fashions. This was true of all the industrially advanced countries where economic communication was rapidly expanding, namely the USA, Britain, France and Germany. From here, the materials were exported to other countries.
It is surprising that the formal culture of Historicism, fought against, mocked and surmounted for a hundred years, should today be seeing a revival, in modernized form, through new techniques in DTP typography (1990) and multi-stylistic typography (1995). Then and now, this style is a frightened revolt of the emotional, artistic individual in the face of a feared, standardized confinement.

**Klassizismus**
**Néo-classicisme**
**Classicism**

**1760–1830** Die Verfeinerung der Technik und der Entwurfswerkzeuge bestimmten die Entwicklung der neuen Schriftformen in der Zeit des Klassizismus. Angeregt von den Kupferstichschriften, die durch die Radiertechnik des Barock und des Rokoko entstanden, wurden die Buchstaben kontrastreicher: die Haarstriche wurden feiner und schärfer profiliert. Die Buchstaben wurden nicht mehr mit der Hand geschrieben, sondern mit Lineal, Zirkel und Rastersystem konstruiert. Durch die sehr dünnen und sehr dicken Linien in den Buchstaben waren klassizistische Schriften schlechter lesbar als andere Antiquaschriften, aber die Eleganz, Präzision und Souveränität der Formen wirkt bis heute ungebrochen.
Klare Gliederungen und einfache Formen nach dem Vorbild der Antike bestimmten die Kultur des Klassizismus und auch das Bild der Alphabete.
Wie in den Jahrhunderten davor wurde auch jetzt nicht die Form der Typographie in Frage gestellt. Hauptformen waren Blocksatz und Mittelachse, Aufgabe der Typographie war, unter den vielen verschiedenen Schriften die geeignetste für einen bestimmte Zweck zu finden und den Text optimal zu strukturieren. Neuerungen wurden in vielen Ländern Europas verwirklicht, die Zentren waren England, Italien, Frankreich und Deutschland. Die bedeutendsten Schriftentwerfer und Typografen waren John Bell, Giambattista Bodoni, die Familie Didot und Justus Erich Walbaum. Mit den Schriften entstanden zahlreiche außergewöhnliche Bücher, die, anders als in der Zeit des Ro-

**1760–1830** Le raffinement des techniques et le matériel déterminent l'élaboration de nouveaux caractères pendant la période classique. Inspirées de l'écriture utilisée en gravure sur cuivre issue du procédé des eaux-fortes baroques et rococo, les lettres deviennent plus contrastées, s'affinent et se précisent. Ces caractères cessent d'être dessinés à la main, ils sont désormais «construits» à la règle et au compas sur un système de grille. Pourtant ces lignes tantôt très fines, tantôt très épaisses, rendent la lecture des écrits classiques plus difficile que celle des textes en Antique, mais l'élégance, la précision et la maîtrise des formes conservent un attrait inchangé jusqu'à nos jours.
La culture de la période néo-classique et l'image de son alphabet se définissent par un sens du rationnel et une sobriété des formes selon les modèles de l'Antiquité.
Comme au cours des siècles précédents, personne ne remet la forme typographiée en question. La composition en placards d'après un axe central prédomine; parmi les écritures les plus diverses, la typographie est considérée comme la plus apte à remplir une mission définie et à structurer le texte de manière optimale. Plusieurs pays d'Europe, comme l'Angleterre, l'Italie, la France et l'Allemagne participent à ces innovations avec des créateurs importants comme John Bell, Giambattista Bodoni, la famille Didot et Justus Erich Walbaum. Ces caractères donnent naissance à des livres surprenants qui, à l'inverse des publications de la période rococo, présentent une grande sobriété décorative. A la place des lettrines ornementées, on trouve des capitales d'un corps légèrement supérieur, dérivées du caractère de base.
Les deux volumes du «Manuale Tipografico» de Giambattista Bodoni, publié à

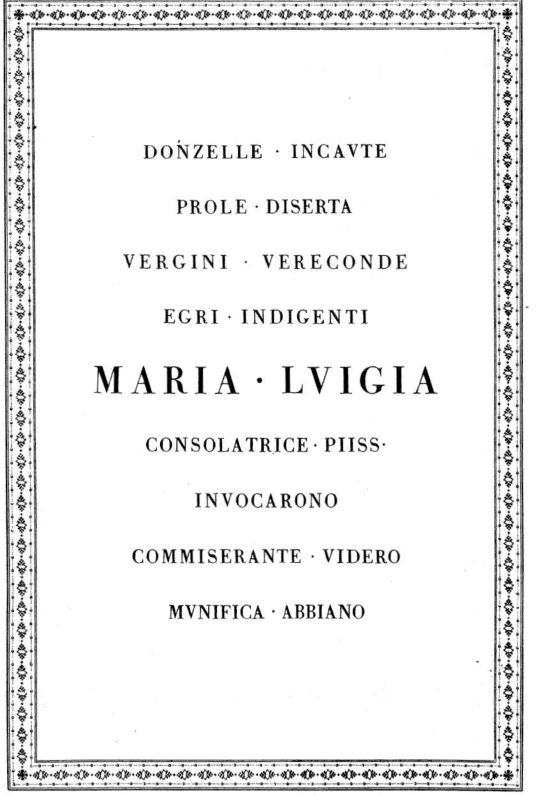
DONZELLE · INCAVTE
PROLE · DISERTA
VERGINI · VERECONDE
EGRI · INDIGENTI
MARIA · LVIGIA
CONSOLATRICE · PIISS·
INVOCARONO
COMMISERANTE · VIDERO
MVNIFICA · ABBIANO

1816 Bodoni

Versuch,
den
Ursprung der Spielkarten,
die
Einführung des Leinenpapieres,
und
den Anfang der Holzschneidekunst
in Europa
zu erforschen.
Von
Joh. Gottl. Immanuel Breitkopf.
Erster Theil,
welcher die Spielkarten und das Leinenpapier enthält.
Mit vierzehn Kupfertafeln.
Leipzig,
bey Joh. Gottl. Immanuel Breitkopf.
1784.

1784 Breitkopf

DIE

Walbaum

DIE
JUNGFRAU VON ORLEANS.
EINE ROMANTISCHE TRAGÖDIE
VON
SCHILLER.
Mit einem Kupfer.
BERLIN.
BEI JOHANN FRIEDRICH UNGER.
1802.

1802 Unger

A a
B b
C c
D d
E e
F f
G g
H h
I i
J j
K k
L l
M m
N n
O o
P p
Q q
R r
S s
T t
U u
V v
X x
Y y
Z z

1811 Didot

**1760–1830** Technological refinements and improved tools characterize the development of new typefaces in the Classical period. Inspired by the letter forms of copperplate engravings, born out of Baroque and Rococo etching techniques, alphabets sported greater contrast: hairlines became finer and more clearly defined. Letters were no longer drawn free-hand, but constructed with rulers, compasses and grids. The very thick and very fine lines make Classical typefaces more difficult to read than other roman styles, but their elegance, precision and supreme ease are still impressive today.
Clear structures and simple forms modeled on Antiquity dominated the culture and also the alphabets of the Classical period.
Typographical layout was never questioned in this era, nor in the eras preceding it. The main layouts used were justification left and right, or about a central axis; the task of typography was to find, from among the many different faces available, the one best suited for a specific requirement and for an optimally-structured text. New ideas were put into practice in many European countries, the main innovative centers being England, Italy, France and Germany. John Bell, Giambattista Bodoni, the Didot family and Justus Erich Walbaum were the major type designers and typographers of the time. With these new fonts, numerous unusual books appeared, which, in contrast to the Rococo, were very sparsely decorated. Ornamental initial letters were re-

koko, sehr sparsam mit Buchschmuck versehen waren. An die Stelle verzierter Initialen traten etwas größere Versalien aus der Grundschrift.
Eines der berühmtesten Bücher der Zeit ist das zweibändige „Manuale Tipografico" von Giambattista Bodoni, das nach seinem Tod erschien und 373 Schriften vorstellte. Bodoni selbst hat über hundert Antiquaschriften, fünfzig Kursivschnitte und fast dreißig griechische Alphabete geschaffen. Die Zeit war bestimmt von Forschung und Suche nach Klarheit. Normbestrebungen führten auch in der Typographie zu neuen Standards: um 1785 entstand aus einer Weiterentwicklung des Maßsystems von Fournier durch François Ambroise Didot das typographische Punktsystem, das von vielen Ländern als verbindlich für Schriftmaterial übernommen wurde.
Im Klassizismus kam die Entwicklung der Antiquaschrift formal an ein vorläufiges Ende, da die Variationsmöglichkeiten der Serifenbildung, der Kontraste der Buchstaben und die Achsenstellung der Rundungen ausgeschöpft waren. Kurz vor dem Abklingen des Klassizismus wurden um 1810 zwei für die gesamte Schriftentwicklung bedeutende und weiterführende Neuschöpfungen geschaffen: die serifenlosen und die serifenbetonten Schriften kündigten eine neue Zeit an.

*A a*
*B b*
*C c*
*D d*
*E e*
*F f*
*G g*
*H h*
*I i*
*J j*
*K k*
*L l*
*M m*
*N n*
*O o*
*P p*
*Q q*
*R r*
*S s ſ*
*T t*
*U u*
*V v v*
*X x*
*Y y*
*Z z*

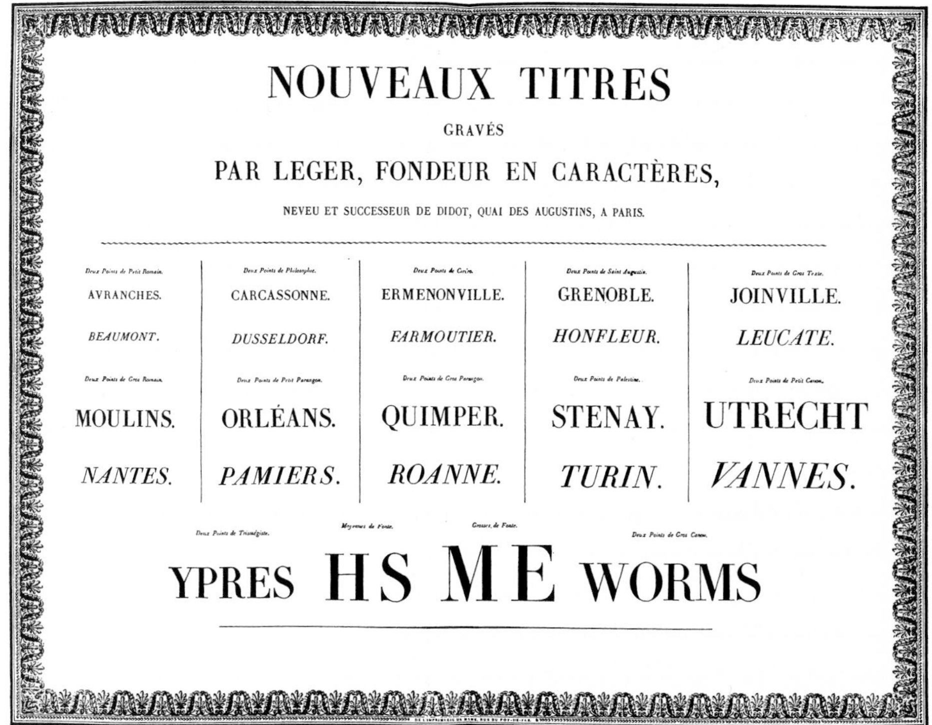

ca. 1806 Léger

titre posthume et présentant 373 types d'écritures, comptent parmi les ouvrages les plus célèbres de l'époque. Bodoni avait lui-même dessiné plus de cent dérivées de l'Antique, cinquante italiques et près de 30 alphabets grecs.
A cette époque déterminée par la recherche de clarté, ces aspirations à une normalisation amènent aussi la typographie à se standardiser. Ainsi vers 1785, le système métrique de Fournier trouve des applications dans le système des points instauré par François Ambroise Didot, système qui sera repris dans bien d'autres pays et deviendra un impératif dans l'imprimerie. D'un point de vue formel, l'évolution de l'Antique connaît alors une interruption provisoire car les possibilités de variations offertes par les sérifs, les contrastes et l'axe des arrondis semblent épuisées. Vers 1810, juste avant que ne commence le déclin du néoclassicisme, deux nouvelles créations essentielles pour le développement de la typographie voient le jour : les caractères sans sérif et les sérifs accentués, qui annoncent les temps nouveaux.

PRIMERA PARTE
DEL INGENIOSO HIDALGO
DON QUIXOTE
DE LA MANCHA.

CAPÍTULO PRIMERO.
*Que trata de la condicion, y exercicio del famoso hidalgo Don Quixote de la Mancha.*

Adria, Città antica d' Italia, che diede il nome al Golfo Adriati.

1811 Didot
1780 Ibarra
1788 Bodoni

placed by larger capitals from the basic body type.
One of the most famous books of this period is the two-volume "Manuale Tipografico" by Giambattista Bodoni, published posthumously and illustrating 373 typefaces. Bodoni himself designed over a hundred roman, fifty italic and almost thirty Greek alphabets.
This was an era marked by research and a quest for clarity. A general desire for norms extended to typography also: in c. 1785, following on from Fournier's system of units, François Ambroise Didot developed the typographical point system, which was adopted by many countries as a standard for printed material.
Where forms were concerned, the Classical period saw a temporary end to any further developments of the roman font, as the possibilities for variation of the serif forms, the contrasts of the letters and the axis position of the curves were exhausted. The fading years of Classicism, c. 1810, saw two major developments which were to have a major influence on typography: sans-serif and slab serifs heralded a new age.

**Barock**
**Baroque**
**Baroque**

**Ab ca. 1590** Die Kultur des Barock spiegelt eine Zeit wieder, in der sich Reichtum und Fülle in leuchtenden Farben und schwellenden Formen darstellten. Es war eine Epoche, in der das Mäzenatentum der Fürsten hervorragende Leistungen in den Bereichen Theater, Musik, Malerei und Architektur hervorbrachte. Namen wie Rubens, van Dyck, Rembrandt oder Bach und Händel genießen bis heute größtes Ansehen. Nach den Entwicklungen von Schrift und Typographie in der Renaissance sind die Neuerungen aus der Zeit des Barock vor allem als eine Zwischenstufe auf dem Weg zum Klassizismus zu sehen. Deshalb werden die Barockschriften häufig als „Übergangsantiqua" bezeichnet. Allerdings gab es Ergebnisse, die sowohl den Weg zur nächsten Stilentwicklung ebneten, als auch als eigenständige Formen die Schriftkultur bereicherten.
Die Buchstaben der Barock-Antiqua waren unbeeinflußt von der Strichführung der Breitfeder wie in der Handschrift. Sie waren abgeleitet von der präzisen Formkunst der Kupferstecherschriften. Die Alphabete wurden kontrastreicher gezeichnet, zwischen Haar- und Grundstrichen wurde deutlicher unterschieden. Dadurch entstand eine geschmeidige Eleganz in den Alphabeten und eine abwechslungsreiche Spannung im Textbild.
Die barocke Schriftentwicklung wurde besonders in Holland, England und Frankreich vorangetrieben. Christoffel van Dyck und Johann Michael Fleischmann schufen in Holland wichtige Alphabete, die die importierten und weit

**A partir d'env. 1590** La culture baroque reflète une période où la richesse et l'abondance se traduisaient dans la luminosité des couleurs et l'arrondi des formes. Ce fut une époque marquée par le mécénat des princes qui généra maints chefs-d'œuvre dans le théâtre et la musique ainsi qu'en architecture. Des grands noms comme Rubens, Van Dyck, Rembrandt, ou Bach et Händel symbolisent ce moment de l'histoire.Après l'évolution des caractères d'imprimerie et de la typographie pendant la Renaissance, on considère plutôt les apports de l'ère baroque comme marquant une transition qui mène au classicisme, ce qui vaut aux caractères baroques d'être fréquemment qualifiés d' «Antique de transition». Toutefois, des événements se produisirent qui anticipaient sur le style de la période suivante tout en enrichissant la culture autonome de l'écrit.
Les caractères de l'Antique baroque n'étaient plus déterminés par le tracé de la calligraphie des manuscrits écrits à la plume. Ils portaient plutôt l'empreinte de l'expression formelle et précise des caractères gravés dans le cuivre. Le dessin des alphabets devenait plus contrasté, on différenciait nettement les pleins et les déliés. Ainsi, les alphabets gagnèrent-ils en souplesse et en élégance et la page en diversité et en intérêt.
C'est surtout en Hollande, en Angleterre et en France que l'on commença à élaborer une typographie de style baroque. Christoffel van Dyck et Johann Michael Fleischmann créèrent en Hollande des alphabets notoires qui prirent le relais d'autres caractères très répandus comme ceux de la Renaissance française. En Angleterre, William Caslon et John Baskerville comptaient parmi les novateurs les plus créatifs. En fait, c'est la typographie hollandaise qui avait influencé les ca-

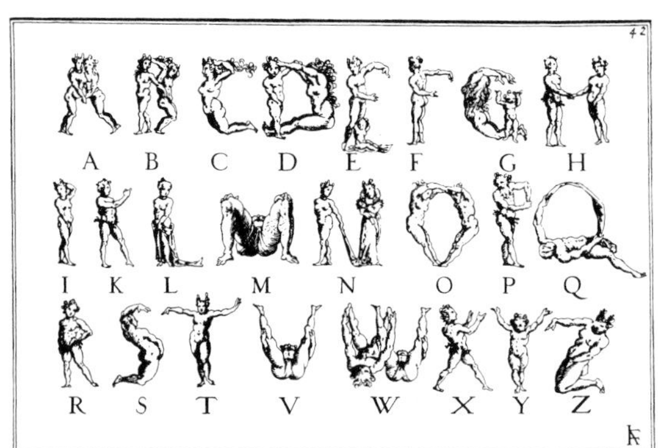

1596 Franco

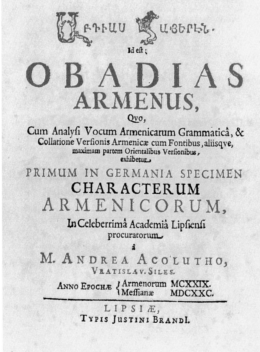

DAs einige Glieder der nach GOttes Wort Reformirten Christl. Gemeine / in hiesigen Churfürstl. Brandenburgischen Residentz-Städten / bey sich erwogen/wie dieselbe durch GOttes Seegen sich eine Zeither durch allerhand Vertriebene / aus der Pfaltz / Franckreich / Schlesien und andern Orten am Ober- und Nieder-Rhein-Strohm / also vermehret / daß in denen von ihnen bißher gebrauchten Kirchen und Tempeln der Raum für Sie und Ihren Gottesdienst zu enge werden wolle / auch viele wegen zerstreueter und entlegener Wohnung / die Zeit zum Gottesdienste nicht wahrnehmen könten / daher Sie dann deßen wider ihren Willen offt entbehren müßen;
So haben sich solche Glieder mit andern mehreren vereiniget / und zusammen Se. Churfürstl. Durchl. zu Brandenburg / dero Gnädigsten Landes-Herren / unterthänigst Supplicando angetre-

1695 Liebpert

Id est;
OBADIAS
ARMENUS,
Quo,
Cum Analysi Vocum Armenicarum Grammaticâ, & Collatione Versionis Armenicæ cum Fontibus, aliisqve, maximam partem Orientalibus Versionibus, exhibetur,
PRIMUM IN GERMANIA SPECIMEN
CHARACTERUM
ARMENICORUM,
In Celeberrimâ Academiâ Lipsiensi procuratorum,
à
M. ANDREA ACOLUTHO,
VRATISLAV. SILES.
ANNO EPOCHÆ { Armenorum MCXXIX. Messianæ MDCXXC.
LIPSIÆ,
TYPIS JUSTINI BRANDI.

1680 Brand

ca. 1719 Losenauer

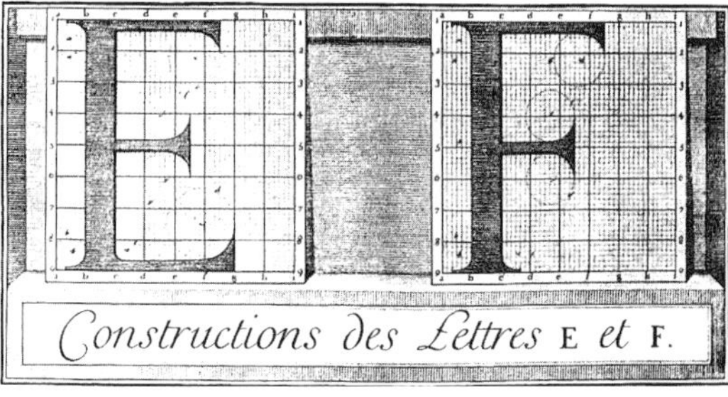

1692 Grandjean de Fouchy

**From c. 1590** The culture of the Baroque mirrors an age, where wealth and abundance were represented by brilliant colors and luxuriant shapes. It was an epoch in which royal patronage gave rise to magnificent achievements in theater, music, art and architecture. Names such as Rubens, van Dyck, Rembrandt, Bach and Händel, who were active during this epoch, are still held in high regard today. After the initial developments in writing and typography in the Renaissance, the innovations of the Baroque are really more of a step towards Classicism, which is why Baroque fonts are often referred to as "transitional roman". There were, however, a number of important events which helped smooth this transition to the next stylistic developments and also made their own contribution to the art of writing with their aesthetic forms.
The letters of Baroque roman were no longer determined by the strokes of the broad-nibbed quill. They were derived from the precise forms used in copperplate engravings. Greater contrasts were worked into the alphabets; there was greater differentiation between hair lines and stems. Fonts thus took on a supple elegance, while the text on the page was marked by tension resulting from the alternation of thick and thin.
Holland, England and France played a major role in the development of Baroque typography. In Holland, Christoffel van Dyck and Johann Michael Fleischmann created important alphabets which succeeded the widespread imported French Renaissance types. In England, William

verbreiteten französischen Renaissance-Alphabete ablösten. In England gehörten William Caslon und John Baskerville zu den kreativen Neuerern. Caslons Schriften waren von den holländischen Schriften beeinflußt. Ihre formalen Qualitäten und die gute Lesbarkeit ließen sie fast zu englischen Nationalschriften werden. Baskerville wurde neben seinen modernen Alphabeten vor allem wegen der richtungsweisenden Einfachheit seiner typographischen Gestaltung berühmt.

Ein für Frankreich einschneidender Vorgang war, daß König Ludwig XIV. 1692 befahl, für die „Imprimerie Royal", die 1639 von Kardinal Richelieu in den Werkstätten des Louvre gegründet wurde, Exklusivschriften entwerfen zu lassen. Um diese neue „königliche" Schrift zu gestalten wurde eine akademische Kommission gebildet. Diese entwickelte Pläne, in denen die Schrift in einer Rasterkonstruktion entstehen sollte: man teilte ein Quadrat in ein 64-teiliges Quadrat-Rastersystem und jedes dieser Quadrate wieder in 36. So entstand ein Netz von insgesamt 2.304 kleinen Quadraten. Philippe Grandjean zeichnete anhand dieses Rastersystems ein Alphabet, mit dem Namen „Romain du Roi". Trotz der logisch-rationalen Vorgehensweise, die aus dem aufklärerischen französischen Denken verständlich wird, wurde die Schrift eine „normale" Barock-Antiqua, denn Grandjean orientierte sich weitgehend an der Harmonie des Auges und nicht an der Geometrie. 1702 waren die ersten Schriftschnitte gefertigt und wenig später erschien das erste Buch mit der neuen Schrift.

1627 Kilian

1710 Shelley

(44)

bracht; hingegen nehmen sie ab, und müssen vergehen, wo ein jeder aus dem Pöbel wil dem Studiren nachgehen: Denn wer sich rühmet, daß er studiret habe, der wird nicht leicht einen Kauffmann abgeben, schreibet und urtheilet Gramond, ein Frantzösischer Historicus.

Mittel Schwabacher.

Außer dem ist des Königs ernster und gnädiger Wille, auch allen weltlichen Obrigkeiten und Geistlichen inclusivè, kund gemacht, per Mandatum, acht zu haben auf die Schulen, die ingenia stupida, oder tölpische Köpffe, die Halßstarrige und Faule nicht zu toleriren, oder zu dulten; sondern zu andern Handthierungen und Manufacturen anzuweisen, damit die Fleißigen nicht gehindert, die Seminaria nicht beschweret, noch die beneficia nichtswürdigen Leuten möchten conferiret werden, und also dem Publico kein Schade oder Schande von ihnen erwachse. Berlin vom 8 September des 1708. Jahres.

Neue Leipziger grobe Mittel Fractur.

Lob der Buchdruckerey.

O Kunst! der nichts zu gleichen ist,
Die Kirche kan zu keiner Frist
Hier ohne dich bestehen:
Was acht ich Rath-Hauß, Canceley,
Was Schöppenstuhl, was Schreiberey,
Wo du dich nicht läst sehen?
Du bist der Künste Königin,
Ja selbst der Weißheit Meisterin:
Daß Advocaten sind gelehrt,
Daß man den Artzt hält hoch und werth,
Daß man die Lehrer liebet:

1710 Pater

B
A
C
H

ca. 1725 Caslon

1722 Andrade

ractères de Caslon, mais leurs qualités formelles et leur excellente lisibilité en firent presque l'écriture nationale anglaise. Quant à Baskerville, il acquit sa notoriété pour ses alphabets modernes, mais surtout à cause de la sobriété de sa mise en page.

La France connut un moment décisif quand, en 1692, Louis XIV ordonna que l'on conçoive des caractères exclusivement destinés à «L'imprimerie Royale» fondée par le cardinal de Richelieu dans les ateliers du Louvre en 1639. Pour élaborer cette nouvelle typographie «royale», on réunit alors une commission académique. Cette dernière réalisa des planches comportant des grilles sur lesquelles devaient s'inscrire les nouvelles lettres: on divisa un carré en 64 cases selon un système de quadrillage, puis chacune d'elles en 32 autres carrés plus petits. C'est ainsi que l'on obtint une trame comportant 2 304 carrés au total. Philippe Grandjean dessina alors, au moyen de ce système, un alphabet qui fut baptisé «Romain du Roi». En dépit de cette démarche logique et rationnelle, pur fruit de la raison et de la pensée française, on obtint une «banale» Antique baroque, car Grandjean ne s'était guère laissé guider par la géométrie, mais davantage par l'harmonie. En 1702, les premières planches de caractères étaient achevées et, peu après, le premier livre avec la nouvelle écriture paraissait.

Caslon and John Baskerville were among the most creative innovators. Caslon's fonts were influenced by Dutch patterns. Their formal qualities and legibility almost made them England's national typefaces. Baskerville became famous, not just for his modern alphabets, but also for the innovative simplicity of his typographical designs.

A radical event for France was Louis XIV's order in 1692 that exclusive fonts be developed for the "Imprimerie Royale," founded by Cardinal Richelieu in the workshops of the Louvre in 1639. An academic commission was put together to design this new "royal" typeface. Plans were drawn up to produce the type using a grid construction, whereby a square was split into a 64-square grid system and each of these squares into 36 smaller squares. A network was thus formed comprising 2,304 tiny squares. Using this grid system, Philippe Grandjean designed an alphabet entitled Romain du Roi. Despite the logical, rational process behind it, a typical manifestation of the French way of thinking at the time, the font turned out to be a "normal" Baroque roman, as Grandjean concentrated more on visual harmony rather than geometry. In 1702 the first type was ready and soon after, the first book was printed in the new face.

**1400–1600** Die Welt der Renaissance knüpfte in allen geistigen und künstlerischen Bereichen bewußt an die Ideale der griechischen und römischen Antike an. Dadurch kam es zu Neuentwicklungen in Kultur und Wissenschaft. Die Humanisten machten den Weg für neue Ansätze frei. Es wurden zahlreiche Universitäten gegründet, es wurden neue Länder entdeckt, viele technischen Entwicklungen ließen eine neue Zeit anbrechen.
Die folgenreichste handwerkliche Neuerung war die Erfindung des Handgießapparats für bewegliche Einzelbuchstaben durch Johannes Gutenberg um 1450. Um 1455 war Gutenbergs 42zeilige Bibel fertiggestellt, für die sechs Handsetzer zwei Jahre lang den Satz für die 1282 Folioseiten herstellten.
Sehr schnell breitete sich die Druckkunst aus, im Jahr 1500 gab es bereits 1100 Druckereien in Europa. In diesen ersten sechs Jahrzehnten entstanden ungefähr vierzigtausend verschiedene Druckwerke, diese Erstdrucke werden als „Inkunabeln" bezeichnet. Damit begann eine neue Zeit der Schriftentwicklung. War die Schrift, die Gutenberg für seine Bibeldrucke verwendete, noch eine Schrift mit gotischem Charakter, um der Qualität der Handschriften nicht nachzustehen, bildete sich um 1465 in Venedig die Antiqua als neue Druckschriftenform heraus.
Die Handschriften der Humanisten, die ihre Werke selbst niederschrieben, lehnten sich an die karolingische Minuskel an. Die Schrift, die dabei entstand, wurde als humanistische Minuskel bezeichnet. Sie

**1400–1600** Dans tous les domaines intellectuels et artistiques, le monde de la Renaissance s'orienta consciemment aux idéaux de l'Antiquité grecque et romaine, ce qui contribua à l'émergence d'un renouveau culturel et scientifique. Les humanistes préparaient la voie pour de nouvelles idées. Partout, d'innombrables universités voyaient le jour, on découvrait de nouveaux territoires, et l'évolution des techniques laissait présager une ère nouvelle.
Dans l'artisanat, l'invention qui aura le plus de conséquences est indéniablement celle de Johannes Gutenberg qui, en 1450 conçut un appareil à fondre des caractères mobiles. En 1455, la Bible de Gutenberg avec ses pages de 42 lignes, était prête; six hommes avaient travaillé pendant deux ans pour composer ses 1282 folios.
L'imprimerie se répandit alors très vite. En l'an 1500, il existait déjà 1100 ateliers en Europe. Environ quarante mille imprimés divers paraissaient au cours des six premières décennies, des ouvrages que l'on désigne aujourd'hui sous le nom d'»incunables». Ainsi s'ouvrait une ère nouvelle dans l'histoire de la typographie. Tandis que les caractères utilisés par Gutenberg dans sa Bible faisaient encore référence aux lettres gothiques, ceci pour n'être en rien inférieurs à la qualité des manuscrits, un caractère d'imprimerie aux formes nouvelles, l'Antique, apparaissait à Venise vers 1465.
Les manuscrits des humanistes, qui rédigeaient eux-mêmes leurs œuvres, s'inspiraient des minuscules carolingiennes, ce qui donna naissance à un caractère appelé «minuscule humaniste». Les imprimeurs italiens comprirent très vite que la typographie permettait d'élaborer des formes différentes de la calligraphie et ne tardèrent pas à la prendre pour modèle.

Beniuolentiam autem a perſonis ducimus:aut a cauſis accipimus:ſed
perſonarum non eſt:ut pleriq; crediderint:triplex ratio:ex litigatore:&
aduerſario:& iudice.Nam exordium duci nonnunq̃ etiã ab actore cau-
ſæ ſol&:q̃q̃ enim pauciora de ſe ipſo dicit:& parcius:plurimũ tamẽ ad
oĩa momenti eſt in hoc poſitũ:ſi uir bonus creditur:ſic enĩ continget:
ut nõ ſtudium aduocati uideatur afferre:ſed pene teſtis fidem.Quare
in primis exiſtimetur ueniſſe ad agendum ductus officio uel cognatio-

1471 Jenson

Non a furib3 neq3 a latronib3 se libe-
rabūt dij lignei 7 lapidei 7 inaurati 7
inargētati: quib3 iniqui fortiores sūt.
Aurū et argentū et vestimentū quo o-
perti sunt auferent illis et abibūt: nec
sibi auxiliū ferent. Itaq3 melius est esse
regem ostentantē virtutē suā aut vas
in domo utile ī quo gloriabitur qui
possidet illud: q̄ falsi dij: uel ostiū ī domo q̄d custodit que in pace sunt: q̄
falsi dij. Sol quidē et luna ac sidera
cum sint splendida et emissa ad utili-

1450–55 Gutenberg

NARRA QVIVI LA DIVA POLIA LA NOBILE ET
ANTIQVA ORIGINE SVA.ET COMO PER LI PREDE
CESSORI SVI TRIVISIO FVE EDIFICATO.ET DI QVEL
LA GENTE LELIA ORIVNDA. ET PER QVALE MO-
DO DISAVEDVTA ET INSCIA DISCONCIAMENTE
SE INAMOROE DI LEI IL SVO DILECTO POLIPHILO.

1499 Manuntius

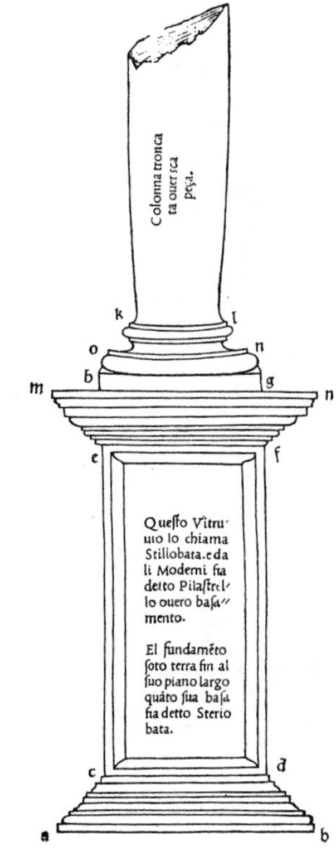


1503 Pacioli

**1400–1600** The world of the Renaissance was consciously linked to Ancient Greek and Roman ideals, both in spiritual and artistical spheres. This led to new developments in culture and science. Humanists forged the way for new ideas, many universities were founded, new countries were discovered; technological advance heralded the dawning of a new age.
The most revolutionary technical innovation was the invention of the manual caster for movable type by Johannes Gutenberg in c.1450. Gutenberg's 42-line Bible was completed in 1455; it took six compositors two years to produce the type for the 1,282 folio pages.
The art of printing spread extremely quickly; by 1500 there were already as many as 1,100 printing workshops in Europe. In these first six decades alone, approximately forty thousand different printed works were produced, termed "incunabula". A new period in the development of writing had begun. In his bibles, Gutenberg used a typeface which was Gothic in character, in order to try and retain the quality of handwritten texts; however, in c. 1465 roman emerged in Venice as the new printing standard.
The handwriting used by the Humanists, who wrote out their works themselves, was based on the Carolingian Minuscule, and came to be known as Humanist Minuscule. It soon became the model for Italian printers, who rapidly recognized that printing allowed forms other than those of hand-written calligraphy.
The roman typeface of 1465 was soon fol-

wurde Vorbild für die italienischen Drucker, die schnell erkannten, daß die Typographie andere Formengestaltung ermöglichte als die handschriftliche Kalligraphie. Schnell folgten der Antiqua von 1465 ausgefeiltere Alphabete von Nicolaus Jenson und Aldus Manutius, die in Venedig neue Maßstäbe setzten. Später war es Claude Garamond in Paris, dessen Schriften zwischen 1530 und 1550 auf der Basis der venezianischen Antiqua erschienen.

Die Renaissance-Antiquaschriften wurden in zwei Gruppen eingeteilt: die venezianischen und die französischen. Beide Gruppen entwickelten sich aus Formen, die mit der schrägangesetzten Breitfeder geschrieben wurden.

Renaissanceschriften haben ein ruhiges, gleichmäßiges Gesamtbild. Sie haben hohe Ober- und Unterlängen. Die offenen Einzelformen lassen das Textbild hell erscheinen.

Um 1600 war die Renaissance-Antiqua die vorherrschende Buchschrift in ganz Europa. Noch heute gehören die Alphabete der Renaissancezeit zu den am häufigsten gebrauchten und am besten lesbaren überhaupt. Die klaren Details, die ruhige harmonische Form dienen der Lesbarkeit des Textes.

ſetz den einen fues in den puncten.n.vnd den anderen auf die lini.l.k.in die ſelb ſtat ſetz ein.v. in diſem centrum laß den cirkel mit dem einen fues ſtil ſten/vnd mit dem anderen reyß auß den puncten.n.ein cirkellini/darnach zeuch auß diſem centrū.v.gerad lini in die puncten der lini.p.q.Darnach nym mit einem cirkel auß den vorgemelten dryangel die leng.a.e.vnd ſetz jn mit dem ein fues in den puncten.o. vnd den anderen auf die lini.k.l.in den ſelben puncten ſetz ein.r. darauß zeuch gerad linien biß in die puncten.r.ſ.Darnach ſetz den cirkel mit dem einen fues in den puncten.r. vnd reyß mit dem anderen ein cirkellini durch den puncten.o.alſo iſt der ober cirkel auß der vorgemelten leng.d.e.gemacht/vnd gehört an die aufrecht mauer zůſtellen/aber der vnder cirkel der auß der leng.a.e.gemacht iſt/gehört auf den orizon oder ebne zůſtellen.Wie das hernach iſt aufgeriſſen/darauß auch die ſeyten horalogia ſind zů nemen.

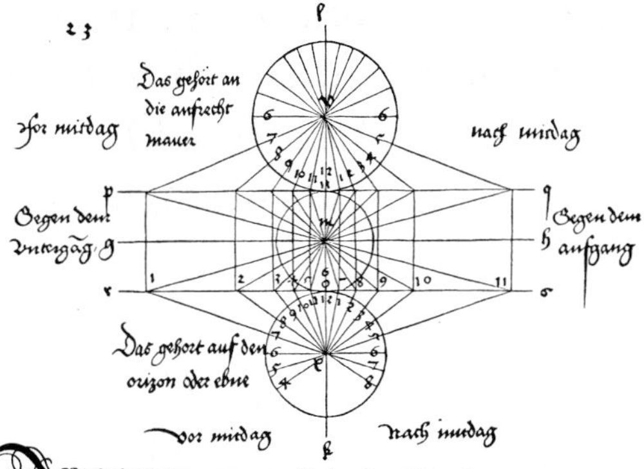


Zu diſen horalogien mag man manicherley ecketter ſtöck machen darauf allerley horalogia zůmachen ſind/aber hie will jch ein gemeynen brauch durch ein cubum anzeygen den man an etlichen orten beſchneiden ſol/vnd darnach auf die ſelben felder die die ſonn beſcheynen mag horalogia ſtellen/thů dem alſo/bezeychen den cubum auf einer ſeyten an ſeinen vier ecken mit.k. l.m.n.vnd ſetz ein puncten.c.mitten darein/daraus reyß ein cirkellini die die vier ſeyten der fierung an rür/vnd teyl die cirkellini mit zweyen geraden linien in vier gleyche teyl zů rechten wincklen/alſo das die zwerchlini.a.c.b.den orizon bedeut/vnnd die aufrecht lini oben ſey das zenit auf dem haubt/vnnd vnden das.e.ſey der fues oder baſis. Darnach teyl ein yetlichs vierteyl des cirkels in neuntzig teyl wie oben gelert iſt/vnd zel von dem.b.vberſich gegen dem.d.ſo vil grad der polus hoch ſtet wo du biſt/aber hie zů Nüremberg wie gemelt iſt die eleuation neun vnd viertzig grad wie vor gemelt/vnd wo ſich diſer grad endet do ſetz einen puncten.f. Darnach reyß ein gerade lini.f.c.vñ gar an die zirkellini do hin ſetz ein.g.diſe lini deuttet axem mundi daran die welt vmb get/nach diſer lini ſollen die ſtangē die den ſchatten an den maurē geben gericht werden/aber auf den verwentē ſeytē/vñ auf der ebne mag man die ſtangē nach dem equinoctial oder zenit richten/die ſtile der vier horalogia im cirkel des equinoctials ſollen yetlicher in ſeinem felt zů gleychen winckelen aufrecht ſten. Darnach reyß durch das centrum. c.ein gerade kreütz lini von einem ende des cirkels zům anderen/alſo das die lini der art vnd diſe vier gleyche winckel machen vnd ſetzen an jre ende.h.i.diſe lini zeyget an den equinoctial zů beden ſeyten

1525 Dürer

ABC
abcde

ca. 1555 de Tournes

A l'Antique de 1465, succédèrent bientôt les alphabets plus raffinés de Nicolaus Jenson et d'Alde Manuce qui fixaient de nouvelles normes à Venise. Puis vint Claude Garamont, à Paris, dont les caractères conçus entre 1530 et 1550 semblaient dérivés de l'Antique vénitienne.

Les caractères renaissance de l'Antique se répartissent en deux familles, vénitienne et française, toutes deux issues des figures que l'on calligraphiait en appliquant la plume en biais. Ils présentent un aspect régulier et harmonieux. Les jambages supérieurs et inférieurs sont élancés, les diverses formes sont franches et confèrent au corps du texte une impression de clarté.

Vers 1600, l'Antique renaissance s'était imposée dans toute l'édition européenne. Aujourd'hui encore, les alphabets de cette époque comptent parmi les plus utilisés et les plus lisibles. La clarté des détails, leurs lignes harmonieuses permettent une lecture aisée du texte.

Medici

ca. 1495 Bembo

COPIE D'VNE LETTRE

DV DVC D'ALVE A

L'EVESQVE · D'ANVERS.

DON Fernando Aluarez de Toledo, Duc d'Alue, &c. Lieutenant, Gouuerneur, & Capitaine general.

TRESREVEREND Pere en Dieu, treſchier & bien amé. Oultre la Bible vniuerſelle, que le Roy auoit donné charge au docteur Arias Montanus de faire imprimer en ceſte ville en quatre langues auec ſes interpretations, lon eſt auſſi apres pour imprimer autres volumes de choſe ſeruāt à l'intelligēce de ladicte Bible & parties d'icelle, leſquelles s'intitulent *Apparato ſacro*; & entre icelles y a pluſieurs eſcriptz compoſez par ledict docteur Arias Montanus, & autres perſonnes doctes, qui ont trauaillé & trauaillent à enrichir ceſt œuure. Mais comme ſemblables choſes nouuelles, ou non imprimees auparauant, ne ſe peuuent imprimer ſans viſitation & approbation precedente, eſtant ceſt œuure plus ſolemnel que les ordinaires; il conuient que ladicte approbation ſe face auec toute auctorité & ſolemnité poſſible. Et d'autant que ledict docteur Arias Montanus ne peult eſtre iuge de ſes propres eſcriptz; nous vous requerons, que quand ledict docteur Arias Montanus vous preſentera leſdicts eſcriptz, & pieces ſeruantes audict œuure, vous les commettiez à la perſonne ou perſonnes du college des viſitateurs, que à noſtre ordonnance & par auctorité de ſa Maieſté vous auez eſleu, afin qu'ils les viſitent & approuuent, pour autant que touche la ſanité de la doctrine; & que icelle approbation ſoit auſſi ſolemnizee auec le ſuffrage de tous ceulx du college, ou la plus grande partie, & l'interuention de voſtre auctorité: afin que ceſte approbation ſe face comme requiert vn œuure ſi principal. Ce que nous vous recommandons tant plus à certes pour auoir entendu par lettres de ſa Maieſté, que telle eſt ſon intētion. Treſreuerend Pere en Dieu treſchier & bien amé, noſtre Seigneur vous ait en ſa ſaincte guarde. D'Anuers, ce dernier iour de Feburier 1570.

ca. 1570 Plantin

lowed by more refined alphabets, such as those by Nicolaus Jenson and Aldus Manutius, who set new standards in Venice. Later, between 1530 and 1550, Claude Garamond in Paris produced typefaces which were also based on the Venetian roman. The Renaissance roman fonts were thus divided into two groups, the Venetian and the French. Both were developed from forms originally hand-written with a broad nib held at an angle.

Renaissance typefaces are smooth and regular in form, with tall ascenders and long descenders. The open forms of the letters give a light appearance to the text on the page.

By 1600, Renaissance roman was the predominant book font throughout Europe. Even today, Renaissance alphabets are among those most frequently used and easiest to read. Clear detail and smooth, harmonious structure provide the legibility of the text.

## Gotik
## Gothique
## The Gothic period

**1230–1500** Zentren der Kultur im Zeitalter der Gotik waren die Burgen und Klöster, in denen die junge Generation zu Kriegern oder Geistlichen erzogen werden sollte. Hier wurden sie im Lesen und Schreiben unterrichtet. Mönche schrieben zur Ehre ihres Gottes alte Texte zur Überlieferung und Verteilung an neu entstandene Kirchen ab. Zum Teil waren es auch Auftragsarbeiten weltlicher Herren, die sich auf diese Art eine standesgemäße Bibliothek zusammenschreiben ließen. Abseits der Burgen und Klöster entwickelte sich langsam eine städtische, bürgerliche Kultur, in der andere Formen und Inhalte gesucht wurden. In manchen Städten entstanden Schulen, allmählich begann das Aufbegehren gegen die Feudalherrschaft. Im 12. Jahrhundert war auf allen künstlerischen und handwerklichen Gebieten ein Aufbruch zu einem neuen kulturellen Stil zu spüren. Sichtbar manifestierte sich dieses Formgefühl zunächst in der Baukunst, bald wurde auch die Schrift von diesem Wandlungsbestreben erfaßt: die Buchstaben streckten sich in die Höhe, sie wurden aneinander gerückt, die Rundungen wurden gebrochen und die gotische Minuskel mit ihrer gitterartigen Struktur entstand. Diese gotischen Kleinbuchstaben entwickelten sich seit dem 11. Jahrhundert langsam in mehreren Übergangsformen aus der spätkarolingischen Minuskel. Es gab zahlreiche Varianten, die alle durch Regelmäßigkeit und Strenge der Formen gekennzeichnet sind. Bei der Textura, die seit dem 13. Jahrhundert verwendet wurde, ist das Schriftbild von geraden,

**1230–1500** A la période gothique, les châteaux et les monastères étaient les centres de la vie culturelle où les jeunes générations se destinaient à devenir des guerriers ou des clercs. C'est là qu'on leur enseignait la lecture et l'écriture. Les moines copiaient des textes anciens qu'ils transmettaient et distribuaient dans les églises nouvellement construites. Il s'agissait aussi parfois de commandes profanes venant de seigneurs qui, de cette manière, se constituaient des bibliothèques dignes de leur rang. Peu à peu, une culture bourgeoise et urbaine se développa à l'écart des châteaux et des monastères, une culture qui recherchait d'autres formes et d'autres contenus. Des écoles étaient créées dans plusieurs villes, tandis que l'on commençait à contester la féodalité. Le 12e siècle connut un bouleversement dans tous les domaines de l'artisanat et de l'art, qui devait déboucher sur un nouveau style. Ce sentiment formel se manifesta d'abord en architecture, puis l'écriture ne tarda pas à obéir à ces aspirations au changement. Les lettres s'étiraient en hauteur, se resserraient les unes contre les autres, des ruptures interrompaient les courbes, et la minuscule gothique apparut avec sa structure qui rappelle une grille. Ces petites lettres gothiques avaient commencé à se constituer progressivement à partir de la minuscule carolingienne et présentaient plusieurs formes transitoires. Il existait de nombreuses variantes, toutes caractérisées par la régularité et la rigueur des formes. Dans la textura utilisée depuis le 13e siècle, l'écriture se définissait par l'aspect des traits droits et verticaux, des angles aigus et acérés. Les formes arrondies avaient disparu, les jambages supérieurs et inférieurs étaient de petite taille. Les lettrines élaborées par la suite s'inscrivaient dans ce système formel ri-

12th century Northern German

1204 England

12th century Minuscule

ca. 1300 Textur

1320 Sienna

ca. 1210 Brandenburg

**1230–1500** The cultural centers of the Gothic age were the numerous castles and monasteries where the younger generations trained to be knights or monks and where they were taught to read and write. Monks meticulously copied ancient texts to the glory of God, for posterity and for distribution to newly-consecrated churches. Some of their work was done for secular lords who made use of the monks' copying skills to build up their own private libraries as their status demanded. Alongside the castles and monasteries, the new towns were slowly beginning to develop their own culture, in which other forms and contents were sought. In some of these towns, schools were set up, and medieval man began to rebel against the rule of feudal lords.
A sense of departure towards a new cultural style began to pervade all areas of the arts and crafts during the 12th century. The first visual signs of this new feeling for form were manifested in architecture, but writing was also soon affected by this desire for change: letters

senkrechten Strichen, von scharfen Ecken und spitzen Winkeln bestimmt, die durch das Schreiben mit der Breitfeder entstanden. Es gab keine gerundeten Formen, Ober- und Unterlängen waren kurz gehalten. Auch die später entwickelten Versalien fügten sich in das strenge formale System ein. Eine weitere Variante war die rundgotische Schrift Rotunda, bei der die sehr engen Formen der gotischen Gitterstruktur geweitet wurden, was ein besser lesbares Schriftbild ergab. Die Schrägstriche in den Buchstaben wurden wieder zu Rundungen. Obwohl die Rotunda in Italien, Spanien und Deutschland häufig verwendet wurde, geriet sie seit Mitte des 16. Jahrhunderts in Vergessenheit.
Nach der völligen Akzeptanz gotischer Schriften im 13. Jahrhundert treten ab dem 14. Jahrhundert formale Veränderungen auf, die einen größeren Reichtum der Schriftformen ankündigen.

goureux. L'autre variante était la rotunda, ou gothique ronde, avec une structure en grille étroite et très stricte; elle conférait à la lettre gothique davantage d'ampleur et procurait une meilleure lisibilité. Les traits obliques des lettres s'arrondissaient de nouveau. Bien que la rotunda ait été fréquemment utilisée en Italie, en Espagne et en Allemagne, elle tomba en désuétude dès le milieu du 16e siècle.
Après avoir été employés tels quels au 13e siècle, les caractères gothiques subirent dès le 14e siècle des modifications qui laissaient présager un grand enrichissement formel des types d'écritures.

1313 Italian Minuscule

1358 Rotunda

1472 Hemmel von Andlau

ca. 1490 Geiler von Kaiserberg

1493 Koberger

were stretched tall and placed closer together, their round form broken, and Gothic Minuscule, with its grid-like structure, was born.
These small Gothic letters had undergone various transitional stages from late Carolingian Minuscule since the 11th century. There were numerous variations, all distinguished by regularity and rigidity of form. Textura, in use since the 13th century, was characterized by straight, vertical strokes, sharp corners and acute angles, the result of writing with a broad nib. There were no rounded forms, and ascenders and descenders were kept short. Even majuscules, developed at a later stage, fitted in with the strict formal system. Another variant was Rotunda, a round Gothic script, where the extremely narrow forms of the Gothic grid system were widened somewhat for easier legibility. The oblique strokes in the letters once again became curved. Although Rotunda was widely used in Italy, Spain and Germany, it gradually faded into oblivion during the 16th century.
Gothic scripts were accepted as the standard of the 13th century, yet during the 14th century formal changes began to materialize which heralded a greater wealth of written forms.

**Mittelalter und Romanik**
**Le moyen-âge et la période romane**
**The Middle Ages and Romanesque**

**Ab ca. 700** Als am Ende des 8. Jahrhunderts durch die Machtübernahme der Karolinger ein riesiges Reich in Europa entstand, wurde mit grundlegenden Reformen in Kultur, Politik und Kirche dieses Gebiet organisiert. Die Notwendigkeit, Beschlüsse, Gesetze und Verlautbarungen für das gesamte Reich verständlich zu machen, führte zu einer gemeinsamen Schrift. Es entstand die karolingische Minuskel, die ihren formalen Ausgangspunkt in den altrömischen Schriften hatte. Es war eine aus Kleinbuchstaben bestehende Schrift, deren einzelne Buchstaben breit und rund gehalten waren. Ihre offenen Formen spiegelten die Offenheit einer Gesellschaft wider, die sich von den Kirchendogmen zu befreien suchte. Diese Entwicklung war ein wichtiger Schritt in der Schriftgeschichte, da erstmals voll ausgebildete Kleinbuchstaben entstanden. Neben der römischen Schrift wurde die karolingische Minuskel zur Grundlage auch unserer heutigen lateinischen Schriften. Durch eine ergonomisch verbesserte Haltung des Schreibgeräts, der Feder, und dank der gerundeten Formen konnte die karolingische Minuskel einfacher und schneller geschrieben werden. Es wurden wenige Ligaturen verwendet, die Silben- und Worttrennung erleichterte das Schreiben. Bedingt durch diese Vereinfachungen wurden neue Gruppen an das Schreiben herangeführt. Das bedeutet auch, daß neue Inhalte aufgeschrieben und uns so überliefert wurden. Bis ins 11. Jahrhundert wurde die karolingische Minuskel als Normalschrift verwendet.

**A partir d'env. 700** Avec l'avènement des Carolingiens, un immense empire se forma en Europe à la fin du 8e siècle. Cette dynastie accomplit des réformes fondamentales qui structuraient la vie culturelle, la politique et l'Eglise. La nécessité de faire connaître dans tout l'Empire les décrets, lois et ordonnances, et de les transmettre de manière à ce qu'ils soient compréhensibles, requérait une uniformisation de l'écriture. C'est alors qu'apparut la minuscule carolingienne qui tirait ses origines des anciens caractères romains. C'était une écriture constituée de minuscules où le corps de chaque lettre était large et arrondi. Les formes ouvertes reflétaient l'esprit d'une société qui cherchait à s'émanciper des dogmes figés de l'Eglise. Cette évolution correspond à une étape importante de l'Histoire de la typographie, car pour la première fois, on dessinait des minuscules aux formes achevées. Outre les lettres latines, les minuscules carolingiennes devinrent les caractères de base, d'où s'inspirent les caractères romains d'aujourd'hui. En raison de l'amélioration d'un outil plus ergonomique, la plume, et grâce à ses courbes, la minuscule carolingienne s'écrivait avec davantage de facilité et de rapidité. On utilisait moins de ligatures; la séparation des syllabes et des mots rendait l'écriture plus aisée. Ces simplifications eurent pour conséquence que de nouveaux groupes furent amenés à l'écriture. Ceci signifiait aussi qu'elle fit l'objet de nouveaux sujets qui ont pu nous être transmis. On employa la minuscule carolingienne comme écriture standard jusqu'au 11e siècle.

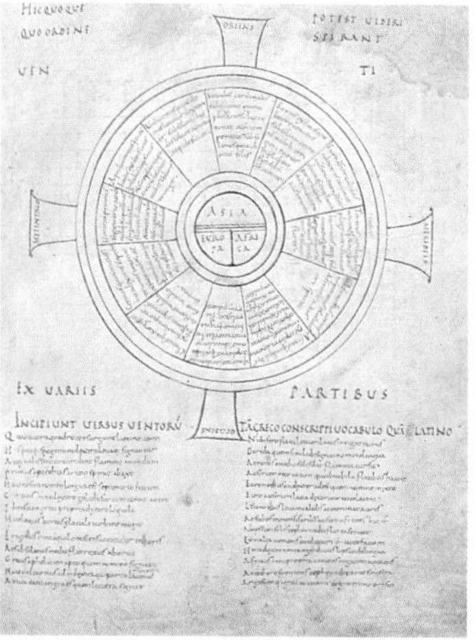

ca. 840 Carolingian Minuscule

Haec dicit dnſ· Ecce ego

Ninth century Half Uncial

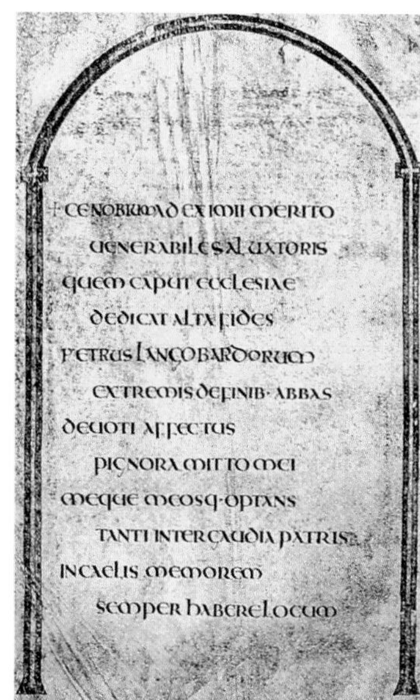

ca. 700 Uncial

Ninth century Initial

Ninth century Carolingian Uncials

12th century Beneventana

**900–1230** The end of the 8th century saw the seizure of power by the Carolingians, who organized their empire, which spanned much of western Europe, by way of basic cultural, political and religious reforms. As it was necessary for the decrees, laws and announcements to be understood by citizens and officials in all parts of the empire, a common style of writing was devised. The result was Carolingian Minuscule, which was based on ancient Roman forms and consisted of small letters broad and round in shape. The open forms reflect the candor of the age, as Carolingians began to liberate themselves from the strict church dogmas of Saint Augustine in his later days. The Carolingian alphabet was an important step in the history of writing, as it was the first system of fully-fledged minuscules. Together with Roman capitals, minuscules formed the basis of later roman alphabets, including our modern-day versions. Its round lettering, along with a new, ergonomically-improved way of holding the quill, simplified the task of writing, as did the general absence of ligatures and the practice of separating words and syllables. This led to new groups being introduced to writing, which in turn produced a new range of subjects to be handed down to us in written form. Carolingian Minuscule was the standard alphabet until well into the 11th century. It then gradually disappeared, as, under the influence of Gothic art, writing took on a new look, and new Gothic alphabets developed out of late Carolingian Minuscule.

**Ab 600 v. Chr.** Zwischen dem Aufstieg der Stadt Rom vom kleinen Bauerndorf zur Weltmacht und ihrem Niedergang entstanden vorbildliche Entwicklungen auf vielen Gebieten. 300 n. Chr. gab es in Rom ca. dreißig öffentliche Bibliotheken.

Bei ihrer Schrift orientierten sich die Römer an den etruskischen und griechischen Buchstaben, die sie zu einem Alphabet aus 21 Zeichen verbanden, das im gesamten römischen Reich verwendet wurde. Die römische Capitalis Monumentalis blieb unverändert in Gebrauch. Ihre Schriftform verband die Nähe zum gesprochenen Laut, wie ihn teilweise die Hieroglyphen kannten, mit der abstrakten Ordnung der Keilschriften. Es war ein neues pragmatisches Konzept. Die römischen Versalien kann man als die Urform der abendländischen Schrift bezeichnen. Sie wurden in Holz geschnitten und in Stein gemeißelt, auf Papyrus und Pergament geschrieben. Sie waren das Privileg der Priester, Gelehrten, Beamten und Kaufleute. In seiner geschichtlichen Entwicklung ging der Schreibfluß zuerst von rechts nach links, dann wechselnd von rechts nach links und wieder zurück von links nach rechts, schließlich nur noch von links nach rechts.

Eines der schönsten Beispiele der römischen Schriftkunst ist auf der Trajanssäule in Rom aus dem Jahr 113 n. Chr. zu sehen. Im Laufe des ersten Jahrhunderts n. Chr. veränderte sich die römische Versalschrift hin zu der ersten Kleinbuchstabenschrift, der römischen Kursiv- oder Minuskelschrift.

Sixth century BC

First century BC Etruscan Latin

Capital Italic

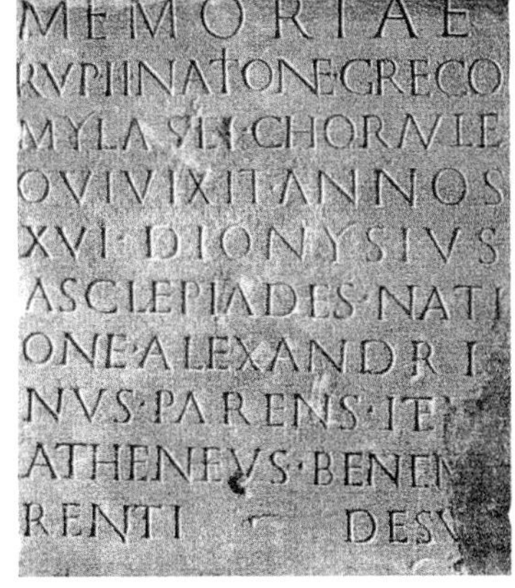

First century Roman Capital

Roman Cursive

Roman Capital Trajan Column

Sixth century Rustika

**A partir de 600 av. J.-C.** Entre la période où Rome n'était qu'un village de paysans, son apogée de puissance mondiale, et son déclin, la cité produisit de nombreuses réalisations exemplaires en maints domaines. En 300 apr. J.-C., il existait environ trente bibliothèques publiques à Rome.

Concernant leur écriture, les Romains s'étaient inspirés des caractères étrusques et grecs qu'ils avaient associés pour en faire un alphabet de 21 lettres, employé dans tout l'Empire romain. La capitale monumentale romaine est restée inchangée quant à son usage. Comme dans le cas des hiéroglyphes, leurs formes alliaient la ressemblance avec le son prononcé à l'ordonnance abstraite des écritures cunéiformes. Il s'agissait d'un concept nouveau et pragmatique.

En fait, nous pouvons considérer la majuscule latine comme la forme originelle de l'écriture occidentale. On la gravait dans le bois, on la taillait dans la pierre, l'inscrivait sur du papyrus ou du parchemin. Elle était le privilège des prêtres, des érudits, des fonctionnaires et des négociants. Aux différents stades de son évolution historique, le flux de l'écriture alla d'abord de droite à gauche, puis s'inversa, revint de droite à gauche, pour passer de nouveau de gauche à droite, jusqu'à ce que les gens n'écrivent plus que de cette manière.

L'un des fleurons de la calligraphie romaine date de l'an 113 apr. J.-C. et se trouve sur la colonne de Trajan à Rome. Au cours des premiers siècles de notre ère, les capitales romaines se modifièrent pour donner les premières minuscules et l'italique romain.

**800 B.C. onwards** From its rise – from small farming village to world power – to its fall, the city of Rome made exemplary progress in all fields of art and science. By 300 A.D. Rome also boasted approximately thirty public libraries.

The Romans based their written system on Etruscan and Greek letters, forming an alphabet of 21 symbols which was employed throughout the empire. Roman Capitals have remained in use in their original form. Their written form closely reflected the spoken sound, as was partly true also of hieroglyphics and the abstract cuneiform scripts. The Roman system was a new, pragmatic concept. Roman capitals can be described as the basic forerunners of western writing. They were carved in wood, chiseled in stone and written on papyrus and parchment. The flow of the lines underwent gradual development, first being written from right to left, then alternately from right to left and then back to left to right, finally resting at left to right. Writing in the Roman Empire was a privilege reserved for priests, scholars, state officials and merchants.

One of the most beautiful examples of Roman writing is preserved on the Trajan Column in Rome, erected in 113 A.D. During the first century A.D., Roman Capitals slowly developed to become the first written system with small letters, called Roman Cursive or Roman Minuscule.

## Phönizien und Griechenland
## Phéniciens et Grèce
## Phoenicia and Greece

**Ab 1700 v. Chr.** Das Land der Phönizier, eines seefahrenden Handelsvolks, lag an der heutigen syrisch-libanesisch-israelischen Mitelmeerküste. Hier wurden nicht nur die damals sichersten und schnellsten Schiffe, sondern auch Leuchttürme als Signale für die Schiffahrt gebaut. Die Herstellung und Bearbeitung von Gegenständen aus Gold, Silber, Elfenbein und aus farbigem Glas bildete die Grundlage für den phönizischen Handel. Zum Färben von Stoffen und zum Schreiben wurde das neu entdeckte Purpur verwendet. Ab ca. 1200 v. Chr. beherrschten die Phönizier den Handel im gesamten Mittelmeerraum. Zum Ausbau ihrer Wirtschaftsmacht gründeten sie Kolonien bis an die Westküste Afrikas. Um 750 v. Chr. wurde Karthago gegründet, wo das erste Münzgeld verwendet wurde und in der Folge die ersten Aktiengesellschaften der Welt entstanden.
Die phönizische Schrift bildete sich um 1400 v. Chr. Sie wurde angeregt durch das ägyptische Hieroglyphen-Schriftsystem und besteht aus 28 Konsonanten. Vokale wurden in diesem System nicht schriftlich festgehalten. Zeugnisse dieser reinen Lautschrift sind bei der Ausgrabung der Stadt Ugarit in einer Tontafel-Bibliothek aufgefunden worden. Die Phönizier entwickelten also das erste belegbare Alphabet der Welt. Es wurde Grundlage für alle weiteren europäischen Schriften.
Um 1200 v. Chr war Griechenland eine blühende Region, die regen Handel mit Afrika, Asien und dem übrigen Europa trieb. Eine für die westliche Zivilisation beeindruckende Kultur entstand mit zahl-

**A partir de 1700 av. J.-C.** Le territoire des Phéniciens, un peuple de marins et de commerçants, s'étendait sur la région côtière de la Méditerranée qui appartient aujourd'hui à la Syrie, au Liban et à Israël. C'est ici que furent inventées les écritures jadis les plus fiables et les plus rapides, et où l'on construisit aussi des phares et des signaux pour la navigation maritime. Les objets en or et en argent, en ivoire et en verre teint qu'on y produisait, constituaient la base du commerce phénicien. Les Phéniciens utilisaient la pourpre, récemment découverte, pour teindre les étoffes ainsi que pour écrire. Vers 1200 av. J.-C., ils détenaient l'hégémonie sur tout le bassin méditerranéen en matière de commerce. Pour étendre leur puissance économique, ils créèrent des comptoirs jusque sur les côtes d'Afrique occidentale. Vers 750 av. J.-C., ils fondèrent Carthage, la première cité à utiliser une monnaie en métal, et où les premières sociétés par actions du monde virent le jour.
L'écriture phénicienne apparut vers 1400 av. J.-C. Inspirée du système des hiéroglyphes égyptiens; elle comprenait 28 consonnes. Des témoignages de cet alphabet phonétique, qui ne comportait aucune voyelle, furent découverts lors des fouilles de la cité d'Ugarit dans une bibliothèque de tablettes de terre cuite. Ce furent donc les Phéniciens qui inventèrent le premier alphabet attesté au monde, et c'est lui qui servit de base aux écritures européennes qui vinrent par la suite.
A partir de 1200 av. J.-C. la Grèce était une région prospère qui pratiquait un commerce intense avec l'Afrique, l'Asie et le reste de l'Europe. Cette culture eut une importance primordiale pour la civilisation occidentale qui s'y développa. Des œuvres innombrables, touchant à

842 BC Phoenician

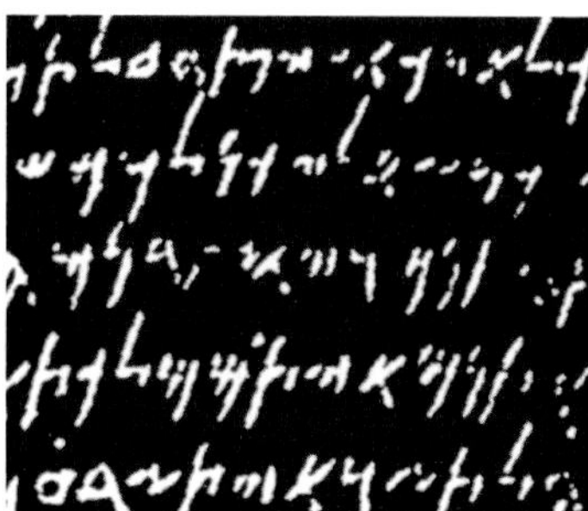
ca. 350 BC Phoenician Sidon

Phoenician

ca. 1300 BC Mykene Linear B

ca. 400 BC Greek Melos

**1700 B.C. onwards** The land of the Phoenicians, a nation of maritime traders, lay on what is today the Mediterranean coastline of Syria, Lebanon and Israel. It was here that the fastest and safest ships were built, and lighthouses signaled the way for the sea-faring community. Objects fashioned from gold, silver, ivory and colored glass formed the basis of Phoenician trading. Tyrian purple, then a new discovery, was used to dye fabric and also as writing ink. By 1200 B.C., the Phoenicians had become the leading traders in the whole of the Mediterranean. They set up colonies stretching to the west coast of Africa to boost their economic power. Around 750 B.C., Carthage was founded, where the first minted coins were used and later the world's first joint-stock companies were established.
Phoenician script began to emerge in around 1400 B.C. It drew its inspiration from Egyptian hieroglyphics and comprised 28 consonants. Vowels were not represented. When the city of Ugarit was excavated, evidence of this purely phonetic system was discovered carved onto a whole library of clay tablets. It is the earliest alphabet of which we have concrete proof and was to become the basis of the writing systems for all the languages of Europe.
Greece in around 1200 B.C. was a prosperous region which pursued extensive trade with Africa, Asia and the rest of Europe. A culture which was to greatly influence the whole of western civilization evolved, with numerous masterpieces in

reichen bis heute bewunderten Arbeiten auf den Gebieten Architektur, Malerei, Plastik, Literatur und Theater. Auch im wissenschaftlichen Bereich erlebten Astronomie, Mathematik, Physik und Landeskunde einen großen Aufschwung. Zwei für das Schreiben und damit auch die Schrift einschneidende Veränderungen erfolgten in den ionischen Kolonien Griechenlands. Hier an der Küste Kleinasiens entstand das Homerische Epos, und Thales von Milet legte mit seinen Lehrsätzen den Grundstein für die modernen Wissenschaften. Um 450 v. Chr. wurde ein verbindliches Maß- und Münzsystem eingeführt, es gab ein geregeltes Bankwesen mit staatlicher Münzprägung.

Im 9. Jahrhundert v. Chr. gaben die Griechen ihre umständliche Silbenschrift auf und übernahmen die systematisch zu verwendenden 22 Zeichen des phönizischen Alphabets, das sie durch die Vokale A, E, I, O, U ergänzten. Diese altgriechische Schrift wurde von der Regierung in Athen durch Gesetz als verbindlich eingeführt. Um 400 v. Chr. bürgerte es sich ein, Texte von links nach rechts zu schreiben. Neben Materialien wie Stein, Ton, Metall und Holz, in die Schrift gemeißelt, getrieben oder geschnitten wurde, schrieb man mit Griffel, Feder und Tinte auf Pergament, Stoff, Papyrus und Leder. Auf öffentlichen Plätzen waren Holztafeln mit allgemeinen Informationen für die Bevölkerung angebracht, die als Vorläufer von Zeitung und Plakat gelten.

ca. 350 BC Greek

263 BC Pergamon

Mykene

ca. 450 BC Melos

ca. 350 Greek

l'architecture, la peinture, la sculpture, la littérature et le théâtre y virent le jour. Dans le domaine scientifique, l'astronomie, les mathématiques, la physique et la géographie y connurent aussi un essor considérable. Deux bouleversements décisifs pour l'écriture, et par conséquent pour le texte, se produisirent dans les colonies grecques de la mer Ionienne. C'est sur la côte d'Asie mineure que naquirent les épopées d'Homère et que Thalès de Milet posa la première pierre de la science moderne avec ses thèses. Un système de mesure ainsi qu'un système monétaire s'instaurèrent vers 450 av. J.-C., et le pays se dota d'une organisation bancaire avec frappe de monnaies.

Au 9e siècle avant J.-C., les Grecs abandonnèrent leur écriture syllabique complexe et adoptèrent les 22 caractères systématiques et faciles d'usage de l'alphabet phénicien qu'ils complétèrent en lui ajoutant les voyelles A, E, I, O, U. Puis le gouvernement d'Athènes promulgua une loi réglementant l'usage de cet ancien alphabet grec. On prit l'habitude d'écrire les textes de gauche à droite vers 400 av. J.-C. Les caractères étaient sculptés, taillés ou gravés dans des matériaux tels que la pierre, l'argile, le métal et le bois; on écrivait au burin, mais aussi à la plume et à l'encre sur du parchemin, des étoffes, du papyrus et du cuir. Dans les lieux publics, on trouvait des panneaux de bois considérés comme les ancêtres des journaux ou des affiches et qui contenaient des informations générales à l'adresse de la population.

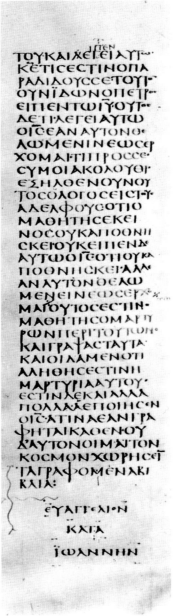

the fields of architecture, painting, sculpture, literature and the theater which we still hold in high regard today. The sciences, in particular astronomy, mathematics, physics and geography, also prospered.

Two important changes which affected writing and lettering occurred in Greece's Ionian colonies. It was here on the coast of Asia Minor that Homer's epics were born and the teachings of Thales of Miletus laid the foundations of modern science. In 450 B.C., the Greeks introduced a compulsory system of measurements and coinage, and there was also a well-organized banking setup with official minted coins.

In the 9th century B.C., the Greeks abandoned their complex syllabary and adopted the 22-symbol system of the Phoenician alphabet, supplementing it with the vowels A, E, I, O and U. This script was made legally binding by the government in Athens. By around 400 B.C., scholars had become accustomed to writing texts from left to right. The written word was chiseled, engraved and cut into stone, clay, metal and wood, and written on parchment, fabric, papyrus and leather with sharp sticks or stones, quills and ink. Wooden tablets carved with important information were hung up in public places for all to see: the ancient forerunners of newspapers and posters.

## Islam

Die bisher gezeigten Beispiele der Moden und Stile in der Entwicklung von Schrift und Typographie basieren auf einer wesentlichen einheitlich formalen Grundlage. Es gibt eine ästhetische Verbindung zwischen Schriftzeichen der Phönizier und Griechen über Romanik und Renaissance bis in unsere Gegenwart. Aber es gab natürlich auch andere formale Ansätze, um Zeichen darzustellen. So gehörte zu der vielfältigen Handelskultur des Islam untrennbar eine reichhaltige Schriftkultur. Während zu Lebzeiten des Propheten Mohammed, der 632 starb, und der ersten ihm nachfolgenden Herrscher, vor allem die mündliche Überlieferung von Texten gepflegt wurde, entstanden seit dem 8. Jahrhundert schriftliche Ergänzungen, die bald von größerer Bedeutung waren, als die ausschließlich verbalen Überlieferungen.
Schon um 800 lernten die Araber von den Chinesen die Papierfabrikation, im 9. Jahrhundert wurden in Bagdad zahlreiche wissenschaftliche Werke aus allen Regionen in die Sprache des Koran übersetzt. Bibliotheken mit mehreren hunderttausend handgeschriebenen Bänden entstanden.
Im eroberten Spanien entstanden Universitäten, durch die sich die islamische Kultur in anderen Ländern ausbreitete. Beispiele für die vielfältige Gestaltung von Büchern waren immer wieder die Koran-Abschriften, denn nach Ansicht der Religionsgelehrten war (und ist) es die eigentliche Funktion der Schrift, die Inhalte des Koran wiederzugeben. Dabei wurde auf bildhafte Darstellungen ver-

Les exemples de modes et de styles présentés jusqu'alors quant à l'évolution de l'écriture et de la typographie se basaient sur des fondements formels qui, pour l'essentiel, sont des constantes. Il existe un lien esthétique qui rattache des caractères phéniciens et grecs, et que l'on retrouve pendant la période romane, et de la Renaissance jusqu'à nos jours. Mais il y eut aussi de toute évidence d'autres démarches formelles de représentation du signe. Ainsi, dans l'Islam, la culture du négoce est indissociable de la richesse scripturale. Tandis que du vivant du prophète Mahomet, décédé en 632, ainsi que des souverains qui lui succédèrent, les textes étaient surtout véhiculés par la tradition orale, on vit apparaître, à partir du 8e siècle, des corollaires écrits qui ne tardèrent pas à prendre davantage d'importance que la transmission par la parole. Vers l'an 800 déjà, les Arabes avaient appris des Chinois la technique de fabrication du papier. Au 9e siècle, à Bagdad, on traduisait dans la langue du Coran d'innombrables ouvrages scientifiques provenant de divers pays. Des bibliothèques contenant plusieurs centaines de milliers de volumes manuscrits virent le jour. Des universités furent créées dans l'Espagne conquise; par ce biais, la culture de l'Islam se répandit dans d'autres pays. Parmi ces livres exemplaires, on trouvait toujours des copies du Coran, car selon l'avis des théologiens, la véritable fonction du livre était (et est) de véhiculer les idées du Coran. En cela, on renonçait aux représentations figuratives, mais on accordait une grande valeur à l'art d'orner les lignes écrites et les citations coraniques. Vers l'an 800, l'écriture hiératique kufi était très répandue, une écriture qui mettait l'accent sur l'horizontalité des lignes. Peu de temps après, ces formes s'enrichirent d'éléments floraux et géométriques,

AD 800 Faijum

ca. AD 850 Iraq

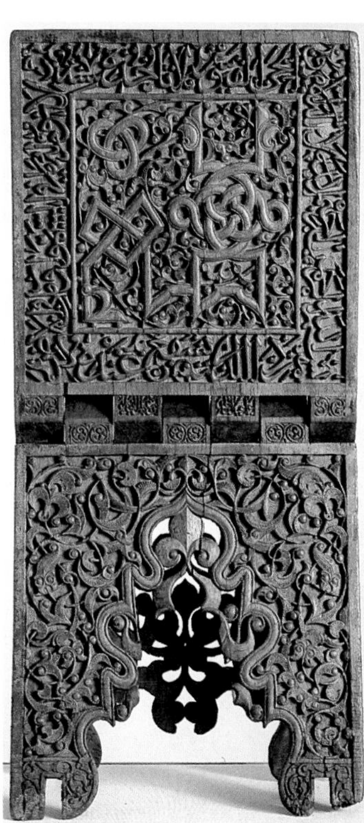

ca. 1250 Turkey

1044 Spain

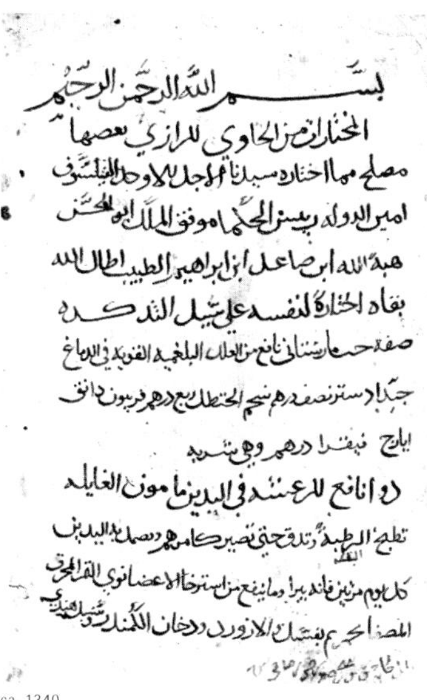

ca. 1340

The examples of fashions and styles running through the development of writing and typography shown up to now have a more or less uniform formal basis. There is an aesthetic link between the written characters of the Phoenicians and the Greeks, those of the Romanesque and Renaissance periods to those of the present day. Yet there were, of course, also other formal attempts to represent characters. Thus Islam's varied range of trading practices, for example, was inextricably linked to a rich written culture. Whereas during the lifetime of the Prophet Mohammed (he died in 632) and his immediate successors it was normal to hand down texts by word of mouth, written elaborations began to surface from the 8th century onwards which soon became more important than the purely oral tradition. The Arabs learned to make paper from the Chinese in around 800, and in the 9th century numerous scientific writings from all religions were translated in Baghdad into the language of the Koran. Libraries were established

zichtet, aber großer Wert auf die kunstvolle ornamentale Ausschmückung von Schriftzeilen und Koranzitaten gelegt. Um 800 war die hieratische Schrift Kufi weitverbreitet, bei der die horizontalen Linien stark betont wurden.
Wenig später wurden die Kufi-Formen mit floralen und geometrischen Elementen oder mit stilisierten Tier- und Menschenköpfen angereichert.
Zusätzlich wurden auch verschiedene dekorative kursive Schriftformen entwickelt. Die ständige hohe Wertschätzung der Schriftkunst im Islam führte dazu, daß die Kalligraphie zum festen Bestandteil jeder künstlerischen Ausbildung gehört.

ca. 1530 Iran

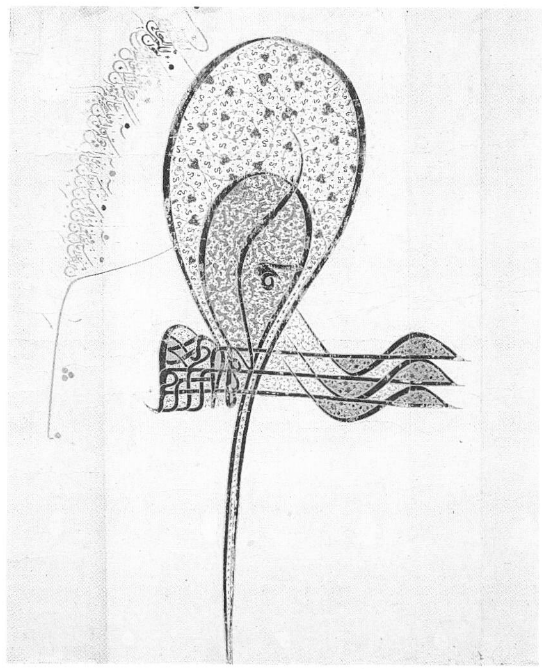

ca. 1550 Istanbul

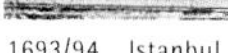

1693/94 Istanbul

ca. 1670 Iran

ou de têtes animales ou humaines stylisées. En complément, diverses formes italiques et décoratives se développèrent. L'estime que l'Islam portait à l'écriture eut pour conséquence que la calligraphie devint l'une des composantes essentielles de la formation artistique.

comprising hundreds of thousands of handwritten volumes. Universities sprang up in Muslim Spain, promoting the spread of Islamic culture to other countries. Transcriptions of the Koran were among the chief subjects of the numerous styles of ornamental calligraphy, for according to religious teachers the true function of writing was (and still is) to reproduce the content of the Koran. Pictorial representations were not used, but great value was attached to the artistic ornamentation of lines of writing and quotations from the Koran. Hieratic Kufic script, with its strongly accentuated horizontal line, was widespread by around 800. Slightly later, the Kufic writing was embellished with floral and geometric elements or with stylized human and animal heads. Various decorative cursive script forms were also developed. Writing was, and is, a matter of great esteem for the Islamic world, which is why calligraphy has always been a component of all forms of artistic training.

**China und Indus-Tal**
**Chine et Vallée de l'Indus**
**China and the Indus Valley**

**Ab 2600 v. Chr.** Mit Hilfe geknüpfter Schnüre begann das System der Zeichen in China. Dieses wurde um 2600 v. Chr. von einer Bilderschrift verdrängt, von der es noch ein Zeugnis gibt, die Inschrift des Kaisers Yü 2278 v. Chr. Gegen 1800 wandelte sich die Bilderschrift zu einer Wortschrift. Jedes Zeichen stellte einen Begriff dar. Neue Zeichen wurden durch Vereinigung vorhandener Zeichen gebildet und durch den ausgeprägten Verwaltungsapparat rasch im ganzen Reich verbreitet. Das Machtmittel einer einheitlichen Schrift war in China so im Bewußtsein verankert, daß die chinesischen Kaiser zu obersten Wächtern der Orthographie wurden und wiederholt Reformen durchführten.
Die chinesische Schrift besteht aus ca. 50.000 Zeichen. Jedes Wortzeichen setzt sich aus einem Bedeutungs- und einem Aussprachezeichen zusammen. Die Schriftzeichen haben eine quadratische Form und sind alle gleich groß. Sie werden von oben nach unten und von rechts nach links angeordnet, in der Volksrepublik China von links nach rechts in horizontalen Zeilen. Die Beherrschung all der Schriftzeichen und deren Praktizieren in der Schönschreibkunst wurde als die höchste Stufe des Gelehrtentums angesehen. Die frühe Entwicklung der Schrift und die hohe Wertschätzung der Schreibkunst in China führte zu wichtigen nachfolgenden Neuerungen. Um 100 n. Chr. wurde in China das Papier erfunden, das unserem heutigen Papier sehr ähnlich war. Es wurde mit Tusche beschrieben, bald auch mit Holzklischees und Tusche

**A partir de 2600 av. J.-C.** Au début, le système des signes chinois était constitué de ficelles nouées. Vers 2600 av. J.-C., ce mode de communication fut ensuite remplacé par des pictogrammes, dont on possède encore un exemplaire, une inscription de l'empereur Yu datant de 2278 av. J.-C. Vers l'année 1800, les pictogrammes évoluèrent pour devenir une écriture où le signe avait valeur de mot. Chaque signe correspondait à un concept. On formait de nouveaux mots en agglutinant des signes existants; mots qu'un puissant appareil administratif répandait rapidement dans tout l'Empire. L'outil de pouvoir que représentait cette uniformisation de l'écriture était tellement ancré dans la conscience chinoise que les empereurs de Chine devinrent les gardiens suprêmes de l'orthographe et procédèrent à plusieurs reprises à des réformes.
L'écriture chinoise comporte environ 50 000 signes. Un idéogramme est composé de signes sémantiques et phonétiques. De forme carrée, les signes ont tous la même dimension. Ils s'écrivent de haut en bas et de droite à gauche, en République populaire de Chine de gauche à droite et en lignes horizontales. En Chine, le fait de maîtriser la totalité de ces caractères ainsi que la calligraphie était considéré comme le stade suprême de l'érudition. L'évolution précoce de l'écriture, tout comme l'admiration que l'on portait à la calligraphie provoqua d'importantes innovations par la suite. Vers l'an 100 apr. J.-C., les Chinois inventèrent un support qui ressemblait fortement à notre papier actuel. On écrivait d'abord à l'encre de Chine, puis très vite, on commença à imprimer au moyen de matrices en bois. C'est vers 860 apr. J.-C. que parut en Chine le «Sutra du diamant», premier livre imprimé qui se présentait sous forme

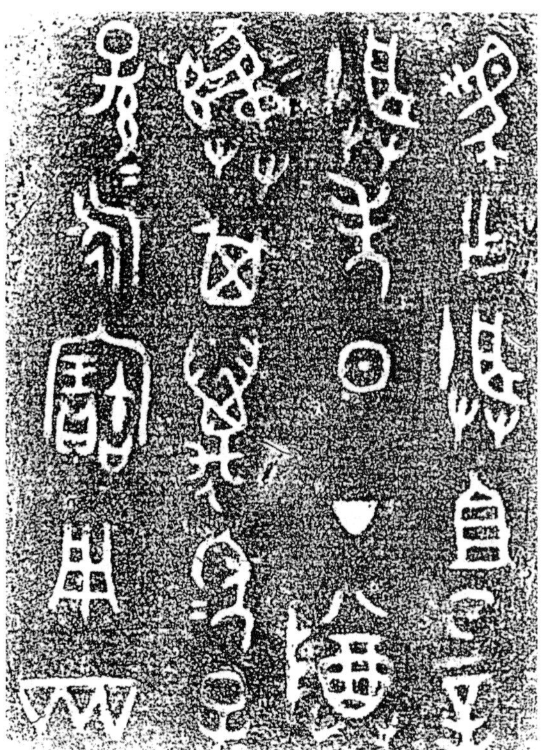
10th century BC China

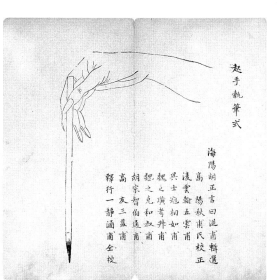

AD 1633 China

ca. 10th century BC China

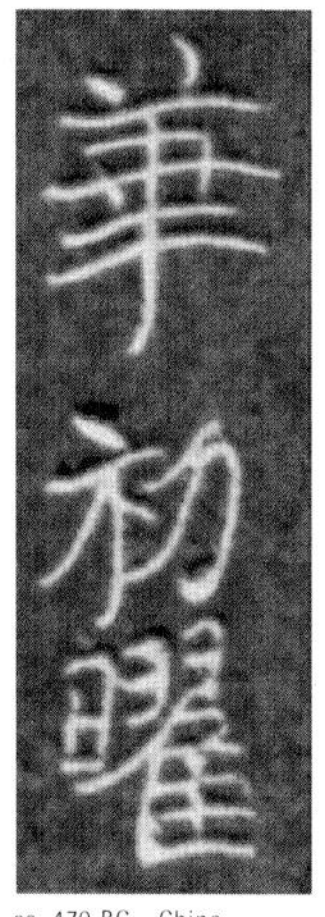
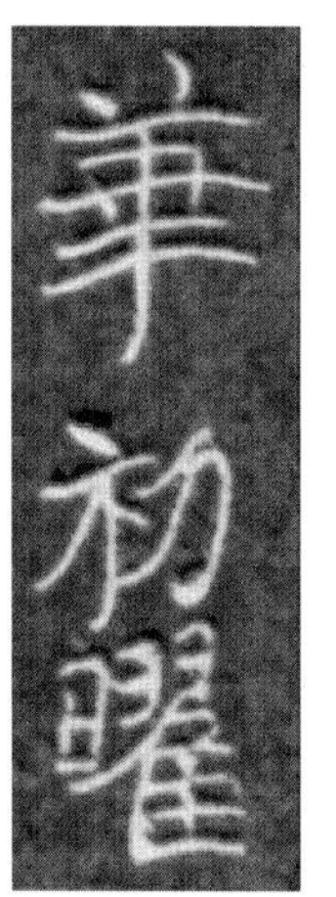
ca. 470 BC China

ca. AD 1610 Japan

**Ab 2600 B.C. onwards** It was with the help of knotted string that a system of signs and symbols began in China. This was replaced by a pictographic writing system in around 2600 B.C., for which we have evidence in the form of an inscription of Emperor Yü dating from 2278 B.C. In around 1800 B.C., this pictographic system developed into one in which the symbols stood for words. Each character represented a single concept. New characters were depicted by combining existing ones and were rapidly introduced to all corners of the empire through China's advanced administrative machine. The awareness that a unified system of writing was also an instrument of power was so deeply rooted in the Chinese consciousness that the emperors became the supreme guardians of their national orthography, implementing many reforms.
The Chinese writing system has c. 50,000 characters. Each is made up of one element indicating the meaning and another indicating the pronunciation. The characters are square in shape and of the same size. They are arranged from top to bottom and from right to left, and, in the People's Republic of China, in horizontal lines from left to right. Scholars were considered to have obtained the highest level of learning in Ancient China when having mastered all the characters, they could practise the art of calligraphy. This early evolution of a written system and the high esteem allotted to calligraphy in China led to important later developments. Paper of a kind very similar to our

bedruckt. Um 860 n. Chr. entstand in China das erste gedruckte Buch der Welt in Form einer Rolle, das „Diamanten-Sutra". Aber auch das erste Buch in gefalteter Form, das unseren Büchern gleicht, wurde in China mit Holzklischees bedruckt. Welchen wichtigen Stellenwert das gedruckte Buch in China hatte, zeigen Enzyklopädien der Ming-Zeit, die 12.000 Bände umfaßten. So ist zu verstehen, daß schon um 1000 n. Chr. hier die ersten beweglichen und wiederverwendbaren Typen aus Ton, später aus Metall, entwickelt und verwendet wurden. Im nordwestlichen Indien entwickelte sich um 2000 v. Chr. eine Zivilisation mit urbaner Kultur, in der gepflasterte Strassen, Kanalisation und Bewässerungssysteme geschaffen wurden. Das Dezimalsystem und die Ziffer Null wurden eingeführt. Ein durchstrukturiertes Regierungssystem ermöglichte ein weitverzweigtes Handelsnetz zu Land und Wasser. Hier entwickelte sich parallel zur ägyptischen und babylonischen Zivilisation eine Kultur, aus der Schriftdenkmäler als Stein- oder Kupfersiegel erhalten sind. Es sind Schriftbildzeichen, die erhaben aus dem Stein gehauen sind oder oder in die Metallplatte getrieben wurden. Insgesamt sind zur Zeit nur etwa 250 solcher Zeichen bekannt, und es ist daher nicht gelungen, diese protoindische Schrift zu entziffern. Nach heutigem Wissensstand nimmt man an, daß diese Schrift eine Mischung aus Wort- und Lautzeichen ist.

AD 1736 China

17th century Japan

ca. 2000 BC

ca. 2000 BC The Indus Valley

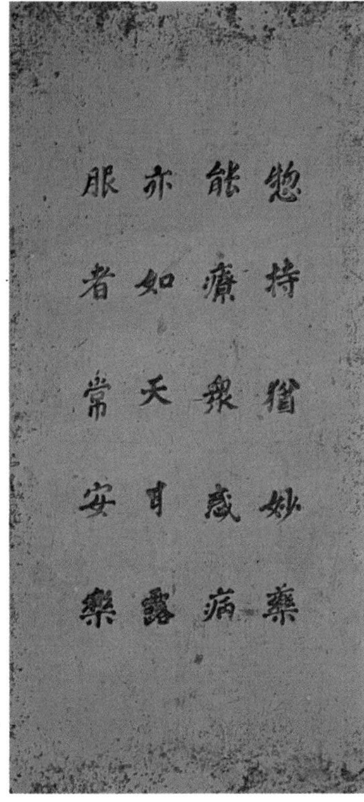


China

18th century India Sanskrit

de rouleau. De même, le premier livre plié dont la forme évoque nos livres actuels avait été imprimé à l'aide de matrices en bois. Les encyclopédies de l'ère Ming, qui comportent quelque 12 000 volumes, montrent bien l'importance accordée au livre imprimé en Chine. On comprend alors aisément que, dès l'an 1000 ap. J.-C., on y inventa, puis utilisa les premiers caractères mobiles, d'abord en terre cuite, puis en métal.
Vers l'an 2000 av. J.-C. une civilisation de culture urbaine capable de construire des rues pavées, des canalisations et des systèmes d'irrigation se développa au nord-ouest de l'Inde. C'est ici que furent instaurés le système décimal et le chiffre zéro. Un régime gouvernemental très structuré permit de créer un réseau d'échanges commerciaux terrestres et fluviaux aux vastes ramifications. Parallèlement aux civilisations égyptienne et babylonienne, une culture, qui nous légua des témoignages de son écriture sous forme de sceaux en pierre ou en cuivre, s'y développa. Ce sont de magnifiques pictogrammes taillés dans la pierre ou gravés dans des plaques de métal. Comme on ne connaît à présent qu'environ 250 caractères de ce type, personne n'a encore réussi à déchiffrer cette écriture proto-indienne. Selon les connaissances actuelles, on suppose qu'il s'agissait d'un amalgame d'idéogrammes et de caractères phonétiques.

paper today was invented in China in c. 100 A.D. It was first written on in Indian ink, and later wooden blocks and ink were used in simple printing processes. The first printed book in the entire world, the "Diamond Sutra", was manufactured in scroll-form in China in c. 860 A.D. The first folded book in a form which resembles modern books was also produced in China, printed with wooden blocks. The great importance attributed to the printed book in China is illustrated by the encyclopedias from the Ming dynasty, which number some 12,000 volumes. We thus assume that as early as 1000 A.D. in China, the first moveable and reusable type was being developed and employed, made first of clay and later of metal.
A civilization had evolved by around 2000 B.C. in the northwest of India whose culture was extremely urbanized, with paved streets, sewerage and irrigation. The decimal system and the figure zero were introduced. A well-structured system of government ensured the successful operation of various branches of trade on land and on water. In parallel with the Egyptians and the Babylonians, the Indus Valley developed a civilization which has left us inscribed artifacts in the form of stone or copper seals. They bear pictograms carved in relief on the stone or engraved into the metal. Only some 250 of these symbols are known to us at present, making it impossible to decipher proto-Indian script. Current scientific opinion assumes that this writing system is a mixture of word-signs and phonograms.

**Ägypten**
**Egypte**
**Egypt**

**Ab 3000 v. Chr.** Im Umfeld hoher technischer und merkantiler Entwicklung entstand um 2800 v. Chr. ein System, in dem Worte ursprünglich durch Bilder dargestellt wurden. Dasselbe Bild konnte zwei oder mehr Bedeutungen haben, wie zum Beispiel Reich und reich oder Arm und arm. Klärung brachten hier kommentierende Zeichen, die die jeweilige Aussprache regelten. In der Folge bildete sich zusätzlich eine Silbenschrift heraus, mit deren Hilfe auch „Fremdworte" dargestellt werden konnten. Die ägyptische Schrift bestand aus ca. 700 Hieroglyphen, zum Teil Wortzeichen, die das Gemeinte bildhaft darstellten, zum Teil Lautzeichen, die nicht bildhaft Darstellbares als Laute ausdrückten. Außerdem gab es 24 Konsonantenzeichen. Vokale kannte die Hieroglyphenschrift nicht. Deutezeichen ohne Lautwert sorgten für eindeutige Auslegung des Geschriebenen. Textzeilen begannen rechts oben und wurden von oben nach unten, von links nach rechts, aber auch von rechts nach links, ohne Wortabstände und ohne Interpunktionen geschrieben. 1799 wurde im Nil-Delta der 196 v. Chr. entstandene „Stein von Rosette" gefunden, der eine dreisprachige Inschrift mit Hieroglyphen, demotischer und griechischer Schrift trägt. Durch diesen Fund konnten die Hieroglyphen 1822 erstmals entziffert werden.

Eine weitreichende Erfindung zur Fixierung von Kommunikationszeichen war die Entwicklung von Papyrus aus dem Mark der Papyrusstaude. Man preßte das in Streifen geschnittene Mark in zwei

**A partir de 3000 av. J.-C.** Vers 2800 av. J.-C., à une époque où se développaient les techniques et le mercantilisme, apparut un système où les mots étaient à l'origine représentés par une image. Celle-ci pouvait avoir une ou plusieurs significations, par exemple le «temps» qu'il fait et le «temps» écoulé, l'or ou or. Dans ce cas, des signes complémentaires réglementaient la prononciation en précisant le sens. Par la suite, une écriture syllabique se constitua qui permit également de représenter des «mots d'origine étrangère». L'écriture égyptienne se composait de 700 hiéroglyphes environ. Une partie d'entre eux avait une valeur figurative, qui représentait par une image ce qu'on voulait exprimer, l'autre une valeur phonétique qui traduisait par des sons ce qu'on ne pouvait représenter par une image. Il existait en outre 24 signes pour les consonnes, l'écriture hiéroglyphique ne connaissant pas les voyelles. Les signes sans valeur phonétique servaient à interpréter clairement le texte. Les lignes de celui-ci commençaient en haut à droite et s'écrivaient de droite à gauche, de haut en bas; mais aussi de gauche à droite, sans espace entre les mots, ni ponctuation. En 1799, on découvrit, dans le Delta du Nil, la fameuse «Pierre de Rosette» gravée en 196 av. J.C. d'une inscription en trois langues : en hiéroglyphes, en démotique et en grec. C'est grâce à cette découverte que l'on réussit enfin à déchiffrer les hiéroglyphes en 1822.

L'autre invention de grande envergure qui servit à fixer la communication écrite fut celle de la fabrication de papyrus à partir de la moelle contenue dans les tiges de la plante. On pressait la moelle coupée en petites bandes superposées par couches croisées pour faire une feuille très fine. Les feuilles étaient coupées à angle droit, les côtés les plus étroits

ca. 2330 BC Saqqara

ca. 1980 BC Assiut

ca. 2050 BC

ca. 1385 BC

196 BC Rosetta Stone

**3000 B.C. onwards** In an environment of intense technological and mercantile development, a system evolved in Egypt in around 2800 B.C. where words were originally represented by pictures. One picture could have two or more meanings, such as with our contemporary words bow, date, or hide, for example. The various meanings were indicated by supplementary symbols which determined the pronunciation of the word. In addition to this, a syllabary gradually evolved with which foreign words could also

kreuzweise übereinander gelegte Schichten zu einem dünnen Blatt zusammen. Die rechteckig beschnittenen Blätter wurden an den Schmalseiten aneinander geklebt und, weil sie beim Falzen gebrochen wären, aufgerollt. Diese Rollen, die man als Bücherrollen bezeichnen kann, waren ca. 20 cm hoch und bis zu 40 Meter lang. Beschrieben wurden sie mit einem Pinsel, der aus einer zerfaserten Binse hergestellt wurde, und mit Tinte, die aus Ruß und Wasser bestand.

étaient ensuite collés les uns aux autres pour que le papyrus ne se brise pas en le pliant, puis on le roulait. Ces livres en forme de rouleaux faisaient environ 20 cm de hauteur et atteignaient jusqu'à 40 m de longueur. Les scribes se servaient d'un pinceau fabriqué à partir de fibres de joncs et employaient une encre constituée d'un mélange de suie et d'eau.

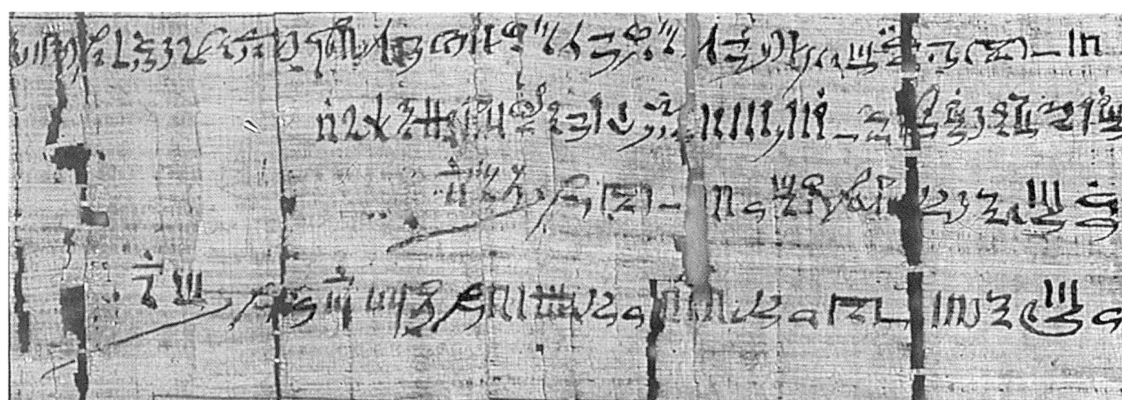

ca. 1150 BC Theben Hieratic

ca. Fourth century BC

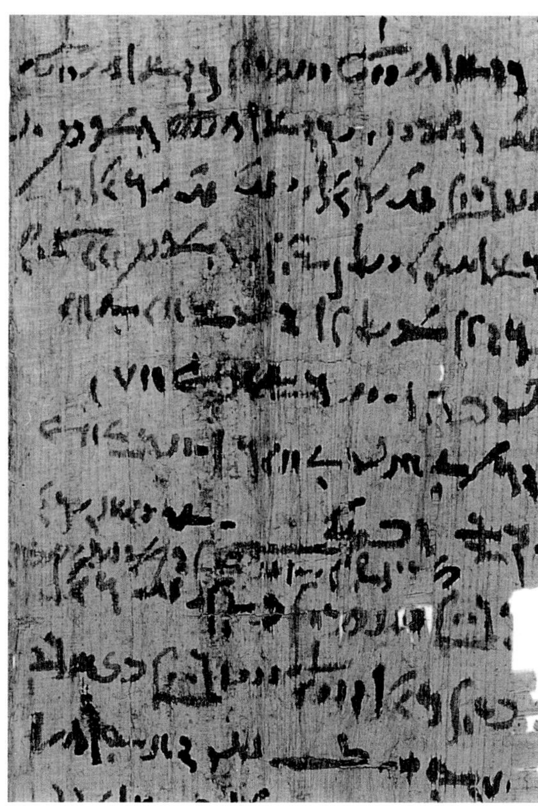

110 BC Demotic

be notated. The Egyptian writing system was made up of c. 700 hieroglyphics. Some of these were word-signs which portrayed the designated word as a picture, some of them phonograms notating non-pictorial concepts as a sound. There were also 24 symbols for consonants. Hieroglyphics had no vowels. Special marks indicating meaning, yet devoid of any phonetic value, enabled a clear interpretation of the written signs. Lines of text started in the top right-hand corner and were written from top to bottom, from left to right and also from right to left, without punctuation or spaces between the words. In 1799, the Rosetta Stone, dating from 196 B.C., was found in the Nile Delta. It bears a trilingual inscription in Egyptian hieroglyphics, demotic script and Greek. This discovery enabled historians to decipher hieroglyphics for the first time in 1822.
A widespread innovation which helped establish this written form of communication was the development of papyrus paper from the stem pith of the papyrus plant. The pith was cut into strips which were arranged criss-cross in two layers before being pressed into a thin sheet. These sheets were cut into rectangles and joined along their shorter edge, then rolled, as the delicate material would have cracked if folded. These scrolls were 8 ins (20cm) high and up to 120 ft (40m) long. Brushes made of frayed rushes and a special ink consisting of soot and water were used to write on the papyrus.

## Mesopotamien / Sumer
## Mésopotamie / Sumer
## Mesopotamia / Sumer

**3000 v. Chr. – 1700 v. Chr.** Im eigentlichen Sinn ist Schrift sichtbar gemachte Sprache. Ihre Entwicklung beginnt ca. 3000 v. Chr. mit dem Auftreten der Keilschrift der Sumerer. Bei der Keilschrift handelt es sich um abstrahierte Wortlautzeichen. Der Weg dorthin umspannte einen Zeitraum von ca. 60.000 Jahren und erfolgte über die Stufen: Bild – Bildzeichen (Piktogramm) – Begriffszeichen (Ideogramm) – Hieroglyphe. Auf jeder dieser Entwicklungsstufen fanden in Jahrtausenden Übereinkünfte statt, die uns als abstrakte Gedankenleistung noch heute beeinflussen und faszinieren.
So wurde der Übergang von Bild- zu Begriffszeichen (ca. 4000 v. Chr.) begleitet von einer Linearität des Aufgezeichneten. An dieser Stelle liegt der Grundstein für das, was wir unter Geschichte verstehen. Was sich bisher vereinzelt ereignete und nur gesprochen mitgeteilt wurde, konnte nun fixiert werden und barg so die Möglichkeit der Geschichtsschreibung. (Konkret wird davon erst ca. 800 v. Chr. von Homer Gebrauch gemacht sowie später von den Verfassern der Bibel.)
Der Übergang von Begriffszeichen zu Hieroglyphen machte es möglich, fremde oder neue Worte ihrem Laut nach durch vorhandene Begriffszeichen auszudrücken. So wurde zum Beispiel der Eigenname einer Person durch vorhandene Begriffszeichen zusammengesetzt. Dieses neue Wort hatte nichts mehr mit der ursprünglichen Bedeutung des Begriffszeichens gemeinsam.
Um 3000 v. Chr. entwickelten die Sumerer die Keilschrift als Wort- und Silben-

**3000 av. J.-C. – 1700 av. J.-C.**
L'écriture est, au sens propre, une visualisation du langage. Son histoire commence vers le troisième millénaire avant J.-C. avec l'émergence des caractères cunéiformes des Sumériens, des caractères ayant fonction de signes phonétiques abstraits. Le chemin parcouru jusqu'à ce stade représente environ 60 000 ans et s'est déroulé par paliers, celui du pictogramme, de l'idéogramme et du hiéroglyphe. Des concordances marquèrent chacune des phases de cette évolution qui s'étendit sur plusieurs millénaires, et des prodiges de la pensée abstraite qui exercent toujours une influence et nous émerveillent aujourd'hui encore.
Le passage du pictogramme à l'idéogramme (env. 4000 av. J.-C.) s'est accompagné d'une linéarité dans la manière de consigner les faits. Or c'est là que reposent les fondements de ce que nous entendons par l'Histoire. Chaque événement survenu jusqu'alors et que l'on transmettait par la parole, pouvait désormais être fixé, permettant ainsi d'écrire l'Histoire. (Concrètement, Homère ne fera usage de cette possibilité que vers 800 av. J.-C., et plus tard aussi les auteurs de la Bible).
Le passage de l'idéogramme au hiéroglyphe ouvrit la possibilité d'exprimer des mots étrangers, ou nouveaux, en fonction de leur prononciation, au moyen des idéogrammes existants. Ces idéogrammes servirent par exemple à composer le nom d'une personne, un nouveau mot qui n'avait plus rien de commun avec la signification initiale de l'idéogramme utilisé.
Vers 3000 av. J.-C., les Sumériens élaborèrent l'écriture cunéiforme, un système d'idéogrammes et d'écriture syllabique comptant d'abord environ 2000 caractères, puis 500 environ, que l'on gravait

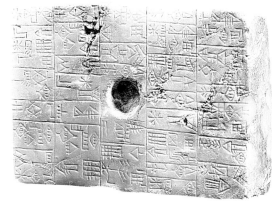

ca. 2600 BC Šuruppak

2430 BC

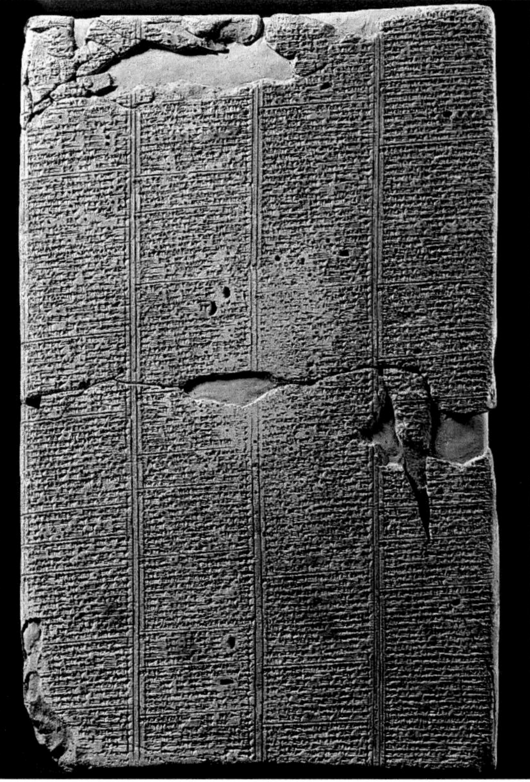

ca. 2100 BC Uruk (front)

ca. 1365 BC Tell el Amarna

ca. 2100 BC Uruk (reverse)

ca. 1200 BC Assur

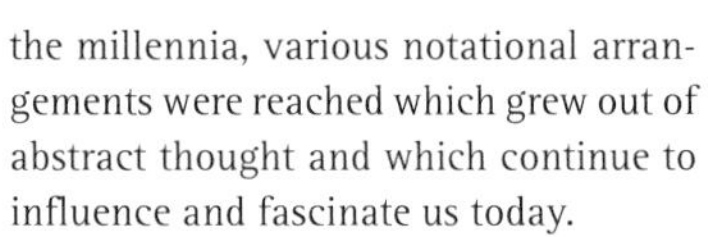

**3000–1700 B.C.** The written word originally stemmed from the desire to achieve a visual representation of speech. Its development began in c. 3000 B.C. with the cuneiform script of the Sumerians. Cuneiform is a series of abstract, phonetic word symbols. It took approximately 60,000 years to reach this stage via a process which went through various phases: pictures, pictorial symbols (pictograms), conceptual symbols (ideograms) and finally hieroglyphics. At each stage in this evolution of writing through the millennia, various notational arrangements were reached which grew out of abstract thought and which continue to influence and fascinate us today.
As pictograms developed into conceptual symbols in c. 4000 B.C., a certain linearity was added to the notation. It was at this time that the foundations were laid for what we understand as history. Past events which prior to this time had only been passed on verbally could now be recorded more permanently, allowing man to begin writing history. (Concrete

schrift, deren zunächst ca. 2000, dann rund 500 Zeichen mit einem kantigen Griffel in den weichen Ton der Schreibtafeln gedrückt wurden. Danach brannte man die Tafeln oder trocknete sie an der Sonne, damit die Schrift haltbar wurde. So konnten Gesetze, Verträge, Befehle oder Briefe schriftlich festgehalten werden. Diese Schrift, die man in Zeilen von links nach rechts schrieb, wurde auch in Vorderasien und Assyrien benutzt. Es gab Kreditgeschäfte, Frachtbriefe, Buchführungen und Quittungen, die gesetzlich vorgeschrieben waren und zur Kontrolle dienten. Anerkannte Zahlungsmittel waren Gold und Silber, was den Beginn des Geldhandels bedeutete. In einer solchen Epoche wurden auch die Kommunikationszeichen verfeinert. Hier entstanden die ältesten Aufzeichnungen, die als Schrift bezeichnet werden können.

1109 BC Assur

First half of first millenium BC

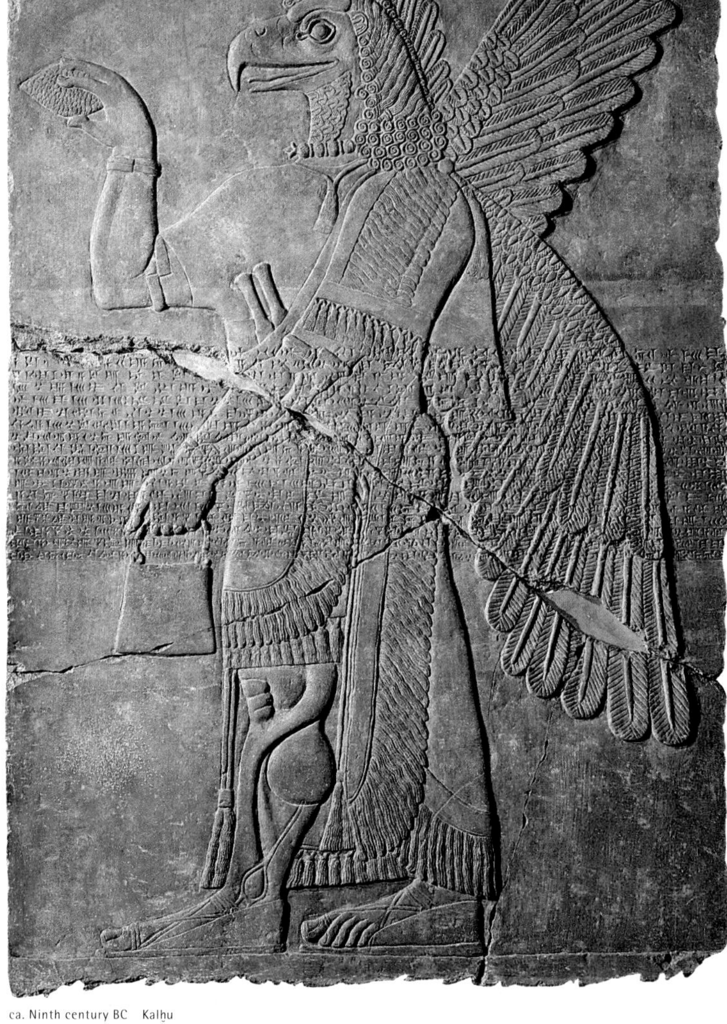

ca. Ninth century BC Kalḫu

au burin sur des tablettes d'argile. Ces tablettes étaient ensuite cuites ou séchées au soleil afin d'être conservées. C'est ainsi que les lois, les contrats, les ordres et la correspondance ont pu être fixés par écrit. Ces caractères écrits en lignes, de gauche à droite, furent utilisés en Asie mineure et par les Assyriens. On découvrit des lettres de crédit, de fret, des comptabilités et des quittances prescrites par la loi à des fins de contrôle. L'or et l'argent, les moyens de paiement courants, marquèrent l'avènement du commerce. C'est à cette époque aussi que les signes servant à la communication s'affinèrent, et que les transcriptions les plus anciennes pouvant être qualifiées d' «écriture», virent le jour.

use was not made of this until about 800 B.C. with Homer and the first writers of what became the Bible).
The transition from ideograms to hieroglyphics enabled new or foreign words to be expressed according to their sound using familiar ideograms. This was how people's names were notated, for example. This new word no longer had anything to do with the ideogram's original meaning.
Around 3000 B.C., the Sumerians began developing cuneiform script as a system of words and syllables, inscribing the over 2000 symbols (later reduced to some 500) into soft clay tablets using a stylus. These tablets were then dried in the sun or fired for durability. The Sumerians were thus able to preserve in writing their laws, contracts, orders and letters. Sumerian cuneiform was written in rows from left to right and was also used in the Near East and Assyria. Letters of credit, consignment notes, ledgers and invoices were prescribed by law and could be used for reference. Gold and silver were the accepted methods of payment, forming the origins of goods-for-money trading practices. This epoch also saw the refinement of symbols used in communication, producing the oldest examples of what could be described as script.

wer

qui

who

# A

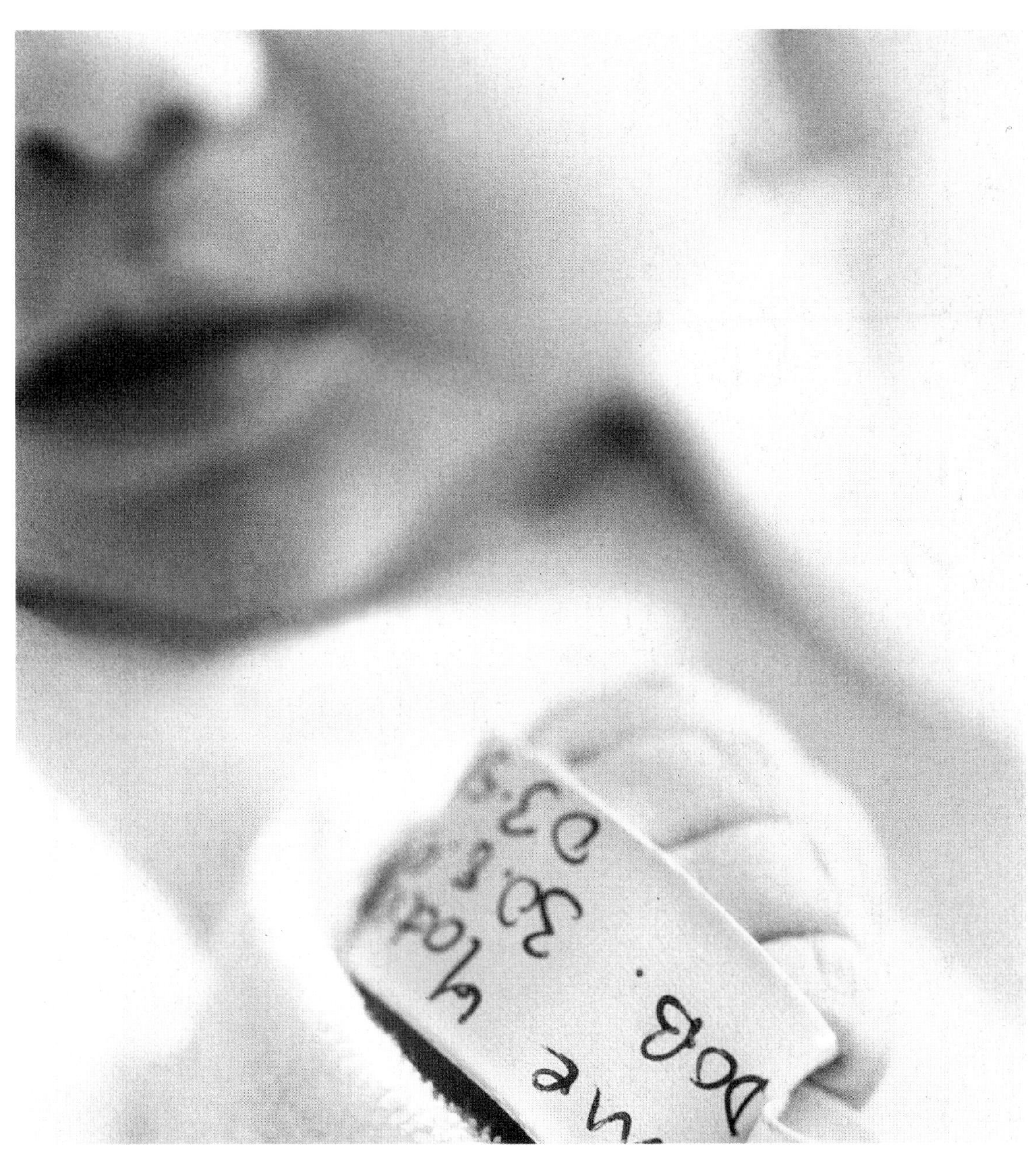

**Achleitner**
**Adobe Systems Inc.**

**Achleitner,** Friedrich – geb. 23.5.1930 in Schalchen, Österreich. – *Schriftsteller, Architekt, Lehrer* – Studium an der Akademie der Bildenden Künste in Wien. 1953–58 freischaffender Architekt in Wien. 1955 Mitglied der „Wiener Gruppe". 1958–64 freiberuflicher Schriftsteller. Seine literarischen Experimente mit dem „Material" Sprache führen zu typographischen Konstruktionen und Bildern. Ab 1963 Dozent an der Akademie der Bildenden Künste in Wien. Seit 1983 Professor für Geschichte und Theorie der Architektur an der Hochschule für angewandte Kunst, Wien. – *Publikationen u.a.:* „hosn rosn baa", Wien 1959; „schwer schwarz", Frauenfeld 1960; „Kinderoper" (mit Konrad Bayer und Oswald Wiener), Wien 1968; „Prosa, Konstellationen, Montagen, Dialektgedichte, Studien", Reinbek 1970; „Quadratroman", Darmstadt, Neuwied 1973; „Österreichische Architektur im 20. Jahrhundert" (4 Bände), Salzburg, Wien 1980–95.

**Adobe Systems Inc.** – *Schriftenhersteller, Schriftenvertrieb* – John Warnock und Charles Geschke gründen 1982 Adobe Systems Inc. in Mountain View, Kalifornien, USA. Adobe stellt 1984 die Software-Technologie für Vektorschriften als Teil des offenen Postscript-Standards vor. Sumner Stone wird Direktor der Typographie bei Adobe. Mit der Adobe-Type-Library erscheinen 1986 die ersten 1.000 Schriften im Adobe PostScript Type 1 Format. Entwicklung und Produktion von zahlreichen Software-Paketen für den Schrift- und Grafik-Design-Bereich: „Adobe Illustrator 88" (1987), „Photoshop

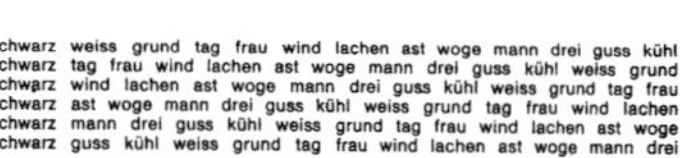
schwarz weiss grund tag frau wind lachen ast woge mann drei guss kühl
schwarz tag frau wind lachen ast woge mann drei guss kühl weiss grund
schwarz wind lachen ast woge mann drei guss kühl weiss grund tag frau
schwarz ast woge mann drei guss kühl weiss grund tag frau wind lachen
schwarz mann drei guss kühl weiss grund tag frau wind lachen ast woge
schwarz guss kühl weiss grund tag frau wind lachen ast woge mann drei

schwarz schwarz schwarz schwarz schwarz schwarz schwarz schwarz schwarz
schwarz schwarz schwarz schwarz schwarz schwarz schwarz schwarz schwarz
schwarz schwarz schwarz schwarz schwarz schwarz schwarz schwarz schwarz
schwarz schwarz schwarz schwarz schwarz schwarz schwarz schwarz schwarz
schwarz schwarz schwarz schwarz schwarz schwarz schwarz schwarz schwarz
schwarz schwarz schwarz schwarz schwarz schwarz schwarz schwarz schwarz

Achleitner 1960 Spread

Achleitner 1995 Cover

Adobe Systems Inc. 1987 Cover

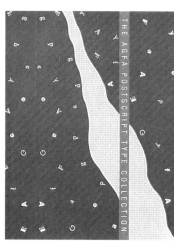

Agfa Compugraphic Corp. 1992 Cover

Aicher 1989 Rotis Sans Serif

**Achleitner,** Friedrich – né le 23.5.1930 à Schalchen, Autriche – *écrivain, architecte, professeur* – études à l'Akademie der Bildenden Künste de Vienne (beaux-arts). 1953–1958, architecte indépendant à Vienne. 1955, membre du «Wiener Gruppe». 1958–1964, auteur indépendant. Ses expériences littéraires, où il utilise la langue comme un «matériau», l'amènent à des constructions et à des images typographiques. A partir de 1963, il enseigne à l'Akademie der Bildenden Künste de Vienne. Professeur d'histoire et de théorie de l'architecture à la Hochschule für angewandte Kunst de Vienne (école supérieure des arts appliqués) depuis 1983. – *Publications, sélection:* «hosn rosn baa», Vienne 1959; «schwer schwarz», Frauenfeld 1960; «Kinderoper» (avec Konrad Bayer et Oswald Wiener), Vienne 1968; «Prosa, Konstellationen, Montagen, Dialektgedichte, Studien», Reinbek 1970; «Quadratroman», Darmstadt, Neuwied 1973; «Österreichische Architektur im 20. Jahrhundert» (4 vol.), Salzbourg, Vienne 1980–1995.

**Adobe Systems Inc.** – *fabricant et distributeur de polices* – Adobe Systems Inc. a été fondée en 1982 par John Warnock et Charles Geschke à Mountain View, Californie, Etats-Unis. En 1984, Adobe présente un logiciel d'écriture numérique faisant partie du standard postscript ouvert. Sumner Stone devient directeur de la typographie. Les 1 000 premiers caractères de format 1 Adobe PostScript Type paraissent en même temps que la Adobe-Type-Library. Développement et production de nombreux logiciels pour le secteur de la composition et du graphisme: «Adobe Illustrator 88» (1987), «Photoshop 1.0» (1989), «Illustrator 1.0» (1989), «Adobe Type Manager ATM» (1989), «Premiere 1.0» (1991), «Adobe Dimension 1.0» (1992), «Adobe Acrobat» (1993). La

**Achleitner,** Friedrich – b. 23.5.1930 in Schalchen, Austria – *author, architect, teacher* – studied at the Akademie der Bildenden Künste in Vienna. 1953–58: freelance architect in Vienna. 1955: member of the Wiener Gruppe. 1958–64: freelance author. His literary experiments with language as material lead to the production of typographical constructions and pictures. From 1963 onwards: lecturer at the Akademie der Bildenden Künste in Vienna. Since 1983: professor of history and architectural theory at the Hochschule für angewandte Kunst in Vienna. – *Publications:* "hosn rosn baa", Vienna 1959; "schwer schwarz", Frauenfeld 1960; "Kinderoper" (with Konrad Bayer and Oswald Wiener), Vienna 1968; "Prosa, Konstellationen, Montagen, Dialektgedichte, Studien", Reinbek 1970; "Quadratroman", Darmstadt, Neuwied 1973; "Österreichische Architektur im 20. Jahrhundert" (4 vols.), Salzburg, Vienna 1980–95.

**Adobe Systems Inc.** – *type manufacturers, font shop* – 1982: John Warnock and Charles Geschke found Adobe Systems Inc. in Mountain View, California, USA. 1984: Adobe introduces software technology for vector fonts as part of the Postscript standard. Sumner Stone becomes director of typography at Adobe. 1986: the first 1,000 fonts are published with the Adobe Type Library in Adobe Post Script Type 1-format. Adobe have developed and produced numerous software packages for type and graphic design: Adobe Illustrator 88 (1987), Photoshop 1.0 (1989), Illustrator 1.0 (1989),

1.0" (1989), „Illustrator 1.0" (1989), „Adobe Type Manager ATM" (1989), „Premiere 1.0" (1991), „Adobe Dimension 1.0" (1992), „Adobe Acrobat" (1993). 1993 erschienen die Schriftkollektionen „Type Basics" und „Wild Type". 1994 Fusion von Adobe und Aldus unter dem Namen „Adobe".

**Agfa Compugraphic Corp.** – *Schriftenhersteller, Schriftenvertrieb* – 1960 Gründung von Compugraphic in Wilmington Massachusetts, USA. 1985 Einführung von Laserschriften. Agfa Gevaert kauft 1988 Compugraphic und nennt sich Agfa Compugraphic. Die Schriftenbibliothek umfaßt über 2.500 Schriften mit vielen Originalschriften, Neuschnitten und lizensierten Schriften.

**Aicher,** Otl – geb. 13. 5. 1922 in Ulm, Deutschland, gest. 1. 9. 1991 in Rotis über Leutkirch, Deutschland. – *Schriftentwerfer, Grafik-Designer, Autor, Lehrer* – 1946–47 Studium an der Akademie der Bildenden Künste in München. 1947 Entwurfsbüro in Ulm. Erscheinungsbilder u. a. für Braun Elektrogeräte (1954), die Deutsche Lufthansa (1969), die Westdeutsche Landesbank (1964), Blohm & Voss (1964), die Bayrische Rückversicherung (1972), das Zweite Deutsche Fernsehen ZDF (1974) und Erco (ab 1976). 1949–54 Mitinitiator und Gründungsmitglied der Hochschule für Gestaltung in Ulm, 1954–66 Dozent an der Abteilung Visuelle Kommunikation, 1956–59 Mitglied des Rektoratkollegiums, 1962–64 Rektor. 1967 Entwurfsbüro in München. 1967–72 Gestaltungsbeauftragter der Olympischen Spiele in München, Entwicklung eines in-

collection de caractères «Type Basics» et «Wild Type» paraît en 1993. En 1994, fusion entre Adobe et Aldus sous le nom «Adobe».

**Agfa Compugraphic Corp.** – *fabricants et distributeurs de polices* – la société Compugraphic a été fondée en 1960 à Wilmington, Massachusetts, Etats-Unis. Introduction de l'écriture laser en 1985. En 1988, Agfa Gevaert achète Compugraphic et la rebaptise Agfa Compugraphic. Sa bibliothèque de polices contient plus de 2 500 écritures y compris de nombreuses polices originales, de nouvelles tailles et des caractères sous licence.

**Aicher,** Otl – né le 13. 5. 1922 à Ulm, Allemagne, décédé le 1. 9. 1991 à Rotis über Leutkirch, Allemagne – *concepteur de polices, graphiste maquettiste, auteur, enseignant* – 1946–1947, études à l'Akademie der Bildenden Künste (beaux-arts) de Munich. 1947, bureau d'études à Ulm. Enseignes pour les appareils électriques Braun (1954), la Deutsche Lufthansa (1969), la Westdeutsche Landesbank (1964), Blohm & Voss (1964), la Bayrische Rückversicherung (1972), la deuxième chaîne de télévision allemande ZDF (1974) et Erco (à partir de 1976), entre autres. 1949–1954, co-initiateur et membre fondateur de la Hochschule für Gestaltung d'Ulm (école supérieure de design); 1954–1966, enseignant au département de communication visuelle; 1956–1959, membre de la direction collégiale, puis recteur en 1962–1964. 1967, bureau d'études à Munich. 1967–1972, chargé du design pour les Jeux Olympiques de Munich, développement d'un

Aicher 1972 Title

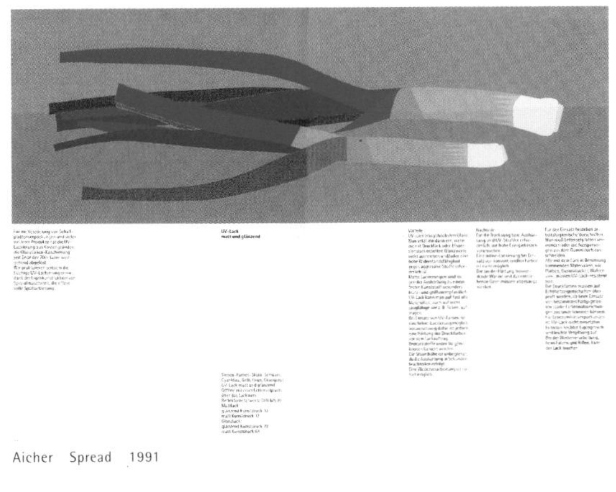

Aicher Spread 1991

|   | n | n |   |   | *n* |
|---|---|---|---|---|---|
| a | b | e | f | g | i |
| o | r | s | t | y | z |
| A | B | C | E | G | H |
| M | O | R | S | X | Y |
| 1 | 2 | 4 | 6 | 8 | & |

Aicher 1989 Rotis Serif

otl aicher

typographie

typographie ist die abbildung des sprechens. die gesprochene sprache aber ist vergänglich. sobald unsere lautzeichen verklungen sind, ist auch die sprache verklungen. in der schrift erhält die sprache eine zeitliche und räumliche dimension. schrift kann man an beliebige stellen transportieren, weit über den rahmen der gesprochenen sprache hinaus. und schrift kann man aufbewahren, an jedem ort, beliebig lange.

typographie ist die bildliche form der sprache. infolgedessen ist die sprache der maßstab der typographie. das maß der sprache ist ihre verständlichkeit. das maß der typographie wäre demnach ihre lesbarkeit. sollte man meinen.

es ist schwieriger, ein wahres bild zu photographieren, als ein unter scheinwerfern inszeniertes objekt darzustellen, sei es eine paprikaschote oder einen akt. und es ist schwieriger, einen text lesbar anzubieten, als daraus eine schöne struktur, ein ästhetisches feld, eine buchstabenlandschaft, ein kunstwerk zu machen.

von diesen schwierigkeiten handelt dieses buch. es ist kein buch über typographie als kunst, sondern über typographie als kommunikation. aber vielleicht besteht die wahre kunst darin, die schwierigkeiten auszuräumen, die einer lesbaren typographie, einer typographie als kommunikation im wege stehen.

Typography is the mirroring of speech. But spoken language is transient. As soon as our tones die away, so too does language. Through writing, language attains a temporal and a spatial dimension. Writing can be randomly moved about, far beyond the confines of spoken language. And it can also be preserved indefinitely.

Typography is the pictorial equivalent of language. By extension, language is what typography is judged by. Language is measured in terms of comprehensibility. Just as typography ought to be measured in terms of legibility. Ought to be.

It is more difficult to take a compelling photograph of an object–be it an apple or a nude–than it is to mount a spotlit display of it. And it is more difficult to produce legible type than to create an alluring textual structure, an aesthetic delight, a landscape of letters, a work of art.

These are the problems this book concerns itself with. It is not a book on typography as art but on typography as communication. But maybe the ability to remove the obstacles to legible typography, viz. typography as communication, is art too.

ernst & sohn
edition druckhaus maack

Aicher 1988 Cover

Adobe Type Manager ATM (1989), Premiere 1.0 (1991), Adobe Dimension 1.0 (1992), Adobe Acrobat (1993). 1993: the font collections Type Basics and Wild Type are produced. 1994: Adobe and Aldus merge under the name Adobe.

**Agfa Compugraphic Corp.** – *type manufacturers, font shop* – 1960: Compugraphic is founded in Wilmington, Massachusetts, USA. 1985: introduction of laser fonts. 1988: Agfa Gevaert buys Compugraphic and renames the company Agfa Compugraphic. Their type library has over 2,500 typefaces with many original fonts, new varieties and licensed fonts.

**Aicher,** Otl – b. 13. 5. 1922 in Ulm, Germany, d. 1. 9. 1991 in Rotis über Leutkirch, Germany – *type designer, graphic designer, author, teacher* – 1946–47: studies at the Akademie der Bildenden Künste in Munich. 1947: design studio in Ulm. Produces graphics for: Braun Elektrogeräte (1954), Deutsche Lufthansa (1969), Westdeutsche Landesbank (1964), Blohm & Voss (1964), Bayrische Rückversicherung (1972), Zweites Deutsches Fernsehen ZDF (1974) and Erco (from 1976), among others. 1949–54: one of the initiators and founder members of the Hochschule für Gestaltung in Ulm. 1954–66: lecturer in the department of visual communication. 1956–59: member of the board of governors. 1962–64: rector. 1967: design studio in Munich. 1967–72: commissioned with the design for the Olympic Games in Munich. Develops an internationally conventional system of pictograms. 1972: moves his

ternational gebräuchlichen Systems von Piktogrammen. 1972 Verlegung des Büros nach Rotis im Allgäu. Aichers umfangreiches praktisches und theoretisches Werk repräsentiert einen verfeinerten und erweiterten gestalterischen Funktionalismus sowie eine ganzheitliche und rationale Einstellung zur visuellen Kommunikation auf der Grundlage ihrer gesellschaftlichen Relevanz. 1984 Gründung des Instituts für analoge Studien. Unterrichtet an der Yale University, USA (1958) und am Museo de Arte Moderna, Rio de Janeiro (1959). Zahlreiche internationale Ausstellungen. – *Schriftentwürfe:* Traffic (für die Münchner Verkehrsbetriebe), Rotis (1988). – *Publikationen u.a.:* „Zeichensysteme" (mit Martin Krampen), München 1980; „Wilhelm von Ockham" (mit Wilhelm Vossenkuhl und Gabriele Greindl), München 1986; „Griffe und Greifen" (mit Robert Kuhn), Köln 1987; „Typographie", Lüdenscheid 1988; „Analog und Digital", Berlin 1991; „Die Welt als Entwurf", Berlin 1991; „Schreiben und Widersprechen", Berlin 1993.

**Albers,** Josef – geb. 19. 3. 1888 in Bottrop, Deutschland, gest. 25. 3. 1976 in New Haven, USA. – *Typograph, Maler, Lehrer* – 1905–08 Ausbildung zum Volksschullehrer in Büren. 1908–13 Volksschullehrer in Bottrop, Weddern und Stadtlohn. 1913–15 Studium an der Königlichen Kunstschule Berlin. 1916–19 Studium an der Kunstgewerbeschule Essen. 1919–20 Studium an der Königlichen bayerischen Akademie der Bildenden Künste München. 1920–23 Studium am Bauhaus in Weimar. 1925 Bauhaus-

système de pictogrammes international. 1972, le bureau s'installe à Rotis dans l'Allgäu. L'importante œuvre pratique et théorique d'Aicher se caractérise par un raffinement et un élargissement du fonctionnalisme, ainsi que par une conception globale et rationnelle de la communication visuelle qui se fonde sur sa pertinence sociale. 1984, fondation de l'Institut für analoge Studien (institut d'études analogiques). A enseigné à la Yale University, Etats-Unis (1958) et au Museo de Arte Moderna, Rio de Janeiro (1959). Nombreuses expositions internationales. – *Polices:* Traffic (pour les transports publics de Munich), Rotis (1988). – *Publications , sélection:* «Zeichensysteme» (avec Martin Krampen), Munich 1980; «Wilhelm von Ockham» (avec Wilhelm Vossenkuhl et Gabriele Greindl), Munich 1986; «Griffe und Greifen» (avec Robert Kuhn), Cologne 1987; «Typographie», Lüdenscheid 1988; «Analog und Digital», Berlin 1991; «Die Welt als Entwurf», Berlin 1991; «Schreiben und Widersprechen», Berlin 1993.

**Albers,** Josef – né le 19. 3. 1888 à Bottrop, Allemagne, décédé le 25. 3. 1976 à New Heaven, Etats-Unis – *typographe, peintre, enseignant* – 1905–1908, formation d'instituteur à Büren. 1908–1913, instituteur à Bottrop, Weddern et Stadtlohn. 1913–1915, études à la Königliche Kunstschule de Berlin (école d'art). 1916–1919, études à la Kunstgewerbeschule d'Essen (école des arts décoratifs). 1919–1920, études à la Königliche bayerische Akademie der Bildenden Künste (beaux-arts), Munich. 1920–1923, études au Bauhaus de Weimar. 1925, enseignant au Bauhaus, y enseigne la typographie, le design de mobilier et la peinture sur verre. Dirige le cycle préliminaire à partir de 1928. Après la fermeture du Bauhaus en 1933, il est appelé au Black Mountain College en Ca-

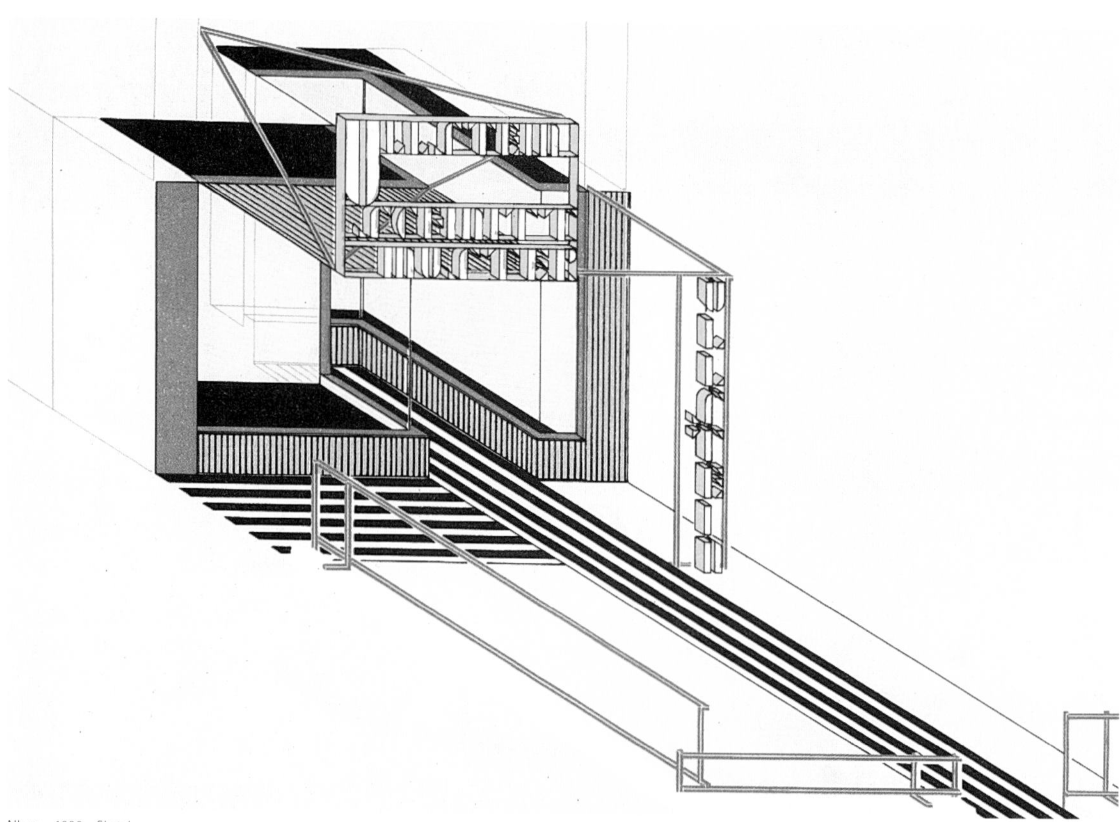

Albers 1926 Sketch

Albers 1928–31 Kombinationsschrift

Albers 1928–31 Kombinationsschrift

POÈME À CRIER ET À DANSER

*Chant III*

Albert-Birot 1967 Page

studio to Rotis in the Allgäu region of Germany. Aicher's extensive practical and theoretical work represents an expanded and refined structural Functionalism, and demonstrates an holistic and rational approach to visual communication on the basis of its social relevance. 1984: founds the Institut für analoge Studien. He has taught at Yale University, USA (1958) and at the Museo de Arte Moderna, Rio de Janeiro (1959). Numerous exhibitions all over the world. – *Fonts:* Traffic (for Munich's public transport services), Rotis (1988). – *Publications include:* "Zeichensysteme" (with Martin Krampen), Munich 1980; "Wilhelm von Ockham" (with Wilhelm Vossenkuhl and Gabriele Greindl), Munich 1986; "Griffe und Greifen" (with Robert Kuhn), Cologne 1987; "Typographie", Lüdenscheid 1988; "Analog und Digital", Berlin 1991; "Die Welt als Entwurf", Berlin 1991; "Schreiben und Widersprechen", Berlin 1993.

**Albers,** Josef – b. 19. 3. 1888 in Bottrop, Germany, d. 25. 3. 1976 in New Haven, USA – *typographer, painter, teacher* – 1905–08: trains as an elementary school teacher in Büren. 1908–13: elementary school teacher in Bottrop, Weddern and Stadtlohn. 1913–15: studies at the Königliche Kunstschule in Berlin. 1916–19: studies at the Kunstgewerbeschule in Essen. 1919–20: studies at the Königliche bayerische Akademie der Bildenden Künste in Munich. 1920–23: studies at the Bauhaus in Weimar. 1925: teaches typography, furniture design and glass painting as a Bauhaus master and from

Meister, unterrichtet Typographie, Möbelentwurf, Glasmalerei. Ab 1928 Leiter des gesamten Vorkurses. Nach der Schliessung des Bauhauses 1933 Berufung an das Black Mountain College in North Carolina, USA. 1948 Rektor des Black Mountain College. 1950 Direktor des Department of Design an der Yale University, New Haven. 1970 Ernennung zum Ehrenbürger der Stadt Bottrop. Zahlreiche Vorträge und Vortragsreihen: Harvard University, Cambridge, Massachusetts (1936–40), Cincinnati Art Academy (1949), Pratt Institute, New York (1949), Hochschule für Gestaltung in Ulm (1953–55), Trinity College, Hartford, Connecticut (1965), University of South Florida, Tampa (1966). – *Schriftentwürfe:* Display (1923), Schablonenschrift (1923–26), Kombinationsschrift aus Glas (1928–31). – *Publikationen:* „Trotz der Geraden", Bern 1961; „Homage to the Square", New Haven 1962; „Interaction of Colors", New Haven 1963. Werner Spies „Albers", Stuttgart, New York 1970; Eugen Gomringer „Josef Albers", Starnberg 1971; Nicholas Fox Weber „Josef Albers, Retrospektive", New York, Köln 1988.

**Albert-Birot,** Pierre – geb. 1876 in Angoulême, Frankreich, gest. 1967 in Paris, Frankreich. – *Schriftsteller, Maler, Bildhauer* – 1916 Gründung der Zeitschrift „SIC" (Sons, Idées, Couleurs), in der Kubisten, Futuristen und Surrealisten veröffentlichen. Bis 1919 erscheinen 41 Hefte. Durch seine verlegerische Tätigkeit und die Form seiner eigenen literarischen Arbeit ist Albert-Birot ein ständiger Anreger bei der Veränderung der Typogra-

Albert-Birot 1916 Signet

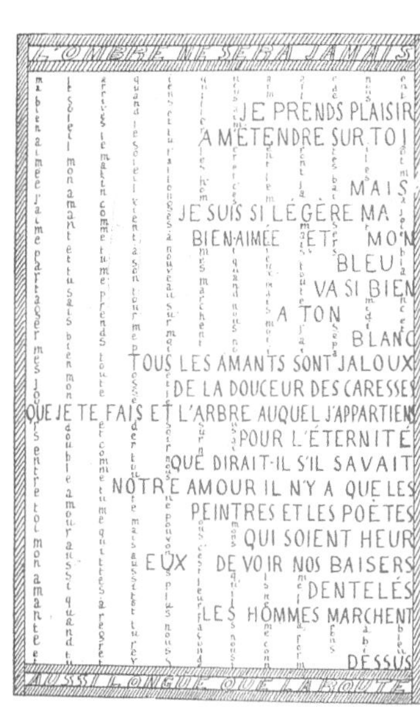


Albert-Birot 1918 Calligram

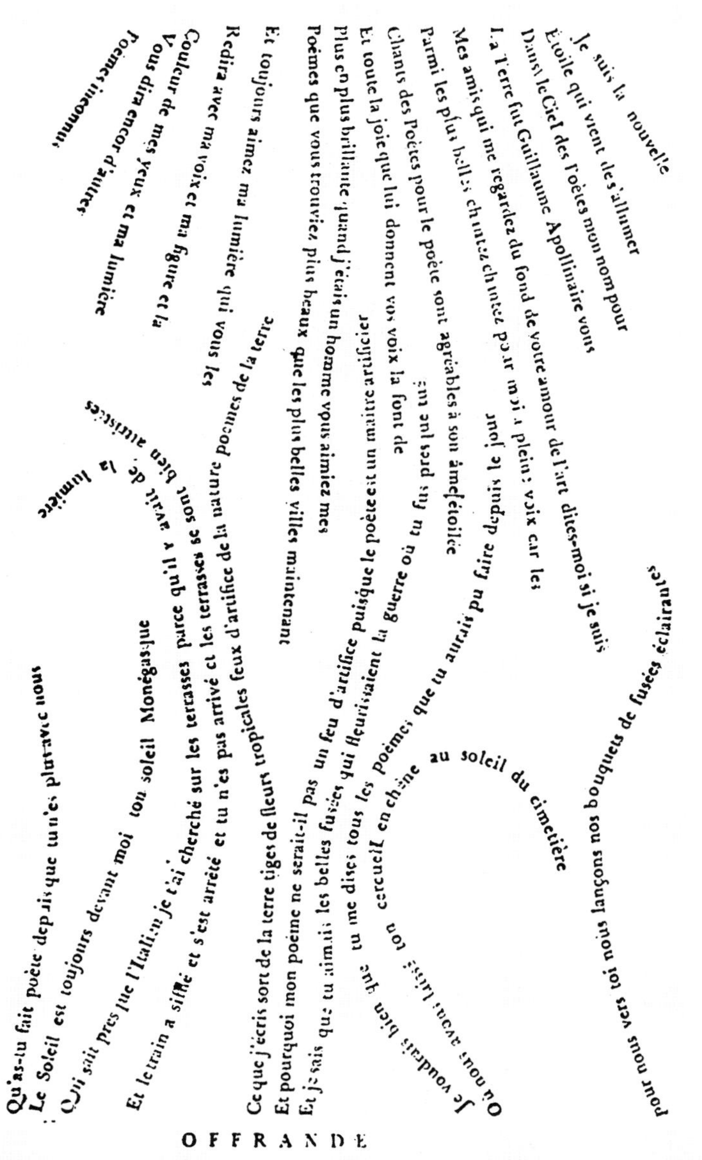


Albert-Birot Calligram

roline du Nord, Etats-Unis. 1948, recteur du Black Mountain College. 1950, directeur du Department of Design à la Yale University, New Haven. En 1970, il est nommé citoyen d'honneur de la ville de Bottrop. Nombreuses conférences et cycles de conférences : Harvard University, Cambridge, Massachusetts (1936–1940), Cincinnati Art Academy (1949), Pratt Institute New York (1949), Hochschule für Gestaltung d'Ulm (école supérieure de design) (1953–1955), Trinity College, Hartford, Connecticut (1965), University of South Florida, Tampa (1966). – *Polices :* Display (1923), Schablonenschrift (1923–1926), Kombinationsschrift aus Glas (1928–1931). – *Publications :* «Trotz der Geraden», Berne 1961; «Homage to the square», New Haven 1962; «Interaction of colors», New Haven 1963. Werner Spies «Albers», Stuttgart, New York 1970; Eugen Gomringer «Josef Albers», Starnberg 1971; Nicholas Fox Weber «Josef Albers, Retrospective», New York, Cologne 1988.

**Albert-Birot,** Pierre – né en 1876 à Angoulême, France, décédé en 1967 à Paris, France – *écrivain, peintre, sculpteur* – fonde en 1916 la revue «SIC» (Sons, Idées, Couleurs) qui publie des cubistes, futuristes et surréalistes. Parution de 41 numéros jusqu'en 1919. En raison de ses activités d'éditeur et de la forme de sa propre création littéraire, Albert-Birot n'a cessé

1928 onwards, heads the entire preparatory course. 1933: after the closure of the Bauhaus, appointment at Black Mountain College in North Carolina, USA. 1948: rector of Black Mountain College. 1950: director of the department of design at Yale University, New Haven. 1970: made an honorary citizen of Bottrop. Numerous lectures and series of lectures at Harvard University, Cambridge, Massachusetts (1936–40), Cincinnati Art Academy (1949), Pratt Institute, New York (1949), Hochschule für Gestaltung, Ulm (1953–55), Trinity College, Hartford, Connecticut (1965) and the University of South Florida, Tampa (1966). – *Fonts:* Display (1923), Schablonenschrift (1923–26), Kombinationsschrift aus Glas (1928–31). – *Publications:* "Trotz der Geraden", Bern 1961; "Homage to the square", New Haven 1962; "Interaction of colors", New Haven 1963. Werner Spies "Albers", Stuttgart, New York 1970; Eugen Gomringer "Josef Albers", Starnberg 1971; Nicholas Fox Weber "Josef Albers, Retrospective", New York, Cologne 1988.

**Albert-Birot,** Pierre – b. 1876 in Angoulême, France, d. 1967 in Paris, France – *author, painter, sculptor* – 1916: launches the magazine "SIC" (Sons, Idées, Couleurs), in which Cubists, Futurists and Surrealists publish work. 41 issues are produced between 1916 and 1919. Albert-Birot, with his publishing and the form of his own literary work, was a constant stimulus for the typographical changes

phie seiner Zeit. – *Publikationen:* „Trente-et-un poèmes de poche", 1917; „Poèmes à hurler et à danser, la joie de sept couleurs", 1919; „La tricotérie", 1920.

**Aldus Corporation** – *Schriftenhersteller* – Paul Brainerd gründet 1984 die Aldus Corporation. Mit dem Layout-Programm „Aldus PageMaker 1.0" für Macintosh beginnt 1985 das Desk Top Publishing (DTP). 1994 Fusion mit Adobe Systems Inc.

**Alessandrini,** Jean Antoine – geb. 3. 8. 1942 in Marseille, Frankreich. – *Schriftentwerfer, Grafik-Designer, Illustrator* – 1959–62 Studium am Collège Technique d'Arts Graphiques, Paris, Mitarbeit in Werbestudios und der Werbeabteilung von „Paris Match". 1963–66 Arbeiten für die Zeitschriften „Lui", „Salut les copains" und „Elle". Ab 1966 Arbeiten für Hollenstein Phototypo, das Comic-Magazin „Pilote" und den Rencontre-Verlag. Zahlreiche Buchumschläge für Gallimard und Mercure de France. – *Schriftentwürfe:* Trombinoscope (1964), Futuriste (1967), Electric-Type (1968), Germain (1969), Hypnos (1969), Akenaton (1969), Showbiz (1969), Vampire (1969), Mirago (1970–71), Alessandrini 7 (1972), Graphicman (1973), Astronef (1976), Combinat (1976), Legitur (1977). – *Publikationen u. a.:* „Henri à l 'Amuseum", Paris 1974; „Sautes d'humour d'Alessandrini", Paris 1977; „Typomanie", Paris 1977.

**Alexiew,** Sheko – geb. 26. 3. 1945 in Stara Sagora, Bulgarien. – *Grafik-Designer, Typograph* – 1971 Studium der Gebrauchsgrafik an der Kunstakademie Sofia. Arbeitet in der Buchgestaltung.

d'œuvrer pour la transformation de la typographie de son temps. – *Publications:* «Trente-et-un poèmes de poche», 1917; «Poèmes à hurler et à danser, la joie de sept couleurs», 1919; «La tricotérie», 1920.

**Aldus Corporation** – *fabricants de polices* – fondée par Paul Brainerd en 1984. En 1985, Desk Top Publishing (DTP) démarre le programme de mise en page «Aldus Pagemaker 1.0» pour Macintosh. 1994, fusion avec Adobe Systems Inc.

**Alessandrini,** Jean Antoine – né le 3. 8. 1942 à Marseille, France – *concepteur de polices, graphiste maquettiste, illustrateur* – 1959–1962, études au Collège technique d'Arts graphiques de Paris, puis collaboration à l'atelier de design publicitaire du département publicité de la revue «Paris Match». 1963–1966, travaux pour les revues «Lui», «Salut les copains» et «Elle». Travaille à partir de 1966 pour Hollenstein Phototypo, le magazine de bandes dessinées «Pilote» et les éditions Rencontre. Nombreuses couvertures de livres pour Gallimard et Mercure de France. – *Polices:* Trombinoscope (1964), Futuriste (1967), Electric-Type (1968), Germain (1969), Hypnos (1969), Akenaton (1969), Showbiz (1969), Vampire (1969), Mirago (1970–1971), Alessandrini 7 (1972), Graphicman (1973), Astronef (1976), Combinat (1976), Legitur (1977). – *Publications, sélection:* «Henri à l'Amuseum», Paris 1974; «Sautes d'humour d'Alessandrini», Paris 1977; «Typomanie», Paris 1977.

**Alexiev,** Sheko – né le 26. 3. 1945 à Stara Sagora, Bulgarie – *graphiste maquettiste, typographe* – 1971, études d'arts graphiques appliqués à l'académie des beaux-arts de Sofia. Spécialisé dans l'illustrations de livres.

**Allner,** Walter – né le 2.1.1909 à Dessau, Allemagne – *graphiste maquettiste, typographe, enseignant* – 1927–1930,

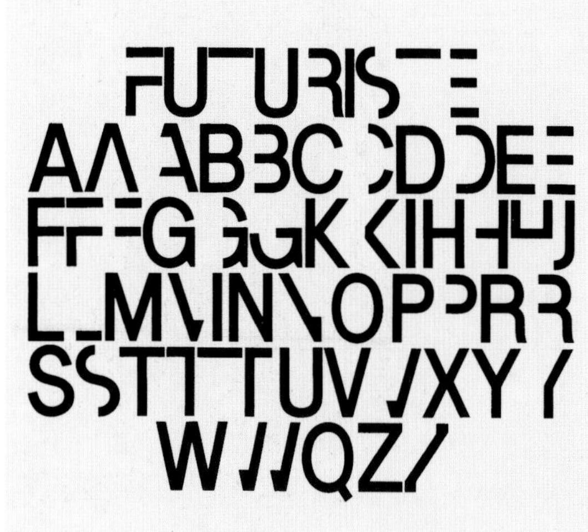

Alessandrini 1967 Futuriste

Alessandrini 1976 Astronef

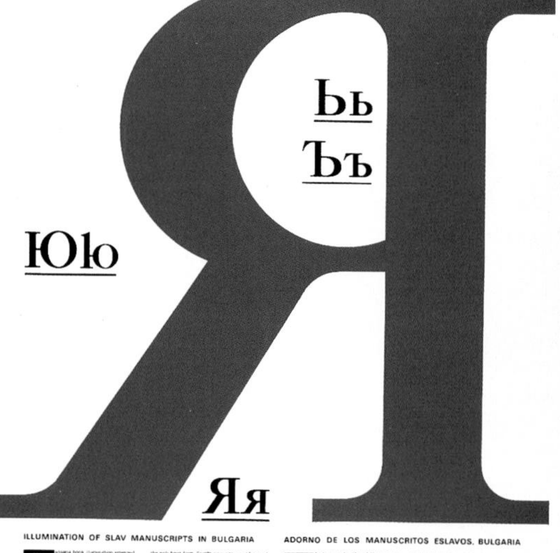

Alexiew 1992 Poster

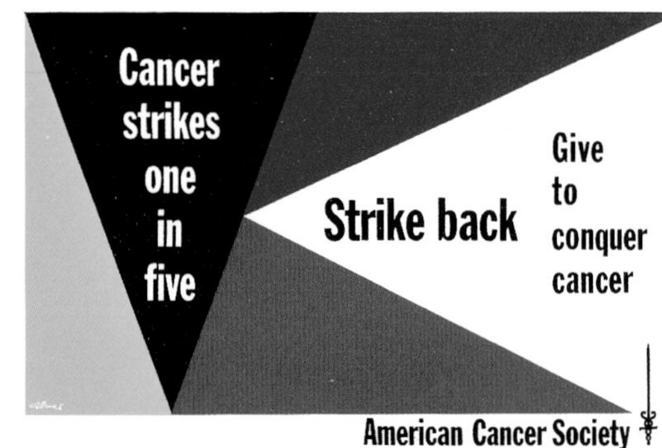


Allner ca. 1948 Poster

of his age. – *Publications:* "Trente-et-un poèmes de poche", 1917; "Poèmes à hurler et à danser, la joie de sept couleurs", 1919; "La tricotérie", 1920.

**Aldus Corporation** – *type manufacturers* – 1984: Paul Brainerd founds the Aldus Corporation. 1985: Desk Top Publishing (DTP) is launched with the layout program Aldus PageMaker 1.0 for Macintosh. 1994: Aldus merges with Adobe Systems Inc.

**Alessandrini,** Jean Antoine – b. 3. 8. 1942 in Marseille, France – *type designer, graphic designer, illustrator* – 1959–62: studies at the Collège Technique d'Arts Graphiques in Paris, then works at various advertising studios and for the advertising department of "Paris Match" magazine. 1963–66: works for the magazines "Lui", "Salut les copains" and "Elle". From 1966 onwards: works for Hollenstein Phototypo, the comic "Pilote" and Rencontre publishers. Produced numerous dust jackets for Gallimard and Mercure de France. – *Fonts:* Trombinoscope (1964), Futuriste (1967), Electric-Type (1968), Germain, Hypnos, Akenaton (1969), Showbiz (1969), Vampire (1969), Mirago (1970–71), Alessandrini 7 (1972), Graphicman (1973), Astronef, Combinat (1976), Legitur (1977). – *Publications include:* "Henri à l'Amuseum", Paris 1974; "Sautes d'humour d'Alessandrini", Paris 1977; "Typomanie", Paris 1977.

**Alexiev,** Sheko – b. 26. 3. 1945 in Stara Sagora, Bulgaria – *graphic designer, typographer* – 1971: studies commercial art at the art academy in Sofia. Works in book design.

**Allner,** Walter – geb. 2. 1. 1909 in Dessau, Deutschland. – *Grafik-Designer, Typograph, Lehrer* – 1927–30 Studium am Bauhaus in Dessau. 1930 Assistent von Piet Zwart in Wassenaar, Niederlande. 1932–33 Assistent von Jean Carlu in Paris, Frankreich. 1933–36 Art Director in Paris. 1945–48 Paris-Redakteur der Zeitschrift „Graphis". 1948–52 Gründer und Herausgeber der Buchreihe „International Poster Annual". Seit 1949 Grafik-Designer in New York. 1963–74 Art Director der Zeitschrift „Fortune" für die er ca. 80 Umschläge entwirft. 1968 Teilnahme an der Ausstellung „50 Jahre Bauhaus". Seit 1974 unterrichtet er an der Parson's School of Design in New York.

**Alsleben,** Kurd – geb. 14. 6. 1928 in Königsberg, Neumark (heute Rußland). – *Computertheoretiker, Lehrer* – 1949 Studium an der Akademie der Bildenden Künste in Karlsruhe. 1959 Studium der Informationswissenschaften und Geschichte. Ab 1960 Computerzeichnungen, Informationstypographie. 1965–69 Gastdozent für Strukturtheorie an der Hochschule für Gestaltung in Ulm. 1969 Gastdozent für Informationsästhetik an der Hochschule der Künste in Berlin. Seit 1970 Professor an der Hochschule der Bildenden Künste in Hamburg, deren Computerabteilung er aufbaut und leitet. 1984 Mitglied der Akademia Internacia de la Sciencj San Marino. – *Publikationen u. a.:* „Ästhetische Redundanz", Quickborn 1962; „Farbwörter", Hamburg 1985.

**American Type Founders Company** (ATF) – Jersey City, USA. – *Schriftgießerei, Schriftmaschinenhersteller* – 1892 nach

This poster was designed by Walter Allner.
It has been produced for his exhibition at
the Bauhaus Dessau
in German Democratic Republic
in November 1989.
WA
Walter Allner
Art Director Designer Lehrer und Maler
in New York, USA. studierte am Bauhaus Dessau
Grunder und Herausgeber
des internationalen Poster-Jahrbuches
Herausgeber des Grafik-Magazins in Paris
Kunstlerischer Direktor des
Fortune Magazines in New York
und Präsident der internationalen
Grafikervereinigung.


Allner 1989 Poster

FORTUNE

Allner 1964 Cover

Kurd Alsleben

Kurd Alsleben

ästhetische Redundanz

ästhetische Redundanz

Abhandlungen über die artistischen Mittel der bildenden Kunst

Vorwort von André Abraham Moles

Verlag Schnelle, Quickborn bei Hamburg

Alsleben 1962 Title

**Allner,** Walter – né le 2. 1. 1909 à Dessau, Allemagne – *graphiste, typographe, enseignant* – 1927–1930, études au Bauhaus de Dessau. 1930, assistant de Piet Zwart à Wassenaar, Pays-Bas. 1932–1933, assistant de Jean Carlu à Paris, France. 1933–1936, directeur artistique à Paris. 1945–1948, rédacteur à Paris de la revue «Graphis». 1948–1952, fondateur et éditeur de la collection «International Poster Annual». Depuis 1949, graphiste à New York. 1963–1974, directeur artistique de la revue «Fortune» dont il réalise la maquette de 80 couvertures environs 1968, participation à l'exposition itinérante «50 Jahre Bauhaus» (50 ans de Bauhaus). Il enseigne à la Parson's School of Design à New York depuis 1974.

**Alsleben,** Kurd – né le 14. 6. 1928 à Königsberg, Neumark (aujourd'hui en Russie) – *informaticien, enseignant* – 1949, études à l'Akademie der Bildenden Künste de Karlsruhe (beaux-arts). 1959, études en sciences de l'information et d'histoire. Dès 1960, dessins sur ordinateur et travaux d'infographie. 1965–1969, maître de conférences en théorie de structure à la Hochschule für Gestaltung d'Ulm (école supérieure de design). 1969, maître de conférence en esthétique de l'information à la Hochschule der Künste de Berlin (école supérieure des beaux-arts). Professeur à la Hochschule der Bildenden Künste de Hambourg à partir de 1970 où il fond le département informatique dont il assure la direction. 1984, membre fondateur de la Akademia Internacia de la Sciencj à Saint-Marin. – *Publications, sélection:* «Ästhetische Redundanz», Quickborn 1962; «Farbwörter», Hambourg 1985.

**American Type Founders Company** (ATF), Jersey City, Etats-Unis – *fonderie de ca-*

**Allner,** Walter – b. 2. 1. 1909 in Dessau, Germany – *graphic designer, typographer teacher* – 1927–30: studies at the Bauhaus in Dessau. 1930: assistant to Piet Zwart in Wassenaar, The Netherlands. 1932–33: assistant to Jean Carlu in Paris, France. 1933–36: art director in Paris. 1945–48: Paris editor of "Graphis" magazine. 1948–52: founder and editor of the "International Poster Annual" series. 1949: starts working as a graphic designer in New York. 1963–74: art director of "Fortune" magazine, for which he designs c. 80 covers. 1968: takes part in the "50 Jahre Bauhaus" traveling exhibition. 1974: starts teaching at the Parson's School of Design in New York.

**Alsleben,** Kurd – b. 14. 6. 1928 in Königsberg, Germany (now Russia) – *computer theoretician, teacher* – 1949: studies at the Akademie der Bildenden Künste in Karlsruhe. 1959: studies information science and history. From 1960 onwards: computerized graphics, information typography. 1965– 69: visiting lecturer of structural theory at the Hochschule für Gestaltung in Ulm. 1969: visiting lecturer of information aesthetics at the Hochschule der Künste in Berlin. Since 1970: professor at the Hochschule der Bildenden Künste in Hamburg, where he sets up and now heads the computer department. 1984: founder member of the Akademia Internacia de la Sciencj San Marino. – *Publications include:* "Ästhetische Redundanz", Quickborn 1962; "Farbwörter", Hamburg 1985.

**American Type Founders Company** (ATF), Jersey City, New Jersey, USA – *type*

dem Zusammenschluß von 23 amerikanischen Schriftgießereien gegründet. 1895 Veröffentlichung des ersten Schriftmusterbuches der ATF. Die Schrift „Cheltenham", die von Bertram Goodhue für die Cheltenham Press entworfen wurde, wird 1904–11 ausgebaut und kommerziell erfolgreich. 1908 Gründung einer typographischen Bibliothek durch ATF unter der Leitung von Henry Lewis Bullen. 1966 wird ATF von White Consolidated Industries gekauft. ATF erwirbt 1970 Lanston Monotype. Durch weitere Fusion entsteht die Kingsley ATF, die inzwischen Software für Computer herstellt. 1993 Liquidation der Firma ATF. Zu den Schriftentwerfern der ATF gehörten F. W. Goudy, M. F. Benton, L. Bernhard, W. Bradley (der die 12 Ausgaben des Magazins „The Chap Book" schrieb, gestaltete und herausgab) und J. W. Phinney.

**Amman,** Jost – geb. 1539 in Zürich, Schweiz, gest. 1591 in Nürnberg, Deutschland. – *Buchillustrator, Holzschneider, Kupferstecher* – Neben zahlreichen Signets (z. B. für Sigmund Feyerabend und Jakob Sabon) und Figurenalphabeten fertigt Amman 1555 die Abbildungen zu Lienhard Fronsbergers „Kriegsbuch" als Holzstiche. Sein bekanntestes Werk ist das „Ständebuch" mit Reimen von Hans Sachs, das S. Feyerabend 1568 in Frankfurt am Main herausgab, in dem zahlreiche Berufe der damaligen Zeit in charakteristischen Tätigkeitsposen erklärend abgebildet sind.

**Andersch,** Martin – geb. 18.11.1921 in München, Deutschland, gest. 22.11.1992 in Hamburg, Deutschland. – *Schrift-*

*ractères, fabricant de machines à écrire* – société fondée en 1892 à la suite de la fusion de 23 fonderies de caractères américaines. 1895, publication du premier catalogue de polices de l'ATF. 1904–1911, la police, appelée «Cheltenham», conçue par Bertram Goodhue pour la Cheltenham Press, est complétée et connaît un grand succès commercial. 1908, fondation d'une bibliothèque spécialisée en typographie alors dirigée par Henry Lewis Bullen. 1966, ATF est reprise par White Consolidated Industries. En 1970, ATF rachète Lanston Monotype. La Kingsley ATF se constitue à la suite de plusieurs fusions, depuis, elle produit des logiciels. 1993, liquidation de la société ATF. Parmi les concepteurs de polices ayant travaillé pour ATF, on compte F .W. Goudy, M. F. Benton, L. Bernhard, W. Bradley (qui a rédigé les articles, conçu la maquette et publié les 12 éditions du magazine «The Chap Book») et J. W. Phinney.

**Amman,** Jost – né en 1539 à Zurich, Suisse, décédé en 1591 à Nuremberg, Allemagne – *illustrateur, graveur sur bois et sur cuivre* – outre de nombreux signets (par ex. pour Sigmund Feyerabend et Jakob Sabon) ainsi que des alphabets figuratifs, Amman réalise en 1555, les illustrations du «Kriegsbuch» de Lienhard Fronsberger, en gravure sur bois. Son œuvre la plus célèbre est le «Ständebuch» sur des poèmes de Hans Sachs, édité par S. Feyerabend à Francfort-sur-le-Main en 1568, on y trouve des représentations et des explications sur les nombreux métiers de l'époque et les gestes liés à leurs activités.

**Andersch,** Martin – né le 18. 11. 1921 à Munich, Allemagne, décédé le 22. 11. 1992 à Hambourg, Allemagne – *écrivain, maquettiste pour l'édition, enseignant* – études d'art et de calligraphie à Munich et à Hambourg. 1962, enseigne la typo-

*Cheltenham Bold Italic*

72 Point 3A 5a
*Regular 9*
60 Point 4A 6a
*2 Complains*
48 Point 4A 6a
*Gigantic Sale 5*

*Cheltenham Bold Italic*

12 Point 15A 32a
*GUIDES DESIGN Sought suggestions wandering through picturesque valleys*
10 Point 16A 34a
*NOTED GROUNDS Children frolic midst wonderful botanical garden and enjoy the exhilarating climate*
8 Point 20A 42a
*COMPLIMENTS ARTIST Many entertaining exhibits of painting are displayed in the main room. The library habitue will be interested in these fascinating art works*
6 Point 20A 42a
*SOLDIERS IN MANEUVERS Numerous troops forming with beautiful accuracy along road passed inspections and started active work. Quickly establish permanent headquarters while zephyrs sharpen their appetite*

24 Point 7A 11a
*REOPEN City folks interested*
18 Point 9A 20a
*SPECIMEN Brochure by old artificer delights boy*
14 Point 14A 28a
*DISTRIBUTER Urchins pleased with many gifts donated weekly by the merchant*

42 Point 5A 7a
*3 MINERALS Specimen Gem*
36 Point 5A 8a
*MORTGAGES 1 Payable Monthly Legal Compound*
30 Point 6A 9a
*4 FINE EFFICIENT Advertisers Prospect Showy Organization*

[ 109 ]

American Type Founders Company 1936 Page

Amman 1568 Page

GOUDY HANDTOOLED
ITALIC
REGULARLY USED
IN DISPLAY
EFFECTIVELY IDENTIFIES
& CALLS ATTENTION
TO THE
MESSAGE OF THE
ADVERTISER

ABCDEFGHIJKLMNOPQRST
UVWXYZ
*abcdefghijklmnopqrstuvwxyz*
$1234567890

American Type Founders Company 1936 Advertisement

*foundry, manufacturers of type machines* – Founded in 1892 by a merger of 23 American type foundries. 1895: publication of the ATF's first book of type models. 1904–11: Cheltenham, a font designed by Bertram Goodhue for Cheltenham Press, is extended and becomes a commercial success. 1908: ATF builds up a typographical library under the supervision of Henry Lewis Bullen. 1966: ATF is bought by White Consolidated Industries. 1970: ATF buys Lanston Monotype. A further merger produces Kingsley ATF, who now make computer software. 1993: ATF is liquidated. Among the type designers who worked for ATF are: F. W. Goudy, M. F. Benton, L. Bernhard, W. Bradley (who wrote, designed and published the 12 issues of the magazine "The Chap Book"), and J. W. Phinney.

**Amman,** Jost – b. 1539 in Zurich, Switzerland, d. 1591 in Nuremberg, Germany – *illustrator of books, woodcut artist, copperplate engraver* – Besides numerous trademarks (e. g. for Sigmund Feyerabend and Jakob Sabon) and figure alphabets, in 1555 Amman finished the illustrations for Lienhard Fronsberger's "Kriegsbuch" as a woodcut. His most famous work is the "Ständebuch" in verse by Hans Sachs, which S. Feyerabend publishes in Frankfurt in 1568, where occupations of the time are explained and depicted in characteristic poses.

**Andersch,** Martin – b. 18.11.1921 in Munich, Germany, d. 22.11.1992 in Hamburg, Germany – *letterer, book designer, teacher* – Studied art and lettering in Munich and Hamburg. 1962: teaches letter-

*schreiber, Buchgestalter, Lehrer* – Studium der Kunst und der Schrift in München und Hamburg. 1962 Lehrer für Schrift und Buchkunst an der Werkkunstschule Hamburg, Professor am Fachbereich Gestaltung der Fachhochschule Hamburg. 1987 Emeritierung. Andersch gestaltet zehn Jahre das „Deutsche Allgemeine Sonntagsblatt". – *Publikationen:* „Spuren Zeichen Buchstaben", Ravensburg 1988; „Manual", Ravensburg 1991.

**Angeluschew,** Boris – geb. 25. 10. 1902 in Plovdiv, Bulgarien, gest. 24. 8. 1966 in Sofia, Bulgarien. – *Buchgestalter, Grafiker, Illustrator* – 1928 Studium an der Kunstakademie Berlin. Arbeitsbereiche Buch- und Schriftgestaltung, Plakat, Karikatur, Briefmarkenentwurf und Malerei.

**Anikst,** Michail A. – geb. 1938 in Moskau, Rußland. – *Typograph, Buchgestalter, Grafik-Designer, Illustrator* – 1962 Studium am Institut für Architektur Moskau. 1986 erhält er den Grand Prix der Biennale Brno, Tschechien.

**Antupit,** Samuel N. – *Grafik-Designer, Art-Director* – Studium an der Yale School of Art and Architecture (u. a. bei Alexej Brodovitch), Abschluß 1956. 1958–1960 Assistant Art Director der Zeitschrift „Harper's Bazaar". 1960–1962 Assistant Art Director der Zeitschrift „Show". 1962 Assistant Art Director beim Condé Nast Verlag, wo er Artikel für die Zeitschriften „Vogue", „Mademoiselle", „Glamour" und „House & Garden" gestaltet. 1963–1964 Arbeiten für das Push Pin Studio, Gestaltung der Zeitschriften „The New York Review of Books" und „Art in America". 1964–1968 Art Director des

Amman 1568 Page

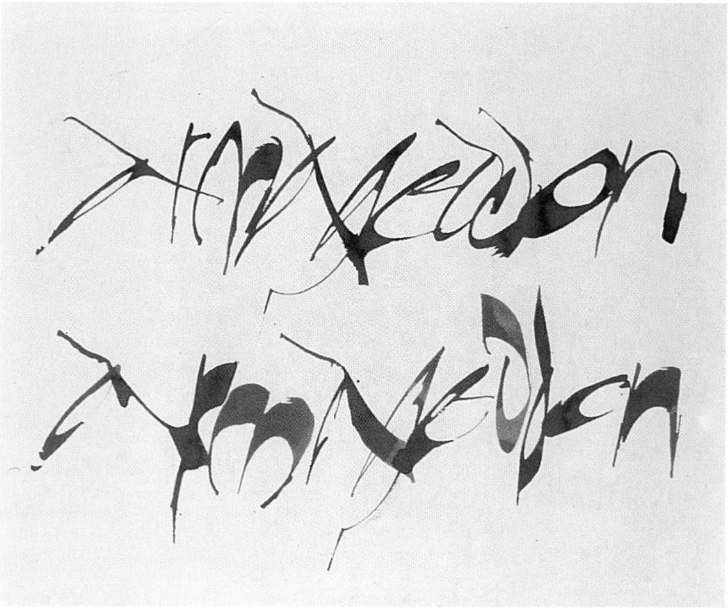

Andersch 1989 Calligraphy

Andersch 1988 Spread

Antupit 1985 Poster

graphie et les arts du livre à la Werkkunstschule de Hambourg (arts appliqués), professeur au département de design à la Fachhochschule de Hambourg. 1987, départ en retraite. Réalise les maquettes du «Deutsche Allgemeine Sonntagsblatt» pendant dix ans. – *Publications:* «Spuren Zeichen Buchstaben», Ravensburg 1988; «Manual», Ravensburg 1991.

**Angeluchev,** Boris – né le 25. 10. 1902 à Plovdiv, Bulgarie, décédé le 24. 8. 1966 à Sofia, Bulgarie – *maquettiste en édition, graphiste, illustrateur* – 1928, études à la Kunstakademie de Berlin (beaux-arts). Spécialisé dans les maquettes de livres et la conception de polices et d'affiches, caricaturiste, dessinateur de timbres-poste et peintre.

**Anikst,** Michaïl A. – né en 1938 à Moscou, Russie – *typographe, maquettiste en édition, graphiste maquettiste, illustrateur* – 1962, études à l'institut d'architecture de Moscou. 1986, Grand prix de la Biennale de Brno, Tchéquie.

**Antupit,** Samuel N. – *graphiste maquettiste, directeur artistique* – études à la Yale School of Art and Architecture (élève d'Alexeï Brodovitch). Diplôme en 1956. 1958–1960 Assistant Art Director de la revue «Harper's Bazaar». 1960–1962, Assistant Art Director de la revue «Show». 1962, Assistant Art Director chez les publications Condé Nast, où il met en page des articles pour les revues «Vogue», «Mademoiselle», «Glamour» et «House & Garden». 1963–1964, travaux pour le Push Pin Studio, maquette des revues

ing and book art at the Werkkunstschule in Hamburg, professor in the Department of Design at the Fachhochschule Hamburg. 1987: retires. Designs the "Deutsches Allgemeines Sonntagsblatt" for a period of ten years. – *Publications:* "Spuren Zeichen Buchstaben", Ravensburg 1988; "Manual", Ravensburg 1991.

**Angelushev,** Boris – b. 25.10.1902 in Plovdiv, Bulgaria, d. 24. 8. 1966 in Sofia, Bulgaria – *book designer, graphic artist, illustrator* – 1928: studies at the Kunstakademie in Berlin. Areas of work: book and type design, posters, caricatures, stamp design and painting.

**Anikst,** Mikhail A. – b. 1938 in Moscow, Russia – *typographer, book designer, graphic designer, illustrator* – 1962: studies at the Institute of Architecture in Moscow. 1986: receives the Grand Prix at the Brno Biennale, Czechoslovakia.

**Antupit,** Samuel N. – *graphic designer, art director* – Studied at the Yale School of Art and Architecture (under Alexei Brodovitch, among others), graduating in 1956. 1958–60: assistant art director of "Harper's Bazaar"magazine. 1960–62: assistant art director of "Show"magazine. 1962: assistant art director for Condé Nast Publications, where he designs articles for the magazines "Vogue", "Mademoiselle", "Glamour" and "House & Garden". 1963–64: works for the Push Pin Studio, does the layout for the magazines "The New York Review of Books"

„Esquire"-Magazins. 1968–1978 eigenes Studio „Antupit & Others Inc.". 1981–98 Design-Vizepräsident des Verlags Harry N. Abrams Inc., New York. Seit 1995 Präsident von CommonPlace Publishing.

**Apeloig,** Philippe – geb. 20. 11. 1962 in Paris, Frankreich. – *Grafik-Designer, Typograph, Lehrer* – 1982–85 Studium an der Ecole Nationale Supérieure des Arts Appliqués und an der Ecole Nationale Supérieure des Arts Décoratifs in Paris. 1985–87 Grafik-Designer des Musée d'Orsay in Paris. 1988 Arbeit bei April Greimann in Los Angeles. 1989 Gründung seines Studios in Paris. Auftraggeber u. a. Imprimerie Nationale, Opéra National, Musée du Louvre, Carré d'Art und die Verlage du Désastre und Odile Jacob. Seit 1992 unterrichtet er Typographie an der ENSAD, Paris. Zahlreiche Auszeichnungen und Ausstellungen, u. a. Galerie Impressions in Paris (1991), „Arc en Rêve" Architekturzentrum in Bordeaux (1992). 1993–94 Stipendiat des französischen Kulturministeriums zur Forschung über Typographie in der Villa Medici in Rom.

**Apollinaire,** Guillaume (d. i. Wilhelm Apollinaris de Kostrowitzki) – geb. 1880 in Rom, Italien, gest. 1918 in Paris, Frankreich. – *Schriftsteller* – 1898 Umzug nach Paris, Leben als Bohemien, Hauslehrer adeliger Familien. Verfasser und Herausgeber erotischer Literatur. Mitarbeit an Zeitungen und Literaturzeitschriften. Viele seiner Gedichte sind in einer freien, bildhaften Form geschrieben und gesetzt. 1904 Zusammentreffen mit Picasso in Paris. 1914 wird Apollinaire Soldat. 1916 schwere Verwundung im Ersten Welt-

«The New York Review of Books» et «Art in America». 1964–1968, Art Director du magazine «Esquire». 1968–1978, travaille dans son propre atelier «Antupit & Others Inc.». 1981–1998 Vice-président du design des éditions Harry N. Abrams Inc., New York. Depuis 1995 président de CommonPlace Publishing.

**Apeloig,** Philippe – né le 20. 11. 1962 à Paris, France – *graphiste maquettiste, typographe, enseignant* – 1982–1985, études à l'Ecole Nationale Supérieure des Arts Appliqués et à l'Ecole Nationale Supérieure des Arts Décoratifs de Paris. Graphiste maquettiste au Musée d'Orsay de 1985 à 1987. Travaille chez April Greiman à Los Angeles en 1988. Fonde son atelier à Paris en 1989. Commanditaires : Imprimerie Nationale, Opéra National, Musée du Louvre, Carré d'Art ainsi que les éditions du Désastre et Odile Jacob. Enseigne la typographie à l'ENSAD, Paris depuis 1992. Nombreuses distinctions et expositions, entre autres à la Galerie Impressions à Paris (1991), «Arc en Rêve», centre d'architecture de Bordeaux (1992). 1993–1994, bourse du Ministère de la Culture pour effectuer des recherches sur la Typographie à la Villa Médicis à Rome.

**Apollinaire,** Guillaume (Wilhelm Apollinaris de Kostrowitzki) – né en 1880 à Rome, Italie, décédé en 1918, Paris, France – *écrivain* – s'installe à Paris en 1898, vie de bohème, précepteur pour des familles d'aristocrates. Auteur et éditeur de littérature érotique. Collabore à des journaux et revues littéraires. Bon nombre de ses poèmes sont écrits en vers libres et composés en utilisant des formes suggestives. Rencontre Picasso à Paris en 1904. Part sous les drapeaux en 1914. Gravement blessé en 1916 pendant la Première Guerre mondiale. 1917, première de sa pièce de théâtre «Les mamelles de Tirésias» à Paris. – *Publications, sélection :*

Apeloig 1991 Poster

Apeloig 1991 Poster

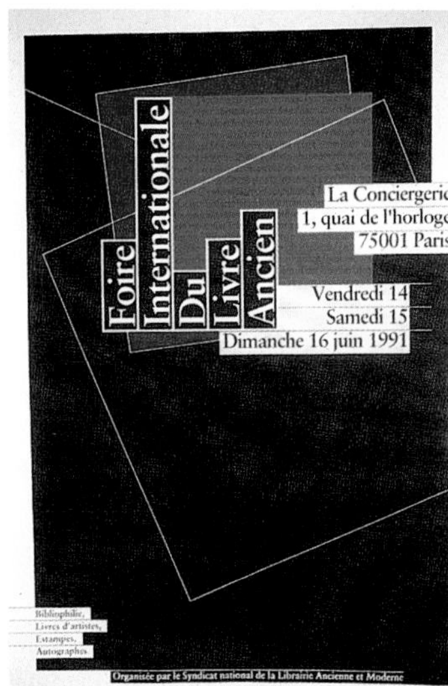

Apeloig 1992 Poster

Apeloig 1989 Poster

Apollinaire 1918 Calligram

and "Art in America". 1964–68: art director of "Esquire" magazine. 1968–78: opens his own studio, Antupit & Others Inc. 1981–98: vice-president of design at the publishing house Harry N. Abrams Inc., New York. From 1995 onwards: president of CommonPlace Publishing.

**Apeloig,** Philippe – b. 20.11.1962 in Paris, France – *graphic designer, typographer, teacher* – 1982–85 : studies at the Ecole Nationale Supérieure des Arts Appliqués and at the Ecole Nationale Supérieure des Arts Décoratifs in Paris. 1985–87 : graphic designer for the Musée d'Orsay in Paris. 1988 : works for April Greimann in Los Angeles. 1989 : opens a studio in Paris. Clients include the Imprimerie Nationale, Opéra National, Musée du Louvre, Carré d'Art and the publishers du Désastre and Odile Jacob. From 1992 onwards : teaches typography at the ENSAD, Paris. Numerous awards and exhibitions, including at the Galerie Impressions in Paris (1991), and at the Arc en Rêve architecture center in Bordeaux (1992). 1993–94 : receives a grant from the French Ministry for the Arts to carry out research on typography at the Villa Medici in Rome.

**Apollinaire,** Guillaume (real name: Wilhelm Apollinaris de Kostrowitzky) – b. 1880 in Rome, Italy, d. 1918 in Paris, France – *writer* – 1898: moves to Paris, lives as a bohemian; private tutor to noble families. Writer and publisher of erotic literature. Works on newspapers and magazines. Many of his poems are written and set in a free, illustrative form. 1904: meets Picasso in Paris. 1914: Apol-

krieg. 1917 Aufführung seines Theaterstücks „Les mamelles de Tirésias" in Paris. – *Publikationen u. a.:* „Les onze mille verges", 1907; „Alcools", 1913; „Antitradition futuriste", 1913; „Les peintres cubistes", 1913; „Le poète assassiné", 1916; „Calligrammes", 1918. R. Taupin, L. Zukofsky „Le style Apollinaire", Paris 1934; C. Giedion-Welcker „Die neue Realität bei Apollinaire", Zürich 1945.

**Arndt,** Alfred – geb. 26. 11. 1898 in Elbing, Deutschland (heute Polen), gest. 7. 10. 1976 in Darmstadt, Deutschland. – *Gebrauchsgrafiker, Architekt* – Zeichenlehre in einem Bau- und Konstruktionsbüro in Elbing. 1920–21 Studium an der Akademie Königsberg, Ostpreußen. 1921–26 Studium am Bauhaus in Weimar und Dessau, Kurse bei J. Itten, P. Klee, W. Kandinsky und O. Schlemmer. 1926–29 freier Architekt in Thüringen. 1929–32 Lehrer am Bauhaus, Leiter der Ausbauabteilung (Metall- und Möbelwerkstatt, Wandmalerei). 1933 Reklamearbeiten und Entwürfe für Industriebauten in Thüringen. 1945–1948 Baurat in Jena. Ab 1948 freier Architekt und Maler in Darmstadt. Arndts Arbeiten sind Beispiele für die frühe Phase der konstruktivistischen, experimentellen Typographieauffassung am Bauhaus. – *Publikation u. a.:* H. M. Wingler „Alfred Arndt. Maler und Architekt", Darmstadt 1968.

**Arp,** Hans – geb. 16. 9. 1886 in Straßburg, Frankreich, gest. 7. 6. 1966 in Basel, Schweiz. – *Maler, Dichter, Bildhauer* – 1900–1908 Studien an der Ecole des Arts et Métiers in Straßburg, Akademie der Schönen Künste, Weimar, Académie Ju-

Arndt 1923 Signet

Arndt 1927 Menu

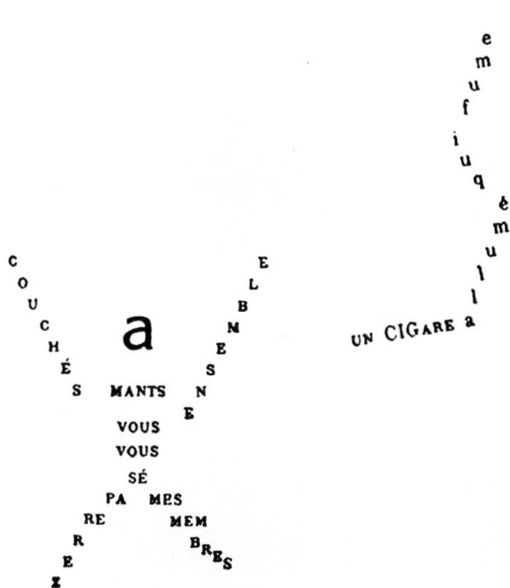

Apollinaire 1918 Calligram

Arndt 1927 Poster

«Les onze mille verges», 1907; «Alcools», 1913; «Antitradition futuriste», 1913; «Les peintres cubistes», 1913; «Le poète assassiné», 1916; «Calligrammes», 1918. R. Taupin, L. Zukofsky «Le style Apollinaire», Paris 1934; C. Giedion Welcker «Die neue Realität bei Apollinaire», Zurich 1945.

**Arndt,** Alfred – né le 26. 11. 1898 à Elbing, Allemagne (aujourd'hui en Pologne), décédé le 7. 10. 1976 à Darmstadt, Allemagne – *graphiste, architecte* – apprentissage du dessin dans une entreprise de bâtiment à Elbing. 1920–1921, études à l'académie de Königsberg, Prusse orientale. 1921–1926 études au Bauhaus de Weimar et de Dessau où il suit les cours de J. Itten, P. Klee, W. Kandinsky et O. Schlemmer. 1926–1929, architecte indépendant en Thuringe. 1929–1932, enseignant au Bauhaus, où il dirige l'atelier d'aménagement intérieur (métal, menuiserie, peinture murale). En 1933, il réalise des publicités et des maquettes pour des bâtiments industriels en Thuringe. 1945–1948, conseiller à l'urbanisme à Iéna. A partir de 1948, architecte indépendant et peintre à Darmstadt. Les travaux de Arndt illustrent la période où le Bauhaus prônait une typographie constructiviste et expérimentale. – *Publication, sélection:* H. M. Wingler «Alfred Arndt. Maler und Architekt», Darmstadt 1968.

**Arp,** Jean – né le 16. 9. 1886 à Strasbourg, France, décédé le 7. 6. 1966 à Bâle, Suisse – *peintre, poète, sculpteur* – 1900–1908, études à l'Ecole des Arts et Métiers

linaire becomes a soldier. 1916: badly wounded in the First World War. 1917: his play, "Les mamelles de Tirésias", is performed in Paris. – *Publications include:* "Les onze mille verges", 1907; "Alcools", 1913; "Antitradition futuriste", 1913; "Les peintres cubistes", 1913; "Le poète assassiné", 1916; "Calligrammes", 1918. R. Taupin, L. Zukofsky "Le style Apollinaire", Paris 1934; C. Giedion-Welcker "Die neue Realität bei Apollinaire", Zurich 1945.

**Arndt,** Alfred – b. 26.11.1898 in Elbing, Germany (today Poland), d. 7. 10. 1976 in Darmstadt, Germany – *commercial artist, architect* – Studied art in a construction and drawing office in Elbing. 1920–21: studies at the Akademie Königsberg, East Prussia. 1921–26: studies at the Bauhaus in Weimar and Dessau, attending courses taught by J. Itten, P. Klee, W. Kandinsky and O. Schlemmer. 1926–29: freelance architect in Thuringia. 1929–32: teacher at the Bauhaus, head of the interior design department (metal and furniture workshop, murals). 1933: advertising work and designs for industrial buildings in Thuringia. 1945–48: director of the planning department and building control office in Jena. From 1948 onwards: freelance architect and painter in Darmstadt. Arndt's work is an example of the early Constructivist, experimental approach to typography at the Bauhaus. – *Publications include:* H. M. Wingler "Alfred Arndt. Maler und Architekt", Darmstadt 1968.

**Arp,** Hans – b. 16. 9. 1886 in Strasbourg, France, d. 7. 6. 1966 in Basle, Switzerland – *painter, poet, sculptor* – 1900–08: stud-

lian in Paris. 1912 Mitarbeit am Almanach „Der Blaue Reiter" von Wassily Kandinsky. 1915 Umzug nach Zürich. 1916 Mitbegründer der Dada-Bewegung und des Cabaret Voltaire in Zürich. Arps anti-klassischer Literaturstil läßt eine experimentelle Auffassung von Typographie entstehen. 1925 Umzug nach Paris. 1927 erste Einzelausstellung in der Galerie Surréaliste Paris. 1930 Teilnahme an der Ausstellung „Cercle et Carré". Eintritt in die Gruppe „abstraction-création" in Paris. 1937 Eintritt in die Gruppe „Allianz". 1942 Flucht in die Schweiz. 1954 Auszeichnung mit dem internationalen Preis für Skulptur der Biennale in Venedig. 1962 Retrospektivausstellung im Musée National d'Art Moderne in Paris. – *Publikationen u. a.:* „Der Pyramidenrock", Erlenbach 1924; „Die Kunst-ismen" (mit El Lissitzky), Zürich, München, Leipzig 1925; „On my way. Poetry and Essays 1912–47", New York 1948; „Unseren täglichen Traum. Erinnerungen 1914–54", Zürich 1955; „Gesammelte Gedichte 1, 2, 3", Zürich, Wiesbaden 1963–84. Carola Giedion-Welcker „Hans Arp", Stuttgart 1957; W. F. Arntz „Das grafische Werk", Haag 1980; Bernd Rau „Hans Arp. Die Reliefs", Stuttgart 1981; Aimée Bleikasten „Bibliographie 1 & 2", London 1981–1983.

**Arpke,** Otto – geb. 16. 10. 1886 in Braunschweig, Deutschland, gest. 4. 12. 1943 in Berlin, Deutschland. – *Grafiker, Illustrator, Maler, Lehrer* – 1900–05 Lithographenlehre, Besuch der Kunstgewerbeschule Hannover. 1905–07 Dekorateur und Maler in Düsseldorf. 1907–08 Mal-

appliqués de Strasbourg, Akademie der Schönen Künste de Weimar (beaux-arts), Académie Julian à Paris. 1912, collaboration à l'almanach «Der Blaue Reiter» (Le Cavalier bleu) de Vassili Kandinsky. S'installe à Zurich en 1915. 1916, cofondateur du mouvement Dada et du Cabaret Voltaire, tous deux à Zurich. Son style littéraire anti-classique permet la naissance d'une typographie expérimentale. S'installe à Paris en 1925. 1927, première exposition personnelle à la Galerie Surréaliste, à Paris. 1930, participation à l'exposition «Cercle et Carré». Entre dans le groupe «abstraction-création» à Paris. 1937, adhère au groupe «Allianz». 1942, s'enfuit en Suisse. 1954, obtient le prix international de sculpture à la Biennale de Venise. 1962, exposition rétrospective au Musée National d'Art Moderne de Paris. – *Publications, sélection :* «Der Pyramidenrock», Erlenbach 1924; «Die Kunst-ismen» (avec El Lissitzky), Zurich, Munich, Leipzig 1925; «On my way. Poetry and Essays 1912–1947», New York 1948; «Unseren täglichen Traum. Erinnerungen 1914–1954», Zurich 1955; «Gesammelte Gedichte 1,2,3», Zurich, Wiesbaden 1963–1984. Carola Giedion-Welcker «Hans Arp», Stuttgart 1957; W. F. Arntz «Das grafische Werk», Haag 1980; Bernd Rau «Hans Arp. Die Reliefs», Stuttgart 1981; Aimée Bleikasten «Bibliographie 1 & 2», Londres 1981–1983.

**Arpke,** Otto – né le 16. 10. 1886 à Braunschweig, Allemagne, décédé le 4. 12. 1943 à Berlin, Allemagne – *graphiste, illustrateur, peintre, enseignant* – 1900–1905, apprentissage de lithographe, fréquente la Kunstgewerbeschule de Hanovre (école des arts décoratifs). 1905–1907, décorateur et peintre à Düsseldorf. 1907–1908, étudie la peinture à Bruxelles. 1911–1912, graphiste et peintre à Hanovre. 1912, s'installe à Berlin dans son

Arp 1929 Poster

Arpke 1928 Signet

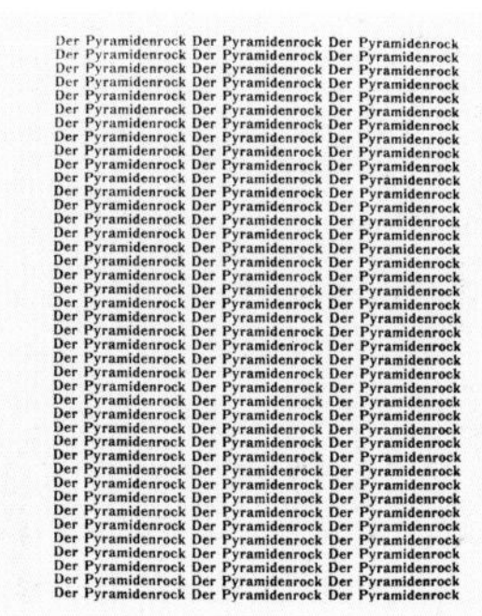

Arp 1924 Cover

Arpke 1928 Poster

ies at the Ecole des Arts et Métiers in Strasbourg, Akademie der Schönen Künste in Weimar and the Académie Julian in Paris. 1912: works on the almanac "Der blaue Reiter" by Wassily Kandinsky. 1915: moves to Zurich. 1916: one of the founders of the Dada movement and Cabaret Voltaire in Zurich. An experimental approach to typography emerges through Arp's anti-classical literary style. 1925: moves to Paris. 1927: first solo exhibition in the Galerie Surréaliste in Paris. 1930: participates in the exhibition "Cercle et Carré". Joins the abstraction-création group in Paris. 1937: joins the group "Allianz". 1942: escapes to Switzerland. 1954: is awarded the international prize for sculpture at the Venice Biennale. 1962: a retrospective exhibition is held in the Musée National d'Art Moderne in Paris. – *Publications:* "Der Pyramidenrock", Erlenbach 1924; "Die Kunst-ismen" (with El Lissitzky), Zurich, Munich, Leipzig 1925; "On my way. Poetry and Essays 1912–47", New York 1948; "Unseren täglichen Traum. Erinnerungen 1914–54", Zurich 1955; "Gesammelte Gedichte 1, 2, 3", Zurich, Wiesbaden 1963–84. Carola Giedion-Welcker "Hans Arp", Stuttgart 1957; W.F. Arntz "Das grafische Werk", Haag 1980; Bernd Rau "Hans Arp. Die Reliefs", Stuttgart 1981; Aimée Bleikasten "Bibliographie 1 & 2", London 1981–83.

**Arpke,** Otto – b. 16.10.1886 in Brunswick, Germany, d. 4.12.1943 in Berlin, Germany – *graphic artist, illustrator, painter, teacher* – 1900–05: studies lithography and attends the Kunstgewerbeschule in

studien in Brüssel. 1911–12 Grafiker und Maler in Hannover. 1912 Umzug nach Berlin, eigenes Atelier. 1919–24 Atelier Stahl-Arpke, Entwürfe für die Filmindustrie (z. B. 1920 Plakat für den Film „Das Kabinett des Dr. Caligari"). 1920 Beteiligung an der Gründung des Bundes Deutscher Gebrauchsgraphiker (BDG). 1928 Berufung als Professor an die Kunst- und Gewerbeschule Mainz. 1928–33 Arbeiten für die Kostümbälle „Im Reich der Reklame" in Berlin. 1929–42 Umschlagsentwürfe für die Zeitschrift „Die neue Linie". 1932–39 Plakatentwürfe für die Schifffahrtslinie „Norddeutscher Lloyd Bremen". 1933 Umzug von Mainz nach Berlin. 1935–43 Leiter der Abteilung Gebrauchsgrafik an der Meisterschule für Buchgewerbe und Grafik Berlin. 1936 Ausstattung des Prospekts für die Olympiade Berlin. Buchillustrationen.

**Arrighi,** Ludovico degli – geb. ca. 1475 in Cornedo Vicentino, Italien, gest. 1527 in Rom, Italien. – *Kalligraph, Drucker, Schriftentwerfer* – Als erster Kalligraph druckt Arrighi Muster seiner Schreibschriften in den zwei Schreibmeisterbüchern „La Operina" (1522) und „Il modo de temperare le penne" (1523). 1524 eröffnet er zusammen mit Lautitius Perugino eine kleine Druckerei, in der Ausgaben zeitgenössischer Autoren erscheinen, die in Arrighis Kursivschriften gesetzt sind.

**Asaba,** Katsumi – geb. 1940 in Yokohama, Japan. – *Grafik-Designer* – 1958 Abschluß des Design-Studiums an der Kanagawa High School of Technology. Eintritt in die Werbeabteilung des Kaufhau-

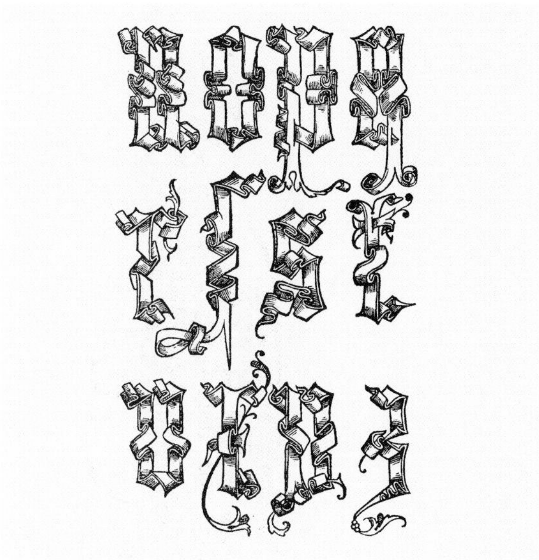

Arrighi 1523

Asaba 1981

ばらばりばるばればばろ
びらびりびるびれびびろ
ぶらぶりぶるぶれぶぶろ
べらべりべるべれべべろ
ぼらぼりぼるぼれぼぼろ

Asaba 1985 Poster

Asaba 1992 Poster

propre atelier. 1919–1924, atelier Stahl-Arpke, maquettes pour l'industrie cinématographique (par ex. 1920, affiche du film «Le cabinet du Docteur Galigari»). 1920, participe à la fondation du «Bund Deutscher Gebrauchsgraphiker» (BDG). 1928, nommé professeur à la Kunst- und Gewerbeschule de Mayence. 1928–1933, travaux pour les bals costumés «Im Reich der Reklame» à Berlin. 1929–1942, maquettes de couvertures pour la revue «Die neue Linie». 1932–1939, maquettes d'affiches pour la compagnie maritime «Norddeutscher Lloyd Bremen». 1933, quitte Mayence pour Berlin. 1935–1943, directeur du département de graphisme industriel à la Meisterschule für Buchgewerbe und Grafik de Berlin (école de graphisme et des arts du livre). 1936, maquette du prospectus des Jeux Olympiques de Berlin. Illustrations de livres.

**Arrighi,** Ludovico degli – né vers 1475 à Cornedo Vicentino, Italie, décédé en 1527 à Rome, Italie – *calligraphe, imprimeur, concepteur de polices* – il est le premier calligraphe à avoir imprimé des modèles de calligraphie dans deux manuels destinés aux scribes, «La Operina» (1522) et «Il modo de temperare le penne» (1523). En 1524, il ouvre une petite imprimerie avec Lautitius Perugino et publie des éditions d'auteurs contemporains composées en italique Arrighi.

**Asaba,** Katsumi – né en 1940 à Yokohama, Japon – *graphiste maquettiste* – 1958, termine ses études de design à la Kanagawa High School of Technology.

Hanover. 1905–07: interior designer and painter in Düsseldorf. 1907– 08: studies painting in Brussels. 1911–12: graphic artist and painter in Hanover. 1912: moves to Berlin and sets up his own studio. 1919–24: runs the Stahl-Arpke studio and designs for the film industry (including a poster for the film "Das Kabinett des Dr. Caligari" in 1920). 1920: helps to found the Bund Deutscher Gebrauchsgraphiker (Association of German Commercial Artists). 1928: professorship at the Kunst- und Gewerbeschule in Mainz. 1928–33: works for the "Im Reich der Reklame" fancy-dress balls in Berlin. 1929–42: designs the covers of the magazine "Die neue Linie". 1932–39: designs the posters for the "Norddeutscher Lloyd Bremen" shipping line. 1933: moves from Mainz to Berlin. 1935–43: head of the department of commercial art at the Meisterschule für Buchgewerbe und Grafik in Berlin. 1936: production of the brochure for the Olympics in Berlin. Book illustrations.

**Arrighi,** Ludovico degli – b. c. 1475 in Cornedo Vicentino, Italy, d. 1527 in Rome, Italy – *calligrapher, printer, type designer* – Arrighi was the first calligrapher to print examples of his script, in the two writing manuals "La Operina" (1522) and "Il modo de temperare le penne" (1523). 1524: he opens a small printery with Lautitius Perugino, where contemporary authors are published in Arrighi's italic type.

**Asaba,** Katsumi – b. 1940 in Yokohama, Japan – *graphic designer* – 1958: graduates in design from the Kanagawa High School of Technology. Joins the publicity

ses Hamatsu. 1960 Studium am Kuwazama Institute, 1960–64 am Institut für Buchdruck von Keinosuke. 1964 Art Director der Light Publicity Werbeagentur. 1975 Eröffnung eines eigenen Design-Studios. Zahlreiche Arbeiten für das Kaufhaus Seibu. 1989 Entwurf für die Europalia Japan in Antwerpen.

**Aschoff,** Eva – geb. 26. 4. 1900 in Göttingen, Deutschland, gest. 22. 9. 1969 in Freiburg, Breisgau, Deutschland. – *Buchkünstlerin, Kalligraphin* – 1921–23 Schriftstudium bei F. H. E. Schneidler an der Stuttgarter Kunstakademie. Danach Ausbildung zur Buchbinderin in München und Hamburg. 1928–64 Buchbinderwerkstatt in Freiburg. 1952–63 Buchumschläge für den Insel Verlag, den Hanser Verlag und den Fischer Verlag. 1964 Arbeit an Schriftblättern und Landschaftsbildern. – *Publikationen:* Jan Kröger „Eva Aschoff – eine ganz stille Künstlerin", in „Die deutsche Schrift" 85/1954. Viviane Engelmann „Eva Aschoff" (2 Bände, Magisterarbeit), Freiburg 1993.

**Ashbee,** Charles Robert – geb. 17. 5. 1863 in Isleworth, England, gest. 1942 in Kent, England. – *Schriftentwerfer, Drucker, Architekt, Formgeber* – Studium am Kings College Cambridge. 1883–1885 Lehre als Architekt. 1888 Gründung der „Guild of Handicraft", einer Vereinigung zur Hebung des Niveaus des Kunstgewerbes. Nach dem Tod von William Morris (1896) kauft Ashbee zwei Druckmaschinen und weiteres Zubehör aus den Beständen der Kelmscott Press und eröffnet 1898 seine „Essex House Press". Der Großherzog von Hessen ruft Ashbee 1899 für den Bau und

Entre dans le service de publicité des grands magasins Hamatsu. 1960, études au Kuwazama Institute. 1960–1964, institut d'édition de Keinosuke. 1964, Art director de l'agence publicitaire Light Publicity. 1975, ouvre son propre atelier de design, nombreux travaux pour les magasins Seibu. 1989, maquette de l'Europalia Japan à Anvers.

**Aschoff,** Eva – née le 26. 4. 1900 à Göttingen, Allemagne, décédée le 22. 9. 1969 à Fribourg-en-Brisgau, Allemagne – *artiste du livre, calligraphe* – 1921–1923, étude de la calligraphie chez F. H. E. Schneidler à la Kunstakademie de Stuttgart (beaux-arts). Puis formation de reliure à Munich et à Hambourg. 1928–1964, atelier de reliure à Fribourg. 1952–1963, couvertures de livres pour les maison d'édition Insel, Hanser et Fischer. 1964, travaux de lettrisme et réalisations de paysages. – *Publications:* Jan Kröger «Eva Aschoff – eine ganz stille Künstlerin» dans «Die deutsche Schrift» 85/1954; Viviane Engelmann «Eva Aschoff» (2 vol., mémoire de maîtrise), Fribourg 1993.

**Ashbee,** Charles Robert – né le 17. 5. 1863 à Isleworth, Angleterre, décédé en 1942 à Kent, Angleterre – *concepteur de polices, imprimeur, architecte, créateur de formes* – études au Kings College de Cambridge. 1883–1885, apprentissage d'architecte. En 1888, il fonde la «Guild of Handicraft», une association visant à relever le niveau de l'artisanat. A la mort de William Morris (1896), Ashbee achète deux presses d'imprimerie et des accessoires du stock de la Kelmscott Press; il ouvre la «Essex House Press» en 1898. En 1899, le grand-duc de Hesse invite Ashbee en Allemagne pour qu'il y réalise des travaux de construction et des aménagements intérieurs au château de Darmstadt. Partisan du mouvement Arts & Crafts, Ashbee prônait une revalorisation

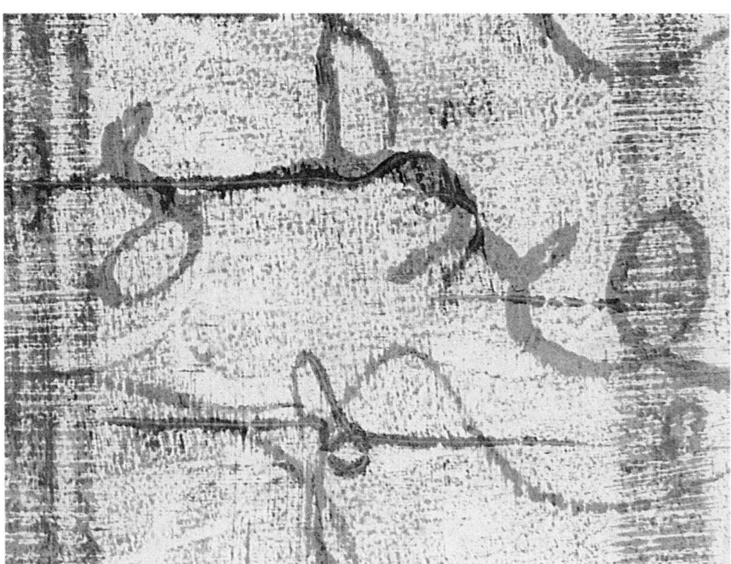

Aschoff

Aschoff

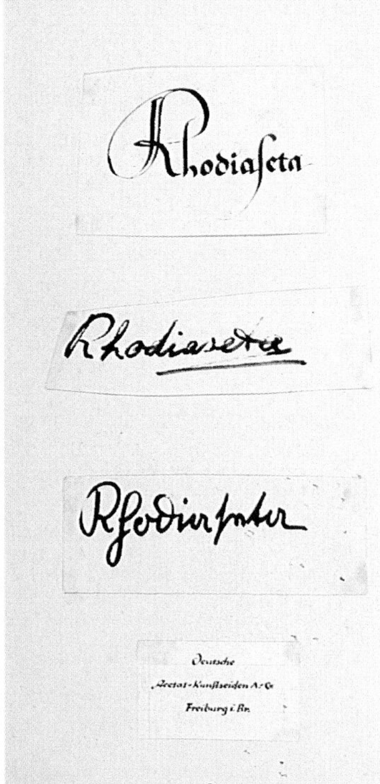


Aschoff ca. 1930 Sketch

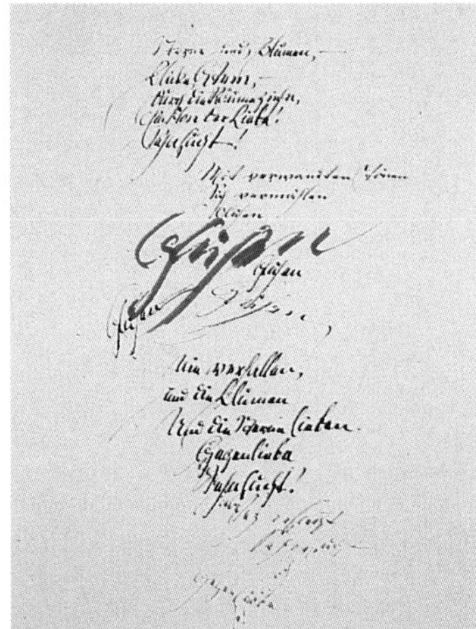

Aschoff 1985 Cover

department of the Hamatsu department store. 1960: studies at the Kuwazama Institute. 1960–64: Institute for Book Printing in Keinosuke. 1964: art director of the Light Publicity advertising agency. 1975: opens his own design studio. Carries out work for the Seibu department store. 1989: works on the design for the Europalia Japan in Antwerp.

**Aschoff,** Eva – b. 26. 4. 1900 in Göttingen, Germany, d. 22. 9. 1969 in Freiburg, Breisgau, Germany – *book artist, calligrapher* – 1921–23: studies lettering with F. H. E. Schneidler at the Stuttgarter Kunstakademie. Then trains as a bookbinder in Munich and Hamburg. 1928–64: bookbinder workshop in Freiburg. 1952–63: dust jackets for the publishers Insel, Hanser and Fischer. 1964: works on original lettering designs and landscape paintings. – *Publications:* Jan Kröger "Eva Aschoff – eine ganz stille Künstlerin" in "Die deutsche Schrift" 85/1954; Viviane Engelmann "Eva Aschoff" (2 vols., MA thesis), Freiburg 1993.

**Ashbee,** Charles Robert – b. 17. 5. 1863 in Isleworth, England, d. 1942 in Kent, England – *type designer, printer, architect, designer* – Studied at King's College, Cambridge. 1883–85: trains as an architect. 1888: founds the Guild of Handicraft, an organization which aims to improve the standard of handicrafts. After the death of William Morris (1896), Ashbee buys up two printing presses and various other equipment from the Kelmscott Press and in 1898 opens his Essex House Press. The Grand Duke of Hesse invites Ashbee to Germany for the build-

die Inneneinrichtung des Darmstädter Schlosses nach Deutschland. Ashbee tritt als Vertreter der „Arts and Crafts"-Bewegung für eine Neubewertung gestalterischer Form im 20. Jahrhundert auf. 1902 Verlegung der Presse nach Chipping Campden. 1915 Professor für englische Literatur in Kairo, Ägypten. 1919–23 Zivilberater der englischen Verwaltung in Palästina. – *Schriftentwürfe:* Endeavour Type (1901), Prayer Book Type (1903). – *Publikationen:* „A Few Chapters on Workshop Reconstruction and Citizenship", London 1894; „An Endeavour towards the Teachings of J. Ruskin and W. Morris", London 1901; „Craftsmenship in Competitive Industry", London 1908; „The Private Press. A Study in Idealism", London 1909; „Should we Stop Teaching Art?", London 1911; „Where the Great City Stands", London 1917. Nikolaus Pevsner „William Morris, C. R. Ashbee und das 20. Jahrhundert" in „Deutsche Vierteljahresschrift für Literaturwissenschaft und Geistesgeschichte" 14/1936.

**Association Typographique Internationale** (ATypI) – 1957 Gründung durch Charles Peignot als Vereinigung von Schriftentwerfern, -herstellern und -anwendern. Ziele der Vereinigung sind die Förderung der Typographie und der internationale Schutz der Schriften. Organisation internationaler Kongresse, die jeweils in anderen Städten der Welt unter wechselnden Themenschwerpunkten mit renommierten Referenten stattfinden.

**Atelier National de Création Typographique** (ANCT), Paris. – 1984 Gründung des Ateliers in den Räumen der Impri-

I. THE FIRST BEGINNINGS OF THE GUILD AND SCHOOL OF HANDICRAFT IN EAST LONDON.

ERHAPS AS THE GUILD of Handicraft has now been in existence for some twelve years, it may not be inexpedient to give some account of it in writing. No record, either of its work & progress, or of the ideas upon which it was founded, has ever yet been given, & I am anxious in reply to the many requests received at Essex House to offer something that may meet this want.
To those who were acquainted with the beginnings of the Guild, the various educational undertakings connected with it between the years 1888 and 1894, the first workshop on the top of a warehouse in Commercial Street, & the classes for the study of design & the reading of Ruskin, carried on by me at Toynbee Hall in the years 1886 and 1887, it will be known that the Guild as originally conceived was the productive side of an undertaking that was also educational. From the volume of the Transactions of the Guild and School of Handicraft published in 1890, the following may be quoted as of interest by comparison with what has been accomplished since that date, and what we at present have in prospect.
"The Guild and School had its beginnings in the years 1886–87 in a small Ruskin class, conducted at Toynbee Hall, and composed of three
b

Ashbee 1901 Page

4. Atypi-Seminar 1981

ATypI 1981 Logo

ATypI | TypeLab

ATypI 1996 Logo

des formes pour le 20e siècle. 1902, la presse s'installe à Chipping Campden. 1915, professeur de littérature anglaise au Caire, Egypte. 1919–1923, conseiller civil de l'administration anglaise en Palestine. – *Polices :* Endeavour Type (1901), Prayer Book Type (1903). – *Publications :* «A Few Chapters on Workshop Reconstruction and Citizenship», Londres 1894; «An Endeavour towards the Teachings of J. Ruskin and W. Morris», Londres 1901; «Craftmenship in Competitive Industry», Londres 1908; «The Private Press. A Study in Idealism», Londres 1909; «Should we Stop Teaching Art?», Londres 1911; «Where the Great City Stands», Londres 1917. Nicolas Pevsner «William Morris, C. R. Ashbee und das 20. Jahrhundert» dans «Deutsche Vierteljahresschrift für Literaturwissenschaft und Geistesgeschichte» 14/1936.

**Association Typographique Internationale** (ATypI) – fondée en 1957 par Charles Peignot comme association de concepteurs, producteurs et utilisateurs de caractères d'imprimerie. Le but de cet organisme est de soutenir la typographie et de protéger les droits sur les polices. Organise des congrès internationaux qui ont lieu dans diverses villes du monde sur des thèmes variés et avec la participation de personnalités célèbres.

**Atelier National de Création Typographique** (ANCT), Paris – atelier fondé en

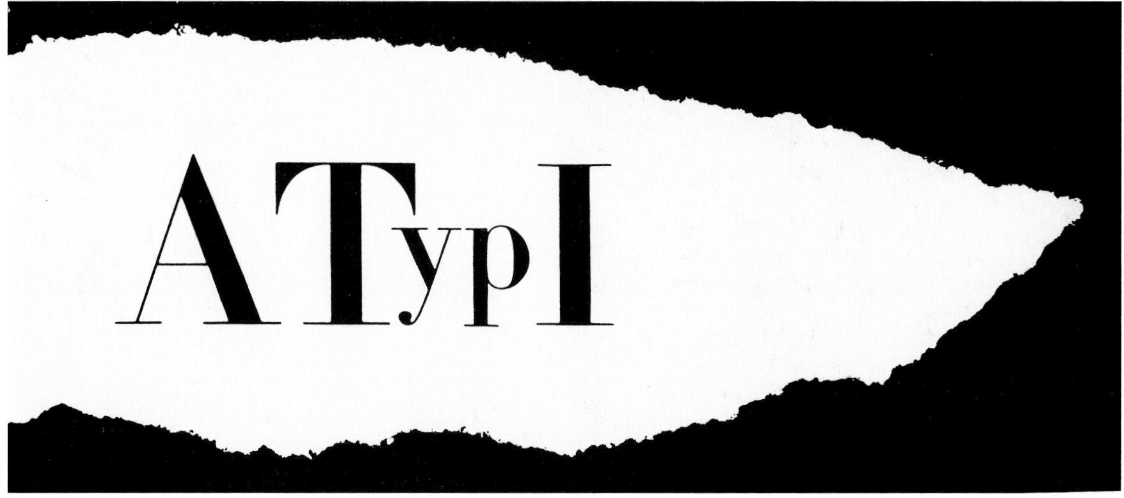


ATypI Logo

ing and interior design of his palace in Darmstadt. Acting as a representative of the Arts and Crafts Movement, Ashbee campaigned for a re-assessment of artistic form in the twentieth century. 1902: the press is moved to Chipping Campden. 1915: professor of English literature in Cairo, Egypt. 1919–23: civilian advisor to the British administration in Palestine. – *Fonts:* Endeavour Type (1901), Prayer Book Type (1903). – *Publications:* "A Few Chapters on Workshop Reconstruction and Citizenship", London 1894; "An Endeavour towards the Teachings of J. Ruskin and W. Morris", London 1901; "Craftsmenship in Competitive Industry", London 1908; "The Private Press. A Study in Idealism", London 1909; "Should we Stop Teaching Art?", London 1911; "Where the Great City Stands", London 1917. Nikolaus Pevsner "William Morris, C. R. Ashbee und das 20. Jahrhundert" in "Deutsche Vierteljahresschrift für Literaturwissenschaft und Geistesgeschichte" 14/1936.

**Association Typographique Internationale** (ATypI) – 1957: founded by Charles Peignot as an association of type designers, manufacturers and users. The aims of the association are to promote typography and to protect type and typefaces at an international level. Organizes international conventions – which always take place in different cities of the world with changing topics and famous speakers.

**Atelier National de Création Typographique** (ANCT), Paris – 1984: the Atelier is founded on the premises of the

Atelier National de Création Typographique
**Auriol**
**Austin**

merie Nationale zur Förderung der Typographie. Es entstand ein Bildungszentrum mit experimentellem Charakter zur Ausbildung auf den Gebieten Schrift, Typographie und Grafik-Design.
**Auriol,** Georges (d. i. Jean Georges Huyot) – geb. 1863 in Beauvais, Frankreich, gest. 1938 in Paris, Frankreich. – *Schriftentwerfer, Grafiker, Maler, Schriftsteller* – 1883 Eintritt in die Redaktion der Zeitschrift „Chat noir", dem Organ des Cabaret „Chat noir" in Paris. Auf Anfrage von Georges Peignot, dem Leiter der Schriftgießerei Peignot & Sons, entwirft Auriol 1900 mehrere Alphabete. Freundschaft mit Th.-A. Steinlen und H. de Toulouse-Lautrec. – *Schriftentwürfe:* Auriol (1901–04), La Française, Champlève, Le clair de lune, Le Robur.
**Austin,** Richard – geb. ca. 1768 in London, England, gest. 1830 in London, England. – *Stempelschneider, Holzstecher, Schriftgießer* – Holzstich-Lehre bei Thomas Bewick. 1788 tätig als Stempelschneider in „The British Letter Foundry" von John Bell, nach dem Verkauf der Schriftgießerei an Simon Stephenson bis 1796 für diese Firma tätig. Danach Stempelschneider für die schottischen Schriftgießereien A. Wilson und W. Miller. 1819 eröffnet Austin seine eigene Schriftgießerei „The Imperial Letter Foundry" in London. – *Schriftentwürfe:* Tooled Roman (1788), Bell (1788), Fry's Ornamented (1796).

1984 dans les locaux de l'Imprimerie Nationale pour promouvoir la typographie. Création d'un centre d'apprentissage utilisant des polices expérimentales pour assurer une formation dans les domaines touchant les caractères d'imprimerie, la typographie et la maquette.
**Auriol,** Georges (Jean Georges Huyot) – né en 1863 à Beauvais, France, décédé en 1938 à Paris, France – *concepteur de polices, graphiste, peintre, écrivain* – 1883, entre à la rédaction de la revue «Chat noir», l'organe du Cabaret du même nom à Paris. 1900, crée plusieurs alphabets à la demande de Georges Peignot, directeur de la fonderie de caractères Peignot & Sons. Ami de Th.-A. Steinlen et de H. de Toulouse-Lautrec. – *Polices:* Auriol (1901–1904), La Française, Champlève, Le clair de lune, Le Robur.
**Austin,** Richard – né vers 1768 à Londres, Angleterre, décédé en 1830 à Londres, Angleterre – *tailleur de types graveur sur bois, fondeur de caractères* – apprentissage de la gravure sur bois chez Thomas Bewick. 1788, travaille comme fondeur de caractères à la «British Letter Foundry» de John Bell. Reste dans cette entreprise jusqu'en 1796 après la vente de la fonderie de caractères à Simon Stephenson. Exerce ensuite comme tailleur de types à la fonderie de caractères écossaise de A. Wilson et W. Miller. En 1819, Austin ouvre sa propre fonderie de caractères «The Imperial Letter Foundry» à Londres. – *Polices:* Tooled Roman (1788), Bell (1788), Fry's Ornamented (1796).

**La production de l'Atelier**
**Projects and works commissioned**

**Qu'il s'agisse d'exercices pratiques collectifs ou de programmes de recherche personnalisés, un travail positif est accompli en permanence au sein de l'Atelier. Au cours des dernières années, les élèves ont ainsi réalisé, sous la conduite de leurs professeurs, l'adaptation de caractères classiques (Romain du Roi, Perrin) ou de caractères faisant partie de la collection protégée de l'Imprimerie nationale (Anisson, Gauthier).**

**Au titre de la commande publique, ils ont conçu et mis au point, dans les mêmes conditions, un caractère de texte et de signalisation pour le Ministère de la Culture dans le cadre de la célébration du Bicentenaire de la Révolution française et un caractère d'édition pour l'Imprimerie nationale. Des stagiaires ont créé leur propre caractère, en participant avec succès à des concours internationaux ou à des compétitions organisées par des associations culturelles.**

**Divers projets sont actuellement à l'étude. Ils répondent à des demandes formulées par des administrations publiques ou des éditeurs de publications professionnelles. A noter également un programme de développement de signalétique routière, des recherches prenant en compte les nouvelles normes européennes de télétexte et l'étude d'un caractère de consultation de haute densité, toujours dans l'esprit de la commande publique.**

Whether it consists of collective practical exercises or personal research programs, the work conducted at the Institute is thorough.
In previous years, and under the guidance of their directors, students explored variations on classic typefaces (Romain du Roi, Perrin) and typefaces included in collections protected by the Imprimerie nationale (Anisson, Gauthier). Under similar circumstances, students conceived and refined a typeface for text and signage systems commissioned by the Ministry of Culture, which was designed for the celebration of the Bicentenary of the French Revolution. Another typeface was designed for editorial purposes at the Imprimerie nationale. Students worked on their individual new typefaces, and successfully participated in international competitions organized by cultural associations.

At present, several different projects are in the works. These projects function in response to requests made by public administrations or publishers. Among other works, students are developing a road signage system and a typeface for teletext, in which the new european telecommunication norms must be considered.

**Mais il est évident que l'impact réel de l'ANCT ne peut être jugé sur ces seules prestations effectuées par les stagiaires dans le cadre de leur programme de formation: Il faut le chercher dans la qualité du travail que les anciens stagiaires produisent ensuite dans leurs différentes affectations professionnelles à travers le monde et, par delà la diversité des talents, dans la communauté de pensée qu'ils continuent à entretenir pour le plus grand rayonnement de l'école française de typographie.** It is evident that the real impact of the Institute is based upon two main factors: the quality of achievement at the Institute and the level of professionalism attained by students once they have moved forward in their careers.
Through the diversity of their talents and their flexibility with a visual language, it is expected that students will utilize their valuable knowledge of French typography while participating in a creative community anywhere in the world.

abcdefghijklmnopqr
stuvwxyz« ¨ ´ ` ˆ ' »–-
çß!?†#–+=%/*:,;.
ABCDEFGHIJKLMN
OPQRSTUVWXYZÇ
&£¥$§[()]
1234567890 1234567890

**Caractère Bi-89**
**pour le Bicentenaire de la**
**Revolution française**

Atelier Nationale de Création Typographique 1992 Spread

| | n | **n** | **n** | | *n* |
|---|---|---|---|---|---|
| a | b | e | f | g | i |
| o | r | s | t | y | z |
| A | B | C | E | G | H |
| M | O | R | S | X | Y |
| 1 | 2 | 4 | 6 | 8 | & |

Auriol 1901–04 Auriol

| | n | **n** | | | *n* |
|---|---|---|---|---|---|
| a | b | e | f | g | i |
| o | r | s | t | y | z |
| A | B | C | E | G | H |
| M | O | R | S | X | Y |
| 1 | 2 | 4 | 6 | 8 | & |

Austin 1788 Bell

Imprimerie Nationale to promote typography. An education center with a view to experiment is set up to teach type, typography and graphic design.
**Auriol,** Georges (real name: Jean Georges Huyot) – b. 1863 in Beauvais, France, d. 1938 in Paris, France – *type designer, graphic artist, painter, author* – 1883: joins the editorial staff of the magazine "Chat Noir", the literary organ of the cabaret of the same name in Paris. 1900: Auriol designs several alphabets at the request of Georges Peignot, director of the type foundry Peignot & Sons. Friends with Th.-A. Steinlen and H. de Toulouse-Lautrec. – *Fonts:* Auriol (1901–04), La Française, Champlève, Le clair de lune, Le Robur.
**Austin,** Richard – b. c. 1768 in London, England, d. 1830 in London, England – *punch cutter, woodcut artist, type founder* – Trained as a wood engraver with Thomas Bewick. 1788: works as a punch cutter in John Bell's British Letter Foundry. After the type foundry is sold to Simon Stephenson, he works for Stevenson's company until 1796. Then works as a punch cutter for the Scottish type foundry A. Wilson and W. Miller. 1819: Austin opens his own type foundry, The Imperial Letter Foundry, in London. – *Fonts:* Tooled Roman (1788), Bell (1788), Fry's Ornamented (1796).

# *B*

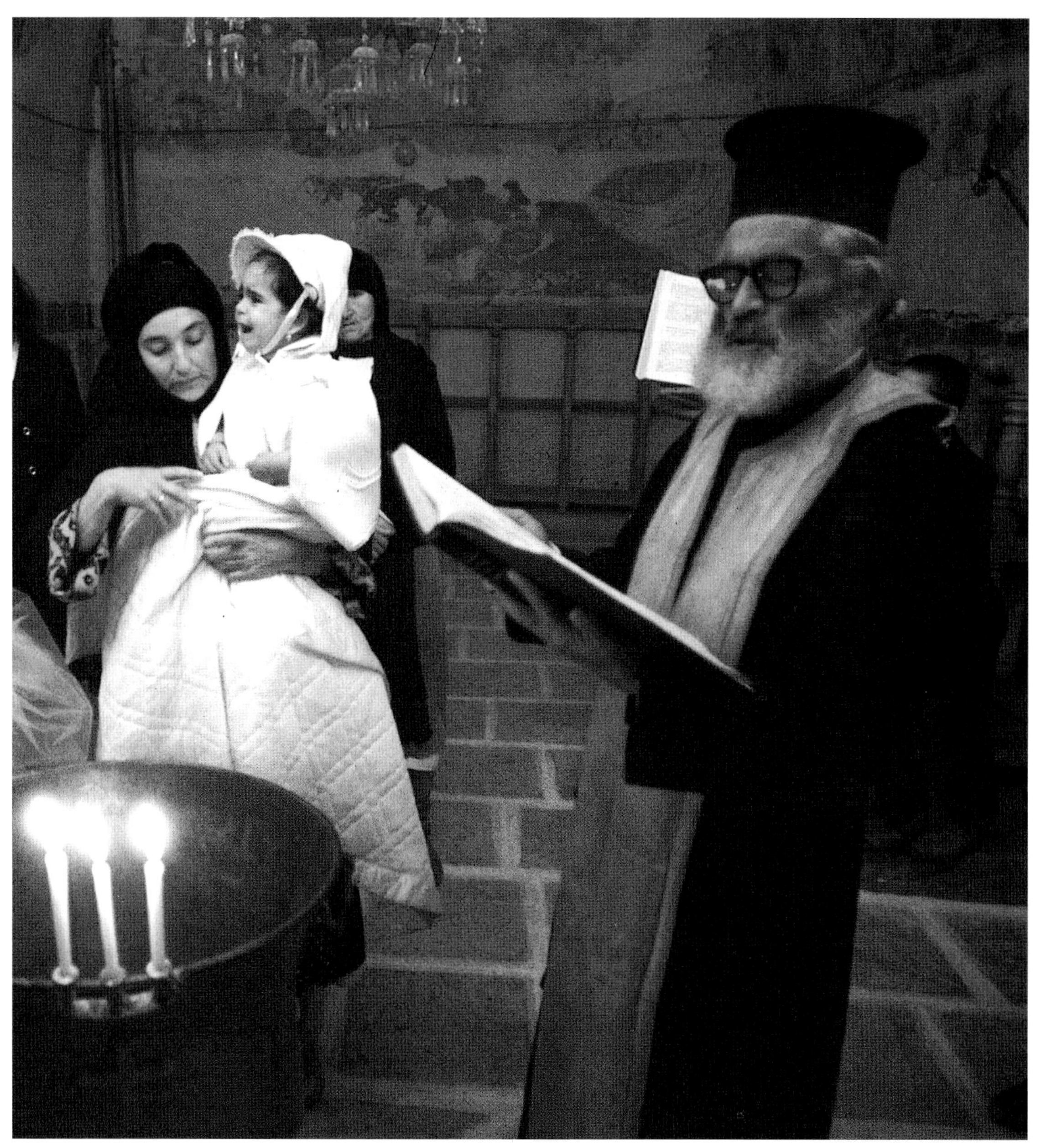

**Baines**
**Baldessari**
**Balla**

**Baines,** Phil – geb. 8. 12. 1958 in Kendal, Westmorland, England. – *Typograph, Schriftentwerfer, Lehrer* – 1985 Abschluß seiner Studien an der St. Martins School of Art, London. 1985–87 Studium am Royal College of Art. Arbeiten als freier Grafik-Designer für Verlage, Kunstorganisationen und das Fernsehen. Beiträge für die digitale Zeitschrift „Fuse" 1989, 1993 und 1995. Seit 1991 unterrichtet Baines Typographie am Central Saint Martins College of Art & Design, London. Die amerikanische Zeitschrift „Details" verwendet 1994 exklusiv seine Schrift „Toulon". – *Schriftentwürfe:* Can You (1989), Ushaw (1993), Toulon (1994), Horncastle (1994), (FF) You Can (Read me) (1991). – *Publikation:* „SToneUTTERS", London 1992.
**Baldessari,** John – geb. 17. 6. 1931 in National City, Kalifornien, USA. – *Maler, Lehrer* – 1949–57 Studium am State College, San Diego. 1953–54 Leiter der Fine Arts Gallery, San Diego. 1962–70 Lehrtätigkeit an der University of California, San Diego. 1971 Lehrtätigkeit am Hunter College, New York. In den siebziger Jahren Hinwendung zu narrativer Kunst mit Fotos und erläuternden Texten.
**Balla,** Giacomo – geb. 18. 7. 1871 in Turin, Italien, gest. 1. 3. 1958 in Rom, Italien. – *Maler, Gestalter* – Studium an der Accademia Albertina, Turin. 1895 Umzug nach Rom. 1910 Mitunterzeichner des Futuristischen Manifests. 1912 erste futuristische Arbeiten. 1915 Verfasser des Manifests „Die futuristische Neukonstruktion des Weltalls" mit Fortunato Depero. In seinen abstrakten futuristischen Gemälden

**Baines,** Phil – né le 8. 12. 1958 à Kendal, Westmorland, Angleterre – *typographe, concepteur de polices, enseignant* – termine ses études à la St. Martins School of Art de Londres en 1985. Etudie au Royal College of Art de 1985 à 1987. Travaille comme graphiste maquettiste indépendant pour des maisons d'édition, des institutions artistiques et la télévision. Ecrit pour la revue digitale «Fuse» en 1989, 1993 et 1995. Baines enseigne la typographie au Central Saint Martins College of Art & Design de Londres depuis 1991. En 1994, la revue américaine «Details» utilise en exclusivité sa police «Toulon». – *Polices:* Can You (1989), Ushaw (1993), Toulon (1994), Horncastle (1994), (FF) You Can (Read me) (1991). – *Publication:* «SToneUTTERS», Londres 1992.
**Baldessari,** John – né le 17. 6. 1931 à National City, Californie, USA – *peintre, enseignant* – 1949–1957, études au State College, San Diego. 1953–1954, directeur de la Fine Art Gallery, San Diego. 1962–1970, enseignant à la University of California, San Diego. 1971, enseignant au Hunter College, New York. Pendant les années 70, il se consacre à l'art narratif avec utilisation de photos et de textes explicatifs.
**Balla,** Giacomo – né le 18. 7. 1871 à Turin, Italie, décédé le 1. 3. 1958 à Rome, Italie – *peintre, créateur* – études à l'Accademia Albertina de Turin. S'installe à Rome en 1895. Cosignataire en 1910 du manifeste futuriste et premiers travaux futuristes en 1912. Rédige le «Manifeste de la reconstruction futuriste de l'univers» avec Fortunato Depero en 1915. Utilise de nombreux éléments typographiques dynamiques dans des peintures futuristes abstraites. Nombreuses maquettes d'affiches scripturales. Après 1925, se consacre aux arts appliqués et crée des motifs pour tissus, des vêtements, du mo-

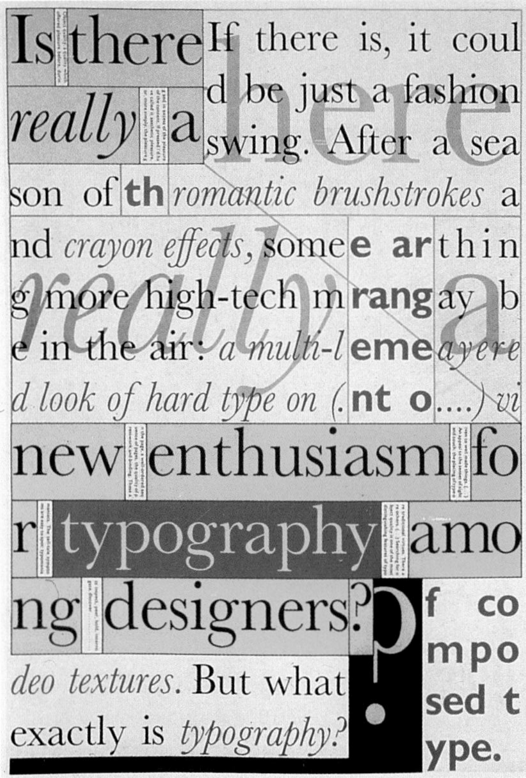


Baines 1989 Illustration

Baines 1985 Cover

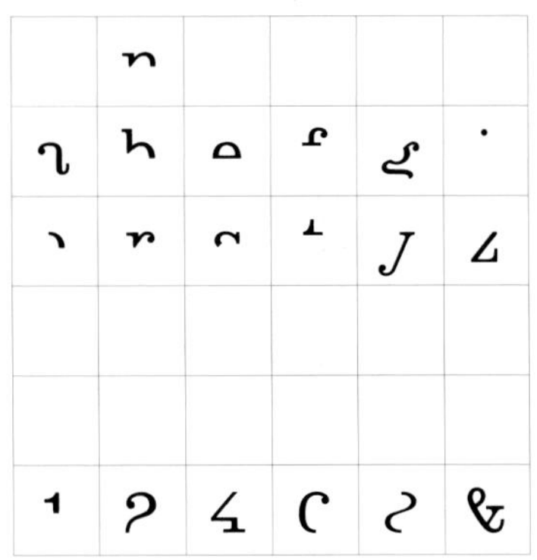

Baines 1991 You Can

Balla 1914 Sketch

**Baines,** Phil – b. 8. 12. 1958 in Kendal, Westmorland, England – *typographer, type designer, teacher* – 1985: finishes his studies at St. Martins School of Art, London. 1985–87: studies at the Royal College of Art. Works as a freelance graphic designer for publishing houses, art organizations and television. Contributes to the digital magazine "Fuse" in 1989, 1993 and 1995. Baines has taught typography at Central Saint Martins College of Art & Design in London since 1991. 1994: the American magazine "Details" exclusively uses his font "Toulon". – *Fonts:* Can You (1989), Ushaw (1993), Toulon (1994), Horncastle (1994), (FF) You Can (Read me) (1991). – *Publication:* "SToneUTTERS", London 1992.
**Baldessari,** John – b. 17. 6. 1931 in National City, California, USA – *painter, teacher* – 1949–57: studies at State College, San Diego. 1953–54: director of the Fine Arts Gallery, San Diego. 1962–70: teaches at the University of California, San Diego. 1971: teaches at Hunter College, New York. In the seventies, turns to narrative art with photos and explanatory texts.
**Balla,** Giacomo – b. 18. 7. 1871 in Turin, Italy, d. 1. 3. 1958 in Rome, Italy – *painter, designer* – Studied at the Accademia Albertina in Turin. 1895: moves to Rome. 1910: signs the Futurist manifesto and in 1912 produces his first Futurist work. 1915: draws up the manifesto "The Futurist Reconstruction of the Universe" with Fortunato Depero. He uses numerous typographic elements in dynamic array in his abstract, Futurist paintings. Numerous scriptural poster designs. After

verwendet er zahlreiche typographische Elemente in dynamischer Anordnung. Zahlreiche skripturale Plakatentwürfe. Nach 1925 Hinwendung zur angewandten Gestaltung: Stoffe, Kleider, Möbel. Teilweise Rückkehr zu figurativer Malerei. Zahlreiche internationale Ausstellungen. – *Publikationen u. a.:* A. Barricelli „Balla", Rom 1967; V. Dortch Dorazio „Balla – an Album of his Life and Works", Venedig, New York 1970; G. Lista „Giacomo Balla Futuriste", Lausanne 1984; M. Fagiolo „Balla, Il Futuristo", Mailand 1987.

**Ballmer,** Theo (Auguste Théophile) – geb. 29. 9. 1902 in Basel, Schweiz, gest. 10. 12. 1965 in Basel, Schweiz. – *Grafiker, Fotograf, Lehrer* – 1920 Lehre als Chromo-Lithograph bei Wassermann in Basel. 1926–28 Grafiker in der Firma Hoffmann-La Roche. 1928–30 Bauhaus Dessau. 1930 Rückkehr in die Schweiz, eigenes Atelier für Grafik und Fotografie. 1931 Leitung der Fachklasse Gebrauchsgrafik an der Allgemeinen Gewerbeschule Basel bis zu seinem Tod 1965. Unterrichtet historische und neue Schriften, Grafik und Fotografie. Gestaltung zahlreicher Plakate gegen Krieg und Faschismus, Plakate für Ausstellungen (u. a. „Neues Bauen"; „Büro 1928"), Ausstellungsgestaltung („Schweizer Landesausstellung", Zürich 1939), Entwurf konstruktiver Schriften und experimentelle Fotografie. Zahlreiche Besprechungen und Ausstellungen seiner Plakate der zwanziger und dreißiger Jahre.

**Ballmer,** Walter – geb. 22. 7. 1923 in Liestal, Schweiz. – *Grafik-Designer, Typograph, Maler, Lehrer* – 1940–1944

Theo Ballmer 1928 Poster

Theo Ballmer 1928 Poster

Theo Ballmer 1929 Sketch

bilier. Retour à la peinture figurative. Nombreuses expositions internationales. – *Publications, sélection:* A. Barricelli «Balla», Rome 1967; V. Dortch Dorazio «Balla – an Album of his Life and Works», Venise, New York 1970; G. Lista «Giacomo Balla Futuriste», Lausanne 1984; M. Fagiolo «Balla, Il Futuristo», Milan 1987.

**Ballmer,** Theo (Auguste Théophile) – né le 29. 9. 1902 à Bâle, Suisse, décédé le 10. 12. 1965 à Bâle, Suisse – *graphiste maquettiste, photographe, enseignant* – 1920, apprentissage de la chromo-lithographie chez Wassermann à Bâle. 1926–1928, graphiste chez Hoffmann-La Roche. 1928–1930, Bauhaus de Dessau. 1930, retour en Suisse, exerce dans son propre atelier de graphiste et de photographe. 1931, responsable du cours de dessin publicitaire à la Allgemeine Gewerbeschule de Bâle jusqu'à sa mort en 1965. Enseigne les caractères anciens et modernes, les arts graphiques et la photographie. Dessinateur de nombreuses affiches contre la guerre et le fascisme et d'affiches d'expositions (par ex. «Neues Bauen»; «Büro 1928»), concepteur d'exposition («Schweizer Landesausstellung», Zurich 1939) et de polices constructives, pratique la photographie expérimentale. Nombreuses expositions et entretiens sur ses affiches des années vingt et trente.

**Ballmer,** Walter – né le 22. 7. 1923 à Liestal, Suisse – *graphiste maquettiste, typographe, peintre, enseignant* – 1940–1944,

1925: turns to applied design for materials, clothes and furniture. Demonstrates a partial return to figurative painting. Has been featured in many international exhibitions. – *Publications include:* A. Barricelli "Balla", Rome 1967; V. Dortch Dorazio "Balla – an Album of his Life and Works", Venice, New York 1970; G. Lista "Giacomo Balla Futuriste", Lausanne 1984; M. Fagiolo "Balla, Il Futuristo", Milan 1987.

**Ballmer,** Theo (Auguste Théophile) – b. 29. 9. 1902 in Basle, Switzerland, d. 10. 12. 1965 in Basle, Switzerland – *graphic artist, photographer, teacher* – 1920: studies chromo-lithography at Wassermann in Basle. 1926–28: graphic artist for the company Hoffmann-La Roche. 1928–30: at the Bauhaus, Dessau. 1930: returns to Switzerland and sets up his own studio as graphic artist and photographer. 1931: heads the senior course in commercial art at the Allgemeine Gewerbeschule in Basle until his death in 1965. Teaches historical and new types, graphic art and photography. Designs numerous posters against war and fascism as well as exhibition posters (including "Neues Bauen"; "Büro 1928"). Designs exhibitions ("Schweizer Landesausstellung", Zurich 1939) and constructive types, works on experimental photography. His posters from the twenties and thirties have been the subject of much discussion and have featured in numerous exhibitions.

**Ballmer,** Walter – b. 22. 7. 1923 in Liestal, Switzerland – *graphic designer, typographer, painter, teacher* – 1940–44: trains

Ausbildung an der Kunstgewerbeschule Basel. 1946 als Grafik-Designer im Studio Boggeri, Mailand, tätig. Freiberuflich tätig, u. a. für Pirelli, Ciba-Geigy, Olivetti und Hoffmann-La Roche. Ab 1956 bis 1981 ausschließlich für Olivetti tätig. Ab 1970 zunehmend künstlerische Arbeiten in konstruktivistischem Stil. 1970 Re-Design des Olivetti-Logos, wofür er 1973 mit einer Goldmedaille in Ljubljana, Jugoslawien ausgezeichnet wird. 1977 Lehrer am Mailänder Politechnikum. 1981 Gründer und Leiter des Studios für Visuelle Kommunikation „Unidesign". Internationale Ausstellungen des freien und angewandten Werks. – *Publikationen u. a.:* in „Art and Graphics", Zürich 1983; in „High Quality" 5/1985, Heidelberg; „Un designer tra arte e grafica", Mailand 1989.

**Bando,** Takaaki – geb. 1957 in Tokushima, Japan. – *Typograph, Grafik-Designer, Lehrer* – 1983 Graduierung an der Musashino Art University, Tokio. 1983–1985 Studium an der Seihokei Corporation Tokio bei dem Grafik-Designer Etsushi Kiyohara. 1985 Gründung der Takaaki Bando Design Inc. in Tokushima. 1990 Beteiligung an der Entwicklung des neuen Design-Magazins „Evolution". Gestalter und Herausgeber von „The International Corporate Identity 1990" im Robundo Verlag Tokio, sowie des 2. Bandes 1992. 1992 Lehrtätigkeit an der Musashino Art University Tokio, 1993 Fachhochschule für Gestaltung Schwäbisch Gmünd; Tokushima University, 1994 Anabuki College Tokushima; Hochschule Bremen. 1995 Gewinner des Kieler Woche-Wettbewerbs für 1996.

Kunstgewerbeschule de Bâle (école des arts décoratifs). 1946, exerce comme maquettiste à l'atelier Boggeri à Milan. Travaille ensuite comme graphiste indépendant et réalise des commandes pour Pirelli, Ciba-Geigy, Olivetti et Hoffmann-La Roche etc. Travaille exclusivement pour Olivetti de 1956 jusqu'en 1981. A partir de 1970, se consacre de plus en plus à son œuvre artistique de style constructiviste. 1970, refonte du logo Olivetti, primé en 1973 d'une médaille d'or à Ljubljana, Yougoslavie.1977, enseignant à l'école polytechnique de Milan. 1981, fondateur et directeur de l'atelier de communication visuelle «Unidesign». Expositions internationales d'arts libres et d'arts appliqués. – *Publications, sélection:* dans «Art and Graphics», Zurich 1983; dans «High Quality»5/1985, Heidelberg; «Un designer tra arte e grafica», Milan 1989.

**Bando,** Takaaki – né en 1957 à Tokushima, Japon – *typographe, graphiste maquettiste, enseignant* – 1983, diplôme à la Musashino Art University, Tokyo. 1983–1985, études à la Seihokei Corporation Tokyo avec le graphiste Etsushi Kiyohara. 1985, Takaaki Bando fonde la Design Inc. à Tokushima. 1990, participe à l'élaboration d' «Evolution», un nouveau magazine de design. Concepteur et éditeur de «The International Corporate Identity 1990» pour les éditions Robundo à Tokyo ainsi que du deuxième volume consacré à l'année 1992. 1992, enseignant à la Musashino Art University de Tokyo; 1993, à la Fachhochschule für Gestaltung (école de design) à Schwäbisch Gmünd; à la Tokushima University; 1994, à l'Anabuki College de Tokushima, à l'université de Brême. 1995, lauréat du concours des Semaines de Kiel pour l'année 1996. – *Publications, sélection:* «Vertical speculation. Mies van der Rohe and his Barcelona Pavillon» dans «Evolution»

Walter Ballmer 1969 Cover

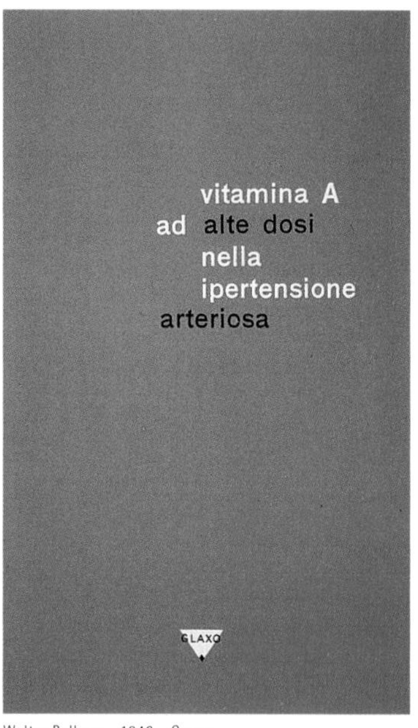

Walter Ballmer 1946 Cover

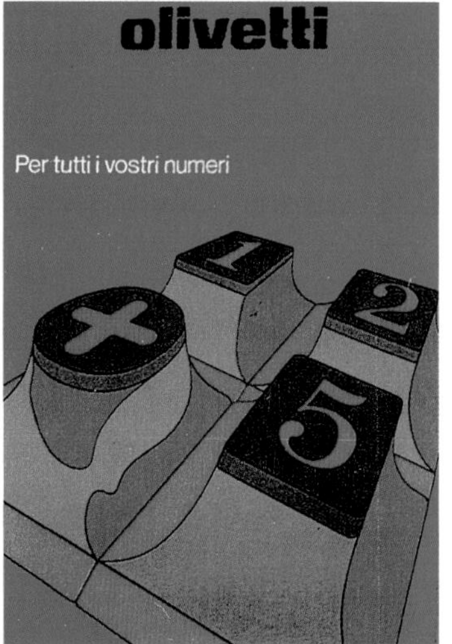

Walter Ballmer 1971 Poster

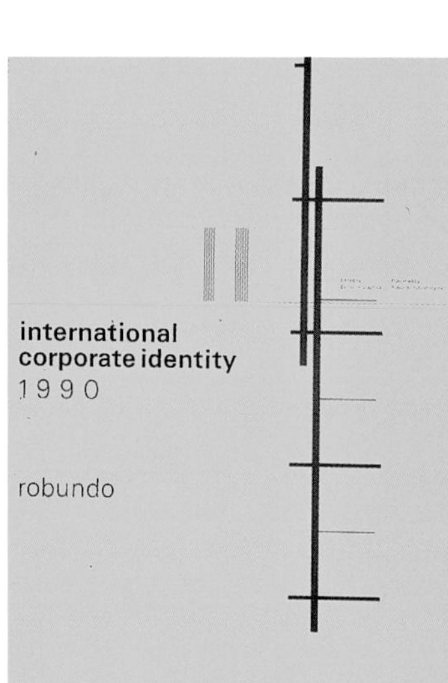

Bando 1990 Cover

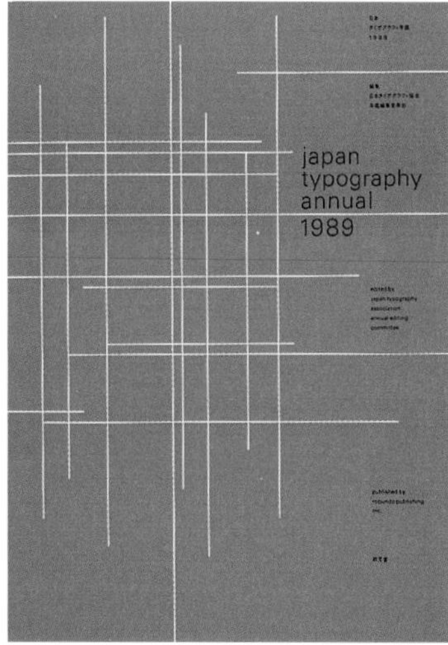

Bando 1989 Cover

at the Kunstgewerbeschule in Basle. 1946: works as a graphic designer for Studio Boggeri in Milan. Freelances for Pirelli, Ciba-Geigy, Olivetti and Hoffmann-La Roche, among others. From 1956–81: works exclusively for Olivetti. Since 1970: increase in artistic work in the Constructivist style. 1970: redesigns the Olivetti logo, for which he is awarded a gold medal in Ljubljana, Yugoslavia in 1973. 1977: teaches at Milan Polytechnic. 1981: founds and manages Unidesign, a studio for visual communication. Exhibitions of his free and applied work world-wide. – *Publications include:* in "Art and Graphics", Zurich 1983; in "High Quality" 5/1985, Heidelberg; "Un designer tra arte e grafica", Milan 1989.

**Bando,** Takaaki – b. 1957 in Tokushima, Japan – *typographer, graphic designer, teacher* – 1983: graduates from the Musashino Art University, Tokyo. 1983–85: studies at the Seihokei Corporation in Tokyo under the graphic designer Etsushi Kiyohara. 1985: founds Takaaki Bando Design Inc. in Tokushima. 1990: participates in the development of the new designer magazine "Evolution". Designs and publishes "The International Corporate Identity 1990" with Robundo publishers in Tokyo, and the second volume thereof in 1992. 1992: teaches at the Musashino Art University in Tokyo, in 1993 at the Fachhochschule für Gestaltung in Schwäbisch Gmünd and at Tokushima University and in 1994 at Anabuki College, Tokushima and the Hochschule in Bremen. 1995: wins the Kieler Woche competition for 1996. –

– *Publikationen u. a.:* „Vertical speculation. Mies van der Rohe and his Barcelona Pavillon" in „Evolution" 1/1990, Tokio; „Takaaki Bando" in „High Quality" 20/1991, Heidelberg; „Takaaki Bando" in „Graphis" 294/1994, New York.

**Banks,** Colin – geb. 16. 1. 1932 in London, England. – *Typograph, Schriftentwerfer, Design-Consultant* – 1948–53 Ausbildung am Mid-Kent College of Printing. Danach Assistent bei Ernst Hoch. Ab 1956 freischaffender Designer. 1958 Gründung des Büros Banks & Miles, London. 1975–82 Büro Banks & Miles in Amsterdam. 1979 Banks erneuert die Hausschrift der London Transport, die „Johnston" von 1916, für das neue Erscheinungsbild des Transportunternehmens. 1990 Gründung von Banks & Miles in Brüssel. 1990–92 Büro Banks & Miles in Hamburg. Design-Consultant für Universitäten und Hochschulen: 1991 Portsmouth University, UNO University Tokio und Helsinki, 1992 University of the South-West, Bristol. Arbeiten für Erscheinungsbilder, u. a. für die British Telecom, den London Transport, The British Council, The Post Office. – *Publikationen u. a.:* „Social Communication", London 1979; „London's Handwriting – a History of Design in London's Transport System", London 1994; „The Development of Edward Johnston's Underground Railway Block-Letter", London 1995.

**Barakow,** Wassil – geb. 4. 9. 1902 in Kazanlik, Bulgarien. – *Buchgestalter, Gebrauchsgrafiker, Maler* – 1937 Studium an der Kunstakademie Sofia, Bulgarien bei Zeno Todorov und Nikola Ganuschev.

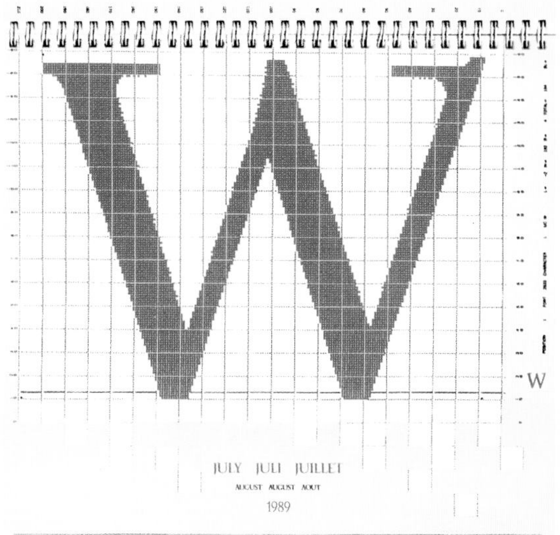

abcdefghijklmnopqrstuvwxyz
ABCDEFGHIJKLMNOPQRSTU
VWXYZÆŒÐŁØ&
1234567890($¢£ƒ%*@)
World Typeface Center Favrile Tom Carnase
JULY JULI JUILLET
1989
10 11 12 13 14 15 16
17 18 19 20 21 22 23
24 25 26 27 28 29 30
31 1 2 3 4 5 6


Banks 1989 Calendar

Banks 1989 Logo British Telecom

Banks 1989 Cover

АБВГДЕЖЗИЙКЛМН
ОПРСТУФХЦЧШЩЪ
ЬЮЯ ЖѢЫЭ
абвгдежзийклмнопрс
тфхцчшщъьюя
[(№1234567890!?VI*%„)]

Barakow 1970 Renaissance Antiqua

1/1990, Tokyo; «Takaaki Bando» dans «High Quality» 20/1991, Heidelberg; «Takaaki Bando» dans «Graphis» 294/1994, New York.

**Banks,** Colin – né le 16. 1. 1932 à Londres, Angleterre – *typographe, concepteur de polices, consultant en design* – 1948–1953, formation au Mid-Kent College of Printing. Puis assistant d'Ernst Hoch. Designer indépendant à partir de 1956. 1958, fondation de l'étude Bank & Miles, Londres. 1975–1982, étude Banks & Miles à Amsterdam. En 1979, Banks refond la «Johnston», la police maison des transports londoniens qui datait de 1916, pour donner une nouvelle image de marque à l'entreprise. 1990, fondation de Banks & Miles à Bruxelles. 1990–1992, agence Banks & Miles à Hambourg. Consultant en design pour diverses universités et écoles supérieures: 1991, Portsmouth University; 1991, UNO University à Tokyo et Helsinki; 1992, University of the South-West, Bristol. Travaux sur l'identité de marque de plusieurs sociétés, entre autres pour British Telecom, London Transport, The British Council, The Post Office. – *Publications, sélection:* «Social Communication», Londres 1979; «London's Handwriting – a History of Design in London's Transport System», Londres 1994; «The Development of Edward Johnston's Underground Railway Block-Letter», Londres 1995.

**Barakov,** Vassil – né le 4. 9. 1902 à Kazanlik, Bulgarie – *maquettiste en édition, designer, peintre* – 1937, études à l'académie des beaux-arts de Sofia, Bulgarie

*Publications include:* "Vertical speculation. Mies van der Rohe and his Barcelona Pavillon" in "Evolution" 1/1990, Tokyo; "Takaaki Bando" in "High Quality" 20/1991, Heidelberg; "Takaaki Bando" in "Graphis" 294/1994, New York.

**Banks,** Colin – b. 16. 1. 1932 in London, England – *typographer, type designer, design consultant* – 1948–53: trains at the Mid-Kent College of Printing; afterwards assistant to Ernst Hoch. From 1956 onwards: freelance designer. 1958: founds the Banks & Miles company, London. 1975–82: Banks & Miles in Amsterdam. 1979: Banks gives London Transport's corporate typeface, Johnston, dating from 1916, a facelift for the corporation's new image. 1990: sets up a branch of Banks & Miles in Brussels. 1990–92: Banks & Miles opens in Hamburg. Acts as a design consultant to various universities and colleges: Portsmouth University (1991), UNO University in Tokyo and Helsinki (1991), University of the South-West in Bristol (1992). Works on logos and designs for various companies, including British Telecom, London Transport, The British Council and The Post Office. – *Publications include:* "Social Communication", London 1979; "London's Handwriting – a History of Design in London's Transport System", London 1991; "The Development of Edward Johnston's Underground Railway Block-Letter", London 1995.

**Barakow,** Vassil – b. 4. 9. 1902 in Kazanlik, Bulgaria – *book designer, commercial artist, painter* – 1937: studies at the art academy in Sofia, Bulgaria with Zeno

Mitglied des Verbands Bulgarischer Künstler. Zahlreiche Auszeichnungen.
**Barnbrook,** Jonathan – geb. 1966 in Luton, England. – *Grafik-Designer, Schriftentwerfer, Typograph* – 1985–88 Studium am Central Saint Martins College of Art and Design in London. 1988–90 Studium am Royal College of Art in London. Zusammenarbeit mit dem Studio „why not associates" in London. Seit 1990 freier Gestalter in London. Zahlreiche Veröffentlichungen. Eine seiner mit der Maschine in Stein geschnittenen Schrifttafeln wird im Victoria & Albert Museum in London gezeigt. 1993 Titel-Gestalter der Firma „Tony Kaye Films". Auftraggeber waren u. a. Volvo, Mazda, Volkswagen, Vidal Sassoon und Lloyd's. Gestaltung von Werbefilmen für Mercury-Prudentials und Hansen's soft drinks. 1996 Gründung seiner eigenen Schriftenhersteller-Firma „Virus". – *Schriftentwürfe:* Prototype (1990), Bastard (1990), Exocet (1991), Mason (1991), Mason sans (1994), Patriot (1994), Nylon (1996), Draylon (1997), Drone (1997), Apocalypso (1997), Prozac (1997).
**Barthes,** Roland – geb. 12. 11. 1915 in Cherbourg, Frankreich, gest. 26. 3. 1980 in Paris, Frankreich. – *Semiologe, Soziologe, Philosoph, Lehrer, Linguist* – 1935 Beginn des Studiums der Klassischen Literatur an der Sorbonne. 1940 Gymnasiallehrer in Paris, Diplomarbeit über die griechische Tragödie. 1941–46 Krankheit und Kur. 1948–49 Bibliothekarsgehilfe, dann Lehrer am Institut Français in Bukarest. 1950 Lektor an der Universität Alexandria, Ägypten. 1977 Lehrstuhl für

auprès de Zeno Todorov et Nikola Ganoucheff. Membre de l'Union des artistes bulgares. Lauréat de nombreux prix.
**Barnbrook,** Jonathan – né en 1966 à Luton, Angleterre – *graphiste maquettiste, concepteur de polices, typographe* – 1985–1988, études au Central Saint Martins College of Art and Design de Londres. 1988–1990, études au Royal College of Art de Londres. Travaille à Londres avec l'atelier „why not associates". Designer indépendant à Londres à partir de 1990. Nombreuses publications. Une plaque de pierre comportant des caractères taillés à la machine est présentée au Victoria & Albert Museum de Londres. En 1993, il crée les bancs-titres de la société «Tony Kaye Films». Commanditaires : Volvo, Mazda, Volkswagen, Vidal Sassoon, Lloyds, etc. Conception de films publicitaires pour Mercury-Prudentials et Hansen's soft drinks. En 1996, il fonde «Virus», une société de production de polices. – *Polices :* Prototype (1990), Bastard (1990), Exocet (1991), Mason (1991), Mason sans (1994), Patriot (1994), Nylon (1996), Draylon (1997), Drone (1997), Apocalypso (1997), Prozac (1997).
**Barthes,** Roland – né le 12. 11. 1915 à Cherbourg, France, décédé le 26. 3. 1980 à Paris, France – *sémiologue, sociologue, philosophe, enseignant, linguiste* – 1935, commence des études de lettres classique à la Sorbonne. 1940, professeur dans un lycée parisien, rédige une thèse de maîtrise sur la tragédie grecque. 1941–1946, maladie et cure. 1948–1949, aide-bibliothécaire, puis enseignant à l'Institut Français de Bucarest. 1950, lecteur à l'université d'Alexandrie, Egypte. 1977, chaire de sémiologie littéraire au Collège de France (sur la proposition de Michel Foucault). Barthes est l'un des principaux représentants du mouvement structuraliste.

Barnbrook Trailer

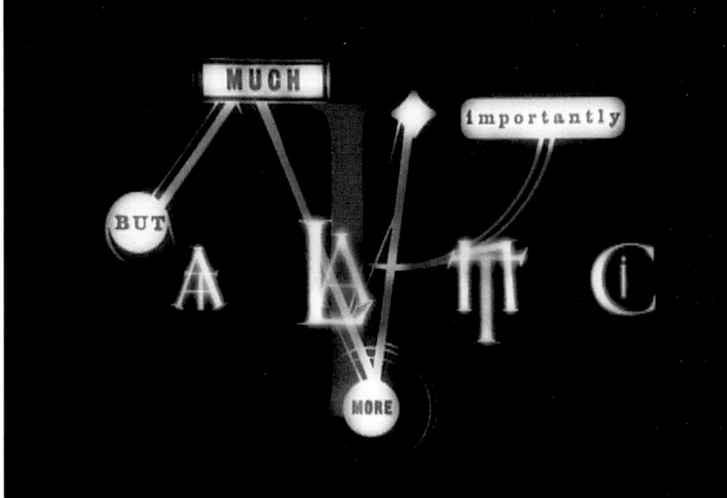

Barnbrook Advertisement

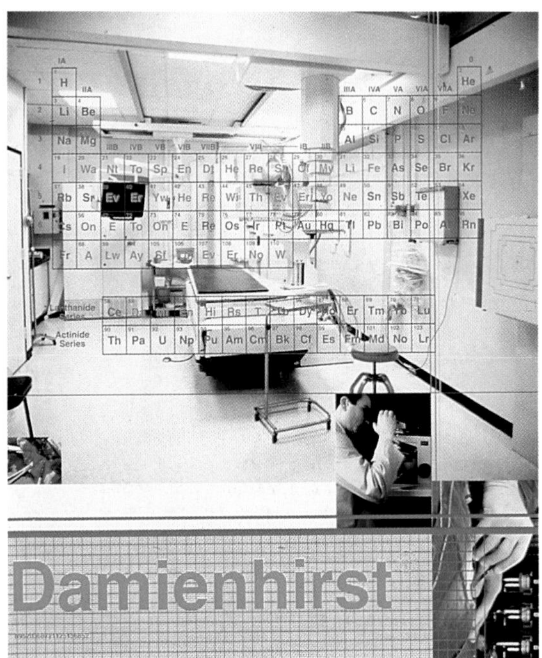

Barnbrook 1997 Cover

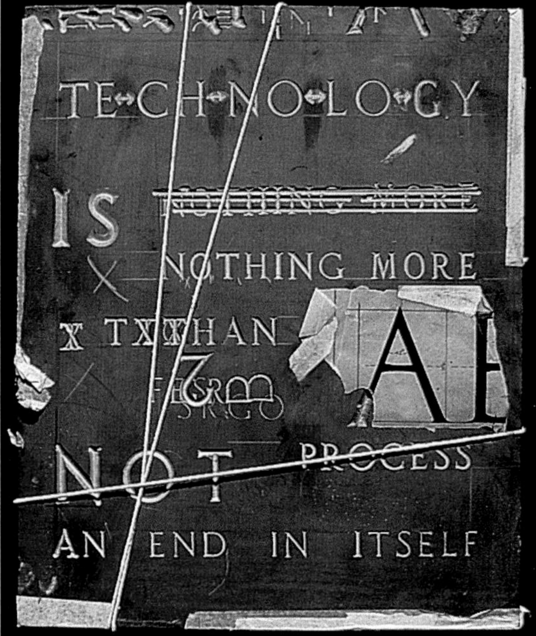

Barnbrook Generated Stonecarving

Todorov and Nikola Ganushev. Member of the Association of Bulgarian Artists. Has received numerous awards.
**Barnbrook,** Jonathan – b. 1966 in Luton, England – *graphic designer, type designer, typographer* – 1985–88: studies at the Central Saint Martins College of Art and Design in London. 1988–90: studies at the Royal College of Art in London. Works with the studio why not associates in London. Since 1990: freelance designer in London. Numerous publications. One of his inscribed slabs with the letters cut into the stone by machine is displayed in the Victoria & Albert Museum in London. 1993: designs titles for Tony Kaye Films. Clients include Volvo, Mazda, Volkswagen, Vidal Sassoon and Lloyd's. Designs promotional films for Mercury-Prudentials and Hansen's soft drinks. 1996: launches his own type manufacturers Virus. – *Fonts:* Prototype (1990), Bastard (1990), Exocet (1991), Mason (1991) and Mason sans (1994), Patriot (1994), Nylon (1996), Draylon (1997), Drone (1997), Apocalypso (1997), Prozac (1997).
**Barthes,** Roland – b. 12. 11. 1915 in Cherbourg, France, d. 26. 3. 1980 in Paris, France – *semiologist, sociologist, philosopher, teacher, linguist* – 1935: begins his studies in classical literature at the Sorbonne. 1940: high-school teacher in Paris. Writes his diploma thesis on Greek tragedies. 1941–46: period of illness and recovery. 1948–49: assistant librarian, then teacher at the Institut Français in Bucharest. 1950: teaches at the University of Alexandria, Egypt. 1977: chair for

Semiologie der Literatur am Collège de France (auf Vorschlag von Michel Foucault). Barthes ist einer der wichtigsten Vertreter strukturalistischer Literaturbetrachtung. Anhand literarischer Texte untersucht er die Mythen der Massenkommunikation, die sich in der unbeabsichtigten Vermischung ihrer Zeichensysteme äußern. – *Publikationen u.a.:* „Le degré zéro de l'écriture", Paris 1953; „Mythologies", Paris 1957; „Eléments de sémiologie", Paris 1964; „Système de la mode", Paris 1967; „L'empire des signes", Genf 1970; „Le plaisir du texte", Paris 1973; „La chambre claire", Paris 1980.

**Baskerville,** John – geb. 28. 1. 1706 in Wolverley, Worcestershire, England, gest. 8. 1. 1775. – *Schriftentwerfer, Schreibmeister, Drucker* – 1725 Umzug nach Birmingham. 1733–37 Schreibmeister in Birmingham. 1750 Gründung einer eigenen Schriftgießerei und Druckerei. 1757 erscheint als sein erstes selbst hergestelltes Buch eine Ausgabe der Schriften Vergils. 1758 erscheint eine Ausgabe von John Miltons „Paradise Lost". 1758 Berufung als Drucker an die Universität Cambridge. Hier entstehen mehrere Ausgaben des „Book of Common Prayer" und 1763 ein Neues Testament in einer von ihm geschnittenen griechischen Type. 1770–73 Herausgabe einer vierbändigen Ausgabe von Ariostos „Orlando Furioso". Baskervilles originale Schriftstempel und Matrizen gehen 1953 als Stiftung in den Besitz der Cambridge University Press über. – *Publikation:* R. Straus und Robert K. Dent „John Baskerville", Cambridge 1907.

PUBLII VIRGILII

MARONIS

BUCOLICA,

GEORGICA,

*ET*

ÆNEIS.

*BIRMINGHAMIAE:*

Typis JOHANNIS BASKERVILLE.

MDCCLVII.

Baskerville 1757 Title

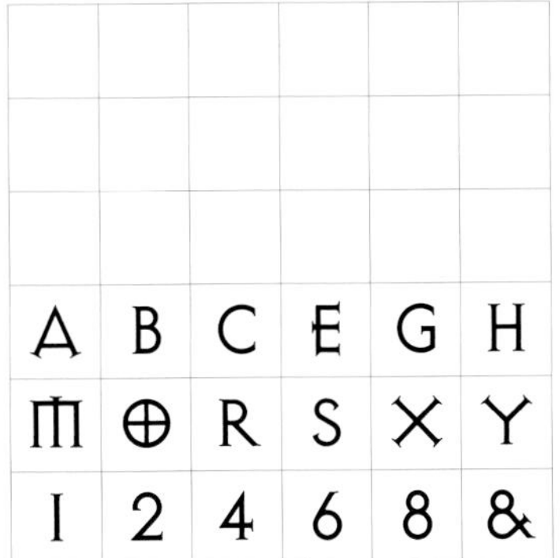


Barnbrook 1991 Exocet

| | n | **n** | | | *n* |
|---|---|---|---|---|---|
| a | b | e | f | g | i |
| o | r | s | t | y | z |
| A | B | C | E | G | H |
| M | O | R | S | X | Y |
| 1 | 2 | 4 | 6 | 8 | & |

Baskerville

*ORLANDO FURIOSO*

DI

LODOVICO ARIOSTO.

*ARGOMENTO.*

*Contra le Donne Rodomonte intende*
*Quanto mal poſſa dir lingua ſallace.*
*Indi verſo il ſuo Regno il cammin prende,*
*Ma luogo trova pria, che al ſuo cor piace.*
*Qui d' Iſabella nuovo amor l' accende;*
*Ma si l' impedimento gli diſpiace*
*Del Frate, ch' ella ha ſeco in compagnia,*
*Che 'l ſellon gli dà morte acerba e ria.*

CANTO VENTESIM'OTTAVO.

I

DONNE, e voi, che le Donne avete in pregio,
Per Dio non date a queſta iſtoria orec- [chia;
A queſta, che l'oſtier dire in diſpregio,
E in voſtra infamia, e biaſmo s' apparecchia;
Benchè nè macchia vi può dar, nè ſregio
Lingua sì vile; e ſia l' uſanza vecchia,
Che 'l volgare ignorante ognun riprenda,
E parli più di quel che meno intenda.

Baskerville 1773 Page

semiology of literature at the Collège de France (at the suggestion of Michel Foucault). Barthes is one of the major representatives of the Structuralist approach to literature. Using literary texts, he examines the myths of mass communication manifested in the unintentional mingling of their systems of signs. – *Publications include:* "Le degré zéro de l'écriture", Paris 1953; "Mythologies", Paris 1957; "Eléments de sémiologie", Paris 1964; "Système de la mode", Paris 1967; "L'empire des signes", Geneva 1970; "Le plaisir du texte", Paris 1973; "La chambre claire", Paris 1980.

**Baskerville,** John – b. 28. 1. 1706 in Wolverley, Worcestershire, England, d. 8. 1. 1775 – *type designer, writing master, printer* – 1725: moves to Birmingham. 1733–37: writing master in Birmingham. 1750: sets up his own type foundry and printing works. 1757: his first printed book is published, an edition of Virgil. 1758: publishes an edition of John Milton's "Paradise Lost". 1758: appointed printer to the University of Cambridge. Here he produces several editions of the "Book of Common Prayer" and in 1763 a New Testament in a Greek type he designs. 1770–73: produces a four-volume edition of Ariosto's "Orlando Furioso". 1953: Baskerville's original letter stamps and matrices are donated to Cambridge University Press. – *Publication:* Ralph Straus and Robert K. Dent "John Baskerville", Cambridge 1907.

A partir de textes littéraires, il analyse les mythes de la communication de masse qui se manifestent dans l'amalgame involontaire de leur système signalétique. – *Publications, sélection:* «Le degré zéro de l'écriture», Paris 1953; «Mythologies», Paris 1957; «Eléments de sémiologie», Paris 1964; «Système de la mode», Paris 1967; «L'empire des signes», Genève 1970; «Le plaisir du texte», Paris 1973; «La chambre claire», Paris 1980.

**Baskerville,** John – né le 28. 1. 1706 à Wolverley, Worcestershire, Angleterre, décédé le 8. 1. 1775 – *concepteur de polices, maître calligraphe, imprimeur* – s'installe à Birmingham en 1725. 1733–1737, maître calligraphe à Birmingham. 1750, crée sa propre fonderie de caractères et son imprimerie. En 1757, il publie les écrits de Virgile, le premier livre de sa fabrication, puis une édition de «Paradise Lost» de John Milton en 1758. Il est nommé imprimeur à l'université de Cambridge en 1758. C'est là que paraissent plusieurs éditions du «Book of Common Prayer», puis en 1763, un nouveau testament en caractères grecs qu'il taille lui-même. 1770–1773, publication d'une édition en quatre volumes de «Orlando Furioso» d'Ariosto. En 1953, les originaux des typons et matrices de Baskerville deviennent la propriété des Cambridge University Press sous forme de fondation. – *Publication:* Ralph Straus et Robert K. Dent «John Baskerville», Cambridge 1907.

**Bass,** Saul – geb. 8. 5. 1920 in New York, USA, gest. 25. 4. 1996 in Los Angeles, USA. – *Typograph, Grafik-Designer, Fotograf, Filmregisseur* – 1936–39 Studium an der Art Students League, New York. 1944–45 Studium am Brooklyn College, New York. 1936–46 freischaffender Designer in New York. 1946–80 Saul Bass & Associates Los Angeles, ab 1981 Bass Yager & Associates Los Angeles. Entwurf zahlreicher Signets und Erscheinungsbilder, z. B. für Bell Systems, AT & T, United Airlines, Warner Communications und Minolta. Entwurf zahlreicher grafischer Symbole für Filme, z. B. „Bonjour tristesse" (1956), „Anatomie eines Mörders" (1959), „Shining" (1980). Filmtitel für über 40 Filme, u. a. „Der Mann mit dem goldenen Arm", „Vertigo", „Psycho" und „Exodus". 1969 Auszeichnung mit einem Oscar und einer Goldmedaille des Filmfestivals Moskau für den Dokumentar-Kurzfilm „Why man creates". 1978 wird Bass in die New York Art Directors Hall of Fame gewählt. 1988 Auszeichnung mit dem „Lifetime Achievement Award" des Art Directors Club of Los Angeles. – *Schriftentwurf:* Rainbow Bass (1982). – *Publikation:* G. Nelson „Saul Bass", New York 1967.

**Baudin,** Fernand – geb. 1918 in Bachte-Maria-Leerne, Ostflandern. – *Buchgestalter, Typograph, Lehrer, Autor* – 1942 Abschluß seiner Ausbildung an der Ecole Nationale Supérieure d'Architecture et des Arts Décoratifs (ENSAAD) in Brüssel. 1943–55 Arbeiten für Verlage, Tageszeitungen, Zeitschriften und Werbeagenturen. 1955–67 Art Director und

**Bass,** Saul – né le 8. 5. 1920 à New York, USA, décédé le 25. 4. 1996 à Los Angeles, USA – *typographe, graphiste maquettiste, photographe, réalisateur de films* – 1936–1939, études à la Art Students League de New York, 1944–1945, études au Brooklyn College de New York. 1936–1946, designer indépendant à New York. 1946–1980, Saul Bass & Associates Los Angeles, puis Bass Yager & Associates Los Angeles à partir de 1981. Crée de nombreux signets et logos, par ex. pour Bell Systems, AT & T, United Airlines, Warner Communications et Minolta. Dessine de nombreux symboles graphiques pour le cinéma, par ex. «Bonjour tristesse» (1956), «Anatomie eines Mörders» (1959), «Shining» (1980). Titres pour plus de 40 films, dont «Der Mann mit dem goldenen Arm», «Vertigo», «Psychose» et «Exodus». 1969, Oscar et médaille d'or du festival du cinéma de Moscou pour son court métrage documentaire «Why man creates». En 1978, Bass est élu à la New York Art Directors Hall of Fame. En 1988, on lui décerne la «Lifetime Achievement Award» d'Art Directors Club of Los Angeles. – *Police:* Rainbow Bass (1982). – *Publication:* G. Nelson «Saul Bass», New York 1967.

**Baudin,** Fernand – né en 1918 à Bachte-Maria-Leerne en Flandre orientale – *maquettiste en édition, typographe, enseignant, auteur* – termine sa formation à l'Ecole Nationale Supérieure d'Architecture et des Arts Décoratifs (ENSAAD) de Bruxelles en 1942. 1943–1945, travaux pour des maisons d'édition, la presse quotidienne, des revues et des agences de publicité. 1955–1967, directeur artistique et rédacteur à la filiale bruxelloise de Lettergieterij Amsterdam. 1965, organisation et design de l'exposition «Stanley Morison et la tradition typographique» à la Bibliothèque royale de Belgique. De-

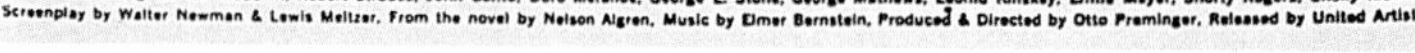

Bass 1955 Poster

BONJOUR TRISTESSE

Bass 1958 Poster

Bass 1961 Poster

Bass 1982 Rainbow Bass

**Bass,** Saul – b. 8. 5. 1920 in New York, USA, d. 25. 4. 1996 in Los Angeles, USA – *typographer, graphic designer, photographer, film director* – 1936–39: studies at the Art Students League in New York. 1944–45: studies at Brooklyn College in New York. 1936–46: freelance designer in New York. 1946–80: runs Saul Bass & Associates, Los Angeles, from 1981 onwards, named Bass Yager & Associates, Los Angeles. Has designed numerous trademarks and logos for Bell Systems, AT & T, United Airlines, Warner Communications and Minolta, for example. Has also designed numerous graphic symbols for films, such as "Bonjour tristesse" (1956), "Anatomy of a Murderer" (1959), "The Shining" (1980) and title sequences for over 40 films, including "The Man with the Golden Arm", "Vertigo", "Psycho" and "Exodus". 1969: awarded an Oscar and also a gold medal by the Moscow Film Festival for the short documentary "Why man creates". 1978: Bass is elected into the New York Art Directors Hall of Fame. 1988: the Art Directors Club of Los Angeles presents him with their Lifetime Achievement Award. – *Font:* Rainbow Bass (1982). – *Publication:* G. Nelson "Saul Bass", New York 1967.

**Baudin,** Fernand – b. 1918 in Bachte-Maria-Leerne, Belgium – *book designer, typographer, teacher, author* – 1942: finishes his training at the Ecole Nationale Supérieure d'Architecture et des Arts Décoratifs (ENSAAD) in Brussels. 1943–55: works for publishing houses, daily newspapers, magazines and advertising agencies. 1955–67: art director and copy-

Texter der Brüsseler Filiale der Lettergieterij Amsterdam. 1965 Organisation und Gestaltung der Ausstellung „Stanley Morison und die typographische Tradition" in der Königlichen Bibliothek von Belgien. Seit 1968 Katalog- und Buchgestaltung für die Königliche Bibliothek und zahlreiche Verlage. 1972 Neugestaltung der belgischen Tageszeitung „Le Soir". Internationale Jurorentätigkeit. Ehren-Vizepräsident der ATypI (Association Typographique Internationale). Lehrtätigkeiten: 1960–66 La Cambre, ENSAAD, Brüssel, 1973–83 NHIBS (van de Velde-Institut), Antwerpen. – *Publikationen u.a.:* „La lettre d'imprimerie", Brüssel 1965; „Stanley Morison et la tradition typographique", Brüssel, Den Haag 1965; „The type-specimen of J.-F. Rosart", Amsterdam 1973; „La typographie au tableau noir", Paris 1984; „How Typography works (and why it is important)", London, New York 1988; „L'effet Gutenberg", Paris 1994.

**Bauer,** Friedrich – geb. 1863 in Droste bei Osterode, Deutschland, gest. 1943 in Schönberg, Taunus, Deutschland. – *Schriftsetzer, Lehrer, Autor, Typograph* – 1877 Schriftsetzerlehre, danach Leiter der Hausdruckerei der Schriftgießerei Schelter & Giesecke, Leipzig. 1890–1891 Mitherausgeber und Redakteur der Zeitschrift „Graphischer Beobachter". 1898–1911 Druckereileiter der Schriftgießerei Genzsch & Heyse, München, später Hamburg. 1911–24 Fachlehrer an der Staatlichen Gewerbeschule Hamburg. – *Publikationen:* „Chronik der Schriftgießereien in Deutschland und den deutschsprachi-

Bass 1963 Poster

Baudin 1988 Cover

puis 1968, maquettes de catalogues et de livres pour la Bibliothèque royale et de nombreux autres éditeurs. 1972, refonte du quotidien belge «Le Soir». Membre de jurys internationaux. Vice-président d'honneur de l'ATypI (Association Typographique Internationale). Enseignements: 1960–1966 La Cambre, ENSAAD, Bruxelles; 1973–1983 NHIBS (van de Velde Institut), Anvers. – *Publications, sélection:* «La lettre d'imprimerie», Bruxelles 1965; «Stanley Morison et la tradition typographique», Bruxelles, La Haye 1965; «The Type-specimen of J.-F. Rosart», Amsterdam 1973; «La typographie au tableau noir», Paris 1984; «How Typography works (and why it is important)», Londres, New York 1988; «L'effet Gutenberg», Paris 1994.

**Bauer,** Friedrich – né en 1863 à Droste près d'Osterode, Allemagne, décédé en 1943 à Schönberg, Taunus, Allemagne – *dessinateur de caractères, enseignant, auteur, typographe* – 1877, apprentissage de la composition, puis directeur de l'imprimerie de la fonderie de caractères Schelter & Giesecke, à Leipzig. 1890–1891, coéditeur et rédacteur de la revue «Graphischer Beobachter». 1898–1911, directeur de l'imprimerie de la fonderie de caractères Schelter & Heyse à Munich, puis à Hambourg. 1911–1924 enseignant à la Staatliche Gewerbeschule de Hambourg. – *Publications:* «Chronik der Schriftgiessereien in Deutschland und

Baudin 1989 Sketch

Baudin 1989 Spread

writer for the Brussels branch of the Lettergieterij Amsterdam. 1965: organizes and plans the exhibition "Stanley Morison and the Typographical Tradition" at the Royal Library of Belgium. From 1968 onwards: designs catalogues and books for the Royal Library of Belgium and many other publishers. 1972: redesigns the Belgian daily "Le Soir". International adjudicator. Honorary vice-president of the Association Typographique Internationale (ATypI). Teaching positions: 1960–66: La Cambre, ENSAAD, Brussels. 1973–83: NHIBS (van de Velde Institute) in Antwerp. – *Publications include:* "La lettre d'imprimerie", Brussels 1965; "Stanley Morison et la tradition typographique", Brussels, The Hague 1965; "The type-specimen of J.-F. Rosart", Amsterdam 1973; "La typographie au tableau noir", Paris 1984; "How Typography works (and why it is important)", London, New York 1988; "L'effet Gutenberg", Paris 1994.

**Bauer,** Friedrich – b. 1863 in Droste near Osterode, Germany, d. 1943 in Schönberg, Taunus, Germany – *letterer, teacher, author, typographer* – 1877: trains as a typesetter, then manages the in-house printing works of the Schelter & Giesecke type foundry in Leipzig. 1890–91: co-publisher and editor of the magazine "Graphischer Beobachter". 1898–1911: manages the in-house printing works of the Genzsch & Heyse type foundry in Munich, later in Hamburg. 1911–24: teacher at the Staatliche Gewerbeschule in Hamburg. – *Publications:* "Chronik der Schriftgießereien in Deutschland und den

gen Nachbarländern", Offenbach 1928; „Die Normung der Buchdruckletter", Leipzig 1929.

**Bauer,** Johann Christian – geb. 1802 in Hanau, Deutschland, gest. 1867 in Frankfurt am Main, Deutschland. – *Stempelschneider, Schriftgießer* – Lehre als Mechaniker und Schlosser. Ab 1827 als Stempelschneider tätig. Leitet 1835–37 mit Christian Nies eine Schriftgießerei in Frankfurt am Main. 1837 Gründung der Bauerschen Gießerei in Frankfurt am Main. 1839–47 Stempelschneider in Edinburgh, Schottland. 1847 Rückkehr nach Frankfurt am Main, Weiterführung der Firma unter dem Namen „Englische Schriftschneiderei und Gravieranstalt".

**Bauer,** Konrad Friedrich – geb. 9. 12. 1903 in Hamburg, Deutschland, gest. 17. 3. 1970 in Schönberg, Deutschland – *Schriftentwerfer, Lehrer* – Schriftsetzerlehre, Studium der Kunstgeschichte. 1928 Beginn seiner Arbeit in der Bauerschen Gießerei, Frankfurt am Main. 1932–36 Redaktion der „Zeitschrift für Bücherfreunde", 1938–40 Redaktion der Bände 7–9 des Jahrbuchs „Imprimatur". 1948 künstlerischer Leiter der Bauerschen Gießerei, Frankfurt am Main. 1952–64 Vorsitzender der Jury des Wettbewerbs „Die schönsten Bücher Deutschlands" der Stiftung Buchkunst, Frankfurt am Main. 1947–48 Lehrauftrag für Buch-, Schrift- und Druckwesen an der Universität Mainz. – *Schriftentwürfe:* (mit Walter Baum) Alpha (1954), Beta (1954), Folio (1956–63), Imprimatur (1952–55), Volta (1956), Verdi (1957), Impressum (1963), alle Bauersche Gießerei, Frankfurt am

den deutschsprachigen Nachbarländern», Offenbach 1928; «Die Normung der Buchdruckletter», Leipzig 1929.

**Bauer,** Johann Christian – né en 1802 à Hanau, Allemagne, décédé en 1867 à Francfort-sur-le-Main, Allemagne – *tailleur de types, fondeur de caractères* – apprentissage de mécanicien et de serrurier. Exerce comme tailleur de types à partir de 1827. Dirige une fonderie de caractères avec Christian Nies à Francfort-sur-le-Main de 1835 à 1837. 1837, crée la fonderie Bauer à Francfort-sur-le-Main. 1839–1847, tailleur de types à Edimbourg, Ecosse. 1847, retour à Francfort-sur-le-Main, où il poursuit les activités de la société sous le nom d'«Englische Schriftschneiderei und Gravieranstalt».

**Bauer,** Konrad Friedrich – né le 9. 12. 1903 à Hambourg, Allemagne, décédé le 17. 3. 1970 à Schönberg, Allemagne – *concepteur de polices, enseignant* – formation de composition, études d'histoire de l'art. Commence à travailler à la fonderie Bauer à Francfort-sur-le-Main en 1928. 1932–1936, emploi à la rédaction de la «Zeitschrift für Bücherfreunde». 1938–1940, rédaction des volumes 7–9 des annales «Imprimatur». 1948, directeur artistique de la fonderie Bauer, à Francfort-sur-le-Main. 1952–1964, président du jury du concours «Die schönsten Bücher Deutschlands» de la fondation pour les arts du livre, à Francfort-sur-le-Main. 1947–1948, chargé de cours sur les arts du livre, de l'écriture et de l'imprimerie à l'université de Mayence. – *Polices:* (avec Walter Baum) Alpha (1954), Beta (1954), Folio (1956–1963), Imprimatur (1952–1955), Volta (1956), Verdi (1957), Impressum (1963), toutes de l'imprimerie Bauer de Francfort-sur-le-Main. – *Publications, sélection:* «Wie eine Druckschrift entsteht», Francfort-sur-le-Main 1931; «Stammbaum der Schrift»,

AVENTUR UND KUNST

Eine Chronik des Buchdruckgewerbes von der Erfindung der beweglichen Letter bis zur Gegenwart, bearbeitet von KONRAD F. BAUER

Privatdruck der Bauerschen Gießerei · Frankfurt a·M im fünfhundertsten Jahre der Buchdruckerkunst 1940

Konrad Bauer 1940 Main Title

Bauersche Gießerei 1916 Specimen

| | n | **n** | | | *n* |
|---|---|---|---|---|---|
| a | b | e | f | g | i |
| o | r | s | t | y | z |
| A | B | C | E | G | H |
| M | O | R | S | X | Y |
| 1 | 2 | 4 | 6 | 8 | & |

Konrad Bauer 1952–55 Imprimatur

| n | n | **n** | **n** | | *n* |
|---|---|---|---|---|---|
| a | b | e | f | g | i |
| o | r | s | t | y | z |
| A | B | C | E | G | H |
| M | O | R | S | X | Y |
| 1 | 2 | 4 | 6 | 8 | & |

Konrad Bauer 1956–63 Folio

Bauersche Gießerei Page

deutschsprachigen Nachbarländern", Offenbach 1928; "Die Normung der Buchdruckletter", Leipzig 1929.

**Bauer,** Johann Christian – b. 1802 in Hanau, Germany, d. 1867 in Frankfurt am Main, Germany – *punch cutter, type founder* – Trained as a mechanic and fitter. From 1827 onwards: works as a punch cutter. 1835–37: runs a type foundry in Frankfurt am Main with Christian Nies. 1837: founds the Bauersche Gießerei in Frankfurt am Main. 1839–47: punch cutter in Edinburgh, Scotland. 1847: returns to Frankfurt am Main and continues to run his company under the name Englische Schriftschneiderei und Gravieranstalt.

**Bauer,** Konrad Friedrich – b. 9. 12. 1903 in Hamburg, Germany, d. 17. 3. 1970 in Schönberg, Germany – *type designer, teacher* – Trained as a typesetter and studied art history. 1928: starts work at the Bauersche Gießerei in Frankfurt am Main. 1932–36: works on the editorial staff for the "Zeitschrift für Bücherfreunde". 1938–40: editor for volumes 7–9 of the yearbook "Imprimatur". 1948: art director for the Bauersche Gießerei in Frankfurt am Main. 1952–64: heads the jury of the competition "Die schönsten Bücher Deutschlands", run by the Stiftung Buchkunst in Frankfurt am Main. From 1947–48 onwards: teaches book design, type and printing at the University of Mainz. – *Fonts:* (with Walter Baum) Alpha (1954), Beta (1954), Folio (1956–63), Imprimatur (1952–55), Volta (1956), Verdi (1957), Impressum (1963), all by the Bauersche Gießerei, Frankfurt.

Main. – *Publikationen u.a.:* „Wie eine Druckschrift entsteht", Frankfurt am Main 1931; „Stammbaum der Schrift", Frankfurt am Main 1937; „Aventur und Kunst", Frankfurt am Main 1940; „Jahreszahlen aus acht Jahrhunderten", Frankfurt am Main 1954; Zahlreiche Aufsätze in „Klimschs Druckerei-Anzeiger", „Börsenblatt des deutschen Buchhandels", „Druckspiegel", „Gebrauchsgraphik". Berthold Hack (Hrsg.) „Konrad F. Bauer", Mainz 1973.

**Bauersche Gießerei,** Frankfurt am Main, Deutschland. – *Schriftgießerei* – 1837 Gründung der Schriftgießerei in Frankfurt am Main durch den Schriftenwerfer und Stempelschneider Johann Christian Bauer. 1880 Eröffnung einer Zweigniederlassung in Stuttgart. Die Bauersche Gießerei erwirbt 1885 die Fundición Tipográfica Neufville (FTN), Barcelona. 1912 Erwerb der Schriftgießerei Numerich, Leipzig. 1916 Erwerb der Schriftgießerei Flinsch & Glock, Frankfurt am Main. 1927 Gründung der Niederlassung „The Bauer Type Foundry" in New York. 1940, im Gutenberg-Jahr, gibt die Bauersche Gießerei das Werk „Aventur und Kunst" heraus. 1972 Schließung des Unternehmens in Frankfurt am Main, der gesamte Schriftguß wird von FTN Barcelona übernommen. Zu den Schriftentwerfern der Bauerschen Gießerei gehören F. W. Kleukens, L. Bernhard, F. H. E. Schneidler, E. R. Weiß, P. Renner, K. F. Bauer und W. Baum.

**Bauhaus,** Weimar, Dessau, Berlin, Deutschland. – *Hochschule für Gestaltung* – 1919 Gründung des Bauhauses in

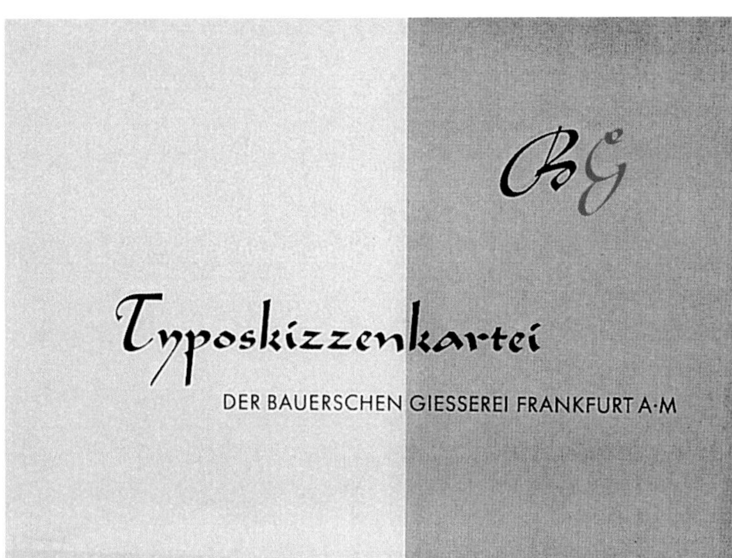


Bauersche Gießerei Cover

Bauersche Gießerei Specimen

Bauersche Gießerei Specimen

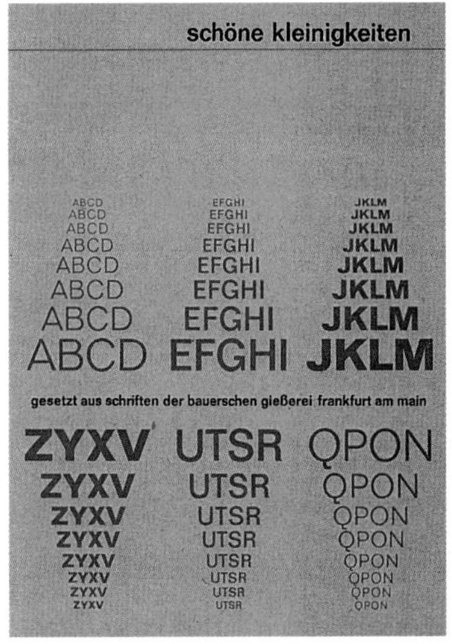


Bauersche Gießerei Cover

Francfort-sur-le-Main 1937; «Aventur und Kunst», Francfort-sur-le-Main 1940; «Jahreszahlen aus acht Jahrhunderten», Francfort-sur-le-Main 1954; de nombreux articles dans le «Klimsch Druckerei-Anzeiger», la «Börsenblatt des deutschen Buchhandels», le «Druckspiegel», «Gebrauchsgraphik». Berthold Hack (éd.) «Konrad F. Bauer», Mayence 1973.

**Bauersche Giesserei** (Fonderie Bauer), Francfort-sur-le-Main, Allemagne – *fonderie de caractères* – la fonderie de caractères Bauer a été créée à Francfort-sur-le-Main en 1837 par le concepteur de polices et tailleur de types Johann Christian Bauer. 1880, création d'une succursale à Stuttgart. 1885, la Bauersche Gießerei achète la Fundición Tipográfica Neufville (FTN) de Barcelone. 1912, acquisition de la fonderie de caractères Numrich, Leipzig. 1916, acquisition de la fonderie de caractères Flinsch & Glock, Francfort-sur-le-Main. 1927, création de la succursale «The Bauer Type Foundry» à New York. 1940, édition de l'ouvrage «Aventur und Kunst» à l'occasion de l'année Gutenberg. 1972, fermeture de l'entreprise de Francfort-sur-le-Main, l'ensemble des activités de fonderie est repris par FTN Barcelone. F. W. Kleukens, L. Bernhard, F. H. E. Schneidler, E. R. Weiss, P. Renner, K. F. Bauer et W. Baum ont compté parmi les concepteurs de polices de la fonderie.

**Bauhaus,** Weimar, Dessau, Berlin, Allemagne – *école supérieure de design et de création* – 1919, Walter Gropius fonde le

– *Publications include:* "Wie eine Druckschrift entsteht", Frankfurt am Main 1931; "Stammbaum der Schrift", Frankfurt am Main 1937; "Aventur und Kunst", Frankfurt am Main 1940; "Jahreszahlen aus acht Jahrhunderten", Frankfurt am Main 1954; various essays in "Klimschs Druckerei-Anzeiger", "Börsenblatt des deutschen Buchhandels", "Druckspiegel" and "Gebrauchsgraphik". Berthold Hack (ed.) "Konrad F. Bauer", Mainz 1973.

**Bauersche Gießerei,** Frankfurt am Main, Germany – *type foundry* – 1837: type designer and punch cutter Johann Christian Bauer founds the type foundry in Frankfurt am Main. 1880: a branch of the foundry is opened in Stuttgart. 1885: the Bauersche Gießerei acquires the Fundición Tipográfica Neufville (FTN) in Barcelona. 1912: acquires the Numerich type foundry in Leipzig. 1916: buys up the Flinsch & Glock type foundry in Frankfurt am Main. 1927: sets up a further branch of the company, The Bauer Type Foundry, in New York. 1940: the Bauersche Gießerei issues "Aventur und Kunst" for the Gutenberg year. 1972: the company in Frankfurt is closed down; the entire type casting equipment goes to FTN Barcelona. F. W. Kleukens, L. Bernhard, F. H. E. Schneidler, E. R. Weiss, P. Renner, K. F. Bauer and W. Baum were among the type designers to work at the Bauersche Gießerei.

**Bauhaus,** Weimar, Dessau, Berlin, Germany – *college of design* – 1919: Walter Gropius founds the Bauhaus in Weimar as a fusion of the Weimar Kunstgewerbeschule and the Hochschule für Bil-

Weimar durch Walter Gropius als Vereinigung der Kunstgewerbeschule und der Hochschule für Bildende Kunst. Die ersten Lehrenden sind L. Feininger, J. Itten und G. Marcks. 1920 Einstellung von G. Muche, P. Klee, O. Schlemmer als weitere Lehrende. 1922 Einstellung von W. Kandinsky. J. Itten verläßt 1923 das Bauhaus, Nachfolger wird L. Moholy-Nagy. Durch die Landtagswahlen in Thüringen 1924 entsteht 1925 ein politischer Rechtsruck, der starke Angriffe gegen das Bauhaus auslöst. Der Meisterrat der Schule beschließt die Auflösung des Bauhauses zum 1. April. Umzug der Schule nach Dessau. Neue Lehrkräfte werden J. Albers, H. Bayer, M. Breuer, H. Scheper, J. Schmidt und G. Stölzl. 1926 Einweihung des Bauhaus-Gebäudes, das nach Entwürfen von W. Gropius gebaut wurde. 1927 beruft Gropius H. Meyer zum Leiter der neugegründeten Bau-Abteilung. Muche verläßt das Bauhaus. 1928 Rücktritt von Gropius als Direktor, Nachfolger wird H. Meyer. Bayer, Breuer, Moholy-Nagy verlassen das Bauhaus. Einstellung von L. Hilberseimer und W. Peterhans. 1929–1930 Wanderausstellung mit Ergebnissen der Bauhaus-Arbeit in Basel, Breslau, Dessau, Essen, Mannheim und Zürich. 1930 Entlassung des Direktors H. Meyer wegen seines linken Engagements. Nachfolger wird L. Mies van der Rohe. 1931 verlassen Klee und Stölzl das Bauhaus. 1932 Schließung des Bauhauses durch die nationalsozialistische Mehrheit im Gemeinderat zum 1. Oktober. Ende Oktober Neueröffnung des privatisierten Bauhauses in Berlin. 1933

«Bauhaus» de Weimar en fusionnant la Kunstgewerbeschule (école des arts décoratifs) et la Hochschule für Bildende Kunst (école des beaux-arts). Les premiers enseignants sont L. Feininger, J. Itten, G. Marcks. 1920, embauche d'autres enseignants dont G. Muche, P. Klee, O. Schlemmer. 1922, arrivée de W. Kandinsky. J. Itten quitte la Bauhaus en 1923 et Moholy-Nagy lui succède. A la suite des élections de 1924, la Thuringe connaît un revirement politique à droite qui provoque de violentes attaques contre le Bauhaus. Le conseil des enseignants de l'école décide de dissoudre le Bauhaus le 1er avril. L'école déménage à Dessau en 1925. De nouveaux enseignants arrivent, dont J. Albers, H. Bayer, M. Breuer, H. Scheper, J. Schmidt, G. Stölzl. 1926, inauguration des bâtiments du Bauhaus conçus par Walter Gropius. 1927, Gropius nomme H. Meyer à la direction du nouveau département d'architecture. Muche quitte le Bauhaus. 1928, Gropius quitte la direction de l'école, H. Meyer lui succède. Bayer, Breuer et Moholy-Nagy quittent le Bauhaus. L. Hilberseimer et W. Peterhans sont embauchés. 1929–1930, exposition itinérante présentant des créations du Bauhaus à Bâle, Breslau, Dessau, Essen, Mannheim et Zurich. 1930, licenciement du directeur H. Meyer à cause de son engagement à gauche. L. Mies van der Rohe lui succède. P. Klee et G. Stölzl quittent l'école. 1932, fermeture le 1er octobre du Bauhaus par un conseil municipal à majorité national-socialiste. Fin octobre, réouverture du Bauhaus sous forme privée à Berlin. 1933, fermeture du Bauhaus de Berlin le 11 avril. Reprise des activités après acceptation de contraintes accablantes. Le Bauhaus est définitivement dissout le 20 juillet sur décision du conseil des enseignants. – *Publications, sélection :* H. Bayer, W. Gropius, I. Gropius

Bauhaus 1921 Cover

Bauhaus 1925 Title

Bauhaus 1922 Advertisement

Bauhaus 1927 Poster

dende Kunst. The first teachers are L. Feininger, J. Itten and G. Marcks. 1920: G. Muche, P. Klee and O. Schlemmer are enrolled as teachers. 1922: W. Kandinsky joins the teaching staff. 1923: J. Itten leaves the Bauhaus; L. Moholy-Nagy succeeds him. Regional elections in Thuringia in 1924 represent a strong political swing to the right, unleashing a wave of critical attacks on the Bauhaus. The college's chief advisor decides to close the Bauhaus on 1st April 1925. The college moves to Dessau. J. Albers, H. Bayer, M. Breuer, H. Scheper, J. Schmidt and G. Stölzl take up teaching positions. 1926: the Bauhaus building is officially opened, built from plans by W. Gropius. 1927: Gropius names H. Meyer head of the new Department of Architecture. Muche leaves the Bauhaus. 1928: Gropius resigns as director; H. Meyer takes his place. Bayer, Breuer and Moholy-Nagy leave the Bauhaus. L. Hilberseimer and W. Peterhans are employed. 1929–30: an exhibition of Bauhaus work tours Basle, Breslau, Dessau, Essen, Mannheim and Zurich. 1930: H. Meyer is dismissed as director because of his left-wing political commitments. L. Mies van der Rohe is his successor. 1931: Klee and Stölzl leave the Bauhaus. 1932: the Bauhaus is closed on 1st October by the National Socialist majority on the district council. At the end of October, a privatized Bauhaus reopens in Berlin. 1933: the Bauhaus is closed on 11th April. Work is continued after an enormous number of conditions are imposed, yet the teaching staff decide to finally close the Bauhaus once and for all

Schließung des Bauhauses Berlin am 11. April. Nach strengen Auflagen Wiederaufnahme der Arbeit; durch Beschluß des Lehrerkollegiums löst sich das Bauhaus am 20. Juli endgültig auf. – *Publikationen u.a.:* H. Bayer, W. Gropius, I. Gropius (Hrsg.) „Bauhaus 1919–1928", New York 1938, Stuttgart 1955; H. M. Wingler „Das Bauhaus", Bramsche 1962; L. Lang „Das Bauhaus 1919–1933, Idee und Wirklichkeit", Berlin 1966; M. Droste „Bauhaus 1919–1933", Berlin 1990.

**Bauhaus-Archiv,** Berlin, Deutschland. – 1960 Gründung des Vereins des Bauhaus-Archivs in Darmstadt, Gründungsdirektor war Hans Maria Wingler. 1964 beginnt Walter Gropius mit Entwurf und Planung neuer Gebäude für das Bauhaus-Archiv auf der Rosenhöhe in Darmstadt, das 1965 im Modell fertig ist. 1967 regt Gropius an, den Neubau nach Berlin zu verlegen. 1971 Verlegung des Bauhaus-Archivs nach Berlin. 1979 Eröffnung des Bauhaus-Archivs Berlin in den neuen Räumen. Das Museum verfügt über die umfangreichste Sammlung zur Geschichte des Bauhauses. Es besitzt die Nachlässe von Walter Gropius, Herbert Bayer, Georg Muche. Die didaktischen Methoden der Lehrenden sind durch zahlreiche Originalarbeiten dokumentiert. Das Museum besitzt eine reichhaltige Architektur-Kollektion und sehr viele Arbeiten, die die Entwicklung der Bauhaus-Werkstätten dokumentieren. Die Vorgeschichte des Bauhauses und die Geschichte des New Bauhaus Chicago sind weitreichend dokumentiert vorhanden. – *Publikationen u. a.:* „Bauhaus Berlin",

Bauhaus 1923 Schablonenschrift

Bauhaus 1929 Poster

(éd.) «Bauhaus 1919–1928», New York 1938, Stuttgart 1955; H. M. Wingler «Das Bauhaus», Bramsche 1962; L. Lang «Das Bauhaus 1919–1933, Idee und Wirklichkeit», Berlin 1966; M. Droste «Bauhaus 1919–1933», Berlin 1990.

**Bauhaus-Archiv,** Berlin, Allemagne – en 1960, fondation du Verein des Bauhaus-Archivs à Darmstadt, dont Hans Maria Wingler devient le directeur. En 1964, Walter Gropius commence à dessiner et à concevoir de nouveaux bâtiments destinés à accueillir le Bauhaus-Archiv sur la Rosenhöhe à Darmstadt; la maquette est achevée en 1965. En 1967, Gropius suggère de construire le bâtiment à Berlin. 1971, transfert du Bauhaus-Archiv à Berlin. 1979, inauguration du Bauhaus-Archiv dans les nouveaux locaux de Berlin. Le musée possède la collection la plus complète concernant l'histoire du Bauhaus. Il détient les successions de Walter Gropius, Herbert Bayer, Georg Muche. Les méthodes pédagogiques des enseignants sont documentées au moyen de nombreux travaux originaux. De plus, ce musée héberge une collection d'architecture d'une grande richesse ainsi que de nombreux travaux témoignant du développement des ateliers du Bauhaus. Il contient également une importante documentation sur les préliminaires du Bauhaus et l'histoire du New Bauhaus de Chicago. – *Publications, sélection:* «Bauhaus Berlin», 1985; «Experiment Bau-

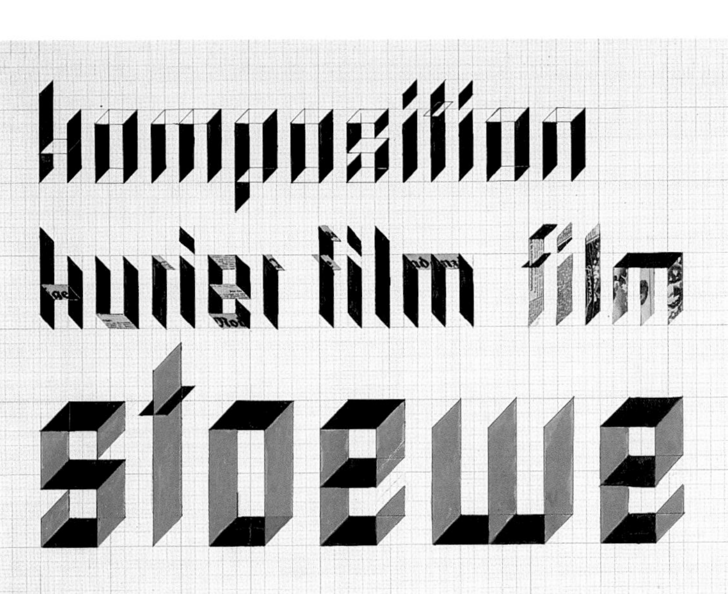

Bauhaus 1930–31

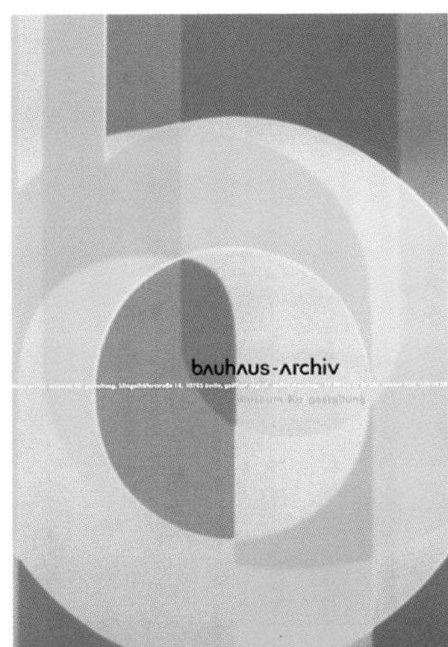


Bauhaus-Archiv 1996 Poster

on 20th July. – *Publications include:* H. Bayer, W. Gropius and I. Gropius (eds.) "Bauhaus 1919–1928", New York 1938, Stuttgart 1955; H. M. Wingler "The Bauhaus: Weimar, Dessau, Berlin and Chicago", Cambridge, Massachusetts 1969; L. Lang "Das Bauhaus 1919–1933, Idee und Wirklichkeit", Berlin 1966; M. Droste "Bauhaus 1919–1933", Berlin 1990.

**Bauhaus Archive,** Berlin, Germany – 1960: Bauhaus Archive is founded in Darmstadt. Founder director is Hans Maria Wingler. 1964: Walter Gropius starts planning and designing new buildings for the Bauhaus Archive on the Rosenhöhe in Darmstadt; the architectural model is completed in 1965. 1967: Gropius suggests moving the new building to Berlin. 1971: the Bauhaus Archive is moved to Berlin. 1979: the Bauhaus Archive is opened in Berlin in the new buildings specially designed for it. The museum's collection is the most extensive in the world on the history of the Bauhaus. It contains unpublished works by Walter Gropius, Herbert Bayer and Georg Muche. The didactic methods of its teachers are documented by numerous original pieces of work. The museum houses an impressive collection of architectural work and documents the development of the Bauhaus workshops. Information on the Bauhaus' past history and the history of the New Bauhaus in Chicago is also part of the museum's display. – *Publications include:* "Bauhaus Berlin", 1985; "Experiment Bauhaus", 1988; "Fotografie am Bauhaus", 1990;

1985; „Experiment Bauhaus", 1988; „Fotografie am Bauhaus", 1990; „Das A und O des Bauhauses", 1995.

**Bauhaus-Museum,** Weimar, Deutschland – 1995 Gründung im Rahmen der Kunstsammlungen zu Weimar. Mit über 500 Exponaten wird die Kunst- und Kunstschulentwicklung von 1900–30 vermittelt, in deren Mittelpunkt das Staatliche Bauhaus in Weimar 1919–25 mit seiner weltweiten Ausstrahlung steht. Umfangreiche Dokumentation der Arbeit Henry van de Veldes in Weimar von 1902–17, besonders als Gründungsdirektor der Großherzoglich Sächsischen Kunstgewerbeschule ab 1907.

**Baum,** Walter – geb. 23.5.1921 in Gummersbach, Deutschland. – *Schriftentwerfer, Grafiker, Lehrer* – 1935–39 Lehre als Schriftsetzer. 1946–48 Studium an der Meisterschule für das gestaltende Handwerk, Offenbach. 1949–72 Leiter des grafischen Ateliers der Bauerschen Gießerei in Frankfurt am Main. 1972–86 Leiter der Kunstschule Westend in Frankfurt am Main. – *Schriftentwürfe:* Volta (1956), Folio (1956–63), Verdi (1957), Impressum (1963), weitere Schriften mit Konrad F. Bauer für die Bauersche Gießerei, Frankfurt am Main.

**Baumann + Baumann** – Barbara Baumann – geb. 4.8.1951 in Niedersachsen, Deutschland. – Studium an der Fachhochschule für Gestaltung Schwäbisch Gmünd. Gerd Baumann – geb. 16.11.1950 in Baden, Deutschland. – Studium an der Fachhochschule für Gestaltung Schwäbisch Gmünd. Während des Studiums gemeinsame gestalterische Arbeiten. 1978

haus», 1988; «Fotografie am Bauhaus», 1990; «Das A und O des Bauhauses», 1995.

**Bauhaus-Museum,** Weimar, Allemagne – fondation en 1995 dans le cadre de la réorganisation des collections d'art de la ville de Weimar. 500 pièces dressent un panorama des mouvements artistiques de 1900 à 1930, la collection est centrée sur le Staatliches Bauhaus de Weimar 1919–1925 et sur son rayonnement dans le monde entier. Elle comprend aussi une documentation étendue sur les travaux de Henry van de Velde à Weimar entre 1902 et 1917, et sur son rôle de directeur et fondateur de la Grossherzoglich Sächsische Kunstgewerbeschule (école des arts décoratifs) en 1907.

**Baum,** Walter – né le 23.5.1921 à Gummersbach, Allemagne – *concepteur de polices, graphiste, enseignant* – 1935–1939, apprentissage de la composition. 1946–1948, études à la Meisterschule für das gestaltende Handwerk (école d'artisanat d'art) d'Offenbach. 1949–1972, dirige l'atelier de graphisme de la fonderie Bauer à Francfort-sur-le-Main. 1972–1986, directeur de la Kunstschule Westend à Francfort-sur-le-Main. – *Polices:* Volta (1956), Folio (1956–1963), Verdi (1957), Impressum (1963); dessine de nombreuses polices pour la fonderie Bauer de Francfort-sur-le-Main, en collaboration avec Konrad F. Bauer.

**Baumann + Baumann** – Barbara Baumann – née le 4.8.1951 en Basse-Saxe, Allemagne – études à la Fachhochschule für Gestaltung (école de design) à Schwäbisch Gmünd. Gerd Baumann – né le 16.11.1950 à Baden, Allemagne – études à la Fachhochschule für Gestaltung à Schwäbisch Gmünd. Réalisent ensemble de nombreux travaux de création pendant leurs études. 1978, fondation de l'agence Baumann + Baumann à Schwä-

Bauhaus-Museum 1994 Card

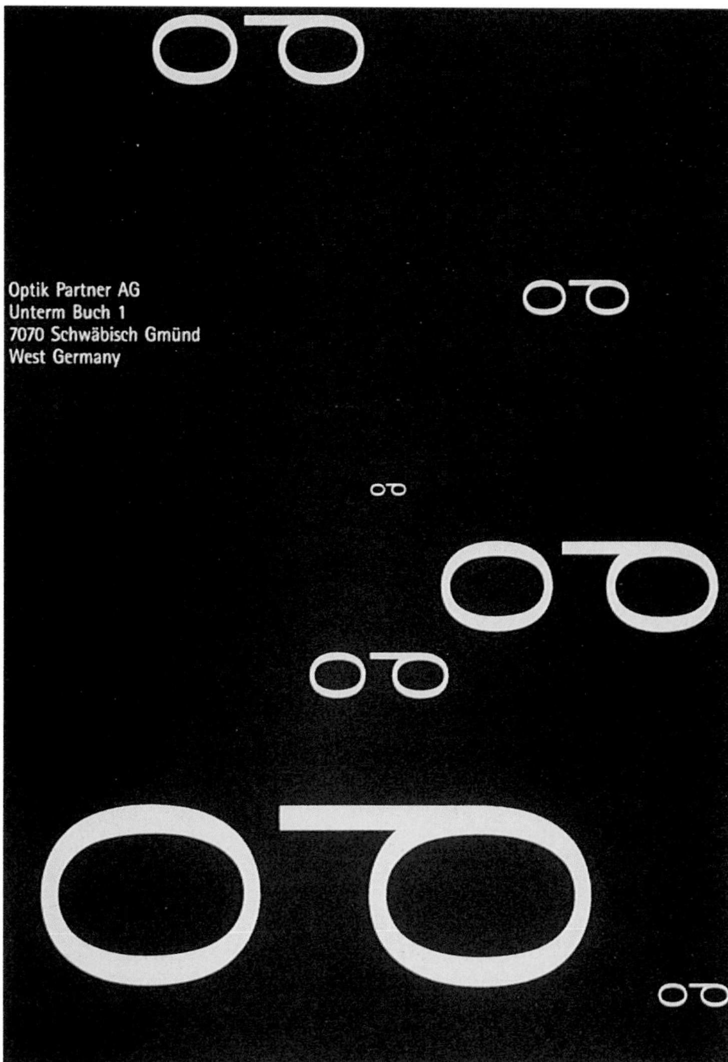

Baumann + Baumann 1989 Poster

| n | n | n | | | |
|---|---|---|---|---|---|
| a | b | e | f | g | i |
| o | r | s | t | y | z |
| A | B | C | E | G | H |
| M | O | R | S | X | Y |
| 1 | 2 | 4 | 6 | 8 | & |

Baum 1956 Volta

"Das A und O des Bauhauses", 1995.

**Bauhaus Museum,** Weimar, Germany – 1995: the museum is founded as part of Weimar's art collection. The development of art and art schools from 1900–30 is illustrated by over 500 exhibits, with the main focus directed at the internationally-influential Staatliches Bauhaus in Weimar from 1919–25. The work of Henry van de Velde in Weimar between 1902–17 is well documented, and especially work performed in his rôle as founder-director of the Großherzoglich Sächsische Kunstgewerbeschule from 1907 onwards.

**Baum,** Walter – b. 23.5.1921 in Gummersbach, Germany – *type designer, graphic artist, teacher* – 1935–39: trains as a typesetter. 1946–48: studies at the Meisterschule für das gestaltende Handwerk, Offenbach. 1949–72: director of the Bauersche Gießerei's graphics studio in Frankfurt am Main. 1972–86: director of the Kunstschule Westend in Frankfurt am Main. – *Fonts:* Volta (1956), Folio (1956–63), Verdi (1957), Impressum (1963), numerous fonts together with Konrad F. Bauer for the Bauersche Gießerei, Frankfurt am Main.

**Baumann + Baumann** – Barbara Baumann – b. 4.8.1951 in Lower Saxony, Germany – Studied at the Fachhochschule für Gestaltung in Schwäbisch Gmünd. Gerd Baumann – b. 16.11.1950 in Baden, Germany – Studied at the Fachhochschule für Gestaltung in Schwäbisch Gmünd. During their studies they worked on designs together. 1978: opening of the Baumann + Baumann studio in Schwä-

Gründung des Büros Baumann + Baumann in Schwäbisch Gmünd. Entwürfe und Gestaltungen für Unternehmen und Kulturinstitutionen. Typographische Arbeiten für die Neubauten des Deutschen Bundestages in Bonn. – *Publikationen u.a.:* „Ein Sechsundzwanzigbuchstabenbuch", München 1987; „lechts und rinks. Orientierungen zwischen Architektur und Parlament", Stuttgart 1995.

**Baumeister,** Willi – geb. 22. 1. 1889 in Stuttgart, Deutschland, gest. 31. 8. 1955 in Stuttgart, Deutschland. – *Maler, Typograph, Werbegestalter, Bühnengestalter, Lehrer* – 1905 Ausbildung als Dekorationsmaler in Stuttgart. Besuch der Zeichenklasse an der Akademie der Bildenden Künste, Stuttgart. 1911 erstes eigenes Atelier in Stuttgart. 1919 Fortsetzung des Studiums, Arbeit als Typograph, Maler und Bühnenausstatter. 1926 verstärkte typographische Arbeit, Vortrag über „Neue Tendenzen in der Typographie" vor dem Grafischen Club, Stuttgart. 1927 Gründungsmitglied des „rings neuer werbegestalter", u. a. mit K. Schwitters, J. Tschichold, R. Michel und W. Dexel. Typographische Arbeiten für die Werkbund-Ausstellung „Die Wohnung" in Stuttgart. 1928 Berufung als Professor für Typographie an die Städelschule, Frankfurt am Main. 1930 Mitglied der Pariser Künstlergruppe „Cercle et Carré". 1933 Entlassung durch die Nationalsozialisten als entarteter Künstler. 1937 in der Ausstellung „Entartete Kunst" in München mit vier Bildern vertreten. 1941 Ausstellungsverbot. 1946 Berufung als Professor an die Kunstakademie Stuttgart. Zahlrei-

Baumann + Baumann 1995 Cover

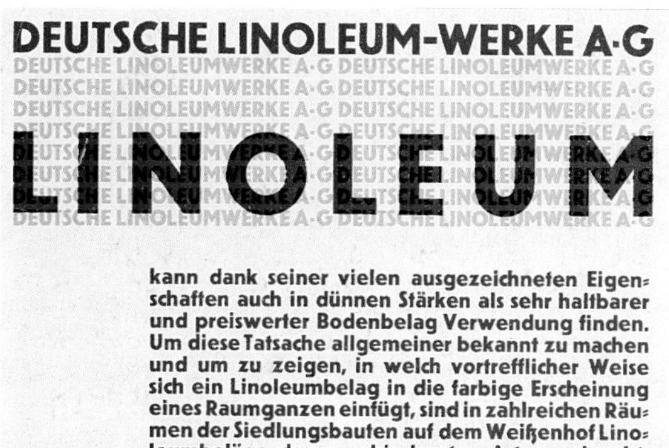


Baumeister 1927 Card

Baumeister 1933 Cover

Baumeister 1930 Cover

l'agence Baumann + Baumann à Schwäbisch Gmünd. Projets et conceptions pour des entreprises et institutions culturelles. Travaux de typographie pour les nouveaux bâtiments du Bundestag à Bonn. – *Publications, sélection:* «Ein Sechsundzwanzigbuchstabenbuch», Munich 1987; «lechts und rinks. Orientierungen zwischen Architektur und Parlament», Stuttgart 1995.

**Baumeister,** Willi – né le 22. 1. 1889 à Stuttgart, Allemagne, décédé le 31.8.1955 à Stuttgart, Allemagne – *peintre, typographe, graphiste publicitaire, scénographe, enseignant* – 1905, formation de peintre décorateur à Stuttgart. Fréquente le cours de dessin de l'Akademie der Bildenden Künste (beaux-arts) de Stuttgart. 1911, premier atelier à Stuttgart. 1919, après une interruption il reprend ses études à la fin de la Première Guerre mondiale et travaille comme typographe, peintre et scénographe. 1926, se consacre davantage aux travaux de typographie, conférence intitulée «Neue Tendenzen in der Typographie» (Nouvelles tendances en typographie) au Grafischer Club de Stuttgart. 1927, membre fondateur du «ring neuer werbegestalter» avec K. Schwitters, J. Tschichold, R. Michel, W. Dexel. Travaux de typographie pour l'exposition «Die Wohnung», organisée par le Werkbund à Stuttgart. 1928, nommé professeur de typographie à la Städelschule de Francfort-sur-le-Main. 1930, membre du groupe d'artistes parisiens «Cercle et Carré». 1933, licencié par les nationalsocialistes qui le considèrent comme un artiste dégénéré. 1937, présent dans l'exposition «Entartete Kunst» (art dégénéré) de Munich avec quatre tableaux. 1941, frappé d'interdiction d'exposer. 1946, nommé professeur à la Kunstakademie (beaux-arts) de Stuttgart. Participe à de nombreuses expositions

bisch Gmünd. The Baumanns have designed for various companies and cultural institutions. They have produced typographical work for the new Bundestag buildings in Bonn. – *Publications include:* "Ein Sechsundzwanzigbuchstabenbuch", Munich 1987; "lechts und rinks. Orientierungen zwischen Architektur und Parlament", Stuttgart 1995.

**Baumeister,** Willi – b. 22. 1. 1889 in Stuttgart, Germany, d. 31. 8. 1955 in Stuttgart, Germany – *painter, typographer, commercial artist, stage designer, teacher* – 1905: trains as an interior decorator in Stuttgart. Attends drawing classes at the Akademie der Bildenden Künste, Stuttgart. 1911: opens his first studio in Stuttgart. 1919: after the First World War, he continues his studies and works as typographer, painter and stage designer. 1926: increases his output of typographical work and holds a lecture on "New Tendencies in Typography" for the Grafischer Club, Stuttgart. 1927: founder member of the ring neuer werbegestalter with K. Schwitters, J. Tschichold, R. Michel and W. Dexel, among others. Produces typographical work for the Deutscher Werkbund experimental housing exhibition, "Die Wohnung", in Stuttgart. 1928: appointed professor of typography at the Städelschule in Frankfurt am Main. 1930: member of the Parisian group Cercle et Carré. 1933: dismissed as a degenerate artist by the National Socialists. 1937: has four of his pictures displayed in the exhibition of "degenerate art" in Munich. 1941: forbidden to exhibit. 1946: professorship at the art academy in Stuttgart.

che internationale Ausstellungen. – *Publikationen u.a.:* „Neue Typographie", in „Form" 10/1926. Heinz Spielmann „Willi Baumeister, das grafische Werk", Hamburg 1972; Wolfgang Kermer (Hrsg.) „Willi Baumeister, Typographie und Werbegestaltung", Stuttgart 1989.

**Baur,** Ruedi – geb. 5. 3. 1956 in Paris, Frankreich. – *Grafik-Designer, Lehrer* – 1975–79 Studium an der Kunstgewerbeschule Zürich. 1979–80 Art Director im Atelier Ballmer, Basel. 1980 Gründung des Ateliers Plus design. 1981 Mitbegründer der Ateliers BBV Lyon, Zürich, Mailand mit M. Baviera und P. Vetter. 1988 Umzug nach Paris, Arbeiten für verschiedene Kulturinstitutionen. Zahlreiche Lehraufträge an europäischen Schulen. Seit 1995 Professor in Leipzig, Deutschland.

**Bayer,** Herbert – geb. 5. 4. 1900 in Haag, Österreich, gest. 30. 9. 1985 in Montecito, USA. – *Grafik-Designer, Typograph, Ausstellungsarchitekt, Maler, Fotograf, Lehrer* – 1919–20 Lehrling im Architekturbüro Schmidthammer, Linz. 1921–25 Studium am Staatlichen Bauhaus in Weimar. 1925 Leiter der Werkstatt für Typographie und Werbegestaltung am Bauhaus in Dessau. Verläßt 1928 das Bauhaus. 1928–29 Art Director der Zeitschrift „Vogue", Berlin, 1928–38 Direktor der Agentur Dorland Berlin, Mitarbeit an der Zeitschrift „Die Neue Linie". 1938 Umzug nach New York, USA. Mit Walter und Ise Gropius Gestaltung der Ausstellung „Bauhaus 1919–28" im Museum of Modern Art in New York. Berater von Verlagen und Industrieunternehmen. 1940

internationales. – *Publications, sélection:* «Neue Typographie» dans «Form» 10/1926. Heinz Spielmann «Willi Baumeister, das grafische Werk», Hambourg 1972; Wolfgang Kermer (éd.) «Willi Baumeister, Typographie und Werbegestaltung», Stuttgart 1989.

**Baur,** Ruedi – né le 5. 3. 1956 à Paris, France – *graphiste maquettiste, enseignant* – 1975– 1979, études à la Kunstgewerbeschule (école des arts décoratifs) de Zurich. 1979–1980, directeur artistique à l'atelier Ballmer, Bâle. 1980, création de l'atelier Plus design. 1981, cofondateur de l'atelier BBV, Lyon, Zurich, Milan avec M. Baviera et P. Vetter. 1988, s'installe à Paris où il travaille pour plusieurs institutions culturelles. De nombreuses interventions dans des écoles européennes. Depuis 1995, professeur à Leipzig, Allemagne.

**Bayer,** Herbert – né le 5. 4. 1900 à Haag, Autriche, décédé le 30. 9. 1985 à Montecito, Etats-Unis – *graphiste maquettiste, typographe, architecte d'expositions, peintre, photographe, enseignant* – 1919–1920, apprenti à l'étude d'architecture Schmidthammer à Linz. 1921–1925, études au Staatliches Bauhaus de Weimar. 1925, responsable de l'atelier de typographie et de publicité au Bauhaus de Dessau. Quitte le Bauhaus en 1928. 1928–1929, directeur artistique de la revue «Vogue», Berlin; 1928–1938 directeur de l'agence Dorland, Berlin, collabore à la revue «Die Neue Linie». S'installe à New York, aux Etats-Unis, en 1938. Conçoit l'exposition «Bauhaus 1919–1928» au Museum of Modern Art New York avec Walter et Ise Gropius. Conseiller de plusieurs éditeurs et entreprises. 1940, chargé de cours pour «The Advertising Guild». 1944, directeur artistique à l'agence Thompson. Conseiller en création pour le centre culturel d'Aspen,

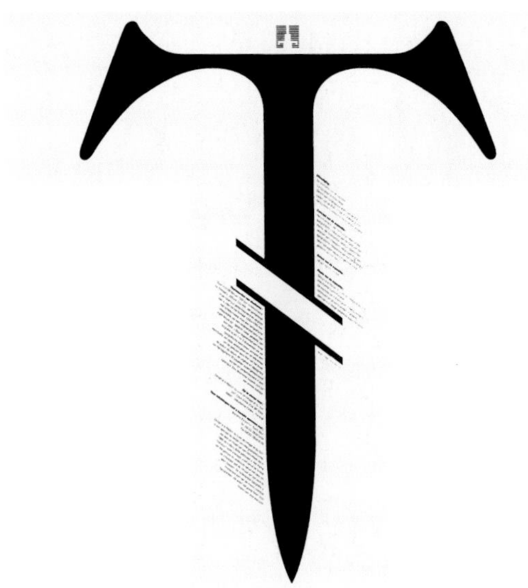

Baur Folder

Bayer 1923 Cover

Baur 1993 Cover

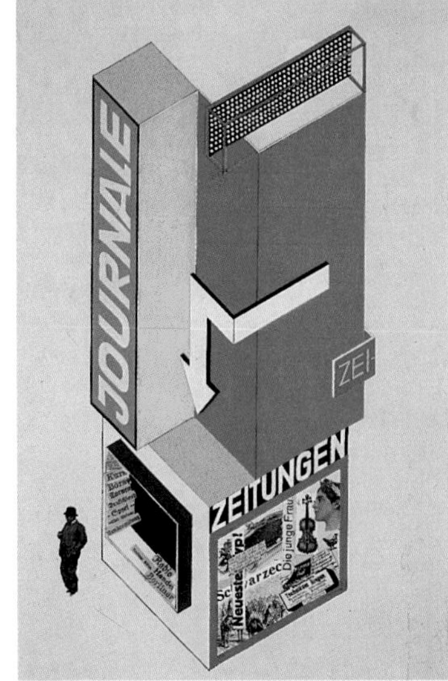


Bayer 1924 Newspaper Stand Design

Numerous exhibitions all over the world. – *Publications include:* "Neue Typographie" in "Form" 10/1926. Heinz Spielmann "Willi Baumeister, das grafische Werk", Hamburg 1972; Wolfgang Kermer (ed.) "Willi Baumeister, Typographie und Werbegestaltung", Stuttgart 1989.

**Baur,** Ruedi – b. 5. 3. 1956 in Paris, France – *graphic designer, teacher* – 1975–79: studies at the Kunstgewerbeschule in Zurich. 1979–80: art director for the Ballmer studio in Basle. 1980: opens the graphics studio Plus design. 1981: sets up the BBV studios in Lyon, Zurich and Milan with M. Baviera and P. Vetter. 1988: moves to Paris and works for various cultural institutions. Numerous teaching positions at European schools. Since 1995: professor in Leipzig, Germany.

**Bayer,** Herbert – b. 5. 4. 1900 in Haag, Austria, d. 30. 9. 1985 in Montecito, USA – *graphic designer, typographer, exhibition architect, painter, photographer, teacher* – 1919–20: apprentice at the Schmidthammer architectural office in Linz. 1921–25: studies at the Staatliche Bauhaus in Weimar. 1925: director of the workshop for typography and commercial design at the Bauhaus in Dessau. 1928: leaves the Bauhaus. 1928–29: art director for "Vogue" magazine in Berlin and 1928–38 director of the Dorland agency in Berlin; works on the magazine "Die neue Linie". 1938: moves to New York, USA. Together with Walter and Ise Gropius, he plans the exhibition "Bauhaus 1919–28" in the Museum of Modern Art in New York. Consultant to publishing houses and industrial concerns. 1940:

Lehrauftrag bei „The Advertising Guild". 1944 Art Director bei der Agentur Thompson. Seit 1946 Gestaltungsberater des Kulturzentrums in Aspen, Colorado. 1946–65 Berater und Leiter der Design-Abteilung der Container Corporation of America. 1948–53 Autor und Gestalter des World Geographic Atlas, 1966–85 Kunst- und Designberater der Atlantic Richfield Company, Los Angeles. 1968 Gestaltung des Katalogs und der Ausstellung „50 Jahre Bauhaus" in Stuttgart. Zahlreiche internationale Ausstellungen und Auszeichnungen. – *Schriftentwürfe:* Universal (1925–30), Bayer-Type (1930–36, Berthold AG). – *Publikationen u.a.:* „Versuch einer neuen Schrift" in „Offset" 7/1926; „Herbert Bayer", Ravensburg 1967. Alexander Dorner „The work of Herbert Bayer", New York 1947; Arthur A. Cohen „Herbert Bayer. The Complete Work", London 1984.

**Beauclair,** Gotthard de – geb. 24. 7. 1907 in Ascona, Schweiz, gest. 31. 3. 1992 in Freiburg, Deutschland. – *Buchgestalter, Lyriker, Verleger* – Ausbildung an der Werkkunstschule in Offenbach (bei Rudolf Koch) und an der Akademie für graphische Künste und Buchgewerbe in Leipzig. 1928 Arbeit als Hersteller im Insel Verlag, Leipzig. 1946–50 Buchgestalter im Scherpe Verlag, Krefeld. 1951 Gründung der Trajanus Presse mit der Schriftgießerei D. Stempel AG in Frankfurt am Main. 1952–62 künstlerischer Leiter des Insel Verlags in Frankfurt am Main. 1962 Gründung seines Verlags „Ars librorum" in Frankfurt am Main. Arbeiten für den Propyläen Verlag in Berlin und die Württembergische

Bayer 1926 Poster

Bayer 1928 Cover

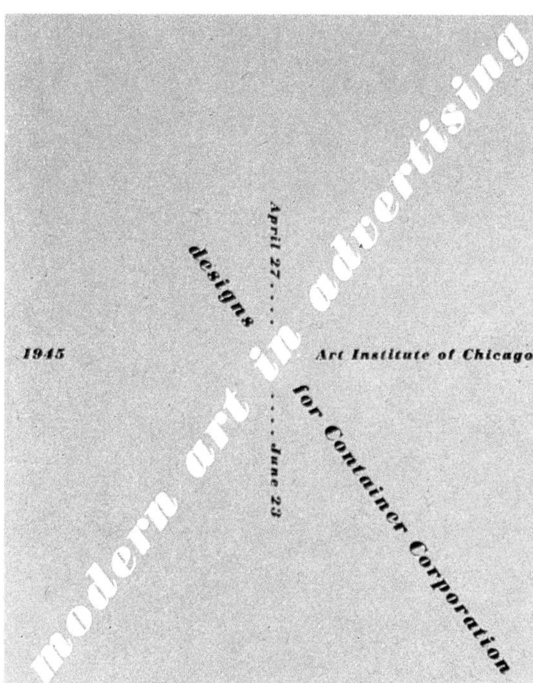

Bayer 1945 Cover

Beauclair 1935 Cover

Colorado, à partir de 1946. 1946–1965, conseiller et directeur du service de design de la Container Corporation of America. 1948–1953, auteur et maquettiste du World Geographic Atlas; 1966–1985, conseiller artistique du design pour la Atlantic Richfield Company, Los Angeles. 1968, maquette du catalogue de l'exposition «50 Jahre Bauhaus» à Stuttgart. De nombreuses expositions internationales et prix. – *Polices:* Universal (1925–30), Bayer Type (1930–1936, Berthold AG). – *Publications, sélection:* «Versuch einer neuen Schrift» dans «Offset» 7/1926; «Herbert Bayer», Ravensburg 1967. Alexander Dorner «The work of Herbert Bayer», New York 1947; Arthur A. Cohen «Herbert Bayer. The Complete Work», Londres 1984.

**Beauclair,** Gotthard de – né le 24. 7. 1907 à Ascona, Suisse, décédé le 31. 3. 1992 à Fribourg, Allemagne – *maquettiste en édition, poète, éditeur* – études à la Werkkunstschule d'Offenbach (chez Rudolf Koch) et à l'Akademie für graphische Künste und Buchgewerbe (arts graphiques et du livre) de Leipzig. 1928, travaille comme producteur pour la maison d'édition Insel á Leipzig. 1946–1950, maquettiste pour les éditions Scherpe à Krefeld. 1951 fonde avec la fonderie de caractères D. Stempel AG le Trajanus Presse à Francfort-sur-le-Main. 1952–1962, directeur artistique des éditions Insel, Francfort-sur-le-Main. Travaux pour la maison d'édition Propyläen à Berlin et pour la Württembergische Bibelanstalt de Stuttgart. 1966–1981, fonde et devient

teaches at The Advertising Guild. 1944: art director for the Thompson agency. From 1946 onwards: design consultant for the arts center in Aspen, Colorado. 1946–65: consultant and director for the design department of the Container Corporation of America. 1948–53: author and designer of the World Geographic Atlas. 1966–85: consultant for art and design to the Atlantic Richfield Company in Los Angeles. 1968: designs the catalogue and plans the exhibition "50 Years of the Bauhaus" in Stuttgart. Bayer staged numerous exhibitions and won many awards worldwide. – *Fonts:* Universal (1925–30), Bayer-Type (1930–36, Berthold AG). – *Publications include:* "Versuch einer neuen Schrift" in "Offset" 7/1926; "Herbert Bayer", Ravensburg 1967. Alexander Dorner "The work of Herbert Bayer", New York 1947; Arthur A. Cohen "Herbert Bayer. The Complete Work", London 1984.

**Beauclair,** Gotthard de – b. 24. 7. 1907 in Ascona, Switzerland, d. 31. 3. 1992 in Freiburg, Germany – *book designer, lyric poet, publisher* – Studied at the Werkkunstschule in Offenbach (under Rudolf Koch) and at the Akademie für graphische Künste und Buchgewerbe in Leipzig. 1928: works as a producer for the Insel publishing house in Leipzig. 1946–50: designer of books for the Scherpe publishing house in Krefeld. 1951: the Trajanus Presse with D. Stempel AG type foundry is founded in Frankfurt am Main. 1952–62: art director of Insel publishers in Frankfurt am Main. 1962: sets up his own publishing house Ars Librorum, in

Bibelanstalt in Stuttgart. 1966–81 Gründung und Leitung der „Edition de Beauclair", in der Mappenwerke und Grafikfolgen zeitgenössischer Künstler erscheinen. 1981 Weiterführung der „Edition de Beauclair" durch Mario Lobmeyr in München. Er wurde mit der Goldmedaille auf der Internationalen Ausstellung Paris 1937 und mit der Goldmedaille auf der Triennale in Mailand 1954 ausgezeichnet. Ausstellungen u.a. in Kopenhagen, Oslo, Stockholm (1960), Helsinki, London (1961), Zürich (1971) und Frankfurt am Main (1987). – *Publikationen u. a.:* „Triangel des Glücks", Leipzig 1929; „Das Buch Sesam", Hamburg 1951; „Zeit, Überzeit", Hamburg 1977; „Sang im Gegenwind", Hamburg 1983. H. Schmoller „Gotthard de Beauclair", Frankfurt am Main 1960; J. A. Kruse (Hrsg.) „Gotthard de Beauclair. Lyriker, Buchgestalter, Verleger", Düsseldorf 1972; „Gotthard de Beauclair. Leben und Werk", Siegburg 1997.

**Beeke,** Anthon – geb. 11. 3. 1940 in Amsterdam, Niederlande. – *Typograph, Grafik-Designer, Lehrer* – Studium an der Kunstschule Amsterdam, danach Assistent verschiedener Designer in Düsseldorf, Brüssel und Amsterdam. 1976–82 Direktor der Agentur „Total Design", Arbeiten für das Stedelijk Museum, Amsterdam und das Van Gogh Museum, Amsterdam. Designer und künstlerischer Leiter der Zeitschriften „Avenue", „Kunstschrift" und „Quote". 1982–85 Büro „Beeke & van Bree BV" mit Rene van Bree. 1985–87 Büro „Anthon Beeke & Associates" mit drei Partnern. Büro „Studio Anthon Beeke BV". Seit 1985 Gastvorle-

directeur des «Edition de Beauclair» qui publient des gravures et séries de gravures d'artistes contemporains. 1981, Mario Lobmeyr continue les activités des «Edition de Beauclair» à Munich. Il reçoit le médaille d'or de l'exposition universelle de Paris en 1937 et le médaille d'or de la Triennale de Milan en 1954. Expositions: Copenhague, Oslo, Stockholm (1960); Helsinki, Londres (1961), Zurich (1971) et Francfort-sur-le-Main (1987). – *Publications, sélection:* «Triangel des Glücks», Leipzig 1929; «Das Buch Sesam», Hambourg 1951; «Zeit, Überzeit», Hambourg 1977; «Sang im Gegenwind», Hambourg 1983. Hans Schmoller «Gotthard de Beauclair», Francfort-sur-le-Main 1960; J. A. Kruse (éd.) «Gotthard de Beauclair. Lyriker, Buchgestalter, Verleger», Düsseldorf 1972; «Gotthard de Beauclair. Leben und Werk», Siegburg 1997.

**Beeke,** Anthon – né le 11. 3. 1940 à Amsterdam, Pays-Bas – *typographe, graphiste maquettiste, enseignant* – études à l'école d'art d'Amsterdam puis assistant de plusieurs designers à Düsseldorf, Bruxelles et Amsterdam. 1976–1982, directeur de l'agence «Total Design», travaux pour le Stedelijk Museum et pour le Van Gogh Museum, tous deux à Amsterdam. Designer et directeur artistique des revues «Avenue», «Kunstschrift» et «Quote». 1982–1985, agence «Beeke & van Bree BV» avec René van Bree. 1985–1987, cabinet «Anthon Beeke & Associates» avec trois associés. Agence «Studio Anthon Beeke BV». Depuis 1985, conférences à la Cranbrook Academy of Arts et à la Rhode Island School of Design. – *Police:* Beeke-Alphabet (1970) – *Publications:* D. Kuilman «Anthon Beeke» dans «Affiche» 3, 4/1992; Pieter Brattinga «Affiches van Anthon Beeke», Amsterdam 1993.

**Behmer,** Marcus – né le 1. 10. 1879 à Weimar, Allemagne, décédé le 6. 9. 1958 à

Beeke 1970 Alphabet

Beeke 1970 Alphabet

Behmer 1920 Page

Beeke 1989 Poster

Behmer 1931 Cover

Frankfurt am Main. Works for the Propyläen publishing house in Berlin and the Württembergische Bibelanstalt in Stuttgart. 1966–81: founder and director of Edition de Beauclair, publishing single-sheet graphic edition and series of graphics by contemporary artists. 1981: Mario Lobmeyr in Munich continues the work of the Edition de Beauclair. He is awarded a gold medal at the International Exhibition in Paris, 1937 and a gold medal at the Triennale Art Festival in Milan, 1954. Exhibitions include: Copenhagen, Oslo, Stockholm (1960); Helsinki, London (1961); Zurich (1971) and Frankfurt am Main (1987). – *Publications include:* "Triangel des Glücks". Leipzig 1929; "Das Buch Sesam", Hamburg 1951; "Zeit, Überzeit", Hamburg 1977; "Sang im Gegenwind", Hamburg 1983. Hans Schmoller "Gotthard de Beauclair", Frankfurt am Main 1960; J. A. Kruse (ed.) "Gotthard de Beauclair. Lyriker, Buchgestalter, Verleger", Düsseldorf 1972; "Gotthard de Beauclair. Leben und Werk", Siegburg 1997.

**Beeke,** Anthon – b. 11. 3. 1940 in Amsterdam, The Netherlands – *typographer, graphic designer, teacher* – Studied at the Amsterdam School of Art, then acted as assistant to various designers in Düsseldorf, Brussels and Amsterdam. 1976–82: director of the agency Total Design. Works for the Stedelijk Museum and the Van Gogh Museum in Amsterdam. Designer and art director of the magazines "Avenue", "Kunstschrift" and "Quote". 1982– 85: runs the Beeke & van Bree BV studio with Rene van Bree. 1985–87: sets

sungen an der Cranbrook Academy of Arts und an der Rhode Island School of Design. – *Schriftentwurf:* Beeke-Alphabet (1970). – *Publikationen u. a.:* D. Kuilman „Anthon Beeke" in „Affiche" 3, 4/1992; Pieter Brattinga „Affiches van Anthon Beeke", Amsterdam 1993.

**Behmer,** Marcus – geb. 1. 10. 1879 in Weimar, Deutschland, gest. 6. 9. 1958 in Berlin, Deutschland. – *Schriftentwerfer, Grafiker, Illustrator* – 1897–1900 Lehre als Dekorationsmaler in München. 1900–01 Aufenthalt in Frankreich. Lebt 1905–09 in Florenz, Italien. Für die Zeitschriften „Insel", „Ver Sacrum" und „Simplizissimus" entwarf Behmer Zeichnungen. Zahlreiche Buchausstattungen. Zahlreiche Ex-Libris-Entwürfe. – *Schriftentwürfe:* Stefan-George-Schrift (1904), Behmer Antiqua (1920), Soni co Hebräisch (1933).

**Behrens,** Peter – geb. 14. 4. 1868 in Hamburg, Deutschland, gest. 27. 2. 1940 in Berlin, Deutschland. – *Architekt, Formgeber, Grafik-Designer, Schriftentwerfer, Lehrer* – 1886–91 Studien zur Malerei an den Akademien Karlsruhe, Düsseldorf und München. 1891–99 als Maler, Typograph und Formgeber in München tätig. 1898 Mitarbeit an der Zeitschrift „Pan". Berufung an die Künstlerkolonie Darmstadt. 1903–07 Direktor der Kunstgewerbeschule Düsseldorf. 1907–14 künstlerischer Beirat der AEG Berlin, Entwicklung des ersten Erscheinungsbildes einer Firma (AEG), vom Briefbogen bis zur Arbeitersiedlung. 1913 Mitbegründer und Vorstandsmitglied des Deutschen Werkbunds. 1922–27 Direktor der Architekturabteilung an der Kunstakademie Wien.

Berlin, Allemagne – *concepteur de polices, graphiste, illustrateur* – 1897–1900, apprentissage de peintre décorateur à Munich. 1900–1901, séjour en France. 1905–1909, vit à Florence, Italie. Dessins pour les revues «Insel», «Ver Sacrum» et «Simplizissimus». De nombreuses maquettes de livres pour divers éditeurs. Maquettes d'ex-libris. – *Polices:* Stefan George-Schrift (1904), Behmer Antiqua (1920), Soni co hébraïque (1933).

**Behrens,** Peter – né le 14. 4. 1868 à Hambourg, Allemagne, décédé le 27. 2. 1940 à Berlin, Allemagne – *architecte, maître de forme, graphiste maquettiste, concepteur de polices, enseignant* – 1886–1891, étudie la peinture aux académies de Karlsruhe, Düsseldorf et Munich. 1891–1899, peintre, typographe et maître de forme à Munich. 1898, collabore à la revue «Pan». Est nommé à la colonie d'artistes de Darmstadt. 1903–1907, directeur de la Kunstgewerbeschule (école des arts décoratifs) de Düsseldorf. 1907–1914, membre du comité artistique d'AEG à Berlin, crée la première ligne graphique et l'identité de la société (AEG) qui se retrouve aussi bien sur le papier à lettres que dans les cités ouvrières. 1913, cofondateur et membre du comité directeur du Deutscher Werkbund. 1922–1927, directeur du département architecture de la Kunstakademie (beaux-arts) de Vienne. 1936, successeur de Hans Poelzig à la direction de la Meisterschule für Architek-

Behmer 1935 Main Title

Behrens 1899 Initial

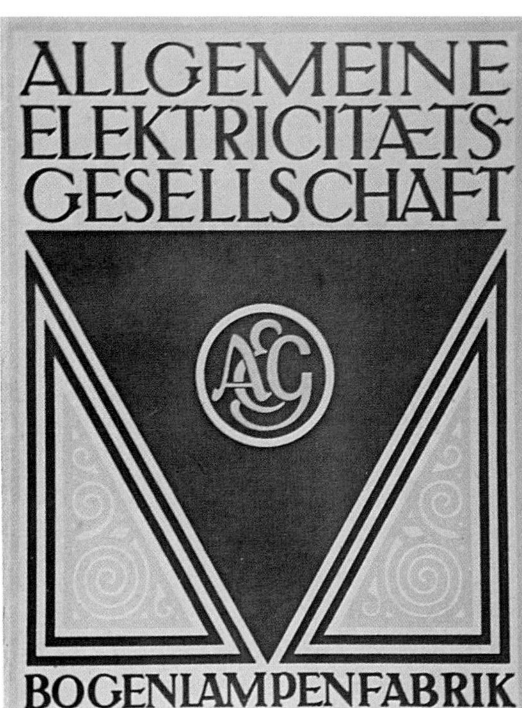

Behrens 1908 Cover

Behrens 1908 Cover

Behrens 1901–07 Behrens-Schrift

up the studio Anthon Beeke & Associates with three partners. Studio Anthon Beeke BV. Since 1985: visiting lecturer at the Cranbrook Academy of Arts and at the Rhode Island School of Design. – *Font:* Beeke-Alphabet (1970). – *Publications include:* D. Kuilman "Anthon Beeke" in "Affiche" 3, 4/1992; P. Brattinga "Affiches van Anthon Beeke", Amsterdam 1993.

**Behmer,** Marcus – b. 1. 10. 1879 in Weimar, Germany, d. 6. 9. 1958 in Berlin, Germany – *type designer, graphic artist, illustrator* – 1897–1900: trains as an interior decorator in Munich. 1900–01: lives in France. 1905–09: lives in Florence, Italy. Behmer drew for the magazines "Insel", "Ver Sacrum" and "Simplizissimus" and produced numerous book designs for publishers and Ex-Libris designs. – *Fonts:* Stefan George-Schrift (1904), Behmer Antiqua (1920), Soni co Hebrew (1933).

**Behrens,** Peter – b. 14. 4. 1868 in Hamburg, Germany, d. 27. 2. 1940 in Berlin, Germany – *architect, designer, graphic designer, type designer, teacher* – 1886–91: studies painting at the academies in Karlsruhe, Düsseldorf and Munich. 1891–99: works as a painter, typographer and designer in Munich. 1898: works on "Pan" magazine. Appointed to the Künstlerkolonie in Darmstadt. 1903–07: director of the Kunstgewerbeschule in Düsseldorf. 1907–14: artistic advisor to AEG in Berlin; develops the first corporate design package (for AEG), from company notepaper to workers' housing estates. 1913: one of the founders and committee members of the Deutscher Werkbund.

1936 Nachfolger Hans Poelzigs als Leiter der Meisterschule für Architektur an der Preußischen Akademie der Künste in Berlin. – *Schriftentwürfe:* Behrens-Schrift (1901–07), Behrens-Antiqua (1907–09), Behrens Mediaeval (1914). – *Publikationen u. a.:* F. Meyer-Schönbrunn „Peter Behrens", Hagen 1913; H. Lanzke „Peter Behrens. 50 Jahre Gestaltung in der Industrie", Berlin 1958; T. Buddensieg, H. Rogge „Industriekultur – Peter Behrens und die AEG", Berlin 1978.

**Beltrán,** Félix – geb. 23. 6. 1938 in Havanna, Kuba. – *Grafik-Designer, Typograph, Maler, Autor, Lehrer* – 1956–62 Studium an der School of Visual Arts, an der American Art School, an der New School for Social Research und am Pratt Institute, New York. 1962 Rückkehr nach Kuba, Arbeit als Grafik-Designer und Design-Consultant in Havanna. 1965–66 Studium am Circulo de Bellas Artes, Madrid. 1967 Design Director des kubanischen Pavillons auf der Expo 67 in Montreal, Kanada. Zahlreiche Lehrtätigkeiten in Havanna. Seit 1984 mexikanischer Staatsbürger, Professor an der Universidad Autonoma Metropolitana und an der Universidad Ibero Americana, Mexiko. Seit 1963 zahlreiche Einzelausstellungen, u. a. in Stockholm, Sofia, Barcelona, Prag, Leipzig, Darmstadt und Caracas. 1974 Entwurf der ersten kyrillischen Schrift in Lateinamerika. – *Publikationen u. a.:* „Desde el Diseño", Havanna 1970; „Letragrafica", Havanna 1973; „Acerca del Diseño", Havanna 1974; „Artes Plasticas"(mit Ramon Cabrera), Havanna 1981.

**Belwe,** Georg – geb. 12. 8. 1878 in Berlin,

tur à la Preussische Akademie der Künste à Berlin (école d'architecture de l'académie des beaux-arts de Prusse). – *Polices:* Behrens-Schrift (1901–1907), Behrens-Antiqua (1907–1909), Behrens Mediaeval (1914). – *Publications, sélection:* F. Meyer-Schönbrunn «Peter Behrens», Hagen 1913; H. Lanzke «Peter Behrens. 50 Jahre Gestaltung in der Industrie», Berlin 1958; T. Buddensieg, H. Rogge «Industriekultur – Peter Behrens und die AEG», Berlin 1978.

**Beltrán,** Félix – né le 23. 6. 1938 à La Havane, Cuba – *graphiste maquettiste, typographe, peintre, auteur, enseignant* – 1956–1962, études à la School of Visual Arts, à la American Art School, à la New School for Social Research et au Pratt Institute de New York. 1962, retour à Cuba où il travaille comme graphiste maquettiste et consultant en design à La Havane. 1965– 1966, études au Circulo de Bellas Artes, Madrid. 1967, directeur du design pour le pavillon cubain de l'exposition universelle de Montréal, Canada. Nombreuses activités d'enseignement à La Havane. Citoyen mexicain depuis 1984, et professeur à l'Universidad Autonoma Metropolitana et à l'Universidad Ibero Americana de Mexico. De nombreuses expositions personnelles depuis 1963 dont Stockholm, Sofia, Barcelone, Prague, Leipzig, Darmstadt et Caracas. 1974, élaboration de la première écriture cyrillique d'Amérique latine. – *Publications , sélection:* «Desde el Diseño», La Havane 1970; «Letragrafica», La Havane 1973; «Acerca del Diseño», La Havane 1974; «Artes Plasticas» (avec Ramon Cabrera), La Havane 1981.

**Belwe,** Georg – né le 12. 8. 1878 à Berlin, Allemagne, décédé en 1954 à Ronneburg, Allemagne – *typographe, concepteur de polices, enseignant* – études à l'institut d'enseignement du Königliches Kunstge-

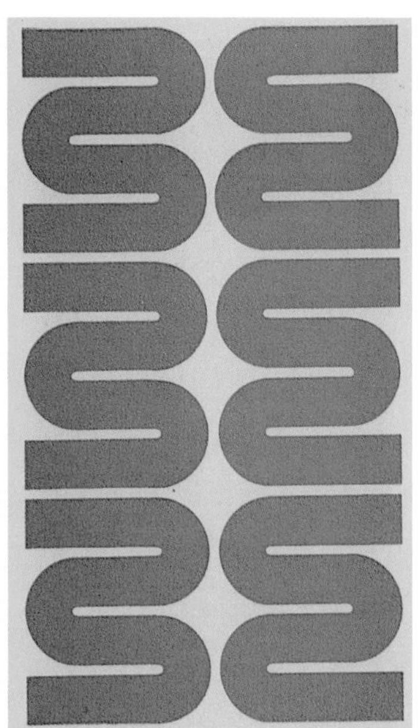

Beltrán 1968 Cover

Beltrán 1969 Poster

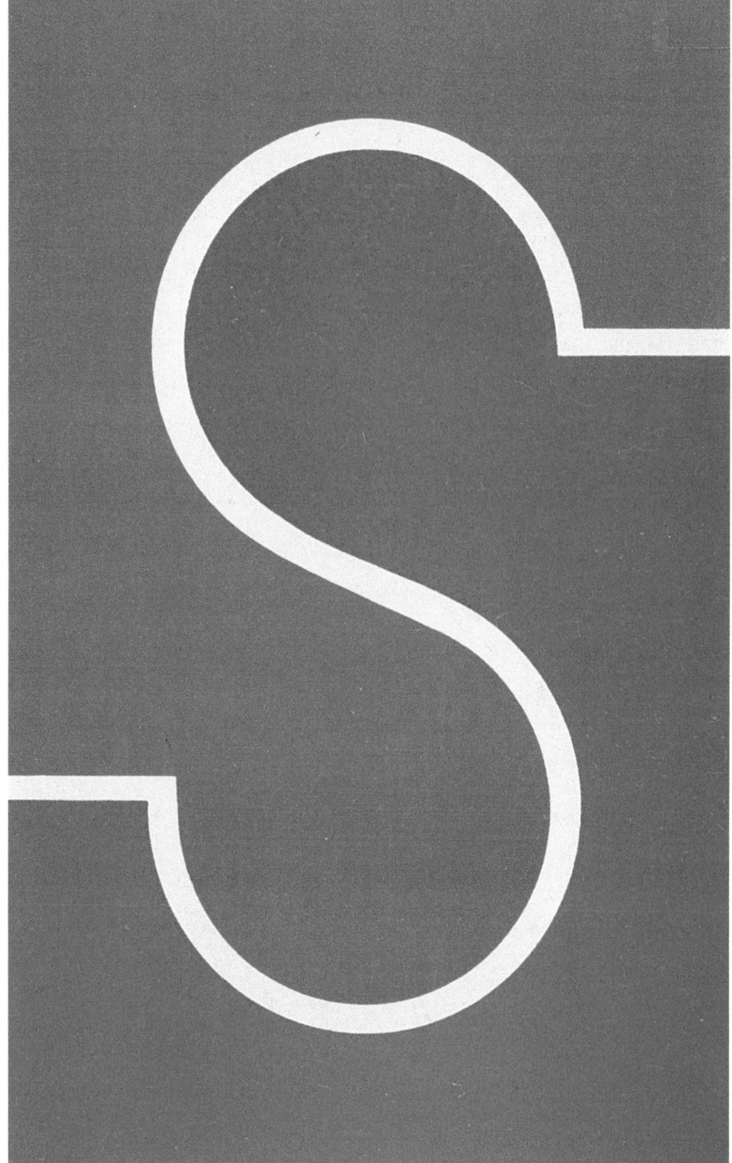

Beltrán 1984 Cover

1922–27: director of the department of architecture at the Kunstakademie in Vienna. 1936: succeeds Hans Poelzig as director of the Meisterschule für Architektur at the Preußische Akademie der Künste in Berlin. – *Fonts:* Behrens-Schrift (1901– 07), Behrens-Antiqua (1907–09), Behrens Mediaeval (1914). – *Publications include:* F. Meyer-Schönbrunn "Peter Behrens", Hagen 1913; H. Lanzke "Peter Behrens. 50 Jahre Gestaltung in der Industrie", Berlin 1958; T. Buddensieg, H. Rogge "Industriekultur – Peter Behrens und die AEG", Berlin 1978.

**Beltrán,** Félix – b. 23. 6. 1938 in Havana, Cuba – *graphic designer, typographer, painter, author, teacher* – 1956–62: studies at the School of Visual Arts, at the American Art School, at the New School for Social Research and at the Pratt Institute in New York. 1962: returns to Cuba and works as a graphic designer and design consultant in Havana. 1965–66: studies at the Circulo de Bellas Artes, Madrid. 1967: design director for the Cuban pavilion at Expo 67 in Montreal, Canada. Numerous teaching positions in Havana. Since 1984: citizen of Mexico, professor at the Universidad Autonoma Metropolitana and at the Universidad Ibero Americana in Mexico. Since 1963, numerous solo exhibitions in Stockholm, Sofia, Barcelona, Prague, Leipzig, Darmstadt and Caracas, among other locations. 1974: designs the first Cyrillic alphabet in Latin America. – *Publications include:* "Desde el Diseño", Havana 1970; "Letragrafica", Havana 1973; "Acerca del Diseño", Havana 1974; "Artes Plasticas" (with

Deutschland, gest. 1954 in Ronneburg, Deutschland. – *Typograph, Schriftentwerfer, Lehrer* – Studium an der Unterrichtsanstalt am Königlichen Kunstgewerbemuseum Berlin. 1900 Gründung der Steglitzer Werkstatt mit F. H. Ehmcke und F. W. Kleukens. Eintritt in die Kunstgewerbeschule Berlin als Lehrender. 1906 Leiter der typographischen Abteilung und der Klasse für exaktes Zeichnen an der Leipziger Akademie für graphische Künste und Buchgewerbe. Zahlreiche Arbeiten für Verlage, u. a. List, Reclam, Eugen Diederichs und Westermann. – *Schriftentwürfe*: Belwe (1907), Belwe Gotisch (1912), Belwe Schrägschrift (1913), Belwe halbfett (1914), Wieland (1926), Schönschrift Mozart (1927). – *Publikation u. a.:* Albert Mundt „Georg Belwe und seine Klasse an der Königlichen Akademie für graphische Künste und Buchgewerbe" in „Archiv für Buchgewerbe" 6/1910.

**Benguiat,** Ed (Ephram Edward) – geb. 27. 10. 1927 in New York, USA. – *Schriftentwerfer, Kalligraph* – Studium an der Columbia University, New York; Workshop School of Advertising Art, New York. 1953 Associate Director des „Esquire". Eröffnung seines Design-Studios in New York. 1962 Eintritt in die „Photo-Lettering Inc.", ist dort Typographic Design Director. 1970 Eintritt in die International Typeface Corporation, Vizepräsident; mit Herb Lubalin Arbeit an der Hauszeitschrift „U&lc". Mitglied in der AGI; Logotypes für „New York Times", „Playboy", „Reader's Digest", „Sports Illustrated", „Esquire", „Look". Unterrichtet seit

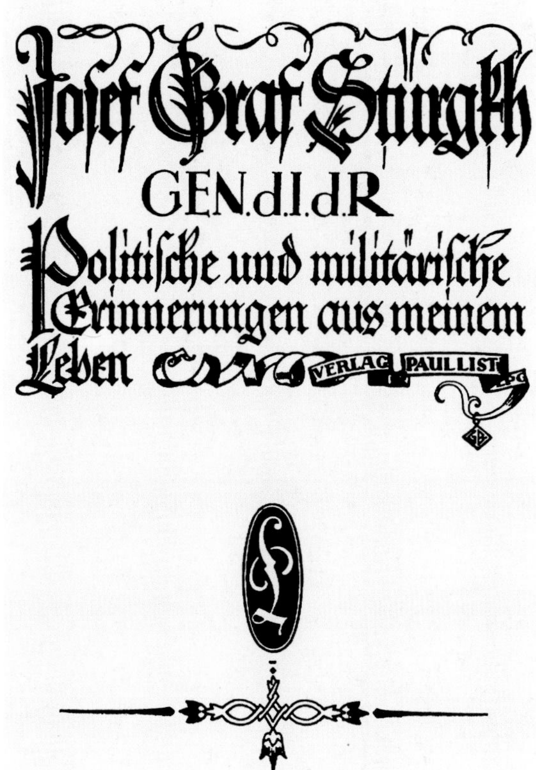


Belwe 1924 Title

Benguiat Logo

Belwe 1930 Logo

| n | n | n | | | n |
|---|---|---|---|---|---|
| a | b | e | f | g | i |
| o | r | s | t | y | z |
| A | B | C | E | G | K |
| M | O | R | S | X | Z |
| 1 | 2 | 4 | 6 | 8 | & |

Belwe 1907 Belwe

| n | n | n | n | | n |
|---|---|---|---|---|---|
| a | b | e | f | g | i |
| o | r | s | t | y | z |
| A | B | C | E | G | H |
| M | O | R | S | X | Y |
| 1 | 2 | 4 | 6 | 8 | & |

Benguiat 1970 ITC Souvenir

| | n | n | | | |
|---|---|---|---|---|---|
| a | b | e | f | g | i |
| o | r | s | t | y | z |
| A | B | C | E | G | H |
| M | O | R | S | X | Y |
| 1 | 2 | 4 | 6 | 8 | & |

Benguiat 1977–79 ITC Bookman

werbemuseum (musée des arts décoratifs) de Berlin. 1900, fonde la Steglitzer Werkstatt avec F. H. Ehmcke et F. W. Kleukens. Commence à enseigner à la Kunstgewerbeschule (école des arts décoratifs) de Berlin. 1906, directeur du département de typographie et responsable du cours de dessin industriel à la Leipziger Akademie für graphische Künste und Buchgewerbe (académie des arts graphiques et du livre). De nombreux travaux pour des éditeurs, dont List, Reclam, Eugen Diederichs et Westermann. – *Polices:* Belwe (1907), Belwe gothique (1912), Belwe italique (1913), Belwe semi-gras (1914), Wieland (1926), Mozart (1927). – *Publications, sélection:* Albert Mundt «Georg Belwe und seine Klasse an der Königlichen Akademie für graphische Künste und Buchgewerbe» dans «Archiv für Buchgewerbe» 6/1910.

**Benguiat,** Ed (Ephram Edward) – né le 27. 10. 1927 à New York, Etats-Unis – *concepteur de polices, calligraphe* – études à la Columbia University, New York; Workshop School of Advertising Art, New York. 1953, Associate Director du magazine «Esquire». Ouverture de son propre atelier de design à New York. 1962, entrée à la «Photo-Lettering Inc.» où il exerce aujourd'hui encore comme Typographic Design Director. 1970, entrée à la International Typeface Corporation dont il est vice-président, collaboration à «U&lc», la revue de la société, avec Herb Lubalin. Membre de l'Alliance Graphique Internationale, logotypes pour le «New York Times», «Playboy», «Reader's Digest», «Sports Illustrated», «Esquire»,

Ramon Cabrera), Havana 1981.

**Belwe,** Georg – b. 12. 8. 1878 in Berlin, Germany, d. 1954 in Ronneburg, Germany – *typographer, type designer, teacher* – Studied at the Königliches Kunstgewerbemuseum in Berlin. 1900: founds the Steglitzer Werkstatt with F. H. Ehmcke and F. W. Kleukens. Joins the Kunstgewerbeschule in Berlin as a teacher. 1906: head of the typography department and the class for accurate drawing at the Leipzig Akademie für graphische Künste und Buchgewerbe. Has produced much work for publishing houses, including List, Reclam, Eugen Diederichs and Westermann. – *Fonts:* Belwe (1907), Belwe Gotisch (1912), Belwe Schrägschrift (1913), Belwe halbfett (1914), Wieland (1926), Schönschrift Mozart (1927). – *Publications include:* Albert Mundt "Georg Belwe und seine Klasse an der Königlichen Akademie für graphische Künste und Buchgewerbe" in "Archiv für Buchgewerbe" 6/1910.

**Benguiat,** Ed (Ephram Edward) – b. 27. 10. 1927 in New York, USA – *type designer, calligrapher* – Studied at Columbia University, New York and the Workshop School of Advertising Art, New York. 1953: associate director of "Esquire" magazine. Opens his own design studio in New York. 1962: joins Photo-Lettering Inc. as typographic design director, a position he still holds today. 1970: joins the International Typeface Corporation and is made vice-president; he works on the in-house magazine "U&lc" with Herb Lubalin. Member of the Alliance Graphique Internationale. He

1961 an der School of Visual Arts, New York. – *Schriftentwürfe u. a.:* Souvenir (1970), Korinna (mit Victor Caruso), Tiffany (1974), Bauhaus (mit Victor Caruso), Bookman (1975), Benguiat (1977– 79), Barcelona (1981), Modern 216 (1982), Caslon 224 (1983), Panache (1988), alle erschienen bei ITC, New York.

**Bense,** Max – geb. 7. 2. 1910 in Straßburg, Frankreich, gest. 29. 4. 1990 in Stuttgart, Deutschland. – *Philosoph, Literat, Semiotiker, Naturwissenschaftler* – Studium der Mathematik, Physik, Geologie und Philosophie an den Universitäten Bonn und Köln. 1946 Professor für Philosophie und Wissenschaftstheorie an der Universität Jena. 1950 Professor an der Technischen Hochschule Stuttgart, Mitbegründer des Studium Generale an der TH Stuttgart, das er von 1954–67 leitet. Unterrichtet 1953–58 an der Hochschule für Gestaltung in Ulm. 1958 Gründung der Studiengalerie an der TH Stuttgart. 1958–60 und 1966–76 Gastprofessur für Ästhetik an der Hochschule für Bildende Künste, Hamburg. 1960 Zusammen mit Elisabeth Walther Gründer und Herausgeber der „edition rot". 1965 Ausstellung „Computergrafik" von Georg Nees in der Studiengalerie der TH Stuttgart. 1975 Gründung der Zeitschrift „Semiosis" mit Gérard Deledalle. In seinem Buch „Theorie der Texte" unternimmt Bense den Versuch, die Literaturtheorie in den Rahmen der Informationstheorie und der Semiotik zu integrieren. – *Publikationen u. a.:* „Raum und Ich", Berlin 1934; „Einleitung in die Philosophie", München 1941; „Plakatwelt", Stuttgart 1952; „Aesthetica",

«Look». Enseigne depuis 1961 à la School of Visual Arts de New York. – *Polices:* plus de 600 polices, entre autres Souvenir (1970), Korinna (avec Victor Caruso, 1974), Tiffany (1974), Bauhaus (avec Victor Caruso, 1975), Bookman (1975), Benguiat (1977–1979), Barcelona (1981), Modern 216 (1982), Caslon 224 (1983), Panache (1988), toutes publiées par ITC, New York.

**Bense,** Max – né le 7. 2. 1910 à Strasbourg, France, décédé le 29. 4. 1990 à Stuttgart, Allemagne – *philosophe, écrivain, sémiologue, spécialiste en sciences de la nature* – études de mathématiques, physique, géologie et philosophie aux universités de Bonn et de Cologne. 1946, professeur de philosophie et d'épistémologie à l'université d'Iéna. 1950, professeur à la Technische Hochschule (université technique) de Stuttgart, cofondateur du cursus d'études générales à la TH de Stuttgart, cursus qu'il dirige de 1954 à 1967. 1953–1958, enseigne à la Hochschule für Gestaltung (école supérieure de design) d'Ulm. 1958, crée la Studiengalerie de la TH de Stuttgart. 1958–1960 et 1966–1976, cycles de conférences sur l'esthétique à la Hochschule für Bildende Künste (école supérieure des beaux-arts) de Hambourg. 1960, fondateur et éditeur de l' «edition rot» avec Elisabeth Walther. 1965, exposition «Computergrafik» de Georg Nees à la Studiengalerie de la TH de Stuttgart. 1975, fonde la revue «Semiosis» avec Gérard Deledalle. Dans son livre «Theorie der Texte», Bense tente d'intégrer la théorie littéraire dans le cadre de la théorie de l'information et de la sémiotique. – *Publications, sélection:* «Raum und Ich», Berlin 1934; «Einleitung in die Philosophie», Munich 1941; «Plakatwelt», Stuttgart 1952; «Aesthetica», Stuttgart 1954; «Programmierung des Schönen», Baden-Baden 1960; «Theorie der Texte»,

max bense
theorie
der texte

kiepenheuer & witsch

Bense 1962 Cover

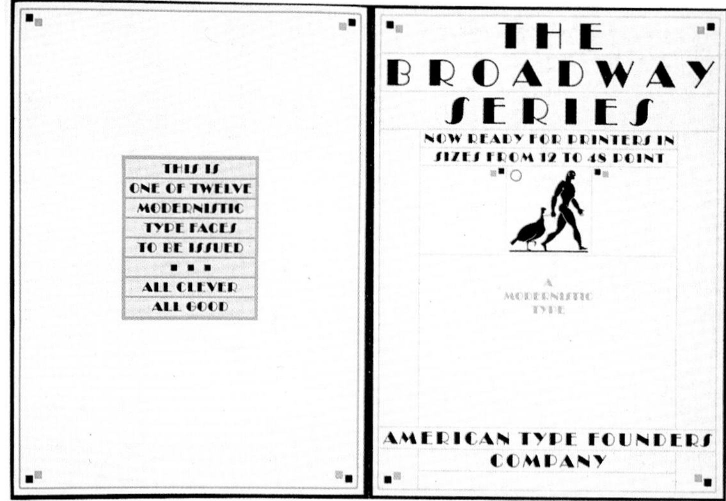


Morris Fuller Benton 1929 Cover

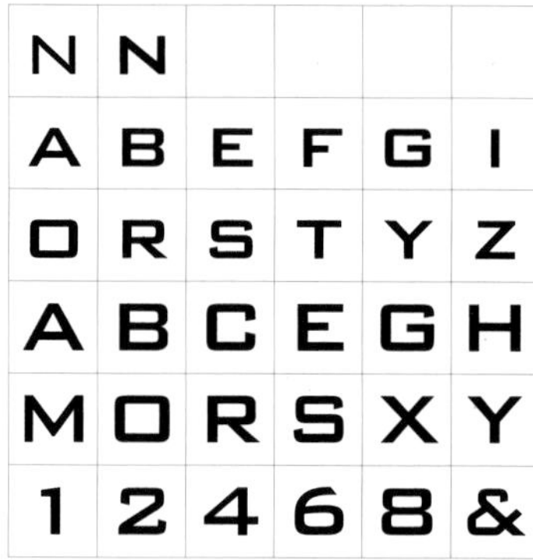


Morris Fuller Benton 1930 Bank Gothic

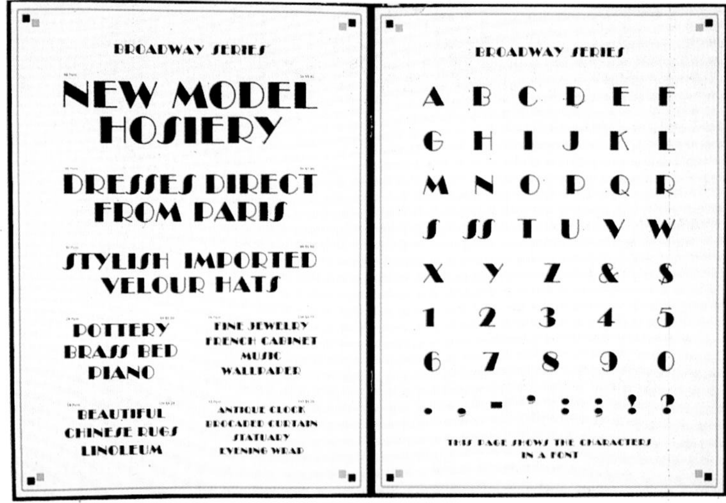


Morris Fuller Benton 1929 Spread

| | n | | | | *n* |
|---|---|---|---|---|---|
| a | b | e | f | g | i |
| o | r | s | t | y | z |
| A | B | C | E | G | H |
| M | O | R | S | X | Y |
| 1 | 2 | 4 | 6 | 8 | & |

Morris Fuller Benton 1931 Stymie

has produced logotypes for the "New York Times", "Playboy", "Reader's Digest", "Sports Illustrated", "Esquire" and "Look". He has taught at the School of Visual Arts in New York since 1961. – *Fonts include:* Souvenir (1970), Korinna (with Victor Caruso, 1974), Tiffany (1974), Bauhaus (with Victor Caruso, 1975), Bookman (1975), Benguiat (1977–79), Barcelona (1981), Modern 216 (1982), Caslon 224 (1983) and Panache (1988), all published by ITC, New York.

**Bense,** Max – b. 7. 2. 1910 in Strasbourg, France, d. 29. 4. 1990 in Stuttgart, Germany – *philosopher, littérateur, semiologist, scientist* – Studied mathemathics, physics, geology and philosophy at the universities of Bonn and Cologne. 1946: professor of philosophy and the theory of science at the University of Jena. 1950: professor at the Technische Hochschule (TH) in Stuttgart. A founder member of the Studium Generale in Stuttgart, which he directed from 1954– 67. 1953–58: teaches at the Hochschule für Gestaltung in Ulm. 1958: founds the Studiengalerie at the TH Stuttgart. 1958– 60 and 1966–76: visiting professor of aesthetics at the Hochschule für Bildende Künste in Hamburg. 1960: sets up and publishes the "edition rot" together with Elisabeth Walther. 1965: Georg Nees holds his exhibition on computer graphics in the Studiengalerie at the TH Stuttgart. 1975: launches the magazine "Semiosis" together with Gérard Deledalle. In his book "Theorie der Texte" ("Theory of Texts"), Bense attempts to integrate literary theory into information theory and semiotics.

Stuttgart 1954; „Programmierung des Schönen", Baden-Baden 1960; „Theorie der Texte", Köln 1962; „Das Auge Epikurs", Stuttgart 1979. E. Walther, L. Harig (Hrsg.) „Muster möglicher Welten", Wiesbaden 1970; E. Walther, U. Bayer (Hrsg.) „Zeichen von Zeichen für Zeichen", Baden-Baden 1990.

**Benton,** Linn Boyd – geb. 1844 in Little Falls, New Jersey, USA, gest. 1932 in Plainfield, New Jersey, USA. – *Schriftentwerfer* – Erhält 1885 ein Patent für eine von ihm entwickelte Maschine zum maschinellen Stempelschnitt. Mitinhaber der Schriftgießerei Benton, Waldo & Co, die 1892 in ATF aufgeht. Ab 1892 Direktor und Hauptberater der ATF. – *Schriftentwurf:* Century (mit de Vinne, 1894).

**Benton,** Morris Fuller – geb. 1872 in Milwaukee, USA, gest. 1948 in Morristown, USA. – *Ingenieur, Schriftentwerfer* – Nach einer Ausbildung als Mechaniker und Ingenieur tritt Benton in die American Type Founders Company (ATF) ein. Er wird bei ATF Schriftentwerfer und Hausdesigner. – *Schriftentwürfe:* Benton entwirft über 200 Alphabete, die alle bei ATF erschienen, u. a. Alternate Gothic (1903), Franklin Gothic (1903–12), Cheltenham (1904), Clearface (1907), News Gothic (1908), Cloister Oldstyle (1913), Souvenir (1914), Century Schoolbook (1919), Broadway (1928), Bulmer (1928), Bank Gothic (1930), Stymie (mit S. Hess und G. Powell, 1931), American Text (1932).

**Berlewi,** Henryk – geb. 30. 10. 1894 in Warschau, Polen, gest. 2. 8. 1967 in Paris, Frankreich. – *Gebrauchsgrafiker, Maler,*

Cologne 1962; «Das Auge Epikurs», Stuttgart 1979. E. Walther, L. Harig (éd.) «Muster möglicher Welten», Wiesbaden 1970; E. Walther, U. Bayer (éd.) «Zeichen von Zeichen für Zeichen», Baden-Baden 1990.

**Benton,** Linn Boyd – né en 1844 à Little Falls, New Jersey, Etats-Unis, décédé en 1932 à Plainfield, New Jersey, Etats-Unis – *concepteur de polices* – 1885, invente une machine capable de tailler automatiquement les caractères et dépose un brevet. Copropriétaire de la fonderie de caractères Benton, Waldo & Co, qui deviendra la American Type Founders Company (ATF) en 1892. A partir de 1892, directeur et principal conseiller de l'ATF. – *Police:* Century (avec Th. L. de Vinne, 1894).

**Benton,** Morris Fuller – né en 1872 à Milwaukee, USA, décédé en 1948 à Morristown, USA – *ingénieur, concepteur de polices* –Benton entre à la American Type Founders Company (ATF) après une formation de mécanicien et d'ingénieur. Très tôt, M. F. Benton exerce comme concepteur de polices et designer maison chez ATF. – *Polices:* Benton a élaboré plus de 200 alphabets, tous publiés chez ATF, dont Alternate Gothic (1903), Franklin Gothic (1903–1912), Chelterham (1904), Clearface (1907), News Gothic (1908), Cloister Oldstyle (1913), Souvenir (1914), Century Schoolbook (1919), Broadway (1928), Bulmer (1928), Bank Gothic (1930), Stymie (avec S. Hess et G. Powell, 1931), American Text (1932).

**Berlewi,** Henryk – né le 30. 10. 1894 à Varsovie, Pologne, décédé le 2. 8. 1967 à Paris, France – *designer, peintre, peintre de dé-*

Morris Fuller Benton 1935 Specimen

Linn Boyd Benton 1894 Specimen

Morris Fuller Benton 1934 Type Sample

Morris Fuller Benton 1932 American Text

Linn Boyd Benton 1895 Century Old Style

– *Publications include:* "Raum und Ich", Berlin 1934; "Einleitung in die Philosophie", Munich 1941; "Plakatwelt", Stuttgart 1952; "Aesthetica", Stuttgart 1954; "Programmierung des Schönen", Baden-Baden 1960; "Theorie der Texte", Cologne 1962; "Das Auge Epikurs", Stuttgart 1979. E. Walter, L. Harig (eds.) "Muster möglicher Welten", Wiesbaden 1970; E. Walter, L. Harig (eds.) "Zeichen von Zeichen für Zeichen", Baden-Baden 1990.

**Benton,** Linn Boyd – b. 1844 in Little Falls, New Jersey, USA, d. 1932 in Plainfield, New Jersey, USA – *type designer* – 1885: obtains a patent for a punch-cutting machine he developed. Joint-owner of the Benton, Waldo & Co. Type Foundry, which is incorporated in the ATF in 1892. From 1892 onwards: director and chief consultant to the ATF. – *Font:* Century (with Th. L. de Vinne, 1894).

**Benton,** Morris Fuller – b. 1872 in Milwaukee, USA, d. 1948 in Morristown, USA – *engineer, type designer* – After training as a mechanic and engineer, Benton joined the ATF, where he became type designer and in-house designer with ATF. – *Fonts:* Benton developed over 200 alphabets, all of which were published by ATF, including Alternate Gothic (1903), Franklin Gothic (1903–12), Cheltenham (1904), Clearface (1907), News Gothic (1908), Cloister Oldstyle (1913), Souvenir (1914), Century Schoolbook (1919), Broadway (1928), Bulmer (1928), Bank Gothic (1930), Stymie (with S. Hess and G. Powell, 1931), American Text (1932).

**Berlewi,** Henryk – b. 30. 10. 1894 in Warsaw, Poland, d. 2. 8. 1967 in Paris, France

*Bühnenmaler, Kunstkritiker* – 1904–09 Gymnasium und Kunstschule in Warschau. 1909–1910 Aufenthalt in Antwerpen, Belgien. 1911–12 Studium an der Ecole des Beaux-Arts, Paris. 1913 Rückkehr nach Warschau, 1922–23 Aufenthalt in Berlin, Erarbeitung seines Gestaltungskonzepts „Mechano-Faktur". 1923 Einer der Mitbegründer der Künstlergruppe „BLOK" in Warschau. 1924 Gründung des Büros „Reklama Mechano" (Mechanische Reklame). Erstellung der ersten Reklamestände. 1926 Entwürfe für Theaterdekorationen. 1928 Übersiedlung nach Paris. Rückkehr zur gegenständlichen Malerei, 1957 Rückkehr zur abstrakten Kunst. 1962 Einladung durch die Akademie der Künste, Berlin. Dokumentarfilm über Berlewi von Hans Cürlis. Erste Versuche der Darstellung von „Mechano-Faktur" im Film. Zahlreiche Ausstellungen, u. a. Galerie „Der Sturm", Berlin 1924; Galerie Creuze, Paris 1957; Helmhaus Zürich, 1960; Grand Palais, Paris 1963. – *Publikationen u. a.:* „Mechano-Faktur", Warschau 1924; „Funktionale Grafik der zwanziger Jahre in Polen" in „Neue Grafik", Olten 9/1961. Zahlreiche Artikel und Essays über alte und neue Gestaltung. Hubert Colleye „Berlewi", Antwerpen 1937; Eckard Neumann „Henryk Berlewi and Mechano-Faktur" in „Typographica", London 9/1961.

**Berlingska Stilgjuteriet,** Lund, Schweden. – *Schriftgießerei* – Frederik Johan Berling gliedert 1837 der 1745 von Carl Gustav Berling in Lund gegründeten Druckerei eine Schriftgießerei an. Von 1874 Berlingska Stilgjuteriet, Lund. Die

*cors, critique d'art* – 1904–1909, lycée et école d'art à Varsovie. 1909–1910, séjours à Anvers, Belgique. 1911–1912 études à l'Ecole des Beaux-Arts, Paris. 1913, retour à Varsovie. 1922–1923, séjour à Berlin et élaboration de son concept plastique appelé «Mechano-Faktur». 1923, cofondateur du groupe d'artistes «BLOK» à Varsovie. 1924, fonde l'atelier «Reklama Mechano» (réclame mécanique). Réalise les premiers stands publicitaires. 1926, dessine des décors de théâtre. 1928, s'installe à Paris. Retour à la peinture figurative. 1957, revient à l'art abstrait. 1962, invitation de l'Akademie der Künste (beaux-arts) de Berlin. Film documentaire de Hans Cürlis sur Berlewi. Premières tentatives de représentation de «Mechano-Faktur» au cinéma. De nombreuses expositions, dont l'une à la galerie «Der Sturm» à Berlin en 1924; Galerie Creuze, Paris, 1957; Helmhaus Zurich, 1960; Grand Palais, Paris, 1963. – *Publications, sélection:* «Mechano-Faktur», Varsovie 1924; «Funktionale Grafik der zwanziger Jahre in Polen» dans «Neue Grafik», Olten 9/1961. De nombreux articles et essais sur le design ancien et nouveau. Hubert Colleye «Berlewi», Anvers 1937; Eckard Neumann «Henryk Berlewi and Mechano-Faktur» dans «Typographica», Londres 9/1961.

**Berlingska Stilgjuteriet,** Lund, Suède – *fonderie de caractères* – en 1837, Frederik Johan Berling incorpore une fonderie de caractères à l'imprimerie fondée en 1745 par Carl Gustav Berling. En 1874, elle prend le nom de fonderie de caractères Berlingska Stilgjuteriet, Lund, Suède. 1888–1943, la fonderie est exploitée sous forme de société anonyme. 1980, dissolution de la fonderie, les matrices sont conservées au Musée d'histoire de l'art de Lund.

**Berlow,** David – né le 9. 4. 1955 à Boston,

Berlewi 1925 Advertisement

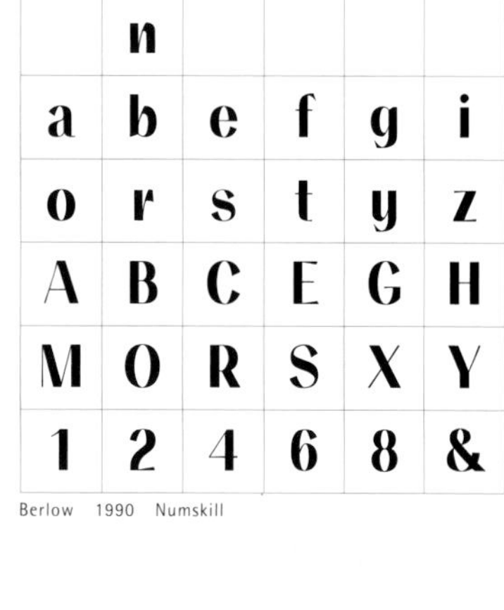


Berlow 1990 Numskill

Berlewi 1924 Cover

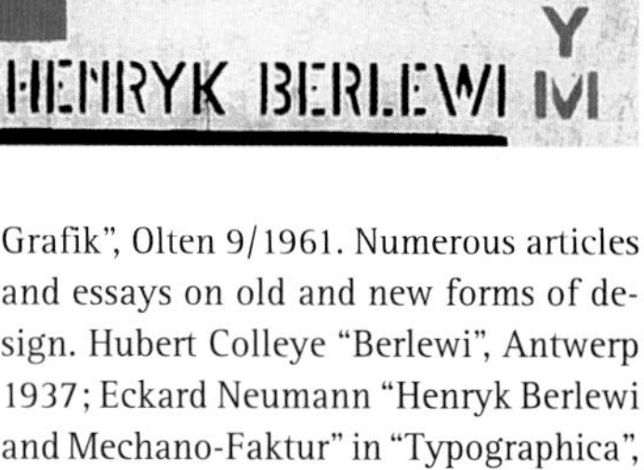


Berlewi 1924 Advertisement

– *commercial artist, painter, stage-set painter, art critic* – 1904–09: attends grammar school and art school in Warsaw. 1909–10: lives in Antwerp, Belgium. 1911–12: studies at the Ecole des Beaux-Arts in Paris. 1913: returns to Warsaw. 1922–23: lives in Berlin and works on his "Mechano-Faktur" design concept. 1923: one of the founder members of the BLOK group in Warsaw. 1924: opens the studio Reklama Mechano ("Mechanical Advertising"). Designs his first advertising stands. 1926: designs stage sets for the theater. 1928: moves to Paris. Returns to representational painting. 1957: returns to abstract art. 1962: invited to the Akademie der Künste in Berlin. Hans Cürlis makes a documentary on Berlewi. First attempts to present "Mechano-Faktur" on film. Numerous exhibitions, including at Galerie Der Sturm, Berlin 1924, Galerie Creuze, Paris 1957, Helmhaus, Zurich 1960, Grand Palais, Paris 1963. – *Publications include:* "Mechano-Faktur", Warsaw 1924; "Funktionale Grafik der zwanziger Jahre in Polen" in "Neue Grafik", Olten 9/1961. Numerous articles and essays on old and new forms of design. Hubert Colleye "Berlewi", Antwerp 1937; Eckard Neumann "Henryk Berlewi and Mechano-Faktur" in "Typographica", London 9/1961.

**Berlingska Stilgjuteriet,** Lund, Sweden – *type foundry* – 1837: Frederik Johan Berling affiliates a type foundry to the printery founded by Carl Gustav Berling in Lund in 1745. From 1874, the concern in Lund is named the Berlingska Stilgjuteriet. 1888–1943: the type foundry is a

Schriftgießerei wird 1888–1943 als Aktiengesellschaft geführt. 1980 Auflösung der Schriftgießerei, die Matrizen werden im Kunstgeschichtlichen Museum Lund aufbewahrt.
**Berlow,** David – geb. 9. 4. 1955 in Boston, Massachusetts, USA. – *Schriftentwerfer, Unternehmer* – 1972–77 Studium von Kunst und Kunstgeschichte an der Universität von Wisconsin in Madison. 1978–82 Arbeit als Schriftentwerfer für Mergenthaler-Linotype. 1982–89 Arbeit als Schriftentwerfer für Bitstream Inc. 1989 zusammen mit Roger Black Gründung von „The Font Bureau". Hier entstehen über 200 verschiedene Schriften und Logotypes für Kunden wie „The Chicago Tribune", „The Wall Street Journal", „Newsweek", „Esquire" und „Rolling Stone" sowie Schriften für Apple Computer, Microsoft und die International Typeface Corporation (ITC). Initiiert die „Font Bureau Retail Library", in der bis heute über 400 Schriften angeboten werden. Vizepräsident des „Interactive Bureau", einem Online-Medien-Projekt. – *Schriftentwürfe:* Neben zahlreichen Überarbeitungen historischer Schriften entstanden Millenium (1989–1996), Numskill (1990), Yernacular (1992), Esperanto (1995), Online Gothic (1995), Nature (1995), Truth (1995), Hitech (1995), Zenobia (1995).
**Bernard,** Pierre – geb. 25. 2. 1942 in Paris, Frankreich. – *Grafik-Designer, Typograph, Kalligraph* – Ausbildung an der Ecole Nationale Supérieure des Arts Décoratifs und am Institut de l'Environnement de Paris. 1970 Mitbegründer der De-

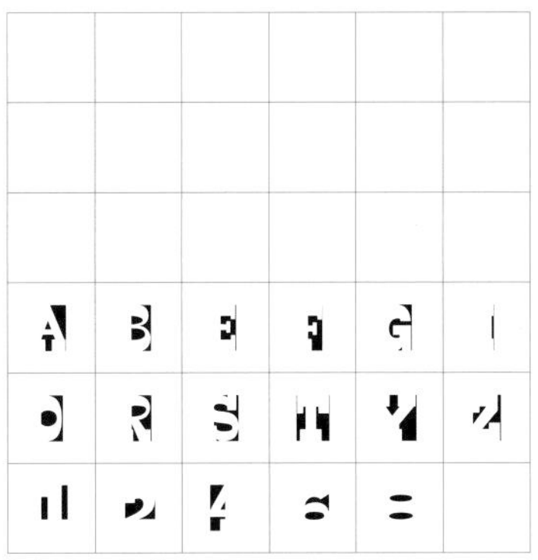
Berlow 1992 Yernacular

Bernard 1992 Poster

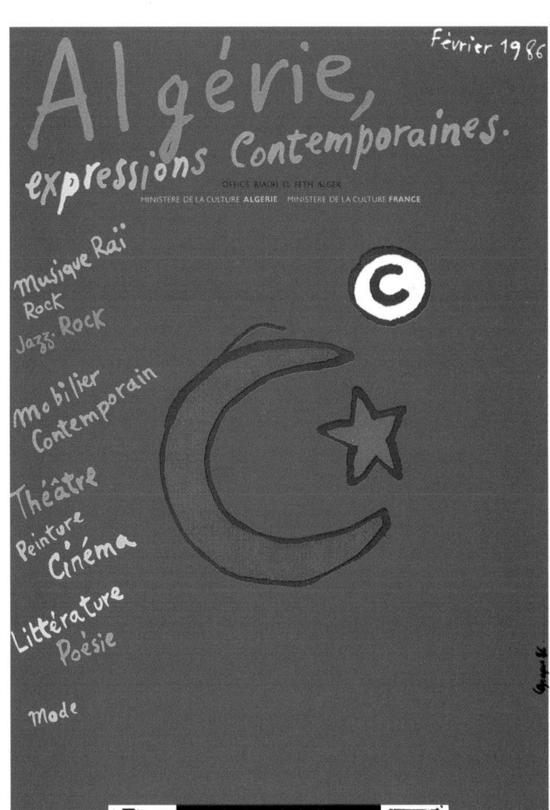

Bernard 1986 Poster

Bernard Logo

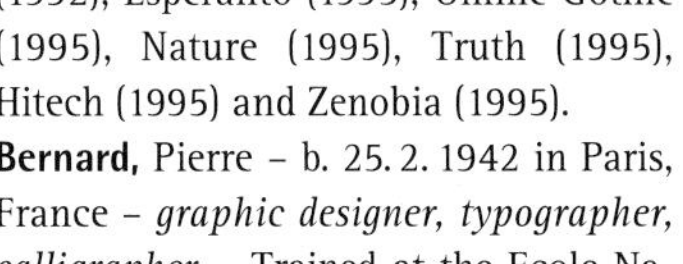

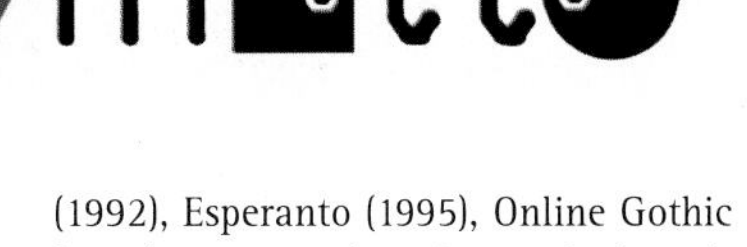
Bernard Logo

Massachusetts, Etats-Unis – *concepteur de polices, chef d'entreprise* – 1972–1977, études d'art et d'histoire de l'art à l'université du Wisconsin à Madison. 1978–1982, exerce comme concepteur de polices pour Mergenthaler-Linotype. 1982–1989, concepteur de polices pour Bitstream Inc. 1989, fonde «The Font Bureau» avec Roger Black, où naissent plus de 200 polices et logos pour des clients tels que «The Chicago Tribune», «The Wall Street Journal», «Newsweek», «Esquire» et «Rolling Stone», ainsi que pour les ordinateurs Apple, Microsoft et la International Typeface Corporation (ITC). Il est à l'initiative de la «Font Bureau Retail Library» qui propose aujourd'hui plus de 400 polices. Vice-président du «Interactive Bureau», un projet de médias online. – *Polices:* outre de nombreuses révisions de polices historiques, création de Millenium (1989–1996), Numskill (1990), Yernacular (1992), Esperanto (1995), Online Gothic (1995), Nature (1995), Truth (1995), Hitech (1995), Zenobia (1995).
**Bernard,** Pierre – né le 25. 2. 1942 à Paris, France – *graphiste maquettiste, typographe, calligraphe* – formation à l'Ecole Nationale Supérieure des Arts Décoratifs et à l'Institut de l'Environnement de Paris. 1970, cofondateur du groupe de de-

stock company. 1980: the type foundry is closed; the matrices are sent to the Lund Museum of Art History.
**Berlow,** David – b. 9. 4. 1955 in Boston, Massachusetts, USA – *type designer, entrepreneur* – 1972–77: studies art and art history at the University of Wisconsin in Madison. 1978–82: works as a type designer for Mergenthaler-Linotype. 1982–89: works as a type designer for Bitstream Inc. 1989: opens The Font Bureau with Roger Black. To date, they have produced over 200 typefaces and logotypes for customers such as "The Chicago Tribune", "The Wall Street Journal", "Newsweek", "Esquire" and "Rolling Stone", as well as fonts for Apple Computers, Microsoft and the International Typeface Corporation (ITC). Berlow initiated the Font Bureau Retail Library, which today offers over 400 alphabets. Vice-president of the Interactive Bureau, an online media project. – *Fonts:* besides reworkings of numerous historical fonts, Berlow has designed Millenium (1989–1996), Numskill (1990), Yernacular (1992), Esperanto (1995), Online Gothic (1995), Nature (1995), Truth (1995), Hitech (1995) and Zenobia (1995).
**Bernard,** Pierre – b. 25. 2. 1942 in Paris, France – *graphic designer, typographer, calligrapher* – Trained at the Ecole Nationale Supérieure des Arts Décoratifs and at the Institut de l'Environnement de

signgruppe Grapus. 1990 Auflösung der Gruppe und Mitbegründer des „Atelier de Création Graphique" mit Dirk Behage und Fokke Draaijer. Entwürfe von Erscheinungsbildern für das Musée du Louvre und die französischen Nationalparks. – *Publikationen u. a.:* „Grapus, différentes tentatives", Paris 1985. Rick Poynor „Pierre Bernard" in „Eye" 3/1991, London.

**Bernhard,** Lucian – geb. 15. 3. 1883 in Stuttgart, Deutschland, gest. 29. 5. 1972 in New York, USA. – *Grafiker, Illustrator, Maler, Schriftentwerfer, Innenarchitekt, Lehrer* – Geboren als Emil Kahn nimmt er 1905 sein Pseudonym an. Autodidakt, kurze Zeit Studium an der Akademie München. 1901 Übersiedlung nach Berlin. 1903 Sieger eines Plakatwettbewerbs für Priester-Streichhölzer. 1904 künstlerischer Leiter der Deutschen Werkstätten für Handwerkskunst. 1910 Zusammenarbeit mit Hans Sachs bei der Herausgabe der Zeitschrift „Das Plakat". 1920 Professor an der Akademie der Künste, Berlin. 1923 Umzug nach New York. 1928 Gründung des Studios „Contempora" mit Rockwell Kent, Paul Poiret, Bruno Paul und Erich Mendelsohn. Arbeit als Grafiker und Innenarchitekt. Nach 1930 vor allem Arbeit als Maler und Plastiker. – *Schriftentwürfe:* Bernhard Antiqua (1912), Bernhard Fraktur (1912–22), Bernhard Privat (1919), Bernhard Schönschrift (1925–28), Bernhard Handschrift (1928), Bernhard Fashion (1929), Bernhard Gothic (1929–31), Negro (1930), Lilli (1930), Lucian (1932), Bernhard Tango (1933), Bernhard Modern (1933–38), Aigrette (1939). – *Publikation u.a.:* F. Plietzsch

signers Grapus. 1990, dissolution du groupe et fondation de l'Atelier de Création Graphique avec Dirk Behage et Fokke Draaijer. Conception du logo du Musée du Louvre et du Parc National français. – *Publications, sélection:* «Grapus différentes tentatives», Paris 1985. Rick Poynor «Pierre Bernard» dans «Eye» 3/1991, Londres.

**Bernhard,** Lucian – né le 15. 3. 1883 à Stuttgart, Allemagne, décédé le 29. 5. 1972 à New York, Etats-Unis – *graphiste, illustrateur, peintre, concepteur de polices, architecte d'intérieur, enseignant* – né sous le nom d'Emil Kahn, il prend son pseudonyme en 1905. Autodidacte, courtes études à l'académie de Munich. S'installe à Berlin en 1901. 1903, remporte un concours d'affiches pour les allumettes Priester. 1904, directeur artistique des Deutsche Werkstätten für Handwerkskunst. 1910, collaboration avec Hans Sachs à la publication de la revue «Das Plakat». 1920, professeur à l'Akademie der Künste (beaux-arts) de Berlin. 1923, s'installe à New York. 1928, fonde l'atelier «Contempora» avec Rockwell Kent, Paul Poiret, Bruno Paul et Erich Mendelsohn. Exerce comme graphiste et architecte d'intérieur. A partir de 1930, travaille surtout comme peintre et plasticien. – *Polices:* Bernhard Antiqua (1912), Bernhard Fraktur (1912–1922), Bernhard Privat (1919), Bernhard Schönschrift (1925– 1928), Bernhard Handschrift (1928), Bernhard Fashion (1929), Bernhard Gothic (1929–1931), Negro (1930), Lilli (1930), Lucian (1932), Bernhard Tango (1933), Bernhard Modern (1933–1938), Aigrette (1939). – *Publications, sélection:* F. Plietzsch «Lucian Bernhard», Deutsches Museum, Hagen 1913. De nombreux articles dans les revues et ouvrages internationaux à propos de son œuvre, nombreuses expositions et participations à des expositions.

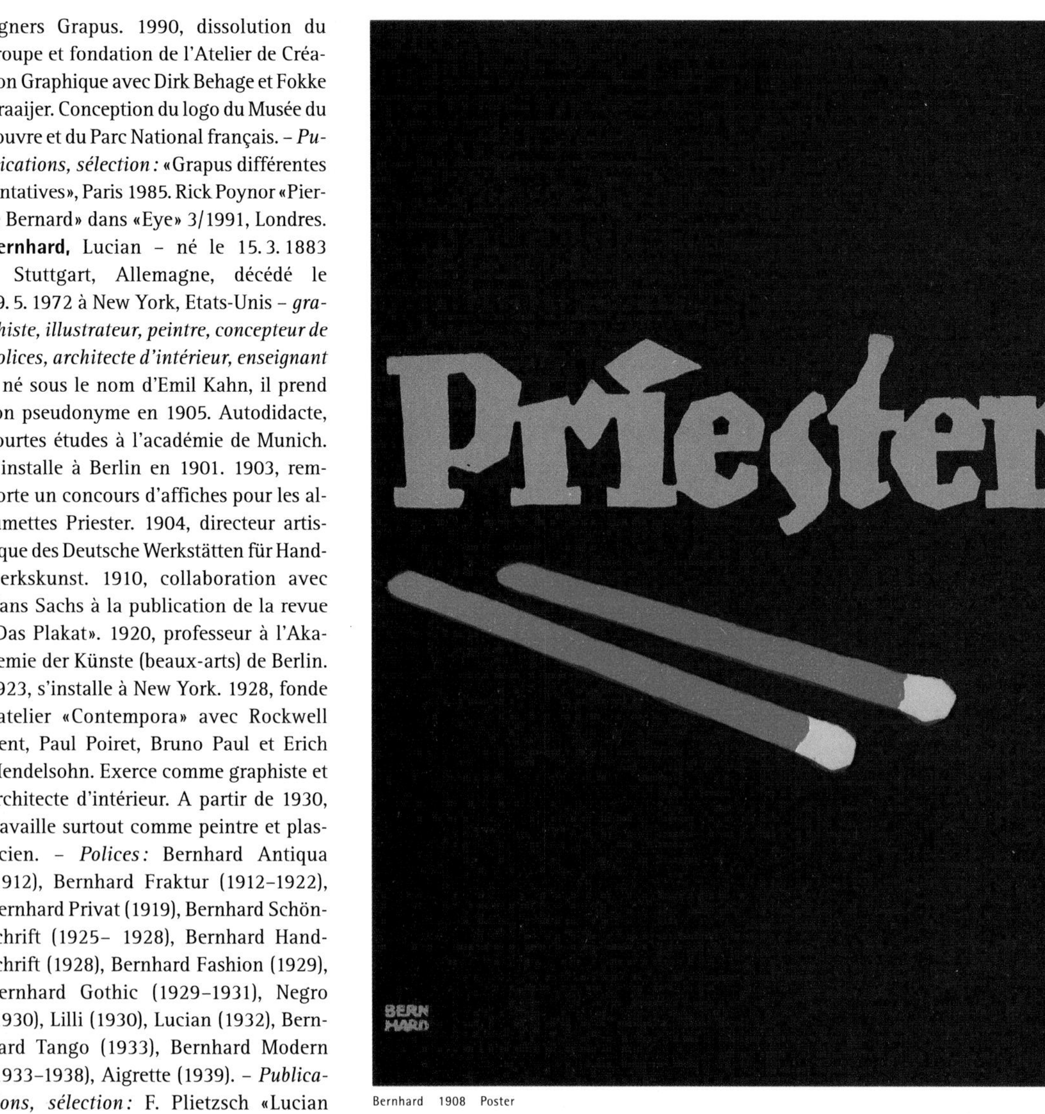


Bernhard 1908 Poster

Bernhard 1915 Poster

Bernhard Type Sample

| | n | **n** | | | *n* |
|---|---|---|---|---|---|
| a | b | e | f | g | i |
| o | r | s | t | y | z |
| A | B | C | E | G | H |
| M | O | R | S | X | Y |
| 1 | 2 | 4 | 6 | 8 | & |

Bernhard 1933–38 Bernhard Modern

Paris. 1970: one of the founders of the Grapus group of designers. 1990: the group is disbanded and Bernard is one of the founders of the Atelier de Création Graphique with Dirk Behage and Fokke Draaijer. Designs for the Musée du Louvre and the French national parks. – *Publications include:* "Grapus, différentes tentatives", Paris 1985. Rick Poynor "Pierre Bernard" in "Eye" 3/1991, London.

**Bernhard,** Lucian – b. 15. 3. 1883 in Stuttgart, Germany, d. 29. 5. 1972 in New York, USA – *graphic artist, illustrator, painter, type designer, interior designer, teacher* – Born as Emil Kahn, he assumes his pseudonym in 1905. Mostly self-taught, he studied briefly at the Akademie in Munich. 1901: moves to Berlin. 1903: wins a poster competition for Priester matches. 1904: art director of the Deutsche Werkstätten für Handwerkskunst. 1910: works with Hans Sachs on the publication of the magazine "Das Plakat". 1920: professor at the Akademie der Künste in Berlin. 1923: moves to New York. 1928: opens the Contempora studio with Rockwell Kent, Paul Poiret, Bruno Paul and Erich Mendelsohn. Works as a graphic artist and interior designer. After 1930, he primarily works as a painter and sculptor. – *Fonts:* Bernhard Antiqua (1912), Bernhard Fraktur (1912–22), Bernhard Privat (1919), Bernhard Schönschrift (1925–28), Bernhard Handschrift (1928), Bernhard Fashion (1929), Bernhard Gothic (1929–31), Negro (1930), Lilli (1930), Lucian (1932), Bernhard Tango (1933), Bernhard Modern (1933–38), Aigrette (1939). – *Publications in-*

„Lucian Bernhard", Deutsches Museum Hagen 1913. Zahlreiche Artikel in internationalen Zeitschriften und Büchern zu seinem Werk, zahlreiche Ausstellungen und Ausstellungsbeteiligungen.

**Berthold AG,** Berlin, Deutschland – *Schriftgießerei und Schriftmaschinenhersteller* – 1858 gründet Hermann Berthold in Berlin ein Institut für Galvanotypie. Der Verband Deutscher Schriftgießer beauftragt Hermann Berthold 1878 mit der Herstellung eines Urmaßes für das typographische Maßsystem. Mit Wilhelm Foerster legt er den typographischen Punkt (= 0,376065 mm) und die deutsche Schrift-Normalhöhe fest. 1893 Filialen in Stuttgart und St. Petersburg. 1918 Leipzig, Riga, Budapest und Wien. Die Berthold AG ist die größte Schriftgießerei der Welt. 1935 Kontakte mit Uher, dem Konstrukteur der Fotosetzmaschine „Uhertype". 1950 Beginn der Tätigkeit von G. G. Lange für die Berthold AG Berlin. 1958 Prototyp des ersten Akzidenz-Fotosatzgerätes „Diatype". 1960 erstes Seriengerät. 1961 wird G. G. Lange künstlerischer Leiter der Berthold AG. 1967 stellt Berthold die erste tastaturgesteuerte Akzidenz-Fotosatzmaschine der Welt, die „Diatronic" vor. Berthold Systeme meldet 1995 Vergleich an. Die Berthold Schriftenbibliothek umfaßte über 2.000 Schriften.

**Bertieri,** Raffaello – geb. 5. 1. 1875 in Florenz, Italien, gest. 30. 5. 1941 in Florenz, Italien. – *Schriftentwerfer, Typograph* – 1886 Lehrling in einer Druckerei. 1902 Redakteur in einem Druck- und Verlagshaus in Mailand. Beginn der Her-

Bernhard 1913 Poster

Bernhard 1913 Poster

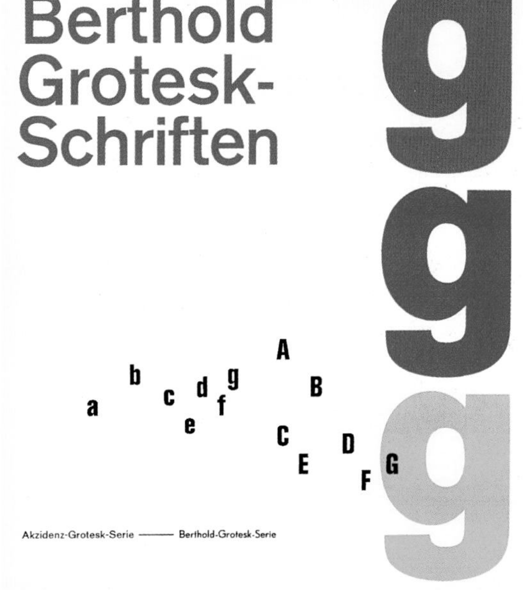

Berthold AG Type Specimen

Berthold AG Cover

Berthold AG Type Specimen

*clude:* F. Plietzsch "Lucian Bernhard", Deutsches Museum, Hagen 1913. Numerous articles have been published on his work in international magazines and books; numerous exhibitions.

**Berthold AG,** Berlin, Germany – *type foundry and manufacturers of type machines* – 1858: Hermann Berthold founds an institute for electrotype in Berlin. 1878: the Association of German Type Founders commissions Berthold with the production of a standard measurement for the typographical system of measures. Together with Wilhelm Foerster he determines the typographical point (= 0.376065 mm) and the standardized German type height. 1893: branches open in Stuttgart and St. Petersburg. 1918: Berthold AG comes to Leipzig, Riga, Budapest and Vienna. It is the largest type foundry in the world. 1935: forms contacts with Uher, the designer of the Uhertype filmsetting machine. 1950: G. G. Lange starts working for Berthold AG in Berlin. 1958: a prototype for the first jobbing filmsetting machine, Diatype, is developed. 1960: production of the first serial instrument. 1961: G. G. Lange is made art director of Berthold AG. 1967: Berthold introduces Diatronic, the first keyboard-controlled jobbing filmsetting machine in the world. 1995: Berthold Systeme comes to a settlement. Berthold's font library contains over 2,000 typefaces.

**Bertieri,** Raffaello – b. 5. 1. 1875 in Florence, Italy, d. 30. 5. 1941 in Florence, Italy – *type designer, typographer* – 1886: apprentice in a printing works. 1902: ed-

**Berthold AG,** Berlin, Allemagne – *fonderie de caractères et fabricant de machines à écrire* – en 1858, Hermann Berthold fonde à Berlin un institut de galvanotypie. 1878, le Verband Deutscher Schriftgiesser (Union allemande des fondeurs de caractères) charge Hermann Berthold de la fabrication d'un étalon pour le système de mesure typographique. Avec Wilhelm Foerster, il définit le point typographique (= 0,376065 mm) et la hauteur normale des lettres allemandes. 1893, filiales à Stuttgart et Saint-Pétersbourg, puis à Leipzig, Riga, Budapest et Vienne en 1918. La Berthold AG est la plus grande fonderie de caractères au monde. En 1935, contacts avec Uher, le constructeur de la photocomposeuse «Uhertype». 1950, début des activités de G. G. Lange au service de la Berthold AG de Berlin. 1958, prototype du premier appareil de photocomposition pour ouvrages de ville, le «Diatype». 1960, premier appareil de série. 1961, G. G. Lange devient directeur artistique de la Berthold AG. 1967, Berthold présente les premières machines au monde commandées par clavier et capable de photocomposer des ouvrages de ville, les «Diatronics». En 1995, Berthold annonce qu'elle fait l'objet d'un règlement judiciaire. La typothèque de Berthold regroupait plus de 2 000 écritures.

**Bertieri,** Raffaello – né le 5. 1. 1875 à Florence, Italie, décédé le 30. 5. 1941 à Florence, Italie – *concepteur de polices, typographe* – 1886, apprenti dans une imprimerie. 1902, rédacteur chez un imprimeur-éditeur à Milan. Commence à pu-

ausgabe der Zeitschrift „Il Risorgimento Grafico". Gründung des Druck- und Verlagshauses Bertieri & Vansetti in Mailand, in dem u. a. die Werke Gabriele d'Annunzios erschienen. Ab 1911 Schriftentwürfe für die Schriftgießerei Nebiolo. 1925 Auszeichnung seiner Bücher auf der Internationalen Ausstellung in Paris. Ausstellung im Museum Plantin-Moretus, Antwerpen, „Die Bücher Raffaello Bertieris". – *Schriftentwürfe:* Inkunabula (1911), Sinibaldi (1926), Paganini (mit A. Butti, 1928), Iliade (1930), Ruano (1933). – *Publikationen u. a.:* „Il numero di pagina nel libro moderno" in „Il Risorgimento Grafico" 22/1925; „Il libro italiano nel novecento", Mailand 1928; „L'arte di G. B. Bodoni", Mailand o. J.; „20 Alfabeti brevemente illustrati", Mailand 1933.

**Bierma,** Wigger – geb. 17. 6. 1958 in Hengelo, Niederlande. – *Grafik-Designer, Lehrer* – 1979–84 Grafik-Design-Studium an der Academie voor Beeldende Kunsten, Arnheim. 1985 Arbeit als Assistent von Walter Nikkels, für den Kunstfonds Amsterdam und für literarische Verlage. Unterrichtet 1986–90 Typographie an der Academie voor Beeldende Kunsten, Arnheim. Zahlreiche der von ihm gestalteten Bücher werden 1986–94 zu den schönsten Büchern der Niederlande gewählt. 1987 Teilnahme an der Ausstellung „Holland in vorm" im Stedelijk Museum in Amsterdam. 1989 Briefmarkenentwürfe für die belgische und niederländische Post. 1994 Unterricht an der Hogeschool voor de Kunsten, Arnheim. 1995 Unterricht an der Gerrit Rietveld Academie, Amsterdam.

blier la revue «Il Risorgimento Grafico». Fondation à Milan de l'imprimerie-éditions Bertieri & Vansetti qui publie entre autres les œuvres de Gabriel d'Annunzio. A partir de 1911, conception de polices pour la fonderie de caractères Nebiolo. 1925, ses livres sont primés à l'exposition internationale de Paris. Exposition au Museum Plantin-Moretus, Anvers: «Les livres de Raffaello Bertieri». – *Polices:* Inkunabula (1911), Sinibaldi (1926), Paganini (avec A. Butti, 1928), Iliade (1930), Ruano (1933). – *Publications, sélection:* «Il numero di pagina nel libro moderno» dans «Il Risorgimento Grafico» 22/1925; «Il libro italiano nel novecento», Milan 1928; «L'arte di G. B. Bodoni», Milan sans date; «20 Alfabeti brevemente illustrati», Milan 1933.

**Bierma,** Wigger – né le 17. 6. 1958 à Hengelo, Pays-Bas – *graphiste maquettiste, enseignant* – 1979–1984, études d'arts graphiques à l'Academie voor Beeldende Kunsten d' Arnhem. 1985, exerce comme assistant de Walter Nikkels pour le Fonds artistique d'Amsterdam et pour des éditeurs de littérature. 1986–1990, enseigne la typographie à l'Academie voor Beeldende Kunsten d'Arnhem. 1986–1994, plusieurs livres dont il réalise la maquette, sont primés plus beaux livres des Pays-Bas. 1987, participation à l'exposition «Holland in vorm» au Stedelijk Museum d'Amsterdam. 1989, dessins de timbres pour les postes belge et néerlandais. 1994, enseigne à la Hogeschool voor de Kunsten, Arnhem. 1995, enseigne à la Gerrit Rietveld Academie, Amsterdam.

**Bierut,** Michael – né le 29. 8. 1957 à Cleveland, Ohio, Etats-Unis – *graphiste maquettiste* – études d'arts graphiques et de design à l'University of Cincinnati's College of Design Architecture, Art and Planning, diplôme en 1980. Vice-président du service d'arts graphiques et de

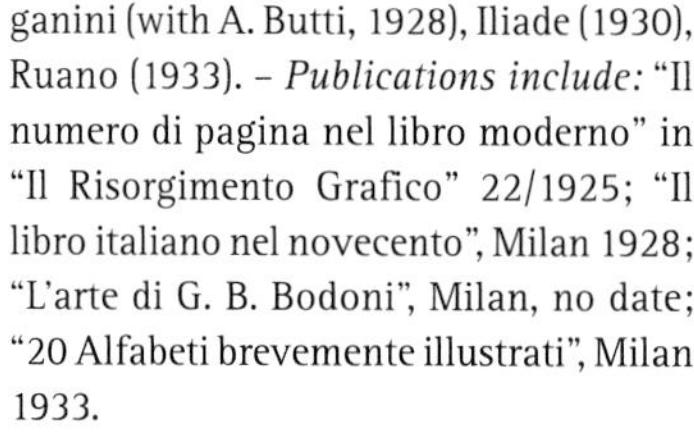


Bertieri 1928 Paganini

Bierma 1988 Cover

Bierut 1988 Poster

Bierut 1989 Logotype

itor in a printing and publishing house in Milan. Begins publishing the magazine "Il Risorgimento Grafico". Founds the Bertieri & Vansetti printing and publishing house in Milan, which publish the works of Gabriele d'Annunzio, among others. From 1911 onwards: designs fonts for the Nebiolo type foundry. 1925: his books are awarded prizes at the International Exhibition in Paris. Exhibition at the Museum Plantin-Moretus, Antwerp: "The Books of Raffaello Bertieri". – *Fonts:* Inkunabula (1911), Sinibaldi (1926), Paganini (with A. Butti, 1928), Iliade (1930), Ruano (1933). – *Publications include:* "Il numero di pagina nel libro moderno" in "Il Risorgimento Grafico" 22/1925; "Il libro italiano nel novecento", Milan 1928; "L'arte di G. B. Bodoni", Milan, no date; "20 Alfabeti brevemente illustrati", Milan 1933.

**Bierma,** Wigger – b. 17. 6. 1958 in Hengelo, The Netherlands – *graphic designer, teacher* – 1979–84: studies graphic design at the Academie voor Beeldende Kunsten in Arnhem. 1985: works as assistant to Walter Nikkels,for the Amsterdam Kunstfonds and for literary publishing houses. 1986–90: teaches typography at the Academie voor Beeldende Kunsten in Arnhem. 1986–94: many of the books he designs are designated the most beautiful books in The Netherlands. 1987: takes part in the exhibition "Holland in vorm" at the Stedelijk Museum in Amsterdam. 1989: designs stamps for the Dutch and Belgian postal services. 1994: teaches at the Hogeschool voor de Kunsten in Arnhem. 1995: teaches at the

**Bierut,** Michael – geb. 29. 8. 1957 in Cleveland, Ohio, USA. – *Grafik-Designer* – Grafik-Design-Studium an der University of Cincinnati's College of Design, Architecture, Art and Planning; Graduierung 1980. Vize-Präsident der Grafik-Design-Abteilung von Vignelli Associates, New York. Seit 1990 Partner von Pentagram Design, New York. Arbeiten für The Council of Fashion Designers of America, Alfred A. Knopf Inc., die Disney Development Company, Nickelodeon, die Princeton University und das American Ballet Theater. Grafik-Design-Consultant der Mohawk Paper Mills, der Brooklyn Academy of Music, des Kindermuseums in St. Paul, Minnesota. Veröffentlichungen zu Fragen des Design in der Zeitschrift „I. D.". Zahlreiche Auszeichnungen für seine Arbeiten. Senior Critic für Grafik-Design an der Yale School of Art.
**Bigelow,** Charles – geb. 29. 7. 1945 in Detroit, Michigan, USA. – *Schriftentwerfer, Lehrer* – 1957–63 Studium an der Cranbrook School, Michigan (graduiert), Sommer 1964 Columbia University, Sommer 1964 L'Università per gli Stranieri, Perugia, Italien, Sommer 1966 Wayne State University, Detroit, 1967 Reed College, Portland, Oregon (graduiert), 1967–69 San Francisco Art Institute, 1972–74, 1976 Portland State University, Oregon, Sommer 1979 Rochester Institute of Technology, New York. 1972–73 Arbeiten als Media-Spezialist für Oregon Environmental Council, Portland. 1974–75 Art Director des „Oregon Times Magazine", Portland. Seit 1976 eigenes Büro zusammen mit Kris Holmes: „Bigelow & Hol-

Bierut 1991 Poster

Bierut 1984 Poster

design de Vignelli Associates, New York. Depuis 1990, partenaire de Pentagram Design, New York; travaux pour le Council of Fashion Designers of America, Alfred A. Knopf Inc., Disney Development Company, Nickelodeon, Princeton University et l'American Ballet Theater. Consultant en graphisme et design pour les Mohawk Paper Mills, la Brooklyn Academy of Music, et le Musée des enfants de St. Paul, Minnesota. Publie des articles traitant des problèmes de design dans la revue «I.D.». Nombreuses distinctions. Senior Critic en graphisme et design à la Yale School of Art.
**Bigelow,** Charles – né le 29. 7. 1945 à Detroit, Michigan, Etats-Unis – *concepteur de polices, enseignant* – études 1957–1963 à la Cranbrook School, Michigan (diplôme); été 1964 à la Columbia University; été 1964 à l'Università per gli Stranieri, Pérouse, Italie; été 1966 à la Wayne State University Detroit; 1967 au Reed College de Portland, Oregon (diplôme); 1967–1969 San Francisco Art Institute; 1972–1974 et 1976, Portland State University, Oregon; été 1979, Rochester Institute of Technology, New York. 1972– 1973, exerce comme spécialiste des médias pour l' Oregon Environmental Council à Portland. 1974–1975, Art Director de l' «Oregon Times Magazine», Portland. Depuis 1976, il

Bierut Logotype

| | n | | | | |
|---|---|---|---|---|---|
| a | b | e | f | g | i |
| o | r | s | t | y | z |
| A | B | C | E | G | H |
| M | O | R | S | X | Y |
| 1 | 2 | 4 | 6 | 8 | & |

Bigelow 1991 Chicago

| | n | | | | |
|---|---|---|---|---|---|
| a | b | e | f | g | i |
| o | r | s | t | y | z |
| A | B | C | E | G | H |
| M | O | R | S | X | Y |
| 1 | 2 | 4 | 6 | 8 | & |

Bigelow 1991 Geneva

Gerrit Rietveld Academie in Amsterdam.
**Bierut,** Michael – b. 29. 8. 1957 in Cleveland, Ohio, USA – *graphic designer* – Studied graphic design at the University of Cincinnati's College of Design, Architecture, Art and Planning; graduated in 1980. Vice-president of Vignelli Associates graphic design department, New York. Since 1990: partner of Pentagram Design, New York. Has produced work for The Council of Fashion Designers of America, Alfred A. Knopf Inc., the Disney Development Company, Nickelodeon, Princeton University and the American Ballet Theater. Graphic design consultant to Mohawk Paper Mills, the Brooklyn Academy of Music and the Children's museum in St. Paul, Minnesota. Has published articles on design issues in the magazine "I.D.". Has won numerous awards for his work. He is senior critic for graphic design at the Yale School of Art.
**Bigelow,** Charles – b. 29. 7. 1945 in Detroit, Michigan, USA – *type designer, teacher* – University education: 1957–63 Cranbrook School, Michigan (where he graduates); summer 1964 Columbia University and L'Università per gli Stranieri, Perugia, Italy; summer 1966 Wayne State University, Detroit; 1967 Reed College, Portland, Oregon (where he graduates); 1967–69 San Francisco Art Institute; 1972–74, 1976 Portland State University, Oregon; summer 1979 Rochester Institute of Technology, New York. 1972–73: works as media specialist for the Oregon Environmental Council in Portland. 1974–75: art director of the "Oregon Times Magazine", Portland. Since 1976:

mes". 1983 organisiert und leitet er das internationale Seminar „Der Computer und die Hand beim Schriftentwurf" an der Stanford University. 1987 mit dem F. W. Goudy-Preis für Typographie des Rochester Institute of Technology ausgezeichnet. 1987 Re-Design der Zeitschrift „Scientific American". Zahlreiche Veröffentlichungen in Fachzeitschriften. – *Schriftentwürfe:* Leviathan (1979), Lucida (1984–95, bisher über 50 verschiedene Schnitte und Variationen), Apple Chicago (1991), Apple Geneva (1991), Microsoft Wingdings 1–3 (1992). 1980 entwickeln Hans Ed. Meier, Charles Bigelow und Kris Holmes phonetische Buchstaben für die Sprachen amerikanischer Indianer auf der Grundlage der Schrift „Syntax" von Hans Ed. Meier.

**Bilibin,** Iwan Jakowlewitsch – geb. 1876 in Tarchowka bei St. Petersburg, Rußland, gest. 1942 in Leningrad, Rußland. – *Maler, Grafiker, Buchillustrator, Lehrer, Bühnenbildner* – 1895–98 Studium an der Schule zur Förderung der Künste in St. Petersburg. 1896–1900 Studium an der Juristischen Fakultät der Universität St. Petersburg. 1898 Studium in München. 1898–1900 Studium an der Privatschule von Prinzessin Teniseva bei Ilja Repin in St. Petersburg. Ab 1900 Mitglied der Künstlergruppe „Welt der Kunst". 1907–08 Arbeiten für das Theater, u. a. Bühnenbild für Diaghilews Inszenierung von Mussorgskijs „Boris Godunow". 1907–17 Lehrer an der Schule zur Förderung der Künste in St. Petersburg. 1920–25 in Ägypten. 1925–36 in Paris. 1936 Rückkehr nach Leningrad, lehft am In-

possède avec Kris Holmes l'agence «Bigelow & Holmes». 1983, organise et dirige un séminaire international sur le thème «L'ordinateur et la main pour concevoir une police» à la Stanford University. 1987, reçoit le prix F. W. Goudy de typographie du Rochester Institute of Technology. 1987, nouveau design de la revue «Scientific American». De nombreuses publications dans des revues spécialisées. – *Polices:* Leviathan (1979), Lucida (1984–1995 en plus de 50 tailles et variantes), Apple Chicago (1991), Apple Geneva (1991), Microsoft Wingdings 1–3 (1992). En 1980, Hans Ed. Meier, Charles Bigelow et Kris Holmes élaborent des lettres phonétiques pour les langues des Indiens d'Amérique sur la base de la police «Syntax» de Hans Ed. Meier.

**Bilibine,** Ivan Jakovlevitch – né en 1876 à Tarkhovka, près de Saint-Pétersbourg, Russie, décédé en 1942 à Leningrad, Russie – *peintre, graphiste, illustrateur de livres, enseignant, scénographe* – 1895–1898, études à l'école pour la promotion des arts à Saint-Pétersbourg. 1896–1900, études à la faculté de droit de l'université de Saint-Pétersbourg. 1898, études à Munich. 1898–1900, fréquente le cours privé de la Princesse Teniseva avec Ilia Répine à Saint-Pétersbourg. A partir de 1900, membre du groupe d'artistes «Monde de l'art». 1907–1908, travaille pour le théâtre et réalise des scénographies pour «Boris Godounov» de Moussorgsky, mis en scène par Diaghilev. 1907–1917, enseignant à l'école pour la promotion des arts à Saint-Pétersbourg. Vit en Egypte de 1920 à 1925, puis à Paris de 1925 à 1936. Retour à Leningrad en 1936 où il enseigne à l'institut de peinture, de sculpture et d'architecture à l'académie des beaux-arts. Les diverses activités artistiques de Bilibine s'inscrivent dans la mouvance de l'Art nou-

Bilibin 1904 Poster

Bilibin 1907 Cover

Bilibin 1931 Cover

Bilibin 1901–03 Cover

Bilibin 1921 Initial

has his own studio, Bigelow & Holmes, together with Kris Holmes. 1983: organizes and chairs the international seminar on "The Computer and the Hand in Type Design" at Stanford University. 1987: wins the F. W. Goudy Prize for typography from the Rochester Institute of Technology. 1987: redesigns "Scientific American" magazine . Has had many articles published in journals. – *Fonts:* Leviathan (1979), Lucida (1984–95, before this over 50 different designs and variations), Apple Chicago (1991), Apple Geneva (1991), Microsoft Wingdings 1–3 (1992). 1980: Hans Ed. Meier, Charles Bigelow and Kris Holmes develop phonetic letters for the languages of the American Indians based on the "Syntax" font by Hans Ed. Meier.

**Bilibin,** Ivan Yakovlevitch – b. 1876 in Tarchovka near St. Petersburg, Russia, d. 1942 in Leningrad, Russia – *painter, graphic artist, book illustrator, teacher, set designer* – 1895–98: studies at the School for the Promotion of the Arts in St. Petersburg. 1896–1900: studies in the law faculty of the University of St. Petersburg. 1898: studies in Munich. 1898–1900: studies at Princess Teniseva's private school with Ilya Repin in St. Petersburg. From 1900 onwards: member of the World of Art group. 1907–08: works in the theater, including designing the set for Diaghilev's production of Mussorgsky's "Boris Godunov". 1907–17: teaches at the School for the Promotion of the Arts in St. Petersburg. 1920–25: lives in Egypt. 1925–36: lives in Paris. 1936: returns to Leningrad and teaches

stitut für Malerei, Bildhauerei und Architektur an der Akademie der Künste. Die vielseitigen gestalterischen Tätigkeiten Bilibins repräsentieren die russische Spielart des Jugendstils. – *Publikation u.a.:* S. W. Golynez „Iwan Bilibin", Leningrad 1981.

**Bill,** Max – geb. 22. 12. 1908 in Winterthur, Schweiz, gest. 8. 12. 1994 in Berlin, Deutschland. – *Architekt, Maler, Typograph, Designer, Plastiker, Lehrer, Politiker* – Ausbildung: 1924–27 Kunstgewerbeschule Zürich, 1927–29 Bauhaus Dessau. 1929 Umzug nach Zürich, Gründung seines Ateliers, tätig als Gestalter und Publizist. 1932–36 Mitglied der Künstlergruppe „abstraction-création", Paris. 1944–45 Lehrer an der Kunstgewerbeschule Zürich. 1951–56 Gründungsrektor der Hochschule für Gestaltung Ulm, Direktor der Abteilungen Architektur und Produkt-Design. 1955 Herausgeber der Gesammelten Schriften von Wassily Kandinsky. 1967–74 Professor für Umweltgestaltung an der Hochschule für Bildende Künste in Hamburg. 1967–74 Mitglied im Schweizer Parlament. Bis zu seinem Tod Vorstandsvorsitzender des Bauhaus-Archivs, Berlin. Zahlreiche internationale Einzel- und Gruppenausstellungen. – *Publikationen u.a.:* „Quinze variations sur un même thème 1935–38", Paris 1938; „Über Typographie" in „Schweizer Grafische Mitteilungen" 4/1946; „Wassily Kandinsky", Paris 1951; „Form", Basel 1952. T. Maldonado „Max Bill", Buenos Aires 1955; Max Bense u. a. „Max Bill", Teufen 1958; Margit Staber „Max Bill", St. Gallen 1971;

Bill 1951 Poster

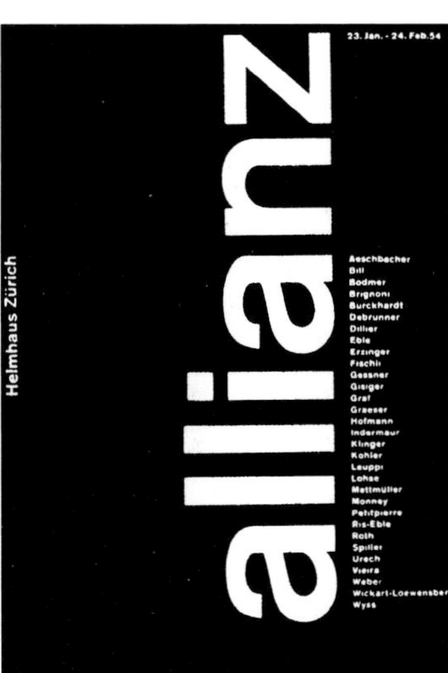


Bill 1954 Poster

Bill 1959 Cover

Bill 1949 Poster

Bill 1960 Poster

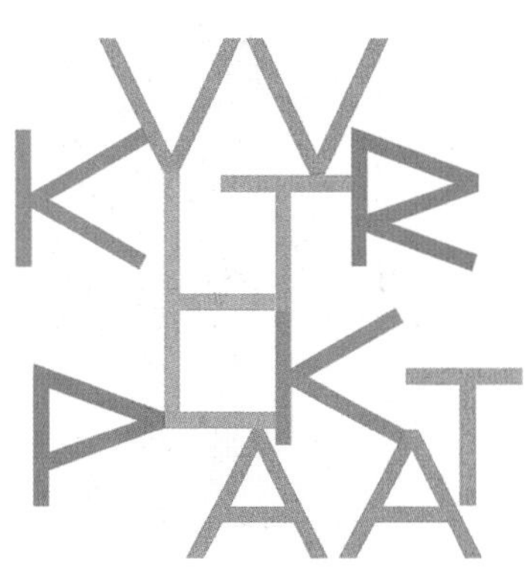


Bill 1974 Title

at the Institute for Painting, Sculpture and Architecture at the Academy of Arts. Bilibin's versatile artistic work is representative of the Russian variation on Art Nouveau. – *Publications include:* S. W. Golynez "Ivan Bilibin", Leningrad 1981.

**Bill,** Max – b. 22. 12. 1908 in Winterthur, Switzerland, d. 8. 12. 1994 in Berlin, Germany – *architect, painter, typographer, designer, sculptor, teacher, politician* – 1924–27: Kunstgewerbeschule in Zurich. 1927–29: at the Bauhaus in Dessau. 1929: moves to Zurich and opens his studio; works as a designer and publicist. 1932–36: member of the "abstraction-création" group, Paris. 1944–45: teaches at the Kunstgewerbeschule in Zurich. 1951–56: first rector of the Hochschule für Gestaltung in Ulm; director of the departments of architecture and product design. 1955: publishes the complete works of Wassily Kandinsky. 1967–74: professor of environmental design at the Hochschule für Bildende Künste in Hamburg. 1967–74: member of Swiss parliament. Chairman of the board of directors of the Bauhaus-Archive in Berlin until his death. Has been featured in numerous solo and group exhibitions all over the world. – *Publications include:* "Quinze variations sur un même thème 1935–38", Paris 1938; "Über Typographie" in "Schweizer Grafische Mitteilungen" 4/1946; "Wassily Kandinsky", Paris 1951; "Form", Basle 1952. Tomas Maldonado "Max Bill", Buenos Aires 1955; Max Bense et al "Max Bill", Teufen 1958; Margit Staber "Max Bill", St. Gallen 1971; Eduard Hüttinger "Max Bill", Zurich 1977; W. Spies "Kon-

veau russe. – *Publications, sélections:* S. W. Golynez «Ivan Bilibine», Leningrad 1981.

**Bill,** Max – né le 22. 12. 1908 à Winterthur, Suisse, décédé le 8. 12. 1994 à Berlin, Allemagne – *architecte, peintre, typographe, designer, plasticien, enseignant, homme politique* – formation: 1924–1927, Kunstgewerbeschule (école des arts décoratifs) de Zurich; 1927–1929, Bauhaus de Dessau. 1929, s'installe à Zurich, il y fonde un atelier où il exerce comme créateur et journaliste. 1932–1936, membre du groupe d'artistes parisiens «abstraction-création». 1944–1945, enseigne à la Kunstgewerbeschule de Zurich. 1951–1956, recteur et fondateur de la Hochschule für Gestaltung (école supérieure de design) d'Ulm; il y dirige le département d'architecture et de design de produits. 1955, édite les œuvres complètes de Vassili Kandinsky. 1967–1974, professeur d'arts de l'environnement à la Hochschule für Bildende Künste (école supérieure des beaux-arts) de Hambourg. 1967–1974, membre du parlement suisse. Président du conseil d'administration du Bauhaus-Archiv jusqu'à sa mort. Nombreuses expositions internationales personnelles et collectives. – *Publications, sélection:* «Quinze variations sur un même thème 1935–38», Paris 1938; «Über Typographie» dans «Schweizer Grafische Mitteilungen» 4/1946; «Vassili Kandinsky», Paris 1951; «Form», Bâle 1952. Tomas Maldonado «Max Bill», Buenos Aires 1955; Max Bense et autres «Max Bill», Teufen 1958; Margit Staber

Eduard Hüttinger „Max Bill", Zürich 1977; Werner Spies „Kontinuität", Frankfurt am Main 1986.

**Birdsall,** Derek – geb. 1. 8. 1934 in Yorkshire, England. – *Grafik-Designer, Art Director, Lehrer, Typograph* – 1949–52 Ausbildung am Wakefield College of Art und 1952–55 an der Central School of Art and Design, London. 1957 Beginn seiner Arbeit als freischaffender Designer. 1960–65 Mitglied der Agentur BDMW Associates, London. 1964 Design des ersten Kalenders für Pirelli-Reifen. Zeitweise Art Director der Zeitschriften „Town", „Nova", „Connoisseur", „The Independent Magazine". Design Consultant der Mobil Corporation, IBM Europa und der United Technologies Corporation. Arbeit für Penguin Books. 1983 Gründungspartner der Omnific Studios, London. Lehrtätigkeiten: ab 1957 an der London School of Printing und an der Central School of Arts and Crafts in London, 1987–88 Royal College of Art in London. – *Publikationen u.a.:* „A Book of Chess", London 1974; „The Technology of Man", London 1978; Veröffentlichungen in „17 Graphic Designers", London 1963. Jeremy Myerson „White Space, Black Hat" in „Eye" 9/1993.

**Bisti,** Dmitri S. – geb. 1925 in Moskau, Rußland. – *Grafik-Designer, Illustrator, Lehrer* – Zahlreiche buchtypographische Arbeiten. 1985 Gutenbergpreisträger der Stadt Leipzig, Deutschland. Professor und Vizepräsident der Akademie der Bildenden Künste der UdSSR in Moskau.

**Bitstream Inc.** – *Schriftenhersteller* – Matthew Carter und Mike Parker gründen

«Max Bill», Saint-Gall 1971; Eduard Hüttinger «Max Bill», Zurich 1977; Werner Spies «Kontinuität», Francfort-sur-le-Main 1986.

**Birdsall,** Derek – né le 1. 8. 1934 dans le Yorkshire, Angleterre – *graphiste maquettiste, directeur artistique, enseignant, typographe* – formation: 1949–1952, Wakefield College of Art; 1952–1955, Central School of Art and Design, Londres. 1957, commence à travailler comme designer indépendant. 1960–1965, membre de l'agence BDMW Associates, Londres. 1964, conçoit le premier calendrier pour les pneus Pirelli. Art Director par intermittence pour les revues «Town», «Nova», «Connoisseur», «The Independent Magazine». Consultant en design pour la Mobil Corporation, IBM Europe et United Technologies Corporation. Travaille pour les Penguin Books. 1983, fonde avec ses partenaires les «Omnific Studios», à Londres. Enseignement: A partir de 1957 à la London School of Printing et à la Central School of Arts and Crafts, Londres; 1987–1988, Royal College of Art, Londres. – *Publications, sélection:* «A Book of Chess», Londres 1974; «The Technology of Man», Londres 1978; publications dans «17 Graphic Designers», Londres 1963. Jeremy Myerson «White Space, Black Hat» dans «Eye» 9/1993.

**Bisti,** Dmitri S. – né en 1925 à Moscou, Russie – *graphiste maquettiste, illustrateur, enseignant* – de nombreux travaux de typographie pour l'édition. 1985, lauréat du Prix Gutenberg de la ville de Leipzig, Allemagne. Professeur et vice-président de l'académie des arts plastiques d'URSS à Moscou.

**Bitstream Inc.** – *fabricant de polices* – en 1981, Matthew Carter et Mike Parker fondent la société à Cambridge, Massachusetts, Etats-Unis dans le but de dévelop-

Birdsall 1961 Folder

Birdsall 1964 Advertisement

Birdsall 1990 Poster

Bitstream Inc. 1989 Type Specimen Book

tinuität", Frankfurt am Main 1986.

**Birdsall,** Derek – b. 1. 8. 1934 in Yorkshire, England – *graphic designer, art director, teacher, typographer* – 1949–52: studies at the Wakefield College of Art and from 1952–55 at the Central School of Art and Design, London. 1957: begins working as a freelance designer. 1960–65: member of the agency BDMW Associates in London. 1964: designs the first calendar for Pirelli Tires. Intermittent art director of the magazines "Town", "Nova", "Connoisseur" and "The Independent Magazine". Design consultant for the Mobil Corporation, IBM Europe and United Technologies Corporation. Works for Penguin Books. 1983: founder partner of Omnific Studios, London. Teaching positions: from 1957 onwards, at the London School of Printing and the Central School of Arts and Crafts, London; 1987–88 at the Royal College of Art, London. – *Publications include:* "A Book of Chess", London 1974; "The Technology of Man", London 1978; publications in "17 Graphic Designers", London 1963. Jeremy Myerson "White Space, Black Hat" in "Eye" 9/1993.

**Bisti,** Dmitri S. – b. 1925 in Moscow, Russia – *graphic designer, illustrator, teacher* – Numerous typographical work for books. 1985: wins the Gutenberg Prize from the City of Leipzig, Germany. Professor and vice-president of the Russian Academy of Fine Arts in Moscow.

**Bitstream Inc.** – *type manufacturers* – 1981: Matthew Carter and Mike Parker found the company in Cambridge, Massachusetts, USA, with a view to developing and marketing digitally saved fonts

1981 in Cambridge, Massachusetts, USA, die Firma, um digital gespeicherte Schriften für DTP zu entwickeln und zu vertreiben. Erste Schrift bei Bitstream: Bitstream Charter.
**Bittrof,** Max – geb. 27. 11. 1890 in Frankfurt an der Oder, Deutschland, gest. 1972 in Frankfurt am Main, Deutschland. – *Typograph, Grafik-Designer, Schriftentwerfer* – Ausbildung in Krefeld und Wuppertal-Elberfeld als Lithograph und Drucker. 1919 Umzug nach Frankfurt am Main, Arbeit als freier Grafiker. 1923 Mitbegründer des Bundes Deutscher Gebrauchsgraphiker (BDG). 1927–29 grafische Arbeiten für den Autohersteller Opel in Rüsselsheim, Deutschland. 1930–68 grafische Arbeiten für Telefonbau & Normalzeit, Frankfurt am Main. 1948 Entwurf der Geldscheine der Bank Deutscher Länder. Ab 1949 zahlreiche Briefmarkenentwürfe für die Deutsche Bundespost. Ab 1957 Lehrauftrag für Gebrauchsgrafik an der Kunstschule Westend, Frankfurt am Main. Zahlreiche Entwürfe für die Tabakindustrie. – *Schriftentwurf:* Element Fraktur (1934). – *Publikationen u. a.:* Frithjof Dahl „Max Bittrof" in „Gebrauchsgraphik" 12/1928; Friedrich Friedl „Max Bittrof" in „Eye" 9/1993.
**Black,** Roger – geb. 14. 8. 1948 in Austin, Texas, USA. – *Art Director, Typograph* – Ausbildung am Poynter Institute, anschließend Art Center Vevey, Schweiz, Art Center Pasadena, School of Visual Art, New York. U. a. Art Director der Zeitschriften „The New York Times", „Newsweek", „Rolling Stone". Re-Design der Zeitschriften „Esquire", „Foreign

Bittrof 1928 Advertisement

Bittrof 1926 Cover

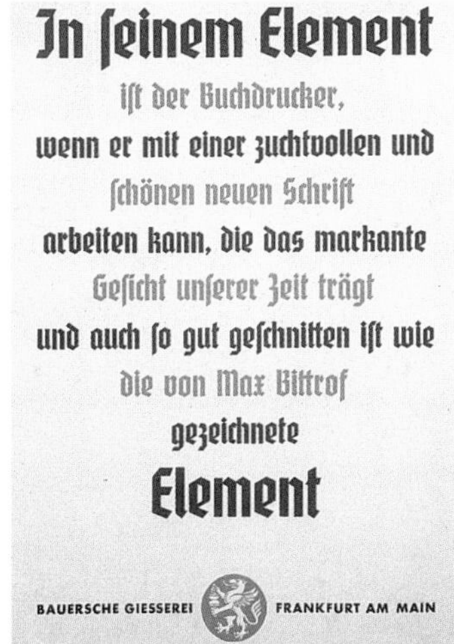
In ſeinem Element
iſt der Buchdrucker,
wenn er mit einer zuchtvollen und
ſchönen neuen Schrift
arbeiten kann, die das markante
Geſicht unſerer Zeit trägt
und auch ſo gut geſchnitten iſt wie
die von Max Bittrof
gezeichnete
Element

BAUERSCHE GIESSEREI FRANKFURT AM MAIN

Bittrof 1934 Specimen

Bittrof 1928 Initials

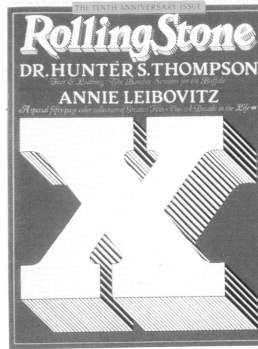

Black 1977 Cover

per et de distribuer des polices numériques pour la DTP. Première police parue chez Bitstream : la Bitstream Charter.
**Bittrof,** Max – né le 27. 11. 1890 à Francfort-sur-l'Oder, Allemagne, décédé en 1972 à Francfort-sur-le-Main, Allemagne – *typographe, graphiste maquettiste, concepteur de polices* – formation de lithographe et d'imprimeur à Krefeld et à Wuppertal-Elberfeld. 1919, s'installe à Francfort-sur-le-Main, où il travaille comme graphiste indépendant. 1923, cofondateur du «Bund Deutscher Gebrauchsgraphiker» (BDG). 1927–1929, travaux graphiques pour le constructeur automobile Opel, Rüsselsheim, Allemagne. 1930–1968, travaux graphiques pour Telefonbau & Normalzeit, Francfort-sur-le-Main. 1948, dessine les billets de la Bank Deutscher Länder. A partir de 1949, nombreux dessins de timbres pour les postes allemandes. A partir de 1957, chargé de cours en graphisme utilitaire à la Kunstschule Westend de Francfort-sur-le-Main. Nombreux dessins pour l'industrie des tabacs. – *Police:* Element Fraktur (1934). – *Publications, sélection:* Frithjof Dahl «Max Bittrof» dans «Gebrauchsgraphik» 12/1928; Friedrich Friedl «Max Bittrof» dans «Eye» 9/1993.
**Black,** Roger – né le 14. 8. 1948 à Austin, Texas, Etats-Unis – *directeur artistique, typographe* – formation : Poynter Institute, puis Art Center Vevey, Suisse, Art Center Pasadena, School of Visual Art New York. Art Director de revues, dont «The New York Times», «Newsweek», «Rolling Stone». Nouveau design des re-

for DTP. Bitstream's first typeface is Bitstream Charter.
**Bittrof,** Max – b. 27. 11. 1890 in Frankfurt an der Oder, Germany, d. 1972 in Frankfurt am Main, Germany – *typographer, graphic designer, type designer* – Trained in Krefeld and Wuppertal-Elberfeld as a lithographer and printer. 1919: moves to Frankfurt am Main and works as a freelance graphic artist. 1923: one of the founders of the Bund Deutscher Gebrauchsgraphiker (BDG or Association of German Commercial Artists). 1927–29: produces graphics for the car manufacturer Opel in Rüsselsheim, Germany. 1930–68: graphic work for Telefonbau & Normalzeit in Frankfurt am Main. 1948: designs bank notes for the Bank Deutscher Länder (the precursor of the Bundesbank). From 1949 onwards: designs numerous stamps for the German Federal Post Office. From 1957 onwards: teaches commercial art at the Westend art school in Frankfurt am Main. Has produced numerous designs for the tobacco industry. – *Font:* Element Fraktur (1934). – *Publications include:* Frithjof Dahl "Max Bittrof" in "Gebrauchsgraphik" 12/1928; Friedrich Friedl "Max Bittrof" in "Eye" 9/1993.
**Black,** Roger – b. 14. 8. 1948 in Austin, Texas, USA – *art director, typographer* – Education: Poynter Institute and then the Art Center in Vevey, Switzerland, the Art Center Pasadena, and the School of Visual Art, New York. Art director of the magazines "The New York Times", "Newsweek" and "Rolling Stone", among others. Redesigns the magazines "Esquire"

Affairs". 1987 Gründung der Firma „Roger Black Inc.", 1989 Gründung der Firma „The Font Bureau" mit David Berlow in Boston. – *Publikation:* „Roger Black's Desktop Design Power", New York 1991.

**Blase,** Karl Oskar – geb. 24. 3. 1925 in Köln, Deutschland. – *Typograph, Grafik-Designer, Lehrer* – Studium der Grafik und der Malerei in Wuppertal. 1950 Atelier mit Felix Müller (Müller-Blase) 1952–58 Leiter des Ateliers für die Gestaltung der Ausstellungen der Amerikahäuser in Deutschland. 1955–88 Briefmarkenentwürfe für die Deutsche Bundespost. 1957–68 Grafik-Design für die Zeitschrift „form". 1958 Dozent, 1966 Professor für Kunst und Visuelle Kommunikation an der Staatlichen Werkkunstschule, Gesamthochschule Kassel. 1966–78 Plakate und visuelle Gestaltung für das Staatstheater Kassel. Seit 1988 Mitglied des Kunstbeirats beim Bundesminister für Post und Telekommunikation, seit 1992 Vorsitzender dieses Gremiums. Zahlreiche Auszeichnungen für typographische und grafische Arbeiten. 1964 Teilnahme an der documenta III, Abteilung Design und Grafik. Zahlreiche Ausstellungen. – *Publikationen u. a.:* „Blase Grafik-Design", Köln 1995; zahlreiche Aufsätze in Fachzeitschriften.

**Bloemsma,** Evert – geb. 8. 7. 1958 in Den Haag, Niederlande. – *Grafik-Designer, Schriftentwerfer, Fotograf* – 1976–81 Grafik-Design-Studium an der Academie voor Beeldende Kunsten in Arnheim. 1981–83 Arbeiten als Designer und Drucker. 1983–85 Grafik-Designer in Den Haag. 1984–87 beteiligt an der Gestal-

vues «Esquire», «Foreign Affairs». 1987, fondation de la société «Roger Black Incorporated». 1989, fonde avec David Berlow, la société «The Font Bureau» à Boston. – *Publication:* «Roger Black's Desktop Design Power», New York 1991.

**Blase,** Karl Oskar – né le 24. 3. 1925 à Cologne, Allemagne – *typographe, graphiste maquettiste, enseignant* – études des arts graphiques et de la peinture à Wuppertal. 1950, partage un atelier avec Felix Müller (Müller-Blase). 1952–1958, dirige l'atelier de conception des expositions des Centres américains en Allemagne. 1955–1988 dessins de timbres pour les postes allemandes. 1957–1968, maquettes pour la revue «form». 1958, enseignant dans le supérieur; 1966, professeur d'art et de communication visuelle à la Staatliche Werkkunstschule, Gesamthochschule de Kassel. 1966–1978, affiches et conception visuelle pour le Staatstheater de Kassel. Depuis 1988, membre du conseil artistique auprès du Ministre des postes et des télécommunications; président de cet organe depuis 1992. De nombreuses distinctions pour ses travaux typographiques et graphiques. 1964, participation à la «documenta III» à Kassel dans la section design et arts graphiques. Nombreuses expositions. – *Publications, sélection:* «Blase Grafik-Design», Cologne 1995; nombreux essais dans les revues spécialisées.

**Bloemsma,** Evert – né le 8. 7. 1958 à La Haye, Pays-Bas – *graphiste maquettiste, concepteur de polices, photographe* – 1976–1981, études d'arts graphiques et de design à l'Academie voor Beeldende Kunsten d'Arnhem. 1981–1983, exerce comme graphiste et imprimeur. 1983–1985, graphiste maquettiste à La Haye. 1984–1987, participe à la mise en page du magazine d'architecture «Forum» et utilise une ancienne version de sa Grotesk.

Blase 1967 Poster

Blase 1963 Poster

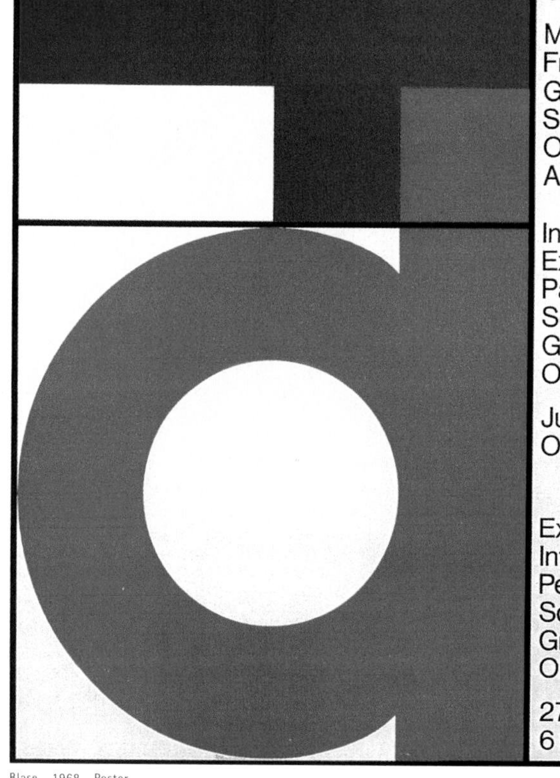

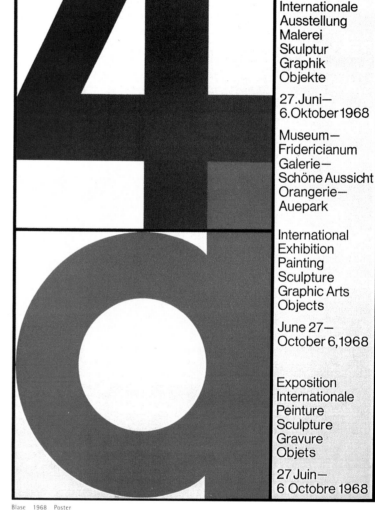

Blase 1968 Poster

and "Foreign Affairs". 1987: sets up Roger Black Incorporated. 1989: founds The Font Bureau with David Berlow in Boston. – *Publications:* "Roger Black's Desktop Design Power", New York 1991.

**Blase,** Karl Oskar – b. 24. 3. 1925 in Cologne, Germany – *typographer, graphic designer, teacher* – Studied graphics and painting in Wuppertal. 1950: opens the Müller-Blase studio with Felix Müller. 1952–58: director of the Studio for the Planning of American Housing Exhibitions in Germany. 1955–88: designs stamps for the German post office. 1957–68: graphic design for "form" magazine. 1958: lecturer and in 1966 professor of art and visual communication at the Staatliche Werkkunstschule, Gesamthochschule in Kassel. 1966–78: produces posters and visual designs for the state theater in Kassel. Since 1988: member of the Art Advisory Committee to the German Minister of Post and Telecommunications; appointed chairman in 1992. Has won many awards for his typographic and design work. 1964: takes part in documenta III in Kassel, in the department of design and graphics. Numerous exhibitions. – *Publications include:* "Blase Grafik-Design", Cologne 1995; numerous essays in journals.

**Bloemsma,** Evert – b. 8. 7. 1958 in The Hague, The Netherlands – *graphic designer, type designer, photographer* – 1976–81: studies graphic design at the Academie voor Beeldende Kunsten in Arnhem. 1981–83: works as a designer and printer. 1983–85: graphic designer in The Hague. 1984–87: helps design the ar-

tung des Architektur-Magazins „Forum", in dem eine frühe Version seiner Groteskschrift verwendet wird. 1987–89 in Hamburg, Deutschland, Arbeit in der typographischen Abteilung der Firma URW. 1990–91 Typograph bei Océ van der Grinten in Venlo, Niederlande. – *Schriftentwurf:* FF Balance (1993).
**Blokland,** Erik van – geb. 29. 8. 1967 in Gouda, Niederlande. – *Schriftentwerfer, Grafik-Designer* – Studiert an der Koninklijke Academie voor Beeldende Kunsten in Den Haag, Abschluß 1989. Praktikum bei Total Design, Amsterdam. Seit 1990 selbständig in Den Haag. Entwickelt 1989 „LettError" mit Just van Rossum in Berlin, zunächst als Zeitschrift, dann als Bezeichnung ihrer gemeinsamen Arbeit. 1989 entsteht der RandomFont Beowolf, eine Schrift, deren Konturen sich ständig über ein „random"-Programm verändern. – *Schriftentwürfe:* Beowolf (1989), Erikrighthand (1991), Trixie (1991), Kosmik (Flipper Font, mit Just van Rossum, 1993), Federal (1996).
**Blumenthal,** Joseph – geb. 4. 10. 1897 in New York, USA, gest. 11. 7. 1990 in West Cornwell, Connecticut, USA. – *Schriftentwerfer, Drucker, Verleger* – 1926 Gründung der Spiral Press. 1930 Aufenthalt in Europa. 1934 Eröffnung seiner Druckerei in New York. 1971 schließt Blumenthal seine Druckerei und schreibt über Typographie. – *Schriftentwurf:* Emerson (vorher Spiral, 1936–39). – *Publikationen:* „The Art of the printed Book", New York 1973; „Typographic Years, a Printer's Journey through half a Century 1925–1975", New York 1982.

Bloemsma 1993 Specimen

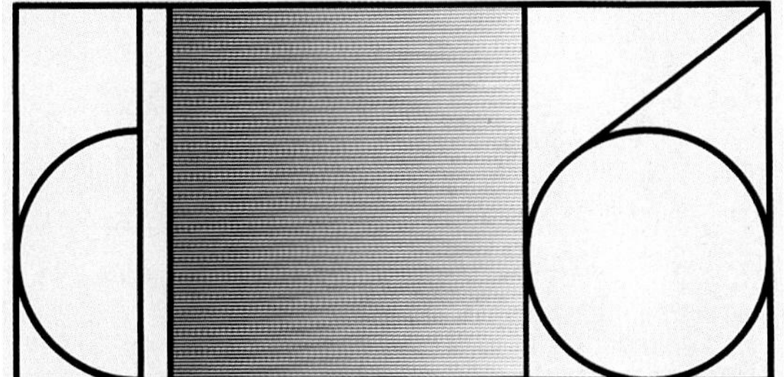

Blokland 1986 Card

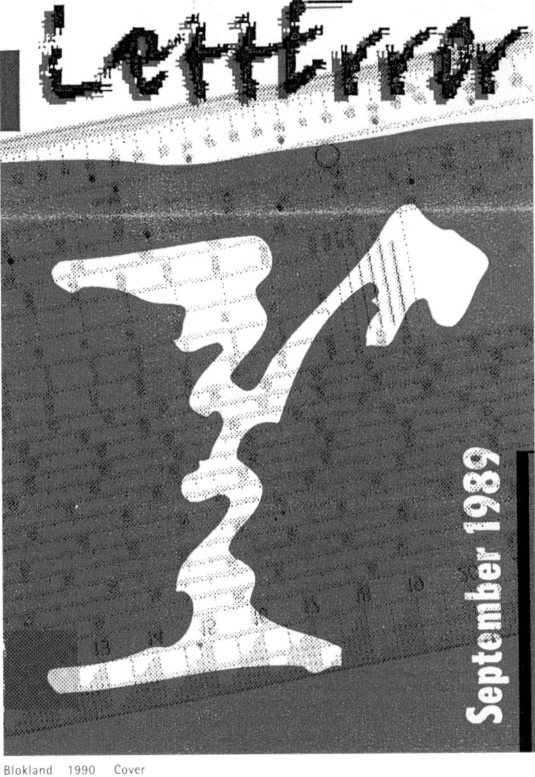

Blokland 1990 Cover

Blase 1987 Poster

| | n | | | | |
|---|---|---|---|---|---|
| a | b | e | f | g | i |
| o | r | s | t | y | z |
| A | B | C | E | G | H |
| M | O | R | S | X | Y |
| 1 | 2 | 4 | 6 | 8 | & |

Blokland 1989 Beowolf

| | n | | | | |
|---|---|---|---|---|---|
| a | b | e | f | g | i |
| o | r | s | t | y | z |
| A | B | C | E | G | H |
| M | O | R | S | X | Y |
| 1 | 2 | 4 | 6 | 8 | & |

Blokland 1991 Erikrighthand

1987–1989, vit à Hambourg, Allemagne et travaille dans le service typographie de la société URW. 1990–1991, typographe chez Océ van der Grinten à Venlo, Pays-Bas. – *Police:* FF Balance (1993).
**Blokland,** Erik van – né le 29. 8. 1967 à Gouda, Pays-Bas – *concepteur de polices, graphiste maquettiste* – études à la Koninklijke Academie voor Beeldende Kunsten de La Haye, diplômé en 1989. Stage chez Total design, Amsterdam. Indépendant à La Haye depuis 1990. En 1989, il publie à Berlin, avec Just van Rossum, «LettError», d'abord sous forme de revue, puis comme image de leur collaboration. 1989, création de RandomFont Beowolf, une police dont les contours se modifient sans cesse sous l'effet d'un programme «random». – *Polices:* Beowolf (1989), Erikrighthand (1991), Trixie (1991), Kosmik (Flipper Font, avec Just van Rossum, 1993), Federal (1996).
**Blumenthal,** Joseph – né le 4. 10. 1897 à New York, Etats-Unis, décédé le 11.7. 1990 à West Cornwell, Connecticut, Etats-Unis – *concepteur de polices, imprimeur, éditeur* – 1926, fondation de Spiral Press. 1930, séjour en Europe. 1934, retour à New York où il ouvre sa propre imprimerie. 1971, Blumenthal ferme son imprimerie et écrit sur la typographie. – *Police:* Emerson (avant Spiral, 1936–1939). – *Publications:* «The Art of the Printed Book», New York 1973; «Typographic Years, a Printer's Journey through Half a Century 1925–1975», New York 1982.

chitecture magazine "Forum", which uses an early version of his sans-serif type. 1987–89: lives in Hamburg and works in the typography department of the URW company. 1990–91: typographer for Océ van der Grinten in Venlo, The Netherlands. – *Font:* FF Balance (1993).
**Blokland,** Erik van – b. 29. 8. 1967 in Gouda, The Netherlands – *type designer, graphic designer* – Studied at the Koninklijke Academie voor Beeldende Kunsten in The Hague, graduating in 1989. Period of practical training with Total Design in Amsterdam. He has been self-employed in The Hague since 1990. 1989: launches his LettError project with Just van Rossum in Berlin, first as a magazine and then as a terminology for their joint work. In 1989 they design the RandomFont Beowolf, a typeface whose contours are constantly altered by a "random" program. – *Fonts:* Beowolf (1989), Erikrighthand (1991), Trixie (1991), Kosmik (a flipper font, with Just van Rossum, 1993), Federal (1996).
**Blumenthal,** Joseph – b. 4. 10. 1897 in New York, USA, d. 11. 7. 1990 in West Cornwell, Connecticut, USA – *type designer, printer, publisher* – 1926: founds the Spiral Press. 1930: spends time in Europe. 1934: returns to New York and opens a printing workshop. 1971: Blumenthal closes down his printing works and writes about typography. – *Font:* Emerson (before Spiral, 1936–39). – *Publications:* "The Art of the Printed Book", New York 1973; "Typographic Years, a Printer's Journey through Half a Century 1925–1975", New York 1982.

**Bodoni,** Giambattista – geb. 26. 2. 1740 in Saluzzo, Piemont, Italien, gest. 30. 11. 1813 in Parma, Italien. – *Graveur, Schriftentwerfer, Typograph, Drucker, Verleger* – 1758–66 Schriftsetzer in der Druckerei von Propaganda Fide, einer Einrichtung des Vatikans. 1766 Einladung des Herzogs von Parma, eine Druckerei aufzubauen und zu leiten. 1768 Beginn seiner Tätigkeit in der Stamperia Reale. 1770 Eröffnung seiner eigenen Schriftgießerei. Veröffentlicht 1771 den ersten eigenen typographischen Beitrag „Fregi e Majuscole". 1782 ernennt Karl III. von Spanien Bodoni zu seinem Kammer-Typographen. 1788 erscheint das Buch „Manuale Tipografico", in dem 100 lateinische, 50 kursive und 28 griechische Minuskeltypen abgedruckt sind. 1790 erlaubt der Herzog von Parma Bodoni die Einrichtung einer eigenen Druckerei, der „Tipi Bodoni". Als erste Bände erscheinen griechische, römische und italienische Klassikerausgaben. 1806 erscheint das Werk „L'Oratio Dominica in CLV linguas versa", das in 215 Schrifttypen gesetzt ist. 1818 gibt die Witwe Bodonis das von ihr vollendete große zweibändige Werk „Manuale Tipografico" heraus, das die Ergebnisse der gesamten Tätigkeit zeigt: es beinhaltet Antiquaschriften, griechische, deutsche, asiatische und russische Schriften sowie Linien, Einfassungen, Zeichen, Ziffern und Notenschriften. 1963 Eröffnung des Bodoni-Museums in Parma.

**Bodoni-Museum,** Parma, Italien – 1843 kauft die Herzogin von Parma die umfangreiche typographische Hinterlassenschaft Bodonis, die den Grundstock für

**Bodoni,** Giambattista – né le 26. 2. 1740 à Saluzzo, Piémont, Italie, décédé le 30. 11. 1813 à Parme, Italie – *graveur, concepteur de polices, typographe, imprimeur, éditeur* – 1758–66, compositeur à l'imprimerie Propaganda Fide, une institution du Vatican. 1766, le duc de Parme l'invite à installer une imprimerie et à la diriger. 1768, début de ses activités à la Stamperia Reale. 1770, inauguration de sa propre fonderie de caractères. 1771, il publie son premier essai sur la typographie «Fregi e Majuscole». 1782, Charles III d'Espagne nomme Bodoni typographe de sa chancellerie. 1788, publication du «Manuale Tipografico», un livre contenant 100 types de basses casses latins, 50 italiques et 28 grecs. 1790, le duc de Parme autorise Bodoni à installer sa propre imprimerie, la «Tipi Bodoni»; les premiers volumes à paraître sont des éditions des classiques grecs, latins et italiens. 1806, publication de «l'Oratio Dominica in CLV linguas versa», composé à l'aide de 215 types d'écriture. 1818, la veuve de Bodoni achève et édite les deux volumes du «Manuale Tipografico» réunissant l'ensemble de l'œuvre de son mari; ils contiennent des Antiqua, des caractères grecs et allemands, asiatiques et russes, ainsi que des filets, des encadrés, des signes, chiffres et notations. 1963, inauguration du Musée Bodoni à Parme.

**Musée Bodoni,** Parme, Italie – en 1843, la duchesse de Parme achète l'imposante succession typographique de Bodoni, celle-ci doit constituer le fonds d'un musée Bodoni. En 1963, inauguration du musée Bodoni dans le Palais Farnèse de la Pilotta, au dernier étage de la Bibliothèque Palatine à Parme. Le musée réunit environ 1000 ouvrages édités par Bodoni ainsi que de la correspondance, des documents, des épreuves d'imprimerie, des modèles d'écriture, des matrices origi-

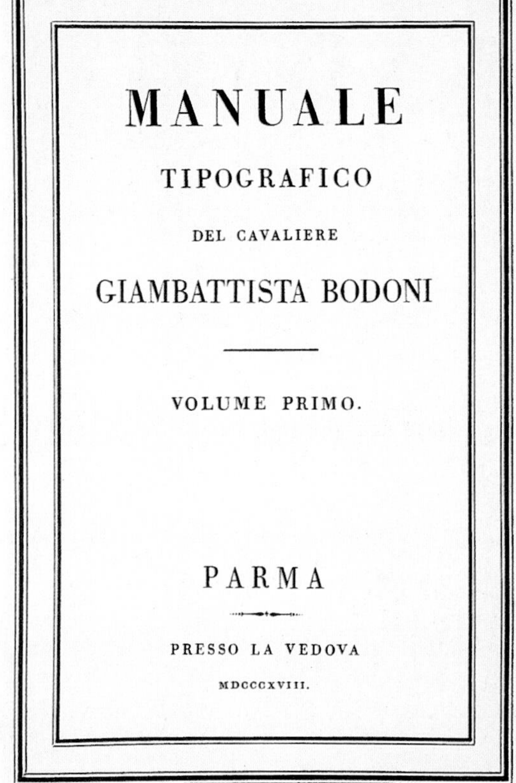
MANUALE
TIPOGRAFICO
DEL CAVALIERE
GIAMBATTISTA BODONI
VOLUME PRIMO.
PARMA
PRESSO LA VEDOVA
MDCCCXVIII.

Bodoni 1818 Book Title

PAPALE
Quousq; tandem abutêre, Catilina,
SALUZZO

Bodoni 1818 Page

| n | n | n | n |  | n |
|---|---|---|---|---|---|
| a | b | e | f | g | i |
| o | r | s | t | y | z |
| A | B | C | E | G | H |
| M | O | R | S | X | Y |
| 1 | 2 | 4 | 6 | 8 | & |

Bodoni

Bodoni 1771 Initial

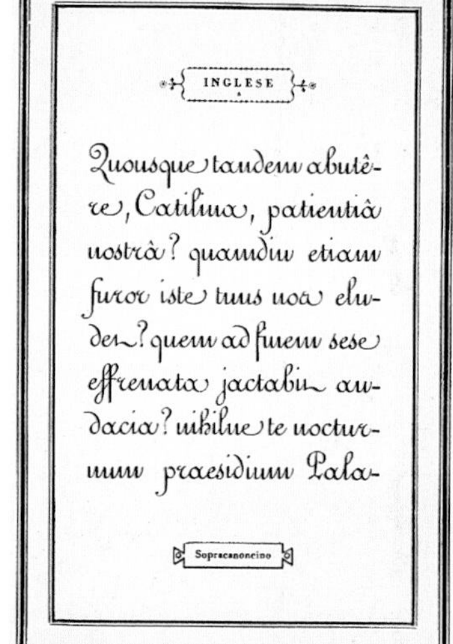
INGLESE
Quousque tandem abutêre, Catilina, patientiâ nostrâ? quamdiu etiam furor iste tuus nos eludet? quem ad finem sese effrenata jactabit audacia? nihilne te nocturnum praesidium Pala-
Soprasanoncino

Bodoni 1818 Page

Bodoni 1771 Initial

Bodoni 1771 Initial

**Bodoni,** Giambattista – b. 26. 2. 1740 in Saluzzo, Piedmont, Italy, d. 30. 11. 1813 in Parma, Italy – *engraver, type designer, typographer, printer, publisher* – 1758–66: typesetter in the Vatican's Propaganda Fide printing works. 1766: the Duke of Parma invites Bodoni to set up and run a printing works. 1768: begins working in the Stamperia Reale. 1770: opens his own type foundry. 1771: publishes his first typographical contribution "Fregi e Majuscole". 1782: Charles III of Spain names Bodoni his court typographer. 1788: the book "Manuale Tipografico" is published, containing 100 roman, 50 italic and 28 Greek minuscule fonts. 1790: the Duke of Parma gives Bodoni permission to open his own printing works, Tipi Bodoni. The first books to be published are volumes of Greek, Roman and Italian classics. 1806: "L'Oratio Dominica in CLV linguas versa" is produced, set in 215 typefaces. 1818: Bodoni's widow completes and publishes her late husband's mighty "Manuale Tipografico" in two volumes, a witness to Bodoni's entire creative working life. It contains roman, Greek, gothic, Asian and Russian fonts, and lines, borders, symbols, numbers and musical notation. 1963: the Bodoni Museum is opened in Parma.

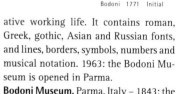

**Bodoni Museum,** Parma, Italy – 1843: the Duchess of Parma buys up Bodoni's extensive typographic estate as the basis for a Bodoni Museum. 1963: the Bodoni Museum is officially opened in the Palatina Library on the top floor of the Farnese Pilotta Palace in Parma. The museum houses approximately 1,000 of Bodoni's printed

ein Bodoni-Museum bilden soll. 1963 Einweihung des Bodoni-Museums im Farnesischen Palast der Pilotta, im obersten Stock der Palatina-Bibliothek in Parma. Im Museum sind etwa 1.000 Buchausgaben Bodonis versammelt sowie Briefwechsel, Dokumente, Probedrucke, Schriftmuster, Originalmatritzen und Arbeitsmaterialien. 1964 Gründung des „Forschungszentrums G. B. Bodoni", das die kulturelle Tätigkeit des Museums anregen und unterstützen soll. Veranstaltung von Ausstellungen, u. a. von Alberto Tallone, Arnoldo Mondadori, Giovanni Mardersteig, Hermann Zapf und Tibar Szanti. Die Präsidenten des Museums: Pietro Trevisani (1963–67), Baldassarre Molossi (1967–73), Angelo Giavarella (seit 1973). Der für italienische Verleger gestiftete Bodoni-Preis „Città di Parma" ging bisher an die Verlage Mondadori, Einaudi, CEI, Jaca Book und Lucini-Cerastico. Forschungszentrum und Museum haben über 20 Werke mit Dokumentationen der Bestände herausgegeben.

**Boehland,** Johannes – geb. 16. 4. 1903 in Berlin, Deutschland, gest. 5. 9. 1964 in Berlin, Deutschland. – *Schriftentwerfer, Kalligraph, Zeichner, Lehrer* – 1920–1926 Ausbildung an der Unterrichtsanstalt am Staatlichen Kunstgewerbemuseum Berlin und an den Vereinigten Staatsschulen für freie und angewandte Kunst in Berlin. 1926 Grafiker an der Staatlichen Porzellanmanufaktur, Berlin. 1929 Lehrer an der Städtischen Kunstgewerbe- und Handwerkerschule Berlin-Charlottenburg. 1931 Lehrer an der Meisterschule für Graphik und Buchgewerbe, Berlin.

Bodoni 1818 Page

Bodoni 1818 Page

Boehland ca. 1934 Poster

Boehland ca. 1932 Logo

Boehland ca. 1930 Sign

nales et des outils de travail. 1964, fondation du «Centre de recherche G. B. Bodoni» destiné à susciter et soutenir les activités culturelles du musée. Organisation d'expositions, entre autres d'Alberto Tallone, Arnoldo Mondadori, Giovanni Mardersteig, Hermann Zapf et Tibar Szanti. Ont été présidents du musée : Pietro Trevisani (1963–1967), Baldassarre Molossi (1967–1973), Angelo Giavarella (depuis 1973). Jusqu'à présent, le prix Bodoni «Città di Parma» destiné aux éditeurs italiens a été décerné aux éditions Mondadori, Einaudi, CEI, Jaca Book et Lucini-Cerastico. Le centre de recherche et le musée ont déjà édité plus de 20 ouvrages de documentation sur le fonds.

**Boehland,** Johannes – né le 16. 4. 1903 à Berlin, Allemagne, décédé le 5. 9. 1964 à Berlin, Allemagne – *concepteur de polices, calligraphe, dessinateur, enseignant.* – 1920–1926, formation à l'institut d'enseignement du Staatliches Kunstgewerbemuseum (musée des arts décoratifs) de Berlin et aux Vereinigten Staatsschulen für freie und angewandte Kunst (écoles des arts appliqués), Berlin. 1926, graphiste pour la Staatliche Porzellanmanufaktur de Berlin (manufacture d'Etat des porcelaines). 1929, enseigne à la Städtische Kunstgewerbe- und Handwerkerschule de Berlin-Charlottenburg. 1931, enseigne à la Meisterschule für Graphik und Buchgewerbe de Berlin (école des arts graphiques et du livre).

books, in addition to letters, documents, trial prints, font samples, original matrices and work materials. 1964: the G.B. Bodoni Research Center is founded, with the aim of initiating and supporting cultural activity in the museum. It has staged exhibitions of the work of Alberto Tallone, Arnoldo Mondadori, Giovanni Mardersteig, Hermann Zapf and Tibar Szanti, among others. Pietro Trevisani (1963–67), Baldassare Molossi (1967–73) and Angelo Giavarella (since 1973) have been presidents of the museum. The endowed Bodoni Prize for Italian publishers, the "Città di Parma", has to date been awarded to the publishers Mondadori, Einaudi, CEI, Jaca Book and Lucini-Cerastico. The research center and museum have published over 20 works documenting the contents of the museum.

**Boehland,** Johannes – b. 16. 4. 1903 in Berlin, Germany, d. 5. 9. 1964 in Berlin, Germany – *type designer, calligrapher, artist, teacher* – 1920–26: trains at the Unterrichtsanstalt am Staatlichen Kunstgewerbemuseum and at the Vereinigte Staatsschulen für freie und angewandte Kunst in Berlin. 1926: graphic artist at the state porcelain factory in Berlin. 1929: teaches at the Städtische Kunstgewerbe- und Handwerkerschule in Berlin-Charlottenburg. 1931: teaches at the Meisterschule für Graphik und Buchgewerbe in Berlin. 1945: teaches at the Hochschule für Bildende Künste in Berlin. 1951: head of the graphics department at the Werkkunstschule in Wiesbaden. 1954: lecturer at the Meisterschule für Graphik und Buchgewerbe in Berlin. Exhibitions:

1945 Lehrer an der Hochschule für Bildende Künste, Berlin. 1951 Leiter der Abteilung Grafik an der Werkkunstschule, Wiesbaden. 1954 Dozent an der Meisterschule für Graphik und Buchgewerbe, Berlin. Ausstellungen: Galerie Gerd Rosen, Berlin (1959), Gutenberg-Museum, Mainz (1963), Städtisches Museum, Wiesbaden (1965). – *Schriftentwurf:* „Balzac" (1951, D. Stempel AG). – *Publikationen u. a.:* „Das Unbegrenzte in der Schrift" in „Archiv für Druck und Papier" 2/1956. Hellwag „Johannes Boehland", Berlin, Leipzig 1939.

**Bohn,** Hans – geb. 23. 12. 1891 in Oberlahnstein, Deutschland, gest. 10. 5. 1980 in Frankfurt am Main, Deutschland. – *Grafiker, Schriftentwerfer, Typograph, Lehrer* – Ausbildung an der Technischen Lehranstalt Offenbach, danach bis 1914 im Ullstein Verlag Berlin tätig. 1919–30 Grafiker in der Schriftgießerei Gebr. Klingspor, Offenbach, enger Mitarbeiter Karl Klingspors. Nach dem Zweiten Weltkrieg Arbeit als Gebrauchsgrafiker. 1946–56 Lehrer an der Meisterschule für das gestaltende Handwerk in Offenbach. Tätig für zahlreiche Verlage, u. a. Rowohlt, Schneekluth, Ullstein, Fischer. – *Schriftentwürfe:* Orplid (1929), Mondial (1936–39), Allegro (1937).

**Boldizar,** Ivan – geb. 6. 11. 1917 in Novi Sad, Jugoslawien, gest. 5. 6. 1986 in Novi Sad, Jugoslawien. – *Schriftentwerfer, Grafiker, Kalligraph, Lithograph, Lehrer* – Lithographenlehre, danach in Druckereien in Deutschland, Jugoslawien und Ungarn tätig. Nach 1945 Leiter der Abteilung Schriftgrafik an der Werkkunst-

1945, professeur à la Hochschule für Bildende Künste, Berlin (école supérieure des beaux-arts). 1951, directeur du département d'arts graphiques à la Werkkunstschule de Wiesbaden. 1954, maître de conférences à la Meisterschule für Graphik und Buchgewerbe de Berlin. Expositions: Galerie Gerd Rosen, Berlin (1959); Gutenberg Museum, Mayence (1963); Städtisches Museum, Wiesbaden (1965). – *Polices:* «Balzac» (1951, D. Stempel AG). – *Publications, sélection:* «Das Unbegrenzte in der Schrift» dans «Archiv für Druck und Papier» 2/1956. Hellwag «Johannes Boehland», Berlin, Leipzig 1939.

**Bohn,** Hans – né le 23. 12. 1891 à Oberlahnstein, Allemagne, décédé le 10. 5. 1980 à Francfort-sur-le-Main, Allemagne – *graphiste, concepteur de polices, typographe, enseignant* – formation au Technische Lehranstalt d'Offenbach (institut d'études techniques), puis activités pour les éditions Ullstein jusqu'en 1914. 1919–1930, graphiste à la fonderie de caractères Gebr. Klingspor à Offenbach, collabore étroitement avec Karl Klingspor. Après la Seconde Guerre mondiale, exerce comme graphiste utilitaire. 1946–1956, enseigne à la Meisterschule für das gestaltende Handwerk (école d'artisanat d'art) à Offenbach. Travaille pour de nombreux éditeurs, dont Rowohlt (Hambourg), Schneekluth (Darmstadt), Ullstein (Berlin), Fischer (Francfort-sur-le-Main). – *Polices:* Orplid (1929), Mondial (1936–1939), Allegro (1937).

**Boldizar,** Ivan – né le 6. 11. 1917 à Novi Sad, Yougoslavie, décédé le 5. 6.1986 à Novi Sad, Yougoslavie – *concepteur de polices, graphiste, calligraphe, lithographe, enseignant* – apprentissage de lithographe, exerce ensuite dans plusieurs imprimeries en Allemagne, Yougoslavie et Hongrie. Après 1945, dirige le dépar-

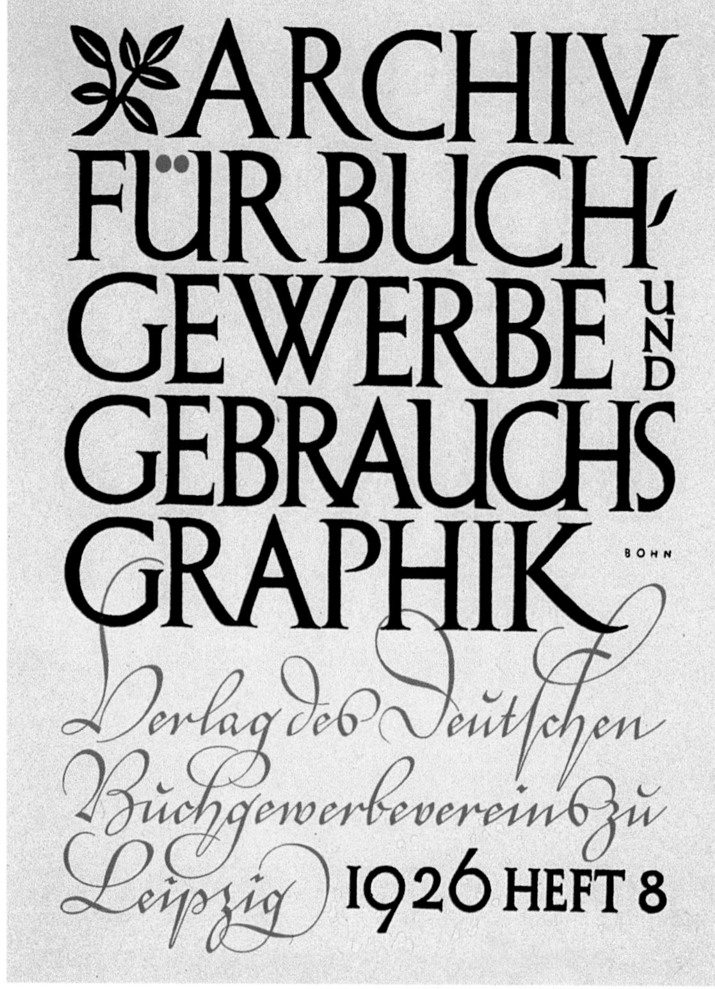


Bohn 1926 Cover

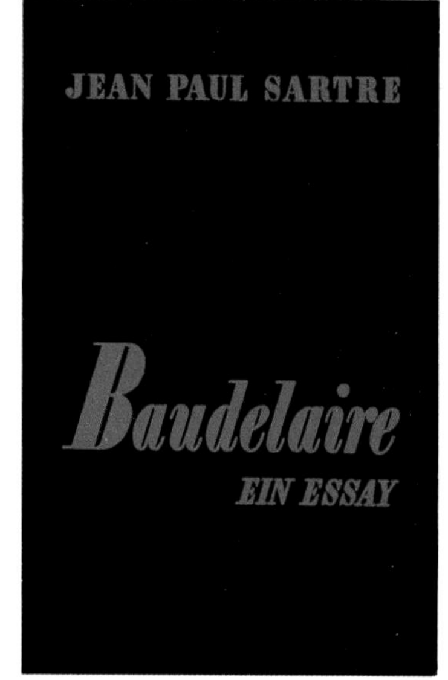


Bohn 1949 Cover

| | n | | | | |
|---|---|---|---|---|---|
| a | b | e | f | g | i |
| o | r | s | t | y | z |
| A | B | C | E | G | H |
| M | O | R | S | X | Y |
| 1 | 2 | 4 | 6 | 8 | & |

Bohn 1937 Allegro

Galerie Gerd Rosen in Berlin (1959), Gutenberg Museum in Mainz (1963), Städtisches Museum in Wiesbaden (1965). – *Font:* Balzac (1951, D. Stempel AG). – *Publications include*: "Das Unbegrenzte in der Schrift" in "Archiv für Druck und Papier" 2/1956. Hellwag "Johannes Boehland", Berlin, Leipzig 1939.

**Bohn,** Hans – b. 23. 12. 1891 in Oberlahnstein, Germany, d. 10. 5. 1980 in Frankfurt am Main, Germany – *graphic artist, type designer, typographer, teacher* – Trained at the Technische Lehranstalt in Offenbach and then worked for the Ullstein publishing house in Berlin until 1914. 1919–30: graphic artist for the Klingspor type foundry in Offenbach, working closely with Karl Klingspor. After the Second World War, he works as a commercial artist. 1946–56: teaches at the Meisterschule für das gestaltende Handwerk in Offenbach. Has worked for numerous publishers, including Rowohlt (Hamburg), Schneekluth (Darmstadt), Ullstein (Berlin) and Fischer (Frankfurt am Main). – *Fonts:* Orplid (1929), Mondial (1936–39), Allegro (1937).

**Boldizar,** Ivan – b. 6. 11. 1917 in Novi Sad, Yugoslavia, d. 5. 6. 1986 in Novi Sad, Yugoslavia – *type designer, graphic artist, calligrapher, lithographer, teacher* – Trained as a lithographer and then worked for various printing workshops in Germany, Yugoslavia and Hungary. After 1945: head of the type graphics department at the Novi Sad Arts and Crafts School. From 1960 onwards: head of the graphic design department at the largest printing and publishing house in Novi

schule Novi Sad. Ab 1960 Leiter der Grafik-Design-Abteilung des größten Druck- und Verlagshauses in Novi Sad. 1966 Auszeichnung mit dem Preis „Die goldene Form" in Bratislava, CSSR. Seit 1975 freischaffender Künstler. Für seine Verdienste um die Kalligraphie erhielt er 1981 den Ehrenpreis der Stadt Novi Sad. 1984 Ausstellung im Klingspor-Museum, Offenbach am Main. – *Schriftentwürfe:* Boldiz (1975), Janus (1975), Triton (1975).

**Bons,** Jan – geb. 2. 4. 1918 in Rotterdam, Niederlande. – *Grafik-Designer, Künstler, Illustrator* – 1936–38 Ausbildung an der Koninklijke Academie voor Beeldende Kunsten in Den Haag und an der Nieuwe Kunstschool in Amsterdam. 1940–41 Arbeiten als Ausstellungsgestalter an der Utrechter Messe. Seit 1943 zahlreiche Buchillustrationen, u. a. zu Kafka (1943), Vernal (1945), Dekker (1960), Melville (1970). Seit 1946 zahlreiche Plakate für das Stedelijk Museum in Amsterdam. Mit dem Architekten Gerrit Rietveld zahlreiche Ausstellungsgestaltungen, Wandbilder und Skulpturen. 1950 Wandbilder für die niederländische Ausstellung in Philadelphia, USA. Für eine Ausstellung der niederländischen Regierung in Mexico City entsteht 1952 ein 450 $m^2$ großes Wandbild. 1958 Wandbild im niederländischen Pavillon der Weltausstellung in Brüssel, Belgien. Seit 1959 künstlerischer Berater der Firma van Ommeren. 1962 Zeichnungen für das Buch „Die Verbindung" der niederländischen Post PTT. 1968 Auszeichnung mit dem H. N. Werkman-Preis für Typographie der Stadt Amsterdam. 1975 Signet für „700 Jahre Am-

Bons 1959 Poster

Bons 1980 Poster

Boldizar Calendar

Bons 1975 Poster

Bons 1992 Poster

tement de graphisme et de calligraphie à l'école d'artisanat d'art de Novi Sad. A partir de 1960, dirige le service d'arts graphiques et de design de la plus grande société d'imprimerie et d'édition de Novi Sad. 1966, lauréat du prix «La forme d'or» à Bratislava (Tchécoslovaquie). Artiste indépendant à partir de 1975. Reçoit le prix d'honneur de la ville de Novi Sad pour son œuvre de calligraphie en 1981. 1984, exposition au Klingspor-Museum d'Offenbach. – *Polices:* Boldiz (1975), Janus (1975), Triton (1975).

**Bons,** Jan – né le 2. 4. 1918 à Rotterdam, Pays-Bas – *graphiste maquettiste , artiste, illustrateur* – 1936–1938, formation au Koninklijke Academie voor Beeldende Kunsten de La Haye et à la Nieuwe Kunstschool d'Amsterdam. 1940–1941, travaille comme décorateur de stands à la foire d'Utrecht. Dès 1943, illustre de nombreux livres de Kafka (1943), Vernal (1945), Dekker (1960) et Melville (1970). A partir de 1946, conçoit des affiches pour le Stedelijk Museum d'Amsterdam. Réalise l'architecture de plusieurs expositions avec Gerrit Rietveld, ainsi que des fresques et des sculptures. 1950, fresques pour l'exposition néerlandaise de Philadelphie, Etats-Unis. 1952, réalisation d'une immense fresque de 450 $m^2$ pour une exposition du gouvernement hollandais à Mexico City. 1958, fresque du pavillon hollandais à l'exposition universelle de Bruxelles, Belgique. Depuis 1959, conseiller artistique de la société van Ommeren. 1962, dessins pour le manuel «Liaison» des postes néerlandaises PTT. 1968, lauréat du prix de typographie H. N. Werkman de la ville d'Amsterdam. 1975, signets pour le «700e an-

Sad. 1966: awarded the prize "The Golden Form" in Bratislava, Czechoslovakia. Freelance artist since 1975. 1981: he receives the prize of honour of the City of Novi Sad for his contributions to calligraphy. 1984: exhibition in the Klingspor Museum in Offenbach am Main. – *Fonts:* Boldiz (1975), Janus (1975), Triton (1975).

**Bons,** Jan – b. 2. 4. 1918 in Rotterdam, The Netherlands – *graphic designer, artist, illustrator* – 1936–38: trains at the Koninklijke Academie voor Beeldende Kunsten in The Hague and at the Nieuwe Kunstschool in Amsterdam. 1940–41: works as an exhibition planner for the Utrecht Trade Fair. Since 1943: has produced numerous book illustrations for books by Kafka (1943), Vernal (1945), Dekker (1960) and Melville (1970), among others. Since 1946: has designed many posters for the Stedelijk Museum in Amsterdam. Designs many exhibitions, murals and sculptures with the architect Gerrit Rietveld. 1950: completes the murals for the Dutch Exhibition in Philadelphia, USA. 1952: produces a mural 450$m^2$ in size for an exhibition held by the Dutch government in Mexico City. 1958: does the mural for the Dutch Pavilion at the World Exhibition in Brussels, Belgium. Since 1959: art consultant to the van Ommeren company. 1962: does the drawings for the Dutch postal service's book "The Connection". 1968: is awarded the City of Amsterdam's H. N. Werkman Prize for Typography. 1975: designs the logo for "700 Years of Amsterdam". – *Publications include:* H. Kuh "Jan Bons" in "Gebrauchsgraphik" 1/1960, Munich; B. Majorick

sterdam". – *Publikationen u. a.:* H. Kuh „Jan Bons" in „Gebrauchsgraphik" 1/1960, München; B. Majorick „Genesis van een Compositie", Amsterdam 1961; „Jan Bons, Affiches", Katalog Museum Fodor, Amsterdam 1975.

**Boom,** Irma – geb. 15. 12. 1960 in Lochem, Niederlande. – *Typographin, Grafik-Designerin, Lehrerin* – Grafik-Design-Studium an der Academie voor Kunst en Industrie, Enschede, Niederlande. Nach Mitarbeit im Studio Dumbar in Den Haag und in der Design-Abteilung des niederländischen Fernsehens 1985–90 Senior-Designer bei der Design-Abteilung der Staatlichen Druckerei (SDU) in Den Haag. Hier entstehen u. a. 1985–86 das Erscheinungsbild des Kulturministeriums (WVC) und 1987–88 Briefmarken-Jahrbücher der niederländischen Post PTT. 1990 Plakat, Prospekt und Drucksachen für das Holland-Festival. Ab 1990 freiberufliche Designerin mit Studio in Amsterdam. Hier entstehen u. a. Kataloge für das Stedelijk Museum Amsterdam, das Centraal Museum Utrecht, die Rijksakademie van Beeldende Kunsten, Amsterdam. 1993 Briefmarken für die niederländische Post PTT. Unterrichtet in den Niederlanden an den Akademien von Arnheim, Amsterdam und Maastricht; in den USA an der Rhode Island School of Design, Providence, der Yale University, New Haven, Cal Arts, Valencia, Kalifornien und an der School of the Art Institute of Chicago. Zahlreiche Preise und Auszeichnungen seit 1989.

**Bosshard,** Hans Rudolf – geb. 25. 1. 1929 in Balm/Lottstetten, Deutschland. –

niversaire d'Amsterdam». – *Publications, sélection:* H. Kuh «Jan Bons» dans «Gebrauchsgraphik» 1/1960, Munich; B. Majoricks «Genesis van een Compositie», Amsterdam 1961; «Jan Bons, Affiches», catalogue Museum Fodor, Amsterdam 1975.

**Boom,** Irma – née le 15. 12. 1960 à Lochem, Pays-Bas – *typographe, graphiste maquettiste, enseignante* – études d'arts graphiques et de design à l'Academie voor Kunst en Industrie, Enschede, Pays-Bas. Après avoir travaillé au Studio Dumbar, La Haye, et dans le service design de la télévision néerlandaise, exerce de 1985 à 1990 comme Senior-Designer pour le service de design de l'imprimerie nationale (SDU) de La Haye. C'est dans ce contexte qu'elle crée entre autres (1985–1986) le logo du ministère de la culture (WVC), (1987–1988) les annales des timbres des postes néerlandaises PTT, ainsi que des prospectus et imprimés pour le Festival de Hollande. A partir de 1990, elle exerce comme indépendante dans son propre atelier à Amsterdam. Elle réalise des catalogues pour le Stedelijk Museum d'Amsterdam, le Centraal Museum d'Utrecht, la Rijksakademie van Beeldende Kunsten d'Amsterdam. 1993, timbres-poste pour les postes néerlandaises PTT. A enseigné aux Pays-Bas dans les académies d'Arnhem, d'Amsterdam et de Maastricht, aux Etats-Unis à la Rhode Island School of Design, à Providence; à la Yale University, New Haven; à la Cal Arts, Valencia, Californie, à la School of the Art Institute of Chicago. De nombreux prix et distinctions depuis 1989.

**Bosshard,** Hans Rudolf – né le 25. 1. 1929 à Balm/Lottstetten, Allemagne – *typographe, graphiste, peintre, écrivain, enseignant* – 1944–1948, apprentissage de typographe à Schaffhausen, Suisse. Fré-

Boom 1990 Poster

Boom 1990 Poster

Bosshard 1980 Cover

"Genesis van een Compositie", Amsterdam 1961; "Jan Bons Affiches", catalogue of the Fodor Museum, Amsterdam, 1975.

**Boom,** Irma – b. 15. 12. 1960 in Lochem, The Netherlands – *typographer, graphic designer, teacher* – Studied graphic design at the Academie voor Kunst en Industrie in Enschede, The Netherlands. 1985–90 After working for Studio Dumbar in The Hague and in the design department of Dutch television she took on a post as senior designer in the design department of the State Printing House (SDU) in The Hague. Here she produced logos for the Dutch Ministry of Education and the Arts (WVC) from 1985–86, stamp yearbooks for the Dutch postal service PTT in 1987 and 1988, and the poster, brochure and other printed matter for the Holland Festival in 1990, among other work. Since 1990: freelance designer with her own studio in Amsterdam. Her work here includes catalogues for the Stedelijk Museum in Amsterdam, the Centraal Museum in Utrecht and the Rijksakademie van Beeldende Kunsten in Amsterdam. 1993: designs stamps for the Dutch postal service PTT. In The Netherlands she has taught at the academies in Arnhem, Amsterdam and Maastricht and in the USA at the Rhode Island School of Design, Providence, Yale University in New Haven, Cal Arts in Valencia, California and at the School of the Art Institute of Chicago. Has won numerous prizes and awards since 1989.

**Bosshard,** Hans Rudolf – b. 25. 1. 1929 in Balm/Lottstetten, Germany – *typographer, graphic artist, painter, littérateur,*

*Typograph, Grafiker, Maler, Literat, Lehrer* – 1944–48 Typographenlehre in Schaffhausen, Schweiz. Besuch von Zeichenkursen an der Schule für Gestaltung Zürich. 1948 Beginn des Werkes in freier Malerei und Grafik sowie literarische Arbeiten. Zahlreiche internationale Einzel- und Gruppenausstellungen, zahlreiche Veröffentlichungen. 1956 Gründung der „Janus-Presse" Zürich, Herausgabe von illustrierten Büchern und Grafikmappen. 1960–63 Lehrer für Typographie an den Gewerbeschulen Winterthur und Weinfelden. 1963–91 Lehrer für Typographie und freies Gestalten an der Schule für Gestaltung, Zürich. 1963–94 Leiter des Weiterbildungslehrgangs für typographische Gestalter. 1967–91 Mitredakteur der internationalen Zeitschrift für Hochdruckgrafik „Xylon". Seit 1993 Buchgestalter und publizistische Tätigkeit für Zeitschriften. – *Publikationen:* „Form und Farbe", Zürich 1961; „Einführung zur Formenlehre", Zürich 1971; „Gestaltgesetze", Zürich 1971; „Proportion", Zürich 1973; „Farbe", Zürich 1974; „Technische Grundlagen zur Satzherstellung", St. Gallen 1980; „Mathematische Grundlagen zur Satzherstellung", St. Gallen 1985; „Die Demontage der Regel", Zürich 1994.

**Bossjatzki,** Ewgeni – geb. 21. 5. 1929 in Sofia, Bulgarien. – *Buchgestalter, Grafiker, Illustrator* – 1954 Studium an der Kunstakademie Sofia. Zahlreiche Auszeichnungen.

**Boton,** Albert – geb. 17. 4. 1932 in Paris, Frankreich. – *Schriftentwerfer, Grafik-Designer, Lehrer* – 1952–55 Ausbildung im Atelier Troy, Paris. Kurs in Kalligraphie an der Ecole Estienne, Paris.

pharaon broadway
voie antique grand voyage
NAVY CUT TANAGRA
BOUEES MARINES COLONNE IONIQUE

Boton 1972 Type Specimen

| | n | | | | |
|---|---|---|---|---|---|
| a | b | e | f | g | i |
| o | r | s | t | y | z |
| A | B | C | E | G | H |
| M | O | R | S | X | Y |
| 1 | 2 | 4 | 6 | 8 | & |

Boton 1973 Chinon

| | n | n | n | | n |
|---|---|---|---|---|---|
| a | b | e | f | g | i |
| o | r | s | t | y | z |
| A | B | C | E | G | H |
| M | O | R | S | X | Y |
| 1 | 2 | 4 | 6 | 8 | & |

Boton 1985 Elan

Hans Rudolf Bosshard
Technische Grundlagen zur Satzherstellung

Bosshard 1980 Spread

| n | n | n | | | n |
|---|---|---|---|---|---|
| a | b | e | f | g | i |
| o | r | s | t | y | z |
| A | B | C | E | G | H |
| M | O | R | S | X | Y |
| 1 | 2 | 4 | 6 | 8 | & |

Boton 1986 Boton

quente le cours de dessin de la Schule für Gestaltung (école de design) de Zurich. 1948, commence à travailler comme peintre et graphiste et écrit des textes littéraires. De nombreuses expositions internationales personnelles et collectives, ainsi que des publications. 1956, création des «Janus-Presse», Zurich, où il édite des livres illustrés et des estampes. 1960–1963, enseigne la typographie dans les écoles techniques de Winterthur et de Weinfelden. 1963–1991, enseigne la typographie et la forme libre à la Schule für Gestaltung de Zurich. 1963–1994, dirige un cycle de formation permanente pour maquettistes typographes. 1967–1991, rédacteur adjoint de la revue internationale de gravure en relief «Xylon». Depuis 1993, exerce comme maquettiste et journaliste pour plusieurs revues. – *Publications:* «Form und Farbe», Zurich 1971; «Einführung zur Formenlehre», Zurich 1971; «Gestaltgesetze», Zurich 1971; «Proportion», Zurich 1973; «Farbe», Zurich 1974; «Technische Grundlagen zur Satzherstellung», Saint-Gall 1980; «Mathematische Grundlagen zur Satzherstellung», Saint-Gall 1985; «Die Demontage der Regel», Zurich 1994.

**Bossjatzki,** Evgueni – né le 21. 5. 1929 à Sofia, Bulgarie – *maquettiste pour l'édition, graphiste, illustrateur* – 1954, études à l'académie des beaux-arts de Sofia. De nombreuses distinctions.

**Boton,** Albert – né le 17. 4. 1932 à Paris, France – *concepteur de polices, graphiste maquettiste, enseignant* – 1952–1955, formation à l'atelier Troy, Paris. Cours de calligraphie à l'Ecole Estienne, Paris.

*teacher* – 1944–48: learns typography in Schaffhausen, Switzerland. Attends art classes at the Schule für Gestaltung in Zurich. 1948: begins to produce work in the field of free painting and graphics as well as literary pieces. Features in numerous solo and group exhibitions all over the world and has written many publications. 1956: founds the Janus Press in Zurich. Publishes illustrated books and graphics folders. 1960–63: teaches typography at the trade schools in Winterthur and Weinfelden. 1963–91: teaches typography and free design at the Schule für Gestaltung in Zurich. 1963–94: directs the further education course for typographical designers. 1967–91: one of the editors of the international magazine for letterpress graphics, "Xylon". Since 1993: book designer and publicist for various magazines. – *Publications:* "Form und Farbe", Zurich 1961; "Einführung zur Formenlehre", Zurich 1971; "Gestaltgesetze", Zurich 1971; "Proportion", Zurich 1973; "Farbe", Zurich 1974; "Technische Grundlagen zur Satzherstellung", St. Gallen 1980; "Mathematische Grundlagen zur Satzherstellung", St. Gallen 1985; "Die Demontage der Regel", Zurich 1994.

**Bossyatzki,** Evgeni – b. 21. 5. 1929 in Sofia, Bulgaria – *book design, graphics, illustrations* – 1954: studies at the Sofia art academy. Numerous awards.

**Boton,** Albert – b. 17. 4. 1932 in Paris, France – *type designer, graphic designer, teacher* – 1952–55: trains at the Troy studio in Paris. Attends a course in calligraphy at the Ecole Estienne in Paris.

phie an der Ecole Estienne, Paris. 1955–57 Schriftentwerfer in der Schriftgießerei Deberny & Peignot, Paris, unter Adrian Frutiger und Ladislas Mendel. 1957–58 Arbeit in der grafischen Abteilung des Designstudios Technès. 1958–66 Schriftentwerfer und Grafik-Designer im Atelier Hollenstein, Paris. 1966–68 Agentur Psycho, Paris. 1968–73 Art Director in der Agentur Delpire, Paris. Entwürfe u. a. für Cacharel, Citroën, Schlumberger. Seit 1968 Lehrauftrag für Kalligraphie und Schriftentwurf an der Ecole Nationale Supérieure des Arts et Design, Paris (ENSAD). 1973 selbständiger Gestalter. 1978–81 Agentur Guépard, Paris, mit Roger Saingt. Seit 1981 Art Director in der Agentur Carré Noir. Seit 1988 Lehrauftrag an der ANCT, Imprimerie Nationale, Paris. – *Schriftentwürfe:* seit 1958 zahlreiche Schriftentwürfe, u. a. für die Collection Hollenstein: Chadking (1958), Roc (1959), Brasilia (1960), Eras (1961), Primavera (1963), Rialto (1964); für Delpire: Black Boton (1970), Zan (1970), Pharaon (1971), Pampam (1974); für Mecanorma: Hillman (1972), Chinon (1973), Hudson (1973), Elan (1985), Boton, Navy Cut (1986) sowie Schriften für die Firmen Agfatype, Berthold AG, Gabor und ITC.

**Bradley,** William H. – geb. 10. 7. 1868 in Boston, Massachusetts, USA, gest. 1962. – *Schriftentwerfer, Typograph, Verleger, Illustrator* – Autodidakt. 1885 erste Gestaltungsarbeiten in Chicago. 1894 beginnt mit seinen Entwürfen von Theaterplakaten und Umschlägen für „The Inland Printer“ und „The Chap Book“ der Jugendstil in Amerika. 1895 gründet er

1955–1957, concepteur de polices à la fonderie de caractères Deberny & Peignot, Paris, sous la direction d'Adrian Frutiger et de Ladislas Mendel. 1957–1958, travaille dans le service de graphisme de l'atelier de design Technès. 1958–1966 concepteur de polices et graphiste maquettiste à l'Atelier Hollenstein, Paris. 1966–1968, Agence Psycho, Paris. 1968–1973, directeur artistique à l'agence Delpire, Paris. Maquettes pour Cacharel, Citroën, Schlumberger. Chargé de cours en calligraphie et en conception de polices à l'Ecole Nationale Supérieure des Arts et Design (ENSAD) à partir de 1968. 1973, designer indépendant. 1978–1981, agence Guépard, Paris avec Roger Saingt. Directeur artistique de l'agence Carré Noir depuis 1981. Chargé de cours à l'ANCT, Imprimerie Nationale, Paris, à partir de 1988. – *Polices :* création de nombreuses polices depuis 1958, dont certaines pour la collection Hollenstein : Chadking (1958), Roc (1959), Brasilia (1960), Eras (1961), Primavera (1963), Rialto (1964); pour Delpire : Black Boton (1970), Zan (1970), Pharaon (1971), Pampam (1974); pour Mecanorma : Hillman (1972), Chinon (1973), Hudson (1973), Elan (1985), Boton, Navy Cut (1986) et d'autres pour les sociétés Agfatype, Berthold AG, Gabor et ITC.

**Bradley,** William H. – né le 10. 7. 1868 à Boston, Massachusetts, Etats-Unis, décédé en 1962 – *concepteur de polices, typographe, éditeur, illustrateur* – autodidacte. 1885, premiers travaux sur la forme à Chicago. 1894, représente l'Art nouveau aux Etats-Unis avec ses affiches de théâtre et enveloppes pour le «Inland Printer» et «The Chap Book». En 1895, il ouvre la «Wayside Press», Springfield, Massachusetts. Il exerce les fonctions de critique, de maquettiste, d'imprimeur et d'éditeur pour sa revue littéraire men-

Bradley 1895 Poster

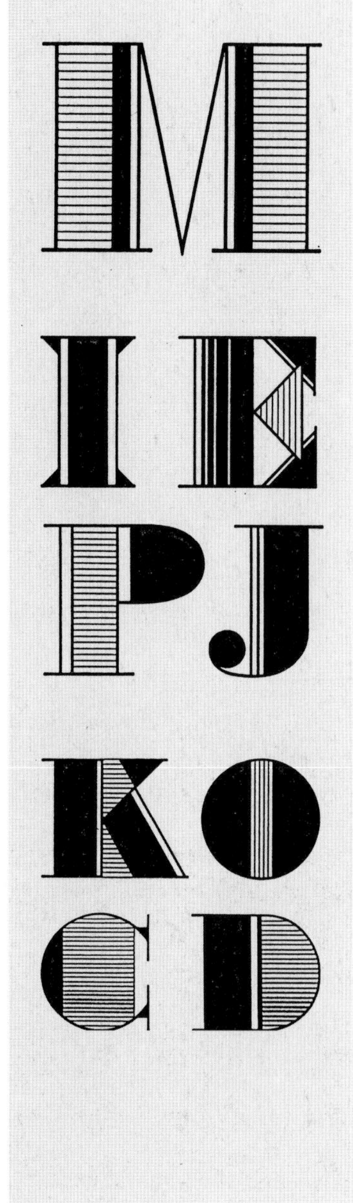

Bradley Specimen

1955–57: type designer for the Deberny & Peignot type foundry in Paris under Adrian Frutiger and Ladislas Mendel. 1957–58: works in the graphics department of the Technès design studio. 1958–66: type designer and graphic designer for the Hollenstein studio in Paris. 1966–68: works for the Psycho agency in Paris. 1968–73: art director of the Delpire agency in Paris. Designs for Cacharel, Citroën and Schlumberger, among others. Since 1968: teaching position for calligraphy and type design at the Ecole Nationale Supérieur des Arts et Design in Paris (ENSAD). 1973: self-employed designer. 1978–81: Guépard agency in Paris with Roger Saingt. Since 1981: art director to the Carré Noir agency. Since 1988: teaching position at the ANCT, Imprimerie Nationale in Paris. – *Fonts:* since 1958 he has produced numerous fonts, including Chadking (1958), Roc (1959), Brasilia (1960), Eras (1961), Primavera (1963) and Rialto (1964) for the Collection Hollenstein, Black Boton (1970), Zan (1970), Pharaon (1971) and Pampam (1974) for Delpire, and Hillman (1972), Chinon (1973), Hudson (1973), Elan (1985), Boton and Navy Cut (1986) for Mecanorma. He has also designed fonts for Agfatype, Berthold AG, Gabor and ITC.

**Bradley,** William H. – b. 10. 7. 1868 in Boston, Massachusetts, USA, d. 1962 – *type designer, typographer, publisher, illustrator* – autodidact. 1885: produces his first designs in Chicago. 1894: Art Nouveau starts in the USA with Bradley's theater posters and his book jackets for "The

die „Wayside Press" in Springfield, Mass. Für seine monatliche Literaturzeitschrift „Bradley, his Book" ist er Herausgeber, Gestalter, Kritiker und Drucker. Nach 1900 Mitarbeiter der „American Type Founders Company", für die er Schriften und Ornamente entwirft sowie eine Serie von 12 Ausgaben der Hauszeitschrift „The American Chap Book". 1907 Arbeit für das „Collier's Magazine". 1920 Typographie-Beauftragter für die Publikationen und Filme der Hearst-Publikationen. Er erhält den „Dean of American Typographers". – *Publikationen u.a.:* „Will Bradley: his Chap Book", New York 1955. Roberta Wong „Bradley: American Artist and Craftsman", New York 1972.

**Braille,** Louis – geb. 4. 1. 1809 in Coupuray, Frankreich, gest. 6. 1. 1852 in Paris, Frankreich. – *Erfinder, Lehrer* – Erblindet mit drei Jahren. 1819 Besuch der Blindenschule in Paris, wo er ab 1826 unterrichtet. 1829 entwickelt Braille das Schriftsystem der Blindenschrift. Die Schrift wird durch Abtasten mit den Fingern gelesen und ist aus leicht erhobenen Punkten aufgebaut. Mit sechs Punkten in zwei parallelen Reihen wie auf einem Würfel angeordneten Punkten werden durch Auslassungen die einzelnen Buchstaben codiert.

**Brainerd,** Paul – geb. 1947. – *Journalist, Unternehmer* – Ausbildung als Schriftsetzer, Studium an der University of Oregon und an der University of Minnesota. Chefredakteur der „Minnesota Daily", Operations-Director des „Minneapolis Star". Vizepräsident beim Software-Entwickler Atex, der spezialisierte Satzsysteme für

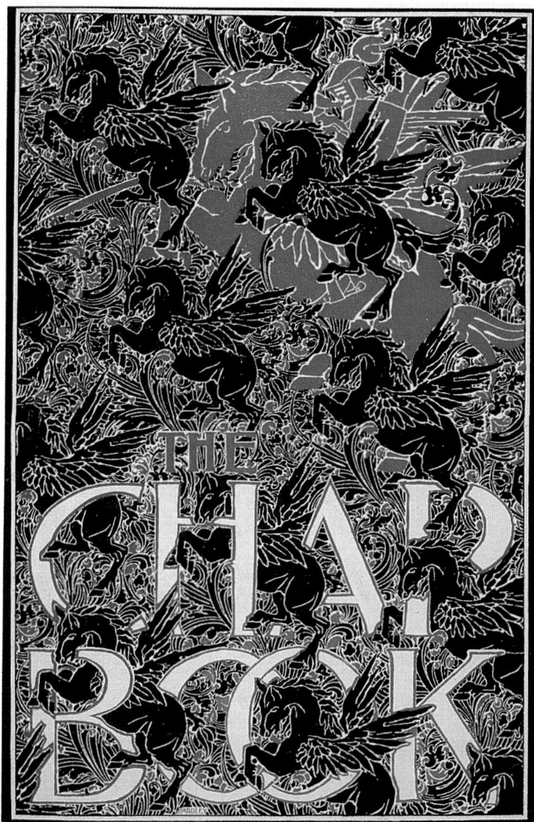

Bradley 1895 Cover

Bradley Initial

| a | b | c | d | e |
|---|---|---|---|---|
| ●○ | ●○ | ●● | ●● | ●○ |
| ○○ | ●○ | ○○ | ○● | ○● |
| ○○ | ○○ | ○○ | ○○ | ○○ |

| f | g | h | i | j |
|---|---|---|---|---|
| ●● | ●● | ●○ | ○● | ○● |
| ●○ | ●● | ●● | ●○ | ●● |
| ○○ | ○○ | ○○ | ○○ | ○○ |

| k | l | m | n | o |
|---|---|---|---|---|
| ●○ | ●○ | ●● | ●● | ●○ |
| ○○ | ●○ | ○○ | ○● | ○● |
| ●○ | ●○ | ●○ | ●○ | ●○ |

| p | q | r | s | t |
|---|---|---|---|---|
| ●● | ●● | ●○ | ○● | ○● |
| ●○ | ●● | ●● | ●○ | ●● |
| ●○ | ●○ | ●○ | ●○ | ●○ |

| u | v | x | y | z |
|---|---|---|---|---|
| ●○ | ●○ | ●● | ●● | ●○ |
| ○○ | ●○ | ○○ | ○● | ○● |
| ●● | ●● | ●● | ●● | ●● |

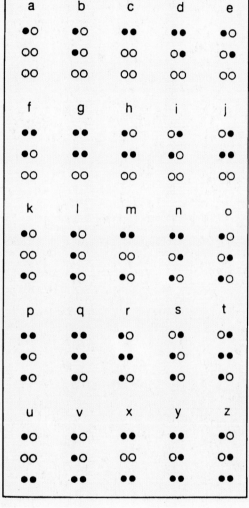

Braille 1809–52 Braille

suelle appelée «Bradley, his Book». Travaille pour la «American Type Founders Company» à partir de 1900; il y crée des écritures et des ornements ainsi que 12 éditions de la revue de la société «The American Chap Book». 1907, travaille pour le «Collier's Magazine». 1920, chargé de la typographie pour les publications et films du groupe Hearst. Il reçoit le «Dean of American Typographers». – *Publications, sélection:* «Will Bradley: his Chap Book», New York 1955. Roberta Wong «Bradley: American Artist and Craftsman», New York 1972.

**Braille,** Louis – né le 4. 1. 1809 à Coupuray, France, décédé le 6. 1. 1852 à Paris, France – *inventeur, enseignant* – devenu aveugle à l'âge de trois ans. 1819, fréquente l'institut des aveugles à Paris, où il enseignera à partir de 1826. 1829, Braille élabore un système d'écriture pour aveugles que l'on appellera Alphabet Braille. Cette écriture se lit du bout des doigts, au toucher, elle est constituée d'éléments et de points légèrement saillants. Les lettres sont codées sur six lignes parallèles comme les points d'un dé à jouer, et sont séparées par un blanc.

**Brainerd,** Paul – né en 1947 – *journaliste, chef d'entreprise* – formation de compositeur, études à la University of Oregon et à la University of Minnesota. Rédacteur en chef du «Minnesota Daily», Operations Director du «Minneapolis Star».

Inland Printer" and "The Chap Book". 1895: opens The Wayside Press in Springfield, Massachusetts. Publisher, designer, critic and printer of his monthly literary magazine "Bradley, his Book". After 1900: works for the American Type Founders Company, for whom he develops fonts and ornaments and a series of 12 issues of their in-house magazine "The American Chap Book". 1907: works for "Collier's Magazine". 1920: typographical representative for the publications and films of Hearst Publications. He is awarded the Dean of American Typographers. – *Publications include:* "Will Bradley: his Chap Book", New York 1955. Roberta Wong "Bradley: American Artist and Craftsman", New York 1972.

**Braille,** Louis – b. 4. 1. 1809 in Coupuray, France, d. 6. 1. 1852 in Paris, France – *inventor, teacher* – Went blind at the age of three. 1819: attends the school for the blind in Paris where he begins teaching in 1826. 1829: Braille develops a writing system for the blind. This alphabet is read with the fingers and based on a system of slightly raised dots arranged as on a dice. Various dots are omitted from the two parallel rows of six dots to produce each individual coded letter.

**Brainerd,** Paul – b. 1947 – *journalist, entrepreneur* – Trained as a typesetter and studied at the University of Oregon and at the University of Minnesota. Chief editor of the "Minnesota Daily". Operations director of the "Minneapolis Star". Vice-president of the Atex company, a software developer which produces specialized type systems for newspapers and

Zeitungen und Zeitschriften herstellt. Gründet 1984 in Seattle das Software-Unternehmen Aldus Corporation. Einführung des Begriffs „Desktop Publishing", (DTP). 1994 Verleihung des Gutenberg-Preises der Stadt Mainz, Deutschland.

**Brand,** Chris – geb. 16. 9. 1921 in Utrecht, Niederlande. – *Typograph, Schriftentwerfer, Kalligraph* – 1940 Kalligraphie-Studium. arbeitet 1948–53 in Brüssel. Ab 1950 Lehrtätigkeit an der Akademie St. Joost in Breda, ab 1955 an der Koninklijke Academie in 's Hertogenbosch. 1974 Auszeichnung mit dem Nassau-Breda-Preis. Ausstellungen in Offenbach (1960), London (1965), Prag (1969), Breda (1974), Brno (1980), Brüssel (1996). *Schriftentwurf:* Albertina (1964). – *Publikation u. a.:* Kurt Löb „Chris Brand" in „Novum" 10/1973.

**Brattinga,** Pieter – geb. 31. 1. 1931 in Hilversum, Niederlande. – *Typograph, Grafik-Designer, Ausstellungsdesigner, Galerist, Autor* – 1951–70 Design-Director in der Druckerei de Jong & Co., Hilversum. 1955–72 Herausgabe der „Quadrat-Prints", in denen er die Arbeiten verschiedener Gestalter vorstellt. 1960 Ausstellung „Das holländische Plakat" im Landesgewerbeamt Stuttgart. 1961–64 Vorsitzender der Abteilung Visuelle Kommunikation am Pratt Institute, New York. 1964 Eröffnung seines eigenen Gestaltungsbüros „Form Mediation International" in Amsterdam. 1966–70 Generalsekretär der ICOGRADA. 1968–71 Präsident des Art Directors Club der Niederlande. 1972 Eröffnung seiner Galerie „Print Gallery Pieter Brattinga" in Amsterdam mit

abcdefghijklmnopqrstuvwxyz
ﬁﬀﬂﬃﬄfbfhfkææçð
1234567890?!&†‡.:,;"—-–/§[]()»«*1234567890
ABCDEFGHIJKLMNOPQRSTUVWXYZÆŒ£

Brand 1964 Albertina Italic

Vice-président d'Atex, une société de développement de logiciels qui fabrique des systèmes de composition spécialisés pour la presse quotidienne et périodique. En 1984, il fonde la société d'informatique Aldus Corporation à Seattle. Introduit le terme «Desktop Publishing» (DTP). 1994, lauréat du Prix Gutenberg de la ville de Mayence, Allemagne.

**Brand,** Chris – né le 16. 9. 1921 à Utrecht, Pays-Bas – *typographe, concepteur de polices, calligraphe* – 1940, études de calligraphie. 1948–1953, travaille à Bruxelles. Enseigne à l'académie St. Joost à Breda à partir de 1950, puis à la Koninklijke Academie de Bois-le-Duc à partir de 1955. 1974, lauréat du Nassau-Breda Prijs. Expositions à Offenbach (1960), Londres (1965), Prague (1969), Breda (1974), Brno (1980), Bruxelles (1996). – *Police:* Albertina (1964). – *Publication, sélection:* Kurt Löb «Chris Brand» dans «Novum» 10/1973.

**Brattinga,** Pieter – né le 31. 1. 1931 à Hilversum, Pays-Bas – *typographe, graphiste maquettiste, architecte d'expositions, galeriste, auteur* – 1951–1970, directeur du design à l'imprimerie de Jong & Co, Hilversum. 1955–1972, publication des «Quadrat-Prints» dans lesquels il présente les travaux de divers créateurs. En 1960, il organise l'exposition «Das holländische Plakat» au Landesgewerbeamt de Stuttgart. 1961–1964, président du département de communication visuelle au Pratt Institute de New York. 1964, ouvre son propre atelier de design «Form Mediation International» à Amsterdam. 1966–1970, secrétaire général de l'ICOGRADA. 1968–1971, président du Art Directors Club des Pays-Bas. En 1972, il inaugure sa galerie la «Print Gallery Pieter Brattinga» à Amsterdam avec l'exposition «15 Signs by Sandberg». De nombreuses conférences et séminaires. Plusieurs ex-

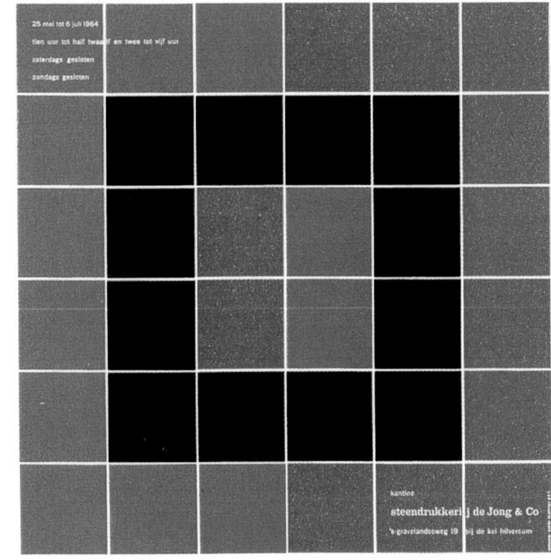


Brattinga 1964 Poster

afdrukken van houten letters uit de verzameling van kelly


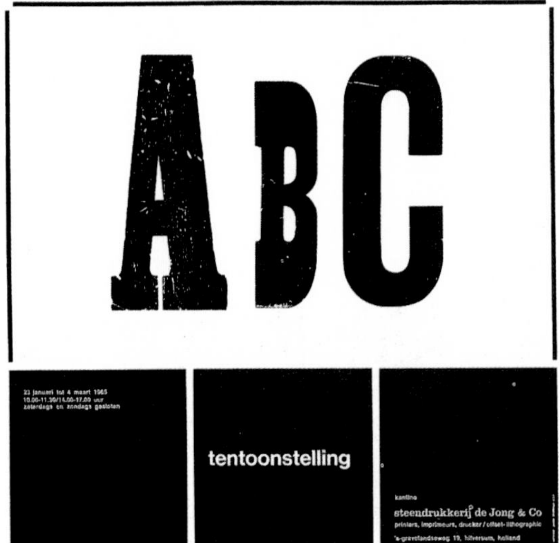


Brattinga 1965 Poster

magazines. 1984: founds the software company Aldus Corporation in Seattle. Introduces the concept of desktop publishing (DTP). 1994: is awarded the Gutenberg prize by the city of Mainz, Germany.

**Brand,** Chris – b. 16.9.1921 in Utrecht, The Netherlands – *typographer, type designer, calligrapher* – 1940: studies calligraphy. 1948–53: works in Brussels. From 1950 onwards: teaches at the Academie St. Joost in Breda. From 1955 onwards: teaches at the Koninklijke Academie in 's Hertogenbosch. 1974: is awarded the Nassau-Breda Prize. Exhibitions in Offenbach (1960), London (1965), Prague (1969), Breda (1974), Brno(1980), Brussels (1996). – *Font:* Albertina (1964). –*Publications include:* Kurt Löb "Chris Brand" in "Novum" 10/1973.

**Brattinga,** Pieter – b. 31. 1. 1931 in Hilversum, The Netherlands – *typographer, graphic designer, exhibition planner, gallery owner, author* – 1951–70: design director for the de Jong & Co. printery in Hilversum. 1955–72: publishes "Quadrat-Print", where he introduces the work of various designers. 1960: organizes the exhibition "Das holländische Plakat" in the Landesgewerbeamt in Stuttgart. 1961–64: head of the department for visual communication at the Pratt Institute in New York. 1964: opens his own design studio, Form Mediation International, in Amsterdam. 1966–70: general secretary of ICOGRADA. 1968–71: president of the Art Directors Club of The Netherlands. 1972: opens his gallery, Print Gallery Pieter Brattinga, in Amsterdam with the

der Ausstellung „15 Signs by Sandberg". Zahlreiche Vorträge und Lehrveranstaltungen. Zahlreiche Ausstellungen, u. a. Frankfurt am Main (1959), Minneapolis (1963), Lodz (1974), Berlin (1976), Otterlo (1980). – *Publikationen u.a.:* „Industrial Design in the Netherlands", Minneapolis 1964; „History of the Dutch Poster" (mit Dick Dooijes), Amsterdam 1968; „Planning for Industry, Art & Education", New York 1970; „60 Plakate, neun holländische Grafiker", München 1972; „Sandberg" (mit Ad Pedersen), Amsterdam 1975; „Influences on Dutch Graphic-Design 1900–45", Burgenstock 1987. Waldmann „The Activities of Pieter Brattinga", Tokio 1989.

**Breitkopf,** Johann Gottlob Immanuel – geb. 23. 11. 1719 in Leipzig, Deutschland, gest. 28. 1. 1794 in Leipzig, Deutschland. – *Schriftentwerfer* – Studium an der Universität Leipzig. 1736 Buchdruckergeselle. 1745 auf Wunsch des Vaters Leiter seiner Druckerei in Leipzig. Entwickelt ein erfolgreiches System des Musiknotendrucks mit Teillinien. – *Schriftentwurf:* Breitkopf-Fraktur (1793). – *Publikationen u. a.:* „Nachricht von der Stempelschneiderei und Schriftgießerei", Leipzig 1777, Nachdruck Berlin 1925; „Über Bibliographie und Bibliophilie", Leipzig 1793; „Buchdruckerei und Buchhandel in Leipzig", Leipzig 1793, Nachdruck Leipzig 1964.

**Breker,** Walter – geb. 12. 4. 1904 in Bielefeld, Deutschland, gest. 16. 9. 1980 in Olsberg, Deutschland. – *Gebrauchsgrafiker, Lehrer* – 1918–22 Lithographenlehre. 1922–25 Studium an der Kunstgewerbe-

Brattinga 1960 Poster

Nachricht von der Stempelschneiderey und Schriftgießerey. Zur Erläuterung der Enschedischen Schriftprobe. Von J. G. J. Breitkopf. Leipzig, 1777.

Breitkopf 1777 Cover

| | n | | | | |
|---|---|---|---|---|---|
| a | b | c | f | g | i |
| d | r | ſ | t | y | z |
| A | B | C | E | G | H |
| M | O | R | S | X | Y |
| 1 | 2 | 4 | 6 | 8 | 9 |

Breitkopf 1793 Breitkopf-Fraktur

positions, par ex. à Francfort-sur-le-Main (1959), Minneapolis (1963), Lodz (1974), Berlin (1976), Otterlo (1980). – *Publications, sélection :* «Industrial Design in the Netherlands», Minneapolis 1964; «History of the Dutch Poster» (avec Dick Dooijes), Amsterdam 1968; «Planning for Industry, Art & Education», New York 1970; «60 Plakate, neun holländische Grafiker», Munich 1972; «Sandberg» (avec Ad Pedersen), Amsterdam 1975; «Influences on Dutch Graphic-Design 1900–1945», Burgenstock 1987. G. Waldmann «The Activities or Pieter Brattinga», Tokyo 1989.

**Breitkopf,** Johann, Gottlob, Immanuel – né le 23. 11. 1719 à Leipzig, Allemagne, décédé le 28. 1. 1794 à Leipzig, Allemagne – *concepteur de polices* – études à l'université de Leipzig. 1736, compagnon imprimeur. 1745, dirige l'imprimerie de son père sur le souhait de ce dernier. Invente un excellent système d'impression des notes de musique avec portées. – *Police :* Breitkopf-Fraktur (1793). – *Publications, sélection :* «Nachricht von der Stempelschneiderei und Schriftgiesserei», Leipzig 1777, réédition Berlin 1925; «Über Bibliographie und Bibliophilie», Leipzig 1793; «Buchdruckerei und Buchhandel in Leipzig», Leipzig 1793, réédition Leipzig 1964.

**Breker,** Walter – né le 12. 4. 1904 à Bielefeld, Allemagne, décédé le 16. 9. 1980 à Olsberg, Allemagne – *graphiste utilitaire, enseignant* – 1918–1922, apprentissage de lithographe. 1922–1925, études à la Kunstgewerbeschule de Bielefeld (école

exhibition "15 Signs by Sandberg". Numerous lectures and courses. Numerous exhibitions, including in Frankfurt am Main (1959), Minneapolis (1963), Lodz (1974), Berlin (1976) and Otterlo (1980). – *Publications include:* "Industrial Design in the Netherlands", Minneapolis 1964; "History of the Dutch Poster" (with Dick Dooijes), Amsterdam 1968; "Planning for Industry, Art & Education", New York 1970; "60 Plakate, neun holländische Grafiker", Munich 1972; "Sandberg" (with Ad Pedersen), Amsterdam 1975; "Influences on Dutch Graphic-Design 1900–45", Burgenstock 1987. G. Waldmann "The Activities of Pieter Brattinga", Tokyo 1989.

**Breitkopf,** Johann Gottlob Immanuel – b. 23. 11. 1719 in Leipzig, Germany, d. 28. 1. 1794 in Leipzig, Germany – *type designer* – Studied at the University of Leipzig. 1736: book printer's apprentice. 1745: at the wish of his father, Breitkopf takes over his father's printing works in Leipzig. Develops a successful system for printing music with staves. – *Font:* Breitkopf-Fraktur (1793). – *Publications include:* "Nachricht von der Stempelschneiderei und Schriftgießerei", Leipzig 1777, reprint Berlin 1925; "Über Bibliographie und Bibliophilie", Leipzig 1793; "Buchdruckerei und Buchhandel in Leipzig", Leipzig 1793, reprint Leipzig 1964.

**Breker,** Walter – b. 12. 4. 1904 in Bielefeld, Germany, d. 16. 9. 1980 in Olsberg, Germany – *commercial artist, teacher* – 1918–22: trains as a lithographer. 1922–25: studies at the Kunstgewerbeschule in Bielefeld. 1925–30: art director

schule Bielefeld. 1925–30 künstlerischer Leiter großer Druckereien in Zwickau, Hamburg, Magdeburg. 1930–34 Lehrassistent bei Prof. Wilhelm Deffke an der Kunstgewerbeschule Magdeburg. 1934–54 Leiter der Abteilung für angewandte Grafik und Buchkunst an der Werkkunstschule Krefeld. 1954–69 Professor an der Staatlichen Kunstakademie Düsseldorf, Leiter der Klasse für angewandte Grafik. Zahlreiche Ausstellungen, u. a. Kunsthalle Bremerhaven (1967), Klingspor-Museum Offenbach (1972), Städtische Kunsthalle Düsseldorf (1963). – *Publikation u.a.:* H. P. Willberg (Hrsg.) „Marken und Marken", Berlin 1984.

**Bremer Presse** – *Pressendruck-Verlag* – Willy Wiegand und Ludwig Wolde gründen 1911 in Bremen die Bremer Presse nach dem Vorbild der englischen Doves Press. Die Schriften, mit denen die Bücher der Bremer Presse gesetzt werden, sind von Louis Höll geschnitten. Zahlreiche Initialen werden von Anna Simons entworfen. 1922 beginnt der „Verlag der Bremer Presse" seine verlegerische Arbeit. 1923–24 zwei Bände mit Werken Homers. 1926–29 fünfbändige Luther-Bibel und der Psalter. 1931 „Missale Romanum", die in der von Willy Wiegand entworfenen Schrift „Liturgica" gesetzt wird. Die Initialen entwirft Frieda Thiersch. – *Publikation u. a.:* Josef Lehnacker, Herbert Post, Rudolf Adolph „Die Bremer Presse", München 1964.

**Brendler,** Wien, Österreich. – *Schriftgießerei* – 1875 Karl Brendler, Sohn des Schriftgießers Johann Gottlieb Brendler aus Wien, übernimmt die Schriftgießerei

des arts décoratifs). 1925–1930, directeur artistique de grandes imprimeries à Zwickau, Hambourg, Magdebourg. 1930–1934, assistant du professeur Wilhelm Deffke à la Kunstgewerbeschule de Magdebourg (école des arts décoratifs). 1934–1954, dirige le département de graphisme appliqué et des arts du livre à la Werkkunstschule de Krefeld (arts appliqués). 1954–1969, professeur à la Staatliche Kunstakademie de Düsseldorf (beaux-arts), où il est responsable de la classe de graphisme appliqué. Nombreuses expositions, par ex. à la Kunsthalle de Bremerhaven (1967), Klingspor-Museum, Offenbach (1972), Städtische Kunsthalle, Düsseldorf (1963). – *Publication, sélection:* H. P. Willberg (éd.) «Marken und Marken», Berlin 1984.

**Bremer Presse** – *éditions et presses* – en 1911, Willy Wiegand et Ludwig Wolde fondent la Bremer Presse à Brême en prenant les Doves Press anglaises pour modèle. Les caractères qui servent à composer les livres de la Bremer Presse sont taillés par Louis Höll. Anna Simons crée de nombreuses initiales. En 1922, le «Verlag der Bremer Presse» démarre ses activités d'édition. 1923–1924, parution des œuvres de Homère en deux volumes. 1926–1929, bible de Luther en cinq volumes et psautier. 1931, le «Missale Romanum» est composé en «Liturgica», une police créée par Willy Wiegand. Les lettrines sont dessinées par Frieda Thiersch. – *Publication, sélection:* Joseph Lehnacker, Herbert Post, Rudolf Adolph «Die Bremer Presse», Munich 1964.

**Brendler,** Vienne, Autriche – *fonderie de caractères* – en 1875, Karl Brendler, fils du fondeur de caractères Johann Gottlieb Brendler de Vienne, reprend la fonderie de caractères Carl Fromme et la gère sous son nouveau nom. Les fils de Brendler, Karl et Joseph, entrent dans la société en

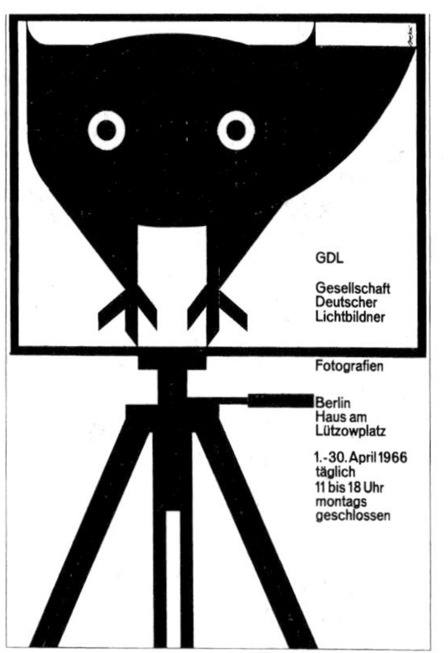


Breker 1966 Poster

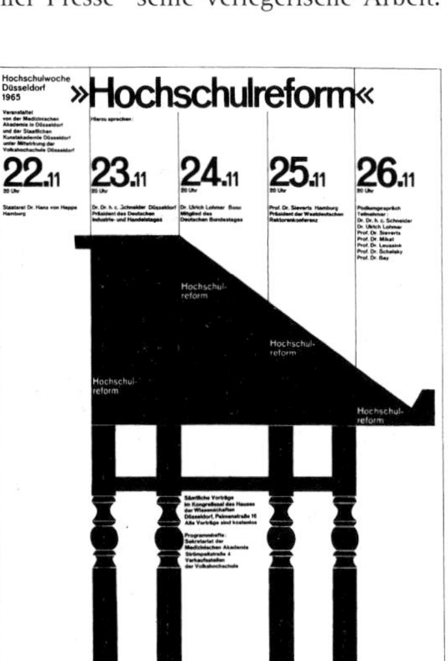


Breker 1965 Poster

Bremer Presse 1926 Page

Breker 1969 Poster

Breker 1963 Trademark

Bremer Presse 1926 Page

for large printing works in Zwickau, Hamburg and Magdeburg. 1930–34: teaching assistant to Prof. Wilhelm Deffke at the Kunstgewerbeschule in Magdeburg. 1934–54: head of the department for applied graphics and book art at the Werkkunstschule in Krefeld. 1954–69: professor at the Staatliche Kunstakademie in Düsseldorf and director of the applied graphics course. Numerous exhibitions, including in the Kunsthalle Bremerhaven (1967), the Klingspor Museum in Offenbach (1972) and the Städtische Kunsthalle in Düsseldorf (1963). – *Publications include:* H. P. Willberg (ed.) "Marken und Marken", Berlin 1984.

**Bremer Presse** – *publishers of fine editions* – 1911: Willy Wiegand and Ludwig Wolde set up the Bremer Presse in Bremen, modelling it on the Doves Press in England. Louis Höll produces the typefaces used in the type setting of Bremer Presse books. Anna Simons designs numerous initials. 1922: the Bremer Presse's publishing house starts work. 1923–24: two volumes of Homer's works are published. 1926–29: publication of a five-volume Luther bible and the Psalter. 1931: "Missale Romanum" is produced, set in Willy Wiegand's Liturgica font. Frieda Thiersch designs the initials. – *Publications include:* Josef Lehnacker, Herbert Post, Rudolf Adolph "Die Bremer Presse", Munich 1964.

**Brendler,** Vienna, Austria – *type foundry* – 1875: Karl Brendler, the son of type founder Johann Gottlieb Brendler from Vienna, takes over the Carl Fromme type foundry and runs it under his own name.

Carl Fromme und führt sie unter eigenem Namen weiter. Die Söhne Brendlers, Karl und Joseph, treten 1896 in die Firma ein, die jetzt Karl Brendler & Söhne heißt. 1914 stirbt Karl Brendler, seine Söhne führen die Firma weiter. 1965 Verkauf des Unternehmens an die Schriftgießerei Johannes Wagner, Ingolstadt, Deutschland.

**Bretteville,** Sheila Levrant de – geb. 4. 11. 1940 in Brooklyn, New York, USA. – *Grafik-Designerin, Lehrerin* – 1962 Abschluß des Kunstgeschichte-Studiums am Bernard College, New York. 1964 Abschluß des Grafik-Design-Studiums an der Yale University, New Haven. 1971 Schaffung des ersten Design-Programms für Frauen am California Institute of the Arts. 1973 Gründung des „Women's Graphic Center" in „The Woman's Building", einem Zentrum für Frauenkultur in Los Angeles. 1975 Plakat zur Konferenz „Women in Design: the Next Decade". 1981–91 Leiterin der Abteilung Kommunikationsdesign und Illustration am Otis Art Institute der Parson's School of Design. Gründung des „Brooklyn Design Workshop", in dem Design-Studierende Alternativprojekte und Non-Profit-Projekte bearbeiten. Gleichzeitig eigenes Design-Studio in Los Angeles, Arbeiten für Auftraggeber wie Olivetti, Warner Bros. und „Los Angeles Times". 1991 Direktor des Graduate-Programm für Grafik-Design an der Yale University.

**Brignall,** Colin – geb. 1940 in Warwickshire, England. – *Fotograf, Schriftentwerfer, Art Director* – Zunächst Arbeit als Fotograf für Mode, Presse und Industrie. 1964 Eintritt in das Letraset Type Studio

Bretteville 1996 Installation

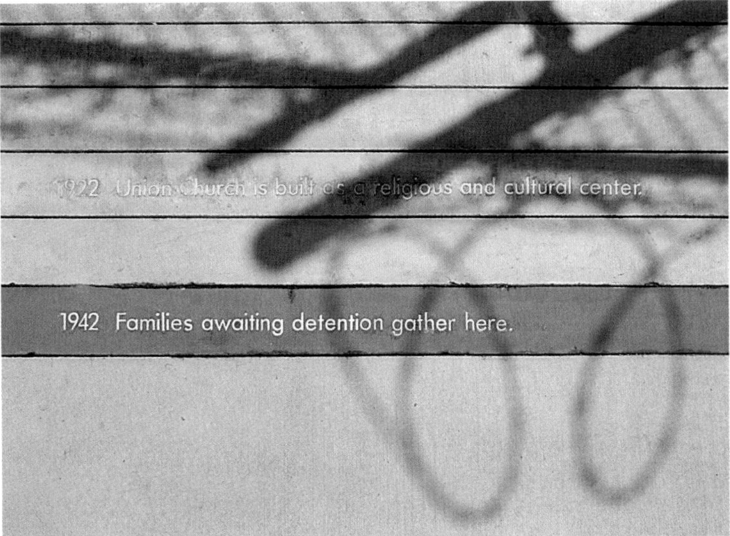


Bretteville 1996 Installation

1896, celle-ci s'appelle désormais Karl Brendler & Söhne. Au décès de Karl Brendler, en 1914, ses fils reprennent la société. 1965, vente de l'entreprise à la fonderie de caractères Johannes Wagner, Ingolstadt, Allemagne.

**Bretteville,** Sheila Levrant de – née le 4. 11. 1940 à Brooklyn, New York, Etats-Unis – *graphiste maquettiste, enseignante* – en 1962, termine ses études d'histoire de l'art au Bernard College de New York. 1964, termine ses études de maquettiste à la Yale University, New Haven. 1971, elle crée le premier programme de design pour femmes au California Institute of Arts. 1973, fondation du «Women's Graphic Center» dans le «Woman's Building», un centre culturel consacré aux femmes à Los Angeles. 1975, affiche pour la conférence «Women in Design : the Next Decade». 1981–1991, dirige le département de design, communication et illustration de l'Otis Art Institute de la Parson's School of Design. Elle fonde le «Brooklyn Design Workshop» où des étudiants en design peuvent travailler sur des projets alternatifs non-lucratifs. Gère en même temps son propre atelier de design à Los Angeles où elle travaille pour des commanditaires tels que Olivetti, Warner Bros. et «Los Angeles Times». 1991, directrice du Graduate Program de maquettistes à la Yale University.

**Brignall,** Colin – né en 1940 dans la Warwickshire, Angleterre – *photographe, concepteur de polices, directeur artistique* – commence par travailler comme photographe de mode et photographe pour la

Bretteville 1995 Installation

| | | | | | |
|---|---|---|---|---|---|
| | n | | | | |
| a | b | e | f | g | i |
| o | r | s | t | y | z |
| A | B | C | E | G | H |
| M | O | R | S | X | Y |
| 1 | 2 | 4 | 6 | 8 | & |

Brignall 1969 Aachen

Brendler's sons, Karl and Joseph, join the company, now called Karl Brendler & Söhne. 1914: Karl Brendler dies; his sons take over the business. 1965: the company is sold to the Johannes Wagner type foundry in Ingolstadt, Germany.

**Bretteville,** Sheila Levrant de – b. 4. 11. 1940 in Brooklyn, New York, USA – *graphic designer, teacher* – 1962: completes her studies in art history at Bernard College in New York. 1964: concludes her studies in graphic design at Yale University, New Haven. 1971: creates the first women's design program at the California Institute of the Arts. 1973: founds the Women's Graphic Center in The Woman's Building, a women's cultural center in Los Angeles. 1975: designs the poster for the conference "Women in Design: the Next Decade". 1981–91: head of the department for communication design and illustration at the Otis Art Institute of the Parson's School of Design. Sets up the Brooklyn Design Workshop, where design students work on alternative and non-profit projects. She also opens her own design studio in Los Angeles, working for customers such as Olivetti, Warner Bros. and the "Los Angeles Times". 1991: director of the graduate program for graphic design at Yale University.

**Brignall,** Colin – b. 1940 in Warwickshire, England – *photographer, type designer, art director* – Started work as photographer for the fashion trade, the press and industry. 1964: joins the Letraset Type Studio as photographic technician.

als Fototechniker. Seit 1974 Type Designer bei Letraset International, seit 1980 Type Director von Esselte Letraset in London. – *Schriftentwürfe:* Aachen, Premier Lightline (1969), Premier Shaded, Superstar, Octopuss (1970), Italia (1974), Romic (1979), für Letraset International entwarf er bisher über 100 Alphabete.
**Brinkmann + Bose** – *Verlag* – 1980 Gründung des Verlags durch Günter Karl Bose und Erich Brinkmann. Günter Karl Bose – geb. 6. 3. 1951 in Debstedt, Deutschland. – Studium der Germanistik, Politik und Kunstgeschichte in Freiburg. 1980–96 Verleger in Berlin. Seit 1993 Professor für Typographie an der Hochschule für Graphik und Buchkunst in Leipzig. Erich Brinkmann – geb. 6. 12. 1947 in Freiburg im Breisgau, Deutschland. – Studium der Germanistik und Politik in Freiburg. Seit 1980 Verleger in Berlin. Neben soziologischen, philosophischen und literarischen Veröffentlichungen erscheinen Publikationen zu Film, Kunst und Typographie, die häufig prämiert wurden. 1987 Herausgabe des Buches Jan Tschichold „Die neue Typographie" (von 1928) mit einem Beiheft zur Geschichte des Buches von W. Doede, Gerd Fleischmann und Jan Tschichold. 1991–92 Herausgabe der „Schriften 1925–74" von Jan Tschichold in zwei Bänden.
**British Letter Foundry,** London, England – *Schriftgießerei* – Der Buch- und Zeitungsverleger John Bell gründet 1788 die British Type Foundry. 1789 tritt Simon Stephenson als Teilhaber in die Firma ein, die nun „Bell and Stephenson's British Letter Foundry" heißt. John Bell

presse et l'industrie. 1964, entre au Letraset Type Studio comme technicien photographe. Type designer chez Letraset International à partir de 1974 et Type Director de Esselte Letraset à Londres à partir de 1980. – *Polices:* Aachen (1969), Premier Lightline (1969), Premier Shaded (1970), Superstar (1970), Octopuss (1970), Italia (1974), Romic (1979) toutes chez Letraset International; il a créé plus de 100 alphabets jusqu'à présent.
**Brinkmann + Bose** – *éditions* – maisons d'édition fondée en 1980 par Günter Karl Bose et Erich Brinkmann. Günter Karl Bose – né le 6. 3. 1951 à Debstedt, Allemagne – études de littérature allemande, de sciences politiques et d'histoire de l'art à Fribourg. Professeur de typographie à la Hochschule für Graphik und Buchkunst (école supérieure des arts graphiques et des arts du livre) de Leipzig depuis 1993. Erich Brinkmann – né le 6. 12. 1947 à Fribourg-en-Brisgau, Allemagne – études de littérature allemande et de sciences politiques à Fribourg. 1980–1996, Editeur à Berlin. Publication de littérature, de textes de sociologie et de philosophie, ainsi que sur le cinéma, l'art et la typographie, souvent primés. Réédition en 1987 de l'ouvrage «Die neue Typographie» (1928) de Jan Tschichold, avec un tiré à part de Werner Doede, Gerd Fleischmann et Jan Tschichold, traitant de l'histoire du livre. Publication en deux volumes de «Schriften 1925–1974» de Jan Tschichold.
**British Letter Foundry,** Londres, Angleterre – *fonderie de caractères* – en 1788, l'éditeur et homme de presse, John Bell fonde la British Type Foundry. En 1789, Simon Stephenson entre dans la société où il détient des parts, celle-ci s'appelle désormais «Bell and Stephenson's British Letter Foundry». John Bell quitte la société. En 1793, la société s'appelle «Simon

Brinkmann + Bose 1990 Poster

Brinkmann + Bose 1994 Cover

Brinkmann + Bose 1990 Poster

Brinkmann + Bose 1990 Main Title

Since 1974: type designer for Letraset International. Since 1980: type director for Esselte Letraset in London. – *Fonts:* Aachen (1969), Premier Lightline (1969), Premier shaded (1970), Superstar (1970), Octopuss (1970), Italia (1974), Romic (1979), all with Letraset International. To date he has designed over 100 alphabets.
**Brinkmann + Bose** – *publishing house* – 1980: the publishing house is founded by Günter Karl Bose and Erich Brinkmann. Günter Karl Bose – b. 6. 3. 1951 in Debstedt, Germany – Studied German studies, politics and art history in Freiburg. 1980–96: publisher in Berlin. Since 1993: professor of typography at the Hochschule für Graphik und Buchkunst in Leipzig. Erich Brinkmann – b. 6. 12. 1947 in Freiburg im Breisgau, Germany – Studied German studies and politics in Freiburg. Since 1980: publisher in Berlin. Besides sociological, philosophical and literary publications, Brinkmann + Bose also publish books on film, art and typography which have won many awards. 1987: Jan Tschichold's book from 1928, "Die neue Typographie", is republished with a supplement on the history of the book by Werner Doede, Gerd Fleischmann and Jan Tschichold. 1991–92: publication of Jan Tschichold's "Schriften 1925–74" in two volumes.
**British Letter Foundry,** London, England – *type foundry* – 1788: book and newspaper publisher John Bell sets up the British Type Foundry. 1789: Simon Stephenson joins the company, now called Bell and Stephenson's British Letter Foundry, as a partner. John Bell leaves

scheidet aus der Firma aus. Die Gießerei heißt 1793 „Simon Stephenson & Co", dann „Simon & Charles Stephenson". 1797 wird die Gießerei auf einer Auktion zum Kauf angeboten. Mitte des 19. Jahrhunderts kommen die Matrizen der Bell Antiqua in den Besitz der Schriftgießerei Stephenson, Blake & Co in Sheffield.

**Brodovitch,** Alexey – geb. 1898 in Ogolitschi, Rußland, gest. 15. 4. 1971 in Le Thor, Frankreich. – *Grafik-Designer, Fotograf, Art Director, Lehrer* – Ausbildung in St. Petersburg. 1920–26 Buchgestalter und Plakatmaler in Paris, Bühnenmaler für Diaghilevs „Ballets Russes". 1930 Umzug in die USA. Designer bei N. W. Ayer, Philadelphia und New York. 1934–58 Art Director bei „Harper's Bazaar", Zusammenarbeit u. a. mit Man Ray, Henri Cartier-Bresson, Richard Avedon, Irving Penn, A. M. Cassandre, Herbert Bayer. Arbeiten für „Saks of Fifth Avenue" (1940–49), „Portfolio" (1949–51). Lehrtätigkeit in New York an der Cooper Union (1940), am Pratt Institute (1940), an der New School for Social Research (1941–49), am American Institute of Graphic Arts, (1964), an der School of Visual Arts (1964–65). 1967 Umzug nach Frankreich. Zahlreiche Ausstellungen, u. a. „Alexey Brodovitch and his Influence", Philadelphia 1972. Zahlreiche Auszeichnungen, u. a. Aufnahme in die Hall of Fame des New York Art Directors Club (1972). – *Schriftentwurf:* Abro Alphabet (1950). – *Publikationen u.a.:* „Alexey Brodovitch and his Influence", Philadelphia 1972. Allen Porter, Georges Tourdjman „Alexey Brodovitch", Paris 1982.

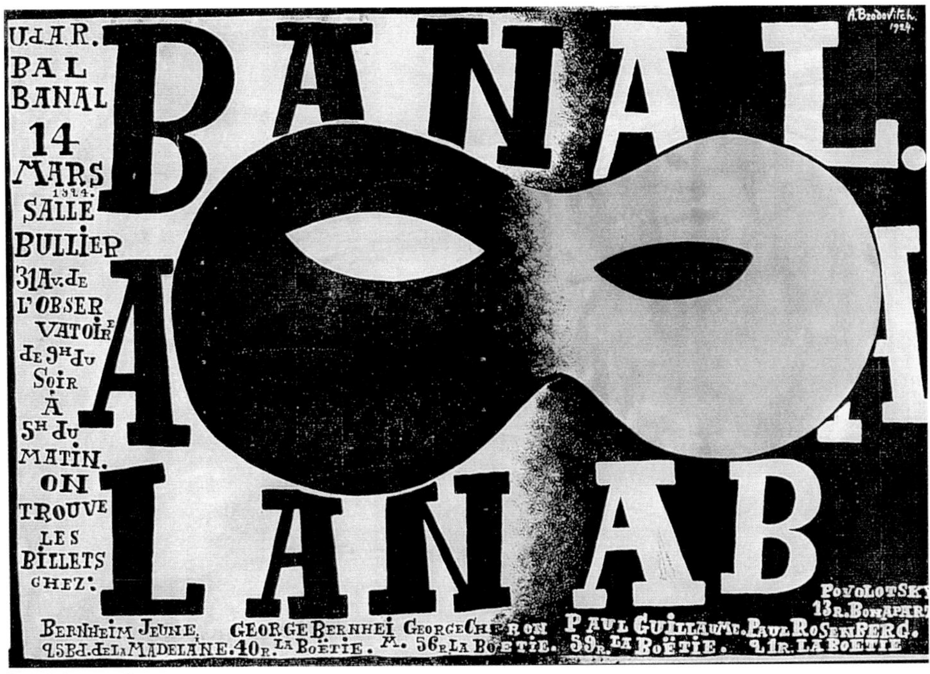

Brodovitch 1924 Poster

Stephenson & Co, puis peu de temps après «Simon & Charles Stephenson». En 1797, la société est vendue aux enchères. Au milieu du 19e siècle, les matrices de la Bell Antiqua deviennent propriété de la fonderie de caractères Stephenson, Blake & Co de Sheffield.

**Brodovitch,** Alexeï – né en 1898 à Ogolitchi, Russie, décédé le 15. 4. 1971 à Le Thor, France – *graphiste maquettiste, photographe, directeur artistique, enseignant* – formation à Saint-Pétersbourg. 1920–1926, maquettiste en édition et peintre affichiste à Paris, peint des décors pour les «Ballets Russes» de Diaghilev. 1930, s'installe aux Etats-Unis. Designer chez N.W. Ayer, Philadelphie et New York. 1934–1958, directeur artistique chez le magazine «Harper's Bazaar» à New York, collabore avec Man Ray, Henri Cartier-Bresson, Richard Avedon, Irving Penn, A. M. Cassandre, Herbert Bayer etc. Travaille en indépendant pour «Saks of Fifth Avenue» (1941–1949), le magazine «Portfolio» (1949–1951). Enseigne à la Cooper Union New York (1940), au Pratt Institute de New York (1940), à la New School for Social Research de New York (1941–1949), à l'American Institute of Graphic Arts, New York (1964), à la School of Visual Arts, New York (1964–1965). 1967, s'installe en France. Nombreuses expositions, par ex. : «Alexeï Brodovitch and his Influence», Philadelphie 1972. Nombreuses distinctions, dont élection au Hall of Fame de l'Art Directors Club de New York (1972). – *Police:* Abro Alphabet (1950). – *Publications, sélection :* «Alexey Brodovitch and his Influence», Philadelphie 1972. Allen Porter, Georges Tourdjman «Alexey Brodovitch», Paris 1982.

Brodovitch 1936 Cover

Brodovitch 1946 Cover

Brodovitch 1952 Cover

the company. 1793: the foundry renames to Simon Stephenson & Co., then to Simon & Charles Stephenson. 1797: the foundry is put up for auction. Mid 19th century: the matrices for Bell Antiqua come under the ownership of the Stephenson, Blake & Co. type foundry in Sheffield.

**Brodovitch,** Alexey – b. 1898 in Ogolitchi, Russia, d. 15. 4. 1971 in Le Thor, France – *graphic designer, photographer, art director, teacher* – Trained in St. Petersburg. 1920–26: book designer and poster artist in Paris. Scene painter for Diaghilev's Ballet Russe. 1930: moves to the USA. Designer for N.W. Ayer, Philadelphia and New York. 1934–58: art director for "Harper's Bazaar" magazine in New York. Works with Man Ray, Henri Cartier-Bresson, Richard Avedon, Irving Penn, A. M. Cassandre and Herbert Bayer, among others. Works freelance for Saks of Fifth Avenue (1941–49) and "Portfolio" magazine (1949–51). Teaches at the Cooper Union (1940), the Pratt Institute (1940), the New School for Social Research (1941–49), the American Institute of Graphic Arts (1964) and the School of Visual Arts (1964–65), all in New York. 1967: moves to France. Numerous exhibitions, including in 1972 "Alexey Brodovitch and his Influence" in Philadelphia. Numerous awards, including acceptance into the New York Art Directors Club Hall of Fame. – *Font:* Abro Alphabet (1950). – *Publications include:* "Alexey Brodovitch and his Influence", Philadelphia 1972. Allen Porter, Georges Tourdjman "Alexey Brodovitch", Paris 1982.

**Brody,** Neville – geb. 23. 4. 1957 London, Großbritannien. – *Grafik-Designer, Art Director, Schriftentwerfer* – 1975 Hornsey College (Malerei), 1976–79 London College of Printing. Zahlreiche Schallplattencover. 1981–86 Art Director des Magazins „The Face". 1983–87 Umschläge für das Londoner Magazin „City Limits". 1987–90 Arbeiten für das englische Magazin „Arena". Brody ist Art Director der Zeitschriften „Per Lui" und „Lei" vom Condé Nast Verlag, Mailand, und der französischen Zeitschrift „Actuel". 1988 Arbeiten und Erscheinungsbilder für Nike, Premiere, den ORF, das Haus der Kulturen der Welt, Berlin, das Deutsche Theater in Hamburg, das Kaufhaus Parco in Tokio und das digitale Publikationsmedium „Fuse". Brodys typographischer Stil verwendet ästhetische Elemente des Art Deco sowie außereuropäische Einflüsse. Seine Gestaltungssprache wird zum internationalen Vorbild des aufbrechenden computerorientierten Designs. Ausstellungen u.a.: „The Graphic Language of Neville Brody", London, Edinburgh, Berlin, Hamburg, Frankfurt am Main, Tokio (1988–90). – *Schriftentwürfe:* Arcadia (1990), Industria (1990), Insignia (1990), Blur (1991), Pop (1991), Gothic (1991), Harlem (1991). – *Publikationen:* J. Wozencroft „The Graphic Language of Neville Brody", London 1988; J. Wozencroft „The Graphic Language of Neville Brody No. 2", London, München 1994.

**Brownjohn,** Robert – geb. 1925 in New Jersey, USA, gest. 1970 – *Maler, Grafik-Designer, Typograph* – Studiert Malerei und Design am Institute of Design in Chi-

**Brody,** Neville – né le 23. 4. 1957 à Londres, Angleterre – *graphiste maquettiste, directeur artistique, concepteur de polices* – formation : 1975, Hornsey College (peinture), 1976–1979, London College of Printing. Nombreuses maquettes de pochettes de disques. 1981–1986, Art Director du magazine anglais «The Face». 1983–1987, couvertures du magazine londonien «City Limits». 1987–1990, travaux pour le magazine anglais «Arena». Brody a été directeur artistique des revues «Per Lui» et «Lei» des publications Condé Nast, Milan, et de la revue française «Actuel». 1988, travaux et études d'identités pour Nike, la chaîne de télévision Premiere, l'ORF, la Maison des cultures du monde de Berlin, le Deutsches Theater de Hambourg, les grands magasins Parco à Tokyo, et le média de publications digitales «Fuse». Le style typographique se caractérise par des emprunts d'éléments Art déco et des influences extra-européennes. Son langage formel devient un modèle international en ce qui concerne le design assisté par ordinateur. Expositions : «The Graphic Language of Neville Brody», Londres, Edimbourg, Berlin, Hambourg, Francfort-sur-le-Main, Tokyo (1988–1990). – *Polices :* Arcadia (1990), Industria (1990), Insignia (1990), Blur (1991), Pop (1991), Gothic (1991), Harlem (1991). – *Publications :* J. Wozencroft «The Graphic language of Neville Brody», Londres 1988 ; J. Wozencroft «The Graphic Language of Neville Brody N° 2», Londres, Munich 1994.

**Brownjohn,** Robert – né en 1925 dans le New Jersey, Etats-Unis, décédé en 1970 – *peintre, graphiste maquettiste, typographe* – études de peinture et de design à l'Institute of Design de Chicago et d'architecture à l'Institute of Design et à l'Illinois Institute of Technology. 1957, fonde l'agence Brownjohn, Chermayeff + Geis-

Brody 1989 Poster

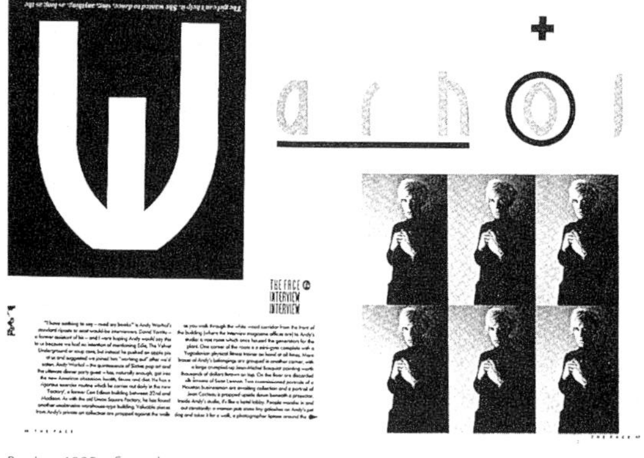


Brody 1985 Spread

Brody 1987 Spread

Brody 1990 Insignia

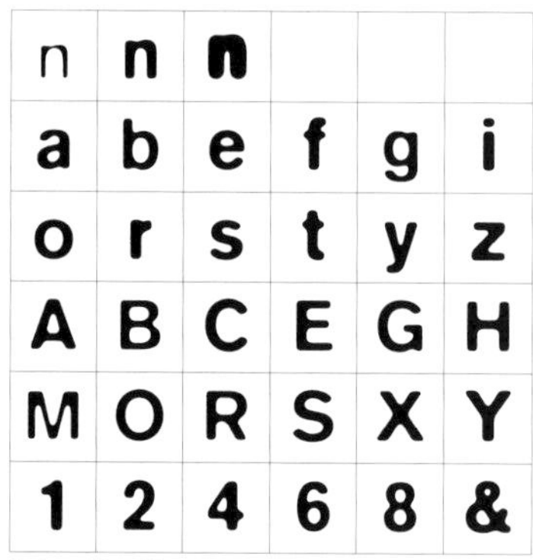


Brody 1991 Blur

Brody Logo

**Brody,** Neville – b. 23. 4. 1957 in London, England – *graphic designer, art director, type designer* – 1975: studies painting at Hornsey College and from 1976–79 at the London College of Printing. Numerous record covers. 1981–86: art director of the English magazine "The Face". 1983–87: covers for the London magazine "City Limits". 1987– 90: works for the "Arena" magazine. Art director for the magazines "Per Lui" and "Lei" of the Condé Nast Publications in Milan and the French magazine "Actuel". 1988: works and designs for Nike, Premiere TV, ORF, the House of World Culture in Berlin, the Deutsches Theater in Hamburg, Parco department store in Tokyo and the digital medium of publication "Fuse". Brody's typographical style uses aesthetic elements from Art Deco and betrays non-European influence. His graphic language has become an international model for the new age of computer-orientated design. Exhibitions: "The Graphic Language of Neville Brody", London, Edinburgh, Berlin, Hamburg, Frankfurt am Main and Tokyo (1988–90). – *Fonts:* Arcadia (1990), Industria (1990), Insignia (1990), Blur (1991), Pop (1991), Gothic (1991), Harlem (1991). – *Publications:* J. Wozencroft "The Graphic Language of Neville Brody", London 1988; J. Wozencroft "The Graphic Language of Neville Brody No. 2", London, Munich 1994.

**Brownjohn,** Robert – b. 1925 in New Jersey, USA, d. 1970 – *painter, graphic designer, typographer* – Studied painting and design at the Institute of Design in Chicago and architecture at the Institute

cago und Architektur am Institute of Design, Illinois Institute of Technology. 1957 Gründung der Agentur Brownjohn, Chermayeff + Geismar in New York. 1958 Gestaltung der Ausstellung „Streetscape“ für den amerikanischen Pavillon auf der Weltausstellung Brüssel. 1960 verläßt Brownjohn Chermayeff + Geismar in New York und wird 1961 Design Director von McCann-Erickson Ltd. in London. Neben zahlreichen Filmtiteln entwirft er 1964 den Titel für den James Bond-Film „Goldfinger“. Lehrt am Pratt Institute und an der Cooper Union in New York.

**Brudi,** Walter – geb. 24. 1. 1907 in Stuttgart, Deutschland, gest. 9. 12. 1987 in Stuttgart, Deutschland. – *Schriftentwerfer, Kalligraph, Illustrator, Maler, Lehrer* – 1923–28 Studium an der Akademie der Bildenden Künste in Stuttgart. 1932 Leiter der Abteilung Typographie und Buchgestaltung an der Höheren graphischen Fachschule, Berlin. 1935 Berufung an die Meisterschule für Deutschlands Buchdrucker in München. 1945 Gründung der „Tübinger Universitätspresse“ mit H. Leins, H. Laupp und F. H. E. Schneidler. 1949 Berufung an die Stuttgarter Akademie als Leiter der Klasse für Buchgrafik und Typographie. 1959 Rektor der Akademie. 1966 Ausbau seiner Klasse zum Institut für Buchgestaltung. – *Schriftentwürfe:* Orbis (1953), Brudi Mediaeval mit kursiv (1953–1954), Pan (1954). – *Publikationen:* „Schriftschreiben und Schriftzeichnen“, Stuttgart 1952; „Das Abenteuer der Handschrift“, Stuttgart 1959; „Schriftzeichen“, Stuttgart 1971; „Grafik, Malerei“, Stuttgart, Zürich 1987.

1971; „Grafik, Malerei“, Stuttgart, Zürich 1987.

Brody 1989 Poster

Brownjohn Invitation

Brownjohn 1970 Poster

Brody 1988 Page

Brudi Cover

Brudi Cover

of Design, Illinois Institute of Technology. 1957: opens the Brownjohn, Chermayeff + Geismar studio in New York. 1958: designs the “Streetscape” display for the American pavilion at the World Exhibition in Brussels. 1960: Brownjohn leaves Chermayeff + Geismar in New York and becomes design director for McCann-Erickson Ltd. in London. He designs the title sequence for the James Bond film “Goldfinger” in 1964, besides titles for numerous other films. Teaches at the Pratt Institute and the Cooper Union in New York.

**Brudi,** Walter – b. 24. 1. 1907 in Stuttgart, Germany, d. 9. 12. 1987 in Stuttgart, Germany – *type designer, calligrapher, illustrator, painter, teacher* – 1923–28: studies at the Akademie der Bildenden Künste in Stuttgart. 1932: head of the department for typography and book design at the Höhere graphische Fachschule in Berlin. 1935: appointed to the Meisterschule für Deutschlands Buchdrucker in Munich. 1945: founds the Tübingen University Press with Hermann Leins, Hilde Laupp and F. H. E. Schneidler. 1949: appointed to the Stuttgarter Akademie as head of the book graphics and typography class. 1959: rector. 1966: his class is enlarged and becomes the Institute for Book Design. – *Fonts:* Orbis (1953), Brudi Mediaeval with italics (1953–54), Pan (1954). – *Publications:* “Schriftschreiben und Schriftzeichnen”, Stuttgart 1952; “Das Abenteuer der Handschrift”, Stuttgart 1959; “Schriftzeichen”, Stuttgart 1971; “Grafik, Malerei”, Stuttgart, Zurich 1987.

mar à New York. 1958, conception de l’exposition «Streetscape» pour le pavillon américain de l’Exposition universelle de Bruxelles. En 1960, Brownjohn quitte Chermayeff + Geismar et devient Design-Director de McCann-Erickson Ltd. à Londres en 1961. Il réalise de nombreux titres de films dont celui de «Goldfinger» (James Bond) en 1964. Il a enseigné au Pratt Institute et à la Cooper Union à New York.

**Brudi,** Walter – né le 24. 1. 1907 à Stuttgart, Allemagne, décédé le 9. 12. 1987 à Stuttgart, Allemagne – *concepteur de polices, calligraphe, illustrateur, peintre, enseignant* – 1923–1928, études à l’Akademie der Bildenden Künste (beaux-arts) de Stuttgart. 1932, dirige le département de typographie et des arts du livre à la Höhere graphische Fachschule de Berlin (école supérieure d’arts graphiques). 1935, nomination à la Meisterschule für Deutschlands Buchdrucker de Munich (école allemande des imprimeurs du livre). En 1945, il fonde la «Tübinger Universitätspresse» avec Hermann Leins, Hilde Laupp et F. H. E. Schneidler. 1949, nomination à la Stuttgarter Akademie (beaux-arts) où il dirige la classe d’arts graphiques et de typographie de l’édition. 1959, recteur du Stuttgarter Akademie. 1966, sa classe prend de l’importance et devient l’Institut für Buchgestaltung (institut des arts du livre). – *Polices :* Orbis (1953), Brudi Mediaeval avec italique (1953–1954), Pan (1954). – *Publications :* «Schriftschreiben und Schriftzeichnen», Stuttgart 1952; «Das Abenteuer der Handschrift», Stuttgart 1959; «Schriftzeichen», Stuttgart 1971; «Grafik, Malerei», Stuttgart, Zurich 1987.

**Büchler**
**Bulmer**
**Burchartz**

**Büchler,** Robert – geb. 23. 4. 1914 in Basel, Schweiz. – *Typograph, Lehrer* – 1929–33 Schriftsetzerlehre in Basel, danach bis 1947 als Typograph tätig. Ausbildung zum Gewerbelehrer und Besuch der Fachschule für Buchdruck an der Gewerbeschule Basel. 1947 Fachlehrer für Schriftsatz, Leiter der Tageskurse für typographisches Gestalten und der Polygraphischen Fachschule Basel. Ab 1965 Vorsteher der Lehrlingsabteilung für kunstgewerbliche Berufe.

**Bulmer,** William – geb. 1757 in Newcastle upon Tyne, England, gest. 1830. – *Typograph, Drucker* – Druckerlehre in Newcastle upon Tyne, danach Arbeiten für John Bell in London. 1792–1802 druckt Bulmer in neun Bänden das dramatische Werk von William Shakespeare für die Verleger Boydell und Nicol. Danach entstehen drei Bände mit Werken von John Milton. 1794–96 druckt Bulmer zwei Foliobände „The History of the River Thames".

**Burchartz,** Max – geb. 28. 7. 1887 in Elberfeld, Deutschland, gest. 31. 1. 1961 in Essen, Deutschland. – *Typograph, Grafik-Designer, Künstler, Lehrer, Theoretiker* – Ausbildung 1906–08 an der Akademie Düsseldorf. Nach dem Ersten Weltkrieg in Hannover. Enge Kontakte mit Kurt Schwitters und El Lissitzky. Verbindung zum Bauhaus Weimar. 1924 Gründung des Studios „Werbe-Bau" mit Johannis Canis in Bochum. 1926–33 Lehrer für Gebrauchsgrafik, Fotografie, Typographie und Werbelehre an der Folkwangschule in Essen. 1927 Gründungsmitglied der Gruppe „ring neuer werbegestalter". 1933

**Büchler,** Robert – né le 23 . 4. 1914 à Bâle, Suisse – *typographe, enseignant* – 1929–1933, apprentissage de compositeur à Bâle où il exerce jusqu'en 1947 comme typographe. Formation d'enseignant, puis fréquente les cours spécialisés pour les imprimeurs du livre à la Gewerbeschule de Bâle. 1947, enseigne la composition, dirige les cours réguliers de mise en forme typographique à la Polygraphische Fachschule de Bâle (école technique de graphisme). A partir de 1965, président de la section apprentissage pour les métiers de l'artisanat.

**Bulmer,** William – né en 1757, à Newcastle upon Tyne, Angleterre, décédé en 1830 – *typographe, imprimeur* – apprentissage d'imprimeur à Newcastle upon Tyne, puis travaux pour John Bell à Londres. 1792–1802, Bulmer imprime le théâtre de Shakespeare en neuf volumes pour les éditeurs Boydell et Nicol. Puis il réalise trois volumes avec des œuvres de John Milton. 1794–1796, Bulmer imprime «The History of the River Thames» en deux volumes folio.

**Burchartz,** Max – né le 28. 7. 1887 à Elberfeld, Allemagne, décédé le 31. 1. 1961 à Essen, Allemagne – *typographe, graphiste maquettiste, artiste, enseignant, théoricien* – formation: 1906–1908 à l'académie de Düsseldorf. Vit à Hanovre après la Première Guerre mondiale. Contacts étroits avec Kurt Schwitters et El Lissitzky. Entretient des relations avec le Bauhaus de Weimar. 1924, création de l'atelier «Werbe-Bau» avec Johannis Canis à Bochum. 1926–1933, enseigne le graphisme utilitaire, la photographie, la typographie et la publicité à la Folkwangschule d'Essen. 1933, licencié par les nazis. 1949, directeur de la Folkwangschule et retour aux arts libres à partir de 1957. Membre fondateur du groupe «ring neuer werbegestalter» en 1927.

a bcd
e fgh
ij kl
mn
o pqr
s
A BCD typographie
E F u v
GH wxyz
IJ K
LMN
O PQ
RST
U VWX
Y
Z

Büchler 1960 Poster

Burchartz 1928 Poster

NINTH DAY. 43

see you in this field of contest, brandishing your unerring lance, or quietly reposing beneath the panoply of your seven-bulls-hide shield!..

Lysander. This must be a very extraordinary champion.

Lisardo. 'In his way' (as they call it) he hath absolutely no compeer; and Magliabecchi yields entirely to his ascendant genius—for Nennius not only loves *bokes* as lustily as did the librarian of the Grand Duke of Tuscany, but he hath something more than a mere *title-page* acquaintance with them. His memory also is equally faithful and well-furnished—and such a gluttonous bibliomaniacal appetite doth he possess, that even *Rymer*, the *Gallia Christiana*, and Bouquet's *Recueil des Historiens des Gaules*, will scarcely suffice him for a twelvemonth's 'victualling.' Mabillon, Montfaucon, and Muratori are his dear delights as foreign authors; while his deal-shelves groan beneath the weight of annotation upon our home historians; such as Gildas, Jeffrey, his namesake, Ingulph, Hoveden, Malmesbury, Matthew Paris, Ralph de Diceto, and Benedictus Abbas, &c. &c.—and then for the 'scribbled margins' (as Warburton used to express it) of his Leland, Camden, Twysden, Gale, Sparke, Hearne, Batteley, Grose, King, and others of the like character—oh, 'twould do your heart good only to have a glimpse of them!

Lorenzo. More and more wondrous!

Lisardo. I have not yet done with Nennius. He hath no small knowledge of the art of design; and brandishes his pencil upon castles, cathedrals, and churches, that it were a marvel to see how his drawers and portfolios are crammed with the same. There is not a church, nor place of worship, nor castle, within the counties of *Sussex*, *Kent*, and *Bedford*,

Bullmer 1817 Page

Burchartz 1930 Poster

**Büchler,** Robert – b. 23. 4. 1914 in Basle, Switzerland – *typographer, teacher* – 1929–33: trains as a typesetter in Basle, then works as typographer until 1947. Trains as a vocational college teacher and attends the Fachschule for Buchdruck at the Gewerbeschule in Basle. 1947: teacher of form, head of the day courses for typographic design and of the Polygraphische Fachschule in Basle. From 1965 onwards: head of the trainee department for trades in arts and crafts.

**Bulmer,** William – b. 1757 in Newcastle upon Tyne, England, d. 1830 – *typographer, printer* – Trained as a printer in Newcastle upon Tyne, then went to London to work for John Bell. 1792–1802: Bulmer prints the dramatic works of William Shakespeare in nine volumes for the publishers Boydell and Nicol. He then produces three volumes of John Milton's works. 1794–96: Bulmer prints two folio volumes on "The History of the River Thames".

**Burchartz,** Max – b. 28. 7. 1887 in Elberfeld, Germany, d. 31. 1. 1961 in Essen, Germany – *typographer, graphic designer, artist, teacher, theoretician* – 1906–08: trains at the Düsseldorf Academy. After the First World War spends time in Hanover. Close contact with Kurt Schwitters and El Lissitzky. Connections with the Bauhaus in Weimar. 1924: sets up the Werbe-Bau studio with Johannis Canis in Bochum. 1926–33: teaches applied graphics, photography, typography and advertising at the Folkwangschule in Essen. 1933: dismissed by the National Socialists. 1949: director of the Folk-

Entlassung durch die Nationalsozialisten. 1949 Direktor der Folkwangschule. Ab 1957 Rückkehr zur freien Kunst. – *Publikationen u. a.:* „Gleichnis der Harmonie", München 1949; „Gestaltungslehre", München 1953; „Schule des Schauens", München 1962. J. Stürzebecher „Max ist endlich auf dem richtigen Weg", Frankfurt am Main, Baden 1993.

**Burke,** Jackson – geb. 1908 in San Francisco, Kalifornien, USA, gest. 1975. – *Schriftentwerfer* – Studium an der University of California in Berkeley. 1949–63 Direktor der Schriftentwicklung bei Mergenthaler-Linotype. Verantwortlich für die Entwicklung des TeleTypesetting-Systems (TTS) für Zeitschriften und für die Entwicklung von Schriften für den indischen Sprachraum. – *Schriftentwürfe:* Trade Gothic (1948–60), Majestic (1953–56), Aurora (1960).

**Burkhardt,** Klaus – geb. 31. 3. 1928 in Bürgel, Deutschland. – *Typograph, Pressendrucker, Verleger, Lehrer* – 1947 Schriftsetzerlehre. 1952 Schriftsetzermeister. 1958 Mitbegründer des „Atelier Rauls" Stuttgart. 1958–90 Lehrer für Typographie an der Grafischen Fachschule Stuttgart. Entwurf und Druck großformatiger Buchstabenbilder im Buchdruck. Seit 1964 Arbeiten mit Fotosatz. Veröffentlichung von Einzeldrucken, Kalendern, Postkarten und Mappenwerken. Zahlreiche Ausstellungen, u. a. in Ulm (1961), Stockholm (1962), Frankfurt am Main (1964), Eindhoven (1967), Offenbach (1995). – *Publikationen u.a.:* „Mein Stundenbuch", 1962; „200 Portraits einer 5", 1966; „Alles was rund ist, ist gut", 1990.

Burchartz 1924 Flyer

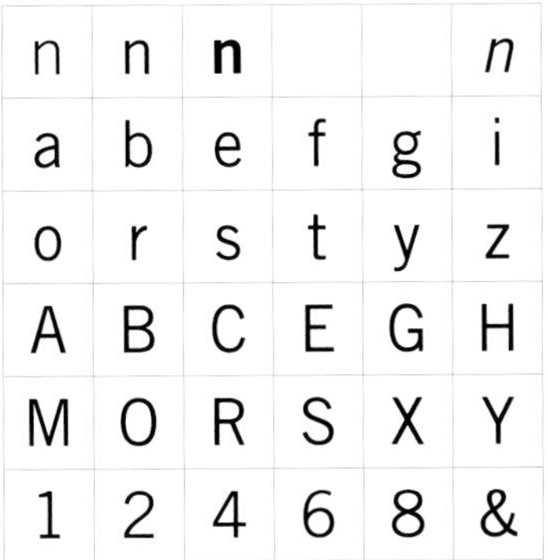


Burke 1948–60 Trade Gothic

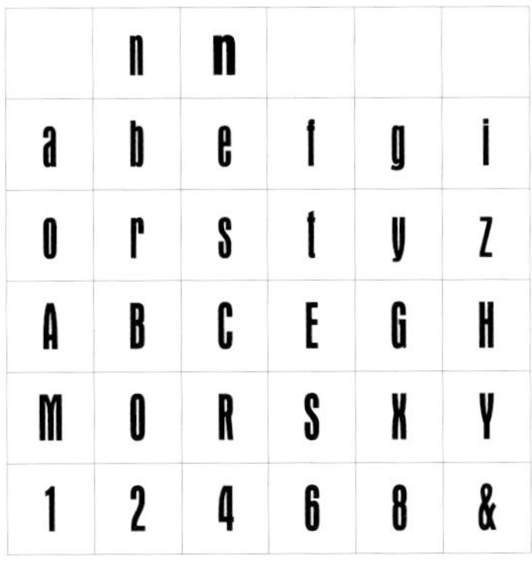


Burke 1960 Aurora

– *Publications, sélection:* «Gleichnis der Harmonie», Munich 1949; «Gestaltungslehre», Munich 1953; «Schule des Schauens», Munich 1962. J. Stürzebecher «Max ist endlich auf dem richtigen Weg», Francfort-sur-le-Main, Baden 1993.

**Burke,** Jackson – né en 1908 à San Francisco, Californie, Etats-Unis, décédé en 1975 – *concepteur de polices* – études à l'University of California à Berkeley. 1949– 1963, directeur du développement des écritures chez Mergenthaler-Linotype. Responsable du développement du TeleTypesetting-System (TTS) pour la presse, et du développement d'écritures pour les territoires de langue indienne. – *Polices:* Trade Gothic (1948–1960), Majestic (1953–1956), Aurora (1960).

**Burkhardt,** Klaus – né le 31. 3. 1928 à Bürgel, Thuringe, Allemagne – *typographe, imprimeur à la presse, éditeur, enseignant* – 1947, apprentissage de composition. 1952, compagnon compositeur. 1958, cofondateur de l' «Atelier Raul» à Stuttgart. Professeur de typographie à la Grafische Fachschule de Stuttgart (école des arts graphiques). 1958–1990, conception et impression de lettrines de grand format pour le livre. Utilise la photocomposition depuis 1964. Publication d'estampes originales, de calendriers, de cartes postales et de folios. Nombreuses expositions, par ex. à Ulm (1961), Stockholm (1962), Francfort-sur-le-Main (1964), Eindhoven (1967), Offenbach (1995). – *Publications, sélection:* «Mein Stundenbuch», 1962; «200 Portraits einer 5», 1966; «Alles was rund ist, ist gut», 1990.

Burchartz 1924 Sketch

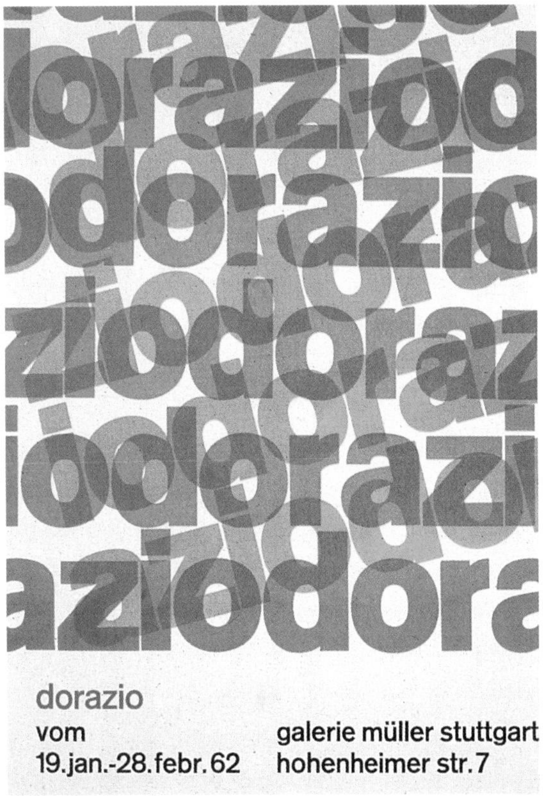


Burkhardt 1962 Poster

wangschule. 1957: returns to free art. 1927: founder member of the group ring neuer werbegestalter. – *Publications include:* "Gleichnis der Harmonie", Munich 1949; "Gestaltungslehre", Munich 1953; "Schule des Schauens", Munich 1962. J. Stürzebecher "Max ist endlich auf dem richtigen Weg", Frankfurt am Main, Baden 1993.

**Burke,** Jackson – b. 1908 in San Francisco, California, USA, d. 1975 – *type designer* – Studied at the University of California in Berkeley. 1949–63: director of type development for Mergenthaler-Linotype. Responsible for the development of the TeleTypesetting System (TTS) for magazines and for the development of fonts for native American languages. – *Fonts:* Trade Gothic (1948–60), Majestic (1953–56), Aurora (1960).

**Burkhardt,** Klaus – b. 31. 3. 1928 in Bürgel, Thuringia, Germany – *typographer, printer of fine editions, publisher, teacher* – 1947: trains as a typesetter. 1952: master typesetter. 1958: one of the founders of the Atelier Raul in Stuttgart. 1958–90: teacher of typography at the Grafische Fachschule in Stuttgart. Has designed and printed large-format illuminated letters. 1964 onwards: starts working with filmsetting techniques. Publishes single prints, calendars, postcards and folder work. Numerous exhibitions, including in Ulm (1961), Stockholm (1962), Frankfurt am Main (1964), Eindhoven (1967) and Offenbach (1995). – *Publications include:* "Mein Stundenbuch", 1962; "200 Portraits einer 5", 1966; "Alles was rund ist, ist gut", 1990.

**Burns,** Aaron – geb. 25. 5. 1922 in Passaic, New Jersey, USA, gest. 16. 7. 1991 in New York, USA. – *Grafik-Designer, Typograph, Lehrer* – 1941–43 Ausbildung an der Newark Evening School of Fine and Industrial Arts, New Jersey. 1948 Assistant Graphic-Designer von Herb Lubalin im Studio Sudler & Hennessy, New York. 1952–60 Director of Design and Typography von „The Composing Room", New York. 1960 Gründer des „International Center for the Typographic Arts", New York. 1970 Gründung der „International Typeface Corporation of New York" zusammen mit Herb Lubalin. Herausgabe der Hauszeitschrift „Upper and lower case (U&lc)". 1976 Gründung der Firma „Design Processing International", New York. 1983 Aufnahme in die Hall of Fame des Art Directors Club . 1985 Ernennung zum Vize-Ehrenvorsitzenden der ATypI. 1985 Auszeichnung mit der Goldmedaille des TDC of New York. 1955–60 Lehrer in Typographie am Pratt Institute, New York. – *Publikationen u. a.:* „Typography", New York 1961. H. Hara „The New Work of Herb Lubalin and Aaron Burns" in „Graphic Design" 9/1971, Tokio; „Aaron Burns" in „Art Directors Club Annual", New York 1983.

**Burtin,** Will – geb. 27. 1. 1908 in Köln, Deutschland, gest. 18. 1. 1972 in New York, USA. – *Typograph, Art Director, Grafik-Designer, Lehrer* – Ausbildung bis 1930 an der Kölner Werkschule, 1938 Emigration in die USA, eigenes Studio in New York. Pionier bei der verständlichen grafischen Umsetzung von Informationen. 1939–43 und 1959–72 Professor am

**Burns,** Aaron – né le 25. 5. 1922 à Passaic, New-Jersey, Etats-Unis, décédé le 16. 7. 1991 à New York, Etats-Unis – *graphiste maquettiste, typographe, enseignant* – 1941– 1943, formation à la Newark Evening School of Fine and Industrial Arts, New Jersey. 1948, assistant graphiste maquettiste d'Herb Lubalin à l'atelier Sudler & Hennessy, New York. 1952–1960, directeur du design et de la typographie de «The Composing Room», New York. 1960, fondateur du «International Center for the Typographic Arts», New York. 1970, fondation du «International Typeface Corporation of New York» avec Herb Lubalin. Edition de la revue de la société «Upper and lower case (U&lc)». 1976, fonde la société «Design Processing International», New York. 1983, entrée au Art Directors Club Hall of Fame. 1985, nommé vice-président d'honneur de l'Association Typographique Internationale (ATypI). 1985, lauréat de la médaille d'or du Type Directors Club of New York. 1955–1960, enseigne la typographie au Pratt Institute, New York. – *Publications, sélection :* «Typography», New York 1961. H. Hara «The New Work of Herb Lubalin and Aaron Burns» dans «Graphic Design» 9/1971, Tokyo; «Aaron Burns» dans Art Directors Club Annual, New York 1983.

**Burtin,** Will – né le 27. 1. 1908 à Cologne, Allemagne, décédé le 18. 1. 1972 à New York, Etats-Unis – *typographe, directeur artistique, graphiste maquettiste, enseignant* – formation à la Kölner Werkschule jusqu'en 1930. Emigre aux Etats-Unis en 1938 et crée son propre atelier à New York. Pionnier de la transposition graphique de l'information. 1939–1943 et 1959–1972, professeur au Pratt Institute de New York. 1945–1949, Art Director de la revue «Fortune». 1949, inauguration de son «Studio de recherche visuelle et de design». Commanditaires : IBM, Her-

Burns 1961 Advertisement

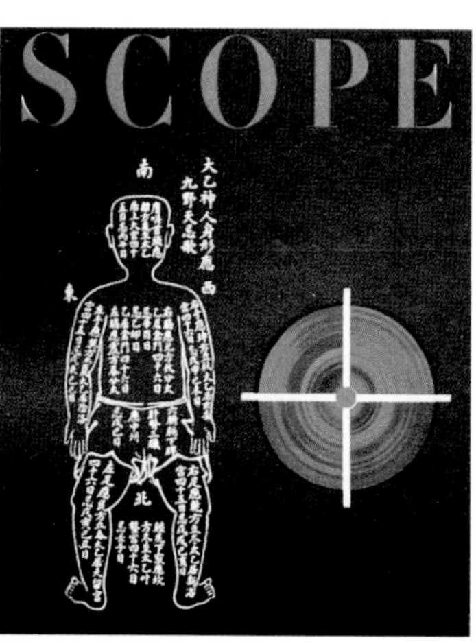

Burtin 1951 Cover

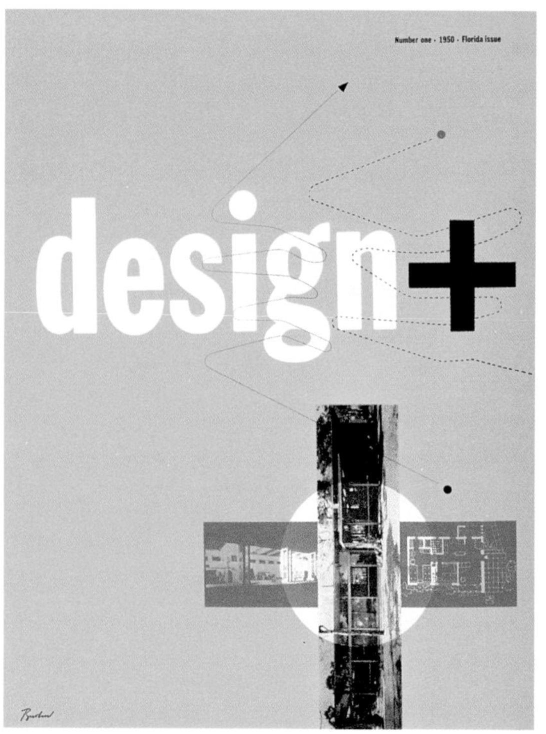

Burtin 1950 Cover

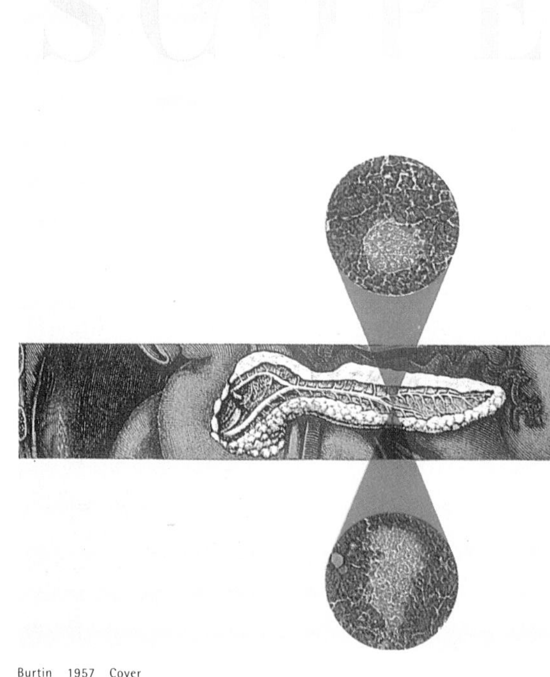

Burtin 1957 Cover

**Burns,** Aaron – b. 25. 5. 1922 in Passaic, New Jersey, USA, d. 16. 7. 1991 in New York, USA – *graphic designer, typographer, teacher* – 1941–43: trains at the Newark Evening School of Fine and Industrial Arts, New Jersey. 1948: assistant graphic designer to Herb Lubalin at the Sudler & Hennessy Studio in New York. 1952–60: director of design and typography for The Composing Room, New York. 1960: founder of the International Center for the Typographic Arts in New York. 1970: founds the International Typeface Corporation of New York with Herb Lubalin. Edits the in-house magazine "Upper and lower case" (U&lc). 1976: sets up the company Design Processing International in New York. 1983: is accepted into the Art Directors Club Hall of Fame. 1985: named honorary vice-president of the Association Typographique Internationale (ATypI). 1985: is awarded a gold medal by the Type Directors Club of New York. 1955–60: teaches typography at the Pratt Institute in New York. – *Publications include:* "Typography", New York 1961. H. Hara "The New Work of Herb Lubalin and Aaron Burns" in "Graphic Design" 9/1971, Tokyo; "Aaron Burns" in the "Art Directors Club Annual", New York 1983.

**Burtin,** Will – b. 27. 1. 1908 in Cologne, Germany, d. 18. 1. 1972 in New York, USA – *typographer, art director, graphic designer, teacher* – Trained at the Kölner Werkschule until 1930. 1938: emigrates to the USA and sets up his own studio in New York. He pioneers the comprehensible graphic presentation of information. 1939

Pratt Institute, New York. 1945–49 Art Director der Zeitschrift „Fortune". 1949 Eröffnung seines „Studios für visuelle Forschung und Design", Auftraggeber u. a. IBM, Hermann Miller, Smithsonian Institution, Union Carbide, Eastman Kodak. 1961 Zusammenarbeit und Heirat mit der Grafik-Designerin Cipe Pineles. Burtin war amerikanischer Präsident der AGI. 1971 Auszeichnung mit der Goldmedaille und einer retrospektiven Ausstellung durch die AIGA.

**Butti,** Allessandro – geb. 1893, gest. 1959 in Turin, Italien. – *Schriftentwerfer, Lehrer* – Künstlerischer Leiter der Schriftgießerei Nebiolo, Turin. Lehrte an der Scuola Vigliandi-Paravia in Turin. – *Schriftentwürfe:* Paganini (mit A. Bertieri, 1928), Quirius (1939), Athenaeum (mit A. Novarese, 1945), Normandia (mit A. Novarese, 1946–49), Rondine (1948), Augustea (mit A. Novarese, 1951), Fluidum (1951), Microgramma (mit A. Novarese, 1952).

**Buzzi,** Paolo – geb. 15. 2. 1874 in Mailand, Italien, gest. 18. 2. 1956 in Mailand, Italien. – *Dichter, Kritiker* – Studiert Jura und Philologie in Pavia. 1898 erster Gedichtband „Rapsodie Leopardiane". 1905 Gründung der Zeitschrift „Poesia" mit Marinetti und Benelli. 1909 Anschluß an die futuristische Bewegung, futuristisch-typographische Inszenierung seiner Gedichtbände. Mit seinen Arbeiten schafft Buzzi eine Verbindung von Form und Inhalt in der Typographie. – *Publikationen u.a.:* „Aeroplani", Mailand 1909; „Versi libri", Mailand 1913; „canto quotidiano", Mailand 1933; „Atomiche", Mailand 1952.

| | n | n | n | | n |
|---|---|---|---|---|---|
| a | b | e | f | g | i |
| o | r | s | t | y | z |
| A | B | C | E | G | H |
| M | O | R | S | X | Y |
| 1 | 2 | 4 | 6 | 8 | & |

Butti 1951 Augustea

| | n | n | | | n |
|---|---|---|---|---|---|
| a | b | e | f | g | i |
| o | r | s | t | y | z |
| A | B | C | E | G | H |
| M | O | R | S | X | Y |
| 1 | 2 | 4 | 6 | 8 | & |

Butti 1945 Athenaeum

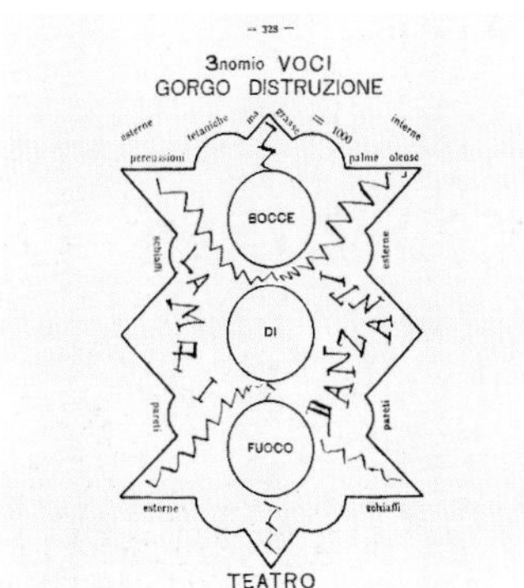


Buzzi 1915 Page

Burtin 1952 Cover

il VoOOOoLOooooo

18h 56.m 56s, 96 — 37.° 15'. 35'', 71

Come dall'IMBYTO vinoso alla boccia d'assenzio

ALI ALI ALI ALI ALI ALI ALI ALI ALI

ALI ALI ALI ALIIIII AAAALI

ALIIIII ALI

ALI ALI AAAALI

ALI ali ali ali ali ali ALI

Buzzi 1915 Page

mann Miller, Smithsonian Institution, Union Carbide, Eastman Kodak. A partir de 1961, il travaille avec son épouse, la graphiste maquettiste Cipe Pineles. Burtin a été président américain de l'AGI. 1971, obtient la médaille d'or de l'AIGA et une exposition rétrospective.

**Butti,** Alessandro – né en 1893, décédé en 1959 à Turin, Italie – *concepteur de polices, enseignant* – directeur artistique de la fonderie de caractères Nebiolo, Turin. Enseigne à la Scuola Vigliandi-Paravia de Turin. – *Polices:* Paganini (avec A. Bertieri, 1928), Quirius (1939), Athenaeum (avec A. Novarese, 1945), Normandia (avec A. Novarese, 1946–1949), Rondine (1948), Augustea (avec A. Novarese, 1951), Fluidum (1951), Microgramma (avec A. Novarese, 1952).

**Buzzi,** Paolo – né le 15. 2. 1874 à Milan, Italie, décédé le 18. 2. 1956 à Milan, Italie – *poète, critique* – études de droit et de philosophie à Pavie, Italie. 1898, premier recueil de poèmes «Rapsodie Leopardiane». 1905, fondation de la revue «Poesia» avec F. T. Marinetti et Sem Benelli. 1909, rejoint le mouvement futuriste, mise en scène typographique et futuriste de ses recueils de poèmes. Buzzi a créé une nouvelle alliance entre forme et contenu en typographie. – *Publications, sélection:* «Aeroplani», Milan 1909; «Versi libri», Milan 1913; «canto quotidiano», Milan 1933; «Atomiche», Milan 1952.

–43 and 1959–72: professor at the Pratt Institute in New York. 1945–49: art director of "Fortune" magazine. 1949: opens his Studio for Visual Research and Design and works for IBM, Hermann Miller, the Smithsonian Institution, Union Carbide and Eastman Kodak, among others. 1961: works with and marries the graphic designer Cipe Pineles. Burtin was the American president of AGI. 1971: AIGA awards him a gold medal and stages a retrospective exhibition of his work.

**Butti,** Alessandro – b. 1893, d. 1959 in Turin, Italy – *type designer, teacher* – Art director of the Nebiolo type foundry in Turin. Taught at the Scuola Vigliani-Paravia. – *Fonts:* Paganini ( with A. Bertieri, 1928), Quirius (1939), Athenaeum (with A. Novarese, 1945), Normandia (with A. Novarese, 1946–49), Rondine (1948), Augustea (with A. Novarese, 1951), Fluidum (1951), Microgramma (with A. Novarese, 1952).

**Buzzi,** Paolo – b. 15. 2. 1874 in Milan, Italy, d. 18. 2. 1956 in Milan, Italy – *poet, critic* – Studied law and philology in Pavia, Italy. 1898: his first volume of poems, "Rapsodie Leopardiane", is published. 1905: launches the magazine "Poesia" with F. T. Marinetti and Sem Benelli. 1909: joins the Futurist movement and produces his works of poetry using a Futurist typography. With his literary works Buzzi created a new am algam of form and content in typography. – *Publications include:* "Aeroplani", Milan 1909; "Versi libri", Milan 1913; "canto quotidiano", Milan 1933; "Atomiche", Milan 1952.

手
祖母
祖父
正方形
三角形

# C

**Caflisch,** Max – geb. 25. 10. 1916 in Winterthur, Schweiz. – *Schriftentwerfer, Lehrer, Autor* – 1932–36 Lehre als Schriftsetzer. 1936–43 erster Akzidenzsetzer in verschiedenen Druckereien. 1941–42 Fachlehrer für Typographie an der Allgemeinen Gewerbeschule Basel. 1943–62 künstlerischer Leiter der Druckerei Benteli in Bern. 1962–81 Leiter der grafischen Abteilung der Kunstgewerbeschule Zürich, gleichzeitig Fachlehrer für Typographie. 1973–78 Dozent an der Technikerschule der grafischen Industrie, Zürich. Berater für Schriftdesign bei IBM New York (1962–66), für die Bauersche Gießerei, Frankfurt am Main (1965–66), für die Firma Dr. Rudolf Hell, Kiel (1972–89), für die Firma Adobe Systems Inc. (seit 1990). – *Schriftentwurf:* Columna (1955). – *Publikationen u. a.:* „William Morris, der Erneuerer der Buchkunst", Bern 1959; „Fakten zur Schriftgeschichte", Zürich 1973; „Schrift und Papier", Grellingen 1973; „Typographie braucht Schrift", Kiel 1978. A. Berlincourt u.a. „Max Caflisch. Typographia practica", Hamburg 1988.

**Calvert,** Margaret – geb. 1936 – *Schriftentwerferin, Typographin, Grafik-Designerin, Lehrerin* – 1953–57 Studium an der Chelsea School of Art, London. 1964 Partnerin von Jock Kinneir bei „Kinneir, Calvert Associates". Hier entstehen Schriften und Zeichensysteme für die Autobahnen, die British Rail, die Flughäfen, Krankenhäuser, die Armee und 1980 für die Tyne & Wear Metro des Vereinigten Königreiches. Unterrichtet am Royal College of Art in London und ist dort 1987–1991 Leiterin der Abteilung Gra-

**Caflisch,** Max – né le 25. 10. 1916 à Winterthur, Suisse – *concepteur de polices, enseignant, auteur* – 1932–1936, apprentissage de composition typographique. 1936–1943, premier compositeur sur Akzidenz dans diverses imprimeries. 1941–1942, enseigne la typographie à la Allgemeine Gewerbeschule (école des arts et métiers) de Bâle. 1943–1962, directeur artistique de l'imprimerie Benteli, à Berne. 1962–1981, directeur de la section d'arts graphiques de la Kunstgewerbeschule (école des arts décoratifs) de Zurich, en même temps qu'il enseigne la typographie. 1973–1978, professeur à la Technikerschule der grafischen Industrie (école technique de l'industrie des arts graphiques) de Zurich. Conseiller en design de polices pour IBM New York (1962–1996), pour la Bauersche Giesserei à Francfort-sur-le-Main (1965–1966), pour la société Dr. Rudolf Hell, à Kiel (1972–1989), et pour la société Adobe Systems Inc. (depuis 1990). – *Police:* Columna (1955). – *Publications, sélection:* «William Morris, der Erneuerer der Buchkunst», Berne 1959; «Fakten zur Schriftgeschichte», Zurich 1973; «Schrift und Papier», Grellingen 1973; «Typographie braucht Schrift», Kiel 1978. A. Berlincourt et autres «Max Caflisch. Typographia practica», Hambourg 1988.

**Calvert,** Margaret – née en 1936 – *conceptrice de polices, typographe, graphiste maquettiste, enseignante* – 1953–1957, études à la Chelsea School of Art, Londres. 1964, partenaire de Jock Kinneir chez «Kinneir, Calvert Associates». C'est dans ce contexte que sont conçus les polices et systèmes signalétiques des autoroutes, de la British Rail, des aéroports, des hôpitaux, de l'armée et, en 1980, de la Tyne & Wear Metro du Royaume-Uni. A enseigné au Royal College of Art de Londres à partir de 1966 et y dirige la sec-

TENTOON
*AMSTERDAM 1949*
STELLING
*DAS SCHWEIZER BUCH*
HET
*LE LIVRE SUISSE*
ZWITSERSE
*IL LIBRO SVIZZERO*
BOEK

Caflisch 1949 Cover

Caflisch 1952 Cover

Calvert 1994 A26

COLUMNA
VERSALIEN

Caflisch 1955 Columna

| n | n | n | | | |
|---|---|---|---|---|---|
| a | b | e | f | g | i |
| o | r | s | t | y | z |
| A | B | C | E | G | H |
| M | O | R | S | X | Y |
| 1 | 2 | 4 | 6 | 8 | & |

Calvert 1980 Calvert

**Caflisch,** Max – b. 25. 10. 1916 in Winterthur, Switzerland – *type designer, teacher, author* – 1932–36: trains as a compositor. 1936–43: first job setter in various printing works. 1941–42: specialist teacher of typography at the Allgemeine Gewerbeschule in Basle. 1943–62: art director of the Benteli printing works in Bern. 1962–81: head of the graphics department at the Kunstgewerbeschule in Zurich; also teacher of typography. 1973–78: lecturer at the Technikerschule der grafischen Industrie in Zurich. Consultant on type design to IBM in New York (1962–66), for the Bauersche Gießerei in Frankfurt am Main (1965–66), for the Dr. Rudolf Hell company in Kiel (1972–89) and for Adobe Systems Inc. (since 1990). – *Font:* Columna (1955). – *Publications include:* "William Morris, der Erneuerer der Buchkunst", Bern 1959; "Fakten zur Schriftgeschichte", Zurich 1973; "Schrift und Papier", Grellingen 1973; "Typographie braucht Schrift", Kiel 1978. A. Berlincourt et al "Max Caflisch. Typographia practica", Hamburg 1988.

**Calvert,** Margaret – b. 1936 – *type designer, typographer, graphic designer, teacher* – 1953–57: studies at the Chelsea School of Art in London. 1964: enters into a partnership with Jock Kinneir at Kinneir, Calvert Associates. Here, typefaces and notation systems are developed for the United Kingdom's Roads, Railways, National Airports, Hospitals, The Army and, in 1980, the Tyne & Wear Metro. Calvert has taught at the Royal College of Art in London since 1966 where she has been head of graphic de-

phic Arts and Design. – *Schriftentwürfe:* Transport (mit Jock Kinneir, 1963), Calvert (1980), A 26 (1994).

**Cangiullo,** Francesco – geb. 1884 in Neapel, Italien, gest. 1977 in Livorno, Italien. – *Maler, Schriftsteller, Journalist* – 1912 Eintritt in die futuristische Bewegung. Mitarbeit an den futuristischen Zeitschriften „Lacerba", „Noi", „L'Italia Futurista". 1914 Teilnahme an der internationalen Futuristen-Ausstellung in Rom mit Bildern und Plastiken. 1916 Ausstellung seiner Gedichte im Cabaret Voltaire, Zürich. Viele seiner literarischen Arbeiten, vor allem „Caffeconcerto", sind mit typographischen Visualisierungen angereichert. 1921 Art Director des „Theater der Überraschungen", das von Rodolfo de Angelis geleitet wird. 1925 Trennung vom Futurismus. – *Publikationen u. a.:* „Le cocottesche", Neapel 1912; „Caffeconcerto, alfabeto a sorpresa", Mailand 1918; „Poesia pentagrammata", Neapel 1923; „Le serate futuriste", Neapel 1930.

**Carboni,** Erberto – geb. 22. 11. 1899 in Parma, Italien, gest. 1984 in Parma, Italien. – *Typograph, Grafik-Designer, Ausstellungsdesigner* – Architekturstudium an der Accademia di belle Arti in Parma, Diplom 1921. Danach Architekturentwürfe, Innenausstattungen und grafische Arbeiten. 1933 Umzug nach Mailand. 1934 Auftrag zum Entwurf der Fassade des Triennale-Gebäudes für die Ausstellung italienischer Flugtechnik. Arbeit als freier Grafik-Designer für Studio Boggeri und die Olivetti-Werbeabteilung. Ständig Gestaltung von Ausstellungen und Werbekampagnen für große italienische Unter-

Cangiullo 1916 Cover

Carboni 1954–56 Advertisement

Carboni Logo

Cangiullo 1914–15 Drawing

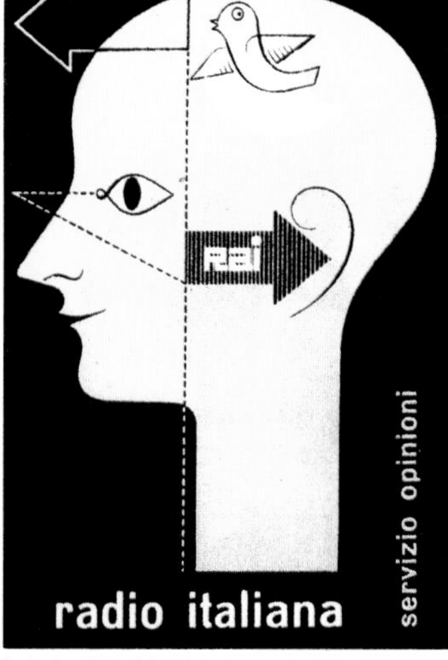

Carboni 1949 Advertisement

tion des arts graphiques et de design de 1987 à 1991. – *Polices:* Transport (avec Jock Kinneir, 1963), Calvert (1980), A 26 (1994).

**Cangiullo,** Francesco – né en 1884 à Naples, Italie, décédé en 1977 à Livourne, Italie – *peintre, écrivain, journaliste* – adhère au mouvement futuriste en 1912. Collabore aux revues futuristes «Lacerba», «Noi», «L'Italia Futurista». 1914, présente des tableaux et des sculptures futuristes à l'exposition futuriste internationale de Rome. 1916, expose ses poèmes au Cabaret Voltaire à Zurich. Bon nombre de ses créations littéraires, surtout «Caffeconcerto» sont enrichies de travaux typographiques. 1921, directeur artistique du «Théâtre des Surprises» dirigé par Rodolfo de Angelis. 1925, se distancie du futurisme. – *Publications, sélection:* «Le cocottesche», Naples 1912; «Caffeconcerto, alfabeto a sorpresa», Milan 1918; «Poesia pentagrammata», Naples 1923; «Le serate futuriste», Naples 1930.

**Carboni,** Erberto – né le 22. 11. 1899 à Parme, Italie, décédé en 1984 à Parme, Italie – *typographe, graphiste maquettiste, architecte d'expositions* – études d'architecture à la Accademia di belle Arti, Parme, diplôme en 1921. Puis, projets d'architecture, de décoration intérieure et travaux graphiques. S'installe à Milan en 1933. Commande d'un projet de façade pour le bâtiment de la Triennale en 1934 en prévision de l'exposition sur les techniques italiennes de l'aéronautique. Travaille comme graphiste maquettiste indépendant pour le Studio Boggeri et le service de publicité d'Olivetti. Conçoit l'architecture d'expositions et des campagnes publicitaires pour de

sign since 1987. – *Fonts:* Transport (with Jock Kinneir, 1963), Calvert (1980), A 26 (1994).

**Cangiullo,** Francesco – b. 1884 in Naples, Italy, d. 1977 in Livorno, Italy – *painter, writer, journalist* – 1912: joins the Futurist Movement. Works on the Futurist magazines "Lacerba", "Noi" and "L'Italia Futurista". 1914: takes part in the international Futurist exhibition in Rome with paintings and sculptures. 1916: an exhibition of his poems is held at the Cabaret Voltaire in Zurich. Many of his literary works are embellished with typographical visualizations, in particular "Caffeconcerto". 1921: art director of the Theater of Surprises, run by Rodolfo de Angelis. 1925: Cangiullo takes leave of Futurism. – *Publications include:* "Le cocottesche", Naples 1912; "Caffeconcerto, alfabeto a sorpresa", Milan 1918; "Poesia pentagrammata", Naples 1923; "Le serate futuriste", Naples 1930.

**Carboni,** Erberto – b. 22. 11. 1899 in Parma, Italy, d. 1984 in Parma, Italy – *typographer, graphic designer, exhibition designer* – Studied architecture at the Accademia di belle Arti in Parma, gaining his diploma in 1921. Architectural designs, interior design and graphic work followed on from this. 1933: moves to Milan. 1934: is commissioned to design the façade of the Triennale building for the exhibition of Italian aircraft technology. Works as a freelance graphic designer for Studio Boggeri and Olivetti's publicity department. Designs frequent exhibitions and advertising campaigns for major Italian companies (Agip, Pirelli,

nehmen (Agip, Pirelli, Motta, Barilla, Campari, Bertolli, Montecatini). 1950 mit dem Nationalpreis für Werbegrafik, 1951 mit dem Nationalpreis für Ausstellungsdesign und 1952 mit dem großen Preis der Mailänder Triennale ausgezeichnet. – *Publikationen u. a.:* „Esposizioni e Nostre", Mailand 1957; „Werbung für den Rundfunk und das Fernsehen", Würzburg, Wien 1959; „Venticinque Campagne Pubblicitarie", Mailand 1961.

**Carnase,** Tom – geb. 1939 – *Typograph, Schriftentwerfer, Lehrer, Grafik-Designer* – Nach dem Studium 1959 Eintritt in die Agentur Sudler & Hennessey Inc., New York. 1964–68 freier Gestalter, Gründung des Studios Bonder & Carnase Inc. (WTC). 1969–79 Vizepräsident und Partner in der Agentur Lubalin, Smith, Carnase Inc. 1979 Gründung des Studios Carnase Computer Typography. 1980 Mitbegründer und Präsident des World Typeface Center Inc. als unabhängige Agentur für Schriftentwurf. Herausgabe der Hauszeitschrift „Ligature". Neben dem Schriftentwurf grafische Gestaltung von Verpackungen, Ausstellungen, Erscheinungsbildern und Signets für Auftraggeber wie ABC, CBS, Coca-Cola, den Condé Nast Verlag, Doubleday Publishing, NBC. Lehrtätigkeit u. a. an der University of Cincinnati, Ohio, am Pratt Institute, New York, an der Herron School of Art, Indiana, an der Parson's School of Design, New York, am Cleveland Institute of Art, Ohio, an der University of Monterrey, Mexiko, am Rochester Institute of Technology, New York. – *Schriftentwürfe:* - Avantgarde Gothic (mit Herb Lu-

Carnase 1973 Logo

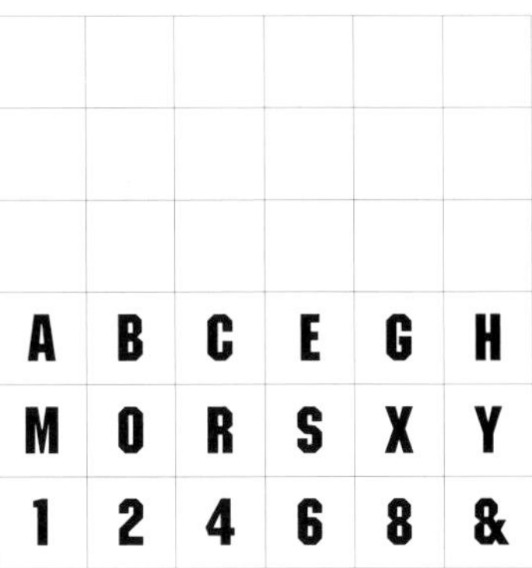


Carnase WTC Carnase

Carnase 1974 Cover

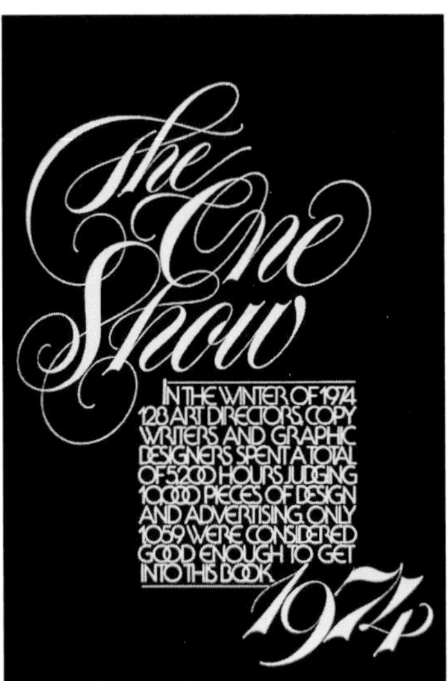


Carnase 1973 Cover

grandes entreprises italiennes (Agip, Pirelli, Motta, Barilla, Campari, Bertolli, Montecatini). 1950, obtient le prix national de Graphisme publicitaire, puis le prix national d'Architecture d'expositions en 1951; 1952, Grand Prix de la Triennale de Milan. – *Publications, sélection:* «Esposizioni e Nostre», Milan 1957; «Werbung für den Rundfunk und das Fernsehen», Würzburg, Vienne 1959; «Venticinque Campagne Pubblicitarie», Milan 1961.

**Carnase,** Tom – né en 1939 – *typographe, concepteur de polices, enseignant, graphiste maquettiste* – 1959, entre après ses études à l'agence Sudler & Hennessey Inc., New York. 1964–1968, travaille comme maquettiste indépendant, fonde le Studio Bonder & Carnase Inc. (WTC). 1969–1979, vice-président et partenaire de l'agence Lubalin, Smith, Carnase Inc. 1979, fonde le Studio Carnase Computer Typography. 1980, cofondateur et président du World Typeface Center Inc., une agence indépendante de conception de polices. Edition de la revue maison «Ligature». Outre ses activités de concepteur de polices, design d'emballages, d'expositions, d'identités et de signets pour plusieurs commanditaires, dont ABC, CBS, Coca-Cola, les publications Condé Nast, Doubleday Publishing, NBC. Enseignement à la University of Cincinnati, Ohio, au Pratt Institute, New York, à la Herron School of Art, Indiana, à la Parson's School of Design, New York, au Cleveland Institute of Art, Ohio, à la University of Monterrey, Mexique, au Rochester Institute of Technology, New York, etc. – *Polices:* Avantgarde Gothic (avec Herb Lubalin), WTC Carnase Text, WTC Favrille, WTC Goudy, WTC Our Bodoni (avec Massimo Vignelli), 223 Caslon, LSC Book, WTC Our Futura, WTC 145; avec R. Bonder: Gorilla, Grizzly, Grouch, Honda, Ma-

Motta, Barilla, Campari, Bertolli, Montecatini). 1950: is awarded the National Prize for Commercial Art. 1951: receives the National Prize for Exhibition Design. 1952: is awarded the Grand Prix at the Milan Triennale. – *Publications include:* "Esposizioni e Nostre", Milan 1957; "Werbung für den Rundfunk und das Fernsehen", Würzburg, Vienna 1959; "Venticinque Campagne Pubblicitarie", Milan 1961.

**Carnase,** Tom – b. 1939 – *typographer, type designer, teacher, graphic designer* – 1959: after completing his studies, Carnase joins the agency Sudler & Hennessey Inc. in New York. 1964–68: freelance designer. Opens the studio Bonder & Carnase Inc. (WTC). 1969–79: vice-president and partner of the agency Lubalin, Smith, Carnase Inc. 1979: opens the Carnase Computer Typography studio. 1980: co-founder and president of the World Typeface Center Inc., an independent type design agency. Publication of the in-house magazine "Ligature". Besides type design, Carnase has designed graphics for packaging, exhibitions, corporate identities and logos for numerous clients, including ABC, CBS, Coca-Cola, Condé Nast Publications, Doubleday Publishing and NBC. He has held teaching positions at the University of Cincinnati in Ohio, the Pratt Institute in New York, the Herron School of Art in Indiana, the Parson's School of Design in New York, the Cleveland Institute of Art in Ohio, the University of Monterrey in Mexico, and the Rochester Institute of Technology in New York, among others. – *Fonts:* Avant-

balin), WTC Carnase Text, WTC Favrille, WTC Goudy, WTC Our Bodoni (mit Massimo Vignelli), 223 Caslon, LSC Book, WTC Our Futura, WTC 145; zusammen mit R. Bonder: Gorilla, Grizzly, Grouch, Honda, Machine, Milano Roman, Tom's Roman und Pioneer.

**Carpenter,** Ron – geb. 1950 in Dorking, England. – *Schriftentwerfer* – 1967 Arbeit als Landkartenzeichner. 1968 Ausbildung in der Schriftenabteilung der Monotype Corporation. 1976 bei der Qualitätskontrolle der Monotype-Schriften tätig. Assistiert Robin Nicholas 1980 beim Entwurf der Kursiven seiner Schrift Nimrod. 1984 Senior Type Designer der Monotype Corporation, entwirft u. a. ergänzende Schnitte der Times New Roman. – *Schriftentwürfe:* Cantoria (1986), Calisto (1987), Amasis (1990).

**Carrà,** Carlo – geb. 11. 2. 1881 in Quargnento, Piemont, Italien, gest. 1966 in Mailand, Italien. – *Maler, Lehrer* – 1893 Lehre bei einem Dekorationsmaler. Lebt 1899–1900 in Paris, Frankreich, Arbeiten für die Dekoration der Weltausstellung. 1906 Studium an der Kunstakademie Brera in Mailand. 1910 Mitunterzeichner des „Manifests der futuristischen Maler" und des „Technischen Manifests der futuristischen Malerei". 1911–12 Zusammentreffen mit Picasso, Braque und Apollinaire in Paris. Verfaßt 1913 das Manifest „Malerei der Töne, Geräusche und Gerüche". 1916 Wechsel von der futuristischen Malerei und Bewegung zur Pittura Metafisica. 1919–22 Mitarbeiter der Zeitschrift „Valori Plastici". Ab 1922 Kunstkritiker der Mailänder Zeitung

| | n | **n** | | | *n* |
|---|---|---|---|---|---|
| a | b | e | f | g | i |
| o | r | s | t | y | z |
| A | B | C | E | G | H |
| M | O | R | S | X | Y |
| 1 | 2 | 4 | 6 | 8 | & |

Carpenter 1987 Calisto

chine, Milano Roman, Tom's Roman, Pioneer.

**Carpenter,** Ron – né en 1950 à Dorking, Angleterre – *concepteur de polices* – 1967, travaille comme cartographe. En 1968, il suit une formation au service des fontes chez Monotype Corporation. Travaille en 1976 au contrôle de qualité des polices Monotype. En 1980, il assiste Robin Nicholas lors de la conception des italiques de sa fonte Nimrod. En 1984, devenu Senior Type Designer de la Monotype Corporation, il dessine et complète les fontes du Times New Roman. – *Polices :* Cantoria (1986), Calisto (1987), Amasis (1990).

**Carrà,** Carlo – né le 11. 2. 1881 à Quargnento, Piémont, Italie, décédé en 1966, à Milan, Italie – *peintre, enseignant* – 1893, apprentissage chez un peintre décorateur. 1899–1900, vit à Paris, et travaille à la décoration de l'exposition universelle. 1906, études à l'académie des beaux-arts Brera à Milan. 1910, cosignataire du «Manifeste des peintres futuristes» et du «Manifeste technique de la peinture futuriste». 1911–1912, rencontre Picasso, Braque et Apollinaire à Paris. 1913, rédige le manifeste «Peinture des sons, bruits et odeurs». 1916, passage de la peinture futuriste à la Pittura Metafisica. 1919–1922, collaborateur de la revue «Valori Plastici». A partir de 1922, cri-

Carrà 1914–15 Cover

Carrà 1915 Cover

garde Gothic (with Herb Lubalin), WTC Carnase Text, WTC Favrille, WTC Goudy, WTC Our Bodoni (with Massimo Vignelli), 223 Caslon, LSC Book, WTC Our Futura, WTC 145; with R. Bonder: Gorilla, Grizzly, Grouch, Honda, Machine, Milano Roman, Tom's Roman, Pioneer.

**Carpenter,** Ron – b. 1950 in Dorking, England – *type designer* – 1967: works as a cartographer. 1968: trains in Monotype Corporation's type department. 1976: works on the quality control of Monotype fonts. 1980: helps Robin Nicholas to design the italics for his Nimrod typeface. 1984: senior type designer of the Monotype Corporation. Designs supplementary weights of Times New Roman. – *Fonts:* Cantoria (1986), Calisto (1987), Amasis (1990).

**Carrà,** Carlo – b. 11. 2. 1881 in Quargnento, Piedmont, Italy, d. 1966 in Milan, Italy – *painter, teacher* – 1893: trains with an interior decorator. 1899–1900: lives in Paris, France and works on the decoration for the World Exposition. 1906: studies at the Brera art academy in Milan. 1910: signs the "Manifesto of Futurist Painters" and the "Technical Manifesto of Futurist Painting". 1911–12: meets Picasso, Braque and Apollinaire in Paris. 1913: writes the manifesto "Painting with Tones, Noise and Smell". 1916: switches from Futurist painting and the Futurist movement to Metaphysical painting. 1919–22: works on the "Valori Plastici" magazine. From 1922 onwards: art critic for the Milan newspaper "L'am-

„L'ambrosiano". 1941–52 Professor für Malerei an der Kunstakademie Brera. Carràs alphabetische Phantasien eröffnen formale und inhaltliche Vorstellungen, die in der experimentellen Typographie, der Fluxus-Bewegung und im Theaterbereich bis heute Wirkung zeigen. – *Publikationen u. a.:* „Guerrapittura", Mailand 1915; „Pittura Metafisica", Florenz 1919.

**Carson,** David – geb. 8. 9. 1957 in Corpus Christi, Texas, USA. – *Typograph, Grafik-Designer, Lehrer* – 1970 Profisurfer. Beendet 1977 seine Soziologie-Studien an der San Diego State University und dem Oregon College of Commercial Art. 1979 Lehrer an der Real Life Private School in Grants Pass, Oregon. 1982–87 Lehrer für Soziologie, Psychologie, Wirtschaft und Geschichte an der Torrey Pines High School in Del Mar. 1983 dreiwöchiger Grafiklehrgang in Rapperswil, Schweiz. 1983–87 Gestaltung der Zeitschrift „Transworld Skateboarding". 1988 Gestaltung der Zeitschrift „Musician". 1989–91 Gestaltung der Zeitschrift „Beach Culture". Seit 1989 Verwendung des Computers für Layoutarbeiten. 1991–92 Neugestaltung der seit über 30 Jahren erscheinenden Zeitschrift „Surfer". 1992–95 Gestaltung der Zeitschrift „Ray Gun". 1993 als Grafik-Designer tätig für Levi's, Nike, Pepsi Cola, American Express, CITIBANK, Coca-Cola, MCI, National Bank, Sega und für die Musiker David Byrne und Prince. Ausstellung in der Neuen Sammlung, München (1995). Zahlreiche Vorträge. Carsons typographische Arbeit führt formale Experimente der „Objets trouvés"-Bewegung fort. Carson wird in

tique d'art pour le journal milanais «L'ambrosiano». 1941–1952, professeur de peinture à l'académie des beaux-arts Brera. L'imagination de Carrà en matière d'alphabets a donné naissance à des idées formelles qui ont eu un impact sur la typographie expérimentale, le mouvement Fluxus et exercent aujourd'hui encore une forte influence. – *Publications, sélection:* «Guerrapittura», Milan 1915; «Pittura Metafisica», Florence 1919.

**Carson,** David – né le 8. 9. 1957 à Corpus Christi, Texas, Etats-Unis – *typographe, graphiste maquettiste, enseignant* – 1970, surfeur professionnel. 1977, termine ses études de sociologie à la San Diego State University et à l'Oregon College of Commercial Art. 1979, enseigne à la Real Life Private School à Grants Pass, Oregon. 1982–1987, enseigne la sociologie, la psychologie, l'économie et l'histoire à la Torrey Pines High School à Del Mar. 1983, stage de graphisme pendant trois semaines à Rapperswil, Suisse. 1983–1987, maquette de la revue «Transworld Skateboarding». 1988, maquette de la revue «Musician». 1989–1991, maquette de la revue «Beach Culture». Utilise l'ordinateur pour ses travaux de maquette depuis 1989. 1991–1992, nouvelle maquette pour la revue «Surfer» qui existe depuis plus de 30 ans. 1992–1995, maquette de la revue «Ray Gun». 1993, graphiste maquettiste pour Levi's, Nike, Pepsi Cola, American Express, CITIBANK, Coca-Cola, MCI, National Bank, Sega et pour les musiciens David Byrne et Prince. Exposition à la Neue Sammlung, Munich (1995). De nombreuses conférences. Le travail typographique de Carson se situe dans le prolongement des expériences formelles du mouvement «Objets trouvés». Les travaux de Carson deviennent un modèle maintes fois copiés pendant les années 1990. – *Publication:* Lewis Blackwell

DAVID
16/21 gennaio
FORTUNY graphic
palazzo
fortuny
19CARSON
95

Carson 1995 Poster

Carson 1992 Cover

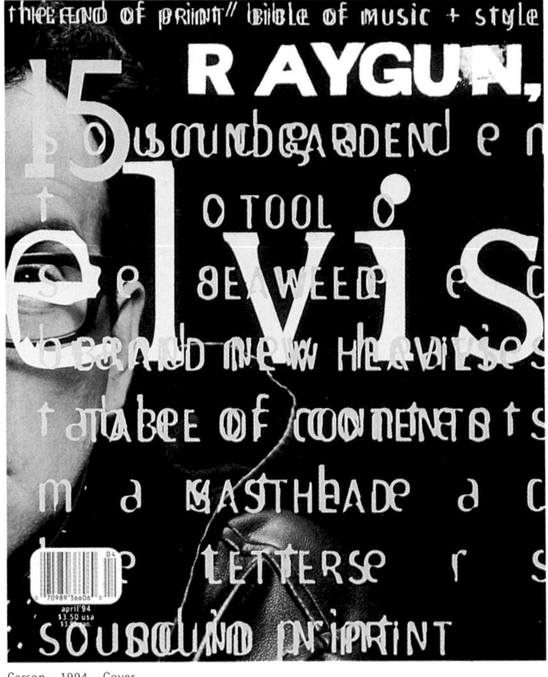


Carson 1994 Cover

brosiano". 1941–52: professor of painting at the Brera art academy. Carrà's alphabet fantasies broadened ideas concerning form and content whose influence, still evident today, extended to experimental typography, the Fluxus movement and the world of the theater. – *Publications include:* "Guerrapittura", Milan 1915; "Pittura Metafisica", Florence 1919.

**Carson,** David – b. 8. 9. 1957 in Corpus Christi, Texas, USA – *typographer, graphic designer, teacher* – 1970: professional surfer. 1977: completes his studies in sociology at the San Diego State University and the Oregon College of Commercial Art. 1979: teaches at the Real Life Private School in Grants Pass, Oregon. 1982–87: teaches sociology, psychology, economics and history at the Torrey Pines High School in Del Mar. 1983: three-week graphics course in Rapperswil, Switzerland. 1983–87: designs "Transworld Skateboarding" magazine. 1988: designs "Musician" magazine. 1989–91: designs "Beach Culture" magazine. 1989: Carson starts using computers for layout work. 1991–92: redesigns the over 30-year-old "Surfer" magazine. 1992–95: designs "Ray Gun" magazine. 1993: works as a graphic designer for Levi's, Nike, Pepsi Cola, American Express, CITIBANK, Coca-Cola, MCI, National Bank and Sega, and for the musicians David Byrne and Prince. Exhibition at the Neue Sammlung, Munich (1995). Numerous lectures. Carson's typographical work continues the formal experiments of the "Objets trouvés" movement. Carson has become a model frequently imitated in the 1990s.

den 90er Jahren zum vielkopierten Vorbild. – *Publikation:* Lewis Blackwell „David Carson – The End of Print", München 1995.

**Carter,** Matthew – geb. 1. 10. 1937 in London, England. – *Schriftentwerfer* – 1956 Lehre als Stempelschneider in den Niederlanden. 1963–65 typographischer Berater von Crossfield Electronics. 1965 Umzug in die USA, Arbeiten für Mergenthaler-Linotype in New York. 1971 Rückkehr nach England, weitere Arbeiten für Linotype. 1980–84 typographischer Berater von „Her Majesty's Stationery Office". Die Royal Society of Arts wählt ihn 1981 zum „Royal Designer for Industry". Gründet 1981 mit Mike Parker die Bitstream Inc. in Cambridge, Massachusetts, als erste unabhängige amerikanische Firma für digitale Schriftproduktion. Gründet 1992 mit Cherie Cone die Carter & Cone Type Inc.; hier entstehen die Alphabete Mantinia, Elephant, Sophia und eine neue Version der Galliard. 1995 Erscheinungsbild für das Walker Arts Center, Minneapolis, mit dem Serifen durch ein Computerprogramm an eine Grundschrift an- oder abgesetzt werden können. – *Schriftentwürfe:* Snell Roundhand (1966), Cascade Script (1966), Gando Ronde (mit H. J. Hunziker, 1970), Olympian (1970), Auriga (1970), CRT Gothic (1974), Video (1977), Bell Centennial (1978), Galliard (1978), V&A Titling (1981), Bitstream Charter (1987) sowie mehrere griechische, koreanische, hebräische und indische (Devanagari-) Schriften.

**Carter,** Sebastian – geb. 1941 in England.

Carson 1990 Cover

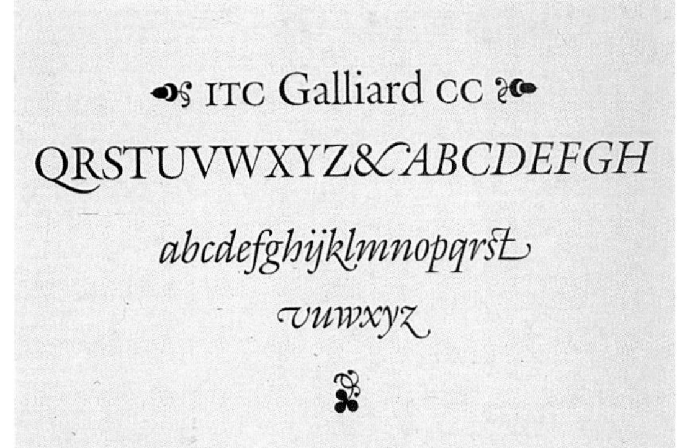

Matthew Carter 1978 Galliard

THE resemblance BETWEEN YOU AND ANDRE IS UNCANNY BECAUSE YOU BOTH ARE WEARING THE NEW AIR CHALLENGE FUTURE tennis SHOE FROM NIKE WITH THE EXOSKELETAL STRAPPING AND THE HUARACHE-FIT™ INNERBOOT system WHICH MOLDS TO YOUR FEET AND YOU BOTH ENJOY THE BETTER LATERAL MOTION BECAUSE OF THE LONGITUDINAL FLEX LINES AND HERRINGBONE OUTSOLE AND YOU SHARE AN INCREDIBLE AMOUNT OF cushioning HATS OFF TO THE NIKE-AIR® CUSHIONING IN THE HEEL AND FOREFOOT AND THERE ARE MYRIAD OTHER THINGS YOU HAVE IN COMMON LIKE THE FOOTFRAME™ DEVICE AND THE MIDFOOT tension STRAP WITH RUGGED HOOK-AND-LOOP CLOSURE FOR INSTANCE AND LET'S FACE IT IF IT WEREN'T FOR THE HAIR AND THE earring AND THE WIMBLEDON CUP YOU GUYS COULD BE, LIKE, TWINS.

NIKE AIR

Carson 1993 Advertisement

| n | n | **n** | | | |
|---|---|---|---|---|---|
| a | b | e | f | g | i |
| o | r | s | t | y | z |
| A | B | C | E | G | H |
| M | O | R | S | X | Y |
| 1 | 2 | 4 | 6 | 8 | & |

Matthew Carter 1978 Bell Centennial

| | n | **n** | **n** | | *n* |
|---|---|---|---|---|---|
| a | b | e | f | g | i |
| o | r | s | t | y | z |
| A | B | C | E | G | H |
| M | O | R | S | X | Y |
| 1 | 2 | 4 | 6 | 8 | & |

Matthew Carter 1987 Charter

«David Carson – The End of Print», Munich 1995.

**Carter,** Matthew – né le 1. 10. 1937 à Londres, Angleterre – *concepteur de polices* – 1956, apprentissage en taille de types aux Pays-Bas. 1963–1965, conseiller typographe chez Crossfield Electronics. 1965, s'installe aux Etats-Unis, travaille pour Mergenthaler-Linotype à New York. 1971, retour en Angleterre où il poursuit des activités pour Linotype. 1980–1984, conseiller typographique de «Her Majesty's Stationery Office». 1981, la Royal Society of Arts l'élit «Royal Designer for Industry». En 1981, il fonde la Bitstream Inc. à Cambridge, Massachusetts, avec Mike Parker; c'est la première société américaine indépendante de production de polices numériques. En 1992, il fonde la Carter & Cone Type Inc. avec Cherie Cone, où sont conçus les alphabets Mantinia, Elephant, Sophia ainsi qu'une nouvelle version du Galliard. 1995, identité pour le Walker Arts Center de Minneapolis où les sérifs peuvent être ajoutés ou supprimés d'une fonte de base au moyen d'un logiciel. – *Polices:* Snell Roundhand (1966), Cascade Script (1966), Gando Ronde (avec H. J. Hunziker, 1970), Olympian (1970), Auriga (1970), CRT Gothic (1974), Video (1977), Bell Centennial (1978), Galliard (1978), V&A Titling (1981), Bitstream Charter (1987), ainsi que plusieurs alphabets grecs, coréens, hébreux et indiens (Devanagari).

**Carter,** Sebastian – né en 1941 en An-

– *Publication:* Lewis Blackwell "David Carson – The End of Print", Munich 1995.

**Carter,** Matthew – b. 1. 10. 1937 in London, England – *type designer* – 1956: trains as a punch cutter in the Netherlands. 1963–65: typography consultant for Crossfield Electronics. 1965: moves to the USA and works for Mergenthaler-Linotype in New York. 1971: moves back to England where he continues to work for Linotype. 1980–84: typography consultant to Her Majesty's Stationery Office. 1981: the Royal Society of Arts makes him a Royal Designer for Industry. 1981: he and Mike Parker set up Bitstream Inc. in Cambridge, Massachusetts, the first independent American company to manufacture digital type. 1992: founds Carter & Cone Type Inc. with Cherie Cone, where the alphabets Mantinia, Elephant, Sophia and a new version of Galliard are produced. 1995: corporate identity for the Walker Arts Center, Minneapolis, where serifs can be added to or removed from the base type using a computer program. – *Fonts:* Snell Roundhand (1966), Cascade Script (1966), Gando Ronde (with H. J. Hunziker, 1970), Olympian (1970), Auriga (1970), CRT Gothic (1974), Video (1977), Bell Centennial (1978), Galliard (1978), V&A Titling (1981), Bitstream Charter (1987), as well as various Greek, Korean, Hebrew and Indian (Devanagari) typefaces.

**Carter,** Sebastian – b. 1941 in England –

– *Grafik-Designer, Buchgestalter, Autor* – Ausbildung am King's College, Cambridge. Ab 1963 Arbeiten für Verlage in Paris und London, Mitarbeit bei der Rampant Lions Press seines Vaters Will Carter in Cambridge. 1971 Francis Minns Award der National Book League. 1983–84 Präsident des Double Crown Club. – *Publikationen u. a.:* „First Principles of Typography in the 1970s", London 1973; „Twentieth-Century Type Designers", London 1987, sowie zahlreiche Aufsätze in Zeitschriften.

**Carter,** Will – geb. 24. 9. 1912 in Slough bei London, England. – *Schriftentwerfer, Drucker* – 1930–32 Ausbildung in der Druckerei Unwin Brothers in Woking. 1938 Tätigkeit in der Werkstatt von Paul Koch in Frankfurt am Main, Freundschaft mit Hermann Zapf. Lernt 1948 das Stempelschneiden bei David Kindersley. 1949 Gründung der Rampant Lions Press in Cambridge. – *Schriftentwürfe:* Klang (1955), Dartmouth (1961), Octavian (mit D. Kindersley, 1961).

**Casey,** Jaqueline S. – geb. 1927 in Quincy, USA. – *Grafik-Designerin, Typographin* – Studium am Massachusetts College of Art. Seit 1955 Tätigkeit als Design-Direktorin des Design Services des Massachusetts Institute of Technology in Cambridge, Boston. 1982 wird das MIT für die Arbeit von J. S. Casey vom American Institute of Graphic Arts (AIGA) mit dem Design Leadership Award ausgezeichnet.

**Caslon I.,** William – geb. 1692 in Cradley, Worcestershire, England, gest. 1766 in Bethnal Green, England. – *Graveur,*

*An aquaduct had the advantage of reaching the high parts of a city while carrying a greater volume of water than a pipe. Pliny shows admirable thrift in suggesting the re-use of much of the previous two unfinished projects, which had been abandoned because of engineering failures and, Pliny suspected, corruption.*

from which it appears that the water must be brought, as was attempted the first time, on an arched structure, so that it may reach not just the flat and low-lying parts of the city. A very few arches are still standing: some can also be built up from the dressed stone which was pulled down from the earlier structure; some part of it, in my judgment, should be made of brick-work, for this would be both easier and cheaper. But what is needed above all is for you to send out a water engineer or an architect, so that what has happened may not recur. This one thing I assert, that both the usefulness and the beauty of the work are worthy of your age.

38. The effort must be made to bring water to the city of Nicomedia. I am truly confident that you will approach this task with the diligence which you ought to show. But, by heaven, it is also your duty diligently to investigate whose fault it is that the people of Nicomedia have until now wasted so much money, in case it was in the course of doing each other favours that they began and abandoned aquaducts. What you thus discover, bring to my attention.

*Nicomedia lay at the head of the first major bay below the Bosphorus, near what is now called Izmit. It was founded in 264 BC by King Nicomedes I as his capital. Under the Romans it was the legal centre of Bithynia, and claimed to be the first city though, as the Roman governors were itinerant, there was no true provincial capital.*

39. A theatre at Nicaea, Sir, most of which has already been built, though it is still incomplete, has swallowed up more than ten million sesterces (so I am informed; for the balance-sheet for the project has not yet been examined); I fear it may have been in vain. For it is sinking and gapes with cracks, because either the soil is wet and spongy, or the stone itself is soft and crumbling:

*Nicaea was founded in 301 BC by King Lysimachus and named after his first wife. It was situated at the head of the present-day lake of Iznik about forty-five kilometres south-west of Nicomedia, with which city it strove for first place in wealth and power.*

Sebastian Carter 1995 Page

Twentieth century TYPE designers
SEBASTIAN CARTER

Sebastian Carter 1987 Cover

Will Carter 1955 Klang

Casey 1980 Poster

Two Lines Great Primer.

Quoufque tandem abutere Catilina, p
*Quoufque tandem a-butere, Catilina, pa-*

Caslon 1763 Specimen

gleterre – *graphiste maquettiste, maquettiste en édition, auteur* – formation au King's College à Cambridge. Travaille pour des éditeurs à Paris et à Londres dès 1963. Employé chez les Rampant Lions Press de son père Will Carter à Cambridge. 1971, Francis Minns Award de la National Book League. 1983–1984, président du Double Crown Club. – *Publications, sélection:* «First Principles of Typography in the 1970s», Londres 1973; «Twentieth-Century Type Designers», Londres 1987 ainsi que de nombreux articles dans plusieurs revues.

**Carter,** Will – né le 24. 9. 1912 à Slough, près de Londres, Angleterre – *concepteur de polices, imprimeur* – 1930–1932, formation à l'imprimerie Unwin Brothers à Woking. 1938, travaille dans l'atelier de Paul Koch à Francfort-sur-le-Main, se lie d'amitié avec Hermann Zapf. Il apprend la taille des types chez David Kindersley en 1948. Création de la Rampant Lions Press à Cambridge en 1949. – *Polices:* Klang (1955), Dartmouth (1961), Octavian (avec D. Kindersley, 1961).

**Casey,** Jaqueline S. – née en 1927 à Quincy, Etats-Unis – *graphiste typographe* – études au Massachusetts College of Art. A partir de 1955, directrice du service design au Massachusetts Institute of Technology à Cambridge, Boston. En 1982, le American Institute of Graphic Arts (AIGA) décerne le Design Leadership Award au MIT pour les travaux de J. S. Casey.

**Caslon I,** William – né en 1692 à Cradley, Worcestershire, Angleterre, décédé en 1766 à Bethnal Green, Angleterre – *graveur, tailleur de types, concepteur de polices* – en 1706, il commence un apprentissage de sept ans comme graveur chez un fabricant de harnais à Londres. 1716, graveur indépendant. En 1721, la «Society for Promoting Christian Know-

*graphic designer, book designer, author* – Educated at King's College, Cambridge. 1963: starts producing work for publishing houses in Paris and London and starts working for his father Will Carter's Rampant Lions Press in Cambridge. 1971: Francis Minns Award from the National Book League. 1983–84: president of the Double Crown Club. – *Publications include:* "First Principles of Typography in the 1970s", London 1973; "Twentieth-Century Type Designers", London 1987 and numerous essays in periodicals.

**Carter,** Will – b. 24. 9. 1912 in Slough near London, England – *type designer, printer* – 1930–32: trains at the Unwin Brothers printing works in Woking. 1938: works at the Paul Koch workshop in Frankfurt am Main; becomes friends with Hermann Zapf. 1948: studies punch cutting with David Kindersley. 1949: founds the Rampant Lions Press in Cambridge. – *Fonts:* Klang (1955), Dartmouth (1961), Octavian (with D. Kindersley, 1961).

**Casey,** Jaqueline S. – b. 1927 in Quincy, USA – *graphic designer, typographer* – studied at the Massachusetts College of Art. 1955: starts working as design director for the Massachusetts Institute of Technology's (MIT) Design Services in Cambridge, Boston. 1982: the MIT is awarded the Design Leadership Award by the American Institute of Graphic Arts (AIGA) for J. S. Casey's work.

**Caslon I,** William – b. 1692 in Cradley, Worcestershire, England, d. 1766 in Bethnal Green, England – *engraver, type founder, type designer* – 1706: begins a seven-year apprenticeship as an engraver

*Schriftschneider, Schriftentwerfer* – Beginnt 1706 eine siebenjährige Lehre als Graveur bei einem Geschirrmacher in London. 1716 selbständiger Graveur. Die „Society for Promoting Christian Knowledge" gibt Caslon 1721 den Auftrag, arabische Alphabete zu schneiden. 1725 Gründung einer eigenen Schriftgießerei. 1734 erscheint seine erste Ein-Blatt-Schriftprobe, in der 47 seiner Schriftschnitte abgedruckt sind. 1737 Umzug der Gießerei in die Chiswell Street in London, wo sie 200 Jahre lang arbeitet.

**Caslon II.**, William – geb. 1720, gest. 1778 – *Schriftschneider, Schriftentwerfer* – 1742 Eintritt in die Firma seines Vaters. Vater und Sohn veröffentlichen 1763 das erste englische Schriftmusterbuch, in dem 56 Alphabete des Vaters und 27 des Sohnes, die er zwischen 1738 und 1763 geschnitten hat, zu sehen sind. 1766 Caslon übernimmt die Geschäfte nach dem Tod des Vaters bis 1778.

**Caslon III.**, William – geb. 1754, gest. 1833 – *Schriftschneider* – Caslon III. verkauft seine Geschäftsanteile an seine Mutter und Schwiegertochter, um mit dem Geld die Schriftgießerei Jackson zu kaufen. Bis 1795 bleibt die Schriftgießerei im Besitz der Familie Caslon.

**Caslon IV.**, William – geb. 1780, gest. 1869 – *Schriftschneider* – 1807 Caslon IV. übernimmt die Geschäfte der Gießerei bis 1819, als die Firma von Blake, Garnett & Co. aufgekauft wird. 1837 Die Schriftgießerei, die noch immer den Namen Caslon trägt, geht in den Besitz der Schriftgießerei Stephenson, Blake & Co. in Sheffield über.

A SPECIMEN

By W. CASLON, Letter-Founder, in Ironmonger-Row, Old-Street, LONDON.

ABCD
ABCDE
ABCDEFG
ABCDEFGHI
ABCDEFGHIJK
ABCDEFGHIJKL
ABCDEFGHIKLMN

French Cannon.

Quousque tandem abutere, Catilina, pati-

*Quousque tandem abutere, Catilina, patientia nostra?*

Two Lines Great Primer.

Quousque tandem abutere, Catilina, patientia nostra? quamdiu nos etiam

*Quousque tandem abutere, Catilina, patientia nostra? quamdiu nos etiam furor*

Two Lines English.

Quousque tandem abutere, Catilina, patientia nostra? quamdiu nos etiam furor iste tuus elu-

*Quousque tandem abutere, Catilina, patientia nostra? quamdiu nos etiam furor*

Double Pica Roman.

Quousque tandem abutere, Catilina, patientia nostra? quamdiu nos etiam furor iste tuus eludet? quem ad finem sese effrenata jactabit audacia?
ABCDEFGHJIKLMNOP

*Double Pica Italick.*

*Quousque tandem abutere, Catilina, patientia nostra? quamdiu nos etiam furor iste tuus eludet? quem ad finem sese effrenata jactabit audacia?*
*ABCDEFGHJIKLMNO*

Great Primer Roman.

Great Primer Italick.

English Roman.

English Italick.

Pica Roman.

Pica Italick.

Small Pica Roman. No 1.

Small Pica Italick. No 1.

Small Pica Roman. No 2.

Small Pica Italick. No 2.

Long Primer Roman. No 1.

Long Primer Italick. No 1.

Long Primer Roman. No 2.

Long Primer Italick. No 2.

Brevier Roman.

Brevier Italick.

Pica Black.

Brevier Black.

Pica Gothick.

Pica Coptick.

Pica Armenian.

English Syriack.

Pica Samaritan.

English Arabick.

Hebrew with Points.

Hebrew without Points.

Brevier Hebrew.

English Greek.

Pica Greek.

Long Primer Greek.

Brevier Greek.

Long Primer Saxon.

Pica Saxon.

Caslon I. 1734 Specimen

Two Lines Great Primer Black.

And be it further hereby enacted, That

Double Pica Black.

And be it further hereby enacted, That the Mayors, Bailiffs, or other head Officers,

Great Primer Black.

And be it further hereby enacted, That the Mayors, Bailiffs, or other head Officers of every Town and place corporate, and City wit-

English Black.

And be it further hereby enacted, That the Mayors, Bailiffs, or other head Officers of every Town and place corporate, and City within this Realm, being Justice or Justices of Peace, shall have the same authority by vertue of this Act, within the limits and precincts of their Ju-

English Black. No 2.

And be it further hereby enacted, That the Mayors, Bailiffs, or other head Officers of every Town and place corporate, and City within this Realm, being Justice or Justices of Peace, shall have the same authority, by vertue of this Act, within the limits and pre-

Caslon II. 1763 Specimen

ledge» commande à Caslon la taille d'un alphabet arabe. Il crée sa propre fonderie de caractères en 1725. En 1734, il publie un premier échantillon d'une fonte sur une page, sur laquelle sont imprimés 47 de ses types. 1737, la fonderie s'installe Chiswell Street à Londres, où elle produira pendant deux siècles.

**Caslon II**, William – né en 1720, décédé en 1778 – *tailleur de types, concepteur de polices* – entre dans l'entreprise paternelle en 1742. 1763, le père et le fils publient le premier catalogue anglais de polices où figurent 56 alphabets du père et 27 du fils, tous taillés entre 1738 et 1763. A la mort de son père en 1766, il reprend l'affaire qu'il dirige jusqu'en 1778.

**Caslon III**, William – né en 1754, décédé en 1833 – *tailleur de types* – Caslon III vend ses parts sociales à sa mère et à sa belle-sœur pour acheter la fonderie de caractères Jackson. Cette fonderie de caractères restera entre les mains de la famille Caslon jusqu'en 1795.

**Caslon IV**, William – né en 1780, décédé en 1869 – *tailleur de types* – Caslon IV reprend la fonderie de caractères en 1807 et la dirige jusqu'en 1819, date à laquelle elle est rachetée par Blake, Garnett & Co. En 1837, la fonderie, qui porte encore le nom de Caslon, est rachetée par la fonderie de caractères Stephenson, Blake & Co. à Sheffield.

| | n | **n** | **n** | | *n* |
|---|---|---|---|---|---|
| a | b | e | f | g | i |
| o | r | s | t | y | z |
| A | B | C | E | G | H |
| M | O | R | S | X | Y |
| 1 | 2 | 4 | 6 | 8 | & |

Caslon

with a London harness-maker. 1716: self-employed engraver. 1721: the Society for Promoting Christian Knowledge commissions Caslon to cast Arabic alphabets. 1725: sets up his own type foundry. 1734: Caslon's first one-page specimen is produced which illustrates 47 of his typefaces. 1737: the type foundry moves to Chiswell Street in London, where it continues to operate for 200 years.

**Caslon II**, William – b. 1720, d. 1778 – *type founder, type designer* – 1742: joins his father's company. 1763: father and son issue the first English book of type specimens, which includes 56 alphabets by Caslon senior and 27 by his son, designed between 1738 and 1763. 1766: after the death of his father, Caslon junior runs the family business until 1778.

**Caslon III**, William – b. 1754, d. 1833 – *type founder* – Caslon III sold his share of the business to his mother and daughter-in-law and used the money to buy the Jackson type foundry. The type foundry remained the property of the Caslon family until 1795.

**Caslon IV**, William – b. 1780, d. 1869 – *type founder* – 1807: Caslon IV takes over the running of the type foundry until 1819, when the foundry is bought by Blake, Garnett & Co. 1837: the type foundry, still under the name of Caslon, becomes the property of the Stephenson, Blake & Co. type foundry in Sheffield.

**Cassandre,** A. M. (d. i. Adolphe Jean-Marie Mouron) – geb. 24. 1. 1901 in Charkow, Ukraine, gest. 17. 6. 1968 in Paris, Frankreich. – *Grafik-Designer, Schriftentwerfer, Maler, Bühnenbildner, Lehrer* – 1918 Studium an der Ecole des Beaux-Arts, an der Académie Julian und im Atelier des Malers Lucien Simon in Paris. Ab 1921 Entwurf von Plakaten. 1922 erstes Atelier am Montparnasse in Paris, signiert seine Arbeiten erstmals mit dem Pseudonym „Cassandre". Ab 1930 Arbeiten für die Firma Nicolas. 1933 erste Arbeiten als Bühnenbildner. Lehrt an der Ecole Nationale des Arts Décoratifs. 1934–35 Lehrt an der Ecole d'Arts Graphiques. 1935 Plakataufträge für französische, schweizer und italienische Auftraggeber. 1936 Vertrag mit der Zeitschrift „Harper's Bazaar", New York, über mehrere Titelbilder. 1939 Rückkehr nach Frankreich, Arbeiten für Theater, Ballett sowie freie Malerei. 1948 Mitbegründer der AGI (Alliance Graphique International). Es entstehen bis 1959 Bühnenbilder für Aix-en-Provence, die Comédie Française, die bayerische Staatsoper München und für Bühnen in Florenz. 1962 Ernennung zum Offizier der Ehrenlegion. 1963 Entwurf des Signets für Yves Saint-Laurent. Cassandres Verbindung von angewandter und freier Gestaltung gab ihm über Jahrzehnte eine Vorbildrolle im Grafik-Design. Ausstellungen u. a. 1936 Retrospektive Plakatausstellung im Museum of Modern Art, New York, 1942 Galerie René Drouin (Malerei), 1966 Galerie Motte, Genf, 1966 Galerie Janine Hao, Paris, 1967 Rijksakademie Amsterdam.

**Cassandre,** A. M. (Adolphe Jean-Marie Mouron) – né le 24. 1. 1901 à Kharkov, Ukraine, décédé le 17. 6. 1968 à Paris, France – *graphiste maquettiste, concepteur de polices, peintre, scénographe, enseignant* – 1918, études à l'Ecole des Beaux-Arts, à l'Académie Julian et à l'atelier du peintre Lucien Simon à Paris. Commence à créer des affiches à partir de 1921. Premier atelier à Montparnasse en 1922; il signe pour la première fois ses travaux de son pseudonyme «Cassandre». Travaille pour la société Nicolas à partir de 1930. Premières scénographies en 1933. Enseigne à l'Ecole Nationale des Arts Décoratifs. 1934–1935, enseigne à l'Ecole d'Arts Graphiques. En 1935, il réalise des affiches pour des commanditaires français, suisses et italiens. En 1936, il signe un contrat avec le magazine «Harper's Bazaar», New York, et crée plusieurs couvertures du magazine. 1939, retour en France, travaux pour le théâtre, le ballet et comme artiste peintre indépendant. 1948, cofondateur de l'AGI (Alliance Graphique International). Jusqu'en 1959, il crée des décors pour Aix-en-Provence, la Comédie Française, le bayerische Staatsoper de Munich et les scènes de Florence. Décoré de la Légion d'honneur en 1962. Dessine le signet d'Yves Saint-Laurent en 1963. La manière dont Cassandre alliait le graphisme publicitaire à la création libre lui a fait jouer un rôle déterminant pendant plusieurs décennies. Expositions: 1936, exposition rétrospective d'affiches au Museum of Modern Art, New York; 1942, Galerie René Drouin (peintures); 1966, Galerie Motte, Genève; 1966, Galerie Janine Hao, Paris; 1967, Rijksakademie, Amsterdam. – *Polices:* Bifur (1929), Acier (1930), Acier Noir (1936), Peignot (1937), Touraine (avec Charles Peignot, 1947), Cassandre (1968). – *Publications, sélection:* R. K.

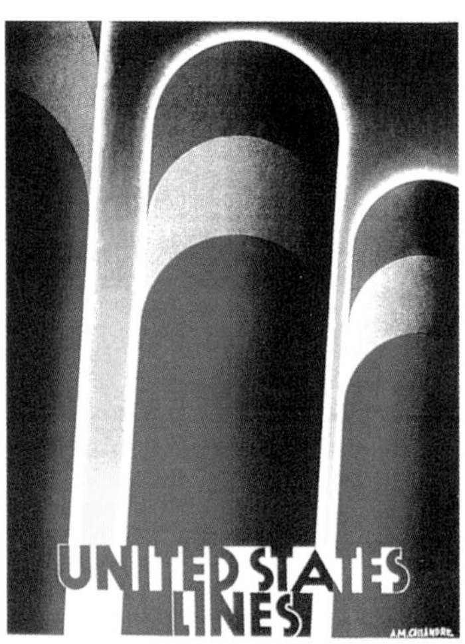

UNITED STATES LINES
LONDON 14 Regent Street
PARIS 10 Rue Auber
BREMEN 43 An der Weide


Cassandre 1954 Poster

Cassandre 1954 Poster

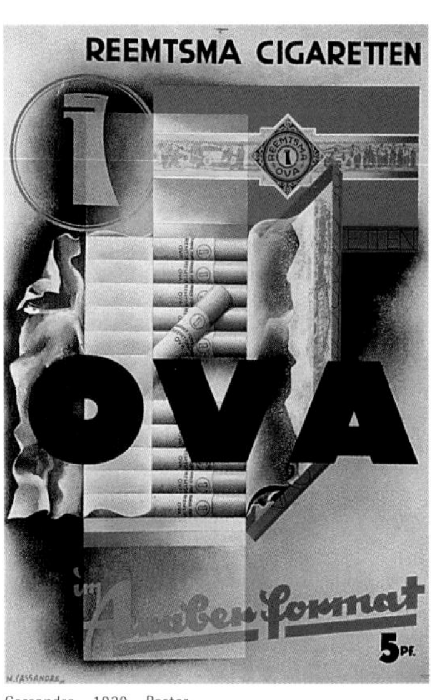


Cassandre 1929 Poster

Cassandre 1929 Bifur

**Cassandre,** A. M. (real name: Adolphe Jean-Marie Mouron) – b. 24. 1. 1901 in Kharkov, Ukraine, d. 17. 6. 1968 in Paris, France – *graphic designer, type designer, painter, set-designer, teacher* – 1918: studies at the Ecole des Beaux-Arts, at the Académie Julian and with painter Lucien Simon in his studio in Paris. 1921: starts designing posters. 1922: has his first studio in Montparnasse in Paris, signs his works for the first time with the pseudonym "Cassandre". 1930: starts producing work for the Nicolas company. 1933: produces his first work as a set-designer. Teaches at the Ecole Nationale des Arts Décoratifs. 1934–35: teaches at the Ecole d'Arts Graphiques. 1935: designs posters for French, Swiss and Italian clients. 1936: contract with "Harper's Bazaar" magazine, New York, for various covers. 1939: returns to France. Works for the theater and the ballet and produces work in the field of free painting. 1948: co-founder of the AGI (Alliance Graphique International). Until 1959 he produces sets for Aix-en-Provence, the Comédie Française, the Bavarian State Opera in Munich and for stages in Florence. 1962: is made an officer of the Legion of Honor. 1963: designs Yves Saint-Laurent's trademark. With his combination of applied and free design, Cassandre was a model in the field of graphic design for several decades. Exhibitions: retrospective poster exhibition at the Museum of Modern Art, New York (1936), Galerie René Drouin (paintings, 1942), Galerie Motte, Geneva (1966), Galerie Janine Hao, Paris (1966), Rijksakademie in Amsterdam (1967). –

– *Schriftentwürfe:* Bifur (1929), Acier (1930), Acier Noir (1936), Peignot (1937), Touraine (mit Charles Peignot, 1947), Cassandre (1968). – *Publikationen u. a.:* R. K. Brown, S. Reinhold „The Poster Art of A. M. Cassandre", New York 1979; Henri Mouron „Cassandre", München 1985.

**Cato,** Ken – geb. 30. 12. 1946 in Brisbane, Australien. – *Typograph, Grafik-Designer* – Studium am Royal Melbourne Institute of Technology, danach Mitarbeit in mehreren Design-Studios. 1970 Gründung der Agentur Cato Hibberd Design mit seinem englischen Partner Terry Hibberd. Umwandlung der Agentur in Cato Design mit Büros in Sydney, Tokio, Singapur, Hong Kong, Auckland und Los Angeles. Arbeitsgebiete u. a., Erscheinungsbilder und Editorial Design für internationale Auftraggeber. 1980 Gründungsmitglied der „Australian Writers and Art Directors Association". – *Publikationen:* „The View from Australia", Tokio 1986; „Design for Business", Tokio 1987; „First Choice", Tokio 1989; „Graphics in the Third Dimension", Tokio 1992.

**Caxton,** William – geb. um 1420 in Kent, England, gest. 1491 in London, England. – *Typograph, Schriftentwerfer, Drucker* – Lehre in einem Londoner Tuchhandelsgeschäft, danach über 30 Jahre in Brügge, Flandern tätig. Später im Hofdienst der Herzogin von York. Bei wiederholten Aufenthalten in Köln lernt er die Kunst des Buchdrucks. Er bringt die Kenntnisse des Setzens mit beweglichen Einzelbuchstaben als Erster nach England. 1475

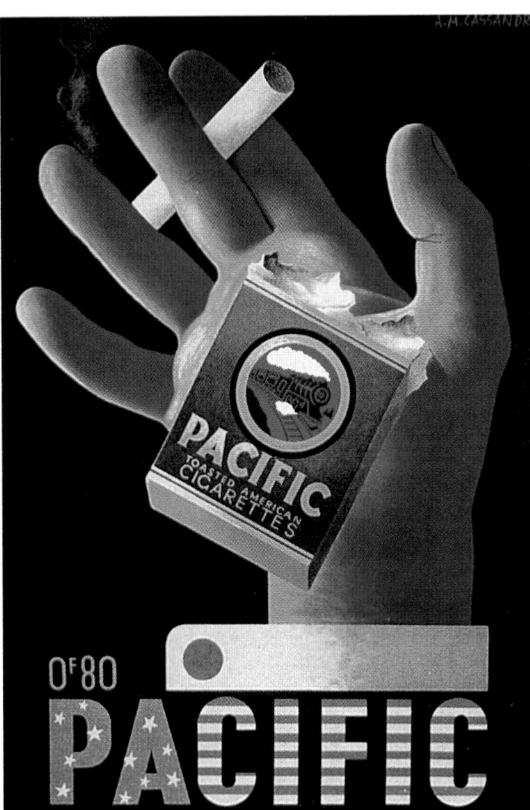


Cassandre 1935 Poster

Cato Logo

| N | N | N | | | |
|---|---|---|---|---|---|
| A | b | E | f | q | i |
| O | R | S | T | y | Z |
| A | B | C | E | G | H |
| M | O | R | S | X | Y |
| 1 | 2 | 4 | 6 | 8 | & |

Cassandre 1937 Peignot

Here begynneth the boke intituled Eracles, and also of Godefrey of Boloyne, the whiche speketh of the Conquest of the holy londe of Iherusalem, conteynyng diuerse warres and noble faytes of Armes made in the same Royāme, and in the contrees adiacent And also many merueyllous werkes happed and fallen as wel on this syde, as in tho partyes this tyme duryng, And how the valyant duc Godefrey of Bolowne conquerd with the swerd the sayd Royamme, And was kynge there,

The ffirst chapitre treateth how Eracles conquerd Perse and slewe Cosdroe, and brought in to Iherusalem the very crosse, capitulo primo,

The Auncyent hystoryes saye that Eracles was a good crysten man and gouernour of thempyre of Rome, But in his tyme Machomet had ben whiche was messager of the deuil And made the peple to vnderstonde, that he was a prophete sente from our lorde, In the tyme of Eracles was the fals lawe of machomet sowen and sprad abrode in many partyes of thoryent, and

Caxton 1481 Page

Brown, S. Reinhold «The Poster Art of A. M. Cassandre, New York 1979; Henri Mouron «Cassandre», Munich 1985.

**Cato,** Ken – né le 30. 12. 1946 à Brisbane, Australie – *typographe, graphiste maquettiste* – études au Royal Melbourne Institute of Technology, puis emplois dans divers ateliers de design. En 1970, il fonde l'agence Cato Hibberd Design avec Terry Hibberd, son partenaire anglais. L'agence devient ensuite Cato Design avec des filiales à Sydney, Tokyo, Singapour, Hong-Kong, Auckland et Los Angeles. Spécialités : identités et design éditorial pour des commanditaires internationaux. 1980, membre fondateur de l'«Australian Writers and Art Directors Association». – *Publications :* «The View from Australia», Tokyo 1986; «Design for Business», Tokyo 1987; «First Choice», Tokyo 1989; «Graphics in the Third Dimension», Tokyo 1992.

**Caxton,** William – né vers 1420 à Kent, Angleterre, décédé en 1491 à Londres, Angleterre – *typographe, concepteur de polices, imprimeur* – apprentissage chez un drapier londonien, puis plus de 30 années d'activités à Bruges, Flandres. Il entre ensuite au service de la duchesse d'York. Apprend l'art du livre au cours de plusieurs séjours à Cologne. Il était le premier à introduire la technique de composition au moyen de caractères mobiles en Angleterre. En 1475, Caxton travaille dans l'imprimerie de Colard Mansion à Bruges et imprime les deux premiers ou-

*Fonts:* Bifur (1929), Acier (1930), Acier Noir (1936), Peignot (1937), Touraine (with Charles Peignot, 1947), Cassandre (1968). – *Publications include:* R. K. Brown, S. Reinhold "The Poster Art of A. M. Cassandre", New York 1979; Henri Mouron "Cassandre", Munich 1985.

**Cato,** Ken – b. 30. 12. 1946 in Brisbane, Australia – *typographer, graphic designer* – Studied at the Royal Melbourne Institute of Technology and then went on to work for several design studios. 1970: opens the agency Cato Hibberd Design with his English partner, Terry Hibberd. The agency is later restructured as Cato Design and opens offices in Sydney, Tokyo, Singapore, Hong Kong, Auckland and Los Angeles. Areas of work are : corporate identities and editorial design for international clients. 1980: founder member of the Australian Writers and Art Directors Association. – *Publications:* "The View from Australia", Tokyo 1986; "Design for Business", Tokyo 1987; "First Choice", Tokyo 1989; "Graphics in the Third Dimension", Tokyo 1992.

**Caxton,** William – b. c. 1420 in Kent, England, d. 1491 in London, England – *typographer, type designer, printer* – Trained with a London cloth merchant and then spent over 30 years working in Bruges, Flanders. He later worked at court for the Duchess of York. Through his various stays in Cologne, he learned the art of book printing. He was the first to introduce the knowledge of how to set print with moveable type to England. 1475: Caxton prints two books at Colard Mansion's printing press in Bruges, the

druckt Caxton in Brügge die ersten zwei Bücher in der Druckerei von Colard Mansion, die ersten Bücher, die in englischer Sprache gedruckt sind. 1476 Einrichtung einer Werkstatt in Westminster, wo er neben der Druckerei eine Schreibwerkstatt betreibt. Hier entstehen 1477 „The Dictes or Sayengis of the Philosophres", das erste in England mit Datum und Ortsangabe gedruckte Buch. – *Publikationen u. a.:* William Blades „The Life and Typography of William Caxton, England's First Printer", 2 Bände, London 1861–63.

**Central Lettering Record**, London – *Schriftarchiv* – Nicolete Gray und Nicholas Biddulph legten am Central Saint Martins College of Art and Design in London eine Sammlung von 10 000 Fotografien und 3 000 Dias über Schrift und angrenzende Bereiche an. Ein Teil der Sammlung umfaßt Schriftprospekte und Schrift-Werbematerial seit den 50er Jahren. In einem zweijährigen Forschungsprogramm, das Phil Baines leitet, wird die Sammlung katalogisiert und zugänglich gemacht.

**Central Saint Martins College of Art and Design,** London – 1845 Gründung der Saint Martins School of Art in London. 1896 Gründung der Central School of Arts and Crafts (ab 1966 Arts and Design) in London. 1984 Zusammenlegung der Grafik-Design-Kurse beider Schulen. 1989 Zusammenlegung beider Schulen, die seit ihrer Gründung großen Einfluß auf die britische Kunst- und Designausbildung ausüben.

**Cerri,** Pierluigi – geb. 21. 3. 1939 in Orta San Guilo, Italien. – *Grafik-Designer.* –

vrages publiés en langue anglaise. 1476, installation d'un atelier à Westminster où il exploite une imprimerie et un atelier d'écriture. C'est là qu'en 1477, il édite «The Dictes or Sayengis of the Philosophres», le premier livre imprimé en Angleterre comportant la mention du lieu et de la date de fabrication. – *Publications, sélection* – William Blades «The Life and Typography of William Caxton, England's First Printer», 2 vol., Londres 1861–1863.

**Central Lettering Record**, Londres – *archive des fontes* – Nicolete Gray et Nicholas Biddulph ont réuni une collection comprenant 10 000 photographies et 3 000 diapositives concernant l'écriture et les domaines annexes au Central Saint Martins College of Art and Design de Londres. Une partie de ce matériel concerne les prospectus et la publicité depuis les années 50. Phil Baines dirige un programme de recherches de deux ans afin d'établir le catalogue de cette collection et de la rendre accessible.

**Central Saint Martins College of Art and Design**, Londres – 1845, fondation de la Saint Martins School of Art à Londres. 1896, fondation de la Central School of Arts and Crafts (Arts and Design à partir de 1966) à Londres. 1984, fusion des cours de design publicitaire de ces deux écoles. 1989, fusion des deux écoles, qui exercent une forte influence sur l'enseignement des arts et du design en Angleterre depuis leur création.

**Cerri**, Pierluigi – né le 21. 3. 1939 à Orta San Guilo, Italie – *graphiste maquettiste* – études d'architecture à l'université de Milan. En 1974, il fonde l'agence «Gregotti Associati» (avec Vittorio Gregotti et Augusto Cagnardi) à Milan. Commanditaires : Ferrari, RAI, Fontana Arte, Unifor, B & B Italia, etc. Dessine le signet de la 18e Triennale de Milan en 1985. Ouvre une agence à Venise en 1986. Architec-

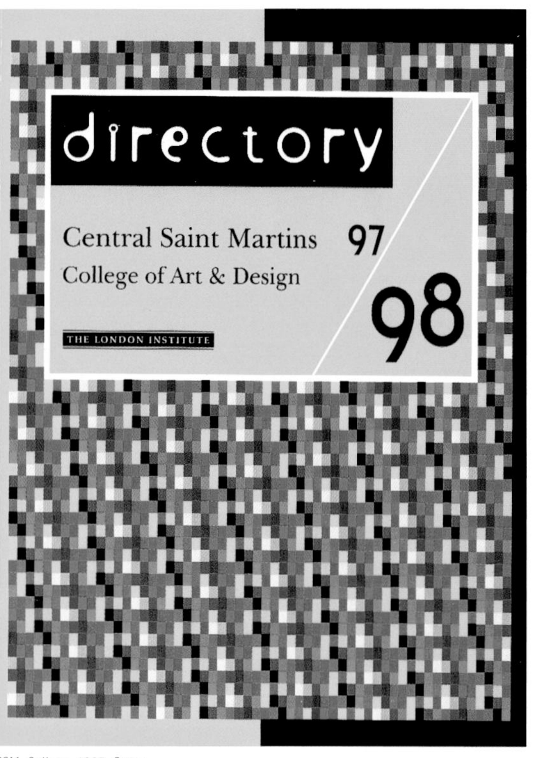

CSM College 1997 Cover

Cerri 1986 Cover

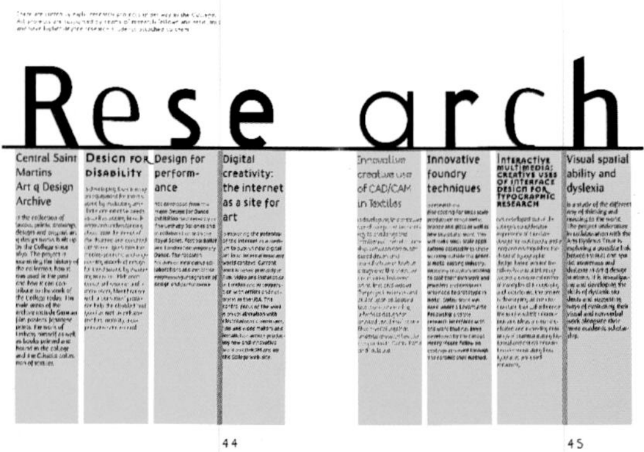

CSM College 1997 Spread

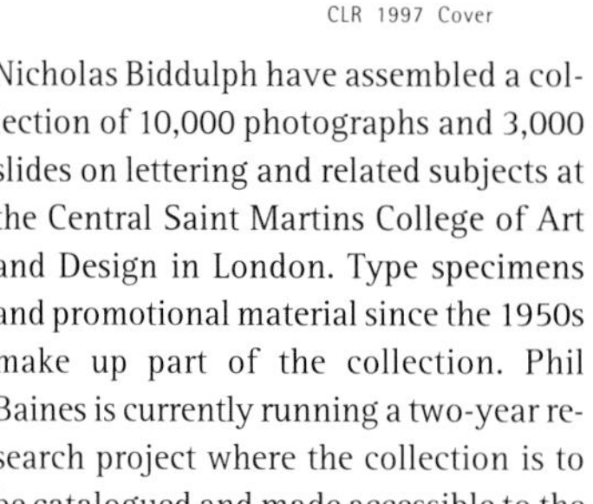

CLR 1997 Cover

n n n
a b e f g i
o r s t y z
A B C E G H
M O R S X Y
1 2 4 6 8 &

Chappell 1938–46 Lydian

first two books to be printed in the English language. 1476: opens up a workshop in Westminster, running a writing workshop as well as a printing press. It is here that "The Dictes or Sayengis of the Philosophres" is produced in 1477, the first printed book in England giving date and place of origin. – *Publications include:* William Blades "The Life and Typography of William Caxton, England's First Printer", 2 vols., London 1861–63.

**Central Lettering Record**, London – *lettering archive* – Nicolete Gray and Nicholas Biddulph have assembled a collection of 10,000 photographs and 3,000 slides on lettering and related subjects at the Central Saint Martins College of Art and Design in London. Type specimens and promotional material since the 1950s make up part of the collection. Phil Baines is currently running a two-year research project where the collection is to be catalogued and made accessible to the public.

**Central Saint Martins College of Art and Design**, London – 1845: Saint Martins School of Art is founded in London. 1896: the Central School of Arts and Crafts (renamed the Central School of Arts and Design in 1966) is founded in London. 1984: the two schools merge their graphic design courses. 1989: the two schools are merged. Since their founding, both schools have exerted great influence on education in Britain in the fields of art and design.

**Cerri**, Pierluigi – b. 21. 3. 1939 in Orta San Guilo, Italy – *graphic designer* – Studied architecture at the University of Milan.

Architekturstudium an der Universität Mailand. 1973 Gründung des Büros „Gregotti Associati" (mit Vittorio Gregotti und Augusto Cagnardi) in Mailand. Auftraggeber waren u. a. Ferrari, RAI, Fontana Arte, Unifor, B & B Italia. 1985 Entwurf des Signets für die 18. Triennale in Mailand. 1986 Eröffnung eines Büros in Venedig. Gestaltung der Ausstellungen „Las Formas de la Industria" (Madrid 1987), „Le Corbusier" (Paris, Turin 1988), „I Goti" (Mailand 1994). Entwicklung des Erscheinungsbildes des Palazzo Grassi in Venedig. Betreuung der Gestaltung der Design-Zeitschrift „Pagina". Ständiger Mitarbeiter der Zeitschriften „Rassegna" und „Casabella". – *Publikationen:* „Gregotti Associati 1973–1988" (Co-Autor), Mailand 1980; „Pubblicità d'autore", Mailand 1983.

**Chappell,** Warren – geb. 1904 in Richmond, Virginia, USA, gest. 1991 in Charlottesville, Virginia, USA. – *Schriftentwerfer, Buchgestalter, Illustrator* – Studium an der Arts Student League, New York. 1931–32 Studium bei Rudolf Koch in Offenbach, Deutschland als Stempelschneider und Schriftentwerfer. 1935 Studium der Illustration am Colorado Springs Fine Art Center. Lehrer an der Arts Student League. Zahlreiche Zusammenarbeiten mit dem Verleger Alfred A. Knopf, Illustrator und typographischer Berater des „Book of the Month Club". – *Schriftentwürfe:* Lydian (1938–46), Trajanus (1940).

**Chermayeff,** Ivan – geb. 6. 6. 1932 in London, England. – *Typograph, Grafik-Designer, Illustrator, Lehrer* – Studium an

Chermayeff 1977 Poster

Chermayeff ca. 1958 Cover

Chermayeff 1986 Poster

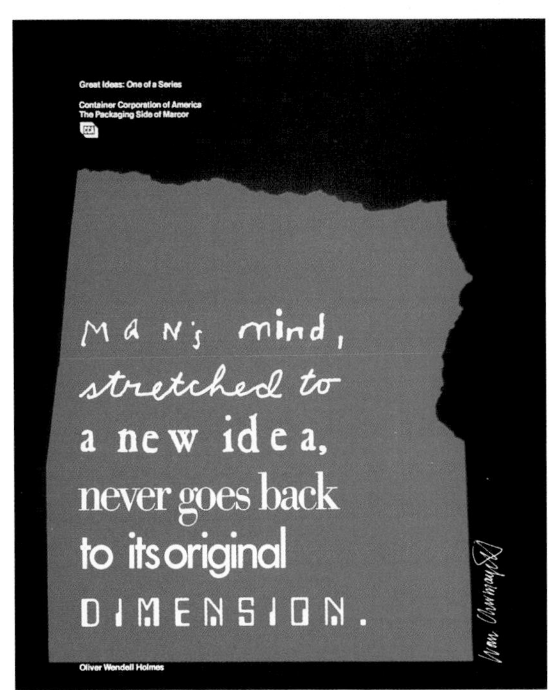

Chermayeff 1986 Poster

ture des expositions «Las Formas de la Industria» (Madrid 1987), «Le Corbusier» (Paris, Turin 1988), «I Goti» (Milan 1994). Réalisation de l'identité du Palazzo Grassi à Venise. Suit la maquette de la revue de design «Pagina». Travaille régulièrement pour les revues «Rassegna» et «Casabella». – *Publications:* «Gregotti Associati 1973–1988» (Co-auteur), Milan 1980; «Pubblicità d'autore», Milan 1983.

**Chappell,** Warren – né en 1904 à Richmond, Virginie, Etats-Unis, décédé en 1991 à Charlottesville, Virginie, Etats-Unis – *concepteur de polices, maquettiste en édition, illustrateur* – études à la Arts Student League à New York. 1931–1932, études de tailleur de types et de concepteur de polices chez Rudolf Koch à Offenbach, Allemagne. 1935, étudie l'art de l'illustration au Colorado Springs Fine Art Center. Enseigne à la Arts Student League. De nombreux travaux pour l'éditeur Alfred A. Knopf, illustrateur et conseiller typographique du «Book of the Month Club». – *Polices:* Lydian (1938–1946), Trajanus (1940).

**Chermayeff,** Ivan – né le 6. 6. 1932 à Londres, Angleterre – *typographe, graphiste maquettiste, illustrateur, enseignant* – études à la Harvard University, Cambridge, Massachusetts (1950–1952),

1973: opens the Gregotti Associati studio in Milan with Vittorio Gregotti and Augusto Cagnardi. Clients include Ferrari, RAI, Fontana Arte, Unifor and B & B Italia. 1985: designs the logo for the 18th Triennale in Milan. 1986: opens a studio in Venice. Plans the exhibitions "Las Formas de la Industria" (Madrid 1987), "Le Corbusier" (Paris, Turin 1988) and "I Goti" (Milan 1994). Develops a corporate identity for the Palazzo Grassi in Venice. Supervises the layout of "Pagina" design magazine. Works for "Rassegna" and "Casabella" magazines on a permanent basis. – *Publications:* "Gregotti Associati 1973–1988" (co-author), Milan 1980; "Pubblicità d'autore", Milan 1983.

**Chappell,** Warren – b. 1904 in Richmond, Virginia, USA, d. 1991 in Charlottesville, Virginia, USA – *type designer, book designer, illustrator* – Studied at the Arts Student League in New York. 1931–32: studies with Rudolf Koch in Offenbach, Germany as a punch cutter and type designer. 1935: studies illustration at the Colorado Springs Fine Art Center. Teaches at the Arts Student League. Chappell produced much work with the publisher Alfred A. Knopf and was also illustrator and typography consultant for the "Book of the Month Club". – *Fonts:* Lydian (1938–46), Trajanus (1940).

**Chermayeff,** Ivan – b. 6. 6. 1932 in London, England – *typographer, graphic designer, illustrator, teacher* – Studied at Harvard University, Cambridge, Massachusetts (1950–52), at the Institute of Design, Illinois, Chicago (1952–54), and at Yale University, New Haven, Connecticut

der Harvard University, Cambridge, Massachusetts (1950–52), am Institute of Design, Illinois, Chicago (1952–54) und an der Yale University, New Haven, Connecticut (1954–55). 1955 Assistent des Grafik-Designers Alvin Lustig, New York. 1956 Art Director bei Columbia Records, New York. 1957 Gründung der Agentur Brownjohn, Chermayeff und Geismar (mit Robert Brownjohn und Thomas Geismar) in New York. 1960 Gründung des Design-Büros Chermayeff & Geismar. Erscheinungsbilder für: Chase Manhattan Bank (1959), Mobil Oil Corporation (1962), Xerox Corporation (1965), Museum of Modern Art (1970), Lincoln Center (1988). Zahlreiche Auszeichnungen, u. a. 1979 Goldmedaille des American Institute of Graphic Arts (AIGA), 1985 Yale Arts Award Medal (mit T. Geismar), 1992 Grand Prix der Biennale von Brno, Tschechien. Illustriert zahlreiche Kinderbücher. Unterrichtet am Brooklyn College, New York (1956–57), an der School of Visual Arts, New York (1959–65), an der Cooper Union und am Kansas City Art Institute. – *Publikationen u. a.:* „First Words" (mit Jane Clark Chermayeff), New York 1989; „Collages", New York 1991; „First Shapes" (mit Jane Clark Chermayeff), New York 1991.

**Chruxin,** Christian – geb. 13. 12. 1937 in Hannover, Deutschland. – *Typograph, Grafik-Designer* – Studium in Kassel und Berlin. Ab 1961 Arbeiten für Verlage u. a., Fietkau Verlag, Verlag Neue Kritik, Voltaire Verlag, Gerhardt Verlag, Rowohlt Verlag. Zahlreiche Auszeichnungen. Herausgeber der Buchreihe „Projekte und

à l'Institute of Design, Illinois, Chicago (1952–1954), à la Yale University, New Haven, Connecticut (1954–1955). 1955, assistant du graphiste maquettiste Alvin Lustig, à New York. 1956, Art Director chez Columbia Records, New York. 1957, fondation de l'agence Brownjohn, Chermayeff et Geismar (avec Robert Brownjohn et Thomas Geismar) à New York. 1960, fondation de l'atelier de design Chermayeff & Geismar. Identités pour: Chase Manhattan Bank (1959), Mobil Oil Corporation (1962), Xerox Corporation (1965), Museum of Modern Art (1970), Lincoln Center (1988). De nombreuses distinctions, par ex., 1979, médaille d'or de l'American Institute of Graphic Arts (AIGA), 1985, Yale Arts Award Medal (avec T. Geismar), 1992, Grand Prix de la Biennale de Brno, Tchéquie. A illustré de nombreux livres pour enfants. Enseignement au Brooklyn College, New York (1956–1957), à la School of Visual Arts, New York (1959–1965), à la Cooper Union et au Kansas City Art Institute. – *Publications, sélection:* «First Words» (avec Jane Clark Chermayeff), New York 1989; «Collages», New York 1991; «First Shapes» (avec Jane Clark Chermayeff), New York 1991.

**Chruxin,** Christian – né le 13. 12. 1937 à Hanovre, Allemagne – *typographe, graphiste maquettiste* – études à Kassel et à Berlin. Travaux pour plusieurs éditeurs à partir de 1961, parmi lesquels le Fietkau Verlag, Verlag Neue Kritik, Voltaire Verlag, Gerhardt Verlag, Rowohlt Verlag. Nombreuses distinctions. Editeur de la collection «Projekte und Modelle» (avec Joachim Krausse). Nombreuses expositions. Enseignant en design audiovisuel à la Hochschule der Künste (école supérieure des beaux-arts) de Berlin. 1971, scénographies, costumes et réalisation de films pour la télévision, décors du stu-

Chruxin 1967 Cover

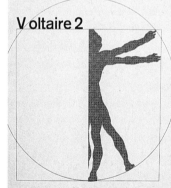

Chruxin ca. 1965 Cover

Chwast 1987 Poster

Chwast 1981 Buffalo

(1954–55). 1955: assistant to the graphic designer Alvin Lustig in New York. 1956: art director for Columbia Records in New York. 1957: sets up the agency Brownjohn, Chermayeff and Geismar (with Robert Brownjohn and Thomas Geismar) in New York. 1960: opens the Chermayeff & Geismar design studio. Designs corporate identities for Chase Manhattan Bank (1959), the Mobil Oil Corporation (1962), the Xerox Corporation (1965), the Museum of Modern Art (1970) and the Lincoln Center (1988). He has won many awards, including a gold medal from the American Institute of Graphic Arts (AIGA, 1979), the Yale Arts Award Medal (with T. Geismar, 1985), and the Grand Prix at the Brno Biennale, Czech Republic (1992). He has illustrated numerous children's books. He has taught at Brooklyn College in New York (1956–57), at the School of Visual Arts in New York (1959–65), at the Cooper Union and at the Kansas City Art Institute. – *Publications include:* "First Words" (with Jane Clark Chermayeff), New York 1989; "Collages", New York 1991; "First Shapes" (with Jane Clark Chermayeff), New York 1991.

**Chruxin,** Christian – b. 13. 12. 1937 in Hanover, Germany – *typographer, graphic designer* – Studied in Kassel and Berlin. 1961: starts working for various publishers, including Fietkau Verlag, Verlag Neue Kritik, Voltaire Verlag, Gerhardt Verlag and Rowohlt Verlag. Numerous awards. Edits a series of books entitled "Projekte und Modelle" with Joachim Krauße. Numerous exhibitions. Teaches

Modelle“ (mit Joachim Krauße). Zahlreiche Ausstellungen. Lehrt Fernsehdesign an der Hochschule der Künste, Berlin. 1971 Szenenbilder, Kostümentwürfe und Regie für Fernsehfilme, Studioarchitektur für den Beat-Club, Radio Bremen. Erscheinungsbild für das Künstlerhaus Bethanien, Berlin.

**Chwast,** Seymour – geb. 18. 8. 1931 in New York, USA. – *Grafik-Designer, Illustrator, Schriftentwerfer, Lehrer* – 1948–51 Studium an der Cooper Union, New York. 1954 Gründung des Push Pin Studios in New York mit Milton Glaser. 1955–81 Art Director der Hauszeitschrift „Push Pin Graphic“. 1982 Direktor von „Push Pin Lubalin Peckolik“ in New York. 1983 in die Hall of Fame des Art Directors Club of New York gewählt. Ab 1985 Direktor von „The Pushpin Group“. Zahlreiche Buchillustrationen, Schallplattencover, Plakate, Headline-Schriften sowie Illustrationen für Zeitschriften. Auftraggeber u. a. Mobil Oil, Büchergilde Gutenberg, „Frankfurter Allgemeine Zeitung“-Magazin, Mohawk Paper Mills. Unterrichtet an der School of Visual Arts, New York (1972–74) und an der Cooper Union, New York (ab 1975). – *Schriftentwurf:* Chwast Buffalo (1981). – *Publikationen u. a.:* „Seymour Chwast. The left-handed Designer“, New York 1985; „Graphic Style“ (mit Steven Heller), New York, London 1988; „The Alphabet Parade“, New York 1991.

**Cissarz,** Johann Vincenz – geb. 22. 1. 1873 in Danzig, Deutschland (heute Gdańsk, Polen), gest. 23. 12. 1942 in Frankfurt am Main, Deutschland. – *Grafiker, Maler* –

dio du Beat-Club pour Radio Bremen. Identité du Künstlerhaus Bethanien, Berlin.

**Chwast,** Seymour – né le 18. 8. 1931 à New York, Etats-Unis – *graphiste maquettiste, illustrateur, concepteur de polices, enseignant* – 1948–1951, études à la Cooper Union, New York. 1954, fonde le Push Pin Studio avec Milton Glaser, à New York. 1951–1981, Art Director de la revue maison «Push Pin Graphic». 1982, directeur de «Push Pin Lubalin Peckolik», à New York. 1983, élu au Hall of Fame de l'Art Directors Club of New York. A partir de 1985, directeur de «The Pushpin Group». Nombreuses illustrations de livres et de pochettes de disques, affiches, gros titres ainsi que des illustrations pour des revues. Travaux pour Mobil Oil, Büchergilde Gutenberg, «Frankfurter Allgemeine Zeitung» Magazin, Mohawk Paper Mills. A enseigné à la School of Visual Arts, New York (1972–1974) et à la Cooper Union, New York (à partir de 1975). – *Police:* Chwast Buffalo (1981). – *Publications, sélection:* «Seymour Chwast. The left-handed Designer», New York 1985; «Graphic Style» (avec Steven Heller), New York, Londres 1988; «The Alphabet Parade», New York 1991.

**Cissarz,** Johann Vincenz – né le 22. 1. 1873 à Danzig, Allemagne (aujourd'hui Gdansk, Pologne), décédé le 23. 12. 1942 à Francfort-sur-le-Main, Allemagne – *graphiste, peintre* – vit à Darmstadt à partir de 1903. En 1906, il est chargé de cours en fabrication du livre au Lehr- und Versuchsanstalt de Stuttgart (Institut d'enseignement expérimental).

Chwast 1989 Poster

KATALOG DER DEUTSCHEN BUCHGEWERBE-AUSSTELLUNG PARIS 1900 HERAUSGEGEBEN VOM DEUTSCHEN BUCHGEWERBE-VEREIN ZU LEIPZIG

Cissarz 1900 Cover

Chwast 1963 Poster

Cissarz 1924 Cover

TV design at the Hochschule der Künste in Berlin. 1971: designs sets and costumes and directs films for television. Does the studio architecture for the Beat-Club, Radio Bremen and designs the corporate identity for the Künstlerhaus Bethanien in Berlin.

**Chwast,** Seymour – b. 18. 8. 1931 in New York, USA – *graphic designer, illustrator, type designer, teacher* – 1948–51: studies at the Cooper Union in New York. 1954: opens the Push Pin Studio in New York with Milton Glaser. 1955–81: art director of the in-house magazine "Push Pin Graphic". 1982: director of "Push Pin Lubalin Peckolik" in New York. 1983: is elected into the New York Art Directors Hall of Fame. 1985: becomes director of The Pushpin Group. Numerous book illustrations, record covers, posters, headline type and illustrations for magazines. Clients include Mobil Oil, Büchergilde Gutenberg, the "Frankfurter Allgemeine Zeitung"-Magazin and Mohawk Paper Mills. He has taught at the School of Visual Arts in New York (1972–74) and the Cooper Union in New York (from 1975 onwards). – *Font:* Chwast Buffalo (1981). – *Publications include:* "Seymour Chwast. The left-handed Designer", New York 1985; "Graphic Style" (with Steven Heller), New York, London 1988; "The Alphabet Parade", New York 1991.

**Cissarz,** Johann Vincenz – b. 22. 1. 1873 in Danzig, Germany (now Gdansk, Poland), d. 23. 12. 1942 in Frankfurt am Main, Germany – *graphic artist, painter* – 1903: moves to Darmstadt. 1906: teaching position for book design at the Lehr-

Seit 1903 in Darmstadt, 1906 Lehrauftrag für Buchausstattung an der Lehr- und Versuchsanstalt in Stuttgart. 1916 Berufung nach Frankfurt am Main als Leiter einer Klasse für Malerei an der Kunstgewerbeschule. 1933 Ehrenvorsitzender des Bundes Deutscher Gebrauchsgraphiker (BDG). Zahlreiche Entwürfe von Buchschmuck, Buchillustrationen, Bucheinbänden und Plakaten.

**Cleland,** Thomas Maitland – geb. 1880 in Brooklyn, New York, USA, gest. 1964 in Danbury, Connecticut, USA. – *Buchgestalter, Maler, Illustrator* – Studium am Artisan Institute in Chelsea, New York. Als Buchgestalter für die Caslon Press und die Merrymount Press tätig. Gründet die Cornhill Press in Boston. 1907–08 Art Director des „McClure's Magazine". 1925 Arbeiten für die Hauszeitschrift des Papierherstellers Westvaco Corporation. Zahlreiche typographische Arbeiten, u. a. Gestaltung der Zeitschrift „Fortune". – *Schriftentwürfe:* Della Robbia (1902), Westminster Oldstyle (1902), Amsterdam Garamont (mit M. F. Benton, 1917).

**Cobden-Sanderson,** Thomas James – geb. 2. 12. 1840 in Alnwick, England, gest. 7. 9. 1922 in London. England. – *Buchbinder, Drucker, Typograph* – 1859– 63 Studium in Cambridge, ab 1871 Advokat und Ingenieur, Lehre als Buchbinder. 1893 Eröffnung seiner Werkstatt in Hammersmith, London. 1900 Gründung der Doves Press mit Emery Walker. Bis zur Trennung von Walker erscheinen 19 Bände, danach unter seiner alleinigen Leitung weitere 30 Bände, u. a. 1903–05 eine 5-bändige Bibel, 1906 Goethes

En 1916, Cissarz est appelé à Francfort-sur-le-Main pour diriger un atelier de peinture à la Kunstgewerbeschule (l'école des arts décoratifs). 1933, président d'honneur du BDG. Nombreuses maquettes de livres, illustrations, reliures et affiches.

**Cleland,** Thomas Maitland – né en 1880 à Brooklyn, New York, Etats-Unis, décédé en 1964 à Danbury, Connecticut, Etats-Unis – *maquettiste en édition, peintre, illustrateur* – études à l'Artisan Institute à Chelsea, New York. Maquettiste pour les Caslon Press et les Merrymount Press. Fonde la Cornhill Press à Boston. 1907–1908, Art Director du «McClure's Magazine». 1925, travaux pour la revue d'entreprise du fabricant de papier Westvaco Corporation. Nombreux travaux typographiques, la maquette de la revue «Fortune». – *Polices:* Della Robbia (1902), Westminster Oldstyle (1902), Amsterdam Garamont (avec M. F. Benton, 1917).

**Cobden-Sanderson,** Thomas James – né le 2. 12. 1840 à Alnwick, Angleterre, décédé le 7. 9. 1922 à Londres, Angleterre – *relieur, imprimeur, typographe* – 1859–1863 études à Cambridge, avocat et ingénieur à partir de 1871, apprentissage de reliure. Il fonde son atelier en 1893 à Hammersmith, Londres. 1900, création des Doves Press avec Emery Walker. 19 ouvrages sont publiés jusqu'à sa séparation de Walker, puis 30 autres sous sa direction, dont une bible en 5 volumes de 1903 à 1905, un «Faust» de Goethe en 1906 et une édition des œuvres de Shakespeare. En 1916, il immerge tout son matériel de composition dans la Tamise en signe de protestation contre la mécanisation de la composition et de l'imprimerie. – *Publications, sélection:* «The ideal book or book beautiful», Londres 1901; «Credo», Hammersmith 1908; «Cosmic Vision», Londres 1922.

Cleland 1902 Della Robbia

Cleland 1917 Amsterdam Garamont

Cobden-Sanderson 1901 Page

Cobden-Sanderson 1909 Page

und Versuchsanstalt in Stuttgart. 1916: appointment as head of a painting class at the Kunstgewerbeschule in Frankfurt am Main. 1933: honorary chairman of the BDG. Numerous designs for book decoration and book illustrations, book covers and posters.

**Cleland,** Thomas Maitland – b. 1880 in Brooklyn, New York, USA, d. 1964 in Danbury, Connecticut, USA – *book designer, painter, illustrator* – Studied at the Artisan Institute in Chelsea, New York. Worked as a book designer for Caslon Press and Merrymount Press. Founded the Cornhill Press in Boston. 1907–08: art director of "McClure's Magazine". 1925: produces work for paper manufacturer Westvaco Corporation's in-house magazine. Numerous typographic work, including designs for "Fortune" magazine. – *Fonts:* Della Robbia (1902), Westminster Oldstyle (1902), Amsterdam Garamont (with M. F. Benton, 1917).

**Cobden-Sanderson,** Thomas James – b. 2. 12. 1840 in Alnwick, England, d. 7. 9. 1922 in London, England – *bookbinder, printer, typographer* – 1859–63: studies in Cambridge. 1871: advocate and engineer, trains as a bookbinder. 1893: opens his workshop in Hammersmith, London. 1900: founds Doves Press with Emery Walker. Until he and Walker part company, 19 volumes are published. Cobden-Sanderson then goes on to supervise the publishing of a further 30 books alone, including a 5-volume bible (1903–05), Goethe's "Faust" (1906) and editions of works by Shakespeare. 1916: in protest against the mechanization of

„Faust“ und Ausgaben mit Werken Shakespeares. 1916 versenkt er als Protest gegen die Mechanisierung des Satz- und Druckhandwerks sein gesamtes Satzmaterial in der Themse. – *Publikationen u. a.:* „The ideal book or book beautiful“, London 1901; „Credo“, Hammersmith 1908; „Cosmic Vision“, London 1922.

**Cochin,** Charles Nicolas – geb. 1715 in Paris, Frankreich, gest. 1790 in Paris, Frankreich. – *Kupferstecher, Illustrator* – Lehre bei seinem gleichnamigen Vater. 1739 Annahme einer hervorragenden künstlerischen Position am Hofe Ludwig XV., wo er u. a. Madame Pompadours jüngeren Bruder ausbildet. 1749–51 ausgedehnte Reise in Italien. Cochin ist ein Hauptvertreter des Linienstiches, entwirft zahlreiche Vignetten und Zierstücke. Seine Arbeiten beeinflussen die Alphabete Baskervilles, Didots und Bodonis.

**Comeriner,** Erich – geb. 3. 6. 1907 in Wien, Österreich, gest. 1978 in Tel Aviv, Israel. – *Grafik-Designer, Fotograf* – 1927–28 Studium am Bauhaus, Kurse bei J. Albers, H. Bayer und J. Schmidt. 1928 Eröffnung des Fotostudios Comofot in Berlin. 1929 Beteiligung an der Ausstellung „Film und Foto“ des Deutschen Werkbunds in Stuttgart. 1929–33 Zusammenarbeit mit L. Moholy-Nagy bei Werbearbeiten und Bühnenbildgestaltungen. Freischaffender Pressefotograf und Gebrauchsgrafiker. 1934 Auswanderung nach Palästina. Ab 1940 freier Fotograf und Grafik-Designer in Tel Aviv.

**Confalonieri,** Giulio – geb. 20. 8. 1926 in Mailand, Italien. – *Grafik-Designer* – Nach dem Studium der Wirtschaftswis-

*abcdefghiklmnopqrstuvwxyz*
Meistersinger *Schopenhauer*
abcdefghiklmnopqrstuvwxyz
1234567890 *1234567890*

Cochin 1918 Cochin-Antiqua

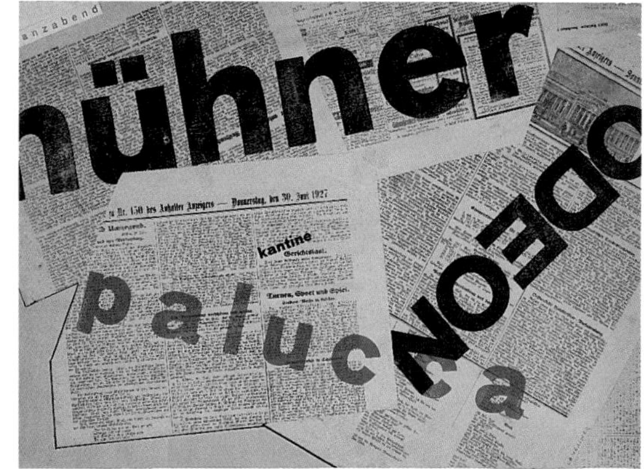


Comeriner 1927 Collage

**Cochin,** Charles Nicolas – né en 1715 à Paris, France, décédé en 1790 à Paris, France – *graveur sur cuivre, illustrateur* – apprentissage avec son père. 1739, occupe une position privilégiée à la cour de Louis XV, où il est, entre autres, chargé de la formation du jeune frère de Madame de Pompadour. Voyage prolongé en Italie de 1749 à 1751. Cochin a été l'un des principaux représentants de la gravure en hachures, il a dessiné de nombreuses vignettes et ornements. Ses travaux ont influencé les alphabets de Baskerville, Didot et Bodoni.

**Comeriner,** Erich – né le 3. 6. 1907 à Vienne, Autriche, décédé en 1978 à Tel Aviv, Israël – *graphiste maquettiste, photographe* – 1927–1928, études au Bauhaus dans les ateliers de J. Albers, H. Bayer et J. Schmidt. 1928, inauguration du studio photographique Comofot à Berlin. 1929, participation à l'exposition «Film und Foto» du Deutscher Werkbund à Stuttgart. De 1929 à 1933, il travaille avec L. Moholy-Nagy et réalise des travaux publicitaires et des scénographies. Photographe de presse indépendant et graphiste publicitaire. 1934, émigration en Palestine. Photographe et graphiste maquettiste indépendant à Tel-Aviv à partir de 1940.

**Confalonieri,** Giulio – né le 20. 8. 1926 à Milan, Italie – *graphiste maquettiste* – étudie les sciences économiques puis fonde un atelier de graphisme. Travaille pour Esso, Mondadori, Pirelli, Westing-

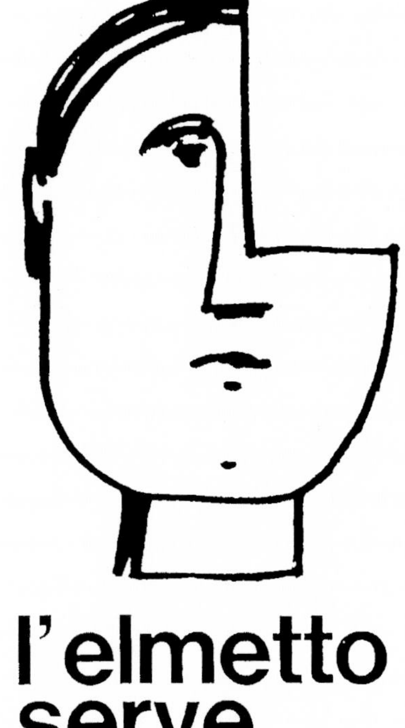


Confalonieri Poster

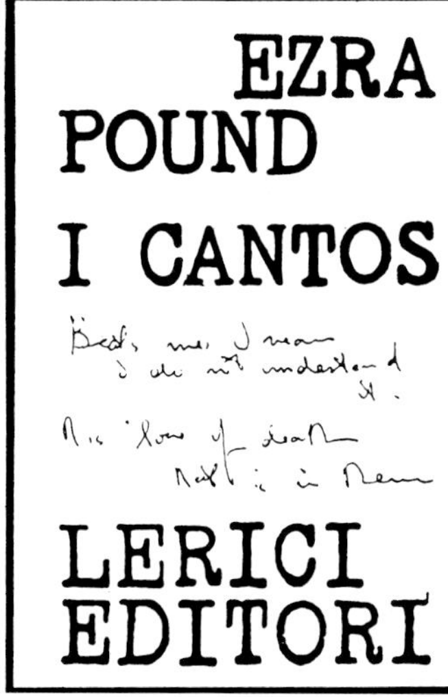


Confalonieri 1970 Cover

the composition and printing trade, he sinks his entire typesetting material in the Thames. – *Publications include:* "The ideal book or book beautiful", London 1901; "Credo", Hammersmith 1908; "Cosmic Vision", London 1922.

**Cochin,** Charles Nicolas – b. 1715 in Paris, France, d. 1790 in Paris, France – *copperplate engraver, illustrator* – Trained with his father, also called Charles Nicolas. 1739: takes up an excellent position in art at the court of Louis XV, where his pupils include Madame Pompadour's younger brother. 1749–51: undertakes a long trip to Italy. Cochin was one of the main representatives of line etching and designed numerous vignettes and ornaments. His work influenced the alphabets of Baskerville, Didot and Bodoni.

**Comeriner,** Erich – b. 3. 6. 1907 in Vienna, Austria, d. 1978 in Tel Aviv, Israel – *graphic designer, photographer* – 1927–28: studies at the Bauhaus, taking courses with J. Albers, H. Bayer and J. Schmidt. 1928: opens the Comofot photographic studio in Berlin. 1929: takes part in the "Film und Foto" exhibition organized by the Deutscher Werkbund in Stuttgart. 1929–33: works with L. Moholy-Nagy on advertising projects and stage set designs. Freelance press photographer and commercial artist. 1934: emigrates to Palestine. From 1940 onwards: freelance photographer and graphic designer in Tel Aviv.

**Confalonieri,** Giulio – b. 20. 8. 1926 in Milan, Italy – *graphic designer* – After completing his studies in economics, Confalonieri set up his own graphics stu-

senschaften Gründung eines grafischen Ateliers. Arbeiten u. a. für Esso, Mondadori, Pirelli, Westinghouse. Regelmäßige Teilnahme an den Mailänder Triennale-Ausstellungen, auf der 15. Triennale Auszeichnung mit einer Goldmedaille. Art Director des Verlags Lerici. Aufsätze in den Zeitschriften „Domus", „Graphis", „Imago". – *Publikationen u. a.:* „Imagine di un Libro", Mailand 1965; „12 Finestre", Mailand 1968.

**Cooper,** Oswald Bruce – geb. 13. 4. 1879 in Mountgilead, Ohio, USA, gest. 17. 12. 1940 in Chicago, USA. – *Schriftentwerfer, Kalligraph, Lehrer* – 1894 Druckerlehre. 1899 Umzug nach Chicago, Ausbildung an der Frank Holme School of Illustration, an der F. W. Goudy Schriftschreiben lehrt. 1902 Lehrer an der Holme School, zeitweise Leiter der Schule. 1904 Gründung des Werbebüros Bertsch & Cooper in Chicago mit dem Illustrator Fred S. Bertsch. 1912 Reise durch England, Frankreich und Deutschland. Zeichnet 1913 die Schrift, die für die Werbung der Packard Motor Company verwendet wird. Sie wird von der American Type Foundry (ATF) produziert. 1921 Umwandlung des Büros in ein Studio für Entwurf, Satz und Fotografie. 1924 Verkauf des Studios, wieder selbständige Arbeit als Grafiker. 1939 Arbeit am Erscheinungsbild der Zeitung „Chicago Daily News". – *Schriftentwürfe:* Cooper Oldstyle (1919), Cooper Black (1920), Cooper Italic (1924), Cooper Black condensed (1925), Cooper Hilite (1925), Cooper Initials (1925), Pompeian Cursive (1927), Cooper Fullface (1928).

house, etc. Participe régulièrement aux expositions de la Triennale de Milan, la médaille d'or lui est décernée lors de la 15e Triennale. Directeur artistique des éditions Lerici. Essais dans les revues «Domus», «Graphis», «Imago». – *Publications, sélection :* «Imagine di un Libro», Milan 1965; «12 Finestre», Milan 1968.

**Cooper,** Oswald Bruce – né le 13. 4. 1879 à Mountgilead, Ohio, Etats-Unis, décédé en 17. 12. 1940 à Chicago, Etats-Unis – *concepteur de polices, calligraphe, enseignant* – 1894, apprentissage d'imprimeur. 1899, s'installe à Chicago, formation à la Frank Holme School of Illustration où F. W. Goudy enseigne la calligraphie. 1902, enseignant à la Holme School qu'il dirige à maintes reprises. En 1904, il fonde l'agence de publicité Bertsch & Cooper avec Fred S. Bertsch, à Chicago. 1912, voyage en Angleterre, en France et en Allemagne. En 1913, il crée une fonte qui sera utilisée pour la publicité de la Packard Motor Company. Cette police est distribuée par la American Type Foundry (ATF). En 1921, l'agence est transformée en studio de création, composition et photographie. Le studio est vendu en 1924, Cooper reprend ensuite ses activités de graphiste indépendant. 1939, création de l'identité du journal «Chicago Daily News». – *Polices :* Cooper Oldstyle (1919), Cooper Black (1920), Cooper Italic (1924), Cooper Black condensed (1925), Cooper Hilite (1925), Cooper Initials (1925), Pompeian Cursive (1927), Cooper Fullface (1928).

**Cranach-Presse,** Weimar – les ateliers d'imprimerie Cranach-Presse ont été fondés en 1912 par Harry Graf Kessler à Weimar en Allemagne. Les imprimeurs Erich Dressler, J. H. Mason et Max Kopp comptent parmi les premiers employés. Au fur et à mesure des années, des illustrateurs tels que Aristide Maillol et Eric Gill, des

Cooper 1925 Cooper Hilite

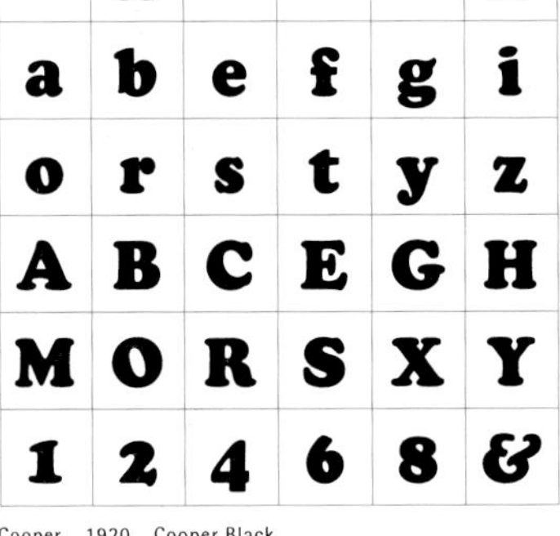


Cooper 1920 Cooper Black

Cranbrook 1977 Page

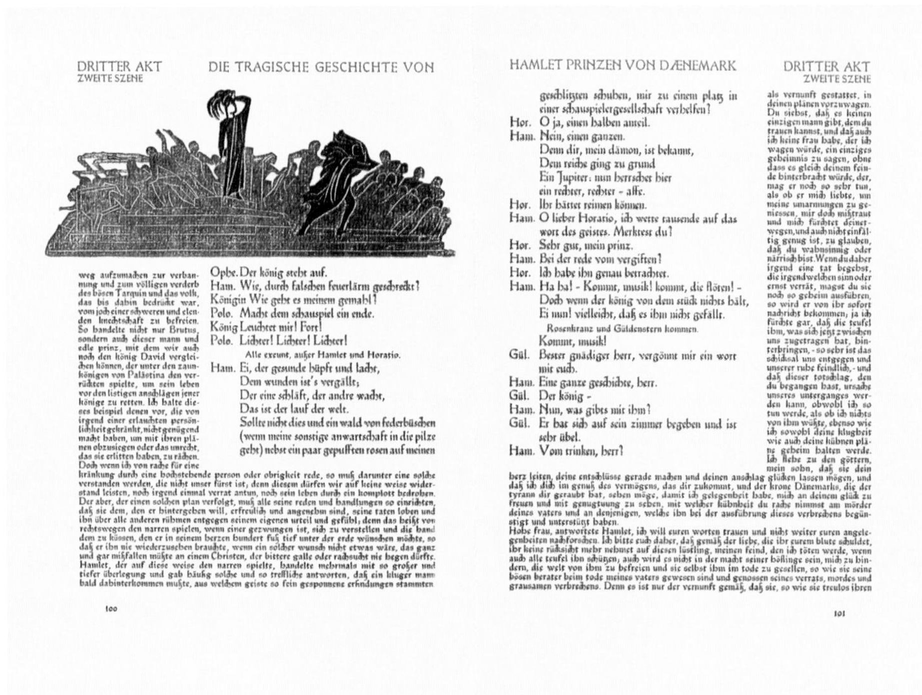


Cranach-Presse 1929 Spread

dio. He has worked for Esso, Mondadori, Pirelli and Westinghouse, among others. Regular participation in the Milan Triennale exhibitions; awarded a gold medal at the 15th Triennale. Art director for Lerici publishers. Has written essays for "Domus", "Graphis" and "Imago" magazines. – *Publications include:* "Imagine di un Libro", Milan 1965; "12 Finestre", Milan 1968.

**Cooper,** Oswald Bruce – b. 13. 4. 1879 in Mountgilead, Ohio, USA, d. 17. 12. 1940 in Chicago, USA – *type designer, calligrapher, teacher* – 1894: trains as a printer. 1899: moves to Chicago. Attends the Frank Holme School of Illustration where F. W. Goudy teaches lettering. 1902: teaches at the Holme School, intermittently also head of the school. 1904: opens the Bertsch & Cooper advertising agency in Chicago with the illustrator Fred S. Bertsch. 1912: travels in England, France and Germany. 1913: draws the lettering which is used in advertisements for the Packard Motor Company and produced by the American Type Foundry (ATF). 1921: turns his agency into a studio for design, typesetting and photography. 1924: sells his studio and resumes work as a freelance graphic artist. 1939: works on the corporate identity for the "Chicago Daily News" newspaper. – *Fonts:* Cooper Oldstyle (1919), Cooper Black (1920), Cooper Italic (1924), Cooper Black condensed (1925), Cooper Hilite (1925), Cooper Initials (1925), Pompeian Cursive (1927), Cooper Fullface (1928).

**Cranach-Presse,** Weimar – 1912: Count Harry Kessler opens the Cranach-Presse,

**Cranach-Presse,** Weimar – 1912 Harry Graf Kessler gründet in Weimar, Deutschland die Cranach-Presse als private Druckwerkstatt. Erste Mitarbeiter sind die Drucker Erich Dressler, J. H. Mason und Max Kopp. Im Laufe der Jahre arbeiten für die Presse als Illustratoren Aristide Maillol und Eric Gill, als Schriftentwerfer Emery Walker und Edward Johnston. Verlegt werden Werke von Henry van de Velde, Maurice Maeterlinck, Theodor Däubler, Wieland Herzfelde, Max Goertz, Hugo von Hofmannsthal, Rainer Maria Rilke und William Shakespeare. Das Papier der Cranach-Presse wird in einer eigenen Papiermühle in Frankreich geschöpft, für Illustrationen und Titel kommen, neben typographisch gestalteten, nur eigenhändig geschnittene Holzschnitte zur Verwendung. Der schon 1913 begonnene Band von Vergils „Eclogen“ erscheint 1927 in drei Ausgaben (Latein mit französischer, englischer und deutscher Übersetzung) mit Holzschnitten von Aristide Maillol. Als zweites der großen illustrierten Bücher der Presse erscheint 1929 die von Edward Gordon Craig illustrierte „Hamlet“-Ausgabe. 1931 Auflösung der Cranach-Presse aus finanziellen Gründen. – *Publikation u. a.:* Renate Müller-Krumbach „Harry Graf Kessler und die Cranach-Presse in Weimar“, Hamburg 1969.

**Cranbrook Academy of Art,** Michigan – Auf dem Gebiet der Ortschaft Bloomfield Hills (einem Vorort von Detroit) in Michigan, gründet der Zeitungsverleger und Mäzen George G. Booth ab 1904 Cranbrook (genannt nach einem Dorf in Kent,

Cranbrook 1977 Page

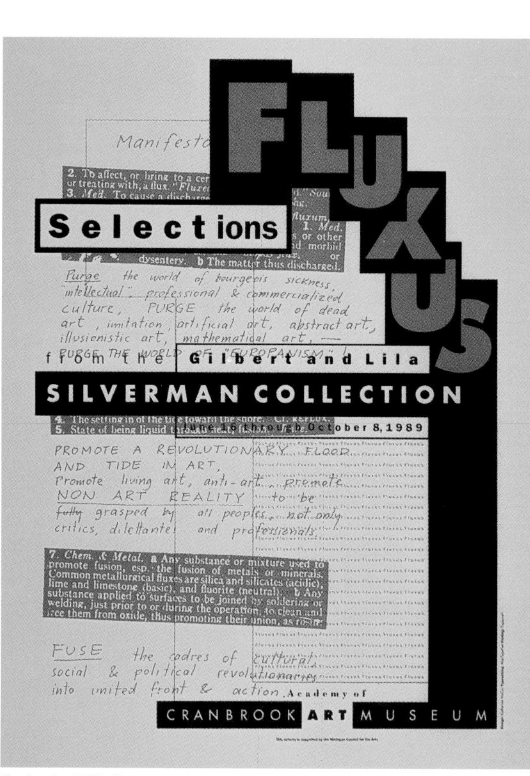


Cranbrook 1989 Poster

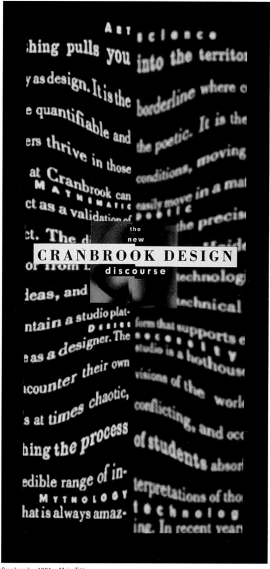

Cranbrook 1991 Main Title

concepteurs de polices comme Emery Walker et Edward Johnston travailleront pour les presses. La maison a édité des œuvres de Henry van de Velde, Maurice Maeterlinck, Theodor Däubler, Wieland Herzfelde, Max Goertz, Hugo von Hofmannsthal, Rainer Maria Rilke et William Shakespeare. Le papier employé à la Cranach-Presse était fabriqué par un moulin que la maison possédait en France. Pour les illustrations, les couvertures ainsi que les créations typographiques, on n'y utilisait que des gravures sur bois réalisées à cet effet. Le volume des «Eclogues» de Virgile, commencé dès 1913 et illustré de gravures sur bois de Maillol, paraît en 1927 en trois éditions (latine, avec des traductions en français, anglais et allemand). Le second grand livre illustré sorti des presses en 1929, est un «Hamlet» avec des planches gravées sur bois par Edward Gordon Craig. En 1931, la Cranach-Presse cesse ses activités pour des raisons financières. – *Publication, sélection:* Renate Müller-Krumbach «Harry Graf Kessler und die Cranach-Presse in Weimar», Hambourg 1969.

**Cranbrook Academy of Art,** Michigan – l'une des cinq institutions fondées à partir de 1904 par George G. Booth, homme de presse et mécène, à Bloomfield Hills, Michigan. L'académie a été ainsi appelée en souvenir d'un village du Kent, Angleterre où vivaient ses ancêtres. Elle comporte une école élémentaire, deux écoles secondaires, une église, un institut de re-

a private printing press, in Weimar, Germany. The first employees are the printers Erich Dressler, J. H. Mason and Max Kopp. Over the years, illustrators such as Aristide Maillol and Eric Gill, and type designers Emery Walker and Edward Johnston work for the press. The press publishes works by Henry van de Velde, Maurice Maeterlinck, Theodor Däubler, Wieland Herzfelde, Max Goertz, Hugo von Hofmannsthal, Rainer Maria Rilke and William Shakespeare. The paper for the Cranach-Presse is made in its own paper mill in France. Besides typographic designs, only woodcuts fashioned by the members of the press are used for title pages and illustrations. 1927: a volume of Virgil's “Eclogues”, begun in 1913, is published in three editions (in Latin with English, French and German translations) with woodcuts by Aristide Maillol. 1929: the second of the press's great illustrated books is published, an edition of “Hamlet”, illustrated by Edward Gordon Craig. 1931: the press is closed for financial reasons. – *Publications include:* Renate Müller-Krumbach “Harry Graf Kessler und die Cranach-Presse in Weimar”, Hamburg 1969.

**Cranbrook Academy of Art,** Michigan – 1904: George G. Booth, newspaper editor and patron, begins building up Cranbrook, one of five establishments, in Bloomfield Hills, Michigan (named after a village in Kent, England, where Booth's ancestors lived). Various institutions evolve here over the years: an elementary

England, in dem seine Vorfahren lebten). Im Laufe der Jahre entstehen eine Elementarschule, zwei Sekundarschulen, eine Kirche, ein wissenschaftliches Institut und eine Kunstakademie mit Museum. 1932 Gründung der Cranbrook Academy of Art mit den Abteilungen Skulptur, Malerei, Architektur. Hinzu kommen Weben und Kostümdesign (1937), Keramik (1938), Design (1938), sowie die Abteilungen Druckgrafik, Fotografie und Metall. Ab 1942 erhält Cranbrook die Erlaubnis, Diplome zu vergeben. Grafik-Design wird 1960 eine eigene Abteilung, deren Leitung 1971 Katherine McCoy übernimmt. Lehrende sind im Laufe der Jahre u. a. die Produktdesigner Charles Eames und Harry Bertoia, der Bildhauer Duane Hanson, die Architekten Eliel und Eero Saarinen und Daniel Libeskind. – *Publikationen u. a.:* „The Cranbrook Vision 1925–50", New York 1983; „Cranbrook Design: The New Discourse", New York 1991.

**Craw,** Freeman – geb. 17. 1. 1917 in East Orange, New York, USA. – *Schriftentwerfer, Typograph, Grafik-Designer* – 1935–39 Studium an der Cooper Union Art School, New York. 1939–43 Grafiker bei der American Colortype Company, New York. Ab 1943 Art Director der Tri-Arts Press, New York, 1958 deren Vizepräsident. Danach Gründung des Studios Freeman Craw Design. – *Schriftentwürfe:* Craw Clarendon (1955–60), Craw Modern (1958–64), Ad Lib (1961), Canterbury, Chaucery, Classic, Cursive, CBS Sans.

**Crous-Vidal,** Enrico – geb. 1908 in Léri-

cherche ainsi qu'une académie des beaux-arts et son musée, toutes ces institutions ont été créées au fur et à mesure. La Cranbrook Academy of Art, fondée en 1932, possédait des sections de sculpture, peinture et architecture. Puis des ateliers de tissage et de costumes (1937), de céramique (1938), de design (1938) ainsi que d'arts graphiques, de photographie et de métaux sont venus s'ajouter. En 1942, l'académie Cranbrook est habilitée à attribuer des diplômes. L'atelier d'arts graphiques devient un département à part entière en 1960; Katherine McCoy en prend la direction en 1971. Les designers de produits Charles Eames et Harry Bertoia, le sculpteur Duane Hanson, les architectes Eliel, Eero Saarinen et Daniel Libeskind ont compté parmi le personnel enseignant. – *Publications, sélection:* «The Cranbrook Vision 1925–1950», New York 1983; «Cranbrook Design : The New Discourse», New York 1991.

**Craw,** Freeman – né le 17. 1. 1917 à East Orange, New York, Etats-Unis – *concepteur de polices, typographe, graphiste maquettiste* – 1935–1939, études à la Cooper Union Art School, New York. 1939–1943, graphiste pour la American Colortype Company, New York. Art Director de la Tri-Arts Press, New York à partir de 1943, puis vice-président en 1958. Il fonde ensuite l'atelier Freeman Craw Design. – *Polices :* Craw Clarendon (1955–1960), Craw Modern (1958–1964), Ad Lib (1961), Canterbury, Chaucery, Classic, Cursive, CBS Sans.

**Crous-Vidal,** Enrico – né en 1908 à Lérida, Espagne, décédé en 1987 à Noyon, France – *concepteur de polices, peintre* – études aux écoles d'art de Lérida et de Barcelone. Il fonde et dirige la revue «Art» de 1933 à 1934. Exil en France en 1939. Travaux de restauration. Graphiste à l'atelier de l'imprimerie Draeger à Mont-

Craw 1961 Poster

Craw 1987 Page

| | n | | | | |
|---|---|---|---|---|---|
| a | b | e | f | g | i |
| o | r | s | t | y | z |
| A | B | C | E | G | H |
| M | O | R | S | X | Y |
| 1 | 2 | 4 | 6 | 8 | & |

Craw 1961 Ad Lib

Crouwel 1967 Cover

school, two high schools, a church, a scientific institute and an art academy with a museum. 1932: the Cranbrook Academy of Art is founded with departments for sculpture, painting and architecture. The departments of weaving and costume design (1937), ceramics (1938), design (1938) and printed graphics, photography and metal are added. 1942: Cranbrook receives permission to issue diplomas. 1960: graphic design becomes a separate department. 1971: Katherine McCoy is made head of the graphic design department. Over the years, the teaching staff has included product designers Charles Eames and Harry Bertoia, sculptor Duane Hanson, and architects Eliel and Eero Saarinen and Daniel Libeskind. – *Publications include:* "The Cranbrook Vision 1925– 50", New York 1983; "Cranbrook Design: The New Discourse", New York 1991.

**Craw,** Freeman – b. 17. 1. 1917 in East Orange, New York, USA – *type designer, typographer, graphic designer* – 1935–39: studies at the Cooper Union Art School, New York. 1939–43: graphic artist for the American Colortype Company, New York. 1943: is made art director of Tri-Arts Press, New York, becoming their vice-president in 1958. He then opens the Freeman Craw Design Studio. – *Fonts:* Craw Clarendon (1955–60), Craw Modern (1958–64), Ad Lib (1961), Canterbury, Chaucery, Classic, Cursive, CBS Sans.

**Crous-Vidal,** Enrico – b. 1908 in Lérida, Spain, d. 1987 in Noyon, France – *type designer, painter* – Studied at the art schools in Lérida and Barcelona. 1933–

da, Spanien, gest. 1987 in Noyon, Frankreich. – *Schriftentwerfer, Maler* – Studium an den Kunstschulen in Lérida und Barcelona. 1933–34 Gründung und Leitung der Zeitschrift „Art". 1939 Exil in Frankreich. Restaurierungsarbeiten. Grafiker im Atelier der Druckerei Draeger in Montrouge. 1950 eigenes grafisches Atelier in Boulogne-Billancourt, Stoff- und Dekorationsentwürfe. Mitarbeiter in der Fonderie Typographique Française (FTF). Ende der fünfziger Jahre Hinwendung zur Malerei. – *Schriftentwürfe:* Champs Elysées (1959), Ile de France (1960).

**Crouwel,** Wim – geb. 21. 11. 1928 in Groningen, Niederlande. – *Typograph, Grafik-Designer, Schriftentwerfer, Lehrer, Museumsleiter* – 1947–49 Studium an der Akademie Minerva in Groningen. 1951–52 Studium am Instituut voor Kunstnijverheidsonderwijs in Amsterdam. 1952 eigenes Design-Büro, 1957–61 Wim Crouwel Design Studio, Amsterdam. Unterrichtet 1957–63 am Instituut voor Kunstnijverheidsonderwijs in Amsterdam. 1958 mit dem H. N. Werkman-Preis der Stadt Amsterdam ausgezeichnet. 1963 Gründung der Agentur „Total Design" mit Friso Kramer, Paul Schwarz und Benno Wissing. 1963–66 Generalsekretär der ICOGRADA. Ab 1964 Grafik-Design für das Stedelijk Museum, Amsterdam. Unterrichtet 1965–85 an der Universität Delft. 1972–80 Direktor von Total Design in Amsterdam. 1985–93 Direktor des Boymans-van Beuningen Museums in Rotterdam. 1987 Stiftungsprofessur für Kunst- und Kulturwissenschaften an der Erasmus-Universität Rot-

Crouwel 1957 Poster

Crouwel 1963 Poster

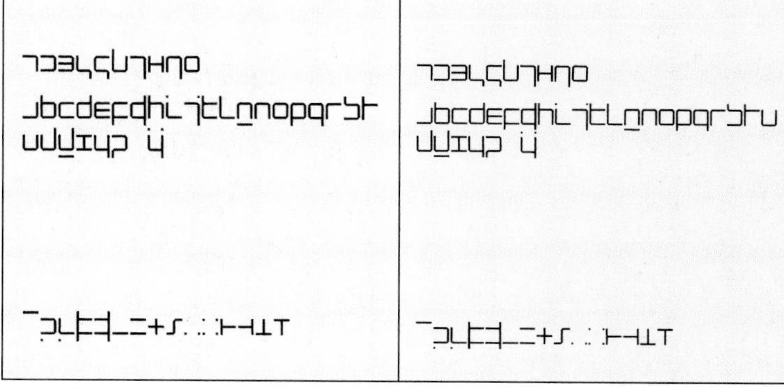

Crouwel 1967 Spread

Crouwel 1959 Poster

rouge. En 1950, il ouvre son propre atelier de graphiste à Boulogne-Billancourt, où il crée des motifs d'imprimés pour tissus et travaille comme décorateur. Collaboration avec la Fonderie Typographique Française (FTF). Il se consacre ensuite à la peinture à la fin des années cinquante. – *Polices:* Champs Elysées (1959), Ile de France (1960).

**Crouwel,** Wim – né le 21. 11 1928 à Groningue, Pays-Bas. – *typographe, graphiste maquettiste, concepteur de polices, enseignant, directeur de musée* – 1947–1949, études à l'Akademie Minerva à Groningue. 1951–1952, études à l'Instituut voor Kunstnijverheidsonderwijs d'Amsterdam. En 1952, il ouvre son propre atelier de design, puis de 1957 à 1961, le Wim Crouwel Design Studio Amsterdam. 1957–1963, enseigne à l'Instituut voor Kunstnijverheidsonderwijs d'Amsterdam. En 1958, il reçoit le prix H. N. Werkman de la ville d'Amsterdam. 1963, création de l'agence «Total Design» avec Friso Kramer, Paul Schwarz et Benno Wissing. 1963–1966, secrétaire général de l'ICOGRADA. A partir de 1964, il réalise des travaux graphiques pour le Stedelijk Museum, Amsterdam. Il enseigne à l'université de Delft de 1965 à 1985. Directeur de Total Design à Amsterdam de 1972 à 1980. Directeur du Boymans-van Beuningen Museum à Rotterdam de 1985 à 1993. Chaire d'art et d'histoire des cultures à l'université Erasmus de Rotterdam en 1987. En 1990, il reçoit le prix Anton Stankowski, puis le prix Piet Zwart en 1991. Travaille comme

34: founder and director of "Art" magazine. 1939: exile in France. Restoration work. Graphic artist for the Draeger Press studio in Montrouge. 1950: opens his own graphics studio in Boulogne-Billancourt, producing fabric and décor designs. Works for the Fonderie Typographique Française (FTF). Turns to painting at the end of the Fifties. – *Fonts:* Champs Elysées (1959), Ile de France (1960).

**Crouwel,** Wim – b. 21. 11. 1928 in Groningen, The Netherlands – *typographer, graphic designer, type designer, teacher, museum curator* – 1947–49: studies at the Minerva Akademie in Groningen. 1951–52: studies at the Instituut voor Kunstnijverheidsonderwijs in Amsterdam. 1952: opens his own design studio. 1957–61: the Wim Crouwel Design Studio in Amsterdam. 1957–63: teaches at the Instituut voor Kunstnijverheidsonderwijs in Amsterdam. 1958: receives the H. N. Werkman Prize from the city of Amsterdam. 1963: founds the studio Total Design with Friso Kramer, Paul Schwarz and Benno Wissing. 1963–66: general secretary of ICOGRADA. From 1964 onwards: produces graphic design for the Stedelijk Museum in Amsterdam. 1965–85: teaches at the University of Delft. 1972–80: director of Total Design in Amsterdam. 1985–93: director of the Boymans-van Beuningen Museum in Rotterdam. 1987: professor of art and cultural studies at the Erasmus University in Rotterdam. 1990: is awarded the Anton Stankowski Prize. 1991: is awarded the Piet Zwart Prize. Since 1993 graphic de-

terdam. 1990 mit dem Anton Stankowski-Preis ausgezeichnet. 1991 Auszeichnung mit dem Piet Zwart-Preis. Seit 1993 Grafik-Designer in Amsterdam. – *Schriftentwürfe:* New Alphabet (1967), Fodor Alphabet (1969). – *Publikationen:* „Nieuw Alfabet", Hilversum 1967; „An International Survey of Packaging" (mit Kurt Weidemann), London 1968; „Ontwerpen en Drukken", Amsterdam 1974.

**Curwen Press** – 1863 John Curwen gründet die Curwen Press in London. Der Enkel Harold Spedding Curwen (1885–1949) übernimmt 1913 die Presse. Er hat ein Jahr in Leipzig, Deutschland und bei Edward Johnston an der Central School of Arts and Crafts studiert. Curwen reduziert die über 200 Schriften der Presse auf die Caslon Old Face, später kommen die Imprint und Goudy Kennerley dazu. Seit 1915 hat die Presse als Emblem ein weißes Einhorn. 1920 tritt Oliver Simon in die Presse ein und ist für die Typographie der meisten Bücher verantwortlich. 1932–39 erscheinen 16 Ausgaben der Informationsbroschüre „Curwen Press Newsletter" sowie zahlreiche bibliophile Publikationen. Für die Presse arbeiten Künstler wie Paul Nash, Edward Bawden, Edward McKnight Kauffer, Graham Sutherland und John Piper, die originalgrafische Suiten schaffen. Harold Curwen beendet 1940 seine aktive Arbeit in der Presse. 1964 Zusammenschluß mit der Firma Harleyprint. Basil Harley wird Managing Director. Seit 1969 erscheinen jährlich die „Curwen's British Wildlife Calendars".

**Cyan** – *Grafik-Design-Studio* – Gegrün-

graphiste maquettiste à Amsterdam depuis 1993. – *Polices:* New Alphabet (1967), Fodor Alphabet (1969). – *Publications:* «Nieuw Alfabet», Hilversum 1967; «An International Survey of Packaging» (avec Kurt Weidemann), Londres 1968; «Ontwerpen en Drukken», Amsterdam 1974.

**Curwen Press** – 1863 fondée par John Curwen à Londres. En 1913, le petit-fils, Harold Spedding Curwen (1885–1949) reprend la presse après avoir passé un an à Leipzig en Allemagne et avoir étudié chez Edward Johnston à la Central School of Arts and Crafts. Curwen réduit les quelque 200 polices des presses à la Caslon Old Face, plus tard, il ajoute la Imprint et la Goudy Kennerley. A partir de 1915, les presses prennent pour l'emblème une licorne blanche. Oliver Simon entre aux presses en 1920 et devient responsable de la typographie pour la plupart des livres. De 1932 à 1939 les Curwen Press publient 16 éditions de la brochure «Curwen Press Newsletter», ainsi que de nombreux ouvrages pour bibliophiles. Des artistes comme Paul Nash, Edward Bawden, Edward McKnight Kauffer, Graham Sutherland et John Piper, ont créé des séries typographiques originales pour les presses. Harold Curwen cesse de travailler pour les presses en 1940. En 1964, fusion des presses avec la société Harleyprint. Basil Harley devient directeur du management. Depuis 1969 ils publient le «Curwen's British Wildlife Calendars».

**Cyan** – *atelier de graphisme* – fondé en 1992 par Daniela Haufe et Detlef Fiedler à Berlin. Daniela Haufe – née le 5. 11. 1966 à Berlin, Allemagne – apprentissage de composition typographique, maquettiste en édition pendant trois ans, puis indépendante pendant une année au cours de laquelle elle se consacre surtout à la pho-

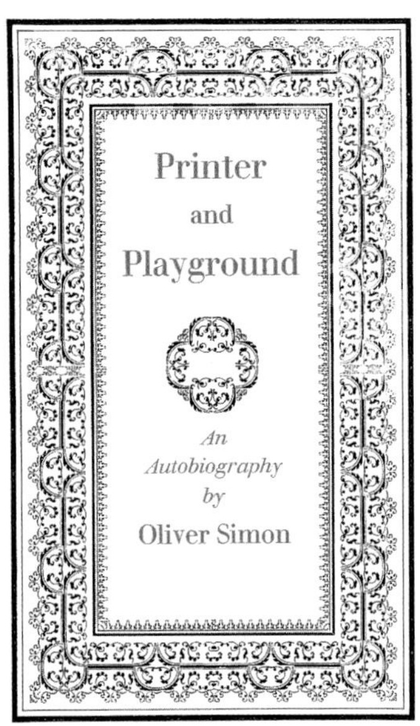


Curwen Press 1956 Cover

Cyan 1997 Program

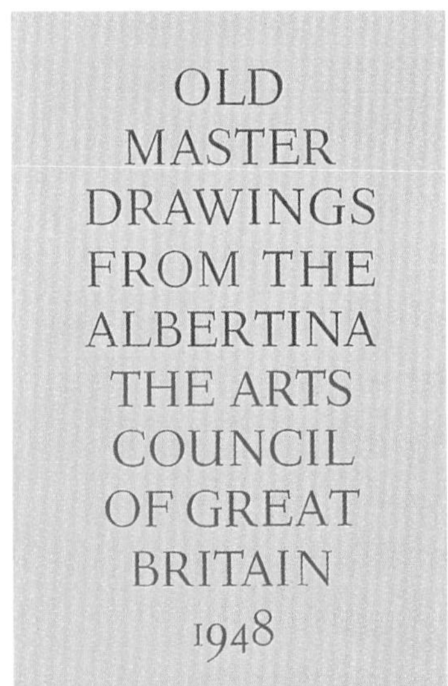

OLD MASTER DRAWINGS FROM THE ALBERTINA
THE ARTS COUNCIL OF GREAT BRITAIN
1948

Curwen Press 1948 Cover

Cyan 1992 Poster

Cyan 1992 Poster

signer in Amsterdam. – *Fonts:* New Alphabet (1967), Fodor Alphabet (1969). – *Publications:* "Nieuw Alfabet", Hilversum 1967; "An International Survey of Packaging" (with Kurt Weidemann), London 1968; "Ontwerpen en Drukken", Amsterdam 1974.

**Curwen Press** – 1863: John Curwen founds the Curwen Press in London. 1913: his grandson, Harold Spedding Curwen (1885– 1949), takes over the press. After spending a year studying in Leipzig, Germany and with Edward Johnston at the Central School of Arts and Crafts, Curwen reduces the number of typefaces used by the press (numbering over 200) to Caslon Old Face, later adding Imprint and Goudy Kennerley. 1915: a white unicorn is made the press's emblem. 1920 Oliver Simon joins the press. He is responsible for the typography of most of the books. 1932–39: 16 issues of the "Curwen Press Newsletter" are published, as well as numerous lavish editions for bibliophiles. Artists such as Paul Nash, Edward Bawden, Edward McKnight Kauffer, Graham Sutherland and John Piper have worked for the press, creating original graphic suites. 1940: Harold Curwen stops working for the press. 1964: the Curwen Press and Harleyprint merge. Basil Harley is made managing director. "Curwen's British Wildlife Calendars" have been published every year since 1969.

**Cyan** – *graphic design studio* – Founded in 1992 by Daniela Haufe and Detlef Fiedler in Berlin. Daniela Haufe – b. 5. 11. 1966 in Berlin, Germany – Trained as a compositor, then spent three years

det 1992 von Daniela Haufe und Detlef Fiedler in Berlin. Daniela Haufe – geb. 5. 11. 1966 in Berlin, Deutschland. – Schriftsetzerlehre, danach drei Jahre gestalterische Verlagsarbeit, ein Jahr freie Arbeiten, vor allem Fotografie. Eintritt in die Berliner Grafik-Design-Gruppe Grappa, Austritt 1992. Detlef Fiedler – geb. 3. 7. 1955 in Schönebeck an der Elbe, Deutschland. – Architekturstudium in Weimar, danach Gartenarchitekt am Schloß Belvedere. Gestalter bei der Deutschen Werbe- und Anzeigengesellschaft (Dewag). Gründungsmitglied der Grafik-Design-Gruppe Grappa, Austritt 1992. Cyan arbeitet vornehmlich für den kulturellen Bereich. Auftraggeber u. a. Bauhaus Dessau sowie die Berliner Staatsoper. Es entstanden ein multimediales Bühnenbild für eine Oper von Kurt Weill, ein Ballettfilm für Copland. Seit 1991 Mitherausgeber und Gestalter der Zeitschrift für Gestaltung „Form + Zweck".

**Cyliax,** Walter – geb. 1899 in Leipzig, Deutschland, gest. 1945 in Wien, Österreich. – *Grafik-Designer, Typograph* – Studium an der Akademie für graphische Künste und Buchgewerbe in Leipzig. Umzug nach Zürich. 1929 künstlerischer Leiter der Druckerei Fretz, Zürich und Herausgeber des umfangreichen Sonderhefts „Schweiz" der Zeitschrift „Archiv für Buchgewerbe und Gebrauchsgrafik", Leipzig. 1929 Beteiligung an der Ausstellung „Film und Foto" des Deutschen Werkbunds in Stuttgart. Kontakte zum „ring neuer werbegestalter". 1933 Beteiligung an der Gründung der Zeitschrift „Typographische Monatsblätter", die der

Cyan 1994 Poster

Cyan 1995 Poster

Cyliax 1929 Poster

tographie. Entre chez Grappa, un groupe de graphistes maquettistes berlinois qu'elle quitte en 1992. Detlef Fiedler – né le 3. 7. 1955 à Schönebeck an der Elbe, Allemagne – études d'architecture à Weimar, puis paysagiste au château du Belvédère. Travaille comme designer pour la Deutsche Werbe- und Anzeigengesellschaft (Dewag). Membre fondateur du groupe Grappa qu'il quitte en 1992. Cyan travaille surtout dans le domaine culturel. Commanditaires : Bauhaus Dessau et le Staatsoper de Berlin. Création d'une scénographie multimédia pour un opéra de Kurt Weill, un ballet filmé pour Copland. Depuis 1991, coéditeur et maquettiste de la revue de design «Form + Zweck».

**Cyliax,** Walter – né en 1899 à Leipzig, Allemagne, décédé en 1945 à Vienne, Autriche – *graphiste maquettiste, typographe* – études à l'Akademie für graphische Künste und Buchgewerbe (arts graphiques et du livre) à Leipzig. S'installe à Zurich. 1929, directeur artistique de l'imprimerie Fretz à Zurich où il dirige les numéraux spéciaux sur la Suisse (Schweiz) de la revue «Archiv für Buchgewerbe und Gebrauchsgrafik» de Leipzig. En 1929, il participe à l'exposition «Film und Foto» organisée par le Deutscher Werkbund à Stuttgart. Contacts avec le «ring neuer werbegestalter». En 1933, il participe à la création de la revue «Typographische Monatsblätter» éditée

designing for various publishers and a year working freelance, concentrating on photography. Joined the graphic design group Grappa in Berlin; left the group in 1992. Detlef Fiedler – b. 3. 7. 1955 in Schönebeck an der Elbe, Germany – Studied architecture in Weimar, then worked as a landscape gardener for Schloß Belvedere. Designer for the Deutsche Werbe- und Anzeigengesellschaft (Dewag). Founder member of the graphic design group Grappa; left in 1992. Cyan primarily works for the arts sector. Clients include Bauhaus Dessau and the Staatsoper in Berlin. Cyan has designed a multimedia set for a Kurt Weill opera and a Copland ballet film. Since 1991: co-editors and designers of the design magazine "Form + Zweck".

**Cyliax,** Walter – b. 1899 in Leipzig, Germany, d. 1945 in Vienna, Austria – *graphic designer, typographer* – Studied at the Akademie für graphische Künste und Buchgewerbe in Leipzig. Moved to Zurich. 1929: art director of the Fretz printing plant in Zurich and editor of a voluminous special edition of the "Archiv für Buchgewerbe und Gebrauchsgrafik" magazine in Leipzig, entitled "Switzerland". 1929: takes part in the "Film und Foto" exhibition held by the Deutscher Werkbund in Stuttgart. Contact with the "ring neuer werbegestalter". 1933: involved in the launching of the "Typographische Monatsblätter" periodical, published by the Schweizer Typographenbund in Bern. 1937: moves to Vienna and runs a large printing works.

**Czeschka,** Carl Otto – b. 22. 10. 1878 in

Schweizer Typographenbund in Bern herausgab. 1937 Umzug nach Wien, Leiter einer Großdruckerei.

**Czeschka,** Carl Otto – geb. 22. 10. 1878 in Wien, Österreich, gest. 30. 7. 1960 in Hamburg, Deutschland. – *Grafiker, Maler, Schriftentwerfer, Lehrer* – 1894–99 Studium an der Akademie der Bildenden Künste in Wien. 1890 Mitglied der Wiener Sezession. 1905 Mitarbeiter der 1903 gegründeten „Wiener Werkstätte". 1907 Ausstattung der Inszenierung „King Lear" von Max Reinhardt am Deutschen Theater in Berlin. 1907–43 Lehrer an der Kunstgewerbeschule in Hamburg. 1909 Ernennung zum Professor. 1911 Ausstellung seiner Arbeit im Museum für Kunst und Gewerbe in Hamburg. 1918–50 grafische Arbeiten für die Zigarrenfabrik L. Wolff. 1920 Signet der Justus-Brinckmann-Gesellschaft in Hamburg. 1947 Titelentwurf für die Wochenzeitung „Die Zeit". 1974 gehen große Teile des Nachlasses an das Museum für Kunst und Gewerbe in Hamburg. – *Schriftentwürfe:* Olympia (1909), Czeschka-Antiqua (1914).

par l'union des typographes suisses à Berne. Il s'installe à Vienne en 1937, où il dirige une grande imprimerie.

**Czeschka,** Carl Otto – né le 22. 10. 1878 à Vienne, Autriche, décédé le 30. 7. 1960 à Hambourg, Allemagne – *graphiste, peintre, concepteur de polices, enseignant* – 1894–1899, études à l'Akademie der Bildenden Künste (beaux-arts) de Vienne. 1890, membre de la Sécession de Vienne. 1905, travaille pour la «Wiener Werkstätte», créée en 1903. En 1907, il conçoit les décors de «King Lear» pour une mise en scéne de Max Reinhardt au Deutsches Theater de Berlin. De 1907 à 1943, enseigne à la Kunstgewerbeschule (école des arts décoratifs) de Hambourg. 1909, est nommé professeur. 1911, expose ses travaux au Museum für Kunst und Gewerbe de Hambourg. 1918–1950, travaux graphiques pour le fabricant de cigarettes L. Wolff. 1920, réalise le signet de la société Justus Brinckmann de Hambourg. En 1947, il dessine la maquette du titre de l'hebdomadaire hambourgeois «Die Zeit». En 1974, une grande partie de la succession entre au Museum für Kunst und Gewerbe de Hambourg. – *Polices:* Olympia (1909), Czeschka-Antiqua (1914).

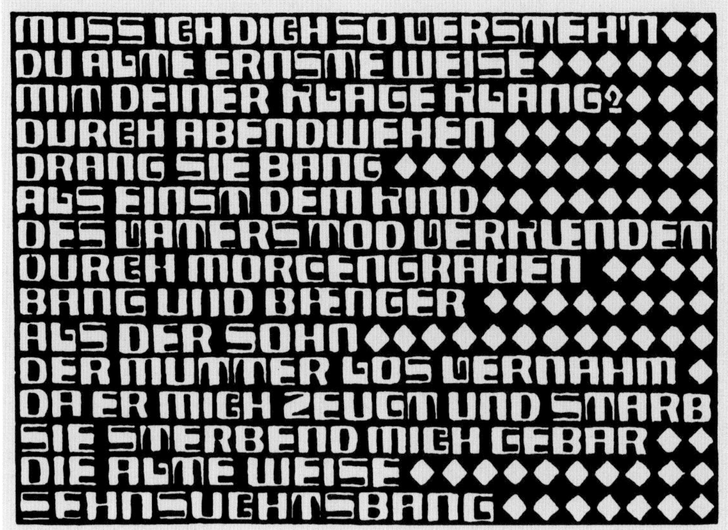

Czeschka 1905 Sketch

Czeschka

Czeschka 1911 Cover

Czeschka 1909 Cover

Vienna, Austria, d. 30. 7. 1960 in Hamburg, Germany – *graphic artist, painter, type designer, teacher* – 1894–99: studies at the Akademie der Bildenden Künste in Vienna. 1890: member of the Vienna Secession. 1905: works for the Wiener Werkstätte, founded in 1903. 1907: designs the décor for Max Reinhardt's production of "King Lear" staged at the Deutsches Theater in Berlin. 1907–43: teacher at the Kunstgewerbeschule in Hamburg. 1909: made a professor. 1911: exhibition of his works in the Museum für Kunst und Gewerbe in Hamburg. 1918–50: produces graphics for the L. Wolff cigar factory. 1920: designs the logo for the Justus-Brinckmann-Gesellschaft in Hamburg. 1947: designs the title for the Hamburg weekly "Die Zeit". 1974: a large proportion of his estate is given to the Museum für Kunst und Gewerbe in Hamburg. – *Fonts:* Olympia (1909), Czeschka-Antiqua (1914).

# D

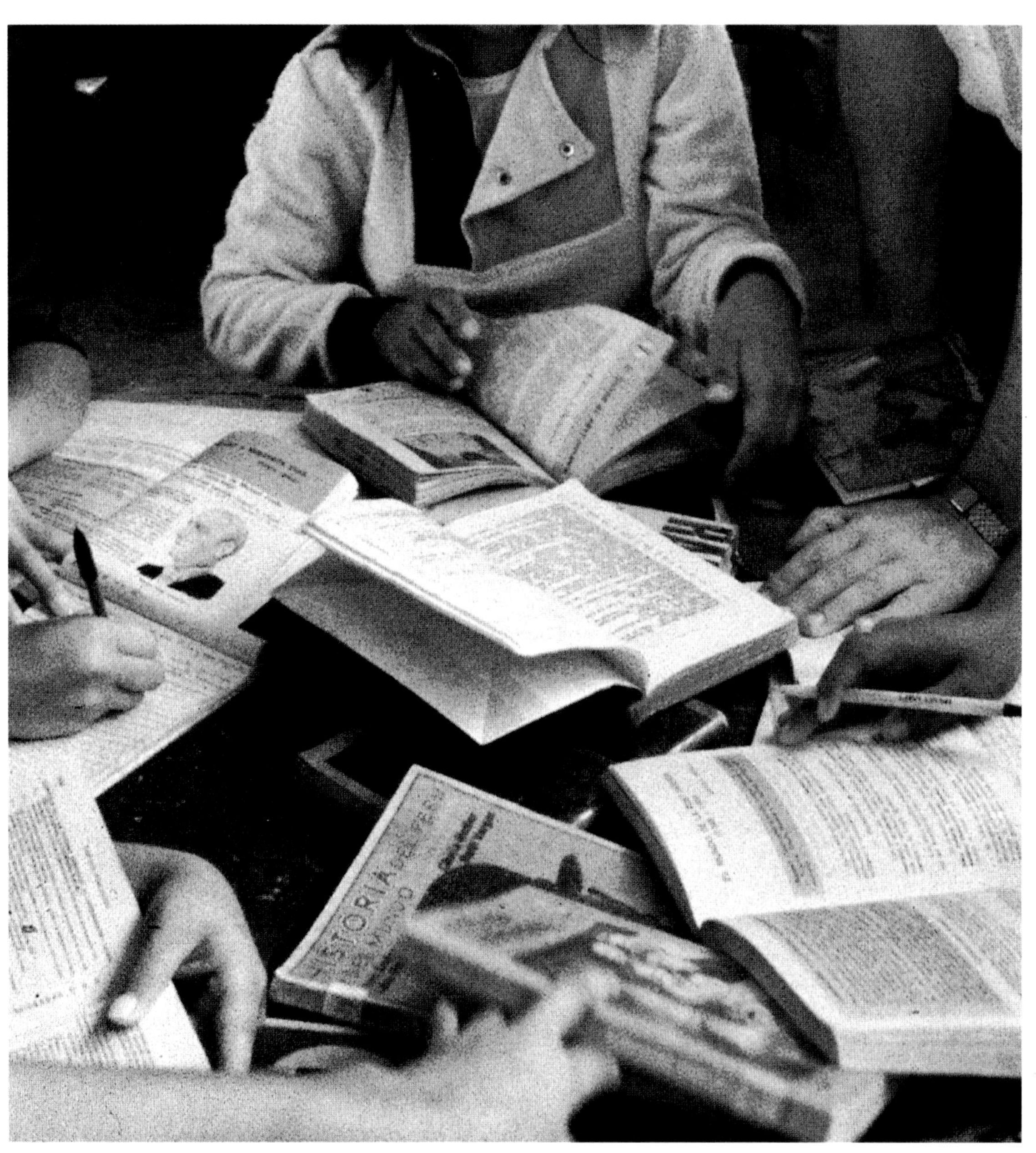

**Damase,** Jacques – geb. 1930 in Brest, Frankreich. – *Verleger, Typograph* – 1947–48 erste verlegerische Arbeiten. 1950–53 Leiter der Zeitschrift „Labyrinthe“. 1957–62 Veröffentlichung von Büchern über Chagall, Braque und Picasso. Mitarbeit an der Zeitschrift „Connaissance des Arts“. Arbeiten für den Condé Nast Verlag, Art Director der Zeitschrift „House and Garden“. 1964 Herausgabe eines Buches über Sonia Delaunay. 1968 Eröffnung seiner Galerie in Paris: Ausstellungen u. a. der Werke von Sonia und Robert Delaunay, Félicien Rops, Alan Davie, Alexander Calder. 1973 Eröffnung seiner Galerie in Brüssel: Ausstellungen u. a. der Werke von Max Ernst, Jean Cocteau, Jim Dine. Herausgabe zahlreicher Katalogbücher. 1979 Ausstellung über seine Verlegertätigkeit im Centre Georges Pompidou, Paris. Ab 1980 Wanderausstellung seiner Bücher. 1980–90 Herausgabe zahlreicher Bücher über Mode und über Typographie. Ernennung zum „Chevalier des arts et des lettres“ durch den französischen Kultusminister. 1989 wird ihm für die Herausgabe seines Buches „Les Ballets suédois 1920–1925“ der Prix Odilon Redon verliehen. 1991 Ausstellung im Gutenberg-Museum, Mainz. Neuausgabe des „Manuel Typographique“ von Fournier mit Bertram Schmidt-Friderichs. – *Publikationen u. a.:* „Révolution Typographique depuis Stéphane Mallarmé“, Genf 1966. Bertram Schmidt-Friderichs „Jacques Damase. Visuelle Leidenschaft“, Mainz 1991.

**Deberny & Peignot** – *Schriftgießerei* – 1923 Zusammenschluß der Schriftgieße-

**Damase,** Jacques – né en 1930 à Brest, France – *editeur, typographe* – 1947–1948, premières activités dans l'édition. Dirige la revue «Labyrinthe» de 1950 à 1953. Entre 1957 et 1962, publication d'ouvrages sur Chagall, Braque et Picasso. Collaborateur à la revue «Connaissance des Arts». Travaux pour les publications Condé Nast, directeur artistique de la revue «House and Garden». 1964, publication d'un livre sur Sonia Delaunay. 1968, inauguration de sa galerie à Paris où il expose des œuvres de Sonia et Robert Delaunay, Félicien Rops, Alan Davie, Alexander Calder. Il ouvre une galerie à Bruxelles en 1973, où il expose des œuvres de Max Ernst, Jean Cocteau, Jim Dine, etc. Edition de nombreux catalogues. 1979, exposition sur ses activités d'éditeur au Centre Georges Pompidou, à Paris, puis exposition itinérante de ses livres à partir de 1980. Edition de nombreux ouvrages sur la mode et la typographie de 1980 à 1990. Il est décoré «Chevalier des arts et des lettres» par le ministre de la culture. En 1989, le prix «Odilon Redon» lui est décerné pour un ouvrage intitulé «Les Ballets suédois 1920–1925». Exposition au Gutenberg-Museum à Mayence en 1991. Nouvelle édition du «Manuel Typographique» de Fournier avec Bertram Schmidt-Friderichs. – *Publications, sélection:* «Révolution typographique depuis Stéphane Mallarmé», Genève 1966. Bertram Schmidt-Friderichs «Jacques Damase. Visuelle Leidenschaft», Mayence 1991.

**Deberny & Peignot** – *fonderie de caractères* – 1923, création de la Fonderie Deberny & Peignot à la suite de la fusion des fonderies de caractères Girard & Cie. et Peignot & Cie. Sur l'initiative de Charles Peignot, les polices créées par Cassandre, Bifur (1929), Acier noir (1936), Peignot (1937) et Touraine (1947) y sont pro-

Damase 1986 Cover

Damase 1991 Cover

Damase 1985 Cover

| | | | | | |
|---|---|---|---|---|---|
| | n | n | | | |
| a | b | e | f | g | i |
| o | r | s | t | y | z |
| A | B | C | E | G | H |
| M | O | R | S | X | Y |
| 1 | 2 | 4 | 6 | 8 | & |

Deck 1990 Template Gothic

**Damase,** Jacques – b. 1930 in Brest, France – *publisher, typographer* – 1947–48: starts working in publishing. 1950–53: director of "Labyrinthe" magazine. 1957–62: publishes books on Chagall, Braque and Picasso. Works for "Connaissance des Arts" magazine. Works for Condé Nast Publications; art director of "House and Garden" magazine. 1964: publishes a book on Sonia Delaunay. 1968: opens his gallery in Paris, exhibiting works by Sonia and Robert Delaunay, Félicien Rops, Alan Davie and Alexander Calder, among others. 1973: opens his gallery in Brussels, where the artists Max Ernst, Jean Cocteau and Jim Dine are exhibited (among others). Publishes many catalogues. 1979: an exhibition is held documenting his work as a publisher in the Centre Georges Pompidou in Paris. From 1980 onwards: touring exhibition of his books. 1980–90: publishes numerous books on fashion and typography. The French Minister of Education and the Arts makes him a Chevalier des arts et lettres. 1989: his book "Les Ballets suédois 1920–1925" is published, for which he is awarded the Prix Odilon Redon. 1991: exhibition in the Gutenberg Museum in Mainz. Publishes a new edition of Fournier's "Manuel Typographique" with Bertram Schmidt-Friderichs. – *Publications include:* "Révolution Typographique depuis Stéphane Mallarmé", Geneva 1966. Bertram Schmidt-Friderichs "Jacques Damase. Visuelle Leidenschaft", Mainz 1991.

**Deberny & Peignot** – *type foundry* – 1923: the Girard & Cie. and Peignot & Cie. type foundries merge to form the Fonderie De-

reien Girard & Cie. und Peignot & Cie. zur Fonderie Deberny & Peignot. Durch Initiative Charles Peignots wurden Cassandres Schriften Bifur (1929), Acier noir (1936), Peignot (1937) und Touraine (1947) produziert. 1950 Entwicklung von Lichtsetzmaschinen mit dem Namen Lumitype, die in den USA den Namen Photon tragen. 1952 Eintritt Adrian Frutigers in die Firma. 1972 Schließung der Firma, das Schriften-Gußprogramm wird von der Haas'schen Schriftgießerei weitergeführt.

**Deck,** Barry – geb. 1. 12. 1962 in Mount Pleasant, Iowa, USA. – *Grafik-Designer, Schriftentwerfer, Lehrer* – 1986 Abschluß des Studiums an der Northern Illinois University. 1986–87 Junior Designer bei Lipman + Simons, Chicago. 1987 Grafik-Designer bei Kym Abrams Design, Chicago. Seit 1987 freischaffender Grafik-Designer in Los Angeles, Chicago, New York. Auftraggeber waren u. a. das American Center for Design, „Emigre", Sony Music, RCA Records, Santa Fe Railroads, „Fuse", Museum of Contemporary Art, Chicago. 1989 Abschluß eines Studiums am California Institute of the Arts. Unterrichtet 1989–90 Typographie und Grafik-Design an der Otis/Parson's School of Design, Los Angeles. Zahlreiche Fernsehspots, u. a. für Crystal Pepsi (1992), Reebok, Nickelodeon (1993), Vox TV, Köln (1995). 1995 Gründung seiner Firma Dysmedia in New York. – *Schriftentwürfe:* Canicopulus Script (1989), Barry Sans Serif (1989), Mutant Industry Roman (1989), Bombadeer, Industry Sans, Template Gothic (1990), Caustic Biomorph (1992), Arbitrary (1992), Cyberotica (1994).

Deck 1990 Advertisement

|  | n | **n** |  |  |  |
|---|---|---|---|---|---|
| a | b | e | f | g | i |
| o | r | s | t | y | z |
| A | B | C | E | G | H |
| M | O | R | S | X | Y |
| 1 | 2 | 4 | 6 | 8 | & |

Deck 1992 Arbitrary

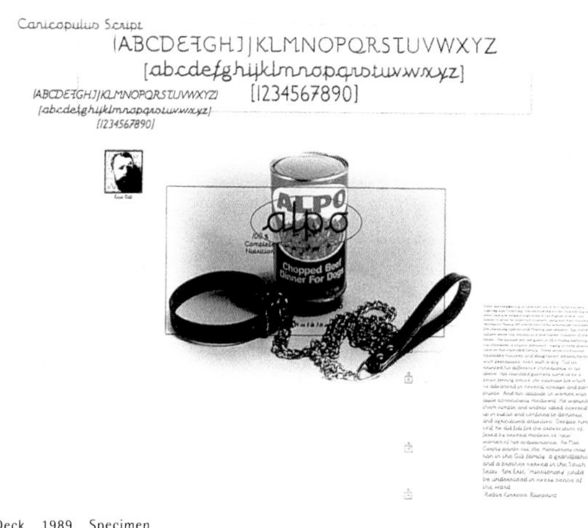


Deck 1989 Specimen

Deck 1988 Poster

duites. 1950, invention d'une machine à photocomposer appelée Lumitype, qui porte le nom de Photon aux Etats-Unis. Adrian Frutiger entre dans la société en 1952. Fermeture de la fonderie en 1972, la fonderie de caractère Haas reprend le catalogue des fontes.

**Deck,** Barry – né le 1. 12. 1962 à Mount Pleasant, Iowa, Etats-Unis – *graphiste maquettiste, concepteur de polices, enseignant* – 1986, termine ses études à la Northern Illinois University. 1986–1987, Junior Designer chez Lipman + Simons, Chicago. 1987, graphiste maquettiste chez Kym Abrams Design, Chicago. Graphiste maquettiste indépendant à Los Angeles, Chicago, New York depuis 1987. Commanditaires: American Center for Design, «Emigre», Sony Music, RCA Records, Santa Fe Railroads, «Fuse», Museum of Contemporary Art, Chicago etc. 1989, termine ses études au California Institute of the Arts. Enseigne la typographie et le design graphique à la Otis/Parson's School of Design, Los Angeles de 1989 à 1990. De nombreuses publicités télévisées, par ex. pour Crystal Pepsi (1992), Reebok (1993), Nickelodeon (1993), Vox TV, Cologne (1995). Fonde sa société, la Dysmedia, à New York, en 1995. – *Polices:* Canicopulus Script (1989), Barry Sans Serif (1989), Mutant Industry Roman (1989), Bombadeer, Industry Sans, Template Gothic (1990), Caustic Biomorph (1992), Arbitrary (1992), Cyberotica (1994).

berny & Peignot. On the initiative of Charles Peignot, Cassandre's typefaces Bifur (1929), Acier noir (1936), Peignot (1937) and Touraine (1947) are produced. 1950: photo-typesetting machines are developed under the name Lumitype (known as Photon in the USA). 1952: Adrian Frutiger joins the foundry. 1972: the foundry closes down. Its type program is continued by the Haas'schen type foundry.

**Deck,** Barry – b. 1. 12. 1962 in Mount Pleasant, Iowa, USA – *graphic designer, type designer, teacher* – 1986: graduates from Northern Illinois University. 1986–87: junior designer for Lipman + Simons, Chicago. 1987: graphic designer for Kym Abrams Design, Chicago. Since 1987: freelance graphic designer in Los Angeles, Chicago and New York. Clients include the American Center for Design, "Emigre" magazine, Sony Music, RCA Records, Santa Fe Railroads, "Fuse" magazine and the Museum of Contemporary Art, Chicago. 1989: completes a course of studies at the California Institute of the Arts. 1989–90: teaches typography and graphic design at the Otis/Parson's School of Design, Los Angeles. Numerous TV commercials, including for Crystal Pepsi (1992), Reebok (1993), Nickelodeon (1993) and Vox TV, Cologne (1995). 1995: founds his own company, Dysmedia, in New York. – *Fonts:* Canicopulus Script (1989), Barry Sans Serif (1989), Mutant Industry Roman (1989), Bombadeer, Industry Sans, Template Gothic (1990), Caustic Biomorph (1992), Arbitrary (1992), Cyberotica (1994).

**Deffke,** Wilhelm – geb. April 1886 im Rheinland, Deutschland, gest. August 1950 in Berlin, Deutschland. – *Grafik-Designer, Lehrer* – Zahlreiche Plakate, Signets, Werbedrucksachen und Architekturentwürfe. Auftraggeber waren u. a. Krupp, Mercedes-Benz, Siemens & Halske, AEG, Bata, Reemtsma, British American Tobacco, BMW. 1912 Leiter der Klasse für Gebrauchsgrafik der Höheren Fachschule für Dekorationskunst Berlin (Reimann-Schule). Bis 1933 Leitung der Kunstgewerbe- und Handwerkerschule in Magdeburg; Entlassung aus dem Amt durch die Nationalsozialisten. 1949 Mitarbeit bei der Neugründung der Landesgruppe Berlin des Bundes Deutscher Gebrauchsgraphiker (BDG). – *Publikation:* Sonderheft „Wilhelm Deffke“ der Zeitschrift „Seidels Reklame“ 1/1923, Berlin.
**de Harak,** Rudolph – geb. 10. 4. 1924 in Culver City, Kalifornien, USA. – *Grafik-Designer, Typograph, Illustrator, Maler, Lehrer* – Autodidakt. Ab 1950 eigenes Designstudio in New York. Gestaltet über 350 Umschläge für Taschenbücher des Verlages McGraw-Hill mit gleichem Rasteraufbau und Grotesk-Typographie. Gestaltet 1967 den Pavillon „Man, his Planet and Space“ auf der Weltausstellung in Montreal. 1969 Gründung des Studios Corchia, de Harak, Inc. mit Al Corchia. Gestaltet 1970 den amerikanischen Pavillon auf der Weltausstellung in Osaka. Entwickelt ein Zeichensystem für das Metropolitan Museum of Art. Lehrtätigkeiten: School of Visual Arts (1962–63), Parson's School of Design (1973–78), Pratt Institute (1973–78), Kent State Uni-

**Deffke,** Wilhelm – né en avril 1886 en Rhénanie, Allemagne, décédé en août 1950 à Berlin, Allemagne – *graphiste maquettiste, enseignant* – nombreux affiches, signets, imprimés publicitaires et projets d'architecture. Commanditaires : Krupp, Mercedes-Benz, Siemens & Halske, AEG, Bata, Reemtsma, British American Tobacco, BMW, etc. En 1912, il dirige l'atelier de graphisme publicitaire à la Höhere Fachschule für Dekorationskunst (école supérieure des arts décoratifs) de Berlin (Reimann-Schule). Dirige le Kunstgewerbe- und Handwerkerschule (école d'artisanat d'art et des arts appliqués) de Magdebourg jusqu'en 1933; puis il est démis de ses fonctions par le régime du national-socialisme. En 1949, il participe à la réorganisation de la section berlinoise du Bund Deutscher Gebrauchsgraphiker (BDG). – *Publication :* Numéro spécial «Wilhelm Deffke» de la revue «Seidels Reklame» 1/1923, Berlin.
**de Harak,** Rudolph – né le 10. 4. 1924 à Culver City, Californie, Etats-Unis – *graphiste maquettiste, typographe, illustrateur, peintre, enseignant* – autodidacte. Dirige son atelier de design à New York à partir de 1950. Conçoit la maquette de plus de 350 couvertures de livres de poche pour les éditions McGraw-Hill en utilisant la même grille et des caractères grotesk. Dessine le pavillon «Man, his Planet and Space» pour l'exposition universelle de 1967 à Montréal. 1969, fondation de l'atelier Corchia, de Harak, Inc. avec Al Corchia. En 1970, il conçoit le pavillon américain pour l'exposition universelle d'Osaka. Création d'un système signalétique pour le Metropolitan Museum of Art. Enseignement à la School of Visual Arts (1962–1963), à la Parson's School of Design (1973–1978), au Pratt Institute (1973–1978), à la Kent State University Ohio (1976), à la Yale University (1979–

Deffke ca. 1920 Logotype

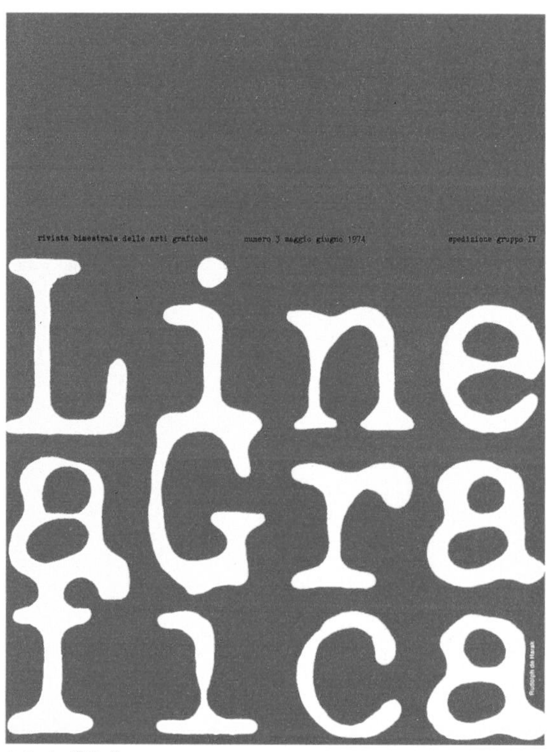

de Harak 1974 Cover

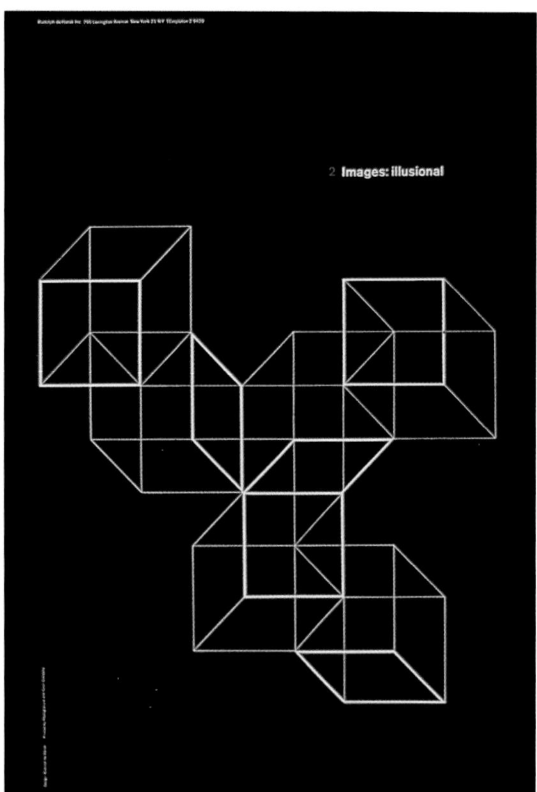

de Harak 1959 Poster

de Harak 1961 Cover

**Deffke,** Wilhelm – b. April 1886 in the Rhineland, Germany, d. 1950 in Berlin, Germany – *graphic designer, teacher* – Produced numerous posters, logos, advertising matter and architectural designs. Krupp, Mercedes-Benz, Siemens & Halske, AEG, Bata, Reemtsma, British American Tobacco and BMW were among his clients. 1912: head of the commercial art course at the Höhere Fachschule für Dekorationskunst in Berlin (Reimann-Schule). Until 1933: head of the Kunstgewerbe- und Handwerkerschule in Magdeburg. Removed from office by the National Socialists. 1949: helps set up the Berlin faction of the Bund Deutscher Gebrauchsgraphiker (BDG). – *Publication:* special issue of "Seidels Reklame" magazine, entitled "William Deffke" 1/1923, Berlin.
**de Harak,** Rudolph – b. 10. 4. 1924 in Culver City, California, USA – *graphic designer, typographer, illustrator, painter, teacher* – Self-taught. 1950: opens his own design studio in New York. Designs over 350 covers for paperback books published by McGraw-Hill with the same grid system and sans-serif typography. 1967: designs the pavilion "Man, his Planet and Space" for the world exhibition in Montreal. 1969: opens the Corchia, de Harak Studio, Inc. with Al Corchia. 1970: designs the American pavilion for the world exhibition in Osaka. Develops a notation system for the Metropolitan Museum of Art. Teaching positions: School of Visual Arts (1962–63), Parson's School of Design (1973–78), Pratt Institute (1973–78), Kent State Uni-

versity, Ohio (1976), Yale University (1979–80), Cooper Union, New York (seit 1979). Zahlreiche Ausstellungen und Auszeichnungen.

**Depero,** Fortunato – geb. 30. 3. 1892 in Fonto, Italien, gest. 29. 11. 1960 in Rovereto, Italien. – *Maler, Werbegrafiker, Möbelentwerfer* – 1911 erste Ausstellung künstlerischer Arbeiten in Rovereto. 1914 Umzug nach Rom. Anschluß an die futuristische Bewegung, Teilnahme an der ersten freien futuristischen Ausstellung in Rom. Verfaßt 1915 mit Giacomo Balla das Manifest „Riconstruzione Futurista dell'Universo". 1916 Zusammenarbeit mit Diaghilews „Ballets Russes"; Bühnenbild für „Le Chant du Rossignol" von Andersen mit Musik von Strawinsky. 1920 erste Wandteppiche und Werbeplakate. 1922 Eröffnung des „Cabaret del Diavolo", dessen gesamte Ausstattung wie Möbel, Dekoration, Beleuchtung Depero entworfen hat. Erhält 1925 auf der Internationalen Ausstellung moderner Raumkunst in Paris zahlreiche Auszeichnungen. Veröffentlicht 1927 das verschraubte Buch „Depero Futurista", eine Selbstdarstellung. Zahlreiche Ausstellungen, Kostümentwürfe für das Ballett „The New Babel". Gestaltung und Umschläge für die Zeitschriften „Vanity Fair", „Vogue" und „The New Yorker". Ab 1926 Gestaltung und Ausstellungsstände für Campari. In „Numero Unico Futurista Campari" verfaßt Depero 1931 das Manifest „Il Futurismo e l'Arte Pubblicataria". 1933 Herausgabe der Zeitschrift „Dinamo Futurista" in Rovereto. Lebt 1948–49 in New York. Gründet 1957 in Rovereto das „Museum Depero",

Depero 1934 Poster

Depero 1927 Cover

Depero 1931 Advertisement

Depero 1930 Advertisement

1980) et à la Cooper Union New York (depuis 1979). Nombreuses expositions et distinctions.

**Depero,** Fortunato – né le 30. 3. 1892 à Fonto, Italie, décédé le 29. 11. 1960 à Rovereto, Italie – *peintre, graphiste publicitaire, concepteur de mobilier* – 1911, première exposition de ses travaux à Rovereto. Il s'installe à Rome en 1914 et adhère au mouvement futuriste. Participe à la première exposition futuriste libre à Rome. En 1915, il écrit le manifeste «Riconstruzione Futurista dell'Universo» avec Giacomo Balla. Travaille pour les «Ballets Russes» de Diaghilev en 1916; scénographie du «Chant du Rossignol» d'Andersen sur une musique de Stravinsky. 1920, premières tapisseries murales et affiches publicitaires. 1922, ouverture du «Cabaret del Diavolo», dont le mobilier, la décoration et les éclairages sont entièrement conçus par Depero. Il obtient de nombreux prix à l'occasion de l'exposition internationale des arts de la décoration intérieure à Paris en 1925. Publie le livre boulonné «Depero Futurista» en 1927, une sorte de statement. Nombreuses expositions, costumes pour le ballet «The New Babel». Maquettes de couvertures pour les revues «Vanity Fair», «Vogue», et «The New Yorker». A partir de 1926, stands d'exposition pour Campari. En 1931, Depero publie le manifeste «Il Futurismo e l'Arte Pubblicataria» dans le «Numero Unico Futurista Campari». 1933, publication de la revue «Dinamo Futurista» à Rovereto. Vit à New York de 1948 à 1949. En 1957, il fonde le «Museum Depero» à Rovereto, musée inau-

versity, Ohio (1976), Yale University (1979–80) and the Cooper Union, New York (since 1979). Numerous exhibitions and prizes.

**Depero,** Fortunato – b. 30. 3. 1892 in Fonto, Italy, d. 29. 11. 1960 in Rovereto, Italy – *painter, commercial artist, furniture designer* – 1911: the first exhibition of his art work is held in Rovereto. 1914: moves to Rome. Joins the Futurist Movement and takes part in the first free exhibition of Futurist art in Rome. 1915: Writes the manifesto "Riconstruzione Futurista dell' Universo" with Giacomo Balla. 1916: works with Diaghilev's Ballets Russe. Designs the set for Andersen's "Le Chant du Rossignol" with music by Stravinsky. 1920: produces his first wall hangings and placards. 1922: the Cabaret del Diavolo opens with interiors designed by Depero, including the furniture, decoration and lighting. 1925: receives numerous prizes at the International Exhibition of Modern Interior Design in Paris. 1927: publishes his bolted book "Depero Futurista", a self-portrayal. Numerous exhibitions; costume designs for the ballet "The New Babel". Designs and covers for "Vanity Fair", "Vogue" and "The New Yorker" magazines. 1926: starts effecting designs and designing exhibition stands for Campari. In "Numero Unico Futurista Campari", Depero issues the manifesto "Il Futurismo e l'Arte Pubblicataria" in 1931. 1933: publishes the "Dinamo Futurista" in Rovereto. 1948–49: lives in New York. 1957: founds the Depero Museum in Rovereto, opened in 1959. – *Publications include:* "Libro Imbullonato", Milan 1927;

das 1959 eröffnet wird. – *Publikationen u. a.:* „Libro Imbullonato", Mailand 1927; „Fortunato Depero nelle Opere e nella Vita", Trento 1940; „So I Think, So I Paint", Trento 1947. G. Gianti „Fortunato Depero Futurista", Mailand 1951; B. Passamani „Fortunato Depero", Rovereto 1981.

**Derrida,** Jacques – geb. 15. 7. 1930 in El Biar, Algerien. – *Philosoph, Lehrer, Linguist* – Lehrt 1960–64 an der Sorbonne. 1964–84 Professor für Philosophiegeschichte an der Ecole Normale Supérieure in Paris. 1983 Gründungsdirektor des Collège International de Philosophie. Seit 1984 Forschungsleiter an der Ecole des Hautes Etudes en Sciences Sociales. In seinen dem Strukturalismus verbundenen Arbeiten wird der Versuch unternommen, die metaphysische Verfassung zu ergründen, die jeglicher systematischen Theorie zugrundeliegt. Schrift ist für ihn ein Modell für die „Differenz", die zwischen dem Zeichen und dem Bezeichneten entsteht. Sie verdankt ihre Kraft nicht der Eindeutigkeit, sondern ihrer offenen Interpretierbarkeit. – *Publikationen u. a.:* „La Voix et le Phénomène", Paris 1967; „De la Grammatologie", Paris 1967.

**Designers Republic** – *Grafik-Design-Gruppe* – 1986 gründet Ian Anderson, der an der Universität Sheffield Philosophie studiert und eine regionale Rock-Band leitet, das Studio. Er beginnt Handzettel (Flyer) für Konzerte zu gestalten. Es entstehen zahlreiche T-Shirts, Plakate, Schallplatten- und CD-Cover für Musikgruppen. Zahlreiche Ausstellungen und Ausstellungsbeteiligungen. 1994 widmet

guré en 1959. – *Publications, sélection:* «Libro Imbullonato», Milan 1927; «Fortunato Depero nelle Opere e nella Vita», Trente 1940; «So I Think, So I Paint», Trente 1947. G. Gianti «Fortunato Depero Futurista», Milan 1951; B. Passamani «Fortunato Depero», Rovereto 1981.

**Derrida,** Jacques – né le 15. 7. 1930 à El Biar, Algérie – *philosophe, enseignant, linguiste* – enseigne à la Sorbonne de 1960 à 1964. Professeur d'histoire de la philosophie à l'Ecole Normale Supérieure de Paris de 1964 à 1984. Fonde le Collège International de Philosophie en 1983 et le dirige. Directeur de recherche à l'Ecole des Hautes Etudes en Sciences Sociales depuis 1984. Dans ses travaux proches du structuralisme, Derrida tente d'explorer la composante métaphysique qui est à la base de toute théorie systématique. Pour lui, l'écriture illustre la «différence» qui apparaît entre le signe et le signifié. Sa force ne relève pas de son univocité mais de son interprétabilité ouverte. – *Publications, sélection:* «La Voix et le Phénomène», Paris 1967; «De la Grammatologie», Paris 1967; «L'Ecriture et la Différence», Paris 1967.

**Designers Republic** – *groupe de graphistes* – en 1986, Ian Anderson, étudiant en philosophie à l'université de Sheffield et leader d'un groupe de rock régional, fonde l'atelier. Il commence à dessiner des prospectus (Flyers) pour les concerts. Conception de nombreux T-shirts, affiches, pochettes de disques et C.D. pour des groupes. Nombreuses expositions. En 1994, le magazine «Emigre» a édité un numéro spécial sur The Designers Republic. – *Publication:* «New & Used – The World Of The Designers Republic», Londres 1997.

**Detterer,** Ernst Frederic – né en 1888 à Lake Mills, Wisconsin, Etats-Unis, décédé en 1947 à Chicago, Etats-Unis – *calli-*

Derrida Cover

Designers Republic 1993 Cover

Designers Republic

Designers Republic 1990 Cover

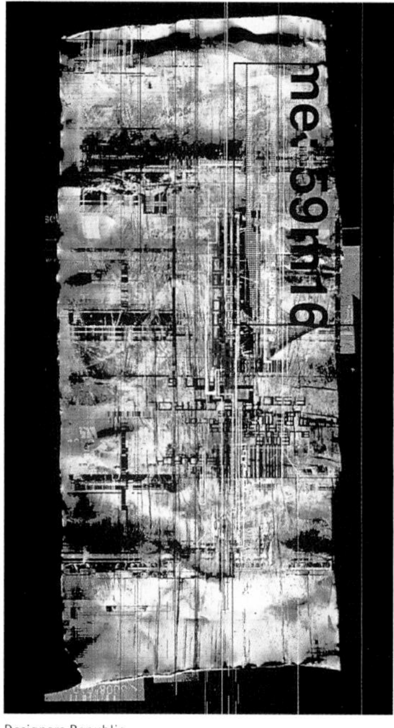

Designers Republic

"Fortunato Depero nelle Opere e nella Vita", Trento 1940; "So I Think, So I Paint", Trento 1947. G. Gianti "Fortunato Depero Futurista", Milan 1951; B. Passamani "Fortunato Depero", Rovereto 1981.

**Derrida,** Jacques – b. 15. 7. 1930 in El Biar, Algeria – *philosopher, teacher, linguist* – 1960–64: teaches at the Sorbonne. 1964–84: history of philosophy professor at the Ecole Normale Supérieure in Paris. 1983: founder director of the Collège International de Philosophie. Since 1984: director of research at the Ecole des Hautes Etudes en Sciences Sociales. With his work, closely linked with Structuralism, Derrida attempts to explain the metaphysical state which forms the basis of all systematic theory. For him, type is a model for the "difference" which arises between the symbol and the denoted object or concept. Its strength does not lie in its clarity but in its capacity to be openly interpreted. – *Publications include:* "La Voix et le Phénomène", Paris 1967; "De la Grammatologie", Paris 1967; "L'Ecriture et la Différence", Paris 1967.

**Designers Republic** – *graphic design group* – 1986: the studio is founded by Ian Anderson, a philosophy student at Sheffield University and leader of a local rock band. Anderson begins by designing flyers for concerts. Produces numerous T-shirts, posters, record and CD covers for various groups. Numerous exhibitions. In 1994 "Emigre" magazine devoted a whole issue to The Designers Republic. – *Publication:* "New & Used – The World Of The Designers Republic", Lon-

„Emigre" den Designers Republic eine gesamte Ausgabe. – *Publikation:* „New & Used – The World of The Designers Republic", London 1997.

**Detterer,** Ernst Frederic – geb. 1888 in Lake Mills, Wisconsin, USA, gest. 1947 in Chicago, USA. – *Kalligraph, Schriftentwerfer, Lehrer* – Kunststudium am Moravian College in Bethlehem, Pennsylvania. 1909–10 Studium an der School of Industrial Art des Pennsylvania Museums in Philadelphia. 1913 Schriftunterricht bei Edward Johnston. Unterrichtet 1912–21 Holzschnitt, Schrift und Satztechnik am Chicago Normal College. 1921–31 Geschichte der Druckkunst am Chicago Art Institute. Zusammenarbeit mit der Ludlow Typograph Company in Chicago, die ihn mit dem Entwurf einer Schrift in Anlehnung an die Formen Anton Jensons beauftragt. Bekanntschaft mit Robert Hunter Middleton. 1928 Middleton ergänzt Detterers Schrift „Eusebius" um mehrere Schnitte. – *Schriftentwurf:* Eusebius (1923).

**Dexel,** Walter – geb. 7. 2. 1890 in München, Deutschland, gest. 8. 6. 1973 in Braunschweig, Deutschland. – *Maler, Typograph, Gebrauchsgrafiker, Lehrer* – 1910–14 Studium der Kunstgeschichte an der Universität München. Besuch der Mal- und Zeichenschule von Prof. Hermann Gröber in München. 1914 Studienaufenthalt in Paris, erste Einzelausstellung in München. Promotion über französische illuminierte Handschriften aus dem 14. und 15. Jahrhundert. 1916–28 Ausstellungsleiter des Jenaer Kunstvereins, Arbeiten als freier Maler und Gebrauchsgrafiker.

ABCDEFGH
IJKLMNOPQR
STUVWXYZ

Detterer 1923 Eusebius Bold

Dexel 1924 Poster

Dexel 1929 Poster

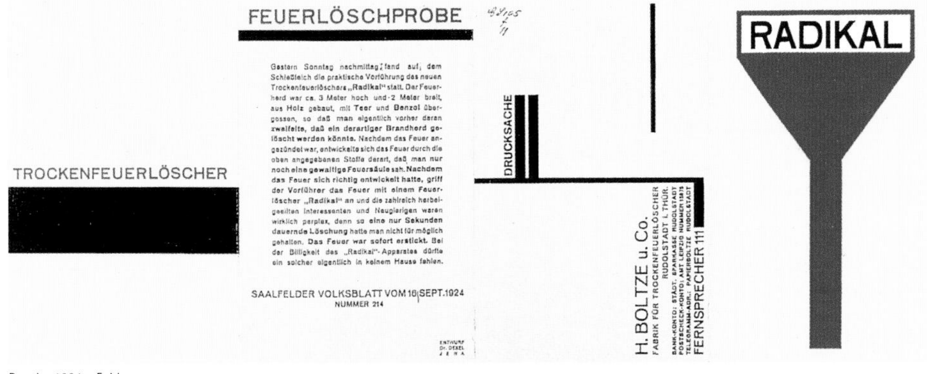


Dexel 1924 Folder

*graphe, concepteur de polices, enseignant* – études d'art au Moravian College de Bethlehem, Pennsylvanie; puis de 1909 à 1910, études à la School of Industrial Art du Pennsylvania Museum de Philadelphie. 1913, suit les cours de calligraphie d'Edward Johnston. De 1912 à 1921, il enseigne la gravure sur bois, l'écriture et les techniques de composition au Chicago Normal College, puis l'histoire de l'imprimerie au Chicago Art Institute de 1921 à 1931. Collabore avec la Ludlow Typograph Company de Chicago qui le charge de dessiner une police s'inspirant des formes créées par Anton Jenson. Rencontre Robert Hunter Middleton. En 1928, Middleton complète «Eusebius», une police de Detterer, et lui ajoute plusieurs tailles. – *Police:* Eusebius (1923).

**Dexel,** Walter – né le 7. 2. 1890 à Munich, Allemagne, décédé le 8. 6. 1973 à Braunschweig, Allemagne – *peintre, typographe, graphiste industriel, enseignant* – 1910–1914, études d'histoire de l'art à l'université de Munich. Fréquente l'école de dessin et de peinture du prof. Hermann Gröber à Munich. 1914, séjour d'étude à Paris et première exposition personnelle à Munich. Thèse sur les enluminures des manuscrits français des 14e et 15e siècles. Dirige les expositions du Kunstverein de Iéna de 1916 à 1928, peintre indépendant et graphiste industriel. Rencontre Kurt Schwitters en 1919,

don 1997.

**Detterer,** Ernst Frederic – b. 1888 in Lake Mills, Wisconsin, USA, d. 1947 in Chicago, USA – *calligrapher, type designer, teacher* – Studied art at Moravian College in Bethlehem, Pennsylvania. 1909–10: studies at the Pennsylvania Museum's School of Industrial Art in Pennsylvania. 1913: takes lettering classes under Edward Johnston. 1912–21: teaches woodcut art, lettering and typesetting techniques at the Chicago Normal College. 1921–31: teaches the history of printing at the Chicago Art Institute. Works with the Ludlow Typograph Company in Chicago who commission him to design a typeface based on the designs of Anton Jenson. Makes the acquaintance of Robert Hunter Middleton. 1928: Middleton draws up several new weights of Detterer's Eusebius typeface. – *Font:* Eusebius (1923).

**Dexel,** Walter – b. 7. 2. 1890 in Munich, Germany, d. 8. 6. 1973 in Braunschweig, Germany – *painter, typographer, commercial artist, teacher* – 1910–14: studies art history at the University of Munich. Attends Prof. Hermann Gröber's painting and drawing classes in Munich. 1914: studies in Paris. First solo exhibition in Munich. Ph.D. on French illuminated manuscripts from the 14th and 15th centuries. 1916–28: exhibition manager for the Jenaer Kunstverein. Works as a freelance painter and commercial artist. 1919: becomes acquainted with Kurt Schwitters and in 1921 with Theo van

1919 Bekanntschaft mit Kurt Schwitters, 1921 mit Theo van Doesburg. Veranstaltet 1923 die „Konstruktivisten"-Ausstellung in Jena. Mitglied der „Novembergruppe". 1925 erste beleuchtete Straßenrichtungsweiser und Reklamelaternen in Jena. 1927 Licht- und Reklamearbeiten in Frankfurt am Main. 1928 Dozent für Gebrauchsgrafik und Kulturgeschichte an der Kunstgewerbe- und Handwerkerschule Magdeburg. Mitglied im „ring neuer werbegestalter". 1935 als „entarteter Künstler" von den Nationalsozialisten aus dem Schuldienst entlassen. 1936–42 Professor für theoretischen Kunst- und Formunterricht an der Staatlichen Hochschule für Kunsterziehung in Berlin. 1942–55 Aufbau und Leitung der „Formensammlung der Stadt Braunschweig – Institut für handwerkliche und industrielle Formgebung". 1961 Wiederaufnahme der freien Malerei. – *Publikationen u. a.:* „Das Wohnhaus heute" (mit Grete Dexel), Leipzig 1928; „Hausgerät, das nicht veraltet", Ravensburg 1938; „Das Hausgerät Mitteleuropas", Braunschweig 1962; „Der Bauhausstil – ein Mythos", Starnberg 1976. Walter Vitt „Dexel-Werkverzeichnis der Druckgrafik", Köln 1971; Friedrich Friedl „Walter Dexel – Neue Reklame", Düsseldorf 1987; R. Wöbkemeier, W. Vitt „Walter Dexel Werkverzeichnis", Heidelberg 1995.

**Didot,** François – geb. 1689 in Paris, Frankreich, gest. 1757 in Paris, Frankreich. – *Buchhändler, Verleger, Drucker, Schriftgießer, Papierhersteller* – Begründer der Verleger- und Druckerfamilie Didot. Lehre bei dem Drucker und Verle-

puis Theo van Doesburg en 1921. En 1923, il organise l'exposition «Constructivistes» à Iéna. Membre du «Groupe Novembre». En 1925, il crée les premiers panneaux de signalisation routière lumineux et des publicités lumineuses à Iéna. Travaux sur les publicités lumineuses à Francfort-sur-le-Main en 1927. Enseigne le graphisme industriel et l'histoire des cultures à la Kunstgewerbe- und Handwerkerschule (école d'artisanat et des arts décoratifs) de Magdebourg en 1928. Membre du «ring neuer werbegestalter» (cercle des nouveaux graphistes publicitaires). En 1935, considéré comme «artiste dégénéré» par les nazis, il est suspendu de ses fonctions d'enseignant. 1936–1942, professeur de pédagogie de l'art et de la forme à la Staatliche Hochschule für Kunsterziehung (école supérieure de pédagogie de l'art) à Berlin. 1942–1955, constitution et direction de la «Formensammlung der Stadt Braunschweig – Institut für handwerkliche und industrielle Formgebung» (collection de design de la ville de Braunschweig – institut de design artisanal et industriel). Se consacre de nouveau à la peinture à partir de 1961. – *Publications, sélection:* «Das Wohnhaus heute» (avec Grete Dexel), Leipzig 1928; «Hausgerät, das nicht veraltet», Ravensburg 1938; »Das Hausgerät Mitteleuropas», Braunschweig 1962; «Der Bauhausstil – ein Mythos», Starnberg 1976. Walter Vitt «Dexel-Werkverzeichnis der Druckgrafik», Cologne 1971; Friedrich Friedl «Walter Dexel – Neue Reklame», Düsseldorf 1987; R. Wöbkemeier, W. Vitt «Walter Dexel Werkverzeichnis», Heidelberg 1995.

**Didot,** François – né en 1689 à Paris, France, décédé en 1757 à Paris, France – *libraire, éditeur, imprimeur, fondeur de caractères, fabricant de papier* – père fondateur de la famille Didot, dynastie d'im-

LES AVENTURES

DE TÉLÉMAQUE,

FILS D'ULYSSE.

PAR M. DE FÉNÉLON.

IMPRIMÉ PAR ORDRE DU ROI

POUR L'ÉDUCATION

DE MONSEIGNEUR LE DAUPHIN.

A PARIS,

DE L'IMPRIMERIE DE FRANÇ. AMBR. DIDOT L'AINÉ.

M. DCC. LXXXIII.

François A. Didot 1783 Type Specimen

Didot-Antiqua

ABCDEFGHIJK

LMNOPQRSTU

VWXYZÆŒ

abcdefghijklmno

pqrſstuvwxyz

äöüæœ

..!ﬂﬁ&ſſﬁﬂſs?:;

KRYSTALLINE

1234567890

171

Pierre Didot 1810 Specimen

Pierre Didot Monogram

Doesburg. 1923: stages the Constructivist exhibition in Jena. Member of the Novembergruppe. 1925: produces his first illuminated road signs and neon signs in Jena. 1927: works in lighting and advertising in Frankfurt am Main. 1928: teaches applied graphics and art history at the Kunstgewerbe- und Handwerkerschule in Magdeburg. Member of the "ring neuer werbegestalter". 1935: dismissed from teaching by the National Socialists, who label him a "degenerate artist". 1936–42: professor of theoretical art and design at the Staatliche Hochschule für Kunsterziehung in Berlin. 1942–55: sets up and runs the Formensammlung der Stadt Braunschweig – Institut für handwerkliche und industrielle Formgebung. 1961: returns to free painting. – *Publications include:* "Das Wohnhaus heute" (with Grete Dexel), Leipzig 1928; "Hausgerät das nicht veraltet", Ravensburg 1938; "Das Hausgerät Mitteleuropas", Braunschweig 1962; "Der Bauhausstil – ein Mythos", Starnberg 1976. Walter Vitt "Dexel – Werkverzeichnis der Druckgrafik", Cologne 1971; Friedrich Friedl "Walter Dexel – Neue Reklame", Düsseldorf 1987; R. Wöbkemeier, W. Vitt "Walter Dexel Werkverzeichnis", Heidelberg 1995.

**Didot,** François – b. 1689 in Paris, France, d. 1757 in Paris, France – *bookseller, publisher, printer, type founder, paper manufacturer* – Founder of the Didot family of printers and publishers. Trained under the printer and publisher André Pralard. 1713: opens his own workshop as a publisher. Publishes the works of Abbé Prévost and editions of the classics by

ger André Pralard. 1713 Eröffnung seiner eigenen Werkstatt als Verleger. Herausgabe der Werke von Abbé Prévost sowie Klassiker-Ausgaben von Sophokles, Aristophanes u. a. Seit 1753 Buchdrucker.

**Didot,** François Ambroise (1. Sohn von François Didot) – geb. 1730 in Paris, Frankreich, gest. 1804 in Paris, Frankreich. – *Drucker, Verleger, Stempelschneider* – 1775 erscheinen die ersten Druckschriften Didots, geschnitten vom Stempelschneider Pierre Louis Waflard. Im Auftrag Ludwig XV. werden die Werke von Corneille, Racine und Fénelon 1783 in Didots Schriften gesetzt und gedruckt. 1785 überarbeitet er das von P. S. Fournier entwickelte typographische Maßsystem. 1789 Übergabe seiner Druckerei an seinen Sohn Pierre, seiner Schriftgießerei an seinen Sohn Firmin.

**Didot,** Pierre François (2. Sohn von François Didot) – geb. 1732 in Paris, Frankreich, gest. 1795 in Paris, Frankreich. – *Verleger, Buchdrucker, Schriftgießer* – 1789 Erwerb der Papeteries de l'Essonne bei Paris. Druckt 1791 die französische Verfassung. Ab 1793 wird in seiner Druckerei die Ausgabe der Schriften Rousseaus hergestellt.

**Didot,** Pierre (1. Sohn von François Ambroise Didot) – geb. 1761 in Paris, Frankreich, gest. 1853 in Paris, Frankreich. – *Drucker, Schriftgießer, Verleger* – Studium am Collège d'Harcourt. Erhält 1797 die Erlaubnis, seine Pressen in den Räumen des Louvre aufzustellen. Hier entstehen die „Editions du Louvre": Virgil (1798), Horaz (1799), La Fontaine (1802), Racine (1801–05). 1798 und 1806 Aus-

primeurs et d'éditeurs. Apprentissage chez André Pralard, imprimeur et éditeur. Ouvre son propre atelier d'édition en 1713. Publie les œuvres de l'Abbé Prévost ainsi que des œuvres classiques de Sophocle, Aristophane etc. Imprimeur à partir de 1753.

**Didot,** François Ambroise (fils aîné de François Didot) – né en 1730 à Paris, France, décédé en 1804 à Paris, France – *imprimeur, éditeur, tailleur de types* – en 1775 paraissent les premières fontes Didot, taillées Pierre Louis Waflard. En 1783, les œuvres de Corneille, Racine et Fénelon sont composées et imprimées en caractères Didot à la demande de Louis XV. En 1785, il remanie le système de mesure typographique élaboré par P. S. Fournier. En 1789, il lègue son imprimerie à son fils Pierre et la fonderie de caractères à son autre fils Firmin.

**Didot,** Pierre François (fils cadet de François Didot) – né en 1732 à Paris, France, décédé en 1795 à Paris, France – *editeur, imprimeur, fondeur de caractères* – 1789, acquisition des Papeteries de l'Essonne, près de Paris. En 1791, il imprime la Constitution française. Fabrication des œuvres de Rousseau dans son imprimerie à partir de 1793.

**Didot,** Pierre (fils aîné de François Ambroise Didot) – né en 1761 à Paris, France, décédé en 1853 à Paris, France – *imprimeur, fondeur de caractères, éditeur* – études au Collège d'Harcourt. En 1797, il est autorisé à installer ses presses dans les locaux du Louvre. Il y crée les «Editions du Louvre»: Virgile (1798), Horace (1799), La Fontaine (1802), Racine (1801–1805). En 1798 et en 1806, ses ouvrages

LIVRE XVIII.

d'un beau prétexte pour contenter leur ambition, et pour se jouer des hommes crédules: ces hommes, qui avoient abusé de la vertu même, quoiqu'elle soit le plus grand don des dieux, étoient punis comme les plus scélérats de tous les hommes. Les enfans qui avoient égorgé leurs pères et leurs mères, les épouses qui avoient trempé leurs mains dans le sang de leurs époux, les traitres qui avoient livré leur patrie après avoir violé tous les sermens, souffroient des peines moins cruelles que ces hypocrites. Les trois juges des enfers l'avoient ainsi voulu; et voici leur raison : c'est que les hypocrites ne se contentent pas d'être méchans comme le reste des impies; ils veulent

Pierre F. Didot 1787 Specimen

PROSPECTUS.

Le recueil des Peintures antiques de Pietro-Sante Bartoli, dont nous annonçons une seconde édition au public, parut pour la premiere fois in-folio, à Paris, en 1757. Deux illustres savants, le Comte de Caylus et M. Mariette, consacrerent les plus grands soins à l'exécution d'une entreprise aussi intéressante, afin qu'elle répondît à la célébrité dont ils jouissoient dans la république des lettres.

A peine ils en ouvrirent la souscription qu'elle fut remplie; leurs noms, qui font seuls leur éloge, étoient trop imposants dans le monde littéraire pour ne pas y obtenir un pareil succès. Nous rappellerons ce qu'en a dit M. le Beau, dans l'Éloge de M. le Comte de Caylus: « M. le « Comte de Caylus, dit-il, voyoit avec regret que les ouvrages des anciens « peintres, dont on a fait de nos jours la découverte sous les ruines

François A. Didot 1782 Specimen

SPECIMEN
DES
NOUVEAUX CARACTÈRES
DE LA FONDERIE ET DE L'IMPRIMERIE
DE P. DIDOT, L'AINÉ,
CHEVALIER DE L'ORDRE ROYAL DE SAINT-MICHEL,
IMPRIMEUR DU ROI ET DE LA CHAMBRE DES PAIRS,
DÉDIÉ
À JULES DIDOT, FILS,
CHEVALIER DE LA LÉGION D'HONNEUR.

À PARIS,
CHEZ P. DIDOT, L'AINÉ, ET JULES DIDOT, FILS
RUE DU PONT DE LODI, N° 6.
MDCCCXIX

Pierre Didot 1819 Specimen

LE QUINZE.

Cette épître se trouve en tête de mon édition in-folio des œuvres de Boileau, en deux volumes, tirée seulement à 125 exemplaires, dont Sa Majesté a daigné agréer la dédicace.

AU ROI.

SIRE,

D'un monarque guerrier, l'un de tes fiers aïeux,
Despréaux a chanté le courage indomptable,
La marche menaçante et le choc redoutable,
Les assauts, les combats, et les faits merveilleux.
LOUIS, applaudis-toi d'un plus heureux partage.
Plus beau, plus fortuné, toujours cher à la paix,
Ton règne ami des lois doit briller d'âge en âge;
Tous nos droits affermis signalent tes bienfaits.
Le ciel t'a confié les destins de la France:
Qu'il exauce nos vœux, qu'il veille sur tes jours!
De ta carrière auguste exempte de souffrance
Que sa bonté pour nous prolonge l'heureux cours!

Pierre Didot 1819 Specimen

Sophocles and Aristophanes, among others. From 1753 onwards: printer.

**Didot,** François Ambroise (François Didot's oldest son) – b. 1730 in Paris, France, d. 1804 in Paris, France – *printer, publisher, punch cutter* – 1775: Didot's first prints are produced, cut by the punch cutter Pierre Louis Waflard. 1783: Louis XV commissions works by Corneille, Racine and Fénelon to be set and printed in Didot's typefaces. 1785: Didot reworks the typographic system of units developed by P. S. Fournier. 1789: Didot's son Pierre takes over his father's printing workshop and his son Firmin takes over the type foundry.

**Didot,** Pierre François (François Didot's second-oldest son) – b. 1732 in Paris, France, d. 1795 in Paris, France – *publisher, printer, type founder* – 1789: acquires the Papeteries de l'Essonne near Paris. 1791: prints the French constitution. From 1793 onwards: his printing works publishes an edition of Rousseau's works.

**Didot,** Pierre (François Ambroise Didot's oldest son) – b. 1761 in Paris, France, d. 1853 in Paris, France – *printer, type founder, publisher* – Studied at the Collège d'Harcourt. 1797: is granted permission to set up his printing presses in the Louvre, where he prints the "Editions du Louvre": Virgil (1798), Horace (1799), La Fontaine (1802) and Racine (1801–05).

Didot, P.
**Didot, F.**
**Didot, H.**
**Didot, L.**

zeichnung seiner Bücher mit einer Goldmedaille auf der Industrieausstellung in Paris. Die Bücher werden bis 1809 ausschließlich mit den Schriften seines Bruders Firmin gesetzt. 1809 Gründung einer eigenen Schriftgießerei. Erwirbt 1818 über 3.000 Originalstempel der Schriften Baskervilles.

**Didot,** Firmin (2. Sohn von François Ambroise Didot) – geb. 1764 in Paris, Frankreich, gest. 1836 in Mesnic-sur-l'Estrée, Frankreich. – *Stempelschneider, Schriftgießer, Buchdrucker, Verleger, Schriftsteller* – Studium alter Sprachen. 1783 erster Schriftschnitt, überarbeitet die Antiqua-Schriften seines Vaters. Erhält 1797 für seine Arbeiten auf dem Gebiet der Stereotypie ein Patent. Seine Schriften wurden bei den Drucken der „Editions du Louvre" seines Bruders Pierre Didot verwendet. 1812 Ernennung zum Direktor der Gießerei der Imprimerie Impériale. Eine der von ihm geschriebenen Tragödien wird 1823 im Théatre de l'Odéon aufgeführt.

**Didot,** Henri (Sohn von Pierre François Didot) – geb. 1765 in Paris, Frankreich, gest. 1852 in Paris, Frankreich. – *Stempelschneider, Schriftgießer* – Entwickelt 1819 ein Wort-Gieß-Gerät „Moule Polyamatype". Schneidet Alphabete in 2 1/2-Punkt-Größe.

**Didot,** Léger (2. Sohn von Pierre François Didot) – geb. 1767 in Paris, Frankreich, gest. 1829 in St. Jean d'Heurs, Frankreich. – *Papierhersteller* – Direktor der Papiermühle in l'Essonne bei Paris, die sein Vater 1789 gekauft hatte. Entwickelt 1798 mit Nicolas-Louis Robert eine

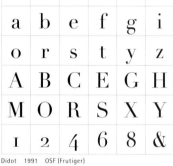

obtiennent une médaille d'or à l'exposition industrielle de Paris. Jusqu'en 1809, ses livres sont exclusivement composés avec des fontes de son frère Firmin. 1809, création d'une fonderie de caractères. En 1818, il achète plus de 3 000 types originaux des fontes Baskerville.

**Didot,** Firmin (fils cadet de François Ambroise Didot) – né en 1764 à Paris, France, décédé en 1836 à Mesnic-sur- l'Estrée, France – *tailleur de types, fondeur de caractères, imprimeur, éditeur, écrivain* – études de langues. Premières tailles de types en 1783; il remanie les antiques de son père. En 1797, il obtient une licence pour ses travaux de stéréotypie. Son frère, Pierre Didot utilise ses caractères pour les impressions réalisées aux «Editions du Louvre». En 1812, il est nommé directeur de la fonderie de l'Imprimerie Impériale. Une tragédie dont il est l'auteur est représentée au Théâtre de l'Odéon en 1823.

**Didot,** Henri (fils de Pierre François Didot) – né en 1765 à Paris, France, décédé en 1852 à Paris, France – *tailleur de types, fondeur de caractères* – en 1819, il élabore une machine à fondre des mots le «Moule Polyamatype». Taille des alphabets de corps 2 1/2.

**Didot,** Léger (Fils de Pierre François Didot) – né en 1767 à Paris, France, décédé en 1829 à St. Jean d'Heurs, France – *fabricant de papier* – directeur des moulins à papier de l'Essonne près de Paris que son père avait achetés en 1789. En 1798, il invente avec Nicolas-Louis Robert une machine à filtrer la pâte à papier qui permet pour la première fois de fabriquer des rouleaux de papier.

**Didot,** Ambroise (fils aîné de Firmin Didot) – né en 1790 à Paris, France, décédé en 1876 à Paris, France – *fondeur de caractères* – en 1827, il reprend la fonderie de caractères de son père avec son frère Hyacinthe. Il est nommé Imprimeur

ODE I.

AD VENEREM.

INTERMISSA, Venus, diu
Rursus bella moves. Parce, precor, precor!
Non sum qualis eram bonæ
Sub regno Cinaræ. Desine, dulcium
Mater sæva Cupidinum,
Circa lustra decem flectere mollibus
Iam durum imperiis. Abi

Pierre Didot 1799 Specimen

| | n | n | | | *n* |
|---|---|---|---|---|---|
| a | b | e | f | g | i |
| o | r | s | t | y | z |
| A | B | C | E | G | H |
| M | O | R | S | X | Y |
| 1 | 2 | 4 | 6 | 8 | & |

Didot 1991 OSF (Frutiger)

Firmin Didot Monogram

QUINTI
HORATII FLACCI
CARMINUM
LIBER QUARTUS.

ODE I.
AD VENEREM.

INTERMISSA, Venus, diu
Rursus bella moves. Parce, precor, precor!
Non sum qualis eram bonæ
Sub regno Cinaræ. Desine, dulcium
Mater sæva Cupidinum,
Circa lustra decem flectere mollibus
Iam durum imperiis. Abi
Quo blandæ iuvenum te revocant preces.

Pierre Didot 1799 Page

ÉPÎTRE
SUR LES PROGRÈS
DE L'IMPRIMERIE.

A MON PERE.

*CET art qui tous les jours multiplie avec grace*
*Et les vers de Virgile et les leçons d'Horace;*
*Qui, plus sublime encor, plus noble en son emploi,*
*Donne un texte épuré des livres de la Loi,*
*Et, parmi nous de Dieu conservant les oracles,*
*Pour la religion fit ses premiers miracles;*
*Des grands événements cet art conservateur,*
*Trop ingrat seulement envers son inventeur,*
*N'a pas su nous transmettre avec pleine assurance*
*Le génie étonnant qui lui donna naissance.*
*Toi qui sus concevoir tant de plans à la fois,*
*A l'immortalité pourquoi perdre tes droits?*

Firmin Didot 1784 Page

1798 and 1806: his books are awarded a gold medal at the industrial exhibition in Paris. Pierre's prints were set exclusively in his brother Firmin's typefaces until 1809. 1809: opens his own type foundry. 1818: acquires over 3,000 original punches of Baskerville's typefaces.

**Didot,** Firmin (François Ambroise Didot's second-oldest son) – b. 1764 in Paris, France, d. 1836 in Mesnic-sur-l'Estrée, France – *punch cutter, type founder, printer, publisher, author* – Studied classical languages. 1783: cuts his first typefaces and reworks his father's roman alphabets. 1797: is granted a patent for his developments in the field of stereotype printing. His typefaces are used in his brother Pierre Didot's "Editions du Louvre" series. 1812: he is made director of the Imprimerie Impériale type foundry. 1823: one of his tragedies is performed at the Théâtre de l'Odéon.

**Didot,** Henri (Pierre François Didot's oldest son) – b. 1765 in Paris, France, d. 1852 in Paris, France – *punch cutter, type founder* – 1819: develops a word-casting machine entitled "Moule Polyamatype". Engraves alphabets in 2 1/2-point.

**Didot,** Léger (Pierre François Didot's second-oldest son) – b. 1767 in Paris, France, d. 1829 in St. Jean d'Heurs, France – *paper manufacturer* – Ran the paper mill in l'Essonne near Paris which his father bought in 1789. 1798: develops the Didot paper-making machine with Nicolas-Louis Robert, the first to produce endless paper for rolls.

**Didot,** Ambroise (Firmin Didot's oldest son) – b. 1790 in Paris, France, d. 1876 in

Langsieb-Papiermaschine, mit der erstmals Maschinenpapier für die Rolle hergestellt wird.

**Didot,** Ambroise (1. Sohn von Firmin Didot) – geb. 1790 in Paris, Frankreich, gest. 1876 in Paris, Frankreich. – *Schriftgießer* – Übernimmt 1827 mit seinem Bruder Hyacinthe die Schriftgießerei des Vaters. Wird 1829 zum Imprimeur du Roi ernannt. Mitglied des L'Institut de France. 1834 Eingliederung der Schriftgießerei in die Fonderie Générale.

**Didot,** Hyacinthe (2. Sohn von Firmin Didot) – geb. 1794 in Paris, Frankreich, gest. 1880 in Paris, Frankreich. – *Schriftgießer* – Übernimmt 1827 mit seinem Bruder Ambroise die Schriftgießerei des Vaters. 1834 Eingliederung der Schriftgießerei in die Fonderie Générale.

**Didot,** Jules (Sohn von Pierre Didot) – geb. 1794 in Paris, Frankreich, gest. 1871 in Caen, Frankreich. – 1828 Verkauf von Teilen der Schriftgießerei. 1830 Verkauf seiner Druckerei nach Brüssel an den belgischen Staat, wo sie den Grundstock für die königliche Druckerei bildet. Zurück in Paris gründet er die Fonderie et Imprimerie Normale. Verliert 1838 den Verstand.

**Diederichs,** Eugen – geb. 22. 6. 1867 in Löbitz, Deutschland, gest. 10. 9. 1930 in Jena, Deutschland. – *Verleger* – Ausbildung als Landwirt. 1888 Lehre als Buchhändler in Halle. 1897 Beginn der Verlagstätigkeit in Leipzig. 1904 Umzug des Verlags nach Jena. 1907 Mitbegründer des Deutschen Werkbunds. Zum 25-jährigen Verlagsjubiläum erscheint 1921 der Almanach „Wille und Gestal-

du Roi en 1829. Membre de l'Institut de France. En 1834, la fonderie de caractère est incorporée à la Fonderie Générale.

**Didot,** Hyacinthe (fils cadet de Firmin Didot) – né en 1794 à Paris, France, décédé en 1880 à Paris, France – *fondeur de caractères* – en 1827, il reprend la fonderie de caractères de son père avec son frère Ambroise. En 1834, la fonderie de caractère est incorporée à la Fonderie Générale.

**Didot,** Jules (fils de Pierre Didot) – né en 1794 à Paris, France, décédé en 1871 à Caen, France – en 1828, il vend des parts de la fonderie de caractères. En 1830, il vend son imprimerie à l'Etat belge qui l'installe à Bruxelles où elle constituera la base de l'Imprimerie Royale. De retour à Paris, il crée la Fonderie et Imprimerie Normale. Perd la raison en 1838.

**Diederichs,** Eugen – né le 22. 6. 1867 à Löbitz, Allemagne, décédé le 10. 9. 1930 à Iéna, Allemagne – *editeur* – formation d'agriculteur. 1888, apprentissage de libraire à Halle. 1897, début de ses activités d'éditeur à Leipzig. En 1904, la maison d'édition s'installe à Iéna. 1907, cofondateur du Deutscher Werkbund. En 1921, il publie l'almanach «Wille und Gestaltung» pour le 25e anniversaire de

VIE DE M. DE FÉNÉLON.

humilier ou nous confondre, ne paroît occupé que de nos intérêts et de notre bonheur.

Fénélon vouloit que toutes les affaires de son diocese lui fussent rapportées, et il les examinoit par lui-même; mais la moindre chose importante dans la discipline ne se décidoit que de concert avec ses vicaires généraux et les autres chanoines de son conseil, qui s'assembloit deux fois la semaine. Jamais il ne s'y est prévalu de son rang ou de ses talents, pour décider par autorité, sans persuasion : il reconnoissoit les prêtres pour ses freres, recevoit leurs avis, et profitoit de leur expérience. *Le pasteur*, disoit-il, *a besoin d'être encore plus docile que le troupeau;* il faut

François A. Didot 1787-92 Specimen

Didot 1991 Roman Initials (Frutiger)

| | n | n | | | *n* |
|---|---|---|---|---|---|
| a | b | e | f | g | i |
| o | r | s | t | y | z |
| A | B | C | E | G | H |
| M | O | R | S | X | Y |
| 1 | 2 | 4 | 6 | 8 | & |

Didot 1991 (Frutiger)

Firmin Didot

ABCDEFGHIJ
KLMNOPQRS
TUVWXYZ
abcdefghijklmno
pqrstuvwxyz
1234567890
fi & fl
Imprimerie

Firmin Didot 1800 Specimen

Paris, France – *type founder* – 1827: he and his brother Hyacinthe take over their father's type foundry. 1829: Ambroise is made an Imprimeur du Roi. Member of the Institut de France. 1834: the type foundry is incorporated into the Fonderie Générale.

**Didot,** Hyacinthe (Firmin Didot's second-oldest son) – b. 1794 in Paris, France, d. 1880 in Paris, France – *type founder* – 1827: he and his brother Ambroise take over their father's type foundry. 1834: the type foundry is incorporated into the Fonderie Générale.

**Didot,** Jules (Pierre Didot's son) – b. 1794 in Paris, France, d. 1871 in Caen, France – 1828: sells parts of the type foundry. 1830: sells his printing workshop to the Belgian state in Brussels where it forms the basis of the Royal printing works. Founds the Fonderie et Imprimerie Normale in Paris. 1838: goes mad.

**Diederichs,** Eugen – b. 22. 6. 1867 in Löbitz, Germany, d. 10. 9. 1930 in Jena, Germany – *publisher* – Trained as a farmer. 1888: trains as a bookseller in Halle. 1897: founds a publishing house in Leipzig. 1904: Diederich's publishing house moves to Jena. 1907: co-founder of the Deutscher Werkbund. 1921: his publishing house prints the "Wille und Gestaltung" almanac to commemorate its 25-year anniversary. 1924: Diederichs is awarded an honorary doctorate by the University of Cologne. The publishing

tung". Die Kölner Universität verleiht Diederichs 1924 die Ehrendoktorwürde. Im Verlag erscheinen zahlreiche Gesamtausgaben, u. a. Novalis (ab 1898), Tolstoi (ab 1900), Tschechow (ab 1900), Giordano Bruno (1904), Bergson (1908), Kierkegaard (ab 1909). Für den Verlag sind als Buchgestalter tätig: Melchior Lechter, Fritz Helmut Ehmcke, Emil Rudolf Weiss, F. H. E. Schneidler. – *Publikationen:* „Aus meinem Leben", Jena 1927. W. G. Oschilewski „Eugen Diederichs und sein Werk", Jena 1936.

**Diethelm,** Walter – geb. 7. 2. 1913 in Zürich, Schweiz, gest. 10. 6. 1986 in Zürich, Schweiz. – *Schriftentwerfer, Typograph, Grafik-Designer* – 1928–32 Lehre als Grafiker in Zürich. 1932–33 Studium an der Kunstgewerbeschule Zürich, danach an der Académie Ranson und der Académie de la Grande Chaumière in Paris. Zehn Jahre künstlerischer Leiter der Druckerei Fretz in Zürich. Ab 1954 eigenes Designbüro in Zürich. Arbeitsbereiche sind Bücher, Leitsysteme, Ausstellungsgestaltung, Signets und Plakate. Zahlreiche Ausstellungen und Auszeichnungen. – *Schriftentwürfe:* Diethelm-Antiqua (1948–50), Sculptura (1957), Arrow, Abacus, Aktiv, Capitol, Gloriette. – *Publikationen u. a.:* „Die Type", Basel 1946; „Signet, Signal, Symbol", Zürich 1970; „Form + Communication", Zürich 1974; „Visual Transformation", Zürich 1982.

**Dijck,** Christoph van (auch van Dyck, Christoffel) – geb. 1601 in Dexheim, Niederlande, gest. 1669 in Amsterdam, Niederlande. – *Schriftschneider, Schriftgießer* – 1640 Goldschmied in Amster-

la maison d'édition. En 1924, l'université de Cologne lui décerne le titre de docteur honoris causa. Les éditions publient de nombreuses œuvres complètes, entre autres Novalis (à partir de 1898), Tolstoï (à partir de 1900), Tchékov (à partir de 1900), Giordano Bruno (1904), Bergson (1908), Kierkegaard (à partir de 1909). Melchior Lechter, Fritz Helmut Ehmcke, Emil Rudolf Weiss, F. H. E. Schneidler ont travaillé pour les éditions comme maquettistes. – *Publications:* «Aus meinem Leben», Iéna 1927. W. G. Oschilewski «Eugen Diederichs und sein Werk», Iéna 1936.

**Diethelm,** Walter – né le 7. 2. 1913 à Zurich, Suisse, décédé le 10. 6. 1986 à Zurich, Suisse – *concepteur de polices, typographe, graphiste maquettiste* – 1928–1932, apprentissage de graphiste à Zurich. 1932–1933, études à la Kunstgewerbeschule (école des arts décoratifs) de Zurich puis à l'Académie Ranson et à l'Académie de la Grande Chaumière à Paris. Directeur artistique de l'imprimerie Fretz à Zurich pendant dix ans. Exploite son propre atelier de design à Zurich à partir de 1954, et se spécialise dans l'édition, les systèmes de signalisation, l'architecture d'exposition, les signets et les affiches. Nombreuses expositions et distinctions. – *Polices:* Diethelm-Antiqua (1948– 1950), Sculptura (1957), Arrow, Abacus, Aktiv, Capitol, Gloriette. – *Publications, sélection:* «Die Type», Bâle 1946; «Signet, Signal, Symbol», Zurich 1970; «Form + Communication», Zurich 1974; «Visual Transformation», Zurich 1982.

**Dijck,** Christoph van (aussi van Dyck, Christoffel) – né en 1601 à Dexheim, Pays-Bas, décédé en 1669 à Amsterdam, Pays-Bas – *tailleur de types, fondeur de caractères* – orfèvre à Amsterdam en 1640, puis indépendant en 1643. Ouvre en

Diethelm 1959 Poster

Diethelm 1948–50 Diethelm Antiqua

PROEVEN
Van
LETTEREN,
*Die gesneden zijn door Wylen*
CHRISTOFFEL VAN DYCK,
*Soo als de selve verkoft sullen werden ten huyse van de Weduwe Wylen*
DANIEL ELSEVIER,
*Op't Water, by de Papenbrugh, in den Olmboom, op Woensdagh, den 5 Martii, 1681.*

Dijck 1681 Specimen

house has printed numerous sets of complete works, including Novalis (from 1898 onwards), Tolstoy (from 1900 onwards), Chekhov (from 1900 onwards), Giordano Bruno (1904), Bergson (1908) and Kierkegaard (from 1909 onwards). The following book designers have worked for Diederich's publishing company: Melchior Lechter, Fritz Helmut Ehmcke, Emil Rudolf Weiss and F. H. E. Schneidler. – *Publications:* "Aus meinem Leben", Jena 1927. W.G. Oschilewski "Eugen Diederichs und sein Werk", Jena 1936.

**Diethelm,** Walter – b. 7. 2. 1913 in Zurich, Switzerland, d. 10. 6. 1986 in Zurich, Switzerland – *type designer, typographer, graphic designer* – 1928–32: trains as a graphic artist in Zurich. 1932–33: studies at the Kunstgewerbeschule in Zurich, then at the Académie Ranson and the Académie de la Grande Chaumière in Paris. Art director of the Fretz printing works in Zurich for ten years. From 1954 onwards: runs his own design studio in Zurich. His areas of work include books, signage systems, exhibition planning, logos and posters. Numerous exhibitions and awards. – *Fonts:* Diethelm-Antiqua (1948–50), Sculptura (1957), Arrow, Abacus, Aktiv, Capitol, Gloriette. – *Publications include:* "Die Type", Basle 1946; "Signet, Signal, Symbol", Zurich 1970; "Form + Communication", Zurich 1974; "Visual Transformation", Zurich 1982.

**Dijck,** Christoph van (also van Dyck, Christoffel) – b. 1601 in Dexheim, The Netherlands, d. 1669 in Amsterdam, The Netherlands – *type cutter, type founder* –

dam, ab 1643 selbständig. 1647 Eröffnung seiner Schriftgießerei in Amsterdam. Schnitt zahlreicher Schriften für die Druckerei Elzevier, die nach van Dijcks Tod sein Schriftmaterial kauft. Van Dijck entwickelte eine niederländische Form der Antiqua mit spitz auslaufenden Serifen und senkrechter Achsenstellung bei den Rundungen.

**DiSpigna,** Tony – geb. 6. 12. 1943 in Forio d'Ischia, Italien. – *Schriftentwerfer, Typograph, Kalligraph, Grafik-Designer, Lehrer* – Studium am New York City Community College und am Pratt Institute, New York. Nach dem Studium erste Arbeiten für das Studio Bonder & Carnase Inc. 1969 Eintritt in das Studio Lubalin, Smith, Carnase. 1973–78 eigenes Studio in New York. 1978 Vize-Präsident des Studio Herb Lubalin Associates. 1980 eigenes Studio Tony DiSpigna Inc. Unterrichtet am Pratt Institute, an der School of Visual Arts und am New York Institute of Technology. – *Schriftentwürfe:* Serif Gothic (mit H. Lubalin, 1972–74), Playgirl, Lubalin Graph (mit H. Lubalin), Fattoni, WNET.

**Does,** Bram de – geb. 19. 7. 1934 in Amsterdam, Niederlande. – *Schriftentwerfer, Typograph* – 1952 Lehre als Setzer und Drucker in der Druckerei seiner Eltern. 1953–56 Studium an der Amsterdamse Grafische School. 1958–88 mit kurzen Unterbrechungen Betreuung der typographischen Gestaltung der Firma Joh. Enschedé en Zonen in Haarlem. 1961 Gründung seiner privaten Presse „Spectatorpers". 1963 Typograph beim Em. Querido Verlag, Amsterdam. Zahlreiche Auszeichnungen, u. a. „Goldenen Letter"

DiSpigna Logo

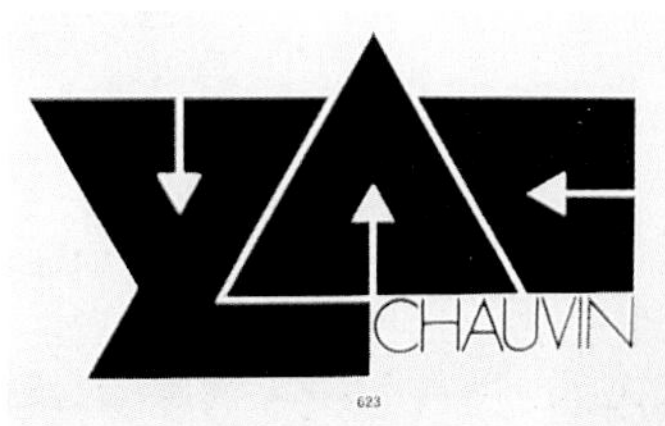


DiSpigna 1972 Logo

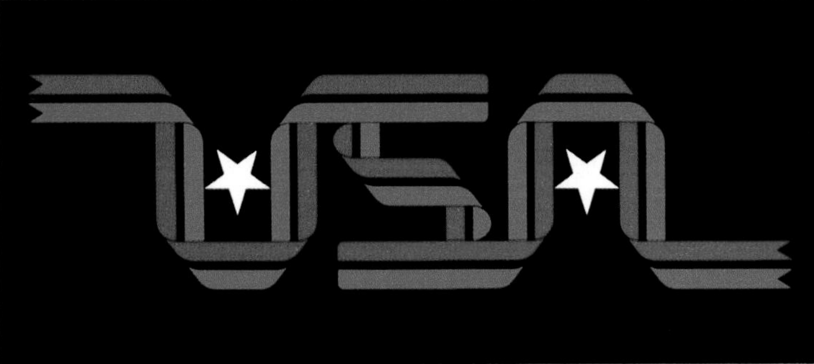


DiSpigna 1976 Sketch

DiSpigna Logo

ABCDEFGHIJKLMNOPQRSTUVXWXYZ
1234567890 1234567890 .,:;!?",„"" ‹›«»()[]*†‡&

ABCDEFGHIJKLMNOPQRSTUVXWXYZ
1234567890 1234567890 .,:;!?",„"" ‹›«»()[]*†‡&

Does 1978–81 Trinité 1, 2, 3

1647 une fonderie de caractères à Amsterdam. Taille de nombreux types pour l'imprimerie Elzevier, qui rachètera les matrices à la mort de van Dijck. Van Dijck a développé une forme d'antique typiquement néerlandaise avec des sérifs se terminant en pointe et des rondeurs avec l'axe vertical.

**DiSpigna,** Tony – né le 6. 12. 1943 à Forio d'Ischia, Italie – *concepteur de polices, typographe, calligraphe, graphiste maquettiste, enseignant* – études au New York City Community College et au Pratt Institute à New York. A la fin de ses études, il réalise ses premiers travaux pour l'atelier Bonder & Carnase Inc. 1969, entre à l'atelier Lubalin, Smith, Carnase. 1973–1978, travaille dans son propre atelier à New York. En 1978, il devient vice-président de l'atelier Herb Lubalin Associates. Création de l'atelier Tony DiSpigna Inc. en 1980. A enseigné au Pratt Institute, à la School of Visual Arts et au New York Institute of Technology. – *Polices:* Serif Gothic (avec Herb Lubalin, 1972–1974), Playgirl, Lubalin Graph (avec Herb Lubalin), Fattoni, WNET.

**Does,** Bram de – né le 19. 7. 1934 à Amsterdam, Pays-Bas – *concepteur de polices, typographe* – 1952, apprentissage de compositeur typographe et d'imprimeur dans l'imprimerie de ses parents. 1953–1956, études à la Amsterdamse Grafische School. De 1958 à1988, il suit avec quelques interruptions la typographie et la mise en page à la société Joh. Enschedé en Zonen à Haarlem. 1961, création de sa propre imprimerie «Spectatorpers». 1963, typographe aux éditions Em. Querido à Amsterdam en 1963. Nombreuses distinctions, entre autres la «Goldene Letter» de Leipzig (1982), le prix

1640: goldsmith in Amsterdam. 1643: self-employed. 1647: opens a type foundry in Amsterdam. Cuts numerous typefaces for Elzevier printers, who buy up van Dijck's type materials after his death. Van Dijck developed a Dutch form of the roman typeface with pointed serifs and vertical axis position of the curves.

**DiSpigna,** Tony – b. 6. 12. 1943 in Forio d'Ischia, Italy – *type designer, typographer, calligrapher, graphic designer, teacher* – Studied at the New York City Community College and at the Pratt Institute in New York. After completing his studies he produced work for the studio Bonder & Carnase Inc. 1969: joins the Lubalin, Smith, Carnase studio. 1973–78: has his own studio in New York. 1978: vice-president of Herb Lubalin Associates. 1980: opens his own studio, Tony DiSpigna Inc. DiSpigna has taught at the Pratt Institute, the School of Visual Arts and at the New York Institute of Technology. – *Fonts:* Serif Gothic (with Herb Lubalin, 1972–74), Playgirl, Lubalin Graph (with Herb Lubalin), Fattoni, WNET.

**Does,** Bram de – b. 19. 7. 1934 in Amsterdam, The Netherlands – *type designer, typographer* – 1952: trains as a typesetter and printer in his parents' printing workshop. 1953–56: studies at the Amsterdamse Grafische School. 1958–88: supervises Joh. Enschedé en Zonen's typographic design in Haarlem more or less continuously over this period. 1961: founds his own private press, Spectatorpers. 1963: typographer for Em. Querido publishers in Amsterdam. Has won nu-

in Leipzig (1982), H. N. Werkman-Preis der Stadt Amsterdam für seine Schrift Trinité (1991), Premio Felice Feliciano in Verona für die Gestaltung des Pressendrucks „The Steadfast Tin Soldier of Joh. Enschedé en Zonen, Haarlem" (1993). – *Schriftentwürfe:* Trinité 1, 2, 3 (1978–81), Lexicon (1990–91).

**Doesburg,** Théo van (eigentlich Christiaan Emil Marie Küpper) – geb. 30. 8. 1883 in Utrecht, Niederlande, gest. 7. 3. 1931 in Davos, Schweiz. – *Maler, Architekt, Schriftsteller, Typograph* – 1899 Studium der Malerei. 1908 erste Einzelausstellung in Den Haag. Erste literarische Texte. Ab 1914 typographische Arbeiten. 1916 Bekanntschaft mit Piet Mondrian. Gründet in Leiden mit Huszar, Mondrian, Oud und van't Hoff die Gruppe De Stijl. Unter der Redaktion von Theo van Doesburg erscheint 1917 die erste Nummer der Zeitschrift „De Stijl". 1922 erscheint das erste von insgesamt vier Heften der von Théo van Doesburg herausgegebenen Zeitschrift „Mécano". 1924 Umzug nach Paris. 1928 Vorlesungen über moderne Malerei an der Universität Utrecht. 1930 erscheint die erste Voraus-Nummer der von Théo van Doesburg herausgegebenen Zeitschrift „Art Concret". Zwischen 1910–31 publiziert van Doesburg etwa 400 Artikel über Malerei, Bildhauerei, Architektur, Typographie, Musik, Literatur und allgemeine Themen in internationalen Zeitschriften. – *Publikationen u. a.:* „Klassiek – Barok – Modern", Antwerpen 1920; „Grundbegriffe der neuen gestaltenden Kunst", München 1924; „Die Scheuche" (mit K. Schwitters

H. N. Werkman de la ville d'Amsterdam pour sa police Trinité (1991), Premio Felice Feliciano à Vérone pour la maquette de «The Steadfast Tin Soldier of Joh. Enschedé en Zonen, Haarlem» (1993). – *Polices:* Trinité, 1, 2, 3 (1978–1981), Lexicon (1990–1991).

**Doesburg,** Théo van (Christiaan Emil Marie Küpper) – né le 30. 8. 1883 à Utrecht, Pays-Bas, décédé le 7. 3. 1931 à Davos, Suisse – *peintre, architecte, écrivain, typographe* – 1899, études de peinture. 1908, première exposition personnelle à La Haye. Premiers textes littéraires. Travaux typographiques à partir de 1914. Rencontre Piet Mondrian en 1916. Il fonde le groupe De Stijl avec Huszar, Mondrian, Oud et van't Hoff à Leyde. 1917, parution du premier numéro de la revue «De Stijl» dirigée par van Doesburg. 1922, parution du premier des quatre numéros de la revue «Mécano», publiée par van Doesburg. S'installe à Paris en 1924. Conférences sur la peinture moderne à l'université d'Utrecht en 1928. Parution en 1930, du numéro d'annonce de la revue «Art Concret», publié par Théo van Doesburg. De 1910 à 1931, il publie environ 400 articles sur la peinture, la sculpture, l'architecture, la typographie, la musique, la littérature et sur des thèmes généraux dans des revues internationales. – *Publications, sélection:* «Klassiek – Barok – Modern», Anvers 1920; «Grundbegriffe der neuen gestaltenden Kunst», Munich 1924; «Die Scheuche» (avec K. Schwitters et K. Steinitz), Munich 1925; «Das Buch und seine Gestaltung», dans «Die Form» 21/1929. Jan Leering «Théo van Doesburg 1883–1931», Eindhoven 1968; Joost Baljeu «Théo van Doesburg», New York 1974.

**Domela,** Cesar – né le 15. 1. 1900 à Amsterdam, Pays-Bas, décédé en 1992 à Paris, France – *peintre, sculpteur, typo-*

Doesburg 1919 Alphabet

Doesburg 1923 Poster

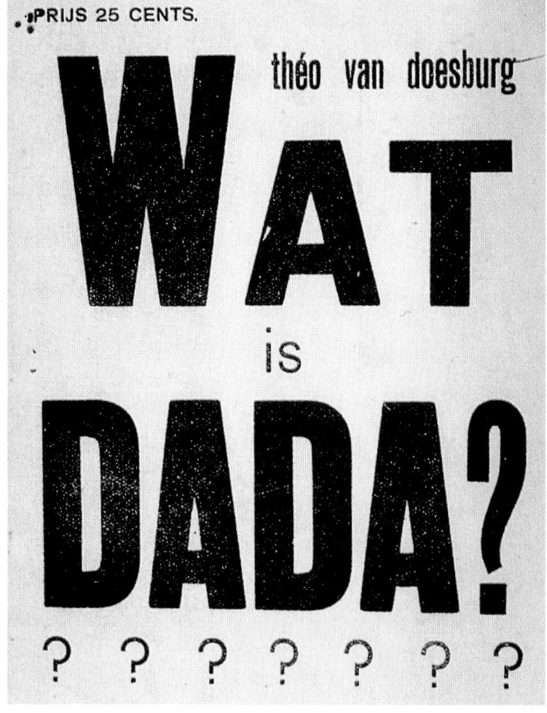

Doesburg 1923 Cover

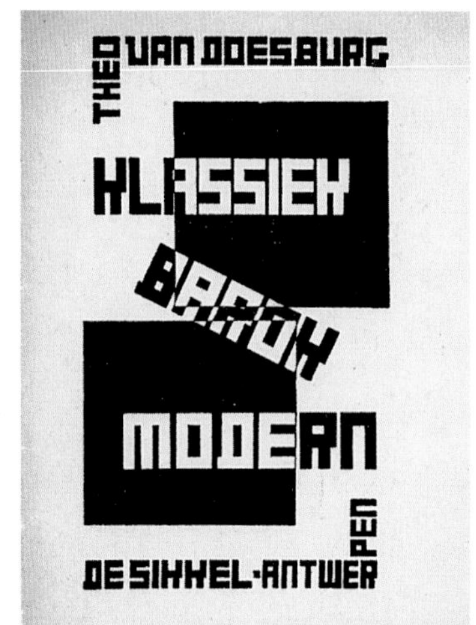

Doesburg 1920 Cover

merous awards, including the Golden Letter in Leipzig (1982), Amsterdam's H. N. Werkman Prize for his Trinité typeface (1991) and the Premio Felice Feliciano in Verona for the design of "The Steadfast Tin Soldier of Joh. Enschedé en Zonen, Haarlem" (1993). – *Fonts:* Trinité 1, 2, 3 (1978–81), Lexicon (1990–91).

**Doesburg,** Théo van (real name: Christiaan Emil Marie Küpper) – b. 30. 8. 1883 in Utrecht, The Netherlands, d. 7. 3. 1931 in Davos, Switzerland – *painter, architect, author, typographer* – 1899: studies painting. 1908: first solo exhibition in The Hague. Produces his first literary texts. From 1914 onwards: typographic work. 1916: becomes acquainted with Piet Mondrian. Founds the De Stijl group in Leiden with Huszar, Mondrian, Oud and van't Hoff. 1917: the first number of "De Stijl" magazine is published, edited by van Doesburg. 1922: the first of four issues of "Mécano" magazine is published by van Doesburg. 1924: moves to Paris. 1928: gives lectures on modern painting at the University of Utrecht. 1930: the first issue of "Art Concret" magazine is published by van Doesburg. Between 1910–31, van Doesburg had c. 400 articles on painting, sculpture, architecture, typography, music, literature and more general topics published in various international magazines. – *Publications include:* "Klassiek – Barok – Modern", Antwerp 1920; "Grundbegriffe der neuen gestaltenden Kunst", Munich 1924; "Die Scheuche" (with K. Schwitters and K. Steinitz), Munich 1925; "Das Buch und seine Gestaltung" in "Die Form"

und K. Steinitz), München 1925; „Das Buch und seine Gestaltung", in „Die Form" 21/1929. Jan Leering „Théo van Doesburg 1883–1931", Eindhoven 1968; Joost Baljeu „Théo van Doesburg", New York 1974.

**Domela,** Cesar – geb. 15. 1. 1900 in Amsterdam, Niederlande, gest. 1992 in Paris, Frankreich. – *Maler, Bildhauer, Typograph, Grafik-Designer* – Beginnt 1919 als Autodidakt zu malen. 1921 Malerei-Studium in Berlin. 1923 Beteiligung an der „Novembergruppe" in Berlin. 1925 Mitglied der Gruppe De Stijl. 1927–33 zahlreiche Entwurfsarbeiten im Stil der Neuen Typographie für Unternehmen und Verlage in Berlin. Einziges auswärtiges Mitglied der Künstlergruppe „die abstrakten hannover". 1928 Kontakte zum Bauhaus. Mitglied der Künstlergruppen „Cercle et Carré" und „abstraction-création". 1933 Umzug nach Paris. 1937 Gründung der Zeitschrift „Plastique" mit Hans Arp und Sophie Täuber-Arp. Gründet 1946 die Künstlergruppe „Centre des Recherches" in Paris. Zahlreiche Ausstellungen, u. a. von Retrospektiven in Paris, Grenoble und Amsterdam (1987). – *Publikationen u. a.:* Christian Zervos „Domela", Amsterdam 1966; Kees Broos, Flip Bool „Domela", Den Haag 1980; G. B. Martini, A. Ronchetti „Cesar Domela", Genua 1981; Marie-Odile Briot „Domela – Schilderijen, reliefs, grafisch Œuvre", Apeldoorn, Paris 1990.

**Dooijes,** Dick – geb. 6. 5. 1909 in Amsterdam, Niederlande. – *Schriftentwerfer, Typograph, Autor, Lehrer* – Ausbildung in der Lettergieterij Amsterdam. 1926–36

Domela 1924 Cover

Domela 1928 Advertisement

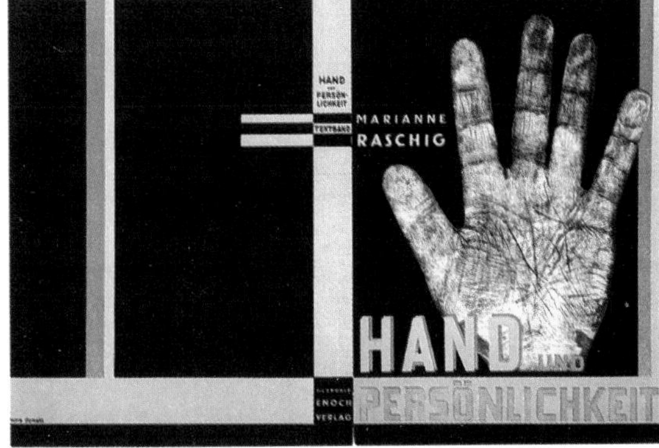


Domela 1931 Cover

Dooijes 1969 Lectura

*graphe, graphiste maquettiste* – 1919, commence à peindre comme autodidacte. 1921, étudie la peinture à Berlin. 1923, membre du «Groupe Novembre» à Berlin. 1925, membre du groupe «De Stijl». Entre 1927 et 1933, il réalise de nombreux travaux dans le style de la nouvelle typographie pour des entreprises et des éditeurs berlinois. Seul membre externe du groupe d'artiste les «Abstraits d'Hanovre». 1928, contacts avec le Bauhaus. Membre des groupes «Cercle et Carré» et «abstraction-création». 1933, s'installe à Paris. En 1937, il fonde la revue «Plastique» avec Jean Arp et Sophie Täuber-Arp. En 1946, il fonde le groupe d'artistes «Centre des Recherches» à Paris. Nombreuses expositions, dont une rétrospective à Paris, Grenoble et Amsterdam (1987). – *Publications, sélection:* Christian Zervos «Domela», Amsterdam 1966; Kees Broos, Flip Bool «Domela», La Haye 1980; G. B. Martini, A. Ronchetti «Cesar Domela», Gêne 1981; Marie-Odile Briot «Domela – Schilderijen, reliefs, grafisch Œuvre», Apeldoorn, Paris 1990.

**Dooijes,** Dick – né le 6. 5. 1909 à Amsterdam, Pays-Bas – *concepteur de polices, typographe, auteur, enseignant* – formation à la Lettergieterij d'Amsterdam. 1926–1936, étudie à l'Instituut voor

21/1929. Jan Leering "Théo van Doesburg 1883–1931", Eindhoven 1968; Joost Baljeu "Théo van Doesburg", New York 1974.

**Domela,** Cesar – b. 15. 1. 1900 in Amsterdam, The Netherlands, d. 1992 in Paris, France – *painter, sculptor, typographer, graphic designer* – 1919: starts painting. Self-taught. 1921: studies painting in Berlin. 1923: becomes involved with the Novembergruppe in Berlin. 1925: member of De Stijl. 1927–33: numerous designs in the New Typographic style for various companies and publishers in Berlin. Sole non-Hanoverian member of the artists' group die abstrakten hannover. 1928: contacts with the Bauhaus. Member of the artists' groups Cercle et Carré and abstraction-création. 1933: moves to Paris. 1937: launches "Plastique" magazine with Hans Arp and Sophie Täuber-Arp. 1946: founds the group Centre des Recherches in Paris. Numerous exhibitions, including retrospectives in Paris, Grenoble and Amsterdam (1987). – *Publications include:* Christian Zervos "Domela", Amsterdam 1966; Kees Broos, Flip Bool "Domela", The Hague 1980; G. B. Martini, A. Ronchetti "Cesar Domela", Genoa 1981; Marie-Odile Briot "Domela – Schilderijen, reliefs, grafisch Œuvre", Apeldoorn, Paris 1990.

**Dooijes,** Dick – b. 6. 5. 1909 in Amsterdam, The Netherlands – *type designer, typographer, author, teacher* – Trained at the Lettergieterij in Amsterdam. 1926–36: studies at the Instituut voor Kunstnijverheidsonderwijs and at the Academie voor Beeldende Kunsten in Amster-

Studium am Instituut voor Kunstnijverheidsonderwijs und an der Academie voor Beeldende Kunsten in Amsterdam. 1940 Nachfolger von S. H. de Roos in der Lettergieterij Amsterdam. 1968–74 Direktor der Gerrit Rietveld Akademie in Amsterdam. – *Schriftentwürfe:* Rondo (mit Stephan Schlesinger, 1948), Mercator (1958), Contura (1966), Lectura (1969). – *Publikationen u. a.:* „A History of the Dutch Poster" (mit Pieter Brattinga), Amsterdam 1968; „Boektypografische verkenningen", Amsterdam 1986; „Wegbereiders van de moderne boektypografie in Nederland", Amsterdam 1988; „Mijn leven met letters", Amsterdam 1991.

**Dorfsman,** Louis – geb. 25. 4. 1918 in New York, USA. – *Grafik-Designer, Typograph* – 1936–39 Studium an der Cooper Union, New York. 1939 Arbeiten für die Weltausstellung in New York. 1946 Eintritt als Designer in die Columbia Broadcasting System (CBS). Gestaltet 1962–66 das visuelle Erscheinungsbild vom Briefpapier bis zum Leitsystem im neuen CBS-Hauptsitz. Berühmt wird sein Wandrelief „Gastro-typographic-Assemblage" in der Cafeteria des Gebäudes. Der Art Directors Club of New York, der ihn im Laufe der Jahre mit 12 Goldmedaillen auszeichnet, wählt ihn 1978 in die Hall of Fame. Das American Institute of Graphic Arts verleiht ihm 1979 eine Goldmedaille. 1988 Präsident des Studios Louis Dorfsman Design New York. – *Publikation u. a.:* Dick Hess, Marion Müller „Dorfsman & CBS", New York 1987.

**Double Crown Club** – Oliver Simon, Holbrook Jackson und Hubert Foss gründen

Kunstnijverheidsonderwijs et à l'Academie voor Beeldende Kunsten d'Amsterdam. En 1940, il succède à S. H. de Roos à la Lettergieterij d'Amsterdam. 1968–1974, directeur de la Gerrit Rietveld Akademie à Amsterdam. – *Polices:* Rondo (avec Stephan Schlesinger, 1948), Mercator (1958), Contura (1966), Lectura (1969). – *Publications, sélection:* «A History of the Dutch Poster» (avec Pieter Brattinga), Amsterdam 1968; «Boektypografische verkenningen», Amsterdam 1986; «Wegbereiders van de moderne boektypografie in Nederland», Amsterdam 1988; «Mijn leven met letters», Amsterdam 1991.

**Dorfsman,** Louis – né le 25. 4. 1918 à New York, Etats-Unis – *graphiste maquettiste, typographe* – 1936–1939, études à la Cooper Union, New York. 1939, travaille pour l'exposition universelle de New York. En 1946, il entre comme designer au Columbia Broadcasting System (CBS). De 1962 à 1966, il réalise l'identité visuelle du nouveau siège CBS, du papier à lettres jusqu'au système signalétique. Son relief mural «Gastro-typographic-Assemblage» réalisé dans la cafétéria de l'immeuble devient célèbre. En 1978, l'Art Directors Club of New York qui lui a déjà décerné douze médailles d'or, l'élit au Hall of Fame. En 1979, il reçoit une médaille d'or de l'American Institute of Graphic Arts. Président de l'atelier Louis Dorfsman Design de New York en 1988. – *Publication, sélection:* Dick Hess, Marion Müller «Dorfsman & CBS», New York 1987.

**Double Crown Club** – en 1924, Oliver Simon, Holbrook Jackson et Hubert Foss fondent ce club à Londres, c'est une dining-club pour les amis des arts du livre et de la typographie. Le dîner d'inauguration a lieu le 31. 10. 1924. Le second dîner conférence se déroule autour du thème «Ecritures du présent», le 25e dîner

Dorfsman 1987 Cover

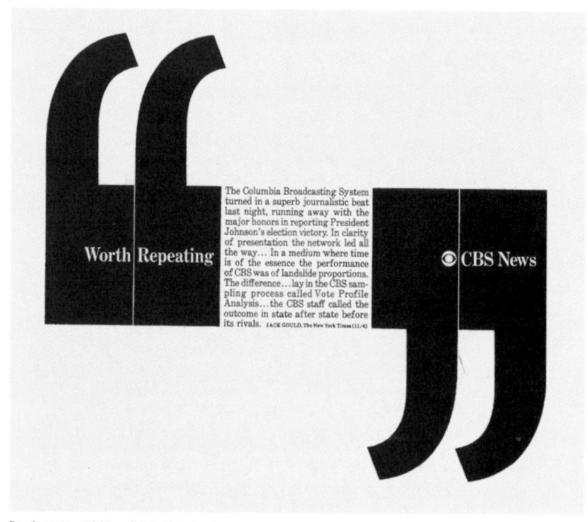

Dorfsman 1964 Advertisement

Dorfsman 1986 Cover

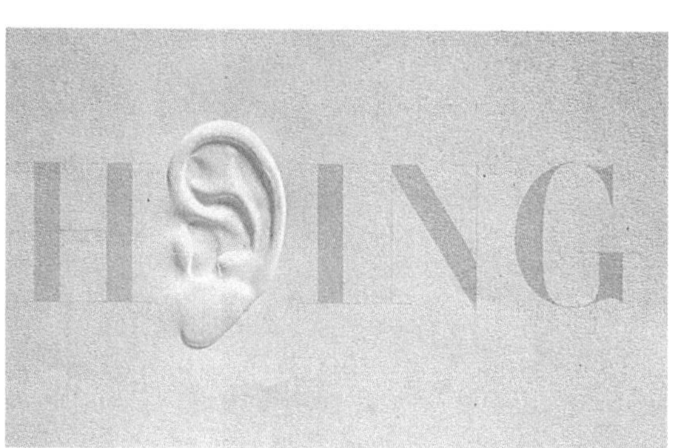
Dorfsman Cover

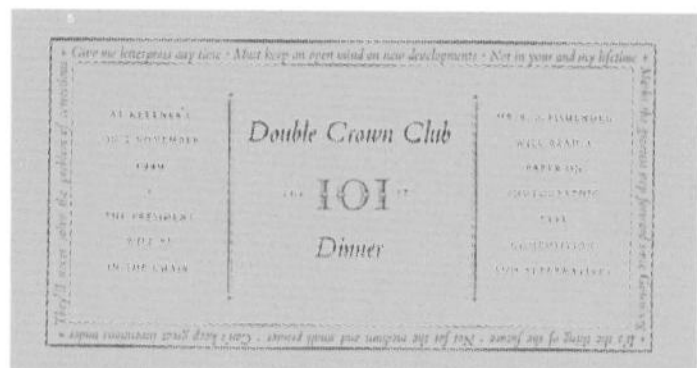

Double Crown Club 1946 Invitation

dam. 1940: succeeds S. H. de Roos at the Lettergieterij in Amsterdam. 1968–74: director of the Gerrit Rietveld Akademie in Amsterdam. – *Fonts:* Rondo (with Stephan Schlesinger, 1948), Mercator (1958), Contura (1966), Lectura (1969). – *Publications include:* "A History of the Dutch Poster" (with Pieter Brattinga), Amsterdam 1968; "Boektypografische verkenningen", Amsterdam 1986; "Wegbereiders van de moderne boektypografie in Nederland", Amsterdam 1988; "Mijn leven met letters", Amsterdam 1991.

**Dorfsman,** Louis – b. 25. 4. 1918 in New York, USA – *graphic designer, typographer* – 1936–39: studies at the Cooper Union, New York. 1939: produces work for the world exhibition in New York. 1946: joins the Columbia Broadcasting System (CBS) as a designer. 1962–66: designs CBS' corporate identity from writing paper to the signage system in the new CBS head office. His "Gastro-Typographic-Assemblage" mural relief in the building's café has become famous. 1978: the Art Directors Club of New York, which has awarded him 12 gold medals over the years, elects him into their Hall of Fame. 1979: the American Institute of Graphic Arts awards him a gold medal. 1988: president of the Louis Dorfsman Designs studio in New York. – *Publications include:* Dick Hess, Marion Müller "Dorfsman & CBS", New York 1987.

**Double Crown Club** – 1924: Oliver Simon, Holbrook Jackson and Hubert Foss found the club in London as a dining club for lovers of books and typography. The founder's dinner takes place on 31. 10.

1924 in London den Club als Dining-Club für die Freunde der Kunst des Buches und der Typographie. Das Gründungsessen findet am 31. 10. 1924 statt. Das zweite Vortragsessen steht unter dem Thema „Schriften der Gegenwart", 1930 findet das 25. Vortragsessen statt, 1944 das 75. Vortragsredner sind im Laufe der Jahre u. a. John Carter, Stanley Morison, Jan Tschichold, Beatrice Warde. 1989 findet das 300. Essen statt.

**Doves Press** – *Verlag, Druckerei* – 1900 Gründung der Presse in Hammersmith durch T. J. Cobden-Sanderson (1840–1922) und Emery Walker (1851–1933), die der „Arts and Crafts"-Bewegung nahestehen. Die Bücher der Presse werden ohne Schmuck und Dekor gestaltet. Als einzige Satzschrift wird die von Edward Prince nach Angaben von Emery Walker geschnittene „Doves Roman" verwendet, deren Form von der Antiqua Nikolaus Jensons aus dem Jahr 1470 abgeleitet wird. 1903–05 Herausgabe der Doves-Bibel in fünf Bänden mit Initialen von Edward Johnston. Emery Walker verläßt 1909 die Doves Press. 1916 Herausgabe von J. W. Goethes „Auserlesene Lieder". Nach dem Druck eines Katalogs über die Erzeugnisse der Doves Press versenkt Cobden-Sanderson den gesamten Schriftenvorrat der Presse in der Themse.

**Doyle,** Stephen – geb. 4. 11. 1956 in Baltimore, Maryland, USA. – *Grafik-Designer, Typograph, Lehrer* – 1978 Abschluß des Studiums an der Cooper Union, New York. Designer bei „Esquire", danach bei „Rolling Stone". Eintritt in das von Tibor Kalman gegründete Studio „M &

¶THE DOVES PRESS was founded in 1900, to attack the problem of pure Typography, as presented by ordinary books in the various forms of prose, verse, & dialogue, & keeping always in view the principle laid down in The Book Beautiful, that "The whole duty of Typography is to communicate to the imagination, without loss by the way, the thought or image intended to be conveyed by the Author," to attempt its solution rather by the arrangement of the whole book, as a whole, with due regard to its parts and the emphasis of its divisions, than by the splendour of ornament, intermittent, page after page.

¶The works selected for this purpose have been chosen, partly for the sake of the particular typographical problem presented by them; but partly also in view of the second object of the Press, viz., to print in a monumental form some of the great literary expressions of man's creative or destructive thought. In this respect, the selection has been mainly accidental; though, as will appear from the Catalogue, the Books Printed have a certain relation to one another and to a plan. The Bible has seemed to the Printer to be the one supreme work of English literature, albeit not of English original thought. So, on the other hand, in Paradise Lost & Paradise Regain'd, Milton has given to English religious emotion its dominant form & immortal expression. Again, Faust, the tragedy of Goethe, has re-opened the eternal debate

2

Doves Press 1908 Page

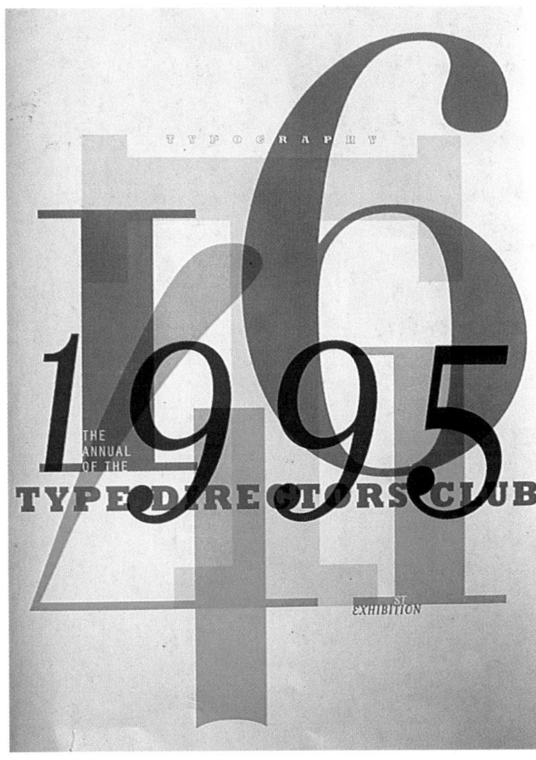


Doyle 1995 Cover

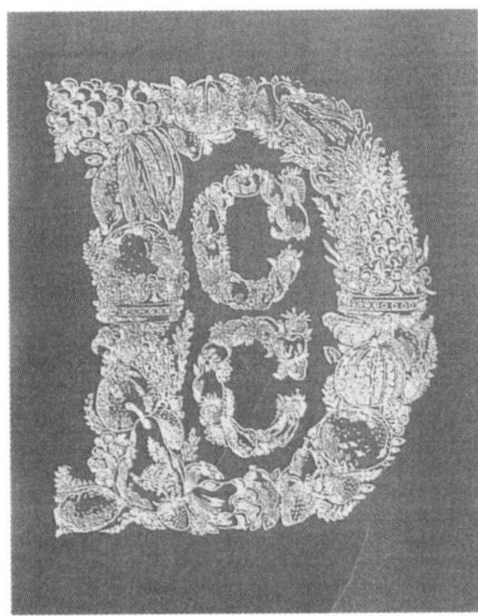

Double Crown Club 1969 Invitation Card

Doyle 1989

a lieu en 1930 et le 75e en 1944. Parmi les intervenants ayant participé au cours de ces années, on compte John Carter, Stanley Morison, Jan Tschichold, Beatrice Warde. Le 300e dîner a eu lieu en 1989.

**Doves Press** – *editions, imprimerie* – en 1900, T. J. Cobden-Sanderson (1840–1922) et Emery Walker (1851–1933), proches du mouvement Arts & Crafts, fondent les presses à Hammersmith. Les livres qui en sortent sont dépourvus d'ornements et de décor. Le seul caractère utilisé en composition est le «Doves Roman» taillé par Edward Prince selon les indications d'Emery Walker; sa forme est dérivée de l'antique de 1470 de Nicolas Jenson. De 1903 à 1905, les presses publient la Bible-Doves en 5 volumes avec des lettrines d'Edward Johnston. En 1909, Emery Walker quitte les Doves Press. Publication des «Auserlesene Lieder» de J. W. Goethe en 1916. Après l'impression d'un catalogue des réalisations des presses, Cobden-Sanderson immerge tout le stock de caractères dans la Tamise.

**Doyle,** Stephen – né le 4. 11. 1956 à Baltimore, Maryland, Etats-Unis – *graphiste maquettiste, typographe, enseignant* – 1978, termine ses études à la Cooper Union, New York. Designer chez «Esquire», puis pour «Rolling Stone». Entre au Studio «M & Co» fondé par Tibor Kalman. 1985, création du Studio Drenttel Doyle

1924. The second lecture evening is held under the motto "Today's Typefaces". The 25th dinner takes place in 1930 and the 75th in 1944. Over the years, invited speakers have included John Carter, Stanley Morison, Jan Tschichold and Beatrice Warde. 1989: the 300th meeting is held.

**Doves Press** – *publishing house, printing works* – 1900: the press is founded in Hammersmith by T. J. Cobden-Sanderson (1840–1922) and Emery Walker (1851–1933), sympathizers of the Arts and Crafts Movement. Books printed by the press are devoid of any decoration or illustration. Doves' sole type is Doves Roman, cut by Edward Prince according to instructions from Emery Walker, and derived from Nicolaus Jenson's roman typeface from 1470. 1903–05: the Doves Bible is published in five volumes with initials by Edward Johnston. 1909: Emery Walker leaves Doves Press. 1916: J. W. Goethe's "Auserlesene Lieder" are published. After printing a catalogue of the items produced by Doves Press over the years, Cobden-Sanderson sinks the press's entire typesetting material in the River Thames.

**Doyle,** Stephen – b. 4. 11. 1956 in Baltimore, Maryland, USA – *graphic designer, typographer, teacher* – 1978: completes his studies at the Cooper Union in New York. Designer for "Esquire" and then "Rolling Stone". Joins the M & Co. studio, founded by Tibor Kalman. 1985: opens the Drenttel Doyle Partners studio in New York with Bill Drenttel and Tom Kluepfel. Their areas of work are designs for books and magazines, corporate identities, packaging and exhibition design.

Co.". 1985 Gründung des Studios Drenttel Doyle Partners in New York zusammen mit Bill Drenttel und Tom Kluepfel. Arbeitsgebiete: Buch- und Zeitschriftengestaltung, Erscheinungsbilder, Verpackungen, Ausstellungsgestaltung. Auftraggeber waren u. a. das Museum of Modern Art, Metropolitan Transit Authority, The World Financial Center, Champion Paper. Unterrichtet an der Cooper Union und an der Yale University.

**Drescher,** Arno – geb. 1882 in Auerbach, Deutschland, gest. 1971 in Braunschweig, Deutschland. – *Grafik-Designer, Schriftentwerfer, Lehrer* – 1896 Ausbildung zum Lehrer. 1902–04 Lehrer. 1905–07 Studium an der Akademie für Kunstgewerbe Dresden, danach Fachlehrer in der Kunsterzieherausbildung. Ab 1920 Professor an der Akademie für Kunstgewerbe Dresden. 1940–45 Direktor der Akademie für Graphische Künste und Buchgewerbe Leipzig. 1945–60 freiberuflicher Grafiker in Leipzig, seit 1960 in Braunschweig. – *Schriftentwürfe:* Arabella (1936–39), Antiqua 505 (1955–56).

**Dreyfus,** John – geb. 15. 4. 1918 in London, England. – *Typograph, Autor* – Ausbildung an der Oundle School und am Trinity College in Cambridge. 1955–82 typographischer Berater der Monotype Corporation, wo er an der Produktion der Schriften Univers (A. Frutiger), Dante (G. Mardersteig), Sabon (J. Tschichold), Spectrum (J. van Krimpen), Mercurius (I. Reiner), Octavian (W. Carter, D. Kindersley) und anderen mitwirkt. 1956–77 europäischer Repräsentant des Limited Edition Club New York, für den er zahlreiche

Partners à New York avec Bill Drenttel et Tom Kluepfel. Activités : maquettes de livres et de revues, identités, emballages et design d'expositions. Commanditaires : le Museum of Modern Art, Metropolitan Transit Authority, The World Financial Center, Champion Paper etc. Enseigne à la Cooper Union et à la Yale University.

**Drescher,** Arno – né en 1882 à Auerbach, Allemagne, décédé en 1971 à Braunschweig, Allemagne – *graphiste maquettiste, concepteur de polices, enseignant* – 1896, formation d'enseignant. 1902–1904, enseignant. 1905–1907, étudie à l'Akademie für Kunstgewerbe (arts décoratifs) de Dresde, puis professeur de pédagogie de l'art. A partir de 1920, professeur à l'Akademie für Kunstgewerbe de Dresde. 1940–1945, directeur de l'Akademie für Graphische Künste und Buchgewerbe (arts graphiques et arts du livre) de Leipzig. 1945–1960, graphiste indépendant à Leipzig, puis à Braunschweig à partir de 1960. – *Polices :* Arabella (1936–1939), Antiqua 505 (1955–1956).

**Dreyfus,** John – né le 15. 4. 1918 à Londres, Angleterre – *typographe, auteur* – formation à la Oundle School et au Trinity College, Cambridge. 1955–1982, conseiller typographique de la Monotype Corporation, où il contribue à la production de polices comme l'Univers (A. Frutiger), Dante (G. Mardersteig), Sabon (J. Tschichold), Spectrum (J. van Krimpen), Mercurius (I. Reiner), Octavian (W. Carter, D. Kindersley). 1956–1977, représentant européen du Limited Edition Club, New York, conception de nombreuses maquettes de livres. 1957, membre fondateur de l'Association Typographique Internationale (ATypI), qu'il préside de 1968 à 1973, et dont il est actuellement président d'honneur. Il organise le Congrès international Caxton en 1976. Obtient le prix Gutenberg de la ville de

Drescher 1936–39 Arabella

Drescher 1956 Antiqua 505 Bold

Dumbar Stamps

Clients include the Museum of Modern Art, the Metropolitan Transit Authority, The World Financial Center and Champion Paper. Doyle has taught at the Cooper Union and at Yale University.

**Drescher,** Arno – b. 1882 in Auerbach, Germany, d. 1971 in Braunschweig, Germany – *graphic designer, type designer, teacher* – 1896: trains as a teacher. 1902–04: teacher. 1905–07: studies at the Akademie für Kunstgewerbe in Dresden and then works as a specialist teacher training art teachers. From 1920 onwards: professor at the Akademie für Kunstgewerbe in Dresden. 1940–45: director of the Akademie für Graphische Künste und Buchgewerbe in Leipzig. 1945–60: freelance graphic artist in Leipzig, and from 1960 onwards in Braunschweig. – *Fonts:* Arabella (1936–39), Antiqua 505 (1955–56).

**Dreyfus,** John – b. 15. 4. 1918 in London, England – *typographer, author* – Studied at the Oundle School and at Trinity College in Cambridge. 1955–82: typography consultant for Monotype Corporation, working on the production of Univers (A. Frutiger), Dante (G. Mardersteig), Sabon (J. Tschichold), Spectrum (J. van Krimpen), Mercurius (I. Reiner), Octavian (W. Carters, D. Kindersley) and other typefaces. 1956–77: European representative of the Limited Edition Club, New York, for whom he designs many books. 1957: cofounder of the Association Typographique Internationale (ATypI), of which he is president from 1968–73, and is currently honorary president. 1976: organizes the International Caxton Congress

Bücher gestaltet. 1957 Mitbegründer der Association Typographique Internationale (ATypI), Präsident von 1968–73, jetzt Ehrenpräsident. Organisiert 1976 den Internationalen Caxton-Kongreß in London. 1996 mit dem Gutenberg-Preis der Stadt Mainz ausgezeichnet. Zahlreiche Essays zu typographischen Themen, u. a. in „Signature", „Fine Print", „Penrose Annual", „The Times Literary Supplement", „Matrix". – *Publikationen u. a.:* „The Survival of Baskerville's Punches", 1949; „The Work of Jan van Krimpen", 1953; „Italic Quartett", 1966; „Aspects of French Eighteenth-Century Typography", 1982; „Morris and the Printed Book", 1989; „A Typographical Masterpiece", 1990; „Into Print", 1994.

**Dumbar,** Gert – geb. 16. 5. 1940 in Batavia, Niederländisch-Indien (heute Djakarta, Indonesien) – *Grafik-Designer, Typograph, Lehrer* – 1959–64 Studium der Malerei und der Visuellen Kommunikation an der Koninklijke Academie voor Beeldende Kunsten in Den Haag. 1964–67 Studium der Typographie am Royal College of Art in London. 1967–76 Mitbegründer und Teilhaber des Studios Tel Graphic Designs. 1977 Gründung des Studio Dumbar, Den Haag. Arbeiten u. a. für Philips, Apple, Zanders Papiere, die niederländische Post PTT, das Rijksmuseum in Amsterdam, die niederländische Eisenbahngesellschaft, Nike und das Holland-Festival. Lehrt 1985–87 als Visiting Professor am Royal College of Art in London. Zeitweilige Lehrtätigkeit an der Universität Bandung, Indonesien, an der Cranbrook University in Detroit und

Dreyfus 1973 Cover

Dumbar 1987 Poster

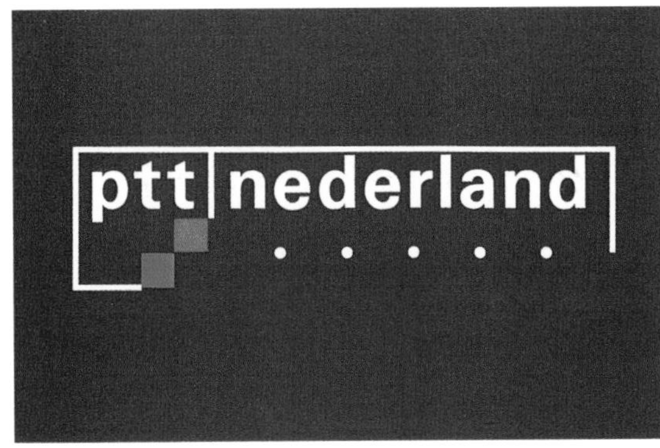


Dumbar 1989 Logo

Dumbar 1990 Poster

Mayence en 1996. De nombreux essais sur la typographie, publiés dans «Signature», «Fine Print», «Penrose Annual», «The Times Literary Supplement», «Matrix» etc. – *Publications, sélection:* «The Survival of Baskerville's Punches», 1949; «The Work of Jan van Krimpen», 1953; «Italic Quartett», 1966; «Aspects of French Eighteenth-Century Typography», 1982; «Morris and the Printed Book», 1989; «A Typographical Masterpiece», 1990; «Into Print», 1994.

**Dumbar,** Gert – né le 16. 5. 1940 à Batavia, L'Inde néerlandais (aujourd'hui Djakarta, Indonésie) – *graphiste maquettiste, typographe, enseignant* – 1959–1964, études de peinture et de communication visuelle à la Koninklijke Academie voor Beeldende Kunsten de La Haye. 1964–1967, études de typographie au Royal College of Art de Londres. 1967–1976, cofondateur et associé du Studio Tel Graphic Design. 1977, fondation du Studio Dumbar, à La Haye. Travaille pour Philips, Apple, Zanders Papiere, les postes des néederlandais PTT, le Rijkmuseum d'Amsterdam, les chemins de fer des Pays-Bas, Nike et le Holland Festival. Enseigne au Royal College of Art de Londres de 1985 à 1987. Enseignement intermittent dans les universités de Bandung, Indonésie, à la Cranbrook University de Detroit et à la Hochschule der Bildenden

in London. 1996: is awarded the Gutenberg Prize by the city of Mainz. Numerous essays on typographical themes, published in "Signature", "Fine Print", "Penrose Annual", "The Times Literary Supplement" and "Matrix", among others. – *Publications include:* "The Survival of Baskerville's Punches", 1949; "The Work of Jan van Krimpen", 1953; "Italic Quartett", 1966; "Aspects of French Eighteenth-Century Typography", 1982; "Morris and the Printed Book", 1989; "A Typographical Masterpiece", 1990; "Into Print" 1994.

**Dumbar,** Gert – b. 16. 5. 1940 in Batavia, Dutch East Indies (now Jakarta, Indonesia) – *graphic designer, typographer, teacher* – 1959–64: studies painting and visual communication at the Koninklijke Academie voor Beeldende Kunsten in The Hague. 1964–67: studies typography at the Royal College of Art in London. 1967–76: co-founder and partner of the Tel Graphic Designs studio. 1977: opens Studio Dumbar in The Hague. Works for various clients, including Philips, Apple, Zanders Papiere, the Dutch post offices PTT, the Rijksmuseum in Amsterdam, the Dutch railways, Nike and the Holland Festival. 1985–87: visiting professor at the Royal College of Art in London. Temporary teaching positions at the University of Bandung, Indonesia, at Cranbrook University in Detroit and since 1995 at the Hochschule der Bildenden Künste in Saarbrücken. 1987–88: president of the British Designers and Art Directors Association. Permanent member of British Rail's advisory committee for design.

seit 1995 an der Hochschule der Bildenden Künste in Saarbrücken. 1987–88 Präsident der British Designers and Art Directors Association. Ständiges Mitglied des Designbeirats der britischen Eisenbahngesellschaft. 1989 werden Michel de Boer (geb. 1954) und Kitty de Jong (geb. 1951) Managing Partner des Studio Dumbar. 1995 Eröffnung eines zweiten Studio Dumbar in Rotterdam, das von Henri Ritzen geleitet wird. 1995 Auszeichnung mit dem Titel „Doctor in Design honoris causa" durch die Nottingham Trent University in England. – *Publikationen:* Hanna Gerken „Studio Dumbar", Mainz 1993; Rick Poynor, Klaus Honnef „Behind the Seen. Studio Dumbar", Mainz 1996.

**Dürer,** Albrecht – geb. 21. 5. 1471 in Nürnberg, Deutschland, gest. 6. 4. 1528 in Nürnberg, Deutschland. – *Maler, Holzschneider, Kupferstecher* – Innerhalb seines umfangreichen freien künstlerischen Werkes befaßt sich Dürer intensiv mit Form, Konstruktion und Proportion von Schrift. Seine Signatur ist Ausdruck dieser gestalterischen Auseinandersetzung. – *Publikationen u. a.:* „Unterweisung der Messung (mit dem Zirkel und Richtscheit in Linien, Ebenen und ganzen Körpern)", Nürnberg 1525; „Befestigungen", Nürnberg 1527; „Proportionen des Menschen", Nürnberg 1528. Ernst Crous „Dürer und die Schrift", Berlin 1933; Ernst Panofsky „Albrecht Dürer", Princeton 1948; H. Rupprich „Dürer. Schriftlicher Nachlaß", Berlin 1956; F. Winkler „Albrecht Dürer. Leben und Werk", Berlin 1957.

**Dwiggins,** William Addison – geb. 1880 in Martinsville, USA, gest. 25. 12. 1956 in

Künste (école supérieure des beaux-arts) de Sarrebruck. 1987– 1988, président de la British Designers and Art Directors Association. Membre permanent du comité consultatif de design des chemins de fer britanniques. En 1989, Michel de Boer (né en 1954) et Kitty de Jong (née en 1951) entrent comme managing partners au Studio Dumbar. 1995, ouverture d'un second Studio Dumbar à Rotterdam dirigé par Henri Ritzen. En 1995, il obtient le titre de «Doctor in Design honoris causa» de la Nottingham Trent University en Angleterre. – *Publications:* Hanna Gerken «Studio Dumbar», Mayence 1993; Rick Poynor, Klaus Honnef «Behind the Seen. Studio Dumbar», Mayence 1996.

**Dürer,** Albrecht – né le 21. 5. 1471 à Nuremberg, Allemagne, décédé le 6. 4. 1528 à Nuremberg, Allemagne – *peintre, graveur sur bois et sur cuivre* – Dürer, dont l'œuvre est considérable s'est consacré intensivement à la forme, la construction et les proportions des lettres. Sa signature est l'expression de ses préoccupations formelles. – *Publications, sélection:* «Unterweisung der Messung (mit dem Zirkel und Richtscheit in Linien, Ebenen und ganzen Körpern)» Nuremberg, 1525; «Befestigungen», Nuremberg 1527; «Proportionen des Menschen», Nuremberg 1528. Ernst Crous «Dürer und die Schrift», Berlin 1933; Ernst Panofsky «Albrecht Dürer», Princeton 1948; H. Rupprich «Dürer. Schriftlicher Nachlass», Berlin 1956; F. Winkler «Albrecht Dürer. Leben und Werk», Berlin 1957.

**Dwiggins,** William Addison – né en 1880 à Martinsville, Etats-Unis, décédé le 25. 12. 1956 à Hingham, Massachusetts, Etats-Unis – *concepteur de polices, imprimeur, typographe, graphiste maquettiste* – études à la Frank Holme School of Illustration à Chicago, chez Frederic W. Goudy. 1903–1904, travaille dans sa

Dürer 1525 Sketch

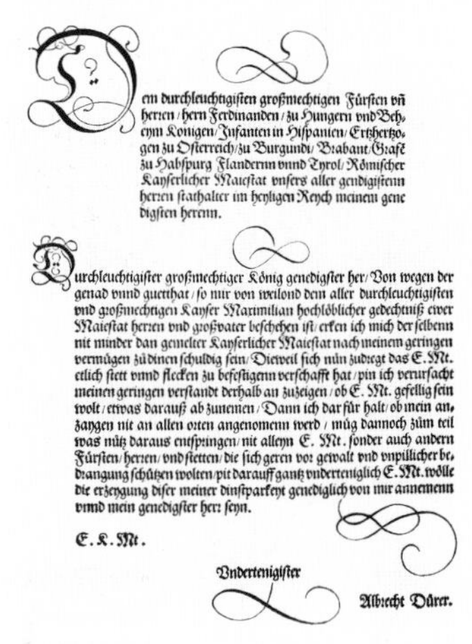

Dem durchleuchtigiſten groſzmechtigen Fürſten vñ herren / hern Ferdinanden / zu Hungern vnd Behemn Konigen / Infanten in Hiſpanien / Ertzhertzogen zu Oſterreich / zu Burgundi / Brabant / Grafe zu Habſpurg Flanderen vnnd Tyrol / Römiſcher Kayſerlicher Maieſtat vnſers aller gendigiſtenn herren ſtathalter im heyligen Reych meinem genedigſten herenn.

Durchleuchtigiſter groſzmechtiger König genedigſter her / Von wegen der genad vnnd guetthat / ſo mir von weilond dem aller durchleuchtigiſten vnd groſzmechtigen Kayſer Maximilian hochlöblicher gedechtniſz ewer Maieſtat herren vnd groſzvater beſchehen iſt / erken ich mich der ſelbenn nit minder dan gemelter Kayſerlicher Maieſtat nach meinem geringen vermügen zů dinen ſchuldig ſein / Dieweil ſich nůn zůtregt das E. Mt. etlich ſtett vnnd flecken zu befeſtigenn verſchafft hat / pin ich verurſacht meinen geringen verſtandt derhalb an zuzeigen / ob E. Mt. gefellig ſein wolt / etwas darauſz ab zunemen / Dann ich dar für halt / ob mein anzaygen nit an allen orten angenomenn werd / mûg dannoch zům teil was nütz daraus entſpringen / nit alleyn E. Mt. ſonder auch andern Fürſten / herren / vnd ſtetten / die ſich geren vor gewalt vnd vnpillicher bedrangung ſchützen wolten / pit darauff gantz vnderteniglich E. Mt. wölle die erzeygung diſer meiner dinſtparkeyt genediglich von mir annemenn vnnd mein genedigſter herr ſeyn.

E. K. Mt.

Vndertenigiſter

Albrecht Dürer.

Dürer

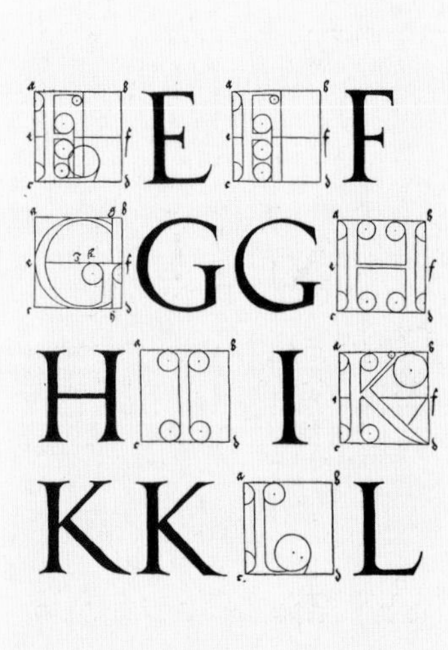

Dürer 1525 Sketch

1989: Michel de Boer (b. 1954) and Kitty de Jong (b. 1951) become managing partners of Studio Dumbar. 1995: a second Studio Dumbar opens in Rotterdam, which is run by Henri Ritzen. 1995: is awarded an honorary Doctorate in Design by Nottingham Trent University in England. – *Publications:* Hanna Gerken "Studio Dumbar", Mainz 1993; Rick Poynor, Klaus Honnef "Behind the Seen. Studio Dumbar", Mainz 1996.

**Dürer,** Albrecht – b. 21. 5. 1471 in Nuremberg, Germany, d. 6. 4. 1528 in Nuremberg, Germany – *painter, woodcarver, copper engraver* – Dürer's extensive, free art work illustrates how intensively he concentrated on the form, construction and proportions of letters. His signature is a clear expression of this structural analysis. – *Publications include:* "Unterweisung der Messung (mit dem Zirkel und Richtscheit in Linien, Ebenen und ganzen Körpern)", Nuremberg 1525; "Befestigungen", Nuremberg 1527; "Proportionen des Menschen", Nuremberg 1528. Ernst Crous "Dürer und die Schrift", Berlin 1933; Ernst Panofsky "Albrecht Dürer", Princeton 1948; H. Rupprich "Dürer. Schriftlicher Nachlaß", Berlin 1956; F. Winkler "Albrecht Dürer. Leben und Werk", Berlin 1957.

**Dwiggins,** William Addison – b. 1880 in Martinsville, USA, d. 25. 12. 1956 in Hingham, Massachusetts, USA – *type designer, printer, typographer, graphic designer* – Studied at the Frank Holme School of Illustration in Chicago under Frederic W. Goudy. 1903–04: runs his own printing workshop in Cambridge,

Hingham, Massachusetts, USA. – *Schriftentwerfer, Drucker, Typograph, Grafik-Designer* – Studium an der Frank Holme School of Illustration in Chicago bei Frederic W. Goudy. Führt 1903–04 seine eigene Druckerei in Cambridge, Ohio. 1905–16 Werbegrafiker in Hingham, Massachusetts. 1917–18 Leiter der Harvard University Press. Gründet 1919 in Boston die „Society of Calligraphers", deren Präsident und einziges Mitglied er ist. Verwendet 1922 erstmals den Begriff „Graphic Designer". Lernt 1923 Alfred A. Knopf kennen und entwirft für dessen Verlag im Laufe der Jahre über 300 Buchumschläge. Ab 1928 Schriftentwürfe für Mergenthaler-Linotype, mit der er 27 Jahre zusammenarbeitet. 1929 Auszeichnung mit der Goldmedaille des American Institute of Graphic Arts (AIGA). – *Schriftentwürfe:* Metro (1929–30), Electra (1935–49), Caledonia (1938), Eldorado (1953), Falcon (1961) sowie zahlreiche Entwürfe, die nicht produziert wurden. – *Publikation:* „Layout in Advertisement", New York 1928.

propre imprimerie à Cambridge, Ohio. 1905–1916, graphiste publicitaire à Hingham, Massachusetts. 1917–1918, dirige les Harvard University Press. En 1919, il fonde la «Society of Calligraphers» à Boston, dont il est le président et l'unique membre. En 1922, il utilise pour la première fois le terme de «Graphic-Designer». Il rencontre Alfred A. Knopf en 1923 et conçoit près de 300 couvertures de livres pour sa maison d'édition au cours des années de leur collaboration. A partir de 1928, il conçoit des polices pour Mergenthaler-Linotype qui l'emploira pendant 27 ans. En 1929, il reçoit la médaille d'or de l'American Institute of Graphic Arts (AIGA). – *Polices:* Metro (1929–1930), Electra (1935–1949), Caledonia (1938), Eldorado (1953), Falcon (1961) ainsi que de nombreux dessins qui n'ont pas été produits. – *Publication:* «Layout in Advertisement», New York 1928.

Alberti Dureri clarissimi picto ris et Geometræ de Symetria partium in rectis formis hũanorum corporum, Libri in latinum conuersi.

Lectori.
Si qui forte leges Germanæ audacia dextræ
Scripta per ausonios currere iussa sonos
Da veniam erratis, neq; enim non esse ueremur,
Et nouitate tibi concilietur opus
Nam labor exemplo caruit nec signa priorum
Vlla pedum per quæ nos graderemur erant
Si qua tamen namq; haud nostra omnia dãno, placebũt
Scripta tuus decorans augeat ista fauor.
Jo.

Dürer 1525 Page

Strange case of Dr. Jekyll and Mr. Hyde. Robert Louis Stevenson. With illustrations by W. A. Dwiggins.

RANDOM HOUSE · *NEW YORK* 1929

Dwiggins 1929 Main Title

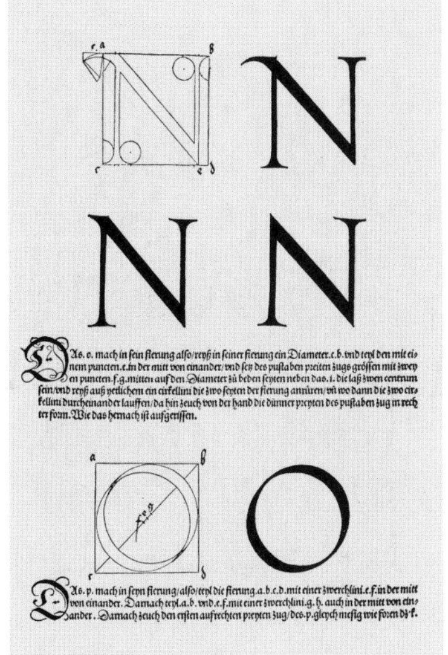

Dürer 1525 Page

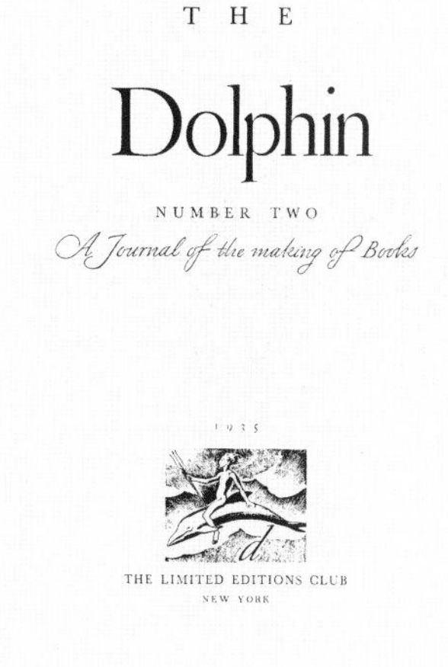

THE
Dolphin
NUMBER TWO
A Journal of the making of Books
THE LIMITED EDITIONS CLUB
NEW YORK


Dwiggins 1935 Cover

| | n | n | | | n |
|---|---|---|---|---|---|
| a | b | e | f | g | i |
| o | r | s | t | y | z |
| A | B | C | E | G | H |
| M | O | R | S | X | Y |
| 1 | 2 | 4 | 6 | 8 | & |

Dwiggins 1935–49 Electra

Ohio. 1905–16: commercial artist in Hingham, Massachusetts. 1917–18: director of Harvard University Press. 1919: founds the Society of Calligraphers in Boston and is their president and sole member. 1922: he is the first to use the term "graphic designer". 1923: over the years, Dwiggins designs over 300 book covers for his friend Alfred A. Knopf's publishing firm. 1928: produces his first type designs for Mergenthaler-Linotype, whom he works with for 27 years. 1929: is awarded a gold medal by the American Institute of Graphic Arts (AIGA). – *Fonts:* Metro (1929–30), Electra (1935–49), Caledonia (1938), Eldorado (1953), Falcon (1961) and numerous other designs which were never manufactured. – *Publication:* "Layout in Advertisement", New York 1928.

# E

**Eckersley,** Tom – geb. 30. 9. 1914 in Lowton, England. – *Grafik-Designer, Illustrator, Typograph* – 1930–34 Studium an der Salford School of Art. 1934–40 Zusammenarbeit mit Eric Lombers, grafische Arbeiten für London Transport und Shell. 1937–39 Dozent an der Westminster School of Art. Während des Zweiten Weltkriegs Kartograph bei der Royal Air Force, Plakatgestaltung für das General Post Office und das Informationsministerium. Seit 1945 freischaffender Grafiker. Auftraggeber u. a. der World Wide Fund for Nature, London Transport, Imperial War Museum, BBC. 1949 Auszeichnung „Officer of the British Empire" (O. B. E.) für sein Plakatschaffen. 1957–76 Leiter der Grafik-Design-Abteilung des London College of Printing. Seit 1970 Gastprofessor an der Yale University. 1977 Wandgestaltung für die U-Bahn-Station des Londoner Flughafens Heathrow. Zahlreiche Auszeichnungen und Ausstellungen. – *Publikationen u. a.:* „Poster Design", London 1954. F. H. K. Henrion „Top Graphic Design", Zürich 1983.

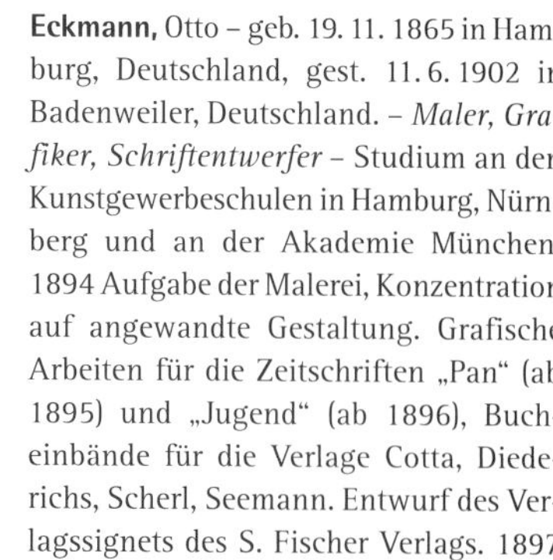

**Eckmann,** Otto – geb. 19. 11. 1865 in Hamburg, Deutschland, gest. 11. 6. 1902 in Badenweiler, Deutschland. – *Maler, Grafiker, Schriftentwerfer* – Studium an den Kunstgewerbeschulen in Hamburg, Nürnberg und an der Akademie München. 1894 Aufgabe der Malerei, Konzentration auf angewandte Gestaltung. Grafische Arbeiten für die Zeitschriften „Pan" (ab 1895) und „Jugend" (ab 1896), Bucheinbände für die Verlage Cotta, Diederichs, Scherl, Seemann. Entwurf des Verlagssignets des S. Fischer Verlags. 1897

**Eckersley,** Tom – né le 30. 9. 1914 à Lowton, Angleterre – *graphiste maquettiste, illustrateur, typographe* – 1930–1934, études à la Salford School of Art. 1934–1940, travaux avec Eric Lombers, plusieurs réalisations pour le London Transport et pour Shell. Enseigne à la Westminster School of Art de 1937 à 1939. Cartographe de la Royal Air Force pendant la Seconde Guerre mondiale, affiches pour le General Post Office et le ministère de l'information. Graphiste indépendant depuis 1945. Commanditaires : le World Wide Fund for Nature, London Transport, Imperial War Museum, BBC. Décoré «Officer of the British Empire» (O. B. E.) pour ses activités d'affichiste en 1949. De 1957 à 1976, il dirige la section d'arts graphiques du London College of Printing. Enseigne à plusieurs reprises à la Yale University à partir de 1970. Décoration murale pour le métro à l'aéroport Heathrow à Londres en 1977. Nombreuses expositions. – *Publications, sélection :* «Poster Design», Londres 1954. F. H. K. Henrion «Top Graphic Design», Zurich 1983.

**Eckmann,** Otto – né le 19. 11. 1865 à Hambourg, Allemagne, décédé le 11. 6. 1902 à Badenweiler, Allemagne – *peintre, graphiste, concepteur de polices* – études aux Kunstgewerbeschule (écoles des arts décoratifs) de Hambourg, de Nuremberg, puis à l'Akademie de Munich. Il renonce à la peinture en 1894 pour se consacrer aux arts appliqués. Travaux graphiques pour les revues «Pan» (à partir de 1895) et «Jugend» (à partir de 1896), reliures pour les éditions Cotta, Diederichs, Scherl, Seemann. Création du signet des éditions S. Fischer. 1897, enseigne la peinture ornementale au Unterrichtsanstalt des Königlichen Kunstgewerbemuseums (institut d'enseignement du musée royal des arts décoratifs) de Berlin. 1899, des-

Eckersley 1957 Poster

Eckersley 1975 Poster

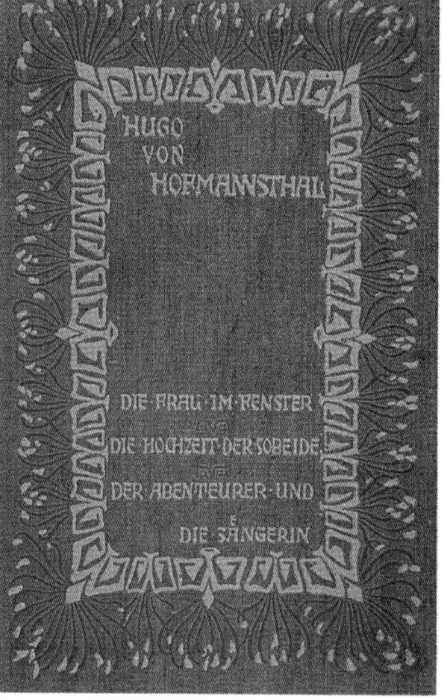


Eckmann 1899 Cover

Eckersley 1969 Poster

Eckmann 1900 Cover

**Eckersley,** Tom – b. 30. 9. 1914 in Lowton, England – *graphic designer, illustrator, typographer* – 1930–34: studies at the Salford School of Art. 1934–40: works with Eric Lombers, producing graphics for London Transport and Shell. 1937–39: lecturer at the Westminster School of Art. Cartographer with the Royal Air Force during the Second World War. Designs posters for the General Post Office and the Ministry of Information. From 1945 onwards: freelance graphic artist, with the World Wide Fund for Nature, London Transport, the Imperial War Museum and the BBC among his clients. 1949: made an Officer of the British Empire (O.B.E.) for his posters. 1957–76: head of the London College of Printing's graphic design department. Since 1970: various periods as a visiting professor at Yale University. 1977: designs the walls for Heathrow Airport's Underground stations. Numerous awards and exhibitions. – *Publications include:* "Poster Design", London 1954. F. H. K. Henrion "Top Graphic Design", Zurich 1983.

**Eckmann,** Otto – b. 19. 11. 1865 in Hamburg, Germany, d. 11. 6. 1902 in Badenweiler, Germany – *painter, graphic artist, type designer* – Studied at the Kunstgewerbeschule in Hamburg and Nuremberg and at the academy in Munich. 1894: gives up painting and concentrates on applied design. Graphic work for the magazines "Pan" (from 1895 onwards) and "Jugend" (from 1896 onwards). Book covers for Cotta, Diederichs, Scherl and Seemann publishers. Designs the logo for S. Fischer publishing house. 1897: teach-

Lehrer für ornamentale Malerei an der Unterrichtsanstalt des Königlichen Kunstgewerbemuseums Berlin. 1899 Entwurf des Signets für die Zeitschrift „Die Woche". 1900–02 grafische Arbeiten für die Allgemeine Elektrizitätsgesellschaft (AEG). – *Schriftentwürfe:* Eckmann-Schrift (1900), Fette Eckmann (1902). – *Publikationen u. a.:* „Schriften und Ornamente nach Entwürfen von O. Eckmann", Offenbach 1902.

**Edelmann,** Heinz – geb. 20. 6. 1934 in Usti Nad Labem, Tschechoslowakei. – *Illustrator, Typograph, Grafik-Designer* – 1953–58 Studium an der Kunstakademie Düsseldorf, Diplom als Zeichenlehrer. Seit 1959 freiberufliche Tätigkeit als Gebrauchsgrafiker in Düsseldorf, Werbeillustrationen, Theater- und Filmplakate, Mitarbeit an der Zeitschrift „Twen". Zahlreiche Buchillustrationen. 1961–66 Dozent für Grafik-Design an der Werkkunstschule Düsseldorf. 1967–68 Art Director des Beatles-Films „Yellow Submarine" in London. Lebt 1970–83 in Den Haag. Zahlreiche Plakate für den Westdeutschen Rundfunk in Köln. Buchgestaltung für die Verlage Carl Hanser und Klett-Cotta. 1971–72 Lehraufträge an der Fachhochschule Köln (1979–81) und an der Koninklijke Academie in Den Haag. 1983–86 Professor an der Fachhochschule Düsseldorf. Seit 1984 regelmäßige Illustrationen für das „FAZ"-Magazin. 1986–96 Professor an der Staatlichen Akademie für Bildende Künste, Stuttgart. – *Publikationen u. a.:* „Die 51 schönsten Buchumschläge von Heinz Edelmann", Stuttgart 1982; „Direct Access. 165 Ideen

Eckmann 1899 Cover

Edelmann 1963 Poster

Edelmann 1981 Cover

sine le signet de la revue «Die Woche». 1900–1902, travaux graphiques pour la Allgemeine Elektrizitätsgesellschaft (AEG). – *Polices:* Eckmann-Schrift (1900), Fette Eckmann (1902). – *Publication:* «Schriften und Ornamente nach Entwürfen von O. Eckmann», Offenbach 1902.

**Edelmann,** Heinz – né le 20. 6. 1934 à Usti Nad Labem, Tchécoslovaquie – *illustrateur, typographe, graphiste maquettiste* – 1953–1958, études à la Kunstakademie (beaux-arts) de Düsseldorf, diplôme de professeur de dessin. Exerce comme graphiste indépendant à Düsseldorf à partir de 1959, réalise des illustrations pour la publicité ainsi que des affiches pour le théâtre et le cinéma. Collaborateur de la revue «Twen». Nombreuses illustrations de livres. Il enseigne les arts graphiques à la Werkkunstschule (école des arts appliqués) de Düsseldorf de 1961 à 1966. Directeur artistique du film des Beatles «Yellow Submarine» à Londres de 1967 à 1968. Vit à La Haye de 1970 à 1983. Nombreuses affiches pour la Westdeutsche Rundfunk, Cologne. Maquettes de livres pour les éditions Carl Hanser et Klett-Cotta. 1971–1972, chargé de cours à la Fachhochschule (arts graphiques) de Cologne, à la Koninklijke Academie de La Haye de 1979–1981, puis à la Fachhochschule de Düsseldorf de 1983 à 1986. Réalise régulièrement des illustrations pour le magazine du «Frankfurter Allgemeine Zeitung» (FAZ) depuis 1984. Professeur à la Staatliche Akademie für Bildende Künste (beaux-arts) de Stuttgart de 1986 à 1996. – *Publications, sélection:* «Die 51 schönsten Buchumschläge von Heinz Edelmann», Stuttgart 1982; «Direct Access. 165 Ideen und 55 Plakate für den

Edelmann 1965 Poster

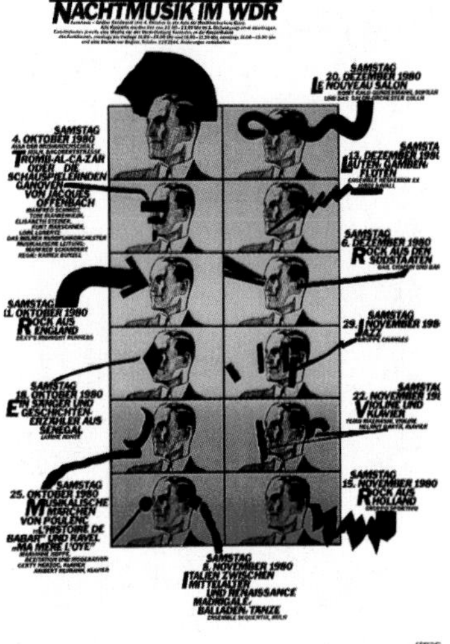


Edelmann 1980 Poster

es ornamental painting at the Unterrichtsanstalt des Königlichen Kunstgewerbemuseums in Berlin. 1899: designs the logo for "Die Woche" magazine. 1900–02: graphic work for the Allgemeine Elektrizitätsgesellschaft (AEG). – *Fonts:* Eckmann-Schrift (1900), Fette Eckmann (1902). – *Publication:* "Schriften und Ornamente nach Entwürfen von O. Eckmann", Offenbach 1902.

**Edelmann,** Heinz – b. 20. 6. 1934 in Usti Nad Labem, Czechoslovakia – *illustrator, typographer, graphic designer* – 1953–58: studies at the Kunstakademie in Düsseldorf, obtaining an art teacher diploma. From 1959 onwards: freelance commercial artist in Düsseldorf. Produces illustrations for advertising and theater and film posters, and works for "Twen" magazine. Numerous illustrations for books. 1961–66: lecturer of graphic design at the Werkkunstschule in Düsseldorf. 1967–68: art director for the Beatles' film "Yellow Submarine" in London. 1970–83: lives in The Hague. Numerous posters for Westdeutscher Rundfunk in Cologne. Designs books for Carl Hanser and Klett-Cotta publishers. 1971–72: teaching positions at the Fachhochschule in Cologne (1979–81) and at the Koninklijke Academie in The Hague. 1983–86: professor at the Fachhochschule in Düsseldorf. From 1984 onwards: regularly produces illustrations for the "Frankfurter Allgemeine Zeitung" magazine. 1986–96: professor at the Staatliche Akademie für Bildende Künste in Stuttgart. – *Publications include:* "Die 51 schönsten Buchumschläge von Heinz Edelmann", Stuttgart 1982;

und 55 Plakate für den Westdeutschen Rundfunk", Mainz 1993.
**Ehmcke,** Fritz Helmut – geb. 16. 10. 1878 in Hohensalza, Deutschland, (heute Polen) gest. 3. 2. 1965 in Widdersberg, Deutschland. – *Grafiker, Illustrator, Typograph, Schriftentwerfer* – 1893–97 Lithographenlehre in Berlin, danach als Lithograph tätig. 1899–1901 Studium an der Unterrichtsanstalt des Kunstgewerbemuseums Berlin. 1900 Gründung der Steglitzer Werkstatt mit G. Belwe und F. W. Kleukens. 1903 Berufung an die Kunstgewerbeschule Düsseldorf. 1907 Gründungsmitglied des Deutschen Werkbundes. 1913–38 Professor an der Kunstgewerbeschule München. 1913–34 Gründung und Leitung der Rupprecht Presse in München, in der 57 Publikationen erscheinen. 1946–48 Professor an der Hochschule der Bildenden Künste, München. Bis 1963 Entwürfe für Verlage, u. a. für Bruckmann, Diederichs, Fischer, Hanser, Insel, Kiepenheuer, Piper, Ullstein. – *Schriftentwürfe:* Ehmcke-Antiqua (1909), Ehmcke-Fraktur (1912), Ehmcke-Rustika (1914), Ehmcke-Schwabacher (1920), Ehmcke-Mediaeval (1923), Ehmcke-Latein (1925), Ehmcke-Elzevier (1927). – *Publikationen u. a.:* „Gildenzeichen", Offenbach 1907; „Ziele des Schriftunterrichts", Jena 1911; „Amtliche Graphik", München 1918; „Kulturpolitik", Frankfurt am Main 1947; „Geordnetes und Gültiges", München 1955.
**Eidenbenz,** Hermann – geb. 4. 9. 1902 in Cannanore, Indien, gest. 25. 2. 1993 in Basel, Schweiz. – *Grafiker, Lehrer* – 1918–22 Studium an der Kunstgewerbe-

Westdeutschen Rundfunk», Mayence 1993.
**Ehmcke,** Fritz Helmut – né le 16. 10. 1878 à Hohensalza, Allemagne (aujourd'hui Pologne), décédé le 3. 2. 1965 à Widdersberg, Allemagne – *graphiste, illustrateur, typographe, concepteur de polices* – 1893–1897, apprentissage de lithographie à Berlin où il exerce ensuite. 1899–1901, études au Unterrichtsanstalt des Königlichen Kunstgewerbemuseums (institut d'enseignement du musée royal des arts décoratifs) de Berlin. Il fonde un atelier à Steglitz en 1900 avec G. Belwe et F. W. Kleukens. Nomination à la Kunstgewerbeschule (école des arts décoratifs) de Düsseldorf en 1903. Membre fondateur du Deutscher Werkbund en 1907. Professeur à la Kunstgewerbeschule de Munich de 1913 à 1938. Fonde la Rupprecht Presse à Munich en 1913, qu'il dirige jusqu'en 1934 et où il publie 57 ouvrages. 1946–1948, professeur à la Hochschule der Bildenden Künste (école supérieure des beaux-arts) de Munich. Travaux pour les éditions Bruckmann, Diederichs, Fischer, Hanser, Insel, Kiepenheuer, Piper, Ullstein, etc. jusqu'en 1963. – *Polices:* Ehmcke-Antiqua (1909), Ehmcke-Fraktur (1912), Ehmcke-Rustika (1914), Ehmcke-Schwabacher (1920), Ehmcke-Mediaeval (1923), Ehmcke-Latein (1925), Ehmcke-Elzevier (1927). – *Publications, sélection:* «Gildenzeichen», Offenbach 1907; «Ziele des Schriftunterrichts», Iéna 1911; «Amtliche Graphik», Munich 1918; «Kulturpolitik», Francfort-sur-le-Main 1947; «Geordnetes und Gültiges», Munich 1955.
**Eidenbenz,** Hermann – né le 4. 9. 1902 à Cannanore, Inde, décédé le 25. 2. 1993 à Bâle, Suisse – *graphiste, enseignant* – 1918–1922, études à la Kunstgewerbeschule (école des arts décoratifs) de Zurich. Graphiste chez Wilhelm Deffke à Berlin de 1923 à 1925 et chez O. H. W.

Ehmcke 1928 Poster

Ehmcke 1911 Cover

Eidenbenz ca. 1948 Logo

Eidenbenz 1948 Logo

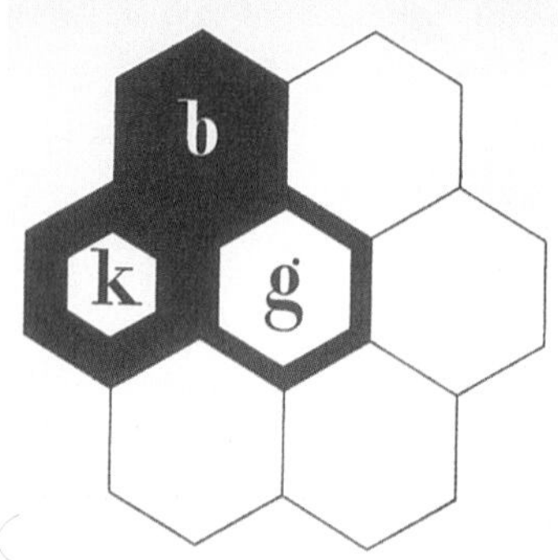

Eidenbenz 1948 Logo

"Direct Access. 165 Ideen und 55 Plakate für den Westdeutschen Rundfunk", Mainz 1993.
**Ehmcke,** Fritz Helmut – b. 16. 10. 1878 in Hohensalza, Germany (today Poland), d. 3. 2. 1965 in Widdersberg, Germany – *graphic artist, illustrator, typographer, type designer* – 1893–97: trains and then works as a lithographer in Berlin. 1899–1901: studies at the Unterrichtsanstalt des Kunstgewerbemuseums in Berlin. 1900: founds the Steglitzer Werkstatt with G. Belwe and F. W. Kleukens. 1903: appointed to the Kunstgewerbeschule in Düsseldorf. 1907: founder member of the Deutscher Werkbund. 1913–38: professor at the Kunstgewerbeschule in Munich. 1913–34: founds and runs the Rupprecht Presse in Munich, which produces 57 publications. 1946–48: professor at the Hochschule der Bildenden Künste in Munich. Designs for various publishers, including Bruckmann, Diederichs, Fischer, Hanser, Insel, Kiepenheuer, Piper and Ullstein until 1963. – *Fonts:* Ehmcke-Antiqua (1909), Ehmcke-Fraktur (1912), Ehmcke-Rustika (1914), Ehmcke-Schwabacher (1920), Ehmcke-Mediaeval (1923), Ehmcke-Latein (1925), Ehmcke-Elzevier (1927). – *Publications include:* "Gildenzeichen", Offenbach 1907; "Ziele des Schriftunterrichts", Jena 1911; "Amtliche Graphik", Munich 1918; "Kulturpolitik", Frankfurt am Main 1947; "Geordnetes und Gültiges", Munich 1955.
**Eidenbenz,** Hermann – b. 4. 9. 1902 in Cannanore, India, d. 25. 2. 1993 in Basle, Switzerland – *graphic artist, teacher* –

schule Zürich. 1923–25 Grafiker bei Wilhelm Deffke in Berlin. 1925–26 Grafiker bei O. H. W. Hadank in Berlin. 1926–32 Lehrer für Schrift und Grafik an der Kunstgewerbeschule Magdeburg. 1932–53 grafisches Büro in Basel mit seinen Brüdern Reinhold und Willy. 1937 Mitarbeit am Schweizer Pavillon der Weltausstellung in Paris. 1940–43 Lehrer an der Allgemeinen Gewerbeschule Basel. 1953–55 Lehrer an der Werkkunstschule und an der Technischen Hochschule in Braunschweig. 1955–67 künstlerischer Leiter und Werbeberater der Firma Reemtsma in Hamburg. Entwurf zahlreicher Plakate, Signets und Banknoten für die Schweiz und Deutschland.

**8vo** visual engineering – *Grafik-Design-Studio* – 1985 Gründung des Design-Studios in London durch Mark Holt, Simon Johnston und Hamish Muir. Mark Holt – geb. 27. 8. 1958 in Leeds, England. – Studium am Newcastle upon Tyne Polytechnic. 1980–84 freier Grafik-Designer in San Francisco, USA. Simon Johnston – geb. 27. 5. 1959 in Leamington Spa, England. – Studium an der Bath Academy of Art. 1981–82 Studium an der Allgemeinen Gewerbeschule in Basel. 1989 Austritt aus 8vo, Gründung seines Grafik-Design-Studios „Praxis" in Los Angeles. Hamish Muir – geb. 30. 3. 1957 in Paisley, Schottland. – Studium an der Bath Academy of Art. 1981–1982 Studium an der Allgemeinen Gewerbeschule in Basel. 1986 Gründung der Zeitschrift „Octavo – Journal of Typography" durch M. Burke, M. Holt, S. Johnston und H. Muir. Michael Burke – geb. 21. 8. 1944 in Leices-

Hadank à Berlin de 1925 à 1926. Enseigne la calligraphie et les arts graphiques à la Kunstgewerbeschule (école des arts décoratifs) de Magdebourg de 1926 à 1932. Atelier de graphisme à Bâle avec ses frères Reinhold et Willy de 1932 à 1953. Travaille pour le pavillon suisse de l'exposition universelle de Paris en 1937. 1940–1943, enseignant à la Allgemeine Gewerbeschule de Bâle. 1953–1955, professeur à la Werkkunstschule (école des arts appliqués) et à l'université technique de Braunschweig. 1955–1967, directeur artistique et conseiller publicitaire de la société Reemtsma à Hambourg. Nombreuses affiches, signets, billets de banque pour la Suisse et la République fédérale d'Allemagne.

**8vo** visual engineering – *atelier de graphisme et de design* – l'atelier de design est fondé à Londres en 1985 par Mark Holt, Simon Johnston et Hamish Muir. Mark Holt – né le 27. 8. 1958 à Leeds, Angleterre – études au Newcastle upon Tyne Polytechnic. 1980–1984, graphiste maquettiste indépendant à San Francisco, Etats-Unis. Simon Johnston – né le 27. 5. 1959 à Leamington Spa, Angleterre – études à la Bath Academy of Art. 1981–1982, études à la Allgemeine Gewerbeschule (école des arts et métiers) de Bâle. Quitte 8vo en 1989 et fonde son propre atelier de graphiste «Praxis» à Los Angeles. Hamish Muir – né le 30. 3. 1957 à Paisley, Ecosse – études à la Bath Academy of Art. 1981–1982, études à la Allgemeine Gewerbeschule de Bâle. En 1986, M. Burke, M. Holt, S. Johnston et H. Muir fondent la revue «Octavo – Journal of Typography». Michael Burke – né le 21. 8. 1944, Leicestershire, Angleterre –

Eidenbenz 1936 Poster

8vo ca. 1990 Poster

10 Barley Mow Passage
Chiswick
London W4 4PH

Telephone 01 994 6477
Fax 01 994 1533
Telex 8811418

John L. Boult
Director
BSc(Hons)

Product first

8vo 1985 Card

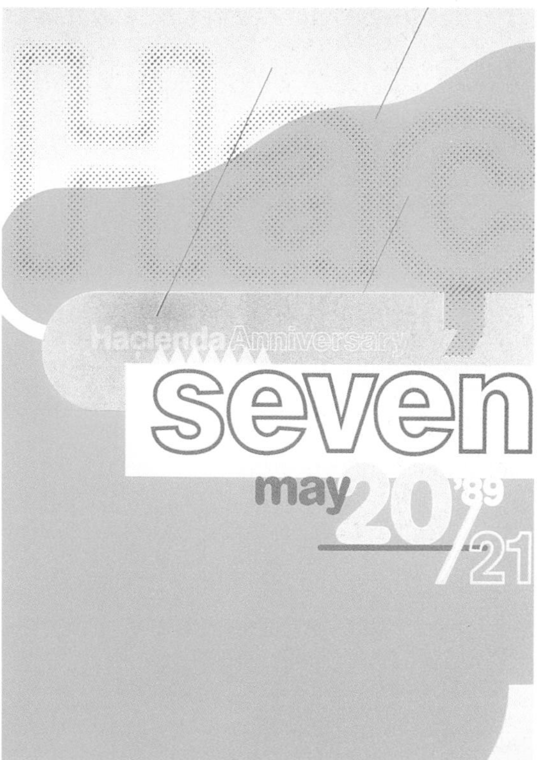


8vo 1990 Poster

1918–22: studies at the Kunstgewerbeschule in Zurich. 1923–25: graphic artist for Wilhelm Deffke in Berlin. 1925–26: graphic artist for O. H. W. Hadank in Berlin. 1926–32: teacher of type and graphics at the Kunstgewerbeschule Magdeburg.1932–53: has a graphics studio in Basle with his brothers Reinhold and Willy. 1937: works on the Swiss pavilion for the world exhibition in Paris. 1940–43: teaches at the Allgemeine Gewerbeschule in Basle. 1953–55: teaches at the Werkkunstschule and at the Technische Hochschule in Braunschweig. 1955–67: art director and advertising consultant for the Reemtsma company in Hamburg. Eidenbenz designed numerous posters, logos, and also bank notes for Switzerland and Germany.

**8vo** visual engineering – *graphic design studio* – 1985: the design studio is founded in London by Mark Holt, Simon Johnston and Hamish Muir. Mark Holt – b. 27. 8. 1958 in Leeds, England – Studied at Newcastle upon Tyne Polytechnic. 1980–84: freelance graphic designer in San Francisco, USA. Simon Johnston – b. 27. 5. 1959 in Leamington Spa, England – Studied at the Bath Academy of Art. 1981–82: studies at the Allgemeine Gewerbeschule in Basle. 1989: leaves 8vo and opens his own design studio, Praxis, in Los Angeles. Hamish Muir – b. 30. 3. 1957 in Paisley, Scotland – Studied at the Bath Academy of Art. 1981–82: studies at the Allgemeine Gewerbeschule in Basle. 1986: the magazine "Octavo – Journal of Typography" is launched by M. Burke, M. Holt, S. Johnston and H.

tershire, England. – Studium am Ravensbourne College of Art and Design. 1969–73 Mitglied des Designteams für die Olympischen Spiele in München 1972 unter der Leitung von Otl Aicher. Unterrichtet 1976–88 am Ravensbourne College of Art and Design. Seit 1988 Professor für Grafik-Design an der Fachhochschule Schwäbisch Gmünd. Seit 1987 Teilhaber an 8vo. Zu den Auftraggebern von 8vo gehören u. a. American Express, das Boymans-van Beuningen Museum in Rotterdam, Christie's, das Design Museum London, London Records, Scandinavian Airline Systems, die Zanders Feinpapiere AG. Zahlreiche Artikel über 8vo in Zeitschriften, u. a. in „Blueprint", „Novum", „High Quality", „The Independent". Zahlreiche Ausstellungen, u. a. in San Francisco (1987), Amsterdam (1988), Offenbach (1989), Rotterdam (1993).

**Eksell,** Olle – geb. 22. 3. 1918 in Dalecarlia, Schweden. – *Grafik-Designer, Typograph* – 1939–41 Schaufensterdekorateur, gleichzeitig Studium bei Hugo Steiner-Prag in Stockholm. 1941–45 Gestalter in der Werbeagentur Ervaco. 1946 Studium an der Art Center School in Los Angeles. Organisiert 1947 die erste amerikanische Grafik-Design-Ausstellung im Nationalmuseum Stockholm. Arbeit als freier Gestalter, zahlreiche Buchgestaltungen und -illustrationen. Seit 1952 Beiträge und Illustrationen für die Tageszeitung „Aftonbladet". 1965 Auszeichnung mit einer Silbermedaille auf der Buchmesse in Leipzig. Gründet 1969 Olle Eksell Design, Stockholm. Zahlreiche Gestaltungsaufgaben für die Industrie,

études au Ravensbourne College of Art and Design. 1969–1973, membre de l'équipe chargée du design pour les Jeux Olympiques de 1972 à Munich sous la direction d'Otl Aicher. Enseigne au Ravensbourne College of Art de 1976 à 1988. Professeur d'arts graphiques et de design à la Fachhochschule (école de design) de Schwäbisch Gmünd depuis 1988. A des participations à 8vo depuis 1987. Parmi les commanditaires de 8vo, on compte : American Express, le Boymans-van Beuningen Museum de Rotterdam, Christie's, le Design Museum de Londres, London Records, Scandinavian Airline Systems, le Zanders Feinpapiere AG. Nombreux articles sur 8vo dans des revues telles que «Blueprint», «Novum», «High Quality», «The Independent». Nombreuses expositions : à San Francisco (1987), Amsterdam (1988), Offenbach (1989), Rotterdam (1993) etc.

**Eksell,** Olle – né le 22. 3. 1918 à Dalecarlia, Suède – *graphiste maquettiste, typographe* – 1939–1941, décorateur de vitrines et études chez Hugo Steiner-Prag à Stockholm. 1941–1945, graphiste à l'agence de publicité Ervaco. 1946, études à la Art Center School de Los Angeles. Il organise la première exposition d'arts graphiques et de design américain au musée national de Stockholm en 1947. Travaille comme designer indépendant, nombreuses maquettes de livres et illustrations. Essais et illustrations pour le quotidien «Aftonbladet» depuis 1952. La médaille d'argent du salon du livre de Leipzig lui est décernée en 1965. En 1969, il fonde Olle Eksell Design à Stockholm. Nombreuses commandes pour l'industrie, emballages, identités, design d'exposition. – *Publications, sélection :* «Per Plex : Tuff und Tuss», 1953; «Design-Ekonomi», 1964; «Corporate Design Programms», New York 1967.

Eksell 1964 Cover

Elffers 1948 Cover

Eksell Cover

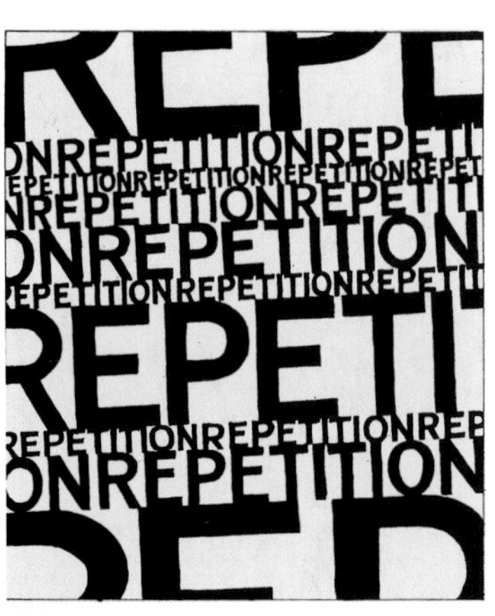


Eksell

Elffers 1932 Cover

Muir. Michael Burke – b. 21. 8. 1944 in Leicestershire, England – Studied at the Ravensbourne College of Art and Design. 1969–73: member of the 1972 Olympic Games design team in Munich under the direction of Otl Aicher. 1976–88: teaches at the Ravensbourne College of Art and Design. Since 1988: professor of graphic design at the Fachhochschule Schwäbisch Gmünd. Partner of 8vo since 1987. 8vo's clients include American Express, the Boymans-van Beuningen Museum in Rotterdam, Christie's, the Design Museum in London, London Records, Scandinavian Airline Systems and Zanders Feinpapiere AG. Numerous articles on 8vo have been published in various magazines, including "Blueprint", "Novum", "High Quality" and "The Independent". Numerous exhibitions, including in San Francisco (1987), Amsterdam (1988), Offenbach (1989) and Rotterdam (1993).

**Eksell,** Olle – b. 22. 3. 1918 in Dalecarlia, Sweden – *graphic designer, typographer* – 1939–41: window-dresser, at the same time studying under Hugo Steiner-Prag in Stockholm. 1941–45: designer for the Ervaco advertising agency. 1946: studies at the Art Center School in Los Angeles. 1947: organizes the first American graphic design exhibition in Stockholm's National Museum. Works as a freelance designer and produces numerous book designs and illustrations. From 1952 onwards: contributions to and illustrations for the "Aftonbladet" daily newspaper. 1965: awarded a silver medal at the book fair in Leipzig. 1969: founds Olle Eksell Design in Stockholm. Numerous design

Verpackungen, Erscheinungsbilder, Ausstellungs- und Display-Design. – *Publikationen u. a.:* „Per Plex: Tuff und Tuss", 1953; „Design-Ekonomi", 1964; „Corporate Design Programs", New York 1967.

**Elffers,** (Dick) Dirk Cornelis – geb. 9. 12. 1910 in Rotterdam, Niederlande, gest. 17. 6. 1990 in Amsterdam, Niederlande. – *Grafik-Designer, Maler* – 1929 Abschluß der Ausbildung an der Academie van Beeldende Kunsten in Rotterdam. 1931–34 Assistent von Paul Schuitema. 1934–37 Assistent von Piet Zwart. 1937 erste Ausstellung seiner Bilder in Rotterdam. Lehrer an der Academie van Beeldende Kunsten in Rotterdam. Zusammenstellung des Grafik-Heftes der Zeitschrift „De 8 en Opbouw" mit Wim Brusse. Bei der Bombadierung Rotterdams 1940 wird sein Atelier völlig zerstört. Umzug nach Amsterdam, wo er seitdem lebt. 1949 Auszeichnung mit dem H. N. Werkman-Preis der Stadt Amsterdam. Gestaltet 1954–67 die Plakate für das Holland-Festival. 1970–76 Dozent an der Koninklijke Academie in s'Hertogenbosch. Gestaltet 1980 die niederländische Gedenkausstellung im Staatsmuseum Auschwitz-Birkenau und 1983 in Westerbork. – *Publikationen u. a.:* „Vorm en Tegenvorm", Amsterdam 1976. N. A. Douwes Dekker „Dick Elffers, Typography and Posters", Den Haag 1968; Max Bruinsma „Een leest heeft drie voeten: Dick Elffers en de Kunst", Amsterdam 1989.

**Elsner&Flake** – *Schriftenhersteller* – Veronika Elsner und Günther Flake gründen 1986 die Firma Elsner&Flake Design-Studios mit dem Ziel, Schriften in

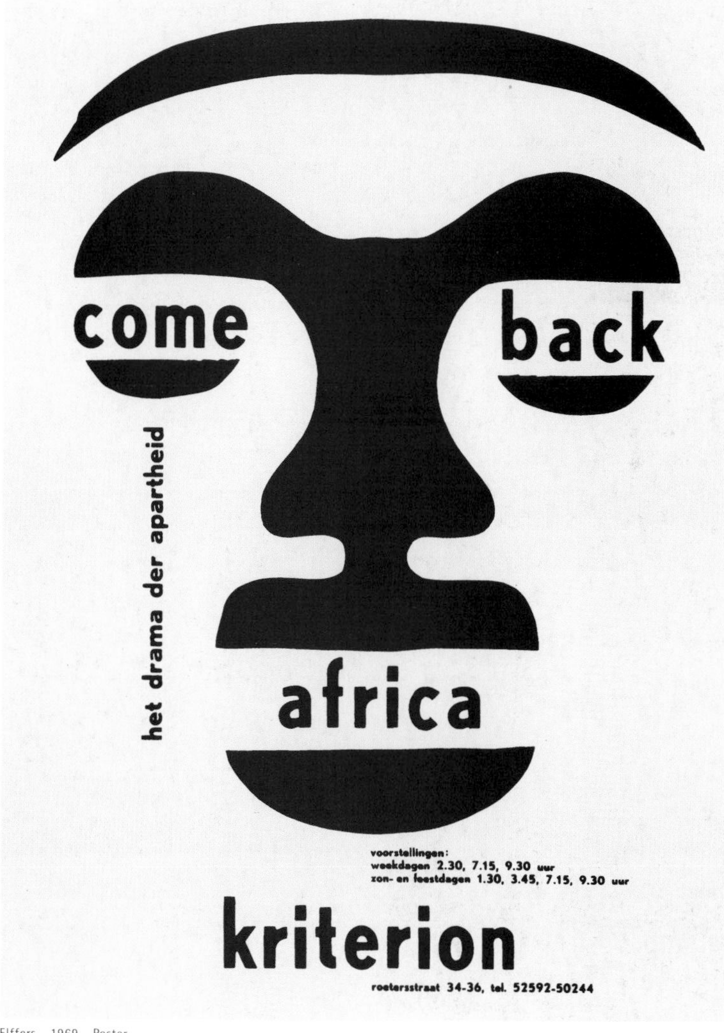


Elffers 1969 Poster

Elffers 1961 Poster

Elsner&Flake 1997 Cover

**Elffers,** (Dick), Dirk Cornelis – né le 9. 12. 1910 à Rotterdam, Pays-Bas, décédé le 17. 6. 1990 à Amsterdam, Pays-Bas – *graphiste maquettiste, peintre* – il termine sa formation à l'Academie van Beeldende Kunsten à Rotterdam en 1929. Assistant de Paul Schuitema de 1931 à 1934, puis de Piet Zwart de 1934 à 1937. Première exposition de ses tableaux à Rotterdam en 1937. Enseigne à l'Academie van Beeldende Kunsten de Rotterdam. Dirige les cahiers graphiques de la revue «De 8 en Opbouw» avec Wim Brusse. Son atelier est entièrement détruit par le bombardement de Rotterdam en 1940, il s'installe alors à Amsterdam où il restera jusqu'à la fin de sa vie. Le prix H. N. Werkman de la ville d'Amsterdam lui est décerné en 1949. De 1954 à 1967, il crée les affiches du Holland-Festival. 1970–1976, professeur à la Koninklijke Academie de s'Hertogenbosch. En 1980, il crée le design de l'exposition commémorative néerlandaise au musée d'Auschwitz-Birkenau, puis à Westerbork en 1983. – *Publications, sélection:* «Vorm en Tegenvorm», Amsterdam 1976. N. A. Douwes Dekker «Dick Elffers, Typography and Posters», La Haye 1968; Max Bruinsma «Een leest heeft drie voeten: Dick Elffers en de Kunst», Amsterdam 1989.

**Elsner&Flake** – *fabricant de fontes* – en 1986, Veronika Elsner et Günther Flake créent la société Elsner&Flake Design-Studio qui vend des fontes numériques.

commissions for industry, packaging, corporate identities, exhibitions and displays. – *Publications include:* "Per Plex: Tuff und Tuss", 1953; "Design-Ekonomi", 1964; "Corporate Design Programs", New York 1967.

**Elffers,** (Dick) Dirk Cornelis – b. 9. 12. 1910 in Rotterdam, the Netherlands, d. 17. 6. 1990 in Amsterdam, the Netherlands – *graphic designer, painter* – 1929: finishes his studies at the Academie van Beeldende Kunsten in Rotterdam. 1931–34: assistant to Paul Schuitema. 1934–37: assistant to Piet Zwart. 1937: first exhibition of his pictures in Rotterdam. Teaches at the Academie van Beeldende Kunsten in Rotterdam. Compiles a graphics volume of the magazine "De 8 en Opbouw" with Wim Brusse. 1940: his studio is completely destroyed during the bombing of Rotterdam. Moves to Amsterdam, where he lives until his death. 1949: is awarded the H. N. Werkman Prize by the city of Amsterdam. 1954–67: designs the posters for the Holland Festival. 1970–76: lecturer at the Koninklijke Academie in 's Hertogenbosch. 1980: designs the Dutch memorial exhibition in the Staatsmuseum Auschwitz-Birkenau and in 1983 in Westerbork. – *Publications include:* "Vorm en Tegenvorm", Amsterdam 1976. N. A. Douwes Dekker "Dick Elffers, Typography and Posters", The Hague 1968; Max Bruinsma "Een leest heeft drie voeten: Dick Elffers en de Kunst", Amsterdam 1989.

**Elsner&Flake** – *type manufacturers* – 1986: Veronika Elsner and Günther Flake found the Elsner&Flake Design Studios

digitaler Form anzubieten. Veronika Elsner – geb. 8. 8. 1952 in Schenefeld, Deutschland. – 1970–74 Studium an der Fachhochschule für Gestaltung in Hamburg. Günther Flake – geb. 8. 7. 1951 in Hamburg, Deutschland. – Mitarbeit bei URW: Projektleiter und Leiter der Schriftenfertigung, 1982–86 künstlerischer Direktor. Zahlreiche Veröffentlichungen zu Schriftfragen. 1992 Gründung der Mail-Order-Firma Elsner&Flake fontinform GmbH. Die Schriftbibliothek umfaßt 1996 ca. 1.400 Schriften im PostScript Format. Elsner&Flake haben sich auf die Gestaltung hebräischer Schriften spezialisiert. Es werden 60 Schnitte aus 12 Schriftfamilien angeboten, daneben osteuropäische Zeichensätze sowie außereuropäische Schriften.

**Emigre** – *Magazin* – 1984 Rudy VanderLans, Menno Meyjes und Marc Susan gründen in Kalifornien das Magazin als Zeitschrift für experimentelles Grafik-Design. In dieser Zeitschrift kommen im Laufe der Jahre die wichtigsten jungen Gestalter zu Wort, es wird eines der einflußreichsten Medien der neuen Typographie im digitalisierten Zeitalter. Der Untertitel der Zeitschrift: "Das Magazin, das Grenzen ignoriert". Einzelne Hefte stehen unter Themenschwerpunkten. An den von Rudy VanderLans gestalteten Heften wirken zahlreiche Gestalter mit, u. a. Allen Hori, Jeffery Keedy, Rick Valicenti, Barry Deck, Pierre Di Sciullo, Zuzana Licko. 1985 Gründung von „Emigre Fonts" mit Alphabeten von Zuzana Licko. – *Publikationen:* „Emigre: Graphic Design into the Digital Realm", New York 1994.

Veronika Elsner – née le 8. 8. 1952 à Schenefeld, Allemagne – 1970–1974 études à la Fachhochschule für Gestaltung (école supérieure de design) de Hambourg. Günther Flake – né le 8. 7. 1951 à Hambourg, Allemagne – travaille chez URW : directeur de projet et directeur de fabrication des fontes. 1982–1986, directeur artistique. Nombreuses publications sur des problèmes d'écritures. En 1992, il fonde l'entreprise Mail-Order Elsner&Flake fontinform GmbH. En 1996, la typothèque contient env. 1400 fontes en format PostScript. Elsner&Flake se sont spécialisés dans la création de caractères hébraïque et proposent 60 types différents pour 12 familles de caractères. Par ailleurs, ils vendent des jeux de caractères cyrilliques et des alphabets non-européens.

**Emigre** – *magazine* – cette revue de design et de graphisme a été créée en 1984 par Rudy VanderLans, Menno Meyjes et Marc Susan en Californie. Au cours de ces dernières années, la revue a permis aux jeunes créateurs de pointe de s'exprimer dans ses colonnes; elle est devenue l'un des médias déterminants de la nouvelle typographie de l'ère du numérique. La revue se définit elle-même à travers son sous-titre : «La revue qui ignore les frontières». Certains numéros sont consacrés à des thèmes spécifiques. La maquette de la revue est conçue par Rudy VanderLans. De nombreux créateur y ont collaboré : Allen Hori, Jeffery Keedy, Rick Valicenti, Barry Deck, Pierre Di Sciullo, Zuzana Licko. 1985, création d'Emigre Fonts avec des alphabets de Zuzana Licko. – *Publication :* «Emigre : Graphic Design into the Digital Realm», New York 1994.

**Enschedé en Zonen,** Haarlem, Pays-Bas – *fonderie de caractères* – en 1703, Izaac Enschedé fonde une imprimerie à Haarlem. En 1743, son fils Johannes Ensche-

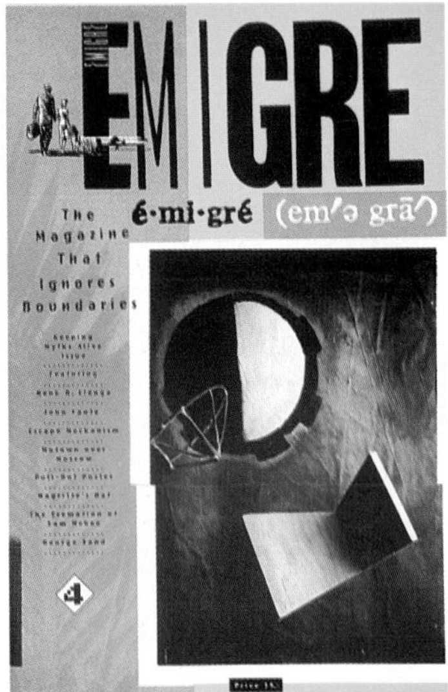

Emigre 1986 Cover

Emigre 1986 Spread

Emigre 1986 Cover

Emigre 1986 Cover

Emigre 1990 Cover

with the aim of producing typefaces in digital form. Veronika Elsner – b. 8. 8. 1952 in Schenefeld, Germany – 1970–74: studies at the Fachhochschule für Gestaltung in Hamburg. Günther Flake – b. 8. 7. 1951 in Hamburg, Germany – Worked for URW as project manager and head of type production and from 1982–86 is art director. Numerous publications on subjects concerning typography. 1992: founds the mail-order company Elsner&Flake fontinform GmbH. 1996: the Elsner&Flake type library has c. 1,400 typefaces in PostScript format. Elsner&Flake specialize in the design of Hebrew typefaces, with 60 weights from 12 type families. They also offer a range of Eastern European fonts and non-European alphabets.

**Emigre** – *magazine* – 1984: Rudy VanderLans, Menno Meyjes and Marc Susan launch the magazine in California as a periodical for popular culture. Over the years, the magazine has acted as a forum for prominent young designers and has been one of the most influential media of new typography in the digital age. The magazine's subtitle is "The magazine that ignores boundaries"; each issue is devoted to a certain subject. Numerous designers have worked for the issues Rudy VanderLans designs, including Allen Hori, Jeffery Keedy, Rick Valicenti, Barry Deck, Pierre Di Sciullo and Zuzana Licko. 1985: Emigre Fonts is established, with alphabets by Zuzana Licko. – *Publications:* "Emigre: Graphic Design into the Digital Realm", New York 1994.

**Enschedé en Zonen,** Haarlem, The Nether-

**Enschedé en Zonen,** Haarlem, Niederlande. – *Schriftgießerei* – Izaac Enschedé gründet 1703 eine Druckerei in Haarlem. Sein Sohn Johannes Enschedé (1708–1780) erweitert die Druckerei seines Vaters 1743 um eine Schriftgießerei. Stempelschneider des Unternehmens sind Johann Michael Fleischmann, Jacques François Rosart und sein Sohn Matthias. Die Söhne von Johannes Enschedé, Johannes II und Jacobus, treten 1773 in die Schriftgießerei ein. Durch Einkäufe gelangt historisches und neues Schriftenmaterial in die Firma: 1815 Matrizen von Pierre und Firmin Didot aus Paris, 1901 Kauf der Lettergieterij Rotterdam mit Schriften, u. a. von Johann Friedrich Unger. Nachdem das Unternehmen in ununterbrochener Folge von Nachkommen Izaac Enschedés geleitet wurde, wird der Schriftguß 1991 eingestellt. Die Bestände werden in einer firmeneigenen Stiftung „Stichting Museum Enschedé" aufbewahrt. 1991 Gründung der „Enschedé Font Foundry", die die Tätigkeit der Schriftgießerei im Computerzeitalter weiterführt.

**Enzensberger,** Hans Magnus – geb. 11. 11. 1929 in Kaufbeuren, Deutschland. – *Schriftsteller* – 1949–54 Studium der Literaturwissenschaft, Sprachen und Philosophie in Erlangen, Hamburg, Freiburg und Paris. Ab 1955 Tätigkeit als Redakteur des Süddeutschen Rundfunks und als Verlagslektor. Gastdozent an der Hochschule für Gestaltung in Ulm und an der Universität Frankfurt am Main. 1963 Auszeichnung mit dem Georg-Büchner-Preis der Deutschen Akademie für Sprache und Dichtung, Darmstadt. 1965–75 Heraus-

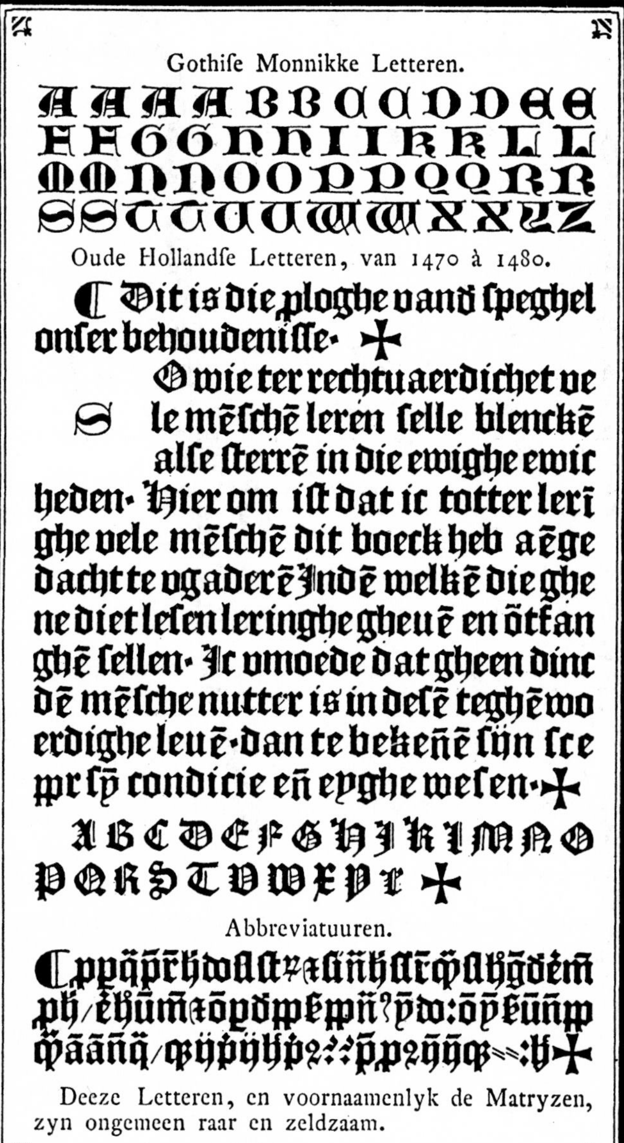


Enschedé en Zonen 1768 Specimen

Dubbelde Garmond
GESCHREVEN SCHRIFT.

De Liefhebbers van Konsten en Weetenschappen zien hier het tweede, voor de Haerlemsche Lettergietery gesneeden, Geschreeven Schrift, door wylen den Heer JOAN MICHAEL FLEISCHMAN, den grootsten en konstigsten Letter-Stempelsnyder, die 'er ooit in de Waereld geweest is, en mogelyk komen zal, in 1768 voleindigt; zynde zyn laatste Konst-Werkstuk voor deeze Lettergietery, en de laatste door hem gejusteerde Matryzen. Zyn Naam en Konst zal, door zyne uitmuntende Letteren, die ten getale van ruim zeventig onderscheidene Schriften zig in de Haarlemsche Lettergietery bevinden, na verloop van veele Eeuwen, nog door de Geleerde Waereld met roem vermeld worden.

Enschedé en Zonen 1768 Specimen

Enzensberger 1961 Cover

dé (1708–1780) adjoint une fonderie de caractères à l'imprimerie de son père. Parmi les tailleurs de types de l'entreprise, on compte Johann Michael Fleischmann, Jacques François Rosart et son fils Matthias. En 1773, les fils de Johannes Enschedé, Johannes II et Jacobus entrent eux aussi à la fonderie. A la suite d'acquisitions, du matériel historique et de nouvelles fontes deviennent la propriété de l'entreprise, en 1815; par ex. des matrices de Pierre et Firmin Didot de Paris; puis en 1901 avec la reprise de la Lettergieterij Rotterdam, des fontes de Johann Friedrich Unger et d'autres. L'entreprise a été constamment dirigée par les descendants d'Izaac Enschedé jusqu'en 1991, date à laquelle la fonderie cesse ses activités. Les stocks sont conservés par la fondation «Stichting Museum Enschedé». La «Enschedé Font Foundry» est créée en 1991, elle perpétue les activités de la fonderie en les adaptant à l'ère de l'ordinateur.

**Enzensberger,** Hans Magnus – né le 11. 11. 1929 à Kaufbeuren, Allemagne – *écrivain* – 1949–1954, études de littérature, de langues et de philosophie à Erlangen, Hambourg, Fribourg et Paris. A partir de 1955, rédacteur à la Süddeutsche Rundfunk et lecteur pour une maison d'édition. Enseigne à la Hochschule für Gestaltung (école supérieure de design) d'Ulm et à l'université de Francfort-sur-le-Main. En 1963, la Deutsche Akademie für Sprache und Dichtung de Darmstadt lui décerne le prix Georg-Büchner. De 1965 à 1975, il édite la revue «Kursbuch»

lands – *type foundry* – 1703: Izaac Enschedé opens a printing workshop in Haarlem. 1743: his son, Johannes Enschedé (1708–80), adds a type foundry to his father's printing works. Johann Michael Fleischmann, Jacques François Rosart and his son Matthias are the company punch cutters. 1773: Johannes Enschedé's sons, Johannes II and Jacobus, join the foundry. The company buys up historical and new type material, including Pierre and Firmin Didot's matrices from Paris in 1815 and the Lettergieterij Rotterdam in 1901 with typefaces by Johann Friedrich Unger, among others. The foundry is run by Izaac Enschedé's descendants in unbroken succession until 1991, when production is stopped. The foundry's equipment is preserved in a company foundation, the Stichting Museum Enschedé. 1991: the Enschedé Font Foundry is set up, which carries on the tradition of the foundry in the age of computers.

**Enzensberger,** Hans Magnus – b. 11. 11. 1929 in Kaufbeuren, Germany – *author* – 1949–54: studies literature, languages and philosophy in Erlangen, Hamburg, Freiburg and Paris. From 1955 onwards: works as an editor for the Süddeutscher Rundfunk and as a publisher's reader. Visiting lecturer at the Hochschule für Gestaltung in Ulm and at the University of Frankfurt am Main. 1963: is awarded the Georg Büchner Prize by the Deutsche Akademie für Sprache und Dichtung in Darmstadt. 1965–75: publishes "Kursbuch" magazine with Karl Markus Michel. 1980–83: publishes

geber der Zeitschrift „Kursbuch" mit Karl Markus Michel. 1980–83 Herausgeber der Zeitschrift „Trans Atlantik". 1982 Auszeichnung mit dem Pasolini-Preis für Poesie der Stadt Rom. Seit 1985 Herausgeber der „Anderen Bibliothek" im Greno Verlag, später im Eichborn Verlag. 1961 typographische Ausstellung der Kinderreimsammlung „Allerleirauh" (Frankfurt am Main). 1981 erscheint in der Wochenzeitung „Die Zeit" sein Artikel über die Veränderung der Typographie durch den Fotosatz. – *Publikationen u. a.:* „Landessprache", Frankfurt am Main 1960; „Blindenschrift", Frankfurt am Main 1964; „Der kurze Sommer der Anarchie", Frankfurt am Main 1972; „Mausoleum", Frankfurt am Main 1975; „Der Untergang der Titanic", Frankfurt am Main 1978. Seit 1957 zahlreiche literarische und politische Veröffentlichungen.

**Erbar,** Jakob – geb. 2. 8. 1878 in Düsseldorf, Deutschland, gest. 1. 7. 1935 in Köln, Deutschland. – *Schriftentwerfer, Lehrer* – Schriftsetzerlehre in Düsseldorf, absolvierte Schriftkurse bei Fritz Helmut Ehmcke und Anna Simons. Akzidenzsetzer in der Druckerei Dumont-Schauberg in Köln. 1908 Lehrer an der Städtischen Berufsschule. 1919–35 Lehrer an den Kölner Werkschulen. – *Schriftentwürfe:* Feder-Grotesk (1908), Erbar-Grotesk mit Varianten Lumina, Lux, Phosphor (1922–30), Koloss (1923), Candida (1936).

**Eremiten-Presse** – *Verlag* – Victor Otto (Vauo) Stomps (1897–1970) gründet 1949 zusammen mit Helmut Knaupp und Ferdinand Möller in Frankfurt die Eremiten-Presse als bibliophilen Verlag für Dich-

avec Karl Markus Michel. De 1980 à 1983, il dirige la revue «Trans Atlantik». En 1982, la ville de Rome lui décerne le prix Pasolini de poésie. Depuis 1985, il dirige la «Andere Bibliothek» d'abord aux éditions Greno, puis aux éditions Eichborn. 1961, exposition typographique de la collection de poèmes pour enfants «Allerleirauh» (Francfort-sur-le-Main). En 1981, publication d'un article sur la modification de la typographie avec l'avènement de la photocomposition, dans l'hebdomadaire «Die Zeit». – *Publications, sélection:* «Landessprache», Francfort-sur-le-Main 1960; «Blindenschrift», Francfort-sur-le-Main 1964; «Der kurze Sommer der Anarchie», Francfort-sur-le-Main 1972; «Mausoleum», Francfort-sur-le-Main 1975; «Der Untergang der Titanic», Francfort-sur-le-Main 1978. Depuis 1957, de nombreux essais sur la littérature et la politique.

**Erbar,** Jakob – né le 2. 8. 1878 à Düsseldorf, Allemagne, décédé le 1. 7. 1935 à Cologne, Allemagne – *concepteur de polices, enseignant* – apprentissage de composition typographique à Düsseldorf, puis cours de calligraphie chez Fritz Helmut Ehmcke et Anna Simons. Compositeur sur Akzidens à l'imprimerie Dumont-Schauberg à Cologne. Enseigne à la Städtische Berufsschule (institut municipal d'enseignement professionnel) en 1908, puis dans plusieurs instituts de Cologne de 1919 à 1935. – *Polices:* Feder-Grotesk (1908), Erbar-Grotesk avec les variantes Lumina, Lux, Phosphor (1922– 1930), Koloss (1923), Candida (1936).

**Eremiten-Presse** – *éditions* – en 1949, à Francfort, Victor Otto (Vauo) Stomps (1897–1970) fonde avec Helmut Knaupp et Ferdinand Möller une maison d'édition, la Eremiten-Presse, qui publiera des ouvrages pour bibliophiles, de la poésie et des livres d'art. De 1952 à 1953, pub-

abcdefghijklmnopqrſstuvw
Grönland xyz Spitzbergen
12345 Magdeburg 67890
AABCDEFGHIJKLMMN
NOPQRSTUVVWWXYZ
UNION PHÖNIX OASE

Erbar 1922–30 Erbar-Grotesk I

Ernst 1938 Cover

Schmale
halbfette
Erbar-Grotesk
ist eine Ergänzung unserer schönen Erbar-Grotesk • Sie ist jetzt in den Graden von 6 bis 84 Punkten vorhanden
Ludwig & Mayer
Schriftgießerei Frankfurt-M

Erbar 1929 Advertisement

Eremiten-Presse 1979 Cover

Ernst 1931 Collage

"Trans Atlantik" magazine. 1982: is awarded the Pasolini Prize for Poetry by the city of Rome. From 1985 onwards: editor of the "Andere Bibliothek" for the Greno publishing house, later for Eichborn publishers. 1961: typographic exhibition of a collection of children's rhymes, "Allerleirauh", in Frankfurt am Main. 1981: his article on the changes in typography brought about by filmsetting is published in the weekly "Die Zeit". – *Publications include:* "Landessprache", Frankfurt am Main 1960; "Blindenschrift", Frankfurt am Main 1964; "Der kurze Sommer der Anarchie", Frankfurt am Main 1972; "Mausoleum", Frankfurt am Main 1975; "Der Untergang der Titanic", Frankfurt am Main 1978. Since 1957 Enzensberger has penned numerous literary and political publications.

**Erbar,** Jakob – b. 2. 8. 1878 in Düsseldorf, Germany, d. 1. 7. 1935 in Cologne, Germany – *type designer, teacher* – Trained as a compositor in Düsseldorf and took courses in type with Fritz Helmut Ehmcke and Anna Simons. Job typesetter for the Dumont-Schauberg printing works in Cologne. 1908: teaches at the Städtische Berufsschule. 1919–35: teaches at the Kölner Werkschule. – *Fonts:* Feder-Grotesk (1908), Erbar-Grotesk with the variants Lumina, Lux and Phosphor (1922–30), Koloss (1923), Candida (1936).

**Eremiten-Presse** – *publishing house* – 1949: with Helmut Knaupp and Ferdinand Möller, Victor Otto (Vauo) Stomps (1897–1970) founds the Eremiten-Presse in Frankfurt, a publishing house for bibliophile editions on poetry and art.

tung und Kunst. 1952–53 Herausgabe der Zeitschrift „Konturen". 1956 Beginn der Herausgabe der „Streit-Zeit-Schrift". 1967 Trennung vom Verlag, Umzug Vauo Stomps nach Berlin, wo er die „Neue Raben Presse" gründet. Weiterführung der Eremiten-Presse durch Hülsmann und Reske. Die Stadt Mainz stiftet 1978 den Vauo-Stomps-Preis für besondere verlegerische Verdienste, der alle zwei Jahre verliehen wird. – *Publikationen u. a.:* Ralf Ruhl „Die Eremiten-Presse und Ihr Gründer V. O. Stomps. Portrait eines Kleinverlags", Wiesbaden 1985; Martin Ebbertz, Friedolin Reske „Vier Jahrzehnte Eremiten-Presse 1949–89", Düsseldorf 1989.

**Ernst**, Max – geb. 2. 4. 1881 in Brühl, Deutschland, gest. 1. 4. 1976 in Paris, Frankreich. – *Maler, Grafiker, Dichter* – 1910–14 Studium der Philosophie an der Universität Bonn. 1913 erste Ausstellung in Bonn. 1919 erste Collagen. Gründung der Dada-Gruppe Köln (mit J. Th. Baargeld). 1920 Einladung durch André Breton zur Ausstellung von Collagen in Paris. 1931 erste Einzelausstellung in den USA. 1934 Begegnung mit James Joyce in Zürich. 1938 Austritt aus der Gruppe der Surrealisten. Zahlreiche Auszeichnungen, u. a. der Preis für Malerei in Venedig (1954). Hauptsächlich grafische Arbeiten. Zahlreiche Ausstellungen, u. a. im Musée Nationale d'Art Moderne, Paris (1959), Museum of Modern Art, New York (1961), Wallraf-Richartz-Museum, Köln (1962), Guggenheim-Museum, New York (1975) und im Grand Palais, Paris (1975). Immer wieder, besonders in dem Werk „Maximiliana ou l'exercice illégal de

Ernst 1920 Poster

lication de la revue «Konturen». 1956, début de la publication de «Streit-Zeit-Schrift». En 1967, Vauo se sépare de la maison d'édition et s'installe à Berlin où il fonde la «Neue Raben Presse». La Eremiten-Presse est reprise par Hülsmann et Reske. En 1978, la ville de Mayence crée le prix «Vauo Stomps», décerné tous les deux ans pour distinguer des activités d'édition. – *Publications :* Ralf Ruhl «Die Eremiten-Presse und Ihr Gründer V. O. Stomps. Portrait eines Kleinverlags», Wiesbaden 1985; Martin Ebbertz, Friedolin Reske «Vier Jahrzehnte Eremiten-Presse 1949–1989», Düsseldorf 1989.

**Ernst,** Max – né le 2. 4. 1881 à Brühl, Allemagne, décédé le 1. 4. 1976 à Paris, France – *peintre, graphiste, poète* – 1910–1914, études de philosophie à l'université de Bonn. 1913, première exposition à Bonn. Premiers collages en 1919. Fonde le groupe Dada de Cologne (avec J. Th. Baargeld). En 1920, André Breton l'invite à exposer ses collages à Paris. Première exposition personnelle aux Etats-Unis en 1931. Rencontre James Joyce en 1934 à Zurich. Quitte le groupe des Surréalistes en 1938. Nombreuses distinctions, dont le prix de peinture de Venise (1954). Réalise surtout des travaux graphiques. Nombreuses expositions, entre autres au Musée Nationale d'Art Moderne, Paris (1959), Museum of Modern Art, New York (1961), Wallraf-Richartz-Museum, Cologne (1962), Guggeheim-Museum, New York (1975), Grand Palais, Paris (1975). Max Ernst n'a cessé de se confronter à l'écriture et à la typographie, surtout dans l'œuvre «Maximiliana ou

Ernst 1924 Cover

Ernst 1964 Page

1952–53: publishes "Konturen" magazine. 1956: begins publishing "Streit-Zeit-Schrift". 1967: Vauo Stomps leaves the company and moves to Berlin, where he establishes the Neue Raben Presse. Hülsmann and Reske continue to run the Eremiten-Presse. From 1978 onwards: every two years, the city of Mainz awards the Vauo Stomps Prize for special achievements in the field of publishing. – *Publications:* Ralf Ruhl "Die Eremiten-Presse und Ihr Gründer V. O. Stomps. Portrait eines Kleinverlags", Wiesbaden 1985; Martin Ebbertz, Friedolin Reske "Vier Jahrzehnte Eremiten-Presse 1949–89", Düsseldorf 1989.

**Ernst**, Max – b. 2. 4. 1881 in Brühl, Germany, d. 1. 4. 1976 in Paris, France – *painter, graphic artist, poet* – 1910–14: studies philosophy at the University of Bonn. 1913: first exhibition in Bonn. 1919: produces his first collages. Founds the Dada group in Cologne with J. Th. Baargeld. 1920: exhibition of collages in Paris at the invitation of André Breton. 1931: first solo exhibition in the USA. 1934: meets James Joyce in Zurich. 1938: leaves the Surrealist group. Numerous awards, including a painting prize in Venice (1954). His output mainly comprised graphic work. Numerous exhibitions, including at the Musée Nationale d'Art Moderne in Paris (1959), the Museum of Modern Art in New York (1961), the Wallraf-Richartz-Museum in Cologne (1962), the Guggenheim Museum in New York (1975) and the Grand Palais in Paris (1975). Max Ernst constantly tackled the subject of type and typography, and es-

l'astronomie" setzt sich Max Ernst mit Schrift und Typographie auseinander, die er in surrealer, individueller Vielfalt interpretiert. – *Publikationen u. a.:* „Répétitions. Poèmes par Paul Eluard", Paris 1922; „Histoire naturelle", Paris 1926; „Beyond Paintings and other Writings", New York 1948; „Maximiliana ou l'exercice illégal de l'astronomie", (mit Iliazd), Paris 1964. Werner Spies (Hrsg.) „Max Ernst. Œuvre-Katalog 1–5", Köln 1975–87; Edward Quinn (Hrsg.) „Max Ernst", Zürich 1977.

**Ernst-Ludwig-Presse** – *Privatpresse des Großherzogs Ernst Ludwig von Hessen Darmstadt* – 1907 Unter dem Patronat des Großherzogs Ernst Ludwig von Hessen wird in Darmstadt die Ernst-Ludwig-Presse gegründet. Als Leiter wird Friedrich Wilhelm Kleukens (1878–1956) berufen, sein Bruder Christian Heinrich Kleukens (1880–1954) wird drucktechnischer Leiter. Insgesamt veröffentlicht die Presse 130 Titel. Neuere Autoren waren Rilke, Verhaeren und Binding. Verlegerisch wird ein Teil der Bände von Anton Kippenbergs Insel Verlag betreut. Bei der Gestaltung verzichtet man weitgehend auf ornamentales oder dekoratives Beiwerk. Lediglich die Titelseiten sind geschmückt und Initialen markieren die Textanfänge. F. W. Kleukens verläßt 1914 die Ernst-Ludwig-Presse. Auf Initiative des Verlegers Kurt Wolff setzt Ch. H. Kleukens 1920 die Arbeit der Ernst-Ludwig-Presse bis 1945 in Mainz fort.

**Excoffon,** Roger – geb. 7. 9. 1910 in Marseille, Frankreich, gest. 1983 in Paris, Frankreich. – *Grafik-Designer, Typo-*

l'exercice illégal de l'astronomie» qu'il a interprétée d'une manière à la fois surréaliste et personnelle. – *Publications, sélection:* «Répétitions. Poèmes par Paul Eluard», Paris 1922; «Histoire naturelle», Paris 1926; «Beyond Paintings and other Writings», New York 1948; «Maximiliana ou l'exercice illégal de l'astronomie» (avec Iliazd), Paris 1964. Werner Spies (éd.) «Max Ernst. Œuvre-Katalog 1–5», Cologne 1975–1987; Edward Quinn (éd.) «Max Ernst», Zurich 1977.

**Ernst-Ludwig-Presse** – *presse privé du grand-duc Ernst Ludwig de Hesse* – la Ernst-Ludwig-Presse a été fondée en 1907 à Darmstadt sous l'égide du grand-duc Ernst Ludwig de Hesse. Celui-ci appelle Wilhelm Kleukens (1878–1956) comme directeur, son frère Christian Heinrich Kleukens (1880–1954) devient directeur technique de l'imprimerie. Les presses Publient au total 130 titres. Parmi les nouveaux auteurs comptent Rilke, Verhaeren et Binding. Un parti du suivi d'édition des volumes est confié à Anton Kippenberg de la maison d'édition Insel. En matière de maquette, on renonce aux ornements et décors. Seules les couvertures dont décorées et des lettrines signalent le début du texte. En 1914, F. W. Kleukens quitte la Ernst-Ludwig-Presse. En 1920, sur l'initiative de l'éditeur Kurt Wolff, Ch. H. Kleukens reprend les activités de la Ernst-Ludwig-Presse jusqu'en 1945 à Mayence.

**Excoffon,** Roger – né le 7. 9. 1910 à Marseille, France, décédé en 1983 à Paris, France – *graphiste maquettiste, typographe, concepteur de polices* – études de droit à l'université d'Aix-en-Provence. Etudes de peinture à Paris. Exerce comme graphiste. Il fonde son atelier en 1947. Cofondateur de l'agence de publicité «U + O», (Urbi et Orbi) à Paris, conseiller artistique de la fonderie de caractères Olive à Marseille (jusqu'en 1959). En 1968, il réalise

H Y P E R I O N   A N   B E L L A R M I N

DER liebe Vaterlandsboden gibt mir wieder Freude und Leid.
Ich bin jetzt alle Morgen auf den Höhn des Korinthischen Isthmus, und, wie die Biene unter Blumen, fliegt meine Seele oft hin und her zwischen den Meeren, die zur Rechten und zur Linken meinen glühenden Bergen die Füße kühlen.
Besonders der eine der beiden Meerbusen hätte mich freuen sollen, wär' ich ein Jahrtausend früher hier gestanden.
Wie ein siegender Halbgott, wallte da zwischen der herrlichen Wildnis des Helikon und Parnaß, wo das Morgenrot um hundert überschneite Gipfel spielt, und zwischen der paradiesischen Ebene von Sicyon der glänzende Meerbusen herein, gegen die Stadt der Freude, das jugendliche Korinth, und schüttete den erbeuteten Reichtum aller Zonen vor seiner Lieblingin aus.
Aber was soll mir das? Das Geschrei des Jakals, der unter den Steinhaufen des Altertums sein wildes Grablied singt, schreckt' ja aus meinen Träumen mich auf.
Wohl dem Manne, dem ein blühend Vaterland das Herz erfreut und stärkt! Mir ist, als würd' ich in den Sumpf geworfen, als schlüge man den Sargdeckel über mir zu, wenn einer an das meinige mich mahnt, und wenn mich

Ernst-Ludwig-Presse 1912 Page

Excoffon 1965 Poster

| | | | | | |
|---|---|---|---|---|---|
| A | B | C | E | G | H |
| M | O | R | S | X | Y |
| 1 | 2 | 4 | 6 | 8 | & |

Excoffon 1951 Banco

| | | | | | |
|---|---|---|---|---|---|
| | n | n | n | | n |
| a | b | e | f | g | i |
| o | r | s | t | y | z |
| A | B | C | E | G | H |
| M | O | R | S | X | Y |
| 1 | 2 | 4 | 6 | 8 | & |

Excoffon 1962–66 Antique Olive

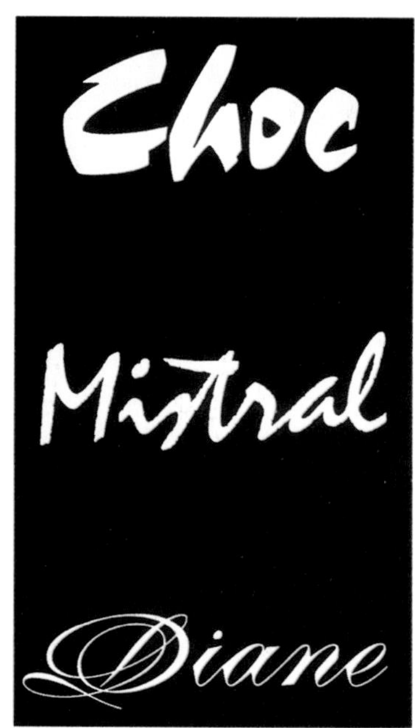


Excoffon 1953–56 Specimen

pecially in his work "Maximiliana ou l'exercice illégal de l'astronomie", interpreting it with a surreal and individual variety. – *Publications include:* "Répétitions. Poèmes par Paul Eluard", Paris 1922; "Histoire naturelle", Paris 1926; "Beyond Paintings and other Writings", New York 1948; "Maximiliana ou l'exercice illégal de l'astronomie" (with Iliazd), Paris 1964. Werner Spies (ed.) "Max Ernst. Œuvre-Katalog 1–5", Cologne 1975–87; Edward Quinn (ed.) "Max Ernst", Zurich 1977.

**Ernst-Ludwig-Presse** – *private press of Ernst Ludwig, Grand Duke of Hesse* – 1907: the Ernst-Ludwig-Presse is founded in Darmstadt under the patronage of Ernst Ludwig, Grand Duke of Hesse. Friedrich Wilhelm Kleukens (1878–1956) is appointed to run the press; his brother, Christian Heinrich Kleukens (1880–1954), is printing manager. The press publishes 130 titles. Rilke, Verhaeren and Binding are new authors featured in the series. The press oversees the publishing of a part of the books from Anton Kippenberg's Insel publishing house. The designs are mostly devoid of ornament or decoration. Only covers boast any decoration, and initials mark the beginning of passages of text. 1914: F. W. Kleukens leaves the Ernst-Ludwig-Presse. 1920: acting on the initiative of publisher Kurt Wolff, Ch. H. Kleukens continues the work of the press in Mainz until 1945.

**Excoffon,** Roger – b. 7. 9. 1910 in Marseille, France, d. 1983 in Paris, France – *graphic designer, typographer, type designer* – Studied law at the University of Aix-en-Provence. Studied painting in

*graph, Schriftentwerfer* – Jura-Studium an der Universität Aix-en-Provence. Studium der Malerei in Paris. Tätigkeit als Grafiker. 1947 Gründung seines Studios. Mitbegründer der Werbeagentur „U + O" (Urbi et Orbi) in Paris, künstlerischer Berater der Schriftgießerei Fonderie Olive in Marseille (bis 1959). 1968 Entwurf der Bildzeichen für die Winterolympiade in Grenoble. 1972 Gründung seiner Agentur Excoffon Conseil. Zahlreiche Plakatentwürfe u. a. für Air France, Bally-Schuhe, Dunlop, Sandoz, SNCF. Mitarbeit an der Zeitschrift „Le Courrier Graphique", „Typographica", „Esthétique Industrielle" und „Techniques Graphiques". – *Schriftentwürfe:* Chambord (1945), Banco (1951), Mistral (1953), Choc (1955), Diane (1956), Calypso (1958), Antique Olive (1962–66).

**Eye** – *Zeitschrift* – 1990 Gründung der Zeitschrift „Eye" als internationales Organ für Grafik-Design. Unter seinem Herausgeber Rick Poynor wird „Eye", die viermal jährlich erscheint, schnell ein leitendes Forum für kritische Diskussionen über Gestaltung. Zahlreiche Vertreter neuer Gestaltung werden erstmals in „Eye" vorgestellt (u. a. Grappa, P. Scott Makela, Bruce Mau), signifikante Repräsentanten früherer Zeit werden neu besprochen (u. a. Alvin Lustig, Karel Teige, Ladislav Sutnar, Max Bittrof). Autoren sind u. a. Steven Heller, Ellen Lupton, J. Abbott Miller, William Owen, Robin Kinross, Jeffery Keedy, Jonathan Barnbrook, Klaus Thomas Edelmann, Erik Spiekermann. Jährlich erscheint ein Sonderheft „Typography".

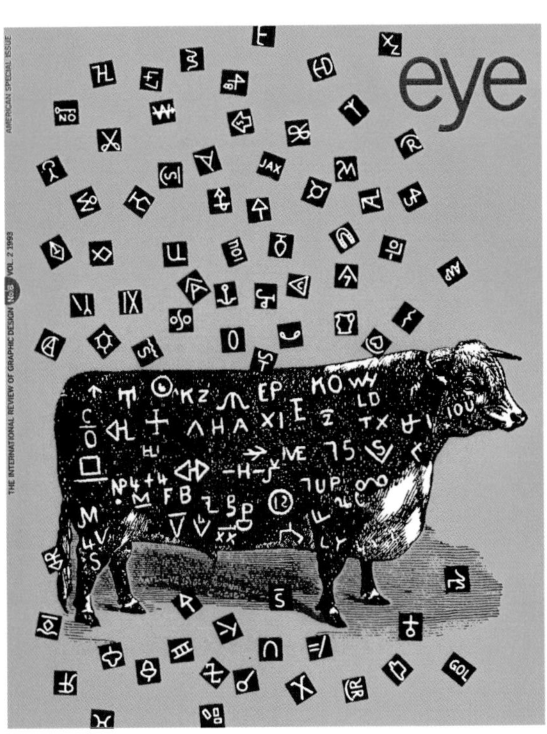

Eye 1993 Cover

eye
TYPE
BITS AND PIECES PUT TOGETHER TO PRESENT A SEMBLANCE OF A WHOLE


Eye 1995 Cover

Excoffon 1973 Logo

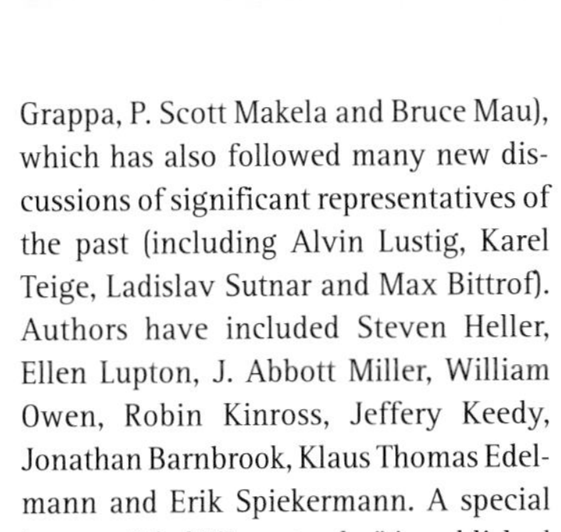


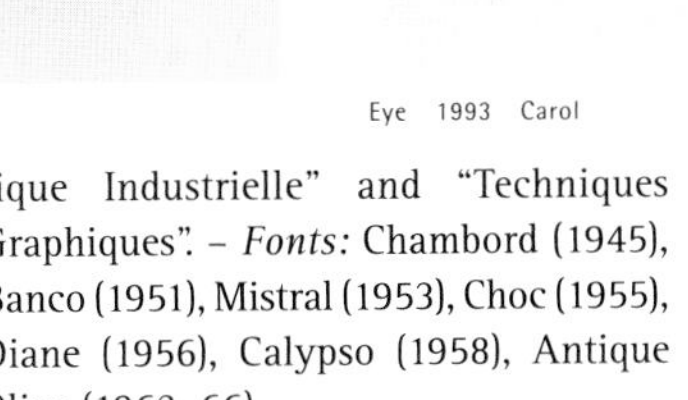

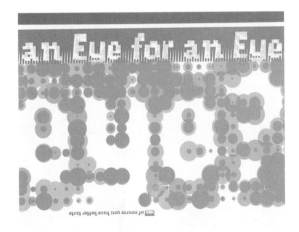

Eye 1993 Carol

les pictogrammes signalétiques des Jeux Olympiques d'hiver de Grenoble. Fondation de l'agence Excoffon Conseil en 1972. Nombreuses affiches pour Air France, Bally, Dunlop, Sandoz, SNCF etc. Collabore aux revues : «Le Courrier Graphique», «Typographica», «Esthétique Industrielle» et «Techniques graphiques». – *Polices :* Chambord (1945), Banco (1951), Mistral (1953), Choc (1955), Diane (1956), Calypso (1958), Antique Olive (1962–1966).

**Eye** – *revue* – fondée en 1990, la revue «Eye» est l'organe international des arts graphiques et du design. Dirigée par Rick Poynor, la revue trimestrielle «Eye» est très vite devenue un grand forum de discussion et de critique du design. De nombreux créateurs des nouvelles tendances y ont été présentés pour la première fois (entre autres Grappa, P. Scott Makela, Bruce Mau), elle a aussi traité de personnalités significatives des périodes antérieures (entre autres Alvin Lustig, Karel Teige, Ladislav Sutnar, Max Bittrof). Parmi les auteurs, on compte Steven Heller, Ellen Lupton, J. Abbott Miller, William Owen, Robin Kinross, Jeffery Keedy, Jonathan Barnbrook, Klaus Thomas Edelmann, Erik Spiekermann. Un numéro spécial «Typography» paraît une fois par an.

Paris, works as a graphic designer. 1947: opens his studio. Co-founder of the U + O (Urbi et Orbi) advertising agency in Paris. Art consultant for the Fonderie Olive type foundry in Marseille (until 1959). 1968: designs the symbols for the Winter Olympics in Grenoble. 1972: founds his agency Excoffon Conseil. Excoffon produced numerous poster designs for Air France, Bally shoes, Dunlop, Sandoz and SNCF, among others and has contributed to the magazines "Le Courrier Graphique", "Typographica", "Esthétique Industrielle" and "Techniques Graphiques". – *Fonts:* Chambord (1945), Banco (1951), Mistral (1953), Choc (1955), Diane (1956), Calypso (1958), Antique Olive (1962–66).

**Eye** – *magazine* – 1990: "Eye" magazine, an international organ for graphic design, is launched. Published quarterly, "Eye" soon establishes itself as a leading forum for the critical discussion of design under its editor Rick Poynor. Numerous representatives of new design concepts have first appeared in "Eye" (including Grappa, P. Scott Makela and Bruce Mau), which has also followed many new discussions of significant representatives of the past (including Alvin Lustig, Karel Teige, Ladislav Sutnar and Max Bittrof). Authors have included Steven Heller, Ellen Lupton, J. Abbott Miller, William Owen, Robin Kinross, Jeffery Keedy, Jonathan Barnbrook, Klaus Thomas Edelmann and Erik Spiekermann. A special issue entitled "Typography" is published once a year.

...ных Курил

*...прав на острова, а Токио — ...*

а обещают вскоре
радио или телевидению

*...м, вступает в опасную фазу*

В воскресенье
семь субъектов Федерации
выберут себе губернатора

# F

**Fairbank**
**Feininger**

**Fairbank,** Alfred John – geb. 12. 7. 1895 in Grimsby, England, gest. 14. 3. 1982 in Hove, England. – *Kalligraph, Schriftentwerfer* – Studium der Kalligraphie an der Central School of Arts and Crafts in London. 1921–39 Entwurf kalligraphischer Arbeiten. Zahlreiche Initialen-Entwürfe für Bücher. 1947 Studien zur Reform der Handschrift. Zeitweilig Präsident der „Society of Scribers and Illuminators" und Vizepräsident der „Society for Italic Handwriting". Fairbanks Wirken trägt zu einem Wiederaufleben der italienischen handgeschriebenen Kursive bei, sein Werk über die Handschrift wird als das kenntnisreichste geachtet. – *Schriftentwurf:* Narrow Bembo (1928). – *Publikationen:* „A Handwriting Manual", London 1932; „Book of Scripts", London 1949; „Humanistic Script" (mit R. W. Hunt), London 1960; „Renaissance Handwriting" (mit B. L. Wolpe), London 1960.
**Feininger,** Lyonel – geb. 17. 7. 1871 in New York, USA, gest. 13. 1. 1956 in New York, USA. – *Künstler, Lehrer* – 1887 Reise nach Deutschland, Zeichen- und Malunterricht an der Tagesschule der Allgemeinen Gewerbeschule in Hamburg. 1888 Umzug nach Berlin, Studium an der Königlichen Akademie der Künste. 1890 erste Karikaturen für die „Humoristischen Blätter". 1894–1914 Karikaturen für die Zeitschriften „Ulk", „Lustige Blätter" und „Narrenschiff". 1907 erste Gemälde. 1913 Atelier in Weimar. Auf Einladung Franz Marcs Teilnahme am „Ersten Deutschen Herbstsalon" in Berlin. Freundschaft mit Kandinsky, Klee und Jawlensky. 1917 erste Einzelausstellung in der Galerie

**Fairbank,** Alfred John – né le 12. 7. 1895 à Grimsby, Angleterre, décédé le 14. 3. 1982 à Hove, Angleterre – *calligraphe, concepteur de polices* – études de calligraphie à la Central School of Arts and Crafts de Londres. 1921–1939, travaux de calligraphie. Nombreux dessins de lettrines pour l'édition. En 1947, il réalise des esquisses pour la réforme de l'écriture manuscrite. Plusieurs fois président de la «Society of Scribers and Illuminators» et vice-président de la «Society for Italic Handwriting». Les activités de Fairbank ont contribué à la renaissance de l'italique manuscrite à l'italienne, son œuvre sur l'écriture manuscrite est considérée comme exemplaire. – *Police:* Narrow Bembo (1928). – *Publications:* «A Handwriting Manual», Londres 1932; «Book of Scripts», Londres 1949; «Humanistic Script» (avec R. W. Hunt), Londres 1960; «Renaissance Handwriting» (avec B. L. Wolpe), Londres 1960.
**Feininger,** Lyonel – né le 17. 7. 1871 à New York, Etats-Unis, décédé le 13. 1. 1956 à New York, Etats-Unis – *artiste, enseignant* – il part pour l'Allemagne en 1887 et prend des cours de dessin et de peinture à l'école de jour de la Allgemeine Gewerbeschule (arts et métiers) de Hambourg. S'installe à Berlin en 1888 et étudie à la Königliche Akademie der Künste. En 1890, il dessine ses premières caricatures pour les «Humoristische Blätter». 1894–1914, caricatures pour les revues «Ulk», «Lustige Blätter» et «Narrenschiff». Peint ses premiers tableaux en 1907. Atelier à Weimar en 1913. Participe au «Erster Deutscher Herbstsalon» (premier salon d'automne) de Berlin sur l'invitation de Franz Marc. Se lie d'amitié avec Kandinsky, Klee et Jawlensky. 1917, première exposition personnelle à la galerie «Der Sturm» à Berlin. 1918, premières gravures sur bois; se rallie au «Novem-

THE moving Moon went up the sky,
And no where did abide:
Softly she was going up,
And a star or two beside—

IN his loneliness and fixedness he yearneth towards the journeying Moon, and the stars that still sojourn, yet still move onward; and everywhere the blue sky belongs to them, and is their appointed rest and their native country and their own natural homes, which they enter unannounced, as lords that are certainly expected & yet there is a silent joy at their arrival.

A Verse and Gloss from The Rime of the Ancient Mariner by Samuel Taylor Coleridge, written out for C. M. W. by A. J. F. 21 June MCMXXXII.

Fairbank 1932 Page

BAUHAUSDRUCKE
NEUE EUROPAEISCHE
GRAPHIK
ERSTE MAPPE
MEISTER D. STAATLICHEN BAUHAUSES
IN WEIMAR

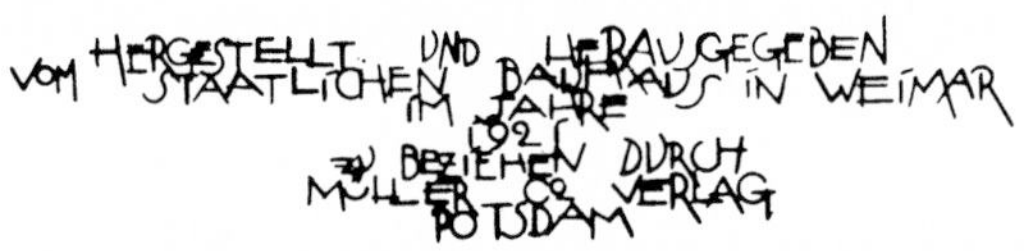


Feininger 1921 Page

Some Famous Englishmen
Sir Philip Sidney, William Shakespeare,
Oliver Cromwell, John Milton,
John Bunyan, Sir Christopher Wren,
Sir Isaac Newton, Henry Purcell,
Captain Cook, Sir James Arkwright,
Lord Nelson, Duke of Wellington,
J. M. W. Turner, George Stevenson,
Michael Faraday, John Keats,
Charles Darwin, Charles Dickens.

ABCDEFGHIJJKLMNOPQR
STUVWXYZ

Fairbank 1932 Page

The introductory episode of the hero's captivity in Valencia, and his consequent entanglement with the Sultana, provide the excuse for his detention outside Teruel till he is just too late to claim his reward, after fulfilling the conditions imposed on him. At the same time, his attitude towards the Sultana serves to enchance his reputation for constancy. The heroine's acceptance of a rival suitor could hardly be explained by the arbitrary power of a father, or his old-fashioned weakness for keeping a promise, though both are used as subordinate motives, and help to give the play the desired air of antiquity. Some more compelling force was required, and that is provided by the rival suitor's hold

Fairbank Page

**Fairbank,** Alfred John – b. 12. 7. 1895 in Grimsby, England, d. 14. 3. 1982 in Hove, England – *calligrapher, type designer* – Studied calligraphy at the Central School of Arts and Crafts in London. 1921–39: calligraphic work. Designs numerous initials for books. 1947: studies on the reform of handwriting. Temporary president of the Society of Scribes and Illuminators and vice-president of the Society for Italic Handwriting. Fairbank contributed to a revival of handwritten Italian italics and his works on handwriting are considered the most knowledgeable in the field. – *Font:* Narrow Bembo (1928). – *Publications:* "A Handwriting Manual", London 1932; "Book of Scripts", London 1949; "Humanistic Script" (with R. W. Hunt), London 1960; "Renaissance Handwriting" (with B. L. Wolpe), London 1960.
**Feininger,** Lyonel – b. 17. 7. 1871 in New York, USA, d. 13. 1. 1956 in New York, USA – *artist, teacher* – 1887: travels to Germany and takes lessons in drawing and painting at the Tagesschule der Allgemeinen Gewerbeschule in Hamburg. 1888: moves to Berlin and studies at the Königliche Akademie der Künste. 1890: produces his first caricatures for the "Humoristische Blätter". 1894–1914: caricatures for "Ulk", "Lustige Blätter" and "Narrenschiff" magazines. 1907: produces his first paintings. 1913: studio in Weimar. Takes part in the Erster Deutscher Herbstsalon in Berlin at the invitation of Franz Marc. Friendship with Kandinsky, Klee and Jawlensky. 1917: first solo exhibition in Der Sturm gallery in Berlin. 1918: produces his first wood-

„Der Sturm“ in Berlin. 1918 erste Holzschnitte. Anschluß an die „Novembergruppe“ in Berlin. Bekanntschaft mit Walter Gropius. 1919 Berufung in den „Meisterrat“ des neu gegründeten „Bauhauses“ in Weimar. Auf dem Titelblatt des Bauhaus-Manifests erscheint Feiningers Holzschnitt „Kathedrale“. Seine grafischen Arbeiten mit integrierten Textzeilen sind Beispiele der frühen, individualistischen Auffassung über Schrift und Typographie am Bauhaus. 1920 erste Museums-Einzelausstellung im Anger-Museum in Erfurt. 1921 Leiter der grafischen Druckerei des Bauhauses. Als erste Veröffentlichung des Bauhauses erscheint eine Mappe mit 12 Holzschnitten Feiningers. 1924 Gründung der Ausstellungsgemeinschaft „Die blaue Vier“ mit Jawlensky, Kandinsky und Klee. 1936 Rückkehr nach New York. 1944 Retrospektive im Museum of Modern Art in New York. 1947 Wahl zum Präsidenten der „Federation of American Painters and Sculptors“. 1955 Mitglied des „National Institute of Arts and Letters“.

**Feitler,** Bea – geb. 5. 2. 1938 in Rio de Janeiro, Brasilien, gest. 1982. – *Grafik-Designerin, Malerin, Lehrerin* – 1956–59 Studium an der Parson's School of Design in New York. Schallplattencover für Atlantic Records. 1959 Rückkehr nach Brasilien, Studium der Malerei am Museu de Arte Moderna in Rio de Janeiro, Arbeit in einer Werbeagentur und für die Zeitschrift „Senhor“. 1960 Gründung des „Studio G“ (mit Sergio Jaguaribe und Glauco Rodrigues) in Rio de Janeiro. Gestaltung von Büchern, Schallplattencov-

Feininger 1919 Cover

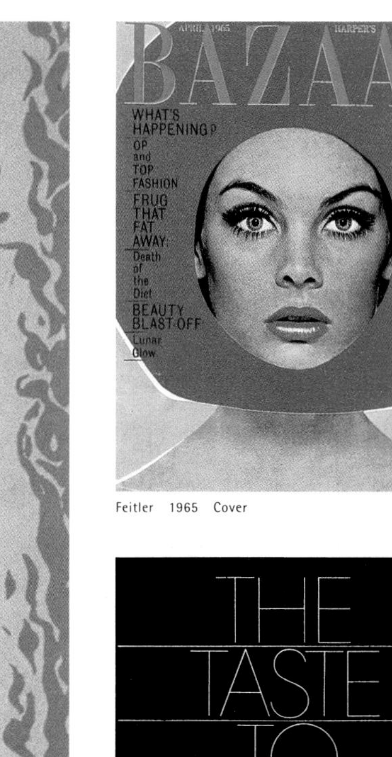


Feitler 1965 Cover

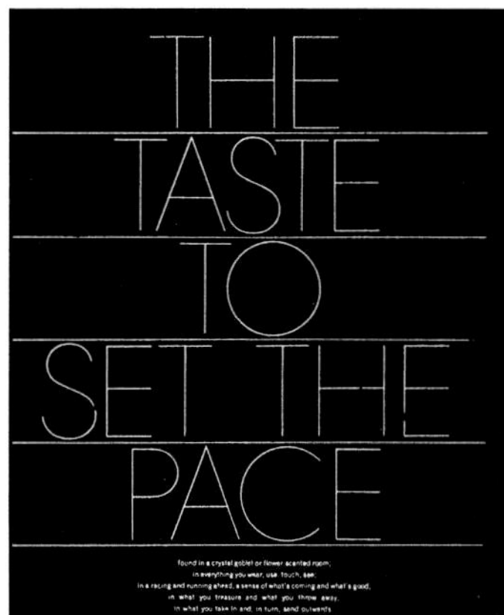


Feitler 1965 Page

bergruppe» (Groupe Novembre) de Berlin. Rencontre Walter Gropius. En 1919, il est appelé au «Meisterrat» (conseil des maîtres) du Bauhaus de Weimar qui vient d'être créé. «Cathédrale», une gravure sur bois de Feininger est publiée sur la couverture du Manifeste du Bauhaus. Ses travaux graphiques, où il intègre des lignes de texte, illustrent les premières conceptions individualistes de la pratique des arts de l'écriture et de la typographie au Bauhaus. Première exposition personnelle dans un musée, au Anger Museum d'Erfurt, en 1920. Dirige l'atelier d'imprimerie et d'arts graphiques au Bauhaus en 1921. La première publication du Bauhaus contient un folio de 12 gravures sur bois de Feininger. En 1924, il fonde le groupe «Die blaue Vier» (Quatre bleu) avec Jawlensky, Kandinsky et Klee. Retour à New York en 1936. 1944, rétrospective au Museum of Modern Art de New York. En 1947, il est élu président de la «Federation of American Painters and Sculptors». 1955, membre du «National Institute of Arts and Letters».

**Feitler,** Bea – née le 5. 2. 1938 à Rio de Janeiro, Brésil, décédée 1982 – *graphiste maquettiste, peintre, enseignante* – 1956–1959, études à la Parson's School of Design de New York. Couvertures de disques pour Atlantic Records. Elle retourne au Brésil en 1959 et étudie la peinture au Museu de Arte Moderna de Rio de Janeiro, travaille pour une agence de publicité et pour la revue «Senhor». 1960, fondation du «Studio G» à Rio de Janeiro (avec Sergio Jaguaribe et Glauco Rodrigues). Maquettes de livres, couvertures de disques, affiches de théâtre. S'installe

cuts. Joins the Novembergruppe in Berlin and makes the acquaintance of Walter Gropius. 1919: asked to join the Meisterrat (council of master craftsmen) of the recently-founded Bauhaus in Weimar. Feininger's "Cathedral" woodcut appears on the cover of the Bauhaus manifesto. His graphic works with their integrated lines of text are examples of Bauhaus' early, individualistic views on type and typography. 1920: stages his first solo museum exhibition in the Anger-Museum in Erfurt. 1921: head of the Bauhaus graphic printing workshop. The first Bauhaus publication is a folder with 12 of Feininger's woodcuts. 1924: founds the exhibition society Die blaue Vier with Jawlensky, Kandinsky and Klee. 1936: moves back to New York. 1944: retrospective exhibition at the Museum of Modern Art in New York. 1947: elected president of the Federation of American Painters and Sculptors. 1955: member of the National Institute of Arts and Letters.

**Feitler,** Bea – b. 5. 2. 1938 in Rio de Janeiro, Brazil, d. 1982 – *graphic designer, painter, teacher* – 1956–59: studies at Parson's School of Design in New York. Produces record cover designs for Atlantic Records. 1959: returns to Brazil and studies painting at the Museu de Arte Moderna in Rio de Janeiro. Works in an advertising agency and for "Senhor" magazine. 1960: founds Studio G with Sergio Jaguaribe and Glauco Rodrigues in Rio de Janeiro. Designs books, record covers and posters for the theater. 1961:

ern, Theaterplakaten. 1961 Umzug nach New York. 1963–72 Co-Art Director der Zeitschrift „Harper's Bazaar". Seit 1976 freischaffende Gestalterin, u. a. Design-Director des „Rolling Stone"-Magazins. Vorlesungen an der School of Visual Arts, New York.

**Feliciano,** Felice – geb. 1433 in Verona, Italien, gest. 1479 in Rom, Italien. – *Dichter, Kunsthistoriker, Drucker* – Als Kenner der römischen Antike beschäftigt sich Feliciano 1460 mit den Proportionen alter Inschriften. Veröffentlicht 1463 eine Abhandlung über die Schrift der römischen Steindenkmäler. Fertigung der ersten geometrischen Konstruktion einer Antiqua-Kapitalschrift. Gründet 1476 in Pojana bei Verona eine Druckerei. Sein berühmtestes Druckwerk ist die italienische Übersetzung von Petrarcas „De viris illustribus".

**Fella,** Edward – geb. 1938 in Detroit, USA. – *Illustrator, Fotograf, Typograph, Lehrer* – 1957–85 Arbeit als Grafik-Designer und Illustrator in Detroit. 1985 Studium am Center for Creative Studies in Detroit. 1985–87 Studium an der Cranbrook Academy of Art. 1987–90 Gestaltung von Flyern, Katalogen und Plakaten für die Detroit Focus Gallery, das Poetry Resource Center, den Detroit Artists' Market. 1994 Entwurf einer Serie von 170 illustrativen Elementen „Fellaparts" für Emigre Fonts. Unterrichtet seit 1987 Grafik-Design am California Institute of the Arts in Valencia, Kalifornien. – *Schriftentwurf:* Out West on a 15 Degree Ellipse (1993).

**Figgins,** Vincent – geb. 1766, gest.

à New York en 1961. Art Director adjoint de la revue «Harper's Bazaar» en 1963–1972. Maquettiste indépendante depuis 1976 et directrice du design du magazine «Rolling Stone». Conférences à la School of Visual Arts, New York.

**Feliciano,** Felice – né en 1433 à Vérone, Italie, décédé en 1479 à Rome, Italie – *poète, historien d'art, imprimeur* – connaissant parfaitement l'antiquité romaine, Feliciano se consacre dès 1460, à l'étude des proportions des anciennes inscriptions. En 1463, il publie un traité sur les caractères gravés dans la pierre des monuments romains. Réalisation de la première construction géométrique de capitales antiques. En 1476, il fonde une imprimerie à Pojana, près de Vérone. L'œuvre la plus célèbre sortie de son atelier est une traduction en italien de «De viris illustribus» de Pétrarque.

**Fella,** Edward – né en 1938 à Detroit, Etats-Unis – *illustrateur, photographe, typographe, enseignant* – 1957–1985, travaux comme graphiste maquettiste et illustrateur à Detroit. 1985, études au Center for Creative Studies à Detroit. 1985–1987, études à la Cranbrook Academy of Art. 1987–1990, maquettes de flyers, catalogues et affiches pour la Detroit Focus Gallery, le Poetry Resource Center, le Detroit Artists' Market. En 1994, il réalise les esquisses pour une série de 170 éléments illustratifs «Fellaparts» pour Emigre Fonts. Enseigne les arts graphiques et le design au California Institute of the Arts à Valencia, Californie. – *Police:* Out West on a 15 Degree Ellipse (1993).

**Figgins,** Vincent – né en 1766, décédé le 29. 2. 1844 à Peckham, Angleterre – *fondeur de caractères* – apprentissage à la fonderie de caractères Joseph Jackson à Londres, en 1782. Crée en 1792 une fonderie de caractères à Londres, qui existera

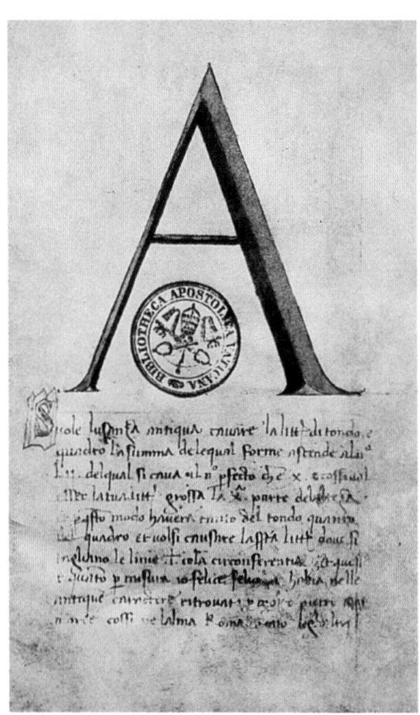

Feliciano 1463 Page

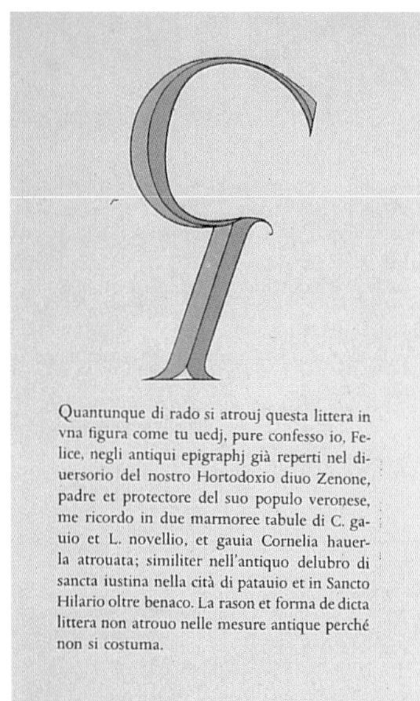

Quantunque di rado si atrouj questa littera in vna figura come tu uedj, pure confesso io, Felice, negli antiqui epigraphj già reperti nel diuersorio del nostro Hortodoxio diuo Zenone, padre et protectore del suo populo veronese, me ricordo in due marmoree tabule di C. gauio et L. novellio, et gauia Cornelia hauerla atrouata; similiter nell'antiquo delubro di sancta iustina nella città di pataiuo et in Sancto Hilario oltre benaco. La rason et forma de dicta littera non atrouo nelle mesure antique perché non si costuma.


Feliciano 1463 Page

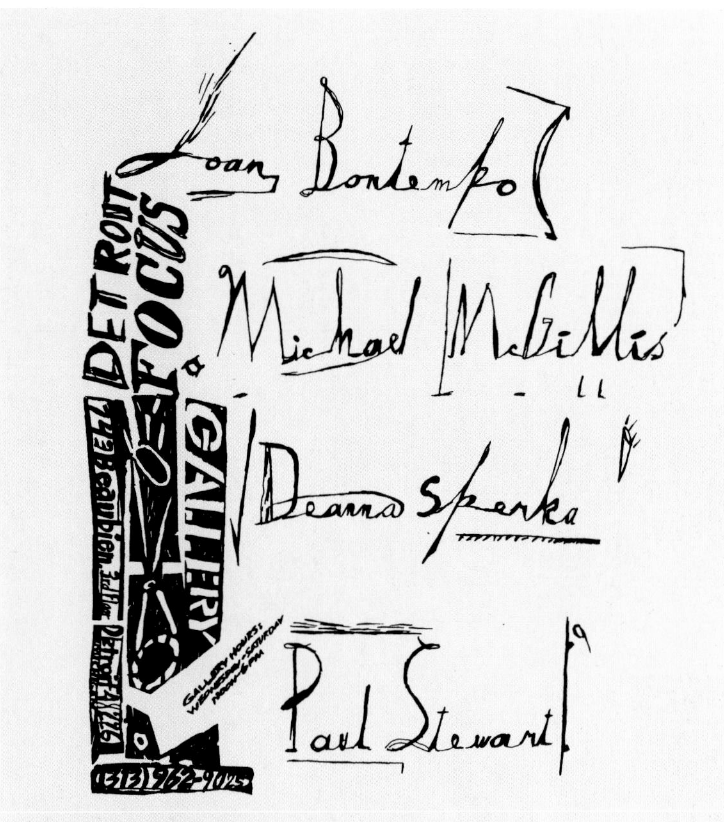


Fella 1987 Flyer

Feliciano 1463 Page

Fella ca. 1987 Logo

moves to New York. 1963–72: joint art director of "Harper's Bazaar". Since 1976: freelance designer, including a post as design director of "Rolling Stone" magazine. Feitler lectures at the School of Visual Arts in New York.

**Feliciano,** Felice – b. 1433 in Verona, Italy, d. 1479 in Rome, Italy – *poet, art historian, printer* – 1460: an authority on Roman antiquity, Feliciano studies the proportions of old inscriptions. 1463: publishes a treatise on the writing on Roman stone slabs. Produces the first geometric construction of a roman majuscule alphabet. 1476: opens a printing workshop in Pojana near Verona. His most famous printed work is the Italian translation of Petrarch's "De viris illustribus".

**Fella,** Edward – b. 1938 in Detroit, USA – *illustrator, photographer, typographer, teacher* – 1957–85: works as a commercial artist in Detroit. 1985: studies at the Center for Creative Studies in Detroit. 1985–87: studies at the Cranbrook Academy of Art. 1987–90: designs flyers, catalogues and posters for the Detroit Focus Gallery, the Poetry Resource Center and Detroit Artists' Market. 1994: designs a series of 170 illustrative elements entitled "Fellaparts" for Emigre Fonts. Since 1987: has taught graphic design at the California Institute of the Arts in Valencia. – *Font:* Out West on a 15 Degree Ellipse (1993).

**Figgins,** Vincent – b. 1766, d. 29. 2. 1844 in Peckham, England – *type founder* – 1782: trains at Joseph Jackson's type foundry in London. 1792: founds a type

29. 2. 1844 in Peckham, England. – *Schriftgießer* – 1782 Lehre in der Schriftgießerei Joseph Jackson in London. 1792 Gründung einer Schriftgießerei in London, die bis 1908 existiert. Die Oxford Press gibt Figgins 1796 den Auftrag, griechische Schriften in fünf Größen herzustellen. 1815 Herausgabe eines Schriftmusterbuchs, in dem unter der Bezeichnung „Antiques" erstmals Schriften gezeigt werden, die später Egyptienne-Schriften (also serifenbetonte Schriften) genannt werden. 1817 Herausgabe einer Schriftprobe mit speziellen Schriften für Zeitungen, 1825 mit irischen, griechischen, hebräischen, indischen, persischen und syrischen Schriften. Nachdem die erste serifenlose Schrift 1816 in einem Schriftmusterbuch von William Caslon IV. vorgestellt wird, veröffentlicht Figgins 1832 eine „Sans Serif"-Schrift. 1833 Herausgabe des Schriftmusterbuches „Specimen of Printing Types". Die Schriftentwürfe von Figgins entsprechen dem neuen Geschmack des neuen Industriezeitalters. – *Schriftentwürfe:* Gresham (1792), Figgins Shaded (1816), Egyptian (1817).

**Fili,** Louise – geb. 12. 3. 1951 in Orange, New Jersey, USA. – *Grafik-Designerin* – Studium am Skidmore College, Saratoga Springs, New York, und an der School of Visual Arts, New York. Danach Designerin bei Herb Lubalin Associates. 1978–89 Art Director bei Pantheon Books. Unter ihrer Leitung entstehen über 2.000 Buchumschläge. 1990 Gründung ihres Studios „Louise Fili Ltd.". Arbeitsgebiete sind Verpackung, Signets-Design und Typographie für Auftraggeber aus Verlagen

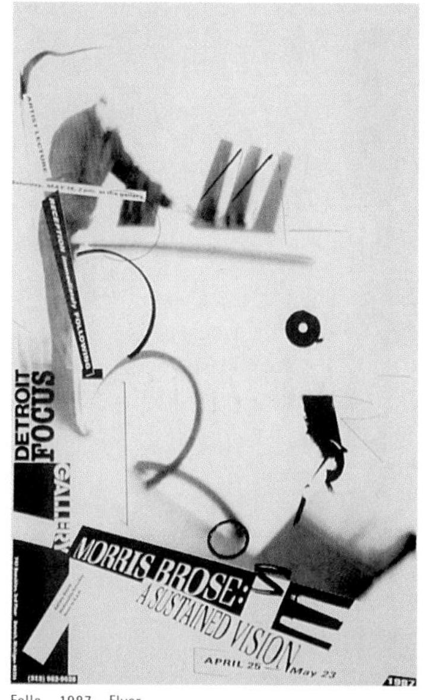

Fella 1987 Flyer

Fella 1992 Invitation

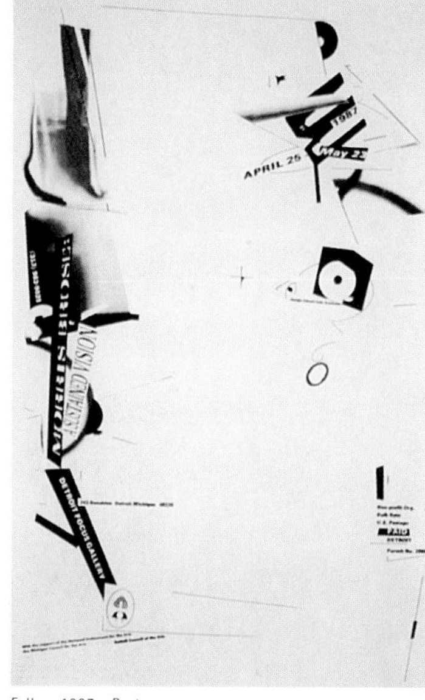

Fella 1987 Poster

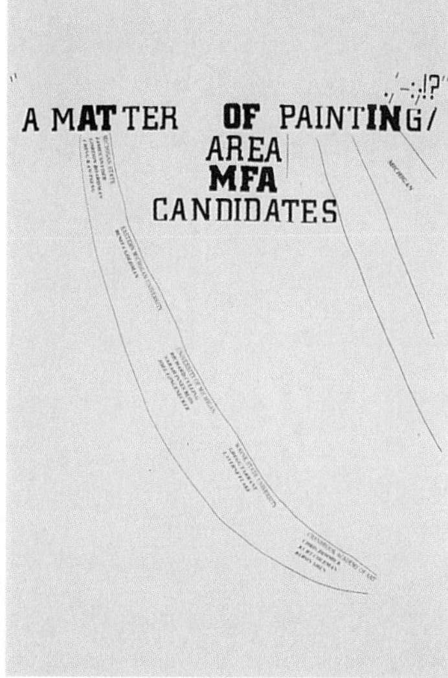

Fella 1985 Poster

Fili 1985 Poster

Fili 1985 Cover

Fili 1988 Logo

jusqu'en 1908. En 1796, la Oxford Press demande à Figgins de fabriquer des caractères grecs en cinq corps. En 1815, il publie un manuel de modèles d'écritures qui présente des caractères appelés pour la première fois «Antiques» et qui prendront ensuite le nom d'Egyptiennes (fontes à sérif accentués). En 1817, il publie des échantillons de fontes spécialement destinées à la presse quotidienne puis en 1825, des caractères irlandais, grecs, hebraïques, indiens, perses et syriaques. Après la présentation de la première fonte sans sérif en 1816 dans le manuel d'échantillons de William Caslon IV, Figgins publie en 1832 une fonte «Sans sérif». En 1833, il édite le catalogue de caractères «Specimen of Printing Types». Les créations de Figgins reflétaient les goûts nouveaux à l'ère de l'industrialisation. – *Polices:* Gresham (1792), Figgins Shaded (1816), Egyptian (1817).

**Fili,** Louise – née le 12. 3. 1951 à Orange, New Jersey, Etats-Unis – *graphiste maquettiste* – études au Skidmore College, Saratoga Springs, New York ainsi qu'à la School of Visual Arts, New York. Exerce ensuite comme designer chez Herb Lubalin Associates. 1978–1989, Art Director chez Pantheon Books, 2 000 couvertures de livres sont réalisées sous sa direction. En 1990, elle fonde son propre atelier, le «Louise Fili Ltd.». Spécialités: emballages, signets, typographie pour des commanditaires des secteurs de l'édition et de l'industrie. – *Publications, sé-*

foundry in London which exists until 1908. 1796: Oxford Press commissions Figgins to produce Greek alphabets in five different sizes. 1815: Figgins publishes a book of type specimens, the first to depict typefaces (labeled "Antiques") which later become known as slab-serif typefaces. 1817: issues a book of type specimens containing special fonts for newspapers and in 1825 a volume containing Irish, Greek, Hebrew, Indian, Persian and Syrian type. 1832: after the introduction of the first sans-serif typeface by William Caslon IV in a specimen book in 1816, Figgins also produces a sans-serif alphabet. 1833: the "Specimen of Printing Types" is published. Figgins' type designs were in keeping with the new tastes of the new industrial age. – *Fonts:* Gresham (1792), Figgins Shaded (1816), Egyptian (1817).

**Fili,** Louise – b. 12. 3. 1951 in Orange, New Jersey, USA – *graphic designer* – Studied at Skidmore College, Saratoga Springs in New York and at the School of Visual Arts in New York. Then designer for Herb Lubalin Associates. 1978–89: art director for Pantheon Books where she supervises the designing of over 2,000 book covers. 1990: opens her studio Louise Fili Ltd. Produces packaging designs, trademarks and typography work for publishers and industry and commerce. – *Publications*

und Wirtschaft. – *Publikationen u. a.:* „Italian Art Deco" (mit Stephen Heller), New York 1993; „Dutch Modern" (mit Stephen Heller), New York 1994.

**Finlay,** Ian Hamilton – geb. 28. 10. 1925 in Nassau, Bahamas. – *Literat, Künstler, Verleger* – Aufgewachsen in Schottland, kurzes Studium an der Glasgow School of Art, Militärdienst, Arbeit als Schäfer. Veröffentlichung von Kurzgeschichten und Theaterstücken. 1961 Gründung von „The Wild Hawthorn Press" (mit Jessie McGuffie), in der Werke zeitgenössischer Dichter und Künstler verlegt werden. 1962 Gründung des Periodikums „Poor. Old. Tired. Horse", einem Forum für Verbales und Visuelles, für Traditionelles und Modernistisches, für Kreatives und Theoretisches. Beginnt 1964 mit der Produktion von Gedichten, die als Skulptur im Raum stehen. 1966 Umzug nach Stonypath, Gartengestaltung. 1987 Ehrendoktorwürde der Universität von Aberdeen. Seit 1968 zahlreiche Ausstellungen mit Katalogpublikationen sowie zahlreiche eigene literarische und künstlerische Publikationen. In Finlays Werk, das von der visuellen Poesie ausging, wird die Welt der Zeichen und der Schrift durch kritische Reflexionen in die Gegenwart vermittelt.

**Fiore,** Quentin – geb. 12. 2. 1920 in New York, USA. – *Grafik-Designer, Schriftzeichner, Typograph, Autor* – Autodidakt als Gestalter. Ende der 30er Jahre kurze Zeit Zeichen- und Malkurse in New York bei George Grosz und Hans Hofmann. Besuch im „New Bauhaus" in Chicago mit einem Empfehlungsschreiben von Grosz

*lection:* «Italian Art Deco» (avec Stephen Heller), New York 1993; «Dutch Modern» (avec Stephen Heller), New York 1994.

**Finlay,** Ian Hamilton – né le 28. 10. 1925 à Nassau, Bahamas – *homme de lettres, artiste, éditeur* – a été élevé en Ecosse, brèves études à la Glasgow School of Arts, service militaire, puis berger. Publie des nouvelles et des pièces de théâtre. En 1961, il fonde «The Wild Hawthorn Press» (avec Jessie McGuffie), où il publie les œuvres de poètes et d'artistes contemporains. 1962, création du périodique «Poor. Old. Tired. Horse», forum d'expression verbale et visuelle, traditionnelle et moderniste, créative et théorique. En 1964, il commence à produire des sculptures – poèmes mis en espaces. S'installe à Stonypath en 1966, devient jardiniste. L'université d'Aberdeen lui décerne le titre de Docteur honoris causa en 1987. Nombreuses expositions et publications de catalogues depuis 1968, éditions de ses œuvres littéraires et plastiques. Dans l'œuvre de Finlay qui puise dans la poésie visuelle, l'univers des signes et de l'écriture est mis en évidence par le biais d'une réflexion critique.

**Fiore,** Quentin – né le 12. 2. 1920 à New York, Etats-Unis – *graphiste maquettiste, dessinateur de caractères, typographe, auteur* – designer autodidacte. A la fin des années 30, il prend quelques cours de dessin à New York avec Georg Grosz et Hans Hofmann. Fréquente le «New Bauhaus» de Chicago grâce à une lettre de recommandation de Grosz adressée à Moholy-Nagy. Dessins de caractères et travaux de typographie pour le designer Lester Beall. A la fin des années 40, il devient Art Director chez Christian Dior et Bonwit Teller. Pendant les années 50, il réalise des maquettes de revues, des films éducatifs, des systèmes signalétiques et des graphismes pour plusieurs comman-

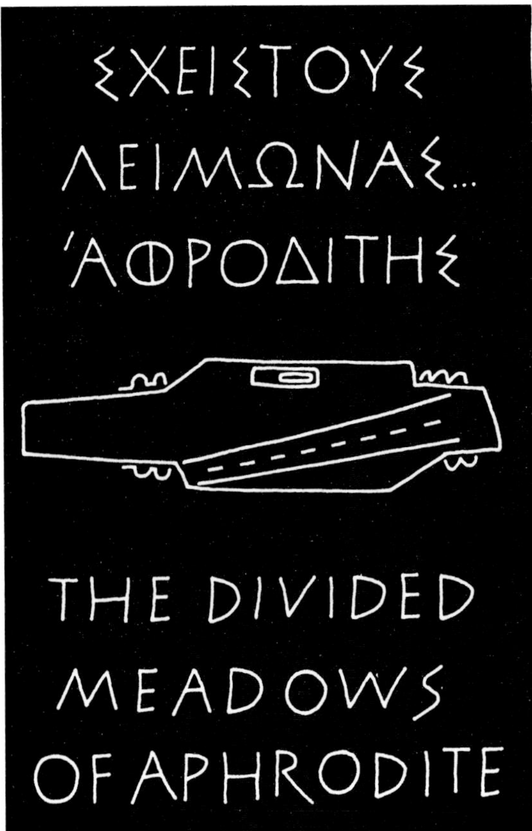
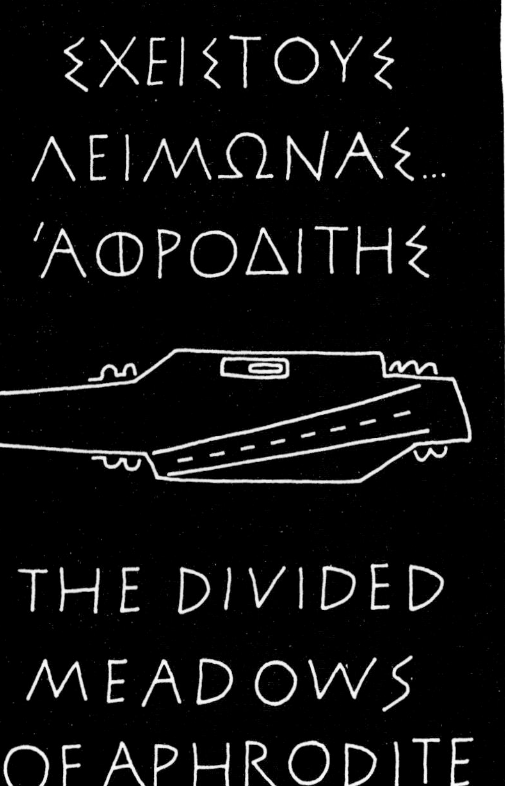

Finlay 1975 Artwork

Finlay 1973 Poem

Finlay 1982 Label

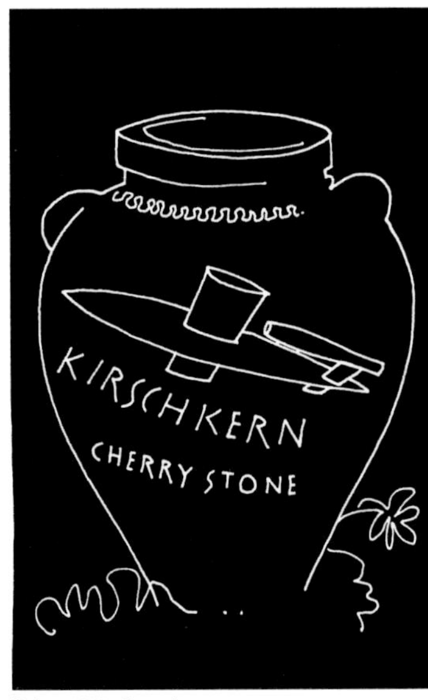


Finlay 1975 Artwork

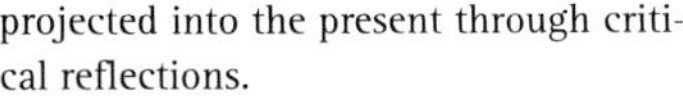

*include:* "Italian Art Deco" (with Stephen Heller), New York 1993; "Dutch Modern" (with Stephen Heller), New York 1994.

**Finlay,** Ian Hamilton – b. 28. 10. 1925 in Nassau, Bahamas – *littérateur, artist, publisher* – Grew up in Scotland; brief period of study at the Glasgow School of Art, military service, work as a shepherd. Published short stories and plays. 1961: founds The Wild Hawthorn Press (with Jessie McGuffie) which publishes works by contemporary poets and artists. 1962: launches the periodical "Poor. Old. Tired. Horse", a forum for the verbal and the visual, the traditional and the modern, the creative and the theoretical. 1964: starts producing poems designed in forms of sculptures. 1966: moves to Stonypath and designs gardens. 1987: is awarded an honorary doctorate from the University of Aberdeen. Since 1968: has staged numerous exhibitions with catalogues and has produced a great number of his own literary and artistic publications. In Finlay's works, which grew out of Visual Poetry, the world of symbols and writing is projected into the present through critical reflections.

**Fiore,** Quentin – b. 12. 2. 1920 in New York, USA – *graphic designer, letterer, typographer, author* – Self-taught designer. End of the 1930s: takes drawing and painting lessons for a brief period with George Grosz and Hans Hofmann in New York. Attends the New Bauhaus in Chicago with a letter of recommendation from Grosz addressed to Moholy-Nagy. Produces type designs and other typographical work for the designer Lester Beall.

an Moholy-Nagy. Schriftzeichnungen und typographische Arbeiten für den Gestalter Lester Beall. Ende der 40er Jahre Art Director bei Christian Dior und Bonwit Teller. In den 50er Jahren Zeitschriftengestaltung, Lehrfilme, Leitsysteme und Grafik-Design für Auftraggeber wie Ford Foundation, RCA, Bell Laboratories, American Medical Association, Gulf + Western Inc. Entwicklung eines Re-Designs der Zeitschrift „Life" der Time-Life Inc. in New York. Für die „Banco de Brasil" in Rio de Janeiro Entwurf des typographischen Konzepts und der Hausschrift „Itabori". Gestaltet 1967 als Co-Autor das Manuskript „The Medium is the Massage" von Marshall McLuhan. Gestaltet 1968 als Co-Autor das Buch „War and Peace in the Global Village" von Marshall McLuhan. Gestaltet 1970 das Buch „Do it! Scenarios of the Revolution" von Jerry Rubin und das Buch „I Seem to Be a Verb" von Richard Buckminster Fuller. 1970 Media-Consultant für „Homefax" (auch „Picturephone" genannt) der Firmen RCA und NBC, die eine der ersten interaktiven Technologien ist, die Fernsehen, Kopiergerät und Telefon verbindet und somit den Ausdruck einer „Elektronischen Zeitung" ermöglicht. Entwurfsarbeiten für die Verlage Random House, Abrams, George Braziller, Houghton-Mifflin, The University of Illinois Press, The University of Southern Illinois Press, The University of Wisconsin Press. Zusammenarbeit mit den Werbeagenturen D'Arcy Advertising Agency, Foote, Cone & Belding Advertising, McCann-Erickson Advertising Agency. Neben sei-

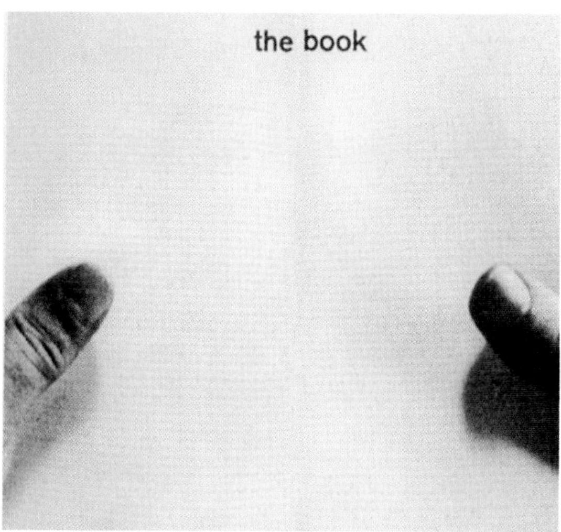

Fiore 1967 Spread

Fiore 1967 Cover

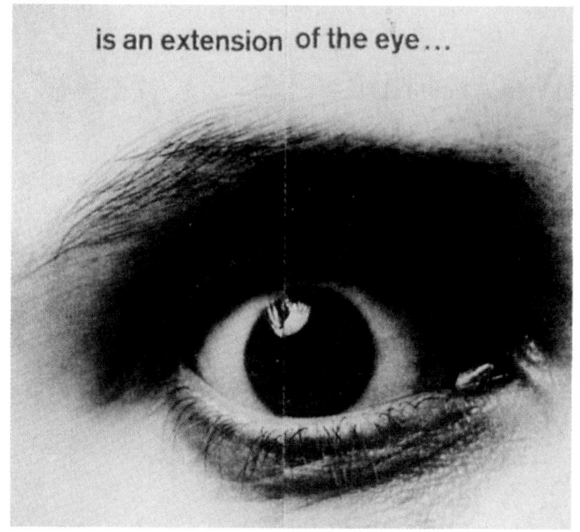


Fiore 1967 Spread

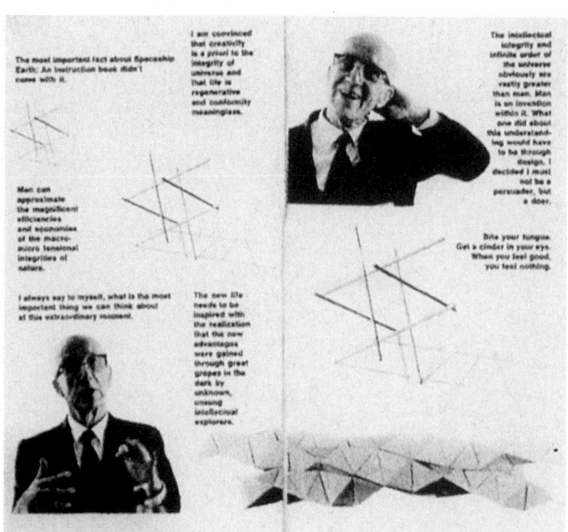

Fiore 1970 Spread

ditaires tels que Ford Foundation, RCA, Bell Laboratories, American Medical Association, Gulf + Western Inc. Refonte de la maquette de la revue «Life» de Time-Life Inc. à New York. Il crée un concept typographique pour la «Banco de Brasil» de Rio de Janeiro ainsi que la maquette de la revue de la société «Itabori». En 1967, mise en page du manuscrit «The Medium is the Massage» de Marshall McLuhan, en qualité de co-auteur. En 1968, mise en page et co-auteur de «War and Peace in the Global Village» de Marshall McLuhan. 1970, mise en page de «Do it! Scenarios of the Revolution» de Jerry Rubin et de «I Seem to Be a Verb» de Richard Buckminster Fuller. Toujours en 1970, Média-Consultant pour le «Homefax» (également appelé «Picturephone») des sociétés RCA et NBC, soit l'une des premières technologies interactives faisant intervenir la télévision, la photocopieuse et le téléphone pour permettre d'imprimer un «journal électronique». Maquettes pour les éditions Random House, Abrams, George Braziller, Houghton-Mifflin, The University of Illinois Press, The University of Southern Illinois Press, The University of Wisconsin Press. Collaboration avec des agences de publicité dont D'Arcy Advertising Agency, Foote, Cone & Belding Advertising, McCann-Erickson Advertising Agency. A côté de ses activités de graphiste ma-

End of the 1940s: art director for Christian Dior and Bonwit Teller. 1950s: designs magazines, makes educational films and signage systems and produces graphic designs for clients who include the Ford Foundation, RCA, Bell Laboratories, the American Medical Association and Gulf + Western Inc. Develops a re-design for Time-Life Inc.'s "Life" magazine in New York. Draws up a typographical concept and designs the in-house magazine "Itabori" for the Banco de Brasil in Rio de Janeiro. 1967: designs and co-authors the manuscript "The Medium is the Massage" by Marshall McLuhan. 1968: designs and co-authors "War and Peace in the Global Village" by Marshall McLuhan. 1970: designs the book "Do it! Scenarios of the Revolution" by Jerry Rubin. 1970: designs the book "I Seem to Be a Verb" by Richard Buckminster Fuller. 1970: media consultant for RCA and NBC's "Homefax" (also called "Picturephone"), one of the first interactive technologies to combine TV, Xerox and the telephone in order to be able to print out an "electronic newspaper". Design work for the publishing houses Random House, Abrams, George Braziller, Houghton-Mifflin, The University of Illinois Press, The University of Southern Illinois Press and The University of Wisconsin Press. Has worked for the D'Arcy Advertising Agency, Foote, Cone & Belding Advertising and McCann-Erickson Advertising Agency. Besides his work as a graphic designer, the results of Fiore's collaboration

ner Arbeit als Grafik-Designer werden die Ergebnisse der Zusammenarbeit mit McLuhan und Buckminster Fuller zu Inkunabeln einer Zeit, die in den 90er Jahren Anerkennung und Nachahmung erfährt. – *Schriftentwurf:* Itabori (o.J.).

**Fischer,** Florian – geb. 5. 9. 1940 in Niesky, Lausitz, Deutschland. – *Typograph, Grafik-Designer, Planer, Lehrer* – 1963–67 Studium der Innenarchitektur, Produktgestaltung und der visuellen Kommunikation an der Staatlichen Hochschule der Bildenden Künste Stuttgart und an der Hochschule für Gestaltung Ulm. Arbeit als Planer und als Dozent für mehrere Projekte im Bildungswesen und in der Gemeinwesenarbeit. 1976 Gründung des Designbüros „Fischer & Spiekermann“ (mit Erik Spiekermann) in Berlin. 1979 Gründung des Designbüros „MetaDesign“ (mit Dieter Heil und Erik Spiekermann) in Berlin. 1984–89 Design-Koordinator von „Sedley Place Design GmbH“ in Berlin. 1990–91 Gastprofessor für Corporate Identity an der Hochschule der Künste Berlin. 1990–96 Partner bei Fischer & Scholz in Berlin. Seit 1996 freier Berater. Zahlreiche und umfangreiche Arbeiten für Erscheinungsbilder, u. a. für die BfG Bank (1979–83), die Deutsche Bundespost (1984–90), die Deutsche Postbank (seit 1991) und die Deutsche Post AG (seit 1994). Zahlreiche Auszeichnungen, Ausstellungsbeteiligungen, Vorträge und Lehraufträge.

**Fleckhaus,** Willy – geb. 21. 12. 1925 in Velbert, Deutschland, gest. 12. 9. 1983 in Castelfranco, Italien. – *Grafik-Designer, Typograph, Lehrer, Art Director* – 1950–53

quettiste, les travaux réalisés en collaboration avec McLuhan et Buckminster Fuller sont devenus des incunables, souvent imités et jouissant d'un grand prestige pendant les années 90. – *Police:* Itabori (sans date).

**Fischer,** Florian – né le 5. 9. 1940 à Niesky, Lausitz, Allemagne – *typographe, graphiste maquettiste, concepteur, enseignant* – 1963–1967, études d'architecture intérieure, de design industriel, de communication visuelle à la Hochschule der Bildenden Künste (école supérieure des beaux-arts) de Stuttgart et à la Hochschule für Gestaltung (design) d' Ulm. Exerce comme concepteur et comme professeur dans le cadre de divers projets éducatifs et d'utilité publique. En 1976, il fonde l'agence de design «Fischer & Spiekermann» (avec Erik Spiekermann) à Berlin, puis en 1979 et toujours à Berlin, l'agence de design «MetaDesign» (avec Dieter Heil et Erik Spiekermann). 1984–1989, coordinateur du design pour «Sedley Place Design GmbH» à Berlin. 1990–1991, séminaires sur les identités d'entreprises à la Hochschule der Künste (beaux-arts) de Berlin. 1990–1996, associé chez Fischer & Scholz à Berlin. Conseiller indépendant depuis 1996. De nombreux travaux sur les identités, par ex. pour la BfG Bank (1979–1983), la Deutsche Bundespost (1984–1990), la Deutsche Postbank (depuis 1991), la Deutsche Post AG (depuis 1994). Nombreuses distinctions et participations à des expositions, conférences et séminaires.

**Fleckhaus,** Willy – né le 21. 12. 1925 à Velbert, Allemagne, décédé le 12. 9. 1983 à Castelfranco, Italie – *graphiste maquettiste, typographe, enseignant, directeur artistique* – 1950–1953, rédacteur pour «Aufwärts», une revue pour la jeunesse publiée par le Bund-Verlag, Co-

Fischer 1981 Display

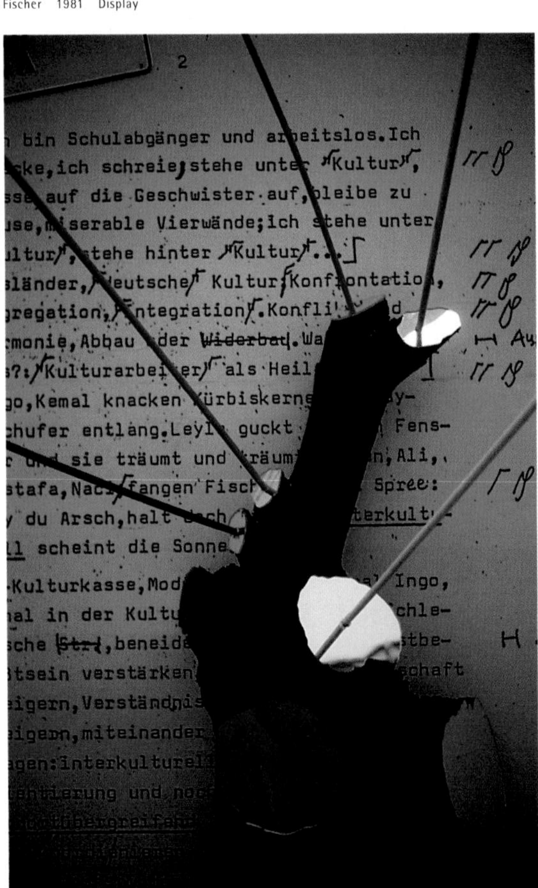

Fischer 1984

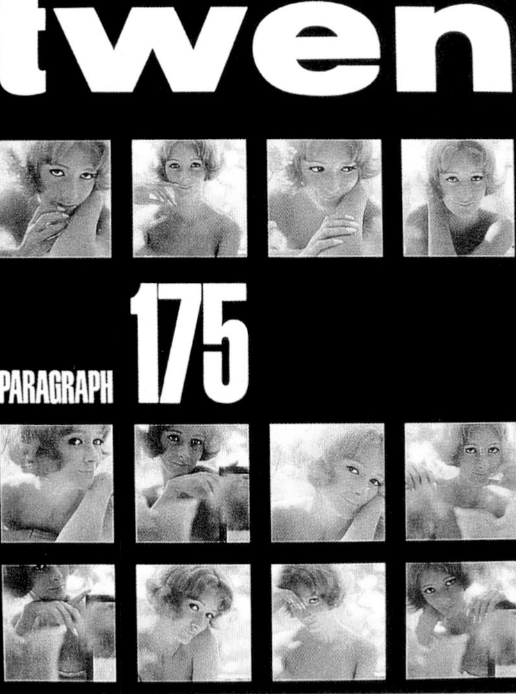


Fleckhaus 1963 Cover

Fleckhaus 1970 Spread

with McLuhan and Buckminster Fuller have become incunabula of an age much praised and copied in the 1990s. – *Font:* Itabori (undated).

**Fischer,** Florian – b. 5. 9. 1940 in Niesky, Lausitz, Germany – *typographer, graphic designer, planner, teacher* – 1963–67: studies interior design, product design and visual communication at the Staatliche Hochschule der Bildenden Künste in Stuttgart and at the Hochschule für Gestaltung in Ulm. Works as a planner and lecturer on numerous educational and community projects. 1976: opens the Fischer & Spiekermann design studio with Erik Spiekermann in Berlin. 1979: opens the MetaDesign studio with Dieter Heil and Erik Spiekermann in Berlin. 1984–89: design coordinator for Sedley Place Design GmbH in Berlin. 1990–91: visiting professor of corporate identity design at the Hochschule der Künste in Berlin. 1990–96: partner of Fischer & Scholz in Berlin. Since 1996: freelance consultant. Has produced extensive work for corporate identities, including for the BfG Bank (1979–83), Deutsche Bundespost (1984–90), Deutsche Postbank (since 1991) and the Deutsche Post AG (since 1994). Numerous awards, participation in exhibitions, lectures and teaching positions.

**Fleckhaus,** Willy – b. 21. 12. 1925 in Velbert, Germany, d. 12. 9. 1983 in Castelfranco, Italy – *graphic designer, typographer, teacher, art director* – 1950–53: editor of Bund-Verlag's teenage magazine, "Aufwärts", in Cologne. 1953: art director of "Aufwärts". Acts as consultant for

Redakteur der Jugendzeitschrift „Aufwärts“ des Bund-Verlags, Köln. 1953 gestalterische Leitung der „Aufwärts“. Beratende Tätigkeit für das Verlagshaus DuMont-Schauberg, Köln. 1956–76 Ausstellungsdesign für die Ausstellung „Photokina“ und Neugestaltung des Katalogs. 1959 Gründung der Zeitschrift „twen“ mit Adolf Theobald und Stephan Wolf. Fleckhaus ist Art Director und zeitweise Chefredakteur. 1959 Beginn der Zusammenarbeit mit dem Suhrkamp Verlag, Entwurf der Umschläge der „Bibliothek Suhrkamp“. 1963 Entwurf der Umschläge und des Erscheinungsbildes der „edition suhrkamp“. Umfassende Gestaltung von Buchprogramm und Werbemitteln des Insel und des Suhrkamp Verlags. 1964 Mitbegründer des Art Directors Club von Deutschland, Präsident von 1972–74. Zahlreiche Entwurfsarbeiten für Unternehmen wie Ilford, Mercedes-Benz, die Deutsche Bundespost und Zeitschriften wie dem „Zeit-Magazin“, „Merian“, „Mode und Wohnen“ und die Tageszeitung „Die Welt“. 1968 erste Arbeiten für den Westdeutschen Rundfunk, Zusammenarbeit mit Heinz Edelmann. 1970 gekündigt als Art Director des „twen“. 1974 Berufung als Professor an die Gesamthochschule Essen. Einrichtung einer Sommerakademie für Studierende in der Toskana. 1980 Wechsel als Professor an die Gesamthochschule Wuppertal. Art Director des von ihm gestalteten „FAZ-Magazins“. – *Publikationen:* Siegfried Unseld „Der Marienbader Korb“, Hamburg 1976; Hansjörg Stulle (Hrsg.) „Nach-Lese“, Stuttgart 1980.

**Fleischmann,** Gerd – geb. 6. 10. 1939 in

logne. 1953, dirige la maquette de «Aufwärts». Conseiller pour les éditions DuMont-Schauberg de Cologne. 1956–1976, design de l'exposition «Photokina» et refonte de la maquette du catalogue. En 1959, il fonde la revue «twen» avec Adolf Theobald et Stefan Wolf. Fleckhaus y exerce la fonction de directeur artistique et de rédacteur en chef à plusieurs reprises. Début de sa collaboration avec les éditions Suhrkamp en 1959 et conception des couvertures de la collection «Bibliothek Suhrkamp». En 1963, il crée la maquette et l'identité de l' «edition suhrkamp». Nombreuses maquettes pour des collections et matériel publicitaire pour les éditions Insel et Suhrkamp. 1964, cofondateur de l'Art Directors Club d'Allemagne qu'il préside de 1972 à 1974. Nombreuses maquettes pour Ilford, Mercedes-Benz, Deutsche Bundespost et des revues telles que le «Zeit-Magazin», «Merian», «Mode und Wohnen», et le quotidien «Die Welt». Premiers travaux pour le Westdeutsche Rundfunk en 1968 et collaboration avec Heinz Edelmann. En 1970, il est licencié de son poste de directeur artistique de «twen». En 1974, il est nommé professeur à la Gesamthochschule d' Essen. Fonde une université d'été pour étudiants en Toscane. Enseigne comme professeur à la Gesamthochschule de Wuppertal à partir de 1980. Directeur artistique du «FAZ-Magazin» dont il a créé la maquette. – *Publications, selection:* Siegfried Unseld «Der Marienbader Korb», Hambourg 1976; Hansjörg Stulle (éd.) «Nach-Lese», Stuttgart 1980.

**Fleischmann,** Gerd – né le 6. 10. 1939 à

Gerd Fleischmann 1990 Cover

Gerd Fleischmann

Fleckhaus 1972 Cover

unkorrigiert, akzeptiert und als Moment des Bildes gedeutet. Don van Vliet hält an seinen Eigenheiten fest, wie er es auch als Musiker tat.

DON VAN VLIET

Dauer der Ausstellung in Bielefeld: 27. November 1993 – 16. Januar 1994

Die Ausstellung des Bielefelder Kunstvereins entstand in Zusammenarbeit mit dem Künstler und wurde als Wanderausstellung konzipiert. Anfang 1994 geht sie nach Odense/Dänemark (Kunsthallen Brandts Klædefabrik) und ist im Herbst 1994 in Brighton/England (The Royal Art Pavilion. Art Gallery and Museums) zu sehen. Weitere Ausstellungsstationen sind in Vorbereitung.

Zur Eröffnung sprechen:
Eberhard David
Oberbürgermeister der Stadt Bielefeld
Dr. Florian Böllhoff
1. Vorsitzender des Bielefelder Kunstvereins
Dr. Andreas Beaugrand
Bielefelder Kunstverein

Gerd Fleischmann 1993 Invitation

DuMont-Schauberg publishers in Cologne. 1956–76: designs the “Photokina” exhibition and re-designs the catalogue. 1959: launches “twen” magazine with Adolf Theobald and Stephan Wolf. Fleckhaus is art director and at various stages also editor-in-chief. 1959: begins working for the publishing house Suhrkamp and designs the covers for the Bibliothek Suhrkamp series. 1963: designs the covers and corporate identity for edition suhrkamp. Extensive design for Insel and Suhrkamp publishers (book programs and advertising). 1964: cofounder of the Art Directors Club in Germany, president from 1972–74. Numerous designs for companies such as Ilford, Mercedes-Benz, Deutsche Bundespost and for various magazines, including “Zeit-Magazin”, “Merian”, “Mode und Wohnen” and “Die Welt” newspaper. 1968: produces his first work for the Westdeutsche Rundfunk and works with Heinz Edelmann. 1970: is dismissed as “twen” art director. 1974: appointed professor at the Gesamthochschule in Essen. Sets up a summer academy for students in Tuscany. 1980: changes his professorial position to the Gesamthochschule in Wuppertal. Designer and art director of the “Frankfurter Allgemeine Magazin”. – *Publications include:* Siegfried Unseld “Der Marienbader Korb”, Hamburg 1976; Hansjörg Stulle (ed.) “Nach-Lese”, Stuttgart 1980.

**Fleischmann,** Gerd – b. 6. 10. 1939 in Nuremberg, Germany – *typographer, teacher* – Studied physics and mathematics in Erlangen. 1960–64: studies art,

Fleischmann, G.
**Fleischmann, J. M.**

Nürnberg, Deutschland. – *Typograph, Lehrer* – Studium der Physik und Mathematik in Erlangen. 1960–64 Studium der Kunsterziehung, Werkerziehung, Fotografie und Pädagogik an der Hochschule für Bildende Künste in Berlin sowie der Mathematik an der Freien Universität in Berlin. 1964 Fotograf und grafischer Gestalter mit Arbeiten, u. a. für die Nationalgalerie Berlin, den Deutschen Akademischen Austauschdienst (DAAD) und private Galerien. Nach mehreren Jahren als Lehrer an Berliner Gymnasien und einem Jahr als Kommunikationsplaner in Bonn seit 1971 Professor am Fachbereich Design der Fachhochschule Bielefeld. Buchgestaltungen und Orientierungssysteme (u. a. für die Universität Bielefeld und den Sächsischen Landtag) sowie Ausstellungen. Zahlreiche Auszeichnungen von der Stiftung Buchkunst. 1980 Bewahrung der letzten Druckerei aus der Zeit der irischen Revolution für den Bunratty Folk Park in Bunratty, Irland. 1988 Organisation des ersten deutschen DTP-Wettbewerbs für die Zeitschrift „Page". Mitbegründer des „Forums Typografie". Gründungsmitglied der Gesellschaft zur Förderung der Druckkunst, Leipzig. – *Publikationen:* „Plakate in der irischen Provinz, Irish Country Posters", Bielefeld, Essen 1982; „Bauhaus. Drucksachen, Typographie, Reklame", Düsseldorf 1984; „Das Stabenbuch", Essen 1986.

**Fleischmann,** Johann Michael – geb. 1701 in Nürnberg, Deutschland, gest. 1768 in Amsterdam, Niederlande. – *Schriftgießer, Stempelschneider* – 1723–24 Lehre als Schriftgießer. Nach Wanderjahren 1727

Nuremberg, Allemagne – *typographe, enseignant* – études de physique et de mathématiques à Erlangen. 1960–1964, études de pédagogie de l'art, des matériaux et de photographie à la Hochschule für Bildende Künste (école supérieure des beaux-arts) de Berlin et de mathématiques à la Freie Universität (université libre) de Berlin. Exerce comme photographe et graphiste pour la Nationalgalerie de Berlin, le Deutscher Akademischer Austauschdienst (DAAD) et des galeries privées à partir de 1964. Au terme de plusieurs années d'enseignement dans divers lycées berlinois et d'une année à Bonn comme chargé de la communication, il est professeur de design à la Fachhochschule (institut universitaire technique) de Bielefeld depuis 1971. Maquettes de livres, systèmes signalétiques (pour l'université de Bielefeld et le parlement de Saxe) ainsi que plusieurs expositions. De nombreuses distinctions de la «Stiftung Buchkunst» (fondation des arts du livre). 1980, conservation de la dernière imprimerie datant de la révolution irlandaise pour le Bunratty Folk Park à Bunratty, Irlande. En 1988, il organise le premier concours allemand de DTP pour la revue «Page». Cofondateur du «Forum Typografie». Membre fondateur de la «Gesellschaft zur Förderung der Druckkunst» (société de promotion des arts de l'imprimerie) de Leipzig. – *Publications:* «Plakate in der irischen Provinz, Irish Country Posters», Bielefeld, Essen 1982; «Bauhaus. Drucksachen, Typographie, Reklame», Düsseldorf 1984; «Das Stabenbuch», Essen 1986.

**Fleischmann,** Johann Michael – né en 1701 à Nuremberg, Allemagne, décédé en 1768 à Amsterdam, Pays-Bas – *fondeur de caractères, tailleur de types* – 1723–1724, apprentissage de fondeur de caractères. Après des années de voyage il

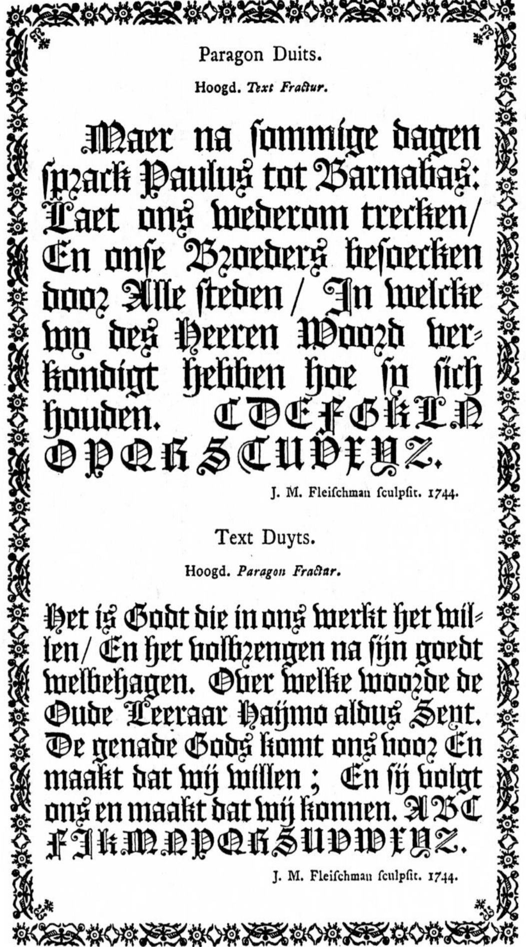

Paragon Duits.

Hoogd. *Text Fractur.*

Maer na sommige dagen sprack Paulus tot Barnabas: Laet ons wederom trecken/ En onse Broeders besoecken door Alle steden / In welcke wy des Heeren Woord verkondigt hebben hoe sy sich houden. CDEFGKLN OPQRSTUVWXYZ.

J. M. Fleischman sculpsit. 1744.

Text Duyts.

Hoogd. *Paragon Fractur.*

Het is Godt die in ons werkt het willen/ En het volbrengen na sijn goedt welbehagen. Over welke woorde de Oude Leeraar Haymo aldus Seyt. De genade Gods komt ons voor En maakt dat wy willen ; En sy volgt ons en maakt dat wy konnen. ABC FIKMNOPQRSTUVWXYZ.

J. M. Fleischman sculpsit. 1744.

Johann Michael Fleischmann 1768 Black Letters

VOORBERICHT.

Reeds eenige Jaaren geleeden, heb ik ten dienste van hun, die zich door my in 't behandelen der Viool lieten onderwyzen, de tegenwoordige Regels opgesteld. Het verwonderde my menigmaal grootlyks, dat 'er tot gebruik van een zo gewoon en by de meeste Concerten byna onontbeerlyk Werktuig, als de Viool met der daad is, geene Handleiding ten voorschyn kwam; daar men echter reeds overlang goede Grondbeginsels, en voornaamlyk eenige Regels aangaande de byzondere Strykmanier had noodig gehad. Het deed my dikwyls niet weinig leed, wanneer ik ondervond, dat de Leerlingen zo slecht onderweezen waren, dat men niet alleen alles van 't begin af weder herhaalen;

Johann Michael Fleischmann 1766 Specimen

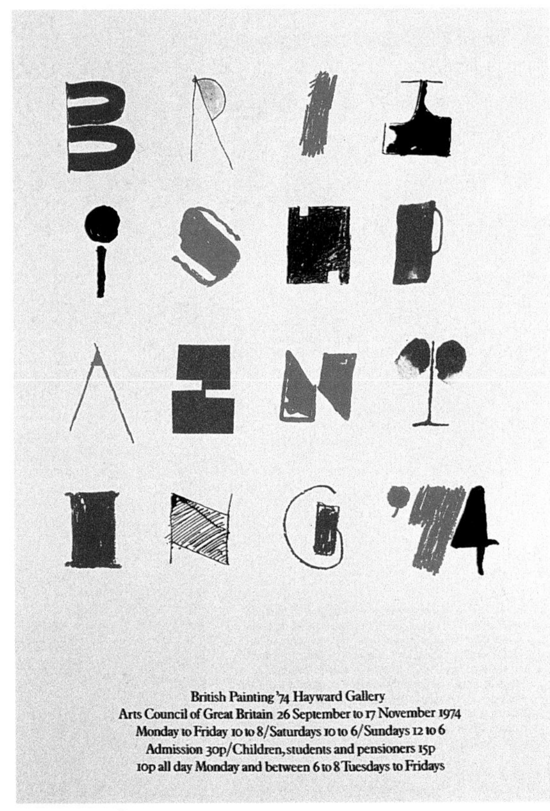


Fletcher 1974 Poster

crafts, photography and education at the Hochschule für Bildende Künste in Berlin and mathematics at the Freie Universität in Berlin. 1964: as photographer and graphic designer, Fleischmann works for clients who include the Nationalgalerie in Berlin, the Deutsche Akademische Austauschdienst (DAAD) and private galleries. After teaching for many years at various high schools in Berlin and after one year as communications planner in Bonn, Fleischmann was made professor in the design department of the Fachhochschule in Bielefeld in 1971, where he still teaches. Has developed book designs and orientation systems (including for the University of Bielefeld and Saxony's state parliament) and exhibitions. Numerous awards from the Stiftung Buchkunst. 1980: conservation of the last printing press from the time of the Irish Rising for the Bunratty Folk Park in Bunratty, Ireland. 1988: organizes the first German DTP competition for "Page" magazine. Co-founder of Forum Typografie. Founder member of the Gesellschaft zur Förderung der Druckkunst in Leipzig. – *Publications:* "Plakate in der irischen Provinz, Irish Country Posters", Bielefeld, Essen 1982; "Bauhaus. Drucksachen, Typographie, Reklame", Düsseldorf 1984; "Das Stabenbuch", Essen 1986.

**Fleischmann,** Johann Michael – b. 1701 in Nuremberg, Germany, d. 1768 in Amsterdam, The Netherlands – *type founder, punch cutter* – 1723–24: trains as a type founder. After years of travel, spends time in Frankfurt am Main in 1727 and

in Frankfurt am Main, 1728 in Amsterdam und Den Haag. 1732 selbständiger Stempelschneider. 1734 selbständiger Schriftgießer. 1735 Verkauf seiner Gießerei an den Amsterdamer Drucker Rudolph Wetstein. Als Stempelschneider arbeitet Fleischmann nun hauptsächlich für die Firma Enschedé in Haarlem.

**Fletcher,** Alan – geb. 27. 9. 1931 in Nairobi, Kenia. – *Grafik-Designer, Typograph* – 1950–51 Studium an der Central School of Arts and Crafts in London. 1952–53 Assistent des Typographen Herbert Spencer. 1953–56 Studium am Royal College of Art in London. 1956 Studium an der Yale University School of Architecture and Design in New Haven, USA. Danach Gestalter, u. a. für Container Corporation of America, IBM und das „Fortune Magazine". 1959–62 Grafiker in London. 1962 Gründung des Studios „Fletcher, Forbes, Gill" mit Colin Forbes und Bob Gill. Auftraggeber waren u. a. Penguin Books und Olivetti. 1965 Austritt von Bob Gill, Eintritt von Theo Crosby in das Studio. Aus dem Studio „Crosby, Fletcher, Forbes" wird 1972 „Pentagram" in London. Auftraggeber u. a. das Victoria & Albert Museum, Olivetti, Kodak, IBM, Lloyd's of London. 1979 Lehrauftrag an der Yale University. 1992 Austritt aus „Pentagram", selbständiger Gestalter, Mitarbeiter der Zeitschrift „Domus" und des „FAZ-Magazins". Berater der „Phaidon Press". 1993 wird er mit dem „Prince Philip Prize for the Designer of the Year" ausgezeichnet. – *Publikationen u.a.:* „Graphic Design: Visual Comparisons" (mit C. Forbes, B. Gill), London 1963; „A Sign

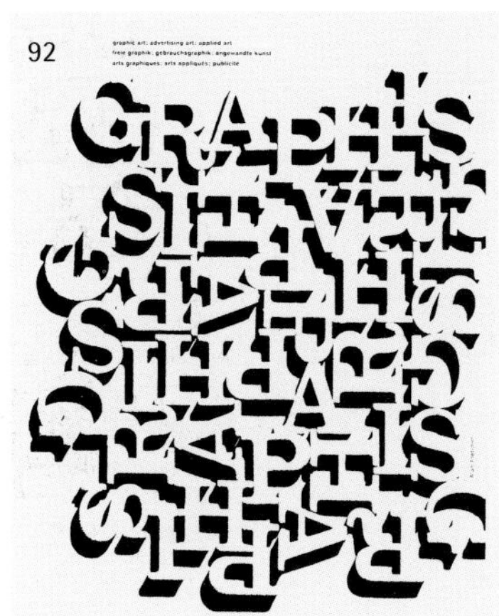


Fletcher 1957 Cover

Fletcher 1992 Cover

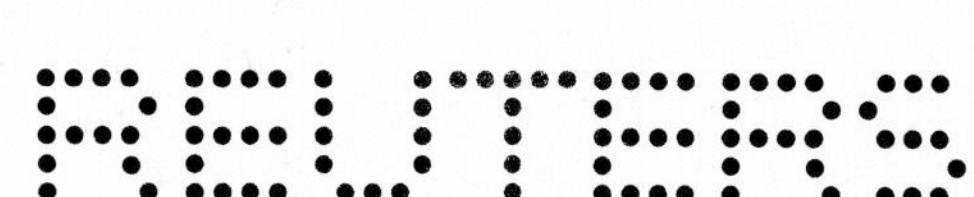


Fletcher 1968 Logo

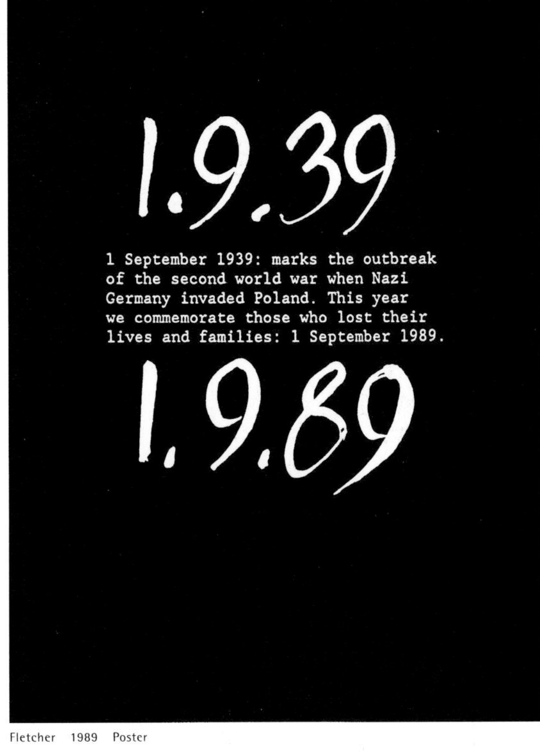


Fletcher 1989 Poster

Fletcher 1989 Signet

s'installe en 1727 à Francfort-sur-le-Main, puis à Amsterdam et à La Haye en 1728. Tailleur de caractères indépendant en 1732, puis fondeur de caractères en 1734. En 1735, il vend sa fonderie à Rudolph Wetstein, un imprimeur d'Amsterdam. Dès lors Fleischmann travaillera principalement comme tailleur de types pour la société Enschedé de Haarlem.

**Fletcher,** Alan – né le 27. 9. 1931 à Nairobi, Kénia – *graphiste maquettiste, typographe* – 1950–1951, études à la Central School of Arts and Crafts de Londres. 1952–1953, assistant du typographe Herbert Spencer. 1953–1956, études au Royal College of Art de Londres. 1956, études à la Yale University School of Architecture and Design à New Haven, Etats-Unis. Puis designer, entre autres pour la Container Corporation of America, IBM et le «Fortune Magazine». 1959–1962, graphiste à Londres. En 1962, il fonde l'agence «Fletcher, Forbes, Gill» avec Colin Forbes et Bob Gill. Commanditaires: Penguin Books Olivetti etc. En 1965, Bob Gill quitte l'agence et est remplacé par Theo Crosby. En 1972, l'agence «Crosby, Fletcher, Forbes» de Londres devient «Pentagram». Commanditaires: Victoria & Albert Museum, Olivetti, Kodak, IBM, Lloyd's of London etc. Chargé de cours à La Yale University en 1979. Il quitte «Pentagram» en 1992 et travaille comme maquettiste indépendant pour des revues telles que «Domus» et le «FAZ-Magazin». Conseiller des «Phaidon Press». En 1993, le prix «Prince Philip Prize for the Designer of the Year» lui est décerné. – *Publications, sélection:* «Graphic Design: Visual Comparisons» (avec C. Forbes, B. Gill), Londres 1963; «A Sign Systems Manual» (avec C. Forbes,

in Amsterdam and The Hague in 1728. 1732: freelance punch cutter. 1734: freelance type founder. 1735: sells his type foundry to the Amsterdam printer Rudolph Wetstein. From then on, Fleischmann works primarily for the Enschedé company in Haarlem as a punch cutter.

**Fletcher,** Alan – b. 27. 9. 1931 in Nairobi, Kenya – *graphic designer, typographer* – 1950–51: studies at the Central School of Arts and Crafts in London. 1952–53: assistant to the typographer Herbert Spencer. 1953–56: studies at the Royal College of Art in London. 1956: studies at the Yale University School of Architecture and Design in New Haven, USA. He then works as a designer for the Container Corporation of America, IBM and "Fortune Magazine", among others. 1959–62: graphic artist in London. 1962: founds the "Fletcher, Forbes, Gill" studio with Colin Forbes and Bob Gill. Clients include Penguin Books and Olivetti. 1965: Bob Gill leaves the studio and Theo Crosby replaces him. 1972: Pentagram evolves out of the Crosby, Fletcher, Forbes studio in London. Clients include the Victoria & Albert Museum, Olivetti, Kodak, IBM and Lloyd's of London. 1979: teaching position at Yale University. 1992: leaves Pentagram and works as a freelance designer on "Domus" and the "FAZ-Magazin". Consultant to Phaidon Press. 1993: awarded the Prince Philip Prize for the Designer of the Year. – *Publications include:* "Graphic Design: Visual Comparisons" (with C. Forbes, B. Gill), London

Systems Manual" (mit C. Forbes, B. Gill), London 1970; „Identity Kits: a pictorial survey of visual arts" (mit G. Facetti), London 1971; „Pentagram: the work of five designers" (mit Pentagram-Partnern), London 1972; „Living by Design" (mit Pentagram-Partnern), London 1978; „Ideas on Design" (mit Pentagram-Partnern), London 1986.

**The Fleuron** – *Zeitschrift* – Nach Vorgesprächen zwischen Holbrook Jackson, Francis Meynell, Bernard Newdigate, Oliver Simon und Stanley Morison in London wird 1922 die Herausgabe einer Zeitschrift beschlossen, die über typographische Themen berichten soll. 1923 erscheint das erste Heft der Zeitschrift „The Fleuron: a Journal of Typography". Die Hefte 1–4 (1923–25) werden von Oliver Simon herausgegeben, von der Curven Press gedruckt und vom Verlag The Fleuron Limited verlegt. Die Hefte 5–7 (1926–30) werden von Stanley Morison herausgegeben, von der Cambridge University Press gedruckt und in England von ihr verlegt, in den USA vom Verlag Doubleday. Die sieben Ausgaben mit insgesamt ca. 1.500 Seiten enthalten Besprechungen typographischer Bücher und Aufsätze zu vielfältigen typographischen Fragen, u. a. von Stanley Morison, A. F. Johnson, Francis Meynell, Bernard Newdigate.

**Flusser,** Vilém – geb. 12. 5. 1920 in Prag, Tschechoslowakei, gest. 27. 11. 1991 in Prag, Tschechien. – *Philosoph, Design- und Medienkritiker, Lehrer, Journalist* – 1939 Beginn des Philosophie-Studiums an der Universität Prag. 1940 Emigration nach London, später nach Brasilien. Fort-

B. Gill), Londres 1970; «Identity Kits: a pictural survey of visual arts» (avec G. Facetti), Londres 1971; «Pentagram: the work of five designers» (avec les associés de Pentagram), Londres 1972; «Living by Design» (avec les associés de Pentagram), Londres 1978; «Ideas on Design» (avec les associés de Pentagram), Londres 1986.

**The Fleuron** – *revue* – en 1922, Holbrook Jackson, Francis Meynell, Bernard Newdigate, Oliver Simon et Stanley Morison se réunissent à Londres et décident de publier une revue traitant de typographie. Le premier exemplaire de la revue «The Fleuron: a journal of typography» paraît en 1923. De 1923 à 1925, les numéros 1–4 sont dirigés par Oliver Simon, imprimées aux Curven Press et publiées par les éditions The Fleuron Limited. Les numéros 5–7 (1926–1930) sont dirigés par Stanley Morison, imprimés par la Cambridge University Press qui l'édite en Angleterre, tandis que les éditions Doubleday la diffuse aux Etats-Unis. Les sept éditions, soit environ 1500 pages, contenaient des critiques de livres sur la typographie et des articles de Stanley Morison, A. F. Johnson, Francis Meynell, Bernard Newdigate etc. traitant de divers problèmes typographiques.

**Flusser,** Vilém – né le 12. 5. 1920 à Prague, Tchécoslovaquie, décédé le 27. 11. 1991 à Prague, Tchéquie – *philosophe, critique de design et des médias, enseignant, journaliste* – commence ses études de philosophie à l'université de Prague en 1939. Emigre à Londres en 1940 puis au Brésil où il poursuit ses études. 1957, premières publications sur des thèmes portant sur la philosophie du langage. Enseigne l'épistémologie en 1959. Professeur de philosophie de la communication et membre du conseil de la Fundaçao Bienal das Artes en 1963. A partir de 1966, il fait de nombreux cycles de conférences

THE FLEURON
A JOURNAL OF TYPOGRAPHY
EDITED
BY
STANLEY MORISON

Cambridge — New York
THE UNIVERSITY PRESS — DOUBLEDAY PAGE & CO
1926

The Fleuron 1926 Cover

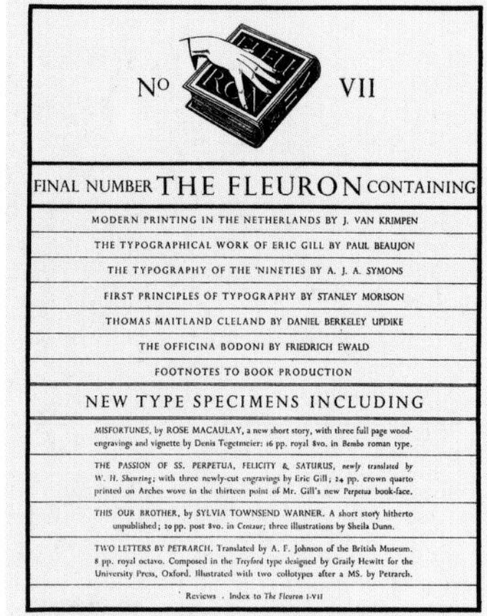
Nº VII

FINAL NUMBER THE FLEURON CONTAINING

MODERN PRINTING IN THE NETHERLANDS BY J. VAN KRIMPEN
THE TYPOGRAPHICAL WORK OF ERIC GILL BY PAUL BEAUJON
THE TYPOGRAPHY OF THE 'NINETIES BY A. J. A. SYMONS
FIRST PRINCIPLES OF TYPOGRAPHY BY STANLEY MORISON
THOMAS MAITLAND CLELAND BY DANIEL BERKELEY UPDIKE
THE OFFICINA BODONI BY FRIEDRICH EWALD
FOOTNOTES TO BOOK PRODUCTION

NEW TYPE SPECIMENS INCLUDING

MISFORTUNES, by ROSE MACAULAY, a new short story, with three full page wood-engravings and vignette by Denis Tegetmeier: 16 pp. royal 8vo. in Bembo roman type.

THE PASSION OF SS. PERPETUA, FELICITY & SATURUS, *newly translated by W. H. Shewring*; with three newly-cut engravings by Eric Gill; 24 pp. crown quarto printed on Arches wove in the thirteen point of Mr. Gill's new *Perpetua* book-face.

THIS OUR BROTHER, by SYLVIA TOWNSEND WARNER. A short story hitherto unpublished; 20 pp. post 8vo. in *Centaur*; three illustrations by Sheila Dunn.

TWO LETTERS BY PETRARCH. Translated by A. F. Johnson of the British Museum. 8 pp. royal octavo. Composed in the *Treyford* type designed by Graily Hewitt for the University Press, Oxford. Illustrated with two collotypes after a MS. by Petrarch.

Reviews . Index to *The Fleuron* I-VII

CAMBRIDGE UNIVERSITY PRESS · 21s. NET

The Fleuron 1930 Cover

Flusser 1987 Cover

1963; "A Sign Systems Manual" (with C. Forbes, B. Gill), London 1970; "Identity Kits: a pictorial survey of visual arts" (with G. Facetti), London 1971; "Pentagram: the work of five designers" (with partners at Pentagram), London 1972; "Living by Design" (with partners at Pentagram), London 1978; "Ideas on Design" (with partners at Pentagram), London 1986.

**The Fleuron** – *magazine* – 1922: after preliminary discussions in London, Holbrook Jackson, Francis Meynell, Bernard Newdigate, Oliver Simon and Stanley Morison decide to launch a magazine dealing with typographical themes. 1923: the first issue of the magazine "The Fleuron: a journal of typography" is published. 1923–25: nos. 1–4 of the magazine are edited by Oliver Simon, printed by Curven Press and published by The Fleuron Limited. Nos. 5–7 (1926–30) are edited by Stanley Morison, printed by Cambridge University Press and published in England by CUP and in the USA by Doubleday. The seven issues with a total of app. 1,500 pages contain discussions on typographical books and essays on a wealth of typographical themes by Stanley Morison, A. F. Johnson, Francis Meynell and Bernard Newdigate, among others.

**Flusser,** Vilém – b. 12. 5. 1920 in Prague, Czechoslovakia, d. 27. 11. 1991 in Prague, Czech Republic – *philosopher, design and media critic, teacher, journalist* – 1939: starts studying philosophy at the University of Prague. 1940: emigrates to London, then to Brazil. Continues his

setzung seiner Studien. 1957 erste Publikationen über sprachphilosophische Themen. 1959 Dozent für Wissenschaftsphilosophie. 1963 Professor für Kommunikationsphilosophie und Ratsmitglied der Fundaçao Bienal das Artes. Ab 1966 zahlreiche Vortragsreisen, Gastprofessor an Universitäten Europas und Nordamerikas. 1972 Umzug nach Italien, später nach Südfrankreich. In seiner Arbeit „Die Schrift" legt er dar, daß unser Begriff der fortschreitenden Historie in der linearen Aufzeichnung der Schrift begründet ist. Das Verlassen dieser Schriftlinie zugunsten der Punktwelt des Computers bezeichnet für Flusser das Ende der Geschichte. – *Publikationen u.a.:* „Lingua e Realidade", São Paulo 1963; „La Force du Quotidien", Paris 1972; „Le Monde codifié", Paris 1972; „Für eine Philosophie der Fotografie", Göttingen 1983; „Die Schrift", Göttingen 1987; „Gesammelte Schriften 1–14", Mannheim ab 1993.

**Fonderie Typographique Française** (FTF) Paris – *Schriftgießerei* – Durch den Zusammenschluß der Schriftgießereien Turlot, Renault, Allainguillaume, Berthier & Durey und Huart entstand 1921 die FTF. Das Unternehmen wird von E. Marcou, dem ehemaligen Besitzer der Gießerei Renault, geleitet. 1969 Verlegung der Firma nach Champigny-sur-Marne, Umbenennung der Firma in Société Nouvelle de la Fonderie Typographique Française. 1974 Verkauf an die Fundición Tipográfica Neufville, Barcelona.

**FontShop** – *Schriftenversandhaus* – 1989 Gründung des Versandhauses für digitalisierte Schriften durch Erik und Joan

FontShop 1993 Advertisement

FontShop 1989 Cover

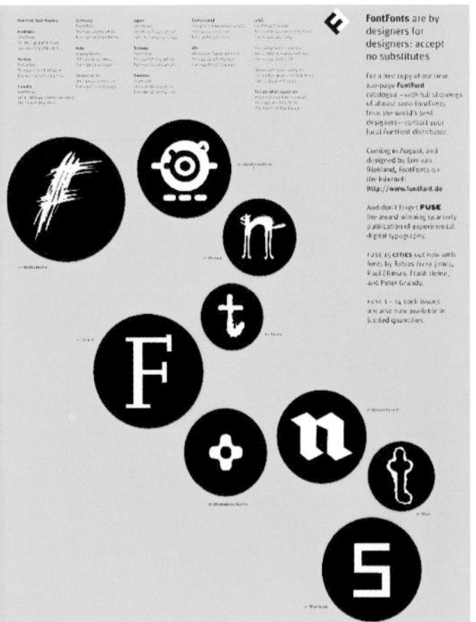


FontShop 1996 Advertisement

dans plusieurs universités européennes et d'Amérique du Nord. S'installe en Italie en 1972, puis dans le Sud de la France. Dans son essai «Die Schrift» (L'Ecriture), il démontre comment notre compréhension du déroulement historique se fonde sur la forme linéaire de l'écriture. Pour Flusser, l'abandon de la ligne écrite au profit de l'univers des points de l'ordinateur marque la fin de l'Histoire. – *Publications, sélection :* «Lingua e Realidade», São Paulo 1963; «La Force du Quotidien», Paris 1972; «Le Monde codifié», Paris 1972; «Für eine Philosophie der Fotografie», Göttingen 1983; «Die Schrift», Göttingen 1987; «Gesammelte Schriften 1–14», Mannheim, à partir de 1993.

**Fonderie Typographique Française** (FTF) Paris – *fonderie de caractères* – fondée en 1921 à Paris à la suite de la fusion des fonderies de caractères Turlot, Renault, Allainguillaume, Berthier & Durey et Huart. L'entreprise a été dirigée par E. Marcou, l'ancien propriétaire de la fonderie Renault. En 1969, la société s'installe à Champigny-sur-Marne où elle prend le nom de «Société Nouvelle de la Fonderie Typographique Française». Elle est cédée à la Fundición Tipográfica Neufville, Barcelona en 1974.

**FontShop** – *société de vente de polices par correspondance* – en 1989, Erik et Joan Spiekermann fondent cette société de vente de polices numériques par correspondance à Berlin. FontShop Internatio-

studies. 1957: produces his first publications on the philosophy of language. 1959: lecturer of philosophy of science. 1963: professor of philosophy of communication and member of the advisory council of the Fundaçao Bienal das Artes. From 1966 onwards: numerous lecture tours. Visiting professor at universities in Europe and North America. 1972: moves to Italy and later to the South of France. In his paper "Die Schrift" he explains that our concept of progressive history is substantiated in the linear notation of writing. Abandoning this written line in favor of the computer's world of dots marks the end of history for Flusser. – *Publications include:* "Lingua e Realidade", São Paolo 1963; "La Force du Quotidien", Paris 1972; "Le Monde codifié", Paris 1972; "Für eine Philosophie der Fotografie", Göttingen 1983; "Die Schrift", Göttingen 1987; "Gesammelte Schriften 1–14", Mannheim from 1993 onwards.

**Fonderie Typographique Française** (FTF) Paris – *type foundry* – 1921: the FTF is born as the result of a merger of the Turlot, Renault, Allainguillaume, Berthier & Durey and Huart type foundries. The company is run by E. Marcou, the former owner of the Renault foundry. 1969: the company moves to Champigny-sur-Marne and is renamed the Société Nouvelle de la Fonderie Typographique Française. 1974: the foundry is sold to Fundición Tipográfica Neufville, Barcelona.

**FontShop** – *font mail order firm* – 1989: a mail order firm for digitized typefaces, FontShop International (FSI), is founded

FontShop
**Forsberg**
**Forum Typografie**

Spiekermann in Berlin. Der FontShop International (FSI) verfügt über Filialen in Europa, Amerika und Australien. Er versendet über 30.000 Schriften von ca. 50 Herstellern sowie osteuropäische, kyrillische und asiatische Zeichensätze, Logos, Symbole und Bildzeichen. Die FontShop-Bibliothek mit eigenen Schriften, die 1990 gegründet wird, umfaßt 750 Fonts. Es werden Alphabete mit neuen Techniken, wie dem Random-Effekt bei FF Beowolf und FF Beo sans, herausgegeben. 1991 wird die erste Ausgabe der digitalisierten Publikation „Fuse" veröffentlicht. Veranstalter der „Fuse"-Konferenzen in London (1994) und in Berlin (1995).

**Forsberg,** Karl Erik – geb. 1914 in Munsö, Schweden. – *Schriftentwerfer, Typograph* – Ausbildung an der Schule für das Buchhandwerk in Schweden. 1938–41 Studium an der Akademie für Graphische Künste und Buchgewerbe in Leipzig. 1942–49 Gestalter im Verlag Almqvist & Wiksell, der Druckerei der Universität von Uppsala. Seit 1950 Gestalter im Verlag P. A. Norstedt & Söner. Neben buchgestalterischen Arbeiten entstehen Zeitschriften, Briefmarkenentwürfe und Briefmarken. 1954 erscheint eine Bibel-Ausgabe mit Reproduktionen von Zeichnungen, Radierungen und Gemälden Rembrandts, die Forsberg gestaltet und die in seiner Schrift Berling gesetzt ist. 1983 Ehrendoktor der Universität Uppsala. – *Schriftentwürfe:* Parad (1936), Lunda (1938), Berling (1951–58), Carolus (1954), Ericus (o.J.).

**Forum Typografie** – 1984 Gründung des Forums Typografie in Berlin, um Schrift und Typographie ins Bewußtsein der Öf-

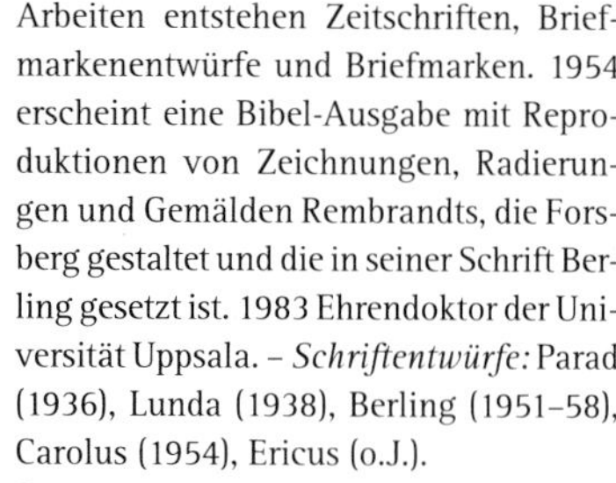

nal (FSI) a des filiales en Europe, en Amérique et en Australie. Elle diffuse plus de 30 000 polices produites par environ 50 sociétés ainsi que des caractères est-européens, cyrilliques et asiatiques, des logos, symboles et pictogrammes. La FontShop-Bibliothek, qui possède ses propres polices, est fondée en 1990, et réunit 750 fontes. Elle diffuse des alphabets produits grâce aux nouvelles technologies, comme les effets random pour la FF Beowolf et la FF Beo sans. 1991, première édition de la publication numérique «Fuse». Organise les «Fuse»-conférences à Londres (1994) et à Berlin (1995).

**Forsberg,** Karl Erik – né en 1914 à Munsö, Suède – *concepteur de polices, typographe* – formation à l'école des arts du livre en Suède. 1938-1941, études à l'Akademie für Graphische Künste und Buchgewerbe (arts graphiques et du livre) de Leipzig. 1942- 1949, maquettiste aux éditions Almqvist & Wiksell de l'imprimerie de l'université d'Uppsala. Maquettiste pour les éditions P. A. Norstedt & Söner à partir de 1950. Outre les maquettes de livres, il crée aussi des maquettes de revues et dessine des timbres. En 1954, il réalise la maquette d'une édition de la Bible comprenant des reproductions de dessins, estampes et tableaux de Rembrandt qui est composée en Berling, une police dessinée par Forsberg. 1983, docteur honoris causa de l'université d'Uppsala. – *Polices:* Parad (1936), Lunda (1938), Berling (1951-1958), Carolus (1954), Ericus (sans date).

**Forum Typografie** – le «Forum Typografie» a été fondé à Berlin en 1984 pour attirer l'attention du public sur l'écriture et la typographie et lui faire comprendre son utilité en matière de communication. Il organise des rencontres annuelles autour de thèmes variés qui font ensuite l'objet de publications.

| | | | | | |
|---|---|---|---|---|---|
| | n | n | | | n |
| a | b | e | f | g | i |
| o | r | s | t | y | z |
| A | B | C | E | G | H |
| M | O | R | S | X | Y |
| 1 | 2 | 4 | 6 | 8 | & |

Forsberg 1951–58 Berling

Forum Typografie 1995 Cover

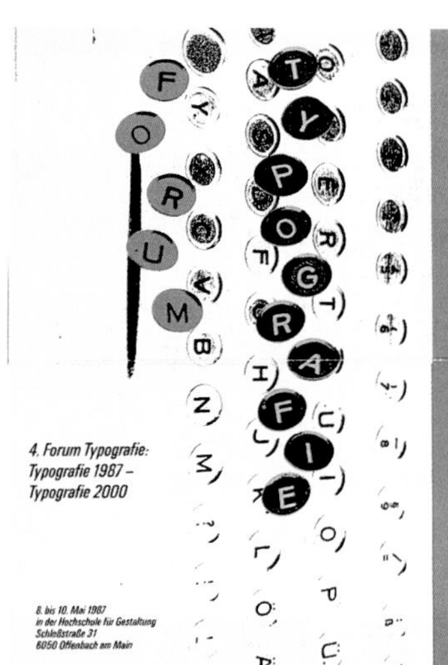

4. Forum Typografie:
Typografie 1987 –
Typografie 2000
8. bis 10. Mai 1987
in der Hochschule für Gestaltung
Schloßstraße 31
6050 Offenbach am Main


Forum Typografie 1987 Poster

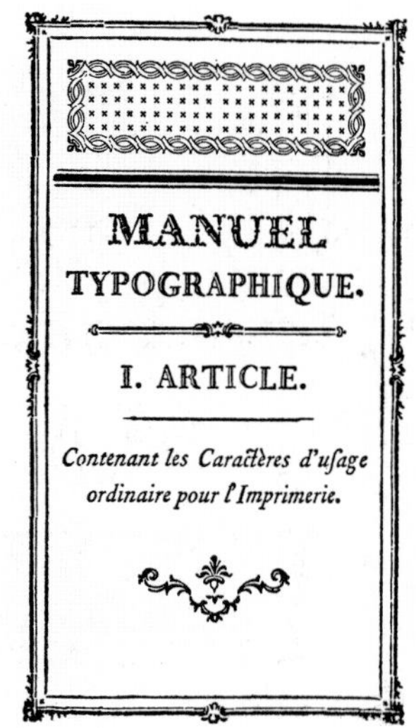

MANUEL
TYPOGRAPHIQUE.
I. ARTICLE.
*Contenant les Caractères d'ufage ordinaire pour l'Imprimerie.*

Fournier 1766 Page

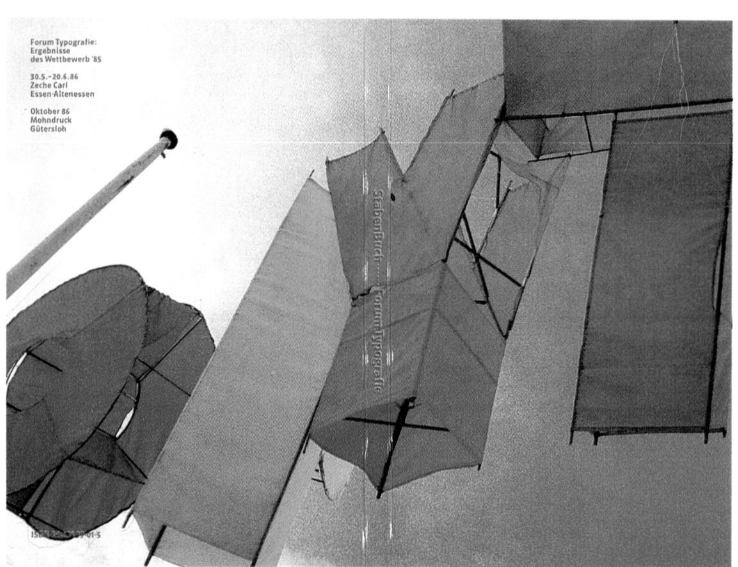

Forum Typografie 1986 Cover

N°. LXXVIII. 76
GROS-CANON.
*L'homme eft toû-jours la dupe des plaifirs.*

Fournier 1766 Page

by Erik and Joan Spiekermann in Berlin, with branches in Europe, America and Australia. Has over 30,000 fonts from c. 50 manufacturers, and Eastern European, Cyrillic and Asian alphabets, logos, symbols and icons. FontShop's library of its own fonts, opened in 1990, has c. 750 typefaces. FontShop has produced alphabets using new techniques, such as the random effect with FF Beowolf and FF Beo sans. 1991: the first number of the digitized publication "Fuse" is issued. FontShop has organized "Fuse" conferences in London (1994) and Berlin (1995).

**Forsberg,** Karl Erik – b. 1914 in Munsö, Sweden – *type designer, typographer* – Trained at the School for Book Crafts in Sweden. 1938–41: studies at the Akademie für Graphische Künste und Buchgewerbe in Leipzig. 1942–49: designer for the publishers Almqvist & Wiksell, the University of Uppsala press. From 1950 onwards: designer for the publishers P. A. Norstedt & Söner. Besides book designs, Forsberg has designed stamps and produced magazines and stamps. 1954: a bible is published with reproductions of drawings, etchings and paintings by Rembrandt, designed by Forsberg and set in his Berling typeface. 1983: is awarded an honorary doctorate by the University of Uppsala. – *Fonts:* Parad (1936), Lunda (1938), Berling (1951–58), Carolus (1954), Ericus (undated).

**Forum Typografie** – 1984: Forum Typografie is founded in Berlin with the aim of introducing type and typography to the public and stressing its usefulness for various means of communication. Year-

fentlichkeit zu tragen und ihren Nutzen für die Kommunikation deutlich zu machen. Jährliche Bundestreffen mit wechselnden Themen, die umfangreich dokumentiert werden.

**Fournier,** Pierre Simon – geb. 15. 9. 1712 in Paris, Frankreich, gest. 8. 10. 1768 in Paris, Frankreich. – *Schriftgießer, Stempelschneider, Schriftentwerfer* – Genannt Fournier le Jeune. Lehre im Betrieb seines Vaters Jean Claude Fournier. Entwickelt 1737 ein typographisches Maßsystem, das von F. A. Didot überarbeitet wird. 1739 Eröffnung seiner eigenen Schriftgießerei. Veröffentlicht 1742 ein Schriftmusterbuch, das von J. J. Barbou gedruckt wird. Insgesamt schneidet Fournier 60.000 Stempel für ca. 150 eigene Alphabete. Beendet 1760 den Entwurf einer Notenschrift, die 1762 patentiert wird und die 1764 in mehreren Größen erscheint. Nach Fourniers Tod 1768 führen seine Witwe und sein Sohn die Schriftgießerei weiter. – *Publikationen u. a.:* „Essai d'un nouveau caractère de fonte pour l'impression de la musique", Paris 1756; „Manuel typographique" (2 Bände), Paris 1764–66.

**Fox,** Mark – geb. 7. 9. 1961 in Covina, USA. – *Grafik-Designer, Lehrer* – 1984 Abschluß seiner Studien an der University of California, Los Angeles, danach freischaffender Gestalter. 1986 Gründung des Ateliers „Black Dog", Auftraggeber waren u. a. Apple Computers, Levi Strauss, Elektra Records, „Esquire", „The Washington Post", Warner Bros. Es entstehen komplette Erscheinungsbilder für Restaurants. Unterrichtet 1993 Grafik-Design am California College of Arts and

MANUEL
TYPOGRAPHIQUE.

PREMIÈRE PARTIE.

LA GRAVURE,
OU TAILLE DES POINÇONS.

POUR être un bon Graveur de Caractères, il faut être Typographe, c'eſt-à-dire, ſavoir tous les détails du méchaniſme de la Fonderie & de l'Imprimerie, afin d'y aſſujétir ſon travail. Maître de l'art, le Graveur doit tout prévoir dans la fonte & dans l'impreſſion. C'eſt par-là que les Simon de Colines, les Garamond, les

Fournier 1766 Page

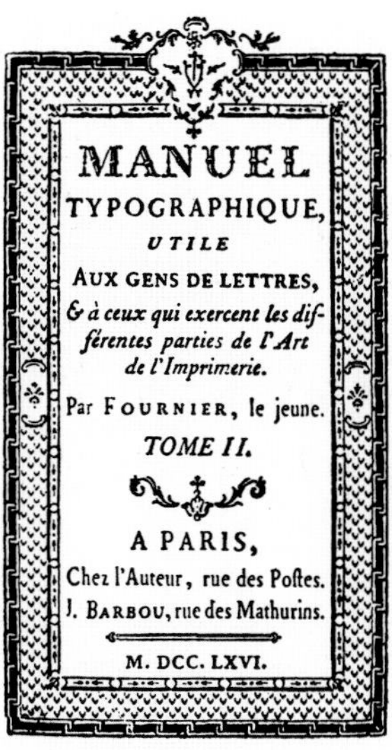

MANUEL
TYPOGRAPHIQUE,
UTILE
AUX GENS DE LETTRES,
& à ceux qui exercent les différentes parties de l'Art de l'Imprimerie.
Par FOURNIER, le jeune.
TOME II.
A PARIS,
Chez l'Auteur, rue des Poſtes.
J. BARBOU, rue des Mathurins.
M. DCC. LXVI.

Fournier 1766 Cover

Fox 1991 Poster

**Fournier,** Pierre Simon – né le 15. 9. 1712 à Paris, France, décédé le 8. 10. 1768 à Paris, France – *fondeur de caractères, tailleur de types, concepteur de polices* – appelé Fournier le Jeune. Apprentissage dans l'entreprise de son père Jean Claude Fournier. En 1737, il invente un système de mesure typographique qui sera remanié par F. A. Didot. Ouvre sa propre fonderie de caractères en 1739. En 1742, il publie un catalogue de fontes qui sera imprimé par J. J. Barbou. Au total, Fournier a taillé 60 000 types pour environ 150 alphabets qu'il a dessinés lui-même. En 1760, il achève la conception de notes de musique, brevetées en 1762, et publiées en plusieurs corps en 1764. En 1768, à la mort de Fournier, sa veuve et son fils reprennent la fonderie de caractères. – *Publications, sélection:* «Essai d'un nouveau caractère de fonte pour l'impression de la musique», Paris 1756; «Manuel typographique» (2 vol.), Paris 1764–1766.

**Fox,** Mark – né le 7. 9. 1961 à Covina, Etats-Unis – *graphiste maquettiste, enseignant* – termine ses études à la University of California de Los Angeles, en 1984, puis exerce comme graphiste indépendant. En 1986, il fonde l'atelier «Black Dog». Commanditaires: Apple Computers, Levi Strauss, Elektra Records, «Esquire», «The Washington Post», Warner Bros, etc. Réalisation d'identités complètes pour des restaurants. Enseigne les arts graphiques et le design au California College of Arts and Crafts de San Francisco en 1993. Elu vice-président de l'AIGA de San Francisco en 1994. – *Pub-*

ly national meetings take place on varying themes, all comprehensively documented.

**Fournier,** Pierre Simon – b. 15. 9. 1712 in Paris, France, d. 8. 10. 1768 in Paris, France – *type founder, punch cutter, type designer* – Known as Fournier le Jeune. Trains at the company of his father, Jean Claude Fournier. 1737: develops a typographical system of measures which F. A. Didot reworks. 1739: opens his own type foundry. 1742: publishes a book of type specimens which is printed by J. J. Barbou. In total, Fournier cuts 60,000 punches for c. 150 of his own alphabets. 1760: completes his design for a system of musical notation which is patented in 1762 and which is produced in various sizes in 1764. 1768: Fournier's widow and son continue to run the type foundry after Fournier's death. – *Publications include:* "Essai d'un nouveau caractère de fonte pour l'impression de la musique", Paris 1756; "Manuel typographique" (2 vols.), Paris 1764–66.

**Fox,** Mark – b. 7. 9. 1961 in Covina, USA – *graphic designer, teacher* – 1984: completes his studies at the University of California in Los Angeles and starts work as a freelance designer. 1986: opens the Black Dog studio. Clients include Apple Computers, Levi Strauss, Elektra Records, "Esquire", "The Washington Post" and Warner Bros. He produces complete corporate imagery packages for restaurants. 1993: teaches graphic design at the California College of Arts and Crafts in San Francisco. 1994: elected vice-president of

Crafts, San Francisco. 1994 Wahl zum Vize-Präsidenten der AIGA San Francisco. – *Publikation:* „The New American Logo" (Co-Autor), New York 1994.

**Franke,** Karl – geb. 28. 1. 1894 in Silberberg, Deutschland, gest. 18. 10. 1952 in Stuttgart, Deutschland. – *Typograph, Autor* – Schriftsetzer-Lehre in Berlin. Teilnahme an den Kursen der Berliner Kunstgewerbe- und Handwerkerschule. Teilnahme im Bildungsverband der Deutschen Buchdrucker in Berlin, Mitglied des Hauptvorstands. Leitung der Skizzier- und Entwurfskurse für Fortgeschrittene. 1947 Arbeiten für die Zeitschrift „Welt der Frau". Seit 1948 Gestaltungsarbeiten für die Büchergilde Gutenberg. 1950 Mitbegründer und 2. Vorsitzender des „Deutschen Typokreises e. V." in Frankfurt am Main. 1950–52 Gestalter und Fachautor für die Zeitschrift „Form und Technik". – *Publikation u. a.:* Willi Mengel u. a. „Karl Franke. Eine Würdigung seines Schaffens", Frankfurt am Main 1957.

**Frej,** David – geb. 10. 1. 1959 in Maquoketa, Iowa, USA. – *Grafik-Designer* – 1977–82 Studium der Malerei und Fotografie an der University of Illinois in Champaign, Urbana. 1982–85 Arbeit als freier Künstler. 1985–87 Studium an der Cranbrook Academy of Art, Michigan. Nach dem Studium sechs Monate Arbeit im Studio Dumbar in Den Haag. 1988 Gründung seiner Grafik-Design-Firma „Influx Co" in Chicago. Auftraggeber aus den Bereichen Finanzen, Computerprodukte, Industrial Design und Einzelhandel. Zwei Jahre unabhängiger Börsenmakler und Warenhändler. 1994 Eintritt

*lication:* «The New American Logo» (coauteur), New York 1994.

**Franke,** Karl – né le 28. 1. 1894 à Silberberg, Allemagne, décédé le 18. 10. 1952 à Stuttgart, Allemagne – *typographe, auteur* – apprentissage de compositeur en typographieà Berlin. Fréquente les cours de la Berliner Kunstgewerbe- und Handwerkerschule (école de l'artisanat d'art et des arts décoratifs). Fait partie du Bildungsverband der Deutschen Buchdrukker (institut de formation de l'imprimerie allemande) à Berlin, dont il devient membre du comité directeur. Responsable de cours d'esquisse et de dessin pour étudiants du second cycle. 1947, travaux pour la revue «Welt der Frau». Maquettes pour la Büchergilde Gutenberg à partir de 1948. En 1950, il est membre fondateur et vice-président du «Deutscher Typokreis e.V.» à Francfort-sur-le-Main. 1950–1952, maquettiste et rédacteur pour la revue «Form und Technik». – *Publication, sélection:* Willy Mengel et autres «Karl Franke. Eine Würdigung seines Schaffens», Francfort-sur-le-Main 1957.

**Frej,** David – né le 10. 1. 1959 à Maquoketa, Iowa, Etats-Unis – *graphiste maquettiste* – 1977–1982, études de peinture et de photographie à l'University of Illinois à Champaign, Urbana. Exerce comme artiste indépendant de 1982 à 1985. Etudes à la Cranbrook Academy of Art, Michigan de 1985 à 1987. Travaille ensuite pendant six mois à l'atelier „Dumbar" à la Haye. Fonde sa société de conception graphiste et de design, „Influx Co", en 1988, à Chicago. Commanditaires dans les secteurs de la finance, de l'informatique, du design industriel et du commerce de détail. Exerce pendant deux ans comme courtier en bourse et commerçant. En 1994, il entre au sevice de Marketing de la société „Lerner Associates Ltd" de sa femme Nancy Lerner. Enseigne la com-

Frej ca. 1990 Poster

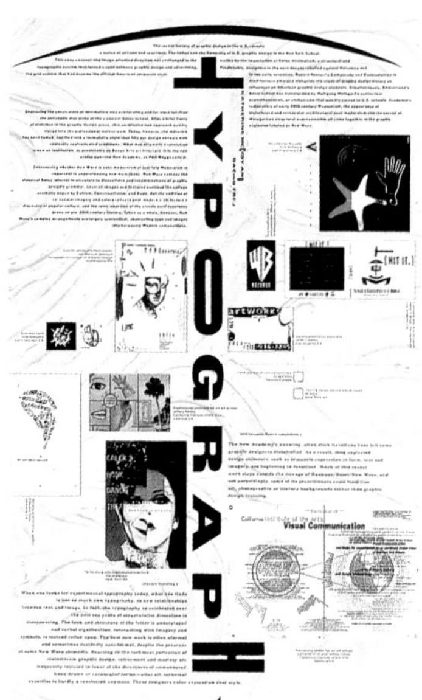

Frej 1988 Page

Franke 1952 Logo

Friedl 1972 Poster

Friedl 1993

Friedl 1982 Advertising

the AIGA San Francisco. – *Publication:* "The New American Logo" (co-author), New York 1994.

**Franke,** Karl – b. 28. 1. 1894 in Silberberg, Germany, d. 18. 10. 1852 in Stuttgart, Germany – *typographer, author* – Trained as a typesetter in Berlin. Attended courses at the Berliner Kunstgewerbe- und Handwerkerschule. Active member of the Bildungsverband der Deutschen Buchdrucker in Berlin, member of the board. Head of the advanced sketching and design courses. 1947: produces work for "Welt der Frau" magazine. From 1948 onwards: designs for the Büchergilde Gutenberg. 1950: co-founder and vice-chairman of the Deutscher Typokreis e.V. in Frankfurt am Main. 1950–52: designer and technical author for "Form und Technik" magazine. – *Publication:* Willi Mengel et al "Karl Franke. Eine Würdigung seines Schaffens", Frankfurt am Main 1957.

**Frej,** David – b. 10.1.1959 in Maquoketa, Iowa, USA – *graphic designer* – 1977–82: studies painting and photography at the University of Illinois in Champaign, Urbana. 1982–85: freelance artist. 1985–87: studies at the Cranbrook Academy of Art, Michigan. After completing his studies, works for six months at the Dumbar studio in The Hague. 1988: launches his graphic design company, Influx Co, in Chicago. Clients are from the fields of finance, computer products, industrial design and retail trade. Works for two years as an independent stockbroker and commodity dealer. 1994: joins Lerner Associates Ltd. marketing services company

in die Marketing-Services-Firma Lerner Associates Ltd. seiner Frau Nancy Lerner. Unterrichtet Visuelle Kommunikation am Art Institute of Chicago.

**Friedl,** Friedrich – geb. 7. 9. 1944 in Fulnek, Sudetenland, Tschechoslowakei. – *Gestalter, Lehrer* – 1961–64 Schriftsetzerlehre und Schriftsetzer in Stuttgart. 1968–72 Grafik-Design-Studium an der Werkkunstschule Darmstadt. 1972–82 Dozent für Typographie an der Fachhochschule Darmstadt. 1982–83 Professor für Typographie an der Fachhochschule Hildesheim. Seit 1983 an der Hochschule für Gestaltung in Offenbach am Main. Herausgeber der Mappenserie „Alphabetisches" (Klaus Basset, 1985; Franz Mon, 1989; Alfons Holtgreve, 1993; Herbert Pfeiffer, 1996). Seit 1964 zahlreiche Ausstellungen freier und angewandter Arbeiten. Veranstaltet Ausstellungen, u. a. Felix Beltran (1973), „Das gewöhnliche Design" (1977), Odermatt & Tissi (1984), Hans Hillmann (1985), 8vo, London (1989), why not associates, London (1991), Herbert Heckmann (1995). 1993 zeigt das Klingspor-Museum Offenbach die Ausstellung „Zehn Jahre Typographie-Ausbildung an der HfG, Offenbach, Prof. F. Friedl". – *Publikationen u.a.:* „Das gewöhnliche Design", Köln 1979; „Thesen zur Typografie" (3 Bände), Eschborn 1985–89; „Typographische Mitteilungen. Sonderheft Elementare Typographie" (Reprint), Mainz 1986; „Walter Dexel. Neue Reklame", Düsseldorf 1987. Artikel in Zeitschriften wie „Graphis", „Octavo", „U&lc", „Eye", „Novum", „Page", „Form".

**Friedlaender,** Henri – geb. 15. 3. 1904 in

munication visuelle à l'Art Institute de Chicago.

**Friedl,** Friedrich – né le 7. 9. 1944 à Fulnek, Sudètes, Tchécoslovaquie – *maquettiste, enseignant* – apprentissage de la composition typographique de 1961 à 1964 et travaux de composition à Stuttgart. 1968–1972, études d'arts graphiques à la Werkkunstschule (école des arts appliqués) de Darmstadt. 1972–1982, enseigne la typographie à la Fachhochschule (institut universitaire technique) de Darmstadt. 1982–1983, professeur de typographie à la Fachhochschule de Hildesheim; puis à la Hochschule für Gestaltung (école supérieure de design) d'Offenbach-sur-le-Main à partir de 1983. Dirige la série de port-folios «Alphabetisches» (Klaus Basset, 1985; Franz Mon, 1989; Alfons Holtgreve, 1993; Herbert Pfeiffer, 1996). A partir de 1964, nombreuses expositions de travaux libres et d'applications, entre autres Felix Beltran (1973), «Das gewöhnliche Design» (1977), Odermatt & Tissi (1984), Hans Hillmann (1985), 8vo, Londres (1989), why not associates, Londres (1991), Herbert Heckmann (1995). En 1993, le Klingspor-Museum d'Offenbach organise l'exposition «Zehn Jahre Typographie-Ausbildung an der HfG Offenbach, Prof. Friedrich Friedl» (dix ans d'enseignement de la typographie à l'école supérieure de design d'Offenbach, prof. Friedrich Friedl). – *Publications, sélection:* «Das gewöhnliche Design», Cologne 1979; «Thesen zur Typografie» (3 vol.), Eschborn 1985–1989; «Typographische Mitteilungen. Sonderheft Elementare Typographie» (reédition), Mayence 1986; «Walter Dexel. Neue Reklame», Düsseldorf 1987; ainsi que des articles dans des revues telles que «Graphis», «Octavo», «U&lc», «Eye», «Novum», «Page», «form».

**Friedlaender,** Henri – né le 15. 3. 1904 à Lyon, France, décédé le 15. 11. 1996 à

Friedlaender Calendar

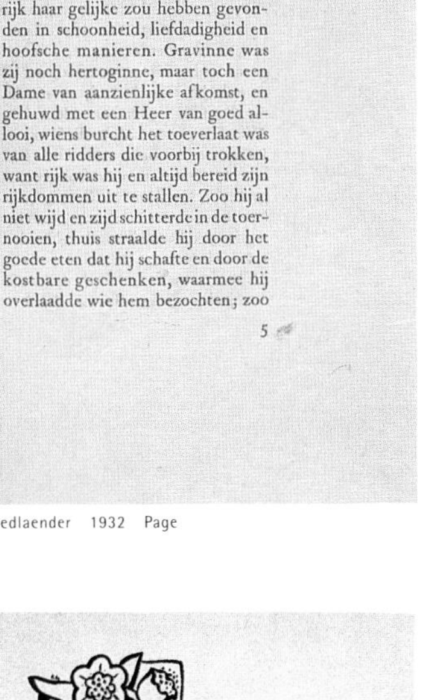

DAAR LEEFDE EENS EEN Edelvrouwe, die binnen geen enkel rijk haar gelijke zou hebben gevonden in schoonheid, liefdadigheid en hoofsche manieren. Gravinne was zij noch hertoginne, maar toch een Dame van aanzienlijke afkomst, en gehuwd met een Heer van goed allooi, wiens burcht het toeverlaat was van alle ridders die voorbij trokken, want rijk was hij en altijd bereid zijn rijkdommen uit te stallen. Zoo hij al niet wijd en zijd schitterde in de toernooien, thuis straalde hij door het goede eten dat hij schafte en door de kostbare geschenken, waarmee hij overlaadde wie hem bezochten; zoo

5

Friedlaender 1932 Page

HEINRICH MANN
DER HASS
DEUTSCHE ZEITGESCHICHTE
QUERIDO VERLAG·AMSTERDAM


Friedlaender 1933 Cover

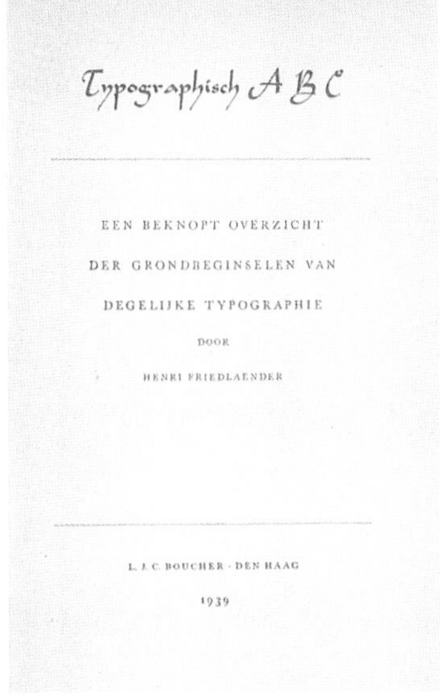


Friedlaender 1939 Cover

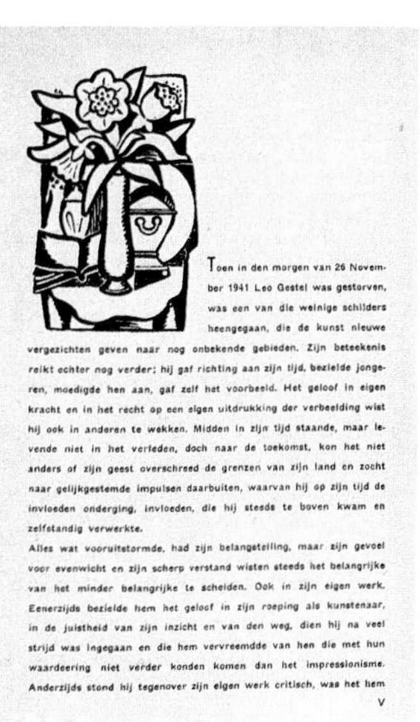

Toen in den morgen van 26 November 1941 Leo Gestel was gestorven, was een van die weinige schilders heengegaan, die de kunst nieuwe vergezichten geven naar nog onbekende gebieden. Zijn beteekenis reikt echter nog verder; hij gaf richting aan zijn tijd, bezielde jongeren, moedigde hen aan, gaf zelf het voorbeeld. Het geloof in eigen kracht en in het recht op een eigen uitdrukking der verbeelding wist hij ook in anderen te wekken. Midden in zijn tijd staande, maar levende niet in het verleden, doch naar de toekomst, kon het niet anders of zijn geest overschreed de grenzen van zijn land en zocht naar gelijkgestemde impulsen daarbuiten, waarvan hij op zijn tijd de invloeden onderging, invloeden, die hij steeds te boven kwam en zelfstandig verwerkte.

Alles wat vooruitstormde, had zijn belangstelling, maar zijn gevoel voor evenwicht en zijn scherp verstand wisten steeds het belangrijke van het minder belangrijke te scheiden. Ook in zijn eigen werk. Eenerzijds bezielde hem het geloof in zijn roeping als kunstenaar, in de juistheid van zijn inzicht en van den weg, dien hij na veel strijd was ingegaan en die hem vervreemdde van hen die met hun waardeering niet verder konden komen dan het impressionisme. Anderzijds stond hij tegenover zijn eigen werk critisch, was het hem

V

Friedlaender 1948 Page

run by his wife, Nancy Lerner. Frej has taught visual communication at the Art Institute of Chicago.

**Friedl,** Friedrich – b. 7. 9. 1944 in Fulnek, Sudetenland, Czechoslovakia – *designer, teacher* – 1961–64: trains and works as a typesetter in Stuttgart. 1968–72: studies graphic design at the Werkkunstschule in Darmstadt. 1972–82: lectures typography at the Fachhochschule in Darmstadt. 1982–83: professor of typography at the Fachhochschule in Hildesheim. Since 1983: has taught at the Hochschule für Gestaltung in Offenbach am Main. Publisher of the "Alphabetisches" series of folders (Klaus Basset, 1985; Franz Mon, 1989; Alfons Holtgreve, 1993; Herbert Pfeiffer, 1996). Since 1964: has organized numerous exhibitions of free and applied work, including by Felix Beltran (1973), "Das gewöhnliche Design" (1977), Odermatt & Tissi (1984), Hans Hillmann (1985), 8vo, London (1989), why not associates, London (1991), and Herbert Heckmann (1995). 1993: the Klingspor Museum in Offenbach stages the exhibition "Zehn Jahre Typographie-Ausbildung an der HfG, Offenbach, Prof. F. Friedl". – *Publications include:* "Das gewöhnliche Design", Cologne 1979; "Theses about Typography" (3 vols.), Eschborn 1985–89; "Typographische Mitteilungen. Sonderheft Elementare Typographie" (reprint), Mainz 1986; "Walter Dexel. Neue Reklame", Düsseldorf 1987; articles in magazines such as "Graphis", "Octavo", "U&lc", "Eye", "Novum", "Page" and "form".

**Friedlaender,** Henri – b. 15. 3. 1904 in

Lyon, Frankreich, gest. 15. 11. 1996 in Motza-Illit, Israel. – *Typograph, Schriftentwerfer, Kalligraph, Lehrer* – 1922–24 Volontär in der Druckerei Simon, Berlin. Abendkurse an der Kunstgewerbe- und Handwerkerschule. 1925–26 Studium an der Akademie für Graphische Künste und Buchgewerbe in Leipzig. 1926–27 Handsetzer bei Jakob Hegner in Hellerau. 1927–28 Handsetzer bei Gebr. Klingspor in Offenbach. 1928 Handsetzer in der Druckerei Hartung in Hamburg. 1929–32 Disponent bei Haag-Drugulin in Leipzig. 1932 Umzug in die Niederlande, Buchgestalter für verschiedene Verlage. Unterrichtet ab 1936 Typographie und Schriftschreiben. 1942–45 taucht er während der deutschen Besatzung der Niederlande unter. 1950–70 Leiter der Hadassah Druckereifachschule, Jerusalem. Seit 1970 unterrichtet er an der Bezalel Kunstakademie in Jerusalem. Buchgestalter und Schriftentwerfer. 1971 Auszeichnung mit dem Gutenberg-Preis der Stadt Mainz. – *Schriftentwurf:* Hadassah Hebräisch (1958) – *Publikationen u. a.:* „Typografisch ABC", Amsterdam 1939; „Über Buchstaben und Ziffern", Jerusalem 1960; „Formen des Buches", Jerusalem 1962; „Die Entstehung meiner Hadassah Hebräisch", Hamburg 1967.

**Friedman,** Daniel – geb. 18. 7. 1945 in Cleveland, USA, gest. 6. 7. 1995 in New York, USA. – *Grafik-Designer, Designer, Lehrer* – Beendet 1967 seine Studien am Carnegie Institute of Technology, Pittsburgh, Pennsylvania. Danach Studium an der Hochschule für Gestaltung in Ulm, Deutschland und an der Allgemeinen Ge-

Motza-Illit, Israël – *typographe, concepteur de polices, calligraphe, enseignant* – 1922–1924, stagiaire à l'imprimerie Simon, Berlin. Cours du soir à la Kunstgewerbe- und Handwerkerschule (école d'artisanat d'art). 1925–1926, études à l'Akademie für Graphische Künste und Buchgewerbe (arts graphique et arts du livre) de Leipzig. 1926–1927, compositeur manuel chez Jakob Hegner à Hellerau. 1927–1928, compositeur manuel chez Gebr. Klingspor à Offenbach. 1928, compositeur manuel à l'imprimerie Hartung à Hambourg. 1929–1932, chargé de la mise en page chez Haag-Drugulin à Leipzig. Il s'installe aux Pays-Bas en 1932 et devient maquettiste pour plusieurs maisons d'édition. Enseigne la typographie et la calligraphie à partir de 1936. De 1942 à 1945, il vit dans la clandestinité pendant l'occupation des Pays-Bas par les Allemands. 1950–1970, directeur de l'école technique de l'imprimerie Hadassah à Jérusalem. Enseigne à l'académie Bezalel, Jérusalem, à partir de 1970. Maquettiste et concepteur de polices. En 1971, le prix Gutenberg de la ville de Mayence lui est décerné. – *Police:* Hadassah hébraïque (1958). – *Publications, sélection:* «Typografisch ABC», Amsterdam 1939; «Über Buchstaben und Ziffern» (Des lettres et des chiffres), Jérusalem 1960; «Formen des Buches», Jérusalem 1962; «Die Entstehung meiner Hadassah Hebräisch», Hambourg 1967.

**Friedman,** Daniel – né le 18. 7. 1945 à Cleveland, Etats-Unis, décédé le 6. 7. 1995 à New York, Etats-Unis – *graphiste maquettiste, designer, enseignant* – en 1967, il termine ses études au Carnegie Institute of Technology, Pittsburgh, Pennsylvanie. Etudie ensuite à la Hochschule für Gestaltung (école supérieure de design) d'Ulm, Allemagne et à l'institut professionnel de Bâle, Suisse. Il retourne aux

Friedman 1987 Logo

Friedman 1973 Poster

Friedman 1973 Logo

Friedman 1971 Cover

Friedman 1989 Logo

Lyon, France, d. 15. 11. 1996 in Motza-Illid, Israel – *typographer, type designer, calligrapher, teacher* – 1922–24: trainee at the Simon printing works in Berlin. Attends evening classes at the Kunstgewerbe- und Handwerkerschule. 1925–26: studies at the Akademie für Graphische Künste und Buchgewerbe in Leipzig. 1926–27: hand compositor for Jakob Hegner in Hellerau. 1927–28: hand compositor for Gebr. Klingspor in Offenbach. 1928: hand compositor for the Hartung printing works in Hamburg. 1929–32: chief clerk for Haag-Drugulin in Leipzig. 1932: moves to The Netherlands and works as a book designer for various publishing houses. From 1936 onwards: teaches typography and calligraphy. 1942–45: goes into hiding during the German occupation of The Netherlands. 1950–70: head of the Hadassah College of the Printing Trade in Jerusalem. From 1970 onwards: teaches at the Bezalel art academy in Jerusalem. Book and type designer. 1971: is awarded the Gutenberg Prize by the city of Mainz. – *Font:* Hadassah Hebräisch (1958). – *Publications include:* "Typografisch ABC", Amsterdam 1939; "Über Buchstaben und Ziffern", Jerusalem 1960; "Formen des Buches", Jerusalem 1962; "Die Entstehung meiner Hadassah Hebräisch", Hamburg 1967.

**Friedman,** Daniel – b. 18. 7. 1945 in Cleveland, USA, d. 6. 7. 1995 in New York, USA – *graphic designer, designer, teacher* – 1967: finishes his studies at the Carnegie Institute of Technology in Pittsburgh, Pennsylvania. He then starts studying at the Hochschule für Gestaltung in Ulm,

werbeschule Basel, Schweiz. 1970 Rückkehr in die USA, unterrichtet an der Yale University, New Haven. Aufbau des New Yorker Büros von Pentagram London. 1972–75 unterrichtet er an der School of Visual Arts, New York. 1975 Eröffnung seines eigenen Studios in New York, Auftraggeber u. a. Alchimia, Alessi, Triade, CITIBANK, Rizzoli, The Rockefeller Foundation. 1994 Professur an der Cooper Union. 1995 Ausstellung mit Arbeiten aus 25 Jahren in Boston. – *Publikationen u. a.:* „Radical modernism", New Haven 1994.

**Fronzoni,** Angiolo Giuseppe (A. G.) – geb. 5. 3. 1923 in Pistoia, Italien. – *Grafik-Designer, Architekt, Produkt-Designer, Verleger, Lehrer* – Autodidakt. Seit 1945 Grafik-Designer in Brescia. Gründet 1947 das Magazin „Punta", das er bis 1965 herausgibt und gestaltet. 1956 Umzug nach Mailand. 1964 Möbelentwürfe. 1965–67 Tätigkeit als Mitherausgeber und Gestalter der Zeitschrift „Casabella", 1966 Restaurierung des Palazzo Balbi in Genua. Seit 1967 unterrichtet er Visuelle Kommunikation in Mailand, Monza und Urbino. 1975–77 Gründung und Leitung des Istituto di Communicazione Visiva in Mailand. 1982 Gründung und Leitung seiner eigenen Gestaltungsschule in Mailand. Ausstellungen: Mailand (1963), London (1965), Buenos Aires (1966), Turin (1969), Brno (1975), New York (1993). – *Publikation:* U. Apollonio „A. G. Fronzoni", Turin 1969.

**Froshaug,** Anthony – geb. 20. 10. 1920 in London, England, gest. 15. 6. 1984 in London, England. – *Typograph, Drucker, Leh-*

10-21 maggio 1979 Lo spazio dei gesti Teatro del falcone 10-11 ore 21 Rolando Mignani I modi della riflessione 12-13 ore 21 Giovanni A Bignone Ipercubo 14-15 ore 18 Giulia no Galletta Fontane 17 ore 21 Angelo Pretolani La non divi na commedia 19 ore 21 Aure lio Caminati Il ritrovamento del corpo di san Marx - Il penultimo degli evangelisti 20-21 ore 17 Beppe Dellepia ne Genesi Comune di Geno va Assessorato alla cultura

Fronzoni 1979 Page

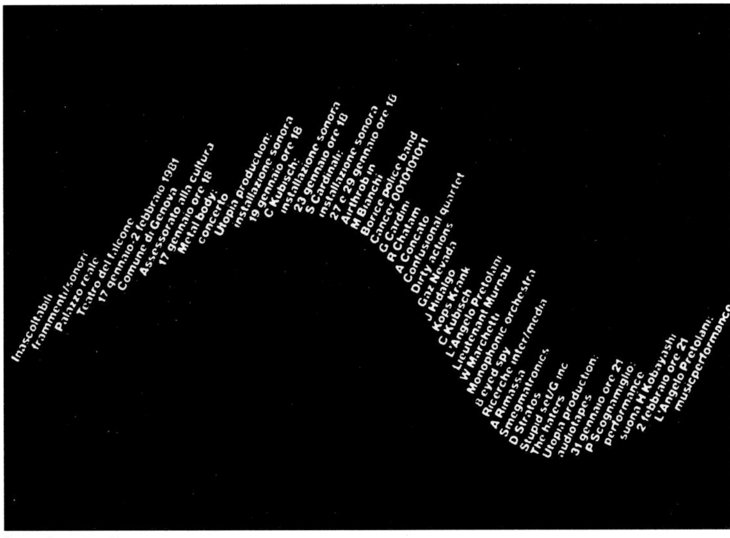

Fronzoni 1981 Poster

Etats-Unis en 1970 et enseigne à la Yale University à New Haven. Fondation de la succursale new-yorkaise de l'agence Pentagram de Londres. Enseigne à la School of Visual Arts de New York de 1972 à 1975. En 1975, il fonde son propre atelier à New York. Commanditaires : Alchimia, Alessi, Triade, CITIBANK, Rizzoli, The Rockefeller Foundation. 1994, chaire de professeur à la Cooper Union. 1995, exposition à Boston des travaux couvrant une période de 25 ans. – *Publication :* «Radical modernism», New Haven 1994.

**Fronzoni,** Angiolo Giuseppe (A. G.) – né le 5. 3. 1923 à Pistoia, Italie – *graphiste maquettiste, architecte, designer industriel, éditeur, enseignant* – autodidacte et graphiste maquettiste à Brescia à partir de 1945. En 1947, il fonde le magazine «Punta» qu'il publie et conçoit jusqu'en 1965. S'installe à Milan en 1956. 1964, design de mobilier. 1965– 1967, coéditeur et maquettiste de la revue «Casabella». Restaure le Palazzo Balbi de Gêne en 1966. Enseigne la communication visuelle à Milan, Monza et Urbino à partir de 1967. Entre 1975 et 1977, il fonde et dirige l'Istituto di Communicazione Visiva à Milan. 1982, il fonde et dirige sa propre école de design à Milan. Expositions : Milan (1963), Londres (1965), Buenos Aires (1966), Turin (1969), Brno (1975), New York (1993). – *Publication :* U. Apollonio «A. G. Fronzoni», Turin 1969.

**Froshaug,** Anthony – né le 20. 10. 1920 à Londres, Angleterre, décédé le 15. 6. 1984 à Londres, Angleterre – *typographe, imprimeur, enseignant* – 1937–1939,

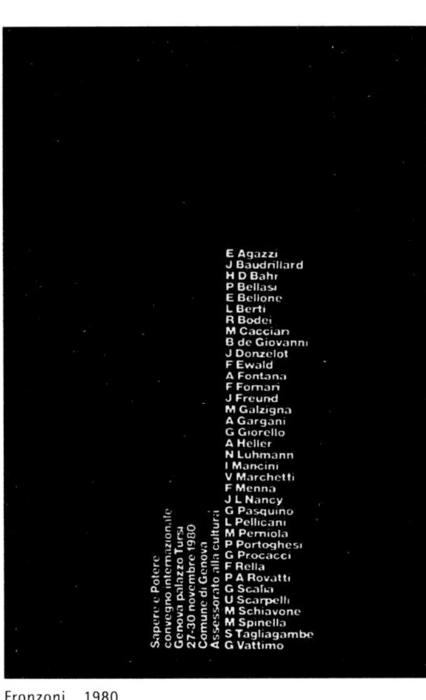


Fronzoni 1980

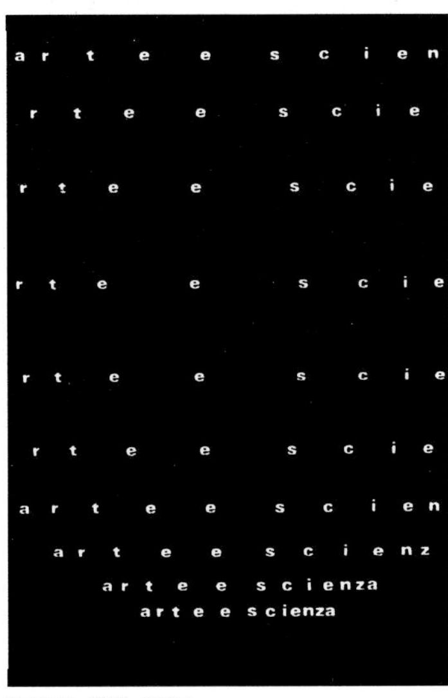


Fronzoni 1979 Poster

Fronzoni 1966 Poster

Germany and at the Allgemeine Gewerbeschule in Basle, Switzerland. 1970: returns to the USA and teaches at Yale University in New Haven. Sets up Pentagram London's New York office. 1972–75: teaches at the School of Visual Arts in New York. 1975: opens his own studio in New York. Clients include Alchimia, Alessi, Triade, CITIBANK, Rizzoli and The Rockefeller Foundation. 1994: professorship at the Cooper Union. 1995: exhibition of 25 years' of work in Boston. – *Publication:* "Radical modernism", New Haven 1994.

**Fronzoni,** Angiolo Giuseppe (A. G.) – b. 5. 3. 1923 in Pistoia, Italy – *graphic designer, architect, product designer, publisher, teacher* – Self-taught. From 1945 onwards: graphic designer in Brescia. 1947: launches "Punta" magazine which he edits and designs until 1965. 1956: moves to Milan. 1964: designs furniture. 1965– 67: works as co-editor and designer of "Casabella" magazine. 1966: restores the Palazzo Balbi in Genoa. From 1967 onwards: teaches visual communication in Milan, Monza and Urbino. 1975–77: founds and runs the Istituto di Communicazione Visiva in Milan. 1982: founds and runs his own school of design in Milan. Exhibitions: Milan (1963), London (1965), Buenos Aires (1966), Turin (1969), Brno (1975) and New York (1993). – *Publication:* U. Apollonio "A. G. Fronzoni", Turin 1969.

**Froshaug,** Anthony – b. 20. 10. 1920 in London, England, d. 15. 6. 1984 in London, England – *typographer, printer, teacher* – 1937–39: studies at the Central

*rer* – 1937–39 Studium an der Central School of Arts and Crafts in London. 1940–43 Medizinstudium an der St. Mary's Hospital School in London. Seit 1940 Arbeit in London. 1949–52 Drucker in Cornwall. 1952–53 Typographie-Lehrer an der Central School of Arts and Crafts in London. 1954–57 Drucker in Cornwall. 1957–61 Professor für Grafik-Design und Visuelle Kommunikation an der Hochschule für Gestaltung in Ulm. Gestaltet hier die Hefte 1–5 der Zeitschrift „ulm". Rückkehr nach England, unterrichtet an mehreren englischen Kunstschulen. 1961– 64 Tutor am Royal College of Art, London. 1967–69 Architektur-Studium an der Architectural Association School in London. Unterrichtet 1970–84 an der Central School of Art and Design in London. – *Publikationen u.a.:* „Typographic norms", Birmingham, London 1964; „Typography 1945–1965", Watford 1965. Zahlreiche Artikel in Zeitschriften wie „ulm" 4/1959, Ulm; „Design 178/1963", London; „Ark" 35/1964, London; „The Designer" 167/1967, London; „Typographic" 7/1975, London.

**Frutiger,** Adrian – geb. 24. 5. 1928 in Unterseen, Schweiz. – *Schriftentwerfer, Grafik-Designer, Illustrator, Typograph, Lehrer* – 1944–48 Schriftsetzerlehre in Interlaken. 1948–51 Studium an der Kunstgewerbeschule Zürich bei Walter Käch und Alfred Willimann. 1952 künstlerischer Leiter der Schriftgießerei Deberny & Peignot in Paris. Unterrichtet 1952–60 an der Ecole Estienne. Unterrichtet 1954– 68 an der Ecole Nationale Supérieure des Arts Décoratifs, Paris. 1962 Gründung des ei-

études à la Central School of Arts and Crafts à Londres. 1940–1943, études de médecine à la St. Mary's Hospital School à Londres. Travaille à Londres à partir de 1940. 1949–1952, imprimeur en Cornouailles. 1952–1953, enseigne la typographie à la Central School of Arts and Crafts de Londres. 1954–1957, imprimeur en Cornouailles. 1957–1961, professeur d'arts graphiques, de design et de communication visuelle à la Hochschule für Gestaltung (école supérieure de design) d'Ulm, Allemagne. Y réalise la maquette des numéros 1–5 de la revue «ulm». Retour en Angleterre où il enseigne dans plusieurs écoles d'art. 1961–1964, tuteur au Royal College of Art à Londres. 1967–1969, études d'architecture à la Architectural Association School de Londres. 1970–1984, enseigne à la Central School of Art and Design de Londres. – *Publications, sélection:* «Typographic norms», Birmingham, Londres 1964; «Typography 1945–1965», Watford 1965. Nombreux articles dans des revues telles que «ulm» 4/1959, Ulm; «Design» 178/ 1963, Londres; «Ark» 35/1964, Londres; «The Designer» 167/1967, Londres; «Typographic» 7/1975, Londres.

**Frutiger,** Adrian – né le 24. 5. 1928 à Unterseen, Suisse – *concepteur de polices, graphiste maquettiste, illustrateur, typographe, enseignant* – 1944–1948, apprentissage de composition typographique à Interlaken. 1948–1951, études à la Kunstgewerbeschule (école des arts décoratifs) de Zurich chez Walter Käch et Alfred Willimann. 1952, directeur artistique de la fonderie de caractères Deberny & Peignot à Paris. 1952–1960, enseigne à l'Ecole Estienne. 1954–1968, enseigne à l'Ecole Nationale Supérieure des Arts Décoratifs, Paris. En 1962, il fonde son propre atelier à Arcueil, près de Paris, avec André Gürtler et Bruno Pfäffli.

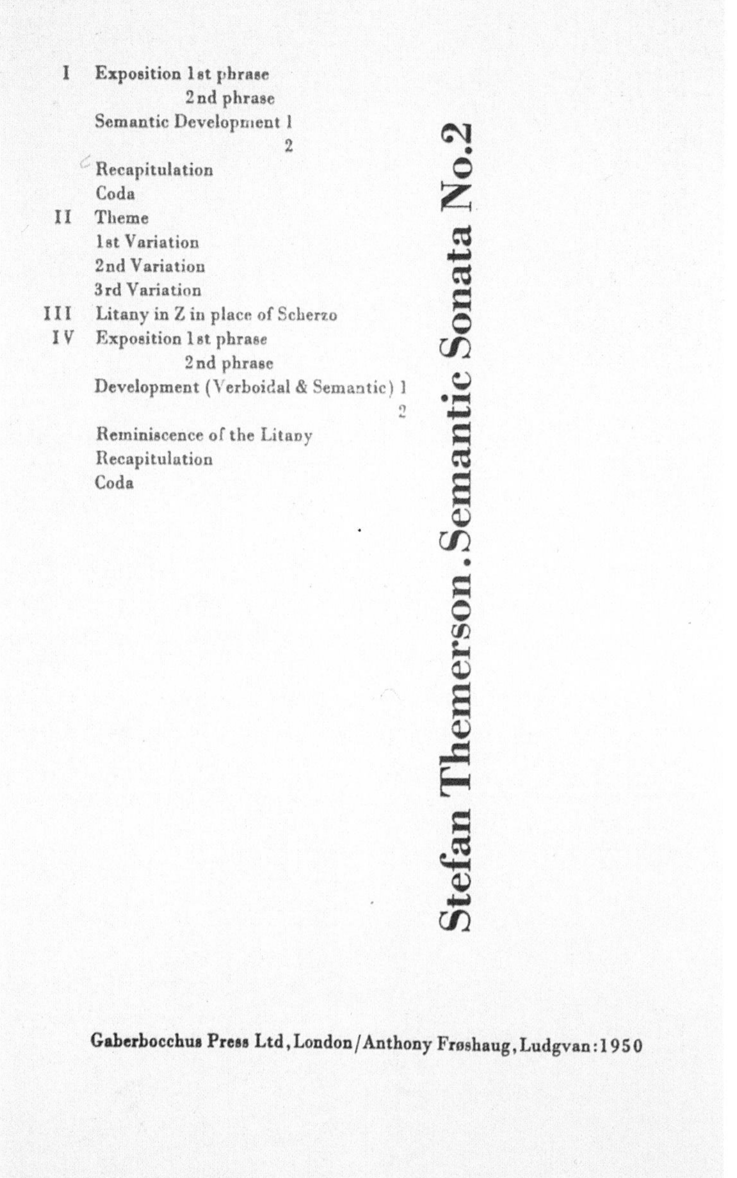


Froshaug 1950 Cover

Froshaug 1946

Froshaug 1951 Page

School of Arts and Crafts in London. 1940–43: studies medicine at St. Mary's Hospital School in London. From 1940 onwards: works in London. 1949–52: printer in Cornwall. 1952–53: teacher of typography at the Central School of Arts and Crafts in London. 1954–57: printer in Cornwall. 1957–61: professor of graphic design and visual communication at the Hochschule für Gestaltung in Ulm. Here he designs numbers 1–5 of "ulm" magazine. Returns to England and teaches at various English art schools. 1961– 64: tutor at the Royal College of Art in London. 1967–69: studies architecture at the Architectural Association School in London. 1970–84: teaches at the Central School of Art and Design in London. – *Publications include:* "Typographic norms", Birmingham, London 1964; "Typography" 1945–1965, Watford 1965. Numerous articles in magazines such as "ulm" 4/1959, Ulm; "Design" 178/ 1963, London; "Ark" 35/1964, London; "The Designer" 167/1967, London and "Typographic" 7/1975, London.

**Frutiger,** Adrian – b. 24. 5. 1928 in Unterseen, Switzerland – *type designer, graphic designer, illustrator, typographer, teacher* – 1944–48: trains as a typesetter in Interlaken. 1948–51: studies at the Kunstgewerbeschule in Zurich under Walter Käch and Alfred Willimann. 1952: art director of the Deberny & Peignot type foundry in Paris. 1952–60: teaches at the Ecole Estienne. 1954–68: teaches at the Ecole Nationale Supérieure des Arts Décoratifs in Paris. 1962: opens his own studio with André Gürtler and Bruno Pfäff-

genen Ateliers mit André Gürtler und Bruno Pfäffli in Arcueil bei Paris. 1963–81 Berater von IBM bei der Gestaltung von Schriften für den Schreibmaschinenbereich. Seit 1968 enge Zusammenarbeit mit der Stempel AG, Frankfurt am Main, die mit der Linotype AG und Linotype-Hell AG fortgesetzt wird. 1968 Ernennung zum Chevalier des Arts et des Lettres. Entwurf der von Computern lesbaren Schrift OCR-B (Optical Character Recognition), die zum internationalen Standard wird. 1986 Auszeichnung mit dem Gutenberg-Preis der Stadt Mainz. 1992 Umzug nach Bremgarten bei Bern. – *Schriftentwürfe:* Phoebus (1953), Ondine (1954), Président (1954), Meridien (1955), Univers (1957), Opéra (1959–60), Egyptienne (1960), Apollo (1964), Serifa (1967), OCR-B (1968), Iridium (1975), Frutiger (1976), Glypha (1979), Breughel (1982), Icone (1982), Versailles (1982), Centennial (1986), Avenir (1988), Vectora (1991). – *Publikationen u. a.:* „Schrift, Ecriture, Lettering", Zürich 1951; „Der Mensch und seine Zeichen", Frankfurt, Echzell 1978–81; „Type Sign Symbol", 1980.

**Fry,** Joseph – geb. 1728 in Birmingham, England, gest. 1787 in London, England. – *Schriftgießer* – Nach seinem Medizinstudium im Schokoladenhandel tätig. 1764 Gründung der Schriftgießerei „Fry & Pine" (mit William Pine) in Bristol, England. Das erste Schriftmusterbuch der Firma erscheint 1766. 1768 Umzug der Schriftgießerei nach London, wo 1770 ein zweites Schriftmusterbuch erscheint. 1778 Umbenennung der Firma in „J. Fry

MIHIQVIDEMNVMQV
AMPERSVADERIPOTVIT
ANIMOSDVMINCORPI
ORIBVSESSENTMORT
ALIBVSVIVERECVMEX

Frutiger 1951 Woodcut

| n | n | **n** | **n** | | *n* |
|---|---|---|---|---|---|
| a | b | e | f | g | i |
| o | r | s | t | y | z |
| A | B | C | E | G | H |
| M | O | R | S | X | Y |
| 1 | 2 | 4 | 6 | 8 | & |

Frutiger 1957 Univers

Frutiger Cover

| | n | | | | |
|---|---|---|---|---|---|
| a | b | e | f | g | i |
| o | r | s | t | y | z |
| A | B | C | E | G | H |
| M | O | R | S | X | Y |
| 1 | 2 | 4 | 6 | 8 | & |

Frutiger 1968 OCR-B

SCIENCES

Frutiger Logo

| | n | | | | |
|---|---|---|---|---|---|
| a | b | e | f | g | i |
| o | r | s | t | y | z |
| A | B | C | E | G | H |
| M | O | R | S | X | Y |
| 1 | 2 | 4 | 6 | 8 | & |

Fry 1768 Fry's Baskerville

1963–1981, conseiller chez IBM en matière de conception de polices pour les services de production de machines à écrire. A partir de 1968, étroite collaboration avec la Stempel AG, Francfort-sur-le-Main, puis avec la Linotype AG et la Linotype-Hell AG. Décoré Chevalier des Arts et des Lettres en 1968. Dessine les caractères OCR-B (Optical Character Recognition) lisibles par ordinateur qui deviennent un standard international. 1986, Prix Gutenberg de la ville de Mayence. Il s'installe en Suisse, à Bremgarten, près de Berne, en 1992. – *Polices:* Phoebus (1953), Ondine (1954), Président (1954), Meridien (1955), Univers (1957), Opéra (1959–1960), Egyptienne (1960), Apollo (1964), Serifa (1967), OCR-B (1968), Iridium (1975), Frutiger (1976), Glypha (1979), Breughel (1982), Icone (1982), Versailles (1982), Centennial (1986), Avenir (1988), Vectora (1991). – *Publications, sélection:* «Schrift, Ecriture, Lettering», Zurich 1951; «Der Mensch und seine Zeichen» (3 vol.), Francfort-sur-le-Main, Echzell 1978–1981; «Type Sign Symbol», Zurich 1980.

**Fry,** Joseph – né en 1728 à Birmingham, Angleterre, décédé en 1787 à Londres, Angleterre – *fondeur de caractères* – étudie la médicine puis ecerce dans le négoce du chocolat. En 1764, il crée la fonderie de caractères «Fry & Pine» (avec William Pine) à Bristol, Angleterre. Publication, en 1766, du catalogue de fontes de la société. En 1768, la fonderie s'installe à Londres, elle publie un second catalogue de fontes en 1770. La société s'appelle «J. Fry & Co» à partir de 1778. Elle achète de nombreuses matrices de polices étrangères provenant de la James Foundry. En 1787, Joseph Fry lègue la fonderie à ses

li in Arcueil, Paris. 1963–81: consultant for IBM, advising on typefaces for typewriters. From 1968 onwards: works in close collaboration with Stempel AG in Frankfurt am Main, and then with Linotype AG and Linotype-Hell AG. 1968: made a Chevalier des Arts et des Lettres. He designs OCR-B (Optical Character Recognition), a font which can be read by computers and which has become an international standard. 1986: is awarded the Gutenberg Prize by the city of Mainz. 1992: moves to Bremgarten near Bern, Switzerland. – *Fonts:* Phoebus (1953), Ondine (1954), Président (1954), Meridien (1955), Univers (1957), Opéra (1959–60), Egyptienne (1960), Apollo (1964), Serifa (1967), OCR-B (1968), Iridium (1975), Frutiger (1976), Glypha (1979), Breughel (1982), Icone (1982), Versailles (1982), Centennial (1986), Avenir (1988) and Vectora (1991). – *Publications include:* "Schrift, Ecriture, Lettering", Zurich 1951; "Der Mensch und seine Zeichen" (3 vols.), Frankfurt am Main, Echzell 1978–81; "Type Sign Symbol", Zurich 1980.

**Fry,** Joseph – b. 1728 in Birmingham, England, d. 1787 in London, England. – *type founder* – Studied medicine and then worked for the chocolate trade. 1764: founds the Fry & Pine type foundry with William Pine in Bristol, England. 1766: the company publishes its first book of type specimens. 1768: the type foundry moves to London where a second book of type specimens is issued in 1770. 1778: the company is renamed J. Fry & Co. It acquires many foreign language matrices from the James Foundry. 1787: Joseph

& Co". Erwerb zahlreicher Fremdsprachen-Matrizen aus der James Foundry. 1787 überläßt Joseph Fry die Firma seinen Söhnen, Umbenennung in „Edmund Fry & Co", 1788 in „Type Street Foundry". Edmund Fry verkauft 1828 die Firma an William Thorowgood in London. – *Schriftentwürfe:* Fry's Baskerville (1768), Old Face Open (1788).

**Fuel** – *Zeitschrift, Grafik-Design-Studio* – 1990 studieren Peter Miles, Damon Murray, Stephen Sorrell am Royal College of Art in London. Sie entwickeln 1991 die Zeitschrift „Fuel" mit einem Zuschuß des Royal College. Der Untertitel der Zeitschrift lautet „a four-letter word". Das erste Heft erscheint unter dem Titel „Girl". Heft 2/1991 mit dem Titel „Hype" hat eine Auflage von 50. Heft 3/1992 mit dem Titel „USSR" ist eine Plakatserie und ein 8mm-Film. Heft 4/1992 mit dem Titel „Cash" hat eine Auflage von 1.000 Exemplaren. Heft 5/1993 wird als T-Shirt ausgeliefert. Das Grafik-Design-Studio entwirft für Kunden wie Diesel Jeans, Virgin und unabhängige Schallplattenfirmen.

**Fugger,** Wolfgang – geb. ca. 1515 in Nürnberg, Deutschland, gest. 4. 1. 1568 in Nürnberg, Deutschland. – *Schreibmeister* – Lehre bei dem Schreibmeister Johann Neudörffer dem Älteren in Nürnberg. Vom Rat der Stadt Nürnberg erhält er die Genehmigung zum Druck mehrerer Bücher. Nach finanziellem Mißerfolg verläßt er die Stadt und zieht nach Passau. 1553 erscheint Fuggers „Schreibbüchlein" mit 207 bedruckten Blättern, die neben Kanzleischriften verschiedene Kurrentschriften („gelegte, geschobene, gerundete, gewölb-

fils et l'entreprise prend le nom de «Edmund Fry & Co», puis en 1788, celui de «Type Street Foundry». En 1828, Edmund Fry vend la société à William Thorowgood de Londres. – *Polices :* Fry's Baskerville (1768), Old Face Open (1788).

**Fuel** – *revue, atelier de design* – en 1990, Peter Miles, Damon Murray, Stephen Sorrell étudient ensemble au Royal College of Art à Londres. En 1991, ils créent la revue «Fuel» avec une subvention du Royal College. La revue a le sous-titre «a four-letter word» (un mot de quatre lettres). Le premier numéro paraît sous le titre «Girl». Le numéro 2/1991, avec «Hype» pour titre, est tiré à 50 exemplaires. Le numéro 3/1992, «USSR», comporte une série d'affiches et un film en 8mm. Le numéro 4/1992, «Cash», est tiré à 1 000 exemplaires. Le numéro 5/1993 est vendu sous forme de t-shirt. L'atelier de design travaille pour des commanditaires tels que Diesel Jeans, Virgin et d'autres sociétés indépendantes de production de disques.

**Fugger,** Wolfgang – né vers 1515 à Nuremberg, Allemagne, décédé le 4. 1. 1568 à Nuremberg, Allemagne – *scribe* – apprentissage chez le scribe Johann Neudörffer l'ancien, à Nuremberg. Le conseil municipal de la ville lui délivre l'autorisation d'imprimer plusieurs livres. A la suite de déboires financiers, il quitte la ville et s'installe à Passau. Le „Schreibbüchlein" (recueil d'écritures) de Fugger qui contient 207 pages imprimées paraît en 1553 ; outre des caractères de chancelleries, on y trouve diverses écritures courantes. Revient à Nuremberg en 1567. – *Publications :* «Ein nutzlich und wolgegrundt Formular, Mancherley schöner schriefften», Nuremberg 1553. Edition d'un fac-similé avec une préface de Fritz Funke, Leipzig 1958.

Fuel 1991 Cover

Fugger 1553 Page

Fuel ca. 1992 Poster

Fugger 1553 Page

Fry leaves the business to his sons. The company is renamed Edmund Fry & Co. and in 1788 becomes the Type Street Foundry. 1828: Edmund Fry sells the foundry to William Thorowgood in London. – *Fonts:* Fry's Baskerville (1768), Old Face Open (1788).

**Fuel** – *magazine, graphic design studio* – 1990: Peter Miles, Damon Murray and Stephen Sorrel study at the Royal College of Art in London. 1991: they develop the magazine "Fuel" using a Royal College grant. The subtitle of the magazine is "a four-letter word". The first number is issued with the title "Girl". 50 copies are printed of no. 2/1991, entitled "Hype". No. 3/1992, entitled "USSR", is a series of posters and an 8mm film. 1,000 copies are printed of no. 4/1992, entitled "Cash". No. 5/1993 is issued as a T-shirt. The graphic design studio has produced designs for clients such as Diesel Jeans and Virgin and for various independent record labels.

**Fugger,** Wolfgang – b. c.1515 in Nuremberg, Germany, d. 4.1.1568 in Nuremberg, Germany – *writing master* – Trained under writing master Johann Neudörffer the Elder in Nuremberg. The council of the city of Nuremberg granted Fugger permission to print several books. But financial failure caused Fugger to leave the city and move to Passau. 1553: Fugger's "Schreibbüchlein" is published with 207 printed pages displaying court hands and several black letter scripts. 1567: returns to Nuremberg. – *Publications:* "Ein nutzlich und wolgegrundt Formular, Mancherley schöner schriefften", Nuremberg

te") zeigen. 1567 Rückkehr nach Nürnberg. – *Publikationen u. a.:* „Ein nutzlich und wolgegrundt Formular,Mancherley schöner schriefften", Nürnberg 1553; Faksimile-Ausgabe mit einer Einführung von Fritz Funke, Leipzig 1958.

**Fundición Tipográfica Neufville,** Barcelona – *Schriftgießerei* – 1885 Kauf der Gießerei Ramirez y Rialp in Barcelona durch die Firma Kramer & Fuchs. Die Leitung der Schriftgießerei hat Jacob de Neufville. Carl Hartmann, der Sohn von Georg Hartmann, dem Besitzer der Bauerschen Gießerei, Frankfurt am Main, übernimmt 1922 die Firma Neufville. 1971 Übernahme der Fundición Tipográfica National durch die Firma Neufville. Nach der Schließung der Bauerschen Gießerei 1972 wird das Gußprogramm von Wolfgang Hartmann, dem Leiter der Gießerei Neufville, fortgesetzt.

**Fuse** – *Digitale Publikation* – 1991 erste Ausgabe, herausgegeben von FontShop und Neville Brody. Alle drei Monate erscheint eine neue Ausgabe, die eine Diskette mit vier neuen Schriften und vier mit den Schriften gestaltete Plakate enthält. Themen der einzelnen Ausgaben sind u. a. Virtual, Pornography, Propaganda, Disinformation, Crash, Religion, Cyber. Teilnehmende Gestalter sind u. a. Ian Swift, Phil Baines, Malcolm Garrett, David Berlow, Thomas Nagel, Barry Deck, Jeffery Keedy, Alexander Branczyk, Rick Valicenti, Pierro di Sciullo, Gerard Unger, David Carson.

**Fust,** Johann – geb. ca. 1400 in Mainz, Deutschland, gest. ca. 1466 in Paris, Frankreich. – *Handschriftenhändler,*

**Fundición Tipográfica Neufville,** Barcelone – *fonderie de caractères* – en 1885, la société Kramer & Fuchs achète la fonderie Ramirez y Rialp de Barcelone dirigée par Jacob de Neufville. En 1922, elle est reprise par Carl Hartmann, le fils de Georg Hartmann, qui est propriétaire de la Bauersche Giesserei de Francfort-sur-le-Main. En 1971, l'entreprise Neufville achète la Fundición Tipográfica National. En 1972, après la fermeture de la Bauersche Giesserei, le catalogue des fontes est repris par Wolfgang Hartmann, le directeur de la fonderie Neufville.

**Fuse** – *publication numérique* – première édition en 1991, publiée par FontShop et Neville Brody. Une nouvelle édition paraît tous les trois mois sous forme de disquette. Elle contient quatre nouvelles polices et quatre affiches créées avec la nouvelle police. Les numéros de la revue ont traité jusqu'à présent de virtuel, de pornographie, de propagande, de désinformation, de crash, de religion, et des cyberspaces. Ian Swift, Phil Baines, Malcolm Garrett, David Berlow, Thomas Nagel, Barry Deck, Jeffery Keedy, Alexander Branczyk, Rick Valicenti, Pierro di Sciullo, Gerard Unger, David Carson, etc. y ont collaboré jusqu'à présent.

**Fust,** Johann – né vers 1400 à Mayence, Allemagne, décédé vers 1466 à Paris, France – *négociant en manuscrits, marchand, imprimeur* – en 1449, Fust et Gu-

Fuse 1991 Poster

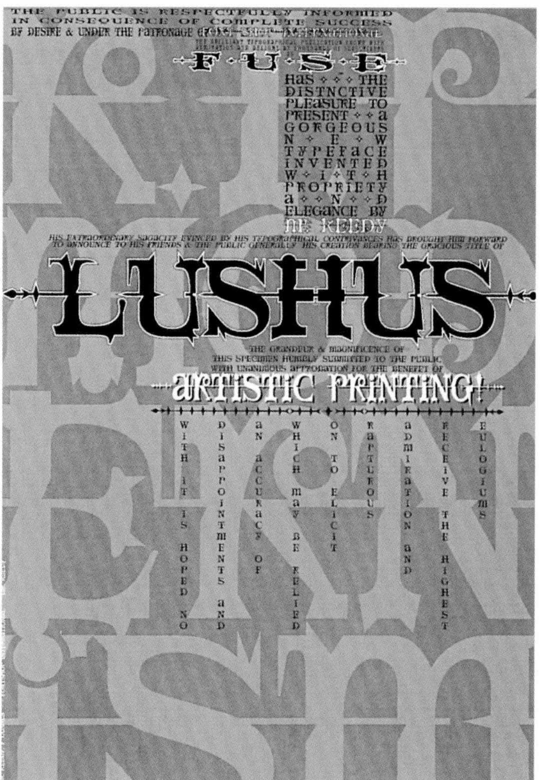

Fuse 1992 Poster

Fundición Tipográfica Neufville 1985 Logo

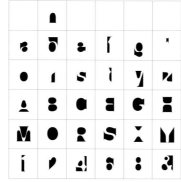

Fuse Brody Auto Suggestion

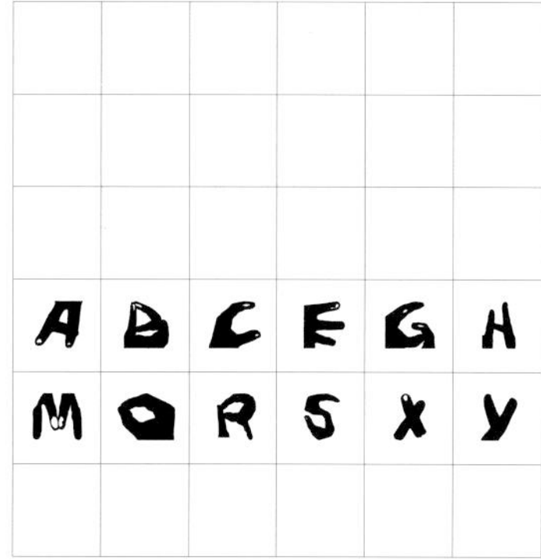
Fuse Carson Fingers

1553; facsimile edition with an introduction by Fritz Funke, Leipzig 1958.

**Fundición Tipográfica Neufville,** Barcelona – *type foundry* – 1885: the Kramer & Fuchs company buy the Ramirez y Rialp type foundry in Barcelona. Jacob de Neufville is the foundry manager. 1922: Carl Hartmann, son of Georg Hartmann, the owner of the Bauersche type foundry in Frankfurt am Main, takes over the Neufville company. 1971: the Neufville foundry takes over the Fundición National. 1972: the type program is continued by Wolfgang Hartmann, head of the Neufville type foundry, after the Bauersche type foundry closes down.

**Fuse** – *digital publication* – 1991: first issue, published by FontShop and Neville Brody. A new issue appears every three months containing a diskette with four new typefaces and four posters designed using these typefaces. Themes dealt with in the various numbers have included "Virtual", pornography, propaganda, disinformation, "Crash", religion and "Cyber". Ian Swift, Phil Baines, Malcolm Garrett, David Berlow, Thomas Nagel, Barry Deck, Jeffery Keedy, Alexander Branczyk, Rick Valicenti, Pierro di Sciullo, Gerard Unger and David Carson have been among the contributing designers.

**Fust,** Johann – b. c.1400 in Mainz, Germany, d. c.1466 in Paris, France – *dealer in manuscripts, merchant, printer* – 1449: Fust and Gutenberg form a business relationship. Gutenberg receives a financial loan for the manufacture of printing machines. 1450: a contract is signed by Fust and Gutenberg concerning the

Fust

*Kaufmann, Buchdrucker* – Fust und Gutenberg gehen 1449 eine geschäftliche Verbindung ein: Gutenberg erhält einen Kredit zur Herstellung von Druckgeräten. 1450 Vertrag zwischen Gutenberg und Fust über den Aufbau einer Druckerei. 1452 Darlehen Fusts an Gutenberg. 1455 Rechtsstreit zwischen Fust und Gutenberg, der zur Auflösung der Geschäftsverbindung führt, und der Gutenberg die Erfindung der Buchdruckkunst streitig macht. Übernahme der Werkstatt Gutenbergs durch Fust. Fust und sein Schwiegersohn Peter Schöffer drucken 1457 den Mainzer Psalter, eine 42-zeilige Bibel mit 340 Folioseiten. In ihm findet sich das älteste bekannte Druckerzeichen (Emblem) der Welt als Kennzeichnung eines Produkts aus ihrer Werkstatt. Daraus entwickelt sich das heutige Verlagssignet. 1459 entsteht ein zweiter Mainzer Psalter, der Benediktiner-Psalter. Als vierter Druck der Werkstatt Fust-Schöffer erscheint das „Rationale divinorum" von Guillemus Durandus. Weiter entstehen das „Catholicon" von Johann Balbus de Janua (1460), die Kreuzzugsbulle für Papst Pius II. (1463) und „De officiis" von Cicero als erste datierte Ausgabe eines klassischen Autors (1465). Ausweisung aus Mainz ca. 1465. Wohnrecht in Frankfurt am Main.

tenberg traitent leurs affaires ensemble : Gutenberg obtient un prêt pour fabriquer une presse. En 1450, Fust et Gutenberg signent un contrat portant sur la création d'une imprimerie. En 1452, Fust prête de l'argent à Gutenberg. En 1455, litige entre Fust et Gutenberg qui entraîne la fin de leur coopération; l'invention de l'imprimerie par Gutenberg est contestée. Fust reprend l'atelier d'imprimerie de Gutenberg. En 1457, Fust et son gendre Peter Schöffer, impriment le psautier de Mayence, une bible en 42 lignes avec 340 folios. Elle contient le plus ancien emblème designant un ouvrage sorti de cet atelier et connu jusqu'alors. Le signet actuel des maisons d'édition en est dérivé. En 1459, parution du second psautier de Mayence, le psautier bénédictin. La 4è publication sortie de l'atelier Fust-Schöffer est le «Rationale divinorum» de Guillemus Durandus. Puis viennent «Catholicon» de Johann Balbus de Janua (1460), la Bulle du pape Pie II sur la croisade (1463) et «De officiis» de Cicéron, la première édition datée (1465) d'un auteur classique. Expulsé de Mayence vers 1465. Puis droit de résidence à Francfort-sur-le-Main.

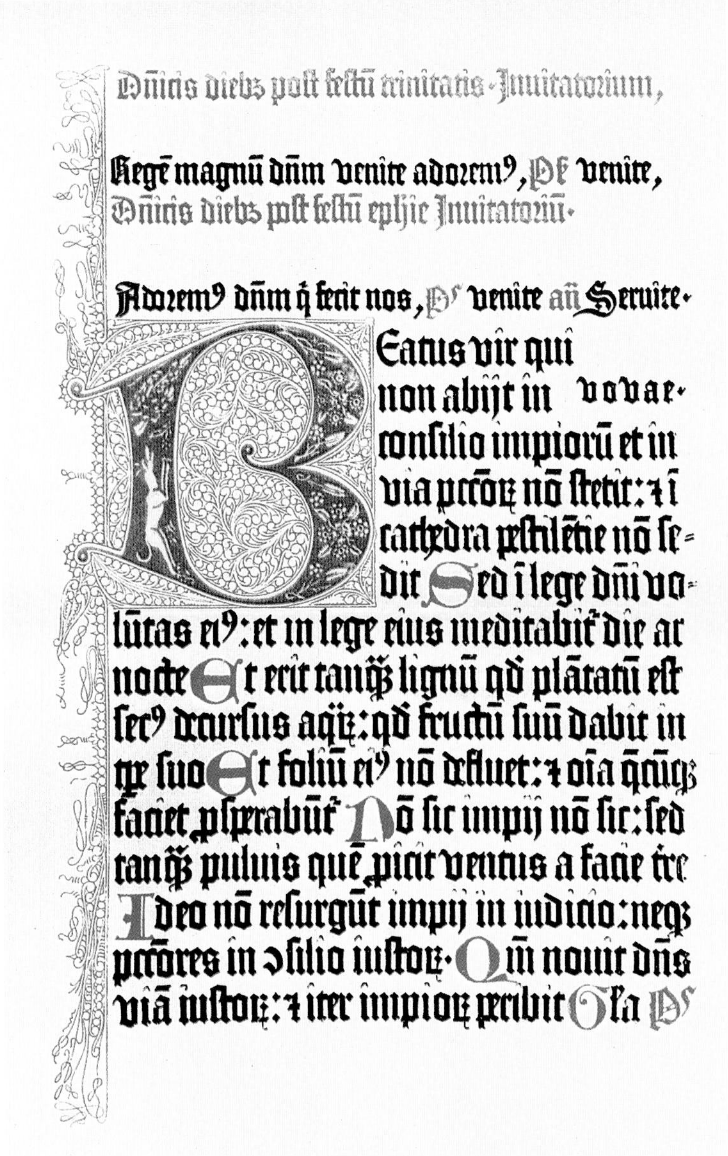

Fust 1457 Page

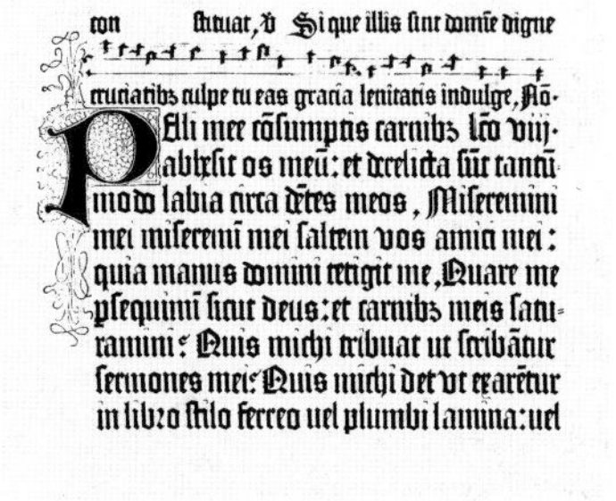

Fust 1457 Page

Fust 1457 Page

Fust 1457 Emblem

setting up of a printing workshop. 1452: Fust lends Gutenberg more money. 1455: lawsuit between Fust and Gutenberg which leads to the breaking off of business ties and which disputes Gutenberg's invention of the art of printing. Fust takes over Gutenberg's workshop. 1457: Fust and his son-in-law Peter Schöffer print the Mainzer Psalter, a 42-line Bible with a folio of 340 pages. The book contains the oldest known printer's mark (emblem) in the world, indicating that it is a product from their workshop. The publishers' current trademark is based on this original. 1459: a second Mainz Psalter is printed, the Benedictine Psalter. The fourth work to be printed at the Fust-Schöffer workshop is the "Rationale divinorum" by Guillemus Durandus. Further publications are "Catholicon" by Johann Balbus de Janua (1460), Pope Pius II's crusade bull (1463) and "De officiis" by Cicero as the first dated edition of a Latin classic (1465). C. 1465: Fust is eventually expelled from Mainz, but permitted to live in Frankfurt am Main.

# G

**Games,** Abram – geb. 29. 7. 1914 in London, England. – *Grafik-Designer, Lehrer* – Autodidakt, kurze Zeit Studium an der St. Martins School of Art in London. 1932–36 Arbeit in einem Reklamestudio. 1936–40 freiberuflich als Plakatmaler tätig. 1940–45 Informationsplakate für das Britische Kriegsministerium. 1946 Briefmarkenentwürfe für England, Irland, Israel, Portugal. Plakatentwürfe u.a. für Shell, London Transport, BEA, BOAC, Guinness, die „Financial Times", Osram. Unterrichtet 1946–53 am Royal College of Art in London. 1951 Entwurf des Signets für das „Festival of Britain" und 1953 für „BBC Television". Zahlreiche Auszeichnungen, u. a. 1991 „President's Award" der Designers and Art Directors Association von England. Zahlreiche internationale Einzel- und Gruppenausstellungen. – *Publikationen:* „Over my Shoulder", London 1960.

**Gan,** Alexei Michailowitsch – geb. 1893, gest. 1940. – *Grafik-Designer, Theoretiker* – Studium am Stroganow-Institut in Moskau. 1918–20 Arbeiten an der Inszenierung des Theaterstücks „Wir" mit Kostümen von A. Rodtschenko. Leiter der Sektion für Massenschauspiele. Bekanntschaft mit K. Malewitsch. 1920–21 Mitglied des Moskauer „Instituts für künstlerische Kultur" (INCHUK). Mitarbeit bei der Ausarbeitung des Programms der „Sektion für praktische Ideologie". Als radikaler Vertreter der Produktionskunst 1922 Gründungsmitglied der „Ersten Arbeitsgruppe der Konstruktivisten". 1922–23 Herausgeber und Gestalter der Zeitschriften „Kino-Fot" und „LEF" (Linke

**Games,** Abram – né le 29. 7. 1914 à Londres, Angleterre – *graphiste maquettiste, enseignant* – autodidacte, brèves études à la St. Martins School of Art de Londres. 1932–1936, travaille dans un atelier de publicité. Exerce comme peintre d'affiches indépendant de 1936 à 1940. Entre 1940 et 1945, il réalise des affiches informatives pour le ministère de la Guerre britannique. 1946, timbres-poste pour l'Angleterre, l'Irlande, l'Israël et le Portugal. Il conçoit des affiches pour Shell, London Transport, BEA, BOAC, Guinness, le «Financial Times», Osram etc. Enseigne au Royal College of Art de Londres de 1946 à 1953. Signet du «Festival of Britain» en 1951, et pour «BBC Television» en 1953. Nombreuses distinctions dont le «President's Award» des Designers and Art Directors Association d'Angleterre en 1991. Nombreuses expositions internationales personnelles et de groupe. – *Publication:* «Over my Shoulder», Londres 1960.

**Gan,** Alexeï Michaïlovitch – né en 1893, décédé en 1940 – *graphiste maquettiste, théoricien* – études à l'Institut Stroganoff de Moscou. Travaille pour la mise en scène de la pièce de théâtre «Nous», costumes de A. Rodtchenko, de 1918 à 1920. Directeur de la section des spectacles pour les masses. Ami de K. Malévitch. 1920–1921, membre de l' «Institut pour la culture artistique» (INCHUK) de Moscou. Participe à l'élaboration du programme de la «Section de l'idéologie pratique». En 1922, ferme partisan de l'art productif, il fait partie des membres fondateurs du «Premier groupe de travail des constructivistes». Il publie et réalise la maquette des revues «Kino-Fot» et «LEF» (front des arts de gauche) de 1922 à 1923. En 1925, il participe à la «Première exposition d'affiches de cinéma». Membre de l'OSA de 1926 à 1930, il participe à la

Games 1943 Poster

Games 1972 Memorial Panel

Gan 1928 Cover

Games 1986 Poster

Garamond 1540 Grec du Roi

**Games,** Abram – b. 29. 7. 1914 in London, England – *graphic designer, teacher* – Self-taught. Short period of study at St. Martins School of Art in London. 1932–36: works in an advertising agency. 1936–40: freelance poster designer. 1940–45: designs informative posters for the British War Office. 1946: designs postage stamps for England, Ireland, Israel and Portugal. Poster designs for Shell, London Transport, BEA, BOAC, Guinness, the "Financial Times" and Osram, among others. 1946–53: teaches at the Royal College of Art in London. 1951: designs the logo for the Festival of Britain and for BBC Television in 1953. Numerous awards, in 1991 e.g. is given the President's Award by the Designers and Art Directors Association of England. Numerous solo and joint exhibitions all over the world. – *Publication:* "Over My Shoulder", London 1960.

**Gan,** Alexei Mikhailovich – b. 1893, d. 1940 – *graphic designer, theorist* – Studied at the Stroganov Institute in Moscow. 1918–20: works on the production of the play "We" with costumes by A. Rodchenko. Head of the section for mass plays. Acquaintance with K. Malevich. 1920–21: member of the Moscow Institute for Artistic Culture (INCHUK). Works on the section for practical ideology's program. 1922: as a radical representative of production art he is founder member of the First Constructivists' Workgroup. 1922–23: editor and designer of the "Kino-Fot" and "LEF" (Left Front of the Arts) magazines. 1925: takes part in the First Exhibition of Film Posters.

Front der Künste). 1925 Beteiligung an der „Ersten Ausstellung des Filmplakats". 1926–30 Mitglied der OSA, Teilnahme an der „Ersten Ausstellung zeitgenössischer Architektur". Gestalter der Zeitschrift „Moderne Architektur". 1928–32 Gründungsmitglied der Gruppe „Oktober" (mit A. Wesnin, G. Kluzis, El Lissitzky). Gestaltet Tribünen und Kioske, Plakate und Bücher. – *Publikation u. a.:* „Konstruktivismus", Twer 1922.

**Ganeau,** François – geb. 1912 in Paris, Frankreich, gest. 1983. – *Maler, Bildhauer, Grafiker, Bühnenbildner, Schriftentwerfer* – Freundschaft mit Maurice Olive, dem Eigentümer der Fonderie Olive in Marseille. Ende der 70er Jahre Ausstattungen für das Theaterfestival in Aix-en-Provence, die Mailander Scala, die Comédie Française und die Pariser Oper. Zusammenarbeit mit Roger Excoffon. – *Schriftentwurf:* Vendôme (1951–54).

**Garamond,** Claude – geb. ca. 1480 in Paris, Frankreich, gest. 1561 in Paris, Frankreich. – *Schriftgießer, Verleger, Stempelschneider, Schriftentwerfer* – 1510 Lehre als Stempelschneider bei Simon de Colines in Paris. 1520 Gehilfe von Geoffroy Tory. Die erste Schrift Garamonds wird in einer Ausgabe des Buches „Paraphrasis in Elegantiarum Libros Laurentii Vallae" 1530 von Erasmus verwendet. Sie beruht auf der 1455 von Aldus Manutius geschnittenen Schrift „De Aetna". König Franz I. gibt 1540 bei Garamond eine griechische Schrift in Auftrag. Sie wird in drei Größen als „Grec du Roi" exklusiv für den Druck griechischer Bücher durch Robert Estienne zur

5

PAVLI IOVII NOVOCOMENſis in Vitas duodecim Vicecomitum Mediolani Principum Præfatio.

ETVSTATEM nobiliſſimæ Vicecomitum familiæ qui ambitioſius à præalta Romanorũ Cæſarum origine, Longobardiſq; regibus deducto ſtemmate, repetere contẽdunt, fabuloſis penè initiis inuoluere videntur. Nos autem recentiora illuſtrioráque, vti ab omnibus recepta, ſequemur: cõtentique erimus inſigni memoria Heriprandi & Galuanii nepotis, qui eximia cum laude rei militaris, ciuiliſque prudentiæ, Mediolani principem locum tenuerunt. Incidit Galuanius in id tempus quo Mediolanum à Federico AEnobarbo deletũ eſt, vir ſumma rerum geſtarum gloria, & quod in fatis fuit, inſigni calamitate memorabilis. Captus enim, & ad triumphum in Germaniam ductus fuiſſe traditur: ſed non multo pòſt carceris catenas fregit, ingentíque animi virtute non ſemel cæſis Barbaris, vltus iniurias, patriã reſtituit. Fuit hic (vt Annales ferunt) Othonis nepos, eius qui ab inſigni pietate magnitudinéque animi, ca nente illo pernobili claſſico excitus, ad ſacrũ bellum in Syriam contendit, communicatis ſcilicet conſiliis atque opibus cũ Guliermo Montiſſerrati regulo, qui à proceritate corporis, Longa ſpatha vocabatur. Voluntariorum enim equitum ac peditum delectæ no-

A.iii.

Garamond 1549 Page

| | n | **n** | | | *n* |
|---|---|---|---|---|---|
| a | b | e | f | g | i |
| o | r | s | t | y | z |
| A | B | C | E | G | H |
| M | O | R | S | X | Y |
| 1 | 2 | 4 | 6 | 8 | & |

Garamond 1917 Garamont Amsterdam

| | n | **n** | | | *n* |
|---|---|---|---|---|---|
| a | b | e | f | g | i |
| o | r | s | t | y | z |
| A | B | C | E | G | H |
| M | O | R | S | X | Y |
| 1 | 2 | 4 | 6 | 8 | & |

Garamond 1925 Stempel Garamond

| | n | **n** | **n** | | *n* |
|---|---|---|---|---|---|
| a | b | e | f | g | i |
| o | r | s | t | y | z |
| A | B | C | E | G | H |
| M | O | R | S | X | Y |
| 1 | 2 | 4 | 6 | 8 | & |

Garamond 1972 Berthold Garamond

«Première exposition d'architecture contemporaine». Réalise la maquette de la revue «Architecture moderne». 1928–1932, membre fondateur du groupe «Octobre» (avec A. Wesnin, G. Kluzis, El Lissitzky), conçoit des tribunes et des kiosques, des affiches et des maquettes de livres. – *Publication, sélection:* «Constructivisme », Twer 1922.

**Ganeau,** François – né en 1912 à Paris, France, décédé en 1983 – *peintre, sculpteur, graphiste, scénographe, concepteur de polices* – ami de Maurice Olive, le propriétaire de la fonderie Olive à Marseille. A la fin des années 70, scénographies pour le festival de théâtre d'Aix-en-Provence, la Scala de Milan, la Comédie Française et l'Opéra de Paris. Collaboration avec Roger Excoffon. – *Police:* Vendôme (1951–1954).

**Garamond,** Claude – né vers 1480 à Paris, France, décédé en 1561 à Paris, France – *fondeur de caractères, éditeur, tailleur de types, concepteur de polices* – 1510, apprentissage de tailleur de types chez Simon de Colines à Paris. 1520, assistant de Geoffroy Tory. La première fonte de Garamond est utilisée en 1530 pour une édition de «Paraphrasis in Elegantiarum Libros Laurentii Vallae» d'Erasme. Celle-ci s'inspire des types «De Aetna» taillés par Alde Manuce en 1455. En 1540, le roi François Ier lui commande une écriture grecque. Elle est publiée par Robert Estienne, en trois corps appelés «Grec du

1926–30: member of the OSA, participation in the First Exhibition of Contemporary Architecture. Designer of the magazine "Modern Architecture". 1928–32: founder member of the October group with A. Wesnin, G. Kluzis, El Lissitzky. Designed rostrums and kiosks, posters and books. – *Publications include:* "Constructivism", Tver 1922.

**Ganeau,** François – b. 1912 in Paris, France, d. 1983 – *painter, sculptor, graphic designer, set-designer, type designer* – Friendship with Maurice Olive, owner of the Fonderie Olive in Marseille. End of the 1970s: décor for the theater festival in Aix-en-Provence, La Scala in Milan, the Comédie Française and the Paris opera. Worked with Roger Excoffon. – *Font:* Vendôme (1951–54).

**Garamond,** Claude – b. c.1480 in Paris, France, d. 1561 in Paris, France – *type founder, publisher, punch cutter, type designer* – 1510: trains as a punch cutter with Simon de Colines in Paris. 1520: trains with Geoffroy Tory. 1530: Garamond's first type is used in an edition of the book "Paraphrasis in Elegantiarum Libros Laurentii Vallae" by Erasmus. It is based on Aldus Manutius' type De Aetna, cut in 1455. 1540: King Francis I commissions Garamond to cut a Greek type. Garamond's ensuing Grec du Roi is used by Robert Estienne in three sizes exclusively for the printing of Greek books. From 1545 onwards: Garamond also works as a publisher, first with Pierre

Verfügung gestellt. Zusammen mit Pierre Gaultier und später mit Jean Barbe ist Garamond ab 1545 auch als Verleger tätig. Als erstes Buch verlegt er „Pia et Religiosa Meditatio“ von David Chambellan. Die Bücher sind in von Garamond entworfenen Schriften gesetzt. Nach Garamonds Tod erwerben Christoph Plantin aus Antwerpen, die Gießerei Le Bé und die Frankfurter Gießerei Egenolff-Bermer einen Großteil der Originalstempel und Originalmatrizen Garamonds. Garamonds Schriften, die zwischen 1530 und 1545 entstanden sind, gelten als Höhepunkte im Schriftschaffen des 16. Jahrhunderts. In vielen Nachschnitten werden seine Schriften bis heute hergestellt und verwendet.

**Garrett,** Malcolm – geb. 2. 6. 1956 in Northwich, England. – *Grafik-Designer* – 1974–75 Studium der Typographie an der Universität Reading. 1975–78 Grafik-Design-Studium am Polytechnikum Manchester. 1977 Entwurfsarbeiten für die Musik-Gruppe „Buzzcocks“. 1978 Art Director von Radar Records. 1982 Art Director der Zeitschrift „New Sounds – New Styles“. Gründet 1983 die Firma „Assorted images“ (Ai, mit Kasper de Graaf) als multimediales Beratungsstudio. Zusammenarbeit mit Jamie Reid, Steven Appleby, Norman Hathaway. Ab 1990 grafische Arbeiten für Bloomingdales in den USA und Parco in Japan.

**Gassner,** Christof – geb. 24. 4. 1941 in Zürich, Schweiz. – *Grafik-Designer, Typograph, Schriftentwerfer, Lehrer* – 1957–62 Studium an der Kunstgewerbeschule Zürich. Danach Arbeiten für die

Roi» réservés à l'impression de livres en grec. Garamond commence ses activités d'éditeur à partir de 1545 d'abord avec Pierre Gaultier, puis avec Jean Barbe. Il publie son premier livre, «Pia et Religiosa Meditatio» de David Chambellan, composé comme les suivants en caractères dessinés par Garamond. A la mort de Garamond, Christoph Plantin d'Anvers, la fonderie Le Bé et la fonderie Egenolff-Bermer de Francfort achètent une grande partie des types et des matrices originales de Garamond. Les caractères qu'il a dessinés entre 1530 et 1545 sont considérées comme l'apogée de la création de fontes du 16e siècle. Ses polices sont encore fabriquées et utilisées aujourd'hui en plusieurs variantes.

**Garrett,** Malcolm – né le 2. 6. 1956 à Northwich, Angleterre – *graphiste maquettiste* – 1974–1975, études de typographie à l'université de Reading. 1975–1978, études de design et d'arts graphiques à Manchester. 1977, graphismes pour le groupe de musique «Buzzcocks». 1978, Art Director de Radar Records. 1982, Art Director de la revue «New Sounds – New Styles». En 1983, il fonde «Assorted images» (Ai, avec Kasper de Graaf), un studio de consultation multimédia. Travaille avec Jamie Reid, Steven Appleby, Norman Hathaway. A partir de 1990, travaux graphiques pour Bloomingdales aux Etats-Unis et Parco au Japon.

**Gassner,** Christof – né le 24. 4. 1941 à Zurich, Suisse – *graphiste maquettiste, typographe, concepteur de polices, enseignant* – 1957–1962, études à la Kunstgewerbeschule (école des arts décoratifs) de Zurich. Exerce ensuite pour la Braun AG à Francfort et les éditions DM-Test à Stuttgart. 1966–1992, graphiste maquettiste indépendant à Francfort-sur-le-Main. Commanditaires : Canton, le ZDF,

Garrett 1991 Cover

Gassner 1980 Advertisement

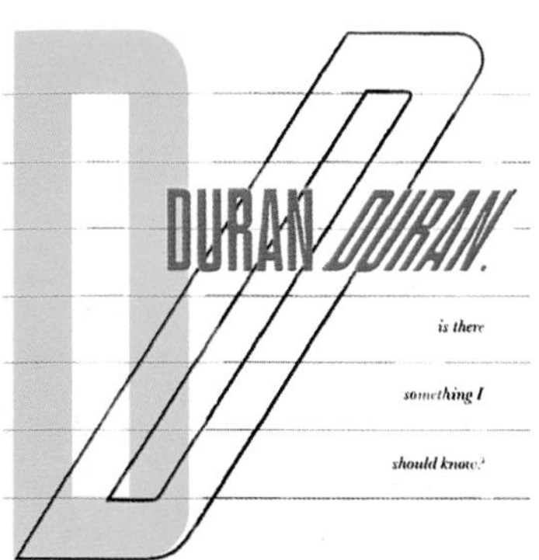

Garrett Cover

Garrett Cover

Gassner 1993 Poster

Gaultier and later with Jean Barbe. The first book he publishes is "Pia et Religiosa Meditatio" by David Chambellan. The books are set using typefaces designed by Garamond. After Garamond's death, Christoph Plantin from Antwerp, the Le Bé type foundry and the Frankfurt foundry Egenolff-Bermer acquire a large proportion of Garamond's original punches and matrices. The typefaces Garamond produced between 1530 and 1545 are considered the typographical highlight of the 16th century. His fonts have been widely copied and are still produced and in use today.

**Garrett,** Malcolm – b. 2. 6. 1956 in Northwich, England – *graphic designer* – 1974–75: studies typography at the University of Reading. 1975–78: studies graphic design at Manchester Polytechnic. 1977: produces designs for the pop group Buzzcocks. 1978: art director of Radar Records. 1982: art director of "New Sounds – New Styles" magazine. 1983: founds the company Assorted images (Ai, with Kasper de Graaf), a multimedia consultancy studio. Works with Jamie Reid, Steven Appleby and Norman Hathaway. From 1990 onwards: graphic work for Bloomingdales in the USA and Parco in Japan.

**Gassner,** Christof – b. 24. 4. 1941 in Zurich, Switzerland – *graphic designer, typographer, type designer, teacher* – 1957–62: studies at the Kunstgewerbeschule in Zurich. Then work for Braun AG in Frankfurt am Main and DM-Test publishing house in Stuttgart. 1966–92: freelance graphic designer in Frankfurt

Braun AG in Frankfurt am Main und den DM-Test Verlag in Stuttgart. 1966–92 freischaffender Grafik-Designer in Frankfurt am Main, Auftraggeber waren u. a. Canton, das ZDF, Letraset. 1985–90 Art Director der Zeitschrift „Öko-Test". 1986–92 Professor an der Fachhochschule Darmstadt. 1992 Art Director der Zeitschrift „Natur", die er neu gestaltet. Seit 1993 Professor an der Universität / Gesamthochschule Kassel. – *Schriftentwürfe:* Vexier (1973), Leopard (1976), Knirsch (1976). – *Publikation:* „Alltag-Ökologie-Design", Mainz 1994.

**Geismar,** Thomas H. – geb. 16. 7. 1931 in Glen Ridge, USA. – *Grafik-Designer* – Studium an der Rhode Island School of Design und an der Yale University, New Haven. 1957 Gründung des Studios „Brownjohn, Chermayeff & Geismar" in New York. 1960 Gründung des Studios „Chermayeff & Geismar" mit Ivan Chermayeff. Gestaltung zahlreicher Firmen-Erscheinungsbilder, u. a. für Mobil Oil, Chase Manhattan Bank, Best Products, Simpson Paper. 1970 Arbeiten für den amerikanischen Pavillon auf der Expo in Osaka, Japan. 1979 Auszeichnung mit einer Goldmedaille des American Institute of Graphic Arts (AIGA). 1985 von Ronald Reagan für die Entwicklung eines nationalen Systems von Standardsymbolen mit dem Presidential Design Award ausgezeichnet. Zahlreiche Ausstellungsgestaltungen.

**Geissbühler,** Karl Domenic – geb. 15. 11. 1933 in Lauperswil, Schweiz. – *Grafik-Designer* – Studium an der Kunstgewerbeschule Zürich (u. a. bei Ernst

Gassner 1979 Poster

Gassner 1983 Poster

Gassner 1979 Advertisement

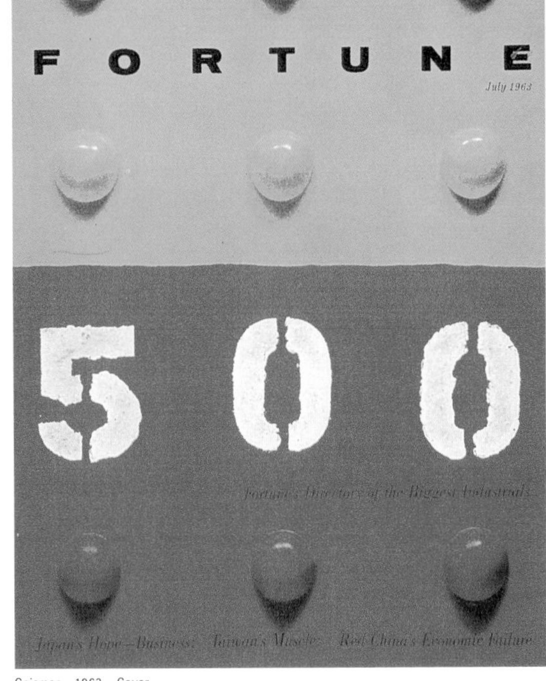

Geismar 1963 Cover

Letraset etc. 1985–1990, directeur artistique de la revue «Öko-Test». 1986–1992, professeur à la Fachhochschule (institut universitaire technique) de Darmstadt. 1992, directeur artistique de la revue «Natur» et refonte de la maquette. Professeur à l'université / Gesamthochschule de Kassel depuis 1993. – *Polices :* Vexier (1973), Leopard (1976), Knirsch (1976). – *Publication :* «Alltag-Ökologie-Design», Mayence 1994.

**Geismar,** Thomas H. – né le 16. 7. 1931 à Glen Ridge, Etats-Unis – *graphiste maquettiste* – études à la Rhode Island School of Design et à la Yale University, New Haven. En 1957, il fonde l'atelier «Brownjohn, Chermayeff & Geismar» à New York. 1960, création de l'agence «Chermayeff & Geismar» avec Ivan Chermayeff. Conception de nombreuses identités d'entreprises, dont Mobil Oil, Chase Manhattan Bank, Best Products, Simpson Paper. 1970, travaux pour le pavillon américain de l'exposition universelle d'Osaka, Japon. En 1979, la médaille d'or de l'American Institute of Graphic Arts (AIGA) lui est décernée. En 1985, il reçoit de Ronald Reagan, le Presidential Design Award pour la conception d'un système national de symboles standards. Nombreuses architectures d'expositions.

**Geissbühler,** Karl Domenic – né le 15. 11. 1933 à Lauperswil, Suisse – *graphiste maquettiste* – études à la Kunstgewerbeschule (école des arts décoratifs) de

am Main. Clients include Canton, ZDF and Letraset. 1985–90: art director of the magazine "Öko-Test". 1986–92: professor at the Fachhochschule in Darmstadt. 1992: re-designs and is art director of "Natur" magazine. Since 1993: professor at the Universität / Gesamthochschule Kassel. – *Fonts:* Vexier (1973), Leopard (1976), Knirsch (1976). – *Publication:* "Alltag-Ökologie-Design", Mainz 1994.

**Geismar,** Thomas H. – b. 16. 7. 1931 in Glen Ridge, USA – *graphic designer* – Studied at the Rhode Island School of Design and at Yale University, New Haven. 1957: the Brownjohn, Chermayeff & Geismar studio in New York is opened. 1960: opens the Chermayeff & Geismar studio with Ivan Chermayeff. Designs numerous corporate identities, including those for Mobil Oil, Chase Manhattan Bank, Best Products and Simpson Paper. 1970: works on the American pavilion for Expo in Osaka, Japan. 1979: is awarded a gold medal by the American Institute of Graphic Arts (AIGA). 1985: Ronald Reagan presents Geismar with the Presidential Design Award for his development of a national system of standard symbols. Has designed numerous exhibitions.

**Geissbühler,** Karl Domenic – b. 15.11.1933 in Lauperswil, Switzerland – *graphic designer* – Studied at the Kunstgewerbeschule in Zurich (tutors included Ernst

Keller und Johannes Itten). Studium an der Hochschule der Künste Berlin. Danach Art Director in einer Werbeagentur. 1963 Gründung seines eigenen Büros, Auftrageber waren u.a. die British Airways und das Opernhaus Zürich. 1981–91 zahlreiche Ausstattungen für verschiedene Opern und Ballette in Basel, Zürich und New York. Zahlreiche internationale Auszeichnungen. – *Publikation:* „Theaterplakate", Zürich 1993.

**Genzsch & Heyse,** Hamburg – *Schriftgießerei* – 1833 Gründung der Schriftgießerei in Hamburg durch Johann August Genzsch. Eintritt des Teilhabers Johann Georg Heyse. 1838 Ankauf der Lampeschen Gießerei in Hamburg. 1866 übernimmt Emil Julius Genzsch, der Sohn des Gründers, die Firma. 1881 Gründung einer Filiale in München. 1906 Tod von E. J. Genzsch. 1909 ist Hermann Genzsch der alleinige Inhaber. 1913 Umwandlung der Firma in eine Aktiengesellschaft. 1930 Schließung der Münchner Filiale. 1943 Zerstörung des Betriebs durch Kriegseinwirkung; Wiederaufbau. 1963 Einstellung des Schriftgusses. – *Schriftentwürfe u. a.:* Neudeutsch (O. Hupp, 1900), Olympia (C. O. Czeschka, 1909), Heraldisch (O. Hupp, 1910), Czeschka-Antiqua (C. O. Czeschka, 1914), Fortuna (F. Bauer, 1930).

**George,** Stefan – geb. 12. 7. 1868 in Büdesheim (Bingen), Deutschland, gest. 4. 12. 1933 in Minusio, Italien. – *Lyriker, Verleger* – 1892 erscheint das erste Heft seiner Zeitschrift „Blätter für die Kunst" für den „George-Kreis". Zu ihm gehören zeitweise Dichter und Wissenschaftler

Zurich (entre autres chez Ernst Keller et Johannes Itten). Etudes à la Hochschule der Künste de Brême (beaux-arts). Exerce ensuite comme directeur artistique dans une agence de publicité. En 1963, il fonde son propre atelier. Commanditaires : British Airways et l'opéra de Zurich. Nombreux décors des opéras et des ballets à Bâle, Zurich et New York de 1981 à 1991. Nombreuses distinctions internationales. – *Publication :* «Theaterplakate», Zurich 1993.

**Genzsch & Heyse,** Hambourg – *fonderie de caractères* – fondée à Hambourg en 1833 par Johann August Genzsch, puis participation de Johann Georg Heyse comme associé. Acquisition de la Lampesche Giesserei de Hambourg en 1838. En 1866, Emil Justus Genzsch, le fils du fondateur de l'entreprise, reprend la société. Création d'une filiale à Munich en 1881. Décès de E. J. Genzsch en 1906. En 1909, Hermann Genzsch est seul propriétaire de l'entreprise qui devient une société anonyme en 1913. Fermeture de la filiale munichoise en 1930. Destruction des bâtiments de l'entreprise pendant la guerre en 1943 et reconstruction. La fonderie cesse ses activités en 1963. – *Polices, sélection :* Neudeutsch (O. Hupp, 1900), Olympia (C. O. Czeschka, 1909), Heraldisch (O. Hupp, 1910), Czeschka-Antiqua (C. O. Czeschka, 1914), Fortuna (F. Bauer, 1930).

**George,** Stefan – né le 12. 7. 1868 à Büdesheim (Bingen), Allemagne, décédé le 4. 12. 1933 à Minusio, Italie – *poète, éditeur* – le premier numéro de sa revue «Blätter für die Kunst» paraît pour le cercle «George-Kreis» en 1892. Des écrivains et intellectuels comme K. Wolfskehl, M. Dauthendey, H. von Hoffmannsthal, G. Simmel, E. Kantorowicz et L. Klages feront partie de ce cercle. En 1903, l'Akzidenz-Grotesk est remaniée,

Geissbühler 1993 Poster

Geissbühler 1995 Poster

Geissbühler 1985 Logo

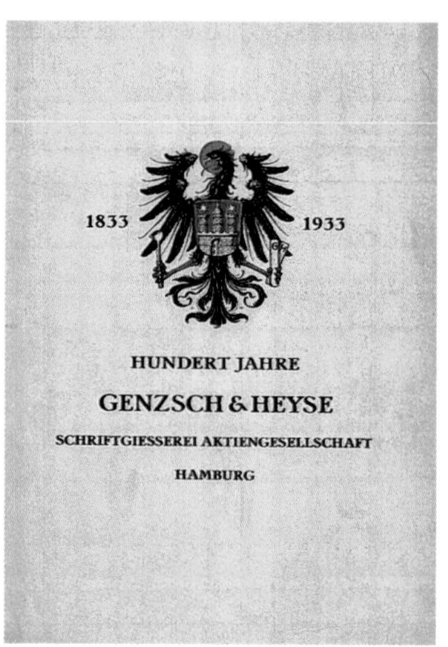

1833
1933
HUNDERT JAHRE
GENZSCH & HEYSE
SCHRIFTGIESSEREI AKTIENGESELLSCHAFT
HAMBURG

Genzsch & Heyse 1933 Cover

Keller and Johannes Itten). Studied at the Hochschule der Künste in Berlin, after which he was art director for an advertising agency. 1963: opens his own studio. Clients include British Airways and the Zurich opera house. 1981–91: produces designs for different operas and ballets in Basle, Zurich and New York. Has won numerous international awards. – *Publication:* "Theaterplakate", Zurich 1993.

**Genzsch & Heyse,** Hamburg – *type foundry* – 1833: Johann August Genzsch opens the type foundry in Hamburg. Johann Georg Heyse joins as a partner. 1838: the Lampesch type foundry in Hamburg is purchased. 1866: Emil Julius Genzsch, the founder's son, takes over the company. 1881: a foundry branch is opened in Munich. 1906: E. J. Genzsch dies. 1909: Hermann Genzsch is the sole owner. 1913: the company becomes a joint-stock company. 1930: the Munich branch is closed. 1943: the foundry is destroyed during the war and rebuilt. 1963: production is stopped. – *Fonts include:* Neudeutsch (O. Hupp, 1900), Olympia (C. O. Czeschka, 1909), Heraldisch (O. Hupp, 1910), Czeschka-Antiqua (C. O. Czeschka, 1914), Fortuna (F. Bauer, 1930).

**George,** Stefan – b. 12. 7. 1868 in Büdesheim (Bingen), Germany, d. 4. 12. 1933 in Minusio, Italy – *lyric poet, publisher* – 1892: the first number of his magazine "Blätter für die Kunst" is published for the George-Kreis (George Circle). The members of this circle have included poets and scientists such as K. Wolfskehl, M. Dauthendey, H. von Hoffmannsthal, G. Sim-

wie K. Wolfskehl, M. Dauthendey, H. von Hoffmannsthal, G. Simmel, E. Kantorowicz, L. Klages. 1903 wird die Akzidenz-Grotesk Georges Handschrift angepaßt und als Stefan-George-Schrift geschnitten und gegossen. Als erster Druck mit dieser Schrift erscheinen 1903 Georges „Das Jahr der Seele", 1906 „Maximin". 1919 erscheint das zwölfte und letzte Heft seiner Zeitschrift „Blätter für die Kunst". Aus Protest gegen die Nationalsozialisten 1933 Emigration in die Schweiz. – *Publikationen u.a.:* G. P. Landmann „Stefan George und sein Kreis", Hamburg 1960; K. Hildebrandt „Das Werk Stefan Georges", Hamburg 1960.

**Gerstner,** Karl – geb. 2. 7. 1930 in Basel, Schweiz. – *Typograph, Grafik-Designer, Maler* – 1945–46 Vorkurs an der Allgemeinen Gewerbeschule Basel. 1946–49 Lehre im Atelier Fritz Bühler in Basel. 1949–52 freier Grafik-Designer. 1953–59 eigenes Studio in Basel, Entwurf sämtlicher grafischer Arbeiten für das 200-jährige Jubiläum von Geigy. 1955–56 Hospitant bei Hans Finsler in der Fotofachklasse, Zürich. Ausgezeichnet mit einer Goldmedaille auf der Triennale 1957 in Mailand. 1959 Gründung der Agentur Gerstner + Kutter (mit Markus Kutter) in Basel. Auftraggeber waren u. a. Citroën, die Holzäpfel KG, die Schwitter AG, das Bech Electronic Center. 1962 Erweiterung der Agentur durch Paul Gredinger zu Gerstner, Gredinger, Kutter (GGK). Auftraggeber waren u. a. Swissair, Volkswagen, IBM, Jägermeister, Burda. Scheidet 1970 aus der exekutiven Geschäftsführung von GGK, um

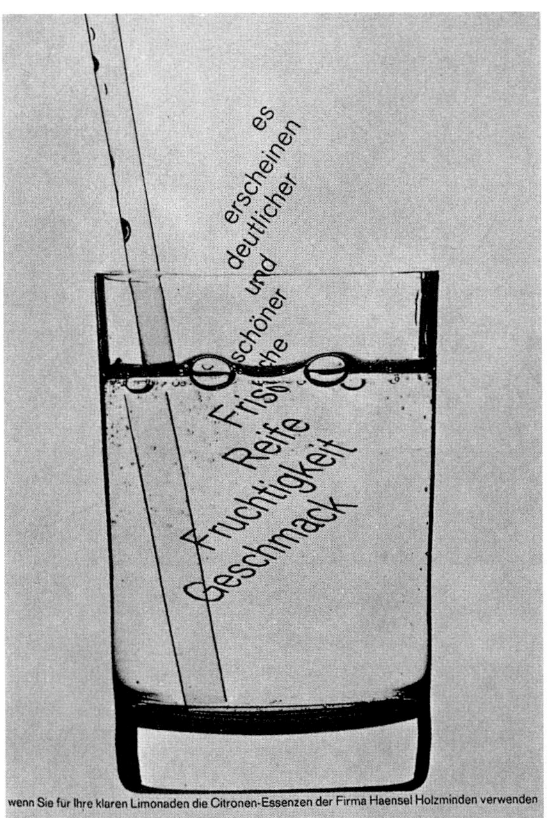


Gerstner Advertisement

Gerstner Advertisement

Gerstner 1959 Cover

Gerstner 1964 Advertisement

taillée et fondue pour correspondre à l'écriture manuscrite de George, puis rebaptisée Ecriture-Stefan-George. «Das Jahr der Seele» de George est le premier texte imprimé au moyen de cette fonte en 1903, suivi de «Maximin» en 1906. Le 12e et dernier numéro des «Blätter für die Kunst» paraît en 1919. En 1933, George émigre en Suisse en signe de protestation contre l'arrivée des nazis au pouvoir. – *Publications, sélection :* G. P. Landmann «Stefan George und sein Kreis», Hambourg 1960; K. Hildebrandt «Das Werk Stefan Georges», Hambourg 1960.

**Gerstner,** Karl – né le 2. 7. 1930 à Bâle, Suisse – *typographe, graphiste maquettiste, peintre* – 1945–1946, cours préliminaire à la Allgemeine Gewerbeschule (école des arts et métiers) de Bâle. 1946–1949, apprentissage à l'atelier Fritz Bühler à Bâle. 1949–1952, graphiste maquettiste indépendant. De 1953 à 1959, il exerce dans son propre atelier à Bâle et crée toute une palette de graphismes pour le 200e anniversaire de Geigy. Stagiaire au cours de photographie de Hans Finsler à Zurich en 1955–1956. Une médaille d'or lui est décernée à la Triennale de Milan en 1957. 1959, fondation de l'agence Gerstner + Kutter (avec Markus Kutter) à Bâle. Commanditaires : Citroën, le Holzäpfel AG, le Schwitter AG, le Bech Electronic Center etc. Avec l'arrivée de Paul Gredinger en 1962, l'agence s'agrandit et devient GGK : Gerstner, Gredinger, Kutter. Commanditaires : Swissair, Volkswagen, IBM, Jägermeister, Burda etc. Il cesse de s'occuper de la gestion de GGK en 1970 pour se consacrer exclusivement

mel, E. Kantorowicz and L. Klages. 1903: the Stefan-George-Schrift Akzidenz-Grotesk is re-designed, cut and cast to match George's handwriting. The first publication to be printed with this font is George's "Das Jahr der Seele" in 1903. 1906: "Maximin". 1919: the 12th and last number of his "Blätter für die Kunst" magazine is produced. 1933: George emigrates to Switzerland in protest against the Nazis. – *Publications include:* G. P. Landmann "Stefan George und sein Kreis", Hamburg 1960; K. Hildebrandt "Das Werk Stefan Georges", Hamburg 1960.

**Gerstner,** Karl – b. 2. 7. 1930 in Basle, Switzerland – *typographer, graphic designer, painter* – 1945–46: preparatory course at the Allgemeine Gewerbeschule in Basle. 1946–49: trains at Fritz Bühler's studio in Basle. 1949–52: freelance graphic designer. 1953–59: opens his own studio in Basle. Designs the graphics for Geigy's 200-year jubilee. 1955–56: sits in on Hans Finsler's photographic classes in Zurich. 1957: is awarded a gold medal at the Triennale in Milan. 1959: founds the Gerstner + Kutter agency (with Markus Kutter) in Basle. Clients include Citroën, Holzäpfel AG, Schwitter AG and Bech Electronic Center. 1962: the agency expands to include Paul Gredinger and is renamed Gerstner, Gredinger, Kutter (GGK). Clients include Swissair, Volkswagen, IBM, Jägermeister and Burda. 1970: Gerstner leaves the executive management of GGK in order to devote

sich ausschließlich der Entwicklung seiner freien Kunst zu widmen. Seit 1957 zahlreiche Ausstellungen der freien und angewandten Gestaltung. – *Schriftentwürfe:* Gerstner Programm (1967), Gerstner Original (1987). – *Publikationen u.a.:* „Kalte Kunst?", Teufen 1957; „Die Neue Graphik" (mit Markus Kutter), Teufen 1959; „Programme entwerfen", Teufen 1964; „Do-it-yourself-Kunst", Köln 1970; „Kompendium für Alphabeten", Teufen 1972; „Typographisches Memorandum", St. Gallen 1972; „Der Geist der Farbe", Stuttgart 1981; „Die Formen der Farbe", Frankfurt am Main 1986.

**Gerz,** Jochen – geb. 4. 4. 1940 in Berlin, Deutschland. – *Künstler, Lehrer* – 1958–60 Studium der Sinologie, Germanistik und Anglistik an der Universität Köln. Seit 1959 Texte und Übersetzungen. 1962 Studium der Urgeschichte an der Universität Basel. Lebt seit 1966 in Paris. Gründung des Autorenverlags „Agentzia". 1968 erste Buchveröffentlichung, erste Aktionen. Seit 1969 Foto-Text-Arbeiten. Seit 1972 Videoarbeiten, Installationen, Performances. Seit 1979 große Wandarbeiten mit Foto-Text-Verbindungen. Seit 1984 Projekte mit Esther Shalev-Gerz. 1990–92 Professor an der Kunsthochschule Saarbrücken. Seit 1968 zahlreiche Ausstellungen. 1972 und 1987 Teilnahme an der documenta V und VII in Kassel. – *Publikationen u.a.:* „Footing", Paris 1968; „Replay", Paris 1969; „Annonceteil", Neuwied 1971; „Die Beschreibung des Papiers", Neuwied 1973; „Die Zeit der Beschreibung" (4 Bände), Spenge 1974–83; „Exit", Hamburg 1974;

à la création artistique libre. Nombreuses expositions d'arts libres et appliqués depuis 1957. – *Polices:* Gerstner Programm (1967), Gerstner Original (1987). – *Publications, sélection:* «Kalte Kunst?», Teufen 1957; «Die Neue Graphik» (avec Markus Kutter), Teufen 1959; «Programme entwerfen», Teufen 1964; «Do-it-yourself-Kunst», Cologne 1970; «Kompendium für Alphabeten», Teufen 1972; «Typographisches Memorandum», Saint-Gall 1972; «Der Geist der Farbe», Stuttgart 1981; «Die Formen der Farbe», Francfort-sur-le-Main 1986.

**Gerz,** Jochen – né le 4. 4. 1940 à Berlin, Allemagne – *artiste, enseignant* – 1958–1960, études à l'université de Cologne de la sinologie, de l'anglais et de la littérature allemande. Textes et traductions à partir de 1959. Etudie la préhistoire à l'université de Bâle en 1962. Vit à Paris depuis 1966. Fonde «Agentzia», éditions gérées par les auteurs. Premières publications et premières actions en 1968. Réalise des œuvres photo-texte depuis 1969. Vidéos, installations et performances depuis 1972. Grandes œuvres murales alliant le texte et la photo depuis 1979. Projets avec Esther Shalev-Gerz depuis 1984. Professeur à la Kunsthochschule (école des beaux-arts) de Sarrebruck de 1990 à 1992. Nombreuses expositions depuis 1968. Participation aux «documenta V» et «VII» de Kassel, en 1972 et 1987. – *Publications, sélection:* «Footing», Paris 1968; «Replay», Paris 1969; «Annonceteil», Neuwied 1971; «Die Beschreibung des Papiers», Neuwied 1973; «Die Zeit der Beschreibung» (4 vol.), Spenge 1974–1983; «Exit», Hambourg 1974; «Die Schwierigkeit des Zentaurs beim vom Pferd steigen», Munich 1976; «Les Livres de Gandelu», Liège 1977; «Von der Kunst», Dudweiler 1985; «Texte», Bielefeld 1985; «Eine Ausstel-

Gerz 1968 Work on Paper

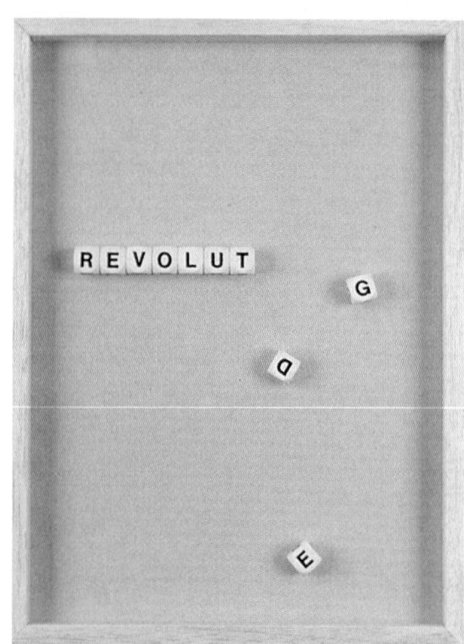


Gerz 1971 Object

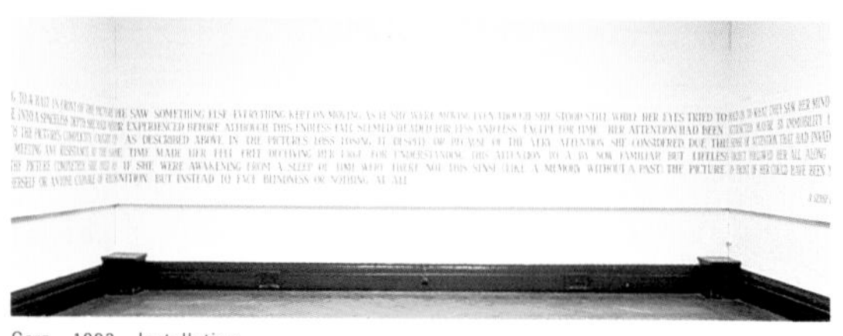

Gerz 1993 Installation

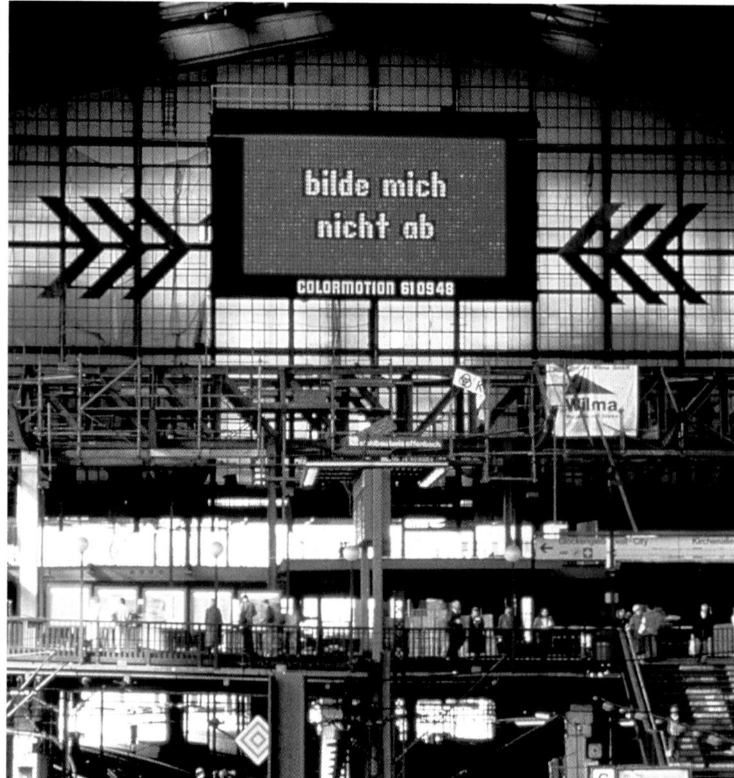


Gerz 1989 Installation

Gerz 1993 Poster

himself entirely to his free art. Since 1957: numerous exhibitions of free and applied design. – *Fonts:* Gerstner Programm (1967), Gerstner Original (1987). – *Publications include:* "Kalte Kunst?", Teufen 1957; "Die Neue Graphik" (with Markus Kutter), Teufen 1959; "Programme entwerfen", Teufen 1964; "Do-it-yourself-Kunst", Cologne 1970; "Kompendium für Alphabeten", Teufen 1972; "Typographisches Memorandum", St. Gallen 1972; "Der Geist der Farbe", Stuttgart 1981; "Die Formen der Farbe", Frankfurt am Main 1986.

**Gerz,** Jochen – b. 4. 4. 1940 in Berlin, Germany – *artist, teacher* – 1958–60: studies Sinology, German and English at the University of Cologne. 1959: first texts and translations. 1962: studies prehistory at the University of Basle. 1966: moves to Paris. Founds the author-publishers "Agentzia". 1968: publishes his first book, first promotions. 1969: starts producing photo-text work. From 1972 onwards: video work, installations, performances. From 1979 onwards: large murals with photographic and textual links. From 1984 onwards: projects with Esther Shalev-Gerz. From 1990–92: professor at the Kunsthochschule in Saarbrücken. Since 1968: numerous exhibitions. 1972 and 1987: participation in documenta V and VII in Kassel. – *Publications include:* "Footing", Paris 1968; "Replay", Paris 1969; "Annonceteil", Neuwied 1971; "Die Beschreibung des Papiers", Neuwied 1973; "Die Zeit der Beschreibung" (4 vols.), Spenge 1974–83; "Exit", Hamburg 1974; "Die Schwierigkeit des Zen-

„Die Schwierigkeit des Zentaurs beim vom Pferd steigen", München 1976; „Les Livres de Gandelu", Lüttich 1977; „Von der Kunst", Dudweiler 1985; „Texte", Bielefeld 1985; „Eine Ausstellung", Köln 1988; „Life after Humanism", Stuttgart 1992.

**Gill,** Bob – geb. 17. 1. 1931 in New York, USA. – *Illustrator, Grafik-Designer, Filmemacher, Lehrer* – Studium an der Philadelphia Museum School of Art (1948–51), an der Pennsylvania Academy of Fine Arts in Philadelphia (1951) und am City College of New York (1952 und 1955). 1954–60 freier Grafiker und Illustrator in New York. Unterrichtet 1955–60 an der School of Visual Arts und am Pratt Institute in New York (1959). 1960–62 Art Director bei der Werbeagentur Charles Hobson in London. 1962 Gründung des Studios „Fletcher, Forbes, Gill" (mit Alan Fletcher und Colin Forbes) in London. 1967–75 Austritt aus dem Studio; freier Grafiker, Illustrator und Filmemacher in London. Unterrichtet 1967–69 an der Central School of Art in London, 1969 an der Chelsea School of Art in London. 1968 Einzelausstellung im Stedelijk Museum Amsterdam. Unterrichtet 1970–75 am Royal College of Art in London, 1972–74 an der Hornsey School of Art in London. Lebt seit 1976 in New York. Zahlreiche Dokumentar- und Industriefilme, u.a. für Olivetti, Pirelli, Holiday Inn, die Singapore Airlines und das Lincoln Center. Zahlreiche Illustrationen für Zeitschriften wie „Fortune", „Esquire", „Architectural Forum", „Town". Zahlreiche Buchillustrationen, Platten-

lung», Cologne 1988; «Life after Humanism», Stuttgart 1992.

**Gill,** Bob – né le 17. 1. 1931 à New York, Etats-Unis – *illustrateur, graphiste maquettiste, réalisateur, enseignant* – études à la Philadelphia Museum School of Art (1948–1951), à la Pennsylvania Academy of Fine Arts de Philadelphie (1951) et au City College of New York (1952 et 1955). Exerce comme graphiste indépendant et illustrateur à New York de 1954 à 1960. Enseigne à la School of Visual Arts (1955–1960) et au Pratt Institute de New York (1959). Art Director à l'agence de publicité Charles Hobson à Londres de 1960 à 1962. En 1962, il fonde l'atelier Fletcher, Forbes, Gill (avec Alan Fletcher et Colin Forbes) à Londres. Quitte l'atelier en 1967, puis exerce à Londres comme graphiste indépendant, illustrateur et réalisateur jusqu'en 1975. Enseigne à la Central School of Art à Londres de 1967 à 1969, puis en 1969 à la Chelsea School of Art à Londres. Exposition personnelle au Stedelijk Museum d'Amsterdam en 1968. Enseigne au Royal College of Art de Londres de 1970 à 1975, et de 1972 à 1974 à la Hornsey School of Art. Vit à New York depuis 1976. Nombreux films documentaires et pour l'industrie, par ex. pour Olivetti, Pirelli, Holiday Inn, les Singapore Airlines et le Lincoln Center. Nombreuses illustrations pour des revues telles que «Fortune», «Esquire», «Architectural Forum», «Town». Nombreuses

Bob Gill 1976 Invitation

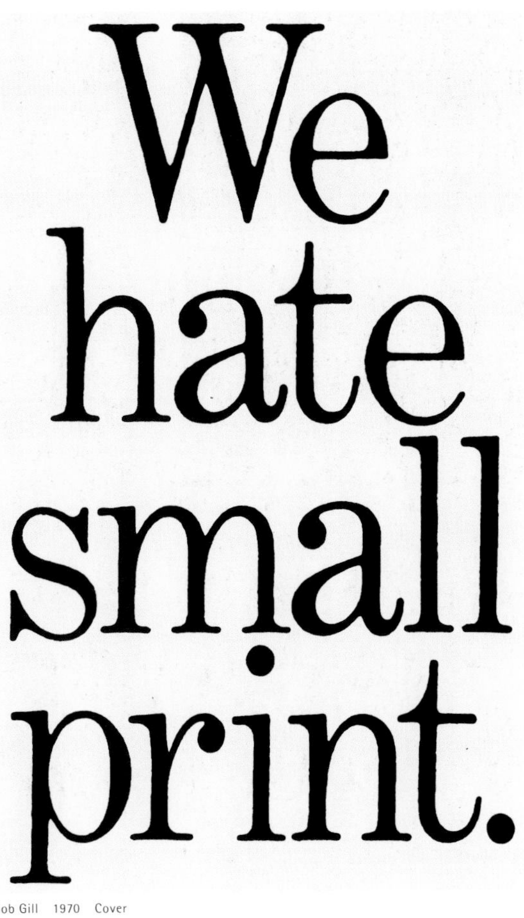

Bob Gill 1970 Cover

Bob Gill 1962 Poster

Bob Gill 1977 Logo

taurs beim vom Pferd steigen", Munich 1976; "Les Livres de Gandelu", Liège 1977; "Von der Kunst", Dudweiler 1985; "Texte", Bielefeld 1985; "Eine Ausstellung", Cologne 1988; "Life after Humanism", Stuttgart 1992.

**Gill,** Bob – b. 17. 1. 1931 in New York, USA – *illustrator, graphic designer, filmmaker, teacher* – Studied at the Philadelphia Museum School of Art (1948–51), at the Pennsylvania Academy of Fine Arts in Philadelphia (1951) and at the City College of New York (1952 and 1955). 1954–60: freelance graphic artist and illustrator in New York. 1955–60: teaches at the School of Visual Arts and the Pratt Institute in New York (1959). 1960–62: art director of the Charles Hobson advertising agency in London. 1962: founds the Fletcher, Forbes, Gill studio (with Alan Fletcher and Colin Forbes) in London. 1967–75: leaves the studio. Freelance graphic artist, illustrator and film-maker in London. 1967–69: teaches at the Central School of Art in London and in 1969 at London's Chelsea School of Art. 1968: solo exhibition at the Stedelijk Museum in Amsterdam. 1970–75: teaches at the Royal College of Art in London and from 1972–74 at London's Hornsey School of Art. He has lived in New York since 1976. Numerous documentaries and industrial films, including for Olivetti, Pirelli, Holiday Inn, Singapore Airlines and the Lincoln Center. Numerous illustrations for magazines such as "Fortune", "Esquire", "Architectural Forum" and "Town". Numerous book illustrations, record covers and logos. – *Publications include:* "A–Z",

cover und Signets. – *Publikationen u.a.:* „A–Z", Boston 1961; „What Colour is your World?", London 1963; „Graphic Design: visual comparisons" (mit A. Fletcher, C. Forbes), London 1963; „Illustration: Aspects and Directions" (mit J. Lewis), London, New York 1964; „Bob Gill's Portfolio", London 1968; „I keep changing", Mailand 1971; „Ups and downs", Mailand 1974; „Forget all the rules you ever learned about graphic design", New York 1981.

**Gill,** Eric – geb. 22. 2. 1882 in Brighton, England, gest. 17. 11. 1940 in Uxbridge, England. – *Bildhauer, Grafiker, Schriftentwerfer* – Studium an der Chichester Technical and Art School. 1899–1903 Arbeit in einem Architekturbüro. Schriftkurse an der Central School of Arts and Crafts in London bei Edward Johnston. 1905–09 buchgestalterische Arbeiten für den Insel Verlag in Leipzig: Initialen und Titelseiten. 1906 Entwurf von Initialen für die Ashedene Press. 1907 Umzug nach Ditchling, Sussex. Hier entstehen Steinskulpturen, u. a. für das BBC-Haus in London. 1914 plastische Arbeiten für die Kreuzstationen der Westminster Cathedral in London. 1924 Umzug nach Capel-y-ffin. 1925–31 Arbeiten für die Golden Cockerell Press: Initialen und Illustrationen, eine exklusiven Textschrift. 1928 Umzug nach Pigotts bei High Wycombe. Arbeiten für den Hauptsitz der Londoner U-Bahn-Verwaltung. Gründung einer eigenen Handpresse für bibliophile Drucke mit seinem Schwiegersohn. 1930 Illustrationen für die letzte Ausgabe der Zeitschrift „The Fleuron". 1937 Entwurf einer

illustrations de livres, pochettes de disques et signets. – *Publications, sélection:* «A–Z», Boston 1961; «What Colour is your World?», Londres 1963; «Graphic Design: visual comparisons» (avec A. Fletcher, C. Forbes), Londres 1963; «Illustration: Aspects and Directions» (avec J. Lewis), Londres, New York 1964; «Bob Gill's Portfolio», Londres 1968; «I keep changing», Milan 1971; «Ups and downs», Milan 1974; «Forget all the rules you ever learned about graphic design», New York 1981.

**Gill,** Eric – né le 22. 2. 1882 à Brighton, Angleterre, décédé le 17. 11. 1940 à Uxbridge, Angleterre – *sculpteur, graphiste, concepteur de polices* – études à la Chichester Technical and Art School. Travaille dans un cabinet d'architecte de 1899 à 1903. Cours de calligraphie chez Edward Johnston, à la Central School of Arts and Crafts de Londres. 1905–1909, maquettes de livres pour les éditions Insel à Leipzig: lettrines et couvertures. Dessine des lettrines pour les Ashedene Press en 1906. S'installe en 1907 à Ditchling, Sussex où il réalise des sculptures sur pierre, entre autres pour l'immeuble de la BBC à Londres. 1914, travaux plastiques pour le chemin de croix de la cathédrale de Westminster à Londres. Il s'installe à Capel-y-ffin en 1924. Travaux pour la Golden Cockerell Press de 1925 à 1931 : lettrines, illustrations et une fonte spéciale pour les textes. 1928, déménagement et installation à Pigotts, près de High Wycombe. Travaux pour le siège de l'administration du métro londonien. Il fonde une presse artisanale avec son gendre pour imprimer des livres pour bibliophiles. Illustrations de la dernière édition de la revue «The Fleuron» en 1930. Dessine, en 1937, un timbre-poste qui restera en circulation pendant 15 ans. En 1936, il est nommé «Royal Designer for

ABCDEFG
HIJKLMNO
PQRSTUV
WXYZ
abcdefghijkl
mnopqrstuv
?æœ&£
1234567890

Eric Gill 1929 Gill sans medium

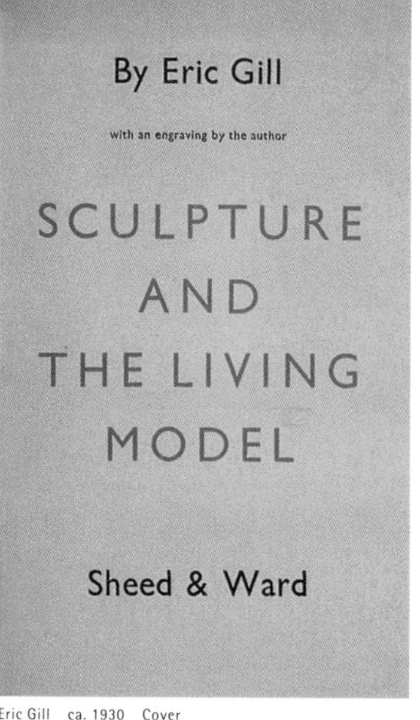

By Eric Gill
with an engraving by the author
SCULPTURE
AND
THE LIVING
MODEL
Sheed & Ward


Eric Gill ca. 1930 Cover

| | n | **n** | | | *n* |
|---|---|---|---|---|---|
| a | b | e | f | g | i |
| o | r | s | t | y | z |
| A | B | C | E | G | H |
| M | O | R | S | X | Y |
| 1 | 2 | 4 | 6 | 8 | & |

Eric Gill 1929–30 Perpetua

| | n | **n** | **n** | | n |
|---|---|---|---|---|---|
| a | b | e | f | g | i |
| o | r | s | t | y | z |
| A | B | C | E | G | H |
| M | O | R | S | X | Y |
| 1 | 2 | 4 | 6 | 8 | & |

Eric Gill 1930–31 Joanna

Eric Gill and
Denis Teg
UNHOLY
TRINITY
Text by Eric Gill • Pictures by Denis
Tegetmeier • Published by Dents for
Hague & Gill Ltd at 2 Shillings net


Eric Gill 1931 Cover

Boston 1961; "What Colour is your World?", London 1963; "Graphic Design: visual comparisons" (with A. Fletcher and C. Forbes), London 1963; "Illustration: Aspects and Directions" (with J. Lewis), London, New York 1964; "Bob Gill's Portfolio", London 1968; "I keep changing", Milan 1971; "Ups and downs", Milan 1974; "Forget all the rules you ever learned about graphic design", New York 1981.

**Gill,** Eric – b. 22. 2. 1882 in Brighton, England, d. 17. 11. 1940 in Uxbridge, England – *sculptor, graphic artist, type designer* – Studied at the Chichester Technical and Art School. 1899–1903: works in an architect's office. Takes lessons in lettering with Edward Johnston at the Central School of Arts and Crafts in London. 1905–09: produces initials and book covers for Insel publishers in Leipzig. 1906: designs initials for Ashedene Press. 1907: moves to Ditchling, Sussex. Here he produces stone sculptures, including for the BBC building in London. 1914: produces sculptures for the stations of the cross in Westminster Cathedral in London. 1924: moves to Capel-y-ffin. 1925–31: works for the Golden Cockerell Press (initials, illustrations and an exclusive text type). 1928: moves to Pigotts near High Wycombe. Works for London Underground's administrative headquarters. With his son-in-law he founds his own hand-press which prints luxury bibliophile editions. 1930: illustrations for the last number of "The Fleuron" magazine. 1937: designs a postage stamp which is in use for 15 years. 1936: made a Royal Designer for

Briefmarke, die 15 Jahre lang verwendet wird. 1936 Auszeichnung als „Royal Designer for Industry". Fertigt 1938 Steintafeln für das Gebäude des Völkerbundes in Genf. – *Schriftentwürfe:* Gill sans (1927–30), Golden Cockerell Roman (1929), Perpetua (1929–30), Solus (1929), Joanna (1930–31), Aries (1932), Floriated Capitals (1932), Bunyan, Pilgrim (1934), Jubilee (1934). – *Publikationen u.a.:* „Essay on Typography", London 1931; „Autobiography", London 1940. R. Speaight „The Life of Eric Gill", London 1966; R. Brewer „Eric Gill, the man who loved letters", London 1973; R. Harling „The letter forms and type design of Eric Gill", Westerham 1976; F. MacCarthy „Eric Gill", New York 1989.

**Glaser,** Milton – geb. 26. 6. 1929 in New York, USA. – *Grafik-Designer, Illustrator, Lehrer* – 1948–51 Studium an der Cooper Union in New York. 1952–53 Studium an der Accademia delle Belle Arti in Bologna, Italien bei Giorgio Morandi. 1954–74 Gründungsmitglied und Präsident des „Push Pin Studio" (mit Seymour Chwast, Reynold Ruffins, Edward Sorel) in New York. 1955–74 Herausgeber und Co-Art Director der Zeitschrift „Push Pin Graphic" (mit Reynold Ruffins und Seymour Chwast). Seit 1961 unterrichtet er am Pratt Institute und an der School of Visual Arts in New York. 1968 Design-Director der Zeitschrift „New York Magazine". 1970 Austritt aus dem „Push Pin Studio". Arbeiten in den Bereichen Innenarchitektur, Möbeldesign, Produktdesign und Grafik-Design. 1974 Gründung und Präsident der „Milton Glaser Inc." in

Glaser Typeface

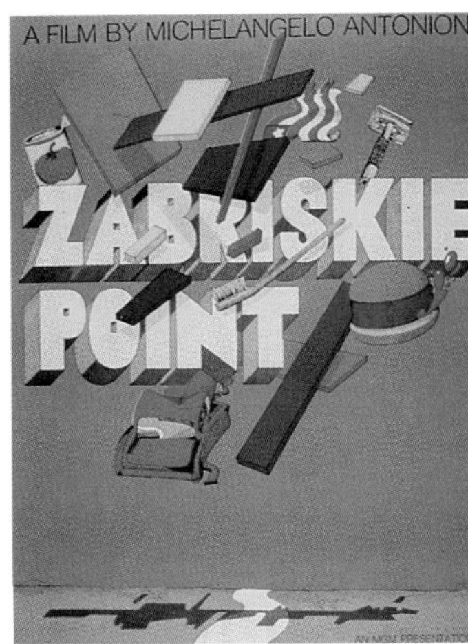


Glaser 1972 Poster

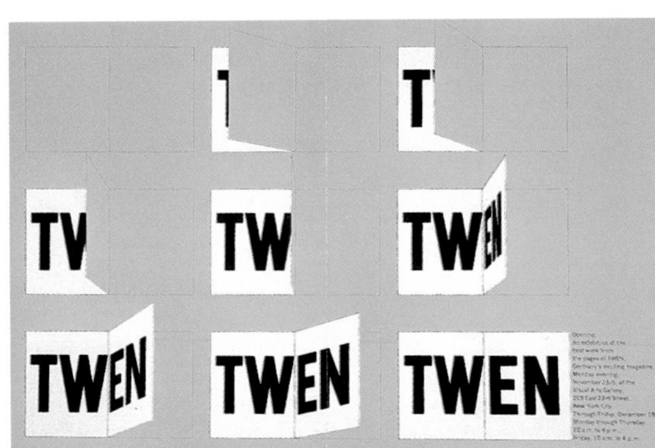
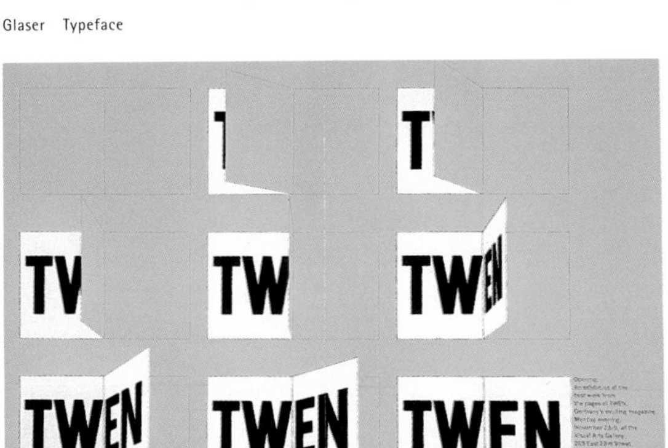


Glaser Poster

Glaser 1968 Poster

Glaser 1973 Logo

Glaser 1973 Glaser Stencil

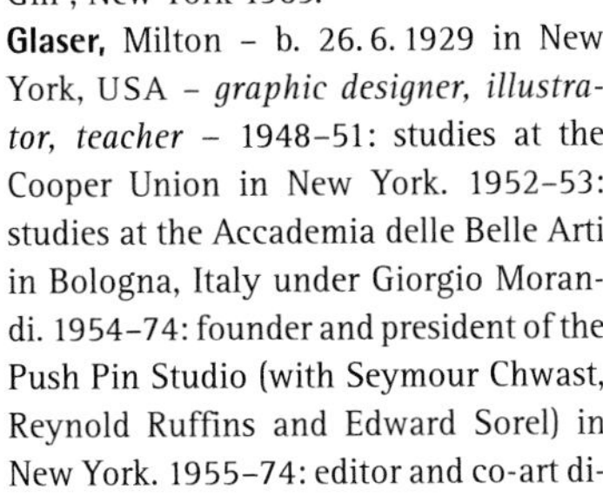

Industry». En 1938, il réalise des stèles pour le bâtiment de la Société des Nations à Genève. – *Polices:* Gill sans (1927–1930), Golden Cockerell Roman (1929), Perpetua (1929–1930), Solus (1929), Joanna (1930–1931), Aries (1932), Floriated Capitals (1932), Bunyan, Pilgrim (1934), Jubilee (1934). – *Publications, sélection:* «Essay on Typography», Londres 1931; «Autobiography», Londres 1940. R. Speaight «The Life of Eric Gill», Londres 1966; R. Brewer «Eric Gill, the man who loved letters», Londres 1973; R. Harling «The letter forms and type design of Eric Gill», Westerham 1976; F. MacCarthy «Eric Gill», New York 1989.

**Glaser,** Milton – né le 26. 6. 1929 à New York, Etats-Unis – *graphiste maquettiste, illustrateur, enseignant* – 1948–1951, études à la Cooper Union à New York. 1952–1953, études chez Giorgio Morandi, à la Accademia delle Belle Arti à Bologne, Italie. Fonde le «Push Pin Studio» en 1954 à New York (avec Seymour Chwast, Reynold Ruffins, Edward Sorel) et le préside jusqu'en 1974. Editeur et codirecteur artistique de la revue «Push Pin Graphic» (avec Seymour Chwast, Reynold Ruffins) de 1955 à 1974. A partir de 1961, enseigne au Pratt Institute et à la School of Visual Arts de New York. Design-Director de la revue «New York-Magazine» en 1968. Il quitte le «Push Pin Studio» en 1970. Travaille comme architecte d'intérieur, designer de mobilier, designer industriel et graphiste. En 1974, il fonde et préside la «Milton Glaser Inc.»

Industry. 1938: produces stone tablets for the League of Nations building in Geneva. – *Fonts:* Gill sans (1927–30), Golden Cockerell Roman (1929), Perpetua (1929–30), Solus (1929), Joanna (1930–31), Aries (1932), Floriated Capitals (1932), Bunyan, Pilgrim (1934), Jubilee (1934). – *Publications include:* "Essay on Typography", London 1931; "Autobiography", London 1940. R. Speaight "The Life of Eric Gill", London 1966; R. Brewer "Eric Gill, the man who loved letters", London 1973; R. Harling "The letter forms and type design of Eric Gill", Westerham 1976; F. MacCarthy "Eric Gill", New York 1989.

**Glaser,** Milton – b. 26. 6. 1929 in New York, USA – *graphic designer, illustrator, teacher* – 1948–51: studies at the Cooper Union in New York. 1952–53: studies at the Accademia delle Belle Arti in Bologna, Italy under Giorgio Morandi. 1954–74: founder and president of the Push Pin Studio (with Seymour Chwast, Reynold Ruffins and Edward Sorel) in New York. 1955–74: editor and co-art director of the "Push Pin Graphic" magazine (with Reynold Ruffins and Seymour Chwast). From 1961 onwards: teaches at the Pratt Institute and the School of Visual Arts in New York. 1968: design director of the "New York Magazine". 1970: leaves the Push Pin Studio. Works in interior, furniture, product and graphic design. 1974: founder and president of Milton Glaser Inc., New York. Re-designs nu-

New York. Re-Design zahlreicher Zeitschriften wie „Paris Match", „L'Express", „Esquire" und „Jardin des Modes". 1975–77 Design-Direktor der Zeitschrift „Village Voice". Für die Supermarktkette „Grand Union Company" entwirft Glaser 1978 das gesamte zwei- und dreidimensionale Erscheinungsbild. Viele seiner Arbeiten werden international berühmt: z.B. das Bob Dylan-Plakat für CBS Records (1966) oder das Signet „I love New York" für das New York State Department of Commerce (1973). 1983 Gründung der Firma „WBMG" (mit Walter Bernard). 1989 Präsident der „Aspen International Design Conference". – *Schriftentwürfe:* Babyfat, Babycurls, Baby Teeth, Houdini, Glaser Stencil. – *Publikationen u.a.:* „If Apples had Teeth" (mit Shirley Glaser), New York 1960; „Graphic Design", New York 1975; „The Milton Glaser Poster Book", New York 1977.

**Goldberg,** Carin – geb. 12. 6. 1953 in New York, USA. – *Grafik-Designerin, Lehrerin* – Studium an der Cooper Union School of Art in New York. Danach Designerin für CBS Television Network, CBS Records und Atlantic Records. 1982 Gründung des Studios „Carin Goldberg Design". Auftraggeber waren u. a. die Verlage Simon & Schuster, Random House, Grove und William Morrow, sowie Nonsuch Records. Unterrichtet seit 1983 an der School of Visual Arts in New York. Teilnahme an zahlreichen Symposien und Fachkonferenzen. Auszeichnungen vom Type Directors Club, vom Art Directors Club und vom American Institute of Graphic Arts (AIGA).

à New York. Refonte de la maquette de nombreuses revues comme «Paris Match», «L'Express», «Esquire», «Jardin des Modes». 1975–1977, Design-Director de la revue «Village Voice». 1978, travaux bi- et tridimensionnels sur l'identité de la chaîne de supermarchés «Grand Union Company». Bon nombre de ses travaux sont célèbres dans le monde entier, par ex. l'affiche Bob Dylan pour CBS Records (1966), ou le signet «I love New York» pour le New York State Department of Commerce (1973). En 1983, il fonde la société «WBMG» (avec Walter Bernard). 1989, président du «Aspen International Design Conference». – *Polices:* Babyfat, Babycurls, Baby Teeth, Houdini, Glaser Stencil. – *Publications, sélection:* «If Apples had Teeth» (avec Shirley Glaser), New York 1960; «Graphic Design», New York 1975; «The Milton Glaser Poster Book», New York 1977.

**Goldberg,** Carin – née le 12. 6. 1953 à New York, Etats-Unis – *graphiste maquettiste, enseignante* – études à la Cooper Union School of Art de New York. Puis designer pour CBS Television Network, CBS Records et Atlantic Records. En 1982, elle fonde l'atelier «Carin Goldberg Design». Commanditaires: Editions Simon & Schuster, Random House, Grove, William Morrow et Nonsuch Records. Enseigne à la School of Visual Arts, New York depuis 1983. Participe à de nombreux symposiums et congrès. Prix du Type Directors Club, de l'Art Directors Club et de l'American Institute of Graphic Arts (AIGA).

**Gomringer,** Eugen – né le 20. 1. 1925 à Cachuela Esperanza, Bolivie – *écrivain, éditeur, enseignant* – études de sciences économiques et d'histoire de l'art à Berne et à Rome. Secrétaire de Max Bill à la Hochschule für Gestaltung (école supérieure de design) d'Ulm de 1954 à 1958. Création en 1960 de la «eugen gomrin-

THE BOOK OF J
TRANSLATED FROM THE HEBREW BY
DAVID ROSENBERG
INTERPRETED BY
HAROLD BLOOM

Goldberg 1990 Cover

Goldberg 1987 Cover

Goldberg 1986 Cover

silencio silencio silencio
silencio silencio silencio
silencio silencio
silencio silencio silencio
silencio silencio silencio

Gomringer 1954 Concrete Poetry

merous magazines, such as "Paris Match", "L'Express", "Esquire" and "Jardin des Modes". 1975–77: design director of "Village Voice" magazine. 1978: Glaser designs the two- and three-dimensional corporate imagery for the Grand Union Company supermarket chain. Much of his work has become internationally famous, e.g. his Bob Dylan poster for CBS Records (1966) or the "I love New York" logo for the New York State Department of Commerce (1973). 1983: founds the company WBMG with Walter Bernard. 1989: president of the Aspen International Design Conference. – *Fonts:* Babyfat, Babycurls, Baby Teeth, Houdini, Glaser Stencil. – *Publications include:* "If Apples had Teeth" (with Shirley Glaser), New York 1960; "Graphic Design", New York 1975; "The Milton Glaser Poster Book", New York 1977.

**Goldberg,** Carin – b. 12. 6. 1953 in New York, USA – *graphic designer, teacher* – Studied at the Cooper Union School of Art in New York. Then designer for CBS Television Network, CBS Records and Atlantic Records. 1982: founds the studio Carin Goldberg Design. Clients include the publishing houses Simon & Schuster, Random House, Grove and William Morrow, and Nonsuch Records. Since 1983: teacher at the School of Visual Arts in New York. Has participated in numerous symposia and specialist conferences. Has received awards from the Type Directors Club, the Art Directors Club and the American Institute of Graphic Arts (AIGA).

**Gomringer,** Eugen – b. 20. 1. 1925 in

**Gomringer,** Eugen – geb. 20. 1. 1925 in Cachuela Esperanza, Bolivien. – *Schriftsteller, Verleger, Lehrer* – Studium der Nationalökonomie und Kunstgeschichte in Bern und Rom. 1954–58 Sekretär von Max Bill an der Hochschule für Gestaltung in Ulm. 1960 Gründung der „eugen gomringer press" in Frauenfeld, in der zahlreiche Publikationen visueller und konkreter Poesie erscheinen. 1962–67 Geschäftsführer des Schweizerischen Werkbundes in Zürich. 1967–85 Kulturbeauftragter der Rosenthal AG in Selb, danach Berater des Unternehmens. Seit 1978 Professor für Theorie der Ästhetik an der Kunstakademie Düsseldorf. – *Publikationen u.a.:* „konstellationen", Bern 1953; „33 konstellationen", St. Gallen 1960; „das stundenbuch", München 1965; „15 konstellationen" (mit Robert S. Gessner), Zürich 1965; „Josef Albers", Starnberg 1968; „Camille Graeser", Niederteufen 1968; „Poesie als Mittel der Umweltgestaltung", Itzehoe 1969; „worte sind schatten", Reinbek 1969; „zur sache der konkreten", St. Gallen 1988; „inversion und öffnung", Piesport 1988.

**Goodhue,** Bertram Grosvenor – geb. 28. 4. 1869 in Pomfret, Connecticut, USA, gest. 23. 4. 1924 in New York, USA. – *Architekt, Typograph* – Ausbildung zum Architekten. In Anlehnung an die Drucke der Kelmscott Press zeichnet Goodhue um 1890 Initialen, Umrahmungen und Dekorationen für Bücher und Zeitschriften. Gestaltet 1892 das Buch „Book of Common Prayer" für D. B. Updike, der 1893 die „Merrymount Press" in Boston gründet. Entwirft 1896 die Schrift



Gomringer 1969 Concrete Poetry



Gomringer 1969 Concrete Poetry

Gomringer 1968 Concrete Poetry



Goodhue 1902 Cheltenham Old Style

**Gomringer,** Eugen – b. 20. 1. 1925 in Cachuela Esperanza, Bolivia – *author, publisher, teacher* – Studied economics and art history in Bern and Rome. 1954–58: secretary to Max Bill at the Hochschule für Gestaltung in Ulm. 1960: founds the eugen gomringer press in Frauenfeld which publishes many volumes of visual and concrete poetry. 1962–67: manager of the Schweizerische Werkbund in Zurich. 1967–85: cultural affairs representative for Rosenthal AG in Selb, then consultant to the company. Since 1978: professor of aesthetic theory at the Kunstakademie in Düsseldorf. – *Publications include:* "konstellationen", Bern 1953; "33 konstellationen", St. Gallen 1960; "das stundenbuch", Munich 1965; "15 konstellationen" (with Robert S. Gessner), Zurich 1965; "Josef Albers", Starnberg 1968; "Camille Graeser", Niederteufen 1968; "Poesie als Mittel der Umweltgestaltung", Itzehoe 1969; "worte sind schatten", Reinbek 1969; "zur sache der konkreten", St. Gallen 1988; "inversion und öffnung", Piesport 1988.

**Goodhue,** Bertram Grosvenor – b. 28. 4. 1869 in Pomfret, Connecticut, USA, d. 23. 4. 1924 in New York, USA – *architect, typographer* – Trained as an architect. C. 1890: Goodhue draws initials, borders and decorations for books and magazines very similar to those in Kelmscott Press prints. 1892: designs a "Book of Common Prayer" for D. B. Updike, who founds Merrymount Press in Boston in 1893. 1896: designs his Merrymount typeface and ornaments the Merrymount Press' edition of the "Altar Book" and various other publications issued by the press.

**Gomringer,** Eugen – né le 20. 1. 1925 à Cachuela Esperanza, Bolivie – *écrivain, éditeur, enseignant* – études d'économie politique et d'histoire de l'art à Berne et à Rome. 1954–1958, secrétaire de Max Bill à la Hochschule für Gestaltung d'Ulm. 1960, fondation de la «eugen gomringer press» à Frauenfeld; il y publie de nombreux ouvrages de poésie visuelle et concrète. 1962–1967, administrateur du Schweizerischer Werkbund à Zurich. 1967–1985, attaché culturel de la Rosenthal AG à Selb, puis conseiller de l'entreprise. Professeur d'esthétique à la Kunstakademie (beaux-arts) de Düsseldorf depuis 1978. – *Publications, sélection:* «konstellationen», Berne 1953; «33 konstellationen», Saint-Gall 1960; «das stundenbuch», Munich 1965; «15 konstellationen (avec Robert S. Gessner), Zurich 1965; «Josef Albers», Starnberg 1968; «Camille Graeser», Niederteufen 1968; «Poesie als Mittel der Umweltgestaltung», Itzehoe 1969; «worte sind schatten», Reinbek 1969; «zur sache der konkreten», Saint-Gall 1988; «inversion und öffnung», Piesport 1988.

**Goodhue,** Bertram Grosvenor – né le 28. 4. 1869 à Pomfret, Conneticut, Etats-Unis, décédé le 23. 4. 1924 à New York, Etats-Unis – *architecte, typographe* – formation d'architecte. Vers 1890, Goodhue dessine des lettrines s'inspirant des réalisations de la Kelmscott Press. Il conçoit des encadrés et des décors pour des livres et des revues. En 1892, il crée la maquette du «Book of Common Prayer» pour D. B. Updike qui fondera la «Merrymount Press» à Boston, en 1893. En 1896, il conçoit la fonte «Merrymount» et le décor de l'»Altar Book» édité par la Merry-

„Merrymount" und den Buchschmuck für die Ausgabe „Altar Book" der Merrymount Press sowie Entwürfe für weitere Publikationen der Presse. Zahlreiche Buchausstattungen für den Verlag Stone & Kimball in Chicago. – *Schriftentwürfe:* Merrymount (1896), Cheltenham Old Style (1902). – *Publikation:* „Book Decorations", New York 1931.

**Gottschalk,** Fritz – geb. 30. 12. 1937 in Zürich, Schweiz. – *Grafik-Designer* – 1956–60 Studium an der Kunstgewerbeschule Zürich. 1962–63 Studium an der Kunstgewerbeschule Basel. 1963 Umzug nach Kanada. Arbeiten am EXPO 67-Projekt für Paul Arthur Associates. 1966 Gründung des Ateliers „Gottschalk + Ash International" (mit Stuart Ash) in Montreal. 1967 Entwurf des Signets für Kanadas Jahrhundert-Feiern. 1975 Beauftragter für Gestaltungsfragen des Olympischen Komitees. 1978 Gründung des Studios „Gottschalk + Ash International" in Zürich. 1985 Entwurf des neuen Schweizer Passes.

**Goudy,** Frederic William – geb. 8. 3. 1865 in Bloomington, USA, gest. 11. 5. 1947 in Marlborough-on-Hudson, USA. – *Schriftentwerfer, Typograph, Verleger, Lehrer* – 1888 Buchhalter für Kredit- und Hypothekengeschäfte. 1889 Umzug nach Chicago, Immobiliengeschäfte. 1892 Gründer der Zeitschrift „Modern Advertising", von der nur wenige Hefte erscheinen. 1895 Gründung einer Druckwerkstatt in Chicago, Druck der Zeitschrift „American Chap-Book". 1897 Entwurf seiner ersten Schrift „Camelot Old Style". Typographische Entwürfe für Verlage und Firmen. 1903 Gründung der „Village Press". Als er-

mount Press, ainsi que d'autres publications de ces éditions. Nombreuses maquettes de livres pour les éditions Stone & Kimball de Chicago. – *Polices:* Merrymount (1896), Cheltenham Old Style (1902). – *Publication:* «Book Decorations», New York 1931.

**Gottschalk,** Fritz – né le 30. 12. 1937 à Zurich, Suisse – *graphiste maquettiste* – 1956–1960, études à la Kunstgewerbeschule (école des arts décoratifs) de Zurich. 1962–1963, études à la Kunstgewerbeschule de Bâle. S'installe au Canada en 1963. Travaille pour Paul Arthur Associates dans le cadre de l'EXPO 67. En 1966, il fonde l'atelier «Gottschalk + Ash International» (avec Stuart Ash) à Montréal. 1967, signet pour les fêtes du centenaire du Canada. 1975, chargé du design et de la conception par le Comité olympique. En 1978, il fonde l'agence «Gottschalk + Ash International» à Zurich. Dessine le nouveau passeport suisse en 1985.

**Goudy,** Frederic William – né le 8. 3. 1865 à Bloomington, Etats-Unis, décédé le 11. 5. 1947 à Marlborough-on-Hudson, Etats-Unis – *concepteur de polices, typographe, éditeur, enseignant* – 1888, comptable dans une société de crédit et d'hypothèque. 1889, s'installe à Chicago et travaille comme agent immobilier. En 1892, il fonde la revue «Modern Advertising», dont peu de numéros paraissent. En 1895, il crée une imprimerie à Chicago et imprime la revue «American Chap-Book». En 1897, il dessine sa première fonte «Camelot Old Style». Créations typographiques pour des maisons d'édition et des sociétés. En 1903, il fonde la «Village Press». La première œuvre publiée est un essai de William Morris. En 1904, ses publications sont primées lors de l'exposition universelle de St. Louis. En 1908, la «Village Press» est détruite par un in-

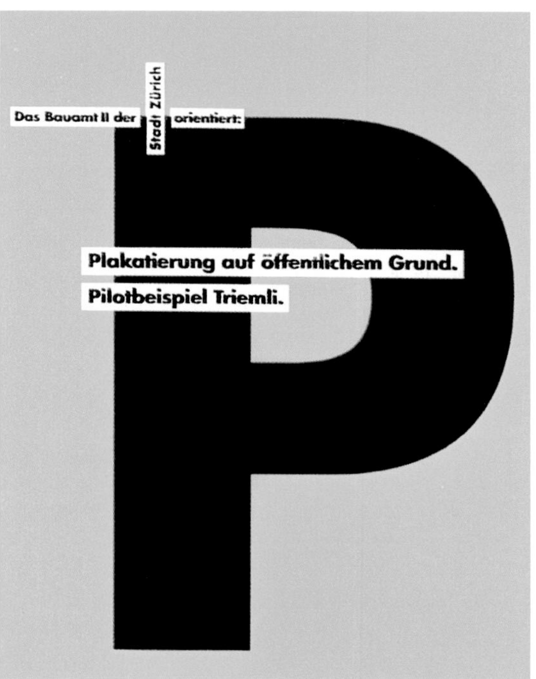


Gottschalk Poster

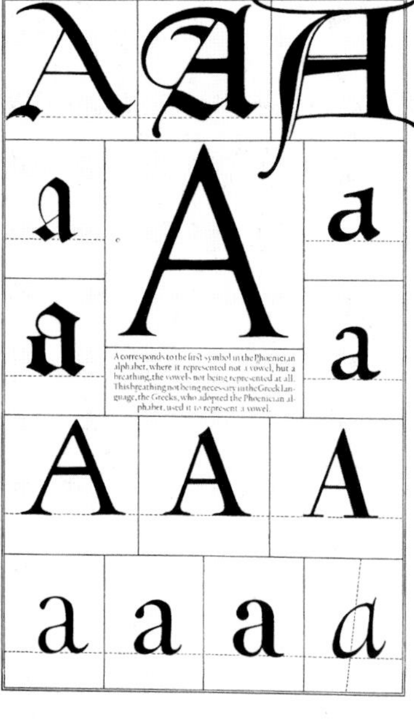

Goudy Page

Gottschalk 1987 Poster

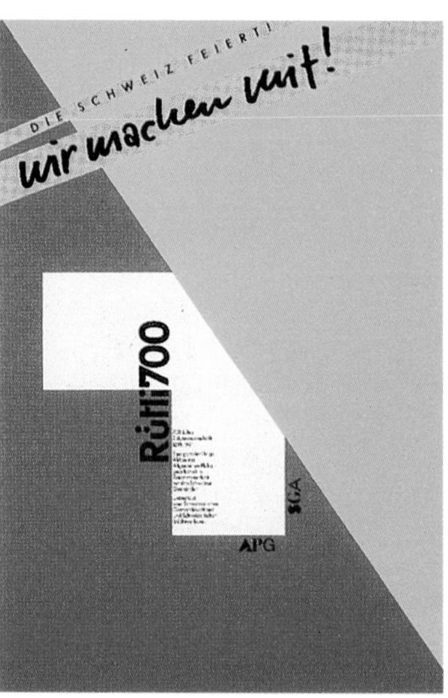


Gottschalk 1987 Poster

TYPOGRAPHICA
AN OCCASIONAL PAMPHLET
TREATING OF PRINTING · LETTER-DESIGN · AND ALLIED ARTS

THE VILLAGE PRESS AND
LETTER FOUNDERY
NEW YORK

Goudy 1903 Cover

Fashions numerous book designs for the Stone & Kimball publishers in Chicago. – *Fonts:* Merrymount (1896), Cheltenham Old Style (1902). – *Publication:* "Book Decorations", New York 1931.

**Gottschalk,** Fritz – b. 30. 12. 1937 in Zurich, Switzerland – *graphic designer* – 1956–60: studies at the Kunstgewerbeschule in Zurich. 1962–63: studies at the Kunstgewerbeschule in Basle. 1963: moves to Canada. Works on EXPO 67 projects for Paul Arthur Associates. 1966: opens his own studio, Gottschalk + Ash International, with Stuart Ash in Montreal. 1967: designs the logo for Canada's Centennial. 1975: representative for design for the Olympics committee. 1978: opens Gottschalk + Ash International in Zurich. 1985: designs the new Swiss passport.

**Goudy,** Frederic William – b. 8. 3. 1865 in Bloomington, USA, d. 11. 5. 1947 in Marlborough-on-Hudson, USA – *type designer, typographer, publisher, teacher* – 1888: book-keeper for credit and mortgage companies. 1889: moves to Chicago, works in real estate. 1892: launches "Modern Advertising" magazine which issues only a few numbers. 1895: opens a print workshop in Chicago and prints the "American Chap-Book". 1897: designs his first type, Camelot Old Style. Produces typographical designs for various publishing houses and companies. 1903: founds the Village Press. The first publication is an essay by William Morris. 1904: his publications are awarded prizes at the world exhibition in St. Louis. 1908: the Village Press is destroyed by fire.

stes Werk verlegt er einen Aufsatz von William Morris. 1904 Auszeichnung seiner Druckwerke auf der Weltausstellung in St. Louis. Die Village Press wird 1908 durch einen Brand vernichtet. 1909–24 Wiedereröffnung und Leitung der Village Press in Forest Hills. 1914 Vertrag mit der American Type Founders Company über die Herstellung und Nutzung seiner Schriften. 1916 Verkauf von acht neuen Schriften an die Caslonsche Gießerei in London. Zahlreiche Firmen bestellen bei Goudy Exklusivschriften. 1920–40 künstlerischer Berater der Lanston Monotype Co. 1924 Umzug mit seinem Verlag nach Marlborough-on-Hudson. 1925 Eröffnung seiner eigenen Schriftgießerei. Die Werkstätten für Schriftenzeichnung, Schriftschneiden, Schriftgießen, Satz, Druck und Buchbinden werden 1939 durch Feuer zerstört. 1940 Lehrstuhl für Schriftkunst an der Universität Syracuse. 1947 wird in der Kongreßbibliothek in Washington in Anwesenheit Goudys die Ausstellung „Goudyana" eröffnet. Insgesamt entwirft Goudy 116 Schriften und veröffentlicht 59 schriftstellerische Arbeiten. – *Schriftentwürfe u. a.:* Copperplate (1905), Kennerley (1911), Goudy Old Style (1915), Deepdene (1927), Remington Typewriter (1929), Californian (1938), Bulmer (1939). – *Publikationen u. a.:* „Typologia. Studies in type design and type making", Berkeley 1940; „A half-century of type design and typography, 1895–1945" (2 Bände), New York 1946. D. J. R. Bruckner „Frederic Goudy", New York 1990.

**Graf,** Carl Bernhard – geb. 13. 12. 1926 in

cendie. Elle ouvre de nouveau en 1909 à Forest Hill, où Goudy la dirige jusqu'en 1924. Contrat avec la «American Type Founders Company» en 1914, sur la fabrication et la diffusion de ses fontes. En 1916, il vend huit nouvelles fontes à la fonderie Caslon de Londres. De nombreuses sociétés commandent des fontes exclusives chez Goudy. 1920–1940, conseiller artistique de Lanston Monotype Co. En 1924, il s'installe à Marlborough-on-Hudson avec sa maison d'édition, il y ouvre sa propre fonderie de caractères en 1925. Les ateliers de graphisme, de taille de types, la fonderie, l'atelier de composition, l'imprimerie et l'atelier de reliure sont détruits par un incendie en 1939. En 1940, il obtient une chaire de calligraphie à l'université de Syracuse. En 1947, l'exposition «Goudyana» est inaugurée en sa présence à la bibliothèque du congrès à Washington. Goudy a créé 116 polices et publié 59 textes. – *Polices, sélection:* Copperplate (1905), Kennerley (1911), Goudy Old Style (1915), Deepdene (1927), Remington Typewriter (1929), Californian (1938), Bulmer (1939). – *Publications, sélection:* «Typologia. Studies in type design and type making», Berkeley 1940; «A half-century of type design and typography, 1895–1945» (2 vol.), New York 1946. D. J. R. Bruckner «Frederic Goudy», New York 1990.

Goudy Page

GOUDY MEDIAEVAL
Drawings begun Aug. 19, 16 pt. finished Sept. 15, 1930
A NEW TYPE is here presented, which, judged by pragmatic standards, may not meet the approval of those critics who demand in their types the elimination of any atavistic tendency. Quite obviously it cannot be judged fairly by the advertising compositor or the job printer, nor must unassuming legibility be made its first criterion. At one time books were entirely written out by hand, but the qualities that made the writing charming defy completely successful reproduction in types. Goudy Mediæval presents a face that in its lower case borrows the freedom of the pen of the Renaissance. Its capitals however, owe less to the pen-hands since they are more or less composites of monastic ms. & painted Lombardic forms. The designer hopes nevertheless, that his capitals will be found in accord with

Goudy 1930 Mediaeval

Goudy Demonstration of Development

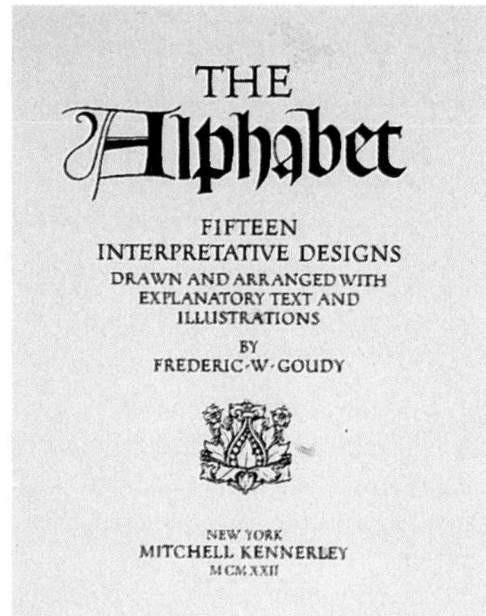

THE
Alphabet
FIFTEEN
INTERPRETATIVE DESIGNS
DRAWN AND ARRANGED WITH
EXPLANATORY TEXT AND
ILLUSTRATIONS
BY
FREDERIC·W·GOUDY
NEW YORK
MITCHELL KENNERLEY
MCMXXII


Goudy 1922 Cover

| | n | n | | | n |
|---|---|---|---|---|---|
| a | b | e | f | g | i |
| o | r | s | t | y | z |
| A | B | C | E | G | H |
| M | O | R | S | X | Y |
| 1 | 2 | 4 | 6 | 8 | & |

Goudy 1915 Goudy Old Style

| | n | | | | n |
|---|---|---|---|---|---|
| a | b | e | f | g | i |
| o | r | s | t | y | z |
| A | B | C | E | G | H |
| M | O | R | S | X | Y |
| 1 | 2 | 4 | 6 | 8 | & |

Goudy 1927 Deepdene

1909–24: the Press is reopened and run under Goudy's management in Forest Hills. 1914: signs a contract with the American Type Founders Company governing the manufacture and use of his typefaces. 1916: sells 8 new typefaces to the Caslon type foundry in London. Numerous companies commission Goudy to design exclusive typefaces for them. 1920–40: art consultant to the Lanston Monotype Co. 1924: he and his publishing house move to Marlborough-on-Houston. 1925: opens his own type foundry. 1939: the workshops for type design, type cutting, type foundry, typesetting, printing and bookbinding are destroyed by fire. 1940: teaching post for calligraphy at the University of Syracuse. 1947: the Goudyana exhibition is opened in Goudy's presence at the Library of Congress in Washington. Goudy designed a total of 116 fonts and published 59 literary works. – *Fonts include:* Copperplate (1905), Kennerley (1911), Goudy Old Style (1915), Deepdene (1927), Remington Typewriter (1929), Californian (1938), Bulmer (1939). – *Publications include:* "Typologia. Studies in type design and type making", Berkeley 1940; "A half-century of type design and typography, 1895– 1945" (2 vols.), New York 1946. D. J. R. Bruckner "Frederic Goudy", New York 1990.

Zürich, Schweiz, gest. 16. 1. 1968 in Zürich, Schweiz. – *Grafik-Designer* – 1942–43 Besuch der Allgemeinen Klasse der Kunstgewerbeschule Zürich. 1943–47 Lehre als Grafiker, anschließend Besuch der Grafik-Fachklasse an der Kunstgewerbeschule Zürich. 1948–53 tätig als Grafiker im Büro A. Wirz in Zürich. 1953–68 eigenes Atelier in Zürich. Aufträge aus der Pharmaindustrie, der Uhren- und Möbelindustrie. Zahlreiche Plakate für das Helmhaus und das Kunstgewerbemuseum Zürich.

**Grandjean de Fouchy,** Philippe – geb. 1666 in Mâcon, Frankreich, gest. 1714 in Paris, Frankreich. – *Schriftgießer, Stempelschneider* – 1692 Berufung in die Kommission zur Erarbeitung einer neuen Antiqua-Schrift, die durch Louis XIV. angeordnet und von der Académie des Sciences ausgerichtet wird. Die Schrift soll ausschließlich in der Imprimerie Royale verwendet werden. Auf der Grundlage der Ergebnisse der Académie des Sciences wird Grandjean 1694 beauftragt, die „Romain du Roi" zu schneiden. Die ersten Schnitte seiner Schrift erscheinen 1702. Sie werden erstmals in dem Werk „Médailles sur les principaux Evénements du Règne de Louis le Grand" verwendet. Nach Grandjeans Tod wird seine Schrift durch Jean Alexandre, der sein Assistent war, erst 1745 in 21 Graden mit Kursivschrift fertiggestellt. Sie war das ganze 18. Jahrhundert in Gebrauch und regte u. a. P. F. Didot und P. S. Fournier zu Weiterführungen und Nachahmungen an.

**Granjon,** Robert – geb. 1513, gest. 16. 11. 1589 in Rom, Italien. – *Schrift-*

**Graf,** Carl Bernhard – né le 13. 12. 1926 à Zurich, Suisse, décédé le 16. 1. 1968 à Zurich, Suisse – *graphiste maquettiste* – 1942–1943, fréquente les cours de formation générale de la Kunstgewerbeschule (école des arts décoratifs) de Zurich. 1943–1947, apprentissage de graphiste, puis cours de graphisme à la Kunstgewerbeschule de Zurich. Exerce de 1948 à 1953 comme graphiste à l'agence A. Wirz, à Zurich, puis dans son propre atelier de 1953 à 1968. Effectue des commandes pour l'industrie pharmaceutique, l'horlogerie et la menuserie. Nombreuses affiches pour le Helmhaus et le musée des arts décoratifs de Zurich.

**Grandjean de Fouchy,** Philippe – né en 1666 à Mâcon, France, décédé en 1714 à Paris, France – *fondeur de caractères, tailleur de types* – en 1692, il est appelé à faire partie de la commission dirigée par l'Académie des Sciences et chargée de créer une nouvelle antique sur ordre de Louis XIV. Cette fonte devait être exclusivement utilisée par l'Imprimerie Royale. Sur la base des résultats obtenus par l'Académie des Sciences, Grandjean a pour tâche de tailler le «Romain du Roi» en 1694. Les premiers types de cette fonte paraissent en 1702 et sont utilisés la première fois pour imprimer l'œuvre «Médailles sur les principaux Evénements du Règne de Louis le Grand». A la mort de Grandjean, sa fonte ne sera achevée et réalisée en 21 corps avec italique qu'en 1745 par son assistant Jean Alexandre. Elle a été employée pendant tout le 18e siècle et a inspiré de nombreuses variantes et imitations à P. F. Didot et P. S. Fournier.

**Granjon,** Robert – né en 1513, décédé le 16. 11. 1589 à Rome, Italie – *fondeur de caractères, tailleur de types, imprimeur, éditeur* – fils présumé de l'imprimeur et éditeur parisien Jean Granjon. Appren-

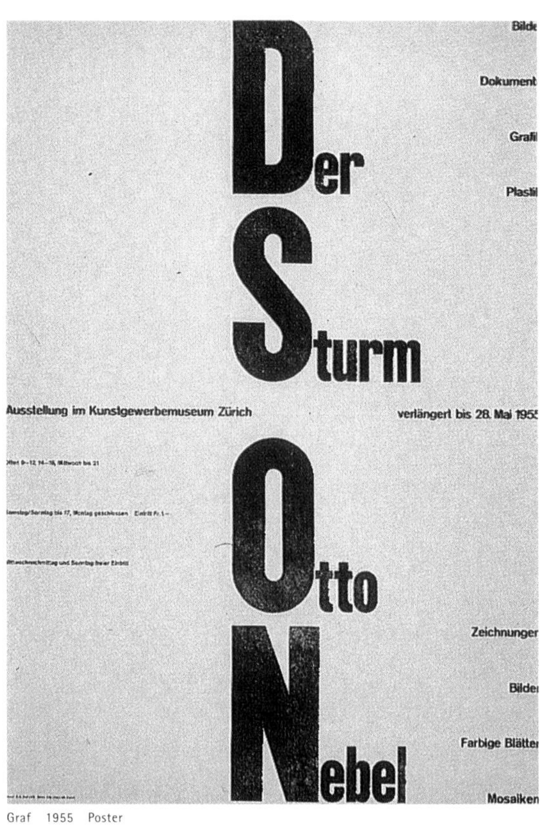

Graf 1955 Poster

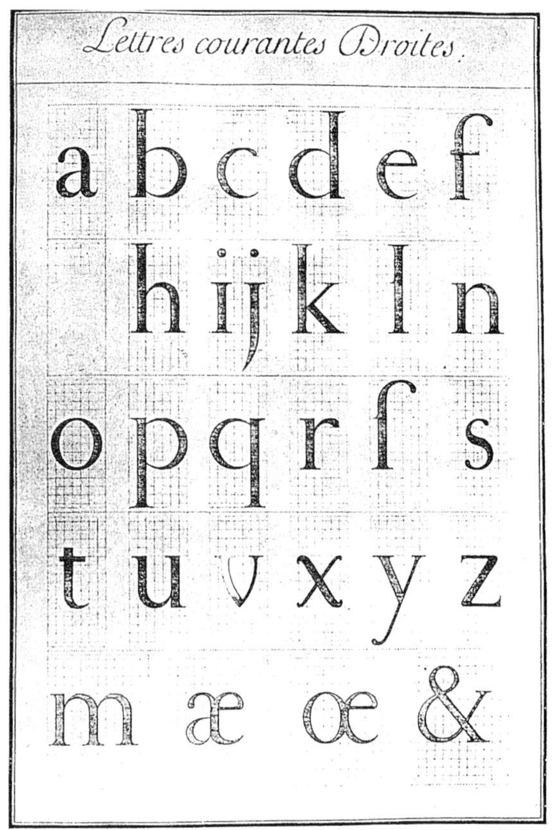

Grandjean 1692 Typeface

*Une deſcription hiſtorique & allégorique aurait accompagné les deſſins, & chaque page de la deſcription devait être entourée d'un cadre différent, également compoſé & gravé par Cochin. Au bas de chaque deſcription ſe trouverait un cul-de-lampe allégorique.*

Grandjean Romain du Roi, italic

**Graf,** Carl Bernhard – b. 13. 12. 1926 in Zurich, Switzerland, d. 16. 1. 1968 in Zurich, Switzerland – *graphic designer* – 1942–43: attends general art classes at the Kunstgewerbeschule in Zurich. 1943–47: trains as a graphic artist, then attends the advanced graphics class at the Kunstgewerbeschule in Zurich. 1948–53: works as a graphic designer for A. Wirz offices in Zurich. 1953–68: has his own studio in Zurich. Receives commissions from the pharmaceutical, clock and furniture industries. Numerous posters for the Helmhaus and the Kunstgewerbemuseum in Zurich.

**Grandjean de Fouchy,** Philippe – b. 1666 in Mâcon, France, d. 1714 in Paris, France – *type founder, punch cutter* – 1692: is asked to join the commission ordered by Louis XIV and organized by the Académie des Sciences for the production of a new roman typeface. The new type is intended for the exclusive use of the Imprimerie Royale. 1694: based on the results of the Académie des Sciences, Grandjean is asked to cut the Romain du Roi. 1702: initial designs are completed. They first appear in the work "Médailles sur les principaux Evénements du Règne de Louis le Grand". Grandjean's typeface is finally completed long after his death in 1745 by his assistant, Jean Alexandre, in 21 sizes with italics. It was in use throughout the 18th century and inspired P. F. Didot and P. S. Fournier, among others, to imitate it and carry on its tradition.

**Granjon,** Robert – b. 1513, d. 16. 11. 1589 in Rome, Italy – *type founder, punch cutter, printer, publisher* – It is assumed that

*gießer, Schriftschneider, Buchdrucker, Verleger* – Vermutlich Sohn des Pariser Druckers und Verlegers Jean Granjon. Lehre als Goldschmied. Seit 1543 Stempelschneider in Paris. 1543–48 als Stahlstempelschneider tätig. Als sein erster Druck erscheint 1549 in Paris das Neue Testament als griechisch-lateinische Ausgabe in Taschenbuchformat. 1550–51 Zusammenarbeit mit dem Schriftgießer Michel Felandat. 1556–57 als Stempelschneider, Schriftgießer, Drucker und Verleger in Lyon tätig. Entwurf der gotischen Kursivschrift „Civilité", die ihm große Anerkennung einbringt. Entwurf und Schnitt eines Musiknotensatzes für ein in Paris veröffentlichtes Gesangbuch. 1562 Rückkehr nach Paris, tätig als Schriftschneider und Schriftgießer. 1563–70 enge Arbeitskontakte zu Christoph Plantin. Als Gegenstück zu Garamonds „Grec du Roi" entsteht 1565 auf Veranlassung von Plantin die Schrift „Parangonne Grecque". 1570–74 bietet Granjon seine Schriften auf der Frankfurter Messe zum Kauf an. Auf Bitten Papst Gregors XIII. geht Granjon 1578 nach Rom, um für die Druckerei des Vatikans exotische Schriften zu schneiden. Bis zu seinem Tod entstehen neun Schriften und ein Musiknotensatz. Während seiner Tätigkeit als Stempelschneider entwirft Granjon ca. 50 verschiedene Alphabete, für die er ca. 6.000 Stempel schneidet.

**Grappa** – *Grafik-Design-Gruppe* – März 1989 : Gründung der Gruppe als Zusammenschluß freischaffender Grafik-Designer. Gründungsmitglieder: Kerstin Baar-

Granjon 1557 Main Title

Grappa 1994

Grappa Poster

Grappa 1990 Page (Detail)

tissage d'orfèvre. Tailleur de types à Paris à partir de 1543, taille des types d'acier de 1543 à 1548. Une édition en latin et en grec du Nouveau Testament en format livre de poche sort en 1549 de son atelier. De 1550 à 1551, il collabore avec le fondeur de caractères Michel Felandat. Il exerce ensuite de 1556 à 1557 comme tailleur de types, fondeur de caractères, imprimeur et éditeur à Lyon. Il dessine «Civilité», une italique gothique qui lui vaut la célébrité. Dessine et taille un jeu de notes de musique pour un livre de chants publié à Paris. Retour à Paris en 1562, où il travaille comme tailleur de types et fondeur de caractères. Contacts étroits avec Christoph Plantin de 1563 à 1570. En 1565, à l'initiative de Plantin, il réalise «Parangonne Grecque», une fonte qui est le contrepoint du «Grec du Roi» de Garamond. De 1570 à 1574, Granjon vend ses fontes à la foire de Francfort. En 1578, Granjon part pour Rome sur l'invitation du Pape Grégoire XIII, afin d'y tailler des fontes non latines pour l'imprimerie du Vatican. Jusqu'à sa mort, il crée neuf fontes et un jeu de notes de musique. Au cours de ses activités de tailleur de types, Granjon a dessiné une cinquantaine d'alphabets pour lesquels, il a taillé environ 6 000 caractères.

**Grappa** – *groupe de graphistes maquettiste* – ce groupe réunissant plusieurs graphistes maquettiste indépendants a

Robert was the son of the Paris printer and publisher Jean Granjon. Trained as a goldsmith. From 1543 onwards: punch cutter in Paris. 1543–48: works as a steel punch cutter. 1549: his first book is published in Paris, a pocket book edition of the New Testament in Greek and Latin. 1550–51: works with type founder Michel Felandat. 1556–57: works as a punch cutter, type founder, printer and publisher in Lyon. Designs the Gothic italic typeface Civilité which earns him great acclaim. Designs and cuts the notes and musical symbols for a song book published in Paris. 1562: returns to Paris where he works as a punch cutter and type founder. 1563–70: works closely with Christoph Plantin. 1565: Granjon produces his Parangonne Greque typeface at the instigation of Plantin as a counterpart to Garamond's Grec du Roi. 1570–74: Granjon offers his typefaces for sale at the Frankfurt fair. 1578: Granjon goes to Rome at the request of Pope Gregory XIII to cut exotic type for the Vatican's printing workshop. During his lifetime Granjon created 9 typefaces and a set of musical symbols. Whilst working as a punch cutter, Granjon designed c. 50 different alphabets, for which he cut c. 6,000 punches.

**Grappa** – *graphic design group* – March 1989: the group is formed by a number of freelance graphic designers. Founder

Grappa
**Grapus**

mann – geb. 1961 in Berlin, Deutschland. – *Grafik-Designerin, Schriftsetzerin* – Schriftsetzerlehre. 1982–86 Studium an der Fachschule für Werbung und Gestaltung in Berlin-Schöneweide. 1986–88 Typographie im Aufbau-Verlag und im Verlag Volk und Welt in Berlin. 1990 Gestaltungskonzept der Zeitschrift „Ypsilon" (mit D. Haufe und H. Grebin). Dieter Fehsecke – geb. 1955 in Salzwedel, Deutschland. – *Grafik-Designer* – Schildermalerlehre. 1981–85 Grafik-Design-Studium an der Fachschule für Werbung und Gestaltung in Berlin-Schöneweide. 1985–88 Grafiker bei der DEWAG. 1988–89 Erscheinungsbild für das Bauhaus Dessau (mit D. Fiedler). Zahlreiche Auszeichnungen. Andreas R. Trogisch – geb. 1959 in Riesa, Deutschland. – *Grafik-Designer, Lehrer* – 1983–88 Studium an der Fachschule für Werbung und Gestaltung in Berlin-Schöneweide. Lehrauftrag an der Hochschule der Künste Berlin. Zahlreiche Auszeichnungen. Weitere Gesellschafterinnen sind Heike Grebin – geb. 1959 in Rostock, Deutschland und Ute Zscharndt – geb. 1964 in Zinnowitz, Deutschland. Grappa arbeitet vorwiegend im Kulturbereich. Auftraggeber waren u.a. die Akademie der Künste in Berlin, das Bauhaus Dessau, die Brandenburgische Kunstsammlungen Cottbus, das Filmmuseum Potsdam und die Internationalen Kurzfilmtage Oberhausen, die Verlage Cantz, Ernst & Sohn, Henschel, Nicolai, Rütten & Loening. Zahlreiche Vorträge an Kunst- und Gestaltungsschulen.

**Grapus** – *Grafik-Design-Studio* – Pierre

été fondé en mars 1989. Les membres fondateurs étaient : Kerstin Baarmann – née en 1961 à Berlin, Allemagne – *graphiste maquettiste, compositeur en imprimerie* – apprentissage de la composition en imprimerie. 1982–1986, études à la Fachschule für Werbung und Gestaltung (école de publicité et de design) de Berlin-Schöneweide. 1986–1988, typographe au Aufbau-Verlag et aux éditions Volk und Welt à Berlin. 1990, conception de la maquette de la revue «Ypsilon» (avec D. Haufe et H. Grebin). Dieter Fehsecke – né en 1955 à Salzwedel, Allemagne – *graphiste maquettiste* – apprentissage de peintre d'enseignes. 1981–1985, études de graphisme et de design à la Fachschule für Werbung und Gestaltung de Berlin-Schöneweide. 1985–1988, graphiste chez DEWAG. 1988–1989, identité pour le Bauhaus Dessau (avec D. Fiedler). Nombreuses distinctions. Andreas R. Trogisch – né en 1959 à Riesa, Allemagne – *graphiste maquettiste, enseignant* – 1983–1988, études à la Fachschule für Werbung und Gestaltung de Berlin-Schöneweide. Chargé de cours à la Hochschule der Künste (école supérieure des beaux-arts) de Berlin. Nombreuses distinctions. Parmi les autres membres de Grappa, on compte Heike Grebin – née en 1959 à Rostock, Allemagne et Ute Zscharndt – née en 1964 à Zinnowitz, Allemagne. Le groupe Grappa travaille surtout pour le secteur culturel. Commanditaires : l'Akademie der Künste à Berlin, le Bauhaus Dessau, le Brandenburgische Kunstsammlungen à Cottbus, le musée de films à Potsdam et les Internationale Kurzfilmtage Oberhausen, les éditions Cantz, Ernst & Sohn, Henschel, Nicolai, Rütten & Loening. Nombreuses conférences dans des écoles d'art et de design.

**Grapus** – *agence de graphisme et de de-*

Grapus 1981 Poster

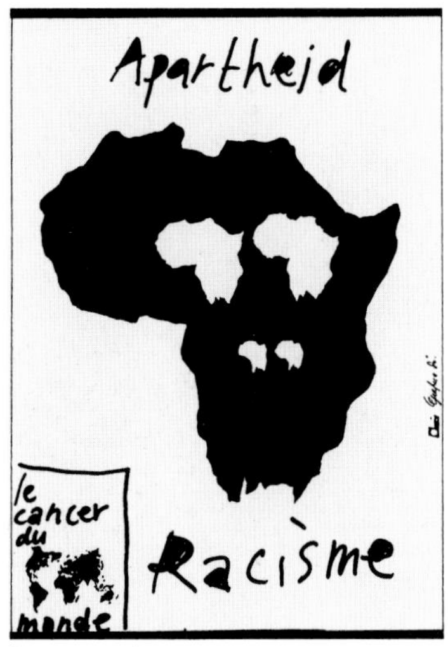


Grapus 1986 Poster

LES TROMPETTES DE LA MORT

Grapus Invitation Card

Grapus 1982 Poster

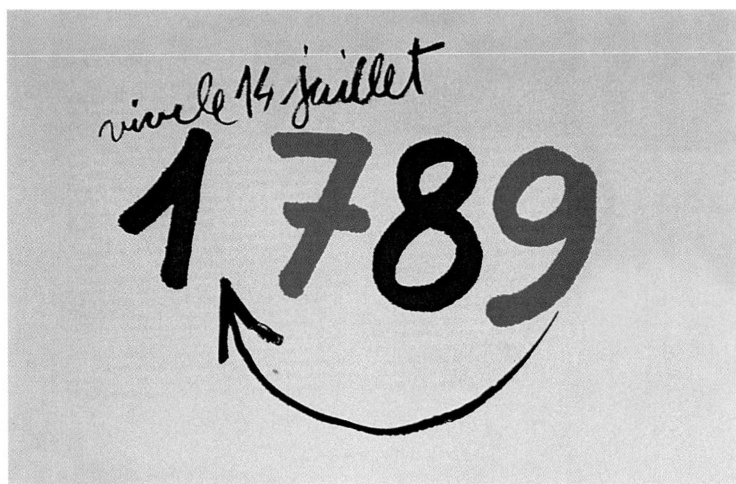


Grapus 1978 Poster

members: Kerstin Baarman – b. 1961 in Berlin, Germany – *graphic designer, typesetter* – Trained as a typesetter. 1982–86: studies at the Fachhochschule für Werbung und Gestaltung in Berlin-Schöneweide. 1986–88: typography for Aufbau-Verlag and for the publishers Volk und Welt in Berlin. 1990: design concept for "Ypsilon" magazine (with D. Haufe and H. Grebin). Dieter Fehsecke – b. 1955 in Salzwedel, Germany – *graphic designer* – Trained as a sign-writer. 1981–85: studies graphic design at the Fachschule für Werbung und Gestaltung in Berlin-Schöneweide. 1985–88: graphic artist for DEWAG. 1988–89: designs the corporate imagery for the Bauhaus Dessau (with D. Fiedler). Numerous awards. Andreas R. Trogisch – b. 1959 in Riesa, Germany – *graphic designer, teacher* – 1983–88: studies at the Fachschule für Werbung und Gestaltung in Berlin-Schöneweide. Teaching position at the Hochschule der Künste in Berlin. Numerous awards. Further partners are Heike Grebin – b. 1959 in Rostock, Germany and Ute Zscharndt – b. 1964 in Zinnowitz, Germany. Grappa works mainly for the arts sector. Clients include the Akademie der Künste in Berlin, the Bauhaus Dessau, the Brandenburgische Kunstsammlungen in Cottbus, the film museum in Potsdam, the Internationale Kurzfilmtage in Oberhausen, and the publishers Cantz, Ernst & Sohn, Henschel, Nicolai and Rütten & Loening. Numerous lectures at various art and design schools.

**Grapus** – *graphic design studio* – Pierre Bernard – b. 25.2.1942 in Paris, France,

Bernard – geb. 25. 2. 1942 in Paris, Frankreich, Gérard Paris-Clavel – geb. 2. 10. 1943 in Paris, Frankreich und François Miehe – geb. 10. 4. 1942 in La Ferté-Alais, Frankreich lernen sich 1968 im Atelier Populaire der Ecole des Arts Décoratifs in Paris kennen. Die drei gründen 1970 das Studio „Grapus" als ein kreatives Kollektiv mit dem Ziel, Gemeinden, Stadträte, Gewerkschaften und die linken Bewegungen bei ihrer Öffentlichkeitsarbeit zu unterstützen. Es entstehen zahlreiche sozialpolitische Kampagnen und Plakate für kulturelle Ereignisse. 1975 Jean-Paul Bachollet wird Geschäftsführer von „Grapus". Entwurf des Erscheinungsbildes für die Gewerkschaft CGT. 1976 Eintritt von Alexander Jordan in die Gruppe. 1978 F. Miehe tritt aus der Gruppe aus, um eine Lehrtätigkeit an der Ecole Nationale Supérieure des Arts Décoratifs (ENSAD) in Paris zu übernehmen. 1989 Entwurf des Erscheinungsbildes für den Louvre in Paris. 1990 tritt P. Bernard aus der Gruppe aus, um das „Atelier de Création Graphique" mitzugründen. 1991 Verleihung des „Grand Prix National de l'Art Graphique en France" an Grapus. Auflösung der Gruppe in drei unabhängige Designbüros. – *Publikation u. a.:* „Grapus 85", Utrecht 1985.

**Greiman,** April – geb. 22. 3. 1948 in New York, USA. – *Grafik-Designerin, Lehrerin* – 1966–70 Studium am Kansas City Art Institute, Missouri. 1970–71 Studium an der Allgemeinen Kunstgewerbeschule Basel (u. a. bei Armin Hofmann und Wolfgang Weingart). 1971–75 Grafik-Designerin in New York und Philadelphia.

Greiman 1986 Poster

Greiman 1997 Stationery

Greiman 1991 Poster

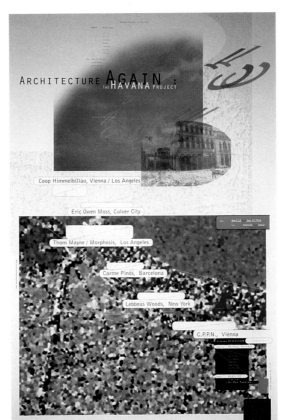

Greiman 1983 Poster

*sign* – Pierre Bernard – né le 25. 2. 1942 à Paris, France, Gérard Paris-Clavel – né le 2. 10. 1943 à Paris, France et François Miehe – né le 10. 4. 1942 à La Ferté-Alais, France après leur rencontre à l'Atelier Populaire de l'Ecole des Arts Décoratifs de Paris. En 1970, ils fondent ensemble l'agence «Grapus», un collectif de créateurs ayant pour objet de soutenir les communes, municipalités, syndicats et les mouvements de gauche dans leur communication avec le public. Ils se chargent de nombreuses campagnes de politique sociale et réalisent des affiches pour des manifestations culturelles. En 1975, Jean-Paul Bachollet devient administrateur de «Grapus». Conception de l'identité de la CGT. 1976, entrée d'Alexander Jordan. Miehe quitte le groupe en 1978 pour enseigner à l'Ecole Nationale Supérieure des Arts Décoratifs (ENSAD) de Paris. Création de l'identité du Louvre en 1989. En 1990, P. Bernard quitte le groupe pour fonder l' «Atelier de Création Graphique». Grapus obtient le «Grand Prix National de l'Art Graphique en France» en 1991. Dissolution du groupe en trois agences de design indépendantes. – *Publication, sélection :* «Grapus 85», Utrecht 1985.

**Greiman,** April – née le 22. 3. 1948 à New York, Etats-Unis – *graphiste maquettiste, enseignante* – 1966–1970, études au Kansas City Art Institute, Missouri. 1970–1971, études à la Allgemeine Kunstgewerbeschule (école générale des arts décoratifs) de Bâle (entre autres, chez Armin Hofmann et Wolfgang Weingart). 1971–1975, graphiste maquettiste à New

Gérard Paris-Clavel – b. 2. 10. 1943 in Paris, France and François Miehe – b. 10. 4. 1942 in La Ferté-Alais, France meet at the Atelier Populaire at the Ecole des Arts Décoratifs in Paris. 1970: the three designers open the Grapus studio, a creative collective which aims to support communities, town councils, labor unions and the political left in their PR work. Grapus organizes many socio-political campaigns and designs numerous posters for cultural events. 1975: Jean-Paul Bachollet is appointed as manager of Grapus. Grapus designs the corporate imagery for the CGT labor union. 1976: Alexander Jordan joins the group. 1978: F. Miehe leaves the group to take on a teaching post at the Ecole Nationale Supérieure des Arts Décoratifs (ENSAD) in Paris. 1989: Grapus designs the corporate imagery for the Louvre in Paris. 1990: P. Bernard leaves the group and is co-founder of the Atelier de Création Graphique. 1991: the Grand Prix National de l'Art Graphique en France is awarded to Grapus. The group disbands and forms three independent design studios. – *Publications include:* "Grapus 85", Utrecht 1985.

**Greiman,** April – b. 22. 3. 1948 in New York, USA – *graphic designer, teacher* – 1966–70: studies at the Kansas City Art Institute in Missouri. 1970–71: studies at the Allgemeine Kunstgewerbeschule in Basle (tutors include Armin Hofmann and Wolfgang Weingart). 1971–75:

Auftraggeber waren u. a. das Museum of Modern Art New York, The Architect's Collaborative Boston, Anspach / Grossman / Portugal in New York. Unterrichtet 1971–75 am Philadelphia College of Art und seit 1994 am Southern California Institute of Architecture (SCI-Arc). 1976 Gründung ihres eigenen Studios in Los Angeles. Zusammenarbeit mit dem Fotografen Jayme Odgers. 1982–84 Leiterin des Visual Communications Program am California Institute of the Arts. Greiman gehört zu der Generation, die das kalifornische Design seit den 80er Jahren repräsentiert. Durch die Verbindung der Lehre der Baseler Schule mit den neuen medialen Entwicklungen (Apple Macintosh) und kalifornischem Zeitgeist entsteht ein reichhaltiger grafischer Stil, der Vorbild wird. – *Publikationen u.a.:* „Hybrid Imagery", New York 1990; „It's Not What You Think It Is", Zürich, München, London 1994.

**Grieshaber,** Helmut Andreas Peter (HAP) – geb. 15. 2. 1909 in Rot, Deutschland, gest. 12. 5. 1981 in Reutlingen, Deutschland. – *Grafiker, Typograph, Künstler, Lehrer* – 1926–27 Schriftsetzerlehre in Reutlingen. 1926–28 Kalligraphie-Kurse im Meisteratelier von F. H. E. Schneidler in Stuttgart. 1928–31 Studium in Paris und London. 1931 erste Ausstellung in London. Lebt 1931–33 in Griechenland und im Vorderen Orient. Ab 1947 freischaffend in Reutlingen (auf der Achalm) tätig. 1951–53 Leiter der Bernsteinschule in Sulz. 1955 Berufung zum Professor an die Kunstakademie Karlsruhe als Nachfolger Erich Heckels. Seit 1957 Her-

York et à Philadelphie. Commanditaires : le Museum of Modern Art, New York ; The Architects' Collaborative, Boston ; Anspach / Grossman / Portugal à New York. Enseigne au Philadelphia College of Art de 1971 à 1975 et depuis 1994 à la Southern California Institute of Architecture (SCI-Arc). Fonde sa propre agence à Los Angeles en 1976. Collaboration avec le photographe Jayme Odgers. De 1982 à 1984 elle dirige le Visual Communications Program à la California Institute of the Arts. Greiman fait partie de la génération qui a fait connaître le design californien dans le monde entier pendant les années 80. En alliant l'enseignement de l'école de Bâle, le développement des nouveaux médias (Apple Macintosh) et l'esprit qui règne en Californie, elle s'est créé un style graphique riche en idées et exemplaire au niveau. – *Publications, sélections :* «Hybrid Imagery», New York 1990 ; «It's Not What You Think It Is», Zurich, Munich, Londres 1994.

**Grieshaber,** Helmut Andreas Peter (HAP) – né le 15. 2. 1909 à Rot, Allemagne, décédé le 12. 5. 1981 à Reutlingen, Allemagne – *graphiste, typographe, artiste, enseignant* – 1926–1927, apprentissage de composition en imprimerie à Reutlingen. 1926–1928, cours de calligraphie à l'atelier de F. H. E. Schneidler à Stuttgart. 1928–1931, études à Paris et à Londres. Première exposition à Londres en 1931. Séjour en Grèce et au Proche Orient de 1931 à 1933. A partir de 1947, il exerce comme indépendant à Reutlingen (auf der Achalm). 1951–1953 dirige la Bernsteinschule à Sulz. En 1955, il est nommé professeur à la Kunstakademie (beaux-arts) de Karlsruhe pour succéder à Erich Heckel. A partir de 1957, il publie «Achalmdrucke», la première édition est nommée «Hommage à Werkman». En 1961, il obtient le «Kunstpreis» de la ville

Grieshaber 1955 Cover

Grieshaber 1948 Poster

Grieshaber 1948 Poster

Grieshaber 1960 Cover

graphic designer in New York and Philadelphia. Clients include the Museum of Modern Art in New York, the Architects' Collaborative in Boston and Anspach / Grossman / Portugal in New York. 1971–75: teaches at the Philadelphia College of Art. Since 1994: teaches at the Southern California Institute of Architecture (SCI-Arc). 1976: opens her own studio in Los Angeles. Works with the photographer Jayme Odgers. Greiman belongs to the generation which has represented design in California at an international level since the 1980s. 1982–1984: head of the Visual Communications Program at the California Institute of the Arts. The combination of Greiman's training at the Basle school with new media developments (Apple Macintosh) and the Californian spirit of the age has produced a varied graphic style which has become a model. – *Publications include:* "Hybrid Imagery", New York 1990; "It's Not What You Think It Is", Zurich, Munich, London 1994.

**Grieshaber,** Helmut Andreas Peter (HAP) – b. 15. 2. 1909 in Rot, Germany, d. 12. 5. 1981 in Reutlingen, Germany – *graphic artist, typographer, artist, teacher* – 1926–27: trains as a typesetter in Reutlingen. 1926–28: takes calligraphy courses at F. H. E. Schneidler's master studio in Stuttgart. 1928–31: studies in Paris and London. 1931: first exhibition in London. 1931–33: lives in Greece and the Near East. From 1947 onwards: works freelance in Reutlingen. 1951–53: head of the Bernsteinschule in Sulz. 1955: professorship at the Kunstakademie in

ausgabe der „Achalmdrucke", erste Ausgabe „Hommage à Werkman". Ausgezeichnet mit dem Kunstpreis der Stadt Düsseldorf 1961. Seit 1964 Herausgeber der Publikationsreihe „Engel der Geschichte". 1968 Auszeichnung mit dem Kulturpreis des Deutschen Gewerkschaftsbundes (DGB). 1971 Auszeichnung mit dem erstmals vergebenen Dürer-Preis der Stadt Nürnberg. 1978 Auszeichnung mit dem Gutenberg-Preis der Stadt Leipzig. Vor allem die Typographie der frühen Plakate ab 1947 und seine experimentellen Pressedrucke machen Grieshaber zu einem für die gestalterische Szene interessanten und unabhängigen Anreger. – *Publikationen u.a.:* „Poesia typographica", Köln 1962; „Malerbriefe", Stuttgart 1967; „Totentanz von Basel", Dresden 1968. W. Boeck „HAP Grieshaber, Holzschnitte", Pfullingen 1959; M. Fuerst „Grieshaber – der Drucker und Holzschneider", Stuttgart 1965; M. Fuerst „Grieshaber – die Plakate 1934–79", Stuttgart 1979; M. Fuerst „HAP Grieshaber. Die Druckgraphik" (2 Bände), Stuttgart 1984–87.

**Griffin,** Rick – geb. in Palos Verdes, USA. – *Grafik-Designer, Illustrator* – Seit 1962 Mitarbeiter bei der Zeitschrift „Surfer Magazine", für die er die Surfer-Cartoon-Figur „Murphy" kreiert („Murph the Surf"). Zahlreiche Airbrush-Bilder. Kurze Zeit Studium an der Chouinard Art School in Los Angeles (heute California Institute of the Arts). Spielt mit der Musik-Gruppe „Jook Savages", für die er 1966 sein erstes psychedelisches Plakat entwirft. Weitere Plakatentwürfe für die

Griffin 1968 Poster

Griffin 1967 Poster

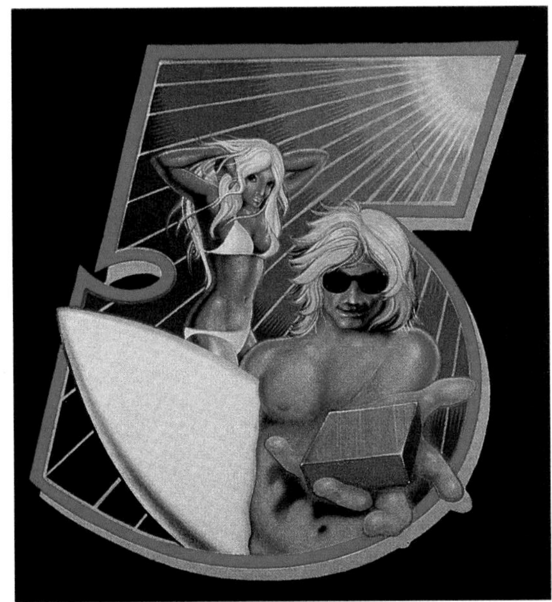
Griffin 1972 Poster

Griffin 1970 Cover

de Düsseldorf. Publie la collection «Engel der Geschichte» à partir de 1964. En 1968, il obtient le Prix de la culture du Deutscher Gewerkschaftsbund (DGB). En 1971, le tout premier Prix Dürer de la ville de Nuremberg lui est décerné, puis le prix Gutenberg de la ville de Leipzig en 1978. La typographie de ses premières affiches et ses publications expérimentales réalisées à partir de 1947 ont fait de Grieshaber l'un des innovateurs les plus intéressants en matière de design. – *Publications, sélection:* «Poesia typographica», Cologne 1962; «Malerbriefe», Stuttgart 1967; «Totentanz von Basel», Dresde 1968. W. Boeck «HAP Grieshaber, Holzschnitte», Pfullingen 1959; M. Fuerst «Grieshaber – der Drucker und Holzschneider», Stuttgart 1965; M. Fuerst «Grieshaber – die Plakate 1934–1979», Stuttgart 1979; M. Fuerst «HAP Grieshaber. Die Druckgraphik» (2 vol.), Stuttgart 1984–1987.

**Griffin,** Rick – né à Palos Verdes, Etats-Unis – *graphiste maquettiste, illustrateur* – collabore à partir de 1962 à la revue «Surfer Magazine» pour laquelle il crée le personnage de bande dessinée «Murphy» («Murph the surf»). Nombreux tableaux à l'aérographe. Courtes études à la Chouinard Art School de Los Angeles (aujourd'hui California Institute of the Arts). Musicien du groupe «Jook Savages» pour qui il dessine sa première affiche psychédélique en 1966. Création

Karlsruhe as Erich Heckel's successor. From 1957 onwards: publishes the "Achalmdrucke", the first being "Hommage à Werkman". 1961: is awarded the city of Düsseldorf's art prize. From 1964 onwards: publisher of the "Engel der Geschichte" series. 1968: is awarded the Deutsche Gewerkschaftsbund's (DGB) arts prize. 1971: the city of Nuremberg's first Dürer Prize is given to Grieshaber. 1978: is awarded the Gutenberg Prize by the city of Leipzig. During his lifetime, Grieshaber provided interesting stimulus for the design scene as an independent, in particular with the typographic work of his early posters from 1947 onwards and his experimental publications. – *Publications include:* "Poesia typographica", Cologne 1962; "Malerbriefe", Stuttgart 1967; "Totentanz von Basel", Dresden 1968. W. Boeck "HAP Grieshaber, Holzschnitte", Pfullingen 1959; M. Fuerst "Grieshaber – der Drucker und Holzschneider", Stuttgart 1965; M. Fuerst "Grieshaber – die Plakate 1934–79", Stuttgart 1979; M. Fuerst "HAP Grieshaber. Die Druckgraphik" (2 vols.), Stuttgart 1984–87.

**Griffin,** Rick – b. in Palos Verdes, USA – *graphic designer, illustrator* – 1962: starts working for "Surfer Magazine", for which he creates the surfing cartoon figure, "Murphy" ("Murph the Surf"). Numerous airbrush pictures. Spends a short period studying at the Chouinard Art School in Los Angeles (now the California Institute of the Arts). Plays in the band Jook Savages and designs his first psychedelic poster for them in 1966. Designs posters

Aktivitäten der „Family Dog" im Avalon Ballroom und Bill Grahams „Fillmore Auditorium". Gründung des Verlags „Berkeley Bonapart", der psychedelische Grafik und Plakate produziert. 1967 Entwurf des Signets der Zeitschrift „Rolling Stone". 1968 Umschlaggestaltung für die Zeitschrift „The Oracle". 1969 Gestaltung des Covers für die Schallplatte „Aoxomoxoa" der Musikgruppe Grateful Dead. Mitarbeit am Sammelband „The Beatles Songbook". 1970 Plakatentwurf zu John Seversons Surfer-Film „Pacific Vibrations". 1976 Ausstellung seiner Arbeiten im „Roundhouse" in London. – *Publikation u. a.:* Gordon McClelland „Rick Griffin", New York 1980.

**Griffo,** Francesco (auch Francesco da Bologna) – geb. 1450, gest. 1518 in Bologna, Italien. – *Schriftgießer, Stempelschneider, Schriftentwerfer* – Lehre als Schriftschneider und Schriftgießer in Bologna. 1495 Zusammenarbeit mit Aldus Manutius, für den er griechische Alphabete und 1501 eine Kursivschrift für eine Vergil-Ausgabe schneidet. Schneidet 1496 das Alphabet für das Werk „De Aetna" von Pietro Bembo. 1502 Trennung von Aldus Manutius. 1503 Schnitt einer Kursivschrift für den Drucker Gershom Soncino in Fano. 1516 eigene Druckerei in Bologna. Für sein erstes Druckwerk schneidet er seine dritte Kursivschrift.

**Grignani,** Franco – geb. 4. 2. 1908 in Pieve Porto Morone, Italien. – *Grafik-Designer, Maler, Lehrer* – 1929–33 Architekturstudium in Turin. Seit 1932 Arbeit als freier Maler, Designer und Fotograf in Mailand. Mitglied der zweiten futuristischen Be-

d'affiches pour les activités de «Family Dog» à l'Avalon Ballroom et pour le «Fillmore Auditorium» de Bill Graham. Fonde les éditions «Berkeley Bonapart» qui produisent des œuvres graphiques psychédéliques et des affiches. En 1967, il dessine le signet de la revue «Rolling Stone». Maquette de la couverture de la revue «The Oracle» en 1968. Affiche et pochette du disque «Aoxomoxoa» de la groupe de la musique Grateful Dead en 1969. Collaboration à l'album «The Beatles Songbook». Dessine l'affiche de «Pacific Vibrations», un film de John Serverson sur le surf, en 1970. Exposition de ses travaux à la «Roundhouse» de Londres en 1976. – *Publications, sélection:* Gordon McClelland «Rick Griffin», New York 1980.

**Griffo,** Francesco (aussi Francesco da Bologna) – né en 1450, décédé en 1518 à Bologne, Italie – *fondeur de caractères, tailleur de types, concepteur de polices* – apprentissage de la taille de types et de la fonte de caractères à Bologne. En 1495, il travaille avec Alde Manuce, pour lequel il taille un alphabet grec et une italique pour une édition de Virgile en 1501. En 1496, il taille l'alphabet utilisé pour «De Aetna» de Pietro Bembo. Griffo quitte Alde Manuce en 1502. En 1503, il taille une italique pour l'imprimeur Gershom Soncino à Fano. Ouvre sa propre imprimerie à Bologne en 1516. Il taille sa troisième italique pour le premier ouvrage qu'il imprimera.

**Grignani,** Franco – né le 4. 2. 1908 à Pieve Porto Morone, Italie – *graphiste maquettiste, peintre, enseignant* – 1929– 1933, études d'architecture à Turin. Travaille comme peintre, designer et photographe indépendant à Milan à partir de 1932. Membre du second mouvement futuriste. 1948–1960, directeur artistique de la revue «Bellezza d'Italia», et membre fondateur du Gruppo Exhibition Design à

| | n | | | | |
|---|---|---|---|---|---|
| a | b | e | f | g | i |
| o | r | s | t | y | z |
| A | B | C | E | G | H |
| M | O | R | S | X | Y |
| 1 | 2 | 4 | 6 | 8 | & |

Griffo 1499 Poliphilus

Grignani 1963 Advertisement

Grignani ca. 1960 Advertisement

Grignani 1967 Advertisement

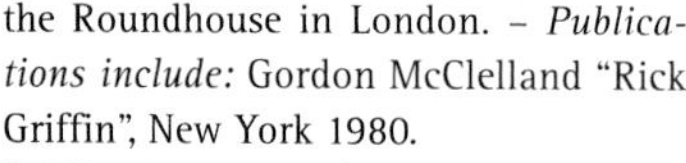

for Family Dog activities in the Avalon Ballroom and Bill Graham's Fillmore Auditorium. Launches Berkeley Bonapart publishing house which produces psychedelic graphics and posters. 1967: designs the logo for "Rolling Stone" magazine. 1968: designs the cover for "The Oracle" magazine. 1969: designs the cover and poster for The Grateful Dead's "Aoxmoxoa" record. Works on "The Beatles Songbook". 1970: designs the poster for John Severson's surfing film "Pacific Vibrations". 1976: his work is exhibited in the Roundhouse in London. – *Publications include:* Gordon McClelland "Rick Griffin", New York 1980.

**Griffo,** Francesco (also Francesco da Bologna) – b. 1450, d. 1518 in Bologna, Italy – *type founder, punch cutter, type designer* – Trained as a punch cutter and type founder in Bologna. 1495: works with Aldus Manutius, for whom he cuts Greek alphabets and in 1501 an italic typeface for a Virgil edition. 1496: cuts the alphabet for Pietro Bembo's "De Aetna". 1502: parts company with Aldus Manutius. 1503: cuts an italic typeface for the printer Gershom Soncino in Fano. 1516: opens his own printing workshop in Bologna. He cuts his third italic alphabet for his first publication.

**Grignani,** Franco – b. 4. 2. 1908 in Pieve Porto Morone, Italy – *graphic designer, painter, teacher* – 1929–33: studies architecture in Turin. 1932: starts work as a freelance painter, designer and photographer in Milan. Member of the second Futurist movement. 1948–60: art director for "Bellezza d'Italia" magazine.

wegung. 1948–60 Art Director der Zeitschrift „Bellezza d'Italia", 1969 Gründungsmitglied der Gruppo Exhibition Design in Mailand. 1969–81 Präsident der italienischen Sektion der Alliance Graphique Internationale (AGI). Seit 1979 unterrichtet er Visuelle Kommunikation an der Nuova Accademia di Belle Arti in Mailand. Jurymitglied des Wettbewerbs „Typomundus 20". Zahlreiche internationale Ausstellungen seiner Malerei, Fotografie und Gebrauchsgrafik. – *Publikationen u.a.:* U. Apollonio, G. Celant „Franco Grignani", Florenz 1965; G. Dorfles „Franco Grignani"; Mailand 1966; G. C. Argan „Franco Grignani", Mailand 1967.

**Groot,** Luc(as) de – geb. 21.6.1963 in Noordwijkerhout, Niederlande. – *Schriftentwerfer, Grafik-Designer, Lehrer* – 1975–82 Gymnasium in Noordwijkerhout, Entwurf zahlreicher Schülerzeitschriften, Plakate und Illustrationen. 1982–87 Studium an der Koninklijke Academie van Beeldende Kunsten in Den Haag, u. a. bei Gerrit Noordzij. Entwickelt 1987 seine Interpolations-Theorie über die optische Erscheinung der Strichstärke von Buchstaben. (Später entwickelt er durch Interpolation einen Zwischenschnitt der Schriften Frutiger 45 und 55, die als neue Schrift für die Design-Zeitschrift „form" dient.) 1988 freier Mitarbeiter in den Studios „Tint", „2D3D", „Studio Dumbar". Neben anderen Aufgaben entwirft er Schriften für Buchumschläge. 1989–93 Arbeiten im Design-Büro „BRS Premsela Vonk" an Corporate-Identity-Projekten, u. a. Entwurf einer Schrift für das niederländische Transport-Ministeri-

Grignani 1949 Advertisement

Grignani 1965 Cover

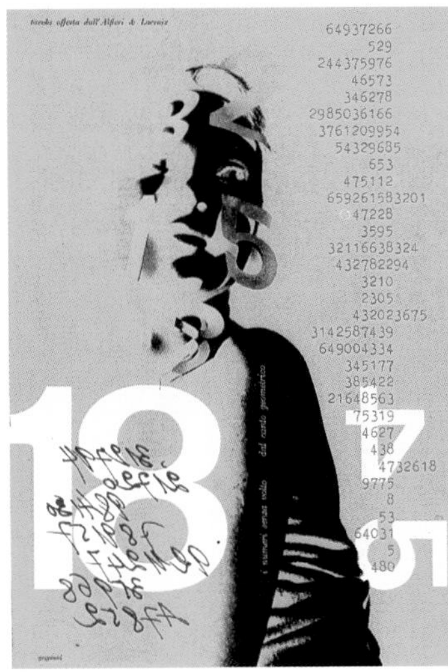
Grignani 1957 Advertisement

Groot 1993 Illustration

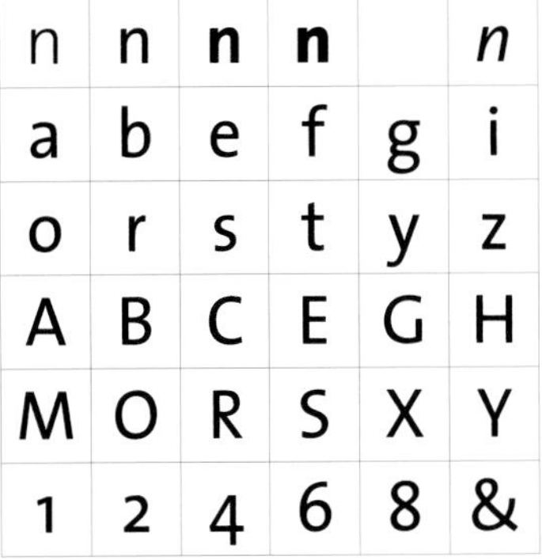

Groot 1994 TheSans

TheSerif specimen:

Groot 1994 TheSerif

Groot 1992 Illustration

Milan en 1969. Président de la section italienne de l'Alliance Graphique Internationale (AGI) de 1969 à 1981. Enseigne la communication visuelle à la Nuova Accademia di Belle Arti de Milan depuis 1979. Membre du jury du concours «Typomundus 20». Nombreuses expositions internationales de ses tableaux, photographies et graphismes industriels. – *Publications, sélection:* U. Apollonio, G. Celant «Franco Grignani», Florence 1965; G. Dorfles «Franco Grignani», Milan 1966; G. C. Argan «Franco Grignani», Milan 1967.

**Groot,** Luc(as) de – né le 21.6.1963 à Noordwijkerhout, Pays-Bas – *concepteur de polices, graphiste maquettiste, enseignant* – études au lycée de Noordwijkerhout de 1975 à 1982; il y conçoit plusieurs journaux scolaires et réalise des affiches et illustrations. 1982–1987, études à la Koninklijke Academie van Beeldende Kunsten de La Haye et chez Gerrit Noordzij. En 1987, il développe sa théorie de l'interpolation sur l'effet optique du corps des bâtons des lettres. (Par la suite, il se servira de l'interpolation pour créer une fonte intermédiaire entre la Frutiger 45 et la Frutiger 55 destinée à la revue de design «form»). En 1988, il collabore comme indépendant avec les studios «Tint», «2D3D», et le «Studio Dumbar». Entre autres il crée des polices pour des couvertures de livres et dans d'autres cadres. De 1989 à 1993, il travaille pour l'agence de design «BRS Premsela Vonk» sur des projets d'identités de sociétés et

1969: founder member of the Gruppo Exhibition Design in Milan. 1969–81: president of the Italian section of the Alliance Graphique Internationale (AGI). Since 1979: has taught visual communication at the Nuova Accademia di Belle Arti in Milan. Member of the jury for the Typomundus 20 competition. There have been numerous international exhibitions of his paintings, photographs and commercial art. – *Publications include:* U. Apollonio, G. Celant "Franco Grignani", Florence 1965; G. Dorfles "Franco Grignani", Milan 1966; G.C. Argan "Franco Grignani", Milan 1967.

**Groot,** Luc(as) de – b. 21.6.1963 in Noordwijkerhout, The Netherlands – *type designer, graphic designer, teacher* – 1975–82: high school in Noordwijkerhout, designs numerous school magazines, posters and illustrations. 1982–87: studies at the Koninklijke Academie van Beeldende Kunsten in The Hague. Teachers include Gerrit Noordzij. 1987: develops his interpolation theory on the optical appearance of the thickness of letter lines. (He later develops a type between Frutiger 45 and 55 using interpolation as a new typeface for "form" design magazine). 1988: freelancer for Tint and 2D3D studios and Studio Dumbar. He designs typefaces for book covers among other tasks. 1989–93: works on corporate identity projects at the design studio BRS Premsela Vonk, including designing a

um. Beginn der Arbeit an seiner Schrift „Thesis". Entwickelt 1991 in London ein Kerning-Programm zum Optimieren der Buchstaben-Abstände bei klassischen Schriften wie Bembo, Garamond, Univers, Futura. Unterrichtet 1992 an der Kunstakademie in Den Bosch. Entwirft von 1992–94 eine Serie von „Trash Faces". 1993 Umzug nach Berlin, Eintritt in das Studio „MetaDesign". Ausbau der Schrift FF Meta von Erik Spiekermann auf 38 Schnitte. Unterrichtet 1995 an der Kunsthochschule Berlin-Weißensee. 1996 an der Kunsthochschule Potsdam. – *Schriftentwürfe:* FF TheMix (1994), FF TheSans (1994), FF TheSerif (1994), FF Nebulae (1994), FF Thesis (1994), FF JesusLovesYouAll (1995), FF MoveMeMultipleMaster (für „Fuse", 1995), FF TheSansMonospace (1996), FF TheSansTypewriter (1996).

**Groothuis,** Rainer – geb. 11. 6. 1959 in Emden, Deutschland. – *Typograph, Grafik-Designer* – Buchhändlerlehre in Emden, 1982 Umzug nach Berlin, Gestalter im Verlag Klaus Wagenbach. 1984 Herstellungsleiter, 1989 Geschäftsführer. 1996 Umzug nach Bremen, Ateliergemeinschaft mit Victor Malsy. Zahlreiche Veröffentlichungen zu Fragen der Buchbranche und der Buchgestaltung. Zahlreiche Auszeichnungen.

**Gürtler,** André – geb. 5. 9. 1936 in Basel, Schweiz. – *Grafik-Designer, Typograph, Schriftentwerfer, Lehrer* – 1952–56 Lehre als Schriftsetzer, erste Kalligraphie-Übungen. 1957 Weiterbildung zum typographischen Gestalter bei Emil Ruder, erste Übungen in Schriftentwurf. 1958–59

dessine p. ex. une police pour le ministère des Transports des Pays-Bas. Commence à travailler sur sa police «Thesis». En 1991, il crée, à Londres, un programme kerning pour optimiser les espaces entre les lettres des polices classiques comme Bembo, Garamond, Univers, Futura. Enseigne à l'académie des beaux-arts de Den Bosch en 1992. De 1992 à 1994, il crée une série de «Trash Faces». S'installe à Berlin en 1993 et entre dans l'agence de design «Meta Design». Complète FF Meta, une police d'Erik Spiekermann pour en faire 38 caractères. Enseigne à la Kunsthochschule (école supérieure des beaux-arts) de Berlin-Weissensee en 1995, puis à la Kunsthochschule de Potsdam en 1996. – *Polices:* FF TheMix (1994), FF TheSans (1994), FF TheSerif (1994), FF Nebulae (1994), FF Thesis (1994), FF JesusLoves-YouAll (1995), FF MoveMeMultipleMaster (pour «Fuse», 1995), FF The SansMonospace (1996), FF The SansTypewriter (1996).

**Groothuis,** Rainer – né le 11. 6. 1959 à Emden, Allemagne – *typographe, graphiste maquettiste* – apprentissage de libraire à Emden. S'installe à Berlin en 1982, maquettiste aux éditions Klaus Wagenbach. 1984, directeur de fabrication, puis administrateur en 1989. En 1996, il s'installe à Brême et travaille dans un atelier collectif avec Victor Malsy. Nombreuses publications sur l'industrie du livre. Nombreuses distinctions.

**Gürtler,** André – né le 5. 9. 1936 à Bâle, Suisse – *graphiste maquettiste, typographe, concepteur de polices, enseignant* – 1952–1956, apprentissage de compositeur en imprimerie, premiers exercices de calligraphie. 1957, formation de maquettiste typographe chez Emil Ruder, premiers exercices de conception de lettres. De 1958 à 1959, il dessine des lettres pour la Monotype Corporation à Salfords,

Groothuis 1992 Cover

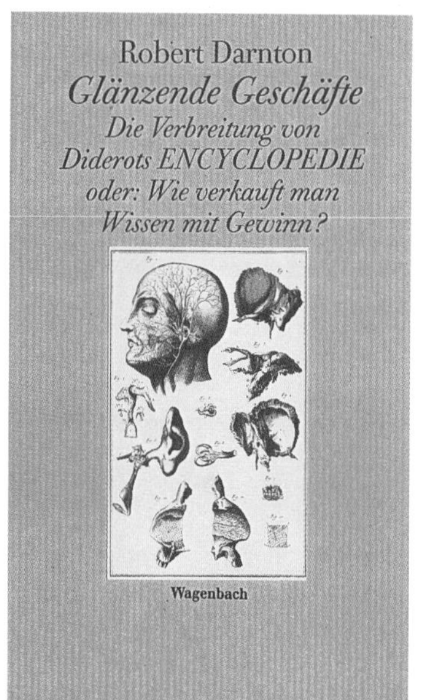


Groothuis 1993 Cover

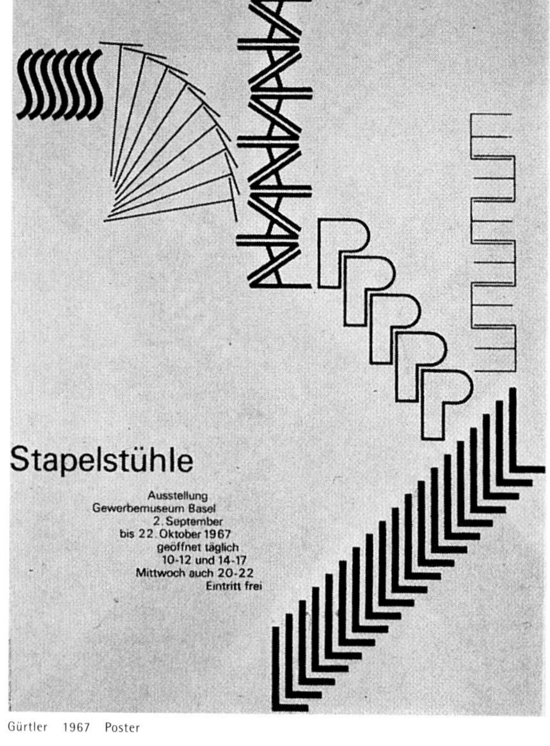


Gürtler 1967 Poster

| n | n | n | n | | |
|---|---|---|---|---|---|
| a | b | e | f | g | i |
| o | r | s | t | y | z |
| A | B | C | E | G | H |
| M | O | R | S | X | Y |
| 1 | 2 | 4 | 6 | 8 | & |

Gürtler 1966 Egyptian 505

| | n | n | n | | *n* |
|---|---|---|---|---|---|
| a | b | e | f | g | i |
| o | r | s | t | y | z |
| A | B | C | E | G | H |
| M | O | R | S | X | Y |
| 1 | 2 | 4 | 6 | 8 | & |

Gürtler 1978 Basilia

font for the Dutch Ministry of Transport. Begins working on his typeface Thesis. 1991: develops a kerning program in London, aiming to optimize the spaces between the letters in classical types such as Bembo, Garamond, Univers and Futura. 1992: teaches at the art academy in Den Bosch. 1992–94: designs a series of "Trash Faces". 1993: moves to Berlin and joins the MetaDesign studio. Works on Erik Spiekermann's type, FF Meta, to produce 38 different designs. 1995: teaches at the Kunsthochschule Berlin-Weißensee. 1996: Kunsthochschule in Potsdam. – *Fonts:* FF TheMix (1994), FF TheSans (1994), FF TheSerif (1994), FF Nebulae (1994), FF Thesis (1994), FF JesusLovesYouAll (1995), FF MoveMeMultipleMaster (for "Fuse", 1995), FF TheSansMonospace (1996), FF TheSansTypewriter (1996).

**Groothuis,** Rainer – b. 11. 6. 1959 in Emden, Germany – *typographer, graphic designer* – Trained as a book-seller in Emden. 1982: moves to Berlin. Designer for the publisher Klaus Wagenbach. 1984: production manager, 1989 managing director. 1996: moves to Bremen and sets up a studio with Victor Malsy. Numerous publications on the book industry and book design. Numerous awards.

**Gürtler,** André – b. 5. 9. 1936 in Basle, Switzerland – *graphic designer, typographer, type designer, teacher* – 1952–56: trains as a typesetter. First exercises in calligraphy. 1957: further training as a typographical designer with Emil Ruder. First exercises in type design. 1958–59: letterer for the Monotype Corporation in

Schriftzeichner bei der Monotype Corporation in Salfords England. 1959–65 Schriftentwerfer und Typograph bei Arian Frutiger in Paris. Zwei Jahre tätig bei Sofratype SA. Ein Jahr Schriftlehrer an der Académie Populaire des Arts Plastiques. Seit 1965 Fachlehrer für Schriftgestaltung an der Schule für Gestaltung Basel: Kalligraphie, Schriftentwurf, Signet- und Logotype-Entwurf, Schriftgeschichte und Schrifttheorie. Daneben Arbeiten für die schriftherstellende Industrie, Buchgestaltung, typographische und grafische Arbeiten. Seit 1969 Gastdozenturen an der Yale University und der Rhode Island School of Design. Gründet 1972 das ATypI-Komitee für Forschung und Ausbildung. Über 20 Jahre Redaktionsmitglied der Zeitschrift „Typografische Monatsblätter“, verschiedene Veröffentlichungen in der Zeitschrift „Visible Language“ (USA) und „tipografica“ (Argentinien). Seit 1982 Gastdozenturen an verschiedenen Universitäten in Mexiko. 1990–94 Mitarbeit am EG COMETT II Programm „Didot“, einem Ausbildungs-Austausch-Programm europäischer Schulen. – *Schriftentwürfe:* Egyptian 505 (1966), Media (mit Chr. Mengelt, E. Gschwind 1976), Basilia (1978), Signa (1978), Haas Unica (1980).

**Gutenberg,** Johannes (d. i. Johannes Gensfleisch zur Laden) – geb. ca. 1400 in Mainz, Deutschland, gest. 3. 2. 1468 in Mainz, Deutschland. – *Drucker, Erfinder* – Gilt als Erfinder der Druckkunst mit beweglichen Lettern (Handsetzgerät). 1434 erste Nennung Gutenbergs in Straßburg. Beauftragt 1436 den Goldschmied Hans

Angleterre. De 1959 à 1965, il travaille comme concepteur de polices et typographe chez Adrian Frutiger à Paris. Exerce pendant deux ans chez Sofratype SA. Enseigne l'art de l'écriture pendant un an à l'Académie Populaire des Arts Plastiques. Professeur de conception scripturale à l'Ecole de design de Bâle depuis 1965 : enseigne la calligraphie, conception de polices, dessin de signets et de logotypes, histoire et la théorie de l'écriture. Il travaille parallèlement pour des fabricants de caractères, réalise des maquettes de livres et des travaux graphiques et typographiques. Cycles de conférences à la Yale University et à la Rhode Island School of Design à partir de 1969. Fonde le comité ATypI pour la recherche et la formation en 1972. Depuis plus de vingt ans, il fait partie de la rédaction de la revue «Typografische Monatsblätter», plusieurs publications dans la revue «Visible Language» (Etats-Unis) et «tipografica» (Argentine). Plusieurs cycles de conférences dans des universités mexicaines depuis 1982. Collabore à la CE COMETT II «Didot», un programme d'échange entre les écoles européennes de 1990 à 1994. – *Polices :* Egyptian 505 (1966), Media (avec Chr. Mengelt, E. Gschwind, 1976), Basilia (1978), Signa (1978), Haas Unica (1980).

**Gutenberg,** Johannes (Johannes Gensfleisch zur Laden) – né vers 1400 à Mayence, Allemagne, décédé le 3. 2. 1468 à Mayence, Allemagne – *imprimeur, inventeur* – est considéré comme l'inventeur de l'imprimerie à caractères mobiles (composition à la main). Première mention du nom de Gutenberg à Strasbourg en 1434. En 1436, il charge l'orfèvre Hans Dünne de lui fabriquer des outils d'im-

Gutenberg 1455 Page

Salfords, England. 1959–65: type designer and typographer with Adrian Frutiger in Paris. Works for Sofratype SA for two years. Teaches lettering for a year at the Académie Populaire des Arts Plastiques. From 1965 onwards: teaches type design at the Schule für Gestaltung in Basle, i. e. calligraphy, type design, trademark and logotype design, history of type and typographical theory. Also works for the type-manufacturing industry and produces book designs and typographical and graphic work. From 1969 onwards: guest lecturer at Yale University and the Rhode Island School of Design. 1972: founder of the ATypI committee for research and education. On the editorial staff of "Typografische Monatsblätter" magazine for over 20 years. Various publications in "Visible Language" (USA) and "tipografica" (Argentina) magazines. From 1982 onwards: guest lecturer at various universities in Mexico. 1990–94: works on the EC COMETT II program Didot, an educational exchange program for schools in Europe. – *Fonts:* Egyptian 505 (1966), Media (with Chr. Mengelt and E. Gschwind, 1976), Basilia (1978), Signa (1978), Haas Unica (1980).

**Gutenberg,** Johannes (real name: Johannes Gensfleisch zur Laden) – b. c.1400 in Mainz, Germany, d. 3. 2. 1468 in Mainz, Germany – *printer, inventor* – Considered the inventor of the art of printing with moveable type (manual caster). 1434: Gutenberg is first mentioned in Strasbourg. 1436: commissions the goldsmith Hans Dünne to make him various print-

Dünne mit der Herstellung von Druckwerkzeugen. Gründet 1438 mit anderen die Druckwerkstatt „Aventur und Kunst". Erhält 1448 in Mainz eine Anleihe von 150 Gulden, errichtet damit die sogenannte „Mainzer Urdruckerei". 1449 Beginn der Zusammenarbeit mit Johannes Fust. 1450 Einrichtung der Gemeinschaftsdruckerei von Fust und Gutenberg im Humbrechthof in Mainz. Ca. 1452–55 Druck einer 42-zeiligen Bibel in zwei Bänden. 1454–61 druckt Gutenberg Ablaßbriefe, den Türkenkalender für 1455, die Türkenbulle in lateinisch und deutsch, den Mainzer Psalter, die 36-zeilige Bibel (in Bamberg), das „Catholicon" und die 49-zeilige Bibel (in Straßburg). 1455 Prozeß Fust gegen Gutenberg. 1461 wird Gutenberg vom Hofgericht in Rottweil mit der Acht belegt, weil er Zinsen für ein Darlehen schuldig geblieben ist. 1465 wird Gutenberg zum Hofmann ernannt. – *Publikationen u.a.:* H. Presser „Johannes Gutenberg in Zeugnissen und Bilddokumenten", Hamburg 1967; A. Ruppel „Johannes Gutenberg. Sein Leben und sein Werk" (Reprint), Nieuwkoop 1967; A. Kapr „Johannes Gutenberg. Tatsachen und Thesen", Leipzig 1977.

**Gutenberg-Gesellschaft** – Internationale Vereinigung für Geschichte und Gegenwart der Druckkunst e.V., Mainz, Deutschland. 1901 Gründung der Gesellschaft in Mainz als Folge der Gutenbergfeiern 1900. Die Gesellschaft verleiht den Gutenberg-Preis. Sie fördert die wissenschaftliche Publikationstätigkeit. Es existieren für 1902–26 Jahresberichte und Beilagen, seit 1926 das von Aloys Rup-

primeur. En 1438, il fonde avec d'autres, un atelier d'imprimerie appelé «Aventur und Kunst». En 1448, grâce à un prêt de 150 florins qu'on lui accorde à Mayence, il crée la (dite «Mainzer Urdruckerei») première imprimerie de Mayence. Début de la coopération avec Johannes Fust en 1449 et création de l'imprimerie Gutenberg et Fust au Humbrechthof à Mayence en 1450. Vers 1452 à 1455, impression des deux volumes de la Bible à 42 lignes. Entre 1454 et 1461, Gutenberg imprime des indulgences, le calendrier turc de 1455, la bulle turque en latin et en allemand, le psautier de Mayence, la Bible à 36 lignes (à Bamberg), le «Catholicon» et la Bible à 49 lignes (à Strasbourg). 1455, procès de Fust contre Gutenberg. En 1461, le tribunal de Rottweil le met au ban parce qu'il doit encore les intérêts sur un prêt contracté antérieurement. En 1465, Gutenberg est appelé à la cour. – *Publications, sélection:* H. Presser «Johannes Gutenberg in Zeugnissen und Bilddokumenten», Hambourg 1967; A. Ruppel «Johannes Gutenberg. Sein Leben und sein Werk» (réédition), Nieuwkoop 1967; A. Kapr «Johannes Gutenberg. Tatsachen und Thesen», Leipzig 1977.

**Gutenberg-Gesellschaft** – Internationale Vereinigung für Geschichte und Gegenwart der Buchkunst e.V. (société internationale pour l'histoire et l'actualité de l'art de l'imprimerie), Mayence, Allemagne – la société a été fondée en 1901 à Mayence à la suite de l'année Gutenberg, célébrée en 1900. La société décerne le Prix Gutenberg. Elle encourage les publications scientifiques. Elle publie des annales et des comptes rendus de 1902 à 1926; puis à partir de 1926, le Gutenberg-Jahrbuch (annales Gutenberg) à l'initiative d'Aloys Ruppel, ainsi que la collection «Kleine Drucke der Gutenberg-Gesellschaft» (petites publications de la société Guten-

Gutenberg 1458 Page

JOSEPH THEELE
Das literarische Denkmal
für Gutenberg

Gutenberg-Gesellschaft 1938 Cover

Das literarische Denkmal
für Gutenberg Festvortrag gehalten in
der Generalversammlung der Gutenberg-Gesellschaft
in Mainz am 27. Juni 1937 von dem Direktor der
Landesbibliothek Fulda Dr. JOSEPH THEELE

Verlag der Gutenberg-Gesellschaft MAINZ 1938

Gutenberg-Gesellschaft 1938 Main Title

ing tools. 1438: founds the printing workshop "Aventur und Kunst" with a few others. 1448: in Mainz he receives a loan of 150 guilders, using this sum to open the first Mainz printing works (Mainzer Urdruckerei). 1449: begins working with Johannes Fust. 1450: Gutenberg and Fust open a joint printing workshop in the Humbrechthof in Mainz. C. 1452–55: Gutenberg's 42-line Bible is printed in two volumes. 1454–61: Gutenberg prints Letters of Indulgence, the 1455 "Turk-Kalendar", the Turkish bulls in Latin and German, the Mainz Psalter, the 36-line Bible (in Bamberg), the "Catholicon" and the 49-line Bible (in Strasbourg). 1455: law suit between Fust and Gutenberg. 1461: the royal court of justice in Rottweil outlaws Gutenberg for failing to pay interest on a loan. 1465: Gutenberg is made a courtier. – *Publications include:* H. Presser "Johannes Gutenberg in Zeugnissen und Bilddokumenten", Hamburg 1967; A. Ruppel "Johannes Gutenberg. Sein Leben und sein Werk" (reprint), Nieuwkoop 1967; A. Kapr "Johannes Gutenberg. Tatsachen und Thesen", Leipzig 1977.

**Gutenberg-Gesellschaft** – Internationale Vereinigung für Geschichte und Gegenwart der Druckkunst e.V., Mainz, Germany – 1901: the society is founded in Mainz as a result of the Gutenberg celebrations in 1900. The society awards the Gutenberg Prize and encourages scholarly publications. There are yearly reports and supplements for the years 1902–1926 and from 1926 onwards also the Gutenberg yearbook, conceived by Aloys Rup-

pel begründete Gutenberg-Jahrbuch, daneben die Reihe „Kleine Drucke der Gutenberg-Gesellschaft", Sonderveröffentlichungen und Festschriften.

**Gutenberg-Museum,** Mainz – *Weltmuseum für Druckkunst* – 1900 Gründung des Gutenberg-Museums durch Mainzer Bürger zum 500. Geburtstag von Johannes Gutenberg. Präsentiert werden die Bereiche Buch, Druck, Schrift, Papier mit allen Nebenbereichen. Im Museum befindet sich ein Tresorraum mit den Drucken Gutenbergs (u.a. seine 42-zeilige Bibel) sowie eine rekonstruierte Gutenberg-Werkstatt. Die Bibliothek umfaßt ca. 70.000 Bände. Das Museum veranstaltet ständig Sonderausstellungen und Vortragsreihen.

**Gutenberg-Preis** – Aus Anlaß des 500. Todestages von Johannes Gutenberg stiftet die Gutenberg-Gesellschaft 1968 in Mainz den Gutenberg-Preis. Erster Preisträger ist 1968 Giovanni Mardersteig. Weitere Preisträger sind u.a. Henri Friedlaender (1971), Hermann Zapf (1974), der Erfinder und Technologe Rudolf Hell (1977), Adrian Frutiger (1986), Lotte Hellinga-Querido (1989), Paul Brainerd (1994) und John Dreyfus (1996). In Leipzig gibt es seit 1959 ebenfalls einen Gutenberg-Preis, der jährlich als Anerkennung maßstabsetzender Leistungen in der Buchkunst verliehen wird. Preisträger waren u. a. H. E. Wolter, Albert Kapr, Werner Klemke, Jan Tschichold, Jürgen Seuss, Hans Peter Willberg, A. D. Gonscharow, HAP Grieshaber, Kurt Loeb, sowie Hochschulen, Verlage und Förderer.

Gutenberg-Museum 1924 Poster

Gutenberg-Gesellschaft 1958 Logo

Gutenberg-Preis 1977 Cover

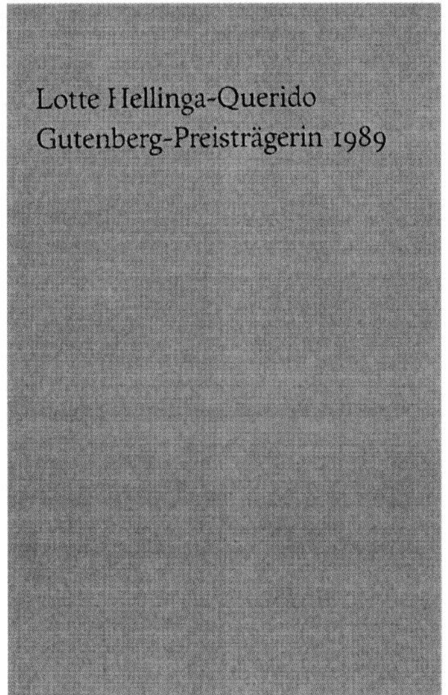


Gutenberg-Preis 1989 Cover

berg), des éditions spéciales et des textes commémoratifs.

**Gutenberg-Museum,** Mayence – *musée mondial de l'art du livre* – fondé en 1900 par des citoyens de la ville de Mayence à l'occasion du 500e anniversaire de la naissance de Johannes Gutenberg. Les domaines du livre, de l'imprimerie, des arts de l'écriture, du papier ainsi que les secteurs annexes y sont représentés. Dans une salle du musée sont conservées les oeuvres de Gutenberg (entre autres sa Bible à 42 lignes) ainsi qu'une reconstitution de son atelier. La bibliothéque contient près de 70 000 volumes. Le musée organise en permanence des expositions temporaires et des cycles de conférences.

**Gutenberg-Preis** (Prix Gutenberg) – en 1968, la Société Gutenberg fonde un prix à l'occasion du 500e anniversaire de la mort de Johannes Gutenberg. En 1968, le premier lauréat était Giovanni Mardersteig. Parmi les autres lauréats, on compte Henri Friedlaender (1971), Hermann Zapf (1974), l'inventeur et spécialiste des technologies Rudolf Hell (1977), Adrian Frutiger (1986), Lotte Hellinga-Querido (1989), Paul Brainerd (1994) et John Dreyfus (1996). Leipzig possédait aussi son Prix Gutenberg depuis 1959 et le décernait pour récompenser des travaux exceptionnels dans le domaine des arts du livre. H. E. Wolter, Albert Kapr, Werner Klemke, Jan Tschichold, Jürgen Seuss, Hans Peter Willberg, A. D. Goncharov, HAP Grieshaber, Kurt Loeb, ainsi que des universités, éditions et mécènes ont été primés.

pel. The society also produces the series "Kleine Drucke der Gutenberg-Gesellschaft", special publications and festschrifts.

**Gutenberg-Museum,** Mainz – *international museum of printing* – 1900: the Gutenberg-Museum is founded by the citizens of Mainz to commemorate Gutenberg's 500th birthday. Subjects covered by the museum include the field of books, printing, typography and paper and all relevant affiliated topics. The museum treasury houses various Gutenberg prints (including his 42-line Bible) and a reconstruction of Gutenberg's workshop. The library has c. 70,000 volumes. The museum stages special exhibitions and lecture series on a regular basis.

**Gutenberg Prize** 1968: on the occasion of the 500th anniversary of Gutenberg's death, the Gutenberg-Gesellschaft in Mainz awards the Gutenberg Prize. The first prize winner in 1968 is Giovanni Mardersteig. Further prize winners have included Henri Friedlaender (1971), Hermann Zapf (1974), the inventor and technologist Rudolf Hell (1977), Adrian Frutiger (1986), Lotte Hellinga-Querido (1989), Paul Brainerd (1994) and John Dreyfus (1996). Leipzig's Gutenberg Prize has been awarded every year since 1959 for model achievements in the field of book art. Prize winners have included H. E. Wolter, Albert Kapr, Werner Klemke, Jan Tschichold, Jürgen Seuss, Hans Peter Willberg, A. D. Gonsharov, HAP Grieshaber, Kurt Loeb and various colleges and universities, publishing houses and sponsors.

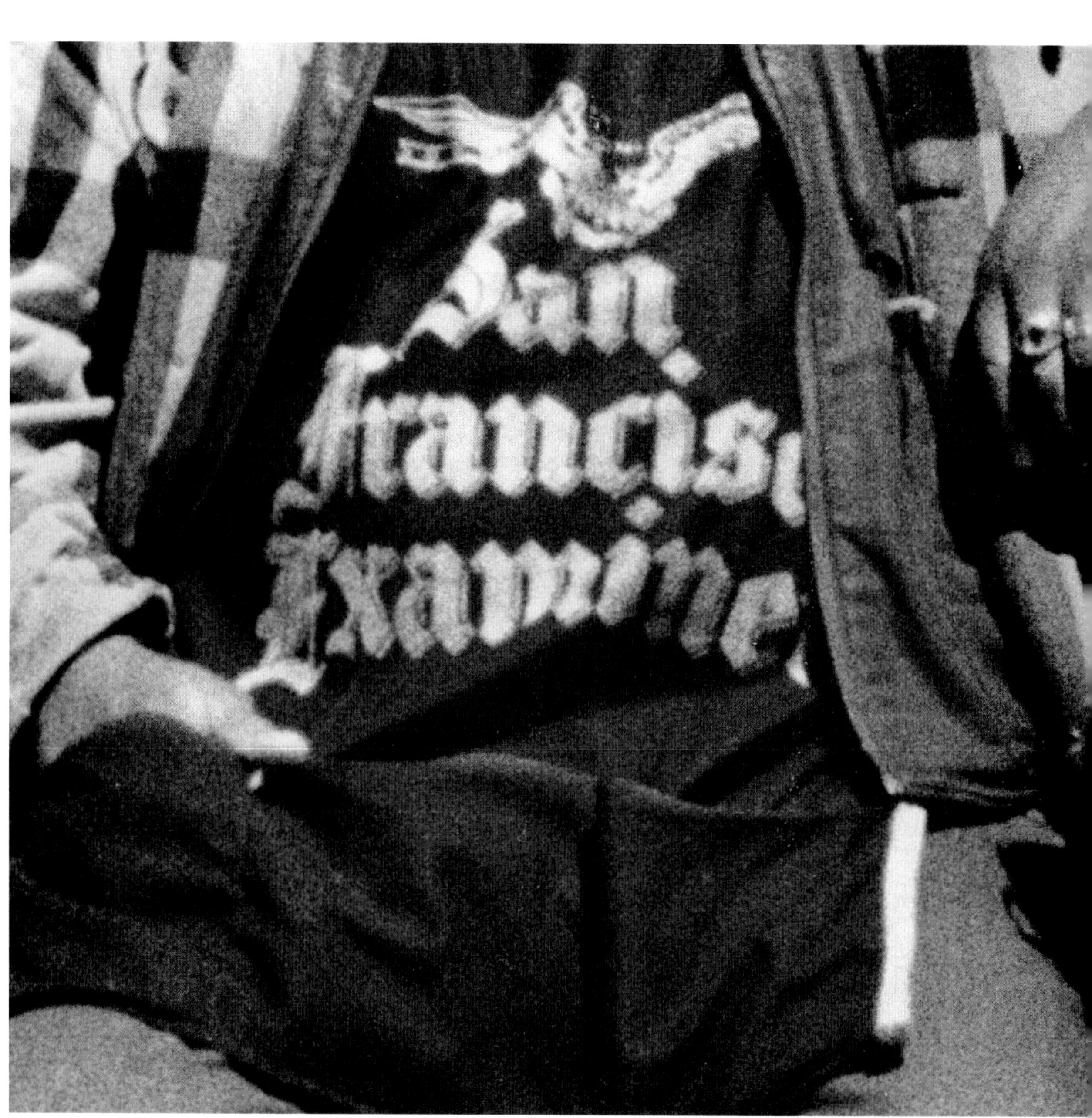

# H

**Haas'sche Schriftgießerei,** Basel – *Schriftenhersteller* – Der Druckermeister Jean Excertier aus Savoyen gründet um 1580 mit seinem Schwager Jacques Foillet aus Lyon in Basel eine Buchdruckerei. Nach dem Tod von Jean Excertier übernimmt Johann Jakob Gerath (1582–1654) die Druckerei. 1718 Eintritt des Nürnberger Stempelschneiders und Schriftgießers Johann Wilhelm Haas, der die Firma übernimmt. Aufschwung unter seinem Sohn Wilhelm Haas-Münch. 1800 übernimmt Wilhelm Haas-Decker das Unternehmen. Nach dem Tod des Vaters übernehmen die Söhne Georg-Wilhelm und Karl-Eduard 1838 das Unternehmen mit wenig Glück. Nach zweimaligem Verkauf übernimmt Fernand Vicarino 1895 das Unternehmen mit Max Krayer, der ab 1904 das Unternehmen leitet. Die Haas'sche Schriftgießerei wird 1927 in eine Aktiengesellschaft umgewandelt. 1944 übernimmt Eduard Hoffmann zusammen mit seinem Sohn Alfred die Leitung der Firma. Das Unternehmen und insbesondere die Abteilung Typographie erfahren einen Aufschwung. Zahlreiche Schriftgießereien werden eingegliedert. 1956 Veröffentlichung der Schrift „Helvetica", die von Max Miedinger entworfen wurde. Weitere Entwerfer der Schriftgießerei waren u. a. Edmund Thiele, Walter Diethelm, André Gürtler. 1980 400-jähriges Jubiläum. 1989 Verkauf aller Rechte an die Linotype AG. Verkauf der Gießwerkstatt an Walter Frutiger.
**Hadank,** O H W (Otto Hans Werner) – geb. 17. 8. 1889 in Berlin, Deutschland, gest. 17. 5. 1965 in Hamburg, Deutschland. –

**Haas'sche Schriftgiesserei,** Bâle – *fonderie de caractères* – vers 1580, le maître imprimeur savoyard Jean Excertier fonde une imprimerie à Bâle avec son beau-frère, Jacques Foillet, de Lyon. Après la mort de Jean Excertier en 1607, l'imprimerie est reprise par Johann Jakob Gerath (1582–1654). Le tailleur de types et fondeur Johann Wilhelm Haas de Nuremberg, entre dans la société en 1718 et la reprend. Celle-ci prospère avec son fils Wilhelm Haas-Münch. Wilhelm Haas-Decker reprend l'entreprise en 1800. A la mort du père, en 1838, les fils Georg-Wilhelm et Karl-Eduard la reprennent mais la chance ne leur sourit guère. L'entreprise est vendue par deux fois, puis reprise par Fernand Vicarino et Max Krayer en 1895, ce dernier la dirige à partir de 1904. La fonderie Haas devient société anonyme en 1927. En 1944, sous Eduard Hoffman, qui la dirige avec son fils Alfred, l'entreprise, et surtout le service des polices, connaissent une période de prospérité. D'autres fonderies de caractères y sont incorporées. En 1956, elle publie l' «Helvetica» dessinée par Max Miedinger. Parmi les concepteurs ayant travaillé pour la fonderie, on compte Edmund Thiele, Walter Diethelm, André Gürtler. En 1980, la société fête son 400e anniversaire. En 1989, tous les droits sont cédés à la Linotype AG. L'atelier de fonte est vendu à Walter Frutiger.
**Hadank,** O H W (Otto Hans Werner) – né le 17. 8. 1889 à Berlin, Allemagne, décédé le 17. 5. 1965 à Hambourg, Allemagne – *graphiste maquettiste, enseignant, concepteur de polices* – 1907 études à l'Unterrichtsanstalt des Kunstgewerbemuseums (institut d'enseignement du Musée des arts décoratifs) de Berlin. 1912–1937, maquettes pour les éditions Cotta, Grote, Phönixdruck, Scherl, Ullstein. 1919, membre fondateur du «Bund

Zum Geleit
Die Haas'ſche Gießerei
in Münchenſtein
bei Baſel
beſorgte den Neuguß der in
dieſer Publikation ange=
wandten Nürnberger
Schwabacher
des
Meiſters
Wilhelm Haas
von Nürnberg
des Gründers obiger Firma

Haas'sche Schriftgießerei 1927 Title

Hadank 1922 Label

Hadank 1926 Emblem

**Haas'sche Schriftgießerei,** Basle – *type manufacturers* – c.1580: master printer Jean Excertier from the Savoy founds a printing works in Basle with his brother-in-law, Jacques Foillet from Lyon. 1607: Johann Jakob Gerath (1582–1654) takes over the printing works after Excertier's death. 1718: Johann Wilhelm Haas, a punch cutter and type founder from Nuremberg, joins the company and soon takes it over. Business prospers under his son, Wilhelm Haas-Münch. 1800: Wilhelm Haas-Decker takes over the company. 1838: after the death of their father, the Haas sons Georg-Wilhelm and Karl-Eduard take over the foundry with little success. After being sold twice, the foundry is taken on in 1895 by Fernand Vicarino and Max Krayer; the latter becomes foundry director in 1904. 1927: the Haas'sche Schriftgießerei becomes a joint-stock company. 1944: Eduard Hoffmann and his son become company directors; the business, and in particular, the typography department, prospers under him and his son Alfred. Numerous type foundries are incorporated. 1956: the Helvetica font is published, designed by Max Miedinger. Further foundry designers have included Edmund Thiele, Walter Diethelm and André Gürtler. 1980: 400-year anniversary. 1989: all rights are sold to Linotype AG. The foundry workshop is sold to Walter Frutiger.
**Hadank,** O H W (Otto Hans Werner) – b. 17. 8. 1889 in Berlin, Germany, d. 17. 5. 1965 in Hamburg, Germany – *graphic designer, teacher, type designer* – 1907: studies at the Unterrichtsanstalt des

*Grafik-Designer, Lehrer, Schriftentwerfer* – 1907 Studium an der Unterrichtsanstalt des Kunstgewerbemuseums Berlin. 1912–37 Entwürfe für Verlage wie Cotta, Grote, Phönixdruck, Scherl, Ullstein. 1919 Gründungsmitglied des Bundes der Deutschen Gebrauchsgraphiker (BDG) in Berlin. 1919–49 Professor für Gebrauchsgrafik an der Hochschule für freie und angewandte Kunst in Berlin. 1921 künstlerischer Berater des Zigarettenherstellers Haus Neuerburg. Zahlreiche Entwürfe für die Marken Overstolz, Ravenklau, Löwenbrück, Gülichplatz und seine eigenen HDK-beschrifteten Zigaretten. Daneben zahlreiche Etiketten-Entwürfe für Spirituosen-Hersteller. 1936 Entwurf des Signets für „Pelikan"-Erzeugnisse der Firma Günther Wagner in Hannover, 1950 Umzug nach Hamburg. 1959 Ehrenmitglied der AGI. Ca. 70 Signets, zahlreiche Verpackungsentwürfe und ein breites gebrauchsgrafisches Werk. Hadank ist Begründer und Neugestalter des Berufsbildes vom Gebrauchsgrafiker zum Grafik-Designer. – *Schriftentwurf:* Ornata (1943).

**Hadders,** Gerard, – geb. 1954 in Rotterdam, Niederlande. – *Grafik-Designer, Fotograf, Lehrer* – 1970–75 Studium an der Akademie van Beeldende Kunsten in Rotterdam, Abteilung Visuelle Kommunikation, und 1977–78 an der Abteilung Kunst. 1979 Mitbegründer des Kulturmagazins „Hard Werken" in Rotterdam. 1980 Mitbegründer des Design-Studios „Hard Werken" in Rotterdam. 1983 Mitbegründer von Tape TV. 1988 Mitbegründer der Zeitschrift „Qwerty". 1992–94 Leiter der Design-Abteilung der Jan van

Hadank ca. 1925 Cover

Hadank ca. 1930 Advertisement

Hadank 1924 Label

Hadank ca. 1930 Package

Kunstgewerbemuseums in Berlin. 1912–37: designs for publishers such as Cotta, Grote, Phönixdruck, Scherl and Ullstein. 1919: founder member of the Bund der Deutschen Gebrauchsgraphiker (BDG) in Berlin. 1919–49: professor of commercial art at the Hochschule für freie und angewandte Kunst in Berlin. 1921: art consultant to Haus Neuerburg cigarette manufacturers. Numerous designs for the brands Overstolz, Ravenklau, Löwenbrück, Gülichplatz and his own cigarettes marked HDK. Also numerous labels for various spirits manufacturers. 1936: designs the logo for the Günther Wagner company's Pelikan products in Hanover. 1950: moves to Hamburg. 1959: honorary member of the Alliance Graphique Internationale (AGI). Approx. 70 logos, numerous packaging designs and a broad spectrum of commercial art work. Hadank was responsible for transforming the commercial art-work profession to that of graphic designer. – *Font:* Ornata (1943).

**Hadders,** Gerard – b. 1954 in Rotterdam, The Netherlands – *graphic designer, photographer, teacher* – 1970–75: studies at the Akademie van Beeldende Kunsten in Rotterdam in the visual communication department and from 1977–78 in the art department. 1979: co-founder of "Hard Werken" art magazine in Rotterdam. 1980: co-founder of the Hard Werken design studio in Rotterdam. 1983: co-founder of Tape TV. 1988: co-founder of

der Deutschen Gebrauchsgraphiker» (union des graphistes industriels allemands) (BDG) de Berlin. De 1919 à 1949, il enseigne le graphisme industriel à la Hochschule für freie und angewandte Kunst (école supérieure des arts appliqués) de Berlin. En 1921, il est conseiller artistique du fabricant de cigarettes Haus Neuerburg. Nombreux graphismes pour les marques Overstolz, Ravenklau, Löwenbrück, Gülichplatz et pour ses propres cigarettes, les HDK. Nombreux dessins d'étiquettes pour des fabricants de spiritueux. En 1936, il dessine le signet pour les produits de «Pelikan» de la société Günther Wagner de Hanovre. Il s'installe à Hambourg en 1950. En 1959, il est membre d'honneur de l' «Alliance Graphique Internationale» (AGI). A réalisé environ 70 signets, de nombreux emballages et laissé une œuvre considérable dans le domaine du graphisme industriel. Hadank a instauré et redéfini l'image de la profession de graphiste industriel qui, avec lui, est devenu un graphiste maquettiste. – *Police:* Ornata (1943).

**Hadders,** Gerard – né en 1954 à Rotterdam, Pays-Bas – *graphiste maquettiste, photographe, enseignant* – 1970–1975, études de communication visuelle à l'Akademie van Beeldende Kunsten de Rotterdam. 1977–1978, études d'art. 1979, cofondateur du magazine culturel «Hard Werken» à Rotterdam. 1980, cofondateur de l'atelier de design «Hard Werken» à Rotterdam. 1983, cofondateur de «Tape TV». 1988, cofondateur de la revue «Qwerty». De 1992 à 1994, il dirige

Eyck-Akademie in Maastricht. 1993 Gründung des Design-Studios „Buro Lange Haven". Auftraggeber waren u. a. das Kunstmuseum Wolfsburg, der Kunstverein Hannover, das Stedelijk Museum Amsterdam, das Stedelijk Museum Schiedam, das Ro Theater Rotterdam, Uitgeverij Bert Bakker, Historische Uitgeverij. Unterrichtet an der Akademie van Beeldende Kunsten in Rotterdam (1984), an der Rietveld Akademie Amsterdam (1987), an der Rhode Island School of Design (1987), der Jan van Eyck-Akademie in Maastricht (1991–95), der St. Joost Akademie in Breda (seit 1995). Umfangreiche Jurytätigkeit.

**Hamersveld,** John van – geb. 1941 – *Grafik-Designer* – Studium am Art Center College of Design und an der Chouinard Art School in Los Angeles. 1963 Entwurf des Plakats zu dem Film „Endless Summer", das als eines der ersten Plakate mit Leuchtfarbe gedruckt wird und von dem bis 1967 800.000 Exemplare verkauft werden. Gründung der Zeitschrift „Surfing Illustrated". Plakatentwürfe für Fillmore-West-Konzerte. Art Director von Capitol Records, Entwürfe für die Plattencover der Beach Boys. Gründung seines Studios „Pinnacle Productions" und „Pinnacle Dance Concerts", einer Multimedia Performance Company. 1967 Kontakte zur Apple Corporation der Beatles in London, Entwurf des Covers zur Platte „Magical Mystery Tour" der Beatles. Weitere Entwürfe für die Musikgruppen The Who, The Rolling Stones („Exile on Main Street", 1972), Jefferson Airplane, Leo Kottke, Dire Straits und Blondie. 1979

la section de design à la Jan van Eyck-Akademie de Maastricht. Il fonde l'atelier de design «Buro Lange Haven» en 1993. Commanditaires : le Kunstmuseum Wolfsburg, le Kunstverein Hannover, le Stedelijk Museum d'Amsterdam, le Stedelijk Museum de Schiedam, le Ro Theater de Rotterdam, Uitgeverij Bert Bakker, Historische Uitgeverij. A enseigné à l'Akademie van Beeldende Kunsten de Rotterdam (1984), à la Rietveld Akademie d'Amsterdam (1987), à la Rhode Island School of Design (1987), à la Jan van Eyck Akademie de Maastricht (1991–1995), à la St. Joost Akademie de Breda (depuis 1995). Membre de nombreux jurys.

**Hamersveld,** John van – né en 1941 – *graphiste maquettiste* – études à l'Art Center College of Design et à la Chouinard Art School à Los Angeles. Dessine en 1963 l'affiche du film «Endless Summer» l'une des premières affiches imprimées avec des couleurs fluorescentes et dont 800 000 exemplaires seront vendus jusqu'en 1967. Fonde la revue «Surfing Illustrated». Dessins d'affiches pour Fillmore West-Konzerte. Directeur artistique des Capitol Records, conception des pochettes de disques des Beach Boys. Fonde son atelier, les «Pinnacle Productions» et les «Pinnacle Dance Concerts», une société de spectacles multimédias. En 1967, il entre en contact avec la Apple Corporation des Beatles à Londres et dessine la pochette du disque «Magical Mystery Tour» des Beatles. Autres travaux pour des groupes de rock tels que The Who, les Rolling Stones («Exile on Main Street», 1972), Jefferson Airplane, Leo Kottke, Dire Straits et Blondie. Travaux pour la revue «Wet» en 1979. Ses travaux pour la culture pop ont été l'emblème de toute une génération. – *Police:* Johnny Deco (1975).

Hadders 1988 Cover

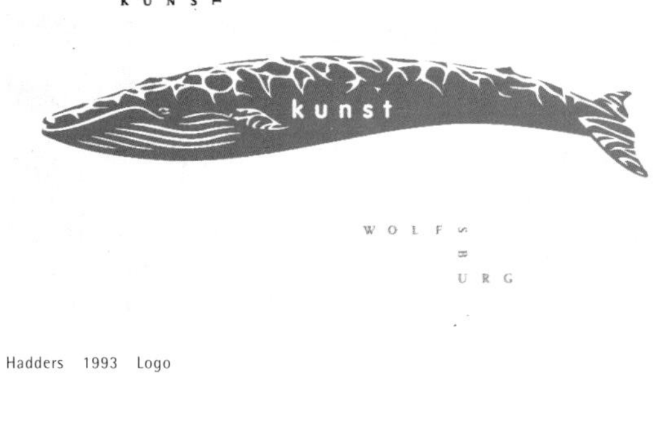


Hadders 1993 Logo

Hadders 1989 Logo

Hadders 1994 Cover

"Qwerty" magazine. 1992–94: head of the design department at the Jan van Eyck Akademie in Maastricht. 1993: opens the design studio Buro Lange Haven. Clients include the Kunstmuseum Wolfsburg, Kunstverein Hannover, the Stedelijk Museum in Amsterdam, the Stedelijk Museum in Schiedam, Ro Theater Rotterdam, Uitgeverij Bert Bakker and the Historische Uitgeverij. Has taught at the Akademie van Beeldende Kunsten in Rotterdam (1984), Rietveld Akademie in Amsterdam (1987), Rhode Island School of Design (1987), the Jan van Eyck Akademie in Maastricht (1991–95) and the St. Joost Akademie in Breda (since 1995). Has judged on many jury panels.

**Hamersveld,** John van – b. 1941 – *graphic designer* – Studied at the Art Center College of Design and at the Chouinard Art School in Los Angeles. 1963: designs the poster for the film "Endless Summer", one of the first posters to be printed in fluorescent colors. 800,000 copies are sold between 1963–67. Launches "Surfing Illustrated" magazine. Designs posters for Fillmore West Concerts. Art director of Capitol Records, producing designs for Beach Boy records. Opens a studio, Pinnacle Productions, and a multimedia performance company entitled Pinnacle Dance Concerts. 1967: contact with the Beatles' Apple Corporation in London. Designs the record cover for "Magical Mystery Tour". Further designs for the groups The Who, The Rolling Stones ("Exile on Main Street", 1972), Jefferson Airplane, Leo Kottke, Dire Straits and Blondie. 1979: produces designs for

Entwürfe für die Zeitschrift „Wet". Mit seinen Arbeiten für die Pop-Kultur entwirft er die Embleme einer ganzen Generation. – *Schriftentwurf:* Johnny Deco (1975).

**Hammer,** Victor – geb. 9. 12. 1882 in Wien, Österreich, gest. 10. 7. 1967 in Lexington, USA. – *Drucker, Maler, Grafiker, Architekt, Schriftentwerfer, Bildhauer, Lehrer, Verleger* – 1897 Lehre bei einem Architekten und Stadtplaner. 1898 Studium an der Kunstakademie Wien. 1922 Umzug nach Florenz. Gründung und Leitung einer Druckerei in Florenz. 1923 Bekanntschaft mit Rudolf Koch. 1929 Umzug mit seiner Druckerei in die Villa Santuccio in Florenz, die Druckerei wird „Stamperia del Santuccio" genannt. Das erste gedruckte Buch ist Miltons „Samson Agonistes" (1931). 1936–39 Professor an der Akademie der Bildenden Künste in Wien. 1939 Emigration in die USA. Unterrichtet bis 1948 am Wells College in Aurora, New York. 1948 Umzug nach Lexington. Herausgabe zahlreicher Bücher, u. a. Hölderlins „Gedichte", gesetzt in seiner Schrift „American Uncial". – *Schriftentwürfe:* Hammer Unziale (1921), Samson (1931), Pindar (1933), American Uncial (1943), Andromaque Uncial (1958). – *Publikationen u. a.:* „The Forms of our Letters", Lexington 1958. Carolyn Hammer „Notes on the Stamperia del Santuccio", Lexington 1963; Carolyn Hammer „Victor Hammer. Artist and Craftsman", Lexington 1981.

**Hard Werken** – *Kulturmagazin, Design-Studio* – 1979 Gründung durch Willem Kars in Rotterdam. Mitarbeiter Gerard Hadders, Rick Vermeulen, Henk Elenga,

Hamersveld 1970 Cover

eine unziale nach zeichnung von victor hammer
geschnitten und herausgegeben von gebr. klingspor in offenbach am main

Hammer 1925 Main Title

| | | | | | |
|---|---|---|---|---|---|
| | n | | | | |
| a | b | e | f | g | í |
| o | r | s | t | y | z |
| A | B | C | E | G | H |
| M | O | R | S | X | Y |
| 1 | 2 | 4 | 6 | 8 | & |

Hammer 1943 American Uncial

**Hammer,** Victor – né le 9. 12. 1882 à Vienne, Autriche, décédé le 10. 7. 1967 à Lexington, Etats-Unis – *imprimeur, peintre, graphiste, architecte, concepteur de polices, sculpteur, enseignant, éditeur* – apprentissage chez un architecte-urbaniste en 1897. Etudes à la Kunstakademie (beaux-arts) de Vienne en 1898. Vit à Florence en 1922 et y fonde et dirige une imprimerie. 1923, rencontre Rudolf Koch. En 1929, son imprimerie s'installe à la Villa Santuccio à Florence et s'appelle désormais «Stamperia del Santuccio». «Samson Agonistes» (1931) est le premier livre qui sort de ses presses. De 1936 à 1939, il est professeur à l'Akademie der Bildenden Künste (beaux-arts) de Vienne. Il émigre aux Etats-Unis en 1939 et enseigne au Wells College d'Aurora, New York jusqu'en 1948. S'installe à Lexington en 1948. Publie de nombreux ouvrages dont les poèmes de Hölderlin composés en «American Uncial», une fonte qu'il avait dessinée. – *Polices:* Hammer Unziale (1921), Samson (1931), Pindar (1933), American Uncial (1943), Andromaque Uncial (1958). – *Publications, sélection:* «The Forms of our Letters», Lexington 1958. Carolyn Hammer «Notes on the Stamperia del Santuccio», Lexington 1963; Carolyn Hammer «Victor Hammer. Artist and Craftsman», Lexington 1981.

**Hard Werken** – *magazine culturel, atelier de design* – fondé en 1979 par Willem Kars à Rotterdam. Collaboration de Gerard Hadders, Rick Vermeulen, Henk

Lionello Venturi (in his History of Art Criticism/New York/1936) considers the theory of Conrad Fiedler «the most important aesthetic thought of the second half of the

Hammer 1949 Hammer Unziale

"Wet" magazine. Hamersveld has created the emblems of an entire generation with his work for the pop industry. – *Font:* Johnny Deco (1975).

**Hammer,** Victor – b. 9. 12. 1882 in Vienna, Austria, d. 10. 7. 1967 in Lexington, USA – *printer, painter, graphic artist, architect, type designer, sculptor, teacher, publisher* – 1897: apprenticeship with an architect and town planner. 1898: studies at the art academy in Vienna. 1922: moves to Florence. Opens and runs a printing workshop in Florence. 1923: makes the acquaintance of Rudolf Koch. 1929: he and his printing works move into the Villa Santuccio in Florence and the business is renamed Stamperia del Santuccio. The first book to be printed is Milton's "Samson Agonistes" (1931). 1936–39: professor at the Akademie der Bildenden Künste in Vienna. 1939: emigrates to the USA and teaches at Wells College in Aurora, New York, until 1948. 1948: moves to Lexington. Publishes numerous books, including a volume of Hölderlin's poems, set in his type American Uncial. – *Fonts:* Hammer Unziale (1921), Samson (1931), Pindar (1933), American Uncial (1943), Andromaque Uncial (1958). – *Publications include:* "The Forms of our Letters", Lexington 1958. Carolyn Hammer "Notes on the Stamperia del Santuccio", Lexington 1963; Carolyn Hammer "Victor Hammer. Artist and Craftsman", Lexington 1981.

**Hard Werken** – *arts magazine, design studio* – 1979: the magazine is launched by Willem Kars in Rotterdam. Contributors are Gerard Hadders, Rick Vermeulen,

Tom van den Haspel, Jan Willem de Kok, die aus der grafischen Werkstatt der „Rotterdamse Kunststichting" hervorgehen. Bis 1982 erscheinen 11 Hefte der Zeitschrift. 1980 Gründung des Design-Studios „Hard Werken" in Rotterdam als Assoziation verschiedener selbständiger Gestalter. Auftraggeber waren u. a. De Kunststichting, das Museum van Landen Volkenkunde, Film International, das „Museum Journal" und die niederländische Post PTT. Gründung der Filiale „Hard Werken Los Angeles Desk" durch Henk Elenga. Auftraggeber waren u. a. Warner Brothers, Esprit. 1989 Auszeichnung mit dem H. N. Werkman-Preis der Stadt Rotterdam. 1993 Zusammenschluß mit dem Büro „Ten Cate Bergmans" zu „Hard Werken – Ten Cate Bergmans Design" (HWTCB). 1993 Gerard Hadders scheidet aus. Die neue Kommunikationsfirma ändert ihren Namen 1994 in „Inizio Design", später „Inizio Multi Media". – *Publikation u. a.:* Paul Hefting, Dirk van Ginkel „Hard Werken – Inizio. From cultural oasis to multimedia", Rotterdam 1995.

**Hassani,** Mir Emad – geb. 1541 in Ghazwin, Iran, gest. 1604 in Isfahan, Iran. – *Kalligraph* – Ausbildung bei seinem Lehrer Mohammad Hossein Tabrizi. Gilt als wichtigster Kalligraph im Bereich der „Nastaliq-Schrift". Die Gestaltungsregeln dieser Schrift werden durch ihn vollkommen erarbeitet, und es ist niemandem gelungen, seinen Neuerungen etwas hinzuzufügen. Diese Schreibart wird zur Nationalform der persischen Kalligraphie. Er hat zahlreiche Schüler und Anhänger. Zu seinen kalligraphischen Werken gehören

Elenga, Tom van den Haspel, Jan Willem de Kok venus de l'atelier graphique «Rotterdamse Kunststichting». 11 numéros de la revue paraissent jusqu'en 1982. L'atelier de design «Hard Werken» est fondé à Rotterdam en 1980 en tant qu'association de designers indépendants. Commanditaires: De Kunststichting, Museum van Land- en Volkenkunde, Film International, le «Museum Journal» et les postes néerlandaises PTT etc. Henk Elenga fonde la filiale «Hard Werken Los Angeles Desk» (HW L.A. Desk). Commanditaires: Warner Brothers, Esprit etc. En 1989, le prix H. N. Werkman de la ville de Rotterdam lui est décerné. 1993, fusion avec l'atelier «Ten Cate Bergmans» et création de «Hard Werken–Ten Cate Bergmans Design» (HWTCB). Gerard Hadders quitte Hard Werken en 1993. La nouvelle société de communication change de nom en 1994 et s'appelle désormais «Inizio Design», puis «Inizio Multi Media». – *Publication, sélection:* Paul Hefting, Dirk van Ginkel «Hard Werken – Inizio. From cultural oasis to multimedia», Rotterdam 1995.

**Hassani,** Mir Emad – né en 1541 à Ghazwin, Iran, décédé en 1604 à Isfahan, Iran – *calligraphe* – formation avec son professeur Mohammad Hossein Tabrizi. Est considéré comme le plus grand calligraphe de l'écriture Nastaliq. Il a défini les règles formelles de cette calligraphie; personne n'a réussi jusqu'alors à ajouter d'autres éléments à ses innovations. Cette façon d'écrire est devenue la forme reconnue au niveau national pour la calligraphie iranienne. Il a eu de nombreux élèves et adeptes. Parmi ses œuvres calligraphiques, on compte «Sobhatolabrar» de «Djami» (1552), «Golestan» de «Saadi» (1578), «Bustan» de «Saadi» (1590), «Tohfatol abrar» de «Djami» (1596) et «Goyet-chogan» de «Arefi» (1599) ainsi que de

Hard Werken 1981 Poster

Hard Werken 1980 Poster

Hassani Calligraphy

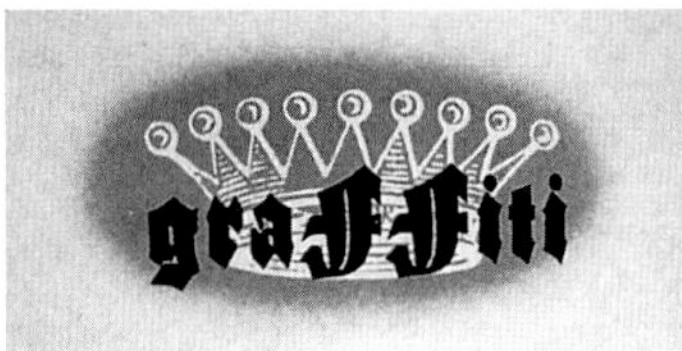

Hard Werken 1983 Logo

Henk Elenga, Tom van den Haspel and Jan Willem de Kok from the Rotterdamse Kunststichting's graphics workshop. 11 numbers of the magazine are published between 1979–82. 1980: the Hard Werken design studio, an association of various freelance designers, is opened in Rotterdam. Clients include De Kunststichting, Museum van Land- en Volkenkunde, Film International, "Museum Journal" and the Dutch post office PTT. Henk Elenga opens the Hard Werken Los Angeles Desk (HW L. A. Desk). Clients include Warner Brothers and Esprit. 1989: awarded Rotterdam's H. N. Werkman Prize. 1993: the studio forms a merger with the Ten Cate Bergmans studio, renaming themselves Hard Werken – Ten Cate Bergmans Design (HWTCB). 1993: Gerard Hadders leaves the studio. 1994: the new communications company changes its name to Inizio Design, and later to Inizio Multi Media. – *Publications include:* Paul Hefting, Dirk van Ginkel "Hard Werken – Inizio. From cultural oasis to multimedia", Rotterdam 1995.

**Hassani,** Mir Emad – b. 1541 in Ghazwin, Iran, d. 1604 in Isfahan, Iran – *calligrapher* – Trained with his teacher, Mohammad Hossein Tabrizi. Considered the most significant calligrapher in the field of nastaliq script. Hassani worked out rules of design for this form of writing to such perfection that no-one has been able to make any additions to his innovations. This script became the national form of Persian calligraphy. Hassani had numerous pupils and followers. His calligraphic works include "Sobhatolabrar"

u. a. „Sobhatolabrar“ von „Djami“ (1552), „Golestan“ von „Saadi“ (1578), „Bustan“ von „Saadi“ (1590), „Tohfatol abrar“ von „Djami“ (1596) und „Goy-und-chogan“ von „Arefi“ (1599) sowie zahlreiche Schrift-Alben, die in Bibliotheken und Museen im Iran, in der Türkei, in Afghanistan, Indien, Rußland und Frankreich aufbewahrt werden.

**Hausmann,** Raoul – geb. 12. 7. 1886 in Wien, Österreich, gest. 1. 2. 1971 in Limoges, Frankreich. – *Maler, Fotograf, Schriftsteller* – 1900 Umzug mit seinen Eltern nach Berlin. 1908–11 Ausbildung an den Studien-Ateliers für Malerei und Plastik in Berlin. 1909–14 Gestaltung von Bucheinbänden für den Eugen Diederichs Verlag in Jena. 1918 Gründung des „Club Dada“, Veröffentlichung des ersten dadaistischen Manifests und Organisation von Dada-Soireen (mit Huelsenbeck, Heartfield, Grosz, Jung, Höch). Entwicklung der Fotomontage, erste „Plakatgedichte“ und „Phonetische Poeme“, 1919 Herausgeber der Zeitschrift „Der Dada“. 1921 Lesung mit Kurt Schwitters in Prag. 1926 Beginn der Niederschrift seines Romans „Hyle“. 1930 erste systematische Kamerafotografien. 1933 Flucht vor den Nationalsozialisten nach Ibiza. 1936 Flucht nach Zürich, 1937 nach Prag, 1938 nach Paris, 1939 nach Peyrat-le-Château. Ließ sich 1944 in Limoges nieder. – *Publikationen u.a.:* „Hurrah! Hurrah! Hurrah!“, Berlin 1921; „Pin and the Story of Pin“ (mit Kurt Schwitters), London 1962; „Sprechspäne“, Flensburg 1962; „Hyle“, Frankfurt am Main 1969; „Am Anfang war Dada“, Gießen 1972; „Schriften bis

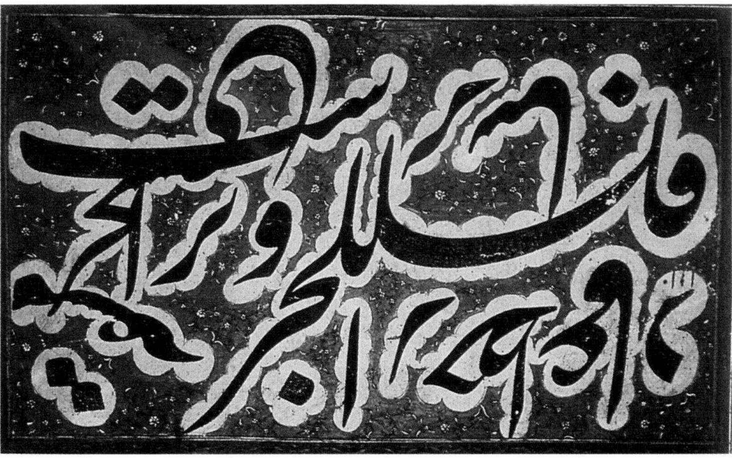
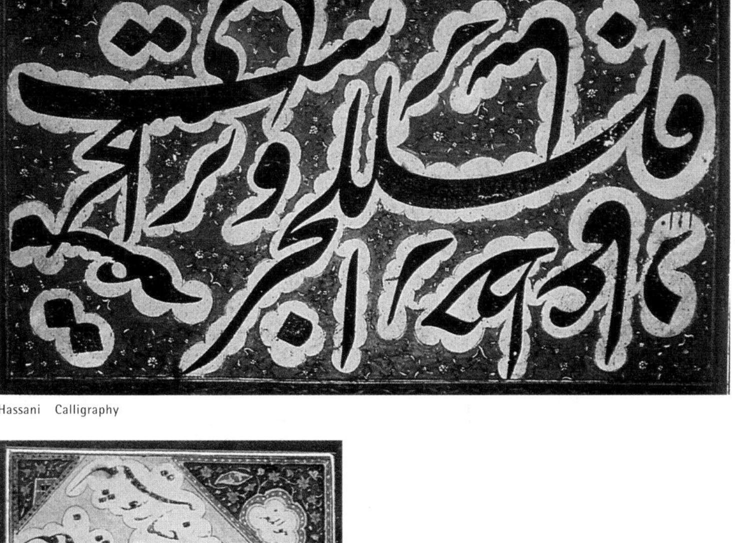

Hassani Calligraphy

Hausmann 1920 Cover

Hassani Calligraphy

Hausmann 1918 Poster

nombreux albums de calligraphie qui sont conservés dans les bibliothèques et musées d'Iran, Afghanistan, Inde, Turquie, Russie et en France.

**Hausmann,** Raoul – né le 12. 7. 1886 à Vienne, Autriche, décédé le 1. 2. 1971 à Limoges, France – *peintre, photographe, écrivain* – en 1900, ses parents viennent s'installer à Berlin. 1908–1911, formation aux ateliers d'apprentissage de la peinture et de la sculpture de Berlin. 1909–1914, dessine des reliures pour les éditions Eugen Diederichs à Iéna. Fonde le «Club Dada» en 1918 et publie le premier manifeste Dada, organise des soirées Dada (avec Huelsenbeck, Heartfield, Grosz, Jung, Höch). Développe le photomontage, premiers «Poèmes affiches» et «Poèmes phonétiques». En 1919, il publie la revue «Der Dada». Lecture publique avec Kurt Schwitters à Prague en 1921. Commence à écrire son roman «Hyle» en 1926. Premières photos systématiques en 1930. Fuit le régime nazi en 1933 et se réfugie à Ibiza, puis à Zurich en 1936, à Prague en 1937. Emigre à Paris en 1938, puis à Peyrat-le-Château en 1939. S'installe à Limoges en 1944. – *Publications, sélection :* «Hurrah! Hurrah! Hurrah», Berlin 1921; «Pin and the Story of Pin» (avec Kurt Schwitters), Londres 1962; «Sprechspäne», Flensburg 1962; «Hyle», Francfort-sur-le-Main 1969; «Am Anfang war Dada», Giessen 1972; «Schriften bis

by Jam'i (1552), "Golestan" by Saadi (1578), "Bustan" by Saadi (1590), "Tohfatol abrar" by Jam'i (1596) and "Goy-and-chogan" by Arefi (1599), as well as numerous albums of script preserved in museums in Iran, Afghanistan, India, Turkey, Russia and France.

**Hausmann,** Raoul – b. 12. 7. 1886 in Vienna, Austria, d. 1. 2. 1971 in Limoges, France – *painter, photographer, author* – 1900: moves to Berlin with his parents. 1908–11: studies at the training studios for painting and sculpture in Berlin. 1909–14: designs book covers for the publishers Eugen Diederichs in Jena. 1918: founds the Club Dada. Publishes the first Dada manifesto and organizes Dada soirées featuring Huelsenbeck, Heartfield, Grosz, Jung and Höch. Develops photomontage and produces his first poster-poems and phonetic poems. 1919: publishes the magazine "Der Dada". 1921: reading with Kurt Schwitters in Prague. 1926: starts writing his novel, "Hyle". 1930: produces his first systematic photography. 1933: flees from the Nazis to Ibiza. 1936: flees to Zurich, Prague (1937), Paris (1938) and Peyrat-le-Château (1939). 1944: finally settles in Limoges. – *Publications include:* "Hurrah! Hurrah! Hurrah!", Berlin 1921; "Pin and the Story of Pin" (with Kurt Schwitters), London 1962; "Sprechspäne", Flensburg 1962; "Hyle", Frankfurt am Main 1969; "Am Anfang war Dada", Gießen 1972; "Schriften bis

1933" (2 Bände), München 1982. A. Haus „Raoul Hausmanns Kameraphotographien", München 1979; A. Koch, „Ich bin immerhin der größte Experimentator Österreichs", Innsbruck 1994.

**Heartfield,** John (d.i. Helmut Herzfelde) – geb. 19.6.1891 in Berlin, Deutschland, gest. 26.4.1968 in Berlin, Deutschland. – *Gebrauchsgrafiker, Verleger* – 1908–11 Studium an der Königlichen Bayrischen Kunstgewerbeschule in München. 1913–14 Werbegrafiker in Mannheim. Fortsetzung des Studiums an der Kunst- und Handwerksschule Berlin. Aus Protest gegen nationalistische Hetze 1916 Umbenennung in John Heartfield. 1917 Gründung der Zeitschrift „Neue Jugend" und des Malik Verlags. 1918 Mitglied der Berliner Dadaisten. 1919–22 Gründung der politisch-satirischen Zeitschrift „Die Pleite" (mit G. Grosz und W. Herzfelde). Arbeiten für Erwin Piscators „Proletarisches Theater" und die Piscator-Bühne. 1923–27 Mitarbeit an der satirischen Wochenzeitschrift „Der Knüppel". 1928 Mitinitiator der „Assoziation Revolutionärer Künstler Deutschlands" (ASSO). 1930 erste Fotomontagen für die „Arbeiter-Illustrierte-Zeitung" (AIZ). 1933 Emigration nach Prag, Fortsetzung der Arbeit für die AIZ. 1938 Flucht nach London, Arbeiten für Verlage und den „Freien Deutschen Kulturbund". 1950 Rückkehr nach Deutschland. Arbeiten als Buchausstatter, Plakatgestalter und Bühnenbildner. Der künstlerische Nachlaß wird 1968 im John-Heartfield-Archiv der Akademie der Künste Berlin aufbewahrt. – *Publikationen u.a.:* „Deutschland, Deutschland über

1933» (2 vol.), Munich 1982. A. Haus «Raoul Hausmanns Kameraphotographien», Munich 1979; A. Koch «Ich bin immerhin der grösste Experimentator Österreichs», Innsbruck 1994.

**Heartfield,** John (Helmut Herzfelde) – né le 19.6.1891 à Berlin, Allemagne, décédé le 26.4.1968 à Berlin, Allemagne – *graphiste publicitaire, éditeur* – 1908–1911, études à la Königliche Bayrische Kunstgewerbeschule (école royale des arts décoratifs) de Munich. 1913–1914, graphiste publicitaire à Mannheim. Reprend ses études à la Kunst- und Handwerksschule (école d'art et d'artisanat) de Berlin. En 1916, il prend le pseudonyme de John Heartfield en signe de protestation contre l'agitation nationaliste. Il fonde la revue «Neue Jugend» en 1917 puis les éditions Malik. Devient membre de Dada Berlin en 1918. Fonde avec G. Grosz et W. Herzfelde la revue politico-satirique «Die Pleite» en 1919, à laquelle il collabore jusqu'en 1922. Travaille pour le «Théâtre prolétarien» d'Erwin Piscator, puis pour le Théâtre Piscator. De 1923 à 1927, il travaille pour l'hebdomadaire satirique «Der Knüppel». En 1928, il fait partie des fondateurs de l' «Assoziation Revolutionärer Künstler Deutschlands» (association des artistes révolutionnaires allemands) (ASSO). Premiers photomontages pour l' «Arbeiter-Illustrierte-Zeitung» (AIZ) en 1930. Emigre à Prague en 1933 où il poursuit sa collaboration avec l'AIZ. S'enfuit à Londres en 1938 et travaille pour des maisons d'édition et pour le «Freie Deutsche Kulturbund». Retour en Allemagne en 1950. Exerce comme maquettiste pour l'édition, affichiste et scénographe. En 1968, sa succession est conservée aux archives John Heartfield de l'Akademie der Künste (beaux-arts) de Berlin. – *Publications, sélection :* «Deutschland, Deutsch-

Heartfield 1927 Cover

Heartfield 1932 Cover

Heartfield 1929 Cover

Heartfield Logo

1933" (2 vols.), Munich 1982. A. Haus "Raoul Hausmanns Kameraphotographien", Munich 1979; A. Koch, "Ich bin immerhin der größte Experimentator Österreichs", Innsbruck 1994.

**Heartfield,** John (real name: Helmut Herzfelde) – b. 19.6.1891 in Berlin, Germany, d. 26.4.1968 in Berlin, Germany – *commercial artist, publisher* – 1908–11: studies at the Königliche Bayrische Kunstgewerbeschule in Munich. 1913–14: commercial artist in Mannheim. Continues his studies at the Kunst- und Handwerksschule in Berlin. 1916: changes his name to John Heartfield in protest against nationalistic propaganda. 1917: launches "Neue Jugend" magazine and Malik publishing house. 1918: member of the Berlin Dada group. 1919–22: launches the political and satirical magazine "Die Pleite" with G. Grosz and W. Herzfelde. Works for Erwin Piscator's Proletarisches Theater and the Piscator-Bühne. 1923–27: works for the satirical weekly magazine "Der Knüppel". 1928: one of the initiators of the Assoziation Revolutionärer Künstler Deutschlands (ASSO). 1930: produces his first photomontages for the "Arbeiter-Illustrierte-Zeitung" (AIZ). 1933: emigrates to Prague and continues to work for the AIZ. 1938: flees to London. Works for various publishers and for the Freie Deutsche Kulturbund. 1950: returns to Germany. Works as a book, poster and set-designer. 1968: his artistic legacy is preserved in the John-Heartfield-Archiv at the Akademie der Künste in Berlin. – *Publications include:* "Deutschland, Deutschland über alles" (with Kurt Tucholsky),

alles" (mit Kurt Tucholsky), 1929. W. Herzfelde „John Heartfield, Leben und Werk", Dresden 1962; E. Siepmann „Montage: John Heartfield", Berlin 1977; M. Töteberg „John Heartfield", Reinbek 1978; R. März (Hrsg.) „John Heartfield. Der Schnitt entlang der Zeit", Dresden 1981; R. März „Heartfield montiert", Leipzig 1993.

**Heiderhoff,** Horst – geb. 27. 8. 1934 in Wülfrath, Deutschland, gest. 10. 7. 1987 in Eisingen, Deutschland. – *Grafik-Designer, Verleger, Lehrer* – 1952–55 Schriftsetzerlehre in Wülfrath. 1956–58 Studium an der Staatlichen Werkkunstschule in Kassel. 1959–62 Studium am Pädagogischen Institut in Weilburg, Universität Gießen. Seit 1960 Verleger und Herausgeber von Büchern. 1963–76 Typograph und Assistent des künstlerischen Leiters der Schriftgießerei D. Stempel AG in Frankfurt am Main. Internationale Vortragstätigkeit. 1969–73 Vorstandsmitglied des International Center for the Typographic Arts (ICTA) in New York. 1970–75 Redakteur für Grafik-Design der Zeitschrift „design international" in Bonn. 1976–81 künstlerischer Leiter der D. Stempel AG. Gastdozent an verschiedenen Fach- und Hochschulen. 1978–80 Lehrauftrag für Grafik-Design an der Fachhochschule Würzburg. 1981–87 Professor an der Fachhochschule Hannover, Studiengang Schrift und Typographie. Heiderhoffs Arbeit hat großen Einfluß auf Auswahl und Qualität der Schriftenbibliothek der Schriftgießerei Stempel. Seine typographische Entwurfsarbeit bestimmt das Erscheinungsbild dieses international wirkenden Unternehmens. –

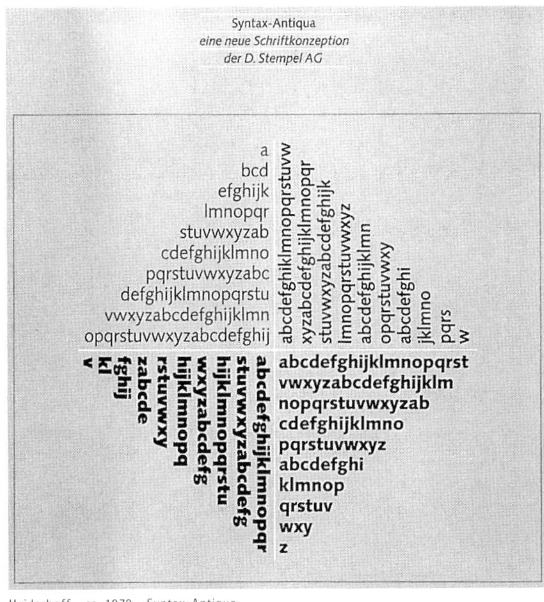

Heiderhoff ca. 1970 Syntax-Antiqua

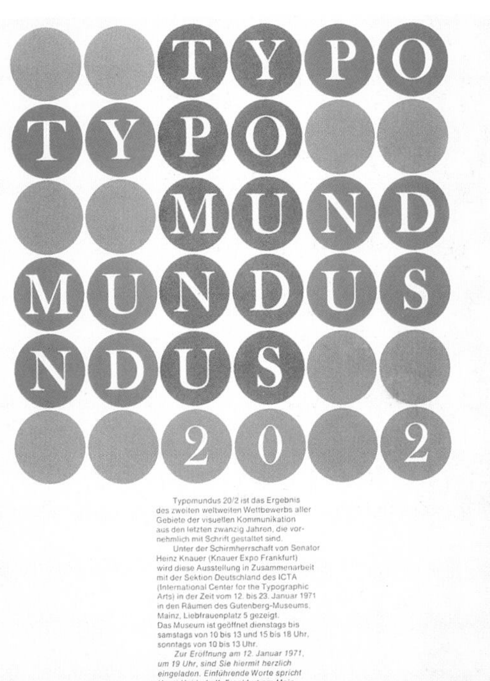

Heiderhoff 1970

Heartfield 1929 Cover

Heiderhoff Specimen

land über alles» (avec Kurt Tucholsky), 1929. W. Herzfelde «John Heartfield, Leben und Werk», Dresde 1962; E. Siepmann «Montage: John Heartfield», Berlin 1977; M. Töteberg «John Heartfield», Reinbek 1978; R. März (éd.) «John Heartfield. Der Schnitt entlang der Zeit», Dresde 1981; R. März «Heartfield montiert», Leipzig 1993.

**Heiderhoff,** Horst – né le 27. 8. 1934 à Wülfrath, Allemagne, décédé le 10. 7. 1987 à Eisingen, Allemagne – *graphiste maquettiste, éditeur, enseignant* – 1952–1955, apprentissage de composition typographique à Wülfrath. 1956–1958, études à la Staatliche Werkkunstschule (école des arts appliqués) de Kassel. 1959–1962, études à l'institut pédagogique de Weilburg, Université de Giessen. Editeur et directeur de collection depuis 1960. Typographe et assistant du directeur artistique de la fonderie D. Stempel AG à Francfort-sur-le-Main, de 1963 à 1976. Conférences dans plusieurs pays. 1969– 1973, membre du comité directeur de l'International Center for the Typographic Arts (ICTA) à New York. 1970–1975, rédacteur chargé des arts graphiques et du design à la revue «design international», Bonn. 1976–1981, directeur artistique de la D. Stempel AG. Cycles de conférences dans plusieurs universités et écoles spécialisées. 1978–1980, chargé de cours de design et d'arts graphiques à l'institut universitaire de Würzburg. 1981–1987, professeur d'arts de l'écriture et de typographie à l'institut universitaire de Hanovre. Le travail d'Heiderhoff a exercé une grande influence sur la sélection et la qualité de la typothèque de la fonderie de caractères Stempel. Ces conceptions typographiques ont déter-

1929. W. Herzfelde "John Heartfield, Leben und Werk", Dresden 1962; E. Siepmann "Montage: John Heartfield", Berlin 1977; M. Töteberg "John Heartfield", Reinbek 1978; R. März (ed.) "John Heartfield. Der Schnitt entlang der Zeit", Dresden 1981; R. März "Heartfield montiert", Leipzig 1993.

**Heiderhoff,** Horst – b. 27. 8. 1934 in Wülfrath, Germany, d. 10. 7. 1987 in Eisingen, Germany – *graphic designer, publisher, teacher* – 1952–55: trains as a typesetter in Wülfrath. 1956–58: studies at the Staatliche Werkkunstschule in Kassel. 1959–62: studies at the Pädagogische Institut Weilburg, University of Gießen. From 1960 onwards: editor and publisher. 1963–76: typographer and assistant to the art director of the D. Stempel AG type foundry in Frankfurt am Main. International lectures. 1969–73: member of the board for the International Center for the Typographic Arts (ICTA) in New York. 1970–75: graphic design editor for "design international" magazine in Bonn. 1976–81: art director of D. Stempel AG. Guest lecturer at various colleges and universities. 1978–80: teaching post for graphic design at the Fachhochschule Würzburg. 1981–87: professor of lettering and typography at the Fachhochschule Hannover. Heiderhoff's work greatly influenced the selection and quality of the type library at the Stempel type foundry. His typographical designs have shaped the image of this international concern. – *Publications include:* "Antiqua oder Fraktur", Frankfurt am Main 1971; "Der Mensch und seine Zeichen" (with

*Publikationen u.a.:* „Antiqua oder Fraktur", Frankfurt am Main 1971; „Der Mensch und seine Zeichen" (mit Adrian Frutiger), Frankfurt am Main, Echzell 1978–81; „Typopictura" (Mitautor), Frankfurt am Main 1981.

**Hell,** Rudolf – geb. 19. 12. 1901 in Eggmühl, Deutschland. – *Erfinder* – 1919–23 Studium der Elektrotechnik an der Technischen Hochschule München. 1927 Promotion. 1931 Entwicklung neuartiger Morsegeräte. 1947 Entwicklung von Geräten für den Übersee-Telegrammdienst. 1950 Entwicklung und Fertigung von Bild-Übertragungsgeräten. 1951–60 Einführung der Klischee-Graviermaschine „Klischograph" in Zeitungsbetrieben. 1964 digitale Schrift- und Bildwiedergabe. Erste elektronische Lichtsetzmaschinen mit digitaler Speicherung („Digiset"). 1990 fusionieren die Linotype AG und die Hell GmbH.

**Henrion,** F. H. K. (Frederick Henri Kay) – geb. 18. 4. 1914 in Nürnberg, Deutschland, gest. 5. 7. 1990 in London, England. – *Grafik-Designer, Typograph, Lehrer* – 1933–34 Ausbildung zum Textildesigner in Paris. 1934–36 Studium Grafik-Design an der Ecole Paul Colin in Paris. 1936 Emigration nach England. Arbeit als freier Designer. 1940–45 Ausstellungsgestaltung für das British Ministry of Information und das United States Office of War Information. Ausgezeichnet mit dem MBE (Member of the British Empire). 1951–73 Gründung und Leitung von Henrion Design Associates, 1973–82 von HDA International und 1982–90 von Henrion, Ludlow and Schmidt. 1955–65

miné l'image de cette entreprise mondialement célèbre. – *Publications, sélection :* «Antiqua oder Fraktur», Francfort-sur-le-Main 1971; «Der Mensch und seine Zeichen» (avec Adrian Frutiger), Francfort-sur-le-Main, Echzell 1978–1981; «Typopictura» (co-auteur), Francfort-sur-le-Main 1981.

**Hell,** Rudolf – né le 19. 12. 1901 à Eggmühl, Allemagne – *inventeur* – 1919–1923, études d'électronique à la Technische Hochschule (institut universitaire des techniques) de Munich. Doctorat en 1927. En 1931, il met au point un nouvel appareil émetteur et récepteur de morse. En 1947, il invente des appareils pour les services télégraphiques intercontinentaux. En 1950, il développe et fabrique des appareils de transmissions d'images. 1951– 1960, introduction de la photograveuse (Klischograph) dans la presse. En 1964, il travaille sur la reproduction numérique de caractères et d'images. Premiers appareils électroniques de photocomposition avec mémoire numérique (Digiset). 1990, Linotype AG et Hell GmbH fusionnent.

**Henrion,** F.H.K. (Frederick Henri Kay) – né le 18.4. 1914 à Nuremberg, Allemagne, décédé le 5. 7. 1990 à Londres, Angleterre – *graphiste maquettiste, typographe, enseignant* – 1933–1934, formation de designer en textiles à Paris. 1934–1936, études d'arts graphiques et de design à l'école Paul Colin à Paris. Emigre en Angleterre en 1936. 1940–1945, architecture d'expositions pour le British Ministry of Information et l'United States Office of War Information. Le titre de MBE (Member of the British Empire) lui est décerné. Fonde «Henrion Design Associates» en 1951, qu'il dirige jusqu'en 1973, puis «HDA International» (1973–1982) et «Henrion, Ludlow and Schmidt» (1982–1990). Enseigne au Royal College

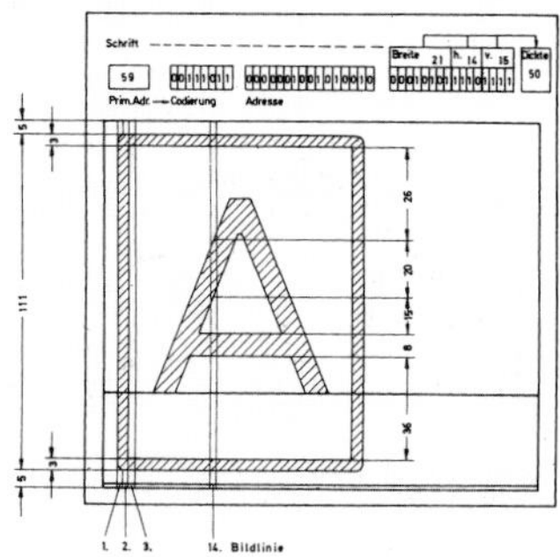

Hell 1985 Construction

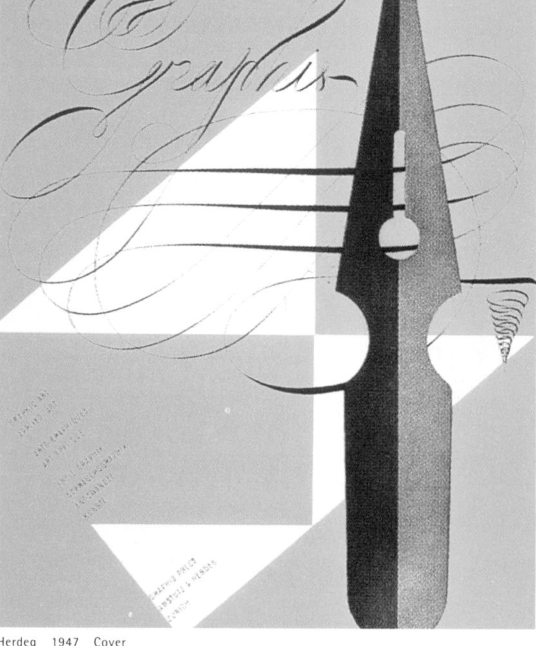


Herdeg 1947 Cover

Henrion ca. 1950 Poster

Henrion ca. 1970 Logo

Adrian Frutiger), Frankfurt am Main, Echzell 1978–81; "Typopictura" (co-author), Frankfurt am Main 1981.

**Hell,** Rudolf – b. 19. 12. 1901 in Eggmühl, Germany – *inventor* – 1919–23: studies electrical engineering at the Technische Hochschule in Munich. 1927: Hell gains his Ph.D. 1931: develops new Morse devices. 1947: develops instruments for the overseas telegram service. 1950: develops and manufactures devices for picture transmission. 1951–60: the Klischograph engraving machine is introduced to various newspaper plants. 1964: digital reproduction of text and pictures. First electronic photo-typesetting machine with digital storage (Digiset). 1990: Linotype AG and Hell GmbH merge.

**Henrion,** F.H.K. (Frederick Henri Kay) – b. 18.4.1914 in Nuremberg, Germany, d. 5. 7. 1990 in London, England – *graphic designer, typographer, teacher* – 1933–34: trains as a textiles designer in Paris. 1934–36: studies graphic design at the Ecole Paul Colin in Paris. 1936: emigrates to England. 1940–45: designs exhibitions for the British Ministry of Information and the United States Office of War Information. Awarded an MBE (Member of British Empire). 1951–73: founds and runs Henrion Design Associates. Director of HDA International from 1973–82 and of Henrion, Ludlow and Schmidt from 1982–90. 1955–65: teaches at the Royal College of Art in London. 1976–79: head of the departement for visual communication at the London College of Printing. President of the Society of Industrial Artists and Designers (1960–62), presi-

Unterrichtet am Royal College of Art in London. 1959 Ausgezeichnet als „Royal Designer for Industry". 1976–79 Leiter der Abteilung Visuelle Kommunikation des London College of Printing. Präsident der Society of Industrial Artists and Designers (1960–62), Präsident der Alliance Graphique Internationale, AGI (1963–66), Präsident des International Council of Graphic Design Associations, ICOGRADA (1968–79). Ausstellungen u.a. Institute of Contemporary Arts in London (1960), Intergraphic Gallery in München (1971). – *Publikationen u. a.:* „Design Coordination and Corporate Image" (mit Alan Parkin) , London, New York 1968.

**Herdeg,** Walter – geb. 3. 1. 1908 in Zürich, Schweiz, gest. 17. 12. 1995 in Meilen, Schweiz. – *Grafiker, Verleger* – Studium an der Kunstgewerbeschule Zürich (bei Ernst Keller) und an der Staatlichen Akademie der Bildenden Künste Berlin (bei O H W Hadank). Seit 1938 Grafisches Atelier und Werbeberatung Amstutz & Herdeg in Zürich. Seit 1942 Herausgeber und Gestalter der Zeitschrift „Graphis", seit 1952 der Jahrbücher „Graphis Annual" und weiterer Publikationen über angewandte Gestaltung. 1986 Verkauf und Rückzug aus dem „Graphis"-Unternehmen. 1987 Auszeichnungen der Parson's School of Design und des American Institute of Graphic Arts (AIGA).

**Hess,** Richard – geb. 27. 5. 1934 in Royal Oak, Michigan, USA. – *Grafik-Designer, Illustrator* – Studium an der Michigan State University. Gründet 1965 die Richard Hess Inc. Zahlreiche Ausstellungen und Auszeichnungen seit 1957.

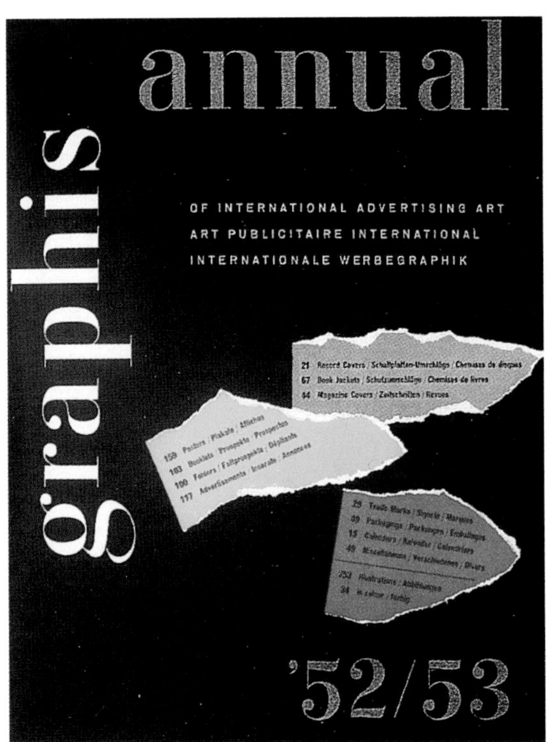


Herdeg 1952 Cover

Herdeg 1948 Logo

Hess ca. 1970 Cover

Herdeg 1948 Logo

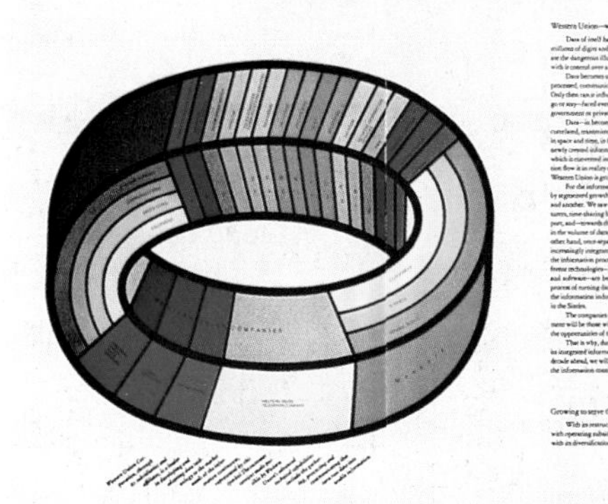

Hess ca. 1970 Spread

of Art à Londres de 1955 à 1965. Dirige la section de communication visuelle de London College of Printing de 1976–1979. Président de la Society of Industrial Artists and Designers (1960–1962), président de l'Alliance Graphique Internationale, AGI (1963–1966), président de l'International Council of Graphic Design Associations, ICOGRADA (1968–1979). Expositions à l'Institute of Contemporary Arts à Londres (1960) et à l'Intergraphic Gallery à Munich (1971). – *Publications, sélection:* «Design Coordination and Corporate Image» (avec Alan Parkin), Londres, New York 1968.

**Herdeg,** Walter – né le 3. 1. 1908 à Zurich, Suisse, décédé le 17. 12. 1995 à Meilen, Suisse – *graphiste, éditeur* – études à la Kunstgewerbeschule (école des arts décoratifs) de Zurich (chez Ernst Keller) puis à la Staatliche Akademie der Bildenden Künste (beaux-arts) de Berlin (chez O H W Hadank). Travaille pour le «Grafisches Atelier und Werbeberatung Amstutz & Herdeg» à Zurich à partir de 1938. Dirige et réalise la maquette de la revue «Graphis» à partir de 1942, puis les annales «Graphis Annual» à partir de 1952, ainsi que d'autres publications de design appliqué. Il vend et quitte la société «Graphis» en 1986. En 1987, il est décoré par la Parson's School of Design et par l'American Institute of Graphic Arts (AIGA).

**Hess,** Richard – né le 27. 5. 1934 à Royal Oak, Michigan, Etats-Unis – *graphiste maquettiste, illustrateur* – études à la Michigan State University. Il fonde la Richard Hess Inc. en 1965. Nombreuses expositions et distinctions depuis 1957.

dent of the Alliance Graphique Internationale, AGI (1963–66) and president of the International Council of Graphic Design Associations, ICOGRADA, between 1968–79. Exhibitions of his Work have been held at the Institute of Contemporary Arts in London (1960) and the Intergraphic Gallery in Munich (1971), among others. – *Publications include:* "Design Coordination and Corporate Image" (with Alan Parkin), London, New York, 1968.

**Herdeg,** Walter – b. 3. 1. 1908 in Zurich, Switzerland, d. 17. 12. 1995 in Meilen, Switzerland – *graphic artist, publisher* – Studied at the Kunstgewerbeschule in Zurich under Ernst Keller and at the Staatliche Akademie der Bildenden Künste in Berlin under O H W Hadank. From 1938 onwards: Grafisches Atelier und Werbeberatung Amstutz & Herdeg in Zurich. From 1942 onwards: editor and designer for "Graphis" magazine and from 1952 for the "Graphis Annual" and further publications on applied design. 1986: the "Graphis" enterprise is sold and Herdeg leaves. 1987: presented with awards by the Parson's School of Design and the American Institute of Graphic Arts (AIGA).

**Hess,** Richard – b. 27. 5. 1934 in Royal Oak, Michigan, USA – *graphic designer, illustrator* – Studied at Michigan State University. 1965: founds Richard Hess Inc. Numerous exhibitions since 1957 and various awards.

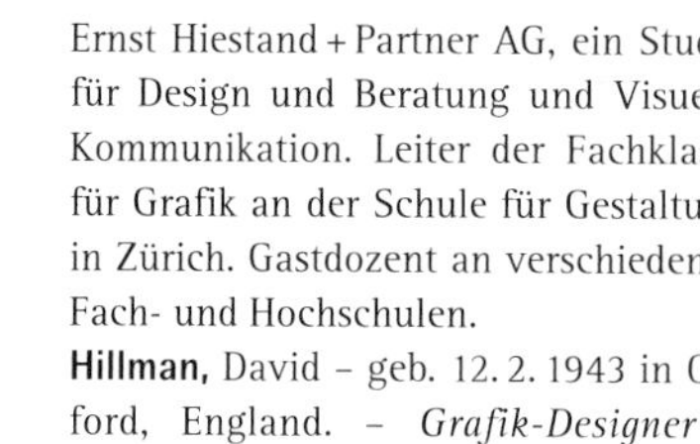

**Hiestand,** Ernst – geb. 16.9.1935 in Zürich, Schweiz. – *Grafik-Designer* – Ausbildung an der Schule für Gestaltung in Zürich. Weiterbildung in Paris, Atelierchef einer Züricher Werbeagentur. 1960–81 Gründung und Leitung des Studios E. + U. Hiestand in Zollikon (mit Ursula Hiestand). Auftraggeber aus Wirtschaft und Kultur, u.a. Entwürfe der Schweizer Banknoten. 1981 Gründung des Studios Ernst Hiestand in Zürich. Seit 1984 Design Consultant von IBM Deutschland. 1986 Gründung des Studios Ernst Hiestand + Partner AG, ein Studio für Design und Beratung und Visuelle Kommunikation. Leiter der Fachklasse für Grafik an der Schule für Gestaltung in Zürich. Gastdozent an verschiedenen Fach- und Hochschulen.

**Hillman,** David – geb. 12.2.1943 in Oxford, England. – *Grafik-Designer* – 1962–65 Studium am London College of Printing and Graphic Arts. Danach Arbeiten als Designer für „The Sunday Times Magazine". 1965–68 Arbeiten für die Zeitschriften „London Life" und „The Sunday Times Magazine". 1968 Art Director der Zeitschrift „Nova". 1975 Gründung seines eigenen Büros, Gestaltung des Layouts der neugegründeten Tageszeitung „Le Matin de Paris". 1978 Eintritt als Partner in das Design-Studio „Pentagram" in London. Entwicklung von Gestaltungsprojekten für The Arts Council of Great Britain, die Tate Gallery, das Design Museum in London, Skandia. Zahlreiche Auszeichnungen, u.a. für das Re-Design der Tageszeitung „The Guardian". – *Publikationen u.a:* „Ideas on Design"

**Hiestand,** Ernst – né le 16.9.1935 à Zurich, Suisse – *graphiste maquettiste* – formation à la Schule für Gestaltung (école de design) de Zurich. Poursuit ses études à Paris, puis chef d'atelier dans une agence de publicité à Zurich. 1960–1981, fonde et dirige l'atelier E. + U. Hiestand à Zollikon (avec Ursula Hiestand). Exécute des commandes pour l'industrie et des organismes culturels et dessine des billets de banque suisses. 1981, fonde le Studio Ernst Hiestand à Zurich. Depuis 1984, Design Consultant pour IBM Allemagne. 1986, création du Studio Ernst Hiestand + Partner AG, atelier de design de conseil et de communication visuelle. Dirige le cours d'arts graphiques à la Schule für Gestaltung de Zurich. Cycles de conférences dans plusieurs universités et instituts universitaires.

**Hillman,** David – né le 12.2.1943 à Oxford, Angleterre – *graphiste maquettiste* – 1962–1965, études au London College of Printing and Graphic Arts. Exerce ensuite comme designer au «The Sunday Times Magazine». De 1965 à 1968, il travaille pour les revues «London Life» et «The Sunday Times Magazine». Art Director de la revue «Nova» en 1968. Il fonde son propre atelier en 1975 et conçoit la maquette d'un nouveau journal «Le Matin de Paris». En 1978, il entre en qualité de partenaire à l'atelier de design «Pentagram» de Londres. Conçoit des projets de design pour The Arts Council of Great Britain, la Tate Gallery, le Design Museum de Londres, Skandia. Nombreuses distinctions, entre autres pour la refonte de la maquette du quotidien «The Guardian». – *Publications, sélection:* «Ideas on Design» (co-auteur), «Pentagames», Londres 1990; Londres 1986; «Pentagram – The Compendium» (co-auteur), Londres 1993; «Nova 1965–1975» (co-auteur), Londres 1993.

Hiestand 1961 Poster

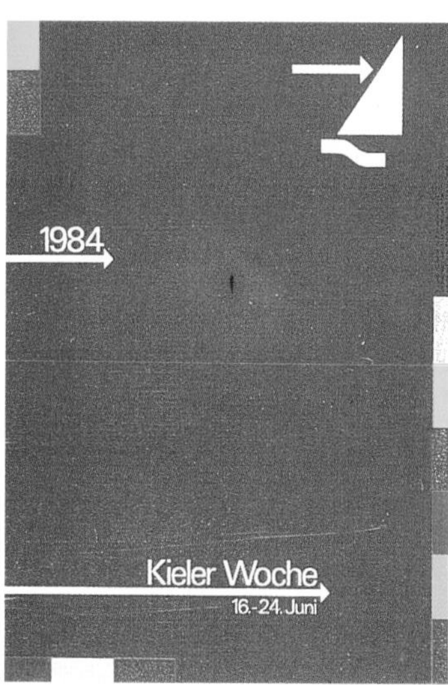

Hiestand 1984 Poster

Hillman 1986 Poster

XIX
ARTICLE 19

Hillman 1991 Logotype

Hillman 1989 Poster

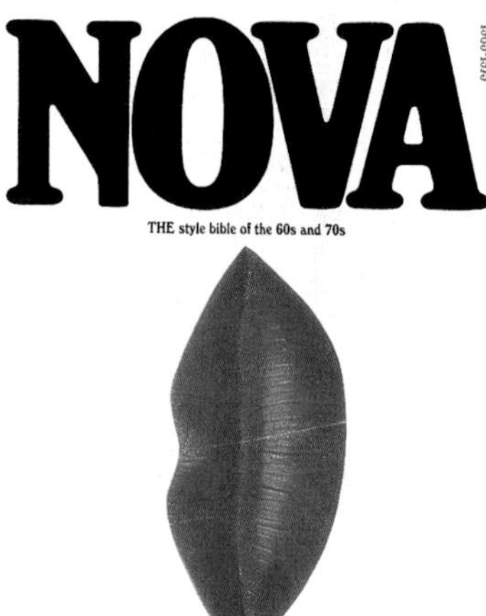


COMPILED BY DAVID HILLMAN & HARRI PECCINOTTI. EDITED BY DAVID GIBBS

Hillman 1993 Cover

**Hiestand,** Ernst – b. 16.9.1935 in Zurich, Switzerland – *graphic designer* – Trained at the Schule für Gestaltung in Zurich. Continued his training in Paris. Art director at a Zurich advertising agency. 1960–81: founds and runs the E. + U. Hiestand studio in Zollikon with Ursula Hiestand. The studio has clients from industry and the arts sector, and also designs Swiss bank notes. 1981: opens the Ernst Hiestand studio in Zurich. 1984: becomes design consultant to IBM Germany. 1986: founds Ernst Hiestand + Partner AG, a studio for design, consultancy and visual communication. Head of the advanced graphics course at the Schule für Gestaltung in Zurich. Guest lecturer at various colleges and universities.

**Hillman,** David – b. 12.2.1943 in Oxford, England – *graphic designer* – 1962–65: studies at the London College of Printing and Graphic Arts. Then works as a designer for "The Sunday Times Magazine". 1965–68: works for "London Life" and "The Sunday Times Magazine". 1968: art director for "Nova" magazine. 1975: opens his own studio and designs the layout for the new daily newspaper "Le Matin de Paris". 1978: becomes a partner at the "Pentagram" design studio in London. Develops design projects for The Arts Council of Great Britain, the Tate Gallery, the Design Museum in London, Skandia. Numerous awards, including for his re-design of "The Guardian" newspaper. – *Publications include:* "Ideas on design" (co-author), London 1986; "Pentagames", London 1990; "Pentagram –

(Co-Autor), London 1986; „Pentagames", London 1990; „Pentagram – The Compendium" (Co-Autor), London 1993; „Nova 1965–75" (Co-Autor), London 1993.

**Hillmann,** Hans – geb. 25. 10. 1925 in Nieder-Mois, Schlesien, Deutschland (heute Polen). – *Grafik-Designer, Illustrator, Lehrer* – 1948–49 Studium an der Staatlichen Schule für Handwerk und Kunst in Kassel. 1949–53 Studium an der Staatlichen Werkakademie Kassel (u. a. bei Hans Leistikow). 1953–56 eigenes Studio in Kassel. 1953–74 zahlreiche Plakatentwürfe für „Neue Filmkunst". Seit 1956 eigenes Studio in Frankfurt am Main. 1959 Gründungsmitglied der Gebrauchsgrafik-Gruppe „Novum". 1961–63 zahlreiche Illustrationen für die Zeitschrift „Twen". 1961–89 Professor für Grafik-Design an der Staatlichen Hochschule für Bildende Künste in Kassel. 1963–64 Gestaltung der Zeitschrift „film". Seit 1980 zahlreiche Illustrationen für die Wochenbeilage der „Frankfurter Allgemeinen Zeitung", das „FAZ-Magazin". – *Publikationen u.a.:* „ABC-Geschichten", Frankfurt am Main 1975; „Fliegenpapier", Frankfurt am Main 1982; „Ein Jogger träumt von der heiligen Monika", Kassel 1989.

**Hinrichs,** Kit – geb. 15. 11. 1941 in Los Angeles, USA. – *Grafik-Designer, Lehrer* – 1963 Abschluß seiner Studien am Art Center College of Design in Los Angeles. Danach Tätigkeit als Designer in New York. Gründung des Büros Russell & Hinrichs in New York. 1972 Gründung des Büros Hinrichs Design Association (mit seiner Frau Linda Hinrichs) in New York. 1976 Umzug nach San Francisco, Grün-

Hillmann 1969 Poster

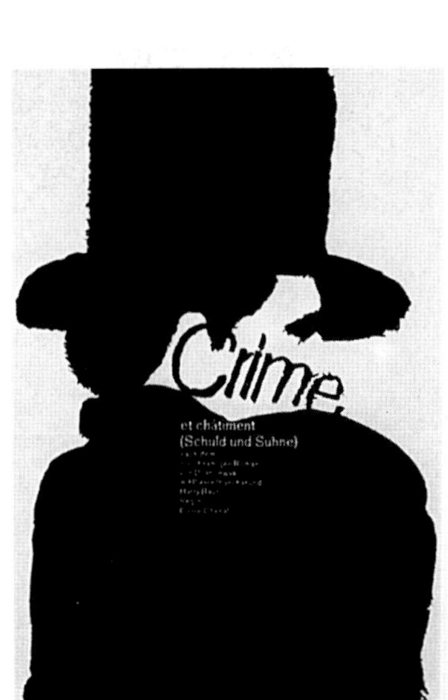

Hillmann 1962 Poster

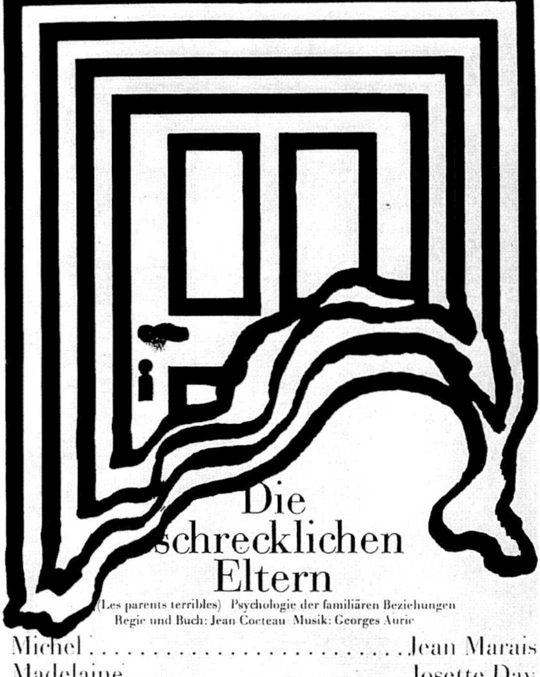

Hillmann 1965 Poster

Hinrichs 1986 Spread

**Hillmann,** Hans – né le 25. 10. 1925 à Nieder-Mois, Silésie, Allemagne, (aujourd'hui en Pologne) – *graphiste maquettiste, illustrateur, enseignant* – 1948–1949, études à la Staatliche Schule für Handwerk und Kunst (école nationale d'art et d'artisanat) de Kassel. 1949–1953, études à la Staatliche Werkakademie (école des arts appliqués) de Kassel (entre autres chez Hans Leistikow). 1953–1956, exerce dans son atelier à Kassel. 1953–1974, nombreuses affiches pour «Neue Filmkunst». Il ouvre un atelier à Francfort-sur-le-Main en 1956. Membre fondateur du groupe de graphistes industriels «Novum» en 1959. Nombreuses illustrations pour la revue «Twen» de 1961 à 1963. Professeur d'arts graphiques et de design à la Staatliche Hochschule für Bildende Künste (école supérieure des beaux-arts) de Kassel de 1961 à 1969. Maquette de la revue «film» de 1963 à 1964. Nombreuses illustrations pour l'hebdomadaire «FAZ-Magazin» de la «Frankfurter Allgemeine Zeitung» depuis 1980. – *Publications, sélection:* «ABC-Geschichten», Francfort-sur-le-Main 1975; «Fliegenpapier», Francfort-sur-le-Main 1982; «Ein Jogger träumt von der heiligen Monika», Kassel 1989.

**Hinrichs,** Kit – né le 15. 11. 1941 à Los Angeles, Etats-Unis – *graphiste maquettiste, enseignant* – termine ses études à l'Art Center College of Design de Los Angeles en 1963. Exerce ensuite comme designer à New York. Fonde l'atelier Russel & Hinrichs à New York. En 1972, il crée l'atelier Hinrichs Design Association (avec sa femme Linda Hinrichs) à New York. S'installe à San Francisco en 1976 et y fonde l'atelier collectif national, Jonson, Pedersen, Hinrichs & Shakery. 1986, fusion de l'atelier de San Francisco et de Penta-

The Compendium" (co-author), London 1993; "Nova 1965–75" (co-author), London 1993.

**Hillmann,** Hans – b. 25. 10. 1925 in Nieder-Mois, Silesia, Germany (now Poland) – *graphic designer, illustrator, teacher* – 1948–49: studies at the Staatliche Schule für Handwerk und Kunst in Kassel. 1949–53: studies at the Staatliche Werkakademie in Kassel (tutors include Hans Leistikov). 1953–56: own studio in Kassel. 1953–74: numerous poster designs for "Neue Filmkunst". 1956: opens his own studio in Frankfurt am Main. 1959: founder member of the commercial art group Novum. 1961–63: numerous illustrations for "Twen" magazine. 1961–89: professor of graphic design at the Staatliche Hochschule für Bildende Künste in Kassel. 1963–64: designs "film" magazine. Since 1980: numerous illustrations for the weekly "FAZ-Magazin" supplement to the "Frankfurter Allgemeine Zeitung". – *Publications include:* "ABC-Geschichten", Frankfurt am Main 1975; "Fliegenpapier", Frankfurt am Main 1982; "Ein Jogger träumt von der heiligen Monika", Kassel 1989.

**Hinrichs,** Kit – b. 15. 11. 1941 in Los Angeles, USA – *graphic designer, teacher* – 1963: completes his studies at the Art Center College of Design in Los Angeles. Then works as a designer in New York. The Russel & Hinrichs studio is opened in New York. 1972: founds the Hinrichs Design Association with his wife Linda Hinrichs in New York. 1976: moves to San Francisco, where a national studio partnership entitled Jonson, Pedersen,

dung der landesweiten Bürogemeinschaft Jonson, Pedersen, Hinrichs & Shakery. 1986 Zusammenschluß des Büros in San Francisco mit Pentagram Design. Zahlreiche Auszeichnungen. Unterrichtet an der School of Visual Arts in New York, an der Academy of Art und am California College of Arts and Crafts in San Francisco. – *Publikationen u.a.:* „Stars & Stripes" (Co-Autor), New York 1987; „TypeWise" (Co-Autor), New York 1991.

**Hlavsa,** Oldrich – geb. 4. 11. 1909 in Náchod, (spätere Tschechoslowakei). – *Typograph* – 1928 Abschluß der Ausbildung als Schriftsetzer, danach Besuch einer grafischen Fortbildungsschule. Seit 1937 Redakteur der Fachzeitschrift „Typografia". Seit 1955 freischaffender Gestalter. Auszeichnungen mit Gold- und Silbermedaillen für Buchgestaltung in Leipzig. Zahlreiche Ausstellungen in Prag (1953), Warschau und Moskau (1955), Leipzig (1960), Montreal (1964), São Paulo (1965), Brno (1966 und 1972). – *Publikationen u.a.:* „Typografická písma latinková", Prag 1957; „A book of type and design", Prag, London 1960; Typographia" (mit Karel Wick, 2 Bände), Prag 1975.

**Hoch,** Hans Peter – geb. 26. 6. 1924 in Aarau, Schweiz. – *Grafik-Designer, Typograph* – Ausbildung an der Höheren Fachschule für das grafische Gewerbe in Stuttgart (1946), der Arbeitsgemeinschaft Bildender Künstler in Bernstein (1946–49) und an der Akademie der Bildenden Künste in Stuttgart (1949–51). Seit 1954 freier Grafik-Designer. Auftraggeber waren u. a. das Landesgewerbeamt Stutt-

gram Design. Nombreuses distinctions. A enseigné à la School of Visual Arts de New York, à l'Academy of Art et au California College of Arts and Crafts de San Francisco. – *Publications, sélection:* «Stars & Stripes» (co-auteur), New York 1987; «TypeWise» (co-auteur), New York 1991.

**Hlavsa,** Oldrich – né le 4. 11. 1909 à Náchod, (plus tard en Tchécoslovaquie) – *typographe* – termine sa formation de compositeur typographe en 1928, fréquente ensuite une école d'arts graphiques. Rédacteur à la revue «Typografia» à partir de 1937. Maquettiste indépendant depuis 1955. Médaille d'or et d'argent pour la meilleure maquette de livre à Leipzig. Nombreuses expositions à Prague (1953), Varsovie et Moscou (1955), Leipzig (1960), Montréal (1964), São Paulo (1965) et Brno (1966 et 1972). – *Publications, sélection:* «Typografická písma latinková», Prague 1957; «A book of type and design», Prague, Londres 1960; «Typographia» (avec Karel Wick, 2 vol.), Prague 1975.

**Hoch,** Hans Peter – né le 26. 6. 1924 à Aarau, Suisse – *graphiste maquettiste, typographe* – formation à la Höhere Fachschule für das grafische Gewerbe (école supérieure des métiers des arts graphiques) à Stuttgart (1946), à la Arbeitsgemeinschaft Bildender Künstler (groupement de travail des artistes peintres et sculpteurs) de Bernstein (1946–1949) et à l'Akademie der Bildenden Künste (beaux-arts) de Stuttgart (1949–1951). Graphiste maquettiste indépendant depuis 1954. Commanditaires: le Landesgewerbeamt Stuttgart, l'Institut für Auslandsbeziehungen, la Deutsche Bundespost, Porsche etc. En 1982, il conçoit l'identité des championnats d'Europe d'athlétisme de Stuttgart. 1983–1989, conception du design et réalisation de

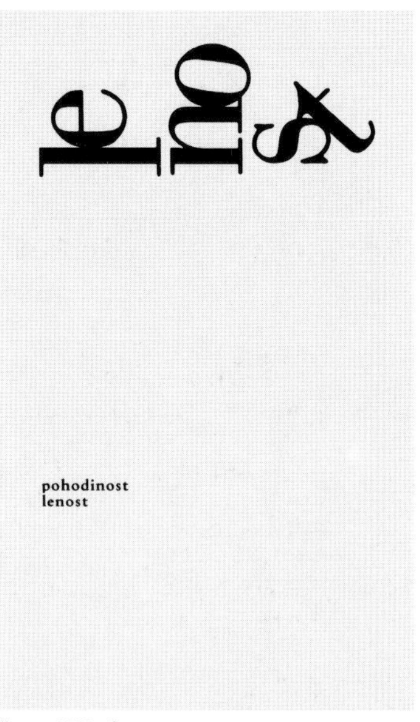

Hlavsa 1969 Cover

Hlavsa 1969 Logo

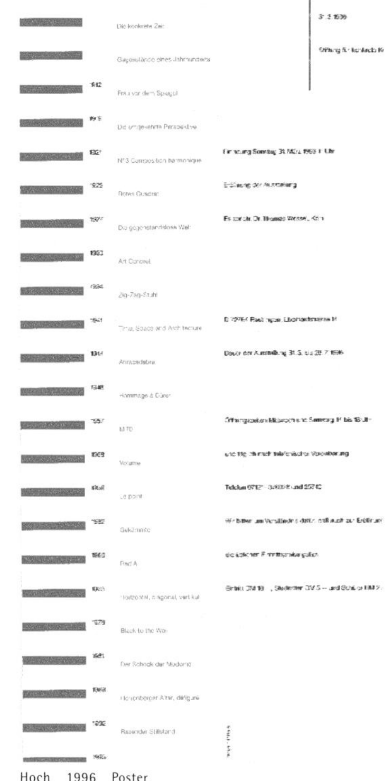
Hoch 1996 Poster

Hlavsa Calendar

Hoch 1994 Stamp

Hinrichs & Shakery is founded. 1986: the San Francisco office merges with Pentagram Design. Numerous awards. Has taught at the School of Visual Arts in New York, at the Academy of Art and at the California College of Arts and Crafts in San Francisco. – *Publications include:* "Stars & Stripes" (co-author), New York 1987; "TypeWise" (co-author), New York 1991.

**Hlavsa,** Oldrich – b. 4. 11. 1909 in Náchod (later in Czechoslovakia) – *typographer* – 1928: completes his training as a typesetter. Attends courses at a continuation school for graphics. From 1937 onwards: editor of the journal "Typografia". Since 1955: freelance designer. Has been awarded gold and silver medals for book design in Leipzig. Numerous exhibitions in Prague (1953), Warsaw and Moscow (1955), Leipzig (1960), Montreal (1964), São Paolo (1965) and Brno (1966 and 1972). – *Publications include:* "Typografická písma latinková", Prague 1957; "A book of type and design", Prague, London 1960; "Typographia" (with Karel Wick, 2 vols.), Prague 1975.

**Hoch,** Hans Peter – b. 26. 6. 1924 in Aarau, Switzerland – *graphic designer, typographer* – Studied at the Höhere Fachschule für das grafische Gewerbe in Stuttgart (1946), the Arbeitsgemeinschaft Bildender Künstler in Bernstein (1946–49) and the Akademie der Bildenden Künste in Stuttgart (1949–51). Since 1954: freelance graphic designer. Clients include the Landesgewerbeamt Stuttgart, Institut für Auslandsbeziehungen, Deutsche Bundespost and Porsche. 1982: cor-

gart, das Institut für Auslandsbeziehungen, die Deutsche Bundespost, Porsche. 1982 Erscheinungsbild der Leichtathletik-Europameisterschaften in Stuttgart. 1983–89 Gestaltungskonzept und Realisation der ständigen Ausstellung „Widerstand im Nationalsozialismus" in der Gedenkstätte Deutscher Widerstand in Berlin. 1988 Preisträger der Stankowski-Stiftung, Stuttgart. Neben dem angewandten Werk entsteht ein reichhaltiges künstlerisches Werk im Sinne von konkret-konstruktiver Gestaltung. – *Publikation u. a.:* „Kunst + Design. Hans Peter Hoch", Stuttgart 1989.

**Höch,** Hannah – geb. 1. 11. 1889 in Gotha, Deutschland, gest. 31. 5. 1978 in Berlin, Deutschland. – *Künstlerin* – 1912 Studium an der Kunstgewerbeschule in Berlin. 1915 Studium an der Lehranstalt des Staatlichen Kunstgewerbemuseums in Berlin. 1917 enge Kontakte zur Dada-Bewegung, erste Collagen. 1918 erste Fotomontagen, Zusammenarbeit mit Raoul Hausmann. 1919 Beteiligung an der ersten Dada-Ausstellung in Berlin. Mitglied der Novembergruppe. 1920 Beteiligung an der „Ersten Internationalen Dada-Messe" in Berlin. 1922 Beteiligung am „Merz-Bau" von Kurt Schwitters in Hannover. 1926–29 Aufenthalt in den Niederlanden, Kontakte zur „De Stijl"-Gruppe. 1929 Beteiligung an der Ausstellung „Film und Foto" in Stuttgart. 1930 Rückkehr nach Deutschland. 1947 erste Montagen aus Farbfotos. 1965 Berufung in die Akademie der Künste Berlin. Buchstabenformen, Texte und Textteile in individueller, künstlerischer Anordnung sind

l'exposition permanente «Widerstand im Nationalsozialismus» (Résistance sous le national-socialisme) au mémorial de la résistance allemande de Berlin. 1988, prix de la fondation Stankowski. Parallèlement à ses travaux en arts appliqués, il crée des œuvres plastiques proche du design concret et constructiviste. – *Publication, sélection:* «Kunst + Design. Hans Peter Hoch», Stuttgart 1989.

**Höch,** Hannah – né le 1. 11. 1889 à Gotha, Allemagne, décédé le 31. 5. 1978 à Berlin, Allemagne – *artiste* – 1912, études à la Kunstgewerbeschule (école des arts décoratifs) de Berlin. 1915, études au Lehranstalt des Staatlichen Kunstgewerbemuseums (institut d'enseignement du musée des arts décoratifs) de Berlin. 1917, contacts étroits avec le mouvement Dada et premiers collages. 1918, premiers photomontages et travail avec Raoul Hausmann. En 1919, elle participe à la première exposition Dada à Berlin. Membre du groupe «Novembre». En 1920, elle participe à la «Erste Internationale Dada-Messe» (première foire internationale Dada) à Berlin. En 1922, elle intervient dans la construction du «Merz-Bau» de Kurt Schwitters à Hanovre. 1926–1929, séjour aux Pays-Bas et contacts avec le groupe «De Stijl». Participe à l'exposition «Film und Foto» à Stuttgart en 1929. Retour en Allemagne en 1930. Premiers montages de photos en couleurs en 1947. Elle est appelée à l'Akademie der Künste

Druckgrafik Ausstellung
Forum für Kulturaustausch
Institut für Auslandsbeziehungen
Stuttgart Charlottenplatz 17
1. Dezember 1976 bis 23. Januar 1977
Di–Fr 11.00–19.00
Sa/So und Feiertage 11.00–17.00
Mo geschlossen
24./25. und 31. Dezember 1976 geschlossen
Holzschnitt am Beispiel HAP Grieshaber
Linolschnitt am Beispiel Karl Rössing
Ätzradierung am Beispiel Horst Janssen
Aquatinta am Beispiel Friedrich Meckseper
Lithografie am Beispiel Paul Wunderlich
Siebdruck am Beispiel Günter Fruhtrunk
Prägedruck am Beispiel Günther Uecker
Mischtechnik am Beispiel Dieter Roth

Hoch 1976 Poster

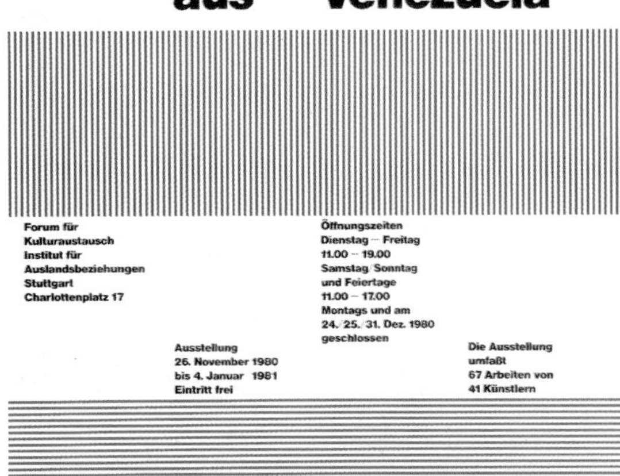


Hoch 1980 Poster

presse ifa
telefax ifa
notiz ifa
75 jahre ifa

Hoch 1991 Signets

Höch Poster

porate imagery for the European light athletics championships in Stuttgart. 1983–89: designs and realizes a permanent exhibition on the German Resistance to National Socialism in the Gedenkstätte Deutscher Widerstand in Berlin. 1988: awarded a prize by the Stankowski-Stiftung in Stuttgart. Besides his applied art, Hoch has produced extensive art work in the field of concrete and constructive design. – *Publications include:* "Kunst + Design. Hans Peter Hoch", Stuttgart 1989.

**Höch,** Hannah – b. 1. 11. 1889 in Gotha, Germany, d. 31. 5. 1978 in Berlin, Germany – *artist* – 1912: studies at the Kunstgewerbeschule in Berlin. 1915: studies at the Lehranstalt des Staatlichen Kunstgewerbemuseums in Berlin. 1917: close contacts with the Dada movement. Produces her first collages. 1918: first photomontages. Works with Raoul Hausmann. 1919: takes part in the first Dada exhibition in Berlin. Member of the Novembergruppe. 1920: takes part in the "first international Dada fair" in Berlin. 1922: takes part in Kurt Schwitters' Merz-Bau project in Hanover. 1926–29: spends time in the Netherlands and has contacts with the De Stijl group. 1929: takes part in the "Film und Foto" exhibition in Stuttgart. 1930: returns to Germany. 1947: first montages made from color photos. 1965: appointed to the Akademie der Künste in Berlin. Individually – and artistically – arranged letter forms, texts and textual components are important pictorial and compositional elements threading through all periods of Hannah

in allen Schaffensperioden Kompositionselemente. – *Publikationen u.a.:* H. Ohff „Hannah Höch", Berlin 1968; H. Remmert, P. Barth „Hannah Höch. Werke und Worte", Berlin 1982; E. Roters (Hrsg.) „Hannah Höch. Eine Lebenscollage" (2 Bände), Berlin 1989; J. Dech, J. E. Maurer (Hrsg.) „Da-da zwischen Reden zu Hannah Höch", Berlin 1991.

**Hochuli,** Jost – geb. 8. 6. 1933 in St. Gallen, Schweiz. – *Grafik-Designer, Schriftsetzer, Typograph, Autor, Lehrer* – 1952–54 Studium an der Kunstgewerbeschule St. Gallen. 1954–55 Volontär bei Rudolf Hostettler in der Druckerei Zollikofer, St. Gallen. 1955–58 Schriftsetzerlehre bei Zollikofer und an der Setzerfachklasse der Kunstgewerbeschule Zürich. 1958–59 Studium an der Ecole Estienne in Paris, Kurse bei Adrian Frutiger. Seit 1959 Atelier für Gebrauchsgrafik und Buchgestaltung in St. Gallen. Unterrichtet 1967–80 Schrift, später Basisausbildung an der Kunstgewerbeschule Zürich, 1980 Schriftunterricht an der Schule für Gestaltung in St. Gallen. Seit 1976 zahlreiche Vorträge und Seminare. 1978 Gründungsmitglied der VGS Verlagsgemeinschaft St. Gallen. Seit 1983 Herausgeber und Gestalter der Broschürenreihe „Typotron" in St. Gallen. 1983 Werkausstellung im Stadtmuseum München. – *Publikationen u. a.:* „Punkt, Cicero und Kaviar. Zum 100. Geburtstag Iron Henry Tschudy", St. Gallen 1982; „Epitaph für Rudolf Hostettler", St. Gallen 1983; „Die Vogelkäfige des Alfons J. Keller", St. Gallen 1985; „Das Detail in der Typographie", Wilmington 1989; „Jost

Berlin en 1965. La forme des lettres, les textes et fragments de textes organisés avec art et originalité ont été des éléments de composition à toutes les époques de sa création. – *Publications, sélection:* H. Ohff «Hannah Höch», Berlin 1968; H. Remmert, P. Barth «Hannah Höch. Werke und Worte», Berlin 1982; E. Roters (éd.) «Hannah Höch. Eine Lebenscollage» (2 vol.), Berlin 1989; J. Dech, J. E. Maurer (éd.) «Da-da zwischen Reden zu Hannah Höch», Berlin 1991.

**Hochuli,** Jost – né le 8. 6. 1933 à Saint-Gall, Suisse. – *graphiste maquettiste, compositeur, typographie, auteur, enseignant* – 1952–1954, études à la Kunstgewerbeschule (école des arts décoratifs) de Saint-Gall. 1954–1955, stagiaire chez Rudolf Hostettler à l'imprimerie Zollikofer, Saint-Gall. 1955–1958, apprentissage de composition typographique chez Zollikofer et à la section de composition typographique de la Kunstgewerbeschule de Zurich. Etudes à l'Ecole Estienne à Paris en 1958–1959, où il suit les cours d'Adrian Frutiger. Atelier de graphiste et de maquettiste à Saint-Gall à partir de 1959. Enseigne les arts de l'écriture, de 1967 à 1980, puis chargé du cours préliminaire à l'école des arts décoratifs de Zurich. En 1980, il enseigne les arts de l'écriture à la Schule für Gestaltung (école de design) de Saint-Gall. A partir de 1976, nombreuses conférences et séminaires. 1978, membre fondateur du VGS Verlagsgemeinschaft Saint-Gall. Depuis 1983, directeur et maquettiste de la collection «Typotron» à Saint-Gall. 1983, exposition de son œuvre au Stadtmuseum de Munich. – *Publications, sélection:* «Punkt, Cicero und Kaviar. Zum 100. Geburtstag von Iron Henry Tschudy», Saint-Gall 1982; «Epitaph für Rudolf Hostettler», Saint-Gall 1983; «Die Vogelkäfige des Alfons J. Keller», Saint-Gall 1985;

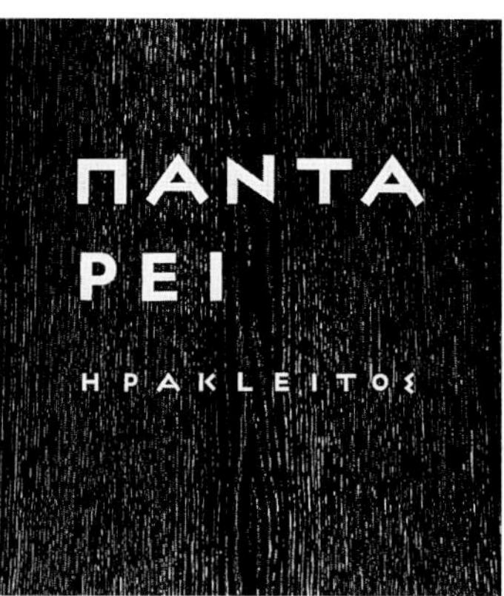

Hochuli

Hochuli Logo

Hochuli Advertisement

Hochuli 1973 Poster

**Bücher machen**

Eine Einführung in die Buchgestaltung, im besonderen in die Buchtypografie von *Jost Hochuli*

Liebe Leserin, lieber Leser Mit dem schmalen Bändchen, das Du in Händen hältst, versucht der Autor, junge Typografen und Grafiker und darüber hinaus alle jene, die an der Buchform interessiert sind, in die Buchgestaltung, insbesondere in die Buchtypografie, einzuführen. ‹Heranzuführen› wäre vielleicht treffender, denn der bescheidene Umfang erlaubt es höchstens, das Thema in großen Zügen zu umreißen.

Auch wenn es sich bei der vorliegenden Drucksache im streng fachlichen Sinn nicht um ein Buch, sondern nur um eine Broschur handelt: mit Ausnahme des Einbandes und des Vorsatzpapiers ist der Gegenstand mit einem ‹richtigen› Buch identisch.

Und wie in Deiner Buchhandlung schon viele Male, hast Du auch jetzt, noch bevor Du es in die Hand genommen hast, gesehen: das Format des Bändchens, seine Dicke, das Material des Deckels. Und hernach – wiewohl unbewußt – abgeschätzt: das Gewicht, die Handlichkeit (mit einer oder mit beiden Händen zu halten?), die Steife oder die Flexibilität und die übrigen taktilen Eigenschaften. Werde ich den Gegenstand gerne in die Hand nehmen, oder ist er meinen Fingern unangenehm? Schließlich hat das auf dem Deckel Gedruckte formale und farbliche Eindrücke hinterlassen und zusammen mit dem Titel und dem Namen des Autors Assoziationen und Erwartungen geweckt.

Du siehst, auch die Gestalt des Buches ist eine komplexe Sache. Sofern Du eine junge Grafikerin oder Typografin, ein junger Grafiker oder Typograf bist, merke Dir vor allem eines: Das Buch als Gebrauchsgegenstand ist ein dreidimensionaler Körper!

Agfa Compugraphic

Hochuli 1989 Cover

Höch's work. – *Publications include:* H. Ohff "Hannah Höch", Berlin 1968; H. Remmert, P. Barth "Hannah Höch. Werke und Worte", Berlin 1982; E. Roters (ed.) "Hannah Höch. Eine Lebenscollage" (2 vols.), Berlin 1989; J. Dech, J. E. Maurer (eds.) "Da-da zwischen Reden zu Hannah Höch", Berlin 1991.

**Hochuli,** Jost – b. 8. 6. 1933 in St. Gallen, Switzerland – *graphic designer, typesetter, typographer, specialist author, teacher* – 1952–54: studies at the Kunstgewerbeschule in St. Gallen. 1954–55: trains under Rudolf Hostettler in the Zollikofer printing works in St. Gallen. 1955–58: trains as a typesetter with Zollikofer and takes advanced type setting courses at the Kunstgewerbeschule in Zurich. 1958–59: studies at the Ecole Estienne in Paris and takes courses run by Adrian Frutiger. 1959: opens a studio for commercial art and book design in St. Gallen. 1967–80: teaches lettering then basic training at the Kunstgewerbeschule in Zurich. 1980: lettering courses at the Schule für Gestaltung in St. Gallen. Since 1976: numerous lectures and seminars. 1978: founder member of the VGS Verlagsgemeinschaft St. Gallen. Since 1983: editor and designer of the "Typotron" brochure series in St. Gallen. 1983: exhibition of his works in the Stadtmuseum in Munich. – *Publications include:* "Punkt, Cicero und Kaviar. Zum 100. Geburtstag Iron Henry Tschudy", St. Gallen 1982; "Epitaph für Rudolf Hostettler", St. Gallen 1983; "Die Vogelkäfige des Alfons J. Keller", St. Gallen 1985; "Das

Hochuli's Alphabugs", Wilmington 1990; „Buchgestaltung als Denkschule", Stuttgart 1991; „Kleine Geschichte der geschriebenen Schrift", St. Gallen 1991.

**Hoefer**, Karlgeorg – geb. 6. 2. 1914 in Schlesisch-Drehnow, Deutschland. – *Schriftentwerfer, Kalligraph, Lehrer* – 1930–35 Schriftsetzerlehre und Schriftsetzer in Hamburg. 1937–39 Ausbildung zum Grafiker an den Technischen Lehranstalten in Offenbach, danach Betriebsassistent in Potsdam. 1946–79 Fachlehrer, später Dozent für Schrift an der Meisterschule für das gestaltende Handwerk in Offenbach, die 1949 in eine Werkkunstschule und 1971 in eine Hochschule für Gestaltung umgewandelt wird. Entwickelt 1950 die Universal-Schreibfeder „Brause 505". Leitet 1981–88 Kalligraphie-Sommer-Workshops in den USA. 1982 Gründung der Offenbacher Schreibwerkstatt für Kalligraphie, 1987 der „Schreibwerkstatt Klingspor Offenbach, Förderkreis internationaler Kalligraphie e. V.". 1994–95 Schriftkurse an der Kunstschule in Basel. Zahlreiche Einzelausstellungen und Auszeichnungen. – *Schriftentwürfe:* Salto (1952), Saltino (1953), Saltarello (1954), Monsun (1954), Prima (1957), Permanent (ab 1962), Zebra (1963), Stereo (1963), Elegance (1964), Programm-Grotesk (1970), Vereinfachte Ausgangsschrift für Schulen (1972), Big Band (1974), Omnia (1990), San Marco (1990), Notre Dame (1991), Sho (1992). – *Publikationen u. a.:* „Das alles mit einer Feder", Iserlohn 1950; „Kalligraphie – gestaltete Handschrift", Düsseldorf 1986. Hans Adolf Halbey „Karlgeorg Hoefer",

Hoefer 1962 Calligraphy

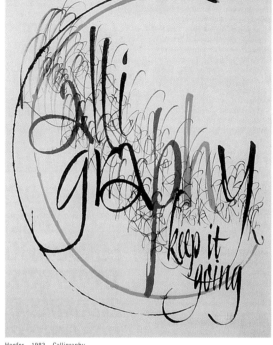

Hoefer 1983 Calligraphy

Hoefer 1988 Calligraphy

| | n | | | | |
|---|---|---|---|---|---|
| a | b | e | f | g | i |
| o | r | s | t | y | z |
| A | B | C | E | G | H |
| M | O | R | S | X | Y |
| 1 | 2 | 4 | 6 | 8 | & |

Hoefer 1990 San Marco

| | | | | | |
|---|---|---|---|---|---|
| | | | | | |
| | | | | | |
| a | b | c | e | g | h |
| m | o | r | s | x | y |
| 1 | 2 | 4 | 6 | 8 | & |

Hoeffer 1990 Omnia

«Das Detail in der Typographie», Wilmington 1989; «Jost Hochuli's Alphabugs», Wilmington 1990; «Buchgestaltung als Denkschule», Stuttgart 1991; «Kleine Geschichte der geschriebenen Schrift», Saint-Gall 1991.

**Hoefer,** Karlgeorg – né le 6. 2. 1914 à Schlesisch-Drehnow, Allemagne – *concepteur de polices, calligraphe, enseignant* – 1930–1935, apprentissage de composition typographique et compositeur à Hambourg. 1937–1939, formation de graphiste aux Technische Lehranstalten (instituts des techniques) d'Offenbach, puis assistant de fabrication à Potsdam. 1946–1979, enseignant puis professeur d'arts de l'écriture à la Meisterschule für das gestaltende Handwerk (école technique d'artisanat et de design) d'Offenbach, qui devient l'école des arts appliqués en 1949 puis l'école supérieure de design en 1971. En 1950, il invente la plume universelle «Brause 505». 1981–1988, dirige l'été des ateliers de calligraphie aux Etats-Unis. En 1982, il fonde la Offenbacher Schreibwerkstatt für Kalligraphie (atelier de calligraphie d'Offenbach), puis la «Schreibwerkstatt Klingspor Offenbach, Förderkreis internationaler Kalligraphie e.V.» (atelier Klingspor, Offenbach. Cercle international de soutien à la calligraphie) en 1987. Cours d'arts de l'écriture à la Kunstschule (école des beaux-arts) de Bâle de 1994 à 1995. Nombreuses expositions personnelles et distinctions. – *Polices:* Salto (1952), Saltino (1953), Saltarello (1954), Monsun (1954), Prima (1957), Permanent (à partir de 1962), Zebra (1963), Stereo (1963), Elegance (1964), Programm-Grotesk (1970), Ecriture de base simplifiée pour les écoles (1972), Big Band (1974), Omnia (1990), San Marco (1990), Notre Dame (1991), Sho (1992). – *Publications, sélection:* «Das alles mit einer Feder»,

Detail in der Typographie", Wilmington 1989; "Jost Hochuli's Alphabugs", Wilmington 1990; "Buchgestaltung als Denkschule", Stuttgart 1991; "Kleine Geschichte der geschriebenen Schrift", St. Gallen 1991.

**Hoefer,** Karlgeorg – b. 6. 2. 1914 in Schlesisch-Drehnow, Germany – *type designer, calligrapher, teacher* – 1930–35: trains and then works as a typesetter in Hamburg. 1937–39: trains as a graphic artist at the Technische Lehranstalten in Offenbach, then assistant in a printing office in Potsdam. 1946–79: specialist teacher, later lettering lecturer at the Meisterschule für das gestaltende Handwerk in Offenbach, which becomes the Werkkunstschule in 1949 and the Hochschule für Gestaltung in 1971. 1950: develops the universal Brause 505 nib. 1981–88: runs summer calligraphy workshops in the USA. 1982: founds the Offenbacher Schreibwerkstatt für Kalligraphie and in 1987 the Schreibwerkstatt Klingspor Offenbach, Förderkreis internationaler Kalligraphie e.V. 1994–95: lettering courses at the Kunstschule in Basle. Numerous solo exhibitions and awards. – *Fonts:* Salto (1952), Saltino (1953), Saltarello (1954), Monsun (1954), Prima (1957), Permanent (from 1962), Zebra (1963), Stereo (1963), Elegance (1964), Programm-Grotesk (1970), Simplified first alphabet for all schools (1972), Big Band (1974), Omnia (1990), San Marco (1990), Notre Dame (1991), Sho (1992). – *Publications include:* "Das alles mit einer Feder", Iserlohn 1950; "Kalligraphie – gestaltete Handschrift", Düsseldorf 1986.

Offenbach 1963; Hermann Zapf u. a. „Schriftkunst. Karlgeorg Hoefer", Hardheim 1989.

**Hoefler,** Jonathan – geb. 22. 8. 1970 in New York, USA. – *Schriftentwerfer* – Autodidakt. Teilnahme an Kursen der Parson's School of Design. 1988–89 Arbeiten im Studio von Roger Black. Seit 1989 freier Gestalter. Eröffnung seines Studios „The Hoefler Type Foundry" in New York. Schriftentwürfe für die Zeitschriften „GQ", „Rolling Stone", „Sports Illustrated", „Harper's Bazaar", „House and Garden", „The New York Times Magazine", das Guggenheim Museum, Apple Computer, American Express, die Musikgruppe „They might be giants" und Alfred A. Knopf Publishers. Seit 1990 zahlreiche Vorträge und Auszeichnungen. – *Schriftentwürfe:* Egiziano Filigree (1989), Champion Gothic (1990), Bodoni Grazia (1990), Hoefler Text (1991–93), Gestalt (1991–93), Ideal Sans (1991), Mazarin (1991), Requiem (1991–96), Ziggurat (1991–93), HTF Didot (1992), Acropolis (1993), HTF Fetish (1993–95), Leviathan (1993), Quantico (1993), Saracen (1993–96), Deseret (1994), Historical Allsorts (1994), William Maxwell Roman (1994), Knockout (1994), Troubadour (1994), HTF Guggenheim (1996), They might be gothic (1996), Hoefler Titling (1996), Pavisse (1996). – *Publikation u. a.:* „Every Art Director needs his own Typeface", New York 1991.

**Höhnisch,** Walter – geb. 14. 2. 1906 in Dresden, Deutschland. – *Schriftentwerfer, Grafiker* – Lehre als Kaufmann. Volontär in einer Werbeagentur. Kurse an der

Iserlohn 1950; «Kalligraphie – gestaltete Handschrift», Düsseldorf 1986. Hans Adolf Halbey «Karlgeorg Hoefer», Offenbach 1963; Hermann Zapf et autres «Schriftkunst. Karlgeorg Hoefer», Hardheim 1989.

**Hoefler,** Jonathan – né le 22. 8. 1970 à New York, Etats-Unis – *concepteur de polices* – autodidacte. Suit les cours de la Parson's School of Design. Travaille à l'atelier Roger Black de 1988 à 1989.Indépendant depuis 1989. Ouvre son atelier «The Hoefler Type Foundry» à New York. Conception de polices pour les revues «GQ», «Rolling Stone», «Sports Illustrated», Harper's Bazaar», «House and Garden», «The New York Times Magazine», le Guggenheim Museum, Apple Computer, American Express, le groupe de musiciens «They might be giants» et Alfred A. Knopf Publishers. Nombreuses conférences et distinctions depuis 1990. – *Polices:* Egiziano Filigree (1989), Champion Gothic (1990), Bodoni Grazia (1990), Hoefler Text (1991–1993), Gestalt (1991–1993), Ideal Sans (1991), Mazarin (1991), Requiem (1991–1996), Ziggurat (1991–1993), HTF Didot (1992), Acropolis (1993), HTF Fetish (1993–1995), Leviathan (1993), Quantico (1993), Saracen (1993–1996), Deseret (1994), Historical Allsorts (1994), William Maxwell Roman (1994), Knockout (1994), Troubadour (1994), HTF Guggenheim (1996), They might be gothic (1996), Hoefler Titling (1996), Pavisse (1996). – *Publication, sélection:* «Every Art Director needs his own Typeface», New York 1991.

**Höhnisch,** Walter – né le 14. 2. 1906 à Dresde, Allemagne – *concepteur de polices, graphiste* – études de commerce. Stagiaire dans une agence de publicité. Cours à l'Abendakademie für Kunstgewerbe (arts décoratifs) de Dresde. En 1927, il suit les cours de Rudolf Koch aux Tech-

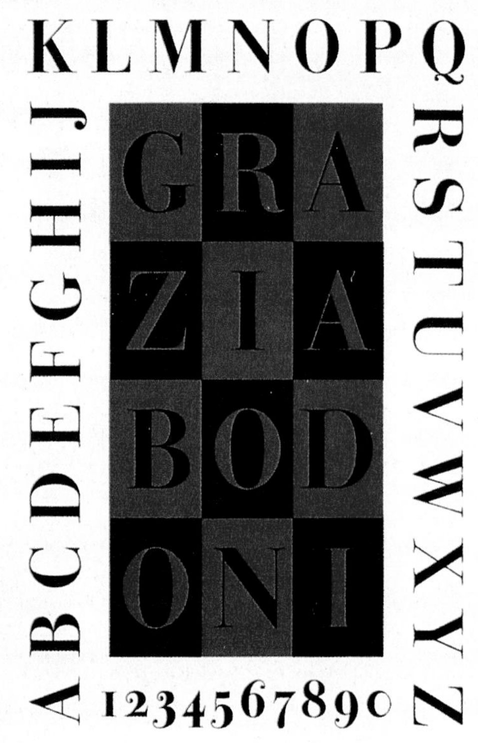


Hoefler 1990 Advertisement

Hofmann 1957 Poster

Frühjahrs-Mode

Höhnisch 1939 Stop

AB CD EF GH
IJ KL
Express mn op
QR ST
uv wx
yz

Höhnisch 1952 Express

National

die straffe organische

Höhnisch 1933 National

Hans Adolf Halbey "Karlgeorg Hoefer", Offenbach 1963; Hermann Zapf et al "Schriftkunst. Karlgeorg Hoefer", Hardheim 1989.

**Hoefler,** Jonathan – b. 22. 8. 1970 in New York, USA – *type designer* – Self-taught. Attends courses at the Parson's School of Design. 1988–89: works at Roger Black's studio. Since 1989: freelance designer. Opens his studio The Hoefler Type Foundry in New York. Designs fonts for "GQ", "Rolling Stone", "Sports Illustrated", "Harper's Bazaar", "House and Garden", "The New York Times Magazine", the Guggenheim Museum, Apple Computer, American Express, the music group They might be giants and Alfred A. Knopf Publishers. Since 1990: numerous lectures and awards. – *Fonts:* Egiziano Filigree (1989), Champion Gothic (1990), Bodoni Grazia (1990), Hoefler Text (1991–93), Gestalt (1991–93), Ideal Sans (1991), Mazarin (1991), Requiem (1991–96), Ziggurat (1991–93), HTF Didot (1992), Acropolis (1993), HTF Fetish (1993–95), Leviathan (1993), Quantico (1993), Saracen (1993–96), Deseret (1994), Historical Allsorts (1994), William Maxwell Roman (1994), Knockout (1994), Troubadour (1994), HTF Guggenheim (1996), They might be gothic (1996), Hoefler Titling (1996), Pavisse (1996). – *Publications include:* "Every Art-Director needs his own Typeface", New York 1991.

**Höhnisch,** Walter – b. 14. 2. 1906 in Dresden, Germany – *type designer, graphic artist* – Training in commerce and at an advertising agency. Took courses at the Abendakademie für Kunstgewerbe in

Abendakademie für Kunstgewerbe in Dresden. 1927 Kurse an den Technischen Lehranstalten in Offenbach bei Rudolf Koch. 1930 Schriftzeichner, von 1949–71 Schriftentwerfer und Grafiker in der Schriftgießerei Ludwig & Mayer. – *Schriftentwürfe:* Tempo (1930), Werbeschrift deutsch (1933), National (1934), Skizze (1935), Stop (1939), Antiqua die Schlanke (1938–39), Express (1952).

**Hoffmeister,** Heinrich – geb. 22. 7. 1857 in Lennep, Deutschland, gest. 21. 9. 1921 in Langen, Deutschland. – *Schriftentwerfer, Grafik-Designer, Lehrer* – Studium an der Kunstakademie Düsseldorf, 1879–80 Studium an der Königlichen Kunstschule Berlin. 1884–98 Zeichenlehrer in Leipzig. 1898–1904 Gründung und Leitung einer Schriftgießerei in Leipzig. 1905 Schriftentwerfer und Hausgrafiker der D. Stempel AG. – *Schriftentwürfe:* Säculum (1907–21), Amts-Antiqua (später Madison genannt, 1909–19), Stempel Fraktur (1914).

**Hofmann,** Armin – geb. 29. 6. 1920 in Winterthur, Schweiz. – *Grafik-Designer, Lehrer* – 1937–39 Studium an der Kunstgewerbeschule Zürich. 1940–43 Lithographenlehre in Winterthur. 1943–48 Lithograph in Basel und Bern. Seit 1948 eigenes Atelier in Basel. 1946–86 Leiter der Fachklasse für Grafik an der Gewerbeschule Basel, später Leiter des Weiterbildungskurses für Visuelle Kommunikation an der Schule für Gestaltung. Die Lehre Hofmanns ist weltweit richtungsweisend für eine experimentelle Herangehensweise bei der Vermittlung gestalterischer Grundlagen und deren Einsatz in

Hofmann 1958 Poster

Hofmann 1967 Poster

Hofmann 1975 Poster

nische Lehranstalten (institut des techniques) d'Offenbach. Dessine des écritures à partir de 1930. Concepteur de polices et graphiste à la fonderie de caractères Ludwig & Mayer de 1949 à 1971. – *Polices:* Tempo (1930), Werbeschrift deutsch (1933), National (1934), Skizze (1935), Stop (1939), Antiqua die Schlanke (1938–1939), Express (1952).

**Hoffmeister,** Heinrich – né le 22. 7. 1857 à Lennep, Allemagne, décédé le 21. 9. 1921 à Langen, Allemagne – *concepteur de polices, graphiste maquettiste, enseignant* – études à la Kunstakademie (beaux-arts) de Düsseldorf, puis de 1879 à 1880, études à la Königliche Kunstschule (école royale des beaux-arts) de Berlin. 1884– 1898, professeur de dessin à Leipzig. Crée une fonderie de caractères à Leipzig en 1898 et la dirige jusqu'en 1904. Concepteur de polices et graphiste maison chez D. Stempel AG en 1905. – *Polices:* Säculum (1907–1921), Amts-Antiqua (appelée Madison par la suite, 1909–1919), Stempel Fraktur (1914).

**Hofmann,** Armin – né le 29. 6. 1920 à Winterthur, Suisse – *graphiste maquettiste, enseignant* – 1937–1939, études à la Kunstgewerbeschule (école des arts décoratifs) de Zurich. 1940–1943, apprentissage de lithographe à Winterthur. 1943–1948, lithographe à Bâle et à Berne. Exerce dans son propre atelier à Bâle dès 1948. 1946–1986, dirige la section d'arts graphiques à la Gewerbeschule de Bâle, puis les cours de formation permanente de communication visuelle à la Schule für Gestaltung (école de design). L'enseigne-

Dresden. 1927: attends courses at the Technische Lehranstalten in Offenbach under Rudolf Koch. 1930: letterer. 1949–71: type designer and graphic artist at the Ludwig & Mayer type foundry. – *Fonts:* Tempo (1930), Werbeschrift deutsch (1933), National (1934), Skizze (1935), Stop (1939), Antiqua die Schlanke (1938–39), Express (1952).

**Hoffmeister,** Heinrich – b. 22. 7. 1857 in Lennep, Germany, d. 21. 9. 1921 in Langen, Germany – *type designer, graphic designer, teacher* – Studied at the Kunstakademie in Düsseldorf. 1879–80: studies at the Königliche Kunstschule in Berlin. 1884–98: teaches art in Leipzig. 1898–1904: founds and runs a type foundry in Leipzig. 1905: type designer and in-house graphic artist for D. Stempel AG. – *Fonts:* Säculum (1907–21), Amts-Antiqua (later called Madison, 1909–19), Stempel Fraktur (1914).

**Hofmann,** Armin – b. 29. 6. 1920 in Winterthur, Switzerland – *graphic designer, teacher* – 1937–39: studies at the Kunstgewerbeschule in Zurich. 1940–43: trains as a lithographer in Winterthur. 1943–48: lithographer in Basle and Bern. 1948: opens his own studio in Basle. 1946–86: head of the advanced graphics course at the Gewerbeschule in Basle, later head of the further training course for visual communication at the Schule für Gestaltung. Throughout the world, Hofmann's theories have pointed the way ahead for an experimental approach to the teaching of design basics and the use thereof in visual communication. From 1955 onwards: guest lecturer at various Ameri-

der Visuellen Kommunikation. Seit 1955 Gastdozent an amerikanischen Universitäten. 1987 Dr. h.c. der Philadelphia University of the Arts. 1988 Ehrenmitglied der Royal Society of Arts in London. – *Publikationen u. a.:* „Graphic Design Manual", Teufen, London 1965; „The Basel School of Design and its Philosophy: the Armin Hofmann years 1946–1986", Philadelphia 1986. H. Wichmann (Hrsg.) „Armin Hofmann, Werk, Wirkung, Lehre", Basel, Boston, Berlin 1989.

**Hollandsworth,** Cynthia – geb. 8. 7. 1956 Washington D.C., USA. – *Schriftentwerferin* – Studium am California College of Arts and Crafts in Oakland, Kalifornien. Managerin der Abteilung Schriftentwurf und -entwicklung bei Agfa Compugraphic in Massachusetts. Beraterin des ITC Typeface Review Board. 1987 Gründung der „Typeface Design Coalition", um den legalen Schutz von Schriftentwürfen und Software in den USA zu sichern. Arbeitet (mit Barbara Gibb) am Neuschnitt einer Schrift von F. H. E. Schneidler, die in den 30er Jahren wegen des Kriegsausbruchs nicht produziert wurde. – *Schriftentwürfe:* Vermeer (1986), Hiroshige (1986), ITC Tiepolo (1987), Agfa Wile Roman (1990), Pompei Capitals (1995), Synthetica (mit Philip Bouwsma, 1996).

**Hollenstein,** Albert – geb. 1930 in Luzern, Schweiz, gest. 10. 8. 1974 in Vernazza, Italien. – *Typograph, Schriftentwerfer* – Vier Jahre Lehre als Typograph. Danach zwei Jahre in der Schweiz tätig. 1955–56 Fabrikationsleiter des Buchclubs „Club du Meilleur Livre". 1956 Eröffnung seines eigenen Studios „Atelier de composition

ment de Hofmann a été déterminant. Il a permis une approche expérimentale dans l'apprentissage des bases formelles et pour leur utilisation en communication visuelle. Depuis 1955, cycle de conférences dans des universités américaines. 1987, Dr. h. c. de la Philadelphia University of the Arts. 1988, membre d'honneur de la Royal Society of Arts de Londres. – *Publications, sélection:* «Graphic Design Manual», Teufen, Londres 1965; «The Basel School of Design and its Philosophy: the Armin Hofmann Years 1946–1986», Philadelphie 1986; H. Wichmann (éd.) «Armin Hofmann, Werk, Wirkung, Lehre», Bâle, Boston, Berlin 1989.

**Hollandsworth,** Cynthia – née le 8. 7. 1956, Washington D.C., Etats-Unis – *concepteur de polices* – études au California College of Arts and Crafts d'Oakland, Californie. Dirige le service de conception de polices et du développement chez Agfa Compugraphic de Massachusetts. Conseillère de la ITC Typeface Review Board. En 1987, elle fonde la «Typeface Design Coalition» dans le but de protéger les droits sur les polices et les logiciels aux Etats-Unis. A travaillé (avec Barbara Gibb) à la refonte d'une police de F. H. E. Schneidler qui n'avait pu être produite dans les années 30 à cause de la déclaration de guerre. – *Polices:* Vermeer (1986), Hiroshige (1986), ITC Tiepolo (1987), Agfa Wile Roman (1990), Pompei Capitals (1995), Synthetica (avec Philip Bouwsma, 1996).

**Hollenstein,** Albert – né en 1930 à Lucerne, Suisse, décédé le 10. 8. 1974 à Vernazza, Italie – *typographe, concepteur de polices* – quatre années d'apprentissage de la typographie, exerce ensuite deux ans en Suisse. 1955–1956, directeur de fabrication du «Club du Meilleur Livre». Il ouvre son propre studio «Atelier de composition typographique à façon» en 1956.

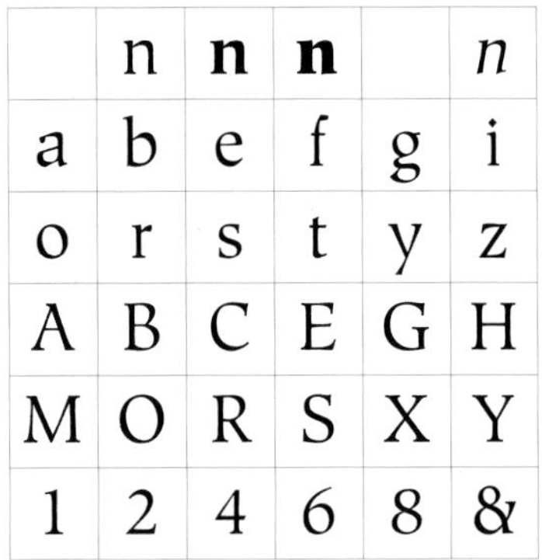

Hollandsworth 1986 Hiroshige

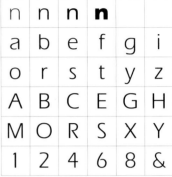

| | n | n | n | | |
|---|---|---|---|---|---|
| a | b | e | f | g | i |
| o | r | s | t | y | z |
| A | B | C | E | G | H |
| M | O | R | S | X | Y |
| 1 | 2 | 4 | 6 | 8 | & |

Hollandsworth 1987 Tiepolo 1

| n | n | n | n | | |
|---|---|---|---|---|---|
| a | b | e | f | g | i |
| o | r | s | t | y | z |
| A | B | C | E | G | H |
| M | O | R | S | X | Y |
| 1 | 2 | 4 | 6 | 8 | & |

Hollenstein 1977 / 78 Eras 1

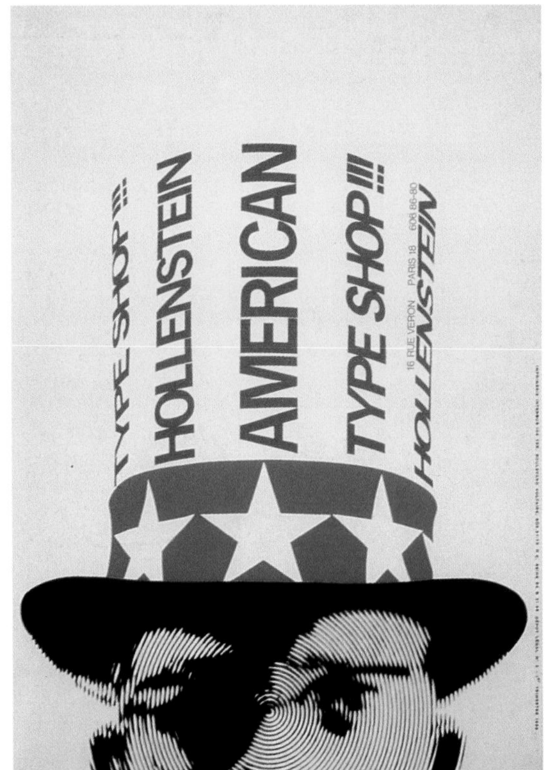


Hollenstein 1968 Advertisement

Hollenstein 1970 Cover

can universities. 1987: honorary doctorate at Philadelphia University of the Arts. 1988: honorary member of the Royal Society of Arts in London. – *Publications include:* "Graphic Design Manual", Teufen, London 1965; "The Basel School of Design and its Philosophy: the Armin Hofmann years 1946–1986", Philadelphia 1986. H. Wichmann (ed.) "Armin Hofmann, Werk, Wirkung, Lehre", Basle, Boston, Berlin 1989.

**Hollandsworth,** Cynthia – b. 8. 7. 1956 in Washington D.C., USA – *type designer* – Studied at the California College of Arts and Crafts in Oakland, California. Manager of the department of type design and development at Agfa Compugraphic in Massachusetts. Consultant to ITC Typeface Review Board. 1987: founds the Typeface Design Coalition to try to secure legal protection for type designs and software in the USA. Has worked with Barbara Gibb on a new version of an F. H. E. Schneidler typeface whose production was prevented in the 1930s by the outbreak of the war. – *Fonts:* Vermeer (1986), Hiroshige (1986), ITC Tiepolo (1987), Agfa Wile Roman (1990), Pompei Capitals (1995), Synthetica (with Philip Bouwsma, 1996).

**Hollenstein,** Albert – b. 1930 in Lucerne, Switzerland, d. 10. 8. 1974 in Vernazza, Italy – *typographer, type designer* – Trained for four years as a typographer. Then worked for two years in Switzerland. 1955–56: production manager at the Club du Meilleur Livre book club. 1957: opens his own studio "Atelier de composition typographique à façon". His

typographique à façon". Der typographischen Abteilung des Studios ist die „Phototypo Hollenstein" angegliedert. Hier werden spezielle Schriften für die Fotobelichtung u. a. von Albert Hollenstein, Albert Boton, André Chante, Jean Alessandrini, Michel Waxman und Jean Larcher entworfen. – *Schriftentwürfe:* brasilia (1958), Eras (mit Albert Boton, 1958), Roc (ca. 1960),Tivi (1968).

**Holmes,** Kris – geb. 19. 7. 1950 in Reedly, Kalifornien, USA. – *Schriftentwerferin* – 1968–71 Studium am Reed College in Portland, Oregon (u. a. Kalligraphie bei Lloyd Reynolds und Robert Palladino). 1975 Studium an der Martha Graham School und an der Alwin Nikolai School in New York (Moderner Tanz). 1976 Studium an der School of Visual Arts in New York (u.a. Schrift bei Ephram Edward Benguiat). 1979 Studium am Rochester Institute of Technology in Rochester, New York (u.a. Kalligraphie und Schriftentwurf bei Hermann Zapf). Kalligraphische und Schriftarbeiten erscheinen in zahlreichen Zeitschriften. 1976 Gründung der Firma Bigelow & Holmes (mit Charles Bigelow). – *Schriftentwürfe:* Insgesamt Entwurf von über 75 Schriften, darunter Leviathan (1979), Shannon (mit Janice Prescott, 1982), Baskerville (Revival, 1982), Caslon (Revival, 1982), ITC Isadora (1983), Sierra (1983), Lucida (mit Charles Bigelow, 1984–95), Galileo (1987), Apple New York (1991), Apple Monaco (1991), Apple Chancery (1994), Kolibri (1994).

**Holzer,** Jenny – geb. 29. 7. 1950 in Gallipolis, USA. – *Künstlerin* – Studium an der Duke University in Ohio, 1970–71

| | *n* | ***n*** | | | |
|---|---|---|---|---|---|
| *a* | *b* | *e* | *f* | *g* | *i* |
| *o* | *r* | *s* | *t* | *y* | *z* |
| *A* | *B* | *C* | *E* | *G* | *H* |
| *M* | *O* | *R* | *S* | *X* | *Y* |
| *1* | *2* | *4* | *6* | *8* | *&* |

Holmes 1985 Isadora

| | n | **n** | | | n |
|---|---|---|---|---|---|
| a | b | e | f | g | i |
| o | r | s | t | y | z |
| A | B | C | E | G | H |
| M | O | R | S | X | Y |
| 1 | 2 | 4 | 6 | 8 | & |

Holmes 1983 Sierra

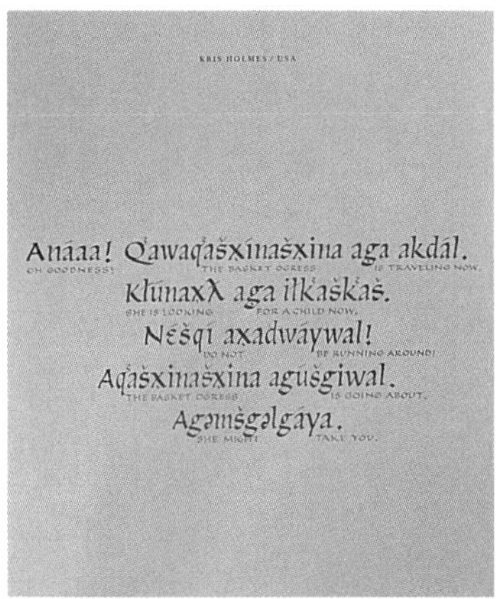

Holmes 1984 Calendar

Holzer 1983 / 84 Installation

Holzer 1987 Installation

Au service de typographie du studio est adjoint la «Phototypo Hollenstein». C'est ici que des polices utilisées pour le photitirage ont été conçues et produites par Albert Hollenstein, Albert Boton, André Chante, Jean Alessandrini, Michel Waxman, Jean Larcher etc. – *Polices :* brasilia (1958), Eras (avec Albert Boton, 1958), Roc (vers 1960), Tivi (1968).

**Holmes,** Kris – née le 19. 7. 1950 à Reedly, Californie, Etats-Unis – *conceptrice de polices* – 1968–1971, études au Reed College à Portland, Oregon (calligraphie chez Lloyd Reynolds et Robert Palladino). 1975, études à la Martha Graham School et à la Alwin Nikolai School de New York (danse moderne). 1976, études à la School of Visual Arts à New York (entre autres, arts de l'écriture chez Ephram Edward Benguiat). 1979, études au Rochester Institute of Technology de Rochester, New York (entres autres, calligraphie et conception de polices chez Hermann Zapf). Ses travaux de calligraphie et ses esquisses sont publiés dans de nombreuses revues. En 1976, elle fonde la société Bigelow & Holmes (avec Charles Bigelow). – *Polices :* a dessiné près de 75 polices, dont Leviathan (1979), Shannon (avec Janice Prescott, 1982), Baskerville (Revival, 1982), Caslon (Revival, 1982), ITC Isadora (1983), Sierra (1983), Lucida (avec Charles Bigelow, 1984–1995), Galileo (1987), Apple New York (1991), Apple Monaco (1991), Apple Chancery (1994), Kolibri (1994).

printing works, Phototypo Hollenstein, is affiliated with the typographic department of the studio. Here typefaces are designed and produced for exposure by designers such as Albert Hollenstein, Albert Boton, André Chante, Jean Alessandrini, Michel Waxman and Jean Larcher. – *Fonts:* brasilia (1958), Eras (with Albert Boton, 1958), Roc (c. 1960), Tivi (1968).

**Holmes,** Kris – b. 19. 7. 1950 in Reedly, California, USA – *type designer* – 1968–71: studies at Reed College in Portland, Oregon (courses include calligraphy under Lloyd Reynolds and Robert Palladino). 1975: studies at the Martha Graham School and the Alwin Nikolai School in New York (modern dance). 1976: studies at the School of Visual Arts in New York (including lettering with Ephram E. Benguiat). 1979: studies at the Rochester Institute of Technology in Rochester, New York (courses include calligraphy and type design with Hermann Zapf). Calligraphic and lettering work is printed in numerous magazines. 1976: founds the Bigelow & Holmes company with Charles Bigelow. – *Fonts:* Holmes has designed over 75 fonts, including Leviathan (1979), Shannon (with Janice Prescott, 1982), Baskerville (Revival, 1982), Caslon (Revival, 1982), ITC Isadora (1983), Sierra (1983), Lucida (with Charles Bigelow, 1984–95), Galileo (1987), Apple New York (1991), Apple Monaco (1991), Apple Chancery (1994) and Kolibri (1994).

**Holzer,** Jenny – b. 29. 7. 1950 in Gallipolis, USA – *artist* – Studies at Duke University in Ohio, University of Chicago

University of Chicago, 1971–72 Ohio University, 1975 Rhode Island School of Design in Providence. 1976–77 Teilnahme am Independent Study Program des Whitney Museums in New York. Erste anonym geklebte Plakate „Truisms“ in Soho. 1979–82 Tätigkeit als Schriftsetzerin. 1982 erste Benutzung elektronischer Zeichen (Leucht-Dioden) bei Schriftbändern über dem Times Square in New York. Teilnahme an der documenta VII in Kassel. 1990 Teilnahme an der Biennale in Venedig als erste weibliche Vertreterin der USA. Die verschiedenen Phasen von Holzers Arbeiten mit Schrift sind „Truisms“ (1977–82), „Inflammatory Essays“ (1979–82), „Living-Series“ (1980–82), „Survival-Series“ (1983–85), „Under a Rock“ (1986–87), „Laments“ (1987–89). – *Publikationen u.a.:* „Black Book“, New York 1980; „Hotel“, New York 1980; „Truisms and Essays“, Halifax 1982; „Black Book Poster“, New York 1988. Michael Auping „Jenny Holzer“, New York 1992; Noemi Smolik „Jenny Holzer“, Köln 1993.

**Hostettler,** Rudolf – geb. 8. 6. 1919 in Zollikofen, Schweiz, gest. 19. 2. 1981 in Engelburg, Schweiz. – *Typograph, Verleger* – Lehre als Schriftsetzer in Zollikofen. 1939 Besuch der London School of Printing. 1941 Arbeit in der Buchdruckerei Feldegg in Zürich, 1943–80 Druckerei Zollikofen in St. Gallen. Aufbau einer betriebseigenen Entwurfsabteilung. Gestaltung des Layouts des „St. Galler Tagblatts“. 1943–52 Leitung der Redaktion der Zeitschrift „Schweizer Graphische Mitteilungen“ (mit Hermann Strehler).

**Holzer,** Jenny – née le 29. 7. 1950 à Gallipolis, Etats-Unis – *artiste* – études à La Duke University dans l'Ohio, de 1970 à 1971 à la University of Chicago, de 1971 à 1972 à la Ohio University, puis en 1975 à la Rhode Island School of Design à Providence. Participe à l'Independant Study Program du Whitney Museum à New York en 1976–1977. Premiers collages d'affiches anonymes «Truisms» à Soho. Travaille comme compositeur en typographie de 1979 à 1982. Utilise pour la première fois des signes électroniques (diodes lumineuses) et des textes défilants situés au-dessus du Times Square à New York en 1982. Participation à la «documenta VII» à Kassel. 1990, participation à la Biennale de Venise comme première femme représentant les Etats-Unis. Les travaux de Holzer basés sur le texte se répartissent sur plusieurs périodes, «Truisms» (1977–1982), «Inflammatory Essays» (1979–1982), «Living-Series» (1980–1982), «Survival-Series» (1983–1985), «Under a Rock» (1986–1987), «Laments» (1987–1989). – *Publications, sélection:* «Black Book», New York 1980; «Hotel», New York 1980; «Truisms and Essays», Halifax 1982; «Black Book Poster», New York 1988. Michael Auping «Jenny Holzer», New York 1992. Noemi Smolik «Jenny Holzer», Cologne 1993.

**Hostettler,** Rudolf – né le 8. 6. 1919 à Zollikofen, Suisse, décédé le 19. 2. 1981 à Engelburg, Suisse – *typographe, éditeur* – apprentissage de composition typographique à Zollikofen. Fréquente la London School of Printing en 1939. Travaille à l'imprimerie Feldegg à Zurich en 1941, puis de 1943 à 1980 à l'imprimerie Zollikofen à Saint-Gall. Développement du service de création de l'entreprise. Conception de la maquette du «Saint-Galler Tagblatt». De 1943 à 1952, il dirige la rédaction de la revue «Schweizer Gra-

Hostettler 1948 Cover

Hostettler 1949 Cover

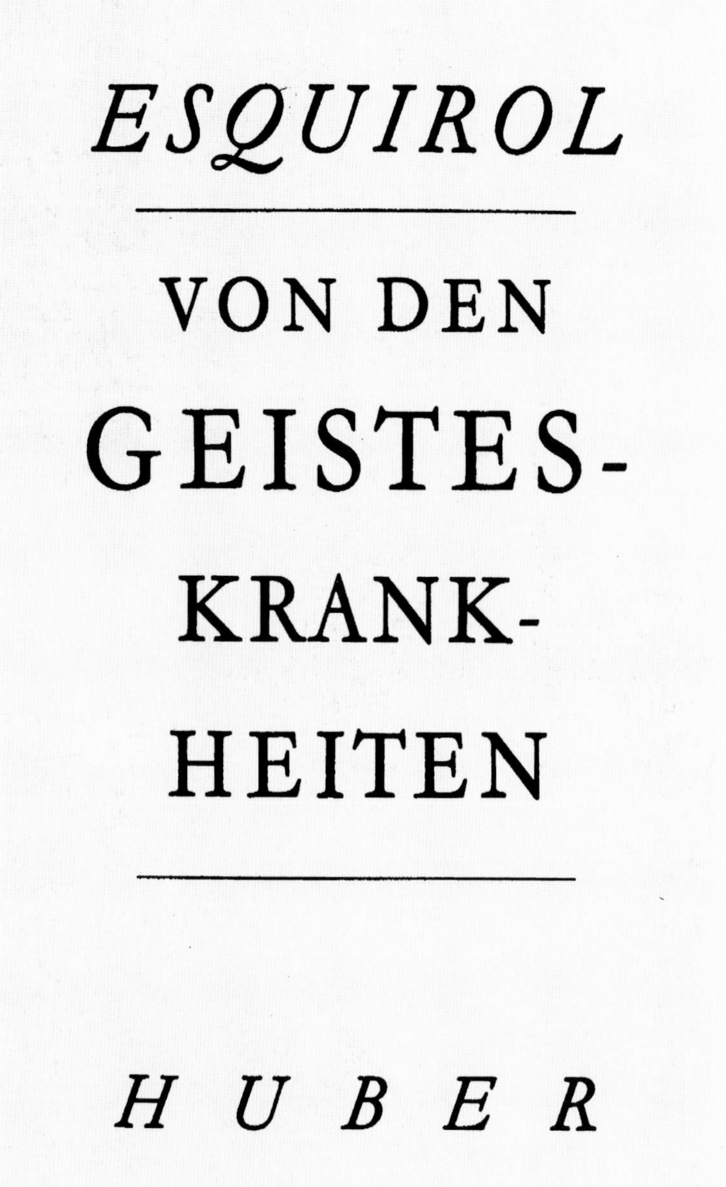

Hostettler 1968 Cover

(1970–71), Ohio University (1971–72) and the Rhode Island School of Design in Providence (1975). 1976–77: takes part in the Whitney Museum's Independent Study Program in New York. Produces her first anonymous posters, "Truisms", in Soho. 1979–82: works as a typesetter. 1982: first uses electronic symbols (light-emitting diodes) in electric banners projected over Times Square in New York. Takes part in documenta VII in Kassel. 1990: is the first woman from the USA to take part in the Venice Biennale. The various stages of Holzer's typographical work are represented by her "Truisms" (1977–82), "Inflammatory Essays" (1979–82), "Living-Series" (1980–82), "Survival-Series" (1983–85), "Under a Rock" (1986–87), "Laments" (1987–89). – *Publications include:* "Black Book", New York 1980; "Hotel", New York 1980; "Truisms and Essays", Halifax 1982; "Black Book Poster", New York 1988. Michael Auping "Jenny Holzer", New York 1992; Noemi Smolik "Jenny Holzer", Cologne 1993.

**Hostettler,** Rudolf – b. 8. 6. 1919 in Zollikofen, Switzerland, d. 19. 2. 1981 in Engelburg, Switzerland – *typographer, publisher* – Trained as a typesetter in Zollikofen. 1939: attends the London School of Printing. 1941: works in the Feldegg printing workshop in Zurich and from 1943–80 at the Zollikofen printing works in St. Gallen. Sets up the company design department. Designs the layout for the "St. Galler Tagblatt". 1943–52: head of the "Schweizer Graphische Mitteilungen" editorial staff (with Hermann Strehler). 1952: the three magazines "Typografische

Nach der Fusion der drei Zeitschriften „Typografische Monatsblätter", „Schweizer Graphische Mitteilungen" und „Revue Suisse de l'Imprimerie" 1952 Hauptredakteur der seitdem unter dem Titel „Typografische Monatsblätter" erscheinenden Zeitschrift. 1963–79 Gestaltung der 15 Bände von „Hubers Klassiker der Medizin und der Naturwissenschaften" für den Verlag Hans Huber in Bern. Drei der Bände werden als „Schönste Bücher der Schweiz" prämiert. 1971 Gründung und Hauptredakteur der 14-tägig erscheinenden Zeitschrift „Druckindustrie" (mit Hermann Strehler). – *Publikationen u. a.:* „Type", St. Gallen, London 1949; „The printers term", St. Gallen, London 1949; „Klassierung der Druckschriften", St. Gallen 1964. Jost Hochuli „Epitaph für R. H.", St. Gallen 1983.

**Hottenroth,** Franz – geb. 1897 in Höxter, Deutschland, gest. 4. 7. 1982 in München, Deutschland. – *Grafiker, Typograph* – 1919 Arbeit in der Reichsdruckerei in Berlin, 1920 in Wohlfeld bei Magdeburg. Nach Arbeiten in Dresden 1925 Umzug nach Frankfurt am Main. Abendkurse bei Rudolf Koch, Paul Renner und Hans Leistikow. Meisterprüfung. Unterrichtet 1927–29 im Bildungsverein der Deutschen Buchdrucker in Berlin Schriftschreiben und Linolschneiden. 1929 Umzug nach München. Buchgestalter in der Druckerei F. Bruckmann. 1949 Mitbegründer der Typographischen Gesellschaft München nach dem Krieg. Veranstaltet zahlreiche Gestaltungs- und Satzkurse. 1954 Leitung der neugegründeten technischen Kommission der Typographische Mitteilungen» (avec Hermann Strehler). En 1952, après la fusion des trois revues «Typografische Monatsblätter», «Schweizer Graphische Mitteilungen» et «Revue Suisse de l'Imprimerie», il devient rédacteur en chef de la nouvelle revue paraissant sous le nom de «Typographische Monatsblätter». De 1963 à 1979, il réalise la maquette des 15 volumes de «Hubers Klassiker der Medizin und der Naturwissenschaften» pour les éditions Hans Huber de Berne. Trois de ces ouvrages seront primés comme «Plus beaux livres suisses». En 1971, il fonde (avec Hermann Strehler) le bimensuel «Druckindustrie» dont il devient rédacteur en chef. – *Publications, sélecion:* «Type», Saint-Gall, Londres 1949; «The printers term», Saint-Gall, Londres 1949; «Klassierung der Druckschriften», Saint-Gall 1964. Jost Hochuli «Epitaph für R. H.», Saint-Gall 1983.

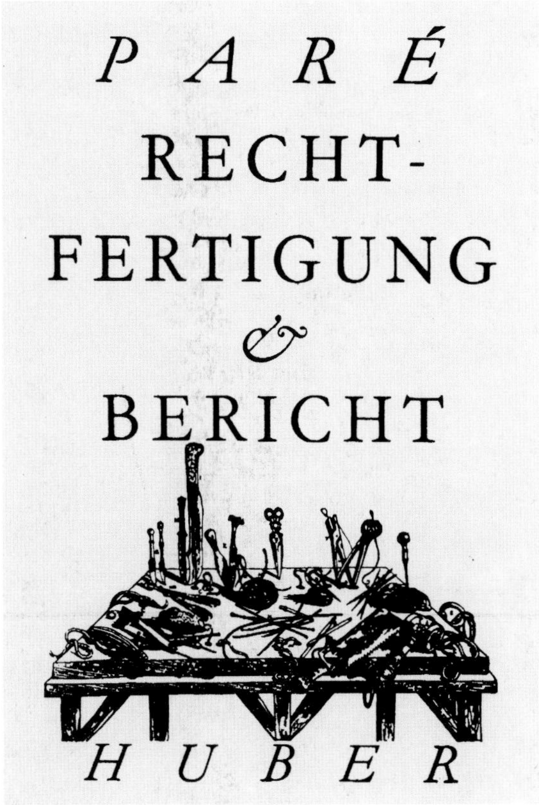

**Hottenroth,** Franz – né le 1897 à Höxter, Allemagne, décédé le 4. 7. 1982 à Munich, Allemagne – *graphiste, typographe* – travaille à la Reichsdruckerei de Berlin en 1919, puis à Wohlfeld près de Magdebourg en 1920. Après avoir exercé à Dresde, il s'installe à Francfort-sur-le-Main en 1925. Suit les cours du soir de Rudolf Koch, Paul Renner et Hans Leistikow. Passe l'examen de maître typographe. Enseigne la calligraphie et la linogravure au Bildungsverein der Deutschen Buchdrucker (association de formation de l'imprimerie allemande du livre) à Berlin de 1927 à 1929. S'installe à Munich en 1929. Maquettiste de livres à l'imprimerie F. Bruckmann. En 1949, après la guerre, il fonde avec d'autres la «Typographische Gesellschaft München» (Société typographique munichoise). Organise de nombreux cours de maquette

PARÉ
RECHT-
FERTIGUNG
&
BERICHT
HUBER

Hostettler 1963 Cover

IDEENWETTBEWERB
Marienplatz
Die Stadt München schreibt zur Klärung
des Neuaufbaues des Marienplatzes einen
städtebaulichen Ideenwettbewerb
mit beschränkter Teilnehmerzahl aus

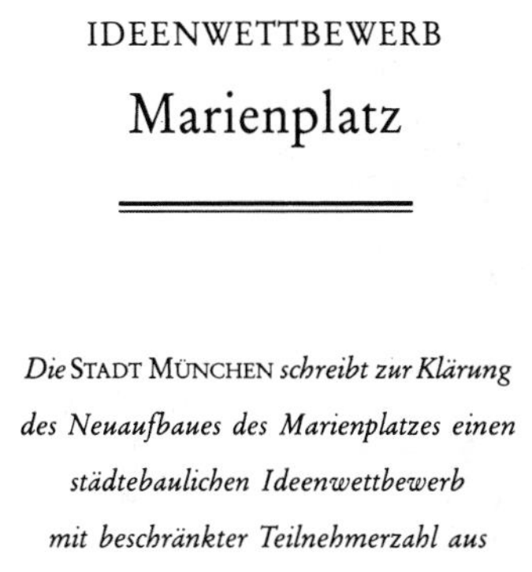

Hottenroth 1952 Main Title

wollen befreit
wirke gutes
schaffe schönes

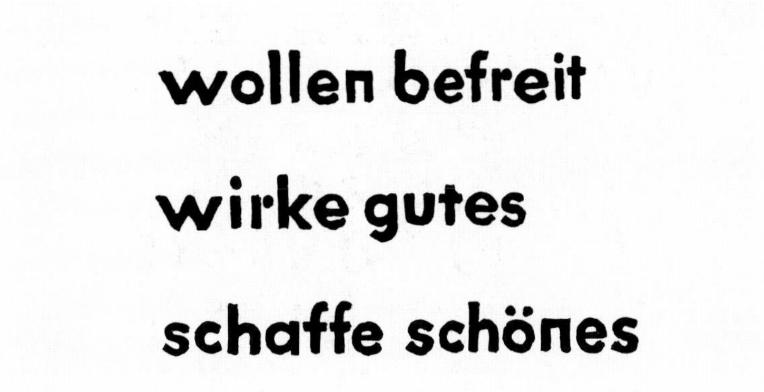

Hottenroth 1931 Sketch

Monatsblätter", "Schweizer Graphische Mitteilungen" and "Revue Suisse de l'Imprimerie" merge and Hostettler is made editor-in-chief of the new journal "Typografische Monatsblätter". 1963–79: designs 15 volumes of "Hubers Klassiker der Medizin und der Naturwissenschaften" for the Hans Huber publishing house in Bern. Three of the volumes receive awards for being the "most beautiful books in Switzerland". 1971: launches and is editor-in-chief of the fortnightly magazine "Druckindustrie" (with Hermann Strehler). – *Publications include:* "Type", St. Gallen, London 1949; "The printers term", St. Gallen, London 1949; "Klassierung der Druckschriften", St. Gallen 1964. Jost Hochuli "Epitaph für R.H.", St. Gallen 1983.

**Hottenroth,** Franz – b. 1897 in Höxter, Germany, d. 4. 7. 1982 in Munich, Germany – *graphic artist, typographer* – 1919: works at the government printing press in Berlin and in 1920 in Wohlfeld near Magdeburg. 1925: after working in Dresden, he moves to Frankfurt am Main. Takes evening classes under Rudolf Koch, Paul Renner and Hans Leistikov. Master craftsman's certificate. 1927–29: teaches lettering and lino-cutting at the Bildungsverein der Deutschen Buchdrucker in Berlin. 1929: moves to Munich. Designs books for the F. Bruckmann printing press. 1949: co-founder of the Typographische Gesellschaft in Munich after the war. Organizes numerous design and typesetting courses. 1954: heads the Typographische Gesellschaft's new technical commission in Munich. 1976: ex-

phischen Gesellschaft München. 1976 Werkausstellung in München, organisiert von der Typographischen Gesellschaft. – *Publikationen u. a.:* G. Trump u. a. „Franz Hottenroth, der Typograph", München 1976.

**Huber,** Max – geb. 5. 6. 1919 in Baar, Schweiz. – *Grafik-Designer, Typograph, Lehrer* – 1935–38 Studium an der Kunstgewerbeschule Zürich. Lehre in einer Werbeagentur. 1938–40 Mitarbeiter von Conzett & Huber bei Zeitschriften-Layouts. 1940 Gestalter und Art Director im Studio Boggeri in Mailand. 1940–42 Studium an der Accademia di Brera in Mailand. Rückkehr in die Schweiz. Freier Grafik-Designer u. a. für die Zeitschrift „Du". Ausstellungsdesign mit Max Bill und Werner Bischof in Zürich. 1944–47 Mitglied der Künstlergruppe „Allianz". Als Grafiker Arbeiten für mehrere sozialistische Verlage und Organisationen. 1946 Umzug nach Mailand, freier Grafik-Designer, Arbeiten für das Kaufhaus Rinascente, Edizioni Einaudi, Legler Stoffe, Rai Radiotelevisione Italiana und den Automobilclub von Italien. 1946–48 Lehrauftrag für Grafik-Design an der Scuola Rinascita in Mailand. 1949–52 Mitglied der Künstlergruppe „Movimento Arte Concreta" in Mailand. 1959–63 Lehrer für Grafik-Design an der Scuola Umanitaria in Mailand. 1971–77 Lehrauftrag für Grafik-Design an der Scuola Politecnica di Design" in Mailand. 1978–84 Lehrer für Grafik-Design und Typographie an der CSIA in Lugano. – *Publikation u.a.:* „Max Huber, progetti grafici 1936–1981", Mailand 1982.

et de composition. En 1954, il dirige la nouvelle commission technique de la «Typographische Gesellschaft München». En 1976, la société typographique organise une exposition de son œuvre à Munich. – *Publication :* G. Trump et autres «Franz Hottenroth, der Typograph», Munich 1976.

**Huber,** Max – né le 5. 6. 1919 à Baar, Suisse – *graphiste maquettiste, typographe, enseignant* – 1935–1938, études à la Kunstgewerbeschule (école des arts décoratifs) de Zurich. Apprentissage dans une agence de publicité. De 1938 à 1940, réalise des maquettes de revues chez Conzett & Huber. En 1940, il devient maquettiste et directeur artistique de l'atelier Boggeri à Milan. 1940–1942, études à l'Accademia di Brera à Milan. Retour en Suisse. Graphiste maquettiste indépendant et travaux pour la revue «Du». Architecte d'expositions avec Max Bill et Werner Bischof à Zurich. 1944–1947, membre du groupe d'artistes «Allianz». Travaille comme graphiste pour plusieurs maisons d'édition et organisations socialistes. S'installe à Milan en 1946, exerce comme graphiste maquettiste indépendant et travaille pour les grands magasins Rinascente, Edizioni Einaudi, Legler Stoffe, Rai Radiotelevisione Italiana et l'Automobile Club d'Italie. 1946–1948, chargé de cours d'arts graphiques et de maquette à la Scuola Rinascita à Milan. 1949–1952, membre du groupe d'artistes «Movimento Arte Concreta»à Milan. Enseigne les arts graphiques et le design à la Scuola Umanitaria de Milan de 1959 à 1963. Chargé de cours d'arts graphiques et de design à la Scuola Politecnica di Design à Milan de 1971 à 1977. Enseigne les arts graphiques, le design et la typographie à la CSIA de Lugano de 1978 à 1984. – *Publications, sélection :* «Max Huber, progetti grafici 1936–1981», Milan 1982.

Huber 1951 Advertisement

Huber 1952 Poster

Huber 1955 Cover

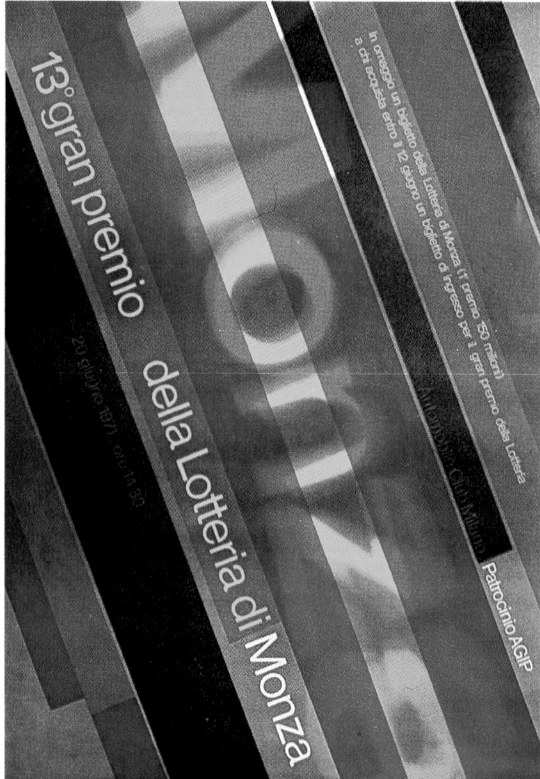

Huber 1974 Poster

hibition of his work in Munich, organized by the Typographische Gesellschaft. – *Publication:* G. Trump et al "Franz Hottenroth, der Typograph", Munich 1976.

**Huber,** Max – b. 5. 6. 1919 in Baar, Switzerland – *graphic designer, typographer, teacher* – 1935–38: studies at the Kunstgewerbeschule in Zurich. Trains at an advertising agency. 1938–40: works for Conzett & Huber on magazine layouts. 1940: designer and art director for Studio Boggeri in Milan. 1940–42: studies at the Accademia di Brera in Milan. Returns to Switzerland. Works as a freelance graphic designer, including for "Du" magazine. Designs exhibitions with Max Bill and Werner Bischof in Zurich. 1944–47: member of the Allianz artists' group. Produces work as a graphic artist for several socialist publishers and organizations. 1946: moves to Milan, freelance graphic designer. Works for the Rinascente department store, Edizioni Einaudi, Legler Stoffe, Rai Radiotelevisione Italiana and Italy's automobile club. 1946–48: teaches graphic design at the Scuola Rinascita in Milan. 1949–52: member of the Movimento Arte Concreta group in Milan. 1959–63: graphic design teacher at the Scuola Umanitaria in Milan. 1971–77: teaches graphic design at the Scuola Politecnica di Design in Milan. 1978–84: graphic design and typography teacher at the CSIA in Lugano. – *Publications include:* "Max Huber, progetti grafici 1936–1981", Milan 1982.

**Huerta,** Gerard – b. 19. 3. 1952 in Los Angeles, USA – *graphic designer* – 1971–74: studies at the Art Center College of De-

**Huerta,** Gerard – geb. 19. 3. 1952 in Los Angeles, USA. – *Grafik-Designer* – 1971–74 Studium am Art Center College of Design. 1974–76 Schallplattencover für CBS Records in New York. 1976 Gründung des Studios Gerard Huerta Design Inc. in New York. Zahlreiche Entwürfe mit gezeichneter Schrift. Signet-Entwürfe u.a. für Swiss Army Brands, MSG Network, CBS Records Masterworks, Waldenbooks, Spelling Entertainment, Nabisco. Entwurf von Schriftzügen für Zeitschriften, u.a. „Times Money", „People", „The Atlantic Monthly", „PC Magazine", „Adweek", „US", „Working Woman", „Word Perfect" und „Architectural Digest". Entwurf von Exklusiv-Alphabeten für das Erscheinungsbild von Waldenbooks, Time-Life und Condé Nast Verlag. Entwurf für „Super Bowl XXVII" (mit Roger Huyssen). Huertas farbenfroh gezeichnete Schrift-Titel im Stil der Plakatmalerei der frühen 20er Jahre sind bemerkenswert als individuelle Stiläußerung und in ihrer technischen Meisterschaft und Vielfalt.

**Hupp,** Otto – geb. 21. 5. 1859 in Düsseldorf, Deutschland, gest. 31. 1. 1949 in Oberschleißheim, Deutschland. – *Graveur, Schriftentwerfer, Heraldiker* – Vier Jahre Ausbildung als Graveur in der Werkstatt seines Vaters. Studien an der Kunstgewerbeschule Düsseldorf. 1878 in München als Ziseleur, Medailleur, Stecher, Lederschnittkünstler und Elfenbeinschnitzer tätig. 1885–1935 Herausgabe seiner „Münchner Kalender". Das erste Heft seines Mappenwerks „Die Wappen und Siegel der deutschen Städte, Flecken und Dörfer" erscheint 1896 in

Huerta 1986 Logo

Huerta Cover

Huerta 1980 Cover

Hupp 1885 Calendar

**Huerta,** Gerard – né le 19. 3. 1952 à Los Angeles, Etats-Unis – *graphiste maquettiste* – 1971–1974, études à l'Art Center College of Design. 1974–1976, pochettes de disques pour CBS Records, New York. 1976, fonde l'agence «Gerard Huerta Design Inc.» à New York. Nombreuses maquettes avec des caractères dessinés à la main. Dessin de signets pour la Swiss Army Brands, MSG Network, CBS Records Masterworks, Waldenbooks, Spelling Entertainment, Nabisco etc. Dessins de paraphes pour des revues, dont «Times Money», «People», «The Atlantic Monthly», «PC Magazine», «Adweek», «US» Working Woman», «Word Perfect» et «Architectural Digest». Dessin d'alphabets exclusifs pour l'identité de Waldenbooks, Time Life et les publications Condé Nast. Conception du «Super Bowl XXVII» (avec Roger Huyssen). Les caractères hauts en couleurs des titres de Huerta, dans le style de l'affiche peinte du début des années 20, caractérisent l'individualité de son style.

**Hupp,** Otto – né le 21. 5. 1859 à Düsseldorf, Allemagne, décédé le 31. 1. 1949 à Oberschleissheim, Allemagne – *graveur, concepteur de polices, héraldiste* – quatre années de formation de graveur dans l'atelier de son père. Etudes à la Kunstgewerbeschule (école des arts décoratifs) de Düsseldorf. En 1878, il exerce à Munich comme frappeur de médailles, poinçonneur, graveur sur cuir et sculpteur sur ivoire. 1885–1935, publie ses «Münchner Kalender» (Calendriers munichois). En 1896, le premier numéro de sa collection «Die Wappen und Siegel der deutschen Städte, Flecken und Dörfer» (les armes et seaux des villes allemandes, terroirs et

sign. 1974–76: designs record covers for CBS Records in New York. 1976: opens the Gerard Huerta Design Inc. studio in New York. Numerous designs with hand-drawn alphabets. Designs logos for Swiss Army Brands, MSG Network, CBS Records Masterworks, Waldenbooks, Spelling Entertainment and Nabisco, among others. Designs the type for various magazines, including "Times Money", "People", "The Atlantic Monthly", "PC Magazine", "Adweek", "US", "Working Woman", "Word Perfect" and "Architectural Digest". Designs exclusive alphabets for Waldenbooks', Time-Life's and Condé Nast Publications corporate imageries. Produces designs with Roger Huyssen for Super Bowl XXVII. Huerta's colorful hand-drawn titles, reminiscent of the posters of the 1920s, are an individual expression of style and remarkable examples of technical mastery.

**Hupp,** Otto – b. 21. 5. 1859 in Düsseldorf, Germany, d. 31. 1. 1949 in Oberschleißheim, Germany – *engraver, type designer, heraldic artist* – Trained as an engraver for four years in his father's workshop. Studied at the Kunstgewerbeschule in Düsseldorf. 1878: works in Munich as a medalist, engraver, leather artist and ivory carver. 1885–1935: his Munich calendar is published. 1896: the first volume of his portfolio, entitled "Die Wappen und Siegel der deutschen Städte, Flecken und Dörfer", is published in an edition of 400. 1898: the second volume is published. 1908–18: designs a great number of postage stamps with heraldic motifs for the Deutsche Post. 1934: his "Wappen-

einer Auflage von 400 Exemplaren, 1898 erscheint das zweite Heft. 1908–18 zahlreiche Briefmarkenentwürfe mit heraldischen Motiven für die Deutsche Post. Das 1911 begonnene Sammelbilder-Einklebewerk „Wappenmarken" der Kaffee-Handels-AG in Bremen (Kaffee Haag) in zehn Heften zu je 16 Seiten ist 1934 vollendet. Es enthält 3.000 von ihm gestaltete Wappen. – *Schriftentwürfe:* Neudeutsch (1900), Liturgisch (1909), Hupp-Antiqua (1909), Hupp-Unziale (1909), Heraldisch (1910), Hupp-Fraktur (1910), Hupp-Antiqua fett (1913), Hupp-Schrägschrift (1922). – *Publikation u. a.:* „Otto Hupp, das Werk eines deutschen Meisters", Leipzig 1940.

**Huszár,** Vilmos (d. i. Vilmos Herz) – geb. 5. 1. 1884 in Budapest, Ungarn, gest. 1960 in Hierden, Niederlande. – *Künstler, Grafik-Designer* – Lehre als Wanddekorateur an der Budapester Kunstschule. 1904 Studium an der Kunstakademie München. Namensänderung in Huszár. Lebt 1909–39 in den Niederlanden. 1915–16 kubistische Experimente, erste Glasfenster-Entwürfe. 1917 erste abstrakte Bilder. Gründungsmitglied der Gruppe „De Stijl". Entwurf des Signets der Zeitschrift „De Stijl", 1918 erste Inneneinrichtungen und Messestände. 1923 Austritt aus der Gruppe „De Stijl". Rückkehr zu gegenständlichen Bildmotiven. Seit 1925 grafische Arbeiten für Werbung und Industrie. Seit 1930 Möbel-Entwürfe für die Firma Metz & Co. – *Publikation u. a.:* Sjarel Ex, Els Hoek „Vilmos Huszár – Schilder en ontwerper 1884–1960", Utrecht 1985.

villages) est publié et tiré à 400 exemplaires, le second numéro paraît en 1898. Nombreux dessins de timbres présentant des blasons pour les postes allemandes de 1908 à 1918. En 1934, l'album de collection d'images «Wappenmarken» (armoiries et emblèmes) commencé en 1911 pour la Kaffee-Handels-AG de Brême (Kaffee Haag) est achevé, il contient dix cahiers de 16 pages et 3 000 blasons dessinés par Hupp. – *Polices:* Neudeutsch (1900), Liturgisch (1909), Hupp-Antiqua (1909), Hupp-Unziale (1909), Heraldisch (1910), Hupp-Fraktur (1910), Hupp-Antiqua fett (1913), Hupp-Schrägschrift (1922). – *Publication, sélection:* «Otto Hupp, das Werk eines deutschen Meisters», Leipzig 1940.

**Huszár,** Vilmos (Vilmos Herz) – né le 5. 1. 1884 à Budapest, Hongrie, décédé en 1960 à Hierden, Pays-Bas – *artiste, graphiste maquettiste* – apprentissage de décoration murale à l'école des beaux-arts de Budapest. 1904, études à la Kunstakademie (beaux-arts) de Munich. Prend le pseudonyme de Huszár. Vit aux Pays-Bas de 1909 à 1939. Expérimente le cubisme de 1915 à 1916, premiers projets de vitraux. Premiers tableaux abstraits en 1917. Membre fondateur de «De Stijl» et création du signet de la revue «De Stijl». 1918, premières décorations d'intérieurs et de stands d'expositions. Quitte le groupe «De Stijl» en 1923 et revient à l'image figurative. Travaux graphiques pour la publicité et l'industrie à partir de 1925. Conçoit des meubles pour la société Metz & Co à partir de 1930. – *Publication, sélection:* Sjarel Ex, Els Hoek «Vilmos Huszár – Schilder en ontwerper 1884–1960», Utrecht 1985.

Huszár 1929 Cover

Huszár 1926 Advertisement

Huszár 1917 Cover

Huszár 1929 Cover

Huszár 1917 Logotype

marken" collector-sticker book for the Kaffee-Handels-AG in Bremen (Kaffee Haag), begun in 1911, is finally completed. There are 10 16-page volumes containing 3,000 coats of arms designed by Hupp. – *Fonts:* Neudeutsch (1900), Liturgisch (1909), Hupp-Antiqua (1909), Hupp-Unziale (1909), Heraldisch (1910), Hupp-Fraktur (1910), Hupp-Antiqua fett (1913), Hupp-Schrägschrift (1922). – *Publications include:* "Otto Hupp, das Werk eines deutschen Meisters", Leipzig 1940.

**Huszár,** Vilmos (originally Vilmos Herz) – b. 5. 1. 1884 in Budapest, Hungary, d. 1960 in Hierden, The Netherlands – *artist, graphic designer* – Trained as an interior decorator at the art school in Budapest. 1904: studies at the Kunstakademie in Munich. Changes his name to Huszár. 1909–39: lives in the Netherlands. 1915–16: Cubist experiments. Produces his first window designs. 1917: produces his first abstract pictures. Founder member of De Stijl. Designs the logo for the "De Stijl" magazine. 1918: first interiors and fair booths. 1923: leaves De Stijl. Returns to representational picture motifs. 1925: starts producing graphics for advertising and industry. From 1930 onwards: designs furniture for the company Metz & Co. – *Publications include:* Sjarel Ex, Els Hoek "Vilmos Huszár – Schilder en ontwerper 1884–1960", Utrecht 1985.

# I

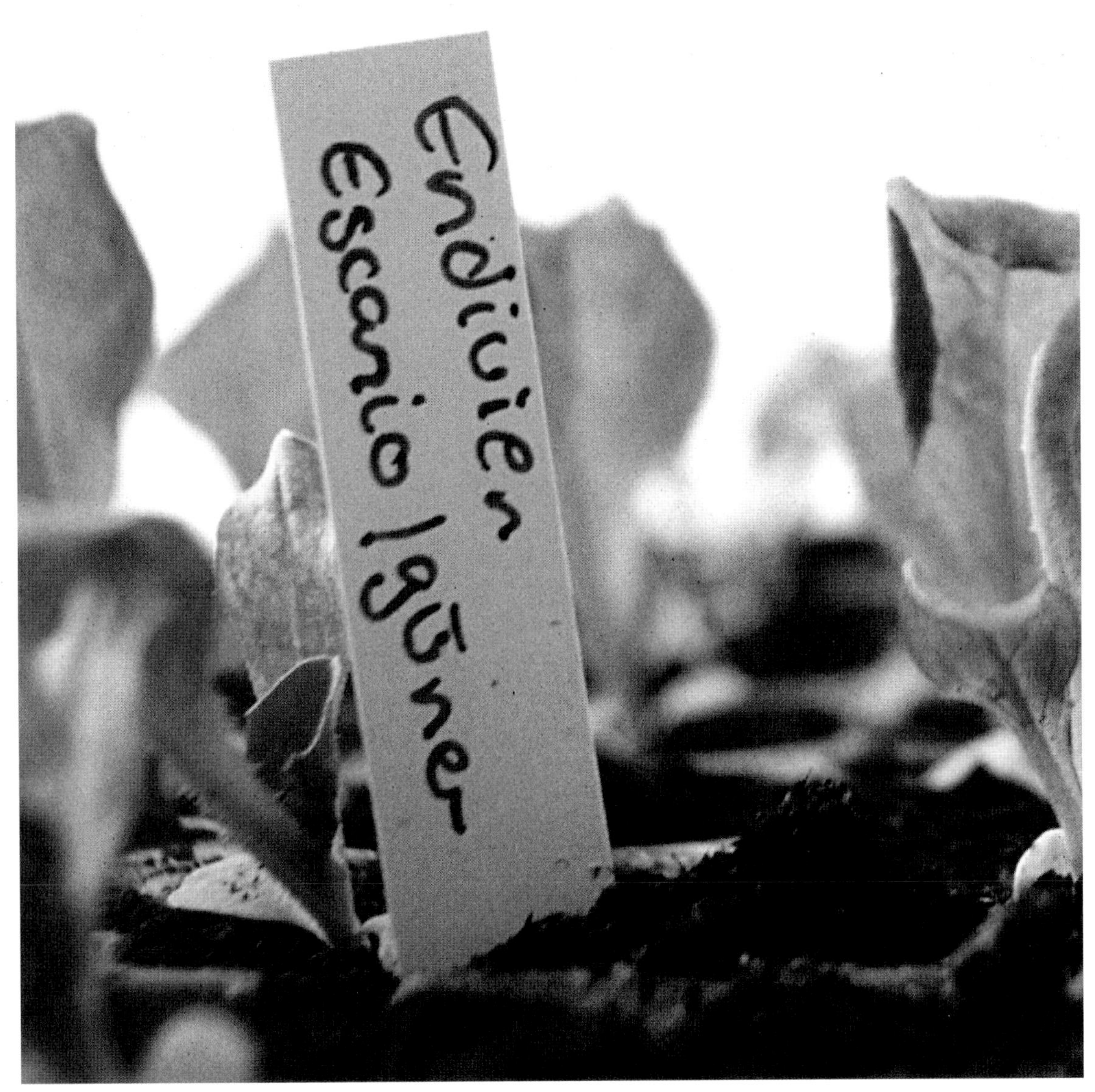

**Ifert,** Gérard – geb. 28.9.1929 in Basel, Schweiz. – *Grafik-Designer, Fotograf, Typograph, Möbel-Designer, Innenarchitekt, Lehrer* – 1945–49 Studium an der Allgemeinen Gewerbeschule Basel. 1950–51 Assistent von Ernst Scheidegger und Lanfranco Bombelli. 1952–53 freier Mitarbeiter bei J. R. Geigy in Basel. Entwurf geometrischer Bilder, die auf mathematischen Systemen beruhen. Entwicklung von stapel- und faltbaren Möbeln. Seit 1960 eigenes Design-Studio in Paris. Konzeption für die Wanderausstellung „Werbung der Steindruckerei de Jong & Co" für das Stedelijk Museum in Amsterdam (mit Wim Crouwel). 1962–64 gestalterische Arbeiten für die Expo 64 in Lausanne. Unterrichtet 1965 an der Ecole Nationale Supérieure des Arts Décoratifs in Paris. Seit 1965 innenarchitektonische Projekte. Ausstellungssystem für das CCI am Centre Georges Pompidou. Leitsystem des Universitätsgeländes „Saint-Martin d'Hères" bei Grenoble (mit Adrian Frutiger). Aus einer von Ifert entworfenen Sitzgruppe für das Haus von Adrian Frutiger entsteht 1978 ein ganzes Möbelprogramm im Zusammensteck-System. Unterrichtet 1980 an der Design-Abteilung der Universität in Cincinnati, Ohio, USA. Ab 1982 an der ESDI-Schule in Paris, 1985–95 an der Ecole Supérieure d'Arts Graphiques et d'Architecture Intérieure in Paris. Zahlreiche internationale Ausstellungen und Publikationen des künstlerischen, typographischen und innenarchitektonischen Werks.

**Igarashi,** Takenobu – geb. 4. 3. 1944 in Takikawa, Japan. – *Grafik-Designer, Bild-*

**Ifert,** Gérard – né le 28.9.1929 à Bâle, Suisse – *graphiste maquettiste, photographe, typographe, designer de meubles, architecte d'intérieur, enseignant* – 1945–1949, études à la Allgemeine Gewerbeschule (école des arts et métiers) de Bâle. 1950–1951, assistant d'Ernst Scheidegger et de Lanfranco Bombelli. 1952–1953, travaux comme prestataire de services pour J.R. Geigy à Bâle. Il conçoit des images géométriques qui reposent sur des systèmes mathématiques. Crée des meubles empilables et pliants. Travaille dans son propre atelier de design à Paris depuis 1960. Conception de l'exposition itinérante «Werbung der Steindruckerei de Jong & Co» (Publicité de l'atelier de lithographie de Jong & Co) pour le Stedelijk Museum d'Amsterdam (avec Wim Crouwel). 1962–1964, travaux de design pour Expo 64 à Lausanne. Enseigne à l'Ecole Nationale Supérieure des Arts Décoratifs de Paris en 1965. Projets d'architecture intérieure depuis 1965. Système d'orientation des expositions du CCI. au Centre Georges Pompidou de Paris. Systèmes de signalisation du campus universitaire de «Saint-Martin d'Hères», près de Grenoble (avec Adrian Frutiger). Création, en 1978, d'une ligne de meubles encastrables à partir de sièges de salons dessinés par Ifert pour la demeure d'Adrian Frutiger. En 1980, il enseigne à la section de design de l'université de Cincinnati, Ohio, Etats-Unis; à partir de 1982, à l'Ecole ESDI à Paris, de 1985 à 1995 à l'Ecole Supérieure d'Arts Graphiques et d'Architecture Intérieure de Paris. Nombreuses expositions internationales. Publications sur son œuvre plastique, typographique et d'architecture d'intérieur.

**Igarashi,** Takenobu – né le 4. 3. 1944 à Takikawa, Japon – *graphiste maquettiste, sculpteur, enseignant* – 1964–1968, étu-

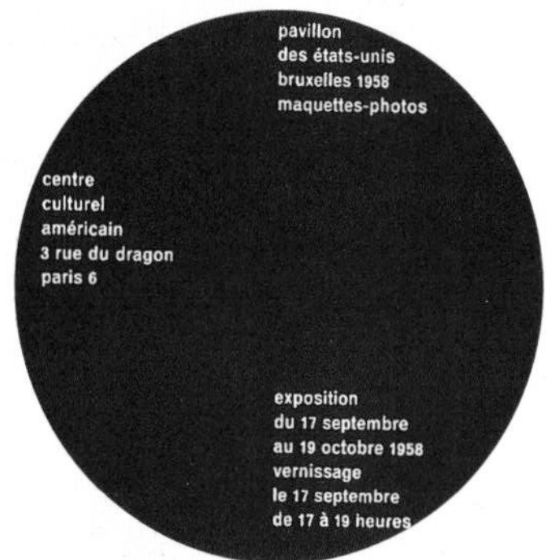


Ifert 1958 Invitation Card

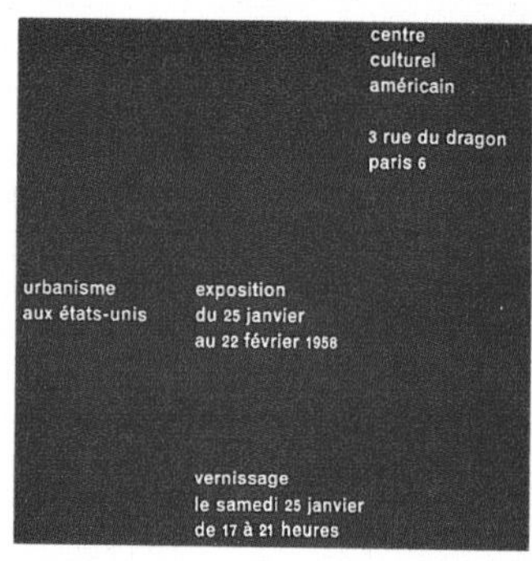


Ifert 1958 Invitation Card

Ifert 1982 Page

Ifert 1987 Invitation Card

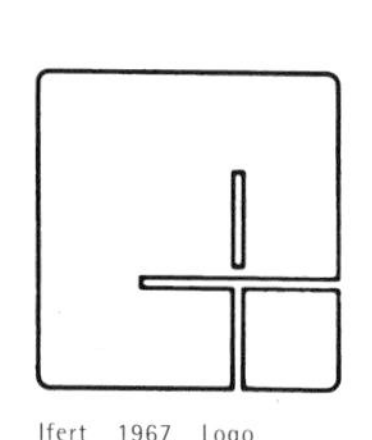

Ifert 1967 Logo

Igarashi 1986 Cover

**Ifert,** Gérard – b. 28.9.1929 in Basle, Switzerland – *graphic designer, photographer, typographer, furniture designer, interior designer, teacher* – 1945–49: studies at the Allgemeine Gewerbeschule in Basle. 1950–51: assistant to Ernst Scheidegger and Lanfranco Bombelli. 1952–53: freelance work for J.R. Geigy in Basle. Designs geometric pictures based on mathematical systems. Develops furniture which can be stacked and collapsed. From 1960 onwards: has his own design studio in Paris. Works with Wim Crouwel on the concept for a traveling exhibition for Amsterdam's Stedelijk Museum "Werbung der Steindruckerei de Jong & Co". 1962–64: design work for Expo 64 in Lausanne. 1965: teaches at the Ecole Nationale Supérieure des Arts Décoratifs in Paris. From 1965 onwards: various interior design projects. Exhibition systems for the CCI at the Centre Georges Pompidou in Paris. Signage system with Adrian Frutiger for the Saint-Martin d'Hères university campus near Grenoble. 1978: a suite of furniture designed by Ifert for Adrian Frutiger's house is developed into an entire range of slot-together furniture. 1980: teaches at the design department of the university in Cincinnati, Ohio, USA, from 1982 onwards at the ESDI school in Paris and at the Ecole Supérieure d'Arts Graphiques et d'Architecture Intérieure in Paris from 1985–95. Numerous international exhibitions and publications of his artistic, typographic and interior design work.

**Igarashi,** Takenobu – b. 4.3.1944 in Takikawa, Japan – *graphic designer,*

*hauer, Lehrer* – 1964–68 Studium an der Tama-Kunsthochschule in Tokio. 1969 Magister in Design an der University of California (UCLA) in Los Angeles. 1970 Gründung des Büros „Takenobu Igarashi Design" in Tokio. 1975–76 Gastdozent an der University of California (UCLA) in Los Angeles. 1979–85 Professor für Design an der Maschinenbauabteilung der Universität in Chiba. 1980 erste „Alphabet-Skulpturen". 1983–88 jährliche Kalender-Entwürfe für das Museum of Modern Art in New York. 1985 Umbenennung seines Büros in „Igarashi Studio". 1989 Gündung eines zweiten Büros in Los Angeles. 1989 Entwurf einer Skulptur, die in 150 Showrooms der Nissan Motor Corporation in den USA aufgestellt wird. 1989 Professor an der Design-Abteilung der Tama Kunstakademie in Tokio. Auftraggeber waren u. a. Polaroid Corporation, Polygon Pictures, Kuboth Computer Inc., Suntory Ltd. Zahlreiche Ausstellungen, u. a. in der Fujie Galerie, Tokio (1973, 1975), Tokyo Designers Space, Tokio (1976, 1979), der Reinhold Brown Gallery, New York (1983), der MacQuaries Gallery, Sydney (1987), dem Deutschen Plakat-Museum, Essen (1993), und der Hosomi Galerie, Tokio (1995). – *Publikationen u.a.:* „Igarashi Alphabets", Zürich 1987; „Rock Scissors Paper", Tokio 1991; „Igarashi Sculptures", Tokio 1992.

**Imprimerie Nationale,** Paris – *Druckerei* – 1640 Gründung der französischen Staatsdruckerei mit einer Schriftgießerei durch Kardinal Richelieu in der Galerie Diane im Louvre. Zunächst „Imprimerie Royale" genannt erhält die Druckerei

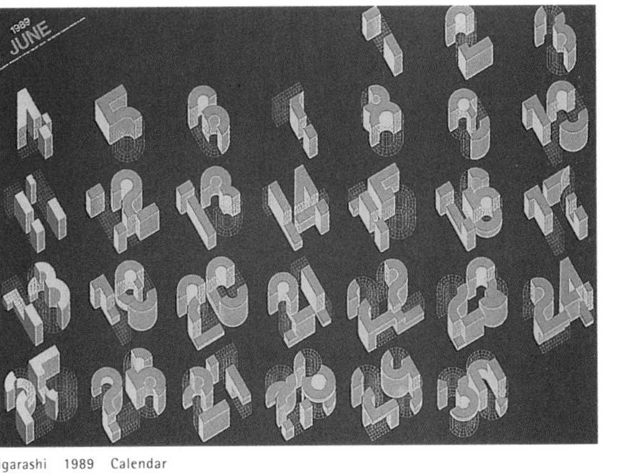

Igarashi 1989 Calendar

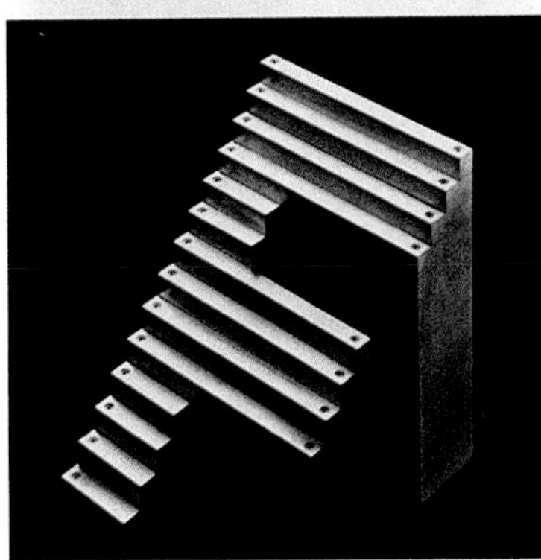

Igarashi 1983 Alphabet

Igarashi 1985 Poster

escargot

Igarashi Logo

Igarashi 1975 Poster

des à l'école supérieure des beaux-arts Tama à Tokyo. 1969, maîtrise de design à la University of California (UCLA) de Los Angeles. 1970, fonde l'atelier «Takenobu Igarashi Design» à Tokyo. 1975–1976, cycle de conférences à l'University of California (UCLA) de Los Angeles. 1979–1985, professeur de design à la section de constructions mécaniques de l'université de Chiba. 1980, premiers «Alphabets sculptures». 1983–1988, dessine les calendriers du Museum of Modern Art de New York. En 1985, son atelier devient l' «Igarashi Studio». 1989, ouverture d'un second atelier à Los Angeles. En 1989, il réalise une sculpture qui sera exposée dans 150 showrooms de la Nissan Motor Corporation aux Etats-Unis. Toujours en 1989, il est professeur à la section de design de l'Académie Tama à Tokyo. Commanditaires : Polaroïd Corporation, Polygon Pictures, Kuboth Computer Inc., Suntory Ltd., etc. Nombreuses expositions, entre autres à la Fujie Galerie, Tokyo (1973, 1975), au Tokyo Designers Space, Tokyo (1976, 1979), à la Reinhold Brown Gallery, New York (1983), à la MacQuaries Gallery, Sydney (1987), au Deutsche Plakat-Museum, Essen (1993), à la Galerie Hosomi, Tokyo (1995). – *Publications, sélection :* «Igarashi Alphabets», Zurich 1987; «Rock Scissors Paper», Tokyo 1991; «Igarashi Sculptures», Tokyo 1992.

**Imprimerie Nationale,** Paris – *imprimerie* – l'imprimerie française d'Etat et sa fonderie de caractères sont fondées en 1640 par le Cardinal de Richelieu dans la Galerie Diane du Louvre. Cette imprimerie d'abord appelée «Imprimerie Royale»

*sculptor, teacher* – 1964–68: studies at the Tama College of Art in Tokyo. 1969: obtains an MA in design at the University of California (UCLA) in Los Angeles. 1970: opens the Takenobu Igarashi Design studio in Tokyo. 1975–76: visiting lecturer at the University of California (UCLA) in Los Angeles. 1979–85: professor of design in the mechanical engineering department of the university in Chiba. 1980: produces his first "alphabet sculptures". 1983–88: designs calendars for the Museum of Modern Art in New York each year. 1985: renames his studio Igarashi Studio. 1989: opens a second studio in Los Angeles. 1989: designs a sculpture which is displayed in 150 Nissan Motor Corporation showrooms in the USA. 1989: professor in the design department of the Tama College of Art in Tokyo. Clients include the Polaroid Corporation, Polygon Pictures, Kuboth Computer Inc. and Suntory Ltd. Numerous exhibitions, including at the Fujie gallery in Tokyo (1973, 1975), Tokyo Designers Space in Tokyo (1976, 1979), Reinhold Brown Gallery in New York (1983), MacQuaries Gallery in Sydney (1987), Deutsche Plakat-Museum in Essen (1993) and the Hosomi Gallery in Tokyo (1995). – *Publications include:* "Igarashi Alphabets", Zurich 1987; "Rock Scissors Paper", Tokyo 1991; "Igarashi Sculptures", Tokyo 1992.

**Imprimerie Nationale,** Paris – *printing workshop* – 1640: Cardinal Richelieu founds the French state printing works with its own type foundry in the Diane gallery in the Louvre. Originally called the Imprimerie Royale, the printing works

1871 den Namen „Imprimerie Nationale". 1925 nach verschiedenen Standorten Umzug in die Rue de la Convention. Gründung einer Filiale in Douai (1973) und Evry (1990). Bis heute arbeiten hier zwei Stempelschneider, und Bleisatz-Buchstaben werden für den eigenen Bedarf gegossen.

**Indiana,** Robert – geb. 13. 9. 1928 in New Castle, Indiana, USA. – *Künstler* – 1945–48 Studium an den Kunstschulen von Indianapolis und Utica, 1949–53 an der School of the Art Institute Chicago und Skowhegan School of Painting and Sculpture, 1953–54 am Edinburgh College of Art und der London University. 1954 Umzug nach New York. 1966 Serie von Gemälden mit dem Wort „Love". Es folgen u. a. „Hug", „Kill", „Yield" und „Eat", die bekannte Werkgruppen der amerikanischen Pop Art werden. In seinem Werk verwendet er Buchstaben, Zahlen und Wortbilder aus der amerikanischen Umgangssprache als abstrakte Grundzeichen.

**Inoue Yu-Ichi** – geb. 14. 2. 1916 in Tokio, Japan, gest. 15. 6. 1985 in Tokio, Japan. – *Kalligraph, Lehrer* – 1935 Volksschullehrer in Tokio, Studium der Malerei an einer Privatschule. 1941 Beginn einer 8-jährigen Lehrzeit in Kalligraphie bei Ueda Sokyu. 1952 Mitbegründer der Kalligraphen-Vereinigung „Bokujin-kai", deren monatliche Zeitschrift „Bokubi" zu einem Forum der Avantgarde-Kalligraphie wird. 1954 Teilnahme an der Ausstellung „Japanese Calligraphy" im Museum of Modern Art in New York. 1955 Teilnahme an der Wanderausstellung der „Bokujin-kai" in Amsterdam, Basel, Paris, Hamburg und

prend le nom d' «Imprimerie Nationale» en 1871. Après plusieurs déménagements, elle s'installe en 1925 dans les locaux de la Rue de la Convention. En 1973, une filiale est construite à Douai, puis une autre à Evry en 1990. Deux tailleurs de types y travaillent encore aujourd'hui et des caractères en plomb y sont encore fondus pour le propre usage de l'imprimerie.

**Indiana,** Robert – né le 13. 9. 1928 à New Castle, Indiana, Etats-Unis – *artiste* – 1945–1948, études dans les écoles d'art d'Indianapolis et d'Utica, puis de 1949 à 1953, à la School of the Art Institute de Chicago et à la Skowhegan School of Painting and Sculpture, en 1953–1954 au Edinburgh College of Art et à la London University. En 1954, il s'installe à New York. En 1966, il réalise une série de tableaux comportant le mot «Love». Puis viennent les séries «Hug», «Kill», «Yield», «Eat» qui s'inscrivent dans le Pop Art américain. Dans ses œuvres, il emploie des lettres, chiffres et pictogrammes issus de l'argot américain utilisés comme abstractions et signes des fondamentaux.

**Inoue Yu-Ichi** – né le 14. 2. 1916 à Tokyo, Japon, décédé le 15. 6. 1985 à Tokyo, Japon – *calligraphe, enseignant* – instituteur au Japon en 1935 et études de peinture dans une école privée. En 1941, il entame huit ans d'apprentissage de la calligraphie chez Ueda Sokyu. Cofondateur en 1952 de l'Union des calligraphes «Bokujin-kai», dont la revue mensuelle «Bokubi» deviendra un forum pour la calligraphie d'avant-garde. En 1954, il participe à l'exposition «Japanese Calligraphy» au Museum of Modern Art de New York. 1955, participation à l'exposition itinérante de la «Bokujin-kai» à Amsterdam, Bâle, Paris, Hambourg et Rome. Expose à la 4. Biennale de São Paulo (1957), participe à «50 Ans d'Art Moderne» à

Indiana 1970 Poster

Indiana 1967 Poster

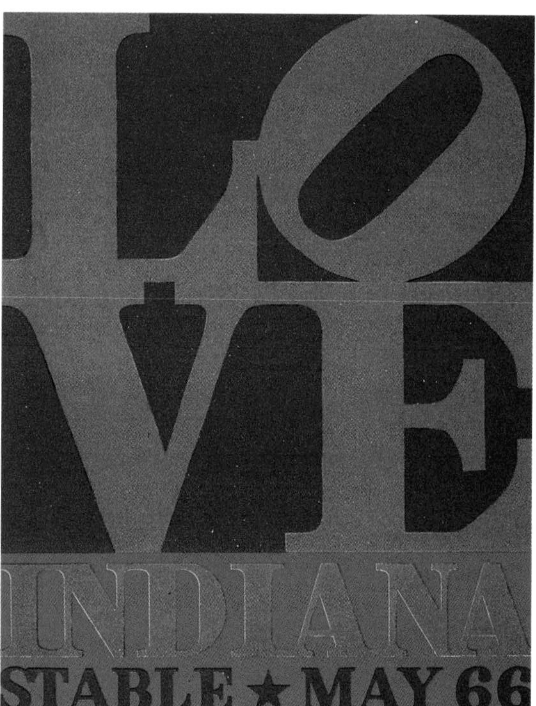

Indiana 1966 Poster

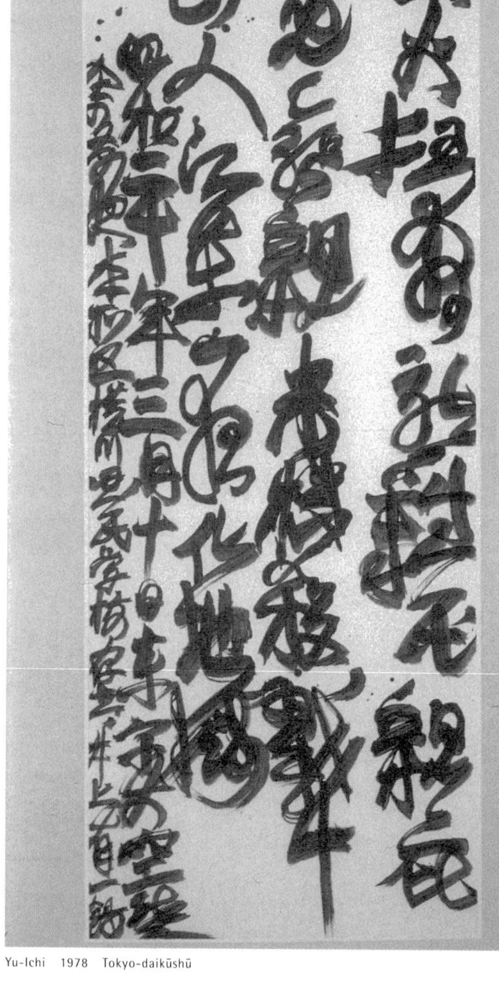
Yu-Ichi 1978 Tokyo-daikūshū

is renamed Imprimerie Nationale in 1871. 1925: after various locations, the printing works moves into rooms in the Rue de la Convention. 1973: a branch is built in Douai. 1990: a branch is built in Evry. Today two punch cutters are still at work here, producing lead composition alphabets for their own use.

**Indiana,** Robert – b. 13. 9. 1928 in New Castle, Indiana, USA – *artist* – 1945–48: studies at the art colleges in Indianapolis and Utica, at the School of the Art Institute in Chicago and the Skowhegan School of Painting and Sculpture from 1949–53 and at the Edinburgh College of Art and London University from 1953–54. 1954: moves to New York. 1966: produces a series of paintings with the word Love. This is followed by the projects Hug, Kill, Yield and Eat, among others, which have became major works of American Pop Art. In his work, Indiana uses letters, numbers and word pictures from American slang as basic abstract symbols.

**Inoue Yu-Ichi** – b. 14. 2. 1916 in Tokyo, Japan, d. 15. 6. 1985 in Tokyo, Japan – *calligrapher, teacher* – 1935: elementary school teacher in Tokyo. Studies painting at a private school. 1941: starts 8 years of training as a calligrapher with Ueda Sokyu. 1952: co-founder of the Bokujin-kai calligraphic association whose monthly magazine "Bokubi" has become a forum for avant-garde calligraphy. 1954: takes part in the "Japanese Calligraphy" exhibition at the Museum of Modern Art in New York. 1955: takes part in the Bokujin-kai traveling exhibition in Amsterdam, Basle, Paris, Hamburg and

Rom. Teilnahme an der 4. Biennale in São Paulo (1957), an der Ausstellung „50 Ans d'Art Moderne" in Brüssel (1958), an der documenta II in Kassel, an der Ausstellung „Schrift und Bild" in Amsterdam und Baden-Baden (1963). – *Publikationen u.a.:* „Gesammelte Schriften von Inoue Yu-Ichi", Tokio 1989; „Der große Luftangriff auf Tokio", Tokio 1995. Günter Aust „Yu-Ichi Inoue", Wuppertal 1965; Masaomi Unagami (Hrsg.) „Inoue Yu-Ichi no sho, Sho by Yu-Ichi 49–77", Tokio 1977.

**International Typeface Corporation** (ITC) – *Schriftenhersteller* – 1970 Gründung der ITC in New York durch Aaron Burns, Herb Lubalin und Edward Rondthaler, um Schriften für Foto- und Computersatz zu entwerfen, lizensieren und weltweit zu vertreiben. Zu den Schriftentwerfern, die für ITC arbeiten, gehören u. a. Herb Lubalin, Tom Carnase, Edward Benguiat, Antonio DiSpigna, Hermann Zapf, Matthew Carter. An Schriftneuschnitten entstehen u. a. Franklin, Century, Souvenir, Berkeley, Korinna. Neue Entwürfe sind u. a. Avantgarde Gothic, Zapf Chancery, Bauhaus, Lubalin Graph. Seit 1974 Herausgabe der Hauszeitschrift „U&lc" (Upper and lower case), die innerhalb kürzester Zeit eine international einflußreiche Typographie-Zeitschrift wird. Herausgeber ist bis zu seinem Tod 1981 Herb Lubalin. 1986 kauft Esselte Letraset ITC, die in New York weiterarbeitet.

**Intertype** – *Setzmaschinenfabrik* – 1912 Gründung der „Intertype Typesetting Company" in Brooklyn durch Hermann Ridder. 1913 Bau der Zeilensetz- und Gießmaschine „Intertype". 1916 Umbenennung

Yu-Ichi 1967 Hana

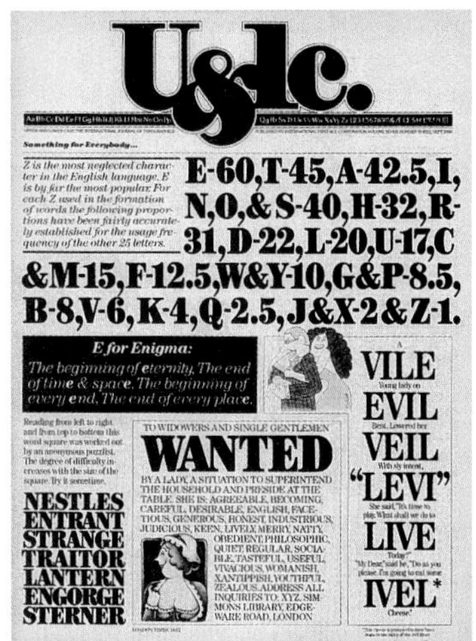


International Typeface Corporation 1980 Cover

Intertype ca. 1935 Cover

International Typeface Corporation 1973 Specimen

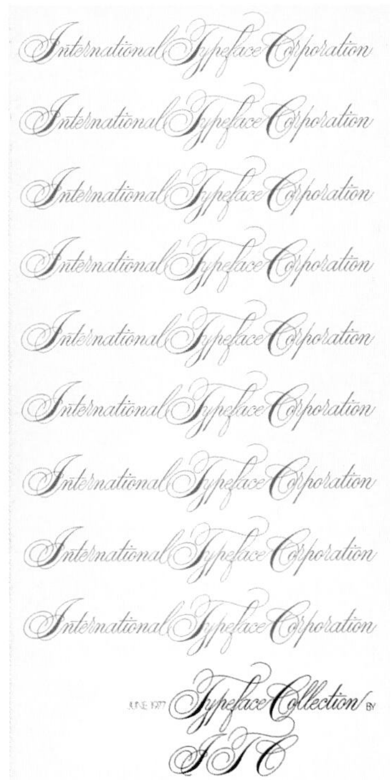


International Typeface Corporation 1977 Specimen

Intertype ca. 1935 Cover

Bruxelles (1958), à la «documenta II» de Kassel, à l'exposition «Schrift und Bild» à Amsterdam et à Baden-Baden (1963). – *Publications, sélection:* «Gesammelte Schriften von Inoue Yu-Ichi» (Œuvres complètes de Inoue Yu-Ichi), Tokyo 1989; «Der grosse Luftangriff auf Tokio» (La grande attaque aérienne sur Tokyo), Tokyo 1995. Günter Aust «Yu-Ichi Inoue», Wuppertal 1965; Masaomi Unagami (éd.) «Inoue Yu-Ichi no sho, Sho by Yu-Ichi 49–77», Tokyo 1977.

**International Typeface Corporation** (ITC) – *fabricant de polices* – ITC a été fondée à New York, en 1970 par Aaron Burns, Herb Lubalin et Edward Rondthaler dans le but de concevoir des polices pour la photocomposition et la PAO, d'exploiter des licences et de distribuer des caractères dans le monde entier. Herb Lubalin, Tom Carnase, Edward Benguiat, Antonio DiSpigna, Hermann Zapf, Matthew Carter comptent parmi les concepteurs de polices ayant travaillé pour ITC. Franklin, Century, Souvenir, Berkeley, Korinna font partie des nouvelles fontes créées dans la société. D'autres, comme Avantgarde Gothic, Zapf Chancery, Bauhaus, Lubalin Graph sont des remaniements. A partir de 1974, ITC publie la revue «U&lc» (Upper and lower case) qui ne tarde pas à devenir une revue de typographie très influente dans le monde entier. Elle a été dirigée par Lubalin jusqu'à sa mort en 1981. En 1986, Esselte Letraset achète ITC qui continue ses activités à New York.

**Intertype** – *fabricant de composeuses* – la société «Intertype Typesetting Company» a été fondée en 1912 par Hermann Ridder à Brooklyn. En 1913, on y construit l'»Intertype», une machine à fondre les caractères et a composer par

Rome. Participation in the 4th Biennale in São Paolo (1957), in the "50 Ans d'Art Moderne" in Brussels (1958), in documenta II in Kassel and in the "Schrift und Bild" exhibition in Amsterdam and Baden-Baden (1963). – *Publications include:* "The Collected Writings of Inoue Yu-Ichi", Tokyo 1989; "The Great Air-Raid on Tokyo", Tokyo 1995. Günther Aust "Yu-Ichi Inoue", Wuppertal 1965; Masaomi Unagami (ed.) "Inoue Yu-Ichi no sho, Sho by Yu-Ichi 49–77", Tokyo 1977.

**International Typeface Corporation** (ITC) – *type manufacturers* – 1970: Aaron Burns, Herb Lubalin and Edward Rondthaler open the ITC in New York with the aim of designing, licensing and marketing typefaces for filmsetting and computer typesetting world-wide. Type designers working for ITC have included Herb Lubalin, Tom Carnase, Edward Benguiat, Antonio DiSpigna, Hermann Zapf and Matthew Carter. ITC has produced many new font versions, including Franklin, Century, Souvenir, Berkeley and Korinna. New designs have included Avantgarde Gothic, Zapf Chancery, Bauhaus and Lubalin Graph. Since 1974: publication of the in-house magazine "U&lc" ("Upper and lower case"), which rapidly becomes an influential typography journal of international acclaim. Herb Lubalin is editor-in-chief until his death in 1981. 1986: Esselte Letraset buys ITC which continues to operate in New York.

**Intertype** – *typesetting machine factory* – 1912: the Intertype Typesetting Company is founded in Brooklyn by Hermann Ridder. 1913: the Intertype line typeset-

der Firma in „Intertype Corporation". 1926 Eröffnung einer Niederlassung in Berlin. 1928 Beginn der eigenen Matrizenfertigung in Berlin. 1944 Übernahme der Linograph Corporation. 1957 Übernahme der Intertype Corporation durch Harris-Seybold in Cleveland unter dem neuen Firmennamen Harris-Intertype Corporation. 1968 Einstellung der Produktion von Matrizen und Bleisatzmaschinen, Fertigung von Fotosatzmaschinen.

**Isfahani,** Mirza Gholamreza – gest. 1884. – *Kalligraph* – Arbeitet als Kalligraph am Hof von Mohammad Schah und Nasseredin Schah aus der Gadjar-Dynastie. Seine Schrift ist die Nastaliq. Fertigt die Schilder der Teheraner Sepahssalar-Schule, „Tohfatol wozara" (1839) und das Album der schönen Schriften (1838–40) an. Die Werke werden in Teheraner Bibliotheken und in der Nationalbibliothek aufbewahrt.

**Isingrin,** Michael – geb. 1500, gest. 1557. – *Drucker, Verleger* – 1531–57 als Drucker in Basel tätig. Entwirft ca. 80 Drucke von Schriftstellern des Altertums, der Renaissance, des Humanismus sowie Kräuter- und Pflanzenbücher. 1557–60 Weiterführung seiner Druckerei durch seine Witwe, 1560–92 durch seinen Schwiegersohn Thomas Guarin.

**Isley,** Alexander – geb. 16.11.1961 in Durham, USA. – *Grafik-Designer, Lehrer* – Studium an der North Carolina State University und an der Cooper Union. Arbeit als Senior Designer der Agentur M&Co von Tibor Kalman. Art Director der Zeitschrift „Spy". 1988 Gründung von „Alexander Isley Design" in New York.

lignes. En 1916, la société est rebaptisée «Intertype Corporation». En 1926, elle ouvre une filiale à Berlin. En 1928, elle commence à fabriquer ses propres matrices. Reprise de Linograph Corporation en 1944. La Intertype Corporation est rachetée en 1957 par Harris-Seybold à Cleveland et prend le nom de Harris-Intertype Corporation. Arrêt de la production de matrices et de composeuses au plomb en 1968, puis production de machines à photocomposer.

**Isfahani,** Mirza Gholamreza – décédé en 1884 – *calligraphe* – a travaillé comme calligraphe à la cour de Mohammad Shah et de Nasseredin Shah de la dynastie Gadjar. Il écrivait en caractères nastaliq. A réalisé les panneaux de l'école Sepahssalar de Téhéran, le «Tohfatol wozara» (1839) et l'album des belles écritures (1838–1840). Ses œuvres sont conservées dans les bibliothèques de Téhéran et à la Bibliothèque Nationale.

**Isingrin,** Michael – né en 1500, décédé en 1557 – *imprimeur, éditeur* – imprimeur à Bâle de 1531 à 1557. Crée environ 80 impressions d'auteurs antiques, de la Renaissance, des humanistes ainsi que des livres de botanique et sur les plantes médicinales. Sa veuve dirige l'imprimerie de 1557 à 1560, puis l'entreprise est reprise par son gendre, Thomas Guarin qui l'exploite de 1560 à 1592.

**Isley,** Alexander – né le 16. 11. 1961 à Durham, Etats-Unis – *graphiste maquettiste, enseignant* – études à la North Carolina State University et à la Cooper Union. Travaille comme Senior Designer à l'agence M&Co de Tibor Kalman. Directeur artistique de la revue «Spy». En 1988, il fonde la «Alexander Isley Design» à New York. Commanditaires : l'American Museum of the Moving Image, l'Archeological Institute of America, Forbes Inc., MTV Networks, Pepsico, Random House,

Isfahani Calligraphy

Isfahani Calligraphy

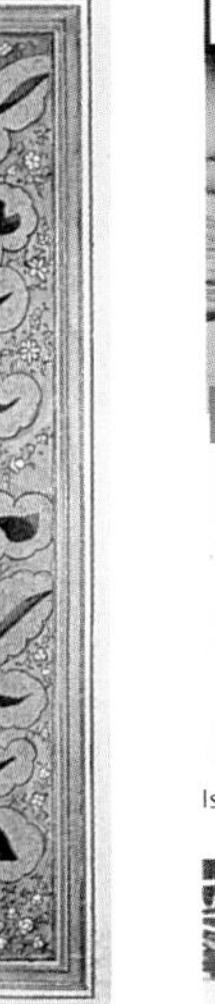

Isley 1989 Cover

Isley 1989 Cover

Von Pappeln. Oleander. Aron. Ye lenger ye lieber. Kölbleskraut.

Spargen. Eibisch. Frawenhar. Haselwurtz. Quendel. Weggraß.

Nachtschatten. Tamarisck. Eniß. Zeiland. Drachen. Groß. Christ.fl

Rosen. Getruckt zů Basell/durch Michael Isingrin. Latte.ff.ff.fü.

Isingrin 1543 Main Title

ting and casting machine is built. 1916: the company is renamed Intertype Corporation. 1926: a branch of the company is opened in Berlin. 1928: matrix production starts in Berlin. 1944: the Intertype Corporation takes over the Linograph Corporation. 1957: the Intertype Corporation is taken over by Harris-Seybold in Cleveland and renamed the Harris-Intertype Corporation. 1968: production of matrices and lead composition machines is stopped. Production of filmsetting machines is begun.

**Isfahani,** Mirza Gholamreza – d. 1884 – *calligrapher* – Worked as a calligrapher at the court of Mohammad Shah and Nasseredin Shah of the Gadjar dynasty. He used nastaliq script. Produced the signs for the Sepahssalar School in Teheran, "Tohfatol wozara" (1839) and an album of beautiful scripts (1838–40). The works are preserved in Teheran libraries and in the National Library.

**Isingrin,** Michael – b. 1500, d. 1557 – *printer, publisher* – 1531–57: works as a printer in Basle. Prints c. 80 books by classic, Renaissance and Humanist authors, and also books on herbs and plants. 1557–60: his widow continues to run the printing workshop. 1560–92: his son-in-law, Thomas Guarin, runs the company.

**Isley,** Alexander – b. 16.11.1961 in Durham, USA – *graphic designer, teacher* – Studied at the North Carolina State University and at the Cooper Union. Works as senior designer for Tibor Kalman's agency M&Co. Art director of "Spy" magazine. 1988: founds Alexander Isley Design in New York. Clients include the

Auftraggeber waren u. a. das American Museum of the Moving Image, das Archaeological Institute of America, Forbes Inc., MTV Networks, Pepsico, Random House, Warner Bros. Records und die Brooklyn Academy of Music. Unterrichtet Design und Typographie an der Cooper Union und an der School of Visual Arts in New York.

**Istlerová,** Clara – geb. 11. 12. 1944 in Prag, Tschechoslowakei. – *Illustratorin, Kalligraphin, Typographin* – Grafik-Design-Studium und Industrie-Design-Studium in Prag und an der Central School of Art in London. Arbeitet als Buchgestalterin. Zahlreiche Auszeichnungen beim Wettbewerb der schönsten Bücher.

**Itten,** Johannes – geb. 11. 11. 1888 in Südern-Linden, Schweiz, gest. 25. 3. 1967 in Zürich, Schweiz. – *Künstler, Lehrer* – 1904–06 Lehrerseminar in Hofwil bei Bern. 1909 Studium an der Ecole des Beaux-Arts in Genf. 1910–12 Sekundarlehrer-Ausbildung an der Universität Bern. 1913–14 Studium an der Akademie Stuttgart. 1916 erste Einzelausstellung in der Galerie „Der Sturm" von Herwarth Walden in Berlin. 1919 Berufung als Meister an das neugegründete Bauhaus in Weimar. Aufbau des Vorkurses. Als früher Vertreter des Bauhauses integriert er skripturale und alphabetische Zeichen in seine Gestaltungslehre. 1922–23 Kündigung und Weggang vom Bauhaus, Umzug in die Schweiz. Eintritt in die internationale Mazdaznan-Tempel-Gemeinschaft in Herrliberg am Zürichsee. Gründung der „Ontos-Werkstätten" für Handweberei und Teppichknüpferei. 1926

Isley 1988 Page

Istlerová 1971 Logo

Istlerová 1979 Cover

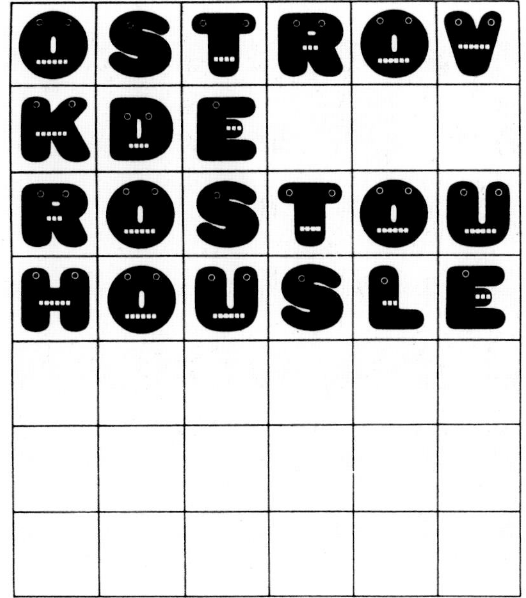


Istlerová 1982

Warner Bros. Records, le Brooklyn Academy of Music etc. A enseigné le design et la typographie à la Cooper Union et à la School of Visual Arts de New York.

**Istlerová,** Clara – née le 11. 12. 1944 à Prague, Tchécoslovaquie – *illustratrice, calligraphe, typographe* – études d'arts graphiques et de design industriel à Prague et à la Central School of Art de Londres. Travaille comme maquettiste de livres. Nombreuses distinctions lors de concours du plus beau livre.

**Itten,** Johannes – né le 11. 11. 1888 à Südern-Linden, Suisse, décédé le 25. 3. 1967 à Zurich, Suisse – *artiste, enseignant* – 1904–1906, études à l'école normale de Hofwil près de Berne. 1909, études à l'école des beaux-arts de Genève. 1910–1912, formation de professeur d'enseignement secondaire à l'université de Berne. 1913– 1914, études à l'académie de Stuttgart. 1916, première exposition personnelle à la galerie «Der Sturm» de Herwath Walden, à Berlin. En 1919, il est appelé comme enseignant au Bauhaus de Weimar à sa création. Conçoit le cours préliminaire. Dans l'esprit de la première phase du Bauhaus, il intègre des signes alphabétiques et des idéogrammes dans son enseignement. Révoqué, il quitte le Bauhaus en 1922–1923 et s'installe en Suisse. Il entre à la communauté internationale du temple Mazdaznan de Herrliberg près du lac de Zurich. Il fonde les «Ontos-Werkstätten» de tissage à la main et de tapisserie. En 1926, il fonde la «Mo-

American Museum of the Moving Image, the Archaeological Institute of America, Forbes Inc., MTV Networks, Pepsico, Random House, Warner Bros. Records and the Brooklyn Academy of Music. Isley has taught design and typography at the Cooper Union and at the School of Visual Arts in New York.

**Istlerová,** Clara – b. 11. 12. 1944 in Prague, Czechoslovakia – *illustrator, calligrapher, typographer* – Studied graphic design and industrial design in Prague and at the Central School of Art in London. Her main area of work is book design. She has won numerous awards in beautiful book competitions.

**Itten,** Johannes – b.11. 11. 1888 in Südern-Linden, Switzerland, d. 25. 3. 1967 in Zurich, Switzerland – *artist, teacher* – 1904–06: attends teachers' college in Hofwil near Bern. 1909: studies at the Ecole des Beaux-Arts in Geneva. 1910–12: trains as a high school teacher at the University of Bern. 1913–14: studies at the academy in Stuttgart. 1916: first solo exhibition in Herwarth Walden's "Der Sturm" gallery in Berlin. 1919: appointed a "Meister" at the new Bauhaus in Weimar. Introduces the Preliminary Course. An early representative of the Bauhaus, Itten integrates script and alphabetic symbols into his designs. 1922–23: leaves the Bauhaus and moves back to Switzerland. Joins the international Mazdaznan Temple Community in Herrliberg on Lake Zurich. Founds the Ontos workshops for hand-weaving and knotted carpets. 1926: founds his own Moderne Kunstschule in Berlin which

Gründung der eigenen „Modernen Kunstschule Berlin", in der Maler, Grafiker, Fotografen und Architekten ausgebildet werden. 1929 Eröffnung des Neubaus der Itten-Schule in Berlin-Wilmersdorf. 1932–37 Direktor der neugegründeten „Höheren Fachschule für textile Flächenkunst" in Krefeld; wechselnder Unterricht in Berlin und Krefeld. 1934 Schließung der Itten-Schule in Berlin durch die Nationalsozialisten. 1937 Werke Ittens sind auf der Ausstellung „Entartete Kunst" in München zu sehen. Kündigung der Direktorenstelle durch die Stadt Krefeld. 1938 Emigration in die Niederlande. 1938–53 Direktor der Kunstgewerbeschule und des Kunstgewerbemuseums Zürich. 1943–60 Leitung der Textilfachschule Zürich. 1949–56 Vorbereitung und Leitung des Rietberg-Museums in Zürich. 1955 Gastdozent an der Hochschule für Gestaltung in Ulm. Zahlreiche Ausstellungen, u.a. im Stedelijk Museum Amsterdam (1957), im Kunsthaus Zürich (1964), und der Biennale Venedig (1966). – *Publikationen u.a.:* „Tagebuch", Berlin 1930, Zürich 1962; „Kunst der Farbe", Ravensburg 1961; „Mein Vorkurs am Bauhaus", Ravensburg 1963; „Werke und Schriften", Zürich 1972; „Tagebücher", Wien 1990.

Wir behaupteten oben, daß alle Stoffe Darstellungsmittel sind.
Nun sind aber alle STOFFE = FORMEN.
ALSO IST DIE
Form =
Darstellungsmittel
Wenn dem so ist und das Wesentliche der FORM in ihrem geistig seelischen Ursprung begründet liegt, welcher nie begriffen werden kann, so ergibt sich ohne Mühe der
SCHLUSS:
Die Darstellungsmittel sind ebensowenig lehrbar wie Form lehrbar ist.
Lehren und lernen heißt begriffen haben und begreifen werden.
Die Behauptung, FORM wäre lehrbar, kann also nur einem niedern Verstande als wahr erscheinen.
Frage: Ist ein Lehren und Begreifen überhaupt
?möglich?
Eines Dinges GRUND ergründen werden wir
nie
Weder Stoff, noch Form, noch Bewegung sind lehrbar, begreifbar,
allein
Wahrnehmung
IST
wahrnehmbar

Itten 1921 Page

Itten 1921 Page

Itten 1919 Cover

derne Kunstschule Berlin» qui forme des peintres, des graphistes, des photographes et des architectes. En 1929, le nouveau bâtiment de la Itten-Schule (école Itten) est inauguré à Berlin-Wilmersdorf. De 1932 à 1937, il dirige la «Höhere Fachschule für textile Flächenkunst» (école supérieure des arts du textile), qui vient d'ouvrir à Krefeld; enseigne tour à tour à Berlin et à Krefeld. En 1934, les nazis ferment la Itten-Schule de Berlin. En 1937, des œuvres de Itten figurent dans l'exposition «Entartete Kunst» (art dégénéré) de Munich. La ville de Krefeld le démet de ses fonctions de directeur. En 1938, il émigre aux Pays-Bas. 1938–1953, directeur de la Kunstgewerbeschule (école des arts décoratifs) et du Kunstgewerbemuseum (musée des arts décoratifs) de Zurich. 1943–1960, dirige l'institut des textiles de Zurich. 1949–1956, préparation et direction du Rietberg-Museum de Zurich. Cycle de conférences à la Hochschule für Gestaltung (école supérieure de design) d'Ulm en 1955. Nombreuses expositions, entre autres au Stedelijk Museum d'Amsterdam (1957), au Kunsthaus de Zurich (1964), à la Biennale de Venise (1966). – *Publications, sélection :* «Tagebuch», Berlin 1930, Zurich 1962; «Kunst der Farbe», Ravensburg 1961; «Mein Vorkurs am Bauhaus», Ravensburg 1963; «Werke und Schriften», Zurich 1972; «Tagebücher», Vienne 1990.

trains painters, graphic artists, photographers and architects. 1929: opens a new building for his school in Berlin-Wilmersdorf. 1932–37: director of the new Höhere Fachschule für textile Flächenkunst in Krefeld; teaches in Berlin and Krefeld. 1934: Itten's school in Berlin is closed down by the National Socialists. 1937: some of Itten's works are included in the exhibition of degenerate art in Munich. The city of Krefeld dismisses him as director. 1938: emigrates to the Netherlands. 1938–53: director of the Kunstgewerbeschule and the Kunstgewerbemuseum in Zurich. 1943–60: director of the Textilfachschule in Zurich. 1949–56: plans and directs the Rietberg Museum in Zurich. 1955: visiting lecturer at the Hochschule für Gestaltung in Ulm. Numerous exhibitions, including at the Stedelijk Museum in Amsterdam (1957), the Kunsthaus in Zurich (1964), and the Biennale in Venice (1966). – *Publications include:* "Tagebuch", Berlin 1930, Zurich 1962; "Kunst der Farbe", Ravensburg 1961; "Mein Vorkurs am Bauhaus", Ravensburg 1963; "Werke und Schriften", Zurich 1972; "Tagebücher", Vienna 1990.

# J

**Jacno,** Marcel – geb. 6. 8. 1904 in Paris, Frankreich. – *Grafik-Designer, Schriftentwerfer, Typograph, Lehrer* – 1923 erste grafische Arbeiten als Autodidakt. 1925–31 Filmplakate für Gaumont und Paramount sowie Zeichentrickfilme. 1934 erste Schriftentwürfe für Deberny & Peignot. Typographische Dekoration für die Halle der grafischen Künste auf der Weltausstellung 1937 in Paris. Unterrichtet 1938 an der School of Fine and Applied Arts in New York. Für den „Club Bibliophile de France" in Marseille entsteht 1946–50 eine vierbändige Ausgabe der Bibel. 1946 Entwurf für die Verpackung der Zigarettenmarke Gauloises. 1950 Entwurf von Buchumschlägen für die Verlage Juillard und Denoël sowie Broschüren für Raffineries Shell. Signet-Entwurf für die Zeitschrift „L'Observateur". Entwirft 1951–71 für das „Théâtre National Populaire" in Paris das Signet, zahlreiche Plakate sowie die Hausschrift „Chaillot". 1952 Entwurf des Plakats zum Theaterfestival in Avignon. Unterrichtet 1953 Typographie an der Ecole Nationale Supérieure des Arts Décoratifs (ENSAD) in Paris. Atelierleiter der Agentur Monteux. Zahlreiche Verpackungsentwürfe für Parfums von Harriet Hubbard Ayer und Révillon, 1954 für Courvoisier Cognac. 1957 Auszeichnung mit der Goldmedaille der „Société d'Encouragement à l'Art et à l'Industrie". Entwirft 1959 das Signet und Plakate für das „Théâtre des Nations", 1977 für das „Théâtre Mémorial Corneille". Entwirft 1968 das Signet, Plakate und eine Zeitschrift für die „Maison la Culture" in Créteil. 1969 Entwurf

**Jacno,** Marcel – né le 6. 8. 1904 à Paris, France – *graphiste maquettiste, concepteur de polices, typographe, enseignant* – 1923, premiers travaux graphiques comme autodidacte. 1925–1931, affiches de cinéma pour Gaumont et Paramount et premiers dessins animés. 1934, premières polices pour Deberny & Peignot. En 1937, il réalise la décoration typographique du hall des arts graphiques de l'exposition universelle de Paris. En 1938, il enseigne à la School of Fine and Applied Arts de New York. Entre 1946 et 1950, à Marseille, il réalise une bible en quatre volumes pour le «Club Bibliophile de France». 1946, dessin du paquet de «Gauloises». En 1950, il conçoit des couvertures de livres pour les éditions Juillard et Denoël ainsi que des brochures pour les Raffineries Shell. Dessin du signet du magazine «L'Observateur». De 1951 à 1971, il conçoit le signet du «Théâtre National Populaire» de Paris, de nombreuses affiches, et dessine «Chaillot», une police spéciale. Conçoit l'affiche du Festival d'Avignon en 1952. Enseigne la typographie à l'Ecole Nationale Supérieure des Arts Décoratifs (ENSAD) de Paris en 1953. Directeur d'atelier à l'Agence Monteux. Nombreuses conceptions d'emballages pour les parfums Harriet Hubbard Ayer et Révillon, puis en 1954 pour les cognacs Courvoisier. En 1957, il reçoit la médaille d'or de la «Société d'Encouragement à l'Art et à l'Industrie». Dessine le signet et les affiches du «Théâtre des Nations» en 1959 et celles du «Théâtre Mémorial Corneille» en 1977. En 1968, il conçoit le signet, les affiches et la maquette de la revue de la Maison de la Culture de Créteil. Dessin du titre du journal «France-Soir» en 1969. Nombreuses expositions de peintures et de travaux graphiques. – *Polices:* Film (1934), Scribe (1936), Jacno (1948), Savoie (1949), Chail-

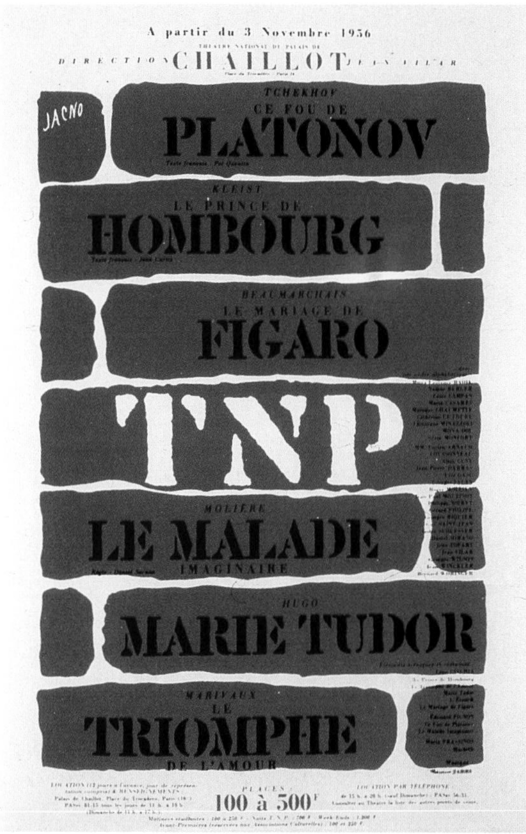


Jacno 1951 Poster

Jacno 1951 Chaillot

Jacno 1946 Packaging

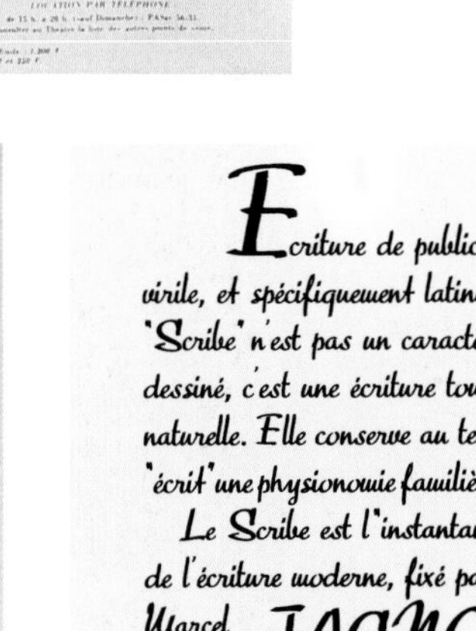

Ecriture de publicité
virile, et spécifiquement latine,
'Scribe' n'est pas un caractère
dessiné, c'est une écriture toute
naturelle. Elle conserve au texte
'écrit' une physionomie familière.
Le Scribe est l'instantané
de l'écriture moderne, fixé par
Marcel JACNO


Jacno 1936 Scribe

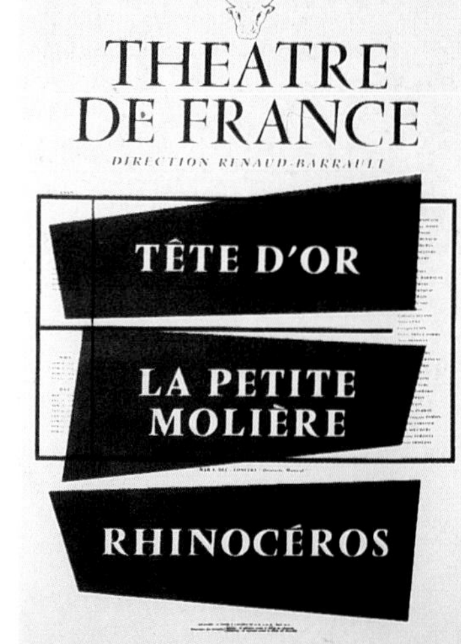


Jacno 1959 Poster

**Jacno,** Marcel – b. 6. 8. 1904 in Paris, France – *graphic designer, type designer, typographer, teacher* – Self-taught. 1923: produces his first graphic work. 1925–31: designs film posters for Gaumont and Paramount and produces animated cartoons. 1934: produces his first typeface designs for Deberny & Peignot. 1937: fashions typographic decorations for the Hall of Graphic Arts at the world exhibition in Paris. 1938: teaches at the School of Fine and Applied Arts in New York. 1946–50: a four-volume edition of the Bible is produced for the Club Bibliophile de France in Marseille. 1946: designs packaging for Gauloises cigarettes. 1950: designs book covers for Juillard and Denoël publishing houses and brochures for Raffineries Shell. Designs the logo for "L'Observateur" magazine. 1951–71: designs the logo for the Théâtre National Populaire in Paris as well as numerous posters and their in-house magazine, "Chaillot". 1952: designs the poster for the theater festival in Avignon. 1953: teaches typography at the Ecole Nationale Supérieure des Arts Décoratifs (ENSAD) in Paris. Art director for the Monteux agency. Produces numerous packaging designs for Harriet Hubbard Ayer and perfumes and Révillon, and in 1954 for Courvoisier cognac. 1957: is awarded a gold medal by the Société d'Encouragement à l'Art et à l'Industrie. 1959: designs the logo and posters for the Théâtre des Nations, and in 1977 for the Théâtre Mémorial Corneille. 1968: designs the logo, posters and magazine for the Maison de la Culture in Créteil. 1969: designs the

der Wortmarke für die Zeitung „France-Soir". Zahlreiche Ausstellungen von Malerei und Gebrauchsgrafik. – *Schriftentwürfe:* Film (1934), Scribe (1936), Jacno (1948), Savoie (1949), Chaillot (1951), Molière (1970), Ménilmontant (1973), Corneille (1978). – *Publikationen u. a.:* „Anatomie de la lettre", Paris 1978. Robert Ranc „Jacno, lettres et images", Paris 1962; Alain Weille „Jacno", Paris 1982.

**Jähn,** Hannes (Johannes Eberhard Hermann) – geb. 28. 4. 1934 in Leipzig, Deutschland, gest. 25. 7. 1987 in Köln, Deutschland. – *Grafik-Designer, Typograph, Fotograf, Lehrer* – 1949–52 Lehre als Schildermaler in Leipzig. 1952 Umzug nach Köln. 1952–55 Studium an der Werkschule Köln. Tätigkeiten als Grafiker und Buchgestalter. 1955–58 Buchumschläge für die Verlage Kiepenheuer & Witsch, Karl Rauch und Limes. 1960 erste Auszeichnungen auf der Ausstellung „Photokina" in Köln. 1972 Gründung der Chihuahua-Press (mit Walther König), in der nur zwei Bücher erscheinen. 1974 Beginn der Arbeit an Stoff-Bildern und freier Malerei. 1979 Ausstellung in der Galerie Zwirner in Köln. 1981 grafische Gesamtgestaltung der Ausstellungen „Westkunst" und „Rausch und Realität" in Köln. 1983 Art Director des Zweitausendeins-Versands. Plakatgestaltung für Oper und Schauspiel in Köln. Ausstellung in der Galerie KK in Essen (mit Roland Topor). 1984–85 Plakatgestaltungen für die Kammerspiele in München und das Thalia-Theater in Hamburg. 1985–87 Professor im Fachbereich Visuelle Kommunikation an der Universität Wuppertal.

lot (1951), Molière (1970), Ménilmontant (1973), Corneille (1978). – *Publications, sélection:* «Anatomie de la lettre», Paris 1978. Robert Ranc «Jacno lettres et images», Paris 1962; Alain Weille «Jacno», Paris 1982.

**Jähn,** Hannes (Johannes Eberhard Hermann) – né le 28. 4. 1934 à Leipzig, Deutschland, décédé le 25. 7. 1987 à Cologne, Allemagne – *graphiste maquettiste, typographe, photographe, enseignant* – 1949–1952, apprentissage de peintre d'enseignes à Leipzig. 1952, s'installe à Cologne. 1952–1955, études à la Werkschule (école des arts et métiers) de Cologne. Exerce comme graphiste et maquettiste en édition. 1955–1958, couvertures de livres pour les éditions Kiepenheuer & Witsch, Karl Rauch et Limes. 1960, premières distinctions à l'exposition «Photokina» à Cologne. Fonde en 1972 les «Chihuahua-Press» (avec Walther König) qui ne publient que deux ouvrages. En 1974, il commence à travailler sur des tableaux-textiles et se consacre à la peinture. 1979, exposition à la Galerie Zwirner de Cologne. En 1981, il réalise l'ensemble de la conception graphique des expositions «Westkunst» et «Rausch und Realität» à Cologne. Directeur artistique de la société de vente par correspondance «Zweitausendeins» en 1983. Conçoit des affiches pour des opéras et spectacles à Cologne. Exposition à la Galerie KK à Essen (avec Roland Topor). En 1984–1985, il réalise les affiches des «Kammerspiele» de Munich et du Thalia-Theater de Hambourg. Professeur au département de communication visuelle de

Jacno 1971 Poster

Jähn 1962 Cover

Kunst hat mit Wissenschaft nichts zu tun. Kunst ist kein Experiment. Es gibt keinen Fortschritt in der Kunst, ebensowenig wie es Fortschritt in der Sexualität gibt. Um es einfach zu sagen: Es gibt nur verschiedene Wege, sie auf die Beine zu stellen.
Man Ray

Westkunst

Jähn 1981 Flyer

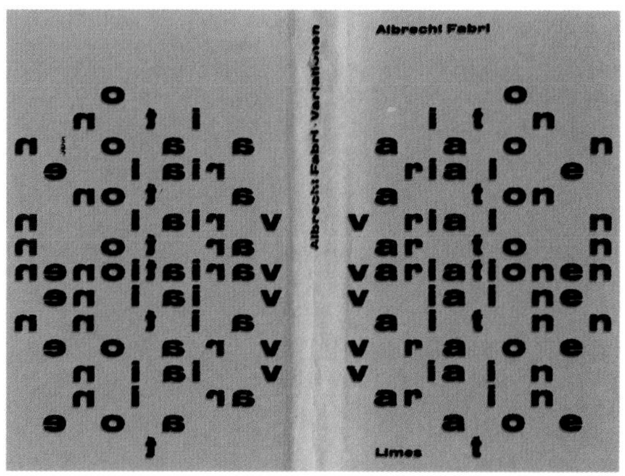


Jähn 1959 Cover

logo for "France-Soir" newspaper. Numerous exhibitions of painting and commercial art. – *Fonts:* Film (1934), Scribe (1936), Jacno (1948), Savoie (1949), Chaillot (1951), Molière (1970), Ménilmontant (1973), Corneille (1978). – *Publications include:* "Anatomie de la lettre", Paris 1978. Robert Ranc "Jacno, lettres et images", Paris 1962; Alain Weille "Jacno", Paris 1982.

**Jähn,** Hannes (Johannes Eberhard Hermann) – b. 28. 4. 1934 in Leipzig, Germany, d. 25. 7. 1987 in Cologne, Germany – *graphic designer, typographer, photographer, teacher* – 1949–52: trains as a sign-writer in Leipzig. 1952: moves to Cologne. 1952–55: studies at the Werkschule in Cologne. Works as a graphic artist and book designer. 1955–58: designs book covers for the publishers Kiepenheuer & Witsch, Karl Rauch and Limes. 1960: is awarded his first prizes at the "Photokina" exhibition in Cologne. 1972: founds the Chihuahua-Press with Walther König; they publish only two books. 1974: starts working with fabric pictures and free painting. 1979: exhibition in the Zwirner gallery in Cologne. 1981: designs the graphics for the "Westkunst" and "Rausch und Realität" exhibitions in Cologne. 1983: art director for the Zweitausendeins mail-order company. Designs posters for the opera and theater in Cologne. Exhibition in the KK gallery in Essen (with Roland Topor). 1984–85: designs posters for the Kammerspiele in Munich and the Thalia-Theater in Hamburg. 1985–87: professor in the visual communication department at

1990 Ausstellung seines Werks im Museum für angewandte Kunst in Köln. 1994 Ausstellung in der Deutschen Bibliothek in Frankfurt am Main. Jähn arbeitet für 49 Verlage und erhält über 100 Auszeichnungen für seine gestalterische Arbeit. In zahlreichen Publikationen werden seine Illustrationen bzw. Vorworte von ihm veröffentlicht. – *Publikationen u. a.:* „Der Turm“, Bad Godesberg 1956; „9 Vista Visions“, Köln 1963; „Universalkalender 1974“, München 1974; „Universalkalender 1975, 1976, 1978, 1979“, alle Frankfurt am Main; „Taschenalmanach 1982, 1984“, alle Frankfurt am Main. Gundel Gelbert (Hrsg.) „Bücher Buchstaben Bilder, Hannes Jähn 1934–1987“, Köln 1990.

**Jandl,** Ernst – geb. 1. 8. 1925 in Wien, Österreich. – *Schriftsteller* – 1946–50 Studium der Germanistik und Anglistik. 1950 Promotion. Langjährige Tätigkeit als Gymnasiallehrer. Seit 1952 Veröffentlichungen. Seit 1970 Mitglied der Akademie der Künste in Berlin. Zahlreiche Ausstellungen seines Werkes, u. a. in der Universitätsbibliothek Frankfurt (1984). Zahlreiche Auszeichnungen, u. a. Hörspielpreis der Kriegsblinden (mit Friederike Mayröcker, 1968), Österreichischer Staatspreis (1978), Georg-Büchner-Preis (1984), Friedrich-Hölderlin-Preis (1995). Jandls Sprachspiele, seine Praxis und Theorie der Literatur und des Schreibens sind für die Typographie Anregungen formaler und inhaltlicher Art. – *Publikationen u. a.:* „Lange Gedichte“, Stuttgart 1964; „Laut und Luise“, Darmstadt 1966; „Sprechblasen“, Darmstadt 1968; „Ge-

l'université de Wuppertal de 1985 à 1987. Exposition de ses œuvres au Museum für angewandte Kunst de Cologne en 1990. Exposition à la Deutsche Bibliothek de Francfort-sur-le-Main en 1994. Jähn a travaillé pour 49 éditeurs et a reçu plus de 100 distinctions pour son travail de création. Ses illustrations et préfaces ont paru dans de nombreuses publications. – *Publications, sélection :* «Der Turm», Bad Godesberg 1956; «9 Vista Visions», Cologne 1963; «Universalkalender 1974», Munich 1974; «Universalkalender 1975, 1976, 1978, 1979», tous à Francfort-sur-le-Main; «Taschenalmanach 1982, 1984», tous à Francfort-sur-le-Main. Gundel Gelbert (éd.) «Bücher Buchstaben Bilder, Hannes Jähn 1934–1987», Cologne 1990.

**Jandl,** Ernst – né le 1. 8. 1925 à Vienne, Autriche – *écrivain* – 1946–1950, études de littérature allemande et d'anglais. Doctorat en 1950. Exerce pendant plusieurs années comme professeur de lycée. Publications depuis 1952. Membre de l'Akademie der Künste de Berlin depuis 1970. Nombreuses expositions de ses œuvres, entre autres à la bibliothèque de l'université de Francfort (1984). Nombreuses distinctions dont le prix de la dramatique radiophonique des aveugles de guerre (avec Friederike Mayröcker, 1968), Österreichischer Staatspreis (Prix d'Autriche) en 1978, Prix Georg Büchner (1984), Prix Friedrich Hölderlin (1995). Les jeux de langage de Jandl, sa pratique ainsi que sa théorie de la littérature et de l'écriture ont inspiré et inspirent des contenus et des formes à la typographie. – *Publications, sélection :* «Lange Gedichte», Stuttgart 1964; «Laut und Luise», Darmstadt 1966; «Sprechblasen», Darmstadt 1968; «Gesammelte Werke» (3 vol.), Darmstadt 1985.

**Jannon,** Jean – né en 1580, décédé en 1658 à Sedan, France – *fondeur de ca-*

Jandl ca. 1968 Concrete Poetry

**film
film
film
film
fi m
f im
fi m
f im
f m
fl m
f im
f m
flim
flim
flim
film
f lm
f lm
fl m
f lm
fl m
f m
f lm
fl m
f m
f lm
f m
fl m
f lm
fl m
fl m
fl m
fl m
fl m
fl m
flim
film
flim
film
flim
film
flim
f m
film
f m
flim
film
flim
film
film
film
film
film**

Jandl 1968

Sur quoy vous me permettrés de vous demander en cette occaſion , ce que, comme i'ay des-ia remarqué, [a] S. Auguſtin demande aux Donatiſtes en vne ſemblable occurrence : *Quoy donc ? lors que nous liſons , oublions nous comment nous auons accouſtumé de parler ? l'eſcriture du grand Dieu deuoit-elle vſer auec nous d'autre langage que le noſtre?*

[a] *Aug. lib. 33. contra Fauſt. c.* 7. Quid ergo? cum legimus , obliuiſcimur quemadmodum loqui ſoleamus? An ſcriptura Dei aliter nobiſcum fuerat quam noſtro more locutura?

Jannon 1642 Antiqua and Italic

the University of Wuppertal. 1990: his work is exhibited in the Museum für angewandte Kunst in Cologne. 1994: exhibition in the Deutsche Bibliothek in Frankfurt am Main. Jähn worked for 49 different publishing houses and won over 100 awards for his designs. Numerous publications have included illustrations and forewords by Jähn. – *Publications include:* "Der Turm", Bad Godesberg 1956; "9 Vista Visions", Cologne 1963; "Universalkalender 1974", Munich 1974; "Universalkalender 1975, 1976, 1978, 1979", all Frankfurt am Main; "Taschenalmanach 1982, 1984", both Frankfurt am Main. Gundel Gelbert (ed.) "Bücher Buchstaben Bilder, Hannes Jähn 1934–1987", Cologne 1990.

**Jandl,** Ernst – b. 1. 8. 1925 in Vienna, Austria – *author* – 1946–50: studies German literature and English. 1950: gains his Ph.D. Spends many years working as a high school teacher. From 1952 onwards: has work published. Since 1970: member of the Akademie der Künste in Berlin. There have been numerous exhibitions of Jandl's work, including at the university library in Frankfurt am Main (1984). He has won numerous awards, including the Hörspielpreis der Kriegsblinden (radio play prize with Friederike Mayröcker, 1968), Austria's state prize (1978), the Georg Büchner Prize (1984) and the Friedrich Hölderlin Prize (1995). Jandl's plays with language, his theories and practice of literature and writing have inspired and still do inspire typography both in form and context. – *Publications include:* "Lange Gedichte", Stuttgart 1964;

sammelte Werke“ (3 Bände), Darmstadt 1985.

**Jannon,** Jean – geb. 1580, gest. 1658 in Sedan, Frankreich. – *Schriftgießer, Stempelschneider, Drucker* – Lehre bei Robert Estienne in Paris. Danach als Drucker und Stempelschneider an der protestantischen Universität in Sedan. Hier entstehen seine Schriften als „Caractères de l'Université“, die er 1621 in einer gedruckten Schriftprobe zusammenstellt. Seine Antiqua-Schrift in 5 Punkt-Größe wird 1625 zum Satz einer Vergil-Ausgabe eingesetzt. Jannons Schriften werden in die 1640 gegründete Imprimerie Royale eingegliedert, wo sie 1642 für den Satz der Edition der Memoiren von Kardinal Richelieu verwendet werden.

**Janson,** Anton – geb. 17. 1. 1620 in Wanden, Friesland, gest. 18. 11. 1687 in Leipzig, Deutschland. – *Stempelschneider, Schriftgießer* – 1635–51 Lehre und Arbeit als Schriftgießer in Amsterdam. 1651–56 Umzug nach Frankfurt am Main, Arbeit in der Schriftgießerei Johann Luther. 1656 Umzug nach Leipzig, Leitung der Schriftgießerei des Buchdruckers Thimotheus Ritzsch. 1658 Kauf der Schriftgießerei des Druckers Johann Erich Hahn, die Janson zu einem selbständigen Betrieb ausbaut. Dadurch gilt er als Begründer des selbständigen Leipziger Schriftgießergewerbes. Neben seinen Antiqua-Schriften entstehen arabische und andere nicht lateinische Schriftarten sowie Frakturschriften. 1688 erscheint von L. Thoma Ittig die Grabrede auf Janson, die seine Verdienste um die Schriftkunst würdigt, in gedruckter Form in

ABCDEFGHIJKLM
abcdefghijklmnop
NOPQRSTUVWXYZ
qrstuvwxyz

*ABCDEFGHIJKLM*
*abcdefghijklmnop*
*NOPQRSTUVWXYZ*
*qrstuvwxyz*

Janson ca. 1690 Janson-Antiqua

La crainte de l'Eternel eſt le chef de ſcience: mais les fols meſpriſent ſapiẽce & inſtruction. Mon fils, eſcoute l'inſtruction de ton

Jannon 1642 Antiqua

SZENT
BIBLIA.
Az-az,
ISTENNEK Ó és UJ
TESTAMEN-
TOMÁBAN
Foglaltatott egész
SZENT IRÁS:
*Magyar nyelvre fordittatott*
*KÁROLYI GASPÁR*
által,
*És moſtan* HATODSZOR *e' kis formában kibotsáttatott a' Belgiumi Academiákban tanúló Magyaroknak fo-goldások által.*
*AMSTELODAMBAN,*
*Nyomt:* M. TÓTFALUSI KIS MIKLÓS
*Költségével 's Betüivel,* 1683-ban.

Janson 1683 Main Title

Mittel Antiqua.
In hilaritate vultus Regis, vita & clementia eius, quaſi imber ſerotinus Poſſide ſapientiam, quia auro melior eſt, & acquire prudentiam, quia precioſior eſt argento. Semita iuſtorum declinat mala, cuſtos animæ ſuæ ſervat viam ſuam. Contritionem præcedit ſuperbia, & ante ruinam exaltabitur ſpiritus. Melius eſt humiliari cum mitibus, quam dividere ſpoliacum ſuperbis.
ÆBCDEFGHIJKLMNOPRSTWUZ-abcdefghikl mnopqrſtuvwxyzſiſſiſt& A I. · 1673

Cicero Antiqua.
Eruditus in verbo, reperiet bona: & qui ſperat in Domino, beatus eſt. Qui ſapiens eſt corde, appellabitur prudens: & qui dulcis eloquio, maiora reperiet. Fons vitæ, eruditio poſſidentis: doctrina ſtultorum fatuitas. Corſapientis, erudiet os eius, & labiis eius addet gratiam. Favus mellis compoſita verba, dulcedo animæ, ſanitas oſſium. Eſt via, quæ videtur homini recta, & noviſſima eius ducunt ad mortem. Anima laborantis laborat ſibi, quia cumpulit eum os ſuum: vir inſipiens, ſodit malum & in labiis eius ignit ardeſcit.
ÆABCDEFGHIJKLMNOPQRSTUVWXYZ
abcdefghiklm nopqrſstuvxyz 1672. A I.

Cicero Arabice.
A I. 1673.

Cicero Samaritaniſch.
A. J. 1674.

Janson 1674 Specimen

ractères, tailleur de types, imprimeur* – apprentissage chez Robert Estienne à Paris. Exerce ensuite comme imprimeur et tailleur de types à l'université protestante de Sedan où il crée les «Caractères de l'Université», une fonte qu'il présente en 1621 sur un échantillon imprimé. En 1625, son antique de corps 5 est utilisée pour composer une édition de Virgile. Les fontes de Jannon ont été employées par l'Imprimerie Royale fondée en 1640, où elles ont servi à la composition des mémoires du Cardinal de Richelieu en 1642.

**Janson,** Anton – né le 17. 1. 1620 à Wanden, Frise, décédé le 18. 11. 1687 à Leipzig, Allemagne – *tailleur de types, fondeur de caractères* – 1635–1651, apprentissage et travail comme fondeur de caractères à Amsterdam. S'installe à Francfort-sur-le-Main en 1651 et y travaille à la fonderie de caractères Johann Luther jusqu'en 1656. S'installe à Leipzig en 1656 et dirige la fonderie de caractères de l'imprimeur Thimotheus Ritzsch. En 1658, il achète la fonderie de caractères de l'imprimeur Johann Erich Hahn qu'il transforme en entreprise indépendante. Ceci lui vaut d'être considéré comme l'initiateur des fonderies de caractères indépendantes de Leipzig. Outre ses Antiques, il a créé des fontes en caractères arabes et non latins ainsi que des caractères allemands. En 1688 paraît une oraison funèbre où L. Thoma Ittig fait l'éloge des mérites et de l'art de Janson. En

"Laut und Luise", Darmstadt 1966; "Sprechblasen", Darmstadt 1968; "Gesammelte Werke" (3 vols.), Darmstadt 1985.

**Jannon,** Jean – b. 1580, d. 1658 in Sedan, France – *type founder, punch cutter, printer* – Trained with Robert Estienne in Paris. He then worked as a printer and punch cutter at the Protestant university in Sedan, where he developed his typefaces as "Caractères de l'Université", collating and printing them in a type specimen in 1621. 1625: his roman typeface in 5-point size is used in an edition of Virgil. Jannon's typefaces are incorporated by the Imprimerie Royal, founded in 1640, and used in an edition of Cardinal Richelieu's memoirs in 1642.

**Janson,** Anton – b. 17. 1. 1620 in Wanden, Friesland, d. 18. 11. 1687 in Leipzig, Germany – *punch cutter, type founder* – 1635–51: trains and works as a type founder in Amsterdam. 1651–56: moves to Frankfurt am Main and works in Johann Luther's type foundry. 1656: moves to Leipzig and runs printer Timotheus Ritzsch's type foundry. 1658: printer Johann Erich Hahn's type foundry is bought. Janson turns it into an independent company, and it is for this that he is considered the founder of the independent type foundry trade in Leipzig. Besides his roman typefaces, Janson also produced Arabic and other non-Latin alphabets and German types. 1688: L. Thoma Ittig's funeral oration to Janson, in which Janson's contribution to typography is praised, is printed in Leipzig. 1919: original matrices for Janson's roman and italic typefaces come into the

Leipzig. Original-Matrizen von Jansons Antiqua- und Kursivschriften gelangen 1919 in den Besitz der Schriftgießerei D. Stempel AG in Frankfurt am Main, wo sie als „Janson-Antiqua" und „Janson-Kursiv" neu herausgegeben werden.

**Jeker,** Werner – geb. 25. 12. 1944 in Mümliswil, Schweiz. – *Grafik-Designer, Typograph, Art Director, Lehrer* – 1960–61 Besuch des Vorkurses an der Kunstgewerbeschule Luzern. 1961–65 Grafikerlehre bei Hugo Welti in Olten. 1965–68 Grafiker bei Designers-Associés in Pully, 1969–71 bei Freddy Huguenin in Lausanne. 1972 Eröffnung eines eigenen Ateliers. Aufträge aus dem Zeitschriftenbereich: Layout und Illustrationen für die „TV-Radio-Zeitung" in Bern (1972–78), Art Director der Zeitschrift „L'Illustré". Zahlreiche prämierte Plakatentwürfe aus dem kulturellen Bereich, u. a. für La Cinémathèque Suisses, Les Musées de l'Art Brut, die Fondation Suisse pour la Photographie und das Kunsthaus Zürich. Ab 1984 Ateliergemeinschaft „Les Ateliers du Nord" in Lausanne mit Antoine Cahen, Claude Frossard und Fairouz Joudié. Seit 1995 Gastprofessor an der Hochschule für Gestaltung in Karlsruhe. – *Publikation:* „Jeker: Les Ateliers du Nord Lausanne", Lausanne 1992.

**Jenson,** Nicolaus – geb. 1420 in Sommevoire, Frankreich, gest. 1480 in Venedig, Italien. – *Stempelschneider, Drucker, Verleger* – Wahrscheinlich Stempelschneider (Münzmeister) der königlichen Münze in Tours oder Paris. Karl VII. sendet Jenson 1458 nach Mainz, um dort die Technik der beweglichen Metallbuchstaben zu lernen

Jeker 1988 Poster

Jeker 1989 Poster

1919, les matrices originales de l'Antique de Janson deviennent la propriété de la fonderie de caractères D. Stempel AG de Francfort-sur-le-Main qui les publie sous le nom de «Janson-Antiqua» et de «Janson-Kursiv».

**Jeker,** Werner – né le 25. 12. 1944 à Mümliswil, Suisse – *graphiste maquettiste, typographe, directeur artistique, enseignant* – fréquente le cours préliminaire de la Kunstgewerbeschule (école des arts décoratifs) de Lucerne en 1960–1961. Apprentissage de graphisme chez Hugo Welti à Olten de 1961 à 1965. Graphiste chez Designers-Associés à Pully de 1965 à 1968, puis chez Freddy Huguenin à Lausanne de 1969 à 1971. Ouvre son propre atelier en 1972. Commandes de revues: maquette et illustrations pour le «TV-Radio-Zeitung» de Berne (1972–1978), directeur artistique de la revue «L'Illustré». Nombreuses affiches primées pour le secteur culturel, entre autres pour la Cinémathèque Suisse, les Musées de l'Art Brut, la Fondation Suisse pour la Photographie, le Kunsthaus Zurich. A partir de 1984, il exerce aux «Ateliers du Nord» à Lausanne avec Antoine Cahen, Claude Frossard et Fairouz Joudié. Invité comme professeur à la Hochschule für Gestaltung (école supérieure de design) de Karlsruhe depuis 1995. – *Publication:* «Jeker: Les Ateliers du Nord Lausanne», Lausanne 1992.

**Jenson,** Nicolay – né en 1420 à Sommevoire, France, décédé en 1480 à Venise, Italie – *tailleur de types, imprimeur, éditeur* – probablement tailleur de types (maître graveur) à l'atelier royal des monnaies à Tours ou à Paris. En 1458, Charles VII délègue Jenson à Mayence pour y apprendre la technique des caractères mobiles afin de l'introduire à Paris. Séjour à Francfort-sur-le-Main. Vit à Venise à partir de 1468, il y fonde et dirige une im-

Te a quartana liberatū gaudeo: itēq; Pyliā. ego dum naues & cætera pa-
rant excurro i Pompeianū. Vectieno uelī gratias q̄ studiosus sit: si quē
quā nactus fueris qui pferat litteras: des āte q̄ discedimus. Vale.

Commodū ad te dederā litteras de pluribus rebus: cū ad me bene
nane Dionysius fuerit: ego quidē non mō placabilē me præbuis-
sē: sed totū remisissē si uenisset qua mēte tu ad me scripseras. Erat eīm
sic in tuis litteris quas Arpini accæperā eū uentur; : facturq; quod ego

Jenson 1470 Specimen

possession of the D. Stempel AG type foundry in Frankfurt am Main where they are reissued as Janson-Antiqua and Janson-Kursiv.

**Jeker,** Werner – b. 25. 12. 1944 in Mümliswil, Switzerland – *graphic designer, typographer, art director, teacher* – 1960–61: attends the preliminary course at the Kunstgewerbeschule in Lucerne. 1961–65: trains as a graphic artist with Hugo Welti in Olten. 1965–68: graphic artist for Designers-Associés in Pully, and from 1969–71 for Freddy Huguenin in Lausanne. 1972: opens his own studio. Receives work from various magazines. Does the layout and illustrations for "TV-Radio-Zeitung" magazine in Bern (1972–78), and is art director of the magazine "L'Illustré". His posters for the arts sector have won numerous awards, including those for La Cinémathèque Suisse, Les Musées de l'Art Brut, Fondation Suisse pour la Photographie and the Kunsthaus in Zurich. From 1984 onwards: opens a joint studio in Lausanne, Les Ateliers du Nord, with Antoine Cahen, Claude Frossard and Fairouz Joudié. Since 1995: visiting professor at the Hochschule für Gestaltung in Karlsruhe. – *Publication:* "Jeker: Les Ateliers du Nord Lausanne", Lausanne 1992.

**Jenson,** Nicolaus – b. 1420 in Sommevoire, France, d. 1480 in Venice, Italy – *punch cutter, printer, publisher* – It is thought Jenson was punch cutter (master of the mint) for the royal mint in Tours or Paris. 1458: Carl VII sends Jenson to Mainz to learn the technique of movable metal type and bring it back to Paris.

und nach Paris zu bringen. Aufenthalt in Frankfurt am Main. Seit 1468 in Venedig, Gründung und Leitung seiner Druckerei, in der ca. 150 Buchausgaben entstehen. Herstellung seiner ersten Antiqua „Cicero, Epistolae ad Brutum", die als vollendet und unerreicht bezeichnet wird. 1471 Herstellung seiner griechischen Schrift als Zitatenschrift. 1473 Herstellung seiner gotischen Schrift, die er in medizinischen und geschichtlichen Büchern verwendet. 1475 Gründung seiner ersten Buchhandelsgesellschaft „Nicolaus Jenson sociique", an der u. a. die Frankfurter Kaufleute Peter Ugelheimer und Johann Rauchfaß beteiligt sind. 1480 Gründung der zweiten Buchhandelsgesellschaft „Johannes de Colonia, Nicolaus Jenson et socii". Lange nach seinem Tod entstehen nach den Formen der Schriften von Jenson neue Alphabete: William Morris entwirft 1890 seine „Golden Type" nach ihrem Aufbau, Cobden-Sanderson entwirft 1900 die Schrift der Doves-Press nach ihrem Vorbild, Bruce Rogers entwirft die „Centaur" in ihrer Nachfolge. 1926 wird die Jenson Antiqua von Morris Fuller Benton als „Cloister Old Style" neu geschnitten.

**Jessen,** Peter – geb. 11. 6. 1858 in Altona, Deutschland, gest. 15. 5. 1926 in Berlin, Deutschland. – *Kunstgeschichtler, Museumsleiter* – Studium der Architektur und Kunstgeschichte in Berlin. 1886–1924 als Nachfolger Alfred Lichtwarks Direktor der Staatlichen Kunstbibliotheken, vormals Bibliothek des Kunstgewerbemuseums in Berlin. 1898 veranstaltet er die große Ausstellung „Die Kunst im

Titulus Decimusquintus.

Tertius priuilegi. §. VI.

Quartum priuile. §. VII.

Jenson 1480 Page

MEISTER DER SCHREIBKUNST

AUS DREI JAHRHUNDERTEN

200 BILDTAFELN

herausgegeben von

PETER JESSEN

JULIUS HOFFMANN VERLAG

STUTTGART

Jessen 1923 Main Title

Quare multarum quoq; gentium patrem diuina oracula futurū:ac in
ipſo benedicēdas oēs gentes hoc uidelic& ipſum quod iam nos uideūs
aperte prædictum eſt:cuius ille iuſtitiæ perfectioēm non moſaica lege
ſed fide cōſecutus eſt:qui poſt multas dei uiſiones legittimum genuit
filium:quem primum omnium diuino pſuaſus oraculo circūcidit:&
cæteris qui ab eo naſcerētur tradidit:uel ad manifeſtum multitudinis
eorum futuræ ſignum:uel ut hoc quaſi paternæ uirtutis iſigne filii re/
tinētes maiores ſuos imitari conaret':aut qbuſcūq; aliis de cauſis.Non
enim id ſcrutādum nobis modo eſt.Poſt Habraam filius eius Iſaac in
pietate ſucceſſit:fœlice hac' hæreditate a parētibus accæpta:q uni uxori

Jenson 1470 Specimen

| | n | | | | n |
|---|---|---|---|---|---|
| a | b | e | f | g | i |
| o | r | s | t | y | z |
| A | B | C | E | G | H |
| M | O | R | S | X | Y |
| 1 | 2 | 4 | 6 | 8 | & |

Jenson 1470–76 Jenson No. 58

primerie où il imprime environ 150 livres. Réalise sa première Antique, la «Cicero Epistolae ad Brutum» considérée comme achevée et inégalée. En 1471, il crée une grecque utilisée pour les citations, puis en 1473, une gothique utilisée dans les livres de médecine et d'histoire. En 1475, il fonde sa première librairie la «Nicolay Jenson sociique» avec la participation de négociants de Francfort comme Peter Ugelheimer et Johann Rauchfaß. En 1480, il fonde sa deuxième librairie la «Johannes de Colonia, Nicolay Jenson et socii». Bien après sa mort, on a réalisé des alphabets s'inspirant des formes des caractères de Jenson. En 1890, William Morris dessine son «Golden Type» selon le même schéma. En 1900, ils servent de modèle pour les caractères des Doves-Press de Cobden-Sanderson, puis vient la «Centaur» de Bruce Rogers. En 1926, Morris Fuller Benton taillera une nouvelle fonte de l'Antique de Jenson et l'appellera «Cloister Old Style».

**Jessen,** Peter – né le 11. 6. 1858 à Altona, Allemagne, décédé le 15. 5. 1926 à Berlin, Allemagne – *historien d'art, directeur de musée* – études d'architecture et d'histoire de l'art à Berlin. En 1886, il succède à Alfred Lichtwark à la direction des Staatliche Kunstbibliotheken (bibliothèques des beaux-arts), autrefois bibliothèque du musée des arts décoratifs de Berlin, et reste en fonction jusqu'en 1924. Organise en 1898 la grande exposition «Die Kunst im Buchdruck» (L'art dans le livre) au

Spends time in Frankfurt am Main. From 1468 onwards: Jenson is in Venice, where he opens and runs his printing workshop which produces c. 150 books. Produces his first roman type "Cicero, Epistolae ad Brutum", which is described as perfect and unequaled. 1471: produces his Greek typeface which is used for quotations. 1473: produces his black letter typeface which he uses in books on medicine and history. 1475: founds his first book trading company, Nicolaus Jenson sociique, whose partners include the Frankfurt businessmen Peter Ugelheimer and Johann Rauchfass. 1480: the second book trading company is launched under the name Johannes de Colonia, Nicolaus Jenson et socii. Even long after his death, Jenson's typefaces have formed the basis for many new alphabets. William Morris based his Golden Type on Jenson's type in 1890, Cobden-Sanderson modeled his typeface for Doves Press on Jenson's alphabets in 1900 and Bruce Rogers emulated them with his Centaur font. 1926: Jenson's roman is recut by Morris Fuller Benton as Cloister Old Style.

**Jessen,** Peter – b. 11. 6. 1858 in Altona, Germany, d. 15. 5. 1926 in Berlin, Germany – *art historian, museum director* – Studied architecture and art history in Berlin. 1886–1924: succeeds Alfred Lichtwark as director of the Staatliche Kunstbibliotheken, formerly the Kunstgewerbemuseum library in Berlin. 1898: stages the large-scale "Die Kunst im Buchdruck" exhibition at the Kunstgewerbemuseum in Berlin. Buys two prints produced by William Morris's Kelmscott-Press for the

Buchdruck" im Kunstgewerbemuseum in Berlin. Ankauf zweier Drucke der Kelmscott-Press von William Morris für das Kunstgewerbemuseum. 1924–30 entwirft Rudolf Koch seine gotische Schrift, die er zu Ehren Peter Jessens „Peter-Jessen-Schrift" nennt. Der Klingspor-Kalender von 1927 wird dem Gedenken Peter Jessens gewidmet. Jessen prägt den Begriff „Buchkunst" und stellt die These vom Buch als Kunstwerk auf. – *Publikationen u.a.:* „William Morris" in „Das Museum" 3/1898; „Die neue Kunst und das Buchgewerbe" in „Archiv für Buchdruckerkunst" 36/1899; „Das Buch als Kunstwerk", Berlin 1900; „Meister der Schreibkunst aus drei Jahrhunderten", Stuttgart 1924.

**Jobin,** Bernard – geb. in Porrentruy, Schweiz, gest. ca. 1597 in Straßburg, Frankreich. – *Drucker, Verleger, Schreibmeister* – In Straßburg zuerst als Holzschneider, später als Drucker tätig. Schuf zwischen 1570–97 ca. 70 Drucke und zahlreiche Einblattdrucke. 1570 Herausgabe des „Fundamentbuchs". Druckt 1573 Johann Fischarts Buch „Flöhhatz, Weibertanz". Verlegt Werke von Paracelsus, Rabelais, Luther, Calvin sowie Musikbücher. 1577 erscheint sein „Neues und Gründliches Canntzleybuech". Seine Druckerei geht 1605 an Johann Carolus über.

**Johnston,** Edward – geb. 11. 2. 1872 in San José, Uruguay, gest. 26. 11. 1944 in Ditchling, England. – *Schriftentwerfer, Kalligraph, Autor, Lehrer* – Medizin-Studium an der Universität in Edinburgh. 1898 Promotion. Umzug nach London.

musée des arts décoratifs de Berlin. Acquisition pour le musée de deux œuvres des Kelmscott-Press de William Morris. Entre 1924 et 1930, Rudolf Koch dessine une gothique qu'il appelle «Peter-Jessen-Schrift» en l'honneur de Peter Jessen. Le Klingspor-Kalender de 1927 est dédié à la mémoire de Peter Jessen. Jessen a marqué le concept d'un «art du livre»de son empreinte et émis la thèse d'un livre qui serait une œuvre d'art. – *Publications, sélection:* «William Morris» dans «Das Museum» 3/1898; «Die neue Kunst und das Buchgewerbe» dans «Archiv für Buchdruckerkunst» 36/1899; «Das Buch als Kunstwerk», Berlin 1900; «Meister der Schreibkunst aus drei Jahrhunderten», Stuttgart 1924.

**Jobin,** Bernard – né à Porrentruy, Suisse, décédé vers 1597 à Strasbourg, France – *imprimeur, éditeur, scribe* – travaille à Strasbourg comme graveur sur bois, puis exerce comme imprimeur. Entre 1570 et 1597, il a réalisé près de 70 travaux imprimés et de nombreuses pages. Publie le «Fundamentbuch» en 1570, puis le «Flöhhatz, Weibertanz» de Johann Fischart en 1573. Edite des œuvres de Paracelse, Rabelais, Luther, Calvin ainsi que des livres de musique. Son «Neues und Gründliches Canntzleybuech» paraît en 1577. L'imprimerie est reprise par Johann Carolus en 1605.

**Johnston,** Edward – né le 11. 2. 1872 à San José, Uruguay, décédé le 26. 11. 1944 à Ditchling, Angleterre – *concepteur de polices, calligraphe, auteur, enseignant* – études de médecine à l'université d'Edimbourg. Thèse de doctorat en 1898. S'installe à Londres. Etude des anciennes techniques d'écriture au British Museum. 1899–1913, enseigne dans la nouvelle section d'arts de l'écriture à la Central School of Arts and Crafts de Londres. 1901–1940, enseigne au Royal College of

Jobin 1579 Title

If I go down to hell,
thou art there also.
If I take the wings of
the morning :
And remain in the uttermost parts of the sea;

Even there also shall thy
hand lead me :
And thy right hand
shall hold me.
If I say, Peradventure
the darkness shall cover me:

Edward Johnston Spread

| | n | **n** | | | *n* |
|---|---|---|---|---|---|
| a | b | e | f | g | i |
| o | r | s | t | y | z |
| A | B | C | E | G | H |
| M | O | R | S | X | Y |
| 1 | 2 | 4 | 6 | 8 | & |

Edward Johnston 1913 Imprint

museum. 1924–30: Rudolf Koch designs his Gothic typeface which he names Peter Jessen-Schrift in honor of Jessen. 1927: the Klingspor calendar for 1927 is dedicated to the memory of Peter Jessen. Jessen shaped the concept of book art and conceived the idea of the book as a work of art. – *Publications include:* "William Morris" in "Das Museum" 3/1898; "Die neue Kunst und das Buchgewerbe" in "Archiv für Buchdruckerkunst" 36/1899; "Das Buch als Kunstwerk", Berlin 1990; "Meister der Schreibkunst aus drei Jahrhunderten", Stuttgart 1924.

**Jobin,** Bernard – b. in Porrentruy, Switzerland, d. c.1597 in Strasbourg, France – *printer, publisher, writing master* – Worked in Strasbourg as a woodcut artist and later as a printer. 1570–97: prints c. 70 volumes and numerous single-page prints. 1570: the "Fundamentbuch" is published. 1573: prints Johann Fischart's book "Flöhhatz, Weibertanz". Publishes works by Paracelsus, Rabelais, Luther and Calvin and various music books. 1577: his "Neues und Gründliches Canntzleybuech" is published. 1605: his printing workshop is passed on to Johann Carolus.

**Johnston,** Edward – b. 11. 2. 1872 in San José, Uruguay, d. 26. 11. 1944 in Ditchling, England – *type designer, calligrapher, author, teacher* – Studied medicine at the University of Edinburgh. 1898: obtains his Ph.D. Moves to London. Studies ancient writing techniques in the British Museum. 1899–1913: teaches at the Central School of Arts and Crafts in London in the new lettering department.

Studien über alte Schreibtechniken im British Museum. 1899–1913 Lehrer an der Central School of Arts and Crafts in London in der neu eingerichteten Klasse für Schrift. 1901–40 Lehrer am Royal College of Art in London. 1906 erscheint sein Buch „Writing and Illuminating and Lettering", das eine Renaissance der Kalligraphie auslöst und als das einflußreichstes Buch gilt, das je über Kalligraphie geschrieben wurde. 1910–30 Schriftentwürfe für die Cranach-Presse von Harry Graf Kessler in Weimar. 1912 Umzug nach Ditchling. 1913 Gründungsmitglied und Herausgeber der Zeitschrift „The Imprint", von der insgesamt neun Ausgaben erscheinen. 1915 gibt Frank Pick, der Direktor von London Transport, bei Johnston eine Schrift für das visuelle Erscheinungsbild der Untergrundbahn in Auftrag. 1916 Schriftentwurf für die Untergrundbahn, Mitarbeit von Eric Gill. Die Zusammenarbeit mit London Transport dauert bis 1940. 1979 wird diese Schrift „London Transport" von Colin Banks zur „New Johnston" überarbeitet. 1928 Druck der Hamlet-Ausgabe mit der von Johnston entworfenen „Hamlet-Type" sowie Holzschnitten von Edward Gordon Craig. 1930 Entwurf einer griechischen Schrift für Harry Graf Kesslers Cranach-Presse in Weimar, von der nur wenige Buchstaben gegossen werden. – *Schriftentwürfe:* Hamlet-Type (1912–27), Imprint-Antiqua (mit G. Meynell und J. H. Mason, 1913), Johnston Sans Serif (1916). – *Publikationen u.a.:* „Writing and Illuminating and Lettering", London 1906; „Manuscript and Inscription Letters

ANMUTHIGE GEGEND

FAUST AUF BLUMIGEN RASEN GEBETTET, ERMÜDET, UNRUHIG, SCHLAFSUCHEND.

Dämmerung.

Geister-Kreis schwebend bewegt, anmuthige kleine Gestalten.

ARIEL

Gesang von Aeolsharfen begleitet.

WENN der Blüthen Frühlings-Regen
Ueber alle schwebend sinkt,
Wenn der Felder grüner Segen
Allen Erdgebornen blinkt,
Kleiner Elfen Geistergrösse
Eilet wo sie helfen kann,
Ob er heilig, ob er böse,
Jammert sie der Unglücksmann.

Die ihr diess Haupt umschwebt im luft'gen Kreise,
Erzeigt euch hier nach edler Elfen Weise,
Besänftiget des Herzens grimmen Strauss,
Entfernt des Vorwurfs glühend bittre Pfeile,
Sein Innres reinigt von erlebtem Graus.
Vier sind die Pausen nächtiger Weile,
Nun ohne Säumen füllt sie freundlich aus.
Erst senkt sein Haupt auf's kühle Polster nieder,
Dann badet ihn im Thau aus Lethe's Fluth;

11

Edward Johnston 1910 Page

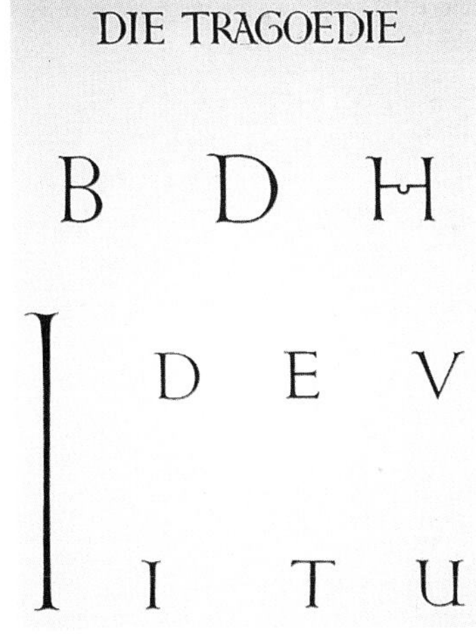


Edward Johnston 1906 Headline, Initials

Edward Johnston 1916 Johnston Sans Serif

Art de Londres. Son livre «Writing and Illuminating and Lettering» qui déclenche une renaissance de la calligraphie paraît en 1906; il est considéré comme étant un ouvrage de référence en matière de calligraphie. Dessine des polices pour la Cranach-Presse de Harry Graf Kessler à Weimar de 1910 à 1930. S'installe à Dichtling en 1912. Membre fondateur et éditeur, en 1913, de la revue «The Imprint» qui publiera neuf numéros au total. En 1915, Frank Pick, le directeur de London Transport lui commande une police pour l'identité visuelle du métro. Il conçoit une police pour le métro avec Eric Gill en 1916. Il collabore avec le London Transport jusqu'en 1940. En 1979, les caractères «London Transport» sont remaniés par Colin Banks qui les appelle «New Johnston». Une édition de Hamlet est imprimée en 1928 utilisant le «Hamlet-Type» de Johnston et illustrée de gravures sur bois d'Edward Gordon Craig. En 1930, il dessine une grecque pour la Cranach-Presse de Harry Graf Kessler à Weimar, mais quelques caractères seulement sont fondus. – *Polices:* Hamlet-Type (1912–1927), Imprint-Antiqua (avec G. Meynell et J. H. Mason, 1913), Johnston Sans Serif (1916). – *Publications, sélection:* «Writing and Illuminating and Lettering», Londres 1906; «Manuscript and Inscription Letters for School and Classes and for Use of Craftsmen», Londres 1909; «Schreibschrift, Zierschrift und angewandte Schrift», Leipzig 1910; «Hand- und Inschrift-Alphabete für Schulen und Fachklassen und für kunstgewerbliche

1901–40: teaches at the Royal College of Art in London. 1906: his book "Writing and Illuminating and Lettering" is published, causing something of a "renaissance" for calligraphy. It is considered the most influential book on calligraphy ever written. 1910–30: designs fonts for Count Harry Kessler's Cranach-Presse in Weimar. 1912: moves to Ditchling. 1913: founder member and editor of "The Imprint" magazine, of which there are a total of nine issues. 1915: Frank Pick, the director of London Transport, commissions Johnston to design a typeface for the London Underground's corporate identity. 1916: Johnston produces a typeface for the Underground. Eric Gill works on the project with him. Johnston works with London Transport until 1940. 1979: Johnston's London Transport type is reworked by Colin Banks to produce New Johnston. 1928: an edition of "Hamlet" is published with Johnston's Hamlet-Type and woodcuts by Edward Gordon Craig. 1930: designs a Greek alphabet for Count Harry Kessler's Cranach-Presse in Weimar, yet only a few of the letters are cast. – *Fonts:* Hamlet-Type (1912–27), Imprint-Antiqua (with G. Meynell and J. H. Mason, 1913), Johnston Sans Serif (1916). – *Publications include:* "Writing and Illuminating and Lettering", London 1906; "Manuscript and Inscription Letters for Schools and Classes and for Use of Craftsmen", London 1909; "Schreibschrift, Zierschrift und angewandte Schrift", Leipzig 1910; "Hand- und Inschrift-Alphabete für Schulen und Fachklassen für kunstgewerbliche Werk-

for School and Classes and for Use of Craftsmen", London 1909; „Schreibschrift, Zierschrift und angewandte Schrift", Leipzig 1910; „Hand- und Inschrift-Alphabete für Schulen und Fachklassen und für kunstgewerbliche Werkstätten", Leipzig 1922. Priscilla Johnston „Edward Johnston", London 1959.

**Johnston,** Simon – geb. 27. 5. 1959 in Leamington Spa, England. – *Grafik-Designer, Lehrer* – 1978 Studium an der School of Graphic-Design der Bath Academy of Art. 1981 Arbeit im Studio Doswald in Zug, Schweiz. 1981–82 Post-Graduate-Studium an der Allgemeinen Gewerbeschule Basel. 1983–84 Arbeiten für Conran Associates in London; Projekte für Olivetti und das Boilerhouse Project am Victoria & Albert Museum in London. 1985 Gründung des Studios „8vo" in London (mit Mark Holt und Hamish Muir). 1986 Herausgabe der Zeitschrift „Octavo, Journal of Typography" (mit Michael Burke, Mark Holt, Hamish Muir). Unterrichtet 1987 am Ravensbourne College of Art and Design. 1989 Trennung von „8vo", Umzug nach Los Angeles, Gründung seines Design-Studios „Praxis". Auftraggeber waren u. a. The Getty Trust, Los Angeles Contemporary Art (LACE), Luna Imaging Inc., Virgin Records, Oxford University Press. Unterrichtet seit 1989 Design und Typographie am Art Center College of Design in Pasadena, USA.

**Jones,** Terry – geb. 2. 9. 1945 in Northhampton, England. – *Grafik-Designer, Art Director* – 1962–65 Grafik-Design-Studium am West of England College of Art in Bristol, England. 1968–70 Assi-

Werkstätten», Leipzig 1922. Priscilla Johnston «Edward Johnston», Londres 1959.

**Johnston,** Simon – né le 27. 5. 1959 à Leamington Spa, Angleterre – *graphiste maquettiste, enseignant* – 1978, études à la School of Graphic-Design de la Bath Academy of Art. Travaille à l'atelier Doswald in Zug, Suisse en 1981. Etudes Post-Graduate à la Allgemeine Gewerbeschule (école des arts et métiers) de Bâle en 1981–1982. Travaille pour Conran Associates à Londres de 1983 à 1984: projets pour Olivetti et le Boilerhouse Project du Victoria & Albert Museum de Londres. En 1985, il fonde l'atelier «8vo» à Londres (avec Mark Holt et Hamish Muir). Publie la revue «Octavo, Journal of Typography» en 1986 (avec Michael Burke, Mark Holt, Hamish Muir). Enseigne au Ravensbourne College of Art and Design en 1987. Quitte «8vo» en 1989 et s'installe à Los Angeles où il fonde son atelier de design «Praxis». Commanditaires: The Getty Trust, Los Angeles Contemporary Art (LACE), Luna Imaging Inc., Virgin Records, Oxford University Press etc. Enseigne le design et la typographie à l'Art Center College of Design de Pasadena, Etats-Unis, depuis 1989.

**Jones,** Terry – né le 2. 9. 1945 à Northhampton, Angleterre – *graphiste maquettiste, directeur artistique* – 1962–1965, études d'arts graphiques et de design au West of England College of Art de Bristol, Angleterre. Assistant directeur artistique de la revue «Good Housekeeping» de 1968 à 1970. Directeur artistique de la revue «Vanity Fair» en 1970–1971, puis de l'édition anglaise de «Vogue» de 1972 à 1977. «Space Age Furniture Collection» en 1976, puis «Sportswear Europe» en 1978–1980. Co-éditeur et directeur artistique de la revue «i-D» depuis 1980, à Londres. Directeur artistique et de

Simon Johnston 1994 Poster

Jones 1981 Cover

Simon Johnston 1996 Cover

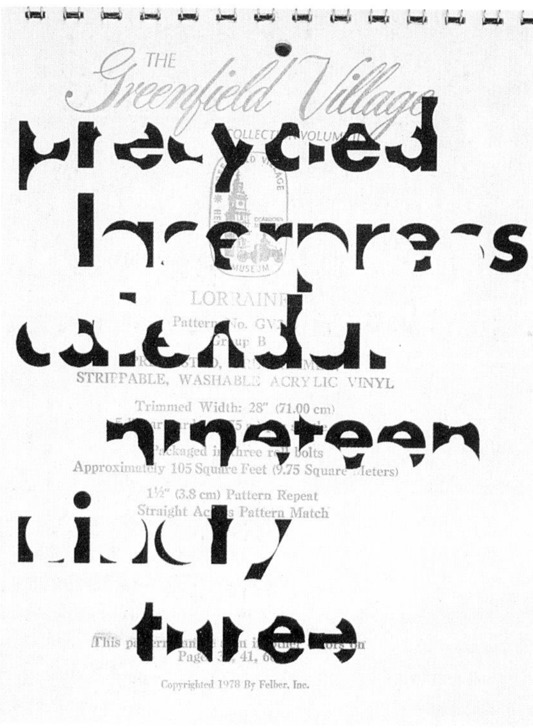


Simon Johnston 1992 Calendar

stätten", Leipzig 1922. Priscilla Johnston "Edward Johnston", London 1959.

**Johnston,** Simon – b. 27. 5. 1959 in Leamington Spa, England – *graphic designer, teacher* – 1978: studies at the Bath Academy of Art's School of Graphic Design. 1981: works for the Doswald studio in Zug, Switzerland. 1981–82: post-graduate studies at the Allgemeine Gewerbeschule in Basle. 1983–84: works for Conran Associates in London. Works on projects for Olivetti and on the Boilerhouse Project at the Victoria & Albert Museum in London. 1985: founds the 8vo studio in London with Mark Holt and Hamish Muir. 1986: publishes "Octavo, Journal of Typography" with Michael Burke, Mark Holt and Hamish Muir. 1987: teaches at Ravensbourne College of Art and Design. 1989: leaves 8vo and moves to Los Angeles, where he opens his design studio, Praxis. Clients include The Getty Trust, Los Angeles Contemporary Art (LACE), Luna Imaging Inc., Virgin Records and Oxford University Press. Since 1989: has taught design and typography at the Art Center College of Design in Pasadena, USA.

**Jones,** Terry – b. 2. 9. 1945 in Northhampton, England – *graphic designer, art director* – 1962–65: studies graphic design at the West of England College of Art in Bristol, England. 1968–70: assistant art director of "Good Housekeeping" magazine. 1970–71: art director of "Vanity Fair" magazine, of the English edition of "Vogue" (1972–77), of "Space Age Furniture Collection" (1976) and "Sportswear Europe" (1978–80). Since 1980: co-edi-

stant Art Director der Zeitschrift „Good Housekeeping". 1970–71 Art Director der Zeitschrift „Vanity Fair", 1972–77 englische Ausgabe der Zeitschrift „Vogue", 1976 „Space Age Furniture Collection", 1978–80 „Sportswear Europe". Seit 1980 Mitherausgeber und Art Director der Zeitschrift „i-D" in London. Art Direction und Imagekampagnen für zahlreiche Modeunternehmen wie Bogner, Esprit, Fiorucci, Hyper Hyper, Mexx, Lorenzini, Swatch. 1986–87 Retrospektiv-Ausstellung seiner Arbeit unter dem Titel „20 Years of Instant Design" in Tokio, London, Genua. Zahlreiche Video- und TV-Arbeiten. Gestaltung des Erscheinungsbildes des Satelliten-Fernsehsenders „Superchannel". – *Publikation:* „A Manual of Graphic Techniques: Instant Design", London 1990.

**Jost,** Heinrich – geb. 13. 10. 1889 in Magdeburg, Deutschland, gest. 27. 9. 1948 in Frankfurt am Main, Deutschland. – *Typograph, Schriftentwerfer, Grafik-Designer* – Buchhändlerlehre, gleichzeitig Besuch der Kunstgewerbe- und Handwerkerschule in Magdeburg. 1908 Umzug nach München. 1911 Abendkurse an der Kunstgewerbeschule in München (bei Paul Renner und Emil Preetorius). Tätigkeit als Buchgestalter und Grafiker. 1923–48 künstlerischer Leiter der Bauerschen Gießerei in Frankfurt am Main. – *Schriftentwürfe:* Fraktur (1925), Atrax (1926), Bauer Bodoni (1926), Aeterna (1927), Beton (1930–36).

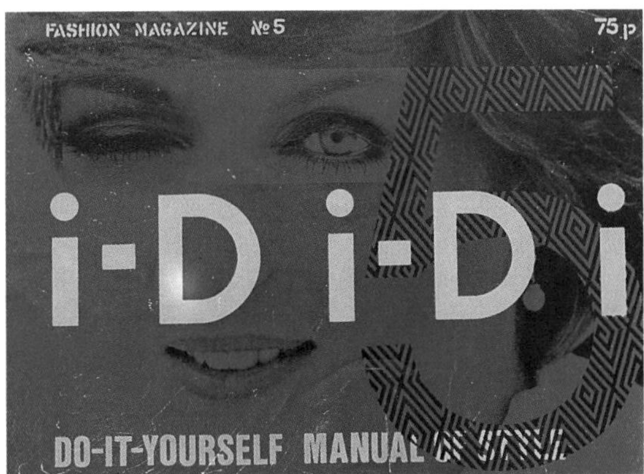


Jones 1981 Cover

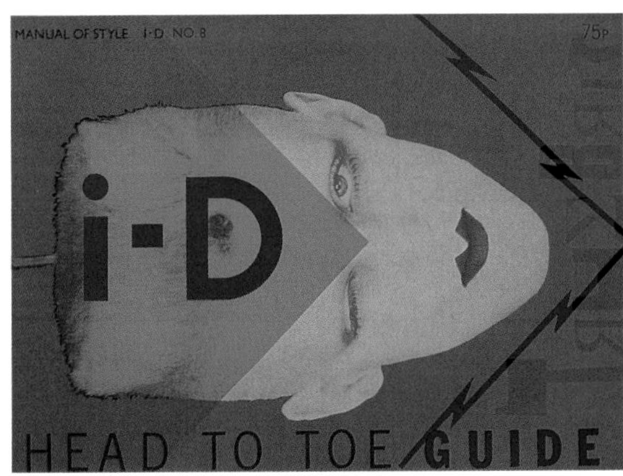


Jones 1981 Cover

Jost ca. 1928 Poster

ABCDEFG
HIJKLMN
OPQRSTU
VWXYZ

Jost 1936 Beton Extra Bold

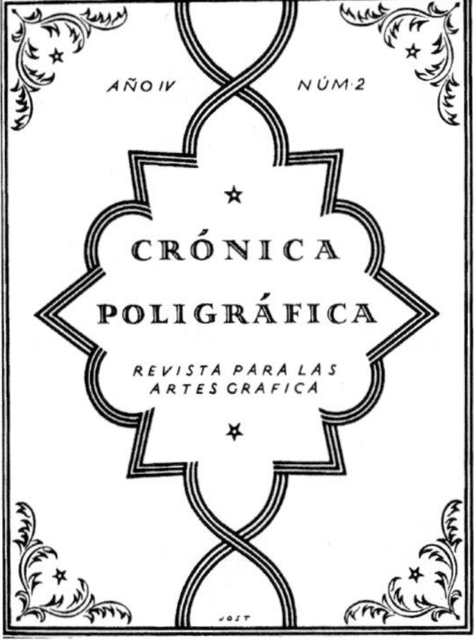


Jost 1923 Cover

tor and art director of "i-D" magazine in London. Has taken on the art direction and organized image campaigns for numerous fashion enterprises, such as Bogner, Esprit, Fiorucci, Hyper Hyper, Mexx, Lorenzini and Swatch. 1986–87: a retrospective exhibition of his work, entitled "20 Years of Instant Design", runs in Tokyo, London and Genoa. Jones has produced much work for video and the TV, including designing the corporate identity for the satellite TV channel Superchannel. – *Publication:* "A Manual of Graphic Techniques: Instant Design", London 1990.

**Jost,** Heinrich – b. 13. 10. 1889 in Magdeburg, Germany, d. 27. 9. 1948 in Frankfurt am Main, Germany – *typographer, type designer, graphic designer* – Trained as a bookseller and also attended courses at the Kunstgewerbe- und Handwerkerschule in Magdeburg. 1908: moves to Munich. 1911: takes evening courses at the Kunstgewerbeschule in Munich under Paul Renner and Emil Preetorius. Works as a book designer and graphic artist. 1923–48: art director of the Bauersche type foundry in Frankfurt am Main. – *Fonts:* Fraktur (1925), Atrax (1926), Bauer Bodoni (1926), Aeterna (1927), Beton (1930–36).

campagnes publicitaires sur l'image de nombreuses maisons de prêt à porter comme Bogner, Esprit, Fiorucci, Hyper Hyper, Mexx, Lorenzini, Swatch. 1986–1987, exposition rétrospective de ses travaux, «20 Years of Instant Design» à Tokyo, Londres, Gêne. Nombreuses vidéos et travaux pour la télévision. Identité de la chaîne par satellite «Superchannel». – *Publication:* «A Manual of Graphic Techniques: Instant Design», Londres 1990.

**Jost,** Heinrich – né le 13. 10. 1889 à Magdebourg, Allemagne, décédé le 27. 9. 1948 à Francfort-sur-le-Main, Allemagne. – *typographe, concepteur de polices, graphiste maquettiste* – apprentissage de libraire en même temps qu'il fréquente la Kunstgewerbe- und Handwerkerschule (école des arts et artisanats décoratifs) de Magdebourg. 1908, s'installe à Munich. En 1911, il suit à Munich les cours du soir de la Kunstgewerbeschule (école des arts décoratifs) (chez Paul Renner et Emil Preetorius). Exerce comme maquettiste de livres et graphiste. Directeur artistique de la Fonderie Bauer de Francfort-sur-le-Main de 1923 à 1948. – *Polices:* Fraktur (1925), Atrax (1926), Bauer Bodoni (1926), Aeterna (1927), Beton (1930–1936).

LAVORA
SAL

# K

**Käch**
**Kalhor**
**Kalman**

**Käch,** Walter – geb. 9. 6. 1901 in Ottenbach, Schweiz, gest. 5. 12. 1970 in Zürich, Schweiz. – *Grafik-Designer, Schriftgrafiker, Lehrer* – Lehre als Lithograph. Ausbildung als Grafiker an der kunstgewerblichen Abteilung der Gewerbeschule in Zürich. 1921–22 Assistent von F. H. Ehmcke an der Kunstgewerbeschule in München. 1925–29 Lehrer für grafisches Entwerfen und Holzschnitt an der kunstgewerblichen Abteilung der Kunstgewerbeschule in Zürich. Ab 1929 freischaffender Grafiker. 1940–67 Lehrer für Schrift an der Kunstgewerbeschule in Zürich.

**Kalhor,** Mohammad Reza – geb. 1825 in Kermanschah, Iran, gest. 1890 in Teheran, Iran. – *Kalligraph* – Erster persischer Kalligraph, der die technischen Schwierigkeiten, die die traditionelle Schönschreibkunst mit dem industriellen Drucken hat, überwindet. Dadurch bekommt diese alte Kunst einen modernen Charakter. Seine Schrift ist die Nastaliq. Zu seinen Arbeitsbereichen gehören u.a. die Zeitung „Scharaf", das Reisebuch von Nasseredin Schah Gadjar Dinestie 1880.

**Kalman,** Tibor – geb. 9. 7. 1949 in Budapest, Ungarn. – *Grafik-Designer* – 1956 Umzug nach New York. 1967–70 Studium der Publizistik an der New York University. 1970–79 Design Director bei Barnes & Noble Bookstore in New York. 1979 Gründung und Leitung des Design-Studios „M&Co" in New York. Aufträge aus den Bereichen Architektur, Musik, Industrie-Design, Film, Video, Fernsehen und Zeitschriftengestaltung. 1994–96 Umzug nach Rom. Herausgabe der Zeitschrift „Colours" der Firma Benetton.

**Käch,** Walter – né le 9. 6. 1901 à Ottenbach, Suisse, décédé le 5. 12. 1970 à Zurich, Suisse – *graphiste maquettiste, dessinateur de lettres, enseignant* – apprentissage de lithographe. Formation de graphiste à la section des arts décoratifs de l'école des arts et métiers de Zurich. Assistant de F. H. Ehmcke à la Kunstgewerbeschule (école des arts décoratifs) de Munich en 1921–1922. Enseigne les arts graphiques et la gravure sur bois à la section des arts décoratifs de l'école des arts et métiers de Zurich de 1925 à 1929. Graphiste indépendant à partir de 1929. Enseigne l'art de l'écriture à l'école des arts et métiers de Zurich de 1940 à 1967.

**Kalhor,** Mohammad Reza – né en 1825 à Kermanshah, Iran, décédé en 1890 à Téhéran, Iran – *calligraphe* – premier calligraphe iranien ayant réussi à surmonter les difficultés techniques que pose l'impression industrielle de l'écriture traditionnelle. Ceci permit de moderniser un art très ancien. Il utilisait les caractères Nastaliq. Kalhor a travaillé pour le journal «Sharaf» et le «Livre de voyage» de Nasseredin Shah Gadjar Dinestie 1880.

**Kalman,** Tibor, – né le 9. 7. 1949 à Budapest, Hongrie – *graphiste maquettiste* – 1956, arrivée à New York. 1967–1970, étudie le journalisme à l'université de New York. 1970–1979, directeur du design chez Barnes & Noble Bookstore à New York. En 1979, il fonde et dirige l'atelier de design «M & Co» à New York. Commandes dans le domaine de l'architecture, de la musique, du design industriel, du cinéma, de la vidéo et de la télévision, maquettes de revues. Vit à Rome de 1994 à 1996. Publie la revue «Colours» de la société Benetton.

**Kamekura,** Yusaku – né le 6. 4. 1915 à Yoshida, Japon, décédé le 11. 5. 1997 à Tokyo, Japon – *graphiste maquettiste, directeur artistique, typographe* – en 1937,

Käch 1927 Poster

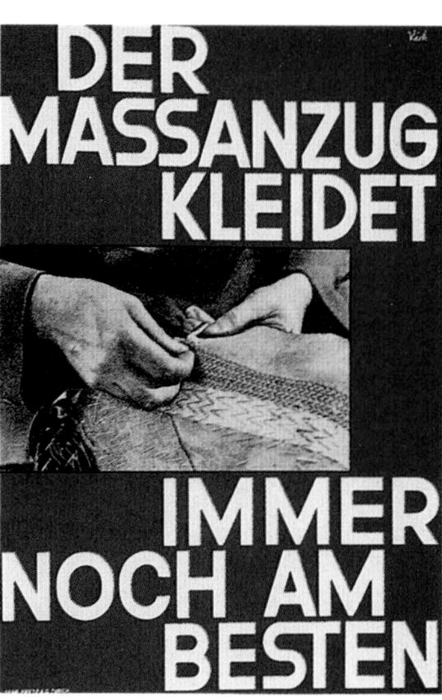

Käch ca. 1928 Poster

Kalhor Calligraphy

**Käch,** Walter – b. 9. 6. 1901 in Ottenbach, Switzerland, d. 5. 12. 1970 in Zurich, Switzerland – *graphic designer, lettering artist, teacher* – Trained as a lithographer, then as a graphic artist in the crafts department of the trade school in Zurich. 1921–22: assistant to F. H. Ehmcke at the Kunstgewerbeschule in Munich. 1925–29: teaches graphic design and woodcarving in the crafts department of the trade school in Zurich. From 1929 onwards: freelance graphic artist. 1940–67: teaches lettering at the Kunstgewerbeschule in Zurich.

**Kalhor,** Mohammed Reza – b. 1825 in Kermansha, Iran, d. 1890 in Teheran, Iran – *calligrapher* – Kalhor was the first Persian calligrapher who was able to overcome the technical difficulty of combining traditional calligraphy with industrial printing. Through this, the ancient art of calligraphy took on a modern character. His script was nastaliq; his works include the "Scharaf" newspaper and the "Travel Book" of Nasseredin Shah Gadjar Dinestie from 1880.

**Kalman,** Tibor – b. 9. 7. 1949 in Budapest, Hungary – *graphic designer* – 1956: moves to New York. 1967–70: studies journalism at New York University. 1970–79: design director for the Barnes & Noble Bookstore in New York. 1979: founds and directs the M&Co design studio in New York. Receives commissions in the following fields: architecture, music, industrial design, film, video, television and magazine design. 1994–96: moves to Rome where he edits Benetton's "Colours" magazine.

**Kamekura,** Yusaku – geb. 6. 4. 1915 in Yoshida, Japan, gest. 11. 5. 1997 in Tokio, Japan. – *Grafik-Designer, Art Director, Typograph* – 1937 Art Director der in englischer Sprache erscheinenden Zeitschrift „Nippon". 1940 Mitarbeiter der Kokusai Bunka Shinkokai, dem Regierungsinstitut für kulturelle Beziehungen Japans zum Ausland. 1951 Mitbegründer des „Japan Advertising Art Club". 1960 Mitbegründer und Leiter des „Japan Design Center" (mit Ikko Tanaka und Katsumasa Nagai). 1961–63 Entwurf des Erscheinungsbilds für die Olympischen Spiele in Tokio 1964. 1964 Entwurf der Innenausstattung des japanischen Pavillons auf der Weltausstellung in New York. 1968 Preisträger der 2. Internationalen Plakat-Biennale in Warschau. 1978 Mitbegründer und Präsident des Japanischen Verbands der Grafik-Designer „JAGDA". Seit 1989 Herausgeber der auf zwanzig Nummern begrenzten Zeitschrift „Creation". – *Publikationen u.a.:* „Paul Rand", Tokio, New York 1959; „Trademarks and Symbols of the World", Tokio, New York 1965; „Yusaku Kamekura: graphic works", Tokio, New York 1971; „The works of Yusaku Kamekura", Tokio 1983.

**Kapitzki,** Herbert W. – geb. 24. 2. 1925 in Danzig (heute Gdańsk Polen) – *Grafik-Designer, Lehrer* – 1941–43 Studium in Danzig bei Prof. Fritz Pfuhle. 1949–52 Studium an der Akademie der Künste in Stuttgart bei Willi Baumeister. 1956–68 für das Landesgewerbeamt Stuttgart tätig. 1964 Teilnahme an der documenta III in Kassel. 1964–68 Dozent an der Hochschule für Gestaltung (HfG) in Ulm.

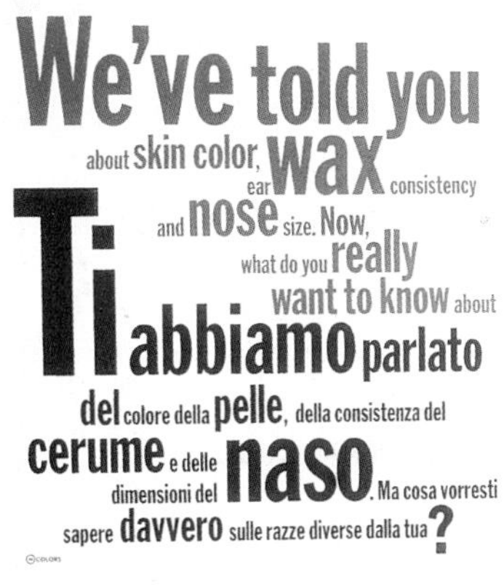

Kalman 1993 Spread

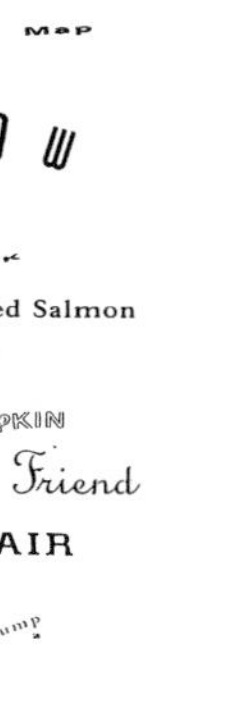


Kalman 1989 Postcard

Kamekura ca. 1968 Logo

Kamekura 1972 Poster

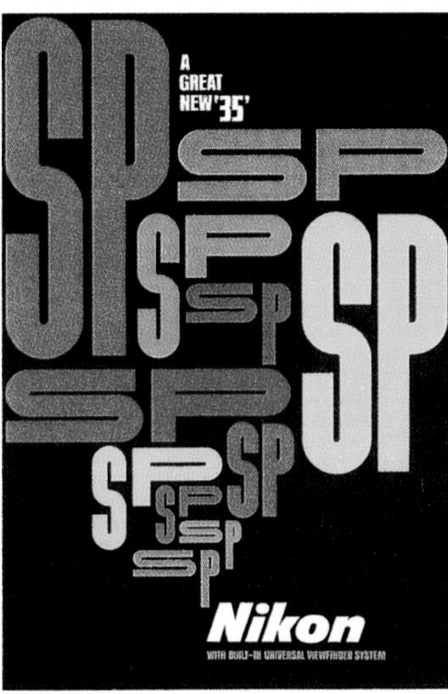


Kamekura 1957 Poster

Kamekura 1962 Poster

il est directeur artistique de l'édition anglaise de la revue «Nippon». En 1940, il travaille pour la Kokusai Bunka Shinkokai, l'institut gouvernemental pour les relations culturelles japonaises avec l'étranger. 1951, cofondateur du «Japan Advertising Art Club». 1960, cofondateur (avec Ikko Tanaka et Katsumasa Nagai) et directeur du «Japan Design Center». 1961–1963, dessine l'identité des Jeux Olympiques de 1964 à Tokyo. En 1964, il conçoit, l'architecture intérieure du pavillon japonais pour l'exposition universelle de New York. Prix de la 2e biennale de l'affiche à Varsovie en 1968. Cofondateur et président de l'Union japonaise des graphistes maquettistes (JAGDA) en 1978. A partir de 1989, il publie la revue «Creation» dont l'édition est limitée à 20 numéros. – *Publications, sélection:* «Paul Rand», Tokyo, New York 1959; «Trademarks and Symbols of the World», Tokyo, New York 1965; «Yusaku Kamekura: graphic works», Tokyo, New York 1971; «The works of Yusaku Kamekura», Tokyo 1983.

**Kapitzki,** Herbert W. – né le 24. 2. 1925 à Danzig (aujourd'hui Gdańsk en Pologne) – *graphiste maquettiste, enseignant* – 1941–1943, études à Danzig avec le professeur Fritz Pfuhle. 1949–1952, études à l' Akademie der Künste (beaux-arts) de Stuttgart avec Willi Baumeister. Exerce de 1956 à 1968 au Landesgewerbeamt (office régional de l'artisanat) à Stuttgart. 1964, participation à la «documenta III» à Kassel. 1964–1968, enseigne à la Hochschule für Gestaltung (école supérieure de

**Kamekura,** Yusaku – b. 6. 4. 1915 in Yoshida, Japan, d. 11. 5. 1997 in Tokyo, Japan – *graphic designer, art director, typographer* – 1937: art director for "Nippon" magazine (published in English). 1940: works for the Kokusai Bunka Shinkokai, the government institute for Japan's cultural relations with countries abroad. 1951: co-founder of the Japan Advertising Art Club. 1960: co-founder and manager of the Japan Design Center (with Ikko Tanaka and Katsumasa Nagai). 1961–63: designs the corporate identity for the 1964 Olympic Games in Tokyo. 1964: is responsible for the interior design of the Japanese pavilion at the world exhibition in New York. 1968: prize-winner at the 2nd International Poster Biennale in Warsaw. 1978: co-founder and president of the Japanese Association of Graphic Designers JAGDA. From 1989 onwards: editor of "Creation" magazine, limited to 20 numbers. – *Publications include:* "Paul Rand", Tokyo, New York 1959; "Trademarks and Symbols of the World", Tokyo, New York 1965; "Yusaku Kamekura: graphic works", Tokyo, New York 1971; "The works of Yusaku Kamekura", Tokyo 1983.

**Kapitzki,** Herbert W. – b. 24. 2. 1925 in Danzig (now Gdańsk, Poland) – *graphic designer, teacher* – 1941–43: studies in Danzig under Prof. Fritz Pfuhle. 1949–52: studies at the Akademie der Künste in Stuttgart under Willi Baumeister. 1956–68: works for the Landesgewerbeamt in Stuttgart. 1964: takes part in documenta III in Kassel. 1964–68: lectures at the Hochschule für Gestaltung

1968 Gründung des „Instituts für Visuelle Kommunikation und Design" (mit Herbert Lindinger und Herbert Ohl). 1970–90 Professor für Visuelle Kommunikation an der Staatlichen Hochschule für Bildende Künste in Berlin. 1970–73 Gestaltungsbeauftragter für das Historische Museum Frankfurt am Main, 1973–77 für das Erscheinungsbild der Schering AG, Berlin. 1993 Gründungsvorstand des Verbandes Deutscher Grafik-Designer (VDGD). Zahlreiche nationale und internationale Ausstellungen. – *Publikationen u.a.:* „Programmiertes Gestalten", Karlsruhe 1980; „Proportionen", Berlin 1981; „Gestaltung: Methode und Konsequenz – ein biographischer Bericht", Stuttgart 1997.

**Kapr,** Albert – geb. 20. 7. 1918 in Stuttgart, Deutschland, gest. 31. 3. 1995 in Leipzig, Deutschland. – *Typograph, Kalligraph, Schriftentwerfer, Autor, Lehrer* – 1933 Schriftsetzerlehre in Stuttgart. 1938–40 Studium an der Kunstakademie in Stuttgart bei F. H. E. Schneidler. 1945–48 Weiterführung des Studiums in Stuttgart. 1947 Dozent an der Freien Kunstschule in Stuttgart. 1951 Berufung an die Hochschule für Graphik und Buchkunst in Leipzig. 1956–78 Gründung und Leitung des Hochschulinstituts für Buchgestaltung, 1959–61 und 1966–73 Rektor der Hochschule für Graphik und Buchkunst in Leipzig. 1961 Nationalpreisträger der DDR und Gutenberg-Preisträger der Stadt Leipzig. 1964–77 künstlerischer Leiter der Dresdner Schriftgießerei VEB Typoart. 1974 Promotion zum Dr. phil. mit dem Thema „Ästhetik der Schriftkunst". – *Schriftent-*

design) d'Ulm. En 1968, il fonde, avec Herbert Lindinger et Herbert Ohl, l' «Institut für visuelle Kommunikation und Design» (institut de communication visuelle et de design). 1970–1990, professeur de communication visuelle à la Hochschule für Bildende Künste (école supérieure des beaux-arts) de Berlin. 1970–1973, chargé du design au musée historique de Francfort-sur-le-Main. 1973–1977, création de l'identité de la Schering AG de Berlin. 1993, membre fondateur et membre du comité directeur du Verband Deutscher Grafik-Designer (VDGD). Nombreuses expositions nationales et internationales. – *Publications, sélection:* «Programmiertes Gestalten», Karlsruhe 1980; «Proportionen», Berlin 1981; «Gestaltung: Methode und Konsequenz – ein biographischer Bericht», Stuttgart 1997.

**Kapr,** Albert – né le 20. 7. 1918 à Stuttgart, Allemagne, décédé le 31. 3. 1995 à Leipzig, Allemagne – *typographe, calligraphe, concepteur de polices, auteur, enseignant* – 1933, apprentissage de composition typographique à Stuttgart. 1938–1940, études à la Kunstakademie (beaux-arts) de Stuttgart avec F. H. E. Schneidler. 1945–1948 reprise des études à Stuttgart. Enseigne à la Freie Kunstschule (école libre des beaux-arts) de Stuttgart en 1947. Nommé à la Hochschule für Graphik und Buchkunst (arts graphiques et arts du livre) de Leipzig en 1951. Fonde l'institut de création bibliophile à la Hochschule für Graphik und Buchkunst de Leipzig en 1956 et le dirige jusqu'en 1978. 1959–1961 et 1966–1973, recteur de la Hochschule für Graphik und Buchkunst de Leipzig. 1961, Prix national de la RDA et lauréat du prix Gutenberg de la ville de Leipzig. 1964–1977, directeur artistique de la fonderie de caractères VEB Typoart de Dresde. 1974,

Kapitzki 1962 Poster

Kapr 1972 Cover

Kapitzki 1967 Logos

Von A bis Z kann
selbst das gute Lexikon
unsere vielgestaltige Welt
nur in großen Zügen erläutern

Kapitzki 1959 Advertisement

(HfG) in Ulm. 1968: launches the Institut für visuelle Kommunikation und Design with Herbert Lindinger and Herbert Ohl. 1970–90: professor of visual communication at the Staatliche Hochschule für Bildende Künste in Berlin. 1970–73: design representative for the Historische Museum in Frankfurt and from 1973–77 for Schering AG's corporate identity (Berlin). 1993: on the founding committee of the Verband Deutscher Grafik-Designer (VDGD). Numerous national and international exhibitions. – *Publications include:* "Programmiertes Gestalten", Karlsruhe 1980; "Proportionen", Berlin 1981; "Gestaltung: Methode und Konsequenz – ein biographischer Bericht", Stuttgart 1997.

**Kapr,** Albert – b. 20. 7. 1918 in Stuttgart, Germany, d. 31. 3. 1995 in Leipzig, Germany – *typographer, calligrapher, type designer, author, teacher* – 1933: trains as a typesetter in Stuttgart. 1938–40: studies at the Kunstakademie in Stuttgart under F. H. E. Schneidler. 1945–48: continues his studies in Stuttgart. 1947: lectures at the Freie Kunstschule in Stuttgart 1951: teaching post at the Hochschule für Graphik und Buchkunst in Leipzig. 1956–78: founds and heads the Institut für Buchgestaltung at the Hochschule für Graphik und Buchkunst in Leipzig. 1959–61 and 1966–73: principal of the Hochschule für Graphik und Buchkunst in Leipzig. 1961: winner of the GDR national award and of Leipzig's Gutenberg Prize. 1964–77: art director of the Dresden type foundry VEB Typoart. 1974: obtains his Dr. phil. with a thesis on the

*würfe:* Faust-Antiqua (1958), Leipziger-Antiqua (1959), Clarendon-Antiqua (1965), Prillwitz-Antiqua (1971), Magna Kyrillisch (1975). – *Publikationen u. a.:* „Deutsche Schriftkunst", Dresden 1955; „Schriftkunst", Dresden 1971; „Gestaltung und Funktion der Typografie" (mit Walter Schiller), Leipzig 1977; „Johannes Gutenberg", Leipzig 1977; „Schrift- und Buchkunst", Leipzig 1982.

**Karow,** Peter – geb. 11. 11. 1940 in Stargard, Pommern, Deutschland (heute Polen). – *Schriftenproduzent, Schriftprogrammentwickler* – 1962–68 Physik-Studium an der Universität Hamburg. 1972 Eintritt bei URW als dritter Partner und Geschäftsführer. 1972 Arbeiten am Ikarus-System. 1975 Entwurf eines Programms zum automatisierten Ausgleichen des Zwischenraums zwischen Buchstaben (Kerning). 1981 Programmierung des Linus-Systems für die Schriftenproduktion von Stempel, Linotype. 1983 Zusammenarbeit von URW und ITC bei der Schriftenherstellung. 1992 Programmierung des hz-Programms (mit H. Zapf) für den Schriftsatz. Entwicklung des Screenmasters für die Grauabstufung von Texten auf Display-Screens. 1995–96 Arbeiten für Adobe in San Francisco. – *Publikationen u.a.:* „Digitale Schriften", Berlin 1992; „Schriftstatistik", Hamburg 1992.

**Kassák,** Lajos – geb. 21. 3. 1887 in Ersekújvár, Ungarn, gest. 22. 7. 1967 in Budapest, Ungarn. – *Schriftsteller, Typograph, Künstler* – Schlosserlehre. 1909 Kontakte zu Apollinaire, Delaunay, Picasso. 1915 in Budapest Herausgabe der Zeitschrift „A Tett" (Die Tat), die verbo-

абвгдеёжзийклмнопрстуфхцчшщъыьэюя
АБВГДЕЁЖЗИЙКЛМНОПРСТУФХЦЧШЩЪЫЬЭЮЯ

Kapr 1975 Magna Kyrillisch

doctorat et thèse sur «l'esthétique de l'art de l'écriture». – *Polices:* Faust-Antiqua (1958), Leipziger-Antiqua (1959), Clarendon-Antiqua (1965), Prillwitz-Antiqua (1971), Magna Kyrillisch (1975). – *Publications, sélection:* «Deutsche Schriftkunst», Dresde 1955; «Schriftkunst», Dresde 1971; «Typoart – Typenkunst» (avec Hans Fischer), Leipzig 1973; «Gestaltung und Funktion der Typografie» (avec Walter Schiller), Leipzig 1977; «Johannes Gutenberg», Leipzig 1977; «Schrift- und Buchkunst», Leipzig 1982.

**Karow,** Peter – né le 11. 11. 1940 à Stargard, Poméranie, Allemagne (aujourd'hui en Pologne) – *fabricant de polices, programmateur* – 1962–1968, études de physique à l'université de Hambourg. Thèse de doctorat sur la production électronique de pions (1971). 1972, entre chez URW en tant que troisième partenaire et administrateur. 1972, programmation pour le système Icare. En 1975, programmation d'équilibrage automatique de l'espace entre les caractères (Kerning). 1981, programmation du système Linus pour la production de fontes de Stempel, Linotype. En 1983, URW et ITC collaborent en matière de production de fontes. 1992, programmation du logiciel-hz (avec Hermann Zapf) pour la composition de textes. Développement du screenmaster pour le réglage de l'intensité des gris dans les textes et écrans défilants. 1995–1996, travaille pour Adobe à San Francisco. – *Publications, sélection:* «Digitale Schriften», Berlin 1992; «Schriftstatistik», Hambourg 1992.

**Kassák,** Lajos – né le 21. 3. 1887 à Ersekújvár, Hongrie, décédé le 22. 7. 1967 à Budapest, Hongrie – *écrivain, typographe, artiste* – apprentissage de serrurerie. Contacts avec Apollinaire, Delaunay, Picasso en 1909. En 1915 il publie la revue «A Tett» (l'action) à Budapest qui

Kapr ca. 1975 Calendar

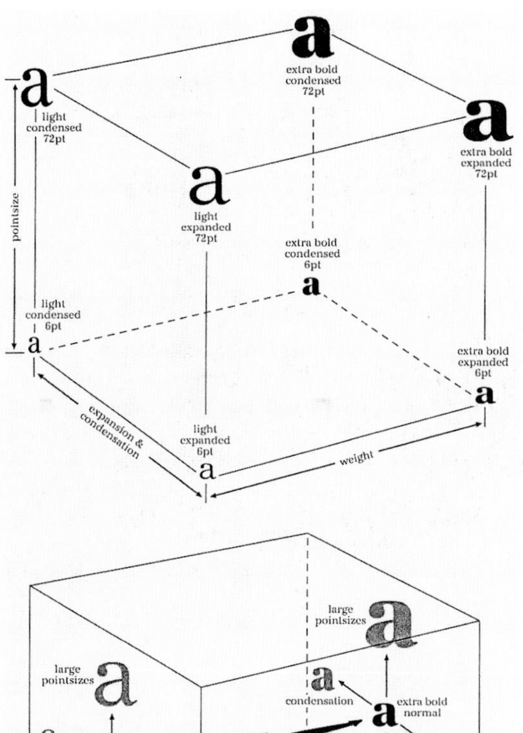


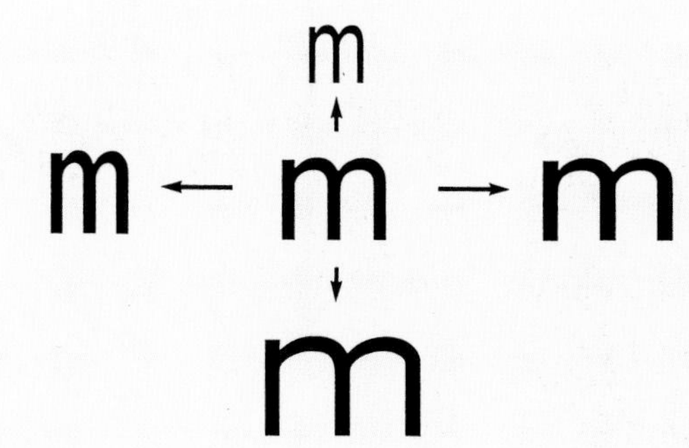

Karow 1992 Demonstration

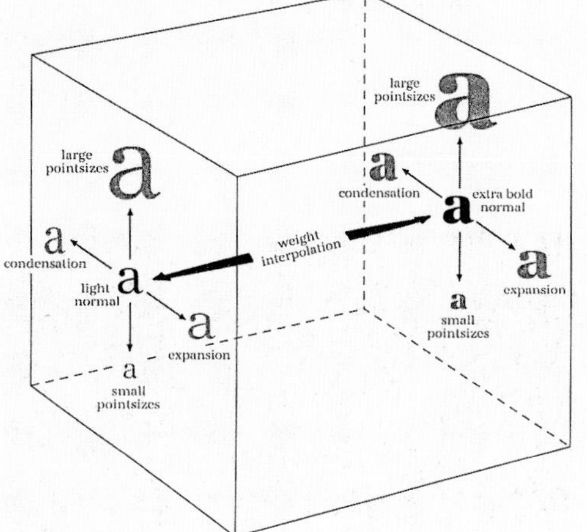


Karow 1992 Demonstration

aesthetics of calligraphy. – *Fonts*: Faust-Antiqua (1958), Leipziger-Antiqua (1959), Clarendon-Antiqua (1965), Prillwitz-Antiqua (1971), Magna Kyrillisch (1975). – *Publications include:* "Deutsche Schriftkunst", Dresden 1955; "Schriftkunst", Dresden 1971; "Gestaltung und Funktion der Typografie" (with Walter Schiller), Leipzig 1977; "Johannes Gutenberg", Leipzig 1977; "Schrift- und Buchkunst", Leipzig 1982.

**Karow,** Peter – b. 11. 11. 1940 in Stargard, Pomerania, Germany (now Poland) – *type producer, developer of type programs* – 1962–68: studies physics at the University of Hamburg. 1972: begins programming the Ikarus system. 1975: develops a program which automatically adjusts the spaces between letters (Kerning). 1981: programs the Linus system for Stempel, Linotype's type production. 1983: URW and ITC work together on the production of typefaces. 1992: programs the hz typesetting program with Hermann Zapf. 1995–96: works for Adobe in San Francisco. – *Publications include:* "Digitale Schriften", Berlin 1992; "Schriftstatistik", Hamburg 1992;

**Kassák,** Lajos – b. 21. 3. 1887 in Ersekújvár, Hungary, d. 22. 7. 1967 in Budapest, Hungary – *author, typographer, artist* – Trained as a metalworker. 1909: has contacts with Apollinaire, Delaunay and Picasso. Publishes "A Tett" (The Deed) magazine which is banned. 1916: launches the magazine and artists' group "Ma" (Today). 1922: spends time in Berlin. Writes his first Dada poems. 1926: returns to Budapest. 1927: co-founder of the ring

ten wird. 1916 Gründung der Zeitschrift und der Künstlergruppe „Ma“ (Heute). 1919 Emigration nach Wien. 1922 Aufenthalt in Berlin. Erste dadaistische Gedichte. 1926 Rückkehr nach Budapest. 1927 Mitbegründer des „rings neuer werbegestalter“ (mit Schwitters, Tschichold u. a.). Leiter der Zeitschrift „Dokumentum“. 1928–38 Herausgeber der Zeitschrift „Munka“ (Arbeit). Nach dem Verbot seiner literarischen Arbeit 1950 Wiederbeginn des künstlerischen Werks mit abstrakten Kompositionen und surrealistischen Collagen. – *Publikationen u.a.:* „Lével Kun Bélához a Müvészet Nevében“ (Ein Brief an Béla Kun im Namen der Kunst), Budapest 1919; „Buch neuer Künstler“ (mit László Moholy-Nagy), Wien 1922; „Az vj Müvészet él“ (Die neue Kunst lebt), Kluj 1926; „Mesterek Köszöntése“ (Gruß an die Meister), Budapest 1965; „Az Izmusok Története“ (Die Geschichte der Ismen), Budapest 1969.

**Käufer,** Josef – geb. 19. 5. 1890 in Beilngries, Deutschland, gest. 9. 3. 1966 in München, Deutschland. – *Typograph, Lehrer* – 1903–07 Schriftsetzerlehre in Regensburg. 1907–11 Wanderjahre. 1909 Eintritt in die „Typographische Gesellschaft München“. 1921–22 Betriebsleiter in Altötting. 1922–26 Akzidenzsetzer in München. 1927 Freundschaft mit Paul Renner. Berufung als Fachlehrer an die neugegründete „Meisterschule für Deutschlands Buchdrucker“ in München. 1951 Ernennung zum Professor. – *Publikationen u.a.:* „Was weißt Du vom Buchdruck“, München 1944; „Der Buchdruck“, Söcking 1949; „Das Setzerlehrbuch“,

est interdite. En 1916, il fonde la revue et le groupe d'artistes «Ma» (Aujourd'hui). Séjour à Berlin en 1922. Premiers poèmes Dada. Retour à Budapest en 1926. Cofondateur du «ring neuer werbegestalter» (Cercle des nouveaux graphistes publicitaires) en 1927 (avec Schwitters, Tschichold, etc.). Dirige la revue «Dokumentum». De 1928 à 1938, il publie la revue «Munka» (Travail). En 1950, après la censure de son œuvre littéraire, il reprend le travail artistique et réalise des compositions abstraites et des collages surréalistes. – *Publications, sélection:* «Lével Kun Bélához a Müvészet Nevében» (Lettre à Béla Kun au nom de l'art), Budapest 1919; «Buch neuer Künstler» (avec László Moholy-Nagy), Vienne 1922; «Az vj Müvészet él» (Le nouvel art vit), Kluj 1926; «Mesterek Köszöntése» (Salut aux Maîtres), Budapest 1965; «Az Izmusok Története» (L'histoire des «ismes»), Budapest 1969.

**Käufer,** Josef – né le 19. 5. 1890 à Beilngries, Allemagne, décédé le 9. 3. 1966 à Munich, Allemagne – *typographe, enseignant* – 1903–1907, apprentissage de composition typographique à Ratisbonne. 1907–1911, années de compagnonnage. Entre à la «Typographische Gesellschaft München» (société typographique de Munich) en 1909. Chef d'atelier à Altötting de 1921 à 1922. Compositeur sur Akzidenz à Munich de 1922 à 1926. Rencontre Paul Renner en 1927. Nommé enseignant à la nouvelle «Meisterschule für Deutschlands Buchdrucker» (école allemande de l'imprimerie) de Munich. Professeur en 1951. – *Publications, sélection:* «Was weißt Du vom Buchdruck», Munich 1944; «Der Buchdruck», Söcking 1949; «Das Setzerlehrbuch», Stuttgart 1956; «Vom Zaubergarten der Typographie», Munich 1960. Erwin Käufer «Vita Educativa. Josef Käufer. Eine Erinnerung», Munich 1991.

Kassak 1922 Cover

Kassak 1932 Cover

Kassak 1922 Invitation Card

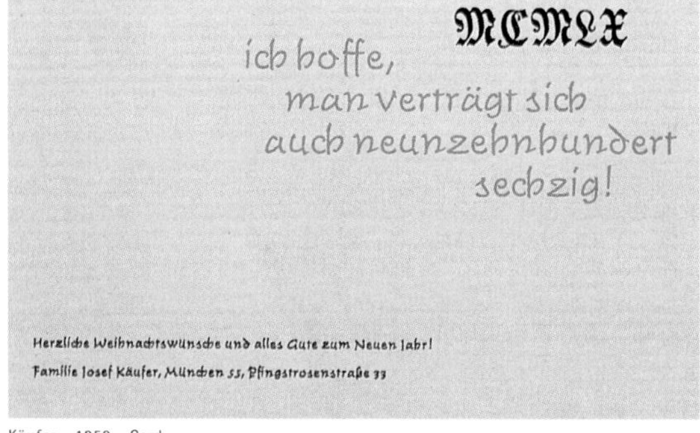


Käufer 1959 Card

neuer werbegestalter with Schwitters and Tschichold, among others. Director of "Dokumentum" magazine. 1928–38: editor of "Munka" (Work) magazine. 1950: after his literary work is banned, Kassák again turns to art, producing abstract compositions and surreal collages. – *Publications include:* "Lével Kun Bélához a Müvészet Nevében" (A Letter to Béla Kun in the Name of Art), Budapest 1919; "Buch neuer Künstler" (with László Moholy-Nagy), Vienna 1922; "Az vj Müvészet él" (New Art Lives), Kluj 1926; "Mesterek Köszöntése" (Greetings to the Masters), Budapest 1965; "Az Izmusok Története" (The History of Isms), Budapest 1969.

**Käufer,** Josef – b. 19. 5. 1890 in Beilngries, Germany, d. 9. 3. 1966 in Munich, Germany – *typographer, teacher* – 1903–07: trains as a typesetter in Regensburg. 1907–11: years of travel. 1909: joins the Typographische Gesellschaft in Munich. 1921–22: production manager in Altötting. 1922–26: job typesetter in Munich. 1927: friendship with Paul Renner. Appointed specialist teacher at the new Meisterschule für Deutschlands Buchdrucker in Munich. 1951: made a professor. – *Publications include:* "Was weißt Du vom Buchdruck", Munich 1944; "Der Buchdruck", Söcking 1949; "Das Setzerlehrbuch", Stuttgart 1956; "Vom Zaubergarten der Typographie", Munich 1960. Erwin Käufer "Vita Educativa. Josef Käufer. Eine Erinnerung", Munich 1991.

**Kauffer,** Edward McKnight – b. 14. 12. 1890 in Great Falls, USA, d. 22. 10. 1954 in New York, USA – *graph-*

Stuttgart 1956; „Vom Zaubergarten der Typographie", München 1960. Erwin Käufer „Vita Educativa. Josef Käufer. Eine Erinnerung", München 1991.

**Kauffer,** Edward McKnight – geb. 14. 12. 1890 in Great Falls, USA, gest. 22. 10. 1954 in New York, USA. – *Grafik-Designer, Bühnenmaler, Innenarchitekt* – 1910–12 Abendkurse in Malerei am Mark Hopkins Institute in San Francisco. 1912 Studium der Malerei, Anatomie und Schriftgestaltung am Art Institute of Chicago. Übernimmt 1912 den Namen „McKnight" von seinem Patron Prof. Joseph McKnight. 1913 Reisen nach Deutschland, Italien und Algerien. 1913–14 Studium der Malerei an der Académie Moderne in Paris. 1914–40 freier Gestalter in London. Auftraggeber waren u. a. London Underground Railways, „Daily Herald", das Museum of London, Shell-Mex und British Petroleum sowie Textilunternehmen. 1925 Buchillustrationen für Burtons „Anatomy of Melancholy" der Nonsuch-Press. Als erster Designer 1936 als „Royal Designer for Industry of the Royal Society of Arts" ausgezeichnet. 1937 Retrospektiv-Ausstellung im Museum of Modern Art in New York. 1941 Umzug nach New York, Auftraggeber waren u. a. die Container Corporation of America, Office of Civilian Defense, American Red Cross, die American Airlines und die Verlage Knopf und Random House. 1955 Gedenkausstellung im Victoria & Albert Museum in London. – *Publikationen u.a.:* „The art of the poster", London 1924; „Advertising art now", New York 1941; „Posters by E. McKnight Kauf-

Kauffer 1935 Cover

Kauffer 1934 Poster

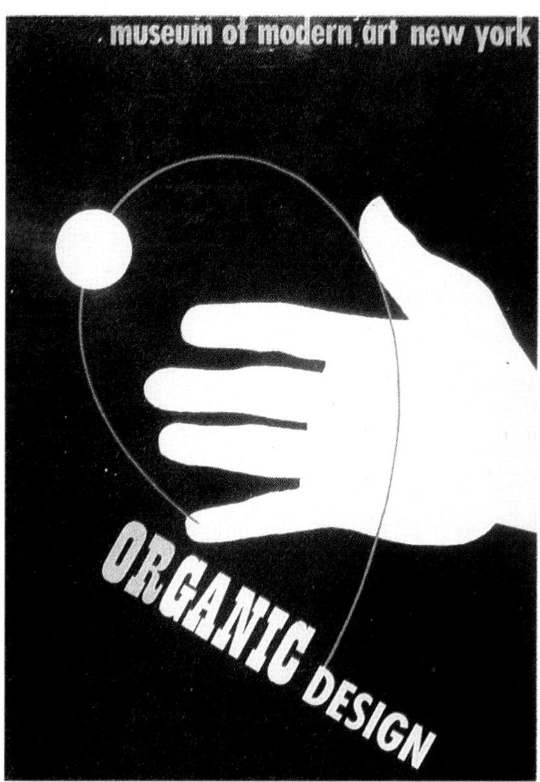

Kauffer 1941 Cover

**Kauffer,** Edward McKnight – né le 14. 12. 1890 à Great Falls, Etats-Unis, décédé le 22. 10. 1954 à New York, Etats-Unis – *graphiste maquettiste, scénographe, architecte d'intérieur* – 1910–1912, cours du soir de peinture au Mark Hopkins Institute de San Francisco. En 1912, il étudie la peinture, l'anatomie et l'art de l'écriture à l'Art Institute of Chicago. En 1912, il prend le nom de McKnight en l'honneur de son professeur Joseph McKnight. 1913, voyage en Allemagne, Italie et Algérie. En 1913–1914, il étudie la peinture à l'Académie Moderne à Paris. Designer indépendant à Londres de 1914 à 1940. Commanditaires : London Underground Railways, «Daily Herald», Museum of London, Shell-Mex, British Petroleum et plusieurs entreprises du textile. En 1925, il réalise des illustrations pour l' «Anatomy of Melancholy» de Burton, paru aux Nonsuch-Press. En 1936, il est le premier designer à recevoir le titre de «Royal Designer for Industry of the Royal Society of Arts». Exposition rétrospective au Museum of Modern Art de New York en 1937. S'installe à New York en 1941. Commanditaires : Container Corporation of America, Office of Civilian Defense, American Red Cross, American Airlines, les éditions Knopf, Random House, etc. Exposition posthume au Victoria & Albert Museum de Londres en 1955. – *Publications, sélection :* «The art of the poster», Londres 1924; «Advertising art now», New York 1941; «Posters by E. McKnight Kauffer», Londres 1925;

*ic designer, scene painter, interior designer* – 1910–12: takes evening classes in painting at the Mark Hopkins Institute in San Francisco. 1912: studies painting, anatomy and type design at the Art Institute of Chicago. 1912: adopts the pseudonym McKnight from his patron, Prof. Joseph McKnight. 1913: travels to Germany, Italy and Algeria. 1913–14: studies painting at the Académie Moderne in Paris. 1914–40: freelance designer in London. Clients include London Underground Railways, the "Daily Herald", the Museum of London, Shell-Mex and British Petroleum, and various textile companies. 1925: is responsible for the book illustrations in Burton's "Anatomy of Melancholy", published by Nonesuch Press. 1936: he is the first designer to be made a Royal Designer for Industry by the Royal Society of Arts. 1937: retrospective exhibition at the Museum of Modern Art in New York. 1941: moves to New York. Clients include the Container Corporation of America, the Office of Civilian Defense, the American Red Cross, American Airlines, and the publishers Knopf and Random House. 1955: commemorative exhibition in the Victoria & Albert Museum in London. – *Publications include:* "The art of the poster", London 1924; "Advertising art now", New York 1941; "Posters by E. McKnight Kauffer", London 1925; "Posters by E. McKnight Kauffer" (intro. by Aldous Huxley), New York 1937; "Memorial Exhibition of the Work of E. McKnight Kauffer", London 1955; "E. McKnight Kauffer: Poster Art 1915–1940" (intro. by Paul Rand), Lon-

fer", London 1925; „Posters by E. McKnight Kauffer" (Einf. von Aldous Huxley), New York 1937; „Memorial Exhibition of the work of E. McKnight Kauffer", London 1955; „E. McKnight Kauffer: Poster Art 1915–1940" (Einf. von Paul Rand), London 1973. Mark Haworth-Booth „E. McKnight Kauffer: A designer and his public", London 1979.

**Kawara,** On – geb. 2. 1. 1933 in Aichi-Ken, Japan. – *Künstler* – Autodidakt. 1952 figurative Gemälde. 1953 Skulpturen. 1958 Architekturstudium in Mexiko. Seit 1965 in New York. 1966 Beginn der Serien „Date Paintings" (Bilder mit dem Datum des Tages, an dem sie entstanden sind) und „I read" (Ausschnitte aus der Tagespresse der Tage, an denen ein „Date Painting" gemalt wurde). 1968 Beginn der Serien „I went" (Fotokopien von Landkarten mit den Markierungen des zurückgelegten Weges am betreffenden Tag), „I met" (Verzeichnis der Namen der am bestimmten Tag getroffenen Personen) und „I got up". – *Publikationen u.a.:* „I am still alive", Berlin 1978; „Continuity, Discontinuity", Stockholm 1980; „On Kawara. Date Paintings in 89 cities", Rotterdam 1991; „Wieder und wieder", Frankfurt am Main 1992.

**Keedy,** Jeffery – geb. 29. 8. 1957 in Battle Creek, USA. – *Grafik-Designer, Schriftentwerfer, Lehrer* – Grafik-Design-Studium und Studium der Fotografie an der Western Michigan University in Kalamazoo, Michigan, Grafik-Design-Studium an der Cranbrook Academy of Art in Bloomfield Hills, Michigan. Arbeit als Grafik-Designer 1981–82 in Boston,

«Posters by E. McKnight Kauffer» (préface d'Aldous Huxley), New York 1937; «Memorial Exhibition of the work of E. McKnight Kauffer», Londres 1955; «E. McKnight Kauffer: Poster Art 1915–1940» (préface de Paul Rand), Londres 1973; Mark Haworth-Booth «E. McKnight Kauffer: A designer and his public», Londres 1979.

**Kawara,** On – né le 2. 1. 1933 à Aichi-Ken, Japon – *artiste* – autodidacte. Tableaux figuratifs en 1952, sculptures en 1953. Etudie l'architecture à Mexico en 1958. S'installe à New York en 1965. Commence en 1966 la série «Date Paintings» (tableaux avec la date du jour où ils ont été réalisés) et «I read» (coupures de la presse quotidienne du jour où il a peint un «Date Painting»). Puis, en 1968, il entame la série «I went» (photocopies de cartes comportant la distance parcourue le jour indiqué), puis «I met» (liste des noms des personnes rencontrées un certain jour) et «I got up». – *Publications, sélection:* «I am still alive», Berlin 1978; «Continuity, Discontinuity», Stockholm 1980; «On Kawara. Date Paintings in 89 cities», Rotterdam 1991; «Wieder und wieder», Francfort-sur-le-Main 1992.

**Keedy,** Jeffery – né le 29.8.1957 à Battle Creek, Etats-Unis – *graphiste maquettiste, concepteur de polices, enseignant* – études d'arts graphiques, de design et de photographie à la Western Michigan University de Kalamazoo, Michigan. Etudes d'arts graphiques et de design à la Cranbrook Academy of Art, à Bloomfield Hill, Michigan. Travaille comme graphiste maquettiste à Boston de 1981 à 1982, puis de 1982 à 1983 chez Clarence Lee Design and Associates Inc. à Honolulu. Graphiste maquettiste indépendant à Chicago et Bloomfield Hill de 1983 à 1985, puis à Los Angeles depuis 1985. A enseigné les arts graphiques et le design à l'Eastern Mich-

| | n | n | | | |
|---|---|---|---|---|---|
| a | b | e | f | g | i |
| o | r | s | t | y | z |
| A | B | C | E | G | H |
| M | O | R | S | X | Y |
| 1 | 2 | 4 | 6 | 8 | & |

Keedy 1989 Keedy Sans

Keedy 1988 Postcard

SOZIO-
LOGIE
UND
LEBEN

Keidel 1952 Cover

JULIUS RODENBERG
GRÖSSE UND
GRENZEN DER
TYPOGRAPHIE
Betrachtungen über typographische Grundfragen, wie sie sich in der Buch-Kunst der letzten sieben Jahrzehnte widerspiegeln
C. E. POESCHEL VERLAG STUTTGART

Keidel 1959 Cover

don 1973. Mark Haworth-Booth "E. McKnight Kauffer: A designer and his public", London 1979.

**Kawara,** On – b. 2. 1. 1933 in Aichi-Ken, Japan – *artist* – Self-taught. 1952: figurative paintings. 1953: sculptures. 1958: studies architecture in Mexico. 1965: moves to New York. 1966: starts executing his series entitled "Date Paintings" (pictures showing the date they were painted) and "I read" (press cuttings from the day a "Date Painting" was painted). In 1968, he began the series "I went" (photocopies of maps with his route of that day marked), "I met" (a list of names of people met on a particular day) and "I got up". – *Publications include:* "I am still alive", Berlin 1978; "Continuity, Discontinuity", Stockholm 1980; "On Kawara. Date paintings in 89 cities", Rotterdam 1991; "Wieder und wieder", Frankfurt am Main 1992.

**Keedy,** Jeffery – b. 29. 8. 1957 in Battle Creek, USA – *graphic designer, type designer, teacher* – Studied graphic design and photography at Western Michigan University in Kalamazoo, Michigan. Studied graphic design at the Cranbrook Academy of Art in Bloomfield Hills, Michigan. 1981–82: works as a graphic designer in Boston, from 1982–83 for Clarence Lee Design and Associates Inc. in Honolulu, from 1983–85 as a freelance graphic designer in Chicago and Bloomfield Hills and from 1985 onwards in Los Angeles. Teaches graphic design at Eastern Michigan University (1984), at the Otis/Parson's School of Design (1987), and at the California Institute of the Arts

1982–83 bei Clarence Lee Design and Associates Inc. in Honolulu, 1983–85 freier Grafik-Designer in Chicago und Bloomfield Hills, seit 1985 in Los Angeles. Unterrichtet 1984 Grafik-Design an der Eastern Michigan University, 1987 an der Otis/Parson's School of Design und seit 1985 am California Institute of the Arts. 1991–96 Programmdirektor des Grafik-Design-Programms. 1995 Gründung seiner Schriftenhersteller-Firma „Cipher". Zusammenarbeit mit Künstlern und Kunstinstitutionen, u.a. dem Museum of Contemporary Art in Los Angeles, dem Santa Monica Museum of Art und dem Pacific Design Center. Veröffentlicht Essays und Interviews in Zeitschriften. – *Schriftentwürfe:* Neo Theo (1989), Keedy Sans (1989), Hard Times (1990).

**Keidel,** Carl – geb. 19. 8. 1902 in Stuttgart, Deutschland, gest. 10. 4. 1981 in Stuttgart, Deutschland. – *Typograph* – Setzer- und Buchdruckerlehre. Zwei Jahre Studium an der Staatlichen Akademie der Bildenden Künste in Stuttgart. 1945 Entwurf des Titels der neugegründeten „Stuttgarter Zeitung". Für seine typographischen Arbeiten erhält Keidel 1950–80 59 Auszeichnungen der Stiftung Buchkunst in Frankfurt am Main. 1967 Verleihung des Professoren-Titels durch die Landesregierung Baden-Württemberg. Herausgabe der von ihm gestalteten großen Lutherbibel, die in der „Original alten Schwabacher" gesetzt wird.

**Keller,** Ernst – geb. 15. 6. 1891 in Villingen, Schweiz, gest. 4. 11. 1968 in Zürich, Schweiz. – *Grafik-Designer, Typograph, Bildhauer, Heraldiker, Lehrer* – 1906

Ernst Keller 1928 Poster

Ernst Keller 1935 Poster

Ernst Keller 1947 Poster

igan University en 1984, à l'Otis/Parson's School of Design en 1987 et au California Institute of the Arts depuis 1985. De 1991 à 1996 il est directeur du Programme d'arts graphiques et de design. Fonde «Cipher», une société de fabrication de fontes, en 1995. Collaboration avec des artistes et des institutions artistiques, entre autres le Museum of Contemporary Art de Los Angeles, le Santa Monica Museum of Art, le Pacific Design Center. Publications d'essais et d'interviews dans diverses revues. – *Polices:* Neo Theo (1989), Keedy Sans (1989), Hard Times (1990).

**Keidel,** Carl – né le 19. 8. 1902 à Stuttgart, Allemagne, décédé le 10. 4. 1981 à Stuttgart, Allemagne – *typographe* – apprentissage de composition typographique et d'imprimerie. Deux années d'études à la Staatliche Akademie der Bildenden Künste (beaux-arts) de Stuttgart. En 1945, il crée le titre d'un nouveau quotidien le «Stuttgarter Zeitung». De 1950 à 1980, Keidel reçoit 59 distinctions de la Stiftung Buchkunst (fondation des arts du livre) de Francfort-sur-le-Main pour ses travaux typographiques. En 1967, le gouvernement de Bade-Wurtemberg lui décerne le titre de professeur. Réalisation de la maquette de la grande Bible de Luther qui est composée en «Original alte Schwabacher».

**Keller,** Ernst – né le 15. 6. 1891 à Villingen, Suisse, décédé le 4. 11. 1968 à Zurich, Suisse – *graphiste maquettiste, typographe, sculpteur, héraldiste, enseignant* – 1906, apprentissage de la lithographie

(from 1985 onwards; director of the graphic design program from 1991–96). 1995: launches his type manufacturing company, Cipher. Works with various artists and art institutes, among them the Museum of Contemporary Art in Los Angeles, the Santa Monica Museum of Art and the Pacific Design Center. Has had essays and interviews published in various magazines. – *Fonts:* Neo Theo (1989), Keedy Sans (1989), Hard Times (1990).

**Keidel,** Carl – b. 19. 8. 1902 in Stuttgart, Germany, d. 10. 4. 1981 in Stuttgart, Germany – *typographer* – Trained as a typesetter and printer. Studied for two years at the Staatliche Akademie der Bildenden Künste in Stuttgart. 1945: designs the title of the new "Stuttgarter Zeitung". 1950–80: Keidel receives a total of 59 awards from the Stiftung Buchkunst in Frankfurt am Main for his typographic work. 1967: made a professor by the state government of Baden-Württemberg. An edition of a large-format Luther bible designed by Keidel is published, set in "Original alte Schwabacher".

**Keller,** Ernst – b. 15. 6. 1891 in Villingen, Switzerland, d. 4. 11. 1968 in Zurich, Switzerland – *graphic designer, typographer, sculptor, heraldist, teacher* – 1906: trains as a lithographer in Aarau.

Lehre als Lithograph in Aarau. Arbeitet 1911–14 in der „Werkstatt für deutsche Wortkunst" in Leipzig. 1914 Rückkehr in die Schweiz. 1916–30 Mitglied des Schweizer Werkbunds. 1917 Weiterbildung bei dem Dekorationsmaler Robert Hunziker in der Gewerbeschule Aarau. 1918 Berufung an die Kunstgewerbeschule in Zürich. Erste Plakatentwürfe. Ab 1920 Fachlehrer für angewandte Grafik an der Kunstgewerbeschule in Zürich. Seit 1925 zahlreiche Entwürfe für die Züricher Stadtverwaltung. 1938 Schrift- und Wappenreliefs für die Schweizer Lebensversicherungs- und Rentenanstalt in Zürich. 1939 Entwurf und Ausführung einer Schriftwand im Raum für protestantische Kirchenkunst auf der Schweizer Landesausstellung. 1956 Pensionierung, Hinwendung zur Bildhauerei. Durch seine Lehrtätigkeit bildet Keller Generationen von Gestaltern aus und wird „Vater der Schweizer Grafik" genannt. Seine Arbeitsbereiche sind das typographische Plakat, Signets und die grafische Gestaltung in der Architektur. – *Publikation:* „Ernst Keller Grafiker. 1891–1968 Gesamtwerk", Zürich 1976.

**Keller,** Peter – geb. 1. 2. 1944 in Basel, Schweiz. – *Typograph, Lehrer* – 1964 Abschluß des Studiums an der Allgemeinen Gewerbeschule Basel. 1965 Arbeit in verschiedenen Druckereien. Kurse bei Emil Ruder in typographischer Konzeption. Unterrichtet 1966 an der Académie Populaire d'Arts Plastiques in Paris. 1968 Gestalter in der Agentur Gerstner, Gredinger & Kutter (GGK) in Basel. 1969 Berufung an die Ecole Nationale Supérieure

à Aarau. 1911–1914, travaille à la «Werkstatt für deutsche Wortkunst» (atelier allemand de l'art du mot) à Leipzig. 1914, retour en Suisse. 1916–1930, membre du Werkbund suisse. En 1917, il suit une formation avec le peintre décorateur Robert Hunziker à la Gewerbeschule (école des arts et métiers) d'Aarau. 1918, nommé à la Kunstgewerbeschule (école des arts décoratifs) de Zurich. Premières affiches. A partir de 1920, professeur d'arts graphiques appliqués à la Kunstgewerbeschule de Zurich. A partir de 1925, nombreux travaux pour la municipalité de Zurich. En 1938, il réalise des reliefs avec textes et armoiries pour les assurances et des maisons de retraite de Zurich. En 1939, élaboration et exécution d'un mur scriptural dans la salle d'art religieux protestant lors de l'exposition nationale de Suisse. Départ en retraite en 1956. Se consacre à la sculpture. Grâce à son enseignement, Keller a formé des générations de créateurs, il est considéré comme le «Père du graphisme suisse». Il a travaillé dans les domaines de l'affiche typographique, du signet et de la création graphique en architecture. – *Publication :* «Ernst Keller Grafiker. 1891–1968 Gesamtwerk», Zurich 1976.

**Keller,** Peter – né le 1. 2. 1944 à Bâle, Suisse – *typographe, enseignant* – termine ses études à la Allgemeine Gewerbeschule (école des arts et métiers) de Bâle en 1964. Travaille dans diverses imprimeries en 1965. Suit les cours de conception typographique d'Emil Ruder. En 1966, il enseigne à l'Académie Populaire d'Arts Plastiques de Paris. Exerce en 1968 comme maquettiste à l'agence Gerstner, Gredinger & Kutter (GGK) à Bâle. 1969, nommé à l'Ecole Nationale Supérieure des Arts Décoratifs (ENSAD) de Paris où il dirige les cours de typographie. 1970, ouvre un atelier à Paris. 1972, cours à



Peter Keller 1975 Cover

Peter Keller 1987 Cover

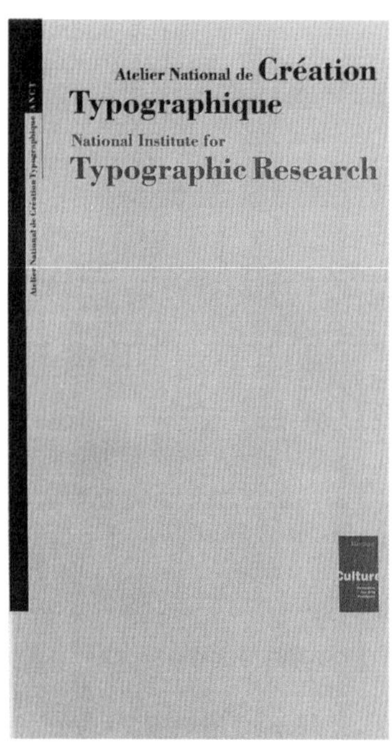


Peter Keller 1992 Cover

Kelley ca. 1969 Logo

1911– 14: works at the Werkstatt für deutsche Wortkunst in Leipzig. 1914: returns to Switzerland. 1916–30: member of the Schweizer Werkbund. 1917: continues his training with the interior decorator Robert Hunziker at the Gewerbeschule in Aarau. 1918: teaching position at the Kunstgewerbeschule in Zurich. Produces his first poster designs. From 1920 onwards: specialist teacher of applied graphics at the Kunstgewerbeschule in Zurich. From 1925 onwards: numerous designs for Zurich's city authorities. 1938: designs relief lettering and trademarks for the Schweizer Lebensversicherungs- und Rentenanstalt in Zurich. 1939: designs and executes a wall of lettering for the room of Protestant church art at the Swiss national exhibition. 1956: retires and turns to sculpture. As a teacher, Keller trained generations of artists and is known as the "father of Swiss graphics". His fields of work were the typographic poster, logos and graphic design in architecture. – *Publication*: "Ernst Keller Grafiker. 1891–1968 Gesamtwerk", Zurich 1976.

**Keller,** Peter – b.1. 2. 1944 in Basle, Switzerland – *typographer, teacher* – 1964: completes his studies at the Allgemeine Gewerbeschule in Basle. 1965: works in various printing workshops. Takes course in typographic conception under Emil Ruder. 1966: teaches at the Académie Populaire d'Arts Plastiques in Paris. 1968: designer for the Gerstner, Gredinger & Kutter (GGK) agency in Basle. 1969: appointed to teach at the Ecole Nationale Supérieure des Arts Dé-

des Arts Décoratifs (ENSAD) in Paris, Leitung der Kurse für Typographie. 1970 Eröffnung seines Studios in Paris. 1972 Kurse an der Ecole Supérieure d'Arts Graphiques (ESAG) in Paris. 1974 Mitarbeit als typographischer Berater bei Roger Tallon in Paris. 1981 Gastvorlesungen an den Fachhochschulen Coventry, Leicester und Stoke-on-Trent in England. 1984 Design-Berater der Ecole Nationale Supérieure des Arts et Métiers (ENSAM) in Paris. Projektberater der Publikationsabteilung der UNESCO. Unterrichtet 1985 am Institut des Arts Visuels (IAV) in Orléans, Mitglied des Ausbildungsgremiums des IAV und Koordinator der Abteilung Grafik. 1986 Berufung in den Wissenschaftsrat der ENSAD in Paris. 1990 Ernennung zum Direktor des Atelier National de Création Typographique (ANCT) durch das Ministerium für Kultur.

**Kelley,** Alton – geb. 17. 6. 1940 in Houlton, USA. – *Grafik-Designer, Illustrator* – Kurzes Industrie-Design-Studium am Philadelphia Museum College of Art und an der Art Students League in New York. Danach zwei Jahre Flugzeugmechaniker in Stratford, USA. 1965 Gründungsmitglied der „Family Dog", einer Gruppe von Freunden in San Francisco, die in der Bay Area Rock & Roll-Veranstaltungen organisieren. Entwurf zahlreicher Plakate, Handzettel und Anzeigen für Aktivitäten von „Family Dog". 1966 Freundschaft mit Stanley Mouse, gemeinsames Studio „Mouse & Kelley". 1968 Teilnahme an der Ausstellung „The Joint Show" in der Moore Gallery in San Francisco (mit Rick Griffin, Victor Moscoso, Stanley Mouse,

Kelley 1969 Poster

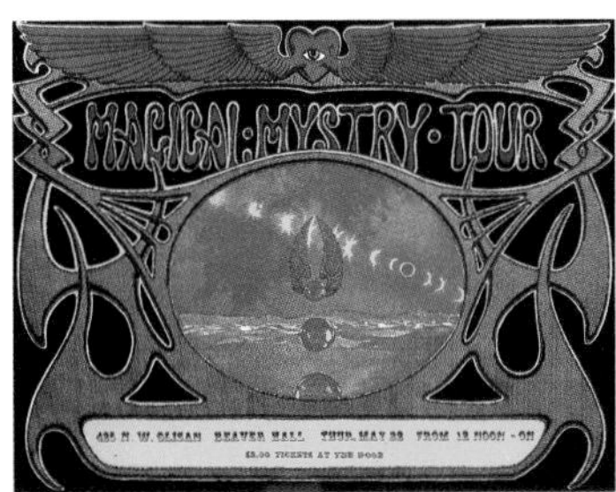


Kelley 1969 Poster

Kelley 1982 Poster

l'Ecole Supérieure d'Arts Graphiques (ESAG) de Paris. 1974, conseiller typographe chez Roger Tallon à Paris. 1981, cycle de conférences dans les écoles spécialisées de Coventry, Leicester et Stoke-on-Trent en Angleterre. Conseiller en design à l'Ecole Nationale Supérieure des Arts et Métiers (ENSAM) à Paris, en 1984. Conseiller du service des publications de l'UNESCO. En 1985, il enseigne à l'Institut des Arts Visuels (IAV) d'Orléans, membre de la commission pédagogique de l'IAV et coordinateur de la section d'arts graphiques. En 1986, il est nommé au conseil scientifique de l'ENSAD à Paris. En 1990, le Ministère de la Culture le nomme directeur de l'Atelier National de Création Typographique (ANCT).

**Kelley,** Alton – né le 17. 6. 1940 à Houlton, Etats-Unis – *graphiste maquettiste, illustrateur* – brèves études de design industriel au Philadelphia Museum College of Art et à l'Art Students League de New York. Travaille ensuite pendant deux ans comme mécanicien en aéronautique à Stratford. Membre fondateur en 1965 de «Family Dog», un groupe d'amis qui organise des concerts de Rock'n Roll dans la Bay Area à San Francisco. Réalise de nombreuses affiches, prospectus et annonces pour les activités de «Family Dog». Rencontre Stanley Mouse en 1966, puis création de l'atelier commun «Mouse & Kelley». Participe à l'exposition «The Joint Show» à la Moore Gallery de San Francisco en 1968 (avec Rick Griffin, Victor Moscoso, Stanley Mouse et Wes Wilson). Kelley séjourne en Oregon et dans le Massachusetts à partir de 1968,

coratifs (ENSAD) in Paris. Runs the typography courses. 1970: opens his studio in Paris. 1972: courses at the Ecole Supérieure d'Arts Graphiques (ESAG) in Paris. 1974: works as typography consultant for Roger Tallon in Paris. 1981: gives lectures at the polytechnics in Coventry, Leicester and Stoke-on-Trent in England. 1984: design consultant to the Ecole Nationale Supérieure des Arts et Métiers (ENSAM) in Paris. Project consultant to the UNESCO publications department. 1985: teaches at the Institut des Arts Visuels (IAV) in Orléans. Member of the IAV's education committee and coordinator of the graphics department. 1986: asked to join ENSAD's scientific and advisory committee in Paris. 1990: made director of the Atelier National de Création Typographique (ANCT) by the Ministry of Culture.

**Kelley,** Alton – b. 17. 6. 1940 in Houlton, USA – *graphic designer, illustrator* – Brief period of study (industrial design) at the Philadelphia Museum College of Art and at the Art Students League in New York, followed by two years as an aeromechanic in Stratford, USA. 1965: founder member of Family Dog, a group of friends in San Francisco who organize rock-and-roll events in the Bay Area. Designs numerous posters, handbills and advertisements for Family Dog activities. 1966: friendship with Stanley Mouse. Joint studio (Mouse & Kelley). 1968: takes part in "The Joint Show" exhibition at the Moore Gallery in San Francisco with Rick Griffin, Victor Moscoso, Stanley Mouse and Wes Wilson. From 1968 Kelley lived

Wes Wilson). Kelley lebt nach 1968 in Oregon und Massachusetts, bevor er sich 1971 endgültig in der San Francisco Bay niederläßt. Elemente aus Comics, Jugendstil und Op Art charakterisieren die Arbeiten Kelleys, der zu den einflußreichsten Plakat-, T-Shirt- und Schallplattencovergestaltern der Flower-Power-Zeit gehört. – *Publikation u. a.:* „Mouse & Kelley", Limpsfield 1979.

**Kelmscott Press** – *Druckerei, Verlag* – Aus seiner Vorliebe für mittelalterliche Drucke und Schriften wächst bei William Morris der Wunsch, eine eigene Handpresse zu gründen. 1891 Gründung der Kelmscott Press durch William Morris in Hammersmith (London). Als erstes Buch erscheint „The story of the glittering plain" von Morris. Anschließend „The stones of Venice" von John Ruskin (1892), „Sonnets and lyrical poems" von Dante Gabriel Rossetti (1894), „The Canterbury Tales" und „The works of Geoffrey Chaucer" mit Holzschnitten von Edward Burne-Jones (1896). Morris verwendet eigene Schriftentwürfe, die er auf der Grundlage von Schriften des 14. und 15. Jahrhunderts entwirft: Golden Type (1889), Troy Type (1891), Chaucer Type (1892). Schriftschneider ist E. P. Prince. Insgesamt werden 53 Publikationen in Auflagen von 200–300 Exemplaren (einmal 500 Exemplaren) in der Kelmscott Press herausgegeben. Die Presse ist Vorbild für Gestalter in Europa und Amerika, sie löst eine Privatpressen-Bewegung aus. 13 Bücher werden von E. Burne-Jones illustriert, weitere Illustratoren sind Walter Crane, L. M. Gere, A. J. Gaskin. Morris entwirft

avant de s'installer définitivement dans la baie de San Francisco en 1971. Des éléments empruntés à la bande dessinée, au style Art Nouveau et au Op Art caractérisent le style de Kelley; il compte parmi les plus importants créateurs d'affiches, de dessins de T-Shirts et de pochettes de disques de la génération «Flower-Power». – *Publication, sélection:* «Mouse & Kelley», Limpsfield 1979.

**Kelmscott Press** – *imprimerie, éditions* – l'intérêt que William Morris portait aux imprimés et aux écrits du Moyen Age a éveillé en lui le désir de créer sa propre presse manuelle. En 1891, William Morris fonde la Kelmscott Press à Hammersmith (Londres). Le premier livre qui paraît est «The story of the glittering plain» de Morris. Puis viennent «The stones of Venice» de John Ruskin (1892), «Sonnets and lyrical poems» de Dante Gabriel Rossetti (1894), «The Canterbury Tales» et «The works of Geoffrey Chaucer» avec des gravures sur bois d'Edward Burne-Jones (1896). Morris utilise ses propres fontes dessinées en s'inspirant de caractères du 14e et du 15e siècle: Golden Type (1889), Troy Type (1891), Chaucer Type (1892). E. P. Prince se charge de la taille des types. Au total, 53 publications tirées à 200–300 exemplaires (une fois 500) sortiront de la Kelmscott Press. Elle fait figure de modèle pour les créateurs européens et américains et déclenche un mouvement de création de presses privées. E. Burne-Jones illustrera 13 ouvrages, d'autres illustrateurs comme Walter Crane, L. M. Gere, A. J. Gaskin y ont également travaillé. Morris a dessiné 384 lettrines, 57 vignettes, 1 082 éléments décoratifs et crée 28 pages de garde et titres pour les publications de ses presses. «Love is enough» a été le dernier livre imprimé et publié par Morris. La Kelmscott Press ferme en mars 1898.

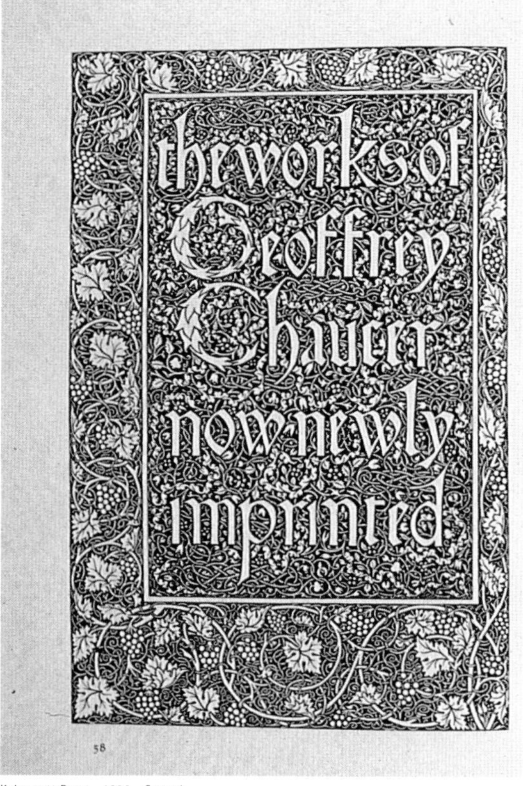

Kelmscott Press 1896 Spread

Kelmscott Press 1892 Trademark

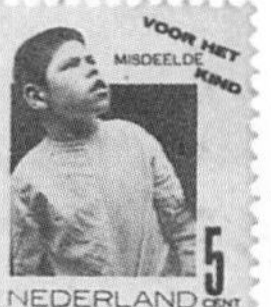

Kiljan 1931 Stamps

in Oregon and Massachusetts before setting in San Francisco Bay in 1971. Kelley's work is characterized by elements from comics, Art Nouveau and Op Art, and Kelley himself is considered one of the most influential poster, T-shirt and album cover designers of the Flower Power era. – *Publications include:* "Mouse and Kelley", Limpsfield 1979.

**Kelmscott Press** – *printing press, publishing house* – William Morris' love of medieval prints and manuscripts prompted him to found his own hand-printing press. 1891: William Morris launches the Kelmscott Press in Hammersmith (London). The first book to be published is "The story of the glittering plain" by Morris. This is followed by "The stones of Venice" by John Ruskin (1892), "Sonnets and lyrical poems" by Dante Gabriel Rossetti (1894), "The Canterbury Tales" and "The works of Geoffrey Chaucer" with woodcuts by Edward Burne-Jones (1896). Morris used his own typefaces which he based on scripts from the 14th and 15th centuries: Golden Type (1889), Troy Type (1891) and Chaucer Type (1892). E.P. Prince was Morris' type founder. The Kelmscott Press produced a total of 53 publications in editions of 200–300 (plus one edition of 500). The press was a model for designers in Europe and America and lead to a wave of private printing presses being established. E. Burne-Jones illustrated 13 books for Morris; further illustrators were Walter Crane, L. M. Gere and A. J. Gaskin. Morris designed a total of 384 initials, 57 borders, 1,082 ornaments and 28 covers and pages for the

insgesamt 384 Initialen, 57 Einfassungen, 1.082 Zierstücke, 28 Titel- und Schriftseiten für die Publikationen seiner Presse. Als letztes Buch wird „Love is enough" von Morris fertig gedruckt und herausgegeben. Im März 1898 wird die Kelmscott Press aufgelöst.

**Kiljan,** Gerard – geb. 26. 10. 1891 in Hoorn, Niederlande, gest. 21. 11. 1968 in Leidschendam, Niederlande. – *Typograph, Fotograf, Lehrer* – 1904–07 Ausbildung an der Kunstnijverheid-Teekenschool Quellinus in Amsterdam. 1909–11 Lithographenlehre in Amsterdam. 1909–16 Abendkurse an der Rijksacademie voor Beeldende Kunsten in Amsterdam. 1911–14 Retuscheur in Amsterdam. 1915–18 Dozent an technischen Schulen in Amsterdam, 1918–30 an der Academie van Beeldende Kunsten in Den Haag, 1920–30 an der Academie voor Beeldende Kunsten in Rotterdam. 1926 typographische Entwürfe für verschiedene Unternehmen. 1928 fotografische Arbeiten. 1929 Teilnahme an der internationalen Ausstellung „Film und Foto" in Stuttgart mit Typo-Foto-Arbeiten. Mitarbeit an der Zeitschrift „De 8 en Opbouw". 1930–59 Leiter der Abteilung Reklame-Entwurf an der Academie van Beeldende Kunsten in Den Haag. Entwürfe für die niederländische Post PTT. 1950 Leiter der Abteilung Industrielle Formgebung an der Academie van Beeldende Kunsten in Den Haag.

**Kindersley,** David – geb. 11. 6. 1915 in Codicote, England, gest. 2. 2. 1995 in Cambridge, England. – *Schriftentwerfer, Kalligraph* – 1933–36 künstlerische Ausbildung durch Eric Gill in Pigotts. 1936–39

Kiljan 1931 Poster

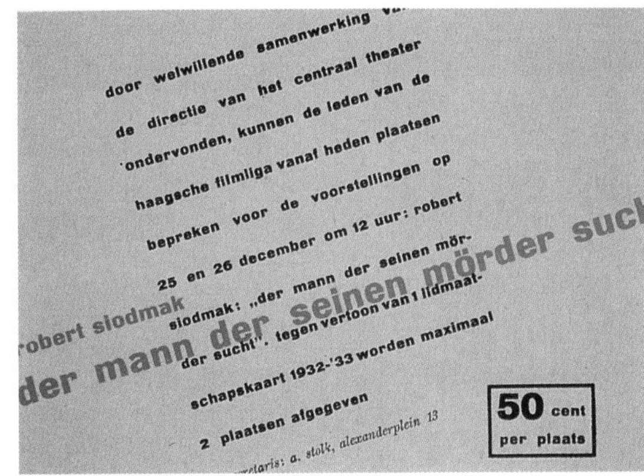

Kiljan 1933 Advertisement

Kiljan 1932 Advertisement

Kiljan 1938 Advertisement

**Kiljan,** Gerard – né le 26. 10. 1891 à Hoorn, Pays-Bas, décédé le 21. 11. 1968 à Leidschendam, Pays-Bas – *typographe, photographe, enseignant* – 1904–1907, formation à la Kunstnijverheid-Teekenschool Quellinus d'Amsterdam. 1909–1911, apprentissage de lithographie à Amsterdam. 1909–1916, cours du soir à la Rijksacademie voor Beeldende Kunsten d'Amsterdam. Retoucheur à Amsterdam de 1911 à 1914. Enseigne dans des écoles techniques d'Amsterdam de 1915 à 1918, à l'Academie van Beeldende Kunsten de La Haye de 1918 à 1930, puis à l'Academie voor Beeldende Kunsten de Rotterdam de 1920 à 1930. Divers travaux typographiques pour des entreprises en 1926. Travaux photographiques en 1928. Participe à l'exposition internationale «Film und Foto» à Stuttgart, en 1929, où il présente des «typo-photos». Il collabore à la revue «De 8 en Opbouw». Directeur du service de la section de graphisme publicitaire à l'Academie van Beeldende Kunsten de La Haye de 1930 à 1959. Travaux pour les postes néerlandaises PTT. Directeur de la section de design industriel à l'Academie van Beeldende Kunsten de La Haye en 1950.

**Kindersley,** David – né le 11. 6. 1915 à Codicote, Angleterre, décédé le 2. 2. 1995 à Cambridge, Angleterre – *concepteur de polices, calligraphe* – 1933–1936, formation artistique auprès d'Eric Gill à Pigotts. Travaille à l'atelier d'Eric Gill de 1936 à

publications of his press. The last book to be produced was "Love is enough", printed and published by Morris. The Kelmscott Press was disbanded in March, 1898.

**Kiljan,** Gerard – b. 26. 10. 1891 in Hoorn, The Netherlands, d. 21. 11. 1968 in Leidschendam, The Netherlands – *typographer, photographer, teacher* – 1904–07: trains at the Kunstnijverheid-Teekenschool Quellinus in Amsterdam. 1909–11: trains as a lithographer in Amsterdam. 1909–16: attends evening classes at the Rijksacademie voor Beeldende Kunsten in Amsterdam. 1911–14: retoucher in Amsterdam. 1915–18: lectures at various technical schools in Amsterdam, from 1918–30 at the Academie van Beeldende Kunsten in The Hague, from 1920–30 at the Academie voor Beeldende Kunsten in Rotterdam. 1926: fashions typographic designs for various companies. 1928: photographic work. 1929: takes part in the international "Film und Foto" exhibition in Stuttgart with his typo-photo work. Works on the magazine "De 8 en Opbouw". 1930–59: head of the advertising design department at the Academie van Beeldende Kunsten in The Hague. Produces designs for the Dutch post office PTT. 1950: head of the department for industrial design at the Academie van Beeldende Kunsten in The Hague.

**Kindersley,** David – b. 11. 6. 1915 in Codicote, England, d. 2. 2. 1995 in Cambridge, England – *type designer, calligrapher* – 1933–36: trains as an artist under Eric Gill in Pigotts. 1936–39: works in Eric Gill's studio. 1939: opens his own

Arbeiten in Eric Gills Studio. 1939 Eröffnung seines Studios in Cambridge. 1946 Gründung einer Werkstatt in Barton bei Cambridge mit Kevin Cribb. 1952 Alphabet für die Straßenbeschilderung für das englische Transportministerium. 1972 Gründung des Verlags „Cardozo Kindersley Editions" mit seiner Frau Lida Lopes Cardozo. Bis 1986 Berater der Firma Letraset in Schriftfragen. Entwicklung des Programms „Logos" zur mechanischen Spationierung von Alphabeten. – *Schriftentwürfe:* Mot Serif (1961), Octavian (mit Will Carter, 1961), Kindersley (1962), Itek Bookface (1976). – *Publikationen u.a.:* „Optical letter spacing and its mechanical application", London 1966; „Mr. Eric Gill", Los Angeles 1967; „Variations on the theme of 26 letters", London 1969; „Optical letter spacing for new printing systems", London 1976; „Letters slate cut" (mit Lida Lopes Cardozo), Cambridge 1981; „David Kindersley 1915–1995", Cambridge 1995.

**Kinneir,** Jock (Richard) – geb. 11. 2. 1917 in Aldershot, England, gest. 23. 8. 1994 in Oxford, England. – *Grafik-Designer, Lehrer* – 1935–39 Studium der Malerei und Gravur an der Chelsea School of Art in London. 1945 Gründung seines Ateliers als Grafik-Designer. Ausstellungsgestalter für das Central Office of Information in London. 1958 gemeinsames Atelier mit Margaret Calvert. Entwürfe für das Leitsystem des Flughafens in Gatwick. 1964 Entwürfe für das Leitsystem aller englischen Straßen. 1964–69 Leiter der Grafik-Design-Abteilung des Royal College of Art in London. Lehrtätigkeit

1939, ouvre ensuite son propre atelier en 1939 à Cambridge. En 1946, il fonde avec Kevin Cribb un atelier à Barton, près de Cambridge. En 1952, il conçoit un alphabet destiné à la signalisation routière pour le ministère britannique des Transports. Fonde les «Cardozo Kindersley Editions» avec sa femme Lida Lopes Cardozo en 1972. Conseiller en écriture de la société Letraset jusqu'en 1986. Elabore le programme «Logos» pour modifier automatiquement l'approche entre les lettres d'alphabets. – *Polices:* Mot Serif (1961), Octavian (avec Will Carter, 1961), Kindersley (1962), Itek Bookface (1976). – *Publications, sélection:* «Mr. Eric Gill», Los Angeles 1967; «Variation on the theme of 26 letters», Londres 1969; «Optical letter spacing for new printing systems», Londres 1976; «Letters slate cut» (avec Lida Lopez Cardozo), Cambridge 1981; «David Kindersley 1915–1995», Cambridge 1995.

**Kinneir,** Jock (Richard) – né le 11. 2. 1917 à Aldershot, Angleterre, décédé le 23. 8. 1994 à Oxford, Angleterre – *graphiste maquettiste, enseignant* – 1935–1939, étudie la peinture et la gravure à la Chelsea School of Art à Londres. Fonde son atelier de graphiste maquettiste en 1945. Architecte d'expositions pour le Central Office of Information de Londres. En 1958, il travaille avec Margaret Calvert dans leur atelier commun. Conception du système de signalisation de l'aéroport de Gatwick. En 1964, il conçoit le système de signalisation pour toutes les routes anglaises. 1964–1969, directeur de la section d'arts graphiques et de design du Royal College of Art de Londres, enseigne jusqu'en 1974. – *Publication, sélection:* «Words and buildings», Londres 1980.

**Kinross,** Robin – né le 19. 2. 1949 à Chipperfield, Angleterre – *écrivain, typographe, éditeur, enseignant* – 1968–1971,

Kindersley 1970 Slate Cut

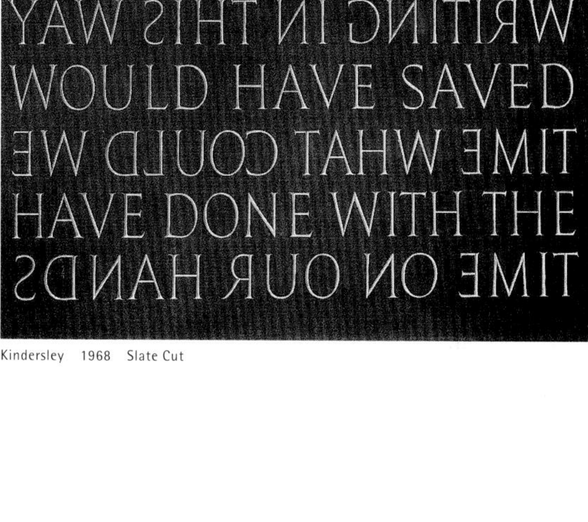


Kindersley 1968 Slate Cut

| | n | | | | *n* |
|---|---|---|---|---|---|
| a | b | e | f | g | i |
| o | r | s | t | y | z |
| A | B | C | E | G | H |
| M | O | R | S | X | Y |
| 1 | 2 | 4 | 6 | 8 | & |

Kindersley 1961 Octavian

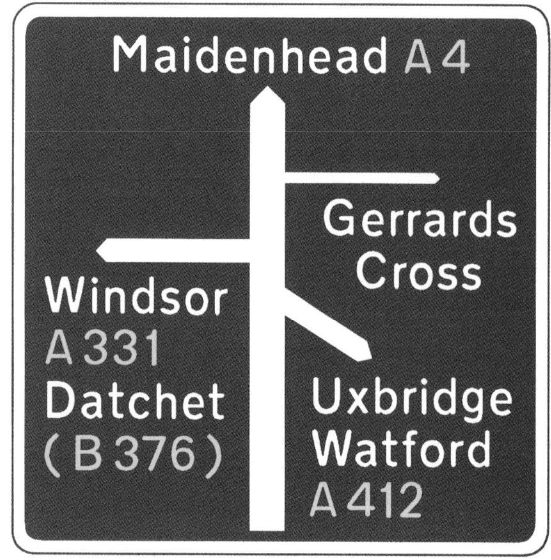


Kinneir ca. 1970 Sign

studio in Cambridge. 1946: opens a workshop in Barton near Cambridge with Kevin Cribb. 1952: produces the alphabet for road signs for the British Ministry of Transport. 1972: founds Cardozo Kindersley Editions publishing house with his wife, Lida Lopes Cardozo. Until 1986: consultant to the Letraset company on typographic matters. Develops the Logos program which mechanically spaces alphabets. – *Fonts:* Mot Serif (1961), Octavian (with Will Carter, 1961), Kindersley (1962), Itek Bookface (1976). – *Publications include:* "Mr. Eric Gill", Los Angeles 1967; "Variations on the theme of 26 letters", London 1969; "Optical letter spacing for new printing systems", London 1976; "Letters slate cut" (with Lida Lopez Cardozo), Cambridge 1981; "David Kindersley 1915–1995", Cambridge 1995.

**Kinneir,** Jock (Richard) – b. 11. 2. 1917 in Aldershot, England, d. 23. 8. 1994 in Oxford, England- *graphic designer, teacher* – 1935–39: studies painting and engraving at the Chelsea School of Art in London. 1945: opens his own studio as a graphic designer. Exhibition planner for the Central Office of Information in London. 1958: opens a joint studio with Margaret Calvert. Produces designs for Gatwick Airport's signage system. 1958–61: graphic consultant for the super-highway sign system. 1964: produces designs for the direction sign systems on all English roads. 1964–69: head of the graphic design department at the Royal College of Art in London. Teaches until 1974. – *Publications include:* "Words and buildings", London 1980.

bis 1974. – *Publikation u. a.:* „Words and buildings", London 1980.

**Kinross,** Robin – geb. 19. 2. 1949 in Chipperfield, England. – *Schriftsteller, Typograph, Verleger, Lehrer* – 1968–71 Studium der englischen Literatur am North-West London Polytechnic. 1971–75 Studium der Typographie und der grafischen Kommunikation an der University of Reading in England. 1979 Dissertation über Otto Neurath und Isotype. Unterrichtet 1977–82 Typographie und typographische Geschichte an der University of Reading. 1980 Gründung der Hyphen Press. Seit 1982 Arbeiten als Verleger und Schriftsteller im Bereich Typographie und Visuelle Kommunikation. Zahlreiche Beiträge, u. a. für „Information Design Journal" (seit 1979), „Blueprint" (1987–90), „Journal of Design History" (1988–92), „Eye" (seit 1990). – *Publikationen u. a.:* „Otto Neurath: Gesammelte bildpädagogische Schriften" (Mitherausgeber), Wien 1991; „Modern typography: an essay in critical history", London 1992.

**Kippenberg,** Anton – geb. 22. 5. 1874 in Bremen, gest. 21. 9. 1950 in Luzern, Schweiz. – *Verleger* – 1905 Leitung des Insel Verlags in Leipzig (gemeinsam mit Carl Ernst Poeschel). 1906 alleiniger Leiter des Insel Verlags. Gründung und erste Ausgabe des jährlich erscheinenden „Insel-Almanach". 1908–16 Kippenbergs Verlag übernimmt den Vertrieb der Drucke der Ernst-Ludwig-Presse sowie der Trajanus-Presse und der Cranach-Presse. Gründet 1912 die „Insel-Bücherei".

**Kis,** Miklós (Nicholas) – geb. 1650 in Alsó-Miszttótfalu, Ungarn, gest. 20. 3.

APOLOGIA
BIBLIORUM
Anno 1684. Amſtelodami impreſſorum,
ut &
ORTHOGRAPHIÆ
in iis obſervatæ:
In Tres Partes diviſa.
I. Epiſtolam Apologeticam, in qua utplurimum tractatur de alterationibus quibusdam, quæ in ea Editione contigerunt, earumque Generibus, cum Exemplis.
II. Catalogum vocum ibi omiſſarum, hic reſtitutarum: ubi interim ſignantur (duntaxat) Loca de neceſſariis mutationibus, vel aliis difficultatibus.
III. Ratiocinationem de Orthographia eo modo inſtituenda.
Ad multorum præjudicia de iis concepta medendum cum coactione à nobis invitis expreſſa.
Per
NICOLAUM KIS de M.TÓTFALU.
CLAUDIOPOLI
Anno 1697.

Kis 1697 Main Title

Kinneir ca. 1970 Sign

SZENT
DAVID
KIRALYNAK
és
PROFETANAK
SZAZÖTVEN
SOLTARI
A' FRANCZIAI Nóćáknak és Verſeknek módjokra
MAGYAR Verſekre fordiaaaak és rendeltseek.
Szentzi MOLNAR ALBERT
által.

Mellyek most ujjolag kinyomtattak
M. TOTFALUSI
KIS MIKLOS
által.
AMSTELODAMBAN.
MDCLXXXVL eſztendöben.

Kis 1665 Main Title

études de littérature anglaise au North-West London Polytechnics. Etudie la typographie et la communication visuelle à l'université de Reading, en Angleterre, de 1971 à 1975. Thèse de doctorat sur Otto Neurath et les Isotypes en 1979. Enseigne la typographie et l'histoire de la typographie de 1977 à 1982 à l'université de Reading. Fonde la Hyphen Press en 1980. Exerce depuis 1982 comme éditeur et écrivain dans le domaine de la typographie et de la communication visuelle. Nombreux articles, entre autres pour «Information Design Journal» (depuis 1979), «Blueprint» (1987–1990), «Journal of Design History» (1988–1992), «Eye» (depuis 1990). – *Publications, sélection:* «Otto Neurath: Gesammelte bildpädagogische Schriften» (co-éditeur), Vienne 1991; «Modern typography: an essay in critical history», Londres 1992.

**Kippenberg,** Anton – né le 22. 5. 1874 à Brême, Allemagne, décédé le 21. 9. 1950 à Lucerne, Suisse – *éditeur* – direction de la maison d'édition Insel à Leipzig (en même temps que Carl Ernst Poeschel) en 1905. Dirige seul le Insel Verlag en 1906. Entre 1908 et 1916, les éditions Kippenberg reprennent la diffusion des publications de la Ernst-Ludwig-Presse, de la Trajanus-Presse et de la Cranach-Presse. En 1912, il crée la «Insel-Bücherei».

**Kis,** Miklós (Nicholas) – né en 1650 à Alsó-Miszttótfalu, Hongrie, décédé le

**Kinross,** Robin – b. 19. 2. 1949 in Chipperfield, England – *author, typographer, publisher, teacher* – 1968–71: studies English literature at North-West London Polytechnic. 1971–75: studies typography and graphic communication at the University of Reading in England. 1979: dissertation on Otto Neurath and Isotype. 1977–82: teaches typography and the history of typography at the University of Reading. 1980: founds Hyphen Press. Kinross has worked as a publisher and author in the field of typography and visual communication since 1982. He has made numerous contributions to various magazines, including "Information Design Journal" (since 1979), "Blueprint" (1987–90), "Journal of Design History" (1988–92) and "Eye" (since 1990). – *Publications include:* "Otto Neurath: Gesammelte bildpädagogische Schriften" (co-editor), Vienna 1991; "Modern typography: an essay in critical history", London 1992.

**Kippenberg,** Anton – b. 22. 5. 1874 in Bremen, Germany, d. 21. 9. 1950 in Lucerne, Switzerland – *publisher* – 1905: management of the Insel publishing house in Leipzig with Carl Ernst Poeschel. 1906: sole director of Insel. 1908–16: Kippenberg's publishing house takes over the sales of prints produced by the Ernst-Ludwig-Presse, Trajanus-Presse and Cranach-Presse. 1912: founds the Insel library.

**Kis,** Miklós (Nicholas) – b. 1650 in Alsó-Miszttótfalu, Hungary, d. 20. 3. 1702 in Kolozsvár, Hungary (now Cluj, Romania) – *punch cutter, type designer, teacher* –

1702 in Kolozsvár, Ungarn (heute Cluj, Rumänien). – *Stempelschneider, Schriftentwerfer, Lehrer* – Theologiestudium in Nagyenyeder. Die reformierte Kirche Ungarns gibt ihm 1680 den Auftrag, in den Niederlanden die erste Bibel in ungarischer Sprache drucken zu lassen. Umzug nach Amsterdam. Für die „Amsterdamer Bibel" schneidet er 1683–85 eine Schrift, die später als die Janson Antiqua berühmt wird. Seit 1683 als Schriftgießer in Amsterdam tätig. 1689 Rückkehr nach Siebenbürgen. 1991 Eröffnung des Misztótfalusi Kis Miklós Museums. – *Publikation u. a.:* György Haiman „Tótfalusi Kis Miklós", Budapest 1972.

**Kisman,** Max – geb. 26. 1. 1953 in Doetinchem, Niederlande. – *Grafik-Designer, Schriftentwerfer, Lehrer* – 1972–77 Studium an der Akademie in Enschede und an der Gerrit Rietveld Akademie in Amsterdam. 1986 Mitbegründer der experimentellen Zeitschrift „Typ, Typografisch Papier", einer alternativen Zeitschrift über Typographie und Kunst. Unterrichtet Grafik-Design an der HKV in Utrecht und an der Gerrit Rietveld Akademie in Amsterdam. 1991 Schriftentwürfe für FontShop International und die digitale Zeitschrift „Fuse". Zusammenarbeit mit der Grafik-Design-Gruppe „Wild Plakken" in Amsterdam. Zahlreiche Auszeichnungen, u.a. mit dem H. N. Werkman-Preis der Stadt Amsterdam. Umzug nach San Francisco. – *Schriftentwürfe:* Tegentonen 88 (1988) und 89 (1989), Ramblas, Cattle Brand (1990), Jacque, Vortex, Rosetta, Scratch, Linear Construct (1991), SSP Quickstep (1994).

20. 3. 1702 à Kolozsvár, Hongrie (aujourd'hui Chij, Romanie) – *tailleur de types, concepteur de polices, enseignant* – études de théologie à Nagyenyeder. En 1680, l'église réformée de Hongrie le charge de faire imprimer en Hollande la première Bible en hongrois. S'installe à Amsterdam. Entre 1683 et 1685, il taille les caractères de la «Bible d'Amsterdam», caractères qui deviendront célèbres sous le nom de Janson Antiqua. Exerce comme fondeur de caractères à Amsterdam à partir de 1683. Retour en Transylvanie en 1689. Le musée Kis Miklós de Misztótfalusi est inauguré en 1991. – *Publication, sélection :* György Haiman «Tótfalusi Kis Miklós», Budapest 1972.

**Kisman,** Max – né le 26. 1. 1953 à Doetinchem, Pays-Bas – *graphiste maquettiste, concepteur de polices, enseignant* – 1972–1977, études à l'académie d'Enschede et à la Gerrit Rietveld Akademie à Amsterdam. Cofondateur, en 1986, de la revue expérimentale «Typ, Typografisch Papier», une revue alternative de typographie et d'art. Enseigne les arts graphiques et le design à la HKV d'Utrecht et à la Gerrit Rietveld Akademie d'Amsterdam. En 1991, il dessine des polices pour FontShop International et la revue numérique «Fuse». Collabore avec le groupe de graphistes-designers «Wild Plakken» à Amsterdam. Nombreuses distinctions, dont le prix H. N. Werkman de la ville d'Amsterdam. S'installe à San Francisco. – *Polices :* Tegentonen 88 (1988) und 89 (1989), Ramblas, Cattle Brand (1990), Jacque, Vortex, Rosetta, Scratch, Linear Construct (1991), SSP Quickstep (1994).

**Klemm,** Walter – né le 18. 6. 1883 à Karlsbad, Allemagne, décédé le 11. 8. 1957 à Weimar, Allemagne – *artiste, graphiste, enseignant* – études d'histoire de l'art à l'université de Vienne en 1901 puis for-

Kisman 1985 Poster

Kisman 1986 Poster

Kisman 1988 Poster

Kisman 1991 Scratch

Kisman 1991 Vortex

Kisman 1987 Stamp

Studied theology in Nagyenyeder. 1680: the Hungarian Reformed Church commissions him to print the first bible in Hungarian in The Netherlands. Moves to Amsterdam. 1683–85: Kis cuts a typeface for his Amsterdam Bible which later becomes famous as Janson Antiqua. From 1683 onwards: type founder in Amsterdam. 1689: returns to Transylvania. 1991: the Misztótfalusi Kis Miklós Museum opens. – *Publications include:* György Haiman "Tótfalusi Kis Miklós", Budapest 1972.

**Kisman,** Max – b. 26. 1. 1953 in Doetinchem, The Netherlands – *graphic designer, type designer, teacher* – 1972–77: studies at the academy in Enschede and at the Gerrit Rietveld Akademie in Amsterdam. 1986: co-founder of the expermental "Typ, Typografisch Papier", an alternative magazine on art and typography. Teaches graphic design at the HKV in Utrecht and at the Gerrit Rietveld Akademie in Amsterdam. 1991: designs type for FontShop International and digital "Fuse" magazine. Works with the Wild Plakken graphic design group in Amsterdam. Numerous awards including Amsterdam's H. N. Werkman Prize. Moves to San Francisco. – *Fonts:* Tegentonen 88 (1988) und 89 (1989), Ramblas, Cattle Brand (1990), Jacque, Vortex, Rosetta, Scratch, Linear Construct (1991), SSP Quickstep (1994).

**Klemm,** Walter – b. 18. 6. 1883 in Karlsbad, Germany, d. 11. 8. 1957 in Weimar, Germany – *artist, graphic artist, teacher* – 1901: studies art history at the university in Vienna and trains at the Kunst-

**Klemm,** Walter – geb. 18. 6. 1883 in Karlsbad, Deutschland, gest. 11. 8. 1957 in Weimar, Deutschland. – *Künstler, Grafiker, Lehrer* – 1901 Studium der Kunstgeschichte an der Universität Wien, Ausbildung an der Wiener Kunstgewerbeschule und Akademie. 1907–08 Aufenthalt in der Künstlerkolonie Dachau. 1913–19 und ab 1921 Professor und Leiter der grafischen Abteilung der Staatlichen Hochschule für Bildende Kunst in Weimar. 1919–21 Formmeister der grafischen Druckerei am Bauhaus Weimar. 1946 Leiter der Fachklasse für freie Grafik in Weimar.

**Kleukens,** Christian Heinrich – geb. 7. 6. 1880 in Achim, Deutschland, gest. 7. 4. 1954 in Darmstadt, Deutschland. – *Drukker, Typograph, Lehrer* – Lehre als Setzer und Buchdrucker in der Steglitzer Werkstatt in Berlin. Typographische Ausbildung an der Akademie für Graphische Künste und Buchgewerbe in Leipzig. 1914–27 Leiter der Ernst-Ludwig-Presse in Darmstadt, 1927–45 in Mainz. 1919–23 Gründung und Leitung der Kleukenspresse (mit Rudolf G. Binding) in Frankfurt am Main. 1929–48 Professor an der Staatsschule für Kunst und Handwerk in Mainz. – *Schriftentwürfe:* Judith-Schrift (1923), Plinius-Schrift (1923), Mainzer Antiqua (1927), Burte-Fraktur (1928), Goethe-Antiqua (1930), Hesiod-Type (1933), Adam Karillon-Schrift (1937). – *Publikation u. a.:* „Die Kunst der Letter", Leipzig 1940.

**Kleukens,** Friedrich Wilhelm – geb. 7. 5. 1878 in Achim, Deutschland, gest. 22. 8. 1956 in Nürtingen, Deutschland. –

Kisman 1987 Poster

Friedrich Wilhelm Kleukens 1926 Scriptura

DA MAN DIE TAPFERKEIT DIESER FÜR UNVERGLEICHLICH ERACHTETE, GAB MAN IHNEN DORT AUF DEM SCHLACHTFELD IHR EIGENES GRAB. WENN SIE ABER DIE ERDE BEDECKT HAT, SPRICHT EIN VON DER STADT HIERZU ERHOBENER MANN, DER IM ANSEHEN STEHT AN EINSICHT UND WÜRDE HERVORZURAGEN, ZU IHREM LOBE WIE ES IHNEN ZUKOMMT. SO WERDEN SIE BESTATTET. DANACH ENTFERNT MAN SICH. WÄHREND DES GANZEN KRIEGES SO OFT ES DAZU KAM, HIELT MAN SICH AN DIESE SITTE. FÜR DIESE ERSTEN NUN WURDE PERIKLES DES XANTHIPPOS SOHN ERWÄHLT ZU REDEN. ALS ER DEN AUGENBLICK GEKOMMEN ERACHTETE TRAT ER, UM WEITHIN IN DER VERSAMMLUNG GEHÖRT ZU WERDEN, VOM GRABE HINWEG AUF EINEN HOCHTRITT DEN MAN ERRICHTET HATTE UND SPRACH ALSO:

DIE vielen, die vor mir von dieser Stätte aus gesprochen haben, preisen den der diese Rede des Gedenkens zum Gesetz erhoben hat; gleich als ob es genug der Ehre sei für die Gefallenen öffentlich zu reden. Mir aber würde es ehrenvoller erscheinen wenn Männer die durch die Tat sich geadelt haben, auch durch die Tat geehrt werden, wie sie ja in Gestalt dieser vom Staate selbst gerüsteten

7

Christian Heinrich Kleukens 1938 Double Spread

Friedrich Wilhelm Kleukens 1910 Initial

HYPERION AN BELLARMIN

DER liebe Vaterlandsboden gibt mir wieder Freude und Leid.

Ich bin jetzt alle Morgen auf den Höhn des Korinthischen Isthmus, und, wie die Biene unter Blumen, fliegt meine Seele oft hin und her zwischen den Meeren, die zur Rechten und zur Linken meinen glühenden Bergen die Füße kühlen.

Besonders der eine der beiden Meerbusen hätte mich freuen sollen ,wär' ich ein Jahrtausend früher hier gestanden.

Wie ein siegender Halbgott, wallte da zwischen der herrlichen Wildnis des Helikon und Parnaß, wo das Morgenrot um hundert überschneite Gipfel spielt, und zwischen der paradiesischen Ebene von Sicyon der glänzende Meerbusen herein, gegen die Stadt der Freude, das jugendliche Korinth, und schüttete den erbeuteten Reichtum aller Zonen vor seiner Lieblingin aus.

Aber was soll mir das? Das Geschrei des Jakals, der unter den Steinhaufen des Altertums sein wildes Grablied singt, schreckt ja aus meinen Träumen mich auf.

Wohl dem Manne, dem ein blühend Vaterland das Herz erfreut und stärkt! Mir ist, als würd' ich in den Sumpf geworfen, als schlüge man den Sargdeckel über mir zu, wenn einer an das meinige mich mahnt, und wenn mich

Friedrich Wilhelm Kleukens 1912 Page

mation à la Kunstgewerbeschule (école des arts décoratifs) et à l'académie de Vienne. 1907–1908, séjour à la colonie d'artistes de Dachau. De 1913 à 1919, puis à partir de 1921, professeur et directeur de la section d'arts graphiques à la Staatliche Hochschule für Bildende Kunst (école nationale supérieure des beaux-arts) de Weimar. 1919–1921, maître de forme à l'imprimerie du Bauhaus de Weimar. En 1946, il dirige le cours d'arts graphiques à Weimar.

**Kleukens,** Christian Heinrich – né le 7. 6. 1880 à Achim, Allemagne, décédé le 7. 4. 1954 à Darmstadt, Allemagne – *imprimeur, typographe, enseignant* – apprentissage de composition typographique et de l'imprimerie à la Steglitzer Werkstatt de Berlin. Formation de typographe à l'Akademie für Graphische Künste und Buchgewerbe (arts graphiques et arts du livre) de Leipzig. Dirige la Ernst-Ludwig-Presse de Darmstadt de 1914 à 1927, et de 1927 à 1945 de Mayence. Fonde et dirige la «Kleukenspresse» (avec Rudolf G. Binding) de 1919 à 1923. Professeur à la Staatsschule für Kunst und Handwerk (école d'art et d'artisanat) de Mayence de 1929 à 1948. – *Polices:* Judith-Schrift (1923), Plinius-Schrift (1923), Mainzer Antiqua (1927), Burte-Fraktur (1928), Goethe-Antiqua (1930), Hesiod-Type (1933), Adam Karillon-Schrift (1937). – *Publication, sélection:* «Die Kunst der Letter», Leipzig 1940.

**Kleukens,** Friedrich Wilhelm – né le 7. 5. 1878 à Achim, Allemagne, décédé le

gewerbeschule and academy in Vienna. 1907–08: spends time in the artist's colony in Dachau. 1913–19 and from 1921 onwards: professor and head of the graphics department at the Staatliche Hochschule für Bildende Kunst in Weimar. 1919–21: form master at the Bauhaus graphic printing workshop in Weimar. 1946: head of the advanced free graphics course in Weimar.

**Kleukens,** Christian Heinrich – b. 7. 6. 1880 in Achim, Germany, d. 7. 4. 1954 in Darmstadt, Germany – *printer, typographer, teacher* – Trained as a typesetter and printer at the Steglitzer Werkstatt in Berlin. Trained as a typographer at the Akademie für Graphische Künste und Buchgewerbe in Leipzig. 1914–27: director of the Ernst-Ludwig-Presse in Darmstadt, 1927–45 in Mainz. 1919–23: founds and runs the Kleukens press in Frankfurt am Main with Rudolf G. Binding. 1929–48: professor at the Staatsschule für Kunst und Handwerk in Mainz. – *Fonts:* Judith-Schrift (1923), Plinius-Schrift (1923), Mainzer Antiqua (1927), Burte-Fraktur (1928), Goethe-Antiqua (1930), Hesiod-Type (1933), Adam Karillon-Schrift (1937). – *Publications include:* "Die Kunst der Letter", Leipzig 1940.

**Kleukens,** Friedrich Wilhelm – b. 7. 5. 1878 in Achim, Germany, d. 22. 8. 1956 in Nürtingen, Germany – *typographer, type designer, illustrator, painter, teacher*

*Typograph, Schriftentwerfer, Illustrator, Maler, Lehrer* – Ausbildung an der Unterrichtsanstalt des Kunstgewerbemuseums in Berlin. 1900 Gründung der „Steglitzer Werkstatt“ (mit G. Belwe und F. H. Ehmcke). 1903 Berufung an die Staatliche Akademie für Buchgewerbe und Graphik in Leipzig. 1906 Berufung an die Künstlerkolonie nach Darmstadt, Ernennung zum Professor. 1907 Gründung der Ernst-Ludwig-Presse in Darmstadt. Trennt sich 1914 von der Ernst-Ludwig-Presse. 1919 Gründung der Ratio-Presse, in der zehn von Kleukens gestaltete Publikationen erscheinen. 1924–31 künstlerischer Beirat der D. Stempel AG in Frankfurt am Main. Ab 1930 mehrjähriger Aufenthalt auf Mallorca. Danach in Darmstadt und Frankfurt am Main als Maler und Illustrator tätig. – *Schriftentwürfe:* Ingeborg-Antiqua (1909), Kleukens-Antiqua (1910), Kleukens-Fraktur (1910), Helga-Antiqua (1911), Gotische Antiqua (1914), Ratio-Latein (1923).

**Klinger,** Julius – geb. 22. 5. 1876 in Wien, Österreich, gest. 1950 in Wien, Österreich. – *Maler, Grafiker, Schriftentwerfer* – Studium am Technologischen Gewerbemuseum in Wien. Ab 1895 Zeichner beim Frauenmagazin „Wiener Mode“, später Illustrator der „Meggendorfer Blätter“. 1896 Umzug nach München. Lebt und arbeitet 1897–1915 in Berlin. Zahlreiche grafische Arbeiten, vor allem Plakatgestaltung. Zusammenarbeit mit der Druckerei Hollerbaum und Schmidt. 1915 Rückkehr nach Wien. Eröffnung seines „Ateliers für Plakatkunst“. Veranstaltung

22. 8. 1956 à Nürtingen, Allemagne – *typographe, concepteur de polices, illustrateur, peintre, enseignant* – formation à la Unterrichtsanstalt des Kunstgewerbemuseums (institut d'enseignement du musée des arts décoratifs) de Berlin. En 1900, il fonde la «Steglitzer Werkstatt» avec G. Belwe et F. H. Ehmcke. 1903, nommé à la Staatliche Akademie für Buchgewerbe und Graphik (arts graphiques et arts du livre) de Leipzig. En 1906, il est muté à la colonie d'artistes de Darmstadt et nommé professeur. Fonde la Ernst-Ludwig-Presse en 1907 à Darmstadt. Se sépare de la Ernst-Ludwig-Presse en 1914. Fonde la Ratio-Presse en 1919 et édite dix ouvrages dont il réalise la maquette. Conseiller artistique de la D. Stempel AG de Francfort-sur-le-Main de 1924 à 1931. Séjourne plusieurs années à Mallorque à partir de 1930. Puis exerce à Darmstadt et à Francfort-sur-le-Main comme peintre indépendant et illustrateur. – *Polices:* Ingeborg-Antiqua (1909), Kleukens-Antiqua (1910), Kleukens-Fraktur (1910), Helga-Antiqua (1911), Gotische Antiqua (1914), Ratio-Latein (1923).

**Klinger,** Julius – né le 22. 5. 1876 à Vienne, Autriche – décédé en 1950 à Vienne, Autriche – *peintre, graphiste, concepteur de polices* – études au Technologisches Gewerbemuseum (Musée des arts et des techniques) de Vienne. Dessinateur au magazine féminin «Wiener Mode» à partir de 1895, puis illustrateur aux «Meggendorfer Blätter». S'installe à Munich en 1896. Vit et travaille à Berlin de 1897 à 1915. Nombreux travaux graphiques, surtout des affiches. Travaille avec l'imprimerie Hollerbaum und Schmidt. S'installe à Vienne en 1915. Ouvre son «Atelier für Plakatkunst». Donne des cours de graphisme industriel. – *Polices:* Klinger-Antiqua (1912), Klinger Type (1925).

Klinger 1919 Poster

Gebr.Klingspor 1925 Karl Klingspor Schrift

Klinger 1923 Poster

Klinger 1937 Poster

– Trained at the Unterrichtsanstalt des Kunstgewerbemuseums in Berlin. 1900: founds the Steglitzer Werkstatt with G. Belwe and F. H. Ehmcke. 1903: teaching position at the Staatliche Akademie für Buchgewerbe und Graphik in Leipzig. 1906: teaching position at the artist's colony in Darmstadt. Made a professor. 1907: the Ernst-Ludwig-Presse is founded in Darmstadt. 1914: leaves the Ernst-Ludwig-Presse. 1919: founds the Ratio-Presse which prints 10 publications designed by Kleukens. 1924–31: artistic advisor to D. Stempel AG in Frankfurt am Main. From 1930 onwards: spends several years in Majorca. He then works in Darmstadt and Frankfurt am Main as a painter and illustrator. – *Fonts*: Ingeborg-Antiqua (1909), Kleukens-Antiqua (1910), Kleukens-Fraktur (1910), Helga-Antiqua (1911), Gotische Antiqua (1914), Ratio-Latein (1923).

**Klinger,** Julius – b. 22. 5. 1876 in Vienna, Austria, d. 1950 in Vienna, Austria – *painter, graphic artist, type designer* – Studied at the Technologisches Gewerbemuseum in Vienna. From 1895 onwards: artist for “Wiener Mode” women's magazine, later illustrator for the “Meggendorfer Blätter”. 1896: moves to Munich. 1897–1915: lives and works in Berlin. Produces much graphic work, especially posters. Works with the Hollerbaum und Schmidt printing works. 1915: moves back to Vienna. Opens a poster studio, Atelier für Plakatkunst. Runs commercial art courses. – *Fonts:* Klinger-Antiqua (1912), Klinger Type (1925).

**Gebr. Klingspor** – *type foundry* – 1892:

von Kursen über Gebrauchsgrafik. – *Schriftentwürfe:* Klinger-Antiqua (1912), Klinger Type (1925).

**Gebr. Klingspor** – *Schriftgießerei* – Der Zigarrenfabrikant Carl Klingspor aus Gießen kauft 1892 die Rudhardsche Gießerei, die 1859 von Johann Peter Nees gegründet wurde. Otto Eckmann wird 1898 mit dem Entwurf einer neuen Schrift beauftragt. Peter Behrens entwirft 1902 seine „Behrens-Schrift" für Klingspor. Nach dem Tod des Vaters 1903 werden die Söhne Eigentümer des Unternehmens. 1906 Umwandlung des Firmennamens in „Gebr. Klingspor". Beginn der Arbeit von Rudolf Koch für das Unternehmen als künstlerischer Berater (bis 1934). Die Schriftgießerei D. Stempel AG in Frankfurt am Main erwirbt 1917 eine Mehrheitsbeteiligung an Gebr. Klingspor. 1956 Ausscheiden der Familie Klingspor aus dem Unternehmen. Künstlerisch-wissenschaftliche Mitarbeiter waren Carl Riedel, Otto Heraeus, Wilhelm H. Lange, Franz Gerhardinger. Stempelschneider sind Louis Hoell, Gustav Eichenauer, Peter Burckhardt. Zu den Schriftentwerfern gehören u. a. Otto Hupp, Rudolf Koch, Walter Tiemann, Rudo Spemann. Verzierungsentwürfe stammen u. a. von Heinrich Vogeler.

**Klingspor,** Karl – geb. 25. 6. 1868 in Gießen, Deutschland, gest. 1. 1. 1950 in Offenbach, Deutschland. – *Unternehmer, Autor* – Der Vater Carl Klingspor erwirbt 1892 die Rudhardsche Gießerei in Offenbach am Main und überträgt seinem Sohn Karl die Leitung. 1895 Eintritt des jüngeren Bruders Wilhelm. 1899 Mit-

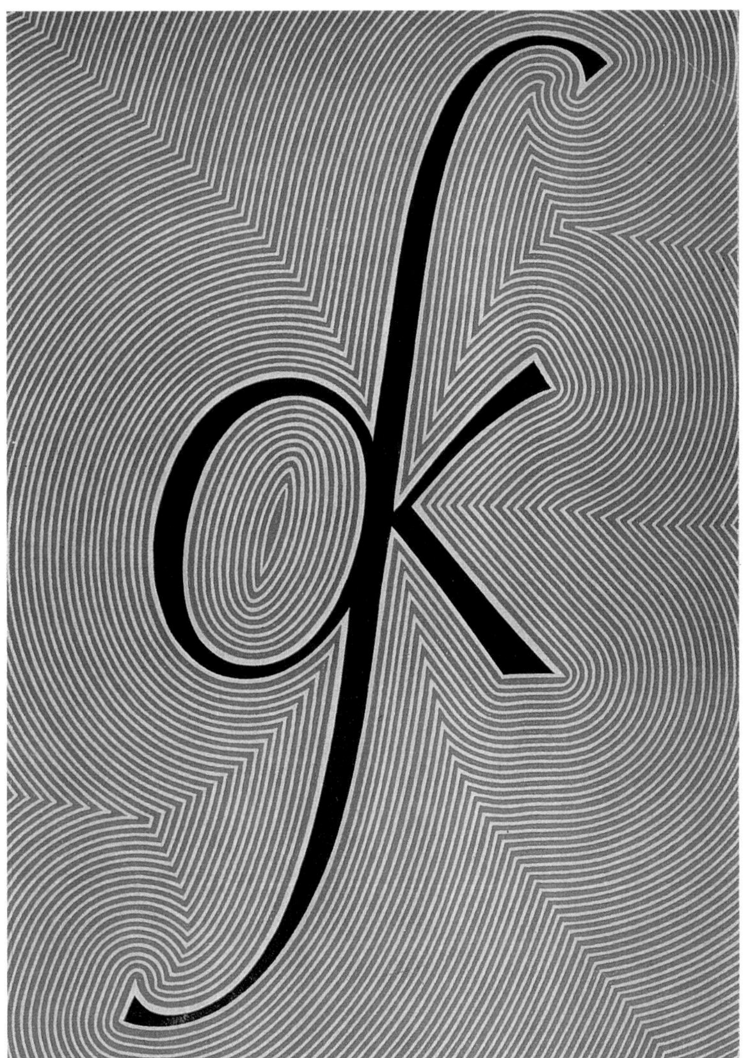

Gebr. Klingspor ca. 1953 Cover

Gebr. Klingspor ca. 1954 Cover

Klingspor 1949 Cover

**Gebr. Klingspor** – *fonderie de caractères* – en 1892, le fabricant de cigarettes Carl Klingspor, de Giessen, achète la fonderie Rudhard qui avait été créée en 1859 par Johann Peter Nees. En 1898, Otto Eckmann est chargé de concevoir une nouvelle fonte. En 1902, Peter Behrens dessine sa «Behrens-Schrift» pour Klingspor. Après la mort du père en 1903, les fils deviennent propriétaires de l'entreprise qui s'appelle «Gebr. Klingspor» à partir de 1906. Rudolf Koch commence à travailler pour l'entreprise en tant que conseiller artistique (jusqu'en 1934). En 1917, la fonderie de caractères D. Stempel AG de Francfort-sur-le-Main acquiert une participation majoritaire dans Gebr. Klingspor. En 1956, la famille Klingspor quitte l'entreprise. Parmi les personnes y ayant travaillé dans les domaines artistiques et scientifiques, on compte: Carl Riedel, Otto Heraeus, Wilhelm H. Lange, Franz Gerhardinger. Louis Hoell, Gustav Eichenauer et Peter Burckhardt y ont été tailleurs de types. Otto Hupp, Rudolf Koch, Walter Tiemann, Rudo Spemann et autres y ont conçu des polices, Heinrich Vogeler y a crée des ornements.

**Klingspor,** Karl – né le 25. 6. 1868 à Giessen, Allemagne, décédé le 1. 1. 1950 à Offenbach, Allemagne – *chef d'entreprise, auteur* – en 1892, son père, Carl Klingspor acquiert la fonderie Rudhard à Offenbach-sur-le-Main et charge son fils Karl de la diriger. En 1895, son frère

Carl Klingspor, cigar manufacturer from Gießen, buys the Rudhard'sche Gießerei which was founded in 1859 by Johann Peter Nees. 1898: Otto Eckmann is commissioned to design a new typeface. 1902: Peter Behrens designs his Behrens-Schrift for Klingspor. 1903: after the death of their father, the ownership of the foundry is passed on to the Klingspor sons. 1906: the company changes its name to Gebr. Klingspor (Klingspor Bros.). Rudolf Koch starts working for the company as art consultant (until 1934). 1917: the D. Stempel AG type foundry in Frankfurt gains majority ownership of Gebr. Klingspor. 1956: the Klingspor family ceases to work for the company. Among those who worked for Klingspor were Carl Riedel, Otto Heraeus, Wilhelm H. Lange and Franz Gerhardinger. Punch cutters were Louis Hoell, Gustav Eichenauer and Peter Burckhardt. Otto Hupp, Rudolf Koch, Walter Tiemann and Rudo Spemann were among the foundry's type designers, with Heinrich Vogeler working on decorations.

**Klingspor,** Karl – b. 25. 6. 1868 in Gießen, Germany, d. 1. 1. 1950 in Offenbach, Germany – *entrepreneur, author* – 1892: father Carl Klingspor acquires the Rudhard'sche Gießerei in Offenbach am Main and appoints his son Karl as director. 1895: Karl's younger brother Wilhelm

glied der Gesellschaft der Bibliophilen. Für seine Verdienste um die künstlerische Buchausstattung und die Entwicklung der Typographie wird ihm 1919 die Würde des Doktor-Ingenieurs (Dr.-Ing. E.h.) der Technischen Hochschule Darmstadt verliehen. 1922 Mitbegründer der Frankfurter Bibliophilen Gesellschaft. Auf Anregung Klingspors wird 1929 der jährlich veranstaltete Wettbewerb der schönsten Bücher eingeführt. 1930 erster Wettbewerb „Die schönsten Bücher" in Leipzig unter seinem Vorsitz. 1937 Ehrenvorsitzender des Vereins der Schriftgießereien. 1940 Verleihung des Gutenberg-Rings durch die Stadt Leipzig. 1948 Ehrenbürger der Stadt Offenbach am Main. – *Publikationen u. a.:* „Über Schönheit von Schrift und Druck", Offenbach 1949. W. H. Lange (Hrsg.) „Karl Klingspor zum 80. Geburtstag", Offenbach 1948.

**Klingspor Museum,** Offenbach am Main. – *Museum für Schriftkunst* – Die Privatsammlung von Dr. Karl Klingspor ist der Grundstock des Museums. 1953 Eröffnung des Klingspor Museums der Stadt Offenbach im Büsing-Palais. Leiter des Museums wird Georg Alexander Mathéy. Der 1938 nach Amerika emigrierte Offenbacher Dr. Siegfried Guggenheim übergibt seine Sammlung 1956 dem Klingspor Museum. 1957 wird Hans A. Halbey Leiter des Museums, 1977 Christian Scheffler. Das Klingspor Museum sammelt und stellt die internationale Buch- und Schriftkunst des 20. Jahrhunderts aus. Es besitzt reichhaltige Bestände zur Geschichte des Buchdrucks, zur Buchillustration, zur Schriftgeschichte,

cadet, Wilhelm, entre dans l'entreprise. 1899, membre de la Société des bibliophiles. En 1919, la Technische Hochschule (école supérieure des technologies) de Darmstadt lui décerne le titre de Docteur-Ingénieur h.c. pour ses mérites en matière de maquettes de livres et de typographie. 1922, cofondateur de la Frankfurter Bibliophilen Gesellschaft (société des bibliophiles de Francfort). En 1929, le concours annuel des plus beaux livres est fondé à l'initiative de Klingspor. Premier concours «Die schönsten Bücher» à Leipzig, en 1930, dont il assure la présidence. En 1937, il est président d'honneur de l'union des fonderies de caractères. En 1940, la ville de Leipzig lui décerne le Gutenberg-Ring. Citoyen d'honneur de la ville d'Offenbach en 1948. – *Publications:* «Über Schönheit von Schrift und Druck», Offenbach 1949. Wilhelm H. Lange (éd.) «Karl Klingspor zum 80. Geburtstag», Offenbach 1948.

**Klingspor Museum,** Offenbach-sur-le-Main – *musée des arts de l'écriture* – la collection privée du dr. Karl Klingspor a constitué la base du musée. Inauguration, en 1953, du Klingspor Museum d'Offenbach dans le Palais Büsing. Georg Alexander Mathéy en prend la direction. En 1956, le dr. Siegfried Guggenheim, ancien Offenbachois émigré aux Etats-Unis en 1938, lègue sa collection au musée. En 1957, Hans A. Halbey devient directeur, puis Christian Scheffler en 1977. Le Klingspor Museum collectionne et présente des travaux originaires de plusieurs pays et des chefs d'œvre de l'art du livre et de l'écriture du 20e siècle. Il détient d'importantes collections sur l'histoire de l'imprimerie du livre, l'illustration, l'histoire de l'écriture, de la reliure, du papier et du graphisme industriel. Il est en possession des successions de Rudolf Koch, Otto Reichert, F. H. E. Schneidler, E. R.

Klingspor Museum 1988 Folder

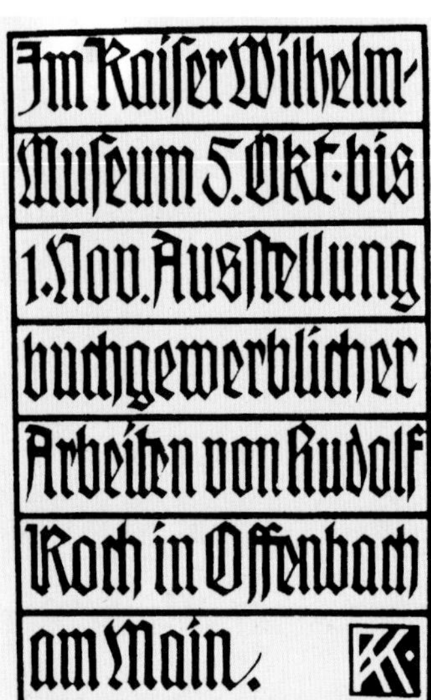


Koch 1910 Poster

Knapp 1961 Spread

Koch 1921 Deutsche Zierschrift

joins him. 1899: member of the Gesellschaft der Bibliophilen. 1919: the Technische Hochschule in Darmstadt awards him an honorary engineering doctorate for his achievements in artistic book design and developments in the field of typography. 1922: co-founder of the Frankfurter Bibliophile Gesellschaft. 1929: Klingspor initiates the establishment of a yearly competition for beautiful books. 1930: the first beautiful books competition ("Die schönsten Bücher") is held in Leipzig with Klingspor as chairman. 1937: honorary president of the Verein der Schriftgießereien. 1940: Leipzig awards him their Gutenber Ring. 1948: made an honorary citizen of Offenbach am Main. – *Publications:* "Über Schönheit von Schrift und Druck", Offenbach 1949. Wilhelm H. Lange (ed.) "Karl Klingspor zum 80. Geburtstag", Offenbach 1948.

**Klingspor Museum,** Offenbach am Main – *museum of type* – Dr. Karl Klingspor's private collection forms the basis of the museum's exhibits. 1953: Offenbach's Klingspor Museum is opened in the Büsing Palais. Georg Alexander Mathéy is museum director. 1956: Dr. Siegfried Guggenheim, who emigrated to the USA in 1938, donates his collection to the Klingspor Museum. 1957: Hans A. Halbey is made museum director and is succeeded by Christian Scheffler in 1977. The Klingspor Museum collects and exhibits items from all over the world pertaining to the art of books and type in the 20th century. The museum's extensive collection includes exhibits on the history of

zum Bucheinband, zur Papierkunde und Gebrauchsgrafik. Das Museum besitzt die künstlerischen Nachlässe u.a. von Rudolf Koch, Otto Reichert, F. H. E. Schneidler, E. R. Weiß, F. H. Ehmcke, Rudolf von Larisch, Heinrich Jost. – *Publikation u. a.:* Christian Scheffler „Klingspor Museum Offenbach", Braunschweig 1981.

**Knapp,** Peter – geb. 5. 5. 1931 in Bäretswil, Schweiz. – *Typograph, Grafik-Designer, Fotograf, Lehrer* – 1947–51 Studien an der Kunstgewerbeschule in Zürich und an der Ecole des Beaux-Arts in Besançon. Danach künstlerischer Leiter von Mövenpick (Zürich) und der Galeries Lafayette (Paris). Art Director der Zeitschriften „Nouveau Féminin", „Elle", „Femme", „Fortune" und „Décoration Internationale". Zahlreiche fotografische Arbeiten und Filme für das französische Fernsehen. Seit 1960 eigenes Studio in Paris. Seit 1983 Lehrer für Fotografie an der Académie Julian (ESAG) in Paris. Zahlreiche Ausstellungen. – *Publikation u. a.:* „Photos d'Elles. Temps de pose 1950–1990", Genf 1993.

**Koch,** Rudolf – geb. 20. 11. 1876 in Nürnberg, Deutschland, gest. 9. 4. 1934 in Offenbach, Deutschland. – *Schriftentwerfer, Typograph, Kalligraph, Lehrer* – 1892–96 Ziseleur-Lehre in Hanau. 1896–97 Zeichenlehrer-Ausbildung an der Kunstgewerbeschule in Nürnberg und an der Technischen Hochschule in München. 1898–1902 Zeichner in einer Buchbinderei in Leipzig. Entwürfe von Buchtiteln, erste Schriftentwürfe. Seit 1906 künstlerischer Berater der Schriftgießerei Gebr. Klingspor in Offenbach. Seit 1908 Lehrer

Koch 1932 Sheet

Koch 1924 Sheet

Weiß, F. H. Ehmcke, Rudolf von Larisch, Heinrich Jost et d'autres. – *Publication, sélection :* Christian Scheffler «Klingspor Museum Offenbach», Braunschweig 1981.

**Knapp,** Peter – né le 5. 5. 1931 à Bäretswil, Suisse – *typographe, graphiste maquettiste, photographe, enseignant* – 1947–1951, études à la Kunstgewerbeschule (école des arts décoratifs) de Zurich et à l'Ecole des Beaux-Arts de Besançon. Il est ensuite directeur artistique de Mövenpick (Zurich) et des Galeries Lafayette (Paris). Directeur artistique des revues «Nouveau Féminin», «Elle», «Femme», «Fortune» et «Décoration Internationale»; nombreux travaux de photographie et films pour la télévision française. A sa propre agence à Paris depuis 1960. Enseigne la photographie à l'Académie Julian (ESAG) à Paris depuis 1983. Nombreuses expositions. – *Publications, sélection :* «Photos d'Elles. Temps de pose 1950– 1990», Genève 1993.

**Koch,** Rudolf – né le 20. 11. 1876 à Nuremberg, Allemagne, décédé le 9. 4. 1934 à Offenbach, Allemagne – *concepteur de polices, typographe, calligraphe, enseignant* – 1892–1896, apprentissage de ciseleur à Hanau. 1896–1897, formation de professeur de dessin à la Kunstgewerbeschule (école des arts décoratifs) de Nuremberg et à la Technische Hochschule (école supérieure des technologies) de Munich. Dessinateur à l'atelier de reliure de Leipzig de 1898 à 1902; dessine des titres de livres et premiers dessins de polices. Conseiller artistique de la fonderie

| n | n | n | | | |
|---|---|---|---|---|---|
| a | b | e | f | g | i |
| o | r | s | t | y | z |
| A | B | C | E | G | H |
| M | O | R | S | X | Y |
| 1 | 2 | 4 | 6 | 8 | & |

Koch 1922 Koch Antiqua

| | n | n | n | | |
|---|---|---|---|---|---|
| a | b | e | f | g | i |
| o | r | s | t | y | z |
| A | B | C | E | G | H |
| M | O | R | S | X | Y |
| 1 | 2 | 4 | 6 | 8 | & |

Koch 1927 Kabel

| | n | | | | |
|---|---|---|---|---|---|
| a | b | e | f | g | i |
| o | r | s | t | y | z |
| A | B | C | E | G | H |
| M | O | R | S | X | Y |
| 1 | 2 | 4 | 6 | 8 | & |

Koch 1929 Zeppelin

printing and type, paper manufacture, book illustration, book covers and applied graphics. It is also in possession of the artistic estates of Rudolf Koch, Otto Reichert, F. H. E. Schneidler, E. R. Weiß, F. H. Ehmcke, Rudolf von Larisch and Heinrich Jost. – *Publications include:* Christian Scheffler "Klingspor Museum Offenbach", Braunschweig 1981.

**Knapp,** Peter – b. 5. 5. 1931 in Bäretswil, Switzerland – *typographer, graphic designer, photographer, teacher* – 1947–51: studies at the Kunstgewerbeschule in Zurich and at the Ecole des Beaux-Arts in Besançon. He is then art director for Mövenpick (Zurich) and Galeries Lafayette (Paris). Art director of "Nouveau Féminin", "Elle", "Femme", "Fortune" and "Décoration Internationale" magazines. Produces much photographic work and numerous films for French television. 1960: opens his own studio in Paris. He has taught photography at the Académie Julian (ESAG) in Paris since 1983. Numerous exhibitions. – *Publications include:* "Photos d'Elles. Temps de pose 1950–1990", Geneva 1993.

**Koch,** Rudolf – b. 20. 11. 1876 in Nuremberg, Germany, d. 9. 4. 1934 in Offenbach, Germany – *type designer, typographer, calligrapher, teacher* – 1892–96: trains as an engraver in Hanau. 1896–97: trains as an art teacher at the Kunstgewerbeschule in Nuremberg and at the Technische Hochschule in Munich. 1898–1902: artist in a bookbindery in Leipzig. Designs book titles and his first typefaces. 1906: starts working as art consultant to the Gebr. Klingspor type foundry in Of-

für Schrift an der Technischen Lehranstalt in Offenbach. 1911–24 Herausgabe der „Rudolfinischen Drucke" (mit Rudolf Gerstung). 1921 Gründung der „Offenbacher Werkgemeinschaft" in der Technischen Lehranstalt in Offenbach. 1930 Ehrendoktor der Evangelischen Theologischen Fakultät der Universität München. 1934 Kochs Werk wird von seinem Sohn Paul (gest. 1943), der in Frankfurt am Main eine Werkstatt führte, fortgesetzt. – *Schriftentwürfe:* Deutsche Schrift (1906–21), Maximilian Antiqua (1913–17), Frühling (1913–17), Wilhelm-Klingspor-Schrift (1920–26), Koch-Antiqua (1922), Deutsche Zierschrift (1921), Neuland (1922–23), Deutsche Anzeigenschrift (1923–34), Peter-Jessen-Schrift (1924– 30), Wallau (1925–34), Kabel (1927), Offenbach (1928), Zeppelin (1929), Marathon (1930–38), Claudius (1931–34), Prisma (1931), Holla (1932), Grotesk-Initialen (1933), Koch Kurrent (1933), Neufraktur (1933–34). – *Publikationen u.a.:* „Die Schriftgießerei im Schattenbild", Offenbach 1918; „Das Schreiben als Kunstfertigkeit", Leipzig 1921; „Das ABC-Büchlein", Leipzig 1934; „Das Schreibbüchlein", Kassel 1939. Georg Haupt „Rudolf Koch, der Schreiber", Leipzig 1936; Wilhelm H. Lange „Rudolf Koch, ein deutscher Schreibmeister", Berlin, Leipzig 1938; Oskar Beyer „Rudolf Koch. Mensch, Schriftgestalter und Erneuerer des Handwerks", Berlin 1949.

**Korger,** Hildegard – geb. 18. 6. 1935 in Reichenberg, Tschechoslowakei (heute Liberec, Tschechische Republik). – *Kalligraphin, Lehrerin* – 1956–59 Studium an

de caractères Klingspor d'Offenbach à partir de 1906. Enseigne les arts de l'écriture au Technische Lehranstalt (institut d'enseignement technique) d'Offenbach à partir de 1908. Publie les «Rudolfinischen Drucke» (avec Rudolf Gerstung) de 1911 à 1924. Fonde la «Offenbacher Werkgemeinschaft» au Technische Lehranstalt d'Offenbach en 1921. Docteur h.c. de la faculté luthérienne de théologie de l'université de Munich en 1930. Son fils Paul (décédé en 1943), qui dirigeait un atelier à Francfort-sur-le-Main, reprend et poursuit l'œuvre de Koch en 1934. – *Polices:* Deutsche Schrift (1906–1921), Maximilian Antiqua (1913–1917), Frühling (1913–1917), Wilhelm-Klingspor-Schrift (1920–1926), Koch-Antiqua (1922), Deutsche Zierschrift (1921), Neuland (1922–1923), Deutsche Anzeigenschrift (1923–1934), Peter-Jessen-Schrift (1924–1930), Wallau (1925–1934), Kabel (1927), Offenbach (1928), Zeppelin (1929), Marathon (1930–1938), Claudius (1931–1934), Prisma (1931), Holla (1932), Grotesk-Initialen (1933), Koch Kurrent (1933), Neufraktur (1933–1934). – *Publications, sélection:* «Die Schriftgiesserei im Schattenbild», Offenbach 1918; «Das Schreiben als Kunstfertigkeit», Leipzig 1921; «Das ABC-Büchlein», Leipzig 1934; «Das Schreibbüchlein», Kassel 1939. Georg Haupt «Rudolf Koch der Schreiber», Leipzig 1936; Wilhelm H. Lange «Rudolf Koch, ein deutscher Schreibmeister», Berlin, Leipzig 1938; Oskar Beyer «Rudolf Koch. Mensch, Schriftgestalter und Erneuerer des Handwerks», Berlin 1949.

**Korger,** Hildegard – née le 18. 6. 1935 à Reichenberg, Tchécoslovaque (aujourd'hui Liberec, République tchèque)– *calligraphe, enseignante* – études à la Fachschule für Angewandte Kunst (école des arts appliqués) de Heiligendamm de 1956 à 1959. Etudes à la Hochschule für

Korger ca. 1980 Calligraphy

ABCDEFGHIJKLMN
OPQRSTUVWXYZ
ÆŒÄÖÜĲÞ §£$&!?*†
abcdefghijklmnopqrstuvwxyz
æœäöü ĳþðßı chckfiflffft
12345678907 1234567890

Korger 1984–89 Kis-Antiqua

Korger 1990 Alphabet

Kosuth 1965 Installation

fenbach. From 1908 onwards: teacher of lettering at the Technische Lehranstalt in Offenbach. 1911–24: publishes the "Rudolfinische Drucke" with Rudolf Gerstung. 1921: founds the Offenbacher Werkgemeinschaft at the Technische Lehranstalt in Offenbach. 1930: is awarded an honorary doctorate by the faculty of Evangelical theology at the University of Munich. 1934: Koch's work is carried on by his son Paul (d. 1943) who runs a workshop in Frankfurt. – *Fonts:* Deutsche Schrift (1906–21), Maximilian Antiqua (1913–17), Frühling (1913–17), Wilhelm-Klingspor-Schrift (1920–26), Koch-Antiqua (1922), Deutsche Zierschrift (1921), Neuland (1922–23), Deutsche Anzeigenschrift (1923–34), Peter-Jessen-Schrift (1924–30), Wallau (1925–34), Kabel (1927), Offenbach (1928), Zeppelin (1929), Marathon (1930–38), Claudius (1931–34), Prisma (1931), Holla (1932), Grotesk-Initialen (1933), Koch Kurrent (1933), Neufraktur (1933–34). – *Publications include:* "Die Schriftgießerei im Schattenbild", Offenbach 1918; "Das Schreiben als Kunstfertigkeit", Leipzig 1921; "Das ABC-Büchlein", Leipzig 1934; "Das Schreibbüchlein", Kassel 1939. Georg Haupt "Rudolf Koch der Schreiber", Leipzig 1936; Wilhelm H. Lange "Rudolf Koch, ein deutscher Schreibmeister", Berlin, Leipzig 1938; Oskar Beyer "Rudolf Koch. Mensch, Schriftgestalter und Erneuerer des Handwerks", Berlin 1949.

**Korger,** Hildegard – b. 18. 6. 1935 in Reichenberg, Czechoslovakia (now Liberec, Czech Republic) – *calligrapher, teacher* – 1956–59: studies at the Fach-

der Fachschule für Angewandte Kunst in Heiligendamm. 1959–63 Studium an der Hochschule für Graphik und Buchkunst in Leipzig. 1963–65 freiberufliche Tätigkeit. 1965–68 Assistentin an der Hochschule für Graphik und Buchkunst in Leipzig. Seit 1976 Beraterin für baugebundene Schriftgestaltung im Büro des Chefarchitekten der Stadt Leipzig. Seit 1979 Dozentin für Schriftgestaltung an der Hochschule für Graphik und Buchkunst in Leipzig. 1992 Ernennung zur Professorin. – *Publikationen u. a.:* „Schriftgestaltung", Dresden 1965; „Schreiben", Leipzig 1971.

**Kosuth,** Joseph – geb. 31. 1. 1945 in Toledo, USA. – *Künstler* – 1963–64 Studium am Cleveland Art Institute. 1965 Studium an der School of Visual Arts in New York. Gründet und leitet 1967 das „Museum of Normal Arts" in New York. Seit 1967 zahlreiche internationale Einzelausstellungen. Seit 1968 Dozent an der School of Visual Arts in New York. 1971–72 Studium der Anthropologie und der Philosophie an der New School for Social Research in New York. 1982 Teilnahmen an der documenta VII und IX (1992) in Kassel. 1988–90 Professor an der Hochschule für Bildende Künste in Hamburg, seit 1991 an der Staatlichen Akademie der Bildenden Künste in Stuttgart. – *Publikationen u.a.:* „Teksten, Textes", Antwerpen 1976; „Bedeutung von Bedeutung", Stuttgart 1981; „Interviews", Stuttgart 1989; „Kein Ausweg, No Exit", Stuttgart 1991; „Kein Ding, kein Ich, keine Form, kein Grundsatz", Stuttgart 1993.

**Krimpen,** Jan van – geb. 12. 1. 1892 in Gouda, Niederlande, gest. 20. 10. 1958 in

Graphik und Buchkunst (école supérieure des arts graphiques et des arts du livre) de Leipzig de 1959 à 1963. Exerce comme indépendante de 1963 à 1965. Assistante à la Hochschule für Graphik und Buchkunst (arts graphiques et arts du livre) de Leipzig de 1965 à 1968. Conseillère en design de caractéres utilisées en architecture au cabinet de l'architecte en chef de la ville de Leipzig à partir de 1976. Enseigne la conception graphique de caractères à la Hochschule für Graphik und Buchkunst (arts graphiques et arts du livre) de Leipzig depuis 1979. Nommée professeur en 1992. – *Publications, sélection:* «Schriftgestaltung», Dresde 1965; «Schreiben», Leipzig 1971.

**Kosuth,** Joseph – né le 31. 1. 1945 à Toledo, Etats-Unis – *artiste* – études au Cleveland Art Institute de 1963 à 1964. Etudes à la School of Visual Arts de New York en 1965. Fonde et dirige le «Museum of Normal Arts» à New York en 1967. Nombreuses expositions internationales depuis 1967. Enseigne à la School of Visual Arts de New York à partir de 1968. De 1971 à 1972, il étudie l'anthropologie et la philosophie à la New School for Social Research à New York. Participe aux «documenta VII» et «IX» en 1982 et 1992 à Kassel. Professeur à la Hochschule für Bildende Künste (école supérieure des beaux-arts) de Hambourg de 1988 à 1990; enseigne à la Staatliche Akademie der Bildenden Künste (beaux-arts) de Stuttgart depuis 1991. – *Publications, sélection:* «Teksten, Textes», Anvers 1976; «Bedeutung von Bedeutung», Stuttgart 1981; «Interviews», Stuttgart 1989; «Kein Ausweg, No Exit», Stuttgart 1991; «Kein Ding, kein Ich, keine Form, kein Grundsatz», Stuttgart 1993.

MPQRS
TUWYZ

Krimpen 1929 Initials

SIG
NAT
URE

Krimpen 1947 Specimen

Krimpen ••JAHR•• Cover

Krimpen 1946 Stamp

| | n | | | | *n* |
|---|---|---|---|---|---|
| a | b | e | f | g | i |
| o | r | s | t | y | z |
| A | B | C | E | G | H |
| M | O | R | S | X | Y |
| 1 | 2 | 4 | 6 | 8 | & |

Krimpen 1936 Van Dijk

schule für Angewandte Kunst in Heiligendamm. 1959–63: studies at the Hochschule für Graphik und Buchkunst in Leipzig. 1963–65: freelance. 1965–68: assistant at the Hochschule für Graphik und Buchkunst in Leipzig. From 1976 onwards: consultant for type design in building at the Leipzig head architect's office. From 1979 onwards: lecturer of type design at the Hochschule für Graphik und Buchkunst in Leipzig. 1992: is made a professor. – *Publications include:* "Schriftgestaltung", Dresden 1965; "Schreiben", Leipzig 1971.

**Kosuth,** Joseph – b. 31. 1. 1945 in Toledo, USA – *artist* – 1963–64: studies at the Cleveland Art Institute. 1965: studies at the School of Visual Arts in New York. 1967: founds and directs the Museum of Normal Arts in New York. Since 1967: numerous solo exhibitions all over the world. From 1968 onwards: lecturer at the School of Visual Arts in New York. 1971–72: studies anthropology and philosophy at the New School for Social Research in New York. 1982: takes part in documenta VII and IX (1992) in Kassel. 1988–90: professor at the Hochschule für Bildende Künste in Hamburg and since 1991 at the Staatliche Akademie der Bildenden Künste in Stuttgart. – *Publications include*: "Teksten, Textes", Antwerp 1976; "Bedeutung von Bedeutung", Stuttgart 1981; "Interviews", Stuttgart 1989; "Kein Ausweg, No Exit", Stuttgart 1991; "Kein Ding, kein Ich, keine Form, kein Grundsatz", Stuttgart 1993.

**Krimpen,** Jan van – b. 12. 1. 1892 in Gouda, The Netherlands, d. 20. 10. 1958

Haarlem, Niederlande. – *Typograph, Kalligraph, Schriftentwerfer* – 1908–12 Studium an der Academie van Beeldende Kunsten in Den Haag. 1912 Typograph und Autor der Zeitschrift „De Witte Mier", die in der Privatpresse „Zilverdistel" herausgegeben wird. Seit 1915 selbständiger Gestalter. 1925–58 typographischer Berater der Schriftgießerei und Druckerei Enschedé en Zonen in Haarlem. 1926 Bekanntschaft mit Stanley Morison. 1929 Entwurf von Initialen für Oliver Simons „Curwen Press". 1946 Entwurf der Briefmarken-Dauerserie („Ziffernserie") für die niederländische Post PTT. 1947 Entwurf der Titelzeile für Oliver Simons Zeitschrift „Signature". Gestaltet „Das Buch der Psalmen" für die Gesellschaft der Holländischen Bibliophilen. – *Schriftentwürfe:* Lutetia (1925), Antigone (1927), Open Roman Capitals (1929), Romanée (1929), Romulus (1931), Van Dijck (1936), Cancelleresca Bastarda (1937), Haarlemmer (1939), Spectrum (1943–45), Sheldon (1947). – *Publikationen u.a.:* „Typography in Holland" in „The Fleuron" 7/1930; „On designing and devising type", New York 1960. John Dreyfus „The work of Jan van Krimpen", London 1952.

**Kriwet,** Ferdinand – geb. 3. 8. 1942 in Düsseldorf, Deutschland. – *Künstler* – Autodidakt. Seit 1961 Seh- und Hörtexte für das Radio. 1962 Teilnahme an der Ausstellung „Skripturale Malerei" in Amsterdam und Berlin sowie „Schrift und Bild" in Amsterdam und Baden-Baden. Seit 1962 Theater-Aufführungen und Mixed-Media-Performances. 1977 Teilnahme an der documenta VI in Kassel.

**Krimpen,** Jan van – né le 12. 1. 1892 à Gouda, Pays-Bas, décédé le 20. 10. 1958 à Haarlem, Pays-Bas – *typographe, calligraphe, concepteur de polices* – 1908–1912, études à l'Academie van Beeldende Kunsten de La Haye. 1912, typographe et auteur pour la revue «De Witte Mier» éditée par les presses privées «Zilverdistel». A partir de 1915, designer indépendant. De 1925 à 1958, il est conseiller en typographie à la fonderie de caractères et imprimerie «Enschedé en Zonen» de Haarlem. Rencontre Stanley Morison en 1926. Dessine des lettrines pour «Curwen Press» d'Oliver Simon en 1929. En 1946, il dessine des timbres-poste pour (»Ziffernserie») utilisée pendant plusieurs années par les postes néerlandaises. En 1947, il conçoit le titre de la revue «Signature» d'Oliver Simon. Maquette du «Livre des Psaumes» pour la société des bibliophiles hollandais. – *Polices:* Lutetia (1925), Antigone (1927), Open Roman Capitals (1929), Romanee (1929), Romulus (1931), Van Dijck (1936), Cancelleresca Bastarda (1937), Haarlemmer (1939), Spectrum (1943- 1945), Sheldon (1947). – *Publications, sélection:* «Typography in Holland» in «The Fleuron» 7/1930; «On designing and devising type», New York 1960. John Dreyfus «The work of Jan van Krimpen», Londres 1952.

**Kriwet,** Ferdinand – né le 3. 8. 1942 à Düsseldorf, Allemagne – *artiste* – autodidacte. «Textes à voir et à entendre» pour la radio depuis 1961. Participe à l'exposition «Skripturale Malerei» à Amsterdam et à Berlin en 1962 ainsi qu'à «Schrift und Bild» à Amsterdam et à Baden-Baden. Représentations de théâtre et performances multimédias depuis 1962. Participe à la «documenta VI» à Kassel en 1977. Nombreux travaux de création et d'art appliqués pour les systèmes d'orientation en intérieur et en extérieur. – *Publications,*

Kriwet 1965 Painting

Kriwet 1975 Text Towers

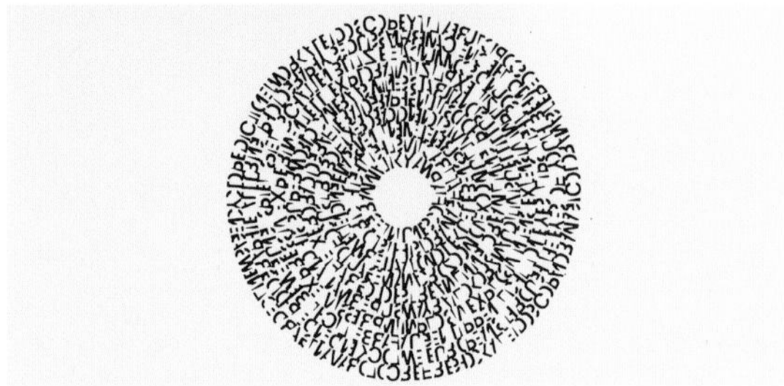
Kriwet 1963 Painting

Kriwet 1965 Painting

in Haarlem, The Netherlands – *typographer, calligrapher, type designer* – 1908–12: studies at the Academie van Beeldende Kunsten in The Hague. 1912: typographer and author of "De Witte Mier" magazine published by the private Zilverdistel press. From 1915 onwards: freelance designer. 1925–58: typography consultant to the Enschedé en Zonen type foundry and printing works in Haarlem. 1926: makes the acquaintance of Stanley Morison. 1929: designs initials for Oliver Simon's Curwen Press. 1946: designs a series of postage stamps (permanent "number series") for the Dutch post office, PTT. 1947: designs the title line for Oliver Simon's "Signature" magazine. Designs "The Book of Psalms" for the Gesellschaft der Holländischen Bibliophilen. – *Fonts:* Lutetia (1925), Antigone (1927), Open Roman Capitals (1929), Romanee (1929), Romulus (1931), Van Dijck (1936), Cancelleresca Bastarda (1937), Haarlemmer (1939), Spectrum (1943–45), Sheldon (1947). – *Publications include:* "Typography in Holland" in "The Fleuron" 7/1930; "On designing and devising type", New York 1960. John Dreyfus "The Work of Jan van Krimpen", London 1952.

**Kriwet,** Ferdinand – b. 3. 8. 1942 in Düsseldorf, Germany – *artist* – Self-taught. From 1961 onwards: texts for the radio. 1962: takes part in the "Skripturale Malerei" exhibition in Amsterdam and Berlin and in "Schrift und Bild" in Amsterdam and Baden-Baden. From 1962 onwards: theater and mixed-media performances. 1977: takes part in docu-

Zahlreiche freie und angewandte Arbeiten zur Orientierung in Gebäuden und Außenanlagen. – *Publikationen u.a.:* „Rotor", Köln 1961; „10 Sehtexte", Köln 1962; „Leserattenfänge", Köln 1965; „Com.mix", Köln 1972; „Kunst und Architektur", Dodenburg 1981.

**Kruger,** Barbara – geb. 1940 in New York, USA. – *Künstlerin, Grafik-Designerin* – Studium an der Syracuse University in New York. 1965 Studium an der Parson's School of Design in New York. Abbruch des Studiums. 1965–69 Layouterin bei Condé Nast in New York. Entwurf von Kleinanzeigen. Arbeiten für die Zeitschriften „Mademoiselle", „House and Garden" und „Seventeen". 1969 Beginn ihrer Kunstproduktion mit gewebten Wandbehängen. Entwurf von Buchumschlägen für die Verlage Harper & Row und Schocken. Gestaltung einer Ausgabe der Zeitschrift „Aperture". Seit 1972 Bild-Text-Montagen aus Werbefotos. 1974 erste Einzelausstellung in New York. 1982 Teilnahme an der documenta VII und VIII (1987) in Kassel und an der Biennale in Venedig. 1989 Entwurf des Plakats mit dem Titel „Your body is a battleground" für den „March on Washington", der für Geburtenkontrolle und Frauenrechte abgehalten wurde. – *Publikationen u.a.:* „My pretty pony", New York 1988; „Love for sale", New York 1990.

Kruger 1989 Poster

Kruger Montage

*sélection:* «Rotor», Cologne 1961; «10 Sehtexte», Cologne 1962; «Leserattenfänge», Cologne 1965; «Com.mix», Cologne 1972; «Kunst und Architektur», Dodenburg 1981.

**Kruger,** Barbara – née en 1940 à New York, Etats-Unis – *artiste, graphiste maquettiste* – études à la Syracuse University à New York. Etudes à la Parson's School of Design à New York en 1965. Interrompt ses études. Maquettiste chez les publications Condé Nast à New York de 1965 à 1969. Conçoit les petites annonces. Travaux pour les revues «Mademoiselle», «House and Garden» et «Seventeen». En 1969, elle commence à réaliser des tapisseries murales. Dessine des couvertures de livres pour les éditions Harper & Row et Schocken. Conçoit une édition de la revue «Aperture». Montages textes/images à partir de photos publicitaires depuis 1972. Première exposition personnelle à New York en 1974. Participe aux «documenta VII» et «VIII» à Kassel en 1982 et 1987 et à la Biennale de Venise. En 1989, elle crée l'affiche «Your body is a battleground» pour la «March on Washington» qui défendait les droits de la femme et le contrôle des naissances. – *Publications, sélection:* «My pretty pony», New York 1988; «Love for sale», New York 1990.

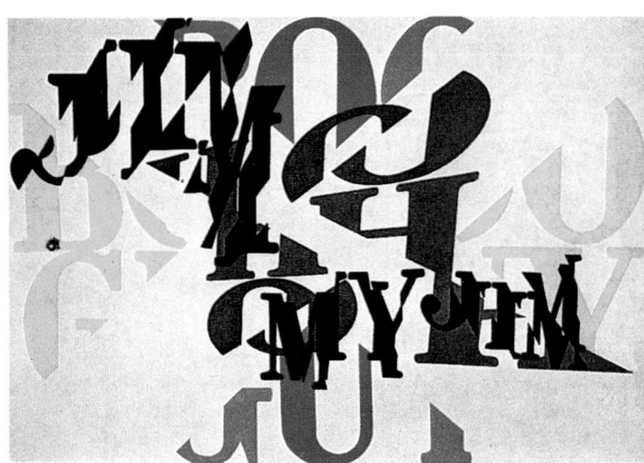
Kriwet 1965 Painting

Kruger 1987

menta VI in Kassel. Has done much free and applied work for orientation systems in buildings and outside. – *Publications include:* "Rotor", Cologne 1961; "10 Sehtexte", Cologne 1962; "Leserattenfänge", Cologne 1965; "Com.mix", Cologne 1972; "Kunst und Architektur", Dodenburg 1981.

**Kruger,** Barbara – b. 1940 in New York, USA – *artist, graphic designer* – Studied at Syracuse University in New York. 1965: studies at the Parson's School of Design in New York. Does not complete her studies. 1965–69: layouter for Condé Nast in New York. Designs classified advertisements. Works for the magazines "Mademoiselle", "House and Garden" and "Seventeen". 1969: starts producing artwork. Her first projects are woven wall hangings. Designs book covers for the publishers Harper & Row and Schocken. Designs a number of "Aperture" magazine. From 1972 onwards: fashions picture-text montages from advertising photos. 1974: first solo exhibition in New York. 1982: takes part in documenta VII and VIII (1987) in Kassel and in the Biennale in Venice. 1989: designs a poster entitled "Your body is a battleground" for the March on Washington for women's rights and birth control. – *Publications include:* "My pretty pony", New York 1988; "Love for sale", New York 1990.

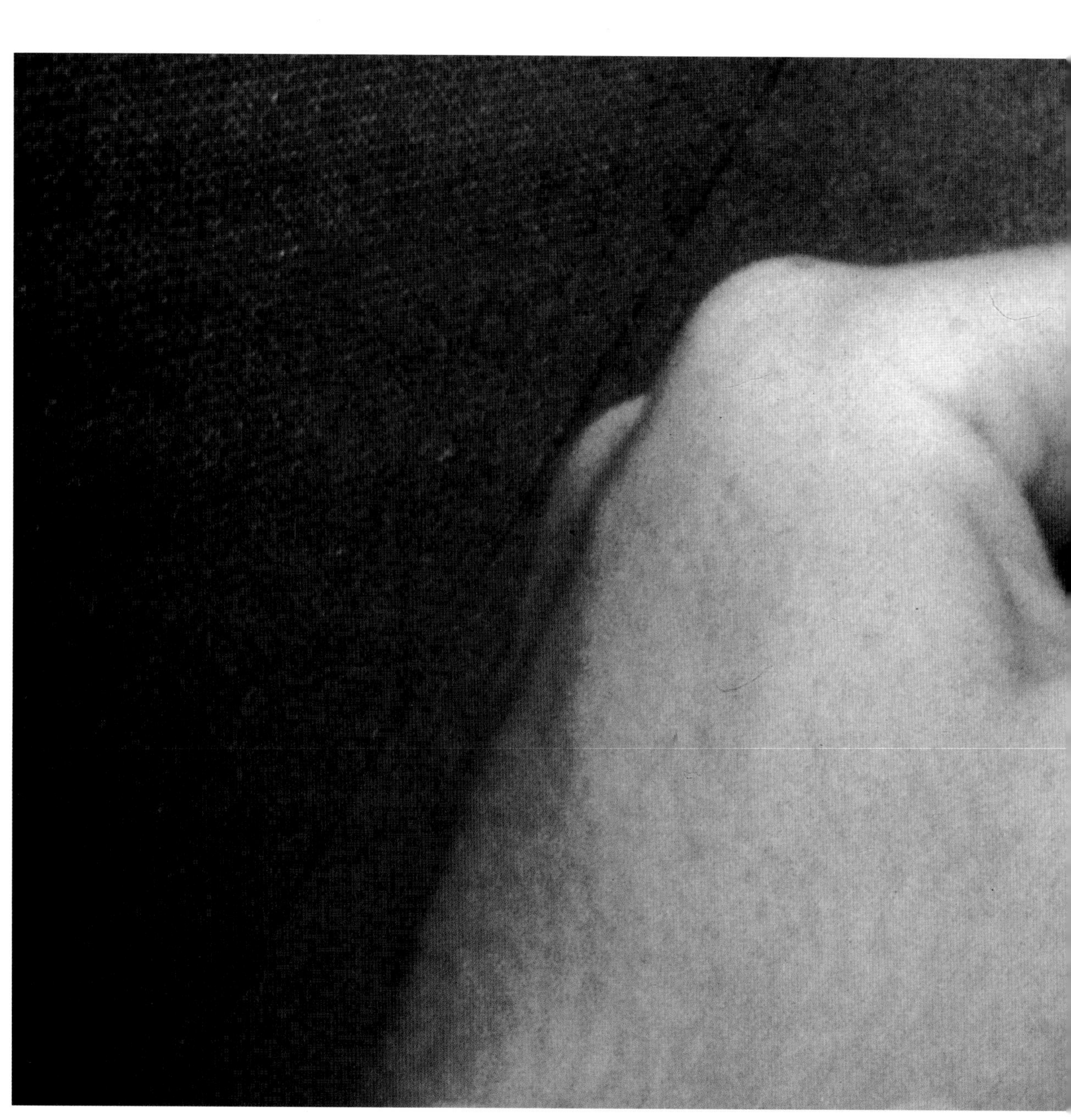

# L

AFRICAN CHRISTIAN DEMOCRATIC PARTY
ACDP
ADM
AMCP
ANC
ANC
DPSA
FP
LUSAP

**Lacan,** Jacques – geb. 13. 4. 1901 in Paris, Frankreich, gest. 9. 9. 1981 in Neuilly, Frankreich. – *Neurologe, Psychiater* – Studium der Medizin. Arbeit an der Klinik Sainte Anne in Paris und Burghölzi in Zürich (mit C. G. Jung). Seit 1932 steht er den Surrealisten und der intellektuellen Avantgarde nahe. Lacan entwickelt die These, daß die Struktur der Sprache das Unbewußte hervorbringt. Mit ihrer Ordnung bildet sie das Gegenstück zum Begehren. Beide zusammen erst lassen das Subjekt in seinem Spannungsfeld entstehen. Lacan arbeitet diesen Gedanken in Seminaren und Vorträgen „Das Drängen des Buchstabens im Unbewußten oder die Vernunft seit Freud" (1957) aus. Damit beinflußt er die strukturalistischen Philosophen und Linguisten wie Barthes und Derrida.

**Lange,** Günter Gerhard – geb. 12. 4. 1921 in Frankfurt an der Oder, Deutschland. – *Schriftentwerfer, Typograph, Lehrer* – 1941–45 Studium an der Akademie für Graphische Künste und Buchgewerbe in Leipzig. Meisterschüler von Prof. Georg Belwe. Danach Assistent von Prof. Walter Tiemann. Arbeit als freier Grafiker und Maler. 1949 Studium an der Hochschule für Bildende Künste in Berlin. Ab 1950 freier Mitarbeiter der H. Berthold AG in Berlin. 1955–60 Dozent für Typographie an der Meisterschule für Grafik und Buchgewerbe in Berlin. Seit 1961 künstlerischer Leiter der H. Berthold AG in Berlin. Seit 1971 für das gesamte Schriftenprogramm verantwortlich. Lehrt u. a. an der Kunstschule Alsterdamm in Hamburg (ab 1972), an der Werblichen Aka-

**Lacan,** Jacques – né le 13. 4. 1901 à Paris, France, décédé le 9. 9. 1981 à Neuilly, France – *neurologue, psychiatre, analyste* – études de médecine. Travaille à la clinique Sainte Anne à Paris, puis au Burghölzi de Zurich (avec C. G. Jung). Affinités avec les surréalistes et l'avant garde intellectuelle depuis 1932. Lacan développe la thèse selon laquelle la structure du langage produit l'inconscient, et son organisation constitue le contrepoint du désir. Ensemble seulement, ces deux aspects font apparaître le sujet dans son champ de tensions. Lacan traite de cette théorie lors de séminaires et de conférences sur la «L'insistance de la lettre dans l'inconscient, ou la raison depuis Freud» (1957). Ces thèses ont influencé les philosophes structuralistes et des linguistes comme Barthes et Derrida.

**Lange,** Günter Gerhard – né le 12. 4. 1921 à Francfort-sur-l'Oder, Allemagne – *concepteur de polices, typographe, enseignant* – 1941–1945, études à l'Akademie für Graphische Künste und Buchgewerbe (arts graphiques et arts du livre) de Leipzig. Elève Maître chez Georg Belwe, puis assistant du professeur Walter Tiemann. Exerce comme peintre et graphiste indépendant. 1949, études à la Hochschule für Bildende Künste (école supérieure des arts plastiques) de Berlin. Travaille comme prestataire de services pour la H. Berthold AG de Berlin à partir de 1950. Enseigne la typographie à la Meisterschule für Grafik und Buchgewerbe (école professionnelle des arts graphiques et des arts du livre) de Berlin. Directeur artistique de la H. Berthold AG depuis 1961 puis responsable de l'ensemble de la gamme de polices à partir de 1971. A enseigné à l'école d'art Alsterdamm à Hambourg (à partir de 1972), à la Werbliche Akademie de Munich (à partir de 1973) au Lehrinstitut für Graphische Ge-

Jacques Lacan
Écrits

Le champ freudien
collection dirigée par Jacques Lacan
aux Éditions du Seuil, Paris

Lacan 1966 Title

ABCDEFGHIJ
KLMNOPQ
RSTUVWXYZ

abcdefghijk
lmnopqrstuvwxyz
chck ff fi fl ft ß

1234567890 &

Lange 1953 Derby

| | n | **n** | | | *n* |
|---|---|---|---|---|---|
| a | b | e | f | g | i |
| o | r | s | t | y | z |
| A | B | C | E | G | H |
| M | O | R | S | X | Y |
| 1 | 2 | 4 | 6 | 8 | & |

Lange 1968–78 Concorde

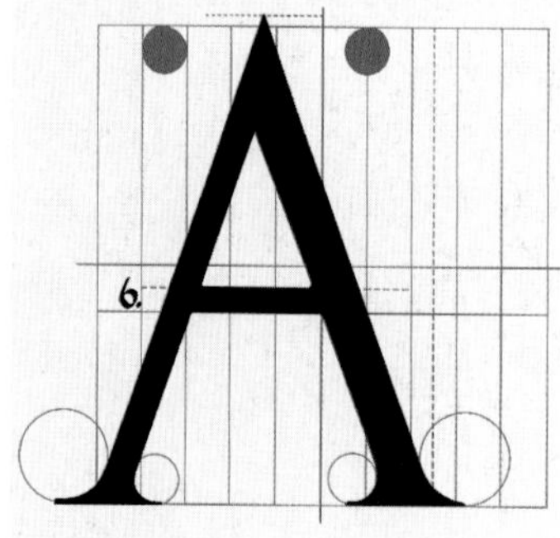


Lange Versalconstruction

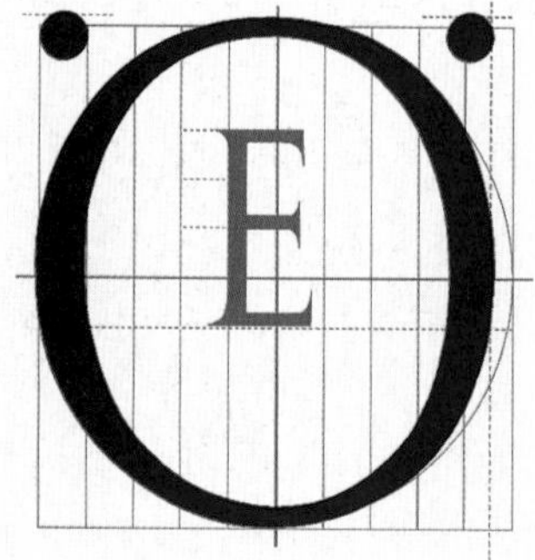


Lange Versalconstruction

**Lacan,** Jacques – b. 13. 4. 1901 in Paris, France, d. 9. 9. 1981 in Neuilly, France – *neurologist, psychiatrist* – Studied medicine. Worked at the Sainte Anne's clinic in Paris and the Burghölzi clinic in Zurich (with C. G. Jung). From 1932 onwards: connections with the Surrealists and the intellectual avant-garde. Lacan developed the theory that the structure of language brings out the unconscious in us. It forms the opposite of desire in its order. Only when both come together can the subject be created in its area of conflict. Lacan elaborated on his idea in a series of seminars and lectures, entitled "The Urge of the Letter in the Unconscious or Reason Since Freud" (1957). His theories influenced Structuralist philosophers and linguists such as Barthes and Derrida.

**Lange,** Günter Gerhard – b. 12. 4. 1921 in Frankfurt an der Oder, Germany – *type designer, typographer, teacher* – 1941–45: studies at the Akademie für Graphische Künste und Buchgewerbe in Leipzig. Student in Prof. Georg Belwe's master class. Then assistant to Prof. Walter Tiemann. Works as a freelance graphic artist and painter. 1949: studies at the Hochschule für Bildende Künste in Berlin. From 1950 onwards: works freelance for H. Berthold AG in Berlin. 1955–60: lecturer of typography at the Meisterschule für Grafik und Buchgewerbe in Berlin. From 1961 onwards: art director for H. Berthold AG in Berlin. From 1971 onwards: Lange is responsible for the company's entire type program. Lange has taught at the Kunstschule Alsterdamm in Hamburg (from 1972 onwards), at the Werbliche Akade-

demie in München (ab 1973), am Lehrinstitut für Graphische Gestaltung in München (ab 1974), an der Hochschule für Graphik und Buchkunst in Leipzig (seit 1995). Zahlreiche Auszeichnungen, u. a. F. W. Goudy-Award des Rochester Institute (1989). 1992 Ehrenmitglied des Art Directors Club von Deutschland. 1996 Ehrenmitglied der AGD Deutschland. – *Schriftentwürfe:* Arena (1951–59), Derby (1953), Regina (1954), Solemnis (1954), Boulevard (1955), Champion (1957), El Greco (1964), Concorde (1968–78), AG Buch (1969–80), Franklin Antiqua (1976), Imago (1979), Bodoni Old Face (1983).

**Larcher,** Jean – geb. 28. 1. 1947 in Rennes, Frankreich. – *Kalligraph, Typograph, Schriftentwerfer, Lehrer* – 1962–65 Ausbildung zum Typographen bei der Handelskammer in Paris. 1965–73 Arbeit als Grafik-Designer für verschiedene Werbeagenturen. 1973 Gründung des eigenen Ateliers für Typographie, Schriftkunst und Kalligraphie. Entwirft mehr als 500 Signets für Presse, Werbung, Industrie, Fernsehen und Verlage. Unterrichtet seit 1984 Typographie und Kalligraphie an Schulen in Frankreich, England, den USA, Belgien und Italien. Nach zahlreichen Schriftentwürfen und Grafik-Design-Arbeiten widmet er sich seit 1983 ausschließlich der Kalligraphie. – *Schriftentwürfe:* Plouf (1970–74), Larcher (1972), Menhir (1973–75), Guapo (1973–75), Soleil (1973–75), Larcher Outline (1974), Optical (1974), Honolulu (1974), Tornade (1974), Digitale (1974), Super Crayon (1976), Vibrator (1976), Castille-

Larcher 1978 Card

Larcher ca. 1980 Logo

Larcher 1988 Poster

Larcher 1978 Poster

Larcher 1977 Logo

staltung (institut d'enseignement des arts graphiques) de Munich (1974) et à la Hochschule für Graphik und Buchkunst (école supérieure des arts graphiques et des arts du livre) de Leipzig à partir de 1995. Nombreuses distinctions dont le F. W. Goudy Award du Rochester Institute (1989). En 1992, il devient membre d'honneur de l'Art Directors Club d'Allemagne, puis membre d'honneur de l'AGD d'Allemagne en 1996. – *Polices:* Arena (1951–1959), Derby (1953), Regina (1954), Solemnis (1954), Boulevard (1955), Champion (1957), El Greco (1964), Concorde (1968–1978), AG Buch (1969–1980), Franklin Antiqua (1976), Imago (1979), Bodoni Old Face (1983).

**Larcher,** Jean – né le 28. 1. 1947 à Rennes, France – *calligraphe, typographe, concepteur de polices, enseignant* – formation de typographe à la Chambre du commerce de Paris de 1962 à 1965. Travaille comme graphiste maquettiste pour plusieurs agences de publicité de 1965 à 1973. En 1973, il fonde son propre atelier de typographie, d'arts de l'écriture et de calligraphie. Dessine plus de 500 signets pour la presse, la publicité, l'industrie, la télévision et des maisons d'édition. A partir de 1984, il enseigne la typographie dans diverses écoles en France, en Angleterre, aux Etats-Unis, en Belgique et en Italie. Après avoir conçu de nombreuses polices et réalise une œuvre importante dans le design et les arts graphiques, il se consacre exclusivement à la calligraphie à partir de 1983. – *Polices:* Plouf (1970–1974), Larcher (1972), Menhir (1973–1975), Guapo (1973–1975), Soleil (1973–1975), Larcher Outline (1974), Optical (1974), Honolulu (1974), Tornade (1974), Digitale (1974), Super Crayon (1976), Vibrator (1976), Castillejo Bauhaus (1980),

mie in Munich (from 1973 onwards), at the Lehrinstitut für Graphische Gestaltung in Munich (from 1974 onwards), and at the Hochschule für Graphik und Buchkunst in Leipzig (since 1995), among others. He has won numerous awards, including the Rochester Institute's F. W. Goudy Award (1989). 1992: honorary member of the Art Directors Club in Germany. 1996: honorary member of the AGD in Germany. – *Fonts:* Arena (1951–59), Derby (1953), Regina (1954), Solemnis (1954), Boulevard (1955), Champion (1957), El Greco (1964), Concorde (1968–78), AG Buch (1969–80), Franklin Antiqua (1976), Imago (1979), Bodoni Old Face (1983).

**Larcher,** Jean – b. 28. 1. 1947 in Rennes, France – *calligrapher, typographer, type designer, teacher* – 1962–65: trains as a typographer at the chamber of commerce in Paris. 1965–73: works as a graphic designer for various advertising agencies. 1973: founds his own studio for typography, type and calligraphy. Designs more than 500 logos for the press, advertising, industry, television and various publishers. From 1984 onwards: teaches typography and calligraphy at various schools in France, England, USA, Belgium and Italy. From 1983 onwards: after designing numerous typefaces and producing much work in the field of graphic design, Larcher devotes himself entirely to calligraphy. – *Fonts:* Plouf (1970–74), Larcher (1972), Menhir (1973–75), Guapo (1973–75), Soleil (1973–75), Larcher Outline (1974), Optical (1974), Honolulu (1974), Tornade (1974), Digitale (1974),

jo Bauhaus (1980), Logoment condensed (1980), New Crayon (1980), Le Lancôme (1981), Latina (1987), Veloz (1987), Gautier Romain (1992). – *Publikationen u.a.:* „Graphismes cinétiques", Paris 1970; „Geometrical Designs and Optical Art", New York 1974; „Ecritures", Paris 1976; „Fantastic Alphabets", New York 1976; „Une Typographie Nouvelle", Paris 1976; „The 3-dimensional Alphabet Coloring Book", New York 1978; „Calligraphies", Paris 1984; „Allover patterns with letterforms", New York 1985; „Typomondo 5", Paris 1985; „Jean Larcher Calligraphe", Rennes 1986.

**Larisch,** Rudolf von – geb. 1. 4. 1856 in Verona, Italien, gest. 24. 3. 1934 in Wien, Österreich. – *Schriftentwerfer, Lehrer* – Studium an der Kunstgewerbeschule in Wien. 1902–10 Schriftlehrer an der Kunstgewerbeschule, 1910–20 an der Graphischen Lehr- und Versuchsanstalt und seit 1920 an der Akademie der Bildenden Künste in Wien. Das Museum für Kunst und Industrie in Wien veranstaltet 1923 eine Ausstellung mit Schriftarbeiten der Schüler Rudolf von Larischs und 1926 (zum 70. Geburtstag von Larisch) eine internationale Ausstellung künstlerischer Schrift. Von Larischs Hauptarbeit galt der Reform des Schriftunterrichts. – *Schriftentwürfe:* Plinius (1904), Wertzeichentype (1906). – *Publikationen u.a.:* „Über Zierschriften im Dienste der Kunst", Wien 1899; „Beispiele künstlerischer Schrift, 1–3", Wien 1900, 1903, 1906; „Über Leserlichkeit von ornamentalen Schriften", Wien 1904; „Unterricht in ornamentaler Schrift", Wien 1905.

Logoment Condensed (1980), New Crayon (1980), Le Lancôme (1981), Latina (1987), Veloz (1987), Gautier Romain (1992). – *Publications, sélection:* «Graphismes cinétiques», Paris 1970; «Geometrical Designs and Optical Art», New York 1976; «Ecritures», Paris 1976; «Fantastic Alphabets», New York 1976; «Une Typographie Nouvelle», Paris 1976; «The 3-dimensional Alphabet Coloring Book», New York 1978; «Calligraphies», Paris 1984; «Allover patterns with letterforms», New York 1985; «Typomondo 5», Paris 1985; «Jean Larcher Calligraphe», Rennes 1986.

**Larisch,** Rudolf von – né le 1. 4. 1856 à Vérone, Italie, décédé le 24. 3. 1934 à Vienne, Autriche – *concepteur de polices, enseignant* – études à la Kunstgewerbeschule (école des arts décoratifs) de Vienne. Enseigne les arts de l'écriture à Vienne, à l'école des arts de 1902 à 1910, à l'institut d'enseignement et de recherche pour les arts graphiques de 1910 à 1920, puis à l'académie des beaux-arts à partir de 1920. En 1923, le Museum für Kunst und Industrie de Vienne organise une exposition des travaux des élèves de Rudolf von Larisch, puis, en 1926, une exposition internationale sur les arts de l'écriture pour son 70è anniversaire. La principale tâche de Larisch a été de réformer l'enseignement de l'écriture. – *Polices:* Plinius (1904), Wertzeichentype (1906). – *Publications, sélection:* «Über Zierschriften im Dienste der Kunst», Vienne 1899; «Beispiele künstlerischer Schrift, 1–3», Vienne 1900, 1903, 1906; «Über Leserlichkeit von ornamentalen Schriften», Vienne 1904; «Unterricht in ornamentaler Schrift», Vienne 1905.

**Leck,** Bart van der – né le 26. 11. 1876 à Utrecht, Pays-Bas, décédé le 13. 11. 1958 à Blaricum, Pays-Bas – *artiste* – de 1900–1904, il fréquente la Rijksschool

BEISPIELE KÜNSTLERISCHER SCHRIFT
HERAUSGEG. V. RUDOLF VON LARISCH
MIT ORIGINALBEITRÄGEN VON

| | | | |
|---|---|---|---|
| GEORGE AURIOL | PARIS | CELDA KLOUČEK | PRAG |
| LEOPOLD BAUER | WIEN | ETHEL LARCOMBE | EXETER |
| R.P.C. DE BAZEL | SANDPOORT | EDMOND LEFEVER | BRÜSSEL |
| H.P. BERLAGE | AMSTERDAM | CH. MACKINTOSH | GLASGOW |
| ADOLPHE CRESPIN | BRÜSSEL | E. MOREAU-NÉLATON | PARIS |
| CHAS. E. DAWSON | LONDON | R. RIEMERSCHMID | MÜNCHEN |
| EMIL DOEPLER D.J. | BERLIN | JAN STUYT | AMSTERDAM |
| FIDUS - HÖPPENER - | BERLIN | HEINR. VOGELER | WORPSWEDE |
| RICHARD GRIMM | CREFELD | OTTO WAGNER | WIEN |
| MAX KLINGER | LEIPZIG | WALTER WEST | NORTHWOOD |

VERLAG: ANTON SCHROLL & C. WIEN. MCMII
ZWEITE FOLGE

Larisch 1903 Main Title

Larisch 1905 Ex Libris

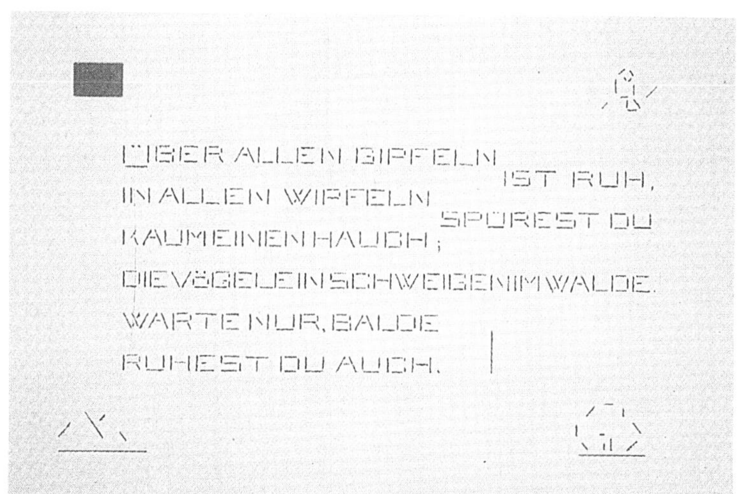


Leck 1932 Sheet

Leck 1919 Poster

Super Crayon (1976), Vibrator (1976), Castillejo Bauhaus (1980), Logoment condensed (1980), New Crayon (1980), Le Lancôme (1981), Latina (1987), Veloz (1987), Gautier Romain (1992). – *Publications include:* "Graphismes cinétiques", Paris 1970; "Geometrical Designs and Optical Art", New York 1974; "Ecritures", Paris 1976; "Fantastic Alphabets", New York 1976; "Une Typographie Nouvelle", Paris 1976; "The 3-dimensional Alphabet Coloring Book", New York 1978; "Calligraphies", Paris 1984; "Allover patterns with letterforms", New York 1985; "Typomondo 5", Paris 1985; "Jean Larcher Calligraphe", Rennes 1986.

**Larisch,** Rudolf von – b. 1. 4. 1856 in Verona, Italy, d. 24. 3. 1934 in Vienna, Austria – *type designer, teacher* – Studied at the Kunstgewerbeschule in Vienna. 1902–10: teaches lettering at the Kunstgewerbeschule, from 1910–20 at the Graphische Lehr- und Versuchsanstalt and from 1920 onwards at the Akademie der Bildenden Künste in Vienna. 1923: the Museum für Kunst und Industrie in Vienna stages an exhibition featuring lettering work by Rudolf von Larisch's students and in 1926, on the occasion of Larisch's 70th birthday, holds an international exhibition on artistic lettering. Von Larisch mainly concentrated on ways of reforming the teaching of lettering. – *Fonts:* Plinius (1904), Wertzeichentype (1906). – *Publications include:* "Über Zierschriften im Dienste der Kunst", Vienna 1899; "Beispiele künstlerischer Schrift, 1–3", Vienna 1900, 1903, 1906; "Über Leserlichkeit von ornamentalen

**Leck,** Bart van der – geb. 26. 11. 1876 in Utrecht, Niederlande, gest. 13. 11. 1958 in Blaricum, Niederlande. – *Künstler* – 1900–04 Besuch der Rijksschool voor Kunstnijverheid und von Abendkursen an der Rijksacademie voor Beeldende Kunsten in Amsterdam. Illustrationen für eine Buchausgabe des „Hohelied Salomons". 1917 Gründungsmitglied der Künstlergruppe „De Stijl". 1917 Bruch mit „De Stijl" über die Rolle der Architektur. 1935 Kontakt zur Künstlergruppe „abstraction-création" in Paris. 1949 Retrospektiv-Ausstellung im Stedelijk Museum in Amsterdam. 1951 Zusammenarbeit mit Gerrit Rietveld. 1954 Rückkehr zur abstrakten Malerei. 1957 Herausgabe autobiographischer Texte. – *Publikation:* W. C. Feltkamp „B. A. van der Leck", Leiden 1956.

**Léger,** Fernand – geb. 4. 2. 1881 in Argentan, Frankreich, gest. 17. 8. 1955 in Gif-sur-Yvette, Frankreich. – *Künstler, Lehrer* – 1897–99 Ausbildung in einem Architektur-Büro in Caen. 1903 Studium an der Ecole des Arts Décoratifs und an der Académie Julian in Paris. 1910 kubistische Farbstudien. Erste Ausstellung (mit Braque und Picasso) in der Galerie Kahnweiler in Paris. 1911 Mitglied der Gruppe „Section d'or" (mit R. Delaunay, Le Fauconnier, Kupka, Villon u. a.). 1921 Illustrationen zu „Lunes en papier" von André Malraux. Verbindung zur Künstlergruppe „De Stijl". Unterrichtet 1929 an der Académie moderne. Mitglied der Künstlergruppe „Cercle et Carré". 1935 Gestaltung des „Saals für Körperkultur" im französischen Pavillon der Weltaus-

Léger 1919 Spread

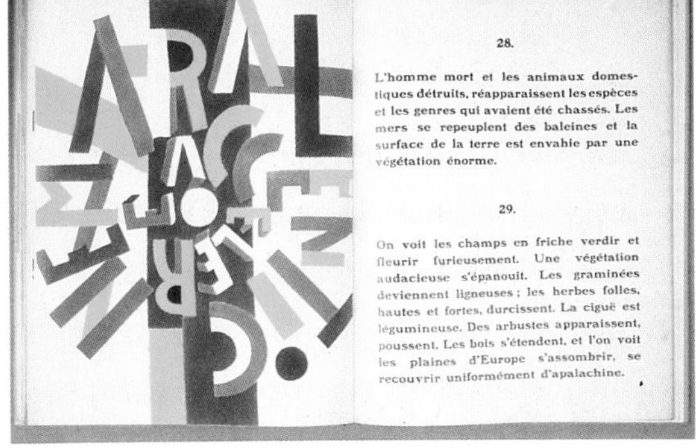
28.

L'homme mort et les animaux domestiques détruits, réapparaissent les espèces et les genres qui avaient été chassés. Les mers se repeuplent des baleines et la surface de la terre est envahie par une végétation énorme.

29.

On voit les champs en friche verdir et fleurir furieusement. Une végétation audacieuse s'épanouit. Les graminées deviennent ligneuses; les herbes folles, hautes et fortes, durcissent. La ciguë est légumineuse. Des arbustes apparaissent, poussent. Les bois s'étendent, et l'on voit les plaines d'Europe s'assombrir, se recouvrir uniformément d'apalachine.

Léger 1919 Spread

voor Kunstnijverheid et les cours du soir de la Rijksacademie voor Beeldende Kunsten à Amsterdam. Illustrations pour une édition du «Cantique des Cantiques». Membre fondateur du groupe d'artistes «De Stijl» en 1917, puis rupture avec le groupe la même année à cause du rôle attribué à l'architecture. En 1935, contacts avec le groupe d'artistes «abstraction-création» de Paris. Rétrospective au Stedelijk Museum d'Amsterdam en 1949. Collaboration avec Gerrit Rietveld en 1951. Retour à la peinture abstraite en 1954. Publie des textes autobiographiques en 1957. – *Publication :* W. C. Feltkamp «B. A. van der Leck», Leiden 1956.

**Léger,** Fernand – né le 4. 2. 1881 à Argentan, France, décédé le 17. 8. 1955 à Gif-sur-Yvette, France – *artiste, enseignant* – formation dans un cabinet d'architecte à Caen de 1897 à 1899. Etudes à l'Ecole des Arts Décoratifs et à l'Académie Julian en 1903. Etudes cubistes de couleurs en 1910. Première exposition (avec Braque et Picasso) à la galerie Kahnweiler à Paris. Membre du groupe «Section d'or» (avec R. Delaunay, Le Fauconnier, Kupka, Villon etc.) en 1911. Illustration des «Lunes en Papier» d'André Malraux en 1921. Relations avec le groupe d'artistes «De Stijl». Enseigne à l'Académie moderne en 1929. Membre du groupe d'artistes «Cercle et Carré». En 1935, il conçoit la «Salle de culture physique» du pavillon français de l'exposition universelle de Bruxelles. Vit aux

Leck 1919 Sketch

Léger 1919

Schriften", Vienna 1904; "Unterricht in ornamentaler Schrift", Vienna 1905.

**Leck,** Bart van der – b. 26. 11. 1876 in Utrecht, The Netherlands, d. 13. 11. 1958 in Blaricum, The Netherlands – *artist* – 1900–04: attends the Rijksschool voor Kunstnijverheid and evening classes at the Rijksacademie voor Beeldende Kunsten in Amsterdam. Produces illustrations for an edition of "Salomon's Song of Songs". 1917: founder member of De Stijl. 1917: breaks with De Stijl over the role of architecture. 1935: has contacts with the abstraction-création artists' group in Paris. 1949: retrospective exhibition in the Stedelijk Museum in Amsterdam. 1951: works with Gerrit Rietveld. 1954: returns to abstract painting. 1957: publishes autobiographical texts. – *Publication:* W. C. Feltkamp "B. A. van der Leck", Leiden 1956.

**Léger,** Fernand – b. 4. 2. 1881 in Argentan, France, d. 17. 8. 1955 in Gif-sur-Yvette, France – *artist, teacher* – 1897–99: trains at an architect's office in Caen. 1903: studies at the Ecole des Arts Décoratifs and at the Académie Julian in Paris. 1910: produces Cubist color studies. First exhibition (with Braque and Picasso) at the Kahnweiler gallery in Paris. 1911: member of the Section d'or group (with R. Delaunay, Le Fauconnier, Kupka and Villon, among others). 1921: produces illustrations for "Lunes en papier" by André Malraux. Contacts with De Stijl. 1929: teaches at the Académie moderne. Member of the Cercle et Carré group. 1935: designs the Hall of Body Culture for the French pavilion at the world exhibi-

stellung in Brüssel. Lebt 1940–45 in den USA, Lehrtätigkeit am Mills College in Oakland. 1945 Rückkehr nach Frankreich. 1949 Illustrationen für „Illuminations" von Arthur Rimbaud. 1950 Herausgabe seines Zyklus „Le cirque" mit 63 Lithographien durch den Verleger Tériade in Paris. 1952 Wandgemälde für das Auditorium der Vereinten Nationen in New York. – *Publikationen u. a.:* „Fonctions de la peinture", Paris 1965; „Mensch, Maschine, Malerei. Aufsätze und Schriften zur Kunst", Bern 1971; „Functions of Painting", New York 1973. Guido Le Noci „Fernand Léger, sa vie, son œuvre, son rêve", Mailand 1971; Poter De Francia „Fernand Léger", New Haven, London 1983; Gilles Néret „Fernand Léger", Paris 1990.

**Leistikow,** Hans – geb. 4. 5. 1892 in Elbing, Deutschland (heute Elblag, Polen), gest. 22. 3. 1962 in Frankfurt am Main, Deutschland. – *Grafiker, Typograph, Lehrer* – 1908–14 Studium an der Kunstakademie in Breslau. 1914–17 freier Maler in Tampadel, Schlesien. Besuch bei Walter Dexel in Jena. 1925–30 Berufung in den Magistrat der Stadt Frankfurt am Main als Leiter des grafischen Büros der Stadtverwaltung. Lehrtätigkeit an der Kunstgewerbeschule in Frankfurt sowie Abendkurse für die Gewerkschaft der Buchdrucker. Entwürfe für das Franfkurter Stadtwappen. 1926–30 Gestaltung der Zeitschrift „Das neue Frankfurt". Typographische Arbeiten für den Verlag Englert und Schlosser in Frankfurt. 1928 Mitglied des „rings neuer werbegestalter". Mit der Gruppe Ernst May gehen Leistikow

Etats-Unis de 1940 à 1945 où il enseigne au Mills College à Oakland. En 1945, retour en France. Illustre les «Illuminations» d'Arthur Rimbaud en 1949. En 1950, il publie aux éditions Tériade, le cycle «Le cirque» qui comprend 63 Lithographies. Fresques pour l'auditorium des Nations Unies à New York en 1952. – *Publications, sélection:* «Fonctions de la peinture», Paris 1965; «Mensch, Maschine, Malerei. Aufsätze und Schriften zur Kunst», Berne 1971; «Functions of Painting», New York 1973. Guido Le Noci «Fernand Léger, sa vie, son œuvre, son rêve», Milan 1971; Poter de Francia «Fernand Léger», New Haven, Londres 1983; Gilles Néret «Fernand Léger», Paris 1990.

**Leistikow,** Hans – né le 4. 5. 1892 à Elbing, Allemagne (aujourd'hui Elblag, Pologne), décédé le 22. 3. 1962 à Francfort-sur-le-Main, Allemagne – *graphiste, typographe, enseignant* – 1908–1914, études à la Kunstakademie (beaux-arts) de Breslau. Peintre indépendant à Tampadel, Silésie de 1914 à 1917. Rend visite à Walter Dexel à Iéna. Engagé par la municipalité de Francfort-sur-le-Main pour diriger le bureau de conception graphique de l'administration municipale de 1925 à 1930. Enseigne à la Kunstgewerbeschule (école des arts décoratifs) de Francfort et donne des cours du soir au Syndicat de l'industrie du livre. Dessine les armoiries de la ville de Francfort. Conçoit la maquette de la revue «Das neue Frankfurt» de 1926 à 1930. Travaux de typographie pour les éditions Englert et Schlosser à Francfort. Membre du «ring neuer werbegestalter» (cercle des nouveaux graphistes publicitaires) en 1928. Leistikow, sa femme Erika et le groupe Ernst May séjournent en Union soviétique de 1930 à 1937. Il y travaille comme dessinateur de plans et maquettiste pour la «Coopérative d'édition des travailleurs étrangers en URSS» et

Leistikow 1929 Cover

Leistikow 1929 Cover

Leistikow 1951 Poster

Leistikow 1930 Poster

Leistikow 1949 Poster

tion in Brussels. Spends 1940–45 in the USA. Teaches at Mills College in Oakland. 1945: returns to France. 1949: produces illustrations for "Illuminations" by Arthur Rimbaud. 1950: his "Le cirque" cycle containing 63 lithographs is published by Tériade in Paris. 1952: paints murals for the United Nations auditorium in New York. – *Publications include:* "Fonctions de la peinture", Paris 1965; "Mensch, Maschine, Malerei. Aufsätze und Schriften zur Kunst", Bern 1971; "Functions of Painting", New York 1973. Guido Le Noci "Fernand Léger, sa vie, son œuvre, son rêve", Milan 1971; Poter De Francia "Fernand Léger", New Haven, London 1983; Gilles Néret "Fernand Léger", Paris 1990.

**Leistikov,** Hans – b. 4. 5. 1892 in Elbing, Germany (today Elblag, Poland), d. 22. 3. 1962 in Frankfurt am Main, Germany – *graphic artist, typographer, teacher* – 1908–14: studies at the Kunstakademie in Breslau. 1914–17: freelance painter in Tampadel, Silesia. Visits Walter Dexel in Jena. 1925–30: appointed director of the city graphics studio by Frankfurt am Main's municipal authorities. Teaches at the Kunstgewerbeschule in Frankfurt and gives evening classes for the printer's labor union. Produces designs for Frankfurt's municipal coat of arms. 1926–30: designs "Das neue Frankfurt" magazine. Typographic work for Englert and Schlosser publishing house in Frankfurt. 1928: member of the ring neuer werbegestalter. 1930–37: Leistikov and his wife Erika go to the Soviet Union with the Ernst May group. Leistikov

und seine Frau Erika 1930–37 in die Sowjetunion. Leistikow arbeitet als Plangrafiker und Buchgestalter für die „Verlagsgenossenschaft ausländischer Arbeiter in der UdSSR“ und fertigt Werbearbeiten für das Mejerhold-Theater an. Bekanntschaft mit Sophie Lissitzky-Küppers und El Lissitzky. 1937 Ausweisung aus der Sowjetunion. Zahlreiche Entwürfe von Buchumschlägen, u. a. für die Verlage Aufbau, B. Schott's Söhne, Chronos, Suhrkamp, Ullstein und List. 1946 Signet-Entwürfe für den neugegründeten Deutschen Werkbund. 1947 Berufung als Stadtgrafiker nach Frankfurt. 1948–59 Leiter der Klasse für freie und angewandte Grafik an der Staatlichen Werkakademie in Kassel. 1957 Verleihung der Goethe-Plakette der Stadt Frankfurt am Main. – *Publikationen u.a.:* „Hans Leistikow“ (Katalog), Frankfurt 1963. Jörg Stürzebecher (Hrsg.) „Exemplarisch: Hans Leistikow“, Kassel 1995.

**Letraset** – *Transferschriftenhersteller* – 1956 hat C. C. J. Davies die Idee, Buchstaben und Zeichen auf eine Folie zu drucken, um sie von dort übertragen zu können. Davies gründet 1959 Letraset Ltd. in London. Erstes Letraset-Erzeugnis sind Schriften, die naß übertragen werden. 1961 Entwicklung und Vermarktung der trocken anzureibenden Schriften mit dem Titel „Instant lettering“, die als vorfabrizierte Buchstaben direkt in den Entwurf eingebracht werden können. 1970 Einführung der „Letragraphica“-Schriften, die von international führenden Gestaltern ausgewählt werden, u. a. von Herb Lubalin, Roger Excoffon, Colin For-

Letraset Transfer Sheet

Letraset 1975 Cover

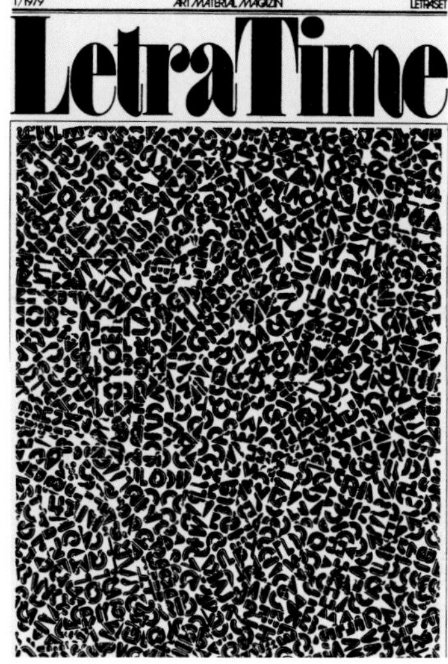

Letraset 1979 Cover

Leistikow 1932 Label

réalise des publicités pour le théâtre de Meyerhold. Rencontre Sophie Lissitzky-Küppers et El Lissitzky. Expulsé d'Union soviétique en 1937. Nombreuses maquettes de couvertures de livres pour les éditions Aufbau, B. Schott's Söhne, Chronos, Suhrkamp, Ullstein et List. En 1946, il réalise des signets pour le «Deutscher Werkbund» nouvellement reconstitué. Engagé comme graphiste de la ville de Francfort en 1947. De 1948 à 1959, il dirige le cours d'arts graphiques et de graphisme appliqué à la Staatliche Werkakademie de Kassel. Il obtient la médaille Goethe de la ville de Francfort-sur-le-Main en 1957. – *Publications, sélection :* «Hans Leistikow» (catalogue), Francfort 1963. Jörg Stürzebecher (éd.) «Exemplarisch : Hans Leistikow», Kassel 1995.

**Letraset** – *fabricant de caractères à décalquer* – en 1956, C. C. J. Davies a l'idée d'imprimer des lettres et des signes sur des films qui permettront de les transférer ensuite sur un autre support. Davies fonde Letraset Ltd. à Londres en 1959. Les premiers produits Letraset se transfèrent en les humidifiant. En 1961, la société invente et distribue des caractères pouvant être décalqués à sec, l' «Instant lettering» est constitué de lettres que l'on transfère directement sur la maquette. En 1970, la société introduit les caractères «Letragraphica», sélectionnés par des créateurs célèbres dans le monde entier comme Herb Lubalin, Roger Excoffon, Colin

works in architectural planning graphics and book design for the Publishing Cooperative of Foreign Workers in the USSR, and produces advertising work for the Mejerhold Theater. Makes the acquaintance of Sophie Lissitzky-Küppers and El Lissitzky. 1937: is deported from the Soviet Union. Designs numerous book covers, including for the publishers Aufbau, B. Schott's Söhne, Chronos, Suhrkamp, Ullstein and List. 1946: designs the logo for the new Deutscher Werkbund. 1947: appointed graphic artist by the city of Frankfurt. 1948–59: head of the free and applied graphics course at the Staatliche Werkakademie in Kassel. 1957: is awarded the Goethe plaque of the city Frankfurt am Main. – *Publications include:* "Hans Leistikow" (catalogue), Frankfurt 1963. Jörg Stürzebecher (ed.) "Exemplarisch: Hans Leistikow", Kassel 1995.

**Letraset** – *manufacturers of transfer type* – 1956: C. C. J. Davies has the idea of printing letters and symbols onto a transparent foil which can then be used to transfer them to paper or another surface. 1959: Davies founds Letraset Ltd. in London. His first Letraset products are letters which have to be wetted before being transferred to paper. 1961: develops and markets "instant lettering", pre-fabricated letters which can be transferred when dry and used directly in designs. 1970: introduces his Letragraphica typefaces selected by leading designers from all over the world, including Herb Lubalin, Roger Excoffon, Colin Forbes and Derek Birdsall. 1977: Letraset products are available

bes und Derek Birdsall. Letraset-Produkte sind 1977 in über 90 Ländern erhältlich. Colin Brignall, der zahlreiche Schriften entwirft, wird 1980 Type Director von Letraset. 1981 Eingliederung von Letraset in die schwedische Unternehmensgruppe Esselte. Die International Typeface Corporation (ITC) wird 1986 eine Abteilung von Esselte Letraset. Letraset produziert ca. 800 Schriften, von denen ca. 350 Exklusivschriften sind.

**Lettergieterij Amsterdam** – *Schriftgießerei* – Nicolaas Tetterode erwirbt 1851 die Schriftgießerei Broese & Comp. Tetterode übernimmt 1856 die Schriftgießerei De Passe en Menne. 1901 Umbenennung in „Lettergieterij Amsterdam voorheen N. Tetterode". 1906–39 Herausgabe der Hauszeitschrift „Typografische Mededeelingen". Die Lettergieterij übernimmt 1913 den Alleinvertrieb der Intertype-Setzmaschinen. 1928–41 Zusammenarbeit mit der Schriftgießerei H. Berthold AG in Berlin. 1985 Einschränkung des Schriftgusses, der zum Teil von der Fundición Typográfica Neufville in Barcelona ausgeführt wird. 1988 Offizielle Einstellung des Schriftgusses in Amsterdam.

**Lettersnijder van Rotterdam,** Hendrik Pieterzoon – *Stempelschneider, Schriftgießer, Buchdrucker* – tätig als Buchdrucker in Antwerpen (1496–1501), in Rotterdam (1501) und in Delft (1508–11). Er ist der erste spezialisierte Schriftgießer der Niederlande, der den Druckern Schriften liefert. Er schneidet ausschließlich gotische Schriften, die „Lettersnijder-Typen" genannt werden.

**Leu,** Olaf – geb. 28. 7. 1936 in Chemnitz,

Forbes et Derek Birdsall. En 1977, les produits Letraset sont distribués dans plus de 90 pays. Colin Brignall, auteur de nombreuses polices, devient directeur de la société en 1980. En 1981, Letraset est incorporé au groupe suédois Esselte. La International Typeface Corporation (ITC) devient un département d'Esselte Letraset en 1986. Letraset a produit près de 800 polices, dont près de 350 en exclusivité.

**Lettergieterij Amsterdam** – *fonderie de caractères* – Nicolaas Tetterode acquiert la fonderie de caractères Broese & Comp. en 1851, puis il reprend la fonderie de caractères De Passe en Menne en 1856. En 1901, la société est rebaptisée «Lettergieterij Amsterdam voorheen N. Tetterode». Elle publie la revue d'entreprise «Typografische Mededeelingen» de 1906 à 1939. En 1913, la Lettergieterij devient distributeur exclusif des machines à composer Intertype. Elle coopère avec la fonderie de caractères H. Berthold AG, de Berlin, de 1928 à 1941. En 1985, les activités de fonderie sont réduites et partiellement reprises par la Fundición Typográfica Neufville de Barcelone. Les activités de fonderie d'Amsterdam cessent officiellement en 1988.

**Lettersnijder van Rotterdam,** Hendrik Pieterzoon – *tailleur de types, fondeur de caractères, imprimeur* – exerce comme imprimeur de livres à Anvers (1496–1501), à Rotterdam (1501) et à Delft (1508–11). Il a été le premier fondeur de caractères des Pays-Bas à livrer des polices aux imprimeurs. Il ne taillait que des lettres gothiques appelés «Types-Lettersnijder».

**Leu,** Olaf – né le 28. 7. 1936 à Chemnitz, Allemagne – *graphiste maquettiste, enseignant* – apprentissage de composition typographique de 1951 à 1954. Designer typographe à l'atelier de la fonderie Bauer à Francfort-sur-le-Main de 1954 à 1957.

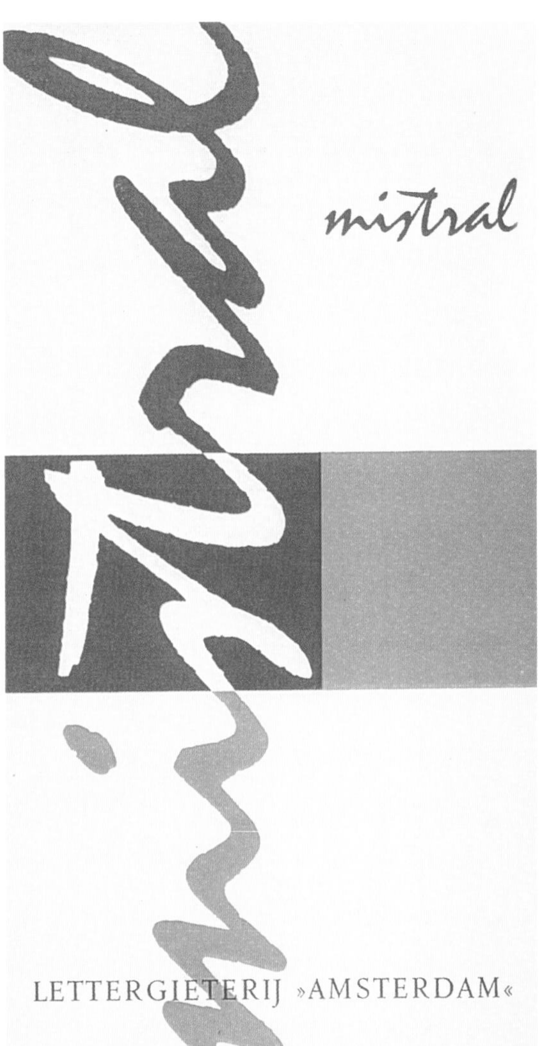


Lettergieterij Amsterdam ca. 1954 Specimen

Item eenē anderē goeden priestere opēbaerde haer ōs lieue vrouwe met haren kinde Jhesu eñ met sinte peter Ende hem werdt ghesept dat hi tot dē kerstenē segghen soude datse hem bekeeren souden tot gode: afleggheнde haer sonden. ende datse god dan vhooren soude.
Itē oec so openbaerde sinte Andries eenē goeden clerc wt Provenciē eñ hy wijsde hem een stede in sinte Peeters kercke tot Anthiochyē: daer den speer ons liefs heeren ihesu cristi begrauen lach. Ende men vantten met grooter blijtschappē. ende met groter deuociē werdt hi gheeert

Lettersnijder van Rotterdam 1495 Specimen

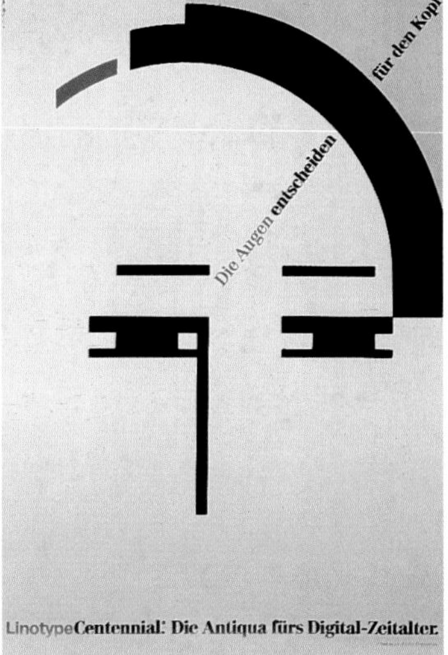


Leu 1985 Poster

Leu 1984 Logo

in over 90 countries. 1980: Colin Brignall, who designed numerous typefaces for the company, is made Letraset's type director. 1981: Letraset is incorporated into the Swedish group Esselte. 1986: the International Typeface Corporation (ITC) becomes part of Esselte Letraset. Letraset has produced c. 800 typefaces, c. 350 of them exclusive Letraset fonts.

**Lettergieterij Amsterdam** – *type foundry* – 1851: Nicolaas Tetterode acquires the Broese & Comp. type foundry. 1856: Tetterode takes over the De Passe en Menne type foundry. 1901: the foundry is renamed Lettergieterij Amsterdam voorheen N. Tetterode. 1906–39: publication of the in-house magazine "Typografische Mededeelingen". 1913: the Lettergieterij takes over the sole marketing rights of Intertype typesetting machines. 1928–41: the Lettergieterij works with the H. Berthold AG type foundry in Berlin. 1985: reduction of the foundry program which is carried out in part by the Fundición Typográfica Neufville in Barcelona. 1988: foundry production in Amsterdam is officially stopped.

**Lettersnijder van Rotterdam,** Hendrik Pieterzoon – *punch cutter, type founder, printer* – Worked as a printer of books in Antwerp (1496–1501), in Rotterdam (1501) and in Delft (1508–11). He was the first specialized type founder in the Netherlands to provide printers with typefaces. He cut only Gothic typefaces which were called "Lettersnijder" types.

**Leu,** Olaf – b. 28. 7. 1936 in Chemnitz, Germany – *graphic designer, teacher* – 1951–54: trains as a typesetter. 1954–57: ty-

Deutschland. – *Grafik-Designer, Lehrer* – 1951–54 Schriftsetzerlehre. 1954–57 typographischer Gestalter im künstlerischen Atelier der Bauerschen Gießerei in Frankfurt am Main. 1959–70 selbständig als Grafik-Designer und Art Director. Unterrichtet 1964 am London College of Printing, 1968 am Manchester College of Art and Design. 1964–72 Generalsekretär des International Center for Typographic Arts (ICTA), Sektion Deutschland. 1971–91 Studio Olaf Leu Design in Frankfurt am Main. Auftraggeber waren u. a. die Gebr. Schmidt Druckfarben, Zanders Feinpapiere, Porsche AG, Reemtsma Cigarettenfabrik, Martin Brinkmann Tabakfabriken und die Linotype AG. Unterrichtet 1975–77 an der State University of Chile. 1977 Mitbegründer der Galerie Intergraphis in München. Unterrichtet 1982–84 an der St. Martins School of Art in London. 1984–90 Chairman of the German Liaison Committee Type Directors Club of New York. Unterrichtet 1985–86 an der Staatlichen Akademie der Bildenden Künste in Stuttgart, seit 1986 an der Fachhochschule in Mainz. Arbeitet seit 1991 weltweit als Referent des Goethe-Instituts. Zahlreiche internationale Ausstellungen. Über 300 nationale und internationale Wettbewerbs-Auszeichnungen. – *Publikationen u.a.:* „Olaf Leu Graphic Design Direction 1955–1990", Mainz 1990; „Stilformen der grafischen Gestaltung", München 1993; „Corporate Design, Corporate Identity", München 1994.

**Lévy,** Jean Benoît – geb. 17. 6. 1959 in Pully, Schweiz. – *Grafik-Designer* –

Lévy 1989 Poster

Leu 1967 Advertisement

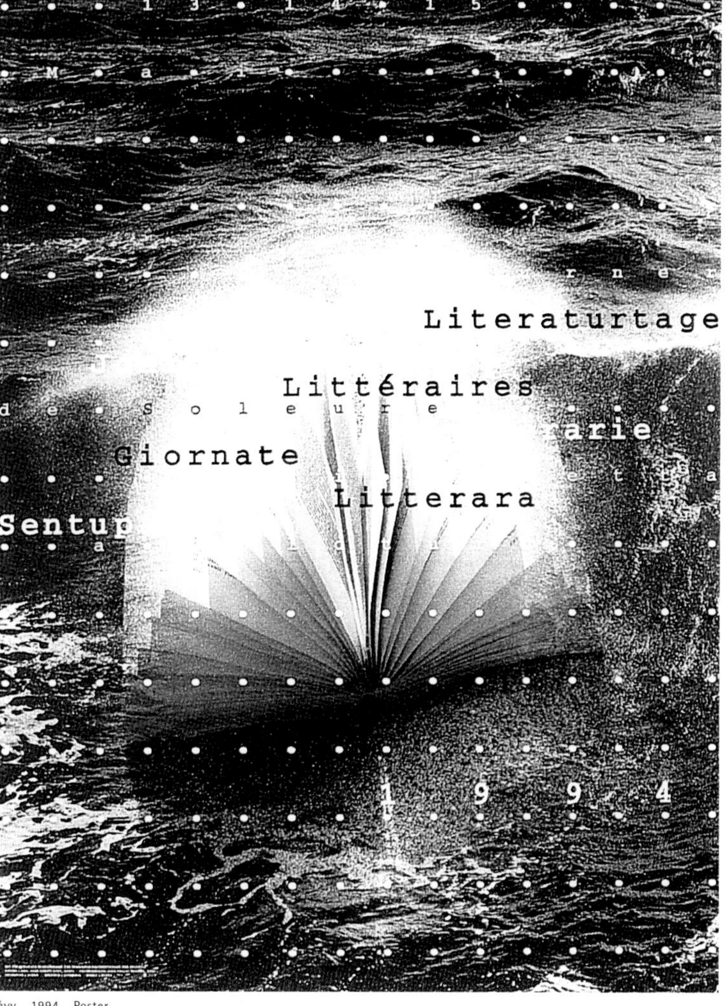


Lévy 1994 Poster

Graphiste maquettiste et directeur artistique indépendant de 1959 à 1970. En 1964, il enseigne au London College of Printing, en 1968 au Manchester College of Art and Design. Secrétaire général de la section allemande de l'International Center for Typographic Arts (ICTA) de 1964 à 1972. Exerce dans sa propre agence, le «Studio Olaf Leu Design» à Francfort-sur-le-Main de 1971 à 1991. Commanditaires : la Gebr. Schmidt Druckfarben, Zanders Feinpapiere, Porsche AG, Reemtsma Cigarettenfabrik, Martin Brinkmann Tabakfabriken et la Linotype AG. Enseigne à la State University of Chile de 1975 à 1977. Cofondateur de la galerie Intergraphis à Munich en 1977. Enseigne à la St. Martins School of Art de Londres de 1982 à 1984. Chairman of the German Liaison Committee Type Directors Club of New York de 1984 à 1990. Enseigne à la Staatliche Akademie der Bildenden Künste (beaux-arts) de Stuttgart de 1985 à 1986, puis à la Fachhochschule de Mayence depuis 1986. Travaille comme porte-parole du Goethe Institut depuis 1991 dans le monde entier. Nombreuses expositions internationales. Plus de 300 distinctions lors de concours nationaux et internationaux. – *Publications, sélection :* «Olaf Leu Graphic Design Direction 1955– 1990», Mayence 1990; «Stilformen der grafischen Gestaltung», Munich 1993; «Corporate Design, Corporate Identity», Munich 1994.

**Lévy,** Jean Benoît – né le 17. 6. 1959 à Pully, Suisse – *graphiste maquettiste* –

pographic designer at the Bauersche Gießerei's art studio in Frankfurt am Main. 1959–70: freelance graphic designer and art director. 1964: teaches at the London College of Printing and in 1968 at the Manchester College of Art and Design. 1964–72: secretary general for the German section of the International Center for Typographic Arts (ICTA). 1971–91: Studio Olaf Leu Design in Frankfurt am Main. Clients include Gebr. Schmidt Druckfarben, Zanders Feinpapiere, Porsche AG, Reemtsma Cigarettenfabrik, Martin Brinkmann Tabakfabriken and Linotype AG. 1975–77: teaches at the State University of Chile. 1977: co-founder of the Intergraphis gallery in Munich. 1982–84: teaches at St. Martins School of Art in London. 1984–90: chairman of the German Liaison Committee Type Directors Club of New York. 1985–86: teaches at the Staatliche Akademie der Bildenden Künste in Stuttgart and from 1986 onwards at the Fachhochschule in Mainz. 1991: Leu starts lecturing for the Goethe Institute on an international scale. Leu has taken part in numerous exhibitions all over the world and has won over 300 national and international awards in various competitions. – *Publications include:* "Olaf Leu Graphic Design Direction 1955–1990", Mainz 1990; "Stilformen der grafischen Gestaltung", Munich 1993; "Corporate Design, Corporate Identity", Munich 1994.

**Lévy,** Jean Benoît – b. 17. 6. 1959 in Pully, Switzerland – *graphic designer* – 1978–83: studies at the Schule für Gestaltung

1978–83 Studium an der Schule für Gestaltung in Basel. 1983–86 Grafik-Designer der Schweizer Mustermesse. Danach freischaffend tätig. 1989 Eröffnung seines Studios „&, Trafic Grafic". Aufträge aus Wirtschaft und Kultur, u. a. für das Kongreßzentrum in Basel. Zahlreiche Auszeichnungen.
**Lewis,** John – geb. 11. 12. 1912 in Rhoose, England. – *Typograph, Illustrator, Autor, Lehrer* – Studium am Goldsmith's College School of Art. 1946 Eintritt in die Druckerei W. S. Cowell Ltd. in Ipswich. 1948–54 grafische Arbeiten für das Aldeburgh Festival. Unterrichtet 1951–64 Grafik-Design am Royal College of Art in London. 1953–64 Gründung und Leitung der Privatpresse „Lion and Unicorn Press Books" des Royal College of Art, die bis 1970 besteht. 1960 Re-Design des grafischen Erscheinungsbildes des Aldeburgh Festivals. 1963–72 Herausgeber und Gestalter zahlreicher Bände der Serie Reinhold Art Paperback für das Studio Vista. Seit 1964 typographischer Berater der Universität Essex. 1969 Präsident des „Double Crown Club" in London. – *Publikationen u.a.:* „A Handbook of Printing Types", London 1947; „Graphic-Design" (mit John Brinkley), London 1954; „Printed Ephemera", London 1962; „Typography: Basic Principles", London 1963; „Illustration: Aspects and Directions" (mit Bob Gill), London 1964; „The 20th Century Book", London 1964; „Typography: Design and Practice" London 1977; „Such Things Happen", Stowmarket 1990.
**Licko,** Zuzana – geb. 2. 10. 1961 in Brati-

études à la Schule für Gestaltung (école de design) de Bâle de 1978 à 1983. Graphiste maquettiste du salon suisse du modèle de 1983 à 1986, exerce ensuite comme indépendant. Ouvre son agence «&, Trafic Grafic» en 1989. Commandes d'entreprises et d'organismes culturels, entre autres pour le Centre des Congrès de Bâle. Nombreuses distinctions.
**Lewis,** John – né le 11. 12. 1912 à Rhoose, Angleterre – *typographe, illustrateur, auteur, enseignant* – études au Goldsmith's College School of Art. Entre à l'imprimerie W. S. Cowell Ltd. d'Ipswich en 1946. Travaux graphiques pour le festival d'Aldeburgh de 1948 à 1954. Enseigne les arts graphiques et le design au Royal College of Art de Londres de 1951 à 1964. En 1953, il fonde la presse privée «Lion and Unicorn Press Books» du Royal College of Art et la dirige jusqu'en 1964, la presse existera jusqu'en 1970. Refonte de l'identité graphique du festival d'Aldeburgh en 1960. Edite et réalise la maquette de nombreux volumes de Reinhold Art Paperback Serie pour l'atelier Vista de 1963 à 1972. Conseiller typographique de l'université d'Essex depuis 1964. Président du «Double Crown Club» de Londres en 1969. – *Publications, sélection:* «A Handbook of Printing Types», Londres 1947; «Graphic-Design» (avec John Brinkley), Londres 1954; «Printed Ephemera», Londres 1962; «Typography: Basic Principles», Londres 1963; «Illustration: Aspects and Directions» (avec Bob Gill), Londres 1964; «The 20th Century Book», Londres 1964; «Typography: Design and Practice», Londres 1977; «Such Things Happen», Stowmarket 1990.
**Licko,** Zuzana – née le 2. 10. 1961 à Bratislava, Tchécoslovaquie – *graphiste maquettiste, conceptrice de polices* – s'installe aux Etats-Unis en 1968. Etudes à l'University of California de Berkeley de

Lewis 1990 Cover

Licko 1988 Lunatix

Licko 1994 Dogma

Licko 1990 Totally Gothic

in Basle. 1983–86: graphic designer for the Schweizer Mustermesse. After this he works freelance. 1989: opens his studio &, Trafic Grafic. Has work commissioned by industry and the arts sector, including work for the congress center in Basle. Numerous awards.
**Lewis,** John – b. 11. 12. 1912 in Rhoose, England – *typographer, illustrator, author, teacher* – Studied at Goldsmith's College School of Art. 1946: joins the W. S. Cowell Ltd. printing works in Ipswich. 1948–54: does graphic work for the Aldeburgh Festival. 1951–64: teaches graphic design at the Royal College of Art in London. 1953–64: founds and runs the Royal College of Art's private press, Lion and Unicorn Press Books, in operation until 1970. 1960: redesigns the Aldeburgh Festival's corporate identity (graphics). 1963–72: publishes and designs numerous books of the Reinhold Art Paperback series for Studio Vista. From 1964 onwards: typography consultant to the University of Essex. 1969: president of the Double Crown Club in London. – *Publications include:* "A Handbook of Printing Types", London 1947; "Graphic-Design" (with John Brinkley), London 1954; "Printed Ephemera", London 1962; "Typography: Basic Principles", London 1963; "Illustration: Aspects and Directions" (with Bob Gill), London 1964; "The 20th Century Book", London 1964; "Typography: Design and Practice", London 1977; "Such Things Happen", Stowmarket 1990.
**Licko,** Zuzana – b. 2. 10. 1961 in Bratislava, Czechoslovakia – *graphic designer,*

slava, Tschechoslowakei. – *Grafik-Designerin, Schriftentwerferin* – 1968 Umzug in die USA. 1981–85 Studium an der University of California in Berkeley. Schriftexperimente mit „Font-Editor". 1983 Heirat und Zusammenarbeit mit Rudy VanderLans, der 1984 die Zeitschrift „Emigre" gründet. Ab Heft 2/1985 werden Lickos Schriften bei der Gestaltung der Zeitschrift eingesetzt. 1985 Gründung von „Emigre Fonts" zur - Vermarktung der Schriften von Licko und anderen Gestaltern (u. a. J. Keedy, B. Deck, J. Downer). 1986 Arbeiten für die neugegründete Zeitschrift „Mac Week" (mit Rudy VanderLans). Arbeiten für Adobe Systems Inc. Kontinuierliche Herausgabe von Emigre-Schriftkatalogen. – *Schriftentwürfe:* Modula (1985), Oakland (1985), Universal (1985), Emigre (1986), Emperor (1986), Citizen (1986), Matrix (1986), Lunatix (1988), Oblong (1988), Senator (1988), Elektrix (1989), Triplex Roman (1989), Variex (mit R. VanderLans, 1989), Journal (1990), Totally Gothic (1990), Quartett (1992), Narly (1993), Dogma (1994), Whirligig (1994), Soda Script (1995), Base-9, 12 (1996).

**Lienemeyer,** Gerhard – geb. 1. 2. 1936 in Bielefeld, Deutschland. – *Grafik-Designer, Typograph, Lehrer* – 1953–56 Lehre als grafischer Zeichner. 1956–57 Höhere Fachschule für das grafische Gewerbe in Stuttgart. 1959–64 Studium der Visuellen Kommunikation an der Staatlichen Werkakademie in Kassel bei Prof. Hans Hillmann. 1960 Beginn der Zusammenarbeit mit Gunter Rambow (Rambow + Lienemeyer, ab 1973 mit van de Sand).

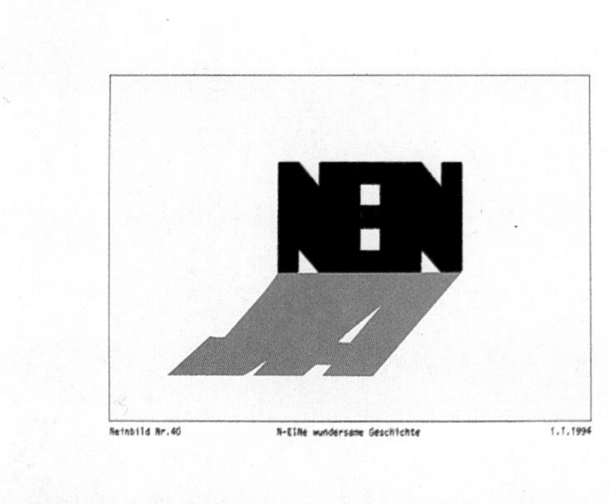

Lienemeyer 1994 Poem

| | n | n | | | |
|---|---|---|---|---|---|
| a | b | e | f | g | i |
| o | r | s | t | y | z |
| A | B | C | E | G | H |
| M | O | R | S | X | Y |
| 1 | 2 | 4 | 6 | 8 | & |

Licko 1986 Citizen

| | n | n | | | n |
|---|---|---|---|---|---|
| a | b | e | f | g | i |
| o | r | s | t | y | z |
| A | B | C | E | G | H |
| M | O | R | S | X | Y |
| 1 | 2 | 4 | 6 | 8 | & |

Licko 1990 Journal

Lienemeyer 1980 Cover

Lienemeyer 1992 Poem

1981 à 1985. Expérience de l'écriture chez «Font-Editor». En 1983, elle épouse Rudy VanderLans qui fondera la revue «Emigre» en 1984 et travaille avec lui. Les polices de Licko sont utilisées par la revue à partir de 2/1985. Création de «Emigre Fonts» en 1985, pour diffuser les polices de Licko et d'autres créateurs (J. Keedy, B. Deck, J. Downer etc.). Travaille pour la nouvelle revue «Mac Week» (avec Rudy VanderLans) en 1986. Travaux pour Adobe Systems Inc., et publication du catalogue de polices d'Emigre. – *Polices :* Modula (1985), Oakland (1985), Universal (1985), Emigre (1986), Emperor (1986), Citizen (1986), Matrix (1986), Lunatix (1988), Oblong (1988), Senator (1988), Elektrix (1989), Triplex Roman (1989), Variex (avec R. VanderLans, 1989), Journal (1990), Totally Gothic (1990), Quartett (1992), Narly (1993), Dogma (1994), Whirligig (1994), Soda Script (1995), Base-9, 12 (1996).

**Lienemeyer,** Gerhard – né le 1. 2. 1936 à Bielefeld, Allemagne – *graphiste maquettiste, typographe, enseignant* – études de graphisme et de dessin de 1953 à 1956. Puis à la Höhere Fachschule für das grafische Gewerbe (école supérieure des métiers des arts graphiques) à Stuttgart de 1956 à 1957. Etudie la communication visuelle à la Staatliche Werkakademie (arts et métiers) de Kassel chez le professeur Hans Hillmann. Commence à travailler avec Gunter Rambow en 1960 (Rambow + Lienemeyer à partir de 1973, avec van de Sand). Travaux pour la Deutsche Lite-

*type designer* – 1968: moves to the USA. 1981–85: studies at the University of California in Berkeley. Experiments with type using Font-Editor. 1983: marries and starts working with Rudy VanderLans, who in 1984 launches "Emigre" magazine. Licko's typefaces are used in the magazine design from 2/1985 onwards. 1985: Emigre Fonts is launched with the aim of marketing typefaces by Licko and other designers (including J. Keedy, B. Deck and J. Downer). 1986: produces work for the new "Mac Week" magazine with Rudy VanderLans and for Adobe Systems Inc. Continuous publication of Emigre fonts type catalogues. – *Fonts:* Modula (1985), Oakland (1985), Universal (1985), Emigre (1986), Emperor (1986), Citizen (1986), Matrix (1986), Lunatix (1988), Oblong (1988), Senator (1988), Elektrix (1989), Triplex Roman (1989), Variex (with R. VanderLans, 1989), Journal (1990), Totally Gothic (1990), Quartett (1992), Narly (1993), Dogma (1994), Whirligig (1994), Soda Script (1995), Base-9, 12 (1996).

**Lienemeyer,** Gerhard – b. 1. 2. 1936 in Bielefeld, Germany – *graphic designer, typographer, teacher* – 1953–56: trains as a graphics draftsman. 1956–57: attends the Höhere Fachschule für das grafische Gewerbe in Stuttgart. 1959–64: studies visual communication at the Staatliche Werkakademie in Kassel under Prof. Hans Hillmann. 1960: starts working with Gunter Rambow (Rambow + Lienemeyer; from 1973 onwards with van de Sand). Produces work for the Deutsche Literaturarchiv, the Deutsche Bibliothek, S. Fi-

Arbeiten u. a. für das Deutsche Literaturarchiv, die Deutsche Bibliothek, den S. Fischer Verlag und das Bauhaus-Archiv in Berlin. Zahlreiche Auszeichnungen. Seit 1988 ein eigenes Büro in Offenbach am Main. Lehrtätigkeiten: Staatliche Werkakademie Kassel (1966–70), Fachhochschule Niederrhein in Krefeld (1973–74), Fachhochschule Würzburg-Schweinfurt in Würzburg (1978–92), Hochschule für Gestaltung in Offenbach (1992), Fachhochschule Darmstadt (1994–96).

**Linotype** – *Schriftenhersteller* – Ottmar Mergenthaler stellt 1886 die erste Zeilensetz- und Gießmaschine mit dem Namen „Blower" vor. 1890 Gründung der Mergenthaler Linotype Company in Brooklyn, USA und der Mergenthaler Linotype & Machinery Ltd. in Manchester, England. 1896 Gründung eines Zweigbetriebs in Berlin. 1900 Vertrag mit der Schriftgießerei D. Stempel AG in Frankfurt am Main über den Schnitt von Schriften für Linotype. Mit einer arabischen Schrift wird 1911 die Entwicklung nichtlateinischer Schriften eingeleitet. Für den Zeitungssatz wird 1922 die Schrift „Ionic" (von C. H. Griffith) hergestellt. 1935 erste Linotype-Maschine mit Lochstreifen-Eingabe. 1958 Einführung des „Linofilm-Systems", dem ersten Fotosatzsystem der Welt. 1976 Einstellung der Bleisatzproduktion. Produktion der elektronischen Fotosatzanlagen „Linotron" und „Linotronic". 1983 Verlegung der Konzernzentrale nach Eschborn. 1984 Produktion des Linotronic 300 Laserbelichters. 1985 Restbeteiligung an der D. Stempel AG. Integration der Fotosatz-Schriftenferti-

raturarchiv, la Deutsche Bibliothek, les éditions S. Fischer et le Bauhaus-Archiv de Berlin. Nombreuses distinctions. Travaille dans son propre atelier à Offenbach depuis 1988. Enseignement: Staatliche Werkakademie, Kassel (1966–1970), Fachhochschule Niederrhein (école supérieure professionnelle du Bas-Rhin) à Krefeld (1973–1974), Fachhochschule Würzburg-Schweinfurt à Würzburg (1978–1992), Hochschule für Gestaltung (école supérieure de design) d'Offenbach (1992), Fachhochschule de Darmstadt (1994–1996).

**Linotype** – *fabricant de polices* – en 1886, Ottmar Mergenthaler présente la première machine «Blower» qui fond et compose des lignes entières. 1890, création de la Mergenthaler Linotype Company à Brooklyn, Etats-Unis et de la Mergenthaler Linotype & Machinery Ltd. à Manchester, Angleterre. Implantation d'une filiale à Berlin en 1896. En 1900, un contrat est conclu avec la fonderie de caractères D. Stempel AG de Francfort-sur-le-Main portant sur la taille de caractères destinés à Linotype. Début de la conception de caractères non latins en 1911, avec l'introduction d'une police arabe. Produit «Ionic» (de C. H. Griffith) en 1922 destinée à la presse quotidienne. Première machines Linotype à bandes perforées en 1935. Introduction du «Linofilm-System», premier système de photocomposition au monde, en 1958. La production de caractères en plomb cesse en 1976. Production d'équipements électroniques de photocomposition, le «Linotron» et le «Linotronic». Le siège du groupe s'installe à Eschborn en 1983. Production du Linotronic 300 Laser en 1984. Participation à la D. Stempel AG en 1985, et intégration de la photocomposition et de la fabrication de polices. Production de la police «Linotype Centennial» (d'A. Frutiger) en

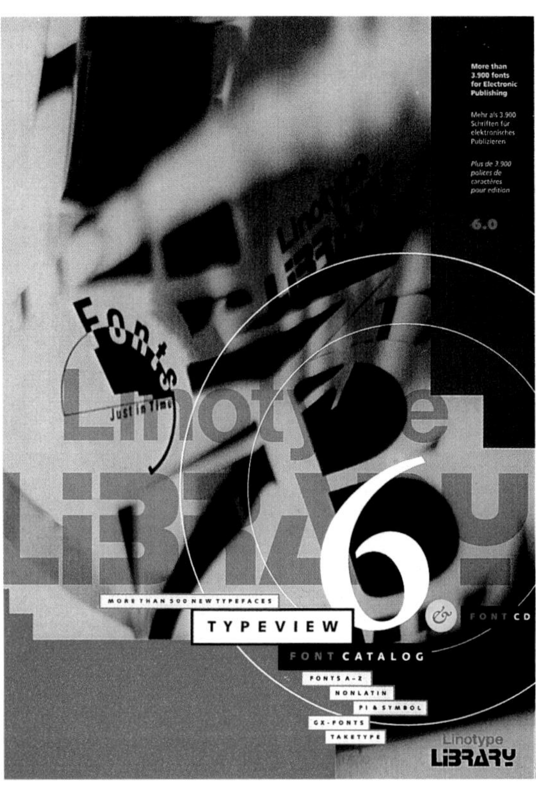


Linotype 1996 Cover

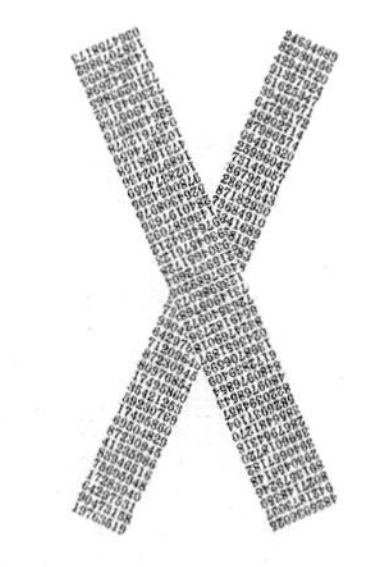

Lionni 1956 Spread

Lionni 1956 Spread

Linotype 1995 Poster

scher Verlag and the Bauhaus-Archiv in Berlin, among others. Numerous awards. From 1988 onwards: has his own studio in Offenbach am Main. Teaching positions: Staatliche Werkakademie in Kassel (1966–70), Fachhochschule Niederrhein in Krefeld (1973–74), Fachhochschule Würzburg-Schweinfurt in Würzburg (1978–92), Hochschule für Gestaltung in Offenbach (1992), Fachhochschule Darmstadt (1994–96).

**Linotype** – *type manufacturers* – 1886: Ottmar Mergenthaler introduces "Blower", the first line composing and casting machine. 1890: the Mergenthaler Linotype Company is founded in Brooklyn, USA. Mergenthaler Linotype & Machinery Ltd. is founded in Manchester, England. 1896: a branch of the company is opened in Berlin. 1900: a contract is signed with the D. Stempel AG type foundry in Frankfurt am Main governing the cutting of type for Linotype. 1911: Linotype starts developing non-Latin alphabets with an Arabic typeface. 1922: Linotype manufactures a typeface for newspaper typesetting Ionic by C. H. Griffith. 1935: the first Linotype machine using punched tape is introduced. 1958: the first filmsetting system in the world, the Linofilm system, is introduced. 1976: the production of lead type is stopped. The Linotron and Linotronic electronic filmsetting machines are produced. 1983: the company moves its head offices to Eschborn. 1984: the Linotronic 300 laser exposer unit is produced. 1985: Linotype buys up the remaining shares in D. Stempel AG. Integrates the production of type using film-

gung. 1986 Produktion der Schrift „Linotype Centennial" (von A. Frutiger). 1990 Zusammenschluß der Linotype AG mit der Dr.-Ing. Rudolf Hell GmbH in Kiel zur Linotype-Hell AG.
**Lionni,** Leo – geb. 5. 5. 1910 in Amsterdam, Niederlande. – *Illustrator, Maler, Grafik-Designer, Schriftsteller, Lehrer* – 1925 Umzug nach Italien. 1928–30 Studium an der Universität Zürich, 1926–28 Universität Genua. Doktor der Nationalökonomie. Als Künstler Autodidakt. 1931–39 Arbeit als Gestalter in Mailand. 1931 Mitglied der futuristischen Bewegung Italiens auf Einladung Marinettis. 1939 Emigration in die USA. 1939–47 Arbeit als Gestalter in Philadelphia, 1949–60 in New York: Art Director der Zeitschrift „Fortune". 1950–57 Design-Director der Firma Olivetti America. Herausgeber der Zeitschrift „Print" (1953–56). 1951 Mitbegründer der Aspen Design Conference in Colorado. Seit 1961 Arbeit als Gestalter in San Bernardo, Italien. Herausgeber der Zeitschrift „Panorama" (1962–63). Zahlreiche internationale Ausstellungen und Auszeichnungen, u. a. Art Directors Hall of Fame (1974). Unterrichtet an der Charles Morris Price School in Philadelphia (1939–47), am Black Mountain College in North Carolina (1946), an der Parson's School of Design in New York (1954), an der University of Illinois (1967). – *Publikationen u.a.:* „Little Blue and Little Yellow", New York 1959; „Designs for the printed page", New York 1960; „The alphabet tree", New York 1968; „Leo Lionni: Plastiken, Ölbilder, Zeichnungen, Druckgrafik", Köln 1974.

Lionni Cover

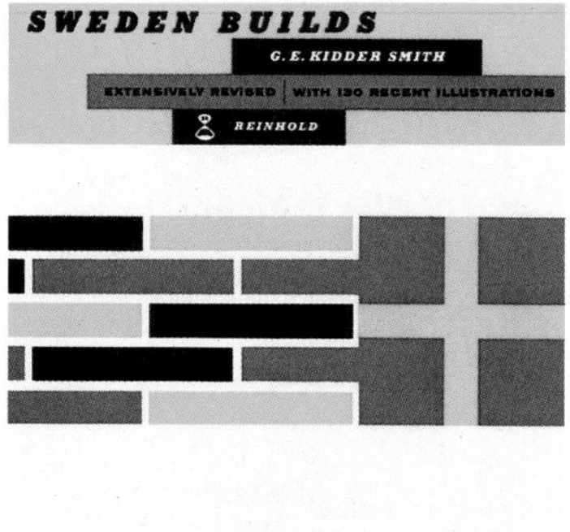

Lionni Cover

Lionni 1960 Cover

1986. Fusion de la Linotype AG et de la Dr.-Ing. Rudolf Hell GmbH de Kiel en 1990 et constitution de la Linotype-Hell AG.
**Lionni,** Léo – né le 5. 5. 1910 à Amsterdam, Pays-Bas – *illustrateur, peintre, graphiste maquettiste, écrivain, enseignant* – s'installe en Italie en 1925. Etudes à l'université de Zurich de 1928 à 1930 et à l'université de Gênes de 1926–28. Docteur en économie nationale. Artiste autodidacte. De 1931 à 1939 il travaille comme designer à Milan. Membre du mouvement futuriste italien sur invitation de Marinetti, en 1931. Emigre aux Etats-Unis en 1939. Travaille comme designer à Philadelphie de 1939 à 1947, puis à New York de 1949 à 1960 directeur artistique de la revue «Fortune». De 1950–1957 Directeur du design d'Olivetti America. Edite la revue «Print» (1953–1956). Cofondateur en 1951 de la Aspen Design Conference au Colorado. Travaille comme designer à San Bernardo, Italie, à partir de 1961. Editeur de la revue «Panorama» (1962–1963). Nombreuses expositions internationales et distinctions, entre autres elu au Art Directors Hall of Fame (1974). A enseigné à la Charles Morris Price School de Philadelphie (1939–1947), au Black Mountain College de Caroline du Nord (1946), à la Parson's School of Design de New York (1954), à l'University of Illinois (1967). – *Publications, sélection:* «Little Blue and Little Yellow», New York 1959; «Designs for the printed page», New York 1960; «The alphabet tree», New York 1968; «Leo Lionni: Plastiken, Ölbilder, Zeichnungen, Druckgrafik», Cologne 1974.

setting. 1986: produces A. Frutiger's Linotype Centennial font. 1990: Linotype AG merges with Dr.-Ing. Rudolf Hell GmbH in Kiel to form Linotype-Hell AG.
**Lionni,** Leo – b. 5. 5. 1910 in Amsterdam, The Netherlands – *illustrator, painter, graphic designer, author, teacher* – 1925: moves to Italy. 1928–30: studies at the University of Zurich and from 1926–28 at the University of Genoa. Doctor of economics and self-taught artist. 1931–39: works as a designer in Milan. 1931: joins Italy's Futurist movement at the invitation of Marinetti. 1939: emigrates to the USA. 1939–47: works as a designer in Philadelphia and from 1949–60 in New York. Art director for "Fortune" magazine. 1950–57 Design director for Olivetti America. Editor of "Print" magazine (1953–56). 1951: co-founder of the Aspen Design Conference in Colorado. From 1961 onwards: works as a designer in San Bernardo, Italy. Editor for "Panorama" magazine (1962–63). Numerous international exhibitions and awards, including being elected into the Art Directors Hall of Fame (1974). Lionni has taught at the Charles Morris Price School in Philadelphia (1939–47), at Black Mountain College in North Carolina (1946), at the Parson's School of Design in New York (1954) and at the University of Illinois (1967). – *Publications include:* "Little Blue and Little Yellow", New York 1959; "Designs for the printed page", New York 1960; "The alphabet tree", New York 1968; "Leo Lionni: Plastiken, Ölbilder, Zeichnungen, Druckgrafik", Cologne 1974.

**Lissitzky,** El (Lasar Markowitsch) – geb. 23. 11. 1890 in Potschinok, Rußland, gest. 30. 12. 1941 in Moskau, UdSSR. – *Maler, Architekt, Grafiker, Typograph, Lehrer* – 1909 Umzug nach Deutschland. Studium an der Polytechnischen Hochschule in Darmstadt. 1914 Rückkehr nach Moskau. 1917–20 zahlreiche Illustrationen in jüdischen Publikationen. 1918 Architektur-Diplom. 1919 Leiter der Werkstatt für Grafik, Druck und Architektur an der Kunsthochschule in Witebsk. Begegnung mit Kasimir Malewitsch. 1921 Leiter der Architekturfakultät an den Höheren künstlerisch-technischen Werkstätten (WChUTEMAS) in Moskau. 1922 Herausgabe der Zeitschrift „Gegenstand" (mit Ilja Ehrenburg) in Berlin. Zahlreiche typographische Arbeiten für den Skythen-Verlag. Entwurf des Buchs „Suprematičeskij skaz pro dva kvadrata v 6ti postrojkach" („Von zwei Quadraten") und „Dlja golosa" („Für die Stimme"). Teilnahme an der 1. Russischen Kunstausstellung in Berlin. Produktion der „1. Kestner-Mappe", der „Figurinen-Mappe" und dem „Proun-Raum". Mitarbeit an der Zeitschrift „ABC – Beiträge zum Bauen". Es entstehen der „Wolkenbügel", die „Lenintribüne", der „Konstrukteur", „Nasci" (mit Kurt Schwitters). 1925 Rückkehr nach Moskau. Bis 1930 Professor an der WChUTEMAS. 1932–40 Mitarbeit an der Zeitschrift „USSR im Bau". 1936–39 mehrmonatige Kuraufenthalte. 1940 Entwurf des Restaurants der sowjetischen Abteilung der Weltausstellung in New York. – *Publikationen u.a.:* „Die Kunstismen" (mit Hans Arp), Zürich 1925. Sophie

**Lissitzky,** El (Lazar Markovitch) – né le 23. 11. 1890 à Potchinok, Russie, décédé le 30. 12. 1941 à Moscou, URSS – *peintre, architecte, graphiste, typographe, enseignant* – s'installe en Allemagne en 1909. Etudes à la Polytechnische Hochschule (école polytechnique) de Darmstadt. Retourne à Moscou en 1914. Réalise de nombreuses illustrations pour des publications juives de 1917 à 1920. Diplôme d'architecte en 1918. Dirige l'atelier d'arts graphiques, d'impression et d'architecture de l'école des beaux-arts de Vitebsk en 1919. Rencontre Kasimir Malévitch. En 1921, il dirige la faculté d'architecture des ateliers des arts et des techniques (VKHOUTEMAS) à Moscou. Edite la revue «Objet» (avec Ilia Ehrenburg) en 1922. Nombreux travaux typographiques pour les éditions Skythes. Maquette du livre «Suprematičeskij skaz pro dva kvadrata v 6ti postrojkach» (Deux carrés) et «Dlja golosa» (Pour la voix). Participe à la première exposition d'art russe de Berlin. Production de la première «Kestner-Mappe» (Folio Kestner), du «Folio aux figurines» et de «L'espace Proun». Travaille à la revue «ABC – Beiträge zum Bauen». Il réalise «L'arc aux nuages», la «Tribune de Lénine», le «Constructeur», «Nasci» (avec Kurt Schwitters). Retour à Moscou en 1925. Professeur au VKHOUTEMAS jusqu'en 1930. Travaille pour la revue «URSS en construction» de 1932 à 1940. Séjour de plusieurs mois en cure entre 1936 et 1939. En 1940, il dessine le restaurant soviétique de l'exposition universelle de New York. – *Publications, sélection:* «Die Kunstismen» (avec Jean Arp), Zurich 1925. Sophie Lissitzky-Küppers «El Lissitzky», Dresde 1967; Sophie Lissitzky-Küppers, El Lissitzky «Schriften, Briefe, Dokumente», Dresde 1977.

**Loesch,** Uwe – né le 23.1. 1943 à Dresde, Allemagne – *maquettiste, typographe,*

El Lissitzky 1922 Spread

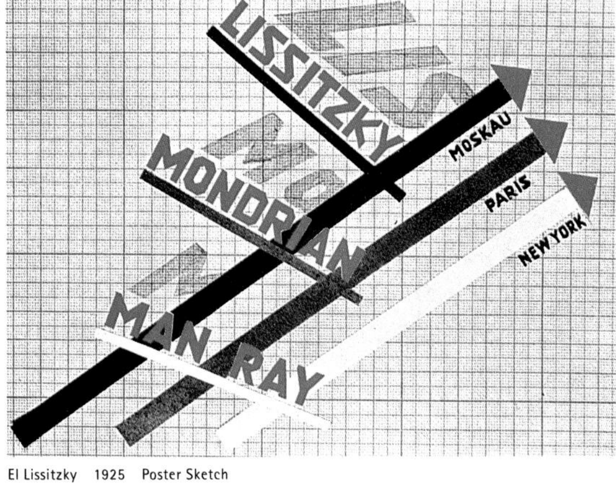

El Lissitzky 1925 Poster Sketch

El Lissitzky 1927 Cover

Loesch 1990 Poster

**Lissitzky,** El (Lazar Markovich) – b. 23. 11. 1890 in Pochinok, Russia, d. 30. 12. 1941 in Moscow, USSR – *painter, architect, graphic artist, typographer, teacher* – 1909: moves to Germany. Studies at the Polytechnische Hochschule in Darmstadt. 1914: returns to Moscow. 1917–20: produces numerous illustrations for Jewish publications. 1918: diploma in architecture. 1919: head of the workshop for graphics, printing and architecture at the art school in Vitebsk. Meets Kasimir Malevich. 1921: head of the faculty of architecture at the Higher Workshops for Art and Technology (VKHUTEMAS) in Moscow. 1922: edits "Weschtsch" (object) magazine with Ilja Ehrenburg in Berlin. Produces much typographic work for Skythen publishers. Designs the books "Suprematičeskij skaz pro dva kvadrata v 6ti postrojkach" ("of two squares") and "Dlja golosa" ("for the voice"). Takes part in the 1st exhibition of Russian art in Berlin. Produces the "1. Kestner-Mappe", the "Figurinen-Mappe" and his "Proun-Raum". Works on the magazine "ABC – Beiträge zum Bauen". He also produces "Cloud Hanger", his "Lenin Rostrum", the "Designer", "Nasci" (with Kurt Schwitters). 1925: returns to Moscow. Professor at the VKHUTEMAS until 1930. 1932–40: works on the magazine "USSR im Bau". 1936–39: spends periods of several months on health cures. 1940: designs the restaurant for the Soviet section of the world exhibition in New York. – *Publications include:* "Die Kunstismen" (with Hans Arp), Zurich 1925. Sophie Lissitzky-Küppers "El Lissitzky", Dresden 1967; S.

Lissitzky-Küppers „El Lissitzky", Dresden 1967; Sophie Lissitzky-Küppers, El Lissitzky „Schriften, Briefe, Dokumente", Dresden 1977.

**Loesch,** Uwe – geb. 23. 1. 1943 in Dresden, Deutschland. – *Grafik-Designer, Typograph, Lehrer* – 1964–68 Studium an der Peter-Behrens-Werkkunstschule Düsseldorf. 1970 Gründung der „Arbeitsgemeinschaft für visuelle und verbale Kommunikation" in Düsseldorf. Er ist Mitglied der Alliance Graphique Internationale, des Art Directors Club Germany und des Type Directors Club New York. Seit 1990 Professor für Kommunikationsdesign an der Universität Wuppertal. Ausstellungen und Veröffentlichungen seiner Plakate weltweit. Seit 1984 in der Sammlung und in diversen Ausstellungen im Museum of Modern Art, New York, vertreten. Zahlreiche Auszeichnungen, u. a. Gewinner der International Invitational Exhibition, Fort Collins, USA (1987), Excellence Award der ICOGRADA (1987 und 1990), Grand Prix, Lahti, Finnland (1993), zehn Auszeichnungen des TDC (1994/95), Goldmedaille des ADC (1995), Silbermedaille des Festival d'Affiches, Chaumont (1996), Goldmedaille der Biennale für Grafik-Design, Brno (1996), Bronzemedaille des Poster Triennale, Toyama, Japan (1997). – *Publikationen u. a.:* „Zeichenzitate", Düsseldorf 1986; „Uwe Loesch: Der Ort, die Zeit und der Punkt", Frankfurt am Main 1991; „Faire le beau – Affiches de Uwe Loesch", Mainz 1993; „Nichtsdestoweniger. Plakate von Uwe Loesch im Museum für Kunst und Gewerbe Hamburg", Mainz 1997.

El Lissitzky 1924 Advertisement

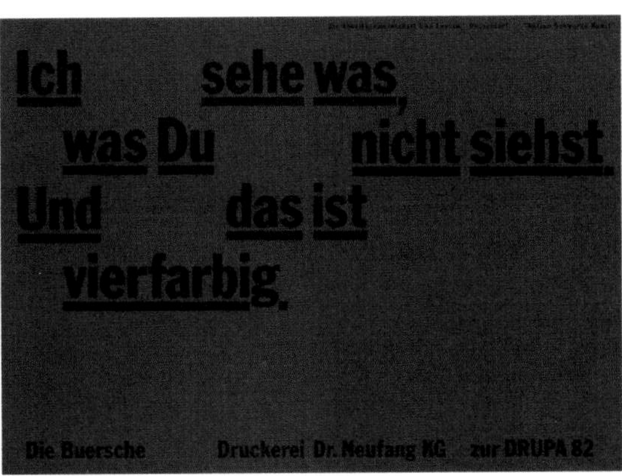


Loesch 1982 Poster

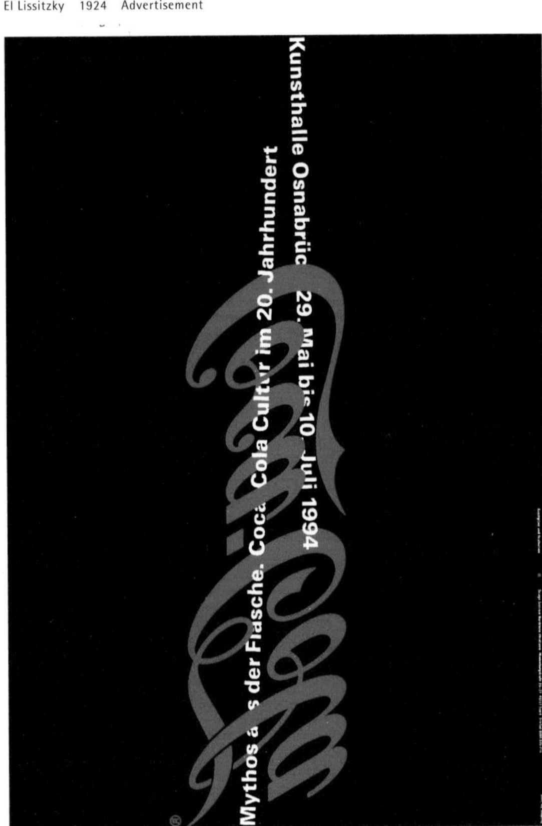


Loesch 1994 Poster

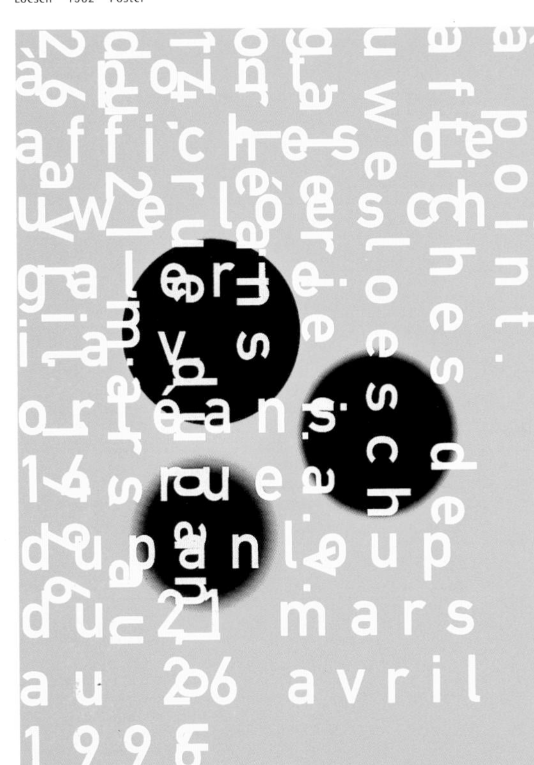


Loesch 1996 Poster

*enseignant* – 1964–1968, études à la Peter-Behrens-Werkkunstschule de Düsseldorf. Il fonde en 1970 la «Arbeitsgemeinschaft für visuelle und verbale Kommunikation» (groupe de travail pour la communication visuelle et verbale) à Düsseldorf. il est membre de l'Alliance Graphique Internationale de Art Directors Club Germany et de Type Directors Club New York. Depuis 1990, professor de design de communication à l'université de Wuppertal. Expositions et publications de ses affiches dans le monde entier. Depuis 1984, il est représenté dans la collection et les diverses expositions du Museum of Modern Art, New York. Nombreuses distinctions, entrs autres lauréat de l'International Invitational Exhibition, Fort Collins, Etats-Unis (1987), Excellence Award ICOGRADA (1987 et 1990), the Grand Prix, Lahti, Finlande (1993), dix récompences de TDC (1994/95), médaille d'or de l'ADC (1995), médaille d'argent du Festivals d'Affiches, Chaumont (1996), médaille d'or de la biennale du design graphique, Brno (1996), médaille bronze de la triennale d'afiches, Toyama, Japan (1997). – *Publications:* «Zeichenzitate», Düsseldorf 1986; «Uwe Loesch : Der Ort, die Zeit und der Punkt», Francfort-sur-le-Main 1991; «Faire le beau – Affiches de Uwe Loesch», Mayence 1993; «Nichtsdestoweniger. Plakate von Uwe Loesch im Museum für Kunst und Gewerbe Hamburg», Mayence 1997.

Lissitzky-Küppers, El Lissitzky "Schriften, Briefe, Dokumente", Dresden 1977.

**Loesch,** Uwe – b. 23.1 1943 in Dresden, Germany – *graphic designer, typographer, teacher* – 1964–68: studies at the Peter-Behrens-Werkkunstschule in Düsseldorf. 1970: founds an association for visual and verbal communication, the Arbeitsgemeinschaft für visuelle und verbale Kommunikation, in Düsseldorf. He is a member of the Alliance Graphique Internationale, of the Art Directors Club Germany and the Type Directors Club New York. From 1990 onwards: professor of communication design at Wuppertal University. His posters are exhibited and published worldwide. From 1984 onwards: his work is included in the museum collection and various exhibitions held at the Museum of Modern Art in New York. Numerous awards, including winning the International Invitational Exhibition, Fort Collins, USA (1987), the ICOGRADA Excellence Award (1987 and 1990), the Grand Prix, Lahti, Finland (1993), ten awards from the TDC (1994/95), the ADC gold medal (1995), the Festivals d'Affiches silver medal, Chaumont (1996), the gold medal at the Biennale for Graphic Design, Brno (1996), and the bronze medal at the Poster Triennale, Toyama (1997). – *Publications include:* "Zeichenzitate", Düsseldorf 1986; "Uwe Loesch: Der Ort, die Zeit und der Punkt", Frankfurt am Main 1991; "Faire le beau – Affiches de Uwe Loesch", Mainz 1993; "Nichtsdestoweniger. Plakate von U. Loesch im Museum für Kunst und Gewerbe Hamburg", Mainz 1997.

**Lohse,** Richard Paul – geb. 13. 9. 1902, in Zürich, Schweiz, gest. 16. 9. 1988 in Zürich, Schweiz. – *Typograph, Grafik-Designer, Maler* – 1918–22 Lehre als Reklamezeichner. 1922–30 Atelier Max Dalang in Zürich. 1930 eigenes Reklameatelier in Zürich. Zahlreiche Arbeiten für Museen, Verlage und Industrie. 1937 Mitbegründer und zweites Vorstandsmitglied der Künstlergruppe „Allianz", Vereinigung moderner Schweizer Künstler. Einrichtung einer Grafik-Ausstellung deutscher und sowjetischer Konstruktivisten in Zürich. 1944–58 Mitarbeit an den Publikationen „abstrakt-konkret", „Spirale" und „Plan". 1947–55 Redakteur und Gestalter der Architektur-Zeitschrift „Bauen und Wohnen". Zahlreiche Ausstellungen und Auszeichnungen, u. a. 1949 mit dem Preis für Schweizer Malerei. 1958–65 Herausgeber und Redakteur der Zeitschrift „Neue Grafik" in Zürich (mit Josef Müller-Brockmann, Hans Neuburg, Carlo Vivarelli). Repräsentierte 1965 die Schweiz auf der 8. Biennale in São Paulo. Teilnahme an der documenta IV (1968) und VII (1982) in Kassel. Repräsentiert 1972 die Schweiz auf der 36. Biennale in Venedig. Gründung der Richard-Paul-Lohse-Stiftung in Zürich. – *Publikationen u.a.:* „Ausstellungsgestaltung", Zürich 1953; „Neue Industriebauten" (mit H. und T. Maurer), Ravensburg 1954; „Modulare und serielle Ordnungen", Köln 1973; „Zeichnungen-Dessins 1935–1985", Baden 1986.

**Lubalin,** Herb (Herbert Frederick) – geb. 17. 3. 1918 in New York, USA, gest. 24. 5. 1981 in New York, USA. – *Grafik-*

**Lohse,** Richard Paul – né le 13. 9. 1902, à Zurich, Suisse, décédé le 16. 9. 1988 à Zurich, Suisse – *typographe, graphiste maquettiste, peintre* – apprentissage de peintre de réclames de 1918 à 1922. Travaille à l'Atelier Max Dalang à Zurich de 1922 à 1930. Ouvre son propre atelier de publicité à Zurich en 1930. Nombreux travaux pour des musées, maisons d'éditions et pour l'industrie. Cofondateur et second membre du comité du groupe d'artistes «Allianz», une association des artistes modernes de la Suisse, en 1937. Organise une exposition de constructivistes allemands et russes à Zurich. De 1944 à 1958, il collabore aux publications «abstrakt-konkret», «Spirale» et «Plan». De 1947 à 1955, il est rédacteur et maquettiste de la revue d'architecture «Bauen und Wohnen». Nombreuses expositions et distinctions, dont le prix de peinture suisse (1949). Editeur et rédacteur de la revue «Neue Grafik» à Zurich (avec Joseph Müller-Brockmann, Hans Neuburg, Carlo Vivarelli) de 1958 à 1965. En 1965, il représente la Suisse à la 8è Biennale de São Paulo. Participe à la «documenta IV» (1968) et «VII» (1982) à Kassel. En 1972, il représente la Suisse à la 36è Biennale de Venise. Crée la fondation Richard Paul Lohse à Zurich. – *Publications, sélection:* «Ausstellungsgestaltung», Zurich 1953; «Neue Industriebauten « (avec H. et T. Maurer), Ravensburg 1954; «Modulare und serielle Ordnungen», Cologne 1973; «Zeichnungen – Dessins 1935– 1985», Baden 1986.

**Lubalin,** Herb (Herbert Frederick) – né le 17. 3. 1918 à New York, Etats-Unis, décédé le 24. 5. 1981 à New York, Etats-Unis – *graphiste maquettiste, typographe, concepteur de polices, enseignant* – 1936–1939, études à la Cooper Union à New York. Travaille pour l'exposition universelle de New York en 1939. Direc-

Lohse ca. 1930 Cover

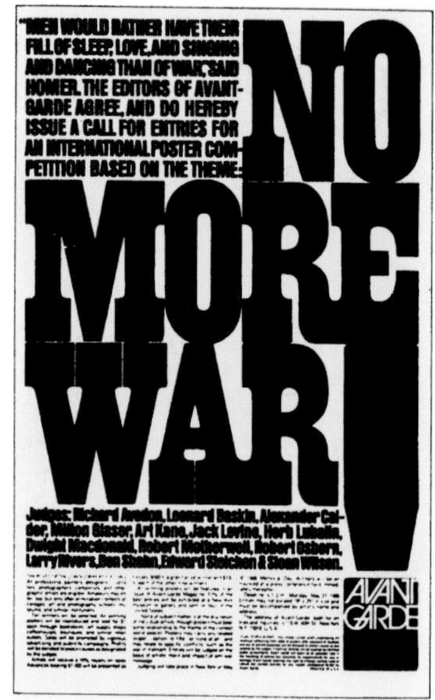

Lubalin 1968 Advertisement

Lohse 1952 Poster

**Lohse,** Richard Paul – b. 13. 9. 1902 in Zurich, Switzerland, d. 16. 9. 1988 in Zurich, Switzerland – *typographer, graphic designer, painter* – 1918–22: trains as an advertising designer. 1922–30: works at the Max Dalang studio in Zurich. 1930: opens his own advertising agency in Zurich. Produces much work for museums, publishers and industry. 1937: co-founder and deputy board member of the artists' group "Allianz", an association of modern Swiss artists. Organizes an exhibition of German and Soviet Constructivist graphics in Zurich. 1944–58: works on the publications "abstrakt-konkret", "Spirale" and "Plan". 1947–55: editor and designer of the architecture magazine "Bauen und Wohnen". Numerous exhibitions and awards, including Switzerland's national painting prize (1949). 1958–65: publisher and editor of "Neue Grafik" magazine in Zurich with Josef Müller-Brockmann, Hans Neuburg and Carlo Vivarelli. 1965: represents Switzerland at the 8th Biennale in São Paolo. Takes part in documenta IV (1968) and VII (1982) in Kassel. 1972: represents Switzerland at the 36th Biennale in Venice. Founds the Richard Paul Lohse-Stiftung in Zurich. – *Publications include:* "Ausstellungsgestaltung", Zurich 1953; "Neue Industriebauten" (with H. and T. Maurer), Ravensburg 1954; "Modulare und serielle Ordnungen", Cologne 1973; "Zeichnungen – Dessins 1935–1985", Baden 1986.

**Lubalin,** Herb (Herbert Frederick) – b. 17. 3. 1918 in New York, USA, d. 24. 5. 1981 in New York, USA – *graphic*

*Designer, Typograph, Schriftentwerfer, Lehrer* – 1936–39 Studium an der Cooper Union in New York. 1939 Arbeiten für die Weltausstellung in New York. Art Director bei Deutsch & Shea Advertising (1941–42), Fairchild Publications (1942–43) und Reiss Advertising (1943–45). 1945 Vizepräsident von Sudler & Hennesey Inc. in New York. 1964–69 Gründung von Herb Lubalin Inc. in New York. 1969–75 Präsident von Lubalin, Smith & Carnase Inc., seit 1975 mit Alan Peckolick. 1970 Gründung der International Typeface Corporation (ITC, mit Aaron Burns) in New York. Unterrichtet 1972 an der Cornell University, 1976–81 an der Cooper Union in New York. 1973 Herausgabe der Hauszeitschrift der ITC „Upper and lower case" (U&lc). Ausgezeichnet mit über 500 Preisen. Zahlreiche Ausstellungen, u. a. Society of typographic Arts in Chicago (1957), im Centre Georges Pompidou in Paris (1979), im ITC Center in New York (1980). Lubalin entwickelt eine assoziative, bildhafte Typographie. – *Schriftentwürfe:* Avantgarde Gothic (mit Carnase, Gschwind, Gürtler, Mengelt, 1970–77), Lubalin Graph (1974), Serif Gothic (mit Tony DiSpigna, 1974). – *Publikation u. a.:* Gertrude Snyder, Alan Peckolick „Herb Lubalin. Art Director, Graphic Designer and Typographer", New York 1985.

**Luidl,** Philipp – geb. 11. 12. 1930 in Diessen, Deutschland. – *Typograph, Grafik-Designer, Autor, Lyriker, Lehrer* – 1945–48 Schriftsetzerlehre in Diessen. 1955 Meisterprüfung als Schriftsetzer in München. Lehrt 1963–91 in München Satz

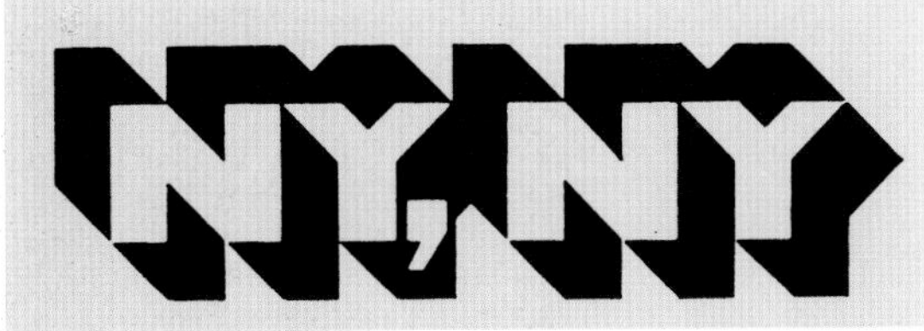


Lubalin 1966 Logo

Lubalin Logo

Lubalin 1966 Logo

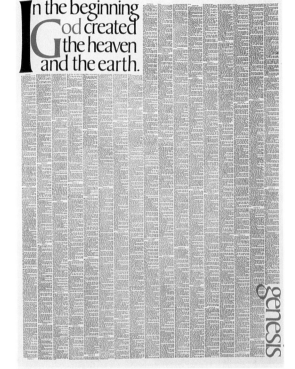

Lubalin Poster

Lubalin

| | n | n | n | | n |
|---|---|---|---|---|---|
| a | b | e | f | g | i |
| o | r | s | t | y | z |
| A | B | C | E | G | H |
| M | O | R | S | X | Y |
| 1 | 2 | 4 | 6 | 8 | & |

Lubalin 1970–77 Avantgarde Gothic

| | n | | | | n |
|---|---|---|---|---|---|
| a | b | e | f | g | i |
| o | r | s | t | y | z |
| A | B | C | E | G | H |
| M | O | R | S | X | Y |
| 1 | 2 | 4 | 6 | 8 | & |

Lubalin 1974 Lubalin Graph 2

teur artistique chez Deutsch & Shea Advertising (1941–1942), Fairchild Publications (1942–1943), Reiss Advertising (1943–1945). Vice-président de Sudler & Hennesey Inc. à New York en 1945. En 1964, il fonde la Herb Lubalin Inc. à New York (1964–1969). Président de Lubalin, Smith & Carnase Inc. de 1969 à 1975 (avec Alan Peckolick à partir de 1975). En 1970, il fonde l'International Typeface Corporation (ITC avec Aaron Burns) à New York. Enseigne à la Cornell University en 1972, puis à la Cooper Union à New York de 1976 à 1981. Publie, en 1973, la revue de l'ITC «Upper and lower case» (U&lc). A reçu plus de 500 prix. Nombreuses expositions, entre autres à la Society of Typographic Arts de Chicago (1957), au Centre Georges Pompidou à Paris (1979), à l'ITC Center de New York (1980). Lubalin a élaboré une typographie associative et figurative. – *Polices:* Avantgarde Gothic (avec Carnase, Gschwind, Gürtler, Mengelt, 1970–77), Lubalin Graph (1974), Serif Gothic (avec Tony DiSpigna, 1974). – *Publication, sélection:* Gertrude Snyder, Alan Peckolick «Herb Lubalin. Art Director, Graphic Designer and Typographer», New York 1985.

**Luidl,** Philipp – né le 11. 12. 1930 à Diessen, Allemagne – *typographe, graphiste maquettiste, auteur, poète, enseignant* – apprentissage de composition typographique à Diessen de 1945–1948. Diplôme de maître artisan compositeur à Munich en 1955. Enseigne la composition et la typographie à l'école professionnelle de Munich, à l'Akademie für das Grafische Gewerbe (métiers des arts graphiques), la

*designer, typographer, type designer, teacher* – 1936–39: studies at the Cooper Union in New York. 1939: produces work for the world exhibition in New York. Art director for Deutsch & Shea Advertising (1941–42), Fairchild Publications (1942–43) and Reiss Advertising (1943–45). 1945: vice-president of Sudler & Hennesey Inc. in New York. 1964–69: founds Herb Lubalin Inc. in New York. 1969–75: president of Lubalin, Smith & Carnase Inc., from 1975 onwards also with Alan Peckolick. 1970: founds the International Typeface Corporation (ITC) with Aaron Burns in New York. 1972: teaches at Cornell University and from 1976–81 at the Cooper Union in New York. 1973: publishes ITC's in-house magazine, "Upper and lower case" (U&lc). Lubalin won over 500 prizes and took part in numerous exhibitions, including at the Society of Typographic Arts in Chicago (1957), at the Centre Georges Pompidou in Paris (1979) and at the ITC Center in New York (1980). Lubalin developed an associative, pictorial typography. – *Fonts:* Avantgarde Gothic (with Carnase, Gschwind, Gürtler, Mengelt, 1970–77), Lubalin Graph (1974), Serif Gothic (with Tony DiSpigna, 1974). – *Publications include:* Gertrude Snyder, Alan Peckolick "Herb Lubalin. Art Director, Graphic Designer and Typographer", New York 1985.

**Luidl,** Philipp – b. 11. 12. 1930 in Diessen, Germany – *typographer, graphic designer, author, lyric poet, teacher* – 1945–48: trains as a typesetter in Diessen. 1955: obtains his master typesetter's diploma in Munich. 1963–91: teaches typesetting

und Typographie an der Berufsschule, der Akademie für das Grafische Gewerbe, der Meisterschule für Mode und an der Fachhochschule. 1966–71 Mitglied des Ausstellungskomitees der Galerie Intergraphis in München. 1972–93 Herausgeber der Publikationsreihen „Werkstattbriefe", „Aus Rede und Diskussion", „Jahresgaben der Typographischen Gesellschaft München". 1979–85 Vorsitzender der Typographischen Gesellschaft München. Zahlreiche Prämierungen im Wettbewerb der „50 schönsten Bücher" der Stiftung Buchkunst in Frankfurt am Main. Zahlreiche Beiträge in Fachzeitschriften. – *Publikationen u.a.:* „Typografie neu", Stuttgart 1971; „Typografie", Hannover 1984; „desktop-knigge", München 1988; „Grundsetzliches 1–6", München 1994; „Typografie, Basiswissen", Ostfildern 1996. Mitarbeit an „Typopictura", Stuttgart 1981; „Bruckmanns Handbuch der Schrift", München 1977; „Bruckmanns Handbuch der Drucktechnik", München 1981.

**Lupton,** Ellen – geb. 1. 12. 1963 in Philadelphia, USA. – *Grafik-Designerin, Autorin, Ausstellungskuratorin, Lehrerin* – 1985 Abschluß des Grafik-Design-Studiums an der Cooper Union in New York. Gründung des Studios „Design Writing Research" (mit J. Abbott Miller) in New York. 1985–92 Kuratorin am „Herb Lubalin Study Center for Design and Typography" an der Cooper Union in New York. Veranstaltet zahlreiche Ausstellungen, u. a. „Global Signage: Semiotics and the Language of International Pictures" (1986), „Numbers" (mit Alan Wolf, 1989),

Meisterschule für Mode (école de la mode), et à la Fachhochschule (institut technique universitaire) de 1963 à 1991. Membre du comité des expositions de la galerie Intergraphis à Munich, de 1966 à 1971. Publie la série «Werkstattbriefe», «Aus Rede und Diskussion», «Jahresgaben der Typographischen Gesellschaft München» entre 1973 et 1993. Président de la Typographische Gesellschaft München de 1979 à 1985. Nombreux prix, dont celui du concours «50 schönsten Bücher» de la Fondation des Arts du livre de Francfort-sur-le-Main. Nombreux articles dans des revues spécialisées. – *Publications, sélection:* «Typografie neu», Stuttgart 1971; «Typografie», Hanovre 1984; «desktop-knigge», Munich 1988; «Grundsetzliches 1–6», Munich 1994; «Typografie, Basiswissen», Ostfildern 1996. Participation à «Typopictura», Stuttgart 1981; «Bruckmanns Handbuch der Schrift», Munich 1977; «Bruckmanns Handbuch der Drucktechnik», Munich 1981.

**Lupton,** Ellen – née le 1. 12. 1963 à Philadelphie, Etats-Unis – *graphiste maquettiste, auteur, architecte d'expositions, enseignante* – termine ses études d'arts graphiques et de design à la Cooper Union à New York; en 1985. Fonde l'atelier «Design Writing Research» (avec J. Abbott Miller), à New York. 1985–1992, conservatrice au «Herb Lubalin Study Center for Design and Typography», de la Cooper Union à New York; organise de nombreuses expositions, dont «Global Signage: Semiotics and the Language of International Pictures» (1986), «Numbers» (avec Alan Wolf, 1989), «Writing and the Body» (1990), «Graphic Design in the Netherlands: A View of Recent Work» (1992). Etudie l'histoire de l'art au Graduate Center de la City University of New York de 1989 à 1992. Conservatrice char-

Luidl 1993 Cover

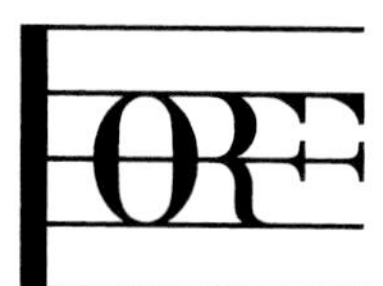

Luidl 1995 Logo

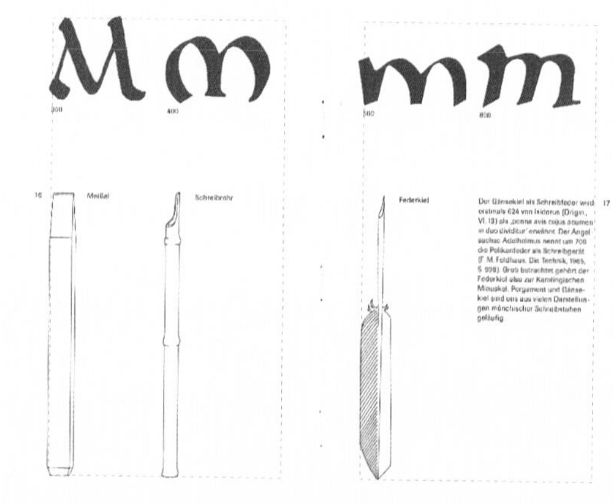


Luidl 1994 Spread

Lupton 1990 Poster

and typography at the Berufsschule in Munich, at the Akademie für das Grafische Gewerbe, at the Meisterschule für Mode and at the Fachhochschule. 1966–71: member of the exhibition committee for the Intergraphis gallery in Munich. 1972–93: editor of the series "Werkstattbriefe", "Aus Rede und Diskussion" and "Jahresgaben der Typographischen Gesellschaft München". 1979–85: chairman of the Typographische Gesellschaft in Munich. Has won numerous prizes in the competition for the "50 most beautiful books" run by the Stiftung Buchkunst in Frankfurt am Main. Numerous contributions to specialist magazines. – *Publications include:* "Typografie neu", Stuttgart 1971; "Typografie", Hanover 1984; "desktop-knigge", Munich 1988; "Grundsetzliches 1–6", Munich 1994; "Typografie, Basiswissen", Ostfildern 1996. Has worked on "Typopictura", Stuttgart 1981; "Bruckmanns Handbuch der Schrift", Munich 1977.

**Lupton,** Ellen – b. 1. 12. 1963 in Philadelphia, USA – *graphic designer, author, exhibition organizer, teacher* – 1985: completes her graphic design studies at the Cooper Union in New York. Opens the Design Writing Research studio with J. Abbott Miller in New York. 1985–92: curator of the Herb Lubalin Study Center for Design and Typography at the Cooper Union in New York. Organizes numerous exhibitions, including "Global Signage: Semiotics and the Language of International Pictures" (1986), "Numbers" (with Alan Wolf, 1989), "Writing and the Body" (1990) and "Graphic Design in the Nether-

„Writing and the Body" (1990), „Graphic Design in the Netherlands: A View of Recent Work" (1992). 1989–92 Studium der Kunstgeschichte am Graduate Center der City University of New York. Seit 1992 Kuratorin für zeitgenössisches Design am Cooper-Hewitt National Museum of Design in New York. Veranstaltet zahlreiche Ausstellungen, u. a. „Living with AIDS: Education through Design" (1993), „Elaine Lustig Cohen, Modern Graphic Designer" (1995), „Mixing Messages: Graphic Design in Contemporary Culture" (1996). Unterrichtet an der Cooper Union (1989–92), an der Yale University School of Art (1992–93), an der Parson's School of Design (1993–95). 1993 Verleihung des „Chrysler Design Award" (mit J. Abbott Miller). Das Studio „Design Writing Research" gründet 1995 innerhalb der Princeton Architectural Press den Verlag „Kiosk" für Publikationen zur Geschichte und Theorie von Industrie- und Grafik-Design. Zahlreiche Aufsätze in Zeitschriften wie „Design Issues", „Design Review", „Print", „ID", „Eye", „Emigre". – *Publikationen u.a.:* „The Bauhaus and Design Theory" (mit J. Abbott Miller), New York, London 1991; „Graphic Design and Typography in the Netherlands: A View of Recent Work", New York 1992; „Letters from the Avant-Garde: Modern Graphic Design" (mit Elaine Lustig Cohen), New York 1996; „Design Writing Research: Writing on Graphic Design" (mit J. Abbott Miller), New York 1996; „Mixing Messages: Graphic Design and Contemporary Culture", New York, London 1996.

**Lustig,** Alvin – geb. 8. 2. 1915 in Denver,

Lupton 1996 Cover

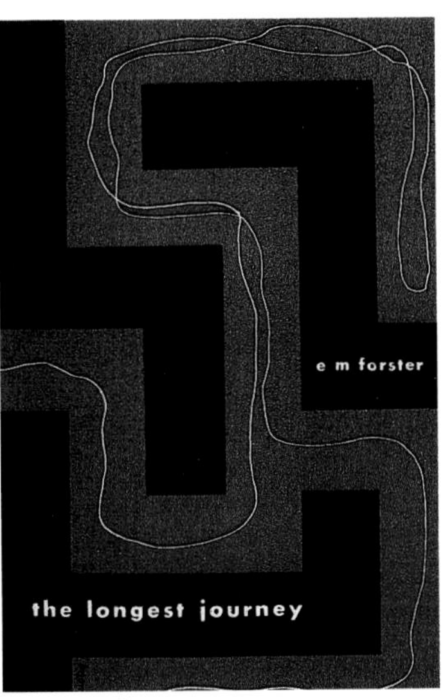
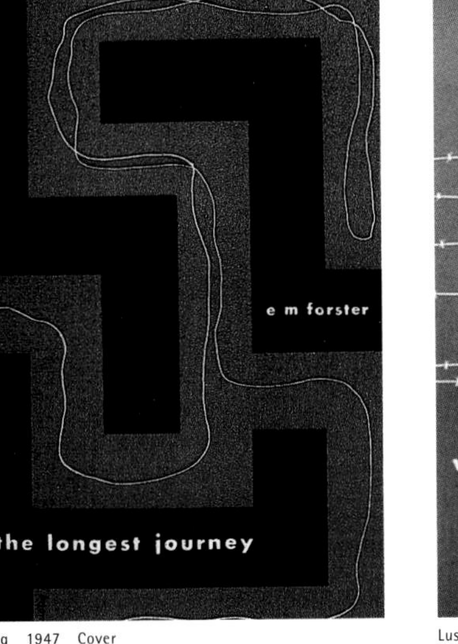

Lustig 1947 Cover

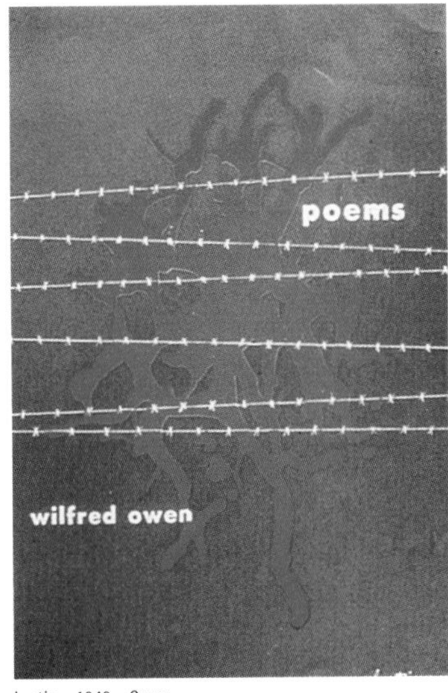

Lustig 1949 Cover

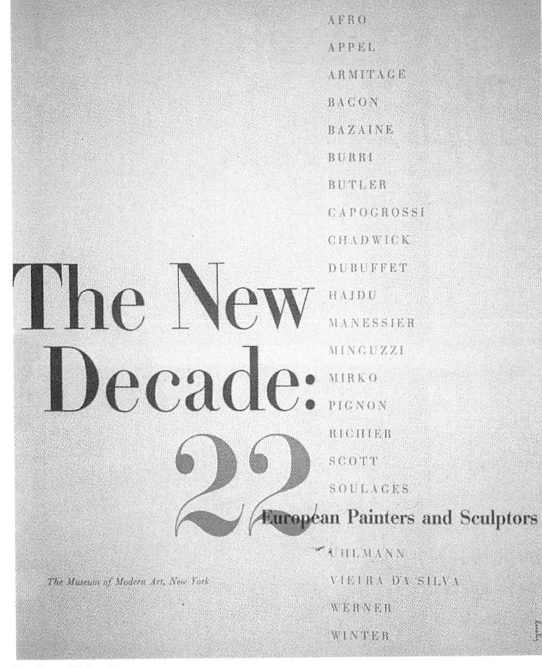

Lustig 1954 Cover

Lustig 1947 Cover

gée du design contemporain au Cooper-Hewitt National Museum of Design à New York; a organisé de nombreuses expositions, dont «Living with AIDS : Education through Design» (1993), «Elaine Lustig Cohen, Modern Graphic Designer» (1995), «Mixing Messages : Graphic Design in Contemporary Culture» (1996). A enseigné à la Cooper Union (1989–1992), à la Yale University School of Art (1992–1993), à la Parson's School of Design (1993–1995). Décorée du «Chrysler Design Award» (avec J. Abbott Miller) en 1993. En 1995, l'atelier «Design Writing Research» fonde les éditions «Kiosk» dépendant de la Princeton Architectural Press, qui publie des ouvrages sur l'histoire et la théorie du design industriel et graphique. De nombreux essais dans des revues telles que «Design Issues», «Design Review», «Print», «ID», «Eye», «Emigre». – *Publications, sélection :* «The Bauhaus and Design Theory» (avec J. Abbott Miller), New York, Londres 1991; «Graphic Design and Typography in the Netherlands : A View of Recent Work», New York 1992; «Letters from the Avant-Garde : Modern Graphic Design» (avec Elaine Lustig Cohen), New York 1996; «Design Writing Research : Writing on Graphic Design» (avec J. Abbott Miller), New York 1996; «Mixing Messages : Graphic Design and Contemporary Culture», New York, Londres 1996.

**Lustig,** Alvin – né le 8. 2. 1915 à Denver, Etats-Unis, décédé le 5. 12. 1955 à New

lands: A View of Recent Work" (1992). 1989–92: studies art history at the City University of New York Graduate Center. From 1992 onwards: curator for contemporary design at the Cooper-Hewitt National Museum of Design in New York. Organizes numerous exhibitions, including "Living with AIDS: Education through Design" (1993), "Elaine Lustig Cohen, Modern Graphic Designer" (1995) and "Mixing Messages: Graphic Design in Contemporary Culture" (1996). Teaches at the Cooper Union (1989–92), at the Yale University School of Art (1992–93) and at the Parson's School of Design (1993–95). 1993: is presented with the Chrysler Design Award (with J. Abbott Miller). 1995: the Design Writing Research studio launches Kiosk as a sub-publishing body of the Princeton Architectural Press. Kiosk specializes in publications on the history and theory of industrial and graphic design. Numerous essays in magazines such as "Design Issues", "Design Review", "Print", "ID", "Eye" and "Emigre". – *Publications include:* "The Bauhaus and Design Theory" (with J. Abbott Miller), New York, London 1991; "Graphic Design and Typography in the Netherlands: A View of Recent Work", New York 1992; "Letters from the Avant-Garde: Modern Graphic Design" (with Elaine Lustig Cohen), New York 1996; "Design Writing Research: Writing on Graphic Design" (with J. Abbott Miller), New York 1996; "Mixing Messages: Graphic Design and Contemporary Culture", New York, London 1996.

**Lustig,** Alvin – b. 8. 2. 1915 in Denver, USA, d. 5. 12. 1955 in New York, USA –

USA, gest. 5. 12. 1955 in New York, USA. – *Grafik-Designer, Lehrer* – 1934 Studium am Art Center College of Design in Los Angeles. 1935 Architekturstudium bei Frank Lloyd Wright in Taliesin, Wisconsin. 1936 Studio (mit John Chariot). 1937–43 Design-Studio und Druckunternehmen in Brentwood, Los Angeles. Gestaltung zahlreicher Erscheinungsbilder, Signets, Plattencover, Buchumschläge. 1944–46 Arbeiten für „Look" in New York. 1945 Bekanntschaft mit Josef Albers. 1946–50 Design-Studio in Los Angeles, 1950–55 in New York. Nach seinem Erblinden 1954 führt seine Frau Elaine das Studio weiter. Zahlreiche Ausstellungen seiner Arbeiten, u. a. im Walker Art Center in Minneapolis (1949) und im Museum of Modern Art in New York (1953). Unterrichtet am Black Mountain College (1945), an der Art Center School in Los Angeles (1947–48), an der Yale University (1951–55). – *Publikation u. a.:* Holland R. Melson (Hrsg.) „The Collected Writings of Alvin Lustig", New York 1958.

**Lustig Cohen,** Elaine – geb. 6. 3. 1927 in Jersey City, USA. – *Grafik-Designerin, Innenarchitektin* – 1945–46 Studium am Sophie Newcomb College, Tulane University. 1947–48 Studium an der University of Southern California. 1948–55 Grafik-Designerin im Studio von Alvin Lustig. Seit 1956 selbständig. 1965 Teilnahme an der Ausstellung „Typomundus 20", 1966 „Fifty years of Graphic Arts in America". Seit 1969 zahlreiche internationale Gruppen- und Einzelausstellungen. Leiterin des Antiquariats „Ex libris" (mit Arthur A. Cohen) in New York. 1996

York, Etats-Unis – *graphiste maquettiste, enseignant* – études à l'Art Center College of Design de Los Angeles en 1934. Etudes d'architecture chez Frank Lloyd Wright à Taliesin, Wisconsin en 1935. Ouvre son propre atelier en 1936 (avec John Chariot). Dirige son propre atelier de design et d'imprimerie à Brentwood, Los Angeles, de 1937 à 1943. Réalise de nombreuses identités, signets, pochettes de disques, couvertures de livres. De 1944 à 1946, il travaille pour la revue «Look» à New York. Rencontre Josef Albers en 1945. Exerce dans son atelier de Design à Los Angeles de 1946 à 1950, puis à New York de 1950 à 1955. Il est atteint de cécité en 1954 et sa femme, Elaine, continue de gérer l'atelier. Nombreuses expositions de ses travaux, entre autres au Walker Art Center de Minneapolis (1949) et au Museum of Modern Art de New York (1953). A enseigné au Black Mountain College (1945), à l'Art Center School de Los Angeles (1947–1948), puis à la Yale University (1951–1955). – *Publication, sélection:* Holland R. Melson (éd.) «The Collected Writings of Alvin Lustig», New York 1958.

**Lustig Cohen,** Elaine – née le 6. 3. 1927 à Jersey City, Etats-Unis – *graphiste maquettiste, architecte d'intérieur* – 1945–1946, études au Sophie Newcomb College, Tulane University. 1947–1948, études à la University of Southern California. Graphiste maquettiste à l'atelier d'Alvin Lustig de 1948 à 1955. Exerce comme indépendante à partir de 1956. Participe à l'exposition «Typomundus 20» en 1965, à «Fifty years of Graphic Arts in America» en 1966. Nombreuses expositions internationales personnelles et collectives depuis 1969. Dirige la librairie de livres anciens «Ex Libris» (avec Arthur A. Cohen) à New York. 1996, commissaire associé de l'exposition «The Avant-Garde

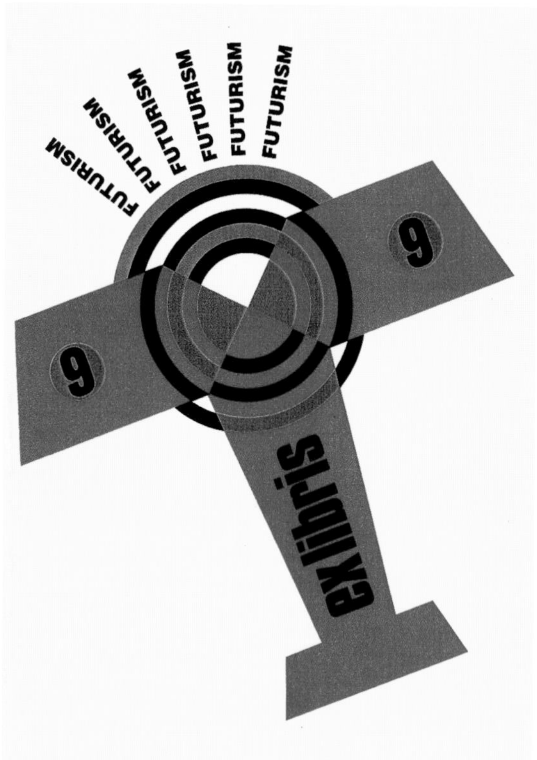

Lustig Cohen 1982 Front Cover

Lustig Cohen 1982 Back Cover

Lustig Cohen 1983 Cover

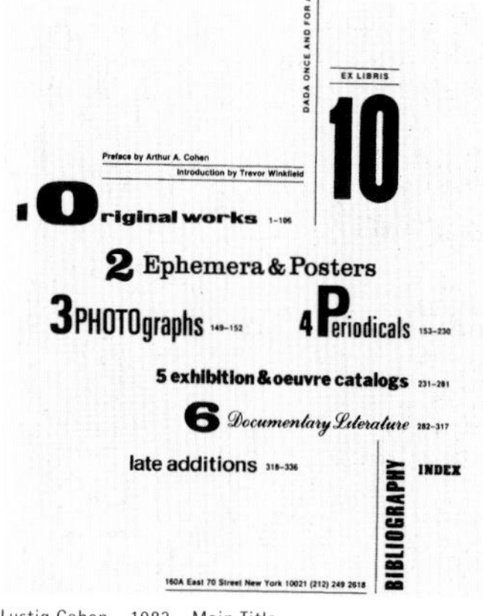


Lustig Cohen 1983 Main Title

*graphic designer, teacher* – 1934: studies at the Art Center College of Design in Los Angeles. 1935: studies architecture under Frank Lloyd Wright in Taliesin, Wisconsin. 1936: opens his own studio with John Chariot. 1937–43: runs his own design studio and printing workshop in Brentwood, Los Angeles. Designs numerous corporate identities, logos and record and book covers. 1944–46: works for "Look" magazine in New York. 1945: makes the acquaintance of Josef Albers. 1946–50: runs a design studio in Los Angeles and from 1950–55 in New York. 1954: Lustig completely loses his sight. His wife, Elaine, continues to run the studio. Numerous exhibitions, including at the Walker Art Center in Minneapolis (1949) and at the Museum of Modern Art in New York (1953). Lustig teaches at Black Mountain College (1945), at the Art Center School in Los Angeles (1947–48) and at Yale University (1951–55). – *Publications include:* Holland R. Melson (ed.) "The Collected Writings of Alvin Lustig", New York 1958.

**Lustig Cohen,** Elaine – b. 6. 3. 1927 in Jersey City, USA – *graphic designer, interior designer* – 1945–46: studies at Sophie Newcomb College, Tulane University. 1947–48: studies at the University of Southern California. 1948–55: graphic designer at Alvin Lustig's studio. From 1956 onwards: works freelance. 1965: participates in the "Typomundus 20" exhibition and in 1966 in "Fifty Years of Graphic Arts in America". Since 1969: numerous group and solo exhibitions worldwide. Runs Ex libris, an antiquari-

Co-Kuratorin der Ausstellung „The Avant-Garde Letterhead". – *Publikationen u. a.:* „Letters from the Avant-Garde: Modern Graphic-Design" (mit Ellen Lupton), New York 1996. Ellen Lupton „Archive: Elaine Lustig Cohen" in „Eye" 17/1995.

**Lutz,** Hans Rudolf – geb. 14. 1. 1939 in Zürich, Schweiz, gest. 17.1.1998 in Zürich, Schweiz. – *Typograph, Verleger, Autor, Lehrer* – 1955–58 Schriftsetzerlehre im Verlags- und Druckhaus Orell Füssli in Zürich. 1959–60 Schriftsetzer in Zürich. 1963 Kurs für typographische Gestaltung bei Emil Ruder und Robert Büchler an der Schule für Gestaltung in Basel. 1964–66 Leiter der Gruppe „expression typographique" im Atelier Albert Hollenstein in Paris. Lehrer an der „Cours 19" Abendschule in Paris. 1965 und 1977 Cover für „Typografische Monatsblätter". 1966 Gründung des Lutz Verlags. 1966–70 Fachlehrer für Schriftsatz an der Schule für Gestaltung in Zürich, seit 1968 an der Schule für Gestaltung in Luzern. 1971–80 Mitbegründer der Verlagsgenossenschaft in Zürich. Beteiligung am Aufbau der Schule für experimentelle Gestaltung „F + F". Seit 1983 grafische Arbeiten für die Multimedia-Gruppe „Unknownmix" in Zürich. Unterrichtet an der Schule für Gestaltung in Zürich. – *Publikationen u.a.:* „Grafik in Kuba", Zürich 1971; „Experiment F + F, 1965–71", Zürich 1971; „Eine Art Geschichte", Zürich 1980; „Ausbildung in typographischer Gestaltung", Zürich 1987; „Die Hieroglyphen von Heute. Grafik auf Verpackungen für den Transport", Zürich 1990; „Typoundso", Zürich 1996.

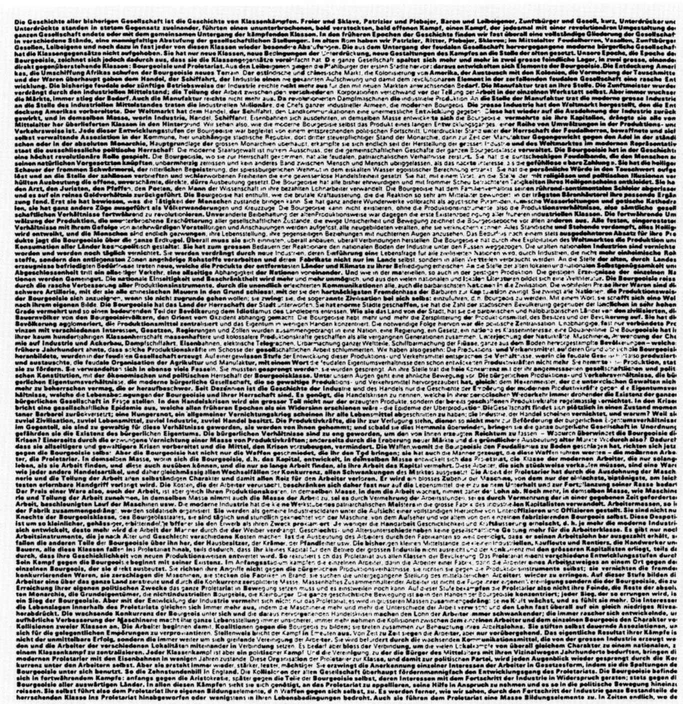

Lutz 1968 Poster

Lutz 1977 Cover

Lutz 1977 Cover

Lutz 1977 Cover

Lutz 1977 Cover

an bookstore in New York with Arthur A. Cohen. 1996: co-organizer of the exhibition "The Avant-Garde Letterhead". – *Publications:* "Letters from the Avant-Garde: Modern Graphic-Design", New York 1996. Ellen Lupton "Archive: Elaine Lustig Cohen" in "Eye" 17/1995.

**Lutz,** Hans Rudolf – b. 14. 1. 1939 in Zurich, Switzerland, d. 17. 1. 1998 in Zurich, Switzerland – *typographer, publisher, author, teacher* – 1955–58: trains as a typesetter at Orell Füssli in Zurich. 1959–60: typesetter in Zurich. 1963: takes a course in typographic design under Emil Ruder and Robert Büchler at the Schule für Gestaltung in Basle. 1964–66: heads the expression typographique group at Albert Hollenstein's studio in Paris. Teaches at the Cours 19 night school in Paris. 1965 and 77: covers for "Typografische Monatsblätter" magazine. 1966: founds Lutz Verlag. 1966–70: specialist typesetting teacher at the Schule für Gestaltung in Zurich, and from 1968 onwards at the Schule für Gestaltung in Lucerne. 1971–80: co-founder of the publishing association in Zurich. Helps to build up the school for experimental design F + F. From 1983 onwards: produces graphic work for the multimedia group Unknownmix in Zurich. Teaches at the Schule für Gestaltung in Zurich. – *Publications include:* "Grafik in Kuba", Zurich 1971; "Experiment F + F, 1965–71", Zurich 1971; "Eine Art Geschichte", Zurich 1980; "Ausbildung in typographischer Gestaltung", Zurich 1987; "Die Hieroglyphen von Heute. Grafik auf Verpackungen für den Transport", Zurich 1990; "Typoundso", Zurich 1996.

Letterhead». – *Publications, sélection:* «Letters from the Avant-Garde: Modern Graphic-Design» (avec Ellen Lupton), New York 1996. Ellen Lupton «Archive: Elaine Lustig Cohen» dans «Eye» 17/1995.

**Lutz,** Hans Rudolf – né le 14. 1. 1939 à Zurich, Suisse, décédé le 17. 1. 1998 à Zurich, Suisse – *typographe, éditeur, auteur, enseignant* – 1955–1958, apprentissage de la composition typographique aux Imprimeries et éditions Orell Füssli à Zurich. Compositeur à Zurich de 1959 à 1960. Cours de maquette et de typographie chez Emil Ruder et Robert Büchler à la l'école de design de Bâle en 1963. Dirige le groupe «expression typographique» de l'Atelier Albert Hollenstein, à Paris, de 1964 à 1966. Donne des cours du soir au «Cours 19» à Paris. En 1965 et 1977, il conçoit les couvertures de la revue «Typografische Monatsblätter» à Saint-Gall. Fonde les éditions Lutz en 1966. Enseigne la composition à l'école de design de Zurich de 1966 à 1970, puis à l'école de design de Lucerne à partir de 1968. Cofondateur de la Verlagsgenossenschaft de Zurich de 1971 à 1980. Participe à la création de la Schule für experimentelle Gestaltung «F + F» (école de design expérimental). Travaux graphiques pour le groupe multimédias «Unknowmix» à Zurich, à partir de 1983. Enseigne à l'école de design de Zurich. – *Publications, sélection:* «Grafik in Kuba», Zurich 1971; «Experiment F + F, 1965– 1971», Zurich 1971; «Eine Art Geschichte», Zurich 1980; «Ausbildung in typographischer Gestaltung», Zurich 1987; «Die Hieroglyphen von Heute. Grafik auf Verpackungen für den Transport», Zurich 1990; «Typoundso», Zurich 1996.

# M

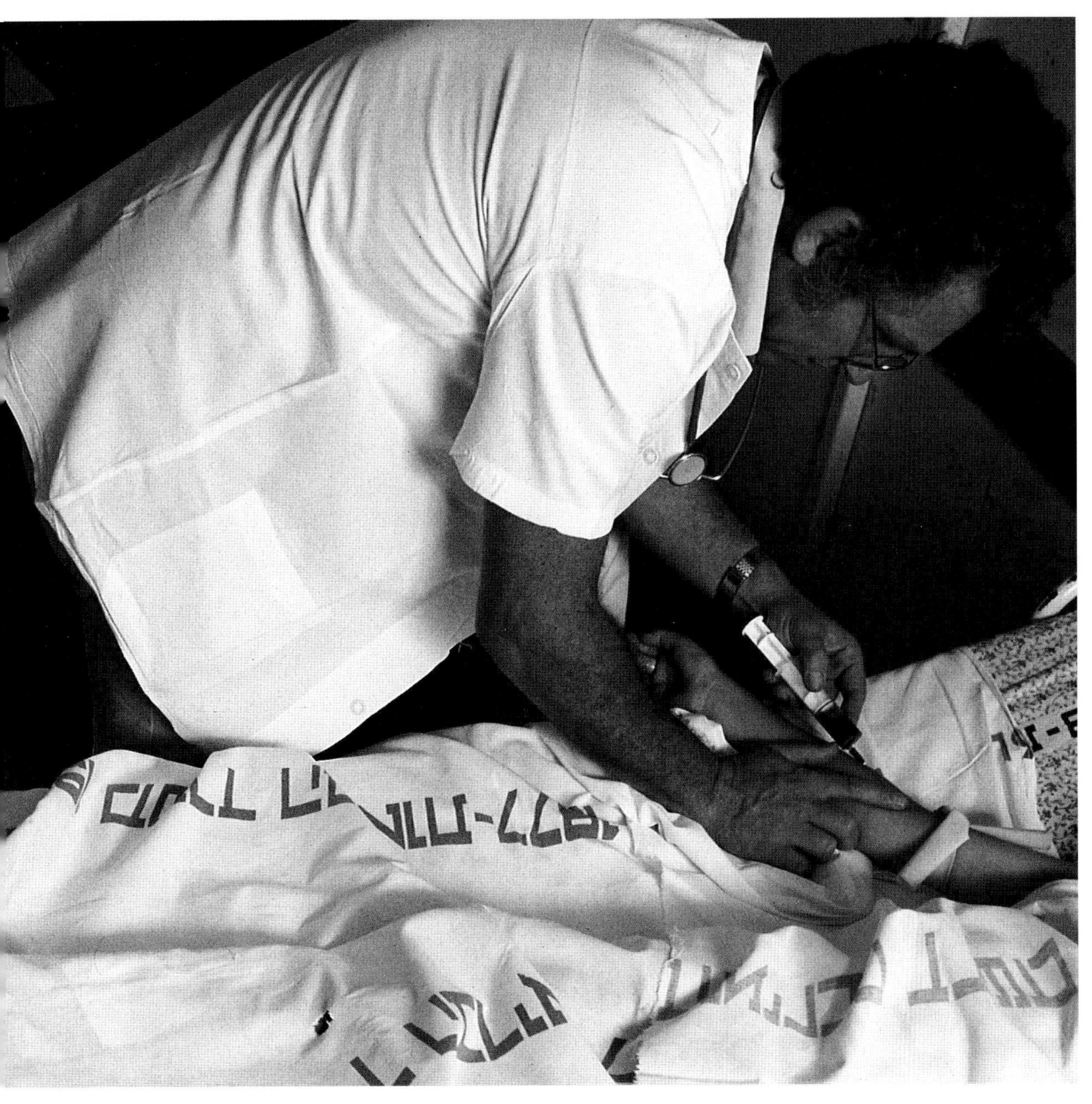

**Maciunas,** George – geb. 1931 in Kaunas, Litauen, gest. 9. 5. 1978 in New York, USA. – *Künstler* – 1949–52 Studium der Kunst und Architektur an der Cooper Union School of Art in New York, 1952–54 der Architektur und Musikwissenschaften am Carnegie Institute of Technology in Pittsburgh, 1955–60 der Kunst der Völkerwanderungszeit am Institute of Fine Arts in New York. Seit 1961 Veranstalter von Fluxus-Konzerten und Aktionen, u.a. die Konzertserie „Musica Antiqua et Nova", „Fluxus Internationale Festspiele neuester Musik" in Wiesbaden (1962), wobei erstmals öffentlich der Begriff „Fluxus" verwendet wird. Seit 1963 Herausgeber der Fluxus-Publikationen, Koordinator und Theoretiker der Fluxus-Bewegung, für die er zahlreiche grafische und typographische Arbeiten entwirft. Die George-Maciunas-Stiftung des Fluxeum Wiesbaden vergibt 1992 erstmals den George-Maciunas-Preis, der alle 47 Monate vergeben wird.

**Mahlow,** Dietrich – geb. 19. 8. 1920 in Seehausen, Altmark, Deutschland. – *Autor, Museumsleiter, Kurator* – 1949–54 Studium Philosophie, Kunstgeschichte und Psychologie in Freiburg. Direktor der Staatlichen Kunsthalle in Baden-Baden, der Städtischen Kunsthalle in Nürnberg. 1967–73 Gründung und Leitung des Instituts für moderne Kunst in Nürnberg. 1975–78 Direktor im Museum für moderne Kunst J. Soto in Venezuela. Veranstaltet ca. 250 Ausstellungen, u. a. „Schrift und Bild" (Design W. Schmidt, 1963), „Die Kunst der Schrift" (1964) und „Auf ein Wort" (1987). Viele der von ihm

**Maciunas,** George – né en 1931 à Kaunas, Lituanie, décédé le 9. 5. 1978 à New York, Etats-Unis – *artiste* – 1949–1952 études d'art et d'architecture à la Cooper Union School of Art de New York; de 1952 à 1954, études d'architecture et de musicologie au Carnegie Institute of Technology de Pittsburgh; puis, de 1955 à 1960, études sur les arts de la période des grandes invasions à l'Institute of Fine Arts de New York. Organise des concerts Fluxus et des actions, dont la série de concerts «Musica Antiqua et Nova», «Fluxus Internationale Festspiele neuester Musik», à Wiesbaden (1962), où le terme «Fluxus» est utilisé pour la première fois en public. A partir de 1963, il édite les publications Fluxus, et a fonction de coordinateur et théoricien du mouvement Fluxus; réalise de nombreux travaux graphiques et typographiques dans ce contexte. En 1992, la fondation George-Maciunas du Fluxeum de Wiesbaden décerne pour la première fois le prix «George Maciunas» attribué tous les 47 mois.

**Mahlow,** Dietrich – né le 19. 8. 1920 à Seehausen, Altmark, Allemagne – *auteur, directeur de musée, conservateur* – 1949–1954, études de philosophie, d'histoire de l'art et de psychologie à Fribourg. Directeur de la Staatliche Kunsthalle de Baden-Baden et de la Städtische Kunsthalle de Nuremberg. En 1967, il fonde l'Institut für moderne Kunst (institut d'art moderne) de Nuremberg qu'il dirige jusqu'en 1973. Directeur du Musée d'art moderne J. Soto au Venezuela. A organisé près de 250 expositions, dont «Schrift und Bild» (Design W. Schmidt, 1963), «Die Kunst der Schrift» (1964) et «Auf ein Wort» (1987). Bon nombre des catalogues qu'il a dirigés sont devenus des ouvrages de références. La croix du mérite de la RFA lui est décernée en 1982.



Maciunas 1963 Poster

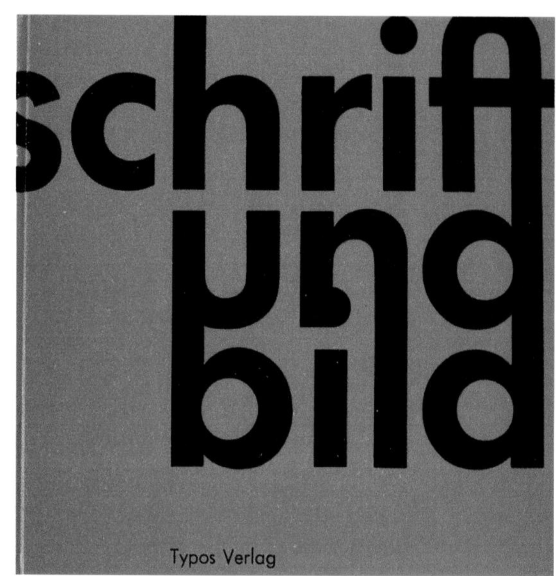


Mahlow 1963 Cover

Majakowski 1920 from Rosta

**Maciunas,** George – b. 1931 in Kaunas, Lithuania, d. 9. 5. 1978 in New York, USA – *artist* – 1949–52: studies art and architecture at the Cooper Union School of Art in New York, architecture and musicology at the Carnegie Institute of Technology in Pittsburgh from 1952–54, and art during the Dark Ages at the Institute of Fine Arts in New York from 1955–60. From 1961 onwards: organizes Fluxus concerts and promotions, including the concert series "Musica Antiqua et Nova" and the "Fluxus Internationale Festspiele neuester Musik" in Wiesbaden (1962), where the word "Fluxus" is used for the first time in public. From 1963 onwards: edits Fluxus publications. Maciunas was the coordinator and theorist of the Fluxus movement, for whom he produced much graphic and typographic work. 1992: the first George Maciunas Prize is awarded by the George-Maciunas-Stiftung des Fluxeum in Wiesbaden. The prize is awarded every 47 months.

**Mahlow,** Dietrich – b. 19. 8. 1920 in Seehausen, Altmark, Germany – *author, museum director, curator* – 1949–54: studies philosophy, art history and psychology in Freiburg. Director of the Staatliche Kunsthalle in Baden-Baden and the Städtische Kunsthalle in Nuremberg. 1967–73: founds and runs the Institut für moderne Kunst in Nuremberg. 1975–78: director of the Museum for modern art J. Soto in Venezuela. Organizes c. 250 exhibitions, including "Schrift und Bild" (Design W. Schmidt, 1963), "Die Kunst der Schrift" (1964) and "Auf ein Wort" (1987). Many of the cat-

erarbeiteten Kataloge wurden Standardwerke. 1982 Verleihung des Bundesverdienstkreuzes der Bundesrepublik Deutschland.
**Majakowski,** Wladimir – geb. 19. 7. 1893 in Bagdadi, Georgien, gest. 14. 4. 1930 in Moskau, UdSSR. – *Dichter, Grafiker, Maler* – 1908–09 Kunststudium am Stroganow-Institut in Moskau, 1911–14 an der Moskauer Schule für Malerei, Bildhauerei und Architektur. 1912 erste Gedichte im Sammelband „Eine Ohrfeige dem öffentlichen Geschmack". 1913–14 Teilnahme an der Werbereise der futuristischen Gruppe „Hylaea" durch Rußland (mit D. Burljuk, W. Kamenskij u.a.). 1913 Aufführung seines ersten Theaterstücks „Wladimir Majakowski" in St. Petersburg. Ab 1919 Gestaltung von Filmplakaten sowie Agitationsplakate für die „Rosta"-Fenster. Werbeplakate für die staatliche Handelsorganisation (mit A. Rodtschenko und W. Stepanowa). 1922–29 zahlreiche Lesungen und Vorträge in Berlin. Teilnahme an der „1. Russischen Kunstausstellung" in Berlin. 1923 Gründungsmitglied der Gruppe „LEF" (Linke Front der Künstler), Herausgeber der Zeitschriften „LEF" und „Nowy LEF". El Lissitzky gestaltete Majakowskis Gedichtband „Dlja golosa" („Für die Stimme"). 1930 Retrospektive seiner Arbeit „20 Jahre Arbeit" in Moskau.
**Majoor,** Martin – geb. 14. 10. 1960 in Baarn, Niederlande. – *Grafik-Designer, Schriftentwerfer, Lehrer* – 1980–86 Grafik-Design-Studium an der Kunstakademie in Arnheim. 1986–88 Arbeit als Typograph bei Océ Niederlande. 1988–90

Majakowski 1918 Poster

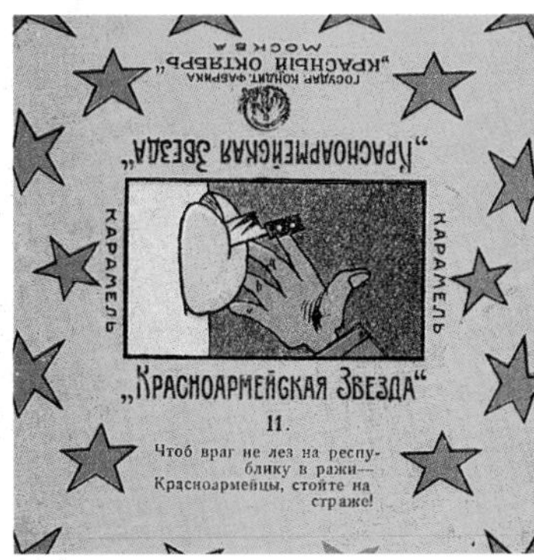


Majakowski 1919 Wrapper

Majakowski 1918 Poster

**Maïakovski,** Vladimir – né le 19. 7. 1893 à Bagdadi, Géorgie, décédé le 14. 4. 1930 à Moscou, URSS – *poète, graphiste, peintre* – 1908–1909, études d'art à l'Institut Stroganov de Moscou; puis de 1911 à 1914, études à l'école de peinture de sculpture et d'architecture de Moscou. Son premier recueil de poèmes, «Gifle au bon goût public», est publié en 1912. De 1913 à 1914, il participe à la tournée du groupe futuriste «Hylaea» à travers la Russie (avec D. Bourlyouk, W. Kamensky etc.). Sa première pièce de théâtre «Vladimir Maïakovski» est représentée à Saint-Pétersbourg en 1913. A partir de 1919, il dessine des affiches de films ainsi que des affiches d'agit-prop pour les «Fenêtres Rosta». Affiches publicitaires pour l'organisme d'état du commerce (avec A. Rodtchenko et W. Stepanova). Nombreuses lectures et conférences à Berlin de 1922 à 1929. Participe à la «1ère Exposition d'art russe» à Berlin. Membre fondateur du groupe «LEF» (Front des artistes de gauche) en 1923, édite les revues «LEF» et «Novy LEF». El Lissitzky réalise la maquette du recueil de poèmes «Dlja golosa» (Pour la voix). 1930, rétrospective de ses travaux «20 ans de travail» à Moscou.
**Majoor,** Martin – né le 14. 10. 1960 à Baarn, Pays-Bas – *graphiste maquettiste, concepteur de polices, enseignant* – 1980–1986, études d'arts graphiques et de design à l'académie des beaux-arts d'Arnhem. Travaille comme typographe chez Océ, Pays-Bas, de 1986 à 1988; exerce en-

alogues he has worked on have become standard works. 1982: is awarded an order of merit by the Federal Republic of Germany.
**Mayakovsky,** Vladimir – b. 19. 7. 1893 in Bagdadi, Georgia, d. 14. 4. 1930 in Moscow, USSR – *poet, graphic artist, painter* – 1908–09: studies art at the Stroganov Institute in Moscow and from 1911–14 at the Moscow School of Painting, Sculpture and Architecture. 1912: his first poems are published in a collection entitled "Boxing the Ears of Public Taste". 1913–14: takes part in a promotional tour of Russia with the Futurist group Hylaea, whose members include D. Burliuk and V. Kamensky, among others. 1913: his first play, "Vladimir Mayakovsky", is performed in St Petersburg. From 1919 onwards: designs film posters and agitation posters for the "Rosta" window. Produces posters for the state trade organization with A. Rodchenko and V. Stepanova. 1922–29: gives numerous lectures and readings in Berlin. Takes part in the 1st exhibition of Russian art in Berlin. 1923: founder member of the "LEF" group (Left Artists' Front). Editor of "LEF" and "Novy LEF" magazines. El Lissitzky designs Mayakovsky's volume of poetry entitled "Dlja golosa" ("for the voice"). 1930: a retrospective exhibition of Mayakovsky's work entitled "20 Years' Work" is held in Moscow.
**Majoor,** Martin – b. 14. 10. 1960 in Baarn, The Netherlands – *graphic designer, type designer, teacher* – 1980–86: studies graphic design at the art academy in Arnhem. 1986–88: works as a typogra-

Arbeit als Grafik-Designer im Musikzentrum „Vredenburg" in Utrecht. Entwurf der Schrift „Scala" für „Vredenburg". Seit 1990 freier Grafik-Designer in Arnheim, verfaßt zahlreiche Artikel. Ausgezeichnet im Wettbewerb „De Best Verzorgde Boeken" der Niederlande. Unterrichtet an der Kunstakademie in Arnheim (1989–94), der Kunstakademie in Breda (1990–95), der Merz Akademie in Stuttgart (1995) und der Rietveld Akademie in Amsterdam (1995). – *Schriftentwürfe:* Serré (1984), Scala (1991), Ocean (1992), Scala Sans (1993), Telefont List (1994), Telefont Text (1994), Scala Jewels (1996). – *Publikationen u.a.:* „FF Scala Sans, a new typeface", Berlin 1993; „A short history of decorated typefaces", Arnheim 1996; „The decorative alphabet from A to Z", Amsterdam 1996.

**Makela,** P. Scott – geb. 1960 in St. Paul, Minnesota, USA. – *Grafik-Designer, Typograph, Schriftentwerfer, Lehrer* – 1984 Abschluß der Studien am Minneapolis College of Art and Design. Danach Gründung seines Studios „Commbine" (mit Paul Knickelbine und Laurie Haycock) in Los Angeles. 1989–90 Studium an der Cranbrook Academy of Art. 1990 Gründung des Studios „Makela + Knickelbine Design" in Los Angeles, 1992 in Minneapolis, Auftraggeber aus Wirtschaft und Kultur, u.a. das Minneapolis College of Art and Design, das American Center for Design in Chicago, die Zeitschrift „Design Quarterly" und Emigre Records. Unterrichtet 1996 (zusammen mit Laurie Haycock Makela) Grafik-Design an der Cranbrook Academy of Art. – *Schriftentwür-*

suite comme graphiste maquettiste au centre de la musique «Vredenburg» à Utrecht. Dessine la police «Scala» pour «Vredenburg». Graphiste maquettiste à Arnhem depuis de 1990, écrit de nombreux articles. Primé lors du concours «De Best Verzorgde Boeken» des Pays-Bas. A enseigné à l'académie des beaux-arts d'Arnhem (1989–1994), à l'académie des beaux-arts de Breda (1990–1995), à la Merz Akademie de Stuttgart (1995) et à l'académie Rietveld d'Amsterdam (1995). – *Polices:* Serré (1984), Scala (1991), Ocean (1992), Scala Sans (1993), Telefont List (1994), Telefont Text (1994), Scala Jewels (1996). – *Publications, sélection:* «FF Scala Sans, a new typeface», Berlin 1993; «A short history of decorated typefaces», Arnhem 1996; «The decorative alphabet from A to Z», Amsterdam 1996.

**Makela,** P. Scott – né en 1960 à St Paul, Minnesota, Etats-Unis – *graphiste maquettiste, typographe, concepteur de polices, enseignant* – termine ses études au Minneapolis College of Art and Design en 1984. Fonde ensuite son propre atelier «Commbine» (avec Paul Knickelbine et Laurie Haycock) à Los Angeles. Etudie à la Cranbrook Academy of Art de 1989 à 1990. En 1990, il fonde l'atelier «Makela + Knickelbine Design» à Los Angeles, puis une filiale à Minneapolis en 1992. Commanditaires de l'industrie et du secteur culturel, entre autres le Minneapolis College of Art and Design, l'American Center for Design de Chicago, la revue «Design Quarterly» et Emigre Records. Enseigne les arts graphiques et le design (avec Laurie Haycock Makela) à la Cranbrook Academy of Art, en 1996. – *Polices:* Carmela (1989), Dead History (1990).

**Manuce,** Alde – né en 1449 à Bassiano, Italie, décédé le 8. 2. 1515 à Venise, Italie

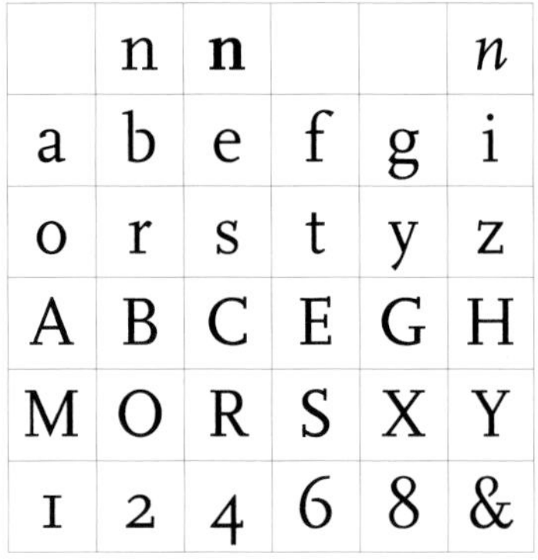


Majoor 1991 Scala

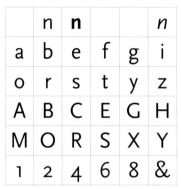

n n n
a b e f g i
o r s t y z
A B C E G H
M O R S X Y
1 2 4 6 8 &

Majoor 1993 Scala Sans

Makela 1992 Poster

Majoor 1995 (Telephone) Page

Majoor

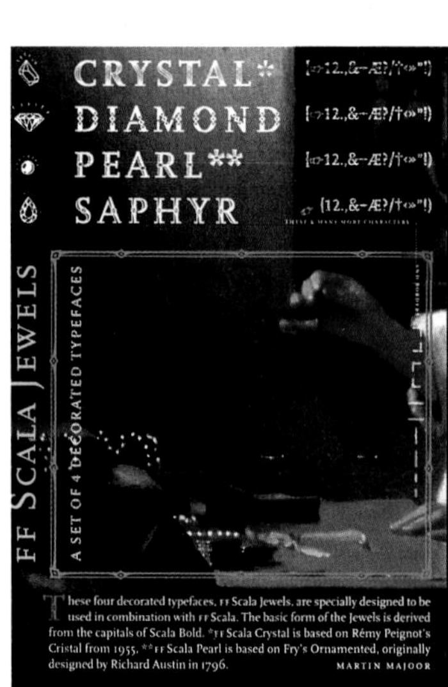


Majoor 1996 Postcard

pher for Océ in The Netherlands. 1988–90: works as a graphic designer at the Vredenburg music center in Utrecht. Designs Scala typeface for Vredenburg. Since 1990: freelance graphic designer in Arnhem. Has written numerous articles. Wins an award in the Dutch "De Best Verzorgde Boeken" competition. Majoor has taught at the art academy in Arnhem (1989–94), at the art academy in Breda (1990–95), at the Merz Akademie in Stuttgart (1995) and at the Rietveld Akademie in Amsterdam (1995). – *Fonts:* Serré (1984), Scala (1991), Ocean (1992), Scala Sans (1993), Telefont List (1994), Telefont Text (1994), Scala Jewels (1996). – *Publications include:* "FF Scala Sans, a new typeface", Berlin 1993; "A short history of decorated typefaces", Arnhem 1996; "The decorative alphabet from A to Z", Amsterdam 1996.

**Makela,** P. Scott – b. 1960 in St Paul, Minnesota, USA – *graphic designer, typographer, type designer, teacher* – 1984: completes his studies at the Minneapolis College of Art and Design. Then opens his own studio Commbine, with Paul Knickelbine and Laurie Haycock in Los Angeles. 1989–90: studies at the Cranbrook Academy of Art. 1990: opens the Makela + Knickelbine Design studio in Los Angeles and in 1992 in Minneapolis. Works for various clients from commerce and industry and the arts sector, including the Minneapolis College of Art and Design, the American Center for Design in Chicago, "Design Quarterly" magazine and Emigre Records. 1996: teaches graphic design at the Cranbrook Academy of Art

*fe:* Carmela (1989), Dead History (1990).
**Manutius,** Aldus – geb. 1449 in Bassiano, Italien, gest. 8. 2. 1515 in Venedig, Italien. – *Verleger, Drucker* – Studium in Rom und Ferrara. Lebt 1482 bei dem Adeligen Giovanni Pico della Mirandola. Erlernt ca. 1490 in Venedig das Drucken, ca. 1494 eigene Druckerei. Veröffentlicht griechische Klassiker-Ausgaben. Diese Ausgaben werden heute als „Aldinen" bezeichnet. 1495–98 Herausgabe der Schriften von Aristoteles in fünf Foliobänden. 1496 Herausgabe seines ersten lateinischen Drucks „De Aetna" von Pietro Bembo, für den Francesco Griffo eine Antiqua-Schrift schneidet. 1499 Herausgabe der „Hypnerotomachia Poliphili" von Francesco Colonia. Diese Ausgabe gilt als der schönste Aldus-Druck. – *Publikationen u.a.:* Antoine Auguste Renouard „Annales d'imprimerie des Aldes", Paris 1803; Julius Schück „Aldus Manutius und seine Zeitgenossen in Italien und Deutschland", Berlin 1862.
**Manwaring,** Michael – geb. 21. 3. 1942 in Palo Alto, USA. – *Grafik-Designer, Lehrer* – 1961–64 Grafik-Design-Studium und Studium der Malerei am San Francisco Art Institute. 1968–76 Gründungsmitglied und Teilhaber im Studio „Reis & Manwaring" (mit Gerald Reis), seit 1976 Gründung und Leitung von „The Office of Michael Manwaring" in San Francisco. Es entstehen Ausstellungsgestaltungen, Leit- und Zeichensysteme, Architekturgrafik und Verpackungen. Zahlreiche Auszeichnungen, u. a. Goldmedaille des Art Directors Club of New York (1980). 1982 Mitbegründer der San Fran-

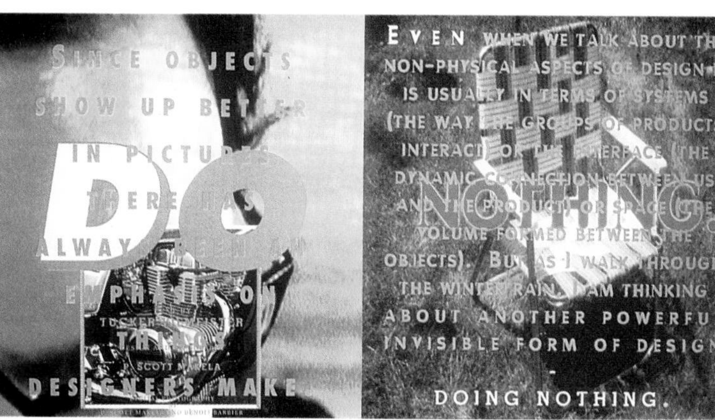

Makela 1992 Spread

Makela 1990 Dead History

Manutius 1499 Page

Manutius 1501 Aldus kursiv

– *éditeur, imprimeur* – études à Rome et à Ferrare. Séjourne chez l'aristocrate Jean Pic de la Mirandole en 1482. Vit à Venise vers 1490 où il apprend l'imprimerie, travaille dans sa propre imprimerie vers 1494. Publie des textes classiques grecs, des ouvrages appellés «Aldes». De 1495 à 1498, il publie les oeuvres d'Aristote en cinq volumes folio; puis en 1496, il édite son premier ouvrage imprimé en caractères latins, «De Aetna» de Pietro Bembo, pour lequel Francesco Griffo taille une antique. Publication de «Hypnerotomachia Poliphili» de Francesco Colonia en 1499. Cet ouvrage est considéré comme le plus beau livre imprimé par Alde. – *Publications, sélection:* Antoine Auguste Renouard «Annales d'imprimerie des Aldes», Paris 1803; Julius Schück «Aldus Manutius und seine Zeitgenossen in Italien und Deutschland», Berlin 1862.
**Manwaring,** Michael – né le 21. 3. 1942 à Palo Alto, Etats-Unis – *graphiste maquettiste, enseignant* – 1961–1964, études d'arts graphiques, de design et de peinture au San Francisco Art Institute. Fonde, avec son partenaire Gerald Reis, l'atelier «Reis & Manwaring» en 1968, atelier qui fonctionnera jusqu'en 1976. En 1976, il crée et dirige «The Office of Michael Manwaring» à San Francisco. Il réalise des architectures d'expositions, des systèmes d'orientation et de signalisation, des dessins d'architecture et des emballages. Nombreuses distinctions, entre autres la médaille d'or de l'Art Directors Club de New York (1980). Cofondateur de la section de l'American Institute of Graphic Arts (AIGA) de San Francisco en 1982. En 1987, il fonde l'atelier

with Laurie Haycock Makela. – *Fonts:* Carmela (1989), Dead History (1990).
**Manutius,** Aldus – b. 1449 in Bassiano, Italy, d. 8. 2. 1515 in Venice, Italy – *publisher, printer* – Studied in Rome and Ferrara. 1482: resides with the nobleman Giovanni Pico della Mirandola. C.1490: Manutius moves to Venice where he learns the art of printing and in c.1494 opens his own printing workshop. Publishes editions of the Greek classics. Today these volumes are known as "Aldines". 1495–98: publishes the writings of Aristotle in five folio volumes. 1496: publishes his first book in Latin, "De Aetna" by Pietro Bembo, with a roman typeface cut by Francesco Griffo. 1499: publishes "Hypnerotomachia Poliphili" by Francesco Colonia. This book is considered the most beautiful product of the Aldine press. – *Publications include:* Antoine Auguste Renouard "Annales d'imprimerie des Aldes", Paris 1803; Julius Schück "Aldus Manutius und seine Zeitgenossen in Italien und Deutschland", Berlin 1862.
**Manwaring,** Michael – b. 21. 3. 1942 in Palo Alto, USA – *graphic designer, teacher* – 1961–64: studies graphic design and painting at the San Francisco Art Institute. 1968–76: is founder and partner of the Reis & Manwaring studio (with Gerald Reis). 1976: founds and runs The Office of Michael Manwaring in San Francisco. Plans exhibitions, designs signage systems and systems of notation, architectural graphics and packaging. Wins numerous awards, including the Art Directors Club of New York's gold medal

cisco-Gruppe des American Institute of Graphic Arts (AIGA). 1987 Gründung des Büros „Environmental Design" (mit David Meckel). Unterrichtet Grafik-Design am California College of Arts and Crafts in Oakland und an der University of California in Berkeley.

**Mardersteig,** Giovanni (Hans) – geb. 8. 1. 1892 in Weimar, Deutschland, gest. 27. 12. 1977 in Verona, Italien. – *Typograph, Schriftentwerfer, Verleger* – 1910–13 Jura-Studium an den Universitäten Bonn, Wien, Kiel und Jena. 1915 Promotion. 1917–22 Mitherausgeber der Halbjahresschrift „Genius" für den Kurt Wolff Verlag. 1922 Gründung und Leitung der Handpresse „Officina Bodoni" in Montagnola, Schweiz. Seit 1927 in Verona, Italien. 1927–36 Herausgabe und Gestaltung der Gesamtausgabe der Werke von Gabriele d'Annunzio in 49 Bänden. Für diese Ausgabe erhält Mardersteig die Erlaubnis der italienischen Regierung, Bodonis Originalmatrizen zum Guß der Satzschrift zu verwenden. 1947–77 Gründung der Setzerei und Druckerei „Stamperia Valdonega" in Verona. 1977 Weiterführung der Druckerei durch seinen Sohn Martino Mardersteig. 1968 Auszeichnung mit dem Gutenberg-Preis der Stadt Mainz. Mardersteig gestaltet und druckt ca. 200 Bücher auf seiner Handpresse und in seiner Druckerei mit einer Gesamtauflage von ca. 40. 000 Exemplaren. – *Schriftentwürfe:* Griffo (1930), Fontana (1936), Zeno (1936), Dante (geschnitten von Charles Malin, 1954), Pacioli (1955). – *Publikationen u.a.:* „Die Officina Bodoni: das Werkbuch einer Hand-

«Environmental Design» (avec David Meckel). A enseigné les arts graphiques et le design au California College of Arts and Crafts d'Oakland et à l'University of California de Berkeley.

**Mardersteig,** Giovanni (Hans) – né le 8. 1. 1892 à Weimar, Allemagne, décédé le 27. 12. 1977 à Vérone, Italie – *typographe, concepteur de polices, éditeur* – 1910–1913, études de droit à Bonn, Vienne, Kiel et Iéna. Doctorat en 1915. De 1917 à 1922, il coédite la revue «Genius» pour les éditions Kurt Wolff. En 1922, il fonde et dirige la presse à main «Officina Bodoni», à Montagnola, Suisse, qui s'installera à Vérone en 1927. De 1927 à 1936, il édite et réalise la maquette des 49 volumes des oeuvres complètes de Gabriel d'Annunzio. Pour ces ouvrages, le gouvernement italien autorise Mardersteig à utiliser les matrices des caractères originaux de Bodoni. En 1947, il fonde un atelier de composition et une imprimerie, la «Stamperia Valdonega» à Vérone où il exerce jusqu'en 1977. En 1977, son fils Martino Mardersteig reprend l'imprimerie. Obtient le Prix Gutenberg de la ville de Mayence en 1968. Mardersteig a conçu et imprimé près de 200 ouvrages sur sa presse à main et dans son imprimerie, ses livres ont atteint un tirage total de 40 000 exemplaires. – *Polices:* Griffo (1930), Fontana (1936), Zeno (1936), Dante (taillé par Charles Malin en 1954), Pacioli (1955). – *Publications, sélection:* «Die Officina Bodoni: Das Werkbuch einer Handpresse in den ersten sechs Jahren ihres Wirkens», Paris 1929. John Dreyfus «Giovanni Mardersteig: an account of his work», New York 1956; John Barr «The Officina Bodoni: Montagnola – Vérone», Londres 1978.

25.1

Manwaring 1985 Logo

FELICE FELICIANO
VERONESE
ALPHABETUM
ROMANUM
Herausgegeben von G. Mardersteig
EDITIONES OFFICINAE BODONI
VERONA


Mardersteig 1960 Cover

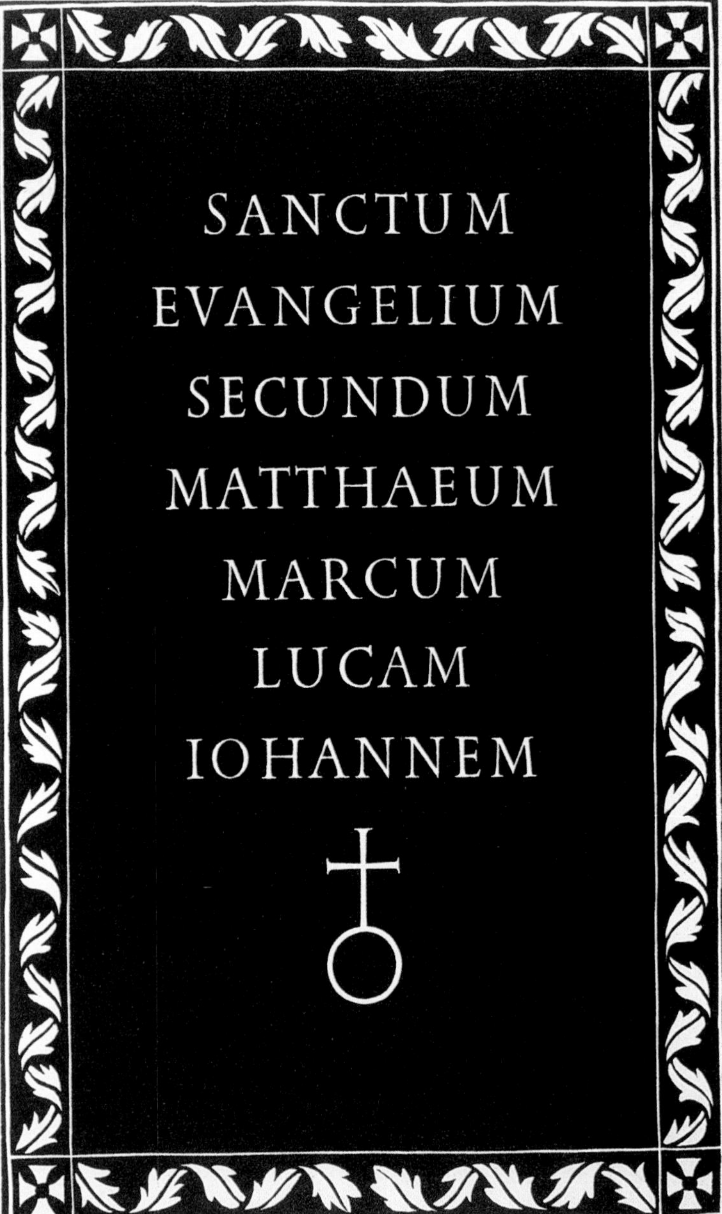


Mardersteig 1963 Cover

(1980). 1982: co-founder of the San Francisco group of the American Institute of Graphic Arts (AIGA). 1987: opens the Environmental Design studio with David Meckel. Manwaring has taught graphic design at the California College of Arts and Crafts in Oakland and at the University of California in Berkeley.

**Mardersteig,** Giovanni (Hans) – b. 8. 1. 1892 in Weimar, Germany, d. 27. 12. 1977 in Verona, Italy – *typographer, type designer, publisher* – 1910–13: studies law at the universities in Bonn, Vienna, Kiel and Jena. 1915: obtains his Ph.D. 1917–22: co-editor of the semiannual magazine "Genius" for Kurt Wolff publishers. 1922: founds and runs the Officina Bodoni hand press in Montagnola, Switzerland. 1927: moves to Verona, Italy. 1927–36: designs and publishes a complete edition of Gabriel d'Annunzio's works in 49 volumes. The Italian government grants Mardersteig permission to use Bodoni's original matrices to cast the type for composition. 1947–77: founds the Stamperia Valdonega typesetting and printing workshop in Verona. 1977: his son Martino Mardersteig takes over the running of the business. 1968: Mardersteig is awarded the Gutenberg Prize by the city of Mainz. Mardersteig designed and printed c. 200 books totaling 40,000 copies on his hand-press and in his printing workshop. – *Fonts:* Griffo (1930), Fontana (1936), Zeno (1936), Dante (cut by Charles Malin, 1954), Pacioli (1955). – *Publications include:* "Die Officina Bodoni: das Werkbuch einer Handpresse in den ersten sechs Jahren

presse in den ersten sechs Jahren ihres Wirkens", Paris 1929. John Dreyfus „Giovanni Mardersteig: an account of his work", New York 1966; John Barr „The Officina Bodoni: Montagnola – Verona", London 1978.

**Marinetti,** Filippo Tommaso – geb. 22. 12. 1876 in Alexandria, Ägypten, gest. 2. 12. 1944 in Bellagio, Italien. – *Schriftsteller, Verleger* – Jurastudium an den Universitäten Pavia und Genua. 1905 Gründung der Zeitschrift „Poesia", die zum Sprachrohr der futuristischen Bewegung wird. 1909 Veröffentlichung seines „Futuristischen Manifests" in der Pariser Zeitung „Figaro". 1911 Kriegsberichterstatter in Libyen. Die von ihm organisierte futuristische Ausstellung wird 1912 in Paris, London, Brüssel, Den Haag, Amsterdam, München, Dresden und Berlin gezeigt. Veröffentlicht „Technisches Manifest der futuristischen Literatur" (1912). Kämpft 1919 mit Mussolini für den Sieg des Faschismus. 1929 Mitglied der Akademie Italiens. Marinettis experimentelle Typographie ist geprägt von dem Bestreben, Inhalte unmittelbar und emphatisch zu visualisieren. – *Publikationen u.a.:* „La Conquête des Etoiles", Paris 1902; „Destruction", Paris 1904; „Zang-Tumb-Tumb", Mailand 1914; „Les mots en liberté futuristes", Mailand 1919; „Poemi simultanei", La Spezia 1933; „Il poema non umano dei meccanismi", Mailand 1940. W. Vaccari „Vita e tumulti di F. T. Marinetti", Mailand 1959; G. Lista „Marinetti", Paris 1976.

**Martens,** Karel – geb. 8. 3. 1939 in Mook en Middelaar, Niederlande. – *Grafik-De-*

| | n | n | n | | n |
|---|---|---|---|---|---|
| a | b | e | f | g | i |
| o | r | s | t | y | z |
| A | B | C | E | G | H |
| M | O | R | S | X | Y |
| 1 | 2 | 4 | 6 | 8 | & |

Mardersteig 1954 Dante

Marinetti 1914 Poem

**Marinetti,** Filippo Tommaso – né le 22. 12. 1876 à Alexandrie, Egypte, décédé le 2. 12. 1944 à Bellagio, Italie – *écrivain, éditeur* – études de droit à Pavie et à Gêne. En 1905, il fonde la revue «Poesia» qui deviendra le forum du mouvement futuriste. Publie son «Manifeste futuriste», en 1909, dans le quotidien le «Figaro». Correspondant de guerre en Libye en 1911. En 1912, il organise une exposition futuriste présentée à Paris, Londres, Bruxelles, La Haye, Amsterdam, Munich, Dresde et Berlin. Il publie le «Manifeste technique de la littérature futuriste» en 1912. En 1919, il combat aux côtés de Mussolini pour la victoire du fascisme. Membre de l'académie italienne en 1929. La typographie expérimentale de Marinetti est marquée par un désir de visualiser des contenus avec emphase. – *Publications, sélection:* «La Conquête des Etoiles», Paris 1902; «Destruction», Paris 1904; «Zang-Tumb-Tumb», Milan 1914; «Les mots en liberté futuristes», Milan 1919; «Poemi simultanei», La Spezia 1933; «Il poema non umano dei meccanismi», Milan 1940. W. Vaccari «Vita e tumulti di F. T. Marinetti», Milan 1959; G. Lista «Marinetti», Paris 1976.

**Martens,** Karel – né le 8. 3. 1939 à Mook en Middelaar, Pays-Bas – *graphiste ma-*

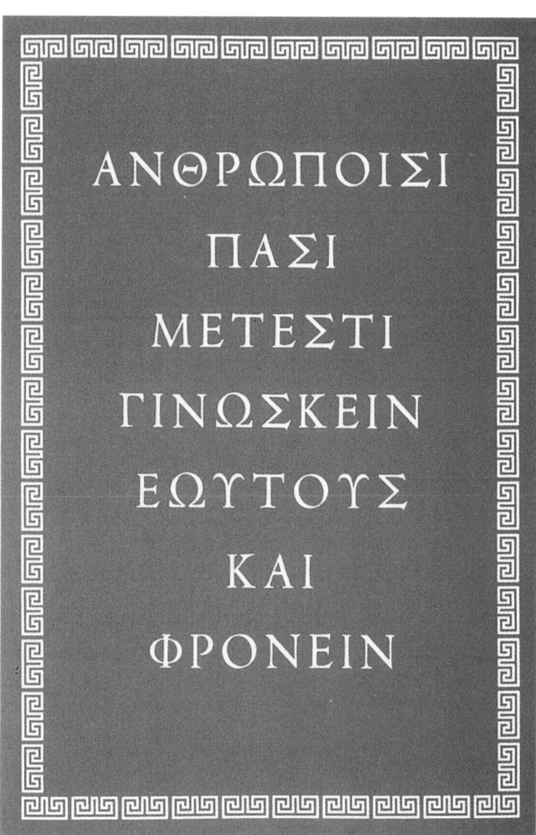


Mardersteig 1966 Cover

Marinetti 1914 Cover

ihres Wirkens", Paris 1929. John Dreyfus "Giovanni Mardersteig: an account of his work", New York 1966; John Barr "The Officina Bodoni: Montagnola – Verona", London 1978.

**Marinetti,** Filippo Tommaso – b. 22. 12. 1876 in Alexandria, Egypt, d. 2. 12. 1944 in Bellagio, Italy – *author, publisher* – Studied law at the universities in Pavia and Genoa. 1905: launches "Poesia" magazine which becomes the mouthpiece of the Futurist movement. 1909: publishes his "Futurist Manifesto" in the Parisian "Figaro" newspaper. 1911: war correspondent in Libya. 1912: he organizes a Futurist exhibition which is shown in Paris, London, Brussels, The Hague, Amsterdam, Munich, Dresden and Berlin. 1912: publishes his "Technical Manifesto of Futurist Literature". 1919: fights with Mussolini for the Fascist cause. 1929: member of the Italian Academy. Marinetti's experimental typography is characterized by his attempts to visualize the contents of the text clearly and emphatically. – *Publications include:* "La Conquête des Etoiles", Paris 1902; "Destruction", Paris 1904; "Zang-Tumb-Tumb", Milan 1914; "Les mots en liberté futuristes", Milan 1919; "Poemi simultanei", La Spezia 1933; "Il poema non umano dei meccanismi", Milan 1940. W. Vaccari "Vita e tumilti di F. T. Marinetti", Milan 1959; G. Lista "Marinetti", Paris 1976.

**Martens,** Karel – b. 8. 3. 1939 in Mook en Middelaar, The Netherlands – *graphic designer, typographer, teacher* – 1957–61: studies at the Academie voor Beeldende

*signer, Typograph, Lehrer* – 1957–61 Studium an der Academie voor Beeldende Kunsten in Arnheim. Seit 1961 selbständiger Gestalter. Auftraggeber waren u.a. Verlage, Regierungs-Institutionen und die niederländische Post PTT, Filmhuis Nijmegen. 1961–69 typographische Arbeiten für den Verlag Van Loghum Slaterus in Arnheim, 1975–81 für den Verlag Socialistiese Uitgeverij Nijmegen (SUN). Unterrichtet 1977–94 Grafik-Design an der Academie voor Beeldende Kunsten in Arnheim. 1985 Ausstellung freier Arbeiten unter dem Titel „Papier" in Nimwegen. Zahlreiche Auszeichnungen, u. a. H. N. Werkman-Preis der Stadt Amsterdam (1993), Designprijs Rotterdam (1995). Ausstellung freier Arbeiten im Stedelijk Museum in Amsterdam. Unterrichtet seit 1994 an der Jan van Eyck Academie in Maastricht. – *Publikationen u. a.:* Jereon Brouwers u. a. „Papier", Nimwegen 1985; Robin Kinnross u.a. „Printed matter, Drukwerk", London 1996.

**Martin,** William – geb. ca. 1765 in Birmingham, England, gest. 1815 in London, England. – *Schriftentwerfer, Schriftschneider* – Ausbildung zum Schriftgießer in Birmingham durch seinen Bruder Robert Martin, der in der Druckerei von Baskerville arbeitet und sich nach deren Auflösung selbständig macht. 1786 Gründung einer Schriftgießerei in London. Stellt u. a. zahlreiche griechische und orientalische Schriften her. Seine Schriften beruhen auf den Alphabeten Baskervilles. Arbeitet für George Nicol, später für William Bulmers „Shakespeare Printing Office". 1795 erscheint die von

*quettiste, typographe, enseignant* – 1957–1961, études à l'Academie voor Beeldende Kunsten à Arnhem. Graphiste indépendant à partir de 1961. Commanditaires : éditeurs, institutions gouvernementales, les postes néerlandaises PTT, Filmhuis Nijmegen. De 1961 à 1969, il réalise des travaux typographiques pour les éditions Van Loghum Slaterus de Arnhem, puis de 1975 à 1981 pour les éditions Socialistiese Uitgeverij Nijmegen (SUN). Enseigne les arts graphiques et le design à l'Academie voor Beeldende Kunsten d'Arnhem de 1977 à 1994. Exposition de travaux libres en 1985 à Nimègue sous le titre de «Papier». Nombreuses distinctions, dont le prix H.N. Werkman de la ville d'Amsterdam (1993), le Designprijs Rotterdam (1995). Expositions de travaux libres au Stedelijk Museum d'Amsterdam. Enseigne depuis 1994 à la Jan van Eyck Akademie de Maastricht. – *Publications, sélection :* Jereon Brouwers et autres «Papier», Nimègue 1985; Robin Kinnross et autres «Printed matter, Drukwerk», Londres 1996.

**Martin,** William – né vers 1765 à Birmingham, Angleterre, décédé en 1815 à Londres, Angleterre – *concepteur de polices, tailleur de types* – son frère, Robert Martin, employé à l'imprimerie de Baskerville et qui exercera à son compte après sa dissolution, le forme comme fondeur de caractères. En 1786, William Martin crée une fonderie de caractères à Londres et produit de nombreux caractères grecs et orientaux. Ses polices s'inspirent des alphabets de Baskerville. Il travaille pour George Nicol, puis pour le «Shakespeare Printing Office» de William Bulmer. En 1795, Bulmer publie, à Londres, une édition des poèmes de Goldsmith et de Parnell. A la mort de Martin, en 1815, W. Bulmer reprend l'entreprise. – *Polices :*

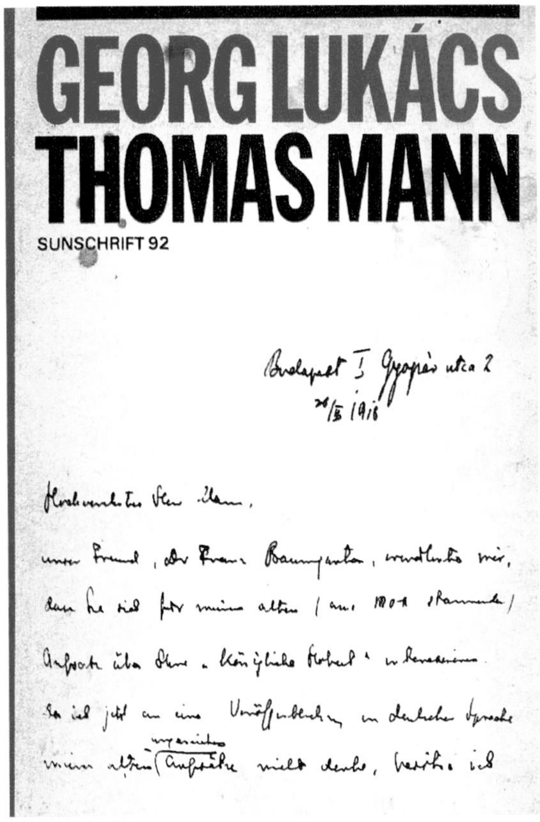

Martens Cover

Martens 1978 Poster

Martin ca. 1790 Bulmer

Kunsten in Arnhem. From 1961 onwards: freelance designer. Clients include various publishing houses, government institutions, the Dutch post office PTT and Filmhuis Nijmegen. 1961–69: produces typographic work for Van Loghum Slaterus publishers in Arnhem, and from 1975–81 for the publishing house Socialistiese Uitgeverij Nijmegen (SUN). 1977–94: teaches graphic design at the Academie voor Beeldende Kunsten in Arnhem. 1985: an exhibition of his free work, entitled "Papier", is held in Nimwegen. Numerous awards, including Amsterdam's H. N. Werkman Prize (1993) and the Designprijs Rotterdam (1995). An exhibition of his free work has been held in the Stedelijk Museum in Amsterdam. Martens has taught at the Jan van Eyck Academie in Maastricht since 1994. – *Publications include:* Jereon Brouwers et al "Papier", Nimwegen 1985; Robin Kinnross et al "Printed matter, Drukwerk", London, 1996.

**Martin,** William – b. c.1765 in Birmingham, England, d. 1815 in London, England – *type designer, type founder* – Trained as a type founder in Birmingham under his brother Robert Martin, who worked at Baskerville's printing works and set up his own business when Baskerville's company closed down. 1786: opens a type foundry in London. Production includes numerous Greek and Oriental alphabets. Martin bases his typefaces on those of Baskerville. Works for George Nicol and later for William Bulmer's Shakespeare Printing Office. 1795: an edition of poems by Goldsmith and

William Bulmer in London gedruckte Ausgabe der Gedichte von Goldsmith und Parnell. Nach Martins Tod 1815 übernimmt William Bulmer das Unternehmen. – *Schriftentwurf:* Bulmer (1790, M. F. Benton 1928).

**Maschadi,** Sultan Ali – geb. 1421 im Iran, gest. 1506 in Masched, Iran. – *Kalligraph* – ausgebildet von Ashar Tabrizi und Hafiz Hadji Mohammad. Arbeitet am Hof von Sultan Hossein Boyoghra mit der Nastaliq-Schrift. Schreibt zahlreiche Gedichtbände, darunter von Khadjo Kermani (1453), Khamsse Nawai (1478), sowie Grabsteine von Sultan Mohammad Bayoghra (1463), Sultan Ebrahim-Ben-Manssour (1494). Seine Arbeiten sind in Bibliotheken und Sammlungen in Frankreich, England, dem Iran, der Türkei, Rußland, Indien, Pakistan, Österreich und Deutschland aufbewahrt.

**Massin,** Robert – geb. 13. 10. 1925 in Boisvillette, Frankreich. – *Typograph, Autor, Grafik-Designer* – Autodidakt. Seit 1948 Grafik-Designer und Art Director in Paris. 1948–52 Gestaltung von Büchern für den „Club Français du Livre". 1952–58 Art Director des „Club du Meilleur Livre", 1958–78 Buchgestaltung für den Verlag Gallimard. 1970 Auszeichnung mit dem „Prix des Graphistes" in Paris. 1979 Leiter von Hachette-Réalités, 1980 von Hachette, Massin. Seit 1981 Herausgeber bei Edition Denoël in Paris. Die von Massin gestalteten Bücher, u.a. „Cent Mille Milliards de Poèmes" von Raymond Queneau (Paris 1961), „Exercice de style" von Raymond Queneau (Paris 1963) gehören zu den außergewöhnlichsten Li-

Maschadi

Massin 1964 Spread

Bulmer (1790, M. F. Benton 1928).

**Maschadi,** Sultan Ali – né en 1421 en Iran, décédé en 1506 à Masched, Iran – *calligraphe* – formation auprès d'Ashar Tabrizi et d'Hafiz Hadji Mohammad. Travaille à la cour du sultan Hossein Boyoghra. Sa calligraphie était de type nastaliq. A transcrit de nombreux recueils de poèmes, dont ceux de Khadjo Kermani (1453), de Khamsse Nawai (1478) et conçu des pierres tombales, dont celle du sultan Mohammad Bayoghra (1463), du sultan Ebrahim-Ben-Manssour (1494). Ses travaux sont conservés dans plusieurs bibliothèques et collections en France, Angleterre, Iran, Turquie, Russie, Inde, Pakistan, Autriche et en Allemagne.

**Massin,** Robert – né le 13. 10. 1925 à Boisvillette, France – *typographe, auteur, graphiste maquettiste* – autodidacte. Graphiste maquettiste, directeur artistique à Paris à partir de 1948. Réalise la maquette d'ouvrages du «Club Français du Livre» de 1948 à 1952. Art Director du «Club du Meilleur Livre» de 1952 à 1958, puis pour les éditions Gallimard de 1958 à 1978. Obtient le «Prix des graphistes» à Paris, en 1970. Directeur de Hachettes-Réalités en 1979, puis de Hachette Massin en 1980. Directeur de collection aux éditions Denoël depuis 1981. Les livres conçus par Massin, dont «Cent Mille Milliards de Poèmes» de Raymond Queneau (Paris 1961) et «Exercice de style», également de Queneau (Paris 1963), font partie des visualisations littéraires les plus extraordi-

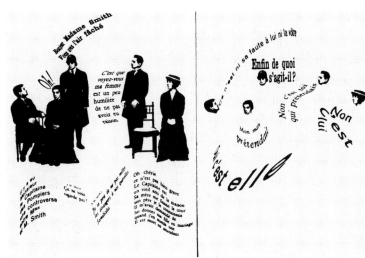

|   | n |   |   |   | *n* |
|---|---|---|---|---|---|
| a | b | e | f | g | i |
| o | r | s | t | y | z |
| A | B | C | E | G | H |
| M | O | R | S | X | Y |
| 1 | 2 | 4 | 6 | 8 | & |

Martin 1790 Bulmer

Massin 1964 Spread

Parnell is published by William Bulmer in London. 1815: William Bulmer takes over the company after Martin's death. – *Fonts:* Bulmer (1790, M. F. Benton 1928).

**Maschadi,** Sultan Ali – b. 1421 in Iran, d. 1506 in Masched, Iran – *calligrapher* – Trained under Ashar Tabrizi and Hafiz Hadji Mohammed. Worked at Sultan Hossein Boyoghra's court. Maschadi worked in nastaliq script. He notated numerous volumes of poetry, including works by Khadjo Kermani (1453) and Khamsse Nawai (1478), and inscribed the tombstones for Sultan Mohammed Bayoghra (1463) and Sultan Ebrahim-Ben-Manssour (1494). His work has been preserved in various libraries and collections in France, England, Iran, Turkey, Russia, India, Pakistan, Austria and Germany.

**Massin,** Robert – b. 13. 10. 1925 in Boisvillette, France – *typographer, author, graphic designer* – Self-taught. From 1948 onwards: graphic designer and art director in Paris. 1948–52: designs books for the Club Français du Livre. 1952–58: art director of the Club du Meilleur Livre and from 1958–78 book-design for Gallimard publishers. 1970: is awarded the Prix des Graphistes in Paris. 1979: director of Hachette-Réalités and in 1980 of Hachette, Massin. From 1981 onwards: editor for Edition Denoël in Paris. Books Massin has designed, including "Cent Mille Milliards de Poèmes" by Raymond Queneau (Paris 1961) and "Exercice de style" by Raymond Queneau (Paris 1963), are counted among the most unusual examples of visualized literature of the past few decades. – *Publications include:*

teratur-Visualisierungen neuerer Zeit. – *Publikationen u.a.:* „Vignettes françaises fin de siècle", Paris 1966; „La lettre et l'image", Paris 1970; Les célébrités de la rue", Paris 1983.

**Materot,** Lucas – geb. um 1600 – *Schreibmeister* – Schreiber in der päpstlichen Kanzlei in Avignon. In Avignon erscheint 1608 sein Schreibbuch „Les œuvres".

**Matsunaga,** Shin – geb. 13. 3. 1940 in Tokio, Japan. – *Grafik-Designer, Lehrer* – Beendet 1964 sein Studium der visuellen Kommunikation an der Nationalen Universität für Kunst und Musik in Tokio. 1964–77 Arbeit in der Werbeabteilung von Shiseido Kosmetik. Seit 1971 eigenes Büro „Shin Matsunaga Design Inc.", Gestaltung von Plakaten, Verpackungen, Erscheinungsbildern. Zahlreiche Auszeichnungen, u. a. Goldmedaille der 12. Plakatbiennale in Warschau (1988). Unterrichtet Grafik-Design an der Nationalen Universität für Kunst und Musik in Tokio. – *Publikationen u. a.:* „The Design Works of Shin Matsunaga", Tokio 1992; „Phollage", Tokio 1993.

**Matter,** Herbert – geb. 25. 4. 1907 in Engelberg, Schweiz, gest. 8. 5. 1984 in Springs, USA. – *Grafik-Designer, Fotograf, Lehrer* – 1924–26 Studium an der Ecole des Beaux-Arts in Genf, 1928–29 Académie moderne in Paris. 1929–32 Zusammenarbeit mit A. M. Cassandre und Le Corbusier. Mitarbeit an der Zeitschrift „Vogue" und in der Werbeabteilung von Deberny & Peignot. Lebt 1932–36 in Zürich. Zusammenarbeit u. a. mit Heiri Steiner, Anton Stankowski, Walter Herdeg, Hans Neuburg. Lebt 1936–46 in New

naires de notre époque. – *Publication, sélection:* «Vignettes françaises fin de siècle», Paris 1966; «La lettre et l'image», Paris 1970; «Les célébrités de la rue», Paris 1983.

**Materot,** Lucas – né vers 1600 – *scribe* – Scribe à la chancellerie pontificale d'avignon. Son livre «Les œuvres» a été publié en 1608 à Avignon.

**Matsunaga,** Shin – né le 13. 3. 1940 à Tokyo, Japon – *graphiste maquettiste, enseignant* – Termine ses études de communication visuelle à l'université nationale d'art et de musique de Tokyo en 1964. Travaille au service de la publicité des cosmétiques Shiseido de 1964 à 1977. Exerce dans son propre atelier, le «Shin Matsunaga Design Inc.» à partir de 1971 où il dessine des affiches, des emballages, des identités. Nombreuses distinctions, entre autres la médaille d'or de la 12è Biennale de l'affiche à Varsovie (1988). Enseigne les arts graphiques et le design à l'université nationale des arts et de la musique de Tokyo. – *Publications, sélection:* «The Design Works of Shin Matsunaga», Tokyo 1992; «Phollage», Tokyo 1993.

**Matter,** Herbert – né le 25. 4. 1907 à Engelberg, Suisse, décédé le 8. 5. 1984 à Springs, Etats-Unis – *graphiste maquettiste, photographe, enseignant* – 1924–1926, études à l'Ecole des Beaux-Arts de Genève, puis à l'Académie moderne de Paris de 1928 à 1929. Travaille avec A. M. Cassandre et Le Corbusier de 1929 à 1932. Collabore à la revue «Vogue» et travaille pour le service publicité de Deberny & Peignot. Vit à Zurich de 1932 à 1936. Travaille avec Heiri Steiner, Anton Stankowski, Walter Herdeg et Hans Neuburg. Vit à New York de 1936 à 1946. Réalisations pour les revues «Harper's Bazaar» et «Vogue». Affiches pour la Container Corporation of America. Architecture in-

Matsunaga 1987 Poster

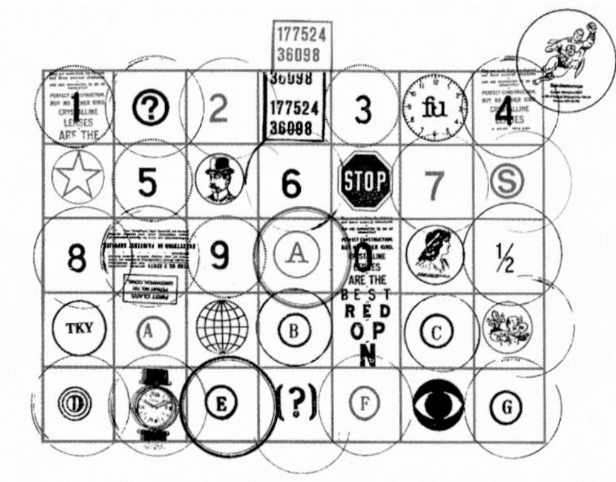

Matsunaga 1978 Card

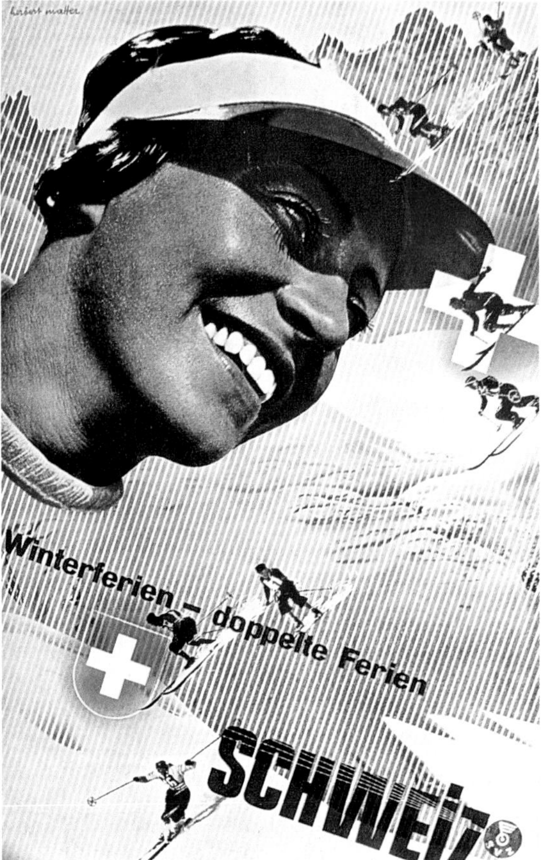


Matter 1934 Poster

Matter 1950 Logo

"Vignettes françaises fin de siècle", Paris 1966; "La lettre et l'image", Paris 1970; "Les célébrités de la rue", Paris 1983.

**Materot,** Lucas – b. c.1600 – *writing master* – Scribe at the papal chancellery in Avignon. 1608: his book on the art of writing, "Les œuvres", is published in Avignon.

**Matsunaga,** Shin – b. 13. 3. 1940 in Tokyo, Japan – *graphic designer, teacher* – 1964: completes his studies in visual communication at the National University of Art and Music in Tokyo. 1964–77: works in the publicity department of Shiseido Cosmetics. 1971: opens his own studio, Shin Matsunaga Design Inc. Designs posters, packaging and corporate identities. Numerous awards, including a gold medal at the 12th Poster Biennale in Warsaw (1988). Teaches graphic design at the National University of Art and Music in Tokyo. – *Publications include:* "The Design Works of Shin Matsunaga", Tokyo 1992; "Phollage", Tokyo 1993.

**Matter,** Herbert – b. 25. 4. 1907 in Engelberg, Switzerland, d. 8. 5. 1984 in Springs, USA – *graphic designer, photographer, teacher* – 1924–26: studies at the Ecole des Beaux-Arts in Geneva and from 1928–29 at the Académie moderne in Paris. 1929–32: works with A. M. Cassandre and Le Corbusier. Works on "Vogue" magazine and for Deberny & Peignot's publicity department. 1932–36: moves to Zurich and works with Heiri Steiner, Anton Stankowski, Walter Herdeg and Hans Neuburg, among others. 1936–46: moves to New York and works for "Harper's Bazaar" and "Vogue"

York. Arbeiten für die Zeitschriften „Harper's Bazaar" und „Vogue". Plakate für die Container Corporation of America. 1939 Innengestaltung des Schweizer Pavillons der Weltausstellung in New York. 1946–66 Design-Berater der Firma Knoll International. 1952–76 Professor für Fotografie an der Yale University in New Haven. 1953–68 Gestaltung aller Drucksachen des Solomon R. Guggenheim Museums in New York. 1954–56 Gestaltung des Erscheinungsbildes der New Haven Railroad. 1968 Realisierung von zehn Unterrichtsfilmen über das Werk von Richard Buckminster Fuller. – *Publikationen u.a.:* „Symbols, Signs, Logos, Trademarks", New York 1977; „H. Matter. Foto-Grafiker. Sehformen der Zeit", Baden 1995.

**Mau,** Bruce – geb. 25. 10. 1959 in Sudbury, Ontario, Kanada. – *Grafik-Designer, Lehrer* – 1978–80 Kunst- und Design-Studium am Ontario College of Art in Toronto. 1980–82 Gestalter bei „Fifty Fingers Inc." in Toronto. 1982–83 Gestalter bei „Pentagram Design" in London. 1983–85 Mitbegründer des Studios „Public Good Design and Communication Inc." in Toronto. 1985 Gründung seines Studios „Bruce Mau Design Inc." in Toronto. Auftraggeber waren u.a. die Art Gallery of Ontario in Toronto, das Wexner Center for the Arts in Columbus, Design Exchange in Toronto, The Whitney Museum of American Art in New York, The Museum of Contemporary Art in Los Angeles, Swatch in New York, das Nederlandse Architectuurinstituut in Rotterdam. Projekt-Zusammenarbeit mit Claes Oldenburg und Coosje van Bruggen (New York),

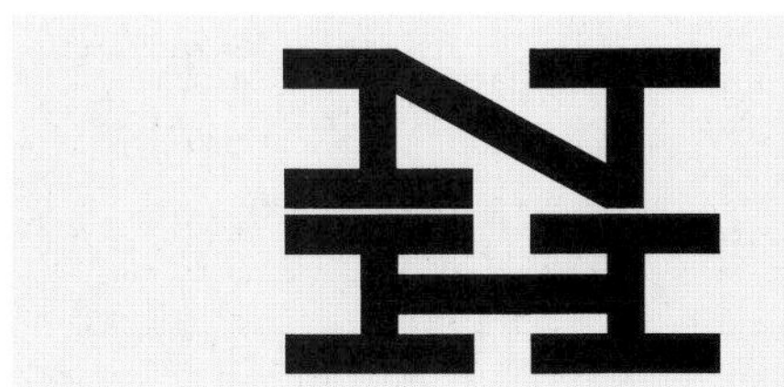

Matter 1954 Signet

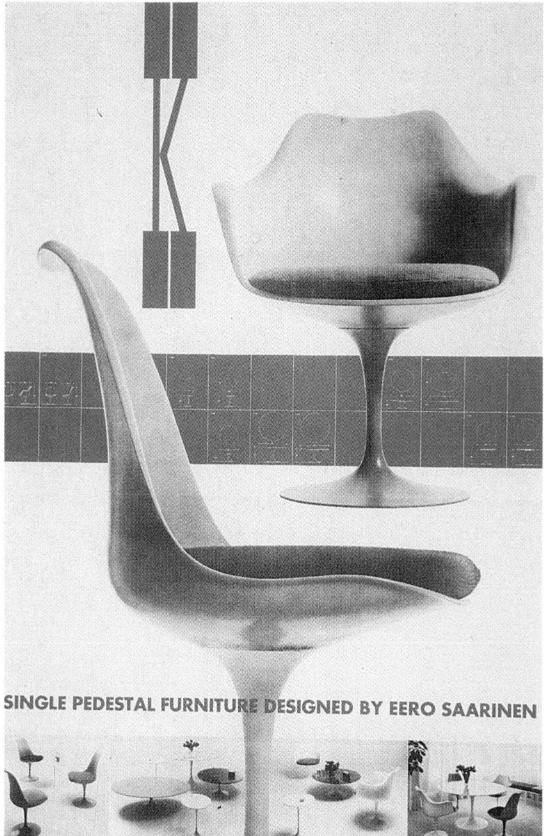


Matter 1957 Poster

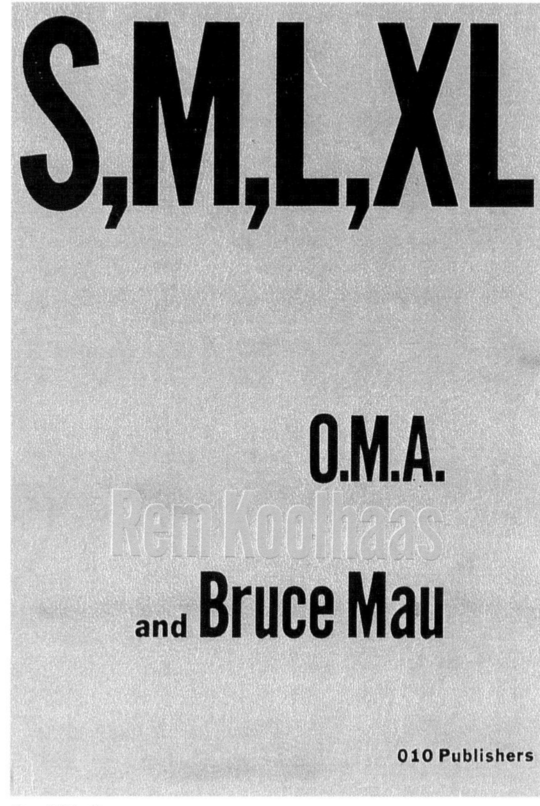

S,M,L,XL
O.M.A.
Rem Koolhaas
and Bruce Mau
010 Publishers


Mau 1995 Cover

Mau 1992 Cover (Detail)

térieure du pavillon suisse de l'exposition universelle de New York en 1939. Conseiller en design de la société Knoll International de 1946 à 1966. Professeur de photographie à la Yale University de New Haven de 1952 à 1976. Maquette de tous les imprimés du Salomon R. Guggenheim Museum de New York de 1953 à 1968. Identité du New Haven railroad en 1954–1956. Réalisation de dix films éducatifs sur l'oeuvre de Richard Buckminster Fuller en 1968. – *Publications, sélection:* «Symbols, Signs, Logos, Trademarks», New York 1977; «Herbert Matter. Foto-Grafiker. Sehformen der Zeit», Baden 1995.

**Mau,** Bruce – né le 25. 10 1959 à Sudbury, Ontario, Canada – *graphiste maquettiste, enseignant* – 1978–1980, études d'art et de design à l'Ontario College of Art de Toronto. Designer chez «Fifty Fingers Inc.» à Toronto de 1980 à 1982. De 1982 à 1983, désigner chez «Pentagram Design» à Londres. Cofondateur, en 1983, de l'atelier «Public Good Design and Communication Inc.» à Toronto où il exerce jusqu'en 1985. En 1985, il fonde son propre atelier le «Bruce Mau Design Inc.» à Toronto. Commanditaires: l'Art Gallery of Ontario à Toronto, le Wexner Center for the Arts à Columbus, Design Exchange à Toronto, The Whitney Museum of American Art à New York, The Museum of Contemporary Art à Los Angeles, Swatch à New York, le Nederlandse Architectuurinstituut à Rotterdam. Collaboration à des projets avec Claes Ol-

magazines. Produces posters for the Container Corporation of America. 1939: does the interior design for the Swiss pavilion at the world exhibition in New York. 1946–66: design consultant to Knoll International. 1952–76: professor of photography at Yale University in New Haven. 1953–68: designs all the printed material for the Solomon R. Guggenheim Museum in New York. 1954–56: designs New Haven Railroad's corporate identity. 1968: completes ten educational films on the work of Richard Buckminster Fuller. – *Publications include:* "Symbols, Signs, Logos, Trademarks", New York 1977; "Herbert Matter. Foto-Grafiker. Sehformen der Zeit", Baden 1995.

**Mau,** Bruce – b. 25. 10. 1959 in Sudbury, Ontario, Canada – *graphic designer, teacher* – 1978–80: studies art and design at the Ontario College of Art in Toronto. 1980–82: designer for Fifty Fingers Inc. in Toronto. 1982–83: designer for Pentagram Design in London. 1983–85: cofounder of the studio Public Good Design and Communication Inc. in Toronto. 1985: opens his own studio, Bruce Mau Design Inc. in Toronto. Clients include the Art Gallery of Ontario in Toronto, the Wexner Center for the Arts in Columbus, Design Exchange in Toronto, The Whitney Museum of American Art in New York, The Museum of Contemporary Art in Los Angeles, Swatch in New York and the Nederlandse Architectuurinstituut in Rotterdam. Works on projects with Claes Oldenburg and Coosje van Bruggen (New York), Frank Gehry (Los Angeles) and Rem Koolhaas (Rotterdam). 1991–93: cre-

Frank Gehry (Los Angeles), Rem Koolhaas (Rotterdam). 1991–93 Creative Director des „I.D. Magazine". 1992 Gast des Getty Research Center for the History of Arts and the Humanities in Los Angeles. Unter dem Titel „S, M, L, XL in Progress: Rem Koolhaas and Bruce Mau" werden 1994 die Ergebnisse der Zusammenarbeit am Southern California Institute of Architecture in Los Angeles veröffentlicht. Unterrichtet u.a. an der Technical University of Nova Scotia in Halifax (1986, 1988), der York University in Toronto (1990), der University of Toronto (1990–92) und der Rice University in Houston. Zahlreiche Auszeichnungen für seine gestalterischen Arbeiten.

**Mavignier,** Almir da Silva – geb. 1. 5. 1925 in Rio de Janeiro, Brasilien. – *Maler, Grafik-Designer, Lehrer* – 1945 Studium der Malerei. Mitwirkung bei der ersten Gruppe abstrakter Maler in Rio de Janeiro. 1953–58 Studium der Visuellen Kommunikation an der Hochschule für Gestaltung in Ulm. 1955 erste Plakatentwürfe. 1958 Mitglied der Künstlergruppe „Zero" in Düsseldorf. 1959 Atelier in Ulm, als Maler und Grafik-Designer tätig. 1965–90 Professor für Malerei an der Staatlichen Hochschule für Bildende Künste in Hamburg. 1968 Teilnahme an der documenta IV in Kassel. 1985 Preisträger der Anton-Stankowski-Stiftung. Zahlreiche Ausstellungen der freien und angewandten Arbeiten. – *Publikationen u.a.:* „Mavignier. Plakate", Hamburg, Essen 1981; „Mavignier. Bilder, Plakate", München, Herning, Hamburg 1990.

**Mayer,** Hansjörg – geb. 1943 in Stuttgart,

denburg et Coosje van Bruggen (New York), Frank Gehry (Los Angeles), Rem Koolhaas (Rotterdam). Directeur de la création pour «I.D. Magazine» de 1991 à 1993. En 1992, il est l'invité du Getty Research Center for the History of Arts and the Humanities à Los Angeles. Publication en 1994 des résultats de sa collaboration avec R. Koolhaas pour le Southern California Institute of Architecture de Los Angeles sous le titre «S, M, L, XL in Progress : Rem Koolhaas and Bruce Mau». A enseigné à la Technical University of Nova Scotia d'Halifax (1986, 1988), à la York University de Toronto (1990), à l'université de Toronto (1990–1992), à la Rice University de Houston. Nombreuses distinctions pour ses travaux de création.

**Mavignier,** Almir da Silva – né le 1. 5. 1925 à Rio de Janeiro, Brésil – *peintre, graphiste maquettiste, enseignant* – étudie la peinture en 1945. Participe au premier groupe de peintres abstraits de Rio de Janeiro. 1953–1958, études de communication visuelle à la Hochschule für Gestaltung (école supérieure de design) d'Ulm. Premiers dessins d'affiches en 1955. Membre du groupe d'artistes «Zero» à Dusseldorf en 1958. Installe son atelier à Ulm en 1959 où il exerce comme peintre et graphiste maquettiste. 1965–1990, professeur de peinture à la Staatliche Hochschule für Bildende Künste (école des beaux-arts) de Hambourg. Participe à la «documenta IV» de Kassel en 1968. Prix de la fondation Anton Stankowski en 1985. Nombreuses expositions de ses travaux graphiques et artistiques. – *Publications, sélection :* «Mavignier. Plakate», Hambourg, Essen 1981; «Mavignier. Bilder, Plakate», Munich, Herning, Hambourg 1990.

**Mayer,** Hansjörg – né en 1943 à Stuttgart, Allemagne – *typographe, éditeur, enseignant* – apprentissage de composition ty-

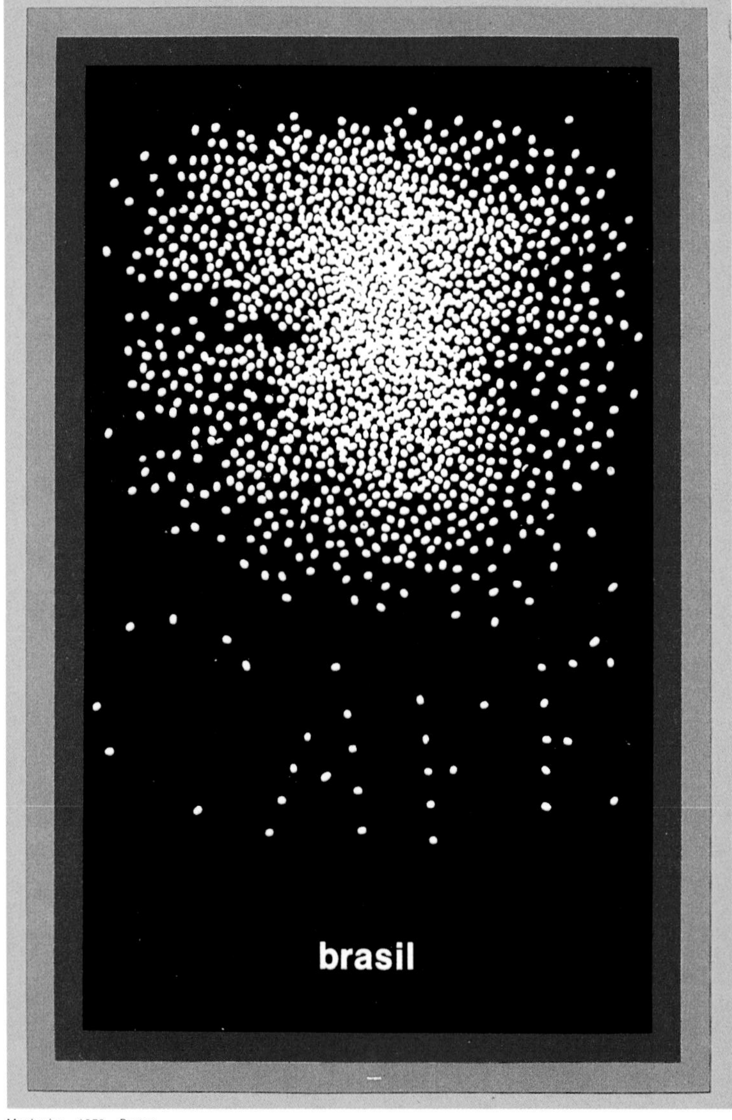

Mavignier 1970 Poster

Mavignier 1961 Poster

Mavignier 1985 Poster

ative director for "I.D. Magazine". 1992: guest at the Getty Research Center for the History of Arts and the Humanities in Los Angeles. 1994: the results of a collaboration project at the Southern California Institute of Architecture in Los Angeles are published under the title "S, M, L, XL in Progress: Rem Koolhaas and Bruce Mau". Mau has taught at the Technical University of Nova Scotia in Halifax (1986, 1988), York University in Toronto (1990), the University of Toronto (1990–92) and Rice University in Houston. He has received numerous awards for his work in design.

**Mavignier,** Almir da Silva – b. 1. 5. 1925 in Rio de Janeiro, Brazil – *painter, graphic designer, teacher* – 1945: studies painting. Becomes involved with the first group of abstract painters in Rio de Janeiro. 1953–58: studies visual communication at the Hochschule für Gestaltung in Ulm. 1955: produces his first poster designs. 1958: member of the group Zero in Düsseldorf. 1959: has a studio in Ulm and works as a painter and graphic designer. 1965–90: professor of painting at the Staatliche Hochschule für Bildende Künste in Hamburg. 1968: takes part in documenta IV in Kassel. 1985: wins the Anton-Stankowski-Stiftung Prize. There have been numerous exhibitions of his free and applied work. – *Publications include:* "Mavignier. Plakate", Hamburg, Essen 1981; "Mavignier. Bilder, Plakate", Munich, Herning, Hamburg 1990.

**Mayer,** Hansjörg – b. 1943 in Stuttgart, Germany – *typographer, publisher,*

Deutschland. – *Typograph, Verleger, Lehrer* – Schriftsetzerlehre. Arbeit in Druckereien in der Schweiz und den USA. 1962–67 Plakatentwürfe für das Studium Generale der Technischen Hochschule in Stuttgart. Gründet die „Edition Hansjörg Mayer" in Stuttgart. Herausgabe zahlreicher Publikationen, u.a. von Franz Mon, Dieter Rot, Emmett Williams, Wolfgang Schmidt und Mary Vieira. 1966 Eröffnung der „Galerie der Edition Hansjörg Mayer" in Stuttgart, in der die Künstler des Verlags ausgestellt werden. 1967 Dozent an der Watford School of Art in Hertfordshire. Durch die Publikationen seines Verlags und die eigene gestalterische Arbeit ist Mayer ein Teil der „Stuttgarter Schule" um Max Bense. – *Publikationen u.a.:* „19 Typographien", Stuttgart 1962; „Alphabet", Stuttgart 1963; „Typoems", Stuttgart 1965; „Alphabetenquadratbuch", Stuttgart 1966; „Typoaktionen", Frankfurt 1967.

**McConnell,** John – geb. 14. 5. 1939 in London, England. – *Grafik-Designer* – Studium am Maidstone College of Art in Kent, England. 1959–60 Arbeit in einer Werbeagentur in London, 1960–61 Werbeabteilung der Air Lingus in Dublin. 1961–63 Grafik-Designer bei Tyndy, Halford & Mills in London. Unterrichtet 1962–63 am Colchester College of Art & Design. 1963–74 Grafik-Design-Büro in London. Gestaltung des Erscheinungsbildes der Modeboutique Biba in London. 1967 Mitbegründer der Firma „Face Photosetting" in London. 1974 Eintritt in das Grafik-Design-Büro „Pentagram" in London. Arbeiten u. a. für Clarks Shoes, den

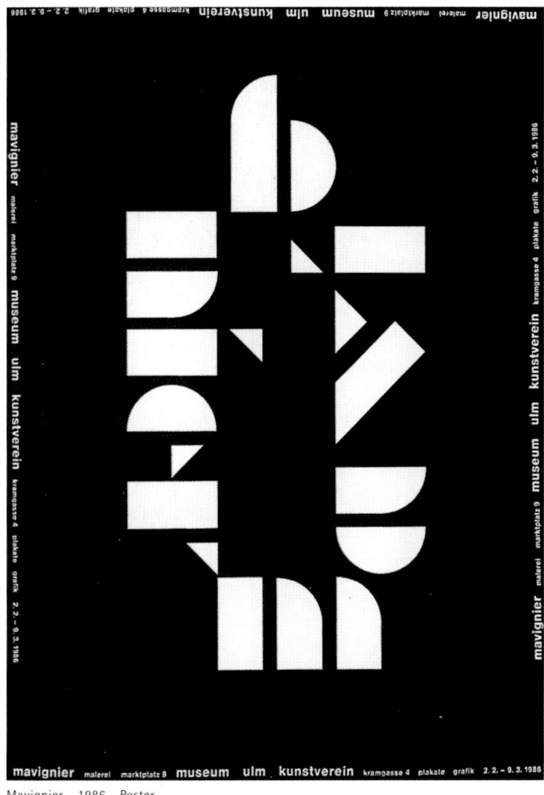

Mavignier 1986 Poster

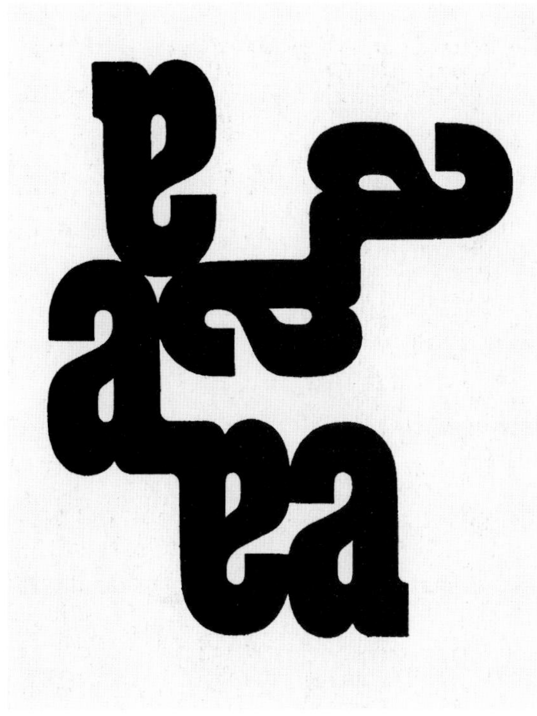
Mayer 1963 Poem

McConnell 1985 Poster

McConnell 1982 Logo

Mayer 1963 Poem

pographique. Travaille dans des imprimeries en Suisse et aux Etat-Unis. 1962–1967, affiches pour les études générales de la Technische Hochschule de Stuttgart. Fondé l'»Edition Hansjörg Mayer» à Stuttgart. Publie de nombreux ouvrages, entre autres des œuvres de Franz Mon, Dieter Rot, Emmett Williams, Wolfgang Schmidt et Mary Vieira. Ouvre la «Galerie der Edition Hansjörg Mayer» à Stuttgart en 1966, où il expose les artistes ayant publié chez lui. Enseigne à la Watford School of Art de Hertfordshire en 1967. Ses publications et son travail de création font de Mayer un représentant de l'«école de Stuttgart» regroupée autour de Max Bense. – *Publications, sélection :* « 19 Typographien», Stuttgart 1962; «Alphabet», Stuttgart 1963; «Typoems», Stuttgart 1965; «Alphabetenquadratbuch», Stuttgart 1966; «Typoaktionen», Francfort 1967.

**McConnell,** John – né le 14. 5. 1939 à Londres, Angleterre – *graphiste maquettiste* – études au Maidstone College of Art du Kent, Angleterre. Travaille dans une agence publicitaire londonienne de 1960 à 1961. Graphiste maquettiste chez Tyndy, Halford & Mills de 1961 à 1963 à Londres. Enseigne au Colchester College of Art & Design de 1962 à 1963. Exerce dans son atelier de graphisme et de design à Londres de 1963 à 1974 et dessine l'identité de la boutique londonienne Biba. En 1967, cofondateur de l'entreprise «Face Photosetting» à Londres. En 1974, il entre à l'atelier de graphisme et de design «Pentagram» à Londres. Travaux pour Clarks Shoes, les éditions Faber &

*teacher* – Trained as a typesetter. Worked at various printing workshops in Switzerland and the USA. 1962–67: designs posters for the Studium Generale courses at the Technische Hochschule in Stuttgart. Founds Edition Hansjörg Mayer in Stuttgart. Issues numerous publications, including by Franz Mon, Dieter Rot, Emmett Williams, Wolfgang Schmidt and Mary Vieira. 1966: opens the Galerie der Edition Hansjörg Mayer in Stuttgart which exhibits works by the artists working for Mayer's publishing house. 1967: lecturer at the Watford School of Art in Hertfordshire. The publications produced by Mayer's publishing company and his own work in design make Mayer part of Max Bense's "Stuttgart School". – *Publications include:* "19 Typographien", Stuttgart 1962; "Alphabet", Stuttgart 1963; "Typoems", Stuttgart 1965; "Alphabetenquadratbuch", Stuttgart 1966; "Typoaktionen", Frankfurt 1967.

**McConnell,** John – b. 14. 5. 1939 in London, England – *graphic designer* – Studied at Maidstone College of Art in Kent, England. 1959–60: works for an advertising agency in London and from 1960–61 for the Air Lingus publicity department in Dublin. 1961–63: graphic designer for Tyndy, Halford & Mills in London. 1962–63: teaches at the Colchester College of Art & Design. 1963–74: runs a graphic design studio in London. Designs the corporate identity for Biba fashion boutique in London. 1967: co-founder of the Face Photosetting company in London. 1974: joins Pentagram graphic design studio in London. Produces work for

Verlag Faber & Faber und Boots Pharmacie. Zahlreiche Auszeichnungen, u. a. Goldmedaille der Grafik-Design-Biennale in Brno (1982), President's Award der „Designers and Art Directors Association" (1985). – *Publikation u. a.:* „Ideas on Design" (Co-Autor), London 1986.

**McCoy,** Katherine – geb. 12. 10. 1942 in Decator, USA. – *Grafik-Designerin, Typographin, Lehrerin* – Industrie-Design-Studium an der Michigan State University. 1967–70 Gestalterin bei Unimark International, Chrysler Corporation, Corporate Design Identity Office und Omnigraphics Inc. 1971 Co-Chairman der Grafischen Abteilung der Cranbrook Academy of Art in Bloomfield Hills, Michigan (mit ihrem Mann Michael). 1972 Gründung des Büros „McCoy & McCoy Association". Auftraggeber waren u.a. Knoll International, MIT Press und Formica Corporation. 1983–85 Präsidentin der „Industrial Designers Society of America". 1987 zusammen mit ihrem Mann mit dem „Society of Typographic Arts Educator Award" ausgezeichnet. 1991 Ausstellungen der Lehrergebnisse unter dem Titel „Cranbrook Design: a New Discourse" in New York und Tokio. Beraterin der Zeitschriften „Design Issues" und „I.D.". – *Publikationen u.a.:* „Projects & Processes", Bloomfield Hills 1976; „Typography", in „I.D." 35/1988; „Cranbrook Design: the New Discourse", New York 1990.

**McLean,** Ruari – geb. 10. 6. 1917 in Schottland. – *Typograph, Autor, Lehrer* – 1936 Eintritt in die Shakespeare Head Press unter Bernard Newdigate. Arbeit als

Faber, Boots Pharmacie, etc. Nombreuses distinctions, entre autres la médaille d'or de la Biennale des Arts graphiques et du Design de Brno (1982), President's Award de la «Designers and Art Directors Association» (1985). – *Publication, sélection:* «Ideas on Design» (coauteur), Londres 1986.

**McCoy,** Katherine – née le 12. 10. 1942 à Decator, Etats-Unis – *graphiste maquettiste, typographe, enseignante* – études de design industriel à la Michigan State University. De 1967 à 1970, designer chez Unimark International, Chrysler Corporation Corporate Design Identity Office et Omnigraphics Inc. 1971, vice recteur du département d'arts graphiques de la Cranbrook Academy of Art à Bloomfield Hills, Michigan (avec son mari Michael). Fondation, en 1972 de l'atelier «McCoy & McCoy Association». Commanditaires: Knoll International, MIT Press, Formica Corporation etc. 1983–1985, présidente de la «Industrial Designers Society of America». En 1987, elle reçoit, avec son mari, le prix «Society of Typographic Arts Educator Award». Exposition de ses matériaux d'enseignement sous le titre «Cranbrook Design: a New Discourse», à New York et à Tokyo en 1991. Conseillère des revues «Design Issues» et «I.D.». – *Publications, sélection:* «Projects & Processes», Bloomfield Hills 1976; «Typography» dans «I.D.» 35/1988; «Cranbrook Design: the New Discourse», New York 1990.

**McLean,** Ruari – né le 10. 6. 1917 en Ecosse – *typographe, auteur, enseignant* – entre à la Shakespeare Head Presse, sous Bernard Newdigate, en 1936. Travaille comme imprimeur en Allemagne. Exerce ensuite pour la revue «The Studio». Auteur et typographe indépendant à partir de 1945. Travaux pour Penguin Books ainsi que pour la revue «Signature». De

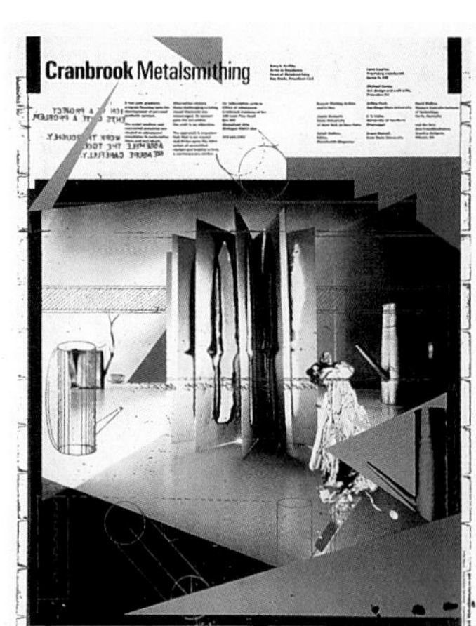

McCoy 1986 Poster

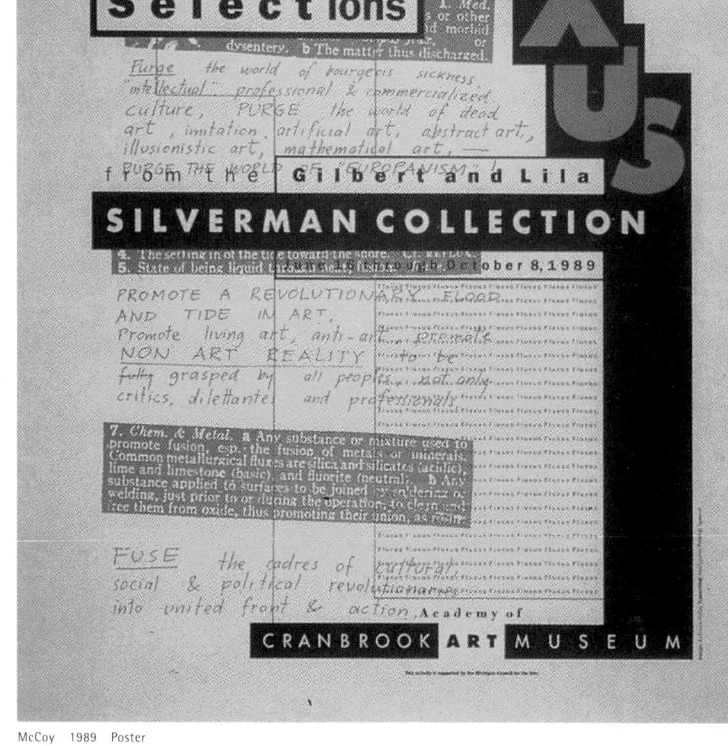


McCoy 1989 Poster

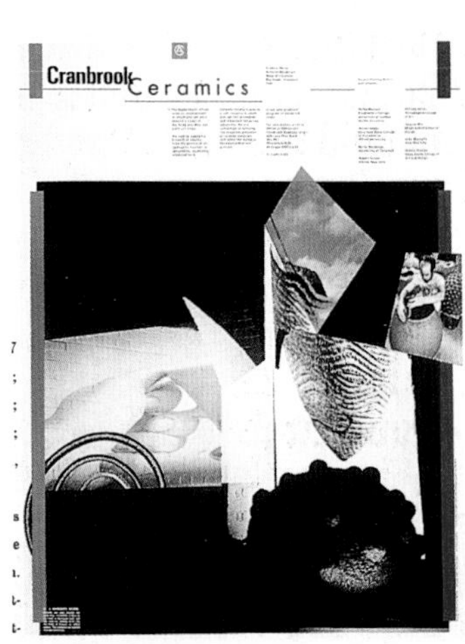


McCoy 1987 Poster

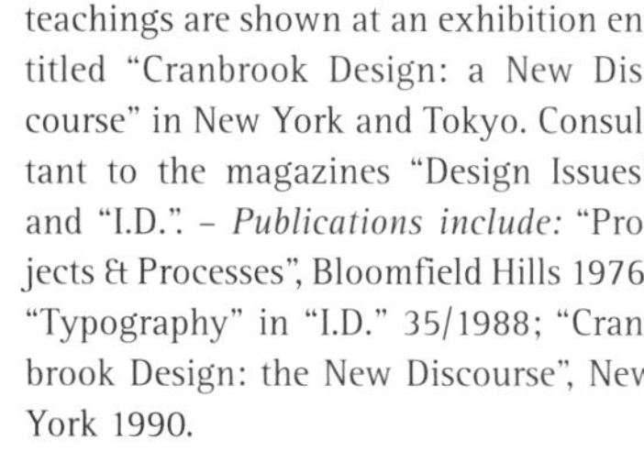

Clarks Shoes, Faber & Faber publishers and Boots Pharmacie, among others. Numerous awards, including a gold medal at the Graphic Design Biennale in Brno (1982) and the Designers and Art Directors Association's President's Award (1985). – *Publications include:* "Ideas on Design" (co-author), London 1986.

**McCoy,** Katherine – b. 12. 10. 1942 in Decator, USA – *graphic designer, typographer, teacher* – Studied industrial design at Michigan State University. 1967–70: designer for Unimark International, the Chrysler Corporation, Corporate Design Identity Office and Omnigraphics Inc. 1971: co-chairperson of the graphics department at the Cranbrook Academy of Art in Bloomfield Hills, Michigan, with her husband, Michael. 1972: founds the McCoy & McCoy Association. Clients include Knoll International, MIT Press and the Formica Corporation. 1983–85: president of the Industrial Designers Society of America. 1987: she and her husband receive the Society of Typographic Arts Educator Award. 1991: the results of her teachings are shown at an exhibition entitled "Cranbrook Design: a New Discourse" in New York and Tokyo. Consultant to the magazines "Design Issues" and "I.D.". – *Publications include:* "Projects & Processes", Bloomfield Hills 1976; "Typography" in "I.D." 35/1988; "Cranbrook Design: the New Discourse", New York 1990.

**McLean,** Ruari – b. 10. 6. 1917 in Scotland – *typographer, author, teacher* – 1936: joins Shakespeare Head Press under Bernard Newdigate. Works as a printer in

Drucker in Deutschland. Danach Arbeit für die Zeitschrift „The Studio". Seit 1945 freier Typograph und Autor. Arbeit für Penguin Books sowie die Zeitschrift „Signature". 1948–51 Tutor für Typographie am Royal College of Art in London. 1958–67 Gründung und Herausgeber der Zeitschrift „Motif", von der 13 Hefte erscheinen. 1966–80 typographischer Berater von „H. M. Stationery Office". Übersetzt Jan Tschicholds Bücher aus dem Deutschen ins Englische: „Typographische Gestaltung" unter dem Titel „Asymmetric Typography" (1967), „Typographische Entwurfstechnik" unter dem Titel „How to draw layouts" (1991), „Die neue Typographie" unter dem Titel „The New Typography" (1995). – *Publikationen u.a.:* „Modern Book Design", London 1951; „Victorian Book Design and Colour Printing", London 1963, 1972; „Magazine Design", London 1969; „Jan Tschichold, typographer", London 1975; „Manual of Typography", London 1980.

**McLuhan,** Marshall – geb. 21. 7. 1911 in Edmonton, Kanada, gest. 31. 12. 1980 in Toronto, Kanada. – *Medientheoretiker* – Studium an der University of Manitoba und in Cambridge. 1953 Herausgabe der Zeitschrift „Explorations", die sich mit Sprache und Medien auseinandersetzt. Seine These, nicht die Botschaft sondern das Wesen der eingesetzten Medien evoziere gesellschaftliche und ästhetische Innovation, markiert den Beginn moderner Medientheorie. Das einflußreiche Taschenbuch „The Medium is the Massage", mit Quentin Fiore als Co-Autor und von Jerome Agel herausgegeben, zeigt eine

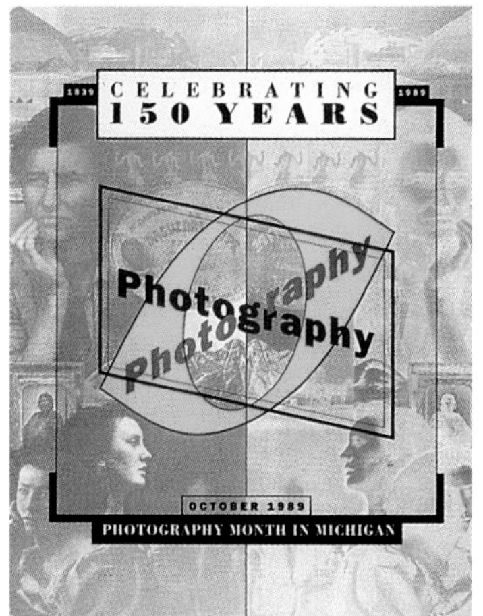


McCoy 1989 Poster

McCoy 1992 Page

McCoy 1994 Logo

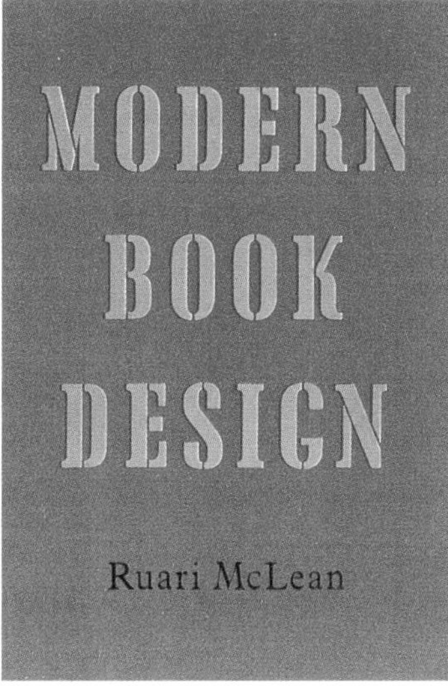


McLean 1951 Cover

50

*Printing, a ditto device*

*Printing, a ditto device*

*Printing, a ditto device* confirmed and extended the new visual stress. It provided the first uniformly repeatable "commodity," the first assembly line—mass production.

It created the portable book, which men could read in privacy and in isolation from others. Man could now inspire—and conspire.

Like easel painting, the printed book added much to the new cult of individualism. The private, fixed point of view became possible and literacy conferred the power of detachment, non-involvement.

*Printing, a ditto device*

*Printing, a ditto device*

*Printing, a ditto device*

*Printing, a ditto device*

*Printing, a ditto device*

*Printing, a ditto device*

*Printing, a ditto device*

*Printing, a ditto device*

McLuhan 1967 Spread

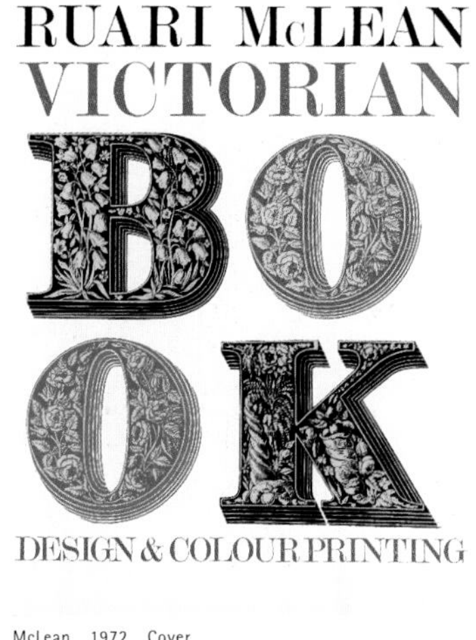


McLean 1972 Cover

1948 à 1951, il est tuteur en typographie au Royal College of Art de Londres. En 1958, il fonde la revue «Motif» qu'il publiera jusqu'en 1967; 13 numéros paraîtront. Conseiller typographique de «H. M. Stationery Office» de 1966 à 1980. A traduit plusieurs ouvrages de Jan Tschichold en anglais: «Typographische Gestaltung» sous le titre de «Asymmetric Typography» (1967); «Typographische Entwurfstechnik», sous le titre de «How to draw layouts» (1991) et «Die neue Typographie» sous le titre «The New Typography» (1995). – *Publications, sélection:* «Modern Book Design», Londres 1951; «Victorian Book Design and Colour Printing», Londres 1963, 1972; «Magazine Design», Londres 1969; «Jan Tschichold, typographer», Londres 1975; «Manual of Typography», Londres 1980.

**McLuhan,** Marshall – né le 21. 7. 1911 à Edmonton, Canada, décédé le 31. 12. 1980 à Toronto, Canada – *théoricien des médias* – études a l'University of Manitoba et à Cambridge. 1953, publie la revue «Explorations», consacrée au langage et aux médias. Sa thèse selon laquelle ce n'est pas le message mais la nature du média utilisé qui suscite l'innovation sociale et esthétique, marque l'avènement de la théorie moderne des médias. Le fameux livre de poche «The Medium is the Massage», écrit en collaboration avec Quentin Fiore et édité par Jerome Agel, présente une foule d'effets visuels. – *Publi-*

Germany. Then works for "The Studio" periodical. From 1945 onwards: freelance typographer and author. Works for Penguin Books and "Signature" magazine. 1948–51: tutor of typography at the Royal College of Art in London. 1958–67: launches and edits "Motif" magazine, of which 13 numbers are published. 1966–80: typography consultant to H. M. Stationery Office. Translates books by Jan Tschichold from German into English: "Typographische Gestaltung" is published as "Asymmetric Typography" (1967), "Typographische Entwurfstechnik" as "How to draw layouts" (1991) and "Die neue Typographie" as "The New Typography" (1995). – *Publications include:* "Modern Book Design", London 1951; "Victorian Book Design and Colour Printing", London 1963, 1972; "Magazine Design", London 1969; "Jan Tschichold, typographer", London 1975; "Manual of Typography", London 1980.

**McLuhan,** Marshall – b. 21. 7. 1911 in Edmonton, Canada, d. 31. 12. 1980 in Toronto, Canada – *media theorist* – Studied at the University of Manitoba and at the University of Cambridge. 1953: editor of "Explorations", a journal which examines the topics of language and media. His hypothesis that it is not the message itself but the nature of the media used to put it across which sparks off social and aesthetic innovation marked the beginnings of modern media theory. His influential paperback, "The Medium is the Massage", co-written with Quentin Fiore (produced by Jerome Agel) contains a wealth of amazing visual effects. – *Publications*

Fülle verblüffender visueller Effekte. – *Publikationen u.a.:* „The Gutenberg Galaxy", Toronto 1962; „Understanding Media", New York 1964; „The Medium is the Massage", New York 1967.

**Megert,** Peter – geb. 28. 3. 1937 in Bern, Schweiz. – *Grafik-Designer, Lehrer* – Grafikerlehre in Bern. 1956–58 Studium an der Kunstgewerbeschule in Bern. 1964–68 eigenes Designbüro in Bern. Studium an der Ohio State University. 1968–70 Grafik-Designer im Westinghouse Corporate Design Center in Pittsburgh. 1970–85 Professor für Visuelle Kommunikation an der Ohio State University in Columbus, Ohio. Seit 1986 eigenes Gestaltungsstudio „Visual Syntax Design Inc." in Columbus, Ohio. Unterrichtet u. a. an der Harvard University in Cambridge, an der Carnegie Mellon University in Pittsburgh und an der Schule für Gestaltung in St. Gallen.

**Meier,** Hans Eduard – geb. 30. 12. 1922 in Horgen, Schweiz. – *Grafiker, Schriftentwerfer, Lehrer* – 1939–43 Schriftsetzerlehre in Horgen. 1943–46 Ausbildung als Grafiker an der Kunstgewerbeschule in Zürich. 1946–48 Mitgestalter der Monatszeitschrift „du", grafische Arbeiten für den Manesse-Verlag. Grafische Arbeiten für Verlage, Agenturen und die UNESCO. Unterrichtet 1950–86 an der Kunstgewerbeschule in Zürich Schriftgestaltung, Kalligraphie, Zeichnen und farbiges Naturstudium. Zahlreiche Plakate für die Tonhalle und das Kunsthaus in Zürich. Kalligraphische Schriftkompositionen. Entwickelt 1968 die „Syntax" als eine serifenlose Schrift mit optisch glei-

*cations, selections:* «The Gutenberg Galaxy», Toronto 1962; «Understanding Media», New York 1964; «The Medium is the Massage», New York 1967.

**Megert,** Peter – né le 28. 3. 1937 à Berne, Suisse – *graphiste maquettiste, enseignant* – apprentissage de graphiste à Berne. 1956–1958, études à la Kunstgewerbeschule (école des arts décoratifs) de Berne. Exerce dans son propre atelier de design à Berne de 1964 à 1968. Etudes à la Ohio State University. Graphiste maquettiste au Westinghouse Corporate Design Center de Pittsburgh de 1968 à 1970. Professeur de communication visuelle à la Ohio State University de Columbus, Ohio de 1970 à 1985. Exerce depuis 1986 dans son atelier de design la «Visual Syntax Design Inc.» à Columbus, Ohio. A enseigné à la Harvard University à Cambridge, à la Carnegie Mellon University de Pittsburgh et à la Schule für Gestaltung (école de design) de Saint-Gall.

**Meier,** Hans Eduard – né le 30. 12. 1922 à Horgen, Suisse – *graphiste, concepteur de polices, enseignant* – apprentissage de la composition typographique à Horgen de 1939 à 1943. 1943–1946, formation de graphiste à la Kunstgewerbeschule (école des arts décoratifs) de Zurich. De 1946 à 1948, il est maquettiste à la revue «du» et réalise des travaux graphiques pour les éditions Manesse. Travaux graphiques pour des éditeurs, des agences et l'UNESCO. De 1950 à 1986, il enseigne le design de caractères, la calligraphie, le dessin et l'esquisse d'après nature avec utilisation de la couleur à la Kunstgewerbeschule de Zurich. Nombreuses affiches pour la Tonhalle et pour le Kunsthaus de Zurich. Compositions calligraphiques. En 1968, il dessine «Syntax», une police sans sérif comportant des lignes d'épaisseur égale qui interprète les caractères humanistes de la Renaissance. Depuis 1984, il a réa-

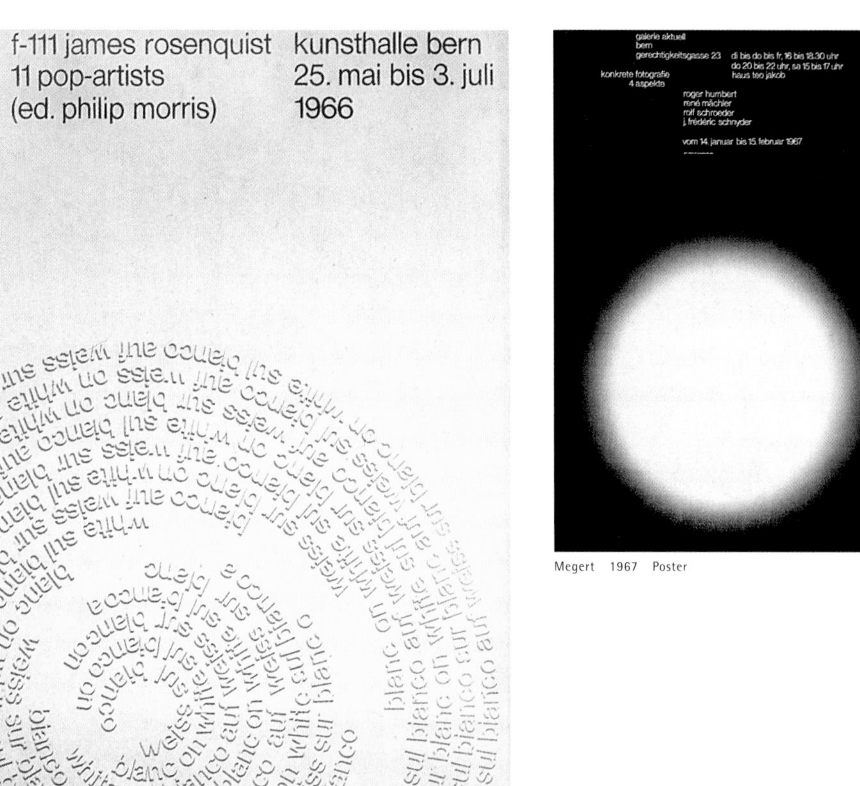

Megert 1966 Poster

Megert 1967 Poster

| | n | **n** | **n** | | *n* |
|---|---|---|---|---|---|
| a | b | e | f | g | i |
| o | r | s | t | y | z |
| A | B | C | E | G | H |
| M | O | R | S | X | Y |
| 1 | 2 | 4 | 6 | 8 | & |

Meier 1968 Syntax

*include:* "The Gutenberg Galaxy", Toronto 1962; "Understanding Media", New York 1964; "The Medium is the Massage", New York 1967.

**Megert,** Peter – b. 28. 3. 1937 in Bern, Switzerland – *graphic designer, teacher* – Trained as a graphic artist in Bern. 1956–58: studies at the Kunstgewerbeschule in Bern. 1964–68: runs his own design studio in Bern. Studies at Ohio State University. 1968–70: graphic designer for the Westinghouse Corporate Design Center in Pittsburgh. 1970–85: professor of visual communication at Ohio State University in Columbus, Ohio. From 1986 onwards: runs his own design studio entitled Visual Syntax Design Inc. in Columbus, Ohio. Has taught at Harvard University in Cambridge, at Carnegie Mellon University in Pittsburgh and at the Schule für Gestaltung in St. Gallen.

**Meier,** Hans Eduard – b. 30. 12. 1922 in Horgen, Switzerland – *graphic artist, type designer, teacher* – 1939–43: trains as a typesetter in Horgen. 1943–46: trains as a graphic artist at the Kunstgewerbeschule in Zurich. 1946–48: one of the designers working on the monthly "du" magazine. Produces graphics for various publishers, including Manesse-Verlag, for various agencies and for UNESCO. 1950–86: teaches type design, calligraphy, drawing and color studies from nature at the Kunstgewerbeschule in Zurich. Produces numerous posters for the Tonhalle and the Kunsthaus in Zurich and calligraphic letter compositions. 1968: develops Syntax, a sans-serif typeface with an optically even thickness of line and

cher Strichstärke, die humanistische Schriften der Renaissance interpretiert. Seit 1984 Arbeiten am Institut für Computersysteme der ETH Zürich, Gestaltung neuer Druckschriften mit dem Computer. 1996–97 Neubearbeitung und Ergänzung seiner Schrift „Syntax": 6 Schnitte, 6 Kursive, alle mit Kapitälchen und Renaissance-Ziffern. – *Schriftentwürfe:* Syntax (1968), Barbedor (1984–86), Barbetwo (1992), Syndor (1992), Oberon (1994), Letter (1995), Lapidar (1995). – *Publikationen u.a.:* „Die Schriftentwicklung", Zürich 1959, Cham 1994; „Schriftgestaltung mit Hilfe des Computers. Typographische Grundregeln mit Gestaltungsbeispielen", Zürich 1993. Erich Schulz-Anker „Formanalyse und Dokumentation einer serifenlosen Linearantiqua auf neuer Basis: Syntax-Antiqua", Frankfurt am Main 1969.

**Mendell + Oberer** – *Grafik-Design-Studio* – 1961 Gründung des Studios in München durch Pierre Mendell und Klaus Oberer. Mendell, Pierre – geb. 17. 11. 1929 in Essen, Deutschland. – 1958–60 Studium an der Kunstgewerbeschule in Basel bei Armin Hofmann. 1960–61 kurze Zusammenarbeit mit Michael Engelmann. 1969 Ausstellung von Plastiken im Modern Art Museum in München. Im Studio verantwortlich für Konzept und Visualisierung. Oberer, Klaus – geb. 5. 5. 1937 in Basel, Schweiz. – 1955–59 Studium an der Kunstgewerbeschule in Basel bei Armin Hofmann. 1959–61 grafische Arbeiten bei Michael Engelmann in München. Im Studio verantwortlich für Realisation und Organisation. Seit 1961 Ent-

Mendell + Oberer 1967 Poster



Mendell + Oberer 1984 Poster

| | n | | | | *n* |
|---|---|---|---|---|---|
| a | b | e | f | g | i |
| o | r | s | t | y | z |
| A | B | C | E | G | H |
| M | O | R | S | X | Y |
| 1 | 2 | 4 | 6 | 8 | & |

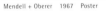

Meier 1992 Syndor 1

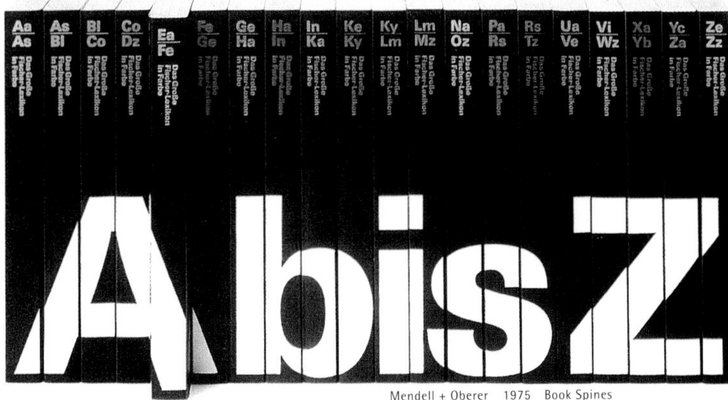


Mendell + Oberer 1975 Book Spines

lisé de nombreux travaux pour l'Institut für Computersysteme de l'ETH à Zurich, et dessiné de nouvelles fontes à l'aide de l'ordinateur. Poursuite du travail sur «Syntax» en 1996–1997 à laquelle il ajoute 6 tailles, 6 italiques, des petites capitales et des chiffres renaissance. – *Polices:* Syntax (1968), Barbedor (1984–1986), Barbetwo (1992), Syndor (1992), Oberon (1994), Letter (1995), Lapidar (1995). – *Publications, sélection:* «Die Schriftentwicklung», Zurich 1959, Cham 1994; «Schriftgestaltung mit Hilfe des Computers. Typographische Grundregeln mit Gestaltungsbeispielen», Zurich 1993. Erich Schulz-Anker «Formanalyse und Dokumentation einer serifenlosen Linearantiqua auf neuer Basis: Syntax-Antiqua», Francfort-sur-le-Main 1969.

**Mendell + Oberer** – *atelier de graphisme et de design* – L'atelier a été fondé à Munich, en 1961, par Pierre Mendell et Klaus Oberer. Mendell, Pierre – né le 17. 11. 1929 à Essen, Allemagne – 1958–1960, études à la Kunstgewerbeschule (école des arts décoratifs) de Bâle chez Armin Hofmann. 1960–1961, brève collaboration avec Michael Engelmann. Exposition de sculptures au musée d'art moderne de Munich en 1969. A l'atelier, c'est lui le responsable de la conception et de la visualisation. Oberer, Klaus – né le 5. 5. 1937 à Bâle, Suisse – 1955–1959, études à la Kunstgewerbeschule de Bâle chez Armin Hofmann. Travaux graphiques chez Michael Engelmann à Munich de 1959 à 1961. Au sein de l'atelier, il est responsable de la réalisation et de l'organisation. Réalisation d'identités, de concepts publicitaires, d'affiches, de ma-

which interprets the Humanist type of the Renaissance. 1984: starts working for the Institute of Computer Systems at the ETH Zurich. Designs new print with computers. 1996–97: reworks and extends his Syntax typeface to include 6 weights and 6 sets of italics, all with small capitals and Renaissance numerals. – *Fonts:* Syntax (1968), Barbedor (1984–86), Barbetwo (1992), Syndor (1992), Oberon (1994), Letter (1995), Lapidar (1995). – *Publications include:* "Die Schriftentwicklung", Zurich 1959, Cham 1994; "Schriftgestaltung mit Hilfe des Computers. Typographische Grundregeln mit Gestaltungsbeispielen", Zurich 1993. Erich Schulz-Anker "Formanalyse und Dokumentation einer serifenlosen Linearantiqua auf neuer Basis: Syntax-Antiqua", Frankfurt am Main 1969.

**Mendell + Oberer** – *graphic design studio* – 1961: Pierre Mendell and Klaus Oberer open the studio in Munich. Mendell, Pierre – b. 17. 11. 1929 in Essen, Germany – 1958–60: studies at the Kunstgewerbeschule in Basle under Armin Hofmann. 1960–61: brief collaboration with Michael Engelmann. 1969: his sculptures are exhibited at the Modern Art Museum in Munich. At Mendell + Oberer Pierre Mendell is responsible for project concepts and visualization. Oberer, Klaus – b. 5. 5. 1937 in Basle, Switzerland – 1955–59: studies at the Kunstgewerbeschule in Basle under Armin Hofmann. 1959–61: produces graphics under Michael Engelmann in Munich. At Mendell + Oberer Klaus Oberer is responsible for project realization and organization. Since 1961 the

wicklung von Erscheinungsbildern, Werbekonzepten, Plakaten, Buch- und Verpackungsgestaltung, Leitsystemen. Zahlreiche Arbeiten für das Museum „Die Neue Sammlung" in München und die Bayrische Staatsoper in München. Zahlreiche Ausstellungen, u. a. in der Galerie Intergraphic (1970) in München, im Design Center Stuttgart (1990). Zahlreiche Auszeichnungen, u. a. Goldmedaille des Art Directors Club Deutschland (1971). – *Publikationen u. a.:* „Studio Mendell + Oberer. Graphik Design", Stuttgart 1976; „L'art pour l'art", Baden 1996. Hans Wichmann „Graphic Design Mendell + Oberer", Basel, Boston, Stuttgart 1987.

**Mendoza y Almeida,** José – geb. 14. 10. 1926 in Sèvres, Frankreich. – *Schriftentwerfer, Typograph, Lehrer* – Ausbildung als Fotograveur bei Cliché Union in Paris. 1953–54 Zusammenarbeit mit Maximilian Vox. 1954–59 Assistent von Roger Excoffon in der Fonderie Olive in Marseille. Seit 1959 freier Typograph und Schriftentwerfer. 1985–90 Lehrer für Schriftentwurf im Atelier National de Création Typographique (ANCT) der Imprimerie Nationale in Paris. – *Schriftentwürfe:* Pascal (1960), Photina (1972), Fidelio (1980), Sully-Jonquières (1980), Mendoza (1990).

**Mengel,** Willi – geb. 3. 10. 1904 in Düsseldorf, Deutschland, gest. 2. 7. 1969 in Würzburg, Deutschland. – *Typograph, Autor, Grafik-Designer* – 1918–21 Schriftsetzerlehre in Düsseldorf. 1929–37 Meisterprüfung in Frankfurt am Main. 1937–39 Schriftsetzer in Wetzlar. 1945–48 Gebrauchsgrafiker in Straubing. 1950–63

quettes de livres et d'emballages, de systèmes de signalisation depuis 1961. Nombreux travaux pour le musée «Die Neue Sammlung» et pour le Bayrische Staatsoper de Munich. Nombreuses expositions, entre autres à la galerie Intergraphic (1970) de Munich, au Design Center de Stuttgart (1990). Nombreuses distinctions, entre autres, la médaille d'or de l'Art Directors Club d'Allemagne (1971). – *Publications, sélection:* «Studio Mendell + Oberer. Graphic Design», Stuttgart 1976; «L'art pour l'art», Baden 1996. Hans Wichmann «Graphic Design Mendell + Oberer», Bâle, Boston, Stuttgart 1987.

**Mendoza y Almeida,** José – né le 14. 10. 1926 à Sèvres, France – *concepteur de polices, typographe, enseignant* – formation de photograveur chez Cliché Union à Paris. Travaille avec Maximilien Vox de 1953 à 1954. Assistant de Roger Excoffon à la Fonderie Olive à Marseille de 1954 à 1959. Typographe et concepteur de polices indépendant à partir de 1959. A enseigné le design de polices de 1985 à 1990 à l'Atelier National de Création Typographique (ANCT) de l'Imprimerie Nationale de Paris. – *Polices:* Pascal (1960), Photina (1972), Fidelio (1980), Sully-Jonquières (1980), Mendoza (1990).

**Mengel,** Willi – né le 3. 10. 1904 à Düsseldorf, Allemagne, décédé le 2. 7. 1969 à Würzburg, Allemagne – *typographe, auteur, graphiste maquettiste* – 1918–1921, apprentissage de la composition typographique à Düsseldorf. Stages de maîtrise à Franfort-sur-le-Main de 1929 à 1937. Compositeur à Wetzlar de 1937 à 1939. Graphiste industriel à Straubing de 1945 à 1948. Directeur de la publicité de la société Linotype GmbH à Francfort-sur-le-Main de 1950 à 1963. Collabore au «Deutscher Typokreis». Publie une série de 13 essais intitulée «Schrift in der Gegenwart» dans la revue «Deutscher

| | n | **n** | | | *n* |
|---|---|---|---|---|---|
| a | b | e | f | g | i |
| o | r | s | t | y | z |
| A | B | C | E | G | H |
| M | O | R | S | X | Y |
| 1 | 2 | 4 | 6 | 8 | & |

Mendoza y Almeida 1972 Photina

| | n | | | | |
|---|---|---|---|---|---|
| *a* | b | *e* | *f* | *g* | i |
| o | r | s | t | *y* | *z* |
| A | B | C | E | G | H |
| M | O | R | S | X | Y |
| 1 | 2 | 4 | 6 | 8 | *&* |

Mendoza y Almeida 1980 Sully Jonquières

| | n | n | **n** | | *n* |
|---|---|---|---|---|---|
| a | b | e | f | g | i |
| o | r | s | t | y | z |
| A | B | C | E | G | H |
| M | O | R | S | X | Y |
| 1 | 2 | 4 | 6 | 8 | & |

Mendoza y Almeida 1990 Mendoza

Mendoza y Almeida Page

studio has developed corporate identities, advertising concepts, posters, book and packaging designs and signage systems. They have done much work for Die Neue Sammlung art museum and the Bayrische Staatsoper in Munich. They have been the subject of numerous exhibitions, including at the Intergraphic gallery (1970) in Munich and at the Design Center in Stuttgart (1990). They have received many awards, including a gold medal from the Art Directors Club in Germany (1971). – *Publications include:* "Studio Mendell + Oberer. Graphik Design", Stuttgart 1976; "L'art pour l'art", Baden 1996. Hans Wichmann "Graphic Design Mendell + Oberer", Basle, Boston, Stuttgart 1987.

**Mendoza y Almeida,** José – b. 14. 10. 1926 in Sèvres, France – *type designer, typographer, teacher* – Trained as a photoengraver at the Cliché Union in Paris. 1953–54: works with Maximilian Vox. 1954–59: assistant to Roger Excoffon at the Fonderie Olive in Marseille. From 1959 onwards: freelance typographer and type designer. 1985–90: teaches type design in the Atelier National de Création Typographique (ANCT) at the Imprimerie Nationale in Paris. – *Fonts:* Pascal (1960), Photina (1972), Fidelio (1980), Sully-Jonquières (1980), Mendoza (1990).

**Mengel,** Willi – b. 3. 10. 1904 in Düsseldorf, Germany, d. 2. 7. 1969 in Würzburg, Germany – *typographer, author, graphic designer* – 1918–21: trains as a typesetter in Düsseldorf. 1929–37: master craftsman's diploma in Frankfurt am Main. 1937–39: typesetter in Wetzlar. 1945–48:

Werbeleiter der Firma Linotype GmbH in Frankfurt am Main. Mitarbeit im „Deutschen Typokreis". Veröffentlichung einer 13-teiligen Aufsatzreihe „Schrift in der Gegenwart" in der Zeitschrift „Deutscher Drucker". – *Publikationen u.a.:* „Zur Typographie des wissenschaftlichen Buches", Frankfurt 1954; „Formprobleme der gegenwärtigen Typographie", Frankfurt 1955.

**Menhart,** Oldřich – geb. 25. 6. 1897 in Prag, Tschechoslowakei, gest. 11. 2. 1962 in Prag, Tschechoslowakei. – *Typograph, Schriftentwerfer, Lehrer* – 1911–14 Setzerlehre in der Druckerei „Politika" in Prag. 1922–29 Faktor der Staatsdruckerei in Prag. Seit 1929 Arbeit als freier Buchgestalter und Kalligraph, seit 1939 für den Verlag Sfinx und den Europäischen Literarischen Klub in Prag. Sein handgeschriebenes Buch „Kytice" von K. J. Erben erscheint 1941. Die Nationalversammlung der CSSR beauftragt Menhart 1950 mit dem Entwurf einer Exklusiv-Schrift für den Druck der Verfassung. Zahlreiche Auszeichnungen, u.a. Staatspreis für Buchgestaltung (1948), Orden der Republik für den Entwurf einer tschechischen Schrift (1952), Goldmedaille der Internationalen Buchausstellung in Leipzig für sein Lebenswerk (1959). – *Schriftentwürfe:* Menhart-Antiqua (1931–36), Menhart Latein (1934–36), Hollar (1939), Figural (1940; M. Gills, 1992), Tschechische Unziale (1948), Manuscript-Antiqua (1949), Monument (1949), Parlament (1950), Grazdanka (1953–55), Triga (1955), Standard (1959–66), Vajgar (1961). – *Publikationen u.a.:* „Vecerni hovory knínomi-

*Willi Mengel*

Zur TYPOGRAPHIE
des wissenschaftlichen
Buches

DEUTSCHER TYPOKREIS EV FRANKFURT AM MAIN

*Gesellschaft zur Höherführung der Setz- und Druckkunst*

Mengel 1954 Main Title

| | n | **n** | | | *n* |
|---|---|---|---|---|---|
| a | b | e | f | g | i |
| o | r | s | t | y | z |
| A | B | C | E | G | H |
| M | O | R | S | X | Y |
| 1 | 2 | 4 | 6 | 8 | & |

Menhart 1940 Figural

ABCDE
FGHIJKLM

Menhart 1949 Monument

karel plicka
PRAG
ein fotografisches bilderbuch

artia
prag·tschechoslowakei
1953

Menhart 1953 Title

Drucker». – *Publications, sélection:* «Zur Typographie des wissenschaftlichen Buches», Francfort 1954; «Formprobleme der gegenwärtigen Typographie», Francfort 1955.

**Menhart,** Oldřich – né le 25. 6. 1897 à Prague, Tchécoslovaque, décédé le 11. 2. 1962 à Prague, Tchécoslovaquie – *typographe, concepteur de polices, enseignant* – 1911-1914, apprentissage de composition typographique à l'imprimerie «Politika» à Prague. Travaille à l'imprimerie nationale de Prague de 1922 à 1929. Exerce comme maquettiste d'édition et calligraphe indépendant à partir de 1929. Travaux pour les éditions Sfinx et le club littéraire européen de Prague à partir de 1939. »Kytice» de K. J. Erben, qu'il écrit à la main, paraît en 1941. En 1950, l'assemblée nationale tchèque charge Menhart de concevoir une police spécifique pour l'impression de la constitution. Nombreuses distinctions, entre autres le «Prix national de la maquette de livre» (1948), l'Ordre de la république pour la conception d'une écriture tchèque (1952), médaille d'or du salon international du livre de Leipzig pour l'ensemble de son œuvre (1959). – *Polices:* Menhart-Antiqua (1931-1936), Menhart-Latine (1934–1936), Hollar (1939), Figural (1940; M. Gills, 1992), Onciale tchèque (1948), Manuscript-Antiqua (1949), Monument (1949), Parlament (1950), Grazdanka (1953–1955), Triga (1955), Standard (1959–1966), Vajgar (1961). – *Publications, sélection:* «Vecerni hovory kninomila Rubricia a starotiskare Tympána»,

commercial artist in Straubing. 1950–63: advertising manager for Linotype GmbH in Frankfurt am Main. Works with the Deutsche Typokreis. Publishes a 13-part essay series entitled "Schrift in der Gegenwart" in the journal "Deutscher Drucker". – *Publications include:* "Zur Typographie des wissenschaftlichen Buches", Frankfurt 1954; "Formprobleme der gegenwärtigen Typographie", Frankfurt 1955.

**Menhart,** Oldřich – b. 25. 6. 1897 in Prague, Czechoslovakia, d. 11. 2. 1962 in Prague, Czechoslovakia – *typographer, type designer, teacher* – 1911–14: trains as a typesetter in the Politika printing works in Prague. 1922–29: composing room foreman at the state printing works in Prague. From 1929 onwards: works as a freelance book designer and calligrapher, and from 1939 onwards for Sfinx publishers and the European Literary Club in Prague. 1941: his hand-written book, "Kytice" by K. J. Erben, is published. 1950: the Czech national assembly commissions Menhart to design an exclusive typeface for the printing of the constitution. Menhart received numerous awards, including the State Prize for Book Design (1948), the Medal of the Republic for his design of a Czech typeface (1952) and a gold medal from the Internationale Buchausstellung in Leipzig for his life's work (1959). – *Fonts:* Menhart Roman (1931–36), Menhart Latin (1934–36), Hollar (1939), Figural (1940; M. Gills, 1992), Czech Uncial (1948), Manuscript Roman (1949), Monument (1949), Parlament (1950), Grazdanka (1953–55), Triga (1955), Standard (1959–66), Vajgar

la Rubricia a starotiskare Tympána", Kroměřiž 1947; „Regule kaligrafu", Prag 1948; „Tvorba typografického pisma", Prag 1956; „Abendgespräche des Buchliebhabers Rubricius und des Alt-Druckers Tympán", Frankfurt 1958; „Schriften", Hamburg 1968. Otto F. Babler „Oldrich Menhart", Antwerpen 1950.

**Mergenthaler,** Ottmar – geb. 11. 5. 1854 in Hachtel, Deutschland, gest. 28. 10. 1899 in Baltimore, USA. – *Erfinder, Unternehmer* – 1868 Uhrmacherlehre bei seinem Onkel Louis Hahl in Bietigheim am Neckar. 1872 Auswanderung nach Amerika. Arbeit, Geschäftsführung und Teilhaber in der Fabrik seines Vetters August Hahl in Washington, 1875 Umzug nach Baltimore. Erhält den Auftrag, die Schreibsetzmaschine für lithographischen Druck von Charles T. Moore zu verbessern und zu vereinfachen. 1879 Entwurf einer Zeilengießmaschine, den er verwirft. 1882 Gründung einer eigenen Werkstatt in der Baltimore Bank Lane. 1884 Fertigung der ersten Stabsetz- und Gießmaschine, erster Guß einer Zeile auf einer von Mergenthaler gebauten Maschine. 1886 Einsatz der ersten „Blower"-Setzmaschine bei der Tageszeitung „New York Tribune". Der Zeitungsverleger Whitelaw Reid gibt der Maschine den Namen „Linotype". 1890 Gründung der Mergenthaler Linotype Company in Brooklyn. 1897 Umzug nach Deming im Staat New Mexico, Niederschrift seiner Memoiren. Vernichtung des gesamten Besitzes durch Feuer. Rückkehr nach Baltimore. 1924 Einrichtung eines Mergenthaler-Museums in seinem Geburtsort Hachtel. – *Publikationen u.a.:*

Kroměřiž 1947; «Regule kaligrafu», Prague 1948; «Tvorba typografického pisma», Prague 1956; «Abendgespräche des Buchliebhabers Rubricius und des Alt-Druckers Tympán», Francfort 1958; «Schriften», Hambourg 1968. Otto F. Babler «Oldrich Menhart», Anvers 1950.

**Mergenthaler,** Ottmar – né le 11. 5. 1854 à Hachtel, Allemagne, décédé le 28. 10. 1899 à Baltimore, Etats-Unis – *inventeur, chef d'entreprise* – apprentissage d'horlogerie chez son oncle Louis Hahl à Bietigheim/Neckar en 1868. Emigre aux Etats-Unis en 1872. Associé et gérant de l'usine de son cousin August Hahl à Washington. S'installe à Baltimore en 1875. Est chargé d'améliorer et de simplifier la machine à composer de l'atelier de lithographie de Charles T. Moore. En 1879, il élabore une machine à fondre des lignes de caractères, puis renonce à son projet. Crée son propre atelier à Baltimore Bank Lane en 1882. En 1884, il achève sa première machine à composer et à fondre les caractères, première fonte d'une ligne sur une machine construite par Mergenthaler. 1886, utilisation des premières machines à composer «Blower» par le quotidien «New York Tribune». L'homme de presse Whitelaw Reid appelle la machine «Linotype». Création de la Mergenthaler Linotype Company à Brooklyn en 1890. En 1897, l'entreprise s'installe à Deming dans l'état du Nouveau Mexique, Mergenthaler rédige ses mémoires. L'ensemble du patrimoine est détruit par un incendie. Retour à Baltimore. En 1924, le musée Mergenthaler est édifié à Hachtel, son lieu de naissance. – *Publications, sélection:* Fritz Schröder «Ottmar Mergenthaler», Berlin 1941; Willi Mengel et autres «Die Linotype erreicht das Ziel», Berlin, Francfort 1955.

**Meyer,** Erich – né en 1898 à Offenbach, Allemagne, décédé en 1983 à Waldshut,

Mergenthaler 1884

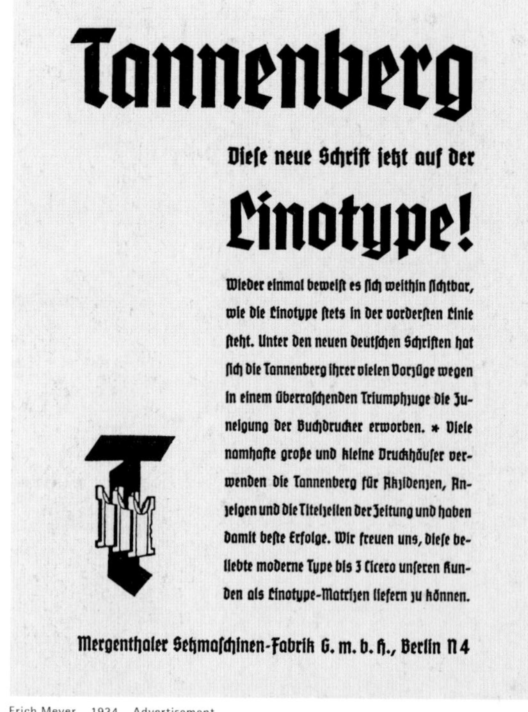


Erich Meyer 1934 Advertisement

Rudolf Meyer 1965 Poster

(1961). – *Publications include:* "Vecerni hovory kninomila Rubricia a starotiskare Tympána", Kroměřiž 1947; "Regule kaligrafu", Prague 1948; "Tvorba typografického pisma", Prague 1956; "Abendgespräche des Buchliebhabers Rubricius und des Alt-Druckers Tympán", Frankfurt 1958; "Schriften", Hamburg 1968. Otto F. Babler "Oldrich Menhart", Antwerp 1950.

**Mergenthaler,** Ottmar – b. 11. 5. 1854 in Hachtel, Germany, d. 28. 10. 1899 in Baltimore, USA – *inventor, entrepreneur* – 1868: trains as a watchmaker under his uncle, Louis Hahl, in Bietigheim am Neckar. 1872: emigrates to America. Works in the factory of his cousin, August Hahl, in Washington and is manager and partner. 1875: moves to Baltimore. Charles T. Moore commissions him to improve and simplify the typesetting machine for his lithographic prints. 1879: produces a design for a line casting machine which he then rejects. 1882: founds his own workshop in Baltimore Bank Lane. 1884: manufactures the first cylinder typesetting and casting machine. The first line is cast on a machine built by Mergenthaler. 1886: the first Blower typesetting machine is put into operation for the "New York Tribune" daily newspaper. The publisher of the newspaper, Whitelaw Reid, gives the machine the name "Linotype". 1890: the Mergenthaler Linotype Company is founded in Brooklyn. 1897: moves to Deming in New Mexico State and writes his memoirs. A fire destroys his property and entire possessions. He returns to Baltimore. 1924:

Fritz Schröder „Ottmar Mergenthaler", Berlin 1941; Willi Mengel u.a. „Die Linotype erreicht das Ziel", Berlin, Frankfurt 1955.

**Meyer,** Erich – geb. 1898 in Offenbach, Deutschland, gest. 1983 in Waldshut, Deutschland. – *Kalligraph, Grafiker, Schriftentwerfer, Sänger, Lehrer* – Ausbildung zum Grafiker an den Technischen Lehranstalten in Offenbach, Schüler von Rudolf Koch. Freier Grafiker in Offenbach. Leiter des Schriftunterrichts an der Kunstgewerbeschule in Frankfurt am Main. Gleichzeitig Ausbildung zum Konzertsänger, Auftritte unter dem Namen Meyer-Stephan. Nach dem Zweiten Weltkrieg als freier Grafiker tätig. – *Schriftentwurf:* Tannenberg (1934).

**Meyer,** Rudolf (Rudi) – geb. 11. 6. 1943 in Basel, Schweiz. – *Grafik-Designer, Typograph, Lehrer* – 1958–63 Studium an der Allgemeinen Gewerbeschule in Basel, u. a. bei Armin Hofmann und Emil Ruder. 1963–64 Zusammenarbeit mit Gérard Ifert in Lausanne. 1964 Umzug nach Paris, Büro als freier Gestalter. 1967 Berufung an die Ecole Nationale Supérieure des Arts Décoratifs in Paris. 1968–74 Gründung des „Bureau d'études Gérard Ifert et Rudolf Meyer". Seit 1974 selbständiger Gestalter in Paris. Auftraggeber waren u.a. LIP Uhren (1974), die SNCF (1976), das RER-Regionalexpressnetz (mit Roger Tallon und Massimo Vignelli, 1977), Renault (1982) und das Opéra Trade Center, Paris (1993–94). 1987 Sieger des Wettbewerbs für das Erscheinungsbild der „Kieler Woche". Unterrichtet seit 1990 am „Atelier National de

Rudolf Meyer 1985 Poster

Rudolf Meyer 1991 Poster

Rudolf Meyer 1986 Poster

Allemagne – *calligraphe, graphiste, concepteur de polices, chanteur, enseignant* – formation de graphiste au Technische Lehranstalt (institut d'études techniques) d'Offenbach, élève de Rudolf Koch. Graphiste indépendant à Offenbach. Directeur du cours d'écriture à la Kunstgewerbeschule (école des arts décoratifs) de Francfort-sur-le-Main. Suit en même temps une formation de chanteur concertiste, se produit sous le nom de Meyer-Stephan. Exerce comme graphiste indépendante après la deuxième guerre mondiale. – *Police:* Tannenberg (1934).

**Meyer,** Rudolf (Rudi) – né le 11. 6. 1943 à Bâle, Suisse – *graphiste maquettiste, typographe, enseignant* – 1958–1963, études à la Allgemeine Gewerbeschule (arts et métiers) de Bâle, chez Armin Hofmann et Emil Ruder. Travaille avec Gérard Ifert à Lausanne de 1963 à 1964. S'installe à Paris en 1964 où il exerce comme designer indépendant. En 1967, il est nommé à l'Ecole Nationale Supérieure des Arts Décoratifs de Paris. Fonde le «Bureau d'études Gérard Ifert et Rudolf Meyer» en 1968. Designer indépendant à Paris depuis 1974. Commanditaires: Montres LIP (1974), la SNCF (1976), le Réseau express régional RER (avec Roger Tallon et Massimo Vignelli, 1977), Renault (1982), l'Opéra Trade Center, Paris (1993–1994). En 1987, il remporte le concours de l'identité de la «Semaine de Kiel». Enseigne à l'Atelier National de Création Typogra-

a Mergenthaler Museum is opened in Hachtel, his place of birth. – *Publications include:* Fritz Schröder "Ottmar Mergenthaler", Berlin 1941; Willi Mengel et al "Die Linotype erreicht das Ziel", Berlin, Frankfurt 1955.

**Meyer,** Erich – b. 1898 in Offenbach, Germany, d. 1983 in Waldshut, Germany – *calligrapher, graphic artist, type designer, singer, teacher* – Trained as a graphic artist at the Technische Lehranstalten in Offenbach. Pupil of Rudolf Koch. Freelance graphic artist in Offenbach. Head of the lettering courses at the Kunstgewerbeschule in Frankfurt am Main. Parallel training as a professional singer, performing under the name of Meyer-Stephan. Freelance graphic artist after the Second World War. – *Font:* Tannenberg (1934).

**Meyer,** Rudolf (Rudi) – b. 11. 6. 1943 in Basle, Switzerland – *graphic designer, typographer, teacher* – 1958–63: studies at the Allgemeine Gewerbeschule in Basle. His tutors include Armin Hofmann and Emil Ruder. 1963–64: works with Gérard Ifert in Lausanne. 1964: moves to Paris and sets up a studio as a freelance designer. 1967: teaching post at the Ecole Nationale Supérieure des Arts Décoratifs in Paris. 1968: the Bureau d'études Gérard Ifert et Rudolf Meyer opens. From 1974 onwards: freelance designer in Paris. Clients include LIP Clocks (1974), SNCF (1976), the RER regional express network (with Roger Tallon and Massimo Vignelli, 1977), Renault (1982) and the Opéra Trade Center, Paris (1993–94). 1987: wins the competition for the design of the

Création Typographique" (ANCT) in Paris. Grafische Arbeiten für Ausstellungen, u.a. „Titien, le siècle d'or de la peinture à Venise" (1993), „Impressionisme, les origines" (1994), „Marianne und Germania" (1995 in Berlin).

**Middleton,** Robert Hunter – geb. 6. 5. 1898 in Glasgow, Schottland, gest. 3. 8. 1985 in Chicago, USA. – *Schriftentwerfer, Drucker* – 1908 Umzug in die USA. Studium der Malerei am Art Institute of Chicago. Seit 1923 Schriftentwerfer der Ludlow Typograph Company. Zusammenarbeit mit Ernst F. Detterer, dessen Schrift „Eusebius" er ab 1923 ausbaut. 1933–71 künstlerischer Leiter der Ludlow Typograph Company. Gründet 1944 die Cherryburn Press. – *Schriftentwürfe:* Eusebius (1923–29), Ludlow Black (1924), Cameo (1927), Record Gothic (1927–60), Delphian Open (1928), Stellar (1929), Tempo (1930–42), Karnak (1931–42), Lafayette (1932), Maifair Kursiv (1932), Eden (1934), Mandate (1934), Umbra (1935), Coronet (1937), Stencil (1938), Radiant (1940), Samson (1940), Flair (1941), Admiral Script (1953), Florentine Kursiv (1956), Formal Script (1956), Wave (1962). – *Publikation u. a.:* Bruce Beck „Robert Hunter Middleton, the man and his letters. Eight essays on his life and career", Chicago 1985.

**Miedinger,** Max – geb. 24. 12. 1910 in Zürich, Schweiz, gest. 8. 3. 1980 in Zürich, Schweiz. – *Schriftentwerfer* – 1926–30 Schriftsetzerlehre in Zürich. Danach Abendkurse an der Kunstgewerbeschule in Zürich. 1936–46 Typograph im Werbeatelier des Kaufhauses „Globus" in Zürich.

phique (ANCT), à Paris, depuis 1990. Travaux graphiques pour diverses expositions, dont «Titien, le siècle d'or de la peinture à Venise» (1993), «Impressionisme, les origines» (1994), «Marianne und Germania» (1995 à Berlin).

**Middleton,** Robert Hunter – né le 6. 5. 1898 à Glasgow, Ecosse, décédé le 3. 8. 1985 à Chicago, Etats-Unis – *concepteur de polices, imprimeur* – s'installe aux Etats-Unis en 1908. Etudes de peinture à l'Art Institute de Chicago. Concepteur de polices à la Ludlow Typograph Company à partir de 1923. Travaille avec Ernst F. Detterer, dont il complète la fonte «Eusebius» à partir de 1923. Directeur artistique de la Ludlow Typograph Company de 1933 à 1971. Fonde la Cherryburn Press en 1944. – *Polices:* Eusebius (1923–1929), Ludlow Black (1924), Cameo (1927), Record Gothic (1927–1960), Delphian Open (1928), Stellar (1929), Tempo (1930–1942), Karnak (1931–1942), Lafayette (1932), Maifair Cursive (1932), Eden (1934), Mandate (1934), Umbra (1935), Coronet (1937), Stencil (1938), Radiant (1940), Samson (1940), Flair (1941), Admiral Script (1953), Florentine Cursive (1956), Formal Script (1956), Wave (1962). – *Publication, sélection:* Bruce Beck «Robert Hunter Middleton, the man and his letters. Eight essays on his life and career», Chicago 1985.

**Miedinger,** Max – né le 24. 12. 1910 à Zurich, Suisse, décédé le 8. 3. 1980 à Zurich, Suisse – *concepteur de polices* – 1926–1930, apprentissage de compositeur en typographie à Zurich, puis cours du soir à la Kunstgewerbeschule (école des arts décoratifs) de Zurich. Typographe à l'atelier de publicité du grand magasin «Globus» à Zurich, de 1936 à 1946. Conseiller clientèle et vendeur de polices à la fonderie Haas à Münchenstein, près de Bâle, de 1947 à 1956. Graphiste indépendant à

Middleton 1928 Delphian Open

Middleton 1937 Coronet

Middleton 1940 Radiant

Light Regular Regular Italic Medium Bold Bold Compact Italic Regular Extended Bold Extended Extra Bold Extended Regular Condensed Bold Condensed Extra Bold Condensed

helvetica helvetica

D. Stempel AG
Typefoundry
Frankfurt am Main

Miedinger ca. 1968 Cover

International Tennis Tournaments in Aix-la-Chapelle
Electro-technical High School of Edinburgh
Container Corporation of America
Pyramids and Temples of Gizeh
Swiss Information Center
Himalaya Mountains
Schiphol Airport
Netherlands

Miedinger 1967 Helvetica Bold Compact Italic

Kieler Woche corporate identity. From 1990 onwards: teaches at the Atelier National de Création Typographique (ANCT) in Paris. Has produced graphics for exhibitions, including "Titien, le siècle d'or de la peinture à Venise" (1993), "Impressionisme, les origines" (1994) and "Marianne und Germania" (1995 in Berlin).

**Middleton,** Robert Hunter – b. 6. 5. 1898 in Glasgow, Scotland, d. 3. 8. 1985 in Chicago, USA – *type designer, printer* – 1908: moves to the USA. Studies painting at the Art Institute of Chicago. From 1923 onwards: type designer for the Ludlow Typograph Company. Works with Ernst F. Detterer and starts producing new weights and versions of Detterer's Eusebius typeface in 1923. 1933–71: art director of the Ludlow Typograph Company. 1944: founds the Cherryburn Press. – *Fonts:* Eusebius (1923–29), Ludlow Black (1924), Cameo (1927), Record Gothic (1927–60), Delphian Open (1928), Stellar (1929), Tempo (1930–42), Karnak (1931–42), Lafayette (1932), Maifair Italic (1932), Eden (1934), Mandate (1934), Umbra (1935), Coronet (1937), Stencil (1938), Radiant (1940), Samson (1940), Flair (1941), Admiral Script (1953), Florentine Italic (1956), Formal Script (1956), Wave (1962). – *Publications include:* Bruce Beck "Robert Hunter Middleton, the man and his letters. Eight essays on his life and career", Chicago 1985.

**Miedinger,** Max – b. 24. 12. 1910 in Zurich, Switzerland, d. 8. 3. 1980 in Zurich, Switzerland – *type designer* – 1926–30: trains as a typesetter in Zurich, after which he attends evening classes at

1947–56 Kundenberater und Schriftenverkäufer der Haas'schen Schriftgießerei in Münchenstein bei Basel. Seit 1956 freier Grafiker in Zürich. Eduard Hoffmann, der Leiter der Haas'schen Schriftgießerei, beauftragt Miedinger 1956 mit der Entwicklung einer neuen Grotesk-Schrift. 1957 Vorstellung der „Haas-Grotesk". 1958 Vorstellung der mageren (bzw. normalen) Haas-Grotesk. 1959 Vorstellung der fetten Haas-Grotesk. 1960 Umbenennung der Schrift „Neue Haas Grotesk" in „Helvetica". 1983 Veröffentlichung der „Neuen Helvetica" auf der Grundlage der früheren „Helvetica" durch die Linotype AG. – *Schriftentwürfe:* Pro Arte (1954), Haas-Grotesk, Helvetica (seit 1957), Horizontal (1965).

**Miller,** J. Abbott – geb. 28. 6. 1963 in Indiana, USA. – *Grafik-Designer, Autor, Lehrer* – 1985 Abschluß des Studiums an der Cooper Union in New York. Arbeit mit Richard Saul Wurman. Gründung des Design-Studios „Design Writing Research" (mit Ellen Lupton) in New York. Auftraggeber waren u. a. Vitra in Weil, Deutschland, das Guggenheim Museum in New York, das National Building Museum in Washington und das American Craft Museum in New York. Konzept und Gestaltung von Ausstellungen, u. a. „The Bauhaus and Design Theory, From Pre-School to Post-Modernism" im Herb Lubalin Study Center der Cooper Union (mit Ellen Lupton, 1991), „Printed Letters: The Natural History of Typography" im Jersey City Museum (1992). Erhält 1993 mit Ellen Lupton den „Chrysler Award". Das Studio „Design Writing Research" grün-

Miller 1991 Cover

Miller 1991 Spread

| n | n | **n** | **n** | | *n* |
|---|---|---|---|---|---|
| a | b | e | f | g | i |
| o | r | s | t | y | z |
| A | B | C | E | G | H |
| M | O | R | S | X | Y |
| 1 | 2 | 4 | 6 | 8 | & |

Miedinger 1957–67 Helvetica

Miller 1991 Spread

Zurich à partir de 1956. En 1956, Eduard Hoffmann, le directeur de la fonderie Haas, charge Miedinger de dessiner une nouvelle Grotesk. Présentation de la Haas-Grotesk en 1957. Présentation de la Haas-Grotesk maigre (normale) en 1958, puis de la Haas-Grotesk gras en 1959. En 1960, la «Neue Haas Grotesk» est appelée «Helvetica». Publication de la «Neue Helvetica» s'inspirant de l'ancienne «Helvetica» par Linotype AG en 1983. – *Polices :* Pro Arte (1954), Haas-Grotesk, Helvetica (depuis 1957), Horizontal (1965).

**Miller,** J. Abbott – né le 28. 6. 1963 à Indiana, Etats-Unis – *graphiste maquettiste, auteur, enseignant* –termine ses études à la Cooper Union, à New York, en 1985. Travaille avec Richard Saul Wurman. Fonde l'atelier de design «Design Writing Research» (avec Ellen Lupton) à New York. Commanditaires : Vitra à Weil, Allemagne; le Guggenheim Museum, New York; le National Building Museum, Washington et l'American Craft Museum, New York. Conception et architecture d'expositions, dont «The Bauhaus and Design Theory, From Pre-School to Post-Modernism» au Herb Lubalin Study Center de la Cooper Union (avec Ellen Lupton, 1991); «Printed Letters : The Natural History of Typography» au Jersey City Museum (1992). Le «Chrysler Award» lui est décerné en 1993 (avec Ellen Lupton). En 1995, l'atelier «Design Writing Re-

the Kunstgewerbeschule in Zurich. 1936–46: typographer for Globus department store's advertising studio in Zurich. 1947–56: customer counselor and typeface sales clerk for the Haas'sche Schriftgießerei in Münchenstein near Basle. From 1956 onwards: freelance graphic artist in Zurich. 1956: Eduard Hoffmann, the director of the Haas'sche Schriftgießerei, commissions Miedinger to develop a new sans-serif typeface. 1957: the Haas-Grotesk face is introduced. 1958: introduction of the roman (or normal) version of Haas-Grotesk. 1959: introduction of a bold Haas-Grotesk. 1960: the typeface changes its name from Neue Haas Grotesk to Helvetica. 1983: Linotype AG publishes its Neue Helvetica, based on the earlier Helvetica. – *Fonts:* Pro Arte (1954), Haas-Grotesk, Helvetica (1957 onwards), Horizontal (1965).

**Miller,** J. Abbott – b. 28. 6. 1963 in Indiana, USA – *graphic designer, author, teacher* – 1985: completes his studies at the Cooper Union in New York. Works with Richard Saul Wurman. Opens a design studio with Ellen Lupton in New York, named Design Writing Research. Clients include Vitra in Weil, Germany, the Guggenheim Museum in New York, the National Building Museum in Washington and the American Craft Museum in New York. Produces concepts and designs for various exhibitions, including "The Bauhaus and Design Theory, From Pre-School to Post-Modernism" in the Herb Lubalin Study Center at the Cooper Union (with Ellen Lupton, 1991) and "Printed Letters: The Natural History of

det 1995 den Verlag „Kiosk" für Publikationen zur Geschichte und Theorie von Industrie- und Grafik-Design. 1996 Herausgeber und Gestaltung der Publikation „Twice", einem Magazin der visuellen Kultur. Unterrichtet an der Parson's School of Design (1988–90), an der Cooper Union (1991), an der Yale University (1992, 1993). – *Publikationen u.a.:* „Printed Letters: The Natural History of Typography", New Jersey 1993; „The Bauhaus and Design Theory" (mit Ellen Lupton), New York 1993; „Design Writing Research: Writing on Graphic Design and Typography" (mit Ellen Lupton), New York 1996.

**Miranda,** Oswaldo – geb. 3. 6. 1949 in Paranaguá, Brasilien. – *Grafik-Designer, Illustrator, Kalligraph* – Beginn seiner Berufstätigkeit im Studio von Casemiro Ambrozewicz. Danach Gestalter im „Griffo Studio" und bei „PAZ Advertising". Herausgeber der Zeitschrift für Grafik-Design und Illustration „Gráfica". Seit 1978 Mitglied des Type Directors Club (TDC) in New York. Veranstaltet die erste Ausstellung des TDC in Brasilien. Veröffentlicht 1981 einen Artikel über das Werk von Herb Lubalin in der Zeitschrift „Raposa". Ausgezeichnet mit zahlreichen Gold- und Silbermedaillen für seine Zeitschriftengestaltung vom Creativity Club of São Paulo. Artikel über sein Werk werden in Zeitschriften wie „Graphis", „Idea" und „Gráfica" veröffentlicht.

**Moholy-Nagy,** László – geb. 20. 7. 1895 in Bácsborsod, Ungarn, gest. 24. 11. 1946 in Chicago, USA. – *Maler, Bildhauer, Fotograf, Typograph* – 1918 Abbruch des Ju-

search» fonde les éditions «Kiosk» qui éditent des ouvrages sur l'histoire et la théorie du design graphique et industriel. En 1996, il édite et réalise la maquette de «Twice», un magasine sur la culture visuelle. A enseigné à la Parson's School of Design (1988–1990), à la Cooper Union (1991), à la Yale University (1992,1993). – *Publications, sélection:* «Printed Letters: The Natural History of Typography», New Jersey 1993; «The Bauhaus and Design Theory» (avec Ellen Lupton), New York 1993; «Design Writing Research: Writing on Graphic Design and Typography» (avec Ellen Lupton), New York 1996.

**Miranda,** Oswaldo – né le 3. 6. 1949 à Paranaguá, Brésil – *graphiste maquettiste, illustrateur, calligraphe* – débute à l'atelier de Casemiro Ambrozewicz, puis exerce comme designer au «Griffo Studio» et chez «PAZ Advertising». Editeur de «Gráfica», une revue spécialisée en graphisme et en illustration. Membre du Type Directors Club (TDC) de New York depuis 1978. A organisé la première exposition de TDC au Brésil. Publie un article sur l'œuvre d'Herb Lubalin dans la revue «Raposa», en 1981. Le Creativity Club de São Paulo lui a décerné plusieurs médailles d'or et d'argent pour ses maquettes de revues. Articles sur son œuvre dans des revues telles que «Graphis», «Idea» et «Gráfica».

**Moholy-Nagy,** László – né le 20. 7. 1895 à Bácsborsod, Hongrie, décédé le 24. 11. 1946 à Chicago, Etats-Unis – *peintre, sculpteur, photographe, typographe* – 1918, interruption de ses études de droit à Budapest. Contacts avec le groupe d'artistes de la revue «Ma» (aujourd'hui) en 1919. S'installe à Berlin en 1920. Première exposition personnelle à la galerie «Sturm» d'Herwarth Walden à Berlin en 1922. Enseigne au Bauhaus de Weimar de

Miranda ca. 1978 Spread

Moholy-Nagy 1925 Cover

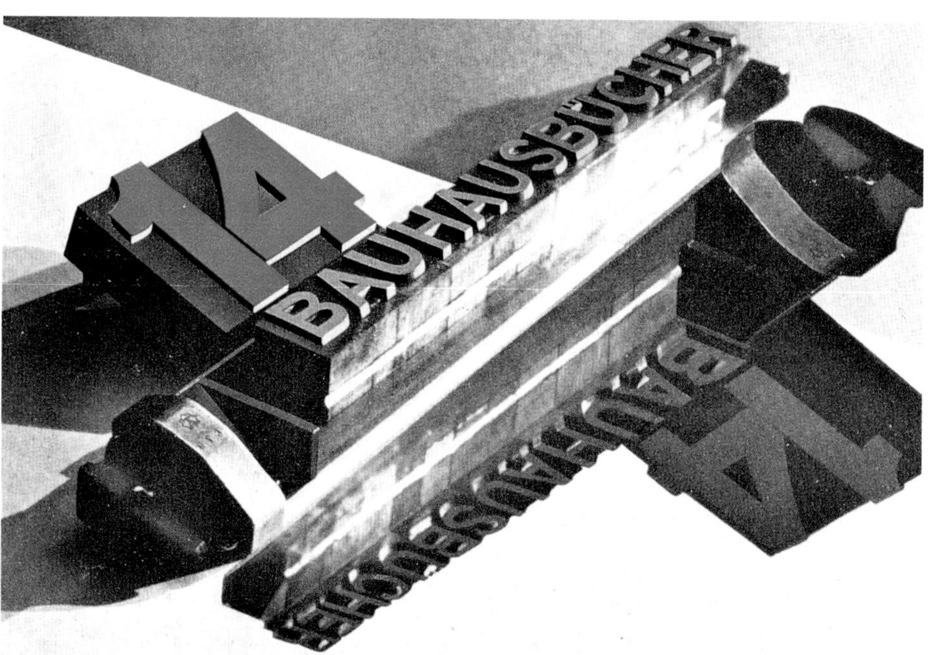

Moholy-Nagy 1929 Cover

Typography" at the Jersey City Museum (1992). 1993: receives the Chrysler Award with Ellen Lupton. 1995: Design Writing Research launches Kiosk, a publishing subsidiary which specializes in publications on the history and theory of industrial and graphic design. 1996: edits and designs "Twice", a magazine for visual culture. Miller has taught at the Parson's School of Design (1988–90), the Cooper Union (1991) and Yale University (1992, 1993). – *Publications include:* "Printed Letters: The Natural History of Typography", New Jersey 1993; "The Bauhaus and Design Theory" (with E. Lupton), New York 1993; "Design Writing Research: Writing on Graphic Design and Typography" (with Ellen Lupton), New York 1996.

**Miranda,** Oswaldo – b. 3. 6. 1949 in Paranaguá, Brazil – *graphic designer, illustrator, calligrapher* – Miranda started his career at Casemiro Ambrozewicz's studio. He then worked as a designer for Griffo Studio and for PAZ Advertising. Editor of "Gráfica", a magazine for graphic design and illustration. From 1978 onwards: member of the Type Directors Club (TDC) of New York. Stages the TDC's first exhibition in Brazil. 1981: publishes an article on the work of Herb Lubalin in "Raposa" magazine. The Creativity Club of São Paolo has awarded Miranda numerous gold and silver medals for his magazine design. Articles on his work have been published in magazines such as "Graphis", "Idea" and "Gráfica".

**Moholy-Nagy,** László – b. 20. 7. 1895 in Bácsborsod, Hungary, d. 24. 11. 1946 in

rastudiums in Budapest. 1919 Kontakte zur Künstlergruppe um die Zeitschrift „Ma" (Heute). 1920 Umzug nach Berlin. 1922 erste Einzelausstellung in Herwarth Waldens Galerie „Sturm" in Berlin. 1923–28 Lehrer am Bauhaus in Weimar, ab 1925 in Dessau. Leiter des Vorkurses. Beschäftigung mit Typographie und Fotografie. Ab 1924 Herausgabe der „Bauhausbücher" (mit Walter Gropius). 1927 Mitbegründer der Zeitschrift „i10" für Film und Fotografie (mit J. J. P. Oud und Willem Pijper). 1928 Austritt aus dem Bauhaus, Rückkehr nach Berlin. 1935 Emigration nach London. Dokumentarfilme, typographische Arbeiten und Werbegestaltung. 1937 Emigration in die USA. Gründung des „New Bauhaus". Nach der Schließung des „New Bauhaus" 1939 (aus finanziellen Gründen), Gründung und Leitung der „School of Design" in Chicago, die 1944 in „Institute of Design" umbenannt wird. 1940 räumliche Arbeiten mit kinetischen und Lichteffekten mit den Materialien Metall und Plexiglas. Seit 1944 Abwendung von plastischen Abeiten, Wiederbeschäftigung mit der Malerei. – *Publikationen u.a.:* „Képeskönyv" (Bilderbuch), Wien 1921; „Buch neuer Künstler" (mit Lajos Kassák), Wien 1922; „Vision in Motion", Chicago 1947. Sybil Moholy-Nagy „Moholy-Nagy, Experiment in Totality", New York 1950; Lucia Moholy-Nagy „Marginalien zu Moholy-Nagy", Krefeld 1972; Andreas Haus „Moholy-Nagy. Fotos und Fotogramme", München 1978.

**Moles,** Abraham A. – geb. 1920 in Paris, Frankreich. – *Philosoph, Lehrer* – Studi-

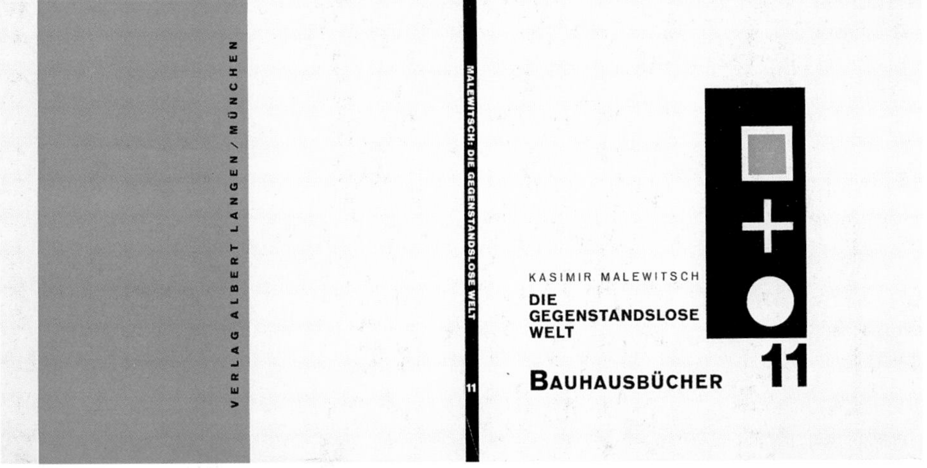


Moholy-Nagy 1927 Cover

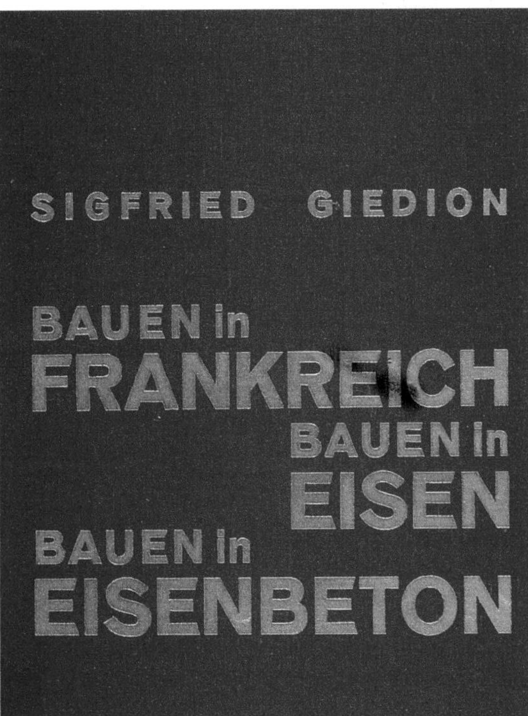


Moholy-Nagy 1928 Cover

Quatrième texte : 35 %

Cinquième texte : 40 %

Sixième texte : 45 %

Septième texte : 50 %

Moles 1969 Scheme

1923 à 1928, et à Dessau à partir de 1925. Dirige le cours préliminaire. Se consacre à la typographie et à la photographie. Edite les «Bauhausbücher» (avec Walter Gropius) à partir de 1924. Cofondateur, en 1927, de la revue «i10» spécialisée en cinéma et en photographie (avec J. J. P. Oud et Willem Pijper). Quitte le Bauhaus en 1928 et retourne à Berlin. Emigre à Londres en 1935. Films documentaires, travaux typographiques et design publicitaire. Emigre aux Etats-Unis en 1937. Il fonde le «New Bauhaus» en 1939. Après la fermeture du «New Bauhaus» (pour des raisons financières), il crée et dirige la «School of Design» de Chicago qui deviendra l' «Institute of Design» en 1944. En 1940, réalisions spatiales avec effets cinétiques et lumineux, où il utilise le métal et le plexiglas. Se détourne de la sculpture en 1944 et retourne à la peinture. – *Publications, sélection:* «Képeskönyv» (Livre d'images), Vienne 1921; «Buch neuer Künstler» (avec Lajos Kassák), Vienne 1922; «Vision in Motion», Chicago 1947. Sybil Moholy-Nagy «Moholy-Nagy, Experiment in Totality», New York 1950; Lucia Moholy-Nagy «Marginalien zu Moholy-Nagy», Krefeld 1972; Andreas Haus «Moholy-Nagy. Fotos und Fotogramme», Munich 1978.

**Moles,** Abraham A. – né en 1920 à Paris, France – *philosophe, enseignant* – études

Chicago, USA – *painter, sculptor, photographer, typographer* – 1918: breaks off his law studies in Budapest. 1919: forms contacts with the artists' group connected with "Ma" ("Today") magazine. 1920: moves to Berlin. 1922: first solo exhibition in Herwarth Walden's Sturm gallery in Berlin. 1923–28: teaches at the Bauhaus in Weimar and from 1925 in Dessau. Director of the "Vorkurs". Works with typography and photography. 1924: publishes the Bauhaus book series with Walter Gropius. 1927: launches "i10", a magazine for film and photography, with J. J. P. Oud and Willem Pijper. 1928: leaves the Bauhaus and returns to Berlin. 1935: emigrates to London. Works in typography and designs advertising and documentaries. 1937: emigrates to the USA. Founds the New Bauhaus. 1939: after the New Bauhaus is forced to close for financial reasons, Moholy-Nagy founds and runs the School of Design in Chicago, renamed the Institute of Design in 1944. 1940: uses metal and Plexiglas to create three-dimensional structures which experiment with kinetic and lighting effects. From 1944 onwards: turns away from sculpture and returns to painting. – *Publications include:* "Képeskönyv" (picture book), Vienna 1921; "Buch neuer Künstler" (with Lajos Kassák), Vienna 1922; "Vision in Motion", Chicago 1947. Sybil Moholy-Nagy "Moholy-Nagy, Experiment in Totality", New York 1950; Lucia Moholy-Nagy "Marginalien zu Moholy-Nagy", Krefeld 1972; A. Haus "Moholy-Nagy. Fotos und Fotogramme", Munich 1978.

**Moles,** Abraham A. – b. 1920 in Paris,

um der Physik und Philosophie. 1945–54 Mitglied des „Centre National de la Recherche Scientifique" (CNRS). 1955–60 Direktor der „Enzyklopädie des Atomzeitalters" in Genf, Schweiz. 1960–69 Dozent an der Hochschule für Gestaltung in Ulm. 1960–68 Dozent an der Ecole supérieure d'organisation in Paris. In dem von Max Bense und Elisabeth Walther herausgegebenen Text „Erstes Manifest der permutationellen Kunst" (Stuttgart 1962) zeigt Moles ein Konzept experimenteller Gestaltung für alle Kulturbereiche auf. Er ist wesentlich beteiligt an den Entwicklungen innerhalb der Informationsästhetik. – *Publikationen u. a.:* „Création scientifique", Genf 1957; „Théorie de l'information et perception esthétique", Paris 1958; „L'affiche dans la société urbaine", Paris 1969.

**Molzahn,** Johannes – geb. 21. 5. 1892 in Duisburg, Deutschland, gest. 31. 12. 1965 in München, Deutschland. – *Maler, Typograph, Lehrer* – Studium an der Großherzoglichen Kunstschule in Weimar. Anschließend Fotografenlehre. 1917 erste Ausstellung in Herwarth Waldens Galerie „Sturm" in Berlin. 1918 Rückkehr nach Weimar. Ab 1922 Werbegestaltung für die Fagus-Werke in Alfeld an der Leine. 1923–28 Professor an der Kunstgewerbeschule Magdeburg als Leiter der Klassen für Werbegrafik, Typographie, Druck und Lithographie. 1925 Gestaltung des Erscheinungsbildes der „Mitteldeutschen Handwerksausstellung" in Magdeburg. 1928–33 Professor für Grafik an der Kunstakademie in Breslau. 1933 Entlassung aus dem Schuldienst

de physique et de philosophie. Membre du «Centre National de Recherche Scientifique» (CNRS) de 1945 à 1954. Dirige l'»Encyclopédie de l'ère atomique» à Genève, Suisse, de 1955 à 1960. Enseigne à la Hochschule für Gestaltung (école supérieure de design) d'Ulm de 1960 à 1969 et à l'Ecole supérieure d'organisation à Paris de 1960 à 1968. Dans le texte «Erstes Manifest der permutationellen Kunst» (premier manifeste de l'art permutationnel) édité par Max Bense et Elisabeth Walther (Stuttgart 1962), Moles présente un concept de création expérimentale valable pour tous les domaines culturels. Il participe activement aux dévelloppements de l'esthétique de la communication. – *Publications, sélection:* «Création scientifique», Genève 1957; «Théorie de l'information et perception esthétique», Paris 1958; «L'affiche dans la société urbaine», Paris 1969.

**Molzahn,** Johannes – né le 21. 5. 1892 à Duisbourg, Allemagne, décédé le 31. 12. 1965 à Munich, Allemagne – *peintre, typographe, enseignant* – études à la Grossherzogliche Kunstschule de Weimar, puis apprentissage de photographe. Première exposition à la galerie «Sturm» de Herwarth Walden à Berlin en 1917. Retour à Weimar en 1918. Publicités pour les Fagus Werke d'Alfeld-sur-la Leine depuis 1922. Professeur à la Kunstgewerbeschule (école des arts décoratifs) de Magdebourg de 1923 à 1928 où il est responsable des cours de graphisme publicitaire, de typographie, d'imprimerie et de lithographie. En 1925, il conçoit l'identité de la «Mitteldeutsche Handwerksaustellung» (salon de l'artisanat) de Magdebourg. Professeur d'arts graphiques à l'académie des beaux-arts de Breslau de 1928 à 1933. Révoqué de l'enseignement par les nazis en 1933. Emigre aux Etats-Unis en 1938. Professeur au département

Molzahn 1927 Cover

Molzahn 1929 Poster

Molzahn 1929 Poster

Momayez 1975 Logo

France – *philosopher, teacher* – Studied physics and philosophy. 1945–54: member of the Centre National de la Recherche Scientifique (CNRS). 1955–60: director of the "Encyclopedia of the Atomic Age" in Geneva, Switzerland. 1960–69: lecturer at the Hochschule für Gestaltung in Ulm. 1960–68: lecturer at the Ecole supérieure d'organisation in Paris. In the "Erstes Manifest der permutationellen Kunst" ("first manifesto of permutational art"), published by Max Bense and Elisabeth Walther in Stuttgart in 1962, Moles sketches out an experimental design concept for all areas of art. Moles has played a major part in developments in information aesthetics. – *Publications include:* "Création scientifique", Geneva 1957; "Théorie de l'information et perception esthétique", Paris 1958; "L'affiche dans la société urbaine", Paris 1969.

**Molzahn,** Johannes – b. 21. 5. 1892 in Duisburg, Germany, d. 31. 12. 1965 in Munich, Germany – *painter, typographer, teacher* – Studied at the Großherzogliche Kunstschule in Weimar and then trained as a photographer. 1917: first exhibition at Herwarth Walden's Sturm gallery in Berlin. 1918: returns to Weimar. Since 1922: designs advertising for the Fagus Werke in Alfeld an der Leine. 1923–28: professor at the Kunstgewerbeschule in Magdeburg. Heads the commercial graphics, typography, printing and lithography courses. 1925: designs the corporate identity for the "Mitteldeutsche Handwerksausstellung" in Magdeburg. 1928–33: professor of graphics at the Kunstakademie in Breslau. 1933: is dis-

durch die Nationalsozialisten. 1938 Emigration in die USA. 1938–41 Professor an der Kunstabteilung der Universität Washington in Seattle. 1943–47 Berufung an die School of Design in Chicago durch László Moholy-Nagy. 1947–52 Professor an der New School of Social Research in New York. 1959 Rückkehr nach Deutschland, lebt in München als freier Künstler. 1965 Aufnahme als Mitglied der Akademie der Künste in Berlin.

**Momayez,** Morteza – geb. 26. 8. 1936 in Teheran, Iran. – *Grafik-Designer, Illustrator, Art Director, Lehrer* – 1952–58 Illustrator und Gestalter für zahlreiche iranische Zeitungen und Zeitschriften. 1962 Art Director der Seven-up Company. 1962–64 Art Director der Kulturzeitschrift „Das Buch der Woche". 1964 Abschluß seiner Studien an der Universität in Teheran. 1965–68 Studien in Paris. Seit 1969 eigenes Grafik-Design-Studio in Teheran. Unterrichtet an der Kunstfakultät der Universität in Teheran. 1972–78 Art Director der Zeitschriften „Rudaki" und „Leben und Kultur". 1973 Leiter der Abteilung Skulpturkunst der UNESCO-Kommission Iran. 1974–78 Gestalter und Art Manager des Internationalen Film-Festivals in Teheran. – *Publikationen u. a.:* „Was ist Grafik-Design", Teheran 1973; „Geschichte der grafischen Kunst im Iran", Teheran 1974; „Zeichen", Teheran 1983; „Plakate", Teheran 1984; „Illustration und Imagination", Teheran 1989.

**Mon,** Franz – geb. 6. 5. 1926 in Frankfurt am Main, Deutschland. – *Schriftsteller, Künstler, Verleger* – Studium der Germa-

Momayez 1974 Poster

Momayez 1975 Poster

Momayez 1976 Logo

Momayez 1993 Logo

des beaux-arts de l'université Washington à Seattle de 1938 à 1941. En 1943, Moholy-Nagy le nomme à la School of Design de Chicago, Molzahn y enseigne jusqu'en 1947. Professeur à la New School of Social Research de New York de 1947 à 1952. Retour en Allemagne en 1959. Vit à Munich comme artiste indépendant. Devient membre de l'académie des beaux-arts de Berlin en 1965.

**Momayez,** Morteza – né le 26. 8. 1936 à Téhéran, Iran. – *graphiste maquettiste, illustrateur, directeur artistique, enseignant* – illustrateur et designer pour plusieurs revues et journaux iraniens de 1952 à 1958. Directeur artistique de la Seven-up Company en 1962. Directeur artistique de la revue culturelle «Le livre de la semaine» de 1962 à 1964. Termine ses études à l'université de Téhéran en 1964. Etudes à Paris de 1965 à 1968. Exerce dans son propre atelier de graphisme et de design à Téhéran à partir de 1969. Enseigne à la faculté des beaux-arts de l'université de Téhéran. Directeur artistique des revues «Rudaki» et «Vie et culture» de 1972 à 1978. En 1973, responsable du département de sculpture auprès de la commission iranienne de l'UNESCO. Designer et directeur artistique du festival international du cinéma de Téhéran de 1974 à 1978. – *Publications, sélection:* «Qu'est ce que le design graphique», Téhéran 1973; «L'histoire des arts graphiques en Iran», Téhéran 1974; «Signes», Téhéran 1983; «Affiches», Téhéran 1984; «Illustration et imagination», Téhéran 1989.

**Mon,** Franz – né le 6. 5. 1926 à Francfort-sur-le-Main, Allemagne – *écrivain, artiste, éditeur* – études de littérature allemande, d'histoire et de philosophie. Pre-

missed from teaching by the Nazis. 1938: emigrates to the USA. 1938–41: professor of art at Washington University in Seattle. 1943–47: László Moholy-Nagy appoints him a teacher at the School of Design in Chicago. 1947–52: professor at the New School of Social Research in New York. 1959: returns to Germany. Lives in Munich as a freelance artist. 1965: made a member of the Akademie der Künste in Berlin.

**Momayez,** Morteza – b. 26. 8. 1936 in Teheran, Iran – *graphic designer, illustrator, art director, teacher* – 1952–58: illustrator and designer for numerous Iranian newspapers and magazines. 1962: art director of the Seven-up Company. 1962–64: art director for the arts magazine "Book of the Week". 1964: finishes his studies at the university in Teheran. 1965–68: studies in Paris. 1969: opens his own graphic design studio in Teheran. Teaches in the art faculty at the university in Teheran. 1972–78: art director for "Rudaki" and "Life and Culture" magazines. 1973: head of the Iranian UNESCO commission's sculpture department. 1974–78: designer and art manager of the International Film Festival in Teheran. – *Publications include:* "What is Graphic Design", Teheran 1973; "The History of Graphic Art in Iran", Teheran 1974; "Symbols", Teheran 1983; "Posters", Teheran 1984; "Illustration and Imagination", Teheran 1989.

**Mon,** Franz – b. 6. 5. 1926 in Frankfurt am Main, Germany – *author, artist, publisher* – Studied German studies, history and philosophy. 1951: first publication of one

nistik, Geschichte und Philosophie. 1951 erste Publikation eines Gedichts in der Zeitschrift „Meta" 7/1951. 1955–91 Tätigkeit als Lektor in einem Schulbuchverlag. 1960 Herausgabe des Sammelbandes „Movens" (mit Walter Höllerer und Manfred de la Motte). 1962–71 Gründung und Leitung des Typos-Verlags. Seit 1963 zahlreiche Gruppen- und Einzelausstellungen der visuellen Arbeiten im In- und Ausland. Für die Biennale in Venedig 1970 entsteht der Textraum „Mortuarium für zwei Alphabete". Seit 1987 Lehraufträge an der Gesamthochschule in Kassel, an der Hochschule für Gestaltung in Karlsruhe und an der Hochschule für Gestaltung Offenbach am Main. – *Publikationen u.a.:* „Artikulationen", Pfullingen 1959; „einmal nur das alphabet gebrauchen", Stuttgart 1967; „Lesebuch", Neuwied 1967; „Texte über Texte", Neuwied 1970; „Knöchel des Alphabets", Offenbach 1989. Eine Ausgabe der gesammelten Texte erscheint seit 1994 im Verlag Gerhard Wolf Janus Press, Berlin.

**Monguzzi,** Bruno – geb. 21. 8. 1941 in Mendrisio, Schweiz. – *Grafik-Designer, Typograph, Lehrer* – 1956–61 Studium an der Ecole des Arts Décoratifs in Genf. Danach Studium an der St. Martins School of Art and Design und an der Central School of Design in London. 1961–63 Gestalter im Studio Boggeri in Mailand. Unterrichtet 1963–65 Typographie an der Fondazione Cini in Venedig. 1965–67 Aufenthalt in den USA. Gestaltung mehrerer Pavillons für die Expo 67 in Montreal. 1968 Rückkehr nach Europa, zunächst Studio in Mailand, danach in Me-

mier poème publié dans la revue «Meta» 7/1951 en 1951. Lecteur dans une maison d'édition spécialisée en livres scolaires de 1955 à 1991. En 1960, il publie le recueil «Movens» (avec Walter Höllerer et Manfred de la Motte). Fonde le Typo-Verlag en 1962 et le dirige jusqu'en 1971. Nombreuses expositions personnelles et collectives de son travail visuel en Allemagne et à l'étranger à partir de 1963. Réalisation de l'espace-texte «Mortuarium für zwei Alphabete» à la Biennale de Venise de 1970. Enseigne à partir de 1987 comme chargé de cours à la Gesamthochschule de Kassel, à la Hochschule für Gestaltung (école supérieure de design) de Karlsruhe et à la Hochschule für Gestaltung d'Offenbach-sur-le-Main. – *Publications, sélection:* «Artikulationen», Pfüllingen 1959; «einmal nur das alphabet gebrauchen», Stuttgart 1967; «Lesebuch», Neuwied 1967; «Texte über Texte», Neuwied 1970; «Knöchel des Alphabets», Offenbach 1989. Une édition des œuvres complètes paraît depuis 1994 aux éditions Gerhard Wolf Janus Press de Berlin.

**Monguzzi,** Bruno – né le 21. 8. 1941 à Mendrisio, Suisse – *graphiste maquettiste, typographe, enseignant* – 1956–1961, études à l'Ecole des Arts Décoratifs de Genève, puis études à la St. Martins School of Art and Design et à la Central School of Design de Londres. Designer au Studio Boggeri à Milan de 1961 à 1963. Enseigne la typographie à la Fondazione Cini à Venise de 1963 à 1965. Séjour aux Etats-Unis de 1965 à 1967. Conçoit plusieurs pavillons pour l'Expo 67 à Montreal. Retour en Europe en 1968, ouvre d'abord un atelier à Milan, puis à Meride, Suisse. Travaux pour l'édition et architecture d'expositions. Conseiller artistique de la revue «Abitare». Professeur à l'école de design de Lugano depuis 1971. Obtient le «Pre-

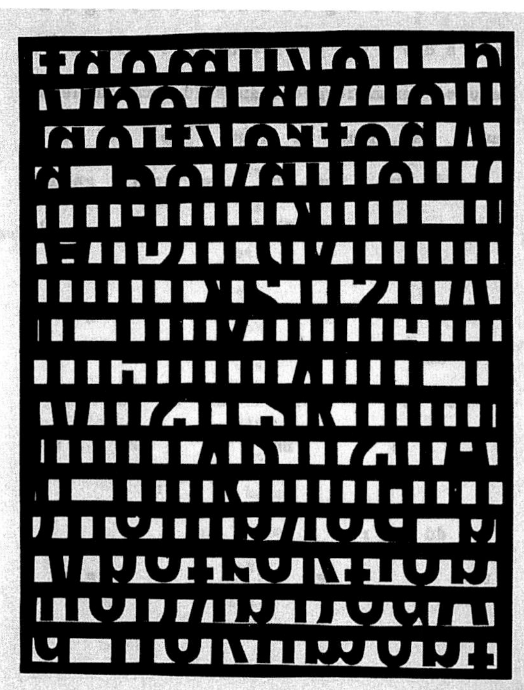

Mon 1963 Collage

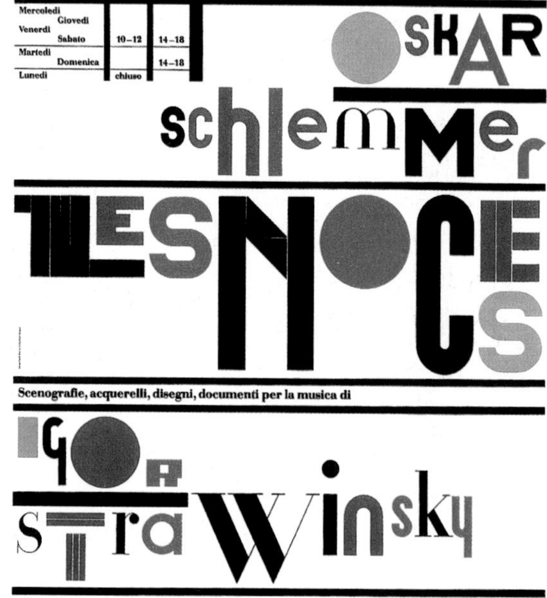

Monguzzi 1988 Poster

Mon 1970

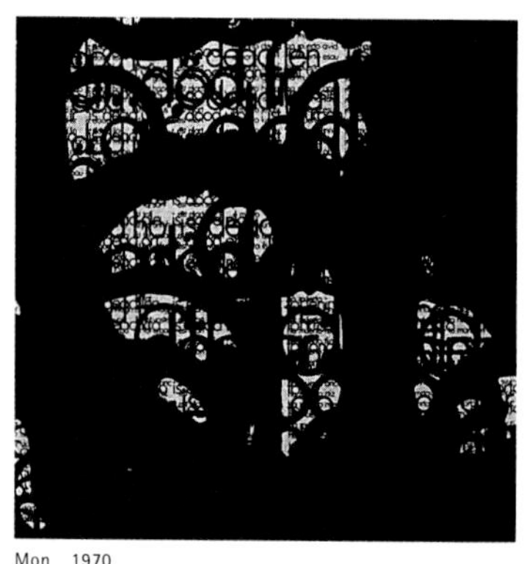

Mon 1970

Monguzzi 1969 Logo

of his poems in the magazine "Meta" 7/1951. 1955–91: works as an editor for publishers of school books. 1960: publishes a volume of collected works entitled "Movens" with Walter Höllerer and Manfred de la Motte. 1962–71: launches and directs the Typos publishing company. From 1963 onwards: numerous joint and solo exhibitions of his visual work in Germany and abroad. 1970: produces a room of texts for the Biennale in Venice entitled "Mortuarium für zwei Alphabete". Since 1987 Mon has held various teaching positions at the Gesamthochschule in Kassel, the Hochschule für Gestaltung in Karlsruhe and the Hochschule für Gestaltung in Offenbach am Main. – *Publications include:* "Artikulationen", Pfullingen 1959; "einmal nur das alphabet gebrauchen", Stuttgart 1967; "Lesebuch", Neuwied 1967; "Texte über Texte", Neuwied 1970; "Knöchel des Alphabets", Offenbach 1989. An edition of his complete works is gradually being published by Gerhard Wolf Janus Press in Berlin (first volume issued in 1994).

**Monguzzi,** Bruno – b. 21. 8. 1941 in Mendrisio, Switzerland – *graphic designer, typographer, teacher* – 1956–61: studies at the Ecole des Arts Décoratifs in Geneva. Then studies at the St Martins School of Art and Design and at the Central School of Design in London. 1961–63: designer for the Boggeri studio in Milan. 1963–65: teaches typography at the Fondazione Cini in Venice. 1965–67: spends time in the USA. Designs several pavilions for Expo 67 in Montreal. 1968: returns to Europe. Runs his own studio first in Milan

ride, Schweiz. Arbeiten für Verlage und Ausstellungs-Design. Künstlerischer Berater der Zeitschrift „Abitare". Seit 1971 Professor an der Schule für Gestaltung in Lugano. 1971 Verleihung des „Premio Bodoni" Preises. Unterrichtet 1982 an der Cooper Union in New York. Gewinnt 1983 den Wettbewerb für das Erscheinungsbild und das Leitsystem des Musée d'Orsay in Paris (mit Jean Widmer). 1987 Leitsystem für das neue „Museo Cantonale d'Arte" in Lugano. Unterrichtet an der Cooper Union in New York. – *Publikationen u.a.:* „Note per una Tipografica informativa", Venedig 1964; „Lo Studio Boggeri 1933–1981", Mailand 1981; „Piet Zwart – the typographical work", Mailand 1987.

**Moniteurs** – *Grafik-Design-Studio* – 1994 Gründung des Studios durch Heike Nehl, Sibylle Schlaich und Heidi Specker in Berlin, Deutschland. Heike Nehl – geb. 27. 2. 1964 in Bünde, Deutschland. – 1984–90 Studium der Visuellen Kommunikation in Bielefeld. 1990–94 Arbeit als Grafik-Designerin bei MetaDesign in Berlin. Sibylle Schlaich – geb. 14. 3. 1964 in Stuttgart, Deutschland. – 1985–87 Studium der Visuellen Kommunikation an der Fachhochschule in Schwäbisch Gmünd. 1988–91 an der Hochschule der Künste in Berlin. 1991–94 Arbeit als Grafik-Designerin bei MetaDesign in Berlin. Heide Specker – geb. 4. 6. 1962 in Damme, Deutschland. – 1984–89 Studium der Visuellen Kommunikation an der Fachhochschule Bielefeld. 1990–94 freischaffende Fotografin. Auftraggeber aus dem kulturellen und wirtschaftlichen Bereich,

Monguzzi 1980 Poster

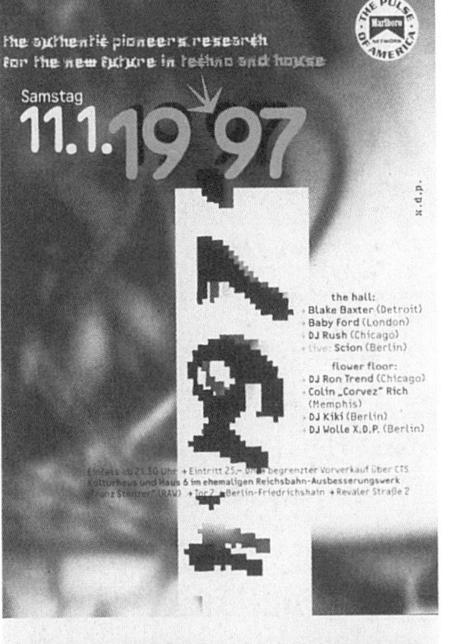


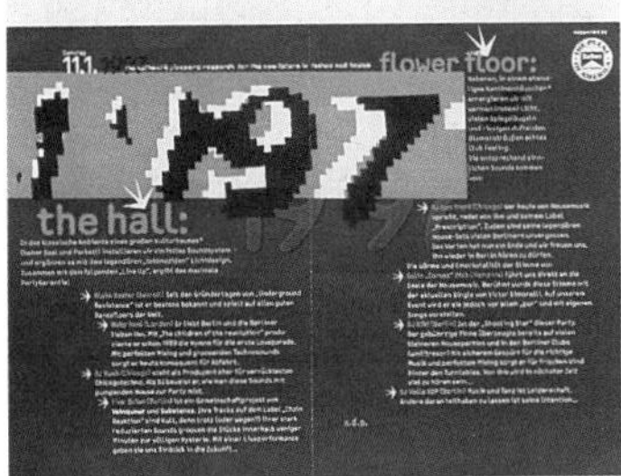


Moniteurs 1997 Folder

Monguzzi 1986 Poster

Moniteurs 1997 Cover

mio Bodoni» en 1971. Enseigne à la Cooper Union de New York en 1982. En 1983, il est lauréat du concours ouvert pour l'identité et le système d'orientation du Musée d'Orsay à Paris (avec Jean Widmer). Système signalétique du nouveau «Museo Cantonale d'Arte» de Lugano en 1987. A enseigné à la Cooper Union de New York. – *Publications, sélection:* «Note per una Tipografica informativa», Venise 1964; «Lo Studio Boggeri 1933–1981», Milan 1981; «Piet Zwart – the typographical work», Milan 1987.

**Moniteurs** – *atelier de graphisme et de design* – fondé en 1994 à Berlin par Heike Nehl, Sibylle Schlaich et Heidi Specker. Heike Nehl – née le 27. 2. 1964 à Bünde, Allemagne – 1984–1990, études de communication visuelle à Bielefeld. 1990–1994, travaille comme graphiste maquettiste chez MetaDesign à Berlin. Sibylle Schlaich – née le 14. 3. 1964 à Stuttgart, Allemagne – 1985–1987, études de communication visuelle à l'école supérieure de Schwäbisch Gmünd. 1988–1991, études à la Hochschule der Künste (école des beaux-arts) de Berlin. Travaille comme graphiste maquettiste chez MetaDesign à Berlin de 1991 à 1994. Heidi Specker – née le 4. 6. 1962 à Damme, Allemagne – 1984–1989, études de communication visuelle à l'institut technique universitaire de Bielefeld. 1990–1994, photographe indépendante. Commanditaires d'organismes culturels et de l'industrie, entre autres: l'Alliierten-Mu-

and then in Meride, Switzerland. Works for various publishers and in exhibition design. Art consultant to "Abitare" magazine. From 1971 onwards: professor at the Schule für Gestaltung in Lugano. 1971: is awarded the Premio Bodoni prize. 1982: teaches at the Cooper Union in New York. 1983: wins the competition for the design of the corporate identity and signage system of the Musée d'Orsay in Paris (with Jean Widmer). 1987: designs the signage system for the new Museo Cantonale d'Arte in Lugano. Teaches at the Cooper Union in New York. – *Publications include:* "Note per una Tipografica informativa", Venice 1964; "Lo Studio Boggeri 1933–1981", Milan 1981; "Piet Zwart – the typographical work", Milan 1987.

**Moniteurs** – *graphic design studio* – 1994: Heike Nehl, Sibylle Schlaich and Heidi Specker open the studio in Berlin, Germany. Heike Nehl – b. 27. 2. 1964 in Bünde, Germany – 1984–90: studies visual communication in Bielefeld. 1990–94: works as a graphic designer for MetaDesign in Berlin. Sibylle Schlaich – b. 14. 3. 1964 in Stuttgart, Germany – 1985–87: studies visual communication at the Fachhochschule in Schwäbisch Gmünd and from 1988–91 at the Hochschule der Künste in Berlin. 1991–94: works as a graphic designer for MetaDesign in Berlin. Heidi Specker – b. 4. 6. 1962 in Damme, Germany – 1984–89: studies visual communication at the Fachhochschule in Bielefeld. 1990–94: freelance photographer. Moniteurs have various clients from the arts and business sector, including the Alliierten Museum

u. a. das Alliierten-Museum Berlin, der Beuth Verlag, das Design-Zentrum Nordrhein-Westfalen, FontShop, die Leipziger Messe und die Telekom. Ausgezeichnet u.a. mit dem „Deutschen Preis für Kommunikationsdesign" 1995 und 1996. Zahlreiche Vorträge zur Typographie im Bereich der neuen Medien.

**Monnerat,** Pierre – geb. 23. 11. 1917 in Paris, Frankreich. – *Grafik-Designer, Lehrer* – Lehre bei einem Architekten. 1936–40 Studium an der Ecole des Beaux-Arts in Lausanne. Seit 1940 selbständiger Grafik-Designer werblicher Arbeiten, u. a. für die Spezialdruckerei Roth & Sauter. Seit 1956 Lehrer an der Ecole des Beaux-Arts in Lausanne. 1957 Gestalter und Organisator des AGI-Pavillons auf der Ausstellung „Graphic 57" in Lausanne. 1958–64 Mitglied des Organisationskomitees und grafischer Berater für die Expo in Lausanne 1964. 1970–74 Leiter der Ecole d'Art Graphique Monnerat in Lausanne. Auswanderung nach Lateinamerika, später Wohnsitz in Barcelona, Spanien, Herausgabe von Kinderbüchern.

**Monotype Corporation** – *Schriftenhersteller, Setz- und Gießmaschinenhersteller* – Tolbert Lanston (1844–1913) entwickelt 1883 seine Idee zur Konstruktion einer Setzmaschine für Mengensatz in Form von Einzelbuchstaben. Bei seiner zweiten Maschine verwendete er 1887 das Heißmetall-Gießverfahren. J. M. Dove beauftragt 1893 die Maschinenfabrik William Sellers in Philadelphia mit dem Bau von 50 Gießmaschinen. 1897 Gründung der Firma Lanston Monotype Corporation in Salfords, England. Die ame-

seum, Berlin, la Beuth Verlag; le Design-Zentrum Nordrhein-Westfalen, FontShop, la Foire de Leipzig et la Telekom. Citée pour le «Deutsche Preis für Kommunikationsdesign» en 1995 et en 1996. Nombreuses conférences sur la typographique dans les nouveaux médias.

**Monnerat,** Pierre – né le 23. 11. 1917 à Paris, France – *graphiste maquettiste, enseignant* – apprentissage chez un architecte. 1936–1940, études à l'Ecole des Beaux-Arts de Lausanne. Graphiste maquettiste indépendant à partir de 1940 et travaux pour l'imprimerie Roth & Sauter. Enseigne à l'Ecole des Beaux-Arts de Lausanne à partir de 1956. Organise et conçoit le pavillon de l'AGI lors de l'exposition «Graphic 57» à Lausanne, en 1957. Membre du comité d'organisation de l'exposition de Lausanne de 1958 à 1964, puis conseiller graphique en 1964. Dirige l'Ecole d'Art Graphique Monnerat à Lausanne de 1970 à 1974. Emigre en Amérique du sud, puis vit à Barcelone, Espagne, publie des livres pour enfants.

**Monotype Corporation** – *fabricant de polices, de machines à fondre les caractères et de composeuses* – en 1883, Tolbert Lanston (1844–1913) a l'idée de construire une composeuse utilisant des caractères indépendants pour la composition de grandes quantités de textes. En 1887, sa deuxième machine emploie le procédé de la fonte de métal à chaud. En 1893, J. M. Dove charge les constructions mécaniques William Sellers de Philadelphie de produire 50 machines. La société Lanston Monotype Corporation est fondée en 1897 à Salfords, Angleterre. Les sociétés Monotype d'Angleterre et des Etats-Unis évoluent dans des directions différentes. «Imprint», la première fonte originale taillée pour la composition mécanique date de 1912. En 1922, Stanley Morison devient conseiller typo-

Monnerat 1961 Poster

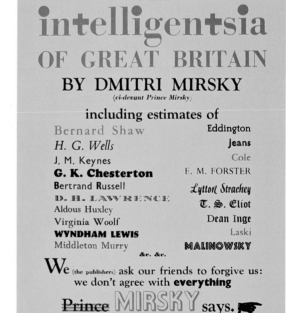

Monotype Corporation 1986 Cover

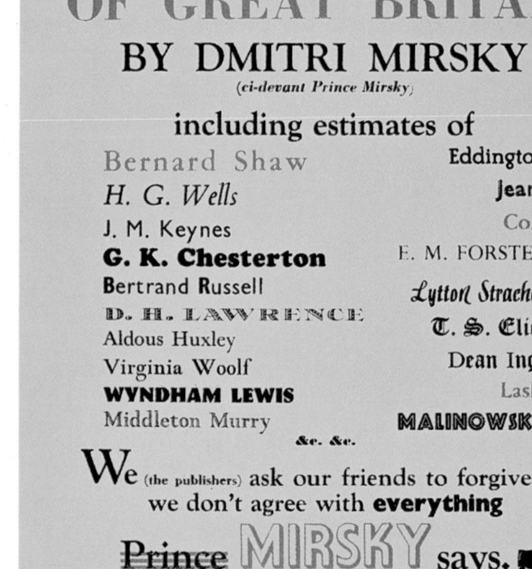

THE
intelligentsia
OF GREAT BRITAIN
BY DMITRI MIRSKY
(ci-devant Prince Mirsky)
including estimates of
Bernard Shaw
H. G. Wells
J. M. Keynes
G. K. Chesterton
Bertrand Russell
D. H. LAWRENCE
Aldous Huxley
Virginia Woolf
WYNDHAM LEWIS
Middleton Murry
Eddington
Jeans
Cole
E. M. FORSTER
Lytton Strachey
T. S. Eliot
Dean Inge
Laski
MALINOWSKY
&c. &c.
We (the publishers) ask our friends to forgive us:
we don't agree with everything
Prince MIRSKY says.

Morison Page

| | *n* | | | | |
|---|---|---|---|---|---|
| *a* | *b* | *e* | *f* | *g* | *i* |
| *o* | *r* | *s* | *t* | *y* | *z* |
| *A* | *B* | *C* | *E* | *G* | *H* |
| *M* | *O* | *R* | *S* | *X* | *Y* |
| *1* | *2* | *4* | *6* | *8* | *&* |

Morison 1923 Blado

in Berlin, Beuth publishers, the Design Zentrum Nordrhein-Westfalen, FontShop, the Leipzig Messe and Telekom. Awards include the Deutsche Preis für Kommunikationsdesign in 1995 and 1996. Numerous lectures on typography in the new media.

**Monnerat,** Pierre – b. 23. 11. 1917 in Paris, France – *graphic designer, teacher* – Trained in an architect's office. 1936–40: studies at the Ecole des Beaux-Arts in Lausanne. From 1940 onwards: he works in commercial art as a freelance graphic designer, producing work for the specialized printers Roth & Sauter among others. From 1956 onwards: teacher at the Ecole des Beaux-Arts in Lausanne. 1957: designs and organizes the AGI pavilion for the "Graphic 57" exhibition in Lausanne. 1958–64: graphics consultant and member of the organizing committee for Expo 1964 in Lausanne. 1970–74: director of the Ecole d'Art Graphique Monnerat in Lausanne. Emigrates to Latin America, and later lives in Barcelona, Spain. Edits children's books.

**Monotype Corporation** – *type manufacturers, manufacturers of typesetting and casting machines* – 1883: Tolbert Lanston (1844–1913) realizes his ideas for a typesetting machine for mass composition in monotype form. 1887: he uses the hot-metal casting method with his second machine. 1893: J. M. Dove commissions William Sellers' machine factory in Philadelphia to build 50 casting machines. 1897: the Lanston Monotype Corporation is founded in Salfords, England. The American and British Monotype Cor-

rikanische und die englische Monotypegesellschaft entwickeln sich in verschiedene Richtungen. Die Schrift „Imprint" wird 1912 als erste Originalschrift für mechanischen Satz geschnitten. 1922 wird Stanley Morison typographischer Berater der Monotype Corp. 1927 Herausgabe der Hauszeitschrift „The Monotype Recorder" unter der Leitung von Beatrice Warde. 1976 Produktion der Lasercomp-Maschine, des ersten Lasersatz-Systems der Welt. The Monotype Corporation wird 1990 von der Firma King Black Associates gekauft. Die Monotype Schriftenbibliothek besteht aus 2.500 Schriften.

**Morison,** Stanley – geb. 6. 5. 1889 in Wanstead, England, gest. 11. 10. 1967 in London, England. – *Typograph, Schriftentwerfer* – 1913–14 Mitarbeiter bei „The Imprint". 1914–18 Kriegsdienstverweigerung und Gefängnis. 1919–21 Mitarbeiter bei Pelican Press, 1921–23 bei Cloister Press in Manchester. 1923–30 Mitherausgeber der Zeitschrift „The Fleuron" (mit Holbrook Jackson, Francis Meynell, B. H. Newdigate, Oliver Simon). 1923–67 typographischer Berater der Monotype Corporation. Seit 1923 Buchgestalter, Gestaltung von Buchumschlägen für den Verlag Victor Gollancz. 1924–30 Mitarbeiter der Cambridge University Press. 1929–60 typographischer Berater der Tageszeitung „The Times". Am 3. 10. 1932 erscheint die „The Times" erstmals in Morisons neuer Schrift „Times New Roman" gesetzt. 1935–51 Herausgeber einer vierbändigen Geschichte der Tageszeitung „The Times". 1945–47 Chefredakteur des

Morison 1932 Cover

Morison 1938 Cover

| | n | **n** | | | *n* |
|---|---|---|---|---|---|
| a | b | e | f | g | i |
| o | r | s | t | y | z |
| A | B | C | E | G | H |
| M | O | R | S | X | Y |
| 1 | 2 | 4 | 6 | 8 | & |

Morison 1932 Times New Roman

The Times
LONDON, WEDNESDAY, SEPTEMBER 28, 1932
THE TIMES
LONDON MONDAY OCTOBER 3 1932

Morison 1932 Headlines

graphique chez Monotype Corp. La revue de la société «The Monotype Recorder» est publié en 1927 sous la direction de Beatrice Warde. Production de la Lasercomp-Machine, le premier système de composition au laser du monde en 1976. En 1990, la Monotype Corporation est rachetée par la société King Black Associates. La bibliothèque des fontes de Monotype contient 2 500 polices.

**Morison,** Stanley – né le 6. 5. 1889 à Wanstead, Angleterre, décédé le 11. 10. 1967 à Londres, Angleterre – *typographe, concepteur de polices* – 1913–1914, travaille sur «Imprint». Incarcéré de 1914 à 1918 comme objecteur de conscience. Employé chez Pelican Press de 1919 à 1921, puis chez Cloister Press à Manchester de 1921 à 1923. Coéditeur de la revue «The Fleuron» (avec Holbrook Jackson, Francis Meynell, B.H. Newdigate, Oliver Simon) de 1923 à 1930. Conseiller typographique de la Monotype Corporation de 1923 à 1967. Réalise des maquettes et des couvertures de livres pour les éditions Victor Gollancz à partir de 1923. Travaille pour la Cambridge University Press de 1924 à 1930. Conseiller typographique du quotidien «The Times» de 1929 à 1960. Le 3. 10. 1932, première édition de «The Times» composé en «Times New Roman», la nouvelle police de Morison. Publication d'une histoire en quatre volumes du quotidien «The Times» de 1935 à 1951. Rédacteur en chef de

porations grow in different directions. 1912: Imprint typeface is the first original alphabet to be cut for mechanical typesetting. 1922: Stanley Morison is made Monotype Corporation's typographic consultant. 1927: publication begins of the in-house magazine, "The Monotype Recorder", under the editorship of Beatrice Warde. 1976: the Lasercomp machine is produced, the first laser typesetting system in the world. 1990: King Black Associates buys Monotype Corporation. Monotype's type library contains 2,500 typefaces.

**Morison,** Stanley – b. 6. 5. 1889 in Wanstead, England, d. 11. 10. 1967 in London, England – *typographer, type designer* – 1913–14: works for "The Imprint". 1914–18: is sent to prison as a conscientious objector. 1919–21: works for Pelican Press and from 1921–23 for Cloister Press in Manchester. 1923–30: co-editor of "The Fleuron" magazine with Holbrook Jackson, Francis Meynell, B. H. Newdigate and Oliver Simon. 1923–67: typography consultant to the Monotype Corporation. From 1923 onwards: book artist. Designs covers for the Victor Gollancz publishing house. 1924–30: works for Cambridge University Press. 1929–60: typography consultant to "The Times" daily newspaper. On 3 October 1932 the first issue of "The Times" set in Morison's new typeface, Times New Roman, is printed. 1935–51: publishes the history of "The Times" in four volumes. 1945–47: editor-in-chief of the "Times Literary Supplement". 1960: made a Royal Designer for Industry. – *Font:* Times New

„Times Literary Supplement“. 1960 Ernennung zum „Royal Designer for Industry“. – *Schriftentwurf:* Times New Roman (1932). – *Publikationen u.a.:* „Four Centuries of fine print“, London 1924; „The Alphabet of Damianus Moyllus“, London 1927; „The calligraphy of Ludovico degli Arrighi“, Paris 1929; „The English newspaper, 1622–1932“, Cambridge 1932; „First Principles of Typography“, Cambridge 1936; „A tally of types“, Cambridge 1953; „Typographic design in relation to photographic composition“, San Francisco 1959. James Moran „Stanley Morison“, London 1971.

**Morris,** William – geb. 24. 3. 1834 in Walthamston, England, gest. 3. 10. 1896 in Hammersmith, England. – *Maler, Designer, Drucker, Verleger, Autor, Typograph, Schriftentwerfer* – 1853–55 Studium am Exeter College in Oxford. 1856 Arbeit im Architekturbüro G. E. Street in Oxford. Hinwendung zur Malerei. 1859 Gründung der Firma Monks, Marshall, Faulkner & Co., Entwurf und Produktion von Schmuck, Kirchenfenstern, Tapeten und ganzen Innendekorationen. 1862 Ausstellung der Produkte der Firma auf der „London International Exhibition of Art and Industry“. 1866–80 Ausstattung von Räumen im St. James Palace, 1867 des Speisesaals im Victoria & Albert Museum. 1875 Neugründung der Firma unter dem Namen „Morris & Co.“. 1877 Ablehnung einer Professur für Literatur an der Universität in Oxford. 1883 Eintritt in die „Democratic Federation“ (später „Social Democratic Federation“). Öffentliche politische Vorträge für den So-

«Times Literary Supplement» de 1945 à 1947. Obtient le titre de «Royal Designer for Industry» en 1960. – *Polices:* Times New Roman (1932). – *Publications, sélection:* «Four Centuries of fine print», Londres 1924; «The Alphabet of Damianus Moyllus», Londres 1927; «The calligraphy of Ludovico degli Arrighi», Paris 1929; «The English newspaper» 1622–1932», Cambridge 1932; «First Principles of Typography», Cambridge 1936; « A tally of types», Cambridge 1953; «Typographic design in relation to photographic composition», San Francisco 1959. James Moran «Stanley Morison», Londres 1971.

**Morris,** William – né le 24. 3. 1834 à Walthamston, Angleterre, décédé le 3. 10. 1896 à Hammersmith, Angleterre – *peintre, designer, imprimeur, éditeur, auteur, typographe, concepteur de polices* – 1853–1855, études à l'Exeter College d'Oxford. En 1856, il est employé au cabinet d'architecte G. E. Street à Oxford. Se consacre à la peinture. Fondation de la société Monks, Marshall, Faulkner & Co. en 1859; conception et production de bijoux, de vitraux, de papier peints et de décorations d'intérieur. En 1862, les produits de la société sont présentés à la «London International Exhibition of Art and Industry». Architecture intérieure des salles du St. James Palace de 1866 à 1880, et du restaurant du Victoria & Albert Museum en 1867. En 1875, la société prend le nom de «Morris & Co.». En 1877, rejet de sa candidature comme professeur de littérature à l'université d'Oxford. Entre à la «Democratic Federation» (plus tard «Social Democratic Federation») en 1883. Prises de parole en public en faveur du socialisme. Fonde la «Art Workers Guild». Crée la Kelmscott Press en 1890. En 1892, il refuse le titre de «Poet Laureat» (Poète de la cour). Derniers dis-

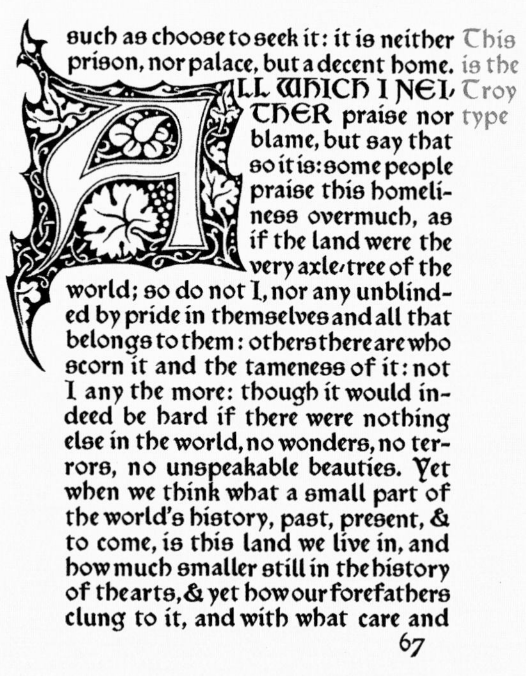
such as choose to seek it: it is neither prison, nor palace, but a decent home. ALL WHICH I NEITHER praise nor blame, but say that so it is: some people praise this homeliness overmuch, as if the land were the very axle-tree of the world; so do not I, nor any unblinded by pride in themselves and all that belongs to them: others there are who scorn it and the tameness of it: not I any the more: though it would indeed be hard if there were nothing else in the world, no wonders, no terrors, no unspeakable beauties. Yet when we think what a small part of the world's history, past, present, & to come, is this land we live in, and how much smaller still in the history of the arts, & yet how our forefathers clung to it, and with what care and

67

This is the Troy type

Morris 1892 Troy Type

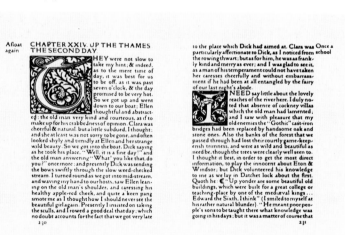

Morris 1890 Golden Type

Morse 1938 Alphabet

Roman (1932). – *Publications include:* "Four Centuries of fine print", London 1924; "The Alphabet of Damianus Moyllus", London 1927; "The calligraphy of Ludovico degli Arrighi", Paris 1929; "The English newspaper, 1622–1932", Cambridge 1932; "First Principles of Typography", Cambridge 1936; "A tally of types", Cambridge 1953; "Typographic design in relation to photographic composition", San Francisco 1959; James Moran "Stanley Morison", London 1971.

**Morris,** William – b. 24. 3. 1834 in Walthamston, England, d. 3. 10. 1896 in Hammersmith, England – *painter, designer, printer, publisher, author, typographer, type designer* – 1853–55: studies at Exeter College, Oxford. 1856: works at G. E. Street's architect's office in Oxford. Starts painting. 1859: Monks, Marshall, Faulkner & Co. is founded. The company designs and manufactures jewelry, stained glass windows, wallpaper and complete interiors. 1862: the company's products are exhibited at the "London International Exhibition of Art and Industry". 1866–80: designs interiors for St James Palace and in 1867 the cafeteria in the Victoria & Albert Museum. 1875: Morris relaunches the company as Morris & Co. 1877: declines a professorship in literature at Oxford University. 1883: joins the Democratic Federation (later called the Social Democratic Federation). Holds public political speeches for the Socialist cause. Founds the Art Workers Guild. 1890: launches the Kelmscott Press. 1892: refuses to be made poet laureate. 1896: gives his last public speech

zialismus. Gründung der „Art Workers Guild". 1890 Gründung der Kelmscott Press. 1892 Ablehnung des Angebots „Poet Laureat" (Hofdichter) zu werden. 1896 letzte öffentliche Ansprache auf einer Versammlung der „Society for Checking the Abuses of Public Advertisement". – *Schriftentwürfe:* Golden Type (1890), Troy Type (1892), Chaucer Type (1893).– *Publikationen u.a.:* „The collected works of William Morris" (24 Bände), London 1910–15, New York 1966. May Morris (Hrsg.) „William Morris, artist, writer, socialist", Oxford 1936; Edward P. Thompson „The communism of William Morris", London 1965.

**Morse,** Samuel – geb. 27. 4. 1791 in Charlestown, Massachusetts, USA, gest. 2. 4. 1872 in New York, USA. – *Maler, Erfinder* – 1827–45 Präsident der National Academy of Design in New York. Seit 1837 Arbeit an einem elektromagnetischen Schreibgerät. 1838 Entwicklung des Telegraphen-Alphabets. Das lateinische Schriftzeichenrepertoire wird durch eine Kombination von Punkten und Strichen ersetzt. Jeder Buchstabe wird durch eine Verbindung von bis zu vier Zeichen dargestellt, jede Ziffer durch fünf Zeichen, jedes Satzzeichen durch die Verbindung von sechs Zeichen.

**Moscoso,** Victor – geb. 1936 in Spanien. – *Grafik-Designer, Lehrer* – Studium an der Cooper Union in New York, danach an der Yale University in New Haven (bei Josef Albers). 1959 Umzug nach San Francisco, Studium am Art Institute of San Francisco, wo er später als Dozent lehrt. Seit 1966 zahlreiche Plakate für

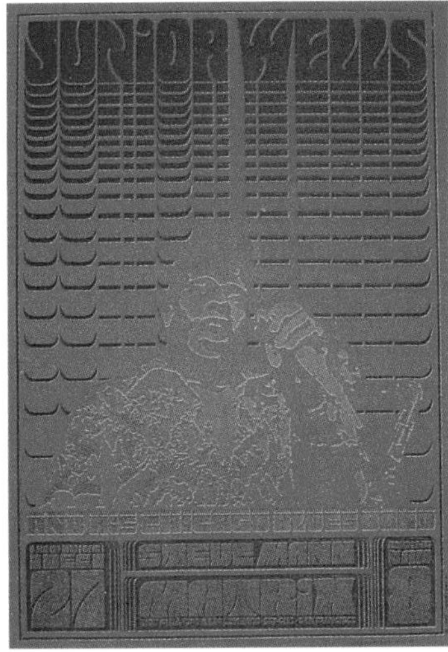

Moscoso 1966 Poster

Moscoso 1967 Poster

Moscoso 1967 Poster

cours politiques en 1896 lors d'une assemblée de la «Society for Checking the Abuses of Public Advertisement». – *Polices:* Golden Type (1890), Troy Type (1892), Chaucer Type (1893). – *Publications, sélection:* «The collected works of William Morris» (24 vol.), Londres 1910–1915, New York 1966. May Morris (éd.) «William Morris, artist, writer, socialist», Londres 1936; Edward P. Thompson «The communism of William Morris», Londres 1965.

**Morse,** Samuel – né le 27. 4. 1791 à Charlestown, Massachusetts, Etats-Unis, décédé le 2. 4. 1872 à New York, Etats-Unis – *peintre, inventeur* – 1827–1845, président de la National Academy of Design de New York. Travaille à partir de 1837 sur un appareil de transcription électromagnétique. Invente l'alphabet télégraphique en 1838. Le répertoire des caractères latins est remplacé par une combinaison de points et de traits. Chaque lettre est représentée par une association de quatre impulsions au maximum, chaque chiffre correspond à cing signes, et la ponctuation à six signes.

**Moscoso,** Victor – né en 1936 en Espagne – *graphiste maquettiste, enseignant* – études à la Cooper Union à New York, puis à la Yale University à New Haven (chez Josef Albers). S'installe à San Francisco en 1959, étudie à l'Art Institute of San Francisco où il enseignera par la suite. Nombreuses affiches pour des concerts au Fillmore West, Avalon Ball-

at a meeting held by the Society for Checking the Abuses of Public Advertisement. – *Fonts:* Golden Type (1890), Troy Type (1892), Chaucer Type (1893). – *Publications include:* "The collected works of William Morris" (24 vols.), London 1910–15, New York 1966. May Morris (ed.) "William Morris, artist, writer, socialist", Oxford 1936; Edward P. Thompson "The communism of William Morris", London 1965.

**Morse,** Samuel – b. 27. 4. 1791 in Charlestown, Massachusetts, USA, d. 2. 4. 1872 in New York, USA – *painter, inventor* – 1827–45: president of the National Academy of Design in New York. From 1837 onwards: works on an electromagnetic writing device. 1838: develops the telegraph alphabet, whereby Latin type is replaced by a system of dots and dashes. Words are represented by various combinations of symbols: a letter has up to four symbols, numbers have five and punctuation marks six symbols.

**Moscoso,** Victor – b. 1936 in Spain – *graphic designer, teacher* – Studied at the Cooper Union in New York, then at Yale University in New Haven under Josef Albers. 1959: moves to San Francisco and studies at the Art Institute of San Francisco, where he later teaches. 1966: starts

Konzerte im Fillmore West, dem Avalon Ballroom und Matrix. 1967 Teilnahme an der Plakat- und Kunstausstellung „Joint Show“ in der Moore Gallery in San Francisco (mit Wes Wilson, Stanley Mouse, Alton Kelley, Rick Griffin, die als die „Big Five“ der Plakat-Szene von San Francisco bezeichnet werden). Seit 1973 zahlreiche Entwürfe von Plattencovern für die CBS (u.a. für Herbie Hancock's „Headhunters“), Round Records (u.a. für Jerry Garcias „Garcia“). Moscoso veröffentlicht seinen Text „Artist Rights Today Information Pamphlet“ am 1. 12. 1986 in Woodacre, Kalifornien. Moscosos verzerrte und ornamentale Schriften in flimmernden Farben werden zum Inbegriff einer Typographie, die bildhaft ausdrückt, was in der populären Musik entsteht.

**Mouse,** Stanley – geb. 10. 10. 1940 in Fresno, USA. – *Grafik-Designer, Illustrator* – 1958–65 Studium an der Art School of the Society of Arts and Crafts in Detroit. T-Shirt-Design. Gründung seines „Mouse Studios“. 1965 Umzug nach San Francisco. 1966 Freundschaft mit Alton Kelley, gemeinsames Studio „Mouse & Kelley“, Entwürfe von Plakaten für die Veranstaltungen von „Family Dog“ im Avalon Ballroom und anderen Aktivitäten in der Bay Area. 1968 Teilnahme an der Ausstellung „The Joint Show“ in der Moore Gallery in San Francisco (mit Rick Griffin, Victor Moscoso, Alton Kelley, Wes Wilson). Aufenthalt in London, grafische Arbeiten für „The Beatles Songbook“ und für „Blind Faith“. 1969 Rückkehr in die USA, weitere Arbeiten mit

room et Matrix à partir de 1966. Participe à l'exposition d'affiches et d'art «Joint Show» à la Moore Gallery de San Francisco en 1967 (avec Wes Wilson, Stanley Mouse, Alton Kelley, Rick Griffin que les milieux de l'affiche à San Francisco appelaient «Big Five»). Réalise à partir de 1973 de nombreuses pochettes de disques pour CBS (entre autres pour Herbie Hancock's «Headhunters») et Round Records (pour Jerry Garcias «Garcia» etc.). Le 1. 12. 1986, Moscoso publie son texte «Artist Rights Today Information Pamphlet» à Woodacre, Californie. Les caractères ornementaux et déformés aux couleurs rutilantes de Moscoso deviennent le symbole d'une typographie qui exprime de manière picturale la nature de la musique pop.

**Mouse,** Stanley – né le 10. 10. 1940 à Fresno, Etats-Unis – *graphiste maquettiste, illustrateur* – études à l'Art School of the Society of Arts and Crafts de Detroit de 1958 à 1965. Design de T-shirts. Fonde le «Mouse Studio». S'installe à San Francisco en 1965. Rencontre Alton Kelley en 1966, puis création de l'atelier commun «Mouse & Kelley», dessine des affiches pour les manifestations de «Family Dog» à l'Avalon Ballroom et pour les autres activités de la Bay Area. Participe, en 1968 à l'exposition «The Joint Show» à la Moore Gallery de San Francisco (avec Rick Griffin, Victor Moscoso, Alton Kelley, Wes Wilson). Séjour à Londres, travaux graphiques pour «The Beatles Songbook» et pour «Blind Faith». Retour aux Etats-Unis en 1969, continue à travailler avec Alton Kelley pour des groupes de musiciens tels que Steve Miller, Journey, Grateful Dead, The Wings, The Rolling Stones. En 1977, il obtient un «Grammy» pour la pochette de «Book of Dreams» de Steve Miller. Les travaux de Mouse ont marqué les milieux de la musique et la

Mouse 1966 Poster

Mouse 1967 Poster

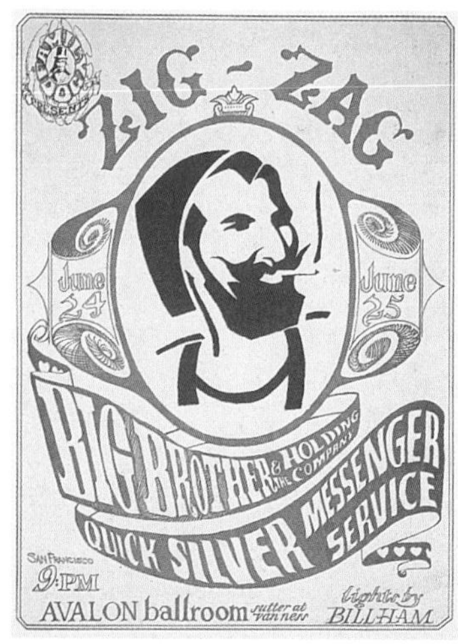


Mouse 1966 Poster

producing posters for concerts in Fillmore West, the Avalon Ballroom and Matrix. 1967: takes part in "The Joint Show" poster and art exhibition at Moore Gallery in San Francisco with Wes Wilson, Stanley Mouse, Alton Kelley and Rick Griffin, known as the "Big Five" of the San Francisco poster scene. From 1973 onwards: produces numerous record covers for CBS (including the cover for Herbie Hancock's "Headhunters") and for Round Records (including the cover for Jerry Garcia's "Garcia"). 1. 12. 1986: Moscoso publishes his "Artist Rights Today Information Pamphlet" in Woodacre, California. Moscoso's distorted and ornamental typefaces in shimmering colors have become the symbol of a typographic style which clearly illustrates the various trends arising from pop music.

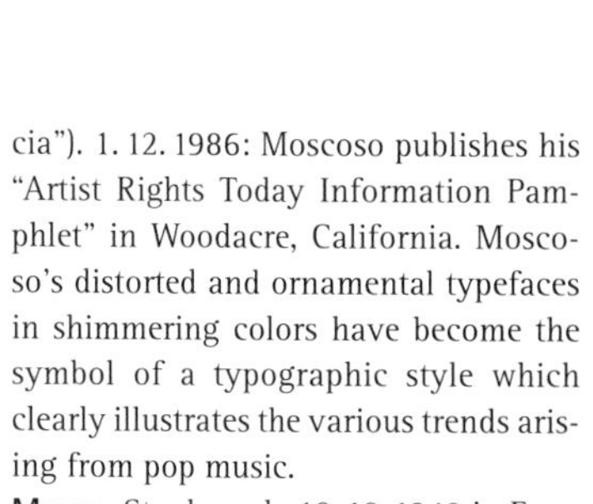

**Mouse,** Stanley – b. 10. 10. 1940 in Fresno, USA – *graphic designer, illustrator* – 1958–65: studies at the Art School of the Society of Arts and Crafts in Detroit. T-shirt design. Opens Mouse Studio. 1965: moves to San Francisco. 1966: founds the Mouse & Kelley joint studio with his friend Alton Kelley. Designs posters for Family Dog events in the Avalon Ballroom and for other Bay Area activities. 1968: takes part in The Joint Show exhibition at the Moore Gallery in San Francisco with Rick Griffin, Victor Moscoso, Alton Kelley and Wes Wilson. Spends time in London. Designs graphics for "The Beatles Songbook" and for Blind Faith. 1969: returns to the USA and continues to work with Alton Kelley for

Alton Kelley für Musikgruppen wie Steve Miller, Journey, Grateful Dead, The Wings, The Rolling Stones. 1977 Auszeichnung mit einem Grammy für die Gestaltung des Umschlags von Steve Millers „Book of Dreams". In den sechziger und siebziger Jahren prägten die Arbeiten von Mouse die Musikszene und die Jugendkultur der USA. – *Publikation:* „Mouse & Kelley", Limpsfield 1979.

**Müller,** Lars – geb. 25. 12. 1955 in Oslo, Norwegen. – *Grafik-Designer, Typograph, Verleger, Lehrer* – 1975–79 Lehre als Grafiker. 1980 Studienaufenthalte in den USA und den Niederlanden. 1981 Junior-Designer bei Total Design in Amsterdam. 1982 Gründung des „Ateliers für Visuelle Kommunikation" in Baden, Schweiz. Auftraggeber aus Kultur und Wirtschaft. 1983 Gründung des Verlags „Lars Müller Publishers", in dem Publikationen zu Architektur, Design und Kunst des 20. Jahrhunderts erscheinen. Lehraufträge u. a. an der Schule für Gestaltung in Zürich (1985–88, 1991), der Höheren Technischen Lehranstalt in Brugg-Windisch (1988), Ecole des Beaux-Arts in Lyon (1993) und der Hochschule für Gestaltung in Offenbach (1996). – *Publikation:* „Josef Müller-Brockmann", Baden 1994.

**Müller,** Rolf – geb. 15. 12. 1940 in Dortmund, Deutschland. – *Grafik-Designer, Typograph* – Studium an der Hochschule für Gestaltung in Ulm. Einjähriger Arbeitsaufenthalt bei Josef Müller-Brockmann in Zürich. 1965 eigenes Gestaltungsbüro in Neu-Ulm, Konzepte für visuelle Erscheinungsbilder. 1967–72 stell-

Mouse 1967 Poster

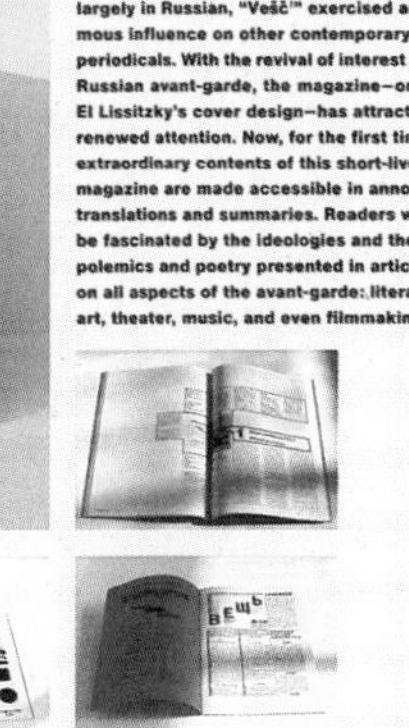
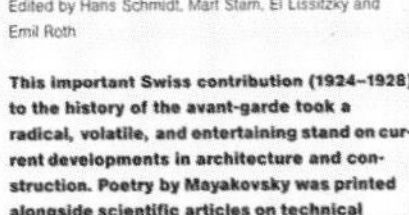

**Vešč' Objet Gegenstand**
Edited by El Lissitzky and Ilya Ehrenburg, Berlin, 1922

**Although it lasted for only two issues, March/April and May, 1922, and was written largely in Russian, "Vešč'" exercised an enormous influence on other contemporary periodicals. With the revival of interest in the Russian avant-garde, the magazine–or rather El Lissitzky's cover design–has attracted renewed attention. Now, for the first time, the extraordinary contents of this short-lived magazine are made accessible in annotated translations and summaries. Readers will be fascinated by the ideologies and theories, polemics and poetry presented in articles on all aspects of the avant-garde: literature, art, theater, music, and even filmmaking.**

35,5×25,7 cm / 14×10⅛", boxed.
Reprint ABC, First Series: 1–6,
Second Series: 1–4.
Separate commentary.
55 pages, G/E.
ISBN 3-906700-63-1

**ABC**
**Contributions on Building–a magazine**
Edited by Hans Schmidt, Mart Stam, El Lissitzky and Emil Roth

**This important Swiss contribution (1924–1928) to the history of the avant-garde took a radical, volatile, and entertaining stand on current developments in architecture and construction. Poetry by Mayakovsky was printed alongside scientific articles on technical innovations. ABC thrived on provocative contradictions!**
In the commentary: Claude Lichtenstein, "Industrialism as a Social and Aesthetic Utopia," Werner Möller, "Useless Beauty," Jörg Stürzebecher, "ABC between Review and Trade Journal," Otaker Máčel, "Stavba and ABC, a Comparison".

33,5×25,5 cm / 13¼×10",
boxed. Nos. 1/2 and 3. Separate commentary, 144 pages,
b+w illustrations. Essay by Roland Nachtigäller and Hubertus Gassner,
annotations by Rita Frommenwiler,
index, G/E
ISBN 3-906700-62-3

18 19

Lars Müller 1996 Spread

culture des jeunes aux Etats-Unis pendant les années 60 et 70. – *Publication:* «Mouse & Kelley», Limpsfield 1979.

**Müller,** Lars – né le 25. 12. 1955 à Oslo, Norvège – *graphiste maquettiste, typographe, éditeur, enseignant* – 1975–1979, apprentissage de graphiste. Séjour d'études aux Etats-Unis et aux Pays-Bas en 1980. Junior Designer chez Total Design à Amsterdam en 1981. Fonde l' «Atelier für Visuelle Kommunikation» à Baden, Suisse en 1982. Commanditaires du secteur culturel et de l'entreprise. En 1983, il crée les éditions «Lars Müller Publishers» qui publient des ouvrages d'architecture, de design et d'art du 20è siècle. Chargé de cours à la Schule für Gestaltung (école de design) de Zurich (1985–1988, 1991), à la Höhere Technische Anstalt (institut d'études techniques) de Brugg-Windisch (1988), à l'Ecole des Beaux-Arts de Lyon (1993), à la Hochschule für Gestaltung (école supérieure de design) d'Offenbach (1996). – *Publication:* «Josef Müller-Brockmann», Baden 1994.

**Müller,** Rolf – né le 15. 12. 1940 à Dortmund, Allemagne – *graphiste maquettiste, typographe* – études à la Hochschule für Gestaltung (école supérieure de design) d'Ulm. Stage d'un an chez Josef Müller-Brockmann à Zurich. Ouvre son propre atelier de design à Neu-Ulm en 1965, conception d'identités visuelles. De

Mouse 1971 Poster

Lars Müller 1988 Poster

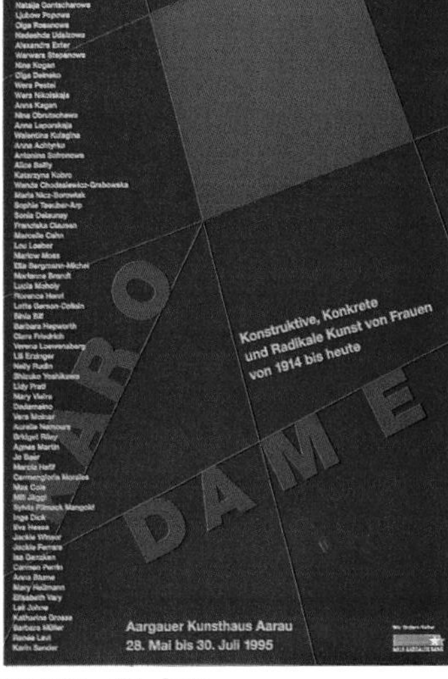

Lars Müller 1995 Poster

groups such as Steve Miller, Journey, Grateful Dead, The Wings and The Rolling Stones. 1977: is awarded a Grammy for his cover design for Steve Miller's "Book of Dreams". In the 1960s and 1970s Mouse's work bore great influence on the music scene and teen and twen culture in the USA. – *Publication:* "Mouse and Kelley", Limpsfield 1979.

**Müller,** Lars – b. 25. 12. 1955 in Oslo, Norway – *graphic designer, typographer, publisher, teacher* – 1975–79: trains as a graphic artist. 1980: studies in the USA and The Netherlands. 1981: junior designer for Total Design in Amsterdam. 1982: founds the Atelier für Visuelle Kommunikation studio in Baden, Switzerland. Works for various clients from industry and commerce and the arts sector. 1983: launches Lars Müller Publishers which issues publications on architecture, design and 20th-century art. Müller has held teaching positions at the Schule für Gestaltung in Zurich (1985–88, 1991), the Höhere Technische Lehranstalt in Brugg-Windisch (1988), the Ecole des Beaux-Arts in Lyon (1993) and at the Hochschule für Gestaltung in Offenbach (1996), among others. – *Publication:* "Josef Müller-Brockmann", Baden 1994.

**Müller,** Rolf – b. 15. 12. 1940 in Dortmund, Germany – *graphic designer, typographer* – Studied at the Hochschule für Gestaltung in Ulm. Spent a year working for Josef Müller-Brockmann in Zurich. 1965: opens his own design studio in Neu-Ulm and works on concepts for visual corporate identities. 1967–72:

vertretender Gestaltungsbeauftragter für die Spiele der XX. Olympiade in München 1972, Mitarbeiter von Otl Aicher. 1970 Gestaltung des Erscheinungsbildes der Stadt Leverkusen. Seit 1972 eigenes Büro für visuelle Kommunikation in München. Visuelle Erscheinungsbilder, Informations- und Orientierungssysteme, Ausstellungsgestaltung. Gestaltet 1972 das Plakat der Kieler Woche. Seit 1985 Chefredakteur und Gestalter der Zeitschrift „High Quality". 1989 Auszeichnung mit einer Goldmedaille des Art Directors Club of New York. 1990 Erscheinungsbild der internationalen Bauausstellung Emscherpark. 1991–93 Präsident der Alliance Graphique Internationale (AGI). Vorlesungen an der Kunstgewerbeschule in Zürich, an der Fachhochschule in Schwäbisch Gmünd, am College of Art and Design in Halifax.

**Müller-Brockmann,** Josef – geb. 9. 5. 1914 in Rapperswil, Schweiz, gest. 30. 8. 1996 in Unterengstringen, Schweiz. – *Grafik-Designer, Typograph, Autor, Lehrer* – zweijährige Lehre als Grafiker. Ein Jahr Hospitant an der Kunstgewerbeschule in Zürich. 1934–36 Assistent bei Walter Diggelmann in Zürich. Seit 1936 selbständig. Ausstellungsgestaltungen im In- und Ausland. 1946–52 Bühnenbilder. Seit 1950 Hinwendung zu sachlich-konstruktiver Gestaltung. Erste Plakate für die Tonhalle in Zürich. Seit 1952 als Grafiker tätig. 1957–60 Fachlehrer für Grafik an der Kunstgewerbeschule in Zürich. 1958–65 Gründer und Mitredakteur der Zeitschrift „Neue Grafik" (mit R. P. Lohse, C. Vivarelli, H. Neuburg). 1962 Berater

1967 à 1972, co-responsable du design pour les XXe Jeux Olympiques de Munich, travaille avec Otl Aicher. Conception de l'identité de la ville de Leverkusen en 1970. Exerce dans son agence de communication visuelle à Munich depuis 1972, où il conçoit des identités, des systèmes de signalisation et d'information, des architectures d'expositions. Dessine l'affiche de la Semaine de Kiel en 1972. Rédacteur en chef et maquettiste de la revue «High Quality» depuis 1985. Médaille d'or de l'Art Directors Club de New York en 1989. Identité de la «internationale Bauausstellung Emscherpark» en 1990. Vice-président de l'Alliance Graphique Internationale (AGI) de 1991 à 1993. Conférences à la Kunstgewerbeschule (école des arts décoratifs) de Zurich, à la Fachhochschule für Gestaltung (école de design) de Schwäbisch Gmünd et au College of Art and Design de Halifax.

**Müller-Brockmann,** Josef – né le 9. 5. 1914 à Rapperswil, Suisse, décédé le 30. 8. 1996 à Unterengstringen, Suisse – *graphiste maquettiste, typographe, auteur, enseignant* – deux années d'apprentissage de graphiste. Auditeur libre pendant un an à la Kunstgewerbeschule (école des arts décoratifs) de Zurich. Assistant de Walter Diggelmann à Zurich de 1934 à 1936. Exerce comme indépendant à partir de 1936. Architectures d'expositions en Suisse et à l'étranger. 1946–1952, scénographies. Se consacre au design objectif et construit à partir de 1950. Premières affiches pour la Tonhalle de Zurich. Exerce comme graphiste à partir de 1952. Professeur d'arts graphiques à la Kunstgewerbeschule de Zurich de 1957 à 1960. Fonde la revue «Neue Grafik» en 1958, exerce comme rédacteur (avec R. P. Lohse, C. Vivarelli, H. Neuburg) jusqu'en 1965. Conseiller et designer pour

Rolf Müller 1972 Poster

Rolf Müller 1987 Cover

Rolf Müller 1981 Cover

Rolf Müller 1981 Cover

Müller-Brockmann 1959 Poster

assistant design consultant to the 20th Olympic Games held in Munich in 1972. Works for Otl Aicher. 1970: designs the corporate identity for the city of Leverkusen. 1972: opens his own studio for visual communication in Munich. Designs visual corporate identities, information and sign-posting systems and exhibitions. 1972: designs the poster for the Kieler Woche. From 1985 onwards: editor-in-chief and designer of "High Quality" magazine. 1989: awarded a gold medal by the Art Directors Club of New York. 1990: designs the corporate identity for the international Emscherpark building exhibition. 1991–93: president of the Alliance Graphique Internationale (AGI). Has given lectures at the Kunstgewerbeschule in Zurich, at the Fachhochschule in Schwäbisch Gmünd and at the College of Art and Design in Halifax.

**Müller-Brockmann,** Josef – b. 9. 5. 1914 in Rapperswil, Switzerland, d. 30. 8. 1996 in Unterengstringen, Switzerland – *graphic designer, typographer, author, teacher* – Spent two years training as a graphic artist. Audited for a year at the Kunstgewerbeschule in Zurich. 1934–36: assistant to Walter Diggelmann in Zurich. 1936: starts working freelance. Designs exhibitions in Switzerland and abroad. 1946–52: designs sets for the theater. 1950: turns to functional, constructive design. Produces his first posters for the Tonhalle in Zurich. From 1952 onwards: works as a graphic artist. 1957–60: specialist teacher for graphics at the Kunstgewerbeschule in Zurich. 1958–65: founder and co-editor of "New Graphic

und Gestalter für die Rosenthal Porzellanwerke in Selb, Deutschland. 1963 Gastdozent an der Hochschule für Gestaltung in Ulm. 1964 Gestalter der Abteilung „Erziehung, Wissenschaft und Forschung" auf der Expo Lausanne. 1965 Gründer der Galerie 58 (später Galerie Seestraße) für konkrete und konstruktive Kunst in Rapperswil. 1967–88 Design-Berater für IBM Europa. 1967 Gründung der Agentur „Müller-Brockmann & Co" mit drei Partnern. Gestaltungsaufträge für Industrie, Verwaltung und Kultur. 1976–84 Trennung von den Partnern der Agentur, Müller-Brockmann führt die Agentur weiter. Beratungs- und Gestaltungstätigkeit, u. a. für Olivetti, die Schweizerischen Bundesbahnen, das Kunsthaus Zürich, das Schweizerische Institut für Kunstwissenschaft. 1978–83 Jury-Mitglied des Deutschen Bundespreises „Gute Form" in Darmstadt. 1980 Gestaltungskonzept für die Zeitschrift „Transatlantik". 1986–93 Wanderausstellung der Plakate in Nord-, Mittel- und Südamerika. Zahlreiche Auszeichnungen, u. a. Goldmedaille des Kantons Zürich für kulturelle Verdienste (1987), „Honorary Designer for Industry" der Royal Society of Arts in London (1988), „Middleton-Award" des American Center for Design in Chicago (1990) und den Design-Preis der Schweiz (1993). 1994 Plakatausstellung (mit Shizuko Müller-Brockmann) in Wiligrad und Cottbus. Mit seinem rationalen Gestaltungsansatz war Müller-Brockmann Wegbereiter einer wissenschaftlichen Auffassung der Visuellen Kommunikation. – *Publikationen*

Müller-Brockmann 1968 Poster

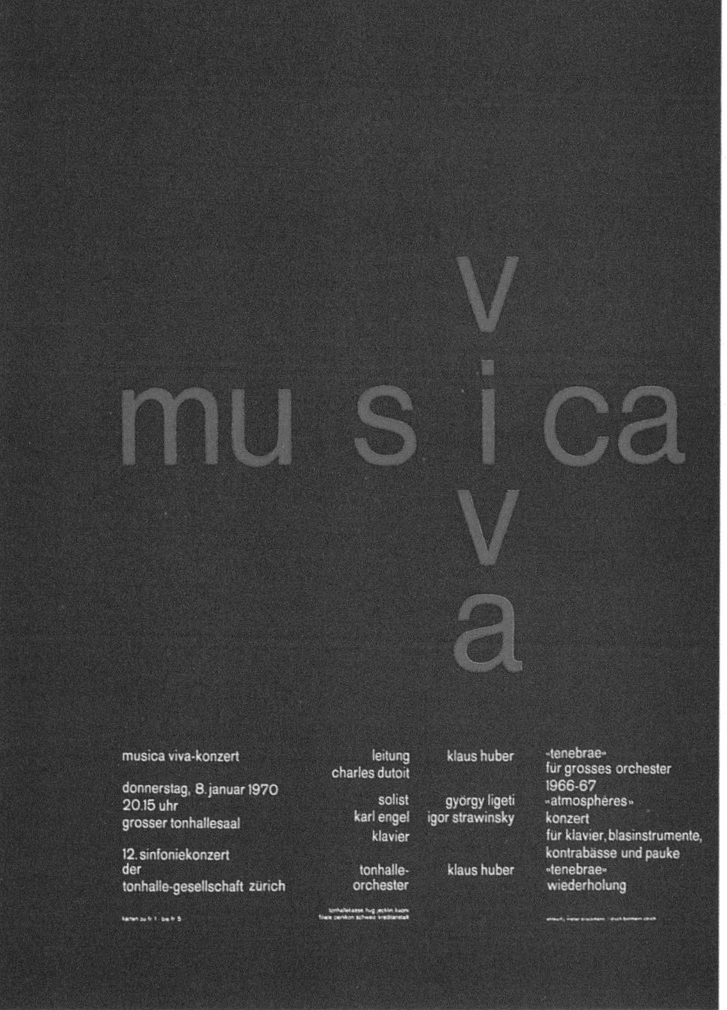

Müller-Brockmann 1969 Poster

les porcelaines Rosenthal à Selb, Allemagne. Séminaires à la Hochschule für Gestaltung (école supérieure de design) d'Ulm en 1963. Design de la section «Education, sciences et recherche» à l'Expo de Lausanne en 1964. En 1965, il fonde la galerie 58 (plus tard Galerie Seestrasse) à Rapperswil, galerie spécialisée en art concret et construit. Conseiller du design chez IBM Europe de 1967 à 1988. En 1967, il crée l'agence «Müller-Brockmann & Co» avec trois autres partenaires. Commandes de l'industrie, des administrations et du secteur culturel. En 1976, ses partenaires quittent l'agence, et Müller-Brockmann continue à y exercer seul jusqu'en 1984. Activités de conseiller et de designer pour Olivetti, les Chemins de fer suisses, le Kunsthaus Zurich, L'institut suisse des arts. Membre du jury du Bundespreis «Gute Form» à Darmstadt en Allemagne de 1978 à 1983. Conception graphique de la revue «Transatlantik» en 1980. Exposition itinérante d' affiches en Amérique du nord, en Amérique centrale et latine de 1986 à 1993. Nombreuses distinctions, dont la médaille d'or du mérite culturel du canton de Zurich (1987), le «Honorary Designer for Industry» de la Royal Society of Arts de Londres (1988), le «Middleton Award» de l'American Center for Design à Chicago (1990), le Prix de Design de Suisse (1993). Exposition d'affiches (avec Shizuko Müller-Brockmann) à Wiligrad et à Cottbus en 1994. Par son design rationnel, Müller-Brockmann a été le précurseur d'une concep-

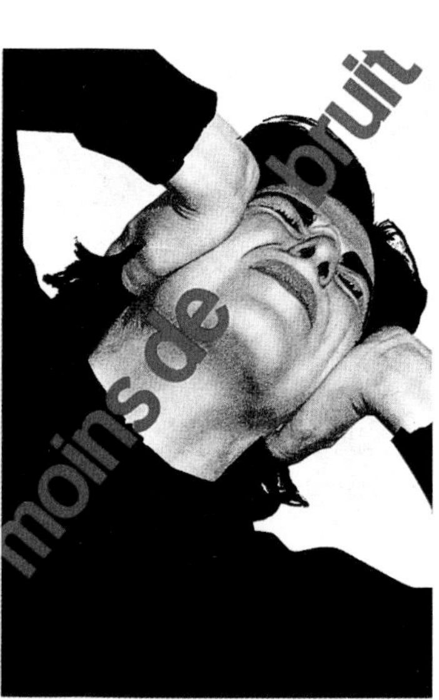

Müller-Brockmann 1960 Poster

Design" magazine with R. P. Lohse, C. Vivarelli and H. Neuburg. 1962: consultant and designer for the Rosenthal porcelain factory in Selb, Germany. 1963: visiting lecturer at the Hochschule für Gestaltung in Ulm. 1964: designs the education, science and research section for Expo Lausanne. 1965: founds his studio, Galerie 58 (later renamed Galerie Seestraße), for concrete and constructive art in Rapperswil. 1967–88: design consultant to IBM in Europe. 1967: founds the Müller-Brockmann & Co. agency with three partners. Has designs commissioned from industry, administrative bodies and the arts. 1976–84: the agency splits. Müller-Brockmann continues to run the agency. Acts as consultant and designer for various companies and institutions, including Olivetti, the Swiss Railways, the Kunsthaus Zürich and the Schweizerisches Institut für Kunstwissenschaft. 1978–83: member of the jury for the German "Gute Form" competition in Darmstadt. 1980: produces the design concept for "Transatlantik" magazine. 1986–93: traveling exhibition of his posters in North, Central and South America. Müller-Brockmann received numerous awards, including a gold medal from the canton of Zurich for his achievements in art (1987), the title of Honorary Designer for Industry from the Royal Society of Arts in London (1988), the Middleton Award from the American Center for Design in Chicago (1990) and Switzerland's design prize (1993). 1994: poster exhibition with Shizuko Müller-Brockmann in Wiligrad and Cottbus. With his rational

*u.a.:* „Gestaltungsprobleme des Grafikers", Teufen 1961; „Geschichte der visuellen Kommunikation", Teufen 1971; „Geschichte des Plakats" (mit Shizuko Müller-Brockmann), Zürich 1971; „Raster-Systeme", Teufen 1981; „Mein Leben: spielerischer Ernst und ernsthaftes Spiel", Baden 1994. Lars Müller (Hrsg.) „Josef Müller-Brockmann", Baden 1994.
**Mumprecht,** Rudolf – geb. 1. 1. 1918 in Basel, Schweiz. – *Künstler, Kalligraph, Illustrator* – 1934–38 Lehre als Kartograph und Lithograph bei Kümmerly & Frey in Bern. Hinwendung zu Grafik und Gebrauchsgrafik. Seit 1944 druckgrafische Arbeiten auf der eigenen Radier-Handpresse. 1960 Fertigung von 13 farbigen Monotypien zu einem Buch von Jacques Prévert, das bei Berggruen in Paris erscheint. Seit 1987 gelangt Mumprecht über konkrete, tachistische und gestische künstlerische Arbeiten in den 70er Jahren zu seinen „Schriftbildern", einer skripturalen Malerei. – *Publikationen u.a.:* Hans Rudolf Schneebeli „Rudolf Mumprecht. Catalogue de l'œuvre gravé sur cuivre 1944–64", Winterthur 1965; Fred Zaugg, Margrit Moser „Rudolf Mumprecht. L'œuvre gravé sur cuivre 1944–80", Bern 1980.
**Munari,** Bruno – geb. 24. 10. 1907 in Mailand, Italien. – *Maler, Fotograf, Grafik-Designer, Industrie-Designer, Lehrer* – 1927–36 Zusammenarbeit mit futuristischen Gruppen in Mailand und Rom. Seit 1929 Zusammenarbeit mit dem „Studio Boggeri". Zahlreiche Artikel und Entwürfe für die Zeitschrift „Campo". Arbeiten als Fotograf und Grafik-Designer, u. a. für

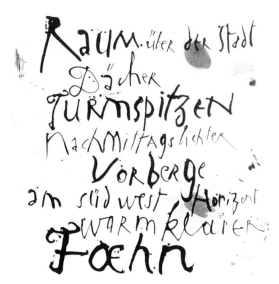

tion scientifique de la communication visuelle. – *Publications, sélection :* «Gestaltungsprobleme des Grafikers», Teufen 1961; «Geschichte der visuellen Kommunikation», Teufen 1971; «Geschichte des Plakats» (avec Shizuko Müller-Brockmann), Zurich 1971; «Raster-Systeme», Teufen 1981; «Mein Leben : spielerischer Ernst und ernsthaftes Spiel», Baden 1994. Lars Müller (éd.) «Josef Müller-Brockmann», Baden 1994.
**Mumprecht,** Rudolf – né le 1. 1. 1918 à Bâle, Suisse – *artiste, calligraphe, illustrateur* – 1934–1938, apprentissage de cartographe et de lithographe chez Kümmerly & Frey à Berne. Se consacre aux arts graphiques et au graphisme industriel. A partir de 1944, il réalise des estampes sur sa propre presse à main. En 1960, série de 13 monotypes en couleur pour un ouvrage de Jacques Prévert édité chez Berggruen à Paris. A partir de 1987, après des travaux concrets, tachistes et gestuels des années 70, il se consacre à une peinture scripturale appelée «Schriftbilder». – *Publications, sélection :* Hans Rudolf Schneebeli «Rudolf Mumprecht. Catalogue de l'œuvre gravé sur cuivre, 1944–1964», Winterthur 1965; Fred Zaugg, Margrit Moser «Rudolf Mumprecht. L'œuvre gravé sur cuivre 1944–1980», Berne 1980.
**Munari,** Bruno – né le 24. 10. 1907 à Milan, Italie – *peintre, photographe, graphiste maquettiste, enseignant* – 1927–1936 travaille avec les groupes futuristes à Milan et à Rome. Employé au «Studio Boggeri» à partir de 1929. Nombreux articles et conceptions graphistes pour la revue «Campo». Travaille comme photographe et graphiste maquettiste pour Olivetti, Pirelli, La Rinascente, IBM, Cinzano et les éditions Mondadori et Einaudi. Ouvre une agence de graphisme et de design en 1930 (avec Ricardo Ricas). Colla-

Mumprecht 1978 Letter Picture

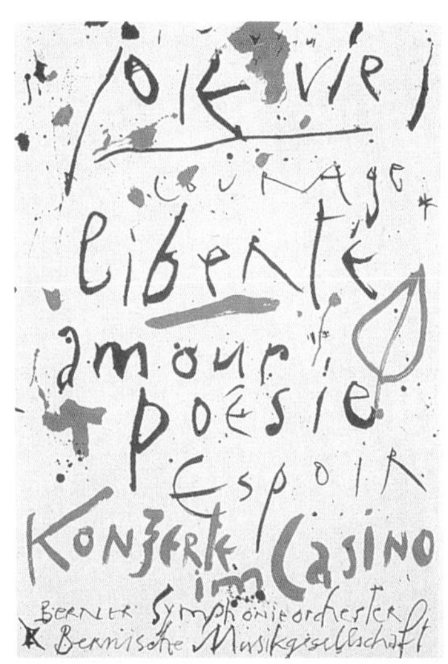

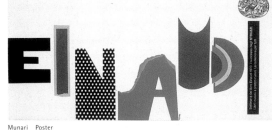

Mumprecht 1984 Poster

Munari Poster

Munari 1960 Poster

Munari 1964 Poster

concept of design, Müller-Brockmann paved the way for a scientific approach to visual communication. – *Publications include:* "Gestaltungsprobleme des Grafikers", Teufen 1961; "Geschichte der visuellen Kommunikation", Teufen 1971; "Geschichte des Plakats" (with Shizuko Müller-Brockmann), Zurich 1971; "Raster-Systeme", Teufen 1981; "Mein Leben: spielerischer Ernst und ernsthaftes Spiel", Baden 1994. Lars Müller (ed.) "Josef Müller-Brockmann", Baden 1994.
**Mumprecht,** Rudolf – b. 1. 1. 1918 in Basle, Switzerland – *artist, calligrapher, illustrator* – 1934–38: trains as a cartographer and lithographer with Kümmerly & Frey in Bern. Turns to graphics and commercial art. From 1944 onwards: produces printed graphics on his own etching press. 1960: produces 13 colored monotypes for a book by Jacques Prévert, published by Berggruen in Paris. From 1987 onwards: Mumprecht's scriptural paintings, called "Schriftbildern", evolve as a result of his experiments with Tachisme and concrete and gestural art in the 70s. – *Publications include:* Hans Rudolf Schneebeli "Rudolf Mumprecht. Catalogue de l'œuvre gravé sur cuivre 1944–64", Winterthur 1965; Fred Zaugg, Margrit Moser "Rudolf Mumprecht. L'œuvre gravé sur cuivre 1944–80", Bern 1980.
**Munari,** Bruno – b. 24. 10. 1907 in Milan, Italy – *painter, photographer, graphic designer, industrial designer, teacher* – 1927–36: works with Futurist groups in Milan and Rome. From 1929 onwards: works with Studio Boggeri. Contributes numerous articles and designs to

Olivetti, Pirelli, La Rinascente, IBM, Cinzano und die Verlage Mondadori und Einaudi. 1930 Eröffnung eines Grafik-Design-Studios (mit Ricardo Ricas). 1939 Mitarbeit an der Wochenzeitung „Tempo". 1948 Gründung der Künstlergruppe „mac" (movimento arte concreta) in Mailand. Seit 1948 Arbeiten als Industrie-Designer, u.a. für die Firmen Danese, Pigomma, Tre-A und Robots. Unterrichtet 1970–71 Design an der Harvard University in Cambridge, Massachusetts. 1984 Plakatentwürfe für Campari. – *Publikationen u.a.:* „I libri Munari", Mailand 1945; „Libro illegibile", Mailand 1949; „Bruno Munari's ABC", New York 1960; „Alfabetiere secondo il metodo attivo", Turin 1960; „Good design", Mailand, New York 1963; „Design e communicazione visive", Bari 1968; „Alfabetiere", Turin 1972. A. Tranchis „L'arte anomalia di Bruno Munari", Bari 1981.

**Museum für Gestaltung,** Zürich – Gründung 1875, 1878 Gründung der Kunstgewerbeschule in Partnerschaft mit dem Kunstgewerbemuseum. Zahlreiche Ausstellungen und Publikationen zu Themen aus Ästhetik, Architektur, Alltagskultur, Design und Visueller Kommunikation. 1984 Umbenennung in „Schule und Museum für Gestaltung". Seit 1986 erwirbt die Designsammlung seriell hergestellte Produkte des 20. Jahrhunderts und hat einen Bestand von ca. 7.000 Objekten und ca. 20.000 Verpackungen. In der öffentlichen Bibliothek sind ca. 70.000 Textdokumente aus den Bereichen Kunst, Architektur, Mode, Produktdesign, Typographie, Grafik, Film, Thea-

Munari 1941

Munari 1965 Poster

Museum für Gestaltung 1993 Invitation Card

Museum für Gestaltung 1996 Invitation Card

borateur à l'hebdomadaire «Tempo» en 1939. Fonde le groupe d'artistes «mac» (movimento arte concreta) en 1948. Exerce comme designer industriel à partir de 1948, entre autres pour les sociétés Danese, Pigomma, Tre-A, Robots. Enseigne les arts graphiques et le design à la Harvard University à Cambridge, Massachusetts de 1970 à 1971. Affiches pour Campari en 1984. – *Publications, sélection:* «I libri Munari», Milan 1945; «Libro illegibile», Milan 1949; «Bruno Munari's ABC», New York 1960; «Alfabetiere secondo il metodo attivo», Turin 1960; «Good design», Milan, New York 1963; «Design e communicazione visive», Bari 1968; «Alfabetiere», Turin 1972. A. Tranchis «L'arte anomalia di Bruno Munari», Bari 1981.

**Museum für Gestaltung** (musée de la création), Zurich – fondé en 1875, puis création en 1878 de la Kunstgewerbeschule (école des arts décoratifs) en tant que partenaire du musée. Nombreuses expositions et publications sur l'esthétique, l'architecture, la culture du quotidien, le design et la communication visuelle. En 1984, le complexe est rebaptisé «Schule und Museum für Gestaltung» (musée et école de design). Depuis 1986, la collection de design acquiert des pièces fabriquées en série au XXè siècle, elle possède près de 7 000 objets et 20 000 emballages. La bibliothèque publique contient environ 70 000 documents sur l'art, l'architecture, la mode, le design de produits, la typographie, les arts graphiques, le cinéma, le théâtre et la joaillerie. La collection graphique, conçue comme modèle

"Campo" magazine. Works as a photographer and graphic designer for various concerns, including Olivetti, Pirelli, La Rinascente, IBM, Cinzano and the publishing houses Mondadori and Einaudi. 1930: opens a graphic design studio with Ricardo Ricas. 1939: works for "Tempo" weekly. 1948: founds the artists' group mac (movimento arte concreta) in Milan. 1948: starts working as a designer for industry for companies which include Danese, Pigomma, Tre-A and Robots. 1970–71: teaches design at Harvard University in Cambridge, Massachusetts. 1984: designs posters for Campari. – *Publications include:* "I libri Munari", Milan 1945; "Libro illegibile", Milan 1949; "Bruno Munari's ABC", New York 1960; "Alfabetiere secondo il metodo attivo", Turin 1960; "Good design", Milan, New York 1963; "Design e communicazione visive", Bari 1968; "Alfabetiere", Turin 1972. A. Tranchis "L'arte anomalia di Bruno Munari", Bari 1981.

**Museum für Gestaltung,** Zurich. 1875: the museum is founded – 1878: the Kunstgewerbeschule is opened in partnership with the Kunstgewerbemuseum, which has numerous exhibitions and publications on various topics pertaining to aesthetics, architecture, design, everyday life and culture and visual communication. 1984: renamed the Schule und Museum für Gestaltung. 1986: the museum's design collection starts to procure 20th-century products manufactured in series. Its collection comprises c. 7,000 objects and 20,000 packaging exhibits. Its public library has amassed c. 70,000

ter und Schmuck gesammelt. Die Grafische Sammlung, die als Beispielsammlung für die angegliederte Schule begonnen wurde, umfaßt Künstlerbücher, Fotografie und Gebrauchsgrafik. Die Plakatsammlung umfaßt 250.000 Objekte.

**Museum für Kunst und Gewerbe,** Hamburg – 1869 auf Initiative Justus Brinckmanns gegründet. Die Sammlungen umfassen über 200.000 Objekte. Vorhanden sind Bestände aus China, Japan,der griechisch-römischen Antike, dem vorderen Orient mit islamischer Kunst, aus Europa vom Mittelalter bis zur Gegenwart, der Volkskunst, der angewandten Grafik und Plakatsammlung und der Fotografie. In zahlreichen Sonderausstellungen gibt es ein weitgefächertes Programm, im Bereich des Grafik-Design u.a. „Neville Brody" (1989), „Ott+Stein" (1993), „David Carson" (1996).

**Museum für Kunsthandwerk,** Frankfurt am Main – 1877 Gründung des Museums als private Initiative Frankfurter Bürger für alle Gebiete der angewandten Kunst. Die Sammlungsgebiete sind: die Europäische Abteilung (Kunsthandwerk vom Mittelalter bis zur Gegenwart), der Vordere Orient (9.–19. Jahrhundert mit islamischer Kunst und Kunsthandwerk), Ostasien (Kunst und angewandte Kunst aus China und Japan vom Neolithikum bis zur Moderne), Buchkunst und Grafik (vom Mittelalter bis zur Gegenwart). 1991 Bildung einer Design-Abteilung und Aufbau einer Design-Sammlung. Ausstellungen u.a. „Uwe Loesch" (1991), „Günther Kieser" (1995).

**Muzika,** František – geb. 26.6.1900 in

pour l'école qui lui est adjointe, réunit des livres d'artistes, des photographies et des graphismes industriels. La collection d'affiche regroupe 250 000 œuvres.

**Museum für Kunst und Gewerbe** (musée des arts et métiers), Hambourg – fondé en 1869 sur l'initiative de Justus Brinkmann. Les collections contiennent plus de 200000 pièces. On y trouve des objets chinois et japonais, et des objets datant de l'antiquité grecque et romaine, venant du Moyen-Orient, de l'art de l'Islam, de l'Europe médiévale jusqu'à nos jours, des arts populaires, du graphisme appliqué et une collection d'affiches et de photographies. Nombreuses expositions temporaires dans le cadre d'un programme très diversifié dans le domaine des arts graphiques, par ex. «Neville Brody» (1989), «Ott + Stein» (1993), «David Carson» (1996).

**Museum für Kunsthandwerk** (musée de l'artisanat d'art), Francfort-sur-le-Main – ce musée destiné à tous les domaines des arts appliqués a été fondé en 1877 en tant qu'institution privée, sur l'initiative de citoyens de Francfort. La collection est répartie en plusieurs sections : l'Europe (artisanat d'art du Moyen-âge jusqu'à nos jours), le Moyen Orient (9–19è siècle, art d'Islam et artisanat d'art), l'Extrême Orient (arts et arts appliqués chinois et japonais du néolithique aux temps modernes), les Arts du livre et arts graphiques (du Moyen-âge jusqu'à nos jours). Constitution d'une section de design et d'une collection de design depuis 1991. Expositions entre autres «Uwe Loesch» (1991) et «Günther Kieser» (1995).

**Muzika,** František – né le 26.6.1900 à Prague, Tchécoslovaquie, décédé le 1.11.1974 à Prague, Tchécoslovaquie – *peintre, typographie, graphiste, scénographe, enseignant* – 1918–1924, études à l'académie des beaux-arts de Prague.

Museum für Kunst und Gewerbe 1981 Page

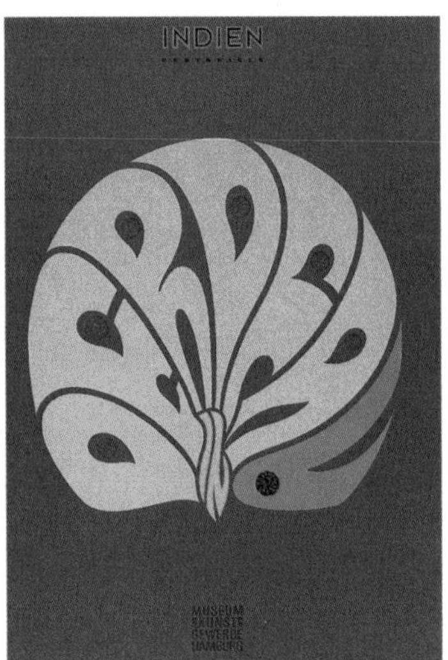

Museum für Kunst und Gewerbe 1991 Invitation Card

Museum für Kunsthandwerk 1996 Poster

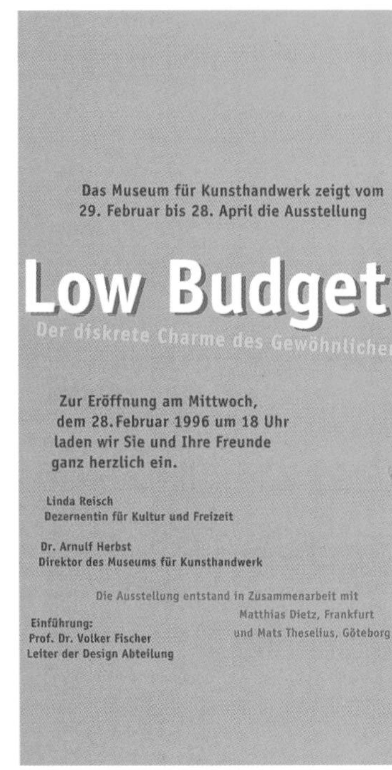

Museum für Kunst und Gewerbe 1996 Invitation Card

texts on art, architecture, fashion, product design, typography, graphics, film, theater and jewelry. The graphics collection, which originated as a collection of study samples for the affiliated art college, has many artist's books, photographs and examples of applied graphics. The poster collection numbers 250,000 items.

**Museum für Kunst und Gewerbe,** Hamburg – 1869: founded on the initiative of Justus Brinckmann. The collection houses over 200,000 objects, including exhibits from China, Japan, Ancient Greece and Rome, the Near East (with examples of Islamic art), Europe (from the Middle Ages to the present day), and examples of folk art, applied graphics, posters and photography. Numerous exhibitions cover a broad range of topics; recent graphic design exhibitions have been devoted to Neville Brody (1989), Ott + Stein (1993) and David Carson (1996).

**Museum für Kunsthandwerk,** Frankfurt am Main – 1877: the museum is founded as a private enterprise by the citizens of Frankfurt for all areas of applied art. The collection covers the following areas: Europe (arts and crafts from the Middle Ages to the present day), the Near East (9th – 19th century with examples of Islamic art and arts and crafts), Eastern Asia (art and applied art from China and Japan from the Neolithic period to the present day) and book art and graphics (from the Middle Ages to the present day). 1991: a design department is installed for which design exhibits are gradually procured. Exhibitions include "Uwe Loesch"

Prag, Tschechoslowakei, gest. 1. 11. 1974 in Prag, Tschechoslowakei. – *Maler, Typograph, Grafiker, Bühnenbildner, Lehrer* – 1918–24 Studium an der Akademie der Künste in Prag. 1921 Mitglied der Künstlergruppe „Devětsil". 1922 Eintritt in die Künstlergruppe „Nová Skupina". 1924–25 Studienaufenthalt an der Académie des Beaux-Arts in Paris. Danach Typograph und Ausstellungsorganisator verschiedener Verlage in Prag. Seit 1929 Bühnen- und Kostümentwürfe für die Theater in Brno und Prag. 1945–70 Professor an der Hochschule für Kunstgewerbe in Prag für Kalligraphie und Grafische Künste. 1949 Auszeichnung mit dem Staatspreis, 1966 mit dem Verdienstkreuz „Orden der Arbeit". – *Publikationen u. a.:* „Die schöne Schrift" (2 Bände), Prag 1965. J. Pečirka „František Muzika", Prag 1947; F. Smejkal „František Muzika, Kresby, scénická knizní tvorba", Prag 1984.

Muzika 1948 Cover

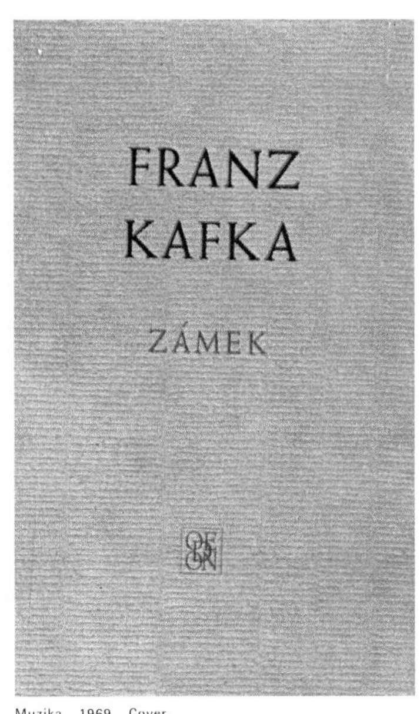

Muzika 1969 Cover

Membre du groupe d'artistes Devětsil en 1921. Fait partie du groupe d'artistes «Nová Skupina» en 1922. Séjour d'études à l'Académie des Beaux-Arts de Paris de 1924 à 1925. Exerce ensuite comme typographe et organise des expositions pour plusieurs maisons d'éditions praguoises. Depuis 1929 création des scénarios et des costumes pour des théâtres à Brno et à Prague. 1945–1970, professeur à l'école des arts décoratifs de Prague, enseigne la calligraphie et les arts graphiques. Obtient le Grand prix d'Etat en 1949 et la Croix du mérite du travail en 1966. – *Publications, sélections:* «Die schöne Schrift» (2 vol.), Prague 1965. J. Pečirka «František Muzika», Prague 1947; F. Smejkal «František Muzika, Kresby, scénická knizní tvorba», Prague 1984.

(1991) and "Günther Kieser" (1995).

**Muzika,** František – b. 26. 6. 1900 in Prague, Czechoslovakia, d. 1. 11. 1974 in Prague, Czechoslovakia – *painter, typographer, graphic artist, set-designer, teacher* – 1918–24: studies at the Academy of Arts in Prague. 1921: member of the artists' group Devětsil. 1922: joins the artists' group Nová Skupina. 1924–25: studies at the Académie des Beaux-Arts in Paris. Then works as a typographer and exhibition organizer for various publishers in Prague. From 1929 onwards: designs sets and costumes for the theaters in Brno and Prague. 1945–70: professor of calligraphy and the graphic arts at the College for Handicrafts in Prague. 1949: is awarded the State Prize and in 1966 an order of merit, a "Work Medal". – *Publications include:* "Die schöne Schrift" (2 vols.), Prague 1965. J. Pečirka, "František Muzika", Prague 1947; F. Smejkal "František Muzika, Kresby, scénická knizní tvorba", Prague 1984.

# N

**Nannucci,** Maurizio – geb. 20. 4. 1939 in Florenz, Italien. – *Künstler* – 1959–62 Studium an den Akademien in Berlin und Florenz. Mitglied der Gruppe „Studio di musica elettronica" in Florenz. Seit 1962 konkrete Poesie, 1965 elektronische Musik, akustische Untersuchungen. 1966–68 Theatermaler in Florenz, Catania und Bratislava. Seit 1966 Arbeiten mit Neonlicht, Verbindung von Schrift und Farbe. 1967 erste Einzelausstellung in Triest. 1970 mit dem Preis „Künstler der Stadt" in Florenz ausgezeichnet. – *Publikationen u.a.:* „Play Text" (mit Jochen Gerz), Paris 1968; „Universum", Florenz 1969; „Nomenclature", Oldenburg 1972; „Poem, Poesie", Neapel 1975; „To cut a long story short", Middelburg 1982; „Sometexts 1962–1972", Frankfurt 1982.

**Nebiolo & Companie,** Turin – *Schriftgießerei, Druckmaschinenhersteller* – Giovanni Nebiolo erwirbt 1878 die Schriftgießerei von G. Narizzano in Turin. Nach Eintritt der Brüder Lazzaro und Giuseppe Levi in das Unternehmen, 1880 Umbenennung in Nebiolo & Companie. 1890 Beginn mit der Herstellung von Druckmaschinen. 1899 Umwandlung in eine Aktiengesellschaft. Zusammenschluß mit der Firma Urania. 1908 Umbenennung in Augustea. Übernahme von Schriftgießereien. 1916 Umbenennung in Nebiolo. Aldo Novarese wird 1952 künstlerischer Leiter. 1978 Übernahme der Maschinenfabrikation durch Fiat. Die Schriftgießerei wird von Italiana Caratteri in Bologna und Turin übernommen.

**Neuburg,** Hans – geb. 20. 3. 1904 in Grulich, Österreich-Ungarn (heute Tschecho-

**Nannucci,** Mauricio – né le 20. 4. 1939 à Florence, Italie – *artiste* – 1959–1962, études aux académies des beaux-arts de Berlin et de Florence. Membre du groupe «Studio di musica elettronica» à Florence. Poésie concrète à partir de 1962, puis musique électronique et expérimentations acoustiques en 1965. Peintre de décors de théâtre à Florence, Catania et Bratislava de 1966 à 1968. Depuis 1966, il utilise des néons associant l'écriture et la couleur. Première exposition personnelle à Trieste en 1967. Prix de «L'artiste de la ville» de Florence en 1970. – *Publications, sélection:* «Play Text» (avec Jochen Gerz), Paris 1968; «Universum», Florence 1969; «Nomenclature», Oldenburg 1972; «Poem, Poesie», Naples 1975; «To cut a long story short», Middelburg 1982; «Sometexts 1962–1972», Francfort 1982.

**Nebiolo & Companie,** Turin – *fonderie de caractères, fabriquants de presses mécaniques* – en 1878, Giovanni Nebiolo acquiert la fonderie de caractères de G. Narizzano à Turin. En 1880, après l'entrée des frères Lazzaro et Guiseppe Levi dans l'entreprise, celle-ci prend le nom de Nebiolo & Companie. En 1890, elle commence à fabriquer des machines pour imprimeries. L'entreprise est transformée en société anonyme en 1899. Fusion avec la société Urania. Elle devient Augustea en 1908. Reprise de fonderies de caractères. S'appelle de nouveau Nebiolo en 1916. En 1952, Aldo Novarese devient directeur artistique. En 1978, Fiat reprend la fabrication des machines. La fonderie de caractères a été rachetée par Italiana Caratteri de Bologne et de Turin.

**Neuburg,** Hans – né le 20. 3. 1904 à Grulich, Autriche-Hongrie (aujourd'hui Tchécoslovaquie), décédé le 24. 6. 1983 à Zurich, Suisse – *graphiste maquettiste, typographe, auteur, peintre, enseignant* – apprentissage à la Orell Füssli AG à Zu-

Nannucci 1992

Nannucci 1969 Poem

Nannucci 1993

Nannucci 1975

Neuburg 1944 Poster

**Nannucci,** Mauricio – b. 20. 4. 1939 in Florence, Italy – *artist* – 1959–62: studies at the academies in Berlin and Florence. Member of the Studio di musica elettronica group in Florence. 1962: starts producing concrete poetry and in 1965 electronic music and investigations into acoustics. 1966–68: theater artist in Florence, Catania and Bratislava. 1966: starts working with neon light, combining text and color. 1967: first solo exhibition in Trieste. 1970: receives Florence's Artist of the City prize. – *Publications include:* "Play Text" (with Jochen Gerz), Paris 1968; "Universum", Florence 1969; "Nomenclature", Oldenburg 1972; "Poem, Poesie", Naples 1975; "To cut a long story short", Middelburg 1982; "Sometexts 1962–1972", Frankfurt 1982.

**Nebiolo & Companie,** Turin – *type foundry, printing machine manufacturers* – 1878: Giovanni Nebiolo acquires G. Narizzano's type foundry in Turin. 1880: the brothers Lazzaro and Giuseppe Levi join the company which is subsequently renamed Nebiolo & Companie. 1890: the company starts manufacturing printing machines. 1899: the business becomes a joint-stock company. Nebiolo & Co. merge with the Urania company. 1908: the enterprise is renamed Augustea and takes over various type foundries. 1916: the enterprise is renamed Nebiolo. 1952: Aldo Novarese is made art director. 1978: Fiat takes over machine production. The type foundry is taken over by Italiana Caratteri in Bologna and Turin.

**Neuburg,** Hans – b. 20. 3. 1904 in Grulich, Austria-Hungary (today Czechoslovakia),

slowakei), gest. 24. 6. 1983 in Zürich, Schweiz. – *Grafik-Designer, Typograph, Autor, Maler, Lehrer* – 1919–22 Lehre bei der Orell Füssli AG in Zürich. 1928–29 Texter in der Werbeagentur Max Dalang in Zürich. 1930–31 freier Gestalter, Arbeiten u. a. mit Max Bill, Herbert Matter, Anton Stankowski. 1932–36 Werbeleiter bei Jean Haecki Import in Basel. 1933–37 Redakteur der Zeitschrift „Industriewerbung" in Basel. Seit 1936 freier Grafiker in Zürich. Auftraggeber waren u. a. die Mustermesse Basel und das Internationale Rote Kreuz. 1952–53 Redakteur der Zeitschrift „Camera". 1958–65 Mitbegründer und -herausgeber der Zeitschrift „Neue Grafik" (mit R. P. Lohse, J. Müller-Brockmann, C. Vivarelli). 1962–64 Direktor des Gewerbemuseums in Winterthur. 1963 Lehrtätigkeit an der Hochschule für Gestaltung in Ulm, 1971 an der Carlton University, School of Industrial Design in Ottawa, Kanada. – *Publikationen u. a.:* „Moderne Werbe- und Gebrauchsgraphik", Ravensburg 1960; „Richard Paul Lohse: 60. Geburtstag" (Hrsg.), Teufen 1962; „50 anni di Grafica costruttiva", Mailand 1983. Max Bill u. a. „Hans Neuburg", Teufen 1964.

**Neudörffer,** Johann der Ältere – geb. 1497 in Nürnberg, Deutschland, gest. 1563 in Nürnberg, Deutschland. – *Schreibmeister, Rechenmeister* – Schüler des Schreib- und Rechenmeisters Caspar Schmidt und des Kanzlisten Paulus Vischer. Veröffentlicht 1519 eine in Holzschnitt ausgeführte Schreibfibel „Fundament", das erste deutsche Vorlagenbuch. 1522–27 Schnitt der „Neudörffer-An-

Neuburg 1948 Advertisement

Neuburg 1967 Poster

Neudörffer 1601 Page

rich de 1919 à 1922. Préparateur de textes à l'agence publicitaire Max Dalang à Zurich de 1928 à 1929. Concepteur graphiste indépendant de 1930 à 1931 ; travaille, entre autres, avec Max Bill, Herbert Matter et Anton Stankowski. 1932–1936, directeur de la publicité chez Jean Haecki Import à Bâle. 1933–1937, rédacteur de la revue «Industriewerbung» à Bâle. Graphiste indépendant à Zurich à partir de 1936 ; commanditaires : la Mustermesse Bâle et le Croix rouge internationale, etc. 1952–1953, rédacteur de la revue «Camera». Cofondateur, en 1958, de la revue «Neue Grafik» qu'il édite jusqu'en 1965 avec R. P. Lohse, J. Müller-Brockmann et C. Vivarelli. 1962–1964, directeur du Gewerbemusem (musée des arts et métiers) de Winterthur. A enseigné à la Hochschule für Gestaltung (école supérieure de design) d'Ulm en 1963, à la Carlton University, School of Industrial Design à Ottawa, Canada, en 1971. – *Publications, sélection :* «Moderne Werbe- und Gebrauchsgraphik», Ratisbonne 1960 ; «Richard Paul Lohse : 60. Geburtstag» (éd.), Teufen 1962 ; «50 anni di Grafica costruttiva», Milan 1983. Max Bill et autres «Hans Neuburg», Teufen 1964.

**Neudörffer,** Johann l'ancien – né en 1497 à Nuremberg, Allemagne, décédé en 1563 à Nuremberg, Allemagne – *scribe, comptable* – élève de Caspar Schmidt, maître scribe et comptable à la chancellerie de Paulus Vischer. En 1519, il publie le premier livre de modèles de caractères d'Allemagne, «Fundament», un manuel d'écriture réalisé en gravure sur bois. De 1522 à 1527, il taille la «Neudörffer-An-

d. 24. 6. 1983 in Zurich, Switzerland – *graphic designer, typographer, author, painter, teacher* – 1919–22: trains at Orell Füssli AG in Zurich. 1928–29: copywriter for the Max Dalang advertising agency in Zurich. 1930–31: freelance designer. Works with Max Bill, Herbert Matter and Anton Stankowski, among others. 1932–36: advertising manager for Jean Haecki Import in Basle. 1933–37: editor of "Industriewerbung" magazine in Basle. 1936: starts freelance graphic work in Zurich. Clients include the Mustermesse Basle and the International Red Cross. 1952–53: editor of "Camera" magazine. 1958–65: co-founder and co-editor of "Neue Grafik" magazine with R. P. Lohse, J. Müller-Brockmann and C. Vivarelli. 1962–64: director of the Gewerbemuseum in Winterthur. 1963: teaches at the Hochschule für Gestaltung in Ulm and in 1971 at Carlton University, School of Industrial Design in Ottawa, Canada. – *Publications include:* "Moderne Werbe- und Gebrauchsgraphik", Ravensburg 1960; "Richard Paul Lohse: 60. Geburtstag" (ed.), Teufen 1962; "50 anni di Grafica costruttiva", Milan 1983. Max Bill et al "Hans Neuburg", Teufen 1964.

**Neudörffer,** Johann, the elder – b. 1497 in Nuremberg, Germany, d. 1563 in Nuremberg, Germany – *writing master, arithmetic master* – Pupil of writing and arithmetic master Caspar Schmidt and the clerk Paulus Vischer. 1519: publishes a writing primer as a woodcut entitled "Fundament", the first German copybook. 1522–27: Nuremberg form cutter Hieronymus Andreä cuts Neudörffer-Andreä-

dreä-Fraktur" in fünf Graden durch den Nürnberger Formschneider Hieronymus Andreä. Diese Schrift gilt als früheste und reinste Frakturschrift und als Ausgangsschrift für alle späteren Schriften. Schreibt 1526 in Albrecht Dürers Werkstatt unter die Gemälde der vier Apostel einen mehrzeiligen religiösen Text in Kurrent. 1538 Veröffentlichung des Schreibbuchs „Anweysung zur gemeynen Handschrift". 1549 „Gesprechbüchlein zweyer Schüler", ein „Initialenbuch". Durch den Enkel Anton Neudörffer wird 1599 eine „Arithmetik" von Johann Neudörffer dem Älteren gedruckt, die in mehreren Auflagen erscheint.

**Neue Sammlung,** München – *Staatliches Museum für angewandte Kunst* – 1925–29 Gründung des staatlichen Instituts „Die Neue Sammlung" als Abteilung für Gewerbekunst des Bayerischen Nationalmuseums in München. Seit 1926 zahlreiche Ausstellungen mit Objekten der eigenen Sammlung aus den Bereichen Keramik, Metalle, Plakate, Bücher, Stoffe und Glas, z. B. „Ehmcke und der Ehmcke-Kreis" (1928), „Deutsche Schrift und ihre Entwicklung" (1933), „Der Heraldiker des Deutschen Reiches Otto Hupp" (1934), „Schweizer Plakate" (1949), „Herbert Bayer" (1956), „Max Bill" (1957), „Peter Behrens" (1958), „Hochschule für Gestaltung Ulm" (1964), „Warenplakate" (1981), „Raymond Savignac" (1982).

**Neumann,** Eckhard – geb. 15. 4. 1933 in Königsberg, Ostpreußen, Polen. – *Grafik-Designer, Autor, Lehrer* – 1956–57 Studium der Werbung und des Grafik-Design an der Werbefachschule in Berlin und an

dreä-Fraktur» en cinq corps pour Hieronimus Andreä à Nuremberg. Cette fonte est considérée comme la Fraktur la plus ancienne et la plus pure; elle a inspiré celles qui lui succéderont. En 1526, dans l'atelier de Dürer, il écrit un texte religieux de plusieurs lignes en caractères manuscrits sous un tableau représentant quatre apôtres. Il publie en 1538 le manuel d'écriture «Anweysung zur gemeynen Handschrift», puis le «Gesprechbüchlein zweyer Schüler», un «Initialenbuch» (livre d'initiales), en 1549. Une «Arithmetik» de Johann Neudörffer l'ancien est imprimée en 1599 par son petit-fils Anton Neudörffer, celle-ci sera éditée plusieurs fois.

**Neue Sammlung,** Munich – *musée d'Etat pour les arts appliqués* – l'institut d'Etat «Die Neue Sammlung» a été fondée entre 1925 et 1929 en tant que département des arts et métiers du Bayerische Nationalmuseum de Munich. Il organise de nombreuses expositions de céramiques, objets en métal, affiches, livres, tissus et travaux de verrerie de la collection à partir de 1926, ainsi que «Ehmcke und der Ehmcke-Kreis» (Ehmcke et le cercle Ehmcke) (1928), «Deutsche Schrift und ihre Entwicklung» (L'écriture allemande et son évolution) (1933), «Der Heraldiker des Deutschen Reiches Otto Hupp» (Héraldique du Reich, Otto Hupp) (1934), «Schweizer Plakate» (Affiches suisses) (1949), «Herbert Bayer» (1956), «Max Bill» (1957), «Peter Behrens» (1958), «Hochschule für Gestaltung Ulm» (école supérieure de design d'Ulm) (1964), «Warenplakate» (Affiches de produits) (1981), «Raymond Savignac» (1982).

**Neumann,** Eckhard – né le 15. 4. 1933 à Königsberg, Prusse orientale, Pologne – *graphiste maquettiste, auteur, enseignant* – étudie la publicité et la conception graphique à la Werbefachschule (école de pu-

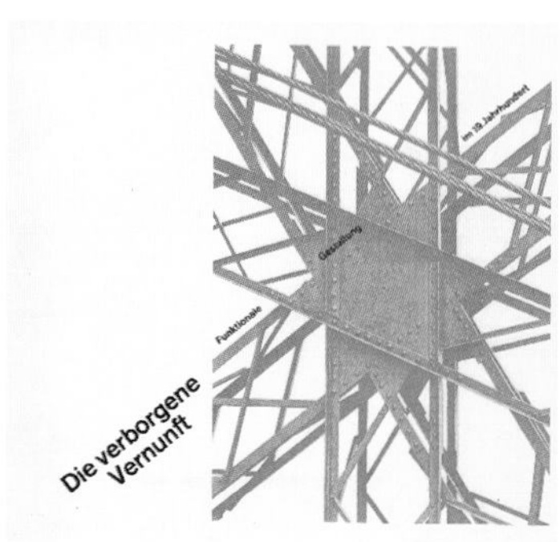

Neue Sammlung 1971 Cover

Neue Sammlung 1993 Poster

Neue Sammlung 1997 Poster

Fraktur in five sizes. This typeface is generally considered to be the first and the purest of all black letter typefaces and the model for all later faces of this kind. 1526: writes several lines of religious text in gothic handwriting under the paintings of the four apostles in Albrecht Dürer's workshop. 1538: publishes the writing manual "Anweysung zur gemeynen Handschrift". 1549: publishes "Gesprechbüchlein zweyer Schüler", a book of initials ("Initialenbuch"). 1599: grandson Anton Neudörffer prints an "Arithmetik" manual by Johann Neudörffer der Ältere, which is reprinted several times.

**Neue Sammlung,** Munich – *state museum of applied art* – 1925–29: Die Neue Sammlung state institute is founded as a commercial art department of the Bayerische Nationalmuseum in Munich. From 1926 onwards: numerous exhibitions display ceramics, metalwork, posters, books, fabrics and glass exhibits from the museum's own collection, including "Ehmcke und der Ehmcke-Kreis" (1928), "Deutsche Schrift und ihre Entwicklung" (1933), "Der Heraldiker des Deutschen Reiches Otto Hupp" (1934), "Schweizer Plakate" (1949), "Herbert Bayer" (1956), "Max Bill" (1957), "Peter Behrens" (1958), "Hochschule für Gestaltung Ulm" (1964), "Warenplakate" (1981) and "Raymond Savignac" (1982).

**Neumann,** Eckhard – b. 15. 4. 1933 in Königsberg, East Prussia, Poland – *graphic designer, author, teacher* – 1956–57: studied advertising and graphic design at the Werbefachschule in Berlin and at the Hochschule für Gestaltung in Ulm.

der Hochschule für Gestaltung in Ulm. 1957–71 Werbeleiter der Swissair Deutschland Direktion in Frankfurt. 1973–75 Leiter der Kommunikations-Gestaltung der Braun AG in Kronberg. 1975–85 Leiter der Design-Promotion beim Rat für Formgebung (German Design Council) in Darmstadt. 1985–88 Dozent für Typographie und Design-Geschichte an der Städtischen Fachhochschule für Gestaltung in Mannheim. Seit 1988 Professor für Kommunikationsdesign an der Fachhochschule Mannheim, Hochschule für Technik und Gestaltung. Gründer und Mitherausgeber des „Jahrbuchs der Werbung" (Düsseldorf 1964–91). Durch seine Ausstellungskonzepte und Publikationen trägt Neumann zur neuen Rezeption der Typographie der 20er und 30er Jahre bei. – *Publikationen u.a.:* „Functional Graphic Design in the 20s", New York 1969; „Bauhaus and Bauhaus People", New York 1970; „Herbert Bayer: Kunst und Design in Amerika 1938–1985", Berlin 1985; „Xanti Schawinsky: Foto", Bern 1989.

**Neurath,** Otto – geb. 10. 12. 1882 in Wien, Österreich, gest. 22. 12. 1945 in Oxford, England. – *Soziologe, Philosoph, Ökonom, Lehrer* – 1901–05 Studium an den Universitäten in Wien und Berlin. Seit 1907 Lehr- und Verwaltungstätigkeiten. 1924–34 Gründung und Leitung des „Gesellschafts- und Wirtschaftsmuseums Wien". Seit 1925 Entwicklung der „Wiener Methode der Bildstatistik", die statistische Fakten in Schaubilder umsetzt. 1934 Emigration nach Den Haag. Gründung des „Mundaneum Instituut Den

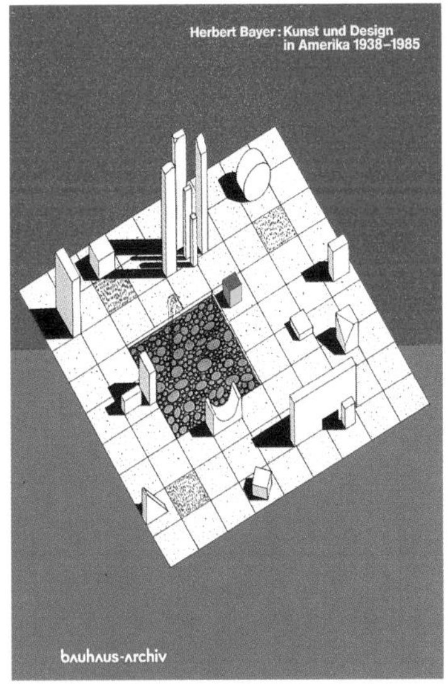

Neumann 1976 Cover

Neumann 1976 Cover



Neumann 1976 Poster

blicité) de Berlin et à la Hochschule für Gestaltung (école supérieure de design) d'Ulm de 1956 à 1957. Directeur de la publicité à la direction de Swissair Allemagne, à Francfort, de 1957 à 1971. Directeur du service design et communication de la Braun AG à Kronberg de 1973 à 1975. Directeur de la promotion du design au conseil allemand pour le désign à Darmstadt de 1975 à 1985. Enseigne la typographie et l'histoire du design de 1985 à 1988 à la Städtische Fachhochschule für Design (école supérieure de design) de Mannheim. Professeur de design et de communication à la Fachhochschule de Mannheim, Hochschule für Technik und Gestaltung (école supérieure des techniques et du design) depuis 1988. Fondateur et coéditeur du «Jahrbuch der Werbung» (annales de la publicité) (Düsseldorf 1964–1991). Par ces conceptions d'expositions et ses publications, Neumann contribue à une nouvelle réception de la typographie des années 20 et 30. – *Publications, sélection :* «Functional Graphic Design in the 20s», New York 1969; «Bauhaus and Bauhaus People», New York 1970; «Herbert Bayer : Kunst und Design in Amerika 1938–1985», Berlin 1985; «Xanti Schawinsky : Foto», Berne 1989.

**Neurath,** Otto – né le 10. 12. 1882 à Vienne, Autriche, décédé le 22. 12. 1945 à Oxford, Angleterre – *sociologue, philosophe, économiste, enseignant* – 1901–1905, études dans les universités de Vienne et de Berlin. Emplois dans l'administration et dans l'enseignement à partir de 1907. Fonde le «Gesellschafts- und Wirtschaftsmuseum» (musée de la société et de l'économie) à Vienne en 1924 et le dirige jusqu'en 1934. Travaille à partir de 1925 à l'élaboration de la «Méthode viennoise de visualisation statistique» qui transpose des données statistiques en schémas.

1957–71: advertising manager for Swissair Deutschland Direktion in Frankfurt. 1973–75: head of communication design for Braun AG in Kronberg. 1975–85: design promotion manager for the German Design Council in Darmstadt. 1985–88: lecturer of typography and history of design at the Städtische Fachhochschule für Gestaltung in Mannheim. Since 1988: professor of communication design at the Fachhochschule Mannheim, Hochschule für Technik und Gestaltung. Founder and co-editor of the "Jahrbuch der Werbung" (Düsseldorf, 1964–91). Through his exhibition concepts and publications, Neumann has done much for the new acceptance of typography from the 20s and 30s. – *Publications include:* "Functional Graphic Design in the 20s", New York 1969; "Bauhaus and Bauhaus People", New York 1970; "Herbert Beyer: Kunst und Design in Amerika 1938–1985", Berlin 1985; "Xanti Schawinsky: Foto", Bern 1989.

**Neurath,** Otto – b. 10. 12. 1882 in Vienna, Austria, d. 22. 12. 1945 in Oxford, England – *sociologist, philosopher, economist, teacher* – 1901–05: studies at the universities in Vienna and Berlin. 1907: starts working in teaching and administration. 1924–34: founds and runs the Gesellschafts- und Wirtschaftsmuseum in Vienna. From 1925 onwards: develops the Viennese Method of Pictorial Statistics which turns statistical facts into graphs. 1934: emigrates to The Hague. Founds the Mundaneum Instituut Den Haag for the development of visual work. 1935: renames his Viennese Method of

Haag" zur Entwicklung visueller Arbeiten. 1935 Umbenennung der „Wiener Methode der Bildstatistik" in „Isotype" (International System of Typographic Picture Education). 1940 Flucht nach England, Internierung. 1941 Fortsetzung der bildpädagogischen Arbeit in Oxford. Gründung des „Isotype Institute" mit seiner Frau, der Wissenschaftlerin und Mathematikerin Marie (geb. Reidemeister). Nach Beendigung der Produktionstätigkeit des „Isotype Institute", 1971 Einrichtung der „Otto und Marie Neurath Isotype-Sammlung" an der Universität Reading. – *Publikationen u.a.:* „Bildstatistik: Führer durch die Ausstellung des Gesellschafts- und Wirtschaftsmuseums in Wien", Leipzig 1927; „Bildstatistik nach Wiener Methode in der Schule", Wien, Leipzig 1933; „International Picture Language – the first rules of Isotype", London 1936; „Basic by Isotype", London 1937. Michael Twyman „The Significance of Isotype", Reading 1975; Marie Neurath (Hrsg.) „Otto Neurath, philosophical papers 1913–1946", Dordrecht 1983.

**Nicholas,** Robin – geb. 25. 4. 1947 in Westerham, Kent, England. – *Schriftentwerfer* – Ausbildung zum Ingenieur. 1965 Eintritt in das „Monotype Type Drawing Office". Seit 1982 Leiter der Schriftzeichenabteilung von Monotype. Entwurf einer serifenlosen Schrift für bit map font-Laserdrucker. Seit 1992 Entwicklungsleiter von Monotype Typography Ltd. und Leiter des Monotype's Typographical Custom Design Service. 1994 Entwurf der Schrift „Plantin Headline" für die Zeitung „Yorkshire Evening Post".

Neurath 1936 Cover

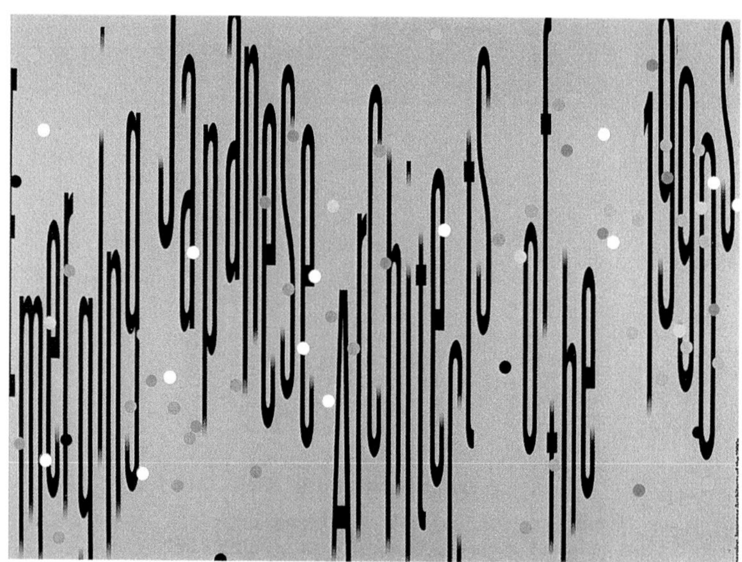

Niijima 1990 Poster

ISOTYPE

International
System
Of
TYpographic
Picture
Education

12

THE QUESTION OF AN
INTERNATIONAL LANGUAGE

The desire for an international language is an old one, and it is more than ever in men's minds at this time of international connections in business and science. But 'debabelization' is a very hard and complex work. The attempt to make one international language has given us a parcel of new languages. The best way out seems to be the use of instruments which are, or have become, international. For this reason this book is in Basic English, because this international language is part of an old language in general use.

The question of an international language has now become important. There are a number of signs pointing to a great development of international organization in the near future—though we are living in a time of warring interests and broken connections. Any work done on the question of international languages—with a view to making a word language, or

13

Neurath 1936 Spread

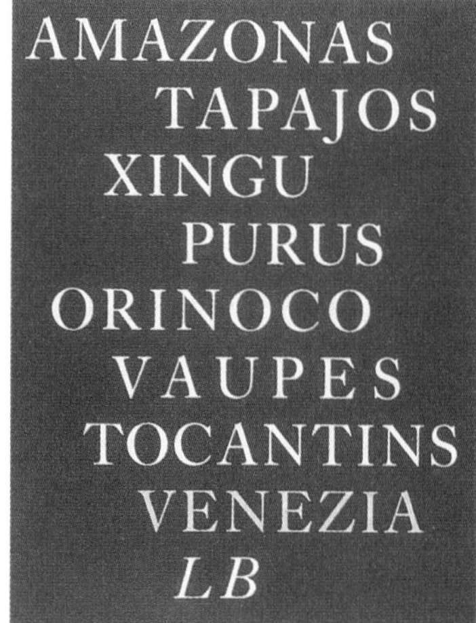

AMAZONAS
TAPAJOS
XINGU
PURUS
ORINOCO
VAUPES
TOCANTINS
VENEZIA
LB


Nikkels 1984 Cover

Emigre à La Haye en 1934. Il fonde le «Mundaneum Instituut Den Haag» pour le développement du travail visuel. En 1935, la «Méthode viennoise de visualisation statistique» est appelée «Isotype» (International System of Typographic Picture Education). Se réfugie en Angleterre en 1940, internement. Poursuit des travaux sur la pédagogie de l'image à Oxford en 1941, il y fonde l' «Isotype Institute» avec sa femme, la mathématicienne Marie (née Reidemeister). En 1971, à la fin des activités de l'»Isotype Institute», la «Collection Isotype Otto et Marie Neurath» s'installe à l'université de Reading. – *Publications, sélection:* «Bildstatistik: Führer durch die Ausstellung des Gesellschafts- und Wirtschaftsmuseums in Wien», Leipzig 1927; «Bildstatistik nach Wiener Methode in der Schule», Vienne, Leipzig 1933; «International Picture Language – the first rules of Isotype», Londres 1936; «Basic by Isotype», Londres 1937. Michael Twyman «The Significance of Isotype», Reading 1975; Marie Neurath (éd.) «Otto Neurath, philosophical papers 1913–1946», Dordrecht 1983.

**Nicholas,** Robin – né le 25. 4. 1947 à Westerham, Kent, Angleterre – *concepteur de polices* – formation d'ingénieur. Entre en 1965 au «Monotype Type Drawing Office». Dirige le service des polices chez Monotype à partir de 1982 et conception d'une police sans sérif pour l'imprimante Laser bit map font. Directeur du développement des polices de la Monotype Typography Ltd à partir de 1992 et directeur de Monotype's Typographical Custom Design Service. En 1994, il dessine la police «Plantin Headline» pour le quotidien «Yorkshire Evening Post». Nicholas a dirigé la production de la police Clarion et suivi la révision de nombreuses polices Monotype autrefois destinées à la

Pictorial Statistics "Isotype" (International System of Typographic Picture Education). 1940: flees to England where he is interned. 1941: continues his work in picture education in Oxford where he founds the Isotype Institute with his scientist and mathematician wife Marie (née Reidemeister). 1971: after the Isotype Institute stops work the Otto and Marie Neurath Isotype Collection is opened at the University of Reading. – *Publications include:* "Bildstatistik: Führer durch die Ausstellung des Gesellschafts- und Wirtschaftsmuseums in Wien", Leipzig 1927; "Bildstatistik nach Wiener Methode in der Schule", Vienna, Leipzig 1933; "International Picture Language – the first rules of Isotype", London 1936; "Basic by Isotype", London 1937. Michael Twyman "The Significance of Isotype", Reading 1975; Marie Neurath (ed.) "Otto Neurath, philosophical papers 1913–1946", Dordrecht 1983.

**Nicholas,** Robin – b. 25. 4. 1947 in Westerham, Kent, England – *type designer* – Trained as an engineer. 1965: joins the Monotype Type Drawing Office. 1982: made head of Monotype's lettering department. Designs a sans-serif typeface for bit map font laser printers. Since 1992: type development manager for Monotype Typography Ltd. and manager of Monotype's Typographical Custom Design Service. 1994: designs the Plantin Headline font for the Yorkshire Evening Post newspaper. Nicholas has supervised the production of the Clarion typeface and also the creation of many Monotype lead composition typefaces destined for

Nicholas leitet die Produktion der Schrift Clarion und die Neuschöpfungen vieler Monotype-Bleisatz-Schriften für neue Technologien u. a.: Bell, Centaur, Janson, Fournier, van Dijck, Bulmer, Walbaum. – *Schriftentwürfe:* Nimrod (1980), Arial (mit Patricia Saunders, 1982), Plantin Headline (1994).

**Niijima,** Minoru – geb. 1948 in Tokio, Japan. – *Grafik-Designer, Typograph, Autor* – 1977 Gründung seines Design-Studios in Tokio. 1980–83 Studium an der Kunstakademie der Yale University. Teilnahme an der Plakatbiennale in Warschau. Zahlreiche Texte zu gestalterischen Themen, u. a. über Odermatt & Tissi in der 5. Ausgabe der Zeitschrift „Creation". Zahlreiche Ausstellungen seiner Arbeiten in Japan und den USA.

**Nikkels,** Walter – geb. 25. 11. 1940 in Herwen en Aerdt, Niederlande. – *Typograph, Lehrer* – 1956–60 Studium an den Kunstakademien in Rotterdam und München sowie an der Accademia Brera in Mailand. Seit 1966 freischaffender Typograph. Gestaltung von Büchern und Ausstellungen. Seit 1976 typographische Gestaltung der Publikationen des Van Abbe Museums in Eindhoven (u. a. für Richard Long, Jannis Kounellis, Lothar Baumgarten, Armando) und der Kataloge des Deutschen Pavillons auf der Biennale in Venedig (Ulrich Rückriem, Anselm Kiefer, Lothar Baumgarten, A. R. Penck). 1979–82 verantwortlich für Grafik-Design und Ausstellungs-Architektur der documenta VII in Kassel. Zahlreiche Auszeichnungen in den Niederlanden und in der Bundesrepublik Deutschland, u. a. H. N. Werk-

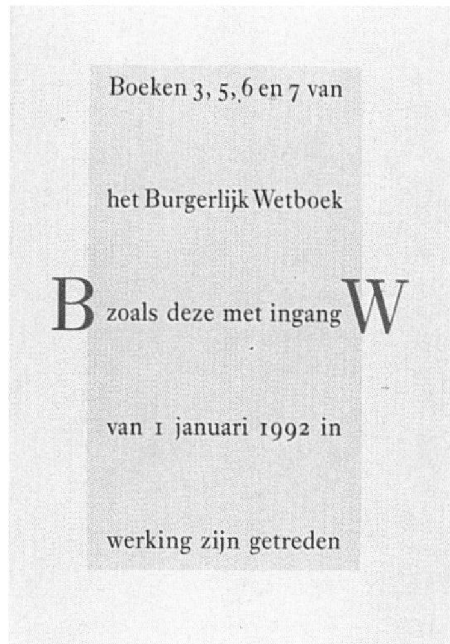


Nikkels 1992 Cover

Nikkels 1982 Logo

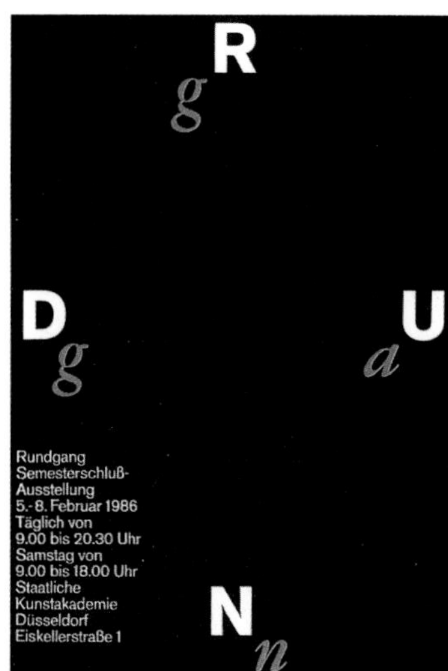


Nikkels 1986 Poster

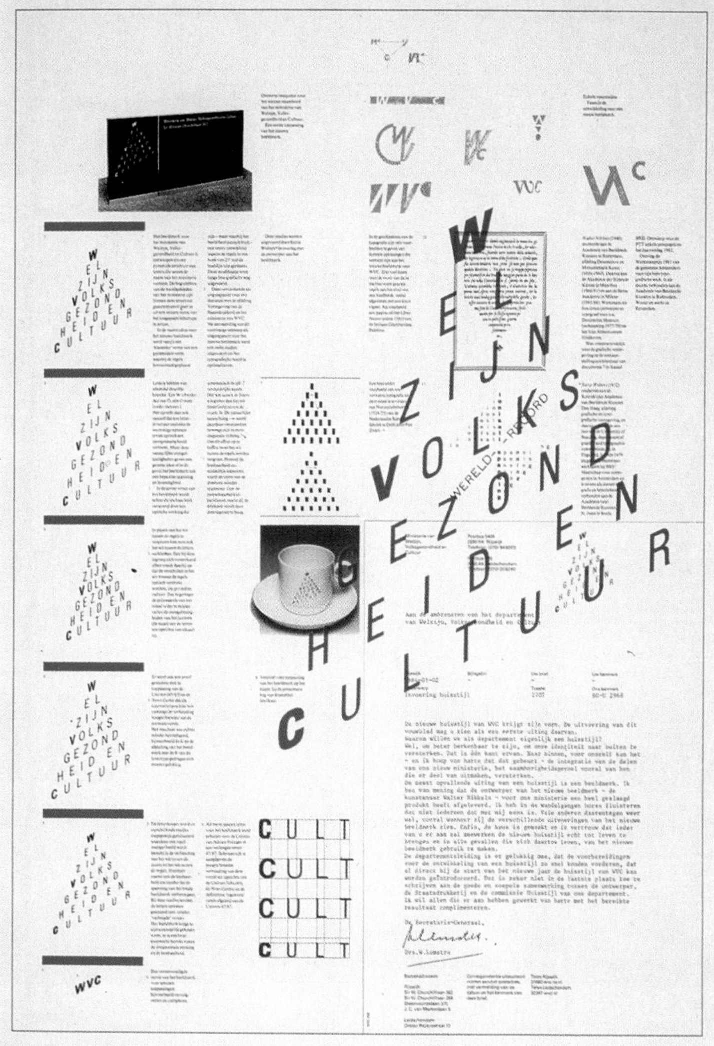


Nikkels 1984 Poster

composition au plomb pour les adapter aux nouvelles technologies; ceci concerne entre autres Bell, Centaur, Janson, Fournier, van Dijck, Bulmer, Walbaum. – *Polices:* Nimrod (1980), Arial (avec Patricia Saunders, 1982), Plantin Headline (1994).

**Niijima,** Minoru – né en 1948 à Tokyo, Japon – *graphiste maquettiste, typographe, auteur* – fonde son atelier de conception graphiste à Tokyo en 1977. Etudes à l'académie des beaux-arts de la Yale University de 1980 à 1983. Participe à la Biennale de l'affiche de Varsovie. Nombreux essais sur le design, entre autres sur Odermatt & Tissi dans numéro 5 de la revue «Creation». Nombreuses expositions de ses travaux au Japon et aux Etats-Unis.

**Nikkels,** Walter – né le 25. 11. 1940 à Herwen en Aerdt, Pays-Bas – *typographe, enseignant* – 1956–1960, études aux académies des beaux-arts de Rotterdam et de Munich ainsi qu'à la Accademia Brera de Milan. Typographe indépendant à partir de 1966. Réalise des maquettes de livres et des expositions. Mise en page typographique des publications du Van Abbe Museum d'Eindhoven à partir de 1976 (entre autres pour Richard Long, Jannis Kounellis, Lothar Baumgarten, Armando), du catalogue du pavillon allemand de la Biennale de Venise (Ulrich Rückriem, Anselm Kiefer, Lothar Baumgarten, A. R. Penck). Responsable de la conception graphique et de l'architecture d'exposition de la «documenta VII» à Kassel de 1979 à 1982. Nombreuses distinctions

new technologies, such as Bell, Centaur, Janson, Fournier, van Dijck, Bulmer and Walbaum. – *Fonts:* Nimrod (1980), Arial (with Patricia Saunders, 1982), Plantin Headline (1994).

**Niijima,** Minoru – b. 1948 in Tokyo, Japan – *graphic designer, typographer, author* – 1977: opens his own design studio in Tokyo. 1980– 83: studies at Yale University's Art Academy. Participates in the poster biennial in Warsaw. Has written numerous texts on design, including on Odermatt & Tissi in No. 5 of "Creation" magazine. There have been numerous exhibitions of his work in Japan and the USA.

**Nikkels,** Walter – b. 25. 11. 1940 in Herwen en Aerdt, The Netherlands – *typographer, teacher* – 1956– 60: studies at the art academies in Rotterdam and Munich and at the Accademia Brera in Milan. From 1966 onwards: freelance typographer. Designs books and exhibitions. From 1976 onwards: designs the typography for the publications issued by the Van Abbe Museum in Eindhoven (including for Richard Long, Jannis Kounellis, Lothar Baumgarten and Armando), and for the catalogues for the German pavilion at the Venice Biennale (Ulrich Rückriem, Anselm Kiefer, Lothar Baumgarten and A. R. Penck). 1979–82: is responsible for the graphic and exhibition design of documenta VII in Kassel. Nikkels has won numerous awards in The Netherlands and Germany, including the H. N. Werkman Prize for his work in typography (1981) and the Charles Nypels Prize (1989). Since 1985: professor of de-

man-Preis für sein typographisches Werk (1981) und Charles-Nypels-Preis (1989). Seit 1985 Professor für Entwurf, Buchgestaltung und Schriftentwicklung an der Kunstakademie Düsseldorf. Ausstellungen u. a. im Stedelijk Museum in Amsterdam und im Centre Georges Pompidou in Paris. – *Publikation u. a.:* Roelof van Gelder u. a. „Rijn, Rhein. Over het werk van Walter Nikkels", Nuth 1989.

**Noorda,** Bob – geb. 15. 7. 1927 in Amsterdam, Niederlande. – *Grafik-Designer, Industrie-Designer* – 1944–47 Studium am Instituut voor Kunstnijverheidsonderwijs in Amsterdam. 1947–50 Grafik-Designer für niederländische Regierungsdrucksachen in Djakarta, Indonesien. 1950–52 freier Grafik-Designer in Amsterdam, seit 1952 in Mailand. Unterrichtet Grafik-Design in Mailand, Venedig und Urbino. 1961–65 Art Director der Firma Pirelli. 1964 ausgezeichnet mit dem „Compasso d'Oro" für das Leitsystem der Mailänder U-Bahn. Weitere Leitsystem-Entwürfe für New York und São Paulo. 1965 Mitbegründer und Senior Vice-President von „Unimark International Design and Marketing Corporation" (mit Massimo Vignelli) in den USA und in Italien. Auftraggeber waren u. a. Alfa Romeo, Philips, Agip, Mitsubishi, Feltrinelli, Mondadori und Olivetti. 1966 Entwurf des Signets der Biennale von Venedig. 1975–78 neues Erscheinungsbild für Agip Petroli, das 1979 mit dem „Compasso d'Oro" ausgezeichnet wird.

**Noordzij,** Gerrit – geb. 2. 4. 1931 in Rotterdam, Niederlande. – *Kalligraph, Typograph, Schriftentwerfer, Lehrer* – 1948

aux Pays-Bas et en République fédérale, dont le Prix H. N. Werkman pour son œuvre typographique (1981) et le prix Charles Nypels (1989). Professeur d'esquisse, de maquette de livre et de conception de polices à la Kunstakademie (beaux-arts) de Düsseldorf depuis 1985. Expositions au Stedelijk Museum d'Amsterdam et au Centre Georges Pompidou à Paris. – *Publication, sélection:* Roelof van Gelder et autres «Rijn, Rhein. Over het werk van Walter Nikkels», Nuth 1989.

**Noorda,** Bob – né le 15. 7. 1927 à Amsterdam, Pays-Bas – *graphiste maquettiste, designer industriel* – 1944–1947, études à l'Instituut voor Kunstnijverheidsonderwijs d'Amsterdam. De 1947 à 1950, il exerce comme graphiste maquettiste à Djakarta, Indonésie, où il conçoit des imprimés pour le gouvernement néerlandais. Graphiste indépendant à Amsterdam de 1950 à 1952, puis à Milan à partir de 1952. A enseigné les arts graphiques et le design à Milan, Venise et Urbin. Directeur artistique de la société Pirelli de 1961 à 1965. Le prix «Compasso d'Oro» lui est décerné en 1964 pour le système de signalisation du métro de Milan. Conçoit d'autres systèmes de signalisation à New York et à São Paulo. Cofondateur et Senior Vice-Président d' «Unimark International Design and Marketing Corporation» (avec Massimo Vignelli), aux Etats-Unis et en Italie en 1965. Commanditaires : Alfa Romeo, Philips, Agip, Mitsubishi, Feltrinelli, Mondadori, Olivetti etc. En 1966, il dessine le signet de la Biennale de Venise. Crée l'identité d'Agip Petroli entre 1975 et 1978; ce logo obtiendra le «Compasso d'Oro» en 1979.

**Noordzij,** Gerrit – né le 2. 4. 1931 à Rotterdam, Pays-Bas – *calligraphe, typographe, concepteur de polices, enseignant* – apprentissage de relieur en 1948. Tra-

Noorda 1966 Poster

Noorda 1974 Logo

Noorda 1963 Poster

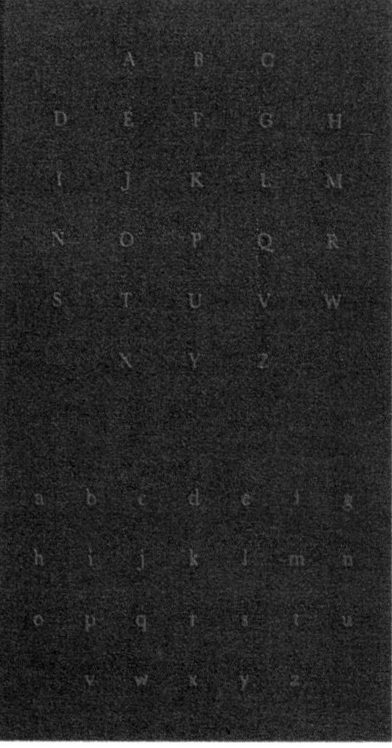

Gerrit Noordzij 1986 Cover

sign, book art and type development at the Kunstakademie Düsseldorf. His work has been exhibited at the Stedelijk Museum in Amsterdam and at the Centre Georges Pompidou in Paris, among other venues. – *Publications include:* Roelof van Gelder et al "Rijn, Rhein. Over het werk van Walter Nikkels", Nuth 1989.

**Noorda,** Bob – b. 15. 7. 1927 in Amsterdam, The Netherlands – *graphic designer, industrial designer* – 1944–47: studies at the Instituut voor Kunstnijverheidsonderwijs in Amsterdam. 1947–50: graphic designer in Jakarta, Indonesia for printed materials issued by the Dutch government. 1950–52: freelance graphic designer in Amsterdam and from 1952 onwards in Milan. Teaches graphic design in Milan, Venice and Urbino. 1961–65: art director for the Pirelli company. 1964: awarded the Compasso d'Oro for his signage system on the Milan subway. Designs further signage systems for New York and São Paolo. 1965: co-founder and senior vice-president of Unimark International Design and Marketing Corporation (with Massimo Vignelli) in the USA and in Italy. Clients include Alfa Romeo, Philips, Agip, Mitsubishi, Feltrinelli, Mondadori and Olivetti. 1966: designs the logo for the Venice Biennale. 1975–78: designs the new corporate identity for Agip Petroli for which he is awarded the Compasso d'Oro in 1979.

**Noordzij,** Gerrit – b. 2. 4. 1931 in Rotterdam, The Netherlands – *calligrapher, typographer, type designer, teacher* – 1948: trains as a bookbinder. 1956–58: production manager for Querido publishers

Buchbinderlehre. 1956–58 Hersteller im Querido-Verlag in Amsterdam. 1960–90 Dozent für Schrift und Typographie an der Koninklijke Academie voor Beeldende Kunsten in Den Haag. Seit 1984 Herausgabe der Ein-Mann-Zeitschrift „Letterletter". 1990 Gestaltung einer niederländischen Bibelausgabe. Noordzijs pädagogisches Konzept und seine ästhetische Qualität hatten großen Einfluß auf Generationen junger Schriftgestalter an der Akademie in Den Haag. – *Schriftentwürfe:* Remer, Tret, Kadmos, Apex, Ruse, Rysia, Burgundica, Algerak, Ruit, Batavian (1980), Dutch Roman (1980). – *Publikationen u. a.:* „The stroke of the pen: fundamental aspects of writing", Den Haag 1982; „Letters en studie", Eindhoven 1983; „Die burgundische Bastarda", München 1983; „De Streek, Theorie van het Schrift", Zaltbommel 1985; „Das Kind und die Schrift", München 1985.

**Noordzij,** Peter Matthias – geb. 27. 1. 1961 in Den Haag, Niederlande. – *Typograph, Schriftentwerfer, Lehrer* – 1980–85 Studium an der Koninklijke Academie voor Beeldende Kunsten in Den Haag und Schriftstudium bei seinem Vater Gerrit Noordzij. 1983 erste Skizzen zu seiner Schrift „PMN Caecilia" während des 3. Studienjahres. 1985–86 Praktikum in einer Druckerei. Seit 1986 Tätigkeit als Typograph. Auftraggeber waren u. a. die Verlage Querido, Arbeiderpers, Meulenhoff sowie die Tageszeitung „de Volkskrant". Unterrichtet 1986–89 Typographie in Arnheim. Seit 1988 Dozent für Schriftdesign an der Kunstakademie in Den Haag. Seit 1991 für die neu gegrün-

Gerrit Noordzij 1985 Cover

Gerrit Noordzij 1972 Stamp

Gerrit Noordzij Specimen

Peter Matthias Noordzij 1991 Caecilia

vaille au service de fabrication des éditions Querido de 1956 à 1958 à Amsterdam. Enseigne les arts de l'écriture et la typographie à la Koninklijke Academie voor Beeldende Kunsten de La Haye de 1960 à 1990. Publie seul la revue «Letterletter» depuis 1984. Maquette d'une édition néerlandaise de la Bible en 1990. Les conceptions pédagogiques de Noordzij et son sens de l'esthétique ont exercé une grande influence sur des générations de jeunes concepteurs de polices à l'Académie de La Haye. – *Polices:* Remer, Tret, Kadmos, Apex, Ruse, Rysia, Burgundica, Algerak, Ruit, Batavian (1980), Dutch Roman (1980). – *Publications, sélection:* «The stroke of the pen: fundamental aspects of writing», La Haye 1982; «Letters en studie», Eindhoven 1983; «Die burgundische Bastarda», Munich 1983; «De Streek, Theorie van het Schrift», Zaltbommel 1985; «Das Kind und die Schrift», Munich 1985.

**Noordzij,** Peter Matthias – né le 27. 1. 1961 à La Haye, Pays-Bas – *typographe, concepteur de police, enseignant* – 1980–1985, études à la Koninklijke Academie voor Beeldende Kunsten de La Haye, et études des arts de l'écriture chez son père Gerrit Noordzij. Premières esquisses de sa police «PMN Caecilia», en 1983, pendant sa troisième année d'études. Stage dans une imprimerie de 1985 à 1986. Travaille comme typographe depuis 1986. Commanditaires: les éditions Querido, Arbeiderpers, Meulenhoff ainsi que le quotidien «de Volkskrant». A enseigné la typo-

in Amsterdam. 1960–90: teaches lettering and typography at the Koninklijke Academie voor Beeldende Kunsten in The Hague. 1984: starts editing the one-man magazine "Letterletter". 1990: designs an edition of a Dutch bible. Noordzij's teaching concepts and his aesthetic quality have made a lasting impression on generations of young type designers at the Academie in The Hague. – *Fonts:* Remer, Tret, Kadmos, Apex, Ruse, Rysia, Burgundica, Algerak, Ruit, Batavian (1980), Dutch Roman (1980). – *Publications include:* "The stroke of the pen: fundamental aspects of writing", The Hague 1982; "Letters en studie", Eindhoven 1983; "Die burgundische Bastarda", Munich 1983; "De Streek, Theorie van het Schrift", Zaltbommel 1985; "Das Kind und die Schrift", Munich 1985.

**Noordzij,** Peter Matthias – b. 27. 1. 1961 in The Hague, The Netherlands – *typographer, type designer, teacher* – 1980–85: studies at the Koninklijke Academie voor Beeldende Kunsten in The Hague. Courses include lettering under his father, Gerrit Noordzij. 1983: produces preliminary sketches for his PMN Caecilia typeface during his third year of study. 1985–86: period of practical training at a printing works. 1986: starts working as a typographer. Clients include the publishers Querido, Arbeiderpers and Meulenhoff and "de Volkskrant" daily newspaper. 1986–89: teaches typography in Arnhem. From 1988 onwards: teaches type design at the art academy in The Hague. Since 1991: responsible for the new Enschedé Font Foundry in Haarlem. – *Fonts:*

dete Enschedé Font Foundry in Haarlem verantwortlich. – *Schriftentwurf:* PMN Caecilia (1991).

**Novarese,** Aldo – geb. 29. 6. 1920 in Pontestura Monferrato, Italien, gest. 16. 9. 1995 in Turin, Italien. – *Schriftentwerfer, Lehrer* – 1931–33 Besuch der Scuola Arteri Stampatori in Turin, wo er Holzschnitt, Kupferstich und Lithographie lernt. 1933–36 Besuch der Scuola Tipografica Guiseppe Vigliandi Paravia in Turin. 1936 Eintritt als Zeichner in die Schriftgießerei Nebiolo in Turin. 1938 Verleihung der Goldmedaille des nationalen Künstlerwettbewerbs „Ludi Juveniles". 1948–58 Lehrer für grafisches Zeichnen an der Scuola Vigliandi Paravia in Turin. Seit 1952 künstlerischer Leiter der Schriftgießerei Nebiolo. 1956 Verleihung der Goldmedaille der Mailänder Messe. 1975 Austritt aus der Schriftgießerei Nebiolo, Arbeit als freier Schriftentwerfer. 1979 Auszeichnung mit dem „Compasso d'Oro". – *Schriftentwürfe:* Landi Linear (1943), Athenaeum (mit A. Butti, 1945), Normandia (mit A. Butti, 1946), Augustea (mit A. Butti, 1951), Microgramma (mit A. Butti, 1952), Eigno (1954), Fontanesi (1954), Egizio (1955–58), Juliet (1955), Ritmo (1955), Garaldus (1956), Slogan (1957), Recto (1958–61), Estro (1961), Eurostile (1962), Forma (1966), Magister (1966), Metropol (1967), Stop (1971), Lapidar (1977), Fenice (1977–80), Novarese (1978), Expert (1983), Colossal (1984), ITC Symbol (1984), ITC Mixage (1985), Arbiter (1989). – *Publikation u. a.:* „Alfabeta", Turin 1964.

graphie à Arnhem de 1986 à 1989. Enseigne la conception de polices à l'académie des beaux-arts de La Haye depuis 1988. Responsable de la nouvelle Enschedé Font Foundry de Haarlem depuis 1991. – *Police:* PMN Caecilia (1991).

**Novarese,** Aldo – né le 29. 6. 1920 à Pontestura Monferrato, Italie, décédé le 16. 9. 1995 à Turin, Italie – *concepteur de polices, enseignant* – formation à la Scuola Arteri Stampatori de Turin de 1931 à 1933, où il apprend la gravure sur bois, sur cuivre et la lithographie. Fréquente la Scuola Tipografica Guiseppe Vigliandi Paravia à Turin de 1933 à 1936. Entre comme dessinateur à la fonderie de caractères Nebiolo de Turin en 1936. Reçoit en 1938 la médaille d'or du concours national des artistes «Ludi Juveniles». Enseigne le dessin graphique à la Scuola Vigliandi Paravia de Turin de 1948 à 1958. Directeur artistique de la fonderie de caractères Nebiolo à partir de 1952. Obtient la médaille d'or du salon de Milan en 1956. Quitte la fonderie de caractères Nebiolo en 1975 et travaille comme concepteur de polices indépendant. Reçoit le «Compasso d'Oro» en 1979. – *Polices:* Landi Linear (1943), Athenaeum (avec A. Butti, 1945), Normandia (avec A. Butti, 1946), Augustea (avec A. Butti, 1951), Microgramma (avec A. Butti, 1952), Eigno (1954), Fontanesi (1954), Egizio (1955– 1958), Juliet (1955), Ritmo (1955), Garaldus (1956), Slogan (1957), Recto (1958– 1961), Estro (1961), Eurostile (1962), Forma (1966), Magister (1966), Metropol (1967), Stop (1971), Lapidar (1977), Fenice (1977–80), Novarese (1978), Expert (1983), Colossal (1984), ITC Symbol (1984), ITC Mixage (1985), Arbiter (1989). – *Publication, sélection:* «Alfabeta», Turin 1964.

**Nypels,** Charles – né le 31. 10. 1895 à Maastricht, Pays-Bas, décédé le 3. 1. 1952

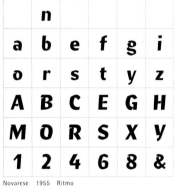

| | n | | | | |
|---|---|---|---|---|---|
| a | b | e | f | g | i |
| o | r | s | t | y | z |
| A | B | C | E | G | H |
| M | O | R | S | X | Y |
| 1 | 2 | 4 | 6 | 8 | & |

Novarese 1955 Ritmo

| n | n | n | n | | n |
|---|---|---|---|---|---|
| a | b | e | f | g | i |
| o | r | s | t | y | z |
| A | B | C | E | G | H |
| M | O | R | S | X | Y |
| 1 | 2 | 4 | 6 | 8 | & |

Novarese 1977–80 Fenice

| | n | n | n | | n |
|---|---|---|---|---|---|
| a | b | e | f | g | i |
| o | r | s | t | y | z |
| A | B | C | E | G | H |
| M | O | R | S | X | Y |
| 1 | 2 | 4 | 6 | 8 | & |

Novarese 1984 Symbol

Nypels ca. 1927 Poster

PMN Caecilia (1991).

**Novarese,** Aldo – b. 29. 6. 1920 in Pontestura Monferrato, Italy, d. 16. 9. 1995 in Turin, Italy – *type designer, teacher* – 1931–33: attends the Scuola Arteri Stampatori in Turin, where he learns woodcut, copper engraving and lithography. 1933–36: attends the Scuola Tipografica Guiseppe Vigliandi Paravia in Turin. 1936: joins the Nebiolo type foundry in Turin as a draftsman. 1938: is awarded a gold medal in the national Ludi Juveniles art competition. 1948–58: teaches graphic drawing at the Scuola Vigliandi Paravia in Turin. From 1952 onwards: art director at the Nebiolo type foundry. 1956: is awarded a gold medal by the Milan trade fair and exhibitions center. 1975: leaves the Nebiolo type foundry and starts working as a freelance type designer. 1979: is awarded the Compasso d'Oro. – *Fonts:* Landi Linear (1943), Athenaeum (with A. Butti, 1945), Normandia (with A. Butti, 1946), Augustea (with A. Butti, 1951), Microgramma (with A. Butti, 1952), Eigno (1954), Fontanesi (1954), Egizio (1955–58), Juliet (1955), Ritmo (1955), Garaldus (1956), Slogan (1957), Recto (1958–61), Estro (1961), Eurostile (1962), Forma (1966), Magister (1966), Metropol (1967), Stop (1971), Lapidar (1977), Fenice (1977–80), Novarese (1978), Expert (1983), Colossal (1984), ITC Symbol (1984), ITC Mixage (1985), Arbiter (1989). – *Publications include:* "Alfabeta", Turin 1964.

**Nypels,** Charles – b. 31. 10. 1895 in Maastricht, The Netherlands, d. 3. 1. 1952 in Groesbeek, The Netherlands – *typogra-*

**Nypels,** Charles – geb. 31. 10. 1895 in Maastricht, Niederlande, gest. 3. 1. 1952 in Groesbeek, Niederlande. – *Typograph, Verleger, Drucker* – 1917 Volontär bei S. H. de Roos in der Druckerei der Lettergieterij Amsterdam. 1920 Drucker und Verleger von klassischen und modernen Schriftstellern. Buchherstellung und -gestaltung für die Verlage La Connaissance (Paris), Le Balancier (Lüttich), The First Editions Club (London), De Gemeenschap (Utrecht), Uitgevers-Maatschappij (Amsterdam). 1938 typographischer Berater der Uitgeverij Het Spectrum in Utrecht. Übersetzt Jan Tschicholds Buch „Typographische Gestaltung" ins Niederländische. 1945 Gründung der „Stichting de Roos" in Utrecht (mit Chr. Leeflang und G. M. van Wees), einer Verlagsgenossenschaft zur Verbreitung der Bibliophilie. Korrespondierendes Mitglied des „Vereins Deutscher Buchkünstler". Für die von ihm gestaltete Ausgabe der Bibel wird ihm 1948 der Staatspreis für Typographie verliehen. Zu seinem Gedenken wird der Charles-Nypels-Preis für Gestaltung eingerichtet und verliehen. Bisherige Preisträger sind: Dieter Roth (1986), Walter Nikkels (1989), Harry N. Sierman (1992), Pierre di Sciullo (1995). Die Typographie von Charles Nypels orientiert sich am klassischen, traditionellen Stil mit ausgleichendem Aufbau und sparsamen Dekorationen. – *Publikationen u. a.:* „Blad, Boek en Band", Amsterdam 1940; „In Memoriam Charles Nypels", Amsterdam 1953. M. und K. van Laar „Charles Nypels, Meester-drukker", Maastricht 1986.

IN DIEN TIJD verscheen een besluit van keizer Augustus tot een volkstelling over geheel de wereld. Deze eerste volkstelling geschiedde onder Quirinius landvoogd van Syrië. En allen gingen om

zich aan te geven, ieder naar zijn eigen stad. Ook Jozef nu ging op uit Galilëa, uit de stad Nazareth, naar Judea, naar de stad van David, Bethlehem geheeten, wijl hij uit het huis en geslacht van David was, om zich aan te geven met Maria zijne verloofde vrouw, die zwanger was. En tijdens

hun verblijf aldaar verliepen de dagen tot zij baren zou, en ze baarde haar eerstgeboren Zoon, en legde hem neder in een kribbe omdat voor hen geene plaats was in de herberg. En daar waren herders in die omstreek, die buiten nachtwacht hielden over hun

kudde. En zie, een Engel des Heeren stond bij hen en de glorie des Heeren omstraalde hen, en ze vreesden met grooten vrees. En de Engel zeide hun: Vreest niet, want zie, ik boodschap U groote vreugde die voor geheel het volk zal zijn: dat U heden een Zaligmaker is geboren, die Christus de Heer is in de stad van David. En dit zij U ten teeken: ge zult een wicht vinden in doeken gewikkeld en in een kribbe gelegen. En plotseling bevond zich bij den Engel eene menigte uit het hemelsche heer, die Gode lofzongen en zeiden:

Eere aan God in den Hooge en op aarde vrede onder de menschen van goeden wil.

Nypels 1923 Spread

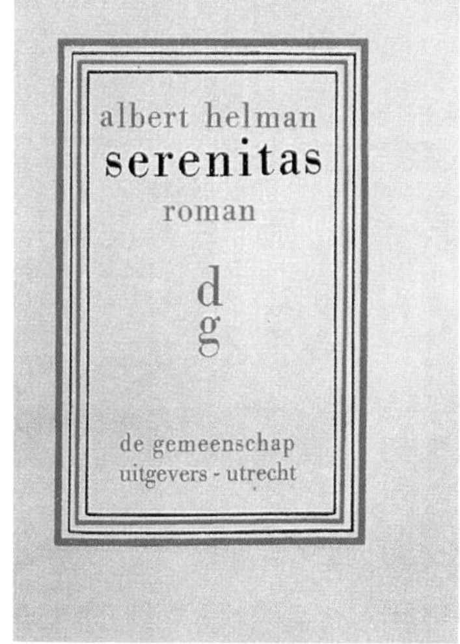


Nypels 1930 Cover

perdue. La vigne eſt confondue, et le figuier languit. Le grenadier, et la palme, et le pomier, et tous les arbres du champ ſont ſechez, pource que la ioye eſt troublee des fils des hommes. Vous preſtres, ceignez vous, et vous plaignez: vous miniſtres de l'autel, lamentez. Vous miniſtres de mon Dieu, entrez dedans, et vous couchez ſur le ſac: car le ſacrifice et la libation eſt oſtee de la maiſon de voſtre Dieu. Sanctifiez le ieuſne, appellez la multitude, aſſemblez les anciens, et tous les habitans de la terre, en la maiſon de voſtre Dieu, et criez au Seigneur. Ah, ah, ah, pour ce iour là: car le iour du Seigneur eſt pres, et viendra comme la deſtruction faicte par le puiſſant. Les vivres ne ſont ilz pas oſtez de devant voz yeulx, auſſi lieſſe et exultation de la maiſon de voſtre Dieu? Les beſtes ſont pourries en leur fiente, les greniers ſont demolis, les granges ſont diſſipees, pource que le froment eſt perdu. Pourquoy a gemi la beſte? Pourquoy ont mugy les vacheries? Pource qu'ilz n'ont que paſturer. Mais auſſi les troupeaux du beſtiail ſont deſtruicts. Seigneur, ie crieray à toy: car le feu a mangé la beaulté du deſert, et la flamme a bruſlé tous les bois de la contree. Mais auſſi les beſtes des champs ont regardé en hault vers toy, comme l'aire deſirant avoir la pluye: pource que les fontaines des eaues ſe ſont ſechees, et le feu a devoré les choſes plaiſantes du deſert.

SONNEZ LA TROMPETTE EN SION, HURLEZ EN MA SAINCTE MONTAGNE, QUE TOUS LES HABITANS DE LA TERRE SOYENT TROUBLEZ, CAR LE IOUR DU SEIGNEUR VIENT, ET LE IOUR DE TENEBRES ET D'OBSCURITÉ EST PRES. LE IOUR DE NUEE ET DE TOURBILLON. UN PEUple grand et fort vient comme le matin qui eſt eſtendu ſur les montagnes. Il n'en y a pas eu de ſemblable à luy depuis le commencement et apres luy, et n'y en aura pas iuſques ès ans de generation et generation. Le feu devorant eſt devant ſa face, et derriere luy eſt la flamme bruſlante. La terre eſt devant luy comme un iardin de volupté, et derriere luy eſt la deſolation du deſert: et n'y a aucun qui le puiſſe eſchapper. Leur regard eſt comme le regard des chevaulx, et courront comme les chevaucheurs. Ilz viendront ſaultans comme le ſon des chariotz ſur les coupeaux des montagnes, comme le ſon de la flamme devorant l'eſteule: comme un peuple fort preparé à la bataille. Les peuples ſeront tormentez pour ſa preſence: toutes faces ſeront reduictes comme un pot de terre. Ilz courront comme hommes forts: et monteront ſur la muraille comme hommes de guerre. Les hommes chemineront en leurs voyes, et ne ſe deſtourneront point de leurs ſentiers. Aucun ne contraindra ſon frere: chaſcun cheminera en ſon ſentier. Mais auſſi ilz cherront par les feneſtres, et ne ſe bleſſeront point. Ilz entreront en la cité, Ilz cour-

Nypels 1923 Spread

à Groesbeek, Pays-Bas – *typographe, éditeur, imprimeur* – 1917, stagiaire chez S. H. de Roos à l'imprimerie de la Lettergieterij Amsterdam. Imprime et édite des écrivains classiques et modernes en 1920. Ouvrages pour les éditions La Connaissance (Paris), Le Balancier (Liège), The First Editions Club (Londres), De Gemeenschap (Utrecht), Uitgevers-Maatschappij (Amsterdam), dont il réalise souvent les maquettes. En 1938, il est conseiller typographique de Uitgeverij Het Spectrum à Utrecht. Traduit l'ouvrage de Jan Tschichold «Typographische Gestaltung» en néerlandais. Fonde en 1945 la «Stichting de Roos» à Utrecht (avec Chr. Leeflang et G. M. van Wees), une coopérative d'édition pour promouvoir la bibliophilie. Membre et correspondant du «Verein Deutscher Buchkünstler» (association des artistes allemands du livre). Le prix national de typographie lui est décerné en 1948 pour une édition de la Bible dont il conçoit la maquette. Le prix Charles Nypels de mise en page est créé en sa mémoire. Parmi les lauréats, on compte jusqu'à présent: Dieter Roth (1986), Walter Nikkels (1989), Harry N. Sierman (1992), Pierre di Sciullo (1995). La typographie de Charles Nypels s'inspirait du style classique et traditionnel, sa mise en page était équilibrée et sobre quant aux décorations. – *Publications, sélection:* «Blad, Boek en Band», Amsterdam 1940; «In Memoriam Charles Nypels», Amsterdam 1953. M. et K. van Laar «Charles Nypels, Meester-drukker», Maastricht 1986.

*pher, publisher, printer* – 1917: trainee under S. H. de Roos in the printing workshop at the Lettergieterij Amsterdam. 1920: prints and publishes works by classic and modern authors. Produces and often also designs books for the publishers La Connaissance (Paris), Le Balancier (Liège), The First Editions Club (London), De Gemeenschap (Utrecht) and Uitgevers-Maatschappij (Amsterdam). 1938: typography consultant to Het Spectrum publishers in Utrecht. Translates Jan Tschichold's book "Typographische Gestaltung" into Dutch. 1945: founds the Stichting de Roos in Utrecht with Chr. Leeflang and G. M. van Wees, a publishing association for the promotion of bibliophile editions. Corresponding member of the Verein Deutscher Buchkünstler. 1948: he is awarded the State Prize for Typography for an edition of the Bible he designs. The Charles Nypels Prize for Design is introduced and awarded in his memory. Prize winners to date are Dieter Roth (1986), Walter Nikkels (1989), Harry N. Sierman (1992) and Pierre di Sciullo (1995). Charles Nypels' typography was based on classic, traditional styles. It demonstrated great balance in its composition and was sparse in its decoration. – *Publications include:* "Blad, Boek en Band", Amsterdam 1940; "In Memoriam Charles Nypels", Amsterdam 1953. M. and K. van Laar "Charles Nypels, Meester-drukker", Maastricht 1986.

# O

**Octavo**
**Odermatt + Tissi**

**Octavo** – *Zeitschrift* – Die Design-Gruppe „8vo" publiziert in London 1986 die erste Ausgabe der auf acht Ausgaben konzipierten Typographie-Zeitschrift „octavo". Die Zeitschrift versteht sich als Organ gegen modische und dekorative Gestaltung. Autoren sind u. a. Robin Kinross, April Greiman, Peter Rea, Wolfgang Weingart, Peter Mayer, Wim Crouwel, Friedrich Friedl und Bridget Wilkins. Die letzte Ausgabe der Zeitschrift erscheint 1992 als CD-Rom.
**Odermatt + Tissi** – *Grafik-Design-Studio* – 1968 Gründung der Ateliergemeinschaft Odermatt + Tissi in Zürich, Schweiz. Siegfried Odermatt – geb. 13. 9. 1926 in Neuheim, Schweiz. – Als Grafiker Autodidakt. Drei Jahre freier Mitarbeiter des Malers und Grafikers Hans Falk. Drei Jahre Tätigkeit in einer Werbeagentur. 1950–68 selbständiger Grafiker. Zahlreiche Auszeichnungen, u. a. Preisträger der 5. Plakat-Biennale in Fort Collins, Colorado, USA (1987), 1. Preis für das Erscheinungsbild der Kieler Woche 1994 (1992). Rosmarie Tissi – geb. 13. 2. 1937 in Schaffhausen, Schweiz. – Studium an der Kunstgewerbeschule in Zürich, danach vierjährige Berufslehre. Zahlreiche Auszeichnungen, u. a. 1. Preis für das beste kulturelle Plakat auf der Plakat-Biennale in Warschau (1986), 1. Preis für das Erscheinungsbild der Kieler Woche 1990 (1988), Henri-de-Toulouse-Lautrec-Medaille in Silber auf der 6. Plakat-Triennale in Essen (1990). 1993 Gastdozentin an der Yale-University in New Haven, USA. – *Schriftentwürfe:* Antiqua Classica (1971), Sonora (1972), Sinaloa

**Octavo** – *revue* – le groupe de designer «8vo» de Londres publie en 1986 le premier des huit numéros prévus de la revue «octavo». La revue se présente comme un organe dirigé contre la mode du design décoratif. Robin Kinross, April Greiman, Peter Rea, Wolfgang Weingart, Peter Mayer, Wim Crouwel, Friedrich Friedl et Bridget Wilkins y ont écrit des articles. La dernière édition de la revue est parue en 1992 sous forme de CD-Rom.
**Odermatt + Tissi** – *atelier de graphisme et de design* – l'atelier Odermatt + Tissi a été fondé en 1968 à Zurich, Suisse. Siegfried Odermatt – né le 13. 9. 1926 à Neuheim, Suisse – graphiste autodidacte. Prestataire de service pour le peintre et graphiste Hans Falk pendant trois ans. Exerce pendant trois ans dans une agence de publicité. 1950–1968, graphiste indépendant. Nombreuses distinctions, dont le prix de la 5e Biennale de l'affiche de Fort Collins, Colorado, Etats-Unis (1987), 1er prix pour l'identité de la semaine de Kiel 1994 (1992). Rosmarie Tissi – née le 13. 2. 1937 à Schaffhausen, Suisse – études à la Kunstgewerbeschule (école des arts décoratifs) de Zurich, puis quatre années d'apprentissage. Nombreuses distinctions, dont le 1er prix de la meilleure affiche culturelle de la Biennale de l'affiche de Varsovie (1986), 1er prix pour l'identité de la Semaine de Kiel de 1990 (1988), médaille d'argent Henri de Toulouse-Lautrec à la 6è Triennale de l'affiche à Essen (1990). Cycle de conférences à la Yale University à New Haven, Etats-Unis, en 1993. – *Polices:* Antiqua classica (1971), Sonora (1972), Sinaloa (1972), Marabu (1972), Mindanao (1975). Commanditaires communs de l'atelier: Univac Computers, le Museum für Gestaltung Zurich, les Editions Arche et le Cimenterie Holderbank. Nombreux articles de presse et expositions, par ex. à

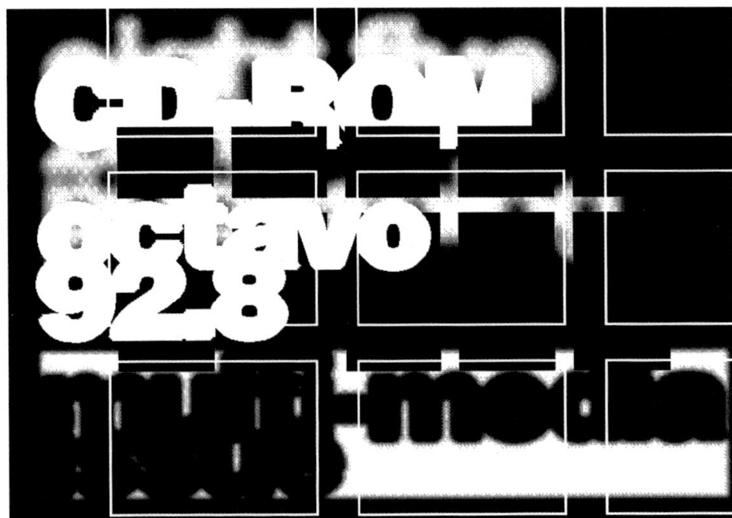

Octavo 1992 Poster

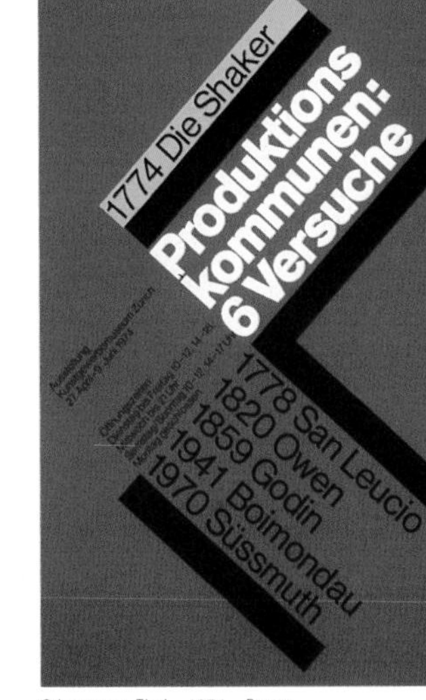

Odermatt + Tissi 1974 Poster

Octavo 1986 Cover

Odermatt + Tissi 1984 Logo

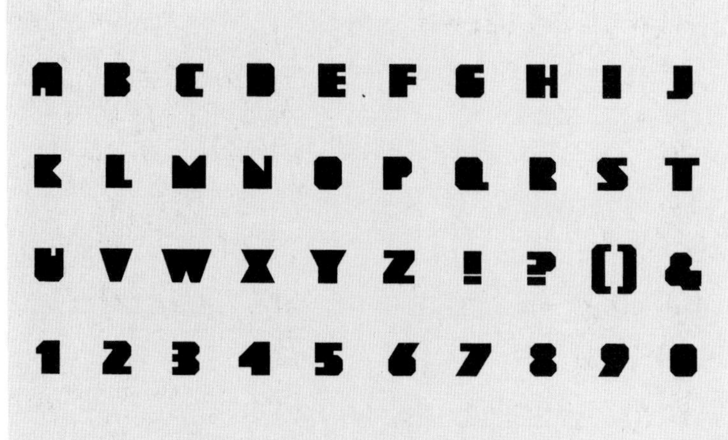

Odermatt + Tissi 1972 Alphabet

**Octavo** – *magazine* – 1986: 8vo design group in London publishes the first number of eight "octavo" typography journals. The magazine sees itself as an organ against fashionable and decorative design. Authors include Robin Kinross, April Greiman, Peter Rea, Wolfgang Weingart, Peter Mayer, Wim Crouwel, Friedrich Friedl and Bridget Wilkins. 1992: the last number of the magazine appears as a CD-ROM.
**Odermatt + Tissi** – *graphic design studio* – 1968: the Odermatt + Tissi studio is founded in Zurich, Switzerland. Siegfried Odermatt – b. 13. 9. 1926 in Neuheim, Switzerland – Self-taught graphic artist. Worked freelance for the painter and graphic artist Hans Falk for three years. Worked in an advertising agency for three years. 1950–68: freelance graphic artist. Has won numerous awards, including at the 5th Poster Biennale in Fort Collins, Colorado, USA (1987) and 1st prize for the corporate identity of the 1994 Kieler Woche (1992). Rosmarie Tissi – b. 13. 2. 1937 in Schaffhausen, Switzerland – Studied at the Kunstgewerbeschule in Zurich, followed by four years' vocational training. Numerous awards, including 1st prize for the best arts poster at the Poster Biennale in Warsaw (1986), 1st prize for the corporate identity of the 1990 Kieler Woche (1988) and the Henri de Toulouse Lautrec silver medal at the 6th Poster Triennale in Essen (1990). 1993: visiting lecturer at Yale University in New Haven, USA. – *Fonts:* Antiqua Classica (1971), Sonora (1972), Sinaloa (1972), Marabu (1972), Mindanao (1975).

(1972), Marabu (1972), Mindanao (1975). Gemeinsame Auftraggeber des Studios u. a. Univac Computers, Museum für Gestaltung Zürich, der Arche Verlag, Zementfabrik Holderbank. Zahlreiche Zeitschriftenartikel und Ausstellungen, u. a. in der Visual Graphics Gallery in New York (1966), der Hochschule für Gestaltung in Offenbach (1984), der Reinhold-Brown Gallery in New York (1992) und dem Deutschen Plakat-Museum in Essen (1996). – *Publikation u. a.:* Siegfried Odermatt + Rosmarie Tissi „Graphic Design", Weingarten, Zürich 1993.

**Olbrich,** Joseph Maria – geb. 22. 12. 1867 in Troppau, Schlesien (heute Opava, Tschechische Republik), gest. 8. 8. 1908 in Düsseldorf, Deutschland. – *Architekt, Kunstgewerbler* – 1890–93 Studium an der Akademie in Wien. 1894–99 Mitarbeiter des Architekten Otto Wagner. 1897 Gründungsmitglied der Wiener Sezession. 1898–99 Entwurf und Ausführung des Ausstellungsgebäudes der Wiener Sezession. 1899 Berufung an die Künstlerkolonie in Darmstadt. Entwürfe für Gegenstände aus allen Gebieten des Kunstgewerbes. 1907 Gründungsmitglied des Deutschen Werkbunds.

**Oliver,** Vaughan – geb. 12. 9. 1957 in Sedgefield, England. – *Grafik-Designer, Illustrator* – 1976–79 Studium an der Newcastle-upon-Tyne Polytechnic. Seit 1980 Entwürfe von Schallplattencovern für „4 AD", seit 1983 unter dem Studionamen „23 Envelope" (mit Nigel Grierson) und ständige Arbeit als Art Director für die Schallplattenfirma „4 AD". Erscheinungsbilder für Musikgruppen.

la Visual Graphics Gallery de New York (1966), à la Hochschule für Gestaltung (école supérieure de design) d'Offenbach (1984), à la Reinhold-Brown Gallery à New York (1992), au Deutsche Plakat-Museum (musée allemand de l'affiche) à Essen (1996). – *Publication, sélection:* Siegfried Odermatt + Rosmarie Tissi «Graphic Design», Weingarten, Zurich 1993.

**Olbrich,** Joseph Maria – né le 22. 12. 1867 à Troppau, Silésie (aujourd'hui Opava, République Tchèque), décédé le 8. 8. 1908 à Düsseldorf, Allemagne – *architecte, artisan d'art* – 1890–1893, études à l'académie des beaux-arts de Vienne. Travaille avec l'architecte Otto Wagner de 1894 à 1899. Membre fondateur de la Sécession de Vienne en 1897. Appelé à la colonie d'artistes de Darmstadt en 1899. Crée des objets dans tous les domaines de l'artisanat d'art. Membre fondateur du Deutscher Werkbund en 1907.

**Oliver,** Vaughan – né le 12. 9. 1957 à Sedgefield, Angleterre – *graphiste maquettiste, illustrateur* – 1976–1979, études au Newcastle-upon-Tyne Polytechnic. A partir de 1980, il crée des pochettes de disques pour «4 AD», travaille ensuite à partir de 1983 comme directeur artistique de la société de productions de disques «4 AD» sous le nom de l'atelier «23 Envelope» (avec Nigel Grierson).

Odermatt + Tissi 1986 Cover

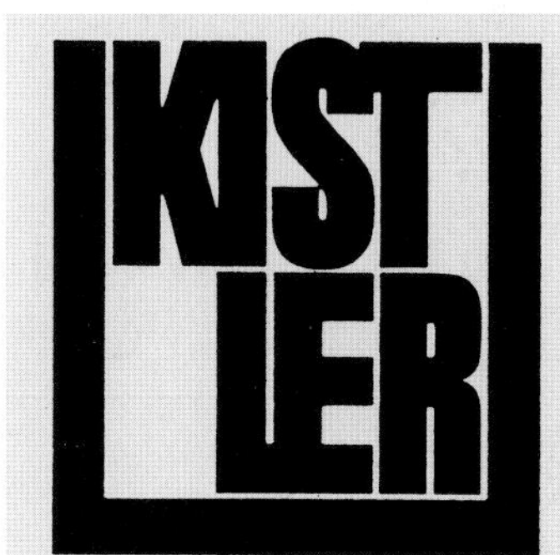
Odermatt + Tissi 1959 Logo

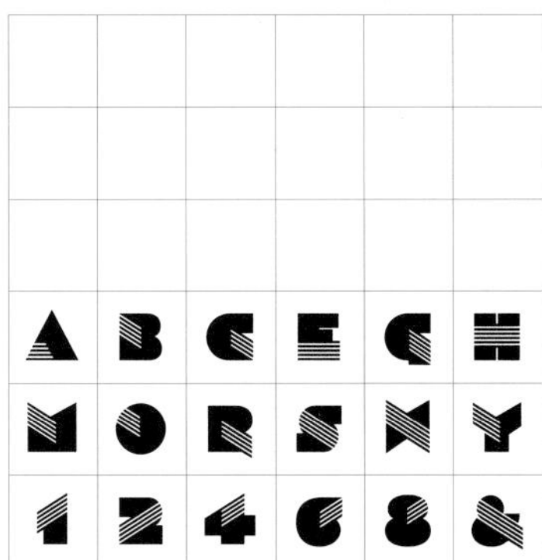
Odermatt + Tissi 1972 Sinaloa

Odermatt + Tissi 1991 Poster

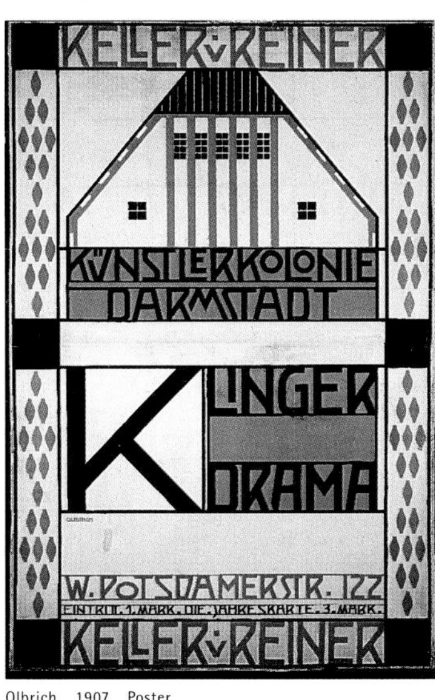

Olbrich 1907 Poster

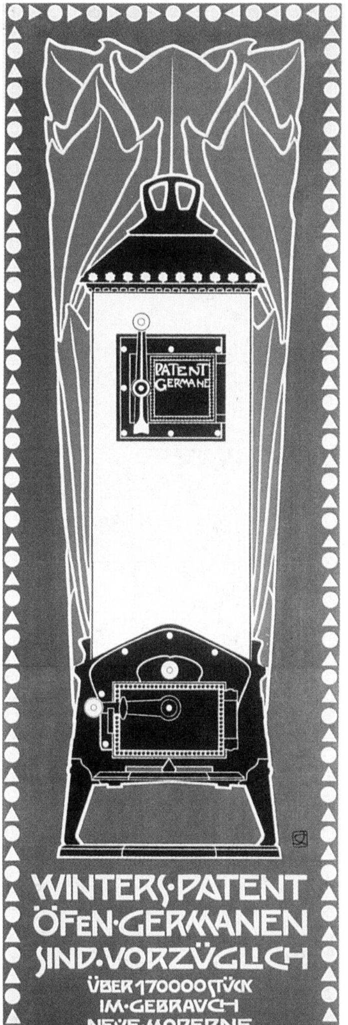

Olbrich ca. 1900 Poster

Clients shared by the studio include Univac Computers, Museum für Gestaltung in Zurich, Arche publishers and Holderbank cement factory. Numerous magazine articles and exhibitions, including at the Visual Graphics Gallery in New York (1966), the Hochschule für Gestaltung in Offenbach (1984), the Reinhold-Brown Gallery in New York (1992) and the Deutsche Plakat-Museum in Essen (1996). – *Publications include:* Siegfried Odermatt + Rosmarie Tissi "Graphic Design", Weingarten, Zurich 1993.

**Olbrich,** Joseph Maria – b. 22. 12. 1867 in Troppau, Silesia (now Opava, Czech Republic) d. 8. 8. 1908 in Düsseldorf, Germany – *architect, craftsman* – 1890–93: studies at the academy in Vienna. 1894–99: works for the architect Otto Wagner. 1897: founder member of the Vienna Secession. 1898–99: designs and builds the Vienna Secession exhibition building. 1899: asked to join the artists' colony in Darmstadt. Designs objects from all areas of the applied arts. 1907: founder member of the Deutsche Werkbund.

**Oliver,** Vaughan – b. 12. 9. 1957 in Sedgefield, England – *graphic designer, illustrator* – 1976–79: studies at Newcastle-upon-Tyne Polytechnic. From 1980 onwards: designs record covers for 4 AD. From 1983 onwards: works as art director for the 4 AD record company under his studio name, 23 Envelope, with Nigel Grierson. Produces corporate identities for music groups. 1987: designs record covers for Virgin Records. 1988: founds v 23 studio with Christopher Bigg after 23 Envelope disbands. 1990: his work is

1987 Entwürfe von Schallplattencovern für die Schallplattenfirma „Virgin". Nach der Auflösung von „23 Envelope", 1988 Gründung des Studios „v 23" (mit Christopher Bigg). 1990 Ausstellung seiner Arbeiten in Nantes, Frankreich, danach in Paris (1991), Amerika und Japan. 1993 Entwurf des Erscheinungsbildes des spanischen Fernsehsenders „Documania". 1994 Plakatentwürfe für das „Young Vic Theatre" in London. – *Publikation u. a.:* Rick Poynor u. a. „Vaughan Oliver", Nantes 1990.

**Oppenheim,** Louis – geb. 1879 in Coburg, Deutschland, gest. 1936 in Berlin, Deutschland. – *Grafiker, Maler, Schriftentwerfer* – 1899–1906 Ausbildung in London. Seit 1910 als freier Grafiker in Berlin tätig. Signiert seine Arbeiten mit der Abkürzung „LO". Auftraggeber waren u. a. die AEG, die Reichsbahn, Persil und Adrema. Seine Plakate galten als wichtige Ergebnisse des „Berliner Plakatstils". 1980 aktualisiert Erik Spiekermann Oppenheims „Lo-Schriften" unter dem Titel „Lo-Type" und zeichnet zusätzliche Schnitte. – *Schriftentwürfe:* Lo-Schrift (1911–14), Fanfare schmal (1927).

**Ortiz-Lopez,** Dennis – geb. 30. 7. 1949 in East Los Angeles, USA. – *Grafik-Designer, Typograph, Schriftzeichner, Lehrer* – 1967–69 Studium der Typographie an der California State University in Long Beach. Danach Grafik-Design-Studium in Los Angeles. 1979–81 Arbeit für die Zeitschrift „Rolling Stone" als Schriftzeichner; zahlreiche Preise und Auszeichnungen. Unterrichtet 1981–84 Schriftzeichnen und Design-Produktions-

identités de groupes de musiciens. En 1987, il réalise des pochettes de disques pour la société «Virgin». Après la dissolution de «23 Envelope», en 1988, il fonde l'atelier «v 23» (avec Christopher Bigg). 1990, expositions de ses travaux à Nantes, France, puis à Paris (1991), aux Etats-Unis et au Japon. En 1993, il dessine l'identité de la chaîne de télévision espagnole «Documania». Affiches pour le «Young Vic Theatre» de Londres en 1994. – *Publication, sélection:* Rick Poynor et autres «Vaughan Oliver», Nantes 1990.

**Oppenheim,** Louis – né en 1879 à Coburg, Allemagne, décédé en 1936 à Berlin, Allemagne – *graphiste, peintre, concepteur de polices* – formation à Londres de 1899 à 1906. Exerce comme graphiste indépendant à Berlin à partir de 1910. Signe ses travaux avec ses initiales «LO». Commanditaires: l'AEG, la Reichsbahn, Persil et Adrema. Ses affiches sont considérées comme des témoignages importants du «style de l'affiche berlinoise». En 1980, Erik Spiekermann actualise les polices «Lo» d'Oppenheim qui paraissent sous le nom de «Lo-Type» et dessine des fontes supplémentaires. – *Polices:* Lo-Schrift (1911–1914), Fanfare schmal (1927).

**Ortiz-Lopez,** Dennis – né le 30. 7. 1949 à East Los Angeles, Etats-Unis – *graphiste maquettiste, calligraphe, enseignant* – 1967–1969, études de typographie à la California State University à Long Beach. Exerce ensuite comme graphiste et designer à Los Angeles. Travaille comme concepteur de police pour la revue «Rolling Stone» de 1979 à 1981; nombreux prix et distinctions. 1981–1984, enseigne les arts de l'écriture et la méthodologie de production de design à la School of Visual Arts de New York, et pendant quelque temps au Montclair State College à New Jersey. Travaille dans son propre atelier à New York depuis 1985, nom-

Oliver 1983 Cover

lightLoTypemager
abcdefghijklmnopqrstuvwxyzß
ABCDEFGHIJKLMNOPQRSTUVWXYZ
äöüæœåøÄÖÜÆŒÅØ 1234567890%
(.,-;:!i?¿–)·[“„”‘»«]+–=/$£†*&§

regularLoTypenormal
abcdefghijklmnopqrstuvwxyzß
ABCDEFGHIJKLMNOPQRSTUVWXYZ
äöüæœåøÄÖÜÆŒÅØ 1234567890%
(.,-;:!i?¿–)·[“„”‘»«]+–=/$£†*&§

mediumLoTypehalbfett
abcdefghijklmnopqrstuvwxyzß
ABCDEFGHIJKLMNOPQRSTUVWXYZ
äöüæœåøÄÖÜÆŒÅØ 1234567890%
(.,-;:!i?¿–)·[“„”‘»«]+–=/$£†*&§

boldLoTypefett
abcdefghijklmnopqrstuvwxyzß
ABCDEFGHIJKLMNOPQRSTUVWXYZ
äöüæœåøÄÖÜÆŒÅØ 1234567890%
(.,-;:!i?¿–)·[“„”‘»«]+–=/$£†*&§

Oppenheim 1980 Lo-Type

Oliver 1990 Poster

Oppenheim 1911 Poster

exhibited in Nantes, France, then in Paris (1991), America and Japan. 1993: designs the corporate identity for the Spanish Documania television channel. 1994: designs posters for the Young Vic Theatre in London. – *Publications include:* Rick Poynor et al "Vaughan Oliver", Nantes 1990.

**Oppenheim,** Louis – b. 1879 in Coburg, Germany, d. 1936 in Berlin, Germany – *graphic artist, painter, type designer* – 1899–1906: trains in London. 1910: starts working as a freelance graphic artist in Berlin. Signs his work with the initials "LO". Clients include AEG, the Reichsbahn, Persil and Adrema. His posters are considered a significant product of the "Berlin poster style". 1980: Erik Spiekermann updates Oppenheim's Lo typefaces under the name of Lo-Type and draws additional weights for them. – *Fonts:* Lo-Schrift (1911–14), Fanfare schmal (1927).

**Ortiz-Lopez,** Dennis – b. 30. 7. 1949 in East Los Angeles, USA – *graphic designer, typographer, letterer, teacher* – 1967–69: studies typography at the California State University in Long Beach, after which he works as a graphic designer in Los Angeles. 1979–81: works as a letterer for "Rolling Stone" magazine; wins numerous prizes and awards. 1981–84: teaches lettering and design production methodology at the School of Visual Arts in New York and for a brief period also at the Montclair State College in New Jersey. 1985: opens his own design studio in New York, working on numerous lettering projects for magazines which include "Sports Illustrated", "People", "U.S.

Methodologie an der School of Visual Arts in New York und kurze Zeit am Montclair State College in New Jersey. Seit 1985 eigenes Entwurfsstudio in New York. Zahlreiche Schriftarbeiten für Zeitschriften, u. a. „Sports Illustrated", „People", „U. S. News & World Report", „PC Magazine". Entwurf des Signets für die Zeitschriften „Spectrum" und „Wirtschaftswoche". Mehrere Entwürfe hebräischer Alphabete.

**Ott + Stein** – *Grafik-Design-Studio* – 1978 Gründung des Studios in Berlin. Nicolaus Ott – geb. 1947 in Göttingen, Deutschland. – 1966 Lehre als grafischer Zeichner. 1969–74 Studium an der Hochschule der Künste in Berlin bei H. W. Kapitzki. Bernard Stein – geb. 1949 in Berlin, Deutschland. – 1972–77 Studium an der Hochschule der Künste in Berlin bei H. Lortz. Ott + Stein gestalten vorwiegend für Kulturinstitutionen und entwerfen typographische Signets. 1988 gemeinsame Gastprofessur an der Hochschule der Künste in Berlin, 1996 an der Gesamthochschule in Kassel. Zahlreiche Ausstellungen, u. a. im IDZ Berlin (1983, 1992), im Museum für Kunst und Gewerbe Hamburg (1993) und im Deutschen Plakat-Museum Essen (1995). – *Publikationen u. a.:* Ulf Erdmann Ziegler „Vom Wort zum Bild und zurück", Berlin 1992; Nils Jockel, Kristin Freireiss „Ott + Stein. Architekturplakat Plakatarchitektur", Hamburg 1996.

**Ovink,** Gerrit Willem – geb. 22. 10. 1912 in Leiden, Niederlande, gest. 4. 2. 1984 in Amsterdam, Niederlande. – *Autor, Wissenschaftler, Lehrer* – Studium der Psy-

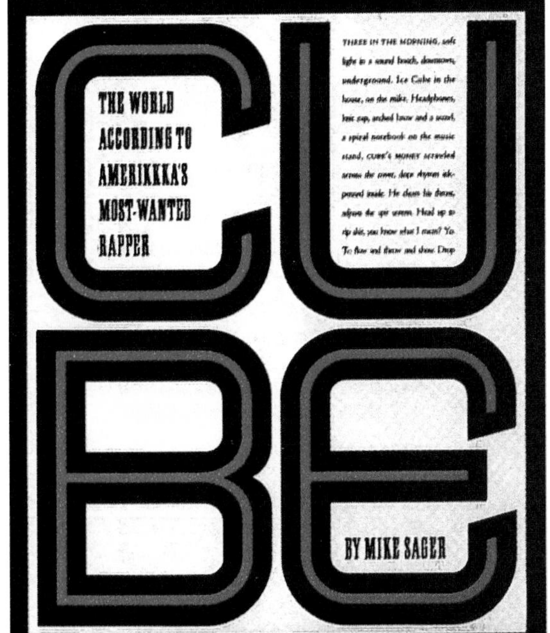

Ortis-Lopez Page

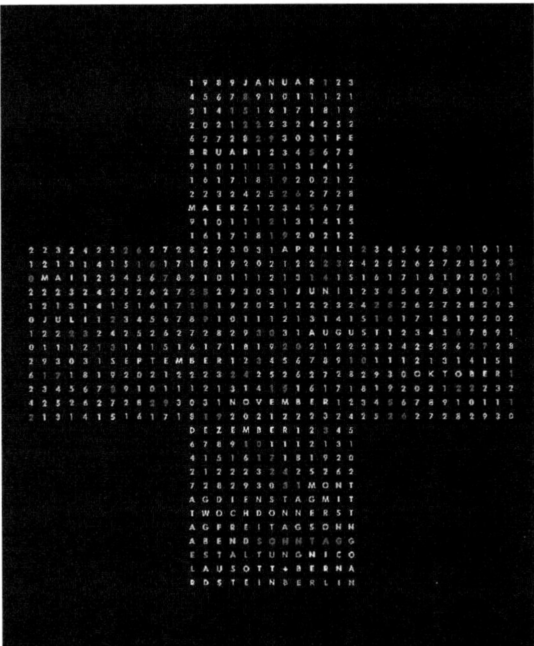
Ott + Stein 1989 Calendar

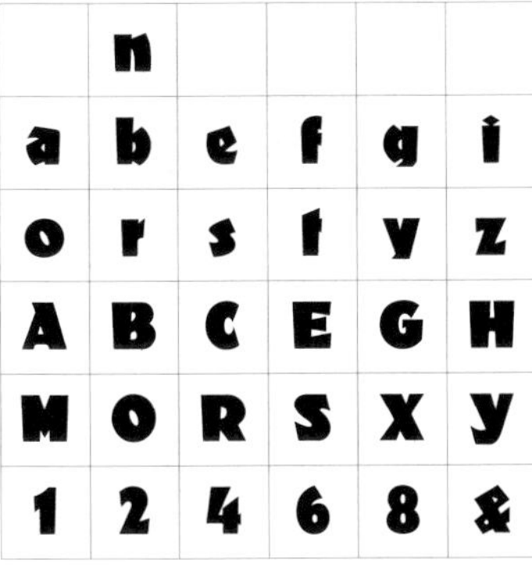
Oppenheim 1927 Fanfare

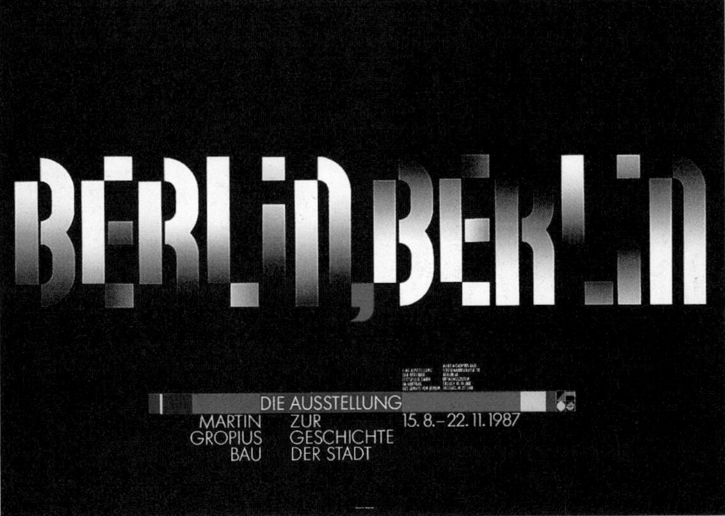

Ott + Stein 1987 Poster

breux travaux calligraphiques pour des revues, dont «Sports Illustrated», «People», «US News & World Report», «PC Magazine». Signets des revues «Spectrum» et «Wirtschaftswoche». A dessiné plusieurs alphabets hébraïques.

**Ott + Stein** – *atelier de graphisme et de design* – atelier fondé en 1978 à Berlin. Nicolaus Ott – né en 1947 à Göttingen, Allemagne – apprentissage de dessinateur graphiste en 1966. Etudes à la Hochschule der Künste (école supérieure des beaux-arts) de Berlin chez H. W. Kapitzki. Bernard Stein, né en 1949 à Berlin, Allemagne – 1972–1977, études à la Hochschule der Künste à Berlin chez H. Lortz. Ott + Stein travaillent surtout pour des institutions culturelles et créent des signets typographiques. Cycles de conférences en commun en 1988 à la Hochschule der Künste de Berlin, puis à la Gesamthochschule de Kassel en 1996. Nombreuses expositions, entre autres, à l'IDZ Berlin (1983 et 1992), au Museum für Kunst und Gewerbe de Hambourg (1993) et au Deutsche Plakat-Museum d'Essen (1995). – *Publications, sélection :* Ulf Erdmann Ziegler «Vom Wort zum Bild und zurück», Berlin 1992; Nils Jockel, Kristin Freireiss «Ott + Stein. Architekturplakat Plakatarchitektur», Hambourg 1996.

**Ovink,** Gerrit Willem – né le 22. 10. 1912 à Leyde, Pays-Bas, décédé le 4. 2. 1984 à Amsterdam, Pays-Bas – *auteur, chercheur, enseignant* – études de psycholo-

News & World Report" and "PC Magazine". Designs the logos for "Spectrum" and "Wirtschaftswoche" magazines. Has designed numerous Hebrew alphabets.

**Ott + Stein** – *graphic design studio* – 1978: the studio is founded in Berlin. Nicolaus Ott – b. 1947 in Göttingen, Germany – 1966: trains as a graphic draftsman. 1969–74: studies at the Hochschule der Künste in Berlin under H. W. Kapitzki. Bernard Stein – b. 1949 in Berlin, Germany – 1972–77: studies at the Hochschule der Künste in Berlin under H. Lortz. Ott + Stein design primarily for cultural institutions and have produced typographical logos. 1988: they share a visiting professorship at the Hochschule der Künste in Berlin and in 1996 at the Gesamthochschule in Kassel. Numerous exhibitions, including at the IDZ Berlin (1983 and 1992), the Museum für Kunst und Gewerbe in Hamburg (1993) and at the Deutsche Plakat-Museum in Essen (1995). – *Publications include:* Ulf Erdmann Ziegler "Vom Wort zum Bild und zurück", Berlin 1992; Nils Jockel, Kristin Freireiss "Ott + Stein. Architekturplakat Plakatarchitektur", Hamburg 1996.

**Ovink,** Gerrit Willem – b. 22. 10. 1912 in Leiden, The Netherlands, d. 4. 2. 1984 in Amsterdam, The Netherlands – *author, scientist, teacher* – Studied psychology, art history and philosophy in Utrecht. Consultant to various institutes. Wrote essays on questions relating to books. 1940–72: art consultant to the Lettergieterij Amsterdam. 1956–82: professor of history and the aesthetics of printing at the university in Amsterdam. 1983: re-

chologie, Kunstgeschichte und Philosophie in Utrecht. Berater von Instituten und Verfasser von Aufsätzen zu Fragen des Buchwesens. 1940–72 künstlerischer Berater der Lettergieterij Amsterdam. 1956–82 Professor für Geschichte und Ästhetik der Druckkunst an der Universität in Amsterdam. 1983 Verleihung des Gutenberg-Preises. – *Publikationen u.a.:* „Legibility, atmosphere-value and forms of printing types" (Phil. Diss.), Leiden 1938; „Hondert jaren lettergieterij in Amsterdam", Amsterdam 1951; „Fitting design into hard business", London 1959; „Von Gutenbergbibel bis Reader's Digest", Mainz 1960; „Anderhalve eeuw boektypografie 1815–1965", Nimwegen, Amsterdam 1965; „Die Gesinnung des Typographen", Mainz 1973.

**Oxenaar,** Otje (R. D. E.) – geb. 1929 in Den Haag, Niederlande. – *Grafik-Designer, Maler, Illustrator, Lehrer* – Studium an der Koninklijke Academie voor Beeldende Kunsten in Den Haag. 1955 Zusammenarbeit mit dem Architekten Gerrit Rietveld an Wandmalereien für die Wiederaufbau-Ausstellung „E 55" in Rotterdam. 1966 erste realisierte Banknotenentwürfe für die niederländische Bank, seit 1967 weitere realisierte Banknotenentwürfe. Entwurf zahlreicher Briefmarken für die niederländische Post PTT. Seit 1976 Direktor der Abteilung Kunst und Design der niederländischen Post PTT. Unterrichtet 1958–70 Grafik-Design an der Koninklijke Academie voor Beeldende Kunsten in Den Haag und visuelle Präsentation an der Technischen Universität in Delft.

gie, d'histoire de l'art et de philosophie à Utrecht. Conseiller de plusieurs instituts et auteur d'essais sur l'édition. Conseiller artistique de la Lettergieterij Amsterdam de 1940 à 1972. Professeur d'histoire de l'esthétique des arts de l'imprimerie à l'université d'Amsterdam. Lauréat du Prix Gutenberg en 1983. – *Publications, sélection:* «Legibility, atmosphere-value and forms of printing types» (thèse de doctorat), Leyde 1938; «Hondert jaren lettergieterij in Amsterdam», Amsterdam 1951; «Fitting design into hard business», Londres 1959; «Von Gutenbergbibel bis Reader's Digest», Mayence 1960; «Anderhalve eeuw boektypografie 1815–1965», Nimègue, Amsterdam 1965; «Die Gesinnung des Typographen», Mayence 1973.

**Oxenaar,** Otje (R.D.E.) – né en 1929 à La Haye, Pays-Bas – *graphiste maquettiste, peintre, illustrateur, enseignant* – études à la Koninklijke Academie voor Beeldende Kunsten de La Haye. En 1955, Oxenaar crée des fresques pour «E 55», l'exposition sur la reconstruction de Rotterdam en collaboration avec l'architecte Gerrit Rietveld. Conçoit ses premiers billets de banque pour la Banque des Pays-Bas en 1966, a dessiné d'autres billets de banque depuis 1967. Dessin de nombreux timbres pour les postes néerlandaises PTT. Directeur du service art et design des postes néerlandaises depuis 1976. A enseigné les arts graphiques et le design à la Koninklijke Academie voor Beeldende Kunsten de La Haye et la présentation visuelle à l'université technique de Delft de 1958 à 1970.

Ovink 1960 Cover

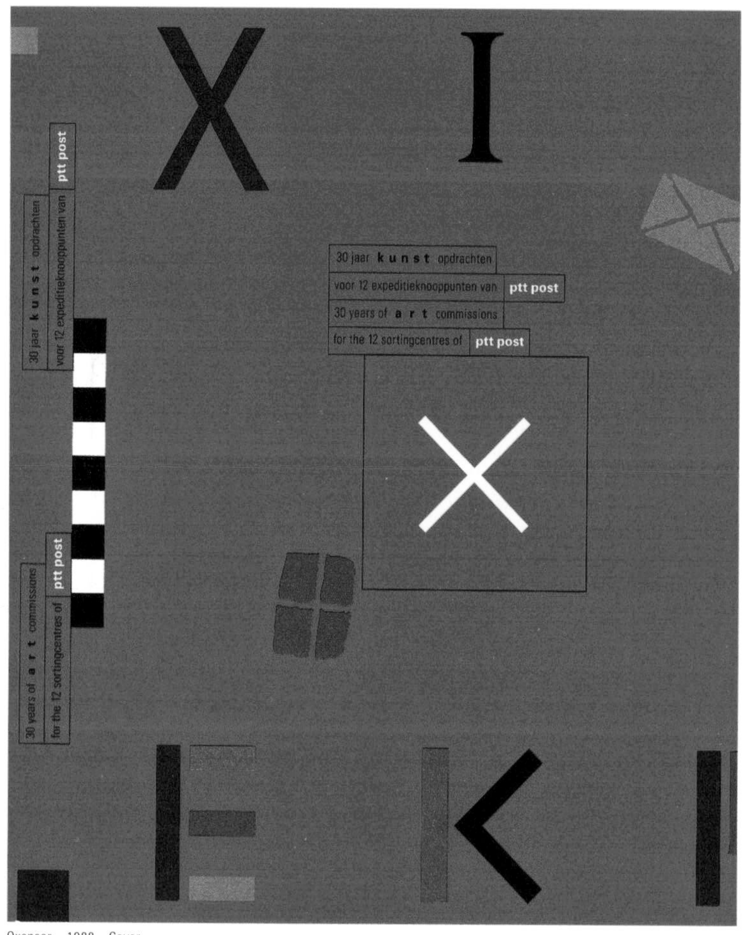

Oxenaar 1988 Cover

Oxenaar 1995 Stamps

ceives the Gutenberg Prize. – *Publications include:* "Legibility, atmosphere-value and forms of printing types" (doctoral dissertation), Leiden 1938; "Hondert jaren lettergieterij in Amsterdam", Amsterdam 1951; "Fitting design into hard business", London 1959; "Von Gutenbergbibel bis Reader's Digest", Mainz 1960; "Anderhalve eeuw boektypografie 1815–1965", Nijmegen, Amsterdam 1965; "Die Gesinnung des Typographen", Mainz 1973.

**Oxenaar,** Otje (R.D.E.) – b. 1929 in The Hague, The Netherlands – *graphic designer, painter, illustrator, teacher* – Studied at the Koninklijke Academie voor Beeldende Kunsten in The Hague. 1955: Oxenaar works with the architect Gerrit Rietveld on murals for the "E 55" city rebuilding exhibition in Rotterdam. 1966: the first of his bank note designs are produced for the Dutch Bank, with further designs issued from 1967 onwards. Has designed numerous postage stamps for the Dutch post offices PTT. 1976: is made director of the PTT's art and design department. 1958–70: teaches graphic design at the Koninklijke Academie voor Beeldende Kunsten in The Hague and visual presentation at the technical university in Delft.

# *PQ*

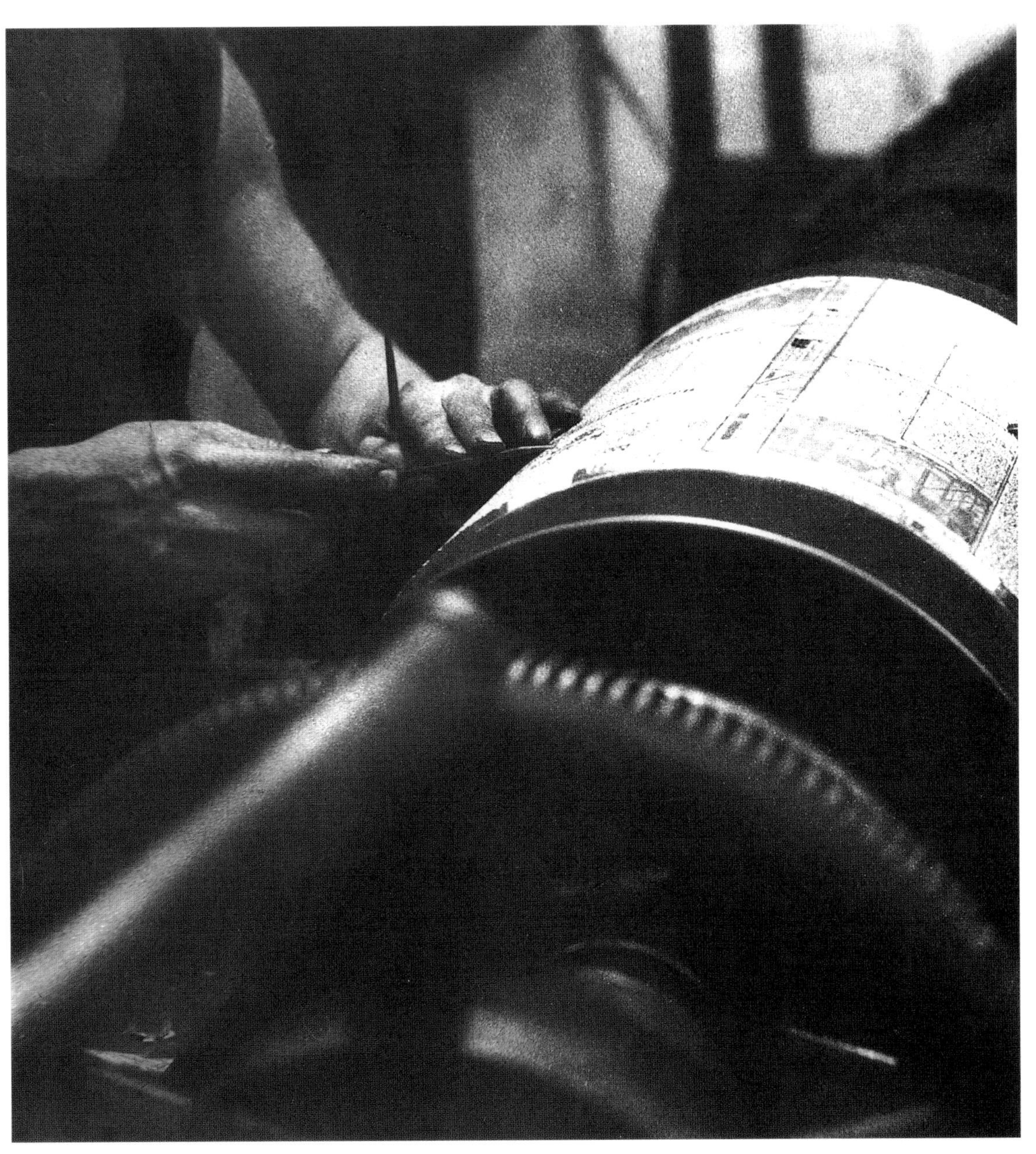

**Pakti,** Bhai – geb. 11. 8. 1925 in Devrukh, Ratnagiri, Indien. – *Grafik-Designer, Typograph, Maler* – 1944–48 Studium am Sir J. J. Institute of Applied Art in Bombay. 1950 Gestalter in der Werbeagentur Shilpi in Bombay. 1955–56 Studium an der Central School of Arts and Crafts in London. Gewinnt einen Plakat-Wettbewerb des National Trust. Repräsentiert 1956 indische Malerei auf der Biennale in Venedig. 1959 künstlerischer Leiter der Calico Mills in Ahmedabad. Arbeit für Benson's Overseas Marketing and Advertising Service.

**Palatino,** Giambattista – geb. ca. 1515 in Brescia, Italien, gest. ca. 1575 in Rom, Italien. – *Schreibmeister* – Tätig in Rom. 1540 Veröffentlichung seines Schreibbuchs „Libro nel qual s'insegna a Scrivere ogni sorte Lettera, Antica, et Moderna", das von Antonio Blado in Rom gedruckt wird. Das Buch erscheint bis 1588 in zahlreichen Auflagen. Fertigt 1544 einen Holzschnitt-Plan von Rom. 1575 Veröffentlichung seines Schreibbuchs „Libro Primo Delle Lettere Maiuscole Antiche Romane".

**Palomares,** Francisco Xavier de Santiago – *Schreibmeister* – Verfaßt 1776 in Madrid das Schreibbuch „Arte nuevo de escribir".

**Pan** – *Zeitschrift* – 1895 Gründung der Zeitschrift in Berlin. Redaktionsleiter der ersten drei Hefte sind der Kunstgeschichtler Julius Meier-Graefe und der Dichter Otto Julius Bierbaum, die nach der Veröffentlichung von Arbeiten Henri de Toulouse-Lautrecs im „Pan" vom Aufsichtsrat entlassen werden. Die Zeitschrift wird zu einem Sprachrohr des deutschen

**Pakti,** Bhai – né le 11. 8. 1925 à Devrukh, Ratnagiri, Inde – *graphiste maquettiste, typographe, peintre* – 1944–1948, études au Sir J. J. Institute of Applied Art de Bombay. 1950, designer à l'agence de publicité Shilpi à Bombay. Etudes à la Central School of Arts and Crafts de Londres de 1955 à 1956. Remporte un concours d'affiches du National Trust. En 1956, il représente la peinture indienne à la Biennale de Venise. Directeur artistique des Calico Mills à Ahmedabad en 1959. Travaux pour Benson's Overseas Marketing and Advertising Service.

**Palatino,** Giambattista – né vers 1515 à Brescia, Italie, décédé vers 1575 à Rome, Italie – *scribe* – a exercé à Rome. Publication en 1540 de son livre d'écriture «Libro nel qual s'insegna a Scrivere ogni sorte Lettera, Antica, et Moderna», imprimé à Rome par Antonio Blado. Cet ouvrage a été réédité plusieurs fois jusqu'en 1588. Réalise un plan de Rome en gravure sur bois en 1544. A publié le livre d'écriture «Libro Primo Delle Lettere Maiuscole Antiche Romane» en 1575.

**Palomares,** Francisco Xavier de Santiago – *scribe* – rédige le manuel d'écrite «Arte nuevo de escribir» à Madrid, en 1776.

**Pan** – *revue* – fondée à Berlin en 1895. L'historien d'art Julius Meier-Graefe et le poète Otto Julius Bierbaum en sont directeurs de rédaction pour les premiers numéros, ils sont licenciés par le conseil d'administration après la publication de travaux d'Henri de Toulouse-Lautrec. La revue fonctionne comme forum de l'Art nouveau en Allemagne. Le signet de la revue, qui représente la tête de Pan est l'œuvre du peintre et illustrateur Franz von Stuck. Parmi les créateurs y ayant travaillé de 1896 à 1900, on compte E. R. Weiß, Otto Eckmann et Peter Behrens.

**Pechey,** Eleisha – né en 1831 à Bury St. Edmunds, Suffolk, Angleterre, décédé en

Palatino 1545 Specimen

Pan ca. 1900 Initials

**Pakti,** Bhai – b. 11. 8. 1925 in Devrukh, Ratnagiri, India – *graphic designer, typographer, painter* – 1944–48: studies at the Sir J. J. Institute of Applied Art in Bombay. 1950: designer for the Shilpi advertising agency in Bombay. 1955–56: studies at the Central School of Arts and Crafts in London. Wins a poster competition staged by the National Trust. 1956: represents Indian painting at the Biennale in Venice. 1959: art director of Calico Mills in Ahmedabad. Works for Benson's Overseas Marketing and Advertising Service.

**Palatino,** Giambattista – b. c.1515 in Brescia, Italy, d. c.1575 in Rome, Italy – *writing master* – Worked in Rome. 1540: publishes his book on the art of writing entitled "Libro nel qual s'insegna a Scrivere ogni sorte Lettera, Antica, et Moderna", which is printed by Antonio Blado in Rome. The book is reprinted many times until 1588. 1544: completes a woodcut plan of Rome. 1575: publishes his writing book, "Libro Primo Delle Lettere Maiuscole Antiche Romane".

**Palomares,** Francisco Xavier de Santiago – *writing master* – 1776: produces his book on the art of writing, "Arte nuevo de escribir", in Madrid.

**Pan** – *magazine* – 1895: the magazine is launched in Berlin. Editors-in-chief of the first three numbers are the art historian Julius Meier-Graefe and the poet Otto Julius Bierbaum, who are dismissed from the journal's advisory board after printing work by Toulouse-Lautrec in "Pan". The magazine soon becomes a mouthpiece for German Jugendstil. The logo of the magazine depicting Pan's head is de-

Jugendstils. Das Signet des Magazins, das den Kopf des Pan zeigt, wird von dem Illustrator und Maler Franz von Stuck entworfen. Gestalterische Mitarbeiter zwischen 1896 und 1900 sind u. a. E. R. Weiß, Otto Eckmann und Peter Behrens.

**Pechey,** Eleisha – geb. 1831 in Bury St. Edmunds, Suffolk, England, gest. 1902 in London, England. – *Drucker, Buchbinder, Schriftentwerfer* – 1846 Lehre als Buchdrucker und Buchbinder in Bury St. Edmunds. Umzug nach London, Arbeit als Korrektor. 1863 Eintritt in die Schriftgießerei Stephenson, Blake & Co. in Sheffield, für die er ab 1873 als Verkaufsleiter in London tätig ist. – *Schriftentwürfe:* Charlemagne (1886), Windsor (1910).

**Peckolick,** Alan – geb. 3. 10. 1940 in New York, USA. – *Grafik-Designer, Typograph, Lehrer* – Studium am Pratt Institute in New York. 1961–63 Art Director bei McCann-Erickson Advertising, 1963–64 bei Kenyon-Eckhardt Advertising. 1964–68 Assistent von Herb Lubalin. 1968–73 eigenes Studio in New York. 1973 Vize-Präsident von Lubalin, Smith, Carnase. 1974 Entwurf der Goldmedaille des Art Directors Club of New York. 1980–82 Präsident von Lubalin, Peckolick Associates. 1982–85 Direktor von Pushpin Lubalin Peckolick. 1986–91 eigenes Studio. 1991 Gründung von Peckolick Inc. mit Büros in Atlanta, New York und Tokio. Unterrichtet u. a. 1975–78 am Pratt Institute in New York und 1981–86 an der Syracuse University in New York. – *Publikation u. a.:* „Herb Lubalin. Art Director, Graphic-Designer and Typographer" (mit Gertrude Snyder), New York 1985.

Pechey 1886 Charlemagne

Peckolick 1982 Poster

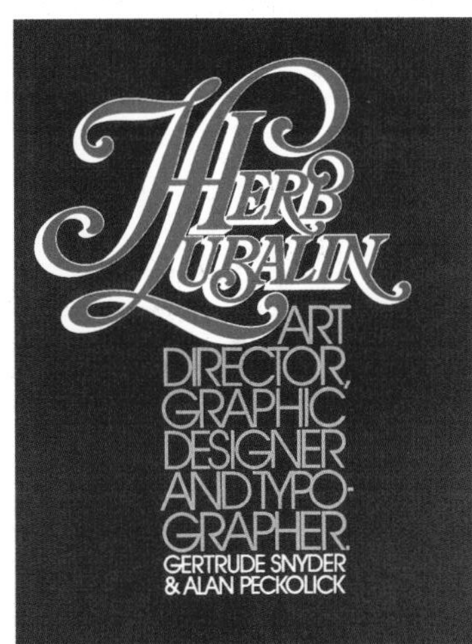

Peckolick 1985 Cover

Peckolick 1984 Logo

1902 à Londres, Angleterre – *imprimeur, relieur, concepteur de polices* – apprentissage d'imprimerie et de reliure à Bury St. Edmunds en 1846. S'installe à Londres où il travaille comme correcteur. Entre à la fonderie de caractères Stephenson, Blake & Co à Sheffield en 1863, pour laquelle il exercera les fonctions de directeur des ventes à Londres à partir de 1873. – *Polices:* Charlemagne (1886), Windsor (1910).

**Peckolick,** Alan – né le 3. 10. 1940 à New York, Etats-Unis – *graphiste maquettiste, typographe, enseignant* – études au Pratt Institute de New York. Directeur artistique chez McCann-Erikson Advertising de 1961 à 1963, puis chez Kenyon-Eckhardt Advertising de 1963 à 1964. Assistant d'Herb Lubalin de 1964 à 1968. Exerce dans son atelier à New York de 1968 à 1973. Vice-président de Lubalin, Smith, Carnase en 1973. Dessine la médaille d'or de l'Art Directors Club of New York en 1974. Président de Lubalin, Peckolick Associates, de 1980 à 1982. Directeur de Pushpin Lubalin Peckolick de 1982 à 1985. Exerce dans sa propre agence de 1986 à 1991. Fonde Peckolick Inc. avec des agences à Atlanta, New York et Tokyo en 1991. A enseigné, entre autres, au Pratt Institute de New York de 1975 à 1978 et à la Syracuse University à New York de 1981 à 1986. – *Publication, sélection:* «Herb Lubalin. Art Director, Graphic-Designer and Typographer» (avec Gertrude Snyder), New York 1985.

signed by illustrator and painter Franz von Stuck. 1896–1900: those working on the design of "Pan" include E. R. Weiß, Otto Eckmann and Peter Behrens.

**Pechey,** Eleisha – b. 1831 in Bury St. Edmunds, Suffolk, England, d. 1902 in London, England – *printer, bookbinder, type designer* – 1846: trains as a printer and bookbinder in Bury St. Edmunds. Moves to London and works as a proof-reader. 1863: joins the Stephenson, Blake & Co. type foundry in Sheffield, working for them as their sales manager in London from 1873 onwards. – *Fonts:* Charlemagne (1886), Windsor (1910).

**Peckolick,** Alan – b. 3. 10. 1940 in New York, USA – *graphic designer, typographer, teacher* – Studied at the Pratt Institute in New York. 1961–63: art director for McCann-Erickson Advertising and from 1963–64 for Kenyon-Eckhardt Advertising. 1964–68: assistant to Herb Lubalin. 1968–73: runs his own studio in New York. 1973: vice-president of Lubalin, Smith, Carnase. 1974: designs the Art Directors Club of New York's gold medal. 1980–82: president of Lubalin, Peckolick Associates. 1982–85: director of Pushpin Lubalin Peckolick. 1986–91: has his own studio. 1991: founds Peckolick Inc. with branches in Atlanta, New York and Tokyo. Peckolick has held teaching positions at the Pratt Institute in New York from 1975 to 1978 and at Syracuse University in New York from 1981 to 1986, among others. – *Publications include:* "Herb Lubalin. Art Director, Graphic-Designer and Typographer" (with Gertrude Snyder), New York 1985.

**Pedersen,** B. Martin – geb. 17.4.1937 in Brooklyn, USA. – *Grafik-Designer, Verleger, Lehrer* – Arbeit in verschiedenen Werbeagenturen und Grafik-Design-Studios. 1966–68 Corporate Design Director der American Airlines. 1968 Gründung seines Büros „Pedersen Design Inc.“. Auftraggeber waren u.a. Dow Jones & Co., die American Airlines, McGraw Hill, Seagrams, Volkswagen, IBM und American Express. Zahlreiche Zeitschriften-Gestaltungen, u.a. für „Business Week“, „Esquire“, „Audio“, „Business Month“. 1978 Gründung der Design Association „Jonson Pedersen Hinrichs & Shakery“ in New York und San Francisco. Erwirbt 1986 den internationalen Verlag „Graphis Press“ mit der Zeitschrift „Graphis“. Ausgezeichnet mit über 300 Preisen für seine gestalterische Arbeit, u.a. durch das American Institute of Graphic Arts (AIGA), den New York Art Directors Club, die Society of Publication Designers und den Type Directors Club of New York.
**Peignot,** Charles Armand – geb. 16.8. 1897 in Paris, Frankreich, gest. 1.11.1983 in Paris, Frankreich. – *Unternehmer, Verleger* – 1919 Eintritt in das Unternehmen seiner Familie Peignot & Cie. Durch die Fusion mit der Firma Girard & Cie., die im Besitz des Schriftmaterials der Firma Deberny ist, wird 1923 die Firma Deberny & Peignot gegründet. 1923–39 künstlerischer Leiter der Firma Deberny & Peignot. Zusammenarbeit mit A. M. Cassandre, dessen Schriften er produziert: Bifur (1928), Acier noir (1936), Peignot (1936). 1927–39 Gründer und Herausgeber der Zeitschrift „Arts & Métiers Graphiques“.

**Pedersen,** B. Martin – né le 17.4.1937 à Brooklyn, Etats-Unis – *graphiste maquettiste, éditeur, enseignant* – travaille dans diverses agences de publicité et ateliers de graphisme et de design. 1966–1968, Corporate Design Director d'American Airlines. Fonde sa propre agence «Pedersen Design Inc.» en 1968. Commanditaires : Dow Jones & Co., American Airlines, McGraw Hill, Seagrams, Volkswagen, IBM, American Express etc. Nombreuses maquettes de revues, dont «Business Week», «Esquire», «Audio» et «Business Month». En 1978, il fonde l'association de design «Jonson Pedersen Hinrichs & Shakery» à New York et San Francisco. En 1986, il rachète les éditions internationales «Graphis Press» et la revue «Graphis». Plus de 300 prix lui ont été décernés pour son travail de création, dont celui de l'American Institute of Graphic Arts (AIGA), le New York Art Directors Club, la Society of Publication Designers, le Type Directors Club de New York.
**Peignot,** Charles Armand – né le 16.8. 1897 à Paris, France, décédé le 1.11.1983 à Paris, France – *chef d'entreprise, éditeur* – entre dans l'entreprise familiale Peignot & Cie en 1919. La société Deberny & Peignot est fondée en 1923 à la suite de la fusion avec la société Girard & Cie qui détenait les matrices de la société Deberny. Directeur artistique de la société Deberny & Peignot de 1923 à 1939. Travaille avec Cassandre dont il produit les fontes : Bifur (1928), Acier noir (1936), Peignot (1936). Fonde la revue «Arts & Métiers Graphiques» en 1927 qu'il publie jusqu'en 1939. Président directeur général de la société de 1939 à 1972. En 1952, Peignot fait entrer Adrian Frutiger dans l'entreprise, comme concepteur de polices et directeur artistique. En 1953, Peignot remet à l'université de Cambridge les ma-

Pedersen 1984 Cover

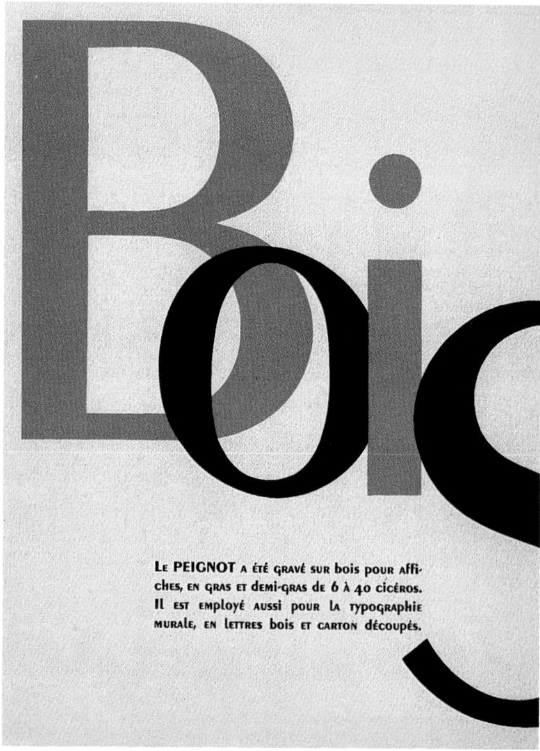

Le PEIGNOT a été gravé sur bois pour affiches, en gras et demi-gras de 6 à 40 cicéros.
Il est employé aussi pour la typographie murale, en lettres bois et carton découpés.

Peignot 1937 Page

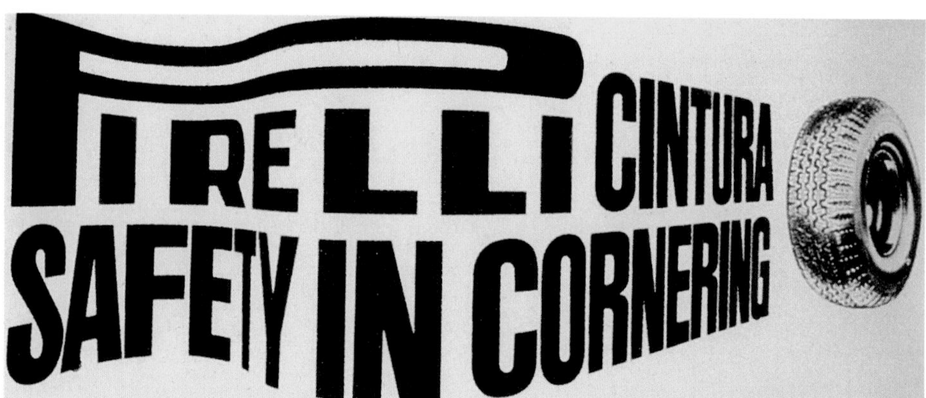

Pentagram 1962 Poster

**Pedersen,** B. Martin – b. 17.4.1937 in Brooklyn, USA – *graphic designer, publisher, teacher* – Work in various advertising agencies and graphic design studios. 1966–68: corporate design director for American Airlines. 1968: opens his own studio, Pedersen Design Inc. Clients include Dow Jones & Co., American Airlines, McGraw Hill, Seagrams, Volkswagen, IBM and American Express. Designs numerous magazines, including "Business Week", "Esquire", "Audio" and "Business Month". 1978: founds the Jonson Pedersen Hinrichs & Shakery design association in New York and San Francisco. 1986: buys up international Graphis Press and their "Graphis" magazine. Pedersen's work in design has won him over 300 prizes, including awards from the American Institute of Graphic Arts (AIGA), the New York Art Directors Club, the Society of Publication Designers and the Type Directors Club of New York.
**Peignot,** Charles Armand – b. 16.8.1897 in Paris, France, d. 1.11.1983 in Paris, France – *entrepreneur, publisher* – 1919: joins the family business, Peignot & Cie. 1923: Peignot merges with Girard & Cie., the company which owns Deberny's type materials, and the Deberny & Peignot company is founded. 1923–39: art director for Deberny & Peignot. Works with A. M. Cassandre and produces the following Cassandre typefaces: Bifur (1928), Acier noir (1936) and Peignot (1936). 1927–39: founds and edits the magazine "Arts & Métiers Graphiques". 1939–72: is the company's general director. 1952: Peignot persuades Adrian Frutiger to join

1939–72 Generaldirektor der Firma. 1952 gewinnt Peignot Adrian Frutiger als Schriftentwerfer und Art Director für sein Unternehmen. 1953 übergibt Peignot die Originalstempel der Baskerville-Schriften, die sich seit 1929 im Besitz von Deberny & Peignot befinden, an die Universität Cambridge. 1954 aktives Engagement bei der Entwicklung des Fotosatzes. Kauft die Rechte zur Produktion der „Lumitype"-Photosatz-Maschinen. 1957–68 Mitbegründer und erster Präsident der Association Typographique Internationale (ATypI).

**Pentagram** – *Design-Studio* – 1962 Gründung des Design-Studios mit Alan Fletcher, Colin Forbes und Bob Gill. 1965 Austritt von Bob Gill, Umbenennung des Studios in Crosby, Fletcher, Forbes (mit Theo Crosby). 1972 Gründung von Pentagram in London (mit T. Crosby, A. Fletcher, C. Forbes, Kenneth Grange, Mervyn Kurlansky). Eintritt von weiteren Partnern: John McConnell (1974), David Hillman (1978), Peter Harrison (1979). 1978 Gründung des Pentagram-Büros in New York, neue Partner sind Etan Manasse (1987), Woody Pirtle (1988), Michael Bierut (1991), Paula Scher (1991) und James Biber (1991). 1986 Gründung des Pentagram Büros in San Francisco, neue Partner werden Kit Hinrichs, Linda Hinrichs, Neil Shakery (1992) und Cowell Williams (1992). Neue Partner von Pentagram in London werden 1990 John Rushworth und Peter Saville. 1991 Gründung des jährlichen „Pentagram Prize" für Studierende in Zusammenarbeit mit der Zeitschrift „Graphis". 1992 Austritt von Alan Fletcher. Auftraggeber waren u. a. Lloyd's

Pentagram 1972 Cover

Pentagram 1962 Logo

Pentagram 1983 Poster

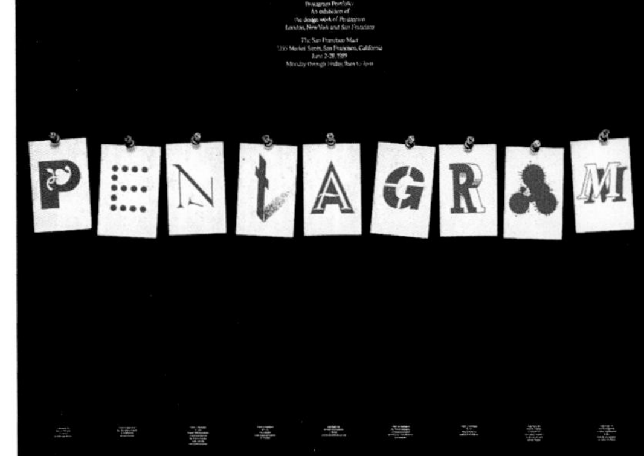

Pentagram 1989 Poster

trices originales des fontes Baskerville dont Deberny & Peignot étaient propriétaires depuis 1929. S'engage activement dans le développement de la photocomposition en 1954. Achète les droits pour produire des machines à photocomposer «Lumitype». Cofondateur de l'Association Typographique Internationale (ATypI) en 1957, dont il est le premier président jusqu'en 1968.

**Pentagram** – *atelier de design* – l'atelier de design réunissant Alan Fletcher, Colin Forbes, Bob Gill est fondé en 1962. Bob Gill quitte l'atelier en 1965, celui-ci s'appelle alors Crosby, Fletcher, Forbes (avec Theo Crosby). Pentagram est fondé à Londres en 1972 (avec T. Crosby, A. Fletcher, C. Forbes, Kenneth Grange, Mervyn Kurlansky). D'autres partenaires y entrent : John McConnell (1974), David Hillman (1978) et Peter Harrison (1979). L'agence Pentragram de New York est créée en 1978 avec de nouveaux partenaires : Etan Manasse (1987), Woody Pirtle (1988), Michael Bierut (1991), Paula Scher (1991) et James Biber (1991). L'agence de San Francisco est fondée en 1986; Kit Hinrichs, Linda Hinrichs, Neil Shakery (1992), Cowell Williams (1992) en sont les nouveaux partenaires. A Londres, de nouveaux collaborateurs entrent chez Pentagram en 1990 : John Rushworth et Peter Saville. Création en 1991 du prix annuel «Pentragram Prize» en coopération avec la revue «Graphis» pour récompenser des étudiants. Alan Fletcher quitte Pentagram en 1992. Commanditaires : Lloyd's of London, Reuters,

the company as their type designer and art director. 1953: Peignot presents the original punches of Baskerville's typefaces, in Deberny & Peignot's possession since 1929, to the University of Cambridge. 1954: active participant in the development of the filmsetting process. Buys the rights to manufacture Lumitype filmsetting machines. 1957–68: cofounder and first president of the Association Typographique Internationale (ATypI).

**Pentagram** – *design studio* – 1962: the design studio is founded by Alan Fletcher, Colin Forbes and Bob Gill. 1965: Bob Gill leaves the studio, which is thus renamed Crosby, Fletcher, Forbes (with Theo Crosby). 1972: Pentagram is launched in London (with T. Crosby, A. Fletcher, C. Forbes, Kenneth Grange and Mervyn Kurlansky). Further partners join the business: John McConnell (1974), David Hillman (1978) and Peter Harrison (1979). 1978: a branch of Pentagram is opened in New York. New partners are Etan Manasse (1987), Woody Pirtle (1988), Michael Bierut (1991), Paula Scher (1991) and James Biber (1991). 1986: a Pentagram studio is set up in San Francisco, with Kit Hinrichs, Linda Hinrichs, Neil Shakery (1992) and Cowell Williams (1992) as new partners. 1990: John Rushworth and Peter Saville join Pentagram in London. 1991: Pentagram launches the yearly Pentagram Prize for students in collaboration with "Graphis" magazine. 1992: Alan Fletcher leaves Pentagram. Clients include Lloyd's of London, Reuters, IBM Europe, Cunard, Nissan, Rank Xerox,

of London, Reuters, IBM Europe, Cunard, Nissan, Rank Xerox, American Express, British Olivetti, die Chase Manhattan Bank, Pirelli, Polaroid, Kodak, Penguin Books. Die Arbeiten der verschiedenen Studiomitglieder als auch die Wirkung von Pentagram als Institution sind von Beginn an ein weltweites Symbol für zeitgemäßes Grafik-Design. – *Publikationen u.a.:* „Pentagram: the work of five designers", London 1972; „Living by Design", London 1978; „Pentagram. The Compendium", Schaffhausen 1993.

**Pfäffli,** Bruno – geb. 20. 3. 1935 in Basel, Schweiz. – *Grafik-Designer, Typograph, Lehrer* – Schriftsetzerlehre in Basel. Besuch der Allgemeinen Gewerbeschule in Basel, Kurse in Typographie und Schrift bei Emil Ruder. 1957–58 Gestalter in einer Großdruckerei in Helsinki, Finnland. Lebt seit 1960 in Paris. 1960–74 Mitarbeiter von Adrian Frutiger, u. a. Gestaltung der „National-Zeitung" in Basel und der „Tribune de Genève" sowie von Kursbüchern und Flugplänen der Air France. Seit 1974 selbständiger typographischer Gestalter, hauptsächlich als Gestalter von Katalogen, Plakaten und Einladungskarten tätig, u. a. für die französischen Nationalmuseen (Louvre, Grand Palais, Schloß Versailles, Musée Picasso, Musée d'Orsay, das Centre Georges Pompidou), die Hayward Gallery und Tate Gallery in London, die National Gallery in Washington, dem Museum of Modern Art und Metropolitan Museum of Art in New York und dem Musée des Beaux-Arts in Ottawa, Kanada. Gestaltung des Erscheinungsbildes der Fondation Cartier

IBM Europe, Cunard, Nissan, Rank Xerox, American Express, British Olivetti, le Chase Manhattan Bank, Pirelli, Polaroid, Kodak, Penguin Books. Dès le début, les travaux des divers membres de l'atelier, ainsi que l'influence de Pentagram devenue une institution, ont été le symbole du graphisme contemporain dans le monde entier. – *Publications, sélection :* «Pentagram : the work of five designers», Londres 1972; «Living by Design», Londres 1978; «Pentagram. The Compendium», Schaffhausen 1993.

**Pfäffli,** Bruno – né le 20. 3. 1935 à Bâle, Suisse – *graphiste maquettiste, typographe, enseignant* – apprentissage de composition typographique à Bâle. Fréquente la Allgemeine Gewerbeschule (arts et métiers) de Bâle et suit les cours de typographie et d'arts de l'écriture d'Emil Ruder. Designer dans une grande imprimerie à Helsinki, Finlande, de 1957 à 1958. Vit à Paris depuis 1960. Travaille avec Adrian Frutiger de 1960 à 1974, réalise, entre autres, la maquette du «National-Zeitung» de Bâle et celle de la «Tribune de Genève» ainsi que des carnets d'horaires pour Air France. Maquettiste et typographe indépendant depuis 1974, se spécialise dans la maquette de catalogues, affiches et cartons d'invitation pour les musées nationaux français (Louvre, Grand Palais, Château de Versailles, Musée Picasso, Musée d'Orsay, le Centre Georges Pompidou), la Hayward Gallery et la Tate Gallery de Londres, la National Gallery de Washington, le Museum of Modern Art et le Metropolitan Museum of Art de New York, et le Musée des Beaux-Arts d'Ottawa au Canada. Dessine l'identité de la «Fondation Cartier pour l'art Contemporain». A enseigné à l'Ecole Nationale Supérieure des Arts Décoratifs de Paris de 1964 à 1968 et au National Institute of Design à Ahmedabad,

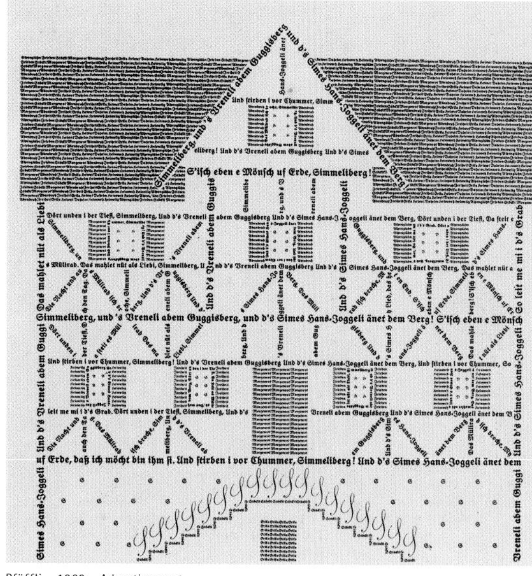

Pfäffli 1969 Advertisement

Pfund 1980 Poster

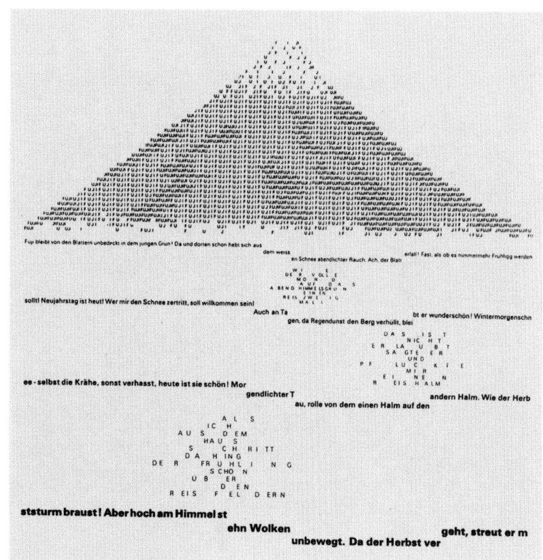

Pfäffli 1971 Advertisement

American Express, British Olivetti, Chase Manhattan Bank, Pirelli, Polaroid, Kodak and Penguin Books. Right from the beginning, the work of the various individual studio members and the influence of Pentagram as an institution were a symbol for contemporary graphic design worldwide. – *Publications include:* "Pentagram: the work of five designers", London 1972; "Living by Design", London 1978; "Pentagram. The Compendium", Schaffhausen 1993.

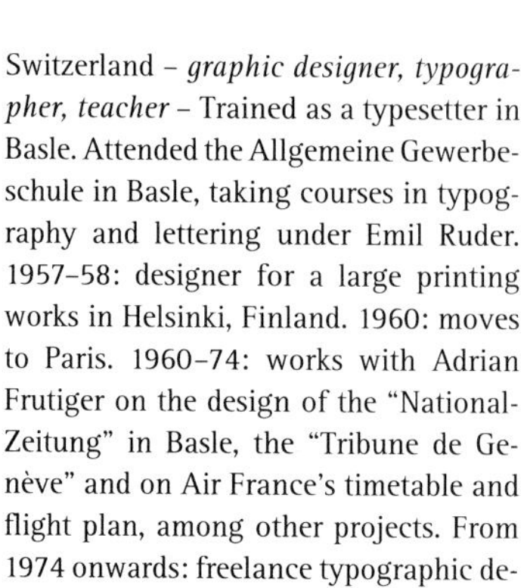

**Pfäffli,** Bruno – b. 20. 3. 1935 in Basle, Switzerland – *graphic designer, typographer, teacher* – Trained as a typesetter in Basle. Attended the Allgemeine Gewerbeschule in Basle, taking courses in typography and lettering under Emil Ruder. 1957–58: designer for a large printing works in Helsinki, Finland. 1960: moves to Paris. 1960–74: works with Adrian Frutiger on the design of the "National-Zeitung" in Basle, the "Tribune de Genève" and on Air France's timetable and flight plan, among other projects. From 1974 onwards: freelance typographic designer, mainly working on catalogue, poster and invitation card designs for a number of clients, including the French national museums (Louvre, Grand Palais, Versailles Palace, Musée Picasso and Musée d'Orsay, the Centre Georges Pompidou), the Hayward Gallery and the Tate Gallery in London, the National Gallery in Washington, the Museum of Modern Art and the Metropolitan Museum of Art in New York, and the Musée des Beaux-Arts in Ottawa, Canada. Designs the corporate identity for the Fondation Cartier

pour l'art Contemporain. Unterrichtet 1964–68 an der Ecole National Supérieure des Arts Décoratifs in Paris und am National Institute of Design in Ahmedabad, Indien.

**Pfund,** Roger – geb. 28. 12. 1943 in Bern, Schweiz. – *Grafik-Designer, Typograph, Maler, Illustrator* – 1963–66 Grafikerlehre in Bern. Studium an der Kunstgewerbeschule in Bern. Seit 1966 eigenes Atelier als Grafiker und Maler. 1969–76 Zusammenarbeit mit Elisabeth Pfund. 1971 Gewinner des Banknoten-Wettbewerbs der Nationalbank der Schweiz. Seit 1972 zahlreiche Plakatentwürfe für politische und kulturelle Institutionen: Amnesty International Bern, Groupo Memoria Genf, das Kunstmuseum Bern, das Cabinet des Estampes in Genf, das Nouveau Théâtre de Poche in Genf und das Musée d'Art et d'Histoire in Genf. Seit 1976 gemeinsames Atelier und zeitweise Zusammenarbeit mit Jean-Pierre Blanchoud. 1979 Beginn der Arbeit an den neuen Banknoten der Nationalbank von Frankreich, von denen die ersten Noten 1993 erscheinen. Umschlagentwürfe für die Editions Zoé in Genf. 1980–83 Organisation und grafische Gestaltung einer Serie von Ausstellungen für Theo Jakob in Genf und Bern. 1982–84 Gestaltung des Buches „L'Art pour l'Aare. Bernische Kunst im 20. Jahrhundert" zum 150. Jubiläum der Kantonalbank in Bern. 1985 Gestaltung des grafischen Erscheinungsbildes des „Centre Culturel Suisse Pro Helvetia" in Paris. 1986–88 Gestaltung des Erscheinungsbildes des Internationalen Museums des Roten Kreuzes in Genf.

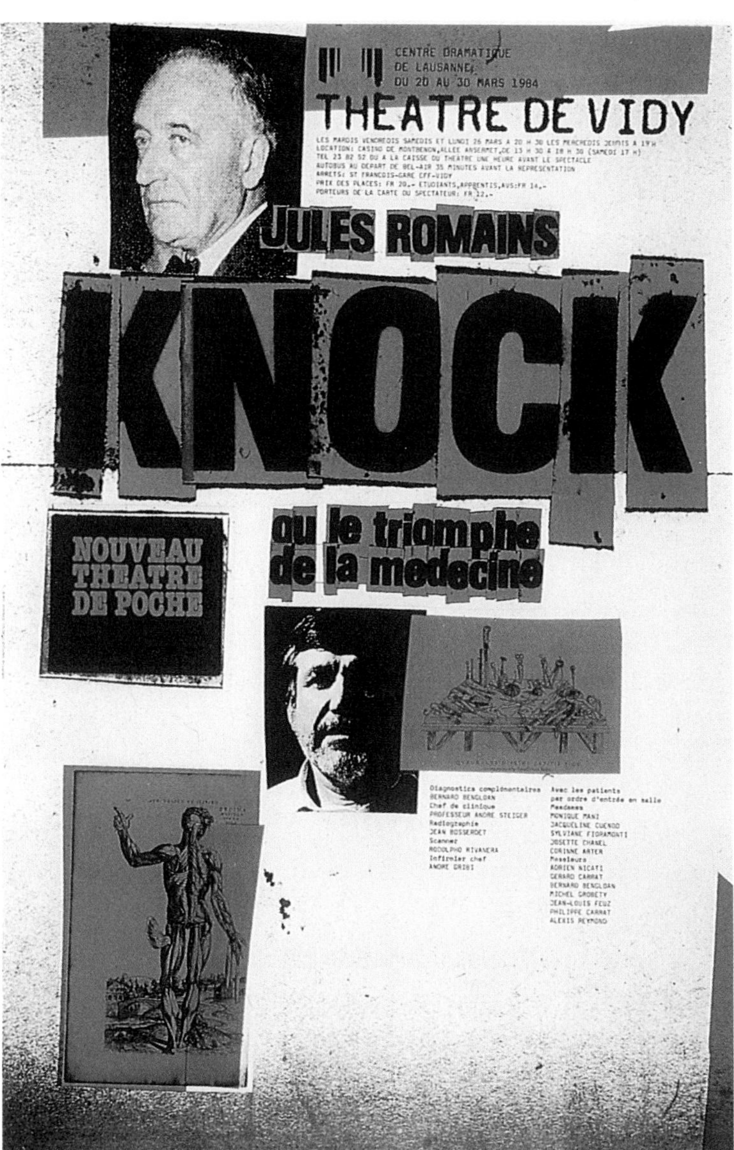


Pfund 1984 Poster

Pfund 1982 Poster

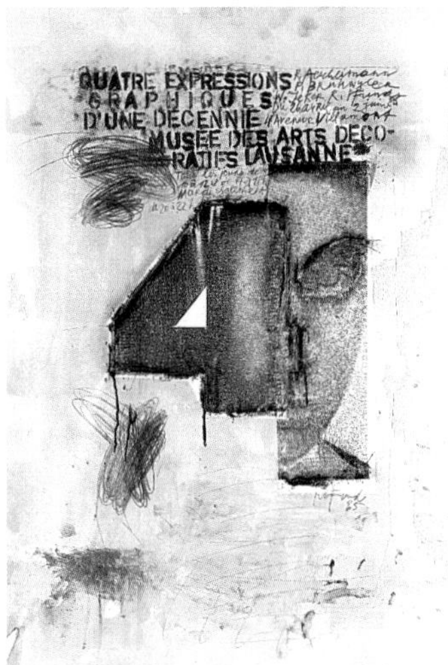


Pfund 1985 Poster

Inde.

**Pfund,** Roger – né le 28. 12. 1943 à Berne, Suisse – *graphiste maquettiste, typographe, peintre, illustrateur* – 1963–1966, apprentissage de graphiste à Berne. Etudes à la Kunstgewerbeschule (école des arts décoratifs) de Berne. Exerce comme graphiste et peintre dans son atelier à partir de 1966. Collaboration avec Elisabeth Pfund de 1969 à 1976. Lauréat du concours de billets de banques de la Banque Nationale Suisse en 1971. Nombreuses affiches pour des organisations politiques et culturelles à partir de 1972 : Amnesty International, Berne; Groupo Memoria, Genève; le Kunstmuseum, Berne; le Cabinet des Estampes, Genève; le Nouveau Théâtre de Poche de Genève; le Musée d'Art et d'Histoire de Genève. Atelier commun avec Jean Pierre Blanchoud avec qui il travaille par intermittence depuis 1976. Commence à travailler sur les nouveaux billets de banque de la Banque de France en 1979, les premières coupures circulent en 1993. Couvertures de livres pour les éditions Zoé à Genève. Organisation et design graphique d'une série d'expositions pour Theo Jakob à Genève et à Berne de 1980 à 1983. Maquette du livre «L'Art pour l'Aare. Bernische Kunst im 20. Jahrhundert» de 1982 à 1984 pour le 150e anniversaire de la Kantonalbank de Berne. Dessine, en 1985, l'identité graphique du «Centre Culturel Suisse Pro Helvetia» à Paris. 1986–1988, identité du Musée de la Croix Rouge Internationale de Genève.

pour l'art Contemporain. Has taught at the Ecole Nationale Supérieure des Arts Décoratifs in Paris from 1964 to 68 and at the National Institute of Design in Ahmedabad, India.

**Pfund,** Roger – b. 28. 12. 1943 in Bern, Switzerland – *graphic designer, typographer, painter, illustrator* – 1963–66: trains as a graphic artist in Bern. Studies at the Kunstgewerbeschule in Bern. 1966: sets up his own graphics and painting studio. 1969–76: works with Elisabeth Pfund. 1971: wins a competition to design the Swiss National Bank's new bank notes. From 1972 onwards: designs numerous posters for political and arts institutions, i.e. Amnesty International in Bern, Groupo Memoria in Geneva, Kunstmuseum Bern, Cabinet des Estampes in Geneva, the Nouveau Théâtre de Poche in Geneva and the Musée d'Art et d'Histoire in Geneva. 1976: opens a joint studio and occasionally works on projects with Jean-Pierre Blanchoud. 1979: starts working on a set of new banknotes for France's national bank. The first notes are issued in 1993. Designs book covers for Editions Zoé in Geneva. 1980–83: organizes and does the graphic design for a series of exhibitions for Theo Jakob in Geneva and Bern. 1982–84: designs the book "L'Art pour l'Aare. Bernische Kunst im 20. Jahrhundert" for the 150th anniversary of the Kantonalbank in Bern. 1985: designs the graphics for the Centre Culturel Suisse Pro Helvetia's corporate identity in Paris. 1986–88: designs the corporate identity for the International Museum of the Red Cross in Geneva.

**Piatti,** Celestino – geb. 5. 1. 1922 in Wangen, Schweiz. – *Grafik-Designer, Typograph, Illustrator* – 1937 Studium an der Kunstgewerbeschule in Zürich. 1938–42 Lehre als Grafiker in Zürich. 1944–48 Grafiker im Atelier Fritz Bühler in Basel. Seit 1948 eigenes Atelier in Riehen bei Basel. Seit 1954 zahlreiche Buchillustrationen. Seit 1959 zahlreiche Briefmarkenentwürfe für die Schweizer Post. 1960 Wandbilder für Ciba AG in Basel. Seit 1960 Gestaltung des Erscheinungsbildes des Deutschen Taschenbuch Verlags (dtv): Umschlaggestaltung, Ausstattung, Typographie, Plakate, Kataloge. Es entstehen über 500 Buchumschläge. 1962 erste Lithographien, Holz- und Linolschnitte. Atelier in Paris. 1967 Wandbilder für den Schweizer Pavillon der Expo in Montreal. 1968 sozialkritische Titelbilder für die Zeitschrift „Nebelspalter". 1968–76 Mitglied der Eidgenössischen Kommission für angewandte Kunst. 1983 Einrichtung des „Celestino-Piatti-Preises für Verlagsgrafik" durch den Deutschen Taschenbuch Verlag und erste Verleihung. 1989 Beginn der Zusammenarbeit mit dem Schweizer Lexikon-Verlag Mengis + Ziehr in Luzern. Gesamtgestaltung von 6 Bänden der Normal- und Luxusausgabe. Zahlreiche Ausstellungen seiner Arbeiten, u. a. in der Kunstbibliothek Berlin (1964), dem Focke-Museum Bremen (1966), dem Kunsthaus Schaller Stuttgart (1976) und dem Stadtmuseum München (1982). – *Publikationen u. a.:* „Reisen mit Pinsel, Stift und Feder", Basel 1962; „Celestino Piatti: Plakate", München 1992. Manuel Gasser „Celestino Piatti", Zürich 1979.

**Piatti,** Celestino – né le 5. 1. 1922 à Wangen, Suisse – *graphiste maquettiste, typographe, illustrateur* – 1937, études à la Kunstgewerbeschule (école des arts décoratifs) de Zurich. 1938–1942, apprentissage de graphiste à Zurich. Graphiste à l'atelier Fritz Bühler à Bâle de 1944 à 1948. Exerce dans son propre atelier à Riehen, près de Bâle, à partir de 1948. Réalise de nombreuses illustrations de livres à partir de 1954. Nombreux timbres pour les postes suisses depuis 1959. Fresques pour Ciba AG à Bâle en 1960. Travaille sur l'identité des éditions Deutscher Taschenbuch Verlag dtv à partir de 1960 : couvertures, maquette, typographie, affiches, catalogues. A réalisé plus de 500 couvertures de livres. Premières lithographies, gravures sur bois et linogravures en 1962. Atelier à Paris. Réalise des fresques pour le pavillon suisse de l'Expo Montréal en 1967. Gros titres critiques pour la revue «Nebelspalter» en 1968. Membre de la Commission helvétique des arts appliqués de 1968 à 1976. Création et première remise du prix «Celestino-Piatti-Preis für Verlagsgrafik» par les éditions dtv en 1983. Début de la collaboration avec les éditions suisses de dictionnaires Mengis + Ziehr à Lucerne, en 1989. Réalise la maquette de 6 volumes de l'édition courante et de l'édition de luxe. Nombreuses expositions de ses travaux, entre autres, à la Kunstbibliothek, Berlin (1964), au Focke Museum, Bremen (1966), au Kunsthaus Schaller, Stuttgart (1976), au Stadtmuseum, Munich (1982). – *Publications, sélection :* «Reisen mit Pinsel, Stift und Feder», Bâle 1962; «Celestino Piatti : Plakate», Munich 1992. Manuel Gasser «Celestino Piatti», Zurich 1979.

**Pierpont,** Frank Hinman – né en 1860 à New Haven, Etats-Unis, décédé en 1937 à Londres, Angleterre – *manager, concep-*

Herausgegeben von Siegfried Thalheimer

Die Affäre Dreyfus

dtv dokumente

Piatti 1963 Cover

das neue Programm April bis September '63: vielseitig anspruchsvoll profiliert

Piatti 1963 Advertisement

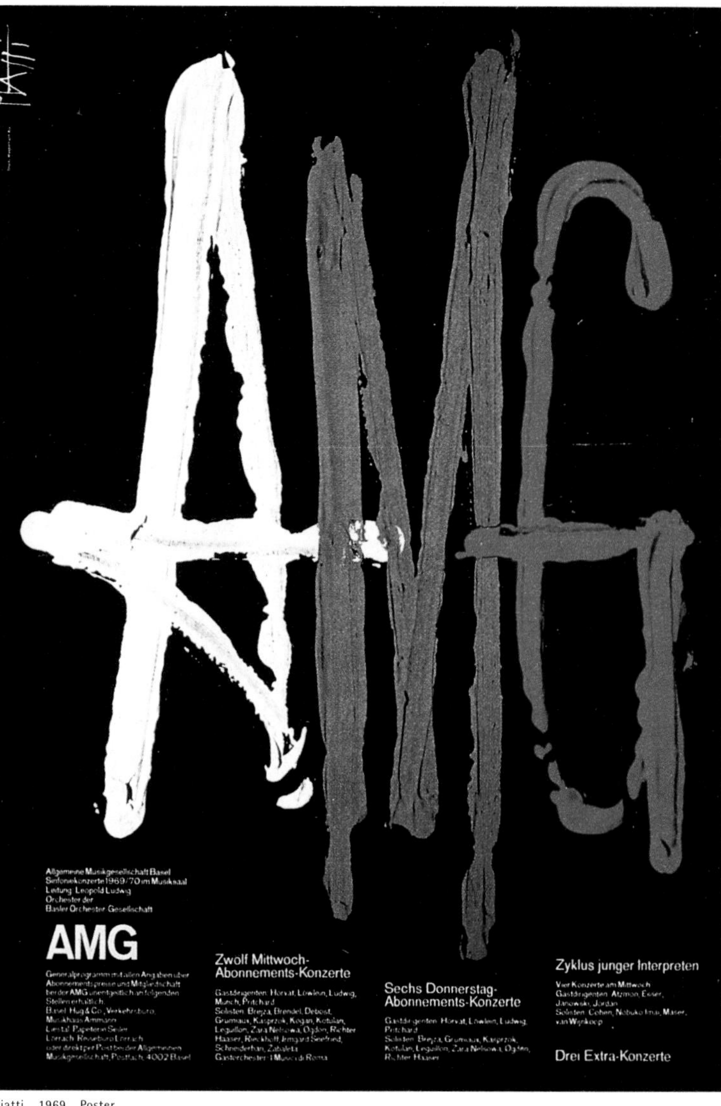


Piatti 1969 Poster

**Piatti,** Celestino – b. 5. 1. 1922 in Wangen, Switzerland – *graphic designer, typographer, illustrator* – 1937: studies at the Kunstgewerbeschule in Zurich. 1938–42: trains as a graphic artist in Zurich. 1944–48: graphic artist at Fritz Bühler's studio in Basle. 1948: opens his own studio in Riehen near Basle. From 1954 onwards: produces numerous book illustrations. From 1959 onwards: designs numerous postage stamps for the Swiss post office. 1960: designs murals for Ciba AG in Basle. From 1960 onwards: creates the corporate identity for Deutscher Taschenbuch Verlag dtv, i.e. cover designs, book designs, typography, posters and catalogues. Piatti designs over 500 book covers. 1962: produces his first lithographs, woodcuts and linocuts. Studio in Paris. 1967: fashions the murals for the Swiss pavilion at Expo in Montreal. 1968: produces sociocritical cover illustrations for "Nebelspalter" magazine. 1968–76: member of the Eidgenössische Kommission für angewandte Kunst (Swiss Federal Commission for Applied Art). 1983: Deutscher Taschenbuch Verlag introduce and award the first Celestino Piatti Prize for graphics in publishing. 1989: starts working for the Swiss encyclopedia publishers Mengis + Ziehr in Lucerne. Designs 6 volumes of their standard and luxury edition in their entirety. There have been many exhibitions of Piatti's works, including at the Kunstbibliothek in Berlin (1964), the Focke Museum Bremen (1966), the Kunsthaus Schaller in Stuttgart (1976) and the Stadtmuseum in Munich (1982). – *Publications include:* "Reisen mit Pinsel, Stift

**Pierpont,** Frank Hinman – geb. 1860 in New Haven, USA, gest. 1937 in London, England. – *Manager, Schriftentwerfer* – 1880 Mechanikerausbildung in Hartford, Connecticut. 1885 Arbeit in einer Patentkanzlei, wo er Zeichnungen einer Setzmaschine für das Patentamt anfertigt. 1894 Umzug nach Europa. 1896 Direktor der Typograph Setzmaschinen-Fabrik in Berlin. 1899–1936 Aufbau und Betriebsleiter des Werks der englischen Monotype in Salfords. Überarbeitung klassischer Schriften für die Monotype-Technik. 1900–12 Weiterentwicklung und Verbesserungen von zahlreichen Monotype-Maschinen zur Schriftherstellung. Als Neuschöpfung einer klassischen Schrift für Monotype entwirft Pierpont 1913 die Plantin (mit dem Berliner Fritz Steltzer). – *Schriftentwurf:* Plantin (1913).

**Pineles,** Cipe – geb. 23.6.1910 in Slovita, Polen – *Typographin, Grafik-Designerin, Art Directorin, Lehrerin* – 1923 Umzug in die Vereinigten Staaten. 1927–31 Studium am Pratt Institute, New York. 1930 Studium an der Louis Comfort Tiffany Foundation, New York. 1931–33 Designerin im Contempora Industrial Design Studio, New York. 1933– 36 Assistentin von Mehemed Fehmy Agha bei der Zeitschrift „Vogue", New York. Art Directorin bei den Zeitschriften „Vogue" London (1936–38), „Glamour" New York (1938–45), „Overseas Woman" Paris (1945–46), „Seventeen" New York (1947–50), „Charm" New York (1950–58), „Mademoiselle" New York (1958–59). 1961 Zusammenarbeit und Heirat mit dem Grafik-Designer Will

Piatti 1969 Poster

| | n | **n** | | | *n* |
|---|---|---|---|---|---|
| a | b | e | f | g | i |
| o | r | s | t | y | z |
| A | B | C | E | G | H |
| M | O | R | S | X | Y |
| 1 | 2 | 4 | 6 | 8 | & |

Pierpont 1913 Plantin

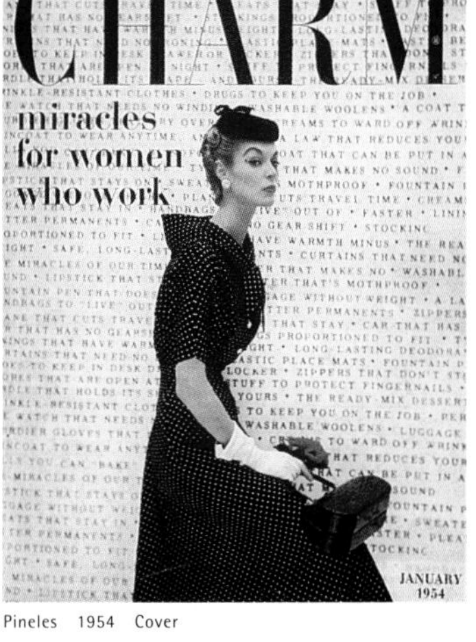


Pineles 1954 Cover

Pineles 1965 Cover

*teur de polices* – 1880, formation de mécanicien à Hartford, Connecticut. Travaille en 1885 au bureau des brevets où il réalise les dessins d'une machine à composer pour l'Office des brevets. S'installe en Europe en 1894. Directeur de la société Typograph Setzmaschinen-Fabrik de Berlin en 1896. Développe et dirige l'usine anglaise Monotype de Salfords de 1899 à 1936. Adapte des fontes classiques à la technologie Monotype. De 1900 à 1912, il développe et améliore de nombreuses machines Monotype destinées à fabriquer des caractères. En 1913, il crée une fonte classique, la Pierpont de Plantin pour Monotype (avec le Berlinois Fritz Seltzer). – *Police:* Plantin (1913).

**Pineles,** Cipe – née le 23.6.1910 à Slovita, Pologne – *typographe, graphiste maquettiste, directeur artistique, enseignante* – s'installe aux Etats-Unis en 1923. Etudes au Pratt Institute, Brooklyn, New York, de 1927 à 1931. Etudes à la Louis Comfort Tiffany Foundation, New York, en 1930. Designer au Contempora Industrial Design Studio, New York de 1931 à 1933. Assistante de Mehemed Fehmy Agha pour la revue «Vogue», New York. Directeur artistique des revues «Vogue», Londres (1936–1938), «Glamour», New York (1938–1945), «Overseas Woman», Paris (1945–1946), «Seventeen», New York (1947–1950), «Charm», New York (1950–1958), «Mademoiselle», New York (1958–1959). Elle travaille à

und Feder", Basle 1962; "Celestino Piatti: Plakate", Munich 1992. Manuel Gasser "Celestino Piatti", Zurich 1979.

**Pierpont,** Frank Hinman – b. 1860 in New Haven, USA, d. 1937 in London, England – *manager, type designer* – 1880: trains as a mechanic in Hartford, Connecticut. 1885: works in a patent office for whom he makes plans of a typesetting machine. 1894: moves to Europe. 1896: director of Typograph Setzmaschinen-Fabrik in Berlin. 1899–1936: helps establish the English branch of Monotype in Salfords and is their factory manager. Adapts classic typefaces to suit Monotype technology. 1900–12: develops and improves many different Monotype machines used in type production. 1913: Pierpont designs Plantin, a new classic typeface creation for Monotype, with Fritz Steltzer from Berlin. – *Font:* Plantin (1913).

**Pineles,** Cipe – b. 23.6.1910 in Slovita, Poland – *typographer, graphic designer, art director, teacher* – 1923: moves to the USA. 1927–31: studies at the Pratt Institute, Brooklyn, in New York. 1930: studies at the Louis Comfort Tiffany Foundation, New York. 1931–33: designer at Contempora Industrial Design Studio, New York. 1933–36: assistant to Mehemed Fehmy Agha at "Vogue" magazine in New York. Art director for the magazines "Vogue" in London (1936–38), "Glamour" in New York (1938–45), "Overseas Woman" in Paris (1945–46), "Seventeen" in New York (1947–50), "Charm" in New York (1950–58) and "Mademoiselle" in New York (1958–59). 1961: works with and marries the graphic

Burtin (gest. 1972). 1961–1972 Grafik-Design-Consultant des Lincoln Center for the Performing Arts, New York. Unterrichtet seit 1963 an der Parson's School of Design in New York, 1977 an der Cooper Union in New York, 1978 an der Harvard University in Cambridge, Massachusetts. Zahlreiche Auszeichnungen, u. a. 1975 als erste Frau in die Art Directors Hall of Fame New York gewählt; 1978 Society of Publication Designers Award of Excellence, New York.

**Pintori,** Giovanni – geb. 14. 7. 1912 in Tresnuraghes, Italien. – *Grafik-Designer, Maler* – 1930–36 Studium am Istituto Superiore per le Industrie Artistiche in Monza. 1936–50 Arbeit als Grafik-Designer in der Werbeabteilung von Olivetti in Mailand, 1950–67 Leiter. Bestimmt durch seine Gestaltung von Kalendern, Plakaten, Anzeigen, Verpackungen, Broschüren und Prospekten das Erscheinungsbild des Unternehmens, das weltweit zu den innovativsten im Bereich der Gestaltung gehört. 1950 Verleihung des „Palma d'Oro"-Preises in Mailand. 1957 mit dem großen Preis der Triennale in Mailand ausgezeichnet. Seit 1968 eigenes Studio als Grafik-Designer in Mailand. Zahlreiche Ausstellungen seines Werks, u. a. im Museum of Modern Art in New York (1952).

**Pirtle,** Woody – geb. 20. 2. 1944 in Corsicana, USA. – *Grafik-Designer* – 1967–69 Studium an der University of Arkansas in Fayetteville. 1969–74 und 1975–78 Grafik-Designer bei „The Richards Group" in Dallas. 1974–75 eigenes Atelier in Houston. 1978–88 Gründung und Leitung

partir de 1961 avec son époux, le graphiste maquettiste Will Burtin (décédé en 1972). Consultante en graphisme et design au Lincoln Center for the Performing Arts, New York, de 1961 à 1972. A enseigné à la Parson's School of Design de New York à partir de 1963, à la Cooper Union, New York, en 1977, à la Harvard University à Cambridge, Massachusetts, en 1978. Nombreuses distinctions. En 1975, elle est la première femme élue à l'Art Directors Hall of Fame, New York; en 1978, elle reçoit le Society of Publication Designers Award of Excellence à New York.

**Pintori,** Giovanni – né le 14. 7. 1912 à Tresnuraghes, Italie – *graphiste maquettiste, peintre* – 1930–1936, études à l'Istituto Superiore per le Industrie Artistiche à Monza. Exerce comme graphiste maquettiste dans le service de publicité d'Olivetti à Milan de 1936 à 1950, qu'il dirige ensuite de 1950 à 1967. Ses calendriers, affiches, annonces, emballages, brochures et prospectus ont marqué l'identité de l'entreprise, leur design comptait parmi les créations les plus innovatrices dans ce domaine. La «Palma d'Oro» lui est décernée à Milan en 1950, il est ensuite lauréat du grand prix de la Triennale de Milan en 1957. Exerce comme graphiste maquettiste dans son atelier de Milan à partir de 1968. Nombreuses expositions de son œuvre, entre autres, au Museum of Modern Art de New York (1952).

**Pirtle,** Woody – né le 20. 2. 1944 à Corsicana, Etats-Unis – *graphiste maquettiste* – 1967–1969, études à l'University of Arkansas à Fayetteville. Graphiste maquettiste pour «The Richards Group» à Dallas de 1969 à 1974, puis de 1975 à 1978. Exerce dans son propre atelier à Houston de 1974 à 1975. Fonde sa propre agence «Pirtle Design» à Dallas en 1978 et la dirige jusqu'en 1988. Commanditaires: Knoll International, Omega Opti-

Pintori 1948 Poster

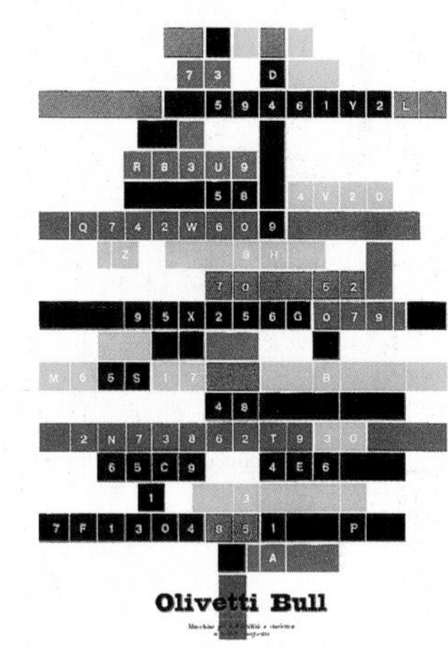

Pintori 1950 Cover

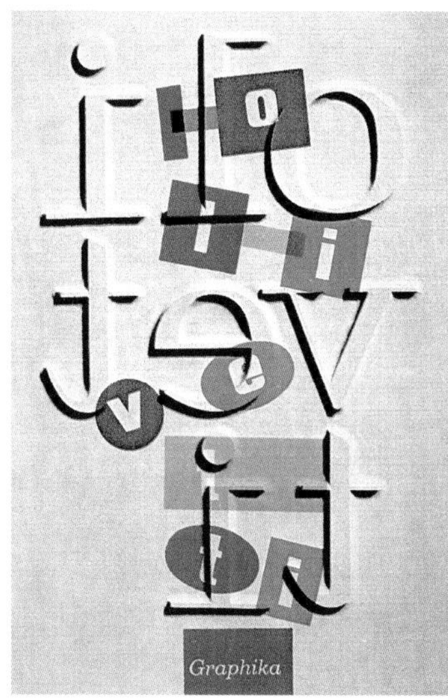

Pintori 1953 Advertisement

designer Will Burtin (d. 1972). 1961–1972: graphic design consultant to the Lincoln Center for the Performing Arts in New York. 1963: starts teaching at the Parson's School of Design in New York. Teaches at the Cooper Union in New York in 1977 and in 1978 at Harvard University in Cambridge, Massachusetts. Has won numerous awards, including in 1975 being the first woman to be elected into the Art Directors Hall of Fame in New York and in 1978 the Society of Publication Designers Award of Excellence.

**Pintori,** Giovanni – b. 14. 7. 1912 in Tresnuraghes, Italy – *graphic designer, painter* – 1930–36: studies at the Istituto Superiore per le Industrie Artistiche in Monza. 1936–50: works as a graphic designer in Olivetti's publicity department in Milan, and heads the department from 1950–67. His calendar, poster, advert, packaging, brochure and pamphlet designs determine the corporate identity of the company, considered one of the most innovative in the field of design. 1950: is awarded the Palma d'Oro Prize in Milan. 1957: awarded the Milan Triennale's grand prix. 1968: opens his own graphic design studio in Milan. There have been numerous exhibitions of his work, including at the Museum of Modern Art in New York (1952).

**Pirtle,** Woody – b. 20. 2. 1944 in Corsicana, USA – *graphic designer* – 1967–69: studies at the University of Arkansas in Fayetteville. 1969–74 and 1975–78: graphic designer for The Richards Group in Dallas. 1974–75: has his own studio in Houston. 1978–88: founds and runs the

seines Büros „Pirtle Design" in Dallas. Auftraggeber waren u. a. Knoll International, Omega Optical Company. Zahlreiche Auszeichnungen durch den Art Directors Club of Houston. Berater der japanischen Zeitschrift „Portfolio". 1982 Wahl in das Direktorium des American Institute of Graphic Arts (AIGA). 1988 Eintritt bei „Pentagram Design" in New York. Auftraggeber waren u. a. Pantone, Rizzoli Publishing, Rockefeller Foundation und die Northern Telecom.

**Plantin,** Christoph – geb. ca. 1520 in Saint-Avertin, Frankreich, gest. 1. 7. 1589 in Antwerpen, Spanische Niederlande (heute Belgien). – *Buchdrucker, Buchbinder* – 1550 Gründung einer Buchdruckerei in Antwerpen. In der angegliederten Schriftgießerei werden Stempel und Matrizen verwendet, die von Claude Garamond, Robert Granjon u.a. angefertigt sind. Jacob Sabon ist in seiner Schriftgießerei tätig. Plantin veröffentlicht 1567 ein Buch mit den Schriften seiner Druckerei, „Index characterum". 1568–72 Herstellung der „Polyglotten-Bibel" in mehreren Bänden: neben dem Urtext werden in parallelen Textspalten andere Übersetzungen abgedruckt (armanisch, hebräisch, griechisch, lateinisch). Erhält 1570 vom spanischen König und vom Heiligen Stuhl Privilegien für die Herstellung aller liturgischen Bücher für Spanien und die Niederlande. Wird Hofdrucker des Königs Philip II. von Spanien. Plantins Schwiegersohn und Haupterbe Jan Moretus setzt das Geschäft ab 1589 fort, das von Nachkommen bis 1871 geführt und dann zum Plantin-Moretus-Museum umgewandelt wird. In

Pirtle 1980 Poster

REGNI NEAPOLITANI
PRIVILEGIVM.

PHILIPPVS DEI GRATIA REX
CASTELLÆ, ARAGONVM, VTRIVSQVE
SICILIÆ, HIERVSALEM, VNGARIÆ, DALMATIÆ, ET CROATIÆ, &c.

ANTONIVS Perrenotus, S.R.C.tit. Sancti Petri ad Vincula Preſbyter, Cardinalis de Granuela, præfatæ Regiæ & Catholicæ Maieſtatis à conſiliis ſtatus, & in hoc Regno locum tenens, & Capitaneus generalis, &c. Mag^co viro Chriſtophoro Plantino, ciui Antuerpienſi, & præfatæ Catholicæ Maieſtatis Prototypographo fideli Regio, dilecto, gratiam Regiam & bonam voluntatem. Cùm ex præclarorum virorum literis certiores facti ſimus, opus Bibliorum quinque linguarum, cum tribus Apparatuum tomis, celeberrimum, reique publicę Chriſtianæ vtiliſſimũ, eiuſdem ſereniſſimæ Maieſtatis iuſſu, ope atque auſpiciis, ad publicam totius Chriſtiani orbis commoditatem & ornamentum, typis longè elegantiſſimis, & præſtantiſſimi viri Benedicti Ariæ Montani præcipua cura & ſtudio. quàm emendatiſſimè à te excuſum eſſe, eiuſdemq; exemplar ſanctiſſimo Domino noſtro PP. Gregorio XIII. oblatum, ita placuiſſe, vt præfatæ Maieſtatis ſanctos conatus, & Regi Catholico in primis conuenientes, ſummopere laudarit, & ampliſſima tibi priuilegia ad hoc opus tuendum Motu proprio conceſſerit; Nos quoque cum naturali genio impellimur ad fouendum præclara quæque ingenia, quæ inſigni quopiam conatu ad publica commoda promouenda atque augenda aſpirant; primùm quidem longè præclariſſimum hoc ſuæ Maieſtatis ſtudium, vt verè Heroicum & Ptolomęi, Eumenis, aliorumque olim conatibus in Bibliothecis inſtruendis eò præſtantius, quòd non vanæ ſtimulo gloriæ, vt illi, ſed rectæ Religionis conſeruandæ & propagandæ zelo ſuſceptum, meritò ſuſpicientes; deinde eximiam operam doctiſſimi B. Ariæ Montani, ac immortali laude dignam admirantes, rebuſque tuis, quemadmodũ tuo nomine expetitur, proſpicere cupientes, ne meritis fraudèris fructibus tantæ operæ, & impenſæ, quæ ſumma ſolicitudine & induſtria in opus ad finem feliciter perducendum à te etiam inſumpta eſſe accepimus; cumque certò conſtet, opus hoc nunquam hactenus hoc in Regno excuſum eſſe, dignumque ipſo S. ſedis Apoſtolicæ ſuffragio ſit iudicatum vt diuulgetur

Plantin 1572 Page

cal Company etc. Nombreuses distinctions de l'Art Directors Club de Houston. Conseiller de la revue japonaise «Portfolio». En 1982, il est élu au directoire de l'American Institute of Graphic Arts (AIGA). Entre, en 1988, chez «Pentagram Design» de New York. Commanditaires: Pantone, Rizzoli Publishing, Rockefeller Foundation, Northern Telecom etc.

**Plantin,** Christophe – né vers 1520 à Saint-Avertin, France, décédé le 1. 7. 1589 à Anvers, les Pays-Bas espagnols (aujourd'hui Belgique) – *imprimeur, relieur* – fonde une imprimerie à Anvers en 1550. La fonderie de caractère qui lui est adjointe utilise des types et des matrices réalisées par Claude Garamond, Robert Granjon etc. Jacob Sabon exerce dans sa fonderie de caractères. En 1567, il publie «Index characterum», un ouvrage présentant les fontes de son imprimerie. Edite la «Bible polyglotte» en plusieurs volumes de 1568 à 1572; des traductions en arménien, hébreu, grec et latin sont imprimées en colonnes parallèles au texte original. En 1570, le roi d'Espagne et le Saint-Siège lui accordent le privilège d'imprimer des ouvrages liturgiques pour l'Espagne et les Pays-Bas. Il devient imprimeur de la cour du roi Philippe II d'Espagne. En 1589, Jan Moretus, gendre et héritier de Plantin reprend l'affaire. Celle-ci sera dirigée par ses descendants jus-

Philippe, par la grace de Dieu Roy de Castille, de Leon, d'Arragon, de Navarre, de Naples, de Secille, de Maillorque, de Sardaine, des isles Indes, & terre ferme de la mer Oceane : Archiduc d'Austrice : Duc de Bourgoingne, de Lothiers, de Brabant, de Lembourg, de Luxembourg, de Geldres, & de Milan : Conte de Habsbourg, de Flandres, d'Arthois, de Bourgoingne : Palatin & de Haynnau, de Hollande, de Zeelande, de Namur, & de Zutphen : Prince de Zwaue : Marquis du saint Empire : Seigneur de Frize, de Salins, de Malines, des cité, Villes, & pays d'Utrecht, d'Overyssel, & Groenin-

Plantin 1572 Civilité

Pirtle Design studio in Dallas. Clients include Knoll International and the Omega Optical Company. Receives numerous awards from the Art Directors Club of Houston. Consultant to the Japanese magazine "Portfolio". 1982: is elected to join the American Institute of Graphic Arts' (AIGA) board of directors. 1988: joins Pentagram Design in New York. Clients include Pantone, Rizzoli Publishing, the Rockefeller Foundation and Northern Telecom.

**Plantin,** Christoph – b. c.1520 in Saint-Avertin, France, d. 1. 7. 1589 in Antwerp, Spanish Netherlands (now Belgium) – *printer, bookbinder* – 1550: opens a printing workshop in Antwerp. The affiliated type foundry uses punches and matrices made by Claude Garamond and Robert Granjon, among others. Jacob Sabon works in Plantin's foundry. 1567: Plantin publishes a book containing typefaces produced by his printing works, entitled "Index characterum". 1568–72: produces a polyglot bible in several volumes, with Armenian, Hebrew, Greek and Latin translations printed in columns parallel to the original text. 1570: the Spanish king and the Holy See grant him the privilege of printing all liturgical books for Spain and the (Spanish) Netherlands. Plantin is made royal printer to King Philip II of Spain. 1589: Plantin's son-in-law and principal heir Jan Moretus takes over the business which is run by the family until 1871, after which it becomes the Plantin Moretus Museum. Plantin was responsible for over 2,000 publications in

Antwerpen und in Leiden ist Plantin für über 2.000 Publikationen verantwortlich. Er ist damit der produktivste Drucker und Verleger seiner Zeit.

**Poeschel,** Carl Ernst – geb. 2. 9. 1874 in Leipzig, Deutschland, gest. 19. 5. 1944 in Scheidegg, Deutschland. – *Drucker, Typograph* – 1892–96 Lehre in der Druckerei seines Vaters in Leipzig (Poeschel & Trepte). 1896–98 Schriftsetzer in Halle, Drucker in Zwickau und München. 1898–1900 Aufenthalt in Amerika. 1900 Eintritt in die Druckerei seines Vaters. 1904–06 Leiter des Insel Verlags in Leipzig. 1904 Reise nach England, Bekanntschaft mit Edward Johnston, Eric Gill. 1907 Gründung der Janus-Presse (mit Walter Tiemann) als erste deutsche Privatpresse. 1918 Übernahme der Janus-Presse durch den Insel Verlag. 1935 Vortragsessen im Double Crown Club in London, wo er Ehrenmitglied ist. 1940 Verleihung der Ehrendoktorwürde der Universität Leipzig und des Gutenbergrings der Stadt Leipzig. – *Schriftentwürfe:* Janus-Presse-Schrift (mit Walter Tiemann, 1907), Winckelmann-Antiqua (1920). – *Publikationen u.a.:* „Antiqua als deutsche Normalschrift" (mit F. L. Habbel), Berlin 1942; „C. E. Poeschel zum 60. Geburtstage am 2. September 1934", Leipzig 1934.

**Polanco,** Juan Claudio Aznar de – *Schreibmeister* – 1719 Herausgabe des Buchs „Arte nueva de escribir" in Mailand.

**Poppl,** Friedrich – geb. 1. 3. 1923 in Soborten, Tschechoslowakei. – *Maler, Grafiker, Typograph, Schriftentwerfer, Kalli-*

qu'en 1871, date à laquelle elle est transformée en Plantin-Moretus Museum. Plantin a réalisé plus de 2 000 publications à Anvers et à Leyde. Il a été l'imprimeur le plus productif de son époque.

**Poeschel,** Carl Ernst – né le 2. 9. 1874 à Leipzig, Allemagne, décédé le 19. 5. 1944 à Scheidegg, Allemagne – *imprimeur, typographe* – 1892–1896, apprentissage dans l'imprimerie de son père (Poeschel & Trepte) à Leipzig. 1896–1898, typographe à Halle, puis imprimeur à Zwikkau et Munich. Séjour aux Etats-Unis de 1898 à 1900. Entre dans l'imprimerie de son père en 1900. Dirige les éditions Insel à Leipzig de 1904 à 1906. Voyage en Angleterre en 1904, où il rencontre Edward Johnston et Eric Gill. Fonde en 1907 la Janus Presse, la première presse privée allemande (avec Walter Tiemann). En 1918, la Janus Presse est reprise par les éditions Insel. En 1935, il donne une conférence lors d'un dîner du Double Crown Club de Londres, dont il est membre d'honneur. En 1940, l'université de Leipzig lui décerne le titre de Docteur honoris causa et le Gutenbergring de la ville de Leipzig. – *Polices :* Janus-Presse-Schrift (avec Walter Tiemann, 1907), Winckelmann-Antiqua (1920). – *Publications, sélection :* «Antiqua als deutsche Normalschrift» (avec F. L. Habbel), Berlin 1942; «C. E. Poeschel zum 60. Geburtstage am 2. September 1934», Leipzig 1934.

**Polanco,** Juan Claudio Aznar de – *scribe* – publie «Arte nueva de escribir», à Milan, en 1719.

**Poppl,** Friedrich – né le 1. 3. 1923 à Soborten, Tchécoslovaquie – *peintre, graphiste, typographe, concepteur de polices, calligraphe, enseignant* – 1939–1941, études à l'école technique d'Etat de Teplitz-Schönau. Exerce comme peintre et graphiste de 1947 à 1949. Etudes à la Werkkunstschule (école des arts appli-

Die Briefe der
Frau Rath Goethe

Gesammelt und herausgegeben
von Albert Köster

Erster Band

Leipzig
Carl Ernst Poeschel
1904

Poeschel 1904 Title

Poppl 1968 Calligraphy

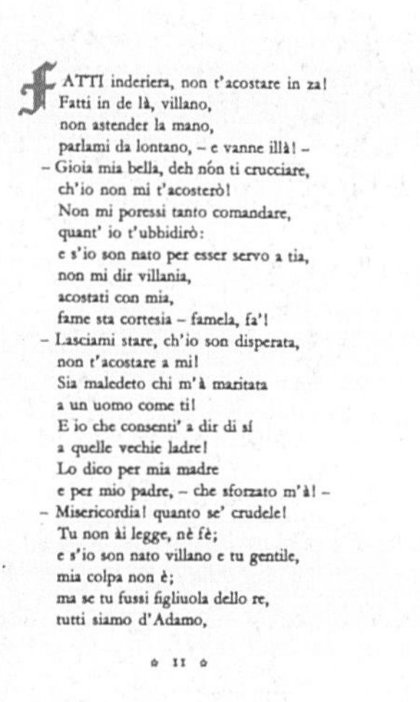

FATTI inderiera, non t'acostare in za!
Fatti in de là, villano,
non astender la mano,
parlami da lontano, – e vanne illà! –
– Gioia mia bella, deh nón ti crucciare,
ch'io non mi t'acosterò!
Non mi poressi tanto comandare,
quant' io t'ubbidirò:
e s'io son nato per esser servo a tia,
non mi dir villania,
acostati con mia,
fame sta cortesia – famela, fa'!
– Lasciami stare, ch'io son disperata,
non t'acostare a mi!
Sia maledeto chi m'à maritata
a un uomo come ti!
E io che consenti' a dir di sí
a quelle vechie ladre!
Lo dico per mia madre
e per mio padre, – che sforzato m'à! –
– Misericordia! quanto se' crudele!
Tu non ài legge, nè fè;
e s'io son nato villano e tu gentile,
mia colpa non è;
ma se tu fussi figliuola dello re,
tutti siamo d'Adamo,

11

GEH, schier dich fort und komm' mir nicht zu nah!
Du Tölpel bleibe dort,
laß deine Hände fort,
bleib' du an deinem Ort und sprich von da! –
„Ach liebes Weibchen, mußt nicht mit mir schmälen,
möcht' meine Lieb' ich stillen.
Soviel kannst du ja gar nicht mir befehlen
wie ich dir bin zu willen.
Bin ich zum Knecht dir gut genug allhier,
so zanke nicht mit mir
und lasse mich zu dir;
tu' den Gefallen mir; – ach tu's doch, – ja?"
– Lass' mich in Ruh', bin außer Rand und Band;
du, rühre mich nicht an!
Fluch denen, die da gaben meine Hand
an einen solchen Mann!
Und ich hab' auch noch zugestimmt alsdann
den Kupplern, den zwei alten;
sollt' sie für Eltern halten,
und sie vergalten mir's mit Zwange, – ha!
„Hergott, wie kann man sich so grausam geben!
Kennst keine Pflicht, – o nein!
Und bin ich garstig, und du fein daneben,
soll ich drum schuldig sein?
Und wärst du selbst des Königs Töchterlein,
von Adam sind wir alle;

11

Poeschel 1926 Spread

| | n | | | | |
|---|---|---|---|---|---|
| a | b | e | f | g | i |
| o | r | s | t | y | z |
| A | B | C | E | G | H |
| M | O | R | S | X | Y |
| 1 | 2 | 4 | 6 | 8 | & |

Poppl 1970 Poppl Exquisit

Antwerp and Leiden. He was thus the most prolific printer and publisher of his day.

**Poeschel,** Carl Ernst – b. 2. 9. 1874 in Leipzig, Germany, d. 19. 5. 1944 in Scheidegg, Germany – *printer, typographer* – 1892–96: trains at his father's printing works (Poeschel & Trepte) in Leipzig. 1896–98: trains as a typesetter in Halle, and as a printer in Zwickau and Munich. 1898–1900: spends time in America. 1900: joins his father's printing shop. 1904–06: director of Insel publishers in Leipzig. 1904: travels to England, where he makes the acquaintance of Edward Johnston and Eric Gill. 1907: launches the Janus-Presse (with Walter Tiemann), the first private press in Germany. 1918: the Janus-Presse is taken over by Insel publishing house. 1935: gives a lecture at the Double Crown Club in London, where he is an honorary member. 1940: is awarded an honorary doctorate by the University of Leipzig and the Gutenberg Ring by the city of Leipzig. – *Fonts:* Janus-Presse-Schrift (with Walter Tiemann, 1907), Winckelmann-Antiqua (1920). – *Publications include:* "Antiqua als deutsche Normalschrift" (with F. L. Habbel), Berlin 1942; "C. E. Poeschel zum 60. Geburtstage am 2. September 1934", Leipzig 1934.

**Polanco,** Juan Claudio Aznar de – *writing master* – 1719: publishes the book "Arte nueva de escribir" in Milan.

**Poppl,** Friedrich – b. 1. 3. 1923 in Soborten, Czechoslovakia – *painter, graphic artist, typographer, type designer, calligrapher, teacher* – 1939–41: studies

*graph, Lehrer* – 1939–41 Studium an der Staatsfachschule Teplitz-Schönau. 1947–49 als Maler und Grafiker tätig. 1950–53 Studium an der Werkkunstschule in Offenbach, u. a. bei Karlgeorg Hoefer, Hans Bohn, Herbert Post. 1953– 55 als Schriftgrafiker in Frankfurt am Main für Industrie und Verlage tätig. 1955–82 Dozent an der Werkkunstschule in Wiesbaden, seit 1973 Professor an der Fachhochschule in Wiesbaden. 1967 Beginn seiner Zusammenarbeit mit der Schriftgießerei H. Berthold AG. Ausstellungen u. a. im Klingspor-Museum Offenbach (1965), im Gutenberg-Museum Mainz (1966), in der Galerie von Oertzen Frankfurt am Main (1984). – *Schriftentwürfe:* Poppl-Antiqua (1967), Poppl-Stretto (1969), Poppl Exquisit (1970), Poppl Heavy (1973), Poppl Pontifex (1976), Poppl Leporello (1977), Poppl Residenz (1977), Poppl Saladin (1979), Poppl College (1981), Poppl Laudatio (1982), Poppl Nero (1982).

**Portfolio** – *Zeitschrift* – Gründung der Zeitschrift „Portfolio" 1950 durch den Verleger George Rosenthal und den Journalisten Frank Zachary als ein Organ für freie und angewandte Künste mit einem Schwerpunkt im Bereich des Grafik-Design. Beauftragung von Alexey Brodovitch zur Gestaltung der Zeitschrift ohne jede Einschränkung. Es erscheinen drei Hefte der Zeitschrift. „Portfolio" spiegelt die Moderne Amerikas wider. Die Gestaltung von Brodovitch regt die nachfolgende Zeitschriften-Gestaltung noch lange an. „Portfolio" wird aus Geldmangel eingestellt.

**Post,** Herbert – geb. 13. 1. 1903 in Mann-

Poppl 1965 Sheet

| n | n | **n** | **n** | | *n* |
|---|---|---|---|---|---|
| a | b | e | f | g | i |
| o | r | s | t | y | z |
| A | B | C | E | G | H |
| M | O | R | S | X | Y |
| 1 | 2 | 4 | 6 | 8 | & |

Poppl 1982 Poppl Laudatio

Poppl 1968 Sketch

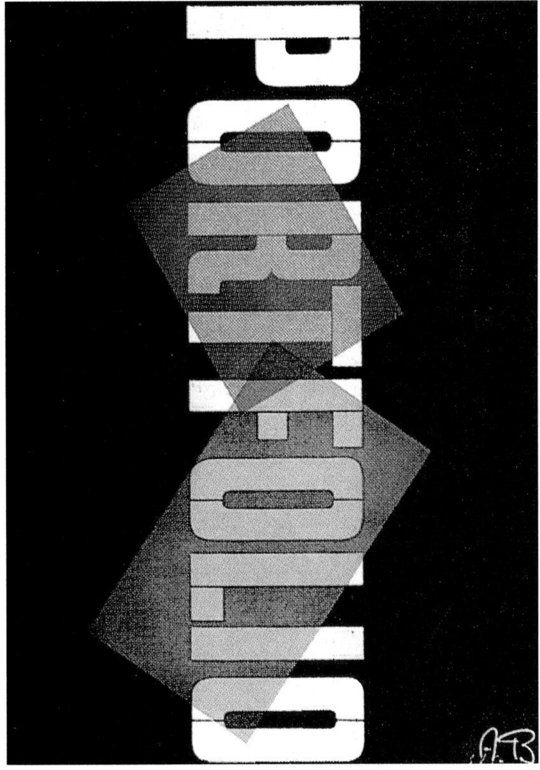


Portfolio 1949 Cover

qués) d'Offenbach de 1950 à 1953, entre autres, chez Karlgeorg Hoefer, Hans Bohn, Herbert Post. Travaille comme graphiste en écriture pour l'industrie et des maisons d'édition, à Francfort-sur-le-Main, de 1953 à 1955. Professeur à la Werkkunstschule (école des arts appliqués) de Wiesbaden de 1955 à 1982, et professeur à la Fachhochschule (institut universitaire technique) de Wiesbaden à partir de 1973. Début de sa collaboration avec la fonderie de caractères H. Berthold AG en 1967. Expositions au Klingspor-Museum d'Offenbach (1965), au Gutenberg-Museum de Mayence (1966), à la Galerie von Oertzen à Francfort-sur-le-Main (1984). – *Polices:* Poppl-Antiqua (1967), Poppl-Stretto (1969), Poppl Exquisit (1970), Poppl Heavy (1973), Poppl Pontifex (1976), Poppl Leporello (1977), Poppl Residenz (1977), Poppl Saladin (1979), Poppl College (1981), Poppl Laudatio (1982), Poppl Nero (1982).

**Portfolio** – *revue* – la revue «Portfolio» a été créée en 1950 par l'éditeur George Rosenthal et le journaliste Frank Zachary pour traiter des arts libres et appliqués en mettant l'accent sur les arts graphiques et le design. Chargé de la maquette, Alexey Brodovitch a toute liberté quant à la conception. Trois numéros paraîtront. Portfolio était un reflet de l'Amérique moderne. La maquette de Brodovitch a inspiré pendant longtemps la conception graphique d'autres revues. Portfolio a cessé de paraître pour des raisons financières.

at the State Technical College in Teplitz-Schönau. 1947–49: works as a painter and graphic artist. 1950–53: studies at the Werkkunstschule in Offenbach. His teachers include Karlgeorg Hoefer, Hans Bohn and Herbert Post. 1953–55: works as a lettering artist in Frankfurt am Main for industry and various publishers. 1955–82: lecturer at the Werkkunstschule in Wiesbaden. From 1973 onwards: professor at the Fachhochschule in Wiesbaden. 1967: starts working with the H. Berthold AG type foundry. Exhibitions have been held at the Klingspor Museum in Offenbach (1965), the Gutenberg Museum in Mainz (1966) and the von Oertzen gallery in Frankfurt am Main (1984), among other venues. – *Fonts:* Poppl-Antiqua (1967), Poppl-Stretto (1969), Poppl Exquisit (1970), Poppl Heavy (1973), Poppl Pontifex (1976), Poppl Leporello (1977), Poppl Residenz (1977), Poppl Saladin (1979), Poppl College (1981), Poppl Laudatio (1982), Poppl Nero (1982).

**Portfolio** – *magazine* – 1950 : "Portfolio" magazine is launched by publisher George Rosenthal and journalist Frank Zachary as a mouthpiece for free and applied art with a special focus on graphic design. Alexey Brodovitch is given free reign with the design of the magazine. Only three numbers of the magazine are published. "Portfolio" mirrored the modernity of America. Brodovitch's designs acted as a stimulus for many future magazine designs. Financial difficulties forced "Portfolio" to cease publication.

**Post,** Herbert – b. 13. 1. 1903 in Mann-

heim, Deutschland, gest. 9. 7. 1978 in Bayersoien, Deutschland. – *Maler, Drucker, Kalligraph, Schriftentwerfer, Lehrer* – Schriftsetzerlehre, Besuch der Fachschule für Buch- und Kunstgewerbe in Frankfurt am Main. Zeichenkurse an der Kunstgewerbeschule in Frankfurt am Main 1921–24 Studium an den Technischen Lehranstalten in Offenbach, Schüler von Rudolf Koch und Mitglied seiner „Offenbacher Werkgemeinschaft". 1926 Fachlehrer für Buchdruck und Schrift an der Kunstgewerbeschule in Halle auf Burg Giebichenstein. Hier entstehen zahlreiche Einblattdrucke, handgeschriebene Bücher und Schriftentwürfe. Unterrichtet 1950 Buch- und Schriftkunst an der Werkkunstschule in Offenbach. Gründung der „Herbert Post Presse", Herausgabe von über 30 bibliophilen Publikationen. 1958 Präsident des Bundes Deutscher Buchkünstler als Nachfolger von F. H. Ehmcke. Seit 1968 Maler und Grafiker in München. 1976–77 Gastprofessuren an den Sommerakademien in Salzburg. Zahlreiche Ausstellungen seines Werks, u. a. im Gutenberg-Museum Mainz (1964), im Klingspor-Museum Offenbach (1978). – *Schriftentwürfe:* Post-Antiqua (1932), Post-Fraktur (1935), Post Mediaeval (1944), Dynamik (1952), Post-Marcato (1961). – *Publikationen u. a.:* „Schrift und Druck", 1941. Rudolf Adolph (Hrsg.) „Herbert Post. Eine Würdigung seines Schriftschaffens zum 60. Geburtstag", Berlin, Stuttgart 1962; Kat. Burg Giebichenstein 1997.

**Poynor,** Rick – geb. 1957 in Surrey, England. – *Autor, Lehrer* – Studium der

**Post,** Herbert – né le 13. 1. 1903 à Mannheim, Allemagne, décédé le 9. 7. 1978 à Bayersoien, Allemagne – *peintre, imprimeur, calligraphe, concepteur de polices, enseignant* – apprentissage de compositeur typographe, fréquente la Fachschule für Buch und Kunstgewerbe (école des arts du livre et des arts décoratifs) de Francfort-sur-le-Main et prend des cours de dessin à école des arts décoratifs de Francfort. Etudie aux Technische Lehranstalten (institut d'enseignement technique) d'Offenbach de 1921 à 1924, où il est l'élève de Rudolf Koch et membre de sa «Offenbacher Werkgemeinschaft». Enseigne l'imprimerie d'édition et les arts de l'écriture à la Kunstgewerbeschule (école des arts décoratifs) de Halle, au château de Giebichenstein, en 1926. Il y réalise des impressions sur une page, des livres calligraphiés et des maquettes d'écritures. En 1950, il enseigne les arts du livre et de l'écriture à la Werkkunstschule (école des arts appliqués) d'Offenbach. Fonde la «Herbert Post Presse», édite plus de trente ouvrages pour bibliophiles. Succède à F. H. Ehmcke comme président du «Bund Deutscher Buchkünstler» en 1958. Peintre et graphiste à Munich à partir de 1968. Cycles de conférences aux académies d'été de Salzbourg de 1976 à 1977. Nombreuses expositions de ses travaux libres et appliqués, entre autres, au Gutenberg-Museum de Mayence (1964), au Klingspor-Museum d'Offenbach (1978). – *Polices :* Post-Antiqua (1932), Post-Fraktur (1935), Post Mediaeval (1944), Dynamik (1952), Post-Marcato (1961). – *Publications, sélection :* «Schrift und Druck», 1941. Rudolf Adolf (éd.) «Herbert Post. Eine Würdigung seines Schriftschaffens zum 60. Geburtstag», Berlin, Stuttgart 1962; Catalogue 1997.

**Poynor,** Rick – né en 1957 à Surrey, Angleterre – *auteur, enseignant* – études

IN JVENGEREN TAGEN
WAR ICH DES MORGENS FROH
DES ABENDS WEINT ICH /
JETZT DA ICH ÄLTER BIN
BEGINN ICH ZWEIFELND
MEINEN TAG / DOCH
HEILIG VND HEITER IST
MIR SEIN ENDE

HÖLDERLIN

Post Calligraphy

Post Cover

| | | | | | |
|---|---|---|---|---|---|
| | n | n | | | |
| a | b | e | f | g | i |
| o | r | s | t | y | z |
| A | B | C | E | G | H |
| M | O | R | S | X | Y |
| 1 | 2 | 4 | 6 | 8 | & |

Post 1932 Post Antiqua

| | | | | | |
|---|---|---|---|---|---|
| | n | n | | | *n* |
| a | b | e | f | g | i |
| o | r | s | t | y | z |
| A | B | C | E | G | H |
| M | O | R | S | X | Y |
| 1 | 2 | 4 | 6 | 8 | & |

Post 1944 Post Mediaeval

Poynor 1991 Cover

heim, Germany, d. 9. 7. 1978 in Bayersoien, Germany – *painter, printer, calligrapher, type designer, teacher* – Trained as a typesetter and attended the Fachschule für Buch- und Kunstgewerbe in Frankfurt am Main. Took art courses at the Kunstgewerbeschule in Frankfurt am Main. 1921–24: studies at the Technische Lehranstalten in Offenbach. Is one of Rudolf Koch's pupils and member of his Offenbacher Werkgemeinschaft. 1926: specialist teacher for printing and lettering at the Kunstgewerbeschule in Halle at Giebichenstein Castle. Here he produces many single-page prints, handwritten books and type designs. 1950: teaches book art and calligraphy at the Werkkunstschule in Offenbach. Founds the Herbert Post Presse and publishes over thirty bibliophile editions. 1958: succeeds F. H. Ehmcke as president of the Bund Deutscher Buchkünstler. From 1968 onwards: painter and graphic artist in Munich. 1976–77: visiting professor at the summer academies in Salzburg. There have been numerous exhibitions of his free and applied work, including at the Gutenberg Museum in Mainz (1964) and at the Klingspor Museum in Offenbach (1978). – *Fonts:* Post-Antiqua (1932), Post-Fraktur (1935), Post Mediaeval (1944), Dynamik (1952), Post-Marcato (1961). – *Publications include:* "Schrift und Druck", 1941. Rudolf Adolph (ed.) "Herbert Post. Eine Würdigung seines Schriftschaffens zum 60. Geburtstag", Berlin, Stuttgart 1962; Catalog 1997.

**Poynor,** Rick – b. 1957 in Surrey, England – *author, teacher* – Studied art history at

Kunstgeschichte an der University of Manchester. Danach für einen akademischen Verlag tätig. Arbeit in der Grafik-Design-Abteilung der amerikanischen Computer Consultancy „CACI" in London. 1986 Assistant Editor des „Designer's Journal". 1988 Deputy Editor der Zeitschrift „Blueprint". 1990 Gründung der Zeitschrift „Eye", einem Forum für Grafik-Design und Typographie. 1991 Herausgabe des Buches „Typography Now: The Next Wave", der ersten Zusammenfassung der typographischen Szene im Computerzeitalter. Beiträge für internationale Zeitschriften, Bücher und Kataloge, u. a. für die Kunstzeitschrift „Frieze" (London), „I.D." (New York). 1994 Gastprofessor an der School of Communication Design am Royal College of Art in London. – *Publikationen u. a.:* „Typography Now: The Next Wave" (mit why not associates), London 1991; „The Graphic Edge", London 1993; „Typography Now Two: Implosion", London 1996.

**Prins,** Ralph – geb. 3. 5. 1926 in Amsterdam, Niederlande. – *Grafik-Designer, Typograph, Bildhauer* – Studium an der Kunstgewerbeschule in Zürich (1944), an der Kunstnijverheidsschool in Amsterdam (1945), an der Koninklijke Academie voor Beeldende Kunsten in Den Haag (1946–49) und der Ecole de Mime Etienne Degroux in Paris (1950). Seit 1955 Entwurf von Plakaten, u. a. für die Vereinten Nationen, das Coordinating Secretariat of the National Unions of Students (COSEC), die PNUD-UNESCO, Amnesty International und verschiedene Kultureinrichtungen. Zahlreiche Auszeichnungen, u. a.

d'histoire de l'art à l'University of Manchester. Exerce ensuite pour des éditions universitaires. Travaille dans le service de graphisme et de design de la société américaine Computer Consultancy «CACI» à Londres. Assistant Editor de la revue «Designer's Journal» en 1986. Deputy Editor de la revue «Blueprint» en 1988. Il fonde la revue «Eye» en 1990, un forum pour les arts graphiques, le design et la typographie. En 1991, il publie l'ouvrage «Typography Now : The Next Wave» qui présente un premier aperçu des milieux de la typographie à l'ère de l'ordinateur. Nombreux essais dans des revues internationales, livres et catalogues, entre autres, pour les revues «Frieze» (Londres), «I.D.» (New York). Cycle de conférence à la School of Communication Design du Royal College of Art de Londres en 1994. – *Publications, sélection :* «Typography Now : The Next Wave» (avec why not associates), Londres 1991; «The Graphic Edge», Londres 1993; «Typography Now Two : Implosion», Londres 1996.

**Prins,** Ralph – né le 3. 5. 1926 à Amsterdam, Pays-Bas – *graphiste maquettiste, typographe, sculpteur* – études à la Kunstgewerbeschule (école des arts décoratifs) de Zurich (1944), à la Kunstnijverheidsschool d'Amsterdam (1945), à la Koninklijke Academie voor Beeldende Kunsten de La Haye (1946–1949), à l'Ecole de Mime Etienne Degroux à Paris (1950). Dessine des affiches pour les Nations Unies, le Coordinating Secretariat of the National Unions of Students (COSEC), la PNUD-UNESCO, Amnesty International et des organismes culturels à partir de

tr e u rende tr o nen
tr ᐅnen n iet
m aa r j ij
b ent n iet van h o u t
j o u w tr o nen
tr EU r e n n lEt
j o u w tr ᐅ nen
zijn van g O U d

Prins 1946 Poem

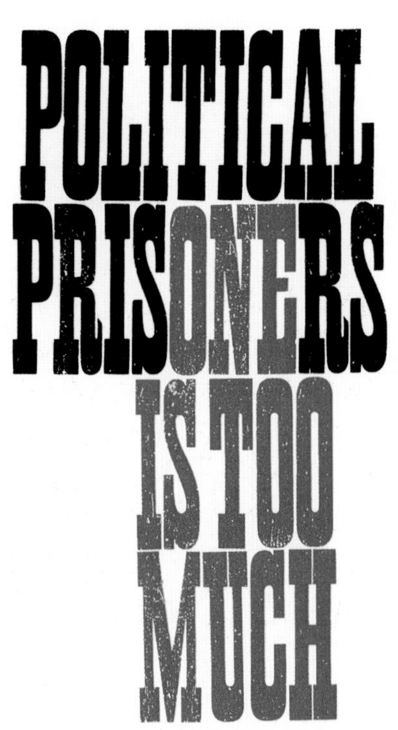


Prins 1967 Poster

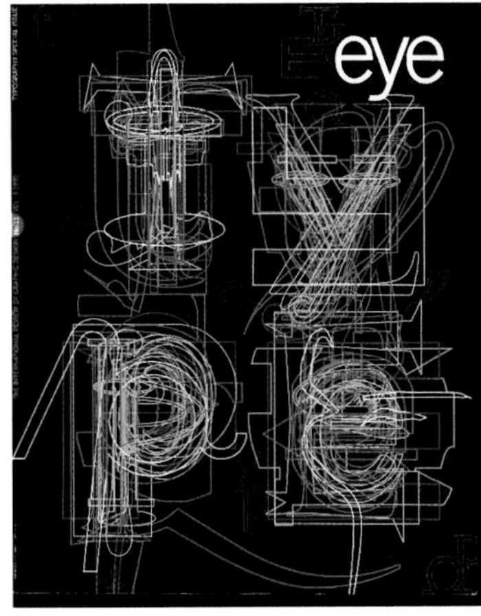


Poynor 1993 Cover

Poynor 1994 Invitation Card

the University of Manchester. He then worked for a university press. Worked in the graphic design department of an American computer consultancy (CACI) in London. 1986: assistant editor of "Designer's Journal". 1988: deputy editor of "Blueprint" magazine. 1990: launches "Eye" magazine, a forum for graphic design and typography. 1991: edits the book "Typography Now: The Next Wave", the first summary of the typography scene in the age of computers. Has made numerous contributions to various international journals, books and catalogues, including "Frieze" art magazine (London) and "I.D." (New York). 1994: visiting professor at the Royal College of Art's School of Communication Design in London. – *Publications include:* "Typography Now: The Next Wave" (with why not associates), London 1991; "The Graphic Edge", London 1993; "Typography Now Two: Implosion", London 1996.

**Prins,** Ralph – b. 3. 5. 1926 in Amsterdam, The Netherlands – *graphic designer, typographer, sculptor* – Studied at the Kunstgewerbeschule in Zurich (1944), at the Kunstnijverheidsschool in Amsterdam (1945), the Koninklijke Academie voor Beeldende Kunsten in The Hague (1946–49) and at the Ecole de Mime Etienne Degroux in Paris (1950). From 1955 onwards: designs posters for various clients who include the United Nations, the Coordinating Secretariat of the National Unions of Students (COSEC), PNUD-UNESCO, Amnesty International and various cultural establishments. He has won numerous awards, including

auf der Plakatbiennale in Warschau (1966, 1968, 1970) und den Henegouwen-Preis für politische Plakate in Mons (1982). Unterrichtet u. a. 1960–63 an der Koninklijke Academie voor Beeldende Kunsten in Den Haag und 1966–86 an der ABK Academie Minerva in Groningen. 1994 mit dem H. N. Werkman-Preis der Stadt Amsterdam ausgezeichnet. – *Publikationen u. a.:* „Humanity. Noodzaak Affiches", Den Haag 1991; „Homo Ludens", Den Haag 1992.

**Puni,** Iwan – geb. 20. 2. 1892 in Kuokkala, Rußland, gest. 28. 12. 1956 in Paris, Frankreich. – *Maler* – 1909 Atelier in St. Petersburg. 1910–11 Studium an der Académie Julian in Paris. Veranstaltet 1916 in St. Petersburg eine Konferenz zur Förderung moderner Kunst (mit K. Malewitsch). 1918 Entwurf von Plakaten zum 1. Mai und zum 1. Jahrestag der Revolution. Professor an der Akademie der Schönen Künste in Leningrad, St. Petersburg. Lehrt auf Einladung von Marc Chagall an der Akademie in Witebsk. Zahlreiche Ausstellungen im In- und Ausland. 1946 französische Staatsbürgerschaft. 1947 Ernennung zum Ritter der Ehrenlegion. 1949 erste Einzelausstellung in New York. Puni gehört zu den freien Gestaltern, die der Typographie durch unorthodoxe Ideen Anregungen geben. – *Publikationen u.a.:* „Schriften zur Kunst (1915–23)", in: „Iwan Puni. Synthetischer Musiker", Berlin 1992. H. Berninger, J. A. Cartier „Pougny, Catalogue de l'œuvre" (2 Bände), Tübingen 1972– 1992; J. Merkert und andere „Iwan Puni", Berlin 1993.

1955. Nombreuses distinctions, entre autres, lors de la Biennale de l'affiche à Varsovie (1966, 1968, 1970), Prix Henegouwen de l'affiche politique à Mons (1982). A enseigné, entre autres, à la Koninklijke Academie voor Beeldende Kunsten de La Haye de 1960 à 1963 et à la ABK Academie Minerva de Groningue de 1966 à 1986. A reçu le Prix H. N. Werkman de la ville d'Amsterdam en 1994. – *Publications, sélection:* «Humanity. Noodzaak Affiches», La Haye 1991; «Homo Ludens», La Haye 1992.

**Pougny,** Ivan – né le 20. 2. 1892 à Kuokkala, Russie, décédé le 28. 12. 1956 à Paris, France – *peintre* – atelier à Saint-Pétersbourg en 1909. Etudes à l'Académie Julian à Paris de 1910 à 1911. Organise à Saint-Pétersbourg une conférence en faveur de l'art moderne en 1916 (avec K. Malévitch). En 1918, il dessine des affiches pour le 1er Mai et le premier anniversaire de la Révolution. Professeur à l'académie des beaux-arts de Leningrad, Saint-Pétersbourg. En 1919, il dessine le sceau du commissariat à l'éducation populaire, département des arts plastiques à Saint-Pétersbourg. Enseigne à l'académie de Vitebsk sur invitation de Marc Chagall. Nombreuses expositions en URSS et à l'étranger. Devient citoyen français en 1946. Est nommé Chevalier de la Légion d'honneur en 1947. Première exposition personnelle à New York en 1949. Pougny a nourri la typographie d'idées peu orthodoxes qui font désormais partie du patrimoine commun bien qu'ayant mis longtemps avant d'être acceptées. – *Publications, sélection:* «Schriften zur Kunst (1915–1923) in «Iwan Puni. Synthetischer Musiker», Berlin 1992. H. Berninger, J. A. Cartier «Pougny, Catalogue de l'œuvre» (2 vol.), Tübingen 1972–1992; J. Merkert et autres «Iwan Puni», Berlin 1993.

Puni 1921 Card

Quay 1988 Cover

| | n | | | | |
|---|---|---|---|---|---|
| a | b | e | f | g | i |
| o | r | s | t | y | z |
| A | B | C | E | G | H |
| M | O | R | S | X | Y |
| 1 | 2 | 4 | 6 | 8 | & |

Quay 1983 Santa Fe

| | n | | | | |
|---|---|---|---|---|---|
| a | b | e | f | g | i |
| o | r | s | t | y | z |
| A | B | C | E | G | H |
| M | O | R | S | X | Y |
| 1 | 2 | 4 | 6 | 8 | & |

Quay 1989 Robotik

| | n | n | n | | n |
|---|---|---|---|---|---|
| a | b | e | f | g | i |
| o | r | s | t | y | z |
| A | B | C | E | G | H |
| M | O | R | S | X | Y |
| 1 | 2 | 4 | 6 | 8 | & |

Quay 1990 Quay Sans

prizes at the Poster Biennale in Warsaw (1966, 1968 and 1970) and the Henegouwen Prize for political posters in Mons (1982). Has held teaching posts at the Koninklijke Academie voor Beeldende Kunsten in The Hague from 1960 to 63 and at the ABK Academie Minerva (1966–86), among others. 1994: is awarded the H. N. Werkman Prize by the city of Amsterdam. – *Publications:* "Humanity. Noodzaak Affiches", The Hague 1991; "Homo Ludens", The Hague 1992.

**Puni,** Ivan – b. 20. 2. 1892 in Kuokkala, Russia, d. 28. 12. 1956 in Paris, France – *painter* – 1909: has a studio in St. Petersburg. 1910–11: studies at the Académie Julian in Paris. 1916: organizes a conference in St. Petersburg to promote modern art (with K. Malevich). 1918: designs posters for the 1st of May and for the first anniversary of the Revolution. Professor at the Academy of Fine Arts in Leningrad, St. Petersburg. 1919: designs a seal for the Commissariat for Adult Education, Fine Arts Department, in St. Petersburg. Is invited by Marc Chagall to teach at the academy in Vitebsk. Numerous exhibitions in Russia and abroad. 1946: is granted French citizenship. 1947: made a knight of the Legion of Honor. 1949: first solo exhibition in New York. Puni was one of the free designers who inspired typography with their unorthodox ideas. – *Publications include:* "Schriften zur Kunst (1915–23)", in "Iwan Puni. Synthetischer Musiker", Berlin 1992. H. Berninger, J. A. Cartier, "Pougny, Catalogue de l'œuvre", Tübingen 1972–1992; J. Merkert et al "Iwan Puni", Berlin 1993.

**Quay,** David – geb. 3. 1. 1948 in London, England. – *Grafik-Designer, Schriftentwerfer, Typograph, Lehrer* – 1963–68 Studium am Ravensbourne College of Art & Design. 1968–75 Arbeit in London als Grafik-Designer. 1983 Mitbegründer von „Quay & Gray Lettering" (mit Paul Gray). 1987 Gründung von „David Quay Design" in London. Auftraggeber waren u. a. Monotype, die H. Berthold AG und die International Typeface Corporation (ITC). Gründet 1988 „Letter Exchange", eine Vereinigung für Aspekte der Schriftkunst. 1990 Mitbegründer von „The Foundry" (mit Freda Sack und Mike Daines), einer Firma, in der PostScript-Schriften entworfen, hergestellt und vermarktet werden. Seit 1992 Mitorganisator des Preises der „Society of Typographic Designers" (STD), seit 1994 der Vortragsreihen, seit 1995 stellvertretender Vorsitzender. Unterrichtet u. a. an der Hochschule für Gestaltung Offenbach (1990), am Bournemouth College of Art (1994), am London College of Printing (1995) und dem Cumbria College of Art & Design (1996). – *Schriftentwürfe:* Santa Fe (1983), Quay (1985), Bordeaux (1988), Helicon, Robotik (1989), ITC Quay Sans (1990).

**Queneau,** Raymond – geb. 21. 2. 1903 in Le Havre, Frankreich, gest. 25. 10. 1976 in Paris, Frankreich. – *Schriftsteller* – Zunächst Arbeit als Bankbeamter und Handelsvertreter. 1924–29 Mitglied der surrealistischen Bewegung. Seit 1936 Generalsekretär des Verlags Gallimard in Paris. Seit 1951 Mitglied der Académie Goncourt. 1954 Leiter der „Encyclopédie de

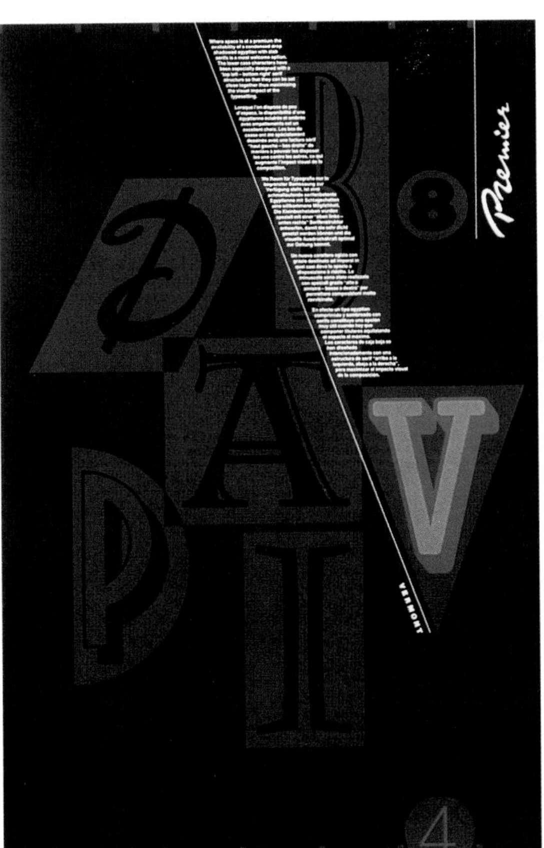

Quay 1987 Cover

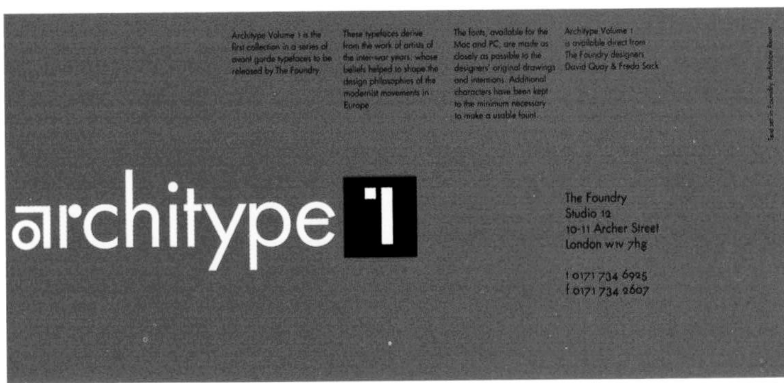

Quay 1991 Card

Queneau 1960 Cover

Raymond Queneau
Die blauen Blumen
Roman / Stahlberg

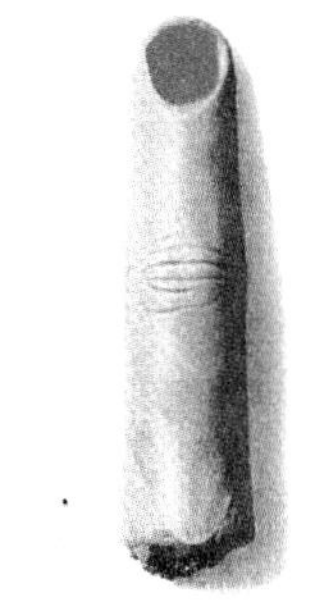
Queneau 1966 Cover

**Quay,** David – né le 3. 1. 1948 à Londres, Angleterre – *graphiste maquettiste, concepteur de polices, typographe, enseignant* – 1963–1968, études au Ravensbourne College of Art & Design. Graphiste maquettiste pour plusieurs ateliers londoniens de 1968 à 1975. Cofondateur de «Quay & Gray Lettering» en 1983 (avec Paul Gray). En 1987, il crée «David Quay Design» à Londres. Commanditaires : Monotype, H. Berthold AG, International Typeface Corporation (ITC), etc. En 1988, il fonde «Letter Exchange», une association de promotion des arts de l'écriture. Parmi sa centaine de membres, on compte des concepteurs de polices, des lapidaires, des typographes et calligraphes qui organisent un programme mensuel de conférences. Cofondateur, en 1990, de «The Foundry» (avec Freda Sack et Mike Daines), une société qui crée, produit et diffuse des polices PostScript. En 1992, il participe à l'organisation du prix de la «Society of Typographic Designers» (STD); séries de conférences à partir de 1994; vice-président de la STD depuis 1995. A enseigné à la Hochschule für Gestaltung (école supérieure de design) d'Offenbach (1990), au Bournemouth College of Art (1994), au London College of Printing (1995), au Cumbria College of Art & Design (1996). – *Polices :* Santa Fe (1983), Quay (1985), Bordeaux (1988), Helicon, Robotik (1989), ITC Quay Sans (1990).

**Queneau,** Raymond – né le 21. 3. 1903 au Havre, France, décédé le 25. 10. 1976 à Paris, France – *écrivain* – commence comme employé de banque et représentant de commerce. Membre du groupe des Surréalistes de 1924 à 1929. Secrétaire général des Editions Gallimard, à Paris, à partir de 1936. Membre de l'Académie Goncourt à partir de 1951. Dirige, en 1954, l'«Encyclopédie de la Pléiade». Les jeux de langage de Queneau, qui allient

**Quay,** David – b. 3. 1. 1948 in London, England – *graphic designer, type designer, typographer, teacher* – 1963–68: studies at Ravensbourne College of Art & Design. 1968–75: works in London as a graphic designer. 1983: co-founder of Quay & Gray Lettering (with Paul Gray). 1987: launches David Quay Design in London. Clients include Monotype, H. Berthold AG and the International Typeface Corporation (ITC). 1988: founds Letter Exchange, an organization for aspects of typography and type. The c. 100 members include type designers, stone cutters, typographers and calligraphers who contribute to a program of monthly lectures. 1990: co-founder of The Foundry with Freda Sack and Mike Daines, a company which designs, manufactures and markets PostScript fonts. From 1992 onwards: co-organizer of the Society of Typographic Designers (STD) prize and from 1994 onwards of the STD's lecture series. 1995: Quay is made vice-chairman of the STD. Quay has taught at the Hochschule für Gestaltung, Offenbach (1990), at Bournemouth College of Art (1994), at the London College of Printing (1995) and at Cumbria College of Art & Design (1996). – *Fonts:* Santa Fe (1983), Quay (1985), Bordeaux (1988), Helicon, Robotik (1989), ITC Quay Sans (1990).

**Queneau,** Raymond – b. 21. 2. 1903 in Le Havre, France, d. 25. 10. 1976 in Paris, France – *author* – Queneau started his working life as a bank employee and sales representative. 1924–29: member of the Surrealist movement. From 1936 onwards: secretary general for Gallimard

la Pléiade". Queneaus schöpferische Sprachspiele, in denen sich Poesie, Träumerei, Ironie und Humor verbinden, sind Anregung für experimentelle typographische Umsetzungen. – *Publikationen u. a.:* „Exercices de style", Paris 1947; „Zazie dans le métro", Paris 1959; „Cent mille milliards de poèmes", Paris 1961.

**Qwer** – *Büro für Kommunikation* – 1994 Gründung des Büros in Düsseldorf durch Iris Utikal und Michael Gais. Iris Utikal – geb. 18. 6. 1966 in Kassel, Deutschland. – 1986–91 Studium der Visuellen Kommunikation an der Fachhochschule in Düsseldorf. 1991–94 selbständige Grafik-Designerin in Düsseldorf. Zahlreiche Projekte mit Uwe Loesch. Michael Gais – geb. 29. 5. 1965 in Hagen, Deutschland. – 1986–92 Studium der Visuellen Kommunikation an der Fachhochschule in Düsseldorf. 1991–94 selbständiger Grafik-Designer in Düsseldorf. Zahlreiche Projekte mit Uwe Loesch. Seit 1995 ist das Büro in Köln. Auftraggeber waren u. a. die Expo 2000 Hannover, die Internationalen Kurzfilmtage Oberhausen, das Wilhelm-Lehmbruck-Museum und bitlab Information Design & Entertainment Köln. Zahlreiche Auszeichnungen, u. a. erster Preis für das Signet der Expo 2000 in Hannover (1994).

la poésie, la rêverie, l'ironie et l'humour ont inspiré des transpositions typographiques expérimentales. – *Publications, sélection:* «Exercices de style», Paris 1947; «Zazie dans le métro», Paris 1959; «Cent mille milliards de poèmes», Paris 1961.

**Qwer** – *agence de communication* – l'agence a été fondée en 1994 par Iris Utikal et Michael Gais à Düsseldorf. Iris Utikal – née le 18. 6. 1966 à Kassel, Allemagne – études de communication visuelle à l'institut universitaire de Düsseldorf de 1986 à 1991. Graphiste maquettiste indépendante à Düsseldorf de 1991 à 1994. Nombreux projets avec Uwe Loesch. Michael Gais – né le 29. 5. 1965 à Hagen, Allemagne – études de communication visuelle à l'institut universitaire de Düsseldorf de 1986 à 1992. Graphiste maquettiste indépendant à Düsseldorf de 1991 à 1994. Nombreux projets avec Uwe Loesch. Exerce à l'agence de Cologne à partir de 1995. Commanditaires : Expo 2000, Hanovre, Festival international du court métrage d'Oberhausen, Wilhelm Lehmbruck Museum, bitlab Information Design & Entertainment, Cologne, etc. Nombreuses distinctions, entre autres, le premier prix pour le signet d'Expo 2000 à Hanovre (1994).

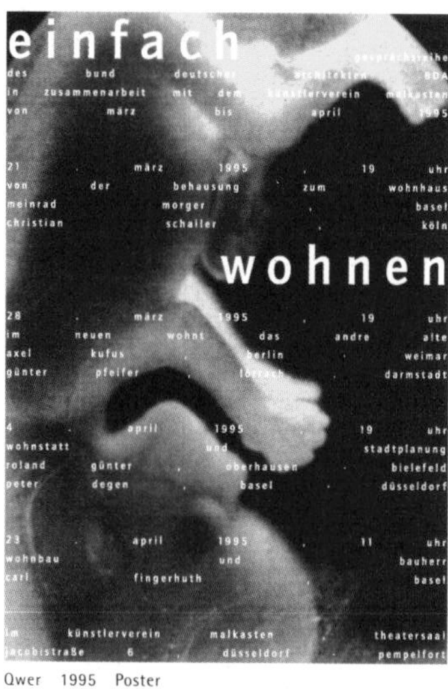

Qwer 1995 Poster

Qwer 1996 Poster

Qwer 1997 Poster

Qwer 1996 Trailer

publishers in Paris. 1951: joins the Académie Goncourt. 1954: head of the "Encyclopédie de la Pléiade". Queneau's linguistic creativity, a melange of poetry, reverie, irony and humor, has inspired many experiments with typography. – *Publications include:* "Exercices de style", Paris 1947; "Zazie dans le métro", Paris 1959; "Cent mille milliards de poèmes", Paris 1961.

**Qwer** – *communication studio* – 1994: the studio is founded in Düsseldorf by Iris Utikal and Michael Gais. Iris Utikal – b. 18.6.1966 in Kassel, Germany – 1986–91: studies visual communication at the Fachhochschule in Düsseldorf. 1991–94: freelance graphic designer in Düsseldorf. Numerous projects with Uwe Loesch. Michael Gais – b. 29.5.1965 in Hagen, Germany – 1986–92: studies visual communication at the Fachhochschule in Düsseldorf. 1991–94: freelance graphic designer in Düsseldorf. Numerous projects with Uwe Loesch. 1995: the studio moves to Cologne. Clients include the Expo 2000 in Hanover, the Internationale Kurzfilmtage Oberhausen, the Wilhelm Lehmbruck Museum and bitlab Information Design & Entertainment in Cologne. Numerous awards, including 1st prize for the Expo 2000 logo in Hanover (1994).

# R

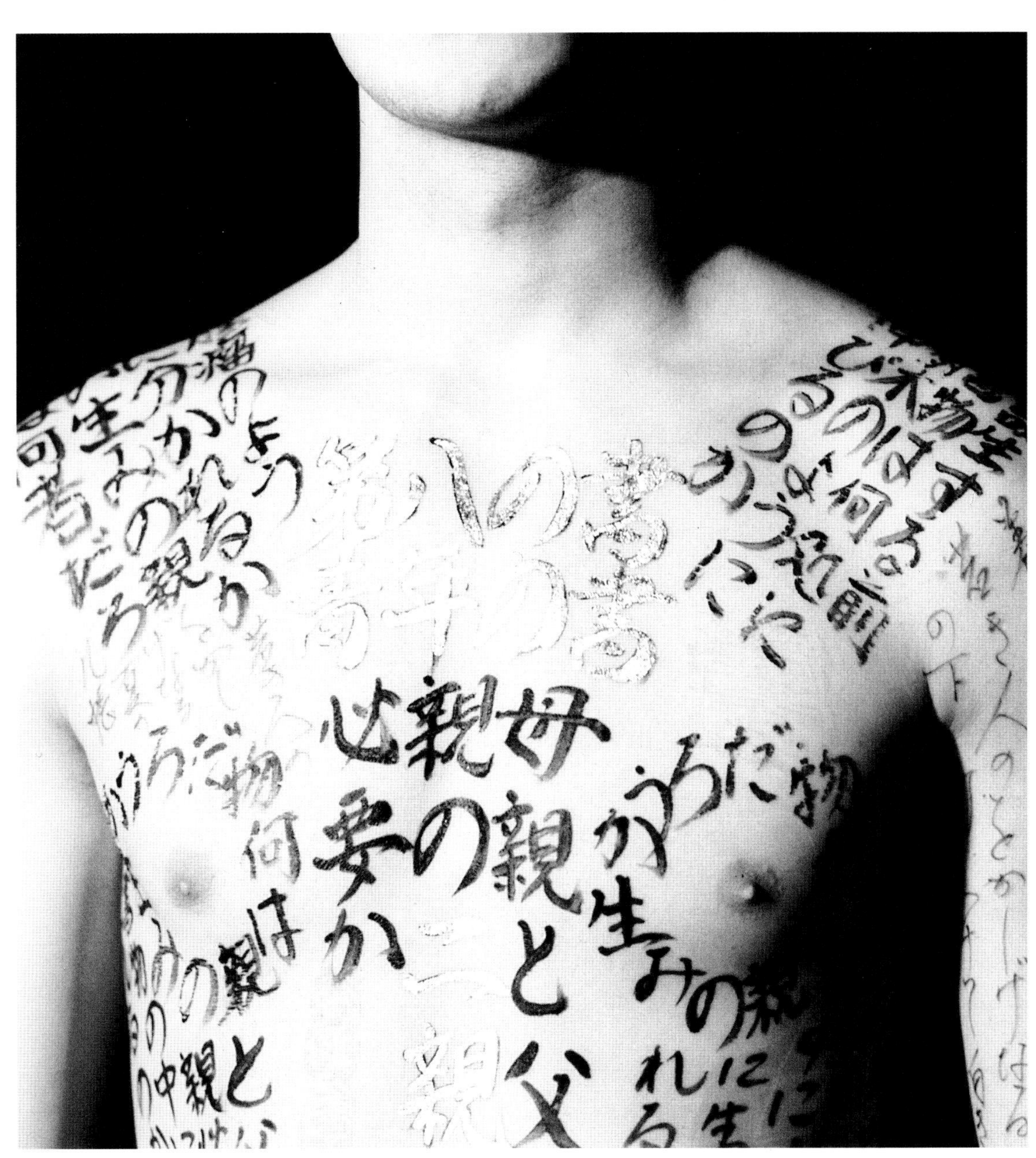

**Rambow**
**Rand**

**Rambow**, Gunther – geb. 2. 3. 1938 in Neustrelitz, Deutschland. – *Grafik-Designer, Lehrer* – 1954–58 Ausbildung zum Glasmaler an der Staatlichen Fachschule in Hadamar. 1958–64 Studium an der Hochschule für Bildende Künste in Kassel. 1960–64 Gründung des Ateliers Rambow + Lienemeyer in Kassel (mit Gerhard Lienemeyer). 1964–68 Arbeit des Ateliers in Stuttgart, 1968–72 in Frankfurt am Main. 1973–88 Ateliergemeinschaft Rambow Lienemeyer van de Sand in Frankfurt. 1988 Nach dem Austritt Lienemeyers Weiterführung des Ateliers Rambow van de Sand. Auftraggeber waren u. a. das Schauspiel Frankfurt, das Theater am Turm und das Städel in Frankfurt, der S. Fischer Verlag, die Galerie René Block in Berlin, der Hessische Rundfunk und das Museum Wiesbaden. 1974–91 Professor für Grafik-Design an der Gesamthochschule in Kassel, seit 1991 an der Hochschule für Gestaltung in Karlsruhe. – *Publikationen u. a.:* „Das sind eben alles Bilder der Straße", Frankfurt am Main 1979; „Traumstoff", Frankfurt am Main 1986; „Fernseh-Design. Modell hessen 3", Berlin 1991. Volker Rattemeyer (Hrsg.) „Rambow: Plakate", Wiesbaden 1988; Heinrich Klotz, Alain Weil „Rambow 1960–96", Stuttgart 1996.

**Rand,** Paul – geb. 14. 8. 1914 in New York, USA, gest. 26. 11. 1996 in Norwalk, Connecticut, USA. – *Grafik-Designer, Illustrator, Typograph, Lehrer* – Studium in New York am Pratt Institute (1930–32), an der Parson's School of Design (1932) und der Art Students League bei George Grosz (1933–34). 1936–41 Art Director der Zeit-

**Rambow,** Gunther – né le 2. 3. 1938 à Neustrelitz, Allemagne – *graphiste maquettiste, enseignant* – 1954–1958, formation de peintre sur verre à la Staatliche Fachschule (institut professionnel d'Etat) de Hadamar. 1958–1964, études à la Hochschule für Bildende Künste (école supérieure des beaux-arts) de Kassel. Fonde, en 1960, l'atelier «Rambow + Lienemeyer» (avec Gerhard Lienemeyer) à Kassel, où il exerce jusqu'en 1964. De 1964 à 1968, il travaille dans son atelier à Stuttgart, puis de 1968 à 1972 à Francfort-sur-le-Main. 1973–1988, Atelier collectif Rambow Lienemeyer van de Sand à Francfort. 1988, après le retrait de Lienemeyer l'atelier continue sous le nom Rambow van de Sand. Commanditaires : le Schauspiel Frankfurt, le Theater am Turm et le Städel à Francfort, les éditions S. Fischer, la Galerie René Block, Berlin, le Hessische Rundfunk, le Musée de Wiesbaden, etc. 1974–1991, professeur d'arts graphiques et de design à la Gesamthochschule de Kassel, puis à la Hochschule für Gestaltung (école supérieure de design) de Karlsruhe, à partir de 1991. – *Publications, sélection:* «Das sind eben alles Bilder der Strasse», Francfort-sur-le-Main 1979; «Traumstoff», Francfort-sur-le-Main 1986; «Fernseh-Design. Modell hessen 3», Berlin 1991. Volker Rattemeyer (éd.) «Rambow: Plakate», Wiesbaden 1988; Heinrich Klotz, Alain Weil «Rambow 1960–96», Stuttgart 1996.

**Rand,** Paul – né le 14. 8. 1914 à New York, Etats-Unis, décédé le 26. 11. 1996 à Norwalk, Connecticut, Etats-Unis – *graphiste maquettiste, illustrateur, typographe, enseignant* – études à New York au Pratt Institute (1930–1932), à la Parson's School of Design (1932), à l'Art Students League, chez George Grosz (1933–1934). Directeur artistique des revues «Esquire» et «Apparel Arts» de 1936 à 1941. Ma-

Rambow 1993 Poster

Rand 1968 Cover

Rambow 1990 Logo

Rand 1962 Logo

**Rambow,** Gunther – b. 2. 3. 1938 in Neustrelitz, Germany – *graphic designer, teacher* – 1954–58: trains as a designer of stained glass at the Staatliche Fachschule in Hadamar. 1958–64: studies at the Hochschule für Bildende Künste in Kassel. 1960–64: opens the Rambow + Lienemeyer studio in Kassel with Gerhard Lienemeyer. 1964–68: the studio operates in Stuttgart and from 1968–72 in Frankfurt am Main. 1973–88: the Rambow Lienemeyer van de Sand joint studio operates from Frankfurt am Main. 1988: after Lienemeyers departure, business continues under Rambow van de Sand. Clients include Schauspiel Frankfurt, the Theater am Turm and the Städel in Frankfurt, S. Fischer publishing house, the René Block gallery in Berlin, the Hessische Rundfunk and Museum Wiesbaden. 1974–91: professor of graphic design at the Gesamthochschule in Kassel and from 1991 onwards at the Hochschule für Gestaltung in Karlsruhe. – *Publications include:* "Das sind eben alles Bilder der Straße", Frankfurt am Main 1979; "Traumstoff", Frankfurt am Main 1986; "Fernseh-Design. Modell hessen 3", Berlin 1991. Volker Rattemeyer (ed.) "Rambow: Plakate", Wiesbaden 1988; Heinrich Klotz, Alain Weil "Rambow 1960–96", Stuttgart 1996.

**Rand,** Paul – b. 14. 8. 1914 in New York, USA, d. 26. 11. 1996 in Norwalk, Connecticut, USA – *graphic designer, illustrator, typographer, teacher* – Studied in New York at the Pratt Institute (1930–32), at the Parson's School of Design (1932) and at the Art Students League under

schriften „Esquire“ und „Apparel Arts“. 1938–45 Gestaltung der Umschläge der Kulturzeitschrift „Direction“. 1941–54 Art Director in der Agentur William H. Weintraub. Seit 1955 freier Grafik-Designer. Auftraggeber waren u. a. IBM, Westinghouse Electric Company, ABC Television, United Parcel Service, Next. Entwarf 1956 für das Erscheinungsbild von Westinghouse die Schrift „Westinghouse Gothic“. 1958 Auszeichnung mit dem Professorentitel der Tama Universität in Tokio. Entwirft 1966 für das Erscheinungsbild von IBM die Schrift „City Medium“. 1958 Verleihung der Goldmedaille des „American Institute of Graphic Arts“ (AIGA). Unterrichtet an der Cooper Union (1938–42), am Pratt Institute (1946), an der Yale University School of Art and Design (seit 1956) und leitet Kurse des „Yale Summer Program“ in Brissago, Schweiz (seit 1977). Rand beeinflußt mit seinen Arbeiten und Texten seit den vierziger Jahren nachfolgende Generationen mit einer Design-Auffassung, die alle erkenntnistheoretischen Gedanken miteinbezieht, jedoch stets der Kommunikation verpflichtet bleibt. – *Schriftentwürfe:* Westinghouse Gothic (1956), City Medium (1966). – *Publikationen:* „Thoughts on Design“, New York 1946; „I know a lot of things“ (mit Ann Rand), New York 1956; „Sparkle and Spin“ (mit Ann Rand), New York 1957; „Advertisement: Ad Vivum or Ad Hominem“, New York 1960; „Little 1“, New York 1962; „Design and the play instinct“, New York 1965; „The Trademarks of Paul Rand“, New York 1966; „Listen, Listen“, New York 1970; „A Designer's Art“,

Rand 1982 Cover

Rand 1986 Logo

Rand 1987 Page

Rand 1981 Poster

quette des couvertures de la revue culturelle «Direction» de 1938 à 1945. Directeur artistique de l'agence William H. Weintraub de 1941 à 1954. Exerce comme graphiste maquettiste indépendant à partir de 1955. Commanditaires : IBM, Westinghouse Electric Company, ABC Television, United Parcel Service, Next, etc. En 1956, il crée «Westinghouse Gothic» pour l'identité de Westinghouse. En 1958, l'université Tama de Tokyo lui décerne le titre de professeur. En 1966, il dessine «City Medium» pour l'identité d'IBM. Médaille d'or de l' «American Institute of Graphic Arts» (AIGA) en 1958. A enseigné à la Cooper Union (1938–1942), au Pratt Institute (1946), à la Yale University School of Art and Design (à partir de 1956), a dirigé les cours du «Yale Summer Program» à Brissago, Suisse (à partir de 1977). Les travaux et écrits de Rand, sa conception du design, qui tient compte de toutes les connaissances et théories sans jamais négliger le rôle de la communication, ont influencé les générations suivantes depuis les années 40. – *Polices :* Westinghouse Gothic (1956), City Medium (1966). – *Publications :* «Thoughts on Design», New York 1946; «I know a lot of things» (avec Ann Rand), New York 1956; «Sparkle and Spin» (avec Ann Rand), New York 1957; «Advertisement : Ad Vivum or Ad Hominem», New York 1960; «Little 1», New York 1962; «Design and the play instinct», New York 1965; «The Trademarks of Paul Rand», New York 1966; «Listen, Listen», New York 1970; «A Designer's Art», New

George Grosz (1933–34). 1936–41: art director of "Esquire" and "Apparel Arts" magazines. 1938–45: designs covers for "Direction" arts magazine. 1941–54: art director for the William H. Weintraub agency. 1955: starts working as a freelance graphic designer. Clients include IBM, Westinghouse Electric Company, ABC Television, United Parcel Service and Next. 1956: designs his Westinghouse Gothic typeface for Westinghouse's corporate identity. 1958: made an honorary professor by Tama University in Tokyo. 1966: designs his City Medium typeface for IBM's corporate identity. 1958: is awarded a gold medal by the American Institute of Graphic Arts (AIGA). Teaches at the Cooper Union (1938–42), the Pratt Institute (1946) and the Yale University School of Art and Design (from 1956 onwards) and heads courses at the Yale Summer Program in Brissago, Switzerland (from 1977 onwards). From the 1940s onwards, Rand greatly influenced the next generation with his artwork, writings and a concept of design which absorbed all kinds of epistemological ideas yet remained dedicated to the process of communication. – *Fonts:* Westinghouse Gothic (1956), City Medium (1966). – *Publications:* "Thoughts on Design", New York 1946; "I know a lot of things" (with Ann Rand), New York 1956; "Sparkle and Spin" (with Ann Rand), New York 1957; "Advertisement: Ad Vivum or Ad Hominem", New York 1960; "Little 1", New York 1962; "Design and the play instinct", New York 1965; "The Trademarks of Paul Rand",

New Haven 1985; „Design Form Chaos", New Haven, London 1993; „Von Lascaux bis Brooklyn", Teufen 1996. Yusaku Kamekura „Paul Rand 1946–1958", Tokio, New York 1959.

**Rapp,** Hermann – geb. am 17. 5. 1937 in Pfullingen, Deutschland. – *Drucker, Grafiker, Holzschneider, Autor* – Schriftsetzerlehre, tätig als Setzer und Lehrlingsausbilder in einer Stuttgarter Druckerei. 1964 Verlagshersteller in Frankfurt am Main, danach künstlerischer Leiter. 1989 Gründung seiner Presse „Offizin Die Goldene Kanne" in Neuweilnau. In dieser druckgrafischen Werkstatt entstehen klassische und experimentelle Drucke in Bleisatz und Holzschnitt.

**Ratdolt,** Erhard – geb. 1447 in Augsburg, gest. 1527 in Augsburg, Deutschland. – *Drucker, Verleger* – Lehre als Formschneider. Verläßt 1474 Augsburg. Betreibt 1476 in Venedig eine Druckerei, die große Erfolge hat. Als erstes Buch erscheint das „Calendarium" des Königsberger Astronomen Johannes Müller. 1480–85 erscheinen weitere fünfzig Bücher, die zu den schönsten der Zeit gehören. 1486 folgt Ratdolt dem Ruf des Bischofs Johann von Werdenberg und zieht mit seiner gesamten Druckerei zurück nach Augsburg. – *Publikationen:* Gilbert E. Redgrave „Erhard Ratdolt and his work in Venice", London 1894; K. Schottenloher „Die liturgischen Druckwerke Ratdolts", Mainz 1922.

**Ray Gun** – *Zeitschrift* – 1992 Gründung von „Ray Gun" in Santa Monica, Kalifornien, als alternative Musikzeitschrift mit dem Untertitel „Music + Style". Kon-

Haven 1985; «Design Form Chaos», New Haven, Londres 1993; «Von Lascaux bis Brooklyn», Teufen 1996. Yusaku Kamekura «Paul Rand 1946–1958», Tokyo, New York 1959.

**Rapp,** Hermann – né le 17. 5. 1937 à Pfullingen, Allemagne – *imprimeur, graphiste, graveur sur bois, auteur* – apprentissage de composition typographique, exerce comme typographe et formateur d'apprentis dans une imprimerie de Stuttgart. Directeur de fabrication dans une maison d'édition à Francfort-sur-le-Main, en 1964, puis directeur artistique. En 1989, il fonde ses propres presses, l'«Offizin Die Goldene Kanne» à Neuweilnau. Il y réalise des impressions classiques et expérimentales en utilisant la composition au plomb et la gravure sur bois.

**Ratdolt,** Erhard – né en 1447 à Augsbourg, Allemagne, décédé en 1527 à Augsbourg, Allemagne – *imprimeur, éditeur* – apprentissage de tailleur de poinçons. En 1474, il quitte Augsbourg. En 1476, à Venise, il exploite une imprimerie qui connaît un grand succès. Son premier livre publié est le «Calendarium» de Johannes Müller, un astronome de Königsberg. 50 autres ouvrages, qui comptent parmi les plus beaux de l'époque, paraissent entre 1480 et 1485. Ratdolt répond à l'appel de l'évêque Johann von Werdenberg en 1486 et retourne à Augsbourg avec son imprimerie. – *Publications:* Gilbert E. Redgrave «Erhard Ratdolt and his work in Venice», Londres 1894; K. Schottenloher «Die liturgischen Druckwerke Ratdolts», Mayence 1922.

**Ray Gun** – *revue* – fondation en 1992 de «Ray Gun» à Santa Monica, Californie, une revue de musique alternative, ayant pour sous-titre «Music + Style». Conception et directeur de publication: Marvin Scott Jarrett; rédaction: Neil Feinemann;

Rapp 1991 Letter

Rapp 1997

CXVI

E igit̃ clemētiſſime
pater per iheſū xp̄z
filiuz tuū dn̄m nr̄m
ſupplices rogam⁹ ⁊
petim⁹: vti accepta
habeas et bn̄dicas
hec do✠na: hec
mu✠nera: hec ſā
✠cta ſacrificia illibata. In p̄mis q̄
tibi offerimus pro eccleſia tua ſancta
catholica quaz pacificare: cuſtodire:
adunare ⁊ regere digneris toto orbe
terrarū vna cū famulo tuo papa nr̄o
N. ⁊ antiſtite nr̄o. N. ⁊ rege nr̄o. N. ⁊
oib⁹ orthodoxis catholice ⁊ apoſto
lice fidei cultorib⁹. Memoria viuorū.
Memēto dn̄e famulorū famularūq͂
tuarū. N. Et oīm circūſtantiū quorū
tibi fides cognita eſt: ⁊ nota deuotio
pro quibus tibi offerimus: vel qui tibi
offerunt hoc ſacrificium laudis: pro ſe
ſuiſq͂ oībus pro redēptione animarū
ſuarū: pro ſpe ſalutis ⁊ incolumitatis

Ratdolt 1496 Page

New York 1966; "Listen, Listen", New York 1970; "A Designer's Art", New Haven 1985; "Design Form Chaos", New Haven, London 1993; "Von Lascaux bis Brooklyn", Teufen 1996. Yusaku Kamekura "Paul Rand 1946–1958", Tokyo, New York 1959.

**Rapp,** Hermann – b. 17. 5. 1937 in Pfullingen, Germany – *printer, graphic artist, woodcut artist, author* – Trained as a typesetter then worked as a typesetter and trained apprentices at a Stuttgart printing works. 1964: production assistant and then art director for a publishing house in Frankfurt am Main. 1989: launches his press, Offizin Die Goldene Kanne, in Neuweilnau. His graphics workshop produces classical and experimental prints from lead composition and woodcuts.

**Ratdolt,** Erhard – b. 1447 in Augsburg, Germany, d. 1527 in Augsburg, Germany – *printer, publisher* – Trained as a form cutter. 1474: leaves Augsburg. 1476: runs a printing workshop in Venice which is extremely successful. The first book to be published is "Calendarium" by the Königsberg astronomer Johannes Müller. 1480–85: fifty books are published, among the most beautiful of their day and age. 1486: at the request of Bishop Johann von Werdenberg Ratdolt and his printing works move back to Augsburg. – *Publications:* Gilbert E. Redgrave "Erhard Ratdolt and his work in Venice", London 1894; K. Schottenloher "Die liturgischen Druckwerke Ratdolts", Mainz 1922.

**Ray Gun** – *magazine* – 1992: "Ray Gun" is launched in Santa Monica, California,

zeption und Herausgeber: Marvin Scott Jarrett. Redaktion: Neil Feineman. Layout: David Carson. Die Zeitschrift wird als eine Collage von Text und Bild ohne Rastervorgaben gestaltet. In der Abteilung „Sound in Print" veröffentlicht „Ray Gun" populär-künstlerische Beiträge der Leser. Mitarbeit zahlreicher bekannter Illustratoren und Fotografen. Bis Heft 30/1995 war David Carson Art Director. Die Hefte 31–32 wurden von Johnson + Wolverton gestaltet. Ab dem Heft 33 ist Robert Hales Art Director.

**Reichert,** Josua – geb. 1937 in Stuttgart, Deutschland. – *Drucker, Künstler, Typograph* – 1959 Studium an der Staatlichen Akademie der Bildenden Künste in Karlsruhe bei HAP Grieshaber. 1960 Atelier in Stuttgart. Neben den Pressendrucken entstehen ab 1960 großformatige Handdrucke für öffentliche Gebäude, u. a. die Marbacher Drucke (1960), Stuttgarter Drucke (1971–73) und für das Gutenberg-Museum Mainz (1986). 1961–72 Atelier in München. 1963 entsteht die Folge „Codex Typographicus" mit 28 Drucken. Für seine typographisch geprägte Grafik wird er 1967 mit dem Preis der Biennale in São Paulo ausgezeichnet. – *Publikationen u.a.:* „Gill", Haidholzen 1989; „Das Werkmanbuch", Bietigheim-Bissingen 1991; „Die Frank-Rühl-Hebräisch und der Haidholzener Psalter", München 1994; „Die Mühldorfer Drucke", Haidholzen 1994.

**Reid,** Jamie – geb. 1940. – *Grafik-Designer, Maler* – 1962–64 Studium an der Wimbledon Art School. 1964–68 Studium an der Croydon Art School. Mitarbeit

maquette : David Carson. La revue est conçue comme un collage de textes et d'images, realisé sans grille de mise en page. Dans sa rubrique «Sound in Print», «Ray Gun» publie les contributions artistiques de ses lecteurs. De nombreux illustrateurs et photographes de renom y collaborent. David Carson y a exercé les fonctions de directeur artistique jusqu'au numéro 30/1995. Les numéros 31–32 ont été conçus par Johnson + Wolverton. Robert Hales est chargé de la direction artistique depuis le numéro 33.

**Reichert,** Josua – né en 1937 à Stuttgart, Allemagne – *imprimeur, artiste, typographe* – 1959, études à la Staatliche Akademie der Bildenden Künste (beaux-arts) de Karlsruhe chez HAP Grieshaber. 1960, atelier à Stuttgart. Outre des impressions traditionnelles, il réalise des impressions manuelles de grand format pour des bâtiments publics à partir de 1960, entre autres les planches de Marbach (1960), planches de Stuttgart (1971–1973) et des planches pour le Gutenberg-Museum de Mayence (1986). Exerce dans son atelier à Munich de 1961 à 1972. Il réalise une série de 28 planches, le «Codex Typographicus», en 1963. Ses travaux graphiques, fortement marqués par la typographie, obtiennent le premier prix de la Biennale de São Paulo en 1967. – *Publications, sélection :* «Gill», Haidholzen 1989; «Das Werkmanbuch», Bietigheim-Bissingen 1991; «Die Frank-Rühl-Hebräisch und der Haidholzener Psalter», Munich 1994; «Die Mühldorfer Drucke», Haidholzen 1994.

**Reid,** Jamie – né en 1940 – *graphiste maquettiste, peintre* – 1962–1964, études à la Wimbledon Art School. 1964–1968, études à la Croydon Art School. Participe,

Ray Gun 1992 Cover

Reichert 1972

Ray Gun 1994 Spread

Reichert Type Picture

as an alternative music magazine, subtitled "Music + Style". Concept, publisher: Marvin Scott Jarrett. Editor: Neil Feineman. Layout: David Carson. The magazine is made up of a collage of text and pictures without a grid system. The "Sound in Print" section in "Ray Gun" prints pop-art contributions from the magazine's readers. Many well-known illustrators and photographers have worked for the magazine. David Carson was art director until no. 30/1995. Nos. 31–32 were designed by Johnson + Wolverton. Robert Hales took over as art director from no. 33 onwards.

**Reichert,** Josua – b. 1937 in Stuttgart, Germany – *printer, artist, typographer* – 1959: studies at the Staatliche Akademie der Bildenden Künste in Karlsruhe under HAP Grieshaber. 1960: studio in Stuttgart. Besides his press prints, Reichert starts producing large-format hand prints for public buildings from 1960 onwards, such as the Marbacher Drucke (1960), the Stuttgarter Drucke (1971–73) and for the Gutenberg Museum in Mainz (1986). 1961–72: studio in Munich. 1963: the "Codex Typographicus" series is produced with 28 prints. 1967: Reichert is awarded a prize for his graphic work in typography at the Biennale in São Paolo. – *Publications include:* "Gill", Haidholzen 1989; "Das Werkmanbuch", Bietigheim-Bissingen 1991; "Die Frank-Rühl-Hebräisch und der Haidholzener Psalter", Munich 1994; "Die Mühldorfer Drucke", Haidholzen 1994.

**Reid,** Jamie – b. 1940 – *graphic designer, painter* – 1962–64: studies at Wimbledon

an einem Film über die Oxford Street mit Malcolm McLaren. 1970–75 Mitbegründer der anarchistischen Zeitschrift „Suburban Press" (mit Jeremy Brook und Nigel Edwards), von der sechs Ausgaben erscheinen. Zahlreiche grafische Arbeiten für die Zeitschrift, Herausgabe von Aufklebern (Stickers) mit anarchistischen Inhalten. 1974 Herausgabe des „Suburban Press Poster Book". 1976–79 Gestaltung aller grafischen Arbeiten für die Punk-Musikgruppe Sex Pistols: T-Shirts, Plakate, Broschüren, Schallplattencover, Anzeigen. Verkauf der „Suburban Press", 1978 grafische Entwürfe für die Musikgruppen Dead Kennedys und Bow Wow Wow. 1979 Art Director des Films „The Great Rock 'n' Roll Swindle" der Sex Pistols. 1980 grafische Entwürfe für Vivienne Westwoods Modeladen „World's End" in London. 1980–82 Arbeit in Paris. Beginn seines Projekts „Leaving the 20th Century". 1982 Rückkehr nach England. Arbeiten für Musik-, Tanz-, Theatergruppen und Verlage. 1986 Retrospektiv-Ausstellung von zwanzig Jahren Arbeit unter dem Titel „Chaos in Cancerland" in der Hamilton's Gallery in Mayfair. – *Publikation:* „Up they rise: the incomplete works of Jamie Reid", London 1987.

**Reiner,** Imre – geb. 18. 8. 1900 in Versec, Ungarn, gest. 21. 8. 1987 in Lugano, Schweiz. – *Grafiker, Typograph, Kalligraph, Autor* – Ausbildung zum Bildhauer in Ungarn und an der Kunstgewerbeschule in Frankfurt am Main. 1921–23 und 1925–27 Studium an der Akademie für Bildende Künste in Stuttgart bei Ernst Schneidler. Aufenthalt

avec Malcolm McLaren, à la réalisation d'un film sur la Oxford Street. 1970–1975, cofondateur de la revue anarchiste «Suburban Press» (avec Jeremy Brook et Nigel Edwards), six numéros paraîtront. Nombreux travaux graphiques pour des revues, éditions d'autocollants (Stickers) avec des slogans anarchistes. En 1974, il publie le «Suburban Press Poster Book». De 1976 à 1979, il crée tous les graphismes de T-shirts, affiches, brochures, pochettes de disques, annonces du groupe Punk Sex Pistols. Vend les «Suburban Press», puis réalise des travaux graphiques pour les groupes Dead Kennedys et Bow Wow Wow en 1978. Directeur artistique du film «The Great Rock'n' Roll Swindle» des Sex Pistols en 1979. Travaux pour la boutique de mode «World's End» de Vivienne Westwoods à Londres, en 1980. Travaille à Paris de 1980 à 1982. Début du projet «Leaving the 20th Century». Retour en Angleterre en 1982. Travaux pour des groupes de musiciens, des compagnies de danse, des troupes de théâtre et des éditeurs. 1986, exposition rétrospective sur vingt années de travail intitulée «Chaos in Cancerland» à la Hamilton's Gallery à Mayfair. – *Publication:* «Up they rise: the incomplete works of Jamie Reid», Londres 1987.

**Reiner,** Imre – né le 18. 8 1900 à Versec, Hongrie, décédé le 21. 8. 1987 à Lugano, Suisse – *graphiste, typographe, calligraphe, auteur* – formation de sculpteur en Hongrie puis à la Kunstgewerbeschule (école des arts décoratifs) de Francfort-sur-le-Main. Etudes à l'Akademie für Bildene Künste (beaux-arts) de Stuttgart de 1921 à 1923, puis de 1925 à 1927 chez Ernst Schneidler. Séjours à New York et à Chicago (1923–1925), à Stuttgart (1927–1930), et à Paris (1930). Vit et travaille à Ruvigliana, Suisse, à partir de 1931. Nombreuses illustrations en gra-

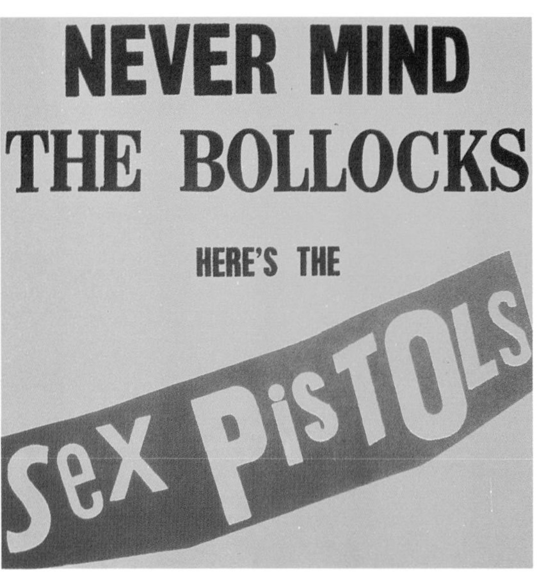

Reid 1977 Cover

Reid 1987 Cover

Reid 1988 Cover

Reiner 1946 Cover

Art School. 1964–68: studies at Croydon Art School. Works with Malcolm McLaren on a film about Oxford Street. 1970–75: co-founder of the anarchy magazine "Suburban Press" with Jeremy Brook and Nigel Edwards. Six numbers of the magazine are published. Reid does much graphic work for the magazine and issues anarchy stickers. 1974: publishes the "Suburban Press Poster Book". 1976–79: designs the graphics for the Sex Pistols punk group, including T-shirts, posters, brochures, record covers and adverts. "Suburban Press" is sold. 1978: does graphic work for the Dead Kennedys and Bow Wow Wow. 1979: art director of the Sex Pistols' film "The Great Rock 'n' Roll Swindle". 1980: produces graphic designs for Vivienne Westwood's World's End fashion store in London. 1980–82: works in Paris. Starts up his "Leaving the 20th Century" project. 1982: returns to England. Works for various music, dance and theater groups and publishing houses. 1986: a retrospective exhibition of twenty years' work entitled "Chaos in Cancerland" is held at Hamilton's Gallery in Mayfair. – *Publication:* "Up they rise: the incomplete works of Jamie Reid", London 1987.

**Reiner,** Imre – b. 18. 8. 1900 in Versec, Hungary, d. 21. 8. 1987 in Lugano, Switzerland – *graphic artist, typographer, calligrapher, author* – Trained as a sculptor in Hungary and at the Kunstgewerbeschule in Frankfurt am Main. 1921–23 and 1925–27: studies at the Akademie für Bildende Künste in Stuttgart under Ernst Schneidler. Spends

in New York und Chicago (1923–25), Stuttgart (1927–30) und Paris (1930). Seit 1931 lebt und arbeitet er in Ruvigliana, Schweiz. Zahlreiche Holzschnitt-Illustrationen für Werke von Homer, Cervantes, Voltaire, Goethe, Dickens, Novalis. 1973 entsteht die Serie der „Ziffernbilder". – *Schriftentwürfe:* Meridian (1930), Gotika, Skyline (1934), Corvinus (1934–35), Matura (1936), Floride (1939), Stradivarius (1945), Reiner Script (1951), Reiner Black (1955), Mustang (1956), Bazaar (1956), London Script (1957), Mercurius (1957), Pepita (1959), Contact (1968). – *Publikationen u.a.:* „Initialen", Basel 1942; „Typo-Graphik I", St. Gallen 1943; „Monogramme", St. Gallen 1947; „Alphabete", St. Gallen 1948; „Schrift im Buch" (mit Hedwig Reiner), St. Gallen 1948; „Typo-Graphik II", St. Gallen 1950. Hans Peter Willberg „Imre Reiner. Die Ziffernbilder", Stuttgart 1975.

**Reisinger,** Dan – geb. 3. 8. 1934 in Kanjiza, Jugoslawien. – *Grafik-Designer, Lehrer* – 1951–54 Studium an der Bezalel Kunstakademie in Jerusalem. 1957–58 freiberuflich in Brüssel tätig. 1958–60 Kunststudium an der Central School of Arts and Design in London. 1960–64 künstlerischer Leiter der Werbeagentur Tab Arieli. 1964 Dozent an der Bezalel Kunstakademie in Jerusalem. 1965–66 Studium für Bühnenbild und dreidimensionales Gestalten an der Central School of Arts and Design in London. Seit 1968 Eröffnung und Leitung seines Grafik-Design-Büros in Tel Aviv. Auftraggeber waren u. a. die Israel Railways, das Tel Aviv Museum, die EL AL Israel Airlines

| | n | | | | |
|---|---|---|---|---|---|
| a | b | e | f | g | i |
| o | r | s | t | y | z |
| A | B | C | E | G | H |
| M | O | R | S | X | Y |
| 1 | 2 | 4 | 6 | 8 | & |

Reiner 1936 Matura

| | n | | | | |
|---|---|---|---|---|---|
| a | b | e | f | g | i |
| o | r | s | t | y | z |
| A | B | C | E | G | H |
| M | O | R | S | X | Y |
| 1 | 2 | 4 | 6 | 8 | & |

Reiner 1959 Pepita

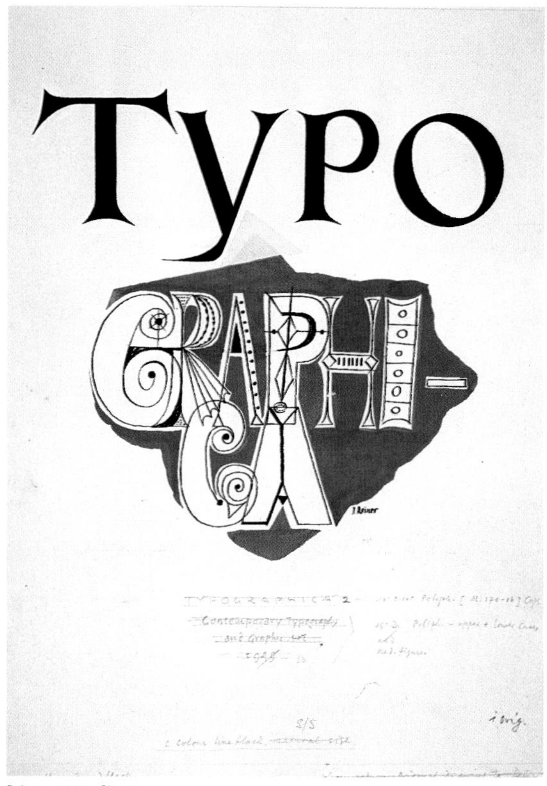


Reiner 1950 Sketch

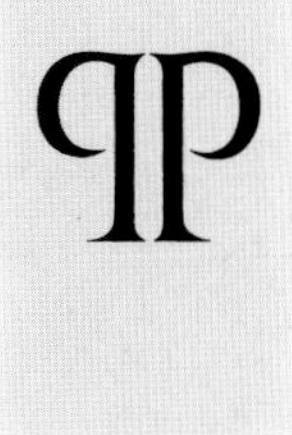

Reiner 1947 Logo

Reisinger ca. 1972 Poster

Reisinger Poster

vure sur bois pour des œuvres d'Homère, Cervantes, Voltaire, Goethe, Dickens, Novalis. En 1973, il réalise la série «Ziffernbilder» (images de chiffres). – *Polices:* Meridian (1930), Gotika, Skyline (1934), Corvinus (1934–1935), Matura (1936), Floride (1939), Stradivarius (1945), Reiner Script (1951), Reiner Black (1955), Mustang (1956), Bazaar (1956), London Script (1957), Mercurius (1957), Pepita (1959), Contact (1968). – *Publications, sélection:* «Initialen», Bâle 1942; «Typo-Graphik I», Saint-Gall 1943; «Monogramme», Saint-Gall 1947; «Alphabete», Saint-Gall 1948; «Schrift im Buch» (avec Hedwig Reiner), Saint-Gall 1948; «Typo-Graphik II», Saint-Gall 1950. Hans Peter Willberg «Imre Reiner. Die Ziffernbilder», Stuttgart 1975.

**Reisinger,** Dan – né le 3. 8. 1934 à Kanjiza, Yougoslavie – *graphiste maquettiste, enseignant* – 1951–1954, études à l'académie Bezalel à Jérusalem. Exerce comme indépendant à Bruxelles de 1957 à 1958. Etudes d'art à la Central School of Arts and Design de Londres de 1958 à 1960. Directeur artistique de l'agence de publicité Tab Arieli de 1960 à 1964. Enseigne à l'académie Bezalel, Jérusalem, en 1964. Etudie la scénographie et la création en trois dimensions à la Central School of Arts and Design de Londres de 1965 à 1966. Ouvre une agence de conception graphique et de design à Tel Aviv en 1968, où il exerce depuis. Commanditaires: Israel Railways, Tel Aviv Museum, EL AL Israel Airlines, Habima National Theater,

time in New York and Chicago (1923–25), Stuttgart (1927–30) and Paris (1930). From 1931 onwards: lives and works in Ruvigliana, Switzerland. Produces numerous woodcut illustrations for works by Homer, Cervantes, Voltaire, Goethe, Dickens and Novalis. 1973: produces his "Ziffernbilder" series. – *Fonts:* Meridian (1930), Gotika, Skyline (1934), Corvinus (1934–35), Matura (1936), Floride (1939), Stradivarius (1945), Reiner Script (1951), Reiner Black (1955), Mustang (1956), Bazaar (1956), London Script (1957), Mercurius (1957), Pepita (1959), Contact (1968). – *Publications include:* "Initialen", Basle 1942; "Typo-Graphik I", St. Gallen 1943; "Monogramme", St. Gallen 1947; "Alphabete", St. Gallen 1948; "Schrift im Buch" (with Hedwig Reiner), St. Gallen 1948; "Typo-Graphik II", St. Gallen 1950. Hans Peter Willberg "Imre Reiner. Die Ziffernbilder", Stuttgart 1975.

**Reisinger,** Dan – b. 3. 8. 1934 in Kanjiza, Yugoslavia – *graphic designer, teacher* – 1951–54: studies at the Bezalel Art Academy in Jerusalem. 1957–58: freelances in Brussels. 1958–60: studies art at the Central School of Arts and Design in London. 1960–64: art director for the Tab Arieli advertising agency. 1964: lecturer at the Bezalel Art Academy in Jerusalem. 1965–66: studies stage set design and three-dimensional design at the Central School of Arts and Design in London. 1968: opens and runs his own graphic design studio in Tel Aviv. Clients include Israel Railways, the Tel Aviv Museum, EL AL Israel Airlines and Habima

und das Habima National Theater. Gestalterischer Berater der Bank of Israel und der Israel Air Force. 1977 Ausstellung seiner Arbeit im Israel Museum in Jerusalem. Seit 1984 Dozent an der Bezalel Kunstakademie in Jerusalem. – *Publikation:* „Symbols", Jerusalem 1996.

**Remington,** Eliphalet – geb. 28. 10. 1793 in Suffield, USA, gest. 12. 8. 1861 in Ilion, USA. – *Erfinder, Unternehmer, Waffenproduzent* – Remington entwickelt in seiner Firma Neuerungen für die Waffenherstellung und eine Schreibmaschine. Sein Sohn Philo führt 1861 die Firma weiter und wird zum Produzenten der ersten kommerziellen Schreibmaschine. 1873 Beginn der Produktion im Remington-Werk in Ilion. W. O. Wyckoff, C. W. Seamans und H. H. Benedict kaufen 1886 die Schreibmaschinen-Abteilung der Firma E. Remington. Umzug der Firma nach New York. Produktion von zahlreichen Schreibmaschinen-Modellen.

**Renner,** Paul – geb. 9. 8. 1878 in Wernigerode, Deutschland, gest. 25. 4. 1956 in Hödingen, Deutschland. – *Grafiker, Maler, Schriftentwerfer, Autor, Lehrer* – Studium der Architektur und Malerei in Berlin, München und Karlsruhe. Danach als Maler in München tätig. 1907–17 Hersteller und Ausstatter im Georg-Müller-Verlag in München. 1911 Mitbegründer einer privaten Schule für Illustration in München. 1925–26 Leitung der Klasse für Werbegrafik und Typographie an der Frankfurter Kunstschule. 1926 Leitung der Grafischen Berufsschulen der Stadt München, ab 1927 der Meisterschule für Deutschlands Buchdrucker. 1933 Beauf-

etc. Conseiller en création de la Bank of Israel et de l'Israel Air Force. Ses travaux sont exposés à l'Israel Museum de Jérusalem en 1977. Enseigne à l'Académie Bezalel depuis 1984. – *Publication:* «Symbols», Jérusalem 1996.

**Remington,** Eliphalet – né le 28. 10. 1793 à Suffield, Etats-Unis, décédé le 12. 8. 1861 à Ilion, Etats-Unis – *inventeur, chef d'entreprise, fabricant d'armes* – dans sa société, Remington invente des techniques innovatrices pour la fabrication d'armes et développe une machine à écrire. En 1861, Philo, son fils reprend l'entreprise et devient le premier fabricant de machines à écrire commercialisées. En 1873, la production débute à l'usine Remington d'Ilion. En 1886, W. O. Wyckoff, C. W. Seamans et H. H. Benedict rachètent le service de production des machines à écrire de la société E. Remington. Implantation de la société à New York. Production de nombreux modèles de machines à écrire.

**Renner,** Paul, né le 9. 8. 1878 à Wernigerode, Allemagne, décédé le 25. 4. 1956 à Hödingen, Allemagne – *graphiste, peintre, concepteur de polices, auteur, enseignant* – études d'architecture et de peinture à Berlin, Munich et Karlsruhe. Exerce ensuite comme peintre à Munich. Responsable de la fabrication aux éditions Georg Müller à Munich, de 1907 à 1917. Cofondateur, en 1911 d'une école d'illustration à Munich. De 1925 à 1926, il dirige les cours de graphisme publicitaire et de typographie à l'école des beaux-arts de Francfort. En 1926, il dirige les écoles professionnelles d'arts graphiques de Munich, puis à partir de 1927, la Meisterschule für Deutschlands Buchdrucker (l'école supérieure de l'imprimerie allemande). En 1933, le gouvernement le charge de la conception de la participation allemande à la Triennale de Milan,

Renner before 1925 Type Specimen

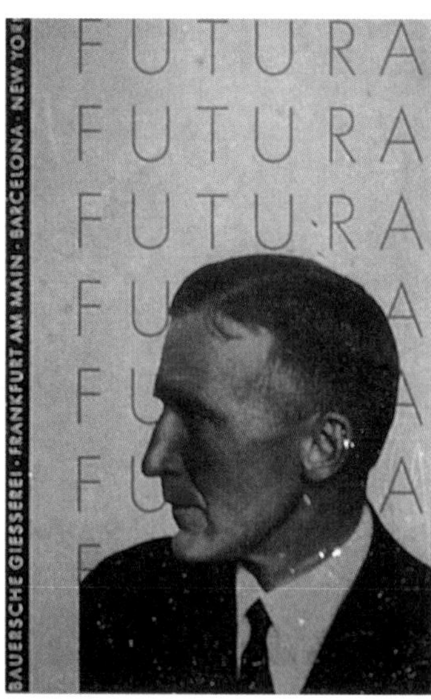


Renner Cover

| n | n | n | n | | n |
|---|---|---|---|---|---|
| a | b | e | f | g | i |
| o | r | s | t | y | z |
| A | B | C | E | G | H |
| M | O | R | S | X | Y |
| 1 | 2 | 4 | 6 | 8 | & |

Renner 1928 Futura

| | n | | | | |
|---|---|---|---|---|---|
| a | b | e | f | g | i |
| o | r | s | t | y | z |
| A | B | C | E | G | H |
| M | O | R | S | X | Y |
| 1 | 2 | 4 | 6 | 8 | & |

Renner 1929 Futura Black

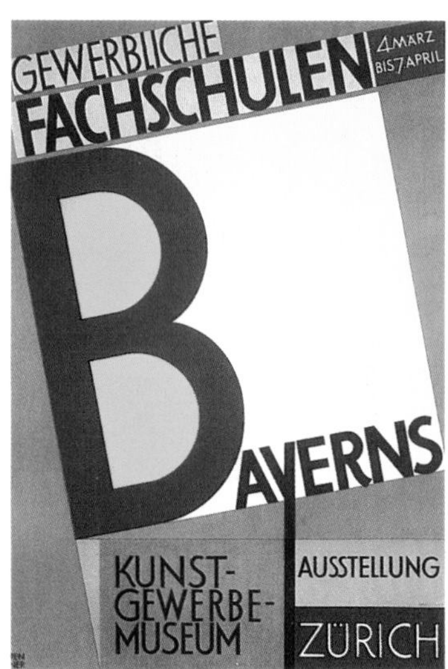


Renner 1928 Poster

National Theater. Design consultant to the Bank of Israel and the Israel Air Force. 1977: his work is exhibited at the Israel Museum in Jerusalem. From 1984 onwards: lecturer at the Bezalel Art Academy in Jerusalem. – *Publication:* "Symbols", Jerusalem 1996.

**Remington,** Eliphalet – b. 28. 10. 1793 in Suffield, USA, d. 12. 8. 1861 in Ilion, USA – *inventor, entrepreneur, weapons manufacturer* – In his company Remington worked on innovations for the weapons industry and also produced a typewriter. 1861: his son Philo takes over the company and manufactures the first commercial typewriter. 1873: production is started at the Remington plant in Ilion. 1886: W. O. Wyckoff, C. W. Seamans and H. H. Benedict buy up the E. Remington company's typewriter department. The business is moved to New York, where it manufactures numerous typewriter models.

**Renner,** Paul – b. 9. 8. 1878 in Wernigerode, Germany, d. 25. 4. 1956 in Hödingen, Germany – *graphic artist, painter, type designer, author, teacher* – Studied architecture and painting in Berlin, Munich and Karlsruhe. Then worked as a painter in Munich. 1907–17: production assistant and presentation manager for Georg Müller Verlag in Munich. 1911: cofounder of a private school for illustration in Munich. 1925–26: head of the commercial art and typography department at the Frankfurter Kunstschule. 1926: director of the city of Munich's Grafische Berufsschulen and from 1927 the Meisterschule für Deutschlands Buch-

tragter des Deutschen Reichs für den Aufbau der Deutschen Abteilung auf der Triennale in Mailand, ausgezeichnet mit dem großen Preis. 1933 Entlassung aus dem Schuldienst. Seit 1934 Maler. Autor zu Themen aus Typographie, Schrift, Grafik und Farblehre. – *Schriftentwürfe:* Futura (1928), Plak (1928), Futura Black (1929), Futura licht (1932), Futura Schlagzeile (1932), Ballade (1937), Renner Antiqua (1939), Steile Futura (1952). – *Publikationen u.a.:* „Typographie als Kunst", München 1922; „Kulturbolschewismus?", Zürich 1932; „Die Kunst der Typographie", Berlin 1948; „Das moderne Buch", Lindau 1946; „Vom Geheimnis der Darstellung", Frankfurt 1955. Philipp Luidl (Hrsg.) „Paul Renner", München 1978

**Ricci,** Franco Maria – geb. 2. 12. 1937 in Parma, Italien. – *Grafik-Designer, Verleger, Typograph* – Studium der Geologie, anschließend Arbeit für Gulf Oil. 1963 Eröffnung seines Grafik-Design-Studios in Parma. 1965 Begegnung mit dem Werk Giambattista Bodonis. Herausgabe des Reprints „Manuale Tipografico" von Bodoni in 900 Exemplaren. Es folgen u. a. Bodonis „Oratio Dominica", Fotos von Lewis Carroll, Gemälde von Tamara de Lempicka, Arbeiten des Art-Deco-Künstlers Demeter Chiparus. 1970–80 Herausgabe eines Reprints der „Encyclopédie de Diderot et d'Alembert". 1981 mit dem Orden „Chevalier des Arts et des Lettres" der französischen Republik ausgezeichnet. 1982 erscheint das erste Heft der von ihm konzipierten und herausgegebenen zweimonatigen Kulturzeitschrift „FMR".

ABCDEFG 1234567890
HIJKLMNO
PQRSTUV
WXYZÄÖÜ abcdefghijklmn
opqrstuvwxyz

Renner 1927 Furura halbfett

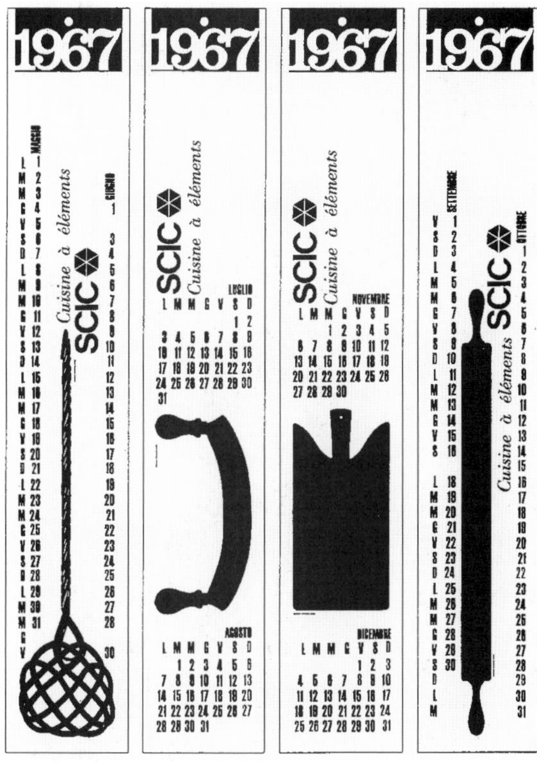


Ricci 1967 Calendar

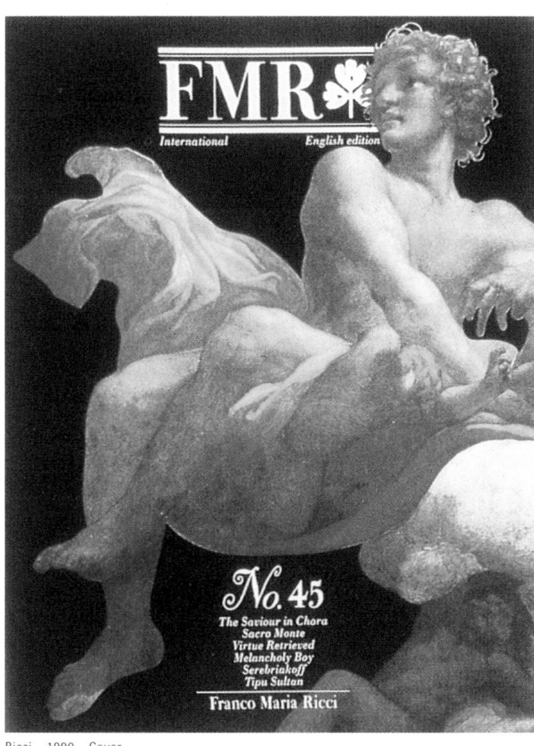


Ricci 1990 Cover

le premier prix lui est décerné. Toujours en 1933, il est révoqué de l'enseignement. Peintre à partir de 1934. Auteur d'ouvrages sur la typographie, l'écriture, les arts graphiques et la théorie des couleurs. – *Polices :* Futura (1928), Plak (1928), Futura Black (1929), Futura licht (1932), Futura Schlagzeile (1932), Ballade (1937), Renner Antiqua (1939), Steile Futura (1952). – *Publications, sélection :* «Typographie als Kunst», Munich 1922; «Kulturbolschewismus ?», Zurich 1932; «Die Kunst der Typographie», Berlin 1948; «Das moderne Buch», Lindau 1946; «Vom Geheimnis der Darstellung», Francfort 1955. Philipp Luidl (éd.) «Paul Renner», Munich 1978.

**Ricci,** Franco Maria – né le 2. 12. 1937 à Parme, Italie – *graphiste maquettiste, éditeur, typographe* – études de géologie, puis emploi chez Gulf Oil. En 1963, il ouvre un atelier de graphisme et de design à Parme. Découvre l'œuvre de Giambattista Bodoni en 1965. Publie une réédition du «Manuale Tipografico» en 900 exemplaires. Puis viennent «Oratio Dominica» de Bodoni, des photos de Lewis Carroll, des tableaux de Tamara de Lempicka, des travaux de l'artiste Art-déco, Demeter Chiparus. Publie une réédition de l' «Encyclopédie de Diderot et d'Alembert» de 1970 à 1980. Le titre de «Chevalier des Arts et des Lettres» de la République française lui est décerné en 1981. Conçoit et édite la revue bimensuelle «FMR», dont le premier numéro paraît en

drucker. 1933: as a representative of the German Reich he is in charge of the design of the German section at the Milan Triennale. Receives the Triennale's Grand Prix. 1933: is dismissed from teaching. Works as a painter from 1934 onwards. Writes on topics pertaining to typography, lettering, graphics and color studies. – *Fonts:* Futura (1928), Plak (1928), Futura Black (1929), Futura licht (1932), Futura Schlagzeile (1932), Ballade (1937), Renner Antiqua (1939), Steile Futura (1952). – *Publications include:* "Typographie als Kunst", Munich 1922; "Kulturbolschewismus?", Zurich 1932; "Die Kunst der Typographie", Berlin 1948; "Das moderne Buch", Lindau 1946; "Vom Geheimnis der Darstellung", Frankfurt 1955. Philipp Luidl (ed.) "Paul Renner", Munich 1978.

**Ricci,** Franco Maria – b. 2. 12. 1937 in Parma, Italy – *graphic designer, publisher, typographer* – Studied geology and then worked for Gulf Oil. 1963: opens a graphic design studio in Parma. 1965: is introduced to the work of Giambattista Bodoni. Publishes a reprint of Bodoni's "Manuale Tipografico" in an edition of 900, followed by Bodoni's "Oratio Dominica", photos by Lewis Carroll, paintings by Tamara de Lempicka and work by the Art Deco artist Demeter Chiparus, among others. 1970–80: publishes a reprint of the "Encyclopédie de Diderot et d'Alembert". 1981: is made a Chevalier des Arts et des Lettres by the French republic. 1982: the first number of the bimonthly arts magazine, "FMR", is issued, which Ricci designs and edits. Ricci

Seit den siebziger Jahren eröffnet Ricci „FMR"-Buchhandlungen in Italien und Mexiko.

**Riedel**, Hubert – geb. 7. 9. 1948 in Berlin, Deutschland. – *Grafik-Designer, Ausstellungsgestalter* – Arbeit im Schwermaschinenbau. 1974 erster Plakatentwurf „Chile wird wieder Chile". 1975–86 Grafiker an der Stadtbibliothek in Berlin, anschließend freiberuflich. Ab 1989 für die Berliner Galerie für Auslandsbeziehungen (ifa) tätig. Einrichtung von Ausstellungen zu Buchthemen u.a.: „Marcus Behmer", „Bücherverbrennung", „Herwarth Walden und der Sturm". Ab 1987 Plakatausstellungen in seiner eigenen Galerie. Zahlreiche Auszeichnungen und Ausstellungen, u. a. bei den Brandenburgischen Kunstsammlungen, Cottbus 1993.

**ring neuer werbegestalter** – *Gestaltergruppe* – 1927 Kurt Schwitters und Robert Michel sahen die Möglichkeit und Notwendigkeit der Gründung einer Gruppe für neue Gestaltung. 1928 Gründung des „rings neuer werbegestalter" als Vereinigung von Gestaltern, die „neuzeitliche" Ausdrucksformen in der Gestaltung der Werbung einsetzen. Die Ziele werden durch Entwurfstätigkeit, Ausstellungen, Vorträge und Mitarbeit an Zeitschriften propagiert. Vorsitzender ist Kurt Schwitters. Bei der Gründung gehören der Gruppe in Deutschland arbeitende Gestalter an: Willi Baumeister, Max Burchartz, Walter Dexel, Cesar Domela, Robert Michel, Kurt Schwitters, Georg Trump, Jan Tschichold, Friedrich Vordemberge-Gildewart. Es folgen Hans Leistikow, Adolf Meyer (Ehrenmitglied),

1982. Ricci a ouvert plusieurs librairies FMR en Italie et au Mexique depuis les années 70.

**Riedel**, Hubert – né le 7. 9. 1948 à Berlin, Allemagne – *graphiste maquettiste, architecte d'expositions* – emploi dans la construction mécanique. Premier dessin d'affiche en 1974 : «Chile wird wieder Chile». Graphiste à la Stadtbibliothek de Berlin de 1975 à 1986, puis indépendant. A partir de 1989, travaille pour la Galerie für Auslandsbeziehungen (ifa) à Berlin. Architecture d'expositions sur le thème du livre, entre autres : «Marcus Behmer», «Bücherverbrennung», «Herwarth Walden und der Sturm». Exposition d'affiches dans sa propre galerie à partir de 1987. Nombreuses distinctions et expositions, entre autres à la Brandenburgische Kunstsammlung de Cottbus, en 1993.

**ring neuer werbegestalter** (cercle des nouveaux graphistes publicitaires) – *groupe de graphistes* – en 1927, Kurt Schwitters et Robert Michel jugent nécessaire de fonder un groupe se consacrant au nouveau design. Le ring neuer werbegestalter naît en 1928, ce groupe de designers applique les formes d'expression «nouvelles» au design publicitaire. Il propage ses idées au moyen de maquettes, expositions, conférences et collaboration à de nombreuses revues. Il est présidé par Kurt Schwitters. Au moment de sa fondation, il réunit des artistes travaillant en Allemagne : Willi Baumeister, Max Burchartz, Walter Dexel, Cesar Domela, Robert Michel, Kurt Schwitters, Georg Trump, Jan Tschichold, Friedrich Vordemberge-Gildenwart. Puis Hans Leistikow, Adolf Meyer (membre d'honneur), Werner Graeff et Hans Richter viennent s'y adjoindre. Le Hollandais Piet Zwart sera le premier membre étranger, Paul Schuitema le second en 1929. Des

Riedel 1974 Poster

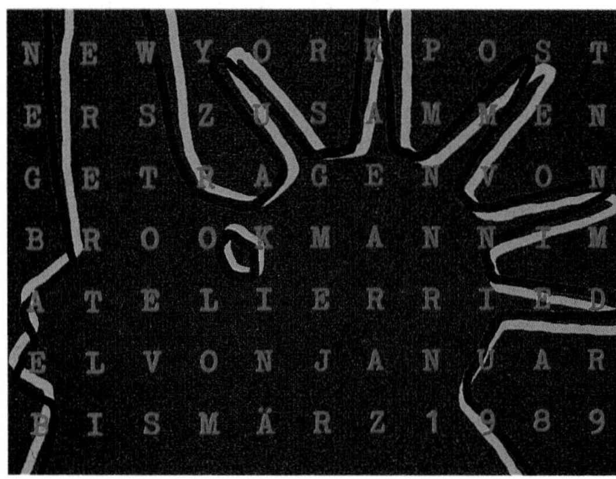


Riedel 1989 Invitation Card

ring neuer werbegestalter 1929 Logo

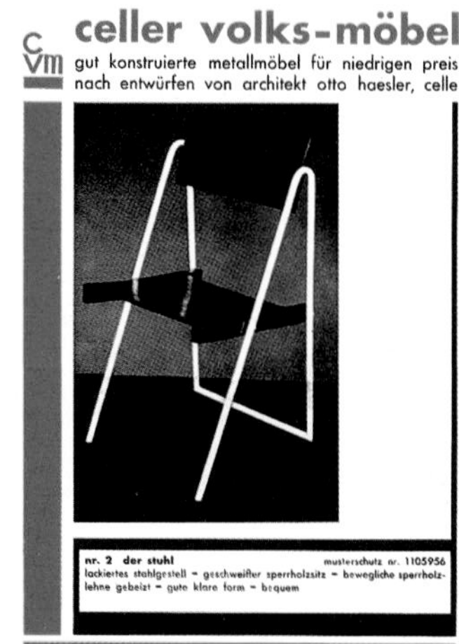


ring neuer werbegestalter ca. 1929 Advertisement

started establishing his chain of FMR bookstores in Italy and Mexico in the 1970s.

**Riedel**, Hubert – b. 7. 9. 1948 in Berlin, Germany – *graphic designer, exhibition planner* – Worked in heavy machinery construction. 1974: produces his first poster, entitled "Chile wird wieder Chile". 1975–86: graphic artist for Berlin's city library, after which he goes freelance. From 1989 onwards: works for the Berliner Galerie für Auslandsbeziehungen (ifa). Designs exhibitions on books, such as "Marcus Behmer", "Bücherverbrennung" and "Herwarth Walden und der Sturm". 1987: starts exhibiting his posters in his own gallery. Riedel has won numerous awards and had many exhibitions of his work held, including at the Brandenburgische Kunstsammlungen in Cottbus in 1993.

**ring neuer werbegestalter** – *design group* – 1927: Kurt Schwitters and Robert Michel recognize the opportunity and necessity for a group to promote new design. 1928: the ring neuer werbegestalter is launched, an association of designers who use "modern" forms of expression in advertising design. The group's aims and objectives are propagated through design work, exhibitions, lectures and work on various magazines. Kurt Schwitters is president of the group. The group's initial members include various German designers, such as Willi Baumeister, Max Burchartz, Walter Dexel, Cesar Domela, Robert Michel, Kurt Schwitters, Georg Trump, Jan Tschichold and Friedrich Vordemberge-Gildenwart. Hans Leis-

Werner Graeff, Hans Richter. Erstes ausländisches Mitglied ist Piet Zwart aus den Niederlanden, 1929 Paul Schuitema. Ausstellungen u. a. in Köln (1928), Barmen (1928), Berlin (1929), Magdeburg (1929), Essen (1929), München (1930) und Stockholm (1931). In dem von Heinz und Bodo Rasch 1930 herausgegebenen Buch „Gefesselter Blick" sind alle „ring"-Mitglieder mit Beiträgen vertreten. Ohne förmliche Auflösung werden die Aktivitäten des „rings" 1933 durch die neue politische Situation beendet. – *Publikationen u. a.:* Heinz und Bodo Rasch (Hrsg.) „Gefesselter Blick", Stuttgart 1930, Reprint Baden 1996; Dieter Ronte, Perdita Lottner „ring neuer werbegestalter. 1928–33. Ein Überblick", Hannover 1990.

**Rockner,** Vincenz – *Schreiber* – Sekretär Kaiser Maximilians. 1513 Entwurf einer Schrift, in der Johann Schönsperger das Gebetbuch Kaiser Maximilians mit Randzeichnungen von Albrecht Dürer, Lucas Cranach, Hans Baldung Grien, Hans Burgkmair und Jörg Breu druckt. 1517 Entwurf der Fraktur, die zum Druck des Buches „Teuerdank" verwendet wird. Es ist eine der ersten Fraktur-Schriften. Sie wird von Hieronymus Andreae geschnitten, das Buch wird von Johann Schönsperger gedruckt. Der „Teuerdank" ist ein gereimter Roman, der das Leben Kaiser Maximilians dichterisch umschreibt.

**Rodenberg,** Julius – geb. 5. 5. 1884 in Bremerhaven, Deutschland, gest. 23. 1. 1970 in Berlin, Deutschland. – *Autor, Bibliothekar* – 1905–08 Studium der Theologie und Kunstgeschichte. 1908 Promotion im Fach Philosophie in Hei-

expositions ont lieu à Cologne (1928), Barmen (1928), Berlin (1929), Magdebourg (1929), Essen (1929), Munich (1930) et Stockholm (1931). En 1930, Heinz und Bodo Rasch publient un ouvrage intitulé «Gefesselter Blick» qui présente des contributions de tous les membres du «ring». En 1933, le «ring» cesse ses activités, sans dissolution formelle, en raison de la nouvelle situation politique. – *Publications, sélection:* Heinz und Bodo Rasch (éd.) «Gefesselter Blick», Stuttgart 1930, réédition Baden 1996; Dieter Ronte, Perdita Lottner «ring neuer werbegestalter. 1928–33. Ein Überblick», Hanovre 1990.

**Rockner,** Vincenz – *scribe* – secrétaire de l'empereur Maximilien. En 1513, il dessine des caractères qui serviront à Johann Schönsperger pour imprimer le Missel de l'empereur Maximilien, décoré de vignettes d'Albrecht Dürer, Lucas Cranach, Hans Baldung Grien, Hans Burgkmair et Jörg Breu. En 1517, il dessine une Fraktur utilisée pour l'impression de «Teuerdank». Ces caractères taillés par Hieronimus Andreae, comptent parmi les premières «Frakturs». L'ouvrage a été imprimé par Johann Schönsperger. Le «Teuerdank» est une biographie romancée et rimée de l'empereur Maximilien.

**Rodenberg,** Julius – né le 5. 5. 1884 à Bremerhaven, Allemagne, décédé le 23. 1. 1970 à Berlin, Allemagne – *auteur, bibliothécaire* – 1905–1908, études de théologie et d'histoire de l'art. 1908, doctorat de philosophie à Heidelberg.

Rockner 1517 Page (Detail)

tikov, Adolf Meyer (honorary member), Werner Graeff and Hans Richter soon join them. The first non-German group member is Piet Zwart from The Netherlands, joined in 1929 by Paul Schuitema. Exhibitions are held in Cologne (1928), Barmen (1928), Berlin (1929), Magdeburg (1929), Essen (1929), Munich (1930) and Stockholm (1931). 1930: all the members of the ring neuer werbegestalter have contributions printed in the book "Gefesselter Blick", edited by Heinz and Bodo Rasch. 1933: the activities of the ring come to an abrupt and informal end as a result of the new political situation in Germany. – *Publications include:* Heinz and Bodo Rasch (eds.) "Gefesselter Blick", Stuttgart 1930, reprinted in Baden in 1996; Dieter Ronte, Perdita Lottner "ring neuer werbegestalter. 1928–33. Ein Überblick", Hanover 1990.

**Rockner,** Vincenz – *scribe* – Secretary to Emperor Maximilian I. 1513: designs a typeface in which Johann Schönsperger prints Emperor Maximilian's prayer book. The book margins contain illustrations by Albrecht Dürer, Lucas Cranach, Hans Baldung Grien, Hans Burgkmair and Jörg Breu. 1517: designs a black letter typeface, one of the first, cut by Hieronymus Andreae. The typeface is used for the book "Teuerdank", printed by Johann Schönsperger. "Teuerdank" is a novel in rhyming verse which describes the life of the Holy Roman Emperor Maximilian.

**Rodenberg,** Julius – b. 5. 5. 1884 in Bremerhaven, Germany, d. 23. 1. 1970 in Berlin, Germany – *author, librarian* – 1905–08: studies theology and art histo-

delberg. 1920–52 Bibliotheksrat an der Deutschen Bücherei in Leipzig, Gründung der Abteilung für künstlerische Drucke. Zahlreiche Aufsätze zu typographischen und künstlerischen Themen. – *Publikationen u.a.:* „Deutsche Pressen", Wien 1925; „Die deutsche Schriftgießerei", Mainz 1927; „In der Schmiede der Schrift", Berlin 1940; „Größe und Grenzen der Typographie", Stuttgart 1959.

**Rodtschenko,** Alexander – geb. 23. 11. 1891 in St. Petersburg, Rußland, gest. 3. 12. 1956 in Moskau, Sowjetunion. – *Maler, Fotograf, Typograph, Lehrer* – 1911–14 Studium an der Kunstschule in Kasan. 1914 Studium der Bildhauerei und Architektur am Stroganow-Institut in Moskau. 1917 Organisator der „Gewerkschaft der Kunstmaler". 1918–22 Leiter des Museumsbüros und Mitglied des Kunstkollegiums im Volkskommissariat für Volksaufklärung. 1919 Lehrer an der Schule des Moskauer „Proletkult". Gewinnt 1920 den 1. Preis bei einem Wettbewerb für einen Zeitungskiosk. 1921–30 Professor an der „Höheren künstlerisch-technischen Werkstätte" (WChUTEMAS). Gestaltung der Zeitschrift „Kino-Fot". 1923 Arbeit mit Dziga Wertow an der Kino-Monatsschau „Kino-Prawda". Typographische Arbeiten für Verlage. Zahlreiche Fotomontagen für Bücher und Plakate. Gestaltet die Zeitschriften „Molodaja Gwardija" und „Juni Kommunist". 1923–25 Redaktionsmitglied der Zeitschrift „LEF". Reklamearbeit mit Wladimir Majakowski. Gestaltet 1925 die „Geschichte der KPdSU (B)" in Plakaten. 1943–45 Leiter der künstlerischen Arbeit

Conseiller bibliothécaire à la Deutsche Bücherei de Leipzig de 1920 à 1952, où il crée le département des impressions d'art. Nombreux essais sur la typographie et l'art. – *Publications, sélection:* «Deutsche Pressen», Vienne 1925; «Die deutsche Schriftgiesserei», Mayence 1927; «In der Schmiede der Schrift», Berlin 1940; «Grösse und Grenzen der Typographie», Stuttgart 1959.

**Rodtchenko,** Alexandre – né le 23. 11. 1891 à Saint-Pétersbourg, Russie, décédé le 3. 12. 1956 à Moscou, Union Soviétique – *peintre, photographe, typographe, enseignant* – 1911–1914, études à l'école des beaux-arts de Kasan. 1914, études de sculpture et d'architecture à l'Institut Stroganoff de Moscou. En 1917, il organise le «Syndicat des artistes peintres». Dirige le Bureau du Musée de 1918 à 1922 et est membre du collège des arts plastiques au commissariat pour l'éducation populaire. Enseigne en 1919 à Moscou, à l'école «Proletkult». En 1920, il remporte le 1er prix lors d'un concours pour le design d'un kiosque à journaux. Professeur aux «Ateliers artistiques et techniques» (VKHOUTEMAS) de 1921 à 1930. Maquette de la revue «Kino-Fot». En 1923, il travaille avec Dziga Vertov pour la revue mensuelle du cinéma «Kino Pravda». Travaux typographiques pour des éditeurs. Nombreux photomontages pour des livres et affiches. Maquette des revues «Molodaja Gvardija» et «Juni Kommunist». Membre de la rédaction de la revue «LEF» de 1923 à 1925. Travaux de publicité avec Vladimir Maïakowski. En 1925, il réalise l'«Histoire du PC soviétique (B)» sous forme d'affiches. Directeur de la création artistique de la «Maison de la technique» à Moscou, de 1943 à 1945. Architecture de l'exposition «L'histoire du PC soviétique (B)» au Musée de la Révolution. En 1948, il dessine une série d'af-

Rodenberg 1940 Cover

Rodtschenko 1919 Title

Rodtschenko 1923 Cover

ry. 1908: gains his Ph.D. in philosophy in Heidelberg. 1920–52: chief librarian at the Deutsche Bücherei in Leipzig. Founds the library's prints department. Rodenberg wrote numerous essays on typography and art. – *Publications include:* "Deutsche Pressen", Vienna 1925; "Die deutsche Schriftgießerei", Mainz 1927; "In der Schmiede der Schrift", Berlin 1940; "Größe und Grenzen der Typographie", Stuttgart 1959.

**Rodchenko,** Alexander – b. 23. 11. 1891 in St. Petersburg, Russia, d. 3. 12. 1956 in Moscow, Soviet Union – *painter, photographer, typographer, teacher* – 1911–14: studies at the art school in Kasan. 1914: studies sculpture and architecture at the Stroganov Institute in Moscow. 1917: organizer of the Union of Painters. 1918–22: head of the Museums Office and member of the arts board at the People's Commissariat for Education. 1919: teaches at the Moscow Proletkult school. 1920: wins first prize in a newsstand competition. 1921–30: professor at the Higher Workshops for Art and Technology (VKHUTEMAS). Designs "Kino-Fot" magazine. 1923: works on "Kino Pravda", a monthly movie theater magazine, with Dziga Vertov. Does typographic work for various publishers and creates numerous photomontages for books and posters. Designs "Molodaya Gvardiya" and "Juni Kommunist" magazines. 1923–25: on the editorial staff of "LEF" magazine. Works on advertising projects with Vladimir Mayakovsky. 1925: presents the history of the CPSU (B) in posters. 1943–45: supervises the artwork at the House of Tech-

im Moskauer „Haus der Technik". Gestaltet die Ausstellung „Die Geschichte der KPdSU (B)" im Revolutionsmuseum. 1948 Entwurf einer Serie monographischer Plakate über Majakowski (mit W. Stepanowa). – *Publikationen u.a.:* Germann Karginov „Rodtschenko", Budapest 1979; David Elliot (Hrsg.) „Rodchenko", Oxford 1979; A. Lawrentiew u. a. „Alexander Rodtschenko. Maler, Konstrukteur, Fotograf", Dresden 1983; S. Chan-Magomedow „Alexander Rodchenko – the complete works", Cambridge 1986.

**Rogers,** Bruce – geb. 14. 5. 1870 in Lynnwood, USA, gest. 21. 5. 1957 in New Fairfield, USA. – *Typograph, Schriftentwerfer, Illustrator, Künstler* – 1885–90 künstlerische Ausbildung am Purdue College bei Lynnwood. Ab 1894 grafische Arbeiten für die Zeitschrift „Modern Art". 1896 Eintritt in die 1888 gegründete „Riverside Press" in Boston. Unter seiner Leitung werden 1900–12 die Vorzugsdrucke der Presse weltberühmt. Rogers gestaltet über 50 Buchausgaben mit z. T. eigens für das Buch geschnittenen Schriften (Montaigne, Riverside Modern, Brimmer). Nach 1912 mehrfach für die Metropolitan Press und die Montague Press tätig. Gestaltet 1915 für die Montague Press das Buch „The Centaur". 1917–19 als Printing Adviser an der Cambridge University Press in England, 1920–28 an der Harvard University Press tätig. Arbeit als Typograph für den Verlag W. E. Rudge in New York. 1929–31 Zusammenarbeit mit der Oxford University Press und Emery Walker. 1935 erscheint die von ihm gestaltete „Oxford Lectern Bible", die in seiner Schrift „Cen-



Rodtschenko 1923 Title

THE BANQUET OF PLATO

APOLLODORUS. I think that the subject of your inquiries is still fresh in my memory; for yesterday, as I chanced to be returning home from Phaleros, one of my acquaintance, seeing me before him, called out to me from a distance, jokingly, 'Apollodorus, you Phalerian, will you not wait a minute?'—I waited for him, and as soon as he overtook me, 'I have just been looking for you, Apollodorus,' he said, 'for I wish to hear what those discussions were on Love, which took place at the party, when Agathon, Socrates, Alcibiades, and some others met at supper. Some one who heard it from Phœnix, the son of Philip, told me that you could give a full account, but he could relate nothing distinctly him-

Rogers 1915 Centaur

Rodtschenko 1924 Package

| | | | | | |
|---|---|---|---|---|---|
| | n | **n** | | | *n* |
| a | b | e | f | g | i |
| o | r | s | t | y | z |
| A | B | C | E | G | H |
| M | O | R | S | X | Y |
| 1 | 2 | 4 | 6 | 8 | & |

Rogers 1915 Centaur

| | | | | | |
|---|---|---|---|---|---|
| | n | | | | *n* |
| a | b | e | f | g | i |
| o | r | s | t | y | z |
| A | B | C | E | G | H |
| M | O | R | S | X | Y |
| 1 | 2 | 4 | 6 | 8 | & |

Rogers 1928–30 Metropolitan

fiches monographiques sur Maïakovski (avec W. Stepanova). – *Publications, sélection:* Germann Karnigov «Rodtschenko», Budapest 1979; David Elliot (éd.) «Rodchenko», Oxford 1979; A. Lawrientev et autres «Alexander Rodtschenko. Maler, Konstrukteur, Fotograf», Dresde 1983; S. Chan-Magomedow «Alexander Rodchenko – the complete works», Cambridge 1986.

**Rogers,** Bruce – né le 14. 5. 1870 à Lynnwood, Etats-Unis, décédé le 21. 5. 1957 à New York, Etats-Unis – *typographe, concepteur de polices, illustrateur, artiste* – 1885–1890, formation artistique au Purdue College près de Lynnwood. Travaux graphiques pour la revue «Modern Art» à partir de 1894. Entrée en 1896 aux «Riverside Press», fondées à Boston en 1888. Sous sa direction (1900–1912), les tirages numérotés des presses deviennent célèbres dans le monde entier. Rogers réalise la maquette de plus de 50 livres avec des caractères spécialement taillés pour certains ouvrages (Montaigne, Riverside Modern, Brimmer). Travaille à plusieurs reprises pour la Metropolitan Press et la Montague Press à partir de 1912. Maquette de «The Centaur», un ouvrage des Montague Press, en 1915. Exerce comme Printing Adviser pour les Cambridge University Press en Angleterre de 1917 à 1919, puis pour la Harvard University Press de 1920 à 1928. Typographe pour les éditions W. E. Rudge à New York. De 1929 à 1931, il travaille avec Emery Walker pour la Oxford University Press. En 1935 paraît la «Oxford Lectern Bible», créee par Rogers et composée en «Cen-

nology. Designs the "History of the CPSU (B)" exhibition in the Revolution Museum. 1948: designs a series of monograph posters on Mayakovsky with V. Stepanova. – *Publications include:* Germann Karginov "Rodtschenko", Budapest 1979; David Elliot (ed.) "Rodchenko", Oxford 1979; A. Lavrentiev et al "Alexander Rodtschenko. Maler, Konstrukteur, Fotograf", Dresden 1983; S. Chan-Magomedow "Alexander Rodchenko – the complete works", Cambridge 1986.

**Rogers,** Bruce – b. 14. 5. 1870 in Lynnwood, USA, d. 21. 5. 1957 in New Fairfield, USA – *typographer, type designer, illustrator, artist* – 1885–90: trains as an artist at Purdue College near Lynnwood. From 1894 onwards: produces graphics for "Modern Art" magazine. 1896: joins Riverside Press in Boston, founded in 1888. 1900–12: Riverside's special prints achieve world acclaim under Rogers' management. Rogers designs over fifty book editions for the press, often using typefaces he has cut exclusively for some of the books (e.g. Montaigne, Riverside Modern and Brimmer). Post-1912: frequent work for Metropolitan Press and Montague Press. 1915: designs the book "The Centaur" for Montague Press. 1917–19: works as printing adviser for Cambridge University Press in England and from 1920–28 for Harvard University Press. Works as a typographer for W. E. Rudge publishers in New York. 1929–31: works with Oxford University

taur" gesetzt ist. – *Schriftentwürfe:* Brimmer (1901), Montaigne (1902), Riverside Modern (1904), Centaur (1915), Metropolitan (1928–30). – *Publikationen u.a.:* „Paragraphs on printing", New York 1943. Joseph Blumenthal „Bruce Rogers, a life in letters", Austin 1989.

**Roller,** Alfred – geb. 2. 10. 1864 in Brünn, Österreich-Ungarn (heute Brno, Tschechien), gest. 22. 6. 1935 in Wien, Österreich. – *Bühnenbildner, Grafiker, Lehrer* – Studium an der Akademie der Bildenden Künste in Wien. 1897–1905 Gründung und Mitglied der Wiener Sezession, für die er zahlreiche Ausstellungsplakate entwirft. 1898 Entwurf des Umschlags zum ersten Heft der Zeitschrift „Ver Sacrum" und weitere Arbeiten für die folgenden Hefte (Initialen, Kalender, Postkarten, Illustrationen). Ab Heft 7 verantwortlicher Redakteur. Ab 1899 Professor an der Kunstgewerbeschule in Wien. 1902 Präsident der Wiener Sezession. 1905 Austritt aus der Wiener Sezession, Mitarbeit beim „Cabaret Fledermaus". Ab 1909 Direktor der Kunstgewerbeschule in Wien. Arbeiten für die Wiener Werkstätten. 1897–1907 Gustav Mahler beauftragt Roller mit dem Entwurf für die Ausstattungen der Wiener Hofoper. Roller entwirft Bühnenbilder (u. a. für Max Reinhardt in Berlin) und Plakate mit ornamentalen und geometrischen Elementen.

**Rondthaler,** Edward – geb. 9. 6. 1905 in Bethlehem, USA. – *Unternehmer* – 1936 gründet Rondthaler den ersten Fotosatz-Servicebetrieb „Photo-Lettering Inc." in New York (mit Harold A. Horman). Dies

taur», une de ses polices. – *Polices:* Brimmer (1901), Montaigne (1902), Riverside Modern (1904), Centaur (1915), Metropolitan (1928–1930). – *Publications, sélection:* «Paragraphs on Printing», New York 1943. Joseph Blumenthal «Bruce Rogers, a life in letters», Austin 1989.

**Roller,** Alfred – né le 2. 10. 1864 à Brünn, Autriche-Hongrie (aujourd'hui Brno, Tschèque) décédé le 22. 6. 1935 à Vienne, Autriche – *scénographe, graphiste, enseignant* – études à l'Akademie der Bildenen Künste (beaux-arts) de Vienne. 1897–1905, fondation et membre de la Sécession de Vienne pour laquelle il dessine de nombreuses affiches d'expositions. En 1898, il conçoit la couverture du premier numéro de la revue «Ver Sacrum» et réalise des travaux pour les autres numéros (lettrines, calendriers, cartes postales, illustrations). Rédacteur en titre à partir du numéro 7. Professeur à la Kunstgewerbeschule (école des arts décoratifs) de Vienne à partir de 1899. Président de la Sécession de Vienne en 1902; il quitte la Sécession en 1905 et travaille pour le «Cabaret Fledermaus». Directeur de la Kunstgewerbeschule de Vienne à partir de 1909. Travaille pour les ateliers de Vienne. Entre 1897 et 1907, Gustav Mahler charge Roller de concevoir des décors pour le Hofoper (opéra) de Vienne. Roller crée des scénographies (entre autres pour Max Reinhardt à Berlin) et des affiches comportant des éléments ornementaux et géométriques.

**Rondthaler,** Edward – né le 9. 6. 1905 à Bethlehem, Etats-Unis – *chef d'entreprise* – en 1936, Rondthaler fonde la première entreprise de photocomposition, la «Photo-Lettering Inc.» à New York (avec Harold A. Horman). Cette date marque le début de la photocomposition industrielle. En 1959, il édite le catalogue des fontes de la maison sous le titre «The-

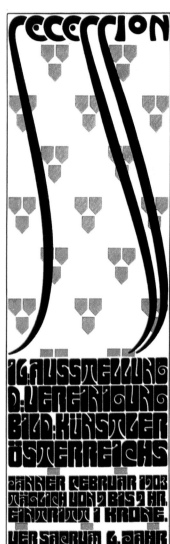

Roller 1902 Poster

Roller 1899 Poster

Rondthaler 1981 Back Cover

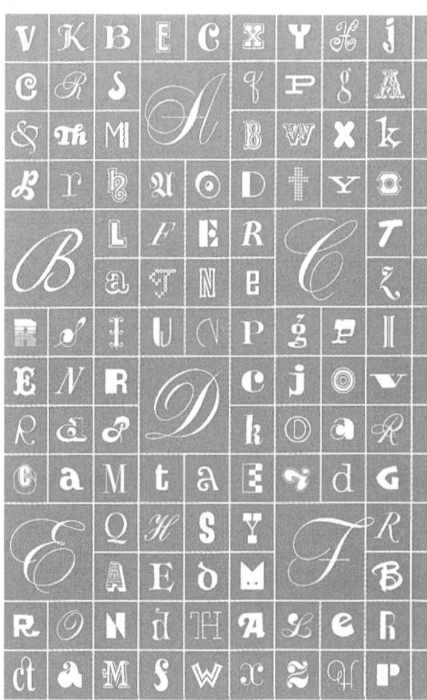
Rondthaler 1981 Front Cover

Press and Emery Walker. 1935: the Oxford Lectern Bible, designed by Rogers and set in his "Centaur typeface", is published. – *Fonts:* Brimmer (1901), Montaigne (1902), Riverside Modern (1904), Centaur (1915), Metropolitan (1928–1930). – *Publications include:* "Paragraphs on printing", New York 1943. Joseph Blumenthal "Bruce Rogers, a life in letters", Austin 1989.

**Roller,** Alfred – b. 2. 10. 1864 in Brünn, Austria-Hungary (today Brno, Czech Republic), d. 22. 6. 1935 in Vienna, Austria – *set-designer, graphic artist, teacher* – Studied at the Akademie der Bildenden Künste in Vienna. 1897–1905: one of the founders and members of the Vienna Secession, for whom he designs numerous exhibition posters. 1898: designs the cover for the first issue of "Ver Sacrum" magazine and contributes initials, calendars, postcards and illustrations to ensuing numbers of the journal. He is responsible editor from number 7 onwards. From 1899 onwards: professor at the Kunstgewerbeschule in Vienna. 1902: president of the Vienna Secession. 1905: leaves the Vienna Secession. Works for Cabaret Fledermaus. From 1909 onwards: director of the Kunstgewerbeschule in Vienna. Produces work for the Wiener Werkstätte. 1897–1907: Gustav Mahler commissions Roller to design the décor for the Hofoper in Vienna. The stage sets Roller designed (for Max Reinhardt in Berlin, among others) and his posters contained ornamental and geometric elements.

**Rondthaler,** Edward – b. 9. 6. 1905 in

war der Beginn des kommerziellen Fotosatzes. 1959 Herausgabe des umfangreichen Schriftenkatalogs seiner Firma unter dem Namen „Thesaurus Vol. 1", dem 1965 der 2. Band und 1971 der 3. Band folgen. 1970 Mitbegründer der „International Typeface Corporation" (ITC) in New York (mit Aaron Burns, Herb Lubalin). – *Publikation:* „Life with letters after they turned photogenic", New York 1981.

**Roos,** Sjoerd Hendrik de – geb. 14. 9. 1877 in Drachten, Niederlande, gest. 2. 4. 1962 in Haarlem, Niederlande. – *Schriftentwerfer, Maler, Grafik-Designer, Typograph* – 1889–92 Ausbildung zum Lithographen. 1892–95 Abendkurse in Amsterdam an der Tekenschool voor Kunstambachten, 1895–98 an der Rijksacademie van Beeldende Kunsten. 1907–42 Mitarbeiter der Lettergieterij Amsterdam als künstlerischer Berater, Typograph und Schriftzeichner. 1925–32 Mitglied der Kommission des Wettbewerbs „De vijftig beste boeken". 1927 Kommissar der niederländischen Abteilung auf der Internationalen Buchkunst-Ausstellung in Leipzig. 1928 Gründung der „Heuvel-Presse" in Hilversum. 1937 mit dem Grand Prix der Weltausstellung in Paris ausgezeichnet. 1942–61 als Landschaftsmaler und typographischer Entwerfer tätig. 1948 Gestaltung einer Sonderausgabe der niederländischen Verfassung für die Krönung von Königin Juliana. – *Schriftentwürfe:* Nieuw Javaans (1909), Hollandse Mediaeval (1912), Zilvertype (1915), Ella Cursief (1916), Erasmus (1923), Grotius-Antiqua (1925), Meidoorn (1927), Nobel (1930), Egmont (1933),

saurus Vol. 1» qui sera suivi, en 1965 d'un 2e volume, puis d'un 3e en 1971. Cofondateur, en 1970, de «International Typeface Corporation» (ITC) à New York (avec Aaron Burns et Herb Lubalin). – *Publication:* «Life with letters after they turned photogenic», New York 1981.

**Roos,** Sjoerd Hendrik de – né le 14. 9. 1877 à Drachten, Pays-Bas, décédé le 2. 4. 1962 à Haarlem, Pays-Bas – *concepteur de polices, peintre, graphiste maquettiste, typographe* – 1889–1892, formation de lithographe. 1892–1895, cours du soir à la Tekenschool voor Kunstambachten à Amsterdam, puis études à la Rijksacademie van Beeldende Kunsten de 1895 à 1898. Employé comme conseiller artistique, typographe et dessinateur de caractères à la Lettergieterij Amsterdam de 1907 à 1942. Membre de la commission du concours «De vijftig beste boeken» de 1925 à 1932. Commissaire de la section néerlandaise lors de l'exposition du livre d'art à Leipzig en 1927. Fonde la «Heuvel-Presse» à Hilversum en 1928. Le Grand Prix de l'exposition universelle de Paris lui est décerné en 1937. Exerce comme peintre paysagiste et concepteur typographe de 1942 à 1961. Maquette d'une édition spéciale de la constitution néerlandaise en 1948 pour le couronnement de la Reine Juliana. – *Polices:* Nieuw Javaans (1909), Hollandse Mediaeval (1912), Zilvertype (1915), Ella Cursief (1916), Erasmus (1923), Grotius-Antiqua (1925), Meidoorn (1927), Nobel (1930) Egmont (1933), Libra (1938), Simp-

Roos 1924 Cover

ZEVEN TEGEN THEBAI

TREURSPEL NAAR HET GRIEKSCH VAN
AISCHYLOS IN NEDERLANDSCHE
VERZEN OVERGEBRACHT
DOOR
P C. BOUTENS

W. L. & J. BRUSSE'S
UITGEVERSMAATSCHAPPIJ
ROTTERDAM MCMXXVIII

Roos 1928 Main Title

| | | | | | |
|---|---|---|---|---|---|
| | n | | | | |
| a | b | e | f | g | i |
| o | r | s | t | y | z |
| A | B | C | E | G | H |
| M | O | R | S | X | Y |
| 1 | 2 | 4 | 6 | 8 | & |

Roos 1912 Hollandse Mediaeval

| | | | | | |
|---|---|---|---|---|---|
| n | n | n | | | n |
| a | b | e | f | g | i |
| o | r | s | t | y | z |
| A | B | C | E | G | H |
| M | O | R | S | X | Y |
| 1 | 2 | 4 | 6 | 8 | & |

Roos 1930 Nobel

| | | | | | |
|---|---|---|---|---|---|
| | | | | | |
| | | | | | |
| | | | | | |
| a | b | c | e | g | h |
| m | o | r | s | x | y |
| 1 | 2 | 4 | 6 | 8 | & |

Roos 1938 Libra

Bethlehem, USA – *entrepreneur* – 1936: Rondthaler founds the first filmsetting service, Photo-Lettering Inc., in New York with Harold A. Horman. This marks the beginning of commercial filmsetting. 1959: publishes a comprehensive catalogue of his company's typefaces entitled "Thesaurus Vol. 1", followed by a second volume in 1965 and a third in 1971. 1970: co-founder of the International Typeface Company (ITC) in New York with Aaron Burns and Herb Lubalin. – *Publication:* "Life with letters after they turned photogenic", New York 1981.

**Roos,** Sjoerd Hendrik de – b. 14. 9. 1877 in Drachten, The Netherlands, d. 2. 4. 1962 in Haarlem, The Netherlands – *type designer, painter, graphic designer, typographer* – 1889–92: trains as a lithographer. 1892–95: takes evening classes at the Tekenschool voor Kunstambachten in Amsterdam and from 1895–98 at the Rijksacademie van Beeldende Kunsten. 1907–42: works for the Lettergieterij Amsterdam as art consultant, typographer and letterer. 1925–32: committee member for the "De vijftig beste boeken" competition. 1927: commissioner of the Dutch section at the Internationale Buchkunst-Ausstellung in Leipzig. 1928: launches Heuvel Press in Hilversum. 1937: awarded the Grand Prix at the world exhibition in Paris. 1942–61: works as a landscape painter and typographic designer. 1948: designs a special edition of the Dutch constitution for the coronation of Queen Juliana. – *Fonts:* Nieuw Javaans (1909), Hollandse Mediaeval (1912), Zilvertype (1915), Ella Cur-

Libra (1938), Simplex (1939), de Roos Roman (1947). – *Publikationen u.a.:* A. A. M. Stols „Het Werk van S. H. de Roos", Amsterdam 1942; J. P. Boterman „Sjoerd H. de Roos, typografische geschriften 1907–1920", Den Haag 1989.

**Rosarivo,** Rául M. – geb. 17. 7. 1903 in Buenos Aires, Argentinien. – *Zeichner, Maler, Typograph, Lehrer* – Arbeit als Künstler, zahlreiche Ausstellungen in Buenos Aires und Mendoza. Illustriert Bücher und spezialisiert sich auf den Entwurf typographischer Diagramme. Unterrichtet an der Escuela Industrial de Artes Gráficas. 1940 Beginn seiner Forschungen über die Proportionen verschiedener Drucke von Gutenberg und seiner Zeitgenossen. 1948 Veröffentlichung unter dem Titel „Divina Proporcion Tipográfica Ternaria". Darin weist er nach, daß die Proportion 2:3 für die Typographie, auf Grund ihrer Gesetzmäßigkeit, die brauchbarsten Gestaltungsverhältnisse ergibt. – *Publikationen u.a.:* „Cómo formar el espiritu en la imprenta", Buenos Aires 1946; „Divina Proporcion Tipográfica Ternaria", Buenos Aires 1948.

**Rosart,** Jacques-François – geb. 1714 in Namur, österreichische Niederlande, gest. 1774 in Brüssel, österreichische Niederlande. – *Stempelschneider, Schriftgießer* – 1740 Beginn seiner Arbeit als Stempelschneider und Schriftgießer in Haarlem. 1741 Herausgabe seines ersten Schriftmusterbuchs mit zwölf Schriften und vierzehn Verzierungen. 1742 Veröffentlichung einer Anzeige in der Zeitung „Oprechte Haarlemse Courant", in der er

lex (1939), de Roos Roman (1947). – *Publications, sélection:* A. A. M. Stols «Het Werk van S. H. de Roos», Amsterdam 1942; J. P. Boterman «Sjoerd H. de Roos, typografische geschriften 1907–1920», La Haye 1989.

**Rosarivo,** Rául M. – né le 17. 7. 1903 à Buenos Aires, Argentine – *dessinateur, peintre, typographe, enseignant* – exerce comme artiste, nombreuses expositions à Buenos Aires et à Mendoza. Illustre des livres et se spécialise dans le dessin de diagrammes typographiques. Enseigne à la Escuela Industrial de Artes Gráficas. Début de ses recherches sur les proportions de diverses impressions de Gutenberg et de ses contemporains en 1940. Publication, en 1948, sous le titre «Divina Proporcíon Tipográfica Ternaria». Il y démontre que la proportion la plus applicable et la plus logique en typographie est le rapport 2 : 3. – *Publications, sélection:* «Cómo formar el espiritu en la imprenta», Buenos Aires 1946; «Divina Proporcion Tipográfica Ternaria», Buenos Aires 1948.

**Rosart,** Jacques-François – né en 1714 à Namur, Basse-Autriche, décédé en 1774 à Bruxelles, Basse-Autriche – *tailleur de types, fondeur de caractères* – commence à travailler comme tailleur de types et fondeur de caractères à Haarlem en 1740. Publie son premier catalogue de caractères en 1741 avec douze polices et quatorze ornements. En 1742, il publie une annonce dans la gazette «Oprechte Haarlemse Courant» où il propose ses caractères et ornements aux libraires et imprimeurs. 1746–1752, réalisation de treize alphabets différents. En 1749, il taille des notes de musique et publie un prospectus de modèles. En 1752, publication d'un nouveau catalogue de caractères. Contrat avec Johannes Enschedé en 1754, portant sur la fabrication de 160 matrices. S'in-

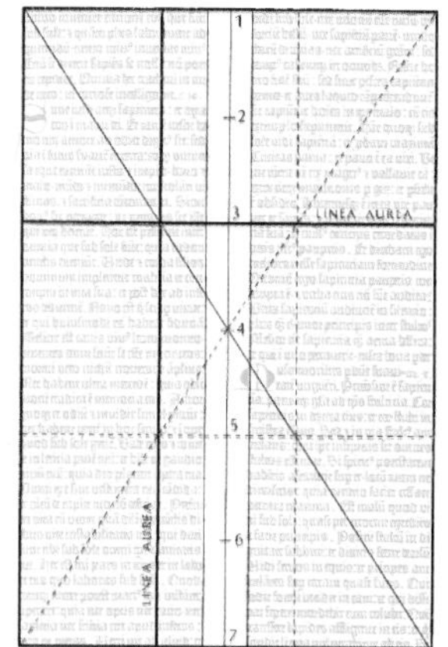


Rosarivo 1948 Page

Rosarivo 1948 Page

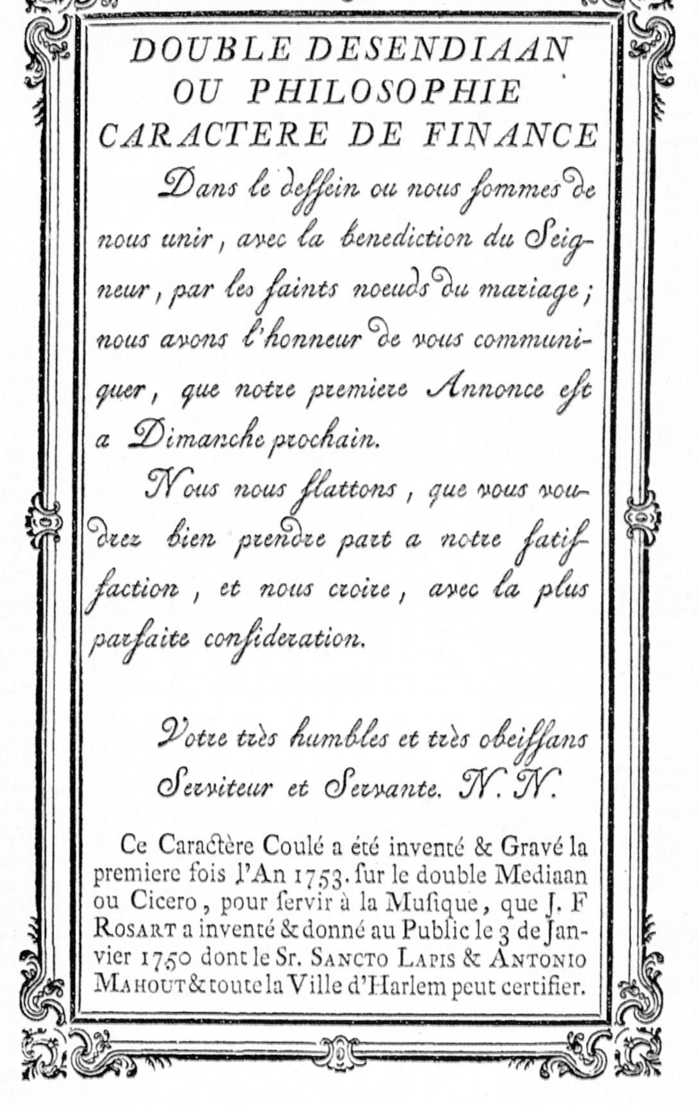
DOUBLE DESENDIAAN
OU PHILOSOPHIE
CARACTERE DE FINANCE

Dans le deſſein ou nous ſommes de nous unir, avec la benediction du Seigneur, par les ſaints noeuds du mariage; nous avons l'honneur de vous communiquer, que notre premiere Annonce eſt a Dimanche prochain.

Nous nous flattons, que vous voudrez bien prendre part a notre ſatiſfaction, et nous croire, avec la plus parfaite conſideration.

Votre très humbles et très obeiſſans Serviteur et Servante. N. N.

Ce Caractère Coulé a été inventé & Gravé la premiere fois l'An 1753. ſur le double Mediaan ou Cicero, pour ſervir à la Muſique, que J. F Rosart a inventé & donné au Public le 3 de Janvier 1750 dont le Sr. Sancto Lapis & Antonio Mahout & toute la Ville d'Harlem peut certifier.

Rosart after 1760 Caractère de Finance

sief (1916), Erasmus (1923), Grotius-Antiqua (1925), Meidoorn (1927), Nobel (1930), Egmont (1933), Libra (1938), Simplex (1939), de Roos Roman (1947). – *Publications include:* A. A. M. Stols "Het Werk van S. H. de Roos", Amsterdam 1942; J. P. Boterman "Sjoerd H. de Roos, typografische geschriften 1907–1920", The Hague 1989.

**Rosarivo,** Rául M. – b. 17. 7. 1903 in Buenos Aires, Argentina – *illustrator, painter, typographer, teacher* – Worked as an artist with numerous exhibitions of his work in Buenos Aires and Mendoza. Illustrated books and specialized in designing typographic diagrams. Taught at the Escuela Industrial de Artes Gráficas. 1940: begins his research on the proportions of various prints by Gutenberg and his contemporaries. 1948: publishes his findings under the title "Divina Proporcion Tipográfica Ternaria". In his thesis he proves that on account of its regularity the proportion 2:3 is the most useful proportion in typographic design. – *Publications include:* "Cómo formar el espiritu en la imprenta", Buenos Aires 1946; "Divina Proporcion Tipográfica Ternaria", Buenos Aires 1948.

**Rosart,** Jacques-François – b. 1714 in Namur, Austrian Low Countries, d. 1774 in Brussels, Belgium – *punch cutter, type founder* – 1740: starts working as a punch cutter and type founder in Haarlem. 1741: publishes his first book of type specimens containing twelve typefaces and fourteen ornaments. 1742: publishes an advertisement in the "Oprechte Haarlemse Courant" newspaper in which he offers

Buchhändlern und Druckern seine Schriften und Verzierungen anbietet. 1746–52 Rosart schneidet dreizehn verschiedene Alphabete. Schneidet 1749 Musiknoten und veröffentlicht davon einen Musterprospekt. 1752 Herausgabe eines neuen Schriftmusterbuchs. 1754 Vertrag mit Johannes Enschedé über die Herstellung von 160 Matrizen. 1759 Umzug nach Brüssel. 1761 Herausgabe seines ersten Schriftmusterbuchs mit 43 Seiten, 1768 seines zweiten Schriftmusterbuchs mit 70 Seiten in Brüssel. Veröffentlicht 1769 in der Zeitung „Gazette des Pays-Bas" eine Anzeige für eine neue Schrift, die wahrscheinlich die erste Werbeanzeige für eine Druckschrift war. 1779 Verkauf seiner Schriftgießerei. – *Publikation:* F. Baudin, N. Hoeflake „The Type Specimen of J. F. Rosart, Brussels 1768", Amsterdam, London, New York 1973.

**Rossum,** Just van – geb. 13. 7. 1966 in Haarlem, Niederlande. – *Schriftentwerfer, Grafik-Designer* – Studium an der Koninklijke Academie van Beeldende Kunsten in Den Haag. 1988 Praktika bei Océ-van der Grinten in Venlo, bei Monotype in Salfords, England und bei MetaDesign in Berlin, Deutschland. 1989 Gründung von „LettError" (mit Erik van Blokland) in Berlin, zunächst Titel einer Zeitschrift, dann Bezeichnung für ihre gemeinsame typographische Arbeit. Gestaltung u. a. für Apple Computer, Adobe Systems, FontShop International, Virtual Valley Inc., Metro Newspapers San José. – *Schriftentwürfe:* Beowolf (mit Erik van Blokland, 1989), Justlefthand (1991), Broken-script (1991), Advert (1991), Beo Sans

Rosart after 1760 Caractère de Musique

| n | n | n | n | | |
|---|---|---|---|---|---|
| a | b | e | f | g | i |
| o | r | s | t | y | z |
| A | B | C | E | G | H |
| M | O | R | S | X | Y |
| 1 | 2 | 4 | 6 | 8 | & |

Rossum 1992 Advert Rough

Rossum 1995 Page

stalle à Bruxelles en 1759. Publication de sa première typothèque de 43 pages en 1761, puis de la seconde, contenant 70 pages, à Bruxelles, en 1768. Publication dans la «Gazette des Pays-Bas» d'une annonce pour une nouvelle police; ce texte de 1769 fut probablement la première publicité pour un caractère d'imprimerie. Vend sa fonderie de caractères en 1779. – *Publication:* F. Baudin, N. Hoeflake «The Type Specimen of J. F. Rosart, Brussels 1768», Amsterdam, Londres, New York 1973.

**Rossum,** Just van – né le 13. 7. 1966 à Haarlem, Pays-Bas – *concepteur de polices, graphiste maquettiste* – études à la Koninklijke Academie van Beeldende Kunsten de La Haye. 1988, stage chez Océ-van der Grinten à Venlo, chez Monotype à Salfords, Angleterre, et chez MetaDesign à Berlin, Allemagne. Fondation, en 1989, de «LettError» (avec Erik van Blokland) à Berlin, d'abord comme revue, puis comme identité de leurs travaux communs en typographie. Designs pour Apple Computer, Adobe Systems, FontShop International, Virtual Valley Inc., Metro Newspapers San José, etc. – *Polices:* Beowolf (avec Erik van Blokland, 1989), Justlefthand (1991), Broken-script (1991), Advert (1991), Beo Sans (1991),

his typefaces and ornaments to booksellers and printers. 1746–52: Rosart cuts thirteen different alphabets. 1749: cuts sets of musical characters which he publishes in a brochure of specimens. 1752: a new book of type specimens is published. 1754: signs a contract with Johannes Enschedé governing the production of 160 matrices. 1759: moves to Brussels. 1761: publishes his first book of type specimens in Brussels, which is 43 pages long, and in 1768 a second specimen book 70 pages long. 1769: his advertisement for a new typeface in the "Gazette des Pays-Bas" newspaper is probably the first ever to market a typeface. 1779: sells his type foundry. – *Publication:* F. Baudin, N. Hoeflake "The Type Specimen of J. F. Rosart, Brussels 1768", Amsterdam, London, New York 1973.

**Rossum,** Just van – b. 13. 7. 1966 in Haarlem, The Netherlands – *type designer, graphic designer* – Studied at the Koninklijke Academie van Beeldende Kunsten in The Hague. 1988: periods of vocational training with Océ-van der Grinten in Venlo, with Monotype in Salfords, England, and with MetaDesign in Berlin, Germany. 1989: founds "LettError" with Erik van Blokland in Berlin, which initially is the title of a magazine but is later used to describe their joint typographic work in general. Rossum has designed for Apple Computers, Adobe Systems, FontShop International, Virtual Valley Inc. and Metro Newspapers in San José, among others. – *Fonts:* Beowolf (with Erik van Blokland, 1989), Justlefthand (1991), Broken-script (1991), Advert

(1991), Schulschrift (1991), Advert Rough (1992), Instant Types (1992).

**Rosta-Fenster** – *Großplakate* – 1919–22 Nach der Oktoberrevolution wurden auf Initiative des sowjetischen Plakatgestalters Michail Tscheremnich, dem sich Wladimir Majakowski und Iwan Maljutin anschlossen, von der Telegrafen-Agentur „Rosta", später von der Hauptverwaltung für politische Aufklärungsinstitutionen des Volkskommissariats für Bildungswesen (GPP) der Sowjetunion, Agitationsplakate herausgegeben. Diese Großplakate werden „Rosta-Fenster", oder auch „Satirefenster der Rosta" genannt. Sie beinhalten die politischen, militärischen und wirtschaftlichen Tagesthemen. In 29 Monaten erscheinen über 1.600 „Rosta-Fenster". Diese sind nicht gedruckt, sondern die gesamte Auflage, die zwischen 50 und 300 Exemplaren beträgt, ist mit Schablonen handgearbeitet. 1929 Ausstellung der „Rosta-Fenster" in der Tretjakow-Galerie. – *Publikation u. a.:* Viktor Duwakin „Rosta-Fenster", Dresden 1975.

**Roth,** Dieter – geb. 21. 4. 1930 in Hannover, Deutschland. – *Dichter, Künstler, Lehrer, Typograph* – 1947–51 Arbeit als Grafik-Designer im Studio F. Wuthrich in Bern. Zusammenarbeit mit den Künstlern Franz Eggenschwiler und Paul Talman in Bern. 1949 erste Holzschnitte und Collagen. 1951 Gründung der Zeitschrift „Spirale" (mit Franz Eggenschwiler und Marcel Wyss). 1951–63 Grafik-Designer, Textilzeichner, Goldschmied in Kopenhagen und Reykjavík. 1957 Umzug nach Reykjavík. Seit 1957 erscheinen zahlreiche

Schulschrift (1991), Advert Rough (1992), Instant Types (1992).

**Rosta Fenêtres** – *grandes affiches* – affiches d'agitation politique publiées de 1919 à 1922, après la Révolution d'Octobre, à l'initiative de l'affichiste soviétique Michaïl Tcheremnich auquel Vladimir Maïakowski et Ivan Malioutine viennent s'associer. Ces affiches sont éditées par l'agence des télégraphes «Rosta», puis par l'administration centrale des institutions d'éducation politique du commissariat à l'éducation populaire (GPP) d'Union Soviétique. Ces grandes affiches étaient appelées «Fenêtres Rosta» ou «Fenêtres satiriques Rosta». Elles traitaient de thèmes actuels: politique, armée et économie. «1 600 Fenêtres Rosta» paraîtront en 29 mois. Elles n'étaient pas imprimées, mais l'ensemble du tirage, soit 50 à 300 exemplaires, était réalisé à la main à partir de pochoirs. Exposition des «Fenêtres Rosta» à la Tretjakov-Galerie, en 1929. – *Publication, sélection:* Viktor Duwakin «Rosta-Fenster», Dresde 1975.

**Roth,** Dieter – né le 21. 4. 1930 à Hanovre, Allemagne – *poète, artiste, enseignant, typographe* – 1947–1951, employé comme concepteur graphiste à l'atelier F. Wuthrich à Berne. Travaille avec les artistes Franz Eggenschwiler et Paul Talman à Berne. Premières gravures sur bois et collages en 1949. Création de la revue «Spirale» (avec Franz Eggenschwiler et Marcel Wyss) en 1951. Graphiste maquettiste, designer en textiles, orfèvre à Copenhague et Reykjavik de 1951 à 1963. S'installe à Reykjavik en 1957. Parution de nombreux livres de Dieter Roth à partir de 1957, le contenu et la forme de ses ouvrages ont une grande importance sur la maquette, surtout concernant les livres d'artistes. A enseigné à la Rhode Island School of Design à Providence (1965–1967), à la Watford School of Graphic Art

Rosta-Fenster 1921 Poster (Detail)

Rosta-Fenster 1919 Poster (Detail)

Rosta-Fenster 1920 Poster

Roth 1981 Poster

(1991), Beo Sans (1991), Schulschrift (1991), Advert Rough (1992), Instant Types (1992).

**Rosta windows** – *large-format posters* – 1919–22: after the October Revolution, Soviet poster designer Mikail Cherenikh, later joined by Vladimir Mayakovsky and Ivan Malyutin, initiates a series of agitprop posters. These are initially issued by the Rosta news agency and later by the main administrative body for political educational institutes of the Soviet Union's People's Commissariat for Education (GPP). These large-format posters were known as "Rosta windows" or "satirical windows of the Rosta". They illustrated the political, military or economic topics of the day. Over the space of 29 months, 1,600 Rosta windows were produced. The individual editions of between 50 and 300 copies were not printed using machines, but stenciled by hand. 1929: an exhibition of Rosta windows is held in the Tretyakov Gallery. – *Publications include:* Viktor Duvakin "Rosta-Fenster", Dresden 1975.

**Roth,** Dieter – b. 21. 4. 1930 in Hanover, Germany – *poet, artist, teacher, typographer* – 1947–51: works as a graphic designer for the F. Wuthrich studio in Bern. Works with artists Franz Eggenschwiler and Paul Talman in Bern. 1949: produces his first woodcuts and collages. 1951: launches "Spirale" magazine with Franz Eggenschwiler and Marcel Wyss. 1951–63: graphic designer, textiles designer and goldsmith in Copenhagen and Reykjavík. 1957: moves to Reykjavík. From 1957 onwards: Roth has many books

Bücher von Dieter Roth, die durch Inhalt und Form große Wirkung auf die Buchgestaltung, besonders auf die Künstlerbücher, haben. Unterrichtet an der Rhode Island School of Design in Providence (1965–67), der Watford School of Graphic Art (1968), der Kunstakademie Düsseldorf (1969–71) und der Kunstakademie in Reykjavík (1982). 1968–80 Teilhaber der Edition Hansjörg Mayer in Stuttgart. Seit 1975 Gründung und Herausgabe der „Zeitschrift für Alles. Review for Everything. Timarit Fyrir Allt“. Gleichzeitig gründet er den „Dieter Roth's Familienverlag“, der ab 1978 als „Dieter Roth's Verlag“ weitergeführt wird. Erhält 1986 den erstmalig verliehenen Charles-Nypels-Preis für sein typographisches Gesamtwerk. – *Publikationen u.a.:* „Mundumculum“, Köln 1967; „Gesammelte Werke“ (40 Bände), Stuttgart, London 1969–1986.

**Ruder,** Emil – geb. 20. 3. 1914 in Zürich, Schweiz, gest. 13. 3. 1970 in Basel, Schweiz. – *Typograph, Lehrer* – 1929–33 Schriftsetzerlehre in Zürich. 1938–39 Studienaufenthalt in Paris. 1941–42 Schriftsatz- und Buchdruck-Studium bei Alfred Willimann und Walter Käch an der Kunstgewerbeschule in Zürich. Seit 1942 freier Typograph in Basel. Fachlehrer für Typographie an der Allgemeinen Gewerbeschule in Basel. Seit 1961 Mitglied der Eidgenössischen Kommission für angewandte Kunst und künstlerischer Berater der Briefmarkenabteilung der Schweizer Post. 1962 Mitbegründer des „International Center for the Typographic Arts“ (ICTA) in New York. Seit 1965 Direktor

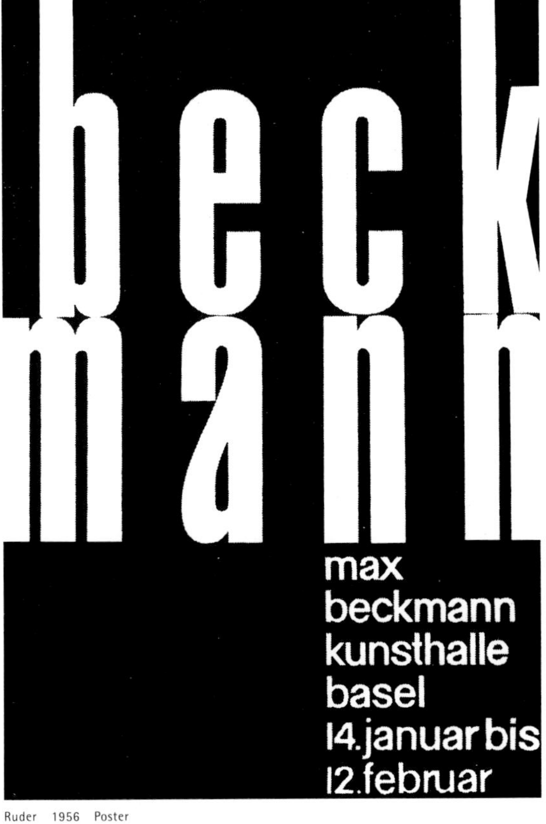


Ruder 1956 Poster

Ruder 1954 Poster

Ruder 1967 Spread

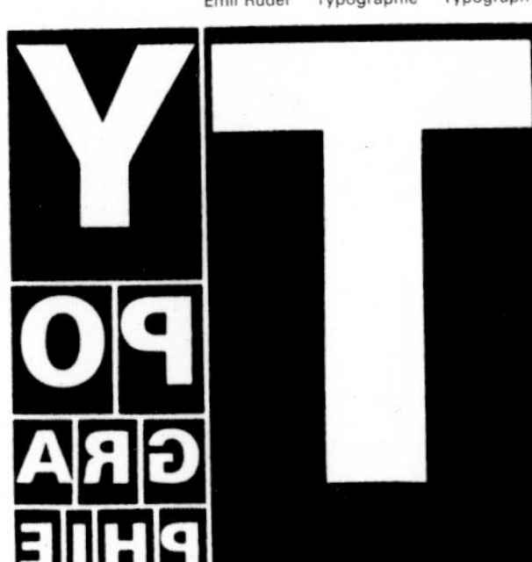


Ruder 1967 Cover

(1968), à la Kunstakademie (beaux-arts) de Düsseldorf (1969–1971), à l'académie des beaux-arts de Reykjavík (1982). Associé de l'Edition Hansjörg Mayer à Stuttgart de 1968 à 1980. Fonde et édite la revue «Zeitschrift für Alles. Review for Everything. Timarit Fyrir Allt» à partir de 1975. Crée en même temps le «Dieter Roth's Familienverlag» qui deviendra le «Dieter Roth's Verlag» en 1978. Lauréat, en 1986, du premier Prix Charles Nypels pour son œuvre typographique. – *Publications, sélection :* «Mundumculum», Cologne 1967; «Gesammelte Werke» (40 vol.), Stuttgart, Londres 1969–1986.

**Ruder,** Emil – né le 20. 3. 1914 à Zurich, Suisse, décédé le 13. 3. 1970 à Bâle, Suisse – *typographe, enseignant* – 1929–1933, apprentissage de composition typographique à Zurich. 1938–1939, séjour d'études à Paris. 1941–1942, études de compositeur et d'imprimeur de livres chez Alfred Willimann et Walter Käch à la Kunstgewerbeschule (école des arts décoratifs) de Zurich. Typographe indépendant à Bâle à partir de 1942. Professeur de typographie à l'Allgemeine Gewerbeschule (école des arts et métiers) de Bâle. Membre de la commission helvétique des arts appliqués et conseiller artistique du département de philatélie des postes suisses à partir de 1961. Cofondateur de l'«International Center for the Typographic Arts» (ICTA) à New York en 1962. Directeur de l'institut d'enseignement du Gewerbemuseum (musée des arts et métiers) de Bâle à partir de 1965.

published whose content and form have greatly influenced book design, and artists' books in particular. Teaches at the Rhode Island School of Design in Providence (1965–67), at the Watford School of Graphic Art (1968), at the Kunstakademie Düsseldorf (1969–71) and at the art academy in Reykjavík (1982). 1968–80: partner of Edition Hansjörg Mayer in Stuttgart. 1975: launches and edits the "Zeitschrift für Alles. Review for Everything. Timarit Fyrir Allt". At the same time he founds Dieter Roth's Familienverlag publishing house, which continues to run under the name of Dieter Roth's Verlag from 1978 onwards. 1986: receives the first Charles Nypels Prize awarded for his work in typography. – *Publications include:* "Mundumculum", Cologne 1967; "Gesammelte Werke" (40 vols.), Stuttgart, London 1969–1986.

**Ruder,** Emil – b. 20. 3. 1914 in Zurich, Switzerland, d. 13. 3. 1970 in Basle, Switzerland – *typographer, teacher* – 1929–33: trains as a typesetter in Zurich. 1938–39: studies in Paris. 1941–42: studies composition and printing under Alfred Willimann and Walter Käch at the Kunstgewerbeschule in Zurich. 1942: starts freelancing as a typographer in Basle. Specialist teacher of typography at the Allgemeine Gewerbeschule in Basle. 1961: becomes a member of the Eidgenössische Kommission für angewandte Kunst and is art consultant to the postage stamp section of the Swiss post office. 1962: co-founder of the International Center for the Typographic Arts (ICTA) in New York. From 1965 onwards: director

der Allgemeinen Gewerbeschule und des Gewerbemuseums Basel. Veröffentlicht zahlreiche Fachartikel in den Zeitschriften „Typographische Monatsblätter", „Schweizer Lehrerzeitung", „Werk", „Form und Technik" und „Druckspiegel". – *Publikation u. a.:* „Typographie", Teufen 1967.

**Rüegg,** Ruedi – geb. 4. 8. 1936 in Zürich, Schweiz. – *Grafik-Designer, Typograph, Lehrer* – Studium an der Kunstgewerbeschule in Zürich. 1960–63 Arbeit als Grafiker im Atelier Josef Müller-Brockmann. 1963–64 Assistent von Paul Rand in Weston, USA. 1964–65 Arbeit als Grafiker in der Nakamoto International Agency in Osaka, Japan. 1965 Rückkehr in die Schweiz, Arbeit im Atelier Müller-Brockmann. 1967–76 Mitinhaber der Werbeagentur Müller-Brockmann & Co. in Zürich. 1976–81 Präsident der Schweizer Abteilung der Association Graphique Internationale (AGI). 1977 Gründung der Bürogemeinschaft Baltis und Rüegg. Unterrichtet an der Eidgenössischen Technischen Hochschule (ETH) in Zürich, an der Cooper Union in New York (1981) und seit 1968 regelmäßig an der Ohio State University in Columbus, USA. – *Publikationen u. a.:* „Typographische Grundlagen", Zürich 1972; „Swiss Posters 1970–1980", Tokio 1982; „Basic Typography: Design with Letters", Zürich 1989.

**Ruscha,** Edward – geb. 16. 12. 1937 in Omaha, USA. – *Künstler, Lehrer* – 1956–60 Studium am Chouinard Art Institute in Los Angeles. 1960–61 Arbeit in einer Werbeagentur. 1962 erstes Ölbild mit dreidimensionalen Buchstaben. 1963

A publié de nombreux articles dans les revues spécialisées «Typographische Monatsblätter», «Schweizer Lehrerzeitung», «Werk», «Form und Technik», «Druckspiegel». – *Publication, sélection:* «Typographie», Teufen 1967.

**Rüegg,** Ruedi – né le 4. 8. 1936 à Zurich, Suisse – *graphiste maquettiste, typographe, enseignant* – études à la Kunstgewerbeschule (école des arts décoratifs) de Zurich. Employé comme graphiste à l'Atelier Josef Müller-Brockmann de 1960 à 1963. Assistant de Paul Rand à Weston, Etats-Unis de 1963 à 1964. Travaille comme graphiste à la Nakamoto International Agency d'Osaka, Japon, de 1964 à 1965. Retour en Suisse en 1965 où il travaille à l'Atelier Müller-Brockmann. Associé de l'agence de publicité Müller-Brockmann & Co à Zurich, de 1967 à 1976. Président de la section suisse de l'Association Graphique Internationale (AGI) de 1976 à 1981. Fonde la «Bürogemeinschaft Baltis und Rüegg» en 1977. A enseigné à la Eidgenössische Technische Hochschule (ETH) de Zurich, à la Cooper Union à New York (1981), et régulièrement depuis 1968 à la Ohio State University à Columbus, Etats-Unis. – *Publications, séliection:* «Typographische Grundlagen», Zurich 1972; «Swiss Posters 1970–1980», Tokyo 1982; «Basic Typography: Design with Letters», Zurich 1989.

**Ruscha,** Edward – né le 16. 12. 1937 à Omaha, Etats-Unis – *artiste, enseignant* – 1956–1960, études au Chouinard Art Institute de Los Angeles. Employé dans une agence de publicité de 1960 à 1961. Premier tableau à l'huile avec lettres tridimensionnelles en 1962. Première exposition individuelle en 1963. Commence la série des bandes de papier et mots dessinés en 1964. Maquette de la revue «Art Forum» sous le pseudonyme d'Eddie Rus-

Rüegg 1975 Poster

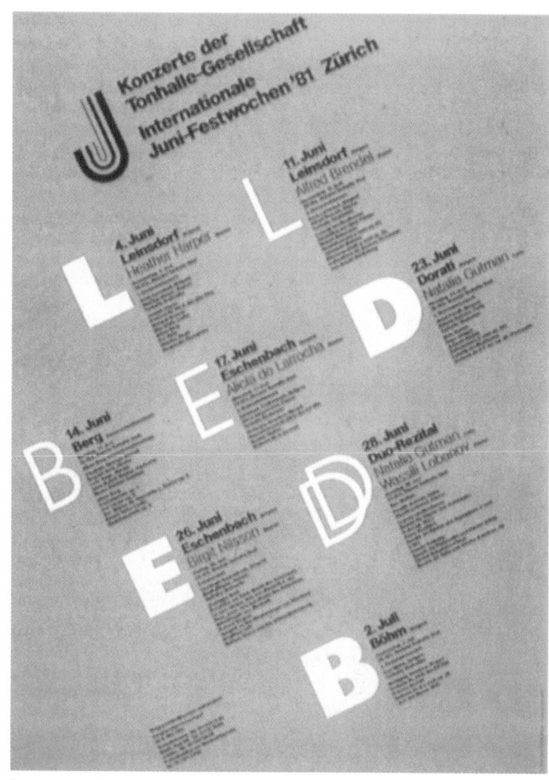


Rüegg 1981 Poster

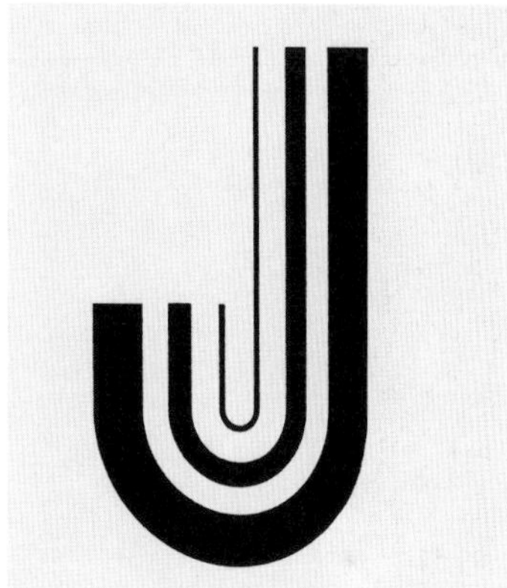

Rüegg ca. 1981 Logo

Ruscha 1963 Painting

of the Allgemeine Gewerbeschule and the Gewerbemuseum in Basle. Ruder published numerous specialist articles in "Typographische Monatsblätter", "Schweizer Lehrerzeitung", "Werk", "Form und Technik" and "Druckspiegel" magazines. – *Publications include:* "Typography", Teufen 1967.

**Rüegg,** Ruedi – b. 4. 8. 1936 in Zurich, Switzerland – *graphic designer, typographer, teacher* – Studied at the Kunstgewerbeschule in Zurich. 1960–63: works as a graphic artist in Josef Müller-Brockmann's studio. 1963–64: assistant to Paul Rand in Weston, USA. 1964–65: works as a graphic artist for the Nakamoto International Agency in Osaka, Japan. 1965: returns to Switzerland and again works for the Müller-Brockmann studio. 1967–76: co-owner of the Müller-Brockmann & Co. advertising agency in Zurich. 1976–81: president of the Swiss section of the Association Graphique Internationale (AGI). 1977: opens the Baltis and Rüegg joint studio. Teaches at the Eidgenössische Technische Hochschule (ETH) in Zurich and at the Cooper Union in New York (1981). Rüegg has taught on a regular basis at Ohio State University in Columbus, USA, since 1968. – *Publications include:* "Typographische Grundlagen", Zurich 1972; "Swiss Posters 1970–1980", Tokyo 1982; "Basic Typography: Design with Letters", Zurich 1989.

**Ruscha,** Edward – b. 16. 12. 1937 in Omaha, USA – *artist, teacher* – 1956–60: studies at the Chouinard Art Institute in Los Angeles. 1960–61: works for an advertising agency. 1962: produces his first

erste Einzelausstellung. Beginnt 1964 die Serie der Papierstreifen-Wörterzeichnungen. Gestaltet 1965–67 unter dem Pseudonym Eddie Russia das Layout der Zeitschrift „Art Forum". Unterrichtet 1969–70 Zeichnen und Druckgrafik an der University of California in Los Angeles. Veröffentlicht 1969 die Mappe „Stains", die mit organischen Substanzen gedruckt wurde und 1970 „News, Mews, Pews, Brews, Stews, Dues", für die er anstelle der üblichen Farben die von Blumen, Schmierfetten und Nahrungsmitteln verwendet. Für die Biennale in Venedig entsteht der „Chocolate Room", der mit 300 Papierbögen, die im Siebdruck mit Schokolade bedruckt wurden, bedeckt ist. Beginnt 1971 die Serie der Gemüse-Bilder. 1972–82 Teilnahme an der documenta VI, VII, VIII in Kassel. 1979 Beginn der Serie der Breitwandbilder. Für die Miami Dade Public Library entsteht 1985 ein achtteiliges Wandbild mit dem Satz „Words without thoughts never to heaven go". Bei der Arbeit von Ruscha sind sowohl im freien Werk (Bilder, Grafik, Fotos), als auch in den ca. 20 Buchpublikationen Schrift, Text und Typographie dominierende Gestaltungselemente. – *Publikationen u.a.:* D. Hickey, P. Plagens, A. Livet, H. T. Hopkins „The Works of Edward Ruscha", San Francisco 1982; Marianne Stockebrand „Ed Ruscha: 4 x 6", Münster 1986.

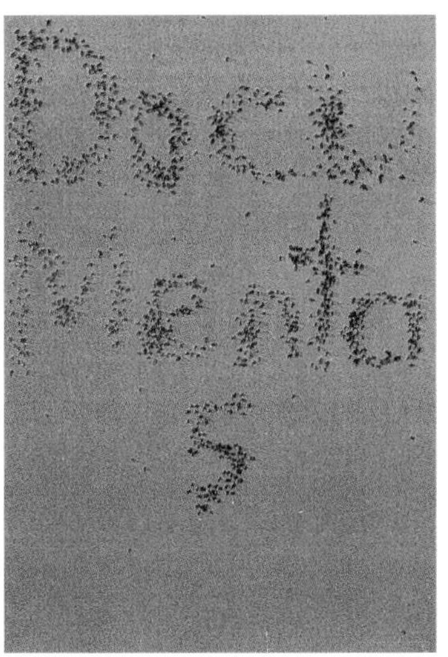
Ruscha 1972 Poster

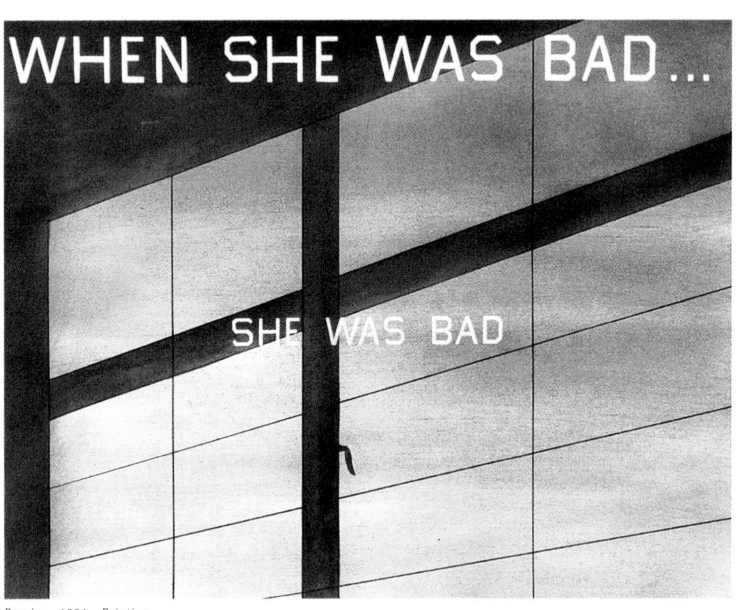

Ruscha 1984 Painting

sia de 1965 à 1967. Enseigne le dessin et les arts graphiques à l'University of California, à Los Angeles de 1969 à 1970. Publie, en 1969, la série de planches «Stains» imprimée avec des substances organiques, puis, en 1970, «News, Mews, Pews, Brews, Stews, Dues», pour laquelle, il remplace la couleur par des fleurs, des lubrifiants et des aliments. Il réalise «Chocolate Room» pour la Biennale de Venise, soit 300 planches de papier imprimées en sérigraphie au moyen de chocolat. En 1971, il commence la série des tableaux-légumes. Participe aux «documenta» VI, VII et VIII à Kassel de 1972 à 1982. Débute une série des grands tableaux muraux en 1979. Réalise en 1985, une fresque en 8 panneaux pour la Miami Dade Public Library, autour de la phrase «Words without thoughts never to heaven go». Dans l'œuvre de Ruscha, qu'il s'agisse de travaux libres (tableaux, estampes, photos) ou de la vingtaine de publications réalisées, l'écriture, le texte et la typographie jouent un rôle prédominant. – *Publications, sélection :* D. Hickey, P. Plagens, A. Livet, H. T. Hopkins «The Works of Edward Ruscha», San Francisco 1982; Marianne Stockebrand «Ed Ruscha : 4 x 6», Münster 1986.

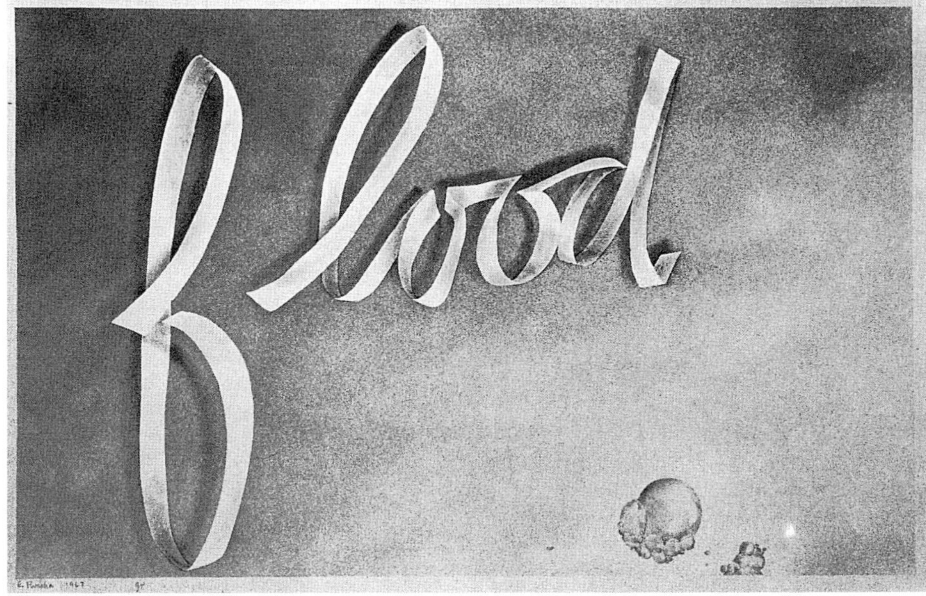

Ruscha 1967 Painting

oil painting with three-dimensional letters. 1963: first solo exhibition. 1964: starts his series of paper-strip-word-drawings. 1965–67: designs the layout for "Art Forum" magazine under the pseudonym Eddie Russia. 1969–70: teaches drawing and printed graphics at the University of California in Los Angeles. 1969: publishes his "Stains" portfolio, printed in organic substances, and in 1970 his "News, Mews, Pews, Brews, Stews, Dues", which he executes using grease, food and colors extracted from flowers instead of normal paint. He creates a "Chocolate Room" for the Biennale in Venice plastered with 300 sheets of paper screen-printed in chocolate. 1971: starts his series of vegetable pictures. 1972–82: takes part in documenta VI, VII and VIII in Kassel. 1979: starts his series of wide-format pictures. 1985: Ruscha creates a mural in eight sections for the Miami Dade Public Library entitled "Words without thoughts never to heaven go". Script, text and typography are a dominant design feature both of Ruscha's free artwork (pictures, graphics and photos) and of his app. 20 book publications. – *Publications include:* D. Hickey, P. Plagens, A. Livet, H. T. Hopkins "The Works of Edward Ruscha", San Francisco 1982; Marianne Stockebrand "Ed Ruscha: 4 x 6", Münster 1986.

CINÈ RADIO
سينما راديو
JOAN CRAWFORD
JEFF CHANDLER
FEMALE ON THE BEACH
JAN STERLING
JOHN PAYNE
COLEEN GRAY
الجمعية التعاونية المنزلية

# S

كوكا كولا
سينما قصر النيل
CINÉ RIVOLI
سينما ريفولى
Walt Disney
The Mightiest Motion Picture of them All!
20000 Leagues
KIRK DOUGLAS · JAMES MASON
PAUL LUKAS · PETER LORRE
Partner
SOCIETE CENTRALE D'ELECTRICITE S.A.E.
Centrelec

**Sabon**
**Salden, G.**

**Sabon,** Jakob – geb. ca. 1535 in Lyon, Frankreich, gest. 10. 9. 1580 in Frankfurt am Main, Deutschland. – *Schriftschneider, Schriftgießer* – 1563 Einrichtung der Schriftgießerei von Chr. Plantin in Antwerpen. 1566–67 Schriftgießer und Stempelschneider in Frankfurt an der Oder. 1572 wieder in Frankfurt am Main tätig, alleinige Übernahme der Schriftgießerei von Chr. Egenolff. Produziert Schriften von Robert Granjon und Claude Garamond, denen er damit zur Verbreitung in Deutschland verhilft. Zum Andenken an Sabons Schriftkunst entwirft Jan Tschichold 1967 eine Schrift, die er „Sabon" nennt. Es ist die erste Schrift, die im Hand-, Einzelbuchstaben- und Zeilenmaschinensatz gleichlaufend ist.
**Salden,** Georg – geb. 28. 8. 1930 in Essen, Deutschland. – *Grafik-Designer, Schriftentwerfer* – 1950–54 Studium an der Folkwangschule in Essen. 1955–72 selbständiger Grafik-Designer. Seit 1972 ausschließlich Entwurf neuer Alphabete. Unter dem Namen GST entstehen Schriften für Layoutsetzereien, die sich 1979 zur „Context Gesellschaft für Typographie und Satztechnik" zusammenschließen. Seit 1989 Versand seiner digitalisierten GST-Schriften. – *Schriftentwürfe:* York (1966–73), Transit (1969), Daphne (1970), Polo (1972–76), Bilbao (1972), Caslon (1972–76), Basta (1972–74), Stresemann (1972–73), Parabella (1972), Mäander (1972), Angular (1973), Brasil (1973–79), Magnet (1973), Hansa (1973), Bonjour (1973), Tandem (1973), Futuranea (1974–77), Congress (1974–75), Ready (1975), Loreley (1977), Gordon

**Sabon,** Jakob – né vers 1535 à Lyon, France, décédé le 10. 9. 1580 à Francfort-sur-le-Main, Allemagne – *tailleur de types, fondeur de caractères* – séjourne à Anvers en 1563, où il développe la fonderie de caractères de Chr. Plantin. Fondeur de caractères et tailleur de types à Francfort-sur-Oder de 1566 à 1567. Exerce de nouveau à Francfort-sur-le-Main en 1572, où il reprend seul la fonderie de Chr. Egenolff. Produit des caractères de Robert Granjon et Claude Garamond, et contribue ainsi à les faire connaître et à les diffuser en Allemagne. En mémoire de son art, Jan Tschichold dessinera en 1967 une fonte qu'il appellera «Sabon». C'est la première police présentant des caractères utilisés autant en composition manuelle qu'en composition automatique.
**Salden,** Georg – né le 28. 8. 1930 à Essen, Allemagne – *graphiste maquettiste, concepteur de polices* – 1950–1954, études à la Folkwangschule d'Essen. 1955–1972, graphiste maquettiste indépendant. A partir de 1972, il se consacre exclusivement à la conception de nouveaux alphabets. De nombreuses polices destinées à des ateliers de composition et de mise en page voient le jour sous le nom de GST, qui en 1979, deviendront la «Context Gesellschaft für Typographie und Satztechnik». Depuis 1989, il vend ses polices numériques GST par correspondance. – *Polices:* York (1966–1973), Transit (1969), Daphne (1970), Polo (1972–1976), Bilbao (1972), Caslon (1972–1976), Basta (1972–1974), Stresemann (1972–1973), Parabella (1972), Mäander (1972), Angular (1973), Brasil (1973–1979), Magnet (1973), Hansa (1973), Bonjour (1973), Tandem (1973), Futuranea (1974–1977), Congress (1974–1975), Ready (1975), Loreley (1977), Gordon (1977), Volante (1978), Loretta

ABCDEFGH

ABCDEFGHIJKLMNOPQRSTUVWXYZ
abcdefghijklmnopqrstuvwxyzß ch ck ff fi fl ft &
ÄÁÂÀÅÃÆÇÉÊËÈĞÌÍÎÏÑÖÓÔÒÕŒŞÜÚÛÙ
äáâàåãæçéêëèğìíîïijñöóôòõœşüúûù
1234567890 £ $ 1234567890
.,:;-!?·"()[]*†§%/„""›‹»«¡¿—

IJKLMNOPQRSTUVWXYZ

*ABCDEFGHIJKLMNOPQRSTUVWXYZ*
*abcdefghijklmnopqrstuvwxyzß ch ck ff fi fl ft &*
*ÄÁÂÀÅÃÆÇÊÉËÈĞÏÍÎÌÑÖÓÔÒÕŒŞÜÚÛÙ*
*äáâàåãæçêéëèğïíîìijñöóôòõœşüúûù*
*1234567890 £ $ 1234567890*
*.,:;-!?·"()[]*†§%/„""›‹»«¡¿—*

abcdefghijklmnopqrstuvwxyz

**ABCDEFGHIJKLMNOPQRSTUVWXYZ**
**abcdefghijklmnopqrstuvwxyzß ch ck ff fi fl ft &**
**ÄÁÂÀÅÃÆÇÊÉËÈĞÏÍÎÌÑÖÓÔÒØÕŒŞÜÚÛÙ**
**äáâàåãæçêéëèğïíîìijñöóôòøõœşüúûù**
**1234567890 £ $ 1234567890**
**.,:;-!?·"()[]*†§%/„""›‹»«¡¿—**

1234567890

Sabon 1967 Sabon Antiqua

| n | n | n | n |  | n |
|---|---|---|---|---|---|
| a | b | e | f | g | i |
| o | r | s | t | y | z |
| A | B | C | E | G | H |
| M | O | R | S | X | Y |
| 1 | 2 | 4 | 6 | 8 | & |

Georg Salden 1972–76 Polo

| n | n | n | n |  | n |
|---|---|---|---|---|---|
| a | b | e | f | g | i |
| o | r | s | t | y | z |
| A | B | C | E | G | H |
| M | O | R | S | X | Y |
| 1 | 2 | 4 | 6 | 8 | & |

Georg Salden 1972–74 Basta

|  | n | n | n |  |  |
|---|---|---|---|---|---|
| a | b | e | f | g | i |
| o | r | s | t | y | z |
| A | B | C | E | G | H |
| M | O | R | S | X | Y |
| 1 | 2 | 4 | 6 | 8 | & |

Georg Salden 1979–81 Tap

**Sabon,** Jakob – b. c.1535 in Lyon, France, d. 10. 9. 1580 in Frankfurt am Main, Germany – *type cutter, type founder* – 1563: spends time in Antwerp setting up Chr. Plantin's type foundry. 1566–67: type founder and punch cutter in Frankfurt an der Oder. 1572: works in Frankfurt am Main. Takes over Chr. Egenolff's type foundry as sole owner. Produces typefaces by Robert Granjon and Claude Garamond, which helps promote their work and gain them recognition in Germany. In homage to Sabon's typographic achievements, Jan Tschichold designs a special typeface in 1967 which he names Sabon. For the first time in history, a typeface of equal quality is produced, whether used in manual casting, monotype or line typesetting.
**Salden,** Georg – b. 28. 8. 1930 in Essen, Germany – *graphic designer, type designer* – 1950–54: studies at the Folkwangschule in Essen. 1955–72: freelance graphic artist. From 1972 onwards: concentrates on designing new alphabets. Numerous typefaces are produced for various layout workshops under the name of GST; these workshops merge in 1979 to form the Context Gesellschaft für Typographie und Satztechnik. 1989: a mail order company is launched selling Salden's digitized GST typefaces. – *Fonts:* York (1966–73), Transit (1969), Daphne (1970), Polo (1972–76), Bilbao (1972), Caslon (1972–76), Basta (1972–74), Stresemann (1972–73), Parabella, Mäander (1972), Angular (1973), Brasil (1973–79), Magnet, Hansa, Bonjour, Tandem (1973), Futuranea (1974–77), Congress (1974–75), Ready (1975), Loreley, Gordon (1977),

(1977), Volante (1978), Loretta (1979–84), Tap (1979–81), Sketchy (1979–80), Galopp (1979), Videon (1981), Deutschkurrent (1983), Turbo (1983), Corvey (1983), Klicker (1984), Dalli (1987), Axiom (1995–97), Carree (1995–97).

**Salden,** Helmut – geb. 20. 2. 1910 in Essen, Deutschland, gest. 2. 1. 1996 in Amsterdam, Niederlande. – *Typograph, Kalligraph, Lehrer* – Studium an der Folkwangschule für Gestaltung in Essen, danach dort bis 1933 Lehrer für Fotografie. 1934 Emigration. Seit 1938 in den Niederlanden als Grafiker tätig. 1939–45 in Haft in Deutschland. 1946 Rückkehr in die Niederlande. Arbeit für Verlage, u. a. für G. A. van Oorschot, Contact, Kiepenheuer & Witsch sowie für De Baak Congrescentrum Noordwijk. Zahlreiche Auszeichnungen, u. a. Niederländischer Staatspreis (1949, 1952), Frieslandpreis (1951), Werkman-Preis der Stadt Amsterdam (1953). Einzelausstellung seiner Arbeit u. a. im Gutenberg-Museum in Mainz (1966), in der Scottish National Gallery of Modern Art in Edinburgh (1970).

**Salisbury,** Mike – geb. 1949 in Provo, Utah, USA. – *Grafik-Designer, Fotograf, Illustrator, Lehrer* – 1967–71 Studium an der University of Southern California. Danach Art Director bei zahlreichen Zeitschriften. Gründet Mike Salisbury Inc. in Torrance, Kalifornien. Auftraggeber waren u.a. Levi Strauss, Lucasfilm, Frances Ford Coppola, Gotcha Sportswear, Disneyland, Michael Jackson, Polygram, Chevrolet sowie die Major-Filmstudios. Grafik-Design für über 300 Filme (u.a. „Alien", „Rocky IV", „Basic Instinct", „Silk-

Helmut Salden 1948 Cover

Salisbury Logo

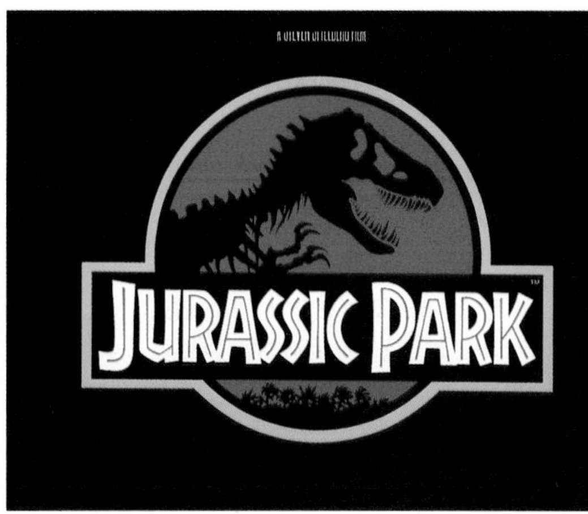

Salisbury 1992 Logo

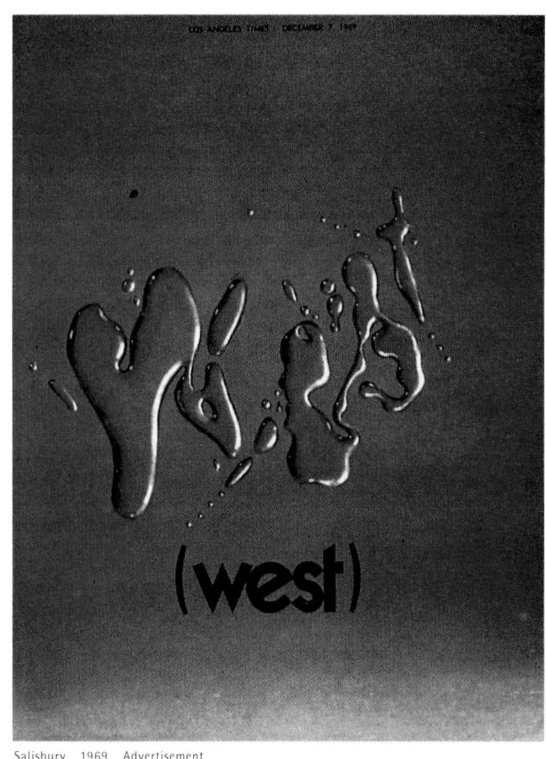

Salisbury 1969 Advertisement

Volante (1978), Loretta (1979–84), Tap (1979–81), Sketchy (1979–80), Galopp (1979), Videon (1981), Deutschkurrent, Turbo, Corvey (1983), Klicker (1984), Dalli (1987), Axiom, Carree (1995–97).

**Salden,** Helmut – b. 20. 2. 1910 in Essen, Germany, d. 2. 1. 1996 in Amsterdam, The Netherlands – *typographer, calligrapher, teacher* – Studied at the Folkwangschule für Gestaltung in Essen and then taught photography there until 1933. 1934: emigrates. From 1938 onwards: works in The Netherlands as a graphic artist. 1939–45: imprisoned in Germany. 1946: returns to The Netherlands. Works for various publishers, including G. A. van Oorschot, Contact and Kiepenheuer & Witsch and for De Baak Congrescentrum in Noordwijk. Salden won numerous awards, including the Dutch State Prize (1949, 1952), the Friesland Prize (1951) and Amsterdam's H. N. Werkman Prize (1953). Exhibitions of his work have been staged at the Gutenberg Museum in Mainz (1966) and the Scottish National Gallery of Modern Art in Edinburgh (1970).

**Salisbury,** Mike – b. 1949 in Provo, Utah, USA – *graphic designer, photographer, illustrator, teacher* – 1967–71: studies at the University of Southern California. Then art director for various magazines. Founds Mike Salisbury Inc. in Torrance, California. Clients include Levi Strauss, Lucasfilm, Frances Ford Coppola, Gotcha Sportswear, Disneyland, Michael Jackson, Polygram, Chevrolet and Major Film Studios. Does the graphic design for over 300 films (including "Alien", "Rocky IV", "Basic Instinct", "Silkwood", "Dead Man

(1979–1984), Tap (1979–1981), Sketchy (1979–1980), Galopp (1979), Videon (1981), Deutschkurrent (1983), Turbo (1983), Corvey (1983), Klicker (1984), Dalli (1987), Axiom (1995–1997), Carree (1995–1997).

**Salden,** Helmut – né le 20. 2. 1910 à Essen, Allemagne, décédé le 2. 1. 1996 à Amsterdam, Pays-Bas – *typographe, calligraphe, enseignant* – études de design à la Folkwangschule d'Essen où il enseignera la photographie jusqu'en 1933. Emigre en 1934. Exerce comme concepteur graphique aux Pays-Bas en 1938. Incarcéré en Allemagne de 1939 à 1945. Retour aux Pays-Bas en 1946. Travaille pour des maisons d'édition, entre autres pour G. A. van Oorschot, Contact, Kiepenheuer & Witsch ainsi que pour De Baak Congrescentrum Noordwijk. Nombreuses distinctions, dont le Prix des Pays-Bas (1949, 1952), le Frieslandpreis (1951), le Prix Werkman de la ville d'Amsterdam (1953). Expositions personnelles au Musée Gutenberg de Mayence (1966); à la Scottish National Gallery of Modern Art à Edimbourg (1970).

**Salisbury,** Mike – né en 1949 à Provo, Utah, Etats-Unis – *graphiste maquettiste, photographe, illustrateur, enseignant* – 1967–71, études à l'University of Southern California. Puis directeur artistique de nombreuses revues. Fonde la Mike Salisbury Inc. à Torrance, Californie. Commanditaires : Levi Strauss, Lucasfilm, Frances Ford Coppola, Gotcha Sportswear, Disneyland, Michael Jackson, Polygram, Chevrolet ainsi que les studios cinématographiques «Major». Concepteur graphique pour plus de 300 films (entre autre pour «Alien», «Rocky IV», «Basic Instinct», «Silkwood», «Dead

wood", „Dead Man Walking", „Jurassic Park", „Dick Tracy"). Unterrichtet am Art Center College of Design (1973, 1974, 1980), am Chouinard Art Institute (1975), an der University of California in Los Angeles (1976) und am Otis College of Art and Design (1996–97). Zahlreiche Auszeichnungen und Ausstellungen.

**Sallwey,** Friedrich Karl – geb. 16. 10. 1918 in Langen, Deutschland. – *Grafiker, Schriftentwerfer* – 1937–39 und 1941–42 Studium an der Kunstgewerbeschule in Offenbach. 1948 Arbeit in der Bauerschen Gießerei als Assistent von Heinrich Jost. Seit 1951 selbständiger Grafiker in Frankfurt am Main. – *Schriftentwürfe:* Information (1955), Present (1974), Sallwey Script (1979), Roundy (1992).

**Sandberg,** Willem – geb. 24. 10. 1897 in Amersfoort, Niederlande, gest. 9. 4. 1984 in Amsterdam, Niederlande. – *Grafik-Designer, Ausstellungsgestalter, Museumsdirektor* – 1919–20 Studium an der Rijksacademie voor Beeldende Kunsten in Amsterdam. 1927 Studium der Psychologie in Wien. Seit 1928 freier Grafiker in Amsterdam. 1930–35 Studium der Psychologie und Assistent in Utrecht. 1938 Beginn seiner Arbeit am Stedelijk Museum in Amsterdam. 1943–45 Schriftgestaltungen unter dem Titel „Experimenta Typographica". 1945–63 Direktor des Stedelijk Museums. 1947 Auszeichnung mit dem H. N. Werkman-Preis. 1949–62 Direktor des Fodor und Willet Museums in Amsterdam. 1964–68 Berater des Israel Museums in Jerusalem. 1968 mit dem Staatspreis für Bildende Kunst und Architektur ausgezeichnet. 1973 Ausstel-

Man Walking», «Jurassic Park», «Dick Tracy»). A enseigné à l'Art Center College of Design (1973, 1974, 1980) et à l'Otis College of Art and Design (1996–1997). Nombreuses expositions et distinctions.

**Sallwey,** Friedrich Karl – né le 16. 10. 1918 à Langen, Allemagne – *graphiste, concepteur de polices* – 1937–1939, 1941–1942, études à la Kunstgewerbeschule (école des arts décoratifs) d'Offenbach. Employé comme assistant de Heinrich Jost à la fonderie Bauer en 1948. Graphiste indépendant à Francfort-sur-le-Main depuis 1951. – *Polices:* Information (1955), Present (1974), Sallwey Script (1979), Roundy (1992).

**Sandberg,** Willem – né le 24. 10. 1897 à Amersfoort, Pays-Bas, décédé le 9. 4. 1984 à Amsterdam, Pays-Bas – *graphiste maquettiste, architecte d'expositions, directeur de musée* – 1919–1920, études à la Rijksacademie voor Beeldende Kunsten à Amsterdam. 1927, études de psychologie à Vienne. Graphiste indépendant à Amsterdam à partir de 1928. Etudes de psychologie et assistant à Utrecht de 1930 à 1935. Commence à travailler au Stedelijk Museum d'Amsterdam en 1938. Dessin de polices sous le titre «Experimenta Typographica» de 1943 à 1945. Dirige le Stedelijk Museum de 1945 à 1963. Lauréat du Prix H. N. Werkman en 1947. Directeur du Fodor und Willet Museum d'Amsterdam de 1949 à 1962. Conseiller de l'Israel Museum à Jérusalem de 1964 à 1968. Staatspreis des Arts plastiques et d'architecture en 1968. Exposition «Sandberg designe le Stedelijk, 1945–1963» au Musée des Arts Décoratifs de Paris, en 1973. Lauréat du Prix Erasmus en 1975. – *Publications, sélection:* «Sleutelwoorden: Piet Zwart», La Haye 1965; «Nu 2», Hilversum 1968. Bibeb «Sandberg, experimenta typographica 1943–68», Nimègue 1969; Ad Pedersen, Pieter Brat-

| | | | | | |
|---|---|---|---|---|---|
| | n | | | | |
| a | b | e | f | g | i |
| o | r | s | t | y | z |
| A | B | C | E | G | H |
| M | O | R | S | X | Y |
| 1 | 2 | 4 | 6 | 8 | & |

Sallwey 1974 Present

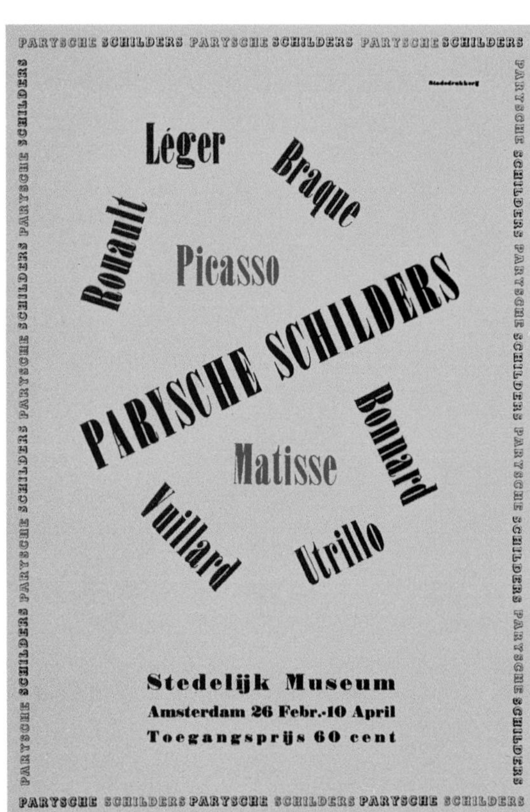


Sandberg 1939 Poster

Sandberg 1952 Poster

Sandberg 1972 Stamps

Walking", "Jurassic Park", "Dick Tracy"). Has taught at the Art Center College of Design (1973, 1974 and 1980), the Chouinard Art Institute (1975), the University of California in Los Angeles (1976) and the Otis College of Art and Design (1996–97). Numerous awards and exhibitions.

**Sallwey,** Friedrich Karl – b. 16. 10. 1918 in Langen, Germany – *graphic artist, type designer* – 1937–39 and 1941–42: studies at the Kunstgewerbeschule in Offenbach. 1948: works at the Bauersche Gießerei as Heinrich Jost's assistant. From 1951 onwards: freelance graphic artist in Frankfurt am Main. – *Fonts:* Information (1955), Present (1974), Sallwey Script (1979), Roundy (1992).

**Sandberg,** Willem – b. 24. 10. 1897 in Amersfoort, The Netherlands, d. 9. 4. 1984 in Amsterdam, The Netherlands – *graphic designer, exhibition planner, museum director* – 1919–20: studies at the Rijksacademie voor Beeldende Kunsten in Amsterdam. 1927: studies psychology in Vienna. 1928: starts freelancing as a graphic artist in Amsterdam. 1930–35: studies psychology and is an assistant in Utrecht. 1938: starts working at the Stedelijk Museum in Amsterdam. 1943–45: designs type under the motto "Experimenta Typographica". 1945–63: director of the Stedelijk Museum. 1947: is awarded the H. N. Werkman Prize. 1949–62: director of the Fodor and Willet Museum in Amsterdam. 1964–68: advisor to the Israel Museum in Jerusalem. 1968: is awarded the State Prize for Fine Art and Architecture. 1973: "Sandberg designe le Stedelijk,

lung „Sandberg designe le Stedelijk, 1945–63" im Musée des Arts Décoratifs in Paris. 1975 mit dem Erasmuspreis ausgezeichnet. – *Publikationen u.a.:* „Sleutelwoorden: Piet Zwart", Den Haag 1965; „Nu 2", Hilversum 1968. Bibeb „Sandberg, experimenta typographica 1943–68", Nijmegen 1969; Ad Pedersen, Pieter Brattinga „Sandberg – een documentaire", Amsterdam 1975; Ad Pedersen u. a. „Sandberg – typograaf als museumman", Amersfoort 1982.

**Sato,** Koichi – geb. 9. 8. 1944 in Takasaki, Japan. – *Illustrator, Kalligraph, Typograph, Lehrer* – 1968 Abschluß seines Kommunikationsdesign-Studiums am Tokyo Art College. 1969 Grafik-Designer bei der Kosmetikfirma Shiseido. 1970 Gründung des „Koichi Sato Design Office". Unterrichtet 1972–87 am Tokyo Art College. Zahlreiche Auszeichnungen, u. a. mit dem 1. Preis des Plakatwettbewerbs des Museums of Modern Art in New York (1988). – *Publikationen u. a.:* „Graphics Japan" (Mitherausgeber), Tokio 1987; „Koichi Sato", Tokio 1990; „Kirei – Plakate aus Japan" (Co-Autor), Schaffhausen 1993.

**Saville,** Peter – geb. 9. 10. 1955 in Manchester, England. – *Grafik-Designer* – 1975–78 Studium an der Manchester Polytechnic. Seit 1979 Schallplattencover-Entwürfe für Factory Records (die er mit gründete), Dindisc, Stiff und Radar. 1983 Gründung von „Peter Saville Associates" in London (mit Brett Wickens). Auftraggeber waren u. a. das Centre Georges Pompidou in Paris, die Modedesigner Yohji Yamamoto und Jil Sander, die

arne jacobsen tot 22 juni
architectuur en meubels
tot 3 juni lucebert
tekeningen
tot 8 juni constant
plastieken
tot 15 juni monticelli
schilderijen
calder tot 22 juni
stabiles en mobiles
stedelijk museum a'dam

Sandberg 1957 Poster

Sato 1988 Poster

tinga «Sandberg – een documentaire», Amsterdam 1975; Ad Pedersen et autres «Sandberg – typograaf als museumman», Amersfoort 1982.

**Sato,** Koichi – né le 9. 8. 1944 à Takasaki, Japon – *illustrateur, calligraphe, typographe, enseignant* – termine ses études de design et de communication au Tokyo Art College en 1968. Employé comme graphiste maquettiste dans la société de cosmétiques Shiseido, en 1969. Fonde le «Koichi Sato Design Office» en 1970. Enseigne au Tokyo Art College de 1972 à 1987. Nombreuses distinctions, entre autres, le 1er prix du concours d'affiches du Museum of Modern Art de New York (1988). – *Publications, sélection:* «Graphics Japan» (co-éditeur), Tokyo 1987; «Koichi Sato», Tokyo 1990; «Kirei – Plakate aus Japan» (co-auteur), Schaffhausen 1993.

**Saville,** Peter – né le 9. 10. 1955 à Manchester, Angleterre – *graphiste maquettiste* – 1975–1978, études au Manchester Polytechnic. Conception de pochettes de disques pour Factory Records (dont il est cofondateur), Dindisc, Stiff et Radar à partir de 1979. Fondation de «Peter Saville Associates» à Londres, en 1983 (en collaboration avec Brett Wickens). Commanditaires: Centre Georges Pompidou, Paris, les stylistes de mode Yoji Yamamoto et Jil Sander, Whitechapel Art Gal-

Sato 1983 Poster

1945–63" exhibition in the Musée des Arts Décoratifs in Paris. 1975: is awarded the Erasmus Prize. – *Publications include:* "Sleutelwoorden: Piet Zwart", The Hague 1965; "Nu 2", Hilversum 1968. Bibeb "Sandberg, experimenta typographica 1943–68", Nijmegen 1969; Ad Pedersen, Pieter Brattinga "Sandberg – een documentaire", Amsterdam 1975; Ad Pedersen et al "Sandberg – typograaf als museumman", Amersfoort 1982.

**Sato,** Koichi – b. 9. 8. 1944 in Takasaki, Japan – *illustrator, calligrapher, typographer, teacher* – 1968: completes his studies in communication design at Tokyo Art College. 1969: graphic designer for Shiseido cosmetics company. 1970: opens the Koichi Sato Design Office. 1972–87: teaches at Tokyo Art College. Sato has won numerous awards, including 1st prize in the poster competition run by the Museum of Modern Art in New York (1988). – *Publications include:* "Graphics Japan" (co-editor), Tokyo 1987; "Koichi Sato", Tokyo 1990; "Kirei – Plakate aus Japan" (co-author), Schaffhausen 1993.

**Saville,** Peter – b. 9. 10. 1955 in Manchester, England – *graphic designer* – 1975–78: studies at Manchester Polytechnic. 1979: starts designing record covers for Factory Records (of which he is co-founder), Dindisc, Stiff and Radar. 1983: founds Peter Saville Associates in London. Brett Wickens works with him. Clients include the Centre Georges Pompidou in Paris, fashion designers Yohji Yamamoto and Jil Sander, the Whitechapel Art Gallery in London and Virgin Records. 1990–93: partner of Pentagram

Whitechapel Art Gallery in London und Virgin Records. 1990–93 Partner bei Pentagram in London, 1996 Austritt und Gründung seines eigenen Studios in London. Partnerschaft mit der Agentur Meiré & Meiré (Köln und Wiesbaden) unter dem Namen „The Apartment".

**Schauer,** Georg Kurt – geb. 2. 8. 1899 in Frankfurt am Main, Deutschland, gest. 11. 12. 1984 in Frankfurt am Main, Deutschland. – *Verleger, Autor* – Studium der Geschichte an den Universitäten Freiburg und Frankfurt am Main. 1922 Volontär im Verlag Rütten & Loening in Frankfurt am Main, danach bei S. Fischer in Berlin. Schriftleiter der Zeitschrift „Philobiblon". 1946–72 Gründung und Leitung der „Verlagsbuchhandlung Georg Kurt Schauer". Herausgabe von Büchern zur Typographie von F. H. Ehmcke, Hermann Zapf, Karl Klingspor und ihm selbst. 1953–59 künstlerischer Leiter der Schriftgießerei D. Stempel AG in Frankfurt am Main. 1958–64 Mitwirkung im Gremium des Deutschen Normenausschusses für Fragen der Klassifikation des Schriftwesens. – *Publikationen u.a.:* „Wege der Buchgestaltung", Stuttgart 1953; „Typographie der Mitte", Frankfurt am Main 1954; „Kleine Geschichte des deutschen Buchumschlags im 20. Jahrhundert", Königstein 1962; „Deutsche Buchkunst 1890–1960", Hamburg 1963.

**Schawinsky,** Xanti (Alexander) – geb. 26. 3. 1904 in Basel, Schweiz, gest. 11. 9. 1979 in Locarno, Schweiz. – *Fotograf, Maler, Grafik-Designer, Lehrer* – 1921–23 Ausbildung im Architekturbüro Merrill in Köln. 1923–24 Studium an der Kunstge-

lery, Londres, Virgin Records, etc. 1990–1993, partenaire de Pentagram à Londres. Quitte Pentagram en 1996 et fonde sa propre agence à Londres. Partenariat avec l'agence Meiré & Meiré (Cologne et Wiesbaden) sous le nom de «The Apartment».

**Schauer,** Georg Kurt – né le 2. 8. 1899 à Francfort-sur-le-Main, Allemagne, décédé le 11. 12. 1984 à Francfort-sur-le-Main, Allemagne – *éditeur, auteur* – études d'histoire dans les l'universités de Freiburg et de Francfort-sur-le-Main. Stagiaire aux éditions Rütten & Loening à Francfort-sur-le-Main en 1922, puis chez S. Fischer à Berlin. Rédacteur en chef de la revue «Philobiblon». Fonde la Verlagsbuchhandlung Georg Kurt Schauer en 1946 et la dirige jusqu'en 1972. Publie des ouvrages sur l'œuvre typographique de F. H. Ehmcke, Hermann Zapf, Karl Klingspor et sur lui-même. Directeur artistique de la fonderie de caractères D. Stempel AG à Francfort-sur-le-Main de 1953 à 1959. Membre du comité allemand des normes, chargé de la classification des caractères et écritures de 1958 à 1964. – *Publications, sélection:* «Wege der Buchgestaltung», Stuttgart 1953; «Typographie der Mitte», Francfort-sur-le-Main 1954; «Kleine Geschichte des deutschen Buchumschlags im 20. Jahrhundert», Königstein 1962; «Deutsche Buchkunst 1890–1960», Hambourg 1963.

**Schawinsky,** Xanti (Alexander) – né le 26. 3. 1904 à Bâle, Suisse, décédé le 11. 9. 1979 à Locarno, Suisse – *photographe, peintre, graphiste maquettiste, enseignant* – 1921–1923, formation au cabinet d'architecte Merill à Cologne. 1923–1924 études à la Kunstgewerbeschule (école des arts décoratifs) de Berlin. 1924–1929, études au Bauhaus à Weimar puis à Dessau. Début de ses expériences photographiques en 1924. Participe, en 1927, à l'exposition «Neue Reklame» au Kunst-

Saville 1979 Cover

Saville 1992 Page

Saville 1985 Logo

Saville 1984/85 Logo

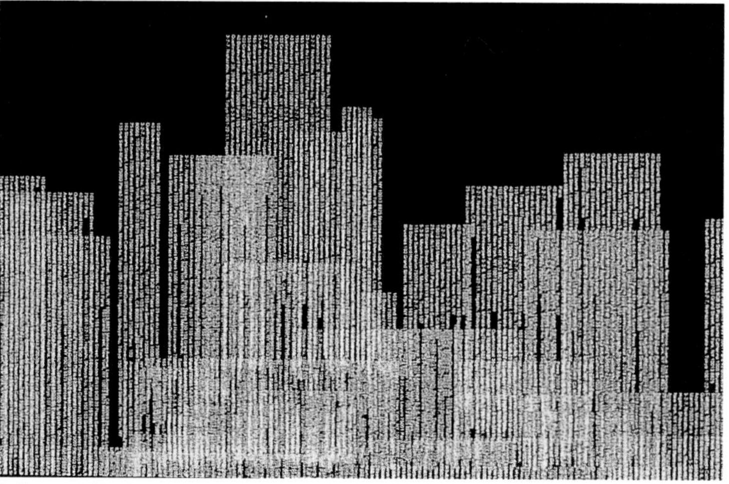
Saville 1994 Advertisement

Schauer 1953 Cover

in London. 1996: leaves Pentagram and opens his own studio in London. Forms a partnership, entitled The Apartment, with the Meiré & Meiré agency in Cologne and Wiesbaden.

**Schauer,** Georg Kurt – b. 2. 8. 1899 in Frankfurt am Main, Germany, d. 11. 12. 1984 in Frankfurt am Main, Germany – *publisher, author* – Studied history at the universities in Freiburg and Frankfurt am Main. 1922: trainee at the Rütten & Loening publishing house in Frankfurt am Main and later at the S. Fischer press in Berlin. Editor of "Philobiblon" magazine. 1946–72: launches and runs the Verlagsbuchhandlung Georg Kurt Schauer. The company publishes books on typography by F. H. Ehmcke, Hermann Zapf, Karl Klingspor and by Schauer himself. 1953–59: art director of the D. Stempel AG type foundry in Frankfurt am Main. 1958–64: member of the German committee for establishing norms in printing types (Deutscher Normenausschuss für Fragen der Klassifikation des Schriftwesens). – *Publications include:* "Wege der Buchgestaltung", Stuttgart 1953; "Typographie der Mitte", Frankfurt am Main 1954; "Kleine Geschichte des deutschen Buchumschlags im 20. Jahrhundert" Königstein 1962; "Deutsche Buchkunst 1890–1960", Hamburg 1963.

**Schawinsky,** Xanti (Alexander) – b. 26. 3. 1904 in Basle, Switzerland, d. 11. 9. 1979 in Locarno, Switzerland – *photographer, painter, graphic designer, teacher* – 1921–23: trains at Merrill architect's office in Cologne. 1923–24: studies at the Kunstgewerbeschule in Berlin. 1924–29:

werbeschule in Berlin. 1924–29 Studium am Bauhaus in Weimar und Dessau. 1924 Fotografische Experimente. 1927 Beteiligung an der Ausstellung „Neue Reklame" im Kunstverein Jena. 1928 Gestaltung des Standes für den Reichsverband der Wohnungsfürsorge-Gesellschaften auf der Ausstellung „bauen und wohnen" in Berlin-Zehlendorf. 1929 Mitgestaltung für den Stand der Firma Junkers auf der Ausstellung „Gas und Wasser" in Berlin. Leiter der Grafik-Abteilung des Städtischen Hochbauamts Magdeburg. 1931 Schautafel-Gestaltung für die „Deutsche Bauausstellung Berlin 1931". 1933–36 Aufenthalt in Italien. Mitarbeit im Grafik-Design-Studio Boggeri; Gestaltung für Auftraggeber wie Motta, Cinzano, Olivetti. 1936 Emigration in die USA. Berufung an das Black Mountain College in North Carolina. Seit 1938 in New York als Ausstellungsgestalter, Grafiker, Maler, Fotograf tätig. 1943–46 Dozent für Grafik-Design und Malerei am New York City College. 1950–54 Lehrauftrag an der New York University. 1961–79 als Maler in Europa und USA tätig. – *Publikationen u.a.:* Barbara Paul u. a. „Xanti Schawinsky, Malerei, Bühne, Grafikdesign, Fotografie", Berlin 1986; E. Neumann, R. Schmid „Xanti Scha-winsky Foto", Bern 1989; Lutz Schöbe u.a. „Xanti Schawinsky. Magdeburg 1929–31. Fotografien", Dessau 1993.

**Schelter & Giesecke** – *Schriftgießerei* – 1819 Gründung der Firma in Leipzig durch den Stempelschneider Johann Andreas Gottfried Schelter (1775–1841) und dem Schriftgießer Christian Friedrich

Saville 1992 Trailer

Schawinsky 1934 Poster

Eine neue Anzeigen- und Auszeichnungsschrift

LICHTE BELWE ANTIQUA

GROßBUCHSTABEN

nach Zeichnung von Professor Georg Belwe In 7 Graden von 16 bis 60 Punkt lieferbar

SCHRIFTGIEßEREI
J. G. SCHELTER & GIESECKE
LEIPZIG C1

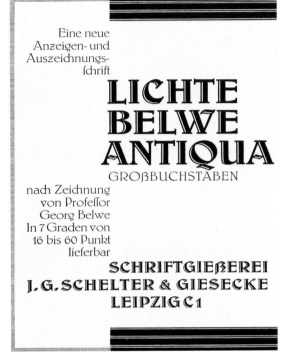

Schelter & Giesecke 1929 Advertisement

KORALLE

Diese Grotesk ergibt mit ihrem einfachen ungekünstelten Schriftbild wirkungsvolle Werbedrucke. Musterblätter zu Diensten

SCHELTER & GIESECKE AG.

Schelter & Giesecke 1931 Advertisement

verein de Iéna. En 1928, il conçoit le stand du Reichsverband der Wohnungsfürsorge-Gesellschaften (Union allemande des sociétés de logements sociaux) «bauen und wohnen» à Berlin-Zehlendorf. Travaux pour le stand de la société Junkers lors de l'exposition «Gas und Wasser» à Berlin, en 1929. Directeur du service du graphisme de l'office municipal de la construction à Magdebourg. En 1931, il conçoit les panneaux informatifs de la «Deutsche Bauausstellung Berlin 1931». Séjour en Italie de 1933 à 1936. Employé à l'agence de design et de conception graphique «Boggeri»; créations pour des commanditaires tels que Motta, Cinzano, Olivetti. Emigre aux Etats-Unis en 1936. Nommé professeur au Black Mountain College de Caroline du Nord. Exerce comme architecte d'exposition, graphiste, peintre et photographe à New York à partir de 1938. Professeur d'arts graphiques, de design et de peinture au New York City College de 1943 à 1946. Chargé de cours à la New York University de 1950 à 1954. Exerce comme peintre aux Etats-Unis et en Europe de 1961 à 1979. – *Publications, sélection:* Barbara Paul et autres «Xanti Schawinsky. Malerei, Bühne, Grafikdesign, Fotografie», Berlin 1986; E. Neumann, R. Schmid «Xanti Schawinsky Foto», Berne 1989; Lutz Schöbe et autres «Xanti Schawinsky. Magdeburg 1929–1931. Fotografien», Dessau 1993.

**Schelter & Giesecke** – *fonderie de caractères* – la société est fondée en 1819 à Leipzig par le tailleur de types Johann Andreas Gottfried Schelter (1775–1841) et par le fondeur de caractères Christian

studies at the Bauhaus in Weimar and Dessau. 1924: starts experimenting with photography. 1927: takes part in the "Neue Reklame" exhibition run by the Kunstverein Jena. 1928: designs a stand for the Reichsverband der Wohnungsfürsorge-Gesellschaften at the "bauen und wohnen" exhibition in Berlin-Zehlendorf. 1929: designs a stand for the Junkers company at the "Gas und Wasser" exhibition in Berlin. Head of the Städtische Hochbauamt Magdeburg graphics department. 1931: designs the display board for the "Deutsche Bauausstellung Berlin 1931" exhibition. 1933–36: spends time in Italy. Works for the Boggeri graphic design studio, designing for clients such as Motta, Cinzano and Olivetti. 1936: emigrates to the USA. Takes up a teaching position at Black Mountain College in North Carolina. From 1938 onwards: works as an exhibition planner, graphic artist, painter and photographer in New York. 1943–46: lectures graphic design and painting at New York City College. 1950–54: teaches at New York University. 1961–79: works as a painter in Europe and the USA. – *Publications include:* Barbara Paul et al "Xanti Schawinsky. Malerei, Bühne, Grafikdesign, Fotografie", Berlin 1986; E. Neumann, R. Schmid "Xanti Schawinsky Foto", Bern 1989; Lutz Schöbe et al "Xanti Schawinsky. Magdeburg 1929–31. Fotografien", Dessau 1993.

**Schelter & Giesecke** – *type foundry* – 1819: the company is founded in Leipzig by punch cutter Johann Andreas Gottfried Schelter (1775–1841) and type

Giesecke (1785–1851). 1839 wird Giesecke alleiniger Betreiber der Firma. Seine Söhne Karl Ferdinand Wilhelm und Bernhard Rudolf erweitern die Firma 1851 um eine Messinglinien-Fabrik, eine galvanoplastische Anstalt und eine Maschinenfabrik. 1946 Verstaatlichung der Firma, Trennung in VEB Buchdruckmaschinenwerk Leipzig und VEB Typoart Dresden.

**Scher,** Paula – geb. 6. 10. 1948 in Washington, USA. – *Grafik-Designerin, Illustratorin* – 1966–70 Studium an der Tyler School of Art (Temple University). 1970 Umzug nach New York. Art Director bei CBS Records, Gestaltung von über 1.000 Schallplattencovern. 1984–91 Gründung des Grafik-Design-Studios „Koppel & Scher" (mit Terry Koppel). 1991 Eintritt als Partner bei Pentagram Design in New York. Arbeiten für Auftraggeber wie Swatch, Champion Paper, die Chase Manhattan Bank, Simon & Schuster und Time Warner. 1993 Vorsitzende des „Graduate Program in Graphic Design" der School of Visual Arts in New York. Als eine der „10 signifikantesten Designerinnen in Amerika" durch die Vereinigung „Women in Design" in Chicago ausgezeichnet. – *Publikationen u.a.:* „The Honeymoon Book", New York 1981; „The Graphic Design Portfolio: How to Make a Good One", New York 1992.

**Schiller,** Walter – geb. 18. 3. 1920 in Hamburg, Deutschland. – *Typograph, Lehrer* – 1938 Volontär als Lithograph und Beginn des Studiums in Wien. 1947–51 Studium an der Hochschule für Graphik und Buchkunst (HGB) in Leipzig. 1955–64 Assistent und Dozent an der HGB, 1964 Pro-

Friedrich Giesecke (1785–1851). Giesecke exploite seul la société à partir de 1839. Ses fils, Karl Ferdinand Wilhelm et Bernhard Rudolf agrandissent l'entreprise en 1851 et lui adjoignent une usine de fabrication de tiges de laiton, un atelier de galvanisation et une usine de constructions mécaniques. La société est nationalisée en 1946, puis est scindée pour former le VEB Buchdruckmaschinenwerk Leipzig et le VEB Typoart, Dresde.

**Scher,** Paula – née le 6. 10. 1948 à Washington, Etats-Unis – *graphiste maquettiste, illustratrice* – 1966–1970, études à la Tyler School of Art (Temple University). S'installe à New York en 1970. Directrice artistique chez CBS Records; maquette de plus de 1 000 pochettes de disques. En 1984, elle fonde (avec Terry Koppel) l'atelier de conception graphique «Koppel & Scher» (1984–1991). Entre comme partenaire chez Pentagram à New York en 1991. Travaux pour Swatch, Champion Paper, Chase Manhattan Bank, Simon & Schuster et Time Warner. Présidente, en 1993, du «Graduate Program in Graphic Design» de la School of Visual Arts à New York. Est considérée comme faisant partie des « 10 meilleurs designers des Etats-Unis» par l'association «Women in Design» de Chicago. – *Publications, sélection:* «The Honeymoon Book», New York 1981; «The Graphic Design Portfolio : How to Make a Good One», New York 1992.

**Schiller,** Walter – né le 18. 3. 1920 à Hambourg, Allemagne – *typographe, enseignant* – stage de lithographie en 1938, puis études d'arts graphiques et d'arts du livre (HGB) à Leipzig. Assistant, puis enseignant à la HGB de 1955 à 1964, professeur à partir de 1964. Lauréat du Prix Gutenberg de la ville de Leipzig en 1968, puis du Nationalpreis der DDR en 1976. Dirige la section de maquette de livres à

Scher 1991 Poster

Scher ca. 1992 Self-portrait

Scher 1992 Poster

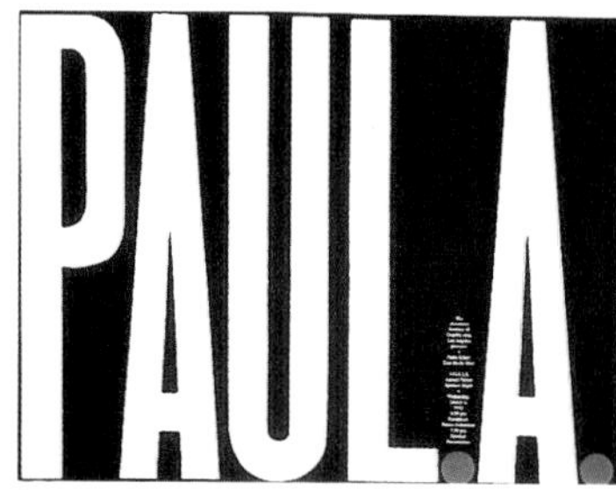


Scher 1989 Poster

founder Christian Friedrich Giesecke (1785–1851). 1839: Giesecke becomes the sole operator of the foundry. 1851: his sons Karl Ferdinand Wilhelm and Bernhard Rudolf expand the foundry, adding to it a brass rule factory, a galvanoplastics unit and a machine shop. 1946: the company is nationalized and split into VEB Buchdruckmaschinenwerk Leipzig (printing machines) and VEB Typoart Dresden.

**Scher,** Paula – b. 6. 10. 1948 in Washington, USA – *graphic designer, illustrator* – 1966–70: studies at the Tyler School of Art (Temple University). 1970: moves to New York. Art director for CBS Records, designing over 1,000 record covers. 1984–91: opens Koppel & Scher graphic design studio with Terry Koppel. 1991: joins Pentagram Design in New York as a partner. Works for clients such as Swatch, Champion Paper, Chase Manhattan Bank, Simon & Schuster and Time Warner. 1993: chairwoman of the Graduate Program in Graphic Design at the School of Visual Arts in New York. Named one of the "10 most significant woman designers in America" by the Women in Design organization in Chicago. – *Publications include:* "The Honeymoon Book", New York 1981; "The Graphic Design Portfolio: How to Make a Good One", New York 1992.

**Schiller,** Walter – b. 18. 3. 1920 in Hamburg, Germany – *typographer, teacher* – 1938: trainee lithographer. Starts studying in Vienna. 1947–51: studies at the Hochschule für Graphik und Buchkunst (HGB) in Leipzig. 1955–64: assistant and

fessor. 1968 mit dem Gutenberg-Preis der Stadt Leipzig und 1976 mit dem Nationalpreis der DDR ausgezeichnet. 1968–73 Leiter der Abteilung Buchgestaltung der HGB. 1981 Leiter des Instituts für Buchgestaltung. – *Publikation:* „Gestalt und Funktion der Typografie" (mit Albert Kapr), Leipzig 1977.

**Schmid,** Helmut – geb. 1. 2. 1942 in Ferlach, Österreich. – *Typograph* – Schriftsetzerlehre in Deutschland, Studium der Typographie an der Allgemeinen Gewerbeschule in Basel bei Emil Ruder. 1964–72 Entwurf von 44 typographischen Umschlägen für die schwedische Fachzeitschrift „Grafisk Revy". 1967–70 Entwurf der japanischen Silbenschrift „Katakana Eru" (zu Ehren Emil Ruders). 1973–76 Entwurf von Drucksachen für das Bundespresseamt in Bonn und für die Sozialdemokratische Partei Deutschlands SPD. Seit 1977 lebt und arbeitet Schmid in Japan. Auftraggeber waren u. a. Taiho Pharmaceutical und Sanyo Electric. 1978 Ausstellung seiner „polittypographien", die zwischen 1974–77 entstanden sind, in der Print Gallery Pieter Brattinga in Amsterdam. 1983–86 Entwurf von Schriftmarken für Produkte von Shiseido Cosmetics. Zahlreiche Aufsätze über Typographie in internationalen Zeitschriften, u. a. in „Typographische Monatsblätter", „Grafisk Revy", „Idea", „Novum". Seit 1992 Herausgabe seiner Broschürenreihe „Typographic Reflections". – *Schriftentwurf:* Katakana Eru (1971). – *Publikation:* „Typography today", Tokio 1980.

**Schmidt,** Hans – geb. 14. 1. 1923 in Leipzig, Deutschland. – *Typograph, Schrift-*

Internationale Beiträge zur Buchgestaltung Herausgegeben vom Institut für Buchgestaltung Leipzig Vierter Band

BUCHKUNST

Inhalt: Théophile Alexandre Steinlen Alexei Krawtschenko Vorformen des Buchdrucks in China und Korea Die chinesische Schriftreform Japanische Grafik Max Schwimmer Horst Erich Wolter José Guadalupe Posada Der Künstler als Reporter Die Dostojewski-Illustrationen von Axl Leskoschek Der Druck mit Metallettern in Korea »Das schönste Gedicht« VEB VERLAG DER KUNST DRESDEN

Schiller 1963 Cover

Schmid 1980 Cover

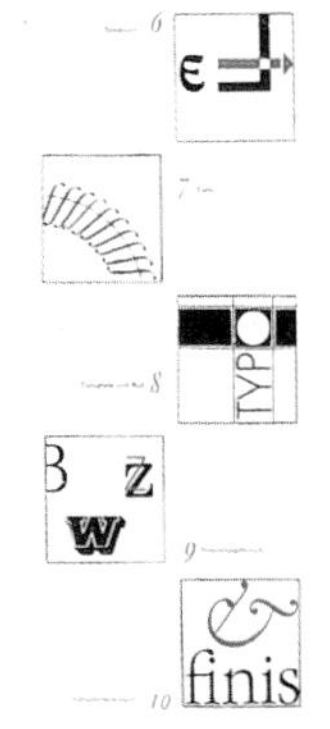

VEB Fachbuchverlag Leipzig

Aus dem Inhalt: Allgemeines zur Lesbarkeit von Schriften und über die Aufgaben der Typografie beim Lesbarmachen von Texten [1] Die Fläche, ihre Möglichkeiten und ihre grafischen Wirkungsmittel [2] Über die Satzschriften [3] Von den Bausteinen und der Bauweise [4] Hervorheben, Auszeichnen, Ordnen [5] Tendenzen der Typografie [6] Typografie und Farbe [7] Typografie und Bild [8] Einige Besonderheiten der wesentlichen Anwendungsbereiche der Typografie [9] Schlußbemerkungen [10]

Gestalt und Funktion der Typografie

Albert Kapr / Walter Schiller

Albert Kapr / Walter Schiller

Gestalt und Funktion der Typografie

Schiller 1977 Cover

la HGB de 1968 à 1973. Dirige l'Institut für Buchgestaltung (institut de design du livre) en 1981. – *Publication:* «Gestalt und Funktion der Typografie» (avec Albert Kapr), Leipzig 1977.

**Schmid,** Helmut – né le 1. 2. 1942 à Ferlach, Autriche – *typographe* – apprentissage de composition typographique en Allemagne, études de typographie à la Allgemeine Gewerbeschule (école des arts et métiers) de Bâle, avec Emil Ruder. De 1964 à 1972, il conçoit 44 couvertures typographiées pour la revue suédoise «Grafisk Revy». De 1967 à 1970, il dessine Katakana Eru, une police syllabique japonaise (en l'honneur d'Emil Ruder). Maquettes d'imprimés pour le Bundespresseamt de Bonn et pour le Parti social démocrate allemand (SPD) de 1973 à 1976. Vit et travaille au Japon depuis 1977. Commanditaires: Taiho Pharmaceutical et Sanyo Electric. En 1978, il expose ses «polittypographien» créées entre 1974 et 1977 à la galerie Pieter Brattinga, à Amsterdam. Dessin de logos pour des produits Shiseido Cosmetics de 1983 à 1986. Nombreux essais sur la typographie dans des revues internationales telles que «Typographische Monatsblätter», «Grafisk Revy», «Idea», «Novum». Publication à partir de 1992 d'une série de brochures «Typographic Reflections». – *Police:* Katakana Eru (1971). – *Publication:* «Typography today», Tokyo 1980.

**Schmidt,** Hans – né le 14. 1. 1923 à Leipzig, Allemagne – *typographe, graphiste,*

lecturer at the HGB. Made professor there in 1964. 1968: awarded the city of Leipzig's Gutenberg Prize and in 1976 the National Prize of the GDR. 1968–73: head of the HGB's department of book design. 1981: head of the institute for book design. – *Publication:* "Gestalt und Funktion der Typografie" (with Albert Kapr), Leipzig 1977.

**Schmid,** Helmut – b. 1. 2. 1942 in Ferlach, Austria – *typographer* – Trained as a typesetter in Germany and studied typography at the Allgemeine Gewerbeschule in Basle under Emil Ruder. 1964–72: designs 44 typographic covers for a Swedish specialist journal, "Grafisk Revy". 1967–70: designs Katakana Eru, a Japanese syllabary, in honor of Emil Ruder. 1973–76: designs stationery for the German government's press and information center in Bonn and for the Germany's Social Democratic Party (SPD). Since 1977 he has lived and worked in Japan. Clients include Taiho Pharmaceutical and Sanyo Electric. 1978: exhibition of his "polit-typographies", created between 1974–77, at the Print Gallery Pieter Brattinga in Amsterdam. 1983–86: designs logotype for Shiseido Cosmetics products. Has written numerous essays on typography for international magazines such as "Typographische Monatsblätter", "Grafisk Revy", "Idea" and "Novum". Since 1992: publication of his "Typographic Reflections" series of brochures. – *Font:* Katakana Eru (1971). – *Publication:* "Typography today" Tokyo 1980.

**Schmidt,** Hans – b. 14. 1. 1923 in Leipzig, Germany – *typographer, lettering artist,*

*grafiker, Künstler, Lehrer* – 1947–51 Studium an der Hochschule für Graphik und Buchkunst (HGB) in Leipzig. 1951–63 Typograph an der Eggebrecht-Presse in Mainz. 1963–83 Dozent und Professor für Schrift an der Werkkunstschule, Hochschule für Gestaltung in Offenbach. Aufstellung von Schrift-Plastiken im öffentlichen Raum in Bremen (1983), Köln (1984), Oberwesel (1996). Zahlreiche Ausstellungen, u. a. im Klingspor Museum Offenbach (1959, 1970, 1988), im Carnegie Institute in Pittsburgh (USA) und im Gutenberg-Museum in Mainz (1963). – *Publikationen u. a.:* Paul Raabe u. a. „Schrift von Hans Schmidt", Wolfenbüttel 1987; Nicola Leffelsend „Schriftliches von Hans Schmidt", Bergisch Gladbach 1992.

**Schmidt,** Johan Michael – geb. ca. 1700 in Frankfurt am Main, Deutschland, gest. 1750 in Berlin, Deutschland. – *Schriftschneider, Schriftgießer* – Lehre in der Lutherischen Schriftgießerei in Frankfurt am Main. Seit 1728 in Den Haag und Berlin tätig. 1742–50 Leiter der Königlich-Preußischen Schriftgießerei in Berlin.

**Schmidt,** Joost – geb. 5. 1. 1893 in Wunstorf, Deutschland, gest. 2. 12. 1948 in Nürnberg, Deutschland. – *Maler, Typograph, Lehrer* – 1910–14 Studium der Malerei an der Hochschule für Bildende Künste in Weimar. 1919–25 Studium der Holzbildhauerei und Typographie am Bauhaus. 1922–23 erste typographische Arbeiten, Firmendrucksachen und Reklame. 1925–32 Lehrer am Bauhaus Dessau. 1928–30 Leiter der plastischen Werkstatt am Bauhaus. 1928–32 Leiter der Re-

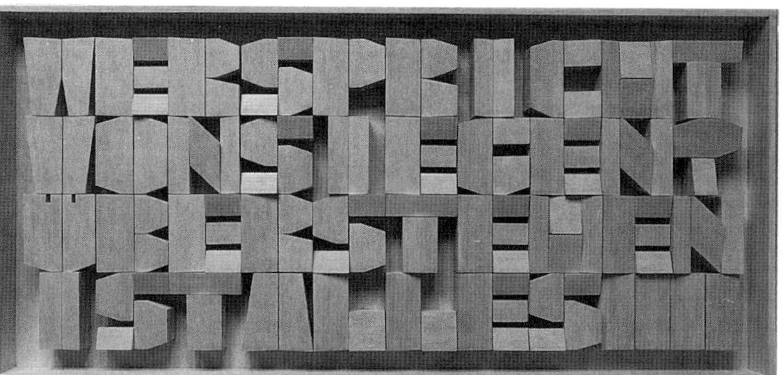
Hans Schmidt 1992 Relief

Hans Schmidt 1971 Page

Joost Schmidt 1926 Cover

Joost Schmidt 1932 Advertisement

*artiste, enseignant* – 1947–1951, études à la Hochschule für Graphik und Buchkunst (école supérieure des arts graphiques et des arts du livre) (HGB) à Leipzig. Typographe chez Eggebrecht-Presse à Mayence de 1951 à 1963. Enseignant puis professeur d'arts de l'écriture à la Werkkunstschule, Hochschule für Gestaltung (école des arts appliqués et de design) d'Offenbach de 1963 à 1983. Expose des lettres-sculptures en espace public à Brême (1983), Cologne (1984), Oberwesel (1996). Nombreuses expositions, entre autres, au Klingspor Museum d'Offenbach (1959, 1970, 1988), au Carnegie Institute de Pittsburgh (Etats-Unis), au Gutenberg Museum de Mayence (1963). – *Publications, sélection:* Paul Raabe et autres «Schrift von Hans Schmidt», Wolfenbüttel 1987; Nicola Leffelsend «Schriftliches von Hans Schmidt», Bergisch Gladbach 1992.

**Schmidt,** Johan Michael – né vers 1700 à Francfort-sur-le-Main, Allemagne, décédé en 1750 à Berlin, Allemagne – *tailleur de types, fondeur de caractères* – apprentissage à la fonderie de caractères Luther, à Francfort-sur-le-Main. Exerce à La Haye et à Berlin à partir de 1728. Dirige la Königlich-Preussische Schriftgiesserei (fonderie royale de Prusse) à Berlin de 1742 à 1750.

**Schmidt,** Joost – né le 5. 1. 1893 à Wunstorf, Allemagne, décédé le 2. 12. 1948 à Nuremberg, Allemagne – *peintre, typographe, enseignant* – 1910–1914, études de peinture à la Hochschule für Bildende Künste (école supérieure des beaux-arts) de Weimar. 1919–1925, études de sculpture sur bois et de typographie au Bauhaus. Premiers travaux typographiques, imprimés de sociétés et publicités entre 1922 et 1923. Enseigne au Bauhaus de Dessau de 1925 à 1932. Dirige l'atelier de formes du Bauhaus de 1928 à 1930. Di-

*artist, teacher* – 1947–51: studies at the Hochschule für Graphik und Buchkunst (HGB) in Leipzig. 1951–63: typographer for the Eggebrecht press in Mainz. 1963–83: lecturer and professor of type at the Werkkunstschule, Hochschule für Gestaltung in Offenbach. His three-dimensional typographic constructions are displayed in public places in Bremen (1983), Cologne (1984) and Oberwesel (1996). Numerous exhibitions, including at the Klingspor Museum in Offenbach (1959, 1970 and 1988), at the Carnegie Institute in Pittsburgh (USA) and at the Gutenberg Museum in Mainz (1963). – *Publications include:* Paul Raabe et al "Schrift von Hans Schmidt", Wolfenbüttel 1987; Nicola Leffelsend "Schriftliches von Hans Schmidt", Bergisch Gladbach 1992.

**Schmidt,** Johan Michael – b. c.1700 in Frankfurt am Main, Germany, d. 1750 in Berlin, Germany – *type cutter, type founder* – Trained at the Lutherische type foundry in Frankfurt am Main. From 1728 onwards: works in The Hague and Berlin. 1742–50: head of the Königlich-Preußische Schriftgießerei in Berlin.

**Schmidt,** Joost – b. 5. 1. 1893 in Wunstorf, Germany, d. 2. 12. 1948 in Nuremberg, Germany – *painter, typographer, teacher* – 1910–14: studies painting at the Hochschule für Bildende Künste in Weimar. 1919–25: studies wood carving and typography at the Bauhaus. 1922–23: produces his first typographic work, printed matter for companies and advertisements. 1925–32: teaches at the Bauhaus in Dessau. 1928–30: head of the

klame-Abteilung und der Druckerei des Bauhauses in Dessau. 1932 Gestaltung des Katalogs für die Bauhaus-Tapeten der Firma Gebr. Rasch in Bramsche. Schriftentwurf und Standgestaltung für Uhertype auf der Leipziger Messe. Nach der Schließung des Bauhauses 1933 in Berlin tätig. Landkartenzeichner. 1934 mit Walter Gropius Gestaltung einer Abteilung der Ausstellung „Deutsches Volk, Deutsche Arbeit" in Berlin. 1935–36 Lehrer an der Privatschule „Kunst und Werk" von Hugo Häring (früher Reimann-Schule). Lehrverbot. 1945 Berufung als Professor an die Hochschule für Bildende Künste in Berlin. 1946 Gestaltung der Ausstellung „Berlin plant". – *Publikation u. a.:* „Joost Schmidt: Lehre und Arbeit am Bauhaus", Düsseldorf 1984.

**Schmidt,** Klaus F. – geb. 25. 5. 1928 in Dessau, Deutschland. – *Grafik-Designer, Typograph* – 1947–48 Journalismus-Studium in Aachen, Deutschland. 1948–50 Schriftsetzerlehre. 1951–56 Schriftsetzer in Detroit und New York. 1952–56 Studium der Betriebswirtschaft in Detroit. 1961 Mitbegründer des „International Center for the Typographic Arts" in New York. Seit 1961 in verschiedenen Positionen in der Agentur Young & Rubicam in New York tätig. Seit 1962 zahlreiche Auszeichnungen durch den Type Directors Club of New York. Veröffentlicht zahlreiche Artikel in amerikanischen und deutschen Fachzeitschriften.

**Schmidt,** Wolfgang – geb. 24. 7. 1929 in Fulda, Deutschland, gest. 8. 3. 1995 in Witzenhausen, Deutschland. – *Grafik-Designer, Typograph, Verleger, Künstler,*

Joost Schmidt 1923 Poster

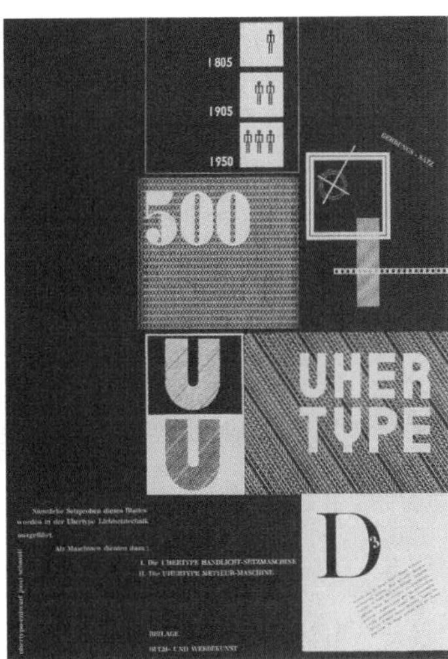


Joost Schmidt 1932 Advertisement

IT IS ALTOGETHER FITTING
that the symbol of our friendship and our appreciation
should be Abraham Lincoln.
For Lincoln and Larmon have much in common.

Lincoln was a worthy leader.
The success and stature of Young & Rubicam stand
as tribute to your twenty years of leadership.

Lincoln was thoughtful of others. The happiness
and many benefits that Young & Rubicam offers its
people are evidence of your similar concern.

Lincoln said: "I do the very best I know how –
the very best I can." You have lived by that principle,
and we are the beneficiaries.

For all of Young & Rubicam, and with the
personal gratitude of each of us, we say sincerely
and very simply, thank you.

DECEMBER 3, 1962

Klaus F. Schmidt 1962 Advertisement

recteur de l'atelier de publicité et d'imprimerie du Bauhaus de 1928 à 1932. Réalise la maquette du catalogue des papiers peints du Bauhaus édité par la société Gebr. Rasch de Bramsche, en 1932. Conception de polices et architecture d'un stand pour Uhertype à la foire de Leipzig. Exerce à Berlin après la fermeture du Bauhaus, en 1933 Cartographe. Avec Walter Gropius, il conçoit l'une des sections de l'exposition «Deutsches Volk, Deutsche Arbeit» à Berlin, en 1934. Enseigne à l'école privée «Kunst und Werk» de Hugo Häring (ancienne Reimann-Schule) de 1935 à 1936. Interdiction d'enseigner. Nommé professeur à la Hochschule für Bildende Künste (école supérieure des beaux-arts) de Berlin, en 1945. Architecture de l'exposition «Berlin plant» en 1946. – *Publication, sélection:* «Joost Schmidt: Lehre und Arbeit am Bauhaus», Düsseldorf 1984.

**Schmidt,** Klaus F. – né le 25. 5. 1928 à Dessau, Allemagne – *graphiste maquettiste, typographe* – 1947–1948, études de journalisme à Aix-la-Chapelle, Allemagne. 1948–1950, apprentissage de composition typographique. Exerce comme compositeur à Detroit et à New York de 1951 à 1956. Etudes d'économie et de gestion à Detroit, de 1952 à 1956. Cofondateur, en 1961, de l'«International Center for the Typographic Arts», à New York. A partir de 1961, il exerce plusieurs fonctions à l'agence «Young & Rubicam» à New York. Nombreuses distinctions du Type Directors Club of New York depuis 1962. A publié de nombreux articles dans des revues spécialisées américaines et allemandes.

**Schmidt,** Wolfgang – né le 24 7 1929 à Fulda, Allemagne, décédé le 8. 3. 1995 à Witzenhausen, Allemagne – *graphiste maquettiste, typographe, éditeur, artiste,*

Bauhaus sculpture workshop. 1928–32: head of the advertising department and printing workshop at the Bauhaus in Dessau. 1932: designs a catalogue of Bauhaus wallpaper for the Gebr. Rasch company in Bramsche. Designs a typeface and exhibition stand for Uhertype at the Leipzig fair. 1933: works in Berlin after the Bauhaus is closed. Cartographer. 1934: designs one of the sectors of the "Deutsches Volk, Deutsche Arbeit" exhibition in Berlin with Walter Gropius. 1935–36: teacher at Hugo Häring's private Kunst und Werk art school (formerly the Reimann school). Then forbidden to teach. 1945: takes up a post as professor at the Hochschule für Bildende Künste in Berlin. 1946: designs the "Berlin plant" exhibition. – *Publications include:* "Joost Schmidt: Lehre und Arbeit am Bauhaus", Düsseldorf 1984.

**Schmidt,** Klaus F. – b. 25. 5. 1928 in Dessau, Germany – *graphic designer, typographer* – 1947–48: studies journalism in Aachen, Germany. 1948–50: trains as a typesetter. 1951–56: works as a typesetter in Detroit and New York. 1952–56: studies business administration in Detroit. 1961: co-founder of the International Center for the Typographic Arts in New York. From 1961 onwards: holds various positions at the Young & Rubicam agency in New York. From 1962 onwards: receives various awards from the Type Directors Club of New York. Has published numerous articles in American and German specialist journals.

**Schmidt,** Wolfgang – b. 24. 7. 1929 in Fulda, Germany, d. 8. 3. 1995 in Witzen-

*Lehrer* – 1950–54 Studium an der Staatlichen Akademie der Bildenden Künste in Stuttgart und an der Werkakademie in Kassel. 1958 Arbeit als Art Director der Zeitschrift „Mobilia" in Kopenhagen. Seit 1959 freischaffend in Frankfurt am Main tätig. Auftraggeber waren u. a. Neue Filmkunst Walter Kirchner, Nils Wiese Vitsoe, Buchhandlung Wendelin Niedlich und Behr Möbel. Mitglied der Gruppe „novumgraphik". 1960 Teilnahme an der Ausstellung „Schrift und Bild" in Amsterdam und Baden-Baden, dazu Gestaltung von Katalog und Plakat. 1957–84 entstehen 21 „Bücher" in Auflagen zwischen 5–100 Exemplaren. 1964 Teilnahme an der documenta III in Kassel. 1977–86 entstehen „Hefte" Nr. 1–4 mit den Titeln „Lebenszeichen Hören", „Mein Horizont", „Drunter und Drüber", „Wolken". Als Lehrer an den Hochschulen in Reykjavik, Island (1957), Ulm (1966, 1968), Bath, England (1969), Offenbach (1970), Kassel (1971–72, 1974), Hamburg (1975) und Berlin (1986–87) tätig. Schmidt unterscheidet bei seiner Arbeit die Bereiche „Visuelle Gestaltung zweckvoll (Design) und zweckfrei (Bild- und Textsysteme)". – *Publikationen u. a.:* Dietrich Mahlow „Arbeiten von Wolfgang Schmidt", Frankfurt 1981; Anke Jaaks „Das doppelte Schmidt-Buch", Mainz 1992.

**Schmidt Rhen,** Helmut – geb. 7. 9. 1936 in Köln, Deutschland. – *Grafik-Designer, Typograph, Lehrer* – 1957–60 Studium an der Staatlichen Werkakademie in Kassel. 1960 Umzug nach Mainz, Arbeit als freier Grafik-Designer, u. a. für den Stadtplaner Ernst May und das Städtische Museum

*enseignant* – 1950–1954, études à la Staatliche Akademie der Bildenden Künste (beaux-arts) de Stuttgart et à la Werkakademie (arts et métiers) de Kassel. Directeur artistique de la revue «Mobilia» à Copenhague, en 1958. Exerce à Francfort-sur-le-Main comme indépendant à partir de 1959. Commanditaires : Neue Filmkunst Walter Kirchner, Nils Wiese Vitsoe, Librairie Wendelin Niedlich, Behr Möbel. Membre du groupe «novumgraphik». Participe à l'exposition «Schrift und Bild» à Amsterdam et à Baden-Baden en 1960, il réalise la maquette du catalogue et l'affiche de l'exposition. De 1957 à 1984, il crée 21 «livres» tirés entre 5 et 100 exemplaires. Participe à la «documenta III» à Kassel en 1964. Réalise les «Hefte» (cahiers) n° 1–4, intitulés «Lebenszeichen Hören», «Mein Horizont», «Drunter und Drüber» et «Wolken» entre 1977 et 1986. Enseigne dans plusieurs écoles d'art, à Reykjavik, Islande (1957), à Ulm, Allemagne (1966, 1968), à Bath, Angleterre (1969), à Offenbach (1970), à Kassel (1971–1972, 1974), à Hambourg (1975) et à Berlin (1986–1987). Dans son travail, Schmidt fait une distinction entre la «Création visuelle fonctionnelle (design) et gratuite (système image et texte)». – *Publications, sélection :* Dietrich Mahlow «Arbeiten von Wolfgang Schmidt», Francfort 1981; Anke Jaaks «Das doppelte Schmidt-Buch», Mayence 1992.

**Schmidt Rhen**, Helmut – né le 7. 9. 1936 à Cologne, Allemagne – *graphiste maquettiste, typographe, enseignant* – 1957–1960, études à la Staatliche Werkakademie (arts et métiers) de Kassel. S'installe à Mayence en 1960 où il travaille comme graphiste et designer indépendant pour l'urbaniste Ernst May et le Städtische Museum de Leverkusen. S'installe à Bâle, en Suisse en 1961; employé

Wolfgang Schmidt 1963 Poster

Wolfgang Schmidt 1966 Poster

Leo Trotzkij
literatur
und
revolution

Wolfgang Schmidt 1968 Cover

Wolfgang Schmidt Cover

hausen, Germany – *graphic designer, typographer, publisher, artist, teacher* – 1950–54: studies at the Staatliche Akademie der Bildenden Künste in Stuttgart and at the Werkakademie in Kassel. 1958: works as art director for "Mobilia" magazine in Copenhagen. 1959: starts freelancing in Frankfurt am Main. Clients include Neue Filmkunst Walter Kirchner, Nils Wiese Vitsoe, Wendelin Niedlich bookstore and Behr Möbel. Member of the novumgraphik group. 1960: takes part in the "Schrift und Bild" exhibitions in Amsterdam and Baden-Baden, for which he designs the catalogue and poster. 1957–84: he produces 21 "books" in editions of between 5 and 100 copies. 1964: takes part in documenta III in Kassel. 1977–86: issues his "Hefte" nos. 1–4, entitled "Lebenszeichen Hören", "Mein Horizont", "Drunter und Drüber" and "Wolken". He taught at various colleges in Reykjavik (1957), Ulm (1966, 1968), Bath (1969), Offenbach (1970), Kassel (1971–72, 1974), Hamburg (1975) and Berlin (1986–87). In his work Schmidt differentiates between "effective" (design) and "pure" (picture and text systems) visual design. – *Publications include:* Dietrich Mahlow "Arbeiten von Wolfgang Schmidt", Frankfurt 1981; Anke Jaaks "Das doppelte Schmidt-Buch", Mainz 1992.

**Schmidt Rhen,** Helmut – b. 7. 9. 1936 in Cologne, Germany – *graphic designer, typographer, teacher* – 1957–60: studies at the Staatliche Werkakademie in Kassel. 1960: moves to Mainz and works as a freelance graphic designer for clients

Leverkusen. 1961 Umzug nach Basel, Schweiz, Arbeit in der Werbeagentur Gerstner + Kutter. Daneben freie künstlerische Arbeit und grafische Aufträge. 1963 Teilnahme an der Ausstellung „Schrift und Bild" in Amsterdam und Baden-Baden. 1965 Umzug nach Köln. 1967–68 Art Director der Zeitschrift „Capital". 1968–73 Dozent für Grafik-Design an der Werkkunstschule Düsseldorf. Gestalterische Arbeiten für Auftraggeber wie den Kunstmarkt Köln, die Bank für Gemeinwirtschaft, die Stadt Köln und den Kunstverein Hannover. 1977–93 Professor an der Fachhochschule in Düsseldorf. 1979–87 Herausgabe der Zeitschrift „Oetz" an der Fachhochschule in Düsseldorf. 1984 Mitbegründer des „Forums Typographie". 1993 Umzug nach Hamburg, Beschäftigung mit konstruktiv-serieller Malerei. Zahlreiche Auszeichnungen und Veröffentlichungen. Ausstellungen seiner freien und angewandten Gestaltung. – *Publikation:* Roland Henss (Hrsg.) „Poesie der Systematik. Design: Schmidt Rhen", Mainz 1996.

**Schmoller,** Hans – geb. 9. 4. 1916 in Berlin, Deutschland, gest. 25. 9. 1985 in London, England. – *Typograph* – 1933–37 Schriftsetzerlehre in Berlin. 1935–36 Studium der Schrift und Kalligraphie an der Höheren grafischen Fachschule in Berlin. 1938 Emigration nach England. 1938–46 Tätigkeit bei Morija Printing Works in Lesotho, Südafrika. 1947–49 Assistent Oliver Simons an der Curwen Press in London. 1949–80 verschiedene Positionen im Verlag Penguin Books (als Nachfolger Jan Tschicholds): 1949–56 Typograph,

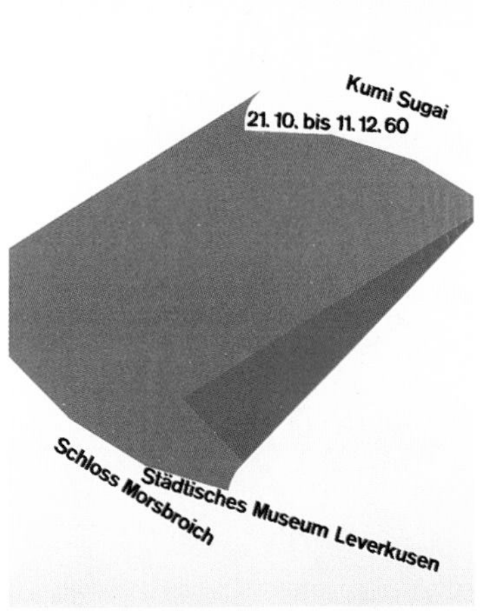

Schmidt Rhen 1960 Poster

Schmidt Rhen 1961 Poster

Schmidt Rhen 1973 Cover

Schmidt Rhen 1973 Poster

à l'agence de publicité Gerstner + Kutter; parallèlement, il réalise des œuvres plastiques et des commandes graphiques. En 1963, il participe à l'exposition «Schrift und Bild» à Amsterdam et à Baden-Baden. S'installe à Cologne en 1965. Directeur artistique de la revue «Capital» de 1967 à 1968. Enseigne les arts graphiques et le design à la Werkkunstschule (école des arts appliqués) de Düsseldorf de 1968 à 1973. Travaux pour le Kunstmarkt de Cologne, la Bank für Gemeinwirtschaft, la ville de Cologne, le Kunstverein de Hanovre. Professeur à la Fachhochschule de Düsseldorf de 1977 à 1993. Publie la revue «Oetz» de 1979 à 1987 à la Fachhochschule de Düsseldorf. Cofondateur de «Forum Typographie» en 1984. S'installe à Hambourg en 1993 et se consacre à une peinture sérielle d'inspiration constructiviste. Nombreuses distinctions et publications. Expositions de ses travaux libres et appliqués. – *Publication :* Roland Henss (éd.) «Poesie der Systematik Design : Schmidt Rhen», Mayence 1996.

**Schmoller,** Hans – né le 9. 4. 1916 à Berlin, Allemagne, décédé le 25. 9. 1985 à Londres, Angleterre – *typographe* – 1933–1937, apprentissage de composition typographique à Berlin. 1935–1936, études d'arts de l'écriture et de calligraphie à l'école supérieure des arts graphiques de Berlin. Emigre en Angleterre en 1938. Employé chez Morija Printing Works au Lesotho, Afrique du Sud, de 1938 à 1946. Assistant d'Oliver Simon aux Curwen Press à Londres de 1947 à 1949. Exerce plusieurs fonctions aux éditions Penguin Books entre 1949 et 1980

who include city planner Ernst May and the Städtische Museum in Leverkusen. 1961: moves to Basle in Switzerland and works for the Gerstner + Kutter advertising agency and as a freelance artist and graphic artist. 1963: takes part in the "Schrift und Bild" exhibition in Amsterdam and Baden-Baden. 1965: moves to Cologne. 1967–68: art director of "Capital" magazine. 1968–73: lecturer of graphic design at the Werkkunstschule in Düsseldorf. Produces designs for clients such as Kunstmarkt Köln, the Bank für Gemeinwirtschaft, the city of Cologne and the Kunstverein in Hanover. 1977–93: professor at the Fachhochschule in Düsseldorf. 1979–87: editor of "Oetz" magazine at the Fachhochschule in Düsseldorf. 1984: co-founder of Forum Typographie. 1993: moves to Hamburg where he concentrates on serial, constructive painting. Numerous awards, publications and exhibitions of his free and applied design. – *Publication:* Roland Henss (ed.) "Poesie der Systematik. Design: Schmidt Rhen", Mainz 1996.

**Schmoller,** Hans – b. 9. 4. 1916 in Berlin, Germany, d. 25. 9. 1985 in London, England – *typographer* – 1933–37: trains as a typesetter in Berlin. 1935–36: studies lettering and calligraphy at the Höhere grafische Fachschule in Berlin. 1938: emigrates to England. 1938–46: works for the Morija Printing Works in Lesotho, South Africa. 1947–49: assistant to Oliver Simons at Curwen Press in London. 1949–80: holds various positions at Penguin Books as Jan Tschichold's successor.

1956–76 Herstellungsleiter, 1960–76 Direktor, 1976–80 Berater. 1955 Mitarbeit am internationalen Bibelprojekt „Liber Librorum". 1968–69 Präsident des Double Crown Club in London. 1969 Gestaltung der Ausgabe „The complete Pelican Shakespeare". 1971 mit einer Goldmedaille auf der Buchmesse in Leipzig für die Shakespeare-Ausgabe, 1976 mit dem Titel „Royal Designer for Industry" ausgezeichnet. 1979–85 Vorstandsmitglied der „Royal Society of Arts". – *Publikationen:* „Giovanni Mardersteig: die Officina Bodoni 1923–77", Hamburg 1979; „Mr. Gladstone's Washi", Newtown 1983.

**Schneider,** Werner – geb. 7. 4. 1935 in Marburg, Deutschland. – *Typograph, Schriftentwerfer, Kalligraph, Lehrer* – 1954–58 Studium, 1959–73 Dozent, seit 1973 Professor für Schrift und Typographie an der Werkkunstschule in Wiesbaden. Freiberuflich als Schriftgestalter tätig. 1988 Lehrauftrag für Paläographie an der Philipps-Universität in Marburg. Zahlreiche Einzelausstellungen seiner schriftgrafischen Arbeit, u. a. in London (1986), Bombay (1987), Wiesbaden (1988), Marburg (1990) und Queensland, Australien (1996). Zahlreiche Auszeichnungen, u. a. Rudo-Spemann-Preis (1957), Type Directors Club of New York (1986, 1989, 1993), Biennale für Grafik-Design in Brno, CSSR (1988, 1992). – *Schriftentwürfe:* Medita (1979), Sublima (1981), Schneider-Antiqua (1987), Schneider Libretto (1995).

**Schneidler,** F. H. Ernst – geb. 14. 2. 1882 in Berlin, Deutschland, gest. 6. 1. 1956 in Gundelfingen, Deutschland – *Typograph,*

(succède à Jan Tschichold) : 1949–1956, typographe, 1956–1976, directeur de fabrication, 1960–1976, directeur, 1976–1980, conseiller. Collaboration, en 1955, au projet international de publication de la Bible «Liber Librorum». Président du Double Crown Club à Londres de 1968 à 1969. Réalise la maquette de l'édition «The complete Pelican Shakespeare», en 1969. Médaille d'or du salon du livre de Leipzig, en 1971, pour son édition de Shakespeare, «Royal designer for Industry» en 1976. Membre du comité directeur de la «Royal Society of Arts» de 1979 à 1985. – *Publications :* «Giovanni Mardersteig : die Officina Bodoni 1923–77», Hambourg 1979; «Mr. Gladstone's Washi», Newtown 1983.

**Schneider,** Werner – né le 7. 4. 1935 à Marbourg, Allemagne – *typographe, concepteur de polices, calligraphe, enseignant* – 1954–1958, études à la Werkkunstschule (école des arts appliqués) de Wiesbaden, y enseigne de 1959 à 1973, professeur titulaire de la chaire d'arts de l'écriture et de typographie à partir de 1973. Concepteur de polices indépendant. Chargé de cours de paléographie à la Philipps-Universität de Marbourg en 1988. Nombreuses expositions personnelles de ses travaux graphiques, entre autres à Londres (1986), Bombay (1987), Wiesbaden (1988), Marbourg (1990), Queensland, Australie (1996). Nombreuses distinctions, dont le Prix Rudo Spemann (1957), prix du Type Directors Club of New York (1986, 1989, 1993), prix de la Biennale d'arts graphiques et de design de Brno, Tchécoslovaquie (1988, 1992). – *Polices :* Medita (1979), Sublima (1981), Schneider-Antiqua (1987), Schneider Libretto (1995).

**Schneidler,** F. H. Ernst. – né le 14. 2. 1882 à Berlin, Allemagne, décédé le 6. 1. 1956 à Gundelfingen, Allemagne – *typo-*

Schmoller 1948 Cover

Schmoller 1951 Label

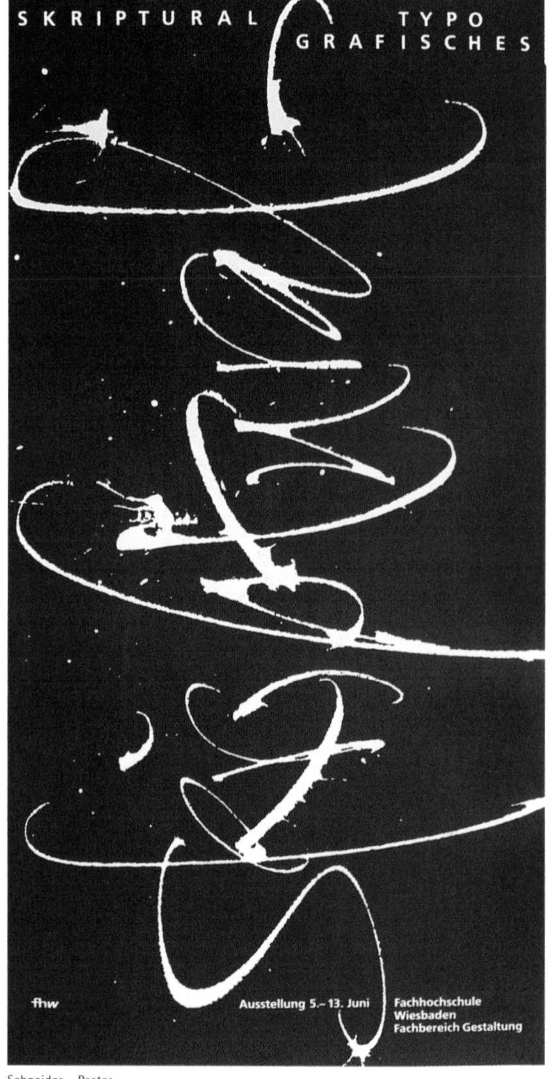


Schneider Poster

From 1949–56 he is typographer, from 1956–76 production manager, from 1960–76 director, and from 1976–80 consultant to Penguin. 1955: works on the international Liber Librorum Bible project. 1968–69: president of the Double Crown Club in London. 1969: designs "The Complete Pelican Shakespeare". 1971: is awarded a gold medal at the Leipzig book fair for his edition of Shakespeare's works. 1976: made a Royal Designer for Industry. 1979–85: one of the board members of the Royal Society of Arts. – *Publications:* "Giovanni Mardersteig: die Officina Bodoni 1923–77", Hamburg 1979; "Mr. Gladstone's Washi", Newtown 1983.

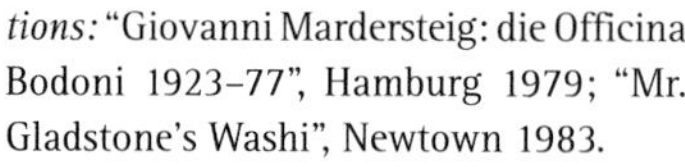

**Schneider,** Werner – b. 7. 4. 1935 in Marburg, Germany – *typographer, type designer, calligrapher, teacher* – 1954–58: studies at the Werkkunstschule in Wiesbaden. 1959–73: lectures at the Werkkunstschule and from 1973 onwards is professor of lettering and typography there. Freelance type designer. 1988: appointed to teach paleography at Philipps University in Marburg. There have been numerous solo exhibitions of his typographic work in London (1986), Bombay (1987), Wiesbaden (1988), Marburg (1990) and Queensland, Australia (1996), among other places. Schneider has won numerous awards, including the Rudo Spemann Prize (1957), awards from the Type Directors Club of New York (1986, 1989, 1993) and at the Biennale for Graphic Design in Brno, Czechoslovakia (1988, 1992). – *Fonts:* Medita (1979), Sublima (1981), Schneider-Antiqua (1987), Schneider Libretto (1995).

*Kalligraph, Schriftentwerfer, Illustrator, Maler, Lehrer* – 1902–03 Studium der Architektur an der Technischen Hochschule in Berlin-Charlottenburg. 1904–05 Studium an der Kunstgewerbeschule in Düsseldorf bei Peter Behrens und F. H. Ehmcke. 1905–06 Lehrer an der Fachschule in Solingen. 1907–20 Professor und Leiter der Graphischen Fachschule an der Kunstgewerbeschule in Barmen. Seit 1912 Buchgestaltung für den Eugen Diederichs - Verlag in Jena. 1920 Berufung an die Württembergische Kunstgewerbeschule in Stuttgart. 1921–25 Gründung und Leitung der Juniperus-Presse an der Akademie in Stuttgart. 1925 Beginn mit den Arbeiten zu „Der Wassermann", einer Sammlung von Studienblättern für Büchermacher in einer Auflage von 70 Exemplaren. Es enthält auf 700 Seiten Buchtitel, Briefköpfe, Ornamente, Signets etc. 1952 Ernennung zum Ehrenmitglied der Staatlichen Akademie der Künste in Stuttgart. 1953 Ausstellung seines Werks in New York, 1957 in Stuttgart. Schneidlers typographisches und pädagogisches Wirken trägt dazu bei, von einer „Stuttgarter Schule der Typographie" zu sprechen. Er bildet zahlreiche Typographen aus, u. a. Eva Aschoff, Walter Brudi, Carl Keidel, Imre Reiner, Rudo Spemann. – *Schriftentwürfe:* Schneidler Schwabacher (1913), Schneidler Fraktur (1914), Schneidler Latein (1914–19), Deutsch-Römisch (1923), Kontrast (1930), Graphik (1934), Schneidler Mediaeval (1936), Legende (1937), Zentenar Fraktur (1937), Amalthea (1956). – *Publikationen u. a.:* „Der Wassermann. Studienblätter für Bücher-

WALLENSTEIN
ZWEITER
TEIL
PICCOLOMINI
DRAMATISCHE
DICHTUNG
IN FÜNF AUFZÜGEN
VON
SCHILLER

Schneidler 1945 Page

Graphik
ABCDEFGH
IKLMNOPQRS
TUVWXYZ
abcdefghijklmnopq
rstuvwxyz
äöüchckffllßtz&.,:;!?
1234567890
Frisch begonnen,
halb gewonnen

Schneidler 1934 Graphik

*graphe, calligraphe, concepteur de polices, illustrateur, peintre, enseignant* – 1902–1903, études d'architecture à la Technische Hochschule (université technique) de Berlin-Charlottenburg. 1904–1905, études à la Kunstgewerbeschule (école des arts décoratifs) de Düsseldorf, chez Peter Behrens et F. H. Ehmcke. 1905–1906, enseigne à l'institut technique de Solingen. 1907–1920, professeur et directeur de l'institut des arts graphiques à la Kunstgewerbeschule (école des arts décoratifs) de Barmen. Maquettes de livres pour les éditions Eugen-Diederichs, à Iéna, à partir de 1912. Nommé à la Württembergische Kunstgewerbeschule (école des arts décoratifs) à Stuttgart, en 1920. Fonde la «Juniperus-Presse» à l'académie de Stuttgart en 1921, il dirigera la presse jusqu'en 1925. Début des travaux sur le «Wassermann» (Verseau), en 1925; réalise une série de planches d'études tirées à 70 exemplaires pour les maquettistes en édition. Cet ouvrage contient 700 pages et présente des titres de livres, des en-têtes de papier à lettre, des ornements, des signets etc. Nommé membre d'honneur de l'académie des beaux-arts de Stuttgart en 1952. Exposition de ses œuvres à New York, en 1953, et à Stuttgart en 1957. L'œuvre typographique et pédagogique de Schneidler a contribué à la constitution d'une «école typographique de Stuttgart». Il a formé de nombreux typographes, comme Eva Aschoff, Walter Brudi, Carl Keidel, Imre Reiner, Rudo Spemann. – *Polices:* Schneidler Schwabacher (1913), Schneidler Fraktur (1914), Schneidler Latein (1914–1919), Deutsch-Römisch (1923), Kontrast (1930), Graphik (1934), Schneidler Mediaeval (1936), Legende (1937), Zentenar Fraktur (1937), Almathea (1956). – *Publications, sélection:* «Der

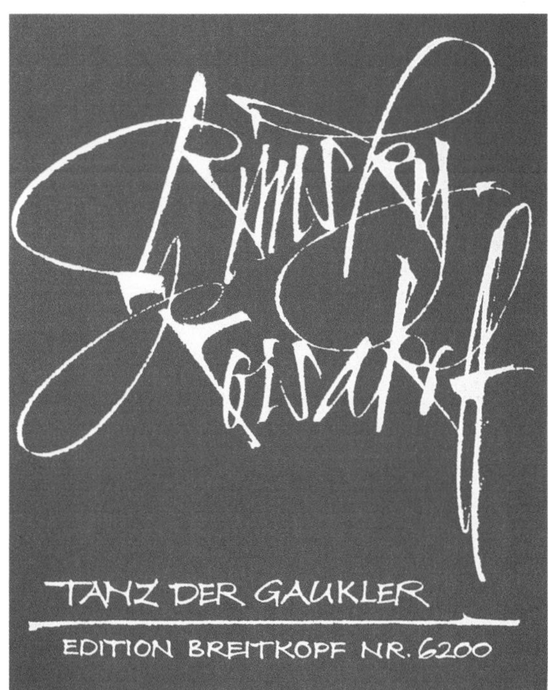


Schneider 1961 Cover

| | n | **n** | | | *n* |
|---|---|---|---|---|---|
| a | b | e | f | g | i |
| o | r | s | t | y | z |
| A | B | C | E | G | H |
| M | O | R | S | X | Y |
| 1 | 2 | 4 | 6 | 8 | & |

Schneidler 1936 Schneidler Mediaeval

**Schneidler,** F. H. Ernst – b. 14. 2. 1882 in Berlin, Germany, d. 6. 1. 1956 in Gundelfingen, Germany – *typographer, calligrapher, type designer, illustrator, painter, teacher* – 1902–03: studies architecture at the Technische Hochschule in Berlin-Charlottenburg. 1904–05: studies at the Kunstgewerbeschule in Düsseldorf under Peter Behrens and F. H. Ehmcke. 1905–06: teaches at the Fachschule in Solingen. 1907–20: professor and head of the Graphische Fachschule at the Kunstgewerbeschule in Barmen. 1912: starts designing books for Eugen Diederichs publishing house in Jena. 1920: takes up a teaching post at the Württembergische Kunstgewerbeschule in Stuttgart. 1921–25: founds and runs the Juniperus-Presse at the academy in Stuttgart. 1925: starts working on "Der Wassermann", a collection of study sheets for bookmakers, published in an edition of 70. The 700 pages include covers, headings, ornaments and logos etc. 1952: made an honorary member of the Staatliche Akademie der Künste in Stuttgart. 1953: his life's work is exhibited in New York and in 1957 in Stuttgart. Schneidler's efforts in typography and teaching contributed to what is often called the "Stuttgart school of typography". He trained numerous typographers, including Eva Aschoff, Walter Brudi, Carl Keidel, Imre Reiner and Rudo Spemann. – *Fonts:* Schneidler Schwabacher (1913), Schneidler Fraktur (1914), Schneidler Latein (1914–19), Deutsch-Römisch (1923), Kontrast (1930), Graphik (1934), Schneidler Mediaeval (1936), Legende (1937), Zentenar Fraktur (1937),

macher", Stuttgart 1945; „Briefe", Stuttgart 1968.

**Schöffer,** Peter – geb. ca. 1425 in Gernsheim, Deutschland, gest. ca. 1503 in Mainz, Deutschland. – *Drucker, Schriftgießer* – 1444–48 Studium an der Universität Erfurt. 1449 Kalligraph und Schönschreiber an der Sorbonne in Paris. 1452 in Mainz tätig, Zusammenarbeit mit Peter Fust. 1457 Herausgabe des „Psalterium Moguntinum", 1459 des „Psalterium Benedictum" und des „Rationale Divinorum". 1462 Herausgabe einer 48-zeiligen Bibel in zwei Bänden. 1463 Druck der „Türkenbulle" von Papst Pius II. 1465–66 Herausgabe von Ciceros „De officiis". 1470 Herausgabe einer Bücheranzeige, die 21 Drucke aufzählt. 1485 Herausgabe des Buchs „Gart der Gesundheit". 1502 Druck seines letzten Buchs.

**Schönsperger,** Johann der Ältere – Geburtsdatum unbekannt, gest. 1520 in Augsburg, Deutschland. – *Drucker* – Seit 1481 Druck von über hundert Werken mit zahlreichen Illustrationen in Augsburg, u. a. „Sachsenspiegel" (1482), „Reformation in Nürnberg" (1488) und „Das Narrenschiff" von Sebastian Brant (1494). 1508 ernennt Kaiser Maximilian Schönsperger zu seinem Diener und Hofdrucker auf Lebenszeit. 1513 beendet Schönsperger den Druck des „Gebetsbuchs", das Kaiser Maximilian für den Sankt-Georgen-Orden in Auftrag gegeben hat. Eine Pergamentausgabe wird mit Randzeichnungen von Albrecht Dürer und Lucas Cranach geschmückt.

**Schraivogel,** Ralph – geb 4. 7. 1960 in Luzern, Schweiz. – *Grafik-Designer, Typo-*

Wassermann. Studienblätter für Büchermacher», Stuttgart 1945; «Briefe», Stuttgart 1968.

**Schöffer,** Peter – né vers 1425 à Gernsheim, Allemagne, décédé vers 1503 à Mayence, Allemagne – *imprimeur, fondeur de caractères* – 1444–1448, études à l'université d'Erfurt. 1449, calligraphe et scribe à la Sorbonne à Paris. Exerce à Mayence en 1452 où il travaille avec Peter Fust. En 1457, il publie le «Psalterium Moguntinum», et en 1459 le «Psalterium Benedictum» puis le «Rationale Divinorum». Publication en 1462 d'une Bible à 48 lignes en deux volumes. Imprime la «Bulle turque» du pape Pie II en 1463. Edition de «De officiis» de Cicéron de 1465 à 1466. Publication, en 1470, d'une annonce sur ses ouvrages comportant une liste de 21 impressions. Edition, en 1485, de l'ouvrage «Gart der Gesundheit». Son dernier livre sera imprimé en 1502.

**Schönsperger,** Johann l'Ancien – date de naissance inconnue, décédé en 1520 à Augsbourg, Allemagne – *imprimeur* – a vécu à Augsbourg, où il a imprimé plus de cent ouvrages avec de nombreuses illustrations à partir de 1481, dont le «Sachsenspiegel» (1482), «Reformation in Nürnberg» (1488), «Das Narrenschiff», de Sebastian Brant (1494). En 1508, l'empereur Maximilien appelle Schönsperger à la cour et le nomme Imprimeur de la cour à vie. En 1513, il achève l'impression du «Gebetsbuch» commandé par l'empereur Maximilien pour l'ordre de Saint-Georges. Les marges d'une édition en parchemin sont ornées de dessins d'Albrecht Dürer et de Lucas Cranach.

**Schraivogel,** Ralph – né le 4. 7. 1960 à Lucerne, Suisse – *graphiste maquettiste, typographe* – 1978–1982, études d'arts graphiques et de design à la Schule für Gestaltung (école de design) de Zurich. Exerce ensuite comme designer indépen-

Schöffer 1459 Catholicon-Type

Schöffer 1485 Page

letabuntur in cubilibus suis.
Exultatiões dei in gutture eorum: et gladij ancipites ĩ ma-
nibus eorum. Ad faciendam
vindictam in nationibus: in-
crepatiões in populis. Ad al
ligandos reges eorum in com-
pedibus: et nobiles eorum in
manicis ferreis. Ut faciant in
eis iudicium conscriptum: glo
ria hec est oĩbus sanctis eius.
Laudate dominum in sanct
eius: laudate eũ in firmamen
to virtutis eius. Laudate eum

Schönsperger 1513 Page

Amalthea (1956). – *Publications include:* "Der Wassermann. Studienblätter für Büchermacher", Stuttgart 1945; "Briefe", Stuttgart 1968.

**Schöffer,** Peter – b. c.1425 in Gernsheim, Germany, d. c.1503 in Mainz, Germany – *printer, type founder* – 1444– 48: studies at the university in Erfurt. 1449: calligrapher and copyist at the Sorbonne in Paris. 1452: works in Mainz and works with Peter Fust. 1457: publishes "Psalterium Moguntinum" and in 1459 "Psalterium Benedictum" and "Rationale Divinorum". 1462: publishes a 48-line Bible in two volumes. 1463: prints Pope Pius II's Turkish bull. 1465– 66: publishes Cicero's "De officiis". 1470: publishes an advertisement for books which lists 21 volumes. 1485: publishes "Gart der Gesundheit". 1502: prints his last book.

**Schönsperger,** Johann, the Elder – date of birth unknown, d. 1520 in Augsburg, Germany – *printer* – From 1481 onwards: prints over a hundred works in Augsburg with numerous illustrations, including the "Sachsenspiegel" (1482), "Reformation in Nürnberg" (1488) and "Das Narrenschiff" by Sebastian Brant (1494). 1508: Emperor Maximilian I makes Schönsperger his life-long servant and court printer. 1513: Schönsperger finishes printing the prayer book ("Gebetsbuch") Emperor Maximilian I had commissioned for the Order of St. George. An edition on parchment is decorated with illustrations in the margin by Albrecht Dürer and Lucas Cranach.

**Schraivogel,** Ralph – b. 4. 7. 1960 in Lucerne, Switzerland – *graphic designer,*

*graph.* – 1978–82 Grafik-Design-Studium an der Schule für Gestaltung in Zürich. Danach freischaffender Gestalter im Bereich Kulturwerbung. Es entstehen Plakate, Kataloge und Signets für Auftraggeber wie das Kunsthaus Zürich, das Museum für Gestaltung Zürich, das Internationale Jazz-Festival, das Theater am Neumarkt, das Filmpodium der Stadt Zürich und Pro Helvetia. Buchgestaltung für mehrere Verlage. Seit 1992 unterrichtet er Visuelle Gestaltung an der Schule für Gestaltung in Zürich. 1992 Ausstellung seiner Plakate im Plakatmuseum Emmerich in Deutschland. 1994 mit dem ersten Preis der 14. Plakatbiennale in Warschau ausgezeichnet.

**Schreyer,** Lothar – geb. 19. 8. 1886 in Dresden-Blasewitz, Deutschland, gest. 18. 6. 1966 in Hamburg-Wohldorf, Deutschland. – *Maler, Autor, Lehrer* – Studium der Kunstgeschichte in Heidelberg, Jura und Kunstgeschichte in Berlin und Leipzig. 1911–18 Dramaturg und Regieassistent am Deutschen Schauspielhaus in Hamburg. 1916–28 Schriftleiter von Herwarth Waldens Zeitschrift „Sturm". 1918–21 Gründung und Leitung der „Sturm"-Bühne in Berlin (mit Herwarth Walden). 1921–23 Lehrer am Bauhaus. 1924–27 Lehrer an der Kunstschule „Der Weg" in Berlin und Dessau. 1928–32 Cheflektor in der Hanseatischen Verlagsanstalt in Hamburg.

**Schrofer,** Jurriaan – geb. 1926 in Den Haag, Niederlande, gest. 1990 in Amsterdam, Niederlande. – *Grafik-Designer, Autor* – Jura-Studium, danach Assistent des Grafikers Dick Elffers. 1954 Gestal-

Schraivogel 1991 Poster

Schraivogel 1995 Poster

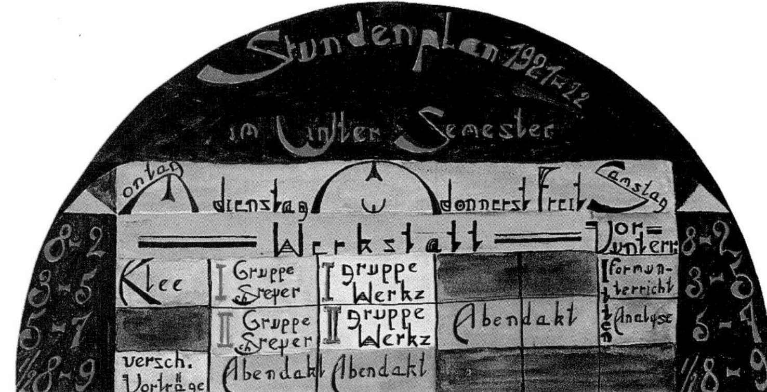


Schreyer 1921–22 Schedule

Schreyer 1923 Letter Picture

dant dans le secteur de la publicité culturelle. Réalise des affiches, catalogues et signets pour des institutions comme le Kunsthaus de Zurich, le Museum für Gestaltung, le Festival international de jazz, le Théâtre du Neumarkt, le Filmpodium de la ville de Zurich et Pro Helvetia. Maquettes de livres pour plusieurs éditeurs. Enseigne la communication visuelle à l'école de design de Zurich depuis 1992. Exposition de ses affiches au Plakatmuseum Emmerich, en Allemagne, en 1992. Premier prix de la 14e Biennale de l'affiche de Varsovie, en 1994.

**Schreyer,** Lothar – né le 19. 8. 1886 à Dresde-Blasewitz, Allemagne, décédé le 18. 6. 1966 à Hambourg-Wohldorf, Allemagne – *peintre, auteur, enseignant* – études d'histoire de l'art à Heidelberg, de droit et d'histoire de l'art à Berlin et à Leipzig. Dramaturge et assistant de régie au Deutsche Schauspielhaus de Hambourg de 1911 à 1918. Rédacteur en chef de la revue «Der Sturm» de Herwarth Walden de 1916 à 1928. Fonde le théâtre du «Sturm» (avec Herwarth Walden) à Berlin, en 1918, et le dirige jusqu'en 1921. Enseigne au Bauhaus de 1921 à 1923. Enseigne à l'école d'art «Der Weg» à Berlin puis à Dessau de 1924 à 1927. Lecteur en chef à la Hanseatische Verlagsanstalt de Hambourg de 1928 à 1932.

**Schrofer,** Jurriaan – né en 1926 à La Haye, Pays-Bas, décédé en 1990 à Amsterdam, Pays-Bas – *graphiste maquettiste, auteur* – études de droit, puis assis-

*typographer* – 1978–82: studies graphic design at the Schule für Gestaltung in Zurich. Then works freelance as a designer in arts advertising. He produces posters, catalogues and logos for clients such as the Kunsthaus and the Museum für Gestaltung in Zurich, the International Jazz Festival, Theater am Neumarkt, city of Zurich's Filmpodium and Pro Helvetia. Designs books for various publishers. From 1992 onwards: teaches visual design at the Schule für Gestaltung in Zurich. 1992: an exhibition of his posters is held in the poster museum in Emmerich in Germany. 1994: is awarded 1st prize at the 14th Poster Biennale in Warsaw.

**Schreyer,** Lothar – b. 19. 8. 1886 in Dresden-Blasewitz, Germany, d. 18. 6. 1966 in Hamburg-Wohldorf, Germany – *painter, author, teacher* – Studied art history in Heidelberg, and law and art history in Berlin and Leipzig. 1911–18: dramaturge and assistant producer at the Deutsche Schauspielhaus in Hamburg. 1916–28: editor of Herwarth Walden's "Sturm" magazine. 1918–21: launches and directs the Sturm-Bühne (theater) in Berlin with Herwarth Walden. 1921–23: teaches at the Bauhaus. 1924–27: teaches at the "Der Weg" art school in Berlin and Dessau. 1928–32: editor-in-chief at the Hanseatische Verlagsanstalt in Hamburg.

**Schrofer,** Jurriaan – b. 1926 in The Hague, The Netherlands, d. 1990 in Amsterdam, The Netherlands – *graphic designer, author* – Studied law and then started working as assistant to graphic artist Dick Elffers. 1954: designs the

tung der Weihnachtsausgabe der Zeitschrift „Drukkersweekblad". 1956 Gestaltung der Broschüre „Soonsbeek Pavilion by Gerrit Rietveld" als „Kwadraatbladen". 1966 gestaltet er mit Willem Sandberg das Buch „Sleutelwoorden" zum 80. Geburtstag von Piet Zwart. 1972 Eintritt in die Leitung des Grafik-Design-Studios „Total Design". Lehrt an der Rietveld Akademie in Amsterdam und an der Akademie in Rotterdam; Direktor der Akademie in Arnheim. – *Publikationen u.a.:* „Letters op maat", Eindhoven 1987. Bibeb „Jurriaan Schrofer", Nijmegen 1972.

**Schuitema,** Paul – geb. 27. 2. 1897 in Groningen, Niederlande, gest. 25. 10. 1973 in Wassenaar, Niederlande. – *Grafik-Designer, Maler, Fotograf, Ausstellungsgestalter, Lehrer* – 1915–20 Studium an der Academie voor Beeldende Kunsten in Rotterdam. 1922–25 Studio in Rotterdam als Maler. Seit 1925 Arbeit als Grafik-Designer und Industrie-Designer, seit 1928 als Fotograf. 1927–31 Werbeleiter bei van Berkel's Patent Company in Rotterdam. 1929 Teilnahme an der „Internationalen Werkbundausstellung Film und Foto" in Stuttgart. Mitglied des „rings neuer werbegestalter". Unterrichtet 1930–63 an der Academie van Beeldende Kunsten in Den Haag. 1932 Briefmarkenentwürfe für die niederländische Post PTT. 1948–67 Gestalter der Zeitschrift „Cement" der Verkoopassociatie Enci-Cemij NV in Wassenaar. – *Publikationen u.a.:* Mark Buchmann (Hrsg.) „Ein Pionier der holländischen Avantgarde, Paul Schuitema", Zürich 1967; D. F. Maan „Paul Schuitema", Rotterdam 1986.

tant du graphiste Dick Elffers. Maquette de l'édition de Noël de la revue «Drukkersweekblad» en 1954. Remaniement de la brochure «Soonsbeek Pavilion by Gerrit Rietveld» en «Kwadraatbladen». En 1966, il crée, avec Willem Sandberg, la maquette de l'ouvrage «Sleutelwoorden» pour le 80e anniversaire de Piet Zwart. Entre à la direction de l'atelier de conception graphique «Total Design» en 1972. A enseigné à la Rietveld Academie à Amsterdam, à l'académie de Rotterdam; directeur de l'académie d'Arnhem. – *Publications, sélection:* «Letters op maat», Eindhoven 1987. Bibeb «Jurriaan Schrofer», Nimègue 1972.

**Schuitema,** Paul – né le 27. 2. 1897 à Groningen, Pays-Bas, décédé le 25. 10. 1973 à Wassenaar, Pays-Bas – *graphiste maquettiste, peintre, photographe, architecte d'expositions, enseignant* – 1915–1920, études à l'Academie voor Beeldende Kunsten de Rotterdam. 1922–1925, peintre dans son atelier de Rotterdam. Exerce comme graphiste maquettiste et designer industriel à partir de 1925, puis comme photographe dès 1928. Directeur de la publicité chez van Berkel's Patent Company à Rotterdam de 1927 à 1931. Participe à la «Internationale Werkbundausstellung Film und Foto» de Stuttgart en 1929. Membre du «ring neuer werbegestalter». Enseigne à l'Academie van Beeldende Kunsten de La Haye de 1930 à 1963. Dessin de timbres pour les postes néerlandaises PTT en 1932. Maquette de la revue «Cement» de la Verkoopassociatie Enci-Cemij NV de Wassenaar de 1948 à 1967. – *Publications, sélection:* Mark Buchmann (éd.) «Ein Pionier der holländischen Avantgarde, Paul Schuitema», Zurich 1967; D. F. Maan «Paul Schuitema», Rotterdam 1986.

**Schule für Gestaltung** – *école des arts et métiers de Bâle* – école de design. En

Schrofer 1955 Advertisement

Schrofer 1963 Poster

Schuitema 1927 Page

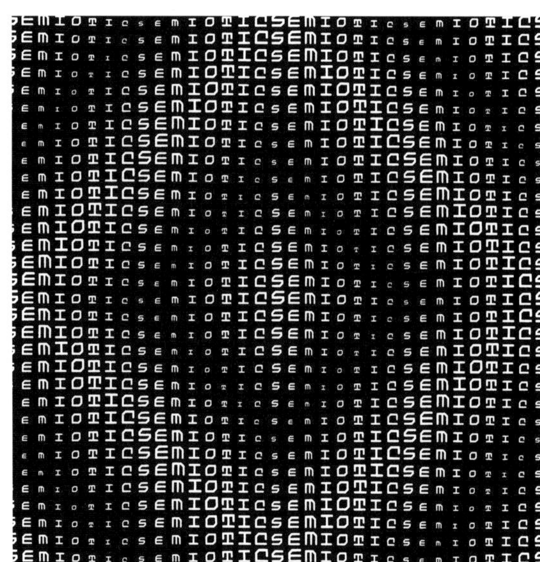

Schrofer ca. 1984 Poster

Schuitema 1928 Card

Christmas edition of "Drukkersweekblad" magazine. 1956: design the "Soonsbeek Pavilion by Gerrit Rietveld" brochure as "Kwadraatbladen". 1966: designs the book "Sleutelwoorden" with Willem Sandberg for Piet Zwart's 80th birthday. 1972: joins the management of Total Design graphic design studio. Teaches at the Rietveld Academie in Amsterdam, at the academy in Rotterdam and is director of the academy in Arnhem. – *Publications include:* "Letters op maat", Eindhoven 1987. Bibeb "Jurriaan Schrofer", Nijmegen 1972.

**Schuitema,** Paul – b. 27. 2. 1897 in Groningen, The Netherlands, d. 25. 10. 1973 in Wassenaar, The Netherlands – *graphic designer, painter, photographer, exhibition planner, teacher* – 1915–20: studies at the Academie voor Beeldende Kunsten in Rotterdam. 1922–25: has a painting studio in Rotterdam. 1925: starts working as a graphic designer and industrial designer and in 1928 as a photographer. 1927–31: publicity manager for van Berkel's Patent Company in Rotterdam. 1929: takes part in the "Internationale Werkbundausstellung Film und Foto" in Stuttgart. Member of the ring neuer werbegestalter. 1930–63: teaches at the Academie van Beeldende Kunsten in The Hague. 1932: designs postage stamps for the Dutch post office PTT. 1948–67: designs "Cement" magazine published by the Verkoopassociatie Enci-Cemij NV in Wassenaar. – *Publications include:* Mark Buchmann (ed.) "Ein Pionier der holländischen Avantgarde, Paul Schuitema", Zurich 1967; D. F. Maan "Paul Schuitema", Rotterdam 1986.

**Schule für Gestaltung** – *Allgemeine Gewerbeschule Basel* – 1796 Gründung einer „Zeichnungsschule" durch die „Gesellschaft zur Förderung des Guten und Gemeinnützigen" (GGG). 1869 Zulassung von Schülerinnen. 1887 Umwandlung der Schule in die „Allgemeine Gewerbeschule Basel" (AGS) als staatliche Institution. 1980 Umbenennung in „Schule für Gestaltung" (SfG). Die Schule hat Abteilungen für Vorbildung, Ausbildung und Weiterbildung. Lehrende und Gestalter wie Emil Ruder, Armin Hofman, Robert Büchler, André Gürtler und Wolfgang Weingart machen die Schule zu einer international einflußreichen Ausbildungsstätte. – *Publikation u. a.:* Manfred Maier „Elementare Entwurfs- und Gestaltungsprozesse. Die Grundkurse an der Kunstgewerbeschule Basel, Schweiz" (vier Bände), Bern, Stuttgart 1977.

**Schüle,** Ilse – geb. 1903 in Vaihingen, Deutschland. – *Typographin, Schriftentwerferin* – 1921–25 Studium an der Kunstgewerbeschule in Stuttgart bei F. H. E. Schneidler, 1925–29 Assistentin. Seit 1929 freie Gestalterin, hauptsächlich für Verlage tätig. – *Schriftentwurf:* Rhapsodie (1951).

**SchumacherGebler,** Eckehart – *Verleger, Unternehmer* – Mitinhaber der Firmen SchumacherGebler, Studio für Typographie und Reprosatz, Buchdruckerei und Verlag SchumacherGebler KG in München. Seit 1974 Herausgabe der jährlichen Publikationsfolge „Bibliothek SG", von der jede Ausgabe einem Autor und einer bestimmten Schrift gewidmet ist. 1988 Gründung des „Journals für Druck-

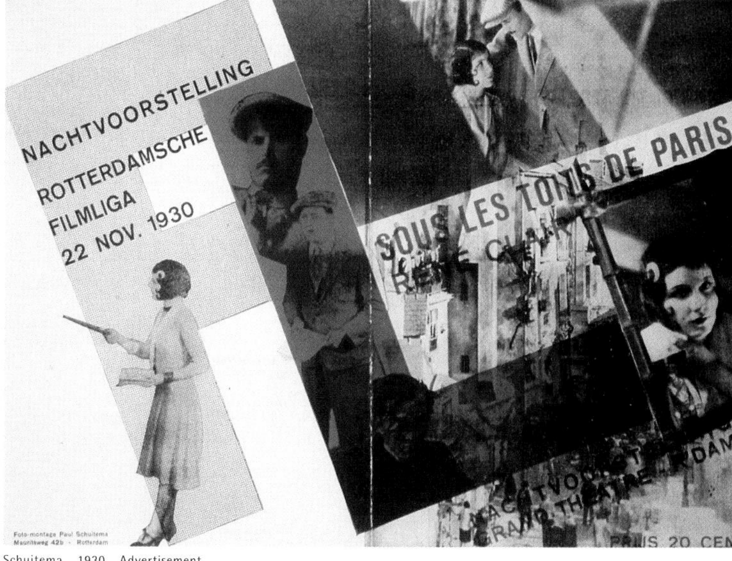


Schuitema 1930 Advertisement

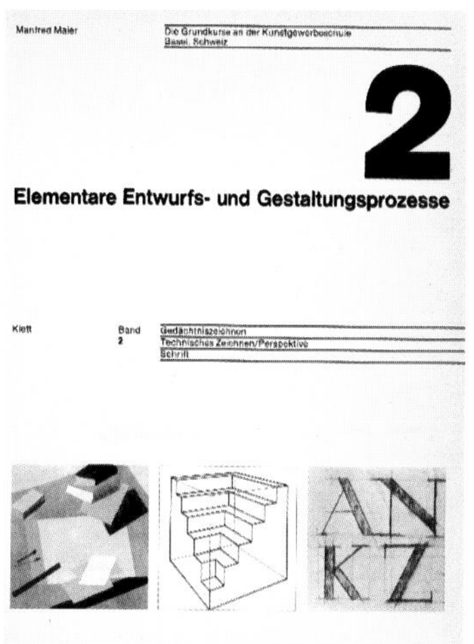


Schule für Gestaltung 1977 Cover

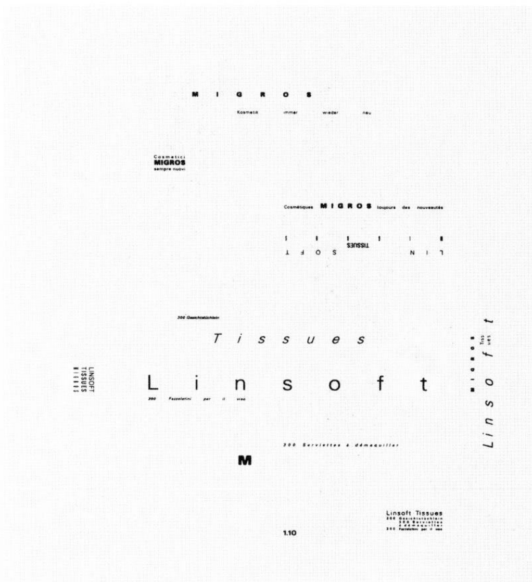


Schule für Gestaltung 1969 Study

1/1234567890 %

Berthold-Schriftweite weit
Berthold-Schriftweite normal
Berthold-Schriftweite eng
Berthold-Schriftweite sehr eng

Schüle 1951 Rhapsodie

1796, la «Gesellschaft zur Förderung des Guten und Gemeinnützigen» (GGG) ou Société d'encouragement du bien et de l'utilité publique, fonde une «Zeichnungsschule» (école de dessin). Première admission d'élèves féminines en 1869. L'école devient une institution d'Etat en 1887 et s'appelle désormais «Allgemeine Gewerbeschule Basel» (école des arts et métiers) (AGS). En 1944, elle est scindée en une section technique, la «Gewerbeschule» (arts et métiers) et une section d'arts décoratifs, la «Kunstgewerbeschule» (AGS) qui deviendra la «Schule für Gestaltung» (SfG) en 1980. L'école dispose de sections préliminaires, de formation normale et de formation permanente. Des enseignants et des créateurs tels qu'Emil Ruder, Armin Hofman, Robert Büchler, André Gürtler et Wolfgang Weingart ont fait la renommé de cette école qui a une influence internationale. – *Publication, sélection:* Manfred Maier "Elementare Entwurfs- und Gestaltungsprozesse. Die Grundkurse an der Kunstgewerbeschule Basel, Schweiz" (4 vol.), Berne, Stuttgart 1977.

**Schüle,** Ilse, née en 1903 à Vaihingen, Allemagne – *typographe, conceptrice de polices* – 1921–1925, études à la Kunstgewerbeschule (école des arts décoratifs) de Stuttgart chez F. H. E. Schneidler, puis assistante de 1925 à 1929. Conceptrice graphique indépendante à partir de 1929, travaille surtout pour des éditeurs. – *Police:* Rhapsodie (1951).

**SchumacherGebler,** Eckehart – *éditeur, chef d'entreprise* – associé des sociétés SchumacherGebler, «Studio für Typographie und Reprosatz, Buchdruckerei und Verlag SchumacherGebler KG» à Munich. Depuis 1974, il édite la série de publications annuelles «Bibliothek SG», dont chaque volume est consacré à un auteur et à une police spécifique. Fonde le «Jour-

**Schule für Gestaltung** – *Allgemeine Gewerbeschule Basel* – 1796: an art school is founded by the Gesellschaft zur Förderung des Guten und Gemeinnützigen (GGG). 1869: the first female students are admitted. 1887: the school becomes the Allgemeine Gewerbeschule Basel (AGS), a state institution. 1980: the school is renamed the Schule für Gestaltung (SfG, school of design). The SfG has departments which offer foundation courses, full courses of study and further education courses. Teachers and designers, such as Emil Ruder, Armin Hofman, Robert Büchler, André Gürtler and Wolfgang Weingart, have helped make the school influential at an international level. – *Publications include:* Manfred Maier "Elementare Entwurfs- und Gestaltungsprozesse. Die Grundkurse an der Kunstgewerbeschule Basel, Schweiz" (4 vols.), Bern, Stuttgart 1977.

**Schüle,** Ilse – b. 1903 in Vaihingen, Germany – *typographer, type designer* – 1921–25: studies at the Kunstgewerbeschule in Stuttgart under F. H.E. Schneidler and is from 1925–29 assistant there. From 1929 onwards: freelance designer, working primarily for publishing houses. – *Font:* Rhapsodie (1951).

**SchumacherGebler,** Eckehart – *publisher, entrepreneur* – Joint holder of the companies SchumacherGebler, Studio für Typographie und Reprosatz and Buchdruckerei und Verlag SchumacherGebler KG in Munich. 1974: starts publishing the yearly "Bibliothek SG" series, where each issue is dedicated to a certain author and typeface. 1988: launches the "Journal

geschichte". 1989 und 1992 Herausgabe des Kalenders „26 Lettern". Erwirbt 1992 den Kern der Druckerei „Offizin Andersen Nexö" in Leipzig (vormals „Offizin Haag-Drugulin"). Gründet 1994 die „Gesellschaft zur Förderung der Druckkunst, Leipzig". Veranstaltet regelmäßig Ausstellungen zu Typographie und Grafik-Design im Typostudio SchumacherGebler in München.

**Schuster,** Horst – geb. 1. 6. 1930 in Dresden, Deutschland. – *Typograph, Lehrer* – 1944–52 Lehre in einem grafischen Großbetrieb in Dresden. Seit 1952 im VEB Verlag der Kunst in Dresden tätig. 1955 erste Auszeichnung im Wettbewerb „Schönste Bücher der DDR", der ca. 80 weitere Auszeichnungen folgen. 1960–65 Fernstudium an der Hochschule für Graphik und Buchkunst in Leipzig bei A. Kapr und W. Schiller. 1966–77 künstlerischer und technischer Leiter des VEB Verlags der Kunst in Dresden. 1977 mit dem Gutenberg-Preis der Stadt Leipzig ausgezeichnet. Seit 1978 Dozent an der Hochschule für Bildende Künste in Dresden. – *Publikation u. a.:* „Giambattista Bodoni", Dresden 1956.

**Schweitzer,** Johannes – geb. 21. 9. 1927 in Frankfurt am Main, Deutschland. – *Grafiker, Kalligraph, Schriftentwerfer* – Kaufmännische Lehre. 1949–51 Studium an der Kunstschule in Offenbach, u. a. bei H. Zapf und Karlgeorg Hoefer. 1951–53 und 1955–57 Studium der Bildhauerei, Grafik und Schriftgestaltung an der Landeskunstschule in Mainz. 1957–87 arbeit als Grafiker. – *Schriftentwürfe:* Dominante (1959), Pensilvania (1982).

nal für Druckgeschichte» en 1988. Publie le calendrier «26 Lettern» en 1989 et en 1992. Il rachète la partie névralgique de l'imprimerie «Offizin Andersen Nexö» (anciennement «Offizin Haag-Drugulin») à Leipzig, en 1992. Fonde la «Gesellschaft zur Förderung der Druckkunst, Leipzig», en 1994. Organise régulièrement des expositions sur la typographie et le design graphique au Typostudio SchumacherGebler à Munich.

**Schuster,** Horst – né le 1. 6. 1930 à Dresde, Allemagne – *typographe, enseignant* – 1944–1952, apprentissage dans une grande entreprise de Dresde, spécialisée dans le graphisme. Employé au VEB Verlag der Kunst de Dresde à partir de 1952. Première distinction en 1955 dans le cadre du concours «Schönste Bücher der DDR» (Plus beaux livres de RDA), il obtiendra ensuite 80 autres prix. 1960–1965, études par correspondance à la Hochschule für Graphik und Buchkunst (école supérieure des arts graphiques et des arts du livre) chez A. Kapr et W. Schiller. Directeur artistique et technique du VEB Verlag der Kunst de Dresde de 1966 à 1977. Lauréat du Prix Gutenberg de la ville de Leipzig en 1977. Professeur à la Hochschule der Bildende Künste (école supérieure des beaux-arts) de Dresde depuis 1978. – *Publication, sélection :* «Giambattista Bodoni», Dresde 1956.

**Schweitzer,** Johannes – né le 21. 9. 1927 à Francfort-sur-le-Main, Allemagne – *graphiste, calligraphe, concepteur de polices* – apprentissage dans le commerce. 1949–1951, études à la Kunstschule (école d'art) d'Offenbach, chez Hermann Zapf et Karlgeorg Hoefer. 1951–1953, puis 1955–1957, études de sculpture, d'arts graphiques et des formes écrites à la Landeskunstschule (école d'art) de Mayence. Exerce comme graphiste de 1957 à 1987.

SchumacherGebler 1984 Invitation Card

SchumacherGebler 1986 Cover

Schweitzer 1959 Dominante

Schuster 1971 Cover

Schwitters 1920 Cover

für Druckgeschichte". 1989 and 1992: publishes the "26 Lettern" calendar. 1992: acquires the main units of the Offizin Andersen Nexö printing workshop in Leipzig (formerly Offizin Haag-Drugulin). 1994: founds the Gesellschaft zur Förderung der Druckkunst, Leipzig. Stages regular exhibitions on typography and graphic design in the Typostudio SchumacherGebler in Munich.

**Schuster,** Horst – b. 1. 6. 1930 in Dresden, Germany – *typographer, teacher* – 1944–52: trains at a major graphics company in Dresden. 1952: starts working for VEB Verlag der Kunst in Dresden. 1955: wins his first prize in the GDR's "most beautiful books" competition, followed by some 80 awards. 1960–65: does a correspondence study course at the Hochschule für Graphik und Buchkunst in Leipzig under A. Kapr and W. Schiller. 1966–77: art director and technical manager at VEB Verlag der Kunst in Dresden. 1977: is awarded Leipzig's Gutenberg Prize. From 1978 onwards: lecturer at the Hochschule für Bildende Künste in Dresden. – *Publications include:* "Giambattista Bodoni", Dresden 1956.

**Schweitzer,** Johannes – b. 21. 9. 1927 in Frankfurt am Main, Germany – *graphic artist, calligrapher, type designer* – Period of business training. 1949–51: studies at the Kunstschule in Offenbach. Tutors include Hermann Zapf and Karlgeorg Hoefer. 1951–53 and 1955–57: studies sculpture, graphics and type design at the Landeskunstschule in Mainz. 1957–87: works as graphic artist. – *Fonts:* Dominante (1959), Pensilvania (1982).

**Schwitters,** Kurt – geb. 20. 6. 1887 in Hannover, Deutschland, gest. 8. 1. 1948 in Kendal, England. – *Maler, Grafiker, Typograph, Schriftsteller* – 1909–14 Studium an der Kunstakademie in Dresden. 1918 Bekanntschaft mit Hans Arp und Raoul Hausmann. Erste „Merz"-Arbeiten. 1918–19 Architekturstudium an der Technischen Hochschule in Hannover. 1920 Beginn des „Merz-Baus" in seinem Haus in Hannover. Erste Einzelausstellung in der Galerie „Der Sturm" in Berlin. 1921 Vortrag mit Raoul Hausmann und Hannah Höch in Prag. 1922 Bekanntschaft mit Theo van Doesburg auf dem Dada-Kongreß in Weimar. 1923–32 Gründer und Herausgeber der Zeitschrift „Merz" in Hannover. Veröffentlicht 1924 in „Merz" Nr. 11 seinen Text „Thesen über Typographie". 1927 Mitbegründer des „rings neuer werbegestalter" und der Künstlergruppe „die abstrakten hannover". 1937 Umzug nach Norwegen, Beginn des zweiten „Merz-Baus" in Oslo. Seine Bilder werden in der Ausstellung „Entartete Kunst" in Deutschland gezeigt. 1940 Emigration nach England. 1947 Beginn der Arbeiten zu „Merzbarn". – *Schriftentwurf:* Phonetische Schrift (1927). – *Publikationen u.a.:* „Anna Blume: Dichtungen", Hannover 1919; „Die Scheuche: ein Märchen" (mit Käte Steinitz, Theo van Doesburg), Hannover 1925. Werner Schmalenbach „Kurt Schwitters", Köln 1967; Ernst Nündel „Kurt Schwitters", Reinbek 1981; John Elderfield „Kurt Schwitters", London 1985.

**Senefelder,** Alois – geb. 6. 11. 1771 in Prag, Böhmen, gest. 26. 2. 1834 in München, Deutschland. – *Schauspieler, Thea-*

Schwitters 1924 Cover

Schwitters 1930 Poster

Schwitters 1927 Poster

– *Polices :* Dominante (1959), Pensilvania (1982).

**Schwitters,** Kurt – né le 20. 6. 1887 à Hanovre, Allemagne, décédé le 8. 1. 1948 à Kendal, Angleterre – *peintre, graphiste, typographe, écrivain* – 1909–1914, études à l'académie des beaux-arts de Dresde. Rencontre Jean Arp et Raoul Hausmann en 1918. Premières œuvres «Merz». Etudes d'architecture à l'université technique de Hanovre de 1918 à 1919. Début du «Merz-Bau» dans sa maison de Hanovre, en 1920. Première exposition personnelle à la galerie «Der Sturm» à Berlin. Conférence avec Raoul Hausmann et Hannah Höch à Prague, en 1921. Rencontre Theo van Doesburg au congrès Dada de Weimar en 1922. Fonde la revue «Merz» en 1923 qu'il publiera jusqu'en 1932 à Hanovre. Parution de son texte «Thesen über Typographie» (Thèses sur la typographie) en 1924, dans le n° 11 de «Merz». Cofondateur, en 1927, du «ring neuer werbegestalter» et du groupe d'artistes «die abstrakten hannover» (les abstraits de Hanovre). S'installe en Norvège en 1937 où il entame le second «Merz-Bau» à Oslo. Ses œuvres sont montrées en Allemagne lors de l'exposition «Entartete Kunst» (art dégénéré). Emigre en Angleterre en 1940. Début des travaux sur le «Merzbarn» en 1947. – *Police :* Phonetische Schrift (écriture phonétique, 1927). – *Publications, sélection :* «Anna Blume : Dichtungen», Hanovre 1919 ; «Die Scheuche : ein Märchen» (avec Käte Steinitz, Theo van Doesburg), Hanovre 1925. Werner Schmalenbach «Kurt Schwitters», Cologne 1967 ; Ernst Nündel «Kurt Schwitters», Reinbeck 1981 ; John Elderfield «Kurt Schwitters», Londres 1985.

**Senefelder,** Alois – né le 6. 11. 1771 à Prague, Bohême, décédé le 26. 2. 1834 à Munich, Allemagne – *comédien, drama-*

**Schwitters,** Kurt – b. 20. 6. 1887 in Hanover, Germany, d. 8. 1. 1948 in Kendal, England – *painter, graphic artist, typographer, writer* – 1909–14: studies at the art academy in Dresden. 1918: meets Hans Arp and Raoul Hausmann. Produces his first "Merz" collages. 1918–19: studies architecture at the Technische Hochschule in Hanover. 1920: starts creating his Merz interior in his house in Hanover. First solo exhibition in the "Der Sturm" gallery in Berlin. 1921: lecture with Raoul Hausmann and Hannah Höch in Prague. 1922: makes the acquaintance of Theo van Doesburg at the Dada Congress in Weimar. 1923–32: launches and edits "Merz" magazine in Hanover. 1924: publishes his text on typography ("Thesen über Typographie") in no. 11 of "Merz". 1927: co-founder of the ring neuer werbegestalter and the artists' group die abstrakten hannover. 1937: moves to Norway and starts the second Merz building in Oslo. His pictures are shown in the exhibition of degenerate art ("Entartete Kunst") in Germany. 1940: emigrates to England. 1947: starts working on his "Merzbarn". – *Font:* Phonetische Schrift (1927). – *Publications include:* "Anna Blume: Dichtungen", Hanover 1919; "Die Scheuche: ein Märchen" (with Käte Steinitz, Theo van Doesburg), Hanover 1925. Werner Schmalenbach "Kurt Schwitters", Cologne 1967; Ernst Nündel "Kurt Schwitters", Reinbek 1981; John Elderfield "Kurt Schwitters", London 1985.

**Senefelder,** Alois – b. 6. 11. 1771 in Prague, Bohemia, d. 26. 2. 1834 in Mu-

*terschriftsteller, Erfinder* – 1797 Erfindung des chemischen Steindrucks (Flachdruck): Er zeichnet mit Seife auf den geschliffenen Stein, gießt Gummiwasser darüber und schwärzt die Zeichnung mit Ölfarbe ein. Die durch Seife fettigen Stellen nehmen die Farbe an; der übrige Stein bleibt weiß. Er verwendet die neue Drucktechnik seit 1806 in München. 1809 Leiter der Steindruckerei für Landkarten und Steuer. Erfindet 1813 das Steinpapier, das den schweren Stein ersetzen soll. Veröffentlicht 1818 sein „Vollständiges Lehrbuch der Steindruckerey", von dem eine französische (1819) und eine italienische (1824) Ausgabe erscheinen. 1826 Herstellung der ersten Mehrfarbendrucke (Mosaikdrucke). Druckt 1833 erstmals auf Stein reproduzierte Ölgemälde auf Leinwand.

**Sichowsky,** Richard von – geb. 28. 5. 1911 in Hamburg, Deutschland, gest. 28. 1. 1975 in Hamburg, Deutschland. – *Typograph, Verleger, Lehrer* – Schriftsetzerlehre. Schüler der Meisterschule für Deutschlands Buchdrucker in München (bei H. Virl, Josef Käufer, Georg Trump). 1937–40 Arbeit im Verlag Heinrich Ellermann in Hamburg. 1946 Berufung an die Landeskunstschule in Hamburg. 1950 Gründung der „Grillen-Presse", bei der er ca. fünfzehn Bücher herausgibt und gestaltet. Seit 1952 im Vorstand der Maximilian-Gesellschaft, für die er 22 Bücher gestaltet. Ausstellungen u. a. im Kunstindustrie-Museum in Oslo (1966), in der Königlichen Bibliothek in Stockholm (1967), im Deutschen Institut in Helsinki (1967) und der Strahov-Bibliothek in

*turge, inventeur* – invente la lithographie chimique (impression à plat) en 1797 : il prépare la pierre en la ponçant, trace son dessin avec du savon, l'enduit ensuite d'une solution de gomme et noircit le dessin à la peinture à l'huile. Les endroits graissés par le savon absorbent la couleur tandis que le reste de la pierre reste blanc. Il applique cette nouvelle technique d'impression à partir de 1806 à Munich. Dirige, en 1809, l'atelier de lithographie du service de cartographie et de l'administration fiscale. En 1813, il invente le papier à lithographie destiné à remplacer les lourdes plaques de pierre. En 1818, il publie son «Vollständiges Lehrbuch der Steindruckerey» qui sera traduit en français (1819) puis en italien (1824). En 1826, il réalise les premières impressions polychromes (impression en mosaïque). 1833, premier tableau à l'huile reproduit sur toile par la technique de lithographie.

**Sichowsky,** Richard von – né le 28. 5. 1911 à Hambourg, Allemagne, décédé le 28. 1. 1975 à Hambourg, Allemagne – *typographe, éditeur, enseignant* – apprentissage de composition typographique. Etudes à la Meisterschule für Deutschlands Buchdrucker (institut professionnel de l'imprimerie allemande) à Munich (chez Hermann Virl, Josef Käufer, Georg Trump). Employé aux éditions Heinrich Ellermann à Hambourg de 1937 à 1940. Nommé à la Landeskunstschule de Hambourg en 1946. Fonde, en 1950, la «Grillen-Presse» pour laquelle il réalise la maquette d'une quinzaine de livres. Membre du comité directeur de la Maximilian-Gesellschaft à partir de 1952; il conçoit la maquette de 22 ouvrages. Expositions au Musée des industries d'art d'Oslo (1966), à la Bibliothèque royale de Stockholm (1967), au Deutsche Institut d'Helsinki (1967), à la bibliothèque Strahov de Prague (1969). – *Publications, sé-*

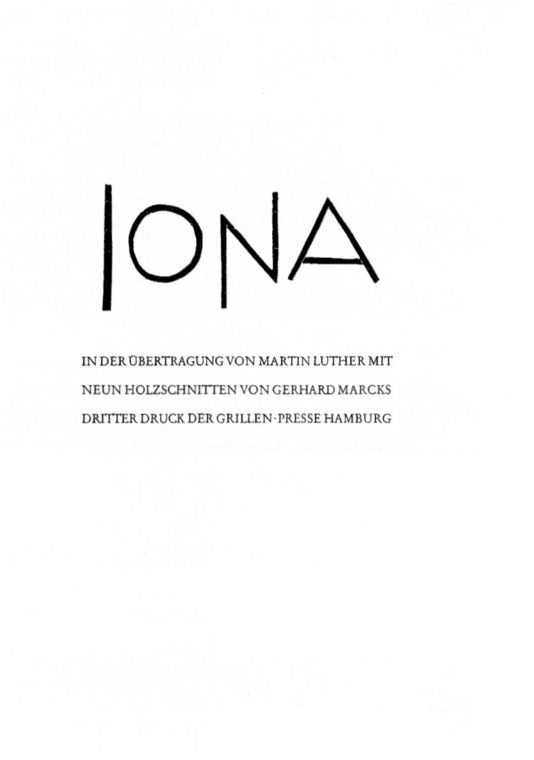

Sichowsky 1951 Page

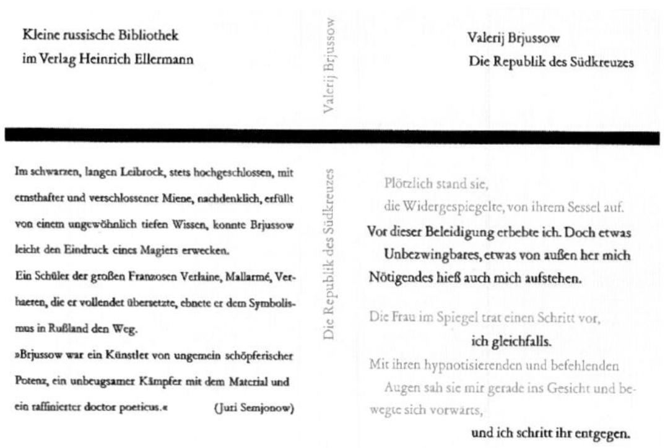

Simon 1945 Cover

Sichowsky 1961 Cover

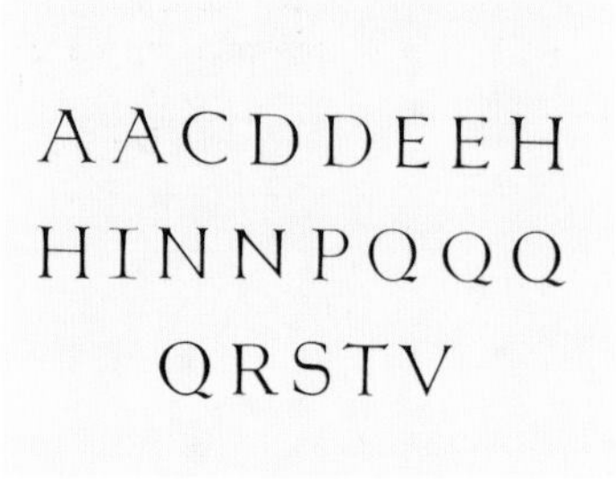

Simons 1921 Initials

nich, Germany – *actor, playwright, inventor* – 1797: invents chemical lithography (surface printing), a process where smooth-grained stone is drawn on in soap, has oily water poured over it and is then blackened with oil paint. The paint adheres only to the greasy, soaped parts of the drawing; the remaining areas stay white. Senefelder starts using this new printing technique in Munich in 1806. 1809: head of a lithographics company (Steindruckerei für Landkarten und Steuer). 1813: invents stone paper to replace heavy stone. 1818: publishes his "Vollständiges Lehrbuch der Steindruckerey", which is also printed in French (1819) and Italian (1824). 1826: produces the first color prints (mosaic prints). 1833: for the first time he prints oil paintings reproduced on stone onto canvas.

**Sichowsky,** Richard von – b. 28. 5. 1911 in Hamburg, Germany, d. 28. 1. 1975 in Hamburg, Germany – *typographer, publisher, teacher* – Trained as a typesetter. Student at the Meisterschule für Deutschlands Buchdrucker in Munich (under Hermann Virl, Josef Käufer and Georg Trump). 1937–40: works for Heinrich Ellermann publishers in Hamburg. 1946: appointed to teach at the Landeskunstschule in Hamburg. 1950: founds Grillen-Presse, where he designs and publishes some fifteen books. From 1952 onwards: committee member of the Maximilian-Gesellschaft, for whom he designs 22 books. His work has been exhibited at the Art Industry Museum in Oslo (1966), the Royal Library in Stockholm (1967), the German Institute in Helsinki (1967) and

Prag (1969). – *Publikationen u. a.:* „Typographie und Bibliophilie. Aufsätze und Vorträge über die Kunst des Buchdrucks aus zwei Jahrhunderten" (mit Hermann Tiemann), Hamburg 1971. Bertold Hack, Otto Rohse (Hrsg.) „Richard von Sichowsky, Typograph", Hamburg 1982.

**Simon,** Oliver – geb. 29. 4. 1895 in Sale, England, gest. 18. 3. 1956 in London, England. – *Typograph, Autor* – Lehre in der Steindruckerei Whittington & Gregg. 1919 Eintritt als Typograph in die Curwen Press in London. 1923–25 Herausgeber der ersten vier Bände der Zeitschrift „The Fleuron". 1924 Mitbegründer und Mitglied des „Double Crown Club" in London. 1935–40 Herausgeber der Zeitschrift „Signature". 1939–56 Präsident der Curwen Press. 1946–54 Herausgabe der neuen Serie der Zeitschrift „Signature". – *Publikationen u. a.:* „Printing of Today", London 1928; „Introduction to Typography", London 1945; „Printer and Playground", London 1956.

**Simons,** Anna – geb. 8. 6. 1871 in Mönchengladbach, Deutschland, gest. 1951 in Prien, Deutschland. – *Grafikerin, Kalligraphin, Schriftkünstlerin, Lehrerin* – 1896–1903 Studium am Royal College of Art in London. 1900 Schülerin von Edward Johnston. 1905 Lehrerin an der Kunstgewerbeschule in Düsseldorf. 1906 mit Gold- und Silbermedaillen auf der Kunstgewerbeausstellung in Dresden ausgezeichnet. Auf Anfrage von Henry van de Velde gibt Simons 1908–14 jedes Jahr einen dreiwöchigen Kurs in Weimar. 1910 Übersetzung von Edward Johnstons Buch „Writing & Illuminating & Letter-

LEVAVI OCULOS MEOS IN MONTES
UNDE VENIET AUXILIUM MIHI
Auxilium meum a Domino,
qui fecit coelum et terram.
Non det in commotionem pedem tuum:
neque dormitet qui custodit te.
Ecce non dormitabit neque dormiet,
qui custodit Israel.
Dominus custodit te. Dominus protectio tua,
super manum dexteram tuam:
Per diem sol non uret te:
neque luna per noctem.
Dominus custodit te ab omni malo:
custodiat animam tuam Dominus.
Dominus custodiat introitum tuum, et exitum tuum:
ex hoc nunc, et usque in saeculum.
AMEN

Simons 1917 Page

WELCHE HAUPTUNTERSCHIEDE sein würden zwischen einem Volke, das in seiner ursprünglichen Sprache sich fortgebildet, und einem solchen, das eine fremde Sprache angenommen, ist in der vorigen Rede auseinandergesetzt. Wir sagten bei dieser Gelegenheit: was das Ausland betreffe, so wollten wir dem eignen Urteile jedweden Beobachters die Entscheidung überlassen, ob in demselben diejenigen Erscheinungen wirklich einträten, die zufolge unsrer Behauptungen darin eintreten müssten; was aber die Deutschen betrifft, machten wir uns anheischig, darzulegen, dass diese sich wirklich also geäussert, wie unsern Behauptungen zufolge das Volk einer Ursprache sich äussern müsse. Wir gehen heute an die Erfüllung unsers Versprechens, und zwar legen wir das zu Erweisende zunächst dar an der letzten grossen und in gewissem Sinne vollendeten Welt-Tat des deutschen Volkes, an der kirchlichen Reformation.
Das aus Asien stammende und durch seine Verderbung erst recht asiatisch gewordene, nur stumme Ergebung und blinden Glauben predigende Christentum war schon für die Römer etwas Fremdartiges und Ausländisches; es wurde niemals von ihnen wahrhaft durchdrungen und angeeignet und teilte ihr Wesen in zwei nicht aneinander passende Hälften, wobei jedoch die Anfügung des fremden Teils durch den angestammten schwermütigen Aberglauben vermittelt wurde. An den eingewanderten Germaniern erhielt diese Religion Zöglinge, in denen keine frühere Verstandesbildung ihr hinderlich war, aber auch kein angestammter Aberglaube sie begünstigte, und so wurde sie denn an dieselben gebracht als ein zum Römer, das sie nun
114

Simons 1922 Page

Simons 1922 Initials

*lection:* «Typographie und Bibliophilie. Aufsätze und Vorträge über die Kunst des Buchdrucks aus zwei Jahrhunderten» (avec Hermann Tiemann), Hambourg 1971. Bertold Hack, Otto Rohse (éd.) «Richard von Sichowsky, Typograph», Hambourg 1982.

**Simon,** Oliver – né le 29. 4. 1895 à Sale, Angleterre, décédé le 18. 3. 1956 à Londres, Angleterre – *typographe, auteur* – apprentissage à l'atelier de lithographie Whittington & Gregg. Entre comme typographe à la Curwen Press, en 1919, à Londres. Publie les quatre premiers volumes de la revue «The Fleuron» entre 1923 et 1925. Cofondateur et membre du «Double Crown Club» de Londres, en 1924. Publie la revue «Signature» de 1935 à 1940. Président de la Curwen Press de 1939 à 1956. Publie la nouvelle édition de la revue «Signature» de 1946 à 1954. – *Publications, sélection:* «Printing of Today», Londres 1928; «Introduction to Typography», Londres 1945; «Printer and Playground», Londres 1956.

**Simons,** Anna – née le 8. 6. 1871 à Mönchengladbach, Allemagne, décédée en 1951 à Prien, Allemagne – *graphiste, calligraphe, scribe, enseignante* – 1896–1903, études au Royal College of Art à Londres. Elève d'Edward Johnston en 1900. Enseigne à la Kunstgewerbeschule (école des arts décoratifs) de Düsseldorf en 1905. Médaille d'or et d'argent à l'exposition des arts décoratifs de Dresde en 1906. A la demande de Henry van de Velde, Simons donne un cours annuel de trois semaines à Weimar de 1908 à 1914. Elle traduit en allemand l'ouvrage d'Edward Johnston «Writing & Illuminating & Lettering» en 1910 sous le titre

at the Strahov Library in Prague (1969), among other venues. – *Publications include:* "Typographie und Bibliophilie. Aufsätze und Vorträge über die Kunst des Buchdrucks aus zwei Jahrhunderten" (with H. Tiemann), Hamburg 1971. Bertold Hack, Otto Rohse (eds.) "Richard von Sichowsky, Typograph", Hamburg 1982.

**Simon,** Oliver – b. 29. 4. 1895 in Sale, England, d. 18. 3. 1956 in London, England – *typographer, author* – Trained at Whittington & Gregg lithographics company. 1919: joins Curwen Press in London as a typographer. 1923–25: edits the first four issues of "The Fleuron" magazine. 1924: co-founder and member of the Double Crown Club in London. 1935–40: editor of "Signature" magazine. 1939–56: president of Curwen Press. 1946–54: editor of the new series of "Signature" magazine. – *Publications include:* "Printing of Today", London 1928; "Introduction to Typography", London 1945; "Printer and Playground", London 1956.

**Simons,** Anna – b. 8. 6. 1871 in Mönchengladbach, Germany, d. 1951 in Prien, Germany – *graphic artist, calligrapher, teacher* – 1896–1903: studies at the Royal College of Art in London. 1900: one of Edward Johnston's students. 1905: teaches at the Kunstgewerbeschule in Düsseldorf. 1906: awarded gold and silver medals at the crafts exhibition in Dresden. 1908–14: at the request of Henry van de Velde, Simons holds a three-week course once a year in Weimar. 1910: translates Edward Johnston's book, "Writing & Illuminating & Lettering", into

ing" unter dem Titel „Schreibschrift, Zierschrift und angewandte Schrift" ins Deutsche. F. H. Ehmcke holt Simons 1914 an die Kunstgewerbeschule nach München. Seit 1918 Mitarbeiterin der Bremer Presse. 1934 Ausstellung ihrer Arbeiten im Schriftmuseum Blanckertz in Berlin. – *Publikation u. a.:* Eberhard Hölscher „Anna Simons", Berlin o. J. (ca. 1935).

**Skolos, Wedell Inc.** – *Grafik-Design-Studio* – 1980 Gründung des Studios Skolos, Wedell Inc. in Boston. Nancy Skolos – geb. 21. 9. 1955 in Cincinnati, USA. – 1975–77 Grafik-Design-Studium an der Cranbrook Academy of Art in Bloomfield Hills. 1977 Studium an der Yale University's School of Art and Architecture. Thomas Wedell – geb. 25. 8. 1949 in Kalamazoo, USA.– Studium der Fotografie an der Cranbrook Academy of Art in Bloomfield Hills. Arbeiten für Auftraggeber wie z.B. EMI Records, Boston Acoustic, Berkeley Type, Digital Equipment Corporation. 1988 Gestaltung einer Ausgabe der Zeitschrift „Design Quarterly" über den Minneapolis Sculpture Garden. Lehraufträge an der Rhode Island School of Design und an der Yale University.

**Society of Scribes and Illuminators (SSI)** – 1921 Gründung der SSI in London durch Lawrence Christie und Graily Hewitt, die Studenten von Edward Johnston sind. Weiteres Gründungsmitglied ist Alfred Fairbank. Die SSI wird als Diskussionsforum für alle am Schreiben und an Kalligraphie Interessierten gegründet. Die SSI veranstaltet 1922 Ausstellungen in Instituten wie dem Victoria & Albert Museum und dem British Crafts Centre. Bei

«Schreibschrift, Zierschrift und angewandte Schrift». En 1914, F. H. Ehmcke fait venir Simons à l'école des arts décoratifs de Munich. Collaboratrice de la Bremer Presse à partir de 1918. Exposition de ses travaux au Schriftmuseum Blanckertz à Berlin, en 1934. – *Publication, sélection :* Eberhard Hölscher «Anna Simons», Berlin sans date (vers 1935).

**Skolos, Wedell Inc.** – *agence de graphisme et de design* – fondation de l'agence Skolos, Wedell Inc. à Boston en 1980. Nancy Skolos – née le 21. 9. 1955 à Cincinnati, Etats-Unis. 1975–1977, études d'arts graphiques et de design à la Cranbrook Academy of Art à Bloomfied Hills. 1977, études à la Yale University's School of Art and Architecture. Thomas Wedell – né le 25. 8. 1949 à Kalamazoo, Etats-Unis. Etudes de photographie à la Cranbrook Academy of Art à Bloomfield Hills. Travaux pour EMI Records, Boston Acoustic, Berkeley Type, Digital Equipment Corporation. Maquette d'une édition de la revue «Design Quarterly» sur le Minneapolis Sculpture Garden en 1988. Chargé de cours à la Rhode Island School et à la Yale University.

**Society of Scribes and Illuminators (SSI)** – La SSI est fondée en 1921 à Londres par Lawrence Christie et Graily Hewitt, étudiants d'Edward Johnston. Alfred Fairbank participe à la création. Le but de la SSI est de servir de podium de discussion pour toutes les personnes s'intéressant à l'écrit et à la calligraphie. En 1922, la SSI organise des expositions dans des institutions telles que le Victoria & Albert Museum et le British Crafts Centre. Lors des rencontres de la SSI, qui compte plus de 1 000 membres, des spécialistes européens interviennent sur des thèmes concernant l'écrit et la calligraphie. La revue de la SSI, «The Scribe», paraît trois fois l'an. La SSI possède une bibliothèque,

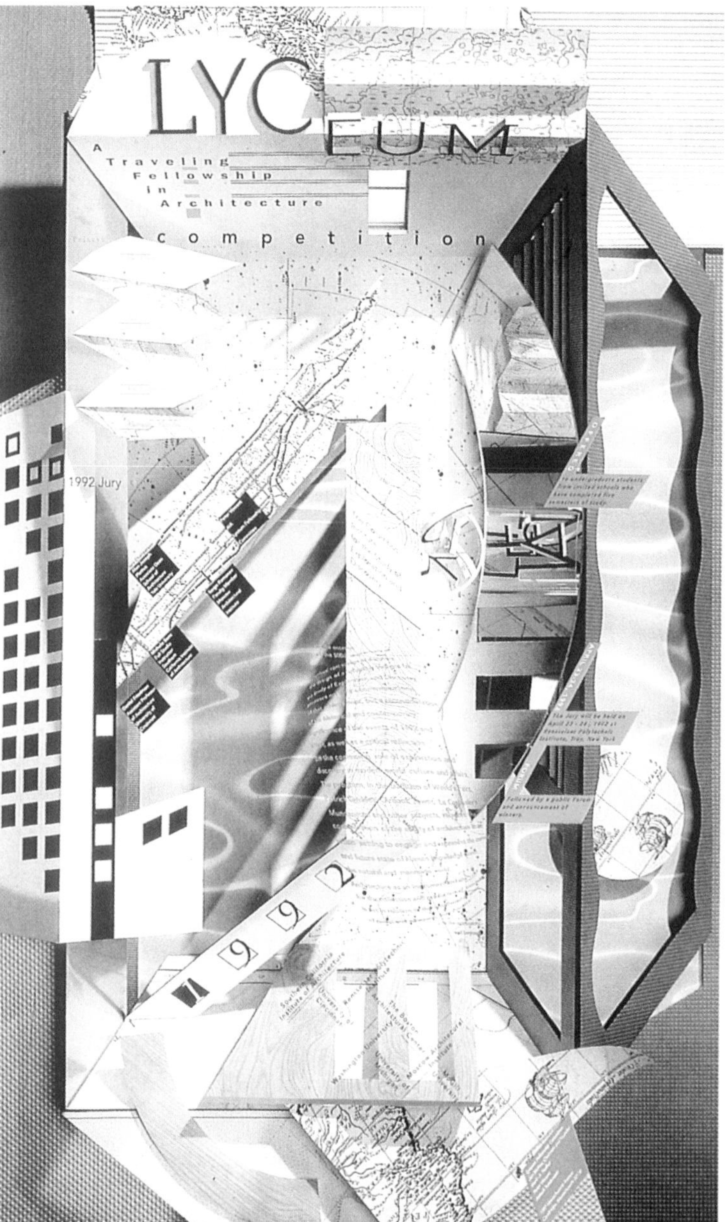

Skolos, Wedell Inc. 1992 Poster

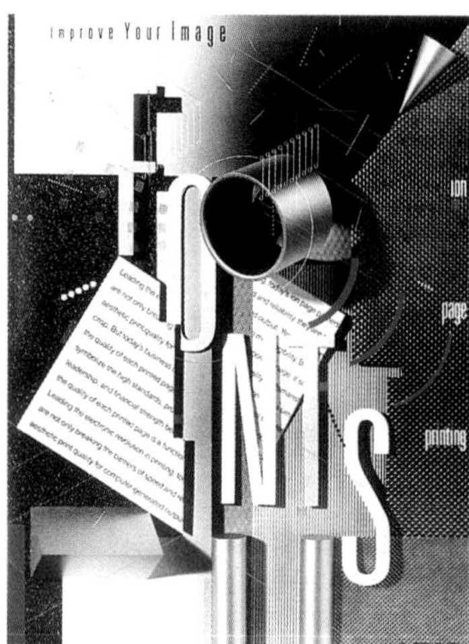

Skolos, Wedell Inc. 1987 Poster

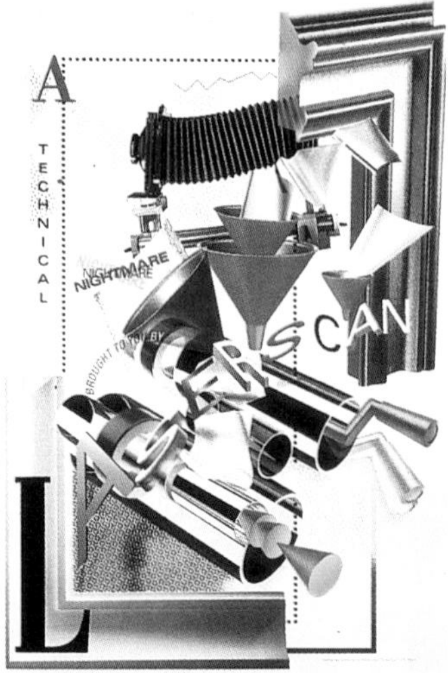

Skolos, Wedell Inc. 1988 Poster

German ("Schreibschrift, Zierschrift und angewandte Schrift"). 1914: F. H. Ehmcke invites Simons to teach at the Kunstgewerbeschule in Munich. From 1918 onwards: works for the Bremer Presse. 1934: an exhibition of her work is held at the Schriftmuseum Blanckertz in Berlin. – *Publications include:* Eberhard Hölscher "Anna Simons", Berlin, no publication date given (c.1935).

**Skolos, Wedell Inc.** – *graphic design studio* – 1980: Skolos, Wedell Inc. is opened in Boston. Nancy Skolos – b. 21. 9. 1955 in Cincinnati, USA. 1975–77: studies graphic design at the Cranbrook Academy of Art in Bloomfield Hills. 1977: studies at Yale University's School of Art and Architecture. Thomas Wedell – b. 25. 8. 1949 in Kalamazoo, USA. Studied photography at the Cranbrook Academy of Art in Bloomfield Hills. Clients include EMI Records, Boston Acoustic, Berkeley Type and the Digital Equipment Corporation. 1988: designs an issue of "Design Quarterly" magazine on the Minneapolis Sculpture Garden. Has taught at Rhode Island School of Design and at Yale University.

**Society of Scribes and Illuminators (SSI)** – 1921: the SSI is founded in London by Lawrence Christie and Graily Hewitt, both students of Edward Johnston. Alfred Fairbank is also one of the founders. The SSI is launched as a forum for discussion for all those interested in writing and calligraphy. 1922: the SSI stages exhibitions in institutes such as the Victoria & Albert Museum and the British Crafts Centre. At meetings held by the SSI, which numbers

den Treffen der SSI, die über 1.000 Mitglieder hat, sprechen internationale Fachleute zu Themen des Schreibens und der Kalligraphie. Die Zeitschrift der SSI, „The Scribe", erscheint dreimal im Jahr. Die SSI besitzt eine Bibliothek, veranstaltet Kurse und veröffentlicht das „Calligraphers Handbook". – *Publikation:* „Present-Day Calligraphy and Illumination", London 1936.

**Society of Typographic Designers (STD)** – 1928 Gründung des Verbands mit dem Namen „British Typographers Guild" durch Vincent Steer, Alfred Vernon, T. Wilson Philip, Stanley Hayter, Charles Hoath, Arnold Jones und Edward Burnett. Das Ziel der BTG ist, das Niveau der englischen Typographie in Theorie und Praxis zu heben. Dazu werden Vorträge und Diskussionen veranstaltet sowie Artikel in Fachzeitschriften publiziert. 1946 Herausgabe der Verbandszeitschrift „The Typographer". 1953 Umbenennung des Verbands in „Society of Typographic Designers" (STD). 1971 Umbenennung der offiziellen STD-Zeitschrift in „Typographic". Der Verband ist Mitglied des „International Council of Graphic Design Associations" (ICOGRADA). Er veranstaltet regelmäßig den Wettbewerb „Typographic Award".

**Solpera,** Jan – geb. 26. 12. 1939 in Jindřichově Hradci, Tschechoslowakei. – *Typograph, Grafik-Designer* – Studium an der Kunsthochschule in Prag bei František Muzika. Als Werbegrafiker und Buchgestalter tätig . Auf der Biennale in Brno (1970, 1978), Premio grafico Fiera di Bologna (1972) und dem Wettbewerb

SSI Sheet

STD 1994 Folder

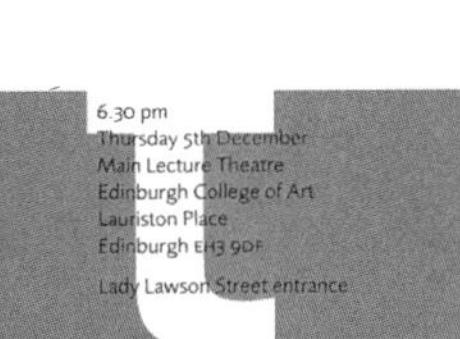

STD 1996 Invitation Card

STD 1996 Cover

organise des cours et publie le «Calligraphers Handbook». – *Publication:* «Present-Day Calligraphy and Illumination», Londres 1936.

**Society of Typographic Designers (STD)** – l'association est fondée en 1928 par Vincent Steer, Alfred Vernon, T. Wilson Philip, Stanley Hayter, Charles Hoath, Arnold Jones et Edward Burnett sous le nom de «British Typographers Guild». L'objectif de la «BTG» était d'élever le niveau de la typographie anglaise en théorie comme en pratique. Elle organise des conférences et des discussions et publie des articles dans les revues spécialisées. En 1946, elle édite la revue de l'association «The Typographer». En 1953, l'association prend le nom de «Society of Typographic Designers» (STD). La revue officielle de la STD est rebaptisée «Typographic» en 1971. L'association est membre de l'«International Council of Graphic Design Associations» (ICOGRADA). Elle organise régulièrement le concours «Typographic Award».

**Solpera,** Jan – né le 26. 12. 1939 à Jindřichově Hrdaci, Tchécoslovaquie – *typographe, graphiste maquettiste* – études à l'école des beaux-arts de Prague chez František Muzika. Exerce comme graphiste publicitaire et maquettiste en édition. Distinctions à la Biennale de Brno (1970, 1978), Premio grafico Fiera di Bologna (1972), concours «Das schön-

over 1,000 members, international specialists give talks on themes related to writing and calligraphy. The SSI's magazine, "The Scribe", is published three times a year. The SSI has a library, runs courses and publishes the "Calligrapher's Handbook". – *Publication:* "Present-Day Calligraphy and Illumination", London 1936.

**Society of Typographic Designers (STD)** – 1928: the society is founded as the British Typographers Guild by Vincent Steer, Alfred Vernon, T. Wilson Philip, Stanley Hayter, Charles Hoath, Arnold Jones and Edward Burnett. The aim of the BTG is to raise standards in English typography in theory and practice. To this end, lectures and discussions are organized and articles published in various journals. 1946: the society magazine, "The Typographer", is published. 1953: the BTG changes its name to the Society of Typographic Designers (STD). 1971: the STD's official magazine is renamed "Typographic". The society is a member of the International Council of Graphic Design Associations (ICOGRADA) and presents its Typographic Award on a regular basis.

**Solpera,** Jan – b. 26. 12. 1939 in Jindřichově Hradci, Czechoslovakia – *typographer, graphic designer* – Studied at the art college in Prague under František Muzika. Works as a commercial artist and designer of books. Has won awards at the Biennale in Brno (1970 and 1978), at the Premio grafico Fiera di Bologna (1972),

„Das schönste Buch" (1975, 1977–1979) ausgezeichnet. Assistent an der Kunsthochschule in Prag. – *Schriftentwurf:* Circo (1971).

**Spemann,** Rudo – geb. 22. 4. 1905 in Würzburg, Deutschland, gest. 11. 7. 1947 in Schepetowka, UdSSR. – *Kalligraph, Künstler, Lehrer* – 1924–30 Studium an der Kunstgewerbeschule in München (bei F. H. Ehmcke und Emil Preetorius) und an der Kunstakademie in Stuttgart (bei F. H. E. Schneidler). 1930–35 Assistent von Schneidler an der Kunstakademie in Stuttgart. 1935–37 freier Grafiker in München. 1937–39 Dozent an der Akademie für Graphische Künste und Buchgewerbe in Leipzig. Seit 1954 vergibt die Stadt Offenbach alle zwei Jahre im Andenken an sein Werk den „Rudo-Spemann-Preis". – *Schriftentwurf:* Gavotte (1942). – *Publikationen u.a.:* Walter Tiemann „Beseelte Kalligraphie", Leipzig 1950; Hans Adolf Halbey „Rudo Spemann 1905–1947. Monographie und Werkverzeichnis seiner Schriftkunst", Offenbach 1981.

**Spencer,** Herbert – geb. 22. 6. 1924 in London, England. – *Grafik-Designer, Fotograf, Maler, Autor, Lehrer* – Seit 1948 eigenes Studio in London; Gestaltung von Büchern, Kunstkatalogen, Signets, Fahrplänen und Geschäftsberichten. 1949–55 Dozent an der Central School of Arts and Crafts in London. 1949–67 Herausgeber der Zeitschrift „Typographica". 1964–73 Herausgeber des Jahrbuchs „Penrose Annual". 1965 Auszeichnung mit dem Titel „Royal Designer for Industry". Seit 1966 unterrichtet er am Royal

ste Buch» (1975, 1977–1979). Assistant à l'école des beaux-arts de Prague. – *Police:* Circo (1971).

**Spemann,** Rudo – né le 22. 4. 1905 à Würzburg, Allemagne, décédé le 11. 7. 1947 à Schepetowka, URSS – *calligraphe, artiste, enseignant* – 1924–1930, études à la Kunstgewerbeschule (école des arts décoratifs) de Munich (chez F. H. Ehmcke et Emil Preetorius) puis à l'académie des beaux-arts de Stuttgart (chez F. H. E. Schneidler). Assistant de Schneidler à l'académie des beaux-arts de Stuttgart de 1930 à 1935. Concepteur graphiste indépendant à Munich de 1935 à 1937. Professeur à l'Akademie für Graphische Künste und Buchgewerbe (académie des arts de l'écriture et du livre) de Leipzig de 1937 à 1939. Tous les deux ans depuis 1954, la ville d'Offenbach décerne, le prix «Rudo-Spemann» en sa mémoire. – *Police:* Gavotte (1942). – *Publications, sélection:* Walter Tiemann «Beseelte Kalligraphie», Leipzig 1950; Hans Adolf Halbey «Rudo Spemann 1905–1947. Monographie und Werkverzeichnis seiner Schriftkunst», Offenbach 1981.

**Spencer,** Herbert – né le 22. 6. 1924 à Londres, Angleterre – *graphiste maquettiste, photographe, peintre, auteur, enseignant* – exerce dans son atelier de Londres à partir de 1948, maquettes de livres, catalogues d'art, signets, horaires de transport et rapports de gestion. Enseigne à la Central School of Arts and Crafts à Londres, de 1949 à 1955. Publie la revue «Typographica» de 1949 à 1967, puis les annales «Penrose Annual» de 1964 à 1973. Obtient le titre de «Royal Designer for Industry» en 1965. Enseigne au Royal College of Art de Londres à partir de 1966, où il occupe la chaire d'arts graphiques de 1978 à 1985. Conseiller du Post Office Stamp Design Committee à Londres, depuis 1968. Exerce comme

Solpera 1985 Poster

Spemann ca. 1937 Cover

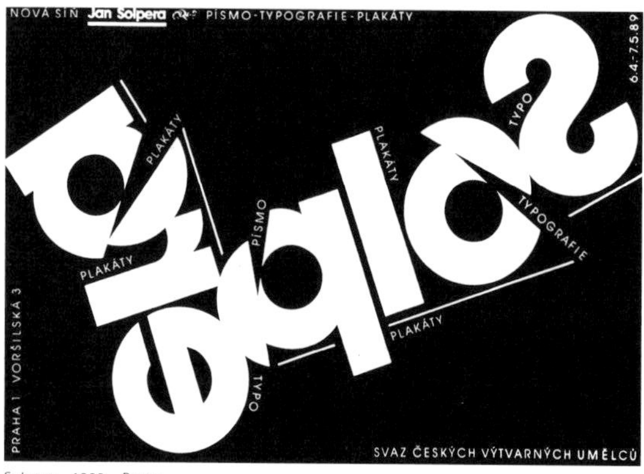


Solpera 1989 Poster

Spemann ca. 1936 Calligraphy

and in the "Das schönste Buch" competition (1975, 1977–1979). Assistant at the Prague art college. – *Font:* Circo (1971).

**Spemann,** Rudo – b. 22. 4. 1905 in Würzburg, Germany, d. 11. 7. 1947 in Shepetovka, USSR – *calligrapher, artist, teacher* – 1924–30: studies at the Kunstgewerbeschule in Munich (under F. H. Ehmcke and Emil Preetorius) and at the Kunstakademie in Stuttgart (under F. H. E. Schneidler). 1930–35: is Schneidler's assistant at the Kunstakademie in Stuttgart. 1935–37: freelance graphic artist in Munich. 1937–39: lecturer at the Akademie für Graphische Künste und Buchgewerbe in Leipzig. The city of Offenbach has awarded the Rudo Spemann Prize in memory of Spemann's work every two years since 1954. – *Font:* Gavotte (1942). – *Publications include:* Walter Tiemann "Beseelte Kalligraphie", Leipzig 1950; Hans Adolf Halbey "Rudo Spemann 1905–1947. Monographie und Werkverzeichnis seiner Schriftkunst", Offenbach 1981.

**Spencer,** Herbert – b. 22. 6. 1924 in London, England – *graphic designer, photographer, painter, author, teacher* – 1948: opens his own studio in London, where he designs books, art catalogues, logos, timetables and business reports. 1949–55: lecturer at the Central School of Arts and Crafts in London. 1949–67: editor of "Typographica" magazine. 1964–73: editor of the "Penrose Annual". 1965: made a Royal Designer for Industry. 1966: starts teaching at the Royal College of Art in London and from 1978–85 is professor of graphic arts. From 1968

College of Art in London, 1978–85 Professor für Graphic Arts. Seit 1968 Berater des Post Office Stamp Design Committee in London. Als Design-Berater für W. H. Smith Ltd., die Tate Gallery, British Rail, das Imperial War Museum, das Royal Institute of British Architects und die University of Leeds tätig. Seit 1970 Direktor von Lund Humphries Publishers in London. – *Publikationen:* „Design in Business Printing", London 1952; „The visible word", London, New York 1969; „Pioneers of Modern Typography", London 1969; „Words, words, words", London, Köln 1972; „New Alphabets A to Z" (mit Colin Forbes), London, New York 1973; „The liberated page", London 1987.

**Spiekermann,** Erik – geb. 30. 5. 1947 in Stadthagen, Deutschland. – *Typograph, Schriftentwerfer, Autor* – 1967 Arbeit als Schriftsetzer und Drucker. 1968 Studium der Kunstgeschichte und Anglistik in Berlin. Unterrichtet 1973–81 am London College of Printing. Freischaffend für die Studios Wolff Olins und Henrion Design Associates tätig. 1979 Gründung von „MetaDesign" in Berlin (mit Florian Fischer und Dieter Heil). Auftraggeber waren u. a. die Bank für Gemeinwirtschaft, Berthold AG, Linotype und Scangraphic. 1989 Gründung von „FontShop", einem Versandhaus für digitalisierte Schriften (mit Joan Spiekermann). 1990 Gründung von „MetaDesign Plus" (mit Uli Mayer und Hannes Krüger). Auftraggeber waren u. a. die Berliner Verkehrsbetriebe, der Westdeutsche Rundfunk, die Stadt Berlin und VW Audi. 1992 Gründung von „MetaDesign West" in San

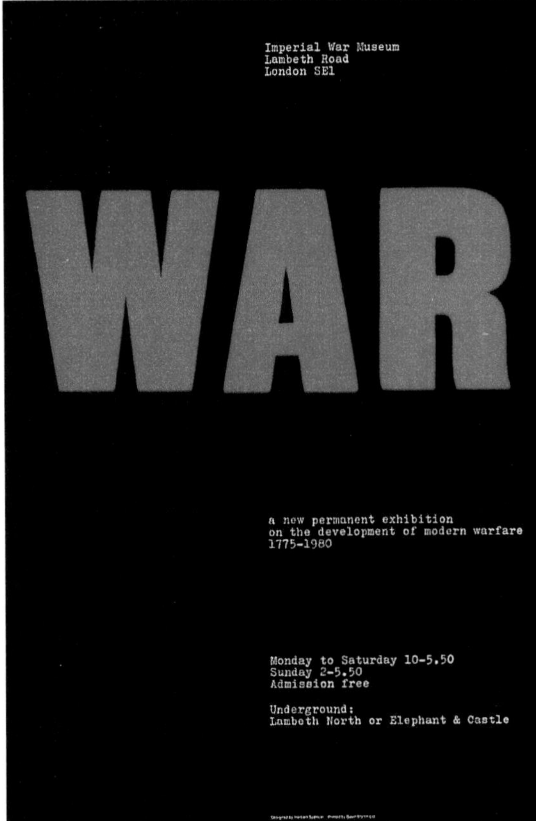


Spencer 1980 Poster

Es gibt eine Regel,
die besagt,
daß Unter- und
Oberlängen
sich nie berühren
dürfen.
Es gibt für
diese Regel die
Ausnahme,
daß Berühren
erlaubt ist, wenn's
besser aussieht.

Spiekermann 1982 Page

| | n | **n** | | | |
|---|---|---|---|---|---|
| a | b | e | f | g | i |
| o | r | s | t | y | z |
| A | B | C | E | G | H |
| M | O | R | S | X | Y |
| 1 | 2 | 4 | 6 | 8 | & |

Spiekermann 1979 Berliner Grotesk

| | n | **n** | | | *n* |
|---|---|---|---|---|---|
| a | b | e | f | g | i |
| o | r | s | t | y | z |
| A | B | C | E | G | H |
| M | O | R | S | X | Y |
| 1 | 2 | 4 | 6 | 8 | & |

Spiekermann 1990 Officina Sans

| | n | **n** | **n** | | *n* |
|---|---|---|---|---|---|
| a | b | e | f | g | i |
| o | r | s | t | y | z |
| A | B | C | E | G | H |
| M | O | R | S | X | Y |
| 1 | 2 | 4 | 6 | 8 | & |

Spiekermann 1991 Meta

conseiller en design auprès de W. H. Smith Ltd., de la Tate Gallery, de British Rail, de l'Imperial War Museum, du Royal Institute of British Architects et de l'University of Leeds. Directeur de Lund Humphries Publishers à Londres depuis 1970. – *Publications :* «Design in Business Printing», Londres 1952; «The visible word», Londres, New York 1969; «Pioneers of Modern Typography», Londres 1969; «Words, words, words», Londres, Cologne 1972; «New Alphabets A to Z» (avec Colin Forbes), Londres, New York 1973; «The liberated page», Londres 1987.

**Spiekermann,** Erik – né le 30. 5. 1947 à Stadthagen, Allemagne – *typographe, concepteur de polices, auteur* – 1967, travaille comme compositeur et imprimeur. 1968, études d'histoire de l'art et d'anglais à Berlin. Enseigne au London College of Printing de 1973 à 1981. Travaille à son compte pour les agences Wolff Olins et Henrion Design Associates. Fonde «MetaDesign» à Berlin en 1979 (avec Florian Fischer et Dieter Heil). Commanditaires : Bank für Gemeinwirtschaft, Berthold, Linotype, Scangraphic. En 1989, il fonde «FontShop» (avec Joan Spiekermann), une société qui vend des polices numériques par correspondance. En 1990, il crée «MetaDesign Plus» (avec Uli Mayer et Hannes Krüger). Commanditaires : Berliner Verkehrsbetriebe, Westdeutscher Rundfunk, Stadt Berlin, VW Audi. Fonde «MetaDesign West» à San Francisco en

onwards: consultant to the Post Office Stamp Design Committee in London. He works for W. H. Smith Ltd., the Tate Gallery, British Rail, the Imperial War Museum, the Royal Institute of British Architects and the University of Leeds as design consultant. 1970: is made director of Lund Humphries Publishers in London. – *Publications:* "Design in Business Printing", London 1952; "The visible word", London, New York 1969; "Pioneers of Modern Typography", London 1969; "Words, words, words", London, Cologne 1972; "New Alphabets A to Z" (with Colin Forbes), London, New York 1973; "The liberated page", London 1987.

**Spiekermann,** Erik – b. 30. 5. 1947 in Stadthagen, Germany – *typographer, type designer, author* – 1967: works as a typesetter and printer. 1968: studies art history and English in Berlin. 1973–81: teaches at the London College of Printing. Works freelance for the Wolff Olins studio and Henrion Design Associates. 1979: founds MetaDesign in Berlin with Florian Fischer and Dieter Heil. Clients include the Bank für Gemeinwirtschaft, Berthold, Linotype and Scangraphic. 1989: opens FontShop, a mail order firm for digitized typefaces, with Joan Spiekermann. 1990: launches MetaDesign Plus with Uli Mayer and Hannes Krüger. Clients include the Berliner Verkehrsbetriebe, Westdeutscher Rundfunk, the city of Berlin and VW Audi. 1992: opens MetaDesign West in San Francisco and

Francisco, 1997 „MetaDesign London". – *Schriftentwürfe:* Berliner Grotesk (1979), Lo Type (1980), Officina (mit J. van Rossum, 1990), Meta (1991), Info (1996). – *Publikationen u. a.:* „Ursache und Wirkung: ein typografischer Roman", Erlangen 1982; „Rhyme and Reason – A Typographic Novel", Berlin 1987; „Studentenfutter", Nürnberg 1989; „Stop stealing sheep" (mit E. M. Ginger), Mountain View 1993. Ulysses Voelker „Zimmermann meets Spiekermann", Bremen 1993.

**Srivastava,** Narendra Nath – geb. 5. 8. 1931 in Delhi, Indien. – *Kalligraph, Typograph, Schriftentwerfer, Lehrer, Illustrator* – Studium der Angewandten Kunst. Seit 1952 Lehrer an der Kunstschule in Delhi. 1967 französisches Staats-Stipendium, seitdem Arbeit in Frankreich und Indien. Zusammenarbeit mit Paul Colin. 1968 Studien über die Modernisierung der Devanagari-Typographie, Entwurf von Devanagari-Alphabeten. 1969 Zusammenarbeit mit Pierre Cardin, Textilentwürfe und Entwurf des Signets „PC". Gewinnt 1972 den Wettbewerb für das Signet des „3rd Asian International Trade Fair". Zahlreiche Ausstellungen seiner Arbeit, u. a. in Paris (1969), Delhi (1972), Bombay (1974), Nürnberg (1974) und Bonn (1975).

**St Bride Printing Library** – 1895 Gründung des St Bride Foundation Institute als technische Schule für junge Drucker. Eröffnung der St Bride Printing Library als Bibliothek des St Bride Foundation Institute. Von Beginn an wird historische Literatur über das Drucken gesammelt. Dazu kommen Schriftkataloge, techni-

1992, puis «MetaDesign London» en 1997. – *Polices:* Berliner Grotesk (1979), Lo Type (1980), Officina (avec J. van Rossum, 1990), Meta (1991), Info (1996) – *Publications, sélection:* «Ursache und Wirkung: ein typografischer Roman», Erlangen 1982; «Rhyme and Reason – A Typographic Novel", Berlin 1987; «Studentenfutter», Nuremberg 1989; «Stop stealing sheep» (avec E. M. Ginger), Mountain View 1993. Ulysses Voelker «Zimmermann meets Spiekermann», Brême 1993.

**Srivastava,** Narendra Nath – né le 5. 8. 1931 à Delhi, Inde – *calligraphe, typographe, concepteur de polices, illustrateur* – études d'arts appliqués. Enseigne à l'école des beaux-arts de Delhi à partir de 1952. Boursier de l'Etat français en 1967. Dès lors, il travaille en France et en Inde. Collaboration avec Paul Colin. 1968, études sur la modernisation de la typographie Devanagari, conception d'alphabets Devanagari. Collabore avec Pierre Cardin en 1969, design de textiles et dessin du logo «PC». En 1972, il remporte le concours du signet de la «3rd. Asian International Trade Fair». Nombreuses expositions, entre autres, à Paris (1969), Delhi (1972), Bombay (1974), Nuremberg (1974) et Bonn (1975).

**St Bride Printing Library** – Création en 1895 du St Bride Foundation Institute, une école professionnelle destinée aux jeunes imprimeurs. Ouverture de la bibliothèque du St Bride Foundation Institute, la St Bride Printing Library. Dès le début, la bibliothèque collectionne des ouvrages historiques sur l'imprimerie. Puis viennent des typothèques, des ouvrages techniques, des manuels, des imprimés anciens et des impressions à la presse manuelle. La bibliothèque possède une importante collection de presses, caractères et objets liés à la composition tra-

St Bride Printing Library

Stankowski 1929 Flyer

Standard Cover

Stankowski 1930 Logo

MetaDesign London in 1997. – *Fonts:* Berliner Grotesk (1979), Lo Type (1980), Officina (with J. van Rossum, 1990), Meta (1991), Info (1996). – *Publications include:* "Ursache und Wirkung: ein typografischer Roman", Erlangen 1982; "Rhyme and Reason – A Typographic Novel", Berlin 1987; "Studentenfutter", Nuremberg 1989; "Stop stealing sheep" (with E. M. Ginger), Mountain View 1993. Ulysses Voelker "Zimmermann meets Spiekermann", Bremen 1993.

**Srivastava,** Narendra Nath – b. 5. 8. 1931 in Delhi, India – *calligrapher, typographer, type designer, teacher, illustrator* – Studied applied art. 1952: starts teaching at the art school in Delhi. 1967: is given a French state grant. From this point on Srivastava works in France and India. Works with Paul Colin. 1968: studies the modernization of Devanagari typography and designs Devanagari alphabets. 1969: works with Pierre Cardin. Designs textiles and the "PC" logo. 1972: wins the competition for the 3rd Asian International Trade Fair logo. Numerous exhibitions of his work, including in Paris (1969), Delhi (1972), Bombay (1974), Nuremberg (1974) and Bonn (1975).

**St Bride Printing Library** – 1895: founding of the St Bride Foundation Institute, a technical school for young printers. The St Bride Printing Library is opened to serve the St Bride Foundation Institute. Even in its early stages the library starts collecting historical literature on printing, and also catalogues of type, technical literature, handbooks, early prints and press prints. The library houses a

sche Literatur, Handbücher, Früh- und Pressendrucke. Die Library besitzt eine große Sammlung von Druckmaschinen, Schriften und Gegenständen, die traditionelles Setzen und Drucken anschaulich machen. Das St Bride Foundation Institute beendet 1922 seine Unterrichtstätigkeit und geht im Laufe der Jahre im London College of Printing auf. 1953 wird die Bibliothek des St Bride Foundation Institute eine eigenständige Institution. 1966 wird die Bibliothek eine öffentliche Bibliothek der City of London. 1976 übergibt die Monotype Corporation der Bibliothek die große Sammlung von Zeichnungen und Arbeitsunterlagen von Eric Gill, die sie 1955 erworben hat.

**Standard,** Paul geb. 19. 5. 1906, gest. 1. 1. 1992. – *Kalligraph, Autor, Lehrer* – Entwickelt 1932 eine kursive Handschrift, die auf Vorlagen von Ludovico degli Arrighi aus dem 15. Jahrhundert aufbaut. Durch seine Initiative entsteht in den angelsächsischen Ländern eine Bewegung zur Erneuerung der Handschrift. Entwurf von Buchumschlägen für mehrere Verlage. Schriftarbeiten für das Museum of Modern Art in New York. Unterrichtet an der Cooper Union, New York. – *Publikationen u. a.:* „Calligraphy's Flowering, Decay & Restauration", Chicago 1947; „Arrighi's Running Hand", New York 1979.

**Stankowski,** Anton – geb. 18. 6. 1906 in Gelsenkirchen, Deutschland. – *Maler, Grafik-Designer, Fotograf, Typograph, Autor* – 1921–26 Lehre als Dekorationsmaler in Düsseldorf. 1927–29 Studium an der Folkwangschule in Essen, u. a. bei Max Burchartz. 1928–29 grafischer Mit-

Stankowski 1956 Poster

Stankowski 1978 Logo

Stankowski 1978 Poster

ditionnelle ainsi qu'à l'imprimerie. En 1922, le St Bride Foundation Institute cesse ses activités d'enseignement et fusionne progressivement avec le London College of Printing. La bibliothèque du St Bride Foundation Institute devient une Institution indépendante en 1953, puis, en 1966, une bibliothèque publique de la City of London. En 1976, la Monotype Corporation remet à la bibliothèque l'importante collection de dessins et de documents de travail d'Eric Gill qu'elle avait acquise en 1955.

**Standard,** Paul – né le 19. 5. 1906, décédé le 1. 1. 1992 – *Calligraphe, auteur, enseignant* – en 1932, il crée une écriture manuscrite cursive inspirée de modèles de Ludovico degli Arrighi et datant du 15e siècle. Grâce à ses initiatives, on assiste à la formation d'un mouvement de rénovation de l'écriture manuscrite dans les pays anglo-saxons. Conception de nombreuses couvertures de livres pour des maisons d'édition. Travaux de graphisme pour le Museum of Modern Art à New York. A enseigné à la Cooper Union à New York. – *Publications, sélection:* «Calligraphy's Flowering, Decay & Restauration», Chicago 1947; «Arrighi's Running Hand», New York 1979.

**Stankowski,** Anton – né le 18. 6. 1906 à Gelsenkirchen, Allemagne – *peintre, graphiste maquettiste, photographe, typographe, auteur* – 1921–1926, apprentissage de peintre décorateur à Düsseldorf. 1927–1929, études à la Folkwangschule d'Essen, entre autres chez Max Burchartz. Employé comme graphiste à l'atelier

large collection of printing machines, typefaces, and objects which illustrate traditional practices of typesetting and printing. 1922: teaching is stopped at the St Bride Foundation Institute which over the years is amalgamated with the London College of Printing. 1953: the library of the St Bride Foundation Institute becomes a separate institution. 1966: the library becomes one of the City of London's public libraries. 1976: Monotype Corporation presents the library with a large collection of Eric Gill's documents and drawings, acquired by the company in 1955.

**Standard,** Paul – b. 19. 5. 1906, d. 1. 1. 1992 – *calligrapher, author, teacher* – 1932: develops an italic script based on 15th-century models by Ludovico degli Arrighi. Individual styles of handwriting experienced something of a revival in Anglo-Saxon countries through Standard's initiative. He designed book covers for various publishers and did typographic work for the Museum of Modern Art in New York. He also taught at the Cooper Union in New York. – *Publications:* "Calligraphy's Flowering, Decay & Restauration", Chicago 1947; "Arrighi's Running Hand", New York 1979.

**Stankowski,** Anton – b. 18. 6. 1906 in Gelsenkirchen, Germany – *painter, graphic designer, photographer, typographer, author* – 1921–26: trains as an interior decorator in Düsseldorf. 1927–29: studies at the Folkwangschule in Essen. Max Burchartz is one of his tutors. 1928–29: works in graphics for Johannes Canis's "werbe-bau" studio in Bochum.

arbeiter im Büro »werbe-bau« von Johannes Canis in Bochum. 1929–36 Arbeit als Grafiker im Reklame-Atelier Max Dalang in Zürich. 1949–51 Fotoreporter und Schriftleiter der Zeitschrift „Stuttgarter Illustrierte". 1951 Gründung seines Gestaltungsbüros, Auftraggeber waren u. a. SEL, der Süddeutsche Rundfunk, Behr-Möbel, Vissmann und die Deutsche Bank. 1965 Entwurf eines Erscheinungsbilds für die Stadt Berlin. 1969–72 Vorsitzender des Ausschusses für visuelle Gestaltung des Organisationskomitees für die Olympischen Spiele 1972. Seit 1977 Zusammenarbeit mit Karl Duschek, Umbenennung des Büros in „Stankowski + Duschek". 1976 mit dem Professoren-Titel durch das Land Baden-Württemberg ausgezeichnet. Zahlreiche Ausstellungen, u. a. im Kunsthaus Zürich (1979). 1983 Gründung der „Stankowski-Stiftung" zur Förderung von Kunst und Design. Bisherige Preisträger u. a. Almir Mavignier (1985), Hans Peter Hoch (1988), Wim Crouwel. – *Publikationen u. a.:* „Funktion und ihre Darstellung in der Werbegrafik", Teufen 1964; „Firmenimage" (mit O. S. Rechenauer), Düsseldorf 1969; „Der Pfeil" (mit J. Stankowski, E. Gomringer), Starnberg 1972; „Anzeige, Inserat" (mit K. Duschek), Stuttgart 1976; „Gucken" (mit E. Gomringer), Leonberg 1979; „Typogramme" (mit H. Heissenbüttel), Leonberg 1985; „Visuelle Kommunikation" (Hrsg. mit K. Duschek), Berlin 1989; „Formfinden", Stuttgart 1991. Günther Wirth, Eugen Gomringer, Stephan von Wiese (Hrsg.) „Anton Stankowski. Das Gesamtwerk", Stuttgart 1983.

«werbe-bau» de Johannes Canis à Bochum de 1928 à 1929. Graphiste à l'atelier de publicité de Max Dalang à Zurich de 1929 à 1936. Reporter photographe et directeur de publication de la revue «Stuttgarter Illustrierte» de 1949 à 1951. Fonde son agence de design en 1951; commanditaires: SEL, Süddeutscher Rundfunk, Behr-Möbel, Vissmann, Deutsche Bank. 1965, dessine l'identité de la ville de Berlin. 1969–1972, président de la commission de la création visuelle du comité d'organisation des Jeux Olympiques de 1972. Travaille avec Karl Duschek à partir de 1977, l'agence s'appelle désormais «Stankowski + Duschek». En 1976, le Land de Bade-Wurtemberg lui décerne le titre le professeur. Nombreuses expositions, entre autres au Kunsthaus de Zurich (1979). Création en 1983 de la «Stankowski Stiftung» (fondation Stankowski) pour promouvoir les arts et le design. Lauréats, sélection: Almir Mavignier (1985), Hans Peter Hoch (1988), Wim Crouwel. – *Publications, sélection:* «Funktion und ihre Darstellung in der Werbegrafik», Teufen 1964; «Firmenimage» (avec O. S. Rechenauer), Düsseldorf 1969; «Der Pfeil» (avec J. Stankowski, E. Gomringer), Starnberg 1972; «Anzeige, Inserat» (avec K. Duschek), Stuttgart 1976; «Gucken» (avec E. Gomringer), Leonberg 1979; «Typogramme» (avec H. Heissenbüttel), Leonberg 1985; «Visuelle Kommunikation» (éd., avec K. Duschek), Berlin 1989; «Formfinden», Stuttgart 1991. Günther Wirth, Eugen Gomringer, Stephan von Wiese (éd.) «Anton Stankowski. Das Gesamtwerk», Stuttgart 1983.

**Staudt,** Rolf – né le 25. 12. 1936 à Coblence, Allemagne – *typographe, enseignant* – apprentissage de composition typographique. 1960–1963, cours du soir de l'école d'arts graphiques de Francfort-sur-

Stankowski 1983 Advertisement

Stankowski 1990 Cover

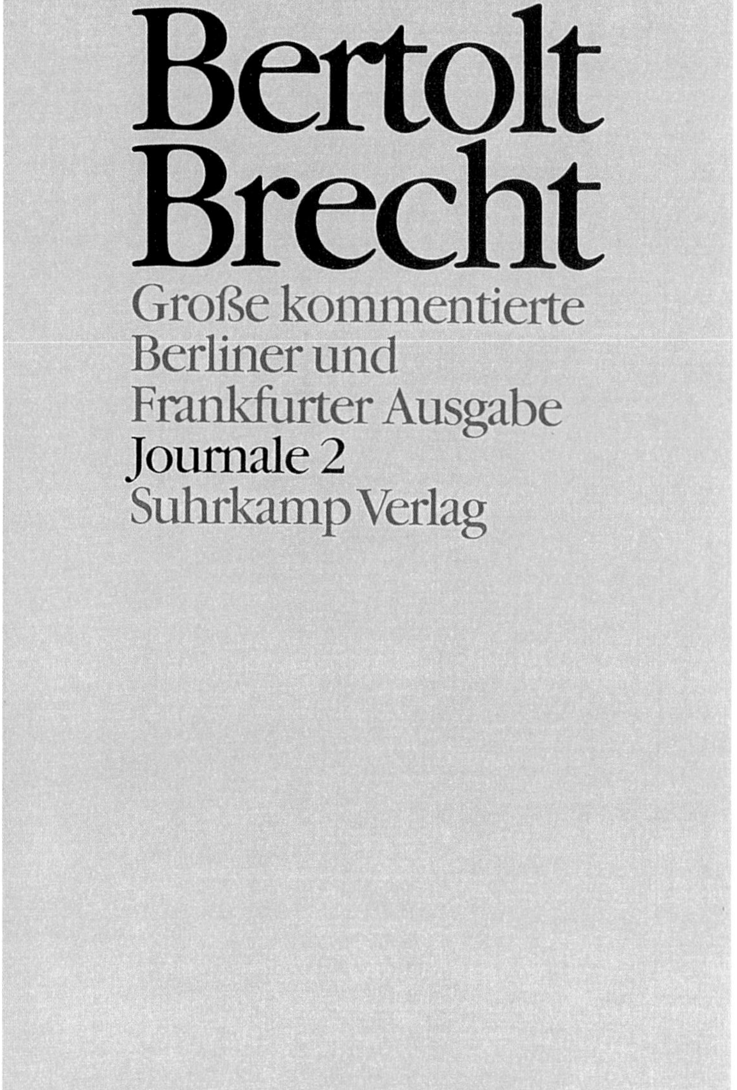

Staudt 1995 Cover

1929–36: works as a graphic artist for the Max Dalang advertising studio in Zurich. 1949–51: press photographer and editor of "Stuttgarter Illustrierte" magazine. 1951: opens his own design studio. Clients include SEL, Süddeutscher Rundfunk, Behr-Möbel, Vissmann and the Deutsche Bank. 1965: designs a corporate identity for the city of Berlin. 1969–72: chairman of the visual design committee for the 1972 Olympic Games. 1977: starts working with Karl Duschek. Renames his studio Stankowski + Duschek. 1976: is made a professor by the state of Baden-Württemberg. Numerous exhibitions, including at the Kunsthaus in Zurich (1979). 1983: launches the Stankowski-Stiftung, a foundation which aims to promote art and design. Prize-winners to date have included Almir Mavignier (1985), Hans Peter Hoch (1988) and Wim Crouwel. – *Publications include:* "Funktion und ihre Darstellung in der Werbegrafik", Teufen 1964; "Firmenimage" (with O. S. Rechenauer), Düsseldorf 1969; "Der Pfeil" (with J. Stankowski, E. Gomringer), Starnberg 1972; "Anzeige, Inserat" (with K. Duschek), Stuttgart 1976; "Gucken" (with E. Gomringer), Leonberg 1979; "Typogramme" (with H. Heissenbüttel), Leonberg 1985; "Visuelle Kommunikation" (co-ed. with K. Duschek), Berlin 1989; "Formfinden", Stuttgart 1991. Günther Wirth, Eugen Gomringer, Stephan von Wiese (eds.) "Anton Stankowski. Das Gesamtwerk", Stuttgart 1983.

**Staudt,** Rolf – b. 25. 12. 1936 in Koblenz, Germany – *typographer, teacher* –

**Staudt,** Rolf – geb. 25. 12. 1936 in Koblenz Deutschland – *Typograph, Lehrer* – Schriftsetzerlehre. 1960–63 Besuch der Abendschule der Grafischen Fachschule in Frankfurt am Main. 1964 Herstellungsleiter des Insel Verlags, erste Kontakte mit Willy Fleckhaus, der die Umschläge der „sammlung insel" gestaltet; Staudt entwirft Typographie und Einbände. 1971 Umschlagentwurf der „Suhrkamp Taschenbücher" (mit W. Fleckhaus). 1972 Konzept für die „insel taschenbücher", der ersten Taschenbuchreihe in Deutschland mit individueller Typographie im Buchinnern. U. a. Gestaltung der Bücher der „Polnischen Bibliothek" im Suhrkamp Verlag (1982) und der „Bibliothek des Deutschen Klassiker Verlags" (1985). Lehrer für Typographie und Buchherstellungstechnik an der Schule des Deutschen Buchhandels in Frankfurt am Main (1964–70) und beim Verlagsfachwirt-Lehrgang der IHK Wiesbaden (1985–94). Zahlreiche Auszeichnungen.

**Stauffacher,** Jack Werner – geb. 19. 12. 1920 in San Francisco, USA. – *Drucker, Verleger, Typograph, Lehrer* – Als Gestalter Autodidakt. 1947 Gründung der „Greenwood Press" in San Francisco, die später einen großen Einfluß auf die amerikanische Privatpressen-Bewegung hat. 1955–58 Studium in Florenz als Fulbright-Stipendiat. 1960–63 Direktor der „New Laboratory Press" am Carnegie Institute of Technology in Pittsburgh. 1974 Berufung als Studienleiter an die University of California in Santa Cruz. 1979 Gründung des „Center for Typographic Language" in San Francisco. Mitglied

Staudt Cover

GOTTHOLD
EPHRAIM
LESSING
WERKE
UND
BRIEFE

in zwölf Bänden

Herausgegeben von Wilfried Barner
zusammen mit Klaus Bohnen, Gunter E. Grimm,
Helmuth Kiesel, Arno Schilson,
Jürgen Stenzel und Conrad Wiedemann

Band 11/2

BRIEFE
VON UND AN
LESSING
1770-1776

Herausgegeben von
Helmuth Kiesel
unter Mitwirkung
von Georg Braungart,
Klaus Fischer
und Ute Wahl

DEUTSCHER
KLASSIKER
VERLAG

Staudt 1988 Main Title

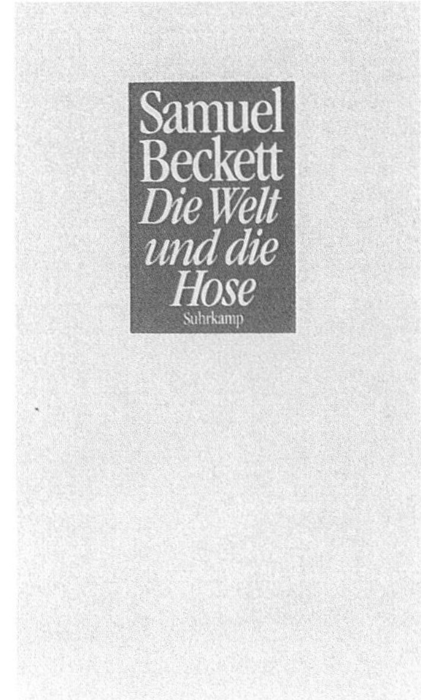

Staudt 1990 Cover

Stauffacher ca. 1950 Emblem

le-Main. Directeur de fabrication au Insel Verlag en 1964, premiers contacts avec Willy Fleckhaus, chargé de la maquette des couvertures de la collection «sammlung insel»; Staudt conçoit la typographie et la reliure. En 1971, il dessine la couverture des «Suhrkamp Taschenbücher» (avec W. Fleckhaus). En 1972, il développe un concept pour les «insel taschenbücher», la première collection de livres de poches allemands présentant une typographie et une maquette individuelle. Maquette des ouvrages de la «Polnische Bibliothek» du Suhrkamp Verlag (1982), de la «Bibliothek des Deutschen Klassiker Verlags» (1985). Enseigne la typographie et les techniques de fabrication du livre à la Schule des Deutschen Buchhandels (école de la librairie allemande) à Francfort-sur-le-Main (1964–1970) et au Verlagsfachwirt-Lehrgang (études de gestion de l'édition) de l'IHK Wiesbaden (1985–1994). Nombreuses distinctions.

**Stauffacher,** Jack Werner – né le 19. 12. 1920 à San Francisco, Etats-Unis – *imprimeur, éditeur, typographe, enseignant* – designer autodidacte. En 1947, il fonde la «Greenwood Press» à San Francisco, cette publication exercera une grande influence sur le mouvement des presses privées américaines. Obtient une bourse Fulbright et étudie à Florence de 1955 à 1958. Directeur de la «New Laboratory Press» du Carnegie Institute of Technology à Pittsburgh de 1960 à 1963. Directeur d'études à l'University of California de Santa Cruz en 1974. Fonde, en 1979, le «Center for Typographic Language» à San Francisco. Membre du comité con-

Trained as a typesetter. 1960–63: attends evening classes at the Grafische Fachschule in Frankfurt am Main. 1964: production manager for Insel publishers. Meets Willy Fleckhaus who designs the jackets for Insel's "sammlung insel". Staudt designs the typography and covers. 1971: designs covers for Suhrkamp Taschenbücher with W. Fleckhaus. 1972: produces a concept for insel taschenbücher, the first series of paperbacks in Germany with text printed in an individual typography. Staudt has designed the books in Suhrkamp's "Polnische Bibliothek" (1982) and the "Bibliothek des Deutschen Klassiker Verlags" (1985), among others. He has taught typography and the techniques of book production at the Schule des Deutschen Buchhandels in Frankfurt am Main (1964–70) and at training courses for publishing management experts at the IHK Wiesbaden (1985–94). Numerous awards.

**Stauffacher,** Jack Werner – b. 19. 12. 1920 in San Francisco, USA – *printer, publisher, typographer, teacher* – Self-taught designer. 1947: launches Greenwood Press in San Francisco, later a major influence on the private press movement in America. 1955–58: studies in Florence on a Fulbright grant. 1960–63: director of New Laboratory Press at the Carnegie Institute of Technology in Pittsburgh. 1974: made principal lecturer at the University of California in Santa Cruz. 1979: founds the Center for Typographic Language in San Francisco. Member of the advisory committee for

des Beirats der Zeitschrift „Visible Language". – *Publikationen u. a.:* „Janson: A Definitive Collection", San Francisco 1954; „Hunt roman: the birth of a type", Pittsburgh 1965; „A printed word has its own measure", San Francisco 1969; „A search for the typographic form of Plato's Phaedrus", San Francisco 1978.

**Steiner,** Albe (Alberto) – geb. 15. 11. 1913 in Mailand, Italien, gest. 7. 8. 1974 in Raffaderi, Italien. – *Grafik-Designer, Typograph, Lehrer* – Ausbildung als Buchhalter, als Gestalter Autodidakt. Seit 1933 Arbeit als Bildredakteur, Ausstellungsgestalter und Grafik-Designer in Mailand. Auftraggeber waren u. a. die Verlage Einaudi, Avanti, Vangelista, Zanichelli und Firmen wie Agfa, Bertelli, Pirelli und Olivetti. 1944 im politischen Widerstand aktiv. Gestaltung der Ausstellung „Mostra della Liberazione". 1945 Art Director der Wochenzeitung „Il Politecnico". Lebt und arbeitet 1946–47 in Mexiko. 1948–58 Lehrer für Geschichte und Technik der grafischen Künste an der Rinascita-Schule in Mailand. 1950–54 Art Director des Kaufhauses „La Rinascente". 1951 Auszeichnung mit dem Grand Prix der Triennale in Mailand. Seit 1954 künstlerischer Berater des Verlags Feltrinelli in Mailand. 1954 mit einer Goldmedaille der Triennale in Mailand ausgezeichnet. 1958–74 Direktor der Scuola dell'Libro Umanitaria. 1962–70 Dozent am Istituto d'Arte in Urbino. – *Publikationen:* „Fotografia", Mailand 1943; „Si fare presto a dire fame", Mailand 1958.

**Steiner,** Heiri (Heinrich) – geb. 1. 10. 1906 in Horgen, Schweiz, gest. 7. 5. 1983 in

sultatif de la revue «Visible Language». – *Publications, sélection:* «Janson: A Definitive Collection», San Francisco 1954; «Hunt roman: the birth of a type», Pittsburgh 1965; «A printed work has its own measure», San Francisco 1969; «A search for the typographic form of Plato's Phaedrus», San Francisco 1978.

**Steiner,** Albe (Alberto) – né le 15. 11. 1913 à Milan, décédé le 7. 8. 1974 à Raffaderi, Italie – *graphiste maquettiste, typographe, enseignant* – formation de comptable, designer autodidacte. Travaille comme rédacteur iconographe, architecte d'expositions, graphiste maquettiste à Milan à partir de 1933. Commanditaires: éditions Einaudi, Avanti, Vangelista, Zanichelli. Travaille aussi pour des sociétés: Agfa, Bertelli, Pirelli, Olivetti. Milite dans la résistance politique en 1944. Architecture de l'exposition «Mostra della Liberazione». Directeur artistique de l'hebdomadaire «Il Politecnico» en 1945. Vit et travaille au Mexique de 1946 à 1947. Enseigne l'histoire et les techniques graphiques de 1948 à 1958 à la l'école Rinascita à Milan. Directeur artistique du grand magasin «La Rinascente» de 1950 à 1954. Obtient le Grand Prix de la Triennale de Milan en 1951. Conseiller artistique des éditions Feltrinelli à Milan à partir de 1954. 1954, médaille d'or de la Triennale de Milan. Directeur de la Scuola dell'Libro Umanitaria de 1958 à 1974. Enseigne à l'Istituto d'Arte à Urbino de 1962 à 1970. – *Publications:* «Fotografia», Milan 1943; «Si fare presto a dire fame», Milan 1958.

**Steiner,** Heiri (Heinrich) – né le 1. 10. 1906 à Horgen, Suisse, décédé le 7. 5. 1983 à Hombrechtikon, Suisse – *graphiste, typographe, illustrateur, architecte d'expositions* – 1924–1927, formation à la Kunstgewerbeschule (école des arts décoratifs) de Zurich chez Ernst Keller, puis

Albe Steiner 1951 Advertisement

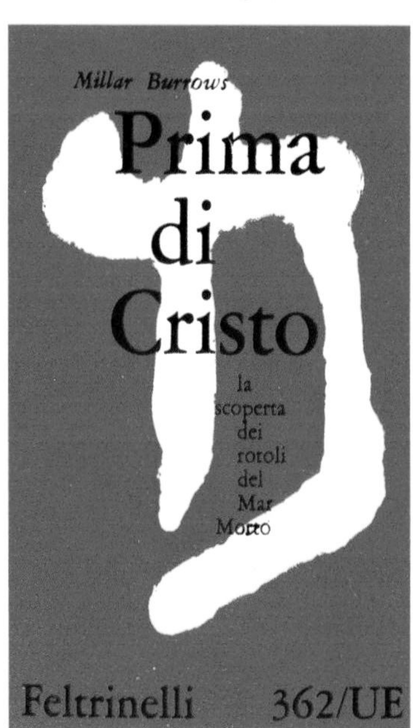


Albe Steiner 1957 Cover

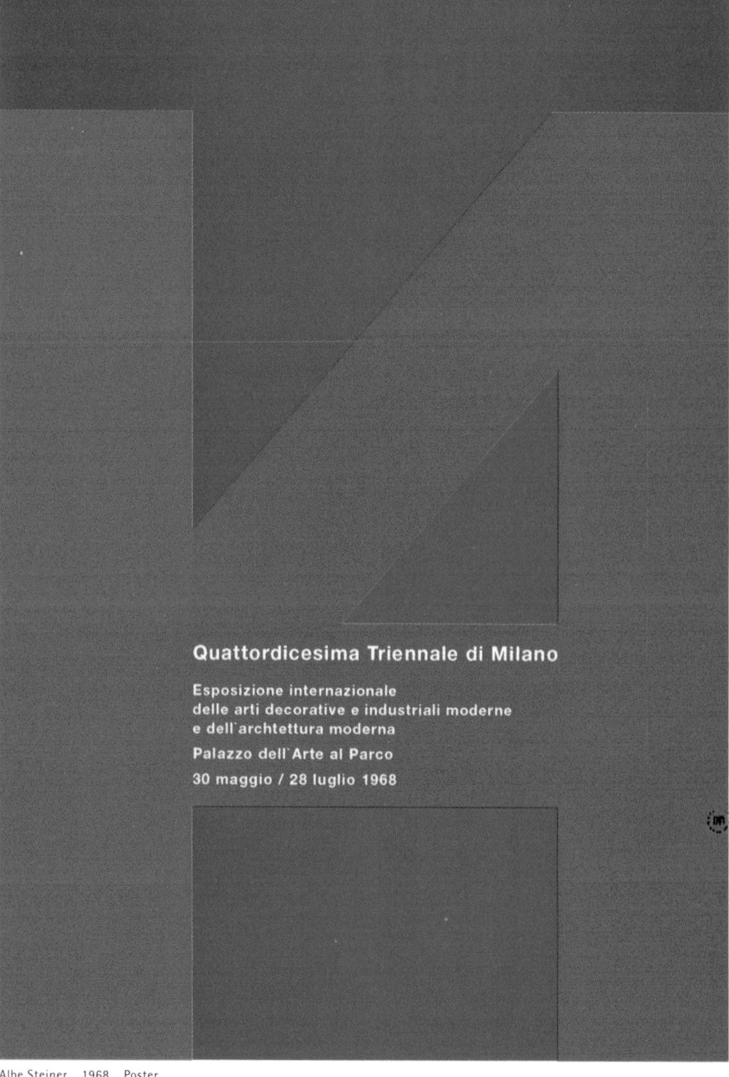


Albe Steiner 1968 Poster

"Visible Language" magazine. – *Publications include:* "Janson: A Definitive Collection", San Francisco 1954; "Hunt roman: the birth of a type", Pittsburgh 1965; "A printed word has its own measure", San Francisco 1969; "A search for the typographic form of Plato's Phaedrus", San Francisco 1978.

**Steiner,** Albe (Alberto) – b. 15. 11. 1913 in Milan, Italy, d. 7. 8. 1974 in Raffaderi, Italy – *graphic designer, typographer, teacher* – Trained as a bookkeeper. Self-taught designer. 1933: starts working as picture editor, exhibition planner and graphic designer in Milan. Clients include Einaudi, Avanti, Vangelista and Zanichelli publishing houses and companies such as Agfa, Bertelli, Pirelli and Olivetti. 1944: active member of the political resistance movement in Italy. Designs the "Mostra della Liberazione" exhibition. 1945: art director of "Il Politecnico" weekly. 1946–47: lives and works in Mexico. 1948–58: teaches history and graphic arts technology at the Rinascita school in Milan. 1950–54: art director of "La Rinascente" department store. 1951: awarded the Grand Prix at the Triennale in Milan. From 1954 onwards: art consultant to Feltrinelli publishers in Milan. 1954: awarded a gold medal at the Triennale in Milan. 1958–74: director of the Scuola dell'Libro Umanitaria. 1962–70: lecturer at the Istituto d'Arte in Urbino. – *Publications:* "Fotografia", Milan 1943; "Si fare presto a dire fame", Milan 1958.

**Steiner,** Heiri (Heinrich) – b. 1. 10. 1906 in Horgen, Switzerland, d. 7. 5. 1983 in Hombrechtikon, Switzerland – *graphic*

Hombrechtikon, Schweiz. – *Grafiker, Typograph, Illustrator, Ausstellungsgestalter* – 1924–27 Ausbildung an der Kunstgewerbeschule in Zürich bei Ernst Keller und an der Vereinigten Staatsschule in Berlin bei O. H. W. Hadank. Ab 1929 eigenes Atelier in Zürich. 1930–35 Lehrer für Grafik an der Kunstgewerbeschule in Zürich. 1934–39 Zusammenarbeit mit dem Fotografen Ernst Heiniger. 1937 Gestaltung einer Abteilung des Schweizer Pavillons auf der Weltausstellung in Paris. 1939 Gestaltung des Pavillons „Vorbeugen und Heilen" auf der Landesausstellung in Zürich. Lebt und arbeitet 1947–58 in Paris. Art Director für die Publikationen der UNESCO, grafische Arbeiten für den U.S. Information Service. 1959 Rückkehr nach Zürich. Arbeiten für Auftraggeber wie Orell Füssli, den Verlag Feltrinelli, das Schauspielhaus Zürich und das Zoologische Museum Zürich. Gestaltung wissenschaftlicher Ausstellungen für die Universität Zürich.

**D. Stempel AG** – *Schriftgießerei* – 1895 Gründung der D. Stempel AG in Frankfurt am Main durch David Stempel (1869–1927). 1898 Eintritt von W. Cunz und P. Scondo als Teilhaber in die Firma. 1900 Vertrag zur alleinigen Herstellung von Linotype-Matrizen in Europa. 1941 wird die Mergenthaler Setzmaschinen-Fabrik in Berlin Mehrheitsaktionärin der D. Stempel AG. Seit 1977 Herstellung von Fotosatzgeräten. 1983 Einstellung der Produktion von Linotype-Matrizen. Linotype erwirbt 1985 den Schriftenbereich der D. Stempel AG. 1986 Beendigung der Tätigkeit der D. Stempel AG. Stiftung der

Heiri Steiner 1928 Poster

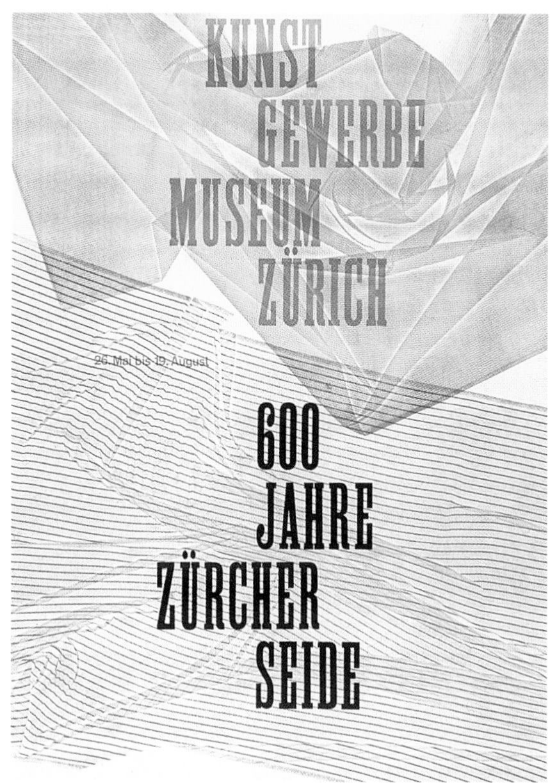


Heiri Steiner 1951 Poster

GOTISCH

Die hohen Dome, die aus alten Städten emporragen, und die gotische Schrift sind der kennzeichnende Ausdruck des Geistes und Wesensgrundes des hohen Mittelalters. Im 13. Jahrhundert hat sich die karolingische Kleinbuchstabenschrift durch Brechung der Rundungen in allmählichem Übergang zu jener eckigen, hochgerechten Schrift entwickelt, die später von den Humanisten die Bezeichnung „gotisch" erhalten hat. Seine Hochblüte hatte dieser Schriftstil zugleich mit der Baukunst im 14. und 15. Jahrhundert. Gutenberg erfand in dieser Zeit die Druckkunst. Die Schriften, die er schnitt und goß, sind daher ihrem Stile nach hochgotisch. Diese ureigene Überlieferung des Buchdrucks hat ihre weiterwirkende Kraft bewahrt. Die gotische Schrift ist in Deutschland lebendig geblieben und in unserer Zeit zu neuer kraftvoller Entfaltung gebracht worden.

Gotische Schriften der Schriftgießerei D. Stempel Akt.-Ges. Frankfurt am Main

Caslon-Gotisch Jaecker-Schrift Tannenberg Gotenburg

D. Stempel AG 1940 Advertisement

Balzac

ABCDEFGHIJKLMN
OPQRSTUVWXYZ
abcdefghijklmno
pqrstuvwxyz
1234567890

D. Stempel AG 1953 Balzac

à la Vereinigte Staatsschule à Berlin, chez O. H. W. Hadank. Travaille dans son propre atelier à Zurich à partir de 1929. Enseigne les arts graphiques à la Kunstgewerbeschule de Zurich de 1930 à 1935. Collaboration avec le photographe Ernst Heiniger de 1934 à 1939. Architecture intérieure d'une section du pavillon suisse lors de l'exposition universelle de Paris, en 1937. Architecture intérieure du pavillon «Vorbeugen und Heilen» (Prévenir et guérir) à l'exposition du Zurich en 1939. Vit et travaille à Paris de 1947 à 1958. Directeur artistique des publications de l'UNESCO, travaux graphiques pour l'U.S. Information Service. Retour à Zurich en 1959. Travaux pour Orell Füssli, les éditions Feltrinelli, le Schauspielhaus et le Jardin zoologique de Zurich. Architecture d'expositions scientifiques pour l'université de Zurich.

**D. Stempel AG** – *fonderie de caractères* – la D. Stempel AG a été fondée en 1895 à Francfort-sur-le-Main par David Stempel (1869–1927). W. Cunz et P. Scondo entrent dans la société en 1898 en qualité d'associés. Contrat d'exclusivité pour la fabrication de matrices Linotype en Europe, en 1900. La Mergenthaler Setzmaschinen-Fabrik de Berlin devient actionnaire majoritaire de la D. Stempel AG, en 1941. Fabrication de machines à photocomposer depuis 1977; la production des matrices Linotype cesse en 1983. Linotype acquiert le secteur polices de la D. Stempel AG en 1985. La société cesse ses activités en 1986. Elle fait don de machines et caractères à la Technische Hoch-

*artist, typographer, illustrator, exhibition planner* – 1924–27: trains at the Kunstgewerbeschule in Zurich under Ernst Keller and at the Vereinigte Staatsschule in Berlin under O. H. W. Hadank. 1929: opens his own studio in Zurich. 1930–35: teaches graphics at the Kunstgewerbeschule in Zurich. 1934–39: works with photographer Ernst Heiniger. 1937: designs a section of the Swiss pavilion at the world exhibition in Paris. 1939: designs the "Vorbeugen und Heilen" ("prevention and cure") pavilion at the national exhibition in Zurich. 1947–58: lives and works in Paris. Is art director for UNESCO's publications and does graphic work for the U.S. Information Service. 1959: returns to Zurich. Works for clients who include Orell Füssli, Feltrinelli publishers and the Schauspielhaus and zoological museum in Zurich. Steiner also designed science exhibitions for Zurich University.

**D. Stempel AG** – *type foundry* – 1895: David Stempel (1869–1927) founds the D. Stempel AG type foundry in Frankfurt am Main. 1898: W. Cunz and P. Scondo join the company as partners. 1900: are granted a contract making them the sole manufacturers of Linotype matrices in Europe. 1941: the Mergenthaler typesetting machine factory in Berlin becomes majority stockholder of D. Stempel AG. 1977: the company starts manufacturing filmsetting machines. 1983: production of Linotype matrices is stopped. 1985: Linotype buys up D. Stempel AG's type department. 1986: D. Stempel AG closes down. The company donates its machines

Maschinen und Schriften an die Technische Hochschule in Darmstadt.

**Stenberg,** Wladimir und Georgij – *Künstler, Illustratoren, Grafiker, Bühnenbildner* – Georgij Stenberg – geb. 20. 3. 1900 in Moskau, Rußland, gest. 15. 10. 1933 in Moskau, UdSSR. Wladimir Stenberg – geb. 4. 4. 1899 in Moskau, Rußland, gest. 1. 5. 1982 in Moskau, UdSSR. – 1912–17 Studium der Email- und Keramikdekoration sowie Bühnenbild am Stroganow-Institut für Kunst in Moskau. 1917–33 Arbeit als Künstlerteam unter dem Signet „2 Stenberg 2". 1918–20 Studium am Institut für Malerei, Bildhauerei und Architektur in Moskau. 1919 Gründungsmitglieder der Künstlergruppe „Obmochu". 1920–24 Mitglieder der „Inchuk". 1922 Teilnahme an der „Ersten Russischen Kunstausstellung" in Berlin. Seit 1923 Entwurf von Filmplakaten; insgesamt entstehen ca. 300 Plakate der Brüder Stenberg und zahlreiche Buchumschläge. 1923 Gestaltung des Pavillons der „Ersten Allrussischen Ausstellung für Landwirtschaft und Heimindustrie" in Moskau. 1925 Verleihung einer Goldmedaille auf der „Exposition Internationale des Art Décoratifs et Industriels Modernes" in Paris. Organisation der ersten Ausstellung von Filmplakaten in Moskau. 1928–33 Gestaltung der Dekorationen auf dem Roten Platz in Moskau zur Feier der Oktoberrevolution. Unterrichten 1929–32 am Institut für Architektur in Moskau. 1933 Unfalltod Georgijs. 1949–61 Entwurf von Dioramen (Durchschaubilder) für Museen. 1952 Verhaftung. 1953 Rehabilitierung. – *Publikation u. a.:*

schule de Darmstadt.

**Stenberg,** Vladimir et Georgy – *artistes, illustrateurs, graphistes, scénographes* – Georgy Stenberg – né le 20. 3. 1900 à Moscou, Russie, décédé le 15. 10. 1933 à Moscou, URSS. Vladimir Stenberg – né le 4. 4. 1899 à Moscou, Russie, décédé le 1. 5. 1982 à Moscou, URSS. Etudient la céramique et les émaux ainsi que la scénographie à l'institut d'art Stroganoff de Moscou, de 1912 à 1917. Travaillent en équipe de 1917 à 1933 sous le logo «2 Stenberg 2». Etudes à l'institut de peinture, sculpture et architecture de Moscou de 1918 à 1920. Membres fondateurs du groupe d'artistes «Obmochu» en 1919. Membres de «Inchuk» de 1920 à 1924. Participent à la «Erste russische Kunstausstellung» (première exposition d'art russe) à Berlin en 1922. Affiches de films à partir de 1923. Les frères Stenberg réaliseront près de 300 affiches et de nombreuses couvertures de livres. Architecture intérieure du pavillon de la «Première exposition sur l'agriculture et l'industrie à domicile en Grande Russie» en 1923 à Moscou. Médaille d'or de l' «Exposition Internationale des Arts Décoratifs et Industriels Modernes» en 1925, à Paris. Organisent la première exposition d'affiches de films à Moscou. Scénographie et décors pour les cérémonies commémoratives de la Révolution d'octobre sur la Place Rouge de 1928 à 1933. Enseignent de 1929 à 1932 à l'institut d'architecture de Moscou. Georgy meurt dans un accident en 1933. Création de dioramas pour les musées de 1949 à 1961. Arrestation en 1952 et réhabilitation en 1953. – *Publication, sélection:* B. Nakow «2 Stenberg 2», Paris, Londres 1975.

**Stephenson & Blake** – *fonderie de caractères* – en 1819, William Garnett, John Stephenson et James Blake achètent la fonderie de William Caslon IV à Londres,

Stenberg 1923 Poster

Stenberg 1928 Poster

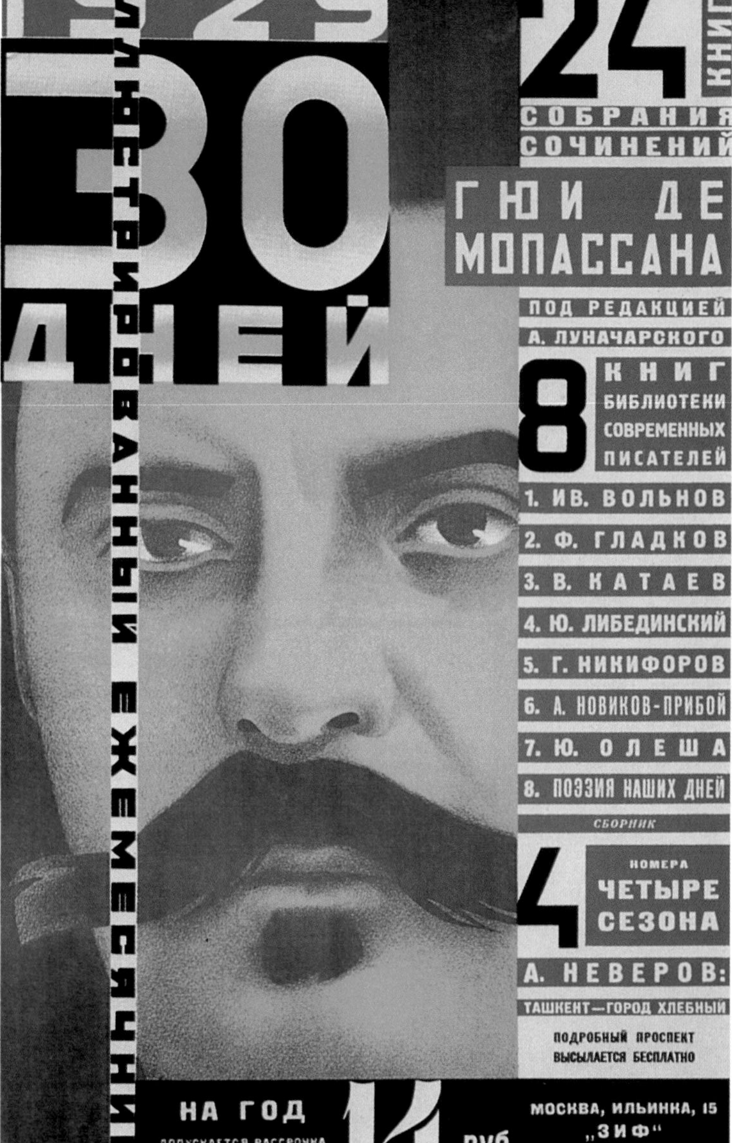


Stenberg 1929 Poster

and typefaces to the Technische Hochschule in Darmstadt.

**Stenberg,** Vladimir and Georgii – *artists, illustrators, graphic artists, set designers* – Georgii Stenberg – b. 20. 3. 1900 in Moscow, Russia, d. 15. 10. 1933 in Moscow, USSR – Vladimir Stenberg – b. 4. 4. 1899 in Moscow, Russia, d. 1. 5. 1982 in Moscow, USSR – 1912–17: they study enamel and ceramic decoration and set design at the Stroganov Art Institute in Moscow. 1917–33: they work together as a team under the name "2 Stenberg 2". 1918–20: study at the Institute of Painting, Sculpture and Architecture in Moscow. 1919: founder members of the artists' group Obmochu. 1920–24: members of Inchuk. 1922: take part in the "First Exhibition of Russian Art" in Berlin. 1923: start designing film posters. The two brothers design a total of c. 300 posters and numerous book covers. 1923: design the pavilion for the "First All-Russian Exhibition for Agriculture and Cottage Industries" in Moscow. 1925: are awarded a gold medal at the "Exposition Internationale des Art Décoratifs et Industriels Modernes" in Paris. Organize the first exhibition of film posters in Moscow. 1928–33: design the decorations for the October Revolution celebrations on Red Square in Moscow. 1929–32: teach at the Institute of Architecture in Moscow. 1933: Georgii is killed in an accident. 1949–61: Vladimir designs dioramas (see-through pictures) for museums. 1952: is arrested. 1953: his name is cleared. – *Publications include:* B. Nakow "2 Stenberg 2", Paris, London 1975.

B. Nakow „2 Stenberg 2", Paris, London 1975.

**Stephenson & Blake** – *Schriftgießerei* – William Garnett, John Stephenson und James Blake kaufen 1819 die Schriftgießerei von William Caslon IV. in London, verlegen sie nach Sheffield und beginnen unter dem Namen Blake, Garnett & Co. die Produktion. Nach dem Austritt von Garnett 1829 Umbenennung der Firma in Blake & Stephenson mit neuen Teilhabern. 1841 Umbenennung in Stephenson, Blake & Co. 1905 Angliederung der Schriftgießerei Charles Reed & Sons. Seit 1914 heißt die Firma Stephenson, Blake & Co. Ltd. 1937 Kauf der Schriftgießerei H. W. Caslon & Co. Im Laufe der Jahre wird die Schriftgießerei eine Tochtergesellschaft der Stephenson Blake (Holdings) Ltd.

**Stiftung Bauhaus Dessau** – Aus dem Bauhaus Dessau DDR hervorgegegangen, hat die Stiftung Bauhaus ihren Sitz im Gebäude des Architekten und Bauhaus-Gründers Walter Gropius. Die Stiftung hat sich als Ort der internationalen Auseinandersetzung mit Problemen des Strukturwandels etabliert. Mit dem Ziel „das Erbe des historischen Bauhauses zu bewahren" arbeiten in der Stiftung die Werkstatt (projektorientiert), die Akademie (pädagogisch) sowie die Sammlung mit Ausstellungen. Das Erscheinungsbild wird seit 1990 von den Gruppen Grappa und Cyan betreut.

**Stoecklin,** Niklaus – geb. 19. 4. 1896 in Basel, Schweiz, gest. 31. 12. 1982 in Riehen, Schweiz. – *Grafiker, Maler* – Nicht beendete Lehre als Maler. 1914 Be-

l'installent à Sheffield et entament la production sous le nom de Blake, Garnett & Co. En 1829, la société prend le nom de Blake & Stephenson après le départ de Garnett. En 1841, elle s'appelle Stephenson, Blake & Co. après l'arrivée de nouveaux associés. Incorporation de la fonderie Charles Reed & Sons en 1905. La société s'appelle Stephenson, Blake & Co. Ltd. à partir de 1914. Elle acquiert la fonderie de caractères H. W. Caslon & Co en 1937. Par la suite, la fonderie deviendra une filiale de Stephenson Blake (Holdings) Ltd.

**Stiftung Bauhaus Dessau** (Fondation du Bauhaus de Dessau) – née du Bauhaus Dessau DDR, la Stiftung Bauhaus a son siège dans le bâtiment de Walter Gropius, architecte et fondateur du Bauhaus. La fondation a été créée en tant que lieu de réflexion internationale sur les problèmes des changements structurels. Dans l'objectif de «conserver l'héritage du Bauhaus historique», la fondation gère l'atelier (travaillant sur des projets), l'académie (pédagogie) ainsi que la collection et les expositions. Depuis 1990, les groupes Grappa et Cyan sont responsables de son identité.

**Stoecklin,** Niklaus – né le 19. 4. 1896 à Bâle, Suisse, décédé le 31. 12 1982 à Riehen, Suisse – *graphiste, peintre* – début d'apprentissage de peintre. Fréquente les cours de l'école des arts décoratifs de Mu-

ABCDEFGHIJKLMN
OPQRSTUVWXYZ
abcdefghijklmnopqrstuvwxyz

Stephenson & Blake 1923 Palace Script

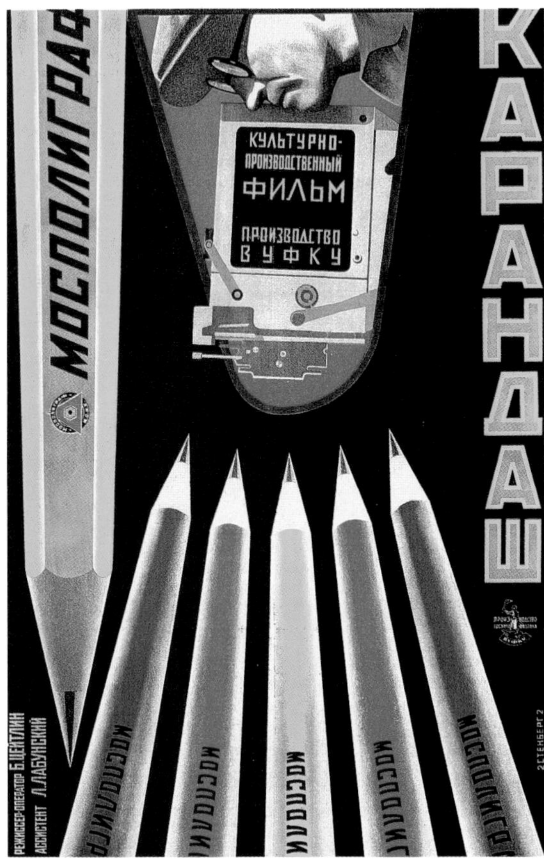


Stenberg 1929 Poster

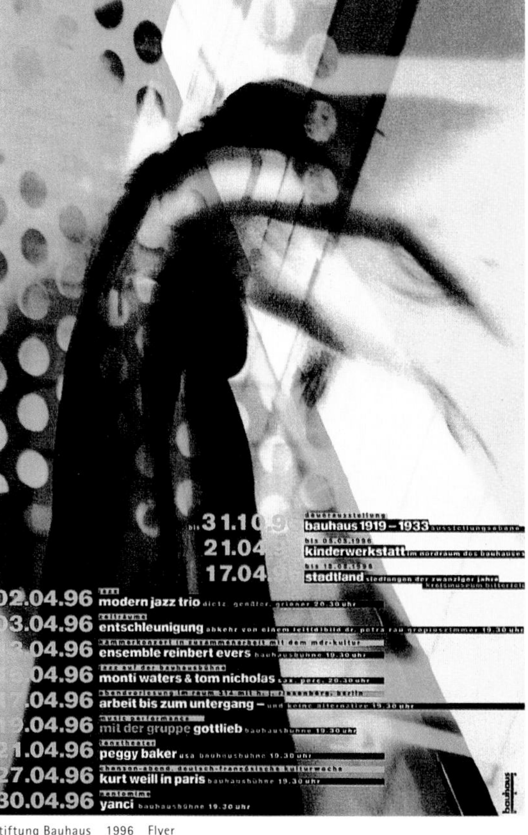


Stiftung Bauhaus 1996 Flyer

**Stephenson & Blake** – *type foundry* – 1819: William Garnett, John Stephenson and James Blake buy up William Caslon IV's type foundry in London, move the business to Sheffield and start production under the name of Blake, Garnett & Co. 1829: the company is renamed Blake & Stephenson after Garnett leaves. 1841: the company name is again changed, this time to Stephenson, Blake & Co., to accommodate new company partners. 1905: affiliation with the Charles Reed & Sons foundry. From 1914 onwards: the company is called Stephenson, Blake & Co. Ltd. 1937: acquisition of the H. W. Caslon & Co type foundry. Over the years the type foundry becomes a subsidiary of Stephenson Blake (Holdings) Ltd.

**Stiftung Bauhaus Dessau** – The Stiftung Bauhaus evolved out of the Bauhaus Dessau DDR and is housed in a building designed by architect and Bauhaus founder Walter Gropius. The foundation has established itself as an organization which undertakes to examine the problems of structural change at an international level. Its aim is to "preserve the legacy of the historic Bauhaus". The foundation has a workshop working on various projects, an academy (education) and a Bauhaus collection with exhibitions. The association's corporate identity has been in the hands of the Grappa and Cyan groups since 1990.

**Stoecklin,** Niklaus – b. 19. 4. 1896 in Basle, Switzerland, d. 31. 12. 1982 in Riehen, Switzerland – *graphic artist, painter* – Incomplete course of studies as a painter. 1914: attends courses at the

such von Kursen an der Kunstgewerbeschule in München. Weiterbildung an der Allgemeinen Gewerbeschule in Basel. Entwurf seines ersten Plakats. 1917 und 1940 Ausstellung seiner Bilder in der Kunsthalle in Basel. 1958 mit dem Kunstpreis der Stadt Basel ausgezeichnet. 1966 Ausstellung seines gebrauchsgrafischen Werks im Gewerbemuseum in Basel. 1971 Entwurf seines letzten Plakats. Stoecklin war der zentrale Vertreter des Stils der „Neuen Sachlichkeit" in der Schweiz. – *Publikationen u. a.:* W. Raeber „N. Stoecklin", Basel 1929; W. George „N. Stoecklin", Paris 1933; H. Birkhäuser „N. Stoecklin", Basel 1943; H. Göhner „Niklaus Stoecklin", Basel 1959; B. Haldner „Niklaus Stoecklin. 1896–1982", Basel 1986.

**Stone,** Reynolds – geb. 13. 3. 1909 in Eton, England, gest. 23. 6. 1979 in Dorchester, England. – *Maler, Illustrator, Kalligraph, Grafik-Designer, Typograph* – 1926–39 Studium der Geschichte an der Cambridge University. 1930–32 Schriftsetzerlehre, 1932–34 Arbeit als Schriftsetzer. Seit 1934 freier Typograph, Buchgestalter, Holzstecher, Steinschneider und Kalligraph. Auftraggeber waren u. a. die Nonesuch Press, die Cambridge University, General Post Office London, British Council, die Linotype Corporation und Royal Mint. Es entstehen zahlreiche Signets, Briefmarken, Buchillustrationen, heraldische Zeichen, Banknoten, Bücher und zwei Alphabete. Für Mitglieder der Königsfamilie entwirft er Ex Libris. Er fertigt den Grabstein für Winston Churchill in der Westminster Abbey. 1953 mit dem „Order of the British Empire", 1956 mit

nich en 1914. Poursuit sa formation à la Allgemeine Gewerbeschule (école des arts et métiers) de Bâle. Première affiche. Ses œuvres sont exposées à la Kunsthalle de Bâle en 1917 et en 1940. Obtient le prix d'art de la ville de Bâle en 1958. Exposition de son œuvre graphique en 1966 au Gewerbemuseum de Bâle. Crée sa dernière affiche en 1971. Stoecklin était le principal représentant du style «Nouvelle objectivité» en Suisse. – *Publications:* W. Raeber «Niklaus Stoecklin», Bâle 1929; W. George «Niklaus Stoecklin», Paris 1933; H. Birkhäuser «Niklaus Stoecklin», Bâle 1943; H. Göhner «Niklaus Stoecklin», Bâle 1959; B. Haldner «Niklaus Stoecklin. 1896–1982», Bâle 1986.

**Stone,** Reynolds – né le 13. 3. 1909 à Eton, Angleterre, décédé le 23. 6. 1979 à Dorchester, Angleterre – *peintre, illustrateur, calligraphe, graphiste maquettiste, typographe* – 1926–1939, études d'histoire de l'art à la Cambridge University. 1930–1932, apprentissage de la composition typographique, travaille ensuite comme compositeur de 1932 à 1934. Typographe indépendant, maquettiste en édition, graveur sur bois, tailleur de pierre et calligraphe à partir de 1934. Commanditaires : Nonesuch Press,Cambridge University, General Post Office London, British Council, Linotype Corporation, Royal Mint. Réalise de nombreux signets, timbres, illustrations de livres, emblèmes héraldiques, billets de banque, maquettes de livres et deux alphabets. Il crée des ex libris pour les membres de la famille royale. Réalise la pierre tombale de Winston Churchill à l'abbaye de Westminster. Décoré de l' «Order of the British Empire» en 1953, obtient le titre de «Royal Designer for Industry» en 1956, puis de «Fellow of the Royal Society of Arts» de Londres en 1964. Le travail de Stone a été

Stoecklin 1922 Poster

Stoecklin 1928 Poster

Stoecklin 1928 Poster

Kunstgewerbeschule in Munich. Continues his training at the Allgemeine Gewerbeschule in Basle. Designs his first poster. 1917 and 1940: his pictures are exhibited at the Kunsthalle in Basle. 1958: awarded the city of Basle's art prize. 1966: an exhibition of his work in commercial graphics is exhibited at the Gewerbemuseum in Basle. 1971: designs his last poster. Stoecklin was the central representative of the New Functionalist style in Switzerland. – *Publications:* W. Raeber "Niklaus Stoecklin", Basle 1929; W. George "Niklaus Stoecklin", Paris 1933; H. Birkhäuser "Niklaus Stoecklin", Basle 1943; H. Göhner "Niklaus Stoecklin", Basle 1959; B. Haldner "Niklaus Stoecklin. 1896–1982", Basle 1986.

**Stone,** Reynolds – b. 13. 3. 1909 in Eton, England, d. 23. 6. 1979 in Dorchester, England – *painter, illustrator, calligrapher, graphic designer, typographer* – 1926–39: studies history at Cambridge University. 1930–32: trains as a typesetter and from 1932–34 works as a typesetter. From 1934 onwards: freelance typographer, book designer, wood engraver, stone cutter and calligrapher. Clients include Nonesuch Press, Cambridge University, the General Post Office in London, the British Council, Linotype Corporation and the Royal Mint. Stone designs numerous logos, postage stamps, book illustrations, heraldic symbols, banknotes, books and two alphabets. He designs an Ex Libris for the members of the royal family. He fashions a tombstone in Westminster Abbey for Winston

dem Titel „Royal Designer for Industry" ausgezeichnet. 1964 Fellow der „Royal Society of Arts" in London. Stones Arbeit ist von den Meistern des 16. Jahrhunderts und von Eric Gill, dessen Schüler er kurze Zeit war, beeinflußt. – *Schriftentwürfe:* Minerva (1954), Janet (1965). – *Publikationen u.a.:* „A Book of Lettering", London 1935. J. W. Goodison „Reynolds Stone: his early development as an engraver on wood", London 1947; Myfanwy Piper „Reynolds Stone", London 1951; J. W. Goodison, R. McLean „Reynolds Stone 1909–1979", Dorchester 1979.

**Stone,** Sumner – geb. 9. 6. 1945 in Venice, Florida, USA. – *Schriftentwerfer, Grafik-Designer* – 1966 Studium am Reed College in Portland, Oregon. Kalligraphie-Unterricht bei Lloyd Reynolds. Danach zwei Jahre Arbeit bei Hallmark Cards in Kansas City als Schriftgestalter. 1972 Eröffnung seines Schrift-Studios in Sonoma, Kalifornien, unter dem Namen „Alpha and Omega Press". Gleichzeitig Studium der Mathematik an der Sonoma State University. 1979 Arbeit bei Autologic Inc. in Boston als Director of Typography, danach in gleicher Position bei Camex Inc. in Boston, 1985–89 bei Adobe Systems Inc. in Kalifornien. Entwirft als erste Originalschrift von Adobe die „Stone-Family"-Schriften, die aus drei Schriftarten besteht: Antiqua, Serifenlos, Informal in insgesamt 18 Schnitten. 1990 Eröffnung seiner Firma „Stone Type Foundry" für digitalen Schriftentwurf in Palo Alto. 1991 Entwurf der „Stone Print" für die amerikanische Grafik-Design-Zeitschrift „Print". – *Schriftentwürfe:*

influencé par celui des maîtres du 16e siècle ainsi que par Eric Gill, dont il fut l'élève pendant une brève période. – *Polices:* Minerva (1954), Janet (1965). – *Publications, sélection:* «A Book of Lettering», Londres 1951. J. W. Goodison «Reynolds Stone : his early development as an engraver on wood», Londres 1947; Myfanwy Piper «Reynolds Stone», Londres 1951; J. W. Goodison, R. McLean «Reynolds Stone 1909–1979», Dorchester 1979.

**Stone,** Sumner – né le 9. 6. 1945 à Venice (Floride), Etats-Unis – *concepteur de polices, graphiste maquettiste* – 1966, études au Reed College à Portland, Oregon. Cours de calligraphie chez Lloyd Reynolds, puis employé pendant deux ans comme dessinateur de caractères chez Hallmark Cards à Kansas City. Ouvre un atelier de graphisme en 1972 à Sonoma, Californie, sous le nom de «Alpha and Omega Press». Etudie parallèlement les mathématiques à la Sonoma State University. Travaille en 1979 chez Autologic Inc. à Boston, comme directeur de la typographie, exerce ensuite la même fonction chez Camex Inc. à Boston, puis chez Adobe Systems en Californie de 1985 à 1989. Crée «Stone Family», la première police originale d'Adobe, composée à partir de trois types de caractères, l'Antiqua, la sans sérif et l'informel dont il réalise 18 fontes. En 1990, il fonde la société «Stone Type Foundry» à Palo Alto, qui crée des polices numériques. Conception de «Stone Print» pour la revue américaine de graphisme «Print». – *Polices:* Stone

Reynolds Stone Logo

Reynolds Stone Page

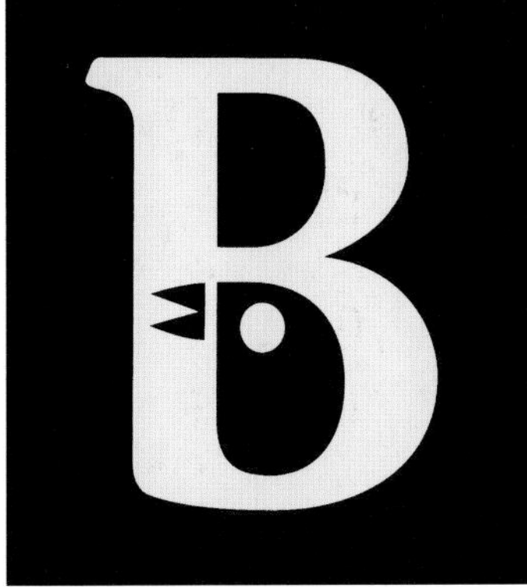

Sumner Stone Illustration

| | n | **n** | **n** | | *n* |
|---|---|---|---|---|---|
| a | b | e | f | g | i |
| o | r | s | t | y | z |
| A | B | C | E | G | H |
| M | O | R | S | X | Y |
| 1 | 2 | 4 | 6 | 8 | & |

Sumner Stone 1987 Stone Sans

| | n | **n** | **n** | | *n* |
|---|---|---|---|---|---|
| a | b | e | f | g | i |
| o | r | s | t | y | z |
| A | B | C | E | G | H |
| M | O | R | S | X | Y |
| 1 | 2 | 4 | 6 | 8 | & |

Sumner Stone 1987 Stone Serif

| | n | | | | |
|---|---|---|---|---|---|
| a | b | e | f | g | i |
| o | r | s | t | y | z |
| ɑ | ʙ | ç | ɛ | ɢ | ʜ |
| ɰ | ɔ | ʀ | ʃ | χ | ʏ |
| | | | | | |

Sumner Stone 1992 Stone Phonetic

Churchill. 1953: he is made an OBE (Officer of the Order of the British Empire) and in 1956 a Royal Designer for Industry. 1964: fellow of the Royal Society of Arts in London. Stone's work was influenced by the great masters of the 16th century and by Eric Gill, his teacher for a brief period. – *Fonts:* Minerva (1954), Janet (1965). – *Publications include:* "A Book of Lettering", London 1935. J.W. Goodison "Reynolds Stone: his early development as an engraver on wood", London 1947; Myfanwy Piper "Reynolds Stone", London 1951; J. W. Goodison, R. McLean "Reynolds Stone 1909–1979", Dorchester 1979.

**Stone,** Sumner – b. 9. 6. 1945 in Venice, Florida, USA – *type designer, graphic designer* – 1966: studies at Reed College in Portland, Oregon. Takes calligraphy lessons with Lloyd Reynolds. Then works as type designer for Hallmark Cards in Kansas City for two years. 1972: opens his type studio, Alpha and Omega Press, in Sonoma, California. At the same time he studies mathematics at Sonoma State University. 1979: works for Autologic Inc. in Boston as director of typography. He later holds the same position at Camex Inc. in Boston and from 1985–89 at Adobe Systems Inc. in California. Abode's first original fonts are Stone's Stone Family typefaces, namely a roman, sans-serif and informal font cut in a total of 18 weights. 1990: opens his Stone Type Foundry company for digital type design in Palo Alto. 1991: designs Stone Print for

Stone Family (1987), Stone Print (1991), Stone Phonetic (mit John Renner, 1992), Silica (1993). – *Publikation:* „Typography on the Personal Computer", San Francisco, London 1991.

**Strzeminski,** Wladyslaw – geb. 21. 11. 1893 in Minsk, Rußland, gest. 26. 12. 1952 in Lodz, Polen. – *Maler, Grafiker, Kunsttheoretiker, Typograph* – 1911–14 Studium an der Technischen Hochschule in St. Petersburg. 1919 Mitglied der Künstlergruppe „Unowis" in Witebsk, Zusammenarbeit mit Malewitsch. 1922 Umzug nach Polen. 1924 Gründung der Künstlergruppe „Blok" in Warschau. Entwurf typographischer Arbeiten. Umschlagsgestaltung für Bücher des Autors Tadeusz Peiper. 1926 Organisation der Künstlergruppe „Praesens" und der gleichnamigen Zeitschrift. 1929 Gründung der Künstlergruppe „a.r." mit K. Kobro, H. Stazewski in Lodz. 1932 Verleihung des Kunstpreises der Stadt Lodz. Mitglied der Künstlergruppe „abstraction-création". Veröffentlicht den Entwurf eines neuen Alphabets „Komunikat". 1945–50 Lehrer an der Hochschule für Bildende Künste in Lodz. – *Schriftentwurf:* Komunikat (1932). – *Publikationen u. a.:* „Teoria widzenia", Krakau 1958. R. Stanislawski u. a. „Wladislaw Strzeminski", Düsseldorf 1980; J. Jedlinski u. a. „Wladyslaw Strzeminski", Bonn 1994.

**Sugiura,** Kohei – geb. 8. 9. 1932 in Tokio, Japan. – *Ausstellungsgestalter, Typograph, Lehrer* – 1951–55 Studium an der Universität für Kunst und Architektur in Tokio. Seit 1955 freier Grafik-Designer, Art Director der Zeitschriften „Space De-

Family (1987), Stone Print (1991), Stone Phonetic (avec John Renner, 1992), Silica (1993). – *Publication:* «Typography on the Personal Computer», San Francisco, Londres 1991.

**Strzemínski,** Vladislav – né le 21. 11. 1893 à Minsk, Russie, décédé le 26. 12. 1952 à Lodz, Pologne – *peintre, graphiste, théoricien d'art, typographe* – 1911–1914, études à l'université technique de Saint-Pétersbourg. 1919, membre du groupe d'artistes «Unowis» à Vitebsk, travaille avec Malévitch. S'installe en Pologne en 1922. Fonde le groupe d'artistes «Blok» à Varsovie, en 1924. Travaux typographiques. Conception de couvertures pour des ouvrages de Tadeusz Peiper. Organise les activités du groupe d'artistes «Praesens» en 1926 et s'occupe de la revue du même nom. Fonde, en 1929, le groupe d'artistes «a.r.» avec K. Kobro et H. Stazewski à Lodz. Membre du groupe d'artistes «abstraction-création». Publie la maquette de «Komunikat», un nouvel alphabet. Enseigne à l'école supérieure des beaux-arts de Lodz de 1945 à 1950. – *Polices:* Komunikat (1932). – *Publications, sélection:* «Teoria widzenia», Cracovie 1958. R. Stanislawski et autres «Wladislaw Strzeminski», Düsseldorf 1980; J. Jedlinski et autres «Wladyslaw Strzeminski», Bonn 1994.

**Sugiura,** Kohei – né le 8. 9. 1932 à Tokyo, Japon – *architecte d'expositions, typographe, enseignant* – 1951–1955, études à l'université des beaux-arts et d'architecture à Tokyo. Graphiste maquettiste indépendant à partir de 1955, directeur artistique des revues «Space Design» (1966–1970), «Toshi Jutaku» (1967–1970), «Ginka» (depuis 1968), «Yu» (1971–1980), «Episteme» (1975–1979). Participe à la conception graphique des Jeux Olympiques de Tokyo en 1964. Cycle de conférences à la Hochschule für Gestaltung

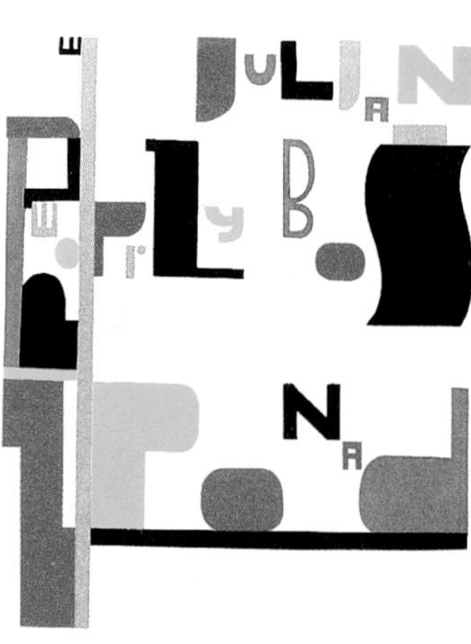

Strzeminski 1930 Cover

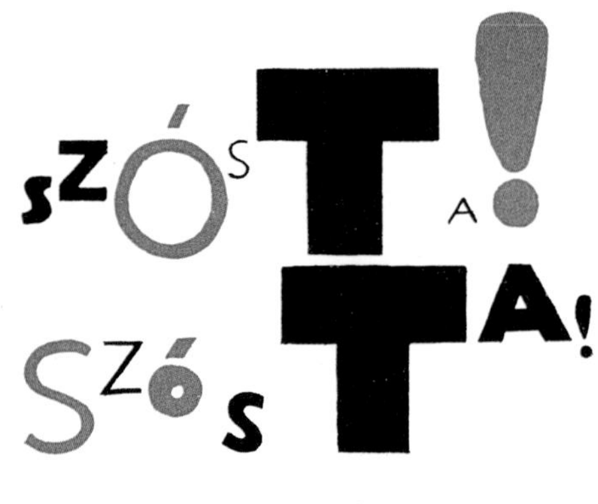

Strzeminski 1926 Cover

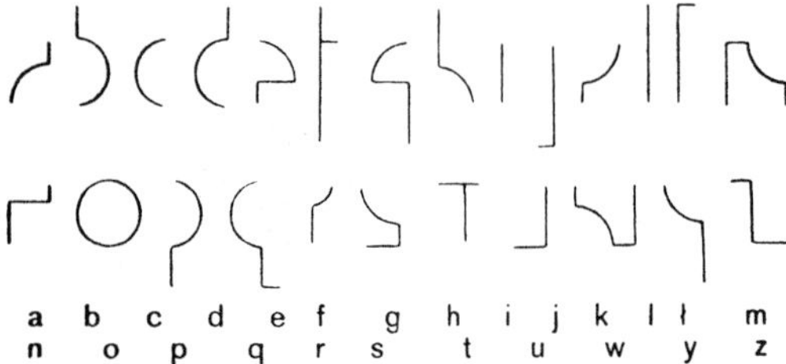


Strzeminski 1932 Komunikat

Strzeminski 1931 Cover

the American graphic-design magazine "Print". – *Fonts:* Stone Family (1987), Stone Print (1991), Stone Phonetic (with John Renner, 1992), Silica (1993). – *Publication:* "Typography on the Personal Computer", San Francisco, London 1991.

**Strzeminski,** Wladyslaw – b. 21. 11. 1893 in Minsk, Russia, d. 26. 12. 1952 in Lodz, Poland – *painter, graphic artist, art theorist, typographer* – 1911–14: studies at the Technical College in St. Petersburg. 1919: member of Unowis artists' group in Vitebsk. Works with Malevich. 1922: moves to Poland. 1924: founds a group of artists (Blok) in Warsaw. Produces typographic work. Designs covers for books by Tadeusz Peiper. 1926: organizes the Praesens artists' group and the magazine of the same name. 1929: founds the artists' group a.r. with K. Kobro and H. Stazevsky in Lodz. 1932: is awarded an art prize by the city of Lodz. Member of the abstraction-création group. Publishes a design for a new alphabet, Komunikat. 1945–50: teaches at the College of Fine Art in Lodz. – *Font:* Komunikat (1932). – *Publications include:* "Teoria widzenia", Cracow 1958. R. Stanislawski et al "Wladislaw Strzemínski", Düsseldorf 1980; J. Jedlinski et al "Wladyslaw Strzeminski", Bonn 1994.

**Sugiura,** Kohei – b. 8. 9. 1932 in Tokyo, Japan – *exhibition planner, typographer, teacher* – 1951–55: studies at the University of Art and Architecture in Tokyo. From 1955 onwards: freelance graphic designer. Art director of "Space Design" (1966–70), "Toshi Jutaku" (1967–70), "Ginka" (from 1968), "Yu" (1971–80) and

sign“ (1966–70), „Toshi Jutaku“ (1967–70), „Ginka“ (seit 1968), „Yu“ (1971–80), „Episteme“ (1975–79). 1964 Mitarbeit bei der grafischen Gestaltung der Olympischen Spiele in Tokio. 1964–65 und 1966–67 Gastdozent an der Hochschule für Gestaltung in Ulm. Zahlreiche Auszeichnungen, u. a. Goldmedaille Leipzig (1978, 1982) und durch das japanische Unterrichtsministerium (1982). Seit 1989 Professor an der Kobe University of Arts and Technology, seit 1993 auch an der Graduate School der Kobe University. – *Publikationen:* „The 3-D Stars“, Tokio 1986; „Asian Cosmology“, Tokio 1989; „Japanese Forms, Asian Forms“, Tokio 1993. T. Tomioka „Kohei Sugiura, his Acts and Arts“, Tokio 1970.

**Sumichrast,** Józef – geb. 26. 7. 1948 in Hobart, USA. – *Illustrator, Grafik-Designer, Typograph* – Studium an der Indiana University, The American Academy of Art und am Illinois Institute of Technology; danach Arbeit in verschiedenen Studios. 1974 Gründung seines Studios. Auftraggeber waren u. a. Reader's Digest, Levi's, New York Botanical Garden. Illustriert 1977 das Buch „Q is for Crazy“ von Ed Leander. Zahlreiche Auszeichnungen, u. a. Type Directors Club (1981, 1985). Sumichrasts illustrative Alphabete finden für Schriftzüge und Zeitschriftencover häufig Verwendung.

**Sutnar,** Ladislav – geb. 9. 11. 1897 in Pilsen, Österreich-Ungarn (heute Tschechien), gest. 17. 11. 1976 in New York, USA. – *Typograph, Ausstellungsgestalter, Maler, Autor, Lehrer* – 1915–16 und 1919–23 Studium der Malerei an der

Strzeminski 1924 Page

Sugiura 1982 Cover

(école supérieure de design) d'Ulm en 1964–1965 et en 1966–1967. Nombreuses distinctions, entre autres, la médaille d'or de Leipzig (1978, 1982), du ministère de l'éducation du Japon (1982). Professeur à l'université des arts et technologies à Kobe depuis 1989, ainsi qu'à la Graduate School de l'université de Kobe depuis 1993. – *Publications:* «The 3-D Stars», Tokyo 1986; «Asian Cosmology», Tokyo 1989; «Japanese Forms, Asian Forms», Tokyo 1993. T. Tomioka «Kohei Sugiura, his Acts and Arts», Tokyo 1970.

**Sumichrast,** Józef – né le 26. 7. 1948 à Hobart, Etats-Unis – *illustrateur, graphiste maquettiste, typographe* – études à l'Indiana University, The American Academy of Art et au Illinois Institute of Technology; travaille ensuite dans divers ateliers. En 1974, il fonde son propre atelier. Commanditaires: Reader's Digest, Levi's, New York Botanical Garden. En 1977, il illustre l'ouvrage «Q ist for Crazy» d'Ed Leander. Nombreuses distinctions, entre autres celles du Type Directors Club (1981, 1985). Les alphabets illustrés de Sumichrast sont souvent utilisés pour les titres et les couvertures de revues.

**Sutnar,** Ladislav – né le 9. 11. 1897 à Pilsen, Autriche-Hongrie (aujourd'hui République Tchèque), décédé le 17. 11. 1976 à New York, Etats-Unis – *typographe, architecte d'expositions, peintre, auteur, enseignant* – études de peinture à Prague, à l'école des arts décoratifs de 1915 à 1916, et de 1919 à 1923, puis à l'école de

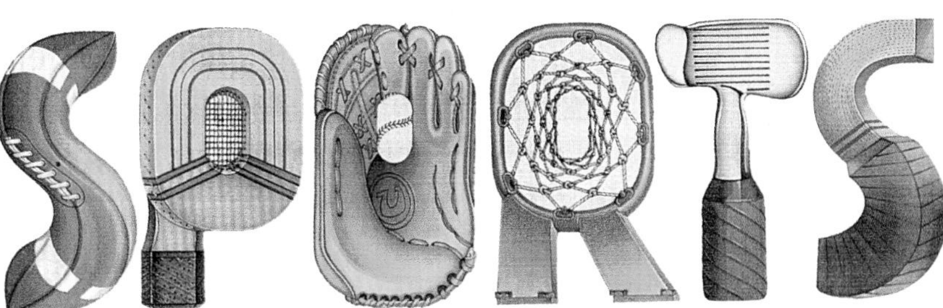


Sumichrast 1989 Illustration

"Episteme" (1975–79) magazines. 1964: works on the graphic design of the Tokyo Olympic Games. 1964–65 and 1966–67: visiting lecturer at the Hochschule für Gestaltung in Ulm. Numerous awards, including the Leipzig gold medal (1978, 1982) and from the Japanese Ministry of Education (1982). From 1989 onwards: professor at Kobe University of Arts and Technology and since 1993 also at Kobe University's Graduate School. – *Publications:* "The 3-D Stars", Tokyo 1986; "Asian Cosmology", Tokyo 1989; "Japanese Forms, Asian Forms", Tokyo 1993. T. Tomioka "Kohei Sugiura, his Acts and Arts", Tokyo 1970.

**Sumichrast,** Józef – b. 26. 7. 1948 in Hobart, USA – *illustrator, graphic designer, typographer* – Studied at the Indiana University, The American Academy of Art and the Illinois Institute of Technology following which he worked for various studios. 1974: opens his own studio. Clients include the Reader's Digest, Levi's and New York Botanical Garden. 1977: illustrates Ed Leander's book, "Q is for Crazy". Numerous awards, including Type Directors Club (1981, 1985). Sumichrast's illustrative alphabets are often used for logos and magazine covers.

**Sutnar,** Ladislav – b. 9. 11. 1897 in Pilsen, Austria-Hungary (today Czech Republic), d. 17. 11. 1976 in New York, USA – *typographer, exhibition planner, painter, author, teacher* – 1915–16 and 1919–23: studies painting at the School of Arts and Crafts in Prague. 1922–23: studies at the School of Applied Graphics in Prague. 1923–39: professor at the State School of

Kunstgewerbeschule in Prag, 1922–23 Studium an der Schule für Gebrauchsgrafik in Prag. 1923–39 Professor an der Staatsschule für Gebrauchsgrafik in Prag, seit 1932 Direktor der Schule. 1929–39 Art Director des Verlags Družxstevni práce in Prag. Gestaltet die Zeitschriften Panorama, Výtvarné snahy und Zijeme. 1929 Auszeichnung mit einer Goldmedaille auf der Weltausstellung in Barcelona. 1939 Gestaltung des Pavillons der Tschechoslowakei auf der Weltausstellung in New York, bleibt in den USA. Seit 1951 eigenes Design-Büro. Art Director der Zeitschrift „Theatre Arts". 1961 Ausstellung im Contemporary Arts Center in Cincinnati, Ohio, unter dem Titel „L. Sutnar: Visual Design in Action". – *Publikationen u. a.:* „Catalog Design Progress: Advancing Standards in Visual Communication" (mit K. Lönberg-Holm), New York 1950; „Design for Point of Sale", New York 1952; „Visual Design in Action: Principles, Purposes", New York 1961.

**Sütterlin,** Ludwig – geb. 23. 7. 1865 in Lahr, Deutschland, gest. 20. 11. 1917 in Berlin, Deutschland. – *Maler, Grafiker, Lehrer* – 1890–95 Studium an der Unterrichtsanstalt des Kunstgewerbemuseums in Berlin. Danach dort als Lehrer tätig. 1915 Entwurf der Reform-Schrift „Sütterlin-Schrift", die 1935 als Schreibschrift an deutschen Schulen eingeführt wird. 1941 wird die Sütterlin-Schrift durch die deutsche Normalschrift, eine lateinische Schrift, ersetzt. – *Publikationen:* „Neuer Leitfaden für den Schreibunterricht", Berlin 1917. O. Schmidt „Im Geiste Sütterlins", 1925.

graphisme appliqué de 1922 à 1923. Professeur à l'école nationale de graphisme appliqué à Prague de 1923 à 1939, il dirigera ensuite l'école à partir de 1932. Directeur artistique des éditions Družxstevni práce de Prague, de 1929 à 1939. Maquette des revues «Panorama», «Výtvarné snahy» et «Zijeme». Médaille d'or de l'exposition universelle de Barcelone en 1929. Conception du pavillon de Tchécoslovaquie pour l'exposition universelle de New York en 1939, reste aux Etats-Unis. Ouvre sa propre agence de design en 1951. Directeur artistique de la revue «Theatre Arts». Exposition de ses œuvres au Contemporary Arts Center de Cincinnati, Ohio, en 1961, sous le titre de «Ladislav Sutnar: Visual Design in Action». – *Publications, sélection:* «Catalog Design Progress: Advancing Standards in Visual Communication» (avec K. Lönberg-Holm), New York 1950; «Design for Point of Sale», New York 1952; «Visual Design in Action: Principles, Purposes», New York 1961.

**Sütterlin,** Ludwig – né le 23. 7. 1865 à Lahr, Allemagne, décédé le 20. 11. 1917 à Berlin, Allemagne – *peintre, graphiste, enseignant* – 1890–1895, études à l'Unterrichtsanstalt des Kunstgewerbemuseums (institut d'enseignement du musée des arts décoratifs) à Berlin. Exerce ensuite comme professeur. Travaille, en 1915, à une réforme de l'alphabet, le «Sütterlin», introduit comme modèle d'écriture courante dans les écoles allemandes en 1935. En 1941, la Sütterlin est remplacée par la «Normalschrift», une écriture latine. – *Publications:* «Neuer Leitfaden für den Schreibunterricht», Berlin 1917. O. Schmidt «Im Geiste Sütterlins», 1925.

Sutnar 1931 Cover

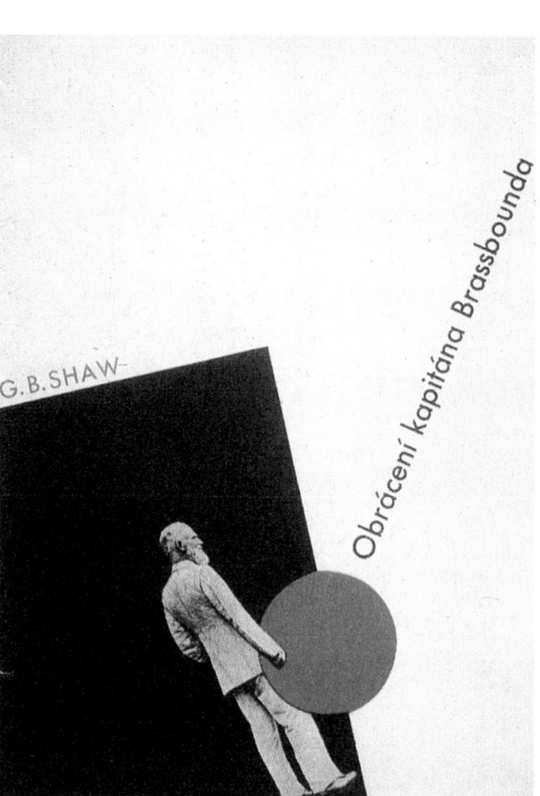

Sutnar 1931 Cover

Sutnar 1947 Cover

Sütterlin ca. 1915 Sütterlin-Schrift

Applied Graphics in Prague and from 1932 is director of the school. 1929–39: art director of Družxstevni práce publishers in Prague. Designs the magazines "Panorama", "Výtvarné snahy" and "Zijeme". 1929: is awarded a gold medal at the world exhibition in Barcelona. 1939: designs the Czech pavilion for the world exhibition in New York. Stays in the USA. 1951: opens his own design studio. Art director for "Theatre Arts" magazine. 1961: an exhibition of his work is held at the Contemporary Arts Center in Cincinnati, Ohio, entitled "Ladislav Sutnar: Visual Design in Action". – *Publications include:* "Catalog Design Progress: Advancing Standards in Visual Communication" (with K. Lönberg-Holm), New York 1950; "Design for Point of Sale", New York 1952; "Visual Design in Action: Principles, Purposes", New York 1961.

**Sütterlin,** Ludwig – b. 23. 7. 1865 in Lahr, Germany, d. 20. 11. 1917 in Berlin, Germany – *painter, graphic artist, teacher* – 1890–95: studies at the Unterrichtsanstalt des Kunstgewerbemuseums in Berlin. Then works there as a teacher. 1915: designs a reformed style of handwriting "Sütterlin-Schrift", introduced in German schools as standard in 1935. 1941: Sütterlin-Schrift is replaced by a standard "Latin" handwriting ("deutsche Normalschrift"). – *Publications:* "Neuer Leitfaden für den Schreibunterricht", Berlin 1917. O. Schmidt "Im Geiste Sütterlins", 1925.

# T

Prazdroj
PLZEŇ
LINDT A.S
Michovský
PUNČOCHY
LYONSKÝ
TOVÁRNA NA KLOBOUKY
Tisíce dárků
pro tisíce přání
KVĚTINY A PÉRA
My Lady
PESSL PESSL
CUKRÁRNA
ZBOŽÍ Z VLASTNÍ TOVÁRNY
NEJVĚTŠÍ ODBORNÝ ZÁVOD
LINE
Štěrba
A. Lindt
MICHOVSKÝ
LYONSKÝ

**Tagliente,** Giovanni Antonio – Daten unbekannt. – *Kalligraph* – ca. 1515–30 in der Staatskanzlei in Venedig tätig. Hilft 1515 Geronimo Tagliente (einem Verwandten) bei den Vorbereitungen zu dem Buch „Libro di abaco“ in Venedig. Veröffentlicht 1524 in Venedig sein verbreitetes Schreibbuch „La vera arte delo excellente scrivere de diversa varie de litere“, 1527 das Stichmusterbuch „Opera nuova...“ in Venedig.

**Tanaka,** Ikko – geb. 13. 1. 1930 in Nara, Japan. – *Grafik-Designer, Typograph, Textildesigner* – 1947–50 Studium an der Universität für Bildende Kunst in Kyoto. 1950–52 Arbeit als Textildesigner bei Kanegafuchi Spinning Mills. 1952–57 Grafik-Designer bei Sankei Press in Tokio. 1960–63 Mitbegründer und Art Director des „Nippon Design Center“ in Tokio. 1963–76 Gründung und Leitung seines Studios „Tanaka Design Atelier“ in Tokio. Auftraggeber waren u. a. das Museum für moderne Kunst in Kyoto, das Victoria & Albert Museum London, die Olympischen Spiele Tokio (1964) und Sapporo (1970). Entwirft 1970 den Pavillon der japanischen Geschichte auf der Expo 70 in Osaka. Seit 1973 Art Director der Seibu-Gruppe, seit 1975 des Seibu Museum of Art in Tokio. 1976 Umbenennung seines Studios in „Ikko Tanaka Design Studio“. Mitbegründer von „Tokio Designers Space“. 1980 Planung, Organisation und Gestaltung der Ausstellung „Japan Style“ in London. Zahlreiche internationale Ausstellungen seiner Arbeiten. – *Publikationen u. a.:* „Modern Design Series: Color Scheme and Design“, Tokio 1969;

**Tagliente,** Giovanni Antonio – dates inconnues – *calligraphe* – exerce vers 1515–1530 à la chancellerie de Venise. En 1515, toujours à Venise, il assiste Geronimo Tagliente (un parent) lors de la préparation de l'ouvrage «Libro di abaco». En 1524, il y publie un livre d'écriture, très diffusé, qui porte le titre de «La vera arte delo excellente scrivere de diversa varie de litere», puis en 1527 le livre de modèles «Opera nuova...» à Venise.

**Tanaka,** Ikko – né le 13. 1. 1930 à Nara, Japon – *graphiste maquettiste, typographe, designer en textiles* – 1947–1950, études à l'université des arts plastiques à Kyoto. 1950–1952, employé comme designer en textiles chez Kanegafuchi Spinning Mills. 1952–1957, graphiste maquettiste pour Sankei Press à Tokyo. 1960–1963, cofondateur et directeur artistique du «Nippon Design Center» à Tokyo. Fonde l'agence «Tanaka Design Atelier» à Tokyo en 1963 et la dirige jusqu'en 1976. Commanditaires : le Musée d'art moderne de Tokyo, le Victoria & Albert Museum, Londres, les Jeux Olympiques de Tokyo (1964) et Sapporo (1970). Conçoit, en 1970, le pavillon de l'histoire du Japon pour Expo 70 à Osaka. Directeur artistique du Groupe Seibu à partir de 1973, et du Seibu Museum of Art de Tokyo, à partir de 1975. En 1976, son agence prend le nom de «Ikko Tanaka Design Studio». Cofondateur du «Tokyo Designers Space». Planification, organisation et conception de l'exposition «Japan Style» en 1980, à Londres. Nombreuses expositions internationales de ses travaux. – *Publications, sélection :* «Modern Design Series : Color Scheme and Design», Tokyo 1969; «The work of Ikko Tanaka», Tokyo 1977; «Surroundings of Designs : Essays», Tokyo 1980.

**Tartakover,** David – né le 29. 1. 1944 à Haïfa, Israël – *graphiste maquettiste, ty-*

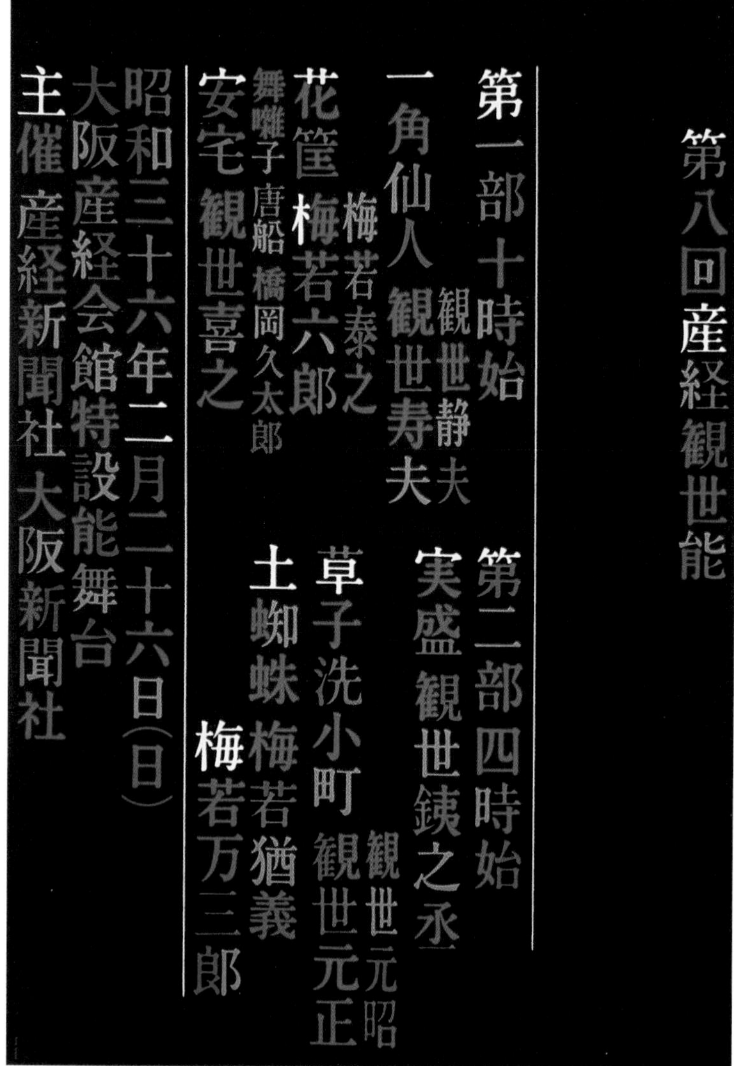

Tanaka 1961 Poster

Tanaka 1977 Poster

Tanaka 1982 Poster

**Tagliente,** Giovanni Antonio – dates unknown – *calligrapher* – c.1515–30: works for the state chancellery in Venice. 1515: assists Geronimo Tagliente (a relation) in the preparation of his book, "Libro di abaco", in Venice. 1524: publishes his popular book on the art of writing, "La vera arte delo excellente scrivere de diversa varie de litere" in Venice and in 1527 a book of engraving specimens "Opera nuova...", also in Venice.

**Tanaka,** Ikko – b. 13. 1. 1930 in Nara, Japan – *graphic designer, typographer, textiles designer* – 1947–50: studies at the University of Fine Arts in Kyoto. 1950–52: works as a textiles designer for Kanegafuchi Spinning Mills. 1952–57: graphic designer for Sankei Press in Tokyo. 1960–63: co-founder and art director of the Nippon Design Center in Tokyo. 1963–76: opens and runs his Tanaka Design Atelier in Tokyo. Clients include the Museum of Modern Art in Kyoto, the Victoria & Albert Museum in London and the Olympic Games in Tokyo (1964) and Sapporo (1970). 1970: designs the Japanese history pavilion for Expo 70 in Osaka. From 1973 onwards: art director of the Seibu group and from 1975 of the Seibu Museum of Art in Tokyo. 1976: renames his studio Ikko Tanaka Design Studio. Cofounder of Tokyo Designers Space. 1980: plans, organizes and designs the "Japan Style" exhibition in London. There have been numerous exhibitions of his work all over the world. – *Publications include:* "Modern Design Series: Color Scheme and Design", Tokyo 1969; "The work of Ikko Tanaka", Tokyo 1977; "Surroundings

„The work of Ikko Tanaka", Tokio 1977; „Surroundings of Designs: Essays", Tokio 1980.

**Tartakover,** David – geb. 29. 1. 1944 in Haifa, Israel. – *Grafik-Designer, Typograph, Lehrer* – Studium an der Bezalel Akademie für Kunst und Design in Jerusalem und am London College of Printing. Seit 1975 Gründung und Leitung seines Studios für Visuelle Kommunikation in Tel Aviv: Sammelt israelisches Design (Plakat, Werbung, Verpackung etc.) und erforscht dessen Geschichte. Unterrichtet seit 1976 an der Bezalel Akademie für Kunst und Design in Jerusalem und zeitweise im Zentrum für Design-Studien „Vital" in Tel Aviv. 1992 Präsident der Grafik-Design-Association von Israel. 1993 Initiator und Kurator der ersten Internationalen Plakat-Biennale im Kunstmuseum in Tel Aviv. Zahlreiche Ausstellungen seiner Arbeit, u. a. in Tokio (1982), Jerusalem (1984), Tel Aviv (1985), Paris (1994) und Chaumont (1995).

**Teige,** Karel – geb. 13. 12. 1900 in Prag, Tschechoslowakei, gest. 1. 10. 1951 in Prag, Tschechoslowakei. – *Grafiker, Typograph, Maler, Schriftsteller* – 1917–21 Studium der Kunstgeschichte in Prag. 1919 Redakteur der Zeitschriften „Kmen", „Čas", „Stavba". 1920 Gründungsmitglied und Leiter des Kunstverbandes „Devětsil". Typographische Gestaltung sämtlicher Publikationen des „Odeon Verlags". Zahlreiche Aufsätze über Typographie. Veröffentlicht 1922 das „Manifest der proletarischen Kunst". Gestaltet 1926 das Buch „ABECEDE" von Vítězlav Nezval, eine Publikation des Konstruktivismus. 1927–

Tartakover 1986 Poster

Tartakover 1988 Poster

Teige 1928 Cover

*pographe, enseignant* – études à l'académie Bezalel d'art et de design à Jérusalem, puis au London College of Printing. Fonde et dirige son agence de communication visuelle à Tel Aviv en 1975 : recherche et collectionne des documents sur l'histoire du design israélien (affiches, publicité, emballages etc.). Enseigne à l'académie Bezalel d'art et de design de Jérusalem à partir de 1976, et sporadiquement au centre d'études de design «Vital» de Tel Aviv. Président de l'association d'arts graphique et de design d'Israël en 1992. Fondateur et commissaire d'exposition de la première Biennale internationale de l'affiche au musée des beaux-arts de Tel Aviv en 1993. Nombreuses expositions de ses travaux, entre autres, à Tokyo (1982), Jérusalem (1984), Tel Aviv (1985), Paris (1994) et Chaumont (1995).

**Teige,** Karel – né le 13. 12. 1900 à Prague, Tchécoslovaquie, décédé le 1. 10. 1951 à Prague, Tchécoslovaquie – *graphiste, typographe, peintre, écrivain* – 1917–1921, études d'histoire de l'art à Prague. 1919, rédacteur pour les revues «Kmen», «Čas», «Stavba». 1920, membre fondateur et directeur de l'organisation «Devětsil». Conception typographique de toutes les publications des édtions «Odeon». Nombreux essais sur la typographie. En 1922, il publie le «Manifeste de l'art prolétarien». En 1926, il réalise la maquette de «ABECEDE», un ouvrage constructiviste de Vítězlav Nezval. De 1927 à 1928, il pub-

Tanaka 1986 Poster

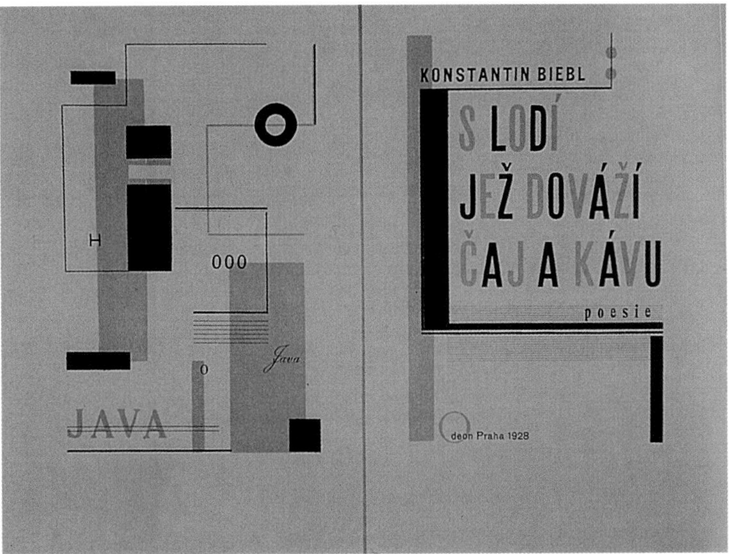


Teige 1928 Cover

of Designs: Essays", Tokyo 1980.

**Tartakover,** David – b. 29. 1. 1944 in Haifa, Israel – *graphic designer, typographer, teacher* – Studied at the Bezalel Academy of Art and Design in Jerusalem and at the London College of Printing. 1975: opens his studio for visual communication in Tel Aviv. He researches and collects items on the history of Israelite design (posters, advertising and packaging etc.). 1976: starts teaching at the Bezalel Academy of Art and Design in Jerusalem and also occasionally at the Vital Center for Design Studies in Tel Aviv. 1992: president of the Graphic Design Association of Israel. 1993: initiator and curator of the first "International Poster Biennale" in Tel Aviv's art museum. There have been numerous exhibitions of his work, including in Tokyo (1982), Jerusalem (1984), Tel Aviv (1985), Paris (1994) and Chaumont (1995).

**Teige,** Karel – b. 13. 12. 1900 in Prague, Czechoslovakia, d. 1. 10. 1951 in Prague, Czechoslovakia – *graphic artist, typographer, painter, writer* – 1917–21: studies art history in Prague. 1919: editor of the magazines "Kmen", "Čas" and "Stavba". 1920: founder member and director of Devětsil art society. Does the typographic design for all the publications issued by Odeon publishers. Writes numerous essays on typography. 1922: publishes the "Manifesto of Proletarian Art". 1926: designs Vítězlav Nezval's book "ABECEDE", a Constructivist publication. 1927–28: edits and designs "RED",

28 Herausgabe und Gestaltung der Zeitschriften „RED", „Pašmo", „Disk". 1929 Mitbegründer der Linken Front der Künste. 1929–30 Gastdozent am Bauhaus, Vorlesungen über „Soziologie und Architektur". Teilnahme an der Ausstellung „Film und Foto" des Deutschen Werkbunds in Stuttgart. 1934 Hinwendung zum Surrealismus. Arbeitet 1935–51 hauptsächlich in den Bereichen Fotocollage und Fotomontage, insgesamt entstehen über 300 Arbeiten. – *Publikationen u. a.:* „Sovětská kultura", Prag 1927; „Film", Prag 1928; „Svét, ktery se sméje", Prag 1928; „Liquidierung der Kunst", Frankfurt am Main 1968.

**Tel Design** – *Design-Studio* – 1962 Gründung von „Truijen en Lucassen" (TeL), Design-Studio, durch Emile Truijen und Jan Lucassen in Den Haag mit dem Schwerpunkt Industrie-Design. 1966 Umbenennung des Design-Studios in „Tel Design Associated". 1966–77 Mitarbeit von Gert Dumbar, Aufbau der Grafik-Design-Abteilung von Tel Design. Seit 1973 Herausgabe der Hauszeitschrift „Telwerk". Seit 1976 ist Tel Design eine GmbH mit vier Designteams, die von Andrew Fallon, Paul Vermijs, Gert Kootstra und Ronald van Lit geleitet werden. 1989 Umbenennung in „Tel Graphic Design, Corporate Design, Environmental Design bv". Auftraggeber u. a. Gasunie, Kadaster, Heineken, Nederlandse Spoorwegen, Boymans-van Beuningen Museum und die Kunsthal Rotterdam. – *Publikation u. a.:* Paul Hefting u. a. „Tel Design 1962–1992", Amsterdam 1992.

**Telingater,** Solomon – geb. 12. 5. 1903 in

lie et conçoit la maquette des revues «RED», «Pašmo» et «Disk». Cofondateur du Front des artistes de gauche en 1929. Cycle de conférences intitulé «Soziologie und Architektur» au Bauhaus de 1929 à 1930. Participe à l'exposition «Film und Foto» du Deutsche Werkbund à Stuttgart. S'oriente vers le surréalisme en 1934. Travaille principalement dans le domaine du photocollage et du photomontage de 1935 à 1951; a réalisé plus de 300 œuvres. – *Publications, sélection:* «Sovětská kultura», Prague 1927; «Film», Prague 1928; «Svét, ktery se sméje», Prague 1928; «Liquidierung der Kunst», Francfort-sur-le-Main 1968.

**Tel Design** – *atelier de design* – l'atelier de design «Truijen en Lucassen» (TeL), spécialisé en design industriel, est fondé en 1962 par Emil Truijen et Jan Lucassen à La Haye. En 1966, l'atelier prend le nom de «Tel Design Associated». Gert Dumbar y travaille de 1966 à 1977 et développe le service de conception graphique et de design de Tel Design. Publie la revue d'entreprise «Telwerk» à partir de 1973. Tel Design a le statut d'une SARL depuis 1976. Quatre équipes de design en font partie, dirigées par Andrew Fallon, Paul Vermijs, Gert Kootstra et Ronald van Lit. Prend le nom de «Tel Grafic Design, Corporate Design, Environmental Design bv» en 1989. Commanditaires: Gasunie, Kadaster, Heineken, Nederlandse Spoorwegen, Boymans-van Beuningen Museum et Kunsthal Rotterdam. – *Publication, sélection:* Paul Hefting et autres «Tel Design 1962–1992», Amsterdam 1992.

**Telingater,** Solomon – né le 12. 5. 1903 à Tiflis, Géorgie, décédé le 1. 10 1969 à Moscou, URSS – *graphiste maquettiste, typographe, concepteur de polices, illustrateur, scénographe* – 1919–1920, études à Bakou, puis au VKHOUTEMAS de Moscou de 1920 à 1921. Participe à la

Tel Design 1993 Poster

Tel Design 1993 Logo

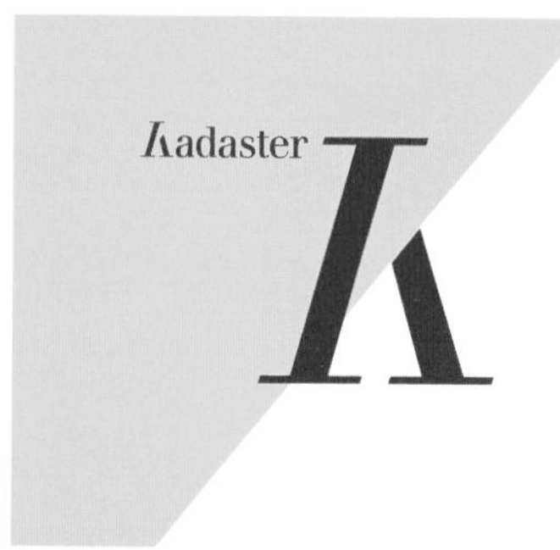
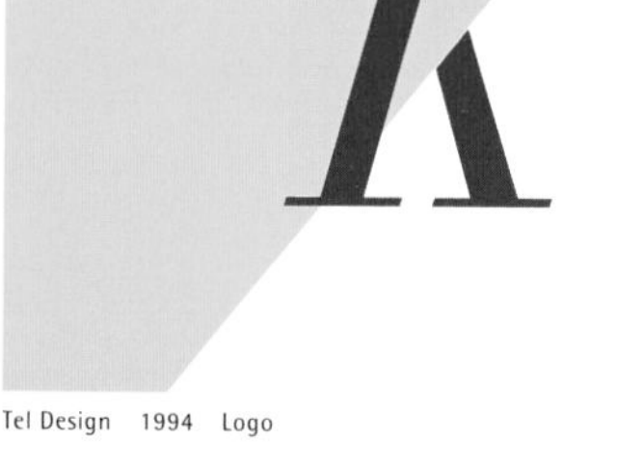


Tel Design 1994 Logo

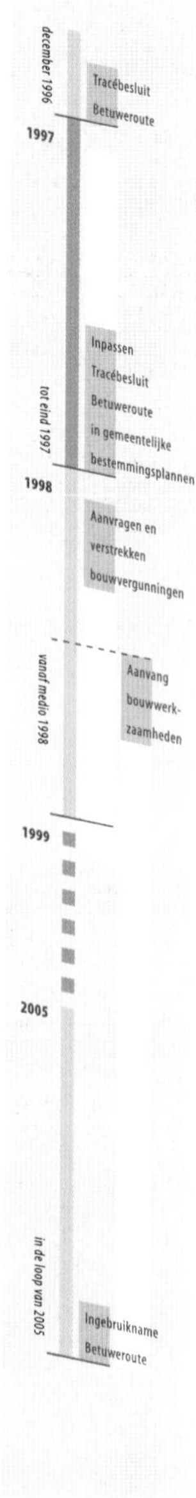


Tel Design 1996 Schedule

Tel Design 1994 Desk

"Pašmo" and "Disk" magazines. 1929: co-founder of the Left Front of the Arts. 1929–30: visiting lecturer at the Bauhaus. Lectures on "Sociology and Architecture". Takes part in the "Film und Foto" photographic exhibition organized by the Deutsche Werkbund in Stuttgart. 1934: turns to Surrealism. 1935–51: Teige concentrates his output on photocollage and photomontage and produces over 300 works in this area. – *Publications include:* "Sovětská kultura", Prague 1927; "Film", Prague 1928; "Svét, ktery se sméje", Prague 1928; "Liquidierung der Kunst", Frankfurt am Main 1968.

**Tel Design** – *design studio* – 1962: the Truijen en Lucassen (TeL) design studio is opened by Emile Truijen and Jan Lucassen in The Hague. Main area of work is industrial design. 1966: the studio is renamed Tel Design Associated. 1966–77: Gert Dumbar works for the studio. Tel Design sets up a graphic design department. 1973: the studio starts publishing its "Telwerk" magazine. Tel Design becomes a limited liability corporation in 1976, with four teams of designers managed by Andrew Fallon, Paul Vermijs, Gert Kootstra and Ronald van Lit. 1989: the company is renamed Tel Graphic Design, Corporate Design, Environmental Design bv. Clients include Gasunie, Kadaster, Heineken, Nederlandse Spoorwegen, the Boymans-van Beuningen Museum and the Kunsthal in Rotterdam. – *Publications include:* Paul Hefting et al "Tel Design 1962–1992", Amsterdam 1992.

**Telingater,** Solomon – b. 12.5.1903 in

Tiflis, Georgien, gest. 1. 10. 1969 in Moskau, UdSSR. – *Grafik-Designer, Typograph, Schriftentwerfer, Illustrator, Bühnenbildner* – 1919–20 Studium in Baku und 1920–21 an der WChUTEMAS in Moskau. 1919 Mitaufbau des Ateliers „Rosta" in Baku. 1921–25 Leiter der künstlerischen Werkstätten des Hauses der kommunistischen Erziehung in Baku. Seit 1925 Arbeit als Grafik-Designer und Typograph in Moskau für den Verlag Gosizdat in konstruktivistischer Typographie und Fotomontage. Mitarbeiter der Architekturzeitschrift „SA". 1927 Zusammenarbeit mit El Lissitzky bei der Organisation der Ausstellung und der Gestaltung des Katalogs „Polygraphische Allunionsausstellung". Telingater entwirft die Typographie. 1928 Gründungsmitglied der Gruppe „Oktober". Teilnahme an der Ausstellung „Pressa" in Köln. 1933–41 künstlerischer Leiter des Verlags der Kommunistischen Partei der Sowjetunion „Partizdat"; diese Arbeiten werden 1937 mit einer Goldmedaille auf der Internationalen Ausstellung in Paris ausgezeichnet. 1959 mit der Goldmedaille der Internationalen Buchausstellung in Leipzig und 1963 mit dem Gutenberg-Preis der Stadt Leipzig ausgezeichnet. Entwurf mehrerer kyrillischer Schriften.

**Tenazas,** Lucille – geb. 17. 12. 1953 in Aklan, Philippinen. – *Grafik-Designerin, Lehrerin* – Studium am College of the Holy Spirit in Manila, danach Grafik-Designerin in einem Pharma-Konzern. 1980–81 Aufbaustudium an der Cranbrook Academy of Art in Bloomfield Hills, Michigan. Arbeitet 1982 als De-

Telingater 1930 Cover

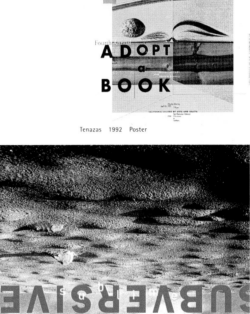

Tenazas 1992 Poster

Tel Design 1996 Cover

Tenazas 1994 Poster

création de l'atelier «Rosta» à Bakou, en 1919. Directeur des ateliers d'art de la maison de l'éducation communiste de Bakou de 1921 à 1925. Travaille, à partir de 1925, comme graphiste maquettiste et typographe pour les éditions Gosizdat de Moscou; il réalise des typographies et des photomontages constructivistes. Collaborateur de la revue d'architecture «SA». En 1927, il collabore, avec El Lissitzky, à l'organisation et à la conception du catalogue de l'exposition «Exposition polygraphique Allunion». Telingater en conçoit la typographie. Membre fondateur du groupe «Octobre» en 1928. Participation à l'exposition «Pressa» à Cologne. 1933–1941, directeur artistique de la «Partizdat», les éditions du parti communiste d'Union soviétique. Ses travaux sont récompensés par une médaille d'or à l'exposition universelle de Paris en 1937. Médaille d'or de la foire internationale du livre de Leipzig (1959) et prix Gutenberg de la ville de Leipzig (1963). A dessiné plusieurs alphabets cyrilliques.

**Tenazas,** Lucille – née le 17. 12 1953 à Aklan, Philippines – *graphiste maquettiste, enseignante* – études au College of the Holy Spirit à Manille, puis graphiste maquettiste pour une grande entreprise pharmaceutique. 1980–1981, études complémentaires à la Cranbrook Academy of Art à Bloomfield Hills, Michigan. Desi-

Tiflis, Georgia, d. 1. 10. 1969 in Moscow, USSR – *graphic designer, typographer, type designer, illustrator, set-designer* – 1919–20: studies in Baku and from 1920–21 at VKHUTEMAS in Moscow. 1919: helps set up the Rosta studio in Baku. 1921–25: head of the art workshop at the House of Communist Education in Baku. 1925: starts working in Constructivist typography and photomontage as a graphic designer and typographer for Gosizdat publishers in Moscow. Works on "SA", a magazine for architecture. 1927: works with El Lissitzky on the organization and catalogue design for the "Polygraphic All-Unions Exhibition". Telingater designs the typography. 1928: founder member of the October group. Takes part in the "Pressa" exhibition in Cologne. 1933–41: art director of the Soviet Communist party publishing house, Partizdat. His work there earns him a gold medal at the International Exhibition in Paris in 1937. Awarded a gold medal at the "Internationale Buchausstellung" in Leipzig (1959) and Leipzig's Gutenberg Prize (1963). Telingater designed several Cyrillic alphabets.

**Tenazas,** Lucille – b. 17. 12. 1953 in Aklan, The Philippines – *graphic designer, teacher* – Studied at the College of the Holy Spirit in Manila and then worked as a graphic designer for a pharmaceuticals company. 1980–81: graduate studies at the Cranbrook Academy of Art in Bloomfield Hills, Michigan. 1982: works as a designer for Harmon Kemp Inc. in New York. 1985: opens the Tenazas Design

signerin bei Harmon Kemp Inc. in New York. 1985 Eröffnung ihres Studios „Tenazas Design" in San Francisco. Auftraggeber waren u. a. das Center for the Arts at Yerba Buena Gardens, das American Institute of Graphic Arts, das San Francisco Museum of Modern Art, Pacific Bell und Esprit. Professorin für Design am California College of Arts and Crafts in San Francisco.

**Thomkins,** André – geb. 11. 8. 1930 in Luzern, Schweiz, gest. 9. 11. 1985 in Berlin, Deutschland. – *Künstler* – 1947–48 Studium an der Kunstgewerbeschule in Luzern. 1950–51 Aufenthalt in Paris, Studium an der Académie de la Grande Chaumière. 1954 Umzug nach Essen. 1967 erste Palindrome und Anagramme. 1971–73 Professor für Malerei an der Kunstakademie in Düsseldorf. 1972 und 1977 Teilnahme an der documenta V und VI in Kassel. 1978 Umzug nach Zürich. 1979 Teilnahme an der 15. Biennale in São Paulo. 1984 Umzug nach München. Zahlreiche Ausstellungen, u. a. im Kunsthaus Zürich (1986). – *Publikation:* „Permanentszene. Zeichnungen, Anagramme, Bilder, Collagen, Objekte und Texte von 1946–77", Stuttgart 1978.

**Thompson,** Bradbury – geb. 25. 3. 1911 in Topeka, USA. – *Grafik-Designer, Typograph* – 1929–34 Studium an der Washburn University in Topeka. 1934–38 Gestaltung von Büchern und Zeitschriften bei Capper Publications in Topeka. 1939–62 Herausgabe und Gestaltung von 59 Ausgaben der Zeitschrift „Westvaco Inspirations" der West Virginia Pulp and Paper Company. 1942–45 Art Director der

gner chez Harmon Kemp Inc. à New York en 1982. Ouvre l'agence «Tenazas Design» à San Francisco en 1985. Commanditaires : le Center for the Arts at Yerba Buena Gardens, l'American Institute of Graphic Arts, le San Francisco Museum of Modern Art, Pacific Bell et Esprit. Professeur de design au California College of Arts and Crafts à San Francisco.

**Thomkins,** André – né le 11. 8. 1930 à Lucerne, Suisse, décédé le 9. 11 1985 à Berlin, Allemagne – *artiste* – 1947–1948, études à la Kunstgewerbeschule (école des arts décoratifs) de Lucerne. 1950–1951, séjour à Paris où il poursuit sa formation à l'académie de la Grande Chaumière. S'installe à Essen en 1954. Premiers palindromes et anagrammes à partir de 1967. Professeur de peinture à l'académie des beaux-arts de Düsseldorf de 1971 à 1973. Participe aux «documenta V» et «VI» à Kassel en 1972 et en 1977. S'installe à Zurich en 1978. Participe à la 15e Biennale de São Paulo en 1979. Vit à Munich à partir de 1984. Nombreuses expositions, entre autres au Kunsthaus de Zurich (1986). – *Publications, sélection :* «Permanentszene. Zeichnungen, Anagramme, Bilder, Collagen, Objekte und Texte von 1946–77», Stuttgart 1978.

**Thompson,** Bradbury – né le 25. 3. 1911 à Topeka, Etats-Unis – *graphiste maquettiste, typographe* – 1929–1934, études à la Washburn University à Topeka. 1934–1938, maquettes de livres et de revues chez Capper Publications à Topeka. Public et réalise la maquette de 59 éditions de la revue «Westvaco Inspirations» de la West Virginia Pulp and Paper Company entre 1939 et 1962. Directeur artistique des revues multilingues «Victory», «USA» et «America» de 1942 à 1945. Directeur artistique de la revue «Mademoiselle» de 1945 à 1959. Maquette des revues «Art News» et «Art News' Annual» de 1945 à

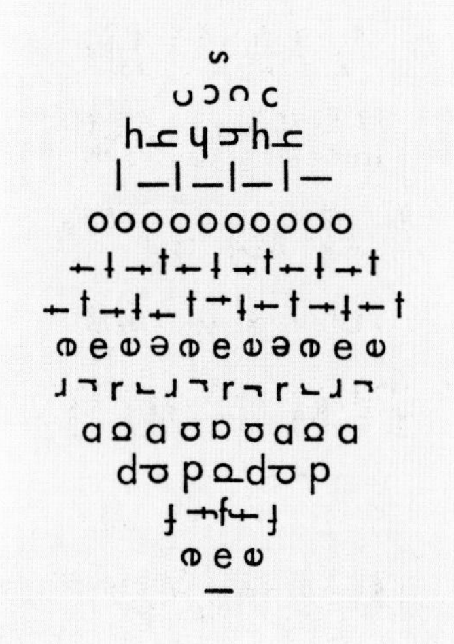

Thomkins 1960 Poem

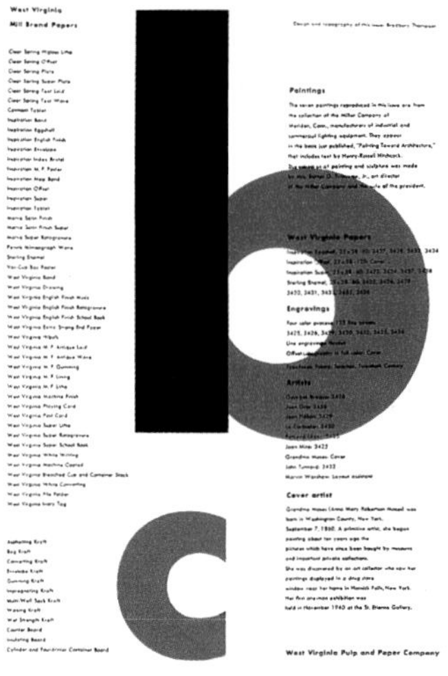

Thompson 1947 Spread

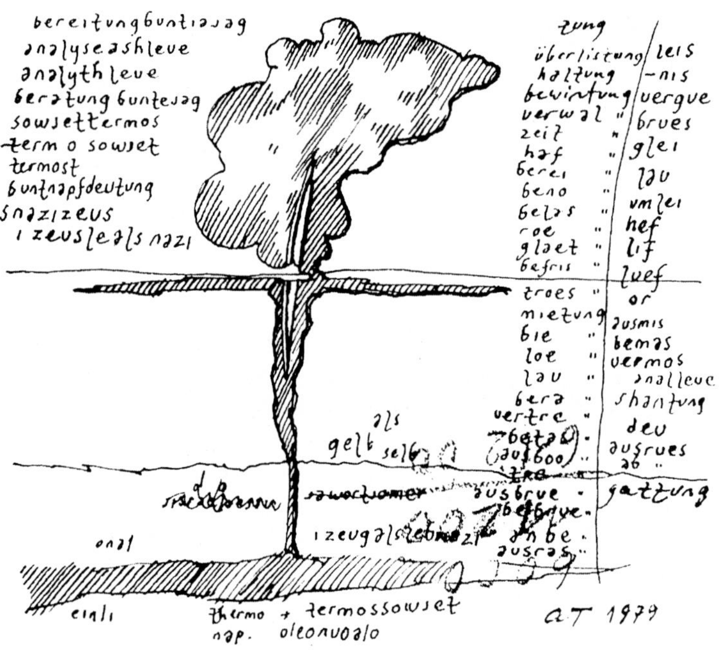

Thomkins 1979 Poem

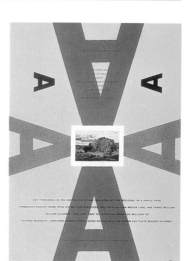

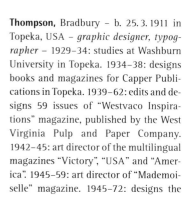

Thompson 1952 Page

studio in San Francisco. Clients include the Center for the Arts at Yerba Buena Gardens, the American Institute of Graphic Arts, the San Francisco Museum of Modern Art, Pacific Bell and Esprit. Professor of design at the California College of Arts and Crafts in San Francisco.

**Thomkins,** André – b. 11. 8. 1930 in Lucerne, Switzerland, d. 9. 11. 1985 in Berlin, Germany – *artist* – 1947–48: studies at the Kunstgewerbeschule in Lucerne. 1950–51: spends time in Paris and studies at the Académie de la Grande Chaumière. 1954: moves to Essen. 1967: produces his first palindromes and anagrams. 1971–73: professor of painting at the art academy in Düsseldorf. 1972 and 1977: takes part in documenta V and VI in Kassel. 1978: moves to Zurich. 1979: takes part in the 15th Biennale in São Paolo. 1984: moves to Munich. Numerous exhibitions, including at the Kunsthaus in Zurich (1986). – *Publications include:* "Permanentszene. Zeichnungen, Anagramme, Bilder, Collagen, Objekte und Texte von 1946–77", Stuttgart 1978.

**Thompson,** Bradbury – b. 25. 3. 1911 in Topeka, USA – *graphic designer, typographer* – 1929–34: studies at Washburn University in Topeka. 1934–38: designs books and magazines for Capper Publications in Topeka. 1939–62: edits and designs 59 issues of "Westvaco Inspirations" magazine, published by the West Virginia Pulp and Paper Company. 1942–45: art director of the multilingual magazines "Victory", "USA" and "America". 1945–59: art director of "Mademoiselle" magazine. 1945–72: designs the

mehrsprachigen Zeitschriften „Victory“, „USA“ und „America“. 1945–59 Art Director der Zeitschrift „Mademoiselle“. 1945–72 Gestaltung der Zeitschriften „Art News“ und „Art News' Annual“. 1975 Auszeichnung mit einer Goldmedaille des American Institute of Graphic Arts (AIGA). 1977 wird er in die Hall of Fame des Art Directors Club of New York gewählt. 1980 erscheint nach einer limitierten Vorauflage die von ihm gestaltete „Washburn Bible“ für die Washburn University Press in Topeka. 1983 Auszeichnung mit der Ehrendoktorwürde der Rhode Island School of Design und mit dem Frederic W. Goudy-Preis des Rochester Institute of Technology. – *Schriftentwürfe:* Monoalphabet (1945), Alphabet 26 (1950). – *Publikationen u. a.:* „The Monoalphabet“, New York 1945; „Modern Painting and Typography“, New York 1947; „Alphabet 26“, New York 1950; „The Design Concept“, New York, London 1982.

**Thorne,** Robert – geb. 1754, gest. 1820. – *Schriftenhersteller* – Thorne erwirbt 1794 die Schriftgießerei des verstorbenen Thomas Cottrell in London. Thorne entwirft und produziert zahlreiche fette, ornamentale Schriften. Sie werden ein Kennzeichen für die Typographie der Zeit der industriellen Revolution.

**Thorowgood,** London – *Schriftgießerei* – William Thorowgood erwirbt 1820 die Schriftgießerei von Robert Thorne nach dessen Tod und 1828 die Type Street Foundry in London. Robert Besley wird 1838 Teilhaber, Umbenennung in W. Thorowgood & Co. W. Thorowgood zieht sich 1849 aus der Firma zurück. Nach

Thompson 1958 Page

ABCD EFGH IJKL MNOP QRST UVW XYZ

Thorne ca. 1810 Thorne Shaded

ABCD EFGH IJKLM NOPQ RSTUV WXYZ abcdef ghijkl mnop qrstuv wxyz

Thorowgood ca. 1820–1836 Thorowgood

And be it hereby enacted, that the Mayors, Ba liffs or other head Officers of every Town or Place corporate, being a Justice

Thorowgood 1824 Open Black, No.1

1972. Obtient la médaille d'or de l'American Institute of Graphic Arts (AIGA) en 1975. Est élu au Hall of Fame de l'Art Directors Club de New York en 1977. Après l'édition d'un tirage limité de la «Washburn Bible» dont il avait réalisé la maquette, publication de l'ouvrage pour la Washburn University en 1980. Docteur honoris causa de la Rhode Island School of Design en 1983 et prix Frederic W. Goudy du Rochester Institute of Technology. – *Polices :* Monoalphabet (1945), Alphabet 26 (1950). – *Publications, sélection :* «The Monoalphabet», New York 1945; «Modern Painting and Typography», New York 1947; «Alphabet 26», New York 1950; «The Design Concept», New York, Londres 1982.

**Thorne,** Robert – né en 1754, décédé en 1820 – *fabricant de fontes* – Robert Thorne acquiert en 1794 la fonderie de caractères du défunt Thomas Cottrell de Londres. Thorne dessine et produit de nombreux caractères gras et ornementaux qui caractériseront la typographe de l'ère de la révolution industrielle.

**Thorowgood,** Londres – *fonderie de caractères* – à la mort de Robert Thorne, William Thorowgood acquiert sa fonderie de caractères en 1820, puis la Type Street Foundry de Londres en 1828. Robert Besley entre comme associé dans l'entreprise en 1838; celle-ci s'appelle désormais W. Thorowgood & Co. W. Thorowgood se retire de la société en 1849.

magazines "Art News" and "Art News' Annual". 1975: is awarded a gold medal by the American Institute of Graphic Arts (AIGA). 1977: is elected into the Art Directors Club of New York's Hall of Fame. 1980: after initial printing of a limited edition, the "Washburn Bible", designed by Thompson, is published by Washburn University Press in Topeka. 1983: is awarded an honorary doctorate by the Rhode Island School of Design and the Frederic W. Goudy Prize by the Rochester Institute of Technology. – *Fonts:* Monoalphabet (1945), Alphabet 26 (1950). – *Publications include:* "The Monoalphabet", New York 1945; "Modern Painting and Typography", New York 1947, "Alphabet 26", New York 1950; "The Design Concept", New York, London 1982.

**Thorne,** Robert – b. 1754, d. 1820 – *manufacturer of type* – 1794: Thorne buys up the late Thomas Cottrell's type foundry in London. Thorne designs and produces numerous bold, ornamental typefaces. They become characteristic of the typographic forms employed at the time of the Industrial Revolution.

**Thorowgood,** London – *type foundry* – 1820: William Thorowgood buys up Robert Thorne's type foundry after Thorne's death and in 1828 the Type Street Foundry in London. 1838: Robert Besley is made a partner and the business is renamed W. Thorowgood & Co. 1849: W. Thorowgood retires from the company. 1905: after various phases of restructuring and renaming, the concern is taken over by Stephenson, Blake & Co. in Sheffield.

mehreren Umwandlungen und Umbenennungen wird 1905 die Nachfolgefirma von Stephenson, Blake & Co. in Sheffield übernommen.

**Tiemann,** Walter – geb. 29. 1. 1876 in Delitzsch, Deutschland, gest. 12. 9. 1951 in Leipzig, Deutschland. – *Schriftentwerfer, Typograph, Maler, Lehrer* – Studium der Malerei an der Kunstakademie in Leipzig. Seit 1898 für Verlage tätig, u. a. S. Fischer, Reclam, Rütten & Loening. 1903 Lehrer an der Staatlichen Akademie für Graphische Künste in Leipzig. Zusammenarbeit mit dem Insel Verlag. 1905 Begegnung mit Karl Klingspor. 1907 Gründung der Janus-Presse (mit C. E. Poeschel). 1920–41 und 1945–46 Direktor der Staatlichen Akademie für Graphische Künste in Leipzig, 1946 mit der Ehrendoktorwürde ausgezeichnet. – *Schriftentwürfe:* Janus-Pressen-Schrift (mit C. E. Poeschel, 1906), Tiemann-Mediaeval (1909), Mediaeval Kursiv (1911), Tiemann Fraktur (1914), Peter-Schlemihl-Schrift (1914), Narziss (1921), Tiemann-Antiqua (1923–26), Tiemann-Gotisch (1924), Kleist-Fraktur (1928), Orpheus (1928), Daphnis (1931), Fichte-Fraktur (1935), Euphorion (1935), Offizin (1952). – *Publikation u. a.:* Georg Kurt Schauer „Walter Tiemann. Ein Vermächtnis", Offenbach 1953.

**Tilp,** Alfred – geb. 25. 3. 1932 in Karlsbad, Böhmen, Tschechoslowakei. – *Typograph, Grafik-Designer, Lehrer* – 1953 Lithographenlehre. 1956–62 Studium des Grafik-Design und der Kunstpädagogik an den Kunstakademien in Stuttgart, Düsseldorf und Berlin. Freiberufliche Tätigkeit für Industrie, Verlage und In-

Après plusieurs transformations et changements de nom, la société est reprise par Stephenson, Blake & Co de Sheffield en 1905.

**Tiemann,** Walter – né le 29. 1. 1876 à Delitzsch, Allemagne, décédé le 12. 9. 1951 à Leipzig, Allemagne – *concepteur de polices, typographe, peintre, enseignant* – études de peinture à l'académie des beaux-arts de Leipzig. Travaille, à partir de 1898 pour des maisons d'édition, dont S. Fischer, Reclam, Rütten & Loening. Enseigne en 1903, à la Staatliche Akademie für Graphische Künste (académie nationale des arts graphiques), de Leipzig. Collaboration avec les éditions Insel. Rencontre Karl Klingspor en 1905. Fondation de la Janus-Presse (avec C. E. Poeschel) en 1907. Dirige la Staatliche Akademie für Graphische Künste de 1920 à 1941, puis de 1945 à 1946. Le titre de docteur honoris causa lui est décerné en 1946. – *Polices:* Janus-Pressen-Schrift (avec C. E. Poeschel, 1906), Tiemann Mediaeval (1909), Mediaeval Kursiv (1911), Tiemann Fraktur (1914), Peter-Schlemihl-Schrift (1914), Narziss (1921), Tiemann-Antiqua (1923–1926), Tiemann-Gotisch (1924), Kleist-Fraktur (1928), Orpheus (1928), Daphnis (1931), Fichte-Fraktur (1935), Euphorion (1935), Offizin (1952). – *Publication:* Georg Kurt Schauer «Walter Tiemann. Ein Vermächtnis», Offenbach 1953.

**Tilp,** Alfred – né le 25. 3. 1932 à Karlsbad, Bohème, Tchécoslovaquie – *typographe, graphiste maquettiste, enseignant* – 1953, apprentissage de lithographe. 1956–1962, études d'arts graphiques, de design et de pédagogie de l'art aux écoles des beaux-arts de Stuttgart, Düsseldorf et Berlin. Exerce comme indépendant pour l'industrie, l'édition et des Institutions. Professeur à l'école supérieure de Rhénanie-Palatinat à Mayence de 1972 à 1982. Re-

Tiemann 1906 Cover

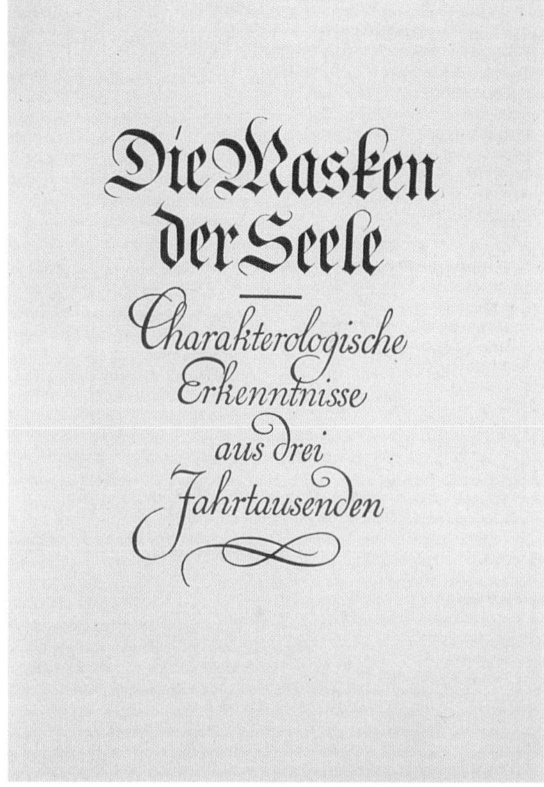


Tiemann 1940 Cover

Tilp 1986 Poster

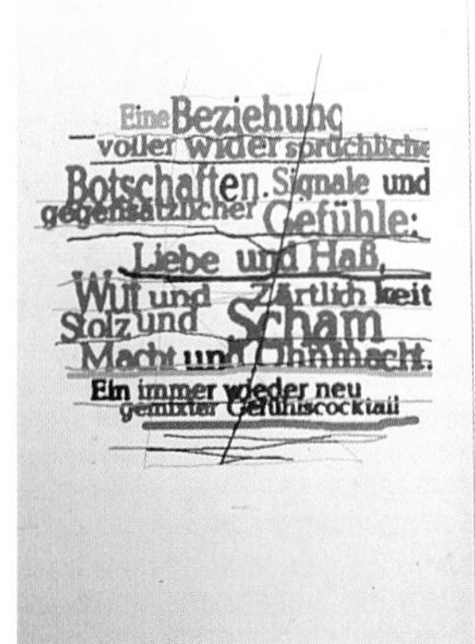

Tilp 1994 Spread

**Tiemann,** Walter – b. 29. 1. 1876 in Delitzsch, Germany, d. 12. 9. 1951 in Leipzig, Germany – *type designer, typographer, painter, teacher* – Studied painting at the art academy in Leipzig. 1898: starts working for various publishers, including S. Fischer, Reclam and Rütten & Loening. 1903: teaches at the Staatliche Akademie für Graphische Künste in Leipzig. Works with the Insel publishing house. 1905: meets Karl Klingspor. 1907: founds the Janus Presse with C.E. Poeschel. 1920–41 and 1945–46: director of the Staatliche Akademie für Graphische Künste in Leipzig. 1946: is awarded an honorary doctorate. – *Fonts:* Janus-Pressen-Schrift (with C.E. Poeschel, 1906), Tiemann-Mediaeval (1909), Mediaeval Kursiv (1911), Tiemann Fraktur (1914), Peter-Schlemihl-Schrift (1914), Narziss (1921), Tiemann-Antiqua (1923–26), Tiemann-Gotisch (1924), Kleist-Fraktur (1928), Orpheus (1928), Daphnis (1931), Fichte-Fraktur (1935), Euphorion (1935), Offizin (1952). – *Publication:* Georg Kurt Schauer "Walter Tiemann. Ein Vermächtnis", Offenbach 1953.

**Tilp,** Alfred – b. 25. 3. 1932 in Karlovy Vary, Bohemia – *typographer, graphic designer, teacher* – 1953: trains as a lithographer. 1956–62: studies graphic design and art education at the art academies in Stuttgart, Düsseldorf and Berlin. Freelances for industry and various publishers and institutions. 1973–82: professor at the Fachhochschule Rheinland-Pfalz in Mainz. 1975: starts producing pictures with letters, collages and scribble art. 1982–96: professor in the design depart-

stitutionen. 1973–82 Professor an der Fachhochschule Rheinland-Pfalz in Mainz. Seit 1975 Buchstabenbilder, Collagen, Scribble-Art. 1982–96 Professor an der Fachhochschule in Würzburg, Fachbereich Gestaltung. Seit 1984 Beschäftigung mit Computerschriften, Herausgabe der Zeitschrift „Tilp". Veröffentlichungen in Zeitschriften wie „Infrarot", „Exit", „Instant", „Hamburger Satzspiegel". Zahlreiche Einzelausstellungen, u. a. in der Print Gallery Pieter Brattinga, Amsterdam (1979), in der Galerie Artfusion, Mainz (1985) und im Rudolf-Alexander-Schröder-Haus, Würzburg (1994). – *Publikationen:* „Tour de France", Würzburg 1983; „Voll ins Leben", Würzburg 1996.

**Tinguely,** Jean – geb. 22. 5. 1925 in Fribourg, Schweiz, gest. 31. 8. 1991 in La Verrerie, Frankreich. – *Künstler* – Lehre als Dekorateur im Kaufhaus „Globus" in Basel. 1941–45 Studium an der Gewerbeschule in Basel. Danach Arbeit als freier Künstler. 1954 erste Einzelausstellung in der Galerie Arnaud in Paris. 1955 erste Konzepte für die Malmaschine „Meta-matics". 1959 Happening „For Static" im Institute of Contemporary Art in London. 1960 Mitglied der in Paris gegründeten Künstlergruppe „Nouveaux Réalistes". Konstruiert die erste sich selbst zerstörende Maschine in New York. Seit 1961 Skulpturen aus Abfalleisen. 1968 Teilnahme an der documenta IV in Kassel. Zu seinen zahlreichen internationalen Ausstellungen zeichnete Tinguely häufig Plakate mit comicartigen Text- und Bildillustrationen. – *Publikationen u. a.:* Pontus Hulten „Jean Tinguely,

Tinguely 1974 Poster

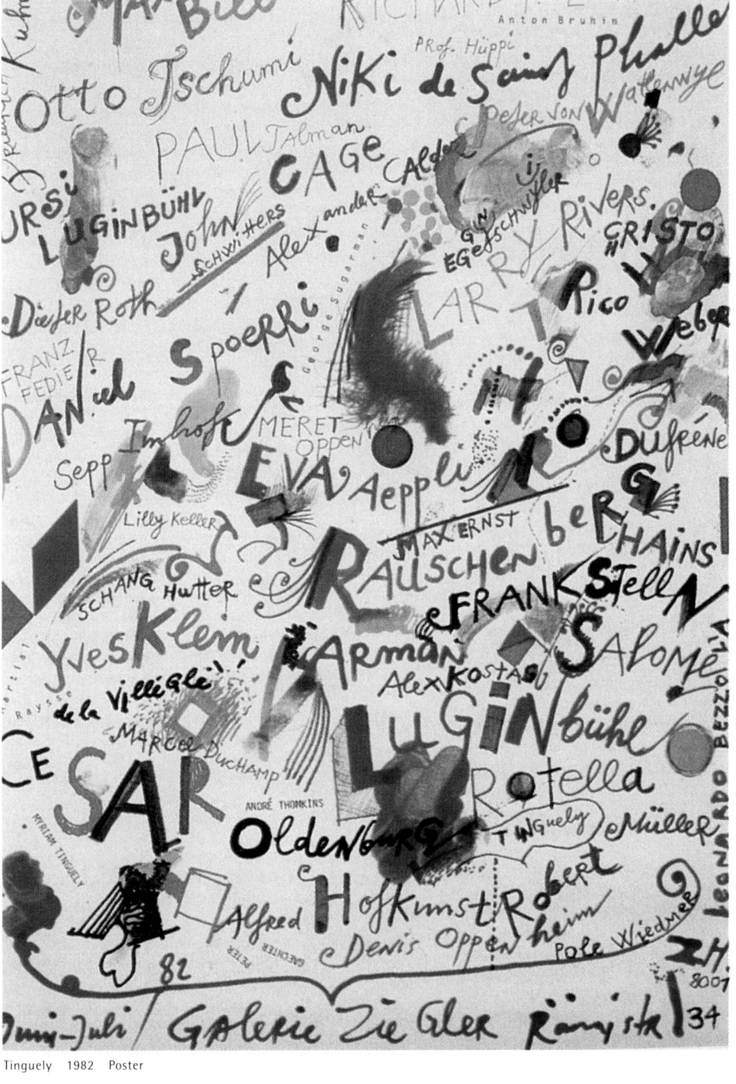


Tinguely 1982 Poster

Tilp 1996 Cover

vient au dessin de lettres, aux collages et au lettrisme à partir de 1975. Professeur de design à l'école supérieure d'arts graphiques de Würzburg de 1982 à 1996. Se consacre depuis 1984 à l'écriture informatique, édite la revue «Tilp». Publications dans des revues comme «Infrarot», «Exit», «Instant», «Hamburger Satzspiegel». Nombreuses expositions personnelles, entre autres, à la Print Gallery Pieter Brattinga, Amsterdam (1979), à la galerie Artfusion, Mayence (1985), à la Rudolf-Alexander-Schröder-Haus, Würzburg (1994). – *Publications:* «Tour de France», Würzburg 1983; «Voll ins Leben», Würzburg 1996.

**Tinguely,** Jean – né le 22. 5. 1925 à Fribourg, Suisse, décédé le 31. 8. 1991 à La Verrerie, France – *artiste* – apprentissage de décorateur au grand magasin «Globus» à Bâle. 1941–1945, études à la Gewerbeschule (arts et métiers) de Bâle. Exerce ensuite comme artiste indépendant. 1954, première exposition personnelle à la galerie Arnaud à Paris. 1955, premières ébauches de sa machine à peindre, la «Meta-matics». 1959, happening «For Static» à l'Institute of Contemporary Art de Londres. 1960, membre du groupe d'artistes parisiens les «Nouveaux Réalistes»; construit la première machine s'autodétruisant à New York. Réalise des sculptures à partir de vieille ferraille dès 1961. Participation à la «documenta IV» à Kassel en 1968. Pour ses nombreuses expositions internationales, Tinguely dessinait fréquemment des affiches comportant des illustrations et des textes rappellant la bande dessinée. – *Publications, sélection:* Pontus Hulten «Jean Tinguely, Meta», Paris 1973; Christina Bi-

ment at the Fachhochschule in Würzburg. 1984: starts studying and working with computer fonts. Produces "Tilp" magazine. Has contributions printed in magazines such as "Infrarot", "Exit", "Instant" and the "Hamburger Satzspiegel". Numerous solo exhibitions, including at the Print Gallery Pieter Brattinga in Amsterdam (1979), the Galerie Artfusion in Mainz (1985) and the Rudolf-Alexander-Schröder-Haus in Würzburg (1994). – *Publications:* "Tour de France", Würzburg 1983; "Voll ins Leben", Würzburg 1996.

**Tinguely,** Jean – b. 22. 5. 1925 in Fribourg, Switzerland, d. 31. 8. 1991 in La Verrerie, France – *artist* – Trained as an interior designer at Globus department store in Basle. 1941–45: studies at the Gewerbeschule in Basle. Then works as a freelance artist. 1954: his first solo exhibition is held at the Galerie Arnaud in Paris. 1955: comes up with his first concepts for the Meta-matics painting machine. 1959: "For Static" happening at the Institute of Contemporary Art in London. 1960: member of Nouveaux Réalistes artist's group founded in Paris. Constructs the first self-destructing machine in New York. 1961: starts producing sculptures made of scrap iron. 1968: takes part in documenta IV in Kassel. Tinguely often drew posters with comic-like text and picture illustrations for his many international exhibitions. – *Publications include:* Pontus Hulten "Jean Tinguely, Meta", Paris 1973; Christina Bischofsberger "Jean Tinguely. Catalogue raisonné. Sculptures and Reliefs 1954–1968", Küs-

Meta", Paris 1973; Christina Bischofsberger „Jean Tinguely. Catalogue raisonné. Sculptures and Reliefs 1954–1968", Küsnacht, Zürich 1982; Stefanie Poley „Jean Tinguely", München 1986.

**Tomaszewski,** Henryk – geb. 10. 6. 1914 in Siedlce, Polen. – *Illustrator, Grafik-Designer, Bühnenbildner, Lehrer* – 1934–39 Studium an der Kunstakademie in Warschau. 1939 Gestaltung des polnischen Industriepavillons auf der Weltausstellung in New York. Seit 1945 Gestalter für Verlage, Theater und kulturelle Institutionen. Seit 1948 Teilnahme an internationalen Plakatausstellungen. Seit 1952 Professor für Plakatgestaltung an der Kunstakademie in Warschau. Durch seine Plakate und sein pädagogisches Wirken entsteht eine polnische Plakatszene, die internationales Ansehen genießt. Zahlreiche Auszeichnungen, u. a. Nationalpreis der polnischen Regierung (1953), erster Preis auf der 7. Biennale in São Paulo, Brasilien (1963), Goldmedaille der Leipziger Buchmesse (1965). 1968 Mitbegründer des Plakatmuseums in Warschau. – *Publikationen u. a.:* „Ksiazka zazalen", Warschau 1961. B. Kwiatkowska „Henryk Tomaszewski", Warschau 1959.

**Tomato** – *Design-Studio* – 1991 Gründung von Tomato in London durch Steve Baker, Dirk van Dooren, Karl Hyde, Rick Smith, Simon Taylor, John Warwicker und Graham Wood. Auftraggeber waren u. a. Channel 4, BBC, Nike, Adidas, Reebok, Levi's, MTV, Philips, Coca Cola und IBM. Auszeichnungen durch den Art Directors Club Deutschland (1993, 1995), Type Directors Club, Tokio (1994, 1995,

schofsberger «Jean Tinguely. Catalogue raisonné. Sculptures and Reliefs 1954–1968», Küsnacht, Zurich 1982; Stefanie Poley «Jean Tinguely», Munich 1986.

**Tomaszewski,** Henryk – né le 10. 6. 1914 à Siedlce, Pologne – *illustrateur, graphiste maquettiste, scénographe, enseignant* – 1934–1939, études à l'académie des beaux-arts de Varsovie. Conception du pavillon de l'industrie polonaise pour l'exposition universelle de New York en 1939. Maquettiste et designer pour des maisons d'édition, le théâtre et des institutions culturelles à partir de 1945. Participe à des expositions internationales d'affiches à partir de 1948. Enseigne depuis 1952 l'art de l'affiche à l'académie des beaux-arts de Varsovie en qualité de professeur. Ses affiches et son engagement pédagogique ont favorisé l'émergence d'une école d'affichistes polonais reconnue dans le monde entier. Nombreuses distinctions, dont le prix du gouvernement polonais (1953), le premier prix de la 7e Biennale de São Paulo, Brésil (1963), la médaille d'or du salon du livre de Leipzig (1965). Cofondateur du musée de l'affiche de Varsovie, en 1968. – *Publications, sélection :* «Ksiazka zazalen», Varsovie 1961. B. Kwiatkowska «Henryk Tomaszewski», Varsovie 1959.

**Tomato** – *atelier de design* – 1991, fondation de l'atelier Tomato à Londres par Steve Baker, Dirk van Dooren, Karl Hyde, Rick Smith, Simon Taylor, John Warwicker et Graham Wood. Commanditaires : Channel 4, BBC, Nike, Adidas, Reebok, Levi's, MTV, Philips, Coca Cola et IBM. Primé par l'Art Directors Club d'Allemagne (1993, 1995), Type Directors Club de Tokyo (1994, 1995, 1997), d & ad, Angleterre (1994, 1995), Golden Globe, Etats-Unis (1994). Participation au «Spoken Word Festival» d'Amsterdam (1995). Expositions à Tokyo (1992, 1997),

Tomaszewski 1959 Poster

Tomaszewski 1975 Poster

Tomaszewski 1963 Poster

Tomato 1992 Poster

nacht, Zurich 1982; Stefanie Poley "Jean Tinguely", Munich 1986.

**Tomaszewski,** Henryk – b. 10. 6. 1914 in Siedlce, Poland – *illustrator, graphic designer, set-designer, teacher* – 1934–39: studies at the art academy in Warsaw. 1939: designs Poland's industrial pavilion for the world exhibition in New York. 1945: starts designing for various publishers, the theater and arts institutions. 1948: first takes part in an international poster exhibition. From 1952 onwards: professor of poster design at the art academy in Warsaw. His posters and teachings have helped establish a Polish poster scene which enjoys recognition worldwide. Numerous awards, including the Polish government's National Prize (1953), 1st prize at the 7th Biennale in São Paolo in Brazil (1963) and a gold medal from the Leipzig book fair (1965). 1968: co-founder of the poster museum in Warsaw. – *Publications include:* "Ksiazka zazalen", Warsaw 1961. B. Kwiatkowska "Henryk Tomaszewski", Warsaw 1959.

**Tomato** – *design studio* – 1991: Tomato is launched in London by Steve Baker, Dirk van Dooren, Karl Hyde, Rick Smith, Simon Taylor, John Warwicker and Graham Wood. Clients include Channel 4, the BBC, Nike, Adidas, Reebok, Levi's, MTV, Philips, Coca Cola and IBM. The studio has received awards from the Art Directors Club of Germany (1993 and 1995), the Type Directors Club of Tokyo (1994, 1995 and 1997), d & ad in England (1994 and 1995) and the Golden Globe, USA (1994). They have taken part in the "Spoken Word Festival" in Amsterdam (1995).

1997), d & ad, England (1994, 1995), Golden Globe, USA (1994). Teilnahme am „Spoken Word Festival“ in Amsterdam (1995). Ausstellungen, u. a. in Tokio (1992, 1997), London (1996, 1997), Tschechien (1996) und Paris (1997). Die von Karl Hyde und Rick Smith gegründete Musikgruppe Underworld ist Teil von Tomato. – *Publikation u. a.:* „Process – a Tomato Project“, London 1996.

**Toorn,** Jan van – geb. 9. 5. 1932 in Tiel, Niederlande. – *Grafik-Designer, Ausstellungsgestalter, Lehrer* – Ausbildung am Instituut voor Kunstnijverheid in Amsterdam. Seit 1957 freier Gestalter. 1960–77 Gestaltung der Kalender der Druckerei Spruyt in Amsterdam. 1965, 1971 Auszeichnung mit dem H. N. Werkman-Preis der Stadt Amsterdam. 1967–73 Grafik-Designer des Stedelijk van Abbe Museums in Eindhoven. Unterrichtet seit 1968 Visuelle Kommunikation an der Gerrit Rietveld Academie in Amsterdam. Einzelausstellungen im Museum Fodor in Amsterdam (1972), in Breda (1987). Seit 1977 Gestaltung der Zeitschrift „DAAT“ (Dutch Art and Architecture Today) des Rijksdienst voor Beeldende Kunsten in Den Haag. 1979 Gestaltung der Ausstellung „Die niederländische Revolte 1559–1609“ in Utrecht. 1979–1980 Briefmarkenserien für die niederländische Post PTT. 1987–88 Gestaltung der Jahresberichte der PTT. Direktor der Jan van Eyck Academie in Maastricht. – *Publikationen:* „Vormgeving in functie van museale overdracht“ (mit J. Leering), Eindhoven 1978; „Professional design, Quad 2“, Maastricht 1980.

Tomato 1992 Cover

Toorn 1979 Stamp

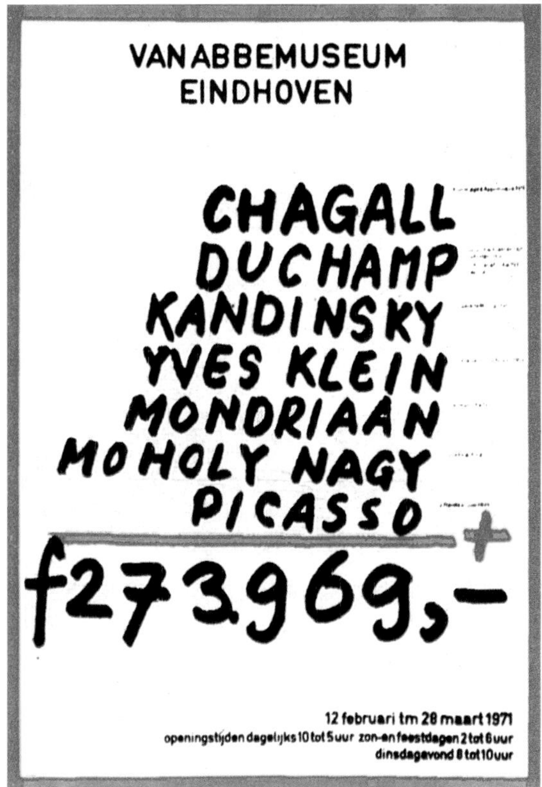

Toorn 1971 Poster



Tomato 1995 Advertisement

Toorn 1973 Calendar

Londres (1996, 1997), en Tchéquie (1996), à Paris (1997). Le groupe de musiciens «Underworld», fondé par Karl Hyde et Rick Smith, fait partie de Tomato. – *Publication, sélection :* «Process – a Tomato Project», Londres 1996.

**Toorn,** Jan van – né le 9. 5. 1932 à Tiel, Pays-Bas – *graphiste maquettiste, scénographe, enseignant* – formation à l'Instituut voor Kunstnijverheid à Amsterdam. Designer indépendant à partir de 1957. Maquette des calendriers de l'imprimerie Spruyt d'Amsterdam de 1960 à 1977. Prix H. N. Werkman de la ville d'Amsterdam en 1965, puis en 1971. Graphiste du Stedelijk van Abbe Museum d'Eindhoven de 1967 à 1973. Enseigne la communication visuelle à la Gerrit Rietveld Academie d'Amsterdam à partir de 1968. Exposition personnelle au musée Fodor à Amsterdam (1972), à Breda (1987). Maquette, depuis 1977, de la revue «DAAT» (Dutch Art and Architecture Today) éditée par le Rijksdienst voor Beeldende Kunsten de La Haye. 1979, conception de l'exposition «La révolte hollandaise 1559–1609» à Utrecht. 1979–1980, série de timbres pour les postes néerlandaises PTT. 1987–1988, maquette des rapports annuels des postes néerlandaises PTT. Directeur de la Jan van Eyck Academie à Maastricht. – *Publications :* Vormgeving in functie van museale overdracht» (avec J. Leering), Eindhoven 1978; «Professional design, Quad 2», Maastricht 1980.

Exhibitions have been shown in Tokyo (1992 and 1997), London (1996 and 1997), the Czech Republic (1996) and Paris (1997). The Underworld group formed by Karl Hyde and Rick Smith is part of Tomato. – *Publications include:* "Process – a Tomato Project", London 1996.

**Toorn,** Jan van – b. 9. 5. 1932 in Tiel, The Netherlands – *graphic designer, exhibition planner, teacher* – Trained at the Instituut voor Kunstnijverheid in Amsterdam. From 1957 onwards: freelance designer. 1960–77: designs a calendar for Spruyt printing works in Amsterdam. 1965 and 1971: is awarded Amsterdam's H. N. Werkman Prize. 1967–73: graphic designer for the Stedelijk van Abbe Museum in Eindhoven. 1968: starts teaching visual communication at the Gerrit Rietveld Academie in Amsterdam. Solo exhibitions at the Museum Fodor in Amsterdam (1972) and in Breda (1987). From 1977 onwards: designs "DAAT" magazine ("Dutch Art and Architecture Today"), published by the Rijksdienst voor Beeldende Kunst in The Hague. 1979: designs the exhibition "The Dutch Revolt 1559–1609" in Utrecht. 1979–1980: produces series of postage stamps for the Dutch post office PTT. 1987–88: designs the PTT's annual reports. Director of the Jan van Eyck Academie in Maastricht. – *Publications:* "Vormgeving in functie van museale overdracht" (with J. Leering), Eindhoven 1978; "Professional design, Quad 2", Maastricht 1980.

**Total Design** – *Design-Studio* – 1963 Gründung von Total Design als multidisziplinärem Design-Studio in Amsterdam durch Wim Crouwel (Grafik-Design), Friso Kramer (Industrie-Design), Benno Wissing (Grafik-Design und Architektur-Design), Dick und Paul Schwarz (Management und Wirtschaft). Für Auftraggeber wie die niederländische Post PTT, den Flughafen Schiphol, das Van Abbe Museum Amsterdam, das Boymans-van Beuningen Museum Rotterdam, das Stedelijk Museum Amsterdam und das Van Gogh Museum Amsterdam entstehen Erscheinungsbilder, Leitsysteme und Ausstellungsgestaltungen. Total Design hat Büros in Amsterdam, Brüssel, Maastricht und seit 1989 in Paris. In den neunziger Jahren Spezialisierung auf Corporate Design (TBI, DSM, Thalys) unter der kreativen Leitung von Hans Paul Brandt, Marcel Speller und Jelle van der Toorn Vrijthoff. – *Publikationen u. a.:* K. Broos „Design: Total Design", Utrecht 1983; Hub. Hubben „Design: Total Design; The Eighties. Entwurf: Total Design; Die Achtziger Jahre", Wormer 1989.

**Tournes,** Jean de – geb. 1504 in Lyon, Frankreich, gest. 1564 in Lyon, Frankreich. – *Verleger, Buchhändler* – Lehre als Drucker in Lyon. 1545 Herausgabe einer Petrarca-Ausgabe, 1547 einer Dante-Ausgabe, 1552 einer Vitruv-Ausgabe, 1559–61 Herausgabe der Chronik von Froissart. Nach seinem Tod 1564 führt sein Sohn (Jean de Tournes II) das Unternehmen weiter. 1585 Umzug der Familie nach Genf wegen Glaubensverfolgung. Weiterführung von Verlag und

**Total Design** – *atelier de design* – 1963, création, à Amsterdam, de Total Design, un atelier de design pluridisciplinaire fondé par Wim Crouwel (conception graphique design) Friso Kramer (design industriel), Benno Wissing (conception graphique et design d'architecture), Dick et Paul Schwarz (management et gestion). Réalise des identités, logos, systèmes d'orientation et architectures d'expositions pour : les postes néerlandaises PTT, l'aéroport Schiphol, le Van Abbe Museum d'Amsterdam, le musée Boymans-van Beuningen, le Stedelijk Museum d'Amsterdam, le musée Van Gogh d'Amsterdam. Total Design a des bureaux à Amsterdam, Bruxelles, Maastricht ainsi qu' à Paris depuis 1989. Dans les années 90 se spécialise dans le Corporate Design (TBI, DSM, Thalys) sous la direction artistique de Hans Paul Brandt, Marcel Speller et Jelle van der Toorn Vrijthoff. – *Publications, sélection :* Kees Broos «Design : Total Design», Utrecht 1983; Hub. Hubben «Design: Total Design; The Eighties. Entwurf: Total Design; Die Achtziger Jahre», Wormer 1989.

**Tournes,** Jean de – né en 1504 à Lyon, France, décédé en 1564 à Lyon, France – *éditeur, libraire* – apprentissage d'imprimeur à Lyon. Publie une édition de Pétrarque en 1545, de Dante en 1547 et de Vitruve en 1552. Edite la chronique de Froissart de 1559 à 1561. Après sa mort, en 1564, son fils (Jean de Tournes II) reprend l'entreprise. La famille s'installe à Genève en 1585 pour fuir les persécutions religieuses. La maison d'édition et l'imprimerie continuent leurs activités jusqu'en 1780.

**Tracy,** Walter – né le 14. 2. 1914 en Angleterre – *concepteur de polices* – 1930–1935, apprentissage de composition typographique et emploi comme compositeur. 1935–1938, typographe chez Bar-

Total Design 1983 Cover

Total Design 1996 Cover

Total Design 1996 Poster

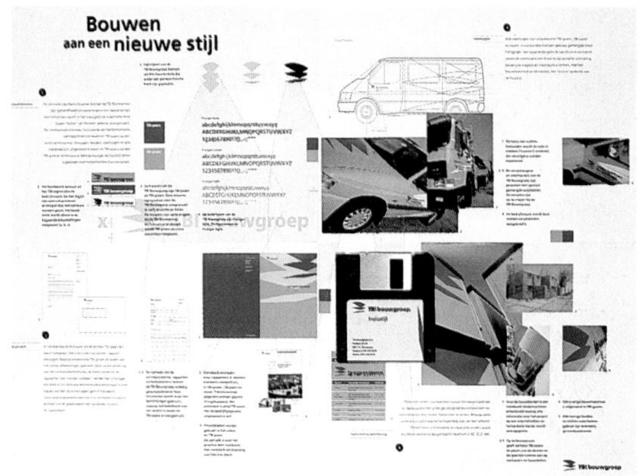

Total Design 1996 Poster

Total Design 1994/95 Train

**Total Design** – *design studio* – 1963: Total Design, a multidisciplinary design studio, is founded in Amsterdam by Wim Crouwel (graphic design), Friso Kramer (industrial design), Benno Wissing (graphic design and architectural design) and Dick and Paul Schwarz (management and finance). They produce corporate identities and signage systems and design exhibitions for clients who include the Dutch post office PTT, Schiphol Airport, the Van Abbe Museum, Stedelijk Museum and Van Gogh Museum (all three in Amsterdam), and the Boymans-van Beuningen Museum in Rotterdam. Total Design has offices in Amsterdam, Brussels and Maastricht, opening a fourth in Paris in 1989. In the nineties they specialize in corporate design (TBI, DSM, Thalys) under the creative direction of Hans Paul Brandt, Marcel Speller and Jelle van der Toorn Vrijthoff. – *Publications include*: Kees Broos "Design: Total Design", Utrecht 1983; Hub. Hubben "Design: Total Design; The Eighties. Entwurf: Total Design; Die Achtziger Jahre", Wormer 1989.

**Tournes,** Jean de – b. 1504 in Lyon, France, d. 1564 in Lyon, France – *publisher, bookseller* – Trained as a printer in Lyon. 1545: publishes an edition of works by Petrarch, in 1547 of works by Dante and in 1552 by Vitruvius. 1559–61: publishes Froissart's chronicles. 1564: Tournes' son Jean de Tournes II takes over the business after his father's death. 1585: religious persecution causes the family to move to Geneva. Tournes' publishing house and printing workshop continues

Druckerei bis 1780.

**Tracy,** Walter – geb. 14. 2. 1914 in England. – *Schriftentwerfer* – 1930–35 Schriftsetzerlehre und Arbeit als Schriftsetzer. 1935–38 Typograph bei der Barnard Press. 1938–46 Arbeit bei Notley Advertising. 1947 freier Gestalter, Mitarbeit im Verlag „Art & Technics", der u. a. die Zeitschrift „Alphabet and Image" produziert. 1948–78 Arbeit bei Linotype England als Leiter der Abteilung Schriftenentwicklung. 1973 Auszeichnung mit dem Titel „Royal Designer for Industry" durch die Royal Society of Arts. Seit 1978 Entwurf mehrerer arabischer Alphabete. – *Schriftentwürfe:* Jubilee (1954), Adsans (1959), Maximus (1967), Telegraph Modern (1969 für die Tageszeitung „The Daily Telegraph"), Times Europa (1972), Telegraph Newface Bold (mit Shelley Winter, 1979), Qadi (1979), Kufics (1980), Oasis (1985), Sharif (1989), Malik (1988), Medina (1989). – *Publikationen u. a.:* „Letters of Credit, a View of Type Design", London 1986; „The Typographic Scene", London 1988.

**Treumann,** Otto – geb. 28. 3. 1919 in Fürth, Deutschland. – *Grafik-Designer* – 1933 Umzug der Familie in die Niederlande. 1935–40 Studium an der Grafisch School und an der Nieuwe Kunstschool in Amsterdam. Seit 1945 freier Grafiker. Auftraggeber waren u. a. die EL AL, Philips, Nederlandse Gasunie, der Verlag Wolters-Noordhoff, die niederländische Post PTT und das Kaufhaus De Bijenkorf. 1946–59 Gestaltung der Hauszeitschrift „Rayon Revue" der Algemene Kunstzijde Unie. 1947 Auszeichnung mit dem H. N.

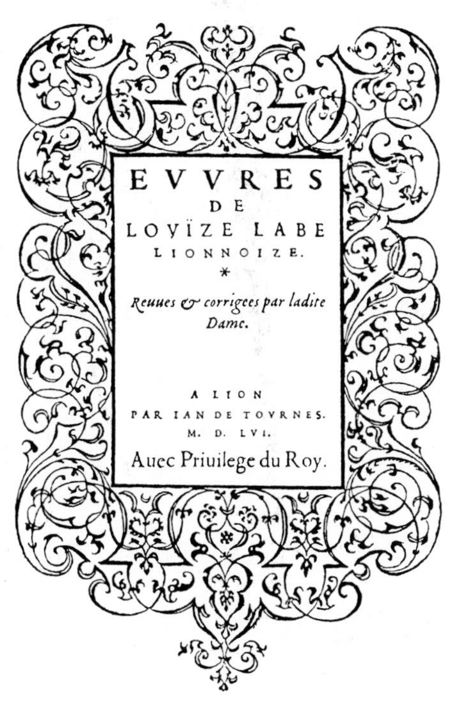


Tournes 1556 Main Title

Treumann 1983 Poster

Treumann 1963 Logo

| | n | | | | |
|---|---|---|---|---|---|
| a | b | e | f | g | i |
| o | r | s | t | y | z |
| A | B | C | E | G | H |
| M | O | R | S | X | Y |
| 1 | 2 | 4 | 6 | 8 | & |

Tracy 1967 Maximus

Treumann 1979 Stamp

to operate until 1780.

**Tracy,** Walter – b. 14. 2. 1914 in England – *type designer* – 1930–35: trains and works as a typesetter. 1935–38: typographer for Barnard Press. 1938–46: works for Notley Advertising. 1947: freelance designer. Works for Art & Technics publishing house which issues "Alphabet and Image" magazine, among others. 1948–78: works for Linotype England as head of the department for type development. 1973: made a Royal Designer for Industry by the Royal Society of Arts. Since 1978: has designed several Arabic alphabets. – *Fonts:* Jubilee (1954), Adsans (1959), Maximus (1967), Telegraph Modern (1969 for "The Daily Telegraph" newspaper), Times Europa (1972), Telegraph Newface Bold (with Shelley Winter, 1979), Qadi (1979), Kufics (1980), Oasis (1985), Sharif (1989), Malik (1988), Medina (1989). – *Publications:* "Letters of Credit, a View of Type Design", London 1986; "The Typographic Scene", London 1988.

**Treumann,** Otto – b. 28. 3. 1919 in Fürth, Germany – *graphic designer* – 1933: the Treumann family moves to The Netherlands. 1935–40: studies at the Grafisch School and at the Nieuwe Kunstschool in Amsterdam. 1945: starts working as a freelance graphic artist. Clients include EL AL, Philips, Nederlandse Gasunie, Wolters-Noordhoff publishers, the Dutch post office PTT, and the De Bijenkorf department store. 1946–59: designs "Rayon Revue", Algemene Kunstzijde Unie's in-house magazine. 1947: is awarded Amsterdam's H. N. Werkman Prize. 1963: de-

nard Press. Employé chez Notley Advertising de 1938 à 1946. Designer indépendant en 1947, travaux pour les éditions «Art & Technics» qui publient, entre autres, la revue «Alphabet and Image». Employé chez Linotype Angleterre comme directeur du service de développement des fontes. En 1973, la Royal Society of Arts lui décerne le titre de «Royal Designer for Industry». A créé plusieurs alphabets arabes depuis 1978. – *Polices:* Jubilee (1954), Adsans (1959), Maximus (1967), Telegraph Modern (1969 pour le quotidien «The Daily Telegraph»), Times Europa (1972), Telegraph Newface Bold (avec Shelley Winter, 1979), Qadi (1979), Kufics (1980), Oasis (1985), Sharif (1989), Malik (1988), Medina (1989). – *Publications:* «Letters of Credit, a View of Type Design», Londres 1986; «The Typographic Scene», Londres 1988.

**Treumann,** Otto – né le 28. 3. 1919 à Fürth, Allemagne – *graphiste maquettiste* – la famille s'installe aux Pays-Bas en 1933. Etudes à la Grafisch School et à la Nieuwe Kunstschool à Amsterdam de 1935 à 1940. Graphiste indépendant à partir de 1945. Commanditaires: l'EL AL, Philips, Nederlandse Gasunie, les éditions Wolters-Noordhoff, les postes néerlandaises PTT et le magasin De Bijenkorf. Maquette de «Rayon Revue», publication de la Algemene Kunstzijde Unie, de 1946 à 1959. Prix H. N. Werkman de la ville

Werkman-Preis der Stadt Amsterdam. 1963 Entwurf des Signets für die Israelische Fluggesellschaft EL AL (mit George Him). Ausstellungen u. a. in Amsterdam (1951), Jerusalem (1957), London (1959). – *Publikation u. a.:* Bibeb „Otto Treumann, grafisch ontwerper", Nimwegen 1970.

**Troxler,** Niklaus – geb. 1. 5. 1947 in Willisau, Schweiz. – *Grafik-Designer, Illustrator, Konzertorganisator* – 1963–67 Schriftsetzerlehre in Luzern. Seit 1966 Organisator der Willisauer Jazzkonzerte, seit 1975 auch Gestalter des Jazz-Festivals in Willisau. 1967–71 Studium an der Schule für Gestaltung in Luzern. 1971–72 Art Director bei Hollenstein Création in Paris. 1973 eigenes Grafik-Design-Studio in Willisau. Auftraggeber waren u. a. das Kleine Theater Luzern und das Kunstgewerbemuseum Zürich. Zahlreiche Auszeichnungen, u. a. Toulouse-Lautrec-Medaille der Plakat-Triennale in Essen (1987), 1. Preis des „Festival des Affiches" in Chaumont, Frankreich (1992), 1. Preis der „Poster Biennale" in Lahti, Finnland (1993). 1996 Gastprofessor an der Gesamthochschule in Kassel. Zahlreiche Plakatausstellungen, u. a. in der Galerie Raeber, Luzern (1978) und in der Reinhold Brown Gallery, New York (1987). – *Publikation u. a.:* „Niklaus Troxler – Jazzplakate", Schaftlach 1991.

**Trump,** Georg – geb. 10. 7. 1896 in Brettheim, Deutschland, gest. 21.12.1985 in München, Deutschland. – *Grafiker, Typograph, Schriftentwerfer, Maler, Lehrer* – 1912–14 Studium an der Kunstgewerbeschule in Stuttgart bei J. V. Cissarz und

d'Amsterdam en 1947. Dessine le logo de la compagnie aérienne israélienne EL AL (avec George Him) en 1963. Expositions à Amsterdam (1951), Jérusalem (1957), Londres (1959). – *Publication, sélection:* Bibeb «Otto Treumann, grafisch ontwerper», Nimègue 1970.

**Troxler,** Niklaus – né le 1. 5. 1947 à Willisau, Suisse – *graphiste maquettiste, illustrateur, organisateur de concerts* – 1963–1967, apprentissage de la composition typographique à Lucerne. Organise des concerts de jazz à Willisau à partir de 1966, puis le Festival de jazz de Willisau à partir de 1975. Etudes à l'école de design de Lucerne de 1967 à 1971. Directeur artistique chez Hollenstein Création à Paris de 1971 à 1972. Exerce dans son propre atelier de conception graphique à Willisau dès 1973. Commanditaires: Le Kleine Theater de Lucerne et le musée des arts décoratifs de Zurich. Nombreuses distinctions, entre autres, la Médaille Toulouse-Lautrec de la Triennale de l'affiche d'Essen (1987), premier prix du «Festival des Affiches» de Chaumont, France (1992), premier prix de la «Poster Biennale» de Lahti, Finlande (1993). Cycle de conférences à l'université de Kassel en 1996. Nombreuses expositions d'affiches à la galerie Raeber, Lucerne (1978), à la Reinhold Brown Gallery, New York (1987). – *Publication, sélection:* «Niklaus Troxler – Jazzplakate», Schaftlach 1991.

**Trump,** Georg – né le 10. 7. 1896 à Brettheim, Allemagne, décédé le 21. 12. 1985 à Munich, Allemagne – *graphiste, typographe, concepteur de polices, peintre, enseignant* – 1912–1914, études à la Kunstgewerbeschule (école des arts décoratifs) de Stuttgart chez J. V. Cissarz, puis, de 1919 à 1923, chez F. H. E. Schneidler, assistant en 1921. Céramiste en Italie de 1923 à 1926. Professeur à la Kunstgewerbeschule (école des arts dé-

Troxler 1980 Poster

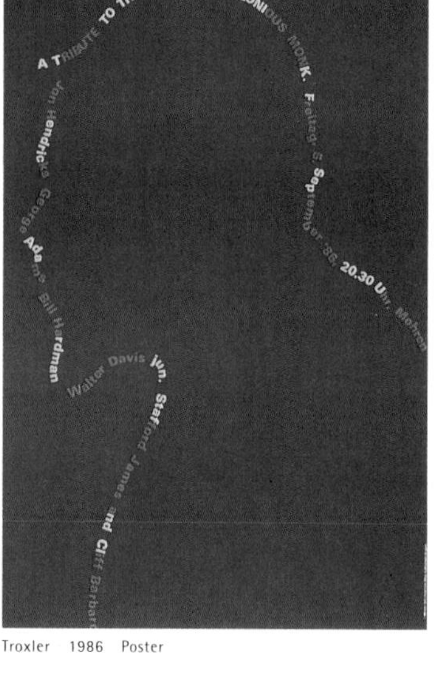
Troxler 1986 Poster

Troxler 1990 Poster

signs the logo for EL AL Israeli airlines with George Him. Various exhibitions in Amsterdam (1951), Jerusalem (1957) and London (1959), among other places. – *Publications include:* Bibeb "Otto Treumann, grafisch ontwerper", Nijmegen 1970.

**Troxler,** Niklaus – b. 1. 5. 1947 in Willisau, Switzerland – *graphic designer, illustrator, concert organizer* – 1963–67: trains as a typesetter in Lucerne. From 1966 onwards: organizes jazz concerts in Willisau and from 1975 onwards also the Jazz Festival in Willisau. 1967–71: studies at the Schule für Gestaltung in Lucerne. 1971–72: art director for Hollenstein Création in Paris. 1973: opens his own graphic design studio in Willisau. Clients include the Kleine Theater in Lucerne and the Kunstgewerbemuseum in Zurich. Troxler has won numerous awards, including the Toulouse Lautrec Medal at the Poster Triennale in Essen (1987), 1st prize at the "Festival des Affiches" in Chaumont, France (1992) and 1st prize at the Poster Biennale in Lahti, Finland (1993). 1996: visiting professor at the Gesamthochschule in Kassel. There have been numerous exhibitions of his posters, including at the Galerie Raeber in Lucerne (1978) and the Reinhold Brown Gallery in New York (1987). – *Publications include:* "Niklaus Troxler – Jazzplakate", Schaftlach 1991.

**Trump,** Georg – b. 10. 7. 1896 in Brettheim, Germany, d. 21. 12. 1985 in Munich, Germany – *graphic artist, typographer, type designer, painter, teacher* – 1912–14: studies at the Kunstgewerbe-

1919–23 bei F. H. E. Schneidler, 1921 dort Assistent. 1923–26 Keramiker in Italien. 1926–29 Professor an der Kunstgewerbeschule in Bielefeld. 1930–35 Zusammenarbeit mit der Schriftgießerei H. Berthold AG in Berlin, seit 1937 mit der Schriftgießerei C. E. Weber in Stuttgart. 1931–34 Direktor der Höheren grafischen Fachschule in Berlin. Als Nachfolger von Paul Renner 1934–53 Leiter der Meisterschule für Deutschlands Buchdrucker in München. 1982 Verleihung der Medaille des Type Directors Club of New York. Ausstellungen seines Werks, u. a. in der Gallery 303 in New York (1966), dem Stadtmuseum München (1981) und dem Klingspor-Museum, Offenbach (1983). Trump arbeitet zunächst im Stil der „Neuen Typographie", später in einem individuell skripturalen Stil. – *Schriftentwürfe:* City (1930), Trump Deutsch (1935), Schadow (1938–52), Forum I (1948), Delphin 1 (1951), Forum II (1952), Amati (1953), Palomba (1954), Delphin 2 (1955), Codex (1955), Signum (1955), Time Script (1956), Trump Mediaeval (1958–60), Trump Gravur (1960), Jaguar (1964), Mauritius (1967). – *Publikationen u. a.:* „Vita Activa: Georg Trump. Bilder, Schriften und Schriftbilder", München 1967. Ph. Luidl, G. G. Lange „Hommage für Georg Trump", München 1981; Ph. Luidl u. a. „Georg Trump. Maler, Schriftkünstler, Grafiker", München 1981.

**Tscherny,** George – geb. 12. 7. 1924 in Budapest, Ungarn. – *Grafik-Designer, Typograph, Lehrer* – 1941 Umzug in die USA. 1946–47 Studium an der Newark School of Fine and Industrial Art. 1947–

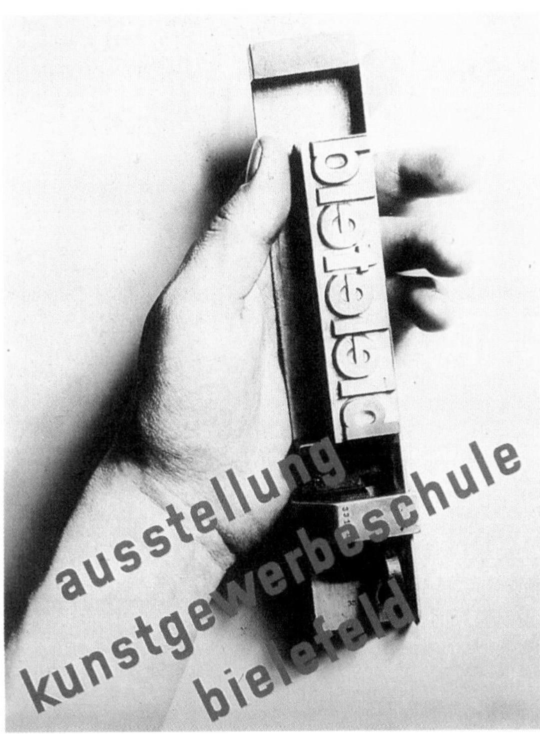


Trump 1927 Poster

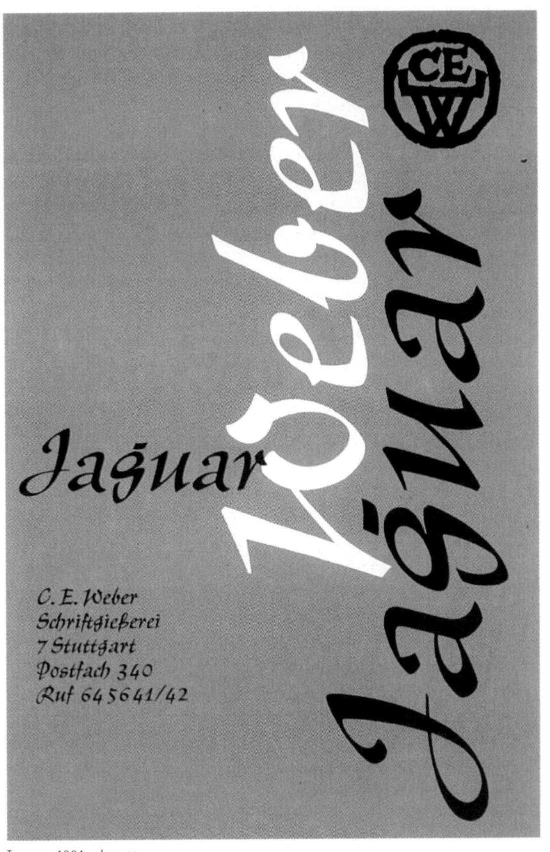


Trump 1964 Jaguar

coratifs) de Bielefeld de 1926 à 1929. Collaboration avec la fonderie de caractères H. Berthold AG à Berlin de 1930 à 1935, puis, à partir de 1937, avec la fonderie C. E. Weber à Stuttgart. 1931–1934, directeur de la Höhere grafische Fachschule (école supérieure des arts graphiques) de Berlin. Succède à Paul Renner à la tête de la Meisterschule für Deutschlands Buchdrucker (école professionnelle de l'imprimerie allemande) qu'il dirige de 1934 à 1953. Médaille du Type Directors Club of New York en 1982. Expositions de ses œuvres à la Gallery 303, New York (1966), au Stadtmuseum München (1981), au musée Klingspor Offenbach (1983). Trump a commencé par adopter le style de la «Nouvelle typographie»; par la suite, il a développé un genre scriptural personnel. – *Polices:* City (1930), Trump Deutsch (1935), Schadow (1938–1952), Forum I (1948), Delphin 1 (1951), Forum II (1952), Amati (1953), Palomba (1954), Delphin 2 (1955), Codex (1955), Signum (1955), Time Script (1956), Trump Mediaeval (1958–1960), Trump Gravur (1960), Jaguar (1964), Mauritius (1967). – *Publications, sélection:* «Vita Activa: Georg Trump. Bilder, Schriften und Schriftbilder», Munich 1967. Ph. Luidl, G. G. Lange «Hommage für Georg Trump», Munich 1981; Ph. Luidl et autres «Georg Trump. Maler, Schriftkünstler, Grafiker», Munich 1981.

**Tscherny,** George – né le 12. 7. 1924 à Budapest, Hongrie – *graphiste maquettiste, typographe, enseignant* – s'installe en 1941 aux Etats-Unis. 1946–1947, études à la Newark School of Fine and Industrial Art. 1947–1950, études au Pratt Institute de New York. Designer d'emballages de

| n | n | **n** | | | *n* |
|---|---|---|---|---|---|
| a | b | e | f | g | i |
| o | r | s | t | y | z |
| A | B | C | E | G | H |
| M | O | R | S | X | Y |
| 1 | 2 | 4 | 6 | 8 | & |

Trump 1930 City

| | n | **n** | | | *n* |
|---|---|---|---|---|---|
| a | b | e | f | g | i |
| o | r | s | t | y | z |
| A | B | C | E | G | H |
| M | O | R | S | X | Y |
| 1 | 2 | 4 | 6 | 8 | & |

Trump 1938–52 Schadow

| | n | **n** | | | *n* |
|---|---|---|---|---|---|
| a | b | e | f | g | i |
| o | r | s | t | y | z |
| A | B | C | E | G | H |
| M | O | R | S | X | Y |
| 1 | 2 | 4 | 6 | 8 | & |

Trump 1958–60 Trump Mediaeval

schule in Stuttgart under J. V. Cissarz and from 1919–23 under F. H. E. Schneidler. 1921: assistant at the school. 1923–26: ceramic artist in Italy. 1926–29: professor at the Kunstgewerbeschule in Bielefeld. 1930–35: works with the H. Berthold AG type foundry in Berlin and from 1937 onwards with the C. E. Weber type foundry in Stuttgart. 1931–34: director of the Höhere grafische Fachschule in Berlin. 1934–53: succeeds Paul Renner as director of the Meisterschule für Deutschlands Buchdrucker in Munich. 1982: awarded a medal by the Type Directors Club of New York. His work has been exhibited in Gallery 303 in New York (1966), in the Stadtmuseum in Munich (1981) and in the Klingspor Museum in Offenbach (1983), among other venues. Trump initially worked in the style of "New Typography" but later adopted an individual, scriptural style. – *Fonts:* City (1930), Trump Deutsch (1935), Schadow (1938–52), Forum I (1948), Delphin 1 (1951), Forum II (1952), Amati (1953), Palomba (1954), Delphin 2 (1955), Codex (1955), Signum (1955), Time Script (1956), Trump Mediaeval (1958–60), Trump Gravur (1960), Jaguar (1964), Mauritius (1967). – *Publications include:* "Vita Activa: Georg Trump. Bilder, Schriften und Schriftbilder", Munich 1967. Ph. Luidl, G. G. Lange "Hommage für Georg Trump", Munich 1981; Ph. Luidl et al "Georg Trump. Maler, Schriftkünstler, Grafiker", Munich 1981.

**Tscherny,** George – b. 12. 7. 1924 in Budapest, Hungary – *graphic designer, typographer, teacher* – 1941: moves to the

50 Studium am Pratt Institute in New York. 1950–53 Verpackungsdesigner. 1953–55 Partner im Büro George Nelson & Associates. 1955 Gründung seiner Firma in New York. Auftraggeber waren u. a. Mobil Oil, die Herman Miller Furniture Company, das Museum of Modern Art New York, die School of Visual Arts New York und IBM. Unterrichtet an der School of Visual Arts (1956–64) und am Pratt Institute (1956–57). 1966 Präsident des American Institute of Graphic Arts (AIGA). 1988 mit der „AIGA's Annual Medal" ausgezeichnet. 1992 Ausstellung im Visual Arts Museum in New York. – *Publikationen u. a.:* „Pull and let go – Design and Communication for the w. c.", New York 1990; „Odd & Even – A Study of House Numbers", New York 1992.

**Tschichold,** Jan – geb 2. 4. 1902 in Leipzig, Deutschland, gest. 11. 8. 1974 in Locarno, Schweiz. – *Typograph, Kalligraph, Autor, Lehrer* – 1919–21 Studium an der Akademie für Graphische Künste und Buchgewerbe in Leipzig. 1921–23 Meisterschüler von Walter Tiemann und Assistent von Hermann Delitsch. 1925 Herausgabe des Sonderhefts „elementare typographie" der Zeitschrift „Typographische Mitteilungen". 1926–33 Lehrer für Typographie und Kalligraphie an der Meisterschule für Deutschlands Buchdrucker in München. 1933 Kündigung des Dienstverhältnisses durch die Nationalsozialisten. Emigration nach Basel, Schweiz. 1933–40 Arbeit im Verlag Benno Schwabe, Lehrauftrag an der Allgemeinen Gewerbeschule in Basel. 1941–46 Typograph im Birkhäuser Ver-

1950 à 1953. Partenaire de l'agence George Nelson & Associates de 1953 à 1955. Fonde sa société à New York en 1955. Commanditaires : Mobil Oil, le Herman Miller Furniture Company, le Museum of Modern Art de New York, la School of Visual Arts de New York et IBM. A enseigné à la School of Visual Arts (1956–1964), puis au Pratt Institute (1956–1957). Président de l'American Institute of Graphic Arts (AIGA) en 1966. Lauréat de la «AIGA Annual Medal» en 1988. Exposition rétrospective au Visual Arts Museum de New York en 1992. – *Publications :* «Pull and let go – Design and Communication for the w. c.», New York 1990; «Odd & Even – A Study of House Numbers», New York 1992.

**Tschichold,** Jan – né le 2. 4. 1902 à Leipzig, Allemagne, décédé le 11. 8. 1974 à Locarno, Suisse – *typographe, calligraphe, auteur, enseignant* – 1919–1921, études à l'Akademie für Graphische Künste und Buchgewerbe (académie des arts graphiques et des arts du livre) de Leipzig. Elève maître de Walter Tiemann et assistant d'Hermann Delitsch de 1921 à 1923. Publie en 1925 le numéro spécial «elementare typographie» de la revue «Typographische Mitteilungen». Enseigne la typographie et la calligraphie à la Meisterschule für Deutschlands Buchdrucker (école professionnelle de l'imprimerie allemande) de Munich, de 1926 à 1933. Radiation de l'enseignement par les national-socialistes en 1933. Emigre à Bâle, Suisse. Travaille aux éditions Benno Schwabe de 1933 à 1940; chargé de cours à la Allgemeine Gewerbeschule (arts et métiers) de Bâle. Typographe aux éditions Birkhäuser à Bâle, de 1941 à 1946. Refonte de la maquette des Penguin Books à Londres de 1946 à 1949. Membre d'honneur du Double Crown Club en 1949. Médaille d'or de l'American Institute of

Tscherny 1958 Cover

Tscherny 1981 Cover

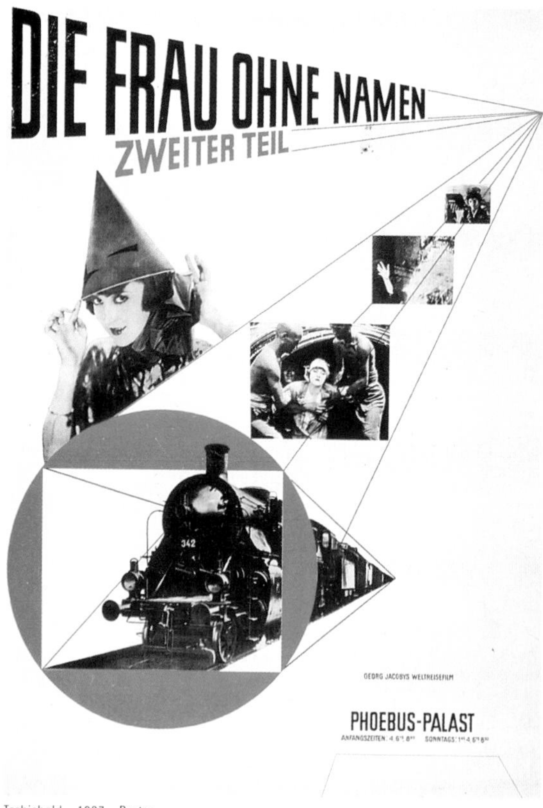


Tschichold 1927 Poster



Tschichold 1927 Invitation Card

USA. 1946–47: studies at the Newark School of Fine and Industrial Art. 1947–50: studies at the Pratt Institute in New York. 1950–53: packaging designer. 1953–55: partner of George Nelson & Associates. 1955: founds his own company in New York. Clients include Mobil Oil, the Herman Miller Furniture Company, the Museum of Modern Art in New York, the School of Visual Arts in New York and IBM. Teaches at the School of Visual Arts from 1956–64 and at the Pratt Institute from 1956–57. 1966: president of the American Institute of Graphic Arts (AIGA). 1988: is awarded the AIGA's Annual Medal. 1992: retrospective exhibition at the Visual Arts Museum in New York. – *Publications:* "Pull and let go – Design and Communication for the w. c.", New York 1990; "Odd & Even – A Study of House Numbers", New York 1992.

**Tschichold,** Jan – b. 2. 4. 1902 in Leipzig, Germany, d. 11. 8. 1974 in Locarno, Switzerland – *typographer, calligrapher, author, teacher* – 1919–21: studies at the Akademie für Graphische Künste und Buchgewerbe in Leipzig. 1921–23: is one of Walter Tiemann's master pupils and assistant to Hermann Delitsch. 1925: publication of Tschichold's special issue of "Typographische Mitteilungen" magazine, entitled "elementare typographie". 1926–33: teaches typography and calligraphy at the Meisterschule für Deutschlands Buchdrucker in Munich. 1933: dismissed from teaching by the National Socialists. Emigrates to Basle, Switzerland. 1933–40: works for Benno Schwabe publishers and teaches at the

lag in Basel. 1946–49 Neugestaltung der Penguin Books in London. 1949 Ehrenmitglied im Double Crown Club. 1954 Auszeichnung mit der Goldmedaille des American Institute of Graphic Arts (AIGA) in New York. 1955–67 Typograph der Firma Hoffmann-La Roche in Basel. 1964 und 1971 Auszeichnung mit der Goldmedaille der Internationalen Buchkunst-Ausstellung in Leipzig. 1965 Auszeichnung mit dem Gutenberg-Preis der Stadt Leipzig. Korrespondierendes Mitglied der Deutschen Akademie der Künste in Berlin. – *Schriftentwürfe:* Transito (1931), Saskia (1931), Zeus (1931), Sabon (1967). – *Publikationen u. a.:* „Die neue Typographie", Berlin 1928; „Typographische Gestaltung", Basel 1935; „Geschichte der Schrift in Bildern", Basel 1941; „Schriftkunde, Schreibübungen und Skizzieren", Basel 1942, Berlin 1951; „Schatzkammern der Schreibkunst", Basel 1946; „Meisterbuch der Schrift", Ravensburg 1953; „Erfreuliche Drucksachen durch gute Typographie", Ravensburg 1960; „Willkürfreie Maßverhältnisse der Buchseite und des Satzspiegels", Basel 1962; „Ausgewählte Aufsätze über Fragen der Gestalt des Buches und der Typographie", Basel 1975; „Jan Tschichold, Leben und Werk", Dresden 1977; „Jan Tschichold. Schriften 1925–1974" (2 Bände), Berlin 1991.

**Twombly,** Carol – geb. 13. 6. 1959 in Concord, USA. – *Schriftentwerferin* – Studium an der Rhode Island School of Design und an der Stanford University. 1984 Auszeichnung mit dem 1. Preis des Morisawa Schriftenwettbewerbs in Japan

Tschichold 1937 Poster

Tschichold 1951 Cover

Tschichold 1938 Poster

| | n | **n** | | | *n* |
|---|---|---|---|---|---|
| a | b | e | f | g | i |
| o | r | s | t | y | z |
| A | B | C | E | G | H |
| M | O | R | S | X | Y |
| 1 | 2 | 4 | 6 | 8 | & |

Tschichold 1967 Sabon

Graphic Arts (AIGA), à New York en 1954. Typographe de la société Hoffmann-La Roche, Bâle, de 1955 à 1967. Médaille d'or du Salon international du livre de Leipzig en 1964, puis en 1971. Prix Gutenberg de la ville de Leipzig en 1965. Membre correspondant de la Deutsche Akademie der Künste (beaux-arts) de Berlin. – *Polices :* Transito (1931), Saskia (1931), Zeus (1931), Sabon (1967). *Publications, sélection :* «Die neue Typographie», Berlin 1928; «Typographische Gestaltung», Bâle 1935; «Geschichte der Schrift in Bildern», Bâle 1941; «Schriftkunde, Schreibübungen und Skizzieren», Bâle 1942, Berlin 1951; «Schatzkammern der Schreibkunst», Bâle 1946; «Meisterbuch der Schrift», Ravensburg 1953; «Erfreuliche Drucksachen durch gute Typographie», Ravensburg 1960; «Willkürfreie Maßverhältnisse der Buchseite und des Satzspiegels», Bâle 1962; «Ausgewählte Aufsätze über Fragen der Gestalt des Buches und der Typographie», Bâle 1975; «Jan Tschichold, Leben und Werk», Dresde 1977; «Jan Tschichold. Schriften 1925– 1974» (2 vol.), Berlin 1991.

**Twombly,** Carol – née le 13. 6. 1959 à Concord, Etats-Unis – *conceptrice de polices* – études à la Rhode Island School of Design et à la Stanford University. Premier prix du concours d'écriture Morisawa au Japon pour sa police Mirarae en 1984. Entre chez Adobe en 1988; elle

Allgemeine Gewerbeschule in Basle. 1941–46: typographer for the Birkhäuser publishing house in Basle. 1946–49: redesigns the typography for Penguin Books in London. 1949: honorary member of the Double Crown Club. 1954: is awarded a gold medal by the American Institute of Graphic Arts (AIGA) in New York. 1955–67: typographer for Hoffmann-La Roche in Basle. 1964 and 1971: is awarded a gold medal at the "International Book Art" exhibition in Leipzig. 1965: awarded a Gutenberg Prize by the city of Leipzig. Corresponding member of the Deutsche Akademie der Künste in Berlin. – *Fonts:* Transito (1931), Saskia (1931), Zeus (1931), Sabon (1967). – *Publications include:* "Die neue Typographie", Berlin 1928; "Typographische Gestaltung", Basle 1935; "Geschichte der Schrift in Bildern", Basle 1941; "Schriftkunde, Schreibübungen und Skizzieren", Basle 1942, Berlin 1951; "Schatzkammern der Schreibkunst", Basle 1946; "Meisterbuch der Schrift", Ravensburg 1953; "Erfreuliche Drucksachen durch gute Typographie", Ravensburg 1960; "Willkürfreie Maßverhältnisse der Buchseite und des Satzspiegels", Basle 1962; "Ausgewählte Aufsätze über Fragen der Gestalt des Buches und der Typographie", Basle 1975; "Jan Tschichold, Leben und Werk", Dresden 1977; "Jan Tschichold. Schriften 1925–1974", Berlin 1991.

**Twombly,** Carol – b. 13. 6. 1959 in Concord, USA – *type designer* – Studied at the Rhode Island School of Design and at Stanford University. 1984: is awarded 1st prize in the Morisawa type compe-

Twombly
**Týfa**
**Type Directors Club**

für die Schrift Mirarae. 1988 Eintritt bei Adobe, für die sie mit Trajan, Charlemagne und Lithos die ersten Adobe-Originalschriften entwirft. – *Schriftentwürfe:* Mirarae (1984), Charlemagne (1989), Lithos (1989), Adobe Trajan (1989), Caslon (1990), Myriad (1992), Viva (1993), Nueva (1994), Chaparral (1997).

**Týfa,** Josef – geb. 5. 12. 1913 in Náchod-Běloves, Österreich-Ungarn (heute Tschechien). – *Typograph, Grafiker, Schriftentwerfer* – Grafikstudium an der Schule Rotter in Prag. Danach künstlerischer Leiter der Exportfirma Centrotex. Entwurf zahlreicher Signets, Plakate, Briefmarken, Buchumschläge und Schriften. Zahlreiche Auszeichnungen, u. a. 1964 auf der 1. Biennale der Gebrauchsgrafik in Brno und 1972 und 1975 für das beste Buch des Jahres. – *Schriftentwürfe:* Kollektiv-Antiqua (1958), Týfa-Antiqua (1959) und Academia (1968).

**Type Directors Club,** New York – 1946 Gründung des „Type Directors Club of New York" durch ca. zwanzig New Yorker Layoutsetzereien, Agenturen, Hochschullehrer und Verlagen. Ziel ist, das Niveau der Typographie und der damit verwandten grafischen Künste zu heben. Dazu werden Vorträge und Seminare veranstaltet, Publikationen herausgegeben. 1955 Veranstaltung des 1. (noch clubinternen) Jahreswettbewerbs. Der 2. Jahreswettbewerb 1956 ist öffentlich, Herausgabe eines Ergebniskatalogs. Seit 1965 Veranstaltung des internationalen Jahreswettbewerbs. Seit 1980 Herausgabe des Jahrbuchs des TDC. 1983 Gründung des Deutschen Komitees des TDC.

y conçoit les premières polices originales Adobe : Trajan, Charlemagne et Lithos. – *Polices :* Mirarae (1984), Charlemagne (1989), Lithos (1989), Adobe Trajan (1989), Caslon (1990), Myriad (1992), Viva (1993), Nueva (1994) et Chaparral (1997).

**Týfa,** Josef – né le 5. 12. 1913 à Náchod-Běloves, Autriche-Hongrie (aujourd'hui République Tschèque) – *typographe, graphiste, concepteur de polices* – études d'arts graphiques à l'école Rotter à Prague. Entre ensuite dans la société d'export Centrotex comme directeur artistique. Conçoit de nombreux signets, affiches, timbres postes, couvertures de livres et caractères. Nombreuses distinctions, entre autres lors de la 1ère Biennale d'arts graphiques appliqués de Brno en 1964, et meilleur livre de l'année en 1972, puis en 1975. – *Polices :* Kollektiv-Antiqua (1958) ; Týfa-Antiqua (1959) et Academia (1968).

**Type Directors Club,** New York – 1946, le «Type Directors Club of New York» a été fondé par une vingtaine d'ateliers de composition, d'agences, de professeurs d'université et de maisons d'édition newyorkais. Son but est de hausser le niveau de la typographie et par là même des arts graphiques apparentés. A ces fins, il organise des conférences et des séminaires, édite des publications et patronne un concours annuel de typographie. Organise en 1955 le premier concours annuel (interne au club). Dès 1956, le second concours est public et s'accompagne d'un catalogue présentant les œuvres. Depuis 1965, le concours annuel est international. Le club publie les annales du TDC depuis 1980. Fondation, en 1983, du comité allemand du TDC. Première conférence multimédias du TDC en 1996. Le club compte environ 300 membres.

**Typoart** Dresde – *fabricant de caractères*

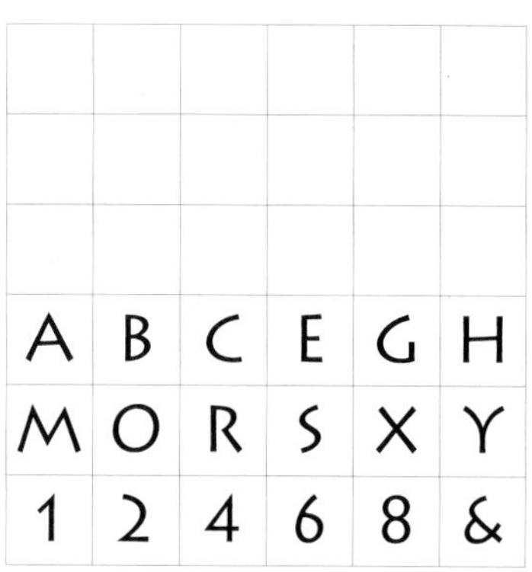

Twombly 1989 Lithos

ABCDEFG
1234567890
abcdefg

Týfa 1959–60 Týfa-Antiqua

| | | | | | |
|---|---|---|---|---|---|
| | n | **n** | | | *n* |
| a | b | e | f | g | i |
| o | r | s | t | y | z |
| A | B | C | E | G | H |
| M | O | R | S | X | Y |
| 1 | 2 | 4 | 6 | 8 | & |

Twombly 1990 Caslon

| | | | | | |
|---|---|---|---|---|---|
| | n | | | | *n* |
| a | b | e | f | g | i |
| o | r | s | t | y | z |
| A | B | C | E | G | H |
| M | O | R | S | X | Y |
| 1 | 2 | 4 | 6 | 8 | & |

Twombly 1992 Myriad (Multiple Master)

Gutenberg & Sons
THE NEWSLETTER OF THE TYPE DIRECTORS CLUB
VOL. 1 NO. 2 MARCH 1984

**Opening Soon Typography 5 (TDC-30)**

As this article is being written, entries for the 30th annual competition sponsored by the Type Directors Club are pouring in. The deadline for submissions is upon us, and Carol Wahler, Coordinator for the competition reports that no let up is in sight. With a small glimpse of the quality of this year's material, this promises to be a signal year for the exhibition.

Working on this year's show has been a labor of love and reward. Our jury is truly a "Blue Ribbon Panel" with such well known designers and typophiles as Photoletterings's Ed Benguiat, Bob Czernyz from Young & Rubicam and John Gibson from MacLaren Advertising in Toronto.

The New York Times is represented by Andy Kner, Olaf Leu of Olaf Leu Design & Partner is coming from Frankfurt, Germany, with Jack Odette of Citibank (last year's exhibition chairman), is serving as alternate juror.

B. Martin Petersen of Pedersen, Hinrichs & Shakery, Inc. and Jessica Weber of the Book of the Month Club round out the jury.

With a panel like this one we should have a top-notch judging. We all think it will be cold days inFebruary well spent.

continued on page 2

- Opening Soon Typography 5 (TDC-30)
- 1983 TDC Medal for Paul Standard and Herb Lubalin
- Losing Sleep
- Have TDC Show Will Travel
- What's New
- Here's Henny! TDC Christmas Party
- Last Impression
- Monster Available
- The Best of the Great
- New Faces
- Editorial

**Losing Sleep**

Being totally involved with typography more than the average person in our graphic community has led me to feel that I have many typographic obligations. I am more than certain we all feel the same way. I'm not referring to the details that we are faced with at our jobs. There's something that goes beyond the daily routines at work. I mean as the saying goes, "losing a lot of sleep over it."

To me designing has never been a job or profession. It's a way of life, like a priest or rabbi. Many things go along

continued on page 4

**1983 TDC Medal for Paul Standard and Herb Lubalin**

One of the many pleasurable duties asked of the president of this club is to award the Medal of the Type Directors Club to an outstanding contributor in our profession. I was most fortunate to be able to do this twice in 1983, giving Paul Standard the TDC medal in the fall and presenting the medal to Herb Lubalin's family the second week of January, 1984.

Many designers like myself who studied calligraphy with Paul Standard

continued on page 2

**Have TDC Show Will Travel**

The 29th annual TDC show exhibits the juried selections of the 1983 Type Directors Club international competition, measuring excellence in the use of type, calligraphy and letterforms, in conjunction with paper, design and printing. Four copies of the winning selections were requested to make up the traveling shows—two shows for the U.S. and Canada, one show for Europe and one show for Japan. TDC-29 contains over 200 selections of diverse sizes, such as logotypes, letterheads, large posters, brochures, books, calendars, and folded or boxed promotional items. Each exhibitor provides space for the show, carefully arranging the selections for best display of the pieces. Most of the pieces are fragile and special attention is given to preserve the selections for as many shows as possible.

As TDC-29 travels, each new exhibitor hosts a gala opening, serving refreshments for their guests. In addition, exhibitors have supplied photos, announcements, invitations and publicity about the show for our archives.

Following are brief descriptions of the places the TDC-29 traveling show has been displayed, with special

continued on page 6

1

Type Directors Club 1984 Title

tition in Japan for her Mirarae typeface. 1988: joins Adobe and designs Adobe's first original display typefaces (Trajan, Charlemagne and Lithos). – *Fonts:* Mirarae (1984), Charlemagne (1989), Lithos (1989), Adobe Trajan (1989), Caslon (1990), Myriad (1992), Viva (1993), Nueva (1994) and Chaparral (1997).

**Týfa,** Josef – b. 5. 12. 1913 in Náchod-Běloves, Austria-Hungary (today Czech Republic) – *typographer, graphic artist, type designer* – Studied graphics at the Rotter School in Prague. Then art director for the Centrotex export company. Týfa has designed logos, posters, postage stamps, book covers and typefaces and won numerous awards, including in 1964 a prize at the 1st Biennale for Graphic-Design in Brno and in 1972 and 1975 for the best book of the year. – *Fonts:* Kollektiv-Antiqua (1958), Týfa-Antiqua (1959) and Academia (1968).

**Type Directors Club,** New York – 1946: the Type Directors Club (TDC) of New York is founded by c. twenty New York layout workshops, agencies, college lecturers and publishing houses. The club aims to elevate standards in typography and its related graphic arts. To this end, the TDC organizes lectures and seminars, issues publications and stages an annual typography competition. 1955: the first competition takes place (for club members only). 1956: the second competition is open to competitors outside the club and a catalogue of results is issued. 1965: the first yearly international competition is held. 1980: the first TDC yearbook is published. 1983: a German TDC commit-

1996 Veranstaltung der 1. Multimedia-Konferenz des TDC. Der Club hat ca. 300 Mitglieder.

**Typoart** Dresden – *Schriftenhersteller* – 1948 Gründung der Firma mit dem Namen VEB Schriftguß Dresden. 1951 Eingliederung der Firmen Schelter & Giesecke (Leipzig) und Schriftguß AG, vorm. Brüder Butter (Dresden). Künstlerischer Leiter wird Herbert Thannhaeuser, nach dessen Tod übernimmt Albert Kapr 1963 die künstlerische Leitung. 1989 Fertigstellung eines Angebots für den digitalen Fotosatz. 1989 Umwandlung von VEB Typoart in Typoart GmbH. Das Schriftenangebot ist auf PostScript- und TrueType-Format umgesetzt und wird auf Diskette angeboten.

**Typographische Gesellschaft München (TGM)** – 1890 Gründung der Typographischen Gesellschaft in München. Seit 1892 Angebot von beruflichen Weiterbildungskursen, seit 1894 Gestaltungswettbewerbe, seit 1898 Vorträge über Kunststile und Entwicklung der Typographie. 1900 Herausgabe einer Broschüre zum zehnjährigen Bestehen der TGM. Veranstaltung einer Ausstellung zum 500. Geburtstag von J. Gutenberg. 1923 Eingliederung der TGM in den Bildungsverband der Buchdruckergewerkschaft. 1933 Auflösung des Bildungsverbands durch die Nationalsozialisten. 1949 Neugründung der TGM. Herausgabe von Publikationen als Jahresgaben. Durch Vortragsangebote und die Herausgabe von Publikationen, die Standardwerke wurden, ist die TGM über ihre Grenzen hinaus einflußreich.

Typoart 1971 Cover

Typographische Gesellschaft München 1900 Cover

TYPOART · FOTOSATZSCHRIFTEN **Prillwitz kursiv mager**

*abcdefghijklmnopqrstuvwxyzß äöü*
*ABCDEFGHIJKLMNOPQRSTUVWXYZ ÄÖÜ*
*1234567890 .,:;'‘-()!?§«»&*

4,5 mm / 12 p
*DIE BESONDEREN BEDINGUNGEN DES FOTOSATZES VERLANGEN NEUE*
***Kriterien für die Gestaltung der Schriftvorlagen und deren Herstellung. Dabei spielt die reproduktionsfähige Reinzeichnung und die Anwendung der Fotografie in der Techno-***

Typoart 1975 Prillwitz Italic

Aus Rede und Diskussion [6]

Gerrit Noordzij
Das Kind
und die Schrift

TGM

Typographische Gesellschaft München 1985 Cover

– fondation de la société en 1948, sous le nom de VEB Schriftguss Dresden. 1951, incorporation des sociétés Schelter & Giesecke (Leipzig) et Schriftguss AG, anciennement Brüder Butter (Dresde). Herbert Thannhaeuser en devient directeur artistique; à sa mort en 1963, Albert Kapr lui succède dans la même fonction. En 1989, la société met au point un système de photocomposition numérique. En 1989, le VEB Typoart devient la Typoart GmbH. Le catalogue des fontes passe en formats PostScript et TrueType, il est diffusé sous forme de disquettes.

**Typographische Gesellschaft München (TGM)** – La société typographique de Munich a été fondée en 1890. Propose des cours de formation permanente à partir de 1892. Organise des concours depuis 1894 et des conférences sur les styles artistiques et le développement de la typographie depuis 1898. En 1900, la TGM publie une brochure à l'occasion de son 10e anniversaire. Organisation d'une exposition pour le 500e anniversaire de J. Gutenberg. En 1923, la TGM est incorporée à l'organisme de formation du syndicat des imprimeurs du livre. En 1933, les nazis dissolvent l'organisme de formation. Nouvelle fondation de la TGM en 1949. Edition de publications annuelles. En raison de la continuité de ses programmes de conférences, ainsi que de ses activités d'édition de publications devenues des ouvrages de référence, la TGM exerce son influence au-delà des frontières.

tee is founded. 1996: the TDC stages its first multimedia conference. The club currently has c. 300 members.

**Typoart** Dresden – *type manufacturers* – 1948: the company is founded under the name of VEB Schriftguß Dresden. 1951: the company affiliates Schelter & Giesecke (Leipzig) and Schriftguß AG, formerly Brüder Butter (Dresden). Herbert Thannhaeuser is made art director. Albert Kapr takes over this position in 1963 after Thannhaeuser's death. 1989: completes a tender for digital filmsetting. 1989: VEB Typoart becomes Typoart GmbH. Typoart's typefaces have been adapted to PostScript and TrueType format and are available on disk.

**Typographische Gesellschaft München (TGM)** – 1890: the TGM is founded in Munich. 1892: starts offering professional training courses. 1894: the society stages its first design competition. 1898: the first lectures on art styles and the development of typography are held. 1900: a brochure is published celebrating 10 years of the TGM. The TGM puts on an exhibition commemorating Johannes Gutenberg's 500th birthday. 1923: the TGM is incorporated into the Bildungsverband der Buchdruckergewerkschaft. 1933: the Bildungsverband is abolished by the National Socialists. 1949: the TGM reforms. Produces yearly publications for its members. The TGM's influence has spread beyond its membership through its continual series of lectures and through its publications, which have become standard works in the field.

# *UV*

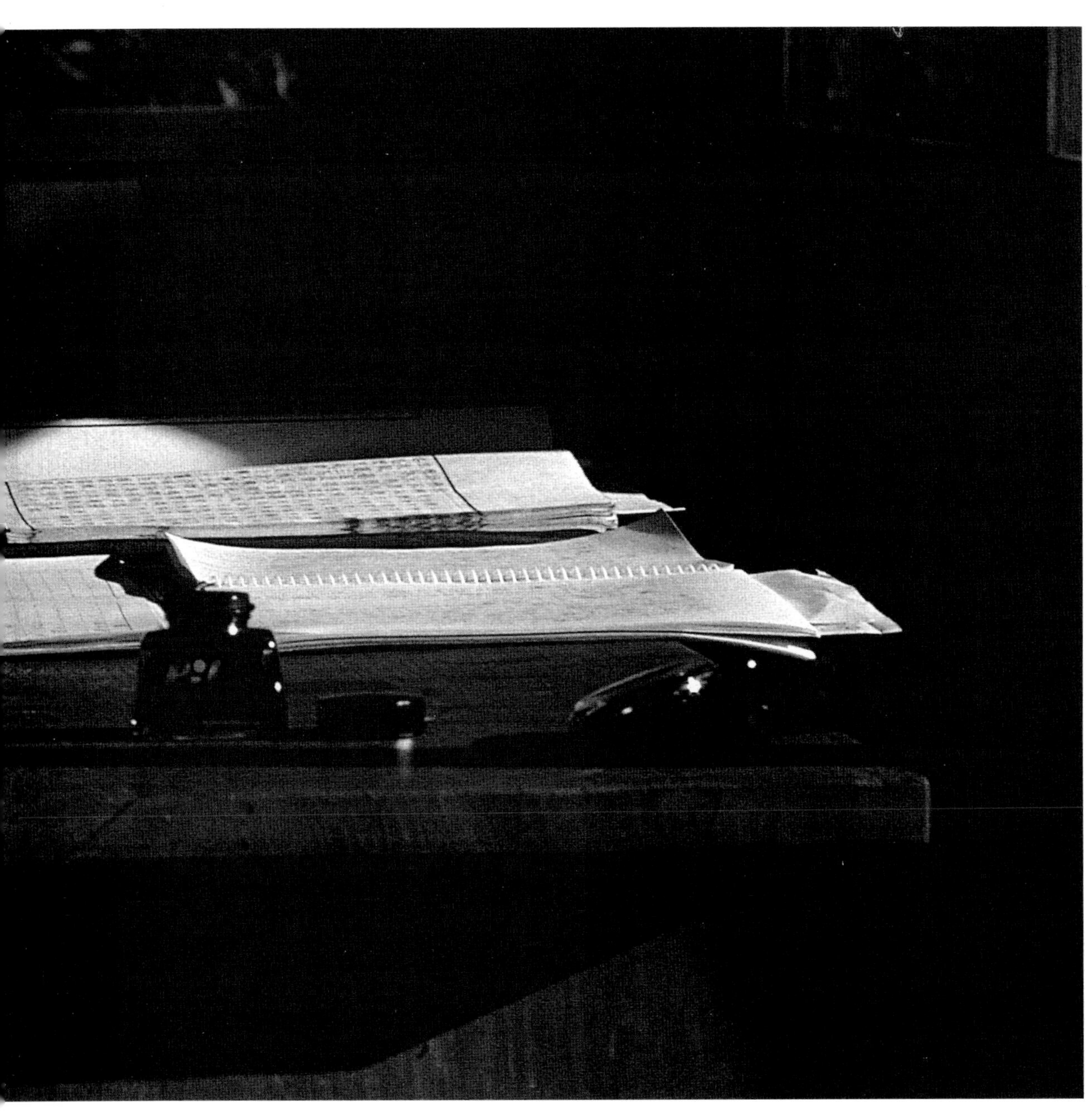

**U&lc**
**Ulm**

**U&lc** (Upper and lower case) – *Zeitschrift* – Die International Typeface Corporation (ITC) in New York gibt seit 1973 vierteljährlich „U&lc" als ihre Hauszeitschrift heraus. Mit dem Untertitel „The international journal of typographics" werden, neben der Präsentation der Schriften der ITC, Artikel über Geschichte und Technik der Typographie sowie Darstellungen einflußreicher typographischer, grafischer und illustrativer Gestalter veröffentlicht. Herausgeber und Art Director von „U&lc" ist 1973–81 Herb Lubalin, der die Zeitschrift inhaltlich und formal in kurzer Zeit zu einem geschätzten und verbreiteten Forum macht. Seit 1986 wird „U&lc" von wechselnden Gestaltern entworfen, u. a. von B. Martin Pedersen, Seymour Chwast, Pentagram, Roger Black Inc. und why not associates. Zu den Autoren gehören u. a. Steven Heller, Marion Muller und Allen Haley.
**Ulm** – *Hochschule für Gestaltung* – Inge Scholl, Otl Aicher und der Schriftsteller Hans Werner Richter (Gründer der Literaten-Vereinigung „Gruppe 47") planen 1949 eine Hochschule für staatsbürgerliche und demokratische Erziehung. Die Stadt Ulm ist bereit, ein Gelände zur Verfügung zu stellen. Inge Scholl errichtet in Ulm die Geschwister-Scholl-Stiftung als Trägerin der neuen Hochschule, deren Rektor Max Bill wird. Mit einem Grundkurs von Walter Peterhans beginnt 1953 der Lehrbetrieb der „Hochschule für Gestaltung, Ulm" (HfG). 1954 beginnt Hans Gugelot mit seiner Arbeit an der HfG, weitere feste Dozenten sind Friedrich Vordemberge-Gildewart und Tomás Mal-

**U&lc** (Upper and lower case) – *revue* – la International Typeface Corporation (ITC) de New York édite «U&lc», la revue trimestrielle de l'entreprise, depuis 1973. Cette revue, qui a pour sous-titre «The international journal of typographics», présente les polices ITC, elle publie en outre des articles sur l'histoire et la technique typographique, et traite aussi de créateurs influents dans le domaine de la typographie, des arts graphiques et de l'illustration. Herb Lubalin, directeur de publication et directeur artistique de 1973 à 1981, l'a transformée en peu de temps en un forum bien diffusé et apprécié autant du point de vue formel que pour son contenu. Depuis 1986, la maquette de «U&lc» est conçue par divers créateurs en alternance, par ex. : B. Martin Pedersen, Seymour Chwast, Pentagram, Roger Black Inc. et why not associates. Steven Heller, Marion Muller et Allen Haley comptent parmi les auteurs.
**Ulm** – *Hochschule für Gestaltung* – école supérieure de design – en 1949, Inge Scholl, Otl Aicher et l'écrivain Hans Werner Richter (fondateur du «Groupe 47») préparent la création d'un institut universitaire ayant pour vocation une éducation citoyenne et démocratique. La ville d'Ulm est disposée à céder un terrain. Inge Scholl crée à Ulm la «Geschwister-Scholl-Stiftung» (Fondation Hans et Sophie Scholl) qui sera l'organisme tutélaire de l'école; Max Bill est nommé recteur. L'enseignement de la «Hochschule für Gestaltung, Ulm» (HfG) débute en 1953 par un cours préliminaire dirigé par Walter Peterhans. Hans Gugelot entre en fonction à la HfG en 1954, d'autres enseignants sont engagés, parmi lesquels Friedrich Vordemberge-Gildewart et Tomás Maldonado. Max Bill démissionne de son poste de recteur en 1956, l'école instaure alors un collège directorial composé

U&lc 1978 Cover

U&lc 1978 Cover

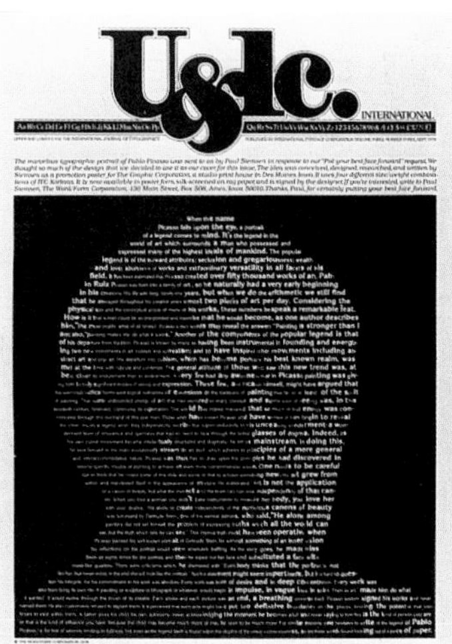

U&lc 1979 Cover

U&lc 1993 Spread

**U&lc** (Upper and lower case) – *magazine* – 1973: the International Typeface Corporation (ITC) in New York starts publishing "U&lc" quarterly as its in-house magazine. Entitled "The international journal of typographics", the magazine includes presentations of ITC typefaces, articles on the history and techniques of typography as well as features on influential typographic artists, graphic designers and illustrators. 1973–81: Herb Lubalin is editor and art director of "U&lc". He works on the content and form of the magazine in such a way that it rapidly becomes a much-revered and widespread forum for typography. Since 1986 "U&lc" has been designed by a number of different artists, including B. Martin Pedersen, Seymour Chwast, Pentagram, Roger Black Inc. and why not associates. Authors have included Steven Heller, Marion Muller and Allen Haley.
**Ulm** – *Hochschule für Gestaltung* – 1949: Inge Scholl, Otl Aicher and the writer Hans Werner Richter (founder of the literary Gruppe 47 association) plan a college for civic and democratic education. The city of Ulm provides suitable premises. Inge Scholl sets up the Geschwister-Scholl-Stiftung in Ulm as the foundation responsible for the new college. Max Bill is designated principal. 1953: teaching starts at the Hochschule für Gestaltung, Ulm (HfG), with a basic course run by Walter Peterhans. 1954: Hans Gugelot starts working at the HfG. Other permanent members of staff are Friedrich Vordemberge-Gildewart and Tomás Maldonado. 1956: Max Bill steps

donado. 1956 Rücktritt von Max Bill als Rektor, Gründung eines Rektoratskollegiums aus Aicher, Gugelot, Maldonado, Vordemberge-Gildewart. Bill verläßt 1957 die HfG. 1958–68 Herausgabe der Zeitschrift „ulm". 1962 wird Aicher Rektor. 1963 Zusammenstellung einer Wanderausstellung mit Studienergebnissen der HfG. Wegen unzureichender finanzieller Voraussetzungen weigern sich 1968 die Dozenten, den Unterricht fortzusetzen. Der baden-württembergische Landtag streicht daraufhin die Zuschüsse für die HfG. Einstellung des Hochschulbetriebs. 1989 Gründung des HfG-Archivs, das seit 1993 zum Ulmer Museum gehört. Erarbeitung zahlreicher Ausstellungen und Publikationen. Die HfG Ulm ist trotz ihrer relativ kurzen Existenz als international richtungsweisendes Modell einer funktionalen und kritischen Gestaltungsausbildung anerkannt. – *Publikationen u. a.:* Herbert Lindinger (Hrsg.) „Hochschule für Gestaltung Ulm. Die Moral der Gegenstände", Berlin 1987; Ch. Wachsmann, E. von Seckendorff, A. Scholz „Hans Gugelot und seine Schüler", Ulm 1990; A. Scholz „Walter Zeischegg", Ulm 1992; I. Albers, Ch. Wachsmann „Bauhäusler in Ulm", Ulm 1993; Ch. Wachsmann u. a. „HfG. Die frühen Jahre", Ulm 1995.

**Ulrichs,** Timm – geb. 31. 3. 1940 in Berlin, Deutschland. – *Künstler, Lehrer* – 1959–66 Architekturstudium an der Technischen Hochschule in Hannover. Seit 1959 entstehen Bilder, Texte, Gegenstände, visuelle Poesie und konzeptionelle Arbeiten. Gründung der „Werbezentrale für Total-

Ulm 1966 Cover

Ulm 1968 Cover

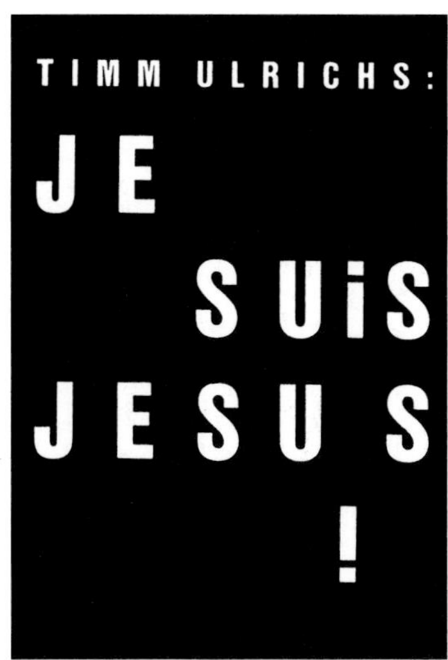

Ulrichs 1968

der anfang vom ende

das beginnende
das beginneende
das beginnenende
das beginnendende
das beginnendeende
das beginnende ende

das ende vom anfang

Ulrichs 1966 Poem

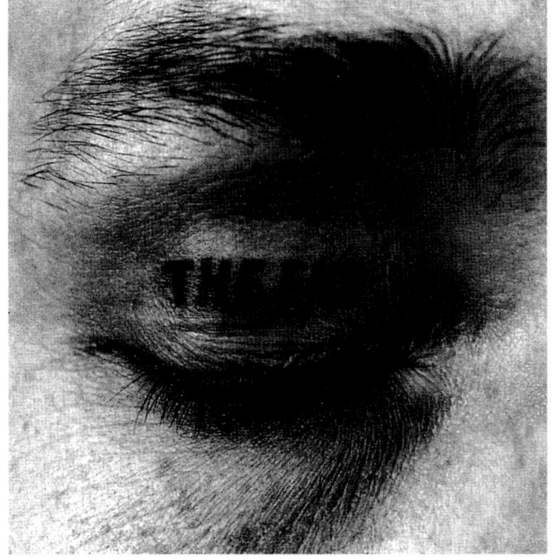
Ulrichs 1970/1981 Tatoo

d'Aicher, Gugelot, Maldonado, Vordemberge-Gildewart. Max Bill quitte la HfG en 1957. De 1958 à 1968, l'école publie la revue «ulm». Aicher devient recteur en 1962. Une exposition itinérante présentant les travaux des étudiants de la HfG est organisée en 1963. En 1968, jugeant le financement insuffisant, les professeurs refusent de continuer à enseigner dans ces conditions. Le gouvernement de Bade-Wurtemberg supprime alors les subventions accordées à la HfG. L'école ferme. La HfG-Archiv est fondée en 1989 et fait partie du musée d'Ulm depuis 1993. Elle se cansacre à l'édition et à l'organisation d'expositions. En dépit de son existence relativement brève, la HfG d'Ulm a été considérée comme un modèle pour l'avenir, elle est reconnue au niveau international pour la formation critique et fonctionnelle qu'elle dispensait. – *Publications, sélection:* Herbert Lindinger (éd.) «Hochschule für Gestaltung Ulm. Die Moral der Gegenstände», Berlin 1987; Ch. Wachsmann, E. von Seckendorff, A. Scholz «Hans Gugelot und seine Schüler», Ulm 1990; A. Scholz «Walter Zeischegg», Ulm 1992; I. Albers, Ch. Wachsmann «Bauhäusler in Ulm», Ulm 1993; Ch. Wachsmann et autres «HfG. Die frühen Jahre», Ulm 1995.

**Ulrichs,** Timm – né le 31. 3. 1940 à Berlin, Allemagne – *artiste, enseignant* – 1959–1966, études d'architecture à la Technische Hochschule (université technique) d'Hanovre. A partir de 1959, il écrit des textes et réalise de nombreux tableaux, objets, travaux conceptuels et poésies visuelles. Fonde la «Werbezentra-

down as principal and Aicher, Gugelot, Maldonado and Vordemberge-Gildewart form a board of rectors. 1957: Bill leaves the HfG. 1958–68: "ulm" magazine is published. 1962: Aicher is made principal. 1963: a traveling exhibition is collated from work by HfG students. 1968: the lecturers refuse to continue teaching due to inadequate funding. Local government in Baden-Württemberg thus cuts all subsidies for the HfG. The college is closed. 1989: the HfG archives are founded, part of the Ulm museum since 1993. Organization of numerous exhibitions and publications. Despite its relatively short existence, the HfG in Ulm was highly acclaimed as an pioneering world model for the functional and critical teaching of design. – *Publications include:* Herbert Lindinger (ed.) "Hochschule für Gestaltung Ulm. Die Moral der Gegenstände", Berlin 1987; Ch. Wachsmann, E. von Seckendorff, A. Scholz "Hans Gugelot und seine Schüler", Ulm 1990; A. Scholz "Walter Zeischegg", Ulm 1992; I. Albers, Ch. Wachsmann "Bauhäusler in Ulm", Ulm 1993; Ch. Wachsmann et al "HfG. Die frühen Jahre", Ulm 1995.

**Ulrichs,** Timm – b. 31. 3. 1940 in Berlin, Germany – *artist, teacher* – 1959–66: studies architecture at the Technische Hochschule in Hanover. 1959: starts producing pictures, texts, objects, visual poetry and Conceptual art works. Founds the Werbezentrale für Totalkunst & Banalismus. 1969–70: constructs an electric phonetic alphabet with the endless text "a tautology is a tautology is a tautology".

Ulrichs
**Unger, G.**

kunst & Banalismus“. Er konstruiert 1969–70 eine elektrische Lautschrift mit dem endlosen Text „Eine Tautologie ist eine Tautologie ist eine Tautologie“. Seit 1972 Professor an der Kunstakademie Münster. Teilnahme an der documenta VI 1977. Zahlreiche Preise und Ausstellungen. – *Publikationen u. a.:* „Klartexte“, Hannover 1966; „Beschriebene Blätter“, Hannover, Göttingen 1967; „Schriftstücke. Collagen 1962, 67“, München 1967; „Lesarten und Schreibweisen“, Stuttgart 1968; „Weiter im Text. Visuelle Texte 1960–65“, Hannover 1969.

**Unger,** Gerard – geb. 22. 1. 1942 in Arnheim, Niederlande. – *Grafik-Designer, Typograph, Schriftentwerfer, Lehrer* – 1963–67 Studium an der Gerrit Rietveld Academie in Amsterdam. Danach Assistent von Wim Crouwel bei Total Design. Seit 1970 Lehrtätigkeit an der Gerrit Rietveld Academie in Amsterdam. Seit 1975 freier Gestalter. 1981 Schrift-Arbeiten für niederländische Münzen und Briefmarken. Zahlreiche Zeitschriften-Neugestaltungen. 1984 Entwurf spezieller Ziffern für die niederländischen Telefonbücher. Berater für digitale Typographie der Firma Océ. Berater der BBC bei der Entwicklung digitaler Schriften. 1984 mit dem H. N. Werkman-Preis ausgezeichnet. 1991 Verleihung des Maurits-Enschedé-Preises. Unterrichtet an der Rhode Island School of Design in Providence, USA (1979) und an der Stanford University (1985). – *Schriftentwürfe:* Markeur (1972), M. O. L. (1975), Demos (1976), Praxis (1976), Flora (1980), Hollander (1985), Swift (1985), Oranda (1992), Amerigo

le für Totalkunst & Banalismus» (centrale publicitaire pour l'art total et la banalité). 1969–1970, conçoit une écriture sonore électronique pour un texte sans fin «Eine Tautologie ist eine Tautologie ist eine Tautologie». Professeur au Kunstakademie de Münster depuis 1972. Participation à la «documenta VI» de Kassel en 1977. Nombreuses distinctions et expositions. – *Publications, sélection:* «Klartexte», Hanovre 1966; «Beschriebene Blätter», Hanovre, Göttingen 1967; «Schriftstücke, Collagen 1962, 67», Munich 1967; «Lesarten und Schreibweisen», Stuttgart 1968; «Weiter im Text. Visuelle Texte 1960–65», Hanovre 1969.

**Unger,** Gerard – né le 22. 1. 1942 à Arnhem, Pays-Bas – *graphiste maquettiste, typographe, concepteur de polices, enseignant* – 1963–1967, études à la Gerrit Rietveld Academie à Amsterdam. Exerce ensuite comme assistant de Wim Crouwel chez «Total Design». Enseigne à la Gerrit Rietveld Academie à partir de 1970. Designer indépendant depuis 1975. Caractères pour les monnaies néerlandaises et timbres-poste à partir de 1981. Nombreuses refontes de maquettes de revues. En 1984, il dessine des chiffres spéciaux pour les bottins de téléphone néerlandais. Conseiller de la société Océ en matière de typographie numérique. Conseiller de la BBC et développement de polices numériques. Lauréat du prix H. N. Werkman en 1984. Prix Maurits Enschedé en 1991. A enseigné à la Rhode Island School of Design à Providence, Etats-Unis (1979), et à la Stanford University (1985). – *Polices:* Markeur (1972), M. O. L. (1975), Demos (1976), Praxis (1976), Flora (1980), Hollander (1985), Swift (1985), Oranda (1992), Amerigo (1987), Cyrano (1989), Argo (1991), Decoder (1993), Gulliver (1993), Swift 2.0 (1995).

**Unger,** Johann Friedrich – né en 1753 à

| | n | **n** | | | |
|---|---|---|---|---|---|
| a | b | e | f | g | i |
| o | r | s | t | y | z |
| A | B | C | E | G | H |
| M | O | R | S | X | Y |
| 1 | 2 | 4 | 6 | 8 | & |

Gerard Unger 1980 Flora

| n | n | **n** | **n** | | *n* |
|---|---|---|---|---|---|
| a | b | e | f | g | i |
| o | r | s | t | y | z |
| A | B | C | E | G | H |
| M | O | R | S | X | Y |
| 1 | 2 | 4 | 6 | 8 | & |

Gerard Unger 1985 Swift

| | n | **n** | | | *n* |
|---|---|---|---|---|---|
| a | b | e | f | g | i |
| o | r | s | t | y | z |
| A | B | C | E | G | H |
| M | O | R | S | X | Y |
| 1 | 2 | 4 | 6 | 8 | & |

Gerard Unger 1992 Oranda

ABCDEFGHIJ
KLMNOPQRS
TUVWXYZ
abcdefghijkl
mnopqrstuv
wxyz-.:,

Gerard Unger 1975 M.O.L.

From 1972 onwards: professor at the Kunstakademie Münster. Takes part in documenta VI in Kassel in 1977. Numerous awards and exhibitions. – *Publications include:* "Klartexte", Hanover 1966; "Beschriebene Blätter", Hanover, Göttingen 1967; "Schriftstücke. Collagen 1962, 67", Munich 1967; "Lesarten und Schreibweisen", Stuttgart 1968; "Weiter im Text. Visuelle Texte 1960–65", Hanover 1969.

**Unger,** Gerard – b. 22. 1. 1942 in Arnhem, The Netherlands – *graphic designer, typographer, type designer, teacher* – 1963–67: studies at the Gerrit Rietveld Academie in Amsterdam. Then assistant to Wim Crouwel at Total Design. 1970: starts teaching at the Gerrit Rietveld Academie in Amsterdam. From 1975 onwards: freelance designer. 1981: does typography work for Dutch coins and postage stamps. Redesigns numerous magazines. 1984: designs special digits for the Dutch telephone directories. Digital typography consultant to Océ. Consultant to the BBC for the development of digital typefaces. Awarded the H. N. Werkman Prize in 1984. 1991: wins the Maurits Enschedé Prize. Has taught at the Rhode Island School of Design in Providence, USA, (1979) and at Stanford University (1985). – *Fonts:* Markeur (1972), M. O. L. (1975), Demos (1976), Praxis (1976), Flora (1980), Hollander (1985), Swift (1985), Oranda (1992), Amerigo (1987), Cyrano (1989), Argo (1991), Decoder (1993), Gulliver (1993), Swift 2.0 (1995).

**Unger,** Johann Friedrich – b. 1753 in Berlin, Germany, d. 1804 in Berlin, Ger-

(1987), Cyrano (1989), Argo (1991), Decoder (1993), Gulliver (1993), Swift 2.0 (1995).

**Unger,** Johann Friedrich – geb. 1753 in Berlin, Deutschland, gest. 1804 in Berlin, Deutschland. – *Drucker, Schriftentwerfer* – Ausbildung und Arbeit als Holzschneider und Drucker. 1780 Gründung einer Druckerei in Berlin, tätig als Verleger von Klassikerausgaben. Erwirbt 1789 das alleinige Recht zum Vertrieb der Schriften Firmin Didots in Deutschland. Seit 1790 Mitglied der Akademie der Künste in Berlin. 1791 Angliederung einer Schriftgießerei an seine Druckerei. Veröffentlicht 1793 seine Ideen zur Erneuerung der Fraktur-Schrift. 1798 Angliederung einer Noten-Gießerei. 1800 Professor für Formschneidekunst an der Königlich-Preußischen Akademie der Künste in Berlin. 1926 wird die Unger-Fraktur von der D. Stempel AG nach den Originalmatrizen neu herausgegeben. – *Schriftentwurf:* Unger-Fraktur (1794, 1926). – *Publikation u. a.:* „Probe einer neuen Art deutscher Lettern, erfunden und in Stahl geschnitten von J. F. Unger", Berlin 1793.

**Valicenti,** Rick – geb. 20. 11. 1951 in Pittsburgh, USA. – *Grafik-Designer, Lehrer* – Bis 1973 Studium der Malerei an der Bowling Green State University. 1973–76 Studium der Fotografie an der University of Iowa. 1978–82 Arbeit als Grafik-Designer im Studio Bruce Beck Design in Evanstone, Illinois. 1982 Gründung seines Studios „R. Valicenti Design" in Chicago. 1988 Gründung des Design-Studios „Thirst" in Chicago, das jetzt in Barrington, Illinois arbeitet. Auftraggeber

Nachricht des Verlegers.

Mein erſter Verſuch neuer deutſcher Druckſchrift, den ich in der Oſtermeſſe 1793 bekannt machte, fand Beifall, wurde aber auch hie und da getadelt. Männer ohne Vorurtheil gegen Neuerungen, und denen guter Geſchmack wohl ſchwerlich abgeſprochen werden kann, munterten mich zu fernerer Vervollkommnung auf, und nun wartete ich nur noch die öffentlichen Urtheile darüber ab. Dieſe ſind jezt wohl größtentheils erſchienen, und lauten dafür und dawider. Einer findet die neuen Lettern den ſchon vorhanden geweſenen ähnlich, mit welchen die großoctav Bibel in Halle gedruckt iſt. Ich verglich ſie, und fand, ſo wie mehrere Perſonen, nicht die geringſte Ähnlich-

A 2

Johann Friedrich Unger 1794 Unger-Fraktur

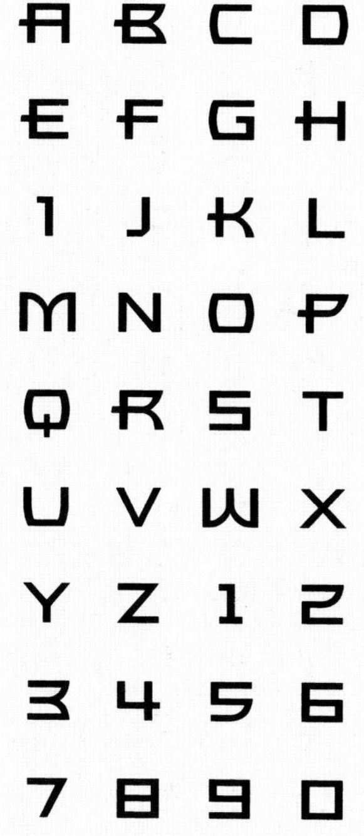


Valicenti 1990 Bronzo

Valicenti 1995 Advertisement

Berlin, Allemagne, décédé en 1804 à Berlin, Allemagne – *imprimeur, concepteur de polices* – formation de graveur sur bois et d'imprimeur. Fonde une imprimerie à Berlin en 1780 et édite des œuvres classiques. En 1789, il acquiert les droits exclusifs de diffusion des fontes Firmin Didot en Allemagne. Membre de l'Akademie der Künste (beaux-arts) de Berlin à partir de 1790. En 1791, il adjoint une fonderie de caractères à son imprimerie. Publie ses conceptions sur le remaniement des caractères Fraktur en 1793. Ajoute à l'entreprise une fonderie de notes de musiques en 1798. Professeur de taille des types à la Königlich-Preussische Akademie der Künste (académie royale des beaux-arts) de Berlin en 1800. La D. Stempel AG diffuse de nouveau la Unger-Fraktur remaniée à partir des matrices originales en 1926. – *Polices:* Unger-Fraktur (1794, 1926). – *Publication:* «Probe einer neuen Art deutscher Lettern, erfunden und in Stahl geschnitten von J. F. Unger», Berlin 1793.

**Valicenti,** Rick – né le 20. 11. 1951 à Pittsburgh, Etats-Unis – *graphiste maquettiste, enseignant* – études de peinture à la Bowling Green State University jusqu'en 1973; puis de 1973–1976, études de photographie à l'University of Iowa. 1978–1982, employé comme graphiste maquettiste à l'atelier Bruce Beck Design à Evanstone, Illinois. Fonde l'atelier «R. Valicenti Design» à Chicago en 1982. Création en 1988 de l'atelier de design «Thirst» à Chicago, l'atelier est actuellement installé à Barrington, Illinois.

many – *printer, type designer* – Trained and worked as a woodcut artist and printer. 1780: opens a printing workshop in Berlin and publishes editions of the classics. 1789: is granted the sole right of sale for Firmin Didot typefaces in Germany. 1790: becomes a member of the Akademie der Künste in Berlin. 1791: a type foundry becomes an affiliate of his printing works. 1793: publishes his ideas on the redesign of black letter typefaces. 1798: a musical notation foundry is affiliated to his business. 1800: professor of form cutting at the Königlich-Preußische Akademie der Künste in Berlin. 1926: D. Stempel AG reissues Unger's black letter typeface, Unger-Fraktur, using the original matrices. – *Font:* Unger-Fraktur (1794, 1926). – *Publication:* "Probe einer neuen Art deutscher Lettern, erfunden und in Stahl geschnitten von J. F. Unger", Berlin 1793.

**Valicenti,** Rick – b. 20. 11. 1951 in Pittsburgh, USA – *graphic designer, teacher* – 1973: ends his studies in painting at the Bowling Green State University. 1973–76: studies photography at the University of Iowa. 1978–82: works as a graphic designer for the Bruce Beck Design Studio in Evanstone, Illinois. 1982: opens his own studio, R. Valicenti Design, in Chicago. 1988: founds the Thirst design studio in Chicago, which now operates from

waren u. a. Gilbert Paper, Cooper Lighting, das American Center for Design und The Arts Club of Chicago. 1992 Gründung von „Thirstype" zum Vertrieb digitalisierter neuer Schriften. 1992 Ausstellung seiner Arbeit in Tokio und Osaka. Unterrichtet u. a. am Chicago Art Institute, an der Cranbrook Academy of Art, an der Stanford University und am California College of Arts and Crafts. – *Schriftentwurf:* Bronzo (mit Mouli Marur, 1990).

**Vanderbyl,** Michael – geb. 9. 2. 1947 in Piedmont, Kalifornien, USA. – *Grafik-Designer, Industrie-Designer, Lehrer* – 1968 Abschluß des Grafik-Design-Studiums am California College of Arts and Crafts. 1973 Gründung von „Vanderbyl Design" in San Francisco. Entwurf von Verpackungen, Leitsystemen, Inneneinrichtungen und Mode. Auftraggeber waren u. a. das Oakland Museum, das San Francisco Museum of Modern Art, Esprit und IBM. Seit 1987 Mitglied der AGI. Zahlreiche internationale Auszeichnungen und Veröffentlichungen. Unterrichtet am California College of Arts and Crafts.

**VanderLans,** Rudy – geb. 10. 10. 1955 in Den Haag, Niederlande. – *Grafik-Designer, Typograph, Verleger* – 1975–79 Grafik-Design-Studium an der Koninklijke Academie voor Beeldende Kunsten in Den Haag. 1980 Arbeit als Grafik-Designer bei „Vorm Vijf" in Den Haag, danach bei „Tel Design". 1981–83 Studium der Fotografie an der University of California in Berkeley. 1983–87 Grafik-Designer bei der Zeitung „San Francisco Chronicle". 1984 Gründung des „Emigre

Commanditaires : Gilbert Paper, Cooper Lighting, l'American Center for Design, et The Arts Club of Chicago. En 1992, il fonde «Thirstype» pour diffuser de nouvelles fontes numériques. Exposition de ses travaux à Tokyo et à Osaka en 1992. A enseigné au Chicago Art Institute, à la Cranbrook Academy of Art, à la Stanford University et au California College of Arts and Crafts. – *Police :* Bronzo (avec Mouli Marur, 1990).

**Vanderbyl,** Michael – né le 9. 2. 1947 à Piedmont, Californie, Etats-Unis – *graphiste maquettiste, designer industriel, enseignant* – 1968, diplôme d'arts graphiques et de design au California College of Arts and Crafts. 1973 fondation de «Vanderbyl Design» à San Francisco. Dessin d'emballages, de systèmes de signalisation, de décorations intérieures et styliste de mode. Commanditaires : Oakland Museum, San Francisco Museum of Modern Art, Esprit et IBM. Membre de l'AGI depuis 1987. Nombreuses distinctions et publications internationales. Enseigne au California College of Arts and Crafts.

**VanderLans,** Rudy – né le 10. 10. 1955 à La Haye, Pays-Bas – *graphiste maquettiste, typographe, éditeur* – 1975–1979, études d'arts graphiques et de design à la Koninklijke Academie voor Beeldende Kunsten de La Haye. Exerce comme graphiste maquettiste chez «Vorm Vijf» à La Haye en 1980, puis chez «Tel Design». Etudie la photographie à l'University of California de Berkeley de 1981 à 1983. Graphiste maquettiste du journal «San Francisco Chronicle» de 1983 à 1987. Fonde «Emigre Magazine» avec Menno Meyjes et Marc Susan en 1984. Fonde l'agence «Emigre Graphics» en 1987. Travaux pour Apple Computer Inc., San Francisco Artspace et GlasHaus, Cairo Cinemafilms. En 1990, il crée la société de production de disques compacts «Emigre

Vanderbyl 1993 Flyer

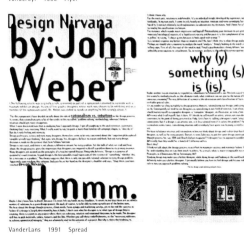

VanderLans 1991 Spread

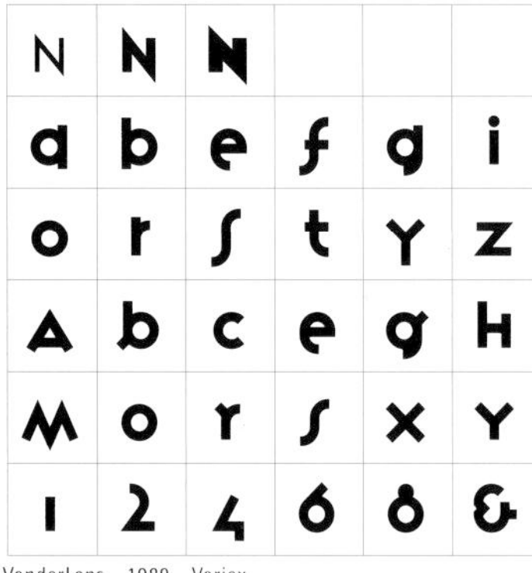

VanderLans 1989 Variex

VanderLans 1992 Cover

Barrington, Illinois. Clients include Gilbert Paper, Cooper Lighting, the American Center for Design and The Arts Club of Chicago. 1992: founds Thirstype which sells and markets new, digitized typefaces. 1992: an exhibition of his work is held in Tokyo and Osaka. Valicenti has taught at the Chicago Art Institute, the Cranbrook Academy of Art, Stanford University and the California College of Arts and Crafts, among others. – *Font:* Bronzo (with Mouli Marur, 1990).

**Vanderbyl,** Michael – b. 9. 2. 1947 in Piedmont, California, USA – *graphic designer, industrial designer, teacher* – 1968: completes his graphic design studies at the California College of Arts and Crafts. 1973: opens Vanderbyl Design in San Francisco. Designs packaging, signage systems, interiors and fashion. Clients include the Oakland Museum, the San Francisco Museum of Modern Art, Esprit and IBM. 1987: becomes a member of AGI. Numerous international awards and publications. Teaches at the California College of Arts and Crafts.

**VanderLans,** Rudy – b. 10. 10. 1955 in The Hague, The Netherlands – *graphic designer, typographer, publisher* – 1975–79: studies graphic design at the Koninklijke Academie voor Beeldende Kunsten in The Hague. 1980: works as a graphic designer for Vorm Vijf in The Hague, then for Tel Design. 1981–83: studies photography at the University of California in Berkeley. 1983–87: graphic designer for the "San Francisco Chronicle". 1984: launches "Emigre Magazine" with Menno Meyjes and Marc Susan. 1987: founds the Emi-

Magazine" mit Menno Meyjes und Marc Susan. 1987 Gründung des Studios „Emigre Graphics". Arbeit für Auftraggeber wie Apple Computer Inc., San Francisco Artspace, GlasHaus und Cairo Cinemafilms. 1990 Gründung der Firma „Emigre Music" für Compact Discs. – *Schriftentwürfe:* Variex (mit Zuzana Licko, 1989), Suburban (1994). – *Publikation:* „Emigre: Graphic Design into the Digital Realm", New York 1994.

**Vanék,** Rostislav – geb. 31. 10. 1945 in Prag, Tschechoslowakei. – *Grafik-Designer, Typograph, Lehrer* – Studium an der Grafikschule und an der Hochschule für angewandte Kunst in Prag. 1971–76 Assistent an der Hochschule. Danach Arbeit als Art Director des Verlags Československý spisovatel in Prag, Gestaltung von Büchern, Zeitschriften und Plakaten. Für seine Arbeit auf der Biennale der Gebrauchsgrafik in Brno (1972–78), auf der Internationalen Buchausstellung in Leipzig (1977) und beim Wettbewerb der schönsten Bücher der CSSR (1970, 1972–79) ausgezeichnet.

**Vardimon,** Yarom – geb. 18. 7. 1942 in Ramat Gan, Israel. – *Grafik-Designer, Typograph, Lehrer* – 1960–64 Grafik-Design-Studium am Chelsea College of Art, an der Regent Polytechnic und am London College of Printing. 1965–67 freiberuflich in Tel Aviv tätig. 1967 Gründung des Studios „Vardimon Design" in Tel Aviv, Arbeit im Bereich Corporate Identity, Typographie, Verpackungsdesign. Seit 1968 Dozent, seit 1977 Präsident der Grafik-Design-Abteilung der Bezalel-Akademie in Jerusalem. Leiter des Expe-

VanderLans 1994 Suburban

Vanék 1978 Cover

Vardimon 1983 Poster

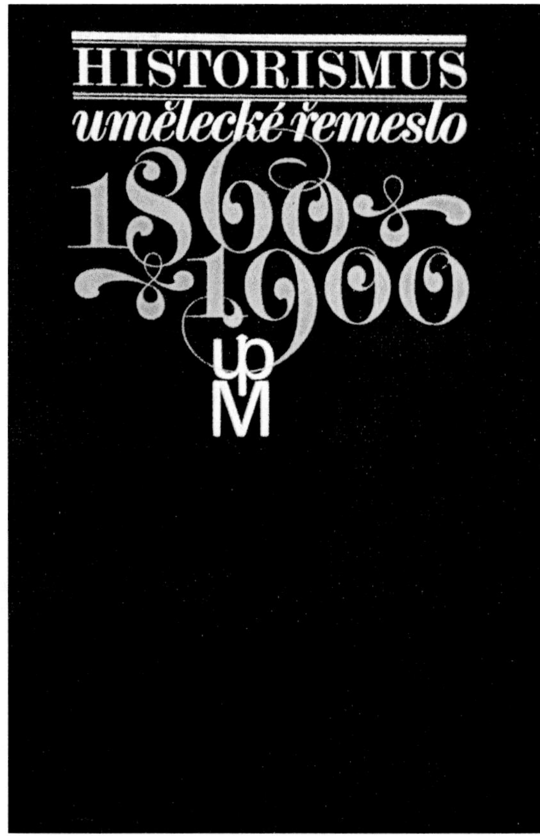


Vanék 1975 Cover

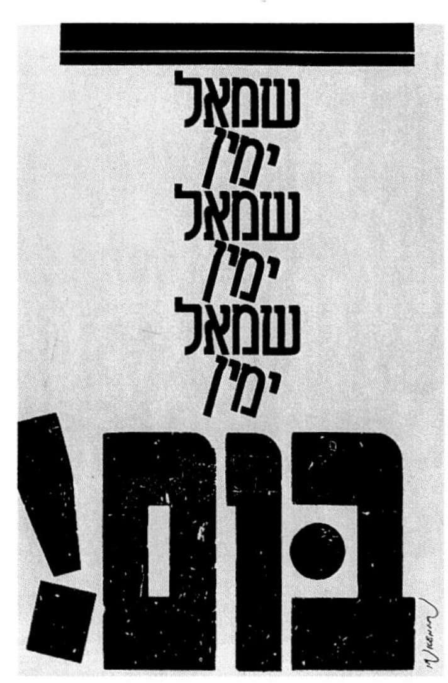


Vardimon 1983 Poster

Music». – *Police:* Variex (avec Zuzana Licko, 1989), Suburban (1994). – *Publication :* «Emigre : Graphic Design into the Digital Realm», New York 1994.

**Vanék,** Rostislav – né le 31. 10. 1945 à Prague, Tchécoslovaquie – *graphiste maquettiste, typographe, enseignant* – études à l'école d'arts graphiques et à l'école supérieure des arts appliqués de Prague, 1971 à 1976 assistant. Travaille ensuite comme directeur artistique aux éditions Československý spisovatel à Prague; il est chargé de la maquette de livres, de revues et d'affiches. Ses travaux sont primés à la Biennale de Brno (1972–1978), au salon international du livre de Leipzig (1977) et à l'occasion du concours des plus beaux livres de Tchécoslovaquie (1970, 1972–1979).

**Vardimon,** Yarom – né le 18. 7. 1942 à Ramat Gan, Israël – *graphiste maquettiste, typographe, enseignant* – 1960–1964, études d'arts graphiques et de design au Chelsea College of Art, au Regent Polytechnic et au London College of Printing. Exerce comme indépendant à Tel Aviv de 1965 à 1967. Fonde l'atelier «Vardimon Design» en 1967 à Tel Aviv où il travaille dans le domaine de l'identité d'entreprises, en typographie et dans le design d'emballages. Enseignant à partir de 1968 et président de la section d'arts graphiques et de design de l'académie Beza-

gre Graphics studio which produces work for clients such as Apple Computer Inc., San Francisco Artspace, GlasHaus and Cairo Cinemafilms. 1990: founds Emigre Music for compact discs. – *Fonts:* Variex (with Zuzana Licko, 1989), Suburban (1994). – *Publication:* "Emigre: Graphic Design into the Digital Realm", New York 1994.

**Vanék,** Rostislav – b. 31. 10. 1945 in Prague, Czechoslovakia – *graphic designer, typographer, teacher* – Studied at the graphics school and at the applied art college in Prague. 1971–76: assistant at the art college. Then works as art director for the Československý spisovatel publishers in Prague, where he designs books, magazines and posters. Vanék has won prizes for his work at the Biennale for Graphic-Design in Brno (1972–78), at the Internationale Buchausstellung in Leipzig (1977) and in the competition for the CSSR's "most beautiful books" (1970, 1972–79).

**Vardimon,** Yarom – b. 18. 7. 1942 in Ramat Gan, Israel – *graphic designer, typographer, teacher* – 1960–64: studies graphic design at the Chelsea College of Art, at Regent Polytechnic and at the London College of Printing. 1965–67: freelances in Tel Aviv. 1967: launches his Vardimon Design studio in Tel Aviv. Works on corporate identities and in typography and packaging design. 1968: lecturer, from 1977 onwards, president of the graphic design department at the Bezalel Academy in Jerusalem. Head of the academy's Experimental Studio. Secretary general (1967), president (1970)

rimental-Studios, 1967 Generalsekretär, 1970 Präsident, 1972 Vorstandsmitglied der „Vereinigung Israelischer Gebrauchsgrafiker". 1974–79 Vizepräsident der ICOGRADA. Zahlreiche Artikel über sein Werk, u. a. in „Idea" und „High Quality".

**Vardjiev,** Todor – geb. 16. 1. 1943 in Blagoevgrad, Bulgarien. – *Grafik-Designer, Schriftentwerfer, Lehrer* – Studium an der Kunstakademie in Sofia, am Herder-Institut in Leipzig (1966–67) und an der Hochschule für Graphik und Buchkunst in Leipzig (1967–72). 1972–76 als Grafik-Designer im Verlag Narodna Mladesh in Sofia tätig. Seit 1976 Dozent an der Kunstakademie in Sofia. 1980, 1982 und 1984 auf der Biennale der Gebrauchsgrafik in Brno ausgezeichnet. – *Schriftentwürfe:* Sredez, Serdika, Balkan.

**Vautier,** Ben – geb. 18. 7. 1935 in Neapel, Italien. – *Künstler* – 1951–55 Arbeit in einer Buchhandlung und als Trödler. 1958–72 Inhaber eines künstlerisch gestalteten Second-Hand-Ladens. Malt Schriftbilder. Seit 1960 zahlreiche internationale Ausstellungen. 1962–70 Zusammenarbeit mit Künstlern der Fluxus-Gruppe in Paris, London, Köln. 1972 Teilnahme an der documenta V in Kassel. 1978 Eröffnung der Galerie „La différence". – *Publikationen u. a.:* „L'art est Prétention", Mailand 1971; „Ecritures de 1958 à 1966", Paris 1971; „My Berlin Inventory", Berlin 1979.

**Velde,** Henry van de – geb. 3. 4. 1863 in Antwerpen, Belgien, gest. 27. 10. 1957 in Zürich, Schweiz. – *Architekt, Maler, Autor, Innenarchitekt, Grafik-Designer, Lehrer* – Studium der Malerei in Ant-

lel de Jérusalem depuis 1977. Secrétaire général de l' «Union des graphistes industriels israéliens» en 1967, président en 1970, et membre du directoire en 1972. Vice-président de l'ICOGRADA de 1974 à 1979. Nombreux articles sur son œuvre dans «Idea», «High Quality», etc.

**Vardjiev,** Todor – né le 16. 1. 1943 à Blagoevgrad, Bulgarie – *graphiste maquettiste, concepteur de polices, enseignant* – études à l'académie des beaux-arts de Sofia, au Herder-Institut de Leipzig (1966–1967) et à la Hochschule für Graphik und Buchkunst (école supérieure des arts graphiques et des arts du livre) de Leipzig (1967–1972). Employé comme graphiste maquettiste aux éditions Narodna Mladesh à Sofia de 1972 à 1976. Enseigne à l'académie des beaux-arts de Sofia depuis 1976. Primé à la Biennale de graphisme industriel de Brno en 1980, 1982 et 1984. – *Polices:* Sredez, Serdika, Balkan.

**Vautier,** Ben – né le 18. 7. 1935 à Naples, Italie – *artiste* – 1951–1955, travaille dans une librairie et comme brocanteur. De 1958 à 1972, il tient une boutique de brocante dont il a conçu le décor. Peint des tableaux comportant des textes. Nombreuses expositions internationales à partir de 1960. Travaille avec des artistes du groupe Fluxus à Paris, Londres et Cologne de 1962 à 1970. Participe à la «documenta V» de Kassel en 1972. Ouvre la galerie «La différence» en 1978. – *Publications, sélection:* «L'art est Prétention», Milan 1971; «Ecritures de 1958 à 1966», Paris 1971; «My Berlin Inventory», Berlin 1979.

**Velde,** Henry van de – né le 3. 4. 1863 à Anvers, Belgique, décédé le 27. 10. 1957 à Zurich, Suisse – *architecte, peintre, auteur, architecte d'intérieur, graphiste maquettiste, enseignant* – études de peinture à Anvers (1881–1884) et à Paris (1884–1885). Conçoit, en 1983, des ornements

АБВГДЕЖЗИЙК
ΛМНОПРСТУФХ
ЦЧШЩЪЬЮЯ
1234567890
абвгдежзийкл
мнопрстуфхцч
шщъьюя
(1234567890-;!?)

Vardjiev Serdika

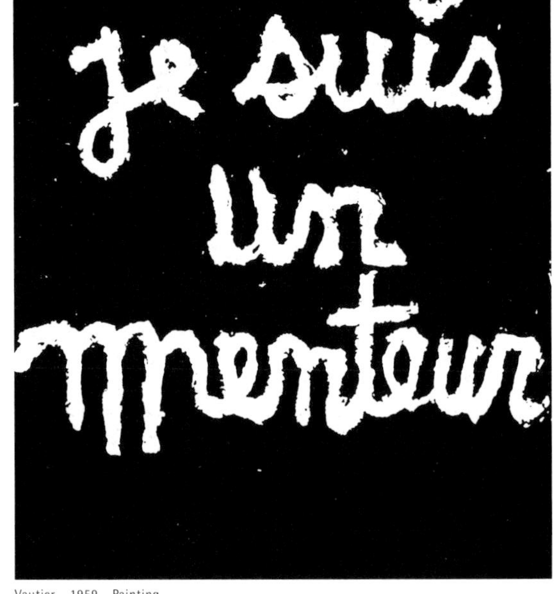


Vautier 1959 Painting

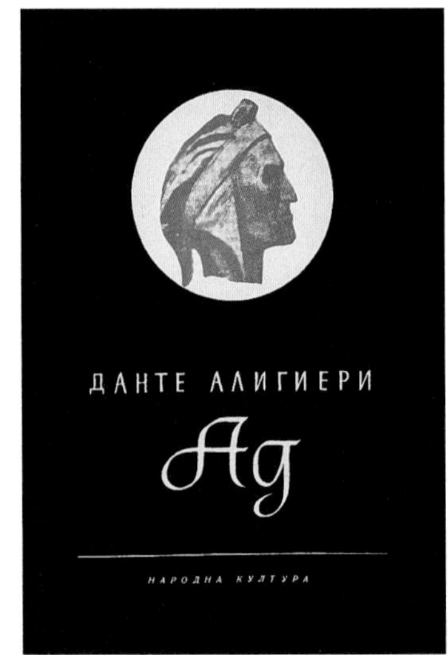


Vardjiev Cover

Vautier 1970 Painting

and board member (1972) of the Graphic Design Association of Israel (GDAI). 1974–79: vice-president of ICOGRADA. There are numerous articles published on his work, including in "Idea" and "High Quality".

**Vardjiev,** Todor – b. 16. 1. 1943 in Blagoevgrad, Bulgaria – *graphic designer, type designer, teacher* – Studied at the art academy in Sofia, at the Herder-Institut in Leipzig (1966–67) and at the Hochschule für Graphik und Buchkunst in Leipzig (1967–72). 1972–76: works as a graphic designer for Narodna Mladesh publishers in Sofia. 1976: starts lecturing at the art academy in Sofia. 1980, 1982 and 1984: awarded prizes at the Biennale for Graphic-Design in Brno. – *Fonts:* Sredez, Serdika, Balkan.

**Vautier,** Ben – b. 18. 7. 1935 in Naples, Italy – *artist* – 1951–55: works in a bookstore and as a junk dealer. 1958–72: owns an artistically designed second-hand store. Paints letter-pictures. From 1960 onwards: numerous exhibitions all over the world. 1962–70: works with Fluxus artists in Paris, London and Cologne. 1972: takes part in documenta V in Kassel. 1978: La différence gallery opens. – *Publications include:* "L'art est Prétention", Milan 1971; "Ecritures de 1958 à 1966", Paris 1971; "My Berlin Inventory", Berlin 1979.

**Velde,** Henry van de – b. 3. 4. 1863 in Antwerp, Belgium, d. 27. 10. 1957 in Zurich, Switzerland – *architect, painter, author, interior designer, graphic designer, teacher* – Studied painting in Antwerp (1881–84) and in Paris (1884–

werpen (1881–84) und in Paris (1884–85). 1893 Entwurf von Ornamenten und Titeln für das belgische Kunstmagazin „Van Nu en Straks". 1893–94 Vorlesungen an der Akademie in Antwerpen. 1901 Berufung zum künstlerischen Berater von Großherzog Wilhelm Ernst in Weimar. Einrichtung des Folkwang-Museums in Hagen, Westfalen. 1907 Gründungsmitglied des Deutschen Werkbundes in München. 1906–14 Bau und Leitung der Kunstgewerblichen Lehranstalten in Weimar. Empfiehlt Walter Gropius als seinen Nachfolger. 1926 Umzug nach Brüssel. 1926–36 Professor an der Universität in Gent. 1926–38 Leiter des „Institut Supérieur d'Architecture et des Arts Décoratifs" in Brüssel. – *Publikationen u. a.:* „Zum neuen Stil", München 1955; „Geschichte meines Lebens", München 1962. Karl Scheffler „Henry van de Velde", Leipzig 1913; K. E. Osthaus „Henry van de Velde. Leben und Schaffen des Künstlers", Hagen 1920; Maurice Casteels „Henry van de Velde", Brüssel 1932.

**Velde,** Jan van de – geb. 1568, gest. 1623, Niederlande. – *Schreibmeister* – 1605 Herausgabe des „Spieghel der Schrijftkonste" in Rotterdam, bis 1621 mehrere Schreibbücher und 1607 Herausgabe des Buchs „Exemplaer-Boec in houdende alderhande Gheschriften", das eine Sammlung von Schreibvorlagen ist. Seine Bücher haben Einfluß auf die zeitgenössischen Schreibmeister Europas.

**Veljović,** Jovica – geb. 1. 3. 1954 in Suvi Do, Jugoslawien. – *Schriftentwerfer, Typograph, Grafik-Designer, Lehrer* – 1974–79 Schrift- und Grafik-Design-Stu-

et des titres pour la revue belge d'art «Van Nu en Straks». Conférences à l'académie d'Anvers de 1893 à 1894. En 1901, le grand-duc Guillaume Ernest le fait venir à Weimar comme conseiller artistique. Conçoit l'aménagement intérieur du Folkwang-Museum à Hagen, Westphalie. Membre fondateur du Werkbund allemand à Munich en 1907. Construction des Kunstgewerblichen Lehranstalten (instituts d'enseignement des arts décoratifs) de Weimar en 1906; il dirigera cette école jusqu'en 1914. Recommande Walter Gropius pour lui succéder. S'installe à Bruxelles en 1926. Professeur à l'université de Gand de 1926 à 1936. Dirige «l'Institut Supérieur d'Architecture et des Arts Décoratifs» de Bruxelles de 1926 à 1938. – *Publications, sélection:* «Zum neuen Stil», Munich 1955; «Geschichte meines Lebens», Munich 1962. Karl Scheffler «Henry van de Velde», Leipzig 1913; K. E. Osthaus «Henry van de Velde. Leben und Schaffen des Künstlers», Hagen 1920; Maurice Casteels «Henry van de Velde», Bruxelles 1932.

**Velde,** Jan van de – né en 1568, décédé en 1623 aux Pays-Bas – *scribe* – 1605, publication du «Spieghel der Schrijftkonste» à Rotterdam. Publie de nombreux manuels d'écriture à Rotterdam et à Haarlem jusqu'en 1621. Parution en 1607 du livre intitulé «Exemplaer-Boec in houdende alderhande Gheschriften» qui présente de nombreux modèles d'écriture. Ses ouvrages ont exercé une forte influence sur les scribes européens de son temps.

**Veljović,** Jovica – né le 1. 3. 1954 à Suvi Do, Yougoslavie – *concepteur de polices, typographe, graphiste maquettiste, enseignant* – 1974–1979, études de calligraphie, d'arts graphiques et de design à

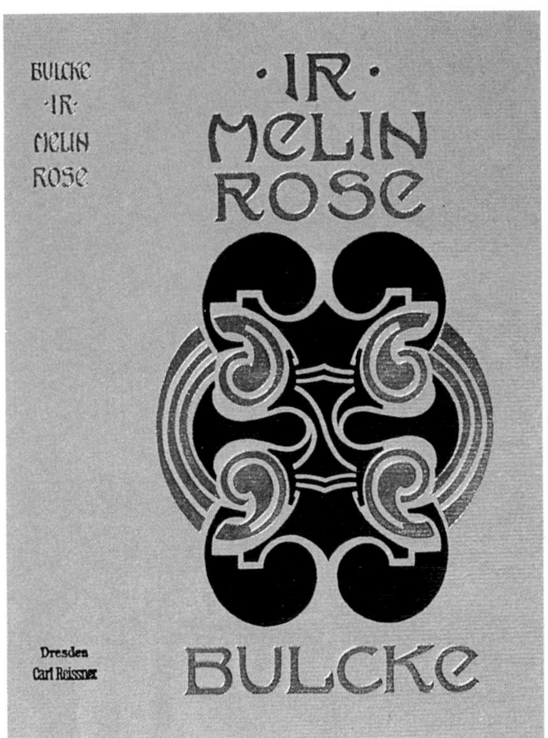

Henry van de Velde 1908 Cover

Veljović Calligraphy

Henry van de Velde ca. 1900 Logo

| | n | **n** | **n** | | *n* |
|---|---|---|---|---|---|
| a | b | e | f | g | i |
| o | r | s | t | y | z |
| A | B | C | E | G | H |
| M | O | R | S | X | Y |
| 1 | 2 | 4 | 6 | 8 | & |

Veljović 1984 Veljović

| *n* | *n* | ***n*** | | | |
|---|---|---|---|---|---|
| *a* | *b* | *e* | *f* | *g* | *i* |
| *o* | *r* | *s* | *t* | *y* | *z* |
| *A* | *B* | *C* | *E* | *G* | *H* |
| *M* | *O* | *R* | *S* | *X* | *Y* |
| *1* | *2* | *4* | *6* | *8* | *&* |

Veljović 1995 Ex Ponto

85). 1893: designs ornaments and covers for the Belgian "Van Nu en Straks" art magazine. 1893–94: lectures at the academy in Antwerp. 1901: made Grand Duke William Ernst's artistic advisor in Weimar. Does the interior design for the Folkwang-Museum in Hagen, Westphalia. 1907: founder member of the Deutsche Werkbund in Munich. 1906–14: builds and directs the Kunstgewerbliche Lehranstalten in Weimar. Recommends Walter Gropius as his successor. 1926: moves to Brussels. 1926–36: professor at the university in Ghent. 1926–38: principal of the Institut Supérieur d'Architecture et des Arts Décoratifs in Brussels. – *Publications include:* "Zum neuen Stil", Munich 1955; "Geschichte meines Lebens", Munich 1962. Karl Scheffler "Henry van de Velde", Leipzig 1913; K. E. Osthaus "Henry van de Velde. Leben und Schaffen des Künstlers", Hagen 1920; Maurice Casteels "Henry van de Velde", Brussels 1932.

**Velde,** Jan van de – b. 1568, d. 1623, The Netherlands – *writing master* – 1605: publishes his "Spieghel der Schrijftkonste" in Rotterdam. Up to 1621: publishes various books on the art of writing in Rotterdam and Haarlem. 1607: publishes his "Exemplaer-Boec in houdende alderhande Gheschriften", a collection of writing models. His books bore great influence on the European writing masters of the time.

**Veljović,** Jovica – b. 1. 3. 1954 in Suvi Do, Yugoslavia – *type designer, typographer, graphic designer, teacher* – 1974–79:

dium an der Akademie der angewandten Künste in Belgrad, 1981–83 Aufbaustudium. Unterrichtet 1987–92 Typographie an der Akademie für angewandte Künste in Belgrad. Seit 1992 Professor für Schrift an der Fachhochschule in Hamburg, Fachbereich Gestaltung. Für seine Arbeit im Bereich Kalligraphie und Typographie ausgezeichnet, u. a. beim XX. Oktobersalon in Belgrad (1979), mit dem „Charles-Peignot-Preis" der Association Typographique Internationale (ATypI) (1985), auf der Internationalen Grafik-Design-Ausstellung „ZGRAF5" in Zagreb (1987), auf der XIII. Biennale der Gebrauchsgrafik in Brno (1988) und auf der Grafik-Biennale in Moskau (1996). – *Schriftentwürfe:* ITC Veljović (1984), ITC Esprit (1985), ITC Gamma (1986) und Ex Ponto multiple master typeface (1995).

**Venezky,** Martin Joel – geb. 20. 8. 1957 in Miami Beach, Florida, USA. – *Grafik-Designer, Typograph, Lehrer* – 1975–79 Studium am Dartmouth College in Hanover, New Hampshire. 1981–85 Grafik-Designer im National Clearinghouse for Alcohol Information in Rockville, Maryland. 1985–86 Grafik-Designer im Design-Studio „Way Out West" in San Francisco, Kalifornien. Entwurf von Erscheinungsbildern. 1986–91 Senior Designer in der „RAM Group" Oakland, Kalifornien. 1990–91 Studium an der University of California Santa Cruz, 1991–93 an der Cranbrook Academy of Art. 1993 tätig im Studio Dumbar in Den Haag. Seit 1993 als freier Gestalter in San Francisco tätig. Auftraggeber waren u. a. Sun Microsystems, Chronicle Books, VH-1, Warner

l'académie des arts appliqués de Belgrade, études complémentaires de 1981 à 1983. Enseigne la typographie à l'académie des arts appliqués de Belgrade de 1987 à 1992. Professeur d'arts de l'écriture à la section de design de la Fachhochschule (école technique supérieure) de Hambourg depuis 1992. Prix du XXe salon d'octobre de Belgrade (1979) pour ses travaux de calligraphie et de typographie, lauréat du prix «Charles Peignot» de l'Association Typographique Internationale (ATypI) en 1985, prix de l'exposition internationale de graphisme et de design «ZGRAF5» de Zagreb en 1987, prix de la XIIIe Biennale du graphisme industriel à Brno en 1988, Biennale du graphisme de Moscou en 1996. – *Polices:* ITC Veljović (1984), ITC Esprit (1985), ITC Gamma (1986) et Ex Ponto multiple master typeface (1995).

**Venezky,** Martin Joel – né le 20. 8. 1957 à Miami Beach, Floride, Etats-Unis – *graphiste maquettiste, typographe, enseignant* – 1975–1979, études au Darmouth College à Hanover, New Hampshire. Exerce comme graphiste maquettiste au National Clearinghouse for Alcohol Information de Rockville, Maryland, de 1981 à 1985. Graphiste maquettiste pour l'agence de design «Way Out West» à San Francisco, Californie, de 1985 à 1986. Conçoit des identités d'entreprises. Senior Designer pour le «RAM Group» à Oakland, Californie, de 1986 à 1991. Etudes à l'University of California Santa Cruz de 1990 à 1991, puis à la Cranbrook Academy of Art de 1991 à 1993. 1993, travaille au Studio Dumbar à La Haye. Exerce depuis 1993 comme indépendant à San Francisco. Commanditaires : Sun Microsystems, Chronicle Books, VH-1, Warner Brothers Records, Island Records, Absolut Vodka et Adobe Systems Inc. Maquette de la revue «Speak». Collaboration

Venezky 1995 Cover

Venezky

Venezky

Venezky

studies lettering and graphic design at the Academy of Applied Arts in Belgrade, with a period of research studies from 1981–83. 1987–92: teaches typography at the Academy of Applied Arts in Belgrade. 1992: is made professor of lettering at the design department of the Fachhochschule in Hamburg. He has won various awards for his work in calligraphy and typography, including the Association Typographiques Internationale (ATypI) Charles Peignot Prize (1985) and prizes at the XX October Salon in Belgrade (1979), the "ZGRAF5" international graphic design exhibition in Zagreb (1987), the XIII Biennale for Graphic-Design in Brno (1988) and the Graphics Biennale in Moscow (1996). – *Fonts:* ITC Veljović (1984), ITC Esprit (1985), ITC Gamma (1986) and Ex Ponto multiple master typeface (1995).

**Venezky,** Martin Joel – b. 20. 8. 1957 in Miami Beach, Florida, USA – *graphic designer, typographer, teacher* – 1975–79: studies at Dartmouth College in Hanover, New Hampshire. 1981–85: graphic designer at the National Clearinghouse for Alcohol Information in Rockville, Maryland. 1985–86: graphic designer for the "Way Out West" design studio in San Francisco, California. Designs corporate identities. 1986–91: senior designer of the RAM Group, Oakland, California. 1990–91: studies at the University of California Santa Cruz and from 1991–93 at the Cranbrook Academy of Art. 1993: works for the Studio Dumbar in Den Haag and starts freelancing as a designer in San Francisco. Clients include Sun

Brothers Records, Island Records, Absolut Vodka und Adobe Systems Inc. Gestaltung der Zeitschrift „Speak". 1993–96 Zusammenarbeit mit Raul Cabra im Büro „Diseño" in San Francisco. Unterrichtet 1993 am California College of Arts & Crafts in San Francisco, 1994–95 an der University of California Santa Cruz in Santa Clara. Zahlreiche Ausstellungen, Auszeichnungen und Vorträge.

**Vespasiano,** Amphiareo – geb. 1501 in Ferrara, Italien, gest. 1563 in Venedig, Italien. – *Schreibmeister* – Seit 1518 ständig in Venedig tätig. Nach dreißig Jahren Tätigkeit als Schreibmeister 1548 Herausgabe seines Werks „Un novo modo d'insegnar a scrivere", das später mit dem Titel „Opera di frate Vespasiano Amphiareo" in weiteren Auflagen erscheint (bis 1620 in 19 Auflagen). Im gleichen Jahr beendet er das Manuskript eines Schreibvorlagenbuchs mit zwanzig Seiten. Ein weiteres Manuskript ist ein Gebetbuch mit dem Titel „Preces piae".

**Vieira,** Mary – geb. 30. 7. 1927 in São Paulo, Brasilien. – *Grafik-Designerin, Künstlerin, Lehrerin* – 1943–47 Studium an der Universität von Minas Gerais und an der Akademie der Schönen Künste in Belo Horizonte. 1947 Ausstellung ihrer Arbeit im „Salon junger brasilianischer Künstler". 1948 erste kinetische Skulpturen. Arbeitet seit 1952 in der Schweiz. 1953 mit dem Skulpturen-Preis der 2. Biennale in São Paulo ausgezeichnet. Arbeiten für das Kunstgewerbemuseum in Zürich (1954) und die Fluglinie Panair do Brasil (1957). Unterrichtet 1966–69 Raumgestaltung an der Kunstgewerbe-

Vespasiano 1554

Vespasiano 1554

Vieira 1996 Poster

Vieira 1996 Poster

avec Raul Cabra á l'agence «Diseño» á San Francisco de 1993 à 1996. Enseigne en 1993 au California College of Arts & Crafts de San Francisco, puis, de 1994 à 1995 à l'University of California Santa Cruz à Santa Clara. Nombreuses expositions, distinctions et conférences.

**Vespasiano,** Amphiareo – né en 1501 à Ferrare, Italie, décédé en 1563 à Venise, Italie – *scribe* – travaille constamment à Venise à partir de 1518. Après trente années d'activités comme scribe, il publie en 1548 la première édition de son ouvrage «Un novo modo d'insegnar a scrivere»; les éditions suivantes seront intitulées: «Opera di frate Vespasiano Amphiareo» (19 rééditions jusqu'en 1620). En cette même année 1548, il achève le manuscrit d'un manuel de modèles d'écritures de vingt pages. Il existe un autre manuscrit d'un autre ouvrage, un livre de prières intitulé «Preces piae».

**Vieira,** Mary – née le 30. 7. 1927 à São Paulo, Brésil – *graphiste maquettiste, artiste, enseignante* – 1943–1947, études à la faculté de pédagogie de l'université de Minas Gerais et à l'académie des beaux-arts de Belo Horizonte. Exposition de ses travaux au «Salon des jeunes artistes brésiliens» en 1947. Premières sculptures cinétiques en 1948. Travaille en Suisse à partir de 1952. Prix de sculpture de la 2e Biennale de São Paulo en 1953. Travaux pour le musée des arts décoratifs de Zurich (1954) et la compagnie aérienne Panair do Brasil (1957). Enseigne la conception spatiale à la Kunstgewerbeschule

Microsystems, Chronicle Books, VH-1, Warner Brothers Records, Island Records, Absolut Vodka and Adobe Systems Inc. Designs "Speak" magazine. 1993–96: works with Raul Cabra at the studio Diseño in San Francisco. 1993: teaches at the California College of Arts & Crafts in San Francisco and from 1994–95 at the University of California Santa Cruz in Santa Clara. Numerous exhibitions, awards and lectures.

**Vespasiano,** Amphiareo – b. 1501 in Ferrara, Italy, d. 1563 in Venice, Italy – *writing master* – From 1518 onwards: works in Venice. 1548: after 30 years' work as a writing master, Vespasiano publishes the first edition of his "Un novo modo d'insegnar a scrivere", which is later reprinted under the title "Opera di frate Vespasiano Amphiareo" (19 impressions up to 1620). In the same year he finishes the manuscript of his 20-page copybook. Another of his manuscripts is a prayer book, entitled "Preces piae".

**Vieira,** Mary – b. 30. 7. 1927 in São Paolo, Brazil – *graphic designer, artist, teacher* – 1943–47: studies in the education department of the University of Minas Gerais and at the Academy of Fine Arts in Belo Horizonte. 1947: exhibits her work at the "Salon of Young Brazilian Artists". 1948: produces her first kinetic sculptures. 1952: starts working in Switzerland. 1953: awarded a sculpture prize at the 2nd Biennale in São Paolo. Produces work for the Kunstgewerbemuseum in Zurich (1954) and the airline Panair do Brasil (1957). 1966–69: teaches interior design at the Kunstgewer-

schule in Basel. – *Publikation:* Alberto Sartoris „1 Polyvolume de Mary Vieira: 50 interactions", Mailand 1974.

**Vignelli,** Massimo – geb. 10. 1. 1931 in Mailand, Italien. – *Grafik-Designer, Typograph, Lehrer* – 1948–50 Studium an der Kunstakademie in Brera, 1950–53 Studium der Architektur in Mailand und 1953–57 in Venedig. 1957–60 Studium am Institute of Technology in Illinois und Massachusetts. 1960–65 Gründung und Leitung seines Studios in Mailand (mit Lella Vignelli). Entwurf der Plakate für die Biennale in Venedig 1962 und 1964. 1964 Verleihung der Goldmedaille der Triennale in Mailand. 1965–71 Umzug in die USA, Mitbegründer und Mitarbeiter von „Unimark International Corporation". Arbeit für Auftraggeber wie Knoll International, die American Airlines, die New York Subway und die Washington Metro. 1971 Gründung von „Vignelli Associates", danach 1978 „Vignelli Design" (beides mit seiner Frau Lella) in New York. Auftraggeber waren u. a. Rizzoli International Publications, Lancia, die United States National Parks Administration und Ciga Hotels. Zahlreiche Auszeichnungen, u. a. Goldmedaille des „American Institute of Graphic Arts" (AIGA) (1983), „Gold Medal for Design" des National Arts Club (1991). Unterrichtet weltweit Design an zahlreichen Schulen. – *Schriftentwurf:* WTC Our Bodoni (mit Tom Carnase, 1989). – *Publikationen u. a.:* „Graphic Design for Nonprofit Organizations", New York 1980; „Knoll Design" (mit Eric Larrabee), New York 1981; „Design: Vignelli", New York 1990.

(école des arts décoratifs) de Bâle de 1966 à 1969. – *Publication:* Alberto Sartoris «1 Polyvolume de Mary Vieira : 50 interactions», Milan 1974.

**Vignelli,** Massimo – né le 10. 1. 1931 à Milan, Italie – *graphiste maquettiste, typographe, enseignant* – 1948–1950, études à l'académie des beaux arts de Brera; 1950–1953, études d'architecture à Milan, puis à Venise de 1953 à 1957. Etudes à l'Institute of Technology de l'Illinois et au Massachusetts de 1957–1960. Fonde un atelier à Milan en 1960 (avec Lella Vignelli) et le dirige jusqu'en 1965. Conception d'affiches pour les biennales de Venise de 1962 et 1964. Médaille d'or de la Triennale de Milan en 1964. Vit aux Etats-Unis de 1965 à 1971, où il participe à la création de «Unimark International Corporation». Il y travaille pour plusieurs commanditaires tels que Knoll International, les American Airlines, le New York Subway et le Washington Metro. Fonde «Vignelli Associates» à New York en 1971, puis «Vignelli Design» en 1978 (toujours avec sa femme Lella). Commanditaires : Rizzoli International Publications, Lancia, United States National Parks Administration et Ciga Hotels. Nombreuses distinctions, entre autres la médaille d'or de l' «American Institute of Graphic Arts» (AIGA) en 1983, et la «Gold Medal for Design» du National Art Club en 1991. Enseigne le design dans plusieurs écoles internationales. – *Police :* WTC Our Bodoni (avec Tom Carnase, 1989). – *Publications, sélection :* «Graphic Design for Nonprofit Organizations», New York 1980; «Knoll Design» (avec Eric Larrabee), New York 1981; «Design : Vignelli», New York 1990.

**Vinne,** Theodore Low de – né le 25. 12. 1828 à Stamford, Connecticut, Etats-Unis, décédé le 16. 2. 1914 à New York, Etats-Unis – *imprimeur, auteur, concep-*

Vignelli 1964 Poster

Vignelli 1992 Poster

Vignelli 1967 Poster

| n | n | **n** | **n** | | *n* |
|---|---|---|---|---|---|
| a | b | e | f | g | i |
| o | r | s | t | y | z |
| A | B | C | E | G | H |
| M | O | R | S | X | Y |
| 1 | 2 | 4 | 6 | 8 | & |

Vignelli 1989 Our Bodoni

beschule in Basle. – *Publication:* Alberto Sartoris "1 Polyvolume de Mary Vieira: 50 interactions", Milan 1974.

**Vignelli,** Massimo – b. 10. 1. 1931 in Milan, Italy – *graphic designer, typographer, teacher* – 1948–50: studies at the art academy in Brera. 1950–53: studies architecture in Milan and from 1953–57 in Venice. 1957–60: studies at the Institute of Technology in Illinois and Massachusetts. 1960–65: opens and runs a studio in Milan with Lella Vignelli. Designs posters for the 1962 and 1964 Biennale in Venice. 1964: is awarded a gold medal by the Triennale in Milan. 1965–71: lives in the USA. Co-founder of Unimark International Corporation, working for clients who include Knoll International, American Airlines, the New York Subway and the Washington Metro. 1971: launches Vignelli Associates and later in 1978 Vignelli Design (both with his wife, Lella) in New York. Clients include Rizzoli International Publications, Lancia, United States National Parks Administration and Ciga Hotels. Numerous awards, including a gold medal from the American Institute of Graphic Arts (AIGA, in 1983) and the Gold Medal for Design from the National Arts Club (1991). Has taught at numerous schools all over the world. – *Font:* WTC Our Bodoni (with Tom Carnase, 1989). – *Publications include:* "Graphic Design for Nonprofit Organizations", New York 1980; "Knoll Design" (with Eric Larrabee), New York 1981; "Design: Vignelli", New York 1990.

**Vinne,** Theodore Low de – b. 25. 12. 1828 in Stamford, Connecticut, USA, d. 16. 2.

**Vinne,** Theodore Low de – geb. 25. 12. 1828 in Stamford, Connecticut, USA, gest. 16. 2. 1914 in New York, USA. – *Drucker, Autor, Schriftentwerfer* – Seit 1850 Arbeit in der Setzerei von Francis Hart in New York, seit 1858 Teilhaber, später mit seinem Sohn alleiniger Inhaber. 1884 Mitbegründer des „Grolier Club", einer privaten Organisation von Bücherfreunden in New York mit eigenem Clubhaus, eigener Bibliothek und eigenen Publikationen. Beauftragt 1894 L. B. Benton mit dem Schnitt der Schrift „Century" für die gleichnamige Zeitschrift. Gründet 1907 „The Theodore de Vinne Press", die zahlreiche, für Typographie und Gestaltung einflußreiche Bücher herausgibt. Zahlreiche Pressendrucke für den „Grolier Club". – *Schriftentwurf:* Century (mit L. B. Benton, 1894). – *Publikationen u. a.:* „Historic Printing Types", New York 1886; „Plain Printing Types", New York 1899; „The Practice of Typography", New York 1902–04. J. H. Brainerd „History of the De Vinne Press", New York 1934.

**Virl,** Hermann – geb. 1903 in München, Deutschland, gest. 2. 8. 1958 in München, Deutschland. – *Holzstecher, Maler, Typograph, Lehrer* – Studium an der Staatsschule für angewandte Kunst in München bei F. H. Ehmcke und Richard Riemerschmid. 1926–31 Lehrer an der grafischen Abteilung der Staatlichen Kunstgewerbeschule in Kassel. 1931–45 Lehrer an der Meisterschule für Deutschlands Buchdrucker in München. 1945–58 freischaffender Grafiker und Maler in München. 1984 Ausstellung seines Werks im Stadtmuseum München. – *Publikatio-*

Vinne 1894 Century Oldstyle Bold

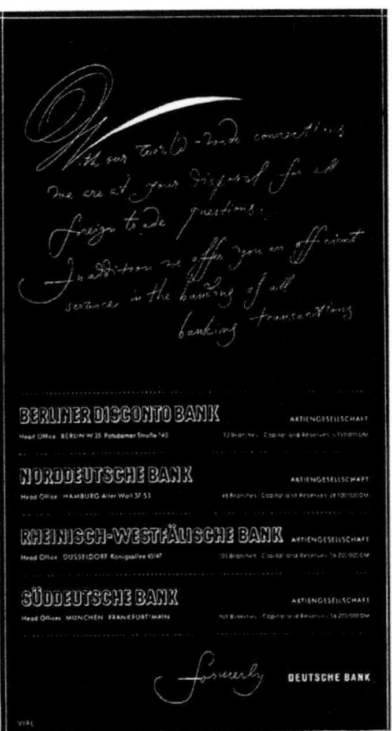


Virl 1951 Advertisement

Virl 1949 Cover

*teur de polices* – employé dans l'atelier de composition de Francis Hart à New York à partir de 1850, sociétaire de l'entreprise en 1858, et propriétaire avec son fils, par la suite. 1884, cofondateur du «Grolier Club» de New York, un organisme privé destiné aux bibliophiles, qui possède un Club House, une bibliothèque et ses propres publications. En 1894, il charge L. B. Benton de tailler la «Century», une police utilisée pour la revue du même nom. En 1907, il fonde «The Theodore de Vinne Press» qui publiera de nombreux ouvrages déterminants pour la typographie et la mise en page. Imprime de nombreuses planches pour le «Grolier Club». – *Polices :* Century (avec L. B. Benton, 1894). – *Publications, sélection :* «Historic Printing Types», New York 1886; «Plain Printing Types», New York 1899; «The Practice of Typography», New York 1902–1904. J. H. Brainerd «History of the De Vinne Press», New York 1934.

**Virl,** Hermann – né en 1903 à Munich, Allemagne, décédé le 2. 8. 1958 à Munich, Allemagne – *graveur sur bois, peintre, typographe, enseignant* – études à la Staatsschule für angewandte Kunst (école nationale des arts appliqués) de Munich chez F. H. Ehmcke et Richard Riemerschmid. Enseigne à la section d'arts graphiques de la Staatliche Kunstgewerbeschule (école nationale des arts décoratifs) de Kassel, de 1926 à 1931. Professeur à la Meisterschule für Deutschlands Buchdrucker (école professionnelle de

1914 in New York, USA – *printer, specialist author, type designer* – 1850: starts working for the Francis Hart typesetting workshop in New York. 1858: is made a partner and is later joint owner of the workshop with his son. 1884: co-founder of the Grolier Club, a private association of book lovers in New York with their own club house, library and publications. 1894: commissions L. B. Benton to cut his Century typeface for the magazine of the same name. 1907: founds The Theodore de Vinne Press. The press publishes numerous books which influence typography and design. Prints several fine editions for the Grolier Club. – *Font:* Century (with L. B. Benton, 1894). – *Publications include:* "Historic Printing Types", New York 1886; "Plain Printing Types", New York 1899; "The Practice of Typography", New York 1902–04. J. H. Brainerd "History of the De Vinne Press", New York 1934.

**Virl,** Hermann – b. 1903 in Munich, Germany, d. 2. 8. 1958 in Munich, Germany – *wood engraver, painter, typographer, teacher* – Studied at the Staatsschule für angewandte Kunst in Munich under F. H. Ehmcke and Richard Riemerschmid. 1926–31: teaches in the graphics department at the Staatliche Kunstgewerbeschule in Kassel. 1931–45: teaches at the Meisterschule für Deutschlands Buchdrucker in Munich. 1945–58: freelance graphic artist and painter in Munich. 1984: an exhibition of his work is held at the Stadtmuseum in Munich. – *Publications:* "Die Entstehung und die Entwicklung der Schrift", Stuttgart 1950. Philipp

*nen:* „Die Entstehung und die Entwicklung der Schrift", Stuttgart 1950. Ph. Luidl u. a. „Hermann Virl", München 1980.

**Vivarelli,** Carlo – geb. 8.5.1919 in Zürich, Schweiz, gest. 12.6.1986 in Zürich, Schweiz. – *Grafik-Designer, Typograph, Künstler* – 1934–39 Grafikerlehre an der Kunstgewerbeschule in Zürich. 1946–47 künstlerischer Leiter im „Studio Boggeri" in Mailand. Besuch der „Accademia di Belle Arti di Brera" in Mailand. Seit 1947 Atelier in Zürich. Zahlreiche international prämierte Plakate. 1958 Teilnahme an der Ausstellung „Konstruktive Grafik" im Kunstgewerbemuseum Zürich. 1958–65 Mitbegründer, Mitherausgeber und Mitredakteur der Zeitschrift „Neue Grafik" (mit R. P. Lohse, J. Müller-Brockmann, H. Neuburg). 1988 Gedenkausstellung in der Stiftung für konstruktive und konkrete Kunst in Zürich. – *Publikation:* Susanne Kappeler „Carlo Vivarelli. Plastik, Malerei, Grafik", Zürich 1988.

**Vogeler,** Heinrich – geb. 12.12.1872 in Bremen, Deutschland, gest. 14.6.1942 in Karaganda, UdSSR (heute Kasachstan). – *Grafiker, Maler, Schriftsteller, Illustrator, Typograph* – 1890–93 Studium an der Kunstakademie in Düsseldorf. 1898 Freundschaft mit R. M. Rilke, für den er Buchillustrationen fertigt. 1899 Mitarbeit an der Zeitschrift „Die Insel". 1906–09 Entwürfe für die Schriftgießerei Gebr. Klingspor (Vogeler-Zierat, 1906). Organisiert 1924 in Berlin die Ausstellung „Sowjetisches Revolutionsplakat". Zahlreiche buchgestalterische Arbeiten für Verlage. 1932 Umzug in die Sowjetunion. 1941 Ausstellung seiner Arbeit in Mos-

l'imprimerie allemande) de Munich, de 1931 à 1945. Graphiste et peintre indépendant à Munich de 1945 à 1958. Exposition de ses œuvres au musée de la ville de Munich en 1984. – *Publications:* «Die Entstehung und Entwicklung der Schrift», Stuttgart 1950. Philipp Luidl et autres «Hermann Virl», Munich 1980.

**Vivarelli,** Carlo – né le 8. 5. 1919 à Zurich, Suisse, décédé le 12. 6. 1986 à Zurich, Suisse – *graphiste maquettiste, typographe, artiste* – 1934–1939, apprentissage de graphiste et études à la Kunstgewerbeschule (école des arts décoratifs) de Zurich. 1946–1947, directeur artistique au «Studio Boggeri» à Milan. Fréquente la «Accademia di Belle Arti di Brera» de Milan. Exerce dans son atelier de Zurich à partir de 1947. Nombreux prix internationaux pour ses affiches. Participe en 1958 à l'exposition «Konstruktive Grafik» au musée des arts décoratifs de Zurich. Cofondateur (avec R. P. Lohse, J. Müller-Brockmann et H. Neuburg) de la revue «Neue Grafik» en 1958, revue à laquelle il collabore comme éditeur et rédacteur jusqu'en 1965. Exposition commémorative à la fondation pour l'art construit et concret en 1988 à Zurich. – *Publication:* S. Kappeler «Carlo Vivarelli. Plastik, Malerei, Grafik», Zurich 1988.

**Vogeler,** Heinrich – né le 12. 12. 1872 à Brême, Allemagne, décédé le 14. 6. 1942 à Karanganda, URSS (aujourd'hui Kazakhstan – *graphiste, peintre, écrivain, illustrateur, typographe* – 1890–1893, études à la Kunstakademie (beaux-arts) de Düsseldorf. Rencontre R. M. Rilke en 1898, dessine des illustrations pour un des ouvrages du poète. Collabore à la revue «Die Insel» en 1899. Dessins pour la fonderie de caractères Gebr. Klingspor (Vogeler-Zierat, 1906) de 1906 à 1909. En 1924, il organise l'exposition «Affiches révolutionnaires soviétiques» à Berlin.

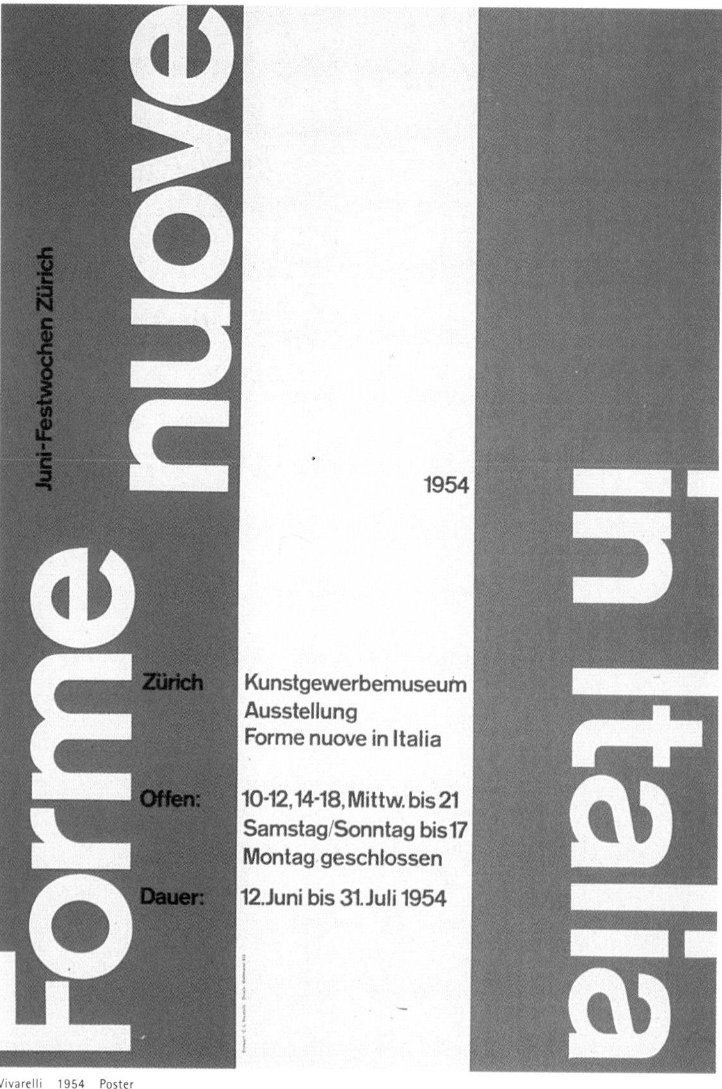

Vivarelli 1954 Poster

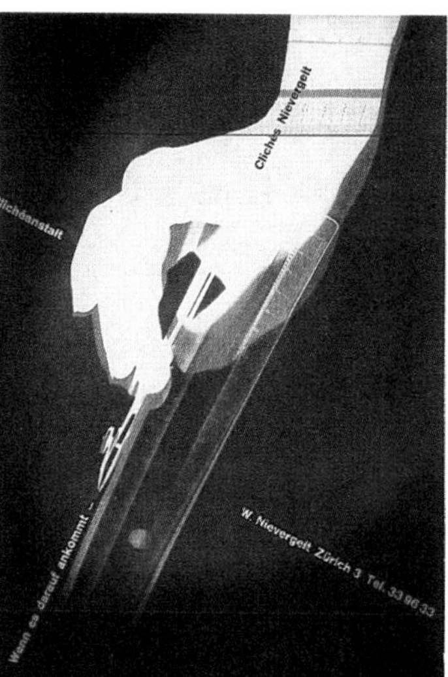

Vivarelli 1953 Advertisement

Vogeler 1903 Cover

Luidl et al "Hermann Virl", Munich 1980.

**Vivarelli,** Carlo – b. 8.5.1919 in Zurich, Switzerland, d. 12.6.1986 in Zurich, Switzerland – *graphic designer, typographer, artist* – 1934–39: trains as a graphic artist and attends the Kunstgewerbeschule in Zurich. 1946–47: art director for Studio Boggeri in Milan. Attends the Accademia di Belle Arti di Brera in Milan. 1947: opens his own studio in Zurich. Many of his posters win international awards. 1958: takes part in the "Konstruktive Grafik" exhibition held at the Kunstgewerbemuseum in Zurich. 1958–65: co-founder, co-publisher and co-editor of "Neue Grafik" magazine with R. P. Lohse, J. Müller-Brockmann and H. Neuburg. 1988: a commemorative exhibition is held at the Stiftung für konstruktive und konkrete Kunst in Zurich. – *Publication:* Susanne Kappeler "Carlo Vivarelli. Plastik, Malerei, Grafik", Zurich 1988.

**Vogeler,** Heinrich – b. 12.12.1872 in Bremen, Germany, d. 14.6.1942 in Karaganda, USSR (now Kazakhstan)– *graphic artist, painter, writer, illustrator, typographer* – 1890–93: studies at the Kunstakademie in Düsseldorf. 1898: makes the acquaintance of R.M. Rilke and illustrates books for him. 1899: works on "Die Insel" magazine. 1906–09: produces designs for the Gebr. Klingspor type foundry (Vogeler-Zierat, 1906). 1924: organizes an exhibition on Soviet Revolution posters in Berlin. Does work in book design for a number of publishers. 1932: moves to the Soviet Union. 1941: his work is exhibited in Moscow. Is

kau. Verbannung nach Kasachstan. – *Publikationen u. a.:* „Das neue Leben", Hannover 1919; „Erinnerungen", Berlin 1952; „Werden. Erinnerungen aus den Jahren 1932–1942", Berlin 1989.

**Vordemberge-Gildewart,** Friedrich – geb. 17. 11. 1899 in Osnabrück, Deutschland, gest. 19. 12. 1962 in Ulm, Deutschland. – *Typograph, Künstler, Autor, Lehrer* – 1919 Studium an der Kunstgewerbeschule und an der Technischen Hochschule in Hannover. 1924 Gründung der Künstlervereinigung „Gruppe k" (mit Hans Nitzschke). Mitglied der Künstlergruppen „Sturm" und „De Stijl". Atelier im Haus der Kestnergesellschaft. Seit 1924 zahlreiche Ausstellungen seines Werks. 1927 Gründung der Künstlergruppe „die abstrakten hannover". 1932 Gründungsmitglied der Künstlergruppe „abstraction-création" in Paris. 1938 auf der Propaganda-Ausstellung „Entartete Kunst" der Nationalsozialisten vertreten. 1942 Gründung und Schriftleitung der „Editions Duwaer" in Amsterdam. 1951–53 Umschlaggestaltung für die niederländische Architekturzeitschrift „Forum". 1952–54 Lehrauftrag an der Academie van Beeldende Kunsten in Rotterdam. 1953 mit dem Preis der 2. Biennale in São Paulo ausgezeichnet. 1954 Berufung an die Hochschule für Gestaltung in Ulm. 1955 Teilnahme an der documenta I in Kassel. – *Publikationen u. a.:* H. L. C. Jaffé „Vordemberge-Gildewart. Mensch und Werk", Köln 1971; Willy Rotzler „Vordemberge-Gildewart", St. Gallen 1979; D. Helms (Hrsg.) „Vordemberge-Gildewart. The Complete Works", München 1990.

Vordemberge-Gildewart 1927 Poster

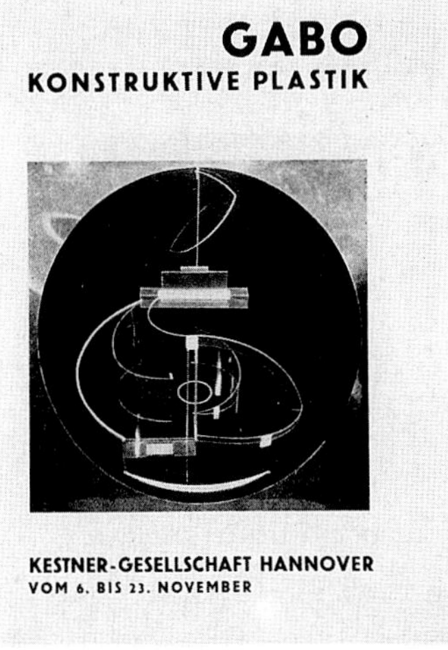

Vordemberge-Gildewart 1930 Cover

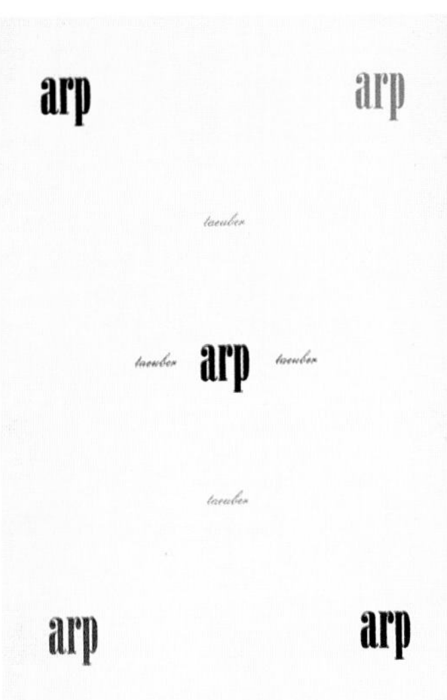

Vordemberge-Gildewart 1944 Cover

Nombreuses maquettes de livres pour des maisons d'édition. S'installe en Union soviétique en 1932. Exposition de ses travaux à Moscou, en 1941. Exilé au Kazakhstan. – *Publications, sélection :* «Das neue Leben», Hanovre 1919; «Erinnerungen», Berlin 1952; «Werden. Erinnerungen aus den Jahren 1932–1942», Berlin 1989.

**Vordemberge-Gildewart,** Friedrich – né le 17. 11. 1899 à Osnabrück, Allemagne, décédé le 19. 12. 1962 à Ulm, Allemagne – *typographe, artiste, auteur, enseignant* – 1919, études à l'école des arts décoratifs et à l'université technique de Hanovre. 1924, création du groupe d'artiste «Gruppe k» (avec Hans Nitzschke). Membre du groupes d'artistes «Sturm» et «De Stijl». Dispose d'un atelier dans le bâtiment de la Kestnergesellschaft. Nombreuses expositions de ses œuvres à partir de 1924. Création, en 1927, du groupe d'artistes «Les abstraits d'Hanovre». Membre fondateur du groupe d'artistes «abstraction-création» à Paris, en 1932. Ses œvres figurent à l'exposition «Art dégénéré» organisée par les national-socialistes en 1938. Fonde les «Editions Duwaer» à Amsterdam en 1942, et dirige la conception graphique. 1951–1953, couvertures pour la revue néerlandaise d'architecture «Forum». Chargé de cours à l'Academie van Beeldende Kunsten de Rotterdam de 1952 à 1954. Prix de la 2e Biennale de São Paulo en 1953. Nommé à la Hochschule für Gestaltung (école supérieure de design) d'Ulm en 1954. Participe à la «documenta I» de Kassel en 1955. – *Publications, sélection :* H. L. C. Jaffé «Vordemberge-Gildewart. Mensch und Werk», Cologne 1971; Willy Rotzler «Vordemberge-Gildewart», Saint-Gall 1979; Dietrich Helms (éd.) «Vordemberge-Gildewart. The Complete Works», Munich 1990.

banished to Kazakhstan. – *Publications include:* "Das neue Leben", Hanover 1919; "Erinnerungen", Berlin 1952; "Werden. Erinnerungen aus den Jahren 1932–1942", Berlin 1989.

**Vordemberge-Gildewart,** Friedrich – b. 17. 11. 1899 in Osnabrück, Germany, d. 19. 12. 1962 in Ulm, Germany – *typographer, artist, author, teacher* – 1919: studies at the Kunstgewerbeschule and at the Technische Hochschule in Hanover. 1924: founds the Gruppe k artists' association with Hans Nitzschke. Member of the Sturm group and De Stijl. Has his own studio in the Kestnergesellschaft building. From 1924 onwards: numerous exhibitions of his work. 1927: founds die abstrakten hannover artists' group. 1932: founder member of the abstraction-création group in Paris. 1938: is featured at the National Socialists' propaganda exhibition of "degenerate art". 1942: founds and is editor of Editions Duwaer in Amsterdam. 1951–53: designs covers for "Forum", a Dutch magazine on architecture. 1952–54: teaches at the Academie van Beeldende Kunsten in Rotterdam. 1953: is awarded a prize at the 2nd Biennale in São Paolo. 1954: appointed head of the visual communication department at the Hochschule für Gestaltung in Ulm. 1955: takes part in documenta I in Kassel. – *Publications include:* H. L. C. Jaffé "Vordemberge-Gildewart. Mensch und Werk", Cologne 1971; Willy Rotzler "Vordemberge-Gildewart", St. Gallen 1979; Dietrich Helms (ed.) "Vordemberge-Gildewart. The Complete Works", Munich 1990.

**Vox,** Maximilien (d. i. Samuel William Théodore Monod) – geb. 16. 12. 1894 in Condé-sur-Noireau, Frankreich, gest. 18. 12. 1974 in Lurs-en-Provence, Frankreich. – *Typograph, Illustrator, Lehrer* – Seit 1913 humoristische Zeichnungen für Zeitungen wie „Humanité" und „Floréale" sowie Illustrationen und Gravuren für Verlage und Zeitschriften. 1924–26 Umschlaggestaltung für Bücher des Verlags Grasset. 1926 Verleihung des Blumenthal-Preises. 1927 Gestaltung der Initialen des „Grand Larousse". Unterrichtet 1928 an der Ecole du Louvre. 1940 Gründung der „Union Bibliophile de France". 1943 Verleihung des Saintour-Preises der Académie française. 1949 Gründung und Herausgabe der Zeitschrift „Caractère". 1952 Gründung und Präsident der „Rencontres de Lurs" in Lurs-en-Provence (mit Jean Garcia, Jean Giono, Lucien Jacques und Robert Ranc). 1954 Entwicklung einer Schriftklassifikation in zehn Gruppen, 1955 Überarbeitung (z. B. neun Gruppen) im Jahrbuch „Caractère Noël". – *Publikationen u. a.:* „Pour une nouvelle classification des caractères", Paris 1955; „Das halbe Jahrhundert 1914–1964" in „Internationale Buchkunst im 19. und 20. Jahrhundert", Ravensburg 1969.

**Vox,** Maximilien (Samuel William Théodore Monod) – né le 16. 12 1894 à Condé-sur-Noireau, France, décédé le 18. 12. 1974 à Lurs-en-Provence, France – *typographe, illustrateur, enseignant* – dessins humoristiques pour la presse, entre autres pour «L'Humanité» et «Floréale» à partir de 1913, illustrations et gravures pour des maisons d'éditions et des revues. Crée des couvertures de livres pour les éditions Grasset de 1924 à 1926. Lauréat du prix Blumenthal en 1926. Dessine le logo du «Grand Larousse» en 1927. Enseigne à l'Ecole du Louvre en 1928. Fonde l' «Union Bibliophile de France» en 1940. Prix Saintour de l'Académie française en 1943. Refonte des directives de mise en page typographique de l'administration du Ministère des finances (1945) et de la radio française (1947). Fonde et édite la revue «Caractère» en 1949. Crée et préside les «Rencontres de Lurs» à Lurs-en-Provence, en 1952 (avec Jean Garcia, Jean Giono, Lucien Jacques et Robert Ranc). En 1954, il établit une classification des caractères en dix groupes, la remanie en 1955 (neuf groupes) dans les annales «Caractère Noël». – *Publications, sélection:* «Pour une nouvelle classification des caractères», Paris 1955; «Das halbe Jahrhundert 1914–1964», in «Internationale Buchkunst im 19. und 20. Jahrhundert», Ravensburg 1969.

CLASSIFICATION MAXIMILIEN VOX

| Groupes | HUMANES | GARALDES | RÉALES | DIDONES | MÉCANES | LINÉALES | INCISES | MANUAIRES | SCRIPTES |
|---|---|---|---|---|---|---|---|---|---|
| Formes des empattements, pieds et extrémités des lettres. | | | | | | | | | |
| Aspects généraux de la lettre capitale | IS | IC | IS | IC | IS | NES NES | ERSI PASC | LAC ESU | [illegible] |
| romain et italique | TS | IC | IS | IE | IN | ICA ICA | AUG NES | POLK BAM | [illegible] |
| Aspects généraux de la lettre bas de casse | ica | aie | ier | ien | icn | sodi sodi | abce abcm | cisel renoi | [illegible] |
| romain et italique | nic | aic | aic | aiu | inc | onde onde | nord nose | lona chou | [illegible] |
| Etudes morphologiques de la lettre | [illegible] | [illegible] | [illegible] | [illegible] | [illegible] | [illegible] | [illegible] | [illegible] | [illegible] |
| Caractères appartenant à chacun des groupes | [illegible] | [illegible] | [illegible] | [illegible] | [illegible] | [illegible] | [illegible] | [illegible] | [illegible] |

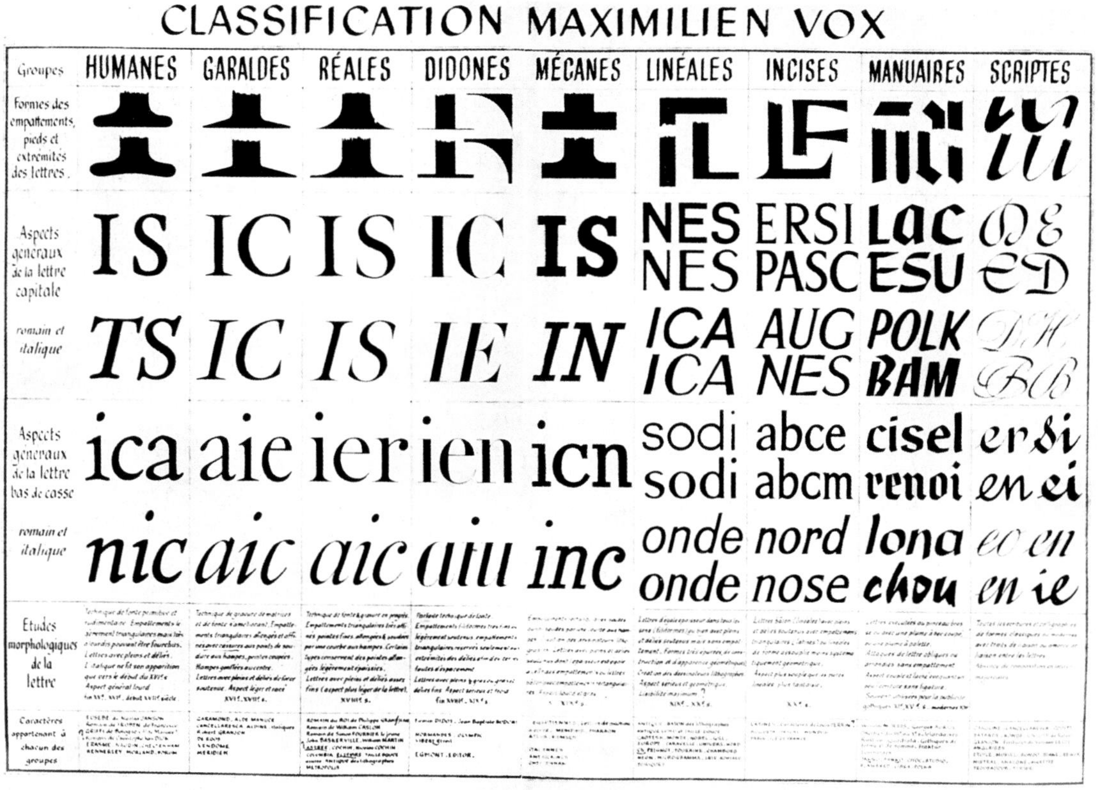

Vox 1955

**Vox,** Maximilien (real name: Samuel William Théodore Monod) – b. 16. 12. 1894 in Condé-sur-Noireau, France, d. 18. 12. 1974 in Lurs-en-Provence, France – *typographer, illustrator, teacher* – 1913: starts producing humorous drawings for magazines such as "Humanité" and "Floréale", and illustrations and engravings for publishing houses and magazines. 1924–26: designs book covers for Grasset publishers. 1926: is awarded the Prix Blumenthal. 1927: designs the initials for the "Grand Larousse". 1928: teaches at the Ecole du Louvre. 1940: founds the Union Bibliophile de France. 1943: is awarded the Prix Saintour by the Académie française. Reworks the guidelines for typographic design for the administration unit of the Ministry of Finance (1945) and the French broadcasting company (1947). 1949: launches and publishes "Caractère" magazine. 1952: founds and is president of the Rencontres de Lurs in Lurs-en-Provence with Jean Garcia, Jean Giono, Lucien Jacques and Robert Ranc. 1954: develops a way of classifying type in ten groups. 1955: reworks his classification (with 9 groups) in the "Caractère Noël" yearbook. – *Publications include:* "Pour une nouvelle classification des caractères", Paris 1955; "Das halbe Jahrhundert 1914–1964" in "Internationale Buchkunst im 19. und 20. Jahrhundert", Ravensburg 1969.

# W

**Wagner,** Johannes – *Schriftgießerei* – Johannes Wagner, der älteste Sohn von Ludwig Wagner, gliedert 1913 der väterlichen Schriftgießerei Ludwig Wagner in Leipzig eine Messinglinienfabrik an. 1921 Gründung der Norddeutschen Schriftgießerei in Berlin durch Johannes Wagner, seinen Bruder Ludwig und seinen Schwager Willy Jahr. 1949 Verlegung des Betriebs nach Ingolstadt. J. Wagner verlegt 1956 den Sitz der Firma Ludwig Wagner nach Berlin West. 1965 Tod von J. Wagner. Arnold Dröse wird 1972 Inhaber der Firma. – *Schriftentwürfe:* Aurora-Grotesk (1912), Steinschrift (1912), Fette Antiqua (1913), Druckhaus-Antiqua (1919), Romana (1930), Neue Aurora Grotesk (1964).

**Walbaum,** Justus Erich – geb. 25. 1. 1768 in Steinlah, Deutschland, gest. 31. 1. 1839 in Weimar, Deutschland. – *Schriftgießer, Schriftentwerfer, Stempelschneider* – Lehre als Gewürzhändler und Konditor in Braunschweig. Weiterbildung zum Formschneider, Notenstecher und Stempelschneider. Übernimmt das Schneiden und Gießen von Gedenkmünzen. 1796 Kauf der Schriftgießerei des Buchdruckers Ernst Wilhelm Kircher in Goslar. 1803 Verlegung der Schriftgießerei nach Weimar. 1828 übergibt Walbaum seine Schriftgießerei an seinen Sohn Theodor, der 1830 tödlich verunglückt. J. E. Walbaum führt das Unternehmen weiter. 1836 verkauft Walbaum seine Schriftgießerei an F. A. Brockhaus in Leipzig. Ein Teil der Schriftgießerei Walbaum und ihrer Matrizen geht 1918 an die H. Berthold AG in Berlin über.

**Wagner,** Johannes – *fonderie de caractères* – en 1913, Johannes Wagner, fils aîné de Ludwig Wagner, étend les activités de la fonderie de caractères Ludwig Wagner en y ajoutant une usine de tiges de laiton. En 1921, Johannes Wagner, son frère Ludwig et Willy Jahr, leur beau-frère, fondent la Norddeutsche Schriftgiesserei à Berlin. En 1949, l'entreprise est transférée à Ingolstadt, puis en 1956, J. Wagner installe de nouveau le siège de la société Ludwig Wagner à Berlin-Ouest. Décès de J. Wagner en 1965. Arnold Dröse reprend la société en 1972. – *Polices :* Aurora-Grotesk (1912), Steinschrift (1912), Fette Antiqua (1913), Druckhaus-Antiqua (1919), Romana (1930), Neue Aurora Grotesk (1964).

**Walbaum,** Justus Erich – né le 25. 1. 1768 à Steinlah, Allemagne, décédé le 31. 1. 1839 à Weimar, Allemagne – *fondeur de caractères, concepteur de polices, tailleur de types* – apprentissage de pâtisserie et d'épicerie à Brunswick. Formation de tailleur de type et de notes. Taille, grave et fond des médailles commémoratives. En 1796, il achète la fonderie de caractères de l'imprimeur Ernst Wilhelm Kircher à Goslar. La fonderie de caractères s'installe à Weimar en 1803. Walbaum lègue la fonderie de caractères à son fils en 1828, mais celui-ci meurt d'un accident en 1830. J. E. Walbaum reprend l'exploitation de la société. En 1836, il cède sa fonderie à F. A. Brockhaus de Leipzig. En 1918, une partie de la fonderie de caractères Walbaum, ainsi que les matrices, est vendue à la H. Berthold AG de Berlin.

**Walker,** Emery – né le 2. 4. 1851 à Londres, Angleterre, décédé le 22. 7. 1933 à Londres, Angleterre – *typographe, concepteur de polices, éditeur* – 1873, apprentissage de dessinateur de caractères, de gravure et d'imprimerie à Londres. En 1886, il devient propriétaire de «Walker

Gruppe 6: Serifenlose Linear-Antiqua

Kanadischer Whisky
Edel (Aurora) -Grotesk mager Größen-Nr. 5/6 – 36

Raststätte Auwaldsee
Edel (Aurora) -Grotesk halbfett Größen-Nr. 5/6 – 48

Flugplatz Salzburg
Edel (Aurora) -Grotesk dreiviertelfett Größen-Nr. 6 – 72

Tiefdruckereien
Edel (Aurora) -Grotesk fett Größen-Nr. 6 – 72

Aluminiumwerk
Edel (Aurora) -Grotesk breitmager Größen-Nr. 4/6 – 36

Eigenwerbung
Edel (Aurora) -Grotesk breithalbfett Größen-Nr. 5/6 – 48

Rasthäuser
Edel (Aurora) -Grotesk breitfett Größen-Nr. 6 – 72

Schriftzeichen sind Formgebilde
Edel (Aurora) -Grotesk enghalbfett Größen-Nr. 6 – 96

Kaufen Sie auch bei Hofman
Edel (Aurora) -Grotesk schmalfett Größen-Nr. 6 – 72

Solitude Rennstreck
Edel (Aurora) -Grotesk engfett Größen-Nr. 6 – 96

Bayrisches Alpenland
Edel (Aurora) -Grotesk kursiv mager Größen-Nr. 5/6 – 36

Egerländer Musiker
Edel (Aurora) -Grotesk, kursiv fett Größen-Nr. 5/6 – 36

Sendung für Bergsteiger
Atlantis Grotesk mager Größen-Nr. 5/6 – 36

Grand Hotel in Moskau
Atlantis Grotesk halbfett Größen-Nr. 5/6 – 48

Industrie - Werbung
Atlantis Grotesk fett Größen-Nr. 6 – 72

Wagner 1978 Specimen

ANTIQUA  KURSIV


ABCDEFGHIJKLMNOP
QRSTUVWXYZ
abcdefghijklmnopqrst
uvwxyz1234567890

*ABCDEFGHIJKLMNOP*
*QRSTUVWXYZ*
*abcdefghijklmnopqrst*
*uvwxyz 1234567890*

Walbaum-Fraktur

ABCDEFGHIJKLMNO
PQRSTUVWXYZ
abcdefghijklmnopqrsſtuv
wxyz 1234567890

Walbaum ca. 1850 Specimen

**Wagner,** Johannes – *type foundry* – 1913: Johannes Wagner, the oldest son of Ludwig Wagner, affiliates a brass rule factory to his father's type foundry in Leipzig. 1921: Johannes Wagner, his brother Ludwig and his brother-in-law Willy Jahr found the Norddeutsche Schriftgießerei in Berlin. 1949: the business moves to Ingolstadt. 1956: J. Wagner moves the Ludwig Wagner company to West Berlin. 1965: J. Wagner dies. 1972: Arnold Dröse is the new company owner. – *Fonts:* Aurora-Grotesk (1912), Steinschrift (1912), Fette Antiqua (1913), Druckhaus-Antiqua (1919), Romana (1930) and Neue Aurora Grotesk (1964).

**Walbaum,** Justus Erich – b. 25. 1. 1768 in Steinlah, Germany, d. 31. 1. 1839 in Weimar, Germany – *type founder, type designer, punch cutter* – Trained as a spice merchant and pastry cook in Braunschweig. Further training as a form cutter, music engraver and punch cutter. Cut and cast commemorative coins. 1796: buys printer Ernst Wilhelm Kircher's type foundry in Goslar. 1803: the type foundry moves to Weimar. 1828: Walbaum hands over the type foundry to his son, Theodor, who is killed in an accident in 1830. J. E. Walbaum continues to run the business. 1836: Walbaum sells his type foundry to F. A. Brockhaus in Leipzig. 1918: H. Berthold AG in Berlin gains possession of part of the Walbaum foundry and some of its matrices.

**Walker,** Emery – b. 2. 4. 1851 in London, England, d. 22. 7. 1933 in London, England – *typographer, type designer, publisher* – 1873: trains as a letterer, engraver

**Walker,** Emery – geb. 2. 4. 1851 in London, England, gest. 22. 7. 1933 in London, England. – *Typograph, Schriftentwerfer, Verleger* – 1873 Lehre als Schriftzeichner, Graveur und Drucker in London. 1886 Inhaber der Reproduktionsanstalt „Walker and Boutall“, seit 1900 „Walker and Cockerell“, seit 1904 „Emery Walker Ltd.“. Gründet 1888 die „Crafts Exhibition Society“. Durch Walkers Vortrag vor der Gesellschaft am 15. 11. 1888 wird William Morris zur Gründung der „Kelmscott Press“ inspiriert. Lehnt das Angebot von Morris ab, Partner zu werden, wird aber sein typographischer Berater. 1889 Beteiligung an der Ausarbeitung der Schrift „Golden Type“ von Morris. 1901 Gründung der „Doves Press“ (mit T. J. Cobden-Sanderson) in Hammersmith. Als erstes Werk wird 1903 die fünfbändige „Doves Bible“ herausgegeben. 1904 Mitarbeit an der Großherzog-Wilhelm-Ernst-August-Ausgabe deutscher Klassiker. Master der „Art Workers Guild“. Seit 1908 Arbeit für die „Ashendene Press“, „Vale Press“, „Eragny Press“ und für die „Cranach-Presse“, für die er alle Schriften entwirft. Austritt aus der „Doves Press“. Arbeit für den Insel Verlag, in den er Edward Johnston und Eric Gill einführt. 1916 Zusammenarbeit mit Bruce Rodgers. 1927 Präsident der „Arts & Crafts Exhibition Society“. 1928 Arbeit an der „Oxford Lectern Bible“ (mit Bruce Rodgers). Wird 1930 in den Adelsstand erhoben. 1933 als „Honorary Fellow of Jesus College“ in Cambridge ausgezeichnet. – *Schriftentwürfe:* Doves Type (1900), Subiaco Type (1901), Cranach Press Roman (1913), Ashendene

| | n | n | | | *n* |
|---|---|---|---|---|---|
| a | b | e | f | g | i |
| o | r | s | t | y | z |
| A | B | C | E | G | H |
| M | O | R | S | X | Y |
| 1 | 2 | 4 | 6 | 8 | & |

Walbaum ca. 1800 Walbaum (Berthold)

| | N | | | | |
|---|---|---|---|---|---|
| A | B | E | F | G | I |
| O | R | S | T | Y | Z |
| A | B | C | E | G | H |
| M | O | R | S | X | Y |
| 1 | 2 | 4 | 6 | 8 | & |

Walbaum ca. 1800 Walbaum Caps

| | n | | | | |
|---|---|---|---|---|---|
| a | b | e | f | g | i |
| o | r | s | t | y | z |
| A | B | C | E | G | H |
| M | O | R | S | X | Y |
| 1 | 2 | 4 | 6 | 8 | & |

Walbaum ca. 1800 Walbaum Fraktur

ODI profanum vulgus & arceo;
Favete linguis: carmina non prius
Audita Musarum sacerdos
Virginibus puerisq; canto.
Regum timendorum in proprios greges,
Reges in ipsos imperium est Iovis,
Clari Giganteo triumpho,
Cuncta supercilio moventis.
Est ut viro vir latius ordinet
Arbusta sulcis, hic generosior
Descendat in Campum petitor,
Moribus hic meliorque fama
Contendat, illi turba clientium
Sit maior: aequa lege Necessitas
Sortitur insignis & imos;
Omne capax movet urna nomen.
Destrictus ensis cui super impia
Cervice pendet, non Siculae dapes
Dulcem elaborabunt saporem,
Non avium citharaeq; cantus
34

Walker 1902 Ashendene Press Type

and Boutall», une société spécialisée dans la reproduction qui s'appellera «Walker and Cockerell» à partir de 1900, puis «Emery Walker Ltd.» en 1904. En 1888, il fonde la «Crafts Exhibition Society» à Londres. Une conférence tenue par Walker devant les membres de la société le 15. 11. 1888 donne à William Morris l'idée de créer la «Kelmscott Press». Il décline l'offre de Morris qui lui propose un partenariat, mais devient son conseiller typographique. En 1889, il participe à l'élaboration du «Golden Type» de Morris. En 1901, il fonde les «Doves Press» (avec J. T. Cobden-Sanderson) à Hammersmith. Le premier ouvrage publié est la «Doves Bible» en cinq volumes qui paraît en 1903. Participe en 1904 à la publication de l'édition de classiques allemands appelée édition du grand-duc Guillaume-Ernest-Auguste. Maître de la «Art Workers Guild». Travaille pour «Ashendene Press», «Vale Press», «Eragny Press» à partir de 1908 et dessine toutes les polices de la «Cranach-Presse» à partir de 1908; quitte les «Doves Press». Exerce pour les éditions Insel où il fait entrer Edward Johnston et Eric Gill. Collaboration avec Bruce Rodgers en 1916. Président de la «Arts & Crafts Exhibition Society» en 1927. Travaille à la «Oxford Lectern Bible» (avec Bruce Rodgers) en 1928. Est anobli en 1930. Titre de «Honorary Fellow of Jesus College» de Cambridge en 1933. – *Polices:* Doves Type (1900), Subiaco Type (1901), Cranach Press Roman (1913), Ashendene Press

and printer in London. 1886: owner of the Walker and Boutall repro plant, renamed Walker and Cockerell in 1900 and Emery Walker Ltd. in 1904. 1888: founds the Crafts Exhibition Society. Walker's speech to the society on 15. 11. 1888 inspires William Morris to launch the Kelmscott Press. Walker turns down Morris's offer of partnership, but instead becomes his typographic advisor. 1889: contributes to the design of Morris's Golden Type. 1901: founds Doves Press with T. J. Cobden-Sanderson in Hammersmith. The first work to be published is the "Doves Bible" in five volumes in 1903. 1904: works on the Grand Duke Wilhelm Ernst August edition of German classics. Master of the Art Workers Guild. 1908: starts working for Ashendene Press, Vale Press, Eragny Press and the Cranach Press, for whom he designs all the typefaces. Leaves Doves Press. Works for Insel Verlag and introduces Edward Johnston and Eric Gill to the publishing house. 1916: works with Bruce Rodgers. 1927: president of the Arts & Crafts Exhibition Society. 1928: works on the "Oxford Lectern Bible" with Bruce Rodgers. 1930: is knighted. 1933: is made an Honorary Fellow of Jesus College in Cambridge. – *Fonts:* Doves Type (1900), Subiaco Type (1901), Cranach Press Roman (1913), Ashendene Press Type, Eragny Press

Press Type, Eragny Press Type, Vale Press Type. – *Publikationen u. a.:* Colin Franklin „Emery Walker", Cambridge 1973; Dorothy A. Harrop „Sir Emery Walker, 1851–1933", London 1986.

**Warde,** Beatrice – geb. 20. 9. 1900 in New York, USA, gest. 14. 9. 1969 in London, England. – *Typographin, Autorin* – 1921–25 Bibliotheksassistentin in der Bibliothek der American Type Founders Company (ATF) in New Jersey. Heiratet 1924 Frederic Warde. Begegnung mit Stanley Morison. 1925 Umzug nach Europa. Beginn der Arbeit für die Monotype Corporation. Entwurf von Plakaten, Anzeigen und Schriftproben. 1926 Beiträge für die Zeitschrift „The Fleuron" unter dem Pseudonym Paul Beaujon. Seit 1927 Herausgeberin der Zeitschrift „The Monotype Recorder" der Monotype Corporation in London. 1929 Leiterin der Publikationsabteilung der Monotype Corp. 1932 Entwurf des Kleinplakats „This is a Printing Office", das in allen Druckereien Englands als Wandschmuck diente und in viele Sprachen übersetzt wurde. Zahlreiche Vortragsreisen in den USA. Als erste Frau wird Beatrice Warde 1967 in den Double Crown Club als Rednerin eingeladen. – *Publikationen:* „Typography in Art Education", London 1946; „The Crystal Goblet: Sixteen Essays on Typography", London 1955.

**Warde,** Frederic – geb. 1894 in Wells, Minnesota, USA, gest. 1939 in New York, USA. – *Drucker, Typograph, Schriftentwerfer* – 1920–22 für das Druck- und Verlagshaus W. E. Rudge unter Bruce Rogers tätig. 1922–24 Leiter der Princeton Uni-

Type, Eragny Press Type, Vale Press Type. – *Publications, sélection:* Colin Franklin «Emery Walker», Cambridge 1973; Dorothy A. Harrop «Sir Emery Walker, 1851–1933», Londres 1986.

**Warde,** Beatrice – née le 20. 9. 1900 à New York, Etats-Unis, décédée le 14. 9. 1969 à Londres, Angleterre – *typographe, auteur* – 1921–1925, assistante bibliothécaire à la bibliothèque de l'American Type Founders Company (ATF) à New Jersey. Epouse Frederic Warde en 1924. Rencontre Stanley Morison. S'installe en Europe en 1925. Commence à travailler pour la Monotype Corporation. Conception d'affiches, d'annonces et de modèles d'écritures. Articles pour la revue «The Fleuron» en 1926, sous le pseudonyme de Paul Beaujon. Edite la revue de la Monotype Corporation de Londres, «The Monotype Recorder», à partir de 1927. Dirige, en 1929, le service des publications de la Monotype Corp. En 1932, elle dessine la petite affiche «This is a Printing Office» qui ornera toutes les imprimeries anglaises et sera traduite en plusieurs langues. Nombreuses conférences aux Etats-Unis. En 1967, Beatrice Warde est la première femme invitée à faire une conférence au Double Crown Club. – *Publications:* «Typography in Art Education», Londres 1946; «The Crystal Goblet: Sixteen Essays on Typography», Londres 1955.

**Warde,** Frederic – né en 1894 à Wells, Minnesota, Etats-Unis, décédé en 1939 à New York, Etats-Unis – *imprimeur, typographe, concepteur de polices* – 1920–1922, employé aux imprimeries éditions W. E. Rudge alors dirigées par Bruce Rogers. 1922–1924, directeur de la Princeton University Press. Epouse Beatrice Becker en 1924. Articles pour la revue «The Fleuron». Edite «The Dolphin, a Journal of the Making of Books» qui pa-

THIS IS
A PRINTING OFFICE

CROSSROADS OF CIVILIZATION
REFUGE OF ALL THE ARTS
AGAINST THE RAVAGES OF TIME
ARMOURY OF FEARLESS TRUTH
AGAINST WHISPERING RUMOUR
INCESSANT TRUMPET OF TRADE

FROM THIS PLACE WORDS MAY FLY ABROAD
NOT TO PERISH ON WAVES OF SOUND
NOT TO VARY WITH THE WRITER'S HAND
BUT FIXED IN TIME HAVING BEEN VERIFIED IN PROOF
FRIEND YOU STAND ON SACRED GROUND
THIS IS A PRINTING OFFICE

Beatrice Warde 1932 Poster

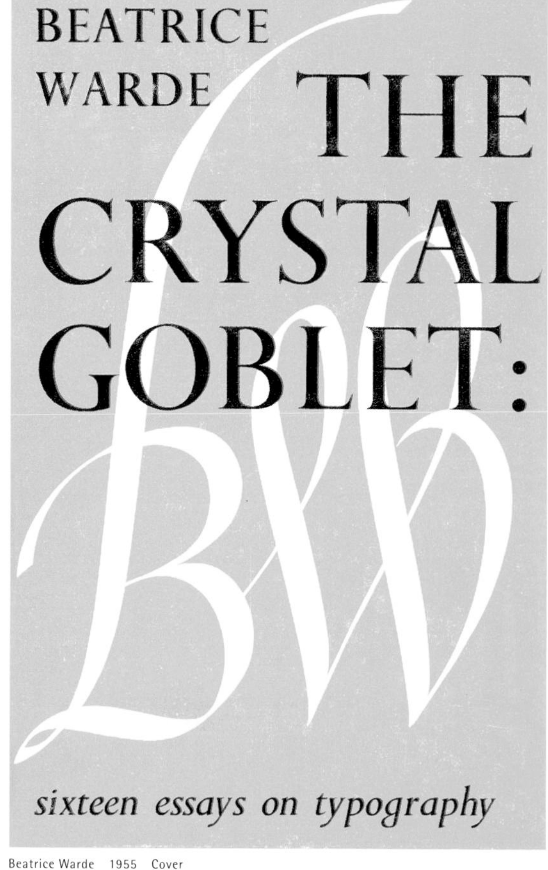


Beatrice Warde 1955 Cover

*ABCDEFGHIJKLMNOPQRSTU*
*VWXYZ abcdefghijklmnopqrstuvwxyz*

Frederic Warde 1929 Arrighi kursiv

Type, Vale Press Type. – *Publications include:* Colin Franklin "Emery Walker", Cambridge 1973; Dorothy A. Harrop "Sir Emery Walker, 1851–1933", London 1986.

**Warde,** Beatrice – b. 20. 9. 1900 in New York, USA, d. 14. 9. 1969 in London, England – *typographer, author* – 1921–25: library assistant at the American Type Founders Company (ATF) library in New Jersey. 1924: marries Frederic Warde. Meets Stanley Morison. 1925: moves to Europe. Starts to work for Monotype Corporation. Designs posters, adverts and type specimens. 1926: contributes to "The Fleuron" magazine under the pseudonym Paul Beaujon. From 1927 onwards: editor of "The Monotype Recorder" magazine, issued by the Monotype Corporation in London. 1929: head of Monotype Corporation's publications department. 1932: designs a small poster, "This is a Printing Office", which adorns the walls of every printing workshop in England and is translated into many languages. Undertakes numerous lecture tours in the USA. 1967: Beatrice Warde is the first woman to be invited to give a talk at the Double Crown Club. – *Publications:* "Typography in Art Education", London 1946; "The Crystal Goblet: Sixteen Essays on Typography", London 1955.

**Warde,** Frederic – b. 1894 in Wells, Minnesota, USA, d. 1939 in New York, USA – *printer, typographer, type designer* – 1920–22: works for the W. E. Rudge printer-publishers under Bruce Rogers. 1922–24: director of Princeton University Press. 1924: marries Beatrice Becker. Writes articles for "The Fleuron" maga-

versity Press. Heiratet 1924 Beatrice Becker. Artikel für die Zeitschrift „The Fleuron". Herausgeber von „The Dolphin, a Journal of the Making of Books", das seit 1932 in Buchform beim Limited Editions Club in New York erscheint. Buchgestalter für die Watch Hill Press, die Morrill Press und für W. E. Rudge. 1937–39 Herstellungsleiter der Oxford University Press in New York. – *Schriftentwürfe:* Arrighi (1925), Arrighi kursiv (1929). – *Publikation:* „Bruce Rogers, designer of books", New York 1926.

**Weber,** C. E. – *Schriftgießerei* – 1856 gründet Christian Emil Weber in Stuttgart eine Schriftgießerei. 1881 Übernahme der Firma durch die Söhne Carl und Adolf Weber. Carl Weber übergibt 1907 die Firma seinem Sohn Eugen und seinem Schwiegersohn Emil Ratzky, dessen Sohn Hermann Ratzky die Firma übernimmt. Das Gußprogramm wird von der D. Stempel AG in Frankfurt am Main und von Johannes Wagner in Ingolstadt übernommen. – *Schriftentwürfe:* Rund-Grotesk (1931), Schadow-Antiqua (1938), Delphin (1951), Trump Mediaeval (1954), Codex (1956).

**Weidemann,** Kurt – geb. 15. 12. 1922 in Eichmedien, Deutschland (heute Polen). – *Typograph, Grafik-Designer, Autor, Lehrer* – 1950–52 Schriftsetzerlehre in Lübeck. 1953–55 Studium der Buchgrafik und Typographie an der Staatlichen Akademie der Bildenden Künste in Stuttgart. Seit 1955 freiberuflich als Grafik-Designer, Werbeberater und Texter tätig. Auftraggeber waren u. a. COOP, Zeiss, die Merck AG, die Mercedes-Benz AG, die

raît sous forme de livre au Limited Editions Club de New York à partir de 1932. Maquettiste de livres pour la Watch Hill Press, la Morrill Press ainsi que pour W. E. Rudge. Directeur de fabrication de la Oxford University Press de New York de 1937 à 1939. – *Polices :* Arrighi (1925), Arrighi italique (1929). – *Publication :* «Bruce Rogers, designer of books», New York 1926.

**Weber,** C. E. – *fonderie de caractères* – Christian Emil Weber crée une fonderie de caractères à Stuttgart en 1856. La société est reprise en 1881 par ses fils Carl et Adolf Weber. En 1907, Carl Weber lègue l'entreprise à son fils Eugen et à son gendre Emil Ratzky dont le fils, Hermann Ratzky, reprendra ensuite la fonderie. Le programme de fontes sera transféré à la D. Stempel AG de Francfort-sur-le-Main et à Johannes Wagner d'Ingolstadt. – *Polices :* Rund-Grotesk (1931), Schadow-Antiqua (1938), Delphin (1951), Trump Mediaeval (1954), Codex (1956).

**Weidemann,** Kurt – né le 15. 12. 1922 à Eichmedien, Allemagne (aujourd'hui en Pologne) – *typographe, graphiste maquettiste, auteur, enseignant* – 1950–1952, apprentissage de composition typographique à Lubeck. 1953–1955, études de graphisme appliqué à l'édition et de typographie à la Staatliche Akademie der Bildenden Künste (beaux-arts) de Stuttgart. Exerce comme graphiste maquettiste, conseiller publicitaire et rédacteur indépendant à partir de 1955. Commanditaires : COOP, Zeiss, la Merck AG, la Mercedes-Benz AG, la Daimler-Benz

CODEX

Die «CODEX» ist eine zum Druck verwandelte Handschrift, der man das Zügige, Hingeworfene, Improvisierte in jeder Figur anmerkt. Sie ist nicht statisch, sondern dynamisch, sie nimmt das Auge mit sich auf den Weg. Die «CODEX» wurde nach Zeichnung von Professor G. Trump geschnitten

C. E. Weber Schriftgießerei Stuttgart

Weber 1956 Codex

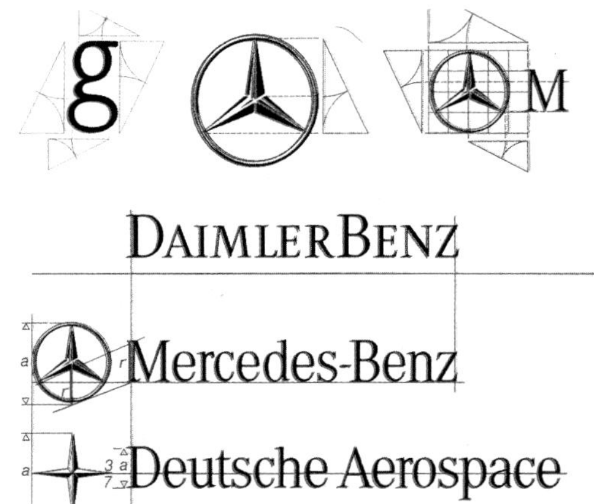


Weidemann 1988–91 Logos

Weidemann 1988 Logo

| | n | **n** | **n** | | *n* |
|---|---|---|---|---|---|
| a | b | e | f | g | i |
| o | r | s | t | y | z |
| A | B | C | E | G | H |
| M | O | R | S | X | Y |
| 1 | 2 | 4 | 6 | 8 | & |

Weidemann 1983 Weidemann

| n | n | **n** | | | *n* |
|---|---|---|---|---|---|
| a | b | e | f | g | i |
| o | r | s | t | y | z |
| A | B | C | E | G | H |
| M | O | R | S | X | Y |
| 1 | 2 | 4 | 6 | 8 | & |

Weidemann 1990 Corporate A

| n | n | **n** | **n** | | *n* |
|---|---|---|---|---|---|
| a | b | e | f | g | i |
| o | r | s | t | y | z |
| A | B | C | E | G | H |
| M | O | R | S | X | Y |
| 1 | 2 | 4 | 6 | 8 | & |

Weidemann 1990 Corporate S

zine. Publishes "The Dolphin, a Journal of the Making of Books", issued in book form by the Limited Editions Club in New York from 1932 onwards. Designs books for the Watch Hill Press, Morrill Press and for W. E. Rudge. 1937–39: production manager for Oxford University Press in New York. – *Fonts:* Arrighi (1925), Arrighi italic (1929). – *Publication:* "Bruce Rogers, designer of books", New York 1926.

**Weber,** C. E. – *type foundry* – 1856: Christian Emil Weber opens a type foundry in Stuttgart. 1881: his sons, Carl and Adolf Weber, assume responsibility for the company. 1907: Carl Weber hands the company over to his son, Eugen, and his son-in-law, Emil Ratzky. Ratzky's son, Hermann Ratzky, takes charge of the company. The foundry type program is taken over by D. Stempel AG in Frankfurt am Main and by Johannes Wagner in Ingolstadt. – *Fonts:* Rund-Grotesk (1931), Schadow-Antiqua (1938), Delphin (1951), Trump Mediaeval (1954) and Codex (1956).

**Weidemann,** Kurt – b. 15. 12. 1922 in Eichmedien, Germany (today Poland) – *typographer, graphic designer, author, teacher* – 1950–52: trains as a typesetter in Lübeck. 1953–55: studies book graphics and typography at the Staatliche Akademie der Bildenden Künste in Stuttgart. 1955: starts freelancing as a graphic designer, advertising consultant and copywriter. Clients include COOP, Zeiss, Merck AG, Mercedes-Benz AG, Daimler-Benz AG, Deutsche Aerospace AG and Deutsche Bahn AG. Designs

Daimler-Benz AG, die Deutsche Aerospace AG und die Deutsche Bahn AG. Als Buchgestalter für Verlage wie Büchergilde Gutenberg, Ullstein, Propyläen, Ernst Klett und Thieme tätig. 1955–64 Schriftleiter der Fachzeitschrift „Der Druckspiegel". 1965–85 Professor an der Staatlichen Akademie der Bildenden Künste in Stuttgart. Seit 1983 Lehrtätigkeit an der Wissenschaftlichen Hochschule für Unternehmensführung in Koblenz. Seit 1987 Berater für das Erscheinungsbild des Daimler-Benz-Konzerns. 1990 Überarbeitung des Erscheinungsbildes der Porsche AG. Seit 1991 Lehrtätigkeit an der Hochschule für Gestaltung im Zentrum für Kunst- und Medientechnologie in Karlsruhe. Zahlreiche Auszeichnungen, u. a. der „Lucky Strike Designer Award" der Raymond-Loewy-Stiftung (1995) und der Verdienstorden der Bundesrepublik Deutschland (1996). – *Schriftentwürfe:* Biblica (1979), ITC Weidemann (1983), Corporate A. S. E. (1985–89). – *Publikationen u. a.:* „Wo der Buchstabe das Wort führt", Stuttgart 1995; „Wortarmut", Karlsruhe 1995. Uta Brandes „Kurt Weidemann. Das Nachbild auf der Netzhaut", Göttingen 1995.

**Weiner,** Lawrence – geb. 10. 2. 1940 in New York, USA. – *Künstler* – Studium der Philosophie und Literatur am Hunter College in New York. Als Künstler Autodidakt. Seit 1960 zahlreiche Einzel- und Gruppenausstellungen, Installationen mit typographischen Elementen und Texten. Seit 1965 Entwurf zahlreicher Plakate für eigene und andere Ausstellungen und Events. 1972 Teilnahme an der do-

FLAGS FOLDED
OUT OF SIGHT
OF THE BARRICADE
1789 (1989)
FAHNEN
AUSSER SICHT
DER BARRIKADEN
GEFALTET

Weiner 1989 Poster

VIEWS FROM ABROAD
EIN EUROPÄISCHER BLICK AUF
DIE AMERIKANISCHE KUNST
DAS WHITNEY MUSEUM OF AMERICAN ART NEW YORK
ZU GAST IM
MUSEUM FÜR MODERNE KUNST
FRANKFURT AM MAIN
31 JANUAR – 4 MAI 1997

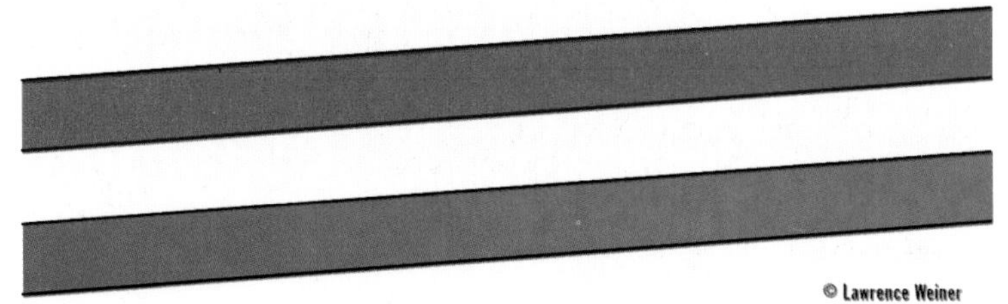


Weiner 1997 Invitation Card

Weingart 1973 Cover

AG, la Deutsche Aerospace AG et la Deutsche Bahn AG. Maquettes de livres pour plusieurs maisons d'édition dont la Büchergilde Gutenberg, Ullstein, Propyläen, Ernst Klett et Thieme. 1955–1964, directeur de publication de la revue de l'industrie du livre «Der Druckspiegel». 1965–1985, professeur à la Staatliche Akademie der Bildenden Künste de Stuttgart. Enseigne à l'école supérieure de management de Coblence depuis 1983. Conseiller de l'identité du groupe Daimler-Benz depuis 1987. Remanie l'identité de Porsche AG en 1990. Enseigne à la Hochschule für Gestaltung (école supérieure de design) du Zentrum für Kunst- und Medientechnologie (centre d'art et des technologies des médias) de Karlsruhe depuis 1991. Nombreuses distinctions, dont le «Lucky Strike Designer Award» de la fondation Raymond Loewy (1995), et l'Ordre du mérite de la République fédérale (1996). – *Polices:* Biblica (1979), ITC Weidemann (1983), Corporate A. S. E (1985–1989). – *Publications, sélection:* «Wo der Buchstabe das Wort führt», Stuttgart 1995; «Wortarmut», Karlsruhe 1995. Uta Brandes «Kurt Weidemann. Das Nachbild auf der Netzhaut», Göttingen 1995.

**Weiner,** Lawrence – né le 10. 2. 1940 à New York, Etats-Unis – *artiste* – études de philosophie et de littérature au Hunter College de New York. Artiste autodidacte. Nombreuses expositions collectives et personnelles à partir de 1960, présente des installations comportant des éléments typographiques et des textes. Depuis 1965, dessine des affiches pour des expositions, les siennes comme d'autres, ainsi que pour des manifestations. Participe à la «documenta V» de Kassel en 1972 et la 36e Biennale de Venise. Dès 1972, les textes qui figurent lors des expositions sont inscrits en capitales

books for publishers who include the Büchergilde Gutenberg, Ullstein, Propyläen, Ernst Klett and Thieme. 1955–64: type manager of the specialist journal "Der Druckspiegel". 1965–85: professor at the Staatliche Akademie der Bildenden Künste in Stuttgart. 1983: starts teaching at the Wissenschaftliche Hochschule für Unternehmensführung in Koblenz. From 1987 onwards: corporate identity consultant to Daimler-Benz. 1990: reworks Porsche AG's corporate identity. 1991: starts teaching at the Hochschule für Gestaltung at the Zentrum für Kunst- und Medientechnologie in Karlsruhe. Weidemann has won numerous awards, including the Lucky Strike Designer Award from the Raymond-Loewy-Stiftung (1995) and Germany's Order of Merit (1996). – *Fonts:* Biblica (1979), ITC Weidemann (1983), Corporate A.S.E. (1985–89). – *Publications include:* "Wo der Buchstabe das Wort führt", Stuttgart 1995; "Wortarmut", Karlsruhe 1995. Uta Brandes "Kurt Weidemann. Das Nachbild auf der Netzhaut", Göttingen 1995.

**Weiner,** Lawrence – b. 10. 2. 1940 in New York, USA – *artist* – Studied philosophy and literature at Hunter College in New York. Self-taught artist. From 1960 onwards: numerous solo and joint exhibitions of his work are held. Produces installations containing typographic elements and texts. From 1965 onwards: designs numerous posters for his own and other exhibitions and events. 1972: takes part in documenta V in Kassel and in the 36th Biennale in Venice. From 1972 onwards: exhibitions display his texts on

cumenta V in Kassel und auf der 36. Biennale in Venedig. Seit 1972 werden bei Ausstellungen seine Texte in Versalbuchstaben auf Wänden angebracht. 1979 Verleihung des Laura-Slobe-Memorial-Prize in Chicago. 1988 Retrospektiv-Ausstellung im Stedelijk Museum in Amsterdam. – *Publikationen u. a.:* „Statements", New York 1968; „10 Words", Paris 1971; „Green As Well As Blue As Well As Red", London 1972; „Once Upon A Time", Mailand 1973. Rudi H. Fuchs „Lawrence Weiner", Eindhoven 1976; Benjamin H. Buchloh „Posters November 1965–April 1986; Lawrence Weiner", Halifax 1986. Dieter Schwarz (Hrsg.) „Lawrence Weiner. Books 1968–1989", Köln 1989.

**Weingart,** Wolfgang – geb. 6. 2. 1941 in Konstanz, Deutschland. – *Typograph, Autor, Lehrer* – 1960 Schriftsetzerlehre in Stuttgart, als Gestalter Autodidakt. Seit 1968 Lehrer für Typographie an der Schule für Gestaltung in Basel als Nachfolger von Emil Ruder. Unterrichtet an zahlreichen Schulen, u. a. an der Yale University und am Philadelphia College of Art & Design sowie jährlich beim „Yale Summer Program in Graphic Design" in Brissago. 1968–85 Mitarbeiter der Zeitschrift „Typographische Monatsblätter", Begründer der periodisch erscheinenden Beilagen „TM/communication" und „Wege zur Typographie". Seit 1972 zahlreiche Vorträge und Vortragsreisen weltweit. Zahlreiche Artikel über seine Arbeit in internationalen Gestaltungszeitschriften. Durch seine Arbeit und Lehre gehört Weingart seit den sechziger Jahren zu den typographisch einflußreichsten Gestaltern

Weingart 1981 Poster

Weingart 1981 Poster

Weingart 1984 Poster

sur les murs. Lauréat du Laura-Slobe-Memorial-Prize de Chicago en 1979. Exposition rétrospective au Stedelijk Museum d'Amsterdam en 1988. – *Publications, sélection :* «Statements», New York 1968; «10 Words», Paris 1971; «Green As Well As Blue As Well As Red», Londres 1972; «Once Upon a Time», Milan 1973. Rudi H. Fuchs «Lawrence Weiner», Eindhoven 1976; Benjamin H. Buchloh «Posters November 1965 – April 1986. Lawrence Weiner», Halifax 1986; Dieter Schwarz (éd.) «Lawrence Weiner. Books 1968–1989», Cologne 1989.

**Weingart,** Wolfgang – né le 6. 2. 1941 à Constance, Allemagne – *typographe, auteur, enseignant* – 1960, apprentissage de composition typographique à Stuttgart, designer autodidacte. Enseigne la typographie à l'école de design de Bâle où il succède à Emil Ruder. Enseigne dans de nombreux établissements, par ex. à la Yale University, au Philadelphia College of Art & Design, participe tous les ans au «Yale Summer Program in Graphic Design» de Brissago. Collaborateur de la revue «Typographische Monatsblätter» de 1968 à 1985, crée les cahiers périodiques «TM/communication» et «Wege zur Typographie». Nombreuses conférences et tournées de conférences internationales à partir de 1972. Nombreux articles sur son travail dans des revues internationales spécialisées dans le design. Depuis les années soixante, le travail et l'enseignement de Weingart font de lui l'un des typographes et créateurs les plus

walls in capital letters. 1979: is awarded the Laura Slobe Memorial Prize in Chicago. 1988: retrospective exhibition of his work at the Stedelijk Museum in Amsterdam. – *Publications include:* "Statements", New York 1968; "10 Words", Paris 1971; "Green As Well As Blue As Well As Red", London 1972; "Once Upon A Time", Milan 1973. Rudi H. Fuchs "Lawrence Weiner", Eindhoven 1976; Benjamin H. Buchloh "Posters November 1965–April 1986. Lawrence Weiner", Halifax 1986; Dieter Schwarz (ed.) "Lawrence Weiner. Books 1968–1989", Cologne 1989.

**Weingart,** Wolfgang – b. 6. 2. 1941 in Constance, Germany – *typographer, author, teacher* – 1960: apprenticeship as a typesetter in Stuttgart. Self-taught designer. 1968: starts teaching typography at the Schule für Gestaltung in Basle as successor to Emil Ruder. Teaches at various universities and colleges, including Yale University and the Philadelphia College of Art & Design, and once a year at the Yale Summer Program in Graphic Design in Brissago. 1968–85: works on "Typographische Monatsblätter" magazine and initiates the regular "TM/communication" and "Wege zur Typographie" supplements. From 1972 onwards: gives numerous talks and undertakes lecture tours all over the world. Numerous articles on his work are published in international design periodicals. Since the 1960s, Weingart has been one of the designers

weltweit. – *Publikationen u. a.:* „Kinder 1 Orient Zeichen", Stuttgart 1964; „Projekte, Projects", Niederteufen 1979.

**Weiß,** Emil Rudolf – geb. 12. 10. 1875 in Lahr, Deutschland, gest. 7. 11. 1942 in Meersburg, Deutschland. – *Grafiker, Maler, Typograph, Schriftentwerfer, Dichter, Lehrer* – 1893–96 Studium an der Kunstakademie Karlsruhe, 1896–97 an der Académie Julian in Paris, 1897–1903 an den Kunstakademien Karlsruhe und Stuttgart. 1894 Beiträge für die Zeitschrift „Pan". Seit 1895 Arbeiten für zahlreiche Verlage, u. a. Cotta, Suhrkamp, S. Fischer, Cassirer, Insel und Rowohlt. Veröffentlicht 1895–1900 fünf Bände mit eigenen Gedichten. 1903 Lehrer an der Malschule des Folkwang-Museums in Essen. 1907–33 Professor an der Unterrichtsanstalt des Königlichen Kunstgewerbemuseums in Berlin. 1910 Wandmalereien für Bruno Pauls Gartenhaus auf der Weltausstellung in Brüssel. 1922–37 Mitglied der Preußischen Akademie der Künste. 1933 Versetzung in den vorzeitigen Ruhestand durch die Nationalsozialisten. 1941 dekorative Malerei für Gebäude in Freiburg im Breisgau. und Dessau. – *Schriftentwürfe:* Weiß Fraktur (1909), Weiß Antiqua (1928), Weiß Kapitale mager (1931), Weiß Lapidar mager (1931), Weiß Kapitale kräftig (1935), Neue Weiß-Fraktur (1935), Weiß Gotisch (1936), Weiß Rundgotisch (1937), Weiß Rundgotisch Initialen (1939). – *Publikationen u. a.:* „Künstler und Buchkünstler gestern, heute und morgen", Frankfurt 1931. Julius Zeitler (Hrsg.) „Sonderheft Emil Rudolf Weiß" 5/1922,

influents au monde. – *Publications, sélection:* «Kinder 1 Orient Zeichen», Stuttgart 1964; «Projekte, Projects», Niederteufen 1979.

**Weiß,** Emil Rudolf – né le 12. 10. 1875 à Lahr, Allemagne, décédé le 7. 11. 1942 à Meersburg, Allemagne – *graphiste, peintre, typographe, concepteur de polices, poète, enseignant* – 1893–1896, études à la Kunstakademie (beaux-arts) de Karlsruhe, à l'Académie Julian de 1896 à 1897 à Paris, puis aux académies de Karlsruhe et de Stuttgart de 1897 à 1903. Articles pour la revue «Pan» en 1894. Travaux pour plusieurs maisons d'édition à partir de 1895, entre autres pour Cotta, Suhrkamp, S. Fischer, Cassirer, Insel et Rowohlt. Publie cinq recueils de poèmes entre 1895 et 1900. Enseigne à la Malschule du Folkwang-Museum d'Essen en 1903. Professeur à l'Unterrichtsanstalt des Königlichen Kunstgewerbemuseums (institut d'enseignement du musée royal des arts décoratifs) de Berlin de 1907 à 1933. Réalise, en 1910, des fresques pour le pavillon de Bruno Paul lors de l'exposition universelle de Bruxelles. Membre de l'académie prussienne des beaux-arts de 1922 à 1937. En 1933, les national-socialistes le forcent à la retraite anticipée. Peintures décoratives pour des bâtiments à Fribourg-en-Brisgau et à Dessau en 1941. – *Polices:* Weiß Fraktur (1909), Weiß Antiqua (1928), Weiß capitale maigre (1931), Weiß lapidaire maigre (1931), Weiß capitale bold (1935), Neue Weiß-Fraktur (1935), Weiß gothique (1936), Weiß Rundgotisch (1937), Weiß Rundgotisch Initialien (1939). – *Publications, sélection:* «Künstler und Buchkünstler gestern, heute und morgen», Francfort 1931. Julius Zeitler (éd.) «Sonderheft Emil Rudolf Weiß» 5/1922, Archiv für Buchgewerbe Leipzig; Herbert Reichner «E. R. Weiß zum fünfzigsten Ge-

Weiß 1899 Poster

Weiß 1912 Cover

Die erste Kursivschrift besaß Aldus Manutius in Venedig. Die meisten Stempelschneider folgten diesem Vorbild, aber schon frühzeitig gab es auch deutsche Kursivschnitte von ausgesprochener Eigenart, wie die formenreiche Schrift, die der aus Schlesien stammende Meister Hieronymus Vietor um das Jahr 1531 verwendete. Ein Meisterwerk der neueren Schriftkunst ist dieser Type gegenübergestellt, die schöne

Weiß-Kursiv

BAUERSCHE GIESSEREI
Frankfurt am Main

Weiß 1940 Weiß-Kursiv

who has exerted great influence on typography all over the world though his work and teaching. – *Publications include:* "Kinder 1 Orient Zeichen", Stuttgart 1964; "Projekte, Projects", Niederteufen 1979.

**Weiß,** Emil Rudolf – b. 12. 10. 1875 in Lahr, Germany, d. 7. 11. 1942 in Meersburg, Germany – *graphic artist, painter, typographer, type designer, poet, teacher* – 1893–96: studies at the art academy in Karlsruhe, from 1896–97 at the Académie Julian in Paris and from 1897–1903 at the art academies in Karlsruhe and Stuttgart. 1894: contributes to "Pan" magazine. 1895: starts producing work for various publishing houses, including Cotta, Suhrkamp, S. Fischer, Cassirer, Insel and Rowohlt. 1895–1900: publishes five volumes of his poems. 1903: teaches at the Folkwang-Museum art school in Essen. 1907–33: professor at the Unterrichtsanstalt des Königlichen Kunstgewerbemuseum in Berlin. 1910: does the murals for Bruno Paul's summer house at the world exhibition in Brussels. 1922–37: member of the Preußische Akademie der Künste. 1933: the National Socialists force him to take early retirement. 1941: does decorative painting for buildings in Freiburg im Breisgau and Dessau. – *Fonts:* Weiß Fraktur (1909), Weiß Antiqua (1928), Weiß Kapitale mager (1931), Weiß Lapidar mager (1931), Weiß Kapitale kräftig (1935), Neue Weiß-Fraktur (1935), Weiß Gotisch (1936), Weiß Rundgotisch (1937), Weiß Rundgotisch Initialen (1939). – *Publications include:* "Künstler und Buchkünstler gestern, heute und

Archiv für Buchgewerbe Leipzig; Herbert Reichner „E. R. Weiß zum fünfzigsten Geburtstag", Leipzig 1925; Kurt Christians, Richard von Sichowski „Emil Rudolf Weiß über Buchgestaltung", Hamburg 1969.

**Werkman,** Hendrik Nicolaas – geb. 29. 4. 1882 in Leens, Niederlande, gest. 10. 4. 1945 in Bakkeveen, Niederlande. – *Drucker, Künstler, Verleger* – Beschäftigt sich seit 1899 mit Fotografie. 1900–03 Arbeit in einer Druckerei mit Verlag und Buchhandlung in Sappemeer und in anderen Betrieben. 1903–07 als Journalist des „Groninger Dagblad" tätig und danach für den „Nieuwe Groninger Courant". 1908–23 Eröffnung und Leitung einer Druckerei in Groningen. 1917 Beginn der malerischen Arbeit. Mitglied der Künstlergruppe „De Ploeg" in Groningen. 1921–22 Herausgabe der Zeitschrift „Blad voor Kunst". 1923 Liquidierung seiner Druckerei, Eröffnung einer neuen, kleineren Druckerei. Herausgabe der ersten Nummer seiner Publikationsfolge „The next call", von der bis 1926 neun Ausgaben erscheinen. Gestaltung und Produktion der ersten Drucke mit Handpresse und Satzmaterial. 1929 entstehen grafische Blätter durch direktes Arbeiten mit der Farbwalze auf Papier und Verwendung der Stempeltechnik. 1939 Ausstellung seiner Arbeit in Amsterdam durch Vermittlung von Willem Sandberg. 1940 Beginn einer Serie von vierzig Ausgaben der Publikation „Blauwe Schuit", die zum Widerstand gegen die deutschen Okkupanten aufruft. 1945 Verhaftung durch die deutschen Okkupationstruppen und

Weiß 1937 Weiß-Rundgotisch

| | n | n | n | | n |
|---|---|---|---|---|---|
| a | b | e | f | g | i |
| o | r | s | t | y | z |
| A | B | C | E | G | H |
| M | O | R | S | X | Y |
| 1 | 2 | 4 | 6 | 8 | & |

Weiß 1928 Weiß Antiqua

Werkman 1925 Cover

Werkman 1926 Calendar

Werkman 1927 Cover

burtstag», Leipzig 1925; Kurt Christians, Richard von Sichowski «Emil Rudolf Weiß über Buchgestaltung», Hambourg 1969.

**Werkman,** Hendrik Nicolaas – né le 29. 4. 1882 à Leens, Pays-Bas, décédé le 10. 4. 1945 à Bakkeveen, Pays-Bas – *imprimeur, artiste, éditeur* – commence à se consacrer à la photographie en 1899. Travaille dans une maison édition et imprimerie de Sappemeer de 1900 à 1903 ainsi que dans plusieurs autres entreprises. Journaliste au «Groninger Dagblad», puis pour le «Nieuwe Groninger Courant» de 1903 à 1907. Ouvre une imprimerie à Groningue en 1908 et la dirige jusqu'en 1923. Commence à peindre en 1917. Membre du groupe d'artistes «De Ploeg» à Groningue. Publie la revue «Blad voor Kunst» en 1921 et 1922. Liquide son imprimerie en 1923 pour en ouvrir une autre plus petite. Edition du premier numéro d'une série de publications appelée «The next call», neuf ouvrages paraîtront jusqu'en 1926. Création et production des premières planches à la presse manuelle utilisant son matériel de composition. En 1929, il réalise des planches graphiques par application directe du rouleau encreur sur le papier et utilise la technique du poinçon. En 1939, Willem Sandberg organise une exposition de ses travaux à Amsterdam. En 1940, il commence à éditer une série de quarante publications, appelées «Blauwe Schuit», qui appellent à la résistance contre l'occupant allemand. Arrêté et fusillé par les Allemands en 1945. – *Publications, sé-*

morgen", Frankfurt 1931. Julius Zeitler (ed.) "Sonderheft Emil Rudolf Weiß" 5/1922, Archiv für Buchgewerbe Leipzig; Herbert Reichner "E. R. Weiß zum fünfzigsten Geburtstag", Leipzig 1925; Kurt Christians, Richard von Sichowski "Emil Rudolf Weiß über Buchgestaltung", Hamburg 1969.

**Werkman,** Hendrik Nicolaas – b. 29. 4. 1882 in Leens, The Netherlands, d. 10. 4. 1945 in Bakkeveen, The Netherlands – *printer, artist, publisher* – 1899: starts working with photography. 1900–03: works for a printing workshop which has its own publishers and bookstore in Sappemeer and for various other businesses. 1903–07: works as a journalist for the "Groninger Dagblad" and later for the "Nieuwe Groninger Courant". 1908–23: opens and runs a printing workshop in Groningen. 1917: starts painting. Member of the De Ploeg artists' group in Groningen. 1921–22: publishes "Blad voor Kunst" magazine. 1923: liquidates his printing works and opens a new, smaller printing workshop. Produces the first number of his "The next call" series, of which 9 issues are published up to 1926. Designs and produces the first prints using a hand-press and typesetting material. 1929: produces graphic prints working with the inking roller directly on paper and using the stamping technique. 1939: Willem Sandberg helps Werkman get his work exhibited in Amsterdam. 1940: starts producing a series of 40 issues of the "Blauwe Schuit" publication which calls for resistance to The Netherlands' German oppressors. 1945: is ar-

Erschießung. – *Publikationen u. a.:* F. R. A. Henkels „Logboek van de Blauwe Schuit", Amsterdam 1946; HAP Grieshaber u. a. „Hommage à Werkman", Stuttgart, New York 1958; H. van Straaten „Hendrik Nicolaas Werkman, de drukker van het Paradijs", Amsterdam 1963; F. Müller, P. F. Althaus, J. Martinet „H. N. Werkman", Teufen 1967; D. Dooijes „Hendrik Werkman", Amsterdam 1970; J. Martinet „Werkman druksels en gebruiksgrafiek", Amsterdam 1977.

**why not associates** – *Grafik-Design-Studio* – 1987 Gründung des Studios in London durch Andrew Altmann, David Ellis und Howard Greenhalgh (der 1992 aus dem Studio ausscheidet und „why not films" gründet). Andrew Altmann – geb. 15. 9. 1962 in Warrington, England. – 1985 Abschluß des Studiums an der St Martins School of Art in London. 1985–87 Studium am Royal College of Art in London. David Ellis – geb. 6. 6. 1962 in Hove, England. – Studium am West Surrey College of Art & Design und an der St Martins School of Art in London. 1984–87 Studium am Royal College of Art in London. Das Studio entwirft für Auftraggeber aus dem Kultur- und Wirtschaftsbereich. Zahlreiche Vorträge und Ausstellungen, u. a. an der Hochschule für Gestaltung Offenbach (1991), GGG Galerie Tokio (1993), Maison du Livre Villeurbanne (1993), St Martins School of Art London (1995). – *Schriftentwurf:* Doddy (1993). – *Publikation u. a.:* „Steelworks" (Hrsg.), London 1990.

**Widmer,** Jean – geb. 31. 3. 1929 in Frauenfeld, Schweiz. – *Grafik-Designer, Art*

*lection:* F. R. A. Henkels «Logboek van de Blauwe Schuit», Amsterdam 1946; HAP Grieshaber et autres «Hommage à Werkman», Stuttgart, New York 1958; H. van Straaten «Hendrik Nicolaas Werkman, de drukker van het Paradijs», Amsterdam 1963; F. Müller, P. F. Althaus, J. Martinet «H. N. Werkman», Teufen 1967; D. Dooijes «Hendrik Werkman», Amsterdam 1970; J. Martinet «Werkman druksels en gebruiksgrafiek», Amsterdam 1977.

**why not associates** – *agence de graphisme et de design* – l'atelier est fondé à Londres en 1987 par Andrew Altmann, David Ellis et Howard Greenhalgh (qui quittera l'entreprise en 1992 pour fonder «why not films»). Andrew Altmann – né le 15. 9. 1962 à Warrington, Angleterre – diplôme de la St Martins School of Art de Londres en 1985. Etudes au Royal College of Art de Londres de 1985 à 1987. David Ellis – né le 6. 6. 1962 à Hove, Angleterre – études au West Surrey College of Art & Design et à la St Martins School of Art de Londres. 1984–1987, études au Royal College of Art de Londres. L'agence travaille pour des institutions culturelles et pour l'industrie. Nombreuses conférences et expositions, entre autres à la Hochschule für Gestaltung (école supérieure de design) d'Offenbach (1991), à la GGG Galerie de Tokyo (1993), à la Maison du Livre de Villeurbanne (1993), à la St Martins School of Art de Londres (1995). – *Police:* Doddy (1993). – *Publication, sélection:* «Steelworks» (éd.), Londres 1990.

**Widmer,** Jean – né le 31. 3. 1929 à Frauenfeld, Suisse – *graphiste maquettiste, directeur artistique, typographe, enseignant* – 1945–1950, études à la Kunstgewerbeschule (école des arts décoratifs) de Zurich, puis à l'école des beaux-arts de Paris, de 1952 à 1955. Directeur artistique de l'agence de publicité SNIP, à Paris de

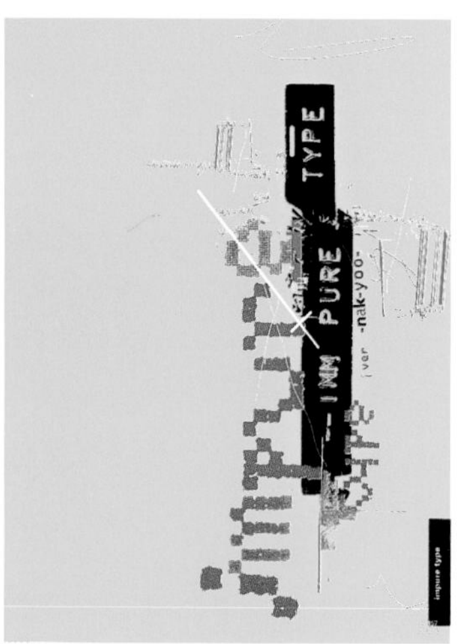

why not associates 1991 Page

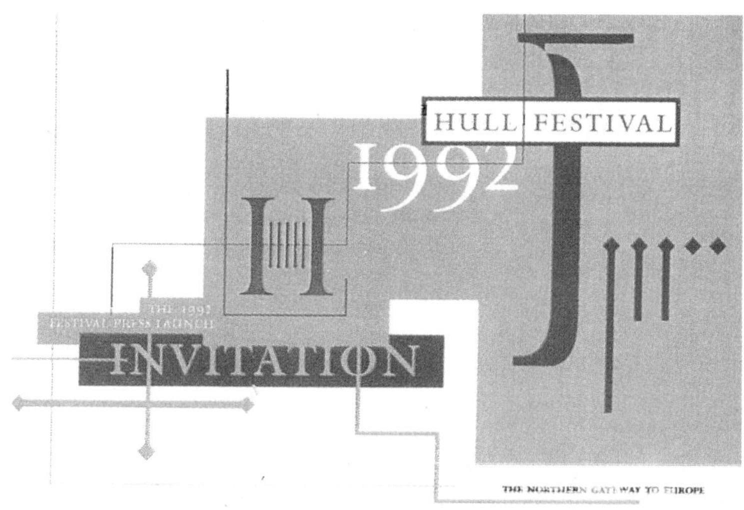


why not associates 1992 Invitation

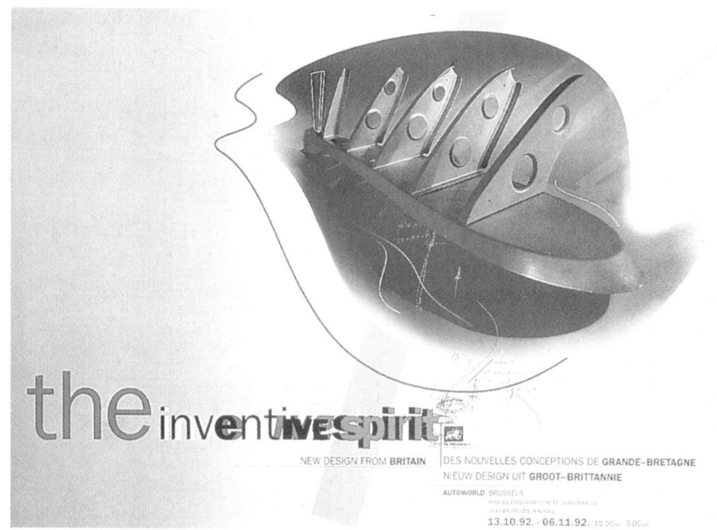


why not associates 1992 Poster

Widmer 1970 Poster

rested and shot by the German occupying forces. – *Publications include:* F. R. A. Henkels "Logboek van de Blauwe Schuit", Amsterdam 1946; HAP Grieshaber et al "Hommage à Werkman", Stuttgart, New York 1958; H. van Straaten "Hendrik Nicolaas Werkman, de drukker van het Paradijs", Amsterdam 1963; F. Müller, P. F. Althaus, J. Martinet "H. N. Werkman", Teufen 1967; D. Dooijes "Hendrik Werkman", Amsterdam 1970; J. Martinet "Werkman druksels en gebruiksgrafiek", Amsterdam 1977.

**why not associates** – *graphic design studio* – 1987: the studio is founded in London by Andrew Altmann, David Ellis and Howard Greenhalgh (who later leaves the studio in 1992 to form his own business, why not films). Andrew Altmann – b. 15. 9. 1962 in Warrington, England – 1985: completes his studies at the St Martins School of Art in London. 1985–87: studies at the Royal College of Art in London. David Ellis – b. 6. 6. 1962 in Hove, England – Studied at the West Surrey College of Art & Design and at the St Martins School of Art in London. 1984–87: studies at the Royal College of Art in London. The studio produces designs for clients from the arts and business sector. The studio has held numerous talks and exhibitions, including at the Hochschule für Gestaltung in Offenbach (1991), the GGG Galerie in Tokyo (1993), the Maison du Livre in Villeurbanne (1993) and at the St Martins School of Art in London (1995). – *Font:* Doddy (1993). – *Publication:* "Steelworks" (ed.), London 1990.

**Widmer,** Jean – b. 31. 3. 1929 in Frauen-

*Director, Typograph, Lehrer* – 1945–50 Studium an der Kunstgewerbeschule in Zürich, 1952–55 an der Ecole des Beaux-Arts in Paris. 1955–59 Art Director in der Werbeagentur SNIP in Paris, 1959–62 des Warenhauses der Galeries Lafayette, 1962–70 der Zeitschrift „Jardin des Modes". 1970 Gründung des Studios „Visual Design" in Paris. Zahlreiche Aufträge für Erscheinungsbilder von Museen: Centre Georges Pompidou (1974), Musée d'Orsay (1983), Musée National d'Art Moderne (1985), Cité de la Musique (1988). Gestaltung des Erscheinungsbildes der Autoroutes du Sud de la France mit über 500 Pictogrammen. Gestaltung der Leitsysteme für das Schloß Versailles und den Louvre. Unterrichtet an der Ecole Nationale Supérieure des Arts Décoratifs in Paris. – *Publikation:* „Jean Widmer", Paris 1991.

**Wiemeler,** Ignaz – geb. 3. 10. 1895 in Ibbenbüren, Deutschland, gest. 25. 5. 1952 in Hamburg, Deutschland. – *Buchbinder, Lehrer* – 1912–13 Buchbinderlehre in Osnabrück. 1914, 1916–21 Studium an der Kunstgewerbeschule in Hamburg. 1921–25 Lehrer an der Kunstgewerbeschule in Offenbach. 1923 Gründungsmitglied des Bundes „Meister der Einbandkunst" in Leipzig, der sich mit der Zusammenarbeit von Druckern, Verlegern, Museen und Bibliotheken beschäftigt. 1925–45 Professor an der Akademie für Graphische Künste und Buchgewerbe in Leipzig. 1935 Ausstellung von fünfzig Einbänden Wiemelers im Museum of Modern Art in New York. 1946–52 Professor an der Landeskunstschule Ham-

why not associates 1989 Poster

1955 à 1959, des «Galeries Lafayette» de 1959 à 1962, puis de la revue «Jardin des Modes» de 1962 à 1970. Fonde l'agence «Visual Design» à Paris en 1970. De nombreux musées lui demandent de créer leur identité: Centre Georges Pompidou (1974), Musée d'Orsay (1983), Musée National d'Art Moderne (1985), Cité de la Musique (1988). Conception de l'identité des autoroutes du Sud de la France pour lesquelles il dessine plus de 500 pictogrammes. Conception du système d'orientation du château de Versailles et du Louvre. Enseigne à l'Ecole Nationale Supérieure des Arts Décoratifs de Paris. – *Publication:* «Jean Widmer», Paris 1991.

**Wiemeler,** Ignaz – né le 3. 10. 1895 à Ibbenbüren, Allemagne, décédé le 25. 5. 1952 à Hambourg, Allemagne – *relieur, enseignant* – 1912–1913, apprentissage de reliure à Osnabrück. 1914 et 1916–1921, études à la Kunstgewerbeschule (école des arts décoratifs) de Hambourg. 1921–1925, enseignant à la Kunstgewerbeschule (école des arts décoratifs) d'Offenbach. Membre fondateur du groupement des «Meister der Einbandkunst» (maîtres artisans relieurs) à Leipzig en 1923, une association qui s'occupe de la collaboration entre les imprimeurs, les éditeurs, les musées et les bibliothèques. 1925–1945, professeur à l'Akademie für Graphische Künste und Buchgewerbe (académie des arts graphiques et des arts du livre) de Leipzig. Exposition de 50 reliures de Wiemeler au Museum of Modern Art de New York en 1935. 1946–1952,

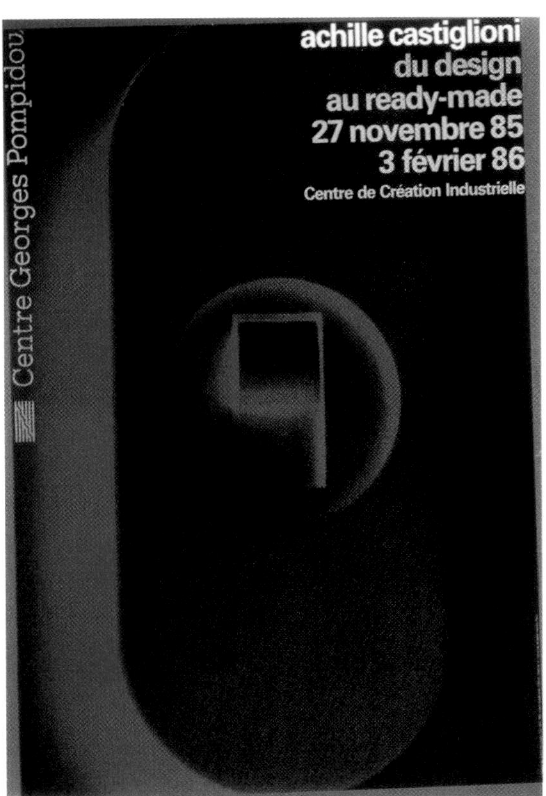

Widmer 1986 Poster

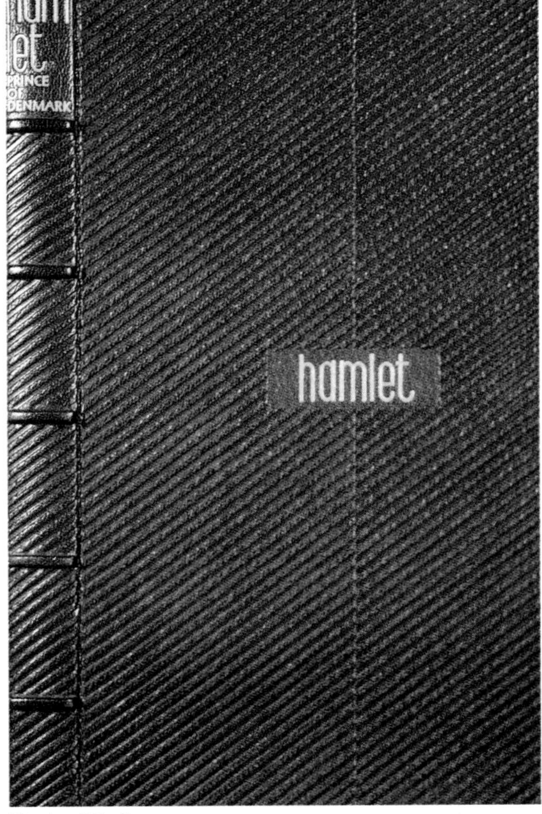

Wiemeler 1925 Cover

feld, Switzerland – *graphic designer, art director, typographer, teacher* – 1945–50: studies at the Kunstgewerbeschule in Zurich, then from 1952–55 at the Ecole des Beaux-Arts in Paris. 1955–59: art director for the SNIP advertising agency in Paris, from 1959–62 for Galeries Lafayette and from 1962–70 for "Jardin des Modes" magazine. 1970: opens the Visual Design studio in Paris. Is commissioned to design numerous corporate identities for museums: Centre Georges Pompidou (1974), Musée d'Orsay (1983), Musée National d'Art Moderne (1985) and Cité de la Musique (1988). Designs the corporate identity for the Autoroutes du Sud de la France with over five hundred pictograms. Designs signage systems for Versailles Palace and the Louvre. Has taught at the Ecole Nationale Supérieure des Arts Décoratifs in Paris. – *Publication:* "Jean Widmer", Paris 1991.

**Wiemeler,** Ignaz – b. 3. 10. 1895 in Ibbenbüren, Germany, d. 25. 5. 1952 in Hamburg, Germany – *bookbinder, teacher* – 1912–13: trains as a bookbinder in Osnabrück. 1914 and 1916–21: studies at the Kunstgewerbeschule in Hamburg. 1921–25: teaches at the Kunstgewerbeschule in Offenbach. 1923: founder member of the Meister der Einbandkunst association in Leipzig which concentrates on the cooperation between printers, publishers, museums and libraries. 1925–45: professor at the Akademie für Graphische Künste und Buchgewerbe in Leipzig. 1935: 50 of Wiemeler's book covers are exhibited at the Museum of Modern Art in New York. 1946–52: pro-

burg. 1972 Veranstaltung des Ignaz-Wiemeler-Symposiums im Gutenberg-Museum der Schweiz in Basel. – *Publikationen u. a.:* Helmut Presser „Ignaz Wiemeler, Buchbinder 1895–1952", Hamburg 1953; Kurt Londenberg „Ignaz Wiemeler, Werkverzeichnis", Hamburg 1990.

**Wiese,** Bruno Karl – geb. 7. 4. 1922 in Berlin, Deutschland. – *Grafik-Designer, Typograph, Lehrer* – 1945–49 Studium an der Staatlichen Hochschule für Bildende Künste in Berlin bei O. H. W. Hadank, 1949–54 Mitarbeiter in dessen grafischem Büro in Hamburg. Seit 1954 eigenes Grafik-Design-Büro in Hamburg. 1978–80 Dozent, 1980–87 Professor für Kommunikations-Design an der Fachhochschule in Kiel. Seit 1982 für das visuelle Programm der Kieler Woche (zusammen mit Fritz Seitz) verantwortlich. Seit 1996 im Kunstbeirat des Bundesministeriums für Post und Telekom. Zahlreiche Ausstellungen und Auszeichnungen. – *Publikation:* H. Köhne, F. Seitz „B. K. Wiese. Visual Design", Kiel 1987.

**Wijdeveld,** Hendricus Theodorus – geb. 4. 10. 1885 in Den Haag, Niederlande, gest. 1987 in Amsterdam, Niederlande. – *Architekt, Bühnenbildner, Verleger* – 1897–99 Lehre im Büro des Architekten J. van Straaten in Amsterdam. 1914 Eröffnung seines Architekturbüros. Entwurf und Bau von Landhäusern, Läden, Theatern und Arbeiterwohnungen. 1917 Gestaltung des Umschlags der Zeitschrift „Architectura". 1918–31 Herausgabe der Zeitschrift „Wendingen, maandblad voor bouwen en sieren" in Amsterdam. Neben seinen eigenen Umschlag-Entwürfen

professeur à la Landeskunstschule (école des beaux-arts) de Hambourg. Symposium Ignaz Wiemeler au Musée Gutenberg suisse, à Bâle en 1972. – *Publications, sélections:* Helmut Presser «Ignaz Wiemeler, Buchbinder 1895– 1952», Hambourg 1953; Kurt Londenberg «Ignaz Wiemeler, Werkverzeichnis», Hambourg 1990.

**Wiese,** Bruno Karl – né le 7. 4. 1922 à Berlin, Allemagne – *graphiste maquettiste, typographe, enseignant* – 1945–1949, études à la Staatliche Hochschule für Bildende Künste (école des beaux-arts) de Berlin, chez O. H. W. Hadank. Travaille à Hambourg pour l'atelier de Hadank de 1949 à 1954. Exerce dans son propre atelier de conception graphique et de design à Hambourg à partir de 1954. Maître de conférence de 1978 à 1980, professor de design et communication visuelle à la Fachhochschule de Kiel de 1980 à 1987. Responsable de la communication visuelle de la semaine de Kiel (avec Fritz Seitz) depuis 1982. Nombreuses expositions et distinctions. – *Publication:* H. Köhne, F. Seitz «B.K. Wiese. Visual Design», Kiel 1987.

**Wijdeveld,** Hendricus Theodorus – né le 4. 10. 1885 à La Haye, Pays-Bas, décédé en 1987 à Amsterdam, Pays-Bas – *architecte, scénographe, éditeur* – 1897–1899, apprentissage chez l'architecte J. van Straaten à Amsterdam. Ouvre son agence d'architecture en 1914. Dessine et construit des maisons de campagne, des boutiques, des théâtres et des logements ouvriers. Conçoit la couverture de la revue «Architectura» en 1917. Fonde, en 1918, la revue «Wendingen, maandblad voor bouwen en sieren» à Amsterdam, qu'il publie jusqu'en 1931. Outre les couvertures qu'il dessine, il publie également des travaux de Vilmos Huszár, El Lissitzky, Jan Toorop, etc. De 1919 à

Wiese 1986 Poster

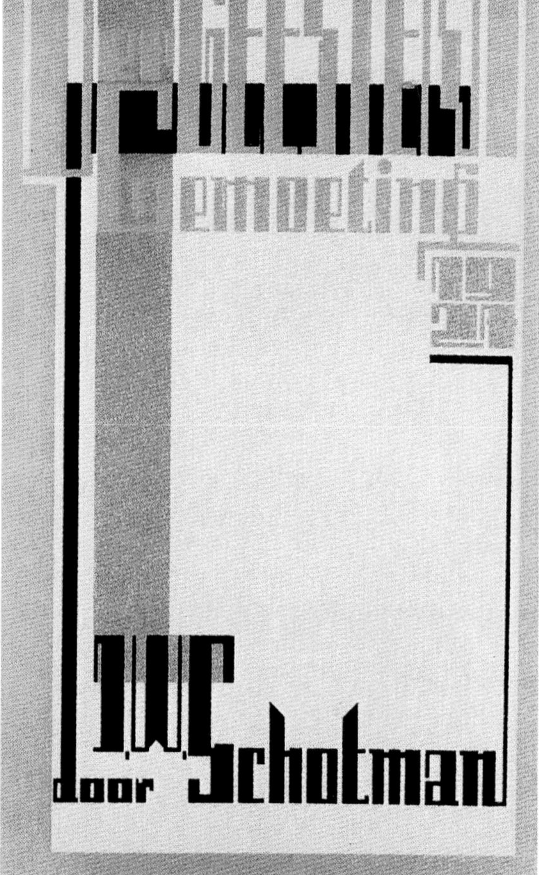

Wijdeveld 1927 Cover

Wiese 1988

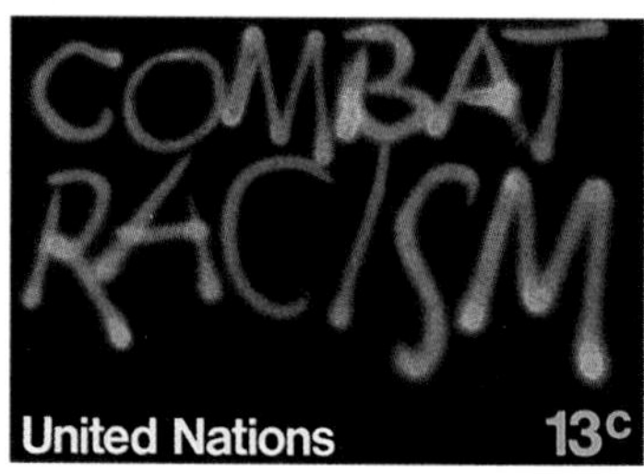

Wiese Stamp

fessor at the Landeskunstschule in Hamburg. 1972: Switzerland's Gutenberg Museum in Basle stages an Ignaz Wiemeler Symposium. – *Publications include:* Helmut Presser "Ignaz Wiemeler, Buchbinder 1895– 1952", Hamburg 1953; Kurt Londenberg "Ignaz Wiemeler, Werkverzeichnis", Hamburg 1990.

**Wiese,** Bruno Karl – b. 7. 4. 1922 in Berlin, Germany – *graphic designer, typographer, teacher* – 1945–49: studies at the Staatliche Hochschule für Bildende Künste in Berlin under O. H. W. Hadank. 1949–54: works for Hadank's graphics studio in Hamburg. 1954: opens his own graphic design studio in Hamburg. 1978–80: lecturer and from 1980–87 professor of communication design at the Fachhochschule in Kiel. From 1982 onwards: Wiese and Fritz Seitz are responsible for the Kieler Woche visual program. Numerous exhibitions and awards. – *Publication*: H. Köhne, F. Seitz "B. K. Wiese. Visual Design", Kiel 1987.

**Wijdeveld,** Hendricus Theodorus – b. 4. 10. 1885 in The Hague, The Netherlands, d. 1987 in Amsterdam, The Netherlands – *architect, set-designer, publisher* – 1897–99: trains at J. van Straaten's architect's office in Amsterdam. 1914: opens his own architect's office. Designs and builds country houses, stores, theaters and workers' apartments. 1917: designs the cover for "Architectura" magazine. 1918–31: launches and publishes the magazine "Wendingen, maandblad voor bouwen en sieren" in Amsterdam. Besides covers designed by himself, Wijdeveld also publishes designs by Vilmos

werden Entwürfe u. a. von Vilmos Huszár, El Lissitzky, Jan Toorop veröffentlicht. 1919–20 Gestaltung von Buchumschlägen für die Verlage Van Loghum Slaterus, Van Holkema und Warendorfin. 1930–31 Entwurf des Kalenders der Amsterdamse Stadsdrukkerij. – *Publikation u. a.:* „Mijn eerste eeuw", Oosterbeek 1985.

**Wild,** Lorraine – geb. 31. 5. 1953 in Ontario, Kanada. – *Grafik-Designerin, Typographin, Lehrerin* – Studium an der Yale University und an der Cranbrook Academy of Art. 1985–92 Leiterin der Grafik-Design-Abteilung des California Institute of the Arts. 1990–95 Mitbegründerin und Partnerin des Design-Studios „ReVerb" in Los Angeles. Arbeit für kulturelle Institutionen, Verlage und Künstler. Unterrichtet an der University of Houston und am California Institute of the Arts. Seit 1995 als freie Gestalterin tätig. Verfaßt Artikel über Design-Geschichte und Design-Theorie.

**Wild Plakken** – *Design-Studio* – Lies Ros und Rob Schröder (zusammen mit Frank Beekers, der das Studio 1988 verläßt) gründen 1977 ein Design-Studio in Amsterdam, das ausschließlich Arbeiten für Auftraggeber aus dem politischen und kulturellen Bereich entwirft. Lies Ros – geb. 8. 4. 1952 in Hengelo, Niederlande. – Studium an der Gerrit Rietveld Academie in Amsterdam. Rob Schröder – geb. 13. 11. 1950 in Oegstgeest, Niederlande. – Studium an der Gerrit Rietveld Academie in Amsterdam. Während des Studiums organisieren sich beide in basisdemokratischen Bewegungen. Gestaltung

Wild Plakken 1991 Poster

Wild 1992 Poster

Wild Plakken 1993 Cover

1920, il crée des couvertures de livres pour les éditions Van Loghum Slaterus, Van Holkema et Warendorfin. Conçoit le calendrier de la Amsterdamse Stadsdrukkerij de 1930 à 1931. – *Publication, sélection:* «Mijn eerste eeuw», Oosterbeek 1985.

**Wild,** Lorraine – née le 31. 5. 1953 en Ontario, Canada – *graphiste maquettiste, typographe, enseignante* – études à la Yale University et à la Cranbrook Academy of Art. Directrice de la section d'arts graphiques et de design du California Institute of the Arts de 1985 à 1992. 1990–1995, Cofondatrice et partenaire de l'agence de design «ReVerb» à Los Angeles. Travaille pour des institutions culturelles, des maisons d'édition et des artistes. Enseigne à l'University of Houston et au California Institute of the Arts. Exerce comme designer indépendante depuis 1995. Ecrit des articles sur l'histoire et la théorie du design.

**Wild Plakken** – *agence de design* – en 1977, Lies Ros et Rob Schröder (ainsi que Frank Beekers qui quittera l'agence en 1988) fondent, à Amsterdam, une agence de design qui travaille exclusivement pour des commanditaires des domaines politiques et culturels. Lies Rob – née le 8. 4. 1952 à Hengelo, Pays-Bas – études à la Gerrit Rietveld Academie d'Amsterdam. Rob Schröder – né le 13. 11. 1950 à Oegstgeest, Pays-Bas – études à la Gerrit Rietveld Academie à Amsterdam. Au cours de leurs études, ils militent tous deux dans des organisations démocra-

Huszár, El Lissitzky and Jan Toorop, among others. 1919–20: designs book covers for Van Loghum Slaterus, Van Holkema and Warendorfin publishing houses. 1930–31: designs calendars for the Amsterdamse Stadsdrukkerij. – *Publications include:* "Mijn eerste eeuw", Oosterbeek 1985.

**Wild,** Lorraine – b. 31. 5. 1953 in Ontario, Canada – *graphic designer, typographer, teacher* – Studied at Yale University and at the Cranbrook Academy of Art. 1985–92: head of the graphic design department at the California Institute of the Arts. 1990–95: co-founder and partner of ReVerb Design Studio in Los Angeles. Produces work for cultural institutions, publishing houses and artists. Has taught at the University of Houston and at the California Institute of the Arts. 1995: starts freelancing as a graphic designer. Wild has written articles on the history and theory of design.

**Wild Plakken** – *design studio* – 1977: Lies Ros and Rob Schröder open a design studio in Amsterdam with Frank Beekers (who leaves the studio in 1988). The studio only works for clients from the political and arts sector. Lies Ros – b. 8. 4. 1952 in Hengelo, The Netherlands – Studied at the Gerrit Rietveld Academie in Amsterdam. Rob Schröder – b. 13. 11. 1950 in Oegstgeest, The Netherlands – Studied at the Gerrit Rietveld Academie in Amsterdam. During their studies, both are members of various grass-roots democracy movements. They design posters, banners, brochures and exhibitions. 1981: the studio is named Wild Plakken. Clients

von Plakaten, Transparenten, Broschüren und Ausstellungen. Das Studio erhält 1981 den Namen „Wild Plakken". Auftraggeber waren u. a. das Museum Fodor in Amsterdam, die Filmzeitschrift „Skrien", die Nederlandse Opera in Amsterdam, das Centraal Museum in Utrecht und Groen Progressief Akkoord. 1992 Ausstellung „Beeld tegen Beeld" im Centraal Museum in Utrecht. 1992–96 ist Max Kisman Mitarbeiter bei „Wild Plakken". – *Publikationen u. a.:* „Dick Elffers and the Arts" (Co-Autoren), Amsterdam 1989; „Beeld tegen Beeld" (Co-Autoren), Amsterdam 1993.

**Willberg,** Hans Peter – geb. 4. 10. 1930 in Nürnberg, Deutschland. – *Typograph, Autor, Lehrer* – 1949–52 Ausbildung als Gebrauchsgrafiker an der Berufsoberschule in Nürnberg. 1952–57 Studium an der Staatlichen Akademie der Bildenden Künste in Stuttgart. 1957–67 freischaffender Buchgestalter. Seit 1962 Vortragstätigkeit sowie journalistische und theoretische Arbeit für Fachzeitschriften. 1968–75 Geschäftsführer der Stiftung Buchkunst in Frankfurt am Main. 1975–96 Professor für Buchgestaltung an der Fachhochschule in Mainz. Zahlreiche Ausstellungen. – *Publikationen u. a.:* „Schrift im Bauhaus, Die Futura von Paul Renner", Neu-Isenburg 1969; „Imre Reiner. Die Ziffernbilder", Stuttgart 1975; „Schriften erkennen" (mit Monika Thomas), Mainz 1988; „Lesetypographie" (mit Friedrich Forssmann), Mainz 1996.

**Williams,** Emmett – geb. 4. 4. 1925 in Greenville, South Carolina, USA. – *Künstler, Autor* – Studium der Poesie am

tiques de base. Ils réalisent des affiches, des banderoles, des brochures et des expositions. En 1981, l'agence prend le nom de «Wild Plakken». Commanditaires: le Museum Fodor d'Amsterdam, la revue de cinéma «Skrien», le Nederlandse Opera d'Amsterdam, le Centraal Museum d'Utrecht, Groen Progressief Akkoord, etc. Exposition «Beeld tegen Beeld» au Centraal Museum d'Utrecht en 1992. Max Kisman travaille avec l'équipe de «Wild Plakken» de 1992 à 1996. – *Publications, sélection:* «Dick Elffers and the Arts» (co-auteurs), Amsterdam 1989; «Beeld tegen Beeld» (co-auteurs), Amsterdam 1993.

**Willberg,** Hans Peter – né le 4. 10. 1930 à Nuremberg, Allemagne – *typographe, auteur, enseignant* – 1949–1952, formation de graphiste industriel au lycée professionnel de Nuremberg. 1952–1957, études à la Staatliche Akademie der Bildenden Künste (beaux-arts) de Stuttgart. Maquettiste de livres indépendant de 1957 à 1967. Conférences, activités de journaliste et travail théorique pour des revues spécialisées à partir de 1962. Administrateur de la fondation des arts du livre à Francfort-sur-le-Main de 1968 à 1975. Professeur de design en édition à la Fachhochschule (école technique supérieure) de Mayence de 1975 à 1996. Nombreuses expositions. – *Publications, sélection:* «Schrift im Bauhaus, Die Futura von Paul Renner», Neu-Isenbourg 1969; «Imre Reiner. Die Ziffernbilder», Stuttgart 1975; «Schriften erkennen» (avec Monika Thomas), Mayence 1988; «Lesetypographie» (avec Friedrich Forssmann), Mayence 1996.

**Williams,** Emmett – né le 4. 4. 1925 à Greenville, Caroline du Sud, Etats-Unis – *artiste, auteur* – études de poésie au Kenyon College. Séjourne à Darmstadt de 1957 à 1959 où il travaille avec Claus Bremer et Daniel Spoerri sur des projets de

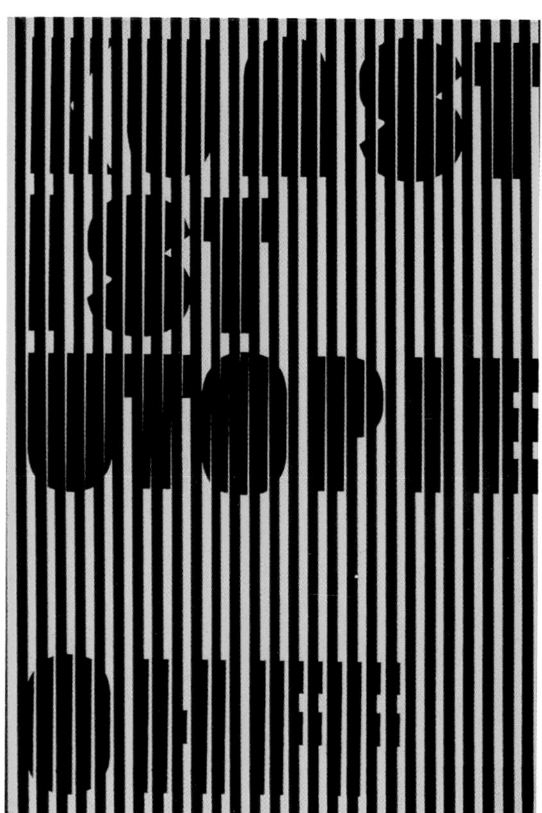

Willberg 1972 Cover

Willberg 1984 Spread

Williams 1976–77 Poem

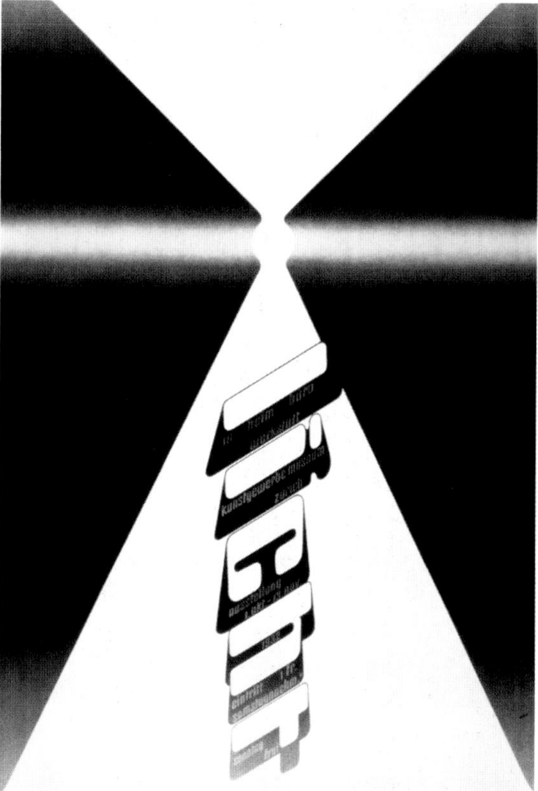

Willimann 1932 Poster

include the Museum Fodor in Amsterdam, "Skrien" film magazine, the Nederlandse Opera in Amsterdam, the Centraal Museum in Utrecht and Groen Progressief Akkoord. 1992: the "Beeld tegen Beeld" exhibition is held at the Centraal Museum in Utrecht. 1992–96: Max Kisman works for Wild Plakken. – *Publications include:* "Dick Elffers and the Arts" (co-authors), Amsterdam 1989; "Beeld tegen Beeld" (co-authors), Amsterdam 1993.

**Willberg,** Hans Peter – b. 4. 10. 1930 in Nuremberg, Germany – *typographer, author, teacher* – 1949–52: trains as a commercial artist at the Berufsoberschule in Nuremberg. 1952–57: studies at the Staatliche Akademie der Bildenden Künste in Stuttgart. 1957–67: freelance designer of books. 1962: starts lecturing and doing journalist and theoretical work for various specialist periodicals. 1968–75: manager of the Stiftung Buchkunst in Frankfurt am Main. 1975–96: professor of book design at the Fachhochschule in Mainz. Numerous exhibitions. – *Publications include:* "Schrift im Bauhaus, Die Futura von Paul Renner", Neu-Isenburg 1969; "Imre Reiner. Die Ziffernbilder", Stuttgart 1975; "Schriften erkennen" (with Monika Thomas), Mainz 1988; "Lesetypographie" (with Friedrich Forssmann), Mainz 1996.

**Williams,** Emmett – b. 4. 4. 1925 in Greenville, South Carolina, USA – *artist, author* – Studied poetry at Kenyon College. 1957–59: works with Claus Bremer and Daniel Spoerri in Darmstadt on concrete poetry and dynamic theater pro-

Kenyon College. 1957–59 in Darmstadt Zusammenarbeit mit Claus Bremer und Daniel Spoerri bei Projekten konkreter Poesie, dynamischem Theater, u. a. Europa-Koordinator der Fluxus-Aktivitäten in Europa. 1966–70 Chefredakteur des Verlags „The Something Else Press" in New York. 1970–72 Professor an der School of Critical Studies des California Institute of the Arts in Valencia, 1972–74 am Nova Scotia College of Art and Design. 1975–77 Gastkünstler am Mount Holyoke College in Massachusetts, 1978–80 am Leverett House der Harvard University, 1980 des Berliner Künstlerprogramms des Deutschen Akademischen Austauschdienstes (DAAD). – *Publikationen u. a.:* „13 Variations on 6 Words of Gertrude Stein", Köln 1965; „Rotapoems«, Stuttgart 1966; „An Anthology of Concrete Poetry" (Hrsg.), New York 1967.

**Willimann,** Alfred – geb. 26. 2. 1900 in Klingnau, Schweiz, gest. 17. 1. 1957 in Zürich, Schweiz. – *Grafiker, Fotograf, Bildhauer, Lehrer* – 1916–17 Studium an der Kunstgewerbeschule in Zürich. Danach als Grafiker und Bildhauer tätig. 1921–24 Studium der Malerei und Bildhauerei an der Akademie der Bildenden Künste in Berlin. 1924–29 Grafiker und Bildhauer in Oerlikon und Zürich. Seit 1930 Lehrer für Zeichnen und Schrift an der Kunstgewerbeschule in Zürich.

**Wilson,** Wes (Robert Wesley Wilson) – geb. 1937. – *Grafik-Designer* – Studium an der San Francisco State University. 1965–66 Entwurf von Handzetteln und Programmheften für Musikfestivals, Plakate für Restaurants. 1966–67 Entwurf

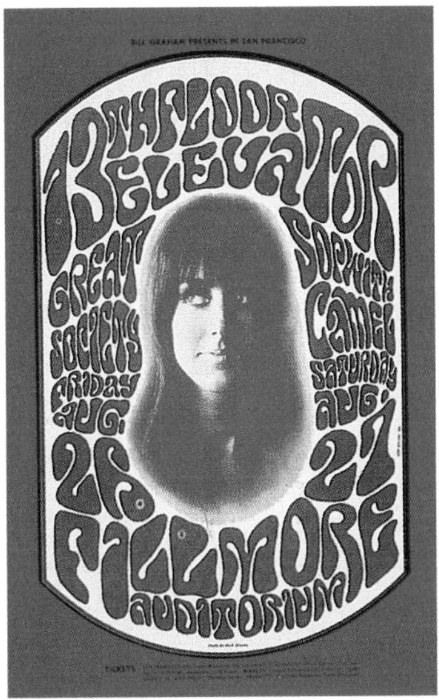

Wilson 1966 Poster

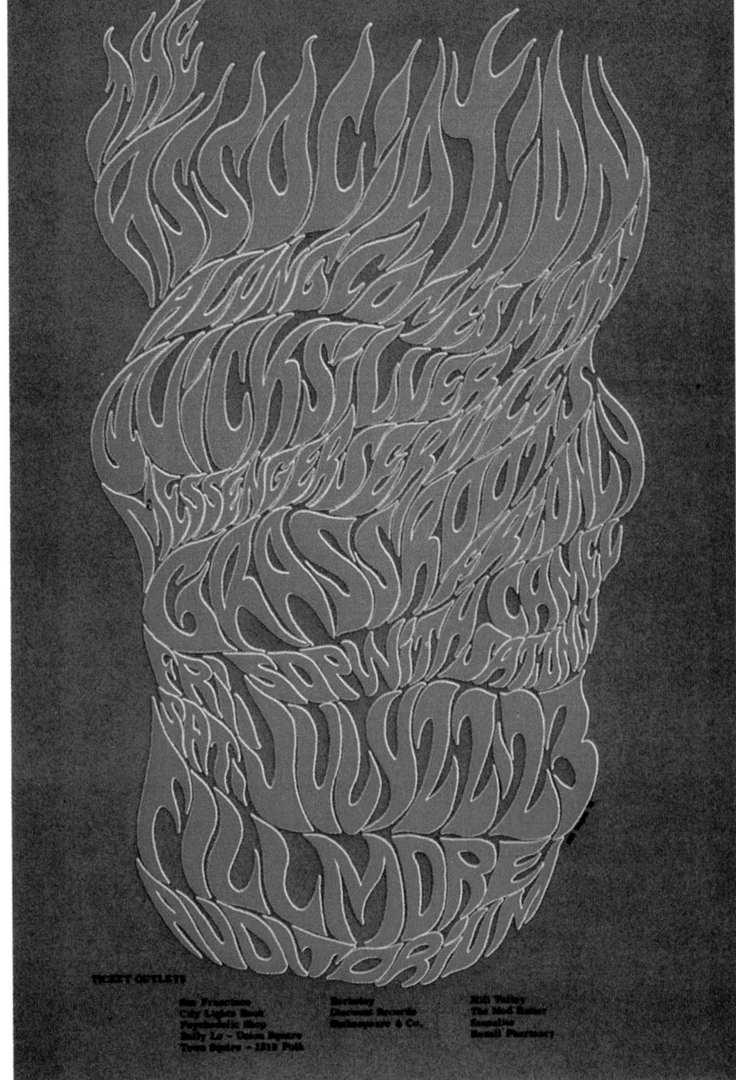

Wilson 1966 Poster

Wilson 1966 Poster

poésie concrète, de théâtre dynamique et coordonne les activités Fluxus en Europe. Rédacteur en chef des éditions «The Something Else Press» à New York de 1966 à 1970. Professeur à la School of Critical Studies du California Institute of the Arts à Valencia de 1970 à 1972, puis au Nova Scotia College of Art and Design de 1972 à 1974. Artiste invité au Mount Holyoke College du Massachusetts de 1975 à 1977, puis à la Leverett House de la Harvard University de 1978 à 1980. Participe au Berliner Künstlerprogramm des Deutschen Akademischen Austauschdienstes (DAAD) (bourses de travail destinées aux artistes) en 1980. – *Publications, sélection:* «13 Variations on 6 Words of Gertrude Stein», Cologne 1965; «Rotapoems», Stuttgart 1966; «An Anthology of Concrete Poetry» (éd.), New York 1967.

**Willimann,** Alfred – né le 26. 2. 1900 à Klingnau, Suisse, décédé le 17. 1. 1957 à Zurich, Suisse – *graphiste, photographe, sculpteur, enseignant* – 1916–1917, études à la Kunstgewerbeschule (école des arts décoratifs) de Zurich. Exerce ensuite comme graphiste et sculpteur. 1921–1924, études de peinture et de sculpture à l'Akademie der Bildenden Künste (beaux-arts) de Berlin. 1924–1929, graphiste et sculpteur à Oerlikon et à Zurich. Enseigne le dessin et les arts de l'écriture à la Kunstgewerbeschule de Zurich à partir de 1930.

**Wilson,** Wes (Robert Wesley Wilson) – né en 1937 – *graphiste maquettiste* – études à la San Francisco State University. 1965–1966, conçoit des prospectus et des programmes pour des festivals de musique ainsi que des affiches pour des restaurants. De 1966 à 1967, il crée des af-

jects, among others. European coordinator of Fluxus activities in Europe. 1966–70: editor-in-chief for The Something Else Press in New York. 1970–72: professor at the California Institute of the Arts' School of Critical Studies in Valencia. 1972–74: professor at the Nova Scotia College of Art and Design. 1975–77: visiting artist at Mount Holyoke College in Massachusetts, from 1978–80 at Harvard University's Leverett House and in 1980 at the Berlin Artists' Program run by the Deutsche Akademische Austauschdienst (DAAD). – *Publications include:* "13 Variations on 6 Words of Gertrude Stein", Cologne 1965; "Rotapoems", Stuttgart 1966; "An Anthology of Concrete Poetry"(ed.), New York 1967.

**Willimann,** Alfred – b. 26. 2. 1900 in Klingnau, Switzerland, d. 17. 1. 1957 in Zurich, Switzerland – *graphic artist, photographer, sculptor, teacher* – 1916–17: studies at the Kunstgewerbeschule in Zurich. Then works as a graphic artist and sculptor. 1921–24: studies painting and sculpture at the Akademie der Bildenden Künste in Berlin. 1924–29: graphic artist and sculptor in Oerlikon and Zurich. 1930: starts teaching art and lettering at the Kunstgewerbeschule in Zurich.

**Wilson,** Wes (Robert Wesley Wilson) – b. 1937 – *graphic designer* – Studied at San Francisco State University. 1965–66: designs handbills and programs for music festivals and posters for restaurants. 1966–67: designs posters for music

von Plakaten für Musikveranstaltungen im Fillmore Auditorium und im Avalon Ballroom in San Francisco. 1967 Teilnahme an der Ausstellung „The Joint Show" (mit Alton Kelley, Victor Moscoso, Rick Griffin, Stanley Mouse) in der Moore Gallery in San Francisco. 1969 Entwurf des Plakats zum Haight-Ashbury-Festival in San Francisco. Zieht sich Anfang der siebziger Jahre in die Ozark Mountains in Missouri zurück.

**Windlin,** Cornel – geb. 1964 in Küsnacht, Schweiz. – *Grafik-Designer, Typograph* – 1983–88 Studium an der Schule für Gestaltung in Luzern. 1987 und 1988–90 Arbeit im Neville Brody Studio in London. 1990–91 Arbeit für die Zeitschrift „The Face" in London. 1991–93 eigenes Studio in London. Seit 1993 eigenes Studio in Zürich, zahlreiche Plakatentwürfe für das Museum für Gestaltung in Zürich.

**Wirth,** Kurt – geb. 12. 9. 1917 in Bern, Schweiz. – *Grafik-Designer, Illustrator, Typograph, Lehrer* – 1933–36 Grafiker-Ausbildung an der Gewerbeschule in Bern. 1937 eigenes Atelier. Illustrationen für Tageszeitungen, grafischer Gestalter für die Pharma-Industrie, Swissair, für Verlage wie Benziger, Haupt und Scherz. Entwirft ca. 200 Buch- und Taschenbuchumschläge für den Fischer Verlag in Frankfurt am Main. Gestaltung von Ausstellungen und Fremdenverkehrswerbung. 1955–60 Mitarbeit in der Arbeitsgemeinschaft von vier Grafik-Designern unter dem Titel „Tim". Gestaltung von Wandbildern und Malerei. 1956–59 Präsident des Verbandes Schweizer Grafiker (VSG). Unterrichtet 1971–87 an der Schu-

fiches pour les concerts du Fillmore Auditorium et l'Avalon Ballroom de San Francisco. Participe à l'exposition «The Joint Show» (avec Alton Kelley, Victor Moscoso, Rick Griffin et Stanley Mouse) à la Moore Gallery de San Francisco en 1967. Dessine l'affiche de l'Haight-Ashbury Festival de San Francisco en 1969. Se retire dans les Ozark Mountains (Missouri) au début des années soixante-dix.

**Windlin,** Cornel – né en 1964 à Küsnacht, Suisse – *graphiste maquettiste, typographe* – 1983–1988, études à la Schule für Gestaltung (école de design) de Lucerne. Travaille pour le Neville Brody Studio de Londres en 1987, puis de 1988 à 1990. Travaux pour la revue «The Face», à Londres, de 1991 à 1993. Exerce dans son propre atelier londonien depuis 1993, nombreuses affiches pour le musée de design de Zurich.

**Wirth,** Kurt – né le 12. 9. 1917 à Berne, Suisse – *graphiste maquettiste, illustrateur, typographe, enseignant* – 1933–1936, formation de graphiste à l'école des arts et métiers de Berne. Exerce dans son atelier à partir de 1937. Illustrations pour des quotidiens, concepteur graphique pour l'industrie pharmaceutique, Swissair et pour des maisons d'édition, entre autres pour Benziger, Haupt et Scherz. Dessine environ 200 couvertures de livres et de livres de poche pour les éditions Fischer de Francfort-sur-le-Main. Design d'expositions et publicités pour le tourisme. De 1955 à 1960, il travaille avec quatre autres graphistes maquettistes dans un atelier commun appelé «Tim». Réalise des fresques et se consacre à la peinture. Président du Verband Schweizer Grafiker (VSG) (Union des graphistes suisses) de 1956 à 1959. Enseigne à la Schule für Gestaltung (école de design) de Berne de 1971 à 1987. – *Publication:* «Drawing, a creative process. Zeichnen,

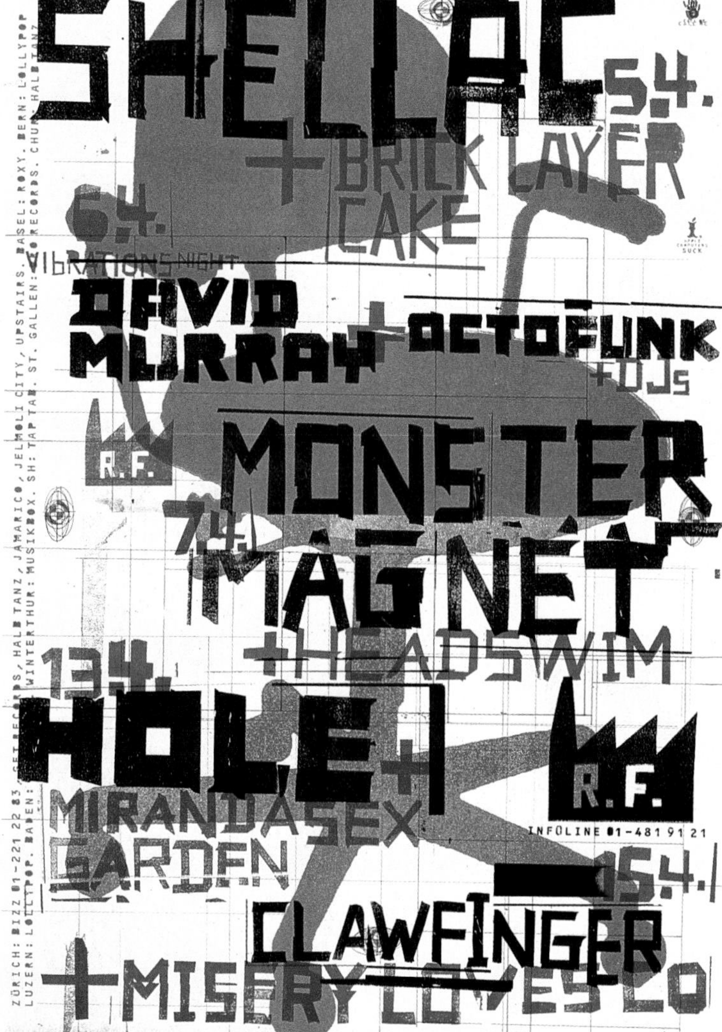
Windlin 1995 Poster

Wirth 1962 Cover

Wirth 1967 Cover

events at Fillmore Auditorium and at the Avalon Ballroom in San Francisco. 1967: takes part in "The Joint Show" with Alton Kelley, Victor Moscoso, Rick Griffin and Stanley Mouse, held at Moore Gallery in San Francisco. 1969: designs the poster for the Haight-Ashbury Festival in San Francisco. Retreats to the Ozark Mountains in Missouri at the beginning of the 1970s.

**Windlin,** Cornel – b. 1964 in Küsnacht, Switzerland – *graphic designer, typographer* – 1983–88: studies at the Schule für Gestaltung in Lucerne. 1987 and 1988–90: works for the Neville Brody Studio in London. 1990–91: works for "The Face" magazine in London. 1991–93: has his own studio in London. 1993: opens his own studio in Zurich. Designs numerous posters for the Museum für Gestaltung in Zurich.

**Wirth,** Kurt – b. 12. 9. 1917 in Bern, Switzerland – *graphic designer, illustrator, typographer, teacher* – 1933–36: trains as a graphic artist at the trade school in Bern. 1937: has his own studio. Does illustrations for daily newspapers. Graphic designer for the pharmaceuticals industry, Swissair and for publishers such as Benziger, Haupt and Scherz. Designs c. 200 hardback and paperback covers for Fischer Verlag in Frankfurt am Main. Designs exhibitions and advertising for tourist bureaus. 1955–60: works in association with four other graphic artists ("Tim"). Murals and art work. 1956–59: president of the Verband Schweizer Grafiker (VSG). 1971–87: teaches at the Schule für Gestaltung in Bern. – *Publi-*

le für Gestaltung in Bern. – *Publikation:* „Drawing, a creative process. Zeichnen, Visualisieren. Dessiner, Création Visuelle", Zürich 1976.

**Wittkugel,** Klaus – geb. 17. 10. 1910 in Kiel, Deutschland, gest. 19. 9. 1985 in Berlin, Deutschland. – *Grafiker, Lehrer* – 1927–29 Lehre als Kaufmann, gleichzeitig Studium an der Staatlichen Kunsthochschule in Hamburg. 1929–32 Studium an der Folkwangschule in Essen bei Max Burchartz. 1932–35 Gebrauchsgrafiker in einem Warenhauskonzern in Berlin. 1935–37 Atelierleiter in einer Berliner Werbeagentur. 1937–39 freischaffend. 1945–49 Gebrauchsgrafiker der Deutschen Zentralverwaltung und Wirtschaftskommission in Berlin. Seit 1946 Arbeiten auf den Gebieten Plakat, Buch, Verpackung, Zeitschrift, Signet, Briefmarke und Ausstellungsgestaltung. 1949–52 Chefgrafiker des Amts für Information der Regierung der DDR. Dozent für Gebrauchsgrafik an der Hochschule für Bildende und Angewandte Kunst in Berlin-Weißensee. Seit 1952 Professor für Grafik an der Kunsthochschule in Berlin. 1958 Verleihung des Nationalpreises der DDR. 1960 und 1963 mit dem Vaterländischen Verdienstorden ausgezeichnet. Seit 1961 ordentliches Mitglied der Deutschen Akademie der Künste in Berlin. – *Publikationen u. a.:* H. Wolf „Klaus Wittkugel, Werke", Dresden 1964; E. Frommhold „Klaus Wittkugel", Dresden 1979.

**Wojirsch,** Barbara – geb. 2. 6. 1940 in Oppeln, Schlesien (heute Opole, Polen). – *Grafik-Designerin, Typographin, Künstlerin* – 1958–60 Ausbildung zur grafi-

Wirth 1973 Poster

Wittkugel 1957 Poster

Wittkugel 1951 Poster

Wittkugel 1973 Cover

Wittkugel 1979 Logo

Visualisieren. Dessiner, Création Visuelle», Zurich 1976.

**Wittkugel,** Klaus – né le 17. 10 1910 à Kiel, Allemagne, décédé le 19. 9. 1985 à Berlin, Allemagne – *graphiste, enseignant* – 1927–1929, apprentissage dans le commerce et simultanément études à la Staatliche Kunsthochschule (école des beaux-arts) de Hambourg. 1929–1932, études à la Folkwangschule d'Essen chez Max Burchartz. 1932–1935, graphiste industriel pour un grand magasin berlinois. 1935–1937, chef d'atelier dans une agence de publicité berlinoise. Indépendant de 1937 à 1939. Graphiste de la Deutsche Zentralverwaltung (administration centrale allemande) et de la Wirtschaftskommission (commission économique) à Berlin de 1945 à 1949. Travaux dans les domaines de l'affiche, du livre, de l'emballage à partir de 1946, crée des signets, des timbres et conçoit des expositions. Chef graphiste de l'Amt für Information (office de l'information) du gouvernement de RDA de la 1949 à 1952. Enseigne le graphisme industriel à la Hochschule für Bildende und Angewandte Kunst (école supérieure des arts plastiques et appliqués) de Berlin-Weissensee de 1949 à 1952. Professeur d'arts graphiques à l'école des beaux-arts de Berlin à partir de 1952. Lauréat du prix national de RDA en 1958. Ordre du mérite pour la patrie en 1960 et en 1963. Membre ordinaire de la Deutsche Akademie der Künste de Berlin à partir de 1961. – *Publications :* H. Wolf «Klaus Wittkugel, Werke», Dresde 1964; E. Frommhold «Klaus Wittkugel», Dresde 1979.

**Wojirsch,** Barbara – née le 2. 6. 1940 à Oppeln, Silésie (aujourd'hui Opole, Pologne) – *graphiste maquettiste, typographe, artiste* – 1958– 1960, formation de graphisme et de dessin. 1960–1964,

*cation:* "Drawing, a creative process. Zeichnen, Visualisieren. Dessiner, Création Visuelle", Zurich 1976.

**Wittkugel,** Klaus – b. 17. 10. 1910 in Kiel, Germany d. 19. 9. 1985 in Berlin, Germany – *graphic artist, teacher* – 1927–29: trains in business and at the same time studies at the Staatliche Kunsthochschule in Hamburg. 1929–32: studies at the Folkwangschule in Essen under Max Burchartz. 1932–35: commercial artist for a large department store in Berlin. 1935–37: art director for a Berlin advertising agency. 1937–39: freelance. 1945–49: commercial artist for the German central administration and economic committee in Berlin. 1946: starts working with posters, books, packaging, magazines, logos, postage stamps and exhibition design. 1949–52: graphics manager at the GDR's government information office. Teaches commercial graphics at the Hochschule für Bildende und Angewandte Kunst in Berlin-Weißensee. From 1952 onwards: professor of graphics at the art college in Berlin. 1958: is awarded the GDR's National Prize. 1960 and 1963: is awarded the Vaterländische Verdienstorden (order of merit). From 1961 onwards: full member of the Deutsche Akademie der Künste in Berlin. – *Publications:* H. Wolf "Klaus Wittkugel, Werke", Dresden 1964; E. Frommhold "Klaus Wittkugel", Dresden 1979.

**Wojirsch,** Barbara – b. 2. 6. 1940 in Oppeln, Silesia (today Opole, Poland) – *graphic designer, typographer, artist* – 1958–60: trains as a graphics draftswoman. 1960–64: studies painting at the

schen Zeichnerin. 1960–64 Studium der Malerei an der Staatlichen Akademie der Bildenden Künste in Stuttgart. Seit 1964 freiberuflich tätig. Grafik-Design für Kulturveranstaltungen, Corporate Design, Plakate sowie Kunst am Bau-Projekte. 1970–96 Gestaltung von Schallplattencovern für ECM Records. Zahlreiche Ausstellungen und Auszeichnungen. – *Publikation:* P. Kemper u. a. „ECM – Sleeves of Desire. A Cover Story", Baden 1996.

**Wolf,** Henry – geb. 23. 5. 1925 in Wien, Österreich. – *Grafik-Designer, Typograph, Lehrer* – 1941 Emigration in die USA. 1941–42 Studium an der School of Industrial Art in New York. 1947–50 Studium an der New School for Social Research in New York. Seit 1946 Arbeit als Grafik-Designer, u. a. für United States Department of State (1951–52), „Esquire" (1952–58), „Harper's Bazaar" (1958–61), McCann Erickson (1964–66) und Trahey-Wolf Advertising (1966–71). Seit 1971 Präsident von Henry Wolf Productions in New York. Zahlreiche Auszeichnungen, u. a. Goldmedaille des „American Institute of Graphic Arts" (1976), Aufnahme in die Art Directors Hall of Fame (1980), Herb Lubalin Award der Society of Publication Designers (1989) und „Honorary Royal Designer for Industry" der Royal Society of Arts in London (1990). 1989–91 Präsident des Art Directors Club of New York. Unterrichtet an der Cooper Union in New York (1955–64), an der School of Visual Arts in New York (1964–74) und an der Parson's School of Design (seit 1974). – *Publikationen u. a.:* „Visual Thinking: Methods for making images memorable",

études de peinture à la Staatliche Akademie der Bildenden Künste (beaux-arts) de Stuttgart. Exerce comme indépendante à partir de 1964. Graphismes et design pour des manifestations culturelles, identités d'entreprises, affiches, projets artistiques et architecturaux. Crée des pochettes de disques pour les ECM Records de Munich de 1970 à 1996. Nombreuses expositions et distinctions. – *Publication :* Peter Kemper et autres «ECM – Sleeves of Desire. A Cover Story», Baden 1996.

**Wolf,** Henry – né le 23. 5. 1925 à Vienne, Autriche – *graphiste maquettiste, typographe, enseignant* – 1941, émigre aux Etats-Unis. 1941–1942, études à la School of Industrial Art à New York. 1947–1950, études à la New School for Social Research de New York. A partir de 1946, il exerce comme graphiste maquettiste pour le United States Department of State (1951–1952), «Esquire» (1952–1958), «Harper's Bazaar» (1958–1961), McCann Erickson (1964–1966), Trahey-Wolf Advertising (1966–1971). Président de Henry Wolf Productions à New York à partir de 1971. Nombreuses distinctions, entre autres la médaille d'or de l' «American Institute of Graphic Arts» (1976), admis au Art Directors Hall of Fame (1980), Herb Lubalin Award de la Society of Publication Designers (1989) et «Honorary Royal Designer for Industry» de la Royal Society of Arts de Londres (1990). Président de l'Art Directors Club de New York, de 1989 à 1991. Enseigne à la Cooper Union de New York (1955–1964), à la School of Visual Arts de New York (1964–1974) et à la Parson's School of Design (à partir de 1974). – *Publications, sélection :* «Visual Thinking: Methods for making images memorable», New York 1988. A. Hawkins «The Art Director at Work», New York 1959.

**Wolf,** Rudolf – né en 1895 à Hechingen,

Wojirsch 1974 Cover

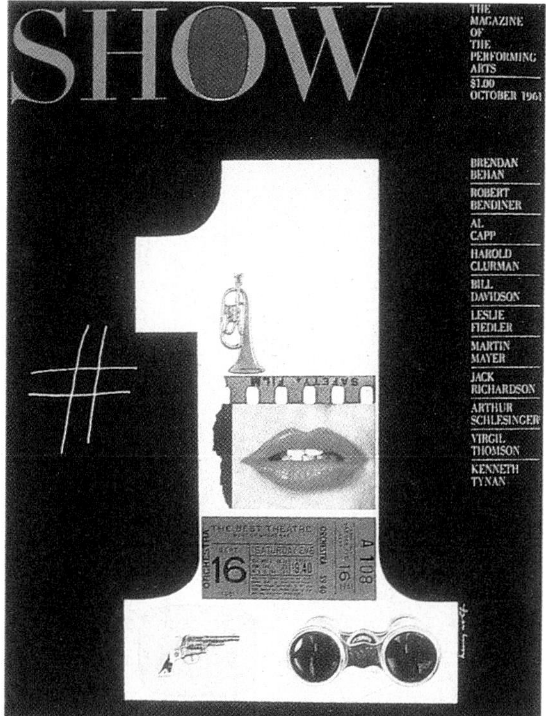

Henry Wolf 1961 Cover

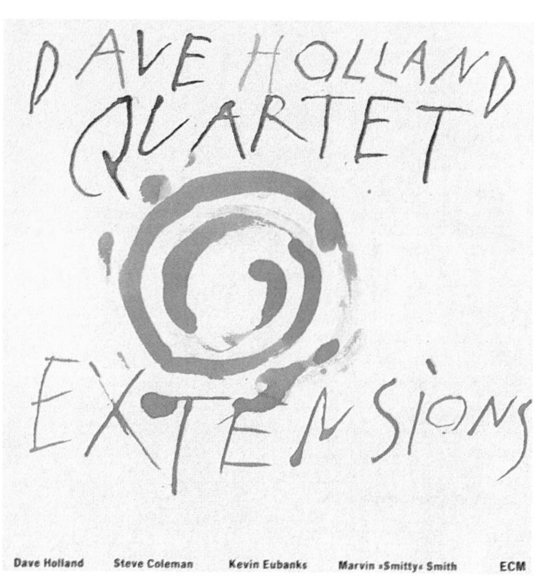

DAVE HOLLAND
QUARTET
EXTENSIONS
Dave Holland Steve Coleman Kevin Eubanks Marvin »Smitty« Smith ECM

Wojirsch 1990 Cover

| n | n | n | n | | n |
|---|---|---|---|---|---|
| a | b | e | f | g | i |
| o | r | s | t | y | z |
| A | B | C | E | G | H |
| M | O | R | S | X | Y |
| 1 | 2 | 4 | 6 | 8 | & |

Rudolf Wolf 1930 Memphis

Staatliche Akademie der Bildenden Künste in Stuttgart. 1964: starts freelancing. Does graphic designs for cultural events, designs posters, and works in corporate design and "art on the building site" projects. 1970–96: designs covers for ECM Records in Munich. Numerous exhibitions and awards. – *Publication:* Peter Kemper et al "ECM – Sleeves of Desire. A Cover Story", Baden 1996.

**Wolf,** Henry – b. 23. 5. 1925 in Vienna, Austria – *graphic designer, typographer, teacher* – 1941: emigrates to the USA. 1941–42: studies at the School of Industrial Art in New York. 1947–50: studies at the New School for Social Research in New York. 1946: starts working as a graphic designer for the United States Department of State (1951–52), "Esquire" (1952–58), "Harper's Bazaar" (1958–61), McCann Erickson (1964–66) and Trahey-Wolf Advertising (1966–71), among others. 1971: is made president of Henry Wolf Productions in New York. Wolf has won numerous awards, including the American Institute of Graphic Arts' gold medal (1976), being elected into the Art Directors Hall of Fame (1980), the Society of Publication Designers' Herb Lubalin Award (1989) and being made an Honorary Royal Designer for Industry by the Royal Society of Arts in London (1990). 1989–91: president of the Art Directors Club of New York. Has taught at the Cooper Union in New York (1955–64), at the School of Visual Arts in New York (1964–74) and at the Parson's School of Design (from 1974 onwards). – *Publications include:* "Visual Thinking: Methods

New York 1988. A. Hawkins „The Art Director at Work“, New York 1959.

**Wolf,** Rudolf – geb. 1895 in Hechingen, Deutschland, gest. 1942 in Frankfurt am Main, Deutschland. – *Werbeleiter, Schriftentwerfer, Lehrer* – Studium an der Universität in Frankfurt am Main, Promotion. 1922–42 als Werbeleiter der D. Stempel AG in Frankfurt am Main tätig, verantwortlich für die Schriftentwicklung. – *Schriftentwurf:* Memphis (1930). – *Publikation:* „Fraktur und Antiqua“, Frankfurt am Main 1934.

**Wolpe,** Berthold – geb. 29. 10. 1905 in Offenbach, Deutschland, gest. 5. 7. 1989 in London, England. – *Typograph, Schriftentwerfer, Lehrer* – 1924–27 Studium an der Kunstgewerbeschule in Offenbach bei Rudolf Koch. Danach Ausbildung zum Goldschmied in Pforzheim. Unterrichtet 1929–33 an der Kunstgewerbeschule in Offenbach und 1930–33 an der Kunstschule in Frankfurt am Main. Entwurf von Gedenkschriften, Warenzeichen, Plakaten, Buchumschlägen und Schriften. 1932 Bekanntschaft mit Stanley Morison. 1935 Emigration nach England. 1935–40 Arbeit bei der „Fanfare Press“ von Ernest Ingham. Gestaltet 1941–75 ca. 1.500 Buchumschläge für den Verlag Faber & Faber. 1955 Teilnahme am internationalen Bibel-Projekt „Liber Librorum“. 1959 Auszeichnung mit dem Titel „Royal Designer for Industry“. 1968 Verleihung des Titels Dr. h. c. des Royal College of Art in London. Unterrichtet am Camberwell College of Art (1948–53), am Royal College of Art in London (1956–75) und an der City & Guilds of London School of Art.

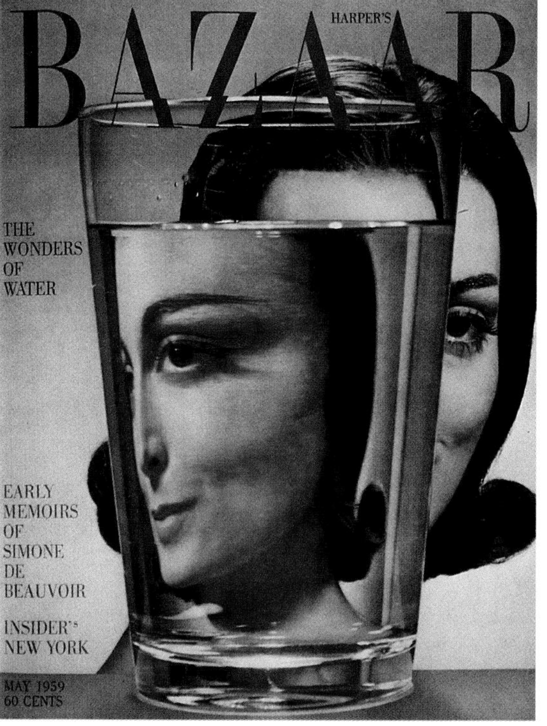


Henry Wolf 1959 Cover

Und wenn mich am Tag die Ferne
Blauer Berge ſehnlich zieht,
Nachts das Übermaß der Sterne
Prächtig mir zu Häupten glüht,
Alle Tag' und alle Nächte
Rühm' ich ſo des Menſchen Los;
Denkt er ewig ſich ins Rechte,
Iſt er ewig ſchön und groß!

Wolpe ca. 1934 Calligraphy

| n | n | **n** | **n** | | *n* |
|---|---|---|---|---|---|
| a | b | e | f | g | i |
| o | r | s | t | y | z |
| A | B | C | E | G | H |
| M | O | R | S | X | Y |
| 1 | 2 | 4 | 6 | 8 | & |

Wolpe 1932–40 Albertus

SCS ALBERTVS MAGNVS
SANCTITATE & DOCTRINA
CELEBRIS
QVEM PIVS PAPA IX
DOCTOREM VNIVERSALIS
ECCLESIAE DECLARAVIT
IPSE PRO NOBIS ORAT

Wolpe 1932–40 Albertus

Allemagne, décédé en 1942 à Francfort-sur-le-Main, Allemagne – *directeur de publicité, concepteur de polices, enseignant* – études à l'université de Francfort-sur-le-Main. Thèse de doctorat. Directeur de la publicité chez D. Stempel AG à Francfort-sur-le-Main de 1922 à 1942, responsable du développement des polices. – *Police:* Memphis (1930). – *Publication:* «Fraktur und Antiqua», Francfort-sur-le-Main 1934.

**Wolpe,** Berthold – né le 29. 10. 1905 à Offenbach, Allemagne, décédé le 5. 7. 1989 à Londres, Angleterre – *typographe, concepteur de polices, enseignant* – 1924–1927, études à la Kunstgewerbeschule (école des arts décoratifs) d'Offenbach chez Rudolf Koch. Puis formation d'orfèvre à Pforzheim. Enseigne à la Kunstgewerbeschule d'Offenbach de 1929 à 1933, puis à la Kunstschule de Francfort-sur-le-Main de 1930 à 1933. Dessine des caractères commémoratifs, des logos, affiches, conçoit des couvertures de livres et des polices. Rencontre Stanley Morison en 1932. Emigre en Angleterre en 1935. Employé au «Fanfare Press» d'Ernest Ingham de 1935 à 1940. Entre 1941 et 1975, il crée environ 1 500 couvertures de livres pour les éditions Faber & Faber. Participe au projet international de Bible «Liber Librorum» en 1955. Titre de «Royal Designer for Industry» en 1959. Titre de Dr. h.c. du Royal College of Art à Londres en 1968. Enseigne au Camberwell College of Art (1948–1953), au Royal College of Art de

for making images memorable", New York 1988. A. Hawkins "The Art Director at Work", New York 1959.

**Wolf,** Rudolf – b. 1895 in Hechingen, Germany, d. 1942 in Frankfurt am Main, Germany – *advertising manager, type designer, teacher* – Studied at the university in Frankfurt am Main. Ph.D. 1922–42: works as advertising manager for D. Stempel AG in Frankfurt am Main, where he is responsible for type design. – *Font:* Memphis (1930). – *Publication:* "Fraktur und Antiqua", Frankfurt am Main 1934.

**Wolpe,** Berthold – b. 29. 10. 1905 in Offenbach, Germany, d. 5. 7. 1989 in London, England – *typographer, type designer, teacher* – 1924–27: studies at the Kunstgewerbeschule in Offenbach under Rudolf Koch. Then trains as a goldsmith in Pforzheim. 1929–33: teaches at the Kunstgewerbeschule in Offenbach and from 1930–33 at the art school in Frankfurt am Main. Designs commemorative reports, trademarks, posters, book covers and typefaces. 1932: meets Stanley Morison. 1935: emigrates to England. 1935–40: works for Ernest Ingham's Fanfare Press. 1941–75: designs c. 1,500 book covers for the publishers Faber & Faber. 1955: takes part in the international Liber Librorum bible project. 1959: made a Royal Designer for Industry. 1968: given an honorary doctorate by the Royal College of Art in London. Wolpe taught at the Camberwell College of Art (1948–53), at the Royal College of Art in London (1956–75) and at the City & Guilds of London School of Art (from 1975 onwards). – *Fonts:* Hyperion (1931), Alber-

– *Schriftentwürfe:* Hyperion (1931), Albertus (1932–40), Tempest (1936), Sachsenwald (1937), Pegasus (1938), Decorata (1950), Johnston's Sans Serif Italic (1973). – *Publikationen u. a.:* „Schriftvorlagen", Kassel 1934; „Marken und Schmuckstücke", Frankfurt am Main 1937; „A Book of Fanfare Ornaments", London 1939; „Renaissance Handwriting" (mit A. Fairbanks), London 1959; „Architectural Alphabet. J. D. Steingruber", London 1972.

**Wolter,** Horst Erich – geb. 26. 8. 1906 in Memel, Deutschland (heute Klaipėda, Litauen), gest. 14. 2. 1984 in Leipzig-Lindenthal, Deutschland. – *Typograph* – 1920–24 Lehre als Schriftsetzer in Memel. 1925–30 Studium an der Abendschule der Akademie für Graphische Künste und Buchgewerbe in Leipzig. 1929–40 Setzereileiter der Offizin Haag-Drugulin in Leipzig. Das von Wolter 1936 gestaltete Buch „Drugulins Schatzkästlein für Bücherfreunde" wird 1937 auf der Pariser Weltausstellung mit einem Ehrendiplom ausgezeichnet. 1948–71 künstlerischer und technischer Leiter der Druckerei VEB Offizin Andersen Nexö in Leipzig. 1959 mit dem neugestifteten „Gutenberg-Preis" der Stadt Leipzig ausgezeichnet. 1962 Verleihung des Professorentitels. 1974 mit dem Nationalpreis für Kunst und Wissenschaft der DDR ausgezeichnet. 1982 Werkausstellung auf der „Internationalen Buchkunst-Ausstellung" (iba) in Leipzig.

**Woodward,** Fred – geb. 13. 4. 1953 in Louisville, Mississippi, USA. – *Grafik-Designer, Art Director, Lehrer* – Als Gestalter Autodidakt. Bis 1987 Art Director der

Londres (1956–1975) et à la City & Guilds of London School of Art (à partir de 1975). – *Polices:* Hyperion, (1931), Albertus (1932–1940), Tempest (1936), Sachsenwald (1937), Pegasus (1938), Decorata (1950), Johnston's Sans Serif Italic (1973). – *Publications, sélection:* «Schriftvorlagen», Kassel 1934; «Marken und Schmuckstücke», Francfort-sur-le-Main 1937; » A Book of Fanfare Ornaments», Londres 1939; «Renaissance Handwriting» (avec A. Fairbanks), Londres 1959; «Architectural Alphabet. J. D. Steingruber», Londres 1972.

**Wolter,** Horst Erich – né le 26. 8. 1906 à Memel, Allemagne (aujourd'hui Klaipėda, Lituanie), décédé le 14. 12. 1984 à Leipzig-Lindenthal, Allemagne – *typographe* – 1920–1924, apprentissage de composition typographique à Memel. 1925–1930, cours du soir de l'Akademie für Graphische Künste und Buchgewerbe (académie des arts graphiques et des arts du livre) de Leipzig. 1929–1940, chef d'atelier de composition à la Offizin Haag-Drugulin de Leipzig. «Drugulins Schatzkästlein für Bücherfreunde», un livre dont Wolter avait conçu la maquette en 1936, obtient un diplôme d'honneur à l'exposition universelle de Paris en 1937. Directeur artistique et technique de l'imprimerie VEB Offizin Andersen Nexö à Leipzig de 1948 à 1971. Lauréat du nouveau «Gutenberg-Preis der Stadt Leipzig» en 1959. Professeur h.c. en 1962. «Nationalpreis für Kunst und Wissenschaft der DDR» (prix des arts et des sciences de RDA) en 1974. Exposition de son œuvre à la «Internationale Buchkunst-Ausstellung» (salon international du livre d'art) de Leipzig en 1982.

**Woodward,** Fred – né le 13. 4. 1953 à Louisville, Mississipi, Etats-Unis – *graphiste maquettiste, directeur artistique, enseignant* – designer autodidacte. Di-

ROSEMARIE SCHUDER

Die Erleuchteten

oder Das Bild des armen Lazarus zu Münster in Westfalen von wenig Furchtsamen auch der Terror der Liebe genannt

UNION VERLAG BERLIN

Wolter 1968 Cover

Woodward 1992 Cover

Woodward 1992 Spread

Woodward 1992 Spread

tus (1932–40), Tempest (1936), Sachsenwald (1937), Pegasus (1938), Decorata (1950), Johnston's Sans Serif Italic (1973). – *Publications include:* "Schriftvorlagen", Kassel 1934; "Marken und Schmuckstücke", Frankfurt am Main 1937; "A Book of Fanfare Ornaments", London 1939; "Renaissance Handwriting" (with A. Fairbanks), London 1959; "Architectural Alphabet. J. D. Steingruber", London 1972.

**Wolter,** Horst Erich – b. 26. 8. 1906 in Memel, Germany (Klaipėda, today Lithuania), d. 14. 2. 1984 in Leipzig-Lindenthal, Germany – *typographer* – 1920–24: trains as a typesetter in Memel. 1925–30: attends evening classes at the Akademie für Graphische Künste und Buchgewerbe in Leipzig. 1929–40: runs the composing room for Offizin Haag-Drugulin in Leipzig. 1937: the book "Drugulins Schatzkästlein für Bücherfreunde", designed by Wolter in 1936, wins an honorary diploma at the world exhibition in Paris. 1948–71: technical and artistic director for the VEB Offizin Andersen Nexö printing works in Leipzig. 1959: awarded Leipzig's new Gutenberg Prize. 1962: made a professor. 1974: wins the GDR's National Prize for Art and Science. 1982: an exhibition of his work is held at the "Internationale Buchkunst-Ausstellung" (iba) in Leipzig.

**Woodward,** Fred – b. 13. 4. 1953 in Louisville, Mississippi, USA – *graphic designer, art director, teacher* – Self-taught designer. Up to 1987: art director for "Texas Monthly" magazine and various regional publications. From 1987 on-

Zeitschrift „Texas Monthly“ und regionaler Publikationen. Seit 1987 Art Director der Zeitschrift „Rolling Stone“, für deren Gestaltung er zahlreiche nationale und internationale Auszeichnungen erhält. 1993 Vizepräsident von Straight Arrow Publications. 1995 Creative Director von Wenner Media. Als jüngstes bisheriges Mitglied wird er 1996 in die Hall of Fame des Art Directors Club of New York gewählt. Unterrichtet Publication Design an der Parson's School of Design in New York und leitet jährliche Workshops am Art Center in Pasadena.

**Wool,** Christopher – geb. 1955 in Boston, USA. – *Künstler* – Studium am Sarah Lawrence College in Bronxville, New York (1973), an der New York Studio School (1974) und an der New York University (1975). Lebt und arbeitet in New York. 1984 erste Einzelausstellung in der Cable Gallery in New York. Weitere Ausstellungen, u. a. in der Galerie Gisela Capitain in Köln (1988, 1992, 1996), im San Francisco Museum of Modern Art (1989, 1996), im Boymans-van Beuningen Museum in Rotterdam (1991), in der Kunsthalle Bern (1991) und der Luhring Augustine Gallery in New York (1990, 1991, 1992, 1995). 1992 Teilnahme an der documenta IX in Kassel. In seinem Werk sind Ornament und Schrift Gegenstand der Malerei, seine Wortbilder sind in Schablonen-Schrift-Charakter ausgeführt. – *Publikationen u. a.:* „Christopher Wool. New York“, San Francisco 1989; „Black Book“, Köln, New York 1989; „Cats in Bag Bags in River“, Rotterdam 1991; „Absent without leave“, Berlin 1993.

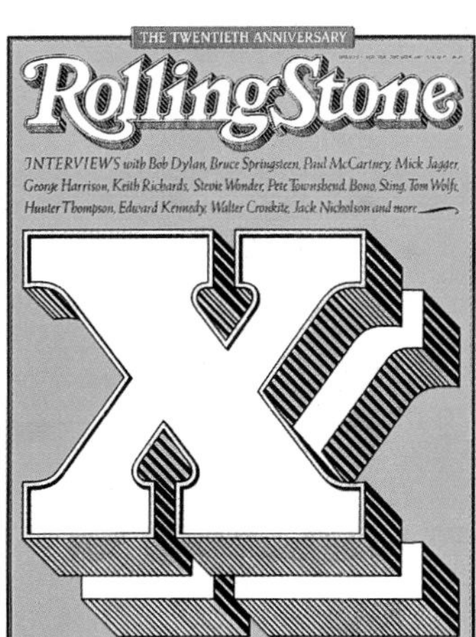


Woodward 1987 Cover

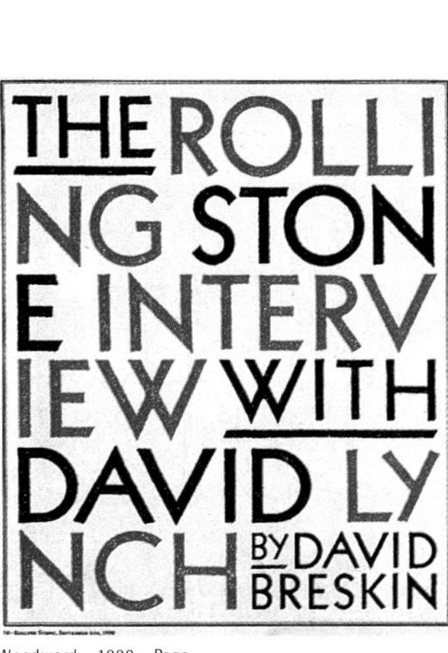


Woodward 1990 Page

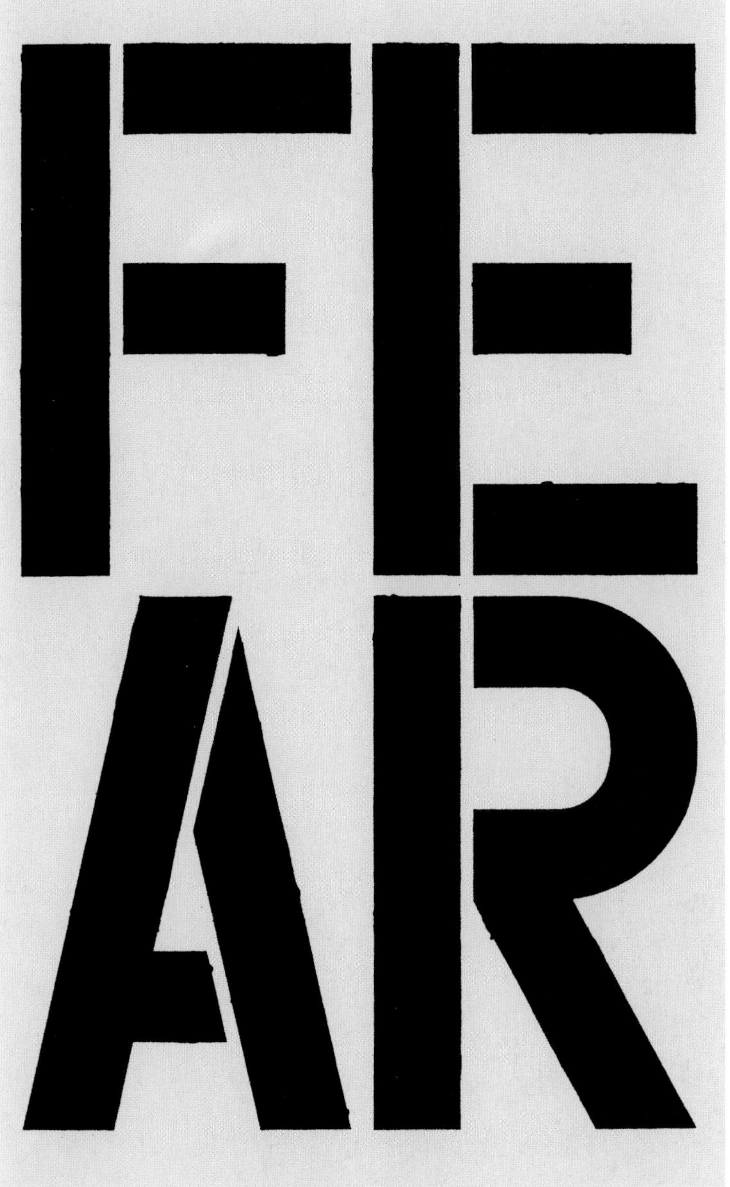

Wool 1990

recteur artistique de la revue «Texas Monthly» et d'autres publications régionales jusqu'en 1987. Directeur artistique de la revue «Rolling Stone» à partir de 1987, de nombreuses distinctions internationales lui sont attribuées pour la maquette de ce magasine. Vice-président de Straight Arrow Publications en 1993. Directeur de la création chez Wenner Media en 1995. En 1996, il est le plus jeune membre élu au Hall of Fame de l'Art Directors Club de New York. Enseigne le design d'édition à la Parson's School of Design de New York et dirige des ateliers annuels à l'Art Center de Pasadena.

**Wool,** Christopher – né en 1955 à Boston, Etats-Unis – *artiste* – études au Sarah Lawrence College de Bronxville, New York (1972), à la New York Studio School (1974) et à la New York University (1975). Vit et travaille à New York. Première exposition personnelle à la Cable Gallery de New York en 1984. Autres expositions, entre autres, à la galerie Gisela Capitain de Cologne (1988, 1992, 1996), au San Francisco Museum of Modern Art (1989, 1996), au Boymans-van Beuningen Museum de Rotterdam (1991), à la Kunsthalle de Berne (1991), à la Luhring Augustine Gallery de New York (1990, 1991, 1992, 1995). Participe à la «documenta IX» de Kassel en 1992. L'écriture et les ornements sont les éléments essentiels de son œuvre peint, il exécute ses pictogrammes au moyen de caractères au pochoir. – *Publications, sélection:* «Christopher Wool. New York», San Francisco 1989; «Black Book», Cologne, New York 1989; «Cats in Bag Bags in River», Rotterdam 1991; «Absent without leave», Berlin 1993.

wards: art director for "Rolling Stone" magazine, winning numerous national and international awards for his magazine design. 1993: vice-president of Straight Arrow Publications. 1995: creative director of Wenner Media. 1996: is the youngest member so far to be elected into the Art Directors Hall of Fame by the Art Directors Club of New York. Teaches publication design at the Parson's School of Design in New York and runs yearly workshops at the Art Center in Pasadena.

**Wool,** Christopher – b. 1955 in Boston, USA – *artist* – Studied at Sarah Lawrence College in Bronxville, New York (1973), at the New York Studio School (1974) and at New York University (1975). Lives and works in New York. 1984: first solo exhibition at Cable Gallery in New York. Further exhibitions have been held at the Gisela Capitain gallery in Cologne (1988, 1992 and 1996), at the San Francisco Museum of Modern Art (1989 and 1996), at the Boymans-van Beuningen Museum in Rotterdam (1991), at the Kunsthalle Bern (1991) and at the Luhring Augustine Gallery in New York (1990, 1991, 1992 and 1995). 1992: takes part in documenta IX in Kassel. In his work, ornament and writing are the subject of his painting; his word pictures are executed in stencil face character. – *Publications include:* "Christopher Wool. New York", San Francisco 1989; "Black Book", Cologne, New York 1989; "Cats in Bag Bags in River", Rotterdam 1991; "Absent without leave", Berlin 1993.

**Wright,** Edward – geb. 1912 in Liverpool, England, gest. 1989 in London, England. – *Grafik-Designer, Typograph, Künstler, Lehrer* – 1930–31 Studium an der Liverpool Art School und 1933–36 an der Bartlett School of Architecture der London University. Lebt und arbeitet 1937–42 in Ecuador. 1948 Einzelausstellung in der Mayor Gallery in London, Mitglied der Künstlergruppe „Independent Group". Veröffentlicht 1956 den Aufsatz „Writing and Environment" in der Zeitschrift „Architectural Design" 26/1956. Zahlreiche Wandgestaltungen und architekturbezogene Gestaltungen mit Schrift, u. a. für die Architects Standard Catalogue Company (1955), der Ideal Home Exhibition „House of the Future" (1956) dem Churchill College in Cambridge (1961) und dem Hinsley House in London (1964). 1968 Gestaltung des räumlichen, drehbaren Signets für New Scotland Yard in London, für den er ein Alphabet entwirft. 1985 Ausstellung des grafischen und malerischen Werks durch den Arts Council in London. 1988 Gestaltung des Schriftzugs der Tate Gallery in Liverpool. Unterrichtet an der Central School of Arts and Crafts in London (1950–55), am Royal College of Art in London, an der Chelsea School of Art und an der Cambridge University School of Architecture. – *Publikation:* Joanna Drew u. a. „Edward Wright, graphic work & painting", London 1985.
**Wunderlich,** Gert – geb. 18. 11. 1933 in Leipzig, Deutschland. – *Grafik-Designer, Typograph, Schriftentwerfer, Lehrer* – 1948–53 Lehre als Schriftsetzer. 1953–58

**Wright,** Edward – né en 1912 à Liverpool, Angleterre, décédé en 1989 à Londres, Angleterre – *graphiste maquettiste, typographe, artiste, enseignant* – 1930–1931, études à la Liverpool Art School, puis à la Bartlett School of Architecture de la London University de 1933 à 1936. Vit et travaille en Equateur de 1937 à 1942. Exposition personnelle à la Mayor Gallery de Londres en 1948, membre du groupe d'artistes «Independent Group». Publie en 1956 l'essai «Writing and Environment» dans la revue «Architectural Design» 26/1956. Nombreuses décorations murales et créations ayant rapport à l'architecture et utilisant l'écriture, entre autres, pour la Architects Standard Catalogue Company (1955), la Ideal Home Exhibition «House of the Future» (1956), le Churchill College de Cambridge (1961) et la Hinsley House à Londres (1964). Dessine le nouveau logo tridimensionnel et rotatif de New Scotland Yard de Londres en 1968 et conçoit un alphabet à cet effet. L'Art Council organise à Londres une exposition de son œuvre peint et graphique en 1985. Dessine le logo de la Tate Gallery de Liverpool en 1988. Enseigne à la Central School of Arts and Crafts de Londres (1950–1955), au Royal College of Art de Londres, à la Chelsea School of Art et à la Cambridge University School of Architecture. – *Publication:* Joanna Drew et autres «Edward Wright, graphic work & painting», Londres 1985.
**Wunderlich,** Gert – né le 18. 11. 1933 à Leipzig, Allemagne – *graphiste maquettiste, typographe, concepteur de polices, enseignant* – 1948–1953, apprentissage de composition typographique. 1953–1958, études à la Hochschule für Graphik und Buchkunst (école supérieure des arts graphiques et des arts du livre) de Leipzig. Travaille ensuite comme ma-

NEW
SCOTLAND
YARD

Wright 1968 Logo

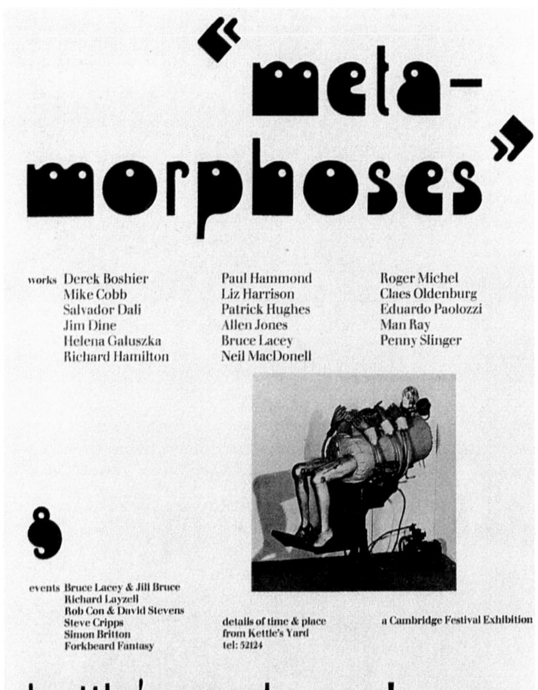


Wright 1978 Poster

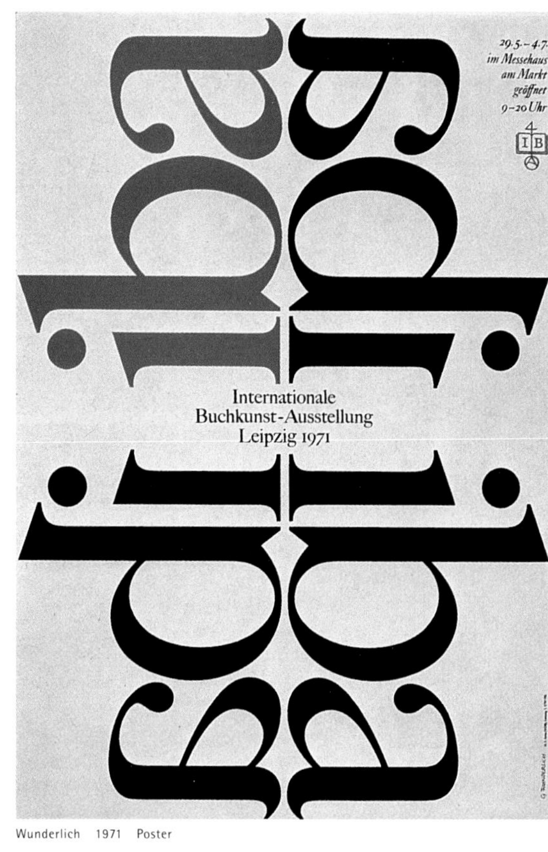


Wunderlich 1971 Poster

ABCDEFGHIJKLMNOP
QRSTUVWXYZ ÄÖÜ
1234567890 /†&!?
abcdefghijklmnopqrstu
vwxyz ch äöü ff ft fi fl ß
1234567890 .,;:„-–*»«

Wunderlich 1964–80 Maxima

**Wright,** Edward – b. 1912 in Liverpool, England, d. 1989 in London, England – *graphic designer, typographer, artist, teacher* – 1930–31: studies at Liverpool Art School and from 1933–36 at London University's Bartlett School of Architecture. 1937–42: lives and works in Ecuador. 1948: solo exhibition at Mayor Gallery in London. Member of the Independent Group of artists. 1956: publishes an essay, "Writing and Environment", in "Architectural Design" 26/1956. Does numerous murals and architectural designs using writing and letters, e.g. for the Architects Standard Catalogue Company (1955), the Ideal Home Exhibition "House of the Future" (1956), Churchill College in Cambridge (1961) and Hinsley House in London (1964). 1968: designs a three-dimensional, rotatable logo for New Scotland Yard in London, for whom he also designs an alphabet. 1985: the Arts Council exhibits his graphic and art work in London. 1988: designs logotype for the Tate Gallery in Liverpool. Wright taught at the Central School of Arts and Crafts in London (1950–55), at the Royal College of Art in London, at the Chelsea School of Art and at Cambridge University School of Architecture. – *Publication:* Joanna Drew et al "Edward Wright, graphic work & painting", London 1985.
**Wunderlich,** Gert – b. 18. 11. 1933 in Leipzig, Germany – *graphic designer, typographer, type designer, teacher* – 1948–53: trains as a typesetter. 1953–58: studies at the Hochschule für Grafik und Buchkunst in Leipzig. He designs books for the Fortschritt printing workshop in

Studium an der Hochschule für Graphik und Buchkunst in Leipzig. Danach Buchgestalter in der Druckerei „Fortschritt" in Erfurt, Zusammenarbeit mit Verlagen. 1966–71 Aspirant, 1971–79 Dozent, seit 1979 Professor für Buchgestaltung, Typographie und Schrift an der Hochschule für Graphik und Buchkunst in Leipzig. 1982–89 Vorsitzender des ICOGRADA-Komitees der DDR. 1986 Gastdozent an der Akademie für Kunst und Design in Peking, China. Ausstellungen seiner Arbeit u.a. in Leipzig (1976, 1981), Warschau (1978) und Amsterdam (1979). U.a. mit dem Kunstpreis der DDR (1976), dem Gutenberg-Preis der Stadt Leipzig (1979) und der Goldmedaille der Biennale der Gebrauchsgrafik in Brno (1984) ausgezeichnet. – *Schriftentwürfe:* Antiqua 58 (1958), Maxima (1964–84).

**Wyss,** Ruedi – geb. 30.9.1949 in Zofingen, Schweiz. – *Grafik-Designer, Typograph, Lehrer* – 1968–72 Grafik-Design-Studium an der Schule für Gestaltung in Biel. Seit 1974 selbständiger Grafik-Designer in Bern. Unterrichtet 1985–87 an der Schule für Gestaltung in Bern, seit 1987 an der Schule für Gestaltung in Zürich.

**Wyss,** Urban – Daten unbekannt. – *Schreibmeister* – 1549 Herausgabe des Schreibbuchs „Libellus valde doctus" in Zürich. 1562 Herausgabe des Buchs „Ein Neuw Fundamentbuch" in Zürich.

Ruedi Wyss 1989 Poster

Ruedi Wyss 1993 Poster

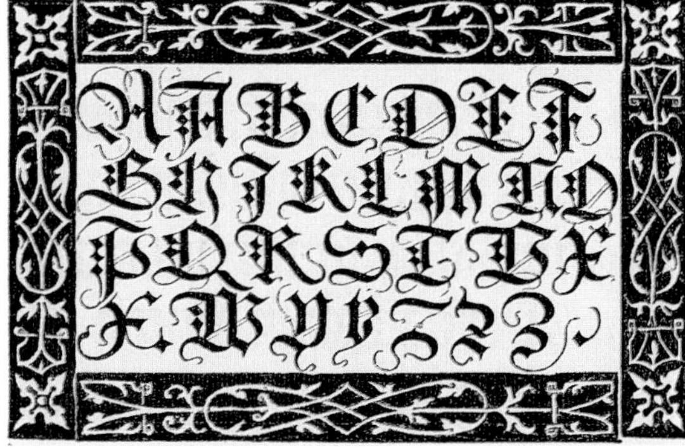
Urban Wyss 1549 Model font

Urban Wyss 1549 Model font

quettiste de livres à l'imprimerie «Fortschritt» d'Erfurt, collabore avec plusieurs maisons d'édition. Enseigne à la Hochschule für Grafik und Buchkunst de Leipzig, d'abord comme assistant de 1966 à 1971, comme maître-assistant de 1971 à 1979, puis comme professeur de design en édition, typographie et arts de l'écriture à partir de 1979. Président du Comité est-allemand de l'ICOGRADA de 1982 à 1989. Cycle de conférences à l'académie des beaux-arts et de design de Pékin, Chine. Exposition de ses travaux à Leipzig (1976, 1981), Varsovie (1978), Amsterdam (1979). Prix d'art de RDA (1976), Prix Gutenberg de la ville de Leipzig (1979), médaille d'or de la Biennale du graphisme industriel de Brno (1984). – *Polices:* Antiqua 58 (1958), Maxima (1964–1984).

**Wyss,** Ruedi – né le 30. 9. 1949 à Zofingen, Suisse – *graphiste maquettiste, typographe, enseignant* – 1968–1972, études d'arts graphiques et de design à la Schule für Gestaltung (école de design) de Bienne. Graphiste maquettiste indépendant à Berne depuis 1974. Enseigne à l'école de design de Berne de 1985 à 1987, enseigne à l'école de design de Zurich depuis 1987.

**Wyss,** Urban – dates inconnues – *scribe* – publication du manuel d'écriture «Libellus valde doctus» à Zurich en 1549. Publication de «Ein Neuw Fundamentbuch" à Zurich en 1562.

Erfurt and works with various publishers. Research assistant (1966–71), lecturer (1971–79) and from 1979 onwards professor of book design, typography and lettering at the Hochschule für Graphik und Buchkunst in Leipzig. 1982–89: chairman of the GDR's ICOGRADA committee. 1986: visiting lecturer at the Academy of Art and Design in Peking, China. Exhibitions of his work have been held in Leipzig (1976 and 1981), Warsaw (1978) and Amsterdam (1979), as well as in other places. Awards include the GDR's Art Prize (1976), Leipzig's Gutenberg Prize (1979) and a gold medal at the Biennale for Graphic-Design in Brno (1984). – *Fonts:* Antiqua 58 (1958), Maxima (1964–84).

**Wyss,** Ruedi – b. 30.9.1949 in Zofingen, Switzerland – *graphic designer, typographer, teacher* – 1968–72: studies graphic design at the Schule für Gestaltung in Biel. 1974: starts freelancing as a graphic designer in Bern. 1985–87: teaches at the Schule für Gestaltung in Bern. 1987: starts teaching at the Schule für Gestaltung in Zurich.

**Wyss,** Urban – dates unknown – *writing master* – 1549: publishes a book on the art of writing, "Libellus valde doctus", in Zurich. 1562: publishes "Ein Neuw Fundamentbuch" in Zurich.

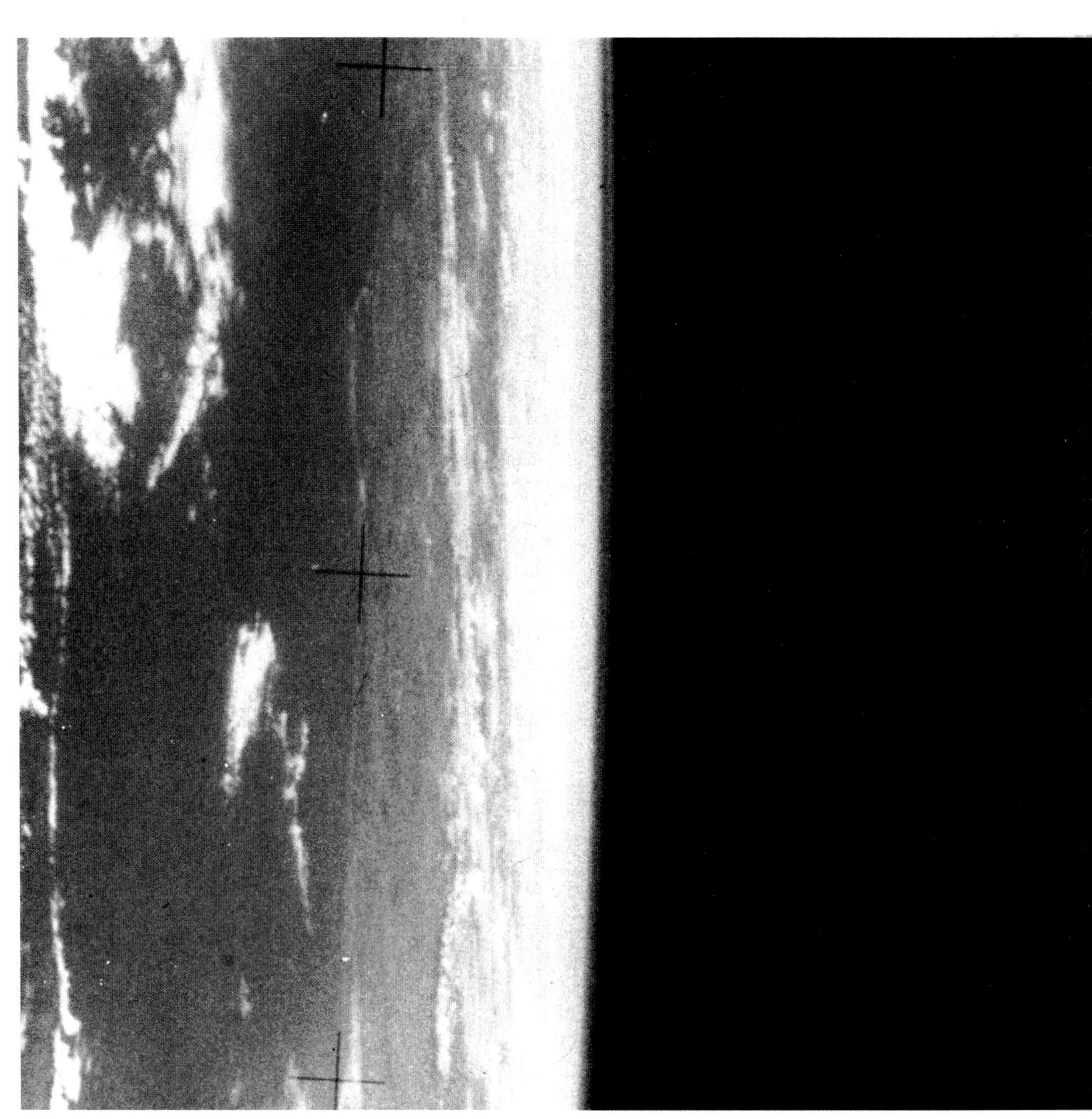

# XYZ

**xplicit** – *Grafik-Design-Studio* – 1994 Gründung in Frankfurt am Main. Alexander Branczyk – geb. 21. 3. 1959 in Oberursel, Deutschland. – 1984–88 Studium an der Hochschule für Gestaltung in Offenbach. 1988–94 Grafik-Designer bei MetaDesign in Berlin. Thomas Nagel – geb 30. 7. 1962 in Frankfurt am Main, Deutschland. – 1985– 88 Studium an der Hochschule für Gestaltung in Offenbach. 1986–89 Grafik-Designer bei Olaf Leu Design in Frankfurt am Main, 1989–94 bei MetaDesign in Berlin. Uwe Otto – geb. 1. 6. 1951 in Braunschweig, Deutschland. – 1969–74 Lehre und Arbeit als Schriftsetzer in Braunschweig und Nürnberg. 1990–92 Produktioner bei Design Pur in Berlin, 1993 bei MetaDesign in Berlin. Hauptarbeitsgebiete von xplicit sind Ausstellungsgestaltung, Corporate Design, Informationsdesign und Zeitschriftengestaltung, u. a. „Frontpage“, „sense“, „Camel Silverpages“ und „ArtKaleidoscope“. Für das eigene Schriftlabel „Face 2 Face“ sind mehr als 200 Alphabete entstanden. – *Schriftentwürfe:* Branczyk: Entebbe (1992), OCR-Alexczyk (1992), Madzine (1992), Czykago (1993), Frontpage (1993), Fuse Synaesthesis (1994), Cryptczyk (1995), Monispace (1995), OCR-Bczyk (1995), Phoneczyk (1996). Nagel: PixMix (1992), ZakkGlobe (1992), HogRoach (1993), WhaleTree (1993).

**Yamashiro,** Ryuichi – geb. 10. 11. 1920 in Osaka, Japan. – *Grafik-Designer, Typograph, Lehrer* – 1938 Abschluß seines Grafik-Design-Studiums an der Kunstgewerbeschule in Osaka. Danach Arbeit als Grafik-Designer in Osaka. Seit 1952 ei-

**xplicit** – *agence de graphisme et de design* – fondée en 1994 à Francfort-sur-le-Main. Alexander Branczyk – né le 21. 3. 1959 à Oberursel, Allemagne – 1984–1988, études à la Hochschule für Gestaltung (école supérieure de design) d'Offenbach. 1988–1994 graphiste maquettiste chez MetaDesign à Berlin. Thomas Nagel – né le 30. 7. 1962 à Francfort-sur-le-Main, Allemagne – 1985–1988, études à la Hochschule für Gestaltung d'Offenbach. 1986–1989 graphiste maquettiste chez Olaf Leu à Francfort-sur-le-Main, 1989–1994, chez MetaDesign à Berlin. Uwe Otto – né le 1. 6. 1951 à Brunswick, Allemagne – 1969–1974, apprentissage de composition typographique à Brunswick et à Nuremberg. 1990–1992, responsable de la production chez Design Plus à Berlin. 1993, graphiste maquettiste chez MetaDesign à Berlin. Les principales spécialités d'xplicit sont l'architecture d'expositions, le corporate design, le design informatif et la maquette de revues «Frontpage», «sense», «Camel Silverpages» et «ArtKaleidoscope». Plus de deux cents alphabets ont été créés pour son label de polices «Face 2 Face». – *Polices:* Branczyk: Entebbe (1992), OCR-Alexczyk (1992), Madzine (1992), Czykago (1993), Frontpage (1993), Fuse Synaesthesis (1994), Cryptczyk (1995), Monispace (1995), OCR-Bczyk (1995), Phoneczyk (1996). Nagel: PixMix (1992), ZakkGlobe (1992), HogRoach (1993), WhaleTree (1993).

**Yamashiro,** Ryuichi – né le 10. 11. 1920 à Osaka, Japon – *graphiste maquettiste, typographe, enseignant* – 1938, diplôme de fins d'études en arts graphiques et en design à l'école des arts décoratifs d'Osaka. Exerce ensuite comme graphiste maquettiste à Osaka. Travaille dans son propre atelier de graphisme à partir de 1952. Crée les fresques qui décorent le pa-

xplicit 1993 Cover

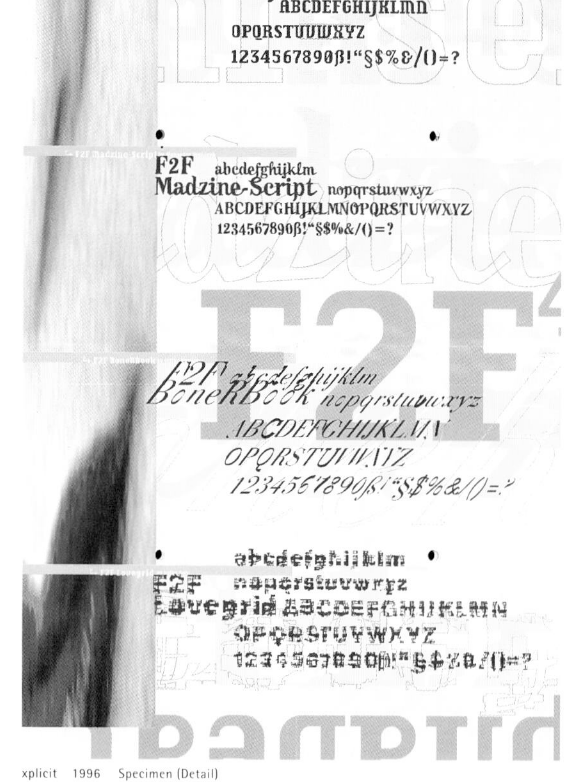


xplicit 1996 Specimen (Detail)

| | n | | | | |
|---|---|---|---|---|---|
| a | b | e | f | g | i |
| o | r | s | t | y | z |
| A | B | C | E | G | H |
| M | O | R | S | X | Y |
| 1 | 2 | 4 | 6 | 8 | & |

xplicit 1992 OCR-Alexczyk

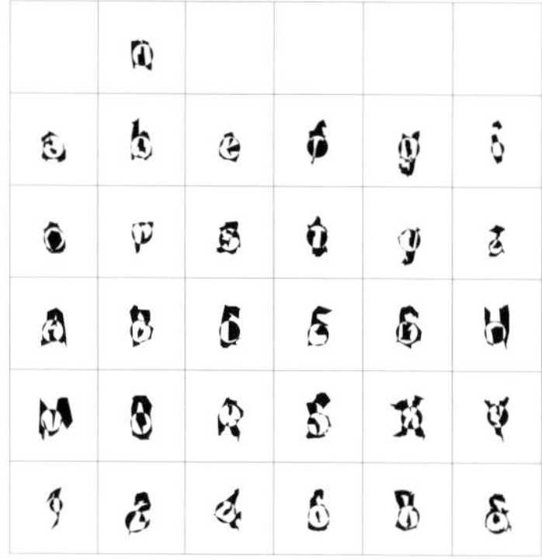

xplicit 1992 ZakkGlobe

| | n | | | | |
|---|---|---|---|---|---|
| a | b | e | f | g | i |
| o | r | s | t | y | z |
| A | B | C | E | G | H |
| M | O | R | S | X | Y |
| 1 | 2 | 4 | 6 | 8 | & |

xplicit 1992 Madzine

**xplicit** – *graphic design studio* – 1994: founded in Frankfurt am Main. Alexander Branczyk – b. 21. 3. 1959 in Oberursel, Germany – 1984–88: studies at the Hochschule für Gestaltung in Offenbach. 1988–94: graphic designer for MetaDesign in Berlin. Thomas Nagel – b. 30. 7. 1962 in Frankfurt am Main, Germany – 1985–88: studies at the Hochschule für Gestaltung in Offenbach. 1986–89graphic designer for Olaf Leu Design in Frankfurt am Main and from 1989–94 for MetaDesign in Berlin. Uwe Otto – b. 1. 6. 1951 in Brunswick, Germany – 1969–74: trains and works as a typesetter in Brunswick and Nuremberg. 1990–92: productioner for Design Pur in Berlin and in 1993 for MetaDesign in Berlin. Xplicit's main areas of work are exhibition, corporate, information and magazine design for "Frontpage", "sense", "Camel Silverpages" and " ArtKaleidoscope". Over 200 alphabets have been created for their own type label, Face 2 Face. – *Fonts*: Branczyk has designed Entebbe (1992), OCR-Alexczyk (1992), Madzine (1992), Czykago (1993), Frontpage (1993), Fuse Synaesthesis (1994), Cryptczyk (1995), Monispace (1995), OCR-Bczyk (1995), Phoneczyk (1996). Nagel has designed PixMix (1992), ZakkGlobe (1992), HogRoach (1993), WhaleTree (1993).

**Yamashiro,** Ryuichi – b. 10. 11. 1920 in Osaka, Japan – *graphic designer, typographer, teacher* – 1938: completes his graphic design studies at the School of Arts and Crafts in Osaka. Then works as a graphic designer in Osaka. 1952: opens his own graphic design studio. 1958: de-

genes Studio. 1958 Gestaltung des japanischen Pavillons auf der Weltausstellung in Brüssel mit Wandgemälden. Gründungsmitglied des „Nippon Design Center". Zeitweise Lehrer an der Kuwazama Design School. 1973 Gründung der „Communication Arts Agentur". Zahlreiche Auszeichnungen für seine Arbeit.

**Yciar,** Juan de – geb. ca. 1523 in Durango, Spanien, gest. ca. 1573 in Logroño, Spanien. – *Schreibmeister, Priester, Autor, Lehrer* – 1548 Herausgabe seines Buchs „Arte subtilissima por la qual se insena a escrevir perfectamente" in Saragossa. Er widmet das Buch Philipp II. von Spanien. Das Buch ist von Jean de Vingles in Holz geschnitten und von Pedro Bernuz gedruckt. Es ist das erste spanische Schreibmeisterbuch und ist von italienischen Vorbildern beeinflußt.

**Zachrisson,** Bror – geb. 12. 12. 1906 in Göteborg, Schweden, gest. 1. 9. 1983 in Stockholm, Schweden. – *Autor, Lehrer* – 1928 Ausbildung am Carnegie Institute of Technology in Pittsburgh, USA. 1932 Promotion zum Doktor der Philosophie an der Universität in Göteborg. 1935–43 Ausbildungsleiter im Verlag Victor Pettersons AG. 1943–53 Rektor der Grafischen Hochschule in Stockholm. 1953–59 Rektor am Institut für höhere Ausbildung im Reklamewesen. Danach Rektor am Institut für Journalismus. Vorsitzender der Vereinigung Nordischer Buchkünstler und Mitglied der Ehrenkommission des Gutenberg-Museums und der Gutenberg-Gesellschaft in Mainz. – *Publikationen u. a.:* „Normalskrift", Stockholm 1943; „Skriftens ABC", Stockholm

Yamashiro 1955 Poster

Yciar 1548 Letra formada Redonda

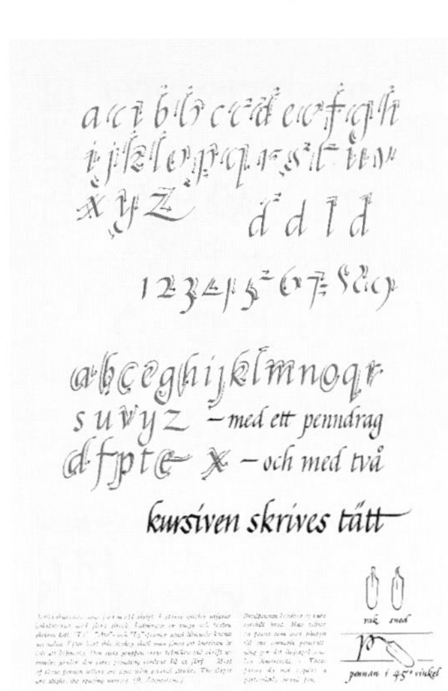


Zachrisson Page

villon japonais de l'exposition universelle de Bruxelles en 1958. Membre fondateur du «Nippon Design Center». Enseigne sporadiquement à la Kuwazama Design School. Fonde la «Communication Arts Agency» en 1973. Son travail a été distingué à maintes reprises.

**Yciar,** Juan de – né vers 1523 à Durango, Espagne, décédé vers 1573 à Logroño, Espagne – *scribe, prêtre, auteur, enseignant* – en 1548, il publie «Arte subtilissima por la qual se insena a escrevir perfectamente» à Saragosse et dédie cet ouvrage à Philippe II d'Espagne. Les caractères de ce livre ont été taillés dans le bois par Jean de Vingles, l'ouvrage a été imprimé par Pedro Bernuz. Cette première publication d'un scribe espagnol a été influencée par les modèles italiens.

**Zachrisson,** Bror – né le 12. 12. 1906 à Göteborg, Suède, décédé le 1. 9. 1983 à Stockholm, Suède – *auteur, enseignant* – 1928, formation au Carnegie Institute of Technology à Pittsburgh, Etats-Unis. 1932, thèse de doctorat de philosophie à l'université de Göteborg. 1935–1943, directeur de la formation aux éditions Victor Pettersons AG. 1943–1953, recteur de l'école supérieure des arts graphiques de Stockholm. 1953–1959, recteur de l'institut d'études supérieures de publicité, puis recteur de l'école supérieure de journalisme. Président de l'union des artistes du livre scandinave et membre de la commission d'honneur du Musée Gutenberg de Mayence et du Gutenberg-Gesellschaft. – *Publications, sélection:* «Normalskrift» Stockholm 1943; «Skrif-

signs the Japanese pavilion with murals for the world exhibition in Brussels. Founder member of the Nippon Design Center. Has taught intermittently at the Kuwazama Design School. 1973: founds the Communication Arts Agency. Yamashiro has received numerous awards for his work.

**Yciar,** Juan de – b. c. 1523 in Durango, Spain, d. c.1573 in Logroño, Spain – *writing master, priest, author, teacher* – 1548: his book, "Arte subtilissima por la qual se insena a escrevir perfectamente", is published in Saragossa. He dedicates the book to Philip II of Spain. The book is carved in wood by Jean de Vingles and printed by Pedro Bernuz. It is the first book by a Spanish writing master and is influenced by Italian models.

**Zachrisson,** Bror – b. 12. 12. 1906 in Gothenburg, Sweden, d. 1. 9. 1983 in Stockholm, Sweden – *author, teacher* – 1928: studies at the Carnegie Institute of Technology in Pittsburgh, USA. 1932: gains his Ph.D. in philosophy at the university in Gothenburg. 1935–43: training manager for the publishing house Victor Pettersons AG. 1943–53: principal of the graphics college in Stockholm. 1953–59: principal of the Institute for Higher Education in Advertising. Then principal of the Institute of Journalism. Chairman of the Society of Nordic Book Artists and member of the Gutenberg Museum and Gutenberg-Gesellschaft honorary commission. – *Publications include:* "Normalskrift", Stockholm 1943; "Skriftens ABC", Stockholm 1943; "Studies in the legibilities of printed text", Stockholm 1965.

1943; „Studies in the legibilities of printed text", Stockholm 1965.

**Zapf,** Hermann – geb. 8. 11. 1918 in Nürnberg, Deutschland. – *Typograph, Schriftentwerfer, Kalligraph, Autor, Lehrer* – 1934–38 Lehre als Retuscheur in der Druckerei Karl Ulrich & Co. in Nürnberg. Als Gestalter Autodidakt. Seit 1938 selbständiger Schriftgrafiker und Kalligraph. 1947–56 künstlerischer Leiter der Schriftgießerei D. Stempel AG in Frankfurt am Main. 1948–50 Lehrer für Schrift an der Werkkunstschule in Offenbach. 1956–74 Berater der Mergenthaler Linotype Company in Brooklyn und Berlin. Arbeiten für Verlage wie Suhrkamp und S. Fischer. 1960 Professor für Grafik-Design am Carnegie Institute of Technology in Pittsburgh, Pennsylvania. 1967–72 Berater von Hallmark Cards. 1969 mit dem F. W. Goudy Award des Rochester Institute of Technology in Rochester, New York ausgezeichnet. 1972–81 Lehrauftrag für Schrift und Typographie an der Technischen Hochschule in Darmstadt. 1974 Verleihung des Gutenberg-Preises der Stadt Mainz. 1977–87 Vize-Präsident von Design Processing International Inc. in New York. 1977–87 Professor für typographische Computerprogramme am Rochester Institute of Technology in Rochester, New York. 1987–91 Vorsitzender der Firma Zapf, Burns & Company in New York für die Entwicklung typographischer Computerprogramme. 1985 Auszeichnung als Honorary Royal Designer for Industry der Royal Society of Arts in London. 1989 Verleihung der Goldmedaille der Internationalen Buchausstel-

tens ABC», Stockholm 1943; «Studies in the legibilities of printed text», Stockholm 1965.

**Zapf,** Hermann – né le 8. 11. 1918 à Nuremberg, Allemagne – *typographe, concepteur de polices, calligraphe, auteur, enseignant* – 1934–1938, apprentissage de retoucheur à l'imprimerie Karl Ulrich & Co. à Nuremberg. Designer autodidacte. Graphiste et calligraphe indépendant à partir de 1938. Directeur artistique de la fonderie de caractères D. Stempel AG à Francfort-sur-le Main de 1947 à 1956. Enseigne les arts de l'écriture à la Werkkunstschule (école des arts appliqués) d'Offenbach de 1948 à 1950. Conseiller de la Mergenthaler Linotype Company à Brooklyn et à Berlin de 1956 à 1974. Travaux pour des maisons d'édition, entre autres, pour Suhrkamp et S. Fischer. Professeur d'arts graphique et de design au Carnegie Institute of Technology à Pittsburgh, Pennsylvanie en 1960. Conseiller de Hallmark Cards de 1967 à 1972. Lauréat du F. W. Goudy Award du Rochester Institute en 1969, à Rochester, New York. Chargé de cours en arts de l'écriture et en typographie à la Technische Hochschule (école supérieure des technologies) de Darmstadt de 1972 à 1981. Prix Gutenberg de la ville de Mayence en 1974. Vice-président de Design Processing International Inc. à New York de 1977 à 1987. Professeur de typographie assistée par ordinateur au Rochester Institute of Technology à Rochester, New York, de 1977 à 1987. Président de "Zapf, Burns & Company" à New York, société pour la mise au point de typographie assisté par ordinateur. Obtient le titre de Honorary Royal Designer for Industry de la Royal Society of Arts à Londres en 1985. Médaille d'or du salon international du livre de Leipzig en 1989. L'œuvre de Zapf allie la connais-

| | n | **n** | | | *n* |
|---|---|---|---|---|---|
| a | b | e | f | g | i |
| o | r | s | t | y | z |
| A | B | C | E | G | H |
| M | O | R | S | X | Y |
| 1 | 2 | 4 | 6 | 8 | & |

Zapf 1958 Optima

| | n | **n** | | | *n* |
|---|---|---|---|---|---|
| a | b | e | f | g | i |
| o | r | s | t | y | z |
| A | B | C | E | G | H |
| M | O | R | S | X | Y |
| 1 | 2 | 4 | 6 | 8 | *&* |

Zapf 1976 Comenius

| | n | | | | *n* |
|---|---|---|---|---|---|
| a | b | e | f | g | i |
| o | r | s | t | y | z |
| A | B | C | E | G | H |
| M | O | R | S | X | Y |
| 1 | 2 | 4 | 6 | 8 | & |

Zapf 1976 Zapf Book 1

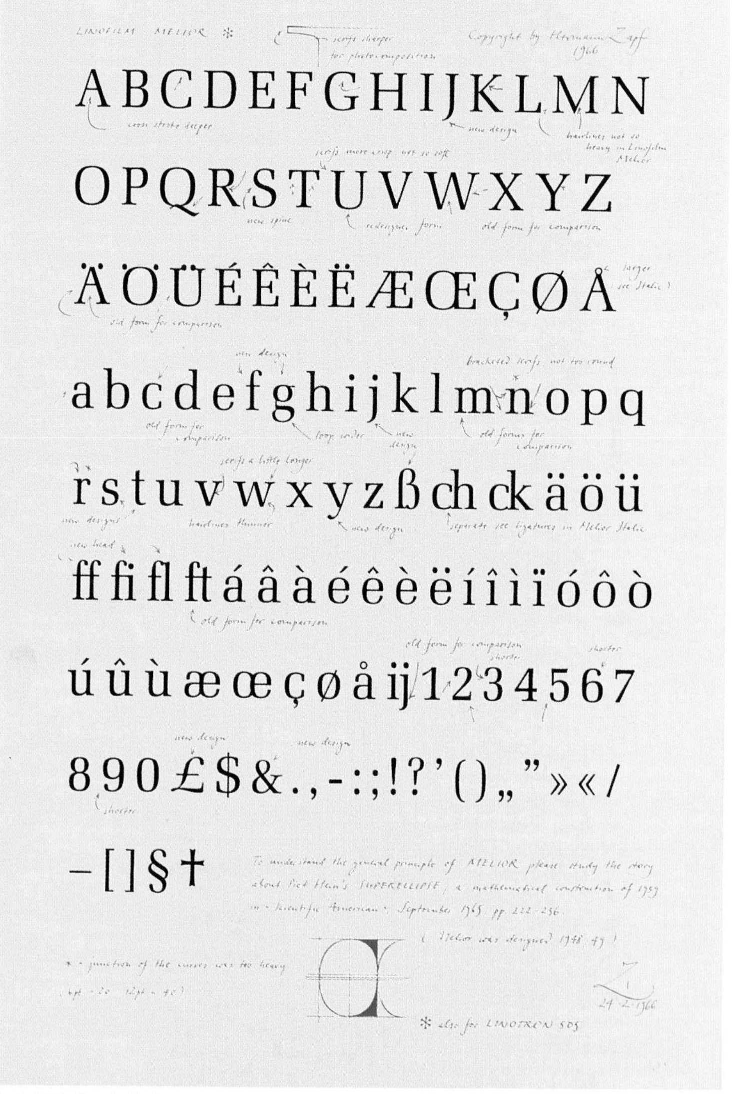


Zapf 1966 Corrected Draft

**Zapf,** Hermann – b. 8. 11. 1918 in Nuremberg, Germany – *typographer, type designer, calligrapher, author, teacher* – 1934–38: trains as a retoucher at the Karl Ulrich & Co. printing works in Nuremberg. Self-taught designer. 1938: starts freelancing as a lettering artist and calligrapher. 1947–56: art director of the D. Stempel AG type foundry in Frankfurt am Main. 1948–50: teaches lettering at the Werkkunstschule in Offenbach. 1956–74: consultant to the Mergenthaler Linotype Company in Brooklyn and Berlin. Works for publishers such as Suhrkamp and S. Fischer. 1960: professor of graphic design at the Carnegie Institute of Technology in Pittsburgh, Pennsylvania. 1967–72: consultant to Hallmark Cards. 1969: is presented with the F. W. Goudy Award by the Rochester Institute in Rochester, New York. 1972–81: teaches lettering and typography at the Technische Hochschule in Darmstadt. 1974: is awarded the Mainz Gutenberg Prize. 1977–87: vice-president of Design Processing International Inc. in New York. 1977–87: professor of typographic computer programming at the Rochester Institute of Technology in Rochester, New York. 1987–91: chairman of Zapf, Burns & Company in New York who develop typographic computer programs. 1985: made an Honorary Royal Designer for Industry by the Royal Society of Arts in London. 1989: awarded a gold medal at the international book exhibition in Leipzig. Zapf's work combines knowledge of the traditional with the possibilities offered by new technologies. –

lung in Leipzig. In Zapfs Schaffen verbinden sich die Kenntnisse der Tradition mit den Möglichkeiten neuer Technologien. – *Schriftentwürfe:* Gilgengart (1938–41), Alkor (1939–40), Novalis (1946), Palatino (1950), Melior (1949), Michelangelo (1950), Sistina (1951), Virtuosa 1 (1952), Saphir (1953), Aldus (1954), Kompact (1954), Alahram Arabisch (1956), Optima (1958), Hunt (1963), Jeanette Script (1967), Firenze (1968), Venture (1969), Textura (1969), Medici (1969), Unical (1970), Optima Griechisch (1971), Missouri (1971), Scriptura (1972), Crown (1972), Orion (1974), Marconi (1976), Comenius (1976), Noris Script (1976), ITC Zapf Book (1976), ITC Zapf International (1977), ITC Zapf Dingbats (1978), Edison (1978), ITC Zapf Chancery (1979), Vario (1982), Aurelia (1983), Euler (1983), Palatino Kyrillisch (1995). – *Publikationen u. a.:* „Das Blumen ABC", Frankfurt am Main 1948; „William Morris", Scharbeutz 1949; „Manuale Typographicum", Frankfurt am Main 1954; „Über Alphabete", Frankfurt am Main 1960; „Ein Arbeitsbericht", Darmstadt, Hamburg 1984; „Kreatives Schreiben", Hamburg 1985; „Poetry through Typography", Rochester 1993; „August Rosenberger", Rochester 1996. John Dreyfus, Knut Erichson „ABC-XYZapf", Offenbach am Main 1989; Martino Mardersteig „L'opera di Hermann Zapf", Verona 1991; Julian Waters „From the Hand of Hermann Zapf", Washington 1993.

**Zapf-von Hesse,** Gudrun – geb. 2. 1. 1918 in Schwerin, Deutschland. – *Buchbinderin, Schriftentwerferin, Lehrerin* –

Zapf 1976 Print

| | n | | | | *n* |
|---|---|---|---|---|---|
| a | b | e | f | g | i |
| o | r | s | t | y | z |
| A | B | C | E | G | H |
| M | O | R | S | X | Y |
| 1 | 2 | 4 | 6 | 8 | & |

Zapf-von Hesse 1952 Diotima

| n | n | **n** | **n** | | *n* |
|---|---|---|---|---|---|
| a | b | e | f | g | i |
| o | r | s | t | y | z |
| A | B | C | E | G | H |
| M | O | R | S | X | Y |
| 1 | 2 | 4 | 6 | 8 | & |

Zapf-von Hesse 1986 Carmina

Zapf-von Hesse ca. 1950 Calligraphy

| | n | **n** | **n** | | *n* |
|---|---|---|---|---|---|
| a | b | e | f | g | i |
| o | r | s | t | y | z |
| A | B | C | E | G | H |
| M | O | R | S | X | Y |
| 1 | 2 | 4 | 6 | 8 | & |

Zapf-von Hesse 1991 Christiana (AB) 20/1

sance de la tradition aux possibilités offertes par les nouvelles technologies. – *Polices:* Gilgengart (1938–1941), Alkor (1939–1940), Novalis (1946), Melior (1949), Palatino (1950), Michelangelo (1950), Sistina (1951), Virtuosa 1 (1952), Saphir (1953), Aldus (1954), Kompact (1954), Alahram arabe (1956), Optima (1958), Hunt (1963), Jeanette Script (1967), Firenze (1968), Venture (1969), Textura (1969), Medici (1969), Unical (1970), Optina grecque (1971), Missouri (1971), Scriptura (1972), Crown (1972), Orion (1974), Marconi (1976), Comenius (1976), Noris Script (1976), ITC Zapf Book (1976), ITC Zapf International (1977), ITC Zapf Dingbats (1978), Edison (1978), ITC Zapf Chancery (1979), Vario (1982), Aurelia (1983), Euler (1983), Palatino cyrillique (1995). – *Publications, sélection:* «Das Blumen ABC», Francfort-sur-le-Main 1948; «William Morris», Scharbeutz 1949; «Manuale Typographicum», Francfort-sur-le-Main 1954; «Über Alphabete», Francfort-sur-le-Main 1960; «Ein Arbeitsbericht», Darmstadt, Hambourg 1984; «Kreatives Schreiben», Hambourg 1985; «Poetry through Typography», Rochester 1993; «August Rosenberger», Rochester 1996. John Dreyfus, Knut Erichson «ABC-XYZapf», Offenbach-sur-le-Main 1989; Martino Mardersteig «L'opera di Hermann Zapf», Vérone 1991; Julian Waters «From the Hand of Hermann Zapf», Washington 1993.

**Zapf-von Hesse,** Gudrun – née le 2. 1. 1918 à Schwerin, Allemagne – *relieuse,*

*Fonts:* Gilgengart (1938–41), Alkor (1939–40), Novalis (1946), Palatino (1950), Melior (1949), Michelangelo (1950), Sistina (1951), Virtuosa 1 (1952), Saphir (1953), Aldus (1954), Kompact (1954), Alahram Arabisch (1956), Optima (1958), Hunt (1963), Jeanette Script (1967), Firenze (1968), Venture (1969), Textura (1969), Medici (1969), Unical (1970), Optima Griechisch (1971), Missouri (1971), Scriptura (1972), Crown (1972), Orion (1974), Marconi (1976), Comenius (1976), Noris Script (1976), ITC Zapf Book (1976), ITC Zapf International (1977), ITC Zapf Dingbats (1978), Edison (1978), ITC Zapf Chancery (1979), Vario (1982), Aurelia (1983), Euler (1983), Palatino Kyrillisch (1995). – *Publications include:* "Das Blumen ABC", Frankfurt am Main 1948; "William Morris", Scharbeutz 1949; "Manuale Typographicum", Frankfurt am Main 1954; "Über Alphabete", Frankfurt am Main 1960; "Ein Arbeitsbericht", Darmstadt, Hamburg 1984; "Kreatives Schreiben", Hamburg 1985; "Poetry through Typography", Rochester 1993; "August Rosenberger", Rochester 1996. John Dreyfus, Knut Erichson "ABC-XYZapf", Offenbach am Main 1989; Martino Mardersteig "L'opera di Hermann Zapf", Verona 1991; Julian Waters "From the Hand of Hermann Zapf", Washington 1993.

**Zapf-von Hesse,** Gudrun – b. 2. 1. 1918 in Schwerin, Germany – *bookbinder, type designer, teacher* – 1934–37: trains as a bookbinder under Prof. Otto Dorfner in Weimar. Then works as a bookbinder in Berlin. 1941: takes lettering courses with

1934–37 Buchbinderlehre bei Prof. Otto Dorfner in Weimar. Danach als Buchbinderin in Berlin tätig. 1941 Schriftunterricht bei Johannes Boehland an der Meisterschule für das grafische Gewerbe in Berlin. 1946–55 Buchbinderwerkstatt in Frankfurt am Main. 1946–54 Lehrerin für Schrift an der Städelschule in Frankfurt am Main. Ausstellung ihrer Arbeit im Klingspor-Museum in Offenbach (1970) und im ITC-Center in New York (1985). 1991 Auszeichnung mit dem Frederic W. Goudy Award des Rochester Institute of Technology in Rochester, New York. – *Schriftentwürfe:* Diotima (1952), Ariadne (1954), Smaragd (1954), Shakespeare (1968), Carmina (1986), Nofret (1987), Christiana (1991), Alcuin (1991), Colombine (1991).

**Zdanewitsch,** Ilia (Iliazd) – geb. 21. 4. 1894 in Tiflis, Georgien, gest. 24. 12. 1975 in Paris, Frankreich. – *Typograph, Maler, Autor, Verleger* – 1911–17 Jura-Studium an der Universität in St. Petersburg. 1912 erste futuristische Vorlesung vor der Künstlergruppe „Bund der Jugend". Veranstaltung von Lesungen und Vorträgen, Herausgabe von Publikationen mit experimentellen Texten und Lautgedichten (Zaum-Sprache genannt), die in experimenteller typographischer Form gestaltet werden. 1921 Publikationen in seinem Verlag „Editions du 41°", experimentelle typographische Tätigkeit in Paris. 1927–33 Stoffentwürfe für Coco Chanel. In Zusammenarbeit mit Pablo Picasso entstehen seit 1940 neun Publikationen mit Originalgrafik, die in der „Edition du 41°" erscheinen. In Zusammenarbeit mit Alber-

*conceptrice de polices, enseignante* – 1934–1937, apprentissage de reliure chez le professeur Otto Dorfner à Weimar. Exerce ensuite comme relieuse à Berlin. Etudie les arts de l'écriture avec Johannes Boehland à la Meisterschule für das grafische Gewerbe (école professionnelle des arts graphiques) de Berlin en 1941. Atelier de reliure à Francfort-sur-le-Main de 1946 à 1955. Enseigne les arts de l'écriture à la Städelschule de Francfort de 1946 à 1954. Exposition de ses travaux au Klingspor-Museum d'Offenbach (1970) et à l'ITC-Center de New York (1985). Lauréate du Frederic W. Goudy Award du Rochester Institute of Technology à Rochester, New York, en 1991. – *Polices:* Diotima (1952), Ariadne (1954), Smaragd (1954), Shakespeare (1968), Carmina (1986), Nofret (1987), Christiana (1991), Alcuin (1991), Colombine (1991).

**Zdanevitch,** Ilia (Iliazd) – né le 21. 4. 1894 à Tiflis, Géorgie, décédé le 24. 12. 1975 à Paris, France – *typographe, peintre, auteur, éditeur* – 1911–1917, études de droit à l'université de Saint-Pétersbourg. Première conférence futuriste devant le groupe d'artistes «Union de la jeunesse» en 1912. Organise des lectures publiques et des conférences, édite des ouvrages contenant des textes expérimentaux et de la poésie sonore (appelée langue «Zaum») avec une mise en page et une typographie expérimentale. 1921, activités d'édition dans sa maison «Editions du 41°»; pratique la typographie expérimentale à Paris. Dessins de textiles pour Coco Chanel de 1927 à 1933. A partir de 1940, il collabore avec Pablo Picasso pour réaliser neuf ouvrages comportant des estampes originales, qu'il publie aux «Editions du 41°». En 1961, il collabore avec Alberto Giacometti au livre «Sentence sans Paroles». En 1964, il publie «Maximiliana ou l'exercice illégal de l'astrono-

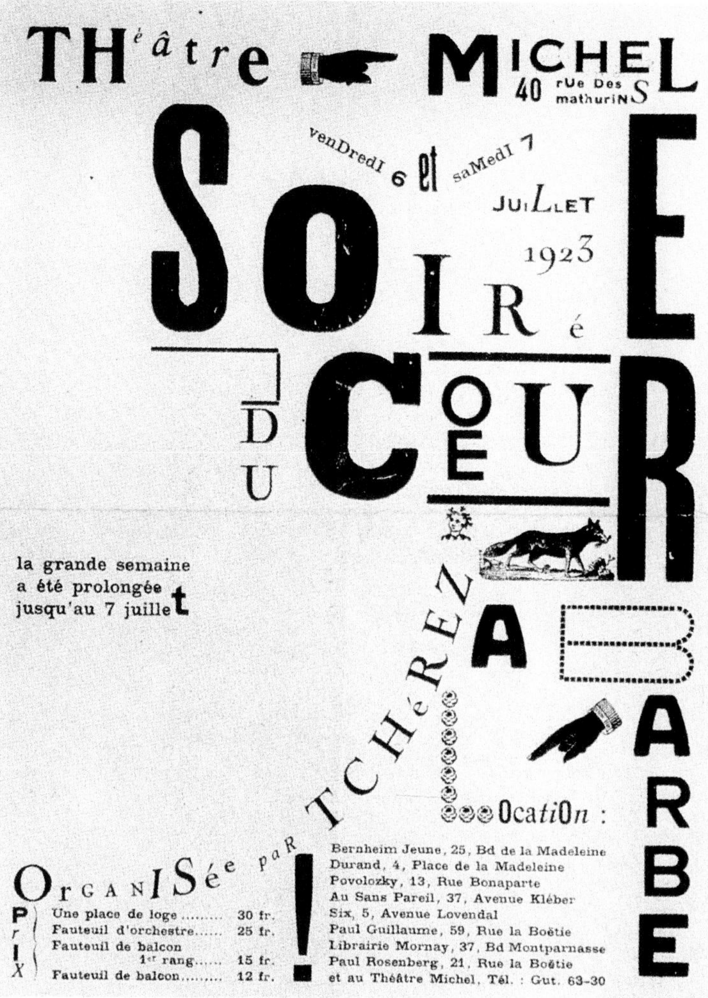


Zdanewitsch 1923 Poster

Zdanewitsch 1923 Page

Zeugin 1960 Poster

Johannes Boehland at the Meisterschule für das grafische Gewerbe in Berlin. 1946–55: bookbinder workshop in Frankfurt am Main. 1946–54: teaches lettering at the Städelschule in Frankfurt am Main. An exhibition of her work is held at the Klingspor Museum in Offenbach (1970) and at the ITC Center in New York (1985). 1991: is presented the F. W. Goudy Award by the Rochester Institute of Technology in Rochester, New York. – *Fonts:* Diotima (1952), Ariadne (1954), Smaragd (1954), Shakespeare (1968), Carmina (1986), Nofret (1987), Christiana (1991), Alcuin (1991), Colombine (1991).

**Zdanevich,** Ilia (Iliazd) – b. 21. 4. 1894 in Tiflis, Georgia, d. 24. 12. 1975 in Paris, France – *typographer, painter, author, publisher* – 1911–17: studies law at the university of St Petersburg. 1912: gives his first Futurist lecture to the Union of Youth artists' group. Organizes lectures and readings and issues publications containing experimental texts and sound poems (called "Zaum Language"), whose design experiments with typographic form. 1921: works in publishing in his Edition du 41° publishing house. Experiments with typography in Paris. 1927–33: designs fabrics for Coco Chanel. 1940: a project with Pablo Picasso produces nine publications with original graphics issued by Edition du 41°. 1961: Iliazd works on the book "Sentence sans Paroles" with Alberto Giacometti. 1964: he and Max Ernst produce "Maximiliana ou l'exercice illégal de l'astronomie" and "L'art de voir de Guillaume Tempel". 1974: the last book Iliazd designs is "Le Cour-

to Giacometti entsteht 1961 das Buch „Sentence sans Paroles". In Zusammenarbeit mit Max Ernst erscheinen 1964 „Maximiliana ou l'exercice illégal de l'astronomie" und „L'art de voir de Guillaume Tempel". Als letztes Buch gestaltet Iliazd 1974 „Le Courtisan grotesque" von Adrian de Monluc mit 16 Farbradierungen von Joan Miró. – *Publikationen u. a.:* „Ostraf Paschi", Tiflis 1919; „Zga jakaby", Tiflis 1920; „Lidantju faram", Paris 1923; „Poésie des mots inconnus", Paris 1949. Olga Djordjadzé „Iliazd", Paris 1978; Françoise Le Gris-Bergmann „Iliazd. Maître d'œuvre du livre moderne", Montreal 1984.

**Zeugin,** Mark – geb. 9. 7. 1930 in Bern, Schweiz. – *Grafik-Designer, Lehrer* – 1948–52 Ausbildung an der Schule für Gestaltung in Luzern und Lehre in einem grafischen Atelier. 1952–54 Grafiker in Basel. 1954–57 Art Director in der Agentur Werner Klapproth in Luzern. 1958 Gründung seines grafischen Ateliers. 1963–72 Präsident des Bundes Grafischer Gestalter der Schweiz. 1965 Umwandlung seines Ateliers in eine Design- und Werbeagentur. 1972–75 Präsident der Arbeitsgemeinschaft Schweizer Grafiker. Seit 1978 Fachlehrer für Grafik an der Schule für Gestaltung in Biel.

**Ziegler,** Zdeněk – geb. 27. 10. 1932 in Prag, Tschechoslowakei. – *Typograph, Grafik-Designer, Lehrer* – 1955–61 Studium an der Technischen Hochschule in Prag. Danach als Gestalter in den Bereichen Plakat, Buch, Ausstellungsgestaltung, Raumausstattung und freier Grafik tätig. 1964 auf der internationalen Ausstellung

mie » et «L'art de voir de Guillaume Tempel» avec Max Ernst. En 1974, Iliazd conçoit la maquette de son dernier ouvrage «Le Courtisan grotesque» d'Adrian de Monluc, qui contient 16 eaux fortes en couleur de Joan Miró. – *Publications, sélection:* «Ostraf Paschi», Tiflis 1919; «Zga jakaby», Tiflis 1920; «Libdantju faram», Paris 1923; «Poésie des mots inconnus», Paris 1949; Olga Djordjadzé «Iliazd», Paris 1978; Françoise Le Gris-Bergmann «Iliazd. Maître d'œuvre du livre moderne», Montréal 1984.

**Zeugin,** Mark – né le 9. 7. 1930 à Berne, Suisse – *graphiste maquettiste, enseignant* – 1948–1952, formation à l'école de design de Lucerne et apprentissage dans un atelier de graphiste. 1952–1954, graphiste à Bâle. 1954–1957, directeur artistique à l'agence Werner Klapproth à Lucerne. Fonde son atelier d'arts graphiques en 1958. Président de l'Union des graphistes suisses de 1963 à 1972. Transforme son atelier en agence de design et de publicité en 1965. Président du Comité de travail des graphistes suisses de 1972 à 1975. Enseigne les arts graphiques à l'école de design de Bienne depuis 1978.

**Ziegler,** Zdeněk – né le 27. 10. 1932 à Prague, Tchécoslovaquie – *typographe, graphiste maquettiste, enseignant* – 1955–1961, études à l'école supérieure des technologies de Prague. Exerce ensuite comme graphiste indépendant, dans le domaine de la conception d'affiches, de livres et comme architecte d'expositions et d'espaces. Participe en 1964 à

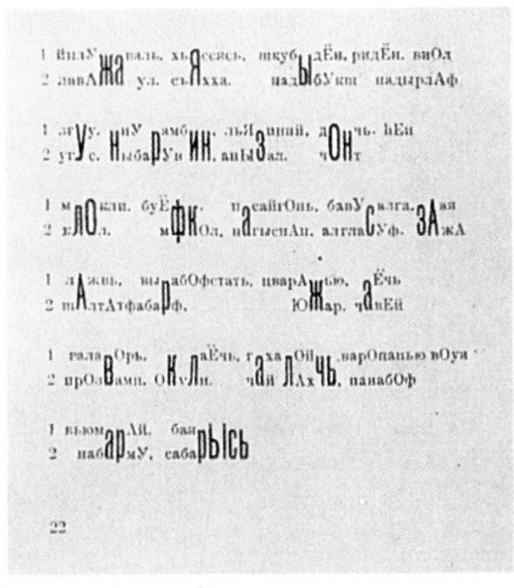
Zdanewitsch 1920 Spread

Ziegler 1990 Poster

Zeugin 1963 Poster

Ziegler 1974 Cover

tisan grotesque" by Adrian de Monluc, with 16 color etchings by Joan Miró. – *Publications include:* "Ostraf Paschi", Tiflis 1919; "Zga jakaby", Tiflis 1920; "Lidantju faram", Paris 1923; "Poésie des mots inconnus", Paris 1949. Olga Djordjadzé "Iliazd", Paris 1978; Françoise Le Gris-Bergmann "Iliazd. Maître d'œuvre du livre moderne", Montreal 1984.

**Zeugin,** Mark – b. 9. 7. 1930 in Bern, Switzerland – *graphic designer, teacher* – 1948–52: trains at the Schule für Gestaltung in Lucerne and at a graphics studio. 1952–54: graphic artist in Basle. 1954–57: art director for the Werner Klapproth agency in Lucerne. 1958: opens his graphics studio. 1963–72: president of the Bund Grafischer Gestalter der Schweiz. 1965: turns his studio into a design and advertising agency. 1972–75: president of the Arbeitsgemeinschaft Schweizer Grafiker. 1978: starts working as a specialist teacher of graphics at the Schule für Gestaltung in Biel.

**Ziegler,** Zdeněk – b. 27. 10. 1932 in Prague, Czechoslovakia – *typographer, graphic designer, teacher* – 1955–61: studies at the technical college in Prague. Then works as a designer of posters and books and in exhibition planning, interior design and free graphics. 1964: takes part and receives an award at the international "Typomundus 64" exhibition. Participates in exhibitions organized by the

„Typomundus 64" ausgezeichnet. Teilnahme an den Ausstellungen der tschechoslowakischen Gruppe „Typo &" u. a. in Prag, Moskau, New York, Berlin und Bratislava. Seit 1990 Lehrer an der Hochschule für angewandte Kunst in Prag. Zahlreiche Auszeichnungen, u. a. „Das schönste Buch des Jahres" (1969, 1970, 1988), „Die besten Plakate des Jahres" (1987, 1988), „Biennale der Gebrauchsgrafik" in Brno (1964, 1978).

**Zimmermann,** Yves José – geb. 17. 4. 1937 in Basel, Schweiz. – *Grafik-Designer, Typograph* – Studium an der Allgemeinen Gewerbeschule in Basel. 1957 Mitarbeit im Studio Will Burtin in New York. Arbeitet 1958 mit dem Architekten Ulrich Franzen, danach Mitarbeit bei Geigy Chemical Corporation in Ardsley (New York), später in Montreal. 1960 Rückkehr nach Basel. 1961 Umzug nach Barcelona. 1961–66 Art Director bei Geigy S. A., Barcelona. 1968 Gründung des eigenen Grafik-Design-Studios. 1975–88 Leitung von „Diseño Integral" (mit André Ricard). 1989 Gründung von „Zimmermann Associados S.L." in Barcelona. Unterrichtete zeitweise an den Gestaltungsschulen Elisava und Eina in Barcelona. – *Publikation u. a.:* „Zimmermann Associados", Barcelona 1993.

**Zwart,** Piet – geb. 28. 5. 1885 in Zaandijk, Niederlande, gest. 24. 9. 1977 in Leidschendam, Niederlande. – *Grafik-Designer, Fotograf, Typograph, Innenarchitekt, Lehrer* – 1902–07 Studium an der Rijksschool voor Kunstnijverheid in Amsterdam. 1908–13 Zeichenlehrer in Leeuwarden. Entwürfe von Schulmöbeln.

l'exposition internationale «Typomundus 64» où un prix lui est décerné. Participe aux expositions du groupe tchèque «Typo &» à Prague, Moscou, New York, Berlin et Bratislava. Enseigne à l'école supérieure des arts appliqués de Prague depuis 1990. Nombreuses distinctions, dont «Le plus beau livre de l'année» (1969, 1970, 1988), «Les meilleures affiches de l'année» (1987, 1988), «Biennale du graphisme publicitaire» de Brno (1964, 1978).

**Zimmermann,** Yves José – né le 17. 4. 1937 à Bâle, Suisse – *graphiste maquettiste, typographe* – études à la Allgemeine Gewerbeschule (école des arts et métiers) de Bâle. 1957, employé à l'agence de Will Will Burtin à New York. Travaille avec l'architecte Ulrich Franzen à partir de 1958, puis chez Geigy Chemical Corporation à Ardsley (New York), puis à Montréal. Retour à Bâle en 1960. S'installe à Barcelone en 1961. Directeur artistique chez Geigy S. A. à Barcelone de 1961 à 1966. Fonde son propre atelier de graphisme et de design en 1968. Dirige «Diseño Integral» (avec André Ricard) de 1975 à 1988. Fonde «Zimmermann Associados S. L.» à Barcelone en 1989. Enseigne par intermittence dans les écoles de design Elisava et Eina à Barcelone. – *Publication, sélection:* «Zimmermann Associados», Barcelone 1993.

**Zwart,** Piet – né le 28. 5. 1885 à Zaandijk, Pays-Bas, décédé le 24. 9. 1977 à Leidschendam, Pays-Bas – *graphiste maquettiste, photographe, typographe, architecte d'intérieur, enseignant* – 1902–1907, études à la Rijksschool voor Kunstnijverheid d'Amsterdam. 1908–1913, professeur de dessin à Leeuwarden. Dessine du mobilier scolaire. Exerce à La Haye comme indépendant de 1913 à 1919. Dessinateur à l'agence de l'architecte Jan Wils à Voorburg de 1919 à 1921. Profes-

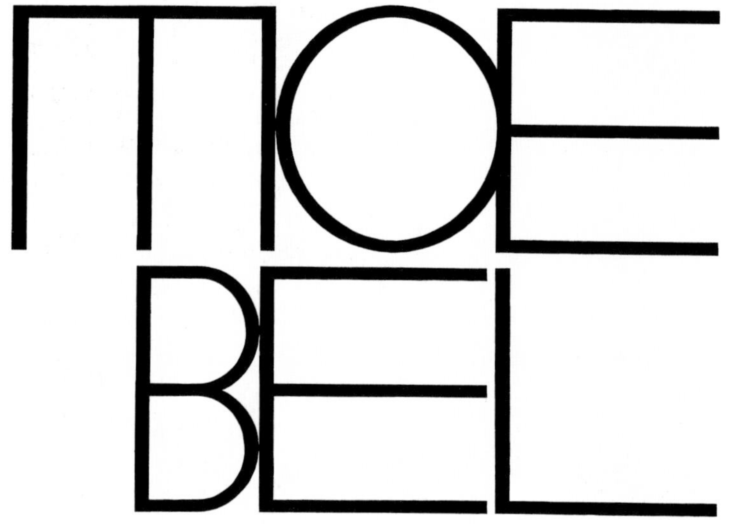


Zimmermann 1957 Poster

Zwart 1925 Page

Zwart 1938 Page

Czechoslovakian Typo & group in Prague, Moscow, New York, Berlin and Bratislava, among other places. 1990: starts teaching at the College of Applied Art in Prague. Ziegler has won numerous awards, including the "Most Beautiful Book of the Year" award (1969, 1970, 1988), the "Best Posters of the Year" award (1987, 1988) and at the "Biennale for Graphic Design" in Brno (1964, 1978).

**Zimmermann,** Yves José – b. 17. 4. 1937 in Basle, Switzerland – *graphic designer, typographer* – Studied at the Allgemeine Gewerbeschule in Basle. 1957: works for the Will Burtin studio in New York. 1958: works with architect Ulrich Franzen and then for the Geigy Chemical Corporation in Ardsley (New York), later in Montreal. 1960: returns to Basle. 1961: moves to Barcelona. 1961–66: art director for Geigy S. A. in Barcelona. 1968: opens his own graphic design studio. 1975–88: runs Diseño Integral with André Ricard. 1989: founds Zimmermann Associados S. L. in Barcelona. Various spells of teaching at the Elisava and Eina design schools in Barcelona. – *Publications include:* "Zimmermann Associados", Barcelona 1993.

**Zwart,** Piet – b. 28. 5. 1885 in Zaandijk, The Netherlands, d. 24. 9. 1977 in Leidschendam, The Netherlands – *graphic designer, photographer, typographer, interior designer, teacher* – 1902–07: studies at the Rijksschool voor Kunstnijverheid in Amsterdam. 1908–13: art teacher in Leeuwarden. Designs school furniture. 1913–19: freelance in The Hague. 1919–21: draftsman for architect Jan

1913–19 freischaffend in Den Haag. 1919–21 Zeichner im Büro des Architekten Jan Wils in Voorburg. 1919–33 Lehrer an der Akademie in Rotterdam. 1921 erste typographische Entwürfe. 1923–27 Assistent im Büro des Architekten H. P. Berlage. 1925–28 Architektur-Korrespondent der Zeitschrift „Het Vaderland" in Den Haag. 1926–28 Kataloggestaltung für die Nederlandsche Kabelfabriek in Delft. 1928 Mitglied im „ring neuer werbegestalter". Beginn des fotografischen Werks. 1928–40 Arbeiten für die niederländische Post PTT: Briefmarken, Bücher, Werbung, Reklamestände. Seit 1929 Entwürfe für die Firma Bruynzeel, Gestaltung einer Anbauküche. Sekretär der niederländischen Abteilung der Ausstellung „Film und Foto" in Stuttgart. 1929 Gestaltung eines Katalogs für die Druckerei Trio in Den Haag. 1930 in der Publikation „Gefesselter Blick" von Heinz und Bodo Rasch in Stuttgart vertreten. 1931 Gastdozent am Bauhaus in Dessau. 1931–37 Präsident der Architekturgruppe „Opbouw". 1942–43 durch die Nationalsozialisten interniert. Seit 1945 Bau und Umbau der Bruynzeel-Büros. Entwurf von Sperrholzmöbeln. Ausgezeichnet mit dem Quellinus-Preis (1959) und dem David-Röell-Preis (1964). – *Publikationen u. a.:* J. Schrofer, W. Sandberg „Sleutelwoorden: Piet Zwart", Den Haag 1965; F. Müller, P. E. Althaus „Piet Zwart", Teufen 1966; Karst Zwart „Piet Zwart en PTT", Den Haag 1968; Kees Broos „Piet Zwart", Den Haag 1973; K. Broos „Retrospective Fotografie: Piet Zwart", Düsseldorf 1981.

Zwart 1928 Advertisement

Zwart 1931 Page

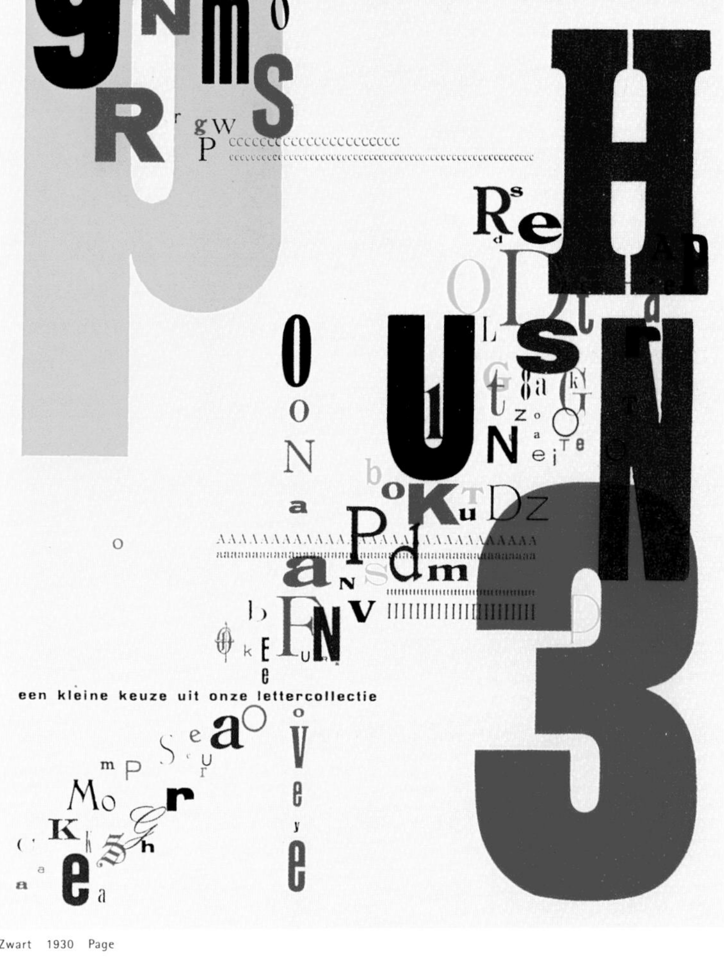


Zwart 1930 Page

seur à l'académie de Rotterdam de 1919 à 1933. Premières esquisses typographiques en 1921. Assistant à l'agence de l'architecte H. P. Berlage de 1923 à 1927. Correspondant en architecture de la revue «Het Vaderland» de 1925 à 1928, à La Haye. Maquette de catalogues pour la «Nederlandsche Kabelfabriek» de Delft, de 1926 à 1928. Membre du «ring neuer werbegestalter» (cercle des nouveaux graphistes publicitaires) en 1928. Début de l'œuvre photographique. 1928–1940, travaux pour les postes néerlandaises PTT : timbres, livres, publicité, stands. Travaux pour la société «Bruynzeel» à partir de 1929, conception d'une cuisine équipée. Secrétaire de la section néerlandaise de l'exposition «Film und Foto» de Stuttgart. Maquette d'un catalogue pour l'imprimerie «Trio» de La Haye en 1929. Présent dans l'ouvrage «Gefesselter Blick» publié à Stuttgart en 1930 par Heinz et Bodo Rasch. Cycle de conférences au Bauhaus de Dessau en 1931. Président du groupe d'architectes «Opbouw» de 1931 à 1937. Interné par les national-socialistes de 1942 à 1943. Construction et réaménagement des «Bruynzeel-Kontore» à partir de 1945. Dessine des meubles en contre-plaqué sans structures en menuiserie. Lauréat du prix Quellinus (1959) et du prix David Röell (1964). – *Publications, sélection :* J. Schrofer, W. Sandberg «Sleutelwoorden : Piet Zwart», La Haye, 1965; F. Müller, P. E. Althaus «Piet Zwart», Teufen 1966; Karst Zwart «Piet Zwart en PTT», La Haye 1968; Kees Broos «Piet Zwart», La Haye 1973; Kees Broos «Retrospective Fotografie : Piet Zwart», Düsseldorf 1981.

Wils in Voorburg. 1919–33: teaches at the academy in Rotterdam. 1921: produces his first typographic designs. 1923–27: assistent to architect H.P. Berlage. 1925–28: architectural correspondent for "Het Vaderland" magazine in The Hague. 1926–28: designs catalogues for the Nederlandsche Kabelfabriek in Delft. 1928: member of the ring neuer werbegestalter. Starts working with photography. 1928–40: works for the Dutch post office PTT, designing postage stamps, books, advertising and advertising stands. 1929: starts designing for the Bruynzeel company. Designs a unit kitchen. Secretary for the Dutch section of the "Film und Foto" exhibition in Stuttgart. 1929: designs a catalogue for Trio printers in The Hague. 1930: is represented in Heinz und Bodo Rasch's publication "Gefesselter Blick" in Stuttgart. 1931: visiting lecturer at the Bauhaus in Dessau. 1931–37: president of the Opbouw architectural group. 1942–43: interned by the National Socialists. 1945: starts building and rebuilding the Bruynzeel offices. Designs plywood furniture. Zwart is awarded the Quellinus Prize in 1959 and the David Röell Prize in 1964. – *Publications include:* J. Schrofer, W. Sandberg "Sleutelwoorden: Piet Zwart", The Hague 1965; F. Müller, P.E. Althaus "Piet Zwart", Teufen 1966; Karst Zwart "Piet Zwart en PTT", The Hague 1968; Kees Broos "Piet Zwart", The Hague 1973; Kees Broos "Retrospective Fotografie: Piet Zwart", Düsseldorf 1981.

## Ergänzungen zu den Bildunterschriften im Wer-Teil
## Compléments des légendes du chapitre Qui
## Additional information to the captions in the Who-section

**Achleitner, Friedrich:** 84 l.o.: aus F. Achleitner „schwer schwarz", Frauenfeld 1960
**Adobe Systems Inc.:** 84 r.o.: Titelblatt von Adobe-Type-Library
**Aicher, Otl:** 85 r.o.: aus der Broschüre „Oberflächiges", Druckhaus Maack, Lüdenscheid 1991, 85 r.u.: Schutzumschlag zu Otl Aicher „typographie", Lüdenscheid 1988
**Albers, Josef:** 86 o.: Entwurf für einen Ladenumbau „Ullstein", 86 l. u. und l.M.: Kombinationsschrift aus Glas , 111 l.o.: Schablonenschrift für Bauhaus aus "Offset, Buch- und Werbekunst" 7/1926
**Albert-Birot, Pierre:** 87 l.o.: Signet für „SIC"
**Allner, Walter:** 89 l.: Ausstellung „Walter Allner", Bauhaus Dessau, 89 r.o.: Titelblatt von „Fortune", Juli 1964, Jubiläum des 500.Heftes. Die Fenster des Hochhauses wurden nach Anweisung Allners erleuchtet.
**American Type Founders Company (ATF):** 90 l.o.: aus „Book of American Types", American Type Founders Company1934. Design: M. F. Benton
**Amman, Jost:** 90 r.o. und 91 l.o.: aus „Ständebuch" herausgegeben von S. Feyerabend, Frankfurt a. M.
**Antupit, Samuel N.:** 91 r.u.: für Buchverlag Harry N. Abrams Inc. Publishers
**Apeloig, Philippe:** 92 l.o.: Ausstellung „Philippe Apeloig, graphiste", Galerie Impressions, Paris, 92 l.u.: „Foire Internationale Du Livre Ancien", Internationales Forum des alten Buches, Paris, 92 M.u.: „Les arts plastique", Press kit cover, Ministry of Culture, France
**Arp, Hans:** 94 l.o.: Ausstellung „Abstrakte und surrealistische Malerei und Plastik", Kunsthaus Zürich. Zusammenarbeit mit W. Cyliax
**Aschoff, Eva:** 96 r.o.: Entwürfe für Wortmarke der Kunstseidenfirma Rhodiaseta
**Ashbee, Charles Robert:** 97 l.o.: aus „An Endeavour towards the Teaching of J. Ruskin and W. Morris", Essex House Press. Buchschmuck: G. Thomson
**Association Typographique Internationale (ATypI):** 97 r.o., r.M.: und u.: Beispiele jährlich wechselnder Eigenlogos, 197 l.o.: Abb. eines Covers für AtyPI
**Atelier National de Création Typographique (ANCT):** 98 o.: aus „Atelier National de Création Typographique", Paris 1992, 322 l.v.: Umschlagentwurf von P. Keller für ANCT
**Baines, Phil:** 100 l.: Illustration für Titelblatt von „graphic world", 100 r.o.: aus: „The BAUHAUS - mistook legibility for communication (it's a man's world)"
**Balla, Giacomo:** 100 r.u.: Bühnenskizze „Printing Machine or Typography"
**Banks, Colin:** 103 l.o.: Design: Zusammenarbeit mit J. Miles, 103 r.o.: Logo für British Telecom, Zusammenarbeit mit J. Miles, 390 r.o.: Cover für Monotype Corporation in Zusammenarbeit mit J. Miles
**Barnbrook, Jonathan:** 104 l.u.: Schutzumschlag zu Damien Hirst „Damienhirst", London 1997. Design: Zusammenarbeit mit D. Hirst, 104 r.u.: machine-generated stonecarving
**Baskerville, John:** 105 r.u.: aus Ariosto „Orlando Furioso", Birmingham, gedruckt von J. Baskerville
**Bass, Saul:** 106 r.o.: mit H. Markowitz
**Bauer, Konrad Friedrich:** 108 l.o.: Design von E. R. Weiß
**Bauhaus:** 110 l.o.: M. Téry-Adler. Titelblatt zu „Utopia", Hrsg.: B. M. Adler, Utopia-Verlag, Weimar 1921, 110 l.u.: A. Arndt, Entwurf für ein Plakat der „Konsumgenossenschaft E.G.M.B.H. Bäckerei Jena", 110 r.o.: L. Moholy-Nagy, Innentitel des Katalogbuches „Staatliches Bauhaus Weimar 1919–1923", 110 r.u.: H. Bayer, Ausstellung „Europäisches Kunstgewerbe", Grassimuseum, Leipzig, 111 l.o.: J. Albers, Schablonenschrift aus der Zeitschrift „Offset, Buch- und Werbekunst" 7/1926, 111 l.u.: K. Kranz, Entwurf für eine Schrift aus Satzmaterial, 111 r.o.: Ausstellung „Bauhaus Dessau im Gewerbemuseum Basel". Design: F. Ehrlich nach einem Entwurf von J. Schmidt
**Bauhaus-Archiv:** 111 r.u.: für „Bauhaus-Archiv Berlin", Design: atelier doppelpunkt (N. Kirschner, C. Trauer, B. Tümmers, P. Reisdorf)
**Bauhaus-Museum:** 112 l.o.: Design: Ott + Stein. 1994 fand die Ausstellung noch in den „Kunstsammlungen zu Weimar" statt. Das Bauhaus-Museum wurde erst 1995 gegründet.
**Baumann+Baumann:** 113 l.o.: zu „lechts und rinks. Orientierungen zwischen Architekur und Parlament", Stuttgart
**Baur, Ruedi:** 114 l.o.: zur Teilnahme an der Ausstellung „La typographie hollandaise", gerichtet an niederländische Mitglieder der AIGA usw.
**Bayer, Herbert:** 110 r.u.: Poster für Bauhaus, 383 l.o.: Cover, gestaltet nach der Schrift Universal (Version 1925) 114 r.u.: Entwurf für einen Zeitungskiosk, 115 l.o.: Ausstellung „Kandinsky" zum 60. Geburtstag, Anhaltischer Kunstverein, 115 l.u.: Titelblatt von „bauhaus" 1/1928, 115 r.o.: Ausstellung „Modern Art in Advertising, Designs for Container Corporation"
**Beeke, Anthon:** 116 l.o.und M.o.: Anthon Beeke-Alphabet. Fotografie: G. Kooiman. Produktion: A. Beeke. Reportage: E. van der Elsken. Ein Kwadraatblad der Steindruckerei de Jong & Co, Hilversum. " 1970: Anthon Beeke und Steindruckerei de Jong & Co, 116 l.u.: Ausstellungsplakat
**Behmer, Marcus:** 116 r.u.: aus „Hebräische Bibel" der Officina Serpentis
**Behrens, Peter:** 117 r.o.: Musterinitiale für Frakturschrift
**Benguiat, Ephram Edward:** 119 r.o.: Logo für „Alphabets & Images", New York
**Benton, Morris Fuller:** 90 l.o.: Design einer Seite für ATF, 120 r.o. und r.u.: für die American Type Founders Company, 121 l.o.: Specimen der Schrift Raleigh Gothic Condensed, 170 l.M.: mit T. M. Cleland, 368 u.: Zusammenarbeit mit W. Martin
**Bernard, Pierre:** 123 r.o.: aus dem Wettbewerb „30 Posters on Environment and Development", United Nations, 123 r.M.: Design: ANCT – Grapus , 123 l.u.: Veranstaltungsreihe zum Thema „ Algérie, expressions contemporaines", 123 r.u.: Logo für „Parc de la Villette", Paris
**Bernhard, Lucian:** 124 r.M.: Bernhard Fraktur
**Berthold AG:** 125 l.u. und r.u.: Schriftprobe Schneider-Antiqua, 125 r.o.: Titelblatt für Berthold Grotesk-Schriften
**Bertieri, Raffaello:** 126 l.o.: Design: A. Butti, Art Director: R. Bertieri
**Bierut, Michael:** 126 r.o.: Poster zur Veranstaltungsreihe im International Design Center New York, „IDCNY", 126 r.u.: Logotype für „Children Museum of Manhattan" , 127 l.o.: für ein Kolloquium zur Revitalisierung New Yorks
**Bilibin, Iwan Jakowlewitsch:** 128 l.o.: Anschlagzettel der „Historischen Ausstellung von Kunstgegenständen in den Sälen des Baron Stieglitz-Museums", 128 M.o.: Cover für „Das lebendige Wort", 128 r.o.: Cover zu „Chrestomathie zur Geschichte der russischen Literatur", 128 l.u.: Cover für die Buchfolge Edition der „Expedition zur Herstellung von staatlichen Wertpapieren"
**Bill, Max:** 129 r.u.: „Kulturplakat", typographische Illustration für ein Plakat
**Black, Roger:** 131 r.u.: Titelblatt zum 10jährigen Bestehen der Zeitschrift „Rolling Stone"
**Blase, Karl Oskar:** 132 r.: Ausstellung „documenta 4", unter Verwendung des Logos „d 4" (von O. Blase), 133 l.u.: Ausstellung „documenta 8", unter Verwendung des Logos „d 8" (von O. Blase)
**Blokland, Erik van:** 133 r.o.: Titelblatt der Zeitschrift „LettError", Zusammenarbeit mit J. v. Rossum
**Bodoni, Giambattista:** 134 r.o. und r.u.M. und 135 l.o. Seiten aus „Manuale Tipografico", Parma, 134 Initialen aus Bodoni „Fregi e Majuscole", Parma
**Boehland, Johannes:** 135 r.: für "Deutsche Keramische Gesellschaft"
**Bons, Jan:** 137 M.u.: Plakat nach Emblem, entworfen zum 700jährigen Bestehen Amsterdams
**Bradley, William H.:** 140 r.: Ultra Modern Initials, 141 l.o.: Titelblatt für die Zeitschrift „The Chap Book"
**Bremer Presse:** 144 r.o.: Anna Simons, Initialen zu Dante „La Divina Commedia", 144 r.u.: Probebogen zu „Missale Romanum", Initialen von F. Thiersch
**Bretteville, Sheila Levrant de:** 145 l.o. und r.o.: „Omoide no Shotkyo", Los Angeles. Colored concrete, stainless steel, brass, 145 l.u.: „Path of Stars". Granite, colored concrete
**Brinkmann + Bose:** 146 l.o.: Ausstellung „Walter Benjamin", Theodor W. Adorno Archiv, Frankfurt a. M., 146 l.u.: Ausstellung „Carlfriedrich Claus", Staatliches Lindenau-Museum, Altenburg
**Brodovitch, Alexey:** 147 o.: Ankündigung für „Bal Banal", 213 l.o.: Design, 433 r.u.: Design
**Brody, Neville:** 148 l.o.: Ausstellung „Die Zeichensprache des N. Brody", Karmeliter Kloster, Frankfurt a. M., 148 r.o.: aus der Zeitschrift „The Face" Nr. 59, März 1985, 148 r.M.: aus der Zeitschrift „Arena" Nr. 5, Herbst 1987, 148 r.u.: Typeface Two, designed for the magazine „The Face" 1984
**Brownjohn, Robert:** 149 r.o.: Ausstellung „Brownjohn, Chermayeff & Geismar", Gallery 303, New York, 167 l.u.: Brownjohn, Chermayeff & Geismar
**Brudi, Walter:** 149 M.u.: Pan-Schrift
**Büchler, Robert:** 150 l.o.: Ausstellung „Typographie. 50 Jahre Handsetzervereinigung Basel", Gewerbemuseum Basel 1960. Plakat und Katalogumschlag
**Bulmer, William:** 150 l.u.: aus „Bibliographical Decameron", Bulmer's London
**Burchartz, Max:** 151 l.o.: „Flugblätter mit dem Buntquadrat", Nr. 2, Eigenwerbung, 151 l.u.: Schriftentwurf, veröffentlicht in: „Imprimatur", 1931
**Burtin, Will:** 429 r.u.: Zusammenarbeit mit C. Pineles
**Buzzi, Paolo:** 153 r.o.: aus „L'Elisse e la Spirale"
**Caflisch, Max:** 156 l.o.: Katalog der Ausstellung „Das Schweizer Buch", Amsterdam
**Cangiullo, Francesco:** 157 l.o.: zu „Parole in libertà", Mailand, 157 l.u.: graue Tinte und Wasserfarbe auf Papier
**Carnase, Tom:** 158 l.u.: Stilisierte Zahl für eine Weihnachtskarte von A. Peckolick, 158 r.u.: Buchtitel, 532 r.u.: Zusammenarbeit mit M. Vignelli
**Carrà, Carlo:** 159 l.u.: Buchumschlag zu „Guerrapittura", 159 r.: Idéogramme de Guerre
**Carter, Matthew:** 213 r.o.: Design , 383 l.o.: Schriftgestaltung
**Carter, Will:** 326 l.u.: mit D. Kindersley
**Casey, Jaqueline S.:** 162 l.u.: Ausstellung „Arts on the Line", Hayden Gallery and Hayden Comdor Gallery, Cambridge, Mass., USA
**Cassandre, A. M.:** 164 r.o.: Ausstellung „Foire de Paris", 165 l.u.: geschnitten von C. Peignot, 424 r.o.: Zusammenarbeit mit C. A. Peignot
**Cato, Ken:** 165 r.o.: für „The Applied Research Corporation Singapore"
**Caxton, William:** 165 r.u.: Caxton's Type, aus „Godfrey of Bologne", Westminster
**Cerri, Pierluigi:** 166 r.o.: Titel des Katalogs zur Ausstellung „Futurismo & Futurismi", Palazzo Grassi, Mailand
**Chermayeff, Ivan:** 167 l.u.: Brownjohn, Chermayeff & Geismar, 167 r.o.: Ausstellung „Mobil Masterpiece Theatre", Museum of Broadcasting, New York, 167 r.u.: für Container Corporation of America
**Chwast, Seymour:** 168 M.u.: Filmfestival Cannes, 169 l.u.: Umzugsankündigung der Filmproduktionsgesellschaft „Elektra"
**Cobden-Sanderson, Thomas James:** 170 r.: aus Cobden-Sanderson „The Ideal Book or Book Beautiful", London
**Comeriner, Erich:** 171 r.o.: aus Stehsatzresten der Bauhaus-Druckerei, auf Zeitungspapier
**Cranach-Presse:** 172 u.: aus William Shakespeares „Hamlet"
**Cranbrook Academy of Art:** 172 r.o. und 173 l.o.: student works, 173 l.u.: Ausstellung „Fluxus selections from the Gilbert and Lia Silverman Collection", Cranbrook Faculty project. Design: K. McCoy, 173 r.: zu H. Aldersley-Williams u. a. „Cranbrook Design. The new discourse", New York 1990
**Craw, Freeman:** 174 l.: Schrift für die American Type Founders Company, 174 r.o.: aus „A 24 Page Book"
**Crouwel, Wim:** 174 r.u.: Cover „new alphabet, eine Möglichkeit für die neue Entwicklung", quadraatbladen, Steindruckerei De Jong & Co, 175 l.o.: Ausstellung „Léger", Stedelijk van Abbe Museum, Eindhoven, 175 r.o.: Ausstellung „Edgar Fornhout", Stedelijk van Abbe Museum, Eindhoven, 175 l.u.: aus Prospekt für „new alphabet", quadraatbladen, Steindruckerei De Jong & Co
**Curwen Press:** 176 l.o.: Umschlagtitel, Faber & Faber. Mit Monotype-Schmuck, Monotype Walbaum-Antiqua, 176 l.u.: Design: H. Schmoller
**Cyan:** 176 r.o.: Programm der Staatsoper Berlin, 176 M.u. und r.u.: Jubiläumsjahr 1993: Kultur Leipzig, 177 l.: Veranstaltung „Film im Gespräch", 177 r.o.: Ausstellung „Lucia Moholy", Bauhaus Dessau, 495 r.: Design
**Cyliax, Walter:** 94 l.o.: Design in Zusammenarbeit mit H. Arp
**Deck, Barry:** 181 l.o.: Anzeige für „Gonz Graphics", USA, Fotografie: Charles Field, 181 r.o.: Canicopulus Script, 181 r.u.: für „California Institute of the Arts"
**Deffke, Wilhelm:** 182 l.o.: Handelsmarke „Foto" der fotografischen Abteilung der Farbenfabriken, vorm. Fr. Bayer & Co. Leverkusen
**Depero, Fortunato:** 183 l.o.: Studio Boggeri
**Derrida, Jacques:** 184 l.o.: Buchtitel, Design von W. Fleckhaus
**Didot, François Ambroise:** 186 l.: Caractère Maigre, auch benutzt von P. Didot, Paris
**Didot, Pierre:** 186 r.o.: Didot-Antiqua, D. Stempel AG (Hrsg.), 187 r.u. und l.u.: aus „ Noveaux Caractères", Paris, 188 o. und l.u.: aus „Horace"
**Didot, Firmin:** 188 r.u.: Didot's Italic, gedruckt von F. Didot, Paris, 189 r.: Deberny & Peignot
**Dijk, Christoph van:** 190 r.: Schriftprobe für den Verkauf von Elsevier-Typen, Amsterdam
**DiSpigna, Tony:** 191 l.o.: Symbol für die Stadt San Antonio, Texas, 191 l.M.u.: Signet für „Büroeinrichtungsunion", Paris, Zusammenarbeit mit H. Lubalin, 191 r.o.: Briefmarkenentwurf für die Zweihundertjahrfeier der USA, 191 M.: Namenszug für ein persönliches Briefpapier
**Does, Bram de:** 191 u.: Trinité, 1. und 2. Zeile: Medium Condensed, 3. und 4 Zeile: Roman Wide
**Doesburg, Théo van:** 192 l.u.: „Dada Soirée"
**Domela, Cesar:** 193 r.o.: Titel einer Mappe, 193 l.u.: Schutzumschlag
**Dorfsman, Louis:** 194 l.u.: Titel einer Broschüre für eine Radiostation, 194 r.o.: für „CBS"-News, 194 r.M.: Titel des Jahrbuchs „AIGA"-Graphic Design USA
**Doyle, Stephen:** 195 r.o.: Jahrbuch des Type Directors Club, Design: Drenttel Doyle Partners, 195 r.u.: Corporate Identity Programm für „The World Financial Center", New York, Zusammenarbeit mit T. Kluepfel
**Dreyfus, John:** 197 l.o.: Cover der Broschüre „ Dossier A-Z 73", Association Typographique Internationale, Design: Zusammenarbeit mit F. Baudin
**Dumbar, Gert:** 197 r.o.: Veranstaltung „Transatlantic Light", Weltpremiere, 197 l.u.: Teil des Erscheinungsbildes der ptt, niederländische Postgesellschaft, 197 r.u.: für Zanders-Papier
**Dürer, Albrecht:** 198 l.: Schriftkonstruktion für den Schildermaler aus „Underweysung der Messung, mit dem Zirckel und richtscheyt" etc., Nürnberg, 198 r.o.: an Erzherzog Ferdinand von Österreich, Brabant und Burgund. Druck: J. Neudörffer, Fraktur: Andreae, 198 r.u. und 199 l.u.: aus „Underweysung der Messung, mit dem Zirckel und richtscheyt" etc., Nürnberg
**Dwiggins, William Addison:** 199 M.u.: Jahrbuch Nr.2
**Eckersley, Tom:** 202 M.o.: Ausstellung „ Tom Eckersley", 202 l.u.: Plakat für „London Transport"
**Edelmann, Heinz:** 203 M.o.: für Atlas Film, 203 r.o.: Buchumschlag, Klett Cotta, 203 l.u.: Musical-Plakat, 203 r.u.: Rundfunkkonzert, WDR
**Ehmcke, Fritz Helmut:** 204 l.o.: Internationale Presseausstellung „Pressa", Köln
**Eidenbenz, Hermann:** 204 l.u.: für „Schweizer Filmarchiv", Lausanne, 204 M.u.: für Hanf- und Leinenverkaufsgesellschaft, 204 r.u.: für „Basler Kulturgemeinschaft", 205 l.o.: Ausstellung „grafa international", Mustermesse Basel
**8vo:** 205 r.o.: Ankündigung eines Plattenalbums, 205 l.u.: Visitenkarte, 205 r.u.: Veranstaltungsplakat
**Elffers, Dirk Cornelius:** 206 r.u.: Titelblatt für die Zeitschrift „de rimpel" Nr. 0, Jan. 1933, 207 l.: Filmplakat
**ElsnerkFlake:** 207 r.u.: Titelseite der Hülle einer Schriften-CD
**Emigre:** 208: Titelblätter der Zeitschrift „Emigre", 208 r.u.: Design: R. VanderLans
**Enschedé en Zonen:** 209 l.: aus „Enschedé 's Proef van Letteren", Haarlem, 209 r.o.: Schreibschrift „Dubbelde Garmond" aus „Enschedé 's Proef van Letteren", Haarlem
**Erbar, Jakob:** 210 l.u.: aus „Archiv für Buchgewerbe und Gebrauchsgraphik" 7/1929
**Ernst, Max:** 210 r.o.: Umschlag zu „Minotaure" Bd. 11, 210 r.u.: Collage zu „Poèmes Visible", 211 o.: Ausstellung „Dada siegt", Zürich, 211 l.u.: Umschlag zu „The Little Review" 8/1924, 211 r.u.: aus „Maximiliana ou l'exercice illégal de l'astronomie" mit der von M. Ernst erfundenen „Geheimschrift"
**Excoffon, Roger:** 212 r.o.: für Air France, 212 r.u.: Mistral(1953) basiert auf Excoffons Handschrift, 213 l.u.: aus einer Serie für „Larousse", Herausgeber von Lexika und Wörterbüchern
**Eye:** 213 l.o.: Titelblatt der Zeitschrift „Eye" 8, Vol. 2/1993. Design: A. Brodovitch, 213 r.o.: Titelblatt der Zeitschrift "Eye" 19, Vol. 5/1995. Design: M. Carter for Walker Art Center, 213 r.u.: Bestellkarte, Design: LettError
**Fairbank, Alfred John:** 216 l.o.: Gedicht von S. T. Coleridge, 216 l.u.:"Woodside Writing Card Nr. 4" für die Dryad Press, 216 r.u.: Detail aus „Lovers of Tervel", Schrift: Narrow Bembo italic
**Feininger, Lyonel:** 216 r.o.: Titelblatt der ersten Mappe „Meister des Staatlichen Bauhauses Weimar", Verlag Müller & Co, Potsdam 1921
**Feitler, Bea:** 217 r.o.: Titelblatt der Zeitschrift „Harper's Bazaar", April 1965, 217 r.u.: aus „Harper's Bazaar"
**Feliciano, Felice:** 218 M.u.: aus „Felix Titling", Monotype 399, in 27 p geschnitten
**Fella, Edward:** 218 r.o., 219 l.o. und l.u.: Handzettel für die „Detroit Focus Gallery", 218 r.u.: Eigenlogo „Design Edward Fella", 219 M.o.: zu einer Art-School-Party
**Fili, Louise:** 219 r.o.: Aufforderung zum Wettbewerb der „Society of Publication Designers SPD", 219 r.M.: Buchumschlag, 219 r.u.: für den Couturier Kubinyi
**Fiore, Quentin:** 221 l.o. und r.o., l.u.: aus Marshall McLuhan und Quentin Fiore „The Medium is the Massage. An Inventory of Effects", produced by Jerome Agel. " 1967 by Jerome Agel und 1996 by Jerome Agel, 221 r.u.: aus R. Buckminster Fuller „I Seem to Be a Verb", 375 r.u.: Entwurf
**Fischer, Florian:** 222 l.o.: Typo-Display „Stulle, Schriftbild" für Layoutsetzerei Stulle, in Zusammenarbeit mit E. Spiekermann, 222 l.u.: Ausschnitt aus „Textbaum", für Internationale Bauausstellung Berlin 1984, Zusammenarbeit mit A. Güler
**Fleckhaus, Willy:** 184 l.o.: Design
**Fleischmann, Gerd:** 223 l.o.: zu „concept art, minimal,

arte povera, land art – farce?“, Marzona, Düsseldorf
**Fleischmann, Johann Michael:** 224 l.: aus „Enschedé's Proef van Letteren“, Haarlem, 224 r.o.: Holländische Schreibschrift
**Fletcher, Alan:** 225 r.o.: 50. Jahrestag: Invasion der Nazis in Polen, Beginn des Zweiten Weltkrieges, 225 r.u.: für Victoria & Albert Museum, London, Zusammenarbeit mit Q. Newark, Pentagram, 425 r.o.: Design
**Flusser, Vilém:** 226 r.u.: Buchtitel, Design: K. Sommer
**FontShop:** 227 l.o.: Titelseite eines Katalogs
**Forum Typografie:** 228 l.M.: Vorder- und Rückseite des Umschlags von „Forum Typografie 0012, Von vorne“, Verlag H. Schmidt, Mainz. Design: R. Wallhäuser, 228 l.u.: Vorder- und Rückseite von „Staben-Buch“, Forum Typografie, Essen 1986, Foto: K. Lammers
**Fournier, Pierre Simon:** 228 r.o., r.u. und 229 l. und r.o.: „Manuel typgraphique“ (Handbuch der Buchdruckerkunst)
**Frej, David:** 230 l.u.: aus Magazin „ID Graphic Design“
**Friedl, Friedrich:** 230 r.o.: Ankündigung eines Vortrags von E. Jandl, 230 M.u.: Ausstellung „10 Jahre Typographie-Ausbildung an der Hochschule für Gestaltung Offenbach“, 230 r.u.: Dokumentation konstruktiver und konkreter Kunst, Edition und Galerie Hoffmann, Friedberg
**Friedländer, Henri:** 231 r.o.: aus Jaques de Baisieux „Het zijden harnas“, 231 r.u.: aus J. Slagter „Leo Gestel“, Den Haag
**Friedman, Daniel:** 232 r.o.: Konzert des Yale Symphony Orchestra, 232 r.M.: für Theater in New Haven, 232 l.u.: Titelblatt der Zeitschrift „Typografische Monatsblätter“ 1/1971, 232 r.u.: für einen Möbelschreiner
**Fronzoni, Angiolo Guiseppe:** 233 r.o.: Konzert-Programm
**Froshaug, Anthony:** 234 l.o.: Katalogtitel, 234 r.o.: Konzert „Stefan Themerson“, 234 l.u.: Geschäftsbericht für „Associated London Properties“
**Frutiger, Adrian:** 235 l.o.: Abschnitt aus „Schrift Ecriture Lettering“, Zürich, 235 M.u.: „Revue sciences“, Edition Herrmann, Paris
**Fugger, Wolfgang:** 236 r.o. und r.u.: aus „Schreibbüchlein“
**Fuse:** 237 l.o.: „Can you and do you want to read me? For typographic magazine. „Fuse“ 1/1991, 237 r.o.: Fuse 4. Design: J. Keedy
**Fust, Johann:** 238 l.: „Psalterium“, J. Fust und P. Schöffer, Mainz, 238 r.M.: Type aus dem „Psalterium“, Zusammenarbeit mit P. Schöffer, 238 r.u.: Druckerzeichen, Zusammenarbeit mit P. Schöffer, 476 l.o.: aus Durandus "Rationale divinorum officiorum", Druck von J. Fust und P. Schöffer
**Games, Abram:** 240 l.o.: Aufruf zum Melden aller Blindgänger, 240 M.o.: Memorial Panel for Holocoust Victims at Hiller House, London
**Gassner, Christof:** 243 l.o.: „ZDF Matinee“, 243 l.u.: Ankündigung von Theatervorstellungen, 349 r.o.: Design
**Geismar, Thomas H.:** 243 r.u.: Titelblatt der Zeitschrift „Fortune“ 7/1963, 167 l.u.: Brownjohn, Chermayeff & Geismar
**Genzsch k Heyse:** 244 l.u.: Festschrift, Hamburg
**Gerstner, Karl:** 245 l.o.: Anzeige für Haensel, Holzminden (Herstellung von Zitronenlimonade). Design: Gerstner + Kutter, 245 r.o.: Design: Gerstner + Kutter, 245 l.u.: Abschlußinserat für Citroen. Design: Gerstner + Kutter, 245 r.u.: Anzeige der „Schwitter AG“, Basel
**Gerz, Jochen:** 246 l.o.: „Der Baum der schweigenden Mehrheit“, Arbeit auf Papier, Privatsammlung, 246 r.o.: „Bilder einer Ausstellung“, Kunst im öffentlichen Raum, Hauptbahnhof Hamburg, 246 M.l.: „Revolution“, Objekt, Privatsammlung, 246 l.u.: „A Sense of Attention“, Installation, Court. Manchester City Art Gallery, 246 r.u.: „People Speak“ " I am You, München/Brüssel
**Gill, Bob:** 247 l.o.: Ankündigung eines Vortrags, 247 l.u.: für die Zeitung „Evening Standard“
**Gill, Eric:** 309 l.o.:Holzschnitt
**Goldberg, Carin:** 250 l.: Buchumschlag, 250 r.o.: Plattencover für „Nonesuch-Records“
**Gomringer, Eugen:** 250 r.u., 251 l. und r.o.: aus „Worte sind Schatten“, 251 r.M.: 4, FÜHR, 4 FÜR VIER
**Gottschalk, Fritz:** 252 l.o.: Ausstellungsplakat, Bauamt Zürich, 252 l.u.: Ausstellungsplakat, Heidi Weber-Haus, Zürich, 252 M.u.: zum Schweizer Fest „Rütli 400“ (Jahre)
**Goudy, Frederic William:** 252 r.o. und 253 l.o.: aus F. W. Goudy „Das Alphabet“, New York 1922, 253 r.M.: Entwicklungsreihe des „G“ von der Unziale des 7. Jahrhunderts bis zum 20. Jahrhundert, aus F. W. Goudy „Das Alphabet“, New York 1922
**Grandjean de Fouchy, Philippe:** 254 r.o.: Entwurf für die neue Druckschrift der Königlichen Druckerei in Paris, 254 u.: Imprimerie Nationale, Paris
**Granjon, Robert:** 255 l.o.: Titel in Civilité
**Grappa:** 255 l.u.: für das Anti-Kriegs-Museum, Berlin, 255 r.o.: für das Filmmuseum Potsdam, 255 r.u.: aus „IDZ-Jahresbericht 1989“
**Grapus:** 256 l.o.: Friedenstage „Mr. Reagan No Neutrons“, 256 r.o.: „Apartheid, Racisme, le cancer du monde“, 256 l.u.: Theaterplakat, 256 r.M.: „Les trompettes de la mort“, Le Theâtre Paris, 256 r.u.: Gedenktag 1978 an den 14.7.1789 (Sturm auf die Bastille)
**Grieshaber, H. A. P. (HAP):** 258 l.o.: Bucheinband K. Leonhard „Gegenwelt“, 258 r.o.: Ausstellung „Zwischen Natur und Abstraktion“, Galerie Herbert Herrmann, Stuttgart, 258 l.u.: Ausstellung „Ritsch!“, Galerie Herbert Herrmann, Stuttgart, 258 r.u.: Umschlag zum Katalog „Expressionismus. Literatur und Kunst 1910–1923“
**Griffin, Rick:** 259 l.o.: Zusammenarbeit mit V. Moscoso
**Grignani, Franco:** 260 l.u., r.o., r.u. und 261 l.o.: für Setzerei „Alfieri & Lacroix“, 261 M.o.: Titelseite zu „Pubblicità in Italia“, L'Ufficio Moderno, Mailand
**Gürtler, André:** 262 r.o.: Ausstellungsplakat, Gewerbemuseum Basel
**Gutenberg, Johannes:** 263: aus der 42-zeiligen Bibel
**Gutenberg-Gesellschaft:** 264 r.o.: Rundgotisch von E. R. Weiß, 264 r.u.: zu J. Theele „Das literarische Denkmal für Gutenberg“
**Gutenberg-Museum:** 265 l.o.: Aufruf zur Unterstützung des Gutenberg-Museums
**Haas'sche Schriftgießerei:** 268 l.: Schriftprobe „Nürnberger Schwabacher“
**Hadank, O. H. W.:** 268 r.M.: Etikett für „Genever“ (Korn), 268 r.u.: Hauswappen „Haus Neuerburg“ (Zigaretten), 269 l.o.: Titelseite eines Werbeheftes für „Haus Neuerburg“, Köln, 269 r.o.: Anzeige für Holzverpackung von Zigaretten „Haus Neuerburg“, 269 l.u.: Etikett für Zigarttenpackung „Haus Neuerburg“
**Hadders, Gerhard:** 270 r.o.: für Kunstmuseum Wolfsburg, 270 r.u.: Back Cover des Ausstellungskatalogs, Kunstmuseum Wolfsburg 1994/95, Zusammenarbeit mit F. Janssens, 272 l.o., l.u. und r.u.: Design
**Hamersveld, Johann van:** 271 l.: Cover von „Art“
**Hammer, Victor:** 271 r.o.: zur Schriftprobe „Hammer-Unziale“
**Hausmann, Raoul:** 273 r.u.: „Plakatgedicht“
**Heartfield, John:** 274 l.o.: Umschlag für die Zeitschrift „Die Arena“, 274 l.u. und r.o.: Schutzumschläge, 274 r.u.: für „Deutsches Theater“, Berlin, 275 l.u.: Schutzumschlag,
**Herdeg, Walter:** 276 r.o.: Cover der Zeitschrift „Graphis“, 277 l.o.: Cover des „Graphis“-Jahrbuches 1952/53
**Hess, Richard:** 277 r.M.: Deckel für ein Schachspiel, 277 r.u.: Doppelseite von „Annual Report of Western Union Company“
**Hillman, David:** 278 r.M.: Symbol für das „International Center Against Censorship“, 278 M.u.: für das „SOS-Kinderdorf“
**Hillmann, Hans:** 279 l.o., r.o. und l.u.: für „Neue Filmkunst Walter Kirchner“
**Hinrichs, Kit:** 279 r.u.: Doppelseite aus von „Skald“
**Hlavsa, Oldrich:** 280 M.o.: Logo für Kunstgewerbemuseum Prag
**Hoch, Hans Peter:** 280 r.o.: Einladung zur Ausstellungseröffnung, 280 r.u.: Briefmarke zum 50. Jahrestag des 20.7.1944, 281 l.o.: Ausstellung „Druckgrafik“, Institut für Auslandsbeziehungen, Stuttgart, 281 r.o.: Ausstellung „Zeitgenössische Grafik aus Venezuela“, Institut für Auslandsbeziehungen, Stuttgart, 281 l.u.: für Institut für Auslandsbeziehungen, Stuttgart
**Hochuli, Jost:** 282 l.o.: Heraklit „Panta Rei“, 282 M.o.: Symbol für „Collection of Natural Science for the Town of St. Gallen“, 282 l.u.: Ausstellung „Walter Käch“, Kunstgewerbemuseum Zürich, 282 r.u.: Schutzumschlag zu Jost Hochuli „Bücher machen“, Agfa Compugraphic
**Hofmann, Armin:** 284 r.o.: Ausstellung „Fernand Léger, Alexander Calder“, Kunsthalle Basel, 285 l.: Ausstellung „Sammlung Cevellini“, Kunsthalle Basel, 285 r.o.: Ausstellung „Formen der Farbe“, Kunsthalle Bern, 285 r.u.: Ausstellung „Plan & Bau“, Gewerbemuseum Basel
**Höhnisch, Walter:** 284 l.u. und r.u.: aus Schriftproben Ludwig & Mayer
**Hollenstein, Albert:** 286 l.u.: für American Type Shop Hollenstein, 286 r.u.: für Hollenstein Fototypo 86
**Holzer, Jenny:** 287 r.o.: aus der Serie „Truisms“
**Hostettler, Rudolf:** 288 l.o.: Titelblatt der Zeitschrift „Schweizer Graphische Mitteilungen“ 10/1948, 288 l.u.: Schutzumschlag
**Hottenroth, Franz:** 289 r.o.: Ausschreibung eines Ideenwettbewerbs der Stadt München
**Huber, Max:** 290 r.o.: für RAI, 290 l.u.: Buchtitel „Enciclopedia del Jazz“, 290 r.u.: „13. Großer Preis der Monza-Lotterie“
**Hupp, Otto:** 291 r.u.: Titelseite des ersten Münchener Kalenders
**Huszár, Vilmos:** 292 l.o.: Titelblatt der Zeitschrift „Wendingen“ (Wohnen), 292 r.o.: aus der Serie „Miss Blanche Cigarettes“, 292 l.u.: Titelblatt der Zeitschrift „De Stijl“ (1917–20), 292 r.M.: Schutzumschlag zu „Jahrbuch für Niederländische Handwerkskunst und Kunstgewerbe“, Rotterdam, 292 r.u.: Detail des Titelblatts „De Stijl“
**Ifert, Gérard:** 294 l.M.: aus einem Prospekt für Panissot & Suhner, Paris, 294 M.u.: Logo für „Omnium Technique“, France
**Igarashi, Takenobu:** 294 r.u.: Titelblatt der Zeitschrift „Graphis“ 245/1986, 295 r.o.: Ausstellung „Expo '85“, 295 l.u.: Aluminium-Alphabet, 295 r.u.: Ausstellung „Die Welt des Shigero Uchida“, Tokio
**Inoue Yu-Ichi:** 296 r.: „Tokyo-daikushu“ = Tokyo bombing, 297 l.o.: „Hana“= mandarava (Sanskrit) = heavenly blossom. " Unac Tokyo
**Isingrin, Michael:** 298 l.u.: aus Leonhard Fuchs „Kreüterbuch“, gedruckt von M. Isingrin, Basel
**Isley, Alexander:** 298 r.o.: Umschlag für den Katalog „Next Wave“, Brooklyn Academy of Music, Zusammenarbeit mit A. Knowlton., 298 r.u.: Umschlag für den Katalog „Ruhr Works, the Arts of a German Region“, 299 l.: aus „26 Poetic Portraits“
**Istlerová, Clara:** 299 r.o.: Buchtitel der Edition „Maj“
**Itten, Johannes:** 300 l.o. und l.u.: 2 Blätter aus „Analysen alter Meister in „Utopia“, Hrsg. B. M. Adler, Utopia-Verlag, Weimar, 300 r.: Einband
**Jacno, Marcel:** 302 l.o. und 303 l.o.: für „Théâtre National Populaire“ , 302 r.u.: für „Théâtre de France“
**Jähn, Hannes:** 303 l.u.: Wortmarke der Ausstellung „Westkunst“, 303 r.o. und r.u.: Buchtitel
**Jandl, Ernst:** 304 l. und r.o.: aus „sprechblasen“, Berlin, Neuwied 1968, 1997 Luchterhand Literaturverlag GmbH, München
**Jannon, Jean:** 304 r.u.: Jannons Antiqua- und Kursivschrift, Imprimerie Royale, Paris 1642
**Janson, Anton:** 305 r.o.: Titelblatt mit Jansons Schrift
**Jeker, Werner:** 306 l.o.: Ausstellung „Man Ray“, Kunsthaus Zürich, 306 r.o.: Ausstellung „Übersichtsausstellung Bildende Kunst 1985–1989“, Kunsthaus Grenchen
**Jenson, Nicolaus:** 306 u.: Detail aus Augustinus „De Civitate Dei“, J. Speyer, Venedig, 307 l.u.: Jenson-Antiqua, wie in Eusebius „De Praeparatione Evangelica“, Venedig
**Johnston, Edward:** 309 l.o.: aus „Faust“, Doves Press. In Holz geschnitten von Eric Gill, 309 l.u.: Überschrift und Initialen für die Doves Press, 309 r.: Schriftentwurf für die Londoner Untergrundbahn London Transport, Mitarbeit von E. Gill
**Johnston, Simon:** 310 l.o.: Ausstellung „Think yourself beyond the box“; Art Center, College of Design, USA, 310 l.u.: Titelblatt des „Bulletin des J.P.Getty Trust“
**Jones, Terry:** 310 r.o.: 4. Umschlagseite der Zeitschrift „i-D“ 4/1981, 311 l.o.: Cover der Zeitschrift „i-D“ 5/1981, 311 l.M.: Cover der Zeitschrift „i-D“ 6/1981
**Jost, Heinrich:** 311 r.o.: Ankündigung des Buchs „Beton als Gestalter“, 311 r.u.: Titelblatt der Zeitschrift „Cr5/1981, 311 l.M.: Cover d**Käch, Walter:** 314 l.: Ausstellungsplakat, Gewerbemuseum Zürich, 314 r.o.: für Zentralverband Schweizer Schneider
**Kalman, Tibor:** 315 l.o.: aus Zeitschrift „Colours“ (Benetton) Nr. 4 , 315 l.u.: für Restaurant „Florent“, Zusammenarbeit mit M. McCarty
**Kamekura, Yusaku:** 315 M.u.: für „ICSID“, Kyoto, 315 r.o.: für 35 mm-Kamera von Nikon
**Kapitzky, Herbert Werner:** 316 l.o.: Ausstellung „photographische essays“, Landesgewerbeamt Baden-Württemberg, Stuttgart, 316 l.u.: TV-Logos zur „Kennzeichnung der Sendestation Südwestfunk“ , 316 r.u.: Werbung für die Reihe „Ölfeuer-Jahrbuch“
**Kapr, Albert:** 316 r.o.: Einband, Verlag der Kunst, Dresden 1966, 519 l.u.: Zusammenarbeit mit W. Schulze
**Karow, Peter:** 317 l.u.: ek Kr-program use only one master font., 317 r.u.: Vergleich: multiple master und hz-Programm
**Kassák, Lajos:** 318 r.o.: Titelblatt der Zeitschrift „Dokumentum“, Sonderbeilage
**Käufer, Josef:** 318 r.u.: Weihnachts-/Neujahrskarte
**Kauffer, Edward Mc Knight:** 319 l.: Umschlag der Broschüre „Orient Line“, 319 r.o.: für Shell, 319 r.u.: Katalogumschlag, Museum of Modern Art, New York
**Keidel, Carl:** 320 l.u. und r.u.: Schutzumschläge
**Keller, Ernst:** 321 l.: Ausstellungsplakat, Kunstgewerbemuseum Zürich
**Keller, Peter:** 321 r.u.: Ankündigung „Züricher Kantonal Schützenfest“, 322 l.u.: Umschlag zu „Atelier National de Création Typographique“
**Kiljan, Gerard:** 324 r.u.: " G. Kiljan, niederländische Briefmarken 1, 5, 6, 12 c, Collection Haags Gemeentemuseum 1931 c/o Beeldrecht Amstelveen, 325 l.o.: für PTT, niederländische Post: „kauft Wohltätigkeitsbriefmarken zu Gunsten mißhandelter Kinder!“. " G. Kiljan, „koopt weldadigheids postzegels... ten bate van het misdeelde kind“, Poster, Collection Haags Gemeentemuseum 1931 c/o Beeldrecht Amstelveen, 325 r.o.: für Haagsche Filmliga. " G. Kiljan „robert slodmak, der mann, der seinen mörder sucht“, Anzeige, Collection Haags Gemeentemuseum 1933 c/o Beeldrecht Amstelveen, 325 r.M.: " G. Kiljan „Per luchtpost, PTT“, Reklamekarte, Collection Haags Gemeentemuseum 1932 c/o Beeldrecht Amstelveen, 325 u.: " G. Kiljan „garage taurus“, Flyer, Collection Haags Gemeentemuseum 1938 c/o Beeldrecht Amstelveen
**Kindersley, David:** 326 l.u.: mit W. Carter, 326 r.o.: aus David Kindersley und Lida Lopes Cardozo „Letters slate cut. Workshop Philosophy and Practise in the Making of Letters“, Cambridge 1981, Abb. 57. Experiment mit Boustrophedon-Schreibweise („as the ox ploughs“)
**Kisman, Max:** 328 l.o.: Veranstaltung im „Paradiso“, Amsterdam, 328 r.o. und r.u. und 329 l.o.: Konzert-Plakate „Paradiso“, Amsterdam
**Kleukens, Friedrich Wilhelm:** 329 r.u.: aus F. Hölderlin „Hyperion“, Ernst Ludwig Presse für den Insel Verlag, Leipzig, 329 l.u.: Initialen aus Ingeborg-Antiqua
**Klinger, Julius:** 330 l.o.: für Tabu-Zigarettenhülsen und -papier
**Klingspor, Gebr.:** 117 r.u.: Schnitt, 330 r.o.: Karl-Klingspor-Schrift, 1925, 331 l.und r.o.: Cover eines Schriftprobenheftes
**Koch, Rudolf:** 332 l.u.: Ausstellung „Arbeiten von Rudolf Koch“, Kaiser-Wilhelm-Museum, Offenbach, 333 r.o.: aus „Archiv für Buchgewerbe“ 1/1924, Beilage S. 64
**Kosuth, Joseph:** 334 r.u.: „One of three chairs“
**Krimpen, Jan van:** 335 l.o.: Initialen für die Curwen Press, 335 l.u.: Schutzumschlag
**Kriwet, Ferdinand:** 336 r.M.: Sehtext
**Larcher, Jean:** 341 l.M.: für ein Audiovisionsunternehmen, 341 r.o.: „200 Jahre Französische Revolution“, 341 l.u.: Ausstellung „Jean Larcher“, Artcurial, Paris
**Leck, Bart van der:** 342 l.u.: J. W. v. Goethe, Gedicht „ Über allen Gipfeln ist Ruh“, 342 r.u.: Ausstellungsplakat, Voor de Kunst, Utrecht, 343 l.u.: „Delftsche Sla Olie, Plantenvet Delfia“ (Pflanzenfett)
**Léger, Fernand:** 343 l.o. und r.o.: aus Blaise Cendrar „La Fin du Monde“
**Leistikow, Hans:** 344 l.o.: Titelseite der Ausstellungskatalogs „Die Wohnung für das Existenzminimum“, 344 M.o.: Titelseite der Broschüre „Städtische Volksbüchereien, Frankfurt a. M.“ mit Stadtwappen von Leistikow (1927), 344 r.: Ausstellung und Gespräch „Mensch und Raum, Baukunst in Deutschland 1901–51“, 344 l.u.: Konzert „Mechanische Musik“ im Rahmen der Ausstellung „Musik im Leben der Völker“, Frankfurt a. M., 344 M.u.: Ausstellung „Städtebau in den Niederlanden“, Städelsches Kunstinstitut, Frankfurt a. M.
**Letraset:** 345 r.o.: Titelseite des Letrasetkatalogs 1975, Design: C. Gassner, 345 r.u.: Titelblatt der Zeitschrift „LetraTime“ 1/1979
**Lettergieterij Amsterdam:** 346 l.o.: Mistral, Design: R. Excoffon, 1953
**Leu, Olaf:** 346 r.u.: für Linotype Centennial
**Lévy, Jean Benoît:** 347 l.o.: für den Züricher Friseur Pierre Gigaud, 347 r.: „Solothurner Literaturtage“
**Lewis, John:** 348 l.u.: Schutzumschlag
**Licko, Zuzana:** 526 r.o.: Zusammenarbeit mit R. VanderLans
**Linotype:** 350 l.o.: Titel der Broschüre "Linotype Library 6", 350 u.: Schriftenübersicht der Linotype Library
**Lionni, Leo:** 351 r.: Titelblatt der Zeitschrift „Fortune“, Februar 1960
**Lissitzky, El:** 352 l.o.: aus W. Majakowski „Für die Stimme“, 352 r.o.: Plakatskizze, 352 l.u.: Katalogumschlag, 353 l.o.: Anzeige für Pelikan-Schreibbänder
**Loesch, Uwe:** 352 r.u.: Kabarettplakat, 353 r.o.: für die Bauersche Druckerei zur Ausstellung „DRUPA“, 353 r.u.: Ausstellung „Uwe Loesch“, Zusammenarbeit mit T. Halbach, 400 r.o.: Ausstellung und Design
**Lubalin, Herb:** 191 l.M.: Zusammenarbeit mit T. DiSpigna, 354 r.o.: Anzeige der Zeitschrift „Avantgarde“, 355 l.o.: Logo „NY“ für den Wirtschaftsrat von New York , 355 l., 2. von o.: für Liaison Newsletter Verlag, 355 l., 3. von o.: für Sandgrenade Gleitschutz, 355 l.u.: aus der Zeitschrift „Avantgarde“, 355 M.u.: mit T. Carnase, Gschwind, Gürtler und Mengelt, 423 M.r. Schutzumschlag des Buches "Herb Lubalin" von A. Peckolick
**Luidl, Philipp:** 356 l.o.: Schutzumschlag, 356 l.M.: für das Carl Orff-Museum, Diessen/Ammersee, 356 l.u.: aus Philipp Luidl „Grundsetzliches“, Heft 1, S. 14/15
**Lupton, Ellen:** 357 l.o.: Schutzumschlag, Zusammenarbeit mit J. Abbott Miller
**Lustig, Alvin:** 357 M.o. und r.o.: Bucheinbände aus der Serie „The New Directions New Classics“, 357 l.u.: Katalogumschlag für das Museum of Modern Art, New York
**Lustig Cohen, Elaine:** 358 l.o. und r.o.: Vorder- und Rückseite von „Ex libris „ 9, 358 l.u.: Umschlag (Vorder- und Rückseite) für „Ex libris“ 10, „Dada-Once for All“. Gestaltende Herausgeberin: E. Lustig-Cohen, Design des Umschlags: T. Cohen, 358 r.u.: Innentitel von „Ex libris“ 10
**Lutz, Hans Rudolf:** 359 l.o.: Karl Marx, gesetzt aus verschiedenen Schnitten der Univers, 359 r.o., l.u., M.u. und r.u.: Titelblätter der Zeitschrift „Typografische Monatsblätter“, St. Gallen, Schweizerische Typo-

grafenbund (Hrsg.)
**Macunias, George:** 362 l.: „Fluxus"-Fest, Staatliche Kunstakademie Düsseldorf
**Mahlow, Dietrich:** 362 r.o.: Bucheinband zu D. Mahlow „Schrift und Bild", Design: W. Schmidt
**Majakowski, Wladimir:** 352 l.o.: Seite aus "Für die Stimme", 362 r.u.: aus „Rosta-Fenster Nr. 555", „willst den Herzenskummer mir nennen?", November 1920, 363 l.o.: Filmplakat „ Nicht für Geld geboren", 363 r.: Theaterplakat „Oktoberrevolution", 451 l.o.: Titel von "Majakowski lächelt Majakowski lacht Majakowski höhnt", 456 l.u.: Text
**Majoor, Martin:** 364 l.u.: Telefonbuchseite
**Makela, P. Scott:** 365 l.o.: aus „Do Nothing", Zusammenarbeit mit Schriftsteller T. Viemeister, in Mohawk Paper's Publikation „Rethinking Design"
**Manutius, Aldus:** 365 l.u.: aus Francesco Colonna „Hypnerotomachia Poliphili", Venedig 1499, 365 r.u.: aus „Virgilius", 1501
**Manwearing, Michael:** 366 l.o.: für ein Spielwarengeschäft
**Mardersteig, Giovanni:** 367 l.u.: aus Heraklit, griechische Schrift nach Griffo-Majuskeln
**Marinetti, Filippo Tommaso:** 367 r.u.: Titelseite von „Zang Tumb Tumb", Edizione Futuriste, Mailand
**Martens, Karel:** 368 r.o.: Programm des „Filmhuis Nijmegen"
**Martin, William:** 368 u.: Zusammenarbeit mit M. F. Benton
**Massin, Robert:** 369 r.o. und r.u.: aus Ionesco „Die kahle Sängerin"
**Matsunaga, Shin:** 370 l.u.: speziell zum Ausstellen entworfenes Plakat
**Mau, Bruce:** 371 r.o.: Katalog zur Ausstellung „S, M, L, XL", Southern California Institute of Architecture, Los Angeles
**Mavignier, Almir da Silva:** 372 r.o.: Ausstellung „Adrian Morellet", 373 l.o.: Ausstellung „Mavignier", Museum Ulm, 400 l.o.: Design
**McConnel, John:** 373 l.u.: Plakatkampagne im Auftrag der "Stiftung Napoli '99" zur Förderung des kulturellen Erscheinungsbildes der Stadt. McConnels Plakat: Neapel nach einem Ausbruch des Vesuvs, 373 M.u.: für den Verlag Faber & Faber, London
**McCoy, Katherine:** 173 l.u.: Design, 374 l.o. und l.u.: Cranbrook Academy of Art, Faculty Project, 374 r.: Programmaufbau des Design-Studiums Cranbrook Academy of Art, 375 l.o.: Ankündigung des Sonderheftes „150 Jahre Fotografie" der Zeitschrift „Photography Monthly", Oktober1989, 375 l.M.: Inhaltsverzeichnis zu „ American Center for Design, Design Year in Review, Fourteenth Annual, 100 Show", Zusammenarbeit mit T. O'Keeffe und M.D. Sylvester, 375 l.u.: für Cranbrook Academy of Art, 400 l.o.: Design
**McLean, Ruari:** 375 M.o.: zu R. McLean „Modern Book Design", 375 r.o.: zu R. McLean „Victorian Book Design"
**McLuhan, Marshall:** 221 l.o., r.o. und l.u.: aus Fiore/McLuhan:"The Medium is the Massage", 375 r.u.: aus „The Medium is the Massage", Entwurf von Q. Fiore
**Megert, Peter:** 376 l.: Ausstellung „Weiß auf Weiß", Kunsthalle Bern, 376 r.o.: Ausstellung „Konkrete Fotografie", Galerie Aktuell, Bern
**Mendell + Oberer:** 377 l.o.: Ausstellung „Gerhard Richter", Neue Bildergalerie H: Friedrich, München, 377 r.o.: für das Museum „Die Neue Sammlung", München, 377 r.u.: Buchrücken der 20bändigen Ausgabe „Großes Fischer-Lexikon in Farbe"
**Miller, J. Abbott:** 357 l.o.: Zusammenarbeit mit E. Lupton, 383 l.o.: Zusammenarbeit mit E. Lupton und M. Mills. Die Schrift ist nach der Universal von H. Bayer (Version 1925) durch M. Carter 1991 gestaltet worden., 383 r.o. und r.u.: Doppelseiten aus „The Abc of the Bauhaus and Design Theory". Gestaltung in Zusammenarbeit mit E. Lupton und M. Mills
**Moholy-Nagy, Lászl. G.:** 110 r.o.: Entwurf eines Innentitels für Bauhaus, 384 r.o. und 385 o.: Schutzumschläge, 384 u.: Titelblatt einer Werbebroschüre für die Bauhausbücher
**Moles, Abraham A.:** 385 r.u.: aus A. A. Moles „Informationstheorie und ästhetische Wahrnehmung", Köln 1971
**Molzahn, Johannes:** 386 l.o.: Bucheinband, 386 r.o. und l.u.: für die Werkbundausstellung „Wohnung und Werkraum", Breslau
**Momayez, Morteza:** 386 r.u.: für ein Jubiläum iranischer Musik, 387 l.o.: „Iranian Theatre", 387 l.u.: für einen Verleger, 387 r.o.: für das deutsche Schauspiel von Peter Weiss „Die Ermittlung" in iranischer Übersetzung, 387 r.u.: Schriftzug für ein Monatsmagazin
**Monguzzi, Bruno:** 388 r.o.: Ausstellung „Oskar Schlemmer/ Les Noces/ Igor Strawinsky", Museo Cantonale d'Arte, Lugano , 388 r.u.: für Gamma-Film, 389 l.o.: Plakat für die Stadträte von Mailand, Aufruf zur Unterstützung nach dem Erdbeben, 389 l.u.: für das Musée d'Orsay, Paris
**Monotype Corporation:** 390 r.o.: Titelblatt einer Schriftprobe Gill Sans. Design: C. Banks, Banks & Miles
**Morison, Stanley:** 390 r.u.: Programm, 391 l.o.: Schutzumschlag, 391 r.u.: Vergleich zwischen altem und neuem Zeitungskopf
**Morris, William:** 393 l.o.: benutzt in Morris' „Historyes of Troye", 393 l.u.: benutzt in Morris' „Golden Legend"
**Moscoso, Victor:** 259 l.o.: Zusammenarbeit mit R. Griffin, 393 l.o.: Konzert „Jumor Wells and his Chicago Blues Band", 393 l.u.: Ausstellung „Poster Show in the Exhibition Hall Dallas", 393 r.: für „Blues Project. Fill More"
**Mouse, Stanley:** 394 l.o., l.u. und 395 l.o.: Konzerte („Family Dog") im Avalon Ballroom, Design: Zusammenarbeit mit A. Kelley, 394 r.: „Cosmic Car Show", 395 l.u.: Veranstaltung, Design: Zusammenarbeit mit A. Kelley
**Müller, Lars:** 395 r.o.: aus Broschüre „Lars Müller Publishers", 395 M.u.: Ausstellung „Skulptur", Aargauer Kunsthaus, 395 r.u.: Ausstellung „ Karo Dame. Konstruktive, Konkrete und Radikale Kunst von Frauen", Aargauer Kunsthaus
**Müller, Rolf:** 396 r.o.: Titelblatt der Zeitschrift „High Quality HQ" 8/1987, Foto: E. Ginoletti, 396 l.u. und M.u.: Publikationen anläßlich des 11. Olympischen Kongresses 1981 (Univers 45), 406 r.o.: Design
**Müller-Brockmann, Josef:** 397 l.o. und r.: Plakate für die Tonhalle Zürich
**Museum für Gestaltung:** 399 M.u.: Design: C. Windlin
**Museum für Kunst und Gewerbe:** 400 l.o.: Ausstellungsplakat, Design: A. Mavignier , 400 l.u.: Ausstellung "Der Pfau in der Wüste". Logo oben: N. Brody, Ausstellungsillustration: C. Berkenhoff und Hauslogo: Ott + Stein
**Museum für Kunsthandwerk:** 400 r.o.: Ausstellung „Uwe Lösch", Design: U. Loesch, 400 r.u.: Einladung zur Ausstellung „Low Budget"
**Muzika, Franti_ek:** 401 l.: Einband für Dilo Vladislava Vancury, vaza, Druzstevi práce, Prag
**Nannucci, Maurizio:** 404 l.o.: für „Europae Lux", Edinburgh, 404 r.o.: Installation für „Lenbachhaus", München, 404 M.u.: „quasi infinito"
**Neudörffer, Johann der Ältere:** 198 r.o.: Druck, 405 r.: aus A. Neudörffer „Schreibkunst", Nürnberg
**Neue Sammlung:** 406 r.o.: Ausstellung „Die verborgene Vernuft", Neue Sammlung, München. Design: R. Müller
**Neumann, Eckhard:** 407 l.o.: Einband für Ausstellungskatalog, Bauhaus-Archiv Berlin , 407 r.o.: Vorder- und Rückseite Ausstellungskatalog, IDZ Berlin, 407 u.: für einen Kongreß, IDZ Berlin
**Neurath, Otto:** 408 l.u.: aus „International Picture Language – the First Rules of Isotype", London
**Niijima, Minoru:** 408 r.o.: Ausstellung „architects of the 1990s"
**Nikkels, Walter:** 408 r.u.: Umschlag zum Katalog „Lothar Baumgarten", Pavillon der BRD, Venedig, 409 l.o.: Buch 3, 5, 6, und 7 des Bürgerlichen Gesetzbuches, 409 l.M.: für „documenta 7", 409 l.u.: Ausstellung „Rundgang", Staatliche Kunstakademie Düsseldorf, 409 r.: Folder „Wohlbefinden, Volksgesundheit und Kultur"
**Noorda, Bob:** 410 l.: Ausstellung „33$^a$ Biennale Internationale d'Arte", Venedig, 410 r.o.: für Pirelli, 410 r.u.: für die U-Bahn São Paulo
**Noordzij, Gerrit:** 410 l.u.: Einband der Schriftprobe „Zetten bij Thieme, Nijmegen", 411 l.o.: Einband „De streek", Van de Garde
**Nypels, Charles:** 412 r.: Ausstellung „H. Jonas", Grandhotel Heerlen, 413 o. und r.u.: aus „La prophétie de Ioël", Leiter-Nypels, Maastricht. Schrift: Plantijn. Holzschnitte: H. Jonas
**Octavo:** 416 l.o.: Plakat für die letzte Ausgabe der Zeitschrift „octavo" als , CD-Rom , 416 l.u.: Titelblatt der Zeitschrift „octavo" Nr. 2
**Odermatt + Tissi:** 416 r.o.: Ausstellung „Die Shaker", Kunstgewerbemuseum Zürich, Design: S. Odermatt, 416 r.M.: Wortmarke für „Küng", Blockflötenbau, Design: R. Tissi, 417 l.o.: Titelblatt für die Zeitschrift „graphis" 241/1996, Design: R. Tissi, 417 M.o.: Tanzplakat, Zürich, Entwurf: R. Tissi, 417 l.M.: Wortmarke für eine Firma, die Schalungstafeln für den Betonbau herstellt. Design: S. Odermatt
**Olbrich, Joseph Maria:** 417 M.u.: Ausstellung „Max Klinger", Künstlerkolonie Darmstadt, 417 r.: für Winter-Patentöfen
**Oliver, Vaughan:** 418 l.o.: Tocsin-Plattenalbum „x mal Deutschland", 418 l.u.: Plakat für ein Symposium über Sowjet-Design
**Oppenheim, Louis:** 418 r.o.: Überarbeitung: E. Spiekermann in Zusammenarbeit mit Berthold-Schriftenatelier, 418 r.u.: für Berlitz
**Ortiz-Lopez, Dennis:** 419 l.o.: aus der Zeitschrift „Rolling Stone", Design: G. Anderson, Satz: D. Ortiz-Lopez
**Ott + Stein:** 112 l.o.: Design einer Karte, 400 l.u.: Ausstellungs- und Hauslogo, 419 r.u.: Ausstellung „Berlin Berlin" zur 750-Jahr-Feier, Martin Gropius Bau, Berlin
**Ovink, Gerrit Willem:** 420 l.o.: Vortrag in der Generalversammlung der Gutenberg-Gesellschaft, 20.6.1958 in Mainz. Gestaltung: D. Dooijes
**Oxenaar, Otje:** 420 r.: Schutzumschlag zu „ptt nederland / ptt post", Zusammenarbeit mit P. C. van der Groen
**Palatino, Giambattista:** 422 l.: Gotische Minuskel des Palatino
**Pan:** 422 r.: Design: O. Eckmann
**Peckolick, Alan:** 423 l.u.: Freier Entritt Dienstags abends im Museum of American Folk Art, Whitney Museum und Guggenheim Museum, 423 M.r.: Schutzumschlag zu Gertrude Snyder/ A. Peckolick „Herb Lubalin"
**Pedersen, B. Martin:** 424 l.o.: Schutzumschlag für „The Annual of the Type Directors Club", Nr. 5
**Peignot, Charles Armand:** 165 l.u.: Entwurf von Cassandre, 424 r.o.: Seite Peignot (in Zusammenarbeit mit A. M. Cassandre) aus Schriftmusterheft
**Pentagram:** 373 l.u.: Entwurf, 425 r.o.: Einladung zum 21. Geburtstag der Tochter Bob Gross', Design: A. Fletcher, 425 l.u.: für „Designers and Art Directors Association", GB, 425 r.u.: Design: N. Shakery
**Pfäffli, Bruno:** 426 l. und l.u.: Anzeigen für die Druckerei Winterthur, Schweiz (Ausschnitte)
**Pfund, Roger:** 426 r.: für die Galerie Numaga, 427 l.: „Théâtre de Vidy", Lausanne, 427 r.o.: „Nouveau Théâtre de Poche", 427 r.u.: Ausstellung von 4 Schweizer Designern, Musée des Arts Décoratifs, Lausanne
**Piatti, Celestino:** 428 r.: für die Konzerte der Allgemeinen Musikgesellschaft (AMG), Basel
**Pierpont, Frank Hinman:** 429 l.u.: Zusammenarbeit mit F. Steltzer. Ursprünglich von R. Granjon entworfen
**Pineles, Cipe:** 429 r.o.: Titelblatt der Zeitschrift „Charm", Januar 1954, 429 r.u.: Titelblatt des Programms zum 50. Geburtstag der AIGA, Zusammenarbeit mit W. Burtin
**Pirtle, Woody:** 523 u.: Design, 431 l.: für das Shakespeare-Festival of Dallas, Juli 1980
**Plantin, Christoph:** 431 r.o.: Antiqua aus „Biblia Polyglotta", Plantin, Antwerpen , 431 u.: Civilité aus der „Biblia Polyglotta"
**Poeschel, Carl Ernst:** 432 l.o.: Schrift: Hamburger Fraktur
**Poppl, Friedrich:** 432 r.o.: Schriftzug mit Ruling Pen: „Don Juan", 433 r.o.: Entwurf für ein Konzertplakat „Bluthochzeit"
**Portfolio:** 433 r.u.: Design: A. Brodovitch
**Post, Herbert:** 434 r.o.: Bucheinband, Insel Verlag
**Poynor, Rick:** 434 r.u.: Schutzumschlag, Design: why not associates, 435 l.u.: Titelblatt der Zeitschrift „eye" 11, Vol.3/1993, 435 r.u.: Einladung der Society of Typographic Designers, zu einem Vortrag von R. Poynor. Design: D. Quay
**Prins, Ralph:** 435 r.o.: für „Amnesty International"
**Quay, David:** 435 r.u.: Design, 437 l.o.: Titelblatt der Schriftprobe Vermont von Premier, Design der Vermont: F. Sack, 437 l.u.: Ankündigung der 1. Serie Architype, enthaltend Architype Renner, Architype Bayer, Architype van der Leck, Architype Doesburg, Architype Tschichold und Architype Bill (The Foundry)., 485 r.o.: Design
**Queneau, Raymond:** 437 r.o.: Bucheinband, Design: P. Zollna, 437 r.u.: Bucheinband, Design: H. Jähn
**Qwer:** 438 l.o.: Zusammenarbeit mit T. Halbach und T. Hagenbucher, 438 r.o.: Ausstellung „Expo 2000", Hannover, 438 u.: Entwurf, nicht umgesetzt
**Rambow, Gunther:** 440 l.o.: Ausstellung „Poster Biennale", Lahti Art Museum, 440 l.u.: für Hessischen Rundfunk, 3. Programm
**Rand, Paul:** 440 r.o.: für American Institute of Graphic Arts, 440 r.u.: American Broadcasting Company, 441 l.o.: Titelseite des Prospekts für „Yale University of Art", Graduate Program, 441 r.: Rebus, von P. Rand auf dem Plakat vermerkt: „an Eye for perception, insight, vision – a Bee for industriousness, dedication, perseverance – an 'M' for motivation, merit, moral strength", 441 l.M.: für die Firma Next, 441 l.u.: aus „A 24 Page"
**Rapp, Hermann:** 442 l.o.: Aquarell, Buchstabe geprägt, 442 l.u.: Buchdruck und Holzschnitt
**Ratdolt, Erhard:** 442 r.: aus „Missale für Augsburg"
**Ray Gun:** 160 r.u.: Titelblatt "Ray Gun" 15/1994, 443 l.o.: Titelblatt der Zeitschrift „Ray Gun. music + style" 1/1992. Design: D. Carson. Marvin Scott Jarrett, Gründer von Ray Gun Publishing, 443 l.u.: Doppelseite der Zeitschrift „Ray Gun. music + style" 14/1995. Marvin Scott Jarrett, Gründer von „Ray Gun Publishing"
**Reichert, Josua:** 443 r.o.: aus Württembergische Landesbibliothek (Hg.) „Die Stuttgarter Drucke von Josua Reichert", Stuttgart-Bad Cannstatt Bogen 7: Hebräisches Alphabet
**Reid, Jamie:** 444 l.o.: Titelblatt der Broschüre „Sex Pistols", 444 l.u.: „für Boy Georgie", Virgin
**Reiner, Imre:** 444 r.u.: Titelblatt der Zeitschrift „graphis" 13/1946, 445 l.u.: Entwurf für „Typographica 2", St. Gallen
**Renner, Paul:** 190 r.o.: Futura, 446 l.o.: Gemeine der Futura, ursprüngliche Form, 446 r.u.: Ausstellung „Gewerbliche Fachschulen Bayerns", Gewerbemuseum Zürich
**Ricci, Franco Maria:** 447 r.u.: Titelblatt der Zeitschrift „FMR" 45/1990
**Riedel, Hubert:** 448 l.o.: Originalgrafik Nr. 1, Grafikzentrum Pankow, 448 l.u.: zu einer Plakatausstellung
**ring neuer werbegestalter:** 448 r.o.: die Schreibweise änderte Schwitters nach 1929 gelegentlich in „neue", Design: K. Schwitters, 448 r.u.: für Möbelfirma, Design: K. Schwitters
**Rockner, Vincenz:** 449: aus „Teuerdank"
**Rodtschenko, Alexander:** 450 l.u.: Titelblatt einer Kostüm-Mappe zu dem Stück „Wir", 450 r.: Titelblatt der Zeitschrift „LEF" Nr. 1, 451 l.o.: Titel von „Majakowski lächelt Majakowski lacht Majakowski höhnt", 451 l.u.: Verpackung von Papirossy (Zigaretten) der Marke Kino
**Roller, Alfred:** 452 l.: Ausstellung „Secession", Vereinigung Bildender Künstler Österreichs, Ver Sacrum, 452 r.o.: Ausstellung „4. Kunstausstellung der Vereinigung Bildender Künstler Österreichs", Ver Sacrum
**Rondthaler, Edward:** 452 u.: Bucheinband „Life with Letters as they Turned Photogenic", New York
**Roos, Sjœrd Hendrik de:** 453 l.o.: Einband für ein Gedicht von A. van Scheltema, W. L. & J. Brusse's Uitgeverij
**Rosarivo, Raul M.:** 454 l.o. und l.u.: aus „Divina Proporcion Tipográfica Ternaria", Buenos Aires
**Rosart, Jacques-François:** 454 r.: aus "Des Caractères de la Fonderie de la Veuve Decellier, Successeur de Jacques-François Rosart", Brüssel, 3. Verm. Aufl. 1779
**Rossum, Just van:** 133 r.o.: Zusammenarbeit mit E. van Blokland, 455 r.u.: Register-Seite aus E. Cleary, J. Sieber, E. Spiekermann „Font Book 1", 487 M.u.: Zusammenarbeit mit E. Spiekermann
**Rosta-Fenster:** 456 l.u.: Nr. 5 Bild links:"Rotarmist, nehmen wir der Bourgeoisie den letzten Halm weg, und sie geht zugrunde". Nr. 5 Bild rechts: „Man muß Frieden machen!", Dezemer 1919, 456 r.o.: Nr. 557 [Bild 1:] „Die Entente hat mal diese ... , ([Bild 2:]...mal jene Macht anerkannt)", Nov. 1920, 456 l.o.: „Work hard and steady, your gun at the ready". Design: W. Lebedev, Text: Majakowski, März/ April 1921
**Roth, Dieter:** 456 r.u.: Ausstellung „Dieter Roth, Bücher und Grafik", Helmhaus Zürich
**Ruder, Emil:** 457 l.o.: Ausstellung „Max Beckmann", Kunsthalle Basel, 457 r.o.: Ausstellung „Die Zeitung", Gewerbemuseum Basel, 457 l.u.: aus E. Ruder "Typographie" Teufen , 457 r.u.: Einband zu E. Ruder "Typographie" Teufen
**Rüegg, Ruedi:** 458 l.o.: Theaterplakat, Opernhaus Zürich, 458 l.u.: für Züricher Juni-Festwochen, 458 r.o.: für Konzerte der Tonhalle-Gesellschaft, Züricher Juni-Festwochen
**Ruscha, Edward:** 459 l.o.: für Ausstellung "documenta V", 19 Kassel
**Sabon, Jakob:** 462 l.: Design: J. Tschichold
**Salisbury, Mike:** 463 l.u.: für die Bekleidungsfirma Gotch, 463 r.o.: für den Film „Jurassic Park", 463 r.u.: für die Zigarrettenmarke West
**Sandberg, Willem:** 464 l.u.: Ausstellung „Parysche Schilders", (Pariser Maler), Stedelijk Museum Amsterdam, 464 r.o.: Ausstellung „Fernand Léger", Museum Fodor, Amsterdam, 465 l.o.: Ausstellung „Arne Jacobsen, Lucebert, Constant, Monticelli, Calder", Stedelijk Museum Amsterdam
**Sato, Koichi:** 465 l.u.: Konzertplakat, 465 r.: Theaterplakat für „Seinenza"
**Saville, Peter:** 466 l.o.: Album-Cover, Zusammenarbeit mit M. Atkins, 466 l.M.: für „Whitechapel Art Gallery, Zusammenarbeit mit B. Wickens, G. Mouat, 466 l.u.: aus „yohi yamamotot 03 magazine", Zusammenarbeit mit S. Wolsteholme, 466 r.o.: Registerseite aus „Font Shop", Flo Motion, Type Design: Zusammenarbeit mit M. Heinz, 466 r.M.: Chrysalis Identity Programm, „y" = Logo von „Butterfly", 467 l.o.: „Channel One", Titelsequenz, Design: Zusammenarbeit mit B. Wickens
**Schawinsky, Xanti:** 467 l.u.: für Illy Caffé, Studio Boggeri
**Scher, Paula:** 468 l.: Werbung für Ambassador Arts's Silk-Screening Capabilities
**Schiller, Walter:** 469 u.: Schutzumschlag (Vorder- und Rückseite) zu A. Kapr/W. Schiller „Gestalt und Funktion der Typografie", Leipzig
**Schmid, Helmut:** 469 r.o.: Titelseite des Schutzumschlages „Typography today", Tokio
**Schmidt, Hans:** 470 l.o.: R. M. Rilke „Wer spricht von Siegen..."
**Schmidt, Joost:** 111 r.o.: Posterentwurf für Bauhaus, 470 l.u.: Titelblatt der Zeitschrift „Form", 470 l.u.: und 471 r.o.: Entwürfe für „Uhertype", Blaupausen, 471 l.: Ausstellung „Staatliches Bauhaus", Weimar
**Schmidt, Wolfgang:** 362 r.o.: Design, 472 l.o.: Ausstellung „Werbegrafik 1920–1930" Göppinger Galerie, Frankfurt a. M., 472 r.o.: für Atlas Film
**Schmidt Rhen, Helmut:** 473 l.o.: Ausstellung „Kumi Sugai", Städtisches Museum Leverkusen, 473 M.o.: Ausstellung „30 junge Deutsche: Architektur, Plastik,

wie

comment

how

| The Tools | Die Werkzeuge | Les Outils |
|---|---|---|
| The Hand-Ax | Faustkeil | Coup de poing |
| The Brush | Pinsel | Pinceau |
| The Calamus or Reed Pen | Binsenhalm oder Rohrfeder | Roseau ou plume de bambou |
| Fishbone | Fischbein | Os de poisson ou fanon |
| The Chisel | Meißel | Burin |
| The Stylus | Griffel | Stylet |
| The Pen | Feder | Plume |
| Movable Type | Letter | Caractères |
| Compasses | Zirkel | Compas |
| The Typewriter | Schreibmaschine | Machine à écrire |
| Light Analog Filmsetting | Licht analoger Fotosatz | Exposition photocomposition analogique |
| Computers Digital Filmsetting | Computer digitaler Fotosatz | Ordinateur photocomposition numérique |

**Faustkeil**

Die Anfänge der Schrift werden unterschiedlich datiert. Wenn die Gravuren auf Kieselsteinen ab ca. 8000 vor unserer Zeitrechnung eine Botschaft darstellen, dann zählen diese Zeichen zu den Vorstufen von Schrift. Die dem Menschen verfügbaren Werkzeuge sind zahlenmäßig gering und spezialisiert. Mit einem Werkzeug werden mehrere Tätigkeiten verrichtet. Hornstein dient zur Bearbeitung von Kiesel, da er härter als der Untergrund ist. Mit ihm können einfache Striche gekerbt werden, die ausreichen, den Darstellungen erkennbare Umrisse zu verleihen.

Die Archäologin Marie König schließt aus Funden aus der mittleren Steinzeit auf eine bewußte Gravur: drei parallele Striche deutet sie als die drei sichtbaren Mondphasen. In vier Strichen sieht sie die Himmelsrichtungen.

Sollte dies zutreffen, handelte es sich um frühe schriftliche Dokumentationen von Zeit und Raum. Nach dem heutigen Stand der mythologischen Forschung ist die Wahrscheinlichkeit groß, daß hier auch bereits ein Zusammenhang zwischen Schrift und Religion besteht, wie er später immer wieder nachgewiesen werden kann.

*Werkzeug:* Stein; *Material:* Stein

**Coup de poing**

La datation des débuts de l'écriture varie considérablement. Si les gravures réalisées sur des galets à partir du 8e millénaire avant notre ère étaient des messages, alors ces signes marquent indéniablement l'une des étapes qui ont précédé l'écrit. Le nombre d'outils dont disposait l'homme était très faible, et chaque instrument servait à plusieurs tâches. Le silex était utilisé pour travailler les galets puisqu'il était plus dur que le support. On l'employait pour graver de simples traits qui suffisaient à tracer des contours rudimentaires permettant d'identifier l'objet représenté.

L'archéologue Marie König déduit à partir d'objets remontant au paléolithique que la gravure était un acte volontaire : elle interprète trois traits parallèles comme les trois phases visibles du cycle lunaire; et quatre traits comme les quatre points cardinaux.

Si cette interprétation s'avérait exacte, il s'agirait de documents très anciens témoignant du temps et de l'espace. Selon l'état actuel des recherches sur les mythologies, il est fort probable qu'il existait déjà à cette époque une relation entre l'écriture et la religion, une relation qui a sans cesse été démontrée par la suite.

*Outil :* pierre; *support :* pierre

**The Hand-Ax**

The origins of writing have been dated back to varying periods. If engravings on pebbles found from c. 8000 BC onwards represent some kind of message, then these could definitely be counted among the first stages of a written form of language.

The tools available to ancient man were few and specialized. Many different tasks were performed with one tool. Hornstone was used to write on pebble, as it was harder than the surface it worked. Simple lines were carved with hornstone which were sufficient to give the object or notion depicted a clear outline.

In her examination of findings from the Mesolithic period, archeologist Marie König concludes that carvings on stone have a certain meaning. She sees three parallel lines as the three visible phases of the moon and four lines as the points of the compass.

If this is true, then these represent an early written documentation of time and space. According to current mythological research, it is highly probable that even at this early stage there was a link between writing and religion, a link which has been proved to exist through many later stages of the history of the written word.

*Tool:* stone; *material:* stone

## Pinsel

Neben dem Faustkeil, mit dem Gravuren in Felsen und Kieselsteine geritzt wurden, zählt der Pinsel zu den ältesten uns bekannten Schreibwerkzeugen.
Diese Pinsel sind dünne Pflanzenstiele, deren Fasern am Ende aufquellen, wenn man sie in Wasser taucht oder mit Speichel anfeuchtet. Damit malen die Ägypter um 3300 vor unserer Zeitrechnung ihre frühen Hieroglyphen. In dieser Zeit ist der Pinsel das Werkzeug für bildliche und schriftliche Darstellungen. Das Bild ist noch Schrift, die Schrift ist noch Bild. Das Werkzeug ist für beide das gleiche.
Der Einfluß des Pinsels auf die späteren Schriften ist nicht genau festzustellen. Vielen der gemeißelten Schriften dient er als vorschreibendes Werkzeug. Wahrscheinlich entstehen die Serifen griechischer und römischer Inschriften durch die Vorzeichnungen mit dem Flachpinsel.
In Ostasien ist der Tierhaarpinsel das dominierende manuelle Schreibwerkzeug. Chinesen wie Japaner haben mit ihm eine künstlerisch hochstehende Schriftkultur entwickelt. Die Kalligraphie gilt als eine eigenständige Kunstrichtung neben der Malerei, der Plastik und dem Holzschnitt. Die chinesischen und japanischen Schriftzeichen verdanken ihre Schönheit und Eigenart ausschließlich dem Pinsel, der hier seine charakteristischen Spuren hinterlassen hat. In der künstlerischen Kalligraphie ist der Pinsel als Ausdrucksmittel ein Synonym für Spontaneität und Persönlichkeit.
In der westlichen Welt spielt der Pinsel als Schreibgerät heute nur noch eine bescheidene Rolle. Pinselschriften gibt es nur wenige. Einige werden für den Buchdruck in Blei geschnitten, tauchen aber erst im zwanzigsten Jahrhundert auf. Heutige Computerprogramme mit ihren Individualisierungsmöglichkeiten erlauben einen unproblematischen Zugang zu Pinselschriften.
*Werkzeug:* Pinsel; *Material:* jegliche glatte Oberflächen

## Pinceau

Outre le coup de poing qui servait à graver la roche et les galets tendres, le pinceau est l'un des plus anciens instruments servant à écrire.
Les premiers pinceaux étaient faits de fines tiges de plantes dont les fibres gonflaient une fois trempées dans l'eau, ou quand on les humidifiait avec la salive. Les Egyptiens les utilisaient pour peindre les premiers hiéroglyphes vers 3 300 avant notre ère. En ces temps reculés, le pinceau servait autant aux représentations scripturales que picturales. L'image avait encore fonction d'écriture, et l'écriture d'image. Toutes deux étaient réalisées avec le même outil.
Il est difficile d'établir l'influence exercée par le pinceau sur les écritures qui suivirent. Pour de nombreuses lettres gravées, il servait à définir préalablement le tracé. D'ailleurs, les sérifs des inscriptions grecques et latines sont sans doute nés d'esquisses au pinceau plat.
En Extrême-Orient, le pinceau en poils d'animaux était l'instrument le plus répandu. Il permit aux Chinois comme aux Japonais de développer cet art supérieur qu'est la calligraphie. Celle-ci était considérée comme autonome et parallèle à la peinture, la sculpture et la gravure sur bois. Les signes chinois et japonais doivent leur beauté et leur élégance exclusivement au pinceau qui donnait aux caractères ce tracé si particulier. Dans la calligraphie d'art, le pinceau est un moyen d'expression, il est synonyme de spontanéité et traduit la personnalité du scribe.
En Occident, le pinceau ne joue plus qu'un rôle mineur dans l'écriture. Il existe peu de caractères au pinceau. Certain d'entre eux, destinés à l'impression de livres, et taillés dans le plomb, ont pourtant resurgi au 20e siècle. Aujourd'hui, les logiciels graphiques et leurs possibilités de programmation individuelle permettent d'accéder facilement aux écritures au pinceau.
*Outil :* pinceau; *support :* toutes surfaces lisses

## The Brush

Besides the hand-ax, which is used to make engravings on rocks and pebbles, the brush is one of the oldest writing tools known to us.
Ancient brushes were made of thin plant stems whose fibers at the end of the stem swelled up when dipped in water or moistened with saliva. This was the tool the Ancient Egyptians used to paint their early hieroglyphics with around c. 3300 B.C. At this point in time, words were still pictures and pictures still words. The tool for both was the same.
The influence of the brush on later written forms has not been clearly ascertained. We do know that many scripts fashioned with a chisel were modeled on writing executed with brushes. The serifs on Greek and Roman inscriptions probably originate from preliminary drafts drawn with a flat brush.
In Eastern Asia the predominant manual writing tool was and still is the brush made of animal hair. The Chinese and the Japanese have developed a lettering culture which is superior in its artistic qualities. Calligraphy is an independent art form alongside painting, sculpture and woodcut art. Chinese and Japanese characters owe their beauty and individuality to the brush, which has left its distinctive mark. In the art of calligraphy, the brush is a form of expression synonymous with spontaneity and personality.
In the modern western world, the brush has been relegated a much more modest rôle as a writing implement. There were and are few brush scripts. Some, cut in lead, were used for printing purposes, but not until the twentieth century. Today's computer programs with their wealth of possibilities for individual font customization give us quick and easy access to brush script.
*Tool:* brush; *material:* any flat surface

**Binsenhalm**
oder Rohrfeder

Zwischen dem zweiten und ersten Jahrtausend v. Chr. löst in Ägypten beim Schreiben von Hieroglyphen der Binsenhalm den Pinsel ab. Es entwickelt sich eine flüssige Priesterschrift, die hieratische Kursive. Aus ihr geht ab dem 7. Jahrhundert eine noch flüssigere Volksschrift hervor, die demotische Schrift. Sowohl die hieratische wie die demotische Schrift lassen deutlich den *Duktus* eines vorne schräg zugeschnittenen Binsenhalms erkennen.
Geschrieben wird auf Papyrus mit einer Flüssigkeit, der Ruß oder Ocker beigemischt ist. Der älteste erhaltene Papyrus wird auf das Jahr 3500 v. Chr. datiert. Aus dem Mark der am Nil wachsenden Papyrusstaude werden dünne Streifen geschnitten, kreuzweise übereinandergelegt und gepreßt oder gehämmert. Der dabei austretende Pflanzensaft verklebt die Streifen, so daß ein beschreibbares Blatt entsteht. Mehrere solcher Blätter werden zu einer Rolle aneinandergeleimt.
Papyrus und Schreibrohr sind leicht zu handhaben. Sowohl Material als auch Werkzeug begünstigen die Veränderung der Zeichen. Ist in der hieratischen Schrift eine Verwandtschaft zu den Hieroglyphen noch erkennbar, so geht dies in der demotischen verloren. Auch in Rom beeinflußt das Schreibrohr, der *calamus*, das gegenüber den in Wachs gedrückten Zeichen durch flüssiges Schreiben Buchstabenverbindungen erlaubt, die Entwicklung von der jüngeren römischen Kursive zu einer fortlaufenden Schrift.
Der Papyrus ist im gesamten Mittelmeerraum verbreitet und wird im arabischen Sprachraum bis ins vierzehnte Jahrhundert gebraucht. Papyri hat man in Kreta, Griechenland, Rom, Palästina, Saudi-Arabien und Mesopotamien gefunden. Für einige Zeit unterliegt die Papyrusausfuhr einem königlich-ägyptischen Monopol. Ein Ausfuhrembargo soll einen Hirten aus Pergamon auf die Idee gebracht haben, Ziegenhaut als Schreibmaterial zu verwenden, das dann als Pergament bekannt wird.
*Werkzeug:* Rohrfeder; *Material:* Papyrus, Pergament, Kalkstein, Topfscherben

**Roseau**
ou plume de bambou

Entre le 1er et le 2e millénaire av. J.-C., les scribes égyptiens remplacèrent progressivement le pinceau par le roseau. C'est alors que s'élabora une écriture plus fluide, la cursive hiératique des prêtres. A partir du 7e siècle, celle-ci donna naissance à une écriture populaire plus fluide encore, appelée démotique. Tout comme les lettres démotiques, les caractères hiératiques sont reconnaissables aux touches laissées par un brin de roseau taillé en biseau.
On écrivait sur du papyrus avec un liquide contenant de la suie et de l'ocre. Le plus ancien papyrus jusqu'alors conservé date de 3500 av. J.-C. Pour fabriquer le papyrus, on coupait de fines lanières dans le cœur des tiges de cette plante répandue sur les berges du Nil. Les bandes étaient tissées entre elles, puis pressées ou martelées et la sève qui s'écoulait pendant le processus les collait ensemble, si bien que l'on obtenait une feuille susceptible d'être écrite. Plusieurs feuilles étaient ensuite collées ensemble pour former un rouleau.
Le papyrus et le roseau étaient d'un maniement facile. Le support comme l'outil favorisèrent alors l'évolution des signes. Si l'écriture hiératique révèle sa parenté avec les hiéroglyphes, la ressemblance disparaît dans l'écriture démotique.
A Rome, le roseau, le calamus, exerça aussi une influence; en effet, à l'inverse des signes imprimés dans la cire, il permettait une écriture plus fluide et des liaisons entre les lettres. Il est à l'origine du passage de l'ancienne cursive romaine à l'écriture courante.
Le papyrus, présent dans tout le bassin méditerranéen, sera également utilisé dans le périmètre linguistique de l'Islam jusqu'au 14e siècle. On en a trouvé en Crète, en Grèce, à Rome, en Palestine, en Arabie Saoudite et en Mésopotamie. Pendant une certaine période, l'exportation du papyrus resta le monopole des pharaons égyptiens. Cet embargo à l'exportation aurait suggéré à un berger de Pergame l'idée d'utiliser de la peau de chèvre comme support écrit, ce procédé donna ensuite naissance au parchemin.
*Outil:* roseau; *support:* papyrus, parchemin, pierre calcaire, poteries

**The Calamus**
or Reed Pen

Between the second and first millennium BC in Egypt, the calamus or hollow rush gradually replaced the brush as a writing implement for hieroglyphics. A fluent form of writing developed, hieratic cursive, which was practiced by priests. Out of this an even more fluid written form, demotic script, evolved from c. 700 BC onwards which was used by the ordinary literate classes. The forms of both hieratic and demotic script clearly indicate that they were executed with a reed pen with a pointed nib.
These new scripts were notated on papyrus paper in liquid mixed with soot or ochre. The oldest sheet of papyrus preserved has been dated to c. 3500 BC The pith of papyrus plants growing along the banks of the Nile was cut into strips which were arranged crisscross in layers and then pressed or hammered together. The ensuing sap stuck the strips together, forming a sheet of primitive paper. These sheets were glued together in a roll.
Papyrus and calamus were easy to handle: materials and tools alike encouraged a development in written symbols. Whereas we can recognize elements of hieroglyphics in hieratic script, these are no longer traceable in demotic forms.
The calamus was also influential in Ancient Rome. In script written with a calamus, letters could be joined up, allowing a more fluid form of writing, something not possible with symbols etched into soft wax. Early roman cursive script thus gradually matured into a flowing alphabet.
Papyrus was found all over the Mediterranean; papyri have been discovered in Crete, Greece, Rome, Palestine, Saudi Arabia and Mesopotamia. The Arab world used papyrus well into the fourteenth century. The Pharaohs of Egypt had a monopoly on the export of papyrus for many years. Legend has it that an export embargo gave a shepherd from Pergamum the idea of using goat's hide as a writing material, a commodity which later became known as parchment (from "Pergament").
*Tool:* calamus; *material:* papyrus paper, parchment, limestone, potsherd

**Fischbein**

In allen uns bekannten Schriftkreisen hat Schrift die Tendenz, sich von bildhaften Umrissen zu lösen. In der mesopotamischen Keilschrift wird dieser Vorgang plastisch vom Werkzeug unterstützt. Die aus der altsumerischen Schrift entwickelten Keile bieten ein anschauliches Beispiel dafür, wie das Schreibwerkzeug auf die Schriftform einwirkt.
Das in der Regel verwendete Fischbein (z.B. Walbarten) wird verkantet in den weichen Ton gedrückt, so daß sich an den Druckstellen keilförmige Vertiefungen bilden.
Die aus solchen Keilen zusammengesetzten Zeichen können schnell geschrieben werden. War eine Tontafel ausgefüllt, wurde sie an der Sonne getrocknet und dann zum Aufbewahren in ein Regal gestellt. Eine große Zahl von Keilschriftdokumenten ist durch Palastbrände erhalten geblieben, da die Tontafeln in Feuer gebrannt und so widerstandsfähiger wurden.
Ein wesentlicher Schritt vom Bild zum abstrakten Zeichen besteht in der immer größeren Vertrautheit des Schreibers mit seinen Zeichen. Hinzu kommt die perfektere Handhabung des Werkzeugs, so daß die Schrift sich unmerklich aber stetig abschleift. Die gewohnten mit dem Fischbein in Ton gedrückten Keilformen wurden auch auf Stein übertragen, diese Übertragung war dem Material Stein aber eigentlich fremd.
Alle damaligen Schriften finden über ihre Sprachgrenze hinaus Verbreitung. Benachbarte Völker schauen sie ab und passen sie an. So fußt beispielsweise die Keilschrift der Hethiter auf der Schrift der Babylonier und Assyrer. Der *Duktus* ihrer Zeichen weist auf das Fischbein als Schreibgerät hin.
*Werkzeug:* Fischbein; *Material:* Ton

**Os de poisson**
ou fanon

Dans toutes les sphères que nous connaissons, l'écriture tendait à se dégager des schémas figuratifs. Dans l'écriture mésopotamienne, ce processus plastique était favorisé par l'outil. L'écriture cunéiforme des premiers alphabets sumériens illustre concrètement la manière dont l'outil influença le dessin de la lettre.
En règle générale, l'os de poisson (par ex. des fanons de baleines) était utilisé par pression dans la terre encore souple si bien que des encoches cunéiformes apparaissaient aux endroits où l'on appuyait.
Les signes composés à partir de ces encoches s'écrivaient rapidement. Dès qu'une tablette d'argile était remplie, on la laissait sécher au soleil, puis elle était rangée dans une étagère. Ainsi, les incendies des palais ont-ils permis la conservation de nombreux documents écrits en caractères cunéiformes. Le feu durcissait l'argile et le rendait plus résistant.
La familiarité croissante du scribe avec le tracé permit une étape décisive, celle du passage de l'image au signe abstrait. A cela s'ajoute un maniement perfectionné de l'outil, de sorte qu'imperceptiblement d'abord, l'écriture devint ensuite de plus en plus fluide. Les lettres cunéiformes, habituellement gravées dans l'argile avec des os de poisson furent transposées sur la pierre, un phénomène qui, en vérité, n'avait rien à voir avec la pierre comme support.
Tous les alphabets anciens se sont répandus au-delà de leurs frontières linguistiques. Les peuples voisins s'en inspiraient et les adaptaient. Ainsi, par exemple, l'écriture cunéiforme des Hittites est dérivée de celle des Babyloniens et des Assyriens; et le tracé des signes indiquait qu'elles avaient été écrites avec des os de poisson.
*Outil:* os de poisson ou fanon; *support:* argile

**Fishbone**

In all areas of writing known to us, the written word has always tended towards a departure from pictorial forms of representation. In the cuneiform of Mesopotamia, the tools used for writing actively supported this process. The wedge-shaped notation developed from Old Sumerian script presents us with a clear example of how writing implements affected the nature of the symbols they formed.
Fishbone (e.g. whalebone) was the most common stylus used to write cuneiform. It was held at an angle and pressed into soft clay to produce wedge-shaped incisions.
Symbols made up of these triangles or wedges could be written at relative speed. Once a clay slab was full, it was left in the sun to dry and then stored on a shelf. Many cuneiform slabs have been preserved through palace fires: hardened by the heat of the flames, the clay became more resistant to the elements and passage of time.
A major aid in the evolution of abstract symbols from pictograms was the writer's growing familiarity with the signs he produced. Scribes also gradually perfected the way they held their writing implements, so that their script slowly but surely lost its rough edges. Cuneiform was also transferred to stone, which proved unsuitable: fishbone and soft clay remained the most common medium.
The scripts common at that time spread far beyond linguistic boundaries. Neighboring peoples copied and adapted forms to suit their languages. Thus Hittite cuneiform has its roots in the cuneiform of the Babylonians and Assyrians. The shapes of their symbols suggest that they also used fishbone as a writing implement.
*Tool:* fishbone; *material:* clay

**Meißel**

Bereits vor siebentausend Jahren sind Meißel aus Stein in Gebrauch. Die ersten Meißel aus Kupfer sind zweitausend Jahre jünger. Eine der ältesten Abbildungen eines Meißels zeigt die „Narmerplatte", eine Schminkpalette vom Anfang des dritten Jahrtausends v. Chr. mit ägyptischen Hieroglyphen. Zu sehen sind dort ein Fisch und ein Meißel. „Nar" bedeutet ägyptisch Fisch, „mer" Meißel, so daß es sich bei dieser Hieroglyphe um eine Phonetisierung des Königsnamens Narmer handelt.

Im Althochdeutschen heißt Meißel „meizan" und bedeutet Schneiden. Das Werkzeug bezeichnet also seine Tätigkeit. Auch das Wort Hieroglyphe, von Herodot gebraucht und mit Heilige Kerbe übersetzt, läßt erkennen, in welcher Form sie häufig anzutreffen war. Ob gekerbt, geschnitten oder gemeißelt, das Ergebnis ist eine Vertiefung in Stein, Ton, Metall oder Holz.

Dem Meißeln der Schrift geht das Schreiben mit dem Pinsel als Vorlage voraus.

In den griechischen Inschriften ist bereits der Ansatz von Serifen zu sehen, die in den römischen Inschriften voll ausgehauen werden. Diese Serifen sind das Merkmal des Flachpinsels. Sie verhindern das Auslaufen der Farbe nach oben oder unten. Beim Meißeln verhindert die Serife außerdem das Aussplittern des Steins an diesen Stellen. Der Meißel ist nicht eigentlich stilbildend, sondern das Ausführungswerkzeug des Pinsels, das sich vom Material her als notwendig erweist. Daß die gemeißelten Schriften häufiger als die geschriebenen überleben, erklärt sich durch das widerstandsfähige Material.

*Werkzeug:* Meißel; *Material:* Stein, Ton, Metall, Holz

**Burin**

Il y a sept mille ans, l'homme utilisait déjà des burins en pierre. Les premiers burins en cuivre apparaîtront deux mille ans plus tard. L'une des premières représentations d'un burin figure sur la «Tablette de Narmer», une palette à maquillage datant du début du 3e millénaire av. J.-C. où figurent des hiéroglyphes égyptiens. On y voit un poisson et un burin; «Nar» signifiant poisson et «mer» burin; ce hiéroglyphe est par conséquent la transcription phonétique de «Narmer», le nom du roi.

En ancien haut allemand, le burin, («Meißel» en allemand moderne), s'appelait «meizan» et signifiait tailler. L'outil désignait donc une activité. De même, le mot «hiéroglyphe» utilisé par Hérodote, et traduit par l'entaille sacrée, rappelle sous quelle forme on le rencontre le plus souvent. Que ce soit une entaille, une encoche ou une engravure, le résultat est toujours un creux gravé dans la pierre, la terre cuite, le métal ou le bois.

L'écriture gravée a précédé l'utilisation du pinceau et lui a servi de modèle. Les inscriptions grecques présentaient déjà les prémisses du sérif, qui arrivera à maturité dans les tailles romaines. Ces sérifs évoquent l'emploi du pinceau pour prévenir les coulures de peinture. Mais en matière de taille au burin, ils évitent les éclats de pierre à certains endroits. En vérité le burin n'était pas un outil susceptible de générer un style, mais plutôt le complément du pinceau; son usage s'avérait nécessaire en fonction du support. Le fait que les lettres gravées avaient une durée de vie plus longue que les lettres écrites, s'explique par la résistance du matériau.

*Outil:* burin; *support:* pierre, terre cuite, métal, bois

**The Chisel**

Stone chisels were in use seven thousand years ago. The first chisels made of copper are not found until two thousand years later. One of the oldest depictions of a chisel is on the Narmer Slab, an ornamental tablet from c. 3000 BC decorated with Egyptian hieroglyphics. The inscription shows a fish (Egyptian *nar*) and a chisel (*mer*), a phonetic interpretation of the name of Pharaoh Narmer.

The word *chisel* entered the English language via Old French *chisel* which has its origins in Latin *caedere*, to cut. The name of the tool thus also describes the action carried out by the tool's user. Even the word *hieroglyph* (from Greek *gluphe*, carving), used by Herodotus and translated as "Holy Carving", shows us in which form hieroglyphics were most frequently found. Whether cut, carved or chiseled, the result was an incision in stone, clay, metal or wood.

Script executed with brushes acted as a model for chiseled letters. Greek inscriptions bear traces of early serifs which are used with clear definition in Roman inscriptions. These serifs are characteristic of the flat brush and were used to prevent the tops and bottoms of letters drawn in paint or ink from running. Chiseling letters with serifs prevented the stone from cracking around the letters. The chisel is not responsible for any major stylistic development; it merely acted as a tool to the brush, a necessary invention for the materials it had to work. The relative permanence of these materials explain why more examples of the chiseled than the written word have survived to the present day.

*Tool:* chisel; *material:* stone, clay, metal, wood

## Griffel

Auf einer Tafel aus Byblos aus dem zweiten Jahrtausend v. Chr. steht folgendes zu lesen: „Die Bronze des Tophet habe ich gewalzt, mit eisernem Griffel habe ich dieses Gerät graviert." Hier sind nicht nur Material und Werkzeug benannt, der Text läßt auch den Arbeitsvorgang erkennen.
Mit einem Griffel haben auch Griechen und Römer auf Täfelchen geschrieben, die ursprünglich aus Holz, später aus Elfenbein oder Metall sind. Die Innenseiten der zusammenklappbaren Täfelchen sind mit Wachs beschichtet, in das die Schrift geritzt wird. Die Griechen nennen sie Diptychon.
Griffel gibt es dazu aus verschiedenen Materialien. Bekannt sind die aus Metall oder Knochen von ca. 15–20cm Länge. Die Römer nennen die Griffel *stilus*. Sie sind an einem Ende zugespitzt und am anderen abgeplattet. Mit der Spitze wird die Schrift eingeritzt, mit dem flachen Ende kann man das Wachs wieder glätten und so die Notiz löschen. Die Wachsschicht kann erneut beschrieben werden. (Die gelöschte Tafel nennt sich *tabula rasa.*)
Diese Art des Schreibens und Löschens hat sich lange erhalten. Bis ins 20. Jahrhundert wird in Schulen mit Griffeln auf Tafeln geschrieben. Beide sind nun aus Schiefer.
Der *Duktus* aller mit dem Griffel geschriebenen Buchstaben ist gleichförmig. Römische Griffelinschriften zeigen, daß die Anmutung der Schrift beim Griffel von den Proportionen der Buchstaben ausgeht. So gibt der Griffel oder *stilus* Anlaß, vom Inhalt des Geschriebenen auf die Art, wie es geschrieben ist, hinzuweisen. Nur wer einen guten *stilus* besitzt, ist imstande, auch eine gute Schrift zu schreiben.
*Werkzeug:* Griffel; *Material:* Holz, Elfenbein, Metall, Ton, Schiefer, Wachs

## Stylet

Sur une tablette datant du 2e millénaire av. J.-C. et originaire de Byblos, on peut lire la phrase suivante : «J'ai laminé le bronze de Tophet, et j'ai gravé cet ustensile avec un stylet en fer.» Ici, le matériau, mais aussi l'outil sont cités; et le texte explique également le procédé. Les Grecs et les Romains écrivaient sur des tablettes à l'aide d'un stylet, d'abord en bois, en ivoire ou en métal. La face intérieure de ces tablettes doubles, que les Grecs appelaient dyptichon, était couverte d'une couche de cire où l'on gravait le texte.
Les stylets existaient en divers matériaux. Les plus répandus étaient en métal ou en os et faisaient 15 à 20 cm de longueur. Les Romains les appelaient stilus. Ils possédaient une extrémité pointue et l'autre aplatie. La pointe servait à graver la lettre, tandis que la spatule permettait de lisser la cire et donc d'effacer l'inscription. La couche de cire pouvait être renouvelée. (Une tablette effacée s'appelait tabula rasa.)
Ce mode d'écriture et d'effacement a subsisté pendant longtemps. Au 20e siècle encore, les enfants écrivaient au tableau avec un crayon d'ardoise. Tableau et crayon étant du même matériau.
Les lettres écrites au stylet présentaient un aspect très uniforme. Les lettres romaines gravées montrent que les proportions des lettres avaient une influence sur le texte. Ainsi le stylet ou la pointe fournissent l'occasion d'établir le rapport entre la teneur du texte et la manière dont il a été écrit. Seul celui qui possède un bon stylet, est capable de bien écrire.
*Outil :* stylet; *support :* bois, ivoire, métal, terre cuite, ardoise, cire

## The Stylus

A tablet from c. 2000–1000 BC found in Byblos bears the following inscription: "I have cast bronze from Tophet, I have engraved this tablet with a stylus made of iron". The artist does not merely name his materials and tools; the text also gives us information as to his working methods.
The Ancient Greeks and the Romans also used styluses for writing purposes. These writing implements (Latin *stilus*) were made of various materials: originally of wood, and later of ivory or metal, we have evidence of styluses made of metal or bone and about 6–8 inches in length. These tools were used to inscribe the inner waxed surfaces of hinged writing tablets, which the Ancient Greeks called diptychs. Styluses were pointed at one end for carving script into the soft wax and blunt at the other for erasing mistakes, whereby the wax surface is smoothed flat making room for new messages. The erased tablet was called a "tabula rasa".
This method of writing and deleting text stayed with us for a long time. School children were still using pointed implements on tablets well into the 20th century, with both "pen" and "paper" made of slate. Letters written with a stylus are uniform in their characteristics. Roman inscriptions made with a stylus illustrate that the elegance of writing in such a way influences the proportions of the letters. Writing with a stylus thus not only records the content of a text but also demonstrates how the text was written. Only he who is possession of a good stylus can produce good writing.
*Tool:* stylus; *material:* wood, ivory, metal, clay, slate, wax

**Feder**

Seit die Vogelfeder das Rohr verdrängt hat, bezeichnet man alle Schreibgeräte als Feder, auch wenn diese vom Material her keine Verwandtschaft zum Federkiel haben. Die Römer kennen bereits Stahlfedern, wie ein Fund bei Avanches beweist, der aus dem ersten Jahrhundert stammt. Feder und Federhalter bestehen aus einem Stück.
Der Gänsekiel wird erstmals im Jahre 624 erwähnt und um 700 berichtet der Angelsachse Adelhelmus von einer Pelikanfeder zum Schreiben. In den Klosterschreibstuben (Scriptorien) des Mittelalters herrscht Arbeitsteilung. An manchen Seiten eines Meßbuches sind Miniaturmaler, Ausschmücker (Illuminatoren) und Schreiber (Kleriki) tätig. Zur Arbeit der Schreiber zählt es, die Gänse- oder Schwanenfeder auszuwählen, zuzuschneiden und sie an der Kielspitze zu spalten. Über diese Kunst erscheint 1544 von Johann Neudörffer d. Ä. (1497–1563) ein eigenes Büchlein. Die von den Römern eingeführte Stahlfeder wird im Mittelalter vom Gänsebad verdrängt und kommt erst im 19. Jahrhundert zur Geltung.
Seit 1808 sind Feder und Federhalter getrennte Teile. Die Feder kann nun ausgetauscht oder erneuert werden. Es gibt seither zwei Grundarten von Schreibfedern: die abgeschrägten und die runden. Die abgeschrägte Feder ergibt einen an- und abschwellenden Strich. Die Anmut dieser Schrift besteht im Wechselzug von dünnen und dicken Strichen. Die runde Feder erzeugt einen gleichbleibend starken Strich und gleicht hierin dem Kugelschreiber.
Die Feder als Schreibgerät, sei es eine Vogel- oder eine Stahlfeder, wird zu einem bedeutenden Träger der Schriftgeschichte.
*Werkzeug:* Vogelfeder, Stahlfeder; *Material:* Pergament, Papier

**Plume**

Depuis que la plume d'oiseau a remplacé le bambou, tous les ustensiles d'écriture s'appellent «plume», même si le matériel ne présente plus aucune parenté avec le tuyau de plume. Une découverte remontant au 1er siècle, faite lors des fouilles d'Avanches, prouve que les Romains connaissaient les plumes en acier. Mais plume et porte-plume étaient alors d'une seule pièce.
Le tuyau de plume est évoqué pour la première fois en 624; puis vers l'an 700, l'Anglo-Saxon Adelhelmus parle d'une plume de pélican servant à écrire. La division du travail régnait chez les copistes des monastères du Moyen Age. La décoration de certaines pages des missels était l'affaire des peintres d'enluminures, et le texte du ressort des scribes (clercs). Le choix des plumes d'oie ou de cygne, leur taille et la manière de fendre le tuyau faisaient partie des tâches du scribe. En 1544, Johann Neudörffer l'Ancien (1497–1563) publiait un petit ouvrage sur cet art. La plume en acier, introduite par les Romains, fut remplacée au Moyen Age par le tuyau de plume. Elle devra attendre le 19e siècle pour être de nouveau appréciée.
Plume et porte-plume sont séparés depuis 1808. Ceci permet de changer la plume ou de la renouveler. Il existe deux catégories de plumes, l'une dont la pointe est taillée en biseau et l'autre arrondie. La plume biseautée donne une écriture riche en pleins et en déliés présentant une succession de traits fins et épais. La plume ronde produit un trait régulier ressemblant au tracé du stylo à bille.
Qu'elle soit tirée d'une plume d'oiseau ou fabriquée en acier, la plume deviendra l'un des éléments essentiels de l'histoire de l'écriture.
*Outil:* plume d'oiseau, plume d'acier; *support:* parchemin, papier

**The Pen**

Writing instruments have been called pens (from Latin *penna*, feather or pen) ever since reeds were replaced by feather quills, even if the material used is not actually feather. The Romans used pens made of steel, as an excavated nib from the 1st century AD found near Avanches has proved. At this stage, nib and holder are still a single item.
The quill was first used in 624 and in 700 the Anglo-Saxon, Adelhelmus, reports of a pelican quill being used for writing. In medieval monastic writing rooms or scriptoria, tasks were divided among the monks. Some pages of the missals were worked on by miniaturists, embellishers (illuminators) and scribes (clerici). One of the scribe's jobs was to select goose and swan feathers, fashion them into pens and slit the nib. Johann Neudörffer the Elder (1497–1563) even published a small volume on this ancient art in 1544. The steel nib was reinvented in the 16th century and has survived to the present day in a variety of forms.
In 1808 the nib and pen holder became two separate entities. This meant that nibs were interchangeable and replaceable. Since this date there have been two kinds of nibs: those with a square (or oblique) writing edge and those which are rounded at the tip. The former produce lines of varying thickness. The beauty of text written with such nibs lies in the elegant contrast of thick and thin lines. Rounded nibs produce lines of equal thickness and are thus very similar to the ball-point pen.
The pen as a writing instrument, whether made of feather or steel, is an important commodity in the history of writing.
*Tool:* quill, steel pen; *material:* parchment, paper

Letter

Auch das „mechanische Schreiben", wie Gutenberg das Drucken nannte, braucht ein Schreibgerät: Letter und Presse. Zwei Tatsachen belegen, daß Johannes Gensfleisch, gen. Gutenberg (etwa 1399–1468), mit der bleiernen Type die geschriebene gotische Textur kopieren wollte. Um der Handschrift mit ihren nie exakt gleichen Buchstaben möglichst nahe zu kommen, hatte er von jeweils einem Buchstaben mehrere verschieden breite Versionen geschnitten. Zum anderen beschäftigt der Goldschmied Gutenberg den Schreiber Peter Schöffer (ca. 1425–1503), der ihn bei der Formgebung der Buchstaben berät.
Alle folgenden Druckschriften bleiben bemüht, die handschriftliche Vorlage möglichst genau nachzuahmen. Ist es anfangs der Gänsekiel, so nimmt man sich später die Stahlfeder zum Vorbild, was im Klassizismus in extrem fetten bzw. feinen Strichen sichtbar wird. Die Erfindung der Lithographie 1796/97 bringt zu Beginn des 19. Jahrhunderts der Druckschrift neue Impulse. Die mit Kreide oder Feder frei auf den Stein gezeichneten Schriften finden ihre Nachahmung in den Pinsel- und Schreibschriften aus Blei. Die Schriftgießereien spornt es immer wieder an, noch das Kurioseste nachzuschneiden, so daß es die vielfältigsten Schriften für den Druck gibt. Die Buchstaben, die lose in den Fächern des Setzkastens liegen, sind nach der Häufigkeit ihres Vorkommens angeordnet. Die meistgebrauchten liegen der rechten Hand am nächsten und nahe zum Körper. Da dies in jeder Sprache andere Buchstaben sind, hat jede Sprache auch eine andere Setzkasteneinteilung.
Der Buchstabe besteht aus einer Legierung von etwa 65 % Blei, 28% Antimon, 5% Zinn und etwas Wismuth. In dieser Legierung gewährleisten Wismuth und Zinn, daß das Metall für den Guß geschmeidig bleibt, und Antimon, daß es für den Druck hart genug ist. Der genaue Prozentsatz der jeweiligen Metalle ist Geheimnis der Schriftgießereien. In der Folgezeit kommt zum Handsatz der Maschinensatz, bei dem sich der Guß in Einzelbuchstaben (Monotype) und Zeilen (Linotype, Intertype, Typograph) unterteilt.
*Werkzeug:* Letter und Presse; *Material:* Pergament und Papier

Caractères

Ce que Gutenberg appelait «écriture mécanique», c'est-à-dire l'imprimerie, exigeait aussi des outils : les caractères et une presse. Deux faits attestent que Johannes Gensfleisch, dit Gutenberg (vers 1399–1468), envisageait de copier les écritures gothiques manuscrites au moyen de caractères mobiles en plomb. Pour se rapprocher le plus possible de l'écriture manuelle et de ses lettres jamais exactement identiques, il avait taillé plusieurs versions différentes de chaque lettre en modifiant la largeur. Par ailleurs, l'orfèvre Gutenberg employait le scribe Peter Schöffer (vers 1425–1503) qui le conseillait sur la forme à donner aux lettres.
Tous les caractères d'imprimerie qui suivront s'efforceront d'imiter le plus possible les écritures manuscrites. Au début, on prenait pour modèle la calligraphie à la plume d'oie, remplacée plus tard par la plume d'acier, ce qui conduisit à des caractères pendant la période néoclassique à montants extrêmement gras ou maigres.
L'invention de la lithographie en 1796–1797 donna de nouvelles impulsions aux caractères d'imprimerie du 19e siècle. Les lettres dessinées librement à la craie ou à la plume sur la pierre trouveront des équivalents dans les caractères en plomb tirés des lettres tracées au pinceau ou à la main. Ce mouvement incita les fonderies de caractères à tailler les types les plus étranges si bien que l'on disposa bientôt d'une multitude de caractères d'imprimerie. Les caractères disposés dans les casiers des casses étaient rangés en fonction de leur fréquence d'utilisation. Les plus employés étaient à droite et près du corps. Comme leur disposition varie selon les langues, chacune d'elles possédait sa propre répartition des caractères.
Ces types étaient constitués d'un alliage de 65% de plomb, de 28% d'antimoine, de 5% de zinc et d'un peu de bismuth. Le bismuth et le zinc permettant au métal de rester souple pendant le moulage, et l'antimoine garantissant la dureté nécessaire à l'impression. Toutefois, les proportions exactes de chaque métal dans l'alliage faisaient partie des secrets des fonderies. Par la suite, la composition à la main fut complétée par la composition mécanique, les machines fondaient d'abord les caractères (Monotype), les lignes (Linotype, Intertype, Typograph).
*Outil* : caractères et Presse; *support :* parchemin et papier

Movable Type

Even "mechanical writing", Gutenberg's term for printing, requires a writing instrument: in this case, movable type and a printing press. Two facts prove that Johannes Gensfleisch, known as Gutenberg (c.1399–1468), was intent on copying handwritten Gothic letters with lead characters. In order to get as close as possible to the handwritten manuscripts which were his model, where the letters are never exactly the same, he cut each letter in varying thicknesses. Goldsmith Gutenberg also employed a scribe, Peter Schöffer (about 1425–1503), who advised him on the design of his alphabets.
All printed matter post-Gutenberg concentrated on copying handwritten texts with the greatest degree of accuracy. Print was initially modeled on manuscripts written with a quill, and later on texts written with a steel nib, which produced a stronger contrast of line in the Baroque period and a mixture of extremely thick and extremely fine lines in the Classical period.
The invention of lithography in 1796/97 injected new impetus into printing at the beginning of the 19th century. Writing executed freehand on stone with chalk or with a quill was emulated in brush scripts and type in lead. Type foundries were constantly trying to imitate the most unusual alphabets, resulting in a great assortment of different typefaces available for printing. The individual letters in the various compartments in the compositor's type case are sorted according to their frequency of use. Those employed most are placed nearest the compositor's right hand and close to the body. As this varies from language to language, each language favors its own particular arrangement of characters in the type case.
Over the course of time, manual setting was gradually joined by machine composition, where a distinction is made between individual letters (Monotype) and lines (Linotype, Intertype, Typograph). The type is cast in an alloy of app. 65% lead, 28% antimony, 5% tin and a small amount of bismuth. The bismuth and tin keep the alloy malleable for casting purposes and antimony ensures that it is hard enough to withstand the printing process. The exact percentages of each of the metals used is a foundry secret.
*Tool:* letters and a printing press; *material:* parchment and paper

**Zirkel** Um 1513 malt Raffael (1483–1520) „Die Schule von Athen" und zeigt dort unter anderem Euklid (um 300 v. Chr.), der den Schülern mit Zirkel und Tafel seine Lehrsätze erklärt. Griechen und Römer kennen den Zirkel. Es ist jedoch nicht überliefert, ob sie mit ihm auch Schrift konstruierten.

Von Leonardo da Vinci (1452–1519) ist bekannt, daß er zur Konstruktion der Antiqua-Majuskeln (Großbuchstaben) einen Zirkel benutzte. Albrecht Dürer (1471–1528) nennt eines seiner Bücher „Unterweysung der Messung mit dem Zirckel und Richtscheydt" (Nürnberg 1525). Darin erklärt er unter anderem die Konstruktion von Antiqua-Majuskeln.

Aus Italien sind Versalkonstruktionen mit dem Zirkel von Damianus Moyllus (1439–1500) in Parma und Vespasiano Amphiareo (1501–1563) in Venedig bekannt.

Fast jeder Renaissance-Künstler und Schreibmeister behilft sich bei der Darstellung von Schrift mit dem Zirkel und dem Lineal. Die den Schülern zugedachten Musterblätter belegen das. Aufgabe ist es dabei, der mit der Feder geschriebenen Form treu zu bleiben, sie zu vermessen und zu korrigieren.

Erst im zwanzigsten Jahrhundert betrachtet man Zirkel und Lineal selbst als Schreibwerkzeuge. Die niederländische Künstlergruppe De Stijl, die russischen Konstruktivisten und vor allem das Bauhaus gelten als geistige Zentren für die Erneuerung der Schrift. Die Bauhaus-Lehrer für Schrift, Josef Albers (1888–1976), Herbert Bayer (1900–1985) und Joost Schmidt (1893–1948), entwickeln die „reine" Form der Schrift, vor allem bei den Minuskeln, durch Zirkel und Lineal. Die so konstruierten Alphabete weichen von der „empfundenen" Geometrie früherer Schriften ab. Sie sind in einem technischen Sinn ideologische Schriften, die in ihrer reinen Form nur geringen Einfluß auf den Mengensatz haben. Die Bauhaus-Alphabete sind in Zeichnungen als Abbild der Gedanken ihrer Zeit überliefert.

*Werkzeug:* Zirkel; zur Schriftvorbereitung

**Compas** Vers 1513, Raphaël (1483–1520) peint «L'Ecole d'Athènes» et représente Euclide (vers 300 av. J.-C.) expliquant ses théorèmes à ses élèves à l'aide d'un tableau et d'un compas. Les Grecs et les Romains connaissaient le compas. Cependant, nous ignorons encore s'ils s'en servaient pour construire leurs lettres. En revanche nous savons que Léonard de Vinci (1452–1519) utilisait un compas pour tracer les antiques capitales (majuscules). Albrecht Dürer (1471–1528) intitula l'un de ses livres «Unterweysung der Messung mit dem Zirckel und Richtscheydt» [Enseignement de la mesure au compas et à l'équerre], (Nuremberg, 1525) où il expliquait, entres autres, comment construire des antiques capitales.

Concernant l'Italie, nous avons connaissance des constructions de majuscules réalisées au compas par Damianus Moyllus (1439–1500) à Parme, et par Vespasiano Amphiareo (1501–1563) à Venise.

Presque tous les artistes et scribes de la Renaissance s'aidaient d'un compas et d'une règle pour tracer leurs lettres, ce dont témoignent les pages de modèles destinés aux élèves. La tâche à accomplir était de recopier à la plume, et avec fidélité, la forme donnée, de la mesurer et de la corriger.

Ce n'est qu'au 20e siècle que l'on traita le compas et la règle comme des instruments d'écriture en soi. Le groupe d'artistes hollandais De Stijl, les constructivistes russes et surtout le Bauhaus devenaient alors les centres intellectuels du renouveau de l'écriture. Les maîtres de typographie du Bauhaus, Joseph Albers (1888–1976), Herbert Bayer (1900–1985) et Joost Schmidt (1893–1948) élaborèrent des formes «pures», surtout des bas de casse, tracées au compas et à la règle. Les alphabets ainsi construits se démarquaient des caractères précédents et de leur géométrie «intuitive». Au sens technique du terme, c'étaient des lettres idéologiques, dont la forme pure n'exerçait qu'une faible influence sur la composition. Par leur dessin, les alphabets du Bauhaus reflétaient la pensée d'une époque.

*Outil:* compas pour réaliser la maquette d'une lettre

**Compasses** In c. 1513, Raphael (1483–1520) painted his "School of Athens" fresco which among others depicts Euclid (in c. 300 BC) teaching his students with the aid of a pair of compasses and a board. Compasses were known to the Ancient Greeks and Romans, yet there is no documentation as to whether they used them in writing or not.

It is known that Leonardo da Vinci (1452–1519) used compasses for the construction of his roman majuscules (capital letters). Albrecht Dürer (1471–1528) gave one of his books the title "Unterweysung der Messung mit dem Zirckel und Richtscheydt"["Teaching Geometry with Compasses and A Rule"], (Nuremberg 1525). His book includes information on how to draw roman majuscules.

From Italy we know that Damianus Moyllus (1439–1500) in Parma and Vespasiano Amphiareo (1501–1563) in Venice used compasses to draw capital letter constructions.

In the Renaissance period, nearly all artists and writing masters used compasses and rules to help them draw letters and alphabets. Proof of this are the specimen sheets intended for their pupils. Writing students had to copy, measure and correct the letter forms from these sheets which had been written with a quill.

It was not until the twentieth century that compasses and rules were considered writing tools. The Dutch group De Stijl, Constructivists in Russia and above all the Bauhaus were the intellectual centers working on new forms of type. Bauhaus lettering teachers Josef Albers (1888–1976), Herbert Bayer (1900–1985) and Joost Schmidt (1893–1948) developed a "pure" type form, especially for minuscule characters, using compasses and rules. The new alphabets constructed in this fashion differ from the "implied" geometry of earlier typefaces. Technically they are ideological alphabets which in their pure form have little influence over mass composition. Sketches of Bauhaus typefaces available to us today are an illustration of the spirit of their age.

*Tool:* compasses for typeface design

## Schreibmaschine

Die Schreibmaschine ist eines der jüngeren Beispiele, wie das Werkzeug auf die Schrift einwirkt und sie verändert.
1864 konstruiert Peter Mitterhofer (1822–1893) eine der ersten Schreibmaschinen aus Holz. Die erste fabrikmäßige Schreibmaschinenproduktion erfolgt 1873 durch die amerikanische Firma Remington. Diese Typenhebelmaschinen besitzen einen Wagen mit Schreibwalze, der nach jedem Anschlag einen Schritt nach links rückt. Da der Wagenschritt immer gleich groß ist (2,54 mm), müssen auch die Buchstaben annähernd gleich breit sein, damit keine zu großen Löcher im Wortbild entstehen. Es werden Schriften entworfen, bei denen die Gesamtbreite des Buchstabens in etwa gleich ist und die als typische Schreibmaschinenschriften bekannt sind. Sie widersprechen der bis dahin gültigen Schriftanatomie, die jedem Buchstaben nur die seinem Bild entsprechende Breite zugesteht.
1961 gelingt es durch die Schreibmaschine mit Kugelkopf, die Schreibmaschinenschrift der Druckschrift wieder anzunähern. Statt des Wagens bewegt sich der Kugelkopf, auf dem die einzelnen Schriftzeichen angeordnet sind. Diese Schreibmaschine kann neun verschieden breite Schreibschritte ausführen und den Text auch in Blockform schreiben. Ein Rechner ermittelt die Wortabstände automatisch, so daß man hier von Schreibsatz spricht. Die spätere Schreibmaschine mit Typenrad funktioniert ähnlich wie die Kugelkopfmaschine. Sie besitzt neben der sogenannten Proportionalschrift einen elektronischen Speicher, der Korrekturen während des Schreibens ermöglicht. Der Anschlag erfolgt nicht mehr wie bei der Typenhebelmaschine mechanisch, sondern elektrisch.
Seit die elektronische Datenverarbeitung in die Büros eingezogen ist, sind aus der Schreibmaschine zwei getrennte Stationen geworden: Die Eingabe-(Tastatur) und die Ausgabeeinheit (Drucker).
*Werkzeug:* Schreibmaschine; *Material:* Papier

## Machine à écrire

L'exemple de la machine à écrire, d'invention récente, montre l'influence exercée par l'outil sur l'écriture et ses modifications.
En 1864, Peter Mitterhofer (1822–1893) construisait l'une des premières machines à écrire en bois. La première production en série de machines à écrire par la société Remington remonte à 1873. Ces machines dont les caractères étaient montés sur des leviers, possédaient un chariot muni d'un rouleau qui se déplaçait d'un cran vers la gauche à chaque frappe. Comme l'avancée du chariot était toujours identique (2,54 mm), les caractères devaient avoir approximativement la même largeur, pour éviter que des espaces n'apparaissent au sein du mot. On dessina alors des polices, dont les lettres présentaient une largeur à peu près égale, connues comme les caractères courrier, typiques de l'écriture à la machine. Ces caractères étaient contraires à l'anatomie de la lettre telle qu'elle était d'usage, et au primat de la relation entre la largeur et la structure de chacune d'elles.
En 1961, la machine à écrire à boule permit de rapprocher l'écriture dactylographiée des caractères d'imprimerie. Au lieu d'un chariot se déplaçant, une boule sur laquelle étaient disposés les divers caractères, s'orientait dans la position requise. Cette machine était capable de sélectionner neuf espaces différents entre les lettres et de réaliser des blocs de textes avec alignement. Un système électronique calculait automatiquement l'espace entre les mots si bien que l'on parle ici de composition dactylographiée. Les machines à écrire qui lui succédèrent étaient équipées d'une marguerite qui fonctionnait comme la boule. Outre un système appelé système proportionnel, elle possédait une mémoire électronique qui permettait de corriger pendant la frappe. Contrairement à la machine à écrire à levier, son fonctionnement n'était plus mécanique mais électrique.
Depuis l'introduction de l'informatique dans les bureaux, deux postes assurent le rôle de la machine à écrire : l'entrée des données a lieu sur le clavier, et l'édition sur l'imprimante.
*Outil :* machine à écrire; *support :* papier

## The Typewriter

The typewriter is one of the more recent examples of how machines affected and changed the process of writing.
In 1864, Peter Mitterhofer (1822–1893) made one of the first typewriters out of wood. The first factory-produced typewriter was introduced in 1873 by the Remington company in America. Their type bar typewriters had a carriage with a roller which sprang one step to the left on every keystroke. As the distance the carriage moved each time was always the same (one-tenth of an inch), characters had to be of similar breadth so that no excessive gaps appear within individual words. Typefaces were developed with letters of more or less equal breadth, easily identifiable as typical typewriter fonts. They were inconsistent with previous type anatomies, where each letter was only as wide as its shape and form required.
The invention of the typewriter with a spherical type head (or "golfball") in 1961 slowly brought typewriter fonts more into line with printing type. With this machine, characters are arranged on the golfball which moves in place of the carriage when a key is struck. This typewriter can perform typing intervals of nine different breadths and can write text in block format. A computer automatically calculates the spaces between the words, a process described as typewriter composition. Later typewriters with daisy wheels function in a similar way to those with golfballs. Besides the ability to print 'proportional' type, this machine also has an electronic memory which allows corrections to be made while typing. The keys are no longer activated mechanically, as with type bar typewriters, but electronically.
Since electronic data processing has swept through offices worldwide, typewriters have been split into two separate components: an input unit (the keyboard) and an output unit (the printer).
*Tool:* typewriter; *material:* paper

**Licht**
Analoger Fotosatz

Der Gedanke, mit Licht nicht nur Bilder herzustellen, sondern auch Schrift zu drucken, ist so alt wie die Fotografie selbst. In den USA werden schon im neunzehnten Jahrhundert Patente auf derartige Verfahren erteilt.
Die erste funktionsfähige Lichtsetzmaschine, die Uhertype, wird aber erst 1925 vorgestellt. Sie besteht aus einer Handlichtsetzmaschine und einer Metteurmaschine. Die Schrift dazu, die Uhertype-Standard-Grotesk, hat Jan Tschichold (1902–1974) gezeichnet. Die Patente der Uhertype werden gekauft, um das weitere Aufkommen des Fotosatzes zu dieser Zeit zu verhindern.
Nach dem Zweiten Weltkrieg wird die Idee des Fotosatzes wieder aufgegriffen. 1949 wird der Fotosetter von Intertype fast zeitgleich mit der Lumitype, der späteren Photon von Deberny und Peignot, beides Mengensatzmaschinen, vorgestellt. 1957 kommt das Diatype-Akzidenz-Fotosetzgerät der H. Berthold AG auf den Markt.
Analoge Fotosatz-Geräte belichten die Buchstaben durch ein stehendes oder bewegtes Negativ mittels Blitzlicht oder Kameraverschluß auf das Fotomaterial (Papier oder Film). Die Schriftbildträger (Typenscheibe und Typenplatte), die mit einem umfangreichen Zeichensatz ausgerüstet sind, werden als Font oder Grid bezeichnet. Ein Zeichensatz (Font) umfaßt neben den Groß- und Kleinbuchstaben, den Ziffern und Interpunktionen sowie den wichtigsten mathematischen Zeichen auch Sonderzeichen wie § und &. Die Zeichenbelegung der Fonts ist von Hersteller zu Hersteller verschieden. Kapitälchen und Minuskelziffern (Mediävalziffern) stehen meist auf einem eigenen Font.
Optomechanische Fotosetzmaschinen sind mit einem System von Linsen, Spiegeln und Prismen ausgerüstet, über das der Lichtstrahl gelenkt wird. Die Zeichen belichtet man einzeln. Die Schriftgröße kann durch eine auswechselbare Optik bestimmt oder stufenlos über ein Zoomobjektiv gewählt werden.
Der Fotosatz erlaubt u. a., die Buchstabenabstände und die Schriftgröße stufenlos zu verändern, die Positiv-Negativ-Montage, Buchstaben zu konturieren. Ultraleichte Schriften werden erst durch den Fotosatz möglich, wie auch jede Form fotografischer Modifikation (Verzerrungen, Überblendungen etc.).
*Werkzeug:* Buchstabennegativ und Lichtquelle; *Material:* Fotopapier oder Film

**Exposition**
photocomposition analogique

L'idée d'utiliser la lumière non seulement pour produire des images, mais aussi pour imprimer des caractères, date de l'invention de la photographie. Dès le 19e siècle, des brevets portant sur des procédés de ce genre ont été déposés aux Etats-Unis.
Pourtant, la première machine à photocomposer capable de remplir cette fonction, la Uhertype, ne fut présentée qu'en 1925; elle utilisait des caractères Uhertype-Standard Grotesk dessinés par Jan Tschichold (1902–1974). Mais à l'époque on racheta tous les brevets portant sur la Uhertype pour ralentir le développement de la photocomposition.
Après la Seconde Guerre mondiale, la photocomposition connut un regain d'intérêt. En 1949, le Fotosetter d'Intertype vit le jour presque en même temps que la Lumitype, puis vint la Photon de Deberny et Peignot, toutes deux étant des machines à composer au kilomètre. En 1957 enfin, H. Berthold AG présentait la photocomposeuse Diatype-Akzidenz.
Pour l'exposition des caractères, les machines de photocomposition analogique opèrent au moyen d'un négatif statique ou mobile, d'un flash ou d'un obturateur et de matériel photographique (papier ou film). Les supports de caractères (disques ou plaques) contiennent un important jeu de caractères. Ils sont appelés fontes ou grilles. Un jeu de caractères comporte les bas de casse et les capitales, mais aussi les chiffres, la ponctuation, les caractères mathématiques les plus courants et des caractères spéciaux comme les «§» et les «&».
La répartition des signes sur le support varie selon les fabricants. Les petites capitales et les chiffres en minuscules se trouvent le plus souvent sur une fonte particulière.
Les machines à photocomposer mécaniques et optiques sont équipées d'un système de lentilles, de miroirs et de prismes qui servent à diriger le rayon lumineux. Chaque caractère est exposé séparément. Le corps de la lettre peut être déterminé au moyen d'objectifs interchangeables, ou être choisi en réglant le zoom.
La photocomposition permet, entre autres, de modifier les corps des caractères et l'espace entre eux. On peut également travailler les contours des lettres à l'aide de montages positif-négatif. Les caractères ultra-light n'ont pu être réalisés qu'avec l'avènement de la photocomposition, au même titre que toutes les autres modifications photographiques (déformation, surexposition, etc.).
*Outil:* négatifs de caractères, source lumineuse; *support:* papier photo ou film

**Light**
Analog Filmsetting

The notion of using light to produce not just pictures but also words is as old as photography itself. Patents for typesetting processes using light were being granted in America as long ago as in the nineteenth century.
The first photo-typesetting machine in operation, the Uhertype, was not introduced until 1925. It was a combination of manual phototypesetting machine and make-up machine. The machine's typefaces, Uhertype, were designed by Jan Tschichold (1902–1974). The patents on Uhertype were bought up at the time to prevent the invention of filmsetting spreading.
After the Second World War the filmsetting process was revived. Two mass composition typesetters were introduced, namely by Intertype with their Fotosetter in 1949, and almost simultaneously by Deberny and Peignot with their Lumitype machine, known as Photon in the USA. In 1957 H. Berthold AG put their Diatype job filmsetting machine on the market.
Analog filmsetting machines expose the letters onto photographic material (paper or film) through a still or moving negative using a flash or camera shutter. The type carriers (type disk and type plate), equipped with an extensive set of characters, are known as fonts or grids. A character set comprises not only upper-case and lower-case letters, numbers and punctuation marks, but also all the major mathematical signs and symbols such as § and &. The assignment of characters within a font varies from manufacturer to manufacturer. Small capitals and minuscule symbols (medieval figures) usually have their own font.
Optomechanical filmsetting machines consist of a system of lenses, mirrors and prisms over which a beam of light is directed. Characters are exposed one at a time. The size of the letters can be adjusted using interchangeable lenses and can be smoothly graded with a zoom lens.
With filmsetting, the spaces between the letters and the font size are infinitely variable. Letters can be outlined using a positive-negative montage. The development of filmsetting has made the production of ultra-light typefaces possible, as well as all kinds of photographic modifications to letters (distortion, superimposition, etc.).
*Tool:* letter negative and light source; *material:* photographic paper or film

**Computer**
Digitaler Fotosatz

Am Beginn der Computerschrift steht das Nachzeichnen der Schwarzweißwerte eines Buchstabens. So werden Buchstaben durch Ja/Nein-Kommandos in digitale Befehle zerlegt. Das Setzen erfolgt über einen gesteuerten Kathodenstrahl (CRT), seit 1971 auch Laserstrahl, der den Buchstaben durch Belichten beziehungsweise durch Nichtbelichten auf das Fotomaterial zeichnet. Digitalisiert werden die Schriftzeichen, indem man ihnen ein Raster unterlegt und die Rasterfelder entsprechend dem Buchstaben schwarz oder weiß markiert. Der so in schwarze und weiße Felder, also in Ja- und Neininformationen (Bits) zerlegte Buchstabe ist für den Belichter der Ein- und Ausschaltplan (Bitmap). Dabei muß jedes Rasterfeld im Rechner notiert werden. Die erste nach diesem Prinzip arbeitende Maschine ist die Digiset der Rudolf Hell GmbH von 1966. Die Auflösung der Zeichen in einzelne digitale Bildpunkte (Pixel) ergibt eine sehr große Menge an Speicherdaten, die man zu reduzieren versucht. Als nächstes wird die Konturenbeschreibung entwickelt. Dabei erhält der Computer Daten über den Buchstabenumriß. Dem Rechner werden in Abständen Orientierungspunkte eingegeben, die er selbständig mit Geraden verbindet. Eine Rundung wird also aus Geraden zusammengesetzt, die bei starker Vergrößerung des Buchstabens zu sehen sind. Zur Zeit werden Zeichen als Vektoren gespeichert. Da der Computer jetzt Teilrundungen selbständig errechnet, wird den Orientierungspunkten nur ihre Abfolge angegeben. Es entstehen kantenfreie Konturen in allen Schriftgrößen.
Durch den Computer ist jede Schrift in ihrer Erscheinung (Schriftgröße, Strichstärke, Neigung etc.) veränderbar geworden. Die Form einer Schrift kann beliebig variiert werden. Programme zum Entwerfen von Schriften wie „Fontographer" erlauben die Entwicklung von Schriften ohne typographische Vorkenntnisse. Über den Zeichenentwurf hinaus gestattet der Computer die Positionierung des Zeichens an jeder Stelle und in jeder Lage. Neben jenen Schriften, die ihre elektronische Anmutung als Zeichen der Fortschrittlichkeit zeigen, sind alle vorhergehenden „historischen" Schriften digital erfaßt und verfügbar. Die Verwendung des Computers hat auch ganz neue Schriften hervorgebracht, die auf sich ständig verändernden Programmen basieren (z.B. Beowolf).
*Werkzeug:* Computer, Laserdrucker und Belichter; *Material:* Papier, Fotopapier oder Film

**Ordinateur**
photocomposition numérique

L'imitation des valeurs noir et blanc d'un caractère marque le début de l'écriture informatisée. Les caractères sont décomposés par les impulsions numériques «oui/non». La composition s'opère par le biais d'un tube cathodique (CRT) commandé par un rayon laser depuis 1971. Ce dispositif permet l'exposition ou la non-exposition des caractères sur un support photographique. Les caractères se superposent sur une trame qui permet de les soumettre à un traitement numérique; les champs de la trame étant marqués en noir ou en blanc, en fonction de la lettre. Le caractère ainsi décomposé en champs blancs et noirs, donc en informations «oui» et «non» (Bits) fonctionne comme un plan d'analyse lors de l'exposition (Bitmap). Chaque champ de la trame doit être analysé par un logiciel. La Digiset de Rudolf Hell GmbH a été la première machine à appliquer ce principe en 1966. Pour convertir les caractères en divers points numériques (pixels), il faut mettre en mémoire une énorme quantité de données que l'on cherche à réduire.
L'étape suivante servira à affiner la résolution des contours. Pour cela, l'ordinateur reçoit des données sur le dessin de la lettre. On fournit à intervalles des points d'orientation au logiciel, qui les relie automatiquement par des lignes. Les courbes sont donc composées de droites que l'on distingue en agrandissant fortement la lettre. Actuellement, les signes sont enregistrés sous forme de vecteurs. Comme l'ordinateur est capable de calculer automatiquement des portions de courbes, on lui fournit leur ordre en même temps que les points d'orientation. Ceci permet de réaliser des contours sans hachures pour tous les corps de caractères.
L'ordinateur est un outil capable de modifier l'aspect de chaque police (corps, graisse, inclinaison etc.) et de générer toutes sortes de variations de forme. Les logiciels de conception de polices, comme le «Fontographer» permettent de créer des caractères sans connaissances préalables en typographie. L'ordinateur sert non seulement à esquisser le caractère, mais aussi à le placer à n'importe quel endroit et dans n'importe quelle position. Parallèlement aux polices, d'inspiration électronique, symboles de progrès, toutes les fontes «historiques» sont disponibles sous forme numérique. L'utilisation de l'ordinateur a donné naissance à des caractères entièrement nouveaux, reposant sur des programmes qui se modifient constamment (par ex. Beowolf).
*Outil :* ordinateur, imprimante laser et photocomposeuse; *support :* papier, papier photo ou film

**Computers**
Digital Filmsetting

The first process whereby characters are produced using computers involves the sketching out of the black-and-white values of a letter. Letters are broken down into unique digital commands using a yes/no command system. This data is then transmitted by beams of electrons controlled by deflecting magnets in a cathode ray tube, or by laser beam, which "draws" the letters onto photographic material by selective exposure. The characters are intricately digitized using a matrix in which the various sections of the matrix are marked black or white according to the shape of the letter. This letter, broken down into its black and white fields, i.e. into yes/no information (bits), acts as an on/off trigger (bitmap) for the imagesetter. Each section of the matrix must be recorded on the computer. The first machine to operate using this principle was the Digiset, produced by Rudolf Hell GmbH in 1966. The resolution of characters into individual digital image points (pixel) produces a vast amount of data, an attempt being subsequently made to reduce this. The next stage is the development of a notation system for letter outlines. Here the computer is fed data which maps out letter contours. A set of orientation points for each letter is entered which the computer joins up with straight lines. A curve is thus actually made up of several straight lines, which become evident when the letter is enlarged. The current computer process stores characters as vectors. As computers are now capable of working out partial curves on their own, the information they receive is limited to a set of orientation points and their sequence. This is sufficient to ensure production of smoothly-rounded letters without any sharp edges in all sizes. Computers also now allow us to change the appearance and the shape of letters (their size, weight, angle of slope, etc.) as we wish. Programs such as "Fontographer" enable us to develop fonts ourselves without any previous typographic knowledge. Computers do not limit our activities to font design; we can also place our customized words, letters and alphabets wherever we wish in our documents and applications.
Besides modern, computer-manufactured alphabets, whose electronic elegance marks the state of progress, older "historic" typefaces have also been digitally recorded and made accessible to us. Through computers, completely new fonts have evolved based on programs which are constantly changing (e.g. Beowolf).
*Tool:* computer, laser printer and imagesetter; *material:* photographic paper or film

## Abgebildete und erwähnte Schriften
## Polices illustrées et mentionées
## Fonts, illustrated and mentioned

## Auswahlbibliographie
## Bibliographie, sélection
## Selected Bibliography

**Aicher, Otl:** *Typographie. Berlin 1988*
**Annenberg, Maurice:** *Type Foundries of America and their Catalogs. Baltimore, Washington 1975*
**Atkins, Kathryn A.:** *Masters of the Italic Letter. Boston 1988*
**Audin, Marius:** *Histoire de l'imprimerie par l'image. Bd. 1–4, Paris 1928–1929*
**ibid.:** *Les Livrets typographiques des Fonderies Françaises créées avant 1800. Paris 1933 Amsterdam 1964*
**Barge, Hermann:** *Geschichte der Buchdruckerkunst. Leipzig 1940*
**Barthel, Gustav/Rahmer, Albert/Stähle, Walter:** *Gestalt und Ausdruck der Antiqua. Stuttgart 1970*
**Barthel, Gustav:** *Konnte Adam schreiben? Köln 1972*
**Baudin, Fernand:** *How typography works and why it is important. London 1989*
**Bauer, Friedrich:** *Chronik der Schriftgießereien in Deutschland und den deutschsprachigen Nachbarländern. Offenbach 2. Aufl. 1928*
**Bauer, Hans:** *Das Alphabet von Ras Schamra. Halle 1932*
**Bauer, Karl Friedrich:** *Aventur und Kunst. Privatdruck der Bauerschen Gießerei, Frankfurt a. M. 1940*
**Bernard, Auguste:** *L'Histoire de l'Imprimerie Royale du Louvre. Paris 1867*
**Biggs, John R.:** *Basic Typography. London 1968*
**Blackwell, Lewis:** *Twentieth-Century Type. London 1992*
**Bohadti, Gustav:** *Die Buchdruckletter. Berlin 1960*
**ibid.:** *Die Walbaum-Schriften und ihre Vorläufer. Berlin 1960*
**Borker, Riekele:** *Handbuch der Keilschriftliteratur. 3 Bde., Berlin 1967–75*
**Bosshard, Hans Rudolf:** *6 Essays zu Typografie, Schrift, Lesbarkeit. Teufen 1996*
**Brandt, Johanns Georg:** *Alphabete und Schriftmuster aus Manuscripten und Druckwerken verschiedener Länder vom 12. bis zum 19. Jahrhundert. Frankfurt a. M. 1859*
**Brasch, Kurt / Senzoku, Takaya:** *Die kalligraphische Kunst Japans. Japanisch-deutsche Gesellschaft e.V., Tokyo 1963*
**Broos, Kees / Hefting, Paul:** *Grafische Formgebung in den Niederlanden. 20. Jahrhundert. Niederländische Ausg. Laren, deutsche Ausg. Basel 1993*
**Bruckner, Albert:** *Schweizer Stempelschneider und Schriftgießer. Geschichte des Stempelschnittes und Schriftgusses in Basel und der übrigen Schweiz von ihren Anfängen bis zur Gegenwart. Basel 1943*
**Bruin, Robert:** *La typographie en France au seizième siècle. Paris 1938*
**Brüning, Ute (ed.):** *Das A und O des Bauhauses. Schriftbilder, Drucksachen, Ausstellungsdesign. Ausstellungskatalog, Bauhaus-Archiv Berlin*
**Caflisch, Max:** *Typographia practica. Hamburg 1988*
**Carter, Rob / Day, B. / Meggs, Philip B.:** *Typographic Design. Form and Communication. New York 2. Aufl. 1993*
**Carter, Rob / Meggs, Philip B.:** *Typographic Specimen. The Great Typefaces. New York 1993*
**Carter, Rob:** *American Typography Today. New York 1993*
**Carter, Sebastian:** *20th Century Type Designers. London, New York 1987*
**Ch'ên Chih-mai:** *Chinese Calligraphers and their Art. London, New York 1966*
**Chappel, Warren:** *A Short History of the Printed Word. New York 1970*
**Cohen, Marcel:** *La grande invention de l'écriture et son évolution. Bd. 1–2, Paris 1958*
**ibid.:** *"Die Kunst der Schrift". Ausstellungskatalog Staatliche Kunsthalle, Baden-Baden. 1964*
**Compton, Susan:** *Russian Avant-Garde Books 1917–34. Cambridge, Mass. 1993*
**Consentius, Ernst:** *Die Typen der Inkunabelzeit. Berlin 1929*
**Corsten, Severin/Füssel, Stephan/Pflug, Günther/Schmidt-Künsemüller, Friedrich Adolf (ed.):** *Lexikon des gesamten Buchwesens. Bd. 1–4: A-Lyser, 2. neu bearb. Aufl. Stuttgart 1987–95*
**Crous, Ernst/Kirchner, Joachim:** *Die gotischen Schriftarten. Leipzig 1928*
**Crous, Ernst:** *Fraktur oder Antiqua? Zwei Beiträge zur Schriftfrage aus dem 18. Jahrhundert. Berlin 1926*
**Dair, Charles:** *Design with Type. Toronto 1967*
**Damase, Jaques:** *Révolution typographique depuis Stéphane Mallarmé. Genf 1966*
**Debes, Dietmar:** *Das Figurenalphabet. Leipzig 1968*
**Degering, Hermann:** *Die Schrift. Atlas der Schriftformen des Abendlandes vom Altertum bis zum Ausgang des 18. Jahrhunderts. Tübingen 3. Aufl. 1952*
**Delitzsch, Friedrich:** *Die Entstehung des ältesten Schriftsystems oder der Ursprung der Keilschriftzeichen. Leipzig 1897*
**Delitzsch, Hermann:** *Geschichte der abendländischen Schreibschriftformen. Leipzig 1928*
**Didot, Pierre / Didot, Jules:** *Spécimen des caractéres de la Fonderie Normale à Bruxelles, provenant de la fonderie de Jules Didot et son père Pierre Didot. Haarlem 1931*
**Döbler, H.:** *Von der Keilschrift zum Computer. Gütersloh*
**Doblhofer, Ernst:** *Die Entzifferung alter Schriften und Sprachen. Stuttgart 1993*
**Doede, Werner:** *Schönschreiben, eine Kunst: Johann Neudörffer und seine Schule im 16. und 17. Jahrhundert. München 1957*
**ibid.:** *Bibliographie deutscher Schreibmeisterbücher von Neudörffer bis 1800. Hamburg 1958*
**ibid.:** *Glanz und Elend der Fraktur. In: Philobiblon. 27. 1983. S. 275–294*
**Dowding, Geoffrey:** *The History of Printing Types. Clerkenwell 1961*
**Driver, Godfrey Rolles:** *Semitic Writing from Pictograph to Alphabet. London 1948. Rev. Aufl. 1954*
**Drucker, Johanna:** *The Alphabetic Labyrinth. The letters in history and imagination. London 1995*
**Dunand, Maurice:** *Byblia Grammata. Documents et recherches sur le développement de l'écriture en Phénicie. Beyrouth 1945*
**Dürer, Albrecht:** *Underweysung der Messung mit dem Zirckel und Richtscheyt. Nürnberg 1525*
**Eason, Ron / Rookledge, Sarah:** *Rookledge's International Handbook of Type Designers. Carshalton Beeches 1991*
**Ehmcke, Fritz Helmut:** *Ziele des Schriftunterrichts. Jena 1911*
**ibid.:** *Schrift: ihre Gestaltung & Entwicklung in neuerer Zeit. Hannover 1925*
**ibid.:** *Die historische Entwicklung der abendländischen Schriftformen. Ravensburg 1927*
**ibid.:** *Persönliches und Sachliches. Berlin 1928*
**ibid.:** *Bestandsaufnahme der deutschen Schrift. – In: Imprimatur. 4. 1933. S. 28–39*
**ibid.:** *Deutsches Schreibbüchlein. Iserlohn 1934*
**Elam, Kimberley:** *Expressive Typography. The word as image. New York 1990*
**Enschedé, Charles:** *Fonderie de caractères et leur matérial dans les Pays-Bas du XVe au XIXe siècle. Haarlem 1908*
**Esposito, Tony/King, J. C.:** *The Postscript Typeface Library. Bd. 1–2, New York 1993*
**Faulmann, Karl:** *Das Buch der Schrift, enthaltend die Schriftzeichen und Alphabete aller Zeiten und aller Völker des Erdkreises. Wien, Pest, Leipzig 1880*
**ibid.:** *: Illustrirte Geschichte der Schrift. 1880*
**ibid.:** *Illustrirte Geschichte der Buchdruckerkunst mit besonderer Berücksichtigung ihrer technischen Entwicklung bis zur Gegenwart. Wien, Pest, Leipzig 1882*
**Feliciano, Felice:** *Alphabetum Romanum. Zürich 1985*
**Finsterer, Alfred (ed.) :** *Hoffmanns Schriftatlas. Ausgewählte Alphabete und Anwendungen aus Vergangenheit und Gegenwart. Stuttgart 1952*
**Fleischmann, Gerd (ed.) :** *Bauhaus. Typografie. Reklame. Düsseldorf, 2. Aufl. 1996*
**Flusser, Vilém:** *Die Schrift. 2. Aufl. Göttingen 1989*
**Foerster, Hans:** *Mittelalterliche Buch- und Urkundenschriften. Berlin 1946*
**Földes-Papp, Karoly:** *Vom Felsbild zum Alphabet. Die Geschichte der Schrift. Stuttgart, Zürich 1966*
**Fournier le Jeune, Pierre S.:** *Manuel Typographique. Bd 1–2, Paris 1764–66*
**Frank, Isaac:** *English Printer's Types of the 16th Century. London 1936*
**Franke, Wolfgang (ed.) :** *China-Handbuch, Düsseldorf 1974. Darin: Hermanová – Novthá, Z.: Schrift und Schriftreform. Und: Franke, Wolfgang: Schriftkunst*
**Freeman, Judi:** *Das Wort-Bild in Dada und Surrealismus. Ausstellungskatalog Frankfurt a. M. . München 1990*
**Friedrich, Johannes:** *Geschichte der Schrift unter besonderer Berücksichtigung ihrer geistigen Entwicklung. Heidelberg 1966*
**Fritz, Georg:** *Geschichte der Wiener Schriftgießereien seit Einführung der Buchdruckerkunst im Jahre 1482 bis zur Gegenwart. Wien 1924*
**Frutiger, Adrian:** *Type, Sign, Symbol. Zürich 1980*
**ibid.:** *: Der Mensch und seine Zeichen. Schriften, Symbole, Signete, Signale. 1. Zeichen erkennen, Zeichen gestalten. 1978. 3. verb. Aufl. Wiesbaden 1991*
**Gaur, Albertine:** *A History of Writing. London 1984*
**ibid.:** *A History of Calligraphy. London 1994*
**Gelb, Ignace Jay:** *Hittite Hieroglyphic Monuments. Chicago 1939*
**ibid.:** *Hittite Hieroglyphs. Bd. 1–3. Chicago 1931, 1935, 1942*
**ibid.:** *Von der Keilschrift zum Alphabet. Grundlagen einer Schriftwissenschaft. (Übersetzung aus dem Amerikanischen), Stuttgart 1958*
**Gerstner, Karl / Kutter, Markus:** *Die neue Grafik. Teufen 1959*
**Gerstner, Karl:** *Programme entwerfen. Programm als Schrift, als Typographie, als Bild, als Methode. Niederteufen 1968*
**ibid.:** *Kompendium für Alphabeten. Teufen 1972*
**Gill, Eric:** *An Essay of Typography. London 1931*
**ibid.:** *An Essay on Typography. London 1936*
**ibid.:** *Autobiography. London 1940*
**Glück, Helmut:** *Schrift und Schriftlichkeit. Stuttgart 1986*
**Goldschmidt, Ernst Philip:** *The Printed Book of the Renaissance. Three lectures on type, illustration, ornament. Cambridge/ Mass. 1950. 2. Aufl. Amsterdam 1966. Reprint: Amsterdam 1974*
**Gottschall, Edward M.:** *Typographic Directions. New York 1964*
**ibid.:** *Typographic Communications Today. New York, Cambridge/Mass., London 1989*
**Goudy, Frederic William:** *The Alphabet. London, New York 1922*
**ibid.:** *The Alphabet: Fifteen Interpretative Designs. New York 1922*
**ibid.:** *American Type Design. Chicago 1924*
**ibid.:** *Typologia. Berkeley, Los Angeles 1940*
**ibid.:** *A half-century of Type Design and Typography 1895–1945. New York 1946*
**Gray, Nicolete:** *Nineteenth Century Ornamented Typefaces. London 1976*
**ibid.:** *A History of Lettering. Creative experiment and letter identity. Oxford 1986*
**Greiman, April:** *A Hybrid Imagerie. The fusion of technology and graphic design. London 1990*
**Gutzwiller, Helmut:** *Die Entwicklung der Schrift vom 12. bis ins 19. Jahrhundert. Solothurn 1981*
**Haab, Armin/Stocker, Alex:** *Lettera 1. Teufen 1954*
**Haettenschweiler, Walter:** *Lettera 2–4. Teufen 1961–1972*
**Hammer, Victor:** *The Forms of our Letters. Lexington 1958*
**Handover, P. M.:** *Geschichtliches über die endstrichlose Schrift, die wir Grotesk nennen. Frankfurt a. M. 1962*
**Hansen, Friedrich:** *Schreibmaschinenschrift, photographische Kunst. München 1918*
**Heiderhoff, Horst:** *Antiqua oder Fraktur. Frankfurt 1971*
**Hermann, Hermann Julius:** *Die Handschriften und Inkunabeln der italienischen Renaissance. Bd. 1–4. Leipzig 1930–33*
**Hochuli, Jost/Kinross, Robin:** *Designing Books. Practise and theory. London 1996*
**Hochuli, Jost:** *Bücher machen. Eine Einführung in die Buchgestaltung im besonderen in die Buchtypografie. München, Berlin 1989*
**ibid.:** *Buchgestaltung in der Schweiz. Zürich 1993*
**Hoffmann, Herbert (ed.):** *Hoffmanns Schriftenatlas. Das Schriftschaffen der Gegenwart in Alphabeten und Anwendungen. Stuttgart 1930*
**Hofstätter, Hans Helmut:** *Jugendstil-Druckkunst. Baden-Baden 1968*
**Holme, C. Geoffrey (ed.):** *Lettering of To-Day. London, special number of The Studio, New York 1937*
**ibid.: (ed.):** *Faces and Figures. London, The Studio 1939*
**Holme, Rathbone/Frost, Kathleen (ed.) :** *Modern Lettering and Calligraphy. A Sequel to 'Lettering of Today'. London, New York 1954*
**Hopkins, Lionel Charles:** *Tai T'ung. The six scripts or the principles of Chinese writing. 1881. Neuauflage Cambridge 1953*
**Hornung, Clarence P.:** *Handbook of Early Advertising Art mainly from American Sources. Typographical and*

*ornamental volume. 3. Aufl. New York, Toronto, London 1956*
**Hostettler, Rudolf:** *The Printer's Terms. 2. Aufl. St. Gallen 153*
**Hrozny, Bedrich:** *Les inscriptions Hittites Hiéroglyphiques. Bd. 1–3, Prag 1933–37*
**Huber, Michel:** *Computergrafik. Das Multi-Media-Lexikon. Augsburg 1994*
**Humboldt, Wilhelm von:** *Über die Buchstabenschrift und ihre Zusammenhänge mit der Sprache. Berlin 1826*
**Igarashi, Takenobu (ed.):** *Designers on Mac. Tokyo, London 1994*
**Imprimerie Nationale:** *Cabinet des Poinçons de l'Imprimerie Nationale de France. Paris 1950*
**ibid.:** *Les Caractères de l'Imprimerie Nationale. Paris 1990*
**Jackson, Donald:** *Alphabet: Die Geschichte vom Schreiben. Frankfurt a. M. 1981*
**Jacno, Marcel:** *Anatomie de la lettre. Paris 1978*
**Jensen, Hans:** *Die Schrift in Vergangenheit und Gegenwart. Berlin 1984*
**Jessen, Peter:** *Meister der Schreibkunst aus drei Jahrhunderten. Stuttgart 1923*
**Johnson, Alfred Forbes:** *Type Designs: Their History and Development. 3. Aufl. London 1966*
**Johnston, Edward:** *Hand- und Inschrift. Leipzig 1912*
**ibid.:** *Schrift Schreibschrift, Zierschrift und angewandte. 5. Aufl. Braunschweig 1955*
**ibid.:** *Lessons in formal writing. New York 1986*
**Jones, T.:** *Wink Instant Design. A manual of graphic techniques. London 1980*
**Kapr, A.:** *Buchgestaltung. Ein Fachbuch für Graphiker, Schriftsetzer, Drucker. Dresden 1963*
**ibid.:** *Schriftkunst. Geschichte, Anatomie und Schönheit der lateinischen Buchstaben. Dresden 1971. Lizenzausg. München, New York, London, Paris 1983*
**Kapr, Albert / Schiller, Walter:** *Gestalt und Funktion der Typografie. Leipzig 1977*
**Käufer, Josef:** *Das Setzerlehrbuch. Die Grundlagen des Schriftsatzes und seiner Gestaltung. Stuttgart 3. Aufl. 1965*
**Kautzsch, Rudolf:** *Die Entstehung der Frakturschrift. Mainz 1921*
**Kinross, Robin:** *Modern Typography. An essay in critical history. London 1993*
**ibid.:** *Fellow Readers. Notes on multiplied language. London 1994*
**Kirchner, Joachim (ed.):** *Lexikon des gesamten Buchwesens. 4 Bde., Stuttgart 1952–56*
**Kleukens, Christian Heinrich:** *Die Kunst der Letter. Leipzig 1942*
**Klingspor, Karl:** *Über Schönheit von Schrift und Druck. Frankfurt a. M. 1949*
**Koch, Rudolf:** *Das Schreiben als Kunstfertigkeit. Leipzig 1921*
**ibid.:** *Das ABC-Büchlein. Leipzig 1934*
**Korger, Hildegard:** *Schrift und Schreiben. Leipzig 1982*
**Krimpen, Jan van:** *On designing and devising types. New York 1960*
**ibid.:** *A letter to Philip Hofer on certain problems connected with the mechanical cutting of punches. Boston 1972*
**ibid.:** *Over het ontwerpen en bedenken van drukletters. Amsterdam 1990*
**Kuckenburg, Martin:** *Die Entstehung von Sprache und Schrift. Köln 1989*
**Larisch, Rudolf von:** *Über Leserlichkeit von ornamentalen Schriften. Wien 1904*
**ibid.:** *Unterricht in ornamentaler Schrift. Wien 1926*
**Lavizzari-Raeuber, Alexandra:** *Arabische Schönschrift. Ausstelllungskatalog Museum Rietberg, Zürich. 1979*
**Lawson, Alexander:** *Printing Types. Boston 1971*
**ibid.:** *Anatomy of a Typeface. Boston 1990*
**Lechner, Herbert:** *Geschichte der modernen Typographie. Von der Steglitzer Werkstatt zum Kathodenstrahl. München 1981*
**Ledderose, Lothar:** *Mi Fu and the Classical Tradition of Chinese Calligraphy. Princeton 1979*
**Lewis, John:** *Typography: Basic Principles. London 1963*
**ibid.:** *Anatomy of Printing. London 1970*
**ibid.:** *The 20th Century Book. Its illustration and design. 2. Aufl., New York, Cincinnati, Toronto, London, Melbourne 1984*
**Lindegren, Erik:** *Våra bokstäver. Göteborg 1959/1960*
**ibid.:** *ABC of Lettering and Printing Typefaces. New York 1982*
**Lommen, Mathieu:** *Letterontwerpers. Haarlem 1987*
**Loubier, Hans:** *Die neue Deutsche Buchkunst. Stuttgart 1921*
**Luidl, Philipp:** *Günter Gerhard Lange. München 1983*
**ibid.:** *Typografie. Ostfildern 1996*
**Lutz, Hans-Rudolf:** *Ausbildung in typographischer Gestaltung. Zürich 1987*
**Massin, Robert:** *La mise en page. Paris 1991*
**ibid.:** *La lettre et l'image. La figuration dans l'alphabet latin du 18e siècle à nos jours. Neuedition Paris 1993*
**McMurtrie, Douglas C.:** *American Type Design in the Twentieth Century with the specimens of the outstanding types procured during this period. Chicago 1924*
**McLean, Ruari (ed.):** *The Thames and Hudson Manual of Typography. London 1980*
**ibid.:** *Typographers on Type. An illustrated anthology from W. Morris to the present day. London 1995*
**Mediavilla, Claude:** *Calligraphie. Paris 1993*
**Meggs, Philip B.:** *History Of Graphic Design. New York 1983*
**ibid.:** *Type and Image. New York 1989*
**Mengel, Willi:** *Druckschriften der Gegenwart, klassifiziert nach DIN 16518. Stuttgart 1966*
**Milchsack, Gustav:** *Was ist Fraktur? 2. Aufl. Braunschweig 1925*
**Morison Stanley:** *Handbuch der Druckkunst: 250 Beispiele mustergültiger Antiquadrucke aus den Jahren 1500 bis 1900. Berlin 1925*
**ibid.:** *Type Designs of the past and present. London 1926*
**ibid.:** *Schrift, Inschrift, Druck. Hamburg 1948*
**ibid.:** *A Tally of Types. Cambridge 1973*
**ibid.:** *Early Italian Writing Books. London 1990*
**Morris, William:** *The Ideal Book. Essays and Lecture on the Arts of the Book. Herausgegeben von William S. Peterson. Berkeley 1982*
**Muess, Johannes:** *Das römische Alphabet. Entwicklung, Form und Konstruktion. München 1989*
**Müller-Brockmann, Josef:** *Rastersysteme, Stuttgart 1981*
**Muzika, Frantisek:** *Die schöne Schrift in der Entwicklung des lateinischen Alphabets. Bd. 1–2 (mit ausführlicher Bibliographie), Hanau 1965*
**Nadolski, Dieter (ed.):** *Didaktische Typografie. Leipzig 1984*
**Nerdinger, Eugen:** *Buchstabenbuch. Schriftentwicklung, Formbedingungen, Schrifttechnik, Schriftsammlung. München 1954*
**Nesbitt, Alexander:** *The History and Technique of Lettering. New York 1957*
**Neudörffer, Antonius:** *Schreibkunst etc. Bd. 1–2, Nürnberg 1631*
**ibid.:** *Schreibkunst etc. Appendix, Nürnberg 1631*
**Novarese, Aldo:** *Lo studio e il disegno des carattere. Turin 1964*
**Ovink, Gerrit Willem:** *Legibility, Atmosphere, Value and Forms of Printing Types. Leiden 1938*
**ibid.:** *Anderhalve eeuw boektypografie in Nederland. Nijmegen 1965*
**Pedersen, B. Martin (ed.) :** *Graphis Typography 1. The international compilation of the best typographic design. Zürich 1994*
**Peignot, Jérôme:** *De l'ecriture à la typographie. Paris 1967*
**Petzet, Erich / Glauning, Otto:** *Deutsche Schrifttafeln des IX. bis XVI. Jahrhunderts. Bd. 1–3 München 1910–1912, Bd. 4–5 Leipzig 1924–1930*
**Plantin, Christoph:** *Index Characterum. Antwerpen 1567. Faksimile-Ausgabe: D. C. McMurtrie, New York 1924*
**Plata, Walter:** *Schätze der Typographie: Gebrochene Schriften. Gotisch, Schwabacher und Fraktur im deutschen Sprachgebiet in der zweiten Hälfte des zwanzigsten Jahrhunderts. Frankfurt 1968*
**ibid.:** *Fraktur, Gotisch, Schwabacher, zeitlos schön. Hildesheim 1982*
**Pollard, Alfred W.:** *Die Doves Press. Wien 1932*
**Poynor, Rick / Booth-Clibborn, Edward / Why Not Associates:** *Typography Now. The Next Wave. London 1991*
**Poynor, Rick:** *The Graphic Edge. London 1993*
**Preetorius, Emil:** *Rudo Spemann. Leinfelden 1955*
**Rahmer, Albert:** *Schriften: Handbuch der Druckschriften. Stuttgart 1974*
**Raible, Werner:** *Zur Entwicklung von Alphabetschrift-Systemen. Heidelberg 1991*
**Rehe, Rolf F.:** *Typographie: Wege zur Besseren Lesbarkeit. St. Gallen 1981*
**Reinecke, Adolf:** *Die deutsche Buchstabenschrift, ihre Entstehung und Entwicklung, ihre Zweckmäßigkeit und völkische Bedeutung. Leipzig 1910*
**Reiner, Imre u. Hedwig:** *Alphabets. St. Gallen 1947*
**ibid.:** *Schrift im Buch. St. Gallen 1948*
**Renner, Paul:** *Mechanisierte Grafik. Schrift, Typo, Foto, Film, Farbe. Berlin 1930*
**ibid.:** *Die Kunst der Typographie. Berlin 1939*
**Rodenberg, Julius:** *Größe und Grenzen der Typographie. Stuttgart 1959*
**Rondthaler, Edward (ed.):** *Alphabet Thesaurus. New York 1965*
**Rosen, Ben:** *Type and Typography. New York 1963*
**Ruder, Emil:** *Typographie. Ein Design-Handbuch. Teufen 1967*
**Ruegg, Ruedi:** *Typografische Grundlagen. Zürich 1972*
**Schauer, Georg Kurt:** *Deutsche Buchkunst 1890–1960. Hamburg 1963*
**ibid.:** *Internationale Buchkunst im 19. Und 20. Jahrhundert. Ravensburg 1969*
**Schlott, Adelheid:** *Schrift und Schreiber im Alten Ägypten. München 1989*
**Schmid, Helmut.:** *Typography Today. (Sonderheft von 'Idea'). Tokyo 1980*
**Schneidler, F. H. Ernst:** *Der Wassermann: Studienblätter für Büchermacher. Stuttgart 1945*
**Seaby, A. Wilfred:** *The Roman Alphabet and Its Derivations. London 1925*
**Soennecken, Friedrich:** *Das deutsche Schriftwesen und die Notwendigkeit seiner Reform. Bonn, Leipzig 1881*
**ibid.:** *Der Werdegang unserer Schrift. Bonn 1911*
**Spencer, Herbert:** *The Visible Word. London 1969*
**ibid.:** *Pioneers of Modern Typography. London 1969, rev. Aufl. 1982*
**ibid.:** *The Liberated Page. An anthology of major typographic experiments of this century as recorded in "Typographica" magazine. London 1. Aufl. 1987, 2. Aufl. 1990*
**Spiekermann, Erik:** *Ursache und Wirkung. Ein typographischer Roman. Berlin 1986. Faksimile der Originalausg.: Mainz 1994*
**Standard, Paul:** *Calligraphy's Flowering, Decay & Restauration. Chicago 1947*
**Steffens, Franz:** *Lateinische Paläographie. 2. Aufl. Berlin, Leipzig 1929*
**Steinberg, Sigfrid Henry:** *Five hundred Years of Printing. Harmondsworth, Middlesex 1955. Dt. Ausg. München 1958*
**Stone, A. Reynolds:** *A Book of Lettering. London 1935*
**Sutton, James / Bartram, Alan:** *An atlas of typeforms. London 1968*
**Thibaudeau, Francis:** *La Lettre d'Imprimerie. Paris 1921*
**Tolmer, A.:** *Mise en Page. The Theory and Practice of Lay-Out. London 1931*
**Tschichold, Jan:** *Typographische Gestaltung. Basel 1935*
**ibid.:** *Schriftkunde, Schreibübungen und Skizzieren für Setzer. Basel 1942*
**ibid.:** *Meisterbuch der Schrift. Ravensburg 1953*
**ibid.:** *Die Neue Typographie. Berlin 1928. Reprint: Berlin 1987*
**Turner, W. / Johnson, Alfred Forbes / Jaspert, W. Pincus:** *The Encyclopaedia of Type Faces. London 1993*
**Ullmann, Berthold Louis:** *The Etruscan Origin of the Latin Alhabet and the Names of the Letters. In: Classical Philology XXII, London 1927*
**ibid.:** *Ancient Writing and its Influences. New York 1963*
**Unger, Ernst:** *Die Keilschrift. Entstehung, System und Ornamentik der Schrift der ältesten Hochkultur. Leipzig 1929*
**Updike, Daniel Berkeley:** *Printing Types. Their history, forms, and use. 3. Aufl. Cambridge, Mass. 1962*
**VanderLans, Rudy / Licko, Zuzana:** *Emigre (The Book) Graphic Design into the Digital Realm. New York 1993*
**Velde, Jan van den:** *Spieghel der Schrijftkonste. Rotterdam 1605*
**Vinne, Theodore Low de:** *Historic Printing Types. New York 1886*
**Waddell, L. A.:** *Egyptian Civilisation, its Sumerian Origin and Real Chronology and Sumerian Origin of Egyptian Hieroglyphs. London 1930*
**Walther, Karl Klaus (ed.) :** *Lexikon der Buchkunst und Bibliophilie. Augsburg 1994*
**Wardrop, James:** *The Script of Humanism. Oxford 1963*
**Weidemann, Kurt:** *Typopictura. 3 Jahrzehnte werbender Typographie. Frankfurt a. M. 1981*
**ibid.:** *Wo der Buchstabe das Wort führt. Ansichten über Schrift und Typographie. Ostfildern 1994*
**Wilkes, Walter:** *Der Schriftguß. Stuttgart 1990*
**Willberg, Hans Peter:** *Buchkunst im Wandel. Frankfurt a. M. 1984*
**Willberg, Hans Peter / Forssman, Friedrich:** *Lesetypographie, Mainz 1997*
**Zapf, Hermann:** *Über Alphabete. Frankfurt a. M. 1960*
**ibid.:** *Manuale Typographicum. Frankfurt a. M., New York 1968*
**Zeller, Otto:** *Der Ursprung der Buchstabenschrift und das Runenalphabet. Osnabrück 1977*

**Periodika**
**Auswahl, sélection, selected**

**Alphabet.** International annual of letterforms. Birmingham, 1964 ff
**Alphabet and Image.** A quarterly of typography and graphic arts. London, 1946–48
**Archiv für Buchgewerbe und Gebrauchsgraphik.** Leipzig, 1861–1943
**Art and Letters.** London, 1948–63
**Arts et métiers graphiques.** Paris, 1927–39
**Arts et techniques graphiques.** Paris, 1957 ff
**Baseline,** London, 1979 ff
**Buch- und Werbekunst.** Leipzig . Vorg. s. Offset-, Buch- und Werbekunst
**Communications.** Paris, 1961 ff
**Der Druckspiegel.** Magazin für Druck und Kommunikation. Heusenstamm, 1946 ff
**Eye.** The international review of graphic design. London, 1991 ff
**Emigre.** Berkeley, 1982ff
The **Fleuron.** A Journal of Typography. Cambridge, London, 1923–30
**Gebrauchsgraphik.** Berlin, 1924/25–72, seit 1972 unter dem Titel **Novum**
**Gutenberg-Jahrbuch.** Mainz, 1926 ff
**High Quality. H Q.** Zeitschrift für das Gestalten, das Drucken und das Gedruckte. Heidelberg, 1985 ff
**Imprimatur.** Ein Jahrbuch für Bücherfreunde. Frankfurt a. M., 1930–1939/40, 1950–1954/55. N.F. 1956/57 ff
**Information Design Journal.** Chester, 1981 ff
The **Journal of Typographic Research.** Cleveland, 1967–1970. cf. Visible Language
**Octavo,** London, 1986–92
**Pagina.** Mailand, 1962–66
The **Penrose Annual.** London, 1898–1940, 1949–82
**Poligrafija.** Moskau, 1924–?
**Print & Publishing.** (Tschechischer Untertitel). Prag, 1992 ff. cf. Typografia
**Print Quarterly.** P. Q. . London, 1984 ff
**Typografia.** Prag, 1893–1940. cf. Print & Publishing
**Typografie.** Fachausgabe der Zeitschrift Papier und Druck. Leipzig, 1952 ff
**Typografische Monatsblätter.** Schweizer Grafische Mitteilungen (wechselnde Untertitel). St. Gallen, 1882/83 ff
Der **Typograph.** Berlin, 1892–1933
**Typographica.** Contemporary typography and graphic art. London, 1949–67
**Typographics.** The art of typography from digital to dyeline. New York, 1995 ff
**Typographische Jahrbücher.** Leipzig, 1880–1929
**Typographische Mitteilungen.** Leipzig, 1903–33
**Typography.** The annual of the Type Directors Club. New York, 1980 ff
**U&lc.** Upper and Lower Case. The international journal of typographics. New York, 1974 ff
**Visible Language.** Ohio, 1970 ff. cf. Journal of Typographic Research
Die **Zeitgemäße Schrift.** Berlin, Leipzig, 1927–43
**Zeitschrift für Bücherfreunde.** Leipzig, 1897/98–1908/09. N.F. 1909/10–36